Contents

II Key to map symbols

III Truckstops in England, Scotland and Wales

VIII Route planning maps

XIV Distances and journey times

1 Road maps of Britain

315 Urban approach maps

- 315 Bristol *approaches*
- 316 Birmingham *approaches*
- 318 Cardiff *approaches*
- 319 Edinburgh *approaches*
- 320 Glasgow *approaches*
- 321 Leeds *approaches*
- 322 London *approaches*
- 326 Liverpool *approaches*
- 327 Manchester *approaches*
- 328 Newcastle *approaches*
- 329 Nottingham *approaches*
- 330 Sheffield *approaches*

331 Town plans

- 331 Aberdeen, Aberystwyth, Ashford, Ayr, Bangor, Barrow-in-Furness, Bath, Berwick-upon-Tweed
- 332 Birmingham, Blackpool, Bournemouth, Bradford, Brighton, Bristol, Bury St Edmunds
- 333 Cambridge, Canterbury, Cardiff, Carlisle, Chelmsford, Cheltenham, Chester, Chichester, Colchester
- 334 Coventry, Derby, Dorchester, Dumfries, Dundee, Durham, Edinburgh, Exeter
- 335 Fort William, Glasgow, Gloucester, Grimsby, Hanley, Harrogate, Holyhead, Hull
- 336 Inverness, Ipswich, Kendal, King's Lynn, Leeds, Lancaster, Leicester, Lewes
- 337 Lincoln, Liverpool, Llandudno, Llanelli, Luton, Macclesfield, Manchester
- 338 London
- 340 Maidstone, Merthyr Tydfil, Middlesbrough, Milton Keynes, Newcastle, Newport, Newquay, Newtown, Northampton
- 341 Norwich, Nottingham, Oban, Oxford, Perth, Peterborough, Plymouth, Poole, Portsmouth
- 342 Preston, Reading, St Andrews, Salisbury, Scarborough, Shrewsbury, Sheffield, Southampton
- 343 Southend-on-Sea, Stirling, Stoke, Stratford-upon-Avon, Sunderland, Swansea, Swindon, Taunton, Telford
- 344 Torquay, Truro, Wick, Winchester, Windsor, Wolverhampton, Worcester, Wrexham, York

345 Index to town plans

361 Index to road maps of Britain

402 County and unitary authority boundaries

www.philips-maps.co.uk

First published in 2009 by Philip's
a division of Octopus Publishing Group Ltd
www.octopusbooks.co.uk
Endeavour House, 189 Shaftesbury Avenue,
London WC2H 8JY
An Hachette UK Company
www.hachette.co.uk

Second edition 2011
First impression 2011

ISBN 978-1-84907-143-7

Cartography by Philip's

Copyright © 2011 Philip's

This product includes mapping data licensed from Ordnance Survey®, with the permission of the Controller of Her Majesty's Stationery Office.

© Crown copyright 2011. All rights reserved. Licence number 100011710

No part of this publication may be reproduced, stored in a retrieval system or transmitted in any form or by any means, electronic, mechanical, photocopying, recording or otherwise, without the permission of the Publishers and the copyright owner.

While every reasonable effort has been made to ensure that the information compiled in this atlas is accurate, complete and up-to-date at the time of publication, some of this information is subject to change and the Publisher cannot guarantee its correctness or completeness.

The information in this atlas is provided without any representation or warranty, express or implied and the Publisher cannot be held liable for any loss or damage due to any use or reliance on the information in this atlas, nor for any errors, omissions or subsequent changes in such information.

The representation in this atlas of any road, drive or track is no evidence of the existence of a right of way.

Data for the speed cameras provided by **PocketGPSWorld.com Ltd**.

Information for National Parks, Areas of Outstanding Natural Beauty, National Trails and Country Parks in Wales supplied by the Countryside Council for Wales.

Information for National Parks, Areas of Outstanding Natural Beauty, National Trails and Country Parks in England supplied by Natural England. Data for Regional Parks, Long Distance Footpaths and Country Parks in Scotland provided by Scottish Natural Heritage.

Information for Forest Parks supplied by the Forestry Commission

Information for the RSPB reserves provided by the RSPB

Gaelic name forms used in the Western Isles provided by Comhairle nan Eilean.

Data for the National Nature Reserves in England provided by Natural England. Data for the National Nature Reserves in Wales provided by Countryside Council for Wales. Darparwyd data'n ymwneud â Gwarchodfeydd Natur Cenedlaethol Cymru gan Gyngor Cefn Gwlad Cymru.

Information on the location of National Nature Reserves in Scotland was provided by Scottish Natural Heritage.

Data for National Scenic Areas in Scotland provided by the Scottish Executive Office. Crown copyright material is reproduced with the permission of the Controller of HMSO and the Queen's Printer for Scotland. Licence number C02W0003960.

Printed in China

Road map symbols

M25	**Motorway**
16 17	**Motorway junctions** – full access, restricted access
	Toll motorway
Pease Pottage Services	**Motorway service area**
	Motorway under construction
S	**Primary route** – dual, single carriageway, services – under construction, narrow
Cardiff	**Primary destination**
25 26	**Numbered junctions** – full, restricted access
	A road – dual, single carriageway – under construction, narrow
	B road – dual, single carriageway – under construction, narrow
	Minor road – dual, single carriageway
	Drive or track
12'9 13'0	**Height restriction, width restriction** – feet and inches
12.5	**Tunnel, weight restriction** – tonnes
2	**Distance in miles**
	Roundabout, multi-level junction,
Toll	**Toll, steep gradient** – points downhill
40 40	**Speed camera** – single, multiple
CLEVELAND WAY	**National trail** – England and Wales
GREAT GLEN WAY	**Long distance footpath** – Scotland
YATTON	**Railway with station, level crossing, tunnel**
ROPLEY	**Preserved railway with level crossing, station, tunnel**
	Tramway
	National boundary
	County or unitary authority boundary
	Car ferry, catamaran
	Passenger ferry, catamaran
CALAIS 1:30	**Ferry destination, journey time** – hours: minutes
	Hovercraft
V P	**Internal ferry** – car, passenger
	Principal airport, other airport or airfield
MENDIP HILLS	**Area of outstanding natural beauty, National Forest** – England and Wales, **Forest park**, **National park**, **National scenic area** – Scotland, **Regional park**
	Woodland
	Beach – sand, shingle
KENNET AND AVON CANAL	**Navigable river or canal**
6	**Lock, flight of locks, canal bridge number**
965	**Viewpoint, spot height** – in metres
	Linear antiquity
P&R	**Park and ride**
29	**Adjoining page number**
SY 70 80	**Ordnance Survey National Grid reference** – see page 402

Tourist information

BYLAND ABBEY	Abbey or priory
WOODHENGE	Ancient monument
SEALIFE CENTRE	Aquarium or dolphinarium
CITY MUSEUM AND ART GALLERY	Art collection or museum
TATE ST IVES	Art gallery
1644	Battle site and date
ABBOTSBURY SWANNERY	Bird sanctuary or aviary
	Camping site
	Caravan site
BAMBURGH CASTLE	Castle
YORK MINSTER	Cathedral
SANDHAM MEMORIAL CHAPEL	Church of interest
SEVEN SISTERS	Country park – England and Wales
LOCHORE MEADOWS	Country park – Scotland
ROYAL BATH & WEST SHOWGROUND	County show ground
MONK PARK FARM	Farm park
HILLIER GARDENS AND ARBORETUM	Garden, arboretum
ST ANDREWS	Golf course – 18-hole
TYNTESFIELD	Historic house
SS GREAT BRITAIN	Historic ship
HATFIELD HOUSE	House and garden
MUSEUM OF DARTMOOR LIFE	Local museum
HOLTON HEATH	National nature reserve
	Marina
NAT MARITIME MUSEUM	Maritime or military museum
SILVERSTONE	Motor racing circuit
CUMBERLAND PENCIL MUSEUM	Museum
	Picnic area
WEST SOMERSET RAILWAY	Preserved railway
THIRSK	Racecourse
LEAHILL TURRET	Roman antiquity
BOYTON MARSHES	RSPB reserve
THRIGBY HALL	Safari park
FREEPORT BRAINTREE	Shopping village
MILLENNIUM STADIUM	Sports venue
ALTON TOWERS	Theme park
	Tourist information centre – open all year – open seasonally
NATIONAL RAILWAY MUSEUM	Transport collection
LEVANT MINE	World heritage site
HELMSLEY	Youth hostel
MARWELL	Zoo
SUTTON BANK VISITOR CENTRE	Other place of interest
GLENFIDDICH DISTILLERY	Other place of interest

Approach map symbols

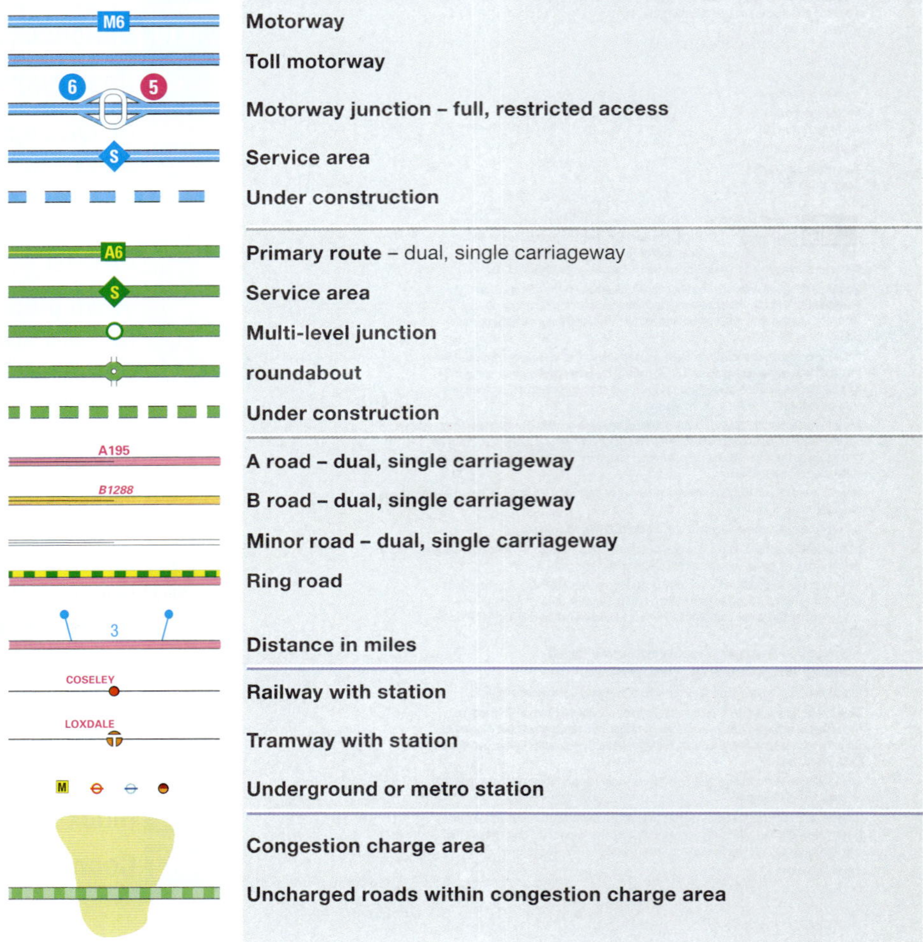

Road map scale 1: 100 000 or 1.58 miles to 1 inch

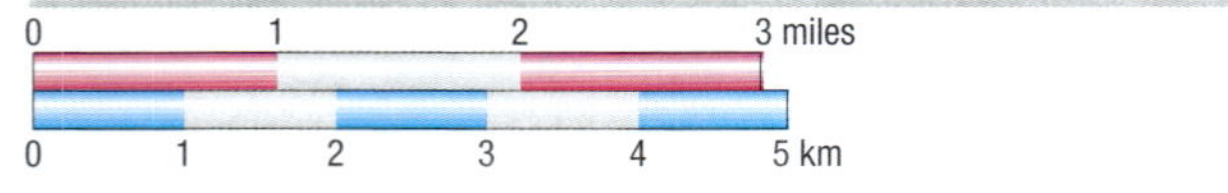

Road map scale (Isle of Man and parts of Scotland)

1: 200 000 or 3.15 miles to 1 inch

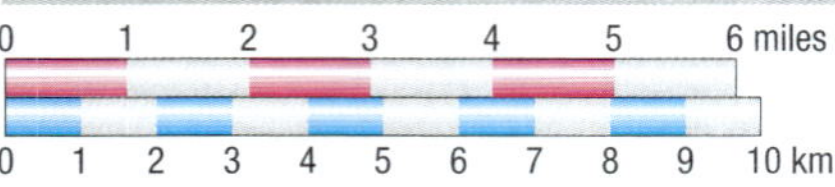

Speed Cameras

Fixed camera locations are shown using the (40) symbol. In congested areas the (40) symbol is used to show that there are two or more cameras on the road indicated.

Due to the restrictions of scale the camera locations are only approximate and cannot indicate the operating direction of the camera. Mobile camera sites, and cameras located on roads not included on the mapping are not shown. Where two or more cameras are shown on the same road, drivers are warned that this may indicate that a SPEC system is in operation. These cameras use the time taken to drive between the two camera positions to calculate the speed of the vehicle. At the time of going to press, some local authorities were considering decommissioning their speed cameras.

Load and vehicle restrictions

Any information on height, width and weight restrictions in the UK as noted on pages 1–314 of this atlas has been derived from the relevant OS material used to compile this atlas. Any information on height, width and weight restrictions on the Isle of Man has been derived from the relevant information as supplied by the Isle of Man Highways Department. Where a warning sign is displayed, any height obstructions, including but not limited to low bridges and overhead cables, are shown in the atlas where such obstructions cross navigable roads selected for inclusion. Height restrictions lower than 16'6" and width restrictions narrower than 13 feet, are all shown in 3 inch multiples and have been rounded down where necessary. Weight restrictions indicate weak bridges and the maximum gross weight which could be supported is shown in tonnes. While every effort has been made to include all relevant and accurate information, due to limitations of scale a single symbol may be used to indicate more than one feature.

Truckstops in England, Scotland and Wales

Thurso Overnight Lorry Park
Skiach Services
Ashgrove Filling Station and Restaurant
Chef's Grill
Stracathro Services
Motorgrill Ballinluig
Tayside Truckstop
Horse Shoe Café
Moto Kinross
Moto Stirling
Muirpark Truckstop
Westway Lorry Park
Roadchef Bothwell (southbound)
Roadchef Hamilton (northbound)
Cedar Café
Redmoss Truckstop
Welcome Break Abington
Hungry Trucker Transport Café
Adderstone Services
Roadchef Annandale Water
Lockerbie Truckstop
Eardley International
Welcome Break Gretna
Nightowl Truckstop Carlisle
Kingstown Truck Park
Moto Washington
Penrith Truckstop
A19 Services
Cleveland Truckstop
Junction 38 Services
Barton Lorry Park
Londonderry Lodge
Quernhow A1 Café and Truckstop
Truckhaven Carnforth
Moto Wetherby
York Lorry Park
Junction 7 Business Park
Andi's Place
Sue's Pitstop Café
Priory Park Lorry Park Truck Stop
Docklands Diner and Truckstop
Whitley Bridge Pallets and Truckstop
Heywood Distribution Park
Redbeck Motel
Ulceby Truckstop
Barney's Café
Diggles Truckstop and Diner
Moto Doncaster North
Poplar 2000 Services – Lymm Truckstop
The Stockyard Truckstop
Caenby Corner Transport Café
Ellesmere Port Truckstop
Lets Eat Café
Junction 29 Truckstop
Ranch Café and Cattle Market Lorry Park
Crewe Truck Stop
Langrick Station Café
Midway Truckstop
Moto Stafford Northbound
The Salt Box Café
Bistro Café
PJ's Transport Café
A1 Truckstop Colsterworth
Dinkys Dinahs
Truckers Rest
Junction 23 Lorry Park
Necton Diner
Standeford Farm Café
Stibbington Diner
Hawkins Transport Village
Kates Cabin Café
Harry Tuffins Transport Café
Birmingham Truckstop
Nightowl Truckstop Rugby
Lincoln Farm Café
Scoffers Café
Hilltop Café
Red Lion Café & Truckstop
Super Sausage Café
Jack's Hill Café
Super Sausage Café
Boss Hoggs Café
Orwell Crossing Lorry Park
The Truckstop Café (Crawley Crossing Bunker Stop)
Moto Cherwell Valley
Ardleigh Truckstop
Watling Street Truckstop
Pont Abraham Services
South Mimms Truckstop
J26 M25 Truckstop
Moto Swansea
Swindon Truckstop
Chris's Café
Crown Road Vehicle Park
Magor Services
Cardiff Gate Services
Square Deal Café
Titan Truckstop
Cardiff West Services
The Avon Lodge
Hideaway Truckstop
Merry Chest Café
Nell's Café
Moto Reading
Moto Medway services
HF Veale and Sons
Nunney Catch Café
Oakdene Café
Ashford International Truckstop
Albion Inn and Truckstop
Hillside Café
Airport Café
The Old Willoughby Hedge Café
The Cabin Café
Gibraltar Club and Café
Portsmouth Truckstop
Café Royal
Pie Stop Café
Smokey Joe's Café

Symbols

- Accommodation
- Advanced booking
- Cafe
- Cash machine
- CCTV
- Fuel
- Truckwash
- Nearest main road. This may be a few miles away
- Security
- Shop
- Showers
- Parking spaces
- Toilets
- TV
- Opening hours

SK93842371
Ordnance Survey National Grid reference

155 E8
Page and grid reference in this atlas

Truckstops near motorways are shown in blue, those near primary routes are green and those on other routes are shown in black.

England

A1 Truckstop Colsterworth

Bourne Road, Colsterworth, Grantham, Lincolnshire NG33 5JN
+44 (0)1476 861543 (day) • +44 (0)1476861901 (eve)
roger@a1truckstop.co.uk
SK93842371 155 E8
At Colsterworth take A151 Bourne. Site is about 200 yards on right.

Mon 24hr **Tue** 24hr **Wed** 24hr **Thu** 24hr **Fri** 24hr **Sat** By appointment **Sun** Closed

Mon 0600-2400 **Tue** 24hr **Wed** 24hr **Thu** 24hr **Fri** 24hr **Sat** Closed **Sun** Closed

A19 Services North

Ron Perry & Son Ltd, Elwick, Hartlepool, Teesside TS27 3HH
+44 (0)1740 644223
sales@ronperry.co.uk
www.ronperry.co.uk
NZ45022902 234 F5
2 miles north of junction with A689

Mon 0600-2100 **Tue** 0600-2100 **Wed** 0600-2100 **Thu** 0600-2100 **Fri** 0600-2100 **Sat** 0600-2100 **Sun** 0600-2100

Mon 0630-1730 **Tue** 0630-1730 **Wed** 0630-1730 **Thu** 0630-1730 **Fri** 0630-1730 **Sat** 0630-1600 **Sun** 0603-1600

A19 Services South

Ron Perry & Son Ltd, Elwick, Hartlepool, Teesside TS27 3HH
+44 (0)1740 644223
sales@ronperry.co.uk
www.ronperry.co.uk
NZ45072914 234 F5
3 miles south junction with A179

Mon 24hr **Tue** 24hr **Wed** 24hr **Thu** 24hr **Fri** 24hr **Sat** 24hr **Sun** 24hr

Mon 0630-1730 **Tue** 0630-1730 **Wed** 0630-1730 **Thu** 0630-1730 **Fri** 0630-1730 **Sat** 0630-1600 **Sun** 0630-1600

Adderstone Services

Belford, Northumberland NE70 7JU
+44 (0)1668 213000
james@purdylodge.co.uk
www.purdylodge.co.uk
NU13213020 **264 C4**
Approximately 13 miles north of A1068 Alnwick turnoff.

Mon 24hr **Tue** 24hr **Wed** 24hr **Thu** 24hr **Fri** 24hr **Sat** 24hr **Sun** 24hr

Mon 24hr **Tue** 24hr **Wed** 24hr **Thu** 24hr **Fri** 24hr **Sat** 24hr **Sun** 24hr

Airport Café

P 20-40
M20 J11 (5 miles)

A20 Main Road, Sellindge, Ashford, Kent TN25 6DA
+44 (0)1303 813185
patanjulie@aol.com
TR11253669 **54 F6**
From M20 J11, Take A20 Sellindge for 5 miles.

Mon 0700-1600 **Tue** 0700-1600 **Wed** 0700-1600 **Thu** 0700-1600 **Fri** 0700-1600 **Sat** 0700-1430 **Sun** 0700-1430

Mon 0700-1600 **Tue** 0700-1600 **Wed** 0700-1600 **Thu** 0700-1600 **Fri** 0700-1600 **Sat** 0700-1430 **Sun** 0700-1430

Albion Inn and Truckstop

P 16
M5 J23 (8 miles)

14 Bath Road, Ashcott, Somerset TA7 9QT
+44 (0)1458 210281
ST42373722 **44 F2**
From M5 J23 take A39 Glastonbury. Site is on right after 8 miles.

Mon 0700-2000 **Tue** 0700-2000 **Wed** 0700-2000 **Thu** 0700-2000 **Fri** 0700-1400 **Sat** 0700-1400 **Sun** 1000-1400

Mon 0700-2000 **Tue** 0700-2000 **Wed** 0700-2000 **Thu** 0700-2000 **Fri** 0700-1400 **Sat** 0700-1400 **Sun** 1000-1400

Andi's Place

Nab Lane, Batley, West Yorkshire WF17 9NG
+44 (0)1924475040
lizzymint08@hotmail.co.uk
SE23122693 **197 B8**
From M62 J27 take A62 Birstall, take 2nd left after roundabout (Pheasant Drive), then right and right again.

Mon 24hr **Tue** 24hr **Wed** 24hr **Thu** 24hr **Fri** 24hr **Sat** 24hr **Sun** 24hr

Mon 0700-1900 **Tue** 0700-1900 **Wed** 0700-1900 **Thu** 0700-1900 **Fri** 0700-1530 **Sat** 0730-1100 **Sun** Closed

Ardleigh Truckstop

Ardleigh, Colchester, Essex CO7 7SL
+44 (0)1543 469183
d.brosnan@virgin.net
TM04842682 **107 F10**
From A12 J29 take A120 Harwich/Clacton for 2 miles.

Mon 0630-1945 **Tue** 0630-1945 **Wed** 0630-1945 **Thu** 0630-1945 **Fri** 0630-1645 **Sat** 0800-1345 **Sun** 0900-1345

Mon 0630-1700 **Tue** 0630-1700 **Wed** 0630-1700 **Thu** 0630-1700 **Fri** 0630-1700 **Sat** 0800-1400 **Sun** 0800-1400

Ashford International Truckstop

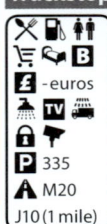

- euros
P 335
M20 J10 (1 mile)

Waterbrook Avenue, Sevington, Ashford, Kent TN24 0LH
+44 (0)1233 502919
ashfordtruckstop@ashfordtruckstop.co.uk
www.ashfordtruckstop.co.uk
TR03273974 **54 F4**
From M20 J10 take A2070 Sevington. Follow signs.

Mon 24hr **Tue** 24hr **Wed** 24hr **Thu** 24hr **Fri** 24hr **Sat** 24hr **Sun** 24hr

Mon 24hr **Tue** 24hr **Wed** 24hr **Thu** 24hr **Fri** 24hr **Sat** 24hr **Sun** 24hr

Barney's Café

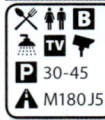

P 30-45
M180 J5

Melton Ross Road (A18), Barnetby, North Lincolnshire DN38 6LB
+44 (0)1652 680966
TA05521065 **200 E5**
From M180/A180 J5 head for Humberside Airport. The site is 300 yards along on the right.

Mon 0600-2100 **Tue** 0600-2100 **Wed** 0600-2100 **Thu** 0600-2100 **Fri** 0600-2000 **Sat** 0600-1400 **Sun** Closed

Mon 0600-2100 **Tue** 0600-2100 **Wed** 0600-2100 **Thu** 0600-2100 **Fri** 0600-2000 **Sat** 0600-1400 **Sun** Closed

Barton Lorry Park

P 40
A1(M) J56

Barton, Richmond, North Yorkshire DL10 6NF
+44 (0)1325 377777
NZ21920799 **224 D4**
From A1(M) J56 head for Barton. The site is on the right after 100 yards.

Mon 24hr **Tue** 24hr **Wed** 24hr **Thu** 24hr **Fri** 24hr **Sat** 24hr **Sun** 24hr

Mon 0600-2200 **Tue** 0600-2200 **Wed** 0600-2200 **Thu** 0600-2200 **Fri** 0600-2100 **Sat** 0700-1500 **Sun** 0800-1600

Birmingham Truckstop

P 100
M42 J6 (5 miles) M6 J5 (6¾ miles)

The Wharf, Wharf Road, Tyseley, Birmingham, B11 2EB
+44 (0)121 6282364
brisklandltd@go.com
SP11598441 **134 G2**
From M42 J6, take A45 westbound 4½ miles to A4040. After ½ mile right into Wharfdale Road, then 1st right. From M42 J5, take A41 westbound 5 miles to A4040. After 1¼ miles, left into Wharfdale Road, then 1st right.

Mon 24hr **Tue** 24hr **Wed** 24hr **Thu** 24hr **Fri** 24hr **Sat** 24hr **Sun** 24hr

Mon 0600-2100 **Tue** 0600-2100 **Wed** 0600-2100 **Thu** 0600-2100 **Fri** 0600-2100 **Sat** 0600-1200 **Sun** Closed

Bistro Café

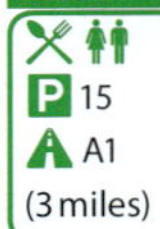

Barrowby View, Nottingham Road, Sedgebrook, Lincolnshire NG32 2EP
+44 (0)1949 843239
frankish687@btinternet.com
SK85503766 **155 B7**
From A1 A52 junction take A52 Nottingham for 3 miles

Mon 0630-1500 **Tue** 0630-1500 **Wed** 0630-1500 **Thu** 0630-1500 **Fri** 0630-1500 **Sat** 0630-1500 **Sun** 0700-1400

Mon 0630-1500 **Tue** 0630-1500 **Wed** 0630-1500 **Thu** 0630-1500 **Fri** 0630-1500 **Sat** 0630-1500 **Sun** 0700-1400

Boss Hoggs Café

London Road, Copdock, Ipswich, Suffolk IP8 3JW
+44 (0)1473 730797
TM10793982 **108 D2**
From A12 J33/A14 J55, take A12 Colchester. Take 1st exit on right, site is 100 yards on.

Mon 0800-1400 **Tue** 0800-1400 **Wed** 0800-1400 **Thu** 0800-1400 **Fri** 0800-1400 **Sat** Closed **Sun** Closed

Mon 0800-1400 **Tue** 0800-1400 **Wed** 0800-1400 **Thu** 0800-1400 **Fri** 0800-1400 **Sat** Closed **Sun** Closed

Caenby Corner Transport Café

Caenby Corner, Glentham, Lincolnshire LN8 2AR
+44 (0)1673 878388
abacuswecan@btinternet.com
SK96718939 **189 D7**
From M180 J4 take A15 Lincoln. After 10 miles, take A361 Market Rasen. Site is immediately on right.

Mon 0700-2100 **Tue** 0700-2100 **Wed** 0700-2100 **Thu** 0700-2100 **Fri** 0700-1800 **Sat** 0700-1600 **Sun** 0700-1600

Mon 0700-2100 **Tue** 0700-2100 **Wed** 0700-2100 **Thu** 0700-2100 **Fri** 0700-1800 **Sat** 0700-1600 **Sun** 0700-1600

Café Royal

Tannery Road, Off West Street, Bridport, Dorset DT6 3QX
+44 (0)1308 422012
SY46309293 **16 C5**
The café is in the centre of Bridport at the bus station on the B3162. Parking is in local authority's West Street Coach Park opposite.

Mon 0600-1900 **Tue** 0600-1900 **Wed** 0600-1900 **Thu** 0600-1900 **Fri** 0600-1900 **Sat** 0600-1900 **Sun** 0600-1900

Mon 0600-1900 **Tue** 0600-1900 **Wed** 0600-1900 **Thu** 0600-1900 **Fri** 0600-1900 **Sat** 0600-1900 **Sun** 0600-1900

Chris's Café

Wycombe Road, Studley Green, Stokenchurch, Buckinghamshire HP14 3XB
+44 (0)1494 482121
SU79069513 **84 F3**
From M40 J5 take the A40 High Wycombe for 5 miles. Site is on right.

Mon 0630-1900 **Tue** 0600-1900 **Wed** 0600-1900 **Thu** 0600-1900 **Fri** 0600-1400 **Sat** 0630-1200 **Sun** Closed

Mon 0630-1900 **Tue** 0600-1900 **Wed** 0600-1900 **Thu** 0600-1900 **Fri** 0600-1400 **Sat** 0600-1200 **Sun** Closed

Cleveland Truckstop

1-5 Puddlers Road, Southbank, Middlesbrough, Cleveland TS6 6TX
+44 (0)1642 465055
NZ53672109 **234 G6**
From A19/A66 junction take A66 (Middlesbrough bypass) Redcar 5¾ miles, exit left onto Normanby Road, 1st right. Site is third left.

Mon 24hr **Tue** 24hr **Wed** 24hr **Thu** 24hr **Fri** 24hr **Sat** 24hr **Sun** 24hr

Mon 0600-2130 **Tue** 0600-2130 **Wed** 0600-2130 **Thu** 0600-2130 **Fri** 0600-1600 **Sat** Closed **Sun** Closed

Crewe Truck Stop

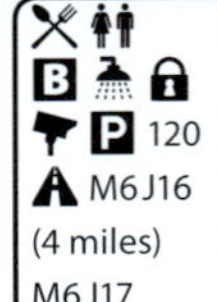

Cowley Way, (off Weston Road), Crewe, Cheshire CW1 6DD
+44 (0)7894 622250
SJ71365432 **168 E2**
From M16 J16 take A500 Crewe. At roundabout take A5020. Cross next roundabout and traffic lights. Take 2nd left. From M6 J17 take A534 5 miles. At roundabout take first exit (A5020). After 400 yards turn right into Cowley Way.

Mon 24hr **Tue** 24hr **Wed** 24hr **Thu** 24hr **Fri** 24hr **Sat** 24hr **Sun** 24hr

Mon 0600-1400 and 1700-2100 **Tue** 0600-1400 and 1700-2100 **Wed** 0600-1400 and 1700-2100 **Thu** 0600-1400 and 1700-2100 **Fri** 0600-1400 and 1700-2100 **Sat** 0600-1400 **Sun** Closed

Crown Road Vehicle Park

Crown Road Vehicle Park, Enfield EN1 1TH
+44 (0)208 443 0602
TQ34769640 **86 F4**
From M25 J25 take A10 London 2¾ miles, then A110 Chingford ½ mile, turn left into Crown Road. Site is on right after 100 yards.

Mon 24hr **Tue** 24hr **Wed** 24hr **Thu** 24hr **Fri** 24hr **Sat** 24hr **Sun** 24hr

Mon 24hr **Tue** 24hr **Wed** 24hr **Thu** 24hr **Fri** 24hr **Sat** 24hr **Sun** 24hr

Diggles Truckstop & Diner

Trafford Wharf Road, Trafford Park, Manchester, Greater Manchester M17 1DJ
+44 (0)7909858532
digglesdiner@gmail.com
SJ79949712 **184 B3**
From J9 M60 head for Trafford Park, follow signs for Imperial War Museum. At Trafford Wharf Road, turn left, site is 1st right.

Mon 24hr **Tue** 24hr **Wed** 24hr **Thu** 24hr **Fri** 24hr **Sat** 24hr **Sun** 24hr

Mon 0700-2200 **Tue** 0700-2200 **Wed** 0700-2200 **Thu** 0700-2200 **Fri** 0700-1400 **Sat** Closed **Sun** Closed
On match days the site is used as a coach park and food hours may vary.

Dinkys Dinahs

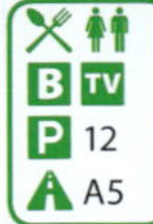

Welshpool Road, Ford, Shrewsbury, Shropshire SY5 9LG
+44 (0)1743 850070
SJ41231323 **149 G8**
From the A5 west of Shrewsbury take the A458 Welshpool for 2 miles. The site is in a layby on the right.

Mon 24hr **Tue** 24hr **Wed** 24hr **Thu** 24hr **Fri** 24hr **Sat** 24hr **Sun** 24hr

Mon 24hr **Tue** 24hr **Wed** 24hr **Thu** 24hr **Fri** 24hr **Sat** 24hr **Sun** 24hr

Docklands Diner and Truckstop

Anderson Road, Goole, E Yorkshire DN14 6UD
+44 (0)1405 766349
SE73412361 **199 C8**
From J36 take A614 Goole at first set of traffic lights (½ miles) turn right into Anderson Road. Site visible on left within 150 yards. Entrance is in A.W. Nielson Road.

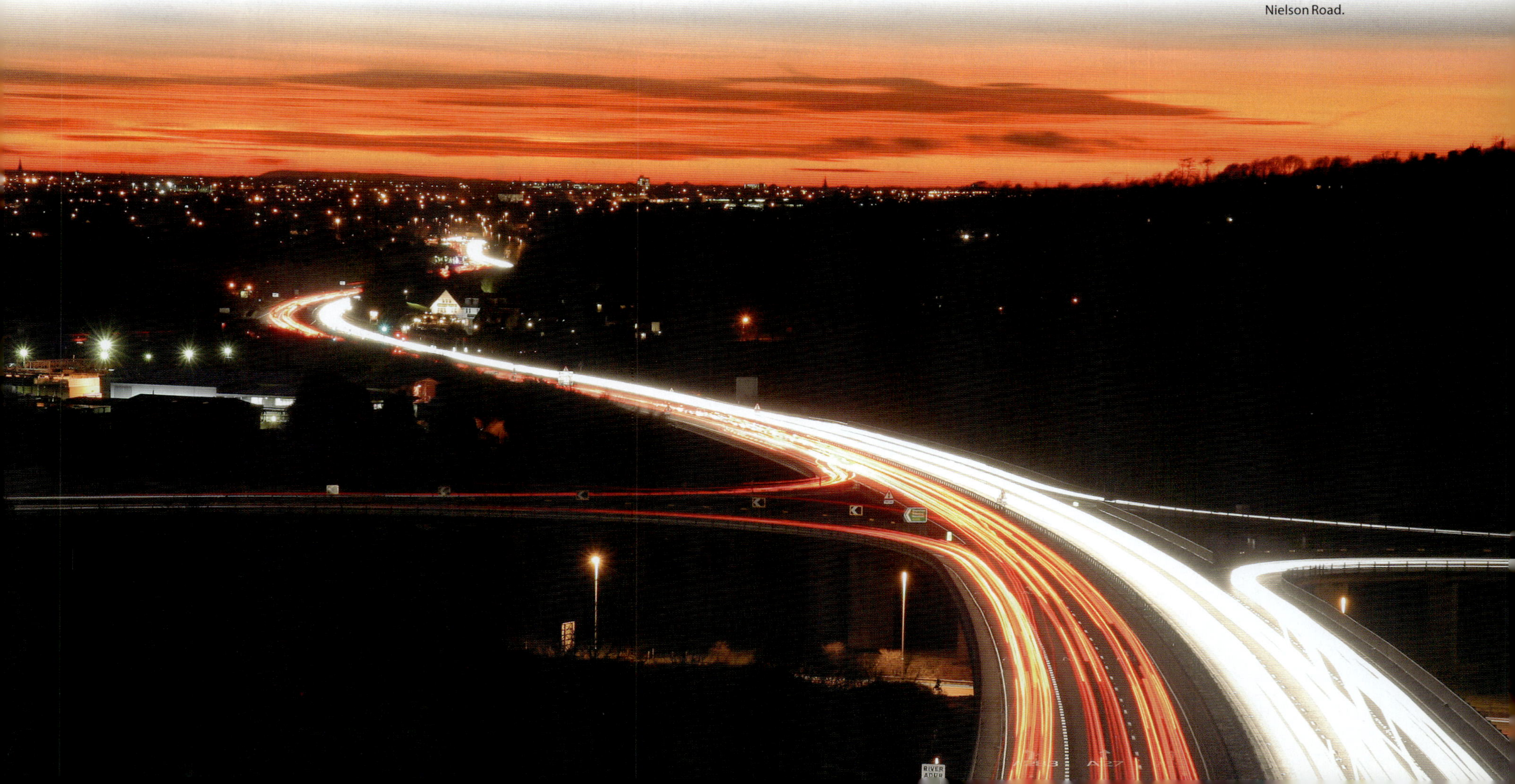

Mon 0600-2200 **Tue** 0600-2200 **Wed** 0600-2200 **Thu** 0600-2200 **Fri** 0600-1400 **Sat** 0800-1300 **Sun** Closed

Mon 0600-2200 **Tue** 0600-2200 **Wed** 0600-2200 **Thu** 0600-2200 **Fri** 0600-1400 **Sat** 0800-1300 **Sun** Closed

Ellesmere Port Truckstop

P 48 M53 J8

Portside North, Merseyton Road, Ellesmere Port, Cheshire CH65 2HQ
+44 (0)151 355 5241
morois@live.co.uk
SJ39987769 **182 F5**

At M53 J8 take A5032 towards the Docks. Take 1st left, go over level crossing and almost immediately take a hard left.

Mon 0600-2200 **Tue** 0600-2200 **Wed** 0600-2200 **Thu** 0600-2200 **Fri** 0600-2200 **Sat** 0600-1300 **Sun** Closed

Mon 0600-2200 **Tue** 0600-2200 **Wed** 0600-2200 **Thu** 0600-2200 **Fri** 0600-2200 **Sat** 0600-1300 **Sun** Closed

Gibraltar Club and Café

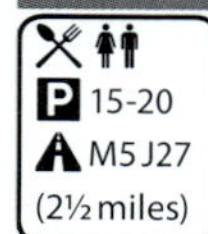

P 15-20 M5 J27 (2½ miles)

Burlescombe, Nr Tiverton, Devon EX16 7JX
+44 (0)1823 672273
ST07891563 **27 D9**

From M5 J27 take the A38 Wellington. Cafe is 2½ miles on left.

Mon 0700-1400 **Tue** 0700-1400 **Wed** 0700-1400 **Thu** 0700-1400 **Fri** 0700-1400 **Sat** 0700-1030 **Sun** Closed

Mon 0700-1400 **Tue** 0700-1400 **Wed** 0700-1400 **Thu** 0700-1400 **Fri** 0700-1400 **Sat** 0700-1030 **Sun** Closed

Hawkins Transport Village

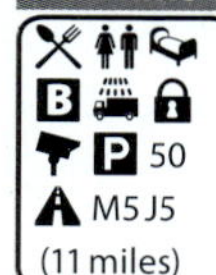

P 50 M5 J5 (11 miles)

Oak Lane/Stallings Lane, Kings Winford, West Midlands DY6 7JS
+44 (0)1384 294949
SO90069014 **133 E8**

M5 J4 take A491 Stourbridge 9 miles. Take B4175 Gornal Sedgley. Site is on left after 2 miles.

Mon 0600-1800 **Tue** 0600-1800 **Wed** 0600-1800 **Thu** 0600-1800 **Fri** 0600-1800 **Sat** 0600-1800 **Sun** Closed

Mon 0600-1800 **Tue** 0600-1800 **Wed** 0600-1800 **Thu** 0600-1800 **Fri** 0600-1800 **Sat** 0600-1800 **Sun** Closed

Heywood Distribution Park

P 200 M66 J3 (1.5 miles)

Pilsworth Road, Heywood, Manchester, Greater Manchester OL10 2TT
+44 (0)1706 368645
david.driver@fsmail.net
www.heywooddistributionpark.com/truckstop/
SD84310948 **195 F10**

From M62, M60 and M66 follow 'Heywood Distribution Park' signs leave M66 J3 (Pilsworth) and follow signs for truck stop.

Mon 24hr **Tue** 24hr **Wed** 24hr **Thu** 24hr **Fri** 24hr **Sat** 24hr **Sun** 24hr

Mon 0700-2200 **Tue** 0700-2200 **Wed** 0700-2200 **Thu** 0700-2200 **Fri** 0700-2200 **Sat** Closed **Sun** Closed

HF Veale & Sons

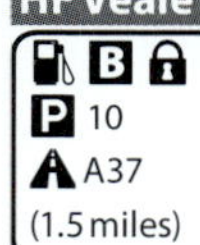

P 10 A37 (1.5 miles)

Broadway Garage, Chilcompton, Radstock, Bath, Somerset BA3 4JW
+44 (0)1761 232298
nfo@hfveale.co.uk
www.hfveale.co.uk
ST64355134 **44 C6**

Take the B3139 off the A 37 at Shepton Mallet.

Mon 0600-2100 **Tue** 0600-2100 **Wed** 0600-2100 **Thu** 0600-2100 **Fri** 0600-2100 **Sat** 0600-2100 **Sun** 0600-2100

Hideaway Truckstop

P 70 M4 J17

Oakley Acres, Dreycott Cerne, Chippenham, Wiltshire SN15 5LH
+44 (0)1249 750645
69094@compass-group.co.uk ST92507938 **62 D2**

From M4 J17 follow 'Services' sign on B4122 Sutton Benger. The site is on the right after ½ miles.

Mon 0600-2400 **Tue** 24hr **Wed** 24hr **Thu** 24hr **Fri** 24hr **Sat** 0000-1100 **Sun** Closed

Mon 0600-2400 **Tue** 24hr **Wed** 24hr **Thu** 24hr **Fri** 24hr **Sat** 0000-1100 **Sun** Closed

Hillside Café

P 10-20 A36

A36 Codford, Warminster, Wiltshire BA12 0JZ
+44 (0)1985 850712
admin@hillsidecafe.co.uk • www.hillsidecafe.co.uk
ST95624058 **46 E3**

The site is on the A36 between Codford St Peter and Upton Lovell

Mon 0600-1700 **Tue** 0600-1700 **Wed** 0600-1700 **Thu** 0600-1700 **Fri** 0600-1700 **Sat** Closed **Sun** Closed

Mon 0600-1700 **Tue** 0600-1700 **Wed** 0600-1700 **Thu** 0600-1700 **Fri** 0600-1700 **Sat** Closed **Sun** Closed

Hilltop Café

P 52 A14, J44

Lorry Park, Rougham Hill, Bury St Edmunds, Suffolk IP33 2RU
+44 (0)7860 170112
TL86926335 **125 E7**

Follow signs from the A14 near to Bury St Edmunds. The site is 200 yards from A14 J44.

Mon 24hr **Tue** 24hr **Wed** 24hr **Thu** 24hr **Fri** 24hr **Sat** 24hr **Sun** 24hr

Mon 0630-2000 **Tue** 0630-2000 **Wed** 0630-2000 **Thu** 0630-2000 **Fri** 0630-1600 **Sat** Closed **Sun** Closed

J26 M25 Truckstop

P 50 J26 M25

Skilletts Hill Farm, Honey Lane, Waltham Abbey, Essex EN9 3QU
+44 (0)7973 913114
dave-cordell@swy.com
www.junction26.org
TQ40489978 **86 E5**

M25 J26 westbound take 4th exit go under M25 and entrance is on left; eastbound take 3rd exit, go right round next roundabout, back under M25 and entrance is on left.

Mon 0600-2100 **Tue** 0600-2100 **Wed** 0600-2100 **Thu** 0600-2100 **Fri** 0600-2000 **Sat** 0600-1400 **Sun** Closed

Mon 0600-2100 **Tue** 0600-2100 **Wed** 0600-2100 **Thu** 0600-2100 **Fri** 0600-2000 **Sat** 0600-1400 **Sun** Closed

Last order for hot food 1 hour before closing

Jacks Hill Café

P 48 M1, J15A (6 miles)

Watling Street, Towcester, Northamptonshire NN12 8ET
+44 (0)1327 351350
capellacaterers@btconnect.com
www.jackshill.co.uk
SP68475007 **120 G3**

From M1 J15, take A43 Oxford, after 6 miles turn right onto A5. Site is 300yds on left.

Mon 24hr **Tue** 24hr **Wed** 24hr **Thu** 24hr **Fri** 24hr **Sat** 24hr **Sun** 24hr

Mon 0600-2130 **Tue** 0600-2130 **Wed** 0600-2130 **Thu** 0600-2130 **Fri** 0600-2030 **Sat** 0600-1430 **Sun** 0730-1400

Junction 23 Lorry Park

P 180 M1 J23

Ashby Road East, Shepshed, Loughborough, Leicestershire LE12 9BS
+44 (0)1509 507480
mail@J23.co.uk
www.j23truckstop.co.uk
SK48631832 **153 F9**

From M1 J23 take A512 Ashby-de-la-Zouch. Site is ¼ miles on right.

Mon 24hr **Tue** 24hr **Wed** 24hr **Thu** 24hr **Fri** 24hr **Sat** 24hr **Sun** 24hr

Mon 0600-2100 **Tue** 0600-2100 **Wed** 0600-2100 **Thu** 0600-2100 **Fri** 0600-2100 **Sat** 0600-1130 **Sun** Closed

Junction 29 Truckstop

P 100 M1, J29

Hardwick View Road, Holmewood Ind Est, Chesterfield, Derbyshire S42 5SA
+44 (0)1246 856536
www.junction29.co.uk
SK43716583 **170 B6**

From M1 J29 take A6175 Clay Cross. Follow 'Lorry Park' signs.

Mon 6am-24hr **Tue** 24hr **Wed** 24hr **Thu** 24hr **Fri** 24hr **Sat** 24hr-12noon **Sun** Closed

Mon 0500-2300 **Tue** 0500-2300 **Wed** 0500-2300 **Thu** 0500-2300 **Fri** 0500-2300 **Sat** 0500-1200 **Sun** Closed

Junction 38 Services

P 90 M6 J38

Old Tebay, Penrith, Cumbria CA10 3SS
+44 (0)1539 624505
feedback@westmorland.com
www.westmorland.com/node/34
NY61550488 **222 E2**

From M6 J38 take A685, at 1st roundabout take 1st exit. Site is on left after about 100 yards.

Mon 24hr **Tue** 24hr **Wed** 24hr **Thu** 24hr **Fri** 24hr **Sat** 24hr **Sun** 24hr

Mon 0630-2230 **Tue** 0630-2230 **Wed** 0630-2230 **Thu** 0630-2230 **Fri** 0630-2230 **Sat** 0630-1930 **Sun** 0630-2230

Junction 7 Business Park

P 20 M65 J7

Blackburn Road, Clayton-le-Moors, Accrington, Lancashire BB5 5JW
+44 (0)1254 398912
SD74233037 **203 G10**

M65 J7 take A6185 Clitheroe. At traffic lights turn left into business park.

Mon 24hr **Tue** 24hr **Wed** 24hr **Thu** 24hr **Fri** 24hr **Sat** 24hr **Sun** 24hr

Mon 0700-1400 **Tue** 0700-1400 **Wed** 0700-1400 **Thu** 0700-1400 **Fri** 0700-1400 **Sat** 0700-1400 **Sun** 0700-1400

Kates Cabin Café

P 36 A1

A1 Northbound, Chesterton, Peterborough, Cambridgeshire PE7 3UJ
+44 (0)1733 235587
katescabin@btconnect.com
TL13219538 **138 D2**

A1 Northbound 1 mile after the end of the A1(M)

Mon 0630-2200 **Tue** 0630-2200 **Wed** 0630-2200 **Thu** 0630-2200 **Fri** 0630-2200 **Sat** 0700-1400 **Sun** Closed

Mon 0630-2200 **Tue** 0630-2200 **Wed** 0630-2200 **Thu** 0630-2200 **Fri** 0630-2200 **Sat** 0700-1400 **Sun** Closed

Northbound only

Kingstown Truck Park

P 30-40 M6 J44

Millbrook Road, Kingstown Ind Est, Carlisle, Cumbria CA3 0EU
+44 (0)777 577 0973
neikor@aol.com
NY39335920 **239 F9**

M6 J44 take A7 Carlisle, right at 2nd lights, 2nd right, 200 yards on right

Mon 24hr **Tue** 24hr **Wed** 24hr **Thu** 24hr **Fri** 24hr **Sat** 24hr **Sun** 24hr

Mon 0700-2000 **Tue** 0700-2000 **Wed** 0700-2000 **Thu** 0700-2000 **Fri** 0700-2000 **Sat** 0700-1700 **Sun** Closed

Langrick Station Café

P 25 A16/A17 (6 miles)

B1192 Main Road, Langrick, Boston, Lincolnshire PE22 7AH
+44 (0)1205 280023
TF26454786 **174 F3**

From A16 take A1121 Boston 3½ miles, then B1192 Brothertoft 2½ miles. From A17 take A1121 Sleaford 3½ miles, then B1192 Brothertoft 2½ miles.

Mon 24hr **Tue** 24hr **Wed** 24hr **Thu** 24hr **Fri** 24hr **Sat** 24hr **Sun** 24hr

Mon 0630-0200 **Tue** 0630-0200 **Wed** 0630-0200 **Thu** 0630-0200 **Fri** 0630-0200 **Sat** 0700-1300 **Sun** 0900-1500

Lets Eat Café

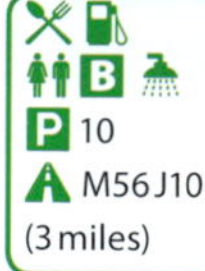

P 10 M56 J10 (3 miles)

A49 Tarporley Road, Lower Whitley, Warrington, Cheshire WA4 4EZ
+44 (0)1928 717322
SJ60607745 **183 F10**

M56 J10 take the A49 Whitchurch. Cafe is on right after 3 miles.

Mon 0800-1800 **Tue** 0730-1800 **Wed** 0730-1800 **Thu** 0730-1800 **Fri** 0730-1800 **Sat** 0730-1400 **Sun** 0830-1400

Mon 0800-1800 **Tue** 0730-1800 **Wed** 0730-1800 **Thu** 0730-1800 **Fri** 0730-1800 **Sat** 0730-1400 **Sun** 0830-1400

Lincoln Farm Café

P 200 M42, J6 (3 miles)

A452 Kenilworth Road, Hampton in Arden, Solihull, Warwickshire B92 0LS
+44 (0)1675 442301
SP21928014 **134 G4**

From M42 J6, take A45 B'ham (E) 1 mile, then A452 Leamington 2 miles. Site is on right.

Mon 0400-2400 **Tue** 0400-2400 **Wed** 0400-2400 **Thu** 0400-2400 **Fri** 0400-2400 **Sat** Closed **Sun** Closed

Mon 0600-2400 **Tue** 0600-2400 **Wed** 0600-2400 **Thu** 0600-2400 **Fri** 0600-2400 **Sat** Closed **Sun** Closed

Londonderry Lodge

P 24 A1

Londonderry, Northallerton, North Yorkshire DL7 9ND
+44 (0)1677 422143
claredalton@btconnect.com
SE30248779 **214 B6**

Just off the A1 about 12 miles south of Scotch Corner. Signposted.

Mon 24hr **Tue** 24hr **Wed** 24hr **Thu** 24hr **Fri** 24hr **Sat** 24hr **Sun** 24hr

Mon 0600-2300 **Tue** 0500-2300 **Wed** 0500-2300 **Thu** 0500-2300 **Fri** 0500-2230 **Sat** 0500-1200 **Sun** Closed

Merry Chest Café

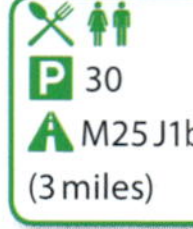

P 30 M25 J1b (3 miles)

Watling Street, Bean, Dartford, Kent DA2 8AH
+44 (0)1474 832371
TQ59297277 **68 E5**

From M25 J1b (via A225 Dartford slip) take A296 Bean 3 miles.

Mon 0630-1500 **Tue** 0630-1500 **Wed** 0630-1500 **Thu** 0630-1500 **Fri** 0630-1500 **Sat** 0630-1200 **Sun** Closed

Mon 0630-1500 **Tue** 0630-1500 **Wed** 0630-1500 **Thu** 0630-1500 **Fri** 0630-1500 **Sat** 0630-1200 **Sun** Closed

Midway Truckstop

P 35 M54

Prees Heath, Whitchurch, Shropshire SY13 4JT
+44 (0)1948 663160
info@midwaytruckstop.co.uk
www.midwaytruckstop.co.uk
SJ55663804 **149 B11**

The site is off the roundabout at the the southern intersection of the A41 the A49 at Whitchurch.

Mon 24hr **Tue** 24hr **Wed** 24hr **Thu** 24hr **Fri** 24hr **Sat** 24hr **Sun** 24hr

Mon 0545-2000 **Tue** 0545-2000 **Wed** 0545-2000 **Thu** 0545-2000 **Fri** 0545-2000 **Sat** 0545-1800 **Sun** 0730-1400

Moto Cherwell Valley

P 747 M40 J10

M40 J10, Northampton Road, Ardley, Bicester, Oxfordshire OX27 7RD
+44 (0)1869 346060
www.moto-way.co.uk
SP55272817 **101 F11**

At M40 J10, signposted from M40 and A43

Mon 24hr **Tue** 24hr **Wed** 24hr **Thu** 24hr **Fri** 24hr **Sat** 24hr **Sun** 24hr

Mon 24hr **Tue** 24hr **Wed** 24hr **Thu** 24hr **Fri** 24hr **Sat** 24hr **Sun** 24hr

Moto Doncaster North

P 30 J5 M18/M180

J5 M18/M180, Doncaster, South Yorkshire DN8 5GS
+44 (0)1302 847700
www.moto-way.co.uk
SE66921112 **199 E7**

Site is at M18 J5, signposted from M18 and M180.

Mon 24hr **Tue** 24hr **Wed** 24hr **Thu** 24hr **Fri** 24hr **Sat** 24hr **Sun** 24hr

Mon 24hr **Tue** 24hr **Wed** 24hr **Thu** 24hr **Fri** 24hr **Sat** 24hr **Sun** 24hr

Moto Stafford Northbound

P 80 M6 J14 (4 miles)

M6 Junction 14/15, Stone, Staffordshire ST15 0EU
+44 (0)1785 810504
www.moto-way.co.uk
SJ88593184 **151 C7**

About 4 miles north of M6 J14

Mon 24hr **Tue** 24hr **Wed** 24hr **Thu** 24hr **Fri** 24hr **Sat** 24hr **Sun** 24hr

Mon 24hr **Tue** 24hr **Wed** 24hr **Thu** 24hr **Fri** 24hr **Sat** 24hr **Sun** 24hr

Moto Medway

137 M2 J4/5

M2 J4/5, Rainhorn, Gillingham, Kent ME8 8PQ
+44 (0)1634 236900
www.moto-way.co.uk
TQ81746347 **69 G10**

On the M2 just east of J4.

Mon 24hr **Tue** 24hr **Wed** 24hr **Thu** 24hr **Fri** 24hr **Sat** 24hr **Sun** 24hr

Mon 24hr **Tue** 24hr **Wed** 24hr **Thu** 24hr **Fri** 24hr **Sat** 24hr **Sun** 24hr

Moto Reading

P 55 M4 J12

M4 Junction 11/12, Burghfield, Reading, Berkshire RG30 3UQ
+44 (0)1189 566966
www.moto-way.co.uk
SU67166976 **65 F7**

Between M4 J11 and J12

Mon 24hr **Tue** 24hr **Wed** 24hr **Thu** 24hr **Fri** 24hr **Sat** 24hr **Sun** 24hr

Mon 24hr **Tue** 24hr **Wed** 24hr **Thu** 24hr **Fri** 24hr **Sat** 24hr **Sun** 24hr

Moto Washington

P 56 A1(M) J64

A1(M), Portobello, Birtley, County Durham DH3 2SJ
+44 (0)191 4103436
www.moto-way.co.uk
NZ28375506 **243 F7**

A1(M) to the North of J64

Mon 24hr **Tue** 24hr **Wed** 24hr **Thu** 24hr **Fri** 24hr **Sat** 24hr **Sun** 24hr

Mon 24hr **Tue** 24hr **Wed** 24hr **Thu** 24hr **Fri** 24hr **Sat** 24hr **Sun** 24hr

Moto Wetherby

P 60 A1(M) J46

A1(M), J46 Kirk Deighton, Wetherby, West Yorkshire LS22 5GT
+44 (0)1937 545 080
www.moto-way.co.uk • SE41425013 **206 C4** A1(M) J46, signposted from A1 and A1(M),

Mon 24hr **Tue** 24hr **Wed** 24hr **Thu** 24hr **Fri** 24hr **Sat** 24hr **Sun** 24hr

Mon 24hr **Tue** 24hr **Wed** 24hr **Thu** 24hr **Fri** 24hr **Sat** 24hr **Sun** 24hr

Necton Diner

P 100 A47

Norwich Road, Necton, Norfolk PE37 8DQ
+44 (0)1760 724180
TF87701010 **159 G7**

On the A47 at the northern edge of Necton.

Mon 0800-1830 **Tue** 0800-1830 **Wed** 0800-1830 **Thu** 0800-1830 **Fri** 0800-1500 **Sat** 0800-1400 **Sun** Closed

Nell's Café

P 30 A2

Marling Cross, A2 Gravesend East, Kent DA12 5UD
+44 (0)1474 362457
lawrence.ys@nell's-cafe.com
www.nells-cafe.com
TQ66127049 **69 E7**

Site is on the A2, 2 miles west of the M2. Take Gravesend (E) exit and follow lorry park signs.

Mon 24hr **Tue** 24hr **Wed** 24hr **Thu** 24hr **Fri** 24hr **Sat** 24hr **Sun** 24hr

Mon 0700-1900 **Tue** 0700-1900 **Wed** 0700-1900 **Thu** 0700-1900 **Fri** 0700-1600 **Sat** 0700-1600 **Sun** 0700-1700

Nightowl Truckstop Carlisle

P 200 M6 J44

Parkhouse Road, Kingstown Industrial Estate, Carlisle, Cumbria CA3 0JR
+44 (0)1288 534192
www.nt-truckstops.com • NY39155969 **239 F9** • M6 J44 take A75 Carlisle. After 150 yards, turn right at traffic lights, cross over the mini-roundabout. The site is 100 yards on the left.

Mon 24hr **Tue** 24hr **Wed** 24hr **Thu** 24hr **Fri** 24hr **Sat** 24hr **Sun** 24hr

Mon 0600-2300 **Tue** 0500-2300 **Wed** 0500-2300 **Thu** 0500-2300 **Fri** 0500-2300 **Sat** 0500-1200 **Sun** 1600-2200

Nightowl Truckstop Rugby

P 240 M1 J18 (2½ miles) or M6 J1 (3½ miles)

(A5) Watling Street, Clifton upon Dunsmore, Rugby, Warwicks, CV23 0AE
+44 (0)1788 535115
www.nt-truckstops.com • SP55317628 **119 B11** From M1 J18, take A5 northbound. The site is on the right after 2½ miles. From M6 J1, take A426 Rugby. After 1 mile take the A5 London. Site is on left after 2½ miles.

Mon 24hr **Tue** 24hr **Wed** 24hr **Thu** 24hr **Fri** 24hr **Sat** Until 1300 **Sun** From 1500

Mon 0600-2230 **Tue** 0600-2230 **Wed** 0600-2230 **Thu** 0600-2230 **Fri** 0600-2130 **Sat** 0700-2315 **Sun** 1600-2130

Nunney Catch Café

A361, Nunney Catch, Nr Frome, Somerset BA11 4NZ
+44 (0)1373 836331
ST73704476 45 E8
From junction of A361 and A359, follow sign for Green Pits Lane directly off roundabout.
Mon 0700-1930 Tue 0700-1930 Wed 0700-1930 Thu 0700-1930 Fri 0700-1600 Sat 0700-1300 Sun Closed
Mon 0730-1900 Tue 0730-1900 Wed 0730-1900 Thu 0730-1900 Fri 0700-1600 Sat 0700-1300 Sun Closed

Oakdene Café

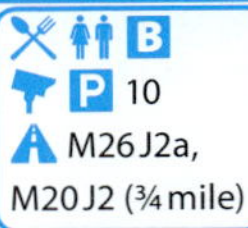

London Road, Wrotham, Sevenoaks, Kent TN15 7RR
+44 (0)1732 884873
janetjevons@yahoo.co.uk
www.oakdenecafe.co.uk
TQ62725869 52 B6
The site is on the A20. From M20 J2 take A20 Wrotham; site is on left after ¾ miles. From M26 J2a take A20 Wrotham. Site is signposted from slip road and is on right after 300 yards.
Mon 0600-1830 Tue 0600-1830 Wed 0600-2200 Thu 0600-1830 Fri 0600-1500 Sat 0600-1500 Sun 0600-1500
Mon 0600-1830 Tue 0600-1830 Wed 0600-2200 Thu 0600-1830 Fri 0600-1500 Sat 0600-1500 Sun 0600-1500

Orwell Crossing Lorry Park

A14 Eastbound, Nacton, Ipswich, Suffolk IP10 0DD
+44 (0)1473 659140
reception@orwellcrossing.com
www.orwellcrossing.com
TM21484120 108 C4
Eastbound on A14, cross Orwell Bridge and take 2nd slip road, site is on the left. Westbound, turn round at J27 and take 1st slip.
Mon 24hr Tue 24hr Wed 24hr Thu 24hr Fri 24hr Sat 24hr Sun 24hr
Mon 24hr Tue 24hr Wed 24hr Thu 24hr Fri 24hr Sat 24hr Sun 24hr

Penrith Truckstop

Penrith Industrial Estate, Penrith, Cumbria CA11 9EH
+44 (0)1768 866995
info@awjtruckstop.co.uk
www.awjtruckstop.co.uk
NY50682954 230 F6
From M6 J40, follow the signs for the truckstop.
Mon 0600-1300 Tue 0600-1300 Wed 0600-1300 Thu 0600-1300 Fri 0600-1300 Sat 0600-1300 Sun 1200-2000
Mon 0600-1300 Tue 0600-1300 Wed 0600-1300 Thu 0600-1300 Fri 0600-1300 Sat 0600-1300 Sun 1200-2000

Pie Stop Café

Pennygillam Industrial Estate, Pennygillam Way, Launceston, Cornwall PL15 7ED
+44 (0)7593 579789
SX32138365 12 E2
From A30 to Bodmin, take A388 (B3254) Launceston, take a sharp right into Western Road. Pennygillam Way is the 4th left. Site is in cash & carry car park.
Mon 0730-1500 Tue 0730-1500 Wed 0730-1500 Thu 0730-1500 Fri 0730-1500 Sat 0730-1200 Sun Closed
Mon 0730-1500 Tue 0730-1500 Wed 0730-1500 Thu 0730-1500 Fri 0730-1500 Sat 0730-1200 Sun Closed

PJ's Transport Café

Sudbury Services, Litchfield Road, Sudbury, Derbyshire DE6 5GX
+44 (0)1283 820669
SK16183025 152 C3
From A5, about 5 miles east of Uttoxeter take A515 Lichfield. Site is on left after 1 mile.
Mon 0700-2200 Tue 0700-2200 Wed 0700-2200 Thu 0700-2200 Fri 0700-1800 Sat Closed Sun Closed
Mon 0700-2200 Tue 0700-2200 Wed 0700-2200 Thu 0700-2200 Fri 0700-1800 Sat Closed Sun Closed

Poplar 2000 Services – Lymm Truckstop

Cliffe Lane, Lymm, Cheshire WA13 0SP
+44 (0)1925 757777
www.moto-way.co.uk
SJ66578481 183 E11
From M6 J20 or M56 J9, follow signs for services.
Mon 24hr Tue 24hr Wed 24hr Thu 24hr Fri 24hr Sat 24hr Sun 24hr
Mon 24hr Tue 24hr Wed 24hr Thu 24hr Fri 24hr Sat 24hr Sun 24hr

Portsmouth truckstop

Railway Triangle, Walton Road, Farlington, Portsmouth PO6 1UJ
+44 (0)23 9237 6000
info@portsmouthtruckstop.co.uk
SU67070451 33 G11
From A27 follow signs for Farlington Services.1st left into Walton Road. Bear left, take first right. Site is just before the next bend.
Mon 24hr Tue 24hr Wed 24hr Thu 24hr Fri 24hr Sat 24hr Sun 24hr
Mon 0700-2030 Tue 0700-2030 Wed 0700-2030 Thu 0700-2030 Fri 0700-1500 Sat 0800-1200 Sun Closed

Priory Park Lorry Park Truck Stop

Henry Boot Way, Priory Way, Hull HU4 7DY
+44 (0)1482 331895
TA04472615 200 B4
From A63 eastbound, pass Humber Bridge and take next exit, follow signs from here. Westbound, follow signs from A63.
Mon 24hr Tue 24hr Wed 24hr Thu 24hr Fri 24hr Sat 24hr Sun 24hr

Quernhow A1 Café and Truckstop

Great North Road, Nr Sinderby, Thirsk, North Yorkshire YO7 4LG
+44 (0)7795 814360
contact@a1cafe.co.uk
www.a1cafe.co.uk
SE33798054 214 C6
On the A1 5 miles north of the end of the A1(M)
Mon 24hr Tue 24hr Wed 24hr Thu 24hr Fri 24hr Sat 24hr Sun Closed
Mon 0600-24hr Tue 24hr Wed 24hr Thu 24hr Fri 24hr Sat 0000-1400 Sun Closed
Northbound only

Ranch Café & Cattle Market Lorry Park

Cattle Market, Old Great North Rd, Newark, Nottinghamshire NG24 1BL
+44 (0)1636 611198
SK79525457 172 E3
From A1 Carlton-on-Trent, take B1164 east, 1st left Great North Rd 1½ m on right.
Mon 24hr Tue 24hr Wed 24hr Thu 24hr Fri 24hr Sat 24hr Sun 24hr
Mon 0800-2100 Tue 0600-2100 Wed 0600-2100 Thu 0600-2100 Fri 0600-2000 Sat Closed Sun Closed

Red Lion Café & Truckstop

120
J16, M1

Weedon Road (A45), Upper Heyford, Northampton, Northamptonshire NN7 4DE
+44 (0)1604 831914
SP68075976 120 F3
From M1 J16, take A45 Northampton. Café in layby of westbound carriageway after 500 yards.
Mon 0700-2330 Tue 0700-2330 Wed 0700-2330 Thu 0700-2330 Fri 0700-2330 Sat 0700-2330 Sun 0700-2330
Mon 0600-2300 Tue 0600-2300 Wed 0600-2300 Thu 0600-2300 Fri 0600-2300 Sat 0600-2300 Sun 0600-2300

Redbeck Motel

Doncaster Road, Crofton, Wakefield, North Yorkshire WF4 1RR
+44 (0)1924862730
www.redbecksite11.com/index.html
SE36611882 197 D11
The site is on the A638, 2 miles east of Wakefield city centre.
Mon 24hr Tue 24hr Wed 24hr Thu 24hr Fri 24hr Sat 24hr Sun 24hr
Mon 24hr Tue 24hr Wed 24hr Thu 24hr Fri 24hr Sat 24hr Sun 24hr

Scoffers Café

A45 Eastbound, Nene Valley Way, Northampton, Northamptonshire NN3 5LU
+44 (0)1604 784500
SP82506258 120 E6
From M1 J15, take A45 eastbound towards Wellingborough. Site on left after Billing roundabout next to Esso.
Mon 24hr Tue 24hr Wed 24hr Thu 24hr Fri 24hr Sat 24hr Sun 24hr
Mon 0730-1530 Tue 0730-1530 Wed 0730-1530 Thu 0730-1530 Fri 0700-2000 Sat 0700-1700 Sun 0700-1700

Smokey Joe's Café

Blackwater, Nr Redruth, Cornwall TR16 5BJ
+44 (0)1209 821810
SW72584486 4 G4
From A30 eastbound take A3047 Scorrier. At roundabout take unmarked road parralled to A30 for ½ miles. Westbound, take A3047 (also marked A307, B397, B398). At first junction turn right and cross over the A30. Bear round to right and at roundabout take unmarked road parallel to A30 for ½ miles.
Mon 0700-2200 Tue 0700-2200 Wed 0700-2200 Thu 0700-2200 Fri 0700-2200 Sat 0700-2200 Sun 0700-2200
Mon 0700-2200 Tue 0700-2200 Wed 0700-2200 Thu 0700-2200 Fri 0700-2200 Sat 0700-2200 Sun 0700-2200

South Mimms Truckstop

St Albans Road, South Mimms, Potters Bar, Hertfordshire EN6 6NE
+44 (0)1707 649998
mimms.truckstop@welcomebreak.co.uk
www.welcomebreak.co.uk
TL22810045 86 E2
Leave the M25 at Junction 23 and follow signs for 'lorry services'
Mon 24hr Tue 24hr Wed 24hr Thu 24hr Fri 24hr Sat 24hr Sun 24hr
Mon 24hr Tue 24hr Wed 24hr Thu 24hr Fri 24hr Sat 24hr Sun 24hr

Square Deal Café

Bath Road, Knowl Hill, Nr Reading, Berkshire RG10 9UR
+44 (0)1628 822426
info@squaredealcafe.com
www.squaredealcafe.com
SU82287948 65 D10
2½ miles east of Twyford on the A4.
Mon 0600-1500 Tue 0600-1500 Wed 0600-1500 Thu 0600-1500 Fri 0600-1500 Sat Closed Sun Closed
Mon 0700-1500 Tue 0700-1500 Wed 0700-1500 Thu 0700-1500 Fri 0700-1500 Sat Closed Sun Closed

Standeford Farm Café

Streamway Hooks, Stafford Road, Standeford, West Midlands WV10 7BN
+44 (0)1902 790389
SJ91160797 133 B8
M54 J2, take A449 Stafford for 2½ miles
Mon 0500-2000 Tue 0500-2000 Wed 0500-2000 Thu 0500-2000 Fri 0500-1400 Sat Closed Sun Closed
Mon 0500-2000 Tue 0500-2000 Wed 0500-2000 Thu 0500-2000 Fri 0500-1400 Sat Closed Sun Closed

Stibbington Diner

2 Old North Road, Stibbington, Peterborough PE8 6LR
+44 (0)1780 782891
stibbycafe@btconnect.com
TL08769829 137 D11
The site is adjacent to the A1 southbound south of Wansford. Follow Nene Valley Railway then services signs from A1.
Mon 24hr Tue 24hr Wed 24hr Thu 24hr Fri 24hr Sat 24hr until 2pm Sun Open from 10pm
Mon 24hr Tue 24hr Wed 24hr Thu 24hr Fri 24hr Sat 0000-1400 Sun 2200-2400

Sue's Pitstop Café

Unit 4 Ledston Luck Enterprise Park, Ridge Road, Kippax, Leeds, West Yorkshire LS25 7BF
+44 (0)113 2863307
SE42953073 206 G4
From M1 J4,6 take A63 Selby. After 3½ miles, take the A656 Castleford. Turn left after ½ miles. From A1(M) J42 or A1, take A63 Leeds. After 1½ miles turn left onto A656 Castleford. Turn left after ½ miles.
Mon 0700-1500 Tue 0700-1500 Wed 0700-1500 Thu 0700-1500 Fri 0700-1500 Sat 0700-1100 Sun Closed
Mon 0700-1500 Tue 0700-1500 Wed 0700-1500 Thu 0700-1500 Fri 0700-1500 Sat 0700-1100 Sun Closed

Super Sausage Café

St Andrews Road, Northampton, Northamptonshire NN1 2SD
+44 (0)1604 636099
SP74876099 120 E4
M1 J6 take A4500 exit, take 1st exit at roundabout (not marked here but A45/A4500) towards town centre. After 6 miles pass The Saints rugby ground on the right, take 1st left (Spencer Bridge Road). About 150 yards after crossing bridge, turn right into St Andrew's Road. Site is immediately on right.
Mon 0600-2100 Tue 0600-2100 Wed 0600-2100 Thu 0600-2100 Fri 0600-1930 Sat 0600-1500 Sun 0830-1300
Mon 0600-2100 Tue 0600-2100 Wed 0600-2100 Thu 0600-2100 Fri 0600-1930 Sat 0600-1500 Sun 0830-1300

Super Sausage Café

A5 Watling Street, Potterspury, Towcester, Northamptonshire NN12 7QD
+44 (0)1908 542964
gail@supersausagecafe.co.uk
www.supersausagecafe.co.uk
SP74944376 102 C4
On A5 between Stony Stratford and Towcester, 1km north of Pottersbury
Mon 0700-1700 Tue 0700-1700 Wed 0700-1700 Thu 0700-1700 Fri 0700-1700 Sat 0800-1500 Sun 0800-1500
Mon 0700-1700 Tue 0700-1700 Wed 0700-1700 Thu 0700-1700 Fri 0700-1700 Sat 0800-1500 Sun 1800-1500
Hours may vary during events at Silverstone.

Swindon Truckstop

A420, Oxford Road, Swindon, Wiltshire SN3 4ER
+44 (0)1793 824812
enquiries@swindontruckstop.co.uk
www.swindontruckstop.co.uk
SU18808642 63 B7
From M4 J15 take A419 north then the A420. Use the Sainsbury service road to enter site.
Mon 0600-2200 Tue 0600-2200 Wed 0600-2200 Thu 0600-2200 Fri 0600-2200 Sat Closed Sun Closed
Mon 0600-2100 Tue 0600-2100 Wed 0600-2100 Thu 0600-2100 Fri 0600-2100 Sat Closed Sun Closed

The Avon Lodge

Third Way, Avonmouth, Bristol, Avon BS11 9YP
+44 (0)117 9827706
avonlodge@btconnect.com
ST52227868 60 D4
From M5 J18, head towards Avonmouth. Third Way is reached via Avonmouth Way.
Mon 0700-2300 Tue 0600-2300 Wed 0600-2300 Thu 0600-2300 Fri 0600-2300 Sat 0600-1100 Sun Closed
Mon 0700-2300 Tue 0700-2300 Wed 0700-2300 Thu 0700-2300 Fri 0700-2300 Sat 0600-1100 Sun Closed

The Cabin Café

Crawley Road, Faygate, Nr Horsham, Sussex RH12 4SE
+44 (0)1293 851575
TQ22953436 51 G8
From M23 J11 take A264 Faygate. Site is on A264 after 2½m, near Faygate roundabout, on left behind Travis Perkins.
Mon 0600-2000 Tue 0600-2000 Wed 0600-2000 Thu 0600-2000 Fri 0600-1400 Sat 0700-1200 Sun Closed
Mon 0600-2000 Tue 0600-2000 Wed 0600-2000 Thu 0600-2000 Fri 0600-1300 Sat 0700-1200 Sun

The Old Willoughby Hedge Café

20-30
A303

A303 Layby, West Knoyle, Salisbury, Wiltshire BA12 6AQ
+44 (0)1963 371099 or 01747 830803 (café)
lcvreg@aol.com ST86313350 45 G11
The site is on the westbound side of the A303, 3 miles to the east of Mere.
Mon 0800-1800 Tue 0800-1800 Wed 0800-1800 Thu 0800-1800 Fri 0800-1800 Sat 0800-1700 Sun 1000-1600
Mon 0800-1800 Tue 0800-1800 Wed 0800-1800 Thu 0800-1800 Fri 0800-1800 Sat 0800-1700 Sun 1000-1600

The Salt Box Café

30
A50
(1 mile)

No.2, Derby Road, Hatton, Derbyshire DE65 5PT
+44 (0)1283 813189
SK21733080 152 C4
From A50, take A511 for Burton-on-Trent. Site is on right after 1 mile. HGVs should access from A511, not main entrance.
Mon 7000-2100 Tue 7000-2100 Wed 7000-2100 Thu 7000-2100 Fri 7000-2100 Sat 7000-1400 Sun Closed Toilets 24hr

The Stockyard Truckstop

250
M18 J1
(1 mile)

Hellaby Lane, Hellaby, Rotherham, South Yorkshire S66 8HN
+44 (0)1709 700200/730083
SK50549331 187 C8
From M18 J1 take A631 Maltby. Left at roundabout, follow for 1 mile site on left.
Mon 0700-2100 Tue 0700-2100 Wed 0700-2100 Thu 0700-2100 Fri 0700-2100 Sat 0600-2100 Sun 0500-1600
Mon 0700-2100 Tue 0700-2100 Wed 0700-2100 Thu 0700-2100 Fri 0700-2100 Sat 0600-2100 Sun 0500-1600
Food limited Saturday evenings

The Truckstop Café (Crawley Crossing Bunker Stop)

Bedford Road, Husborne Crawley, Bedfordshire MK43 0UT
+44 (0)1908 281086
SP95823696 103 D9
From M1 J13 take A4012 Woburn. Site is 300 yards on left.
Mon 24 hours Tue 24 hours Wed 24 hours Thu 24 hours Fri 24 hours Sat 24 hours Sun 24 hours
Mon 0700-2200 Tue 0600-2200 Wed 0600-2200 Thu 0600-2200 Fri 0600-2000 Sat 0700-1200 Sun Closed

Titan Truckstop

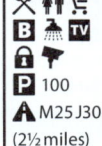

100
M25 J30
(2½ miles)

Stoneness Road, Thurrock, Essex RM20 3AG
+44 (0)1708 258500
ttp@icgl.co.uk
TQ58907729 68 D5
From M25 J30 take A13 Thurrock (Lakeside) W Thurrock. At next junction, take A126 Thurrock (Lakeside) W Thurrock. Continue straight across all junctions. When A126 goes left, carry straight on. Titan Truckstops is at next corner.

Mon 24hr Tue 24hr Wed 24hr Thu 24hr Fri 24hr Sat 24hr Sun 24hr

Mon 0600-1400, 1700-2000 Tue 0600-1400, 1700-2000 Wed 0600-1400, 1700-2000 Thu 0600-1400, 1700-2000 Fri 0600-1400 Sat Closed Sun closed

Truckers Rest

A5 Watling Street, Cannock, Four Crosses, Staffordshire WS11 1SF
+44 (0)1543 469183
SJ95740933 133 B9
From M6 J12 take A5 Cannock for 1½ miles, site is on the right.

Mon 0600-2330 Tue 0600-2330 Wed 0600-2330 Thu 0600-2330 Fri 0600-2330 Sat 0600-1530 Sun 0700-1730

Mon 0600-2330 Tue 0600-2330 Wed 0600-2330 Thu 0600-2330 Fri 0600-2330 Sat 0600-1530 Sun 0700-1730

Truckhaven Carnforth

Scotland Road, Warton, Carnforth, Lancashire LA5 9RQ
+44 (0)1524 736699
www.truckhavencarnforth.co.uk
SD50777171 211 E10
M6 J35 take A601(M), then A6 Carnforth. Site is 600 yards on left.

Mon 24hr Tue 24hr Wed 24hr Thu 24hr Fri 24hr Sat 24hr Sun 24hr

Mon 24hr Tue 24hr Wed 24hr Thu 24hr Fri 24hr Sat 24hr Sun 24hr

Ulceby Truckstop

Ulceby Road, Ulceby, Immingham, Lincolnshire DN40 3JB
+44 (0)1469 540606
www.truck-stop.co.uk
TA13711546 200 D6 At junction of A160 and A1077 Ulceby Road

Mon 24hr Tue 24hr Wed 24hr Thu 24hr Fri 24hr Sat 24hr Sun 24hr

Mon 24hr Tue 24hr Wed 24hr Thu 24hr Fri 24hr Sat 24hr Sun 24hr

Watling Street Truckstop

London Road, Flamstead, St Albans, Hertfordshire AL3 8HA
+44 (0)1582 840270
TL08531508 85 B9
From M1 J9, take A5 Dunstable. The site on the right after 200yds.

Mon 0600-2200 Tue 0600-2200 Wed 0600-2200 Thu 0600-2200 Fri 0600-2000 Sat 0700-1300 Sun Closed

Mon 0600-2200 Tue 0600-2200 Wed 0600-2200 Thu 0600-2200 Fri 0600-2200 Sat 0700-1300 Sun Closed

Whitley Bridge Pallets and Truckstop

Unit 6 The Malting Industrial Estate, Whitley Bridge, Goole, East Yorkshire DN14 0HH
+44 (0)1977 662881
SE55442285 198 C5
At M62 J34, follow sign for 'local traffic', turn left to ind estate before level crossing

Mon 24hr Tue 24hr Wed 24hr Thu 24hr Fri 24hr Sat 24hr Sun 24hr

York Lorry Park

York Auction Centre, Murton, York, North Yorkshire YO19 5GF
+44 (0)1904 489731
SE65175219 207 C9 At Junction of A64, A1070 and A166, take latter, then take first left (Murton Lane) and follow signs to lorry park. HGVs should not try direct approach from A64 southbound; weight limit.

Mon 1600-0900 Tue 1600-0900 Wed 1600-0900 Thu 1600-0900 Fri 1600-0900 Sat Closed Sun Closed

Mon 0700-0900 then 1600-2100 Tue 0700-0900 then 1600-2100 Wed 0700-0900 then 1600-2100 Thu 0700-0900 then 1600-2100 Fri 0700-0900 then 1600-2100 Sat Closed Sun Closed

Scotland

Ashgrove Filling Station and Restaurant

Cairnie, Aberdeenshire. AB54 4TL
01466 760223
NJ48724416 302 E4
On the A96 about 2 miles north-west of Huntly.

Mon 24hr Tue 24hr Wed 24hr Thu 24hr Fri 24hr Sat 24hr Sun 24hr

Cedar Cafe

Granthouse, Berwickshire TD11 3RP
01361 850371
NT81746571 272 B6
In a layby off the southbound carriageway of the A1 ½ miles south of Grantshouse.

Mon 0800-2000 Tue 0800-2000 Wed 0800-2000 Thu 0800-2000 Fri 0800-2000 Sat 0800-1700 Sun 0800-2000
Parking is in the adjacent layby.

Chef's Grill

Perth Road, Newtonmore, Invernesssshire PH20 1BB
01540 673702
NN70999853 291 D9
From the A9 16 miles west of Aviemore, take the B9150 Newtonmore (Perth Road). Site is on the left just before you reach the village.

Mon 0700-2200 Tue 0630-2200 Wed 0630-2200 Thu 0630-2200 Fri 0630-2200 Sat 0700-2000 Sun 0800-2000

Eardley International

Old Burnswark Station, Ecclefechan, Lockerbie, Dumfriesshire DG11 3JD
01576 300500
www.eardleyinternational.com
NY18477526 238 B5
From A74(M) J19, take the B7076 Ecclefechan, then follow the signs.

Mon 24hr Tue 24hr Wed 24hr Thu 24hr Fri 24hr Sat 24hr Sun 24hr

Mon 0600-2100 Tue 0600-2100 Wed 0600-2100 Thu 0600-2100 Fri 0600-2100 Sat 0600-2100 Sun 0600-2100
Last food orders taken 2030. To order food in advance, call 07841 996 607.

Hungry Trucker Transport Cafe

30
A74(M) J14

20, Carlisle Rd, Crawford, Biggar, S Lanarkshire ML12 6TW
01864 502227
NS94842064 259 E10
At A74(M) J14 take A702 Crawford. Southbound take 1st exit at roundabout (Carlisle Road). Site is on right after ¾ miles. Northbound after 1½ miles on A702, turn right into Carlisle Road. Site is on left after ½ mile.

Mon 0700-late Tue 0600-late Wed 0600-late Thu 0600-late Fri 0600-late Sat 0700-1300 Sun 1300-late

Mon 0700-late Tue 0600-late Wed 0600-late Thu 0600-late Fri 0600-late Sat 0700-1300 Sun 1300-late

Horse Shoe Cafe

20
A90 Inchture Junction (1 mile)

Abernyte Road, Inchture, Perthshire PH14 9RS
01828 686283 • NO27952910 286 E6
From the Inchture Junction of the A90 between Perth and Dundee, take B983 Abernyte for 1 mile. Take 1st exit at roundabout; site is on left.

Mon 0800-2100 Tue 0800-2100 Wed 0800-2100 Thu 0800-2100 Fri 0800-2100 Sat 0800-1500 Sun 0800-1900

Lockerbie Lorry Park / Truck Stop

nr Dinwoodie Mains, Johnstonebridge, Lockerbie, Dumfries and Galloway, DG11 2SL
7768654663
NY10468956 248 F4
Northbound A74(M) J17 take B7068 Lockerbie. Cross over the motorway. At the roundabout, take the 1st exit and follow the signs. Southbound, take the B7068 Lockerbie, turn left. At the roundabout take the 1st exit and follow the signs.

Mon 24hr Tue 24hr Wed 24hr Thu 24hr Fri 24hr Sat 24hr Sun 24hr

Mon 0600-2200 Tue 0600-2200 Wed 0600-2200 Thu 0600-2200 Fri 0600-2200 Sat 0700-1200 Sun Closed

Moto Stirling

Pirnhall, Stirling FK7 8EU
01786 813614
www.moto-way.com
NS80388866 278 D6
From M9/M80 J9 follow signs to services.

Mon 24hr Tue 24hr Wed 24hr Thu 24hr Fri 24hr Sat 24hr Sun 24hr

Mon 24hr Tue 24hr Wed 24hr Thu 24hr Fri 24hr Sat 24hr Sun 24hr

Moto Kinross

Turfhills Tourist Centre, Kinross, Perth and Kinross KY13 7NQ
01577 863123
www.moto-way.com
NO10770276 286 G5
Junction 6 M90

Mon 24hr Tue 24hr Wed 24hr Thu 24hr Fri 24hr Sat 24hr Sun 24hr

Mon 24hr Tue 24hr Wed 24hr Thu 24hr Fri 24hr Sat 24hr Sun 24hr

Motorgrill Ballinluig

Ballinluig Services, Pitlochry, Perthshire PH9 0LG
01796 482212
NN97715266 286 B3
Off A9 at the A827 junction – 20 miles north of Perth

Mon 0700-2200 Tue 0700-2200 Wed 0700-2200 Thu 0700-2200 Fri 0700-2200 Sat 0700-2200 Sun 0700-2200

Mon 0800-2030 Tue 0800-2030 Wed 0800-2030 Thu 0800-2030 Fri 0800-2030 Sat 0800-2030 Sun 0800-2030

Muirpark Truckstop

Falkirk Road, Bannockburn, Stirling FK7 8AL
01786 818866
NS82038916 278 D6
From M9 J9/M80 J9, take the A91 Stirling. After 1½ miles, take 3rd exit at roundabout (Falkirk Road). Site is approximately ¾ mile on left.

Mon 24hr Tue 24hr Wed 24hr Thu 24hr Fri 24hr Sat 24hr Sun 24hr

Mon 0600-2000 Tue 0600-2000 Wed 0600-2000 Thu 0600-2000 Fri 0630-1400 Sat Closed Sun Closed

Redmoss Truckstop

35 M74 J11 (5 miles); A74(M) J13 (4 miles)

Carlisle Road (Old A74) by Crawfordjohn, S Lanarkshire ML12 6SX
07831 571856
www.redmoss.co.uk
NS87412704 259 D9
Southbound, from M74 J11 take the B7078 (A70) Edinburgh. After 1½ miles, take the A70 Douglas, pass under the M74, cross over the roundabout and take the 1st left onto the B7078 for 3½ miles. Northbound, from M74 J13 take the A702 Edinburgh, B7078 Douglas for 4 miles.

Mon 0500-2300 Tue 0500-2300 Wed 0500-2300 Thu 0500-2300 Fri 0500-2300 Sat 1300-2300 Sun 1300-2300

Mon 0500-2300 Tue 0500-2300 Wed 0500-2300 Thu 0500-2300 Fri 0500-2300 Sat 1300-2300 Sun 1300-2300

Roadchef Annandale Water

35
A74(M)J16

Johnstonebridge Lockerbie, Dumfriesshire DG11 1HD
01576 470870
www.roadchef.com/motorway-service-area-annandale-water.html
NY10349257 248 E4
Follow signs from A74M J16.

Mon 24hr Tue 24hr Wed 24hr Thu 24hr Fri 24hr Sat 24hr Sun 24hr

Mon 24hr Tue 24hr Wed 24hr Thu 24hr Fri 24hr Sat 24hr Sun 24hr

Roadchef Bothwell (southbound)

M74 Southbound, Bothwell, South Lanarkshire G71 8BG
01698 854123
http://www.roadchef.com/motorway-service-area-bothwell.html
NS70915978 268 D4
On M74 between M74 J4 and J5.

Mon 24hr Tue 24hr Wed 24hr Thu 24hr Fri 24hr Sat 24hr Sun 24hr

Mon 24hr Tue 24hr Wed 24hr Thu 24hr Fri 24hr Sat 24hr Sun 24hr

Roadchef Hamilton (northbound)

M74 Northbound, Hamilton, S Lanarkshire ML3 6JW
01698 282176
www.roadchef.com/motorway-service-area-hamilton.html
NS72495675 268 D4
Signposted from M74, about 1 mile north of J6

Mon 24hr Tue 24hr Wed 24hr Thu 24hr Fri 24hr Sat 24hr Sun 24hr

Mon 24hr Tue 24hr Wed 24hr Thu 24hr Fri 24hr Sat 24hr Sun 24hr

Skiach Services

4D Industrial Est, Evanton, Dingwall, Highland IV16 9XJ
01349 830888
NH62946773 300 C6
From the A9 north of Inverness, take the B9716 Ardross 2 miles east of the Cromarty Firth Bridge. Take the first left.

Mon 24hr Tue 24hr Wed 24hr Thu 24hr Fri 24hr Sat 24hr Sun 24hr

Mon 0700-2200 Tue 0700-2200 Wed 0700-2200 Thu 0700-2200 Fri 0700-2200 Sat 0800-2030 Sun 0800-2030

Stracathro Services

Near Brechin, Angus DD9 7PX
01674 840234
NO62746476 293 G8
On the A90 between Stonehaven and Dundee.

Mon 24hr Tue 24hr Wed 24hr Thu 24hr Fri 24hr Sat 24hr Sun 24hr

Mon 0600-2100 Tue 0600-2100 Wed 0600-2100 Thu 0600-2100 Fri 0600-2100 Sat 0600-2000 Sun 0700-2100
Showers are available during the cafe's opening hours.

Tayside Truckstop

Smeaton Road Dundee DD2 4UT
01382 621941 • NO35713220 287 D7
Wester Gourdie, signposted from A90 Kingsway

Thurso Overnight Lorry Park

Riverside Road, Thurso, Highland
ND12006841 310 C5
From A9 northbound, cross the traffic lights, go over the bridge and take the first right. No facilities on site but the town-centre shops, cafés and takaways are only a short walk away

Mon 24hr Tue 24hr Wed 24hr Thu 24hr Fri 24hr Sat 24hr Sun 24hr

Welcome Break Abington

M74. Abington, Biggar, South Lanarkshire ML12 6RG
01864 502637
www.welcomebreak.co.uk
NS93052482 259 E10
Signposted from M74/A74(M) J13

Mon 24hr Tue 24hr Wed 24hr Thu 24hr Fri 24hr Sat 24hr Sun 24hr

Mon 24hr Tue 24hr Wed 24hr Thu 24hr Fri 24hr Sat 24hr Sun 24hr
Restaurant

Welcome Break Gretna

40
A74(M) J22

A74(M) Trunk Road, Gretna Green, Dumfries DG16 5HQ
01461 337567
www.welcomebreak.co.uk/motorway-service/gretna-green
NY30576877 239 D8
Signposted from A74(M) J22

Mon 24hr Tue 24hr Wed 24hr Thu 24hr Fri 24hr Sat 24hr Sun 24hr

Mon 24hr Tue 24hr Wed 24hr Thu 24hr Fri 24hr Sat 24hr Sun 24hr
Restaurant

Westway Lorry Park

24
M8 J26 (3 miles)
M8 J27 (1 mile)

Westway, Porterfield Road, Paisley, Renfrewshire PA4 8DJ
0141 8866373
NS49656704 267 B9
From M8 J26 (westbound), take A8 Renfrew. At 1st roundabout take 3rd exit; at second roundabout take 2nd exit (Glasgow Road); at 3rd roundabout take second exit (Glebe St); after 1 mile, turn left into Paisley Road; after 1 mile, turn right into Porterfield Road. From M8 J27 (eastbound), take A741 Paisley. Take second exit at roundabout. After 1 mile, turn left into Porterfield Road. Site is on left.

Mon 24hr Tue 24hr Wed 24hr Thu 24hr Fri 24hr Sat 24hr Sun 24hr

Wales

Cardiff Gate Services

20
M4 J30

Cardiff Cardiff Gate Business Park, Pontprenau, Cardiff CF23 8RA
02920 549564
www.welcomebreak.co.uk/motorway-service/cardiff
ST21668290 59 C8
Signposted from M4 J30

Mon 24hr Tue 24hr Wed 24hr Thu 24hr Fri 24hr Sat 24hr Sun 24hr

Mon 24hr Tue 24hr Wed 24hr Thu 24hr Fri 24hr Sat 24hr Sun 24hr

Cardiff West

30
M4, J33

Pontyclun,, Rhondda CF72 8SA
02920 891141
www.moto-way.com
ST09337964 58 D5
M4, J33

Mon 24hr Tue 24hr Wed 24hr Thu 24hr Fri 24hr Sat 24hr Sun 24hr

Mon 24hr Tue 24hr Wed 24hr Thu 24hr Fri 24hr Sat 24hr Sun 24hr

Harry Tuffins Transport Café

Crosslike Supermarket, Churchstoke, Montgomery, Powys SY15 6AR
01588 620226 www.harrytuffin.co.uk
SO27949381 130 E5
From A483 1 mile south of Welshpool take A490 From 8½ miles, then take A489 Craven Arms. Site is on right after ¾ mile.

Mon 0700-1900 Tue 0700-1900 Wed 0700-1900 Thu 0700-2000 Fri 0700-2000 Sat 0700-1930 Sun 0700-1800

Mon 0800-1600 Tue 0800-1600 Wed 0800-1600 Thu 0800-1600 Fri 0800-1600 Sat 0800-1600 Sun 0800-1600
No hot meals after 14.30; there are no dedicated truck bays but the car park is large.

Magor Services

25
M4 J23a

Magor, Caldicot, Monmouthshire NP26 3YL
01633 881515
www.firstmotorway.co.uk
ST42118805 60 B2
Signposted off the M4 J23a

Mon 24hr Tue 24hr Wed 24hr Thu 24hr Fri 24hr Sat 24hr Sun 24hr

Mon 24hr Tue 24hr Wed 24hr Thu 24hr Fri 24hr Sat 24hr Sun 24hr

Moto Swansea

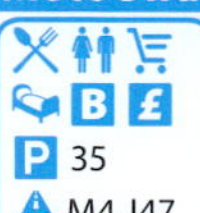

Penllergaer, Swansea. SA4 1GT
01792 896222
www.moto-way.com
SS62099958 75 F10
Signposted from M4 J47.

Mon 24hr Tue 24hr Wed 24hr Thu 24hr Fri 24hr Sat 24hr Sun 24hr

Mon 24hr Tue 24hr Wed 24hr Thu 24hr Fri 24hr Sat 24hr Sun 24hr

Pont Abraham Services

20
M4 J49

Llanedi, Pontarddulais, Carmarthenshire SA4 0FU
01792 884663
www.roadchef.com/motorway-service-area-pont-abraham.html
SN57500734 75 D9
M4 J49

Mon 24hr Tue 24hr Wed 24hr Thu 24hr Fri 24hr Sat 24hr Sun 24hr

Mon 24hr Tue 24hr Wed 24hr Thu 24hr Fri 24hr Sat 24hr Sun 24hr

Cardigan Bay
St. George's Channel
Bristol Channel
ENGLISH
Rosslare 3:30
Rosslare 3:45
Cork 10:00
Roscoff 6:00
Santander 20:00
Guernsey 2:10
Jersey 3:25
St. Malo 5:15
Isles of Scilly
Tremadog Bay
St. Brides Bay
Carmarthen Bay
Swansea Bay
Barnstaple or Bideford Bay
Bridgwater Bay
Bude Bay
Whitesand Bay
Tor Bay
Lyme Bay
Mount's Bay
Falmouth Bay
Wolverhampton
Telford
Shrewsbury
Dudley
Halesowen
Kidderminster
Worcester
Hereford
Gloucester
Aberystwyth
Swansea
Port Talbot
Neath
Merthyr Tydfil
Rhondda
Pontypridd
Caerphilly
Newport
Cardiff
Bristol
Kingswood
Bath
Weston-super-Mare
Cwmbran
Barry
Penarth
Carmarthen
Llanelli
Haverfordwest
Milford Haven
Pembroke
Tenby
Fishguard
St. David's
Barnstaple
Ilfracombe
Minehead
Taunton
Wellington
Tiverton
Exeter
Exmouth
Torquay
Paignton
Plymouth
Newquay
Truro
Falmouth
Penzance
Bodmin
St. Austell
Yeovil
Dorchester
Weymouth
Bridgwater

IX
CHANNEL
LONDON
Birmingham
Nottingham
Derby
Leicester
Coventry
Northampton
Peterborough
Norwich
Cambridge
Ipswich
Oxford
Reading
Southampton
Portsmouth
Brighton
Calais
Boulogne-sur-Mer
Esbjerg 18:00
Hoek van Holland 6:15
Oostende 4:00
Boulogne 1:15
Dunkerque 1:45
Dieppe 4:00
Santander 24:00
Caen 6:00
Le Havre 5:30
Guernsey 2:40
Jersey 3:00
Cherbourg 2:15
St. Malo 4:35
Guernsey 7:00
Jersey 10:30
Cherbourg 3:00
St. Malo 9:00
Strait of Dover
Motorway
junctions - full, restricted
Toll motorway
Services
Primary route
single/dual carriageway
A Road
B Road
Distances - in miles
major
minor
Railway
National boundary
Ferry route
Airport
Scale 1:1 000 000 1cm = 10km 1 inch = 15.78 miles
0 5 10 15 20 25 30 35 40 45 50 miles
0 10 20 30 40 50 60 70 80 km

Route Planner

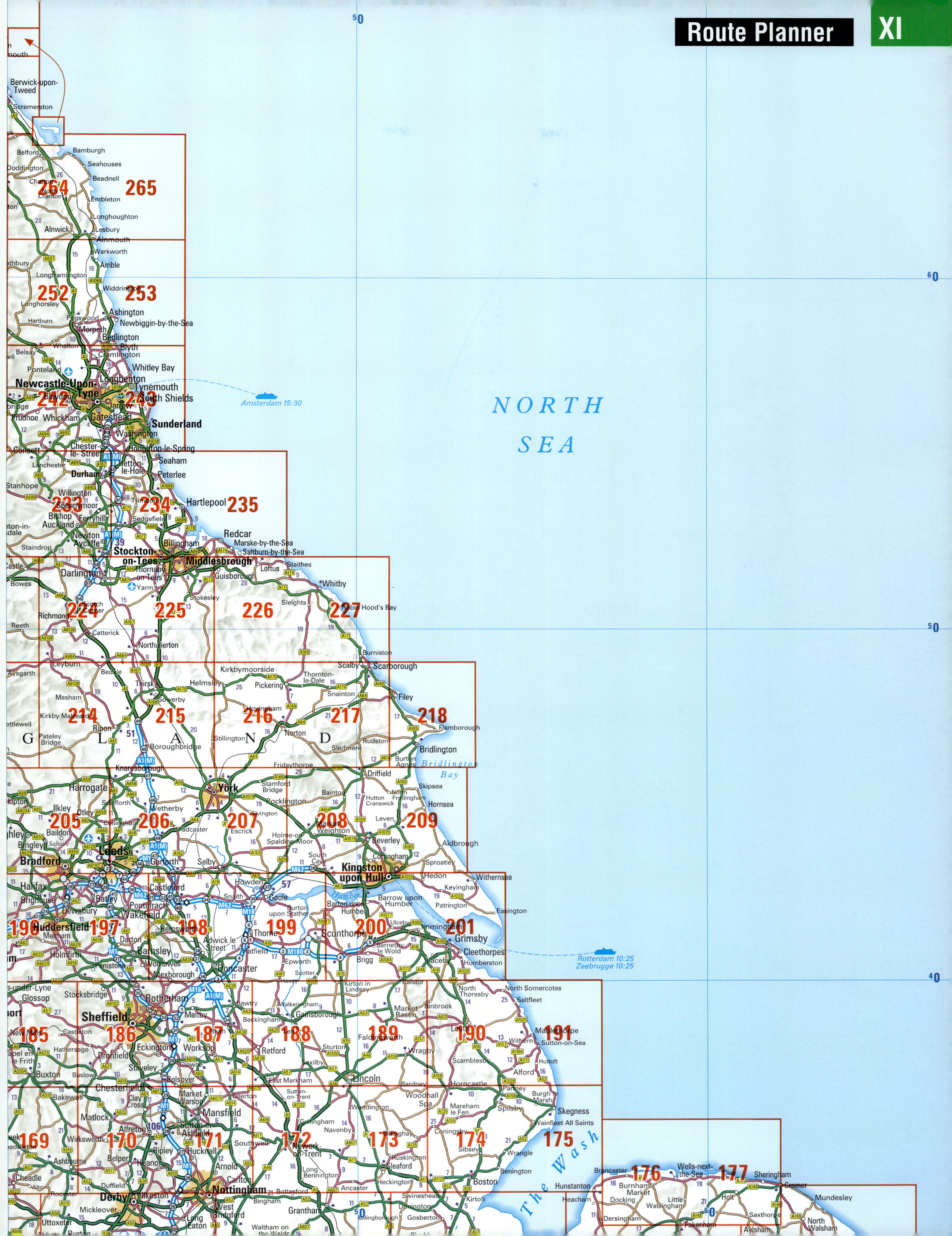
NORTH SEA
Amsterdam 15:30
Rotterdam 10:25
Zeebrugge 10:25
Berwick-upon-Tweed
Newcastle-Upon-Tyne
Sunderland
Middlesbrough
Hartlepool
Whitby
Scarborough
Bridlington
York
Leeds
Bradford
Kingston upon Hull
Grimsby
Doncaster
Sheffield
Lincoln
Nottingham
Derby
Mansfield
Skegness
The Wash
265
264
253
252
243
242
235
234
233
227
226
225
224
218
217
216
215
214
209
208
207
206
205
201
200
199
198
197
196
191
190
189
188
187
186
185
177
176
175
174
173
172
171
170
169

304
305
306
307
308
309
296
297
298
299
300
301
294
295
290
291
288
289
284
285
274
275
276
277
278
254
255
266
267
268
269
Outer Hebrides
Inner Hebrides
North Minch
Little Minch
Lewis
Harris
North Uist
South Uist
Barra
Skye
Rum
Eigg
Coll
Tiree
MULL
Jura
Islay
SCOTLAND
Port Nis
Stornoway
Tairbeart
Loch nam Madadh
Loch Baghasdail
Bagh a Chaisteil
Uig
Portree
Dunvegan
Kyle of Lochalsh
Mallaig
Fort William
Tobermory
Oban
Inverness
Ullapool
Durness
Thurso
Dornoch
Kingussie
Aviemore
Pitlochry
Callander
Stirling
Glasgow
Paisley
Greenock
Dumbarton
Inveraray
Lochgilphead
Campbeltown
Port Ellen
Bowmore
Port Askaig
Gairloch
Lochinver
Kinlochbervie
Tongue
Lairg
Brora
Golspie
Tain
Dingwall
Nairn
Fort Augustus
Spean Bridge
Glencoe
Crianlarich
Killin
Dunoon
Rothesay
Tarbert
Kilmarnock
Hamilton
Motherwell
East Kilbride
Falkirk
Airdrie
Cumbernauld

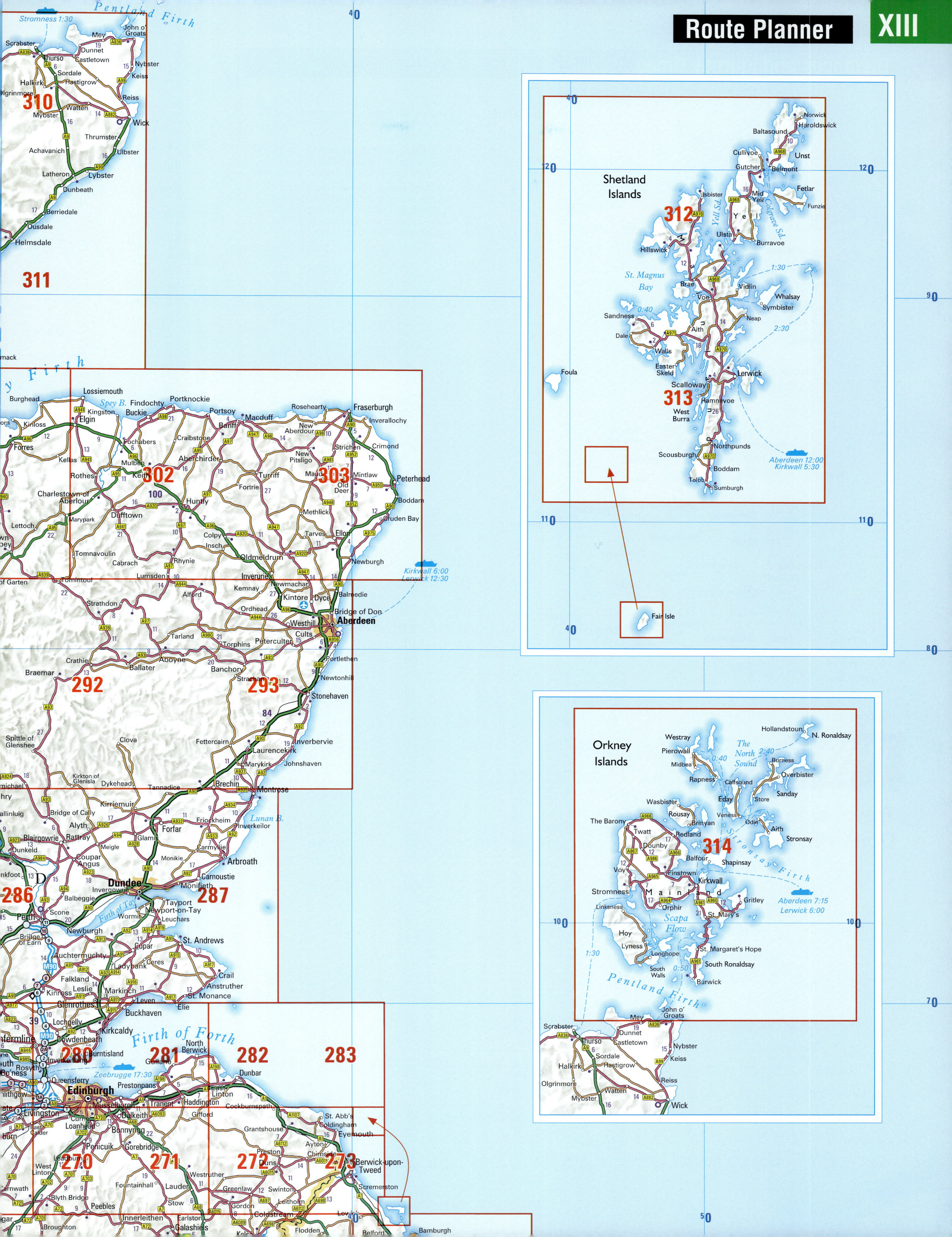
Shetland Islands
312
313
Orkney Islands
314
Baltasound
Haroldswick
Unst
Cullivoe
Gutcher
Belmont
Fetlar
Yell
Mid Yell
Ulsta
Burravoe
Hillswick
St. Magnus Bay
Brae
Voe
Vidlin
Whalsay
Symbister
Sandness
Aith
Walls
Dale
Easter Skeld
Lerwick
Scalloway
Hamnavoe
West Burra
Foula
Scousburgh
Northpunds
Boddam
Sumburgh
Fair Isle
Aberdeen 12:00
Kirkwall 5:30
Westray
Pierowall
Hollandstoun
N. Ronaldsay
The North Sound
Sanday
Stronsay
Rousay
The Barony
Twatt
Dounby
Kirkwall
Stromness
Mainland
Orphir
St. Mary's
Shapinsay
Hoy
Lyness
South Walls
St. Margaret's Hope
South Ronaldsay
Burwick
Pentland Firth
Aberdeen 7:15
Lerwick 6:00
John o' Groats
Scrabster
Thurso
Dunnet
Castletown
Halkirk
Wick
Reiss
Keiss
Nybster
Watten
Mybster
Olgrinmore
Stromness 1:30
310
311
Thrumster
Ulbster
Lybster
Latheron
Dunbeath
Berriedale
Helmsdale
Lossiemouth
Elgin
Forres
Buckie
Portknockie
Portsoy
Banff
Macduff
Fraserburgh
Peterhead
Huntly
Keith
Dufftown
Turriff
Ellon
Newburgh
302
303
Inverurie
Oldmeldrum
Kintore
Dyce
Aberdeen
Westhill
Banchory
Aboyne
Ballater
Braemar
Stonehaven
292
293
Kirkwall 6:00
Lerwick 12:30
Laurencekirk
Inverbervie
Montrose
Brechin
Kirriemuir
Forfar
Arbroath
Carnoustie
Dundee
Perth
286
287
St. Andrews
Cupar
Glenrothes
Kirkcaldy
Firth of Forth
Dunfermline
North Berwick
Dunbar
Edinburgh
Musselburgh
Dalkeith
Livingston
Zeebrugge 17:30
280
281
282
283
270
271
272
273
Penicuik
Peebles
Galashiels
Coldstream
Eyemouth
Berwick-upon-Tweed
Bamburgh

Distances and journey times

Carn Naun Pt
Pen Enys Pt
SOUTH WEST COAST PATH
Mussel Pt
Wicca Pool
Treveal
Trevalgan
Hellesveor
BARBARA HEPWORTH MUS
Porthzennor Cove
Zennor Head
Pendour Cove
Porthglaze Cove
Gurnard's Head
WAYSIDE FOLK MUS
Tremedda
Wicca
Trendrine Fm
Trevessa Fm
Rosewall Hill
Ind Est
Penbeagle
Trendrine Hill
Beagletodn Downs
Breja Fm
Higher Bussow Fm
Balnoon
Halsetown
Trewey
Zennor
ZENNOR CROMLECH
Towednack
Ebal Rocks
Robin's Rocks
Porthmeor Pt
Porthmeor Cove
Great Zawn
Halldrine Cove
Poniou
Boswednack
Treen
Kerrowe Fm
Foage Fm
Amalveor Downs
Amalveor
Amalwhidden
Amalebra
Trink Hill
Cripplesease
Brunnion
Polpeor
Porthmoina Cove
Porthmeor
Bosporthennis
Lit Galver
Lady Downs
Mill Downs
Georgia
Nancledra
Ninnes Bridge
The Wra or Three Stone Oar
Greeb Pt
Whirl Pool
Rosemergy
White Downs
Mulfra Hill
Trenowin Downs
Ashtown
Porthcras Cove
Carn Clough
The Enys
Morvah
Watch Croft
Tredinnick
Trye Fm
CHYSAUSTER ANCIENT VILLAGE
Castle Gate
Tregenden Manor
Levant Zawn
LEVANT MINE
Carn Vellan
Portheras Fm
Trevowhan
Bosullow Common
Lower Ninnes
Mulfra
Carn Du
Lower Boscaswell
Pendeen
Bojewyan
Higher Bojewyan
Boskednan
GEEVOR TIN MINE MUS
Higher Boscaswell
Chun Castle
Bosiliack
Carfury
Newmill
Boscreege Fm
Hellangove Fm
Nanceddan
Botallack Head
Trewellard
Woon Gumpus Common
Little Bosullow
LANYON QUOIT
Crankan
Noongallas
Treassowe Manor
Trezelah
Badger's Cross
Wheal Edward Zawn
Botallack
Wheal Bal Hill
Great Bosullow
Boswarva Carn
Boswarthen
Vellanoweth
Ludgvan
Carnyorth
Carn Kenidjack
Boslow Farm Ho
Bowens Common
Lower Bodinnar
Boswarva
Bosoljack
Rose morran Fm
Tolver
Cape Cornwall
Boscean
Tregeseal
Trengwainton Carn
Madron
Gulval
Trevarrack
Priest's Cove
New Downs
ST JUST MINING DISTRICT
Jericho Fm
Newbridge
TRENGWAINTON GARDEN
Trythogga
Heamoor
Chyandour
Longrock
MARAZION MARSH
The Brisons
BALLOWALL BARROW
St Just
Bosvennan
Bosavern
Crugkern
Derval Fm
Trannack Fm
Bosence Fm
Rosehill
Castle Horneck
PENZANCE
PENLEE HOUSE
Carn Leskys
LAND'S END
Gribba Pt
Kelynack
Dowran
Sellan
Tremethick Cross
Penzance
TRINITY HOUSE NATIONAL LIGHTHOUSE CENTRE
ST MICHAELS MOUNT
Polpry Cove
Trevegean
Bartinney Downs
Grumbla
The Beacon Fm
Treganhoe Fm
Tregavarah
Sancreed
Wherry Town
Toldarne
NEWLYN ART GALLERY
Newlyn
Gwavas Lake
Nanquidno
Land's End
CARN EUNY ANCIENT VILLAGE
Drift Reservoir
Buryas Br
Tredavoe
Gurland Fm
Carn Aire
Carn Brea
Brea Fm
Carn Euny
Brane
Trevean Fm
Drift
Lower Drift
Aire Pt
Crows-an-wra
Catchall
Chywoone Grove
Chywoone Hill
Skilly
Roskilly
Whitesand Bay
Trevedra Fm
Carn Towan
Higher Leah Fm
Bunker's Hill Fm
Trelew Fm
Kerris
Paul
Penlee Pt
Sheffield
Mousehole
The Tribbens
Trevorian Fm
Bosanketh Fm
Banns Fm
Bellowal Fm
Rosevale Fm
Trevithal
St Clement's Isle
Shark's Fin
Sennen Cove
Trevear Fm
Trevorrian Fm
Toldavas Fm
Gamper
Mayon Cliff
Dr Syntax's Head
Mayon
Sennen
Alsia Fm
Penzance Road
Tregadgwith
Trewoofe
Lamorna
Trevelloe Fm
Raginnis
Merlyn Rock
The Mousehole
Pt Spaniard
Longships
Land's End
Trevescan Fm
Skewjack Fm
Trengothal Fm
Bosfranken Fm
Crean
St Buryan
Moorcroft
Choone Fm
Boleigh Fm
Castallack
Penzer Pt
Carn Kez
Carn Greeb
Pordenack Pt
SOUTH WEST COAST PATH
Polgigga
Bosistow Fm
Bottoms
Sparnon
Bown's Barn Fm
Lamorna Valley
Zawn Organ
Kemyel Pt
Mill Bay/Nanjizal
Raftra Fm
Trethewey
Silena
Boskenna
Trevedran
TREGIFFIAN BURIAL CHAMBER
Boscawen Rose
Lamorna Cove
Carn Lês Boel
LOGAN ROCK
Treen
Penberth
Inner Pendower Cove
Zawn Kellys
Porthcurno
MUS OF SUBMARINE TELEGRAPHY
St Levan
Merthen Pt
Boscawen Pt
Tater-du
ST MARY'S 2:40 (ISLES OF SCILLY) (APR-NOV)
Folly Cove
Porth Loe
Porthgwarra
Porth Curno
Cribba Head
Penberth Cove
Porthguarnon
Gwennap Head
Hella Pt
Logan Rock
MINACK OPEN AIR THEATRE
Carn Barges
Round Island
Men-a-vaur
Plumb Island
White Island
St Helen's
Didley's Pt
Porth Morran
Shipman Head
Gimble Pt
Northwethel
St Martin's
Murr Rock
St Martin's Head
CROMWELL'S CASTLE
KING CHARLES CASTLE
St Helen's Pool
Tean
Middle Town
St Martin's Bay
Hell Bay
Castle Down
Northward
Old Grimsby
Higher Town
Hard Lewis Rocks
Scilly Rock
Bryher
The Town
New Grimsby
BLOCKHOUSE
St Martin's Flats
Chimney Rocks
English Island
Gweal
Black Rocks
Crow Island
Samson Hill
Great Pool
Lizard Pt
Borough
Cruther's Pt
Maiden Bower
Merrick Island
Pentle Bay
Great Ganilly
Garden of Maiden Bower
VALHALLA FIGUREHEAD COLLECTION
TRESCO ABBEY GARDENS
Guther's Island
Little Ganilly
Great Innisvouls
Picket Rock
Illiswilgig
Tresco Flats
Skirt Island
Great Ganinick
Eastern Isles
Menawethan
Tresco
Ragged Island
Northern Rocks
Ballard Pt
Puffin Island
Crow Pt
Little Ganinick
Crow Sound
Mincarlo
White Island
Samson
Bar Pt
Stoney Island
The Road
INNISIDGEN BURIAL CHAMBER
Green Island
Halangy Pt
BANTS CARN BURIAL CHAMBER
Watermill Cove
Toll's Island
Carn Morval Pt
Maypole
Taylor's Island
St Mary's
Newford
Holy Vale
Pelistry Bay
ISLES OF SCILLY
St Mary's Pool
Porthloo
LONGSTONE HERITAGE CENTRE
Gap Pt
Rocky Hill
Deep Pt
Hugh Town
A3110
HARRY'S WALLS
St Mary's
PORTH HELLICK DOWN BURIAL CHAMBERS
PENZANCE 2:40 (APR-NOV)
GARRISON WALLS
Old Town
Newfoundland Pt
Porth Cressa
Tolman Pt
Church Pt
Woolpack Pt
Peninnis Head
St Mary's Sound
Inner Head
Outer Head
Haycocks
Annet Head
Porth Conger
Annet Neck
Gugh
Annet
Burnt Island
Dropnose Pt
Smith Sound
The Cove
ISLES OF SCILLY
Lethegus Rocks
Hoe Pt
Codnors Rocks
Wingletang Bay
Great Crebawethan
Isinvrank
Little Crebawethan
Horse Pt
St Agnes
Muncoy Neck
Jacky's Rock
Western Rocks
Melledgan
Rosevear
Rosevean
Daisy
Gorregan

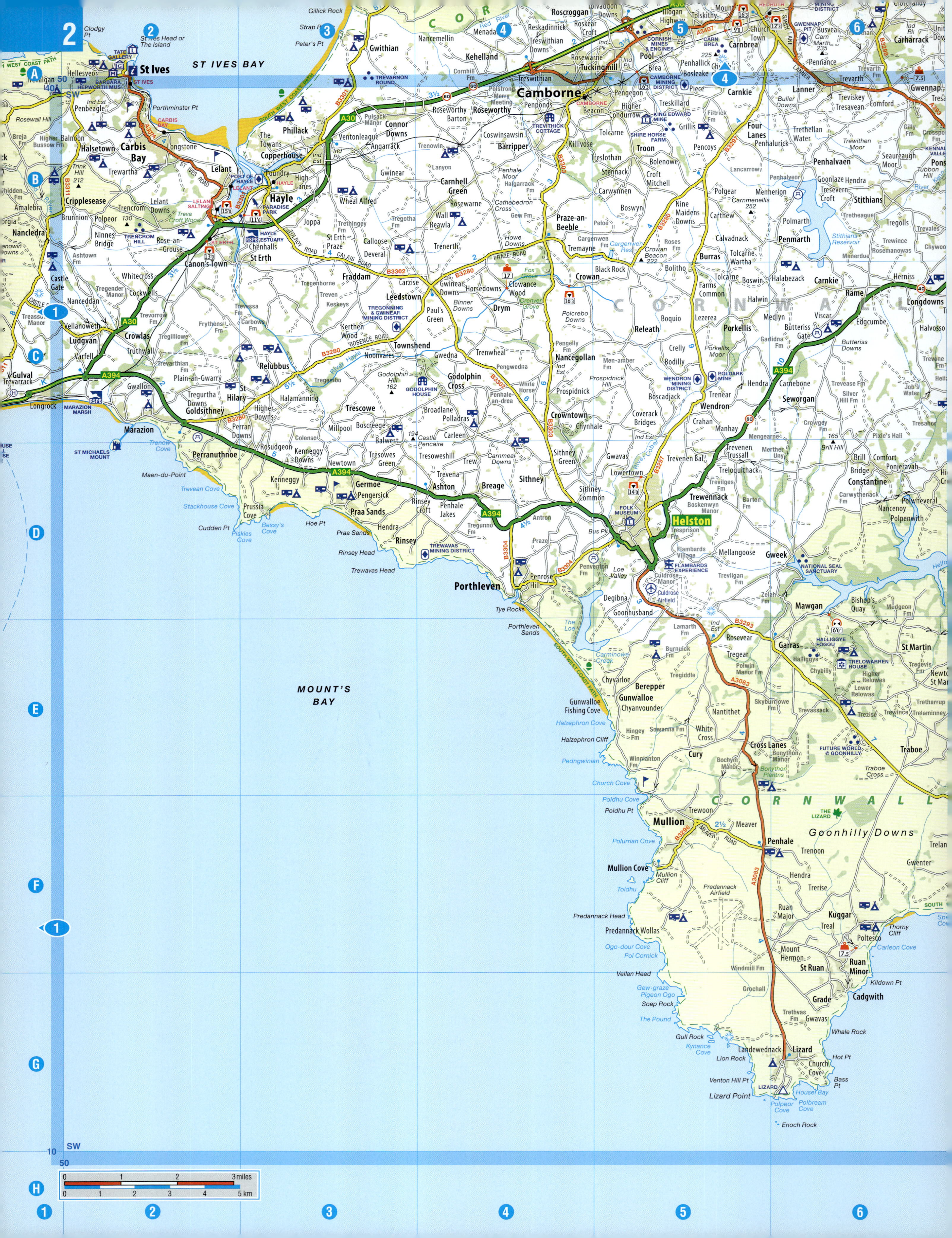

2
ST IVES BAY
St Ives
Carbis Bay
Lelant
Hayle
Phillack
Copperhouse
Camborne
Pool
Carnbrea
Tuckingmill
Roseworthy
Barripper
Connor Downs
Gwithian
Kehelland
Menadarva
Nancemellin
Reskadinnick
Roscroggan
Treswithian
Praze-an-Beeble
Crowan
Leedstown
Fraddam
St Erth
Canon's Town
Crowlas
Ludgvan
Nanceddan
Castle Gate
Nancledra
Cripplesease
Halsetown
Lelant Downs
Rose-an-Grouse
Gulval
Trevarrack
Longrock
Marazion
St Michael's Mount
Perranuthnoe
Goldsithney
St Hilary
Relubbus
Townshend
Godolphin Cross
Trescowe
Germoe
Praa Sands
Rinsey
Ashton
Breage
Porthleven
Sithney
Helston
Crowntown
Nancegollan
Releath
Wendron
Porkellis
Carnkie
Four Lanes
Troon
Penhalvean
Stithians
Longdowns
Edgcumbe
Seworgan
Constantine
Gweek
Mawgan
Garras
St Martin
Trewennack
Gunwalloe
Berepper
Cury
Cross Lanes
Traboe
Mullion
Mullion Cove
Penhale
Goonhilly Downs
Kuggar
Poltesco
Ruan Minor
Cadgwith
Grade
St Ruan
Lizard
Lizard Point
Landewednack
Predannack Wollas
MOUNT'S BAY
CORNWALL
A30
A394
A3083
B3301
B3302
B3280
B3303
B3304
B3297
B3296
B3293
SOUTH WEST COAST PATH
0 1 2 3 miles
0 1 2 3 4 5 km
A B C D E F G H
1 2 3 4 5 6

3
Heads
Wheal Baddon
Fernsplatt
Coombe
Helston Water
Quenchwell
Playing Place
Old Kea
St Michael Penkevil
Ruan River
Caerhays
Hick's Mill
Greenwith Common
Carnon Downs
Killiganoon
Cowlands
Coombe
Higher Trelease Fm
West Portholland
East Portholland
Treveor
Boswinger
Cusgarne
Frogpool
Tregye
Penelewey
Polgerran Wood
Treworga
Trenona
Tretheake Manor
Veryan Green
Treburthes
May's Rock
Porthluney Cove
Greeb Pt
Penare
Penveor Pt
Perranwell
Perranwell Station
Perranarworthal
Devoran
Goon Piper
Penpol
Trevilla
Trelissick
Trelissick Garden
Borlase Wood
Philleigh
Ruan High Lanes
Trewartha
Hartriza Pt
Shag Rock
Veryan Bay
High Pt
Lizard Pool
Dodman Pt
Devoran Mining District
Chycoose
Restronguet Creek
Point
Feock
Harcourt
Treworthal
Crugsillick Manor
Lower Mill
Melinsey
Veryan
Portloe
Tregamenna Manor Fm
Manare Pt
Pennare Fm
Perran Wharf
Devichoys Wood
Carclew
Restronguet Passage
Tregairewoon Fm
Treworlas
Carne
Rosen Cliff
Gerrans Bay
Kiberick Cove
Burnthouse
Laity Moor
Roskrow
Treluswell
Angarrick
Commerrans Fm
Curgurrell
Pibyah Rock
Shannick Pt
Lemoria Rock
Gull Rock
Nare Head
Trewithian
Rosevine
Weir Pt
Carrick Roads
Messack Fm
Mylor Bridge
Mylor Creek
Messack Pt
Trethewell
St Just in Roseland
St Just Pool
Gerrans
Portscatho
Trelever
Penarrow Pt
Trelew
Mylor Churchtown
Penryn
Bissom
Carveth
Mabe Burnthouse
Ind Est
St Gluvias
P&R
Tregew
Flushing
Percuil
Antron Fm
College Res
Trefusis Pt
St Mawes
St Mawes Castle
Greeb Pt
Argal Manor
Mongleath
Falmouth
Nat Maritime Mus Cornwall
Falmouth Docks
Castle Pt
St Mawes Harbour
Porthmellin Head
Porth Fm
Towan Beach
Budock Water
Penmere
Bus Pk
Pendennis Castle
St Anthony
Bohortha
Killigerran Head
Porthbear Beach
Argal Res
Lamanva
Boslowick
Falmouth Town
Treverva
Trewen Fm
Gyllyngvase Beach
Pendennis Pt
St Anthony Head
Place Barton
Zone Pt
Swan Pool
Penjerrick
Penrose Fm
Pennance Pt
Treworval Fm
Carlidnack
Bareppa
Maenporth
High Cliff
Falmouth Bay
Drift Fm
Boskensoe Fm
Mawnan Smith
Bream Cove
Porth Navas
Glendurgan Gardens
Trerose
Rosemullion Head
Helford Passage
Trebah Gdns
Mawnan
Durgan
South West Coast Path
Toll Pt
Helford River
Helford
Treath
St Anthony-in-Meneage
Manaccan
Flushing
Gillan Harbour
Gillan
Carne
Nare Pt
Polnare Cove
Nare Head
Nare Cove
Beacon 112
Roskorwell
Porthallow
Porthkerris Pt
Tregidden
Tregarne
Porthoustock
Polpidnick Fm
Treleage Mill
Trelease Fm
Laddenvean
Rosenithon
St Keverne
Godrevy Cove
Carn-du Rocks
Grugwith
Roskilly
Crousa Common
Roskilly's Open Fm
B3294
Zoar
Lowland Pt
Polcoverack Fm
North Corner
Penhallick
Coverack
Ponsongath
Perprean Cove
Treleaver Cliff
Ebber Rocks
Black Head
A39
A394
A3078
B3289
B3289
B3292
SW
40
10
00
A
B
C
D
E
F
G
H
7
8
9
10
11

5
Wadebridge
Bodmin
St Austell
Lostwithiel
Mevagissey
St Columb Major
Porthcothan
St Issey
St Breock
Egloshayle
Helland
St Mawgan
Newquay Cornwall Airport
Indian Queens
Fraddon
Summercourt
Mitchell
St Enoder
Roche
Bugle
Stenalees
Luxulyan
Lanivet
St Dennis
Nanpean
St Stephen
Grampound
Grampound Road
Probus
Tregony
Sticker
Polgooth
Pentewan
Charlestown
Par
Tywardreath
St Blazey
Fowey
Polkerris
Gorran Haven
Portmellon
St Ewe
Veryan
Portloe
Ruan Lanihorne
Tresillian
Ladock
New Mills
Victoria
Belowda
Tregoss
Goss Moor
Lockengate
Bilberry
Penwithick
Carclaze
Boscoppa
Holmbush
Trewoon
St Mewan
London Apprentice
Heligan
The Lost Gardens of Heligan
Caerhays Castle
St Michael Caerhays
Tregonetha
St Wenn
Withiel
Ruthernbridge
Nanstallon
St Lawrence
Cardinham
Trebyan
Redmoor
Lanlivery
Penpillick
Lanescot
CORNWALL
ST AUSTELL BAY
MEVAGISSEY BAY
VERYAN BAY
Dodman Pt
Black Head
Gwingeas/Gwineas
Chapel Pt
Greeb Pt
Nare Head
Hensbarrow Beacon
Castle Downs
Eden Project
Wheal Martin China Clay Heritage Centre
Lappa Valley
A39
A30
A38
A390
A391
A392
A3058
A3059
A3075
A3078
A3082
A389
B3274
B3275
B3279
B3287
B3269
B3268
St Austell Brewery
Shipwreck & Heritage Centre
Trewithen
World of Model Railways
Spirit of the West Theme Park
Westworld Raceway
Screech Owl Sanctuary
Springfields Fun Park & Pony Centre
Dairyland Farm World
Crealy Great Adventure Park
Cornish Birds of Prey Centre
Pencarrow House
Lanhydrock House
Restormel Castle
St Catherine's Castle
Bodmin & Wenford Railway
Courtroom Experience
Royal Cornwall Showground
SX
70
40
10
A
B
C
D
E
F
G
H
7
8
9
10
11
3
6

6
Bodmin
Cardinham
Millpool
Maidenwell
Warleggan
St Neot
Minions
Upton Cross
Pensilva
Golberton
Trekelvesteps
Commonmoor
Darite
Tremar
St Cleer
St Ive
Gang
Turfdown
Fletchersbridge
Drawbridge
Doublebois
Dobwalls
Moorswater
Liskeard
Merrymeet
Pengover Green
Quethiock
West Taphouse
Middle Taphouse
East Taphouse
St Pinnock
Bofarnel
Braddock
Trewindle
Menheniot
Lostwithiel
Sweethouse
Bridgend
Couch's Mill
Herodsfoot
St Keyne
Trewidland
Tideford
Tideford Cross
Bocaddon
Lanreath
Duloe
Widegates
Hessenford
Trerulefoot
Milltown
Lerryn
St Winnow
Muchlarnick
Pelynt
Sandplace
Morval
St Martin
Seaton
Downderry
Polbathic
St Veep
Golant
Tywardreath
Penpoll
Lanteglos Highway
Trenewan
Watergate
Looe
East Looe
West Looe
Bodinnick
Fowey
Polkerris
Menabilly
Polruan
Lansallos
Porthallow
Talland
Polperro
Crumplehorn
CORNWALL
WHITSAND
A38
A390
A387
B3360
B3359
B3254
B3252
B3253
B3251
B3415
B3269
B3247
SOUTH WEST COAST PATH
Looe Bay
Looe Island
SX
A
B
C
D
E
F
G
H
1
2
3
4
5
6
0
1
2
3 miles
0
1
2
3
4
5 km

7
Kelly Bray
Chilsworthy
Dimson
Gunnislake
Monkscross
St Ann's Chapel
Callington
Bowling Green
Albaston
Drakewalls
Harrowbarrow
Metherell
Norris Green
Morwellham Quay
Calstock
Grenofen
Higher Walreddon
West Down
Horrabridge
Walkhampton
Yelverton
Dousland
Sheepstor
Meavy
Harrowbeer
Crapstone
Buckland Monachorum
THE GARDEN HOUSE
BUCKLAND ABBEY
COTEHELE HOUSE
DUPATH WELL CHAPEL
St Dominick
Bohetherick
Tuckermarsh
Bere Alston
Newbridge
TAMAR VALLEY
TAMAR VALLEY MINING DISTRICT
Milton Combe
Halton Barton
St Mellion
Weir Quay
Cotts
Lopwell
Bere Ferrers
Pillaton
Polborder
Paynter's Cross
Clearbrook
Hoo Meavy
DARTMOOR NATIONAL PARK
DEVON
Shaugh Prior
Wotter
Lee Moor
Roborough
Bickleigh
Woolwell
Cargreen
Landulph
River Tamar
River Tavy
Hatt
Botusfleming
Landrake
Notter
Carkeel
Tamerton Foliot
Widewell
Whitleigh
Southway
Glenholt
Estover
Derriford
Crownhill
PLYMOUTH CITY
Saltash
Trematon
Latchbrook
Burraton
South Pill
Forder
Ernesettle
St Budeaux
Honicknowle
Manadon
Riverside
Wearde
St Stephens
St Erney
St Germans
Antony Passage
Ham
Pennycross
Eggbuckland
Leigham
Longbridge
Woodford
Plympton
Chaddlewood
Underwood
Drakeland Corner
Sparkwell
DARTMOOR ZOOLOGICAL PARK
Cornwood
Lutton
Hemerdon
Lee Mill
Smithaleigh
Langage
Blackpool
Weston Mill
Barne Barton
Peverell
Mannamead
Compton
Efford
Crabtree
Mutley
Ford
Stoke
Laira
Mount Gould
Plymouth
PLYMOUTH
Devonport
Stonehouse
Torpoint
Wilcove
Maryfield
ANTONY HOUSE
Antony
Sheviok
Crafthole
Portwrinkle
St John
Cremyll
MOUNT EDGCUMBE
Millbrook
Insworke
Southdown
Anderton
Freathy
Kingsand
Cawsand
Rame
Rame Head
Penlee Pt
Pier Cellars
Drake's Island
THE SOUND
Mount Batten Pt
Jennycliff Bay
ROYAL CITADEL
NATIONAL MARINE AQUARIUM
CITY MUS & ART GALLERY
MERCHANT'S HOUSE
Cattedown
Turnchapel
Oreston
Pomphlett
Plymstock
Hooe
Elburton
Billacombe
Staddiscombe
Down Thomas
Bovisand Bay
Heybrook Bay
Wembury
Knighton
SOUTH DEVON
Brixton
Yealmpton
Yealmbridge
Torr
Holbeton
Newton Ferrers
Noss Mayo
Worswell
Bridgend
Membland
Battisborough Cross
Mothecombe
Luson
ROSCOFF 6:00
SANTANDER 20:00
(MAR-NOV)
SOUTH WEST COAST PATH
BAY
SX
7
8
9
10
11
A
B
C
D
E
F
G
H

Peat Cott
Whiteworks
Down Ridge
Ter Hill 480
Skir Hill
Foxtor Mires
Nun's Cross Fm
Crane Hill 471
Petre's Bound Stone 515
Naker's Hill
Great Gnats' Head
Green Hill 473
Snowdon
Hartor Tors
Ditsworthy Warren
Giant's Hill
Huntingdon Warren
Pupers Hill
Lud Gate
Hayford Hall
Scorriton Down
Michelcombe
Scorriton
Combe
Holne
Poundsgate
Bench Tor
West Stoke
Middle Stoke
Hannaford Manor
Holne Chase
Venford Reservoir
Holne Moor
Aish Tor
Ausewell Rocks
Ausewell Wood
Boro Wood
Place
Rew
Hele
Ashburton
Caton
Goodstone
Mead
Whistley Hill
Chuley Cross
Ashburton Road
Lower Holbeam Fm
Herebere Fm
Burne Fm
Chercombe Br
Bradley Manor
Newton Abbot
Emblett Hill
Broadridge Wood
Valley Fm
West Ogwell
East Ogwell
Heathfield Fm
HM Prison
Norden Fm
Denbury
Prickly Ball Hedgehog Hospital
Rydon Hill
Halwell Fm
Woodland
Forder Green
Furcombe
Lake
Waye Fm
Quarry Fm
Creek Beacon
Tor Fm
Currington Ho
Well Fm
Waytown
Beacon Hill
Coppa Dolla Fm
Torbryan
Ipplepen
Gullaford
Bickaton
Lee
Barton Ho
Broadhempston
Woolston Green
Landscove
Beara
Baddaford
Buckfast
Buckfast Abbey
Buckfastleigh
Wallaford
Dartmoor Otters & Buckfast Butterflies
South Devon Railway
Bigadon Ho
Hill Fm
King's Wood
Brook Manor
Brook Wood
Holy Brook
Hawson Court
Blackmoor
Priestaford Ho
Southpark Wood
D A R T M O O R
N A T I O N A L   P A R K
Langcombe Hill
Petre's Cross 480
Brown Heath
Bishop's Meads
Avon Dam Reservoir
Water Oak Corner
Gripper's Hill
Lambs Down
Wallaford Down
Lee Moor
Stall Moor
Erme Plains
Quickbeam Hill
Ryder's Rocks
Brent Moor
Black Tor
Shell Top 475
Penn Moor
Penn Beacon 427
Harford Moor
Three Barrows 461
Ugborough Moor
Old Hill
Dockwell Fm
Gidley Br
Dean Wood
Upper Dean
Lower Dean
Dean
Dean Prior
Eden Fm
Harbourneford
White-oxen Manor
Brent Hill 311
Shipley Br
Bloody Pool
Didworthy
Badworthy
Ball Gate 317
Brent Fore Hill
Ball Gate
Sharp Tor
Piles Hill 387
Corringdon Ball
Aish Ridge
Lutton
Leigh Grange
Hillside
Aish
Lydia Br
South Brent
Crowder Park
Ind Est
Brent Mill
Dendles Waste
Dendles Wood
Rook Tor
Hele Cross
Yadsworthy Fm
Torr
Wisdome Fm
Hall Fm
Harford
Hall Cross
Hanger Down
Tor Rocks
Ugborough Beacon 371
Owley
Eastern Beacon
Butterdon Hill 364
Weatherdon Hill
Cheston
Wrangaton
Moorhaven Village
Bittaford
Ivybridge
Henlake Down 212
Cornwood
Lutton
Moor Cross
Yondertown
Delamore
Lower Venton Fm
Hawk's Br
Cadleigh Lodge
Woodland
St Austin's Priory
Lee Mill
Smithaleigh
Bus Pk
Filham
Godwell
Ind Est
Trad Est
Hillhead Cross
Ugborough
Wood Fm
Venn Ho
Kitterford Cross
Well Cross
Fowlescombe Gate
Hunsdon Fm
Southwood Wood
Popple's Br
Marjery Cross
Penquit
Ludbrook
Higher Spriddlescombe
Lotherton Br
Worston
Westlake
Ludbrook Manor
Weeke Fm
Shilston Br
Ermington
Ermington Wood
Mary Cross
Hollowcombe Fm
Yealmbridge
Dunstone
Butland Wood
Modbury
Yarnacombe Fm
East Leigh
Brownston
California Cross
Coldharbour Cross
Hazelwood
Churchland Green
Polston Parks
Babland Fm
Heathfield Manor
Blackwell Park Fm
Higher Wizaller Fm
Woolston Ho
Hole Fm
Holbeton
Ford
Stoliford
Whympston
Luson Borough Fm
Creacombe Fm
Wastor Tor Rock
Langston
Seven Stones Cross
Highland Fm
Combe Fm
Tetwell
Wakeham
Battisborough Cross
Haye Fm
Mothecombe
Kingston
Malthouse Pt
Wonwell Court Fm
Okenbury
South Langston
Ashford
Aveton Gifford
St Ann's Chapel
Waterhead
Easton
Icy Park
Bigbury Court
Bridge End
Venn
Hatch
Bus Pk
Churchstow
Stadbury Fm
Mary's Rocks
Beacon Pt
South West Coast Path
Erme Mouth
Wadham Rocks
Castle Fm
Bigbury
Ringmore
Toby's Pt
Aymer Cove
Challaborough
Warren Pt
Bigbury-on-Sea
Burgh Island
Folly Hill
BIGBURY BAY
Butter Cove
Murray's Rock
Bantham
Buckland
Thurlestone
Kerse Fm
Upton
South Milton
Huxton Cross
Warren Pt
Thurlestone Rock
Horswell Ho
South Huish
Sutton
Davey Park Fm
Beacon Pt
Woolman Pt
Outer Hope
Hope Cove
Galmpton
Burleigh Fm
Withymore Fm
Bolt Tail
Inner Hope
Hope
Bolberry
Redrot Cove
Ferny Hole Pt
Slippery Pt
Southdown Fm
West Cliff
Cathole Cliff
Combe
Rew
Soar
Salcombe
Collaton
Malborough
Salcombe Road
Batson
Salcombe Maritime Museum
Mill Bay
East Portlemouth
Southpool Creek
Chivelstone
South Allington
Kellaton
Bickerton
Hallsands
Hollowcombe Head
Nestley Pt
Middlecombe Fm
Beesands
South Pool
Dunstone
Ford
Beeson
Kernborough
North Pool Fm
Cotmore
Carehouse Cross
Torcross
Limpet Rocks
Stokenham
Chillington
Frittiscombe
Coleridge Ho
Slapton
Slapton Ley
Lower Ley
France Wood
Slapton Sands
South West Coast Path
Start
Harleston
Sherford
Marepark Cross
Frogmore
Mill Fm
East Charleton
West Charleton
Charleton
Ham Pt
Frogmore Creek
Gerston Pt
Kingsbridge Estuary
Kingsbridge
Southville
Dodbrooke
Shop Cen
Cookworthy Mus
Westville
Ind Est
West Alvington
Ticket Wood
Park Fm
Oldaway Fm
Gerston Fm
Woolston
Alston Fm
Wareham Pt
Tosnos Pt
Coombe Fm
Sorley
Sorley Tunnel Adventure Worlds
Ledstone
Slade
Goveton
Warcombe
New Br
Loddiswell
Yanston Fm
Ham Fm
Heath Fm
Fernhill Ho
Woodleigh
Torr
Ind Est
Nutcombe
Yetsonais Fm
Higher Poole Fm
Barnston Fm
East Allington
Lower Coombe Fm
Wallaton Cross
Millcombe
Larcombe Fm
Abbotsleigh
Sheplegh Court
Flear Fm
Cross
Cole's Cross
Binney's Cross
Higher Norton Fm
Malston Mill
Stancombe Cross
Valley Springs
Torr Fm
Gara Mill
Thorn
Dearswell
Hansel
Merrifield
Whitestone Cross
Strete
Lower Green Cross
Pilchard
Burleston
West Combe Fm
Combe Cross
Eastdown
Ash
Bowden
S O U T H
D E V O N
Woolcombe Fm
Morecombe Fm
Pitt Fm
Cuttery Fm
Green Lane End
Hutcherleigh
Woodford
Pasture Fm
Blackawton
Cotterbury Cross
Woodlands Leisure Park
Oldstone
Wadstray
Hillfield
Woodbury Fm
Dreyton Cross
Stone Fm
Capton
Allaleigh
Black Down
Halwell
Boreston Fm
Stanborough Ho
Stanborough Hundred
Bus Pk
Moreleigh
Moreleigh Mount
Stoneleigh Manor
Curtisknowle
Gara Br
Lupridge
Storridge Wood
Topsham Br
Wotton
Hendham Ho
Coarsewell
Bickham Br
Combehead Cross
Coombe Ho
Crabadon Cross
Crabadon
Greyhills Fm
Ball Hill
Hernaford
Higher Poulston
Middle Washbourne
Washbourne
Stonehills
Beenleigh
Tuckenhay
Bow Creek
Bow
Cornworthy
East Cornworthy
Southills Fm
River Wash
Harbertonford
Rolster Br
Harberton
Bennett's Cross
Maryland Ho
Prowse's Luscombe
Stancombe
Coombe Park Fm
Ashprington
Stoke Gabriel
Aish
Totnes Motor Museum
St Mary's Church
Totnes Castle
Bridgetown
Totnes
Totnes Littlehempston
True Street
Berry Pomeroy
Berry Pomeroy Castle
Castle Lodge
Littlehempston
Garcombe Brook
Loventor Manor
Tally Ho
Upshempston
Shadrack
Old Manor
Lillisford Fm
Red Post
Combe Fishacre
Combefish Ho
Lomans Fm
Mount Barton
Moothill Cross
Staverton
South Devon Railway
Hoxham's Cross
Dartington
Dartington Hall
Dartington Cider Press Centre
Shinner's Bridge
Longcause
Cott
Week
Billany Fm
Riverford Br
Hood Ball
Hole
Barkingdon Manor
Kingston Ho
Rattery
Willing Fm
Brownston Fm
Yelland Fm
Pennywell Farm & Wildlife Centre
South Downs Fm
Yarner Beacon
Brook Ho
Whiteley Fm
Tigley
Fork Cross
Culverlane
Mill Cross
Syon Abbey
Harbourne River
Lisburne Fm
Yeo Br
Sandwell Old Manor
Belsford
Tristford Ho
Langford Fm
East Moore Fm
Lincombe
Bradridge Ho
Thorn Fm
Murtwell
Diptford
Christone Cross
Avonwick
Horsebrook
Charford Manor
Webland Fm
River Avon
Black Hall
Baron's Hill Fm
North Huish
Carnamore Fm
D E V O N
SOUTH DEVON
A38
A385
A384
A381
A3122
A379
A3121
B3213
B3372
B3196
B3392
A38
SX
60
70
40
0 1 2 3 miles
0 1 2 3 4 5 km

BABBACOMBE BAY
TOR BAY
START BAY
TORBAY
SOUTH DEVON
Torquay
Paignton
Brixham
Dartmouth
Salcombe
Lower Netherton
Combeinteignhead
Netherton
Knowles Hill
Newton Abbot
Buckland
Wolborough
Milber
Newtake
Haccombe
Trade Est
Higher Charlecombe Fm
Plant World
Middle Rocombe
Lower Gabwell
Higher Gabwell
Aller Park
Langford Br
Abbotskerswell
Aller
Coffinswell
Daccombe
Kingskerswell
Higher Rocombe Barton
Maidencombe
Rock Ho
Bell Rock
Watcombe Head
Stoneycombe
Barton
Watcombe
Combe Pafford
St Marychurch
Model Village
Bygones
Dainton
Blair Hill
North Whilborough
Kerswell Gardens
Edginswell
Riviera Way
Retail Pk
Hele
Babbacombe
Bus Pk
Plainmoor
Moles Cross
Shiphay
Compton Castle
Compton
Chelston
Torre
Ellacombe
Black Head
Kent's Cavern
Widdicombe Fm
Cockington
Torre Abbey
Wellswood
Anstey's Cove
Stantor Barton
Marldon
Cockington
Lead Stone or Flat Rock
Living Coasts
Peaked Tor Cove
Daddyhole Cove
Thatcher Rock
Livermead
Churscombe
Beacon Hill 196
Shorton
Preston
Oldway Mansion
Oldway
Hollicombe
Blagdon
Collaton St Mary
Kirkham House
Primley Ho
St Michaels
Roundham Head
Windmill Hill 166
Paignton Zoo
Goodrington Sands
Quaywest
Clennon Hill
Bus Pk
Goodrington
Yalberton
Saltern Cove
Port Bridge
Elberry Cove
Fishcombe Pt
Broadsands
Berry Head
Brixham Heritage Museum
The Golden Hind
Waddeton
Galmpton
Churston
Churston Ferrers
Alston Fm
Cod Rock
Durl Head
Higher Brixham
St Mary's Bay
Sharkham Pt
Cott Fm
Dittisham
Maypool
River Dart
Paignton & Dartmouth Steam Railway
Southdown
Hillhead
Southdown Cliff
Man Sands
Crabrock Pt
Old Mill Creek
Woodhuish Fm
Cod Rocks
Long Sands
South West Coast Path
Newcomen Engine Ho
Scabbacombe Sands
Scabbacombe Head
Kingston
Norton
St Saviour's Church
Bayard's Cove Fort
Kingswear
Coleton Fishacre Gdn
Ivy Cove
Cotton
Wheatland
P&R
Warfleet
Dartmouth Castle
Venn
Pudcombe Cove
Kelly's Cove
Eastern Black Rock
Newfoundland Cove
Blackstone Pt
Outer Froward Pt
Compass Cove
Stoke Fleming
Combe Pt
Meg Rocks
Redlap Cove
Leonard's Cove
Blackpool
Matthew's Pt
Shaldon Wildlife Trust
The Ness Ho
The Beacon 169
Labrador Bay
Blackaller's Cove
(APR - SEPT)
South Huish
Davey Park Fm
Woolston
Gerston Fm
Gerston Pt
Charleton
Kingsbridge Estuary
Ham Pt
Frogmore Creek
North Pool Fm
Torcross
Limpet Rocks
Cotmore
Kernborough
Burleigh Fm
Alston Fm
Wareham Pt
Ford
Beeson
Beesands
Galmpton
Withymore Fm
Tosnos Pt
South Pool
Dunstone
Salcombe Road
Middlecombe Fm
Bolberry
Malborough
Collaton
Batson
Southpool Creek
Kellaton
Southdown Fm
East Portlemouth
Chivelstone
Hallsands
West Cliff
Combe
Rew
Salcombe Maritime Museum
South Allington
Bickerton
Hollowcombe Head
Mill Bay
Nestley Pt
Catholé Cliff
Soar
South Sands
Start Pt
Lantern Rock
Overbecks Museum & Garden
Black Stone
Shag Rock
Deckler's Cliff
Gorah Rocks
Sleadon Rocks
East Prawle
Ballsaddle Rock
Bolt Head
Little Mew Stone
Ball Rock
Gammon Head
Langerstone Pt
Prawle Pt
Sharpers Head
SY
SX
A
B
C
D
E
F
G
H
7
8
9
10
11
14
15

Padstow
Wadebridge
Padstow Bay
Port Isaac Bay
New Polzeath
Polzeath
Trebetherick
Rock
St Minver
Port Isaac
Port Gaverne
Port Quin
St Endellion
Trelights
Pendoggett
Chapel Amble
St Kew
St Kew Highway
Egloshayle
St Breock
Whitecross
Little Petherick
St Issey
St Merryn
St Ervan
St Eval
Trevose Head
Constantine Bay
Treyarnon
Porthcothan
Harlyn
Trevone
Shop
Towan
Windmill
Crugmeer
Pentire Pt
Rumps Pt
The Mouls
Newland
Gulland Rock
Stepper Pt
Tintagel Head
Tintagel
Trelill
St Mabyn
Washaway
Sladesbridge
Burniere
Bodieve
Trevanson
Edmonton
Tredinnick
St Jidgey
River Camel
CORNWALL
SOUTH WEST COAST PATH
PRIDEAUX PLACE
NATIONAL LOBSTER HATCHERY
PADSTOW MUSEUM
ROYAL CORNWALL SHOWGROUND
LONG CROSS VICTORIAN GDNS
CREALY GREAT ADVENTURE PARK
PENCARROW HOUSE
A39
A389
B3276
B3274
B3314
B3267
0 1 2 3 miles
0 1 2 3 4 5 km

Bay
Black Rock
Wanson Mouth
Wanson
Widemouth
Box's Shop
Budd's Titson
Titson
Marhayes Manor
Merrifield
Bradford Manor
Strawberry Bank
Foxhole Pt
Millook Haven
Millook
Penhalt Fm
Coppathorne
Burracott
Bakesdown
Poundstock
Dizzard Pt
Dizzard
Bangors
Whitstone
Whitstone Head
Treskinnick Cross
Tregole
Trewint
Week Green
Week St Mary
Penhallym
PENHALLAM
Newpark
Street
North Tamerton
Pencannow Pt
St Genny's
Crackington Haven
Bray's Pt
Tremoutha Haven
Chilpark
Coxford
Trelay
Rosecare
Jacobstow
Southcott
Samphire Rock
Middle Crackington
Higher Crackington
Mineshope
Sweets
Wainhouse Corner
Kents
Tobarn
Rosecare Villa Fm
High Cliff
Pengold
Pencuke
Roundhayes Fm
Gull Rock
SOUTH WEST COAST PATH
Newton Fm
Tresparrett Posts
Collamoor Head
Trengune
Headon
Copper Fm
Exe
Middle Wheatley
Higher Whiteleigh
West Curry
Curry Lane
Fire Beacon Pt
Beeny Cliff
Seals Hole
Beeny
Tresparrett
Tresparrett Down
Marshgate
Cansford Fm
Cockport
Treworld
South Wheatley
Maxworthy
Maxworthy Cross
Clubworthy
Darracott Park
Canworthy Water
Canworthy
Copthorne
Bennacott
BOSCASTLE
Penally Pt
MUSEUM OF WITCHCRAFT
Forrabury
Boscastle
Short Island
Grower Rock
Long Island
Lye Rock
Bossiney Haven
Trevalga
Lesnewth
Otterham
Otterham Down
Otterham Station
Hallgarden
Warbstow
Warbstow Cross
Fonston
Trelash
Downinney
Troswell
Brazacott
Billacott
Trillacott
Langdon
North Petherwin
Petherwin Gate
TAMAR OTTER & WILDLIFE CENTRE
TINTAGEL CASTLE
Tintagel
Bossiney
Halgabron
Trethevy
St Nectan's Glen
OLD POST OFFICE
Treven
Tregatta
Truas
Trewarmett
Trebarwith
Hendraburnick
Hendraburnick Down
Davidstow
Hallworthy
Trewassa
Tremail
Treneglos
Three Hammers
Splatt
Tremaine
Tresmeer
Badgall
Trebeath
Tregeare
Egloskerry
Badharlick
Langore
Cold Northcott
Kitttow's Moor
BRITISH CYCLING MUSEUM
THE ARTHURIAN CENTRE
Slaughterbridge
Penpethy
Treligga
Delabole
Rockhead
Pengelly
Camelford
Trevia
Hendra
Tregoodwell
Davidstow Moor
Davidstow Moor Airfield (disused)
Crowdy Reservoir
St Clether
Laneast
Piper's Pool
Tregadillett
HIDDEN VALLEY DISCOVERY FARM
TRETHORNE LEISURE PARK
LAUNCESTON STEAM RAILWAY
Truscott
Newmills
Kennards House
Daw's House
Valley Truckle
Lanteglos
Tramagenna
Trewalder
NORTH CORNWALL MUS & GALLERY
Pencarrow
Watergate
Highertown
Tresinney
Rough Tor
Little Rough Tor
High Moor
Bowithick
Buttern Hill
Bray Down
Trewen
Lower Tregunnon
Polyphant
Lewannick
Two Bridges
South Petherwin
Trewennan
Treforda
Helstone
Knightsmill
Tregreenwell
St Teath
Treveighan
Roughtor Moors
Maiden Tor
Brown Willy
Leskernick Hill
South Carne
Trewint
Fivelanes
Altarnun
ST NONNA'S CHURCH
WESLEY'S COTTAGE
Trebant
Plusha
Trevadlock
Michaelstow
Fentonadle
Trevenning
Harpur's Downs
Candra Hill
Alex Tor
Casehill Downs
Garrow Downs
Butters Tor
Catshole Tor
Codda Tor
Codda
The Beacon
Fox Tor
Tolcarne Tor
Trenhorne
Congdon's Shop
Illand
Newtown
B O D M I N
M O O R
Treswallock Downs
King Arthur's Downs
King Arthur's Hall
Emblance Downs
Garrow Tor
Scribble Downs
Priddacombe Downs
Tolborough
Palmersbridge
East Moor
Carey Tor
Ridge
Bastreet Downs
Trebartha
North Hill
Coad's Green
Tuckingmill
Churchtown
Gam
St Breward
Penrose
Row
St Tudy
Lady Down
Penwood
Carkees Tor
Shallow Water Common
Butterstor Downs
Roughlands
Jamaica Inn (Hotel)
MUSEUM OF SMUGGLING
Bolventor
Toddy Park
Rushyford Gate
Withey Brook
Hawk's Tor
Lank
Bradford
Delford
Carbilly Tor
Hawkstor Downs
STRIPPLE STONES
Brockabarrow Common
Stanning Hill
Minzies Downs
Dozmary Hill
Trezibbett
Smallacoombe Downs
Twelve Men's Moor
Kilmar Tor
Bearah Tor
Berriowbridge
Middlewood
Bathpool
Wenfordbridge
Penpont
Pendrift
Keybridge
Merry Meeting
Blisland
Kerrow
Metherin
Carwen
Menacrin Downs
Temple Tor
Temple
A30
Bois Ho
COLLIFORD LAKE
Dozmary Pool
Gillhouse Downs
Harrowbridge Hill Fm
Ninestones
Hill Tor
Newel Tor
Sharptor
Kingbeare
Botternell
Henwood
North Darley
Billaton
Rilla Mill
Longstone
Trequites
Trehudreth Downs
Waterloo
Blacktor Downs
Simon's Stone
Colliford Lake
Brown Gelly
Browngelly Downs
Bunning's Park
Lord's Park Fm
Westerlake Fm
Stowe's Hill
Witheybrook Marsh
Darleyford
STERTS ARTS CENTRE
Upton Cross
Caradon Town
Helligan Barton
Tredethy
Hellandbridge
Helland
Colvannick Tor
Cardinham Moor
Hard Head
Redhill Downs
Colliford Downs
Sparretts Fm
THE HURLERS STONE CIRCLES
Craddock Moor
Minions
CARADON MINING DISTR CT
Caradon Hill
Gonamena
Higher Carblake
Praze
Trago
Millpool
St Bellarmin's Tor
Maidenwell
Carburrow Tor
Letter Moor
Redhill Downs
Draynes Common
Siblyback Lake
Cryllia Fm
Pensilva
Racecourse Downs
Cardinham Downs
Fore Downs
Warleggan Down
Hobb's Hill
Mutton's Downs
Whitebarrow Downs
Wortha
Northwood
Draynes
GOLITHA FALLS
KING DONIERT'S INSCRIBED STONE
Trekeivesteps
Common Moor
Darite
Tokenbury Corner
Middlehill
Higher Tremarcoombe
Tremarcoombe
TRETHEVY QUOIT
Crow's Nest
Charaton Cross
Clerkenwater
ST PETROC'S CHURCH
Ind Est
Cardinham
Milltown
Old Cardinham Castle
Treslea Downs
Pantersbridge
Warleggan
Mount
Berry Down
Tremaddock
St Neot
Trenant
Treworrick
St Cleer
Tremar
Woolston
Gang
St Ive
A39
A395
A30
B3263
B3266
B3314
B3262
B3254
B3257
B3267

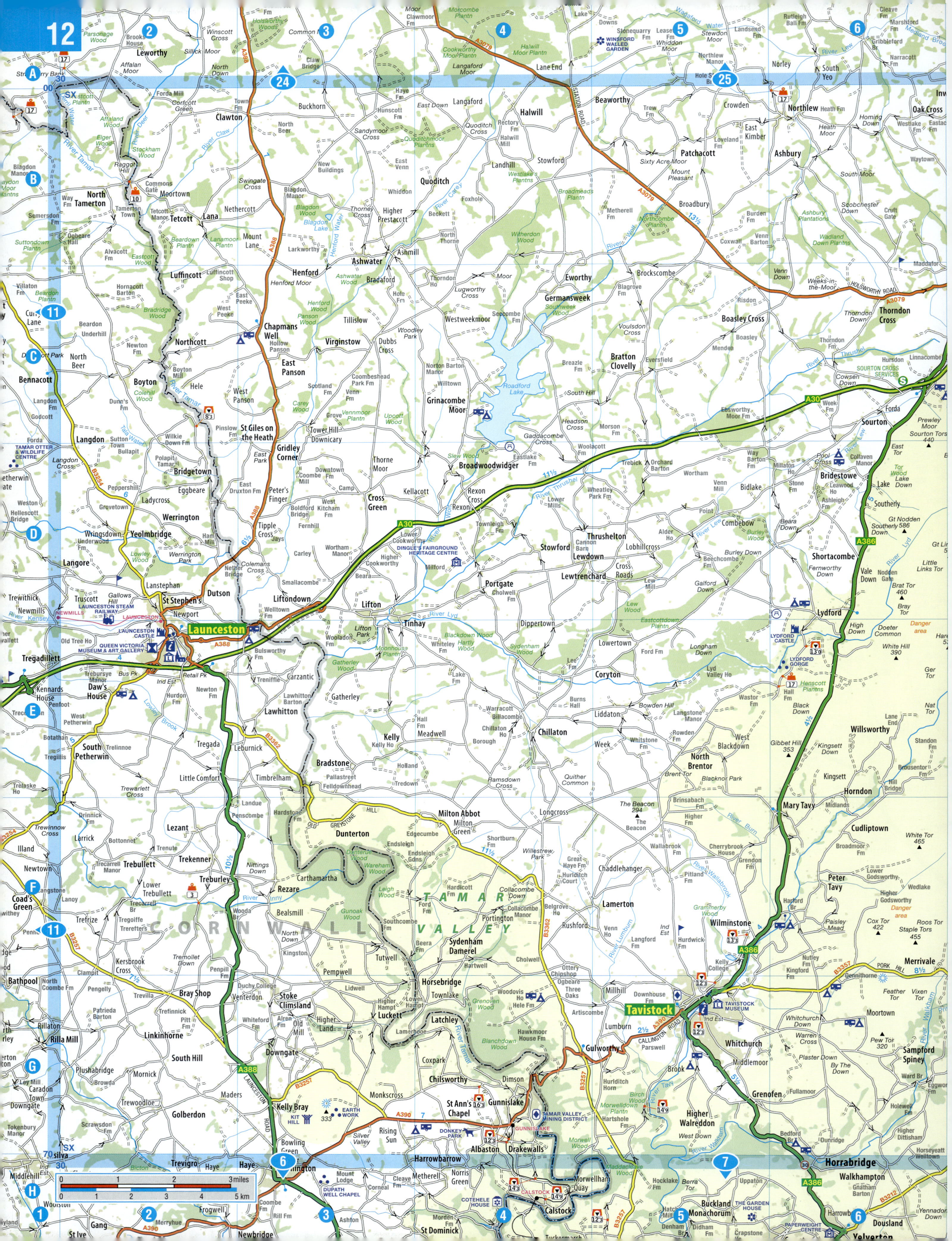

12
Parsonage Wood
Brooks House
Leworthy
Affalan Moor
Winscott Cross
Sillick Moor
North Down
Holsworthy Woods
Common Moor
Claw Bridge
Moor Clawmoor Fm
Morcombe Plantn
Cookworthy Moor Plantn
Langaford Moor
Halwill Moor Plantn
Lane End
Lake
Downs
Winsford Walled Garden
Stonequarry Fm
Lease Fm
Whiddon Moor
Stewdon Moor
Landsend Fm
Northwick Manor
Rutleigh Ball Fm
Cleave Fm
Marshford Fm
Gribbleford
Norley
South Yeo
Narracott
Hole St
Forda Mill
Corkcott Green
Clawton
Town Fm
Buckhorn
North Beer
Hayes Fm
Hunscott Fm
East Down
Langaford
Halwill
Quoditch Cross
Rectory Fm
Halwill Mill
Beaworthy
Trew Fm
Crowden
East Kimber
Loveland Fm
Northlew
Heath Fm
Heath Moor
Oak Cross
Westlake Fm
Homing Down
Eastacott Fm
Sandymoor Cross
Quoditchmoor Plantns
River Tamar
River Deer
River Claw
Affaland Wood
Elgar Wood
Stockham Wood
Raggot Hill
Commons Gate
Moortown
North Tamerton
Tamerton Town
Blagdon Manor
Swingate Cross
New Buildings
East Venn
Whiddon
Quoditch
Landhill
Stowford
Westlake's Plantns
Sixty Acre Moor
Mount Pleasant
Patchacott
Ashbury
Waytown
South Moor
Broadbury
Broadmeads Plantn
Metherell Fm
Northcombe Plantns
Ashbury Plantations
Sobchester Down
Cruft Gate
Burden Fm
Venn Barton
Wadland Down Plantns
Coxwall
Tetcott Manor
Tetcott
Lana
Nethercott
Blagdon Wood
Blagdon Lake
Thorney Cross
Higher Prestacott
Beckett
Foxhole
River Carey
Henford Water
Semersdon Fm
Way Fm
Ogbeare Hall
Suttondown Plantn
Beardown Plantn
Lanamoor Plantn
Mount Lane
Larkworthy
Ashmill
North Thorne
Witherdon Wood
Alvacott Fm
Eastcott Wood
Luffincott
Luffincott Shop
Henford
Henford Moor
Ashwater
Ashwater Wood
Bradaford
Thorndon Ho
Moor
Eworthy
Brockscombe
Venn Down
Weeks-in-the-Moor
Holsworthy Road
Maddaford
Villaton
Beardon Plantn
Hornacott Barton
Bradridge Wood
Curry Lane
East Peeke
West Peeke
Henford Wood
Panson Wood
Hole Fm
Lugworthy Cross
Germansweek
Southweek Wood
Blagrove Fm
Risdon
Thorndon Down
Thorndon Cross
Beardon
Underhill
Northcott
Chapmans Well
Hollow Panson
Tillislow
Westweekmoor
Seccombe Fm
Voulsdon Cross
Boasley Cross
Boasley
North Beer
Newton Fm
Virginstow
Dubbs Cross
Woodley Park
Mendon
Thornbury Fm
Linnacombe
Bennacott
Boyton
Boyton Mill
Hele
East Panson
Norton Barton Manor
Willtown
Breazle Fm
Bratton Clovelly
Eversfield
River Thrushel
Cowsen Down
Sourton Cross Services
Langdon Fm
Godcott
Colehill Wood
Dunn's Fm
West Panson
Scotland Fm
Coombeshead Park Fm
Venn Fm
Carey Wood
Grinacombe Moor
Roadford Lake
South Hill
Ebsworthy Moor Fm
Week Fm
Forda
Sourton
Prewley Moor
Sourton Tors
Langdon
Sutton Town
Bullapit
Wilkie Down Fm
Pinslow Fm
St Giles on the Heath
Gridley Corner
Grove
Vennmoor Plantn
Upcott Wood
Tower Hill
Downicary
Headson Cross
Morson Fm
Gaddacombe Cross
Woolacott Fm
Eastlake Fm
Pool Cross
Colleven Manor
East Tor
Tamar Otter & Wildlife Centre
Langdon Cross
Polapit Tamar
Bridgetown
East Park
Thorne Moor
Slew Wood
Broadwoodwidger
Trebick
Orchard Barton
Wortham
Way Barton Fm
Millaton Ho
Bridestowe
Tor Wood
Lake Down
Lake
Peppershill
Eggbeare
East Druxton Fm
Peter's Finger
Coombe Mill
Downtown
Camp
Cross Green
Kellacott
Rexon Cross
Rexon
River Thrushel
Lower Mills
Wheatley Park Fm
Venn Hill
Bidlake
Stone Fm
Leawood Ho
Ashleigh Ho
Southerly
Weston
Hellescott Bridge
Gravetown
Ladycross
Werrington
Boldford Bridge
West Kitcham Fm
Fernhill
A30
Townleigh Fm
Point
Combebow
Burley Wood
Beara Down
Gt Nodden
Southerly Down 586
Tipple Cross
Jays
Carley
Wortham Manor
Lower Cookworthy
Dingle's Fairground Heritage Centre
Stowford
Thrushelton
Cannon Barn
Lewdown
Lobhillcross
Alder Ho
River Lew
Burley Down
Beechcombe Fm
Shortacombe
Fernworthy Down
Gt Links Tor
Little Links Tor
Wingsdown
Underwood
Yeolmbridge
Werrington Park
Ham Mill
Higher Cookworthy
Beara
Milford
Portgate
Lewtrenchard
Cross Roads
Lew Mill
Galford Down
Vale Down
Nodden Gate
Brat Tor 460
Bray Tor
Langore
Lowley Wood
Nether Bridge
Colemans Cross
Smallacombe
Cholwell Fm
Lew Wood
Eastcottdown Plantn
Trewithick
Newmills
Gallows Hill
Truscott
Launceston Steam Railway
Lanstephan
St Stephen's
Newport
Dutson
Liftondown
Lifton
Welltown
River Lyd
Tinhay
Dippertown
Lydford
High Down
Doeter Common
Danger area
Launceston
Launceston Castle
Queen Victoria Museum & Art Gallery
Old Tree Ho
Tregadillett
Trebursye Manor
Bus Pk
Ind Est
Retail Pk
Bulsworthy Fm
Lifton Park
Wooladon Fm
Moorshole Plantn
Whiteley Fm
Blackdown Wood
Hartly Wood
Sydenham Wood
Lowertown
Lee Fm
Ford Fm
Longham Down
Lydford Castle
Lydford Gorge
White Hill 390
Kennards House
Daw's House
Hurdon Fm
Treniffle
Carzantic
Gatherley Wood
Lake Fm
Coryton
Lyd Valley Ho
Hall Fm
Henscott Plantns
Ger Tor
Trecarne
Penfoot
West Petherwin
Newton Fm
Lawhitton Barton
Lawhitton
Gatherley
Burns Hall
Liddaton
Bowden Hill
Wastor Fm
Black Down
Nat Tor
Botathan
Lowley Brook
Hall Fm
Meadwell
Warracott
Billacombe
Chillaton Ho
Chillaton
Langstone Manor
Willsworthy
Lane End
Standon Fm
South Petherwin
Trelinnoe
Tregilliss
Tregada
Leburnick
Kelly
Kelly Ho
Borough
Week
Whitstone Fm
Rowden Fm
West Blackdown
Gibbet Hill 353
Kingsett Down
Brousentor Fm
North Brentor
Brent Tor
Blacknor Park
Trelaske Ho
Trewarlett Cross
Little Comfort
Timbrelham
Bradstone
Pallastreet
Felldownhead
Holland
Tredown
Ramsdown Cross
Quither Common
Kingsett
Horndon
Hill Bridge
Drinnick Fm
Lezant
Landue
Penscombe
Hardstone Fm
Greystone
Hill
Milton Abbot
Milton Green
Longcross
The Beacon 294
The Beacon
Brinsabach Fm
Higher Fm
River Burn
Mary Tavy
Midlands
Cudliptown
Trewinnow Cross
Larrick
Bottonnet
Trenute
Dunterton
Edgecumbe
Endsleigh
Shortburn Fm
Willestrew Park
Great Haye Fm
Hurdwick Court
Wallabrook Fm
Cherrybrook House
Grendon Fm
White Tor 465
Broadmoor Fm
Illand
Newtown
Trebullett
Trekenner
Treburley
Nittings Down
Carthamartha
Wareham Wood
Endsleigh Gdns
Chaddlehanger
Pitland Fm
River Wallabrook
Lower Godsworthy
Peter Tavy
Wedlake
Lanoy
Lower Trebullett
Trecarrell Manor
Rezare
River Inny
Leigh Wood
Hardicott
Collacombe Down
Tamar Valley
Portington
Collacombe Manor
Belgrove Ho
Rushford
Lamerton
Grammerby Wood
Wilmstone
Higher Godsworthy
Danger area
Cox Tor 422
Roos Tor
Staple Tors 455
Coad's Green
Trefrize
Tregoiffe
Trereiffe
Bealsmill
Gunoak Wood
Ford Fm
Southcombe Fm
Sydenham Damerel
Venn Ho
Langford Fm
Ind Est
Hurdwick Fm
Paisley Mead
Harford Br
Penn
Cornwall
North Down
Kingston
Tutwell
Beera Fm
Hartwell
Cholwell
River Lumburn
Kelly College
Nutley Fm
Kingford Fm
Merrivale
Pork Hill
Bathpool
North Coombe Fm
Clampit
Kersbrook Cross
Tremollet Down
Penpill Fm
Pempwell
Horsebridge
Townlake
Ottery Chipshop
Ogbeare
Three Oaks
Dennithorne
Feather Tor
Vixen Tor
Pengelly
Trevilla
Bray Shop
Duchy College
Venterdon
Stoke Climsland
Lidwell
Higher Hampt
Lower Hampt
Grenoven Wood
Woodovis Ho
Hele Fm
Artiscombe
Millhill
Downhouse Fm
Tavistock
Tavistock Museum
Whitchurch Down
Moortown
Patrieda Barton
Trefinnick
Pitt Fm
Whiteford
Alton Fm
Old Mill
Higher Land
Luckett
Latchley
Lamerhooe
River Tamar
Hawkmoor House Fm
Blanchdown Wood
Lumburn
Callington Road
Parswell
Ind Est
Whitchurch
Warren Cross
Rillaton
Rilla Mill
Linkinhorne
South Hill
Downgate
Gulworthy
Brook
Middlemoor
Plaster Down
By The Down
Pew Tor 320
Sampford Spiney
Plushabridge
Browda
Mornick
Coxpark
Chilsworthy
Dimson
Hurlditch Horn
Birch Wood
River Tavy
Grenofen
Fullamoor
Ward Br
Eggworthy
Caradon Town
Downgate
Trewoodloe
Maders
Launceston Road
Kelly Bray
Kit Hill 333
Earth Work
Monkscross
St Ann's Chapel
Gunnislake
Tamar Valley Mining District
Morwellodown Plantn
Hartshole Fm
Higher Walreddon
Holewell Fm
Tokenbury Manor
Scrawsdon Fm
Golberdon
Silver Valley
Rising Sun
Donkey Park
Albaston
Gunnislake
Drakewalls
Morwell Wood
West Down
River Walkham
Bedford Br
Dunridge
Higher Dittisham
Horseyeatt
Welltown
Silva
Trevigro
Haye
Callington
Bowling Green
Harrowbarrow
Norris Green
Metherell
Cleave Fm
Mount Lodge
Morwellham Quay
Hocklake Fm
Berra Tor
Uppaton
Horrabridge
Walkhampton
Gnatham Barton
Middlehill
Dupath Well Chapel
Cotehele House
Calstock
Buckland Monachorum
The Garden House
Paperweight Centre
Harrowbeer
Dousland
Yennadon Down
Woolston
Coombe Fm
Rill Fm
Ashton
Morden Fm
St Dominick
Tuckermarsh
Denham Br
Didham Fm
Crapstone
Yelverton
Gang
St Ive
Merrylue
Frogwell
Newbridge
Lanxland
A388
A3079
A30
A386
A390
A3072
A3362
B3254
B3257
B3362
B3357
B3212
24
25
SX 30
00
70
1
2
3
4
5
6
7
11
A
B
C
D
E
F
G
H
0 1 2 3 miles
0 1 2 3 4 5 km

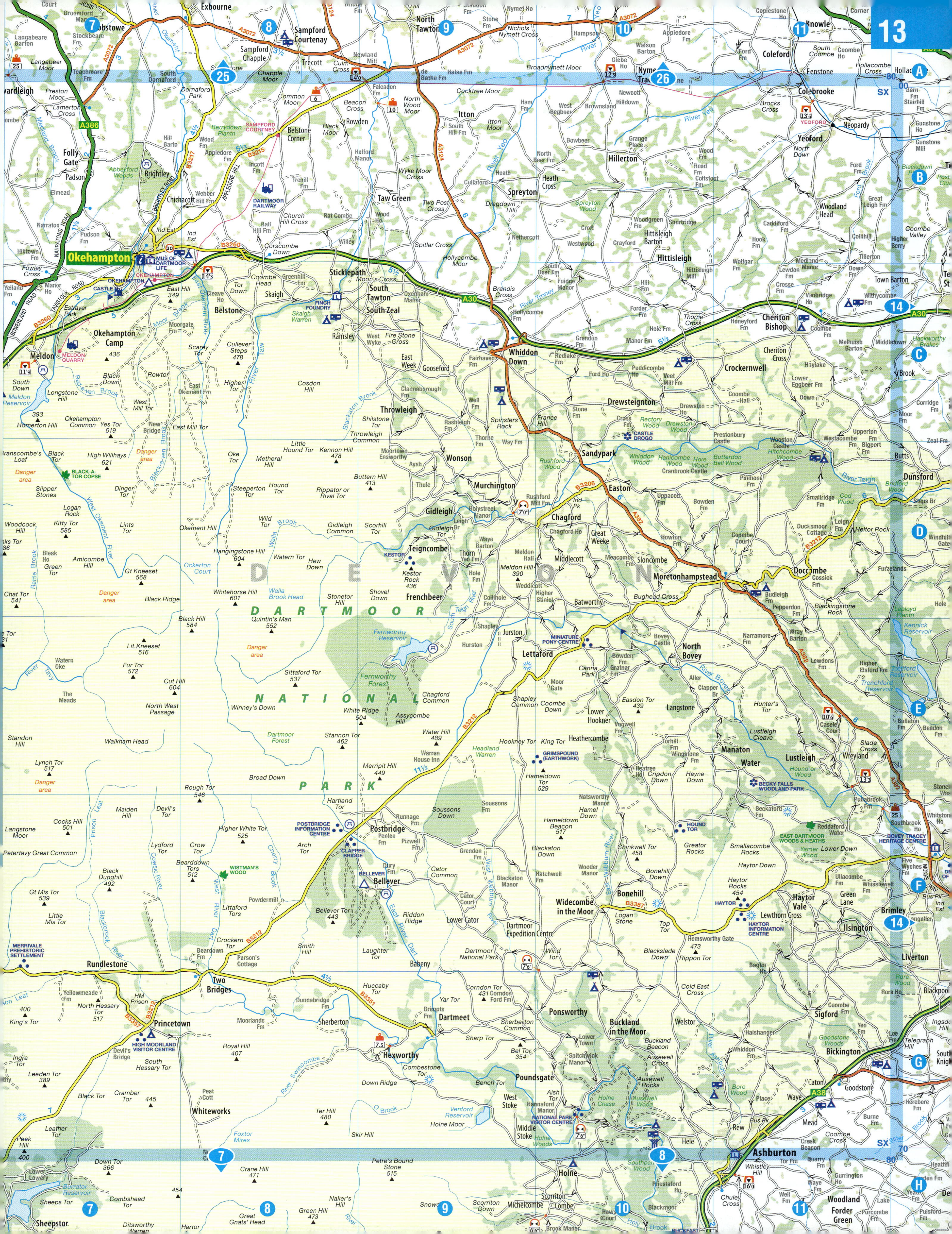
Exbourne
Sampford Courtenay
Sampford Chapple
Trecott
North Tawton
Nichols Nymett Cross
Hampson
Copplestone
Knowle
Coleford
Bobstowe
Stockbeare Fm
Langabeare Barton
Langabeare Moor
Chapple Moor
South Dornaford
Culm Cross
Newland Mill
Bathe Fm
Halse Fm
Broadnymett Moor
Glebe Ho
Nymet Tracey
Walson Barton
Appledore Fm
Ford Ho
South Coombe Fm
Coombe Ho
Fenstone
Hollacombe Cross
Colebrooke
Barn Fm
Stairhill Fm
Preston Moor
Lamerton Cross
Dornaford Park
Common Moor
Beacon Cross
Falcadon Fm
North Wood Moor
Cocktree Moor
Itton
Itton Moor
South Hill Fm
Ham Fm
West Begbeer
Brownsland
Newcott
Hilldown
Brocks Cross
Neopardy
Yeoford
North Dowr
Gunstone Ho
Gunstone Mill
Folly Gate
Padson
Brightley
Abberford Woods
Berrydown Plantn
Sampford Courtenay
Belstone Corner
Black Moor
Rowden
Hill Barton
Wood Fm
Appledore Fm
Incott Fm
Halford Manor
Wyke Moor Cross
Bowbeer
North Beer Fm
Grange Place
Hillerton
Wood Fm
Road Fm
Cottsfoot Fm
Ford Fm
Blackdown Fm
Elmead
Chichacott
Webber Hill Fm
Dartmoor Railway
Trehill Fm
Taw Green
Two Post Cross
Heath
Heath Cross
Spreyton
Dragdown Hill
Spreyton Wood
Woodland Head
Great Leigh Fm
Narraton
Padson Fm
Ind Est
Church Hill Cross
Rat Combe
Wood Ho
Cullaford Fm
Croft
Woodgreen Fm
Shortridge
Hittisleigh Barton
Caddiford Fm
Coombe Valley
Higher Berry
Ball Hill Fm
Corscombe Down
Willey
Spitlar Cross
Hollycombe Moor
Nethercott
Crayford
Westwood
Hook Fm
Collihill
Hittisleigh
Hilltown
Okehampton
Mus of Dartmoor Life
Fowley Cross
Greenhill Fm
Coombe Head
Sticklepath
Moon's Cross
South Beer Fm
Fuidge Manor
Hittisleigh Mill
Wolfgar Fm
Medland Manor
Lewdon Fm
Tillerton
Town Barton
Yellands Fm
Manor Ho
Okehampton Castle
East Hill
Tor Down
Skaigh
South Tawton
Oxenham Manor
Brandis Cross
Hill Fm
Crosse Fm
Vanbridge Ho
Withycombe Fm
Estrayer Park
Cleave Ho
Belstone
Finch Foundry
South Zeal
Hollycombe Fm
Forder Fm
Thorne Cross
Honeyford Fm
Cheriton Bishop
Coombe Fm
Skaigh Warren
Ramsley
Okehampton Camp
Moorgate Fm
Scarey Tor
West Wyke
Fire Stone Cross
Hole Fm
Manor Fm
Grendon Fm
Melhuish Barton
Middletown
Hackworthy Brakes
Meldon
Meldon Quarry
Culliver Steps
East Week
Gooseford
Fairhaven Fm
Whiddon Down
Redlake Fm
Cheriton Cross
Crockernwell
Haylake
Brook
South Down
Black Down
Rowtor
Higher Tor
Cosdon Hill
Ford Ho
Puddicombe Ho
West Mill Fm
Lower Eggbeer Fm
Coombe Hall
Down Fm
Meldon Reservoir
Longstone Hill
West Mill Tor
East Okement Fm
Clannaborough Fm
Throwleigh
Well Fm
Stone Fm
Drewsteignton
Drewston Fm
Corridge Fm
Moor Fm
Homerton Hill
Okehampton Common
Yes Tor
New Bridge
East Mill Tor
Shilstone Tor
Throwleigh Common
Rashleigh Fm
Spinsters Rock
France Hill
Cross Fm
Rectory Wood
Drewston Wood
Castle Drogo
Prestonbury Castle
Westacombe Fm
Upperton Fm
Bigport Fm
Zeal Fm
Branscombe's Loaf
Black Tor
High Willhays
Danger area
Little Hound Tor
Kennon Hill
Oke Tor
Moortown Ensworthy
Thorne Fm
Way Fm
Sandypark
Whiddon Wood
Hanicombe Wood
Hore Wood
Butterdon Ball Wood
Cranbrook Castle
Wooston Castle
Hitchcombe Wood
Butts
Black-a-Tor Copse
Metherall Hill
Aysh
Wonson
Rushford Wood
Pinmoor Fm
River Teign
Dunsford
Bridford Wood
Slipper Stones
Dinger Tor
Steeperton Tor
Hound Tor
Rippator or Rival Tor
Buttern Hill
Thule
Murchington
Rushford Mill Fm
Easton
Uppacott Fm
Bowden Fm
Smallridge Fm
Cod Wood
Steps Br
Logan Rock
Kitty Tor
Lints Tor
Gidleigh
Holystreet Manor
Chagford
Ind Pk
Chagford Ho
Great Weeke
Coombe Court
Ducksmoor Cottage
Leign Fm
Heltor Rock
Woodcock Hill
Okement Hill
Wild Tor
Gidleigh Common
Scorhill Tor
Gidleigh Tor
Waye Barton
Howton Fm
Bleak Ho
Green Tor
Amicombe Hill
Hangingstone Hill
Watern Tor
Hew Down
Teigncombe
Kestor
Kestor Rock
Thorn
Meldon Hall
Middlecott
Meacombe
Sloncombe
Moretonhampstead
Doccombe
Furzelands
Gt Kneeset
Ockerton Court
Chat Tor
Black Ridge
Whitehorse Hill
Walla Brook Head
Stonetor Hill
Shovel Down
Frenchbeer
Hole Fm
Meldon Hill
Weddicott
Collihole Fm
Higher Stiniel
Batworthy
Bughead Cross
Budleigh Fm
Pepperdon Fm
Blackingstone Rock
Cossick Fm
Hole
Black Hill
Dartmoor
Quintin's Man
Shapley
Hurston
Jurston
Miniature Pony Centre
Bovey Castle
North Bovey
Wray Barton
Narramore Fm
Laployd Plantn
Kennick Reservoir
Fernworthy Reservoir
Devon
Lettaford
Bowden Fm
Gratnar
Canna Park
Lewdons Fm
Higher Elsford Fm
Tottiford Reservoir
Trenchford Reservoir
Lit. Kneeset
Watern Oke
Fur Tor
Cut Hill
Sittaford Tor
Fernworthy Forest
Danger area
Moor Gate
Aller
Clapper Br
River Bovey
Hunter's Tor
The Meads
North West Passage
National
Chagford Common
Shapley Common
Coombe Down
Easdon Tor
Langstone
Bullaton Fm
Beadon Fm
Winney's Down
White Ridge
Assycombe Hill
Lower Hookner
Casely Court
Standon Hill
Walkham Head
Dartmoor Forest
Stannon Tor
Water Hill
Hookney Tor
King Tor
Heathercombe
Vogwell Fm
Torhill Fm
Lustleigh Cleave
Slade Cross
Wreyland
Headland Warren
Warren House Inn
Wingstone Fm
Manaton
Lustleigh
Lynch Tor
Grimspound (Earthwork)
Water
Hound or Wood
Broad Down
Merripit Hill
Hameldown Tor
Heatree Ho
Cripdon Down
Hayne Down
Becky Falls Woodland Park
Rough Tor
Park
Natsworthy Manor
Hamel Down
Pullabrook Fm
Maiden Hill
Devil's Tor
Cocks Hill
Hartland Tor
Soussons Down
Soussons Fm
Hameldown Beacon
Hound Tor
Beckaford Fm
Reddaford Water
Southbrook Ho
Bovey Tracey Heritage Centre
Langstone Moor
Higher White Tor
Postbridge Information Centre
Postbridge
Runnage Fm
Penlee Fm
Pizwell Fm
Grendon Fm
Blackaton Down
Chinkwell Tor
Greator Rocks
Smallacombe Rocks
East Dartmoor Woods & Heaths
Yarner Wood
Lower Down
Petertavy Great Common
Lydford Tor
Crow Tor
Beardown Tors
Arch Tor
Clapper Bridge
Cator Common
Hatchwell Fm
Wooder Manor
Bonehill Down
Haytor Down
Ullacombe Fm
Whisselwell Fm
Five Wyches Cross
Black Dunghill
Wistman's Wood
Bellever
Dury Fm
Blackaton Manor
Widecombe in the Moor
Bonehill
Haytor Rocks
Haytor
Haytor Vale
Green Lane
Bus Pk
Gt Mis Tor
Powdermill
Cator Court
Logan Stone
Top Tor
Lewthorn Cross
Haytor Information Centre
Ilsington
Brimley
Little Mis Tor
Littaford Tors
Bellever Tor
Riddon Ridge
Lower Cator
Dartmoor Expedition Centre
Hemsworthy Gate
Merrivale Prehistoric Settlement
Crockern Tor
Parson's Cottage
Smith Hill
Laughter Tor
Babeny
Dartmoor National Park
Wind Tor
Blackslade Down
Rippon Tor
Bagtor Ho
Liverton
Rundlestone
Beardown Fm
Two Bridges
Huccaby Tor
Corndon Tor
Corndon Ford Fm
Cold East Cross
Rora Ho
Blackpool
Yellowmeade Fm
HM Prison
North Hessary Tor
Dunnabridge Fm
Yar Tor
Brimpts Fm
Dartmeet
Ponsworthy
Buckland in the Moor
Welstor
Sigford
Coombe Fm
King's Tor
Princetown
High Moorland Visitor Centre
Moorlands Fm
Sherberton
Sherberton Common
Sharp Tor
Lower Town
Buckland Beacon
Halshanger
Goodstone Woods
Bickington
Telegraph Hill
Devil's Bridge
South Hessary Tor
Royal Hill
Hexworthy
Bel Tor
Spitchwick Manor
Ausewell Cross
Whiddon Fm
Ingra Tor
Leeden Tor
Combestone Tor
Poundsgate
Ausewell Rocks
Holne Chase
Caton Fm
Goodstone
Black Tor
Cramber Tor
Peat Cott
Down Ridge
Bench Tor
Aish Tor
West Stoke
Hannaford Manor
Boro Wood
Place
Waye
Herebere Fm
Whiteworks
Ter Hill
Venford Reservoir
National Park Visitor Centre
Burne Fm
Leather Tor
Skir Hill
Holne Moor
Middle Stoke
Holne Woods
Hele
Rew
Bus Pk
Mead
Coombe Cross
Peek Hill
Foxtor Mires
Holne
Ashburton
Quarry Fm
Creek Beacon
Down Tor
Crane Hill
Petre's Bound Stone
Southpark Wood
Whistley Hill
Gurrington Ho
Lower Lowery
Burrator Reservoir
Combshead Tor
Green Hill
Naker's Hill
Scorriton
Priestaford Fm
Chuley Cross
Wayе Fm
Woodland
Forder Green
Sheeps Tor
Great Gnats' Head
Snowdon
Scorriton Down
Michelcombe
Combe
Hawson Court
Blackmoor
Well Fm
Purcombe Fm
Pulsford Fm
Sheepstor
Ditsworthy Warren
Hartor
Brook Manor
Buckfast
A3072
A386
A3124
A30
B3260
B3215
B3219
B3206
A382
B3212
B3357
B3344
B3387
A38
B3212
25
26
7
8
9
10
11
12
13
14
A
B
C
D
E
F
G
H
SX
80
00
70

Crediton
Shobrooke
Thorverton
Ellerhayes
Nether Exe
Rewe
Bradninch
Sweetham
Newton St Cyres
Brampford Speke
Stoke Canon
Poltimore
Broadclyst
Dog Village
Budlake
Clyst Hydon
Clyst St Lawrence
Westwood
Venny Tedburn
Cowley
Jack in the Green
Rockbeare
Tedburn St Mary
Whitestone
Pinhoe
Monkerton
Clyst Honiton
Marsh Green
Exwick
Exeter
Whipton
Heavitree
Sowton
Aylesbeare
Wheatley
St Thomas
Ide
Longdown
Alphington
Countess Wear
Clyst St Mary
Farringdon
Dunsford
Reedy
Shillingford Abbot
Shillingford St George
Topsham
Clyst St George
Woodbury Salterton
Ebford
Exminster
Bridford
Bridfordmills
Doddiscombsleigh
Woodbury
Exton
Christow
Kennford
Kenn
Lympstone
Lower Ashton
Higher Ashton
Powderham
Kenton
Trusham
Starcross
Withycombe Raleigh
Exmouth
Littleham
Hennock
Cockwood
Chudleigh
Ashcombe
Dawlish Warren
Bovey Tracey
Chudleigh Knighton
Ideford
Luton
Dawlish
Heathfield
Liverton
Coldeast
Holcombe
Preston
Bishopsteignton
Kingsteignton
Teignmouth
Shaldon
Highweek
Newton Abbot
Combeinteignhead
Stokeinteignhead
Netherton
Kingskerswell
Abbotskerswell
Maidencombe
River Exe
River Teign
Cockle Sand
Langstone Rock
Shag Rock
Sprey Pt
Labrador Bay
Straight Pt
Orcombe Rocks
Conger Rocks
Maer Rocks
A377
A30
A38
A380
A376
A379
A381
A383
M5
SX
0
1
2
3miles
4
5 km

Payhembury
Awliscombe
Honiton
Colestocks
Buckerell
Weston
Feniton
Fenny Bridges
Alfington
Gittisham
Hamlet
Littletown
Offwell
Wilmington
Cleave
Dalwood
Widworthy
Axminster
Kilmington
Shute
Weycroft
Millbrook
Talaton
Talewater
Fairmile
Coombelake
Taleford
Woodford
Ottery St Mary
Aller Grove
West Hill
Wiggaton
Fluxton
Metcombe
Venn Ottery
Tipton St John
Southerton
Harpford
Bowd
Sidbury
Sidford
Salcombe Regis
Sidmouth
Cotmaton
Bulverton
Stowford
Newton Poppleford
Burrow
Hawkerland
Colaton Raleigh
Otterton
East Budleigh
Knowle
Little Knowle
Budleigh Salterton
Church Green
Farway
Northleigh
Woodbridge
Southleigh
Harcombe
Trow
Weston
Branscombe
Street
Vicarage
Beer
Seaton
Seaton Bay
Colyford
Colyton
Axmouth
Musbury
Whitford
Seaton Junction
Combpyne
Rousdon
Holcombe
Trill
Abbey Gate
Wyke Green
Smallridge
Ham
Lower Ridge
Hawley Bottom
A30
A35
A3052
A375
B3174
B3172
B3161
B3178
Beer Head
Haven Cliff
Seaton Hole
Dowlands Cliffs and Landslips
Culverhole Pt
Humble Pt
Axmouth to Lyme Regis Undercliffs
Dorset & East Devon Coast
South West Coast Path
Chapman's Rocks
Salcombe Hill Cliff
Dunscombe Cliff
Weston Cliff
Littlecombe Shoot
Coxe's Cliff
Chit Rocks
Tortoiseshell Rocks
Big Picket Rock
Hern Point Rock
Ladram Bay
Smallstones Pt
Chiselbury Bay
Brandy Head
Black Head
Danger Pt
Fairlynch Arts Centre & Museum
Bicton Park
Otterton Mill
North Star
Bicton
River Otter
River Sid
River Coly
River Axe
Seaton Tramway
Pecorama
Beer Quarry Caves
Blackbury Camp
Farway
Burrow Farm
Shute Barton
Cadhay House
Escot Park
All Hallows Museum
Marine House
Museum
Colyton
Axminster
Honiton
Seaton
E A S T   D E V O N
D E V O N
E A S T
7
8
9
10
11
16
27
28
A
B
C
D
E
F
G
H
SY
30
70
00

Axminster
Smallridge
Weycroft
Millbrook
Sector
Tillworth
Hawkchurch
Wadbrook
Tudhay
Castle
Brimley
Holditch
Blackpool Corner
Marshwood
Birdsmoorgate
Marshalsea
Bettiscombe
Pilsdon
Pilsdon Pen
Lewesdon Hill
Blackney
Broadwindsor
Templeman's Ash
Monkwood
Beaminster
Stoke Abbott
Stoke Water
Netherbury
Whitecross
Mapperton
Melplash
Loscombe
North Poorton
South Poorton
Hooke
Salwayash
Waytown
Oxbridge
Kingsland
Filford
Broad Oak
DORSET
Wootton Fitzpaine
Champernhayes Marsh
Monkton Wyld
Raymond's Hill
Wyke Green
Abbey Gate
Penn
Harcombe Bottom
Uplyme
Holcombe
Combpyne
Rousdon
Yawl
Catherston Leweston
Whitchurch Canonicorum
Ryall
Morcombelake
North Chideock
Chideock
Seatown
Golden Cap
Moorbath
Atrim
Dottery
Pymore
Symondsbury
Allington
Bridport
Coneygar
Eype
Higher Eype
Lower Eype
Watton
Skilling
West Bay
Bothenhampton
Walditch
Bradpole
Loders
Yondover
Uploders
Powerstock
Nettlecombe
West Milton
Askerswell
Spyway
Shipton Gorge
Chilcombe
Burton Bradstock
Southover
Puncknowle
Swyre
West Bexington
Charmouth
Lyme Regis
Cobb
Ware
Seven Rock Pt
Pinhay Bay
Humble Pt
DINOSAURLAND
AQUARIUM
CHARMOUTH HERITAGE COAST CENTRE
SOUTH WEST COAST PATH
St Gabriel's Mouth
Wear Cliffs
Ridge Cliff
East Ebb Cove
Eype's Mouth
West Cliff
East Cliff
Burton Cliff
Burton Mere
DORSET & EAST DEVON COAST
AXMOUTH TO LYME REGIS UNDERCLIFFS
Dowlands Cliffs and Landslips
A35
A3052
B3165
B3162
B3163
B3157
A3066
L Y M E
B A Y
15
28
29
SY
30
70
00
0 1 2 3 miles
0 1 2 3 4 5 km
A
B
C
D
E
F
G
H
1
2
3
4
5
6

Dorchester
Weymouth
Portland
Fortuneswell
Easton
Weston
Southwell
Bill of Portland
Chiswell
Grove
Portland Harbour
Weymouth Bay
Chesil Beach
Wyke Regis
Rodwell
Westham
Melcombe Regis
Radipole
Preston
Overcombe
Littlemoor
Chickerell
Charlestown
Broadwey
Upwey
Nottington
Sutton Poyntz
Osmington
Osmington Mills
Poxwell
Upton
Winfrith Newburgh
Chaldon Herring or East Chaldon
Chaldon Down
Warmwell
Owermoigne
Broadmayne
West Knighton
Whitcombe
West Stafford
Crossways
Moreton
Woodsford
Tincleton
Puddletown
Athelhampton
Tolpuddle
Affpuddle
Burleston
Dewlish
Cheselbourne
Piddlehinton
Piddletrenthide
Plush
Charminster
Stinsford
Kingston Maurward
Lower Bockhampton
Higher Bockhampton
Poundbury
Maiden Castle
Winterborne Monkton
Winterborne Herringston
Martinstown or Winterbourne St Martin
Winterbourne Abbas
Winterbourne Steepleton
Bradford Peverell
Stratton
Grimstone
Frampton
Maiden Newton
Cerne Abbas
Nether Cerne
Godmanstone
Forston
Charlton Down
Sydling St Nicholas
Up Sydling
Cattistock
Chilfrome
Toller Porcorum
Toller Fratrum
Wynford Eagle
West Compton
Compton Valence
Kingston Russell
Long Bredy
Littlebredy
Litton Cheney
Abbotsbury
Portesham
Rodden
Langton Herring
Fleet
East Fleet
Coryates
Bincombe
Ringstead Bay
Redcliff Pt
Black Head
Swyre Head
Bat's Head
Dorset
Chesil Beach
Guernsey 2:10
Jersey 3:25
St Malo 5:15
A35
A37
A352
A353
A354
A356
B3157
B3159
B3143
B3142
B3150
B3147
B3390
A
B
C
D
E
F
G
H
7
8
9
10
11
18
29
30

Milton Abbas
Higher Whatcombe
Whatcombe
Lower Whatcombe
Shapwick
Winterborne Whitechurch
Winterborne Kingston
Winterborne Zelston
Almer
Sturminster Marshall
Newton Peveril
Wimborne Minster
Colehill
Canford Bottom
Oakley
Merley
Canford Magna
Lambs Green
Ashington
Hillbutts
Henbury
Corfe Mullen
Broadstone
Bearwood
POOLE
Milborne St Andrew
Anderson
Mapperton
Bere Regis
Tolpuddle
Affpuddle
Briants Puddle
Turners Puddle
Bloxworth
Morden
East Morden
West Morden
Lytchett Matravers
Lytchett Minster
Upton
Hamworthy
Poole
Parkstone
Lilliput
Sandbanks
Waterloo
Creekmoor
Newtown
Branksome
Oakdale
Sherford
Slepe
Organford
Woodlake
Wareham Forest
Lane End
Pallington
Moreton
Bovington Camp
Clouds Hill
DORSET
Sandford
Holton Heath
Hamworthy
Lower Hamworthy
Poole Harbour
Brownsea Island
Arne
Northport
Wareham
Stokeford
Wool
East Burton
East Stoke
Rushton
Worgret
Stoborough
Stoborough Green
Ridge
Holmebridge
East Holme
West Holme
Tadnoll
Blacknoll
Giddy Green
Braytown
Portway
East Knighton
Winfrith Newburgh
Chaldon Herring or East Chaldon
Coombe Keynes
Whiteway
East Lulworth
West Lulworth
Lulworth Camp
Tyneham
Steeple
Kimmeridge
Creech
East Creech
Furzebrook
Norden
Norden Heath
Corfe Castle
Church Knowle
Kingston
Harman's Cross
Woolgarston
Studland
Ulwell
Swanage
New Swanage
Langton Matravers
Herston
Acton
Worth Matravers
St Aldhelm's or St Alban's Head
Purbeck Hills
Worbarrow Bay
Kimmeridge Bay
Studland Bay
Swanage Bay
Durlston Bay
SOUTH WEST COAST PATH
A35
A31
A350
A351
A352
B3075
B3070
B3071
B3351
B3069
0 1 2 3 miles
0 1 2 3 4 5 km
A B C D E F G H
1 2 3 4 5 6
SY
17
30
31

19
Ferndown
Trickett's Cross
West Parley
Parley Cross
Longham
Dudsbury
Bear Cross
Kinson
Ensbury
Northbourne
BOURNEMOUTH
Moordown
Winton
Talbot Village
Wallisdown
Rossmore
Bourne Valley
Talbot Heath
Talbot Woods
Charminster
Strouden
Queen's Park
Springbourne
Boscombe
Pokesdown
Southbourne
Westbourne
West Cliff
Bournemouth
Branksome Park
Canford Cliffs
Canford Cliffs Chine
Flag Head Chine
Poole Head
Branksome Chine
POOLE BAY
BOURNEMOUTH AIRPORT
ADVENTURE WONDERLAND
Hurn
Merritown
West Hurn
Blackwater
Holdenhurst
Muscliff
Mill Throop
Littledown
Iford
Tuckton
West Southbourne
DORSET
Jumpers Green
St Catherine's Hill
Jumpers Common
Fairmile
Purewell
Christchurch
CASTLE & NORMAN HOUSE
RED HOUSE
Stanpit
Stanpit Marsh
Christchurch Harbour
Wick
Mudeford
Friars Cliff
Sandhills
Hengistbury Head
Somerford
Highcliffe
HIGHCLIFFE CASTLE
CHRISTCHURCH BAY
Walkford
Hinton
Burton
Winkton
Sopley
Avon
Ripley
North Ripley
Kingston
Bisterne
Sandford
Shirley
Bransgore
Thorney Hill
Neacroft
Godwinscroft
North Bockhampton
Middle Bockhampton
South Bockhampton
Waterditch
HANTS
Beckley
SAMMY MILLER MOTORCYCLE MUSEUM
New Milton
Ashley
Old Milton
Barton on Sea
Golden Hill
Bashley
Wootton
Ossemsley
Tiptoe
Sway
Hordle
Downton
Everton
Lymington
Pennington
Lymore
Keyhaven
Milford on Sea
Hordle Cliff
Hurst Beach
Battramsley
Setley
Buckland
Mount Pleasant
APPLE COURT
Leagreen
Newlands Manor
Taddiford Fm
Hatherwood Pt
Alum Bay
CHAIR LIFT
THE NEEDLES PARK
The Needles
THE NEEDLES OLD BATTERY
Scratchell's Bay
Main Bench
Totland Bay
Handfast Point or The Foreland
The Pinnacles
DORSET & EAST DEVON COAST
Ballard Pt
GUERNSEY 2:40 (APR-OCT)
JERSEY 3:00 (APR-OCT)
ST MALO 4:35 (MAY-SEPT)
CHERBOURG 2:15 (MAY-SEPT)
A31
A338
A348
A35
A337
B3073
B3059
B3058
B3055
A B C D E F G H
7 8 9 10 11
31 32 20
SZ

Setley
Lymington
Pennington
Everton
Milford on Sea
Keyhaven
KEYHAVEN & PENNINGTON MARSHES NATURE RESERVE
Lower Pennington
Pennington Marshes
Woodside
Walhampton
Waterford
Buckland
Mount Pleasant
St Austins
Boldre
Pilley
Battramsley
Bull Hill
Portmore
Norleywood
South Baddesley
East End
Sowley Pond
Beaulieu Heath
Hatchet Gate
Hatchet Pond
East Boldre
Bucklers Hard
MARITIME MUSEUM
Exbury
EXBURY GARDENS
STEAM RAILWAY
Blackfield
Langley
West Common
Lepe
Stone Pt
Stansore Pt
Calshot
Inchmery Ho
Needs Ore Pt
NORTH SOLENT
Park Shore
Colgrims
Pitts Deep
Pylewell Pt
Lymington Spit
H A N T S
T H E
Hill Head
Cowes Roads
Old Castle Pt
Cowes
East Cowes
MARITIME MUSEUM
OSBORNE HOUSE
Osborne Bay
King's Quay
Gurnard Bay
Gurnard
Gurnard Ledge
Thorness Bay
Rew Street
Northwood
Hillis Corner
Whippingham
Dallimores
Woodside
Wootton
Wootton Bridge
Wootton Common
Newtown Bay
Newtown
NEWTOWN HARBOUR
OLD TOWN HALL
Hamstead
Porchfield
Locksgreen
COLEMANS FARM PARK
Little Whitehouse
Mark's Corner
Parkhurst
Parkhurst Forest
HM Prisons
Fairlee
CLASSIC BOAT MUS
Hunny Hill
Newport
Barton
Staplers
Shide
Pan
Carisbrooke
CARISBROOKE CASTLE & MUSEUM
Gunville
Forest Side
Cranmore
Ningwood
Shalfleet
Bouldnor
Bouldnor Cliff
Yarmouth
YARMOUTH CASTLE
Hurst Beach
HURST CASTLE
Sconce Pt
FORT VICTORIA & MARINE AQUARIUM
Cliff End
Norton
Norton Green
Colwell Bay
Warden Point
Totland
Totland Bay
Freshwater
Freshwater Bay
Pound Green
Weston Manor
Hatherwood Pt
Alum Bay
THE NEEDLES PARK
THE NEEDLES OLD BATTERY
Scratchell's Bay
Main Bench
Afton
DIMBOLA LODGE
Thorley Street
Thorley
Wellow
Newbridge
CALBOURNE WATER MILL
Calbourne
Five Houses
Shalcombe
CHESSELL POTTERY BARNS
Compton Bay
Shippards Chine
Hanover Pt
Brook
Brook Bay
Brookgreen
Hulverstone
Mottistone
MOTTISTONE MANOR GARDEN
The Undercliff
Brighstone
Brighstone Bay
Brighstone Down
Chilton Chine
Hardman Rock
Grange Chine
ISLE OF WIGHT PEARL
Moortown
Limerstone
Shorwell
Yafford
Thorncross
DINOSAUR FARM MUSEUM
Shepherd's Chine
Atherfield Pt
Little Atherfield
Kingston
Whale Chine
Chale
Chale Bay
Chale Green
Blackgang
BLACKGANG CHINE
Watershoot Bay
St Catherine's Pt
St Catherine's Down
ST CATHERINE'S ORATORY
Niton
Whitwell
Bierley
Kingates
St Lawrence
Binnel Pt
Binnel Bay
Puckaster Cove
Godshill
Sandford
Roud
Wroxall
Southford
Nettlecombe
Rookley
Rookley Green
Merstone
Horringford
Arreton
Downend
ROBIN HILL ADVENTURE PARK
Blackwater
Bowcombe
Clatterford
Gatcombe
Chillerton
Calbourne
I S L E O F W I G H T
Newtown River
River Medina
Needles
0:30
0:23
0:55
32
33
19
SZ
30
70
0 1 2 3 miles
0 1 2 3 4 5 km
A
B
C
D
E
F
G
H
1
2
3
4
5
6

21
Lee-on-the-Solent
Rowner
Elson
Hardway
Brockhurst
Gosport
Privett
Ann's Hill
Browndown Pt
Stokes Bay
Alverstoke
Gilkicker Pt
Fort Monckton
Newtown
Clayhall
Burrow Island
Portsmouth
Portsea
Landport
North End
Kingston
Baffins
Portsea Island
Fratton
Milton
Eastney
Somers Town
Southsea
SOUTHSEA CASTLE
BLUE REEF AQUARIUM
NATURAL HISTORY MUSEUM
CLARENCE PIER
ROYAL MARINES MUSEUM
HARBOUR
Langstone Channel
West Winner
East Winner
Fleet
Manor Ho
Hayling Island
Westfield
West Town
South Hayling
Gable Head
Mengham
Selsmore
Eastoke
Hayling Bay
EAST HAYLING LIGHT RLY
Black Pt
Eastoke Pt
Emsworth Channel
Chichester Channel
Longmere Pt
PILSEY ISLAND
Pilsey Sand
Pilsey Island
East Head
West Wittering
Acre Street
Somerley
East Wittering
Bracklesham
Bracklesham Bay
SOLENT
SPITHEAD
Horse Sand Fort
No Mans Land Fort
St Helen's Fort
RYDE PIER HEAD
RYDE ESPLANADE
Ryde
Pelhamfield
QUARR ABBEY
Binstead
Kite Hill
Quarr Hill
Appley
St John's Park
Puckpool Pt
SEAVIEW WILDLIFE ENCOUNTER
Seagrove Bay
Seaview
Elmfield
Pondwell
Oakfield
Haylands
Swanmore
Weeks
Havenstreet
Upton
Ashey
Nettlestone
The Priory
Priory Bay
St Helens
Bembridge Harbour
Bembridge Pt
Ethel Pt
Bembridge
Steyne Cross
Lane End
Foreland
Foreland Fields
Hillway
Long Ledge
Whitecliff Bay
Culver Cliff
Culver Down
Yaverland
Brading
Yarbridge
Morton
Adgestone
Alverstone
Newchurch
Queen's Bower
Winford
Branstone
Merrie Gardens
Apse Heath
Landguard Manor
Lake
Sandown
SANDOWN BAY
Shanklin
Whiteley Bank
Upper Hyde
SHANKLIN CHINE
Horse Ledge
Shanklin Down
Luccombe Village
Luccombe Bay
Dunnose
Upper Bonchurch
Bonchurch
Ventnor
Horseshoe Bay
Ventnor Bay
Lowtherville
COASTAL VISITOR CENTRE
GUERNSEY 7:00 JERSEY 10:30
SANTANDER 24:00 (MAR-NOV)
CAEN 3:45 (APR-SEPT) CHERBOURG 3:00 (APR-NOV)
CAEN 6:00 LE HAVRE 5:30 ST MALO 9:00
22
SZ
7
8
9
10
11
A
B
C
D
E
F
G
H

22
Havant
Leigh Park
Stockheath
Bedhampton
Langstone
Warblington
Emsworth
Denvilles
Brockhampton
Rowlands Castle
Red Hill
Durrants
Stubbermere
West Leigh
Southleigh Park
Southleigh Forest
Westbourne
Woodmancote
Funtington
Hambrook
Southbourne
Hermitage
Lumley
Inlands
Nutbourne
Prinsted
Chidham
Bosham
Broadbridge
Fishbourne
Chichester
Apuldram
Donnington
Stockbridge
Whyke
Portfield
Walderton
Stoughton
East Ashling
West Ashling
Woodend
West Stoke
Mid Lavant
East Lavant
Summersdale
Westhampnett
Westerton
Strettington
Boxgrove
Halnaker
Tangmere
Oving
Aldingbourne
Westergate
Norton
Nyton
Eartham
Crockerhill
Merston
Runcton
North Mundham
South Mundham
Colworth
Lidsey
Shripney
Elbridge
North Bersted
South Bersted
Rose Green
Aldwick
Bognor Regis
Pagham
Nyetimber
Sidlesham
Sidlesham Common
Hunston
Street End
Highleigh
Birdham
Earnley
Somerley
Bracklesham
East Wittering
West Wittering
Acre Street
Shipton Green
West Itchenor
Bosham Hoe
West Thorney
Thorney Island
Church Norton
Norton
East Beach
Selsey
Selsey Bill
Pagham Harbour
Bracklesham Bay
WEST SUSSEX
CHICHESTER HARBOUR
LANGSTONE HARBOUR
Hayling Island
Northney
North Hayling
Stoke
Fleet
Tye
Westfield
West Town
South Hayling
Gable Head
Mengham
Eastoke
Selsmore
Hayling Bay
East Winner
Black Pt
Eastoke Pt
East Head
Pilsey Island
Pilsey Sand
Chichester Channel
Bosham Channel
Thorney Channel
Emsworth Channel
Sweare Deep
North Binness Island
South Binness Island
Long Island
Baker's Island
Langstone Channel
Goodwood
Goodwood Aerodrome & Motor Racing Circuit
Goodwood House
Barn Rocks
Inner Owers
CAEN 6:00
LE HAVRE 5:30
ST MALO 9:00
CAEN 3:45 (APR-SEPT)
CHERBOURG 3:00 (APR-NOV)
ANDER 24:00 (MAR-NOV)
A3(M)
A27
A259
A286
33
34
21
SU
SZ
70
80
90
A
B
C
D
E
F
G
H
1
2
3
4
5
6
0
1
2
3 miles
0
1
2
3
4
5 km

Uckfield
Rocks Park
Manor Park
Little Streele
Blackboys
Pounsley
Waldron Down
Cross in Hand
Mutton Hall
Broad Oak
Cade Street
Heathfield
Old Heathfield
Punnett's Town
Three Cups Corner
Brightling
Oxley's Green
Hollingrove
Twelve Oaks
Cackle Street
Dallington
Darwell Hole
Netherfield
Ridgewood
New Town
Framfield
Possingworth Park
Tinkers Hatch
Tilsmore
Little London
Sandy Cross
Maynard's Green
Chapel Cross
Rushlake Green
Warbleton
Vines Cross
Horam
Foxhunt Green
Waldron
Lions Green
Sharp's Corner
Palehouse Common
Mount Ephraim
Little Horsted
Horsted Green
Hawkhurst Common
Eason's Green
Davis's Town
Halland
Terrible Down
East Hoathly
Burlow
Marle Green
Foul Mile
Trolliloes
Cowbeech
Cowbeech Hill
Bodle Street Green
Penhurst
Ponts Green
Brownbread Street
Ashburnham Forge
Steven's Crouch
Shortgate
Laughton Common
Laughton
Chiddingly
Whitesmith
Muddles Green
Holmes's Hill
Golden Cross
Gun Hill
Hale Green
Grove Hill
Stunts Green
Chilsham
Herstmonceux
Windmill Hill
Ninfield
Lower Street
Hazard's Green
Boreham Street
Hooe Common
Hooe
Whydown
Gotham
Little Common
Cooden
Ashton Green
Glyndebourne
Mark Cross
Ripe
Chalvington
Lower Horsebridge
Upper Horsebridge
Hellingly
Magham Down
Hailsham
Chapel Row
Flowers Green
Golden Cross
Lower Dicker
Upper Dicker
Wickstreet
Arlington
Caneheath
Summer Hill
Wartling
Hall's Cross
Pevensey Levels
Rickney
Hankham
Mill Hill
Westham
Pevensey
Pevensey Bay
Beachlands
Norman's Bay
PEVENSEY BAY
West Firle
Firle
Selmeston
Alciston
Berwick
Polegate
Wilmington
Folkington
Wilmington Green
Milton Street
Alfriston
Willingdon
Lower Willingdon
Wannock
Hampden Park
Stone Cross
Friday Street
Langney
St Anthony's Hill
Roselands
Sovereign Harbour
Langney Pt
Denton
Norton
Bishopstone
East Blatchington
Seaford
Seaford Bay
Seaford Head
Sutton
Litlington
Westdean
Exceat
Friston
East Dean
Jevington
Upperton
Downside
Old Town
Eastbourne
Meads
Holywell
Birling Gap
Seven Sisters
Cuckmere Haven
Beachy Head
Cow Gap
Wish Tower
The Pound
EAST SUSSEX
Pevensey Levels
SOUTH DOWNS WAY
HERSTMONCEUX CASTLE GDNS
OBSERVATORY SCIENCE CENTRE
INTERNATIONAL STUDY CENTRE
MICHELHAM PRIORY
CHIDDINGLY PARISH CHURCH
BENTLEY WILDFOWL & MOTOR MUSEUM
GLYNDE PLACE
FIRLE PLACE
CHARLESTON FARMHOUSE
DRUSILLAS PARK
LONG MAN OF WILMINGTON
ALFRISTON CLERGY HOUSE
FILCHING MANOR MOTOR MUSEUM
LULLINGTON HEATH
CHARLESTON MANOR
SEVEN SISTERS
SEVEN SISTERS SHEEP CENTRE
TOWNER GALLERY & MUSEUM
HOW WE LIVED THEN MUSEUM
LIFEBOAT MUSEUM
REDOUBT FORTRESS MUSEUM
MINIATURE RAILWAY
PEVENSEY CASTLE
OLD MINT HOUSE
A22
A27
A259
A267
A271
A2290
A2270
A26
B2102
B2192
B2124
B2104
B2103
B2096
B2095
37
36
38
7
8
9
10
11
A
B
C
D
E
F
G
H
TQ
TV
50
70
20
90

BARNSTAPLE
OR
BIDEFORD BAY
40
Westward Ho!
Bideford
Abbotsham
Northam Burrows
Sandymere
Mermaid's Pool
Buckleigh
Silford
Rickard's Down
Orchard Hill
Greencliff
THE BIG SHEEP
Knotty Corner
Cockington
ATLANTIC VILLAGE
Handy Cross
Babbacombe Mouth
Babbacombe Cliff
SOUTH WEST COAST PATH
Fairy Cross
Ford
Woodtown
Peppercombe
Northway
Alwington
Yeo Vale
Littleham
Goldworthy
Horns Cross
Hoops
Buck's Cross
Buck's Mills
Gauter Pt
Parkham
Cabbacott
Buckland Brewer
Bocombe
Sedborough
Parkham Ash
Alminstone Cross
Woolfardisworthy
Cranford
West Town
Clifford
Higher Clovelly
Clovelly
Burscott
Bight a Doubleyou
Wood Rock
Gallant Rock
CLOVELLY COURT
Blackchurch Rock
Windbury Pt
Mouth Mill
Beckland Bay
Exmansworthy Cliff
Chapman Rock
Gawlish Cliff
Eldern Pt
Shipload Bay
Barley Bay
Hartland Pt
Titchberry
Blagdon
Damehole Pt
Upright Cliff
Blegbury
Cheristow
Hartland Abbey
HARTLAND ABBEY
HARTLAND QUAY MUS
Stoke
Hartland
Norton
Highdown Cotts
Yapham
Velly
Wrinkleberry
Hugglepit
Burnstone
MILKY WAY ADVENTURE PARK
Slade Fm
Speke's Mill Mouth
DOCTON MILL & GARDENS
NORTH
DEVON
Philham
Natcott
Milford
Elmscott
Eddistone
Tosberry
Welsford
Baxworthy Cross
Burford
Mansley Cliff
Nabor Pt
Gull Rock
Hardisworthy
South Hole
Golden Park
Embury Beacon
Ramtor Rock
Bursdon
Lutsford
Welsford Moor
Summerwell
Huddisford
Duerdon
Stroxworthy
Welcombe
Welcombe Mouth
Tredown
Henaford
Hendon Moor
Meddon
Ashmansworthy
Dipple
Marsland Mouth
Mead
Darracott
Upcott
Gull Rock
Marsland Cliff
Gooseham Mill
Gooseham
Woolley
Eastcott
Cornakey Cliff
Henna Cliff
Vicarage Cliff
Morwenstow
Crosstown
Higher Sharpnose Pt
ST JOHN THE BAPTIST CHURCH
Shop
Woodford
Stanbury Mouth
Lower Sharpnose Pt
Youlstone
West Youlstone
East Youlstone
Redmonsford
Dinworthy
West Putford
East Putford
GNOME RESERVE & WILD FLOWER GARDEN
Bountis Thorne
Colscott
Haytown
Bulkworthy
Abbots Bickington
Newton St Petrock
Stibb Cross
Withacott
Bradworthy
Kimworthy
Bradworthy Cross
Alfardisworthy
Thurdon
Stowford
Upper Tamar Lake
Lower Tamar Lake
Sutcombe
Sutcombemill
Venn Green
Milton Damerel
Brendon
Soldon Cross
Kilkhampton
BROCKLANDS ADVENTURE PARK
Penstowe Park
ST JAMES'S CHURCH
Stibb
Steeple Pt
Duckpool
Coombe
Stowe Cliffs
Long Rock
CORNWALL
Sandy Mouth
Menachurch Pt
Northcott Mouth
Dunsmouth
Northcott
Poughill
Maer
Flexbury
Wrangle Pt
Bude Haven
Compass Pt
BUDE-STRATTON MUSEUM
Bude
Stratton
Launcells
Hersham
Grimscott
Launcells Cross
Red Post
Kingsford
Pancrasweek
DUNSDON
Lana
Woodsdown Hill
Holsworthy Beacon
Woodacott
Thornbury
Bradford
Cookbury
Holemoor
Priestacott
Chilsworthy
Holsworthy
Derril
Pyworthy
Derriton
Whimble
Chasty
Hollacombe
Brandis Corner
Phillips's Pt
Upton
Thorne
Marhamchurch
Helebridge
Widemouth Bay
Widemouth Sand
Black Rock
Wanson Mouth
Box's Shop
Budd's Titson
Titson
Bridgerule
Hopworthy
Lewdown
Lewtrenchard
Merrifield
Bakesdown
Coppathorne
Millook
Poundstock
Treskinnick Cross
Whitstone
Churchtown
Jewell's Cross
Strawberry Bank
Lewersworthy
Clawton
Buckhorn
Halwill
Langaford
Lane End
Chilla
Stowford
11
12
20
30
SS
A39
A388
A3072
A3079
A386
B3254
B3248
B3237
A3073
1
2
3
4
5
6
A
B
C
D
E
F
G
H
0
miles
km

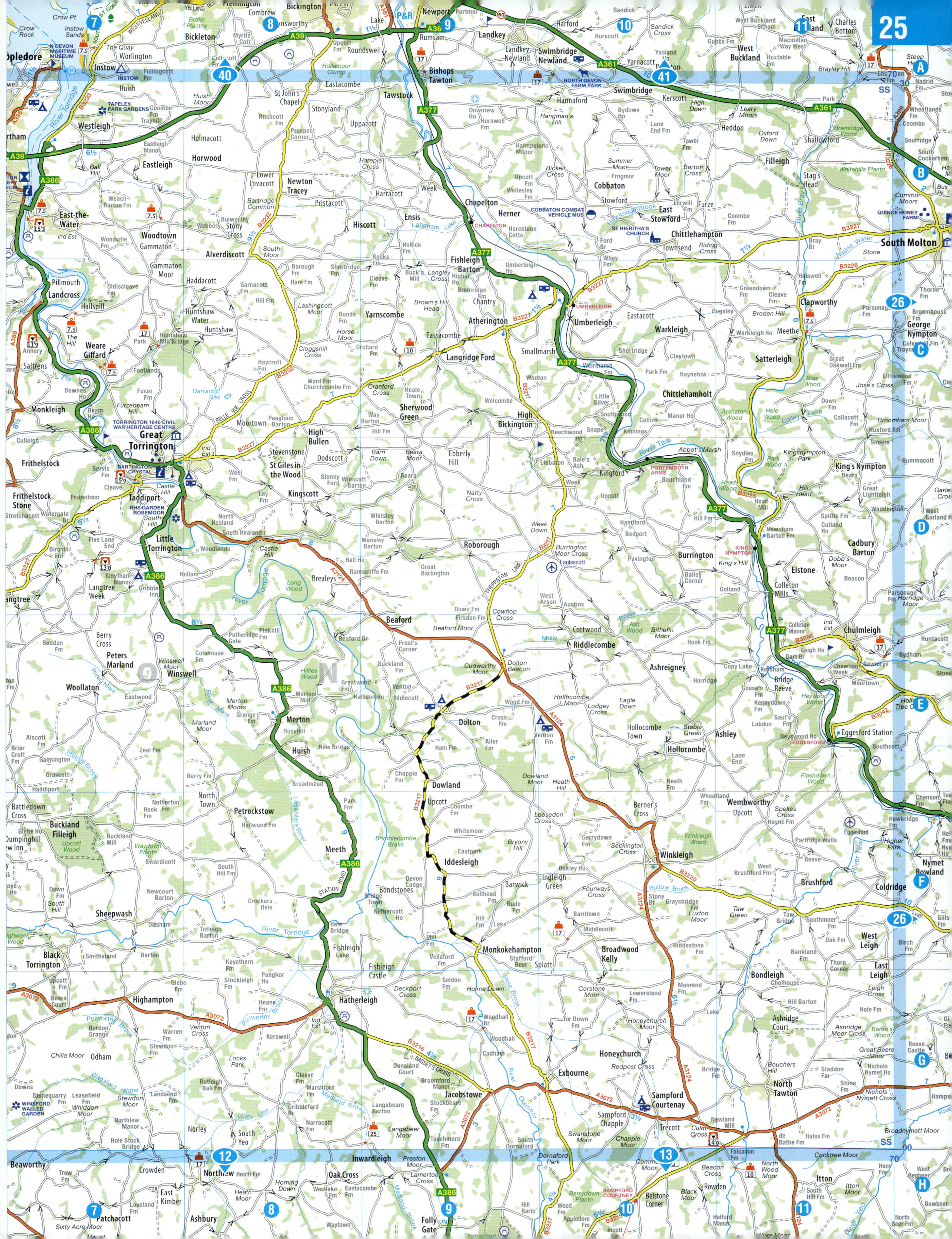
Appledore
Instow
Westleigh
Bideford
Bickleton
Fremington
Bickington
Newport
Landkey
Swimbridge
Swimbridge Newland
Harford
West Buckland
East Buckland
Charles Bottom
Bishops Tawton
Tawstock
Eastacombe
Roundswell
St John's Chapel
Stonyland
Uppacott
Horwood
Eastleigh
Holmacott
Newton Tracey
Lovacott
Hiscott
Ensis
Harracott
Chapelton
Herner
Cobbaton
East Stowford
Chittlehampton
South Molton
Filleigh
Heddon
Shallowford
Stag's Head
Hannaford
Kerscott
Landcross
East-the-Water
Woodtown
Gammaton
Alverdiscott
Huntshaw
Huntshaw Water
Haddacott
Yarnscombe
Fishleigh Barton
Atherington
Umberleigh
Eastacott
Warkleigh
Clapworthy
Satterleigh
George Nympton
Weare Giffard
Annery
Saltrens
Monkleigh
Langridge Ford
Eastacombe
Smallmarsh
Chittlehamholt
Sherwood Green
High Bickington
Great Torrington
Taddiport
Little Torrington
Frithelstock
Frithelstock Stone
Stevenstone
St Giles in the Wood
Kingscott
High Bullen
Dodscott
Ebberly Hill
Kingford
Burrington
King's Nympton
Cadbury Barton
Elstone
Roborough
Beaford
Brealeys
Langtree
Langtree Week
Riddlecombe
Cottwood
Ashreigney
Chulmleigh
Colleton Mills
Bridge Reeve
Eggesford Station
Peters Marland
Winswell
Woolaton
Merton
Huish
Dolton
Dowland
Upcott
Iddesleigh
Hollocombe
Ashley
Wembworthy
Winkleigh
Petrockstow
North Town
Buckland Filleigh
Sheepwash
Meeth
Monkokehampton
Broadwood Kelly
Brushford
Coldridge
Nymet Rowland
West Leigh
East Leigh
Bondleigh
Black Torrington
Highampton
Hatherleigh
Fishleigh Castle
Honeychurch
Exbourne
Sampford Courtenay
North Tawton
Jacobstowe
Inwardleigh
Beaworthy
Northlew
Ashbury
Patchacott
Folly Gate
Itton
River Torridge
River Taw
River Mole
River Lew
A39
A361
A377
A386
A3124
A3072
B3227
B3232
B3217
B3220
B3226
B3216
B3096
B3042
B3233
A388
40
41
12
13
26
A
B
C
D
E
F
G
H
SS
70
30
00

Heasley Mill
Barcombe Down
Balls
Wheatclose Cross
Mounsey Hill 374
Hawkridge Common
South Hill
Steep Head
Twitchen Mill
Twitchen
Brinsworthy
Millbrook
North Molton
Molland Common
Cussacombe Gate
Hawkridge
Hawkridge Ridge
Anstey Gate
West Anstey Common
Draydon
Mounsey Castle
Marsh Hill Ho
Whiterocks Down
Court Down
Louisa Gate
Chilly Bridge
River Exe
Halscombe
Molland
Champson
Bickingcott
Luckworthy
Smallacombe
Great Woods
Abbotts Park
West Anstey
Two Moor Ways
Five Cross Ways
Dulverton
Battleton
Three Gates Cross
Hele Br
Jury Road
Pixton Park
Bury Castle
South Molton
Quince Honey Farm
Common Moors
Bish Mill
Garliford
Newtown
Veraby
River Yeo
Bottreaux Mill
Yeo Mill
East Anstey
Waddicombe
Oldways End
Nightcott
Brushford
Combsland Cross
Wester Bullaford Moor
New Park Fm
Wester New Moor
Cuckoo Fm
Higher Radnidge Moor
Smallacombe Moors
White Moor
Exebridge
Stone
Gorton Hill
Grilstone
Little Hill
Paul's Moor
Luckett Moor
Whitmoor Fm
Whitefield Fm
Shapcott Barton
Tucker's Moor
Den Brook
Hele Manor
Bishops Nympton
Parsonage Fm
Ash Mill
Knowstone
East Knowstone
Roachill
Combe Water
Little Oak Fm
George Nympton
Alswear
Marlansleigh
Rose Ash
Pearchay Fm
Beaple's Moor
Beaple's Hill
Side Moor
Middle Hill Fm
Hares Down
Bickham Moor
South Esworthy Fm
Oakfordbridge
Oakford
Westbrook Wood
Iron Mill Stream
Chain Br
Jose's Cross
Cleave Fm
Westacott Fm
Rowcliffe
Munson Fm
Cherridge
Northcott Manor Fm
Fanny's Cross
Nutcombe Fm
Batsworthy Cross
Batsworthy
Manor Fm
Hares Down Cross
Rackenford Moor
Spurway Mill
Nethercott
Chample's Fm
Hangman's Hill Cross
Broomham Moor
Huxford Fm
Romansleigh
Odam Barton
Langley
Meshaw
Meshaw Barton
Narracott
Burcombe
Ash Moor
Frankhill Fm
Creacombe
Rackenford Manor
Nutcott
Rackenford Fm
Sydeham Fm
Stoodleigh Beacon 301
Wasley
Broadmead Fm
Coleford Bottom
Quoit-at-Cross
Stoodleigh
Stoodleighmoor
East Stoodleigh Barton
Carscombe
Dryhill
Exe Valley
King's Nympton
Hummacott
Great Lightleigh
Garland Cross
West Garland Fm
Waddington
Great Odham Moor Plantn
Batsworthy
Huntacott Water
Burrow Cross
Gidley Cross
Heath Fm
Parsonage Cross
Crowdhole
Wilson Fm
Rackenford
Bulworthy Fm
Mogworthy
Willicroft Moor
Stoneland Fm
Rifton Moor
Diptford Fm
Slade Fm
Cadbury Barton
Waterloo Fm
Week
Lutworthy
Wheadon Fm
Broadridge Fm
Rowden Fm
Bradford Mill
Queen Dart
Ford Down
Colleton Hall
Gibbett Moor Fm
Horestone
Haydon
Churchill
Emerleigh
Hatherland
Moorhayes
Wixon
Hole Fm
Kempland Fm
Beacon
Parsonage Fm
Horridge Moor
Holland Fm
Benley Cross
Bealy Court
Cuddenhay Fm
Horseford Fm
Hellinghayes Fm
Stone Fm
Dart Raffe Fm
Hole Fm
Down Cross
Rose Moor
Witheridge Moor
Great Moor
Holmead Fm
Deepaller
Loxbeare
Courtenay Pitt
Stanterton
Washfield
Marsh Fm
Chulmleigh
Edgiford Cross
Winswood Moor
Affeton Moor
Town Moor
New Br
Ind Est
Witheridge
Charnaford Fm
South Coombe
Witheridge Moor Fm
Templeton Bridge
Templeton
West Bradley
Lurley
Harpridge
Calverleigh
Palmers
Velvains
Huntacott
Sydham
Stonemill
Cheldon
Cheldon Fm
East Worlington
West Worlington
Drayford
Summer Fm
Chapner Fm
Nomansland
Westway Fm
Stretch Down
Moor Barton Fm
Gogland Manor
Mudford Gates
Kelly Fm
Partridge Fm
The Old Rectory
Bradleigh Down
Higher Farleigh
Coombehayes
Longbrag Hill
Cotteylands
Moortown
Hollow Tree Cross
Chawleigh
Leigh
East Leigh
Cobley Fm
Coombe Fm
Thelbridge Barton
Stourton Barton
Washford Moor
Henceford Moor
Deptford Fm
Wood Fm
Cruwys Morchard
Withleigh
Wayland
Hensleigh Fm
St Peter's Church
Eggesford Station
Labbett's Cross
North Lake Fm
Yeatheridge Fm
Thelbridge Cross
Washford Pyne
Littleborough
Puddington
Beer Fm
Pennymoor
Little Heath
Old Ford
Thorne Fm
Down Fm
East Ruckham
Mogridge Fm
Lodge Hill Fm
Southcott
Handsford Fm
Filleigh
Denridge Fm
Forches Cross
River Dalch
Billhole Fm
Cann's Mill Bridge
Pyne Fm
Yowlestone Ho
Hill Fm
Stickeridge Fm
Westway
Way Village
Coombeland
Ashley
Chenson
Tonyfield
Nymet Br
Hawkridge
Bowerthy Wood
Eggesford
Higher Park
Hole
Broadridge Fm
Eastington
Mear Fm
Black Dog
Puddington Bottom
Sunnybrook Fm
Edbury Fm
Trundlemoor Fm
Marshay Fm
Cotton Fm
Meadhayes Fm
Well Town
Little Silver
Well Fm
Way
Nymet Rowland
Lapford
Lapford Cross
Calves Br
Hill Barton
Lane End Fm
Berry Castle
Beech Hill
Emlett Hill
Tridley Foot
Woolfardisworthy
Partridge Hole
Park
Greenhill Cross
Poughill
Upham Fm
Court Place
Upham
Haydon Fm
Higher Pitt Fm
Captain's Fm
Cadeleigh
Cadeleigh Court
East Court
East Barton
Exeter Road
Coldridge
Rudge
Rudge Rew
Middlecott
Redhill
Morchard Bishop
Barton Ho
Farthing Park
South Emlett
Kennerleigh
Hollyford Fm
New House Fm
Marsh Fm
Redyeates Fm
Hayne
Bickleigh Mill
Bickleigh
West Leigh
East Leigh
Allerby
Gillscott Fm
Weeke
Frost
Carrhouse
Oldborough
Scotland
Moor Fm
Welland Down
Down Fm
Rookbeare Fm
Stockleigh English
Wolland Fm
Round Hill
Bickleigh Castle
Birch Fm
Bradiford
Stopgate Cross
Sharland Fm
Venn
Rolstone Barton
Burridge Fm
Splitwell Cross
Stockleigh Court
White Cross
Cheriton Fitzpaine
Forde Cross
Pitt Fm
Way Fm
Perry Fm
Morchard Road
Middle Yeo Fm
Thorne Fm
Dowrich Ho
Vellake
Prowse
Perry
Cadbury
Cadbury Castle
Leigh Cross
Zeal Monachorum
Down St Mary
Ash Bullayne
Gays Fm
Beacon Cross
Newbuildings
Thelbridge Br
Dowrich Br
East Village
Heath Br
Cheriton Cross
North Coombe
East Coombe
Higher Coombe
Fursdon House
Chitterley
Hole Fm
Oak Tree Fm
Gillhouse
Copplestone
Upton Hellions
Lower Creedy
Raddons Hills 235
Bidwell
Bus Cen
Barn's Wood
Reeve Castle
Burston
Lammacott Fm
Clannaborough
Eppletons Fm
Serstone
Sutton Fm
Beers Hill
Bewsley Fm
Elston
Woolsgrove
Brandirons Corner
Clampitt
Bawdenhayes
West Sandford
Sturridge
Sandford
River Creedy
Westwood Fm
Stockleigh Pomeroy
Chapel Fm
Lynch Fm
Ford Cross
Great Beere
Nichols Nymet Ho
Copplestone Ho
Knowle
Priestcombe
Bradley Fm
Uppincott
Baddon Court
Up Exe
Stone Fm
Nichols Nymett Cross
Hampson
Bow
Appledore Fm
Walson Barton
Ford Fm
Coleford
South Coombe
Coombe Fm
Pounds Hill
Long Barn
Creedy Br
Shobrooke Park
Shobrooke
Wood Fm
Efford
Raddon
Thorverton
Berrysbridge
Broadnymett Moor
Glebe Ho
Nymet Tracey
Thorne
Penstone
Hollacombe Cross
Hollacombe
The Collegiate Church of the Holy Cross
Downes Head
Westacott Cotts
Shobrooke Mill Fm
Yendacott Manor
Shute
Yellowford
Heathfield
Willowpark
Nether Exe
Rewe
Newcott
Colebrooke
Barn Fm
Stairhill Fm
Crediton
Park
Trad Est
Fordton
Lake Fm
Sweetham
Norton
Nettacott
Winscott Barton
Jackmoor
Brampford Speke
Burrow Fm
Heath Moor
Yeoford
Neopardy
Gunstone Ho
Denbury Fm
Salmonhutch
Hookway
Newton St Cyres
Ley
Bidwell Barton
Pye Corner
Langford
Stoke Canon
Hunxham
Hillerton
Bowbeare
North Beer Fm
Gunstone Mill
Venny Tedburn
Culvery Bridge
Trobridge Ho
Smallbrook
Rudge Fm
Blackdown Plantn
Ford Fm
Wood Fm
Devon
A361
A377
A396
A399
A3072
B3227
B3226
B3137
B3042
B3096
B3222
B3221
SS
0 1 2 3 miles
0 1 2 3 4 5 km

Brompton Regis
Clatworthy
Clatworthy Reservoir
Wimbleball Lake Res
Brompton Ralph
Tolland
Gaulden Manor
Lawrence
East Combe
Cothelstone
Combe Florey
Pyleigh
Tarr
West Leigh
Huish Champflower
Whitefield
Langley
Langley Marsh
Hoccombe
Chapel Leigh
Northway
Sandhill Park
Ash Priors
Bishops Lydeard
Cushuish
East Lydeard
Yarford
Fulford
Pickney
Nailsbo
Maundown
Northgate
Ford
Fitzhead
Halse
Pond Close
Cotford St Lukes
Upton
Skilgate
Chipstable
Wiveliscombe
Castle
Hartswell
Preston Bowyer
Heathfield
Norton Fitzwarren
Staple
Langford
Hillcommon
Milverton
Houndsmoor
Oake
Oake Green
Blackwell
Raddington
Waterrow
Screedy
Weekmoor
Allerford
Roughmoor
Barr
Upcott
Bishop's Hull
Claypits
Quartley
Petton
Shillingford
Bathealton
Poleshill
Bindon
S O M E R S E T
Hillfarance
Hele
Rumwell
Comeytrowe
Chipley
Sandylands
Gundenham
Langford Budville
East Nynehead
Nynehead
Tone Green
Bradford-on-Tone
Sheppy's Cider
Bampton
Clayhanger
Stawley
Kittisford
Appley
Wellisford
Runnington
Tone
Tonedale
Ham
Hockholler
Middle Stoford
Silver Street
Dipford
Sweethay
Canonsgrove
Ashbrittle
Tracebridge
Cothay Manor
Thorne St Margaret
Payton
Rockwell Green
Wellington
Chelston
Chelston Heathfield
Sawyer's Hill
West Buckland
Huntsham
Greenham
Holywell Lake
Westford
Bagley Green
Stape Cross
Hockworthy
Beam Bridge
Pinksmoor
Budgett's Cross
Bradleigh
Angersleigh
Cove
White Ball
Sampford Arundel
Middle Green
Ford Street
Holcombe Rogus
Whipcott
Pleamore Cross
Voxmoor
Blackmoor
Ruggin
Chelmsine
Lowton
Sampford Moor
Wrangway
Wellington Monument
Red Ball
Westleigh
Burlescombe
North End
Nicholashayne
Wiltown
Leigh Hill
Widcombe Moor
East Mere
Whitnage
Ayshford
Simonsburrow
Clayhidon
Knightshayes Court
Chevithorne
Uplowman
Appledore
Woodgate
Culmstock Beacon
Brownheath
Burnworthy
Bolham
Chettiscombe
Sampford Peverell
Tiverton Parkway
Prescott
Culm Davy
Whitehall
Ashculme
Millhayes
Rosemary Lane
Churchstanton
Cowleymoor
Halberton
Culmstock
Westown
Hemyock
Hemyock Castle
Stapley
Biscombe
Willand
Tiverton
Tiverton Museum of East Devon Life
Grand Western Canal
Uffculme
Yodercott
Coldharbour
Coldharbour Mill
Craddock
Northcott
Churchinford
Lower Town
Manley
Ash Thomas
Smithincott
Ashill
Bolham Water
Madford
Abbey
Smeatharpe
Luxton
Collipriest
Willand
Bradfield
Diggerland
Stenhill
Blackborough
Saint Hill
Stentwood
Bodmiscombe
Ullcombe
Butterleigh
East Butterleigh
St George's Well
Collumpton Services
Stoneyford
Kentisbeare
Hollis Green
Slade
Sheldon
D E V O N
B L A C K D O W N
H I L L S
Cullompton
St Andrew's Church
Dulford
Dunkeswell
Dunkeswell Memorial Museum
Upottery
Rawridge
Colebrook
Luppitt
Bradninch
Mutterton
Westcott
Kerswell
Broadhembury
Wolverstone
Shelvin
Beacon
Monkton
Tiverton
Silverton
Langford
Plymtree
Norman's Green
Colliton
Hembury Fort
Wick
Cotleigh
Ellerhayes
Hele
Payhembury
Tale
Lower Cheriton
Awliscombe
Combe Raleigh
All Hallows Museum
Northcote
Holmsleigh Green
Killerton House
Clyst Hydon
Aunk
Colestocks
Buckerell
Weston
Honiton
Mount Pleasant
Cleave
Columbjohn
Budlake
Clyst St Lawrence
Talaton
Feniton
Fenny Bridges
Gittisham
Hamlet
Littletown
Offwell
Wilmington
Westwood
Brookleigh
Higher Cheriton
Alfington
Broadclyst
Escot Park
Taunton Deane Services
Upottery
Uppacott
M5
A38
A361
A373
A303
A30
A35
B3227
B3190
B3188
B3187
B3181
B3391
B3440
Taunton Road
Hemyock Road
Honiton Road
Uffculme Road
Five Bridges
Milverton Road
Nowers Lane
Oldway Road
Norton Fitzwarren
River Tone
River Culm
River Lowman
River Otter
Grand Western Canal

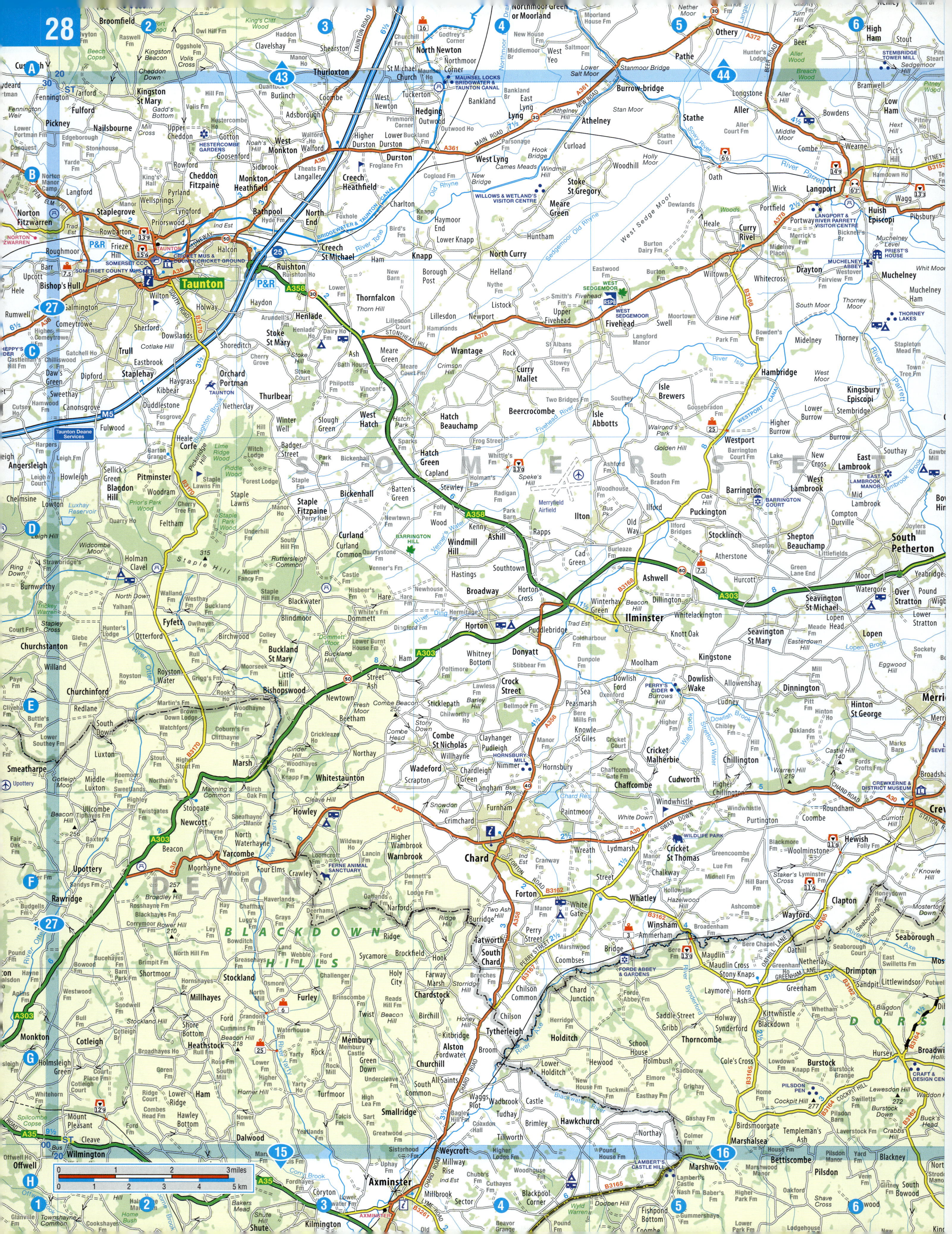
Taunton
Chard
Axminster
Ilminster
Langport
Curry Rivel
South Petherton
Crewkerne
Wellington
SOMERSET
DEVON
BLACKDOWN HILLS
M5
A358
A303
A30
A35
A378
A361
A3088
A370
A372
Hatch Beauchamp
Ilton
Isle Abbotts
Ashill
Broadway
Donyatt
Winsham
Forton
Tatworth
Chardstock
Membury
Stockland
Yarcombe
Upottery
Churchinford
Staple Fitzpaine
Bishopswood
Buckland St Mary
Whitestaunton
Combe St Nicholas
Wadeford
Cricket St Thomas
Whitelackington
Dowlish Wake
Kingstone
Barrington
Shepton Beauchamp
Seavington St Michael
Hinton St George
Merriott
Chillington
Cudworth
Thorncombe
Hawkchurch
Drayton
Muchelney
Huish Episcopi
Aller
High Ham
Othery
Burrow Bridge
Stoke St Gregory
North Curry
Creech St Michael
Ruishton
Henlade
Thurlbear
Orchard Portman
Pitminster
Corfe
Blagdon Hill
Stoke St Mary
West Hatch
Hatch Green
Fivehead
Beercrocombe
Curry Mallet
Isle Brewers
Puckington
Stocklinch
Kingsbury Episcopi
West Lambrook
East Lambrook
Kingston St Mary
Cheddon Fitzpaine
Monkton Heathfield
Bathpool
Staplegrove
Norton Fitzwarren
Bishop's Hull
Wilton
Trull
Staplehay
North Petherton
Thurloxton
West Monkton
Durston
West Lyng
Athelney
Lyng
Stathe
Wick
Pathe
Weycroft
Kilmington
Millbrook
Smallridge
All Saints
Tytherleigh
Chilson
South Chard
Whatley
Fyfett
Holman Clavel
Staple Hill
Newcott
Stockland
Millhayes
Heathstock
Rawridge
Cotleigh
Monkton
Wilmington
Offwell
Mount Pleasant
Dalwood
Shute
Chilson Common
Churchill
Holditch
Birstock
Broadwindsor
Drimpton
Seaborough
Clapton
Wayford
Marshalsea
Blackney
Pilsdon
Bettiscombe
Birdsmoorgate
Marshwood

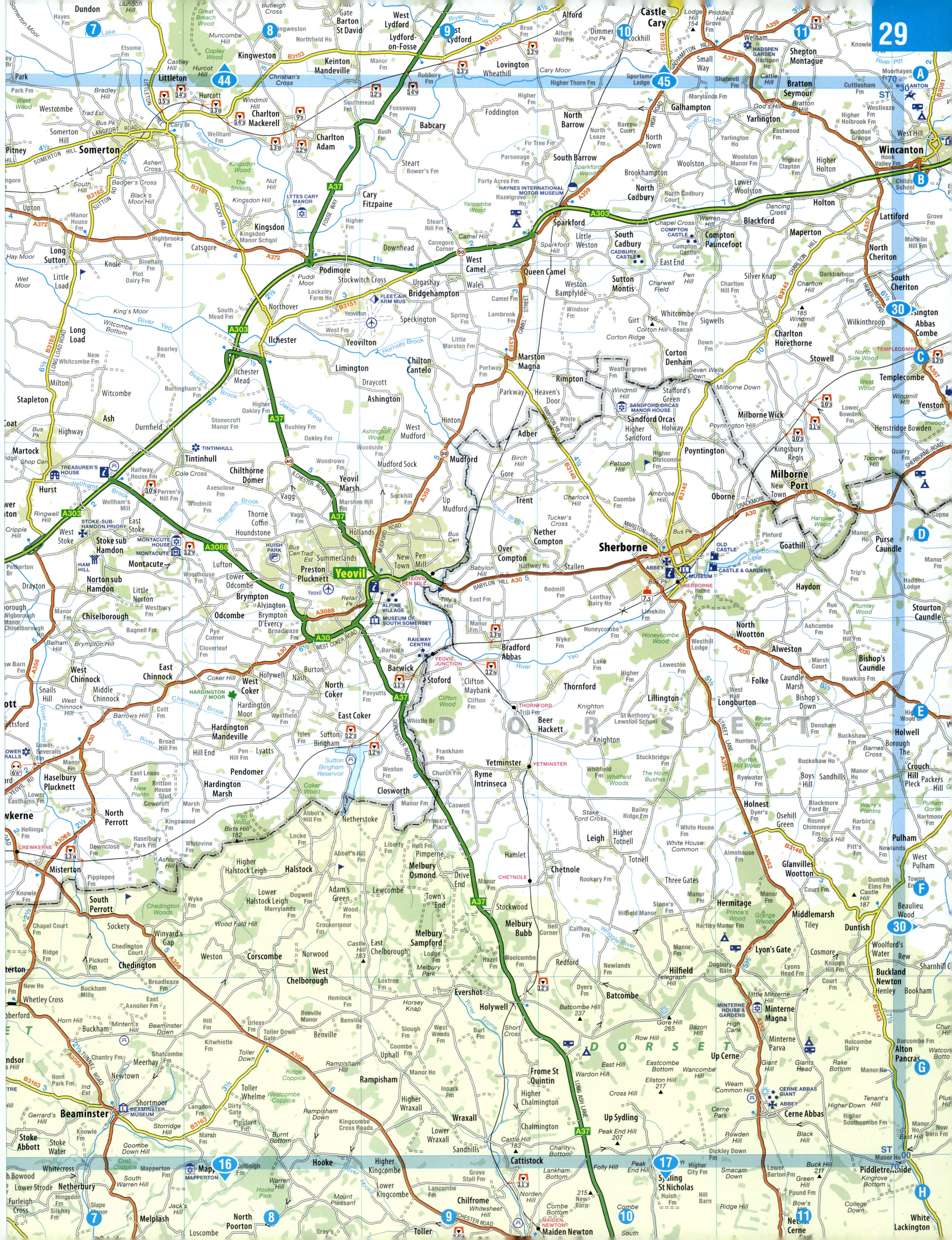

Dundon
Somerton Moor
Butleigh Cross
Gate Barton
St David
West Lydford
Lydford-on-Fosse
West Lydford
Alford
Castle Cary
Dimmer
Cockhill
Shepton Montague
Knowle
Welham
Hadspen Garden
Kingweston
Keinton Mandeville
Lovington
Wheathill
Cary Moor
Higher Thorn Fm
Sportsman's Lodge
Bratton Seymour
Moorhayes
Park
Littleton
Hurcot
Charlton Mackerell
Charlton Adam
Babcary
Foddington
North Barrow
Galhampton
Yarlington
Wincanton
Somerton
Pitney
Langport Road
Somerton Hill
Westcombe
Bradley Hill
Wellham
Kingsdon Wood
South Barrow
North Town
Woolston
Higher Holton
Steart
Bower's Fm
Forty Acres Fm
Haynes International Motor Museum
Brookhampton
North Cadbury
Lower Woolston
Holton
Upton
Kingsdon
Lytes Cary Manor
Cary Fitzpaine
Sparkford
Blackford
Lattiford
Long Sutton
Catsgore
Downhead
Camel Hill
Little Weston
South Cadbury
Compton Castle
Compton Pauncefoot
Maperton
North Cheriton
Knole
Podimore
West Camel
Queen Camel
Cadbury Castle
East End
Silver Knap
South Cheriton
Hay Moor
Little Load
Puddi Moor
Stockwitch Cross
Urgashay
Bridgehampton
Fleet Air Arm Mus
Wales
Weston Bampfylde
South Montis
Charlton Hill
Wet Moor
King's Moor
River Yeo
Northover
South Mead Fm
Yeovilton
Speckington
Spring Fm
Girt
Whitcombe
Sigwells
Charlton Horethorne
Horsington
Abbas Combe
Long Load
Ilchester
Yeovilton
Little Marston Fm
Corton Ridge
Corton Denham
Stowell
Templecombe
Witcombe Bottom
Limington
Chilton Cantelo
Marston Magna
Rimpton
Seven Wells Down
Milton
Ilchester Mead
Draycott
Portway
Parkway
Heaven's Door
Weathergrove
Milborne Down
Stapleton
Witcombe
Ashington
Hinton
Adber
Windmill Hill
Stafford's Green
Sandford Orcas Manor House
Sandford Orcas
Milborne Wick
Yenston
Ash
Higher Oakley Fm
Oakley Brook
Rushley Fm
West Mudford
Higher Sandford
Holway
Poyntington
Henstridge Bowden
Highway
Durnfield
Stonecroft Manor Fm
Oakley Fm
Mudford
Kingsbury Regis
Milborne Port
Martock
Tintinhull
Chilthorne Domer
Woodrows Fm
Yeovil Marsh
Mudford Sock
Birch Hill
Patson Hill
Higher Clatcombe
Oborne
New Town
Hurst
Treasurer's House
Vagg
Gore
Trent
Sherborne
Purse Caundle
West Stoke
Stoke sub Hamdon
Montacute House
Montacute
Lufton
Thorne Coffin
Houndstone
Hollands
Up Mudford
Charlock Hill
Coombe Fm
Ambrose Hill
Old Castle
Castle & Gardens
Goathill
Ham Hill
Norton sub Hamdon
Little Norton
Lower Odcombe
Brympton
Preston Plucknett
Summerlands
Yeovil
New Town
Pen Mill
Babylon Hill
Nether Compton
Over Compton
Stallen
Abbey
Museum
Haydon
Drayton
Chiselborough
Odcombe
Alvington
Brympton D'Evercy
Alpine Village
Museum of South Somerset
Railway Centre
Bradford Abbas
Wyke Fm
Bedmill Fm
Lenthay Dairy Ho
Honeycombe Wood
Limekiln Hill
North Wootton
Stourton Caundle
Balham Hill
Brympton Hill
Pye Corner
Cloverleaf Fm
Barwick
Stoford
Yeovil Junction
Clifton Maybank
Thornford
Lake Fm
Westhill Lodge
Alweston
Bishop's Caundle
West Chinnock
East Chinnock
Coker Hill
West Coker
Holywell
Nash
Burton
North Coker
Payvotts
Clifton Wood
Higher Fm
Lillington
Folke
Caundle Marsh
Bishop's Down
Middle Chinnock
Hardington Moor
Harrington
East Coker
Beer Hackett
Knighton
St Anthony's Leweston School
Longburton
Holwell
Hardington Mandeville
Westfield
Isles Fm
Sutton Bingham
Whistle Br
D O R S E T
Haselbury Plucknett
Hill End
Lyatts
Pendomer
Hardington Marsh
Sutton Bingham Reservoir
Closworth
Frankham
Church Fm
Yetminster
Ryme Intrinseca
Whitfield Fm
Stockbridge
Crouch Hill
Pleck
Packers Hill
Crewkerne
North Perrott
Haselbury
Birts Hill
Netherstoke
Manor Fm
Caswell Fm
Prince's Place
Leigh
Higher Totnell
Holnest
Osehill Green
Misterton
Ashland Hill
Higher Halstock Leigh
Halstock
Liberty Fm
Pimperne
Melbury Osmond
Hamlet
Chetnole
Totnell
White House Common
Glanvilles Wootton
Pulham
West Pulham
South Perrott
Wyke Fm
Lower Halstock Leigh
Adam's Green
Lewcombe
Town's End
Stockwood
Three Gates
Hermitage
Middlemarsh
Tiley
Duntish
Beaulieu Wood
Winyard's Gap
Chedington
Weston
Corscombe
Norwood
Castle Hill
East Chelborough
Melbury Sampford
Melbury Bubb
Hell Corner
Redford
Hilfield Manor
Hilfield
Lyon's Gate
Cosmore
Buckland Newton
Henley
Bookham
Chedington Court
West Chelborough
Evershot
Holywell
Batcombe
Batcombe Hill
Minterne House & Gardens
Minterne Magna
Whetley Cross
Buckham
Benville Manor
Benville
Horsey Knap
Woolcombe Fm
Gore Hill
Minterne Parva
Alton Pancras
Kitwhistle Fm
Toller Down Gate
Uphall
West Woods Fm
Short Cross
D O R S E T
Up Cerne
Giant Hill
Giants Head
Meerhay
Newtown
Rampisham
Frome St Quintin
East Hill
Eastcombe Bottom
Wancombe Hill
Weam Common Hill
Cerne Abbas Giant
Abbey
Cerne Park
Cerne Abbas
Beaminster
Shortmoor
Beaminster Museum
Toller Whelme
Westcombe Coppice
Rampisham Down
Higher Wraxall
Lower Wraxall
Wraxall
Higher Chalmington
Chalmington
Cross Hill
Up Sydling
Peak End Hill
Stoke Abbott
Stoke Water
Storridge Hill
Burnt Bottom
Kingcombe Cross Roads
Sandhills
Castle Hill
Charity Bottom
Cattistock
Whitecross
Mapperton
Hooke
Higher Kingcombe
Lower Kingcombe
Grove Stall Fm
Folly Hill
Peak End Hill
Sydling St Nicholas
Higher City Fm
Piddletrenthide
Netherbury
Warren Hill
Hooke Park
Mount Pleasant
Lancombe Fm
Chilfrome
Whitesheet Hill
Norden Hill
Combe Bottom
Lankham Bottom
Smacam Down
Kingrove Bottom
Melplash
North Poorton
Loscombe
Toller
Maiden Newton
South
Nether Cerne
White Lackington
A37
A303
A359
A3088
A30
A372
A371
A357
A352
A356
B3151
B3153
B3165
B3148
B3145
B3146
B3143

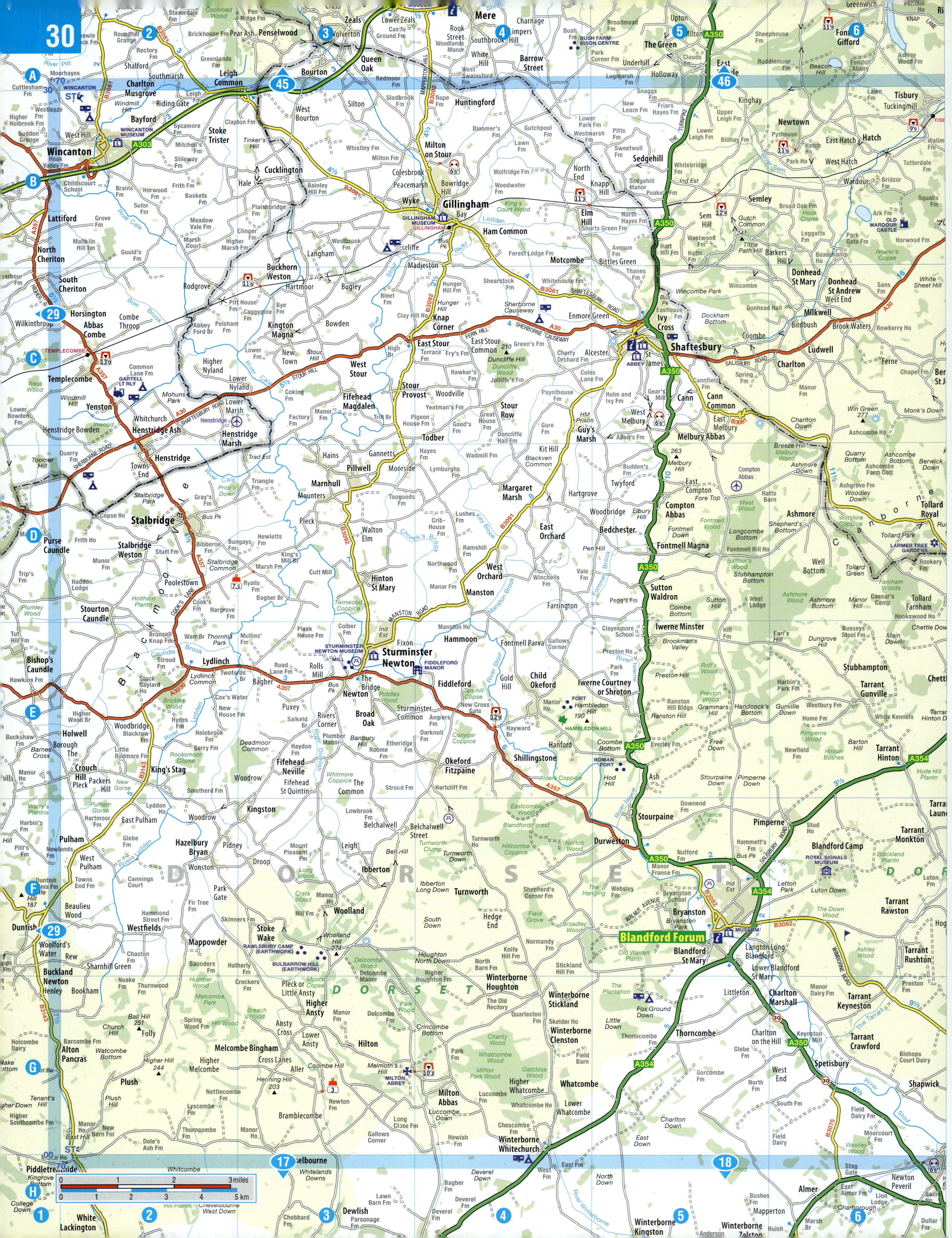

Wincanton
Zeals
Mere
Penselwood
Bourton
Queen Oak
Charlton Musgrove
Bayford
Stoke Trister
Cucklington
Leigh Common
Milton on Stour
Gillingham
Huntingford
Barrow Street
The Green
Upton
East Knoyle
Fonthill Gifford
Tisbury
Newtown
Hatch
West Hatch
Semley
Sedgehill
Ham Common
Motcombe
Madjeston
Kington Magna
Buckhorn Weston
Lattiford
North Cheriton
South Cheriton
Horsington
Abbas Combe
Templecombe
Yenston
Henstridge
Henstridge Ash
Henstridge Marsh
Whitchurch
Stalbridge
Stalbridge Weston
Purse Caundle
Stourton Caundle
Bishop's Caundle
Lydlinch
Poolestown
East Stour
West Stour
Stour Provost
Stour Row
Fifehead Magdalen
Todber
Marnhull
Pillwell
Margaret Marsh
Twyford
East Orchard
West Orchard
Hinton St Mary
Manston
Fontmell Magna
Compton Abbas
Melbury Abbas
Cann
Cann Common
Shaftesbury
Ludwell
Charlton
Donhead St Mary
Donhead St Andrew
Birdbush
Ashmore
Tollard Royal
Sturminster Newton
Fiddleford
Hammoon
Child Okeford
Iwerne Courtney or Shroton
Sutton Waldron
Iwerne Minster
Shillingstone
Okeford Fitzpaine
Broad Oak
Fifehead Neville
King's Stag
Holwell
Kingston
Hazelbury Bryan
Woolland
Ibberton
Turnworth
Stourpaine
Durweston
Pimperne
Tarrant Gunville
Tarrant Hinton
Tarrant Launceston
Tarrant Monkton
Tarrant Rawston
Tarrant Rushton
Tarrant Keyneston
Tarrant Crawford
Blandford Forum
Blandford St Mary
Blandford Camp
Bryanston
Pimperne
Charlton Marshall
Spetisbury
Shapwick
Winterborne Stickland
Winterborne Houghton
Winterborne Clenston
Winterborne Whitechurch
Whatcombe
Milton Abbas
Hilton
Ansty
Higher Ansty
Melcombe Bingham
Mappowder
Pulham
Buckland Newton
Alton Pancras
Plush
Dewlish
Bramblecombe
Piddletrenthide
White Lackington
Almer
Newton Peveril
Winterborne Kingston
Winterborne Zelston
D O R S E T
Cranborne Chase
0 1 2 3 miles
0 1 2 3 4 5 km

W I L T S H I R E
C R A N B O U R N E   C H A S E   A N D
W E S T   W I L T S H I R E   D O W N S
H A M P S H I R E
D O R S E T
Salisbury
Wilton
Ringwood
Wimborne Minster
Fordingbridge
Verwood
Ferndown
Cranborne
Alderholt
Sixpenny Handley
Damerham
Rockbourne
Coombe Bissett
Bishopstone
Broad Chalke
Fovant
Sutton Mandeville
Compton Chamberlayne
Barford St Martin
Burcombe
Quidhampton
Bemerton
Netherhampton
West Harnham
East Harnham
Britford
Odstock
Nunton
Bodenham
Alderbury
Whaddon
Downton
Redlynch
Woodfalls
Breamore
Woodgreen
Hale
Martin
Tidpit
Woodyates
Pentridge
Farnham
Tollard Royal
Cashmoor
Gussage All Saints
Gussage St Michael
Wimborne St Giles
Edmondsham
Knowlton
Witchampton
Hinton Martell
Holt
Colehill
Stapehill
West Moors
St Leonards
St Ives
Ashley Heath
Bisterne
Kingston
Sopley
Hurn
Avon
Ripley
Burley
Linwood
Ibsley
Blashford
Rockford
Hyde
Stuckton
Frogham
Godshill
Sandleheath
Ashford
Bickton
Harbridge
Ellingham
Three Legged Cross
Horton
Chalbury
Pamphill
Sturminster Marshall
Hampreston
Longham
Parley
Dudsbury
Trickett's Cross
Alvediston
Ebbesbourne Wake
Bowerchalke
Woodminton
Swallowcliffe
Ansty
Dinton
Baverstock
Teffont Magna
Ditchampton
Stratford Tony
Homington
Whitsbury
Fifield Bavant
Tarrant Hinton
A354
A338
A31
A36
A30
A348
B3078
B3081
B3080

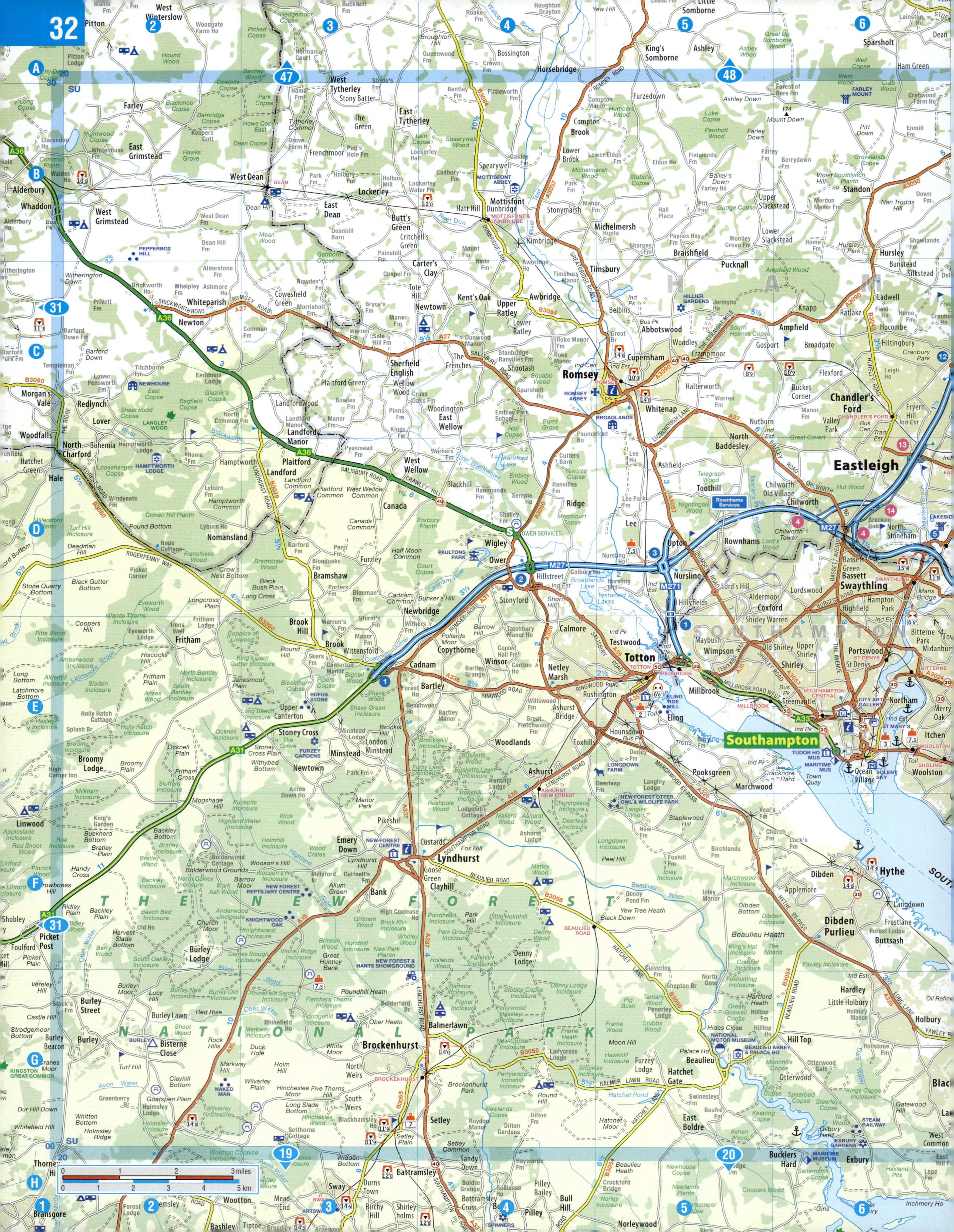
Pitton
West Winterslow
Woodgate Farm Ho
Hedgemoor Copse
Picked Copse
Norman Court
Buckholt Fm
Broughton Hill
Queenwood Fm
Bossington
Houghton Drayton
Yew Hill
Little Somborne
King's Somborne
Ashley
Sparsholt
Lainston Ho
Dean
Ham Green
Pitton Lodge
Hound Wood
West Tytherley
Stony Batter
Horsebridge
Farley Mount
Crab Wood
Forest of Bere Fm
Ashley Down
Mount Down
Crabwood Farm Ho
Farley
Blackmoor Copse
Barnridge Copse
Keepers Cott
Howe Cop East
Park Copse
Tytherley Common
East Tytherley
The Green
Bentley Fm
Pittleworth Fm
Compton Manor
Compton
Brook
Furzedown
Luke Copse
Parnholt Wood
Farley Down
Pitt Down
Enmill Fm
Clarendon Ho
Nightwood Copse
Whitehouse Fm
East Grimstead
Hawks Grove
Dean Copse
Frenchmoor
Pug's Hole Fm
Lockerley Hall
Spearywell Wood
Spearywell
Mottisfont Abbey
Lower Brook
Michelmersh Wood
Eldon Ho
Fishponds
Bailey's Down Farley Ho
Farley Fm
Berrydown
Grovelands Copse
Violet Southlynch Plantn
Standon
Alderbury
Whaddon
West Grimstead
West Dean
Dean Hill
East Dean
Lockerley
Butt's Green
Critchell's Green
Mottisfont
Dunbridge
Hatt Hill
Kimbridge
Stonymarsh
Michelmersh
Pitt Fm
Gudge Copse
Upper Slackstead
Lower Slackstead
Merdon Manor Fm
Nan Tradds Hill
Hursley Park
Hursley
Shawlands Fm
Bunstead Fm
Silkstead
Pepperbox Hill
West Dean Fm
Dean Hill Fm
Mean Wood
Deanhill Barn
Painshill Fm
Carter's Clay
Mount Fm
Hyde Fm
Awbridge
Timsbury Manor
Timsbury
Sharpes Fm
Paynes Hay Fm
Braishfield
Pucknall
Ampfield Wood
Witherington Down
Privett Fm
Brickworth Ho
Whelpley Fm
Ashmore Fm
Alderstone Fm
Cowesfield Green
Morrisholt
Whiteparish
Newton
Rowden's Fm
Bryce's Fm
Chapel Fm
Tote Hill
Newtown
Kent's Oak
Upper Ratley
Lower Ratley
Awbridge
Belbins
Hillier Gardens
Jermyns Ho
Abbotswood
Knapp
Ampfield
Ladwell
Ratlake
Hocombe
Hiltingbury
Barford Down Fm
Barford Park Fm
Templeman's Fm
Barford Down
Titchborne Ho
Common Fm
Warren Fm
Sole Hill Fm
Dunwood Manor
Roke Manor Fm
Roke
Woodley
Cupernham
Crampmoor
Gosport
Broadgate
Cranbury Park
Morgan's Vale
Redlynch
Lower Pensworth Fm
Newhouse
Earldoms Lodge
Glazier's Copse
Sherfield English
Plaitford Green
The Frenches
Stanbridge Ranvilles Fm
Shootash
Squabb Wood
Romsey
Romsey Abbey
Whitenap
Halterworth
Bucket Corner
Flexford
Chandler's Ford
Leigh Ho
Lover
Woodfalls
Sherwood Copse
Langley Wood
Bagfield Copse
Landford Wood
Landford Manor
Bowles Fm
Cross Oaks Fm
Pinns Fm
Woodington
East Wellow
Embley Park School
Burnt Grove
Broadlands
Pauncefoot Ho
Warren Fm
Nutburn
Valley Park
Fryern Hill
North Charford
Bohemia
Hamptworth Lodge
Hatchet Green
Hale
Loosehanger Copse
Hamptworth
Home Fm
Plaitford
Landford
Pyesmead Fm
West Wellow
Kentford Lake
Cutters Barn
Yewtree Copse
Ranvilles Fm
Ridge
Lee Ho
Ashfield
Toothill
North Baddesley
Chilworth Old Village
Chilworth
Eastleigh
Windyeats Fm
Lyburn Ho
Hamptworth Common
Cloven Hill Plantn
Pound Bottom
Plaitford Common
West Wellow Common
Canada
Blackhill
Hammonds Fm
Embley Wood
Semple Ho
Moorcourt Copse
Lee
Nightingale Wood
Rownhams Services
Rownhams
Chilworth Tower
North Stoneham
Bassett
Turf Hill Inclosure
Nomansland
Hope Cottage
Franchises Wood
Barford Fm
Canada Common
Foxbury Plantn
Penn Fm
Half Moon Common
Wigley
Ower
Paultons Park
Nursling Ho
Upton
Nursling
Lord's Hill
Bassett Green
Swaythling
Deadman Hill
Rogerpenny Way
Picket Corner
Crow's Nest Bottom
Bramshaw Wood
Bloodoaks Fm
Bramshaw
Furzley
Court Copse
Hillstreet
Stonyford
Colbury Ho
Broadlands Lake
Testwood Lakes
Hillyfields
Aldermoor
Lordswood
Hampton Park
Mans Bridge
Black Gutter Bottom
Stone Quarry Bottom
Black Bush Plain
Long Cross
Blenman's Fm
Cadnam Common
Bunker's Hill
Newbridge
Shorn Hill
Coxford
Highfield
Longcross Plain
Islands Thorns Inclosure
Fritham Lodge
Brook Hill
Warren's Fm
Storm's Fm
Withers Fm
Pollards Moor
Barrow Hill Fm
Tatchbury Manor Ho
Calmore
Testwood
Shirley Warren
Maybush
Wimpson
Old Shirley
Upper Shirley
Shirley
Portswood
St Denys
Bitterne Park
Midanbury
Pitts Wood Inclosure
Coopers Hill
Irons Well
Fritham
Eyeworth Lodge
Brook
Copythorne
Winsor
Totton
Ashlett Hill
Hiscocks Hill
Janesmoor Plain
Round Hill
Skers Fm
Wittensford
Cadnam
Copied Hall Fm
Bartley Grange
Carlton Ho
Netley Marsh
Rushington
Millbrook
Freemantle
Northam
Long Bottom
Latchmore Bottom
Amberwood Inclosure
Sloden Inclosure
Fritham Plain
North Bentley Inclosure
Blackthorn Copse
Rufus Stone
Upper Canterton
Shave Wood
Shave Green Inclosure
Forest Bar
Bartley
Beechwood Ho
Bartley Manor
Ringwood Road
Willswood Fm
Great Fletchwood Fm
Ashurst Bridge
Eling
Eling Tide Mill
Southampton
Southampton Central
City Art Gallery
Itchen
Merry Oak
Holly Hatch Cottage
Splash Br
Ocknell Plain
Stoney Cross
Stoney Cross Plain
Furzey Gardens
Minstead
Minstead Lodge
London Minstead
Brockis Hill
Woodlands
Foxhills
Hounsdown
Durley
Tudor Ho Mus
Maritime Mus
Town Quay
Ocean Village
Solent Sky
Woolston
Broomy Lodge
Broomy Plain
Withybed Bottom
Newtown
Park Fm
Ashurst
Longdown Farm
Pooksgreen
Marchwood
Crackmore Hard
High Corner Inn
Milkham Inclosure
Fritham Plain
Slufters Inclosure
Acres Down Ho
Manor Park
Lodgehill Cottage
New Forest Otter, Owl & Wildlife Park
Deerleap
Veal's Fm
Linwood
King's Garden
Buckherd Bottom
Bratley Plain
Mogshade Hill
Puckpits Inclosure
Wick Wood
Pikeshill
Mallard Wood
Ashurst Lodge
Staplewood Fm
Church Fm
Cock's Fm
Appleslade Inclosure
Red Shoot Wood
Backley Bottom
Holmhill Inclosure
Emery Down
New Forest Centre
Lyndhurst
Custards
Fox Hill
Longdown Inclosure
Peel Hill
Foxhill Fm
Ipley Inclosure
Birchlands Fm
Hythe
Dibden
Applemore
Handy Cross
Bolderwood Cottage
Bolderwood Grounds
Wooson's Hill
Millyford Br
Cuffnell's Fm
Goose Green
Clayhill
Beaulieu Road
Matley Wood
Decoy Pond Fm
Yew Tree Heath
Ipley Manor
Dibden Bottom
Dibden Inclosure
Langdown
Frostlane
Forest Lodge
Dibden Purlieu
Buttsash
Ridley Plain
Backley Plain
Barrow Moor
New Forest Reptiliary Centre
Ash Wood
Allum Green
Bank
High Coxlease
Pondhead Inclosure
Park Hill
Little Holmhill Inclosure
Denny Wood
Black Down
Beaulieu Heath
King's Hat Inclosure
The Noads
Picket Post
Foulford
Harvest Slade Bottom
Old Ho
Church Moor
Knightwood Oak
Knightwood Inclosure
Gritnam Wood
Brick Kiln Inclosure
Park Ground Inclosure
Denny Lodge
Culverley Fm
Crabtree Inclosure
Fawley Inclosure
Picket Plain
Burley Lodge
Dames Slough Inclosure
Vinney Ridge Inclosure
Brinken Wood
Great Huntley Bank
Hurst Hill Inclosure
New Park Plantn
New Forest & Hants Showground
Hollands Wood
Parkhill Inclosure
North Gate
Shepton Br
Foxhunting Inclosure
Hardley
Little Holbury
Vereley Hill
Lucy Hill
Burley New Inclosure
Burley Old Inclosure
Rhinefield Sandys Inclosure
Poundhill Heath
Rhinefield Inclosure
Fletchers Thorns Inclosure
Bolderford Br
Oberwater
Stubby Copse Inclosure
Denny Lodge Inclosure
Pig Bush
Tantany Wood
Penerley Lodge
Hartford Heath
Hilltop Fm
Holbury Manor
Holbury
Burley Street
Burley Lawn
Red Rise
Markway Inclosure
Aldridgehill Inclosure
Oberwater
Balmerlawn
Frame Wood
Stubbs Wood
Hides Close
National Motor Museum
Beaulieu Abbey & Palace Ho
Hill Top
Castle Hill
Burley
Bisterne Close
Shoot Wood
Rock Hills
Duck Hole
Brockenhurst
Moon Hill
Palace Ho
Beaulieu
Rollstone Fm
Strodgemoor Bottom
Burley Beacon
Turf Hill
Holm Hill
North Weirs
Frame Heath Inclosure
Ladycross Lodge
Hawkhill Inclosure
Furzey Lodge
Hatchet Gate
Moonhills Copse
Otterwood Gate
Otterwood
Kingston Great Common
Avon Water
Clayhill Bottom
Naked Man
Wilverley Plain
Hincheslea Five Thorns Moor
Brockenhurst Park
Parrywood Inclosure
Stockley Inclosure
Round Hill
Balmer Lawn Road
Hatchet Pond
Swinesleys Fm
Gatewood Hill
Dur Hill Down
Greenberry Br
Goatspen Plain
Holmsley Lodge
Wilverley Inclosure
Long Slade Bottom
South Weirs
Setley
Roydon Manor
Dilton Fm
Dilton Gardens
Hatchet Moor
East Boldre
Beufre Fm
Keeping Copse
Gilbury Hard
Exbury Gardens
Steam Railway
West Common
Whitefield Hill
Whitten Bottom
Holmsley Ridge
Sethorns Cottage
Blackhamsley Ho
Setley Plain
Setley Common
Sandy Down
Haywards Fm
Beaulieu Heath
Crockford Bridge
Norley Inclosure
Newhouse Copse
Newlands Plantn
Bucklers Hard
Maritime Museum
Exbury
Haxland Pits
Lepe Fm
Thorney Hill
Whitten Bottom
Sway
Durns Town
Battramsley
Boldre Grange
Battramsley Cross
Pilley Bailey
Pilley
Bull Hill
Coopers Wood
Inchmery Ho
Bransgore
Forest Lodge
Wootton
Tiptoe
Bashley
Mead End
Birchy Hill
Shirley Holms
Spinners
Norleywood
Gins
New Forest National Park
Southampton
SU
A36
A31
A27
A337
A35
M27
M271
A33
A326
B3080
B3078
B3079
B3084
B3057
B3055
B3056
B3054
Scale: 0 1 2 3 miles; 0 1 2 3 4 5 km
19
20
31
47
48

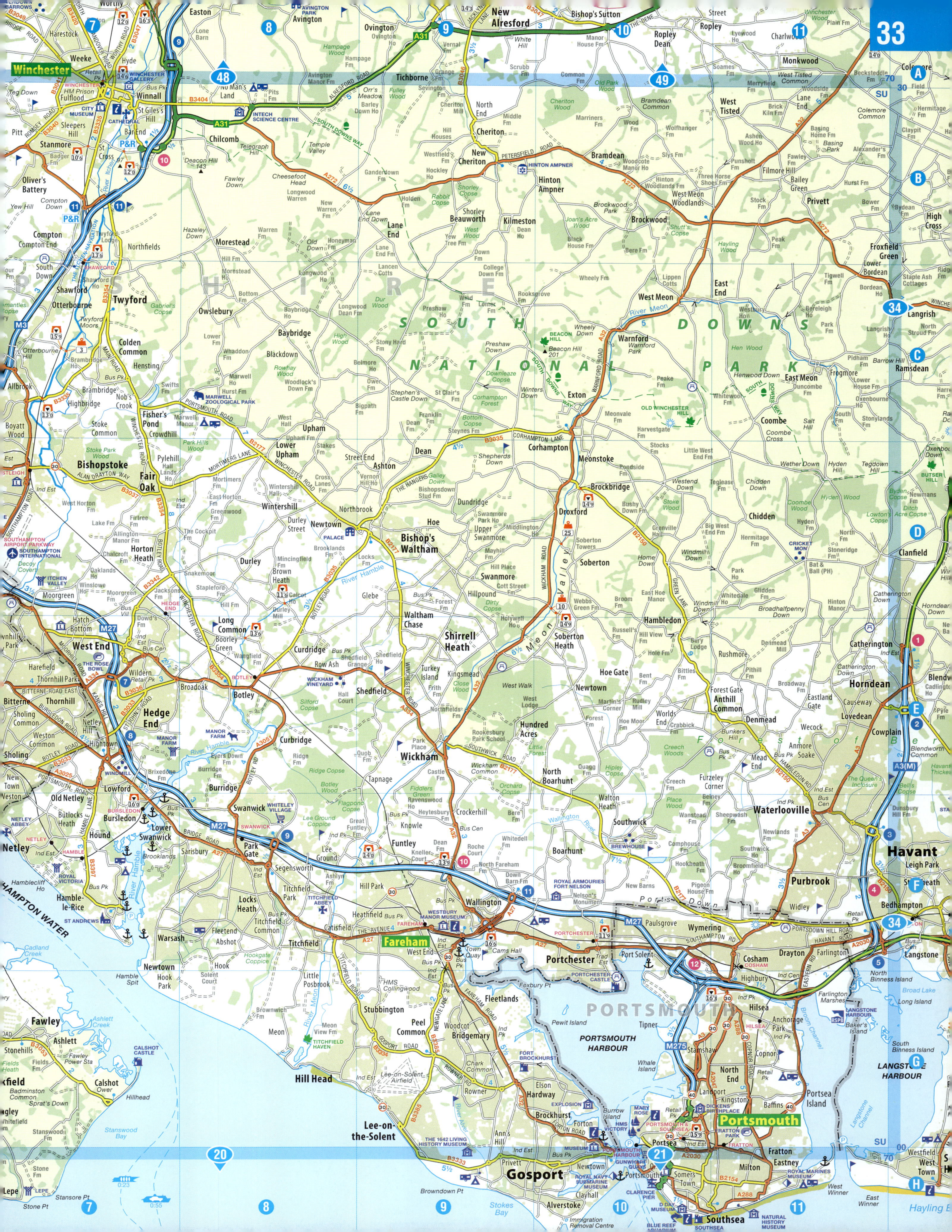

33
Winchester
New Alresford
Bishop's Sutton
Ropley
Ropley Dean
Ovington
Avington
Easton
Tichborne
Cheriton
New Cheriton
Hinton Ampner
Bramdean
West Tisted
Monkwood
Chilcomb
Winnall
Twyford
Owslebury
Morestead
Colden Common
Kilmeston
Beauworth
Lane End
Brockwood
Privett
West Meon
Warnford
East Meon
Langrish
Ramsdean
Exton
Meonstoke
Corhampton
Droxford
Soberton
Soberton Heath
Hambledon
Clanfield
Catherington
Horndean
Denmead
Waterlooville
Purbrook
Havant
Bedhampton
Bishop's Waltham
Upham
Lower Upham
Durley
Bishopstoke
Fair Oak
Horton Heath
West End
Hedge End
Botley
Curdridge
Shedfield
Shirrell Heath
Waltham Chase
Swanmore
Wickham
Knowle
Funtley
Fareham
Portchester
Cosham
Drayton
Paulsgrove
Wymering
Hilsea
Portsmouth
Southsea
Fratton
Eastney
Milton
Gosport
Lee-on-the-Solent
Hill Head
Stubbington
Titchfield
Locks Heath
Warsash
Swanwick
Bursledon
Netley
Hound
Hamble-le-Rice
Fawley
Ashlett
Calshot
Botley
Stoke Common
Otterbourne
Shawford
Compton
Southsea
SOUTH DOWNS NATIONAL PARK
PORTSMOUTH HARBOUR
LANGSTONE HARBOUR
SOUTHAMPTON WATER
Portsea Island
M3
M27
M275
A31
A32
A3
A27

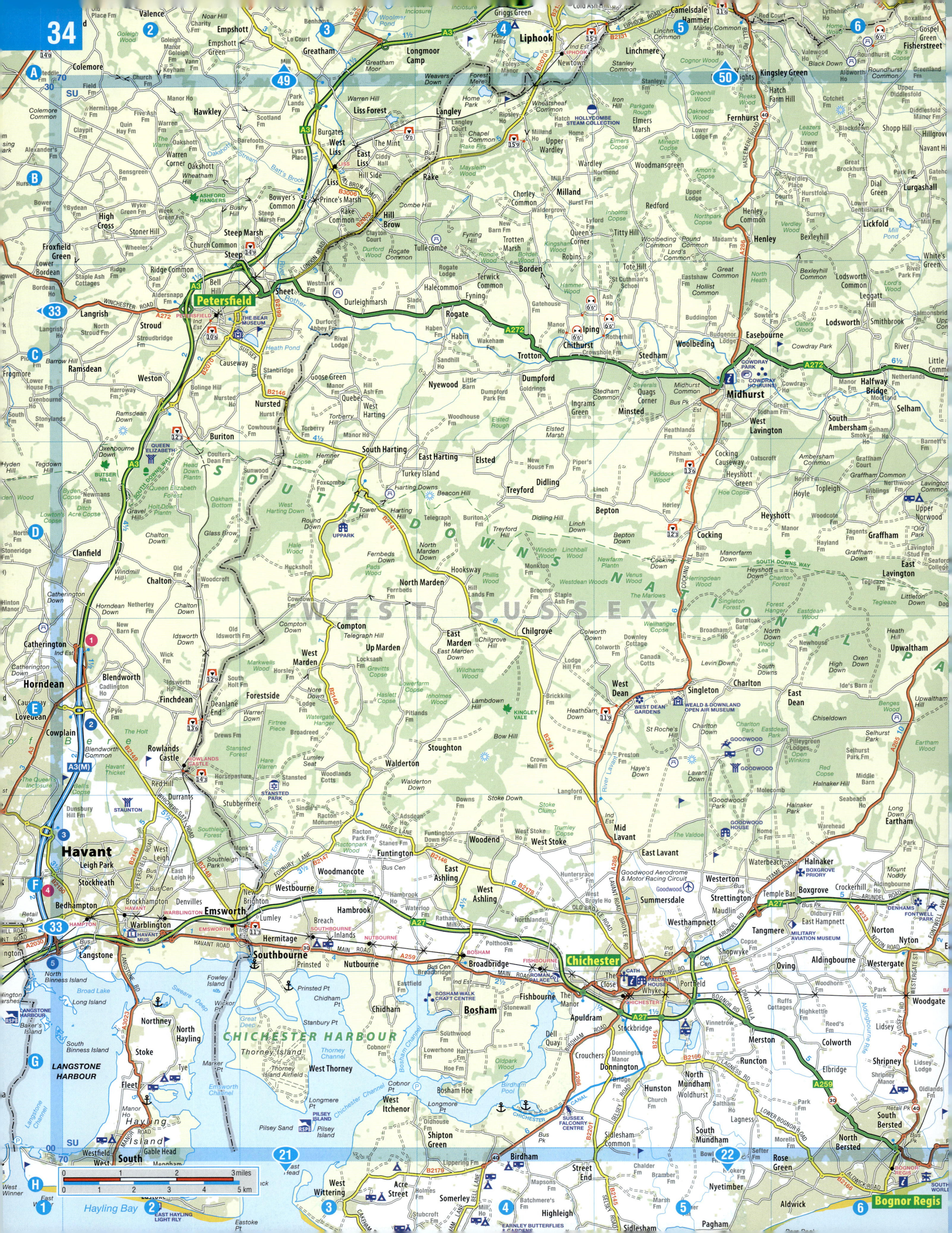

Liphook
Griggs Green
Greatham
Longmoor Camp
Liss Forest
Liss
West Liss
East Liss
Rake
Hill Brow
Hawkley
Empshott
Colemore
Langley
Milland
Linchmere
Camelsdale
Hammer
Kingsley Green
Fernhurst
Henley
Lurgashall
Lickfold
Lodsworth
Woodmansgreen
Redford
Steep
Steep Marsh
Petersfield
Sheet
Langrish
Stroud
Ramsdean
Froxfield Green
High Cross
Durleighmarsh
Rogate
Fyning
Terwick Common
Trotton
Chithurst
Iping
Stedham
Woolbeding
Easebourne
Midhurst
Selham
South Ambersham
West Lavington
Cocking
Heyshott
Graffham
East Lavington
Upwaltham
Nyewood
Dumpford
Elsted
Treyford
Didling
Bepton
South Harting
East Harting
Nursted
Buriton
Weston
Chalton
Clanfield
Catherington
Horndean
Lovedean
Cowplain
Blendworth
Finchdean
Rowlands Castle
Forestside
West Marden
Compton
Up Marden
North Marden
East Marden
Chilgrove
West Dean
Singleton
Charlton
East Dean
Stoughton
Walderton
Woodmancote
Westbourne
Funtington
West Stoke
Woodend
East Ashling
West Ashling
Mid Lavant
East Lavant
Westerton
Strettington
Boxgrove
Halnaker
Summersdale
Westhampnett
Tangmere
Norton
Aldingbourne
Westergate
Oving
Chichester
Havant
Leigh Park
Bedhampton
Warblington
Emsworth
Southbourne
Hermitage
Nutbourne
Bosham
Broadbridge
Fishbourne
Apuldram
Donnington
Hunston
North Mundham
South Mundham
Merston
Runcton
Colworth
Shripney
North Bersted
South Bersted
Bognor Regis
Rose Green
Birdham
Sidlesham
Itchenor
West Itchenor
Chidham
West Thorney
Thorney Island
CHICHESTER HARBOUR
LANGSTONE HARBOUR
Hayling Island
North Hayling
Northney
Stoke
Langstone
West Wittering
Somerley
Highleigh
Nyetimber
Pagham
Aldwick
Hayling Bay
SOUTH DOWNS NATIONAL PARK
WEST SUSSEX
A3
A3(M)
A27
A272
A259
A286
A285
A29
B2070
B2146
B2141
B2145
B2166
33
49
50
21
22

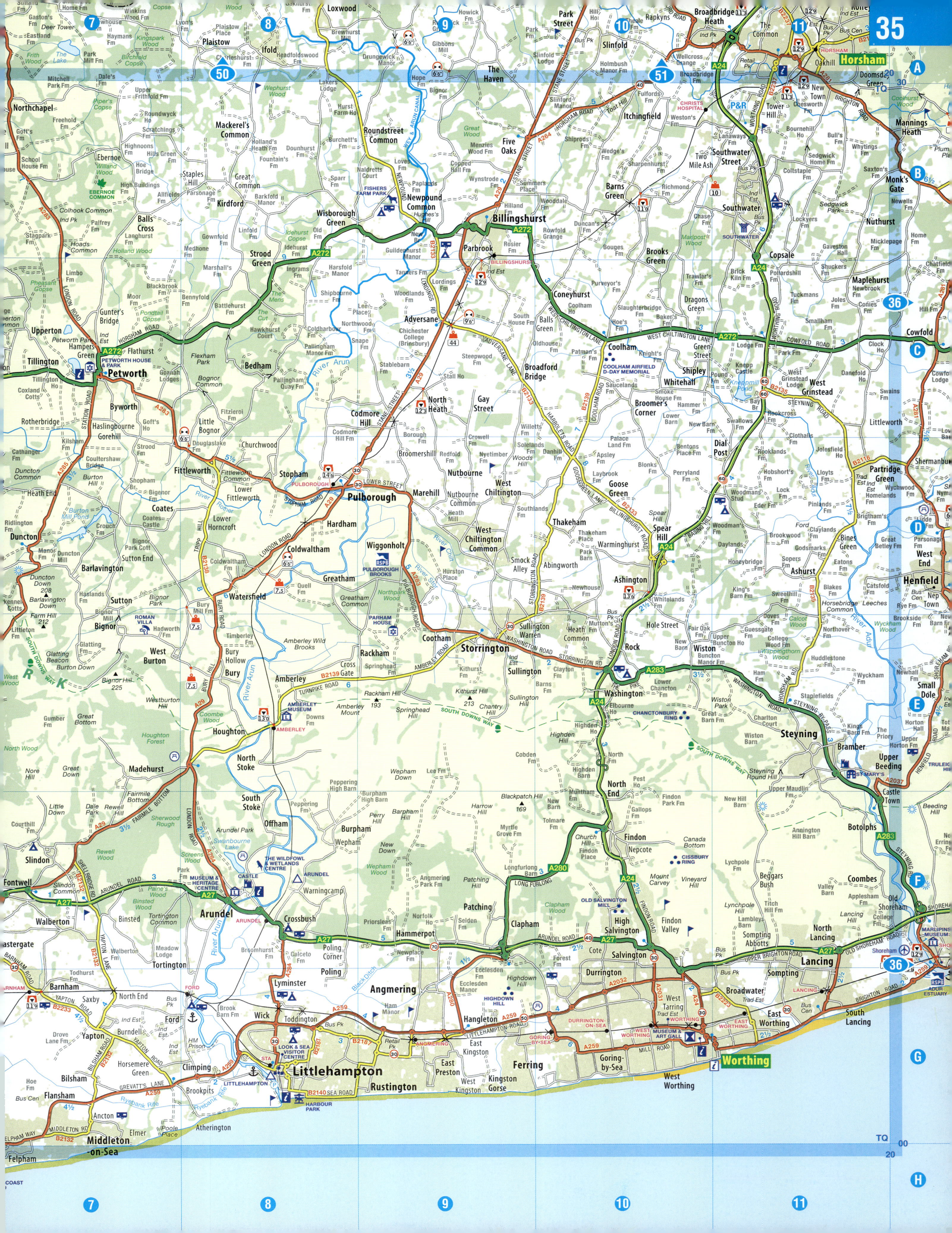
Horsham
Worthing
Littlehampton
Rustington
Arundel
Petworth
Pulborough
Storrington
Billingshurst
Steyning
Henfield
Lancing
Middleton-on-Sea
Felpham
Angmering
Ferring
Goring-by-Sea
Durrington
Salvington
Findon
Washington
Ashington
West Chiltington
Coolham
Cowfold
West Grinstead
Partridge Green
Southwater
Slinfold
Itchingfield
Loxwood
Plaistow
Kirdford
Wisborough Green
Fittleworth
Bury
Amberley
Houghton
Slindon
Fontwell
Walberton
Barnham
Yapton
Climping
Ford
Wick
Lyminster
Crossbush
Poling
Clapham
Patching
Sompting
Bramber
Upper Beeding
Shoreham
Small Dole
Tillington
Northchapel
Duncton
Sutton
Bignor
Madehurst
Burpham
Wepham
Coldwaltham
Hardham
Greatham
Wiggonholt
Thakeham
Heath Common
Sullington
Rock
Wiston
Shipley
Dial Post
Nutbourne
Marehill
Codmore Hill
Adversane
Five Oaks
Barns Green
Mannings Heath
Nuthurst
Copsale
Maplehurst
Monk's Gate
Broadbridge Heath
Christ's Hospital
Tower Hill
Ifold
Newpound Common
Balls Cross
Ebernoe
Stopham
Byworth
A24
A27
A29
A272
A283
A280
A259
A264
A281
SOUTH DOWNS WAY
COAST
TQ
7
8
9
10
11
A
B
C
D
E
F
G
H
36
50
51

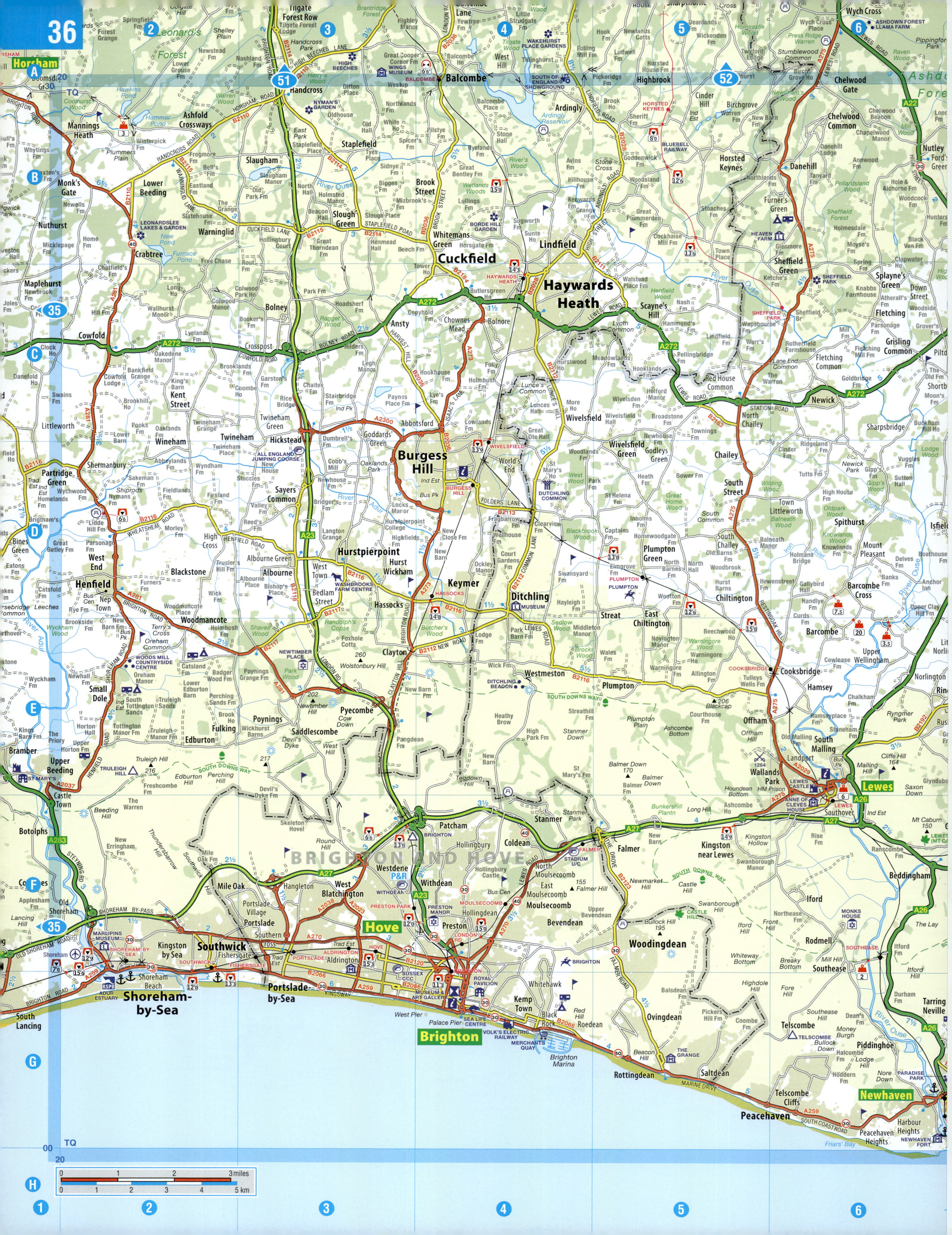
Horsham
Handcross
Balcombe
Highbrook
Wych Cross
Chelwood Gate
Mannings Heath
Ashfold Crossways
Slaugham
Staplefield
Ardingly
Lower Beeding
Nuthurst
Monk's Gate
Warninglid
Brook Street
Slough Green
Whitemans Green
Cuckfield
Lindfield
Haywards Heath
Horsted Keynes
Danehill
Chelwood Common
Crabtree
Maplehurst
Cowfold
Bolney
Crosspost
Ansty
Chownes Mead
Bolnore
Scayne's Hill
Sheffield Green
Fletching
Newick
North Chailey
Chailey
South Chailey
Kent Street
Wineham
Twineham
Twineham Green
Hickstead
Sayers Common
Abbotsford
Goddards' Green
Burgess Hill
Wivelsfield
Wivelsfield Green
World's End
Ditchling Common
Partridge Green
Shermanbury
West Grinstead
Littleworth
Henfield
West End
Blackstone
Albourne
Albourne Green
Hurstpierpoint
Hurst Wickham
Keymer
Ditchling
Hassocks
Clayton
Streat
East Chiltington
Plumpton
Plumpton Green
Chiltington
Barcombe
Barcombe Cross
Spithurst
Cooksbridge
Hamsey
Offham
Woodmancote
Small Dole
Upper Beeding
Bramber
Poynings
Fulking
Edburton
Pyecombe
Saddlescombe
Westmeston
South Malling
Lewes
Wallands Park
Kingston near Lewes
Falmer
Stanmer
Coldean
Patcham
Hollingbury
Westdene
Withdean
Mile Oak
Hangleton
West Blatchington
Portslade
Hove
Preston
Moulsecoomb
Bevendean
Woodingdean
Ovingdean
Rottingdean
Saltdean
Telscombe
Telscombe Cliffs
Peacehaven
Piddinghoe
Newhaven
Rodmell
Southease
Iford
Beddingham
Brighton
Kemp Town
Whitehawk
Southwick
Kingston by Sea
Shoreham-by-Sea
Portslade-by-Sea
South Lancing
Botolphs
Steyning
BRIGHTON AND HOVE
South Downs Way
A23
A27
A272
A259
A275
A26
A22
A281
A283
A2037
A270
35
51
52
0 1 2 3 miles
0 1 2 3 4 5 km

Crowborough
Uckfield
Heathfield
Mayfield
Hailsham
Polegate
Eastbourne
Seaford
Pevensey
Pevensey Bay
PEVENSEY BAY
Herstmonceux
Wadhurst
Rotherfield
Burwash
Etchingham
Cross in Hand
Horam
Willingdon
Westham
Langney
Alfriston
Wilmington
East Dean
Friston
Jevington
E A S T S U S S E X
P e v e n s e y L e v e l s

Flimwell
Gill's Green
Beal's Green
Hawkhurst
Woodsden
Iden Green
Dingleden
Beacon Hill
West Cross
Halden Lane Fm
C M BOOTH COLLECTION OF HISTORIC VEHICLES
Rolvenden
Rolvenden Layne
Small Hythe
SMALLHYTHE PLACE
Reading Street
Appledore Heath
Leg Cross
Ticehurst
Berner's Hill
Dale Hill
Seacox Heath
High Street
Highgate
Pipsden
Four Throws
The Moor
Standen Street
Gun Green
Pashley Manor
PASHLEY MANOR GDNS
Horns Corner
Swiftsden
Sandhurst
Sandhurst Cross
Linkhill
Lamberden
Newenden
Great Job's Cross
KENT & EAST SUSSEX RAILWAY
Peening Quarter
Wittersham
Isle of Oxney
The Stocks
Stone
Stone in Oxney
Moon's Green
Ham Green
Hurst Green
Burgh Hill
Etchingham
ETCHINGHAM
Silverhill
Bodiam
BODIAM CASTLE
Kent Ditch
Northiam
GREAT DIXTER
BRICKWALL HOUSE
Rother Levels
River Rother
FARM WORLD
Northbridge Street
ROTHER VALLEY RLY CENTRE
ROBERTSBRIDGE
Robertsbridge
Salehurst
Snagshall
Ewhurst Green
Mill Corner
Clayhill
Beckley
Four Oaks
Flackley Ash
Peasmarsh
Iden
Houghton Green
Playden
Rye
Rye Foreign
Bowler's Town
East Guldeford
Oxley's Green
Collier's Green
Staplecross
Horns Cross
Broad Oak
Beckley Furnace
John's Cross
Swaile's Green
Cripp's Corner
Chitcombe
Brede
Broadland Row
Udimore
Cock Marling
Rye Harbour
TREASURY OF MECHANICAL MUSIC
YPRES TOWER
HERITAGE CENTRE & TOWN MODEL
CAMBER CASTLE
Mountfield
Vinehall Street
Darwell Reservoir
Darwell Hole
Netherfield
Sedlescombe
Sedlescombe Street
Woodmans Green
Canadia
Whatlington
Powdermill Res
Goatham Green
Cackle Street
Winchelsea
WINCHELSEA
WINCHELSEA COURT MUSEUM
Nook Beach
Penhurst
Battle
BATTLE ABBEY
Starr's Green
Telham
Kent Street
Westfield
Lidham Hill
Guestling Thorn
Icklesham
Broad Street
Pett Level
Winchelsea Beach
Ashburnham Forge
Steven's Crouch
Catsfield
Three Oaks
Guestling Green
Pett
Cliff End
Fairlight Cove
Ninfield
Lower Street
Henley's Down
Crowhurst
Baldslow
Hollington
Silverhill
Ore
St Helen's
Fairlight
Hooe Common
Lunsford's Cross
Sidley
Bulverhythe
Hastings
St Leonards
West Marina
Glyne Gap
UNDERWATER WORLD
SHIPWRECK HERITAGE CENTRE
Old Town
Covehurst Bay
Little Common
Cooden
Bexhill
DELAWARR PAVILION
My Lord's Rock
Hooe Level
Norman's Bay
EAST SUSSEX
TQ
00
70
0 1 2 3 miles
0 1 2 3 4 5 km

Great Heron Wood
Park Hill
Sly Corner
Smith's Fm
Thrift Cott
Higham Fm
HORNE'S PLACE CHAPEL
The Dowels
Bridge Fm
Johnson's Corner
Wey Street Fm
Stockbridge Ho
Poplar Ho
54
Snave
Will's Fm
Brooker Fm
Manor Ho
Norwood Fm
New Barn Fm
Newchurch
Gammon's Fm
Forty Acre Cott
Burmarsh
Donkey Street
Orgarswick Fm
Chapel Cottage Fm
HYTHE ROAD
Dymchurch Wall
55
Sutton Fm
A259
DYMCHURCH MARTELLO TOWER
Dymchurch
DYMCHURCH
Sellinge Fm
Blackmanstone Br
Pickney Bush Fm
Haffenden Fm
Romney Marsh
Lodgeland Fm
Willow Fm
B2080
APPLEDORE
Bridge Fm
Ham Fm
Whitehall Fm
Brenzett Green
Hope Fm
Moat Ho
New House Fm
Poplar Fm
Snargate
North Fording Bungalow
Melon Fm
ROMNEY, HYTHE & DYMCHURCH RAILWAY
St Mary in the Marsh
ST MARY'S BAY
St Mary's Bay
Spring Fm
Ivychurch
A2070
CANAL
Highknock Channel
Priory Fm
Fairfield Court
Becket Barn Fm
Fairfield
Brenzett
Brenzett Place
AERONAUTICAL MUSEUM
Blue House Fm
Yoakes Court Fm
Beechcroft Fm
Honeychild Manor
Brodnyx
ROMNEY WARREN COUNTRY PARK
DYMCHURCH ROAD
ROMNEY WARREN HALT
Brattle Ho
Old Hall Fm
Poplar Hall
Rhee Wall Fm
New Sewer
Hope Fm
Brodnyx Fm
Warren Fm
Dean Court
A259
Brookland
Bush Fm
Sycamore Fm
Old Romney
New Romney
NEW ROMNEY
B2071
Phoenix Caisson
New Buildings Fm
Whitehouse Fm
Hook Ho
Coldharbour Fm
Court Lodge
LYDD ROAD
Hammonds Corner
Kemp's Hill
Littlestone-on-Sea
Romney Sands
White Kemp Sewer
Blue House Fm
Old Cheyne Court
Midley Cotts
Hawthorn Corner
GULDEFORD LANE
Offen's Fm
Guldeford Lane Corner
Walland Marsh
Baynham Fm
Coldicott Fm
Westbrook Fm
ROMNEY ROAD
B2075
Belgar Fm
Footway Fm
Greatstone-on-Sea
ROMNEY SANDS
Newland Fm
Kent Ditch
Little Cheyne Court
Barn Fm
Little Scotney
Westbroke Ho
Jack's Court
Lydd (London Ashford)
Lydd
Pigwell
Red Ho
Lade
Scotney Court
Denge Marsh
Lydd-on-Sea
LYDD INTERNATIONAL RACEWAY
DUNGENESS
RSPB
Boulderwall Fm
Halfway Bush
Point Fm
Danger area
West Ripe
Holmstone
Lydd Ranges
Manor Fm
DUNGENESS
Camber
Jury's Gap
Broomhill Level
Open Pits
Camber Sands
South Brooks
Brickwall Fm
Broomhill Sands
RYE BAY
Dungeness Power Sta
DUNGENESS
THE OLD LIGHTHOUSE
Dungeness
A
B
C
D
E
F
G
H
TR
20
30
00
7
8
9
10
11

56
Hen & Chickens
Seals' Rock
North West Pt
North East Pt
Gannets' Rock
Gannets' Bay
St James's Stone
Tibbetts Hill 138
Tibbett's Pt
LUNDY MARINE NATURE RESERVE
Jenny's Cove
Lundy
Dead Cow Pt
Ackland's Moor 142
Lundy Roads
BIDEFORD 2:00
ILFRACOMBE 2:00
Halftide Rock
Beacon Hill
Castle Hill
Rat Island
Surf Pt
South West Pt
SWANSEA 0:50
LUNDY 2:00
Capstone Pt
Samson's Bay
Water Mouth
WATERMOUTH CASTLE
Hele Bay
Ilfracombe
Hele
MUSEUM
Chambercombe
OLD CORN MILL
Hole Fm
Goosewell
CHAMBERCOMBE MANOR
Kitstone Hill
Berrynarbor
Lee
NORTH DEVON
Sterridge
Hangman Pt
Rawn's Rocks
Blackstone Pt
Little Hangman 218
Gt Hangman 318
Elwill Bay
Trentishoe
SOUTH WEST COAST PATH
Holdstone Down 349
South Dean Fm
Trentishoe Down
Lester Cliff
Girt Fm
Girt Down
Holdstone Fm
Trentishoe Manor
Walner Fm
River Umber
Combe Martin
Knap Down
Verwill Fm
Tattiscombe
Stony Corner
Nutcombe Fm
Westleigh Fm
Truckham Fm
Dean
LONG LANE
Stoneditch Hill
Henstridge
WILDLIFE & DINOSAUR PARK
Cowley Wood
Bull Pt
Pensport Rock
Shag Pt
Flat Pt
Lee Bay
Lee
Lincombe
Higher Slade
Rockham Bay
North Morte Fm
Higher Warcombe
Whitestone
Morte Pt
Mortehoe
Shaftsboro Fm
Campscott Fm
Slade Resrs
Lower Slade
Oakridge Fm
Warmscombe Fm
Ruggaton Fm
Bowden Fm
Smythen Fm
Hempster Fm
Two Pots
Ettiford Fm
Grunta Pool
Borough Cross
Borough Fm
Little Shelfin Fm
Woolacombe
Mill Rock
Manor Fm
Trimstone
B3343
Ind Est
A3123
Hore Down Fm
Stapleton Fm
Berry Down
Berry Down Cross
South Ley
Kentisbury
Higher Week Fm
Kentisbury Down
Bugford
Stonecombe
Preston Ho
A399
Morte Bay
Woolacombe Sand
Ossaborough
Willingcott
Cheglinch
Outer Narracott Fm
River Yeo
Cleave Fm
Highlands Fm
Patchole
Northcote Fm
Kentisbury Ford
Halls Cross
Bridwick Fm
Ivycott
Bradwell
Dean
Dean Cross
Hillcrest Fm
Centery Fm
Higher Aylescott
Bittadon
Collacott Fm
Dingles Fm
Wigmore Fm
Ford Fm
Wistlandpound Reservoir
Roadway
Black Rock
Putsborough Sand
Pickwell Down
Spreacombe Manor
North Downs
West Down
Fullabrook
Burland Fm
Clifton
East Down
Hallsdown Fm
Arlington Beccott
Arlington
Arlington Court
ARLINGTON COURT
Churchill
Churchill Down
Huckham Fm
Besshill Fm
Baggy Pt
SOUTH WEST COAST PATH
Vention
Pickwell
Putsborough
Castle Street Fm
Buckland Down
Stoneyard Wood
Fullabrook Down
Metcombe Down
Little Silver
Hewish Down
Bowden Corner
Deerpark Wood
White Cawsey
Tidicombe Fm
Croyde Bay
Ora Hill
Croyde
Georgeham
North Buckland
Winsham Down Ho
Halsinger Down
Beara Down
Patsford
Swindon Down
Gipsy Corner
Whiddon
Milltown
Whitefield Down
Okewill Cross
Viveham Fm
Plaistow Barton
Garman's Down
Woolley Wood
South Woolley Fm
Rye Park
DEVON
Forda
Darracott
Nethercott
Upcott
Incledon Fm
Halsinger
Winsham
Cross
South Hole Fm
Buckland Manor
Knowle
Beara
Middle Marwood
Crockers
Higher Muddiford
Muddiford
The Warren
Chilbridge Fm
Loxhore
Loxhore Cott
CROYDE ROAD
SAUNTON ROAD
B3231
Lobb
Boode
Marwood
Guineaford
Plaistow Mill
Shirwell Cross
Shirwell
Lower Loxhore
Saunton
Sandy Lane Fm
Braunton
Pippacott
Whitehall
MARWOOD HILL
Kingsheanton
BROOMHILL SCULPTURE GARDENS
Shop Cen
Braunton Down
Luscott Barton
Waterlake
Prixford
Varley Fm
South Hill
Waytown Fm
Youlston Wood
Bratton Cross
Saunton Sands
SOUTH WEST COAST PATH
ELLIOT GALLERY
Velator
Knowl Water
Heanton Punchardon
Mainstone
West Ashford
Springfield Cross
Sepscott Fm
SHIRWELL ROAD
Brightlycott
Chelfham
Horridge
Stoke Rivers
Braunton Burrows
Wrafton
A361
Ashford
Upcott Ho
Burridge
Kingdon's Gardens
Hakeford
Braunton Marsh
Chivenor
Penhill Pt
Bradiford
Raleigh
Pitt Fm
Snapper
Northleigh
Goodleigh
Middle Dean Fm
Airy Pt
Danger area
Horsey Island
Chivenor
Allen's Rock
Pilton
Pottington
Derby
Waytown
Youlden Ho
Dean Head
Hutcherton Down
Stone Cross
River Taw
Saltpill Duck Pond
Penhill
ST ANNE'S CHAPEL & MUSEUM
Barnstaple
MUSEUM OF NORTH DEVON
Coombe Fm
Willesleigh Fm
Gunn
LUNDY 2:00
Sticklepath
Westacott
Broad Sands
Lower Yelland
Yelland
B3233
Muddlebridge
BICKINGTON ROAD
A3125
Bus Pk
East Acland
Birch
Accott
Fremington
Combrew
Bickington
Ind Est
BARNSTAPLE
Newport
Portmore
Sandick
Crow Pt
YELLAND ROAD
Brake Plantn
Brynsworthy
Lake
P&R
A361
Rumsam
Landkey
Harford
Sandick Cross
Crow Rock
Instow Sands
N DEVON MARITIME MUSEUM
WESTYELLAND
Bickleton
Myrtle Cott
A39
Upcott Fm
Roundswell
Hurscott
Sandymere
Appledore
NORTHAM BURROWS
The Quay
Worlington
Collacott Fm
Hollamoor Clump
Bishops Tawton
Landkey Newland
Swimbridge Newland
Yarnacott
Yeoland Ho
Riverton
Instow
INSTOW
Fullingcott
NORTH DEVON FARM PARK
Diddywell
24
Huish
Eastacombe
Tawstock
25
Swimbridge
Kerscott
River Torridge
TAPELEY PARK GARDENS
Coombe Fm
Huish Moor
St John's Chapel
Stonyland
Downrew Ho
Horswell Fm
Hannaford
Bydown Ho
High Down
Westleigh
Trayhill
Rushcott Fm
Hangman's Hill
Lane End Fm
Northam
Eastleigh Manor
Holmacott
Prospect Corner
Uppacott
Halmpstone Manor
Tower Fm
Rickard's
Silford
Fastleigh
Horwood
Harepie
Summer
A377
0 1 2 3 miles
0 1 2 3 4 5 km
SS

7
8
9
10
11
57
A
B
C
D
E
F
G
H
42
25
26
SS
90
60
30
Foreland Pt
Countisbury Cove
Woody Bay
Wringcliff Bay
Lee Bay
Crock Pt
Martinhoe
Lynton
CLIFF RLY
LYN & EXMOOR MUSEUM
Lynmouth
Lynbridge
Countisbury
Wind Hill
Kipscombe Hill 343
WATERSMEET HOUSE
Barton Wood
West Lyn
Dean
Caffyns Fm
Kemacott
Martinhoe Common
LYNTON & BARNSTAPLE RAILWAY
WOODY BAY
Killington
KILLINGTON LANE
WILDNER TOP
Barbrook
East Ilkerton
West Ilkerton
Stock Castle
Stock Common
Martinhoe Cross
Beacon Castle
Bodley
Parracombe
Churchtown
Paracombe Common
South Common
Rowley Down
Blackmoor Gate
Ilkerton Ridge
Shallowford
Furzehill
Cannon Hill
Butter Hill
Saddle Gate
Hillsford Br
Combe Park Ho
Bridge Ball
Cheriton
Roborough Castle
Scob Hill
Shilstone Hill 405
Dry Br
Rockford
Tippacott
Brendon
Leeford
Wilsham
Coombe Fm
Ashton Fm
Southern Wood
Malmsmead
Fellingscott
Slocombeslade
Tippacott Ridge
Malmsmead Hill 389
Great Black Hill
Withycombe Ridge
Brendon Common 412
Doone Country
Badgworthy Hill
Pig Hill
Furzehill Common
Cheriton Ridge
Hoaroak
Exmoor Forest
Hoccombe Water
Brendon Two Gates
Thorn Hill
Winaway
Hoaroak Hill 475
The Chains 487
Pinkery Pond
Exe Plain
Hoar Tor
Blackpitts Gate
Old Burrow Hill 345
Sugarloaf Hill
Embelle Wood
Culbone Wood
Yenworthy Fm
Broomstreet Fm
North Common
Lillycombe Ho
Oare
Oareford
Cloud Fm
Robber's Br
Stowey Ridge
Mill Hill
Black Hill
Chalk Water
Weir Water
Manor Allotment
Madacombe
Kittuck Meads
Larkbarrow
Exmoor
Trout Hill
East Pinford
Swap Hill
Swap Hill 444
Alderman's Barrow 447
Lucott Cross 465
Almsworthy Common
Exford Common
CULBONE CHURCH
Worthy
Yarner Fm
Yearner Wood
Gore Pt
Porlock Bay
Hurlstone Pt
Selworthy Sand
Porlock Weir
Worthy Wood
Porlockford
Smalla Combe
Westcott Brake
Pitt Fm
West Porlock
Toll
Porlock
Doverhay
Bossington
EXMOOR FALCONRY
Lynch
Allerford
Brandish Street
West Luccombe
Weirwood Common
Porlock Common
Shillett Wood
Homebush Wood
Hawk Combe
Glen Lodge
Horner
Tarr Ball Hill
Ley Hill 317
Horner Wood
Babe Hill
The Mound 373
THE DUNKERY & HORNER WOOD
Wilmersham Common
Luccombe Hill
Robin How
Dunkery Beacon 519
Dunkery Gate 387
Hoar Moor
Codsend Moors
River Quarme
Kitnor Heath
Higher Ho
Codsend
Higher Down
Challacombe Common
Challacombe Res
Barton Gate
Swincombe
North Regis 417
Friendship
Stowford
Yelland Cross
Challacombe
Barton Town
Challacombe Mill
NORTH LANE
Old Close Bottom
Weirs Combe
Shoulsbarrow Fm
Sloley Stone
Pinkery Fm
PINKERY
EXMOOR NATIONAL PARK
River Exe
Warren Fm
Prayway Head
Ashcombe Bottom
Duredon Fm
Cornham Fm
Barle Water
Simonsbath
Cloven Rocks Br
Gallon Ho
Higher Riscombe Fm
Pitsworthy Fm
Hill Fm
Alcombe Water
Edgcott
Coombe Fm
Honeymead Fm
Winstitchen Fm
White Water
Pennycombe Water
Ashott Barton
White Cross
Ley
Exford
EXFORD
CHURCH HILL
Larcombe Fm
Horsecombe Fm
Luckwell Bridge
Knightacott
Northland Corner
Lecham Ford Br
Wallover Down
Fullaford Down
Leworthy
Little Bray Cross
Berry Hill
Squallacombe
Rockley Fm
Bray Common
Silkwood Top
Horcombe
Blue Gate
SOMERSET
Thornemead
Chibbet Post
North Court
ROOM HILL ROAD
Road Hill
Staddon Hill
Torre Fm
Haxton Down
Benton
Mockham Down Gate
Down Fm
Wort Wood
River Bray
Lydcott Hall
East Down Wood
Whitefield Down
Five Barrows Cross
Kinsford Gate
Emmett's Grange
Kinsford Water
Wintershead Fm
Horsen Fm
Great Ferny Ball
Pickedstones
Buckworthy
Blackland Fm
Hillway Fm
Landacre Br
Withypool
Nethercote
Kemps
Withycombe
Mockham Down
Lydcott
Little Bray
Cross Gate
Holewater
Beara Hill Head
Poltimore Arms (PH)
Lyddicombe Bottom
Fyldon Common
Gravel Pit Cross
Long Holcombe
Horsen Hill 443
River Barle
Waterhouse Fm
Coret's Gate
Winn Brook
Burrow Wood
Winsford
Winsford Hill 426
Barnacott Fm
Kimbland Fm
Brayfordhill
Brayford
High Bray
Slade Fm
Yarde Gate Fm
North Radworthy
Fyldon
Shortacombe
Sherdon Fm
Woolcombe Allot
Woolcombe Fm
Withypool Common
Withypool Hill 398
North Batsom
Great Bradley
Edbrooke Ho
Goodwells Head
The Old Rectory
Charles
Hudley Mill
Higher Molland
Molland Cross
Popham
Bentwichen
North Heasley
Long Wood
Darlick Moors
Withypool Cross
Sportsman's Inn (PH)
Willingford Fm
Halscombe Allot
Porchester's Post 388
West Water
Westwater Allot
Westwater
Spire Cross
The Allotment
Upcott Fm
Newtown Br
Humewins Fm
Radworthy Down
South Radworthy
Longstone Wells
Litton Water
Twitchen Ridge
Danes Brook
Hawkridge Plain
Hill Fm
Tarr Fm
Liscombe
Spire
Old Ashway
Draydon Knap
Elwell
West Buckland School
East Buckland
Charles Bottom
Macmillan Way West
Bampfylde Hill
Heasley Mill
South Wood
Tabor Hill
North Molton Ridge
Balls Cross
Barcombe Down
Wheatclose Cross
Moorhouse Ridge
Hawkridge Common
Parsonage Fm
TARR STEPS
Ashway Side
Mounsey Hill 374
326
South Hill
West Buckland
Huxtable Fm
Brayley Hill
Steep Head
Litchadon
Holywell Cross
Holywell Res
Brinsworthy
Twitchen Mill
Twitchen
Cussacombe Gate
Molland Common
Cloggs Fm
Lyshwell Fm
Shircombe Fm
Hawkridge
Hawkridge Ridge
Hollowcombe
High Bullen Fm
Millbrook
Pitt Fm
Pulham Fm
Pulsworthy Fm
Kerswell Fm
West Anstey Common
Whiterocks Down
Draydon Fm
Mounsey Castle
Marsh Hill Ho
Park
A361
Leary Moors
Heddon
Oxford Down
Filleigh
Wheatlands Fm
Coombe Fm
Snurridge
South Cockerham
Burcombe
West Park
North Molton
River Mole
Sannacott Fm
Limeslake Fm
Waterbrooks Fm
Bickingcott
Luckworthy
Champson
Molland
Smallacombe
Woodland Fm
West Anstey
Two Moor Ways
Higher
Five Cross Ways
Old Berry Fm
Dulverton
A39
A399
A39
B3358
B3223
B3224
B3225
A396
B3223
A361

BRISTOL CHANNEL
EXMOOR NATIONAL PARK
SOMERSET
58
41
26
27
Hurlstone Pt
Selworthy Sand
Selworthy Beacon 308
Greenaleigh Pt
SOUTH WEST COAST PATH
Bossington
EXMOOR FALCONRY
Lynch
Allerford
Porlock
Selworthy
Woodcombe
Minehead
Madbrain Sands
Warren Pt
BUTLINS
Higher Town
Bratton
Periton
Alcombe
Holnicote
Brandish Street
West Luccombe
Horner
Blackford
Tivington
Great Headon Plantn
Dunster Beach
Luccombe
Tivington Knowle
Periton Hill 295
Penny Hill
Ellicombe
Marsh Street
DUNSTER
Blue Anchor Bay
Huntscott
Wootton Courtenay
Knowle Hill
DOLL MUSEUM
Dunster
DUNSTER CASTLE GARDENS
DUNSTER WATERMILL
Blue Anchor
Luccombe Hill
Brockwell
DUNKERY VINEYARD
Robin How 428
Dunkery Beacon 519
Cowbridge
Timberscombe
Bickham
Carhampton
WEST SOMERSET RAILWAY
Chapel Cleeve
Old Cleeve
Withycombe
Bilbrook
Dunkery Gate 387
Croydon Hill
Black Hill 353
Washford
CLEEVE ABBEY
Hungerford
Torre
Watchet
Warren Bay
Doniford
DONIFORD BEACH HALT
St Decumans
Five Bells
Williton
TROPIQUARIA ANIMAL & ADVENTURE PARK
Sampford Brett
Quantock's Head
East Quantoxhead
West Quantoxhead
Staple
Weacombe
Bicknoller
Cutcombe
Wheddon Cross
Luckwell Bridge
Luxborough
Kingsbridge
Rodhuish
Golsoncott
Roadwater
Beggearn Huish
Yarde
Woodford
Monksilver
COMBE SYDENHAM
Stogumber
Vellow
Newton
Kingswood
Halsway
Crowcombe
Lawford
Churchtown
Pooltown
Stamborough
Treborough
Leighland Chapel
Chidgley
Sticklepath
Elworthy
Higher Vexford
Lower Vexford
Willett
Coleford Water
Winsford
Coppleham
West Howetown
Exton
Bridgetown
Gupworthy
Withiel Florey
Brompton Regis
Clatworthy Reservoir
Clatworthy
Brompton Ralph
Tolland
Pitsford Hill
Tarr
Lydeard St Lawrence
Huish Champflower
Maundown
Langley Marsh
Upton
Hartford
Dulverton
Fitzhead
Halse

43
Weston-super-Mare
Weston Bay
Rudder Rock
Steep Holm
Tower Rock
Calf Rock
Brean Down
Uphill
Hutton
Locking
Banwell
Bleadon
Brean
Berrow Flats
Lympsham
Eastertown
Loxton
Christon
Webbington
Compton Bishop
Biddisham
Badgworth
Rooks Bridge
BRIDGWATER
BAY
Berrow
Brent Knoll
East Brent
Burnham-on-Sea
Edithmead
Highbridge
Mark
Mark Causeway
Watchfield
Stert Island
Stert Flats
Steart
Stolford
Huntspill
East Huntspill
Lilstock
Kilton
Kilve
Stringston
Holford
Burton
Shurton
Stogursey
Stockland Bristol
Otterhampton
Combwich
Pawlett
Puriton
Woolavington
Cossington
Bawdrip
Edington
Catcott
Nether Stowey
Over Stowey
Fiddington
Cannington
Chilton Trinity
Bridgwater
Wembdon
Durleigh
Spaxton
Charlynch
Enmore
Goathurst
North Petherton
Huntworth
Chedzoy
Westonzoyland
Sutton Mallet
Moorlinch
Stawell
Middlezoy
Othery
Northmoor Green or Moorland
North Newton
Thurloxton
Broomfield
Cothelstone
Bishops Lydeard
Kingston St Mary
Burrow-bridge
Athelney
Durston
West Monkton
West Lyng
Q U A N T O C K H I L L S
S O M E R S E T
M5
A38
A39
A370
A371
A372
A361
A358
B3139
B3140
B3141
B3339

Brinsea
Cowslip Green
Butcombe
Strode
Chew Stoke
Stanton Drew
Stanton Wick
Chelwood
BATH
NORTH EAST
Churchill
Lower Langford
Langford
Blackmoor
Redhill
Nempnett Thrubwell
Chew Valley Lake
Bishop Sutton
Sutton Wick
Banwell
Sandford
Dinghurst
Burrington
Rickford
Blagdon
Ubley
Clutton
Temple Cloud
Cameley
Hallatrow
Winscombe
Sidcot
Shipham
Rowberrow
Star
Compton Martin
West Harptree
East Harptree
Hinton Blewett
Farrington Gurney
Compton Bishop
Axbridge
Cheddar
Charterhouse
Mendip Forest
MENDIP HILLS
Priddy
Litton
Chewton Mendip
Ston Easton
Chilcompton
Emborough
Binegar
Gurney Slade
Oakhill
Lower Weare
Weare
Draycott
Rodney Stoke
Westbury-sub-Mendip
Easton
Wookey Hole
Green Ore
West Horrington
East Horrington
Wedmore
Blackford
Cocklake
Clewer
Nyland
Theale
Latcham
Panborough
Mudgeley
Wookey
Wells
Dulcote
Dinder
Croscombe
Shepton Mallet
Doulting
Burtle
Westhay
Meare
Godney
Polsham
Coxley
Southway
North Wootton
Pilton
Cannard's Grave
Prestleigh
Evercreech
Glastonbury
Street
West Pennard
East Pennard
Pylle
Walton
Ashcott
Pedwell
Moorlinch
Greinton
Catcott
Edington
Shapwick
Butleigh
Baltonsborough
West Bradley
Ditcheat
Alhampton
Ansford
Castle Cary
Compton Dundon
Dundon
Somerton
Kingweston
Keinton Mandeville
Barton St David
West Lydford
East Lydford
Lovington
Alford
Charlton Mackerell
Charlton Adam
Babcary
Littleton
Hurcott
Henley
High Ham
Pitney
Othery
Bramwell
Galhampton
North Barrow
South Barrow
SOMERSET
0 1 2 3 miles
0 1 2 3 4 5 km

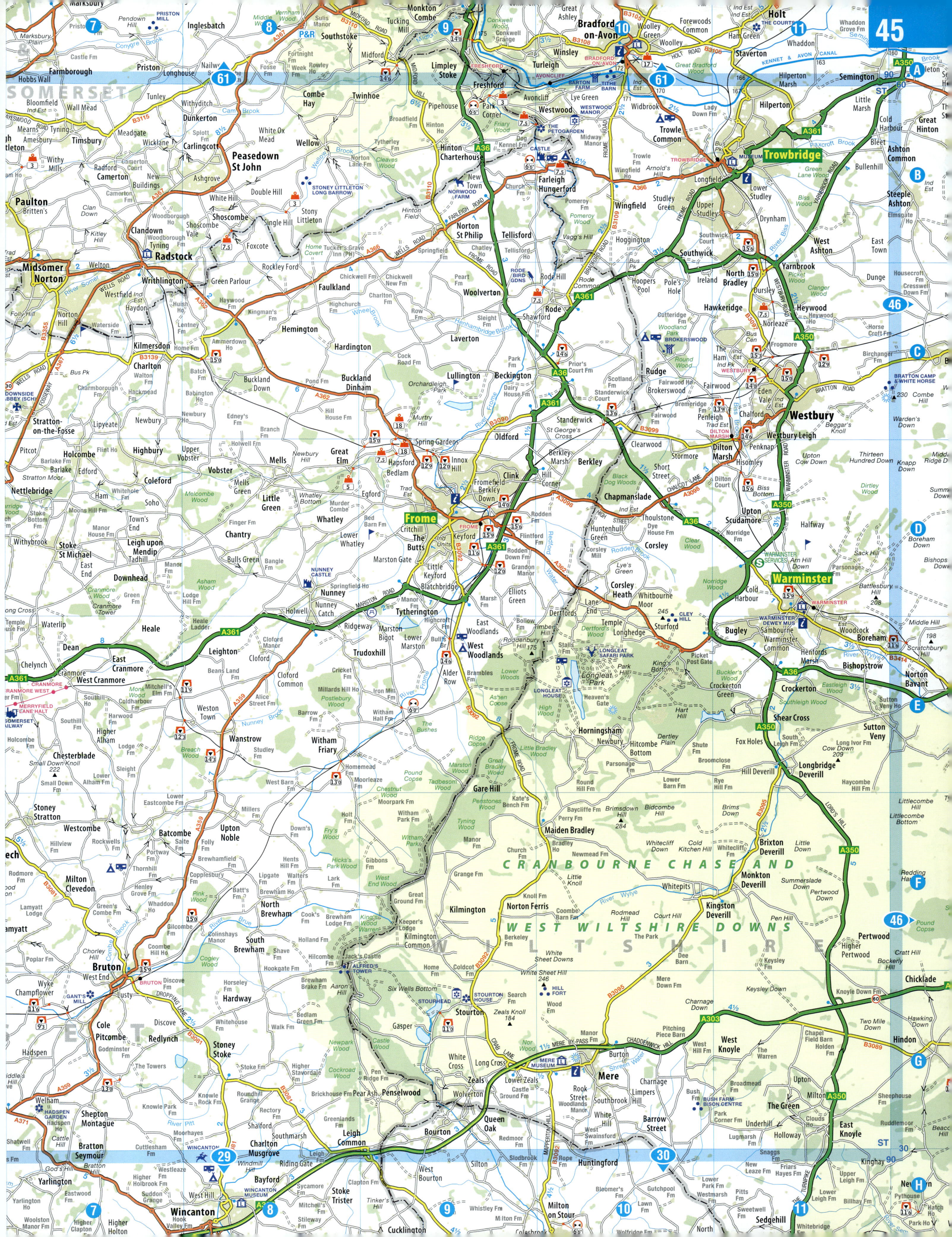

45
SOMERSET
WILTSHIRE
CRANBOURNE CHASE AND
WEST WILTSHIRE DOWNS
Bradford on-Avon
Trowbridge
Westbury
Warminster
Frome
Radstock
Midsomer Norton
Paulton
Peasedown St John
Bruton
Mere
Wincanton
Holt
Staverton
Hilperton
Semington
Melksham
Farleigh Hungerford
Norton St Philip
Beckington
Rode
Woolverton
Laverton
Lullington
Buckland Dinham
Mells
Great Elm
Nunney
Witham Friary
Upton Noble
Batcombe
Milton Clevedon
Evercreech
Shepton Montague
Stourton
Zeals
Kilmington
Maiden Bradley
Horningsham
Longleat Park
Corsley
Chapmanslade
Berkley
Dilton Marsh
Heytesbury
Bishopstrow
Norton Bavant
Sutton Veny
Longbridge Deverill
Brixton Deverill
Monkton Deverill
Kingston Deverill
East Knoyle
West Knoyle
Chicklade
Hindon
Silton
Bourton
Penselwood
Charlton Musgrove
Stoke Trister
Cucklington
Yarlington
Huntingford
Milton on Stour
Sedgehill
Wingfield
Southwick
North Bradley
Heywood
Westwood
Freshford
Limpley Stoke
Southstoke
Midford
Tunley
Timsbury
Camerton
Dunkerton
Wellow
Hinton Charterhouse
Tellisford
Writhlington
Kilmersdon
Hemington
Holcombe
Coleford
Stoke St Michael
Downhead
Leigh upon Mendip
Chantry
Whatley
Tytherington
West Woodlands
Marston Bigot
Trudoxhill
Leighton
East Cranmore
Wanstrow
Chesterblade
A36
A361
A350
A303
A362
A366
A359
A371
A3098
B3092
B3095
B3414
B3108
B3109
30
29
61
46

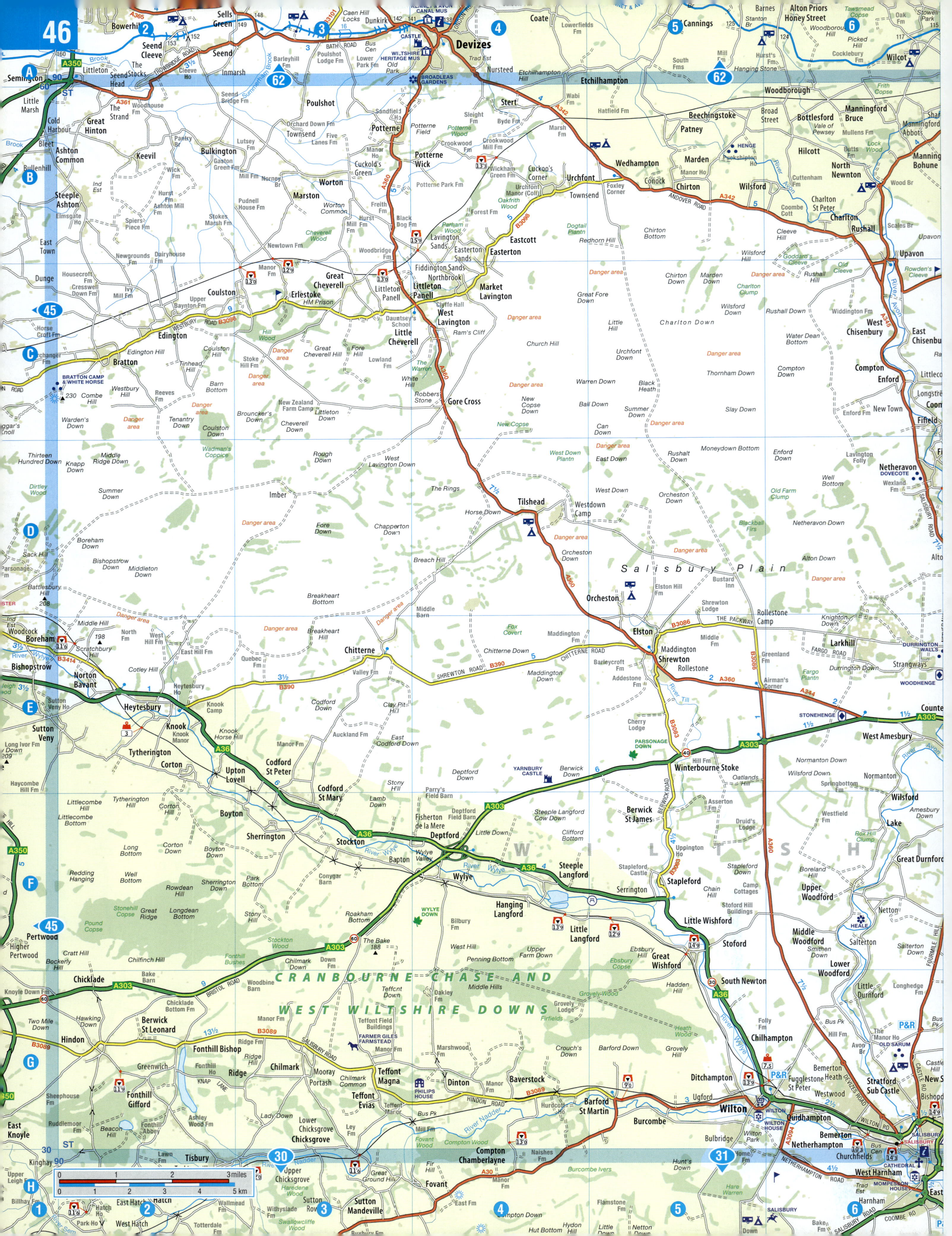
Devizes
Seend
Seend Cleeve
Sells Green
Bowerhill
Semington
Littleton
Poulshot
Potterne
Potterne Wick
Worton
Marston
Bulkington
Keevil
Great Hinton
Steeple Ashton
Ashton Common
Stert
Urchfont
Wedhampton
Etchilhampton
Cannings
Alton Priors
Honey Street
Woodborough
Beechingstoke
Patney
Marden
Chirton
Wilsford
Manningford Bruce
Bottlesford
Hilcott
North Newnton
Charlton
Rushall
Upavon
Coate
Great Cheverell
Little Cheverell
Erlestoke
Coulston
Edington
Bratton
Littleton Panell
Market Lavington
West Lavington
Eastcott
Easterton
Gore Cross
Tilshead
Imber
Salisbury Plain
Orcheston
Elston
Shrewton
Maddington
Rollestone
Larkhill
Chitterne
Heytesbury
Norton Bavant
Bishopstrow
Boreham
Sutton Veny
Knook
Tytherington
Corton
Upton Lovell
Codford St Peter
Codford St Mary
Boyton
Sherrington
Stockton
Deptford
Wylye
Steeple Langford
Hanging Langford
Little Langford
Winterbourne Stoke
Berwick St James
Stapleford
Little Wishford
Great Wishford
Stoford
South Newton
West Amesbury
Wilsford
Lake
Great Durnford
Upper Woodford
Middle Woodford
Lower Woodford
Chilhampton
Stratford Sub Castle
Wilton
Quidhampton
Bemerton
Netherhampton
Barford St Martin
Burcombe
Ugford
Ditchampton
Baverstock
Dinton
Teffont Magna
Teffont Evias
Chilmark
Ridge
Fonthill Bishop
Fonthill Gifford
Hindon
Berwick St Leonard
Chicklade
Pertwood
East Knoyle
Tisbury
Chicksgrove
Compton Chamberlayne
Fovant
Sutton Mandeville
West Harnham
West Hatch
Salisbury Plain
CRANBOURNE CHASE AND WEST WILTSHIRE DOWNS
WILTSHIRE
STONEHENGE
A303
A36
A360
A342
A350
A30
B3098
B390
B3086
B3083
B3089
0 1 2 3 miles
0 1 2 3 4 5 km

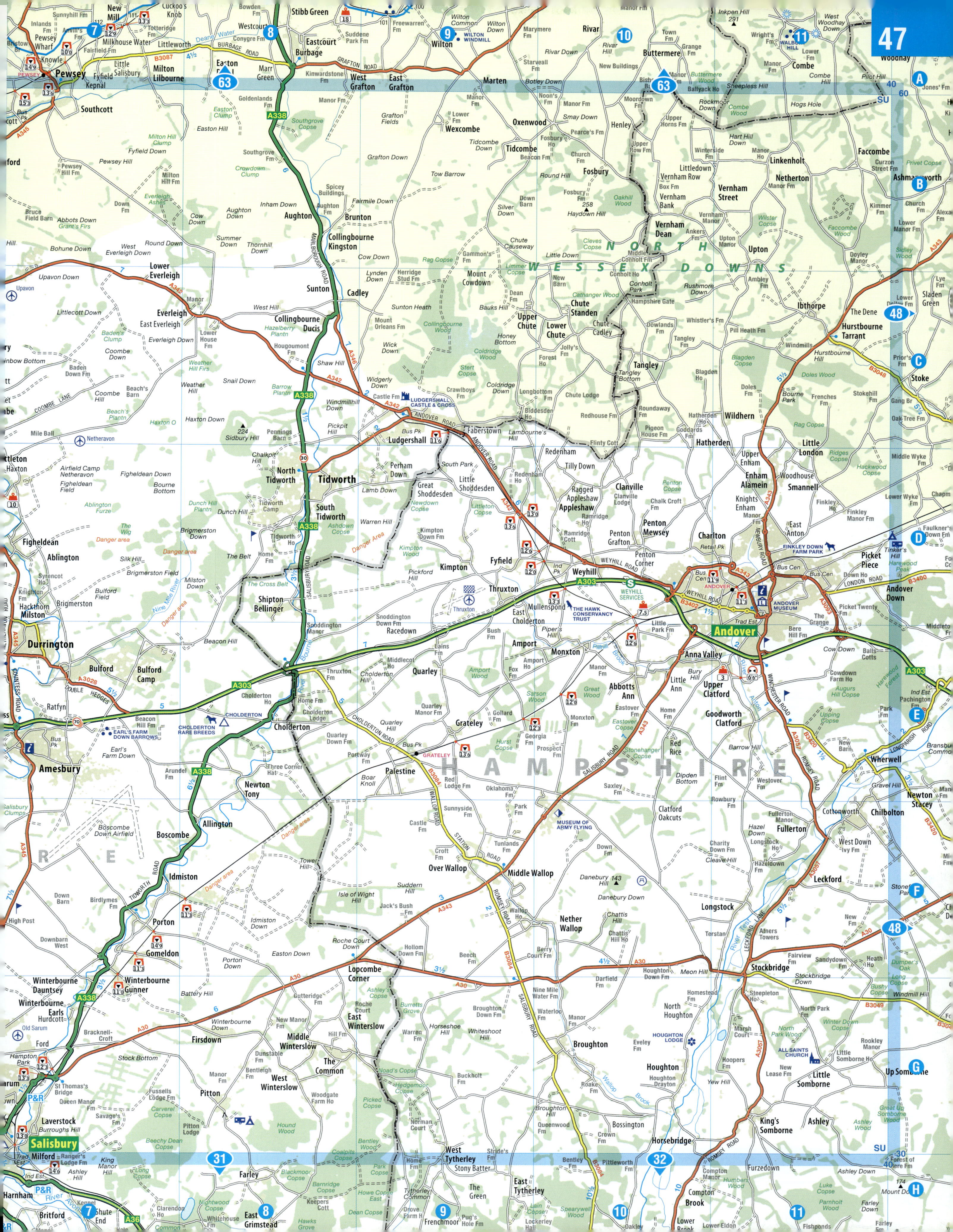
Pewsey
Milton Lilbourne
Easton
Burbage
Stibb Green
Eastcourt
Wilton
Marten
West Grafton
East Grafton
Rivar
Buttermere
Combe
Woodhay
Southcott
Oxenwood
Wexcombe
Tidcombe
Fosbury
Henley
Linkenholt
Netherton
Faccombe
Ashmansworth
Vernham Dean
Vernham Street
Upton
Aughton
Collingbourne Kingston
Brunton
Cadley
Sunton
Collingbourne Ducis
NORTH WESSEX DOWNS
Lower Everleigh
Everleigh
East Everleigh
Upper Chute
Lower Chute
Chute Standen
Chute Cadley
Tangley
Ibthorpe
Hurstbourne Tarrant
Stoke
Ludgershall
Faberstown
Wildhern
Hatherden
Little London
North Tidworth
Tidworth
South Tidworth
Perham Down
Great Shoddesden
Little Shoddesden
Clanville
Appleshaw
Upper Enham
Enham Alamein
Smannell
Knights Enham
Penton Grafton
Penton Mewsey
Charlton
Picket Piece
Andover Down
Figheldean
Ablington
Milston
Durrington
Shipton Bellinger
Kimpton
Fyfield
Thruxton
Weyhill
Andover
East Cholderton
Amport
Monxton
Anna Valley
Bulford
Bulford Camp
Ratfyn
Quarley
Abbotts Ann
Little Ann
Upper Clatford
Goodworth Clatford
Amesbury
Cholderton
Grateley
Palestine
Newton Tony
HAMPSHIRE
Wherwell
Chilbolton
Newton Stacey
Allington
Boscombe
Idmiston
Porton
Gomeldon
Over Wallop
Middle Wallop
Nether Wallop
Fullerton
Leckford
Longstock
Stockbridge
Winterbourne Dauntsey
Winterbourne Gunner
Winterbourne Earls
Lopcombe Corner
East Winterslow
Middle Winterslow
West Winterslow
The Common
Firsdown
Pitton
Broughton
Houghton
Little Somborne
King's Somborne
Ashley
Laverstock
Salisbury
Milford
West Tytherley
East Tytherley
Farley
Horsebridge
Brook
East Grimstead
Britford
Harnham
The Green
Frenchmoor

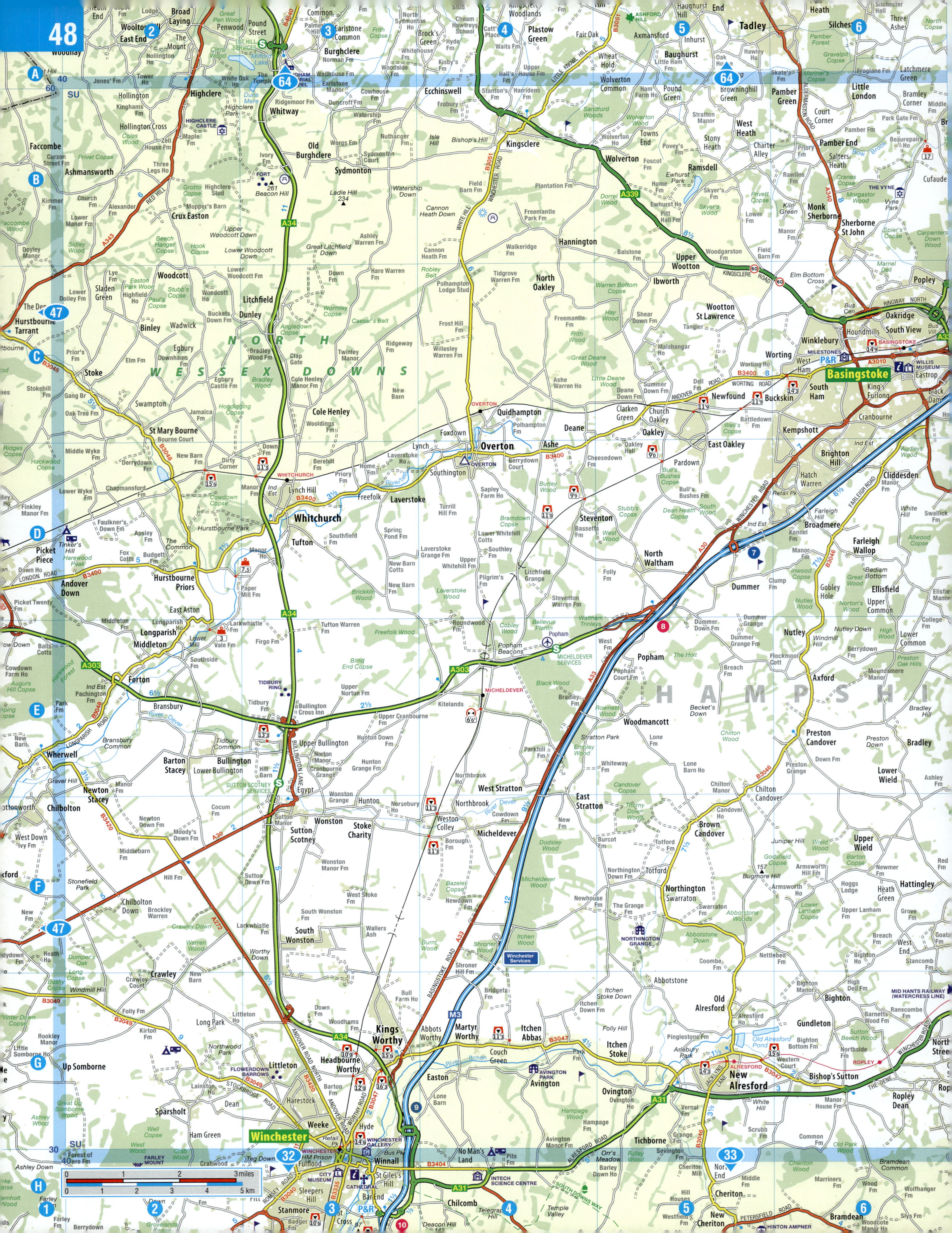

Highclere
Whitway
Faccombe
Ashmansworth
Crux Easton
Old Burghclere
Sydmonton
Burghclere
Ecchinswell
Kingsclere
Wolverton
Tadley
Baughurst
Axmansford
Pamber Green
Little London
Pamber End
Ramsdell
Monk Sherborne
Sherborne St John
Hannington
Upper Wootton
Ibworth
Wootton St Lawrence
North Oakley
Woodcott
Litchfield
Dunley
Binley
Hurstbourne Tarrant
Stoke
St Mary Bourne
NORTH WESSEX DOWNS
Cole Henley
Quidhampton
Overton
Ashe
Deane
Oakley
Church Oakley
East Oakley
Newfound
Buckskin
Worting
Winklebury
South Ham
Basingstoke
Kempshott
Brighton Hill
Cliddesden
Whitchurch
Tufton
Laverstoke
Freefolk
Southington
Steventon
North Waltham
Dummer
Farleigh Wallop
Ellisfield
Hurstbourne Priors
Longparish
Middleton
Andover Down
Picket Piece
Forton
Bransbury
Barton Stacey
Bullington
Upper Bullington
Whitchurch
Chilbolton
Newton Stacey
Wherwell
Popham
Woodmancott
HAMPSHIRE
Preston Candover
Brown Candover
Chilton Candover
Bradley
Axford
Nutley
East Stratton
West Stratton
Micheldever
Northbrook
Sutton Scotney
Wonston
Stoke Charity
Hunton
South Wonston
Crawley
Chilbolton Down
Northington
Swarraton
Upper Wield
Hattingley
Abbotstone
Old Alresford
Itchen Stoke
Ovington
Itchen Abbas
Avington
Martyr Worthy
Abbots Worthy
Kings Worthy
Headbourne Worthy
Littleton
Up Somborne
Sparsholt
Winchester
Easton
Chilcomb
Winnall
New Alresford
Bishop's Sutton
Tichborne
Cheriton
New Cheriton
Bramdean
Ropley Dean
Gundleton
Bighton
Lower Wield
Stanmore
Weeke
Harestock
Tadley
M3
A34
A303
A30
A339
A31

Eversley
Eversley Centre
Eversley Cross
Little Sandhurst
Sandhurst
Yateley
Blackwater
Camberley
York Town
Frimley
Frimley Green
Hawley
Farnborough
Fleet
Church Crookham
Aldershot
Ash
Ash Vale
Mytchett
Deepcut
Heckfield
Hook
Hartley Wintney
Winchfield
Bramshill
Hazeley
Mattingley
Rotherwick
Newnham
Old Basing
Hatch
Mapledurwell
Greywell
Up Nately
North Warnborough
Odiham
Crondall
Ewshot
Hale
Heath End
Badshot Lea
Tongham
Runfold
Seale
Farnham
Wrecclesham
Rowledge
Frensham
Tilford
Elstead
Millbridge
Dockenfield
Churt
Thursley
Hindhead
Grayshott
Haslemere
Shottermill
Camelsdale
Liphook
Bordon
Lindford
Headley
Whitehill
Bramshott
Conford
Greatham
Empshott
Selborne
Oakhanger
Kingsley
Binsted
Bentley
Froyle
Lower Froyle
Upper Froyle
Alton
Holybourne
Chawton
Four Marks
Medstead
Beech
Lasham
Shalden
Bentworth
Herriard
Weston Patrick
Weston Corbett
Upton Grey
Tunworth
Winslade
South Warnborough
Long Sutton
Well
East Worldham
West Worldham
Hartley Mauditt
Upper Farringdon
Lower Farringdon
Newton Valence
East Tisted
West Tisted
Hawkley
Liss Forest
Longmoor Camp
Langley
Kitwood
Monkwood
Colemore
Sleaford
Passfield
Standford
Arford
Hollywater
Griggs Green
Linchmere
Hammer
SURREY

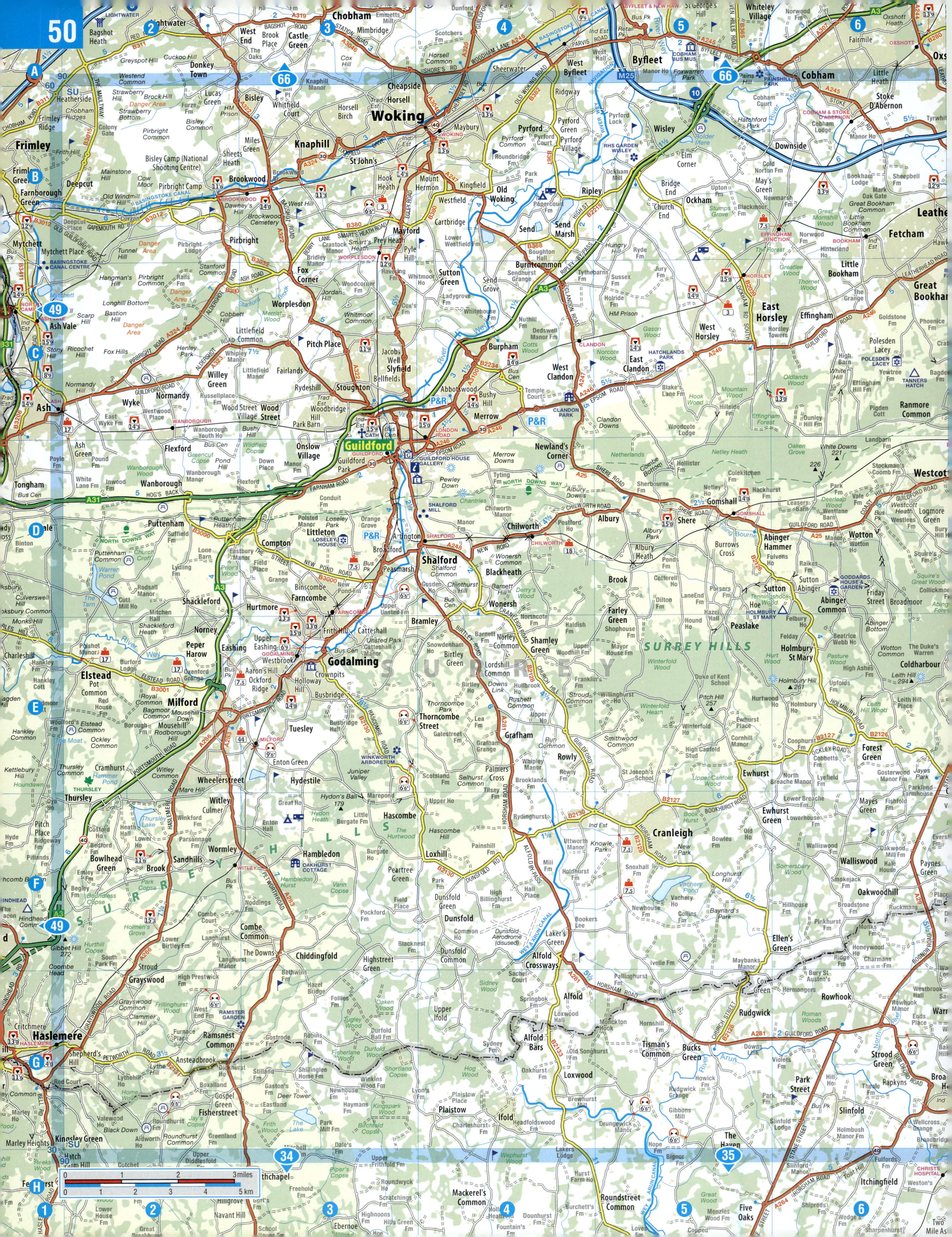
Bagshot Heath
Lightwater
West End
Chobham
Castle Green
Mimbridge
Byfleet
West Byfleet
Cobham
Oxshott Heath
Fairmile
Frimley
Frimley Green
Deepcut
Bisley
Knaphill
Woking
St John's
Pyrford
Wisley
Downside
Farnborough Green
Mytchett
Brookwood
Pirbright
Mayford
Old Woking
Send
Ripley
Ockham
Fetcham
Ash Vale
Worplesdon
Fox Corner
Sutton Green
Send Marsh
Burntcommon
East Horsley
West Horsley
Effingham
Little Bookham
Great Bookham
Ash
Normandy
Wyke
Pitch Place
Jacobs Well
Burpham
West Clandon
East Clandon
Stoughton
Merrow
Guildford
Flexford
Onslow Village
Wanborough
Tongham
Newlands Corner
Puttenham
Compton
Littleton
Artington
Shalford
Chilworth
Albury
Shere
Gomshall
Abinger Hammer
Wotton
Westcott
Farncombe
Hurtmore
Shackleford
Norney
Peper Harow
Eashing
Godalming
Bramley
Wonersh
Blackheath
Brook
Farley Green
Peaslake
Sutton Abinger
Abinger Common
Holmbury St Mary
Coldharbour
SURREY HILLS
Elstead
Milford
Busbridge
Hascombe
Thorncombe Street
Grafham
Shamley Green
Rowly
Ewhurst
Ewhurst Green
Forest Green
Cranleigh
Walliswood
Oakwoodhill
Wheelerstreet
Witley
Wormley
Hydestile
Hambledon
Sandhills
Brook
Thursley
Loxhill
Dunsfold
Alfold Crossways
Alfold
Alfold Bars
Loxwood
Ellen's Green
Rudgwick
Rowhook
Bucks Green
Tisman's Common
Grayswood
Haslemere
Chiddingfold
Combe Common
Ramsnest Common
Fisherstreet
Plaistow
Ifold
Kingsley Green
Shottermill
Fernhurst
Fivehurst
Fivehurst
Slinfold
Itchingfield
Broadbridge Heath
Roundstreet Common
Five Oaks
Mackerel's Common
Strood Green
Rapkyns
SURREY
A3
A31
A25
A281
A246
A247
A320
A322
A323
A324
A286
A283
A264
A29
A24
A245
A248
A3100
B3000
B2126
B2127
B2128
B2130
B2133
B2130
M25
10
66
49
34
35
A
B
C
D
E
F
G
H
1
2
3
4
5
6
0 1 2 3 miles
0 1 2 3 4 5 km

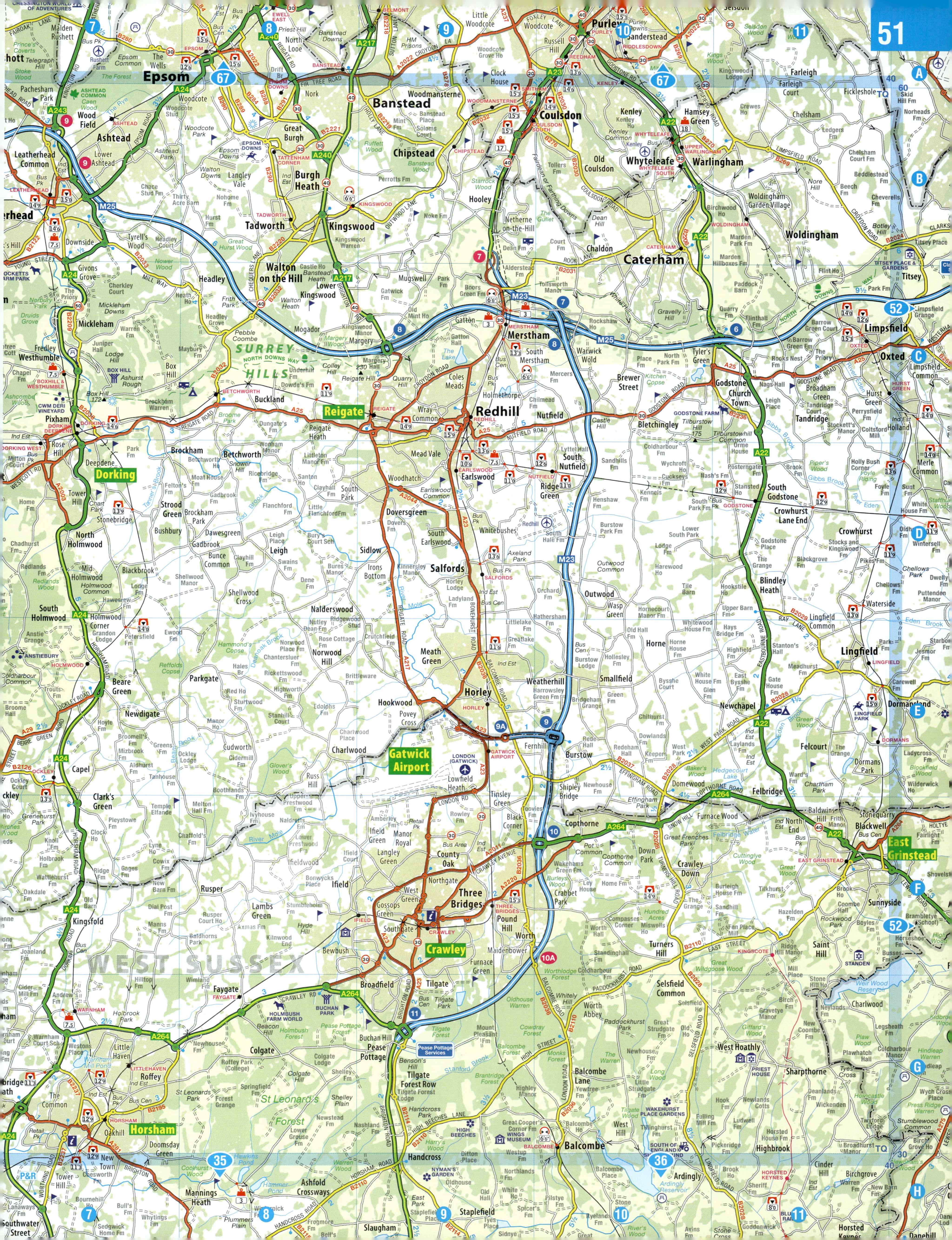
Epsom
Banstead
Coulsdon
Purley
Warlingham
Whyteleafe
Ashtead
Chipstead
Burgh Heath
Tadworth
Kingswood
Caterham
Woldingham
Walton on the Hill
Merstham
Reigate
Redhill
Limpsfield
Oxted
Godstone
Bletchingley
Nutfield
Dorking
Brockham
Betchworth
Buckland
Leigh
Salfords
Earlswood
South Godstone
Horley
Gatwick Airport
Charlwood
Newdigate
Capel
Lingfield
Dormansland
Felbridge
East Grinstead
Copthorne
Crawley
Three Bridges
Ifield
Faygate
Colgate
Horsham
Handcross
Balcombe
Turners Hill
West Hoathly
Sharpthorne
Ardingly
Crawley Down
Pease Pottage
Tilgate
Pound Hill
Rusper
Kingsfold
Warnham
Mannings Heath
Slaugham
Staplefield
SURREY HILLS
WEST SUSSEX
M25
M23
A24
A25
A217
A23
A22
A264
A240
Pease Pottage Services

Biggin Hill
Westerham
Limpsfield
Oxted
Sevenoaks
Tonbridge
Royal Tunbridge Wells
East Grinstead
Edenbridge
Lingfield
Borough Green
Hildenborough
Southborough
Pembury
Crowborough
KENT DOWNS
HIGH WEALD
KENT
EAST SUSSEX
DOWNS
Ashdown Forest
Chiddingstone
Penshurst
Hever
Leigh
Tatsfield
Chevening
Knockholt
Halstead
Shoreham
Otford
Kemsing
Seal
Wrotham
Ightham
Plaxtol
Shipbourne
Hadlow
Sundridge
Brasted
Riverhead
Dunton Green
Ide Hill
Four Elms
Crockham Hill
Marsh Green
Dormansland
Cowden
Hartfield
Withyham
Groombridge
Langton Green
Rusthall
Speldhurst
Bidborough
Fordcombe
Ashurst
Forest Row
Wych Cross
Nutley
Rotherfield
Frant
Mark Cross
Wadhurst
Town Row
Jarvis Brook
Eridge Green
Boarshead
Lye Green
Chuck Hatch
Coleman's Hatch
Blackham
Holtye
Hammerwood
Sunnyside
Fairseat
Stansted
Vigo Village
Platt
Crouch
Tudeley
Capel
Five Oak Green
Matfield
Lamberhurst
Sevenoaks Weald
Underriver
Fletcher's Green
Bough Beech
Markbeech
Mowshurst
Chiddingstone Causeway
Penshurst Place
Hever Castle & Gardens
Chartwell
Emmetts
Knole
Squerryes Court
Ightham Mote
Tonbridge Castle
Groombridge Place Gdns
M25
M26
A25
A21
A26
A22
A264
A233
A227
51
68
36
37
0 1 2 3 miles
0 1 2 3 4 5 km

53
Snodland
Burham
Eccles
Aylesford
Ditton
Larkfield
Leybourne
East Malling
West Malling
Royal British Legion Village
Allington
Sandling
Boxley
Detling
Thurnham
Bearsted
Maidstone
Tovil
Shepway
Otham
Leeds
Hollingbourne
Harrietsham
Lenham
Broomfield
Kingswood
Langley
Sutton Valence
Headcorn
Smarden
Biddenden
High Halden
Tenterden
Rolvenden
Benenden
Sissinghurst
Cranbrook
Hawkhurst
Goudhurst
Lamberhurst
Kilndown
Flimwell
Bewl Water
Hartley
Staplehurst
Marden
Linton
Loose
Coxheath
Yalding
Nettlestead
East Peckham
Paddock Wood
Brenchley
Horsmonden
Teston
Wateringbury
West Farleigh
East Farleigh
Barming Heath
Chart Sutton
Ulcombe
Egerton
Pluckley
Frittenden
Iden Green
Sandhurst
Benenden
KENT
NORTH DOWNS
HIGH WEALD
K E N T
Maidstone Services
Bedgebury Forest
Hemsted Forest
Sissinghurst Castle
Leeds Castle
Headcorn

Rodmersham
Rodmersham Green
Upper Rodmersham
Lynsted
Lewson Street
Norton Ash
Ospringe
Davington
Faversham
Goodnestone
Preston
Whitehill
Culmers
Fostall
Hernhill
Dargate
Dargate Common
Staplestreet
Boughton Street
Dunkirk
Honey Hill
Blean
Tyler Hill
Rough Common
Harbledown
Upper Harbledown
Woodstock
Loyt
Newbury
Kingsdown
Dungate
Erriottwood
Milstead
Bluetown
Hollybushes
Newnham
Doddington
West End
Eastling
North Downs
Plumford
Selgrove
North Street
Gosmere
Sheldwich
Wilgate Green
Sheldwich Lees
Hockley
Throwley
Throwley Forstal
Badlesmere
Leaveland
Crouch
South Street
Gushmere
Hogben's Hill
Neames Forstal
Oversland
Selling
Perrywood
Rhode Common
Chartham Hatch
Thanington
Chartham
Shalmsford Street
Old Wives Lees
Chilham
Shottenden
Bagham
Garlinge Green
Petham
Street End
Lower Hardres
Wichling
Frith
Yewhedges
West Street
Payden Street
Otterden Place
Slade
Woodside Green
Warren Street
The Valley
Kent
Downs
Stalisfield Green
Lenham
Sandway
Lenham Heath
Charing Heath
Charing
Charing Hill
Stocker's Head
Paddock
Pested
Molash
Challock
Godmersham
Bilting
Crundale
Sole Street
Waltham
Eastwell Park
Westwell
Westwell Leacon
Dunn Street
Boughton Aluph
Boughton Lees
Boughton Corner
Wye
Hastingleigh
Elmsted
Ram Lane
Little Chart
Little Chart Forstal
Pluckley
Pluckley Thorne
Egerton
Egerton Forstal
Stone Hill
Mundy Bois
Newland Green
Hothfield
Potters Corner
Goat Lees
Kennington
Bybrook
The Warren
Worten
Fir Toll
Chamber's Green
Ripper's Cross
Maltman's Hill
Biddenden Green
Brook
Troy Town
Naccolt
Hinxhill
Willesborough Lees
Ashford
South Ashford
Willesborough
Sevington
Great Chart
Daniel's Water
Yardhurst
Wissenden
Bethersden
Brissenden Green
Tanden
Shadoxhurst
Chilmington Green
Stanhope
Kingsnorth
Cheeseman's Green
Mersham
Smeeth
Stubbs Cross
Bromley Green
Henghurst
Shirkoak
Orlestone
Hamstreet
Ruckinge
Bilsington
Aldington
Aldington Frith
Bonnington
Clap Hill
Sellindge
Stanford
Westenhanger
Postling
Newingreen
Pedlinge
Lympne
Court-at-Street
West Hythe
Palmarsh
Burmarsh
Newchurch
Dymchurch
St Mary's Bay
St Mary in the Marsh
Tenterden
Leigh Green
Woodchurch
Kenardington
Warehorne
Appledore
Appledore Heath
Reading Street
Small Hythe
Snargate
Ivychurch
Snave
Brenzett Green
Romney Marsh
The Dowels
Isle of Oxney
Stone
High Halden
Bethersden
M20
M2
A2
A20
A28
A252
A251
A2070
A259
0 1 2 3 miles
0 1 2 3 4 5 km

SANDWICH BAY
Shell Ness
Richborough Port
Sandwich Flats
Canterbury
Sturry
Fordwich
Westbere
Stodmarsh
Grove
Preston
Elmstone
Hoaden
Ash
Wingham
Littlebourne
Bramling
Bekesbourne
Patrixbourne
Nackington
Bridge
Bishopsbourne
Adisham
Aylesham
Goodnestone
Chillenden
Nonington
Knowlton
Eastry
Sandwich
Stone Cross
Woodnesborough
Worth
Finglesham
Betteshanger
Northbourne
Tilmanstone
Eythorne
Sholden
Deal
Walmer
Kingsdown
Ringwould
Ripple
Great Mongeham
Sutton
East Studdal
Shepherdswell or Sibertswold
Coldred
Lydden
Whitfield
Temple Ewell
Guston
Martin Mill
West Cliffe
St Margaret's at Cliffe
SOUTH FORELAND
Dover
Buckland
Maxton
Aycliff
Farthingloe
Capel-le-Ferne
Hougham
Alkham
Densole
Hawkinge
Swingfield Minnis
Swingfield Street
Selsted
Denton
Wootton
Barham
Kingston
Womenswold
Woolage Green
Elham
Lyminge
Paddlesworth
Etchinghill
Newington
Cheriton
Channel Tunnel
Folkestone
Sandgate
Saltwood
Hythe
K E N T
Channel Tunnel
CALAIS 1:30
DUNKERQUE 1:45
BOULOGNE 1:45
S T R A I T   O F   D O V E R
A B C D E F G H
7 8 9 10 11
TR 60
TR 30
40

Burry Port
Pembrey
Llanelli
Gorseinon
Loughor
Gowerton
River Loughor
Afon Llwchwr
Llanrhidian Sands
Whiteford Sands
Llanmadoc
Cheriton
Landimore
Llangennith
Llanrhidian
THE GOWER
Gower Peninsula
Reynoldston
Llanddewi
Knelston
Scurlage
Rhossili
RHOSSILI BAY
Middleton
Pitton
Port Eynon
PORT EYNON BAY
Horton
Oxwich
OXWICH BAY
BAE OXWICH
Penmaen
Parkmill
Nicholaston
Ilston
Southgate
Pennard
Bishopston
Kittle
Upper Killay
Killay
Dunvant
Three Crosses
Waunarlwydd
Sketty
Mayals
West Cross
Black Pill
Newton
Oystermouth
The Mumbles
(Y Mwmbwls)
Mumbles Head
Langland
Caswell
BRISTOL CHANNEL
MÔR HAFREN
0 1 2 3 miles
0 1 2 3 4 5 km

SWANSEA BAY
BAE ABERTAWE
Clydach
Ynystawe
Pant-lasau
Morriston (Treforys)
Llansamlet
Landore
Port Tennant
St Thomas
Foxhole
Kilvey Hill
Bon-y-maen
Cwm
Winsh-wen
Pentre-chwyth
Ty-draw
Cefn-hengoed
Pentre-dwr
Llandarcy
Jersey Marine
Crymlyn Burrows
Skewen
Birchgrove
Heol-las
Glais
Lon-las
Trallwn
Neath Abbey
Neath
Castell-Nedd
Cimla
Mount Pleasant
Melincryddan
Briton Ferry (Llansawel)
Baglan
Aberavon
Sandfields
Port Talbot
Taibach
Margam
Groes
Goytre
Cwmafan
Pontrhydyfen
Pwll-y-glaw
Bryn
Brynbryddan
Caerhendy
Pen-y-cae
Corlannau
Tonna (Tonnau)
Aberdulais
Cilfrew
Cadoxton-Juxta-Neath
Rhydding
Bryn-coch
Melincourt
Clyne
AFAN ARGOED
FOREST PARK
Cymer
Croeserw
Duffryn
Blaengwynfi
Abergwynfi
Glyncorrwg
Abercregan
Caerau
Spelter
Nantyffyllon
Maesteg
Llwydarth
Garth
Cwmfelin
Llangynwyd
Blaengarw
Pontycymer
Coytrahen
Tondu
Aberkenfig
Bryncethin
Brynmenyn
Sarn
Kenfig Hill
Pyle (Y Pil)
Cefn Cribbwr
Cefn Cross
Mawdlam
Kenfig
North Cornelly
South Cornelly
Nottage
Porthcawl
Newton
Laleston
Tythegston
Merthyr Mawr
Ogmore
Ogmore-by-Sea (Aberogwr)
St Brides Major (Saint-y-Brid)
Southerndown
Pen-y-fai
Pendre
BRIDGEND
PEN-Y-BONT AR OGWR
Kenfig Sands
Margam Sands
Aberavon Sands
Sker Pt
Porthcawl Pt
Rest Bay
Black Rocks
Tusker Rock
Dunraven Bay
CORK 10:00
ILFRACOMBE 0:50
M4
A48
A4107
A465
7
8
9
10
11

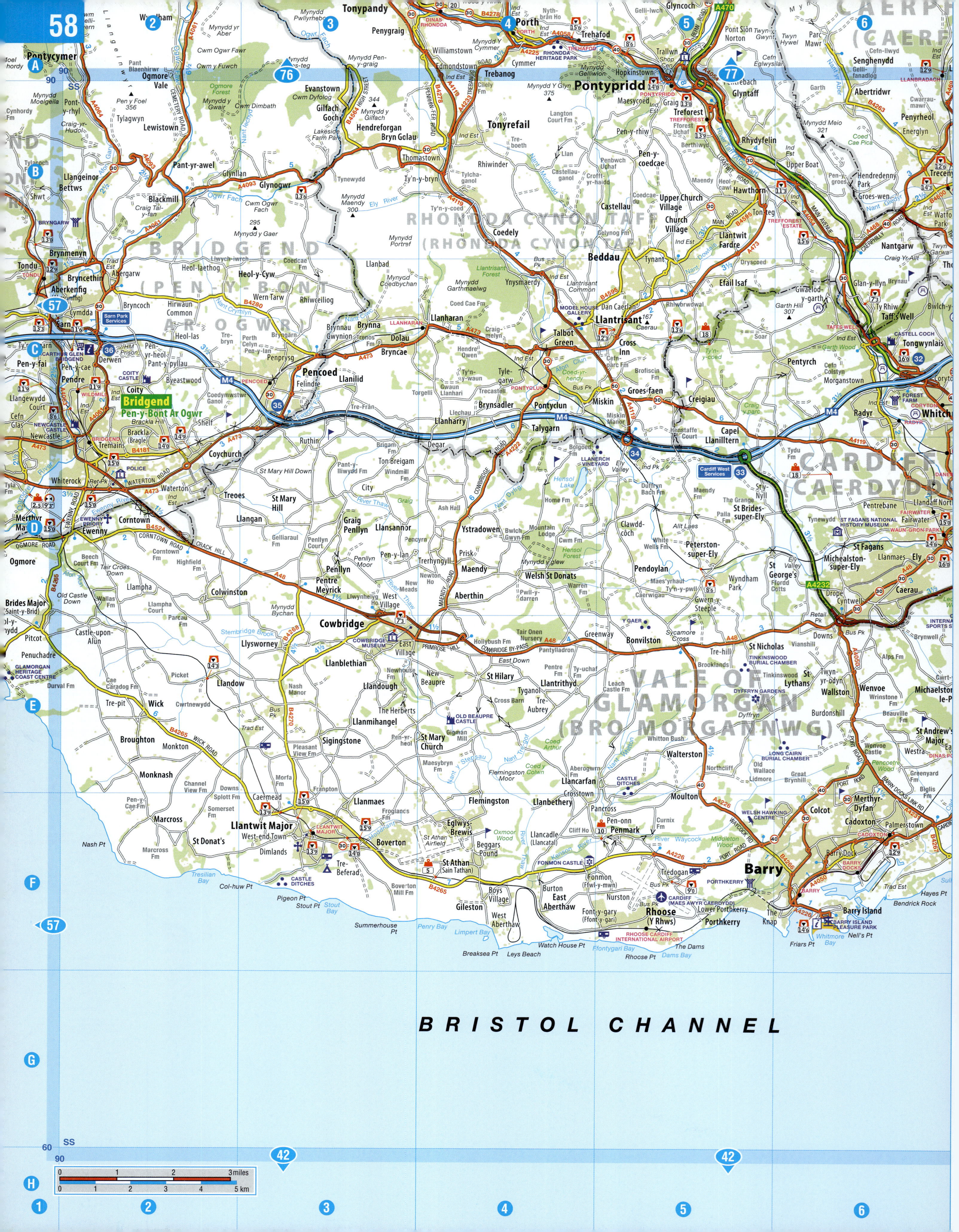

BRISTOL CHANNEL
BRIDGEND (PEN-Y-BONT AR OGWR)
RHONDDA CYNON TAFF (RHONDDA CYNON TAF)
VALE OF GLAMORGAN (BRO MORGANNWG)
CARDIFF (CAERDYDD)
Pontycymer
Ogmore Vale
Tonypandy
Penygraig
Porth
Trehafod
Pontypridd
Williamstown
Trebanog
Tonyrefail
Evanstown
Gilfach Goch
Hendreforgan
Bryn Golau
Thomastown
Glynogwr
Blackmill
Llangeinor
Bettws
Tondu
Brynmenyn
Bryncethin
Aberkenfig
Sarn
Bryncoch
Heol-y-Cyw
Pen-y-fai
Pendre
Coity
Bridgend
Pen-y-Bont Ar Ogwr
Brackla
Newcastle
Tremains
Coychurch
Pencoed
Felindre
Llanilid
Brynna
Dolau
Bryncae
Llanharan
Llanharry
Brynsadler
Pontyclun
Talygarn
Talbot Green
Llantrisant
Beddau
Coedely
Tynant
Llantwit Fardre
Church Village
Upper Church Village
Efail Isaf
Cross Inn
Groes-faen
Miskin
Creigiau
Pentyrch
Capel Llanilltern
Radyr
Tongwynlais
Taff's Well
Nantgarw
Treforest
Hawthorn
Rhydyfelin
Glyntaff
Abertridwr
Senghenydd
Penyrheol
Energlyn
Ruthin
Treoes
St Mary Hill
Llangan
Waterton
Ewenny
Corntown
Ogmore
Merthyr Mawr
Colwinston
Llampha
Brides Major
Pitcot
Castle-upon-Alun
Wick
Broughton
Monknash
Marcross
St Donat's
Llantwit Major
Boverton
Llysworney
Cowbridge
Llandow
Llanblethian
Llandough
Llanmihangel
Sigingstone
Llanmaes
St Mary Church
Flemingston
Eglwys-Brewis
St Athan
Gileston
Aberthaw
East Aberthaw
Penmark
Rhoose (Y Rhws)
Porthkerry
Barry
Barry Island
Cadoxton
Colcot
Merthyr-Dyfan
Palmerstown
Wenvoe
Walterston
Moulton
Llancarfan
Llantrithyd
Bonvilston
St Nicholas
St Lythans
Wallston
Tre-hill
St Hilary
Tyganol
Peterston-super-Ely
St Brides-super-Ely
St George's
Pendoylan
Welsh St Donats
Aberthin
Ystradowen
Maendy
Prisk
Penllyn
Pentre Meyrick
Llansannor
Graig Penllyn
City
Clawdd-côch
Michaelston-super-Ely
St Fagans
Fairwater
Caerau
Pentrebane
Sully
M4
A48
A473
A4119
A4226
A4232
A470
CARDIFF (RHOOSE) CARDIFF INTERNATIONAL AIRPORT
Nash Pt
Col-huw Pt
Summerhouse Pt
Breaksea Pt
Leys Beach
Rhoose Pt
Nell's Pt
Friars Pt
0 1 2 3 miles
0 1 2 3 4 5 km

Cwm felin fach
Crosskeys
Wattsville
Risca
Newport
Casnewydd
Caerleon
Llanbeder
Langstone
Llanmartin
Bedwas
Trethomas
Machen
Upper Ochrwyth
Lower Ochrwyth
Lower Machen
Rogerstone
Bettws
Malpas
Brynglas
Christchurch
Ringland
Llanwern
Underwood
Bishton
Wilcrick
Llandevenny
Caerphilly
(Caerffili)
Lansbury Park
Rudry
Draethen
Rhiwderin
Bassaleg
Pentre-poeth
Glasllwch
Stelvio
Gaer
Maindee
Lliswerry
Duffryn
Maes-glas
Tredegar House
Michaelston-y-Fedw
Pen-y-lan
Cleppa Park
Coedkernew
Castleton
Marshfield
Blacktown
St Brides Wentlooge
Peterstone Wentlooge
Peterstone Gout
Pye Corner
Nash
Goldcliff
Whitson
Saltmarsh
Broadstreet Common
West Usk Lighthouse
Lisvane
Llanishen
Rhiwbina
Birchgrove
Cyncoed
Pentwyn
Llanedeyrn
St Mellons
Trowbridge
Llanrumney
Rumney
Heath
Roath Park
Gabalfa
Llandaff
Pontcanna
Canton
Cathays
Roath
Tremorfa
Splott
Cardiff
Riverside
Butetown
Grangetown
Leckwith
Llandough
Penarth Moors
Cardiff Bay
Cardiff Flats
Orchard Ledges
Dinas Powys
Cogan
Penarth Head
Penarth
Lower Penarth
Cosmeston
Lavernock
Lavernock Pt
Sully
Swanbridge
Sully Island
Ynys Sili
Flat Holm
North West Pt
Lighthouse Pt
Steep Holm
Rudder Rock
Tower Rock
Calf Rock
Monkstone
SEVERN ESTUARY
ABER HAFREN
Clevedon
Kingston Seymour
St Thomas's Head
Woodspring Bay
Middle Hope
Sand Pt
Sand Bay
Wick St Lawrence
Bourton
Hewish
Puxton
Worle
Kewstoke
Milton
Birnbeck Island
Weston-super-Mare
Weston Bay
Uphill
Locking
Banwell
Hutton
Oldmixon
Elborough
Knightcott
NORTH SOMERSET
M4
M5
A48(M)
A370
A371

Llanvaches
Penhow
Parc Seymour
Discoed
Newton Green
Mathern
Beachley
Slimeroad Sand
Pullens Green
Lower Morton
Thornbury
Llanfeder
Llanmartin
Carrow Hill
St Bride's Netherwent
Five Lanes
Highmoor Hill
Caerwent
Crick
Caldicot
Mount Ballan
Leechpool
Portskewett
Sudbrook
Charston Sands
Severn Road Bridge
Old Passage
Aust
Littleton-on-Severn
Elberton
Alveston
Ingst
Olveston
Tockington
Rudgeway
Wilcrick
Bishton
Llandevenny
Magor
Magor Services
Vinegar Hill
Rogiet
Llanfihangel Rogiet
Undy
Whitewall Common
Severn Tunnel Junction
Caldicot Pill
Second Severn Crossing
Severn Tunnel
New Passage
Redwick
Northwick
Awkley
Pilning
Severn Beach
Severn Bridges Visitor Centre
Collister Pill
Summerleaze
Caldicot Level
Redwick
Broad Mead
Porton Grounds
Cold Harbour Pill
Gravel Banks
Marsh Common
Almondsbury
Easter Compton
Over
Patchway
Little Stoke
Bradley Stoke
Frogland Cross
Bedwin Sands
Welsh Grounds
Crook's Marsh
Hallen Marsh
Hallen
Compton Greenfield
Cribbs Causeway
Catbrain
Filton
Stoke Gifford
Winterbourne
King Road
Avonmouth
Lawrence Weston
Henbury
Brentry
Southmead
Harry Stoke
Hambrook
Frenchay
Woodhill Bay
Black Nore
Woodhill
West Hill
Portishead
Redcliff Bay
Sheepway
Shirehampton
Sea Mills
Westbury on Trym
Horfield
Coombe Dingle
Henleaze
Golden Hill
Fishponds
Stapleton
Charlcombe Bay
Walton Bay
Pigeon House Bay
North Weston
Middle Bridge
Lodway
Pill
Easton-in-Gordano
Portbury
Ham Green
Abbots Leigh
Stoke Bishop
Redland
Bishopston
Eastville
Weston-in-Gordano
Clapton-in-Gordano
Walton-in-Gordano
Farley
Margaret's Bay
Ladye Bay
Norton's Wood
Walton St Mary
Clevedon
Tickenham
Middletown
Stone-edge Batch
Nailsea
Wraxall
Tyntesfield
Failand
Leigh Woods
Clifton
Long Ashton
Bower Ashton
Bedminster
Knowle
Brislington
Bristol
Hanham
Kenn
Kenn Moor
West End
Backwell
Flax Bourton
Cambridge Batch
Yanley
Ashton Vale
Bedminster Down
Hengrove
Keynsham
North End
Chelvey
Brockley
Claverham
Cleeve
Barrow Gurney
Barrow Common
Bishopsworth
Withywood
Hartcliffe
Whitchurch
Queen Charlton
Kingston Seymour
Yatton
Felton
Upper Town
Dundry
East Dundry
North Wick
Congresbury
Hewish
Puxton
Wrington
Redhill
Winford
Norton Hawkfield
Norton Malreward
Pensford
Chew Magna
Stanton Drew
Brinsea
Churchill
Langford
Butcombe
Ridgehill
Chew Stoke
Stanton Wick
Chelwood
Blagdon
Ubley
Burrington
Rickford
Nempnett Thrubwell
Chew Valley Lake
Sutton Wick
Clutton
Temple Cloud
North Somerset
City and County of Bristol
Bath and North East Somerset

Whitfield
Kingswood
Charfield
Wortley
Tresham
Leighterton
Doughton
Alderley
Cromhall
Tytherington
Wickwar
Hillesley
Oldbury on the Hill
Westonbirt
Easton Grey
Hawkesbury Upton
Hawkesbury
Didmarton
Sopworth
Sherston
Rangeworthy
Engine Common
Iron Acton
Horton
Little Badminton
Luckington
Norton
Frampton Cotterell
Nibley
Yate
Chipping Sodbury
Old Sodbury
Little Sodbury
Badminton
Alderton
Hullavington
Coalpit Heath
Westerleigh
Wapley
Dodington
Acton Turville
Littleton Drew
Grittleton
Stanton St Quintin
Winterbourne Down
Codrington
Tormarton
Burton
Nettleton
Leigh Delamere
Kington St Michael
Mangotsfield
Pucklechurch
Hinton
Dyrham
West Kington
Castle Combe
Yatton Keynell
Siston
Abson
West Littleton
Marshfield
North Wraxall
Ford
Biddestone
Kingswood
Warmley
Bridge Yate
Doynton
Pennsylvania
Upper Wraxall
Slaughterford
Hardenhuish
Wick
Cold Ashton
Thickwood
Colerne
Corsham
Oldland
Upton Cheyney
Langridge
St Catherine
Box
Easton
Bitton
North Stoke
Woolley
Swainswick
Charlcombe
Northend
Batheaston
Kingsdown
Hawthorn
Neston
Gastard
Saltford
Kelston
Weston
Bath
Bathford
Bathampton
Monkton Farleigh
Atworth
Whitley
Shaw
Beanacre
Corston
Newton St Loe
Twerton
Claverton
Bradford Leigh
Melksham
Burnett
Englishcombe
Combe Down
Monkton Combe
Winsley
Bradford-on-Avon
Holt
Marksbury
Stanton Prior
Inglesbatch
Southstoke
Midford
Limpley Stoke
Turleigh
Staverton
Semington
Farmborough
Priston
Freshford
Westwood
Hilperton
Combe Hay
Twinhoe
Trowbridge
Peasedown St John
Wellow
Dunkerton
GLOUCESTERSHIRE
SOUTH
WILTSHIRE
SOMERSET
COTSWOLDS
M4
M5
A46
A4
A36
A420
A433
A350
A363
A431
A432
A4174
B4040
B4039
B4465
B3108
B3109

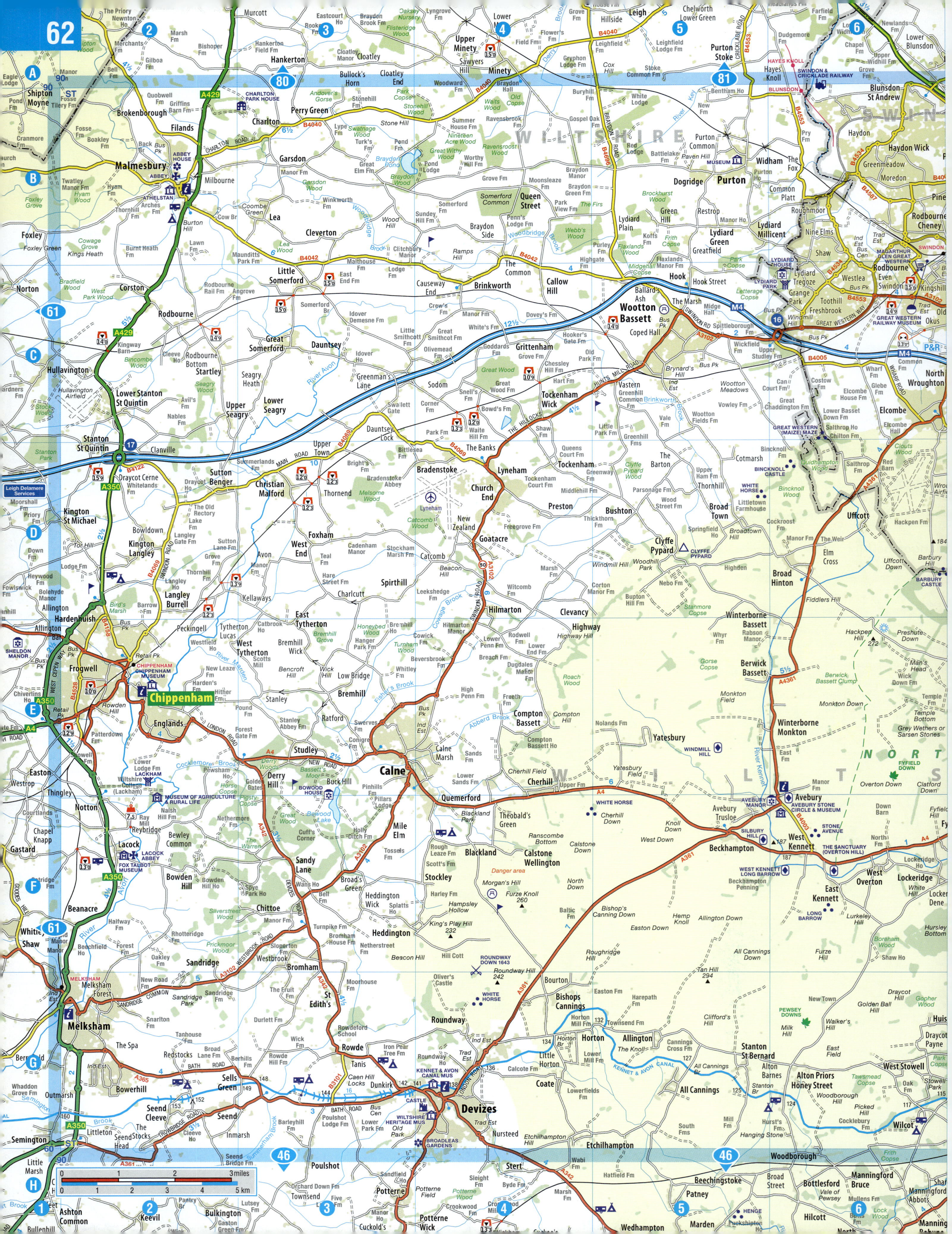

Malmesbury
Chippenham
Calne
Melksham
Devizes
Wootton Bassett
Purton
Lyneham
Avebury
Hankerton
Minety
Upper Minety
Leigh
Purton Stoke
Shipton Moyne
Brokenborough
Charlton
Perry Green
Garsdon
Lea
Cleverton
Little Somerford
Great Somerford
Brinkworth
Dauntsey
Callow Hill
Braydon Side
Somerford Common
Dogridge
Widham
Lydiard Millicent
Lydiard Green
Lydiard Plain
Green Hill
Restrop
Greenfield
Common Platt
Haydon Wick
Rodbourne Cheney
Nine Elms
Shaw
Westlea
Toothill
Freshbrook
Grange Park
Okus
Hook
The Marsh
Coped Hall
Grittenham
Tockenham
Tockenham Wick
Bradenstoke
Church End
Preston
Bushton
Clyffe Pypard
Broad Town
Broad Hinton
Uffcott
Elcombe
North Wroughton
Hullavington
Corston
Rodbourne
Rodbourne Bottom
Startley
Norton
Foxley
Lower Stanton St Quintin
Stanton St Quintin
Clanville
Draycot Cerne
Sutton Benger
Christian Malford
Seagry Heath
Upper Seagry
Lower Seagry
Kington St Michael
Kington Langley
Langley Burrell
Foxham
West End
Thornend
Kellaways
East Tytherton
West Tytherton
Tytherton Lucas
Spirthill
Charlcutt
Catcomb
Goatacre
New Zealand
Hilmarton
Clevancy
Highway
Winterborne Bassett
Berwick Bassett
Winterborne Monkton
Hardenhuish
Frogwell
Allington
Englands
Stanley
Studley
Derry Hill
Bremhill
Bremhill Wick
Ratford
Compton Bassett
Cherhill
Quemerford
Yatesbury
Beckhampton
West Kennett
East Kennett
West Overton
Lockeridge
Avebury Trusloe
Blackland
Calstone Wellington
Heddington
Heddington Wick
Stockley
Mile Elm
Sandy Lane
Chittoe
Bowden Hill
Lacock
Notton
Reybridge
Bewley Common
Easton
Westrop
Thingley
Chapel Knapp
Gastard
Beanacre
Whitley
Shaw
Sandridge
Bromham
Westbrook
St Edith's
Roundway
Rowde
Sells Green
Seend
Seend Cleeve
Bowerhill
Semington
Littleton
Outmarsh
Bishops Cannings
Horton
Allington
All Cannings
Stanton St Bernard
Alton Barnes
Alton Priors
Honey Street
Coate
Etchilhampton
Nurssteed
Poulshot
Potterne
Potterne Wick
Stert
Beechingstoke
Patney
Marden
Woodborough
Manningford Bruce
Bottlesford
Wilcot
Hilcott
Wedhampton
Keevil
Bulkington
Ashton Common
Little Marsh
Bullenhill
Cuckold's Green
WILTSHIRE
WILTSHIRE
M4
A429
A350
A4
A342
A3102
A361
A365
B4040
B4042
B4069
B4005
Leigh Delamere Services
Lyneham
Clyffe Pypard
Kennet & Avon Canal
Great Western (Maize) Maze
Swindon & Cricklade Railway
0 1 2 3 miles
0 1 2 3 4 5 km

63
Swindon
Stratton St Margaret
Upper Stratton
Walcot
Lawn
Park
Liden
Coate
Nythe
Covingham
Dorcan
Eldene
Wroughton
Chiseldon
Badbury
Badbury Wick
Hodson
Wanborough
Liddington
Bishopstone
Hinton Parva
Idstone
Ashbury
Bourton
South Marston
Shrivenham
Watchfield
Longcot
Fernham
Baulking
Uffington
Woolstone
Kingston Lisle
Knighton
Compton Beauchamp
Kingstone Winslow
Sparsholt
West Challow
East Challow
Childrey
Wantage
Letcombe Regis
Letcombe Bassett
Charlton
Highworth
Sevenhampton
Stanton Fitzwarren
Broad Blunsdon
Swanborough
Redlands
Goosey
Denchworth
Lambourn
Upper Lambourn
Eastbury
East Garston
Great Shefford
Shefford Woodlands
Woodlands St Mary
Fawley
South Fawley
Baydon
Aldbourne
Ramsbury
Axford
Mildenhall
Marlborough
Ogbourne St George
Ogbourne St Andrew
Ogbourne Maizey
Southend
Rockley
Upper Upham
Woodsend
Preston
Whittonditch
Chilton Foliat
Leverton
Hungerford
Hungerford Newtown
Wickham
Welford
Weston
Eddington
Froxfield
Little Bedwyn
Great Bedwyn
Chisbury
Savernake Forest
Cadley
Manton
Clench Common
Durley
Crofton
Wilton
Burbage
Easton Royal
Pewsey
Milton Lilbourne
Oare
Wootton Rivers
Shalbourne
Ham
Buttermere
Inkpen
Kintbury
Avington
Upper Green
West Woodhay
Combe
Oxenwood
Tidcombe
Fosbury
Marten
East Grafton
West Grafton
Wexcombe
Bagshot
Templeton
Sanham Green
Anvilles
Halfway
Hoe Benham
Elcot
Clapton
Wickham Heath
Faccombe
Linkenholt
Henley
WESSEX DOWNS
NORTH WESSEX DOWNS
SWINDON
OXON
WEST BERKSHIRE
THE RIDGEWAY
M4
A4
A338
A346
A419
A420
B4000
B4001
B4192
B4507
B4508
Membury Services
UFFINGTON WHITE HORSE & CASTLE
WAYLAND'S SMITHY
ASHDOWN HOUSE
VALE & DOWNLAND MUSEUM CENTRE
CROFTON BEAM ENGINES
WILTON WINDMILL
River Kennet
Kennet and Avon Canal
15
14
81
82
64
47

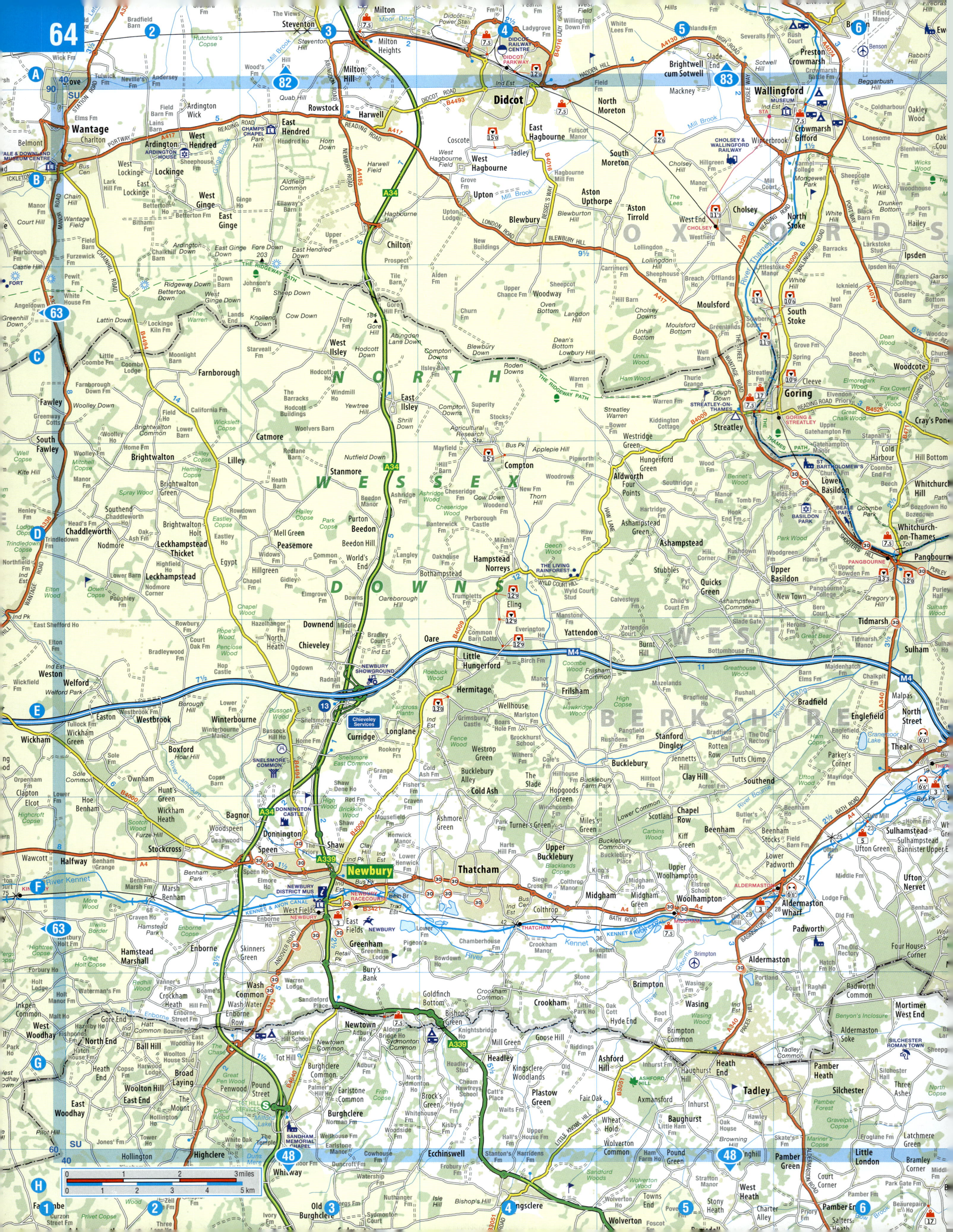

Wantage
Didcot
Wallingford
Newbury
Thatcham
Goring
Streatley
Pangbourne
Theale
Tadley
Kingsclere
Hungerford
Milton
Milton Heights
Milton Hill
Steventon
Harwell
Rowstock
Chilton
Blewbury
Upton
East Hagbourne
West Hagbourne
Aston Tirrold
Aston Upthorpe
North Moreton
South Moreton
Cholsey
Moulsford
North Stoke
South Stoke
Crowmarsh Gifford
Brightwell cum Sotwell
Ardington
Ardington Wick
East Hendred
West Hendred
Lockinge
East Lockinge
West Lockinge
East Ginge
West Ginge
Farnborough
West Ilsley
East Ilsley
Catmore
Lilley
Brightwalton
Brightwalton Green
Brightwalton Holt
Chaddleworth
Leckhampstead
Leckhampstead Thicket
Peasemore
Beedon
Beedon Hill
Stanmore
Compton
Aldworth
Ashampstead
Ashampstead Green
Hampstead Norreys
Chieveley
Downend
Curridge
Hermitage
Frilsham
Yattendon
Bucklebury
Bradfield
Southend
Stanford Dingley
Upper Basildon
Lower Basildon
Tidmarsh
Sulham
Englefield
Donnington
Shaw
Speen
Stockcross
Boxford
Welford
Weston
Wickham
Westbrook
Winterbourne
Bagnor
Enborne
Hamstead Marshall
Marsh Benham
Halfway
Greenham
Crookham
Midgham
Woolhampton
Aldermaston
Aldermaston Wharf
Brimpton
Wasing
Padworth
Headley
Ecchinswell
Burghclere
Highclere
Whitway
Woolton Hill
Ball Hill
East Woodhay
West Woodhay
Newtown
Baughurst
Ashford Hill
Silchester
Pamber Heath
Little London
NORTH WESSEX DOWNS
WEST BERKSHIRE
OXFORDSHIRE
THE RIDGEWAY PATH
THAMES PATH
River Thames
River Kennet
KENNET & AVON CANAL
River Lambourn
River Enborne
River Pang
M4
A34
A4
A339
A417
A329
A340
B4009
B4494
B4493
B4016
B4526
B3051
82
83
63
48
13
12
A
B
C
D
E
F
G
H
1
2
3
4
5
6
40
60
90
SU
DIDCOT RAILWAY CENTRE
NEWBURY RACECOURSE
NEWBURY DISTRICT MUS
DONNINGTON CASTLE
SANDHAM MEMORIAL CHAPEL
BASILDON PARK
THE LIVING RAINFOREST
Chieveley Services
0 1 2 3 miles
0 1 2 3 4 5 km

Marlow
Henley-on-Thames
Maidenhead
Twyford
Caversham
Reading
Tilehurst
Woodley
Earley
Whitley
Wokingham
Bracknell
Binfield
Crowthorne
Sandhurst
Camberley
Yateley
Frimley
Bagshot
Hawley
Flackwell Heath
Bourne End
Cookham
Taplow
Holyport
Burchett's Green
Pinkneys Green
Hurley
Medmenham
Hambleden
Fingest
Turville
Skirmett
Stonor
Nettlebed
Nuffield
Checkendon
Stoke Row
Sonning Common
Shiplake
Wargrave
Charvil
Sonning
Hurst
Winnersh
Sindlesham
Shinfield
Arborfield
Finchampstead
Eversley
Riseley
Swallowfield
Spencers Wood
Three Mile Cross
Burghfield
Mapledurham
Purley on Thames
Calcot
Pangbourne
Emmbrook
Warfield
Easthampstead
Owlsmoor
Little Sandhurst
Blackwater
Hartley Wintney
Hazeley
Heckfield
Stratfield Turgis
Sherfield on Loddon
Hartfordbridge
WOKINGHAM
MAIDENHEAD
WINDSOR
BRACKNELL FOREST
READING
BUCKS
River Thames
Blackwater River
M4
M40
A4
A404
A329(M)
A33
A30
A321
A4155
A4130
A322

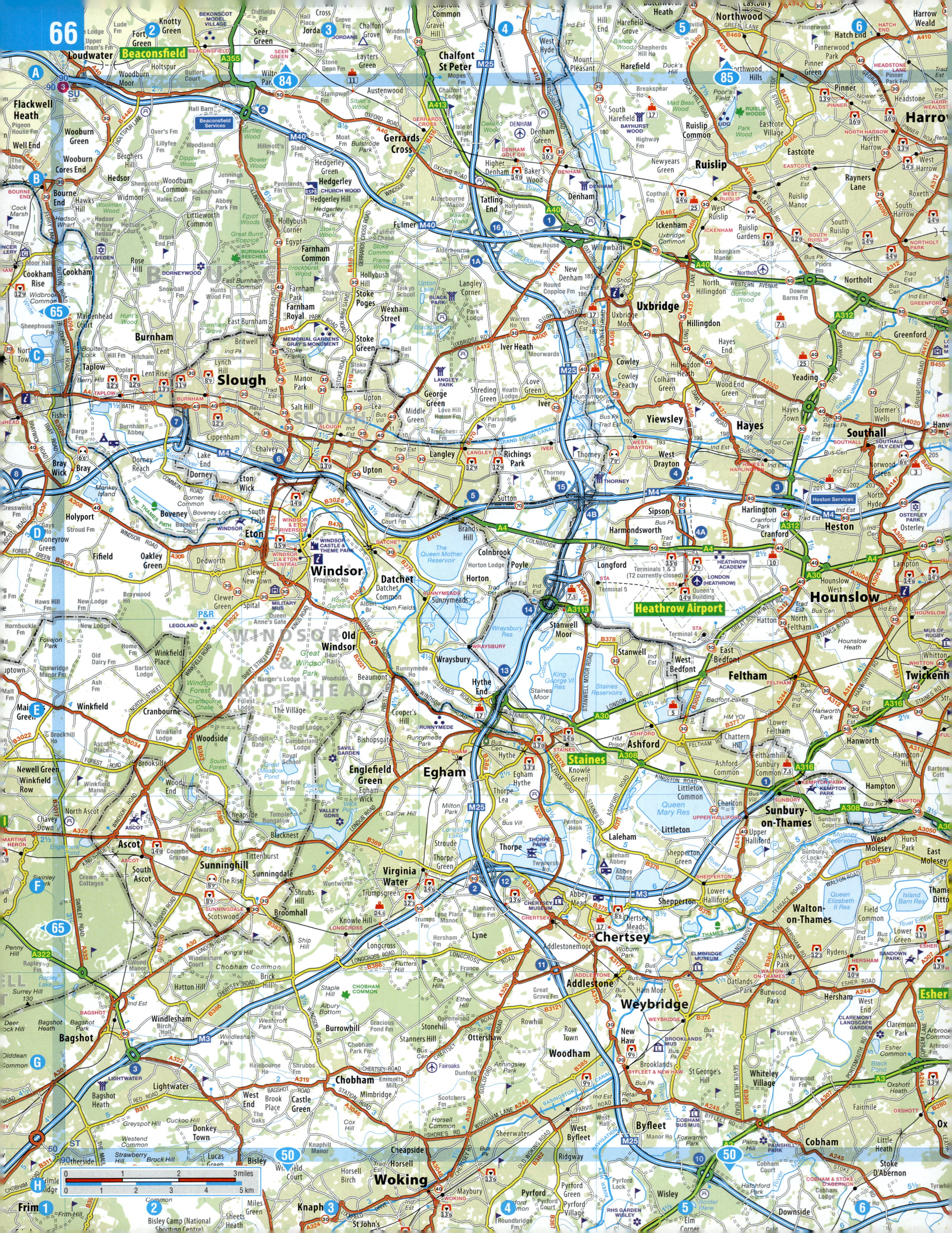
Loudwater
Beaconsfield
Knotty Green
Seer Green
Jordans
Chalfont St Peter
Gerrards Cross
Denham
Harefield
Northwood
Ruislip
Harrow
Eastcote
Rayners Lane
South Harrow
Northolt
Ickenham
Uxbridge
Hillingdon
Flackwell Heath
Wooburn Green
Hedsor
Bourne End
Cookham
Maidenhead
Burnham
Farnham Royal
Stoke Poges
Wexham Street
Fulmer
Iver Heath
Iver
Cowley
Yiewsley
West Drayton
Hayes
Southall
Yeading
Greenford
Slough
Taplow
Langley
George Green
Richings Park
Thorney
Sipson
Harlington
Cranford
Heston
Hounslow
Bray
Dorney
Eton Wick
Eton
Holyport
Windsor
Datchet
Colnbrook
Poyle
Horton
Harmondsworth
Longford
Heathrow Airport
Stanwell
Stanwell Moor
Bedfont
East Bedfont
Feltham
Twickenham
Wraysbury
Old Windsor
Hythe End
Staines
Ashford
Egham
Englefield Green
Winkfield
Cranbourne
Woodside
Ascot
Sunninghill
Sunningdale
Virginia Water
Thorpe
Chertsey
Laleham
Littleton
Shepperton
Sunbury-on-Thames
Hampton
West Molesey
East Molesey
Walton-on-Thames
Addlestone
Weybridge
Hersham
Esher
Bagshot
Lightwater
Windlesham
Chobham
Ottershaw
Woodham
Byfleet
West Byfleet
Cobham
Woking
Knaphill
Horsell
Pyrford
Wisley
Bisley
M4
M25
M40
M3
A4
A30
A40
A308
A316
A3113
Windsor & Maidenhead
Bucks
Slough
The Queen Mother Reservoir
Wraysbury Res
King George VI Res
Staines Reservoirs
Queen Mary Res
Queen Elizabeth II Res
0 1 2 3 miles
0 1 2 3 4 5 km

Stanmore
Finchley
Wood Green
Tottenham
Woodford
Wealdstone
Hendon
Hornsey
Walthamstow
Kenton
Kingsbury
Golders Green
Highgate
Leyton
Leytonstone
Wembley
Hampstead
Stoke Newington
Islington
Sudbury
Willesden
Camden
Hackney
Stratford
Alperton
Kilburn
Shoreditch
Bethnal Green
West Ham
Perivale
Harlesden
Paddington
Finsbury
Ealing
Acton
Holborn
City
Poplar
Blackwall Tunnel
Hammersmith
Kensington
Westminster
Southwark
Bermondsey
Chiswick
Kew Br
Chelsea
Lambeth
Greenwich
Brentford
Fulham
Battersea
Deptford
Camberwell
Peckham
Isleworth
Richmond
Putney
Wandsworth
Clapham
Brixton
Lewisham
Catford
Teddington
Wimbledon
Streatham
Kingston upon Thames
Merton
Mitcham
Penge
Beckenham
Bromley
New Malden
Morden
Thornton Heath
South Norwood
Surbiton
Tolworth
Sutton
Carshalton
Croydon
Chessington
Ewell
Cheam
Wallington
Beddington
Purley
Epsom
Banstead
Coulsdon
Kenley
Whyteleafe
Warlingham
Ashtead
Leatherhead
Chipstead
LONDON
85
86
68
51
TQ

Woodford
Woodford Bridge
Grange Hill
Hainault
Harold Hill
Harold Wood
Great Warley
Little Warley
Childerditch
Herongate
Ingrave
Collier Row
Chase Cross
Rise Park
Gidea Park
Romford
Clayhall
Barkingside
Aldborough Hatch
Fullwell Cross
Marks Gate
Chadwell Heath
Little Heath
Upper Walthamstow
Snaresbrook
Wanstead
Leytonstone
Redbridge
Gants Hill
Cranbrook
Newbury Park
Seven Kings
Goodmayes
Ilford
Loxford
Becontree
Hornchurch
Emerson Park
Ardleigh Green
Upminster
Cranham
West Horndon
Rush Green
Elm Park
Hacton
Corbets Tey
North Ockenden
Bulphan
Forest Gate
Plashet
Barking
Dagenham
Upton Park
Stratford
West Ham
East Ham
Plaistow
Beckton
Canning Town
Creekmouth
Castle Green
South Hornchurch
Rainham
Wennington
South Ockendon
THURROCK
Baker Street
North Stifford
Aveley
Purfleet
Little Thurrock
Chafford Hundred
Grays
South Stifford
West Thurrock
Blackwall Tunnel
Silvertown
North Woolwich
Woolwich Ferry
Woolwich
Thamesmead
Erith Marshes
Belvedere
Erith
Coldharbour
Crayford Ness
Greenwich
Charlton
Plumstead
Abbey Wood
East Wickham
West Heath
Lessness Heath
Northumberland Heath
Slade Green
Dartford Crossing
Greenhithe
Northfleet
Gravesend
Blackheath
Kidbrooke
Shooters Hill
Welling
Bexleyheath
Barnehurst
Barnes Cray
Crayford
Dartford
Temple Hill
Stone
Swanscombe
Knockhall
Eltham
Avery Hill
New Eltham
Blackfen
Bexley
Old Bexley
Coldblow
Bowmans
Hither Green
Lee
Mottingham
Longlands
Sidcup
Lamorbey
Albany Park
Maypole
Wilmington
Fleet Downs
Bean
Betsham
Southfleet
Perry Street
Grove Park
Downham
Chislehurst
Chislehurst West
Foots Cray
North Cray
Ruxley
Hextable
Sutton at Hone
Hawley
Darenth
Lane End
Green Street Green
Longfield
New Barn
Hartley
Istead Rise
Nash Street
Sundridge
Bromley
Bickley
St Paul's Cray
St Mary Cray
Poverest
Swanley
Swanley Village
Hockenden
Kevingtown
Derry Downs
Horton Kirby
South Darenth
Farningham
Crockenhill
Petts Wood
Orpington
Ramsden
Southborough
Bromley Common
Hayes
Keston
Keston Mark
Locksbottom
Farnborough
Green Street Green
Chelsfield
Goddington
Well Hill
Eynsford
Maplescombe
West Kingsdown
Fawkham Green
New Ash Green
Ash
Ridley
Hodsoll Street
Culverstone Green
Vigo Village
Coney Hall
Nash
Leaves Green
Downe
Pratt's Bottom
Badger's Mount
Halstead
Knockholt Pound
Shoreham
Romney Street
Woodlands
Knatts Valley
Stansted
Fairseat
Trottiscliffe
Cudham
Knockholt
Otford
Kemsing
Heaveham
Wrotham
Crowdleham
KENT
DOWNS
Skid Hill Fm
Fickleshole
South
M25
M20
M26
A2
A12
A13
A20
A21
A127
A282
A406

ESSEX
KENT
Rayleigh
Rochford
Hockley
Hawkwell
Basildon
Laindon
Pitsea
Vange
North Benfleet
Thundersley
South Benfleet
Hadleigh
Leigh-on-Sea
Eastwood
Prittlewell
Southchurch
Westcliff-on-Sea
Southend-on-Sea
Canvey Island
Stanford-le-Hope
Corringham
Fobbing
Coryton
Thames Haven
Mucking
Linford
East Tilbury
West Tilbury
River Thames
Isle of Grain
Allhallows
Allhallows-on-Sea
St Mary Hoo
High Halstow
Cliffe
Cooling
Higham
Lower Stoke
Stoke
Grain
Queenborough
Hoo St Werburgh
Chattenden
Wainscott
Frindsbury
Strood
Rochester
Chatham
Gillingham
Brompton
Rainham
Walderslade
Wigmore
Hempstead
Upchurch
Lower Halstow
Newington
Sittingbourne
Kemsley
Milton Regis
Snodland
Birling
Aylesford
Ditton
Leybourne
Larkfield
Cuxton
Halling
Wouldham
Shorne
Cobham
Chalk
Luddesdown
MEDWAY
River Medway
KENT DOWNS
Medway Services

Paglesham Eastend
Horseshoe Corner
Rugwood Fm
Eastwick Head
Island
Great Burwood Fm
Rugwood Head
88
89
Asplins Head
Roper's Fm
Barling Marsh
Barling Hall Fm
Potton Island
New England Island
The Broomway
Barling
Fleet Head
Shelford Head
Stonebridge
Little Wakering Wick
Rushley Island
Havengore Island
Little Wakering
Oxenham Fm
Sharpness Head
Maplin Sands
Havengore Head
Haven Pt
Great Wakering
Samuel's Corner
Wakering Stairs
Bournes Green
Southchurch
North Shoebury
Black Grounds
Pig's Bay
Thorpe Bay
Cambridge Town
Shoeburyness
SEA LIFE CENTRE
Southend Pier
69
Shoebury Ness
West Knock
East Knock
THAMES ESTUARY
Roas Bank
Grain Spit
Grain
Garrison Pt
Jacob's Bank
Cheyney Rock
Blue Town
Sheerness
Scrapsgate
Marine Town
Minster Marshes
The Lappel
Mile Town
West Minster
Queenborough Spit
Halfway Houses
Deadmans Island
Minster
East End
Trouts
Warden Pt
Queenborough
Scocles Fm
Barnland Fm
Warden
Neats Court
Norwood Manor
Eastchurch
Rushenden
Parsonage Fm
Leysdown-on-Sea
Bay View
New Hook Fm
HM Prison
Old Rides Fm
Minster Marshes
Southlees Marshes
Groves Fm
Isle of Sheppey
Capel Hill Fm
Muswell Manor
Stray Marshes
Leysdown Marshes
Ferry Marshes
The Dray
Eastchurch Marshes
Elmley Island
Harty Marshes
Ridham Dock
Iwade
Kings Hill Fm
Spitend Marshes
Shell Ness
TANKERTON BAY
Elmley Marshes
Elliots Fm
THE SWALE
Whitstable
Swalecliffe
Coldharbour Marshes
Dutchman's Island
Isle of Harty
Tankerton
Lower Island
The Swale
Peg Fleet
Spitend Pt
Mocketts
WHITSTABLE BAY
Chestfield
Fowley Island
Sayes Court
Kemsley
South Deep
The Ferry Inn
Horse Sands
Seasalter
Duncan Down
South Street
Tonge Corner
Blacketts
The Oaze
South Oaze
Milton Regis
Teynham Level
Cleve Marshes
Graveney Marshes
Conyer
Seasalter Level
Murston
Teynham Street
Uplees
Nagden Marshes
Clapham Hill
Radfall
Poplar Hall
Court Lodge
Graveney Marshes
Sittingbourne
Snipeshill
Bax
Barrow Green
Luddenham Court
Oare
Ham Marshes
Nagden
Broom Street
Graveney Hill
Monkshill Fm
Yorkletts
Burgess Fm
Bapchild
Radfield
Teynham
Deerton Street
Nash's Fm
Ham Fm
Sandbanks
Graveney
Waterham
Highstreet
Pean Hill
Clowes Fm
Tunstall
Little Dully
Cellarhill
The Old Farm Ho
Four Oaks
The Brents
Faversham
Culmers
Dargate Common
Red Lion Ho
Well Court
Rodmersham
Davington
Denstroude
Highsted
Rodmersham Green
Batteries Fm
Lewson Street
Beacon Hill
Norton Ash
Goodnestone
Langdon Court
Fostall
Hernhill
Honey Hill
Amery Court
Tyler Hill
Woodstock
Upper Rodmersham
Lynsted
Ospringe
Ewell Fm
Dargate
Blean
Loyterton
Judd's Hill
Maison Dieu
Preston
Fairbrook Fm
Staplestreet
Hothe Court Fm
Newbury
Perry Court
Kingsdown
Lynsted Park
54
Mount Ephraim
Boughton Street
Dunkirk
Rough Common
Whitehill
Copton
Brenley Corner
Bexon
Plumford
Sole Street Ho
Upper Harbledown
Selgrove
Crouch
South Street
Hickmans Green
Bluetown
Newnham
North Street
Harbledown
Hollybushes
0 1 2 3 miles
0 1 2 3 4 5 km

7
8
9
10
11
89
A
B
C
D
E
F
G
H
40
90
TR
60
N O R T H
S E A
Margate
Nayland Rock
Fulsam Rock
Walpole Rocks
Palm Bay
Long Nose Spit
Forness Pt
Botany Bay
White Ness
Kingsgate Bay
North Foreland
North Cliff
South Cliff
Dumpton Gap
East Cliff
Cliftonville
Kingsgate
Northdown
Reading Street
St Peters
Broadstairs
Dumpton
Ramsgate
Westgateon Sea
St Mildred's Bay
Westbrook
Garlinge
Twenties
Birchington
Minnis Bay
Grenham Bay
Reculver
RECULVER TOWER & ROMAN FORT
River Wantsum
Plumpudding Island
Herne Bay
Hampton
Studd Hill
Eddington
Beltinge
Bishopstone
Hillborough
Brook Fm
Hawthorn Corner
Greenhill
Broomfield
Herne
Herne Common
Hunters Forstal
Lower Herne
Bullockstone
Hicks Forstal
Maypole
Old Tree
Hoath
Chislet
Boyden Gate
Marshside
Chitty
Upstreet
Sarre
St Nicholas at Wade
Potten Street
Hale
Monkton
Gore Street
Acol
Isle of Thanet
Quex Park
Minster
Manston
KENT INTERNATIONAL
Mount Pleasant
Way
Cliffs End
Pegwell
Pegwell Bay
St Lawrence
Newington
Northwood
Westwood
Haine
Lydden
Upton
Bromstone
Chilton
West Cliff
MOTOR MUSEUM
OOSTENDE 4:00
BOULOGNE 1:15
Shell Ness
SANDWICH & PEGWELL BAY
Richborough Port
SANDWICH BAY
Sandwich Flats
Sandwich
Sandwich Bay Estate
Ebbsfleet Ho
Minster Marshes
Ash Level
River Stour
Stourmouth
West Stourmouth
East Stourmouth
Grove
Westmarsh
Lower Goldstone
Upper Goldstone
Paramour Street
Ware
Elmstone
Preston
Hoaden
Cop Street
Cooper Street
Weddington
Ash
East Street
Guilton
Durlock
Wingham
WINGHAM BIRD PARK
Shatterling
Seaton
Wickhambreaux
Stodmarsh
Westbere Marshes
Westbere
Hersden
Sturry
Fordwich
Canterbury
Calcott
Broad Oak
West Blean Wood
Wildwood Discovery Park
Blaxland Fm
Woodlands Fm
Tile Lodge Fm
Rushbourne Manor
Knave's Ash
Chislet Forstal
Shelvingford
Wagtail
Wade Marsh
Great Brooksend Fm
Brooks End
Monkton Marshes
Plucks Gutter
Sheriffs Court
MINSTER ABBEY (REMAINS)
Sevenscore
ST AUGUSTINE'S CROSS
RICHBOROUGH CASTLE
A299
A28
A253
A256
A255
A254
A257
B2050
B2190
B2051
B2052
B2014
B2054
55

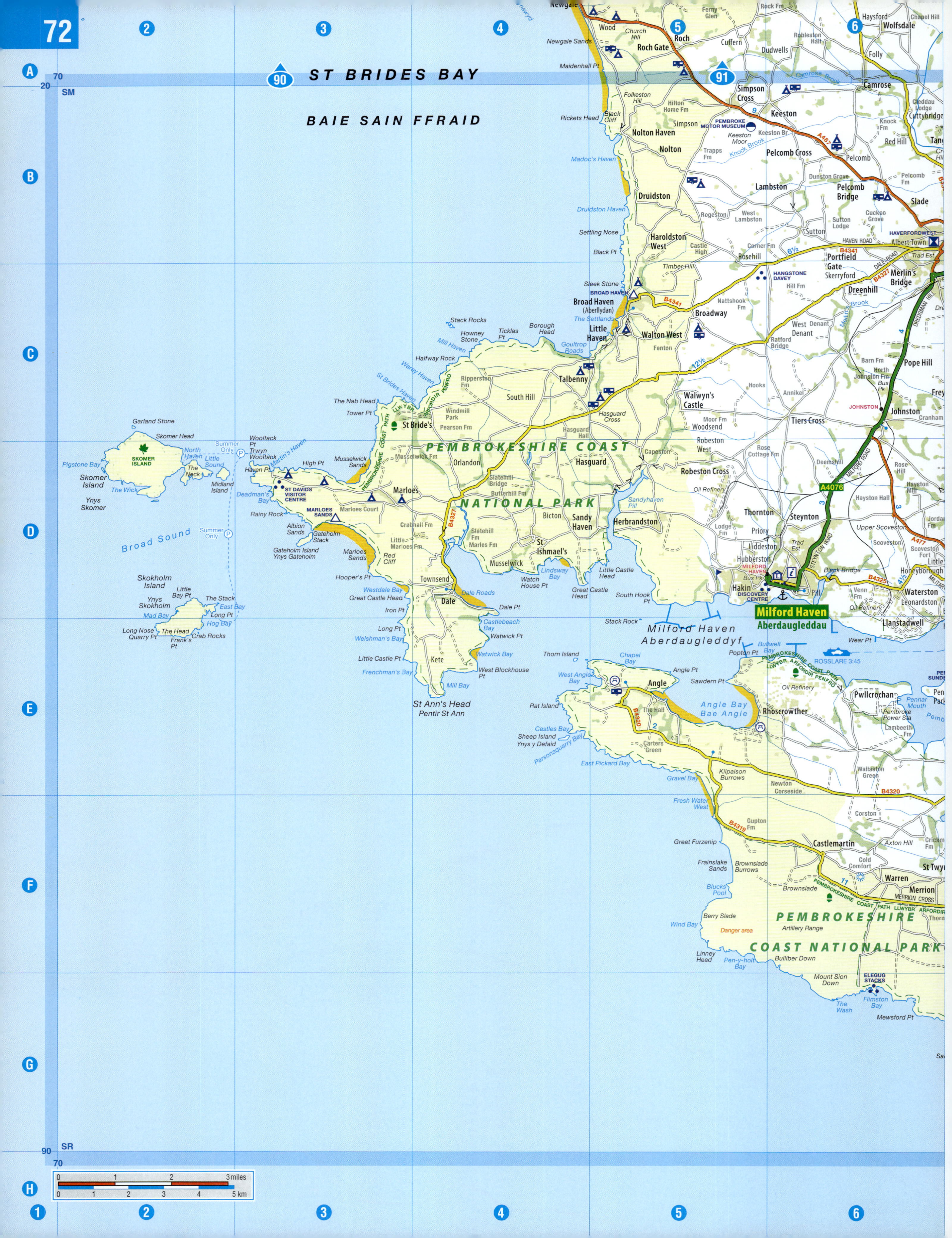

72
ST BRIDES BAY
BAIE SAIN FFRAID
90
91
SM
SR
PEMBROKESHIRE COAST
NATIONAL PARK
PEMBROKESHIRE
COAST NATIONAL PARK
Newgale
Wood
Church Hill
Newgale Sands
Roch Gate
Roch
Ferny Glen
Cuffern
Rock Fm
Robleston Hall
Dudwells
Haysford
Wolfsdale
Chapel Hill
Folly
Maidenhall Pt
Simpson Cross
Camrose
Camrose Brook
Cleddau Lodge
Cuttybridge
Folkeston Hill
Hilton Home Fm
Rickets Head
Black Cliff
Nolton Haven
Nolton
Simpson
PEMBROKE MOTOR MUSEUM
Keeston Moor
Keeston
Keeston Br
Knock Brook
Knock Fm
Red Hill
Trapps Fm
Pelcomb Cross
Pelcomb
Madoc's Haven
Druidston
Dunston Grove
Lambston
Pelcomb Bridge
Pelcomb Fm
Slade
Druidston Haven
Rogeston
West Lambston
Sutton Lodge
Cuckoo Grove
Settling Nose
Haroldston West
Castle High
Sutton
HAVERFORDWEST
Albert Town
HAVEN ROAD
Black Pt
Corner Fm
Rosehill
Portfield Gate
Skerryford
Trad Est
Merlin's Bridge
Timber Hill
HANGSTONE DAVEY
Hill Fm
Dreenhill
Sleek Stone
BROAD HAVEN
Broad Haven (Aberllydan)
The Settlands
Nattshook Fm
Broadway
Little Haven
Walton West
Fenton
West Denant
Denant
Ratford Bridge
Stack Rocks
Howney Stone
Ticklas Pt
Borough Head
Goultrop Roads
Mill Haven
Halfway Rock
Talbenny
Barn Fm
North Johnston Fm
Pope Hill
Wancy Haven
St Brides Haven
Ripperston Fm
South Hill
Walwyn's Castle
Hooks
Annikel
The Nab Head
Tower Pt
Windmill Park
Pearson Fm
St Bride's
Hasguard Cross
Hasguard Hall
Moor Fm
Woodsend
Tiers Cross
JOHNSTON
Johnston
Garland Stone
Skomer Head
North Haven
SKOMER ISLAND
Pigstone Bay
Skomer Island
The Wick
Little Sound
The Neck
Midland Island
Ynys Skomer
Summer Only
Wooltack Pt
Trwyn Wooltack
Martin's Haven
Haven Pt
High Pt
ST DAVIDS VISITOR CENTRE
Deadman's Bay
Musselwick Sands
Musselwick Fm
Orlandon
Hasguard
Capeston
Robeston West
Robeston Cross
Rose Cottage Fm
Deemshill
Rose Hill
Slatemill Bridge
Butterhill Fm
Marloes
MARLOES SANDS
Marloes Court
Rainy Rock
Albion Sands
Gateholm Stack
Gateholm Island
Ynys Gateholm
Marloes Sands
Crabhall Fm
Little Marloes Fm
Red Cliff
Slatehill Fm
Marles Fm
Bicton
Sandy Haven
Sandyhaven Pill
Herbrandston
Oil Refinery
Thornton
Steynton
Lodge Fm
Priory
Liddeston
Hubberston
MILFORD HAVEN
Hayston Hall
Upper Scoveston
Scoveston
Scoveston Fort
Broad Sound
Skokholm Island
Ynys Skokholm
Little Bay Pt
The Stack
East Bay
Mad Bay
Long Pt
Hog Bay
Long Nose
Quarry Pt
The Head
Frank's Pt
Crab Rocks
Hooper's Pt
Townsend
Musselwick
St Ishmael's
Lindsway Bay
Watch House Pt
Little Castle Head
Great Castle Head
South Hook Pt
Hakin
DISCOVERY CENTRE
Pill
Black Bridge
Venn Fm
Waterston
Leonardston
Honeyborough
Westdale Bay
Great Castle Head
Dale Roads
Dale
Dale Pt
Iron Pt
Milford Haven
Aberdaugleddau
Milford Haven
Aberdaugleddyf
Stack Rock
Llanstadwell
Long Pt
Welshman's Bay
Castlebeach Bay
Watwick Pt
Watwick Bay
Wear Pt
Little Castle Pt
Frenchman's Bay
Kete
West Blockhouse Pt
Mill Bay
St Ann's Head
Pentir St Ann
Thorn Island
Chapel Bay
Bullwell Bay
Popton Pt
ROSSLARE 3:45
West Angle Bay
Angle
Angle Pt
Sawdern Pt
Oil Refinery
Pwllcrochan
Pennar Mouth
Pembroke Power Sta
Lambeeth Fm
Rat Island
The Hall
Angle Bay
Bae Angle
Rhoscrowther
Castles Bay
Sheep Island
Ynys y Defaid
Parsonsquarry Bay
Carters Green
East Pickard Bay
Gravel Bay
Kilpaison Burrows
Newton Corseside
Wallaston Green
Fresh Water West
Gupton Fm
Corston
Great Furzenip
Castlemartin
Axton Hill
Frainslake Sands
Brownslade Burrows
Cold Comfort
Warren
St Twynnells
Merrion
MERRION CROSS
Blucks Pool
Brownslade
Artillery Range
Berry Slade
Wind Bay
Danger area
Linney Head
Pen-y-holt Bay
Bulliber Down
Mount Sion Down
ELEGUG STACKS
Flimston Bay
The Wash
Mewsford Pt
A487
A4076
A477
B4341
B4327
B4320
B4319
B4325
B4321
MILFORD ROAD
STEYNTON ROAD
DALE ROAD
PEMBROKESHIRE COAST PATH
LLWYBR ARFORDIR PENFRO
A B C D E F G H
1 2 3 4 5 6
0 1 2 3 miles
0 1 2 3 4 5 km

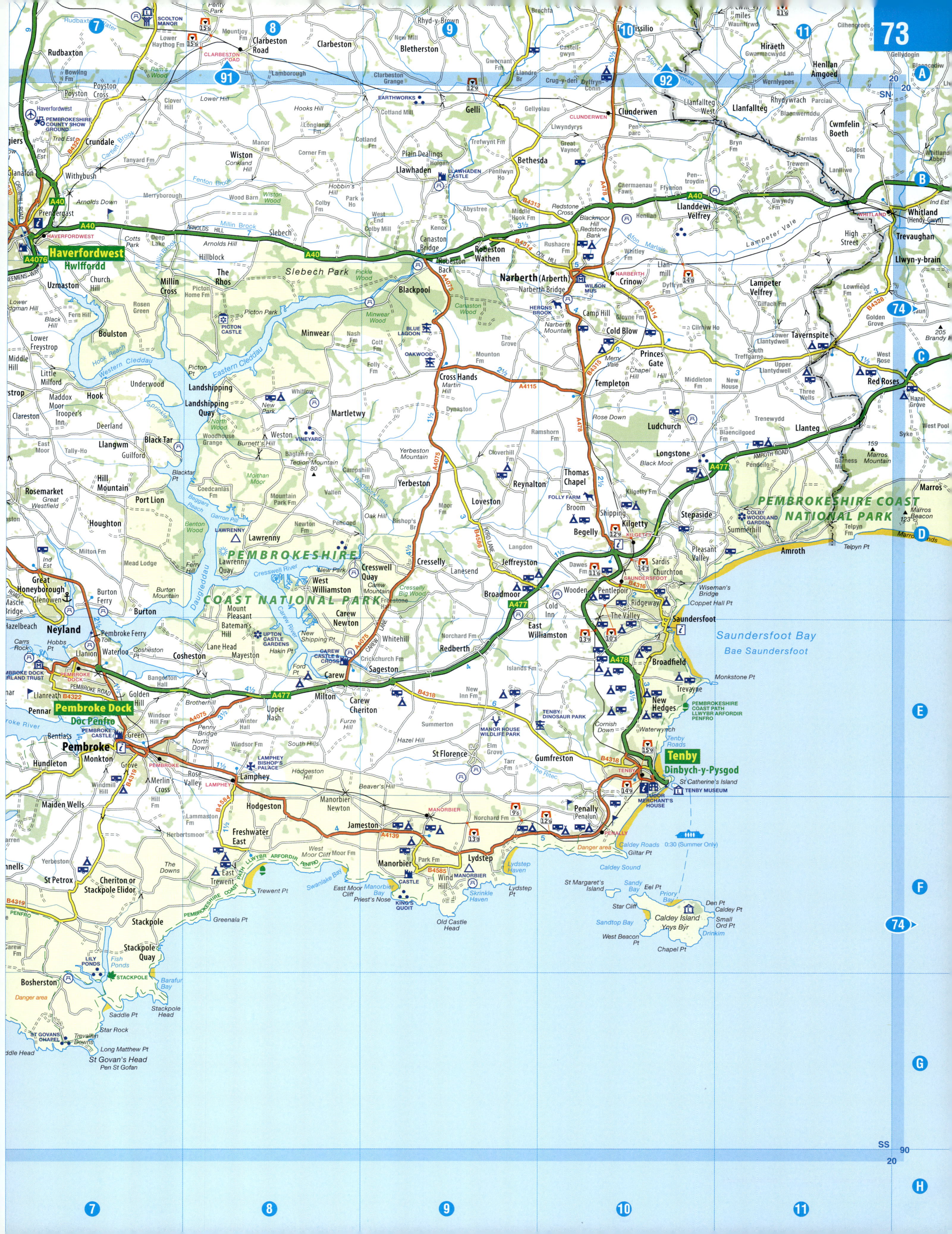

Rudbaxton
Scolton Manor
Clarbeston Road
Clarbeston
Rhyd-y-Brown
Bletherston
Lissilio
Hiraeth
Henllan Amgoed
91
92
Lamborough
Clarbeston Grange
Earthworks
Gelli
Clunderwen
Llanfallteg West
Llanfallteg
Cwmfelin Boeth
Haverfordwest
Pembrokeshire County Show Ground
Crundale
Withybush
Wiston
Plain Dealings
Llawhaden
Llawhaden Castle
Bethesda
Whitland
Prendergast
Haverfordwest
Hwlffordd
Slebech
Slebech Park
Robeston Wathen
Narberth (Arberth)
Narberth Bridge
Wilson Mus
Crinow
Llanddewi Velfrey
Trevaughan
Llwyn-y-brain
High Street
Lampeter Velfrey
Uzmaston
Millin Cross
The Rhos
Picton Castle
Blackpool
Blue Lagoon
Oakwood
Herons Brook
Camp Hill
Cold Blow
Princes Gate
Tavernspite
Red Roses
Boulston
Minwear
Cross Hands
Templeton
Ludchurch
Llanteg
Landshipping
Landshipping Quay
Martletwy
Vineyard
Hook
Llangwm
Black Tar
Rosemarket
Port Lion
Yerbeston
Reynalton
Thomas Chapel
Loveston
Longstone
Kilgetty
Stepaside
Amroth
Pembrokeshire Coast National Park
Colby Woodland Garden
Summerhill
Houghton
Lawrenny
Lawrenny Quay
Pembrokeshire Coast National Park
Cresswell Quay
West Williamston
Cresselly
Jeffreyston
Broadmoor
Begelly
Saundersfoot
Saundersfoot Bay
Bae Saundersfoot
Great Honeyborough
Neyland
Burton
Carew Newton
Carew
Carew Castle & Cross
Sageston
Redberth
East Williamston
Pentlepoir
Broadfield
Pembroke Dock
Doc Penfro
Pembroke Ferry
Cosheston
Upton Castle Gardens
Milton
Carew Cheriton
New Hedges
Pembroke
Monkton
Lamphey
Lamphey Bishop's Palace
St Florence
Manor House Wildlife Park
Tenby Dinosaur Park
Gumfreston
Tenby
Dinbych-y-Pysgod
Tenby Museum
Tudor Merchant's House
Hundleton
Maiden Wells
Hodgeston
Manorbier Newton
Jameston
Penally
Freshwater East
Manorbier
Lydstep
Caldey Island
Ynys Byr
St Petrox
Cheriton or Stackpole Elidor
Stackpole
Stackpole Quay
Bosherston
Lily Ponds
St Govan's Head
Pen St Gofan

CARMARTHEN BAY
BAE CAERFYRDDIN
Carmarthen
Caerfyrddin
St Clears
Sanclêr
Whitland
Laugharne
Pendine
Kidwelly
Ferryside
Llansteffan
Pembrey
Pembrey Forest
Pendine Sands
Laugharne Sands
Cefn Sidan Sands
RHOSSILI BAY
Llangennith
Llanmadoc
Rhossili
Whiteford Sands
Burry Holms
River Taf
Afon Taf
River Towy
Afon Tywi
Gwendraeth
Marros Sands
Marros
Red Roses
Llanddowror
Meidrim
Bancyfelin
Sarnau
Johnstown
Llangain
Llanybri
Pwll-trap
Broadway
Halfpenny Furze
Llandawke
Llansadurnen
Brook
Plashett
Idole
Bancycapel
Llandyfaelog
Mynyddygarreg
Pinged
Penybedd
Morfa Bach
Llangynog
Llanllwch
Merthyr
Trevaughan
Llwyn-y-brain
Llangynin
Backe
Broadlay
Llansaint
Llangunnor
Pensarn
Abergwili
Cwmffrwd
Croesyceiliog
Danger area
MUSEUM OF SPEED
DYLAN THOMAS BOATHOUSE
LAUGHARNE CASTLE
LLANSTEFFAN CASTLE
KIDWELLY CASTLE
WEST WALES CENTRE FOR THE CRAFTS
UNITED COUNTIES SHOWGROUND
CORS GOCH LLANLLWCH
CARMARTHEN MUSEUM
A40
A477
A4066
A484
B4312
B4314
0 1 2 3 miles
0 1 2 3 4 5 km

Llandeilo
Nantgaredig
Pont-ar-gothi
Llanegwad
Felindre
Dryslwyn
Cilsan
Ffairfach
White Mill
Capel Dewi
Llanarthne
Golden Grove
Trapp
Carreg Cennen Castle & Farm Centre
Black Mountain
Nantycaws
The National Botanic Garden of Wales
Penrhiwgoch
Temple Bar
Milo
Llandyfan
Llanddarog
Porthyrhyd
Maesybont
Carmel
Pentre-Gwenlais
Dre-fach
Cwmisfael
Foel-gastell
Llandybie
Llangendeirne
Y Goeden Bach Bonsai Centre
Cefneithin
Gorslas
Cae'r-bryn
Twynmynydd
Brynamman
Crwbin
Drefach
Cwm-y-glo
Cross Hands
Morfa
Pen-y-groes
Saron
Ammanford (Rhydaman)
Glanaman
Garnant
Cwm-mawr
Tumble (Y Tymbl)
Betws
Pontamman
Bancffosfelen
Pontyberem
Pentre
Capel Hendre
Penybanc
Tycroes
Pantyffynnon
Cwmgwili
Greynor
Cwmgors
Four Roads
Pontyates (Pont-iets)
Pont-Henri
Ynys-tawelog
Garnswllt
Llannon
Llwyn-Têg
Pont Abraham Services
Cynheidre
Sylen
Llanedi
Carway
Sardis
Ffos Las
Horeb
Five Roads
Trimsaran
Fforest
Pentrebach
SWANSEA (ABERTAWE)
Hendy
Pontarddulais
Cwm-Dulais
Rhyd-y-gwin
Pontardawe
Trebanos
Swiss Valley
Felinfoel
Allt
Felindre
Rhyd-y-pandy
Craig-cefn-parc
Cwmbach
Pwll
Furnace
Cae-gors Llangennech
Waungron
Pontlliw
Clydach
Sandy
Dafen
Bryn
Ynystawe
Llanelli
Millennium Coastal Park Discovery Centre
Llwyn-hendy
Gorseinon
Penllergaer
Swansea Services
Pant-lasau
Glais
Morfa
Bynea
Yspitty
Loughor
Llangyfelach
Heol-las
Machynys
Wildfowl & Wetlands Centre
Garden Village
Morriston (Treforys)
Birchgrove
River Loughor
Afon Llwchwr
Loughor Castle
Gowerton (Tre-Gwyr)
Cadle
Llansamlet
Fforest-fach
Landore
Pen-clawdd
Crofty
Llanmorlais
Blue Anchor
Three Crosses
Waunarlwydd
Cwmdu
Manselton
Pentre-chwyth
Wernffrwd
Landimore
Llanrhidian Sands
Dunvant (Dynfant)
Killay
Sketty
Glyn Vivian Art Gallery
Port Tennant
St Thomas
Cheriton
Llanrhidian
Upper Killay
Brynmill
Swansea Abertawe
Stembridge
Oldwalls
Llethrid
Burry Green
The Gower
Gower Peninsula
Blackhills
Mayals
Black Pill
SWANSEA BAY
BAE ABERTAWE
Reynoldston
Lunnon
Ilston
Parkmill
Kittle
West Cross
Knelston
Penmaen
Nicholaston
Pennard
Bishopston
Oldway
Mansellfield
The Mumbles (Y Mwmbwls)
Penrice
Southgate
Newton

Gwynfe
Carreg Yr Ogof 585
Pen-rhiw-goch
Bannau Sir Gaer
Brycheiniog 802
Llyn y Fan Fach
Llyn y Fan Fawr
Waun Leuci
Y Gelli
603 Fan Bwlch Chwyth
Ffforest Fawr
94
95
NATIONAL PARK
BANNAU BRYCHEINIOG
Mynydd Du
CARMARTHENSHIRE
(SIR GAERFYRDDIN)
POWYS
BRECON BEACONS
NATIONAL PARK
BANNAU BRYCHEINIOG
Glyntawe
NATIONAL SHOWCAVES CENTRE FOR WALES
DAN-YR-OGOF
CRAIG-Y-NOS
Craig-y-nos
Penwyllt
OGOF FFYNNON DDU
Brynamman
75
Rhosaman
Lower Brynamman
Garnant
Gwaun-Cae-Gurwen
Cwmllynfell
Ystradowen
Tairgwaith
Cwmgors
Upper Cwm-twrch
Lower Cwm-twrch
Cwmgiedd
Abercraf
Caehopkin
Coelbren
Penrhos
Ystradgynlais
Onllwyn
Dyffryn Cellwen
Ystradfellte
Gurnos
Ystalyfera
Godre'r-graig
Seven Sisters
SEVEN SISTERS COLLIERY
Nant-y-cafn
Pontneddfechan
Glyn-neath (Glynedd)
Morfa Glas
Blaengwrach
Cwmgwrach
Rhigos
NEATH PORT TALBOT
(CASTELL-NEDD PORT TALBOT)
Ynysmeudwy
Pontardawe
Rhyd-y-fro
Cilybebyll
Gellinudd
Trebanos
Alltwen
Rhos
Crynant (Creunant)
Rheola Forest
Resolven
Glyn Castle
Clydach
Glais
CEFN COED COLLIERY MUS
Melincourt
Clyne
Cilfrew
Aberdulais
ABERDULAIS FALLS
Tonna (Tonnau)
Cadoxton-Juxta-Neath
Bryn-coch
Rhyddings
Birchgrove
Heol-las
Neath Abbey
Skewen
Llansamlet
Neath Castell-Nedd
Cimla
Melincryddan
Briton Ferry (Llansawel)
Jersey Marine
AFAN ARGOED
SOUTH WALES MINERS MUSEUM
FOREST PARK
Cymer
Croeserw
Duffryn
Caerau
Blaengwynfi
Tynewydd
Treherbert
Blaenrhondda
Glyncorrwg
Abercregan
Cwm-parc
Pontrhydyfen
Pwll-y-glaw
Baglan
Cwmafan
Bryn
Maesteg
Nantyffyllon
Spelter
Blaengarw
Nant-y-moel
Price Town
Wyndham
Pontycymer
Ogmore Vale
Garth
Cwmfelin
Aberavon
Sandfields
Port Talbot
57
58
Goybre
Taibach
Llangynwyd
Lewistown
BRIDGEND

Brecon Beacons
95
96
Craig Cerrig-gleisiad
Craig y Fro
Fan Fawr
Pant Gilfach-wen
Beacons Reservoir
Cantref Reservoir
Nant-ddu
Coed Taf Fawr
Garwnant Visitor Centre
Llwyn-on Reservoir
Llwyn-on Village
Darren Fach
Pen-twyn-isaf
Penderyn
Cwm Cadlan
Pant Sychbant
Mynydd-y-glog
Penmoelallt
Onllwyn
Hirwaun
Penywaun
Tre-Ifor
Tre-Gibbon
Llwydcoed
Robertstown
Aber-nant
Trecynon
Cwmdare
Aberdare
Dare Valley
Gadlys
RHONDDA CYNON TAFF
Cwmbach
Aberaman
Cwmaman
Mountain Ash (Aberpennar)
Penrhiwceiber
Perthcelyn
Abercwmboi
Fernhill
Maerdy
Treorchy (Treorci)
Pentre
Ferndale
Rhondda
Ystrad
Ton-Pentre
Gelli
Tylorstown
Stanleytown
Pont-y-gwaith
Wattstown
Ynyshir
Llwynypia
Clydach Vale
Tonypandy
Trealaw
Penygraig
Porth
Trehafod
Rhondda Heritage Park
Williamstown
Edmondstown
Cymmer
Trebanog
Evanstown
Gilfach Goch
Hendreforgan
Bryn Golau
Tonyrefail
Thomastown
Pontypridd
Hopkinstown
Maesycoed
Graig
Treforest
Trallwn
Glyncoch
Cilfynydd
Pont Siôn Norton
Abercynon
Carnetown
Ynysybwl
Ynysboeth
Tyntetown
Newtown
Edwardsville
Quaker's Yard
Treharris
Trelewis
Taff Merthyr Garden Village
Nelson
Llancaiach
Llancaiach Fawr Manor
Gelligaer
Penpedairheol
Bedlinog
Cwmfelin
Deri
Aberfan
Merthyr Vale
Mount Pleasant
Troedyrhiw
Pentrebach
Abercanaid
MERTHYR TYDFIL (MERTHYR TUDFUL)
Merthyr Tydfil
Merthyr Tudful
Twynyrodyn
Gellideg
Pen-yr-Heolgerrig
Ynysfach Engine House Heritage Centre
Cyfarthfa Castle & Museum
Gurnos
Galon Uchaf
Pen-y-Darren
Dowlais
Dowlais Top
Pant
Trefechan
Cefn-coed-y-cymmer
Vaynor
Pontsticill
Brecon Mountain Railway
Pontsticill Reservoir
Pentwyn Reservoir
Talybont Reservoir
Abercynafon
Danywenallt
Aber-Village
Llangynidr
Coed-yr-ynys
Cwmcrawnon
Mynydd Llangynidr
Mynydd Llangatwg
Llangattock
Crickhowell
Trefil
Rassau
Garnlydan
Brynmawr
Beaufort
Nantyglo
Blaina
BLAENAU GWENT
Ebbw Vale (Glyn Ebwy)
Tredegar
Dukestown
Sirhowy
Tafarnau-bach
Princetown
Rhymney (Rhymni)
Pontlottyn
Fochriw
Abertysswg
New Tredegar (Tredegar Newydd)
Tirphil
Phillip's Town
Bedwellty Pits
Markham
Aberbargoed
Bargoed (Bargod)
Gilfach
Pengam
Blackwood (Coed Duon)
Fleur-de-lis
Pontllanfraith
Hengoed
Ystrad Mynach
Cefn Hengoed
Penybryn
Maesycwmmer
Llanbradach
Senghenydd
Abertridwr
Penyrheol
Bedwas
Trethomas
Caerphilly (Caerffili)
CAERPHILLY (CAERFFILI)
Argoed
Oakdale
Abercarn
Cwm
Waunlwyd
Festival Park
Llanhilleth
Aberbeeg
Crumlin
Newbridge
Ynysddu
Wyllie
Trinant
Wattsville
Cwmfelinfach

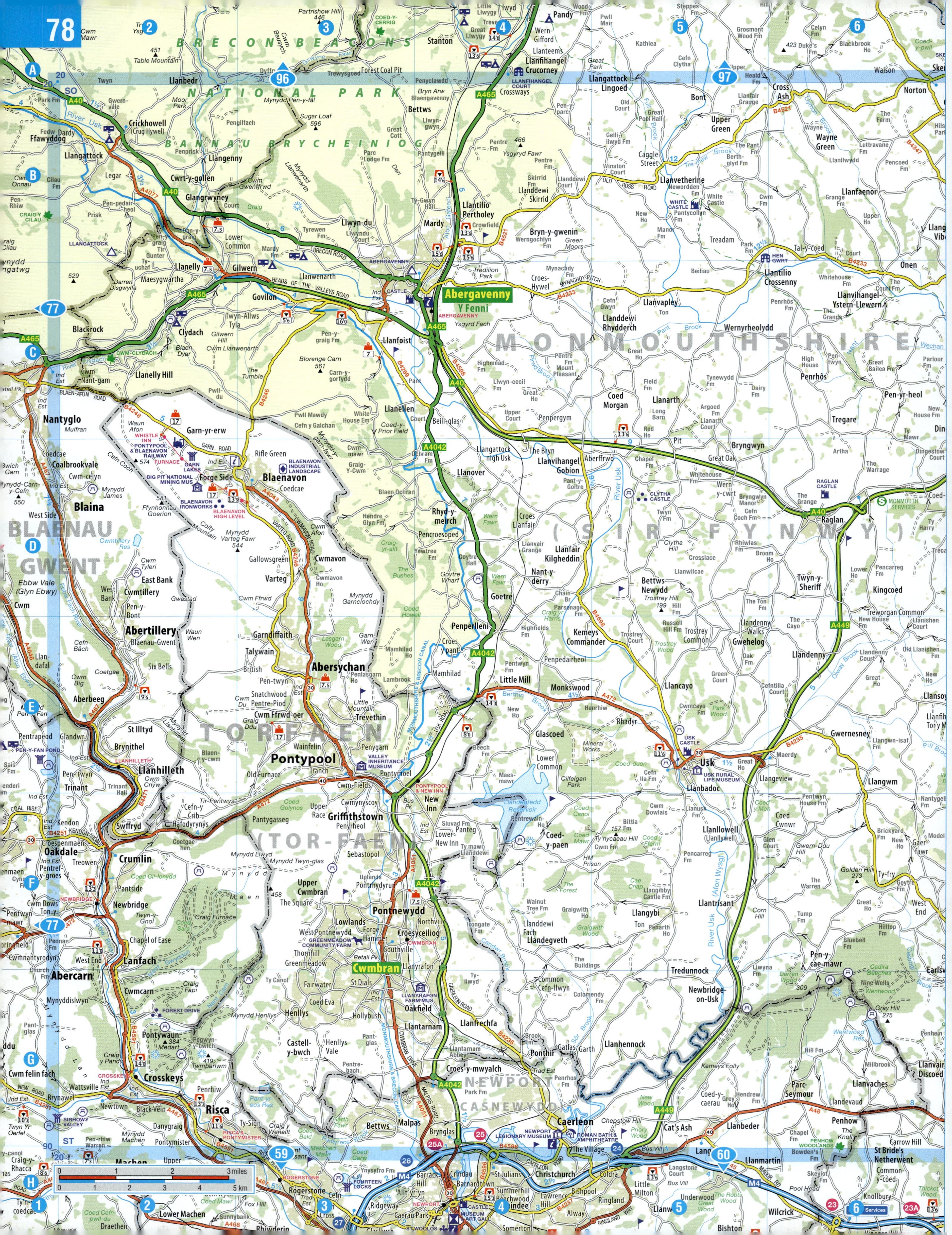
BRECON BEACONS
NATIONAL PARK
BANNAU BRYCHEINIOG
MONMOUTHSHIRE
(SIR FYNWY)
BLAENAU GWENT
TORFAEN
(TOR-FAEN)
NEWPORT
CASNEWYDD
Abergavenny
Y Fenni
Crickhowell
(Crug Hywel)
Llangattock
Llangenny
Cwrt-y-gollen
Glangrwyney
Gilwern
Llanelly
Maesygwartha
Govilon
Clydach
Blackrock
Llanelly Hill
Llanfoist
Llanellen
Nantyglo
Blaina
Brynmawr
Garn-yr-erw
Forge Side
Blaenavon
Varteg
Cwmavon
Abertillery
Abersychan
Talywain
Garndiffaith
Pontypool
Griffithstown
Sebastopol
Pontnewydd
Cwmbran
Croesyceiliog
Llantarnam
Oakfield
Henllys
Risca
Crosskeys
Abercarn
Newbridge
Crumlin
Oakdale
Llanhilleth
Aberbeeg
Six Bells
Cwmtillery
Usk
Llanbadoc
Raglan
Llanarth
Llanover
Goetre
Penperlleni
Little Mill
Monkswood
Llangybi
Caerleon
Christchurch
Malpas
Bettws
Rogerstone
Llantilio Crossenny
Llanvetherine
Llanvapley
Llantilio Pertholey
Mardy
Pandy
Llanfihangel Crucorney
Bettws Newydd
Kemeys Commander
Llandegveth
Llanfrechfa
Llanhennock
Tredunnock
Newbridge-on-Usk
Llanvihangel Gobion
Llanddewi Rhydderch
Coed Morgan
Bryngwyn
Clytha
Tregare
Pen-yr-heol
Llandenny
Llangwm
Gwernesney
Llansoy
Kingcoed
Llantrisant
Pen-y-cae-mawr
Parc-Seymour
Llanvaches
Penhow
Llanbeder
Llanmartin
Langstone
Cat's Ash
Blorenge Cairn 561
Sugar Loaf 596
Table Mountain
Llandegfedd Reservoir
Abergavenny
Pontypool
Usk
A40
A465
A4042
A472
A449
A4051
A467
B4598
B4233
B4246
B4235
M4
River Usk
Monmouthshire & Brecon Canal
Twyn-y-Sheriff
Treadam
Onen
Wernyrheolydd
Penrhos
Bont
Upper Green
Wayne Green
Llanfaenor
Cross Ash
Norton
Llangattock Lingoed
Bryn-y-gwenin
Croes-Hywel
Llwyn-du
Trevethin
Penygarn
Cwmynyscoy
Upper Cwmbran
Fairwater
St Dials
Pontywaun
Cwmcarn
Lanfach
Pantside
Trinant
Swffryd
Wattsville
Ty-Sign
Danygraig
Machen
Lower Machen
Crindau
St Julians
Barnardtown
Summerhill
Beechwood
Coldra
Ringland
Little Milton
Bishpool
Alway
Llanwern
Somerton
0 1 2 3 miles
0 1 2 3 4 5 km

Monmouth
Trefynwy
Chepstow
Coleford
Cinderford
Lydney
Thornbury
FOREST OF DEAN
FOREST PARK
GLOUCESTERSHIRE
SOUTH GLOUCESTERSHIRE
River Severn
River Wye
Sheperdine Sands
Oldbury Sands
Lydney Sand
Saniger Sands
Frampton Sand
Beachley
Tidenham
Tutshill
Sedbury
St Briavels
Tintern Parva
Brockweir
Llandogo
Trellech
Whitebrook
Redbrook
Penallt
Goodrich
Walford
Coughton
Pontshill
Lea
Mitcheldean
Drybrook
Ruardean
Ruspidge
Upper Soudley
Blakeney
Awre
Newnham
Arlingham
Sharpness
Berkeley
Newport
Woodford
Stone
Falfield
Tortworth
Whitfield
Rockhampton
Oldbury-on-Severn
Shepperdine
Aust
Alveston
Elberton
Littleton-on-Severn
Caldicot
Caerwent
Crick
Mathern
Shirenewton
Itton Common
St Arvans
Devauden
Wolvesnewton
Llangovan
Mitchel Troy
Wonastow
Dixton
Welsh Newton
Llangarron
Llanrothal
Marstow
Whitchurch
Symonds Yat
English Bicknor
Lower Lydbrook
Joyford
Christchurch
Berry Hill
Staunton
Newland
Clearwell
Bream
Parkend
Yorkley
Whitecroft
Pillowell
Aylburton
Alvington
Woolaston
Hewelsfield
Coldharbour
Stroat
Broadwell
Milkwall
Ellwood
Cannop
Nailbridge
Littledean
Elton
Blaisdon
Flaxley
Longhope
Westbury-on-Severn
Huntley
Viney Hill
Nibley
Purton
Halmore
Hinton
Ham
Bevington
Tytherington

GLOUCESTERSHIRE
COTSWOLDS
Gloucester
Stroud
Tayton
Tibberton
Rudford
Maisemore
Twigworth
Longford
Longlevens
Parton
Churchdown
Badgeworth
Hucclecote
Barnwood
Brockworth
Bentham
Shurdington
Leckhampton
Huntley
Bulley
Birdwood
Churcham
Highnam
Oakle Street
Minsterworth
Blaisdon
Northwood Green
Flaxley
Elton
Westbury-on-Severn
Elmore Back
Farleys End
Elmore
Hempsted
Quedgeley
Hardwicke
Tuffley
Whaddon
Matson
Brookthorpe
Upton St Leonard's
Cranham
Great Witcombe
Birdlip
Newnham
Arlingham
Longney
Epney
Rodley
Boxbush
Upper Framilode
Fretherne
Saul
Moreton Valence
Whitminster
Frampton on Severn
Haresfield
Harescombe
Painswick
Sheepscombe
Whiteway
Miserden
Edge
Pitchcombe
Slad
Randwick
Whiteshill
Stonehouse
Ebley
Eastington
Cambridge
Slimbridge
King's Stanley
Leonard Stanley
Rodborough
Chalford
Bisley
Thrupp
Brimscombe
Frampton Mansell
Frocester
Woodchester
Minchinhampton
Amberley
Nailsworth
Avening
Cherington
Rodmarton
Coaley
Nympsfield
Uley
Dursley
Cam
Stinchcombe
Berkeley
Woodford
Stone
North Nibley
Wotton-under-Edge
Kingswood
Charfield
Tetbury
Westonbirt
Malmesbury
Hillesley
Leighterton
Long Newnton
Beverston
Kingscote
Horsley
Tetbury Upton
Ashley
Culkerton
River Severn
River Leadon
Frampton Sand
Slimbridge

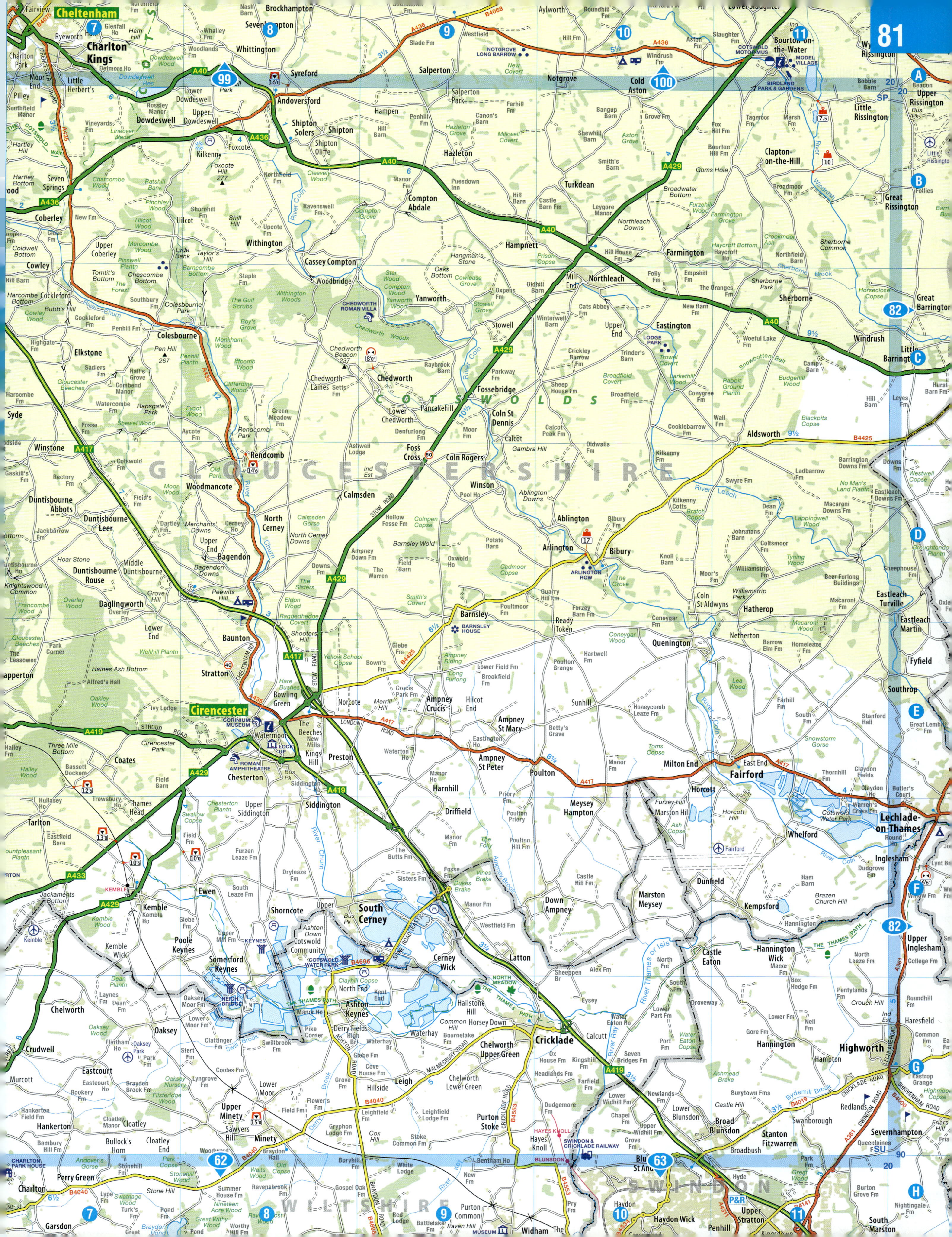

Cheltenham
Charlton Kings
Fairview
Ryeworth
Charlton Park
Moor End
Little Herbert's
Pilley
Southfield Manor
Dowdeswell
Lower Dowdeswell
Upper Dowdeswell
Whittington
Brockhampton
Sevenhampton
Syreford
Andoversford
Salperton
Notgrove
Cold Aston
Bourton-on-the-Water
Little Rissington
Upper Rissington
Great Rissington
Clapton-on-the-Hill
Shipton Solers
Shipton
Shipton Oliffe
Foxcote
Kilkenny
Hazleton
Turkdean
Compton Abdale
Hampnett
Northleach
Farmington
Sherborne
Great Barrington
Windrush
Little Barrington
Seven Springs
Coberley
Upper Coberley
Hilcot
Withington
Cassey Compton
Woodbridge
Yanworth
Stowell
Eastington
Cowley
Cockleford
Colesbourne
Elkstone
Chedworth
Chedworth Laines
Fossebridge
Coln St Dennis
Calcot
Aldsworth
Syde
Winstone
Rendcomb
Woodmancote
Calmsden
Foss Cross
Coln Rogers
Winson
Ablington
Arlington
Bibury
Duntisbourne Abbots
Duntisbourne Leer
Duntisbourne Rouse
Middle Duntisbourne
North Cerney
Bagendon
Daglingworth
Baunton
Barnsley
Ready Token
Quenington
Coln St Aldwyns
Hatherop
Eastleach Turville
Eastleach Martin
Netherton
Fyfield
Southrop
Stratton
Cirencester
Ampney Crucis
Ampney St Mary
Ampney St Peter
Poulton
Sunhill
Milton End
Fairford
Horcott
Lechlade-on-Thames
Whelford
Inglesham
Coates
Chesterton
Preston
Siddington
Harnhill
Driffield
Meysey Hampton
Tarlton
Kemble
Ewen
Shorncote
South Cerney
Poole Keynes
Somerford Keynes
Cerney Wick
Latton
Down Ampney
Marston Meysey
Dunfield
Kempsford
Castle Eaton
Hannington Wick
Hannington
Highworth
Chelworth
Oaksey
Crudwell
Ashton Keynes
Waterhay
Cricklade
Calcutt
Chelworth Upper Green
Chelworth Lower Green
Leigh
Eastcourt
Murcott
Hankerton
Upper Minety
Minety
Purton Stoke
Hayes Knoll
Broad Blunsdon
Lower Blunsdon
Stanton Fitzwarren
Sevenhampton
Perry Green
Charlton
Garsdon
Purton
Widham
Haydon Wick
Penhill
Upper Stratton
South Marston
Blunsdon St Andrew
GLOUCESTERSHIRE
COTSWOLDS
WILTSHIRE
SWINDON
A40
A436
A429
A435
A417
A419
A433
A361
B4425
B4040
B4696
B4019
B4553
River Churn
River Coln
River Thames or Isis
The Thames Path
Cotswold Water Park

82
COTSWOLDS
OXFORDSHIRE
Little Rissington
Great Rissington
Great Barrington
Little Barrington
Windrush
Taynton
Burford
Fulbrook
Westhall Hill
Upton
Signet
Westwell
Holwell
Bradwell Grove
Cotswold Wildlife Park
Shilton
Carterton
Brize Norton
Alvescot
Kencot
Filkins
Broadwell
Langford
Broughton Poggs
Eastleach Turville
Eastleach Martin
Fyfield
Southrop
Little Faringdon
Lechlade-on-Thames
Inglesham
Upper Inglesham
Buscot
Kelmscott
Kelmscott Manor
Grafton
Clanfield
Little Clanfield
Black Bourton
Bampton
Aston
Cote
Brighthampton
Standlake
Shifford
Chimney
Duxford
Radcot
Faringdon
Littleworth
Buckland
Carswell Marsh
Pusey
Gainfield
Hatford
Stanford in the Vale
Shellingford
Great Coxwell
Little Coxwell
Coleshill
Highworth
Watchfield
Longcot
Fernham
Baulking
Uffington
Woolstone
Kingston Lisle
Goosey
Denchworth
West Hanney
East Hanney
Charney Bassett
Lyford
Grove
Wantage
West Challow
Childrey
Garford
Frilford
Kingston Bagpuize
Southmoor
Longworth
Hinton Waldrist
Fyfield
Tubney
Appleton
Netherton
Northmoor
Moreton
Newbridge
Stanton Harcourt
Sutton
Hardwick
Ducklington
Witney
Cogges
High Cogges
South Leigh
Eynsham
Freeland
Barnard Gate
Curbridge
Lew
Yelford
Minster Lovell
Crawley
Hailey
New Yatt
North Leigh
Church Hanborough
Long Hanborough
Combe
Stonesfield
Finstock
Fawler
Ramsden
Leafield
Asthall Leigh
Asthall
Swinbrook
Shipton-under-Wychwood
Milton under Wychwood
Ascott-under-Wychwood
Lyneham
Idbury
Fifield
Chilson
Charlbury
Spelsbury
Chadlington
Taston
Glympton
Kiddington
Wootton
Shorthampton
Walcot
Langley
Fordwells
Whiteoak Green
Wychwood
Blenheim Palace
River Windrush
River Evenlode
Swinbrook
Upper Milton
Westcote
Church Westcote
Nether Westcote
Foscot
Bould
Kingham
Cornbury Park
A40
A361
A415
A417
A420
A4095
A44
B4437
B4449
B4020
B4477
B4508
B4019
B4000
B4047
B4022
Thames Path
Severnhampton
Sevenhampton
3 miles
5 km

BUCKINGHAMSHIRE
OXFORDSHIRE
Oxford
Abingdon
Didcot
Wallingford
Thame
Kidlington
Woodstock
Headington
Cowley
Wheatley
Tackley
Kirtlington
Bletchingdon
Weston-on-the-Green
Chesterton
Wendlebury
Ambrosden
Blackthorn
Merton
Lower Arncott
Upper Arncott
Piddington
Ludgershall
Grendon Underwood
Kingswood
Wotton Underwood
Murcott
Fencott
Charlton on Otmoor
Oddington
Islip
Noke
Boarstall
Brill
Dorton
Ashendon
Horton-cum-Studley
Oakley
Little London
Chilton
Woodeaton
Beckley
Elsfield
Stanton St John
Forest Hill
Holton
Waterperry
Ickford
Worminghall
Shabbington
Long Crendon
Easington
Yarnton
Begbroke
Bladon
Thrupp
Hampton Poyle
Water Eaton
Cutteslowe
Wolvercote
Wytham
Summertown
Marston
Barton
Botley
Cumnor
Chawley
North Hinksey
South Hinksey
New Hinksey
Iffley
Rose Hill
Littlemore
Kennington
Horspath
Garsington
Cuddesdon
Denton
Great Milton
Little Milton
Tiddington
Great Haseley
Little Haseley
Tetsworth
Sandford on Thames
Toot Baldon
Marsh Baldon
Nuneham Courtenay
Radley
Wootton
Sunningwell
Boars Hill
Chislehampton
Stadhampton
Drayton St Leonard
Berinsfield
Dorchester
Clifton Hampden
Culham
Sutton Courtenay
Drayton
Steventon
Milton
Long Wittenham
Little Wittenham
Shillingford
Warborough
Benson
Ewelme
Newington
Chalgrove
Watlington
Cuxham
Pyrton
Shirburn
Stoke Talmage
Adwell
Postcombe
Brightwell Baldwin
Britwell Salome
Crowmarsh Gifford
Cholsey
South Moreton
North Moreton
East Hagbourne
Harwell
Rowstock
East Hendred
Brightwell cum Sotwell
Marcham
Caldecott
Shippon
Cothill
Appleford
Kassam Stadium
Oxford Services
M40
A34
A40
A41
A420
A415
A4074
A329
B4011
B4027

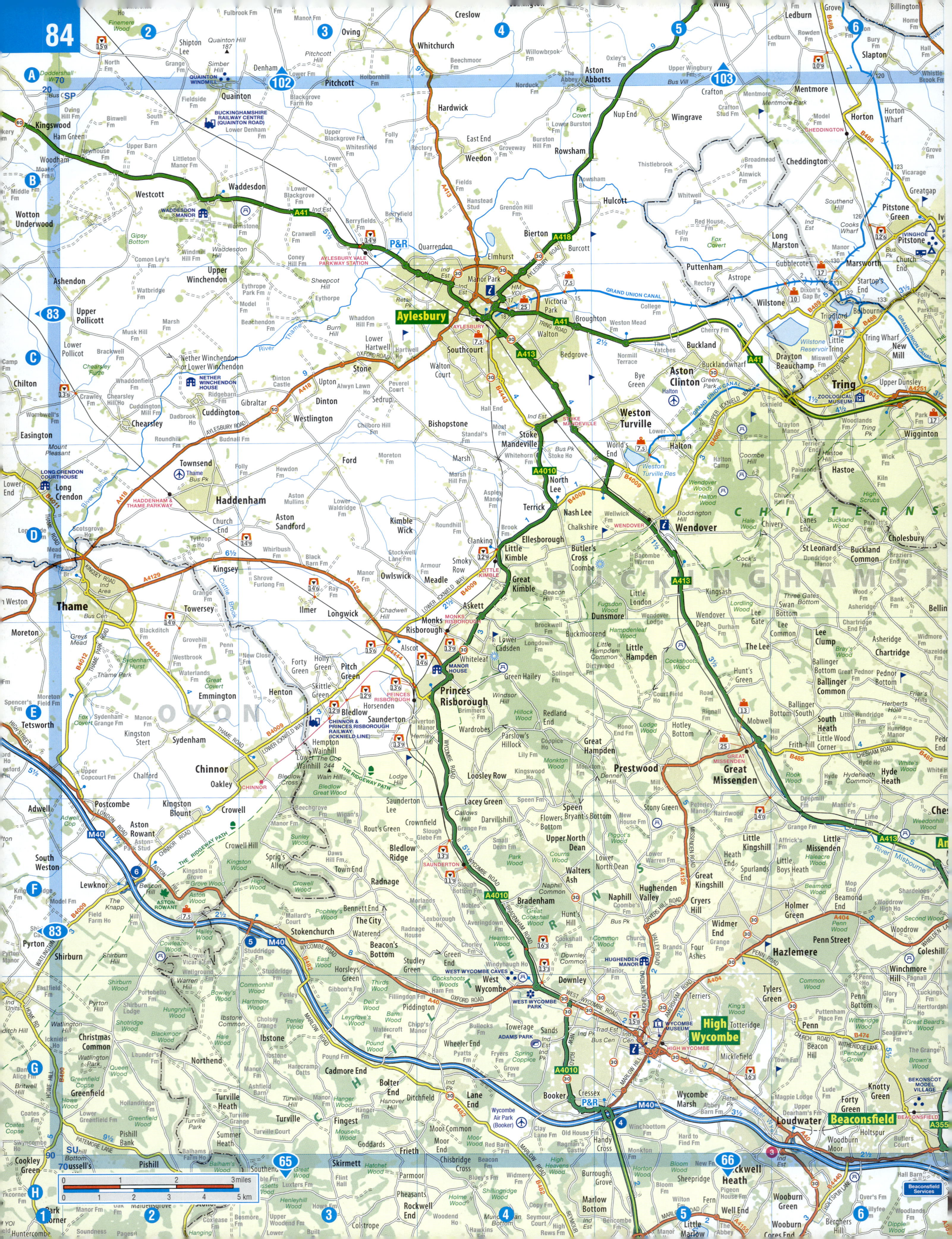

Creslow
Whitchurch
Oving
Quainton
Pitchcott
Hardwick
Aston Abbotts
Wingrave
Ledburn
Mentmore
Slapton
Horton
Cheddington
Kingswood
Westcott
Waddesdon
Wotton Underwood
Weedon
Rowsham
Hulcott
Bierton
Quarrendon
Aylesbury
Upper Winchendon
Ashendon
Upper Pollicott
Lower Pollicott
Chilton
Easington
Long Crendon
Stone
Dinton
Cuddington
Chearsley
Gibraltar
Westlington
Southcourt
Bishopstone
Bedgrove
Broughton
Buckland
Aston Clinton
Tring
Long Marston
Puttenham
Wilstone
Marsworth
Pitstone
Weston Turville
Halton
Stoke Mandeville
Haddenham
Townsend
Ford
Aston Sandford
Kimble Wick
Terrick
Nash Lee
Wendover
Ellesborough
Little Kimble
Great Kimble
Butler's Cross
Chalkshire
Dunsmore
Little Hampden
The Lee
Lee Clump
Chartridge
Cholesbury
Hastoe
Wigginton
Thame
Kingsey
Towersey
Owlswick
Meadle
Askett
Longwick
Ilmer
Monks Risborough
Princes Risborough
Whiteleaf
Alscot
Pitch Green
Forty Green
Henton
Emmington
Bledlow
Saunderton
Horsenden
Sydenham
Chinnor
Oakley
Kingston Blount
Crowell
Tetsworth
Postcombe
Aston Rowant
Adwell
South Weston
Lewknor
Pyrton
Shirburn
Stokenchurch
Radnage
The City
Bledlow Ridge
Lacey Green
Loosley Row
Speen
Great Hampden
Prestwood
Great Missenden
Little Missenden
Little Kingshill
Great Kingshill
Hughenden Valley
Naphill
Walters Ash
Bradenham
Downley
West Wycombe
High Wycombe
Hazlemere
Penn
Tylers Green
Beaconsfield
Loudwater
Wooburn Green
Booker
Lane End
Frieth
Fingest
Turville
Ibstone
Northend
Christmas Common
Cadmore End
Piddington
Wheeler End
Skirmett
Hambleden
Marlow Bottom
Flackwell Heath
Holmer Green
Penn Street
Winchmore Hill
Knotty Green
Forty Green
Chiltern Hills
BUCKINGHAMSHIRE
OXON
CHILTERNS
M40
A41
A418
A413
A4010
A40
A4129
B4009

Luton
Dunstable
Harpenden
Wheathampstead
Kimpton
Redbourn
St Albans
Hemel Hempstead
Berkhamsted
Kings Langley
Abbots Langley
Bricket Wood
London Colney
Radlett
Shenley
Borehamwood
Elstree
Watford
Bushey
Oxhey
South Oxhey
Rickmansworth
Chorleywood
Chesham
Amersham
Little Chalfont
Chalfont St Giles
Chalfont St Peter
Northwood
Stanmore
Edgware
Harrow
Wealdstone
Pinner
Ruislip
Denham
Gerrards Cross
Markyate
Flamstead
Studham
Whipsnade
Kensworth
Caddington
Dagnall
Aldbury
Tring
Great Gaddesden
Nettleden
Bovingdon
Sarratt
Flaunden
Chenies
Latimer
Boxmoor
Apsley
Leverstock Green
Sandridge
Colney Heath
Kinsbourne Green
Batford
Ayot St Lawrence
Chilterns
Hertfordshire
London Gateway Services
M1
M25
A41
A5
A414
A405

Knebworth
Codicote
Welwyn
Ayot St Lawrence
Ayot St Peter
Digswell Park
Oaklands
Burnham Green
Datchworth
Woolmer Green
Bragbury End
Watton at Stone
Whempstead
Dane End
Sacombe
High Cross
Wadesmill
Thundridge
Tonwell
Stapleford
Chapmore End
Waterford
Ware
Latchford
Barwick
Much Hadham
Hadham Cross
Hadham Ford
Wareside
Widford
Hunsdon
Welwyn Garden City
Panshanger
Hertford
Hertingfordbury
Cole Green
Letty Green
Hertford Heath
Great Amwell
Stanstead Abbotts
St Margarets
Eastwick
Hailey
Rye Park
Roydon
Hatfield
Hatfield Garden Village
Stanborough
Lemsford
Essendon
Little Berkhamsted
Bayford
Brickendon
Hoddesdon
Broxbourne
Wormley
Wormley West End
Turnford
Lee Valley Regional Park
Lower Nazeing
Nazeing
Roydon Hamlet
Broadley Common
Welham Green
Brookmans Park
Newgate Street
Cuffley
Goff's Oak
Cheshunt
Flamstead End
Northaw
Potters Bar
South Mimms
Shenley
Shenleybury
Waltham Cross
Waltham Abbey
Upshire
Bulls Cross
Enfield
Enfield Wash
Forty Hill
Crews Hill
Botany Bay
Borehamwood
Barnet
East Barnet
New Barnet
Cockfosters
Hadley
Monken Hadley
Arkley
Totteridge
Southgate
Winchmore Hill
Edmonton
Ponders End
Chingford
Buckhurst Hill
Loughton
Sewardstone
High Beach
Friern Barnet
Finchley
Whetstone
Mill Hill
Edgware
Stanmore
Kingsbury
Hendon
Wood Green
Hornsey
Tottenham
Walthamstow
Woodford
Woodford Bridge
Highams Park
HERTFORDSHIRE
LONDON
A1(M)
A10
A414
A602
M25
M11
Scale
0 1 2 3 miles
0 1 2 3 4 5 km

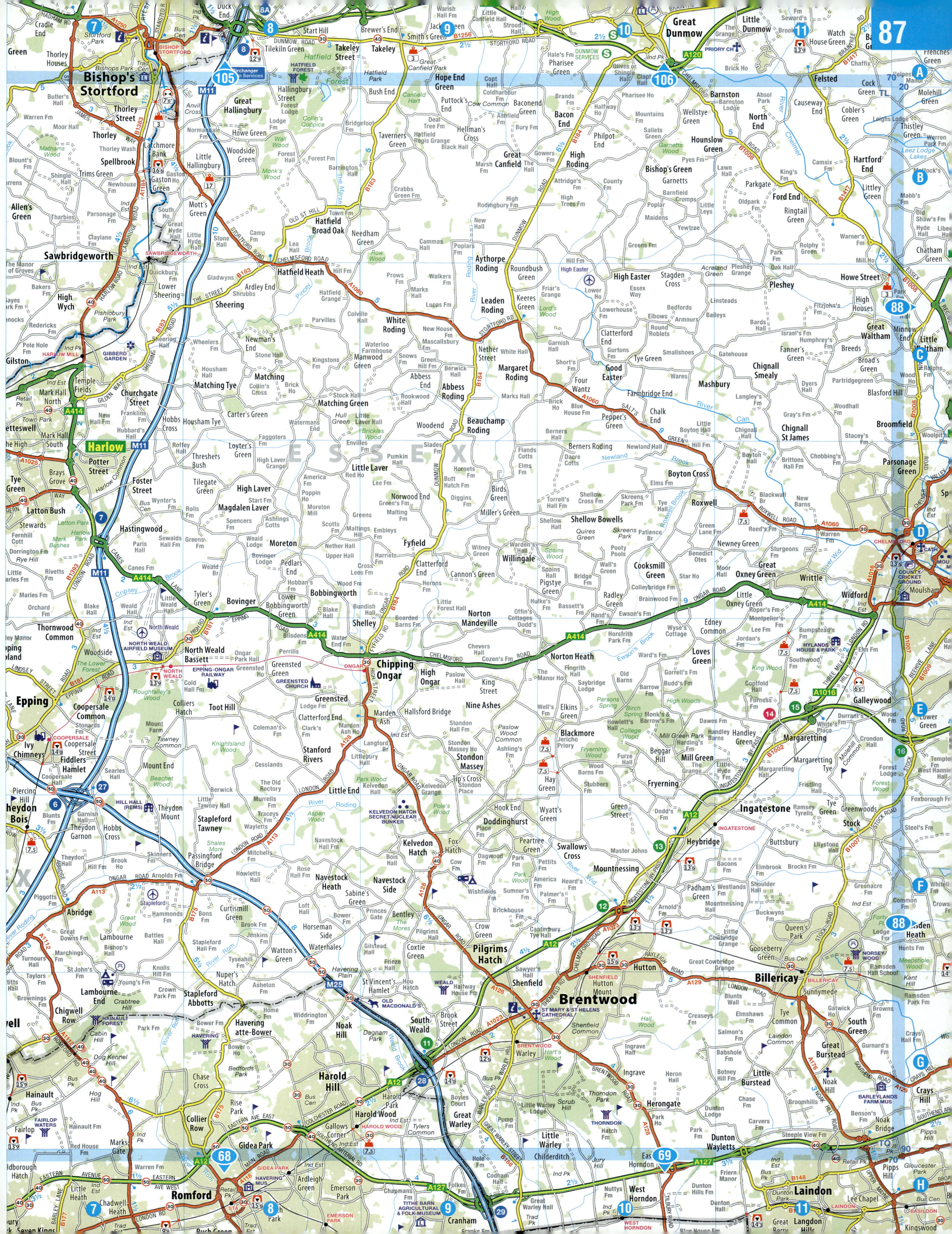
Bishop's Stortford
Great Dunmow
Little Dunmow
Felsted
Takeley
Hatfield Forest
Great Hallingbury
Hatfield Broad Oak
Sawbridgeworth
Hatfield Heath
Sheering
High Wych
Harlow
Great Canfield
High Roding
Aythorpe Roding
Leaden Roding
White Roding
Margaret Roding
Good Easter
High Easter
Pleshey
Barnston
Great Waltham
Broomfield
Chignall St James
Roxwell
Writtle
Chelmsford
Matching
Matching Green
Matching Tye
Abbess Roding
Beauchamp Roding
E S S E X
High Laver
Little Laver
Magdalen Laver
Moreton
Fyfield
Willingale
Shellow Bowells
Boyton Cross
Bobbingworth
North Weald Bassett
Chipping Ongar
High Ongar
Norton Mandeville
Norton Heath
Epping
Coopersale
Theydon Bois
Theydon Mount
Toot Hill
Stanford Rivers
Greensted
Blackmore
Ingatestone
Mill Green
Fryerning
Margaretting
Stondon Massey
Kelvedon Hatch
Doddinghurst
Navestock Heath
Navestock Side
Stapleford Tawney
Stapleford Abbotts
Abridge
Lambourne End
Chigwell Row
Havering-atte-Bower
Pilgrims Hatch
Brentwood
Shenfield
Hutton
Billericay
Noak Hill
Harold Hill
Harold Wood
Gidea Park
Romford
Great Warley
Little Warley
Herongate
Ingrave
Dunton Wayletts
Laindon
West Horndon
Cranham
Hainault
Chadwell Heath
M11
M25
A414
A12
A127
A1060
A120
Dunmow Services
105
106
69
68
88

Great Notley
Silver End
Cressing
Kelvedon
Feering
Coggeshall Hamlet
Tiptree
White Notley
Faulkbourne
Fairstead
Terling
Witham
Chipping Hill
Rivenhall End
Little Braxted
Great Braxted
Great Totham
Little Totham
Tolleshunt Major
Tolleshunt D'Arcy
Tolleshunt Knights
Wickham Bishops
Hatfield Peverel
Nounsley
Boreham
Broomfield
Little Waltham
Great Waltham
Great Leighs
Little Leighs
Springfield
Chelmer Village
Chelmsford
Great Baddow
Sandon
Danbury
Little Baddow
Woodham Walter
Woodham Mortimer
Langford
Heybridge
Heybridge Basin
Maldon
Goldhanger
Osea Island
Northey Island
Mundon
Maylandsea
Mayland
Steeple
Latchingdon
Purleigh
Cold Norton
Stow Maries
Woodham Ferrers
Bicknacre
East Hanningfield
West Hanningfield
South Hanningfield
Hanningfield Reservoir
Rettendon
South Woodham Ferrers
North Fambridge
South Fambridge
Hullbridge
Battlesbridge
Runwell
Wickford
Shotgate
Rawreth
Rayleigh
Hockley
Hawkwell
Ashingdon
Canewdon
Rochford
Creeksea
Burnham-on-Crouch
Althorne
Ostend
Paglesham Churchend
Paglesham Eastend
Great Stambridge
Stroud Green
Eastwood
Thundersley
North Benfleet
Pitsea
Nevendon
Crays Hill
Ramsden Heath
Ramsden Bellhouse
Downham
Stock
Galleywood
Widford
Moulsham
Howe Green
Rudley Green
Hazeleigh
Cock Clarks
Runsell Green
ESSEX
A12
A130
A131
A132
A127
A129
A414
B1018
B1010
B1012
0 1 2 3 miles
0 1 2 3 4 5 km

Colchester Zoo
Wivenhoe
Rowhedge
Alresford
Great Bentley
Weeley
Weeley Heath
Thorpe-le-Soken
Thorpe Hall
Thorpe Park
Kirby Cross
Great Holland
Holland-on-Sea
Frinton Road
Malting Green
Layer de la Haye
Layer Hall
Abberton Reservoir
Abberton
Langenhoe
Fingringhoe
South Green
Fingringhoe Marsh
Fingringhoe Ranges
Geedon Saltings
Aldboro Pt
Moverons
Thorrington
Alresford Hall
Thors Fm
Plumpton's Fm
Gatehouse Fm
Lowermarsh Fm
Greatmarsh Fm
Aingers Green
South Heath
Hollybush Hill
Row Heath
Little Clacton
Cook's Green
Clacton Village
Burrsville Park
Bovill's Hall
St Osyth Heath
Brightlingsea
Hurst Green
Lower Fm
Lamb Fm
Wellwick Fm
Park Fm
Earls Hall
Elm Fm
Bocking's Elm
Great Clacton
Clacton-on-Sea
Rush Green
Ind Est
St Osyth Priory
St Osyth
Mill Dam Lake
St Osyth Lodge Fm
Dalites Fm
Rouses Fm
Clacton
Jaywick
Point Clear
Sandy Pt
St Clere's Hall
Whyers Hall Fm
Cockett Wick Fm
Lee Wick Fm
Wigboro Wick Fm
Seawick
St Osyth Marsh
Lee-over-Sands
St Osyth Beach
Colne Bar
Colne Pt
Brightlingsea Reach
St Osyth Stone Pt
Mersea Stone
Westmarsh Pt
River Colne
Pyefleet Channel
Langenhoe Marsh
Reeveshall Marsh
Pewit Island
Rat Island
Maydays Marsh
Maydays Fm
Mersea Island
East Mersea
Broman's Fm
Cudmore Grove
Colne Estuary
Mersea Flats
Blue Row
Rewsalls Fm
Waldegraves Fm
The Cross
West Mersea
Barrow Hill
Bocking Hall
Ray Island
Strood Channel
The Strood
Peldon
Peldon Lodge
Haxells Fm
Pete Hall
Kemps Fm
Moor Fm
Stafford's Corner
Little Wigborough
Great Wigborough
Copthall Grove
Sampson's Fm
New Hall
Abbotts Hall
Copt Hall Marshes Nature Reserve
Abbot's Hall Saltings
Salcott Channel
Feldy Marshes
Old Hall Marshes
Blackwater Estuary
Pennyhole Fleet
Mersea Fleet
Besom Fleet
Cobmarsh Island
Virley Channel
Old Hall Creek
North Channel
South Channel
Great Cob Island
Tollesbury
Woodrolfe Creek
Tollesbury Wick Marshes
Shinglehead Pt
The Nass
Bohuns Hall
Mell Fm
Decoy Fm
Mill Creek
Mill Pt
Pewet Island
River Blackwater
St Lawrence Bay
Bradwell Creek
The Stone
Ramsey Island
Bradwell Waterside
East End
Bradwell Lodge
Bradwell on Sea
Bradwell Marshes
Sales Pt
Tip Head
St Peter's on the Wall
St Peter's Flat
East Hall Fm
Eastlands
Munkins Fm
Sandbeach
Marshhouse Outfall
Dengie Flat
Dengie
Marshhouse Decoy Pond
Bradwell Brook
Bradwell Hall
Bradwell Wick
Beacon Hill Fm
Mott's Fm
St Lawrence
Stows Fm
Tillingham
Tillingham Marshes
Black House Court Fm
Brick House Fm
Hill Fm
Howe Fm
Glebe Fm
Grange Fm
Grange Outfall
Moynes Fm
Badnocks Fm
Asheldham
Dengie
Manor
Landwick Fm
East Ware Fm
Round Barn
New Hall Fm
Asheldham Brook
Northend
North Wycke
Brook Fm
Southminster
Ray Sand
Wraywick Fm
Court Fm
Turncole Fm
Goldsand Br
Old Turncole
Newman's Fm
Twizzlefoot Br
Montsale
Deal Hall
Brook Fm
Dammer Wick
West Wick
East Wick
Coney Hall
Redward
Holliwell Fm
Holliwell Pt
Burnham-on-Crouch
Burnham Wick
Foulness Sands
Foulness Pt
Ringwood Bar
River Crouch
Wallasea Ness
Crouch Corner
East Newlands
Wallasea Wetlands
Wallasea Island
Brankfleet
Nase Wick
The Quay
Churchend
Lodge Fm
Courtsend
Fishermans Head
River Roach
Quay Reach
Foulness Island
East Wick
Horseshoe Corner
Rugwood Fm
Eastwick Head
Great Burwood Fm
Rugwood Head
Potton Island
Shelford Creek
Asplins Head
New England Island
New England Creek
Shelford Head
Rushley Island
Havengore Island
Sharpness Head
Havengore Head
Oxenham Fm
Haven Pt
Maplin Sands
TM
TR
A133
B1027
B1025
B1021
B1029
B1414
B1441
B1442
B1032
B1033
A
B
C
D
E
F
G
H
7
8
9
10
11
70
71
107
108

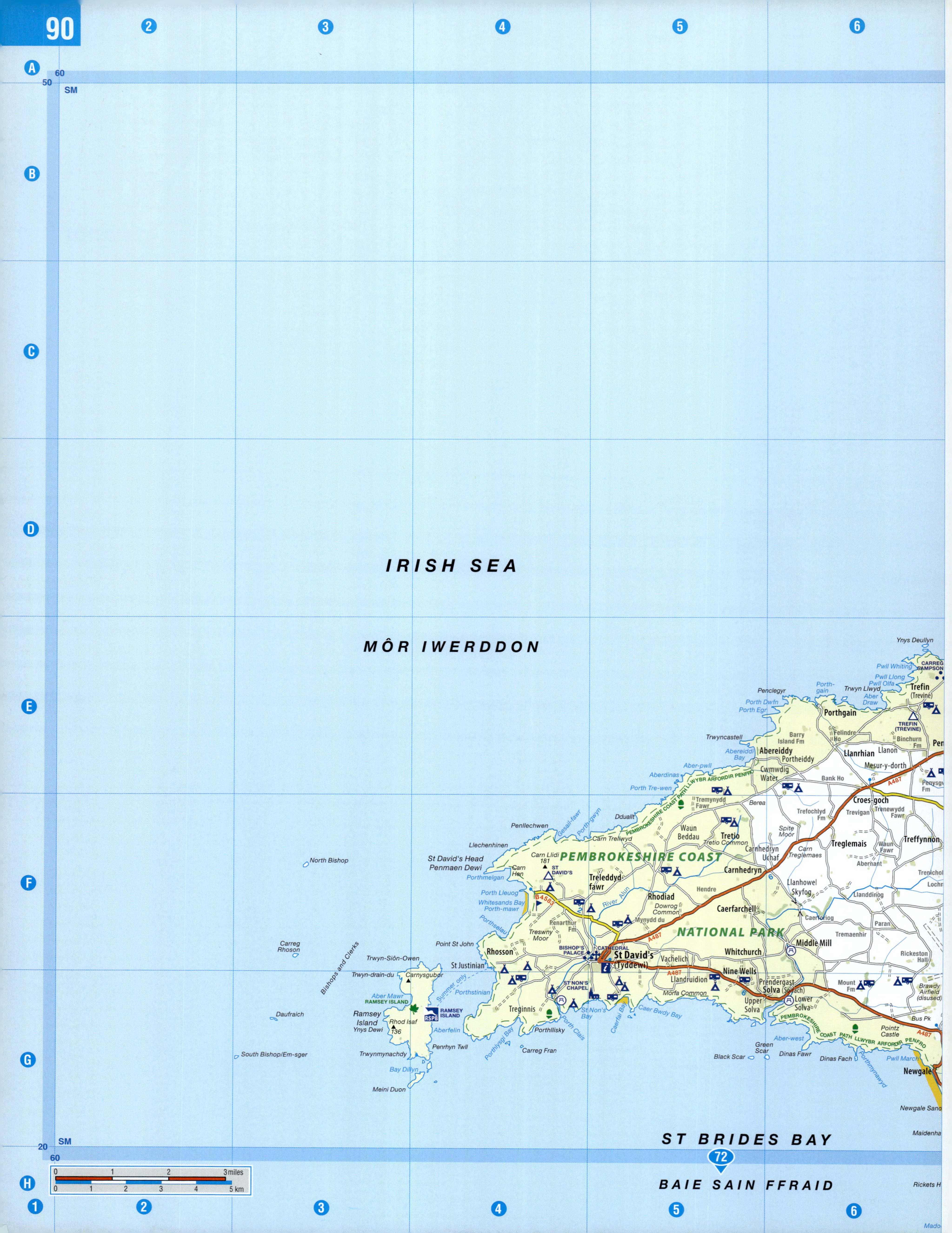

IRISH SEA
MÔR IWERDDON
ST BRIDES BAY
BAIE SAIN FFRAID
PEMBROKESHIRE COAST
NATIONAL PARK
PEMBROKESHIRE COAST PATH LLWYBR ARFORDIR PENFRO
St David's (Tyddewi)
St David's Head
Penmaen Dewi
Ramsey Island
Ynys Dewi
RAMSEY ISLAND
North Bishop
South Bishop/Em-sger
Bishops and Clerks
Carreg Rhoson
Daufraich
Meini Duon
Bay Diliyn
Trwynmynachdy
Penrhyn Twll
Aberfelin
Rhod Isaf
136
Carnysgubor
Trwyn-drain-du
Trwyn-Siôn-Owen
Aber Mawr
St Justinian
Porthstinian
Point St John
Rhosson
Treginnis
Porthlisky
Carreg Fran
Porthlysgi Bay
St Non's Bay
Porth Clais
Caerfai Bay
Caer Bwdy Bay
ST NON'S CHAPEL
BISHOP'S PALACE
CATHEDRAL
Treswny Moor
Penarthur Fm
Porthselau
Whitesands Bay
Porth-mawr
Porth Lleuog
Porthmelgan
Carn Hen
Carn Llidi
181
Llechenhinen
Penllechwen
Gesail-fawr
Porth-gwyn
Dduallt
Carn Trelwyd
Treleddyd-fawr
River Alun
Rhodiad
Dowrog Common
Mynydd du
Vachelich
Llandruidion
Morfa Common
Nine Wells
Whitchurch
Caerfarchell
Hendre
Carnhedryn
Carnhedryn Uchaf
Waun Beddau
Tretio
Tretio Common
Tremynydd Fawr
Porth Tre-wen
Aberdinas
Aber-pwll
Abereiddi Bay
Abereiddy
Barry Island Fm
Porthiddy
Cwmwdig Water
Trwyncastell
Penclegyr
Porth Dwfn
Porth Egr
Porthgain
Porth-gain
Trwyn Llwyd
Aber Draw
Pwll Olfa
Pwll Llong
Pwll Whiting
CARREG SAMPSON
Ynys Deullyn
Trefin (Trevine)
TREFIN (TREVINE)
Binchurn Fm
Felindre Ho
Llanrhian
Llanon
Mesur-y-dorth
Bank Ho
Croes-goch
Trevigan
Trenewydd Fawr
Berea
Trefochlyd Fm
Spite Moor
Carn Treglemaes
Treglemais
Waun Fawr
Abernant
Treffynnon
Llanhowel
Skyfog
Llanddinog
Caerforiog
Paran
Tremaenhir
Middle Mill
Rickeston Hall
Prendergast
Solva (Solfach)
Upper Solva
Lower Solva
Mount Fm
Brawdy Airfield (disused)
Bus Pk
Pointz Castle
Green Scar
Black Scar
Aber-west
Dinas Fawr
Dinas Fach
Porthmynawyd
Pwll March
Newgale
Newgale Sands
Maidenhall
Rickets Hd
A487
B4583
SM
72
0 1 2 3 miles
0 1 2 3 4 5 km

91
CARDIGAN BAY
BAE CEREDIGION
FISHGUARD BAY
BAE ABERGWAUN
PEMBROKESHIRE COAST NATIONAL PARK
PEMBROKESHIRE (SIR BENFRO)
Mynydd Preseli
Strumble Head
Carreg Onnen
Tresinwen
Llanwnda
Cilau
Harbour Village
Goodwick (Wdig)
Dyffryn
Stop-and-Call
Fishguard
Abergwaun
Lower Town
Pen Brush
Trefasser
Penbwchdy
Tregyddulan
St Nicholas
Granston (Treopert)
Abercastle
Manorowen
Scleddau
Mathry
Castlemorris
Letterston (Treletert)
Welsh Hook
Wolf's Castle
Treffgarne
Spittal
Camrose
Simpson Cross
Keeston
Roch
Roch Gate
Nolton Haven
Pelcomb Cross
Tangiers
Rudbaxton
Poyston Cross
Penycwm
Hayscastle
Hayscastle Cross
Mountain Water
Brimaston
Trefgarn Owen
Llandeloy
Llanreithan
Treddiog
Square and Compass
Dinas Head
Dinas Island
Dinas Cross
Brynhenllan
Llanychaer
Newport (Trefdraeth)
Parrog
Nevern
Felindre Farchog
Berry Hill
Cilgwyn
Pontfaen
Brynberian
Tafarn-y-bwlch
Trecwn
Newbridge
Little Newcastle
Puncheston
Castlebythe
Tufton
Henry's Moat
Maenclochog
Rosebush
New Inn
Greenway
Woodstock
Wallis
Rinaston
Ambleston
Triffleton
Walton East
Llys-y-frân
Llys-y-fran Reservoir
Penffordd
Bletherston
Clarbeston
Clarbeston Road
Llanycefn
Gelli
Wiston
Bethesda
A40
A487
B4313
B4329
B4330
B4331
92
72
73

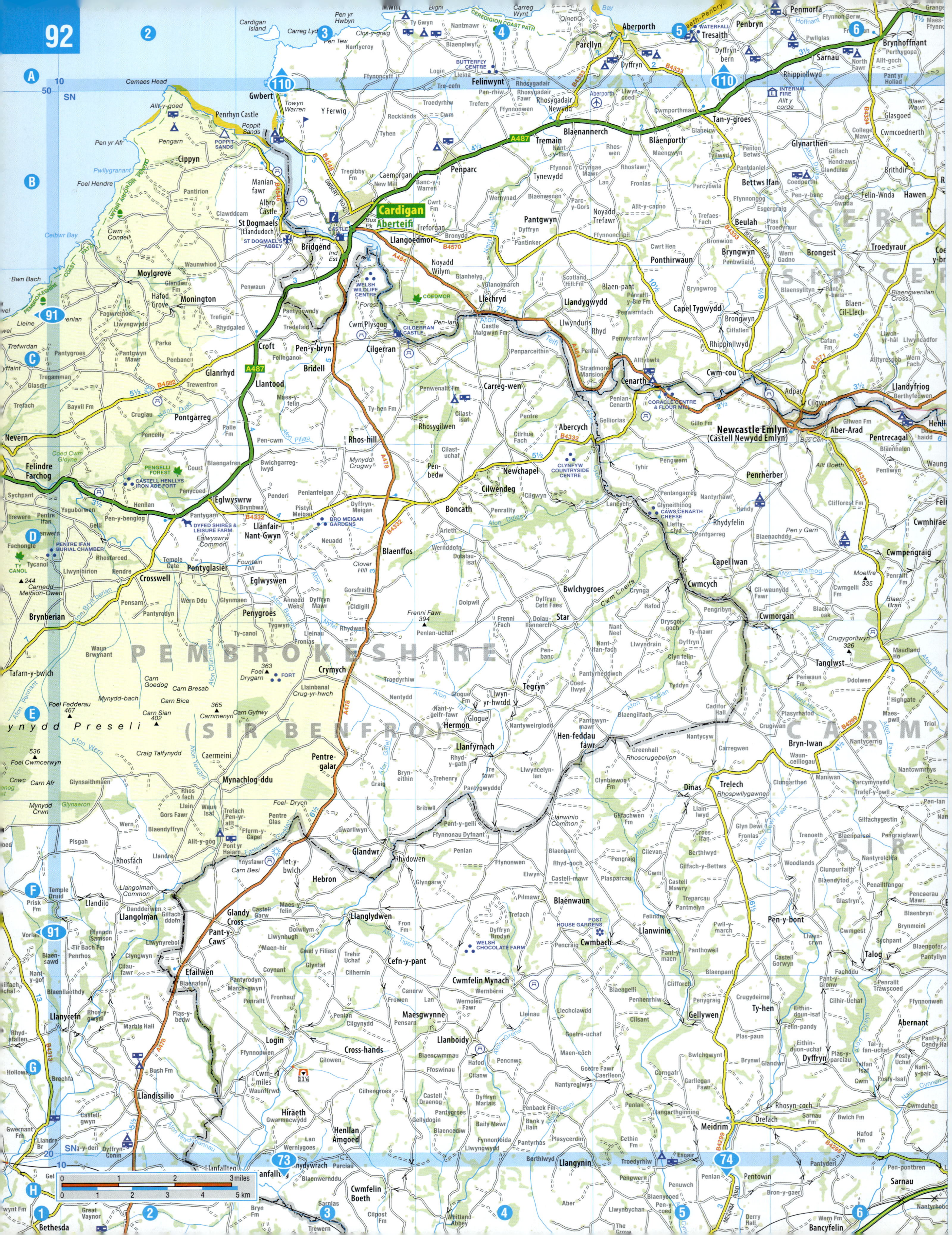
Cardigan Island
Cemaes Head
Gwbert
Penrhyn Castle
Poppit Sands
Pen yr Afr
Pengarn
Cippyn
Pwllgranant
Foel Hendre
Pantirion
Manian fawr
Albro Castle
St Dogmaels (Llandudoch)
St Dogmael's Abbey
Bridgend
Ceibwr Bay
Cwm Connell
Moylgrove
Bwn Bach
Hafod Grove
Monington
Waunwhiod
Penwaun
Croft
Pen-y-bryn
Cilgerran
Cilgerran Castle
Welsh Wildlife Centre
Bridell
Llantood
Glanrhyd
Pontgarreg
Nevern
Felindre Farchog
Castell Henllys Iron Age Fort
Eglwyswrw
Llanfair-Nant-Gwyn
Dyfed Shires & Leisure Farm
Bro Meigan Gardens
Pontyglasier
Crosswell
Brynberian
Eglwyswen
Penygroes
Crymych
Blaenffos
Boncath
Pembrokeshire
(Sir Benfro)
Mynydd Preseli
Foel Cwmcerwyn
Foel Drygarn
Carn Goedog
Carn Sian
Caermeini
Mynachlog-ddu
Pentre-galar
Llanfyrnach
Hermon
Glogue
Tegryn
Hen-feddau fawr
Glandwr
Rhydowen
Hebron
Llangolman
Llandilo
Glandy Cross
Pant-y-Caws
Efailwen
Llanglydwen
Cefn-y-pant
Cwmfelin Mynach
Maesgwynne
Llanboidy
Login
Cross-hands
Llandissilio
Hiraeth
Henllan Amgoed
Cwmfelin Boeth
Llanfallteg
Llangynin
Whitland Abbey
Bethesda
Blaenwaun
Welsh Chocolate Farm
Cwmbach
Llanwinio
Gellywen
Meidrim
Drefach
Trelech
Dinas
Bryn-Iwan
Tanglwst
Cwmorgan
Cwmcych
Capel Iwan
Penrherber
Newcastle Emlyn (Castell Newydd Emlyn)
Aber-Arad
Pentrecagal
Adpar
Cilgwyn
Cenarth
Coracle Centre & Flour Mill
Abercych
Clynfyw Countryside Centre
Newchapel
Cilwendeg
Bwlchygroes
Star
Carreg-wen
Rhos-hill
Llechryd
Cardigan
Aberteifi
Llangoedmor
Penparc
Felinwynt
Butterfly Centre
Ceredigion Coast Path
Aberporth
Parcllyn
Tresaith
Penbryn
Penmorfa
Sarnau
Brynhoffnant
Tan-y-groes
Blaenporth
Blaenannerch
Tremain
Rhosygadair Newydd
Glynarthen
Bettws Ifan
Beulah
Brongest
Troedyraur
Bryngwyn
Ponthirwaun
Pantgwyn
Capel Tygwydd
Llandygwydd
Rhippinllwyd
Cwm-cou
Llandyfriog
Ceredigion
Carmarthenshire
Cwmhiraeth
Cwmpengraig
Pen-y-bont
Talog
Ty-hen
Abernant
Dyffryn
Sarnau
Bancyfelin
Pentowin
Esgair
1 2 3 miles
1 2 3 4 5 km

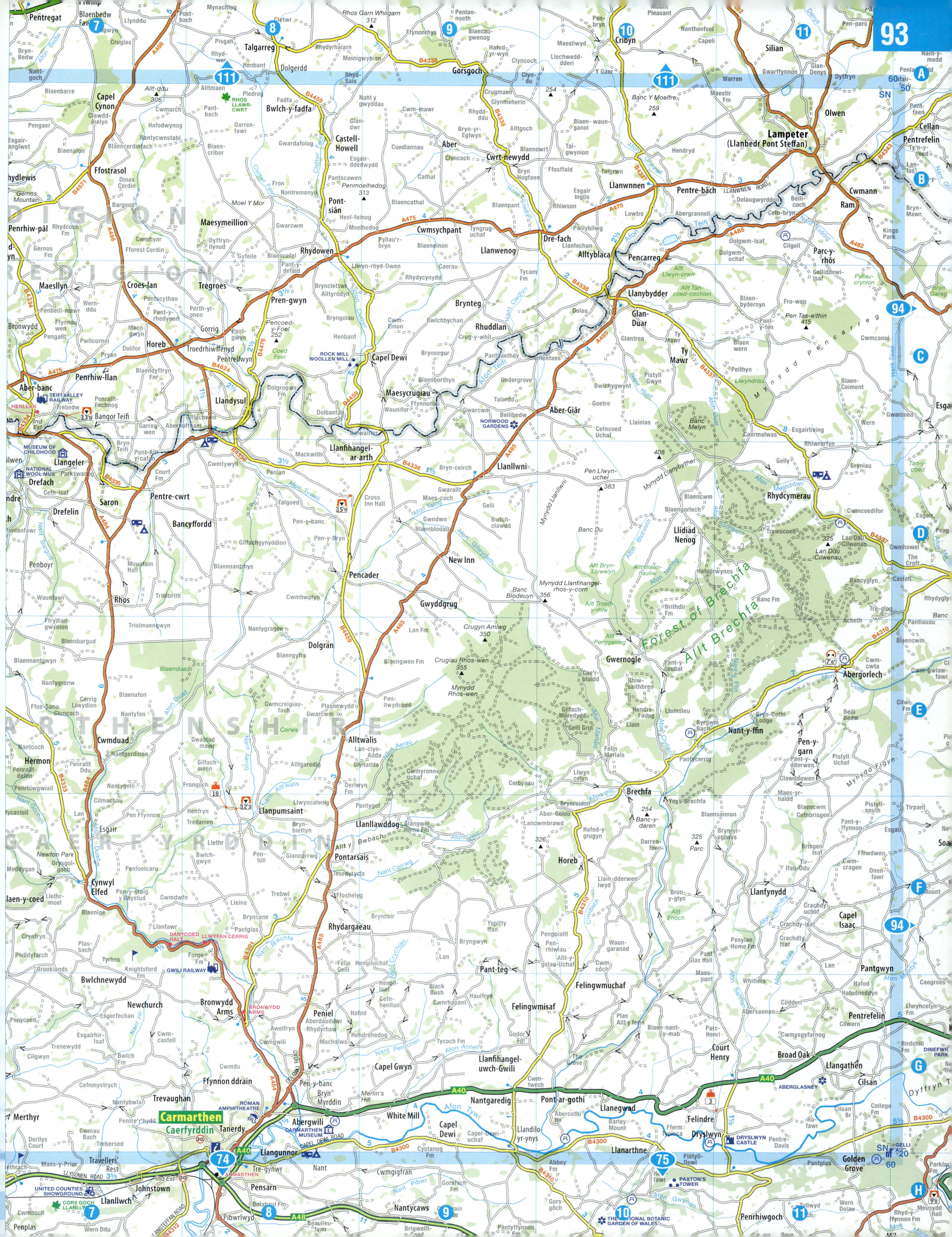
Pentregat
Blaenbedw Fawr
Talgarreg
Dolgerdd
Gorsgoch
Cribyn
Silian
Lampeter
(Llanbedr Pont Steffan)
Olwen
Cellan
Pentrefelin
Capel Cynon
Bwlch-y-fadfa
Castell-Howell
Aber
Cwrt-newydd
Ffostrasol
Llanwnnen
Pentre-bach
Cwmann
Ram
Pont-siân
Maesymeillion
Penrhiw-pâl
Rhydowen
Cwmsychpant
Llanwenog
Dre-fach
Alltyblaca
Pencarreg
Parc-y-rhôs
Maesllyn
Croes-lan
Tregroes
Pren-gwyn
Llanybydder
Glan-Duar
Brynteg
Rhuddlan
Horeb
Troedrhiwffenyd
Capel Dewi
Ty Mawr
Aber-banc
Penrhiw-llan
Llandysul
Maesycrugiau
Aber-Giâr
Bangor Teifi
Llangeler
Drefach
Llanfihangel-ar-arth
Llanllwni
Pentre-cwrt
Saron
Bancyffordd
Rhydcymerau
Drefelin
New Inn
Llidiad Nenog
Pencader
Rhos
Gwyddgrug
Forest of Brechfa
Allt Brechfa
Dolgran
Gwernogle
Abergorlech
Alltwalis
Cwmduad
Hermon
Nant-y-ffin
Llanpumsaint
Brechfa
Llanllawddog
Pontarsais
Horeb
Cynwyl Elfed
Llanfynydd
Rhydargaeau
Capel Isaac
Pant-teg
Bwlchnewydd
Felingwmuchaf
Pantgwyn
Newchurch
Bronwydd Arms
Peniel
Felingwmisaf
Pentrefelin
Court Henry
Broad Oak
Capel Gwyn
Llanfihangel-uwch-Gwili
Llangathen
Cilsan
Ffynnon ddrain
Trevaughan
Carmarthen
Caerfyrddin
Merthyr
Tanerdy
Abergwili
White Mill
Nantgaredig
Pont-ar-gothi
Llanegwad
Felindre
Dryslwyn
Capel Dewi
Llanarthne
Llangunnor
Golden Grove
Travellers' Rest
Johnstown
Llanllwch
Pensarn
Nantycaws
Penrhiwgoch
Pibwrlwyd
CARMARTHENSHIRE
SIR GAERFYRDDIN
CEREDIGION
Mynydd Pencarreg
GWILI RAILWAY
TEIFI VALLEY RAILWAY
NATIONAL WOOL MUS
MUSEUM OF CHILDHOOD
ROCK MILL WOOLLEN MILL
NORWOOD GARDENS
DRYSLWYN CASTLE
PAXTON'S TOWER
THE NATIONAL BOTANIC GARDEN OF WALES
ABERGLASNEY
CARMARTHEN MUSEUM
UNITED COUNTIES SHOWGROUND
CORS GOCH LLANLLWCH
A40
A48
A475
A485
A484
B4459
B4338
B4336
B4310
B4300

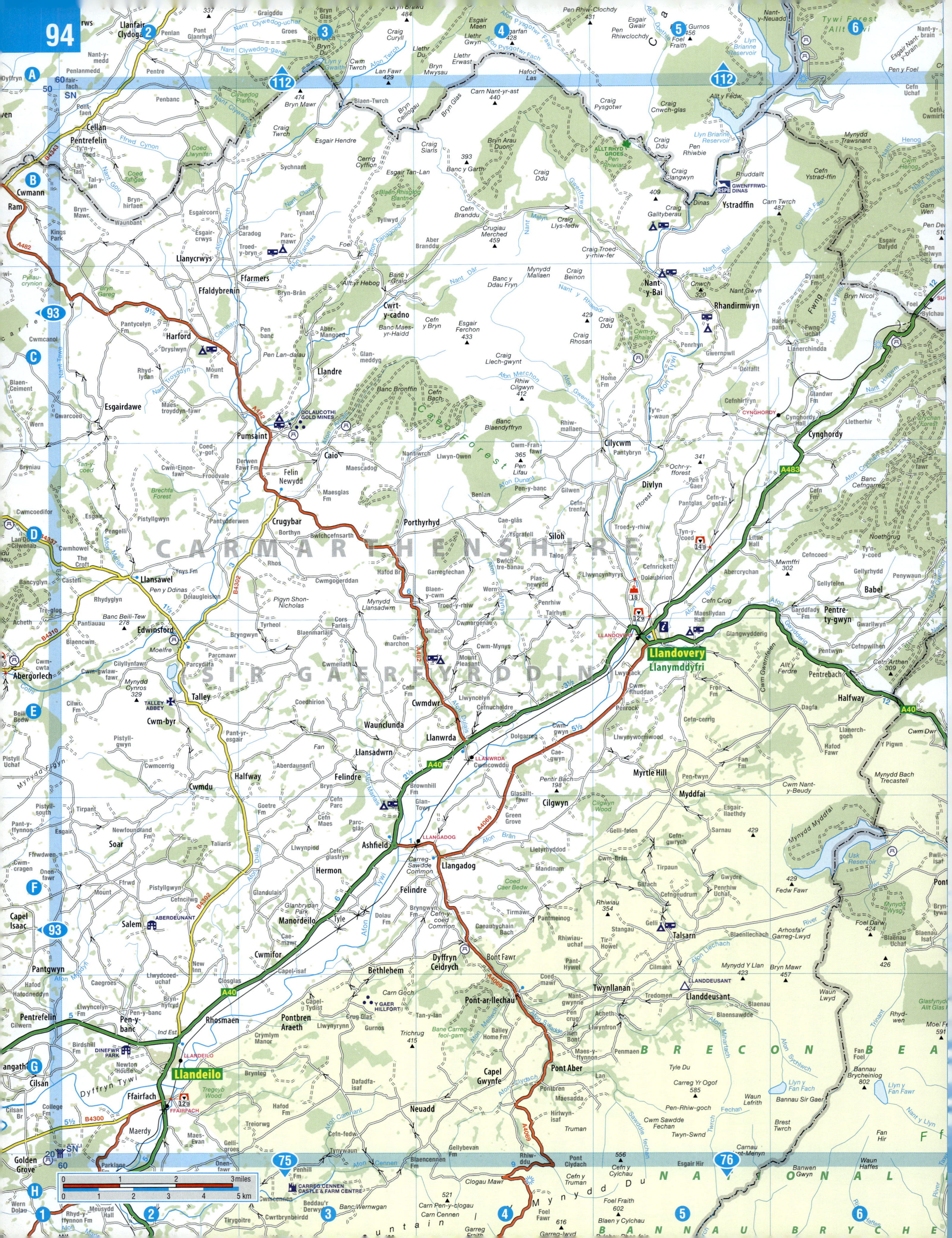

CARMARTHENSHIRE
SIR GAERFYRDDIN
BRECON BEACONS
NATIONAL
BANNAU BRYCHEINIOG
Llandovery
Llanymddyfri
Llandeilo
Llangadog
Llanwrda
Pumsaint
Caio
Cilycwm
Rhandirmwyn
Ystradffin
Nant-y-Bai
Cynghordy
Halfway
Myddfai
Myrtle Hill
Talsarn
Llanddeusant
Twynllanan
Pont-ar-llechau
Capel Gwynfe
Pont Aber
Bethlehem
Ffairfach
Maerdy
Salem
Talley
Edwinsford
Llansawel
Crugybar
Porthyrhyd
Harford
Llanycrwys
Ffarmers
Ffaldybrenin
Llandre
Esgairdawe
Abergorlech
Cwm-byr
Halfway
Cwmdu
Soar
Manordeilo
Cwmifor
Rhosmaen
Pentrefelin
Cilsan
Pantgwyn
Capel Isaac
Llansadwrn
Wauncluda
Llanwrda
Felindre
Ashfield
Hermon
Cwmdwr
Cilgwyn
Llanfair Clydogau
Cwmann
Cellan
Pentrefelin
Cwrt-y-cadno
Siloh
Divlyn
Pentre-ty-gwyn
Babel
Caeo Forest
Llyn Brianne Reservoir
Usk Reservoir
DOLAUCOTHI GOLD MINES
TALLEY ABBEY
DINEFWR PARK
ABERDEUNANT
CARREG CENNEN CASTLE & FARM CENTRE
Y GAER HILLFORT
GWENFFRWD-DINAS
A40
A482
A483
A4069
B4302
B4337
B4310
B4343
B4300
93
75
76
112
A
B
C
D
E
F
G
H
1
2
3
4
5
6
0 1 2 3miles
0 1 2 3 4 5 km

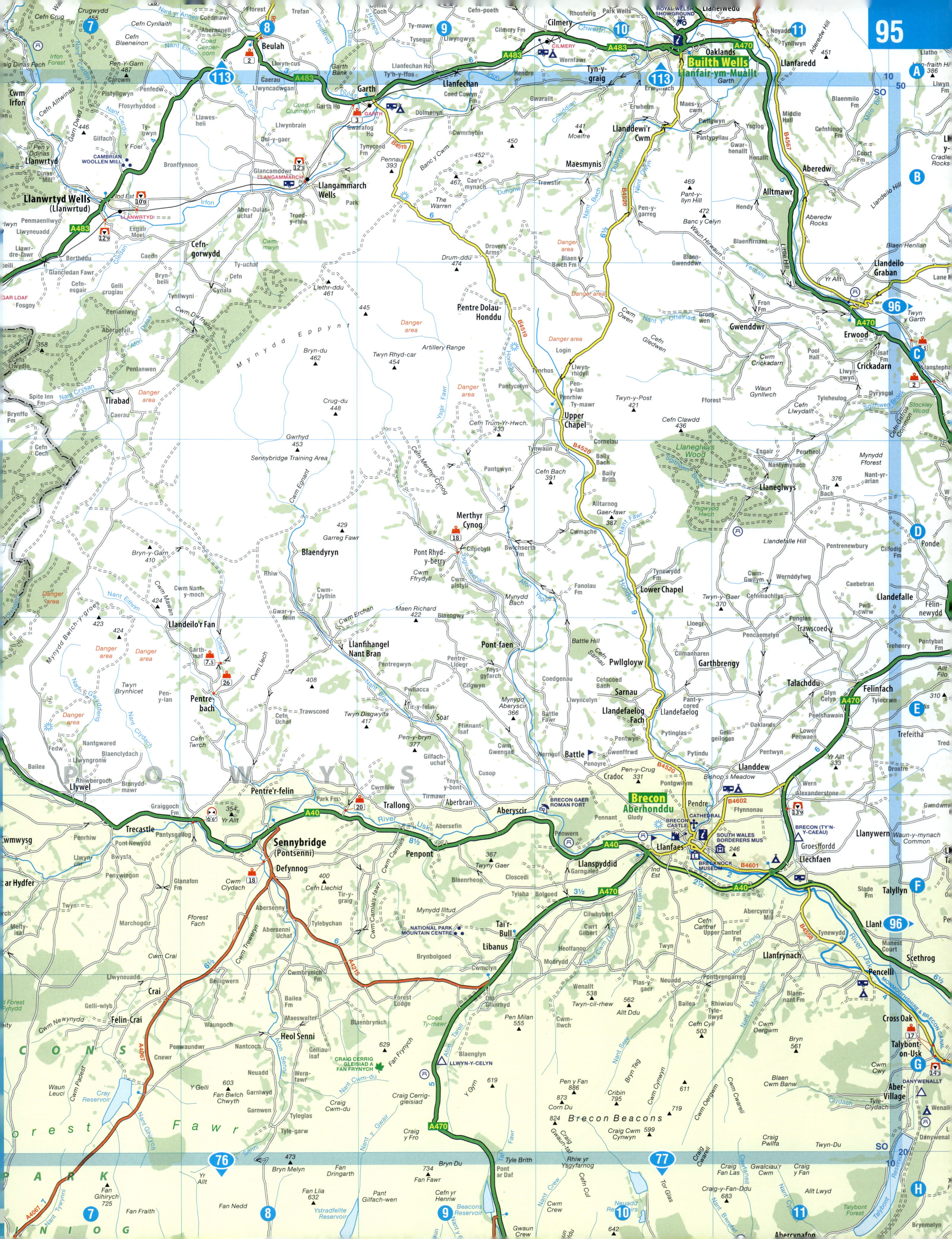
Builth Wells
Llanfair-ym-Muallt
Brecon
Aberhonddu
Llanwrtyd Wells
(Llanwrtud)
Sennybridge
(Pontsenni)
Beulah
Garth
Llangammarch Wells
Llanfechan
Cilmery
Tyn-y-graig
Oaklands
Llanfaredd
Aberedw
Alltmawr
Erwood
Crickadarn
Gwenddwr
Llandeilo Graban
Maesmynis
Llanddewi'r Cwm
Upper Chapel
Lower Chapel
Merthyr Cynog
Pentre Dolau-Honddu
Tirabad
Llandeilo'r Fan
Llanfihangel Nant Bran
Pentre bach
Pont-faen
Pwllgloyw
Sarnau
Llandefaelog Fach
Garthbrengy
Llaneglwys
Llandefalle
Talachddu
Felinfach
Llanddew
Battle
Cradoc
Aberyscir
Trallong
Penpont
Pentre'r-felin
Trecastle
Llywel
Defynnog
Llanspyddid
Llanfaes
Libanus
Tai'r-Bull
Llanfrynach
Pencelli
Scethrog
Talyllyn
Llanywern
Llechfaen
Groesffordd
Cross Oak
Talybont-on-Usk
Aber-Village
Heol Senni
Felin-Crai
Crai
Mynydd Eppynt
Sennybridge Training Area
Brecon Beacons
P O W Y S
Forest Fawr
A483
A470
A40
A4067
A4215
B4519
B4520
B4602
B4601
B4558
B4567
113
96
76
77
7
8
9
10
11
A
B
C
D
E
F
G
H

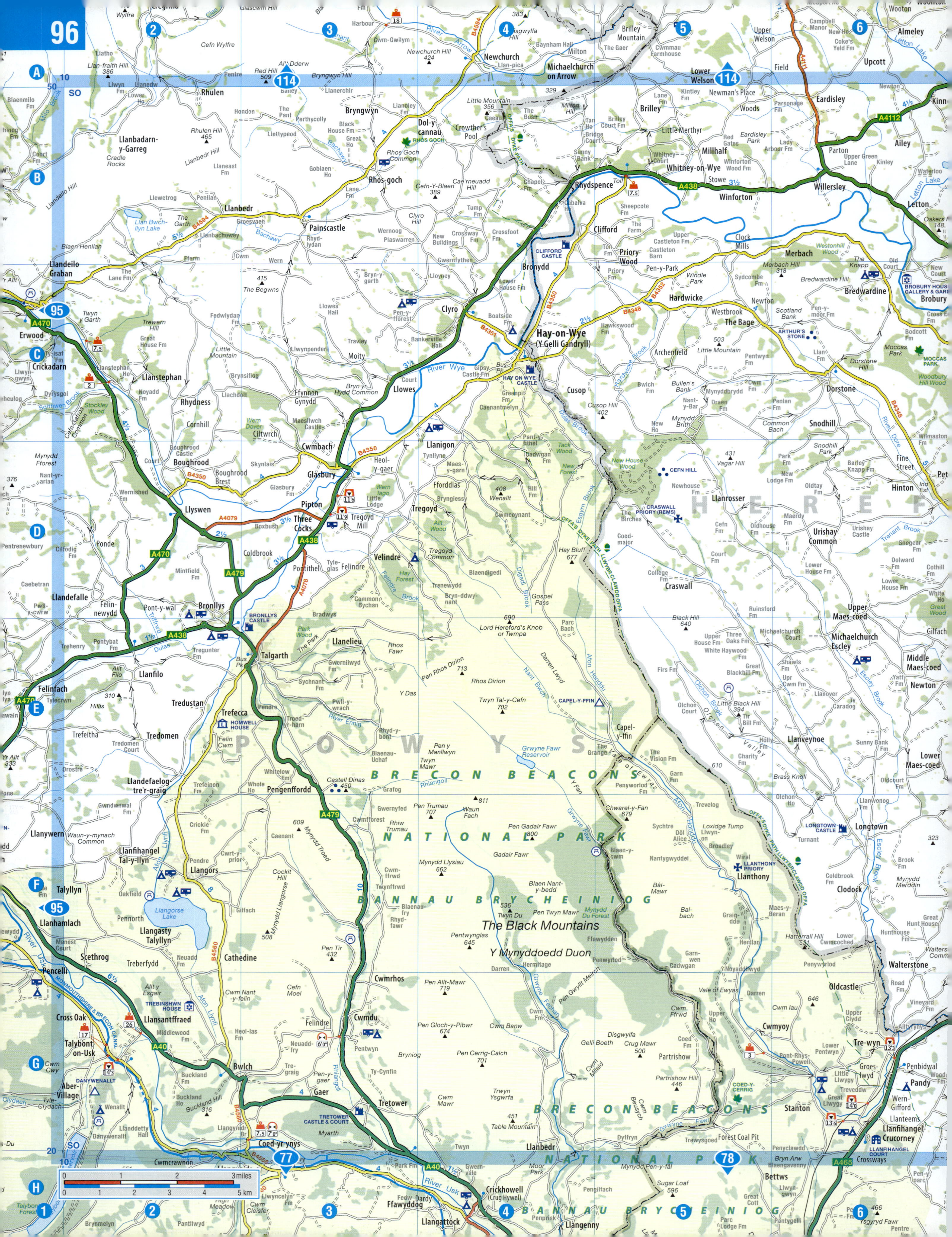

Rhulen
Llanbadarn-y-Garreg
Bryngwyn
Newchurch
Michaelchurch on Arrow
Brilley Mountain
Upper Welson
Almeley
Lower Welson
Eardisley
Brilley
Dol-y-cannau
Rhos-goch
Painscastle
Llanbedr
Rhydspence
Whitney-on-Wye
Winforton
Willersley
Letton
Clifford
Priory Wood
Merbach
Bredwardine
Brobury
Llandeilo Graban
Erwood
Crickadarn
Llanstephan
Rhydness
Clyro
Hay-on-Wye (Y Gelli Gandryll)
Hardwicke
The Bage
Dorstone
Cusop
Llowes
Llanigon
Cwmbach
Glasbury
Boughrood
Llyswen
Three Cocks
Tregoyd
Velindre
Felindre
Bronllys
Talgarth
Llanelieu
Llanfilo
Tredustan
Trefecca
Tredomen
Llandefalle
Felinfach
Llandefaelog tre'r-graig
Pengenffordd
Llanfihangel Tal-y-llyn
Llangors
Llangorse Lake
Talyllyn
Llanhamlach
Llangasty Talyllyn
Pennorth
Scethrog
Cathedine
Pencelli
Cross Oak
Llansantffraed
Talybont-on-Usk
Aber-Village
Bwlch
Coed-yr-ynys
Cwmdu
Tretower
Gaer
Crickhowell (Crug Hywel)
Llangattock
Llangenny
Llanbedr
Ffawyddog
Cwmcrawnon
Craswall
Llanrosser
Michaelchurch Escley
Upper Maes-coed
Urishay Common
Snodhill
Hinton
Newton
Capel-y-ffin
Llanthony
Llanveynoe
Longtown
Clodock
Walterstone
Oldcastle
Cwmyoy
Tre-wyn
Pandy
Stanton
Llanfihangel Crucorney
Bettws
Lord Hereford's Knob or Twmpa
Hay Bluff
Black Hill
Grwyne Fawr Reservoir
BRECON BEACONS NATIONAL PARK
BANNAU BRYCHEINIOG
The Black Mountains
Y Mynyddoedd Duon
Waun Fach 811
Pen y Manllwyn
Pen Allt-Mawr 719
Pen Cerrig-Calch 701
Table Mountain
Sugar Loaf 596
Mynydd Troed 609
Mynydd Llangorse 508
Castell Dinas 450
Pen Tir 432
River Wye
River Usk
Offa's Dyke Path
POWYS
HEREF
A470
A438
A479
A4079
A40
A465
A4078
B4350
B4351
B4348
B4352
B4594
B4560
B4558
95
114
77
78
SO

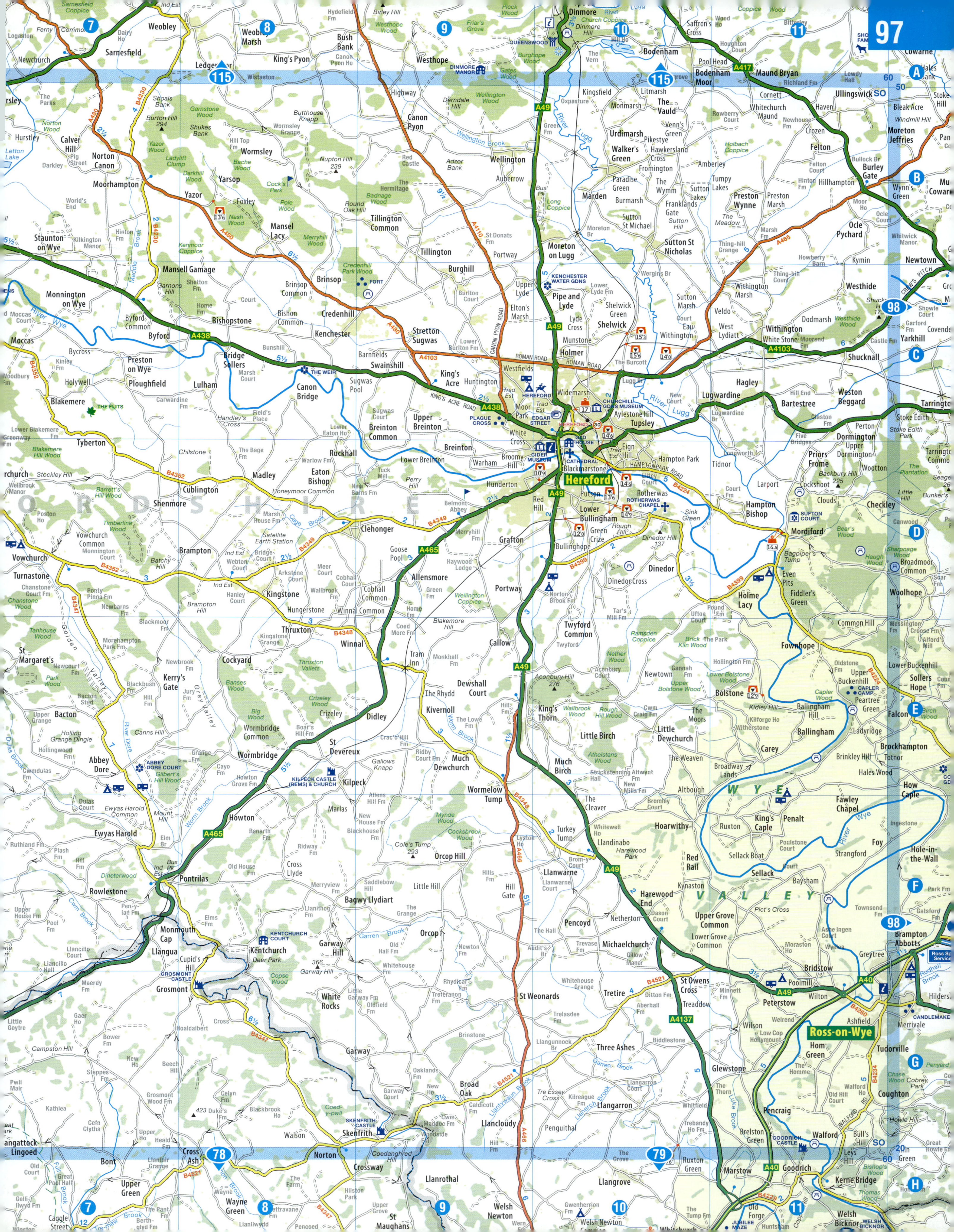
Weobley
Weobley Marsh
Sarnesfield
Ledgemoor
115
King's Pyon
Bush Bank
Birley Hill
Westhope
Canon Pyon
Dinmore
Dinmore Manor
Queenswood
Bodenham
Bodenham Moor
Maund Bryan
Ullingswick
Calver Hill
Norton Canon
Moorhampton
Yazor
Yarsop
Wormsley
Foxley
Mansel Lacy
Wellington
Auberrow
Kingsfield
Monmarsh
Urdimarsh
Pikestye
Litmarsh
The Vauld
Felton
Moreton Jeffries
Burley Gate
Hillhampton
Preston Wynne
Preston Marsh
Ocle Pychard
Staunton on Wye
Tillington Common
Tillington
Burghill
Marden
Sutton Lakes
Sutton St Michael
Sutton St Nicholas
Moreton on Lugg
Portway
Mansell Gamage
Brinsop
Credenhill
Kenchester Water Gdns
Pipe and Lyde
Shelwick
Withington Marsh
Westhide
Newtown
Monnington on Wye
Byford
Bishopstone
Kenchester
Stretton Sugwas
Lyde Cross
Holmer
Withington
Veldo
West Lydiatt
Dodmarsh
Yarkhill
Moccas
Bridge Sollers
Preston on Wye
Swainshill
King's Acre
Huntington
Westfields
Widemarsh
Hagley
Shucknall
Holywell
Blakemere
Ploughfield
Lulham
Canon Bridge
The Weir
Breinton Common
Upper Breinton
Breinton
Plague Cross
Edgar Street
Churchill Gdns Museum
Aylestone Hill
Tupsley
Lugwardine
Bartestree
Weston Beggard
Stoke Edith
Tarrington
Tyberton
Eaton Bishop
Ruckhall
Warham
Broomy Hill
Hunderton
White Cross
Cider Museum
Cathedral
Hereford
Blackmarstone
Eign Hill
Hampton Park
Tidnor
Dormington
Priors Frome
Madley
Cublington
Shenmore
Clehonger
Belmont Abbey
Red Hill
Putson
Lower Bullingham
Rotherwas
Hampton Bishop
Checkley
Mordiford
Vowchurch
Turnastone
Brampton
Kingstone
Allensmore
Grafton
Bullinghope
Dinedor Hill
Dinedor
Portway
Holme Lacy
Even Pits
Fiddler's Green
Woolhope
Hungerstone
Thruxton
Winnal
Cobhall Common
Winnal Common
Callow
Twyford Common
Common Hill
Fownhope
St Margaret's
Kerry's Gate
Cockyard
Tram Inn
Dewshall Court
Newtown
Bolstone
Upper Buckenhill
Peartree Green
Sollers Hope
Falcon
Bacton
The Rhydd
Kivernoll
Crizeley
Didley
King's Thorn
Little Birch
Little Dewchurch
Ballingham Hill
Ballingham
Brockhampton
Abbey Dore
Wormbridge
St Devereux
Kilpeck
Much Dewchurch
Much Birch
Carey
The Weaven
How Caple
Ewyas Harold
Howton
Wormelow Tump
The Cleaver
Altbough
Hoarwithy
King's Caple
Penalt
Fawley Chapel
Foy
Pontrilas
Orcop Hill
Turkey Tump
Red Rail
Sellack
Strangford
Hole-in-the-Wall
Rowlestone
Bagwy Llydiart
Little Hill
Llanwarne
Harewood End
Monmouth Cap
Llangua
Kentchurch
Garway Hill
Orcop
Pencoyd
Upper Grove Common
Brampton Abbotts
Grosmont
White Rocks
St Weonards
Michaelchurch
Tretire
St Owens Cross
Bridstow
Peterstow
Ross-on-Wye
Garway
Broad Oak
Three Ashes
Glewstone
Tudorville
Llangarron
Llancloudy
Pencraig
Coughton
Skenfrith
Walford
Cross Ash
Norton
Crossway
Llanrothal
Llangrove
Marstow
Goodrich
Kerne Bridge
Bont
Upper Green
Wayne Green
St Maughans
Welsh Newton
Welsh Bicknor
Wye Valley
A438
A49
A465
A480
A4103
A417
A466
A4137
A40
B4349
B4352
B4348
B4399
B4224
B4521
B4347
B4230
78
79
98

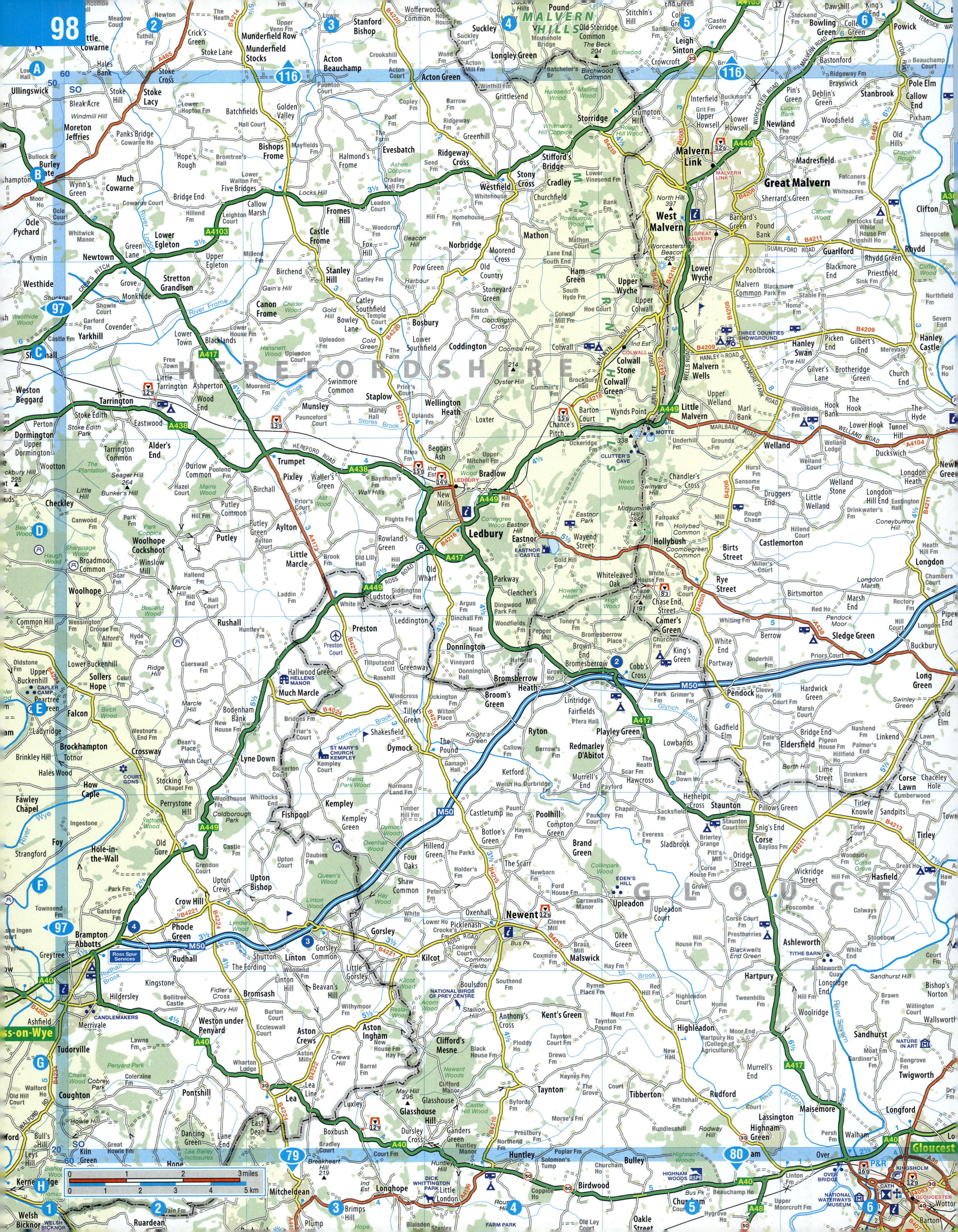

HEREFORDSHIRE
MALVERN HILLS
Great Malvern
Malvern Link
West Malvern
Little Malvern
Malvern Wells
Upper Wyche
Colwall
Ledbury
Newent
Ross-on-Wye
Gloucester
Ullingswick
Stoke Lacy
Bishops Frome
Castle Frome
Acton Beauchamp
Suckley
Stanford Bishop
Munderfield Row
Munderfield Stocks
Burley Gate
Much Cowarne
Moreton Jeffries
Ocle Pychard
Lower Eggleton
Stretton Grandison
Canon Frome
Stanley Hill
Fromes Hill
Evesbatch
Ridgeway Cross
Cradley
Stifford's Bridge
Storridge
Mathon
Westfield
Norbridge
Bosbury
Coddington
Wellington Heath
Staplow
Munsley
Tarrington
Trumpet
Pixley
Aylton
Putley
Little Marcle
Yarkhill
Westhide
Newtown
Dormington
Woolhope
Preston
Much Marcle
Donnington
Dymock
Kempley
Ryton
Redmarley D'Abitot
Bromsberrow Heath
Eastnor
Welland
Castlemorton
Birtsmorton
Longdon
Pendock
Eldersfield
Staunton
Corse
Ashleworth
Hasfield
Tirley
Hartpury
Highleadon
Tibberton
Taynton
Clifford's Mesne
Gorsley
Linton
Upton Bishop
Brampton Abbotts
Weston under Penyard
Aston Ingham
Kilcot
Upleadon
Bromesberrow
Hollybush
Sledge Green
Welland
Hanley Swan
Guarlford
Madresfield
Newland
Clifton
Powick
Leigh Sinton
Kent's Green
Huntley
Bulley
Birdwood
Maisemore
Twigworth
Sandhurst
Lea
Pontshill
Coughton
Kerne Bridge
Rudhall
Hole-in-the-Wall
Fownhope
Sollers Hope
Upper Buckenhill
Falcon
How Caple
Brockhampton
Crossway
Perrystone Hill
Phocle Green
Crow Hill
Kingstone
Bromsash
Glasshouse Hill
Longhope
Mitcheldean
Ruardean
M50
A449
A417
A438
A4103
A40
A4172
B4215
B4216
B4221
B4224
B4211
B4208
B4209
B4218
B4219
B4214
B4220
B4213
116
97
79
80

WORCESTERSHIRE
COTSWOLDS
Evesham
Pershore
Tewkesbury
Cheltenham
Broadway
Winchcombe
Bishop's Cleeve
Charlton Kings
Upton upon Severn
Bengeworth
Fairfield
Hampton
Offenham
Badsey
Bretforton
Aldington
Wickhamford
Childswickham
Willersey
Stanton
Stanway
Toddington
Gretton
Greet
Alderton
Teddington
Ashchurch
Northway
Kinsham
Bredon
Eckington
Defford
Besford
Birlingham
Strensham
Ripple
Twyning
Deerhurst
Apperley
Tredington
Fiddington
Stoke Orchard
Southam
Prestbury
Leckhampton
Churchdown
Parton
Uckington
Boddington
Staverton
Down Hatherley
Norton
Sedgeberrow
Hinton on the Green
Ashton under Hill
Beckford
Kemerton
Overbury
Conderton
Dumbleton
Wormington
Little Comberton
Great Comberton
Elmley Castle
Netherton
Kersoe
Cropthorne
Charlton
Fladbury
Wyre Piddle
Wick
Pinvin
Throckmorton
Bishampton
Peopleton
Drake's Broughton
Wadborough
Stoulton
Norton
Lenchwick
Chadbury
Harvington
Salford Priors
Abbots Salford
Church Lench
Atch Lench
Ab Lench
Bredon's Norton
Westmancote
Kinnersley
Dunstall Common
Earl's Croome
Baughton
Holly Green
Naunton
Upper Strensham
Pirton
High Green
Kempsey
Hatfield
Littleworth
Egdon
Hailes
Didbrook
Buckland
Laverton
Snowshill
Cutsdean
Ford
Temple Guiting
Kineton
Barton
Guiting Power
Hawling
Farmcote
Charlton Abbots
Sevenhampton
Brockhampton
Whittington
Syreford
Andoversford
Shipton
Salperton
Dowdeswell
Woodmancote
Gotherington
Oxenton
Dixton
Alstone
Aston Cross
Walton Cardiff
Forthampton
Chaceley
Coombe Hill
Leigh
Sandfields
Arle
Golden Valley
Shurdington
Badgeworth
Barnwood
Hucclecote
Cleeve Hill
Vale of Evesham
Strensham Services North
Strensham Services South
A46
A44
A38
A435
A40
A436
M5
M50
B4084
B4080
B4077
B4078
B4632

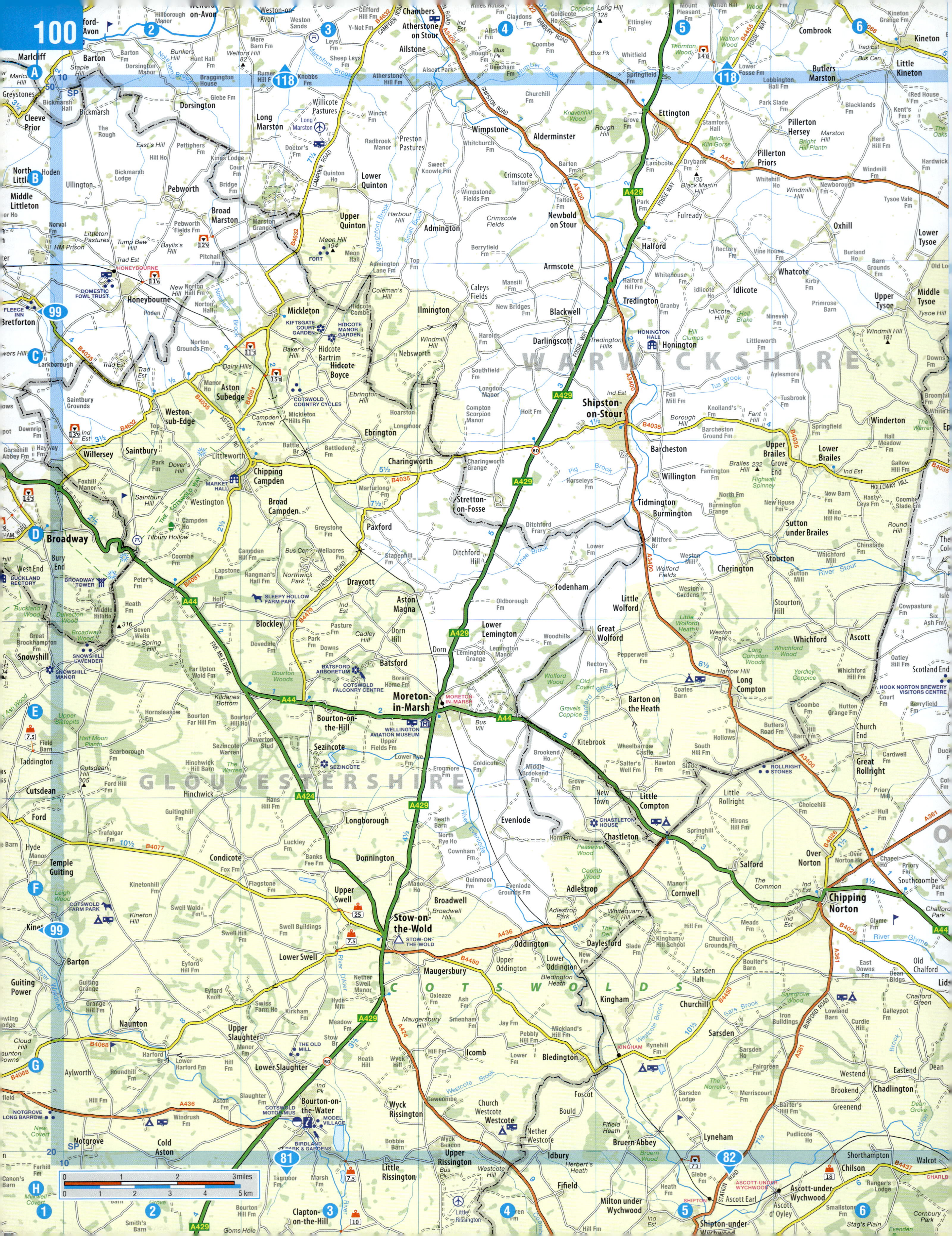
WARWICKSHIRE
GLOUCESTERSHIRE
COTSWOLDS
Marlcliff
Barton
Dorsington
Long Marston
Willicote Pastures
Chambers
Atherstone on Stour
Ailstone
Clifford Chambers
Weston Sands
Combrook
Kineton
Little Kineton
Butlers Marston
Ettington
Pillerton Hersey
Pillerton Priors
Wimpstone
Alderminster
Cleeve Prior
Bickmarsh
North Littleton
Middle Littleton
Ullington
Pebworth
Broad Marston
Upper Quinton
Lower Quinton
Admington
Newbold on Stour
Halford
Fulready
Oxhill
Lower Tysoe
Honeybourne
Mickleton
Hidcote Bartrim
Hidcote Boyce
Ilmington
Armscote
Blackwell
Tredington
Idlicote
Whatcote
Middle Tysoe
Upper Tysoe
Caleys Fields
Darlingscott
Honington
Bretforton
Nebsworth
Aston Subedge
Weston-sub-Edge
Ebrington
Shipston-on-Stour
Willersey
Saintbury
Chipping Campden
Charingworth
Barcheston
Willington
Upper Brailes
Lower Brailes
Winderton
Broad Campden
Westington
Paxford
Stretton-on-Fosse
Tidmington
Burmington
Sutton under Brailes
Broadway
Draycott
Ditchford Hill
Todenham
Cherington
Stourton
Blockley
Aston Magna
Batsford
Lower Lemington
Little Wolford
Great Wolford
Whichford
Ascott
Snowshill
Moreton-in-Marsh
Bourton-on-the-Hill
Sezincote
Barton on the Heath
Long Compton
Kitebrook
Little Rollright
Great Rollright
Taddington
Cutsdean
Hinchwick
Longborough
Evenlode
Little Compton
Chastleton
Ford
Condicote
Donnington
Salford
Over Norton
Chipping Norton
Hyde
Temple Guiting
Upper Swell
Broadwell
Stow-on-the-Wold
Adlestrop
Cornwell
Kineton
Lower Swell
Oddington
Upper Oddington
Lower Oddington
Daylesford
Barton
Maugersbury
Guiting Power
Kingham
Churchill
Sarsden
Naunton
Upper Slaughter
Lower Slaughter
Icomb
Bledington
Aylworth
Bourton-on-the-Water
Wyck Rissington
Church Westcote
Westcote
Nether Westcote
Foscot
Bould
Bruern Abbey
Lyneham
Chadlington
Notgrove
Cold Aston
Upper Rissington
Little Rissington
Idbury
Fifield
Milton under Wychwood
Shipton-under-Wychwood
Ascott-under-Wychwood
Ascott Earl
Ascott d'Oyley
Shorthampton
Chilson
Walcot
Clapton-on-the-Hill
A429
A44
A3400
A424
A436
A361
A422
B4035
B4081
B4632
B4077
B4068
B4450
B4026
B4450
99
118
81
82
0 1 2 3 miles
0 1 2 3 4 5 km

Temple Herdewyke
Burton Dassett
Compton
Fenny Compton
Wormleighton Res
Cedars Fm
Boddington
Aston le Walls
West Farm
Eydon
Canons Ashby House
Canons Ashby
Avon Dassett
Farnborough
Farnborough Hall
Claydon
Granary Museum of Bygones
Edgehill 1642
Radway
Ratley
Arlescote
Warmington
Mollington
Chipping Warden
Edgcote
Culworth
Moreton Pinkney
Canada
Plumpton
Woodend Green
Weedon Lois
Weston
Catanger Llama Trekking
Cropredy
Wardington
Upper Wardington
Williamscott
Great Bourton
Little Bourton
Sulgrave
Sulgrave Manor
Lower Thorpe
Thorpe Mandeville
Upton House
Hornton
Shotteswell
Banbury (Shotteswell)
Horley
Hanwell
Chacombe
Helmdon
Falcutt
Shenington
Alkerton
Brook Cottage
Wroxton
Drayton
Hardwick
Ruscote
Neithrop
Banbury
Grimsbury
Nethercote
Middleton Cheney
Thenford
Marston St Lawrence
Greatworth
Halse
Radstone
Balscote
Shutford
North Newington
Overthorpe
Warkworth
Lower Middleton Cheney
NORTHANTS
Farthinghoe
Great Purston
Whitfield
Brackley
Sibford Gower
Swalcliffe Barn
Tadmarton
Broughton
Broughton Castle
Easington
Calthorpe
Bodicote
Sibford Ferris
Swalcliffe
Lower Tadmarton
Bloxham
Upper Astrop
King's Sutton
Astrop
Newbottle
Charlton
Hinton-in-the-Hedges
Hinton in the Hedges
Turweston
Evenley
Wigginton Heath
Milton
Adderbury
West Adderbury
East Adderbury
Milcombe
Walton Grounds
Aynho
Croughton
Mixbury
East End
Hook Norton
Wigginton
South Newington
River Swere
Barford St John
Barford St Michael
Deddington
Southrop
Hempton
Deddington Castle
Clifton
Souldern
Cottisford
Swerford
Nether Worton
Great Tew
Little Tew
OXFORDSHIRE
Over Worton
North Aston
The Green
Somerton
Fritwell
Baynard's Green
Hardwick
Hethe
Fringford
Newton Morrell
Fewcott
Stoke Lyne
Cherwell Valley Services
Ardley
Dunthrop
Heythrop
Ledwell
Duns Tew
Middle Aston
Upper Heyford
Upper Heyford Airfield (disused)
Sandford St Martin
Middle Barton
Westcott Barton
Steeple Aston
Church Enstone
Enstone
Gagingwell
Steeple Barton
Bucknell
Caversfield
Stratton Audley
Bainton
Cleveley
Hopcroft's Holt
Rousham
Rousham Park House & Garden
Lower Heyford
Caulcott
Bicester
Woodfield
Highfield
Bicester Village
Bicester Town
Bicester North
Middleton Stoney
King's End
Fulwell
Radfordbridge
Kiddington
Over Kiddington
Glympton
Northbrook
Chesterton
Little Chesterton
Graven Hill
Taston
Spelsbury
Ditchley Park
Wootton
Tackley
Nethercott
Kirtlington
Weston-on-the-Green
Wendlebury
Ambrosden
Charlbury
Woodleys
Enslow
Gibraltar
Merton
Lower Arncott
M40
A423
A422
A361
A4260
A43
A41
A44
A4095
B4100
B4035
B4031
B4030
B4525
B4027
B4022
B4437

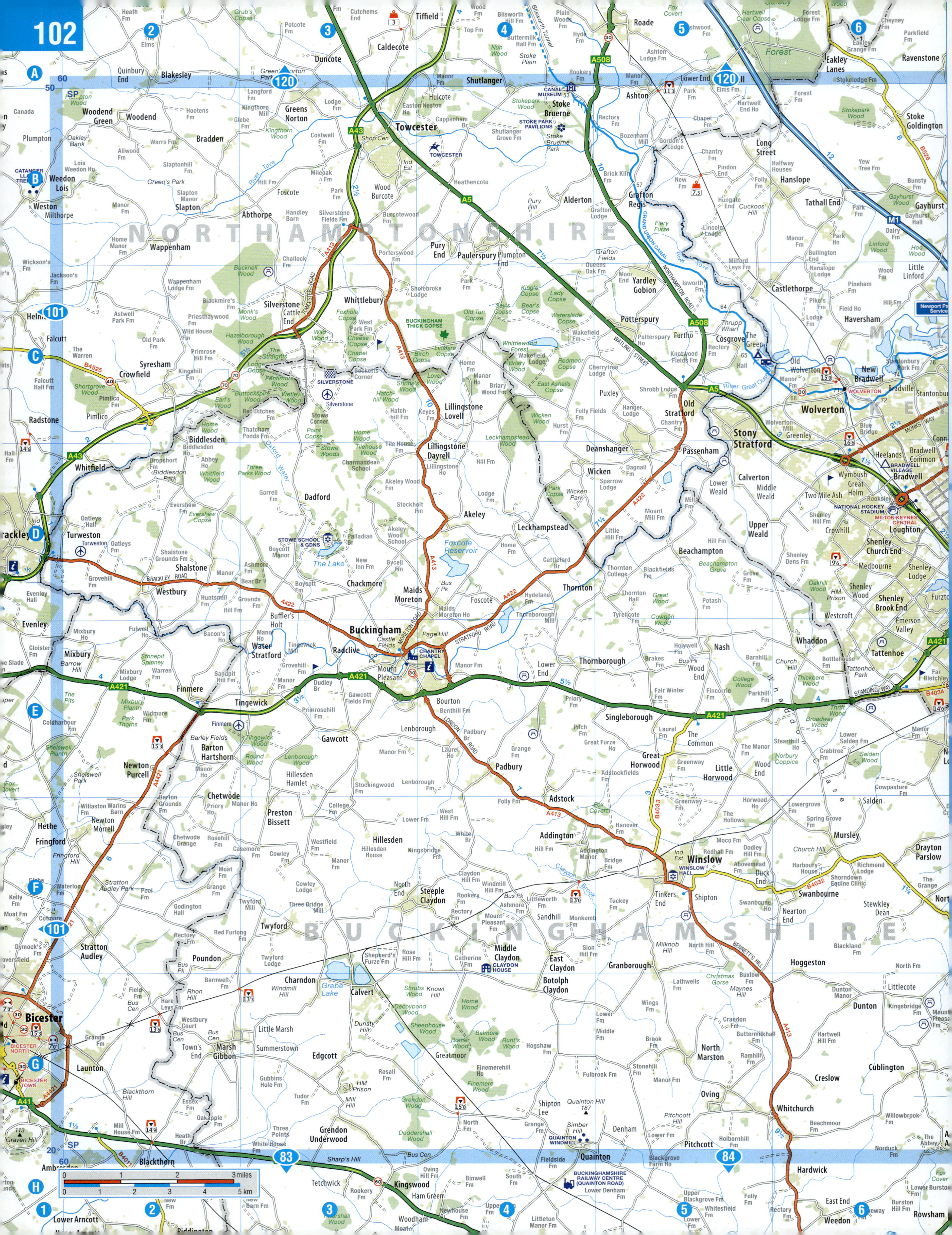
NORTHAMPTONSHIRE
BUCKINGHAMSHIRE
Blakesley
Duncote
Caldecote
Tiffield
Shutlanger
Roade
Ashton
Stoke Bruerne
Towcester
Woodend
Greens Norton
Bradden
Plumpton
Weedon Lois
Weston
Slapton
Abthorpe
Foscote
Wood Burcote
Alderton
Grafton Regis
Hanslope
Long Street
Stoke Goldington
Gayhurst
Tathall End
Wappenham
Pury End
Paulerspury
Plumpton End
Yardley Gobion
Castlethorpe
Little Linford
Haversham
Silverstone
Whittlebury
Potterspury
Furtho
Cosgrove
New Bradwell
Syresham
Crowfield
Radstone
Pimlico
Old Stratford
Wolverton
Lillingstone Lovell
Lillingstone Dayrell
Biddlesden
Stony Stratford
Deanshanger
Passenham
Whitfield
Wicken
Calverton
Bradwell
Turweston
Brackley
Dadford
Akeley
Leckhampstead
Beachampton
Loughton
Shalstone
Westbury
Chackmore
Maids Moreton
Foscote
Thornton
Evenley
Buckingham
Mixbury
Water Stratford
Radclive
Nash
Whaddon
Tattenhoe
Finmere
Tingewick
Thornborough
Singleborough
Gawcott
Lenborough
Newton Purcell
Barton Hartshorn
Padbury
Great Horwood
Little Horwood
Chetwode
Preston Bissett
Adstock
Hethe
Fringford
Hillesden
Addington
Winslow
Mursley
Drayton Parslow
Steeple Claydon
Swanbourne
Shipton
Twyford
Stratton Audley
Poundon
Middle Claydon
East Claydon
Botolph Claydon
Granborough
Hoggeston
Charndon
Calvert
Bicester
Little Marsh
Marsh Gibbon
Edgcott
North Marston
Dunton
Launton
Cublington
Creslow
Oving
Whitchurch
Grendon Underwood
Shipton Lee
Denham
Pitchcott
Quainton
Blackthorn
Kingswood
Hardwick
Lower Arncott
Weedon
Rowsham
A5
A43
A413
A421
A422
A508
A4421
A41
B4525
B4033
B4032
M1
miles
km

Olney
Weston Underwood
Emberton
Filgrave
Tyringham
Sherington
Chicheley
Hardmead
Astwood
North Crawley
Newport Pagnell
Lathbury
Great Linford
Milton Keynes
Bletchley
Fenny Stratford
Far Bletchley
Woughton on the Green
Walton
Simpson
Broughton
Wavendon
Woburn Sands
Aspley Guise
Aspley Heath
Salford
Hulcote
Brogborough
Ridgmont
Husborne Crawley
Cranfield
Moulsoe
Kempston
Biddenham
Bedford
Great Denham
Queen's Park
Fenlake
Cardington
Harrowden
Elstow
Shortstown
Cople
Wootton
Keeley Green
Kempston Hardwick
Stewartby
Marston Moretaine
Lidlington
Millbrook
Ampthill
Maulden
Clophill
Flitwick
Flitton
Silsoe
Steppingley
Westoning
Tingrith
Harlington
Sharpenhoe
Barton-le-Clay
Hexton
Streatley
Toddington
Woburn
Woburn Abbey
Eversholt
Milton Bryan
Potsgrave
Battlesden
Hockliffe
Tebworth
Chalgrave
Chalton
Wingfield
Houghton Regis
Dunstable
Luton
Leagrave
Limbury
Sundon Park
Upper Sundon
Lower Sundon
Leighton Buzzard
Linslade
Heath and Reach
Stoke Hammond
Soulbury
Great Brickhill
Little Brickhill
Bow Brickhill
Eggington
Stanbridge
Tilsworth
Billington
Eaton Bray
Totternhoe
Edlesborough
Slapton
Wing
Ledburn
Mentmore
Cheddington
Wingrave
Ivinghoe Aston
Whipsnade
Kensworth
Caddington
Slip End
Chiltern Hills
Central Bedfordshire
M1
A5
A6
A421
A4146
A507
A428
A418
A505
A4012
Toddington Services
Woburn Safari Park
Wild Animal Kingdom & Leisure Park
Marston Vale Millennium
Stockgrove Park
Leighton Buzzard Narrow Gauge Railway

104
CENTRAL BEDFORDSHIRE
CHILTERNS
LUTON
Sandy
Biggleswade
Potton
Shefford
Letchworth
Hitchin
Stevenage
Baldock
Stotfold
Arlesey
Clophill
Barton-le-Clay
Ashwell
Royston Road
Great Barford
Blunham
Willington
Moggerhanger
Cardington
Girtford
Beeston
Northill
Ickwell Green
Upper Caldecote
Old Warden
Southill
Stanford
Langford
Henlow
Clifton
Campton
Meppershall
Upper Stondon
Lower Stondon
Shillington
Pirton
Ickleford
Holwell
Lilley
Great Offley
King's Walden
St Ippolyts
Gosmore
Preston
Langley
Knebworth
Codicote
Kimpton
Walkern
Benington
Weston
Graveley
Clothall
Wallington
Sandon
Kelshall
Therfield
Bygrave
Newnham
Caldecote
Hinxworth
Edworth
Dunton
Eyeworth
Wrestlingworth
Cockayne Hatley
Tadlow
Guilden Morden
Steeple Morden
Abington Pigotts
Bassingbourn
Litlington
Croydon
Arrington
Hatley St George
Everton
Sutton
Wendy
Shingay
Whaddon
Cottered
Ardeley
Rushden
Weston Hills
Radwell
Norton
Willian
Great Wymondley
Little Wymondley
Hexton
Pegsdon
Higham Gobion
Whitwell
St Paul's Walden
Cromer
Aston
Datchworth
Watton at Stone
Woolmer Green
A1(M)
A505
A507
A600
A602

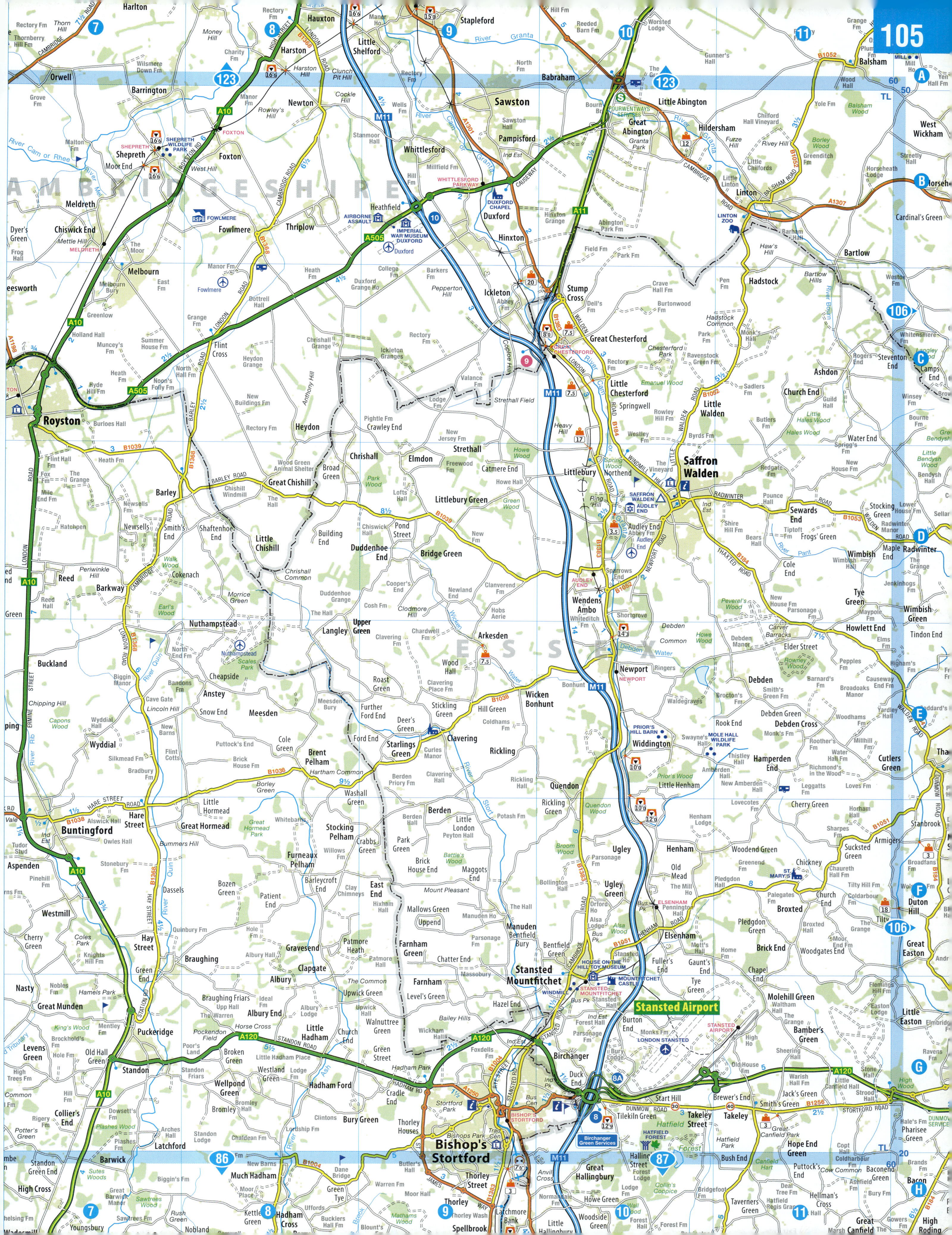
Harlton
Hauxton
Stapleford
Little Shelford
Harston
Orwell
Barrington
Newton
Sawston
Babraham
Little Abington
Great Abington
Hildersham
Balsham
West Wickham
Pampisford
Whittlesford
Shepreth
Foxton
Meldreth
Linton
CAMBRIDGESHIRE
Heathfield
Duxford
Hinxton
Fowlmere
Thriplow
Chiswick End
Melbourn
IMPERIAL WAR MUSEUM DUXFORD
LINTON ZOO
Bartlow
Hadstock
Ickleton
Stump Cross
Great Chesterford
Little Chesterford
Ashdon
Church End
Royston
Flint Cross
Heydon
Chrishall
Elmdon
Strethall
Littlebury
Saffron Walden
Audley End
Great Chishill
Barley
Littlebury Green
Duddenhoe End
Bridge Green
Little Chishill
Sewards End
Radwinter
Wimbish
Reed
Barkway
Nuthampstead
Langley
Upper Green
Arkesden
Wendens Ambo
Newport
Debden
ESSEX
Buckland
Anstey
Meesden
Wicken Bonhunt
Clavering
Rickling
Widdington
Starlings Green
Brent Pelham
Wyddial
Quendon
Buntingford
Hare Street
Great Hormead
Little Hormead
Berden
Stocking Pelham
Furneaux Pelham
Henham
Ugley
Ugley Green
Aspenden
Westmill
Broxted
Thaxted
Elsenham
Manuden
Braughing
Hay Street
Farnham
Stansted Mountfitchet
Stansted Airport
Great Munden
Puckeridge
Albury
Little Hadham
Birchanger
Takeley
Standon
Hadham Ford
Bishop's Stortford
Thorley
Barwick
Much Hadham
Great Hallingbury
High Cross
Hatfield Forest
Great Canfield
Little Easton
Great Easton
Duton Hill
Molehill Green
Bamber's Green
Cutlers Green
Hempstead
Tye Green
Howlett End
Wimbish Green
Great Sampford

Balsham
Weston Green
West Wratting Park
Carlton Green
Temple End
Little Thurlow
Great Thurlow
Little Bradley
Little Thurlow Green
Sowley Green
Harlica Fm
Assington Green
Upper Street
Thurston End
Boxted
Townsend
Hundon
Poslingford
Chilton Street
Stansfield
Gosland Green
Chipley Abbey
Wales End
Glemsford
West Wickham
Burton End
Streetly End
Horseheath
Withersfield
Great Wratting
Little Wratting
Barnardiston
Kedington
Cardinal's Green
Bartlow
Haverhill
Hanchett Village
Calford Green
Shudy Camps
Nosterfield End
Hazel Stub
Castle Camps
Sturmer
Boyton End
Stoke by Clare
Clare
Cavendish
Pentlow
Foxearth
Hickford Hill
Wixoe
Steventon End
Helions Bumpstead
Pale Green
Wiggens Green
Baythorne End
Ashen
Ovington
Belchamp St Paul
Belchamp Otten
Borley Green
Borley
Steeple Bumpstead
Birdbrook
New England
Ridgewell
Tilbury Green
Tilbury Juxta Clare
Little Yeldham
Belchamp Walter
Bulmer
Bulmer Tye
Radwinter
Hempstead
Great Sampford
Little Sampford
Wimbish
Wimbish Green
Tye Green
Howlett End
Stambourne Green
Stambourne
Great Yeldham
Toppesfield
Gestingthorpe
Audley End
Wickham St Paul
Cornish Hall End
ESSEX
Little London
Gainsford End
Castle Hedingham
Sible Hedingham
Great Maplestead
Little Maplestead
Highstreet Green
Howe Street
Finchingfield
Duck End
Morris Green
Thaxted
Cutlers Green
Great Bardfield
Little Bardfield
Wethersfield
Halstead
Gosfield
Monk Street
Lindsell
Shalford
Shalford Green
Beazley End
Bardfield Saling
Great Saling
Jasper's Green
Panfield
Bocking Churchstreet
High Garrett
Duton Hill
Great Easton
Little Easton
Stebbing
Stebbing Green
Bocking
Braintree
Bradwell
Stisted
Churchend
Great Dunmow
Little Dunmow
Bannister Green
Rayne
Great Notley
Black Notley
Cressing
Tye Green
Coggeshall Hamlet
Felsted
Barnston
Silver End
White Notley
High Roding
Hartford End
North End
A120
A131
A1017
A1092
A1307
B1053
B1054
B1057
B1256
105
124
87
88
0 1 2 3 miles
0 1 2 3 4 5 km

107
SUFFOLK
DEDHAM VALE
Sudbury
Colchester
Hadleigh
Lavenham
Long Melford
Coggeshall
Kelvedon
Wivenhoe
Manningtree
Lawford
Capel St Mary
East Bergholt
Dedham
Stratford St Mary
Nayland
Boxford
Bures
Great Cornard
Ballingdon
Brantham
Cattawade
Higham
Holton St Mary
Layham
Raydon
Shelley
Polstead
Stoke-by-Nayland
Assington
Newton
Chilton
Acton
Great Waldingfield
Little Waldingfield
Edwardstone
Groton
Milden
Monks Eleigh
Brent Eleigh
Chelsworth
Bildeston
Kettlebaston
Hitcham
Wattisham
Nedging
Naughton
Semer
Whatfield
Aldham
Elmsett
Offton
Nettlestead
Somersham
Flowton
Burstall
Hintlesham
Chattisham
Washbrook
Copdock
Bentley
Great Wenham
Little Wenham
Wormingford
Little Horkesley
Great Horkesley
Horkesley Heath
Boxted
Boxted Cross
Langham
Ardleigh
Fordham
West Bergholt
Mile End
Eight Ash Green
Aldham
Fordstreet
Chappel
Wakes Colne
White Colne
Colne Engaine
Earls Colne
Great Tey
Marks Tey
Copford
Copford Green
Stanway
Lexden
Abbey Field
The Hythe
Old Heath
Rowhedge
Alresford
Elmstead Market
Great Bromley
Little Bromley
Frating Green
Thorrington
Fingringhoe
Langenhoe
Abberton
Layer de la Haye
Birch
Heckfordbridge
Feering
Messing
Easthorpe
Middleton
Henny Street
Twinstead
Lamarsh
Alphamstone
Pebmarsh
Mount Bures
Little Cornard
Leavenheath
Harrow Street
Honey Tye
Wissington
Abberton Reservoir
River Stour
River Colne
Colne Valley
A12
A120
A134
A1071
A1141
A133
A131
A137
A1124
125
108
88
89

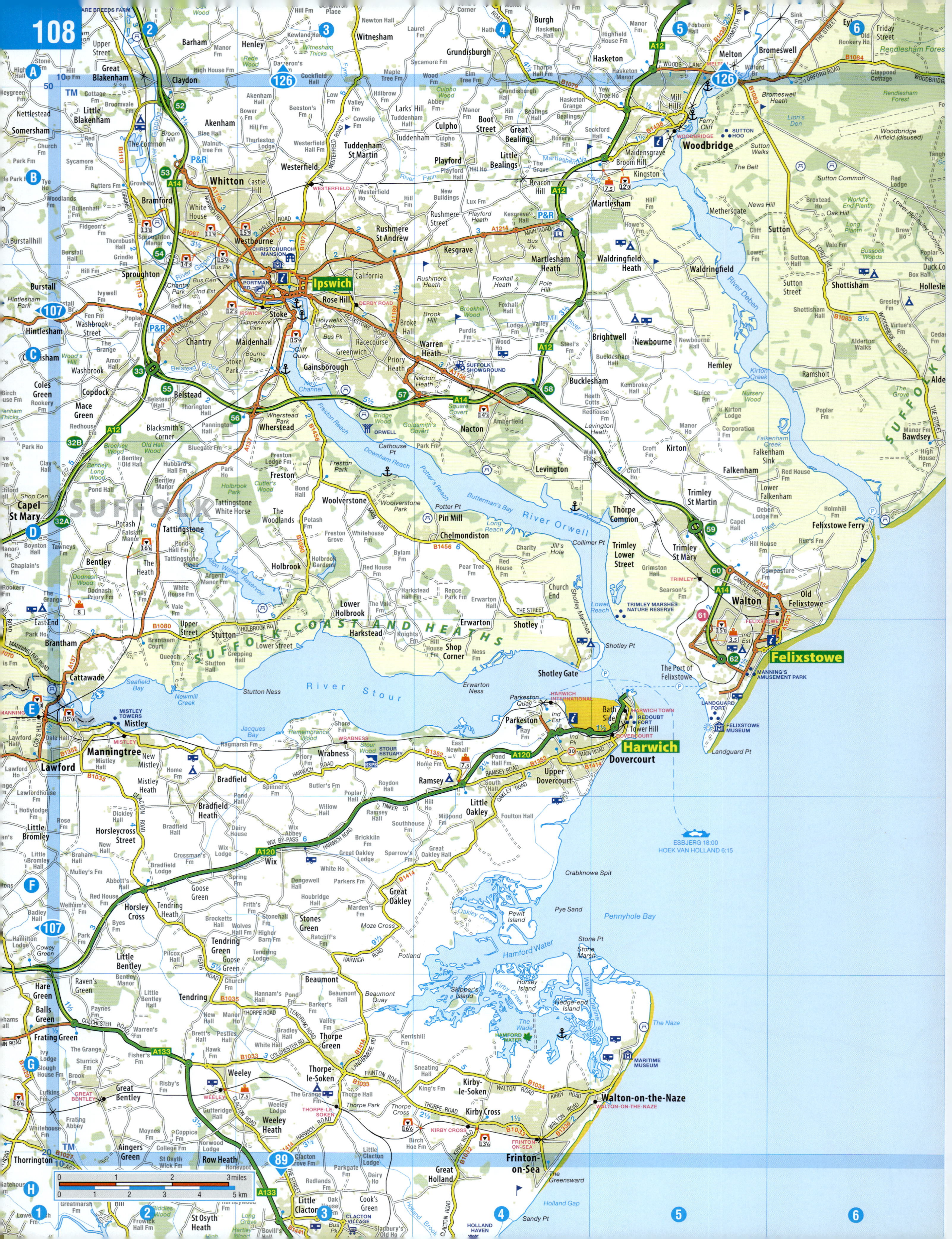

Barham
Henley
Witnesham
Grundisburgh
Burgh
Hasketon
Melton
Bromeswell
Great Blakenham
Claydon
Akenham
Westerfield
Tuddenham St Martin
Culpho
Playford
Little Bealings
Great Bealings
Woodbridge
Martlesham
Kingston
Sutton
Whitton
Bramford
Sproughton
Westbourne
Ipswich
Rushmere St Andrew
Kesgrave
Martlesham Heath
Waldringfield Heath
Waldringfield
Shottisham
California
Rose Hill
Stoke
Chantry
Maidenhall
Gainsborough
Warren Heath
Priory Heath
Brightwell
Newbourne
Hemley
Ramsholt
Bucklesham
Belstead
Wherstead
Nacton
Levington
Kirton
Falkenham
Bawdsey
Freston
Woolverstone
Pin Mill
Chelmondiston
Trimley St Martin
Trimley St Mary
Felixstowe Ferry
Capel St Mary
Tattingstone
Bentley
Holbrook
Lower Holbrook
Harkstead
Erwarton
Shotley
Shotley Gate
Walton
Old Felixstowe
Felixstowe
The Port of Felixstowe
Brantham
Stutton
Cattawade
River Orwell
River Stour
River Deben
SUFFOLK
SUFFOLK COAST AND HEATHS
Mistley
Manningtree
Lawford
Wrabness
Ramsey
Parkeston
Harwich
Dovercourt
Upper Dovercourt
Little Oakley
Bradfield
Wix
Great Oakley
Horsley Cross
Tendring Green
Stones Green
Beaumont
Tendring
Thorpe Green
Thorpe-le-Soken
Weeley
Kirby-le-Soken
Kirby Cross
Walton-on-the-Naze
Frinton-on-Sea
Great Holland
Little Clacton
Great Bentley
Aingers Green
Row Heath
St Osyth Heath
Thorrington
Frating Green
Hamford Water
Pennyhole Bay
ESBJERG 18:00
HOEK VAN HOLLAND 6:15
3 miles
5 km

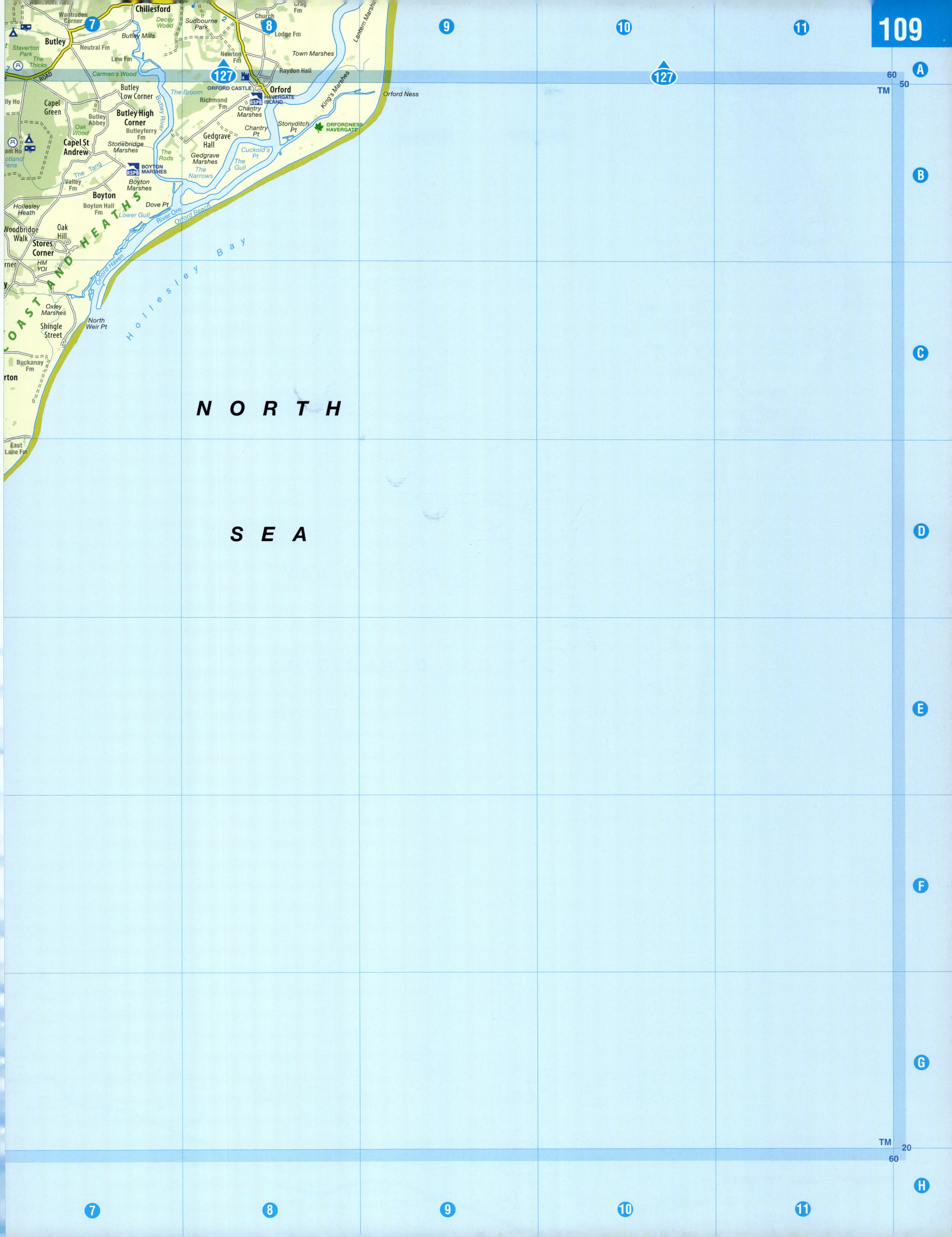

Chillesford
Wantisden Corner
Decoy Wood
Sudbourne Park
Church
Crag Fm
Lodge Fm
Lantern Marshes
Butley
Neutral Fm
Butley Mills
Staverton Park
The Thicks
Low Fm
Newton Fm
Town Marshes
Raydon Hall
127
Carmen's Wood
ORFORD CASTLE
Orford
HAVERGATE ISLAND
King's Marshes
Orford Ness
Butley Low Corner
The Broom
Richmond Fm
Chantry Marshes
Capel Green
Butley Abbey
Butley High Corner
Butley River
Oak Wood
Butleyferry Fm
Gedgrave Hall
Chantry Pt
Stonyditch Pt
ORFORDNESS-HAVERGATE
Capel St Andrew
Stonebridge Marshes
The Rods
Gedgrave Marshes
Cuckold's Pt
The Gull
The Tang
BOYTON MARSHES
The Narrows
Valley Fm
Boyton Marshes
Boyton
Hollesley Heath
Boyton Hall Fm
Dove Pt
Lower Gull
River Ore
Orford Beach
Woodbridge Walk
Oak Hill
Stores Corner
HM YOI
Hollesley Bay
Orford Haven
COAST AND HEATHS
Oxley Marshes
Shingle Street
North Weir Pt
Buckanay Fm
East Lane Fm
NORTH
SEA
TM
60
50
20
7
8
9
10
11
A
B
C
D
E
F
G
H

CARDIGAN BAY
BAE CEREDIGION
Llwyngwril
Rhoslefain
Llanegryn
Peniarth
Bryncrug
Rhyd-yr-onen
Tywyn
Pendre
Caethle
Aberdovey
(Aberdyfi)
Ynyslas
Ynys Tachwedd
Borth Sands
Borth
(Y Borth)
Broad Water
Tonfanau
Narrow Gauge Railway Museum
Neptune Hall
Aberdovey Bar
Beacon 178
Cardigan Island
Carreg Lydan
Mwnt
Aberporth
Parcllyn
Tresaith
Penbryn
Llangrannog
Pontgarreg
Penmorfa
Sarnau
Brynhoffnant
Tan-y-groes
Blaenporth
Blaenannerch
Tremain
Felinwynt
Rhosygadair Newydd
Penparc
Y Ferwig
Gwbert
Cippyn
Cemaes Head
Glynarthen
Ceredigion Coast Path
Cardigan Bay
Traeth-Penbryn
Carreg y Nodwydd
Ynys-Lochtyn
Trwyn Crou
A493
A487
B4333
B4334
SN
SH
128
92

Aberystwyth
Trefechan
Penparcau
Southgate
Fawr
Rhydyfelin
Llanfarian
Chancery
(Rhydgaled)
Blaenplwyf
Llanddeiniol
Llangwyryfon
Llanrhystud
Ty-Newydd
Rhyd-Rosser
Llansantffraed
Llanon
(Llan-non)
Rhydfudr
Trawsnant
Nebo
Rhos Haminiog
Cross Inn
Bethania
Penuwch
Aberarth
Aberaeron
Pennant
Monachty
Cilcennin
Crynfryn
Bwlch-Llan
Capel Betws
Lleucu
New Quay
(Ceinewydd)
New Quay Head
Birds Rock
Traethy Coubal
Cei-bach
Gilfachrheda
Maen-y-
groes
Llanarth
Oakford
Llwyncelyn
Ffos-
y-ffin
Henfynyw
Ciliau
Aeron
Newbridge
Trefilan
Tal-sarn
Llundain-fach
Abermeurig
Gartheli
Llwyn-
y-groes
Cross Inn
Nanternis
Caerwedros
Llwyndafydd
Hafodiwan
Synod Inn
Mydroilyn
Dihewyd
Ystrad Aeron
Felinfach
Felindre
Temple
Bar
CEREDIGION
(SIR CEREDIGION)
Caledrhydiau
Ffynnon-oer
Llangybi
Betws Bledrws
Pentregat
Plwmp
Talgarreg
Gorsgoch
Cribyn
Silian
Capel
Cynon
Bwlch-y-fadfa
Castell-
Howell
Cwrt-newydd
Lampeter
(Llanbedr Pont Steffan)
Olwen
Pentrefelin
A487
A482
A485
A486
B4337
B4342
B4577
B4576
B4338
B4459
B4343
Ceredigion Coastal Path
Fantasy Farm Park
Llanerchaeron

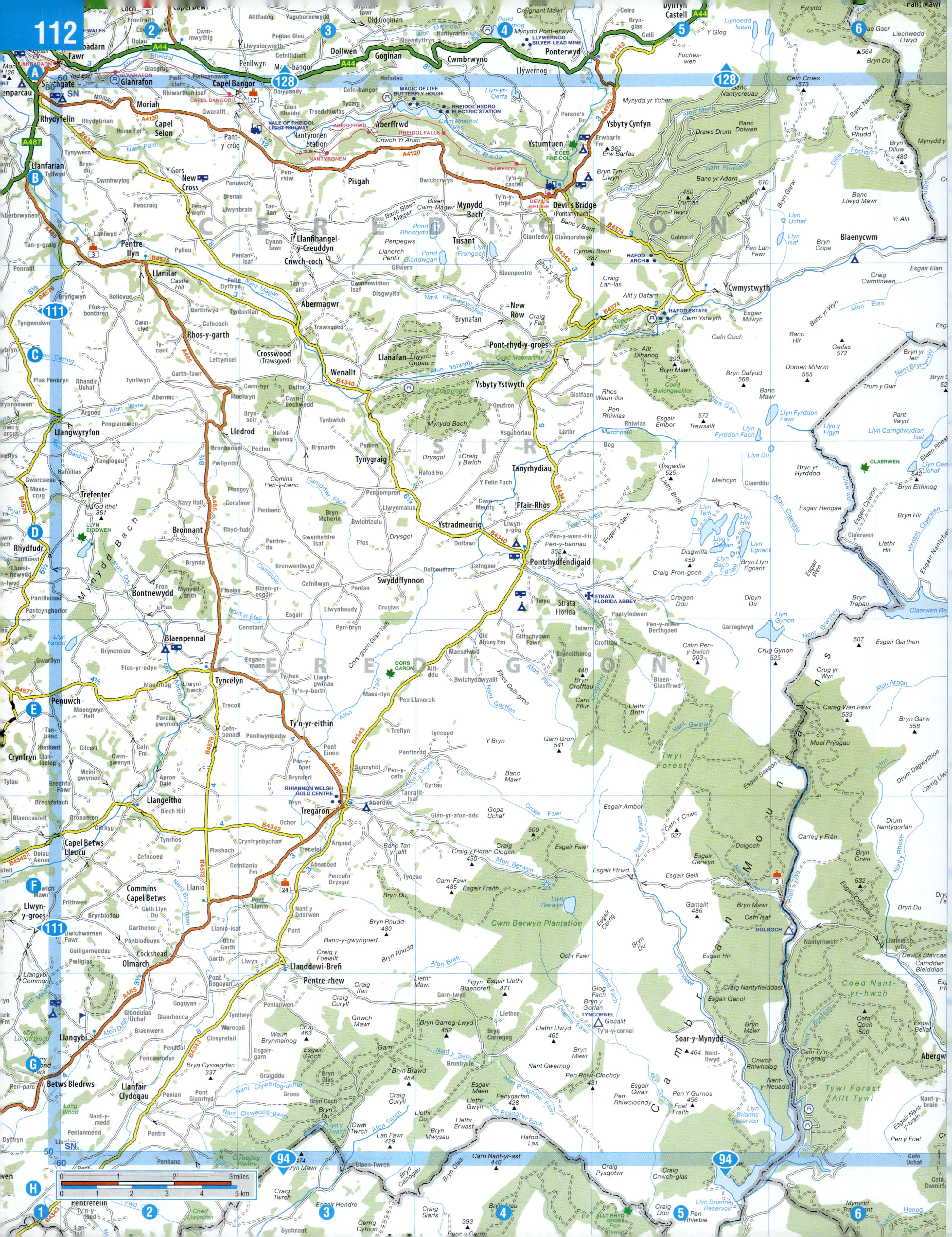

Llanbadarn Fawr
Southgate
Glanrafon
Capel Bangor
Goginan
Dollwen
Cwmbrwyno
Ponterwyd
Devil's Bridge
(Pontarfynach)
Ysbyty Cynfyn
Ystumtuen
Rhydyfelin
Moriah
Capel Seion
Llanfarian
Pisgah
Aberffrwd
Nantyronen Station
Vale of Rheidol Light Railway
Rheidol Falls
Magic of Life Butterfly House
Rheidol Hydro Electric Station
Llywernog Silver-Lead Mine
Pentre-llyn
Llanilar
Llanfihangel-y-Creuddyn
Trisant
Mynydd Bach
New Cross
Cnwch-coch
Abermagwr
New Row
Pont-rhyd-y-groes
Cwmystwyth
Hafod Estate
Hafod Arch
Rhos-y-garth
Crosswood
(Trawsgoed)
Wenallt
Llanafan
Ysbyty Ystwyth
Lledrod
Llangwyryfon
Tynygraig
Tanyrhydiau
Ffair-Rhos
Ystradmeurig
Pontrhydfendigaid
Strata Florida Abbey
Strata Florida
Swyddffynnon
Trefenter
Bronnant
Rhydfudr
Bontnewydd
Blaenpennal
Tyncelyn
Ty'n-yr-eithin
Penuwch
Cors Caron
Rhiannon Welsh Gold Centre
Tregaron
Llangeitho
Capel Betws Lleucu
Commins Capel Betws
Llanio
Llanddewi-Brefi
Pentre-rhew
Olmarch
Llangybi
Betws Bledrws
Llanfair Clydogau
Soar-y-Mynydd
Dolgoch
Tyncornel
Cwm Berwyn Plantation
Twyi Forest
Tywi Forest Allt Tywi
Coed Nant-yr-hwch
Cambrian Mountains
CEREDIGION
SIR CEREDIGION
Claerwen
Llyn Brianne Reservoir
Allt Rhyd y Groes
Llyn Teifi
Llyn Egnant
Llyn Fyrddon Fawr
Llyn Gynon
Llyn Berwyn
Mynydd Bach
Crug Gynon 525
Garn Gron 541
Esgair Fawr
Esgair Gerwyn
Esgair Hir
Blaenycwm
Nant y Maen
Afon Ystwyth
Afon Rheidol
Afon Teifi
Afon Brefi
Afon Tywi
Afon Camddwr
128
111
94
3 miles
5 km

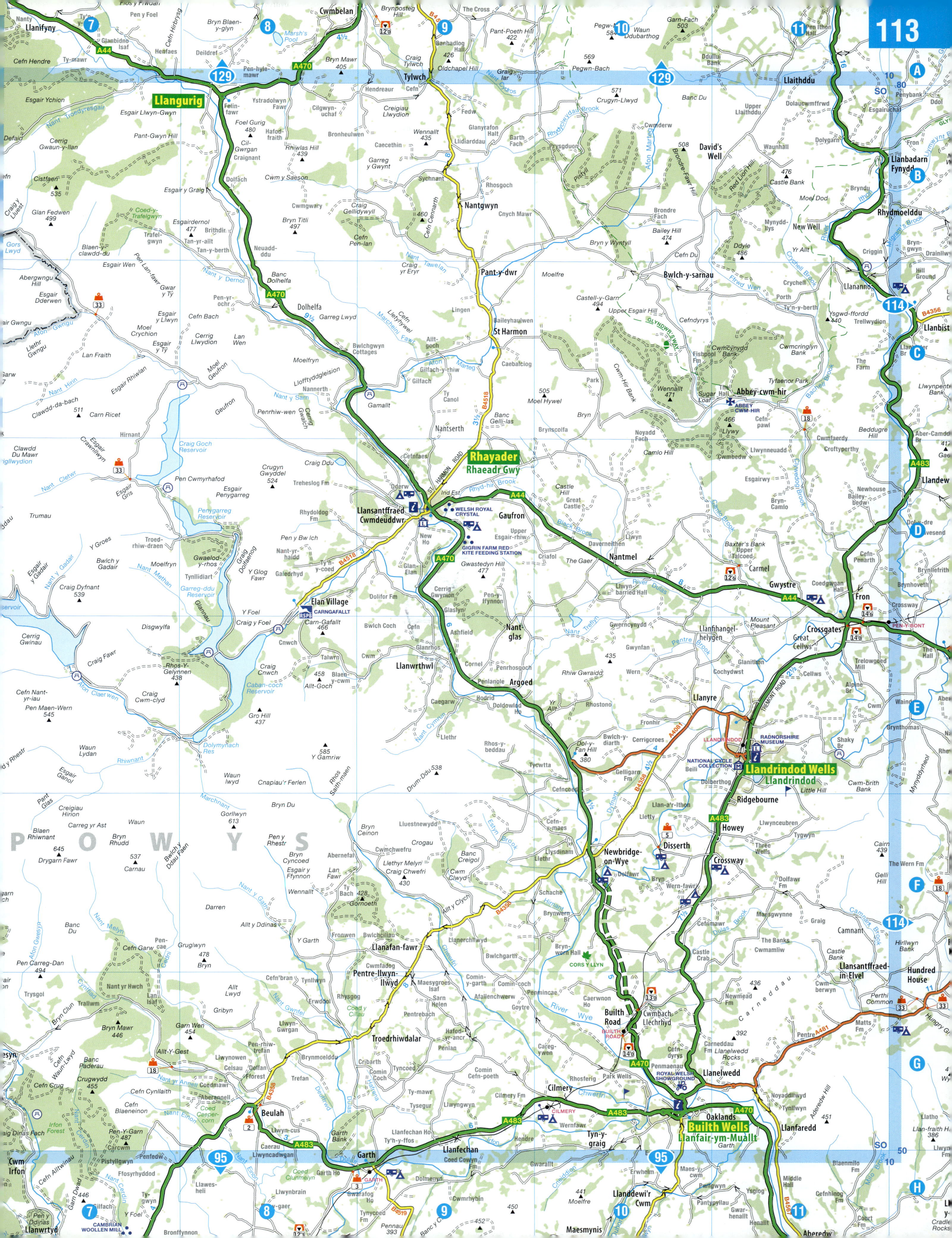
Llanifyny
Cwmbelan
Brynposteg Hill
The Cross
Pegwn-Bach
Garn-Fach 503
Waun Ddubarthog
Pen Ithon Hall
Marsh's Pool
Bryn Blaen-y-glyn
Pen y Foel
Glanbidno Isaf
Cefn Hirbrysg
Deildref
Henfaes
Cefn Hendre
Ty-mawr
A44
A470
Pen-hyle-mawr
Bryn Mawr 405
Tylwch
Craig Tylwch
Banhadlog Hall
426
Oldchapel Hill
Pant-Poeth Hill 422
Graig-lâs
569
Pegwn-Bach
Llaithddu
Dolau Bank
Penybank
Esgairuchaf
Ddol
Llangurig
Esgair Ychion
Felin-fawr
Ystradolwyn Fawr
Hendreaur
Cefn
Cilgwyn-uchaf
Creigiau Llwydion
Fedw
571
Crugyn-Llwyd
Banc Du
Upper Llaithddu
Dolaucwmffrwd
Esgair Llwyn-Gwyn
Nant Troedyresgair
Foel Gurig 480
Cil-Gwrgan
Craignant
Hafod-fraith
Rhiwlas Hill 439
Bronheulwen
Caecethin
Wennallt 435
Glanyrafon Hall
Llidiarddau
Garth Fach
Prysgduon
Cwmderw
508
David's Well
Waunhall
Dolygarn
Fron
Llanbadarn Fynydd
Pant-Gwyn Hill
Cerrig Gwaun-y-llan
Defaid
Garreg y Gwynt
Cwm y Saeson
Sychnant
Rhosgoch
Dolfach
Cistfaen 535
Esgair y Graig
Coed-y-Trafelgwyn
Cwmgwary
Craig Gellidywyll
Cefn Cenarth 460
Nantgwyn
Cnych Mawr
Brondre Fach
Castle Bank
476
Moel Dod
Bryndu
Rhydmoelddu
New Well
Glan Fedwen 499
Trafel-gwyn
Esgairdernol 477
Brithdir
Tan-yr-allt
Tan-y-berth
Bryn Titli 497
Cefn Pen-lan
Bailey Hill 474
Bryn y Wyntyll
Cefn Du
Dyle 486
Yr Allt
Criggin
Bryn-gwyn
Drainllwyn
Gors Lwyd
Blaen-y-clawdd-du
Esgair Wen
Pen Lan-fawr
Neuadd-ddu
Banc Dolhelfa
Craig yr Eryr
Nant Tawelan
Pant-y-dwr
Moelfre
Bwlch-y-sarnau
Crychell
Llananno
Hill Ground
Abergwngu Hill
Esgair Dderwen
Gwar y Tŷ
Pen-yr-ochr
Nant y Dernol
Dolhelfa
Garreg Lwyd
Cefn Lletywel
Lingen
Castell-y-Garn 494
Upper Esgair Hill
Cefndyrys
Porth
Ty'n-y-berth
Ysgwd-ffordd 440
Trellwydion
114
B4356
Llanbister
Esgair Gwngu
Llethr Gwngu
Aton Gwngu
Cefn Bach
Esgair y Llwyn
Moel Crychion
Cerrig Llwydion
Lan Wen
Esgair y Tŷ
Baileyhaulwen
St Harmon
Glyndŵr's Way
Fishpool Fm
Cwmcynydd Bank
Cwmcringlyn Bank
The Farm
Lan Fraith
Bwlchgwyn Cottages
Allt-goch
Gilfach-y-rhiw
Caebalciog
Moelfryn
Moel Geufron
Esgair Rhiwlan
Nant Hirin
Llofftyddgleision
Nannerth
Gilfach
Park
Cwm-Hir Bank
Wennallt 471
Sugar Loaf
Tyfaenor Park
Bachell Brook
Llwynpentre Bank
Clawdd-da-bach
511
Carn Ricet
Nant y Sarn
Gamallt
Ty Canol
505
Moel Hywel
Bryn
Abbey-cwm-hir
Abbey Cwm-Hir
466
Llywy
Cefn-pawl
Beddugre Hill
Aber-Camddwr
Geufron
Penrhiw-wen
Cerrig Gwalch
Nantserth
B4518
Banc Gelli-las
Brynscolfa
Noyadd Fach
Camlo Hill
Cwmfaerdy
Croftyperthy
Hirnant
Craig Goch Reservoir
Clawdd Du Mawr
Esgair Crawnllwyn
33
Crugyn Gwyddel 524
Craig Ddu
Treheslog Fm
Cefnfaes
Rhayader
Rhaeadr Gwy
Llwynneuadd
Cwmbedw
Esgairwy
Nant Clettwr
Pen Cwmyrhafod
Esgair Penygarreg
Esgair Gris
Dderw
Ind Est
Rhyd-hir Brook
Castle Hill
Great Castle
Bryn-Camlo
Newhouse
Bailey-bedw
Llandewi
A483
Trumau
Penygarreg Reservoir
Rhydoldog Fm
Llansantffraed-Cwmdeuddwr
Welsh Royal Crystal
Gaufron
Upper Esgair-rhiw
Black Brook
Daverneithen
Cwm-dre
Gravesend
Troed-rhiw-draen
Y Groes
Pen y Bw Ich
Nant-yr-haidd
New Ho
Gigrin Farm Red Kite Feeding Station
Criafol
The Gaer
Nantmel
Baxter's Bank
Upper Talcoed
Carmel
Cefn-Penarth
Bwlch y Gadair
Moelfryn
Gwaelod-y-rhos
Craig Dolfaenog
Ty'n-y-coed
Glan-Elan
A470
Gwastedyn Hill 477
Esgair y Gadair
Nant y Gadair
Craig Dyfnant 539
Tynllidiart
Y Glog Fawr
Galedrhyd
Garreg-ddu Reservoir
Elan Village
Carngafallt
Dolifor Fm
Cerrig Gwynion
Glaslyn
Pen-y-ffynnon
Llwyn-barried Hall
River Dulas
Gwystre
Coedgwgan Hall
Fron
Brynllefrith
Brynhoveth
Glannau
Y Foel
Craig y Foel
Carn-Gafallt 466
Bwlch Coch
Cefn
Ashfield
Nant-glas
Gwerncynydd
Llanfihangel-helygen
Mount Pleasant
Crossgates
Crossway
Pen-y-Bont
Cerrig Gwinau
Disgwylfa
Cnwch
Cwm
Glanrhos
Llanwrthwl
Nant Treflyn
Gwynfan
Pentre Brook
Great Cellws
Glanithon
Cellws
Trelowgoed Mill
The Hall
Craig Fawr
Rhos-Y-Gelynnen 438
Craig Cwm-clyd
Craig Cnwch
458
Blaen-y-cwm
Allt-Goch
Caban-coch Reservoir
Cornel
Penlanole
Penrhosgoch
Argoed
Rhiw Gwraidd
435
Wern
Cochydwst
Alpine Br
Waina
Cefn Nant-yr-iau
Pen Maen-Wern 545
Aton Claerwen
Gro Hill 437
Caegarw
Hodrid
Doldowlod Ho
Yr Allt
Rhostono
Llanyre
Fronhir
Llandrindod
Radnorshire Museum
Shaky Br
Brynthomas
Waun Lydan
Rhiwnant
Dolymynach Res
585
Y Gamriw
Llethr
Rhos-y-beddau
Dol-y-Fan Hill 380
Bwlch-y-diarth
Cerrigcroes
A4081
National Cycle Collection
Beili
Llandrindod Wells
Llandrindod
Esgair Ganol
Waun lwyd
Cnapiau'r Ferlen
Rhos Saith-maen
Drum-Ddu 538
Tycwtta
Cefncoed
Gelligarn Fm
B4358
Dolberthog
Little Hill
Cwm-brith Bank
Mynyddtheol
Pant Glas
Marchnant
Gorllwyn 613
Bryn Du
Llan-a'r-ithon
Ridgebourne
Creigiau Hirion
Carreg yr Ast
Waun
Bryn Ceinon
Lluestnewydd
Crogau
Cefn-y-maes
Dyfnant
Lletty
Howey
Llwynceubren
Tygwyn
Blaen Rhiwnant
645
Drygarn Fawr
Bryn Rhudd
537
Carnau
Bwlch y Ddau Faen
Pen y Rhestr
Bryn Cyncoed
Esgair y Ffynnon
Abernefal
Cwmchwefru
Lethyr Melyn
Craig Chwefri 430
Banc Creigol
Cwm Clwyd
Llysdinam
Newbridge-on-Wye
Disserth
Crossway
Three Wells
Cairn 439
The Wern Fm
Gelli Hill
POWYS
Nant y Garth
Wennallt
Lan Fawr
Ty Bach
428
Gornoeth
Allt y Clych
B4358
Schache
Dolfawr
Bryn
Wern-fawr
Dolfawr Fm
Maesgwynne
Graig
Camnant
Darren
Allt y Ddinas
Banc Du
Nant Melyn
Y Garth
Fronwen
Bwlchcilau
Llanafan-fawr
Llanerchllwydd
Brynwern Br
Cefnmawr
Dulas Brook
The Banks
Cwmamliw
Castle Bank
Hirllwyn Bank
Pen-cae
Cefn Garw
Gruglwyn
478
Bryn
Bwlchgarth
Bryn-wern Hall
Cors y Llyn
Castle Crab
Llansantffraed-in-Elvel
Hundred House
Pen Carreg-Dan 494
Cefn'bran
Tynllwyn
Cwmfadog
Pentre-llwyn-llwyd
Maesygroes Isaf
Comin-y-garth
Comin-coch
Penmincae
436
Cwm-berwyn
Trysgol
Nant yr Hwch
Lan Isaf
Allt Lwyd
Gribyn
Erwdd
Rhysgog
Coed y Cilau
Sarn Helen
Afallenchwerw
Goytre
Caerwnon Ho
River Wye
Builth Road
Cwmbach Llechrhyd
Newmead Fm
Carneddau
Perthi Common
Matts Fm
Trallwm
Bryn Mawr 446
Garn Wen 454
Llwyn-Gwrgan
Pentrebach
Hafod-yr-ancr
Penlan
Careg-ywen
Cefn-dyrys
392
Carneddau Fm
Llanelwedd Rocks
Pentre
A481
Bryn Clun
Cefn Waun-Lwyd
Allt-Y-Gest
Llwynowen
Pen-rhiw-trefan
Celsau
Dolfan Fforest
Brynmoelddu
Troedrhiwdalar
Cribarth
Comin Coch
Tyncoed
Comin Cefn-poeth
Rhosferig
Park Wells
Penmaenau
Royal Welsh Showground
Llanelwedd
Banc Paderau
Crugwydd 455
Cefn Cynllaith
Aberannell
Coed Caeper-corn
Trefan
Ty-mawr
Tysegur
Llwyngwyn
Cilmery Fm
Cilmery
Chwefri
Noyaddllwyd
Tynllwyn
Aderedw Hill
Cefn Crug
Cefn Blaeneinon
Irfon Forest
Nant Cerdin
Beulah
Llwyn-cus
Garth Bank
Llanfechan Ho
Ty'h-y-ffos
A483
Wernfawr
Tyn-y-graig
Oaklands
A470
Builth Wells
Llanfair-ym-Muallt
Garth
451
Llanfaredd
Latho
Lan-fraith Hill 386
Pen-Y-Garn 487
Carcwm
Pistyllgwyn
Penfedw
Llwyncadwgan
Garth
Llanfechan
Coed Cowyn
Hendre
Gwaralit
Erwhelm
Maes-y-cwm
Blaenmilo Fm
Milo Brook
Cwm Irfon
Cefn Alltwinau
Garn Dwad
446
Gilfach
Pen y Ddinas
Llanwrtyd
Ty-gwyn
Y Foel
Cambrian Woollen Mill
Ffosyrhyddod
Llawes-heli
Llwynbrain
Coed Clunmelyn
Garth Ho
Gwarafog Ho
Tynycoed Fm
Pennau 393
Dolmenyn
Cwmrhybin
Banc y Cwm
452
450
441
Moelfre
Llanddewi'r Cwm
Pwllgwyn
Pantypyllau
Ysgiog
Gwar-henallt
Henallt
Middle Hall
Cefnhinog Fm
Court Fm
Aberedw
B4567
Bronffynnon
Maesmynis
Cradle Rocks

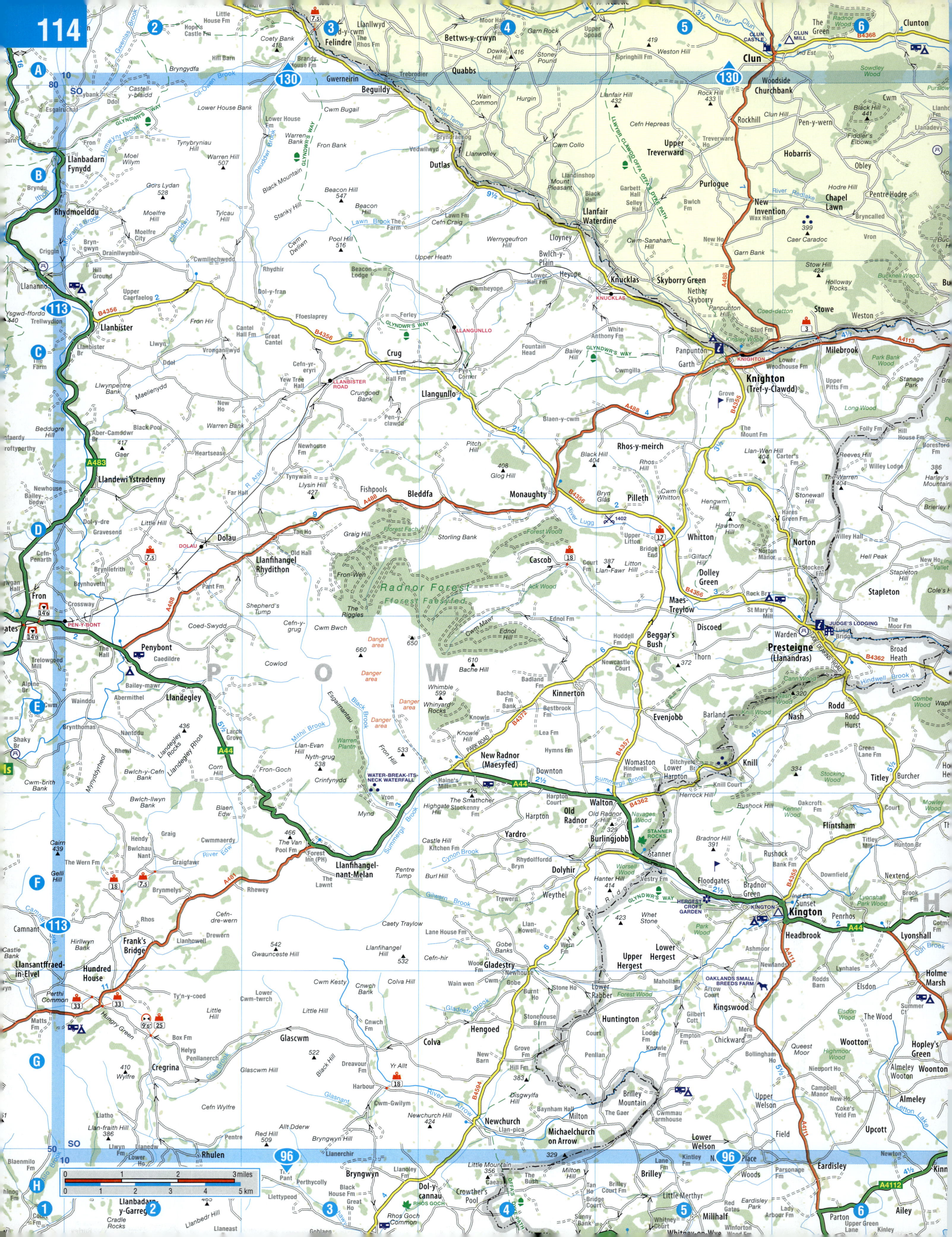
Felindre
Bettws-y-crwyn
Quabbs
Clun
Clunton
Gwerneirin
Beguildy
Woodside
Churchbank
Llanbadarn Fynydd
Dutlas
Upper Treverward
Hobarris
Purlogue
New Invention
Chapel Lawn
Rhydmoelddu
Llanfair Waterdine
Lloyney
Knucklas
Skyborry Green
Llananno
Llanbister
Crug
Llangunllo
Stowe
Panpunton
Garth
Knighton (Tref-y-Clawdd)
Milebrook
Rhos-y-meirch
Llandewi Ystradenny
Fishpools
Bleddfa
Monaughty
Pilleth
Whitton
Norton
Dolau
Llanfihangel Rhydithon
Cascob
Radnor Forest
Fforest Faesyfed
Dolley Green
Stapleton
Maes-Treylow
Discoed
Fron
Penybont
Beggar's Bush
Presteigne (Llanandras)
POWYS
Llandegley
Kinnerton
Evenjobb
Rodd
Nash
New Radnor (Maesyfed)
Knill
Titley
Walton
Old Radnor
Burlingjobb
Flintsham
Yardro
Llanfihangel-nant-Melan
Dolyhir
Kington
Cregrina
Frank's Bridge
Hundred House
Llansantffraed-in-Elvel
Gladestry
Upper Hergest
Lower Hergest
Huntington
Kingswood
Hengoed
Glascwm
Colva
Wootton
Newchurch
Michaelchurch on Arrow
Rhulen
Bryngwyn
Brilley
Eardisley
Upcott
Dol-y-cannau
Crowther's Pool
Little Merthyr
Millhalf
Llanbadarn-y-Garreg
3 miles
5 km

SHROPSHIRE HILLS
HEREFORDSHIRE
Aston on Clun
Little Brampton
Purslow
Clunbury
Twitchen
Hopton Castle
Hoptonheath
Bedstone
Bucknell
Leintwardine
Buckton
Walford
Brampton Bryan
Adforton
Letton
Birtley
Wigmore
Lingen
Limebrook
Kinsham
Byton
Combe Moor
Lower Lye
Aymestrey
Mortimer's Cross
Shobdon
Ledicot
Lugg Green
Kingsland
Staunton on Arrow
Pembridge
Eardisland
Marston
Weston
Broxwood
Luntley
Dilwyn
Sollers Dilwyn
Weobley
Sarnesfield
Ledgemoor
King's Pyon
Canon Pyon
Wormsley
Calver Hill
Norton Canon
Stokesay
Stokesay Castle
Craven Arms
Broome
Clungunford
Abcott
Marlow
Todding
Kinton
Downton on the Rock
Burrington
Elton
Leinthall Starkes
Leinthall Earls
Yatton
Croft
Lucton
Yarpole
Luston
Eyton
Cobnash
Shirl Heath
Lawton
Monkland
Ivington
Brierley
Aulden
Birley
Knapton Green
Westhope
Upper Hill
Onibury
Bromfield
Culmington
Vernolds Common
Stanton Lacy
Ludlow
Ludford
Overton
Richards Castle
Orleton
Comberton
Bircher
Moreton
Eye
The Broad
Leominster
Barons Cross
Ebnall
Stagbatch
Eaton
Stoke Prior
Wharton
Ford
Marlbrook
Newton
Hope under Dinmore
Bodenham
Bodenham Moor
Maund Bryan
Ashford Bowdler
Ashford Carbonell
Woofferton
Brimfield
Ashton
The Hundred
Kimbolton
Stockton
Middleton on the Hill
Middleton
Gosford
Little Hereford
Tenbury Wells
Berrington
St Michaels
Leysters
Leysters Pole
Whyle
Pudleston
Hatfield
Grafton
Docklow
Steen's Bridge
Humber
Risbury
Marston Stannett
Hegdon Hill
Pencombe
Bowley
Little Cowarne
Grendon Green
Bredenbury
Ullingswick
Felton
Moreton Jeffries
Henley
Snitton
Farden
Knowbury
Caynham
Whitton
Greete
Burford
Nash
Cleehill
Hints
Bitterley
Middleton
Hopton Cangeford
Cleedownton
Cleeton St Mary
Knowle
Hope Bagot
Rockgreen
Sheet
Steventon
Hayton's Bent
Upper Hayton
Titterhill
The Hope
Wootton
Whettleton
Norton
Bache
Leominster
Ludlow Castle
Berrington Hall
Croft Castle
Hampton Court
Queenswood
Burford House & Meadow Gallery
Monkland Cheese Dairy
Dunkertons Cider
Mortimer's Cross Water Mill & Battle Centre
A49
A44
A456
A4110
A4113
A4117
A4112
A417
A4214
B4361
B4362
B4360
B4529
B4368
B4367
B4385
B4530
B4365
B4204
116
131
97
60
50
SO
A
B
C
D
E
F
G
H
7
8
9
10
11

SHROPSHIRE
Wyre Forest
MALVERN HILLS
Cleedownton
Silvington
Cleeton St Mary
Farlow
Oreton
Prescott
Bagginswood
Kinlet
Catherton
Neen Savage
Cleobury Mortimer
Hollywaste
Doddington
Hopton Wafers
Foxwood
Cleehill
Hints
Coreley
Knowle
Hope Bagot
Milson
Tilsop
Nash
Neen Sollars
Mamble
Bayton
Clows Top
Rock
Pound Bank
Lem Hill
Far Forest
Fingerpost
Callow Hill
Bliss Gate
Buttonoak
Pound Green
Upper Arley
Bewdley
Wribbenhall
Ribbesford
Stourport-on-Severn
Areley Kings
Astley Cross
Dunley
Heightington
Shatterford
Kingsford
Blakeshall
Trimpley
Low Habberley
Blakebrook
Kidderminster
Franche
Fairfield
Wolverley
Hartlebury
Waresley
Lincomb
Crossway Green
Tenbury Wells
Burford
Boraston
Newnham Bridge
Lindridge
Eardiston
Frith Common
Pensax
Stockton on Teme
Abberley
Great Witley
Astley
Yarhampton
Noutard's Green
Shrawley
Holt Heath
Holt
Ombersley
Little Witley
Orleton
Stanford on Teme
Stanford Bridge
Hanley William
Hanley Child
Broad Heath
Kyre
Sapey Common
Upper Sapey
Shelsley Walsh
Shelsley Beauchamp
Clifton upon Teme
Martley
Wichenford
Sinton Green
Grimley
Hallow
Harpley
High Lane
Collington
Thornbury
Edwyn Ralph
Tedstone Wafer
Tedstone Delamere
Berrow Green
Horsham
Whitbourne
Knightwick
Lulsley
Broadwas
Bromyard
Bromyard Downs
Grendon Green
Hatfield
Bredenbury
Worcester
Lower Broadheath
Upper Broadheath
Rushwick
St John's
Powick
Leigh
Alfrick
Suckley
Linley Green
Stanford Bishop
Acton Beauchamp
Munderfield Row
Munderfield Stocks
Little Cowarne
Pencombe
Hegdon Hill
Longley Green
Storridge
Malvern Link
Newland
Leigh Sinton
Bransford
Ullingswick
Stoke Lacy
Bishops Frome
Evesbatch
Ridgeway Cross
A456
A443
A44
A4103
A442
A449
B4214
B4204
B4202
B4197
132
115
98
SO
60
50
80
0 1 2 3miles
0 1 2 3 4 5 km

Bromsgrove
Redditch
Droitwich Spa
Longbridge
Catshill
Aston Fields
Alcester
Studley
Bournville
Northfield
King's Norton
Hagley
Blakedown
WORCESTERSHIRE
Feckenham
Inkberrow
Alvechurch
Hopwood
Cofton Hackett
Wychbold
Hanbury
Stoke Prior
Astwood Bank
Harvington
Evesham Road
M5
M42
A38
A441
A435
A422
A44
A46
A448
A449
A491
133
118
99

Olton
Solihull
Shirley
Elmdon Heath
Bickenhill
Hampton in Arden
Meriden
Berkswell
Balsall Common
Coventry
Allesley
Coundon
Whoberley
Earlsdon
Stivichall
Baginton
WEST MIDLANDS
Dorridge
Knowle
Bentley Heath
Monkspath
Cheswick Green
Earlswood
Hockley Heath
Packwood
Lapworth
Kingswood
Baddesley Clinton
Temple Balsall
Fen End
Honiley
Kenilworth
Crackley
Stoneleigh
Ashow
Stareton
Cubbington
Lillington
Tanworth-in-Arden
Rowington
Shrewley
Wroxall
Beausale
Haseley
Hatton
Leek Wootton
Old Milverton
Milverton
Warwick
Royal Leamington Spa
Whitnash
Ullenhall
Henley-in-Arden
Lowsonford
Claverdon
Budbrooke
Hampton on the Hill
Myton
Bishop's Tachbrook
Mappleborough Green
Oldberrow
Morton Bagot
Wootton Wawen
Shelfield
Langley
Wolverton
Norton Lindsey
Sherbourne
Barford
Longbridge
Edstone
Bearley
Snitterfield
Aston Cantlow
Little Alne
Great Alne
Alne End
Wilmcote
Pathlow
Ingon
Wasperton
Ashorne
Newbold Pacey
Hampton Lucy
Charlecote
Alveston
Tiddington
Stratford-upon-Avon
Shottery
Bishopton
Drayton
Binton
Temple Grafton
Exhall
Ardens Grafton
Luddington
Welford-on-Avon
Wellesbourne
Walton
Loxley
Moreton Morrell
Moreton Paddox
Lighthorne
Alcester
Oversley Green
Arrow
Wixford
Broom
Bidford-on-Avon
Marlcliff
Barton
Weston-on-Avon
Clifford Chambers
Atherstone on Stour
Ailstone
Alderminster
Wimpstone
Ettington
Pillerton Hersey
Pillerton Priors
Butlers Marston
Kineton
Little Kineton
Combrook
Long Marston
Preston Pastures
W A R W I C K
M42
M40
A46
A429
A452
A3400
A4189
A439
A422
A4177
1
2
3
4
5
6
A
B
C
D
E
F
G
H
134
117
100
0 1 2 3 miles
0 1 2 3 4 5 km

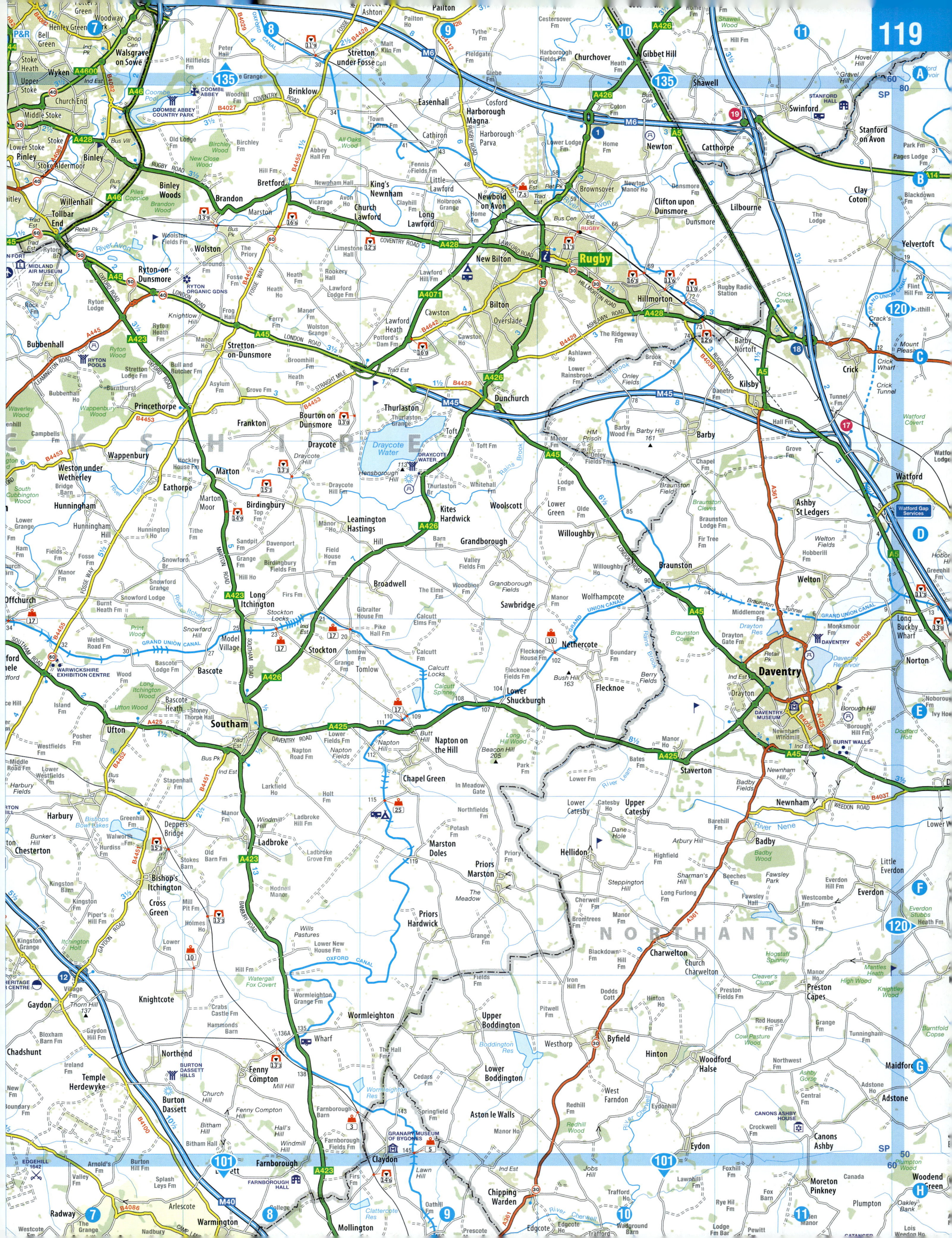

Woodway
Henley Green
Walsgrave on Sowe
Wyken
Stoke Heath
Upper Stoke
Church End
Middle Stoke
Lower Stoke
Pinley
Stoke Aldermoor
Binley
Binley Woods
Willenhall
Tolbar End
Brandon
Bretford
Brinklow
Stretton under Fosse
Easenhall
Harborough Magna
Harborough Parva
Churchover
Gibbet Hill
Shawell
Swinford
Stanford on Avon
Catthorpe
Newton
Clifton upon Dunsmore
Lilbourne
Clay Coton
Yelvertoft
Brownsover
Newbold on Avon
Long Lawford
Church Lawford
King's Newnham
Little Lawford
Cathiron
Coombe Abbey Country Park
Wolston
Marston
Ryton-on-Dunsmore
Midland Air Museum
New Bilton
Rugby
Bilton
Overslade
Cawston
Hillmorton
Barby Nortoft
Kilsby
Crick
Bubbenhall
Stretton-on-Dunsmore
Princethorpe
Frankton
Bourton on Dunsmore
Thurlaston
Dunchurch
Draycote
Draycote Water
Barby
Weston under Wetherley
Wappenbury
Eathorpe
Hunningham
Hunningham Hill
Marton
Birdingbury
Leamington Hastings
Kites Hardwick
Woolscott
Willoughby
Grandborough
Ashby St Ledgers
Watford
Watford Gap Services
Welton
Braunston
Broadwell
Long Itchington
Offchurch
Sawbridge
Wolfhampcote
Nethercote
Flecknoe
Daventry
Drayton
Long Buckby Wharf
Norton
Model Village
Stockton
Bascote
Tomlow
Lower Shuckburgh
Southam
Ufton
Napton on the Hill
Staverton
Newnham
Chapel Green
Harbury
Chesterton
Deppers Bridge
Ladbroke
Marston Doles
Priors Marston
Hellidon
Upper Catesby
Lower Catesby
Badby
Little Everdon
Everdon
Bishop's Itchington
Cross Green
Priors Hardwick
Charwelton
Church Charwelton
Preston Capes
Knightcote
Gaydon
Wormleighton
Upper Boddington
Byfield
Hinton
Woodford Halse
Maidford
Chadshunt
Northend
Fenny Compton
Lower Boddington
Temple Herdewyke
Burton Dassett
Aston le Walls
West Farndon
Adstone
Canons Ashby
Eydon
Farnborough
Claydon
Radway
Arlescote
Warmington
Mollington
Chipping Warden
Edgcote
Moreton Pinkney
Plumpton
Woodend Green
NORTHANTS
Grand Union Canal
Oxford Canal
River Avon
River Nene

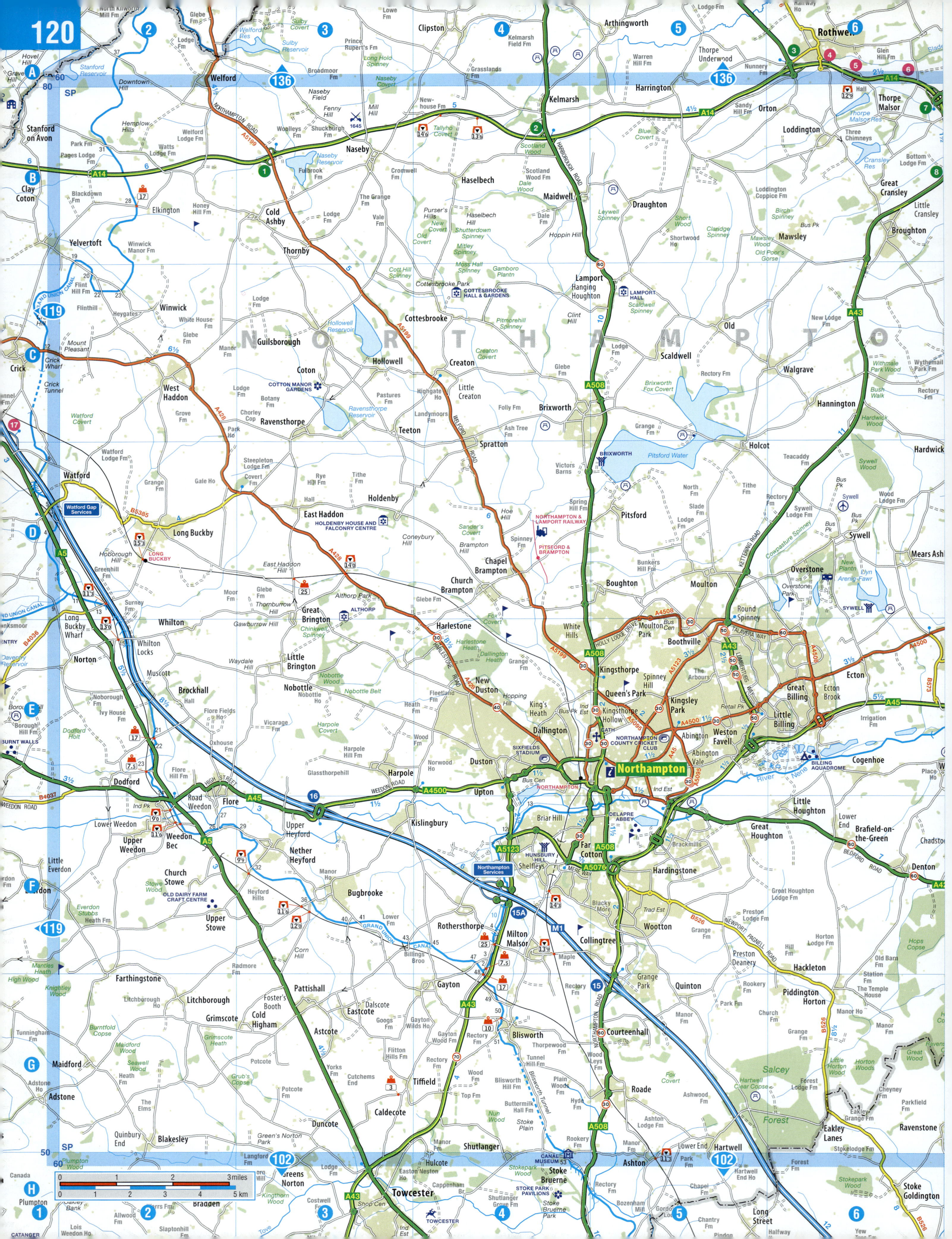
N O R T H A M P T O
Welford
Clipston
Arthingworth
Rothwell
Kelmarsh
Harrington
Thorpe Underwood
Orton
Thorpe Malsor
Loddington
Stanford on Avon
Naseby
Haselbech
Maidwell
Draughton
Great Cransley
Little Cransley
Broughton
Mawsley
Clay Coton
Elkington
Cold Ashby
Thornby
Yelvertoft
Lamport
Hanging Houghton
Winwick
Cottesbrooke
Guilsborough
Old
Scaldwell
Walgrave
Crick
Hollowell
Creaton
Coton
West Haddon
Ravensthorpe
Little Creaton
Brixworth
Hannington
Teeton
Spratton
Holcot
Hardwick
Watford
Watford Gap Services
Holdenby
East Haddon
Pitsford
Pitsford Water
Long Buckby
Chapel Brampton
Church Brampton
Boughton
Moulton
Overstone
Sywell
Mears Ashby
Whilton
Great Brington
Althorp
Harlestone
Long Buckby Wharf
Whilton Locks
Norton
Little Brington
Boothville
Kingsthorpe
Moulton Park
Round Spinney
Ecton
Muscott
Brockhall
Nobottle
New Duston
Queen's Park
Kingsley Park
Great Billing
Little Billing
Dallington
Abington
Weston Favell
Duston
Harpole
Upton
Northampton
Cogenhoe
Dodford
Flore
Weedon
Kislingbury
Upper Heyford
Nether Heyford
Little Houghton
Great Houghton
Brafield-on-the-Green
Denton
Lower Weedon
Upper Weedon
Weedon Bec
Church Stowe
Upper Stowe
Far Cotton
Hunsbury Hill
Shelfleys
Hardingstone
Bugbrooke
Rothersthorpe
Milton Malsor
Collingtree
Wootton
Farthingstone
Gayton
Quinton
Hackleton
Piddington
Horton
Litchborough
Grimscote
Cold Higham
Pattishall
Eastcote
Astcote
Blisworth
Courteenhall
Maidford
Adstone
Tiffield
Roade
Salcey Forest
Caldecote
Duncote
Blakesley
Shutlanger
Hartwell
Ravenstone
Eakley Lanes
Greens Norton
Towcester
Stoke Bruerne
Ashton
Stoke Goldington
Long Street
Plumpton
Bradden
M1
A5
A14
A43
A45
A428
A508
A4500
A5199
SP
136
119
102
3 miles
5 km

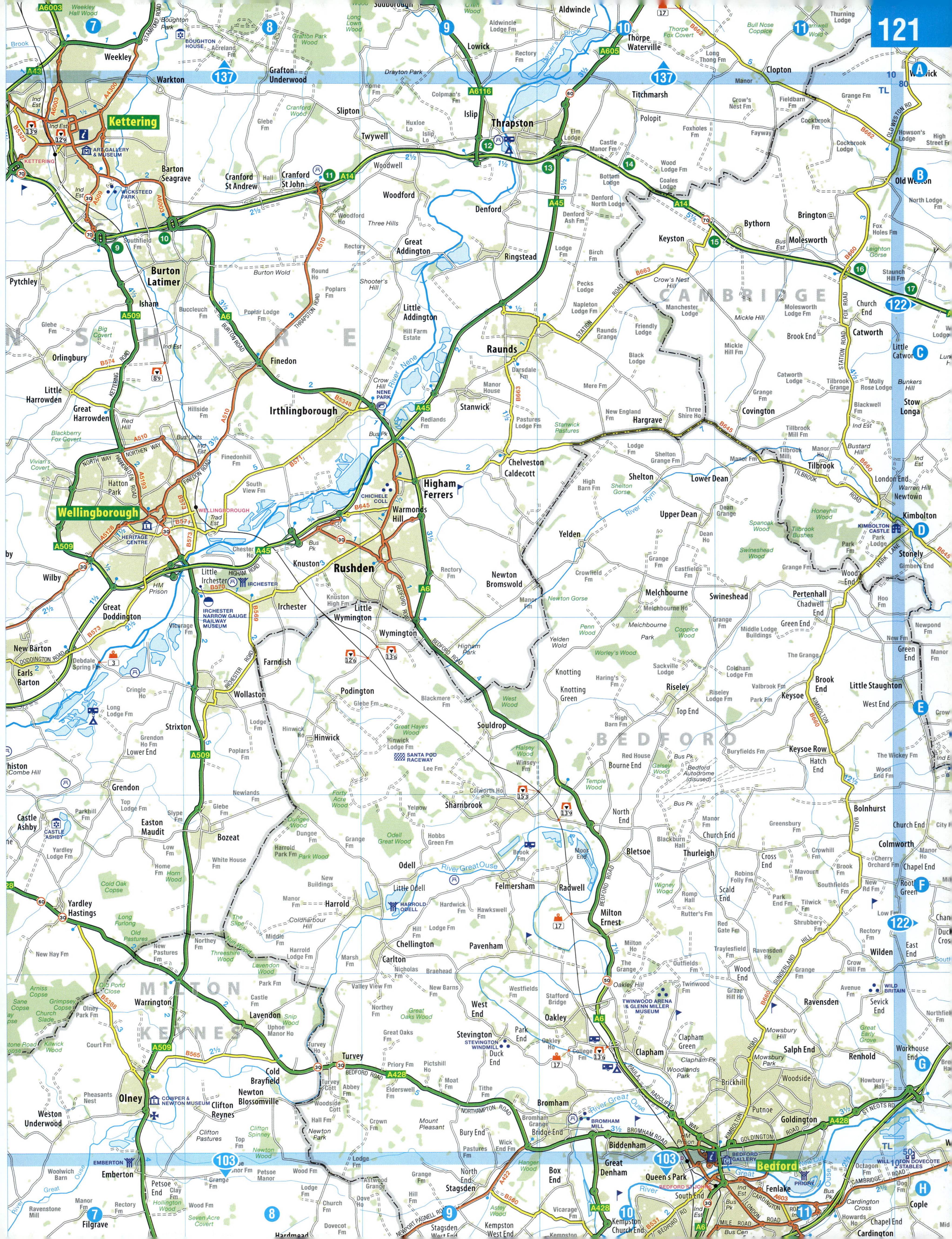
Kettering
Weekley
Warkton
Grafton Underwood
Slipton
Twywell
Lowick
Islip
Thrapston
Aldwincle
Thorpe Waterville
Titchmarsh
Clopton
Polopit
Barton Seagrave
Cranford St Andrew
Cranford St John
Woodford
Denford
Great Addington
Ringstead
Keyston
Bythorn
Brington
Molesworth
Old Weston
Burton Latimer
Isham
Pytchley
Little Addington
Raunds
Finedon
Orlingbury
Little Harrowden
Great Harrowden
Irthlingborough
Stanwick
Hargrave
Covington
Catworth
Higham Ferrers
Chelveston
Caldecott
Shelton
Lower Dean
Upper Dean
Tilbrook
Kimbolton
Stonely
Wellingborough
Rushden
Knuston
Irchester
Little Irchester
Great Doddington
Wilby
Yelden
Newton Bromswold
Melchbourne
Swineshead
Pertenhall
Little Wymington
Wymington
Farndish
New Barton
Earls Barton
Wollaston
Podington
Knotting
Knotting Green
Riseley
Keysoe
Brook End
Little Staughton
Souldrop
Strixton
Hinwick
Grendon
Sharnbrook
Bolnhurst
Colmworth
Castle Ashby
Easton Maudit
Bozeat
Odell
Felmersham
Radwell
Bletsoe
Thurleigh
Milton Ernest
Yardley Hastings
Harrold
Chellington
Carlton
Pavenham
Ravensden
Wilden
Warrington
Lavendon
Olney
Turvey
West End
Stevington
Oakley
Clapham
Cold Brayfield
Newton Blossomville
Clifton Reynes
Bromham
Biddenham
Goldington
Renhold
Salph End
Weston Underwood
Emberton
Great Denham
Queen's Park
Bedford
Fenlake
Kempston
Stagsden
Box End
Filgrave
Cople
Cardington
NORTHAMPTONSHIRE
CAMBRIDGE
BEDFORD
MILTON KEYNES
A6
A14
A45
A509
A428
A6116
A605
A603
B645
B660
River Nene
River Great Ouse
River Kym
Santa Pod Raceway
Irchester Narrow Gauge Railway Museum
Twinwood Arena & Glenn Miller Museum
Cowper & Newton Museum
Kimbolton Castle
Harrold Odell
Stevington Windmill
Bromham Mill
Wicksteed Park
Boughton House

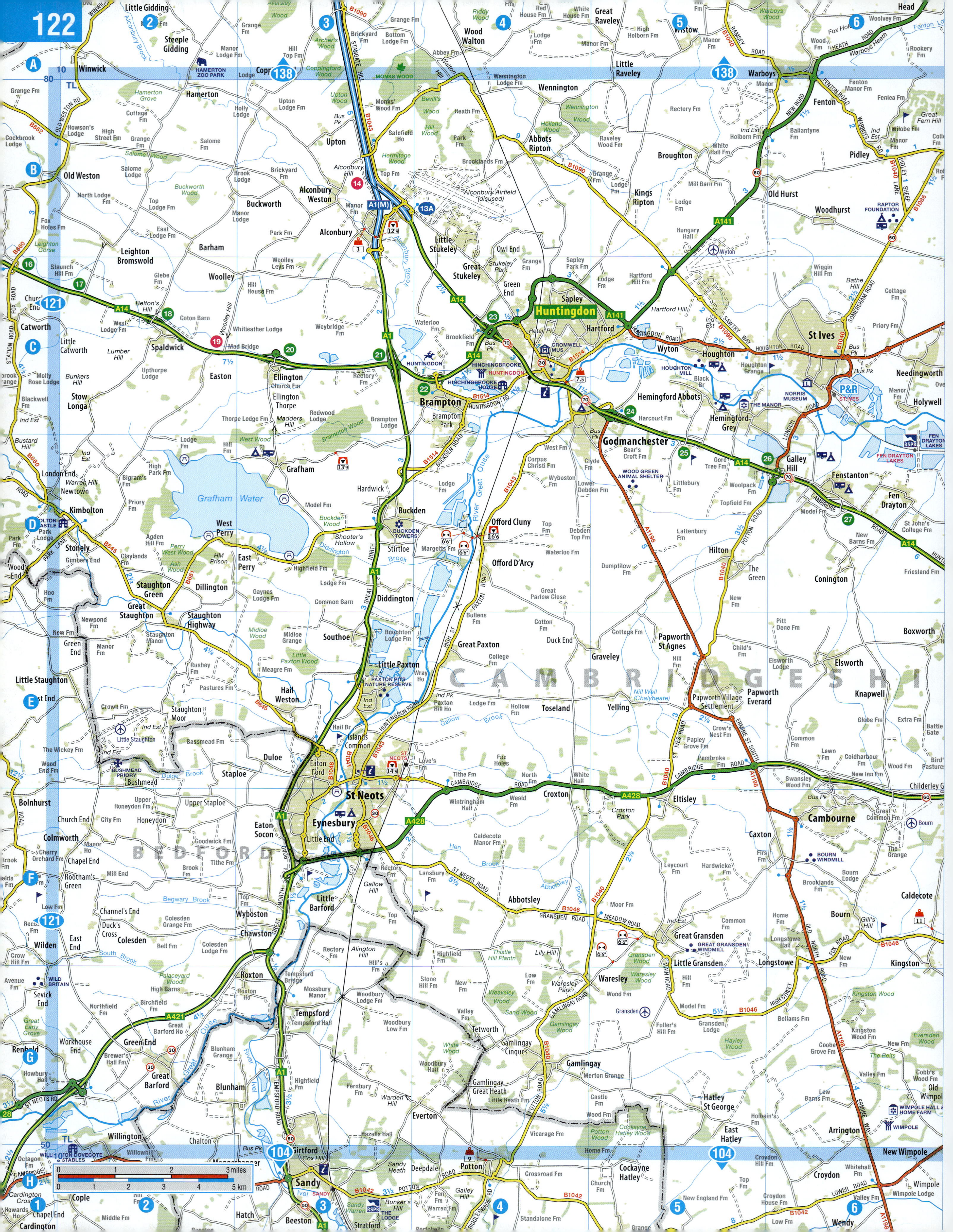

Little Gidding
Steeple Gidding
Winwick
Hamerton
Old Weston
Alconbury Weston
Buckworth
Alconbury
Upton
Leighton Bromswold
Barham
Woolley
Catworth
Spaldwick
Easton
Ellington
Stow Longa
Grafham
Grafham Water
Brampton
Huntingdon
Godmanchester
Hartford
Wyton
Houghton
Hemingford Abbots
Hemingford Grey
St Ives
Fenstanton
Kimbolton
Stonely
West Perry
East Perry
Dillington
Staughton Green
Great Staughton
Staughton Highway
Buckden
Diddington
Southoe
Offord Cluny
Offord D'Arcy
Great Paxton
Little Paxton
Hail Weston
St Neots
Eaton Socon
Eynesbury
Croxton
Eltisley
Papworth Everard
Cambourne
Caxton
Abbotsley
Great Gransden
Waresley
Gamlingay
Tempsford
Roxton
Great Barford
Blunham
Sandy
Potton
Cockayne Hatley
Warboys
Fenton
Pidley
Old Hurst
Woodhurst
Kings Ripton
Abbots Ripton
Wennington
Little Stukeley
Great Stukeley
Hilton
Conington
Elsworth
Knapwell
Boxworth
Graveley
Toseland
Yelling
Papworth St Agnes
Hatley St George
East Hatley
Arrington
Croydon
Wendy
Longstowe
Bourn
Caldecote
Kingston
Wyboston
Chawston
Colmworth
Wilden
Willington
Beeston
Everton
Little Barford
Duloe
Staploe
Bushmead
CAMBRIDGESHIRE
BEDFORD
A1(M)
A1
A14
A141
A428
A1198
A421
A603
B1043
B1040
B1046
B1042
B1090
B645
B660
B661
B1514
TL
1 2 3 miles
1 2 3 4 5 km

123
Ely
Cathedral & Mus
Oliver Cromwell's House
Queen Adelaide
Coveney
Wardy Hill
Mepal
Witcham
Sutton
Wentworth
Witchford
Stuntney
Somersham
Colne
Bluntisham
Earith
The Hermitage
Haddenham
Wilburton
Stretham
Little Thetford
Barway
Soham Cotes
Soham
Aldreth
South Fen
Old Bedford River
River Great Ouse
Chatteris Road
Willingham
Over
Swavesey
Boxworth End
Longstanton
Rampton
Cottenham
Chittering
Upware
Wicken
Wicken Fen
Wicken Corn Windmill
Burwell
Reach
Swaffham Prior
Commercial End
Lode
Anglesey Abbey
Lode Mill
Swaffham Bulbeck
Bottisham
Stow cum Quy
Little Wilbraham
Great Wilbraham
Six Mile Bottom
Fulbourn
Teversham
Cherry Hinton
Cambridge
King's College Chapel
Fitzwilliam Museum
University Botanic Garden
Museum of Technology
Kettle's Yard
Chesterton
Arbury
Kings Hedges
Science Park
Fen Ditton
Horningsea
Milton
Landbeach
Waterbeach
Denny End
Clayhithe
Farmland Mus & Denny Abbey
Histon
Impington
Girton
Oakington
Westwick
Dry Drayton
Bar Hill
Lolworth
Madingley
American Forces Cemetery
Hardwick
Highfields
Coton
Newnham
Grantchester
Barton
Comberton
Toft
Great Eversden
Little Eversden
Harlton
Haslingfield
Trumpington
Great Shelford
Little Shelford
Stapleford
Hauxton
Harston
Newton
Orwell
Barrington
Shepreth
Shepreth Wildlife Park
Foxton
Whittlesford
Sawston
Pampisford
Babraham
Great Abington
Abington
Hildersham
Balsham
West Wratting
Weston Colville
Wandlebury
Gog Magog Hills
Fleam Dyke
River Granta
River Cam
A10
A14
A142
A1123
A1303
A1307
A1309
A1134
A1301
A603
A428
A11
M11
B1049
B1050
B1102
139
105
124

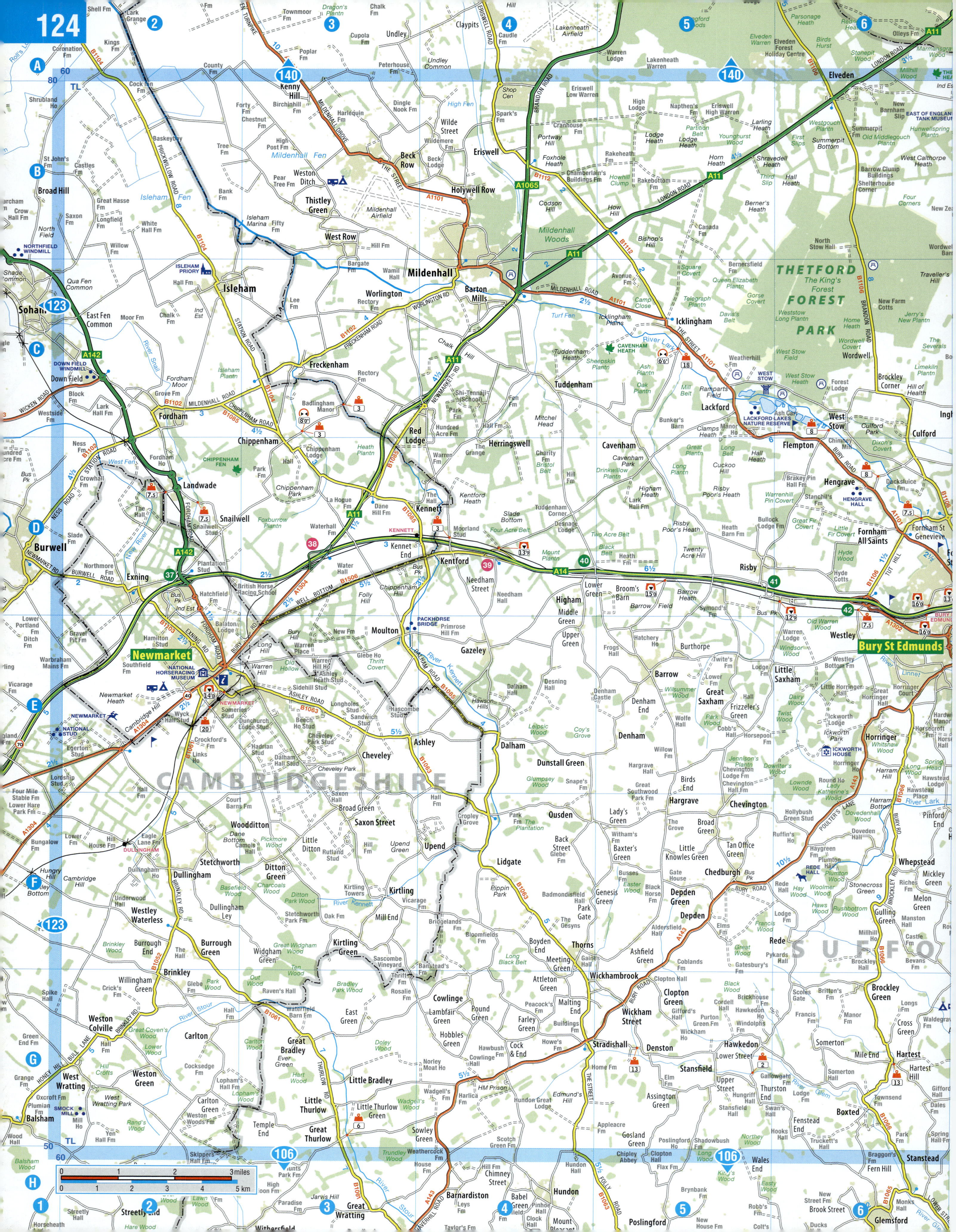

Shell Fm
Lark Grange
Coronation Fm
Kings Fm
County Fm
Townmoor Fm
Dragon's Plantn
Chalk Fm
Cupola Fm
Undley
Claypits
Caudle Fm
Lakenheath Airfield
Warren Lodge
Lakenheath Warren
Elveden Warren
Elveden Forest Holiday Centre
Parsonage Heath
Olleys Fm
Elveden
Poplar Fm
Peterhouse Fm
Undley Common
Kenny Hill
Shop Cen
Eriswell Low Warren
Napthen's Fm
Eriswell High Warren
Larling Heath
Shrubland Ho
Cock Fm
Birchinhill Fm
Forty Fm
Chestnut Fm
Harlequin Fm
Dingle Nook Fm
High Fen
Spark's Fm
Wilde Street
Cranhouse Fm
Lodge Heath
Portway Hill
Foxhole Heath
Westgouch Plantn
Summerpit Bottom
Old Middlegouch Plantn
Hunwellspring Plantn
West Calthorpe Heath
Baskeybay
Tree Fm
High Post Fm
Mildenhall Fen
Beck Row
Beck Lodge
Wildemere Fm
Eriswell
Rakeheath Fm
Chamberlain's Buildings Fm
Horn Heath
Shravedell Heath
St John's Fm
Castles Fm
Broad Hill
Great Hasse Fm
Longfield Fm
Isleham Fen
Pear Tree Fm
Weston Ditch
Thistley Green
Mildenhall Airfield
Holywell Row
Codson Hill
Rakebottom Fm
How Hill
Barrow Clump Buildings
Sheltethouse Corner
Four Corners
New Zealand
Crow Hall Fm
Saxon Fm
North Field
White Hall Fm
Bank Fm
Isleham Marina
Fifty Fm
West Row
Hill Fm
Mildenhall Woods
Canada Fm
Bishop's Hill
North Stow Hall
Northfield Windmill
Willow Fm
Isleham Priory
Hall Fm
Bargate Fm
Wamil Hall
Mildenhall
Barton Mills
Avenue Fm
Square Covert
Bernersfield Fm
Queen Elizabeth Plantn
THETFORD FOREST PARK
The King's Forest
Traveller's Hill
Qua Fen Common
Soham
Isleham
Lee Fm
Worlington
Rectory Fm
Camp Close
Telegraph Plantn
Gorse Covert
Davis's Belt
Icklingham
Icklingham Plains
Turf Fen
New Farm Cotts
Jerry's New Plantn
East Fen Common
Moor Fm
Chalk Fm
Ind Est
Chalk Hill
Westow Long Plantn
Home Heath
Wordwell Covert
Wordwell
West Stow Field
Freckenham
Rectory Fm
Tuddenham Heath
Sheepskin Plantn
Cavenham Heath
Tuddenham
Weatherhill Fm
West Stow
Limekiln Plantn
The Severals
Down Field Windmill
Down Field
Fordham Moor
Grove Fm
Block Fm
Lark Hall Fm
Westside Fm
Fordham
Badlingham Manor
Shi-Tennoji (School)
Fen Fm
Ash Plantn
Oak Plantn
Mill Belt
Ramparts Field
Lackford
Brockley Corner
Hill of Health
Chippenham
Chippenham Fen
Park Fm
Hundred Acre Fm
Red Lodge
Hall Fm
Herringswell
Mitchel Head
Bunker's Barn
Clamps Heath
Lackford Lakes Nature Reserve
West Stow
Culford Park
Culford
Ness Fm
West Fen
Crowhall Fm
Landwade
Fordham Ho
Chippenham Lodge
Chippenham Park
Heath Plantn
Warren Fm
The Grange
Cavenham
Cavenham Park
Charity Fm
Bristol Belt
Great Plantn
Long Belt
Cuckoo Hill
Hall Heath
Flempton
Dixon's Covert
Hengrave
Hengrave Hall
Duckslucice Fm
Burwell
Snailwell
Snailwell Stud
Foxburrow Plantn
La Hogue Fm
Dane Hill Fm
Kennett
Kentford Heath
Slade Bottom
Four Acre Belt
Tuddenham Corner
Desnage Lodge
Two Acre Belt
Higham Heath
Lark Hall Fm
Risby Poor's Heath
Brakey Pin Hall Fm
Warrenhill Pin Covert
Stanchil's Fm
Bullock Lodge Fm
Great Fir Covert
Little Fir Covert
Fornham St Genevieve
Fornham All Saints
Slade Fm
Northmore Fm
Exning
Plantation Stud
Waterhall Fm
Moorland Stud
Kennet End
Kentford
Mount Plantn
Black Belt
Heath Fm
Risby Poor's Heath
Heath Barn Fm
Twenty Acre Hill
Risby
Hyde Wood
Hyde Cotts
Hatchfield Fm
British Horse Racing School
Water Hall
Folly Hill
Chippenham Hill
Needham Street
Needham Hall
Higham
Lower Green
Broom's Barn
Barrow Field
Barrow Heath
Symond's Fm
Old Warren Wood
Westley
Bury St Edmunds
Lower Portland Fm
Gravel Pit Fm
Ditch Fm
Hamilton Stud
Balaton Lodge
Bury Hill
Warren Place
New Fm
Packhorse Bridge
Primrose Hill Fm
Moulton
Middle Green
Upper Green
Frogs Hall
Hatchery Ho
Burthorpe
Warren Lodge
Windsor Wood
Westley Bottom Fm
Little Saxham
Warbraham Mains Fm
Newmarket
Southfield Fm
National Horseracing Museum
Long Hill
Warren Hill
Warren Hill Ho
Ashley Heath Stud
Glebe Ho
Thrift Covert
Gazeley
Desning Hall
Denham Castle
Denham End
Barrow
Willsummer Wood
Great Saxham
Frizzeler's Green
Little Horringer Hall
Horringer Court
Vicarage Fm
Newmarket Heath
Newmarket
Somerles Stud
Sidehill Stud
Longholes Stud
Sandwich Stud
Hascombe Stud
Dalham Hall
Hawson Hills
Wolfe Hall
Park Wood
Cobb's Hall
Horsepool Fm
Ickworth Lodge
Ickworth Park
Ickworth House
Horringer
National Stud
Egerton Stud
Cambridge Hill
Wyck Hall Stud
Dunchurch Lodge Stud
Beech Ho Stud
Cheveley Park Stud
Ashley
Dalham
Leipsic Wood
Coy's Grove
Denham
Willow Fm
Jennison's Plantn
Chevington Lodge Fm
Whitshaw Wood
Crockford's Fm
Links Ho
Hadrian Stud
Dalham Hall Stud
Cheveley
Cheveley Park
Dunstall Green
Glumpsey Wood
Snape's Fm
Hargrave Hall
Birds End
Lowne Wood
Round Ho
Lady Katherine's Wood
Harram Hill
Hawstead Place
Lordship Stud
Four Mile Stable Fm
Lower Hare Park Fm
Hall Fm
CAMBRIDGESHIRE
Court Barns Fm
Saxon Hall
Broad Green
Saxon Street
Hall Fm
Cropley Grove
Park Fm
The Plantation
Ousden
Great Southwood Park Fm
Hargrave
Lady's Green
Chevington
Hollybush Green Stud
Doveden Hall
Harram Bottom
River Lark
Pinford End
Woodditton
Dane Bottom
Camois Hall
Pickmore Wood
Little Ditton
Rutland Stud
Hill Fm
Upend Green
Upend
Witham's Fm
Baxter's Green
Back Street
Glebe Fm
The Grove
Broad Green
Little Knowles Green
Tan Office Green
Ruffin's Ho
Haygreen Fm
Plumton Ho
Whepstead
Bungalow Fm
Lower Fm
Hill House Fm
Eagle Lane Fm
Dullingham
Stetchworth
Ditton Green
Charcoals Wood
Lidgate
Busses Fm
Easter Wood
Gate House
Chedburgh
Plumton Wood
Stonecross Green
Mickley Green
Riches Fm
Hungry Hill
Cambridge Hill
Bottom
Dullingham Ho
Basefield Wood
Dullingham Ley
Kirtling Towers
Kirtling
Vicarage Fm
Pippin Park
Badmondisfield Hall
Park Gate
Genesis Green
Black Horse Fm
Depden Green
Depden
Rede Hall
Hay Wood
Woolmer Wood
Melon Green
Gulling Green
Underwood Fm
Westley Waterless
Ditton Park Wood
Stetchworth Park Fm
River Kennett
Oak Fm
Mill End
Bridgelands Fm
The Gesyns
Haws Wood
Rushbottom Wood
Lodge Fm
Manston Hall
Brinkley Wood
Burrough End
The Hall
Burrough Green
Widgham Green
Great Widgham Wood
Kirtling Green
Sascombe Vineyard
Bloomfields Fm
Boyden End
Long Black Belt
Thorns
Gaines Hall
Meeting Green
Ashfield Green
Aldersfield Hall
Elms Fm
Great Wood
Gatesbury's Fm
Pykards Hall
Rede
Millhill Ho
Castle Fm
Bevans Fm
SUFFOLK
Willingham Green
Brinkley
Ten Wood
Banstead's Fm
Thrift Fm
Rosalie Fm
Attleton Green
Wickhambrook
Clopton Hall
Coblands Fm
Brockley Hall
Crick's Fm
Spike Hall
Glebe Fm
Park Wood
Out Wood
Raven's Hall
Bradley Park Wood
Cowlinge
Lambfair Green
Pound Green
Peacock's Fm
Malting End
Clopton Green
Gifford's Hall
Black Wood
Cordell Hall
Brickhouse Fm
Hawkedon Hall
Windolphs Fm
Scoles Gate
Britton's Fm
Manor Fm
Brockley Green
Longs Fm
Weston Colville
Hall Fm
Carlton
Waterfield Barn Fm
East Green
Hobbles Green
Farley Green
Buildings Fm
Wickham Street
Purton Green Fm
Wickham Ho
Francis Fm
Cross Green
Waldegrave Fm
Green End Fm
Lower Wood
Carlton Wood
Great Bradley
Ever Green
Doley Wood
Hawbush Fm
Cock & End
Howe's Fm
Stradishall
Denston
Hawkedon
Lower Street
Somerton
Mile End
Hartest
Grange Fm
Hill Crofts
Weston Green
Cocksedge Fm
Hart Wood
Norley Moat Ho
Cowlinge Hall
Home Fm
Stansfield
Somerton Hall
Hartest Hill
West Wratting
Oxcroft Fm
Plumian Fm
Smock Mill
West Wratting Park
Lopham's Hall Fm
Carlton Green
Lophams Wood
Westons Woods Fm
Little Bradley
Wadgell's Fm
HM Prison
Harlica Fm
Edmund's Hill
Hundon Great Lodge
Elm Fm
Upper Street
Assington Green
Hungriff Hall
Stansfield Hall
Gallowgate Fm
Thurston End
Lodge Fm
Glem Swan's Hall
Townsend Fm
Gifford Hall
Dales Fm
Balsham
Mill Ho
Rand's Wood
Hill Fm
Temple End
Little Thurlow
Little Thurlow Green
Great Thurlow
Sowley Green
Scotch Green Fm
Appleacre Fm
Gosland Green
Poslingford Ho
Shadowbush Fm
Hooks Hall
Fenstead End
Northey Wood
Boxted
Truckett's Hall
Park Fm
Spring Hall Fm
Wood Hall
Yen Hall Fm
Balsham Wood
Skippers Hall Fm
Weathercock Fm
House Fm
Chipley Abbey
Clopton Hall
Flax Fm
Long Wood
Wales End
Braggon's Fm
Fern Hill
Stanstead
Hunts Park Fm
High Hoon Fm
Jarvis Hill
Paradise Fm
Great Wratting
Barnardiston
Leys Fm
Hill Fm
Chimney Street
Babel Green
Pinhoe Hall
Clock Hall
Hundon Hall
Hundon
Mount Pleasant
Poslingford
Brynbank Fm
New House Fm
Easty Wood
Robb's Fm
Colt's Fm
New Street Fm
Brook Street
Monks Hall
Glemsford
Hare Wood
Streetly Hall
Streetly End
Horseheath
Withersfield
Taylor's Fm
Ducks Hall
Lawn Wood
0 1 2 3 miles
0 1 2 3 4 5 km

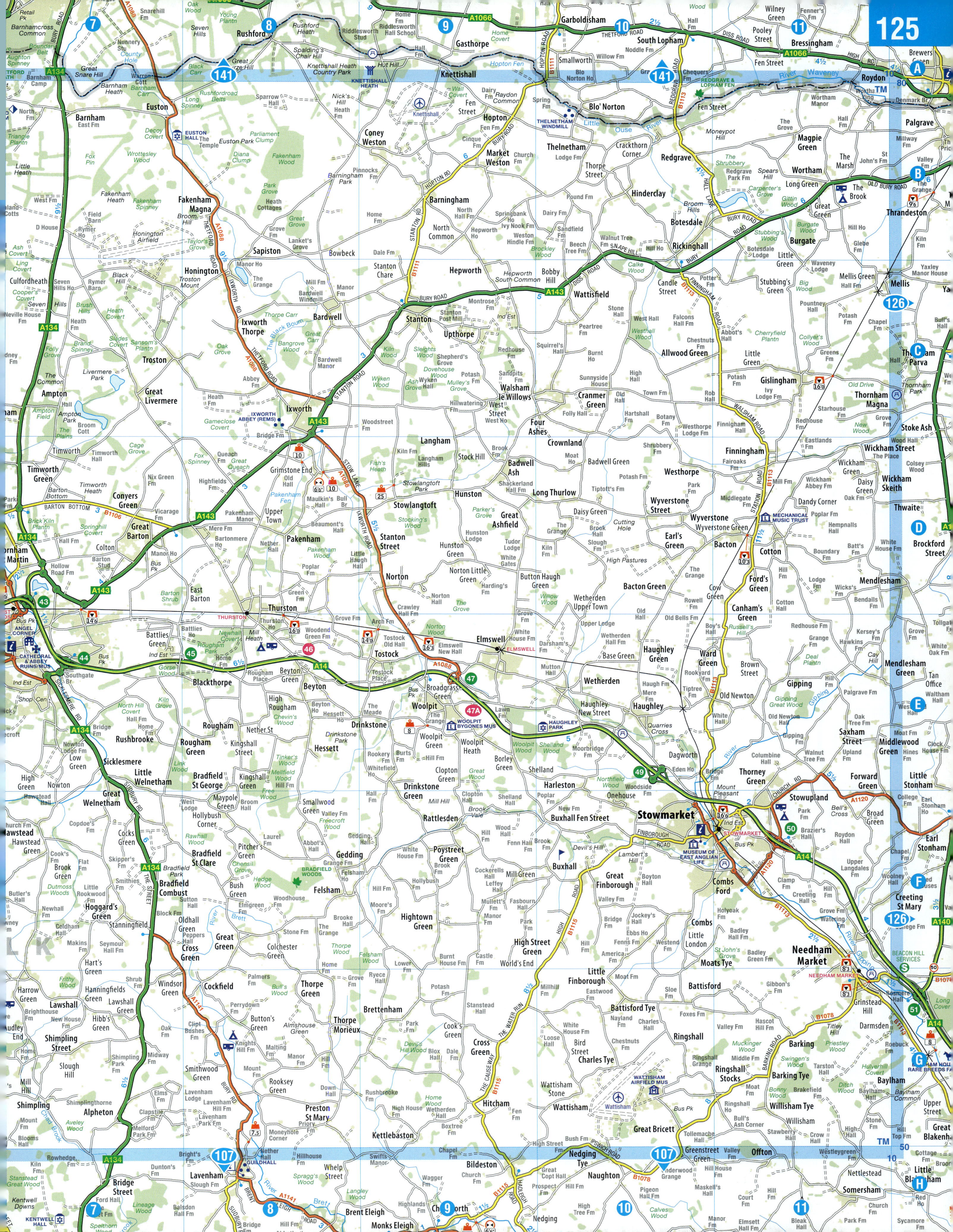
125
Rushford
Garboldisham
Gasthorpe
South Lopham
Bressingham
Roydon
Knettishall
Euston
Barnham
Hopton
Blo' Norton
Coney Weston
Market Weston
Thelnetham
Redgrave
Palgrave
Fakenham Magna
Sapiston
Honington
Barningham
Hepworth
Hinderclay
Botesdale
Rickinghall
Wortham
Burgate
Thrandeston
Mellis
Bardwell
Stanton
Wattisfield
Ixworth Thorpe
Troston
Great Livermere
Ampton
Ixworth
Walsham le Willows
Gislingham
Finningham
Thornham Magna
Langham
Badwell Ash
Westhorpe
Stowlangtoft
Pakenham
Hunston
Great Ashfield
Wyverstone
Bacton
Cotton
Norton
Thurston
Elmswell
Tostock
Beyton
Woolpit
Haughley
Wetherden
Drinkstone
Rougham
Hessett
Shelland
Harleston
Stowmarket
Onehouse
Gedding
Felsham
Rattlesden
Buxhall
Great Finborough
Combs
Needham Market
Battisford
Hitcham
Ringshall
Barking
Bildeston
Kettlebaston
Preston St Mary
Lavenham
Cockfield
Thorpe Morieux
Brettenham
Naughton
Nedging Tye
Offton
Somersham
Little Blakenham
Great Bricett
Sicklesmere
Great Welnetham
Little Welnetham
Bradfield St George
Bradfield St Clare
Bradfield Combust
Stanningfield
Alpheton
Shimpling
Lawshall
Great Barton
Timworth Green
Culford

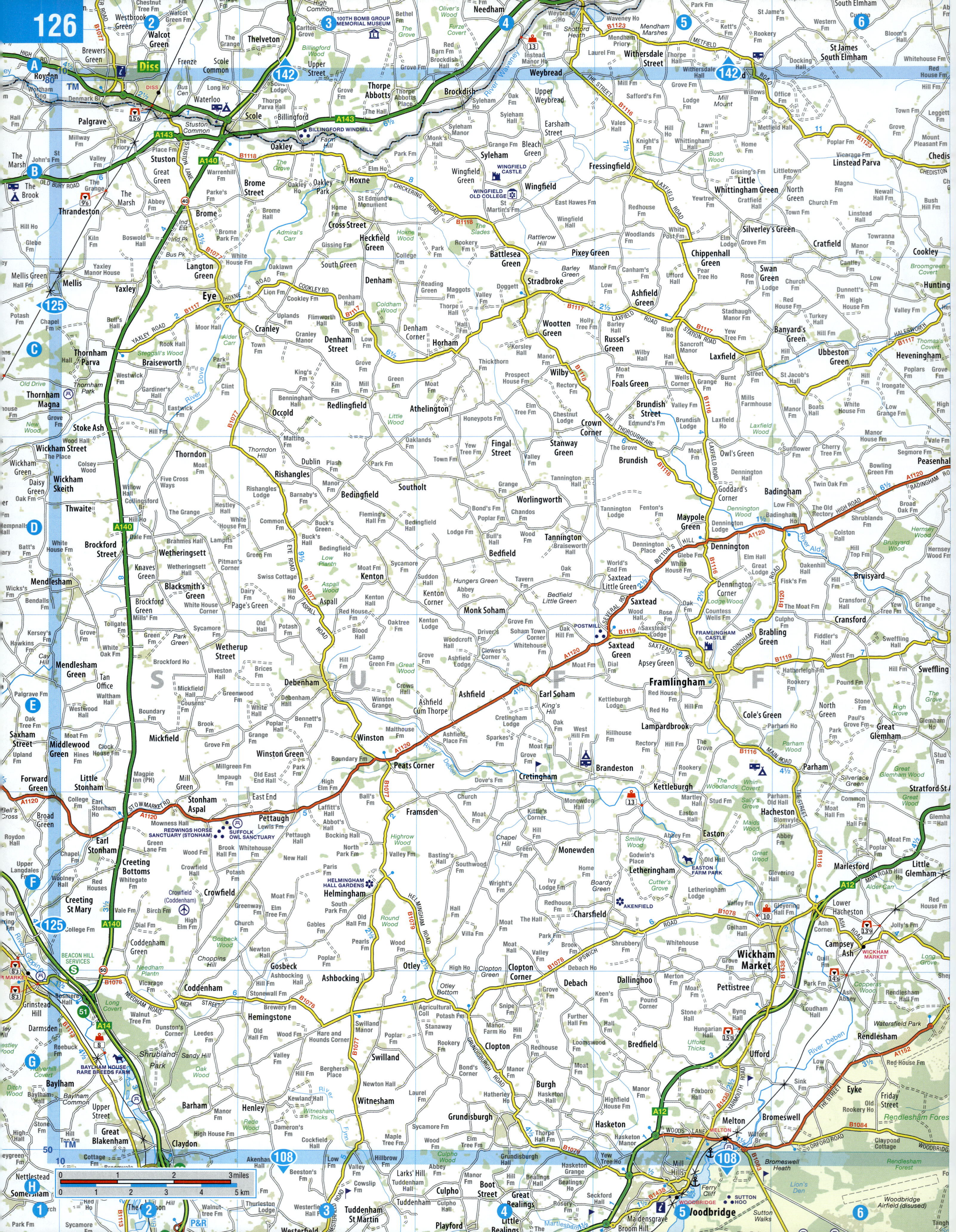

Diss
Roydon
Palgrave
Scole
Thelveton
Billingford
Thorpe Abbotts
Brockdish
Needham
Weybread
Upper Weybread
Withersdale Street
Mendham
St James South Elmham
Oakley
Stuston
Brome
Brome Street
Hoxne
Syleham
Wingfield
Fressingfield
Linstead Parva
Chediston
Thrandeston
Cross Street
Heckfield Green
Battlesea Green
Stradbroke
Pixey Green
Chippenhall Green
Silverley's Green
Cratfield
Cookley
Mellis
Yaxley
Eye
Langton Green
Denham
Cranley
Denham Street
Horham
Wootten Green
Russel's Green
Ashfield Green
Laxfield
Ubbeston Green
Heveningham
Thornham Parva
Braiseworth
Thornham Magna
Occold
Redlingfield
Athelington
Wilby
Foals Green
Brundish Street
Crown Corner
Stanway Green
Fingal Street
Brundish
Owl's Green
Peasenhall
Stoke Ash
Wickham Street
Thorndon
Rishangles
Dublin
Bedingfield
Southolt
Worlingworth
Badingham
Wickham Skeith
Thwaite
Goddard's Corner
Maypole Green
Dennington
Brockford Street
Wetheringsett
Tannington
Bedfield
Kenton
Monk Soham
Saxtead Little Green
Saxtead
Bruisyard
Mendlesham
Blacksmith's Green
Aspall
Kenton Corner
Cransford
Brabling Green
Mendlesham Green
Wetherup Street
Debenham
Ashfield
Earl Soham
Saxtead Green
Framlingham
Sweffling
Palgrave Fm
Saxham Street
Middlewood Green
Mickfield
Winston
Winston Green
Ashfield Cum Thorpe
Peats Corner
Cretingham
Brandeston
Kettleburgh
Lampardbrook
Cole's Green
Great Glemham
Forward Green
Little Stonham
Stonham Aspal
Pettaugh
Framsden
Monewden
Letheringham
Easton
Hacheston
Parham
Stratford St A
Earl Stonham
Creeting Bottoms
Crowfield
Helmingham
Charsfield
Marlesford
Little Glemham
Creeting St Mary
Coddenham Green
Gosbeck
Otley
Clopton Corner
Debach
Dallinghoo
Wickham Market
Pettistree
Campsey Ash
Coddenham
Hemingstone
Ashbocking
Swilland
Clopton
Bredfield
Rendlesham
Ufford
Shrubland Park
Witnesham
Grundisburgh
Burgh
Hasketon
Melton
Bromeswell
Eyke
Baylham
Barham
Henley
Great Blakenham
Claydon
Culpho
Boot Street
Great Bealings
Little Bealings
Tuddenham
Tuddenham St Martin
Playford
Woodbridge
Sutton Hoo
Nettlestead
Somersham
A140
A143
A1120
A1077
B1077
A12
A14
B1078
B1079
B1116
B1117
B1118
B1119
B1120
B1123
A1152
B1084
B1083
B1438
A1120
142
125
108
100TH BOMB GROUP MEMORIAL MUSEUM
BILLINGFORD WINDMILL
WINGFIELD CASTLE
WINGFIELD OLD COLLEGE
FRAMLINGHAM CASTLE
REDWINGS HORSE SANCTUARY (STONHAM)
SUFFOLK OWL SANCTUARY
HELMINGHAM HALL GARDENS
EASTON FARM PARK
BAYLHAM HOUSE RARE BREEDS FARM
BEACON HILL SERVICES
River Waveney
River Dove
River Deben
River Alde
River Fynn
S
U
F
F
0
1
2
3 miles
0
1
2
3
4
5 km

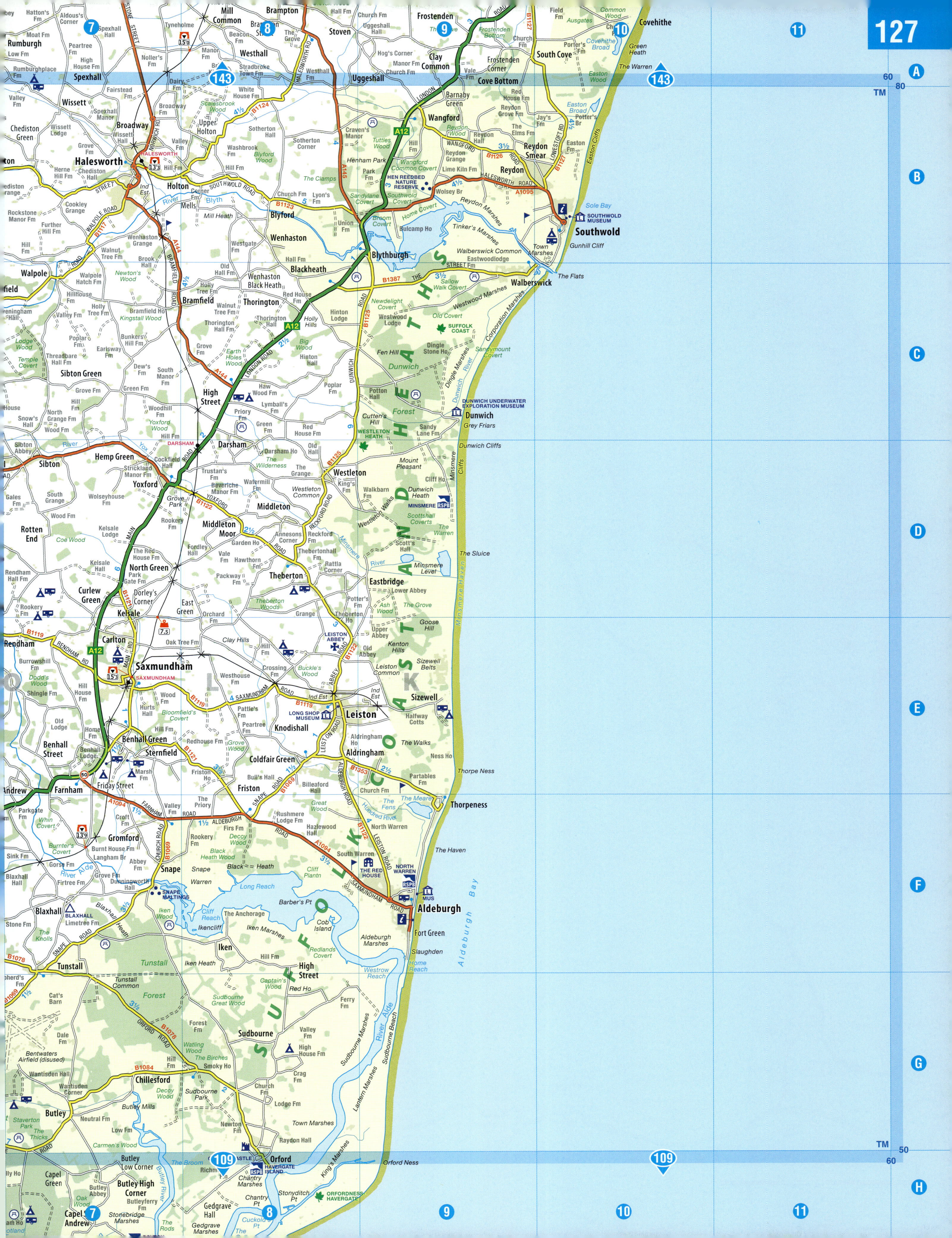

Rumburgh
Spexhall
Wissett
Chediston Green
Halesworth
Holton
Mells
Walpole
Bramfield
Thorington
Sibton Green
Sibton
Hemp Green
Yoxford
Darsham
High Street
Middleton
Middleton Moor
Theberton
Rotten End
Kelsale
North Green
Curlew Green
Carlton
Rendham
Saxmundham
Leiston
Knodishall
Benhall Green
Benhall Street
Sternfield
Friday Street
Farnham
Friston
Coldfair Green
Aldringham
Sizewell
Thorpeness
Gromford
Snape
Blaxhall
Tunstall
Iken
High Street
Aldeburgh
Fort Green
Slaughden
Sudbourne
Chillesford
Butley
Orford
Capel Green
Butley High Corner
Capel St Andrew
Mill Common
Brampton
Stoven
Uggeshall
Frostenden
Clay Common
Covehithe
South Cove
Cove Bottom
Wangford
Reydon Smear
Reydon
Southwold
Blyford
Wenhaston
Blythburgh
Blackheath
Walberswick
Westleton
Eastbridge
Dunwich
Dunwich Forest
Tunstall Forest
Minsmere
Aldeburgh Bay
Orford Ness
Sole Bay
Thorpe Ness
The Haven
The Sluice
The Flats
SUFFOLK
COAST AND HEATHS
A12
A144
A145
A1094
A1095
B1122
B1125
B1119
B1069
B1078
B1084
B1123
B1124
B1126
B1127
B1353
B1387
SOUTHWOLD MUSEUM
DUNWICH UNDERWATER EXPLORATION MUSEUM
LONG SHOP MUSEUM
LEISTON ABBEY
SNAPE MALTINGS
THE RED HOUSE
NORTH WARREN
MINSMERE
HEN REEDBED NATURE RESERVE
ORFORDNESS HAVERGATE
HAVERGATE ISLAND
SUFFOLK COAST
143
109
7
8
9
10
11
A
B
C
D
E
F
G
H
TM
60
80
50

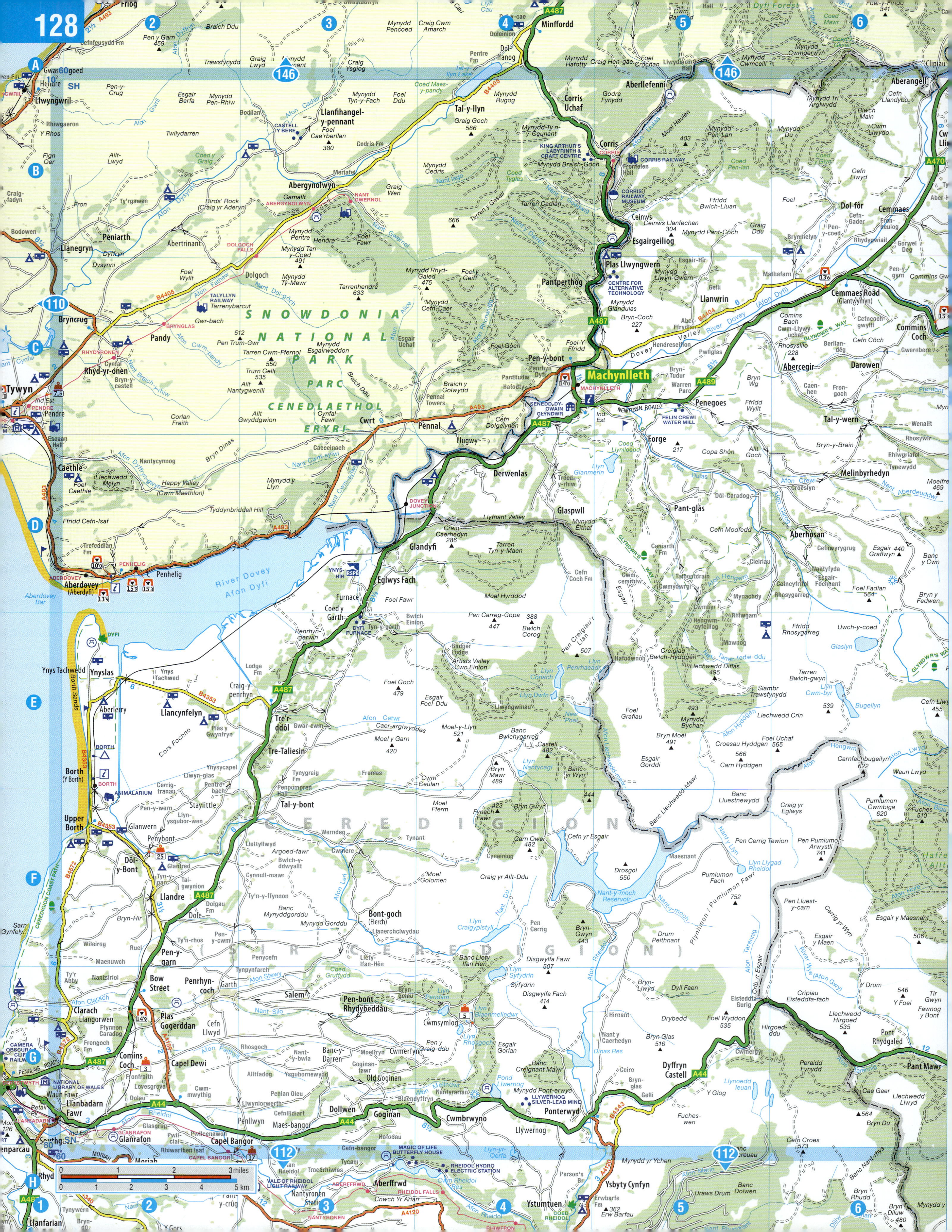

Frlog
Braich Du
Pen y Garn
459
Afon Dyffryn
Mynydd Pencoed
Craig Cwm Amarch
Pentre Ffwrrndan
Minffordd
A487
Dolgellau
Llyn Cau
Dyfi Forest
Cwm Hall
Coed Mawr
Clipiau
Gwasgoed
Cefnfeusydd Fm
Trawsfynydd
Graig Lwyd
Mynydd Pentre
Craig Ysgolg
Mynydd Hafotty
Craig Hen-gae
Foel Crochan
Llwydiarth Hall
Mynydd Cwmcelli
Mynydd Cwmgerwyn
Aberangell
Hendre
Llwyngwril
SH
Pen-y-Crug
Egair Berfa
Mynydd Pen-Rhiw
Mynydd Tyn-y-Fach
Foel Ddu
Coed Maes-y-pandy
Tal-y-llyn Lake
Dol-ffanog
B4405
Mynydd Rugog
Corris Uchaf
Godre Fynydd
Aberllefenni
Mynydd Tri Arglwydd
Cefn Llandybo
Rhiwgaeron
Y Rhos
Twllydarren
Bodilan
Afon Cadair
Castell y Bere
Llanfihangel-y-pennant
Foel Caer'berllan
380
Tal-y-llyn
Graig Goch
586
Mynydd-Ty'n-y-Ceunant
King Arthur's Labyrinth & Craft Centre
Mynydd Braich-Goch
Corris
Corris Railway
Moel Heulen
403
Mynydd Pen-Lan
Mynydd Du
Blwch Main
Cwm Llwydo
Fign Oer
Allt-Lwyd
Coed y Graig
Cedris Fm
Mynydd Cedris
Nant Iago
Coed Tyglas
Fronfelen Hall
Coed Pen-lan
Cefn Llwyd
A470
Craig-fadyn
Ty'rgawen
Fron
Afon Dysynni
Birds' Rock (Craig yr Aderyn)
Abergynolwyn
Gamallt
Nant Gwernol
Graig Wen
666
Tarren y Gesail
Tarren Cadian
Corris Railway Museum
Ffridd Bwlch-Lluan
Foel
Dol-fôr
Cemmaes
Bodowen
Peniarth
Llanegryn
Dyffryn
Dysynni
Abertrinant
Mynydd Pentre
Hendre
Foel Fawr
Cwm Cadian
Ceinws
Ceinws Llanfechan 304
Mynydd Pant-Coch
Graig Ddu
Brynmelyn
Pen-y-coed
Esgairgeiliog
Rhydygwiail
Gorwel Deg
Dolgoch Falls
Mynydd Tan-y-Coed
491
Foel Wyllt
Dolgoch
Mynydd Ty-Mawr
Mynydd Rhyd-Galed
475
Foel y Geiff
Plas Llwyngwern
Centre for Alternative Technology
Esgair-Hir
Mynydd Llwyn-Gwern
Gelli
Mathafarn
Pen-y-gym
Commins Gw
110
Bryncrug
B4405
Talyllyn Railway
Tarrenybarcut
Tarrenhendre
633
Mynydd Cefn-Caer
Pantperthog
Mynydd Glandulas
Llanwrin
Cemmaes Road (Glantwymyn)
Snowdonia National Park
Parc Cenedlaethol Eryri
Brynglas
Gwr-bach
Esgair Uchaf
Bryn-Coch
227
A487
B4404
River Dovey
Afon Dyfi
Commins Bach
Cwm-Llywy-uchaf
Glyndŵr's Way
Cefncoch-gwyllt
Commins Coch
Tywyn
Rhydyronen
Cyntal
Rhyd-yr-onen
Bryn-y-castell
Pandy
Pen Trum-Gwr
512
Mynydd Esgairweddon
Tarren Cwm-Ffernol
550
Trum Gelli
535
Foel Goch
Foel-Y-Ffridd
Dovey Valley
Hendreselfion
Pwllglas
Rhosyslllo 228
Berllan-deg
Cefn Côch
Gwernbere
Allt Nantygwenlli
Braich Du
Braich y Golwydd
Pennal Towers
Pen-y-bont
Penrhyn Dyfi
Pantlludw
Hafodty
Machynlleth
Brynn-Tudur
Warren Parc
A489
Abercegir
Darowen
Ind Est
Pendre
Corlan Fraith
Allt Gwyddgwion
Cynfal-Fawr
Cwrt
Pennal
Cefn Dolgelynen
Senedd-Dy-Owain Glyndwr
Ind Est
Newtown Road
Penegoes
Felin Crewi Water Mill
Ffridd Wyllt
Caen-hen
Fron-goch
Tal-y-wern
Escuan Hall
Bryn Dinas
Caeceinach
Llugwy
A493
Forge
217
Copa Shôn
Allt Goch
Bryn-y-Brain
Rhosywir
Rhiwgriafol
Tynewydd
Caethle
Foel Caethle
Llechwedd Melyn
Afon Dyffryn-gwyn
Nantycynnog
Happy Valley (Cwm Maethlon)
Mynydd y Llyn
Nant Cwm-sylwi
Derwenlas
Coed Llynlloedd
Llyn Glanmerin
Treod-y-rhiw
Afon Dulas
Afon Crewi
Croeslyn
Melinbyrhedyn
Nant Aberdeuddwr
Moelfre 469
Tyddynbriddell Hill
Dovey Junction
Glaspwll
Dôl-Caradog
Pant-glâs
Ffridd Cefn-Isaf
Llyfnant Valley
Mynydd Eithaf
Cefn Modfedd
Aberhosan
Trefeddian
Penhelig
Craig Caerhedyn 286
Tarren Tyn-y-Maen
Glandyfi
Ceniarth Fm
Cefnwyrygrug
Esgair Grafiwyn 440
Banc y Cwn
Aberdovey (Aberdyfi)
Aberdovey Bar
Ynys-hir RSPB
Eglwys Fach
Cefn Coch Fm
Cwm-cemrhiw
Cwmydwr
Talbontdrain
Nant Hengwm
Cleiriau
Nantyfyda
Esgair-Fochnant
Cefncyfrifol
Rhosygarreg
Foel Fadian 564
Bryn y Fedwen
Moel Hyrddod
Furnace
Foel Fawr
Coed y Garth
Dyfi Furnace
Tyn-y-garth
Bwlch Einion
Pen Carreg-Gopa 447
388 Bwlch Corog
Mynachdy
Cwmbyr
Rhiwgam
Hengwm-cyfeiliog
Ffridd Rhosygarreg
Uwch-y-coed
Dyfi
Penrhyn-gerwin
Badger Lodge
Artists Valley
Cwm Einion
Pen Creigiau'r Llan
507
Llyn Penrhaeadr
Hafodwnog
Creigiau Bwlch-Hyddgen
Llechwedd Ddu 495
Afon Tarren-ledw-ddu
Glaslyn
Glyndŵr's Way
Ynys Tachwedd
Ynyslas
Ynys Tachwedd
Borth Sands
Lodge Fm
Craig-y-penrhyn
B4353
A487
Foel Goch 479
Esgair Foel-Ddu
Llyn Conach
Llyn Dwfn
Tarren Bwlch-gwyn
Llyn Cwm-byr
Siambr Trawsfynydd
539
Bugeilyn
Cefn Llwy 455
Aberleri
Llancynfelyn
Plas y Gwynfryn
Tre'r-ddôl
Gwar-cwm
Afon Cetwr
Caer-arglwyddes
Moel y Garn 420
Moel-y-Llyn 521
Llywyngwinau
New Pool
Foel Grafiau
Mynydd Bychan 493
Afon Hyddgen
Llechwedd Crin
Cors Fochno
Borth
Tre-Taliesin
Banc Bwlchygarreg
Castell 482
Llyn Nantycagl
Banc-yr Wyn
Bryn Moel 491
Croesau Hyddgen 566
Foel Uchaf 565
Carn Hyddgen
Hengwm
Carnfachbugeilyn 622
Waun Lwyd
Borth (Y Borth)
Animalarium
Ynysycapel
Llwyn-glas
Tynygraig Fm
Fronlas
Cwm Ceulan
Bryn Mawr 489
Esgair Gorddi
Banc Llechwedd-Mawr
Banc Lluestnewydd
Craig yr Eglwys
Pumlumon Cwmbiga 620
Fuches 510
Cerrig-tranau
Pentre-bach
Staylittle
Pen-y-wern
Llyn-ysgubor-wen
Penpompren Hall
Tal-y-bont
Moel Fferm
Fynach Fawr
423
Bryn Gwyn
444
Ceredigion
Upper Borth
B4353
Glanwern
Penybont
Werndeg
Tynant
Carn Owen 482
Cefn yr Esgair
Maesnant
Pen Cerrig Tewion
Pen Pumlumon-Arwystli 741
Llettyllwyd
Argoed-fawr
Bwlch-y-ddwyallt
Cwmere
Cyneiniog
Drosgol 550
Llyn Llygad Rheidol
Dôl-y-Bont
Glanfred
Tyn-y-garth
Tai-gwynion
Cynnull-mawr
Moel Golomen
Craig yr Allt-Ddu
Pumlumon Fach
Pumlumon Fawr 752
Llandre
A487
Ty'n-y-ffynnon
Dolgau Fm
Afon Leri
Nant Du
Nant-y-moch Reservoir
Nant-y-moch
Plynlimon
Pen Lluest-y-carn
Cerrig yr Wyn
Ceredigion Coast Path
Dole
Banc Mynyddgorddu
Mynydd Gorddu
Bont-goch (Elerch)
Llanerchclwydau
Llyn Craigypistyll
Pen Cerrig
Bryn-Gwyn 443
Drum Peithnant
Esgair y Maesnant
506
Sarn Gynfelyn
Bryn-Hir
Wileirog
Ruel
Ty'n-rhos
Pen-y-cwm
Pen-y-garn
Penycefn
Sir Ceredigion
Lletty-Ifan-Hên
Banc Lletty Ifan Hen
Disgwylfa Fawr 507
Afon Rheidol
Afon Tarennig
Esgair y Maen
Maenuwch
Nantsiriol
Ty'r Abby
Bow Street
Penrhyn-coch
Garth
Tynpynfarch
Afon Stewy
Coed Gruffydd
Llyn Syfydrin
Syfydrin
Disgwylfa Fach 414
Bryn-Llwyd
Dyll Faen
Crib yr Esgair
Afon Wye (Afon Gwy)
Y Drum
546
Y Foel
Tir Gwyn
Fawnog y Bont
Afon Clarach
Clarach
Llangorwen
Ffynnon Caradog
Plas Gogerddan
Cefn Llwyd
Nant Silo
Salem
Pen-bont Rhydybeddau
Bryn-goleu
Llyn Pendam
Cwmsymlog
Llyn Blaenmelindwr
Hirnant
Eisteddfa Gurig
Cripiau Eisteddfa-fach
Llechwedd Hirgoed 535
Drybedd
Foel Wyddon 535
Hirgoed-ddu
Pont Rhydgaled
Frongoch Fm
Comins Coch
Capel Dewi
Rhosgoch
Nant-y-bwla
Banc-y-Darren
Moelfryn
Cwmerfyn
Pen y Graig-ddu
Llyn Rhosgoch
Esgair Gorlan
Nant y Caerhedyn
Dinas Res
Bryn Glas 516
Cwmergyr
Peraidd Fynydd
Pant Mawr
Camera Obscura
Cliff Railway
Penglais Road
A487
Aberystwyth
National Library of Wales
Waun Fawr
Lovesgrove
Alltfadog
Ysguborneywdd
Goginan-fawr
Old Goginan
Afon Melindwr
Nantyrarian
Pond Llwernog
Banc Creignant Mawr
Mynydd Pont-erwyd
Llywernog Silver-Lead Mine
Ceiro
Bryn-glas
Gelli
Dyffryn Castell
A44
Llynoedd Ieuan
Y Glog
Cae Gaer
Llechwedd Llwyd
Llanbadarn Fawr
Cwm-mwythig
Penlan Oleu
Llwyniorwerth
Cefnllidiart
Dolau
Blaendyffryn
Ponterwyd
A44
Fuches-wen
Bryn Du
564
Retail Park
Llanbadarn
Afon Rheidol
Glasgrug
Glanrafon
Pwll-clai
Pwllcenawon
Penllwyn
Maes-bangor
Dollwen
Goginan
Cwmbrwyno
Llywernog
A4120
Cefn Croes 573
Penparcau
Southgate
SN
Capel Bangor
Rhiwarthen Isaf
Halfodau
Cefn-bangor
Magic of Life Butterfly House
Llyn-yr-Oerfa
Mynydd yr Ychen
Banc Bwlch yr Rhosau
Moriah
Capel Bangor
Rheidol Hydro Electric Station
Vale of Rheidol Light Railway
Troedrhiwlas
Tycam
Aberffrwd
Aberffrwd
Cwm Rheidol Res
Parson's Bridge
Ysbyty Cynfyn
Rhyd
Nantyronen Station
Cnwch Yr Arian
Rheidol Falls
Ystumtuen
Coed Rheidol
Erwbarfe Fm
362
Erw Barfau
Draws Drum
Banc Dolwen
Bryn Rhudd
Bryn Diluw 480
Mynydd y
Llanfarian
A487
Tynywern
Y Gors
A4120
Rhiwfron
0 1 2 3 miles
0 1 2 3 4 5 km
A
B
C
D
E
F
G
H
1
2
3
4
5
6
146
112

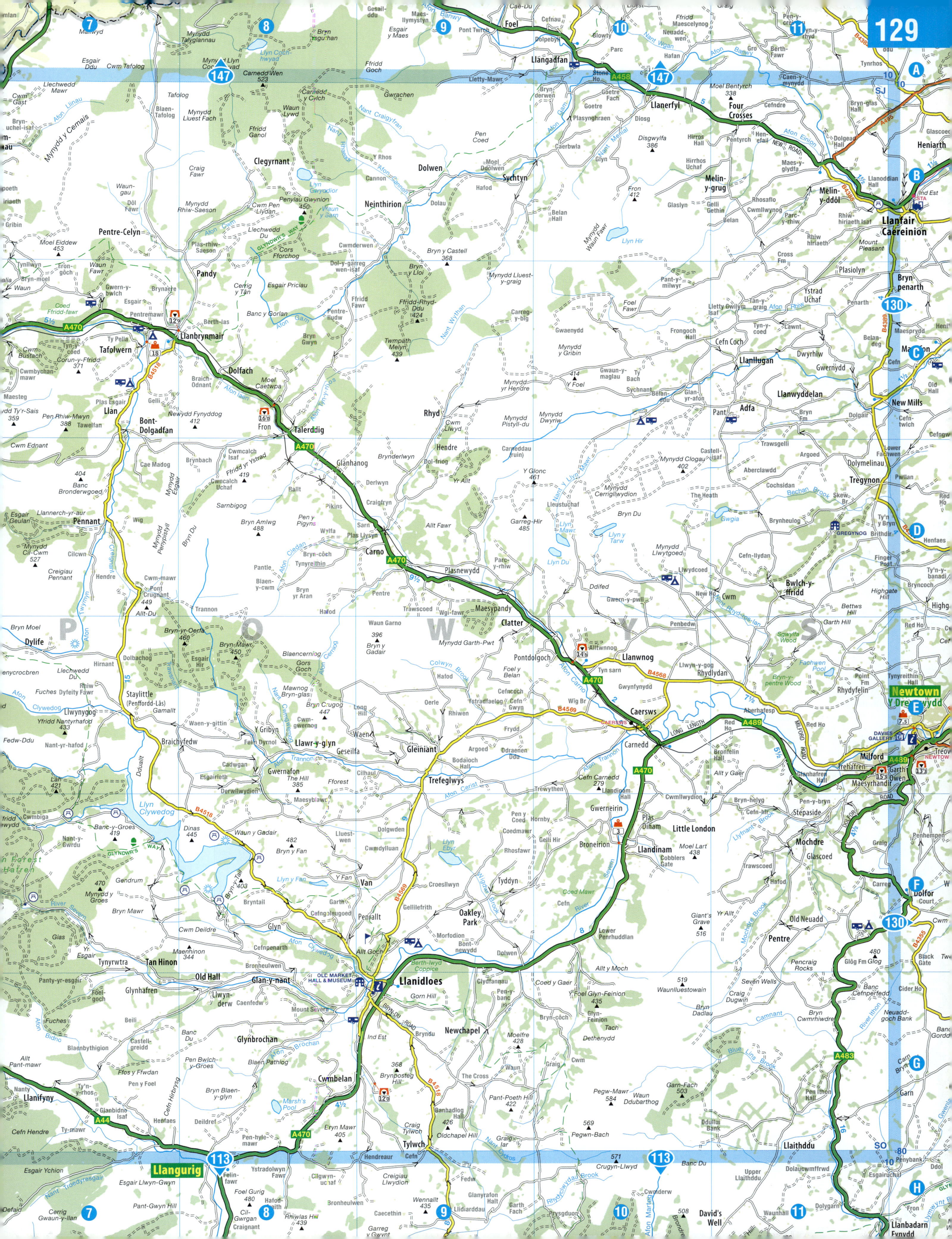
Foel
Llangadfan
Llanerfyl
Four Crosses
Clegyrnant
Dolwen
Sychtyn
Neinthirion
Melin-y-grug
Melin-y-ddôl
Llanfair Caereinion
Pentre-Celyn
Pandy
Llanbrynmair
Tafolwern
Dolfach
Talerddig
Rhyd
Llanllugan
Adfa
Llanwyddelan
New Mills
Llan
Bont-Dolgadfan
Pennant
Carno
Tregynon
Bwlch-y-ffridd
Dylife
Clatter
Maesypandy
Pontdolgoch
Llanwnog
Caersws
Rhydlydan
Newtown
Y Drenewydd
Staylittle
Llawr-y-glyn
Gleiniant
Trefeglwys
Carnedd
Milford
Maesyrhandir
Llyn Clywedog
Van
Little London
Mochdre
Llandinam
Oakley Park
Dolfor
Pentre
Tan Hinon
Old Hall
Glan-y-nant
Llanidloes
Newchapel
Cwmbelan
Llanifyny
Tylwch
Llangurig
Llaithddu
David's Well
Llanbadarn Fynydd
P O W Y S
A470
A458
A489
A483
B4518
B4569
B4568
B4389
B4355
147
113
130

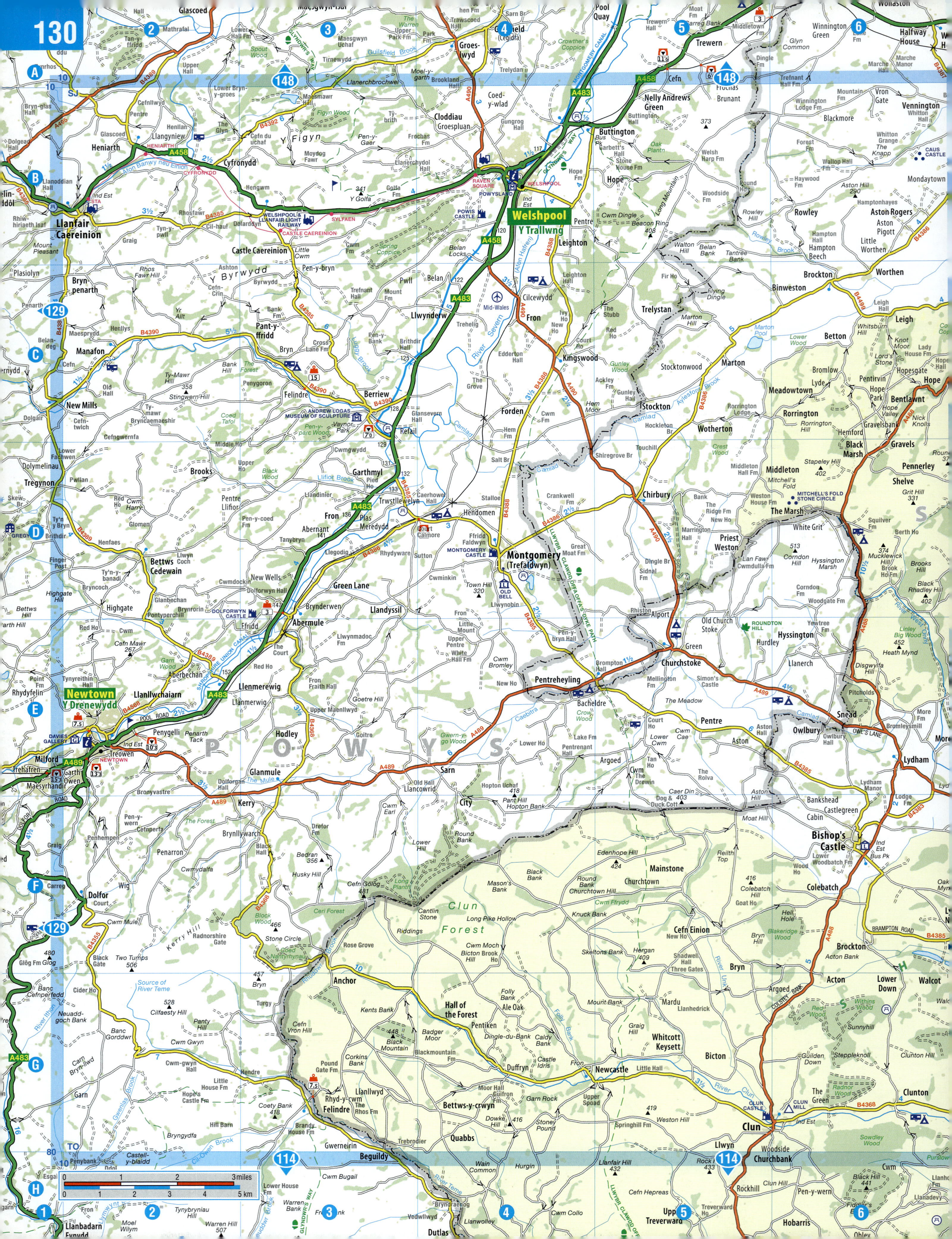

148
Llanfair Caereinion
Heniarth
Cyfronydd
Castle Caereinion
Y Figyn
Y Golfa
Welshpool
Y Trallwng
Leighton
Buttington
Groes-lwyd
Cloddiau
Groespluan
Trewern
Nelly Andrews Green
Hope
Beacon Ring
Long Mountain
Rowley
Aston Rogers
Aston Pigott
Little Worthen
Worthen
Brockton
Binweston
Leigh
Trelystan
Marton
Betton
Stocktonwood
Kingswood
Fron
Cilcewydd
Mid-Wales
River Severn
Llwynderw
Belan
Pwll
Pen-y-bryn
Y Byrwydd
Bryn-penarth
Manafon
Pant-y-ffridd
Bryn
Berriew
Felindre
Refail
Andrew Logan Museum of Sculpture
Forden
Stockton
Wotherton
Meadowtown
Rorrington
Middleton
The Marsh
Black Marsh
Bentlawnt
Hope
Gravels
Pennerley
Shelve
Chirbury
Priest Weston
Hyssington
Marrington
Hendomen
Montgomery (Trefaldwyn)
Montgomery Castle
Old Bell
Town Hill
Garthmyl
Brooks
Tregynon
Dolymelinau
New Mills
Fron
Plas Meredydd
Trwstllewelyn
Abernant
Bettws Cedewain
New Wells
Highgate
Green Lane
Llandyssil
Brynderwen
Abermule
Dolforwyn Castle
Glanbechan
Pontypercyll
Aberbechan
Newtown
Y Drenewydd
Llanllwchaiarn
Llenmerewig
Penygelli
Treowen
Milford
Maesyrhandir
Glanmule
Hodley
Kerry
Sarn
Goltre
City
Pentreheyling
Bacheldre
Churchstoke
Old Church Stoke
Roundton Hill
Green
Brompton Hall
Mellington
Simon's Castle
Pentre
Aston
Snead
Owlbury
Lydham
Bishop's Castle
Colebatch
Brockton
Acton
Lower Down
Walcot
Clunton
Clun
Clun Castle
Woodside
Churchbank
Rockhill
Hobarris
Bicton
Whitcott Keysett
Newcastle
Mainstone
Churchtown
Cefn Einion
Bryn
Argoed
Clun Forest
Hall of the Forest
Anchor
Rose Grove
Cefn Golog
Kerry Hill
Dolfor
Penarron
Brynllywarch
Betws-y-crwyn
Quabbs
Felindre
Rhyd-y-cwm
Beguildy
Dutlas
Llanbadarn Fynydd
114
129
P O W Y S
S
A458
A483
A489
A490
A488
B4385
B4386
B4388
B4389
B4390
B4368
B4355
0 1 2 3 miles
0 1 2 3 4 5 km

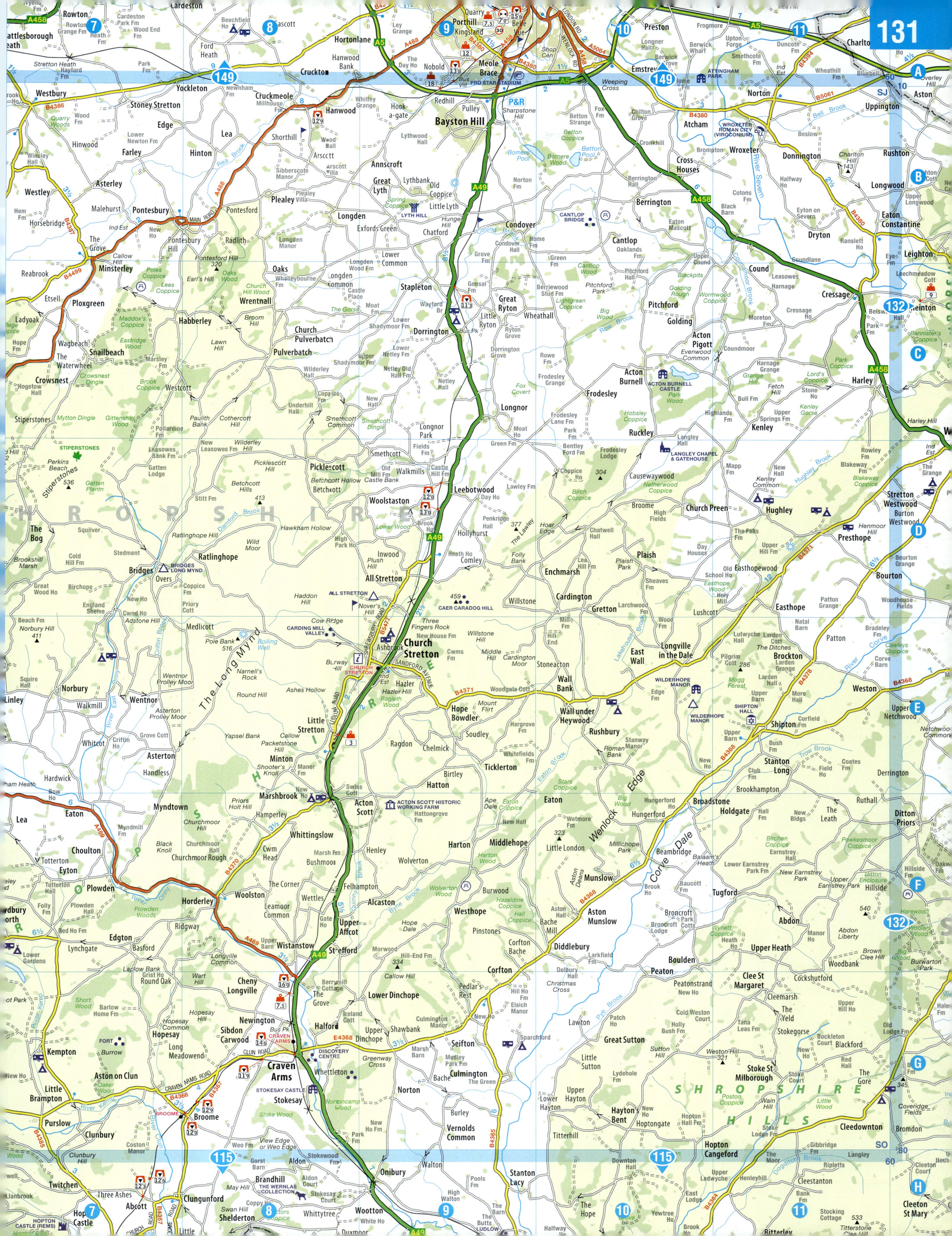
Rowton
Cardeston
Hortonlane
Hanwood Bank
Crutton
Porthill
Kingsland
Meole Brace
PRO STAR STADIUM
Preston
Emstrey
ATTINGHAM PARK
Charlton
Westbury
Yockleton
Cruckmeole
Hanwood
Redhill
Pulley
Bayston Hill
Norton
Uppington
Aston
Stoney Stretton
Edge
Lea
Hook-a-gate
Atcham
WROXETER ROMAN CITY (VIROCONIUM)
Wroxeter
Donnington
Rushton
Hinwood
Farley
Hinton
Shorthill
Arscott
Annscroft
Betton Strange
Cross Houses
Charlton Hill
Westley
Asterley
Great Lyth
Lythbank
Old Coppice
Little Lyth
Berrington
Longwood
Eaton Constantine
Pontesbury
Pontesford
Pleaney
Longden
Exfords Green
Chatford
Condover
CANTLOP BRIDGE
Cantlop
Eyton on Severn
Dryton
The Grove
Minsterley
Radlith
Oaks
Longden Common
Lower Common
Stapleton
Cound
Leighton
Cressage
Reabrook
Etsell
Ploxgreen
Habberley
Wrentnall
Great Ryton
Little Ryton
Wheathall
Pitchford
Golding
Ladyoak
Snailbeach
Wagbeach
The Waterwheel
Church Pulverbatch
Pulverbatch
Dorrington
Acton Pigott
Crowsnest
Westcott
Acton Burnell
ACTON BURNELL CASTLE
Harley
Frodesley
Longnor
Stiperstones
Ruckley
Kenley
STIPERSTONES
Longnor Park
Smethcott
Pickle Scott
Walkmills
Leebotwood
LANGLEY CHAPEL & GATEHOUSE
Causewaywood
Church Preen
Hughley
Stretton Westwood
Woolstaston
Hollyhurst
Presthope
The Bog
Ratlinghope
Bridges
BRIDGES LONG MYND
Overs
All Stretton
ALL STRETTON
CAER CARADOG HILL
Enchmarsh
Cardington
Gretton
Plaish
Easthope
Bourton
Norbury
Medlicott
Church Stretton
CARDING MILL VALLEY
Ashbrook
Hazler
Longville in the Dale
East Wall
Wall Bank
Brockton
Weston
Wentnor
The Long Mynd
Hope Bowdler
Wall under Heywood
WILDERHOPE MANOR
SHIPTON HALL
Shipton
Little Stretton
Ragdon
Chelmick
Soudley
Rushbury
Stanton Long
Asterton
Minton
Tickleton
Hatton
Birtley
Myndtown
Marshbrook
Acton Scott
ACTON SCOTT HISTORIC WORKING FARM
Eaton
Broadstone
Holdgate
Lea
Choulton
Eyton
Whittingslow
Harton
Middlehope
Wenlock Edge
Hungerford
Little London
Corve Dale
Plowden
Horderley
Woolston
Alcaston
Westhope
Munslow
Aston Munslow
Tugford
Abdon
Edgton
Wistanstow
Upper Affcot
Strefford
Corfton
Diddlebury
Peaton
Boulden
Upper Heath
Clee St Margaret
Cheny Longville
Lower Dinchope
Halford
Newington
Sibdon Carwood
Hopesay
Kempton
Craven Arms
Seifton
Culmington
Great Sutton
Stoke St Milborough
Blackford
Aston on Clun
Stokesay
STOKESAY CASTLE
Norton
Vernolds Common
Hayton's Bent
Little Brampton
Purslow
Broome
Clunbury
Cleedownton
Aldon
Onibury
Walton
Stanton Lacy
Hopton Cangeford
Twitchen
Brandhill
Clungunford
Abcott
Shelderton
Wootton
Hopton Castle
Cleeton St Mary
SHROPSHIRE
SHROPSHIRE HILLS

Wellington
Hadley
Ketley
Oakengates
Crackleybank
Weston under Lizard
Arleston
Priorslee
Telford Central
Telford Services
Uppington
Aston
The Rock
Lawley
Telford
Haughton
Shifnal
Tong Norton
Tong
Rushton
Dawley
The Wyke
Longwood
Little Wenlock
Horsehay
Stirchley
Eaton Constantine
Little Dawley
Leighton
Buildwas
Coalbrookdale
Madeley
Halesfield
Kemberton
Albrighton
Ironbridge
Sheinton
Brockton
Grindle
Ryton
Boningale
Benthall
Coalport
Wyke
Broseley
Sutton Maddock
Beckbury
Harley
Homer
Much Wenlock
Norton
Burnhill Green
Stockton
Badger
Ackleton
Pattingham
Willey
Stableford
Stretton Westwood
Callaughton
Nordley
Oldington
Chesterton
Presthope
Astley Abbotts
Worfield
Hilton
Shipley
Trescott
Bourton
Acton Round
Haughton
Cross Lane Head
Muckley Cross
Aston Eyre
Morville
Tasley
Roughton
Hopstone
Seisdon
Monkhopton
Bridgnorth
Claverley
Weston
Upper Netchwood
Upton Cressett
Oldbury
Danesford
Quatford
Heathton
Halfpenny Green
Lower Netchwood
Middleton Baggot
Chetton
Eardington
Bobbington
Middleton Priors
Ditton Priors
Oldfield
Glazeley
Chelmarsh
Quatt
Six Ashes
Neenton
Middleton Scriven
Deuxhill
Sutton
Hampton
Hampton Loade
Four Ashes
Enville
Cleobury North
Sidbury
Billingsley
Birdsgreen
Burwarton
SHROPSHIRE
STAFFORDS
Woodhill
Alveley
Aston Botterell
Highley
Kinver
Chorley
Stottesdon
Loughton
Wheathill
Kingsford
Blakeshall
Prescott
Farlow
Oreton
Silvington
Kinlet
Upper Arley
Pound Green
Holbeache
Trimpley
Fairfield
Wolverley
Cleeton St Mary
Crumpsbrook
Buttonoak
Cleedownton
M54
A5
A41
A442
A458
A454
A464
B4176
B4363
B4364
B4555
150
131
116
0 1 2 3 miles
0 1 2 3 4 5 km

Stretton
Cannock Wood
Gentleshaw
Farewell
Wimblebury
Oldfallow
Horsebrook
Hatherton
Heath Hayes
Cannock
Bridgtown
Norton East
Chase Terrace
Burntwood
Kiddemore Green
Brewood
Norton Canes
Chasetown
Leomansley
Hammerwich
Standeford
Churchbridge
Great Wyrley
Cheslyn Hay
Muckley Corner
Wall
Coven
Shareshill
Little Wyrley
Brownhills
Featherstone
Hilton Park Services
Norton Canes Services
Codsall Wood
Codsall
Essington
Springhill
Pelsall
Clayhanger
Shire Oak
Stonnall
Shenstone
Oaken
Bilbrook
Fordhouses
Bushbury
Bloxwich
Pelsall Wood
Highbridge
Walsall Wood
Shelfield
Vigo
Oxley
Wood Hayes
Ashmore Park
New Invention
Short Heath
Harden
Rushall
Leighswood
Aldridge
Wergs
Perton
Tettenhall
Wednesfield
Willenhall
Bentley
Walsall
Streetly
Little Aston
Four Oaks
Wolverhampton
Compton
Castlecroft
Merry Hill
Blakenhall
Moseley
Bilston
Darlaston
Palfrey
Caldmore
Bradmore
Penn
Lower Penn
Ettingshall
Goldthorn Park
Wednesbury
The Delves
Great Barr
Kingstanding
Sedgley
Coseley
Princes End
Hill Top
Hamstead
Trysull
Wombourne
Tipton
Upper Gornal
The Black Country
Dudley
Gornalwood
Tividale
Oldbury
Great Bridge
Churchfield
West Bromwich
Handsworth
Sandwell
Smethwick
Aston
Birmingham
Swindon
Himley
Kingswinford
Brierley Hill
Netherton
Rowley Regis
Blackheath
Bearwood
Edgbaston
Wordsley
Amblecote
Stourbridge
Lye
Cradley Heath
Old Hill
Halesowen
Harborne
Quinton
Small Heath
Balsall Heath
Sparkhill
Moseley
Pedmore
Wollescote
Hasbury
Selly Oak
Bournville
Kings Heath
Hagley
Romsley
Frankley
Northfield
Kings Norton
Blakedown
Churchill
Broome
Clent
Rubery
Highgate
Stechford
M6
M5
M6 TOLL
M54
A5
A34
A38(M)
A41
A449
A454
A456
A458
A461
A4123
West Midlands
Shropshire
151
117
134

Lichfield
Streethay
Elford
Edingale
Harlaston
Haunton
Clifton Campville
Chilcote
Measham
152
Appleby Magna
Snarestone
Thorpe Constantine
No Man's Heath
Appleby Parva
Whittington
Hademore
Newton Regis
Seckington
Austrey
Norton-Juxta-Twycross
Twycross
Wall
Wigginton
Shenstone
Weeford
Hopwas
Tamworth
Leyfields
Amington
Shuttington
Alvecote
Warton
Orton-on-the-Hill
Glascote
Polesworth
Little Hay
Hints
Bonehill
Fazeley
Wilnecote
Birchmoor
Dordon
St Helena
Grendon
Sheepy Magna
Sheepy Parva
Watford Gap
Drayton Bassett
Dosthill
Freasley
Whittington
Pinwall
Little Aston
Roughley
Streetly
Four Oaks
Middleton
Piccadilly
Grendon Common
Baddesley Ensor
Ratcliffe Culey
Atherstone
Witherley
Mancetter
Fenny Drayton
Sutton Coldfield
Kingsbury
Hurley
Baxterley
Bentley
New Oscott
Wylde Green
Wishaw
Bodymoor Heath
Marston
Lea Marston
Ridge Lane
Oldbury
Hartshill
Erdington
Minworth
Curdworth
Whitacre Heath
Nether Whitacre
Church End
Galley Common
Camp Hill
Ansley
Stockingford
Castle Vale
Water Orton
Tyburn
Gravelly Hill
Castle Bromwich
Coleshill
Shustoke
Furnace End
Over Whitacre
Old Arley
New Arley
Astley
Ward End
Kitt's Green
Tile Cross
Chelmsley Wood
Maxstoke
Fillongley
Corley Ash
Corley
Corley Moor
Bedworth Woodlands
Birmingham
Yardley
Sheldon
Marston Green
Kinwalsey
Chapel Green
Keresley
Ash Green
Sparkhill
Moseley
Acock's Green
Elmdon
Bickenhill
Little Packington
Butler's End
Meriden
Hawkes End
Eaves Green
Olton
Solihull
Hampton in Arden
Berkswell
Allesley
Coundon
Coventry
Whoberley
Balsall Common
Hall Green
Shirley
Barston
Earlsdon
Kingshurst
Castle Bromwich
WEST
MIDLANDS
WARWICKSHIRE
STAFFS
M6 TOLL
M42
M6
A5
A38
A446
A452
A45
Tamworth Services
Corley Services
0 1 2 3 miles
0 1 2 3 4 5 km
133
118

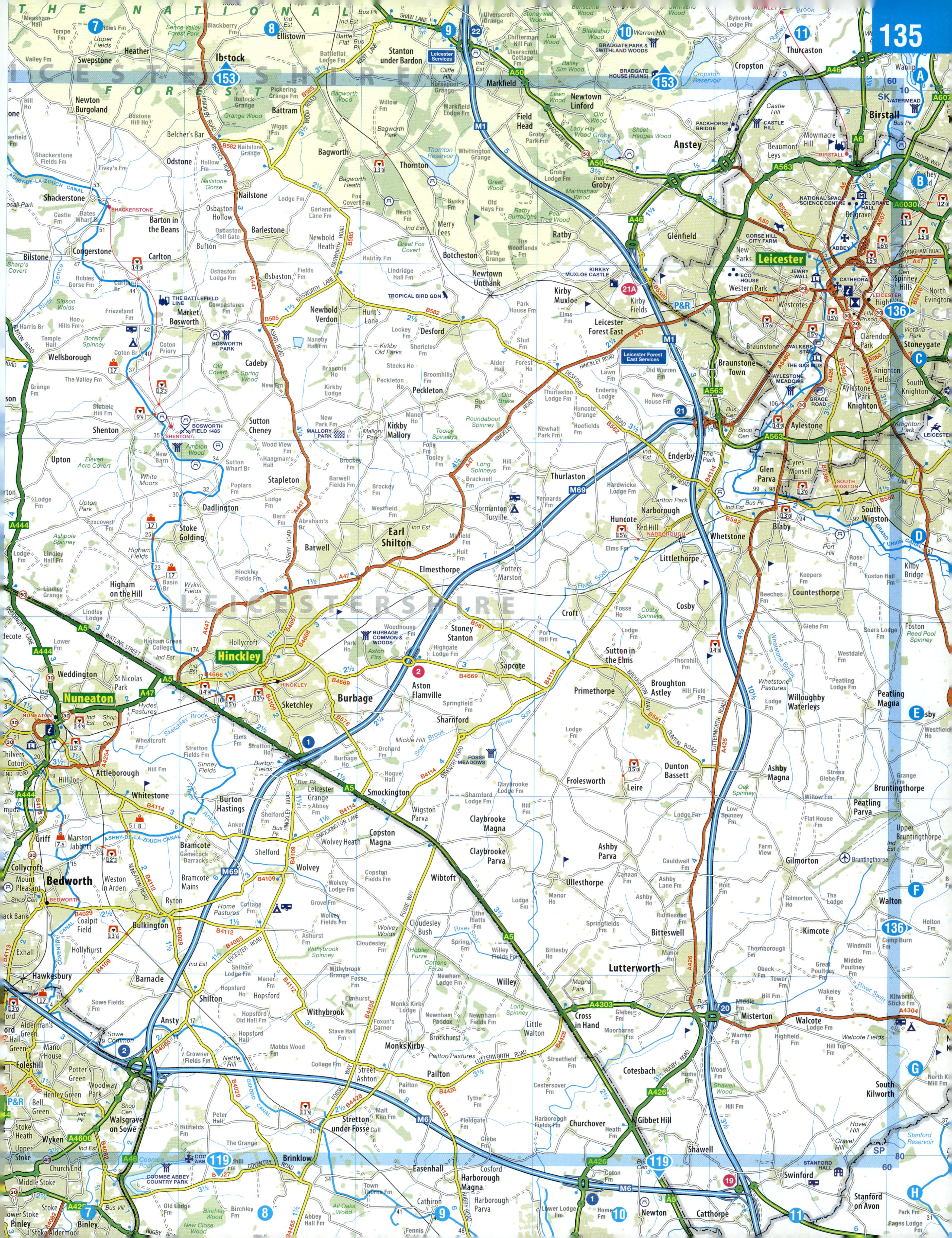
THE NATIONAL FOREST
LEICESTERSHIRE
Ibstock
Ellistown
Stanton under Bardon
Markfield
Thurcaston
Cropston
Anstey
Birstall
Leicester
Newton Burgoland
Heather
Swepstone
Odstone
Nailstone
Bagworth
Thornton
Field Head
Newtown Linford
Groby
Ratby
Glenfield
Shackerstone
Barton in the Beans
Congerstone
Carlton
Barlestone
Osbaston
Market Bosworth
Newbold Verdon
Desford
Kirby Muxloe
Kirby Fields
Leicester Forest East
Braunstone Town
Aylestone
Knighton
Stoneygate
Wellsborough
Cadeby
Sutton Cheney
Shenton
Kirkby Mallory
Peckleton
Enderby
Glen Parva
Upton
Stapleton
Dadlington
Stoke Golding
Thurlaston
Narborough
Huncote
Blaby
South Wigston
Whetstone
Earl Shilton
Barwell
Elmesthorpe
Higham on the Hill
Littlethorpe
Cosby
Countesthorpe
Croft
Stoney Stanton
Sapcote
Sutton in the Elms
Hinckley
Nuneaton
Weddington
Burbage
Sketchley
Aston Flamville
Sharnford
Primethorpe
Broughton Astley
Willoughby Waterleys
Peatling Magna
Attleborough
Whitestone
Burton Hastings
Smockington
Frolesworth
Leire
Dunton Bassett
Ashby Magna
Bruntingthorpe
Bedworth
Bramcote
Shelford
Wolvey
Copston Magna
Claybrooke Magna
Claybrooke Parva
Ashby Parva
Ullesthorpe
Gilmorton
Peatling Parva
Walton
Bulkington
Wibtoft
Kimcote
Bitteswell
Lutterworth
Barnacle
Shilton
Withybrook
Willey
Misterton
Walcote
Ansty
Monks Kirby
Pailton
Little Walton
Cotesbach
South Kilworth
Stretton under Fosse
Churchover
Gibbet Hill
Shawell
Brinklow
Easenhall
Harborough Magna
Swinford
Stanford on Avon
Binley
Catthorpe
Newton
M1
M69
M6
A5
A46
A47
A50
A426
A444
A447
153
136
119

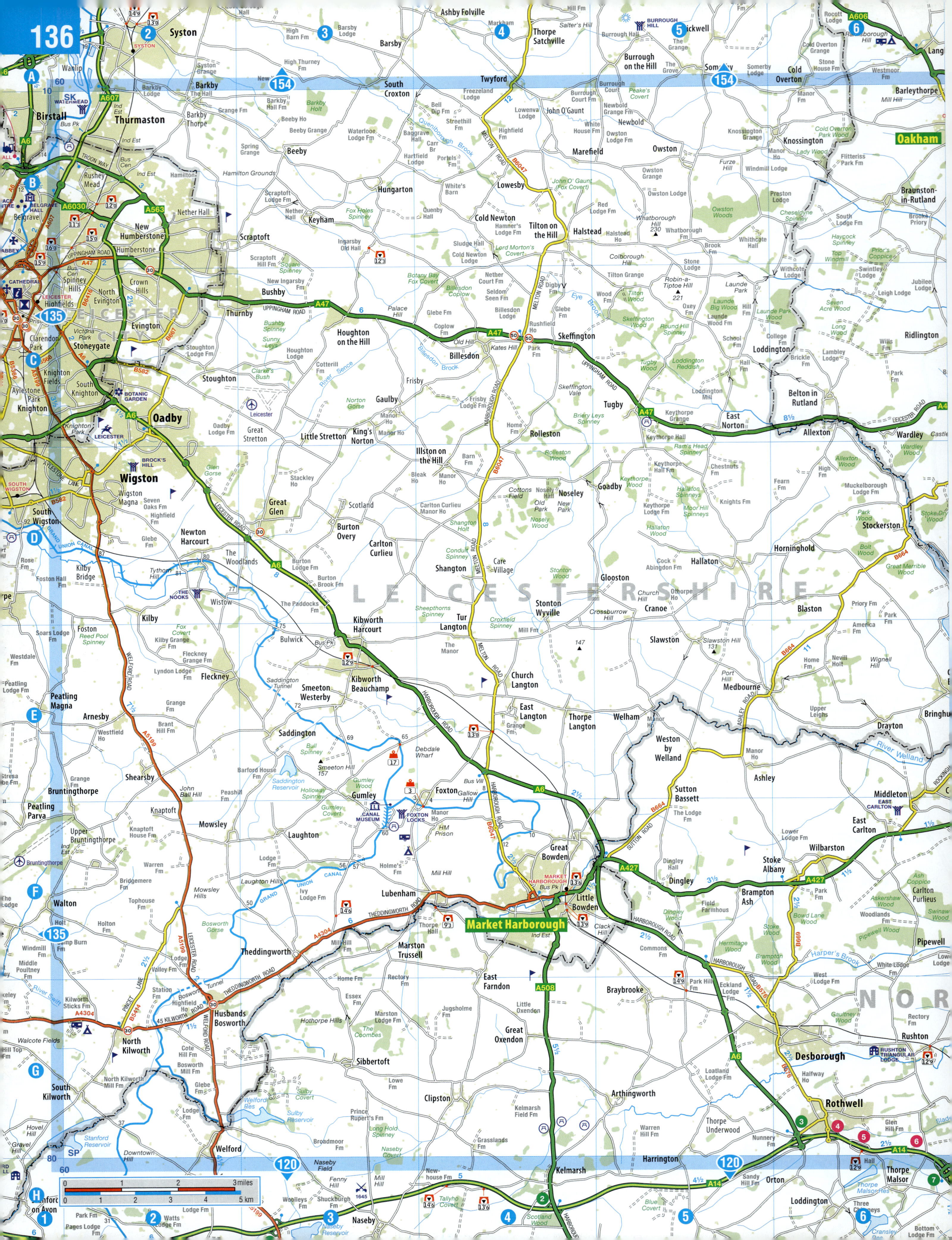

Syston
Ashby Folville
Thorpe Satchville
Burrough on the Hill
Somerby
Cold Overton
Barleythorpe
Oakham
Barsby
South Croxton
Twyford
John O'Gaunt
Newbold
Owston
Knossington
Birstall
Thurmaston
Barkby
Barkby Thorpe
Beeby
Hungarton
Lowesby
Marefield
Keyham
Cold Newton
Tilton on the Hill
Halstead
Braunston-in-Rutland
New Humberstone
Humberstone
Scraptoft
Bushby
Thurnby
Houghton on the Hill
Billesdon
Skeffington
Loddington
Ridlington
LEICESTER CITY
Crown Hills
North Evington
Evington
Highfields
Stoneygate
Clarendon Park
Knighton Fields
South Knighton
Aylestone Park
Knighton
Oadby
Stoughton
Great Stretton
Little Stretton
King's Norton
Gaulby
Frisby
Illston on the Hill
Rolleston
Tugby
East Norton
Belton in Rutland
Allexton
Wardley
Wigston
Wigston Magna
South Wigston
Great Glen
Burton Overy
Carlton Curlieu
Noseley
Goadby
Stockerston
Newton Harcourt
The Woodlands
Shangton
Glooston
Hallaton
Horninghold
LEICESTERSHIRE
Kilby
Wistow
Kibworth Harcourt
Tur Langton
Stonton Wyville
Cranoe
Blaston
Slawston
Medbourne
Fleckney
Kibworth Beauchamp
Smeeton Westerby
Church Langton
East Langton
Thorpe Langton
Welham
Weston by Welland
Drayton
Peatling Magna
Arnesby
Saddington
Shearsby
Bruntingthorpe
Gumley
Foxton
Sutton Bassett
Ashley
Middleton
East Carlton
Peatling Parva
Mowsley
Laughton
Great Bowden
Dingley
Stoke Albany
Wilbarston
Brampton Ash
Carlton Purlieus
Walton
Lubenham
Market Harborough
Little Bowden
Theddingworth
Marston Trussell
East Farndon
Braybrooke
Pipewell
Husbands Bosworth
North Kilworth
Great Oxendon
Desborough
Rushton
South Kilworth
Sibbertoft
Clipston
Arthingworth
Rothwell
Welford
Thorpe Underwood
Harrington
Kelmarsh
Naseby
Orton
Loddington
Thorpe Malsor

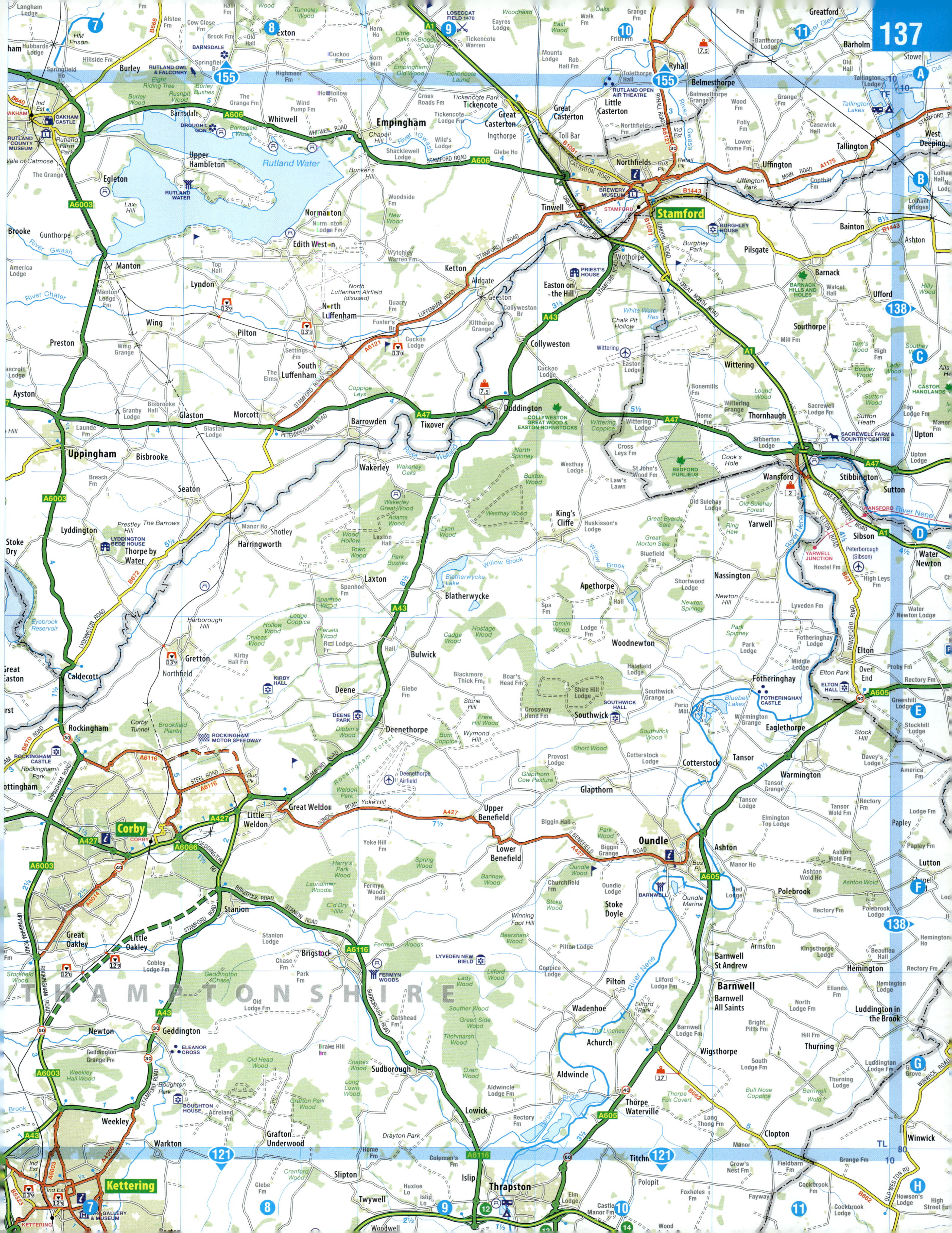
Exton
Empingham
Tickencote
Great Casterton
Little Casterton
Ryhall
Belmesthorpe
Greatford
Barholm
Stowe
Tallington
West Deeping
Uffington
Stamford
Burghley House
Bainton
Ashton
Pilsgate
Barnack
Ufford
Southorpe
Wittering
Thornhaugh
Wansford
Stibbington
Sutton
Sibson
Water Newton
Yarwell
Nassington
Elton
Fotheringhay
Eaglethorpe
Tansor
Cotterstock
Warmington
Oundle
Ashton
Polebrook
Barnwell
Hemington
Luddington in the Brook
Thurning
Wigsthorpe
Clopton
Winwick
Titchmarsh
Thrapston
Islip
Twywell
Slipton
Lowick
Aldwincle
Achurch
Wadenhoe
Pilton
Stoke Doyle
Lower Benefield
Upper Benefield
Glapthorn
Southwick
Apethorpe
Woodnewton
King's Cliffe
Blatherwycke
Bulwick
Deene
Deenethorpe
Laxton
Harringworth
Wakerley
Duddington
Tixover
Collyweston
Easton on the Hill
Ketton
Tinwell
Aldgate
Geeston
North Luffenham
South Luffenham
Morcott
Barrowden
Glaston
Pilton
Wing
Lyndon
Manton
Edith Weston
Normanton
Whitwell
Upper Hambleton
Rutland Water
Egleton
Burley
Oakham
Gunthorpe
Brooke
Preston
Ayston
Uppingham
Bisbrooke
Seaton
Lyddington
Thorpe by Water
Stoke Dry
Caldecott
Gretton
Rockingham
Corby
Great Oakley
Little Oakley
Great Weldon
Little Weldon
Stanion
Brigstock
Sudborough
Geddington
Newton
Weekley
Warkton
Grafton Underwood
Kettering
Northfields
Wothorpe
Great Casterton
NORTHAMPTONSHIRE
A1
A43
A47
A6003
A606
A6121
A427
A6116
A605
A6086
A4300
A1175
B1443
B1081
B672
B670
B662
B671
TL
10
80

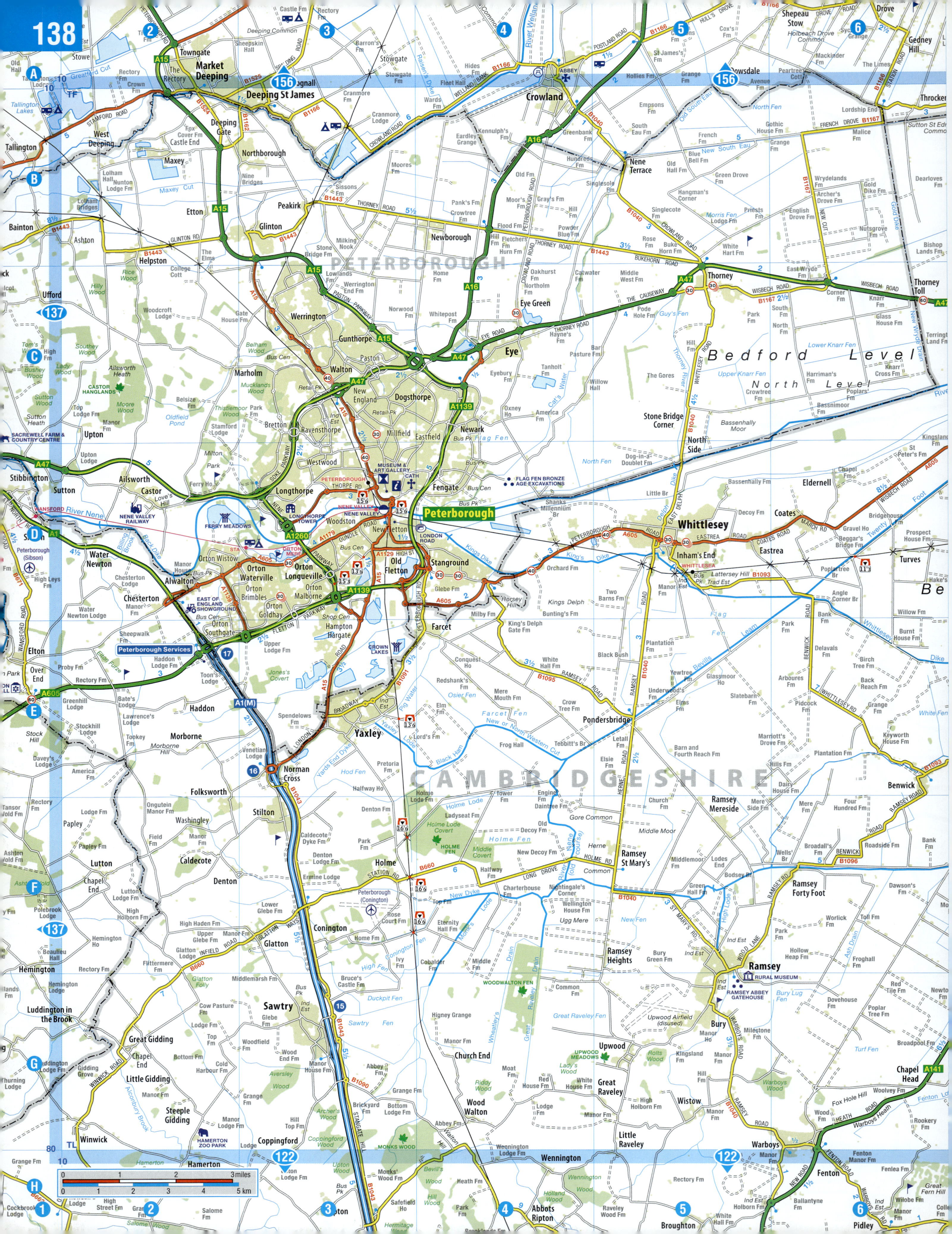
Market Deeping
Deeping St James
Towngate
Deeping Gate
West Deeping
Tallington
Maxey
Northborough
Etton
Peakirk
Glinton
Helpston
Bainton
Ashton
Ufford
Marholm
Upton
Ailsworth
Castor
Sutton
Stibbington
Water Newton
Alwalton
Chesterton
Elton
Crowland
Nene Terrace
Newborough
Eye
Eye Green
Thorney
Thorney Toll
Bedford Level
North Level
PETERBOROUGH
Peterborough
Werrington
Gunthorpe
Walton
Paston
Dogsthorpe
New England
Millfield
Eastfield
Newark
Ravensthorpe
Bretton
Westwood
Longthorpe
Woodston
Fengate
Old Fletton
New Fletton
Stanground
Orton Waterville
Orton Longueville
Orton Wistow
Orton Brimbles
Orton Goldhay
Orton Malborne
Orton Southgate
Hampton Hargate
Farcet
Yaxley
Norman Cross
Haddon
Morborne
Folksworth
Stilton
Caldecote
Denton
Washingley
Lutton
Papley
Holme
Conington
Glatton
Sawtry
Hemington
Luddington in the Brook
Great Gidding
Little Gidding
Steeple Gidding
Winwick
Coppingford
Hamerton
Wennington
Abbots Ripton
Broughton
Whittlesey
Coates
Eastrea
Turves
Eldernell
Inham's End
Pondersbridge
Stone Bridge Corner
North Side
Ramsey
Ramsey St Mary's
Ramsey Mereside
Ramsey Heights
Ramsey Forty Foot
Bury
Upwood
Great Raveley
Little Raveley
Wistow
Warboys
Fenton
Pidley
Benwick
Chapel Head
CAMBRIDGESHIRE
FERRY MEADOWS
NENE VALLEY RAILWAY
MUSEUM & ART GALLERY
FLAG FEN BRONZE AGE EXCAVATIONS
Peterborough Services
CROWN LAKES
EAST OF ENGLAND SHOWGROUND
CASTOR HANGLANDS
SACREWELL FARM & COUNTRY CENTRE
Peterborough (Conington)
HOLME FEN
WOODWALTON FEN
UPWOOD MEADOWS
MONKS WOOD
HAMERTON ZOO PARK
RURAL MUSEUM
RAMSEY ABBEY GATEHOUSE
Upwood Airfield (disused)
A1(M)
A1
A15
A16
A47
A605
A1139
A1260
A1179
B1040
B1443
B1091
B1095
B660
B1043
B1093
B1166
B1167
156
137
122
0 1 2 3 miles
0 1 2 3 4 5 km

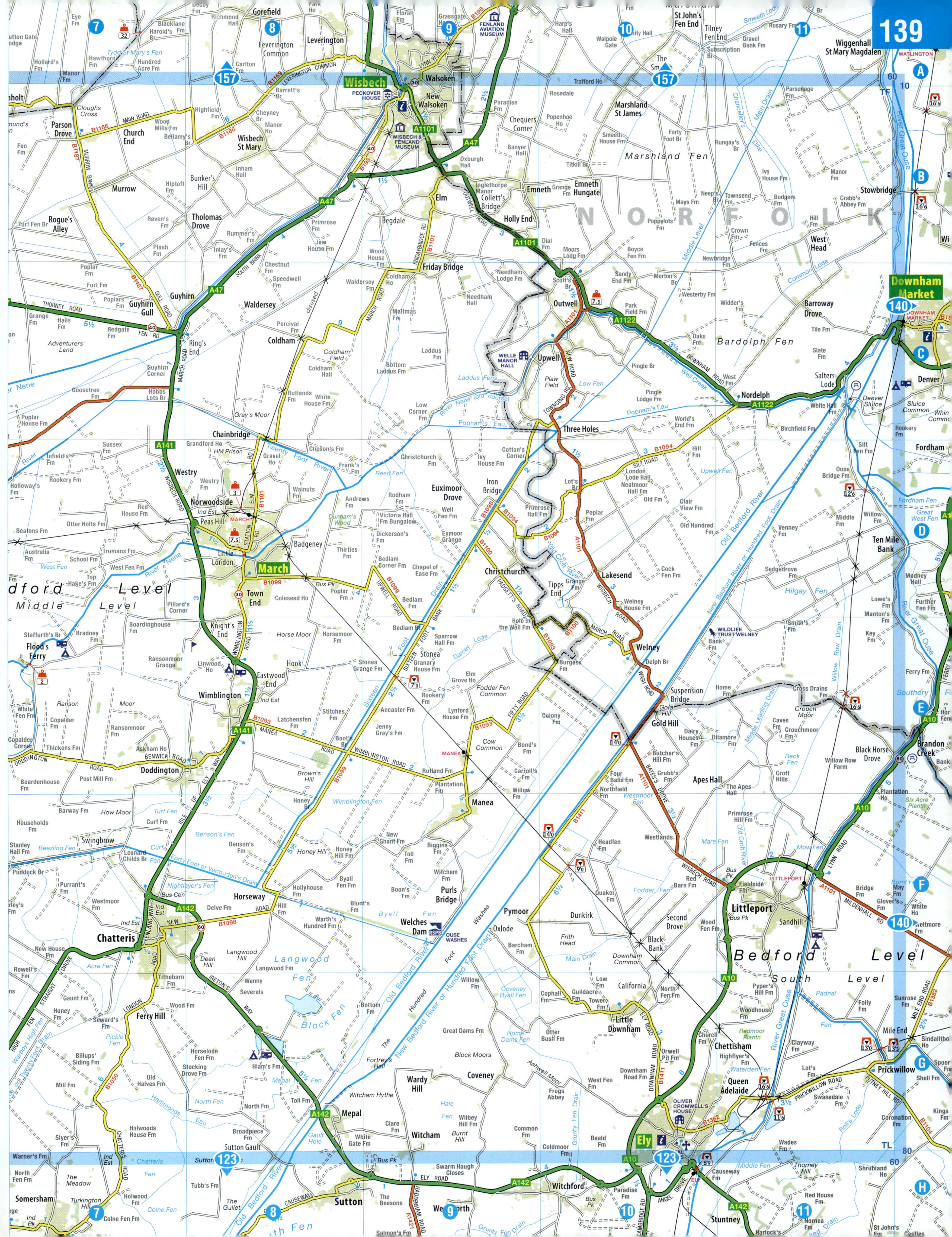

Wisbech
Peckover House
Walsoken
New Walsoken
Wisbech & Fenland Museum
Fenland Aviation Museum
Leverington
Leverington Common
Gorefield
Wisbech St Mary
Parson Drove
Church End
Murrow
Rogue's Alley
Tholomas Drove
Guyhirn
Guyhirn Gull
Ring's End
Elm
Emneth
Emneth Hungate
Marshland St James
Marshland Fen
St John's Fen End
Tilney Fen End
Wiggenhall St Mary Magdalen
Stowbridge
NORFOLK
Holly End
Outwell
Upwell
Welle Manor Hall
Friday Bridge
Waldersey
Coldham
Begdale
Downham Market
Denver
Barroway Drove
Bardolph Fen
Salters Lode
Nordelph
Three Holes
Fordham
Chainbridge
Westry
Norwoodside
March
Town End
Little London
Badgeney
Euximoor Drove
Christchurch
Tipps End
Lakesend
Welney
WWT Welney
Ten Mile Bank
Hilgay Fen
Middle Level
Bedford Level
Flood's Ferry
Knight's End
Wimblington
Eastwood End
Doddington
Stonea
Manea
Gold Hill
Suspension Bridge
Brandon Creek
Apes Hall
Littleport
Sandhill
Bedford Level South Level
Horseway
Purls Bridge
Welches Dam
Ouse Washes
Pymoor
Oxlode
Dunkirk
Black Bank
Little Downham
Chettisham
Queen Adelaide
Prickwillow
Chatteris
Ferry Hill
Langwood Fen
Block Fen
Coveney
Wardy Hill
Witcham
Mepal
Sutton
Sutton Gault
Witchford
Ely
Oliver Cromwell's House
Stuntney
Somersham
Wentworth
Old Bedford River
New Bedford River or Hundred Foot Drain
River Great Ouse
Sixteen Foot Drain
Forty Foot or Vermuden's Drain
Twenty Foot River
River Nene
A47
A141
A142
A1101
A1122
A10
B1099
B1093
B1098
B1100
B1094
B1411
157
140
123

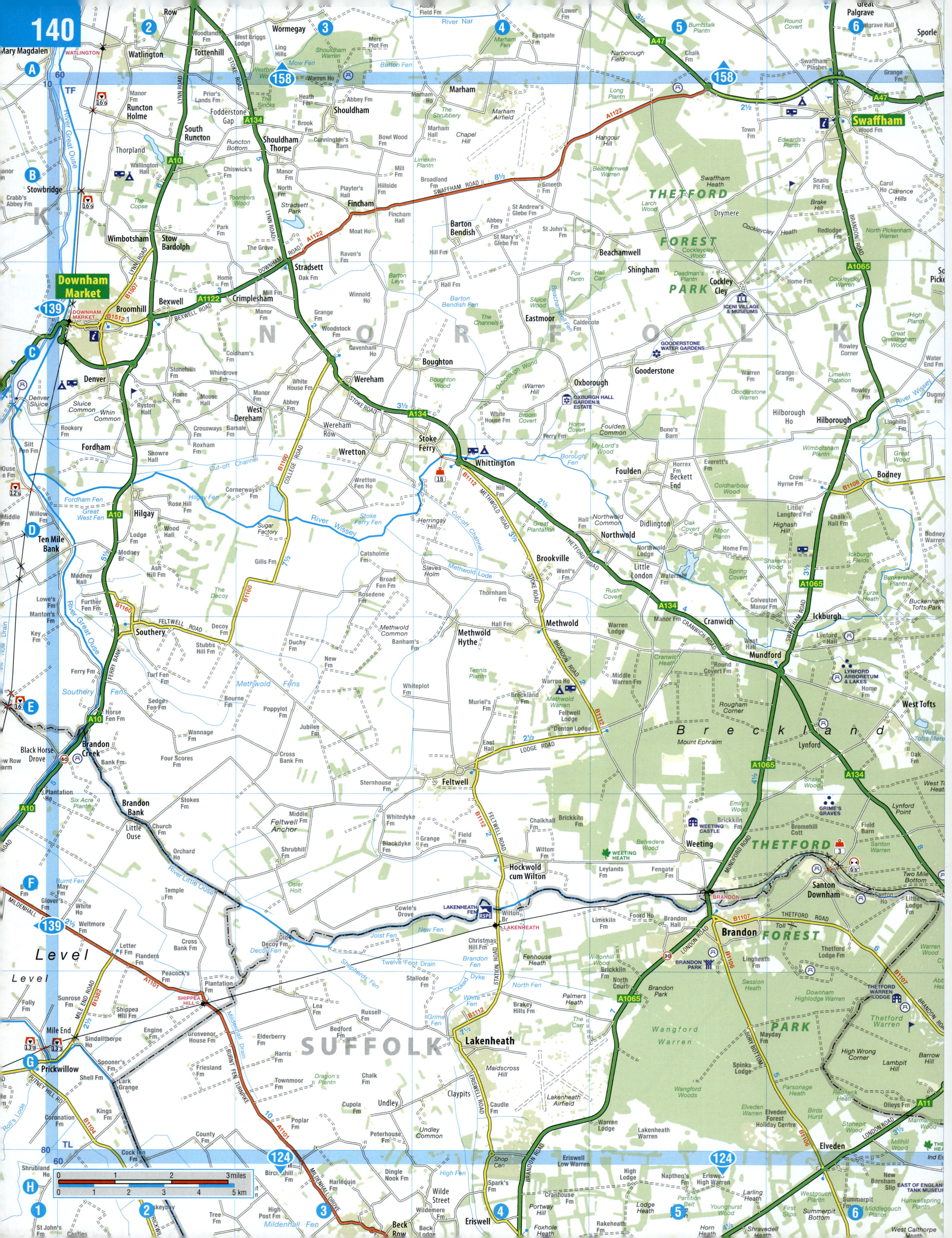

Wormegay
Tottenhill
Watlington
Marham
Swaffham
Runcton Holme
Shouldham
Shouldham Thorpe
South Runcton
Fodderstone Gap
Thorpland
Stowbridge
Fincham
Barton Bendish
Beachamwell
Shingham
Cockley Cley
Wimbotsham
Stow Bardolph
Stradsett
Downham Market
Bexwell
Crimplesham
Broomhill
Eastmoor
N O R F O L K
THETFORD FOREST PARK
ICENI VILLAGE & MUSEUMS
GOODERSTONE WATER GARDENS
Gooderstone
Boughton
Wereham
Denver
Oxborough
OXBURGH HALL GARDEN & ESTATE
West Dereham
Hilborough
Fordham
Wretton
Stoke Ferry
Whittington
Foulden
Bodney
Hilgay
Northwold
Didlington
Ten Mile Bank
River Wissey
Brookville
Little London
Cranwich
Ickburgh
Southery
Methwold
Methwold Hythe
Mundford
LYNFORD ARBORETUM & LAKES
West Tofts
Breckland
Mount Ephraim
Lynford
Brandon Creek
Feltwell
Brandon Bank
Little Ouse
GRIME'S GRAVES
WEETING CASTLE
Weeting
WEETING HEATH
THETFORD
Hockwold cum Wilton
Santon Downham
LAKENHEATH FEN
Brandon
FOREST
BRANDON PARK
Level
Lakenheath
SUFFOLK
Mile End
Prickwillow
PARK
THETFORD WARREN LODGE
Elveden
Eriswell
Wilde Street
Beck Row
Undley
Claypits
Shippea Hill
River Great Ouse
River Little Ouse
Lakenheath Airfield
Wangford Warren
Elveden Forest Holiday Centre
A10
A134
A1122
A47
A1065
A1101
A11
B1160
B1112
B1106
B1107
B1108
158
139
124
TF
TL
0 1 2 3 miles
0 1 2 3 4 5 km

Tottwood
Yaxham
Mattishall
East Tuddenham
Necton
West End
Bradenham
Shipdham
Garvestone
Thuxton
Reymerston
Letton Green
Crowshill
Cranworth
Southburgh
Kimberley
Hardingham
Woodrising
Wymondham
Saham Toney
Ovington
Carbrooke
Watton
Scoulton
Hingham
Deopham
Little Cressingham
Griston
Merton
Northacre
Caston
Little Ellingham
Rockland St Peter
Great Ellingham
Thompson
Stow Bedon
Rockland All Saints
Attleborough
Besthorpe
Spooner Row
Morley St Botolph
Suton
Breckles
Lower Stow Bedon
Great Hockham
Shropham
Puddledock
Old Buckenham
Carleton Rode
Wretham
Stonebridge
Illington
Snetterton
Eccles Road
Wilby
Quidenham
New Buckenham
Banham
Harling Road
East Harling
Bridgham
Kenninghall
Winfarthing
Croxton
Thetford
Brettenham
Shadwell
West Harling
North Lopham
Fersfield
Shelfanger
Rushford
Garboldisham
Gasthorpe
South Lopham
Bressingham
Knettishall
Blo' Norton
Hopton
Euston
Barnham
Coney Weston
Thelnetham
Redgrave
Fen Street
Roydon
A11
A1066
A1075
A134
A47
B1108
B1077
B1111
B1113
B1135
NORFOLK

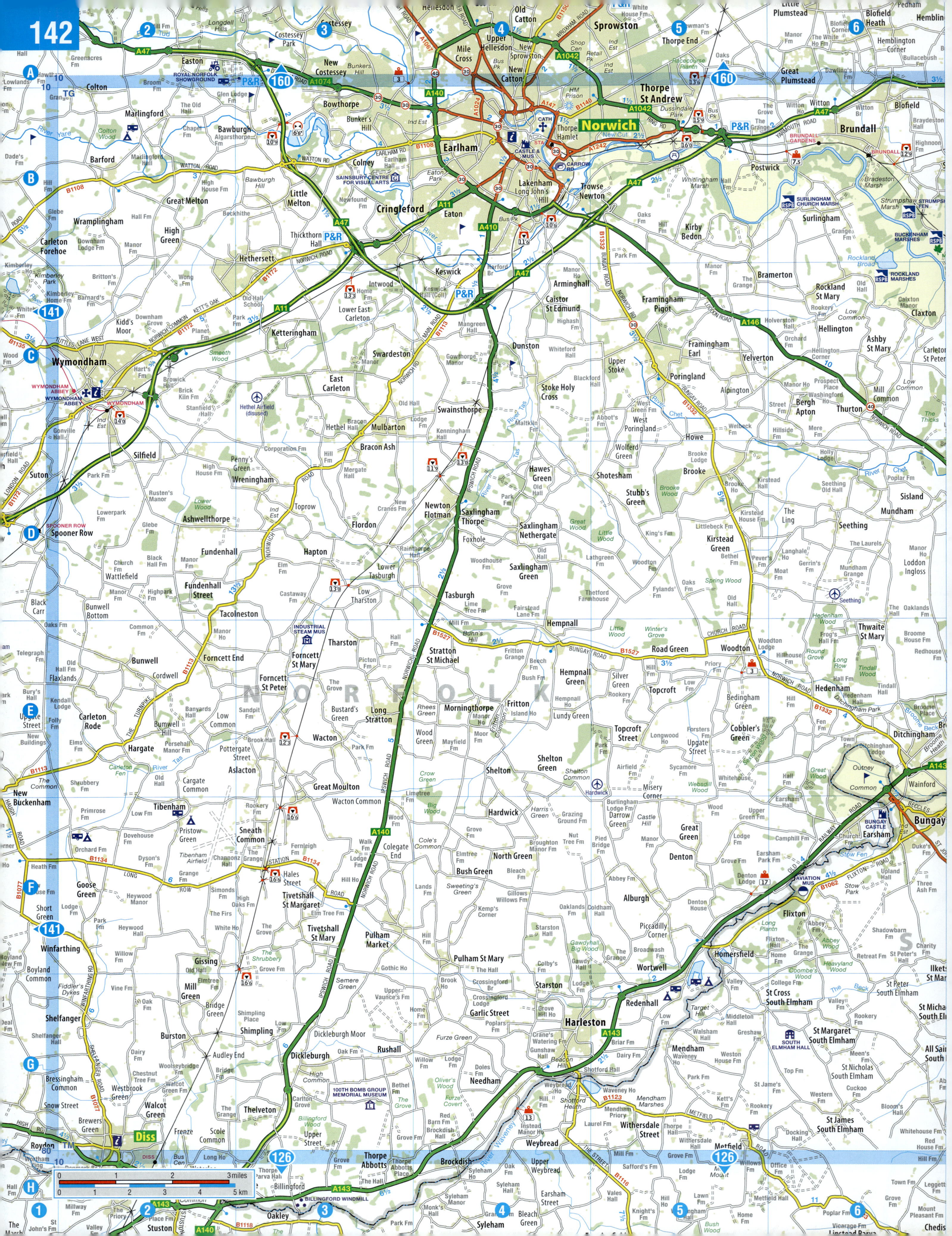
Norwich
Wymondham
Diss
Bungay
Harleston
NORFOLK
Earlham
Cringleford
Sprowston
Thorpe St Andrew
Brundall
Long Stratton
Hethersett

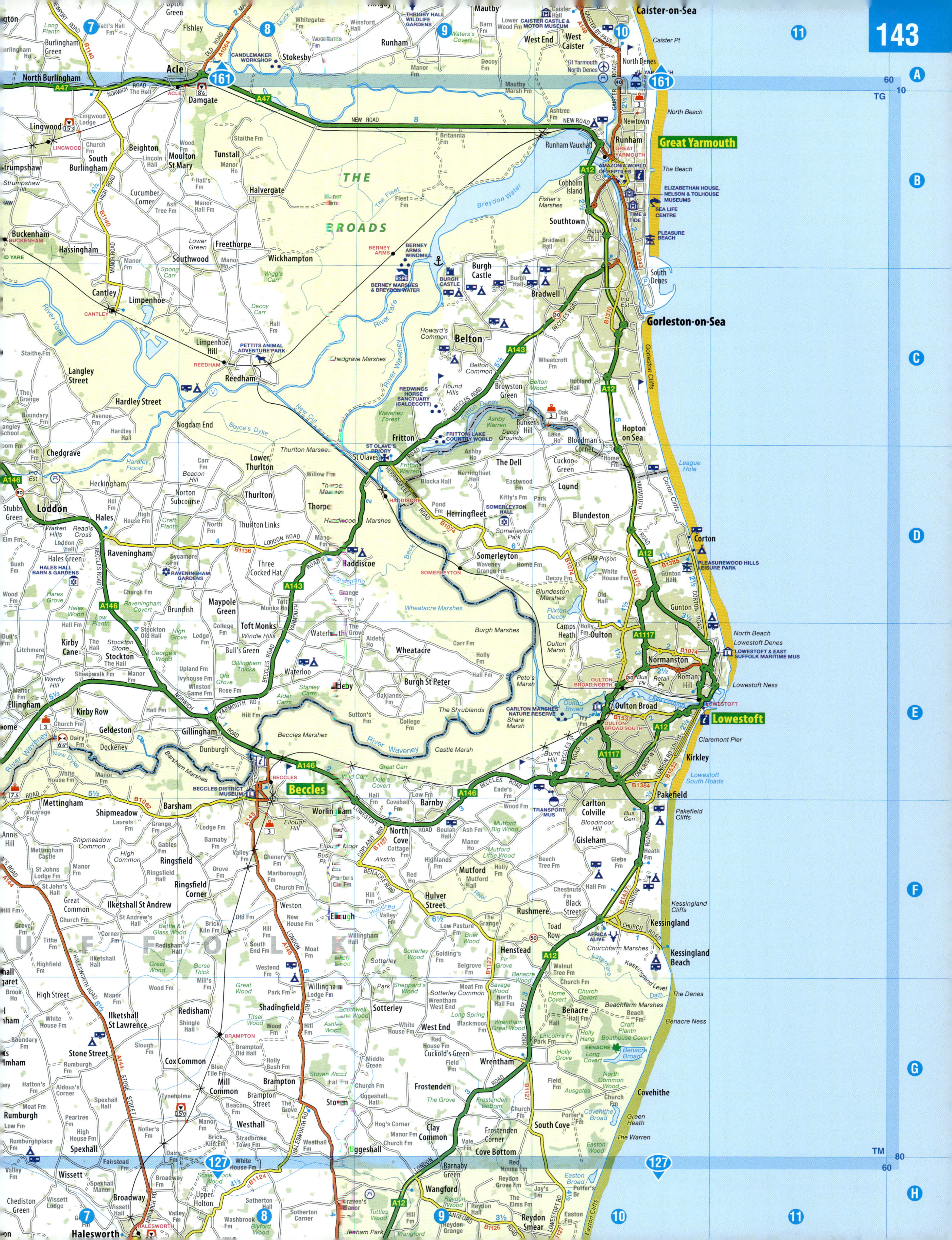
Caister-on-Sea
Caister Pt
North Denes
161
North Beach
Newtown
Great Yarmouth
The Beach
ELIZABETHAN HOUSE, NELSON & TOLHOUSE MUSEUMS
SEA LIFE CENTRE
PLEASURE BEACH
South Denes
Gorleston-on-Sea
Gorleston Cliffs
Hopton on Sea
League Hole
Corton Cliffs
Corton
PLEASUREWOOD HILLS LEISURE PARK
North Beach
Lowestoft Denes
LOWESTOFT & EAST SUFFOLK MARITIME MUS
Lowestoft Ness
Lowestoft
Claremont Pier
Kirkley
Lowestoft South Roads
Pakefield
Pakefield Cliffs
Kessingland Cliffs
Kessingland
Kessingland Beach
The Denes
Benacre Ness
Covehithe
Green Heath
The Warren
Easton Cliffs
Acle
North Burlingham
Lingwood
Beighton
South Burlingham
Moulton St Mary
Tunstall
Halvergate
THE BROADS
Breydon Water
Runham
Stokesby
Fishley
Mautby
West Caister
West End
Runham Vauxhall
Southtown
Burgh Castle
Bradwell
Belton
Fritton
St Olaves
Herringfleet
Somerleyton
Blundeston
Lound
Oulton
Oulton Broad
Carlton Colville
Gisleham
Rushmere
Mutford
Henstead
Wrentham
Benacre
South Cove
Frostenden
Reydon
Wangford
Uggeshall
Sotterley
Brampton
Westhall
Redisham
Barsham
Beccles
Worlingham
North Cove
Barnby
Gillingham
Geldeston
Kirby Row
Ellingham
Mettingham
Shipmeadow
Ringsfield
Ilketshall St Andrew
Ilketshall St Lawrence
Spexhall
Wissett
Halesworth
Holton
Stone Street
Loddon
Hales
Heckingham
Chedgrave
Reedham
Cantley
Limpenhoe
Freethorpe
Wickhampton
Southwood
Hassingham
Buckenham
Strumpshaw
Langley Street
Hardley Street
Thurlton
Haddiscoe
Toft Monks
Aldeby
Wheatacre
Burgh St Peter
Raveningham
Stockton
Kirby Cane
Norton Subcourse
Thorpe
Waveney
River Waveney
River Yare
A47
A12
A146
A143
A1117
B1074
B1136
B1127
B1062
Great Yarmouth
Lowestoft
127
TG
TM

Morfa Nefyn
Nefyn
Pistyll
Edern
Tudweiliog
Llangwnnadl
Aberdaron
Abersoch
Botwnnog
Sarn Meyllteyrn
Llanbedrog
Bardsey Island
Ynys Enlli
Bardsey Sound
Swnt Enlli
PORTH NEIGWL OR HELL'S MOUTH
Aberdaron Bay
St Tudwal's Road
G W Y N E D D
L L E Y N
LLEYN HISTORICAL MARITIME MUSEUM
ST MARY'S ABBEY (RUINS) OLD CROSSES
NEOLITHIC AXE FACTORY
CARN FADRUN (FORT)
PLAS-YN-RHIW
MYNYDD-Y-CRAIG HEADLAND
MYNYDD CILAN
0 1 2 3 miles
0 1 2 3 4 5 km

Lleyn Peninsula
Penrhyn Llyn
BREMADOG BAY
BAE TREMADOG
BARMOUTH BAY
BAE BERMO
Traeth Bach
Morfa Harlech
Morfa Dyffryn
Porthmadog
Pwllheli
Criccieth
Harlech
Barmouth
(Abermaw)
Llanbedr
Tremadog
Penrhyndeudraeth
Minffordd
Portmeirion
Borth-y-Gest
Morfa Bychan
Pentrefelin
Penmorfa
Llanystumdwy
Chwilog
Afon Wen
Abererch
Llanarmon
Llangybi
Pencaenewydd
Rhoslan
Llwyndyrys
Fron
Pentreuchaf
Rhos-fawr
Y Ffôr
Rhyd-y-gwystl
Llannor
Efailnewydd
Denio
Penrallt
South Beach
West End
Marian-y-mor
Traeth Crugan
Talsarnau
Ynys
Glan-y-wern
Glyn-warch
Llanfair
Llandanwg
Pen-sarn
Pentre Gwynfryn
Coed Ystumgwern
Llanenddwyn
Dyffryn Ardudwy
Llanddwywe
Tal-y-bont
Plas-canol
Llanaber
Fairbourne
Llwyngwril
Hendre
Black Rock Sand
Harlech Pt
Shell Island
Pen-ychain
Penrhyn
Cerrig y Barcdy
Carreg yr Imbill
LLOYD GEORGE MUSEUM
CRICCIETH CASTLE
HARLECH CASTLE
SLATE CAVERNS
DYFFRYN BURIAL CHAMBER
RNLI LIFEBOAT MUSEUM
FAIRBOURNE RAILWAY
RAILWAY MUSEUM
MARITIME MUS
PORTHMADOG HARBOUR
BOSTON LODGE HALT
HAFAN Y MOR
PENYCHAIN
ABERERCH
PWLLHELI
CRICCIETH
PORTHMADOG
MINFFORDD
LLANDECWYN
TALSARNAU
TYGWYN
HARLECH
LLANDANWG
PENSARN
LLANBEDR
DYFFRYN ARDUDWY
TALYBONT
LLANABER
BARMOUTH
LLWYNGWRIL
A497
A499
A487
A496
B4354
B4573
A493
163
146
SH
60
40
10
7
8
9
10
11
A
B
C
D
E
F
G
H

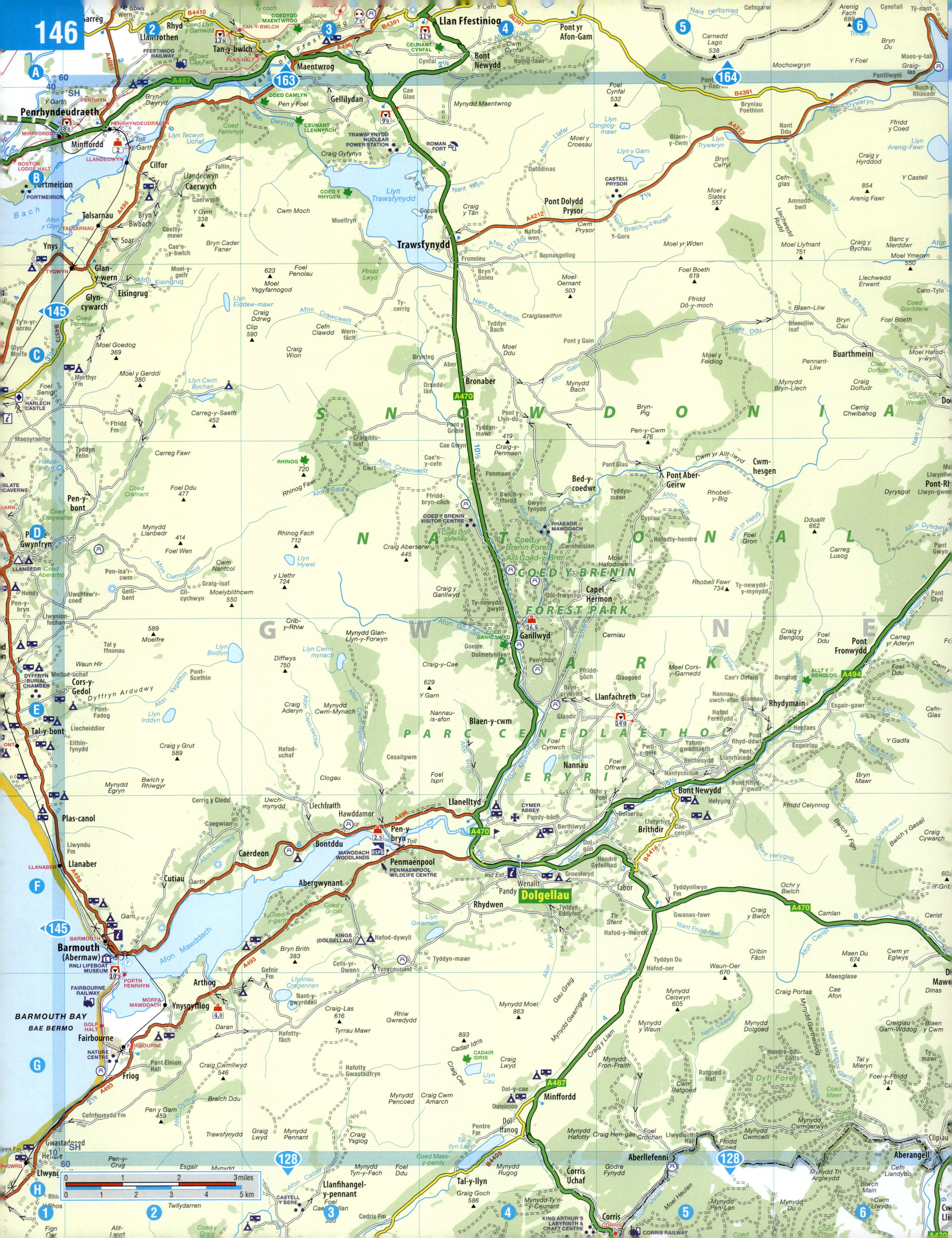
Llan Ffestiniog
Rhyd
Llanfrothen
Tan-y-bwlch
Maentwrog
Gellilydan
Penrhyndeudraeth
Minffordd
Portmeirion
Talsarnau
Cilfor
Caerwych
Ynys
Glan-y-wern
Eisingrug
Glyn-cywarch
Harlech Castle
Pen-y-bont
Gwynfryn
Llanbedr
Dyffryn Ardudwy
Cors-y-Gedol
Tal-y-bont
Plas-canol
Llanaber
Barmouth (Abermaw)
Barmouth Bay
Bae Bermo
Fairbourne
Friog
Arthog
Ynysgyffiog
Caerdeon
Bontddu
Abergwynant
Penmaenpool
Pen-y-bryn
Llanelltyd
Cymer Abbey
Dolgellau
Rhydwen
Pandy
Wenallt
Tabor
Brithdir
Bont Newydd
Rhydymain
Llanfachreth
Nannau
Ganllwyd
Capel Hermon
Pont Aber-Geirw
Cwm-hesgen
Bed-y-coedwr
Bronaber
Trawsfynydd
Llyn Trawsfynydd
Trawsfynydd Nuclear Power Station
Pont Dolydd Prysor
Pont yr Afon-Gam
Bont Newydd
Buarthmeini
Pont Fronwydd
Minffordd
Corris Uchaf
Corris
Aberllefenni
Aberangell
Tal-y-llyn
Llanfihangel-y-pennant
Cadair Idris
Afon Mawddach
Coed y Brenin Forest Park
Dyfi Forest
S N O W D O N I A
N A T I O N A L
P A R K
G W Y N E D D
P A R C C E N E D L A E T H O L
E R Y R I
A470
A487
A494
A496
A493
A4212
B4391
B4410
B4405
B4416
B4573
163
164
145
128

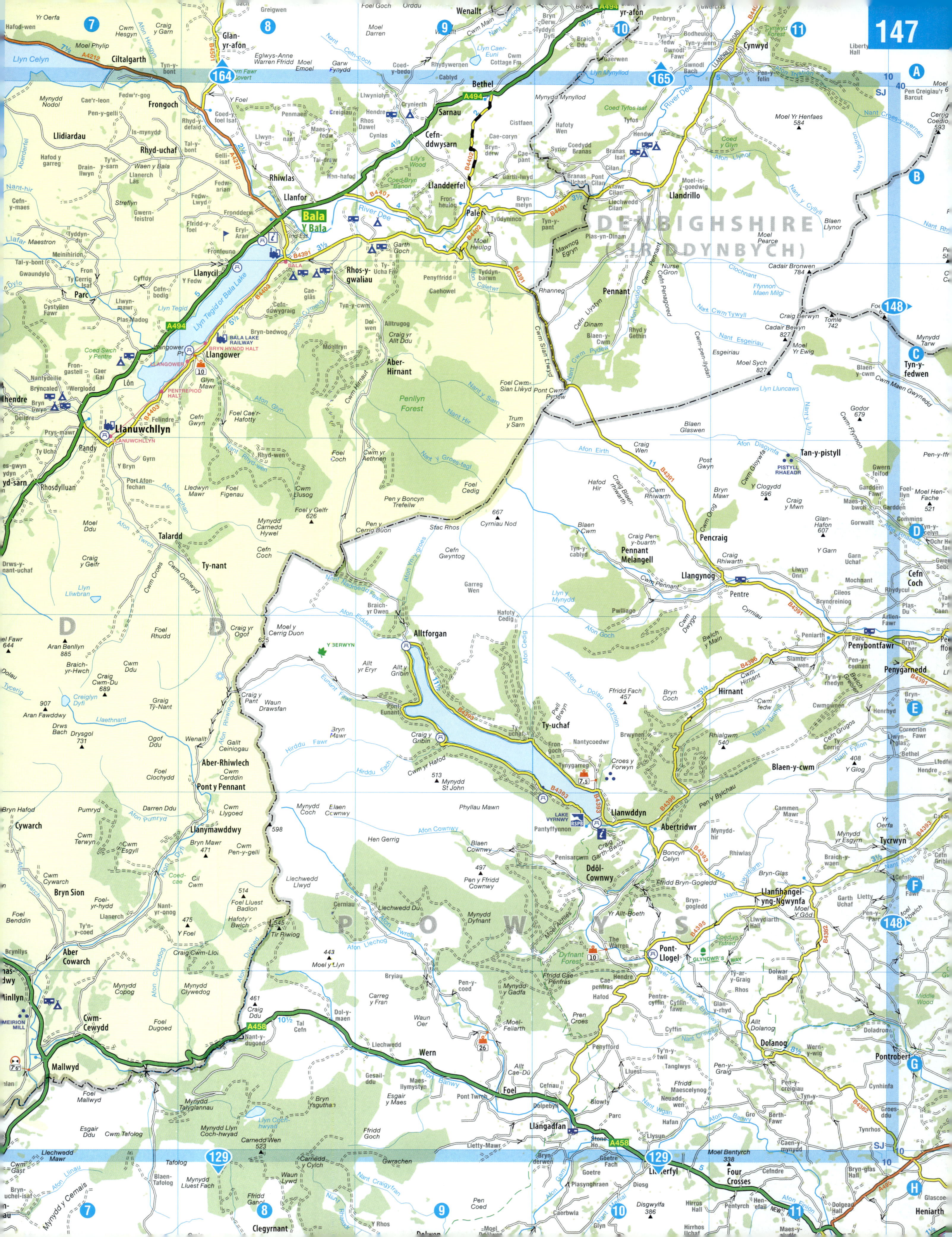

Llyn Celyn
A4212
Ciltalgarth
Moel Phylip
Yr Oerfa
Hafod-wen
Cwm Hesgyn
Craig y Garn
Glan-yr-afon
B4501
Greigwen
Eglwys-Anne Warren Ffridd
Moel Emoel
Garw Fynydd
Foel Goch
Orddu
Moel Darren
Wenallt
Cwm Main
Bryn-Derw
Tyddyn Dyfi
Braich Ddu
yr-afon
Penbryn
Bodheulog
Cynwyd
Liberty Hall
Llyn Caer-Euni
Cwm Cottage Fm
Rhydywernen
Cablyd
Bethel
A494
Sarnau
Cefn-ddwysarn
Frongoch
Mynydd Nodol
Cae'r-leon
Fedw'r-gog
Pen-y-gelli
Llidiardau
Rhyd-uchaf
Is-mynydd
Tal-y-bont
Hafod y garreg
Drain-llwyn
Ty'n-y-sarn
Waen y Bala
Llanerch Las
Rhiwlas
Llanfor
Bala
Y Bala
B4391
River Dee
Llandderfel
Fron-heulog
B4401
B4402
Palé
Tyddyntco
Llyn Mynyllod
Mynydd Myllod
Cistfaen
Cae-coryn
Bryn-derw
Cae-pant
Gartli-lwyd
Hafoty Wen
Syrior
Coedydd Branas
Branas Isaf
Branas Uchaf
Pont Cilan
Cilan
Llawr Cilan
Llechwedd Cilan
Coed Tyfos Isaf
Tyfos
Hendwr
Coed y Glyn
Afon Llynor
Moel-is-y-goedwig
Llandrillo
Moel Yr Henfaes 584
Pen Creigiau'r Barcut
Cerrig Coedio 593
Blaen Llynor
Moel Pearce
DENBIGHSHIRE
(SIR DDYNBYCH)
Plas-yn-Dinam
Mawnog Egryn
Nurse Gron
Cefn Penagored
Clochnant
Cadair Bronwen 784
Ffynnon Maen Milgi
Craig Berwyn
Cadair Berwyn 827
Tomle 742
Mynydd Tarw
Moel Yr Ewig
Nant Esgeiriau
Esgeiriau
Moel Sych 827
Tyn-y-fedwen
Cwm Maen Gwynedd
Llyn Lluncaws
Pennant
Dinam
Blaen-y-Cwm
Rhyd y Gethin
Cwm-pen-llydan
Cwm Sian Llwyd
Rhanneg
Foel Cwm-Sian Llwyd
Pont Cwm Pydew
Tyddyn-barwn
Moel Heulog
Afon Caletwr
Penyfridd
Caehowel
Alltrugog
Craig yr Allt Ddu
Aber-Hirnant
Penllyn Forest
Nant y Sarn
Nant Hir
Trum y Sarn
Rhos-y-gwaliau
Garth Goch
Ty-Ucha Fm
Llanycil
Llyn Tegid or Bala Lake
B4403
Cae-glas
Cefn-ddwygraig
Bryn-bedwog
Moelfryn
Bala Lake Railway
Bryn Hynod Halt
Llangower
Glyn Mawr
Pentrepiod Halt
Llanuwchllyn
Parc
Cefn-bodig
Cyffdy
Llwyn-mawr
Plas Madog
Cystyllen Fawr
Fron-gastell
Caer Gai
Lôn
Werglodd
Nantydeilie
Bryncaled
Bryn Gwyn
Deildre
Prys-mawr
Pandy
Ty Ucha
Y Bryn
Gyrn
Rhosdylluan
Port Afon-fechan
Lledwyn Mawr
Foel Figenau
Cwm Llusog
Foel Cae'r-Hafotty
Cefn Gwyn
Rhyd-wen
Foel Coch
Cwm yr Aethnen
Foel y Gelfr 626
Mynydd Carnedd Hywel
Pen y Boncyn Trefeilw
Foel Cedig
Moel Ddu
Talardd
Ty-nant
Cefn Coch
Craig y Gelfr
Llyn Lliwbran
Cwm Croes
Cwm Cynllwyd
Stac Rhos
Cyrniau Nod 667
Cefn Gwyntog
Garreg Wen
Blaen Glaswen
Craig Wen
Afon Eirth
B4391
Post Gwyn
Tan-y-pistyll
Pistyll Rhaeadr
Y Clogydd 596
Gwern Feifod
Hafod Hir
Cwm Rhiwarth
Bryn Mawr
Craig y Mwn
Glan-Hafon 607
Moel Hen-Fache 521
Blaen y Cwm
Tyn-y-cablyd
Craig Pen-y-buarth
Pennant Melangell
Pencraig
Craig Rhiwarth
Y Garn
Garn Uchaf
Llangynog
Pentre
Cwm Pennant
Mochnant
Cefn Coch
Pwlliago
Afon Goch
Hafoty Cedig
Llyn y Mynydd
Penybontfawr
Penygarnedd
Peniarth
Cyrniau
B4396
Hirnant
Foel Fawr 644
Aran Benllyn 885
Braich-yr-Hwch
Craig Cwm-Du 689
Cwm Ddu
Foel Rhudd
Craig yr Ogof
625
Moel y Cerrig Duon
Y Berwyn
Alltforgan
Braich-yr Owen
Afon Eiddew
Allt yr Eryr
Allt y Gribin
Craig y Pant
Waun Drawsfan
Creiglyn Dyfi
907
Aran Fawddwy
Drws Bach
Drysgol 731
Craig Ty-Nant
Laethnant
Ogof Ddu
Wenallt
Gallt Ceiniogau
Aber-Rhiwlech
Cwm Cerddin
Pont y Pennant
Foel Clochydd
Bryn Mawr
Hirddu Fawr
Craig y Gribin
Cwm yr Hafod
513 Mynydd St John
B4393
Ty-uchaf
Pwll Bwyn
Fron-goch
Nantycoedwr
Croes y Forwyn
Tynygarreg
Llanwddyn
Lake Vyrnwy
RSPB
Pantyffynnon
Ffridd Fach 457
Bryn Coch
Brwynen
Rhialgwm 540
Blaen-y-cwm
Pen Y Bylchau
Abertridwr
Cammen Mawr
Yr Oerfa
Mynydd yr Esgyrn
Tycrwyn
Bryn Hafod
Cywarch
Pumryd
Darren Ddu
Afon Pumryd
Cwm Llygoed
Llanymawddwy
Cwm Terwyn
Cwm Esgyll
Bryn Mawr 471
Cwm Pen-y-gelli
Mynydd Coch
Elaen Cownwy
598
Phyllau Mawn
Afon Cownwy
Hen Gerrig
Blaen Cownwy
497 Pen y Ffridd Cownwy
Penisarcwm
Craig Garth-Bwlch
Ddôl-Cownwy
Boncyn Celyn
Rhiwlas
Bryn-Glas
Llanfihangel-yng-Ngwynfa
Cwm Cywarch
Bryn Sion
Foel-yr-hydd
Llanerch
Foel Benddin
Ty'n-y-coed
Cil Cwm
514 Foel Lluest Badion
Nant-yr-onog
Hafoty'r Bwlch
475 Y Foel
545 Tir Rwiog
Llechwedd Llwyd
Lechwedd Du
Mynydd Dyfnant
P O W Y S
Yr Allt-Boeth
Ffridd Bryn-Gogledd
Bryn-gogledd
Moel Y Golfa
Llwydiarth Hall
Brynllys
Aber Cowarch
Craig Cwm-Llo
443 Moel y Llyn
Afon Llechog
Dyfnant Forest
The Warren
Pont-Llogel
Glyndŵr's Way
B4395
Ty-ar-y-Graig
Rhos
Dolwar Hall
Mynydd Copog
Mynydd Clywedog
461 Craig Ddu
Bryiau
Pen-y-coed
Carreg y Fran
Dol-y-maen
Mynydd y Gadfa
Ffridd Cae Penfras
Cae-penfras
Hafod
Hendre
Pentre-cyffin
Cyffin-fawr
Pren Croes
Glan-y-rhyd
Allt Dolanog
Dolanog
Doladron
Cwm-Cewydd
Foel Dugoed
A458
Nant-y-dugoed
Tal Cefn
Waun Oer
Moel-Feliarth
Cyffin
Llechwedd
Wern
Mallwyd
Foel Mallwyd
Mynydd Talyglannau
Bryn Ysgutha
Gesail-ddu
Maes-llymystyn
Esgair y Maes
Afon Banwy
Pont Twrch
Allt Cae-Dû
Foel
Cefnau
Dolpebyll
Blowty
Penyffordd
Lluest
Ty'n-y-twll
Tangiwys
Ffridd Maeselycnog
Neuadd-wen
Hafan
Pen-y-Graig
Pen-y-creigiau
Berth-Fawr
Pontrobert
Cynhinfa
Groes-ddu
Tynrhos
Esgair Ddu
Cwm Tafolog
Mynydd Llyn Coch-hwyad
Carnedd Wen 523
Ffridd Goch
Llanfgadfan
Lletty-Mawr
Stone Ho
Llysun
Caen-y-mynydd
Moel Bentyrch 338
Llechwedd Mawr
Tafolog
Blaen-Tafolog
Mynydd Lluest Fach
Carnedd y Cylch
Waun Lywd
Gwrachen
Bryn-derwen
Goetre Fach
Goetre
Llanerfyl
Four Crosses
Cefndre
Bryn-glas Hall
Heniarth
Cwm Gast
Bryn-uchel-isaf
Mynydd y Cemais
Ffridd Gander
Clegyrnant
Pen Coed
Plasynghraen
Diosg
Disgwylfa 386
Caerbwla
Glyn
Hirros Hall
Pentyrch
Hen-efail
Hirrhos Uchaf
Dolgead Hall
Glascoed
164
165
129
148
SJ

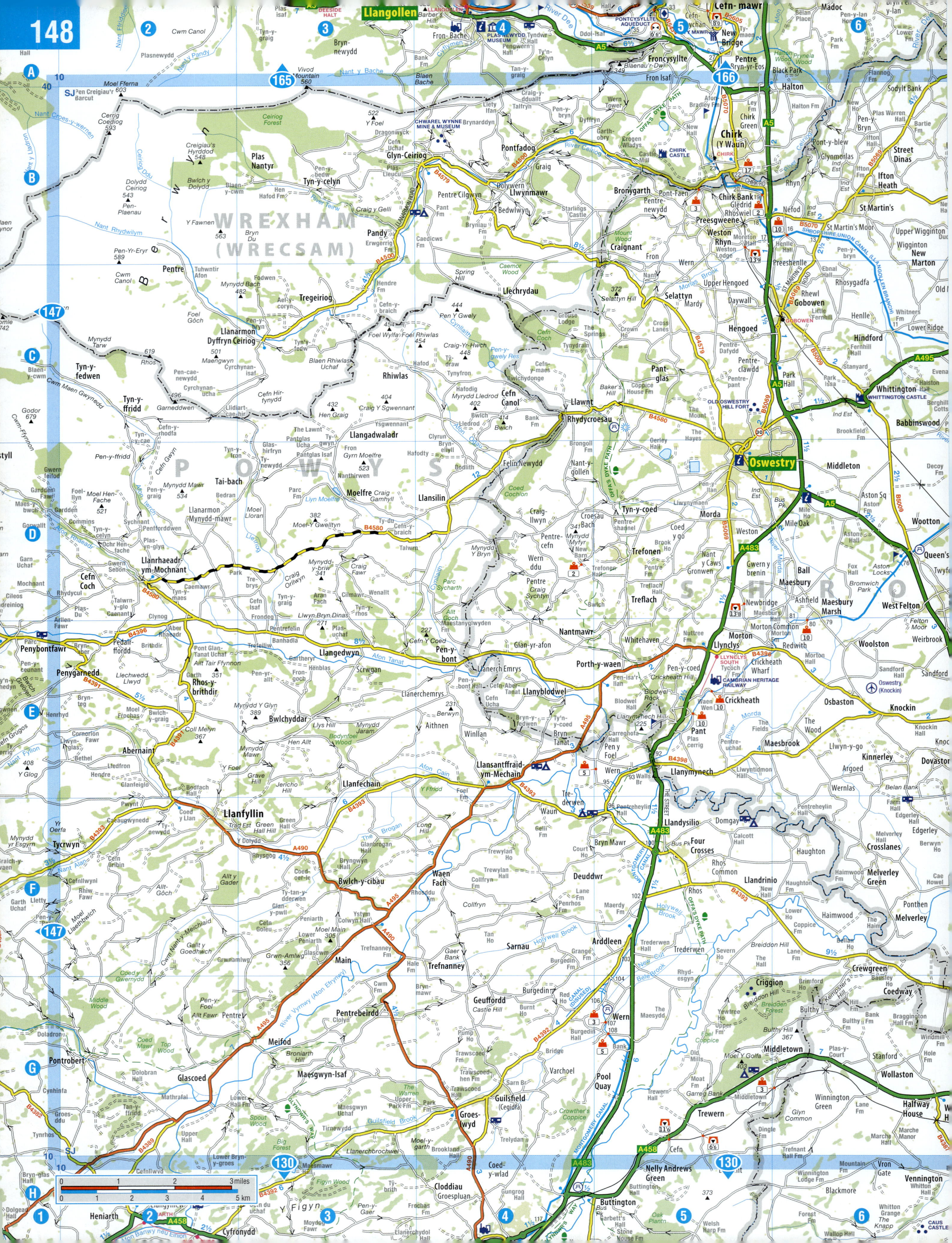

Llangollen
Cefn-mawr
Froncysyllte
Chirk
(Y Waun)
CHIRK CASTLE
Glyn-Ceiriog
Pontfadog
WREXHAM
(WRECSAM)
Llanarmon Dyffryn Ceiriog
Tregeiriog
Pandy
Llechrydau
Selattyn
Weston Rhyn
Preesgweene
St Martin's
Gobowen
Whittington
Oswestry
Llansilin
Rhydycroesau
Llangadwaladr
Moelfre
Llanrhaeadr-ym-Mochnant
Penybontfawr
Llangedwyn
Pen-y-bont
Llanyblodwel
Porth-y-waen
Morda
Trefonen
Maesbury
West Felton
Llynclys
Pant
Llanymynech
Knockin
Kinnerley
Llansantffraid-ym-Mechain
Llanfechain
Llanfyllin
Bwlch-y-cibau
Four Crosses
Llandrinio
Arddleen
Meifod
Pontrobert
Guilsfield
(Cegidfa)
Pool Quay
Middletown
Trewern
Buttington
Welshpool
Cyfronydd
Heniarth
P O W Y S
S H R O
Berwyn
165
166
147
130
0 1 2 3 miles
0 1 2 3 4 5 km

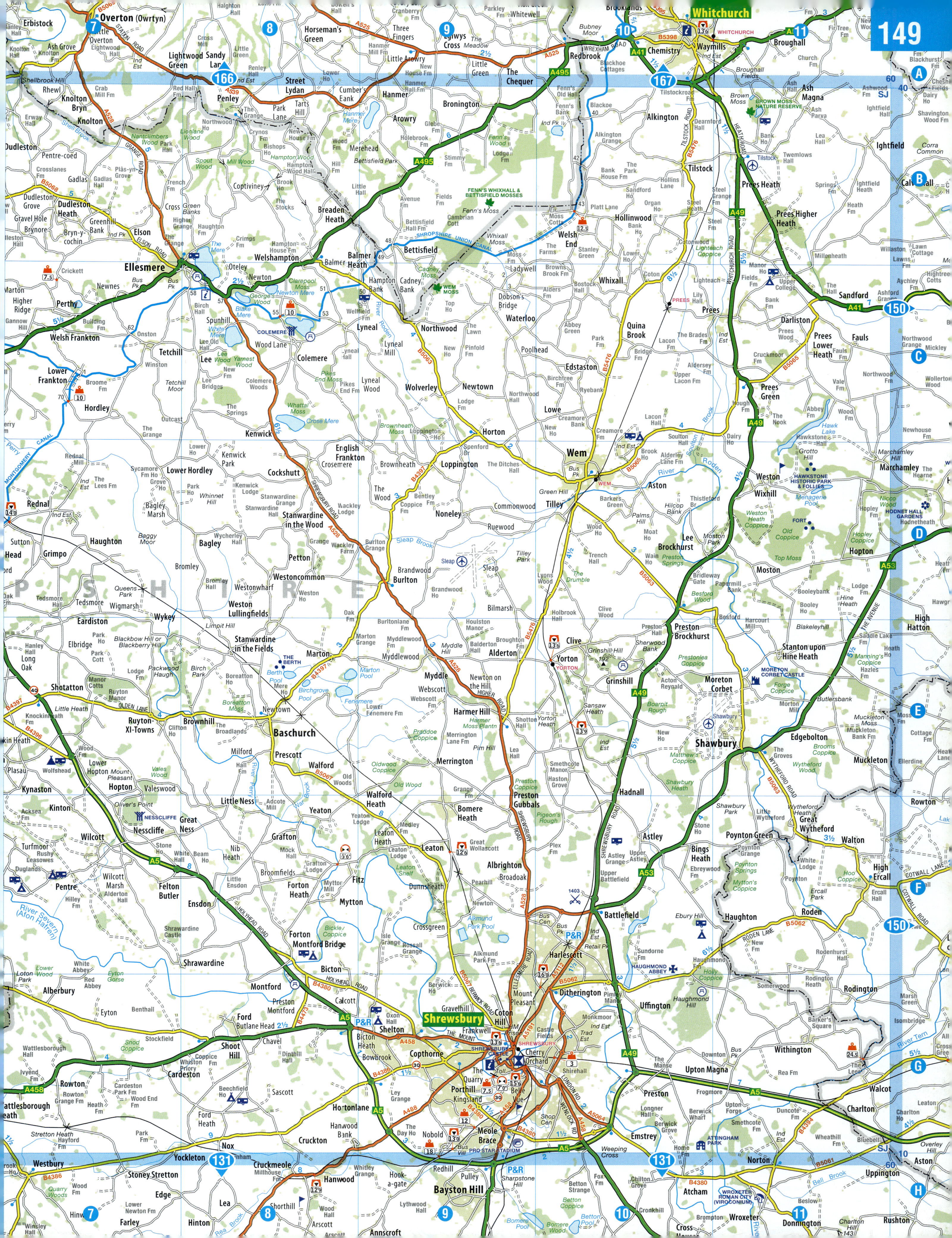
Erbistock
Overton (Owrtyn)
Horseman's Green
Three Fingers
Eglwys Cross
Whitchurch
Broughall
Ash Grove
Lightwood Green
Sandy Lane
Little Arowry
Little Green
The Chequer
Redbrook
Chemistry
Waymills
Street Lydan
Cumber's Bank
Hanmer
Bronington
Penley
Park Lane
Tarts Hill
Arowry
Ash Magna
Rhewl
Knolton Bryn
Knolton
Alkington
Tilstock
Dudleston
Pentre-coed
Gadlas
Breaden Heath
Coptiviney
Welshampton
Prees Heath
Prees Higher Heath
Dudleston Heath
Elson
Ellesmere
Balmer Heath
Bettisfield
Welsh End
Hollinwood
Whixall
Perthy
Newnes
Hampton Bank
Cadney Bank
Dobson's Bridge
Waterloo
Prees
Sandford
Darliston
Fauls
Welsh Frankton
Tetchill
Lyneal
Northwood
Quina Brook
Prees Lower Heath
Lower Frankton
Lee
Colemere
Wood Lane
Wolverley
Newtown
Poolhead
Edstaston
Prees Green
Hordley
Kenwick
Lowe
Horton
Wem
Lower Hordley
Cockshutt
English Frankton
Loppington
Aston
Weston
Wixhill
Marchamley
Rednal
Noneley
Tilley
Commonwood
Haughton
Grimpo
Bagley
Stanwardine in the Wood
Ruewood
Lee Brockhurst
Hopton
Petton
Westoncommon
Burlton
Sleap
Moston
Eardiston
Wykey
Weston Lullingfields
Stanwardine in the Fields
Marton
Myddle
Alderton
Clive
Yorton
Preston Brockhurst
Stanton upon Hine Heath
High Hatton
Shotatton
Ruyton-XI-Towns
Brownhill
Newtown
Baschurch
Harmer Hill
Grinshill
Moreton Corbet
Shawbury
Edgebolton
Muckleton
Prescott
Walford
Merrington
Hadnall
Kynaston
Hopton
Valeswood
Little Ness
Walford Heath
Preston Gubbals
Bomere Heath
Kinton
Nesscliffe
Great Ness
Yeaton
Wilcott
Grafton
Leaton
Astley
Poynton Green
Great Wytheford
Walton
Pentre
Felton Butler
Albrighton
Fitz
Bings Heath
High Ercall
Ensdon
Forton Heath
Mytton
Battlefield
Haughton
Roden
Crossgreen
Forton
Montford Bridge
Harlescott
Shrawardine
Bicton
Montford
Ditherington
Uffington
Rodington
Alberbury
Calcott
Shrewsbury
Coton Hill
Frankwell
Ford
Shelton
Copthorne
Withington
Shoot Hill
Upton Magna
Cardeston
Porthill
Kingsland
Preston
Walcot
Rowton
Hanwood
Charlton
Meole Brace
Emstrey
Westbury
Yockleton
Cruckton
Nox
Cruckmeole
Norton
Stoney Stretton
Edge
Hanwood
Bayston Hill
Atcham
Uppington
Lea
Farley
Hinton
Wroxeter
Donnington
Rushton
Cross Houses
Annscroft
A5
A41
A49
A53
A458
A495
A528
A525
A5112
A5064
A488
B5476
B5063
B5065
B5062
B4397
B4380
B5067
B5068
B5069
B4386
SHROPSHIRE
SHROPSHIRE UNION CANAL
MONTGOMERY CANAL
River Severn (Afon Hafren)
River Roden
River Perry
River Tern
FENN'S WHIXALL & BETTISFIELD MOSSES
BROWN MOSS NATURE RESERVE
HAWKSTONE HISTORIC PARK & FOLLIES
MORETON CORBET CASTLE
HAUGHMOND ABBEY
ATTINGHAM PARK
WROXETER ROMAN CITY (VIROCONIUM)
PRO STAR STADIUM
SJ
7
8
9
10
11
131
150
166
167

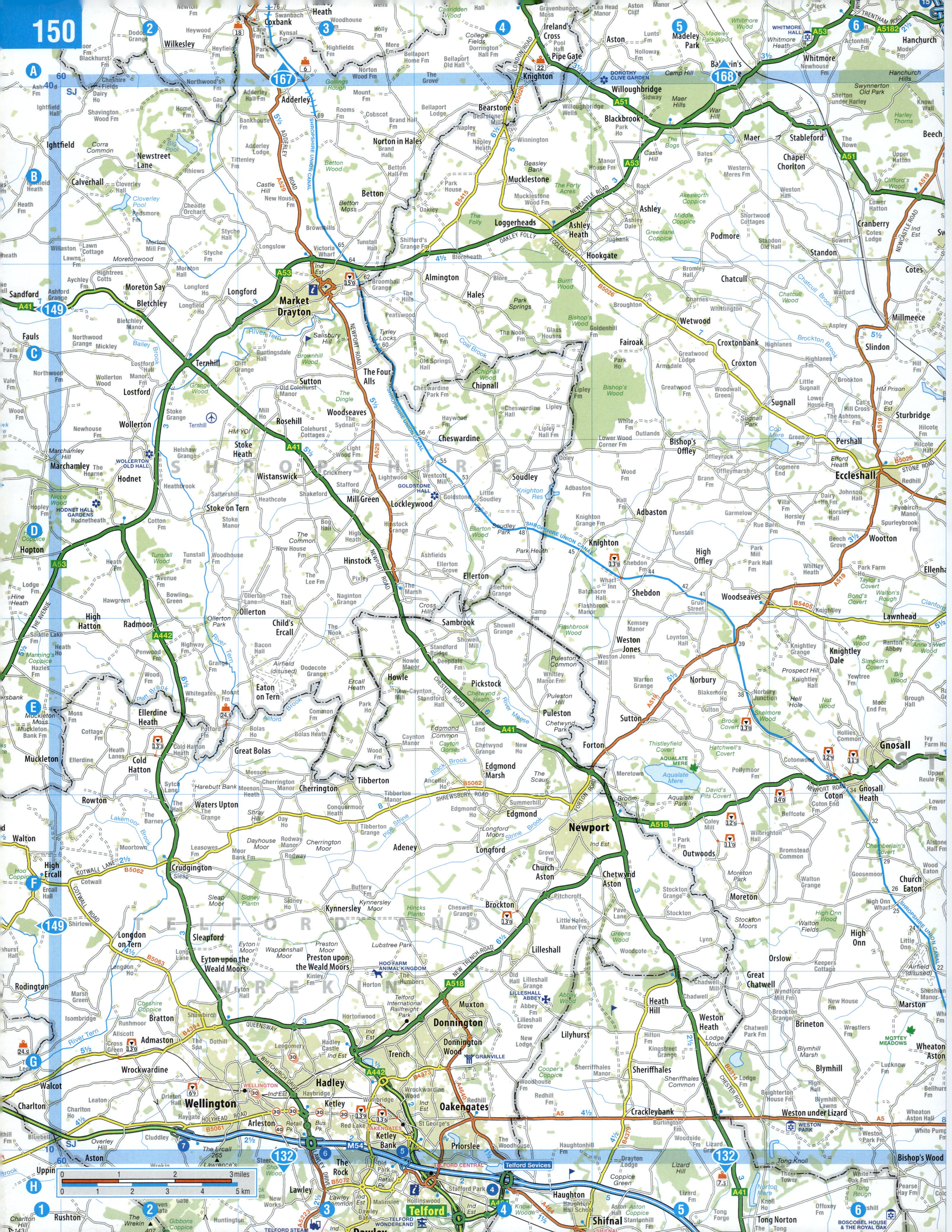

167
168
149
132
SJ
Coxbank
Heath
Wilkesley
Adderley
Knighton
Ireland's Cross
Pipe Gate
Aston
Madeley Park Wood
Baldwin's Gate
Whitmore
Hanchurch
Willoughbridge
DOROTHY CLIVE GARDEN
Bearstone
Blackbrook
Maer
Stableford
Chapel Chorlton
Beech
Ightfield
Newstreet Lane
Calverhall
Norton in Hales
Mucklestone
Betton
Loggerheads
Ashley
Ashley Heath
Hookgate
Podmore
Cranberry
Standon
Cotes
SHROPSHIRE UNION CANAL
Market Drayton
Sandford
Moreton Say
Bletchley
Longford
Almington
Hales
Chatcull
Wetwood
Croxtonbank
Croxton
Millmeece
Slindon
Fauls
Ternhill
Sutton
The Four Alls
Chipnall
Fairoak
Sugnall
Sturbridge
Lostford
Wollerton
Rosehill
Woodseaves
Cheswardine
Bishop's Offley
Pershall
Eccleshall
Marchamley
Hodnet
Stoke Heath
Wistanswick
SHROPSHIRE
Soudley
Stoke on Tern
Mill Green
Lockleywood
Adbaston
GOLDSTONE HALL
HODNET HALL GARDENS
Knighton
High Offley
Wootton
Hopton
Hinstock
Ellerton
Shebdon
Woodseaves
High Hatton
Radmore
Ollerton
Child's Ercall
Sambrook
Weston Jones
Lawnhead
Knightley Dale
Norbury
Howle
Pickstock
Puleston
Sutton
Eaton on Tern
Ellerdine Heath
Muckleton
Cold Hatton
Great Bolas
Edgmond Marsh
Forton
Gnosall
Gnosall Heath
Coton
Aqualate Mere
Tibberton
Cherrington
Waters Upton
Rowton
Edgmond
Newport
Walton
High Ercall
Crudgington
Adeney
Longford
Church Aston
Chetwynd Aston
Outwoods
Moreton
Church Eaton
Kynnersley
Brockton
TELFORD AND WREKIN
Longdon on Tern
Sleapford
Eyton upon the Weald Moors
Preston upon the Weald Moors
Lilleshall
Orslow
High Onn
Great Chatwell
Rodington
HOO FARM ANIMAL KINGDOM
LILLESHALL ABBEY
Heath Hill
Weston Heath
Brineton
Marston
Bratton
Admaston
Donnington
Muxton
Lilyhurst
Blymhill
Wheaton Aston
Wrockwardine
Hadley
Sheriffhales
Walcot
Charlton
Wellington
Ketley
Oakengates
Crackleybank
Weston under Lizard
WESTON PARK
Arleston
Ketley Bank
Priorslee
Aston
Uppington
Telford Services
Bishop's Wood
Lawley
Telford
Stafford Park
Haughton
Shifnal
Tong Norton
Rushton
BOSCOBEL HOUSE & THE ROYAL OAK
0 1 2 3 miles
0 1 2 3 4 5 km

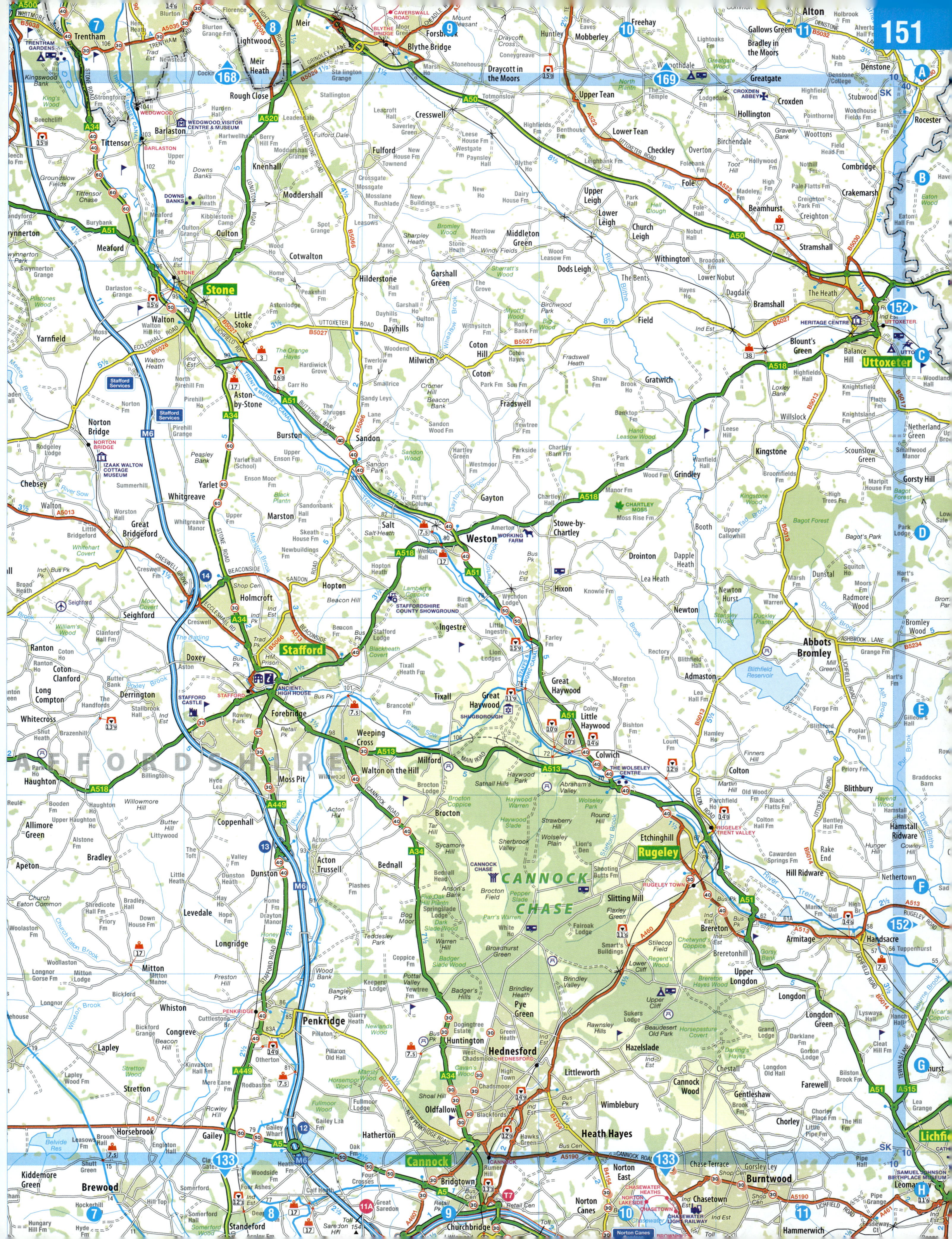
Trentham
Meir
Blythe Bridge
Forsbrook
Draycott in the Moors
Mobberley
Freehay
Alton
Denstone
Rocester
Lightwood
Meir Heath
Rough Close
Barlaston
Tittensor
Cresswell
Fulford
Upper Tean
Lower Tean
Checkley
Hollington
Croxden
Greatgate
Knenhall
Moddershall
Oulton
Meaford
Cotwalton
Hilderstone
Garshall Green
Middleton Green
Dods Leigh
Upper Leigh
Lower Leigh
Church Leigh
Withington
Fole
Beamhurst
Stramshall
Bramshall
The Heath
Uttoxeter
Stone
Little Stoke
Walton
Yarnfield
Dayhills
Milwich
Coton Hill
Coton
Fradswell
Field
Gratwich
Blount's Green
Balance Hill
Aston-by-Stone
Burston
Sandon
Norton Bridge
Chebsey
Great Bridgeford
Yarlet
Whitgreave
Marston
Salt
Weston
Gayton
Stowe-by-Chartley
Grindley
Kingstone
Gorsty Hill
Hopton
Holmcroft
Seighford
Ingestre
Hixon
Drointon
Newton
Abbots Bromley
Stafford
Doxey
Derrington
Ranton
Coton Clanford
Long Compton
Whitecross
Tixall
Great Haywood
Little Haywood
Colwich
Forebridge
Weeping Cross
Milford
Walton on the Hill
Moss Pit
Haughton
STAFFORDSHIRE
Brocton
Coppenhall
Acton Trussell
Bednall
Dunston
Allimore Green
Bradley
Apeton
Levedale
Longridge
Mitton
Whiston
Congreve
Penkridge
Lapley
Stretton
Huntington
Hednesford
Littleworth
Wimblebury
Heath Hayes
CANNOCK CHASE
Etchinghill
Rugeley
Slitting Mill
Brereton
Colton
Blithbury
Hamstall Ridware
Hill Ridware
Armitage
Handsacre
Upper Longdon
Longdon
Longdon Green
Hazelslade
Cannock Wood
Gentleshaw
Chorley
Farewell
Gailey
Horsebrook
Brewood
Kiddemore Green
Oldfallow
Hatherton
Cannock
Bridgtown
Churchbridge
Norton East
Norton Canes
Burntwood
Chasetown
Hammerwich
Lichfield
Standeford

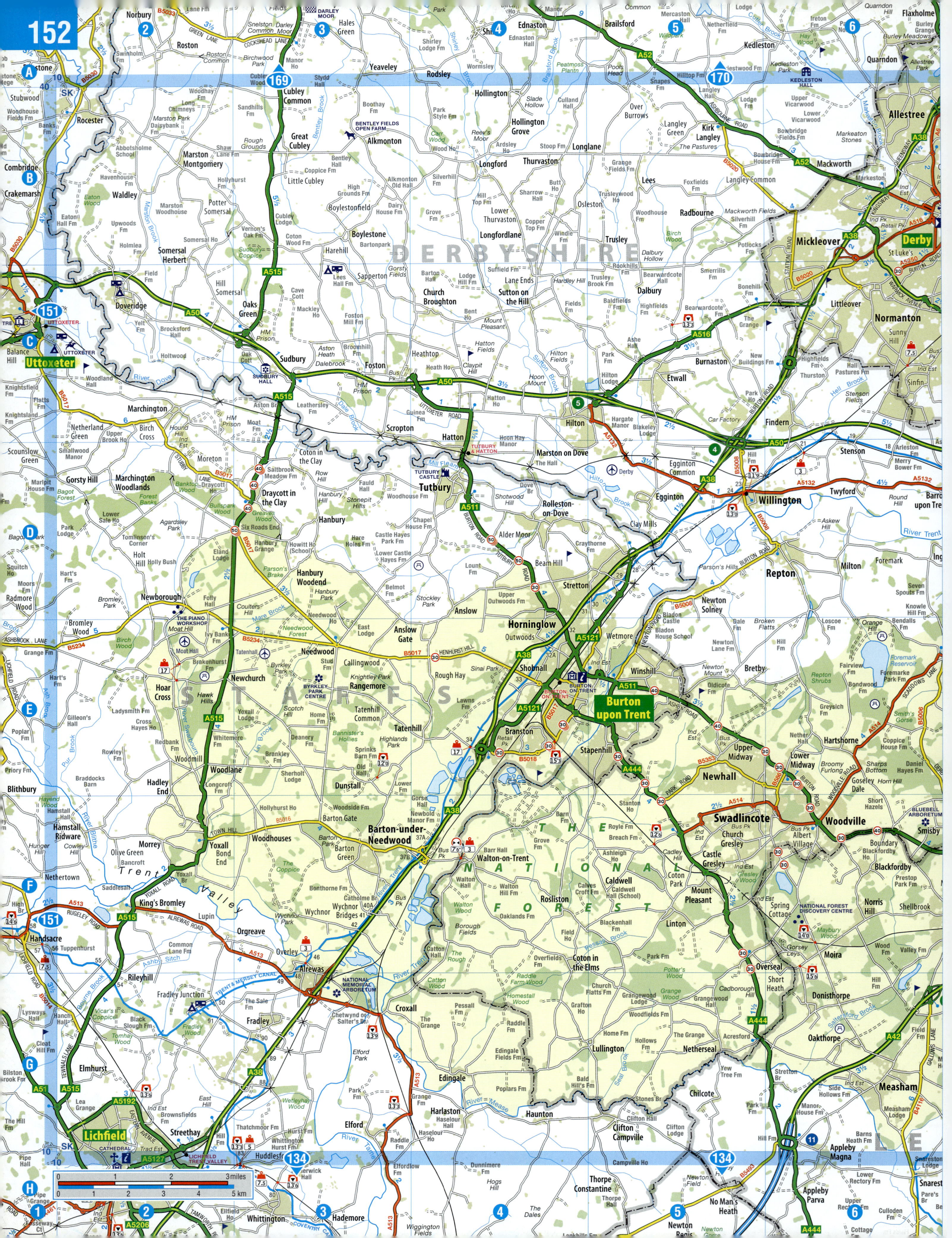
DERBYSHIRE
S T A F F S
THE NATIONAL FOREST
Derby
Burton upon Trent
Uttoxeter
Lichfield
Norbury
Roston
Rocester
Stubwood
Combridge
Crakemarsh
Waldley
Marston Montgomery
Cubley Common
Great Cubley
Little Cubley
Yeaveley
Rodsley
Alkmonton
Ednaston
Hollington
Brailsford
Kedleston
Quarndon
Flaxholme
Allestree
Mackworth
Kirk Langley
Langley Green
Langley Common
Longford
Thurvaston
Longlane
Lees
Radbourne
Mickleover
Littleover
Normanton
Sinfin
Boylestone
Harehill
Sapperton
Church Broughton
Sutton on the Hill
Trusley
Dalbury
Somersal Herbert
Doveridge
Oaks Green
Sudbury
Foston
Heathtop
Etwall
Burnaston
Findern
Hilton
Egginton
Egginton Common
Willington
Stenson
Twyford
Repton
Milton
Foremark
Marchington
Marchington Woodlands
Scropton
Hatton
Tutbury
Tutbury Castle
Marston on Dove
Rolleston-on-Dove
Clay Mills
Coton in the Clay
Draycott in the Clay
Moreton
Hanbury
Hanbury Woodend
Newborough
Anslow
Anslow Gate
Stretton
Horninglow
Outwoods
Wetmore
Winshill
Newton Solney
Bretby
Needwood
Callingwood
Rangemore
Tatenhill
Rough Hay
Shobnall
Branston
Stapenhill
Newhall
Swadlincote
Woodville
Midway
Upper Midway
Hartshorne
Smisby
Hoar Cross
Newchurch
Woodmill
Woodlane
Hadley End
Dunstall
Barton-under-Needwood
Barton Gate
Walton-on-Trent
Rosliston
Caldwell
Linton
Castle Gresley
Church Gresley
Mount Pleasant
Overseal
Moira
Donisthorpe
Oakthorpe
Measham
Netherseal
Lullington
Coton in the Elms
Chilcote
Clifton Campville
Haunton
Harlaston
Elford
Edingale
Croxall
Alrewas
Orgreave
Fradley
Fradley Junction
King's Bromley
Yoxall
Morrey
Woodhouses
Wychnor
Wychnor Bridges
Blithbury
Hamstall Ridware
Handsacre
Rileyhill
Elmhurst
Streethay
Whittington
Huddlesford
Hademore
Thorpe Constantine
No Man's Heath
Appleby Magna
Appleby Parva
National Memorial Arboretum
River Trent
River Dove
Trent Valley
Trent & Mersey Canal
3 miles
5 km

Nottingham
NOTTINGHAM
West Bridgford
Beeston
Stapleford
Chaddesden
Long Eaton
Sandiacre
Risley
Borrowash
Spondon
Ockbrook
Breadsall
Kirk Hallam
Stanton-by-Dale
Chilwell
Attenborough
Toton
Clifton
Wilford
Edwalton
Ruddington
Tollerton
Plumtree
Keyworth
Bradmore
Bunny
Gotham
Barton in Fabis
Thrumpton
Ratcliffe on Soar
Kegworth
Kingston on Soar
West Leake
East Leake
Sutton Bonington
Costock
Wysall
Rempstone
Wymeswold
Hoton
Prestwold
Cotes
Burton on the Wolds
Walton on the Wolds
Barrow upon Soar
Loughborough
Quorn or Quorndon
Sileby
Mountsorrel
Rothley
Cossington
Thurcaston
Cropston
Anstey
Birstall
Woodhouse Eaves
Swithland
Woodthorpe
Shelthorpe
Nanpantan
Hathern
Dishley
Thorpe Acre
Normanton on Soar
Stanford on Soar
Zouch
Long Whatton
Shepshed
Belton
Osgathorpe
Castle Donington
Hemington
Lockington
Diseworth
Isley Walton
Breedon on the Hill
Tonge
Worthington
Melbourne
Kings Newton
Wilson
Shardlow
Cavendish Bridge
Aston-on-Trent
Weston-on-Trent
Elvaston
Ambaston
Thulston
Alvaston
Osmaston
Chellaston
Swarkestone
Stanton by Bridge
Ticknall
Calke
Ashby-de-la-Zouch
Coleorton
Thringstone
Whitwick
Coalville
Swannington
Griffydam
Newbold
Packington
Ravenstone
Normanton le Heath
Hugglescote
Ellistown
Ibstock
Heather
Bardon
Greenhill
Stanton under Bardon
Markfield
Copt Oak
Field Head
Newtown Linford
Newton Burgoland
Battram
Bagworth
Thornton
Charnwood Forest
DERBY CITY
NOTTINGHAMSHIRE
LEICESTERSHIRE
THE NATIONAL FOREST
Derby
Trowell
Stanley Common
West Hallam
Morley
Little Eaton
Strelley
Wollaton
Lenton
Radford
Basford
Bramcote
Dale Abbey
Draycott
Breaston
Sawley
Church Wilne
Shelton Lock
Sinfin Moor
East Midlands Airport
Donington Park
M1
A6
A50
A52
A42
A453
A511
A512
A60
A46
153
154
135
170
171

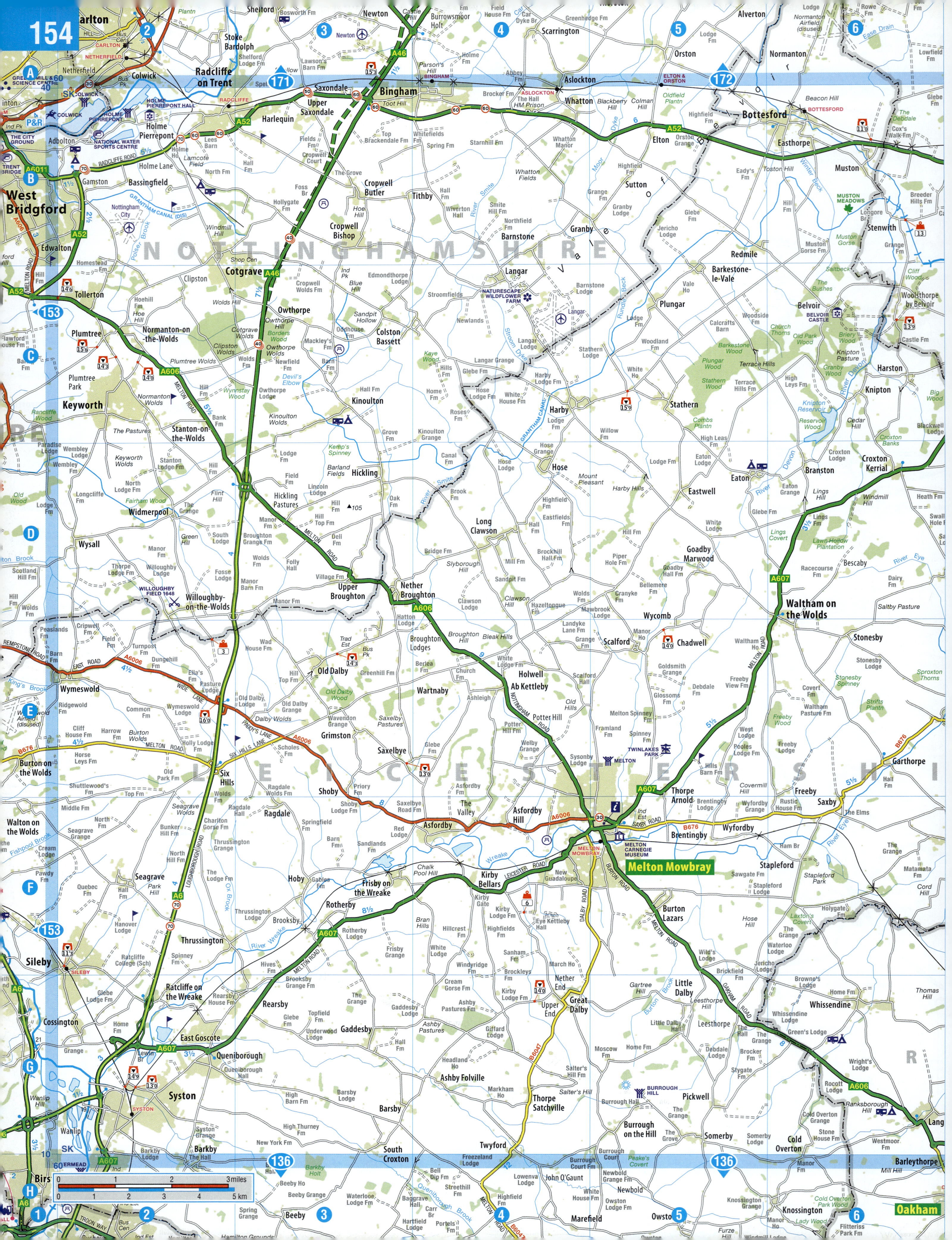

Carlton
Shelford
Newton
Burrowsmoor Holt
Scarrington
Alverton
Orston
Normanton
Radcliffe on Trent
Stoke Bardolph
Colwick
Netherfield
Holme Pierrepont
Saxondale
Upper Saxondale
Bingham
Aslockton
Whatton
Bottesford
Easthorpe
Elton
Muston
Harlequin
Gamston
Bassingfield
Holme Lane
West Bridgford
Edwalton
Tollerton
Cropwell Butler
Cropwell Bishop
Tithby
Barnstone
Granby
Sutton
Redmile
Barkestone-le-Vale
Stenwith
Belvoir
Woolsthorpe by Belvoir
NOTTINGHAMSHIRE
Cotgrave
Clipston
Owthorpe
Colston Bassett
Langar
Plungar
Plumtree
Normanton-on-the-Wolds
Keyworth
Stanton-on-the-Wolds
Kinoulton
Harby
Stathern
Harston
Knipton
Branston
Croxton Kerrial
Eaton
Eastwell
Hose
Long Clawson
Hickling
Widmerpool
Wysall
Upper Broughton
Nether Broughton
Willoughby-on-the-Wolds
Goadby Marwood
Waltham on the Wolds
Wycomb
Chadwell
Scalford
Stonesby
Old Dalby
Wymeswold
Holwell
Ab Kettleby
Wartnaby
Grimston
Saxelbye
Burton on the Wolds
Six Hills
Shoby
Ragdale
Asfordby
Asfordby Hill
Thorpe Arnold
Brentingby
Wyfordby
Freeby
Saxby
Garthorpe
Melton Mowbray
Stapleford
Walton on the Wolds
Seagrave
Hoby
Frisby on the Wreake
Kirby Bellars
Rotherby
Burton Lazars
Thrussington
Sileby
Ratcliffe on the Wreake
Rearsby
Gaddesby
Great Dalby
Little Dalby
Whissendine
Cossington
East Goscote
Queniborough
Syston
Ashby Folville
Thorpe Satchville
Pickwell
Barsby
Burrough on the Hill
Somerby
Cold Overton
Twyford
South Croxton
Barkby
Beeby
John O'Gaunt
Marefield
Newbold
Owston
Knossington
Oakham
Barleythorpe
LEICESTERSHIRE
A46
A52
A606
A607
A6006
A6
B676
171
172
153
136
0 1 2 3 miles
0 1 2 3 4 5 km

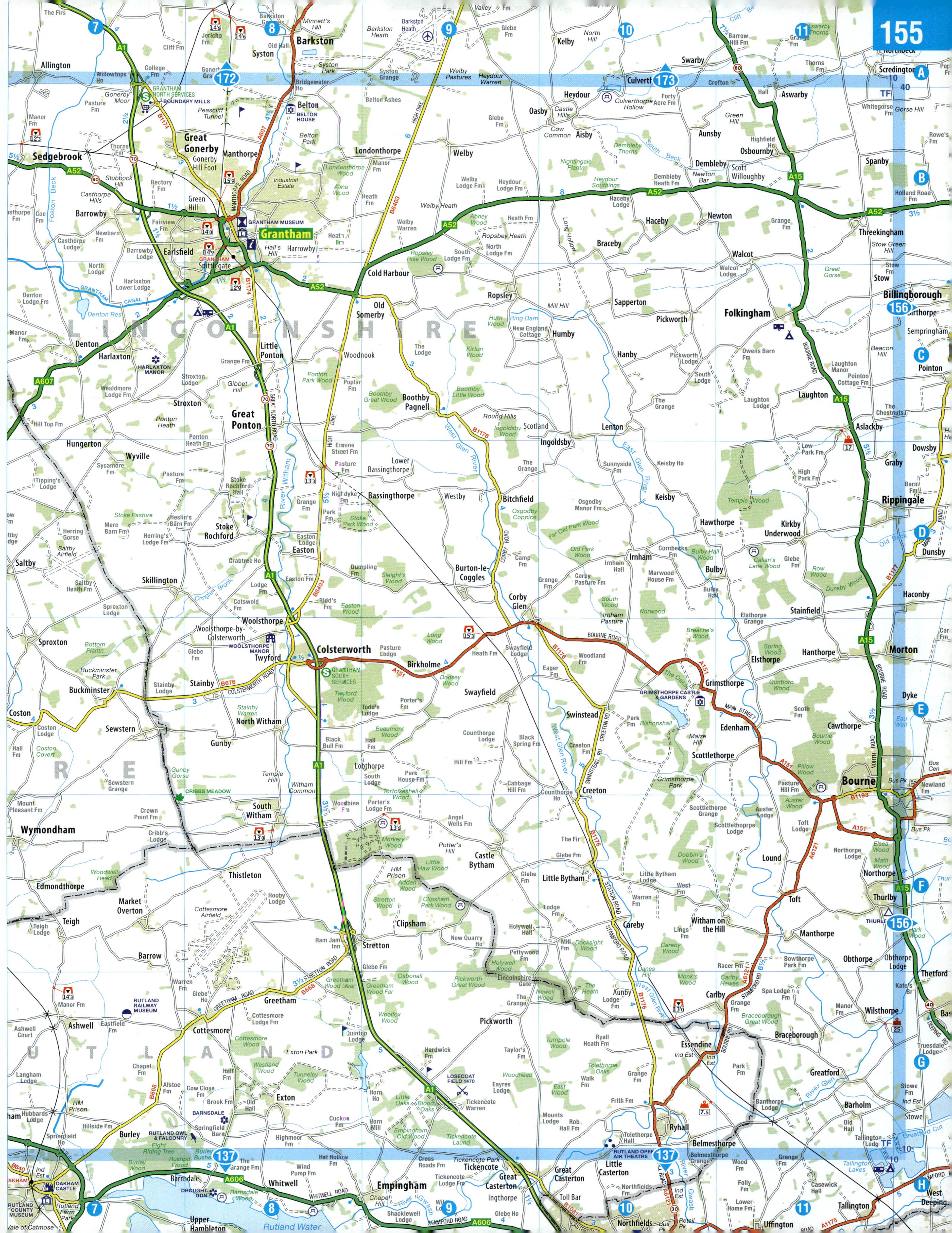
Grantham
Barkston
Syston
Belton
Great Gonerby
Manthorpe
Londonthorpe
Sedgebrook
Allington
Barrowby
Earlsfield
Spittlegate
Harrowby
Cold Harbour
Old Somerby
Denton
Harlaxton
Little Ponton
Great Ponton
Stroxton
Hungerton
Wyville
Stoke Rochford
Skillington
Saltby
Sproxton
Woolsthorpe
Colsterworth
Twyford
Stainby
Buckminster
Coston
Sewstern
North Witham
Gunby
South Witham
Wymondham
Edmondthorpe
Thistleton
Market Overton
Teigh
Barrow
Greetham
Cottesmore
Ashwell
Exton
Burley
Oakham
Whitwell
Empingham
Upper Hambleton
Rutland Water
Tickencote
Great Casterton
Little Casterton
Ryhall
Belmesthorpe
Essendine
Carlby
Pickworth
Stretton
Clipsham
Castle Bytham
Little Bytham
Creeton
Swinstead
Corby Glen
Burton-le-Coggles
Bitchfield
Ingoldsby
Lenton
Ropsley
Boothby Pagnell
Bassingthorpe
Westby
Swayfield
Birkholme
Irnham
Bulby
Hawthorpe
Keisby
Edenham
Grimsthorpe
Scottlethorpe
Elsthorpe
Bourne
Toft
Manthorpe
Thurlby
Obthorpe
Braceborough
Greatford
Barholm
Uffington
Tallington
Folkingham
Pickworth
Sapperton
Walcot
Newton
Haceby
Braceby
Welby
Oasby
Heydour
Aisby
Culverthorpe
Kelby
Swarby
Aswarby
Aunsby
Osbournby
Dembleby
Spanby
Threekingham
Stow
Billingborough
Pointon
Laughton
Aslackby
Dowsby
Graby
Rippingale
Kirkby Underwood
Dunsby
Haconby
Stainfield
Hanthorpe
Morton
Dyke
Cawthorpe
Northorpe
Thurlby
Wilsthorpe
LINCOLNSHIRE
RUTLAND
A1
A52
A607
A151
A15
A606
A6121
B1176
B6403
B676
B668
172
173
137
156
7
8
9
10
11
A
B
C
D
E
F
G
H
GRANTHAM MUSEUM
BELTON HOUSE
HARLAXTON MANOR
WOOLSTHORPE MANOR
GRIMSTHORPE CASTLE & GARDENS
RUTLAND RAILWAY MUSEUM
OAKHAM CASTLE
RUTLAND COUNTY MUSEUM
BARNSDALE
RUTLAND OPEN AIR THEATRE
LOSECOAT FIELD 1470
CRIBBS MEADOW
GRANTHAM SOUTH SERVICES
GRANTHAM NORTH SERVICES
BOUNDARY MILLS

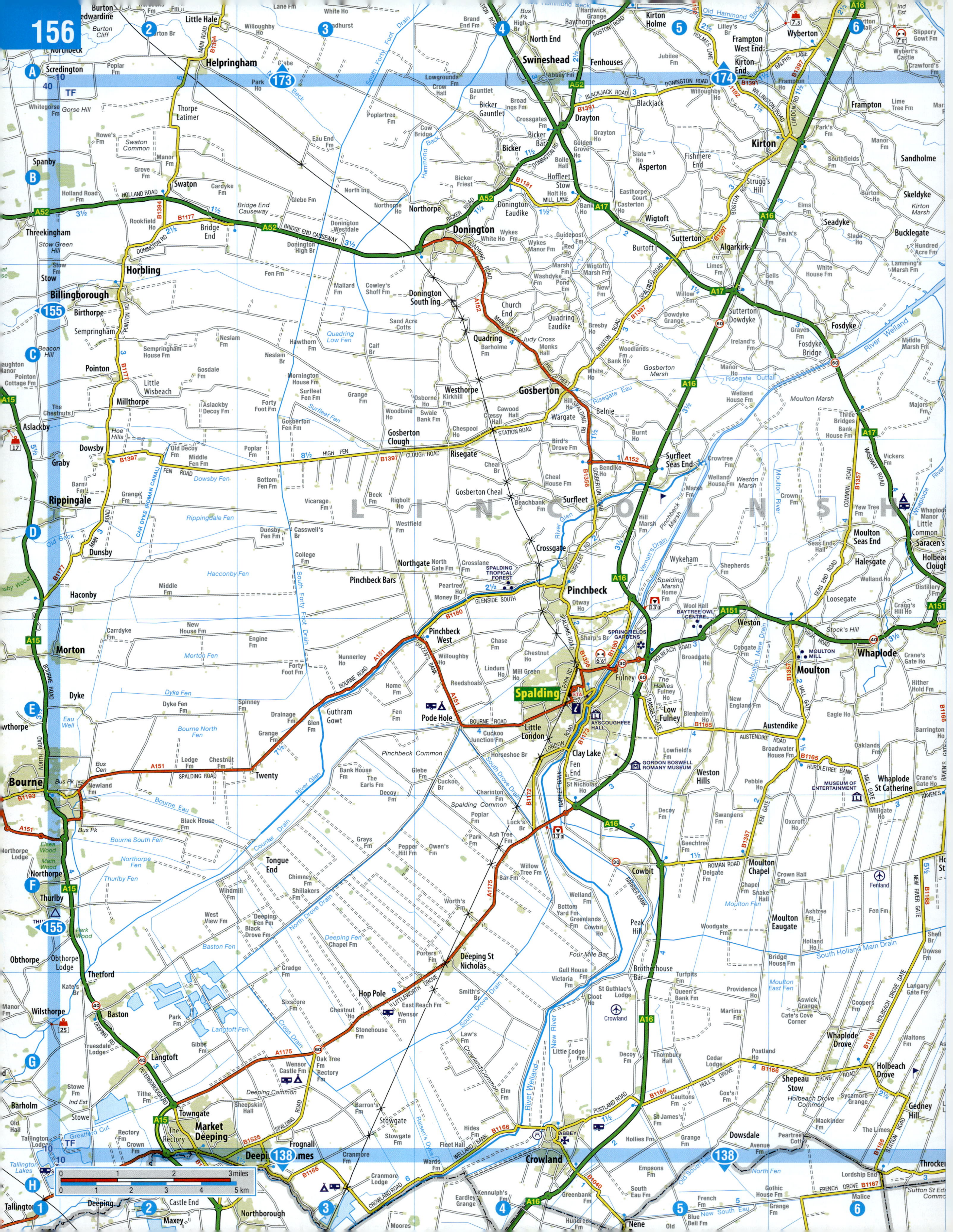

173
174
155
138
Helpringham
Swineshead
Kirton
Frampton
Wyberton
Kirton Holme
Fenhouses
Drayton
Bicker
Swaton
Donington
Horbling
Billingborough
Threekingham
Spanby
Scredington
Thorpe Latimer
Sempringham
Pointon
Millthorpe
Aslackby
Dowsby
Rippingale
Dunsby
Haconby
Morton
Dyke
Bourne
Thurlby
Baston
Langtoft
Market Deeping
Deeping St James
Towngate
Northborough
Tallington
Castle End
Maxey
Frognall
Deeping St Nicholas
Hop Pole
Tongue End
Twenty
Guthram Gowt
Pinchbeck Bars
Pinchbeck West
Pinchbeck
Spalding
Little London
Clay Lake
Fulney
Weston
Moulton
Whaplode
Moulton Chapel
Cowbit
Peak Hill
Crowland
Holbeach Drove
Gedney Hill
Shepeau Stow
Whaplode Drove
Whaplode St Catherine
Weston Hills
Austendike
Surfleet
Surfleet Seas End
Gosberton
Gosberton Clough
Risegate
Quadring
Quadring Eaudike
Westhorpe
Wigtoft
Sutterton
Algarkirk
Fosdyke
Sutterton Dowdyke
Asperton
Fishmere End
Blackjack
Northorpe
Bicker Gauntlet
Donington Eaudike
Church End
Crossgate
Northgate
Wykeham
Moulton Seas End
Holbeach Clough
Saracen's Head
Halesgate
Loosegate
Skeldyke
Sandholme
Bucklegate
Seadyke
Kirton End
Frampton West End
Hoffleet Stow
Dowsdale
Thetford
Wilsthorpe
Obthorpe
Nene
Throckenholt
LINCOLNSHIRE
River Welland
South Forty Foot Drain
Counter Drain
River Glen
South Holland Main Drain
New River
SPALDING TROPICAL FOREST
SPRINGFIELDS GARDENS
AYSCOUGHFEE HALL
GORDON BOSWELL ROMANY MUSEUM
BAYTREE OWL CENTRE
MOULTON MILL
MUSEUM OF ENTERTAINMENT
ABBEY
A52
A17
A16
A15
A151
A1175
A1073
A152
B1397
B1177
B1394
B1180
B1356
B1391
B1166
B1165
B1357
B1168
B1525
B1172
0 1 2 3 miles
0 1 2 3 4 5 km

THE WASH
174
175
158
139
Fishtoft
Scrane End
Freiston Shore
Freiston Low
HM Prison
The Scalp
River Witham
Wyberton Marsh
The Haven
FRAMPTON MARSH
Frampton Marsh
The Cots
Fosdyke Wash
Fleet Haven Outfall
Dawsmere Creek
Crab's Hole
Bull Dog Sand
Lynn Channel
Hospital Cotts
Leadenhall Fm
Trevethoe Fm
Holbeach St Matthew
Holbeach St Marks
Holbeach Marsh
Dawsmere
Gedney Drove End
Gedney Marsh
Black Barn
Welby Ho
Sutton Corner
Lutton Marsh
Lutton Grange
Lutton Leam
Leamlands
Guy's Head
THE WASH
Holbeach Bank
Holbeach Hurn
Gedney Dyke
Holbeach
Fleet Hargate
Chapelgate
Gedney
Fleet
Lutton
Lutton Gowts
Little London
Long Sutton
Garnsgate
Gedney Broadgate
BUTTERFLY & WILDLIFE PARK
Little Sutton
Sutton Crosses
Sutton Bridge
Port Sutton Bridge
Nene Outfall Cut
Wingland Marsh
Lighthouse
Nene Lodge Fm
Terrington Marsh
Walkers Marsh
Ongar Hill
Old Gate
Sutton St James
Tydd St Mary
Tydd Gote
Tydd St Mary's Marsh
Old Inclosed Marsh
Walpole Cross Keys
Walpole Marsh
Walpole St Andrew
Terrington St Clement
AFRICAN VIOLET CENTRE
Orange Row
Bellmount
South Green
Emorsgate
Clenchwarton
West Lynn
ST GEORGE'S GUILDHALL
CUSTOM HOUSE
KING'S LYNN
The Elms
Tilney All Saints
Tilney High End
Hay Green
Wash Dyke
Walpole St Peter
Church End
Four Gotes
Foul Anchor
River Nene
Tydd St Giles
Newton
Ingleborough
NORFOLK
CAMBRIDGE
Sutton St Edmund
Terrington St John
St John's Highway
Walpole Highway
Tilney St Lawrence
Tilney cum Islington
Wiggenhall St Germans
Wiggenhall St Mary the Virgin
Wiggenhall St Peter
Saddle Bow
Eau Brink
West Walton
West Walton Highway
Walton Highway
FENLAND AVIATION MUSEUM
Fitton End
Gorefield
Leverington
Leverington Common
Wisbech
PECKOVER HOUSE
WISBECH & FENLAND MUSEUM
Walsoken
New Walsoken
Marshland St James
Marshland St John's Fen End
Tilney Fen End
The Smeeth
Wiggenhall St Mary Magdalen
WATLINGTON
Parson Drove
Church End
Wisbech St Mary
Marshland Fen
River Great Ouse
Tydd St Mary's Fen
North Level Main Drain
A17
A47
A1101
B1359
B1165
B1166
B1169
B1515
B1390
B198
MARSH ROAD
WASHWAY ROAD
GEDNEY RD
ST JAMES ROAD
WISBECH RD
SUTTON RD
CHURCH LANE
HIGH ROAD
DRAW DIKE
MAIN ROAD
LEVERINGTON COMMON
LYNN ROAD

THE WASH
NORFOLK COAST
NORFOLK
Hunstanton
Ringstead
Heacham
Sedgeford
Docking
Stanhoe
Summerfield
Snettisham
Shepherd's Port
Ingoldisthorpe
Shernborne
Fring
Bircham Newton
Great Bircham
Bircham Tofts
Bagthorpe
Barmer
Syderstone
Blenheim Park
Wicken Green Village
Tattersett
Dersingham
Anmer
Sandringham
Wolferton
West Newton
Babingley
Flitcham
Houghton Hall
New Houghton
East Rudham
West Rudham
Broomsthorpe
Pockthorpe
Harpley
Hillington
Castle Rising
North Wootton
Congham
Roydon
Grimston
Pott Row
South Wootton
Little Massingham
Great Massingham
Gaywood
King's Lynn
West Lynn
Clenchwarton
Fairstead
Hardwick
Bawsey
Leziate
Gayton
Ashwicken
Gayton Thorpe
Rougham
Weasenham All Saints
West Lexham
Middleton
West Winch
North Runcton
East Winch
West Bilney
East Walton
Walton Warren
West Acre
Castle Acre
Newton
South Acre
Fair Green
Tower End
Blackborough End
Setchey
Saddle Bow
Wiggenhall St Germans
Wiggenhall St Peter
Wiggenhall St Mary Magdalen
Tottenhill Row
Tottenhill
Watlington
Wormegay
Pentney
Narborough
Great Palgrave
Sporle
Marham
Shouldham
Shouldham Thorpe
Runcton
Swaffham
Stubborn Sand
Ferrier Sand
Peter Black Sand
Bull Dog Sand
Lynn Channel
River Great Ouse
River Nar
Babingley River
Gaywood River
PEDDARS WAY & NORFOLK COAST PATH
A148
A149
A47
A10
A1065
A1122
B1454
B1153
B1155
B1145
B1440
B1439
0 1 2 3 miles
0 1 2 3 4 5 km

Fakenham
Dereham
Holt
Reepham
Briston
Melton Constable
Guist
Foulsham
North Elmham
Swanton Morley
Little Walsingham
Great Walsingham
Binham
Langham
Letheringsett
Wighton
Warham
New Holkham
North Creake
Sculthorpe
Hempton
Shereford
Helhoughton
East Raynham
West Raynham
South Raynham
Wellingham
Tittleshall
Mileham
Litcham
Stanfield
Brisley
Gressenhall
Beetley
East Bilney
Bittering
Longham
Wendling
Scarning
Toftwood
Yaxham
Mattishall
Welborne
Hockering
North Tuddenham
Lyng
Elsing
Sparham
Bawdeswell
Foxley
Bintree
Twyford
Hindolveston
Thurning
Wood Dalling
Guestwick
Stibbard
Great Ryburgh
Little Ryburgh
Colkirk
Oxwick
Horningtoft
Whissonsett
Gateley
Hindringham
Thursford
Barney
Kettlestone
Fulmodeston
Croxton
Sharrington
Thornage
Brinton
Hunworth
Stody
Edgefield
Hempstead
Saxlingham
Field Dalling
Bale
Gunthorpe
Briningham
Swanton Novers
Wood Norton
Corpusty
Heydon
Salle
Whitwell Street
Lenwade
Weston Longville
Honingham
Colton
Barford
Necton
Ivy Todd
Daffy Green
Westfield
Hoe
Woodgate
Worthing
Billingford
Mill Street
River Wensum
A148
A1067
A1065
A47
B1146
B1110
B1145
NORFOLK

Cromer
Sheringham
Sheringwood
Regis
East Runton
Foulness
Overstrand
Sidestrand
Northrepps
Trimingham
Mundesley
NORFOLK COAST
High Kelling
Bodham
West Beckham
East Beckham
Aylmerton
Felbrigg
Crossdale Street
Roughton
Gresham
Baconsthorpe
Hempstead
Bessingham
Sustead
Metton
Thorpe Market
Southrepps
Gimingham
Lower Street
Trunch
Paston
Knapton
Bacton
Hanworth
Thurgarton
Alby Hill
Aldborough
Plumstead
Matlaske
Edgefield
Barningham Green
Little Barningham
Wickmere
Antingham
Bradfield
Edingthorpe
Swafield
North Walsham
Spa Common
Crostwight
Edgefield Street
Wolterton
Calthorpe
Erpingham
Colby
Suffield
Itteringham
Saxthorpe
Corpusty
Little London
Ingworth
Banningham
Felmingham
Tungate
Blickling
Oulton
Aylsham
Silvergate
Dunkirk
Millgate
Tuttington
Skeyton
Honing
Bengate
Briggate
Worstead
Dilham
Heydon
Oulton Street
Westwick
Swanton Abbott
Burgh next Aylsham
Sloley
Smallburgh
Salle
Southgate
Cawston
Eastgate
Fengate
Marsham
Little London
Brampton
Oxnead
Scottow
Lamas
Little Hautbois
Buxton
Reepham
Booton
Whitwell Street
Brandiston
The Heath
Hevingham
Westgate Street
Stratton Strawless
Sco Ruston
Tunstead
Ashmanhaugh
Neatishead
Great Witchingham
Lenwade
Alderford
Swannington
Upgate
Waterloo
Hainford
Horstead
Coltishall
Belaugh
Hoveton
Wroxham
Attlebridge
Morton
Felthorpe
Newton St Faith
Frettenham
Weston Longville
Horsford
Crostwick
Thorpe Marriot
Horsham St Faith
Spixworth
Rackheath
Salhouse
Woodbastwick
Ringland
Taverham
Drayton
New Rackheath
Little Plumstead
Blofield Heath
Honingham
Costessey
Hellesdon
Old Catton
Sprowston
Thorpe End
Easton
New Costessey
Mile Cross
Upper Hellesdon
Great Plumstead
Bowthorpe
Thorpe St Andrew
Norwich
Brundall
Earlham
Colney
Barford
Thorpe Hamlet
Blofield
A140
A149
A148
A1067
A47
A1042
A1074
B1145
B1149
B1354
B1150
B1140
0 1 2 3 miles
0 1 2 3 4 5 km

NORTH SEA
NORFOLK COAST
THE BROADS
Keswick
Walcott
Ostend
Happisburgh
Whimpwell Green
Eccles on Sea
Bush Estate
Happisburgh Common
Lessingham
Hempstead
Castle Fm
Sea Palling
Hampstead Marshes
WAXHAM GREAT BARN
Waxham
Ingham Corner
Ingham
Stalham
Stalham Green
Chapel Field
MUSEUM OF THE BROADS
EAST RUSTON OLD VICARAGE GDN
East Ruston Hall
Brunstead Grange
Brunstead Common
Hickling
Hickling Green
Hickling Heath
Sutton
Catfield
Catfield Common
HICKLING BROAD
Horsey
HORSEY WINDMILL
Horsey Corner
Blackfleet Broad
WINTERTON DUNES
Winterton Holmes
Somerton Holmes
MARTHAM BROAD
East Somerton
West Somerton
Winterton-on-Sea
Damgate
Potter Heigham
River Thurne
Fritton
Ludham
LUDHAM MARSHES
Bastwick
Martham
Thunder Hill
Mustard Hyrn
Hemsby
Newport
Repps
Rollesby
Thurne
ST BENET'S ABBEY (REMS)
Ormesby St Michael
Ormesby St Margaret
Scratby
California
Sand Cliffs
Clippesby
Burgh St Margaret (Fleggburgh)
Filby
Filby Heath
Filby Broad
Billockby
Thrigby
THRIGBY HALL WILDLIFE GARDENS
Mautby
CAISTER CASTLE & MOTOR MUSEUM
Caister-on-Sea
Caister Pt
West Caister
West End
North Denes
Runham
Stokesby
CANDLEMAKER WORKSHOP
Upton
Upton Green
Fishley
Acle
Damgate
River Bure
Ranworth
Ranworth Broad
South Walsham
FAIRHAVEN WOODLAND & WATER GARDEN
Pilson Green
Upper Street
Johnson Street
Irstead
Barton Turf
Wood Street
ANT BROADS & MARSHES
Sharp Street
How Hill
North Burlingham
Burlingham Green
Lingwood
Beighton
South Burlingham
Moulton St Mary
Tunstall
Runham Vauxhall
Newtown
North Beach
The Beach
AMAZONIA WORLD OF REPTILES
Great Yarmouth
143
A47
A149
A1064
A1062
B1159
B1152
A12

CAERNARFON
BAY
BAE CAERNARFON
Llangwyfan-isaf
Aberffraw
Llangadwaladr
Hermon
Bodorgan
Malltraeth
Newborough
(Niwbwrch)
Pen-lon
Dwyran
ISLE OF ANGLESEY
YNYS MÔN
Newborough Forest
Newborough Warren
NEWBOROUGH WARREN & YNYS LLANDDWYN
Malltraeth Bay
Bae Malltraeth
Malltraeth Sands
Ynys Llanddwyn
Aberffraw Bay
Porth Cwyfan
Porth China
Bethel
Trefdraeth
Capel Mawr
Llangaffo
BARCLODIAD Y GAWRES BURIAL CHAMBER
TACLA TAID TRANSPORT MUSEUM
MODEL VILLAGE
BIRD WORLD
South Sands
Morfa Dinlle
Dinas Dinlle
CAERNARFON AIR MUSEUM
Pontllyfni
Aberdesach
Clynnog-fawr
Capel Uchaf
ST BEUNO'S WELL
Gyrn-goch
Bryn-yr-eryr
Trefor
Llanaelhaearn
Llithfaen
Pistyll
Nefyn
Morfa Nefyn
Edern
Tan-y-graig
LLEYN HISTORICAL MARITIME MUSEUM
TRE'R-CEIRI (FORT)
Yr Eifl 564
Pencaenewydd
Llangybi
Llwyndyrys
Fron
Pentreuchaf
Rhos-fawr
Y Ffôr
Chwilog
Llanarmon
Trwyn y Tâl
Trwyn y Gorlech
Trwyn Porth Dinllaen
Porth Dinllaen
Lleyn Peninsula
Penrhyn Llŷn
178
144
145
SH
0 1 2 3 miles
0 1 2 3 4 5 km

Bangor
Gaerwen
Llanfairpwll
Menai Suspension Bridge
Glan Adda
Maesgeirchen
Llandegai
Minffordd
Coed Mawr
Capel-y-graig
Llanddaniel Fab
Penrhyn Castle
Crymlyn
179
SNOWDONIA
NATIONAL PARK
PARC CENEDLAETHOL
ERYRI
Moel Wnion 580
Llanllechid
Glasinfryn
Tregarth
Rachub
Bethesda
Gerlan
Braichmelyn
Vaynol Hall
Llanedwen
Brynsiencyn
Y Felinheli (Port Dinorwic)
ANGLESEY
(YS MÔN)
Anglesey Sea Zoo
Foel Farm Park
Menai Strait
Pentir
Seion
Llanddeiniolen
Saron
Bethel
Rhiwlas
Carreg y Garth
Mynydd Llandegai
Waterloo Port
Caernarfon
Caernarfon Castle & Town Wall
The Royal Welsh Fusiliers
Segontium Roman Fort
Pont-rug
Llanrug
Bryn Bras Castle
Cwm-y-glo
Clwt-y-bont
Deiniolen
Brynrefail
Penisa'r Waun
Waun
Caeathro
Ceunant
Fachwen
Llanberis Lake Railway
Padarn
Dinorwic
Llanberis
Welsh Slate Museum
Dinorwig Power Station
Dolbadarn Castle (Rems)
Pen-gilfach
Llyn Padarn
Llanfaglan
Bontnewydd
Groeslon
Croesywaun
Waunfawr
Saron
Llanwnda
Dinas
Rhostryfan
Rhosgadfan
Penyffridd
Moel Tryfan
Betws-Garmon
Welsh Highland Railway
Snowdon Mountain Railway
Halfway
Clogwyn
Nant Peris (Old Llanberis)
Gwastadnant
Pass of Llanberis
Snowdon Summit Visitor Centre (Hafod Eryri)
Snowdon (Yr Wyddfa) 1085
Crib-y-Ddysgl 1065
Crib-Goch 921
Glaslyn
Llyn Llydaw
Y Lliwedd 898
Pen-y-Pass
Pen-y-Gwryd Hotel
Pont y Gromlech
Dinas Mot
Glyder Fawr 999
Glyder Fach 994
Tryfan 915
Llyn Ogwen
Pont Pen-y-benglog
Rhaeadr Ogwen
Idwal Cottage
Cwm Idwal
Llyn Idwal
Y Garn 947
Elidir Fawr 795
Foel-Goch 831
Marchlyn Mawr Res
Carnedd y Filiast 821
Mynydd Perfedd 812
Ty'n-y-maes
Carnedd Dafydd 1044
Carnedd Llewelyn 1064
Yr Elen 962
Foel Grach
Foel-Fras 942
Drum 770
Foel-Ganol 533
Pen yr Ole Wen
Pen yr Helgi Du 833
Pont Rhyd-goch
Helyg
Capel Curig
CONWY
Y Foel Goch
Moel Siabod 872
Cefnycerrig
Ceunant Mawr
Y Cribau 591
Craig Ddu
Garnedd
Blaenau Dolwyddelan
Roman Bridge
Moel Meirch 607
Moel Fleiddiau
Moel Dyrnogydd 524
Moel Farlwyd 577
Llechwedd Slate Caverns
Blaenau Ffestiniog
Bethania
Tanygrisiau
Ffestiniog Power Station
Tanygrisiau Reservoir
Moel Ystradau
Congl-y-wal
Rhyd-y-sarn
Llan Ffestiniog
Maentwrog
Gellilydan
Trawsfynydd Nuclear Power Station
Craig Gyfynys
Tan-y-bwlch
Plas Tan-y-bwlch
Llanfrothen
Rhyd
Croesor
Cnicht 689
Moelwyn Mawr 770
Moelwyn Bach 710
Carreg Blaen-Llym
Tan-lan
Garreg
Plas Brondanw
Penrhyndeudraeth
Minffordd
Cilfor
Porthmadog
Tremadog
Penmorfa
Pentrefelin
Criccieth
Llanystumdwy
Lloyd George Museum
Wern
Prenteg
Moel-Ddu 553
Cwm Cloch
Beddgelert
Gelerts Grave
Sygun Copper Mine
Nantmor
Aberglaslyn Hall
Pass of Aberglaslyn
Llyn Dinas
Craflwyn Hall
Bethania
Llyn Gwynant
Penmaen-brith
Hafod Rhisgl
Bryn Gwynant
Moel Hebog 782
Moel Lefn 638
Beddgelert Forest
Pitt's Head
Rhyd-Ddu
Llyn Cwellyn
Snowdon Ranger
Yr Aran 747
Moel Eilio 726
Foel Goch
Moel Cynghorion
Nantlle
Talysarn
Llyn Nantlle Uchaf
Penygroes
Llanllyfni
Tan-yr-allt
Nebo
Nasareth
Pant-glas
Dafarn Faig
Bryncir
Garndolbenmaen
Dolbenmaen
Golan
Brynkir Woollen Mill
Llanfihangel-y-pennant
Craig Cwm Silyn 734
Garnedd-Goch 700
Mynydd Graig Goch 609
Mynydd Tal-y-Mignedd 653
Trum y Ddysgl 709
Mynydd Drws-y-coed
Y Garn
Carmel
Cilgwyn
Fron
Moel Smytho
Rhos Isaf
Groeslon
Llanllyfni
GWYNEDD
Glan-Dwyfach
Llecheiddior
Rhoslan
Dolydd
Glynllifon
Pen-y-Bryn
Llyn Peninsula
Tyddyn-mawr
Cwm Pennant
Cwm Ciprwth
Llyn Cwmystradllyn
Ynys-Wen
Braich-y-ddinas
Llyn Llagi
Yr Arddu 389
Moel y Dyniewyd 382
Mynydd Llyndy
Blaen Nanmor
Hafodydd Brithion
Llyn Edno
Yr Arddu
Ffestiniog Railway
Coed Tremadog
Hafod y Llyn
Hafod Garregog
163
164
145
146
A487
A4085
A498
A4086
A5
A55
A470
A496
A4244
B4418
B4409
B4410

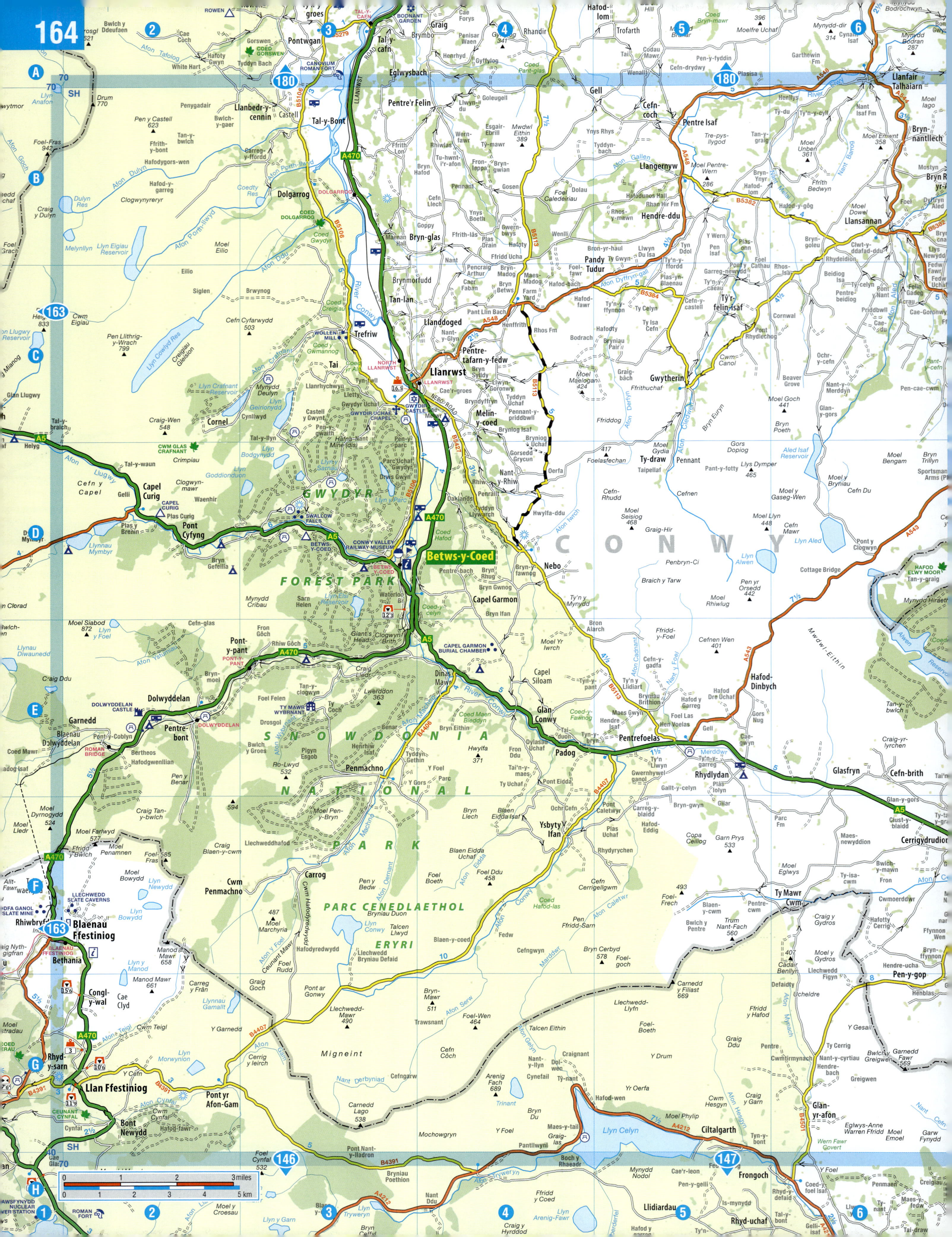
Llanrwst
Betws-y-Coed
Blaenau Ffestiniog
Llan Ffestiniog
Dolwyddelan
Capel Curig
Trefriw
Dolgarrog
Tal-y-Bont
Llangernyw
Pentrefoelas
Cerrigydrudion
Llansannan
Gwytherin
Padog
Ysbyty Ifan
Penmachno
Pentre-bont
Rhydlydan
Glasfryn
Frongoch
Ciltalgarth
Llidiardau
Rhyd-uchaf
C O N W Y
GWYDYR
FOREST PARK
S N O W D O N I A
N A T I O N A L
P A R K
PARC CENEDLAETHOL
ERYRI
Llyn Celyn
Migneint
Mwdwl-Eithin
A5
A470
A548
A543
B5106
B5113
B5427
B4406
B4407
B4391
A4212
B5384
B5382
B4501
180
163
146
147
0 1 2 3 miles
0 1 2 3 4 5 km

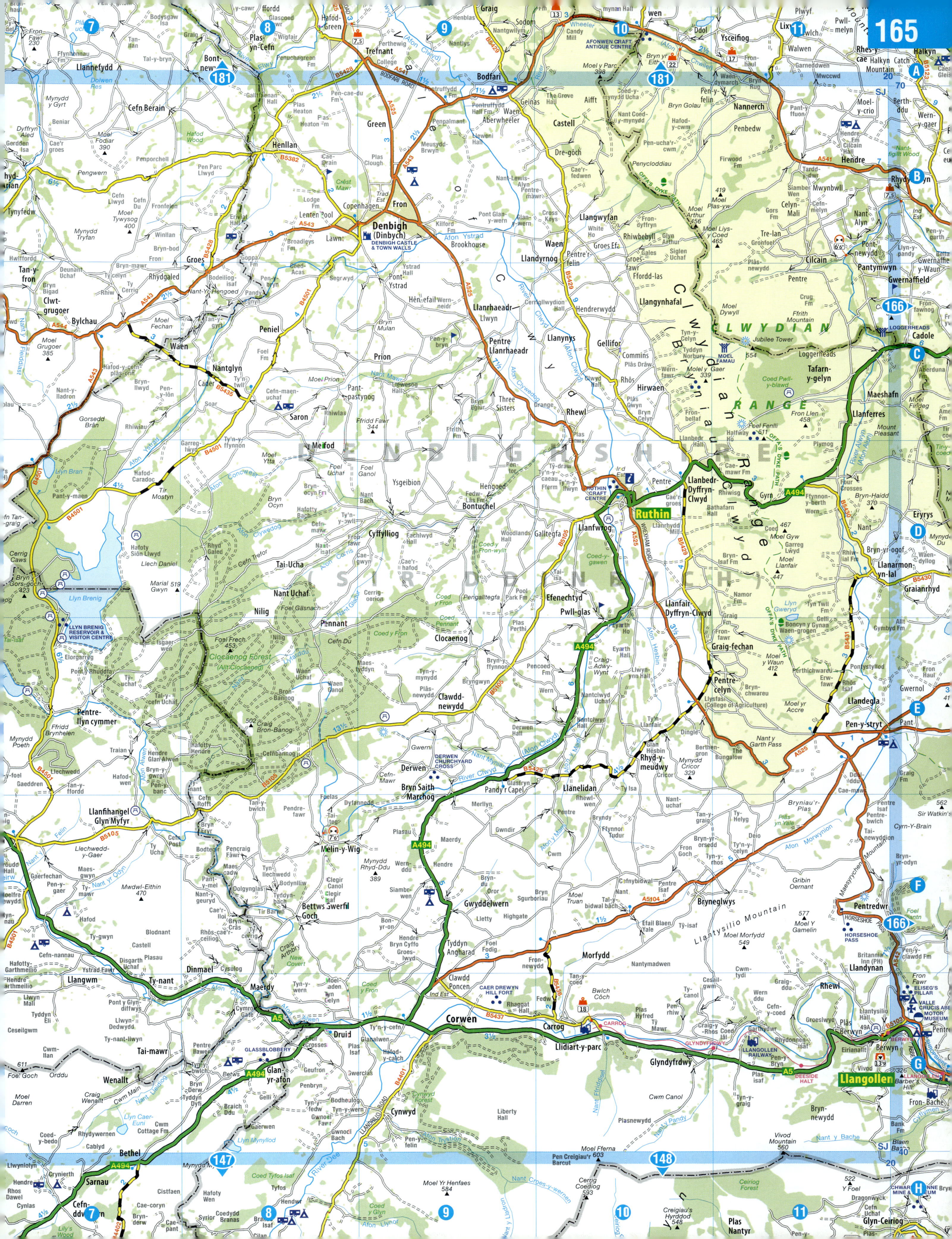

Llannefydd
Trefnant
Bodfari
Ysceifiog
Nannerch
Halkyn
Cefn Berain
Henllan
Green
Castell
Aifft
Denbigh (Dinbych)
DENBIGH CASTLE & TOWN WALLS
Copenhagen
Fron
Llangwyfan
Cilcain
Pantymwyn
Gwernaffield
Rhydymwyn
Groes
Tan-y-fron
Clwt-grugoer
Bylchau
Waen
Peniel
Nantglyn
Llanrhaeadr
Pentre Llanrhaeadr
Llandyrnog
Llangynhafal
Gellifor
Llanynys
CLWYDIAN RANGE
Clwydian Range (Bryniau Clwyd)
Jubilee Tower
MOEL FAMAU
Loggerheads
Cadole
Tafarn-y-gelyn
Maeshafn
Llanferres
Prion
Saron
Rhewl
Hirwaen
Three Sisters
DENBIGHSHIRE (SIR DDINBYCH)
Meifod
Ysgeibion
Bontuchel
Ruthin
RUTHIN CRAFT CENTRE
Llanbedr-Dyffryn-Clwyd
Gyrn
Eryrys
Llanarmon-yn-Ial
Graianrhyd
Llyn Brenig
LLYN BRENIG RESERVOIR & VISITOR CENTRE
Cyffylliog
Tai-Ucha
Llanfwrog
Nant Uchaf
Nilig
Pennant
Efenechtyd
Pwll-glas
Llanfair-Dyffryn-Clwyd
Clocaenog
Clocaenog Forest
Graig-fechan
Pentre-celyn
Llandegla
Pentre-llyn cymmer
Clawdd-newydd
Pen-y-stryt
Derwen
DERWEN CHURCHYARD CROSS
Bryn Saith Marchog
Pandy'r Capel
Llanelidan
Rhyd-y-meudwy
Nant y Garth Pass
Llanfihangel Glyn Myfyr
Melin-y-Wig
Gwyddelwern
Bettws Gwerfil Goch
Bryneglwys
Llantysilio Mountain
Pentredwr
HORSESHOE PASS
Llangwm
Ty-nant
Dinmael
Maerdy
Tyddyn Angharad
Morfydd
Clawdd Poncen
CAER DREWYN HILL FORT
Corwen
Carrog
Llidiart-y-parc
Glyndyfrdwy
Druid
Rhewl
Llandynan
Llangollen
LLANGOLLEN RAILWAY
Tai-mawr
Glan-yr-afon
Wenallt
Cynwyd
Bethel
Sarnau
Cefn-ddwysarn
Glyn-Ceiriog
Plas Nantyr

Halkyn
Flint Mountain (Mynydd Fflint)
Oakenholt
Shotwick
Saughall
Mollington
Backford
Wervin
Picton
Mickle Trafford
Upton
Hoole
Chester
Queensferry
Connah's Quay
Shotton
Garden City
Sealand
Blacon
Northop (Llan-eurgain)
Northop Hall
Rhosesmor
Hendre
Rhydymwyn
Soughton (Sychdyn)
New Brighton
Ewloe
Ewloe Green
Aston
Mancot
Hawarden (Penarlâg)
Sandycroft
River Dee
Handbridge
Vicarscross
Littleton
Christleton
Rowton
Saltney
Lache
Buckley (Bwcle)
Mold Yr Wyddgrug
Mynydd-isa
Pantymwyn
Gwernaffield
Cadole
Spon Green
Drury
Broughton
Bretton
Balderton
Huntington
Eccleston
Lower Kinnerton
Higher Kinnerton
Dodleston
Penyffordd
Pontblyddyn
Llong
Nercwys
Gwernymynydd
Maeshafn
Llanferres
FLINTSHIRE (SIR Y FFLINT)
Leeswood (Coed-Llai)
Treuddyn
Coed-Talon
Hope (Yr Hôb)
Caergwrle
Pulford
Aldford
Bruera
Eryrys
Llanarmon-yn-Ial
Graianrhyd
Rhydtalog
Llanfynydd
Ffrith
Cefn-y-bedd
Llay
Marford
Rossett
Trevalyn
Lavister
Gresford
Churton
Farndon
Holt
Four Crosses
Bwlchgwyn
Brymbo
Gwersyllt
Borras Head
Crewe-by-Farndon
Crewe
Llandegla
Pen-y-stryt
Gwynfryn
Minera
Coedpoeth
New Broughton
Rhosrobin
Acton
Llan-y-pwll
Wrexham Wrecsam
Rhosnesni
Ridleyway
Castletown
Bersham
Rhostyllen
King's Mills
Marchwiel
Isycoed
Shocklach
Bowling Bank
Sutton Green
Talwrn
Rhosllanerchrugog
Johnstown
Sontley
Cross Lanes
Worthenbury
Cock Bank
Penycae
Gyfelia
Eyton
Bangor-is-y-coed
Threapwood
Ruabon (Rhiwabon)
WREXHAM (WRECSAM)
Overton Bridge
Cloy
Tallarn Green
Pentrefelin
Llangollen
Garth
Trevor
Cefn-mawr
Pontcysyllte
Rhos-y-Madoc
Erbistock
Overton (Owrtyn)
Froncysyllte
Pentre
Horseman's Green
Lightwood Green
Sandy Lane
Halton
Chirk (Y Waun)
Rhewl
Knolton Bryn
Knolton
Street Dinas
Dudleston
Penley
Street Lydan
Glyn-Ceiriog
Pandy
Penylan
Bryn-y-Eos
Llandynan
Pentredwr
Horseshoe Pass
Cyrn-Y-Brain
Sir Watkin's Tower
Llandegla Forest
River Dee
Fron-Bache
3 miles
5 km

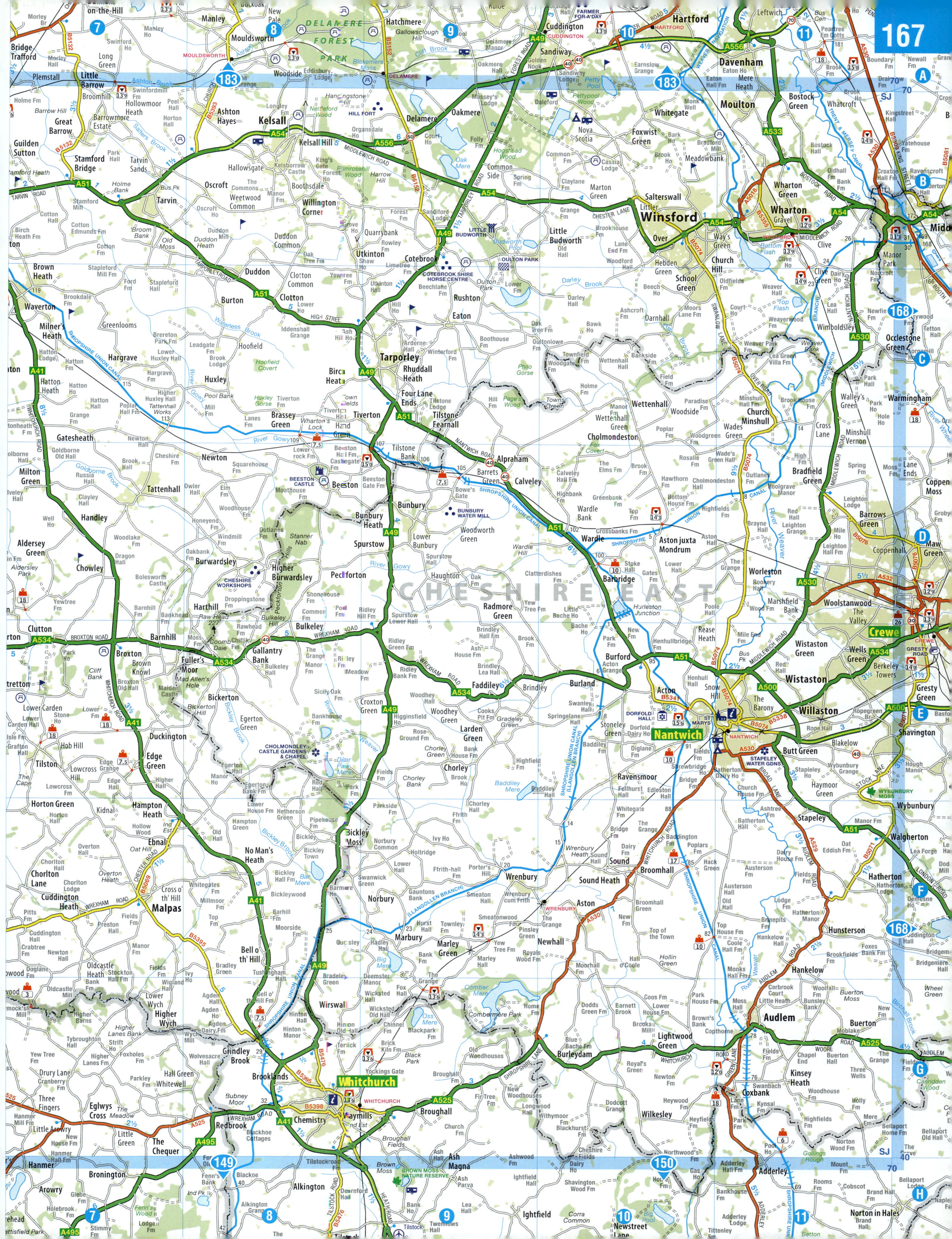
CHESHIRE EAST
DELAMERE FOREST PARK
Winsford
Nantwich
Crewe
Whitchurch
Tarporley
Malpas
Kelsall
Davenham
Moulton
Hartford
Cuddington
Sandiway
Delamere
Oakmere
Tarvin
Wharton
Over
Audlem
Wistaston
Willaston
Wrenbury
Bunbury
Beeston
Burford
Acton
Barbridge
Wardle
Calveley
Alpraham
Faddiley
Burland
Stapeley
Hankelow
Buerton
Adderley
Chemistry
Broughall
Ash Magna
Alkington
Bronington
Redbrook
Grindley Brook
Brooklands
Ightfield
Whitegate
Little Budworth
Cotebrook
Rushton
Eaton
Utkinton
Clotton
Duddon
Burton
Tiverton
Spurstow
Peckforton
Bulkeley
Bickerton
Broxton
Barnhill
Clutton
No Man's Heath
Hampton Heath
Edge Green
Duckington
Tilston
Handley
Chowley
Aldersey Green
Tattenhall
Newton
Huxley
Hargrave
Waverton
Milner's Heath
Gatesheath
Milton Green
Oscroft
Ashton Hayes
Mouldsworth
Manley
Hatchmere
Kingsley
Great Barrow
Stamford Bridge
Guilden Sutton
Little Barrow
Plemstall
Woodside
Darnhall
Church Minshull
Wettenhall
Cholmondeston
Aston juxta Mondrum
Worleston
Marshfield Bank
Woolstanwood
Wells Green
Gresty Green
Shavington
Wybunbury
Walgherton
Hatherton
Hunsterson
Coxbank
Kinsey Heath
Lightwood Green
Burleydam
Wilkesley
Newhall
Aston
Sound Heath
Broomhall
Ravensmoor
Baddiley
Chorley
Marbury
Norbury
Wirswall
Bickley Moss
Cholmondeley Castle Gardens & Chapel
Beeston Castle
Cheshire Workshops
Bunbury Water Mill
Cotebrook Shire Horse Centre
Oulton Park
Hill Fort
Stapeley Water Gdns
Dorfold Hall
Brown Moss Nature Reserve
Wybunbury Moss
Shropshire Union Canal
Llangollen Branch
River Weaver
River Gowy
A49
A51
A54
A41
A534
A530
A500
A525
A529
A556
A533
A495
183
168
149
150
SJ
70

Withington Green
Siddington
Lower Pexhill
Gurnett
Langley
Sutton Lane Ends
Warren
Gawsworth
Oakgrove
Allostock
Rudheath Woods
Blackden Heath
Goostrey
Lower Withington
Hodgehill
Marton
Highlane
Twemlow Green
Cranage
Holmes Chapel
Gladsmoss
Byley
Puddinglake
Sproston Green
Middlewich
Swettenham
Sandlow Green
Rodeheath
North Rode
Bosley
Somerford
Davenport
Newsbank
Hulme Walfield
Eaton
Havannah
Buglawton
Key Green
Congleton
Timbersbrook
Rushton Spencer
Brereton Heath
Brereton Green
Bradwall Green
Hollinsgreen
Ocleston Green
Brownedge
Smethwick Green
West Heath
Lower Heath
Astbury
Mossley
Hightown
Dane in Shaw
Whitemore
Sandbach
Elworth
Arclid
Brookhouse Green
Brownlow
Brownlow Heath
Warmingham
Moston Green
Ettiley Heath
Wheelock
Fourlanes End
Congleton Edge
Newtown
Poolfold
Marsh Green
Gillow Heath
Biddulph
Biddulph Moor
Horton
Crowborough
Coppenhall Moss
Hassall Green
Wheelock Heath
Winterley
Hassall
Thurlwood
Rode Heath
Scholar Green
Mount Pleasant
Mow Cop
Knypersley
Lask Edge
Maw Green
Haslington
Alsager
Church Lawton
Kidsgrove
Newchapel
Crewe
Radway Green
Talke
Goldenhill
Sandyford
Oxford
Endon
Stanley
Barthomley
Shavington
Weston
Englesea-brook
Audley
Red Street
Tunstall
Norton-in-the-Moors
Baddeley Green
Bradeley
Hough
Gorsteyhill
Balterley
Chesterton
Longbridge Hayes
Burslem
Milton
Wybunbury
Walgherton
Betley
Halmer End
Alsagers Bank
Bradwell
Wolstanton
Hanley
Bucknall
Abbey Hulton
Hatherton
Blakenhall
Knutton
May Bank
Checkley
Wrinehill
Madeley Heath
Silverdale
Keele
Newcastle-under-Lyme
Stoke-on-Trent
STOKE-ON-TRENT
Fenton
Madeley
Clayton
Trent Vale
Longton
Dresden
Woore
Onneley
Hanford
Meir
Ireland's Cross
Aston
Madeley Park
Baldwin's Gate
Whitmore
Hanchurch
Trentham
Lightwood
Meir Heath
Pipe Gate
Knighton
Willoughbridge
Blackbrook
Maer
Stableford
Chapel Chorlton
Beech
Barlaston
Rough Close
CHESHIRE EAST
M6
A500
A34
A50
A52
A53
A54
A51
0 1 2 3 miles
0 1 2 3 4 5 km

185
170
151
152
PEAK DISTRICT NATIONAL PARK
STAFFORDSHIRE
DERBYSHIRE
Ashford in the Water
Taddington
Chelmorton
Blackwell
Cowdale
King Sterndale
Harpur Hill
Earl Sterndale
Sterndale Moor
Sheldon
Flagg
Monyash
Over Haddon
Youlgrave
Middleton
Parsley Hay
Pilsbury
Crowdicote
Longnor
Glutton Bridge
Hollinsclough
Flash
Allgreave
Wildboarclough
Wincle
Danebridge
Heaton
Meerbrook
Upper Hulme
Fawfieldhead
Newtown
Sheen
Brund
Hartington
Heathcote
Newhaven
Biggin
Pikehall
Hulme End
Warslow
Upper Elkstone
Lower Elkstone
Thorncliffe
Leek
Ball Haye Green
Poolend
Longsdon
Bradnop
Onecote
Butterton
Wetton
Alstonefield
Hope
Milldale
Alsop en le Dale
Parwich
Ballidon
Grindon
Grindonmoor Gate
Ford
Waterfall
Winkhill
Bottom House
New Street
Waterhouses
Cauldon
Calton
Ilam
Thorpe
Tissington
Bradbourne
Fenny Bentley
Kniveton
Blore
Swinscoe
Mapleton
Ashbourne
Mayfield
Church Mayfield
Clifton
Snelston
Osmaston
Wyaston
Yeaveley
Rodsley
Cubley Common
Great Cubley
Alkmonton
Marston Montgomery
Rocester
Denstone
Quixhill
Alton
Farley
Ramshorn
Wootton
Stanton
Ellastone
Norbury
Roston
Upper Ellastone
Prestwood
Cotton
Oakamoor
Threapwood
Cheadle
Kingsley Holt
Kingsley
Froghall
Foxt
Ipstones
Consall
Wetley Rocks
Cellarhead
Blakeley Lane
Cheddleton
Basford Green
Horse Bridge
Leekbrook
Dilhorne
Caverswall
Forsbrook
Blythe Bridge
Draycott in the Moors
Freehay
Mobberley
Greatgate
Croxden
Hollington
Checkley
Cresswell
Fulford
Lower Tean
Upper Tean
Stubwood
Gallows Green
Bradley in the Moors
Spitalhill
A515
A53
A523
A520
A52
A50
A521
A522
A6
A54
A537
A5012
B5053
B5054
B5032
B5417

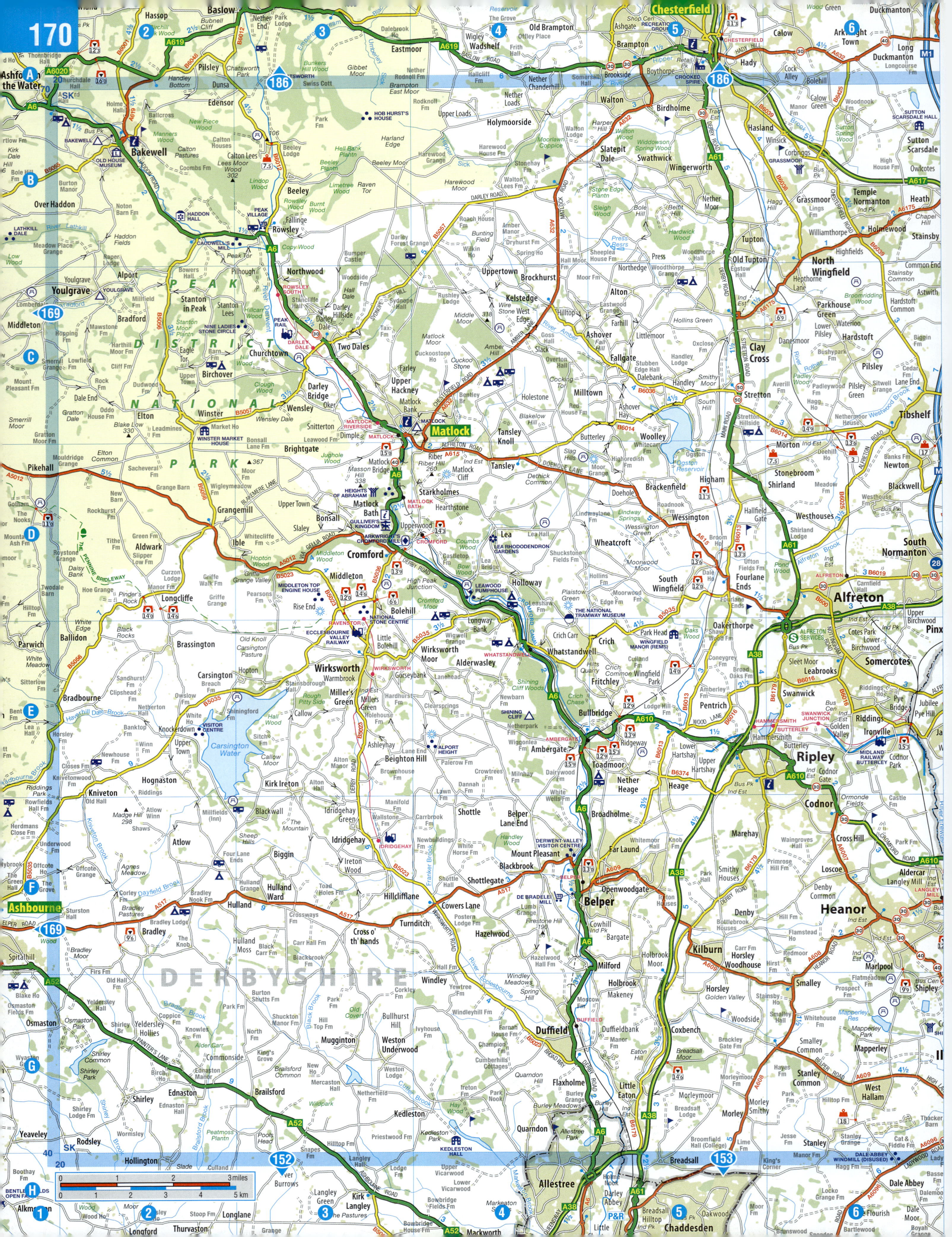

187
172
153
154
NOTTINGHAMSHIRE
SHERWOOD FOREST
SHERWOOD PINES FOREST PARK
CITY OF NOTTINGHAM
Bolsover
Whaley Thorns
Nether Langwith
Langwith
Cuckney
Norton
Budby
Thoresby
Perlethorpe
Walesby
Kirton
Boughton
Ollerton
New Ollerton
Edwinstowe
Market Warsop
Meden Vale
Church Warsop
Shirebrook
Langwith Junction
Scarcliffe
Palterton
Glapwell
Stony Houghton
New Houghton
Pleasley
Doe Lea
Bramley Vale
Ault Hucknall
Rowthorne
Hardwick Hall
Mansfield Woodhouse
Clipstone
Kings Clipstone
Forest Town
Mansfield
Radmanthwaite
Teversal
Stanley
Skegby
Fackley
Stanton Hill
Huthwaite
Sutton in Ashfield
Kirkby in Ashfield
Rainworth
Blidworth
Ravenshead
Papplewick
Linby
Hucknall
Annesley Woodhouse
Annesley
Newstead
Selston
Underwood
Brinsley
Eastwood
Kimberley
Nuthall
Watnall
Bulwell
Bestwood Village
Arnold
Gedling
Carlton
Calverton
Woodborough
Lowdham
Burton Joyce
Epperstone
Oxton
Farnsfield
Edingley
Kirklington
Halam
Southwell
Bilsthorpe
Eakring
Wellow
Ompton
Kneesall
Maplebeck
Winkburn
Normanton
Easthorpe
Westhorpe
Halloughton
Thurgarton
Hoveringham
Gonalston
Caythorpe
Gunthorpe
Bulcote
East Bridgford
Shelford
Newton
Stoke Bardolph
Radcliffe on Trent
Bingham
Saxondale
Netherfield
Colwick
Holme Pierrepont
Mapperley
Sherwood
Basford
Aspley
Cinderhill
Strelley
Bilborough
Trowell
Cossall
Awsworth
Ilkeston
Kirk Hallam
Wollaton
Lenton
Radford
Nottingham
Stapleford
Bramcote
Lidgett
Gorsethorpe
Rufford Abbey
White Post
Lambley
Arnold
Woodthorpe
Carrington
Sneinton
Adbolton
Saxondale
Harlequin

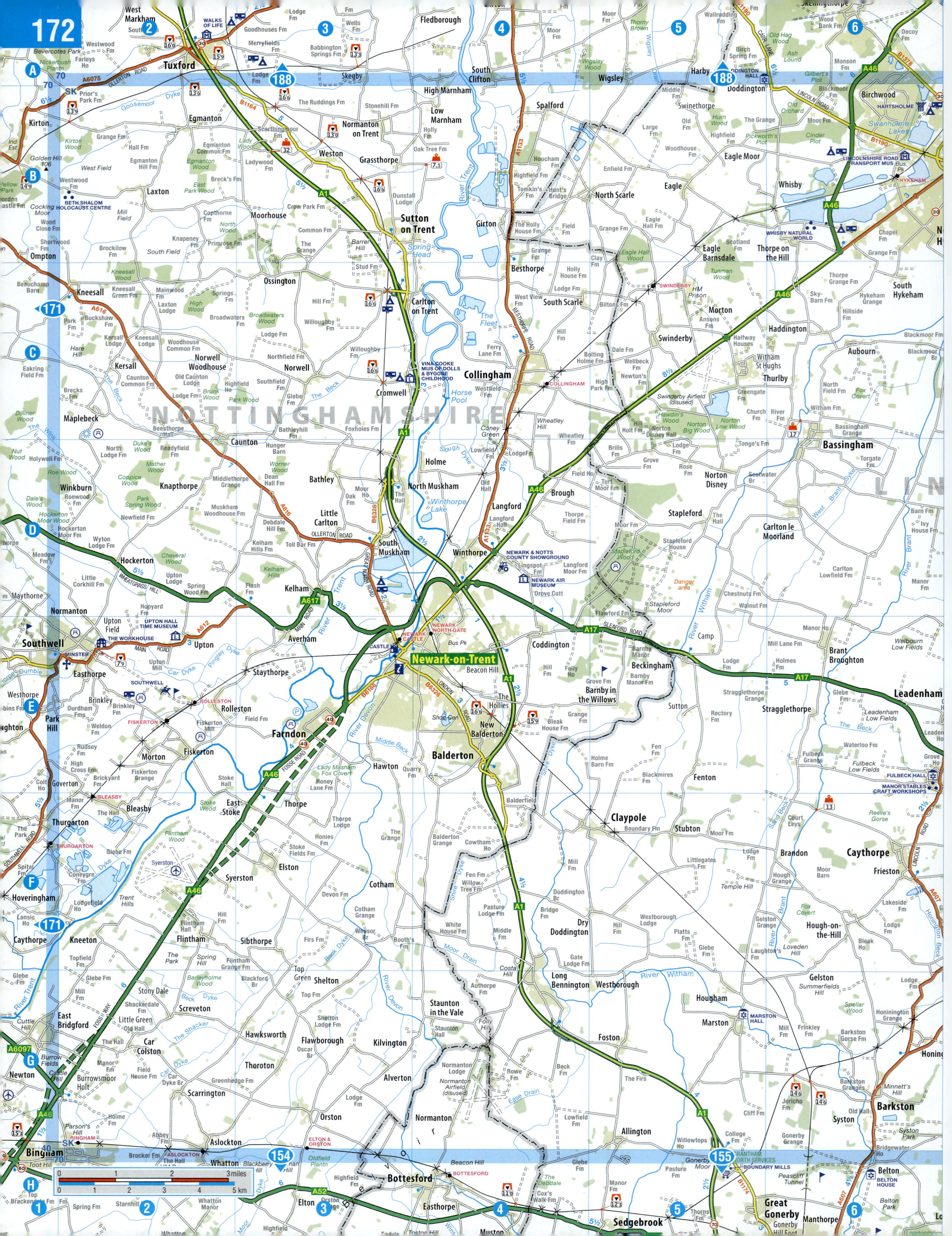
NOTTINGHAMSHIRE
Newark-on-Trent
Tuxford
Southwell
Collingham
Sutton on Trent
Carlton on Trent
Balderton
Farndon
Claypole
Long Bennington
Bottesford
Bingham
Bassingham
Caythorpe
Barkston
Great Gonerby
Leadenham
Brant Broughton
Swinderby
North Scarle
Eagle
Whisby
Harby
Doddington
Birchwood
South Hykeham
Wigsley
Spalford
Girton
Cromwell
Kelham
Averham
Upton
Rolleston
Fiskerton
Thurgarton
Hoveringham
Syerston
Elston
Cotham
Flintham
Sibthorpe
Shelton
Screveton
Hawksworth
Kilvington
Alverton
Orston
Whatton
Aslockton
Elton
Sedgebrook
Allington
Foston
Marston
Hougham
Westborough
Staunton in the Vale
Stubton
Fenton
Beckingham
Coddington
Winthorpe
Langford
Brough
Stapleford
Norton Disney
Carlton le Moorland
Thurlby
Haddington
Morton
Kirton
Egmanton
Laxton
Ossington
Kneesall
Caunton
Maplebeck
Winkburn
Hockerton
Norwell
Bathley
Little Carlton
South Muskham
North Muskham
Holme
Besthorpe
South Scarle
A1
A46
A17
A617
A616
A612
A6097
A52
A1133
A6075
188
171
154
155
0 1 2 3 miles
0 1 2 3 4 5 km

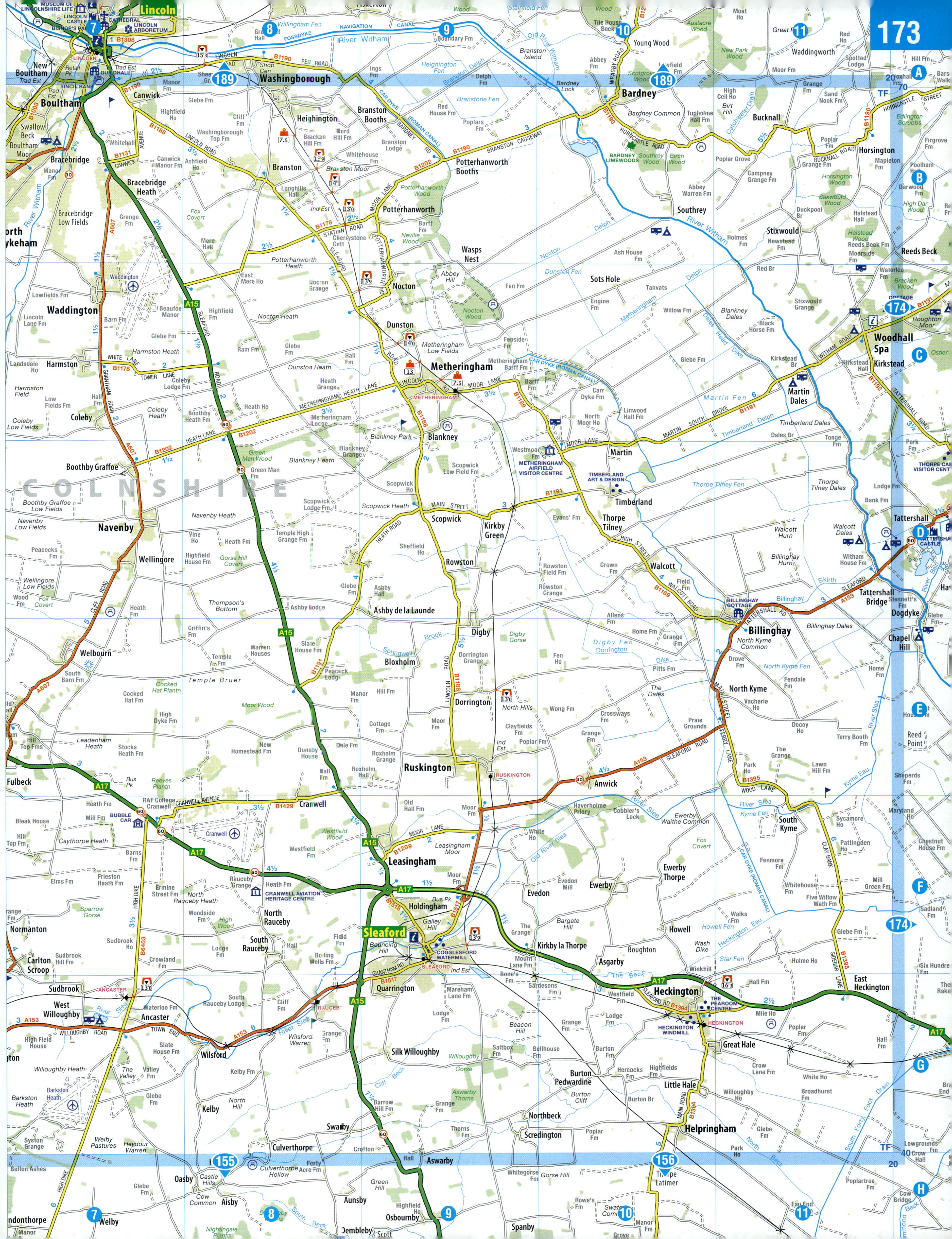
Lincoln
Lincoln Cathedral
Lincoln Castle
Lincoln Arboretum
Museum of Lincolnshire Life
Bishop's Palace
River Witham
Navigation
Fossdyke
Canal
Washingborough
Heighington
Branston
Branston Booths
Bracebridge
Bracebridge Heath
Bracebridge Low Fields
Boultham
New Boultham
Boultham Moor
Swallow Beck
Canwick
Potterhanworth
Potterhanworth Booths
Potterhanworth Heath
Nocton
Nocton Heath
Wasps Nest
Dunston
Dunston Heath
Metheringham
Metheringham Airfield Visitor Centre
Blankney
Blankney Heath
Scopwick
Scopwick Heath
Kirkby Green
Rowston
Digby
Dorrington
Ruskington
Anwick
Ashby de la Launde
Bloxholm
Waddington
Harmston
Coleby
Boothby Graffoe
Navenby
Wellingore
Welbourn
Fulbeck
Leadenham
Temple Bruer
Cranwell
Cranwell Aviation Heritage Centre
RAF College Cranwell
Leasingham
Sleaford
Holdingham
Quarrington
Cogglesford Watermill
North Rauceby
South Rauceby
Wilsford
Ancaster
Sudbrook
West Willoughby
Carlton Scroop
Normanton
Kelby
Culverthorpe
Oasby
Aisby
Aunsby
Swarby
Silk Willoughby
Osbournby
Dembleby
Aswarby
Welby
Evedon
Ewerby
Ewerby Thorpe
Kirkby la Thorpe
Asgarby
Howell
Heckington
Heckington Windmill
The Pearoom Centre
Great Hale
Little Hale
Burton Pedwardine
Helpringham
Scredington
Northbeck
Threekingham
Threckingham Latimer
Spanby
East Heckington
South Kyme
North Kyme
Billinghay
Walcott
Timberland
Thorpe Tilney
Martin
Martin Dales
Timberland Art & Design
Woodhall Spa
Kirkstead
Tattershall
Tattershall Castle
Tattershall Bridge
Dogdyke
Chapel Hill
Stixwould
Southrey
Bardney
Bardney Common
Bardney Limewoods
Bucknall
Horsington
Young Wood
Waddingworth
Reeds Beck
Sots Hole
Kyme Eau
River Slea
Car Dyke (Roman Canal)
Lincolnshire
A15
A17
A153
A607
B1188
B1191
B1202
B1178
B1190
B1189
B1394
B1395
B1517
B1209
B1429
B1192

West Ashby
Edlington
Thimbleby
Horncastle
High Toynton
Scrafield
Winceby
Greetham
Ashby Puerorum
Bag Enderby
Harrington
Sutterby
Fordington
Skendleby Psalter
Aswardby
Langton
Hagworthingham
Sausthorpe
Skendleby
Grebby
Partney
Horsington
Langton
Thornton
Old Woodhall
Martin
Mareham on the Hill
Hameringham
Asgarby
Lusby
Raithby by Spilsby
Hundleby
Spilsby
Ashby by Partney
Halton Holegate
Monksthorpe
Reeds Beck
Scrivelsby
Dalderby
Hareby
Mavis Enderby
Old Bolingbroke
Moorby
Miningsby
High Barn
West Keal
East Keal
Toynton All Saints
Toynton St Peter
Great Steeping
Little Steeping
Toynton Fen Side
Roughton
Roughton Moor
Haltham
Wood Enderby
Wilksby
Woodhall Spa
Kirkstead
Kirkby on Bain
Toft Hill
Mareham le Fen
Revesby
Revesby Bridge
East Kirkby
Hagnaby
LINCOLNSHIRE
Keal Cotes
Tattershall Thorpe
Tumby
Coningsby
Tattershall
Tumby Woodside
New Bolingbroke
West Fen
Fen Side
Stickford
Midville
New Leake
Eastville
Moor Side
Moorhouses
Medlam
Stickney
East Fen
Hawthorn Hill
Tattershall Bridge
Dogdyke
Reedham
Scrub Hill
New York
Sandy Bank
Carrington
Chapel Hill
Bunker's Hill
Hundle Houses
Northlands
Lade Bank
Wrangle Bank
Haven Bank
Sibsey Fen Side
Leake Commonside
Wrangle Lowgate
Wrangle
Sibsey
Frithville
Newham
Gipsey Bridge
Holland Fen
Langrick
Fishtoft Drove
Old Leake
Anton's Gowt
Hill Dyke
Leverton Highgate
Leverton Outgate
Leverton
Leverton Lucasgate
Langrick Bridge
Cowbridge
Frith Bank
Boston Long Hedges
Amber Hill
Brothertoft
Boston West
Haltoft End
Benington
Benington Sea End
Butterwick
East Heckington
Boston
Swineshead Bridge
Hubbert's Bridge
Freiston
Skirbeck
Chain Bridge
Skirbeck Quarter
Kirton Holme
Wyberton
Fishtoft
Freiston Shore
Scrane End
Bicker
Swineshead
North End
Fenhouses
Frampton West End
Kirton End
Frampton
THE
Kirton
Sandholme
A158
A16
A153
A155
A17
A1121
A52
B1183
B1184
B1192
B1195
B1191
B1397
B1391
190
173
156
157
0 1 2 3 miles
0 1 2 3 4 5 km

Willoughby
Claxby
Listoft
Hogsthorpe
Poplar Fm
Burlands Beck
Willoughby Wood
Hogsbeck Ho
Slootby
Welton Low Wood
Welton High Wood
Thwaite Hall
Hasthorpe
Howlet Ho
Slackholme End
Hope Fm
Beeches
Welbourne Fm
HARDY'S ANIMAL FARM
Ingoldmells
FANTASY ISLAND CHILDREN'S PLAYDROME & THE MILLENNIUM ROLLERCOASTER
Ingoldmells Pt
Highfield Fm
Habertoft
North Drain
Addlethorpe
Candlesby Hill
Rookery Fm
Welton le Marsh
Boothby Hall
Boothby Grange
Orby Beck
Manor Fm
Whitehouse Fm
Corner Fm
Marsh House Fm
Candlesby
BAKER'S LANE
Gunby
GUNBY HALL
Orby
Elmtree Ho
Fir Tree Fm
Teapot Hall
FUNCOAST WORLD
Field Fm
Field House Fm
Ashington End
Glebe Fm
The Grange
A158
Home Fm
Black House Fm
Seathorne
Winthorpe
ROMAN BANK
A52
Moat House Fm
Hunger Hill
Elmstead Fm
Willow Lodge
Mill Hill
Burgh le Marsh
BURGH LE MARSH WINDMILL
SKEGNESS ROAD
Bratoft
Moat Fm
Bratoft Corner
South View
BURGH ROAD
LINCOLN ROAD
NATURELAND SEAL SANCTUARY
Firsby
Irby in the Marsh
End House Fm
Croft End
Lloyd's Fm
CHURCH FARM MUS
Middlemarsh Fm
Skegness
WAINFLEET ROAD
SKEGNESS
Grove House Fm
Catchwater Drain
THE LIFEBOAT STATION MUSEUM
B1195
Rookery Fm
The Hundreds
Rivulet Ho
Marsh Fm
Church Fm
Lymn Bank Fm
Oak Br
Croft
CROFT BANK
Croft Marsh
Croft Grange
Thorpe Culvert
THORPE CULVERT
Thorpe St Peter
Poplar Fm
Bank Ho
Seacroft
Kitchen's Yard
Cow Bank Drain
rpe Fendykes
WAINFLEET
Croft Ho
Clough House Fm
Bramble Hills
Crown Fm
HAVENHOUSE
Steeping River
Wainfleet Clough
Wainfleet Common
Crow's Br
New England
New Yard Fm
Wainfleet Haven
Wainfleet Bank
MAGDALEN MUSEUM
White House Fm
White Cross Clough
WAINFLEET
Wainfleet All Saints
Merrifield's Fm
Marsh Farm East
GIBRALTAR POINT
eet St Mary Fen
Wainfleet Tofts
Wainfleet St Mary
Decoy Fm
Decoy Wood
Pepperthorpe Hall
Toft House Fm
BOSTON ROAD
Outmarsh Yard
Gibraltar Point
Ivy Ho
Hall Fm
Wainfleet Harbour
Inner Knock
Avenue Fm
Friskney Eaudyke
Wainfleet Sand
Friskney
Fold Hill
New Marsh
MAIN ROAD
Friskney Tofts
Greens Marsh
Whitehouse Fm
East Toft Fm
Friskney Flats
Tofthouse Fm
The Horseshoe
Home
Wrangle Flats
WASH
Gore Pt
HOLME DUNES
HOLME BIRD OBSERVATORY
NORFOLK COAST PATH
Holme next the Sea
PEDDARS WAY
Old Hunstanton
Manor Ho
St Edmund's Pt
B1161
CLIFF PARADE
Hall
Hunstanton Park
Hunstanton
Bluestone Fm
Lodge Fm
SEA LIFE SANCTUARY
Ringstead
Downs Fm
Ringstead Downs
Manor Fm
Church Fm
NORFOLK LAVENDER
191
157
158
TF
70
40
A B C D E F G H
7 8 9 10 11

2
3
4
5
6
A
B
C
D
E
F
G
H
70
60
TF
HOLKHAM BAY
BRANCASTER BAY
SCOLT HEAD ISLAND
Harbour Channel
Brancaster Harbour
Brancaster Marsh
Norton Creek
Trowland Creek
Burnham Harbour
West Sands
High Cape
The Run
Bob Hall's Sand
Big Gap
Lodge Marsh
Holkham Gap
Holkham Meals
Burrow Gap
Gun Hill
HOLME DUNES
Gore Pt
HOLME BIRD OBSERVATORY
TITCHWELL MARSH
RSPB
BRANCASTER MILLENNIUM ACTIVITY CENTRE
Mow Creek
Deepdale Marsh
Norton Marsh
Overy Marsh
Dale Hole
Warham Salt Marshes
East Fleet
Stonemeal Creek
HOLKHAM
WELLS HARBOUR MINIATURE RAILWAY
PEDDARS WAY & NORFOLK COAST PATH
Holme next the Sea
Thornham
Titchwell
Marsh Side
Brancaster
Brancaster Staithe
Burnham Deepdale
Burnham Norton
Burnham Overy Staithe
Marsh House Fm
Holkham
Wells-Next-The-Sea
STIFFKEY ROAD
COAST ROAD
WELLS-NEXT-THE-SEA
THE MIDDEN
A149
HIGH STREET
NORFOLK COAST
Manor Ho
Barrow Common
The Downs
Hawker's Hill Fm
Westgate Fm
Westgate
Burnham Market
Burnham Overy Town
WELLS ROAD
Peterstone Fm
B1155
Leath Ho
Howe Hill
Bone's Belt
Lucas Hill Wood
HOLKHAM HALL
Holkham Park
The Temple
Old Common Plantn
Golden Gates
Gallow Hill
Northgate Hall Fm
Warham
WARHAM ST MARY HALT
WELLS & WALSINGHAM LIGHT RAILWAY
Field Barn
Wighton
WIGHTON HALT
Hall
Hunstanton Park
Beacon Hill
Lyng Fm
Bluestone Fm
Choseley Fm
Field Ho
B1153
Sussex Fm
Beacon Hill
Gallow Hill
Crow Hall Fm
B1355
Burnham Thorpe
Gravelpit Hill
Manor Ho
Scarboro Wood
Longlands
The Belt
Branthill Fm
Branthill Cotts
B1105
Ringstead
Courtyard Fm
Lugden Hill Fm
Sunderland Fm
Friar's Thorne Fm
Cradle Hall Fm
Muckleton
Field Barn
CREAKE ABBEY (REMS)
Abbey Fm
River Burn
New Holkham
Blunt's Corner
Crabbe Castle Fm
Setons Halt
River Stiffkey
Downs Fm
Ringstead Downs
Neat's Ling
Thornham Corner
Summerfield
Church Fm
High Barn
BURNHAM ROAD
North Creake
Chantry Hills
Quarles Fm
Bunker's Hill
Edgar Fm
NORFOLK LAVENDER
East Hall
St Catherine's Cross
BRANCASTER ROAD
High House Fm
Shammer Ho
Wringate Fm
High Hills
Shepherd's Hill
East Common
Haggard's Lodge
Egmere Fm
Egmere
Berry Hall
Great Walsingham
WALSINGHAM
Little Walsingham
DOCKING RD
SEDGEFORD ROAD
B1454
Docking
The Park
Stanhoe
Sly's Barn
Manor Ho
South Creake
SOUTH CREAKE MAIZE MAZE
Compton Hall
Waterden Fm
Waterden
North Barsham Fm
Eaton
158
Sedgeford
FAKENHAM ROAD
Docking Common
Docking Lodge
Barwick Ho
Barwick Hall Fm
Bloodgate Hill
Southgate
North Barsham
Houghton St Giles
Canister Hall Fm
Great Snoring
Glover's Fm
Church Fm
Fring
Honey Hills
HONEY HILLS
Bircham Newton
Hyde Park
Buildings Fm
Hubbards Fm
The Gorse
West Barsham
Snettisham
Inmere Fm
SNETTISHAM PARK
Snettisham Ho
Ingoldisthorpe
Red Barn Fm
Moor Fm
Great Bircham
Barmer
Leicester Square Fm
White Hall
Syderstone
East Barsham
Shernborne Hall
Shernborne
BIRCHAM MILL
Bircham Tofts
Bagthorpe
Bircham Common
Syderstone Common
Blenheim Park
Sculthorpe Airfield (disused)
Cranmer Hall
Manor Fm
Sculthorpe
Waterhouse Fm
Thorpland Hall
Little Snoring
Thorpland Lodge Fm
Lodge Fm
The Grange
Grove Fm
Mill Ho
LYNN RD
Dersingham
Ling Ho
Frizzleton Fm
Coxford Wood
Wicken Green Village
Tofts Hill
Coxford Heath
Tofts Dale
A148
Fakenham
A1067
Heath Fm
Manor Doddshill Fm
Heath House Fm
Tattersett
Dunton
Dunton Patch
Manor Fm
Sculthorpe Moor
The Heath
Anmer
Dersingham Wood
Gatesend Hill
Coxford
159
Shereford
Hempton
Fakenham Common
FAKENHAM
Pensthorpe Hall
Starmoor Wood
Park Ho
SANDRINGHAM HOUSE
Sandringham
Commodore Wood
Cross Belt
HOUGHTON HALL
Houghton Park
The Mayflower
Bunker's Hill
The Blackgrove
Harpley Common
New Houghton
North Pole Fm
New Buildings
East Rudham
LYNN ROAD
Broomsthorpe
West Rudham
Pynkney Hall
Tatterford
Doughton
Southmill Fm
Toftrees
DEREHAM ROAD
A1065
Manor Ho
Big Wood
The Carr
0 1 2 3 miles
0 1 2 3 4 5 km

Blakeney Pit
Warham Hole
West Sand
Cabbage Creek
Stiffkey Freshes
South Side
Pits Pt
The Marrams
BLAKENEY
Morston Salt Marshes
Great Barnett
Agar Creek
New Cut
CLEY MILL
Stiffkey Salt Marshes
Stiffkey Greens
NORFOLK COAST PATH
PEDDARS WAY AND NORFOLK COAST PATH
Warborough Hill
Stiffkey
Morston
BLAKENEY GUILDHALL
Blakeney
Cley next the Sea
Salthouse
Newgate
Gravelpit Hill
Cockthorpe Common
Battledore Hill
Cockthorpe
Sparrow Hill
Joe's Hill
Wiveton
Wiveton Downs
NORFOLK COAST
Langham Lodge
Oulton Hill
Glandford
Summer House Hill
Swan Lodge
Gallow Hill
Lowes Fm
The Hangs
MUCKLEBURGH COLLECTION
Weybourne
Kelling
NORTH NORFOLK RAILWAY
KELLING HEATH PARK
Kelling Heath
WEYBOURNE
Sheringham Hall
Oak Wood
SHERINGHAM PARK
Weybourne Heath
The Dales
Upper Sheringham
Sheringham
SHERINGHAM
Sheringwood
Pretty Corner
NORFOLK SHIRE HORSE CENTRE
West Runton
WEST RUNTON
Beeston Regis
East Runton
The Roman Camp
The Valley
Stone Hill
CROMER
Langham
Westgate
Long Lane Fm
Short Lane Fm
Old Barn Fm
BINHAM PRIORY & WAYSIDE CROSS
Binham
Ellis Fm
Saxlingham
Manor House Fm
Smoker's Hole
Field Dalling
Letheringsett
PICTURECRAFT GALLERY
LETHERINGSETT WATERMILL
Little Thornage
Holt
HOLT
Gresham's School
Holt Hall (Coll)
Perer's Hills
Cley Park
Lawn Fm
Bayfield Hall
High Kelling
Hundred Acre Wood
Highborough Fm
Bodham
West Beckham
East Beckham
Aylmerton
Felbrigg
FELBRIGG HALL
Roundwood Hill
Great Wood
Barn Plantn
Bounce's Covert
Common Plantn
Gresham
Stonepit Hill
Rookery Fm
Hill Ho
Manor Ho
Lower Bodham
BACONSTHORPE CASTLE (REMS)
Bodham Hill
Heath Fm
Red Ho
Beckett's Fm
Baconsthorpe
Hell Hole
Up Wood
Abbot Fm
Foxburrow Fm
County Fm
Little Marsh
Eastmoor Fm
Clipstreet Fm
Bale Hall
Bale
Stowe Ollands
Sharrington
Common Hill
Ingmote Hill
Hill Ho
Lower Green
Hindringham
Hall Fm
Grange Fm
Bullfer Grove
Brinton
Thornage
Heath Ho
Dam Hill
King's Hills
Edgefield Woods
Edgefield Hall
Lowes Fm
Bunker's Hill
Hempstead
The Dales
Pond Hills
Hole Fm
Manor Ho
Plumstead Green
Plumstead
Barningham Hall
Bessingham
Thurgarton Old Hall
Sustead
Metton
Hanworth
Vinepark Fm
Thursford Green
THURSFORD COLLECTION
Thursford
Old Coach House
Frog Hall Fm
Gunthorpe Park
Gunthorpe
Lobb's Valley
Stody
Hunworth
The Green
Starlings' Hill
Edgefield
Little Wood
Range Fm
Barningham Green
Matlaske
Hall Fm
Lower Street
Aldborough Hall
Aldborough
Thurgarton
Hanworth Hall
Alby Hill
Little Snoring
The Lings
Pigg's Grave
Briningham
Bush Hall
Breck Fm
White Ollands Fm
Fuel Fm
Thwaite Hill Fm
Wood Fm
Forty Acre Plantn
Neat's Close
Melton Constable
FAKENHAM ROAD
Briston
Fir Patch
Edgefield Street
Old Covert
Little Barningham
Wickmere
Erpingham Ho
Barney
Swanton Novers
Park Fm
Melton Hall
Dairy Fm
New Covert
Shrub Fm
Mere Fm
Mannington Hall
MANNINGTON GARDENS
Park Fm
WOLTERTON PARK
Wolterton
Thwaite Hall
Thwaite Common
Erpingham
Calthorpe
Kettlestone
Croxton
Fulmodestone
Swanton Great Wood
Raymere Beck
Briston Common
Roper Fm
Moor Hall
Holly Fm
Mossymere Wood
Itteringham
White House Fm
Clipstone Ho
Common End
Brown's Covert
Tipples' Fm
Raw Hall
Wood Severals
Holme's Wood
Washpit Plantn
Rookery Fm
Blackwater Br
Little London
Saxthorpe
Corpusty
Holly Heath Fm
Moorgate
Common Fm
Hindolveston
Burnthouse Fm
Irmingland Hall
Itteringham Common
Great Wood
Ingworth
Manor Fm
Park Fm
Roundabout Fm
Nethergate
Thurning
Thurning Hall Fm
Foundry Hill
Red Pits
Oulton
Oulton Lodge
The Tower
Blickling
BLICKLING HALL
Abbot's Hall
Church Fm
PENSTHORPE NATURE RESERVE & GARDENS
Little Ryburgh
Stibbard
Fulmodeston Severals
Holly Hill
Field Barn
Wood Norton
Ashcroft Fm
Hindolveston Wood
Black Water
Norton Corner
Oak Grove
Crofton Hall
Hercules Wood
Silvergate
Flash Pit Fm
Drabblegate
Coldham Hall
River Stiffkey
River Glaven
Scarrow Beck
HOLT RD
CROMER ROAD
HOLT ROAD
THORNAGE ROAD
LANGHAM ROAD
SHERINGHAM ROAD
WEYBOURNE RD
LODGE HILL
HOLWAY ROAD
A149
A148
A140
B1156
B1354
B1110
B1149
B1157
B1436
TG
159
160

HOLYHEAD BAY
BAE CAERGYBI
The Skerries
Carmel Head
Pen Carmel
Porth y Dyfn
Porth y Bribys
Church Bay
Porth Swtan
Porth Trwyn
Porth Tywyn-mawr
Porth Penrhyn-mawr
Hen Borth
Harry Furlough's Rocks
Wylfa Head
Wylfa Power Sta
WYLFA POWER STATION & OBSERVATION TOWER
Cemlyn Bay
Porth-y-pistyll
Porth Llanlleiana
Hell's Mouth/Porth Cynfor
Porth Wen
Porthllechog (Bull Bay)
Bull Bay/Porth Llechog
ISLE OF ANGLESEY
YNYS MÔN
ISLE OF ANGLESEY
(SIR YNYS MÔN)
Llanbadrig
Cemaes
Tregele
Burwen
Amlwch
Llanfairynghornwy
Llanfechell
Rhosbeirio
Bodewryd
Llanrhyddlad
Rhydwyn
Llanfflewyn
Carreglefn
Rhosgoch
Rhosybol
Llanfaethlu
Llanbabo
Llyn Alaw
Gwredog
Llanddeusant
Llandyfrydog
Llanfwrog
Elim
Llanerchymedd
Llantrisant
Bachau
Llanfachraeth
Pen-llyn
Carmel
Llanynghenedl
Llechcynfarwy
Trefor
Bodedern
Llanfaelog
Rhosneigr
Caergeiliog
Llanfihangel yn Nhowyn
Bryngwran
Engedi
Gwalchmai
Bodffordd
Heneglwys
Cerrigceinwen
Llangristiolus
Capel Mawr
Bethel
Trefdraeth
Llangadwaladr
Aberffraw
Hermon
Malltraeth
Llangaffo
Llangwyfan-isaf
Pencarnisiog
Bryn Du
Ddrydwy
Dothan
Soar
Holyhead
Caergybi
Holy Island
Ynys Gybi
Goferydd
Llaingoch
Mountain
Porth-y-felin
Môrawelon
Kingsland
Penrhosfeilw
Trearddur
Rhoscolyn
Valley (Y Fali)
Newlands Park
Penrhos
Four Mile Bridge
Llanfairyneubwll
Rhoscolyn Head
Silver Bay
Borthwen
Ynys Feirig
Porth Nobla
Porth Trecastell
Porth Cwyfan
Caethle
Danger area
BARCLODIAD Y GAWRES BURIAL CHAMBER
TY-NEWYDD BURIAL CHAMBER
DIN DRYFOL BURIAL CHAMBER
PRESADDFED BURIAL CHAMBER
TREGWEHELYDD STANDING STONE
LLYNNON WINDMILL
LLYN ALAW VISITOR CENTRE
VALLEY WETLANDS
HOLYHEAD MARITIME MUSEUM
SOUTH STACK CLIFFS
HOLYHEAD MTN HUT CIRCLES
PENRHOS FEILW STANDING STONES
TREFIGNATH BURIAL CHAMBER
TY MAWR STANDING STONE
CAER Y TWR (FORT)
HENBLAS
DUN LAOGHAIRE 1:59
DUBLIN 1:49
DUBLIN 3:15
A5
A55
A5025
A4080
B5109
B5111
B5112
B4422
B4545
162
0 1 2 3 miles
0 1 2 3 4 5 km

7
8
9
10
11
A
B
C
D
E
F
G
H
SH
70
00
180
Amlwch Port
Porthyrychen
Porth Eilian
Llaneilian
Fresh Water Bay
Pengorffwysfa
Cerrig-mân
Porthygwichiaid
Porth Helygen
Porth yr Aber
Ynys Dulas
Garreg Allan
Penysarn
Nebo
Gadfa
Dulas
Gwlybycoed
Llaneuddog
Traeth yr Ora
City Dulas
Derz-Fawr
Tyddyn Isaf
Lligwy Bay
Brynrefail
Moelfre
MOELFRE SEAWATCH CENTRE
Y Swnt
Rhôs Lligwy
DIN LLIGWY
LLIGWY BURIAL CHAMBER
Tal-y-bontan
Ty-mawr
Mynydd Bodafon
Plas Bodafon
Llanallgo
Afon Lligwy
Caerhosllligwy
Maenaddwyn
Parciau
Marianglas
Penciau
Frigan
Borth-wen
Plas Llanfihangel
Tynygongl
Capel Coch
Cefn-du Mawr
Brynteg
Dinas
Benllech
Nant Isaf
Nant-uchaf
CORS ERDDREINIOG
Ynys Isaf
Bettws
Nant Newydd
CORS GOCH
Benllech Sand
Red Wharf Bay
Red Wharf Bay Traeth-coch
Llanbedrgoch
Tyn Coed
Erddreiniog
Maen Eryr
Prysan
Hendre
Glan Gors
Gwenfro Isaf
Tan-y-graig
Pentraeth Forest
STANDING STONE
Llanddyfynan
Gwenfro Uchaf
Pencefn
Benc Fadog
Pentraeth
STONE SCIENCE
Carreg Uchaf
Trefollwyn
Rhosmeirch
Tyn Beudy
CORS BODEILIO
Hendre Fm
Clegyrdy-bach
Talwrn
ORIEL YNYS MÔN
Clegyrdy-mawr
Afon Ceint
Ty-coch
Dyffryn
The Dingle
Ty'n-Coed
Pencraig
Gylched
Rhoscefnhir
Plas Penmynydd
Tai-hirion
Ind Est
Ind Pk
Ceint
Hirdre-faig
Penmynydd
Cefn-poeth
Tyddyn
Y Dolydd
Nant
Twll-y-clawdd
Cefn Du Isaf
Castell
Tyddyn Mawr
Rhyd-yr-arian
Bwlch-gwyn
Ty'n Buarth
Tyn Cae
PILI PALAS
Pentre Berw
Llanfairpwll-gwyngyll
Gaerwen
LLANFAIRPWLL
Ind Est
Bryn-y-fedwen
Brynhyfryd
Llanddaniel Fab
Chwipin
Bus Pk
BRYN CELLI DDU BURIAL CHAMBER
Pont Crug
PLAS NEWYDD
Ty'n-y-Coed
Glanyrafon
Bodlew
Bodowyr-isaf
BODOWYR BURIAL CHAMBER
Dinam
Ysgubor Wen
Lew
Gwydryn Newydd
Pont-Dic
Llanedwen
Bodrida
CAER LEB
ISLE OF ANGLESEY
YNYS MÔN
Trefor
Pen-y-garnedd
Hendrefor
Llansadwrn
Wern Fm
Llanddona
Ffordd Deg
Rhos Newydd
Rhôs
Hafotty Covert
Llyn Bodgylched
Bodgylched
Carwad
Pen-y-bryn
Bryn Cogail
Fedw Fawr
Mariandyrys
Penhwnllys
Bryn Ddol
Cefn
Glan-yr-afon
Caim
PENMON PRIORY
Penmon
Outer Road
Llangoed
Bryn Celyn
Lavan Sands
Llanfaes
BEAUMARIS CASTLE
GAOL & COURTHOUSE
ALLT GOCH FAWR
Coed Cefn
Cefn
Beaumaris
Gallows Point
Trwyn y Penrhyn
Fryars Road
Puffin Island or Priestholm Ynys Seirio
CONWY BAY
BAE CONWY
Hen Bentref Llandegfan
Llandegfan
BEAUMARIS ROAD
Afon Cadnant
Menai Bridge (Porthaethwy)
Garth
Upper Bangor
Hirael
Bangor
ORIEL TEGFRYN
MENAI SUSPENSION BRIDGE
Maesgeirchen
CATHEDRAL
BANGOR
PENRHYN CASTLE
Glan Adda
Minffordd
Coed Mawr
Llandegai
Tal-y-bont
Bryn
Ind Est
Capel-y-graig
Vaynol Hall
Old Hall
Afon Menai
BANGOR SERVICES
Brithdir
Caerhun
Tyddyn-Heilyn
Glasinfryn
Coed Hywel
Halfway Br
Tregarth
Gatehouse
Pen-y-bryn
COCHWILLAN OLD HALL
Rhiw Goch
Rachub
Llanllechid
Marianywinllan
Bronydd Isaf
Abergwyngregyn
Crymlyn
Glyn
Ffridd Ddu
COEDYDD ABER
ABER FALLS
LLANFAIRFECHAN
Llanfairfechan
Garizim
Nant-y-felin
Nant-y-pandy
Garreg Fawr 356
Gorddinog
Cammarnaint
Afon Dhu
Foel-Ganol 533
Afon Anafon
Pen Bryn-du
Drosgl 621
Bwlch-y Ddeufaen
Penmaenan
Moel Wnion 580
SNOWDONIA
NATIONAL PARK
PARC CENEDLAETHOL ERYRI
Gyrn 542
Llefn
Moel Faban 408
Parc
Afon Ffrydlas
Gyrn Wigau
Drosgl 758
Llwytmor
Llyn Anafon
Bera Mawr
Drum 770
Foel-Fras 942
Afon Goch
Moelf 435
163
A5025
A5108
A5025
B5110
B5108
B5109
B5110
B5420
B5109
A5025
A55
A5
A4080
B4419
A4087
A5
A545
B5109
A55
FELINHELI ROAD
B4547
B4409
TALWRN ROAD
PENMYNYDD ROAD
A5114
BRITANNIA ROAD
ASE5

IRISH SEA
MÔR IWERDDON
Llandudno
Conwy
Colwyn Bay (Bae Colwyn)
Old Colwyn
Abergele
Penmaenmawr
Llanfairfechan
SNOWDONIA NATIONAL PARK
PARC CENEDLAETHOL ERYRI
CONWY
Great Ormes Head Pen-y-Gogarth
Llandudno Bay or Ormes Bay
Little Ormes Head
Penrhyn Bay (Bae Penrhyn)
Rhôs-on-Sea
Llandrillo-yn-Rhos
Deganwy
Llandudno Junction Cyffordd Llandudno
Mochdre
Dolwyd
Llansanffraid Glan Conwy
Pentrefelin
Eglwysbach
Tal-y-Bont
Llangernyw
Betws-yn-Rhos
Llanddulas
Llanfair Talhaiarn
Bryn-y-maen
Dwygyfylchi
Capelulo
Henryd
Rowen
Pontwgan
Tal-y-cafn
Gell
Dawn
Pentre Isaf
Dolwen
Llysfaen
Llanelian-yn-Rhos
Penmaenan
179
164

Rhyl
(Y Rhyl)
Prestatyn
Rhuddlan
Dyserth
Kinmel Bay
Towyn
Talacre
Point of Ayr
Gronant
Gwespyr
Llanasa
Mostyn
Mostyn Quay
Ffynnongroyw
Whitford
Holywell
(Treffynon)
St Asaph
Bodelwyddan
St George
Tremeirchion
Bodfari
Caerwys
Halkyn
Pentre Halkyn
Nannerch
West Kirby
Hilbre Island
Little Hilbre Island
DENBIGHSHIRE
FLINTSHIRE
Halkyn Mountain
A55
A548
A547
A525
A5151
A5026
182
165

LIVERPOOL BAY
Crosby Channel
Crosby
Little Crosby
Thornton
Sefton
Lunt
Maghull
Melling
Netherton
Great Crosby
Blundellsands
Brighton le Sands
Waterloo
Seaforth
Litherland
Orrell
Aintree
Fazakerley
Kirkby
Southdene
Bootle
Walton
Norris Green
Clubmoor
West Derby
Tuebrook
Anfield
Everton
Kirkdale
Sandhills
Vauxhall
Liverpool
Knotty Ash
Roby
Huyton
Knowsley
MERSEYSIDE
Stockbridge Village
Longview
Croxteth
Gillmoss
Kensington
Edge Hill
Wavertree
Childwall
Princes Park
Toxteth
Dingle
Mossley Hill
Aigburth
Allerton
Woolton
Gateacre
Netherley
Hunt's Cross
Grassendale
Garston
Speke
Liverpool Airport
BELFAST 8:00
DOUGLAS 4:15 (WINTER ONLY)
DUBLIN 7:00
DOUGLAS 2:30 (MARCH-NOV)
New Brighton
Wallasey
Egremont
Liscard
Seacombe
Poulton
Birkenhead
Claughton
Tranmere
Rock Ferry
New Ferry
Bebington
Bromborough
Bromborough Pool
Port Sunlight
Eastham
Eastham Ferry
River Mersey
Hoylake
Meols
Dove Point
East Hoyle Bank
West Kirby
Hilbre Island
Grange
Caldy
Frankby
Greasby
Saughall Massie
Moreton
Leasowe
Bidston
Upton
Woodchurch
Arrowe Hill
Noctorum
Oxton
Prenton
Landican
Irby
Thurstaston
Thingwall
Pensby
Barnston
Storeton
Brimstage
Heswall
Gayton
Dawpool Bank
Thornton Hough
Raby
Parkgate
Neston
Little Neston
Ness
Willaston
Hinderton
Childer Thornton
Hooton
Little Sutton
Overpool
Great Sutton
Ellesmere Port
Whitby
Stanlow
Stanlow Banks
CHESHIRE WEST & CHESTER
Burton
Puddington
Ledsham
Capenhurst
Two Mills
Woodbank
Shotwick
Saughall
Mollington
Backford
Little Stanney
Stoak
Thornton-le-Moors
Wervin
Picton
Mickle Trafford
Upton
Chester
Hoole
Newton
Sealand
Queensferry
Shotton
Connah's Quay
Garden City
Flint (Y Fflint)
Oakenholt
Flint Mountain
Holywell (Treffynnon)
Bagillt
Greenfield
Mostyn
Halkyn
Rhes-y-cae
Pentre Halkyn
Walwen
Whelston
The Marsh
Flint Marsh
Gayton Sands
Danger area
White Sands
Rockcliffe
M53
M57
M62
A59
A41
A55
A550
A540
A5117

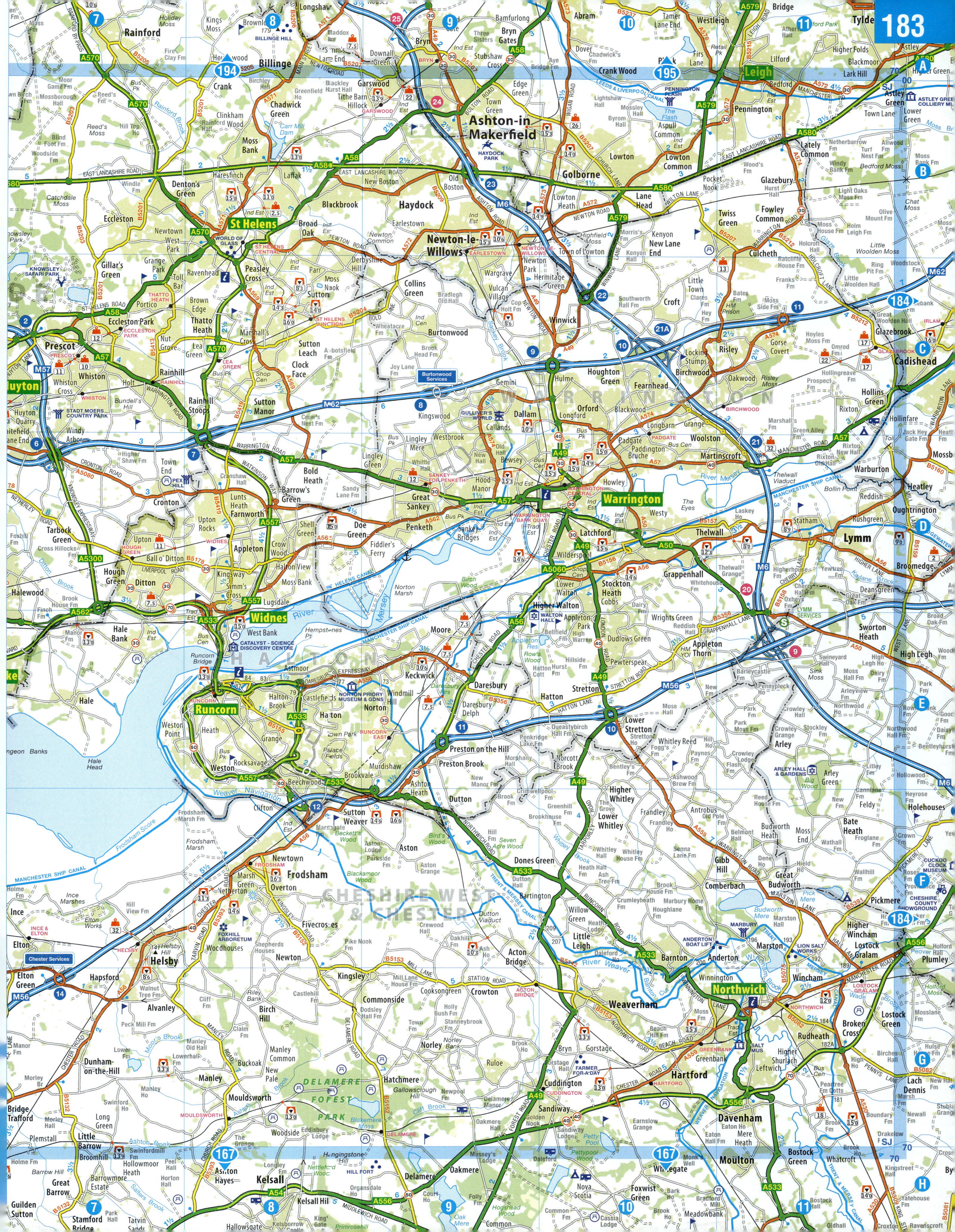
Rainford
Billinge
Crank
Garswood
Bryn
Ashton-in Makerfield
Abram
Golborne
Leigh
Tyldesley
Pennington
Lowton
Lowton Common
Lane Head
Lately Common
Glazebury
Culcheth
St Helens
Eccleston
Haydock
Blackbrook
Newton-le-Willows
Earlestown
Winwick
Croft
Risley
Glazebrook
Cadishead
Prescot
Whiston
Huyton
Rainhill
Rainhill Stoops
Sutton
Sutton Manor
Burtonwood
Kingswood
Houghton Green
Fearnhead
Birchwood
Orford
Dallam
Callands
Warrington
Padgate
Woolston
Martinscroft
Hollins Green
Rixton
Warburton
Great Sankey
Penketh
Bold Heath
Barrow's Green
Farnworth
Cronton
Widnes
Ditton
Hough Green
Halewood
Hale
Hale Bank
Latchford
Thelwall
Lymm
Statham
Heatley
Oughtrington
Grappenhall
Stockton Heath
Appleton
Appleton Thorn
Walton
Moore
Daresbury
Keckwick
Runcorn
Halton
Norton
Weston
Weston Point
Preston on the Hill
Preston Brook
Stretton
Lower Stretton
Hatton
Arley
Dutton
Higher Whitley
Lower Whitley
Antrobus
Great Budworth
Comberbach
Frodsham
Helsby
Elton
Hapsford
Alvanley
Kingsley
Crowton
Acton Bridge
Barnton
Anderton
Marston
Wincham
Northwich
Weaverham
Hartford
Cuddington
Sandiway
Davenham
Moulton
Rudheath
Lostock Gralam
Pickmere
Plumley
Manley
Mouldsworth
Delamere
Delamere Forest Park
Kelsall
Ashton Hayes
Tarvin
Great Barrow
Oakmere
Whitegate
WARRINGTON
HALTON
CHESHIRE WEST & CHESTER
River Mersey
River Weaver
Manchester Ship Canal
M6
M62
M56
A49
A57
A50
A533
A556
A580
A579
A58
A570
A56
A54
A557
A562
A5300
A574
194
195
184
167

Worsley
Swinton
Pendlebury
Salford
Eccles
Manchester
Trafford Park
Stretford
Old Trafford
Hulme
Irlam
Urmston
Flixton
Davyhulme
Cadishead
Partington
Carrington
Sale
Ashton upon Mersey
Chorlton-cum-Hardy
Didsbury
West Didsbury
Withington
Fallowfield
Levenshulme
Gorton
Denton
Droylsden
Failsworth
Dukinfield
Audenshaw
Reddish
Stockport
Bredbury
Romiley
Heaton Moor
Heaton Norris
Cheadle
Cheadle Hulme
Gatley
Wythenshawe
Northenden
Timperley
Altrincham
Bowdon
Hale
Broadheath
Dunham Town
Heatley
Lymm
Warburton
Oughtrington
Manchester Airport
Bramhall
Hazel Grove
Poynton
Handforth
Styal
Wilmslow
Alderley Edge
Prestbury
Bollington
Macclesfield
Hurdsfield
Knutsford
Mobberley
Ashley
Rostherne
Tatton Park
Plumley
Ollerton
Chelford
Marthall
Henbury
Siddington
Gawsworth
Goostrey
Lower Withington
Holmes Chapel
Monks Heath
Nether Alderley
Over Peover
Lower Peover
Twemlow Green
Highlane
Marton
Oakgrove
Sutton Lane Ends
Langley
Adlington
Woodford
Middlewood
Higher Poynton
Kitt's Moss
Heald Green
Woodhouse Park
Baguley
Brooklands
Urmston
Manchester Ship Canal
Bridgewater Canal
GREATER MANCHESTER
0 1 2 3 miles
0 1 2 3 4 5 km
183
195
196
168

Mossley
Stalybridge
Hyde
Marple
New Mills
Glossop
Hadfield
Tintwistle
Charlesworth
Chisworth
Hayfield
Low Leighton
Chinley
Whaley Bridge
Chapel-en-le-Frith
Buxworth
Furness Vale
Disley
Newtown
Strines
Kettleshulme
Taxal
Combs
Dove Holes
Peak Dale
Buxton
Burbage
Harpur Hill
King Sterndale
Cowdale
Chelmorton
Taddington
Sterndale Moor
Flagg
Sheldon
Ashford in the Water
Great Longstone
Little Longstone
Tideswell
Litton
Wardlow
Miller's Dale
Wormhill
Tunstead
Hargatewall
Wheston
Peak Forest
Sparrowpit
Perryfoot
Castleton
Hope
Bradwell
Brough
Aston
Bamford
Thornhill
Great Hucklow
Little Hucklow
Foolow
Eyam
Edale
Barber Booth
Nether Booth
Hathersage
Howden Moors
Bleaklow
Kinder Scout
Crowden
Longdendale
Mottram in Longdendale
Broadbottom
Gamesley
Hollingworth
Padfield
Charlestown
Chunal
Little Hayfield
Birch Vale
Thornsett
Rowarth
Mellor
Marple Bridge
Compstall
Strines
Macclesfield Forest
Wildboarclough
Rainow
Pott Shrigley
PEAK DISTRICT NATIONAL PARK
High Peak
Hope Forest
DERBYSHIRE
Derwent Moors
Ladybower Reservoir
Howden Reservoir
Derwent Reservoir
Mam Tor
Vale of Edale
Edale Moor
Kinder Reservoir
Woodhead Reservoir
Torside Reservoir
Rhodeswood Reservoir
Valehouse Reservoir
Bottoms Reservoir
Arnfield Reservoir
Fernilee Reservoir
Errwood Reservoir
Combs Reservoir
Lamaload Reservoir
Trentabank Reservoir
Snake Road
Pennine Way
Pennine Bridleway
A57
A628
A6
A623
A624
A6024
A515
A53
A54
A537
A5004
B6062
B6049
M67
A6018
A616
196
197
186
169

Langsett
Midhopestones
Stocksbridge
Deepcar
Bolsterstone
Ewden Village
Wortley
Howbrook
Tankersley
Hoyland
Elsecar
Wath upon Dearne
Brampton
Harley
Wentworth
Chapeltown
Burncross
Ecclesfield
Grenoside
Thorpe Hesley
Greasbrough
Kimberworth
Rotherham
Oughtibridge
Wharncliffe Side
High Bradfield
Low Bradfield
Worrall
Wadsley Bridge
Shiregreen
Parson Cross
Wincobank
Tinsley
Brinsworth
Moorgate
Ughill
Stacey Bank
Dungworth
Stannington
Malin Bridge
Hillsborough
Owlerton
Pitsmoor
Attercliffe
Darnall
Catcliffe
Treeton
Sheffield
Walkley
Crookes
Ranmoor
Fulwood
Hollow Meadows
Handsworth
Woodhouse
Manor Estate
Intake
Arbourthorne
Highfield
Ecclesall
Whirlow
Millhouses
Norton Woodseats
Gleadless
Hackenthorpe
Beighton
Ringinglow
Abbeydale
Dore
Beauchief
Jordanthorpe
Highlane
Mosborough
Eckington
Bradway
Totley
New Totley
Dronfield
Dronfield Woodhouse
Coal Aston
Apperknowle
Marsh Lane
Middle Handley
West Handley
Holmesfield
Unstone
Unstone Green
DERBYSHIRE
Bamford
Thornhill
Aston
Hathersage
Upper Padley
Nether Padley
Grindleford
Owler Bar
Eyam
Foolow
Stoney Middleton
Froggatt
Curbar
Calver
Baslow
Bubnell
Hassop
Rowland
Great Longstone
Pilsley
Edensor
Bakewell
Barlow
Wildaw Green
Moorhall
Common Side
Old Whittington
New Whittington
Whittington Moor
Staveley
Brimington
Newbold
Chesterfield
Cutthorpe
Old Brampton
Overgreen
Wadshelf
Eastmoor
Brampton
Walton
Birdholme
Hasland
Calow
Hady
Duckmanton
Long Duckmanton
Arkwright Town
Sutton Scarsdale
PEAK DISTRICT NATIONAL PARK
SOUTH YORKSHIRE
M1
A61
A57
A616
A629
A623
A619
A6

Mexborough
Old Denaby
Swinton
Conisbrough
Denaby Main
Warmsworth
Balby
Bessacarr
Branton
Auckley
Lidget
Blaxton
Finningley
Cadeby
Sprotbrough
High Melton
Loversall
New Edlington
Old Edlington
Wadworth
Wilsic
New Rossington
Rossington
Hay Field
Hayfield Green
Austerfield
Misson
Bawtry
Tickhill
Spital Hill
Harworth
Bircotes
Scaftworth
Everton
Harwell
Newington
Hooton Roberts
Clifton
Thrybergh
Ravenfield
Dalton
Braithwell
Micklebring
Stainton
Maltby
Hellaby
Bramley
Wickersley
Hooton Levitt
Whiston
Morthen
Thurcroft
Brampton en le Morthen
Ulley
Aughton
Aston
Swallownest
Laughton en le Morthen
Laughton Common
Throapham
Dinnington
Letwell
Firbeck
Stone
Oldcotes
Langold
Styrrup
Serlby
Scrooby
Mattersey
Mattersey Thorpe
Blyth
Ranskill
Torworth
Hodsock
Barnby Moor
Sutton
Lound
Carlton in Lindrick
North Carlton
Wallingwells
Gildingwells
North Anston
South Anston
Todwick
Kiveton Park
Wales
Woodsetts
Wigthorpe
Gateford
Shireoaks
Rhodesia
Worksop
Kilton
Manton
Netherthorpe
Thorpe Salvin
Harthill
Killamarsh
Norwood
Spinkhill
Barlborough
Whitwell
Clowne
Creswell
Elmton
Hodthorpe
Belph
Holbeck
Norton
Cuckney
Whaley
Whaley Thorns
Nether Langwith
Langwith
Bolsover
Palterton
Scarcliffe
Stanfree
Oxcroft
Shuttlewood
Woodside
Mastin Moor
Welbeck Abbey
Clumber Park
Hardwick Village
Carburton
Thoresby
Perlethorpe
Budby
Meden Vale
Market Warsop
Church Warsop
Sherwood Forest
Elkesley
Bothamsall
Haughton
Walesby
Ollerton
Boughton
Kirton
Gamston
West Drayton
Eaton
Ordsall
Retford
West Retford
Babworth
Ranby
Scofton
Bilby
Retford (Gamston) Airport
Robin Hood Airport Doncaster Sheffield
South Yorkshire
Nottinghamshire
Long Valley
Great Lake
River Poulter
River Ryton
Chesterfield Canal
M1
M18
A1(M)
A1
A57
A60
A61
A614
A616
A619
A631
A632
A634
A638
A6182

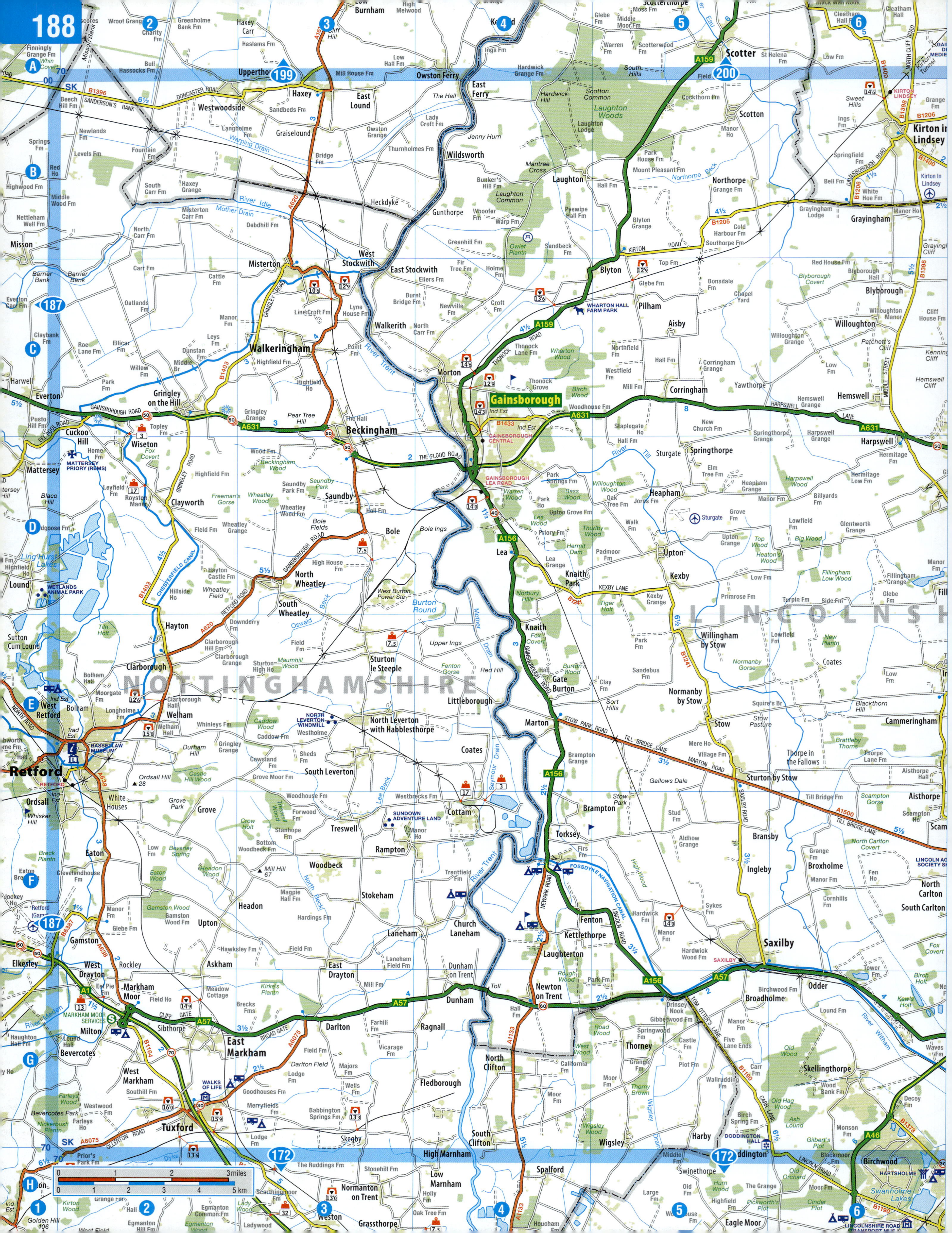

NOTTINGHAMSHIRE
LINCOLNSH
Gainsborough
Retford
Tuxford
Saxilby
Kirton in Lindsey
Scotter
Scotton
Haxey
Westwoodside
Graiselound
East Lound
Owston Ferry
East Ferry
Wildsworth
Laughton
Laughton Woods
Northorpe
Grayingham
Blyborough
Willoughton
Hemswell
Harpswell
Corringham
Springthorpe
Heapham
Upton
Kexby
Willingham by Stow
Normanby by Stow
Stow
Sturton by Stow
Cammeringham
Aisthorpe
Brattleby
Thorpe in the Fallows
Scampton
North Carlton
South Carlton
Broxholme
Ingleby
Bransby
Odder
Broadholme
Skellingthorpe
Birchwood
Harby
Eagle Moor
Swinethorpe
Wigsley
Thorney
Saxilby
Drinsey Nook
Fenton
Kettlethorpe
Laughterton
Newton on Trent
Torksey
Brampton
Marton
Gate Burton
Knaith
Knaith Park
Lea
Blyton
Pilham
Aisby
Morton
Walkerith
East Stockwith
West Stockwith
Misterton
Walkeringham
Gringley on the Hill
Beckingham
Saundby
Bole
North Wheatley
South Wheatley
Hayton
Clarborough
Welham
Clayworth
Wiseton
Mattersey
Everton
Misson
Harwell
Lound
Sutton Cum Lound
Ordsall
West Retford
Bolham
North Leverton with Habblesthorpe
South Leverton
Sturton le Steeple
Littleborough
Coates
Cottam
Treswell
Rampton
Woodbeck
Stokeham
Laneham
Church Laneham
East Drayton
Dunham on Trent
Dunham
Ragnall
Darlton
Fledborough
North Clifton
South Clifton
Spalford
High Marnham
Low Marnham
Normanton on Trent
Weston
Grassthorpe
Skegby
East Markham
West Markham
Markham Moor
Sibthorpe
Milton
Bevercotes
West Drayton
Gamston
Elkesley
Eaton
Headon
Upton
Askham
Grove
White Houses
Heckdyke
Gunthorpe
Wharton Hall Farm Park
Gainsborough Central
Gainsborough Lea Road
West Burton Power Sta
Burton Round
River Trent
River Idle
Chesterfield Canal
Fossdyke Navigation Canal
Sundown Adventure Land
Bassetlaw Museum
North Leverton Windmill
Wetlands Animal Park
Markham Moor Services
Walks of Life
A631
A159
A156
A57
A620
A1
A638
A1133
A1500
A46
A6075
B1403
B1241
B1205
B1206
B1398
B1400
B1433
B1164
SK
199
200
187
172
1 2 3 4 5 6
A B C D E F G H
0 1 2 3 miles
0 1 2 3 4 5 km

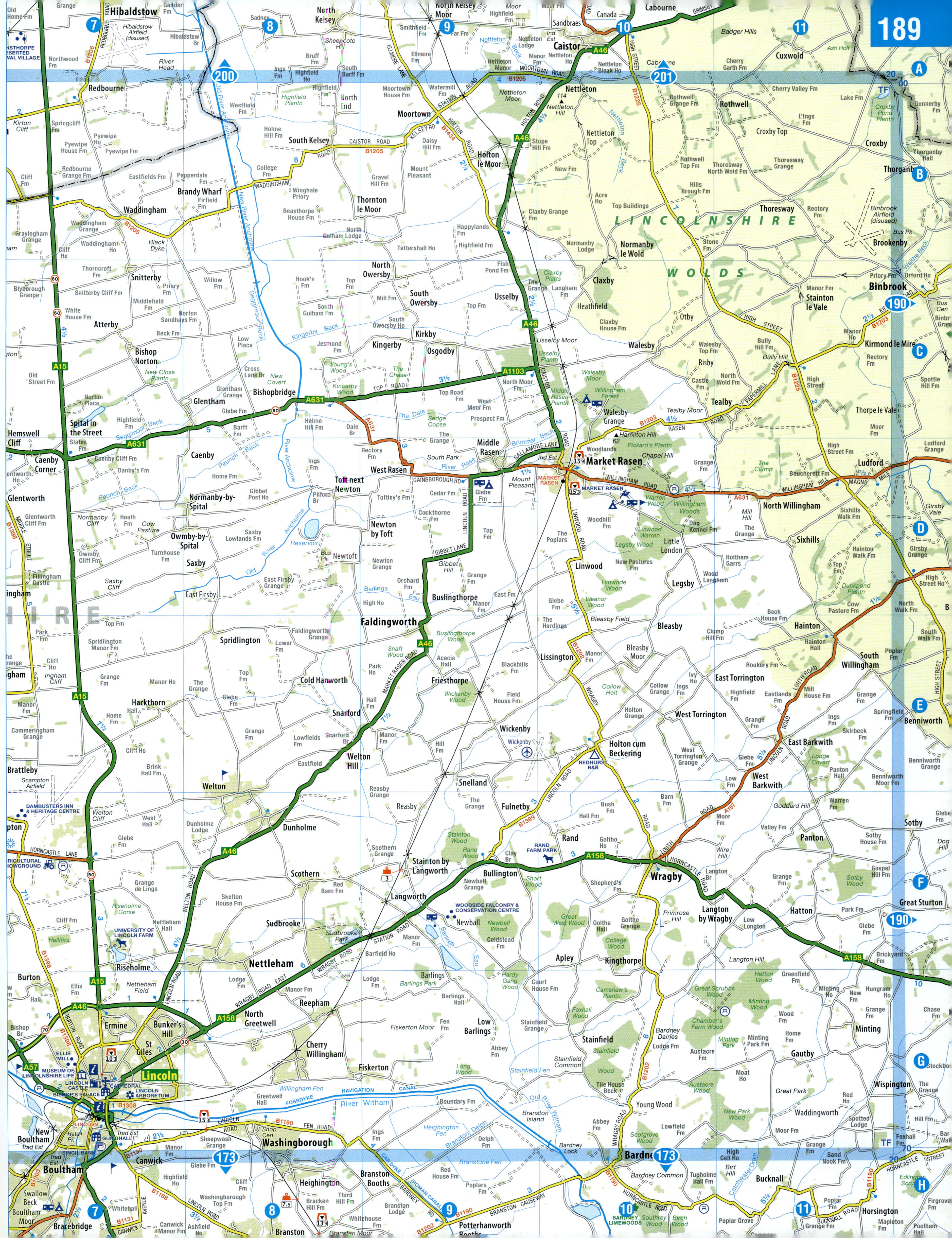

Hibaldstow
North Kelsey
North Kelsey Moor
Canada
Cabourne
Sandbraes
Caistor
Cuxwold
Badger Hills
Hibaldstow Airfield (disused)
Redbourne
Nettleton
Moortown
South Kelsey
Holton le Moor
Rothwell
Croxby Top
Croxby
Thorganby
Waddingham
Brandy Wharf
Thornton le Moor
Nettleton Top
Normanby le Wold
Thoresway
Brookenby
Binbrook Airfield (disused)
LINCOLNSHIRE
WOLDS
Binbrook
Stainton le Vale
Snitterby
North Owersby
South Owersby
Claxby
Usselby
Walesby
Atterby
Kingerby
Osgodby
Kirkby
Kirmond le Mire
Bishop Norton
Bishopbridge
Glentham
Otby
Risby
Tealby
Spital in the Street
Hemswell Cliff
Middle Rasen
Market Rasen
West Rasen
Toft next Newton
Caenby Corner
Caenby
Glentworth
Normanby-by-Spital
Ownby-by-Spital
Newton by Toft
Newtoft
Saxby
North Willingham
Ludford
Little London
Linwood
Sixhills
Legsby
Buslingthorpe
Faldingworth
Spridlington
Lissington
Bleasby
Hainton
South Willingham
Cold Hanworth
Friesthorpe
East Torrington
Hackthorn
Snarford
West Torrington
Wickenby
Holton cum Beckering
East Barkwith
Benniworth
Welton Hill
West Barkwith
Brattleby
Welton
Snelland
Fulnetby
Dunholme
Rand
Sotby
Panton
Scothern
Stainton by Langworth
Bullington
Wragby
Langworth
Langton by Wragby
Hatton
Great Sturton
Newball
Sudbrooke
Nettleham
Apley
Kingthorpe
Riseholme
Burton
Reepham
Barlings
North Greetwell
Ermine
Bunker's Hill
Low Barlings
Stainfield
Minting
Gautby
St Giles
Cherry Willingham
Fiskerton
Lincoln
Wispington
Waddingworth
New Boultham
Washingborough
Young Wood
Canwick
Heighington
Branston Booths
Bardney
Bucknall
Boultham
Horsington
Bracebridge
Branston
Potterhanworth
River Witham
Fossdyke Navigation Canal
A46
A15
A631
A1103
A158
A157
A57
B1205
B1203
B1225
B1202
B1190
200
201
190
173

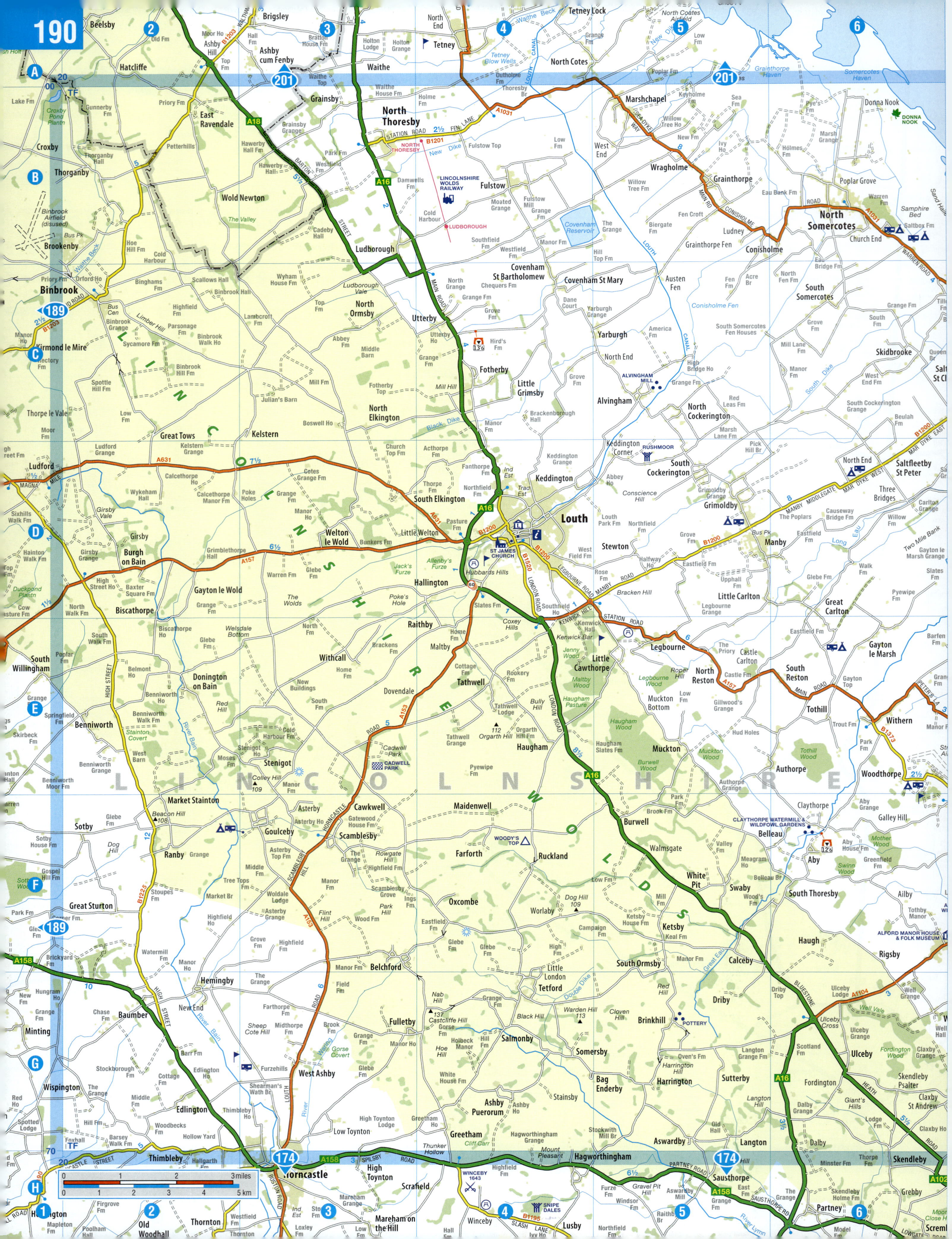

Beelsby
Brigsley
North End
Tetney Lock
North Coates Airfield
Ashby cum Fenby
Holton Le Clay
Tetney
Hatcliffe
Waithe
North Cotes
Marshchapel
Grainsby
North Thoresby
East Ravendale
Croxby
Thorganby
Fulstow
Wragholme
Grainthorpe
Wold Newton
Binbrook Airfield (disused)
The Valley
Brookenby
Covenham Reservoir
Ludney
Conisholme
North Somercotes
Church End
Donna Nook
Binbrook
Ludborough
Covenham St Bartholomew
Covenham St Mary
Austen Fen
South Somercotes
Kirmond le Mire
Utterby
North Ormsby
Yarburgh
Skidbrooke
Fotherby
Little Grimsby
Alvingham
North Cockerington
Thorpe le Vale
North Elkington
Great Tows
Kelstern
Keddington Corner
South Cockerington
Saltfleetby St Peter
Ludford
South Elkington
Keddington
Louth
Grimoldby
Three Bridges
Girsby
Burgh on Bain
Welton le Wold
Little Welton
St James Church
Hubbard's Hills
Stewton
Manby
Gayton le Wold
Hallington
Little Carlton
Great Carlton
Biscathorpe
Raithby
Legbourne
Little Cawthorpe
Castle Carlton
Gayton le Marsh
South Willingham
Witchcall
Maltby
North Reston
South Reston
Donington on Bain
Tathwell
Tothill
Withern
Benniworth
Dovendale
Haugham
Muckton
Authorpe
Woodthorpe
Stenigot
Cadwell Park
Market Stainton
Cawkwell
Maidenwell
Burwell
Claythorpe
Belleau
Claythorpe Watermill & Wildfowl Gardens
Sotby
Goulceby
Asterby
Scamblesby
Farforth
Ruckland
Walmsgate
Aby
Ranby
Woody's Top
White Pit
Swaby
South Thoresby
Great Sturton
Oxcombe
Worlaby
Ketsby
Haugh
Alford Manor House & Folk Museum
Rigsby
Belchford
South Ormsby
Calceby
Hemingby
Tetford
Driby
Baumber
Fulletby
Brinkhill
Pottery
Minting
Salmonby
Somersby
Ulceby
Wispington
West Ashby
Bag Enderby
Harrington
Sutterby
Skendleby Psalter
Claxby St Andrew
Edlington
Ashby Puerorum
Stainsby
Greetham
Langton
Aswardby
Dalby
Thimbleby
Horncastle
High Toynton
Hagworthingham
Sausthorpe
Skendleby
Scrafield
Mareham on the Hill
Winceby
Snipe Dales
Lusby
Partney
Grebby
Thornton
Old Woodhall
LINCOLNSHIRE
LINCOLNSHIRE WOLDS
Lincolnshire Wolds Railway
North Thoresby
Ludborough
A18
A16
A1031
A631
A157
A153
A158
A1104
B1200
B1201
B1203
B1225
B1520
B1373
B1195
201
189
174
0 1 2 3 miles
0 1 2 3 4 5 km

7
8
9
10
11
A
B
C
D
E
F
G
H
NORTH SEA
Toby's Hill
Saltfleet
Saltfleet Haven
Sea View Fm
Rimac
SALTFLEETBY THEDDLETHORPE
Saltfleetby All Saints
Lodge Fm
Theddlethorpe St Helen
Manor Ho
Hall Fm
Theddlethorpe All Saints
Gayton Engine
High Gate
Will Row
Gas Terminal
North End
THE SEAL SANCTUARY & NATURE CENTRE
Westfield Fm
Stain Hill
Meers Bank
Mablethorpe Hall
FUN FAIR
Mablethorpe
Trusthorpe
Strubby Grange
Earl's Br
Grange Fm
Poplar Fm
Willow Fm
Bamber's Br
Strubby
Thorpe
Trusthorpe Hall
Maltby le Marsh
Manor Ho
Sutton on Sea
Sandilands
Mill Hill
Poplar Lodge Fm
Beesby
Abbey Fm
Beesby Grange
Hagnaby
Manor Fm
Washdyke Br
Hannah
Sea Bank Fm
Saleby
Glebe Fm
Markby
Priory Fm
America Fm
Cob Hill
Saleby Manor
The Grange
College Fm
Asserby Turn
Asserby
Willow Fm
Black House Fm
Thoresthorpe
Bilsby
Dryby Fm
Wold Sea Fm
Moat Ho
Huttoft
Manor Fm
Anderby Creek
Alford
The Grange
Thurlby
The Manor
Anderby
Bilsby Field
Farlesthorpe Fen
ON YOUR MARQUES
Wolla Bank
Langham Fm
Mumby
Manor Fm
Chapel Six Marshes
Well Beck Fm
Farlesthorpe
School Fm
Mill Hill
Manor Ho
Authorpe Row
Elsom Fm
Cumberworth
Cherry Fm
Mickleberry Hill
Chapel Pt
Mawthorpe
Bonthorpe
Helsey
Croft Fm
Chapel St Leonards
Manor Fm
Listoft
Hogsthorpe
Willoughby
Poplar Fm
Burlands Beck
Willoughby High Drain
Claxby
Willoughby Wood
Hogsbeck Ho
Sloothby
Howlet Ho
Slackholme End
Hope Fm
Beeches Fm
Welton Low Wood
Hasthorpe
175
Welbourne Fm
HARDY'S ANIMAL FARM
Welton High Wood
Thwaite Hall
Habertoft
Ingoldmells
Highfield Fm
Candlesby Hill
Rookery Fm
Welton le Marsh
Boothby Hall
Addlethorpe
FANTASY ISLAND CHILDREN'S PLAYDROME & THE MILLENNIUM ROLLERCOASTER
Ingoldmells Pt
Manor Fm
Boothby Grange
Whitehouse Fm
Corner Fm
Marsh
A1031
A1104
A157
A52
A1111
B1449
B1196
SALTFLEET ROAD
MABLETHORPE RD
THEDDLETHORPE RD
ALFORD ROAD
SUTTON ROAD
BEESBY ROAD
HUTTOFT ROAD
SUTTON ROAD
EAST ST
THURLBY ROAD
LONG LANE
MUMBY ROAD
MAWTHORPE HILL ROAD
HANBY LANE
SKEGNESS ROAD
BAKER'S LANE
Main Drain
North Drain
Orby Beck
Great Eau
TF
70
00

ISLE OF MAN
Scale 1:200,000
POINT OF AYRE
The Ayres
Rue Pt.
The Lhen
Glentruan
Dhowin
Cranstal
Bride
Andreas
Jurby Head
MANX CROSSES
JURBY SOUTH
Jurby East
Jurby West
Sandygate
Ballasalla
Regaby
The Cronk
St Judes
Dhoor
RAMSEY BAY
Ramsey
GROVE MUSEUM
CURRAGHS WILDLIFE PARK
Sulby
Churchtown
MANX ELECTRIC RAILWAY
Port e Vullen
Orrisdale
Ballaugh
T.T. Course
Glen Auldyn
Maughold
Maughold Head
Rhenculen
Ravensdale
Dreemskerry
Kirk Michael
NORTH BARRULE
Ballajora
CELTIC CRAFT CENTRE
Corrany
Cornaa
Ballaleigh
SNAEFELL
Glen Mona
Barregarrow
Druidale
Dhoon
MURRAYS MOTORCYCLE MUSEUM
Agneash
LAXEY WHEEL AND MINES
Bulgham Bay
MANX TRANSPORT MUSEUM
Knockshary
Cronk-y-Voddy
SNAEFELL MOUNTAIN RAILWAY
Ballaquine
Laxey
St Patrick's I.
PEEL
Peel
COLDEN
LAXEY WOOLLEN MILLS
Old Laxey
Laxey Head
HOUSE OF MANANNAN
BALLALHEANNAGH GARDENS
Contrary Head
TYNWALD CRAFT CENTRE
Fairy Cottage
Laxey Bay
KIPPER MUSEUM
TYNWALD HILL
Patrick
St John's
Greeba
Baldwin
Creg-ny-Baa
Ballacannel
Baldrine
Glenmaye
Clay Head
Lower Foxdale
Crosby
MANX CROSSES
Dalby Pt.
Dalby
Glen Vine
Strang
Onchan
HEYSHAM 3:30
Niarbyl
Foxdale
Union Mills
Tromode
GROUDLE GLEN RAILWAY
Niarbyl Bay
Eairy
Braaid
Spring Valley
ONCHAN PLEASURE PARK
HEYSHAM 2:00 (TT race period only)
SOUTH BARRULE
Cooil
Douglas
Douglas Bay
Close Clark
Ellenbrook
Douglas Head
CAMERA OBSCURA
Ballamodha
St Mark's
Ballaveare
LIVERPOOL 2:30 (March-Nov)
Newtown
Lingague
Ronague
Little Ness
Fleshwick Bay
Grenaby
ISLE OF MAN STEAM RAILWAY
LIVERPOOL 4:15 (Winter only)
Surby
Bradda
Colby
Ballabeg
RUSHEN ABBEY
Santon Head
Bradda Head
Port Erin
Ballasalla
Port Greenaugh
RAILWAY MUS
BILLOWN
The Howe
Four Roads
ISLE OF MAN
Castletown
Derbyhaven
Cregneash
CASTLE RUSHEN
NAUTICAL MUS
St Michael's I.
SCARLETT VISITOR CENTRE
Port St Mary
OLD HOUSE OF KEYS
CREGNEASH VILLAGE FOLK MUSEUM
Scarlett Point
Calf of Man
Spanish Head
Dreswick Pt.
BELFAST 2:55
DUBLIN 2:55
(April-Sept)
Chicken Rock
I S L E O F M A N
NX
SC
miles
km

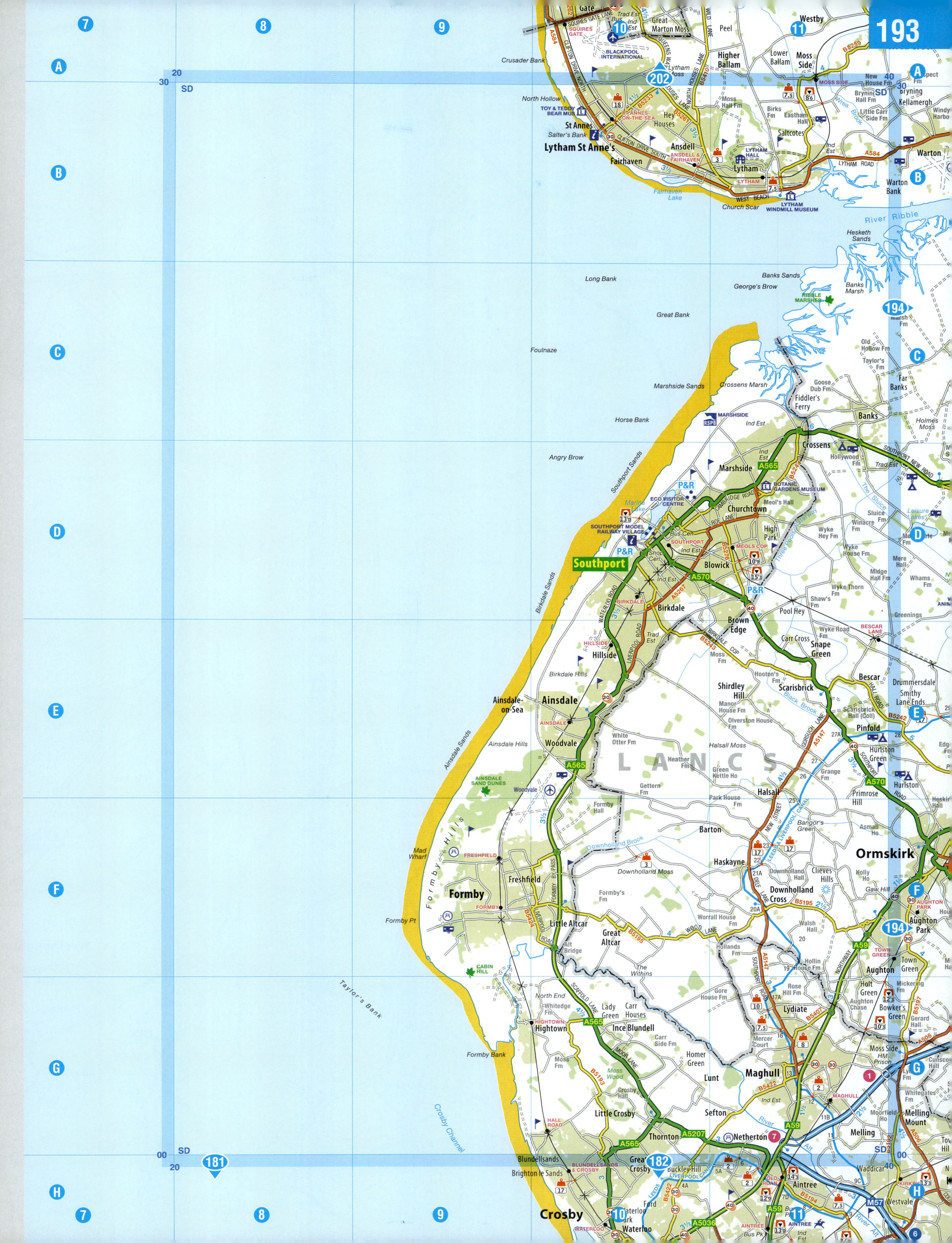

Lytham St Anne's
St Annes
Fairhaven
Ansdell
Lytham
Warton
River Ribble
Banks
Crossens
Marshside
Churchtown
Southport
Birkdale
Hillside
Ainsdale
Ainsdale-on-Sea
Woodvale
Formby
Freshfield
Little Altcar
Great Altcar
Hightown
Ince Blundell
Little Crosby
Thornton
Crosby
Blundellsands
Scarisbrick
Halsall
Haskayne
Downholland Cross
Ormskirk
Aughton
Maghull
Lydiate
Netherton
Aintree
Melling
LANCS

Preston
Fulwood
Ribbleton
Freckleton
Warton
Lytham Road
River Ribble
Longton
Hutton
Bottom of Hutton
Penwortham
Higher Penwortham
Lostock Hall
Bamber Bridge
Walton-le-Dale
Farington
Leyland
Clayton-le-Woods
Whittle-le-Woods
Samlesbury
Samlesbury Bottoms
Hoghton
Brindle
Higher Wheelton
Abbey Village
Withnell
Wheelton
Brinscall
White Coppice
Great Knowley
Chorley
Euxton
Croston
Bretherton
Tarleton
Hesketh Bank
Mere Brow
Holmeswood
Rufford
Mawdesley
Eccleston
Heskin Green
Coppull
Adlington
Horwich
Standish
Blackrod
Aspull
Wigan
Westhoughton
Hindley
Ince in Makerfield
Orrell
Shevington
Appley Bridge
Parbold
Newburgh
Burscough
Burscough Bridge
Ormskirk
Aughton
Skelmersdale
Up Holland
Rainford
Billinge
Ashton-in Makerfield
Pemberton
Abram
Platt Bridge
LANCASHIRE
River Douglas
River Yarrow
Charnock Richard Services
Bolton West Services
M6
M61
M65
M58
A59
A6
A49
A577
A570
A565
A5209
1 2 3 miles
1 2 3 4 5 km

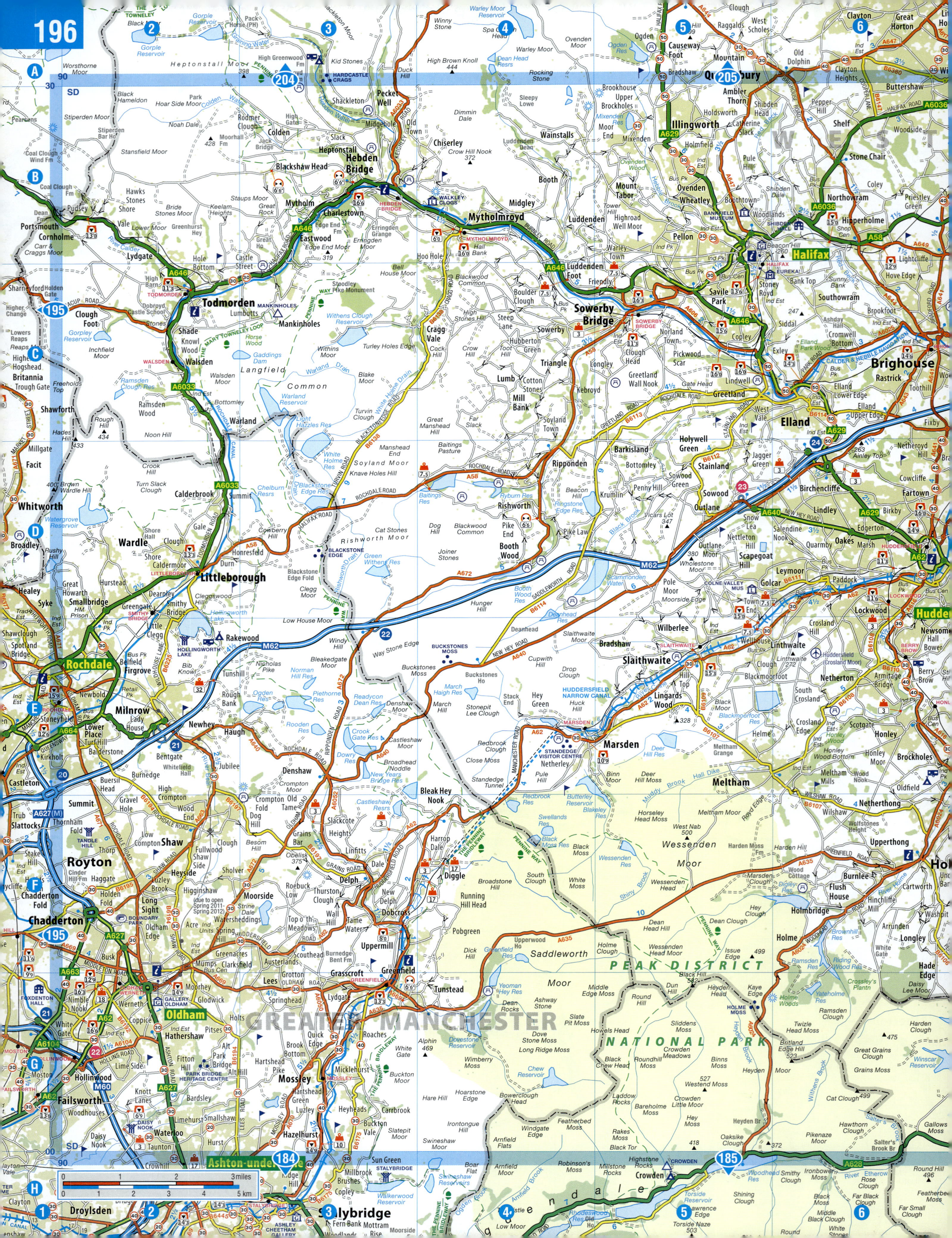

Hebden Bridge
Heptonstall
Mytholmroyd
Todmorden
Walsden
Littleborough
Rochdale
Milnrow
Royton
Chadderton
Oldham
Shaw
Delph
Diggle
Marsden
Slaithwaite
Meltham
Holmfirth
Halifax
Sowerby Bridge
Elland
Brighouse
Ripponden
Rishworth
Greetland
Stainland
Lindley
Golcar
Linthwaite
Uppermill
Greenfield
Mossley
Stalybridge
Ashton-under-Lyne
Failsworth
Hollinwood
Lees
Heckmondwike
Whitworth
Wardle
Hipperholme
Queensbury
Illingworth
Midgley
Luddenden
Cragg Vale
Mankinholes
Sladen
Saddleworth
Moor
Wessenden
PEAK DISTRICT
NATIONAL PARK
GREATER MANCHESTER
WEST
M62
A646
A58
A62
A635
A627(M)
3 miles
5 km

197
Bradford
Bowling
West Bowling
Holme Wood
Bierley
Tong
New Farnley
Beeston
Hunslet
Stourton
Swillington
Oulton
Rothwell
Rothwell Haigh
Woodlesford
Middleton
Morley
Tingley
Gildersome
Drighlington
Adwalton
Birkenshaw
Gomersal
Birstall
Cleckheaton
Liversedge
Heckmondwike
Batley
Dewsbury
Mirfield
Ossett
Horbury
Wakefield
Outwood
Stanley
Lofthouse
Carlton
Methley
Mickletown
Whitwood
Normanton
Altofts
Snydale
Ackton
Sharlston
Crofton
Walton
Sandal Magna
Crigglestone
Middlestown
Overton
Netherton
Flockton
Grange Moor
Lepton
Kirkburton
Shepley
Skelmanthorpe
Clayton West
Denby Dale
Emley
Darton
Mapplewell
Royston
Notton
Woolley
West Bretton
Haigh
High Hoyland
Kexbrough
Barugh
Cawthorne
Dodworth
Barnsley
Worsbrough
Wombwell
Hoyland
Penistone
Silkstone
Thurlstone
Millhouse Green
Ingbirchworth
Holmfirth
Stocksbridge
Langsett
Midhopestones
Thurlstone Moors
Dunford Bridge
Pilley
Tankersley
Wortley
Cudworth
Ryhill
Havercroft
Felkirk
Shafton
Carlton
Monk Bretton
Ardsley
Kendray
Fitzwilliam
Kinsley
South Hiendley
Wragby
Foulby
Crigglestone
YORKSHIRE
SOUTH YORKSHIRE
198
205
206
185
186

Kippax
Micklefield
Ledsham
Fairburn
Monk Fryston
Hillam
South Milford
Lumby
Hambleton
Thorpe Willoughby
Brayton
Gateforth
Burn
Burton Salmon
Birkin
West Haddlesey
Chapel Haddlesey
Camblesforth
Castleford
Brotherton
Byram
Ferrybridge
Knottingley
Kellingley
Kellington
Eggborough
Hensall
Temple Hirst
Hirst Courtney
Carlton
Gowdall
Snaith
West Cowick
Normanton
Whitwood
Pontefract
Featherstone
Darrington
Cridling Stubbs
Whitley
Great Heck
Pollington
Balne
Purston Jaglin
West Hardwick
East Hardwick
Wentbridge
Stapleton
Womersley
Little Smeaton
Walden Stubbs
Fenwick
Sykehouse
Eskholme
Sharlston
Foulby
Wragby
Ackworth Moor Top
Low Ackworth
High Ackworth
Thorpe Audlin
Kirk Smeaton
Badsworth
Norton
Campsall
Askern
Moss
Fitzwilliam
Kinsley
Hemsworth
Upton
Wrangbrook
Skelbrooke
Burghwallis
Sutton
Haywood
Braithwaite
Kirk Bramwith
South Bramwith
Brierley
South Kirkby
South Elmsall
Moorthorpe
Hampole
Skellow
Owston
Carcroft
Thorpe in Balne
Barnby Dun
Cudworth
Grimethorpe
Clayton
Hooton Pagnell
Adwick le Street
Woodlands
Toll Bar
Almholme
Arksey
Kirk Sandall
Dunsville
Great Houghton
Little Houghton
Thurnscoe
Brodsworth
Highfields
Bentley
Edenthorpe
Armthorpe
Billingley
Goldthorpe
Marr
Scawsby
Scawthorpe
Wheatley
Intake
Darfield
Bolton upon Dearne
Barnburgh
Harlington
Doncaster
Hyde Park
Newton
Hexthorpe
Wombwell
Brampton
High Melton
Sprotbrough
Balby
Bessacarr
Cantley
Cadeby
Warmsworth
Wath upon Dearne
Mexborough
Denaby Main
Conisbrough
Swinton
Old Denaby
New Edlington
Loversall
Rossington
NORTH YORKSHIRE
SOUTH YORKSHIRE
A1(M)
A1
M62
M18
A19
A63
A162
A645
A628
A638
A639
A6201
A6021
A635
A630
A18
A6182
A6195
A6023
A633
206
207
197
186
187
miles
km

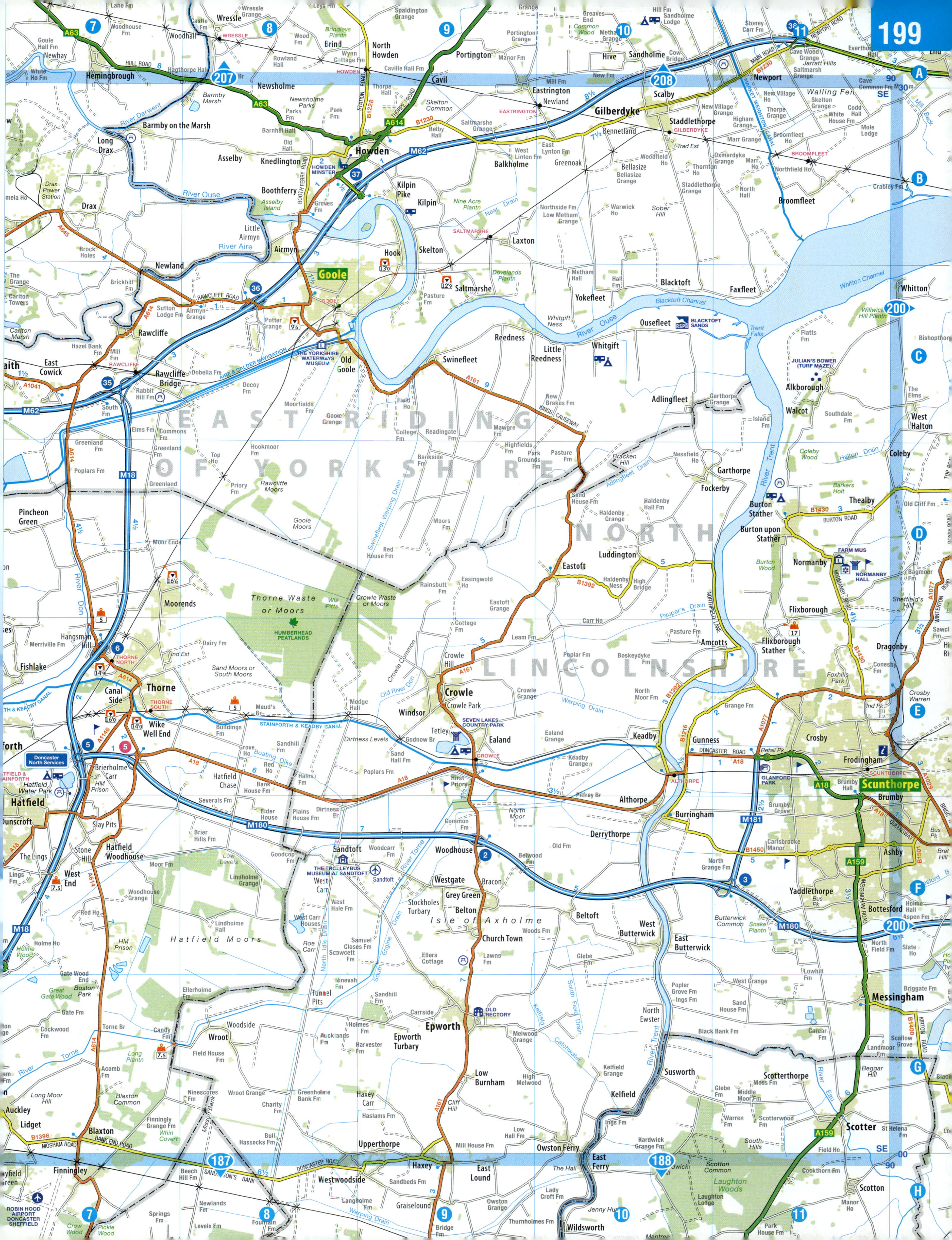

Wressle
Woodhall
Hemingbrough
Newsholme
North Howden
Portington
Eastrington
Newland
Gilberdyke
Scalby
Sandholme
Hive
Newport
Walling Fen
Broomfleet
Barmby on the Marsh
Long Drax
Drax
Asselby
Knedlington
Howden
Cavil
Balkholme
Greenoak
Bennetland
Staddlethorpe
Boothferry
Kilpin Pike
Kilpin
Little Airmyn
Airmyn
Hook
Skelton
Laxton
Newland
Goole
Saltmarshe
Yokefleet
Blacktoft
Faxfleet
Whitton
Rawcliffe
East Cowick
Rawcliffe Bridge
Old Goole
Swinefleet
Reedness
Little Reedness
Whitgift
Ousefleet
Adlingfleet
Alkborough
Walcot
West Halton
Coleby
Garthorpe
Fockerby
Burton Stather
Burton upon Stather
Thealby
Normanby
Luddington
Eastoft
Flixborough
Flixborough Stather
Amcotts
Dragonby
EAST RIDING OF YORKSHIRE
NORTH LINCOLNSHIRE
Thorne Waste or Moors
HUMBERHEAD PEATLANDS
Moorends
Pincheon Green
Fishlake
Thorne
Canal Side
Wike Well End
Crowle
Crowle Park
Windsor
Ealand
Keadby
Gunness
Crosby
Frodingham
Scunthorpe
Brumby
Ashby
Althorpe
Burringham
Derrythorpe
Hatfield
Hatfield Chase
Hatfield Woodhouse
Dunscroft
Stone Hill
West End
The Lings
Sandtoft
Woodhouse
Westgate
Grey Green
Belton
Isle of Axholme
Church Town
Beltoft
West Butterwick
East Butterwick
Yaddlethorpe
Bottesford
Hatfield Moors
Epworth
Epworth Turbary
Wroot
Messingham
North Ewster
Susworth
Scotterthorpe
Kelfield
Scotter
Low Burnham
Blaxton
Auckley
Lidget
Finningley
Upperthorpe
Haxey
East Lound
Owston Ferry
East Ferry
Westwoodside
Graiselound
Wildsworth
Scotton
Laughton Woods
River Ouse
River Trent
River Don
River Aire
River Derwent
Blacktoft Channel
Whitton Channel
M62
M18
M180
M181
A63
A614
A161
A18
A159
A1077
B1230
B1392
B1396
ROBIN HOOD AIRPORT DONCASTER SHEFFIELD
Doncaster North Services
THE YORKSHIRE WATERWAYS MUSEUM
THE TROLLEYBUS MUSEUM AT SANDTOFT
SEVEN LAKES COUNTRY PARK
OLD RECTORY
JULIAN'S BOWER (TURF MAZE)
BLACKTOFT SANDS
NORMANBY HALL
FARM MUS
207
208
200
187
188

KINGSTON UPON HULL
Kingston upon Hull
Humber Bridge
River Humber
NORTH LINCOLNSHIRE
LINCOLNSHIRE
Scunthorpe
South Cave
West End
Riplingham
Eppleworth
Inglemire
Newland
Sculcoates
Stoneferry
Sutton Ings
Summergangs
Marfleet
Willerby
Kirk Ella
West Ella
Anlaby
East Ella
Anlaby Park
Gipsyville
Pickering Park
Hessle
Northfield
Swanland
North Ferriby
Melton
Welton
Elloughton
Brough
Brantingham
Ellerker
Whitton
Whitton Ness
Winteringham
West Halton
Coleby
Thealby
Winterton
Roxby
Dragonby
Appleby
Santon
Frodingham
Brumby
Ashby
Bottesford
Holme
Messingham
Manton
Scotter
Kirton in Lindsey
Hibaldstow
Redbourne
Cadney
Howsham
North Kelsey
South Kelsey
Moortown
Nettleton
Caistor
Cabourne
Clixby
Grasby
Owmby
Searby
Somerby
Bigby
Brigg
Scawby
Sturton
Broughton
Wrawby
Elsham
Worlaby
Bonby
Saxby All Saints
Horkstow
South Ferriby
Barton-upon-Humber
Barton Waterside
Barrow upon Humber
Barrow Haven
New Holland
Goxhill
Goxhill Haven
Thornton Curtis
Wootton
Burnham
Ulceby
Ulceby Skitter
Croxton
Kirmington
Melton Ross
New Barnetby
Barnetby le Wold
Habrough
Great Limber
Little Limber
East Halton
North Killingholme
South End
Kingsforth
Humberside International
Thornton Abbey
Baysgarth House Museum
Elsham Hall
Normanby Hall
Gainsthorpe Deserted Medieval Village
A63
A164
A15
A1077
A18
A180
M180
A1084
A46
A159
A160
A165
A1033
199
188
189
208
209
0 1 2 3 miles
0 1 2 3 4 5 km

Neat Marsh
Foghall Fm
Preston Field
Preston
Somerdon Ho
Longsight Fm
Field Ho
Westfield Fm
West End
Primrose Hill Fm
The Limes
Ind Est
Salt End
Hedon
Haven Side
East End
Rands Fm
209
Daisy Hill
North Park
Elstronwick
Burton Pidsea
Chatt Ho
Manor House
Carr Fm
Prospect Hill
Elmtree Fm
Lelley Grange
Dairy Ho
Magdalen Ho
Old Hall Fm
Forkerleys Fm
Burstwick Drain
Stockholm Hill Fm
Burstwick
Burstwick Grange
Totleys Fm
Wadworth Hill
Eastholme Fm
Halsham Grange
Halsham
Halsham Old Hall
West End
Thorp's Fm
Hill Top Fm
Bunker's Hill
Fox Covert
The Bog
Roos Drain
Sunderland Fm
Glebe Fm
Tunstall
North End
Roos
Redhouse Fm
Renish
Waxholme
North Fm
Carr Fm
Poplar Fm
Hall Fm
Batty's Corner
Rimswell
Rimswell Valley
Rimswell Lodge
Owthorne
Withernsea
Little England Hill
East End
Great Newsome Fm
Frodingham Hall
Burgany Hall
Willow House Fm
Weldon's Plantn
Hollym
Intack Fm
Nevilles Fm
Toffling Fm
Mile Ho
Winestead
Eastfield Ho
Holmpton
Trinity Ho Fm
Cliff Fm
Out Newton
North Fm
Beacon Hill
Welwick
Haverfield Ho
Southfield Fm
Weeton
Gas Terminal
Easington
Skeffling
Humber Fm
Winsetts
South End
Firtholme Fm
Paull
Rose Hill
Hill Fm
Villa Fm
Thorn Hall
Thorngumbald
Camerton
Ryehill
Keyingham
Green Marsh Fm
Low Paull
Fort Paull
Auster Grange
Oxgoddes Fm
Thorn Marsh Fm
Marsh Cott
Berrygate Hill
Ottringham
Keyingham Drain
Hariff Lane Fm
Kelsey Ho
Tower Ho
North End Fm
Carr House Fm
North End Fm
Whinhill Fm
Churchlands Fm
Thorp's Plantn
EAST RIDING OF YORKSHIRE
Paull Holme Sands
Little Humber
Thorney Crofts
New House Fm
Marsh Ho
White House Fm
Keyingham Grange
Nearmarsh Fm
Coate's House Fm
Park Fm
White Hall
Enholmes Hall
Enholmes Fm
Patrington
Patrington Haven
Eastgrowths Fm
Ottringham Grange
Winestead Grange
Sands Ho
Salthaugh Grange
Far Marsh Fm
Long Plantn
Sands Fm
Cherry Cobb Sands
Foulholme Sands
ROTTERDAM 10:25
ZEEBRUGGE 10:25
North Channel
Ivy Ho
Shrubbery Fm
Winestead Fm
Patrington Fm
Winestead Drain
West Fm
Church Fm
Sunk Island
Stone Creek Fm
Crown Fm
Middle Fm
Fisherman's Channel
Channel Fm
Bleak House Fm
Outstray Fm
White House Fm
Wood Fm
South Fm
East Bank Fm
Stone Creek Ho
Sunk Island Sands
The Plumbs
Skeffling Clays
Easington Clays
Kilnsea
SPURN HEAD VISITOR CENTRE
Kilnsea Clays
Trinity Sands
Hawkin's Pt
Halton Marshes
North Killingholme Haven
Oil Terminal
Killingholme Marshes
South Killingholme Haven
Oil Refinery
Immingham Dock
River Humber
Houlton's Covert
Immingham
Luxmore Fm
Works
Ind Est
Trad Est
Power Sta
Immingham Grange
Mauxhall Fm
Roxton Fm
Eleanor Ho
Greenlands Fm
Stallingborough
Newstead Fm
Grimsby Roads
Works
Europarc
Pyewipe
Healing
Healing Covert
Great Coates
Grimsby
GRIMSBY DOCKS
Cleethorpes
Old Clee
Weelsby
Grant Thorold
East Marsh
West Marsh
The Willows
Yarborough
Little Coates
Littlefield
Wellow
Holme Hill
Mouth of the Humber
Spurn Head
Humber Pilot Station
Stony Binks
Bull Sand Fort
Keelby
Primrose Hill
Grange Fm
Wybers Wood
Pyewipe Fm
The Lindens
Riby Cross Roads
Riby
Riby Grange
Grove Wold Fm
Manor Fm
Limber Hill
Great Limber Grange
Aylesby
Laceby
Laceby Acres
Laceby Manor Fm
Nunsthorpe
Bradley
Scartho
Weelsby Woods
Hall Fm
Peaks Tunnel Fm
DISCOVERY CENTRE
PLEASURE ISLAND THEME PARK
Humberston Fitties
Humberston
New Waltham
NORTH EAST LINCOLNSHIRE
Washing Dales Fm
Riby Grove Fm
Irby Dales Wood
Hog Pit Hill
Manor Top Fm
Irby upon Humber
New Close Wood
Swallow Mount
Walk Fm
Irby Holmes Wood
Netherwood Fm
Grove Fm
New Fm
Waltham
WALTHAM WINDMILL
Barnoldby le Beck
Bedlam Hill
Manor Ho
Norman Corner
Brigsley Top
Brigsley
Waltham House Fm
Eastfield Fm
Holton le Clay
Highfield Ho
Bishopthorpe Fm
Low Fm
Tetney High Sands
Northcoates Pt
Tetney Haven
Stonebridge Fm
Horse Shoe Pt
Tetney Lock
North Coates Airfield
North End
Tetney
North Cotes
Tetney Blow Wells
Grange Fm
Low Fm
Silver Hill
The Grange
Swallow
Rookery Fm
Odessa Fm
Beelsby
Old Fm
Badger Hills
Cuxwold
Ash Holt
Cherry Garth Fm
Hatcliffe
Moor Ho
Ashby Hill
Hall Fm
Ashby cum Fenby
Bratton House Fm
Holton Lodge
Holton Grange
Waithe
Outholme Fm
Poplar Fm
The Fitties
Cherry Valley Fm
189
Lake Fm
Croxby Pond Plantn
Priory Fm
East Ravendale
Grainsby
Waithe House Fm
Waithe Top
190
Marshchapel
Keyholme
Sea Fm
Rothwell
L'Ings Fm
Croxby Top
Croxby
Thorganby Hall
Petterhills
Grainsby Grange
Hawerby Hall Fm
North Thoresby
Fulstow Top
Westfield
West End
Willow Tree Ho
Holmes Fm
Ivy Ho
Thoresway
Immingham
Grimsby

Heysham
Lower Heysham
Higher Heysham
Half Moon Bay
HEYSHAM PORT
Heysham Nuclear Power Sta
Heysham Lake
Middleton Sands
Middleton
Overton
Sunderland Bank
Sunderland
Sunderland Point
Shoulder of Lune
Plover Scar
Cockerham Sands
Cockerham Marsh
Pilling Sands
Preesall Sands
North Wharf
DOUGLAS 3:30
DOUGLAS 2:00 (TT RACE PERIOD ONLY)
LARNE 8:00
Lancaster
Heaton
Aldcliffe
Scotforth
Galgate
Glasson
Conder Green
Thurnham
Lower Thurnham
Upper Thurnham
Cockerham
Forton
Hollins Lane
Braides
Sand Side
Pilling
Stake Pool
Dam Side
Preesall
Knott End-on-Sea
Fleetwood
Stalmine
Hambleton
Staynall
Thornton
Cleveleys
Trunnah
Little Thornton
Skippool
Carleton
Poulton-Le-Fylde
Bispham
Norbreck
Anchorsholme
Warbreck
North Shore
Queenstown
Layton
Blackpool
BLACKPOOL
Great Marton
Marton Moss Side
Hawes Side
Common Edge
Squires Gate
South Shore
BLACKPOOL INTERNATIONAL
Little Marton
Mereside
Staining
Normoss
Newton
Weeton
Singleton
Little Singleton
Thistleton
Elswick
Great Eccleston
Little Eccleston
Out Rawcliffe
Eagland Hill
Winmarleigh
Garstang
Catterall
Bowgreave
Churchtown
St Michael's on Wyre
Nateby
Cabus
Scorton
Claughton
Calder Vale
Oakenclough
Dolphinholme
Quernmore
Ortner
Street
Bilsborrow
Myerscough
Barton
Inskip
Crossmoor
Roseacre
Wharles
Catforth
Woodplumpton
Broughton
Esprick
Medlar
Greenhalgh
Corner Row
Wesham
Kirkham
Treales
Westby
Wrea Green
Peel
Higher Ballam
Moss Side
Ribby
Newton
Clifton
Lea Town
Cottam
Ingol
Tanterton
Higher Bartle
Lower Bartle
Preston
Fulwood
Sharoe Green
Cadley
Freckleton
Warton
Lytham St Anne's
Ansdell
Fairhaven
Saltcotes
Penwortham
Lancaster Services
M55
M6
A583
A585
A586
A588
A6
210
211
193

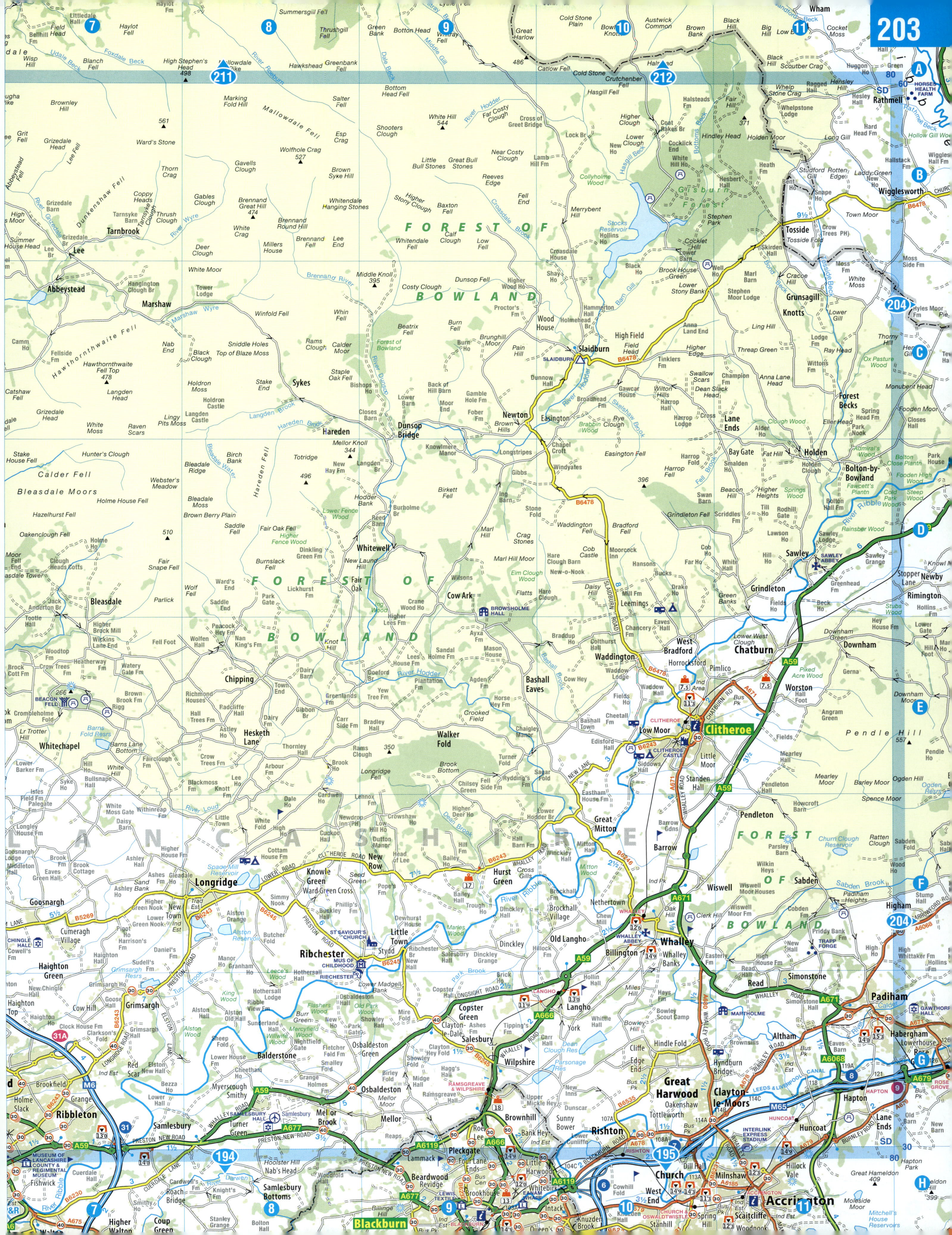
FOREST OF BOWLAND
FOREST OF BOWLAND
LANCASHIRE
FOREST OF BOWLAND
Slaidburn
Newton
Dunsop Bridge
Hareden
Sykes
Whitewell
Chipping
Hesketh Lane
Longridge
Ribchester
Clitheroe
Whalley
Waddington
West Bradford
Chatburn
Grindleton
Sawley
Bolton-by-Bowland
Holden
Tosside
Wigglesworth
Rathmell
Knotts
Grunsagill
Forest Becks
Lane Ends
Easington
Great Mitton
Barrow
Wiswell
Sabden
Higham
Padiham
Simonstone
Read
Billington
Old Langho
Langho
Wilpshire
Salesbury
Clayton-le-Dale
Copster Green
Osbaldeston
Mellor
Mellor Brook
Samlesbury
Ribbleton
Grimsargh
Goosnargh
Haighton Green
Whitechapel
Bleasdale
Abbeystead
Marshaw
Tarnbrook
Lee
Great Harwood
Clayton-le-Moors
Rishton
Church
Accrington
Huncoat
Altham
Blackburn
Pleckgate
Brownhill
Beardwood
Walker Fold
Bashall Eaves
Cow Ark
Low Moor
Pendleton
Pendle Hill
Worston
Downham
Little Town
Hurst Green
Knowle Green
Balderstone
Red Scar
Beacon Fell
Gisburn Forest
Stocks Reservoir
Calder Fell
Bleasdale Moors
Hawthornthwaite Fell
Mallowdale Fell
Cold Stone
Wham
Sawley Abbey
Whalley Abbey
Clitheroe Castle
Browsholme Hall
Stonyhurst
Hoghton Tower
Samlesbury Hall
Museum of Lancashire
Preston New Road
Whalley Road
Longridge Road
Slaidburn Road
M6
M65
A59
A671
A666
A680
A6119
A677
A678
B6478
B6243
B6245
B6246
B5269
B6535
211
212
194
195
204
7
8
9
10
11
A
B
C
D
E
F
G
H

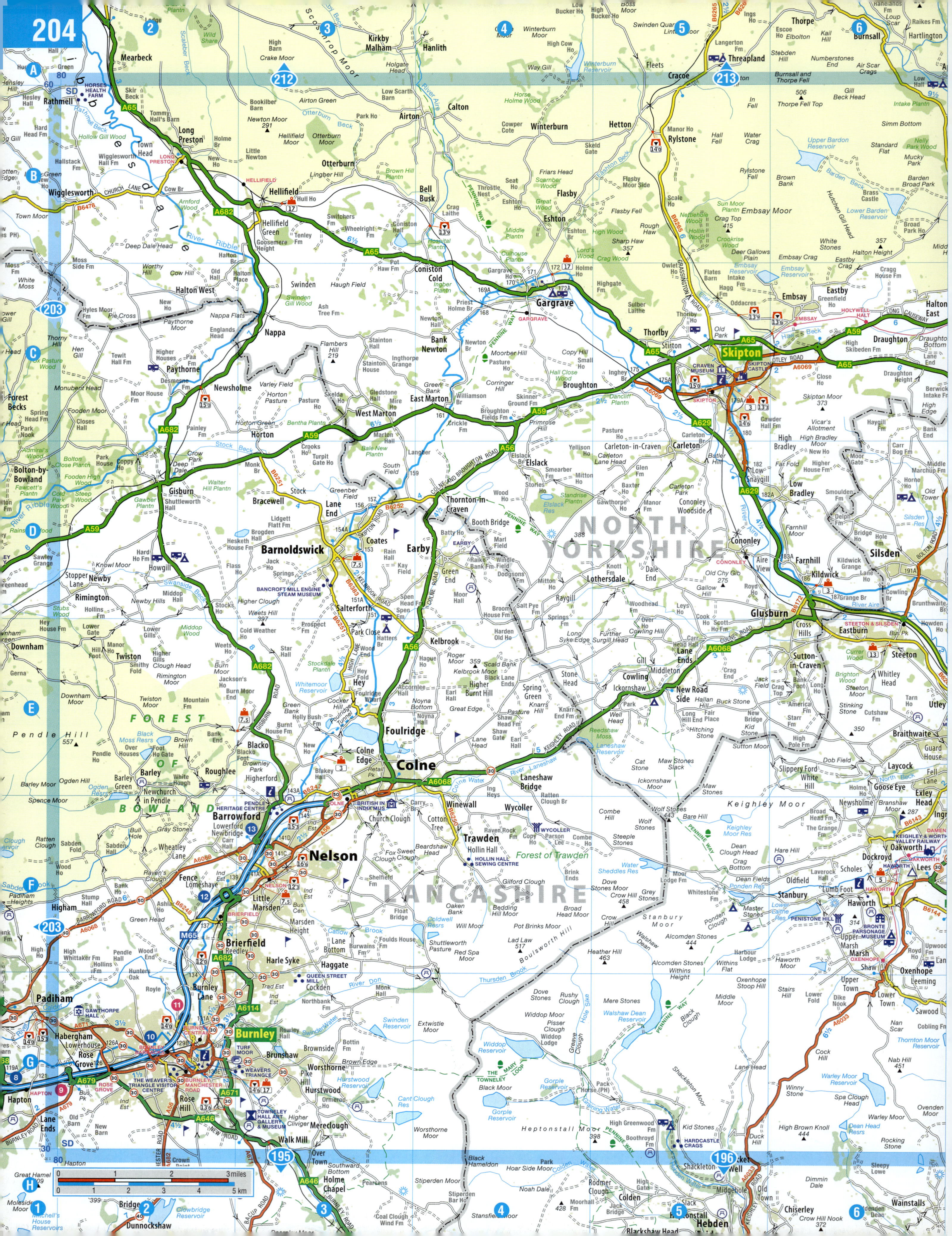
Mearbeck
Rathmell
Long Preston
Wigglesworth
Hellifield
Otterburn
Airton
Calton
Kirkby Malham
Hanlith
Winterburn
Hetton
Cracoe
Threapland
Rylstone
Burnsall
Thorpe
Hartlington
Flasby
Eshton
Bell Busk
Coniston Cold
Gargrave
Embsay
Embsay Moor
Eastby
Draughton
Halton East
Skipton
Thorlby
Stirton
Broughton
Carleton
Carleton-in-Craven
Elslack
Thornton-in-Craven
Earby
Cononley
Farnhill
Kildwick
Glusburn
Cross Hills
Eastburn
Steeton
Silsden
Sutton-in-Craven
Cowling
Lothersdale
Kelbrook
Salterforth
Barnoldswick
West Marton
East Marton
Gisburn
Bracewell
Nappa
Halton West
Swinden
Paythorne
Newsholme
Horton
Rimington
Downham
Twiston
Howgill
Stock
Coates
Foulridge
Colne
Laneshaw Bridge
Trawden
Winewall
Wycoller
Forest of Trawden
Barrowford
Nelson
Brierfield
Burnley
Padiham
Habergham
Rose Grove
Hapton
Higham
Fence
Barley
Roughlee
Blacko
Pendle Hill 557
FOREST OF BOWLAND
Newchurch in Pendle
Worsthorne
Hurstwood
Cliviger
Walk Mill
Haworth
Oxenhope
Oakworth
Stanbury
Oldfield
Laycock
Braithwaite
Keighley Moor
NORTH YORKSHIRE
LANCASHIRE
PENNINE WAY
Boulsworth Hill
Heptonstall Moor
Hebden
Widdop Reservoir
Gorple Reservoir
River Ribble
River Aire
A65
A59
A682
A56
A6068
A629
A646
A671
A679
M65
203
212
213
195
196
0 1 2 3 miles
0 1 2 3 4 5 km

NORTH YORKSHIRE
NIDDERDALE
WEST YORKSHIRE
Harrogate
Keighley
Leeds
Bradford
Skyreholme
Pock Stones Moor
Barden Fell
Rocking Moor
Barden Scale
Bolton Abbey
Hazlewood
Pace Gate
Blubberhouses
Fewston
Timble
Kettlesing
Kettlesing Head
Kettlesing Bottom
Darley
Dacre
Dacre Banks
Bridge
Hartwith
Birstwith
Clint
Hampsthwaite
Killinghall
Burnt Yates
Bedlam
Ripley
Menwith Hill
Beckwithshaw
Pannal
Oatlands
Rossett Green
Spacey Houses
North Rigton
Huby
Weeton
Dunkeswick
Stainburn
Braythorn
Lindley Green
Farnley
Clifton
Askwith
Newall
Otley
Pool
Arthington
Bramhope
Eccup
Adel
Addingham
Middleton
Ilkley
Burley in Wharfedale
Menston
Guiseley
Yeadon
Rawdon
Horsforth
Riddlesden
East Morton
Micklethwaite
Crossflatts
Bingley
Baildon
Saltaire
Shipley
Esholt
Idle
Calverley
Rodley
Bramley
Kirkstall
Headingley
Burley
Armley
Farsley
Stanningley
Pudsey
Thornbury
Undercliffe
Manningham
Heaton
Frizinghall
Cottingley
Wilsden
Harden
Cullingworth
Denholme
Allerton
Thornton
Clayton
Great Horton
Bowling
Laisterdyke
Tong
Farnley
Wortley
Holbeck
Beeston
Hunslet
Middleton
Queensbury
Illingworth
Shelf
Drighlington
Gildersome
Adwalton
Birkenshaw
Oakenshaw
Stone Chair
LEEDS BRADFORD INTERNATIONAL AIRPORT
A59
A65
A61
A658
A660
A629
A650
A6120
A647
A6177
A58
A6110
A643
M621
A64(M)

Brearton
Farnham
Ferrensby
Arkendale
Great Ouseburn
Little Ouseburn
Linton-on-Ouse
Killinghall
Scotton
Knaresborough
Coneythorpe
Flaxby
Allerton Mauleverer
Whixley
Thorpe Underwood
Newton on Ouse
Beningbrough
Nun Monkton
Bilton
Starbeck
Forest Lane Head
Goldsborough
Hopperton
Green Hammerton
Moor Monkton
Harrogate
Harlow Hill
High Harrogate
Woodlands
Calcutt
Little Ribston
Walshford
Hunsingore
Cattal
Kirk Hammerton
Marston Moor
Hessay
Oatlands
Rossett Green
Pannal
Follifoot
Spofforth
North Deighton
Kirk Deighton
Cowthorpe
Tockwith
Rufforth
Long Marston
Hutton Wandesley
Spacey Houses
Ingmanthorpe
Bickerton
Bilton in Ainsty
Kirkby Overblow
Sicklinghall
Wetherby
Walton
Angram
Askham Richard
Dunkeswick
Weeton
Netherby
Linton
Healaugh
Wighill
Bilbrough
Catterton
Collingham
Boston Spa
Thorp Arch
Harewood
Weardley
East Keswick
Clifford
Newton Kyme
Toulston
Colton
Bramham
Bardsey
East Rigton
Tadcaster
Oxton
Stutton
Bolton Percy
Scarcroft
Thorner
Kirkby Wharfe
Ulleskelf
Alwoodley
Moor Allerton
Shadwell
Towton
Chapel Allerton
Roundhay
Wellington Hill
Potterton
Sandhills
Barwick in Elmet
Scholes
Aberford
Saxton
Church Fenton
Leeds
Potternewton
Seacroft
Barkston
Sheepscar
Harehills
Manston
Parlington
Little Fenton
Whitkirk
Halton
Colton
Garforth
Micklefield
Sherburn in Elmet
Hunslet
Beeston
Swillington
Newthorpe
South Milford
Kippax
Lumby
Ledsham
Monk Fryston
Rothwell
Oulton
Middleton
Great Preston
Ledston
Fairburn
Hillam
Burton Salmon
WEST YORKSHIRE
A1(M)
A59
A61
A58
A64
A659
A168
A162
A63
M1
A6120
205
214
215
197
198

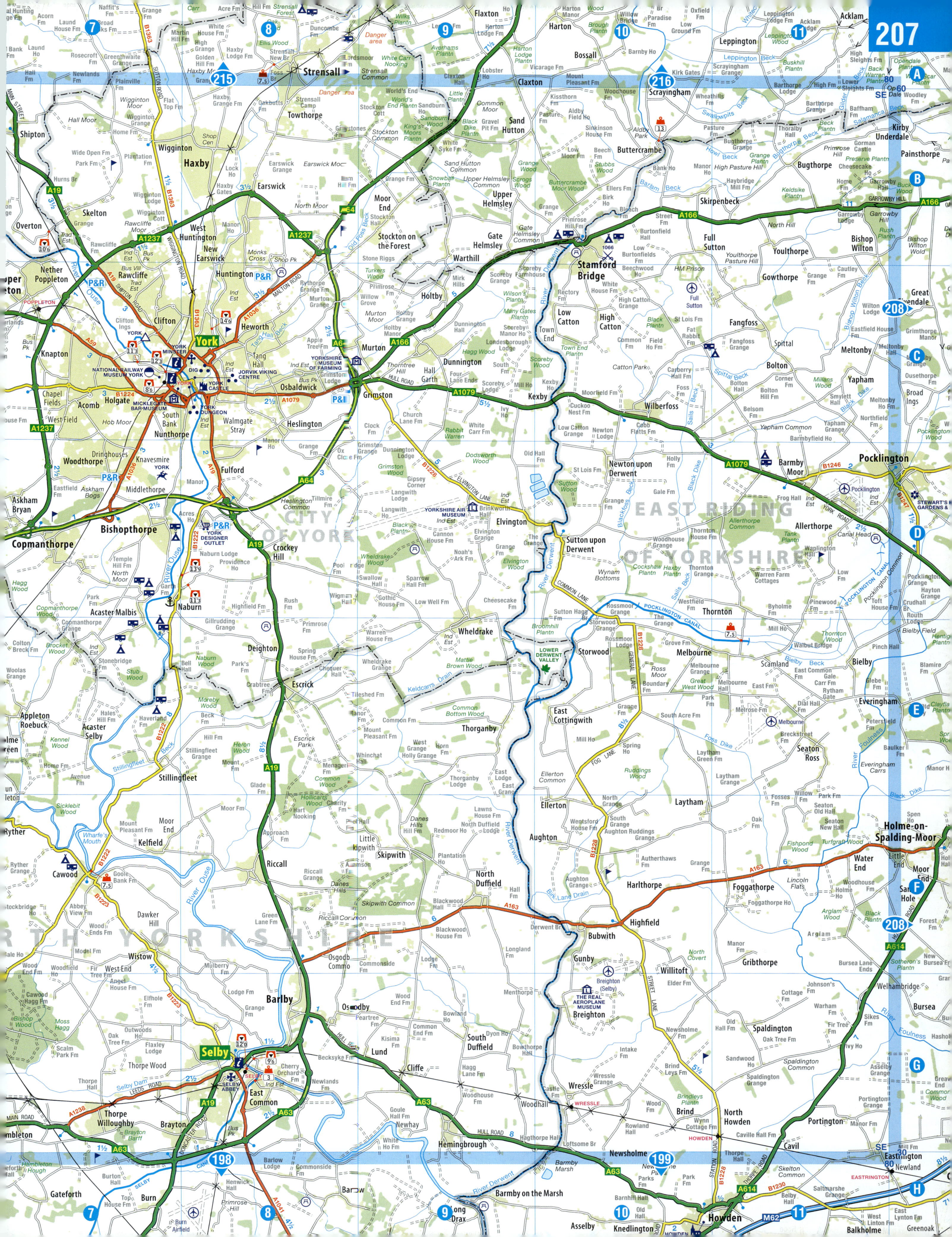

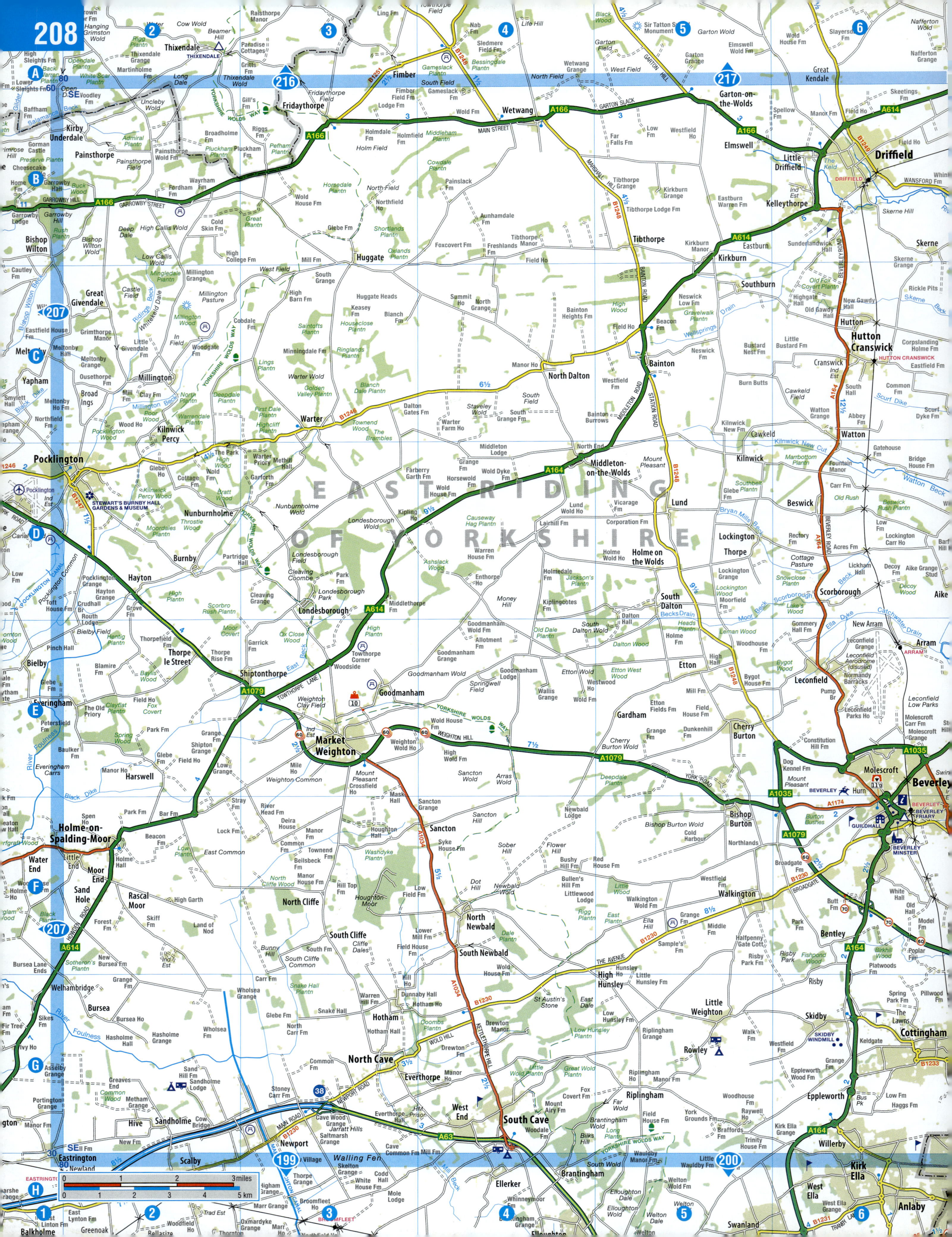
EAST RIDING OF YORKSHIRE
Thixendale
Fimber
Wetwang
Fridaythorpe
Garton-on-the-Wolds
Driffield
Great Kendale
Kirby Underdale
Painsthorpe
Elmswell
Little Driffield
Kelleythorpe
Bishop Wilton
Huggate
Tibthorpe
Kirkburn
Eastburn
Southburn
Skerne
Great Givendale
Hutton Cranswick
Hutton
Millington
North Dalton
Bainton
Yapham
Warter
Kilnwick Percy
Pocklington
Watton
Middleton-on-the-Wolds
Kilnwick
Beswick
Lund
Nunburnholme
Lockington
Burnby
Holme on the Wolds
Hayton
Londesborough
South Dalton
Scorborough
Arram
Bielby
Thorpe le Street
Shiptonthorpe
Goodmanham
Etton
Leconfield
Everingham
Market Weighton
Gardham
Cherry Burton
Molescroft
Beverley
Harswell
Holme-on-Spalding-Moor
Sancton
Bishop Burton
Water End
Moor End
Sand Hole
Rascal Moor
North Cliffe
North Newbald
South Newbald
Walkington
South Cliffe
Bentley
High Hunsley
Bursea
Little Weighton
Skidby
Hotham
Cottingham
Rowley
North Cave
Everthorpe
Riplingham
Eppleworth
West End
South Cave
Hive
Sandholme
Newport
Eastrington
Scalby
Willerby
Kirk Ella
Brantingham
Ellerker
West Ella
Anlaby
Swanland
Balkholme
Broomfleet
Yorkshire Wolds Way
Walling Fen
Stewart's Burnby Hall Gardens & Museum

Foldgate Fm
Corney Fell
Sele Bottom
Swallowhurst
Corney
Buck Barrow 549
Plough Fell
Littlecell Bottom
Prior Park
Ulpha Park
Penn
Frith Hall
Dunnerdale Fells
Scrithwaite Fm
Broughton Mills
Lind End
Hesketh Hall
Bracelet Hall
Troughton Hall
Rosthwaite
Woodland Grove
Yew Bank
Pool Scar
Stable Harvey Moss
Sunny Bank
Thrang Crag Wood
Brown Howe
Grass Paddocks
Top o Selside
Long Crags
Force Mills
Force Forge
Bethecar Moor
Birk Knott
Tarn Riggs
Beacon Tarn
Wool Knott
Greenholme Fm
Water Yeat
Woodland Fell
Green Moor
Woodland
Woodland Hall
Longfield Wood
Moss Side Fm
Lower Hawthwaite
Lane End
Black Moss
Fell End
Fell Gate
Far Houses
Kirkby Park Wood
Well Wood
Ashlack Hall
Heathwaite
Subberthwaite Common
Beck Stones
Great Burney
Blawith Knott
Houkler Hall
Birch Bank
Lin Crag Fm
Blawith
High Nibthwaite
Ickenthwaite
Whitestock Hall
Crosslands
Rusland
Rushland Cross
Rusland Moss
Furness Fells
Arklid Fm
Stock Wood
Nibthwaite Grange
Side Ho
Kirkthwaite Cottage
Low Hay Bridge
High Hay Bridge
Bandrake Head
Hill Park
Oxen Park
Burn Knott
Colton
Lowick Bridge
Sayles Fm
Bank Wood
Beckside
Bouth
Old Hall Fm
Abbots Reading Fm
Lowick
Lowick Hall
Esps Fm
Stone Dykes
Lowick Green
Lowick Common
Wood Gate
Spark Bridge
Tottlebank
Causeway End
Lane End
Haverthwaite
Artcrystal
Beck Bottom
Gawthwaite
Penny Bridge
Greenodd
Legbarrow Pt
Rowsdea Wood
Mearness Fm
Mearness Point
Nab Point
Arrad Foot
Newland Bottom
Plumpton Cottage Fm
Great Oath Hill
Ashes Pt
Hazelhurst Pt
Broughton Beck
Mansriggs
Mansriggs Hall
Bowstead Gates
Newland
Barrow Monument
Lakes Glass Centre
Leven Viaduct
Plumpton Bight
Old Park Wood
South Ulverston
Canal Foot
Hammerside Pt
Cartmel Sands
Sand Gate Marsh
Chapel Island
Lenibrick Point
Cowpren Point
Ulverston
Laurel & Hardy Mus
Stone Cross
Dragley Beck
Outcast
Sandside
Swarthmoor Hall
Swarthmoor
Croftlands
Conishead Priory
Bardsea
Wellhouse Fm
Ulverston Sands
Ulverston Channel
Hycemoor
Bootle
Seaton Hall
Low Kinmont
Cross House
Fellgreen
Fellside
Broomhill
Kiskin
Annaside
Barfield Tarn
Monk Foss Fm
Little Fell
Black Combe
Whitbeck
Gutterby Spa
Stoneside Hill 422
Mere Crags
Bootle Fell
Swinside Fell
Windy Slack
Stoupdale Crags
Gray Stones
Knott Hill
Baystone Bank Res
Fore Slack
Whicham Mill
Hallbeck Ho
Mire Ho
Black Crags
Hill End
Whicham Valley
Whicham Hall
Whicham
Silecroft
Kirksanton
Hestham Hall Fm
Millom
Millom Folk Museum
Haverigg Pool
HM Prison
Haverigg
Haverigg Point
Wortbrig Scar
Steel Green
Hodbarrow
Hodbarrow Point
Borwick Rails
Borwick Rails Harbour
Duddon Estuary
Beck Wood
Millom Park
Woods
Park Ho
The Hill
Low Ho
Millom Marsh
Arnaby Strands
Dunningwell
The Green
Hallthwaites
Brockwood Hall
Lady Hall
Hazel Mount
Duddon Bridge
Greystone Ho
Ash House
Bank End
Broughton in Furness
Eccle Riggs
Coal Gate
Wreaks End
Duddon Mosses
White Moss
Foxfield
Grizebeck
Beanthwaite
Dove Bank
Chapels
Kirkby Moor
Wall End
Angerton Marsh
Kirkby-in-Furness
Sand Side
Beck Side
Gunson Height
Out Park
Low Hall
Soutergate
Duddon Sands
Duddon Channel
Dunnerholme
Marsh Grange
Bank House Moor
Harlock Hill
Harlock Res
Shooting House Hill
Hollowmire
Osmotherley Moor
Higher Lath Fm
Copse Hill
Pennington Res
Ewe Dale
Stewner Park Fm
Soaka Beck Res
Ireleth
Askam in Furness
Far Old Park Fm
Marton
Loppergarth
Pennington
Rowe Head
Lindal Moor
Trinkeld Fm
Lindal in Furness
Flat Woods
Edge Hill
Greenscoe Fm
High Haume Fm
Mouzel Fm
Tytup Hall
South Lakes Wild Animal Park
Roanhead Fm
Sandscale Haws
Thwaite Flat
Dalton Castle
Dalton-in-Furness
Skells Lodge
Great Urswick
Urswick Tarn
Little Urswick
Bolton Heads
Holme Bank
Sunbrick
Sea Wood
Baycliff
Scales
Stainton with Adgarley
Mere Tarn
Wind Hill
Maskel Point
Gleaston Watermill
Gleaston
Colt Park
Aldingham
Dendron
Stank
New Close Fm
North Stank Fm
Newton
Bow Bridge
Furness Abbey
Crosslands Park
Billincoat Fm
Sinkfall
Ind Est
Hawcoat
Ormsgill
Newbarns
Beacon Hill
Barrow-in-Furness
Holbeck
Yarlside
Scarbarrow Hill
Leece
Newbiggin
Sea Mill
Roosecote
Moor Head Cottages
Page Bank Fm
Roosebeck
Peasholmes Fm
Roosebeck Sands
Rampside
Rampside Sands
Westfield Point
Roosecote Sands
Power Sta
Hindpool
Dock Mus
South Newbarns
Salthouse
Barrow Island
Bus Pk
North Walney
North Scale
Vickerstown
Water Garth Nook
Walney Channel
Tummer Hill Marsh
Biggar Sands
Biggar
Earnse Pt
Barrow (Walney Island)
North Walney
North End Marsh
Walney Meetings
Shope Tree Scar
Scarth Channel
Scarth Bight
Lowsy Point
Duddon Sands
Isle of Walney
Long Rein Point
Snab Sands
Wylock Marsh
Snab Pt
Sheep Island
Piel Island
Piel Castle
Piel Harbour
Piel Channel
Roa Island
Foulney Island
South End
Scar End Point
Haws Point
South East Point
Piel Bar
South Channel
Lancaster Sound
MC
A595
A5093
A593
A5092
A590
B5281
A5084
A5087
A5278
220
202
SD
90
60
10
0 1 2 3 miles
0 1 2 3 4 5 km
DOUGLAS 3:30
DOUGLAS 2:00 (TT RACE PERIOD ONLY)
A B C D E F G H
1 2 3 4 5 6

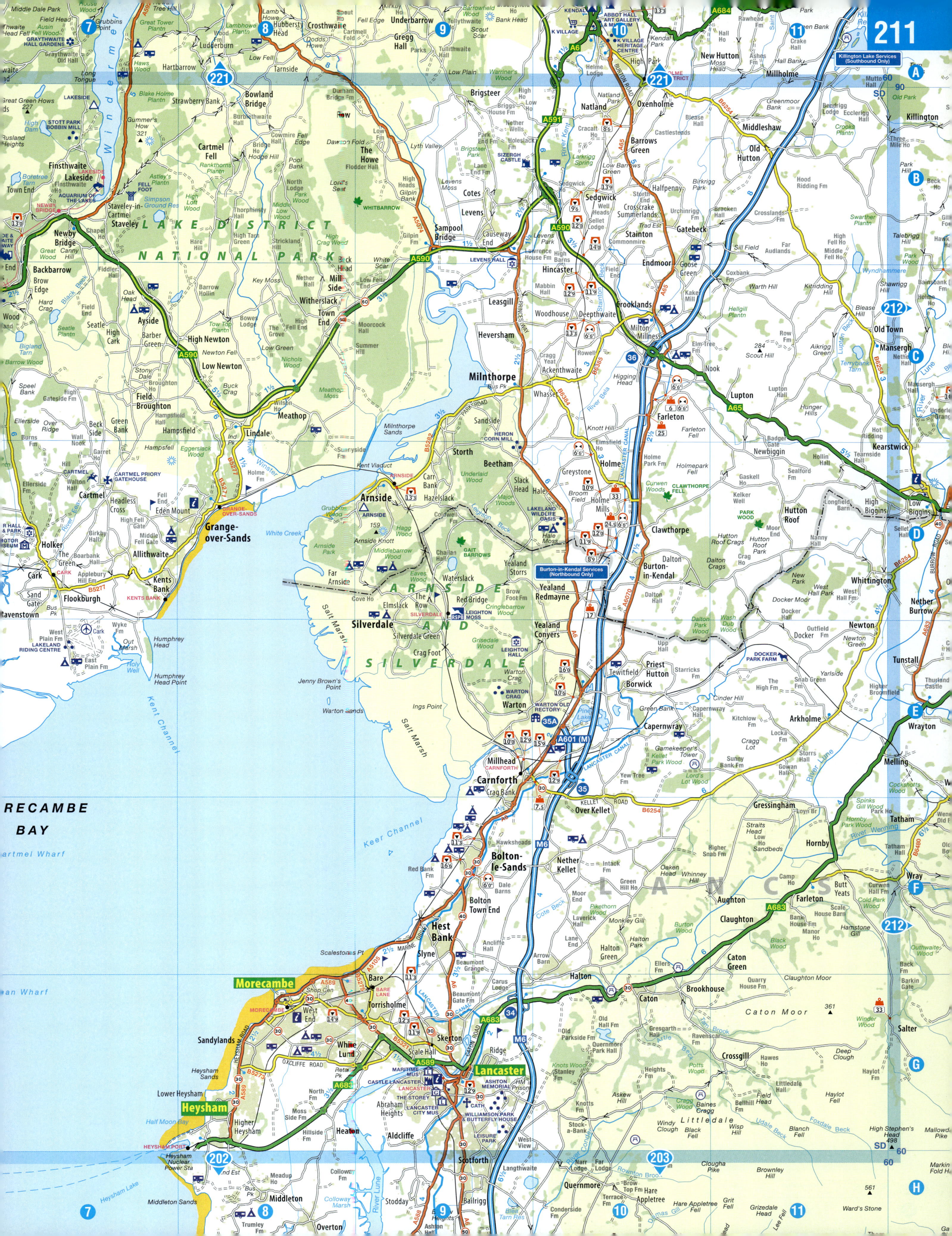

Killington Lake Services (Southbound Only)
Burton-in-Kendal Services (Northbound Only)
LAKE DISTRICT NATIONAL PARK
ARNSIDE AND SILVERDALE
MORECAMBE BAY
LANCS
Windermere
Kent Channel
Keer Channel
Salt Marsh
Newby Bridge
Lakeside
Finsthwaite
Staveley-in-Cartmel
Backbarrow
Cartmel
Cartmel Fell
Bowland Bridge
Crosthwaite
Underbarrow
Gregg Hall
Brigsteer
Natland
Oxenholme
Barrows Green
Sedgwick
Levens
Cotes
Sampool Bridge
Causeway End
Hincaster
Stainton
Endmoor
Gatebeck
Old Hutton
Middleshaw
Killington
New Hutton
Millholme
Leasgill
Heversham
Woodhouse
Deepthwaite
Crooklands
Milnthorpe
Ackenthwaite
Whasset
Farleton
Lupton
Old Town
Mansergh
Kearstwick
Sandside
Storth
Beetham
Holme
Clawthorpe
Hutton Roof
Burton-in-Kendal
Whittington
Nether Burrow
Newton
Tunstall
Arnside
Silverdale
Far Arnside
Yealand Redmayne
Yealand Conyers
Priest Hutton
Borwick
Warton
Capernwray
Arkholme
Melling
Wrayton
Carnforth
Millhead
Over Kellet
Nether Kellet
Bolton-le-Sands
Bolton Town End
Hest Bank
Slyne
Gressingham
Hornby
Tatham
Wray
Farleton
Aughton
Claughton
Caton Green
Caton
Brookhouse
Halton
Salter
Crossgill
Littledale
Caton Moor
Morecambe
Torrisholme
Bare
Skerton
Lancaster
Scotforth
Heysham
Higher Heysham
Lower Heysham
Sandylands
Heaton
Aldcliffe
Middleton
Overton
Quernmore
Grange-over-Sands
Kents Bank
Allithwaite
Flookburgh
Cark
Holker
Lindale
Meathop
Field Broughton
High Newton
Low Newton
Ayside
Witherslack
Town End
Hampsfield
Humphrey Head
Ravenstown
A590
A591
A6
A65
A683
A589
A601 (M)
M6
B6254
221
212
202
203

Killington Lake Services (Southbound Only)
Sedbergh
Millthrop
Garsdale
Garsdale Head
Killington
Gawthrop
Dent
Cowgill
Stone House
Dentdale
Middleton
C U M B R I A
Rise Hill
Deepdale
Whernside
Old Town
Mansergh
Barbon
Casterton
Kirkby Lonsdale
Whittington
Cowan Bridge
Leck
Ireby
Masongill
Westhouse
Thornton in Lonsdale
Ingleton
Chapel-le-Dale
Ribblehead
Ribblehead Viaduct
Y O R K S H I
N A T I O N
Horton in Ribblesdale
Brackenbottom
Helwith Bridge
Stainforth
Langcliffe
Settle
Giggleswick
Austwick
Lawkland
Eldroth
Wharfe
Feizor
Clapham
Newby
Keasden
Mewith Head
High Bentham
Low Bentham
Burton in Lonsdale
Cantsfield
Tunstall
Wrayton
Melling
Wennington
Tatham
Wray
Lowgill
Salter
FOREST OF BOWLAND
Tatham Fells
Burn Moor
Goodber Common
Wham
Rathmell
Mearbeck
Long Preston
Studfold
Selside
Ingleborough
Horton Moor
Dodd Fell
Widdale Fell
Kingsdale
Yordas Cave
White Scar Caves
Gaping Gill
Ingleborough Show Cave
Thornton Force
Pecca Falls
Snow Falls
Stainforth Force
Watershed Mill Visitor Centre
The Yorkshire Dales Falconry & Conservation Centre
Horses Health Farm
Fairfield Mill Arts & Heritage Cent
A683
A684
A65
A687
B6480
B6255
B6254
B6479
222
211
203
204
0 1 2 3 miles
0 1 2 3 4 5 km
SD

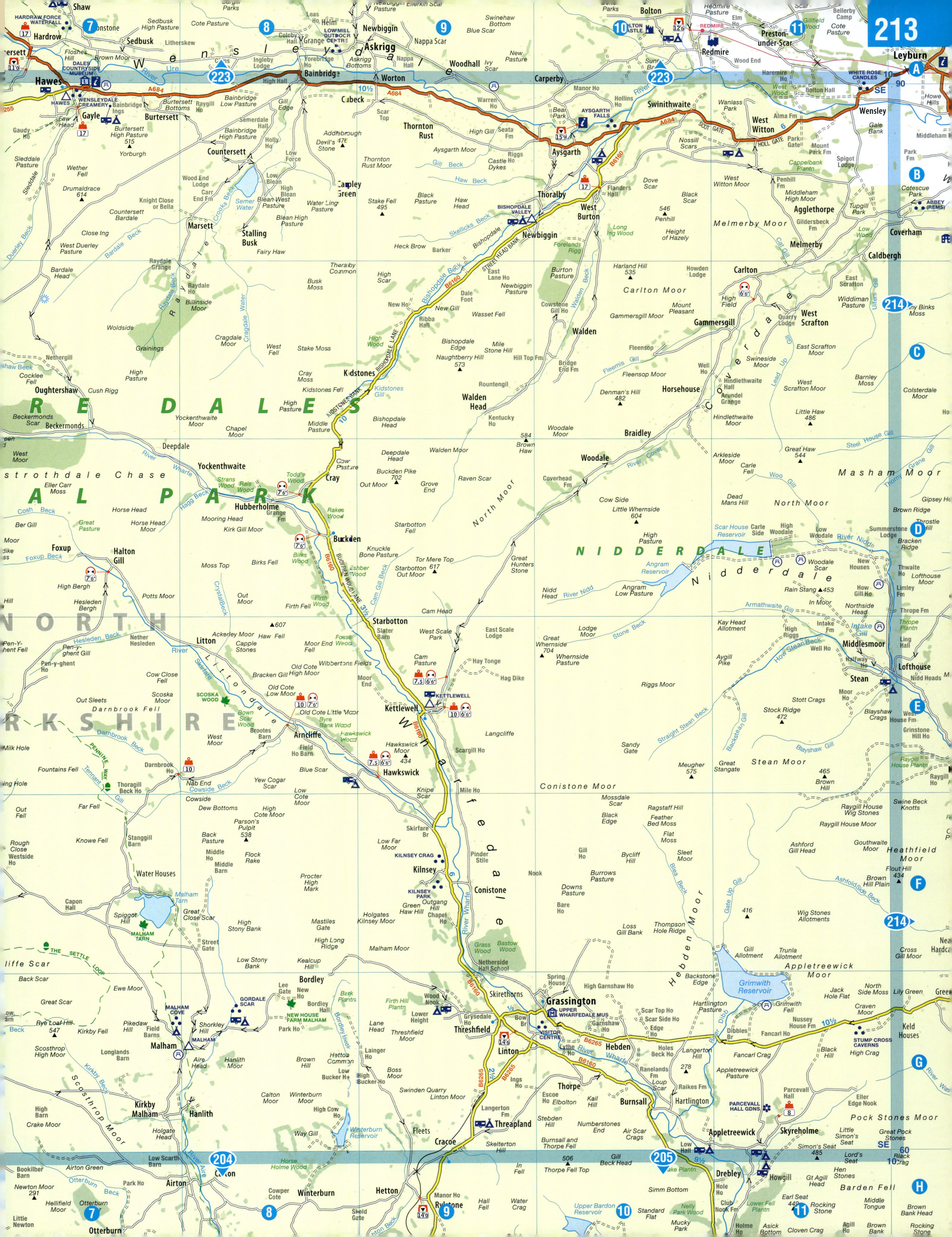

Leyburn
Wensley
Harmby
Spennithorne
Middleham
Constable Burton
Finghall
Patrick Brompton
Newton-le-Willows
Little Crakehall
Great Crakehall
Kirkbridge
Langthorne
Leeming Bar
Leeming
Aiskew
Bedale
Exelby
Londonderry
Gatenby
Morton-on-Swale
Warlaby
Maunby
Coverham
Caldbergh
East Witton
Thornton Steward
Jervaulx Abbey
Thirn
Cowling
Burrill
Firby
Thornton Watlass
Snape
Theakston
Burneston
Carthorpe
Ellingstring
High Ellington
Low Ellington
Well
Sinderby
Colsterdale
Fearby
Healey
Masham
Low Burton
Kirklington
Nosterfield
Binsoe
West Tanfield
Thornborough
Sutton Howgrave
Middleton Quernhow
Leighton
Warthermarske
Swinton
Ilton
Masham Moor
Mickley
North Stainley
Wath
Melmerby
Grewelthorpe
Kirby Malzeard Moor
Middlesmoor
Lofthouse
Stean
Carlesmoor
Kirkby Malzeard
Laverton
Azerley
Nunwick
Hutton Conyers
North Lees
Swetton
Galphay
Greygarth
Dallow
Winksley
Ure Bank
Sharow
Copt Hewick
Bouthwaite
Ramsgill
Bishopton
Ripon
Studley Roger
Studley Royal
Bridge Hewick
Dallow Moor
Low Grantley
High Grantley
Grantley
Aldfield
Littlethorpe
Heathfield Moor
Eavestone
Risplith
Sawley
Wath
Pateley Moor
Nidderdale
Pateley Bridge
Bewerley
Markington
Bishop Monkton
Wilsill
Glasshouses
Smelthouses
Wormald Green
Burton Leonard
Greenhow
Greenhow Hill
Low Laithe
South Stainley
Copgrove
Keld Houses
New York
Summer Bridge
Dacre Banks
Bishop Thornton
Shaw Mills
Heyshaw Moor
Burnt Yates
Bedlam
Pock Stones Moor
Dacre
Ripley
Nidd
Brearton
Darley
Birstwith
Clint
Killinghall
Hampsthwaite
Padside
Thornthwaite
Darley Head
Scotton
Scriven
Thruscross
NORTH YORKSHIRE
0 1 2 3 miles
0 1 2 3 4 5 km

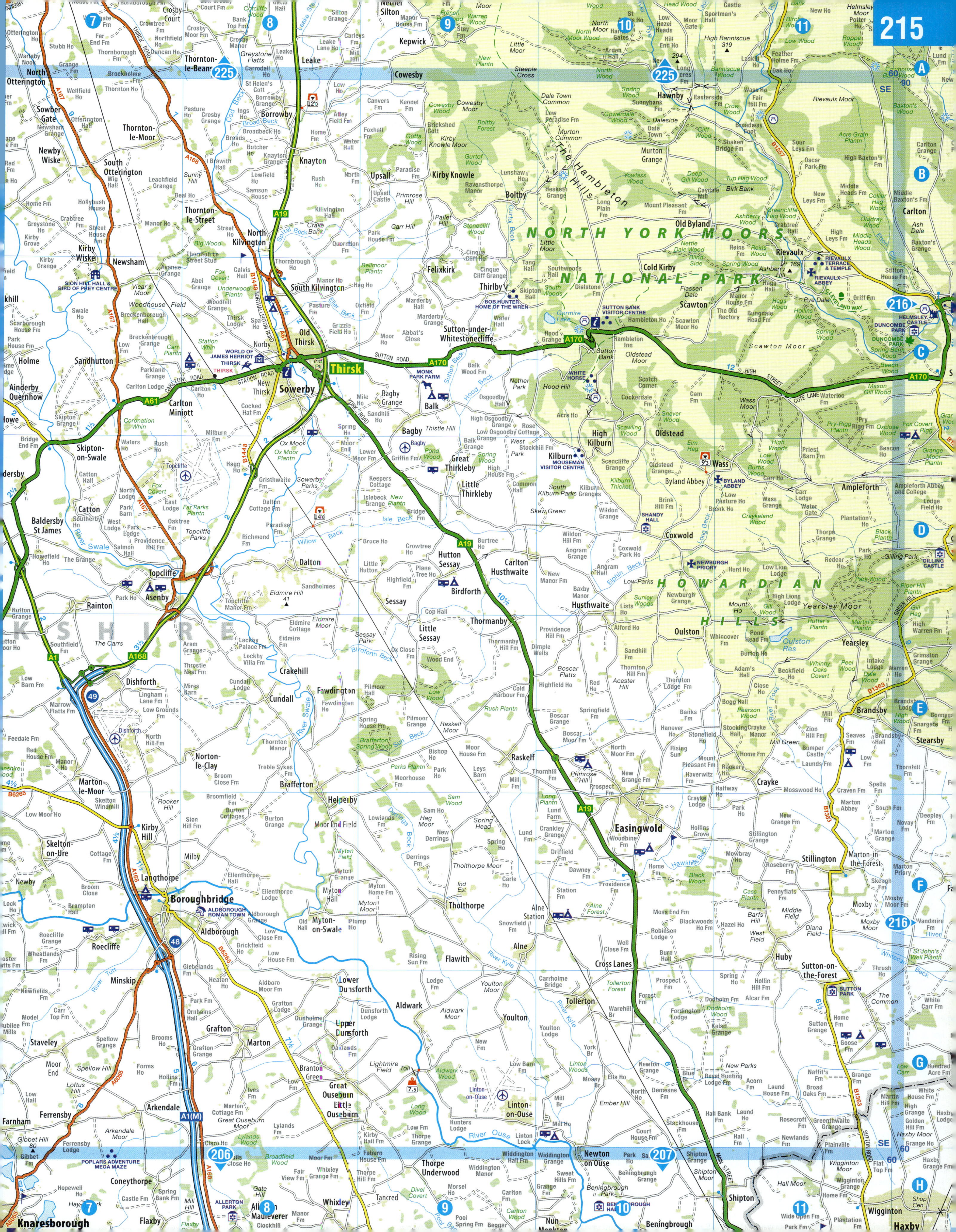
NORTH YORK MOORS
NATIONAL PARK
HOWARDIAN
HILLS
The Hambleton Hills
Thirsk
Sowerby
Old Thirsk
South Kilvington
North Kilvington
Thornton-le-Street
Thornton-le-Moor
Thornton-le-Beans
North Otterington
South Otterington
Newby Wiske
Sowber Gate
Kirby Wiske
Newsham
Sandhutton
Carlton Miniott
Skipton-on-Swale
Baldersby St James
Catton
Topcliffe
Asenby
Rainton
Dishforth
Marton-le-Moor
Kirby Hill
Skelton-on-Ure
Langthorpe
Boroughbridge
Aldborough
Roecliffe
Minskip
Staveley
Ferrensby
Arkendale
Farnham
Knaresborough
Coneythorpe
Allerton Mauleverer
Whixley
Great Ouseburn
Little Ouseburn
Upper Dunsforth
Lower Dunsforth
Aldwark
Grafton
Marton
Helperby
Brafferton
Myton-on-Swale
Cundall
Norton-le-Clay
Crakehill
Fawdington
Dalton
Sessay
Little Sessay
Hutton Sessay
Birdforth
Thormanby
Carlton Husthwaite
Husthwaite
Raskelf
Easingwold
Tholthorpe
Flawith
Alne
Alne Station
Tollerton
Youlton
Linton-on-Ouse
Newton on Ouse
Thorpe Underwood
Shipton
Beningbrough
Cross Lanes
Huby
Sutton-on-the-Forest
Stillington
Marton-in-the-Forest
Crayke
Brandsby
Stearsby
Yearsley
Oulston
Coxwold
Byland Abbey
Wass
Ampleforth
Oldstead
High Kilburn
Kilburn
Great Thirkleby
Little Thirkleby
Bagby
Balk
Felixkirk
Thirlby
Sutton-under-Whitestonecliffe
Kirby Knowle
Boltby
Upsall
Knayton
Borrowby
Leake
Kepwick
Cowesby
Hawnby
Murton Grange
Old Byland
Cold Kirby
Scawton
Rievaulx
Carlton
Wigginton
Haxby
Moxby
Skewsby
Bagby
Topcliffe
A19
A170
A168
A61
A1
A1(M)
A167
B1448
B6265
B1363
SUTTON BANK VISITOR CENTRE
WORLD OF JAMES HERRIOT
MONK PARK FARM
MOUSEMAN VISITOR CENTRE
BOB HUNTER HOME OF THE WREN
SION HILL HALL & BIRD OF PREY CENTRE
ALDBOROUGH ROMAN TOWN
POPLARS ADVENTURE MEGA MAZE
NEWBURGH PRIORY
SHANDY HALL
BYLAND ABBEY
RIEVAULX ABBEY
RIEVAULX TERRACE & TEMPLE
DUNCOMBE PARK
HELMSLEY CASTLE
GILLING CASTLE
SUTTON PARK
BENINGBROUGH HALL
ALLERTON PARK
WHITE HORSE
River Swale
River Ouse
River Kyle
Cod Beck
YORKSHIRE
225
206
207
216
7
8
9
10
11
A
B
C
D
E
F
G
H

Lastingham
Spaunton
Hutton-le-Hole
RYEDALE FOLK MUSEUM
Gillamoor
Fadmoor
Appleton-le-Moors
Kirkbymoorside
Keldholme
Kirkby Mills
Cropton
Wrelton
Sinnington
Aislaby
Middleton
Pickering
PICKERING CASTLE
NORTH YORKSHIRE MOORS RAILWAY
Newton-on-Rawcliffe
Levisham
Lockton
Ellerburn
Thornton-le-Dale
Newbridge
Helmsley
HELMSLEY CASTLE
DUNCOMBE PARK
Carlton
Pockley
Beadlam
Nawton
Wombleton
Welburn
Great Edstone
Marton
Normanby
Harome
Sproxton
Salton
Muscoates
Nunnington
NUNNINGTON HALL
West Ness
East Ness
South Holme
Little Barugh
Great Barugh
Kirby Misperton
FLAMINGO LAND THEME PARK
Brawby
Little Habton
Great Habton
Butterwick
Ryton
Wykeham
High Marishes
Low Marishes
Oswaldkirk
Ampleforth
Ampleforth Abbey and College
Gilling East
Stonegrave
Cawton
Hovingham
HOVINGHAM HALL
Wath
Fryton
Slingsby
Barton-le-Street
Appleton-le-Street
Amotherby
Swinton
Broughton
Malton
Old Malton
Norton-on-Derwent
EDEN CAMP MODERN HISTORY THEME MUSEUM
Rillington
Scagglethorpe
Settrington
Yearsley
Coulton
Scackleton
Howardian Hills
HOWARDIAN HILLS
Coneysthorpe
CASTLE HOWARD
Terrington
Ganthorpe
Dalby
Skewsby
Stearsby
Whenby
Brandsby
Welburn
Bulmer
Whitwell-on-the-Hill
Crambeck
Firby
Kirkham
KIRKHAM PRIORY
Westow
Crambe
Howsham
High Hutton
Low Hutton
Huttons Ambo
Menethorpe
Langton
Kennythorpe
Birdsall
Burythorpe
Leavening
Acklam
Birdsall Brow
Thixendale
Farlington
Sheriff Hutton
SHERIFF HUTTON
West Lilling
Thornton-le-Clay
Foston
Barton-le-Willows
Flaxton
Harton
Bossall
Leppington
Scrayingham
Strensall
Claxton
Sand Hutton
Stamford Bridge
Kirby Underdale
Painsthorpe
Bugthorpe
Haxby
Moxby
Marton-in-the-Forest
A170
A64
A169
B1257
B1363
B1248
226
215
207
208
0 1 2 3 miles
0 1 2 3 4 5 km

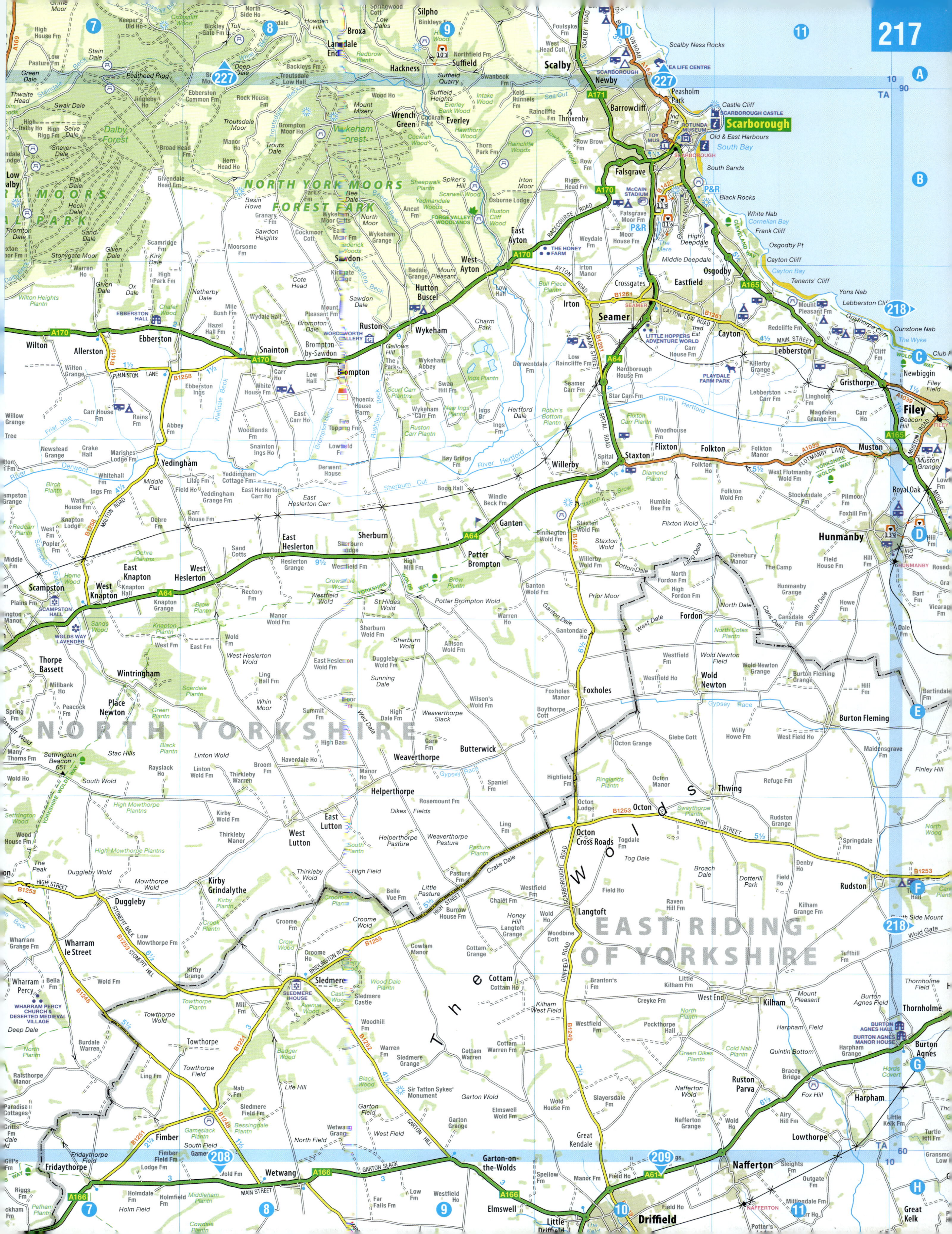

NORTH YORK MOORS FOREST PARK
NORTH YORKSHIRE
EAST RIDING OF YORKSHIRE
The Wolds
Scarborough
Scarborough Castle
Castle Cliff
Rotunda Museum
Old & East Harbours
South Bay
South Sands
Scalby Ness Rocks
Sea Life Centre
Scalby
Newby
Peasholm Park
Barrowcliff
Throxenby
Falsgrave
Seamer
Crossgates
Eastfield
Osgodby
Cayton
Cayton Bay
Black Rocks
White Nab
Cornelian Bay
Frank Cliff
Osgodby Pt
Cayton Cliff
Tenants' Cliff
Yons Nab
Lebberston Cliff
Lebberston
Gristhorpe
Cunstone Nab
The Wyke
Newbiggin
Filey
Muston
Royal Oak
Hunmanby
Hackness
Broxa
Langdale End
Silpho
Suffield
Everley
Wrench Green
Dalby Forest
East Ayton
West Ayton
Irton
Hutton Buscel
Wykeham
Ruston
Sawdon
Brompton-by-Sawdon
Brompton
Snainton
Ebberston
Allerston
Wilton
Yedingham
Staxton
Flixton
Folkton
Willerby
Ganton
Potter Brompton
Sherburn
East Heslerton
West Heslerton
East Knapton
West Knapton
Scampston
Thorpe Bassett
Wintringham
Place Newton
Foxholes
Fordon
Wold Newton
Burton Fleming
Thwing
Octon
Octon Cross Roads
Butterwick
Weaverthorpe
Helperthorpe
East Lutton
West Lutton
Kirby Grindalythe
Duggleby
Wharram le Street
Wharram Percy
Sledmere
Sledmere House
Cottam
Langtoft
Kilham
Rudston
Thornholme
Burton Agnes
Harpham
Ruston Parva
Lowthorpe
Nafferton
Driffield
Little Driffield
Elmswell
Garton-on-the-Wolds
Wetwang
Fimber
Fridaythorpe
Great Kendale
Great Kelk
Sir Tatton Sykes' Monument
Settrington Beacon 651
Yorkshire Wolds Way
Cleveland Way
River Derwent
River Hertford
A170
A171
A165
A64
A1039
A166
A614
B1249
B1253
B1251
B1252
B1248
B1258
B1415
B1261
218
227
208
209

NORTH SEA
FILEY BAY
BRIDLINGTON BAY
EAST RIDING OF YORKSHIRE
217
209
Yons Nab
Lebberston Cliff
Gristhorpe Cliff
Cunstone Nab
The Wyke
Club Pt
North Cliff
Filey Brigg
Brigg End
Filey Sands
Gristhorpe
Newbiggin
Filey Field
Filey
Beacon Hill
Muston
Muston Grange
Muston Sands
Lowfield Fm
Royal Oak
Hunmanby Sands
Primrose Valley
Hunmanby Gap
Hunmanby Moor
Airy Hill Fm
Moor Fm
Reighton Sands
Moor Ho
Reighton Gap
Speeton Sands
Rosedale Fm
Graffitoe Fm
Barf Fm
Vicarage Fm
Reighton
Speeton Hills
Speeton Cliffs
Speeton
Speeton Grange
Speeton Moor
Buckton Cliffs
Buckton Hall
Bempton Cliffs
Scale Nab
Cat Nab
Gull Nook
Standard Hill
Bempton Grange
Wandale Fm
Buckton
Bempton
Dykes Plantn
Dane's Dyke
Thornwick Bay
North Cliff
North Landing
North Moor
Cradle Head
Stottle Bank Nook
Selwicks Bay
Flamborough Head
Flamborough Head Lighthouse
High Stacks
Flamborough
Old Fall Plantn
Flatmere Plantn
Highcliffe Manor
Beacon Fm
Beacon Hill
South Landing
The Crofts
Lynhams
Butterwicks Fm
Newsham Field
High Barn
Field Ho
Sewerby
Sewerby Hall & Garden
Sewerby Rocks
Bondsville Model Village
North Sands
Bridlington
Old Town
Priory
Bayle Mus
The Spa
Old Penny Memories
Hilderthorpe
West Hill
Bessingby
South Sands
Wilsthorpe
Carnaby Moor
Park Rose Bird of Prey Centre
Brackendale Fm
Auburn Fm
Fraisthorpe Sands
Fraisthorpe
Demming Fm
Low Stonehills
High Stonehills
Hamiltonhill Fm
Barmston Sands
Barmston
Barmston Main Drain
Allison Lane End
Ulrome Sands
Lissett
Great Kelk
Park Ho
Gransmoor
Gransmoor Low Ho
Burtoncarr Ho
Woodside Fm
Turtle Hill Fm
Little Kelk Fm
Thornholme Moor
Gransmoor Wood
Harpham
Harpham Grange
Hords Covert
Burton Agnes
Burton Agnes Hall
Burton Agnes Manor House
Burton Agnes Stud Fm
Oak Wood Fm
Thornholme
Burton Agnes Field
Thornholme Field
Haisthorpe
Haisthorpe Field
Carnaby
Carnaby Temple
Hallowkiln Wood
Temple Fm
High Wood
Tufthill Fm
Wold Gate
South Side Mount
Thorpe Hall
Carr Plantn
Sands Wood
West Lawn Wood
Fish Ponds Wood
Wandale Fm
Boynton
Binsdale Fm
North Wood
East Crags Wood
Eastfield Fm
High Easton Fm
Springdale Fm
Rudston
Grindale
Grindale Field
Fox Covert Plantn
Charlestone Fm
Finley Hill
Maidensgrave Fm
Burton Fleming
Hill Fm
Bartindale Fm
Wasters Plantn
North Dale
East Leys Fm
North Mount
Field Greenlands Fm
High Huntow Fm
Reighton Field
Dale Fm
Howe Fm
Pilmoor Fm
Foxhill Fm
Carr Ho
Cliff Fm
Hunmanby
Ind Est
Hill Fm
A165
A1039
A614
B1253
B1229
B1255
B1259
MUSTON ROAD
MOOR ROAD
HUNMANBY ROAD
SPEETON GATE
SCARBOROUGH ROAD
FLAMBOROUGH ROAD
BEMPTON LANE
NORTH MARINE ROAD
LIGHTHOUSE ROAD
MARTON ROAD
MARTON GATE
KINGSGATE
BRIDLINGTON ROAD
WOLDS WAY
TA
10
30
60
90
0 1 2 3 miles
0 1 2 3 4 5 km

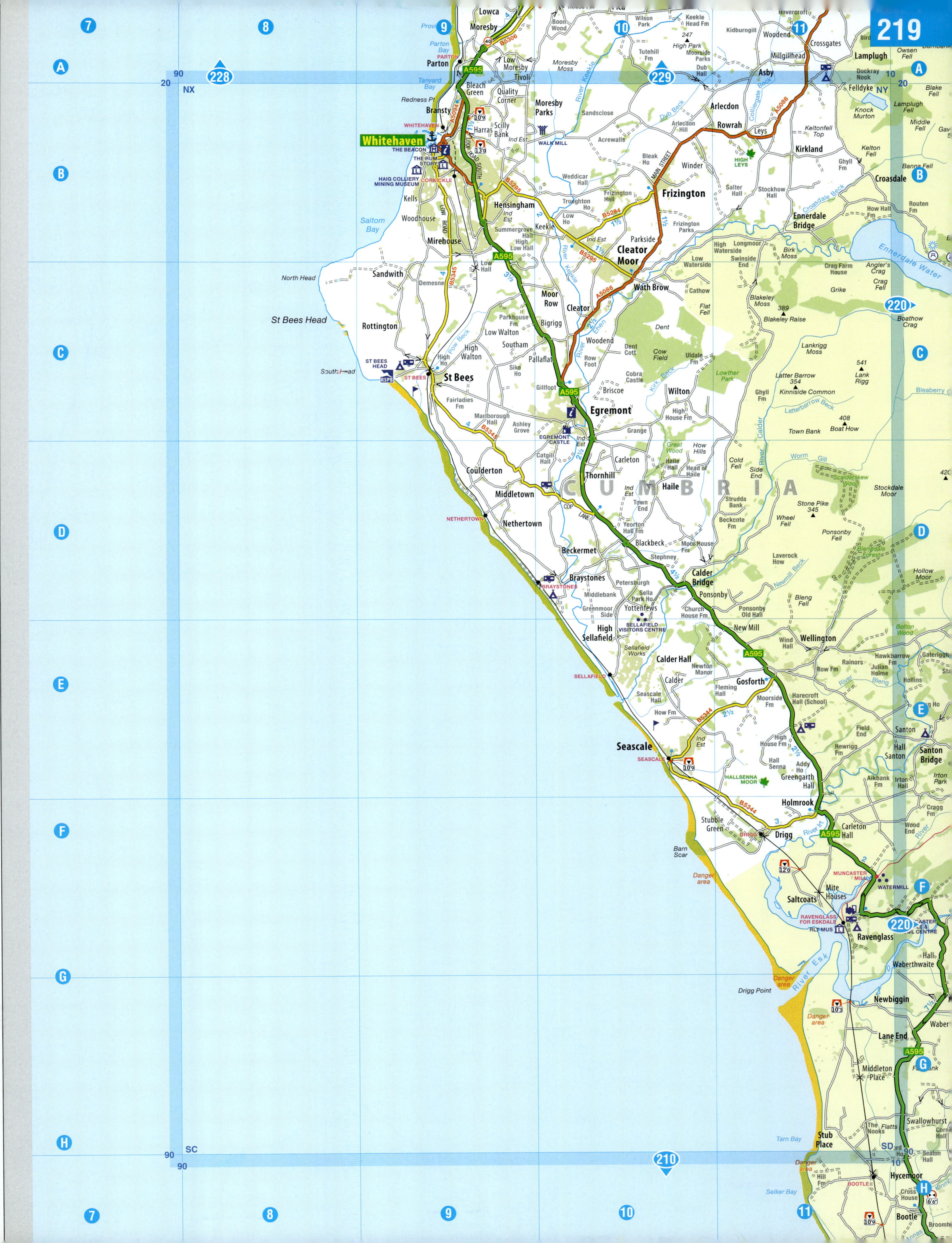

Whitehaven
Parton
Moresby
Lowca
Bransty
Hensingham
Frizington
Cleator Moor
Cleator
Egremont
St Bees
St Bees Head
Sandwith
Rottington
Coulderton
Middletown
Nethertown
Beckermet
Braystones
Calder Bridge
Seascale
Gosforth
Drigg
Holmrook
Ravenglass
Saltcoats
Newbiggin
Bootle
Lamplugh
Kirkland
Ennerdale Bridge
Ennerdale Water
Croasdale
Arlecdon
Rowrah
Asby
C U M B R I A
Sellafield
Calder Hall
Wellington
Santon Bridge
Drigg Point
Tarn Bay
Selker Bay
Saltom Bay
Danger area
A595
B5345
B5344
A5086
B5295

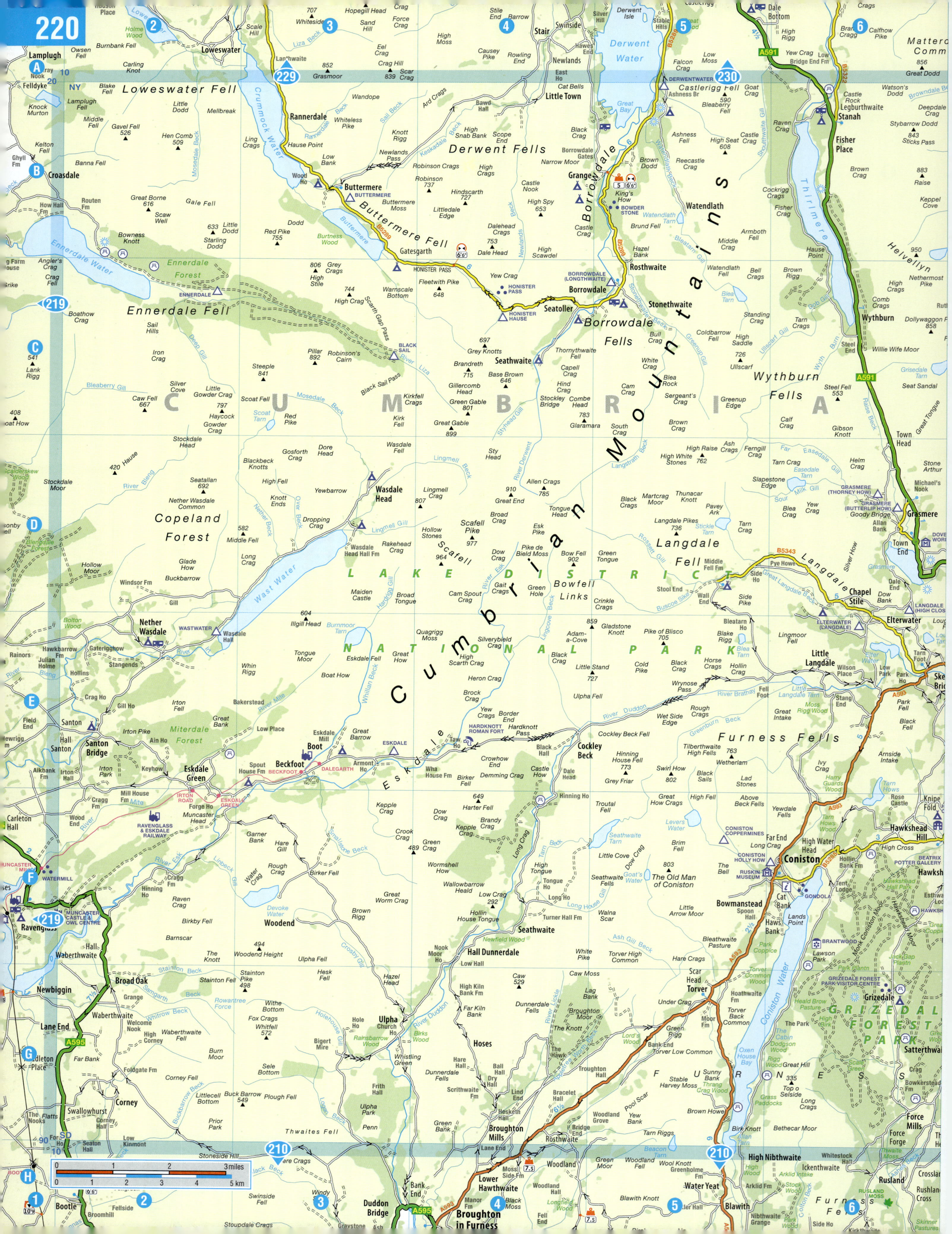
Lamplugh
Loweswater
Loweswater Fell
Crummock Water
Rannerdale
Buttermere
Buttermere Fell
Gatesgarth
Honister Pass
Seatoller
Borrowdale
Rosthwaite
Grange
Derwent Water
Derwent Fells
Castlerigg Fell
Little Town
Watendlath
Stonethwaite
Borrowdale Fells
Thirlmere
Wythburn
Wythburn Fells
Helvellyn
Legburthwaite
Stanah
Fisher Place
Croasdale
Ennerdale Water
Ennerdale Forest
Ennerdale Fell
Pillar 892
Great Gable 899
Kirk Fell
Seathwaite
Scafell Pike 977
Scafell 964
Bowfell 902
Bowfell Links
Langdale Pikes 736
Langdale Fell
Chapel Stile
Elterwater
Grasmere
Town End
Wasdale Head
Wast Water
Nether Wasdale
Copeland Forest
C U M B R I A
Cumbrian Mountains
L A K E   D I S T R I C T
N A T I O N A L   P A R K
Eskdale
Eskdale Green
Beckfoot
Boot
Santon Bridge
Irton Pike
Miterdale Forest
Hardknott Pass
Cockley Beck
Wrynose Pass
Little Langdale
Furness Fells
Coniston
The Old Man of Coniston
Coniston Water
Grizedale
Grizedale Forest Park
Satterthwaite
Hawkshead
Hawkshead Hill
Torver
Seathwaite
Hall Dunnerdale
Ulpha
Broughton Mills
Woodend
Ravenglass
Waberthwaite
Broad Oak
Newbiggin
Lane End
Corney
Bootle
Duddon Bridge
Lower Hawthwaite
Broughton in Furness
Water Yeat
Blawith
High Nibthwaite
Rusland
Force Mills
A591
A593
A595
B5289
B5343
229
230
219
210
0 1 2 3 miles
0 1 2 3 4 5 km

221
Dockray
Knotts
Ullswater
Whale
Shap
Little Strickland
Sleagill
Reagill
Howtown
Sandwick
Martindale
Martindale Common
Bampton
Bampton Grange
Bomby
Rosgill
Glenridding
Patterdale
Bridgend
Hartsop
Grisedale
Deepdale
Dovedale
LAKE DISTRICT
NATIONAL PARK
Haweswater Reservoir
Haweswater
Keld
Shap Abbey
Hardendale
Oddendale
Shap Fells
Kirkstone Pass
Rydal
Ambleside
Clappersgate
Waterhead
Troutbeck
Town End
CUMBRIA
Kentmere
Sadgill
Long Sleddale
Sleddale Forest
Bretherdale Head
Borrowdale
Roundthwaite
Forest Hall
Lowbridge House
Selside
Watchgate
Garnett Bridge
Windermere
Bowness-on-Windermere
Ferney Green
Ings
Staveley
Reston
Bowston
Burneside
Garth Row
Meal Bank
Patton Bridge
Grayrigg
Crook
Kendal
Winster
Storrs
Crosthwaite
Underbarrow
Gregg Hall
Far Sawrey
Near Sawrey
High Wray
Low Wray
Outgate
Colthouse
Lake Bank
New Hutton
Millholme
Brigsteer
Natland
Oxenholme
Barrows Green
Middleshaw
Old Hutton
Killington
Bowland Bridge
Cartmel Fell
Ludderburn
Hartbarrow
Strawberry Bank
The Howe
Row
Windermere
Killington Reservoir
Killington Lake Services (Southbound Only)

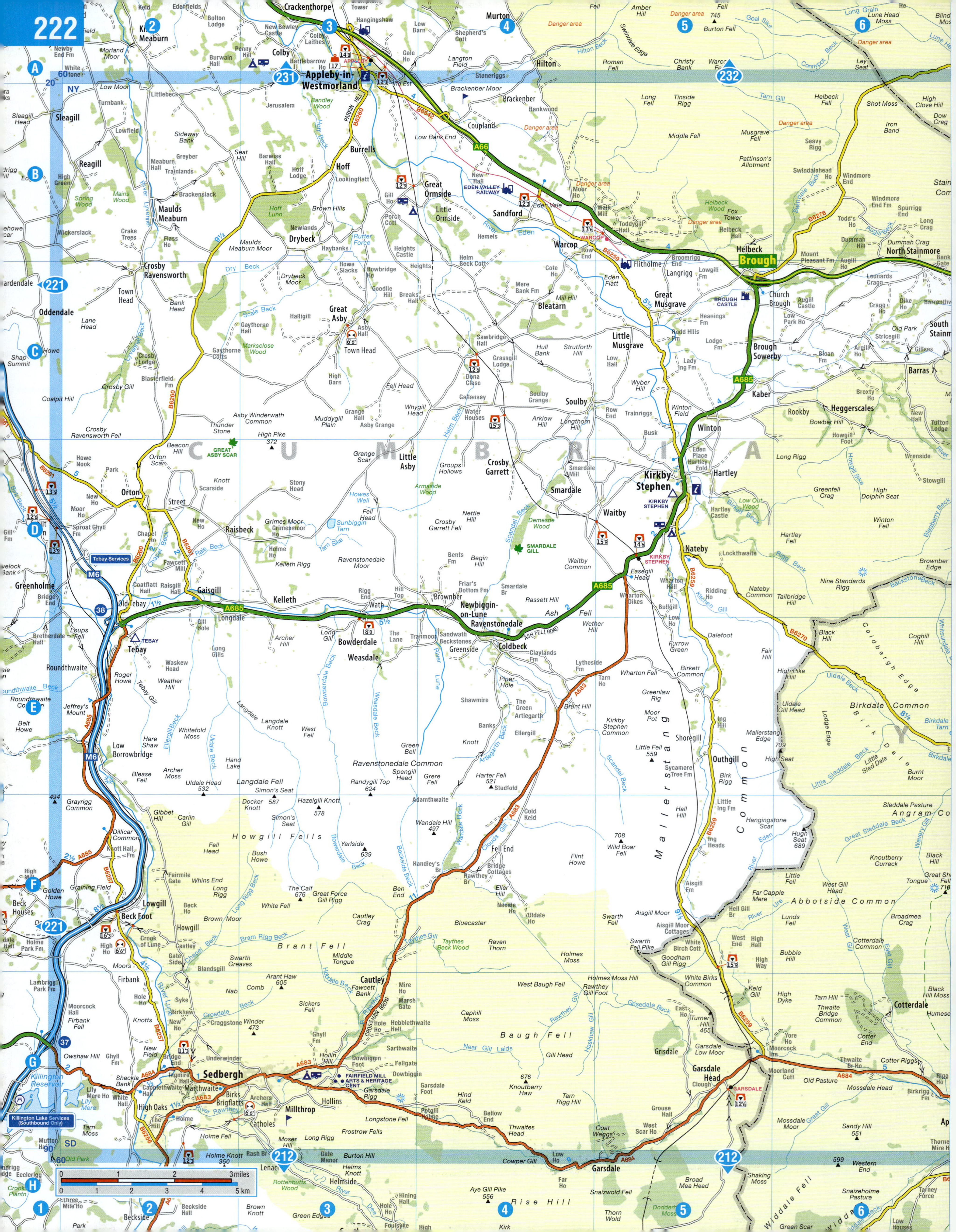

Appleby-in-Westmorland
Brough
Kirkby Stephen
Sedbergh
Tebay
Orton
Great Asby
Crosby Ravensworth
Warcop
Ravenstonedale
Newbiggin-on-Lune
Garsdale Head
Garsdale
Cautley
Howgill Fells
Baugh Fell
Mallerstang Common
C U M B R I A
A66
A685
A683
A684
M6
B6260
B6259
Tebay Services
Killington Lake Services (Southbound Only)

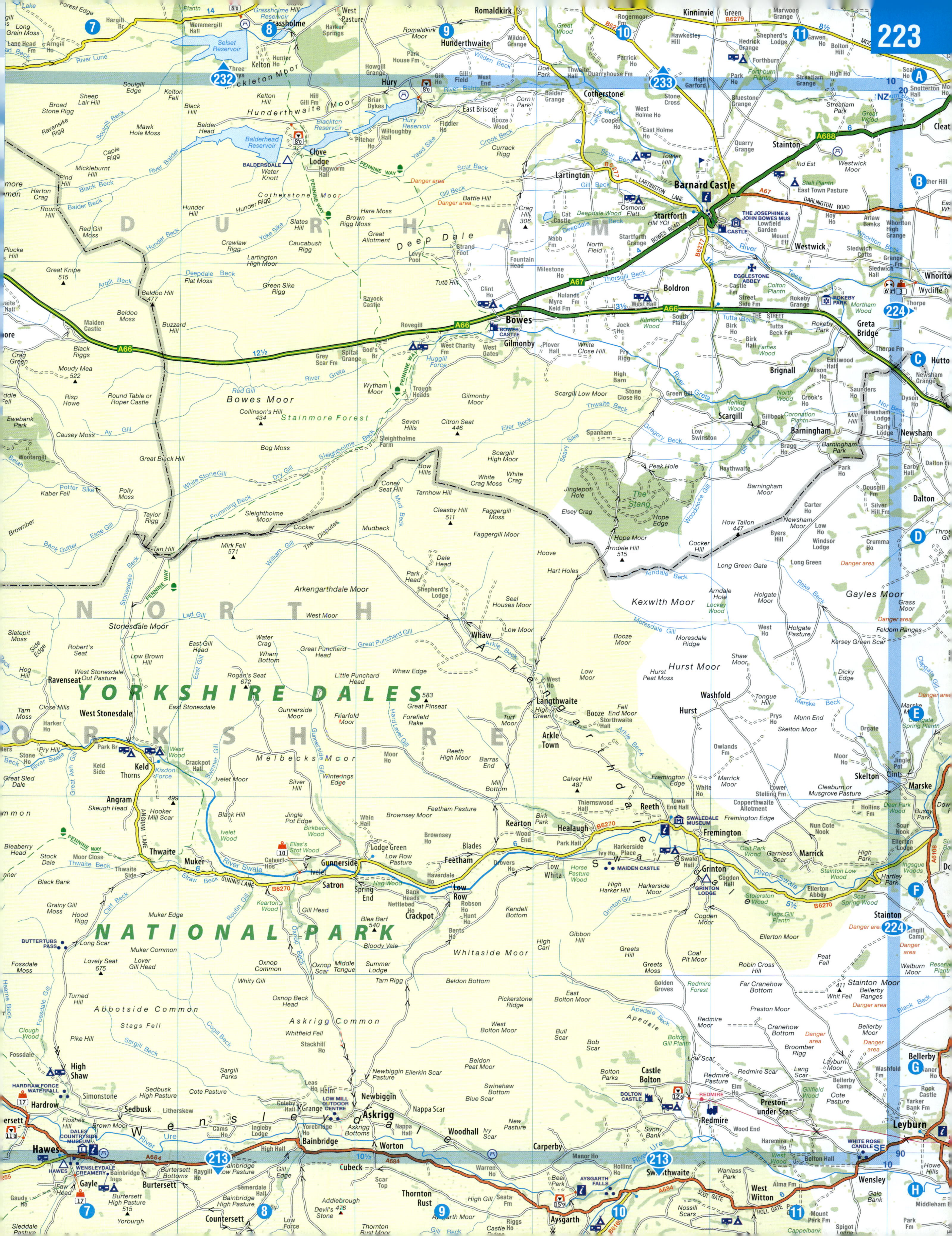

223
Romaldkirk
Kinninvie
Hunderthwaite
Selset Reservoir
Grassholme
Hury
Hury Reservoir
Blackton Reservoir
Balderhead Reservoir
Cotherstone
Barnard Castle
Startforth
Lartington
Stainton
Westwick
Whorlton
Wycliffe
Boldron
Bowes
Gilmonby
Greta Bridge
Brignall
Scargill
Barningham
Newsham
Dalton
Hutton
Bowes Moor
Stainmore Forest
Deep Dale
Cotherstone Moor
Hunderthwaite Moor
D U R H A M
N O R T H
Y O R K S H I R E
YORKSHIRE DALES
NATIONAL PARK
PENNINE WAY
Arkengarthdale
Arkengarthdale Moor
Whaw
Langthwaite
Arkle Town
Booze
Reeth
Healaugh
Fremington
Grinton
Marrick
Marske
Skelton
Hurst
Washfold
Hurst Moor
Kexwith Moor
Gayles Moor
Keld
Thwaite
Muker
Angram
Gunnerside
Ivelet
Satron
Low Row
Feetham
Kearton
Crackpot
Swaledale
River Swale
Ravenseat
West Stonesdale
Stonesdale Moor
Melbecks Moor
Askrigg Common
Abbotside Common
Whitaside Moor
Castle Bolton
Redmire
Preston-under-Scar
Leyburn
Bellerby
Wensley
West Witton
Aysgarth
Carperby
Woodhall
Askrigg
Newbiggin
Bainbridge
Worton
Hawes
Gayle
Burtersett
Hardrow
Simonstone
Sedbusk
High Shaw
Countersett
Thornton Rust
Wensleydale
River Ure
A66
A67
A684
A688
B6270
B6277
B6279
THE JOSEPHINE & JOHN BOWES MUS
EGGLESTONE ABBEY
SWALEDALE MUSEUM
MAIDEN CASTLE
BOLTON CASTLE
HARDRAW FORCE WATERFALL
DALES COUNTRYSIDE MUSEUM
WENSLEYDALE CREAMERY
AYSGARTH FALLS

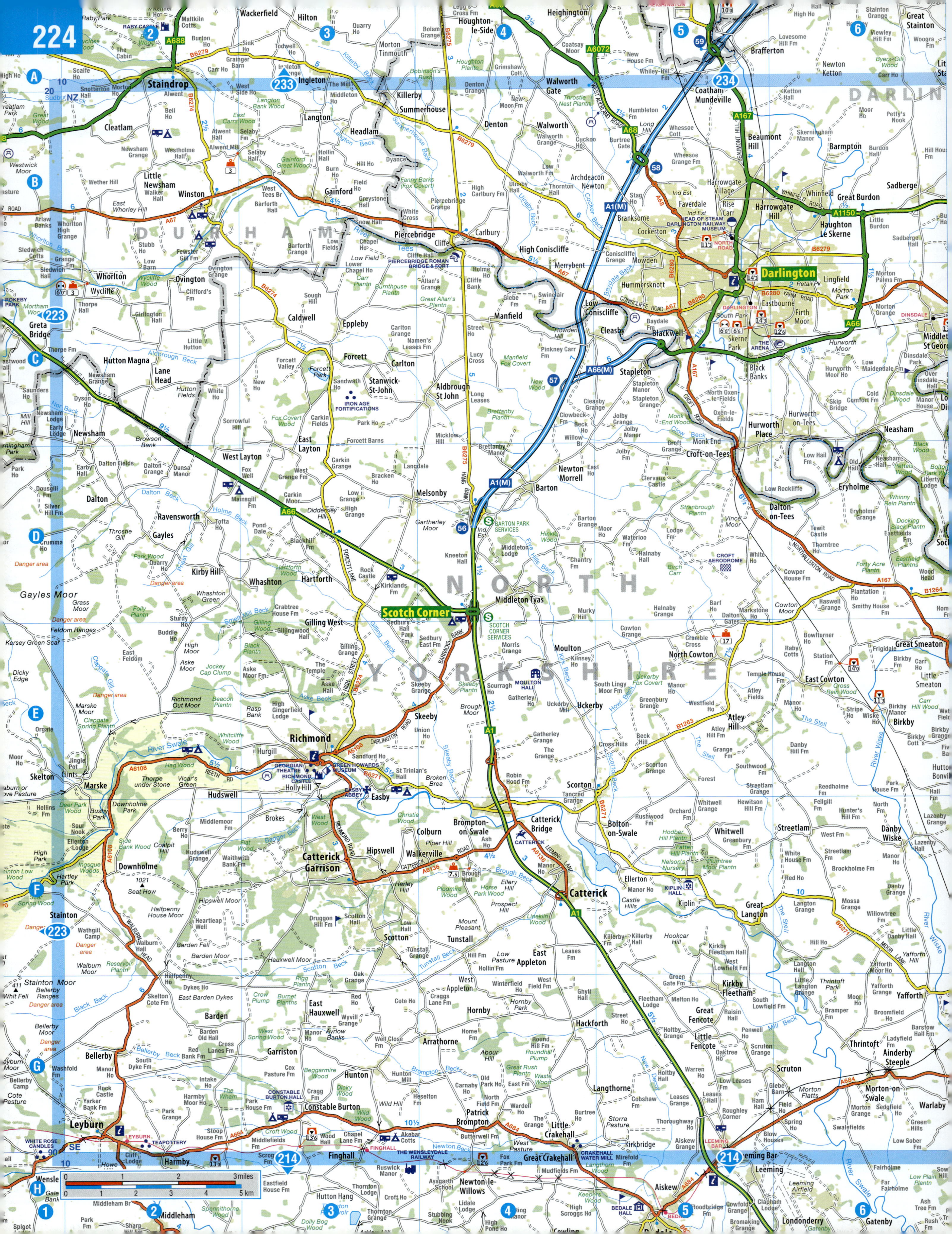

DURHAM
NORTH
YORKSHIRE
DARLINGTON
Darlington
Staindrop
Ingleton
Headlam
Gainford
Winston
Piercebridge
High Coniscliffe
Merrybent
Low Coniscliffe
Cleasby
Blackwell
Stapleton
Manfield
Caldwell
Eppleby
Forcett
Carlton
Stanwick-St-John
Aldbrough St John
Melsonby
Barton
Newton Morrell
Croft-on-Tees
Hurworth Place
Neasham
Eryholme
Dalton-on-Tees
Middleton Tyas
Scotch Corner
Gilling West
Richmond
Skeeby
Moulton
North Cowton
East Cowton
Great Smeaton
Atley Hill
Uckerby
Scorton
Brompton-on-Swale
Colburn
Catterick Bridge
Catterick
Catterick Garrison
Hipswell
Easby
Marske
Hudswell
Downholme
Ravensworth
Gayles
Kirby Hill
Whashton
Hartforth
Dalton
Newsham
Hutton Magna
Greta Bridge
Whorlton
Wycliffe
Ovington
Cleatlam
Little Newsham
Gayles Moor
Skelton
Stainton
Bellerby
Leyburn
Harmby
Finghall
Constable Burton
Hunton
Garriston
East Hauxwell
Barden
Arrathorne
Hornby
Hackforth
Langthorne
Great Crakehall
Little Crakehall
Patrick Brompton
Newton-le-Willows
Kirkby Fleetham
Great Fencote
Little Fencote
Scruton
Ainderby Steeple
Morton-on-Swale
Thrintoft
Yafforth
Danby Wiske
Streetlam
Great Langton
Kiplin
Whitwell
Bolton-on-Swale
Tunstall
East Appleton
Scotton
Birkby
Little Smeaton
Sadberge
Great Burdon
Harrowgate Hill
Haughton Le Skerne
Beaumont Hill
Barmpton
Coatham Mundeville
Brafferton
Heighington
Walworth Gate
Walworth
Denton
Summerhouse
Killerby
Hilton
Wackerfield
Leeming Bar
Leeming
Londonderry
Gatenby
Aiskew
Bedale
Middleham
Wensley
A1(M)
A1
A66
A66(M)
A67
A167
A68
A684
A6108
A6136
B6274
B6275
B6279
B6271
B6280
B1263

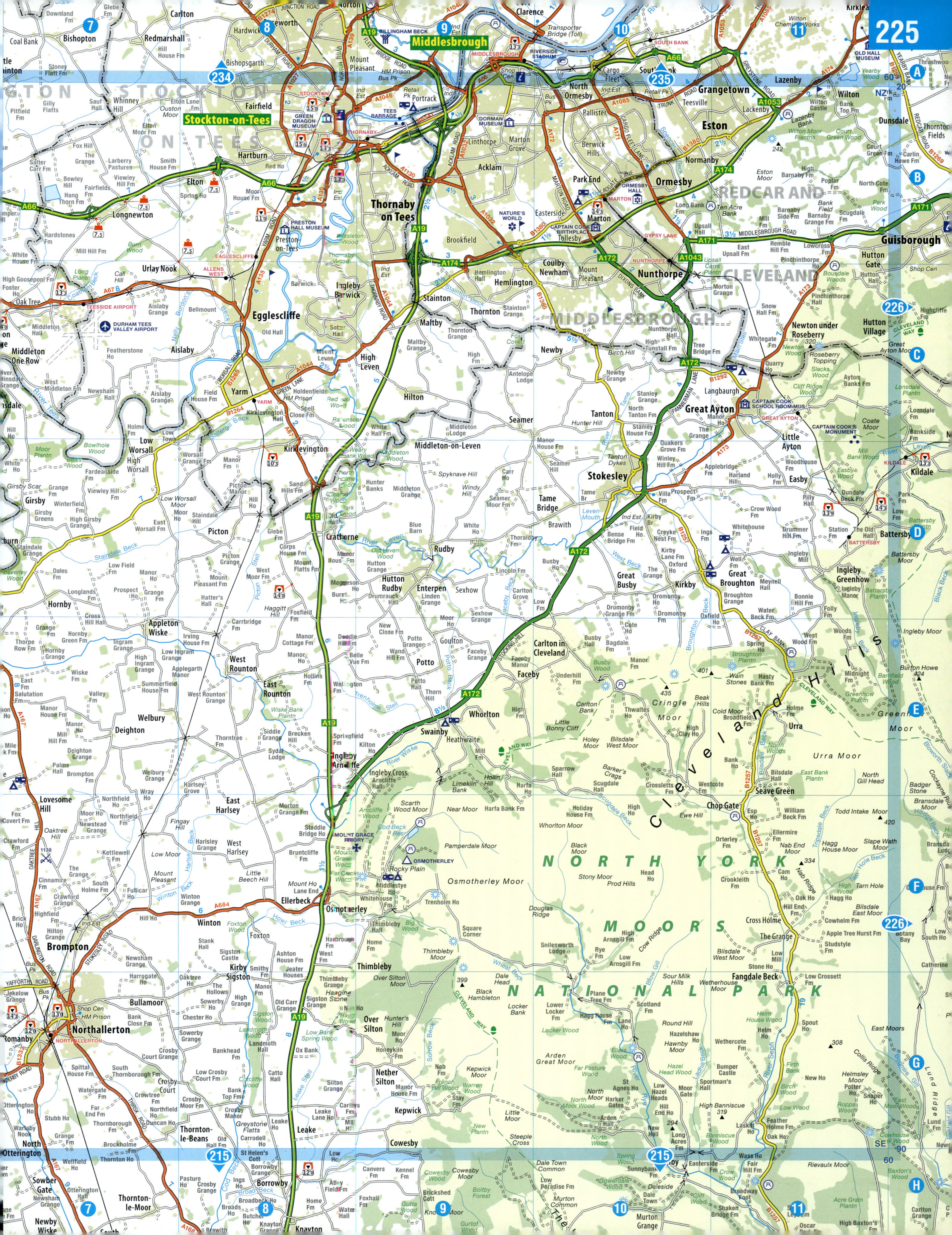
Middlesbrough
Stockton-on-Tees
STOCKTON ON TEES
REDCAR AND CLEVELAND
MIDDLESBROUGH
NORTH YORK MOORS NATIONAL PARK
Cleveland Hills
Billingham Beck
Transporter Bridge (Toll)
Riverside Stadium
Bishopton
Redmarshall
Carlton
Hardwick
Norton
Clarence
Bishopsgarth
Fairfield
Hartburn
Elton
Longnewton
Thornaby on Tees
Preston-on-Tees
Grangetown
Eston
Teesville
Lackenby
Lazenby
Wilton
Dunsdale
North Ormesby
Ormesby
Normanby
Linthorpe
Acklam
Marton Grove
Berwick Hills
Pallister
Park End
Marton
Easterside
Tollesby
Brookfield
Hemlington
Coulby Newham
Nunthorpe
Guisborough
Hutton Gate
Hutton Village
Newton under Roseberry
Great Ayton
Little Ayton
Kildale
Easby
Battersby
Ingleby Greenhow
Urlay Nook
Egglescliffe
Yarm
Aislaby
Stainton
Maltby
Thornton
Newby
High Leven
Hilton
Seamer
Tanton
Stokesley
Middleton-on-Leven
Kirklevington
Low Worsall
High Worsall
Picton
Crathorne
Rudby
Hutton Rudby
Enterpen
Sexhow
Tame Bridge
Great Busby
Kirkby
Great Broughton
Carlton in Cleveland
Faceby
Potto
Whorlton
Swainby
Ingleby Arncliffe
Appleton Wiske
West Rounton
East Rounton
Welbury
Deighton
Hornby
Girsby
Middleton One Row
Teesside Airport
Durham Tees Valley Airport
Lovesome Hill
East Harlsey
West Harlsey
Winton
Ellerbeck
Osmotherley
Thimbleby
Brompton
Northallerton
Romanby
Bullamoor
Kirby Sigston
Over Silton
Nether Silton
Kepwick
Leake
Cowesby
Thornton-le-Beans
Thornton-le-Moor
Borrowby
Knayton
Sowber Gate
North Otterington
Newby Wiske
Chop Gate
Urra
Seave Green
Fangdale Beck
Bilsdale
Urra Moor
Osmotherley Moor
Black Hambleton
Cringle Moor
Cold Moor
Hasty Bank
Carlton Bank
Scarth Wood Moor
Pamperdale Moor
Arden Great Moor
Cleveland Way
Mount Grace Priory
Captain Cook Birthplace Museum
Captain Cook's Monument
Captain Cook Schoolroom Museum
Preston Hall Museum
Green Dragon Museum
Dorman Museum
Nature's World
Ormesby Hall
Stockton
Thornaby
Eaglescliffe
Yarm
Marton
Nunthorpe
Great Ayton
Kildale
Battersby
Northallerton
A19
A66
A172
A174
A171
A1053
A1085
A167
A684
B1257
B1264
234
235
226
215

Saltburn-By-The-Sea
Brotton
Skinningrove
Loftus
Easington
Staithes
Port Mulgrave
Hinderwell
Runswick Bay
Kettleness
Goldsborough
Lythe
Guisborough
Skelton
Upleatham
Lingdale
Kilton Thorpe
Moorsholm
Liverton
Boosbeck
Margrove Park
REDCAR AND CLEVELAND
Hutton Village
Commondale
Castleton
Danby
Ainthorpe
Houlsyke
Lealholm
Glaisdale
Egton
Egton Bridge
Grosmont
Goathland
Kildale
Battersby
Ingleby Greenhow
Westerdale
Church Houses
Low Mill
Thorgill
Rosedale Abbey
Lastingham
Hutton-le-Hole
Gillamoor
Fadmoor
Spaunton
Cropton
Newton-on-Rawcliffe
Levisham
Lockton
Stape
Hartoft End
Ugthorpe
Mickleby
West Barnby
East Barnby
Scaling
Roxby
Newton Mulgrave
Ellerby
Stonegate
Hutton Gate
Wilton
Dunsdale
Gisborough Moor
Kildale Moor
Baysdale Moor
Westerdale Moor
Farndale Moor
Danby High Moor
Glaisdale Moor
Egton High Moor
Rosedale Moor
Wheeldale Moor
Pickering Moor
NORTH YORK MOORS
NORTH YORKSHIRE NATIONAL PARK
NORTH YORK MOORS FOREST PARK
Esk Dale
Farndale
Rudland Ridge
Blakey Ridge
Spaunton Moor
NORTH YORKSHIRE MOORS RAILWAY
RYEDALE FOLK MUSEUM
MOORS CENTRE
A171
A174
B1366
B1266
A173
B1269
B1268
CLEVELAND WAY
235
225
216
0 1 2 3 miles
0 1 2 3 4 5 km

Whitby
Sandsend
Sandsend Ness
East Row
Dunsley
Newholm
Ruswarp
Aislaby
Briggswath
Sleights
Ugglebarnby
Sneaton
Sneatonthorpe
Stainsacre
High Hawsker
Low Hawsker
Raw
Robin Hood's Bay
Fylingthorpe
Littlebeck
Ravenscar
Staintondale
Cloughton
Cloughton Newlands
Burniston
Scalby
Newby
Scarborough
Harwood Dale
Silpho
Broxa
Langdale End
Hackness
Suffield
Everley
Wrench Green
Barrowcliff
Throxenby
Flask Inn
Fylingdales Moor
NORTH YORK MOORS
FOREST PARK
Saltwick Bay
Black Nab
Maw Wyke Hole
White Stone Hole
Castle Chamber
North Cheek or Ness Pt
Robin Hood's Bay
Boggle Hole
Old Peak or South Cheek
Blea Wyke Pt
Beast Cliff
Hayburn Wyke
Rodger Trod
Cloughton Wyke
Hundale Pt
Long Nab
Crook Ness
Cromer Pt
Scalby Ness Rocks
Sea Life Centre
Castle Cliff
Scarborough Castle
Old & East Harbours
South Bay
South Sands
Staintondale Shire Horse Farm
Cleveland Way
A171
A169
A165
B1460
B1410
B1416
B1447
217
TA
OV

EAST
STEWARTRY
COAST
Auchencairn
Auchencairn Bay
Hestan Island
Balcary Bay
Balcary Pt
Airds Pt
Rascarrel
Rascarrel Bay
Castle Muir Pt
Barlocco Bay
Maryport
Flimby
Seaton
Workington
Harrington
Distington
Whitehaven
Parton
Moresby
Lowca
Moss Bay
237
238
219

Blitterlees
Abbey Town
Kingside Hill
Raby
Kelsick
Wedholme Hill
Lessonhall
Oulton
Parton
Micklethwaite
Beckfoot
Highlaws
Pelutho
Dundraw
Waverbridge
Wigton
Newtown
Holme St Cuthbert
Mawbray
Blencogo
Waverton
Old Carlisle
Brackenthwaite
Bromfield
Langrigg
Red Dial
Salta
Edderside
New Cowper
Westnewton
Waterside
Westward
Bolton Low Houses
Allonby
Allonby Bay
Crookdake
Watchhill
Fletchertown
Aspatria
Mealsgate
Hayton
Harriston
Blennerhasset
Prospect
Oughterside
Boltongate
Sandale
Allerby
Crosscanonby
Crosby Villa
Threapland
Plumbland
Parsonby
Bothel
Torpenhow
Ireby
Whitrigg
Bullgill
Gilcrux
Crosby
Birkby
Aughertree
High Ireby
Uldale
Branthwaite
Dearham
Ruthwaite
Sunderland
Tallentire
Blindcrake
Bewaldeth
Orthwaite
Dovenby
Bridekirk
Isel
Redmain
Broughton Moor
Little Broughton
Great Broughton
Kilnhill
Bassenthwaite
Papcastle
Cockermouth
Brigham
Embleton
Wythop Mill
Bridgefoot
Greysouthen
Little Clifton
Eaglesfield
Deanscales
Dean
High Lorton
Low Lorton
Millbeck
Applethwaite
Braithwaite
Portinscale
Keswick
Thornthwaite
Mockerkin
Lamplugh
Loweswater
Thackthwaite
Mosser
Pardshaw
Ullock
Asby
Rowrah
Loweswater Fell
Derwent Fells
Derwent Water
Castlerigg Fell
Skiddaw Forest
Skiddaw
Bassenthwaite Lake
LAKE DISTRICT NATIONAL PARK
WHINLATTER FOREST PARK
SOLWAY COAST
CUMBRIA

Baldwinholme
Cardewlees
Thursby
Dalston
Crofton
Micklethwaite
West Curthwaite
East Curthwaite
Cumdivock
Chalkfoot
Broomfield
Buckabank
Hawksdale
Raughton
Gaitsgill
Unthank
Sprunston
Durdar
Brisco
Burthwaite
Wreay
Foulbridge
Cotehill
Low Cotehill
Hornsby
Holmwrangle
Aiketgate
Armathwaite
Ainstable
Nunclose
Low Hesket
Southwaite
High Hesket
Old Town
Mellguards
Southwaite Services
Raughton Head
Stockdalewath
Skiprigg
Rosley
Welton
Nether Welton
Grassgarth
Brothybeck
Brackenthwaite
Sebergham
Churchtown
Sour Nook
Middlesceugh
Ivegill
Low Braithwaite
Highbridge
Sowerby Row
Hurst
Thomas Close
Calthwaite
Morton
Hutton End
Ruckcroft
Townhead
Lazonby Fell
Plumpton
Brockleymoor
Plumpton Head
New Rent
Unthank
Unthank End
Skelton
Lamonby
Ellonby
Johnby
Little Blencow
Great Blencow
Laithes
Catterlen
Newton Reigny
Newton Rigg
Fair Hill
Penrith
Carleton
Eamont Bridge
Pategill
Castletown
Newbiggin
Stainton
Sockbridge
Tirril
Clifton
Dacre
Soulby
Pooley Bridge
Askham
Lowther
Hackthorpe
Helton
Whale
Greystoke
Greystoke Forest
Berrier
Motherby
Penruddock
Hutton
Beckces
Troutbeck
Whitbarrow Village
Hutton Roof
Mosedale
Mungrisdale
Scales
Threlkeld
Matterdale End
Dockray
Longthwaite
Watermillock
Knotts
Wreay
Howtown
Sandwick
Martindale
Bampton
Bampton Grange
Ullswater
Caldbeck
Hesket Newmarket
Whelpo
Upton
Millhouse
Haltcliff Bridge
High Row
Heggle Lane
Fell Side
Caldbeck Fells
High Pike
Carrock Fell
Skiddaw Forest
LAKE DISTRICT NATIONAL PARK
CUMBRIA
Inglewood Forest
Matterdale Common
Dowthwaitehead
Great Dodd
Castlerigg Stone Circle
Dale Bottom
Birkett Mire
M6
A6
A66
A591
A592
A595
A596
B5305
B5288
B5320

Cumrew
Cumrew Fell
Geltsdale Middle
New Water
Three Pikes 585
Hartchyside
Knarsdale Common
Shawhead
Dewley Fell
Dewley Field
Willyshaw Rigg
Ouston Fell
NINEBANKS
Hawkuplee
Spartywell
Farneyside
Butt Hill
Geil Burn
Lintley
Far Town
Ayle Common
Long Cross
White Hill
Mohope Head
Broadlee
Kevelin Moor
240
241
NY
Newbiggin Fell
Blotting Raise 591
Croglin Fell
The Combs
Lawyer's Cross
Farlam Currick
Green Hill
Broad Mea
High Shield
Thornhope Fell
Grey Nag 656
Black Hill
Castle Nook
Whitley Castle
Kirkhaugh
Kirkside Wood
Ayle
Clarghyll Hall
White Ho
Newshield Moss
Hard Rigg
546
Mohope Moor
Halley Moor
Limestone Brae
Black Cleugh
The Nook
New Ho
The Meres
Broad Meadows
Hesleywell Moor
Middle Rigg
SOUTH TYNEDALE RAILWAY
Wanwood Hill
Blagill
Newbiggin
Bove Wood
Sumburgh Ho
Fieldhead
Croglin Water
Scarrowmanwick Fell
Scarrowmanwick
Hazelgill
Croglin
Caber Fm
Croglin High Hall
Davygill
Thack Moor 609
Tom Smith's Stone
637
Woldgill Burn
Raise
ALSTON
GOSSIPGATE GALLERY
Alston
River Nent
Foreshield
East Cocklake
Browngill
Smallburns Moor
The Dodd
614
Carrshield
Carrshield Moor
Middle Moor
Scale Houses
Lowhall Building
Long Moor
Renwick Fell
Watch Hill
Woldgill Moss
Horse Edge
Park Fell
Black Ho
Hill Ho
Whitewell Fm
A689
Bayle Hill Fm
Middle Fell
Nenthall
Galligill
Whimsey Hill
Greenends
Coalcleugh
Bank Wood
Whin Fell
Crindledyke Fm
Lincowell
Dyrah
Green Rigg
Long Tongue
664
Black Fell
The Dod
Watcher's Hill
Gilderdale Burn
Gilderdale Forest
Brownside Moss
Bayles
Ghyll Ho
Rise
Middle Fell
575
Nettle Hall
Wellgill
Black Hill
Coalcleugh Moor
Scales Moor
Skelling Moor
Haresceugh Fell
Leadgate
Rotherhope Tower
Meadow Flatt
Rice Head
Alston Moor
Nenthead
Old Mill
NENTHEAD MINES HERITAGE CENTRE
Slate Hill
Shaw Side
232
Renwick
Haresceugh
How Moor
Ravenbridge Mill
Loo Gill
HARTSIDE HEIGHT
Benty Hill
Meathaw Hill 486
A686
Hartside Cotts
Blackburn Bank
Dryburn
Cropshall
High Skyreholme
Nunnery Hill
Flinty Fell 614
Westgarth Hill
High Bankhill
Scales
Huddlesceugh Hall
Busk
Low Mill
Gamblesby Fell
Cawood Hill
Rotherhope Fell
Garrigill
Loaning Head
Knoutberry Hill
Nag's Head
Kirkoswald
Raven Beck
Park Head
Viol Moor
Twotop Hill
Fiend's Fell 634
Black Burn
Staneshaw Rigg
Long Grain
Crossgill
Pasture Houses
Ash Gill
Mount Pleasant
Lamb's Head 648
Eden Br
The Mains
Lazonby
Old Parks
Unthank
Hazel Rigg
Hazelrigg Beck
Long Crags
Smittergill Burn
High Ashgill
Little Gill
709
Dead Stones
White Edge
Gamblesby
Glassonbybeck
Little Knapside Hill 657
Aglionby Beck
Smittergill Head
Hound Hill 596
Alston Moor
Lee Hos
West Ashgill
Paper Hill
Pasture Ho
Hole Ho
River South Tyne
Windy Brow
Sally Grain
Eden Lacy
Glassonby
Lacy's Caves
Maughanby Fm
Tarn Ho
Melmerby Mire
Todhills
Hause in the Scar
685
Knapside Hill
Melmerby Fell
Melmerby High Scar
Rigg End
Ousby Fell
Cash Burn
Pikeman Hill
Duffergill Burn
Hill Ho
Tynehead
Yad Moss
Burnhope Seat 746
LONG MEG & HER DAUGHTERS
Marian Lodge
Great Salkeld
Melmerby
Stony Rigg
Bullman Hills
PENNINE WAY
Long Man Hill
Cross Gill
B6277
Slack's Rigg
Scaud Hill
Robberby Fm
Robberby Water
Colstons Fm
Gale Hall
Cuns Fell
Green Fell 740
Rowtling Burn
Raehow End
Lambgreen Hills
Round Hill 686
Tyne Head
Little Salkeld
Nunwick Hall
Dale Beck
Brown Hill
N O R T H
Skirwith Fell
Fallow Hill
Crossgill Head
Calvert End
Tynehead Fell
Yad Moss
Harwood Common
Ashgill Head
Hunsonby
Ousby Moor
Jardine's Fm
Muska Hill
Thack Moor
Bellbeaver Rigg 620
Winskill
Crewgarth
Ousby
Row
Rayson Hall
Sunnygill Beck
High Cap
Kirkdale
Cross Fell 893
Tees Head
River Tees
Frog Hall
Langwathby
LANGWATHBY
Beck Mill
Hole Sike
Townhead
Kirkland Fell
Little Dun Fell
Hard Hill 678
Metalband Hill
Viewing Hill
Greencomb Sike
Herdship Fell
Edenhall
High Barn
Langwathby Moor
Appleside Hill
Whamthorn Plantn
Skirwith Abbey
Skirwith
Skirwith High Moor
Bank Hall
Kirkland
Hanging Walls of Mark Anthony
Wildboar Scar
Middle Tongue
Great Dun Fell 848
Milburn Forest
MOOR HOUSE UPPER TEESDALE
Trout Beck
Moor Ho
Dodgen Pot
West Common
Home Fm
Moorlands Plantn
Primrose Bank
Blencarn Beck
Blencarn
Grumply Hill
Crowdundle Beck
Iron Howe
Green Castle
Moss Burn
Backside Fell
Dolphenby
Udford
River Eamont
Moorside
High Slack
P E N N I N E S
Knock Fell
Rough Sike
Green Burn
Force Burn
Cow Green Reservoir
Highbarn Cott
Hornby Hall
Cross How
Lounthwaite
Red Carle
Burney Hill
Knock Ore Gill
Lodgegill Sike
B6412
Underwood
Loscars
Culgaith
Woodside
Green Fell
Milburn
Howgill Castle
Dufton Fell
Meldon Hill 767
CAULDRON SNOUT WATERFALL
A66
Whinfell Ho
Mill Rigg Fm
Newbiggin
Moorlands Head
Kirk Ho
Gullom Holme
Milburn Beck
Milburn Grange
Slakes Fm
Knock Pike
Swindale Beck
High Scald Fell
Great Mattinson Sike
Haversheaf Hall
Sawmill Cotts
Whinfell Forest
Ash Hill
Oglebird Plantn
ACORN BANK GARDEN
Temple Sowerby
Houtsay Park
Halefield Fm
Low Abbey
Close Ho
Knock Moor
Knock
Brownber Hill
Grain Beck
Leacet Hill
OASIS LAKELAND FOREST VILLAGE
Salter Wood
WETHERIGGS COUNTRY POTTERY
Skygarth Fm
Lowmoor Row
Hale Grange
Kirkby Thore
Stamp Hill Fm
Knock Cross
Great Rundale Beck
Dufton Pike
Bluethwaite Hill
Backstone Edge
Narrowgate Beacon
HIGH CUP NICK
Maize Beck
Rasp Hill
CLIBURN MOSS
Brown Howe
Town Head
Melkinthorpe
Town End
Town Head
Marton Moor
Birks Head
Coatsike Fm
Pusgill Ho
Danger area
Swarth Beck
Maize Beck
Shaw Ho
Ousen Stand
Bridge End
River Eden
Sleastonhow
Marton Ho
DUFTON
Dufton
Bow Hall
503
Peeping Hill
High Cup Gill
Murton Fell
Mickle Fell
Wood Ho
Leith
Melkinthorpe Wood
Cliburn
Long Marton
Park Ho
Middle Tongue
Street Ho
Leith Bank
Crossing Hall
Winter Ho
Edenbank Fm
Trout Beck
Eden Grove
Redlands Bank
Town Head
Dufton Wood
Town End
Stanley Bank Plantn
Glenton Vale
Town Wood
Broom
Brampton
Croft Ends
Scalehowe Wood
Harbour Flatt
594
Murton Pike
Hilton Fell
Keekham Beck Head
Great Strickland
Morland
Bolton
Roger Head
Crackenthorpe
Brampton Tower
Brackenthwaite
Mell Fell
Amber Hill
Little Fell 745
Burton Fell
Greystone Ho
Dallan Bank
Syke Ho
Town Head
Keld
Edenfields
Bolton Lodge
New Bewley Castle
Colby Laithes
Hangingshaw
Low Barn
Shepherd's Cott
Murton
Swindale Edge
Goal Sike
Sheriff Park
Sandriggs Fm
Strickland Lodge Fm
Newby Head
High Field
King's Meaburn
Penny Hill
Colby
Battlebarrow Ho
APPLEBY
Gale Ho
Langton Field
Hilton Beck
Roman Fell
Christy Bank
Warcop Fell
Thrimby Hall
Newby
Newby End Fm
Morland Moor
Burwain Hall
Hilton
Little Strickland
White Stone
White Stone
Dedra Banks
Appleby-in-Westmorland
Stoneriggs
Brackenber Moor
Brackenber
221
222
Thrimby Grange
Bedlands Gate
Littlebeck
Jerusalem
Bandley Wood
Bankwood
Long Fell
Tinside Rigg
Brown Howe
Sleagill Head
Sleagill
Lowfield
Sideway Bank
B6260
B6542
Coupland
Danger area
Musgrave Fell
Seat Hill
Barwise Hall
Burrells
Low Bank End
Middle Fell
Seavy Rigg
Reagill
Meaburn
Greyber
Hoff
Pattinson's Allotment

232
Acton Moor
Broad Gate
Sinderhope
Hangman Hill
Pennypie Ho
Hotburn Hill
Carterway Heads
Brownley Hill
533
Allendale Common
Pikeley Rigg
Hope Fell
Birkside Fell
Pennypie Fell
Blanchland
Berry Bank
Edmundbyers
Kevelin Moor
Sipton Shield
241
Harwoodshield Fell
Newbiggin Fell
Baybridge
242
Edmundbyers Common
Longwell Moor
Riddlehamhope Fell
Newbiggin
Muggleswick
Limestone Brae
Tedham Moss
Hunstanworth
Muggleswick Park
Spartylea
Nookton Back Fell
Townfield
Stoterley Hill
Knockshield Moor
Halleywell Fell
Western Meadows
Uzzles Hill
Buckshott Fell
Black Cleugh
Elpha Green
White Hill
Heatheryburn Moor
Boltshope Park
Ramshaw
Green Hill
Nookton West Fell
Ellers Hill
Sandyford
Cross Rigg
Middle Rigg
White Ridge
Byerhope Moss
Harehope Hill
Carrshield
NORTHUMBERLAND
Byerhope
Boltshope
Middles
Smallburns Moor
Carrshield Moor
Swinhope Moor
Meadow Cott
Quickcleugh Moss
Hunstanworth Moor
Black Hill
421
Muggleswick
Common
The Dodd
Dirt Pot
Edmondbyers Common
614
Boll's Law
540
Coalcleugh
Allendale Common
West End
Allenheads
559
Redburn Edge
Cuthbert's Hill
Horseshoe Hill
519
Feldon Plain
Hisehope Head
409
Skayiock Hill
Middlehope Moor
Killhope Law
Coalcleugh Moor
Westend Moor
Allenheads Park
Redburn Common
561
Bell's Hill
Waskerley Reservoir
Black Hill
Rimey Law
Steward Shield Meadow
231
Killhope Cross
A689
Cleugh Ho
Holy Well
NORTH OF ENGLAND LEAD MINING MUSEUM
Wolfcleugh Common
Rookhope Chimney
Waskerley Park
Bank Foot
ROOKHOPE NURSERIES
Puddingthorn Moor
Bell's Bridge
Lintzgarth
Rookhope
Stanhope Common
Cowhorse Hill
Gold Hill
Silt Foot
Weardale Forest
Sedling Fell
Wolf Cleugh Fm
Lintzgarth Common
Long Hill
512
Weatherhill Engine
Collier Law
516
Weather Law
Mount Pleasant
Middlehope Moor
Smailsburn Common
Smailsburn
Pasture Ho
Knoutberry Hill
High Greenfield
White Edge
Scarsike Head
539
Millstone Rigg
Tunstall
Nag's Head
Lanehead
Black Hill
604
Park Plantn
High Ho
Wolsingham Park Moor
Cornriggs
Whitestone Ho
Middlehope Lodge
Hanging Wells
Ambling Gate
Cowshill
Copthill
Race Head
586
Northgate Fell
Mount Pleasant
Crawleyside
Lamb's Head
648
Moss Moor
Halliwell Ho
Newhouse Moor
709
Dead Stones
Bents Head
Wearhead
Sunderland Fm
Ashy Bank
Hill Crest
Fatherley Hill
459
White Edge
Middle Rigg
West Blackdene
Carr Brow Moor
White Hall
Stanhope
DURHAM DALES CENTRE
Burnhope Reservoir
High Rigg
Ireshopeburn
New Meadows
Rose Hill
Eastgate
Weardale
White Ho
Aller Gill
STANHOPE
Burnhope Moor
Slack Ho
New Ho
East Blackdene
High Kitty Crag
Westgate
Park Ho
Intake Fm
High House
St John's Chapel
Thimbleby Hill
Frosterley Cott
Burnhope Seat
746
Ireshope Plains
Hawkwell Head
Daddry Shield
Hill Hos
Brotherlee
Newlandside Fm
Rogerley Hall
Frosterley
Newlands Hall
Glenwhelt
Littlewood Fm
Snowhope Close
Snape Gate
Bridge End
FROSTERLEY
Coves Ho
Scaud Hill
Great Stony Hill
High Wood Meadows
New Close
Catterick Moss
Ireshope Moor
Noon Hill
High Pike
586
Swinside Ho
NORTH
Hill End
White Kirkley
Harvey Hill
320
Harwood Common
Ashgill Head
Grass Hill Fm
Coldberry End
Causeway Hill
Harthope Moor
Chapel Fell
Snowhope Moor
540
Bollihope Carrs
Bollihope Ho
Harehope
22
Langdon Head
576
Black Hill
Frog Hall
Grasshill Common
Three Pikes
Chapelfell Top
Swinhopehead Ho
607
Snowhope Hill
Juniper Banks
Bollihope
Harthope Moss
Swinhope Moor
Bollihope Common
Force Foot
Harwood
Langdon Fell
Westernhope Moor
Low Black Hill
Allotment Ho
Lingy Hill
Langdon Common
696
Fendrith Hill
Dora's Seat
Swinhope Head
PENNINES
Wager Head
Herdship Fell
Five Pikes
478
Bleak Law
Low End
Black Hill
Harnisha Hill
589
Raven Seat
Pawlaw Pike
488
Pikeston Fell
Red Wing
Church Bowers
White Mere
Binks Ho
Newbiggin Common
Middleton Common
West Common
Touting Hill
B6277
Pike Law
Backside Fell
Peghorn Lodge
Valence Lodge
Langdon Beck
Flushiemere Ho
Carrs Hill
601
South Grain Beck
Hamsterley
Cow Green Reservoir
Cow Green
Widdybank Fell
Hunt Hall
LANGDON BECK
New Ho
Wool Pits Hill
High Beck Head
DURHAM
White Hill
490
Black Hill
393
Sharnberry Flat
Widdy Bank Fm
Forest-in-Teesdale
Ettersgill
Watson's Br
Broadley's Gate
Moor Ho
Red Grooves Ho
Coldberry
Hardberry Hill
529
492 Islington Hill
Middle End Plantn
Hill End
Moor Riggs
Cronkley
Ash Hill
Field Head
Hamsterley Forest
CAULDRON SNOUT WATERFALL
Thistle Green
White Well Green
HIGH FORCE
BOWLEES VISITOR CENTRE
Bowlees
Stable Edge
Club Gill
Monk's Moor
Neighbour Moor Ho
Grain Beck
PENNINE WAY
Hield Ho
Wynch Br
Newbiggin
Hardberry Fm
Snaisgill
Raven Hills
521
Pennington Cott
Pennington Plantn
Merrygill Moss
Cronkley Fell
Noon Hill
Holwick Lodge
Holwick
Eggleston Common
Danger area
Holwick Fell
Holwick Scars
Middle Side
Aukside
Stotley Grange
Knotts Plantn
Lang Hurst
Green Fell End
Swinket Mease Rigg
Dent Bank
High Dyke
Cowlake
339
231
Howden Moss
Rowton Beck
Millstone How Hill
Park End Wood
Park End
Middleton-in-Teesdale
Blackton Head
Woodland Fell
Long Crag
Rotten Rigg
Stotley Hall
Laverock Hall
Stone Houses
Crossthwaite Scars
Teesdale
Mickle Fell
Hagworm Hill
Crossthwaite Common
Step Ends
Egglesburn
Hill Top
Danger area
Bink Moss
Buck Riggs
Harter Fell
481
Laithkirk
Ornella Fm
Staple Moss
Moor Rigg
Bowbank
Mickleton
Bogg Ho
Eggleston
Stobgreen Plantn
B6282
Keekham Beck Head
High Crag
Low Bink Moss
Lune Moor
EGGLESTON HALL
Lune Forest
Cock Lake
520
Thringarth
East Park
Cote Ho
Stoop Hill
Icaron Lodge
Folly Ho
White Ho
Pallet Crag Ho
Philip Reed Moss
Close House Crags
Close Ho
Fish Lake
Sleight Edge
Wythes Hill
Bail Hill
Hollin Side Wood
Paddock Wood
Kinninvie
Forest Edge
High Plantn
Grassholme Reservoir
Stake Hill
Romaldkirk
Rogermoor Fm
Long Grain
Lune Head Moss
Blindgill Moss
Hargill Br
Wemmergill Hall
Grassholme
West Pasture
Harker Springs
Romaldkirk Moor
Great Wood
Hawkesley Hill
Danger area
Long Grain Moss
Lane Head Fm
Arngill Ho
Selset Reservoir
Hunderthwaite
Wilden Grange
River Lune
Hunter Ho
Park House Fm
Parrick Ho
Ley Seat
Three Chimneys
Kelton Ho
Mickleton Moor
Howgill Grange
Hury
Gill Ho
Gill Field
West End
Doe Park
Thwaite Hall
Quarryhouse Fm
High Ho
High Garford
Helbeck
Soulgill Edge
Kelton Fell
222
Kelton Hill
Hill Ho
Gill Fm
223
Balder Grange
Cotherstone
Stone Cross
Road Stone Rigg
Sheep Lair Hill
Black Hill
Hunderthwaite Moor
Briar Dykes
East Briscoe
Corn Park
West Holme Ho
Ravensike Rigg
Balder Head
Blackton Reservoir
Hury Reservoir
Fiddler Ho
Booze Wood
Cooper Ho
East Holme Ho
Musgrave
Seavy Rigg
Mawk Hole Moss
Balderhead Reservoir
Clove Lodge
Pitcher Ho
Willoughby Hall
Currack Rigg
Towler Hill
Pattinson's Allotment
Caple Rigg
BALDERSDALE
0 1 2 3 miles
0 1 2 3 4 5 km

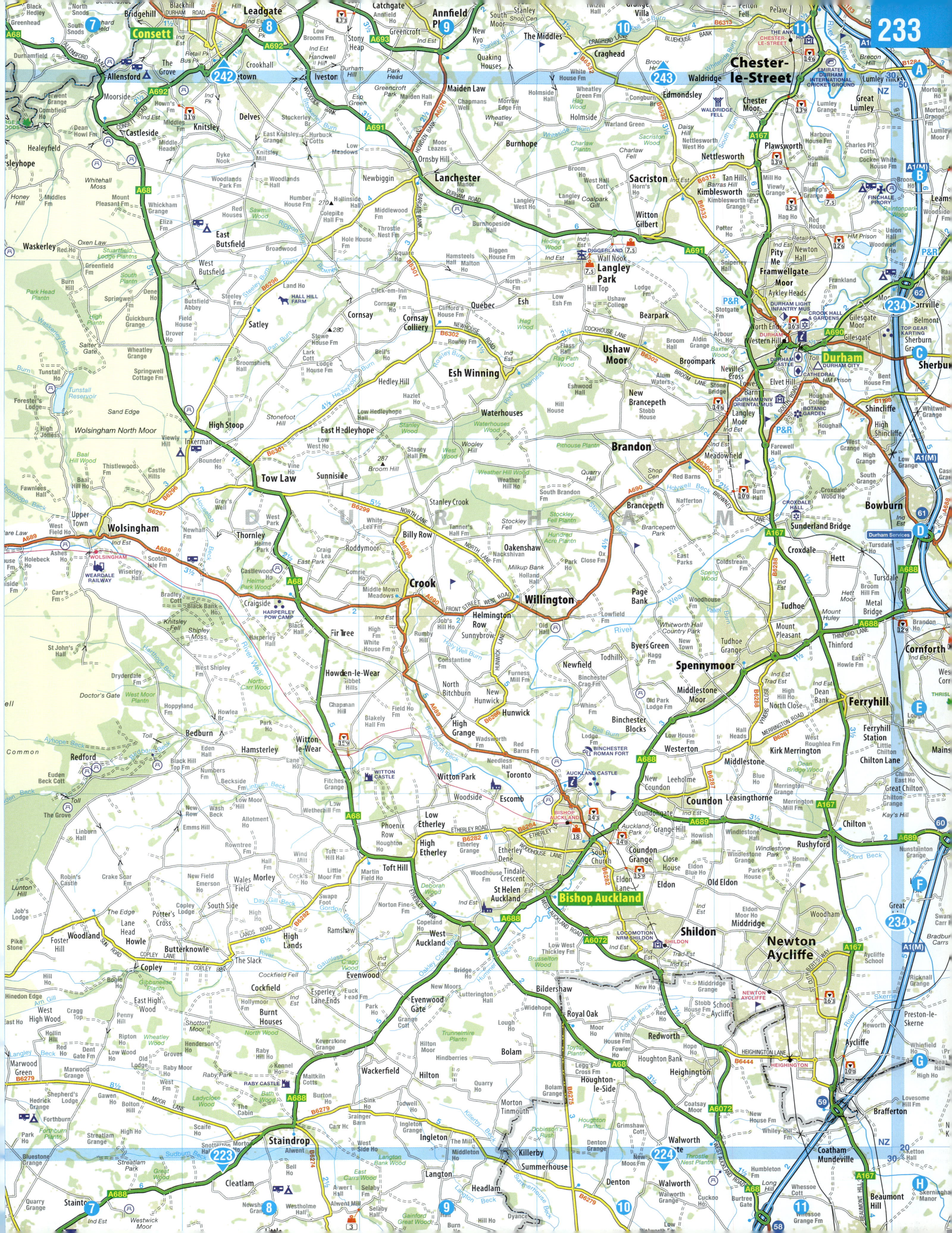
Consett
Leadgate
Annfield Plain
Chester-le-Street
Allensford
Castleside
Lanchester
Sacriston
Kimblesworth
Witton Gilbert
Langley Park
Esh Winning
Cornsay Colliery
Ushaw Moor
Brandon
Durham
Tow Law
Wolsingham
Crook
Willington
Brancepeth
Spennymoor
Ferryhill
Bowburn
Sunderland Bridge
Howden-le-Wear
Witton-le-Wear
Hunwick
Escomb
Bishop Auckland
Coundon
Shildon
Newton Aycliffe
Middlestone
Kirk Merrington
Chilton
Cornforth
West Auckland
Evenwood
Cockfield
Copley
Woodland
Staindrop
Ingleton
Heighington
Redworth
Walworth
Aycliffe
Coatham Mundeville
Bedburn
Hamsterley
Toft Hill
Etherley Dene
St Helen Auckland
High Etherley
Toronto
Wolsingham North Moor
Weardale Railway
Durham Services
Auckland Castle
Binchester Roman Fort
Raby Castle
DURHAM

Houghton-Le-Spring
Seaham
Hetton-le-Hole
Murton
Dalton-le-Dale
Cold Hesledon
Easington
Easington Colliery
Peterlee
Horden
Blackhall Colliery
Shotton Colliery
Wheatley Hill
Wingate
Thornley
Sherburn
Shincliffe
Bowburn
Coxhoe
Cornforth
Ferryhill
Trimdon
Trimdon Colliery
Fishburn
Sedgefield
Bishop Middleham
Chilton
Bradbury
Mordon
Wynyard Village
Thorpe Thewles
Stillington
Great Stainton
Bishopton
Redmarshall
Carlton
Elwick
High Hesleden
Hart
Sheraton
Dalton Piercy
Hartlepool
Seaton Carew
Greatham
Newton Bewley
Wolviston
Billingham
Cowpen Bewley
Haverton Hill
Port Clarence
Norton
Roseworth
Middlesbrough
Stockton-on-Tees
South Bank
North Ormesby
Hartburn
Fairfield
Aycliffe
Preston-le-Skerne
Brafferton
Newton Ketton
Little Stainton
Whitton
Foxton
Pittington
Hallgarth
Littletown
Carrville
Belmont
Leamside
West Rainton
East Rainton
Easington Lane
South Hetton
Haswell
Haswell Plough
Ludworth
Shadforth
Cassop
Quarrington Hill
Kelloe
Old Quarrington
Castle Eden
Hesleden
Station Town
Hutton Henry
Deaf Hill
Hartlepool Bay
Durham Coast
Feathered Rocks
Nose's Pt
Chourdon Pt
Hawthorn Hive
Shippersea Bay
Shot Rock
Fox Holes
Horden Pt
Dene Mouth
Crimdon Park
North Sands
Parton Rocks
The Headland
Middleton
West Park
Rift House
Owton Manor
Graythorp
Seal Sands
Teesmouth
Tees Mouth
D U R H A M
H A R T L E P O O L
TYNE AND WEAR
DARLINGTON
STOCKTON ON TEES
A1(M)
A19
A689
A181
A179
A177
A167
A688
A182
A1086
B1283
B1278
B1280
243
233
224
225
0 1 2 3 miles
0 1 2 3 4 5 km

7
8
9
10
11
A
B
C
D
E
F
G
H
NZ
NORTH SEA
TEES BAY
Coatham Sands
Bran Sands
West Scar
Salt Scar
Redcar Rocks
The Flashes
Grangetown Works
Coatham
Warrenby
REDCAR CENTRAL
Redcar
Mill Howle
REDCAR EAST
BRITISH STEEL REDCAR
TRUNK ROAD
Westfield
Redcar Racecourse
REDCAR
COAST ROAD
Dormanstown
Scanbeck Howle
Marske-by-the-Sea
Stone Gap
Kirkleatham
Wilton Chemical Works
LONGBECK
MARSKE
Grewgrass Fm
Saltburn Scar
Hunt Cliff
SMUGGLERS HERITAGE CENTRE
SALTBURN
Winey Hill Fm
Tofts Fm
Warsett Hill
166
OLD HALL MUSEUM
Yearby
Fell Briggs Fm
Pontac Fm
Saltburn-By-The-Sea
Brough House Fm
Shepherds Ho
New Brotton
INTERNATIONAL RALLY SCHOOL
Thrushwood Fm
New Marske
Horse Close Fm
Lazenby
Yearby Wood
YEARBY BANK
New Buildings Fm
Saltburn Grange
Brotton
Skinningrove
Hummersea Scar
White Stones
ngetown
A1053
Wilton
Park Fm
225
Upleatham
Wand Hills
Gripps Fm
226
Lackenby
Wilton Castle
Bank Top Fm
SKELTON CASTLE
Ind Est
Carlin How
Spring House Fm
Upton
Rockhole Hill
213
Bias Scar
Boulby
Cowbar
Cowbar Nab
Old Nab
Lazenby Bank
Dunsdale
Thornton Fields
Skelton
Craggs Hall
Kilton
Loftus
Grange Fm
Ings Fm
Staithes
Brackenberry Wyke
Wilton Moor Plantns
Court Green Wood
242
REDCAR ROAD
Skelton Green
Trout Hall
New Skelton
East Pastures
East Loftus
Boulby Mine
Court Green Fm
Carlin Howe Fm
Tocketts Fm
Liverton
Easington
Easington Woods
Lingrow Knock
REDCAR AND

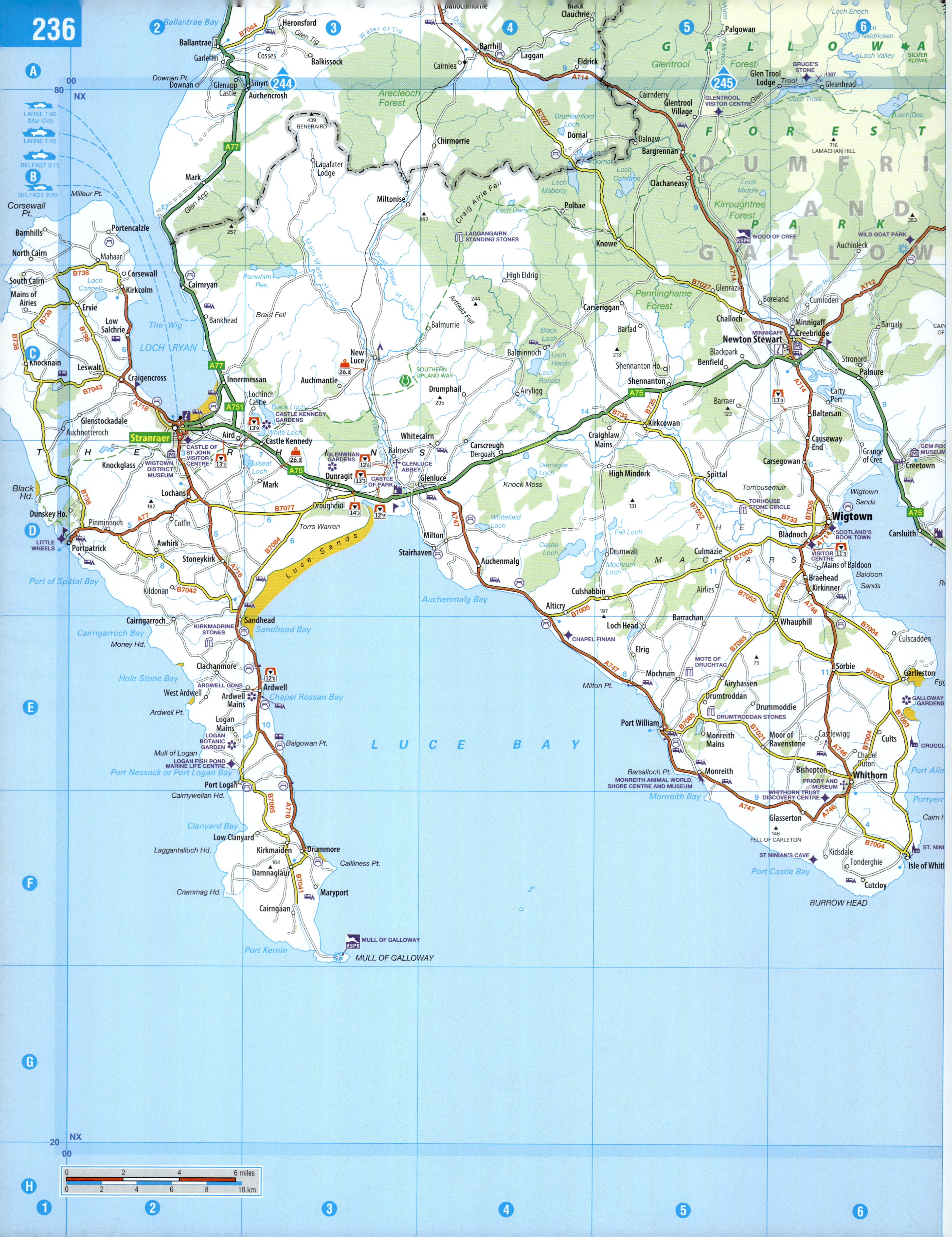
Ballantrae Bay
Ballantrae
Garleffin
Heronsford
Glen Tig
Cosses
Balkissock
Water of Tig
Downan Pt.
Downan
Glenapp Castle
Smyrt
244
Auchencrosh
Barrhill
Laggan
Cairnlea
Eldrick
A714
Black Clauchrie
Palgowan
GALLOWAY
Glentrool Forest
245
Glen Trool Lodge
BRUCE'S STONE
Loch Trool
Gleanhead
Loch Valley
SILVER FLOWE
Loch Enoch
Loch Neldricken
Cairnderry
Glentrool Village
GLENTROOL VISITOR CENTRE
Arecleoch Forest
439 BENERAIRD
Lagafater Lodge
Chirmorrie
B7027
Drumlamford Loch
Dornal
Dalnaw
Bargrennan
Loch Maberry
Loch Ochiltree
Clachaneasy
Loch Middle
FOREST
716 LAMACHAN HILL
DUMFRIES AND GALLOWAY
LARNE 1:00 (Mar-Oct)
LARNE 1:45
BELFAST 3:15
BELFAST 2:20
Milleur Pt.
Corsewall Pt.
Mark
Glen App
A77
Miltonise
Craig Airie Fell
Loch Derry
Polbae
Kirroughtree Forest
WOOD OF CREE
WILD GOAT PARK
Auchinleck
Knowe
Barnhills
Portencalzie
North Cairn
Mahaar
South Cairn
Mains of Airies
B738
Loch Connell
Corsewall
Kirkcolm
Ervie
Low Salchrie
B798
Cairnryan
Penwhirn Res.
Main Water of Luce
Cross Water of Luce
LAGGANGAIRN STANDING STONES
High Eldrig
Artfield Fell
Penninghame Forest
Glenrazie
A714
Boreland
Cumloden
A712
Carseriggan
Barfad
Challoch
Minnigaff
Creebridge
Bargaly
Newton Stewart
Blackpark
Benfield
Bankhead
Braid Fell
The Wig
LOCH RYAN
Knocknain
Leswalt
Craigencross
B7043
Innermessan
Auchmantle
New Luce
SOUTHERN UPLAND WAY
Balmurrie
Balminnoch
Black Loch
Loch Heron
Loch Ronald
Shennanton Ho.
Shennanton
Stronord
Palnure
Drumphail
Airyligg
Tarf Water
A75
B733
B735
Barraer
Carty Port
Baltersan
Glenstockadale
Auchnotteroch
A718
Lochinch Castle
Black Loch
CASTLE KENNEDY GARDENS
White Loch
Stranraer
Aird
Castle Kennedy
Whitecairn
Carscreugh
Dergoals
Craighlaw Mains
Kirkcowan
Causeway End
Grange of Cree
Creetown
GEM ROCK MUSEUM
Knockglass
WIGTOWN DISTRICT MUSEUM
CASTLE OF ST JOHN VISITOR CENTRE
Soulseat Loch
Mark
GLENWHAN GARDENS
Balmesh
GLENLUCE ABBEY
CASTLE OF PARK
Dunragit
Glenluce
Demaglar Loch
Knock Moss
High Mindork
Spittal
Carsegowan
Torhousemuir
TORHOUSE STONE CIRCLE
Wigtown Sands
Wigtown
Black Hd.
Lochans
B7077
Droughduil
Torrs Warren
Whitefield Loch
Bladnoch
SCOTLAND'S BOOK TOWN
Carsluith
Dunskey Ho.
Pinminnoch
Colfin
Awhirk
Stoneykirk
LITTLE WHEELS
Portpatrick
B7084
Luce Sands
Milton
Stairhaven
A747
Fell Loch
Drumwalt
Mochrum Loch
Castle Loch
Culmazie
B7005
THE MACHARS
VISITOR CENTRE
Mains of Baldoon
Baldoon Sands
Port of Spittal Bay
Kildonan
B7042
A716
Auchenmalg
Auchenmalg Bay
Culshabbin
Airlies
B7052
Braehead
Kirkinner
Alticry
Barrachan
Whauphill
A746
Cairngarroch
Sandhead
Sandhead Bay
KIRKMADRINE STONES
Cairngarroch Bay
Money Hd.
CHAPEL FINIAN
Loch Head
Elrig
B7004
Culscadden
Clachanmore
Hole Stone Bay
ARDWELL GDNS
West Ardwell
Ardwell
Ardwell Mains
Chapel Rossan Bay
Ardwell Pt.
Milton Pt.
Mochrum
MOTE OF DRUCHTAG
B7085
Airyhassen
Drumtroddan
Sorbie
Garlieston
GALLOWAY GARDENS
Logan Mains
LOGAN BOTANIC GARDEN
Balgowan Pt.
LUCE BAY
Port William
DRUMTRODDAN STONES
Drummoddie
Monreith Mains
Moor of Ravenstone
Castlewigg
Cults
CRUGGLETON
Mull of Logan
LOGAN FISH POND MARINE LIFE CENTRE
Port Nessock or Port Logan Bay
Port Logan
Cairnywellan Hd.
Barsalloch Pt.
MONREITH ANIMAL WORLD, SHORE CENTRE AND MUSEUM
Monreith
Chapel Outon
Bishopton
PRIORY AND MUSEUM
Whithorn
Port Allen
Monreith Bay
WHITHORN TRUST DISCOVERY CENTRE
Portyerrock
B7065
Clanyard Bay
Low Clanyard
Laggantalluch Hd.
Kirkmaiden
Drummore
Damnaglaur
Cailliness Pt.
B7041
Glasserton
FELL OF CARLETON
ST NINIAN'S CAVE
Kidsdale
Tonderghie
ST. NINIAN'S CHAPEL
Isle of Whithorn
Port Castle Bay
Cutcloy
BURROW HEAD
Crammag Hd.
Maryport
Cairngaan
MULL OF GALLOWAY
Port Kemin
NX
0 2 4 6 miles
0 2 4 6 8 10 km

DUMFRIES AND GALLOWAY
SOLWAY FIRTH
SOLWAY
NITH ESTUARY
Blackshaw Bank
Lockerbie
Halleaths
Lochmaben Castle
Heck
Hightae
Smallholm
Woodlands
Greenhill
Blackford
Castlemilk
Kettleholm
Middleshaw
Kirkwood
Dalton
Ecclefechan
Hoddom Mains
Hoddomcross
Burnswark
Middlebie
Kirtlebridge
Eaglesfield
Brydekirk
Creca
Charlesfield
Carrutherstown
Langdyke
Plans
Clarencefield
Ruthwell
Duncan Savings Bank Museum
Mouswald
Cleughbrae
Collin
Greenlea
Woodside
Elizafield
Racks
Torthorwald
Georgetown
Isle of Man
Greystone
The Grove
Kelton
Glencaple
Bankend
Longbridgemuir
Horseholm
Shearington
Blackshaw
Eastpark
Bowhouse
Caerlaverock Castle
Caerlaverock
Wildfowl & Wetland Centre
Southfield
Cummertrees
Powfoot
Annan
Welldale
Newbie
Shawhill
Watchill
Dornock
Priestside
Bowness-on-Solway
Port Carlisle
Campfield Marsh
Cardurnock
Cardurnock Flats
Anthorn
Longcroft
Whitrigg
Kirkbride
Newton Arlosh
Skinburness
Silloth
Greenrow
Causewayhead
Calvo
Seaville
Blitterlees
Blackdyke
Kingside Hill
Abbey Town
Kelsick
Dundraw
Waverbridge
Beckfoot
Newtown
Pelutho
Highlaws
Aldoth
Blencogo
Carsethorn
Arbigland Gardens
John Paul Jones Cottage Museum
Overton
Burnfoot
Grune Pt
Channel of River Wampool
Channel of River Waver
South Solway Mosses
Thornhill Moss & Meadow
A75
A74(M)
A709
A701
B725
B724
B7020
B723
B7076
B722
B721
B6357
B5300
B5301
B5302
B5307
237
248
228
229
NY
0 1 2 3 miles
0 1 2 3 4 5 km

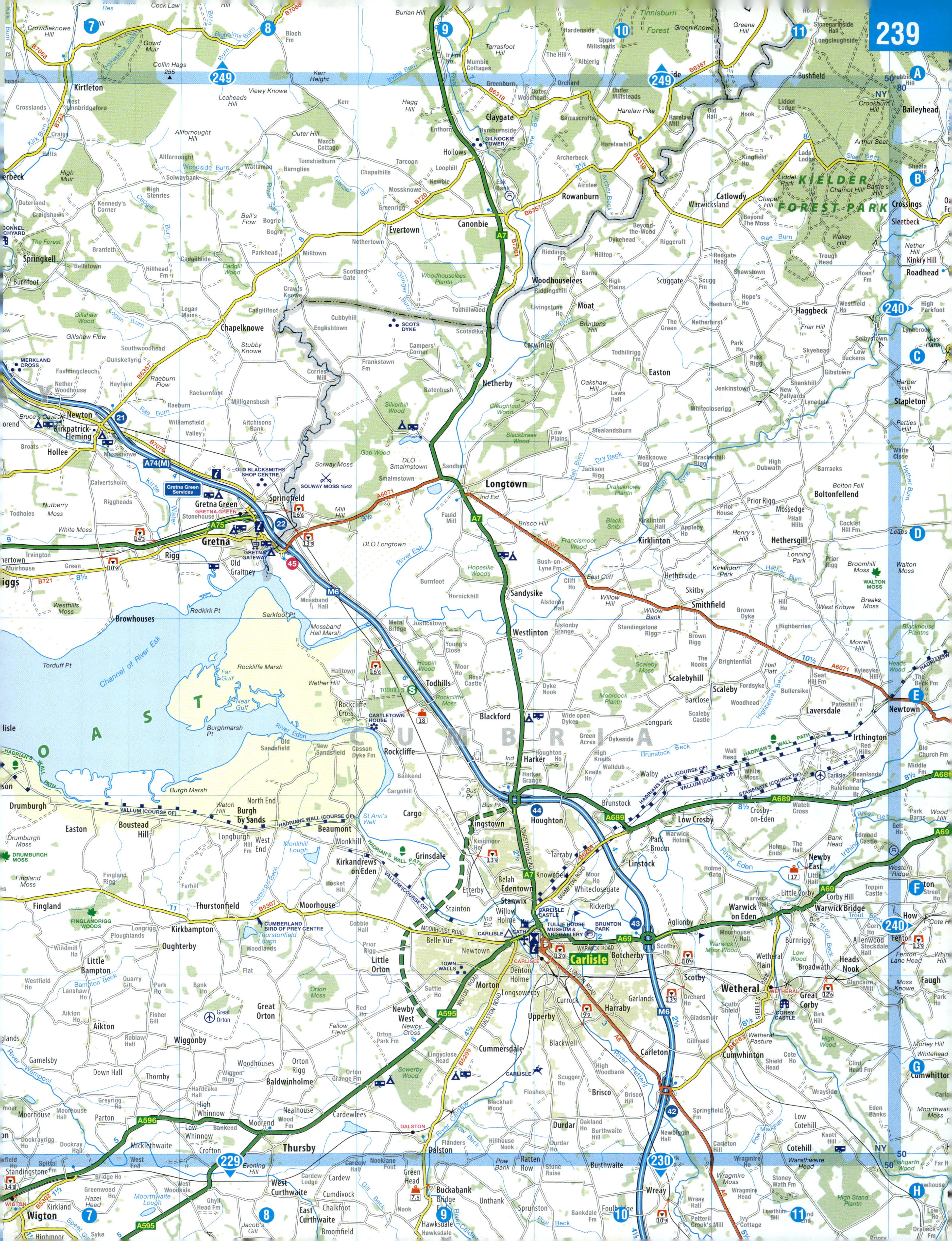

Kirtleton
Crowdieknowe Hill
Collin Hags
Leaheads Hill
Viewy Knowe
Kerr Height
Burian Hill
Tinnisburn Forest
Bushfield
Baileyhead
Greenburn
Claygate
Gilnockie Tower
Hollows
Canonbie
Rowanburn
Catlowdy
Warwicksland
Kielder Forest Park
Crossings
Sleetbeck
Roadhead
Kirkby Hill
Evertown
Springkell
Burnfoot
Chapelknowe
Woodhouselees
Moat
Haggbeck
Netherby
Easton
Stapleton
Scots Dyke
Merkland Cross
Newton
Kirkpatrick-Fleming
Hollee
Gretna Green
Gretna Green Services
Old Blacksmiths Shop Centre
Springfield
Gretna
Gretna Gateway
Rigg
Old Graitney
Solway Moss 1542
Longtown
Boltonfellend
Hethersgill
Kirklinton
Sandysike
Smithfield
Westlinton
Brackenhill
Redkirk Pt
Sarkfoot Pt
Browhouses
Torduff Pt
Channel of River Esk
Rockliffe Marsh
Burghmarsh Pt
River Eden
Rockcliffe
Castletown House
Todhills
Blackford
Harker
Scaleby
Scalebyhill
Laversdale
Newtown
Irthington
Walby
Brunstock
Cumbria
Solway Coast
Hadrian's Wall Path
Drumburgh
Easton
Boustead Hill
Burgh by Sands
Beaumont
Cargo
Kingstown
Houghton
Low Crosby
Crosby-on-Eden
Linstock
Kirkandrews on Eden
Grinsdale
Monkhill
Moorhouse
Thurstonfield
Fingland
Kirkbampton
Oughterby
Little Bampton
Cumberland Bird of Prey Centre
Etterby
Stanwix
Belah
Edentown
Whiteclosegate
Rickerby
Tullie House Museum & Art Gallery
Brunton Park
Carlisle
Botcherby
Belle Vue
Newtown
Denton Holme
Morton
Little Orton
Great Orton
Newby West
Currock
Harraby
Upperby
Scotby
Wetheral
Warwick on Eden
Warwick Bridge
Great Corby
Corby Castle
Aglionby
Aikton
Wiggonby
Cummersdale
Blackwell
Carleton
Cumwhinton
Baldwinholme
Brisco
Durdar
Cotehill
Dalston
Thursby
Crofton
Micklethwaite
Parton
Burthwaite
Wreay
Buckabank
Cardew
Cardewlees
Wigton
Kirkland
Curthwaite
Standingstone
A74(M)
A75
A7
A6071
A689
A69
A595
A596
A6
M6
B6318
B6357
B7068
B721
B5307
B5299
B6263
B5302

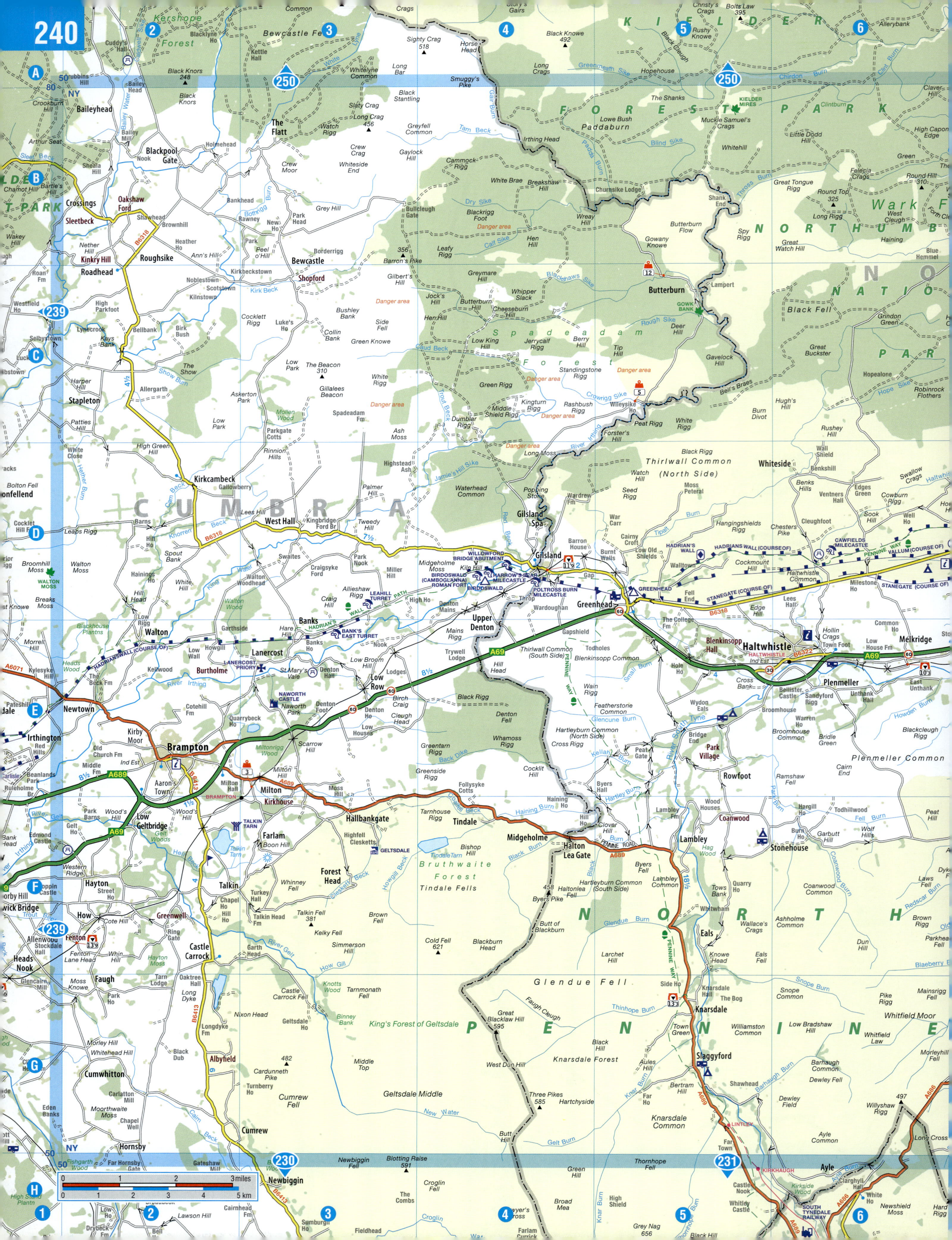

Kershope Forest
Bewcastle Fell
Kielder Forest Park
Northumberland National Park
Cumbria
Spadeadam Forest
Thirlwall Common (North Side)
Baileyhead
Blackpool Gate
The Flatt
Crossings
Sleetbeck
Kinkry Hill
Roughsike
Roadhead
Bewcastle
Shopford
Stapleton
Kirkcambeck
West Hall
Gilsland
Gilsland Spa
Greenhead
Haltwhistle
Melkridge
Plenmeller
Banks
Lanercost
Walton
Burtholme
Newtown
Irthington
Brampton
Milton
Kirkhouse
Hallbankgate
Tindale
Midgeholme
Halton Lea Gate
Lambley
Coanwood
Stonehouse
Eals
Knarsdale
Slaggyford
Ayle
Hayton
Talkin
Farlam
Castle Carrock
Cumwhitton
Cumrew
Albyfield
Newbiggin
Upper Denton
Low Row
Rowfoot
Park Village
Whiteside
Butterburn
Paddaburn
Hadrian's Wall (course of)
Pennine Way
Plenmeller Common
Glendue Fell
King's Forest of Geltsdale
Geltsdale Middle
Knarsdale Forest
Knarsdale Common
Bruthwaite Forest
Tindale Fells
Danger area
A69
A689
B6318
B6413
250
239
230
231
0 1 2 3 miles
0 1 2 3 4 5 km

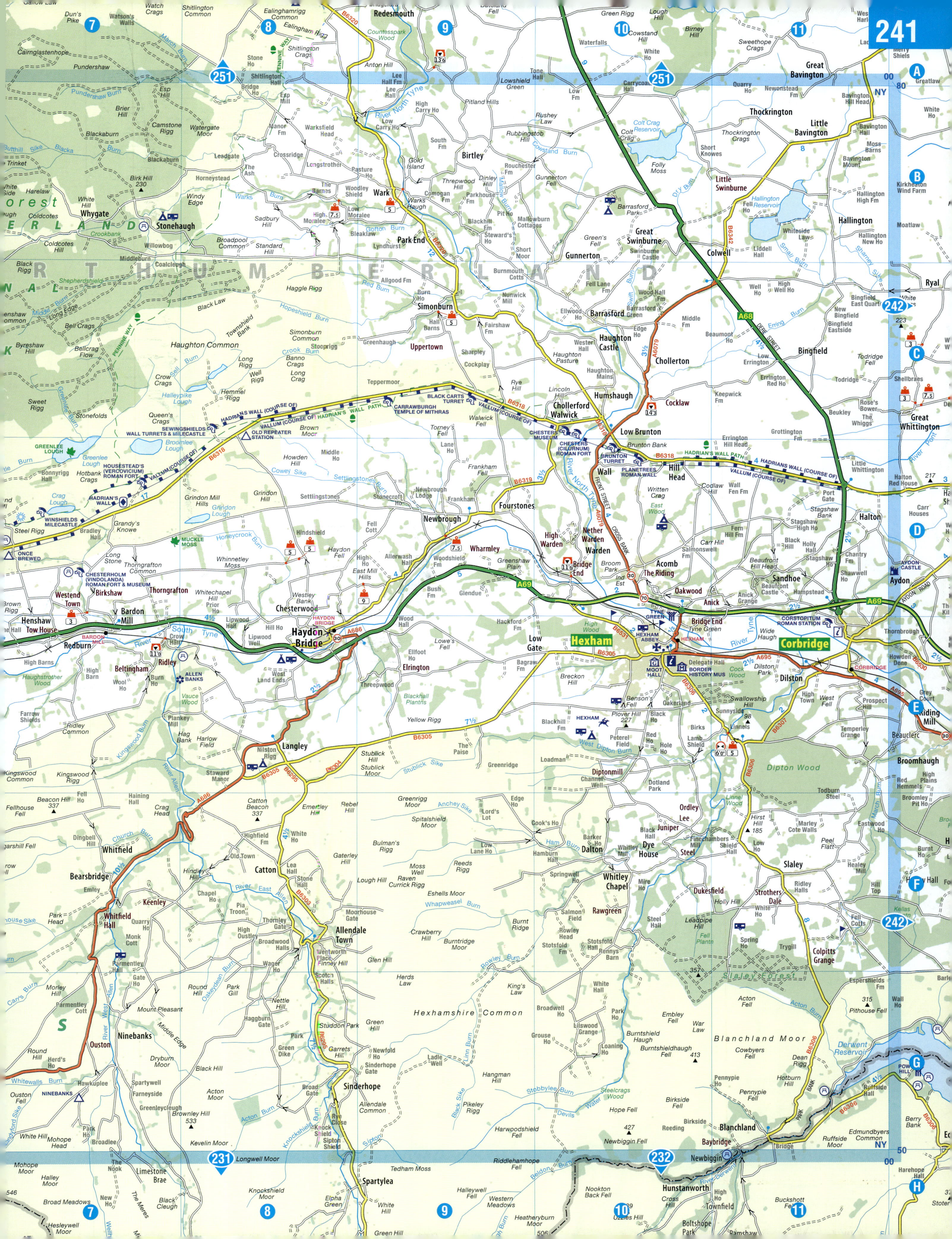

Dun's Pike
Watson's Walls
Watch Crags
Shittlington Common
Ealinghamrigg Common
Redesmouth
Countesspark Wood
Green Rigg
Lough Ho
Cowstand Hill
Birney Hill
Sweethope Crags
Waterfalls
White Ho
Cairnglastenhope
Pundershaw
Stone Ho
Shittlington Crags
Anton Hill
Great Bavington
251
Shittlington Hall
Lee Hall Fm
Lowshield Green
Tone Hall
Carrycoats Hall
Quarry Ho
Newstead Fm
Bavington Hill Head
Greatlaw
Esp Hill
Esp Mill
Pitland Hills
Low Fm
Thockrington
Little Bavington
Brier Hill
Camstone Rigg
Watergate Moor
Manor Fm
Warksfield Head
High Carry Ho
Low Carry Ho
Rushey Law
Colt Crag Reservoir
Colt Crag
Thockrington Crags
Bavington Hall
Blackaburn
Rubbingstob Hill
South Fm
Short Knowes
Moss Barns
Leadgate
Crossridge
Gold Island
Birtley
Coastand Burn
Bavington Mount
Blackaburn
Trinket
Longstrother
Rouchester Fm
Folly Moss
Birk Hill 230
Horneystead
The Ash
Pasture Ho
Threepwood Hill
Dinley Hill
Gunnerton Fell
Little Swinburne
Kirkheaton Wind Farm
White Hill
Whygate
Windy Edge
The Barnes
Woodley Shield
Wark
Comogan Fm
Parkhouse Fm
Barrasford Park
Hallington Reservoirs
Hallington High Fm
Stonehaugh
Sadbury Hill
High Moralee
Low Moralee
Warks Haugh
Blackhill Fm
Pit Ho
Mallowburn Cottages
Fell Ho
Hallington
Moatlaw
Coldcotes Hill
Broadpool Common
Standard Hill
Bleaklaw
Lyndhurst
Park End
Steward's Fm
Short Moor
Green's Fell
Great Swinburne
Swinburne Castle
Colwell
Liddell Hall
Whiteside Law
Hallington New Ho
Black Rigg
Middleburn
Coalcleugh
N O R T H U M B E R L A N D
Burnmouth Cotts
Gunnerton
Allgood Fm
Haggle Rigg
Burn Ho
Simonburn
Numwick Mill
Fell Lane Fm
Wood Hall Fm
Well Ho
High Well Ho
Ryal
Black Law
Long Edge
Hall Barns
Fairshaw Fm
Elwood Ho
Barrasford
Barrasford Green
Middle Fm
Bingfield East Quarter
New Bingfield
Bingfield Eastside
Bell Crags
Townshield Bank
Simonburn Common
Edge Ho
Beaumont Ho
Byreshaw Hill
Haughton Common
Stoopridge
Greenhaugh
Uppertown
Sharpley
Wester Hall
Haughton Castle
Bingfield
Bellcrag Flow
Banno Crags
Long Rigg
Well Rigg
Long Crag
Cockplay
Haughton Pasture
Haughton Mains
Chollerton
Low Errington
Todridge Fell
Crow Crags
Hemmel Rigg
Teppermoor
Rye Hill
Lincoln Hill
Humshaugh
Errington Red Ho
Todridge
Shellbraes
Sweet Rigg
Halleypike Lough
Queen's Crags
Stonefolds
Black Carts Turret
Carrawburgh Temple of Mithras
Hadrian's Wall (Course of)
Hadrian's Wall Path
Vallum (Course of)
Chollerford
Walwick
Cocklaw
Keepwick Fm
Beukley
Rose's Bower
The Whiggs
Great Whittington
Sewingshields Wall Turrets & Milecastle
Old Repeater Station
Brown Moor
Torney's Fell
Walwick Fell
Chesters Museum
Chesters (Cilurnum) Roman Fort
Low Brunton
Grottington Fm
Greenlee Lough
Broomlee Lough
Lane Ho
Brunton Bank
Errington Hill Head
Brunton Turret
Hill Head
Planetrees Roman Wall
Little Whittington
Bonnyrigg Hall
Hotbank Crags
Housesteads (Vercovicium) Roman Fort
Howden Hill
Middle Ho
Frankham Fell
Wall
Codlaw Hill
Wall Fen Fm
Halton Red House
Cowey Sike
Settlingstones
Front Street
Hadrian's Wall
Newbrough Lodge
Frankham
Fourstones
Written Crag
Port Gate
Stagshaw Bank
Crag Lough
Grindon Mill Hills
Grindon Hill
Settlingstones
Stonecroft Ho
East Wood
Halton
Carr Houses
Winshields Milecastle
Grandy's Knowe
Grindon Lough
Newbrough
Fell Cott
Fern Hill Fm
Stagshaw High Ho
Steel Rigg
Bradley Hall
Hindshield
High Warden
Nether Warden
Warden
Carr Hill
Black Hill
Holly Hall
Once Brewed
Muckle Moss
Honeycrook Burn
Haydon Fell
Wharmley
Salmonswell Fm
Chantry Fm
Long Stone
Thorngrafton Common
Whinnetley Moss
High Ho
Allerwash Hall
Woodshield Fm
Greenshaw Plain
Bridge End
Broom Park
Acomb
The Riding
Beaufront Hill Head
Stagshaw Ho
Aydon Castle
Chesterholm (Vindolanda) Roman Fort & Museum
East Mill Hills
Shawwell Ho
Aydon
Westend Town
Birkshaw
Thorngrafton
Whitechapel Hill
Prior Ho
Bush Fm
Glendue
Ind Est
Oakwood
Sandhoe
Beaufront Castle
Hampstead
Brown Rigg
Henshaw
Bardon Mill
Westley Bank
Chesterwood
Haydon Bridge
Wood Hall
Hackford Fm
Tyne Green
Bridge End
Tyne Green
Anick
Anick Grange
Corstopitum Roman Station
Thornbrough
Tow House
Redburn
Lipwood Hall
Lipwood Well
Hill Ho
Lowe's Fell
Low Gate
High Wood
Hexham
Hexham Abbey
Wide Haugh
Corbridge
Howden Dene
High Barns
Beltingham
Ridley
Allen Banks
West Land Ends
Ellfoot Ho
Elrington
Bagraw Fm
Breckon Hill
Moot Hall
Border History Mus
Delegate Hall
Cock Wood
Dilston Park
Dilston
Haughstrother Wood
High Barn
Wool Ho
Burn Ho
Vauce Wood
Threepwood
Blackhall Plantns
Benson's Fell
Oakerland
Swallowship Hill
Sunnyside
High Town
West Fell
Prospect Hill
Grey Court
Farrow Shields
Ridley Common
Plankey Mill
Yellow Rigg
Blackhill Fm
Plover Hill 227
Black Ho
Kinnels
Riding Mill
Beauclerc
Hag Bank
Harlow Field
Langley
The Paise
Hexham
Peterel Field
Red Ho
Hole Ho
Lamb Shield
Temperley Grange
Broomhaugh
Kingswood Common
Kingswood Rigg
Staward Manor
Nilston Rigg
Stublick Hill
Stublick Moor
Greenridge
Loadman
West Dipton Burn
Diptonmill
Channel Well
Dipton Wood
High Plains
Red Hemmels
Broomley Pit Ho
Beacon Hill 337
Fell Ho
Haining Hall
Catton Beacon 337
Emerley Hill
Rebel Hill
Greenrigg Moor
Edge Ho
Dotland Park
Todburn Steel
Fellhouse Fell
Crag Head
Spitalshield Moor
Lord's Lot
Ordley
Linnel Wood
Lee
Hirst Hill 185
Marley Cote Walls
Eastwood Ho
Dingbell Hill
Highfield Fm
White Ho
Cook's Ho
Juniper
Black Hall
Whitley Mill
Finechambers Shield Hall
Peel Flatt
Garshill Fell
Whitfield
Old Town
Bulman's Rigg
Low Lane Ho
Barker Ho
Hamburn Hall
Dalton
Dye House
Steel
Low Ho
Burnt Ho
Hindley Hill
Catton
Lea Hall
Gaterley Hill
Moss Well
Reeds Rigg
Slaley
Healey Mill
Bearsbridge
Emley
Chapel Ho
Stone Hall
Lough Hill
Raven Currick Rigg
Springwell Ho
Whitley Chapel
Mire Ho
Dukesfield
Holly Hill
Strothers Dale
Ridley Halls
Hill Top
Keenley
Pia Troon
Eshells Moor
Whapweasel Burn
Salmon Field
Rawgreen
Leadpipe Hill
Park Head
Whitfield Hall
Quarry Ho
Thornley Gate
Moorhouse Gate
Burnt Ridge
Steel Hall
White Ho
High Oustley
Allendale Town
Crawberry Hill
Rowley Head
Stotsfold Hall
Spring Ho
Trygill
Colpitts Grange
House Sike
Monk Cott
Broadwood Halls
Wentworth Place
Burntridge Moor
Stotsfold Fm
Rennys Barn
Fell Plantn
Parmentley Hall
Wager Ho
Finney Hill
Glen Hill
Slaley Forest
357
Gate Ho
Scotch Halls
Herds Law
King's Law
White Hall
Espershields Fm
Morley Hill
Round Hill
Oakeydean Burn
Park Gill
Acton Fell
315 Pithouse Fell
Parmentley Cott
Mount Pleasant
Nettle Hill
Hexhamshire Common
Broadwell Ho
Park Ho
Embley Fell
War Law
Wall Ho
Middle Edge
Haggburn Gate
Studdon Park
Green Hill
Allsgood Grange
Burntshield Haugh
Burntshieldhaugh Fell
413
Blanchland Moor
Derwent Reservoir
Ninebanks
Ouston
Green Dike
Park
Newfold Ho
Garrets Hill
Grouse Ho
Loaning Ho
Cowbyers Fell
Round Hill
Herd's Ho
Dryburn Moor
Sinderhope Gate
Ladle Well
Black Hill
Hangman Hill
Pennypie Ho
Dean Rigg
Pennypie Fell
Hotburn Hill
Whitewalls Burn
Spartywell
Farneyside
Broad Gate
Sinderhope
Acton Moor
Stobbylee Burn
Steelcrags Wood
Birkside Fell
Ouston Fell
Ninebanks
Hawkuplee
Greenleycleugh
Brownley Hill 533
Allendale Common
Pikeley Rigg
Hope Fell
Ruffside Hall
Knock Shield
Sipton Shield
Rye Close
Harwoodshield Fell
427
Reeding
Birkside
Blanchland
Baybridge
Park Bank
Berry Bank
White Hill
Mohope Head
Broadlee
Park Ho
Kevelin Moor
Newbiggin Fell
Ruffside Moor
Edmundbyers Common
Mohope Moor
The Nook
Limestone Brae
231
Longwell Moor
Riddlehamhope Fell
232
Newbiggin
Bridge
Halley Moor
Tedham Moss
Spartylea
Hunstanworth
Harehope Hall
546
Broad Meadows
New Ho
Black Cleugh
The Meres
Knockshield Moor
Elpha Green
Halleywell Fell
Western Meadows
Nookton Back Fell
Cross Hill
High Ho
Townfield
Buckshott Fell
Stotley
Hesleywell Moor
White Hill
Green Hill
Heatheryburn Moor
506
Ozzies Hill
Boltshope Park
Ramshaw
A68
A69
A686
A695
A6079
B6318
B6319
B6320
B6305
B6306
B6307
B6303
B6304
B6295
B6342
B6531
B6530
NY
80
50
00
A
B
C
D
E
F
G
H
7
8
9
10
11
242

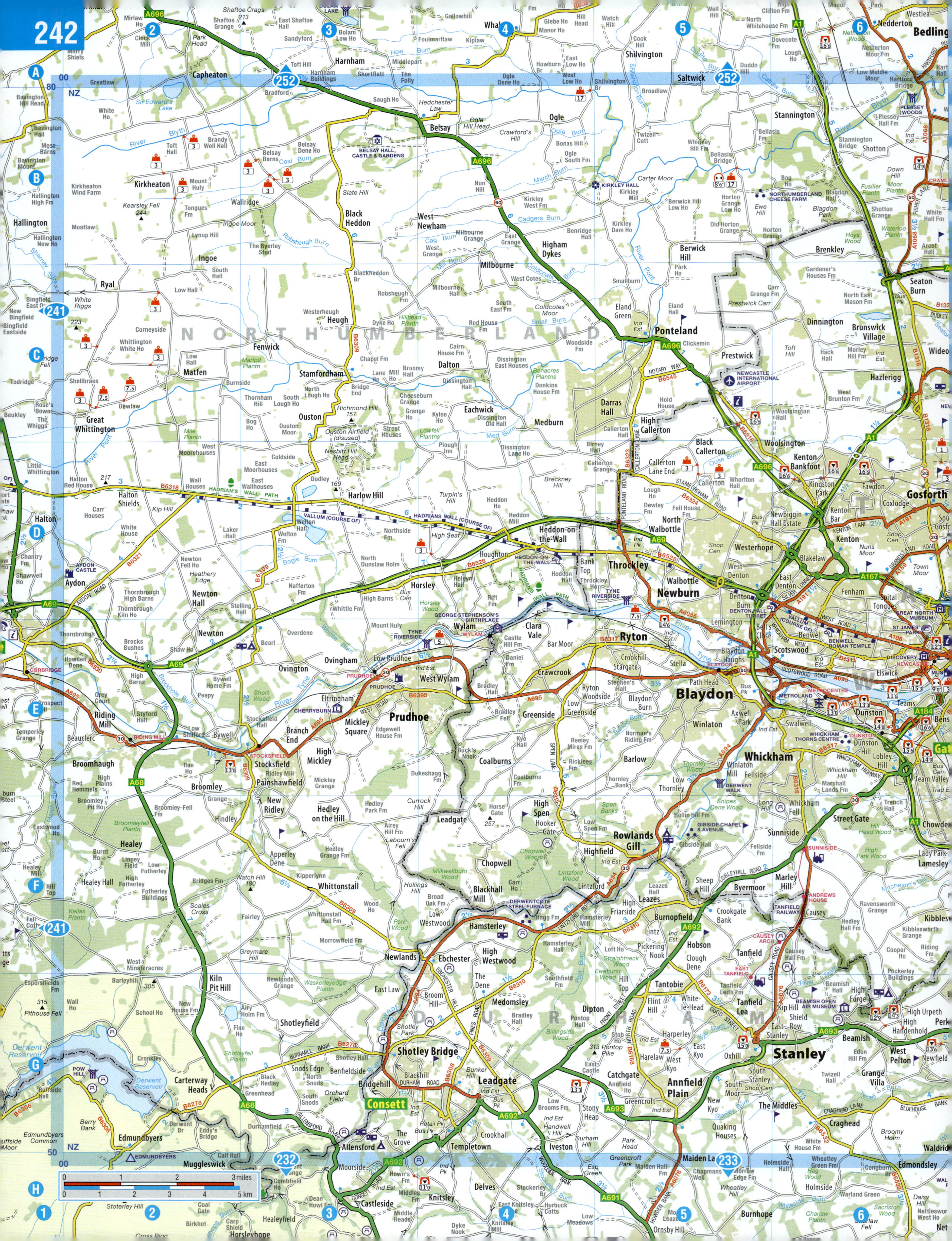

Capheaton
Harnham
Belsay
Ogle
Shilvington
Saltwick
Stannington
Bedlington
Kirkheaton
Wallridge
Hallington
Black Heddon
West Newham
Milbourne
Higham Dykes
Berwick Hill
Brenkley
Ingoe
Ryal
Fenwick
Matfen
Stamfordham
Heugh
Dalton
Eachwick
Medburn
Ponteland
Prestwick
Dinnington
Brunswick Village
Hazlerigg
Great Whittington
Ouston
Harlow Hill
Darras Hall
High Callerton
Black Callerton
Woolsington
Kenton Bankfoot
Gosforth
Halton
Aydon
Newton Hall
Newton
Horsley
Heddon-on-the-Wall
Throckley
Walbottle
Newburn
North Walbottle
Westerhope
Kenton
Fenham
Corbridge
Ovington
Ovingham
Wylam
Clara Vale
Ryton
Crawcrook
Blaydon
Scotswood
Elswick
Newcastle
Riding Mill
Prudhoe
Mickley Square
West Wylam
Greenside
Winlaton
Whickham
Dunston
Broomhaugh
Stocksfield
Painshawfield
High Mickley
Coalburns
Barlow
Broomley
New Ridley
Hindley
Hedley on the Hill
Leadgate
High Spen
Rowlands Gill
Sunniside
Healey
Chopwell
Highfield
Lintzford
Byermoor
Marley Hill
Whittonstall
Blackhall Mill
Hamsterley
Burnopfield
Newlands
Ebchester
High Westwood
Hobson
Tanfield
Kiln Pit Hill
Tantobie
Medomsley
Shotleyfield
Dipton
Stanley
Shotley Bridge
Annfield Plain
Catchgate
Carterway Heads
Leadgate
Consett
Bridgehill
Edmundbyers
Muggleswick
Allensford
Templetown
Iveston
Craghead
Castleside
Knitsley
Delves
Burnhope
Edmondsley
N O R T H U M B E R L A N D
D U R H A M
Derwent Reservoir
A696
A69
A68
A695
A692
A693
A694
A691
A1
252
241
232
233
0 1 2 3 miles
0 1 2 3 4 5 km

Blyth
Cowpen
Isabella Pit
Newsham
South Beach
New Delaval
South Newsham
East Hartford
Shankhouse
Nelson Village
Cramlington
East Cramlington
Seaton Sluice
Seaton Delaval Hall
New Hartley
Seaton
Hartley
Seaton Terrace
Seaton Delaval
Holywell
St Mary's Lighthouse
Southfield
Seghill
Annitsford
Dudley
Burradon
Backworth
Earsdon
Whitley Sands
Whitley Bay
Monkseaton
West Monkseaton
Shiremoor
Cullercoats
Blue Reef Aquarium
Marden
Killingworth
Palmersville
West Allotment
Preston
Tynemouth
Longbenton
Stephenson Railway Mus
Castle & Priory
North Shields
Amsterdam 15:30
Arbeia Roman Fort & Museum
Newcastle upon Tyne
Wallsend
Howdon
Willington
South Shields
Frenchman's Bay
Heaton
Jesmond
Tyne Tunnel
Jarrow
Segedunum Roman Fort
River Tyne
Bede's World
St Paul's Monastery
Westoe
The Leas & Marsden Rock
Horsley Hill
Harton
Marsden
Souter Lighthouse
Hebburn
Byker
Walker
Gateshead International Stadium
Cleadon
Whitburn Colliery
Whitburn
Felling
Boldon
East Boldon
West Boldon
Whitburn Bay
Gateshead
Heworth
Washington
Castletown
Sunderland
Seaburn
Fulwell
Roker
Monkwearmouth
National Glass Centre
Stadium of Light
Deptford
Pallion
Millfield
Springwell
Usworth
Birtley
Hendon
Wildfowl & Wetland Trust
Washington Old Hall
Washington Services
Penshaw Monument
Herrington Country Park
Grangetown
Ryhope
Ryhope Engines Museum
Shiney Row
Philadelphia
Newbottle
Chester-le-Street
Emirates Durham International Cricket Ground
Houghton-Le-Spring
Seaham
Hetton-le-Hole
Dalton-le-Dale
Murton
A1(M)
A19
A1018
A1231
A184
A194(M)
233
234
253

FIRTH OF CLYDE
Ailsa Craig
Swine Cave
Foreland Pt
Stranny Pt
Culzean Bay
CULZEAN CASTLE
Glasson Rock
Barwhin Pt
Maidenhead Bay
Port Murray
Castle Port
Maidens
Turnberry Pt
Turnberry Bay
Turnberry
Brest Rocks
Kirkoswald
Balkenna Isle
Townhead
Dowhill
Wright's Island
Dipple
Chaperdonan
Girvan Mains
Girvan
Glendoune
Horse Rock
Woodland Bay
Old Dailly
Penkill
Burnhead
Houdston
Ardwell
Kennedy's Pass
Pinminnoch
Tormitchell
Kirkland
Merkland
Pinmore
Currarie
Straid
Carleton Bay
Lendalfoot
Whilk Isle
CARLETON CASTLE
Games Loup
Balcreuchan Port
Port Vad
Bennane Head
Bennane Lea
Pinmore Mains
Pinwherry
Poundland
Colmonell
Knockdolian
Alticane
Liglartrie
Bellamore
Ballochmorrie
Heronsford
Ballantrae
Park End
Garleffin
Cosses
Balkissock
Barrhill
Cairnlea
Sgavoch Rock
Downan Pt
Downan
Glenapp Castle
Smyrton
Auchencrosh
Arecleoch Forest
Water of Tig
Water of App
Strawarren Fell
Wee Fell
Knockshin
Currarie Fm
A77
A714
B734
B7044
B741
256
236
NS
NX
0 1 2 3 miles
0 1 2 3 4 5 km

EAST AYRSHIRE
SOUTH AYRSHIRE
Carrick
GALLOWAY FOREST PARK
DUMFRIES AND GALLOWAY
Maybole
Whitefaulds
Collegiate Church
Crossraguel Abbey
Kirkmichael
Aitkenhead
Crosshill
Straiton
Patna
Polnessan
Burnfoot
Waterside
Dalmellington
Cathcartson Visitor Centre
Burnton
Bellsbank
Dalmellington Road
Kilmein Hill
Benquhat Hill
Benbain
Benbeoch
Headmark Moss
Benbraniachan
Craigmark Hill
Minnivey
Cutler
Laight
Cockclay
Snabb
Mossdale
Auchenroy Hill
Grimmet Hill
Turgeny
Dalnean Hill
Bogton Loch
Town's Common
Dalcairney
Pennyarthur Rigg
Craigengillan Home Fm
Bellsbank Plantn
Glenmuck Craig
Muck Water
Ness Glen
Bryan's Heights
Rowantree Craig
Glessel Hill
Muckle Eriff Hill
Eriff Plantn
Little Eriff Hill
Eriff
Loch Muck
Loch Doon
Loch Finlas
Herd's Hill
Wee Hill of Glenmount
Big Hill of Glenmount
Glessel
Cullendoch Hill
Craigencolon
Black Craig
Craiglee
Wee Hill of Craigmulloch
Loch Doon Castle
Craigmalloch
Coran of Portmark
Loch Head
Craigmawhannel
Craigfionn
Loch Macaterick
Macaterick
Black Garpel
Cairnsgarroch
Meaul
Craigchessie
Goat Craigs
Craigneld
Carlin's Cairn
Riders Rig
Craigrine
Meikle Craigtarson
Little Craigtarson
Corserine
Hoodens Hill
Mullwharchar
Millfire
Rhinns Of Kells
Round Loch of the Dungeon
Backhill of Bush (Bothy)
Tops of Craigeazle
Craigeazle
Silver Flowe
Long Loch of the Dungeon
Craignaw
Craignairy
Craig Neldricken
Loch Enoch
Redstone Rig
Loch Neldricken
Loch Valley
Snibe Hill
Little Millyea
Mc Whanns Stone
Darrou
Ellergower Knowe
Round Loch of Glenhead
Long Loch of Glenhead
Rig of the Jarkness
Craiglee
Glenhead
Mid Lump
Loch Dee
Kirrerroch
Craigencallie
Mulldonoch
Loch Trool
Southern Upland Way
Bruce's Stone
Glen Trool Lodge
Glen Trool
Glentrool Visitor Centre
Stroan Br
Glentrool Village
Caldons Wood
Black Craig
Jenny's Hill
Craiget Stone
Glencaird
Lion Ho
White Brae Top
Buchan Hill
Buchan Burn
Meaul
Rig of Loch Enoch
Gloon Burn
Scab Craigs
Benyellary
Rig of the Gloon
Kirn Brae
Benmore
Lagwholt Moor
Corse Craig
Bennan
Craigenjig
Craignacraddoch
Pulnagashel Burn
Palgowan
Knockcravie Burn
Water of Minnoch
Glentrool Forest
Arroch Hill
Sheil Hill
Mullmein Hill
Glencaird Hill
Ealunton Hill
Corraflecknoch Hill
Garwall Hill
Kentie Hill
Breckloch Hill
Black Island
Loch Moan
Suie Hill
Kirkennan Burn
Kirriemore Hill
Kirriereoch Loch
Rig of Kirriereoch
Carnirock Stone
Kirriereoch Hill
Little Spear
Merrick
Black Gairy
Cross Burn
Tarfessock
Carmaddie Brae
Castle on Oyne
Rig of Millmore
The Tores
Shalloch on Minnoch
Cow Craigs
Knocklach Burn
Craiglaigh Hill
Mount Shellie
Waterhead on Minnoch
Cairnfore Hill
Knapps
Ferter
Pindonnan Craigs
Loch Hill
Muck Water
White Clachrie
Black Clauchrie
South Ho
Darnaconnar
Feoch Bank Wood
Feoch Burn
Loch Crongart
Whitefell Plantn
Standard
Shell Hill
Drumjohn
Eldrick
Boreland
Killantringan
Corwar Mains
Corwar Ho
Knockcroon
Lochton
Lochend
Garrick
Corwar Fm
Cairnderry
Barjarg Hill
Dornal Hill
Drumlamford
River Cree
Laniewee Burn
Eldrick Hill
Shalloch
Rig of the Shalloch
Pinmore Burn
Craigmasheenie
Shiel Hill
Cornish Loch
Loch Riecawr
Maccallum
Whitespout Lane
Loch Goosie
Balloching Loch
Craigbrock Hill
Carrick Forest
Loch Brechowie
Craiglure
Loch Skelloch
Loch Bradan Reservoir
Linnshalloch
Stinchar Br
Tairlaw Plantn
Tairlaw Toll
Craig Dhu
Doon of Waterhead
Berclach Loch
Pinverains
Knockbuckle
Tallaminnoch
Stob Hill
Eldrick Hill
Linfern Loch
Lennie
Dunamoddie
The Tappins
River Stinchar
Balloch Plantn
White Row
Pinvalley
South Balloch
Dalwyne
Crs g of Darvine
Meowl of Blackrow
Meowl of Aldinna
Aldinna Loch
Witches Br
Corn Roy
Brandy Well
Glengap Hill
Haggis Hill
Craigenreoch
Polmaddie Hill
Pinbreck Hill
Cairn Hill
Fell Hill
Mid Hill
Drumneillie Hill
Larg Hill
Blackberry Knowe
Balloch Hill
Balskaig Hill
Shangue Plantation
Barr
Alton Albany Fm
Auchensoul
Auchensoul Hill
Doularg Hill
Water of Assel
Milton Br
Milton Hill
Mull of Miljoan
Green Hill
Doon Hill
Penwhapple Reservoir
Maxwellston Hill
Hadyard Hill
Pheelie
Delamford Hill
Craiginmoddie
Doughty Hill
Ferly Burn
Dalquhairn Burn
Auchengairn
Auchengairn Burn
Daljedburgh Hill
Craig Hill
Maxwelston
Craig
Dobbingstone
Dobbingstone Burn
Knockrochar
The Pilot
Knockincullech
Black Hill of Garleffin
Garleffin Fell
Knockoner
Glenalla Fell
Palmullan Burn
Back Fell
Knockskae
Clauchrie Hill
Barony Hill
Cairn Hill
Drumwhirn Fm
Whitehill Fm
Merkland
Meadow Main
Daily
Dalquharran Castle
Wallacetown
Drumbuie
Dalquharran Mains
Drummochreen
Pathhead
Glenton
Kilkerran
Ruglen
Tradunnock
Glengee Wood
Chapelton Wood
Auchalton Fm
Kirkbride Ho
Drumyork Hill
Black Hill of Knockgardner
Cawin Burn
Balbeg Burn
Bennan Hill
Kildoach Hill
Black Hill
Dersalloch Hill
Shalloch Burn
Chapel Burn
Baing Loch
Big Hill of The Baing
Baing
Craig Hill
Craigencallie
Linfairn
Genoch Inner Hill
Water of Girvan
Glenauchie
Clashwinne
Knockdon
Waterhead
Fence of Knockdon
Maratz Hill
Auldcraigoch Hill
Barbellie Wood
Milton
Balminnoch
Blairquhan
High Barbeth
Barskelly
Threave
Kings Hill
Lambdoughty Hill
Sclenteuch Moor Plantn
Gall Hill
Cloncaird Castle
Cloncaird Mains
Cloncaird Moor
Loch Spallander Reservoir
Glenside Hill
Troquhain
Carskeoch
Carskeoch Hill
Keirs Hill
Green Hill
Dalvennan
Dallowie
Whitehill
Patna Hill
Glastron
Shankston Loch
Dyrock Hill
Orchard Fm
Guiltree Hill
Glenhowe
Crawfordston
Barwheys
Drumellan
Chapelton
Lochlands
Barclay
Auchenwynd
Barmoor Wood
Craigfin Wood
Glenshalloch Wood
Drumochill
Longhill
Knockbrake
Craigdow
Craigdow Loch
Altewan
Middleridge
South Mains
Leffinwyne
North Mains
Baltersan Mains
Mochrum Loch
Mochrum
Mochrum Wood
Glenluie
Kirklandhill
Cargilston
Brockloch
East Brockloch
Nether Culzean
Howmoor Cott
Old Trees
Trees
Quarrel Hill
High Mains
Kirk Hill
Craigens Hill
Craig Hill

EAST AYRSHIRE
GALLOWAY FOREST PARK
DUMF
GALLO
Kirkconnel
Kelloholm
Burnside
Dalleagles
CATHCARTSON VISITOR CENTRE
Loch Doon
Carsphairn Forest
Carsphairn
Marbrack
Garryhorn
Lamloch
Brockloch
Bardennoch
Knowehead
Corlae
Benbuie
Cornharrow
Carroch
Craigdarroch
Dundeugh
Glenhoul
Kendoon
Castlefairn
Holmhead
Duchrae
Corriedoo
Drumbuie
Garroch
Glenlee
St John's Town of Dalry
Bogue
Balmaclellan
Blackcraig
Knocklea
Gibbshill
The Glenkens
Afton Reservoir
Southern Upland Way
CAIRN AVEL
KENDOON
BLOWPLAIN
Lochinvar Lodge
245
258
237
NS
NX
50
80
A713
B729
B7000
A702
A712
B741
3 miles
5 km

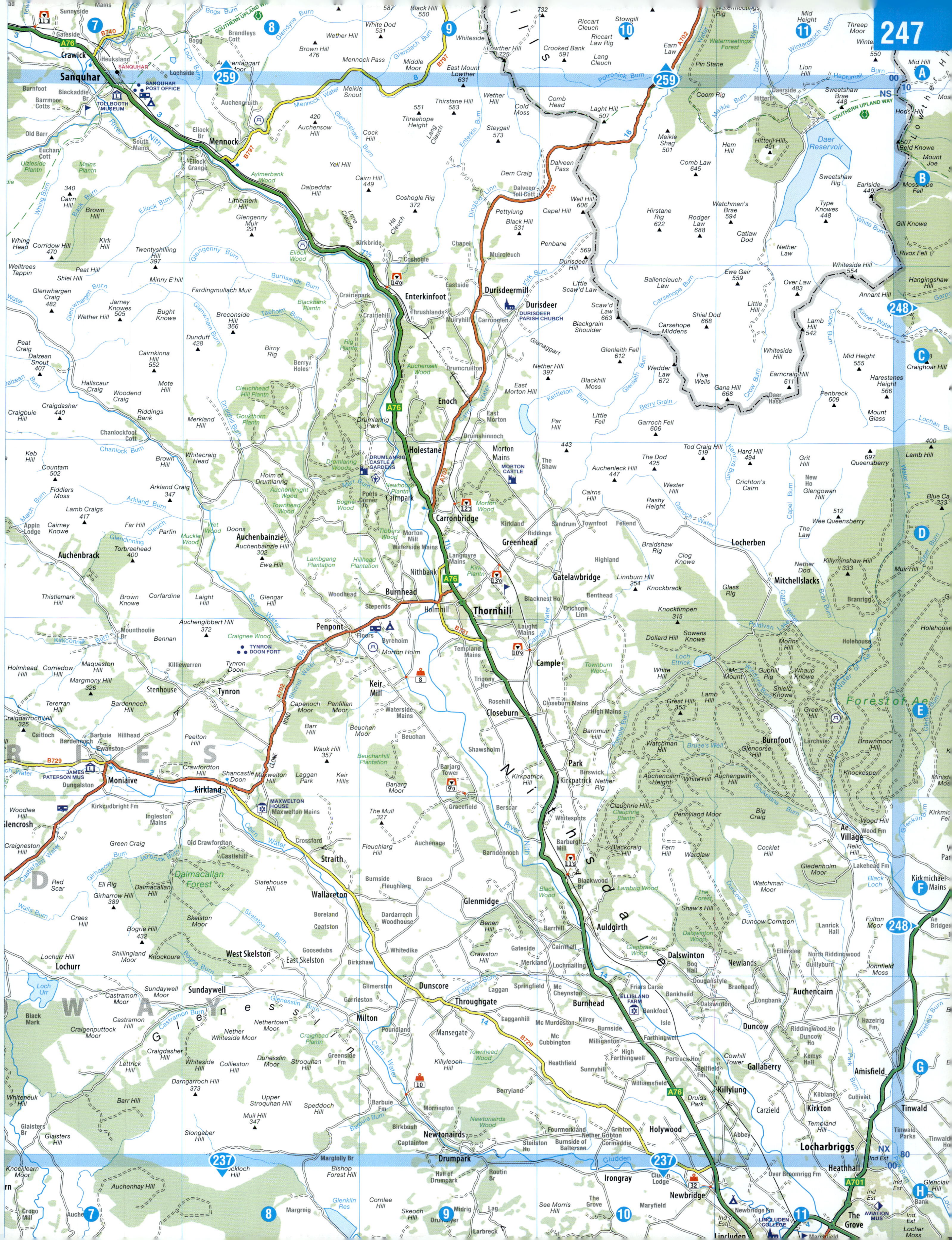

Sanquhar
Crawick
Sanquhar Post Office
Tollbooth Museum
Southern Upland Way
Mennock
Mennock Pass
Lowther Hill
Wether Hill
East Mount Lowther
Thirstane Hill
Enterkinfoot
Durisdeermill
Durisdeer
Durisdeer Parish Church
Dalveen Pass
Daer Reservoir
Enoch
Holestane
Drumlanrig Castle & Gardens
Morton Castle
Carronbridge
Greenhead
Gatelawbridge
Locherben
Mitchellslacks
Burnhead
Thornhill
Penpont
Tynron Doon Fort
Tynron
Cample
Keir Mill
Closeburn
Forest of Ae
Moniaive
James Paterson Mus
Kirkland
Maxwelton House
Park
Kirkpatrick
Nithsdale
Burnfoot
Ae Village
Straith
Wallaceton
Glenmidge
Auldgirth
Dalmacallan Forest
West Skelston
Lochurr
Dalswinton
Sundaywell
Dunscore
Throughgate
Milton
Glenesslin
Ellisland Farm
Duncow
Auchencairn
Killylung
Gallaberry
Amisfield
Kirkton
Tinwald
Holywood
Newtonairds
Drumpark
Locharbriggs
Heathhall
Irongray
Newbridge
Lincluden College
Aviation Mus
The Grove
A76
A702
A701
B729
B797
B740
B731
259
248
237

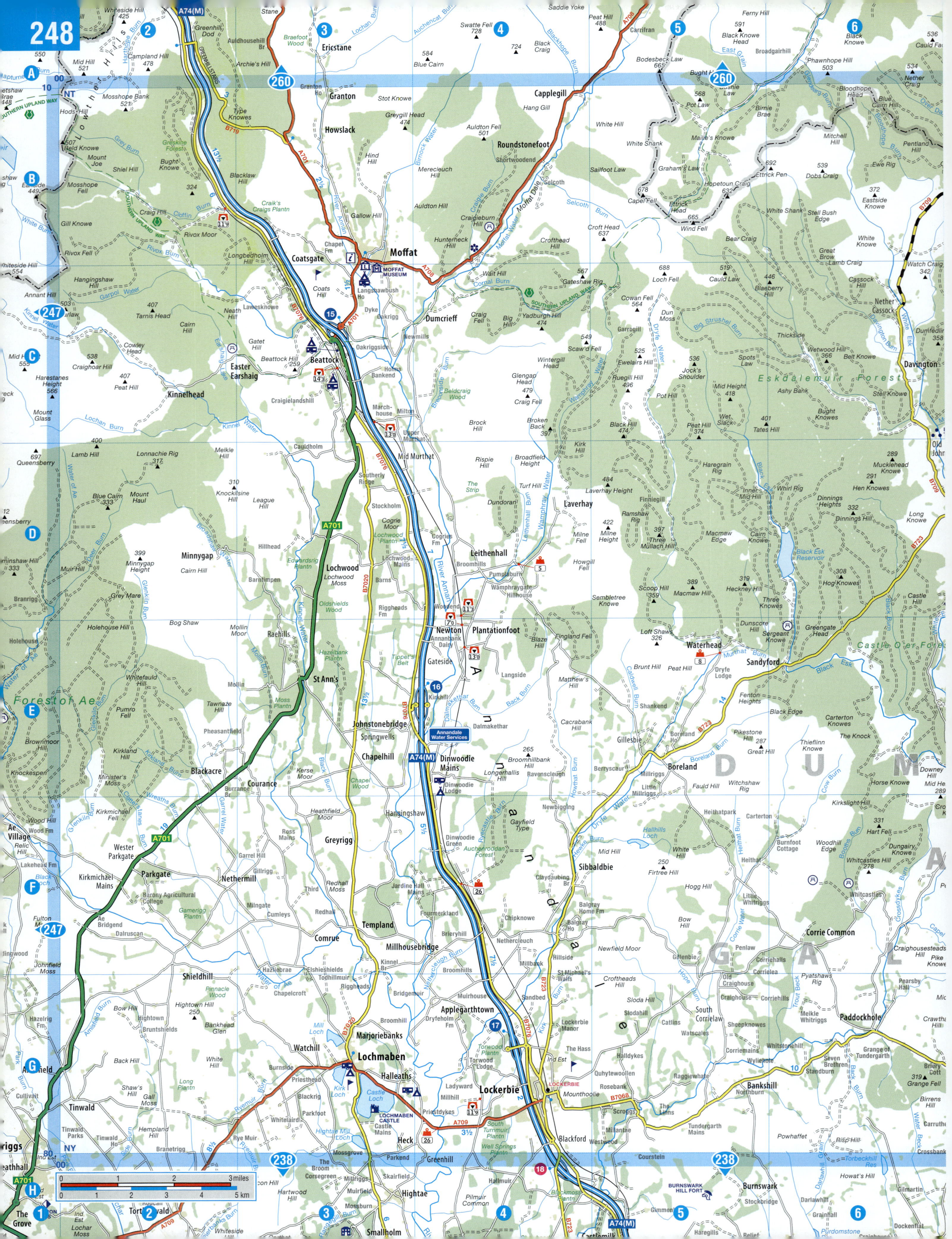

A74(M)
260
247
238
Moffat
MOFFAT MUSEUM
Ericstane
Granton
Howslack
Capplegill
Roundstonefoot
Coatsgate
Beattock
Dumcrieff
Easter Earshaig
Kinnelhead
Craigielandshill
Eskdalemuir Forest
Davington
Laverhay
Leithenhall
Minnygap
Lochwood
Newton
Plantationfoot
Waterhead
Sandyford
St Ann's
Johnstonebridge
Springwells
Chapelhill
Dinwoodie Mains
Annandale Water Services
Boreland
Blackacre
Courance
Forest of Ae
Greyrigg
Parkgate
Kirkmichael Mains
Western Parkgate
Nethermill
Sibbaldbie
Templand
Comrue
Millhousebridge
Corrie Common
Shieldhill
Applegarthtown
Paddockhole
Watchill
Lochmaben
Marjoriebanks
Halleaths
Lockerbie
LOCKERBIE
Bankshill
Tinwald
LOCHMABEN CASTLE
Heck
Blackford
Greenhill
Hightae
Burnswark
BURNSWARK HILL FORT
Smallholm
Tortorwald
The Grove
Queensberry
Lowther Hills
SOUTHERN UPLAND WAY
Annandale
DUMFRIES AND GALLOWAY
Ettrick Pen
Black Esk Reservoir
Castle O'er Forest
A701
A708
A709
B7020
B7076
B7068
B723
B719
B709
A723
0 1 2 3 miles
0 1 2 3 4 5 km

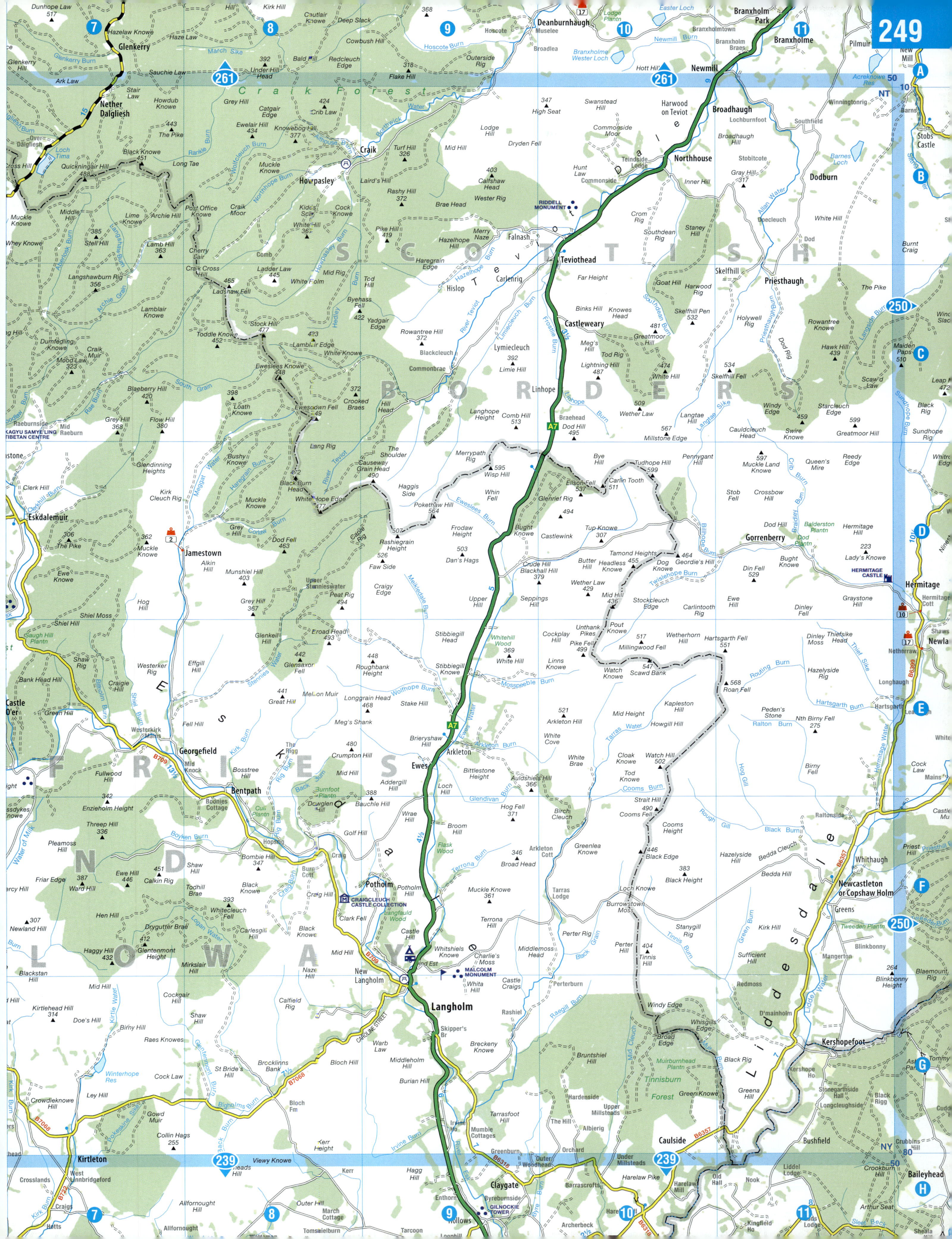

Dunhope Law
Glenkerry
Hazelaw Knowe
Haze Law
Kirk Hill
Coutlair Knowe
Deep Slack
Cowbush Hill
Deanburnhaugh
Muselee
Hoscote
Broadlea
Branxholm Park
Branxholmtown
Branxholme
Branxholm Braes
Pilmuir
New Mill
Newmill
Hott Hill
Easter Loch
Branxholme Wester Loch
Hoscote Burn
March Sike
Bald Hill
Redcleuch Edge
Outerside Rig
Under Hill Head
Flake Hill
Sauchie Law
Ark Law
Glenkerry Burn
Glenkerry Hill
Nether Dalgliesh
Stair Law
Howdub Knowe
Grey Hill
Catgair Edge
Crib Law
Craik Forest
Ewelair Hill
Knowebog Hill
The Pike
High Seat
Swanstead Hill
Harwood on Teviot
Broadhaugh
Lochburnfoot
Winningtonrig
Southfield
Barns
Stobs Castle
Acreknowe Res
NT
Loch Tima
Over Dalgliesh
Black Knowe
Long Tae
Craik
Turf Hill
Mid Hill
Lodge Hill
Dryden Fell
Commonside Moor
Teindside Lodge
Northhouse
Broadhaugh Hill
Stobitcote
Barnes Loch
Cross Hill
Quickningair Hill
Muckle Knowe
Howpasley
Laird's Hill
Calfshaw Head
Hunt Law
Commonside
Inner Hill
Gray Hill
Dodburn
Rashy Hill
Wester Rig
Riddell Monument
Allan Water
Doecleuch
White Hill
Muckle Knowe
Middle Hill
Post Office Knowe
Craik Moor
Kids Scar
Cock Knowe
Brae Head
Crom Rig
Lime Knowe
Archie Hill
White Hill
Pike Hill
Hazelhope Hill
Merry Naze
Falnash
Southdean Rig
Staney Hill
Dod
Burnt Craig
Whey Knowe
Stell Hill
Lamb Hill
Cherry Gair
Comb
Haregrain Edge
Teviothead
Skelfhill
Priesthaugh
Ladder Law
Mid Rig
Carlenrig
Far Height
Goat Hill
Harwood Rig
The Pike
Langshawburn Rig
Craik Cross Hill
Ladshaw Fell
White Folm
Tod Hill
Hislop
Byehass Fell
Binks Hill
Knowes Head
Skelfhill Pen
Holywell Rig
Rowantree Knowe
Lamblair Knowe
Stock Hill
Yadgair Edge
Castleweary
Toddle Knowe
Lamblair Edge
White Knowe
Rowantree Hill
Blackcleuch
Lymiecleuch
Meg's Hill
Greatmoor Hill
Hawk Hill
Maiden Paps
Dumfedling Knowe
Craik Muir
Mood Law
Eweslees Knowe
Commonbrae
Limie Hill
Tod Rig
Lightning Hill
White Hill
Skelfhill Fell
Dod Rig
Scaw'd Law
Blaeberry Hill
Loath Knowe
Ewesdown Fell
Crooked Braes
Linhope
Black Rig
Starcleuch Edge
Windy Edge
Wether Law
Langtae Hill
Grey Hill
Flow Hill
Hill Head
Langhope Height
Comb Hill
Braehead
Dod Hill
Sundhope Rig
Raeburnside
Mid Raeburn
Kagyu Samye Ling Tibetan Centre
Cauldcleuch Head
Swire Knowe
Greatmoor Hill
Millstone Edge
Long Rig
The Shoulder
Causeway Grain Head
Merrypath Rig
Wisp Hill
Bye Hill
Tudhope Hill
Pennygant Hill
Muckle Land Knowe
Queen's Mire
Reedy Edge
Glendinning Heights
Bushy Knowe
Black Burn Head
Haggis Side
Whin Fell
Ellson Fell
Carlin Tooth
Stob Fell
Crosbow Hill
Clerk Hill
Kirk Cleuch Rig
Muckle Knowe
White Hope Edge
Pokethaw Hill
Glenrief Rig
Eskdalemuir
Grey Hill
Dod Fell
Caple Rig
Frodaw Height
Bught Knowe
Castlewink
Tup Knowe
Gorrenberry
Dod Hill
Balderston Plantn
Dod Plantn
Hermitage Hill
The Pike
Muckle Knowe
Jamestown
Rashiegrain Height
Dan's Hags
Tamond Heights
Headless Knowe
Dog Knowe
Geordie's Hill
Lady's Knowe
Alkin Hill
Faw Side
Crude Hill
Blackhall Hill
Butter Hill
Din Fell
Bught Knowe
Hermitage Castle
Hermitage
Ewe Knowe
Munshiel Hill
Upper Stenniswater
Craigy Edge
Wether Law
Twislehope Burn
Grey Hill
Peat Rig
Upper Hill
Seppings Hill
Mid Hill
Stockcleuch Edge
Carlintooth Rig
Ewe Hill
Dinley Fell
Graystone Hill
Hog Hill
Shiel Moss
Shiel Hill
Glenkeil Hill
Broad Head
Stibbiegill Head
Whitehill Wood
Cockplay Hill
Unthank Pikes
Pout Knowe
Wetherhorn Hill
Hartsgarth Fell
Dinley Moss
Thiefsike Head
Newlands
Netherraw
Saugh Hill Plantn
Shaw Rig
Westerker Rig
Effgill Hill
Glensaxon Fell
Roughbank Height
Stibbiegill Knowe
White Hill
Pike Fell
Millingwood Fell
Linns Knowe
Watch Knowe
Scawd Bank
Roan Fell
Routing Burn
Hazelyside Rig
Longhaugh
Bank Head Hill
Craigie Hill
Great Hill
Mel on Muir
Longgrain Head
Wolfhope Burn
Stake Hill
Mossrpeeble Burn
Arkleton Hill
Mid Height
Kapleston Hill
Howgill Hill
Peden's Stone
Nth Birny Fell
Ralton Burn
Hartsgarth Burn
Castle O'er
Green Hill
Fell Hill
Meg's Shank
Westerkirk Mains
Crumpton Hill
Brieryshaw Hill
Arkleton
White Cove
Cloak Knowe
Watch Hill
Tod Knowe
Birny Fell
Cock Law
Georgefield
Mid Knock
The Rigg
Bosstree Hill
Mid Hill
Ewes
Bittlestone Height
Auldshiels Hill
White Brae
Fullwood Hill
Bentpath
Boonies Cottage
Burnfoot Plantn
Adderaill Hill
Loch Hill
Glendivan Burn
Cooms Burn
Enzieholm Height
Bauchle Hill
Hog Fell
Strait Hill
Cooms Fell
Raltonside
Threep Hill
Cull Plantn
Wrae Hill
Birch Cleuch
Cooms Height
Black Burn
Pleamoss Hill
Boyken Burn
Hopsrig
Golf Hill
Broom Hill
Priest Hill
Shaw Hill
Bombie Hill
Calkin Rig
Flask Wood
Arkleton Cott
Broad Head
Greenlea Knowe
Black Edge
Hazelyside Hill
Bedda Cleuch
Whithaugh
Friar Edge
Ward Hill
Ewe Hill
Black Knowe
Craig
Burn Cott
Potholm
Potholm Hill
Terrona Burn
Bedda Hill
Black Height
Newcastleton or Copshaw Holm
Todhill Brae
Whitecleuch Fell
Craig Hill
Craigcleuch Castle Collection
Langfauld Wood
Muckle Knowe
Tarras Lodge
Loch Knowe
Burrowstown Moss
Greens
Tweeden Plantn
Hen Hill
Drygutter Brae
Carlesgill Hill
Black Knowe
Clark Fell
Castle Hill
Terrona Hill
Perter Rig
Stanygill Rig
Kirk Hill
Newland Hill
Glentenmont Height
Haggy Hill
Mirkslair Hill
Naze Hill
Mid Hill
Whitshiels Knowe
Charlie's Moss
Malcolm Monument
Middlemoss Head
Perter Hill
Tinnis Hill
Sufficient Hill
Blinkbonny
Mangerton
Blackstan Hill
Mid Hill
New Langholm
Whita Hill
Castle Craigs
Perterburn
Redmoss
Blinkbonny Height
Blaemount Rig
Kirtlehead Hill
Cockgair Hill
Calfield Rig
Langholm
Rashiel
Windy Edge
D'mainholm
Doe's Hill
Shaw Hill
Birny Hill
Raes Knowes
Skipper's Br
Whisgills Edge
Broad Edge
Kershopefoot
Warb Law
Breckeny Knowe
Bruntshiel Hill
Muirburnhead Plantn
Black Rig
Kershope Ho
Winterhope Res
Cock Law
St Bride's Hill
Brocklinns Bank
Bloch Hill
Middleholm Hill
Burian Hill
Tinnisburn Forest
Hardenside
Upper Millsteads
Greena Hill
Green Knowe
Stonegarthside Hall
Longcleughside
Black Rigg
Crowdieknowe Hill
Ley Hill
Bighoms Burn
Bloch Fm
Gowd Muir
Collin Hags
Kerr Height
Irvine Burn
Irvine Ho
Mumbie Cottages
Tarrasfoot Hill
The Hill
Albierig
Caulside
Bushfield
Crubbins Hill
Kirtleton
Viewy Knowe
Heads Hill
Greenburn
Outer Woodhead
Orchard
Under Millsteads
Harelaw Pike
Liddel Lodge
Crookburn Hill
Baileyhead
West Linnbridgeford
Crosslands
Kerr
Hagg Hill
Claygate
Barrascrofts
Harelaw Mill
Old Hall
Nook
Allfornought Hill
Outer Hill
March Cottage
Enthorn
Byreburnside
Gilnockie Tower
Hollows
Hare
Archerbeck
Kingfield Ho
Arthur Seat
Craigs
Hatts
Allfornought
Tomshielburn
Tarcoon
Loophill
Shealla
SCOTTISH BORDERS
DUMFRIES AND GALLOWAY
Teviotdale
Eskdale
Ewes Water
Liddesdale
A7
B709
B7068
B7068
B722
B6318
B6357
B6399
B6319
NY
50
80
10
261
239
250
7
8
9
10
11
A
B
C
D
E
F
G
H

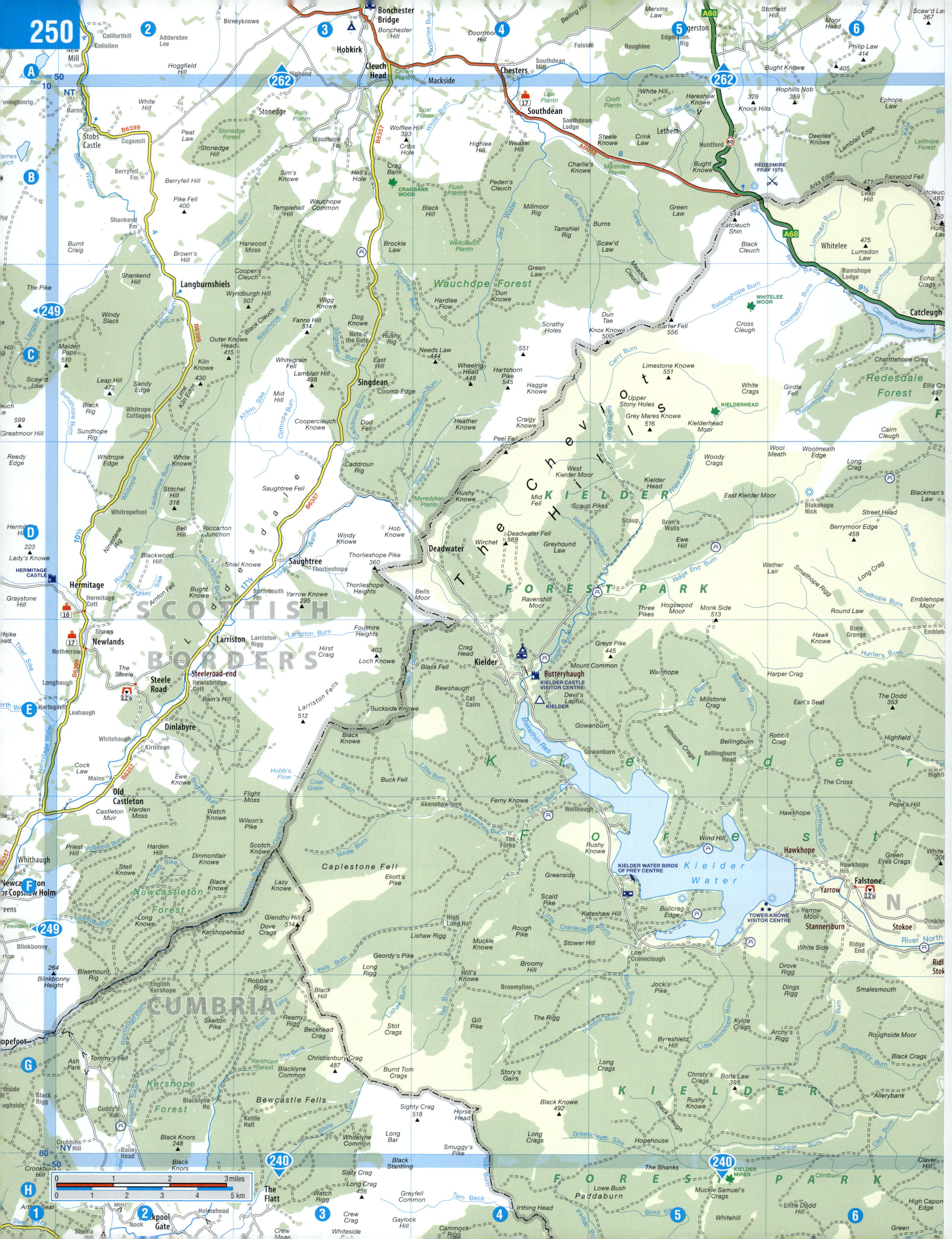

Bonchester Bridge
Hobkirk
Cleuch Head
Mackside
Chesters
Southdean
262
Stonedge
Stobs Castle
Cogsmill
B6399
Langburnshiels
249
Wauchope Forest
Singdean
Carter Fell
Catcleugh
A68
Redesmire
Whitelee
Redesdale Forest
Kielderhead
The Cheviots
Kielder
Deadwater
Saughtree
Liddesdale
Hermitage
Newlands
Steele Road
Larriston
Dinlabyre
SCOTTISH BORDERS
CUMBRIA
Old Castleton
Whithaugh
Newcastleton Forest
Kershope Forest
Bewcastle Fells
Butteryhaugh
Kielder Castle Visitor Centre
Kielder Forest Park
Kielder Water
Kielder Water Birds of Prey Centre
Tower Knowe Visitor Centre
Falstone
Hawkhope
Yarrow
Stannersburn
Caplestone Fell
Christianbury Crag
Paddaburn
240
The Flatt
Kershopefoot
Bailey Head
0 1 2 3 miles
0 1 2 3 4 5 km

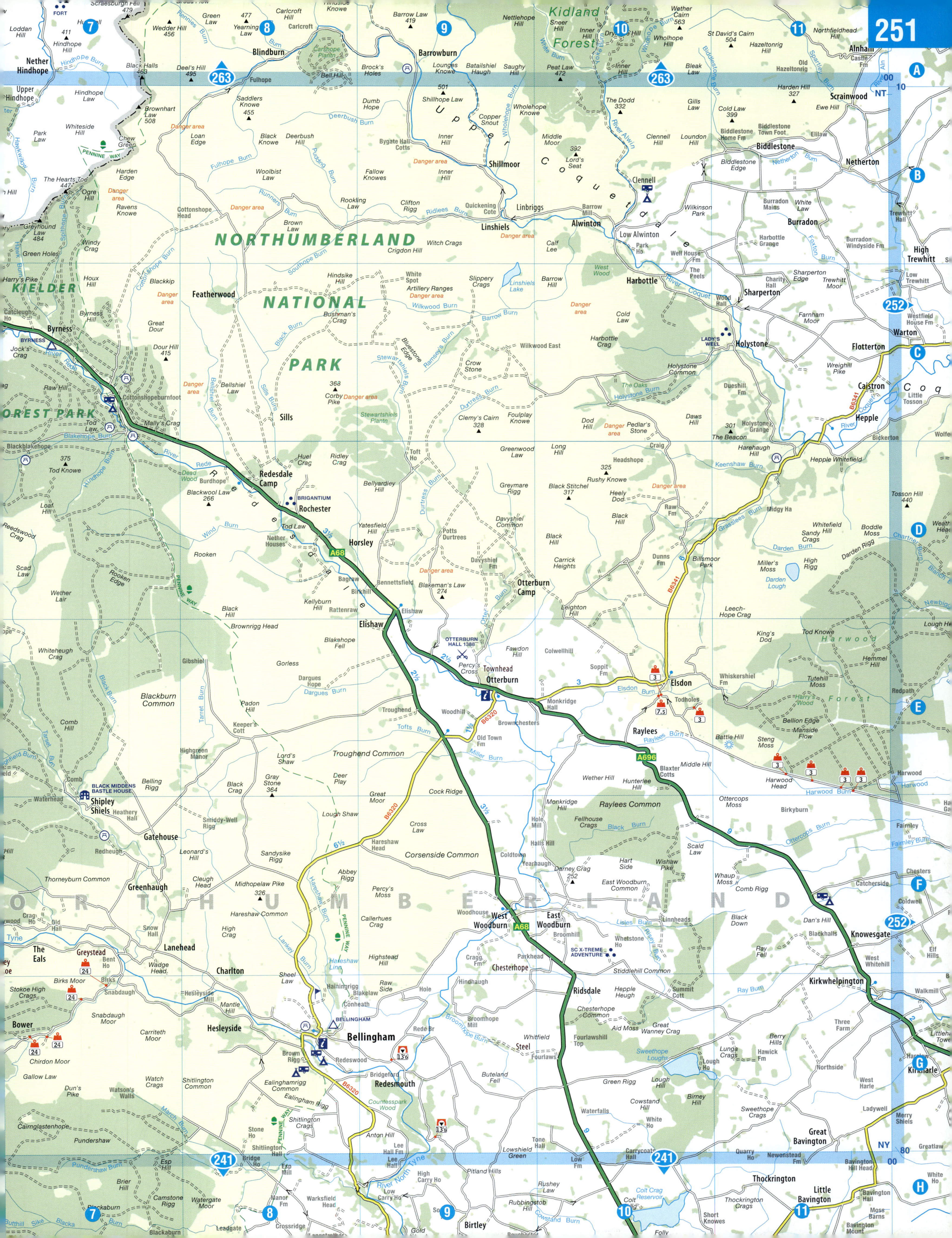

FORT
Scraesburgh Fell 479
Lodden Hill
Nether Hindhope
Upper Hindhope
Hindhope Hill
Hindhope Burn
Hindhope Law
Wedder Hill 456
Green Law
477 Yearning Law
Carlcroft Hill
Carlcroft
Blindburn
Carshope Plantn
Barrow Law 419
Barrowburn
Kidland Forest
Nettlehope Hill
Sneer Hill
Inner Hill
Wether Cairn 563
Wholhope Hill
St David's Cairn 504
Hazeltonrig Hill
Northfieldhead Hill
Alnham
Black Halls 463
Deel's Hill 495
Fulhope
Brock's Holes
Lounges Knowe
Bataishiel Haugh
Saughy Hill
Peat Law 472
Inner Hill
Bleak Law
Old Hazeltonrig
Harden Hill 327
Scrainwood
NT
Brownhart Law 508
Saddlers Knowe 455
Dumb Hope
501 Shillhope Law
Copper Snout
Wholehope Knowe
The Dodd 332
Gills Law
Cold Law 399
Ewe Hill
Whiteside Hill
Park Law
Chew Green
PENNINE WAY
Loan Edge
Black Knowe
Deerbush Hill
Deerbush Burn
Bygate Hall Cotts
Inner Hill
Middle Moor
Clennell Hill
Lounden Hill
Biddlestone Home Fm
Biddlestone Town Foot
Ellilaw
Biddlestone
Biddlestone Edge
Netherton
Danger area
Upper Coquetdale
Shillmoor
392 Lord's Seat
River Alwin
Clennell
Harden Edge
The Hearts Toe 447
Ogre Hill
Fulhope Burn
Woolbist Law
Fallow Knowes
Inner Hill
Burradon Mains
White Law
Burradon
Ravens Knowe
Cottonshope Head
Rookling Law
Clifton Rigg
Quickening Cote
Linbriggs
Barrow Mill
Alwinton
Wilkinson Park
Trewhitt Hall
Greyhound Law 484
Brown Law
Linshiels
Low Alwinton
Harbottle Grange
Burradon Windyside Fm
High Trewhitt
NORTHUMBERLAND NATIONAL PARK
Witch Crags
Crigdon Hill
Calf Lee
Park Ho
Well House Fm
Windy Crag
Green Holes
Southope Burn
The Peels
Harry's Pike
KIELDER
Houx Hill
Blackkip
Featherwood
Hindsike Hill
White Spot
Artillery Ranges
Slippery Crags
Linshiels Lake
Barrow Hill
West Wood
Harbottle
River Coquet
Charity Hall
Sharperton Edge
Sharperton
Trewhitt Moor
Low Trewhitt
Wood Hall
Wilkwood Burn
Barrow Burn
Cold Law
Farnham Moor
Westfield House Fm
Byrness Hill
Byrness
Bushman's Crag
Great Dour
Dour Hill 415
Bluestone Edge
Wilkwood East
Harbottle Crag
LADY'S WELL
Holystone
Flotterton
Warton
Jock's Crag
BYRNESS
River Rede
Stewartshiels Burn
Crow Stone
Holystone Common
Weygill Pike
Caistron
Raw Hill
Cottonshopeburnfoot
Bellshiel Law
368 Corby Pike
The Oaks
Holystone Burn
Dueshill Fm
Little Tosson
FOREST PARK
Tod Law
Mally's Crag
Sills
Stewartshiels Plantn
Clemy's Cairn 328
Foulplay Knowe
Dod Hill
Pedlar's Stone
Daws Hill
301 The Beacon
Holystone Grange
Hepple
Blakehope Burn
Blackblakehope
Bickerton
375 Tod Knowe
Huel Crag
Ridley Crag
Toft Ho
Greenwood Law
Long Hill
Craig
Headshope
Harehaugh Hill
Hepple Whitefield
Dead Wood
Burdhope
Redesdale Camp
325 Rushy Knowe
Keenshaw Burn
Blackwood Law 266
BRIGANTIUM
Rochester
Bellyardley Hill
Greymare Rigg
Black Stitchel 317
Heely Dod
Tosson Hill 440
Loaf Hill
Durtress Burn
Raw Fm
Midgy Ha
Reedswood Crag
Wind Burn
Tod Law
Yatesfield Hill
Davyshiel Common
Black Hill
Grasslees Burn
Whitefield Hill
Sandy Crags
Boddle Moss
Nether Houses
Horsley
Potts Durtrees
Black Hill
Darden Burn
Darden Rigg
Rooken
A68
Davyshiel Fm
Carrick Heights
Dunns Fm
Billsmoor Park
High Rigg
Scad Law
Rooken Edge
Bagraw
Bennettsfield
Danger area
Miller's Moss
Darden Lough
Wether Lair
Birkhill
Blakeman's Law 274
Otterburn Camp
B6341
Black Hill
Kellyburn Hill
Rattenraw
Elishaw
Leighton Hill
Leech-Hope Crag
Brownrigg Head
Elishaw
OTTERBURN HALL 1388
Tod Knowe
King's Dod
Harwood
Lough Hill
Whiteheugh Crag
Blakehope Fell
Fawdon Hill
Colwellhill
Gibshiel
Gorlesss
Percy's Cross
Townhead
Otterburn
Soppit Fm
Elsdon
Whiskershiel Fm
Tutehill Moss
Hemmel Hill
Redpath
Blackburn Common
Dargues Hope
Dargues Burn
Padon Hill
Troughend
Woodhill
Monkridge Hall
Elsdon Burn
Todholes
Harry Wood
Forest
Tarret Burn
Keeper's Cott
Tofts Burn
B6320
Brownchesters
Raylees
Raylees Burn
Bellion Edge
Manside Flow
Comb Hill
Old Town Fm
Battle Hill
Steng Moss
Highgreen Manor
Lord's Shaw
Troughend Common
Miller Burn
A696
Middle Hill
Blaxter Cotts
Deer Play
Wether Hill
Hunterlee Hill
Harwood Head
Harwood Burn
Harwood
BLACK MIDDENS BASTLE HOUSE
Shipley Shiels
Belling Rigg
Black Crag
Gray Stone 364
Great Moor
Cock Ridge
Monkridge Hill
Raylees Common
Ottercops Moss
Birkyburn
Waterhead
Heathery Hall
Smiddy-Well Rigg
Lough Shaw
Fellhouse Crags
Black Burn
Ottercops Burn
Fairnley
Gatehouse
Cross Law
Hole Mill
Scald Law
Fairnley Burn
Redheugh
Hareshaw Head
Halls Hill
Leonard's Hill
Sandysike Rigg
Corsenside Common
Coldtown
Hart Side
Wishaw Pike
Whaup Moss
Thorneyburn Common
Abbey Rigg
Yearhaugh
Darney Crag 252
East Woodburn Common
Comb Rigg
Chesters
Catcherside
Greenhaugh
Cleugh Head
Midhopelaw Pike 326
Percy's Moss
NORTHUMBERLAND
Coldwell
Hareshaw Common
Woodhouse
West Woodburn
East Woodburn
Linnheads
Black Down
Dan's Hill
Knowesgate
Crag
Old Hall
Callerhues Crag
Lisles Burn
Broomhill
Whetstone Ho
Blackhalls
Tyne
Snow Hall
High Crag
Parkhead
SC X-TREME ADVENTURE
Ray Fell
The Eals
Greystead
Bent Ho
Lanehead
Highstead Hill
Hareshaw Linn
Cragg Fm
Chesterhope
Stiddlehill Common
West Whitehill
Kirkwhelpington
Wadge Head
Charlton
Sheel Law
Raw Side
Hindhaugh
Ridsdale
Summit Cott
Birks Moor
Birks
Stokoe High Crags
Snabdaugh
Heslevside Mill
Mantle Hill
Hainingrigg
Blakelaw
Conheath
Hole
Hepple Heugh
Ray Burn
Walkmill
Chesterhope Common
Snabdaugh Moor
BELLINGHAM
Broomhope Mill
Three Farm
Bower
Carriteth Moor
Hesleyside
Bellingham
Rede Br
Aid Moss
Great Wanney Crag
Whitfield
Fourlawshill Top
Steel
Fourlaws
Berry Hills
Hawick Fm
Chirdon Moor
Brown Rigg
Redeswood
Sweethope Loughs
Lunga Crags
Lough Ho
Northside
Kirkharle
Gallow Law
Watch Crags
Shitlington Common
Ealinghamrigg Common
Bridgeford
Redesmouth
Buteland Fell
Green Rigg
Lough Hill
Birney Hill
West Harle
Dun's Pike
Watson's Walls
Ealingham Rigg
Countesspark Wood
Cowstand Hill
Waterfalls
Sweethope Crags
Ladywell
March Burn
Merry Shiels
Cairnglastenhope
Shitlington Crags
White Ho
Pundershaw
Stone Ho
Shitlington Hall
Anton Hill
Lee Hall Fm
Tone Hall
Great Bavington
Carrycoats Hall
Quarry Ho
Newonstead Fm
Greatlaw
NY
Esp Hill
Bridge Ho
Lee Hall
Low Fm
Lowshield Green
Pundershaw Burn
Esp Mill
High Carry Ho
Pitland Hills
Bavington Hill Head
Thockrington
Brier Hill
Carry Ho
Rushey Law
Colt Crag Reservoir
Little Bavington
Bavington Hall
White Ho
Camstone Rigg
Watergate Moor
Manor Fm
Warksfield Head
Rubbingstob Hill
Short Knowes
Thockrington Crags
Moss Barns
Blackaburn
Birtley
Leadgate
Crossridge
Blackaburn
Gold
Folly
Bavington Mount

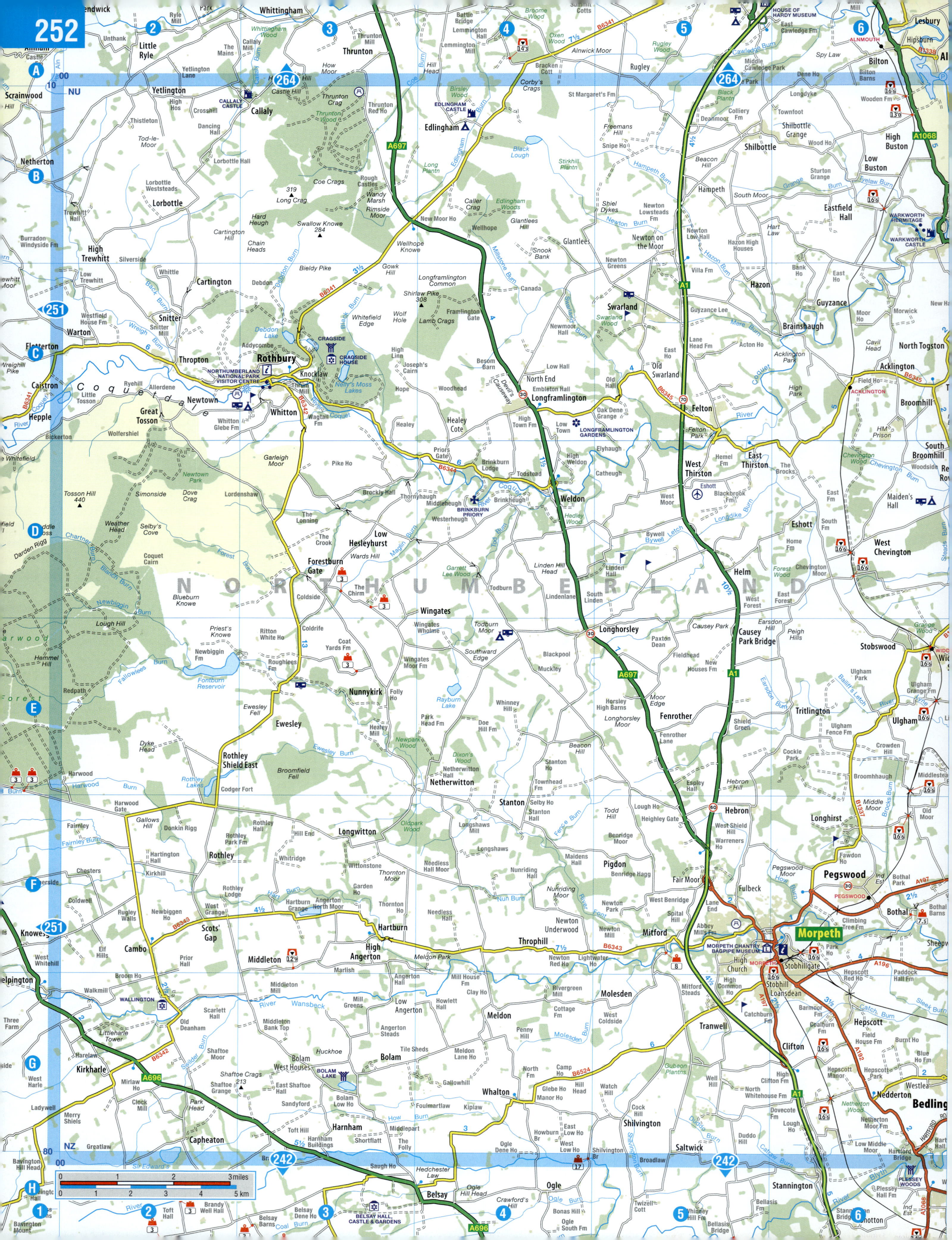

NORTH SEA
Marden Rocks
Alnmouth Bay
Birling
Warkworth
Warkworth Harbour
Pan Pt
Wellhaugh Pt
Coquet Island
Gloster Hill
Amble
Moorhouse Fm
High Hauxley
Togston Hall
Radcliffe
Low Hauxley
Togston Barns
Togston East Fm
A1068
Danger area
Ladyburn Lake
Hadston
DRURIDGE BAY
Druridge Bay
Whitefield Ho
Chibburn Fm
High Chibburn
Widdrington
Hemscott Hill
Highthorn
Cresswell
Warkworthlane Cott
Hagg House
Ellington
Cresswell Home Fm
Linton
Lynemouth
East Moor Fm
Potland Fm
Works
Woodhorn
QUEEN ELIZABETH II
WOODHORN COLLIERY MUS
WOODHORN CHURCH MUS
A189
Bus Cen
Woodbridge
Ashington
Newbiggin-by-the-Sea
Newbiggin Bay
Hirst
North Seaton
WANSBECK
North Seaton Colliery
Stakeford
West Sleekburn
Guide Post
Scotland Gate
Bomarsund
Cambois
Choppington
East Sleekburn
Bedlington Station
Mount Pleasant Fm
North Blyth
Bebside
Cowpen
Blyth
Humford Mill
Isabella Pit
East Hartford
Low Horton Fm
Newsham
South Beach
New Delaval
Laverock Hall
South Newsham
Gloucester Lodge Fm
Shankhouse
Lysdon Fm
Stickley Fm
Meggie's Burn
265
243
NU
NZ

Coul Pt.
Coul
Sunderland
Kilchoman
Machir Bay
Gortan
Conisby
Lyrabus
Blackrock
Redhouse
Esknish
Daill
Bridgend
Islay Ho
267 BEINN DUBH
Camas an Staca
Cabrach
Strone
JURA HOUSE WALLED GARDEN
Am Fraoch Eilean
Rubha na Tràille
Brosdale I.
Bruichladdich
Kilchiaran Bay
Bowmore
BOWMORE ROUND CHURCH
ISLAY LIFE MUSEUM
RHINNS
Tormisdale
Port Charlotte
LOCH INDAAL
Gartbreck
Mulindry
Cattadale
McArthur's Hd.
PORT ASKAIG 2:05
Lossit
Lossit Pt.
Kelsay
OF
Nerabus
ISLAY
Laggan Pt.
Laggan
Bridge Ho
471 BEINN BHAN
491 BEINN BHEIGEIR
Carraig Mhór
Ardtalla
Rubha na Faing
Claddach
Easter Ellister
Wester Ellister
Portnahaven
Port Wemyss
Orsay
Rinns Pt.
I S L A Y
Duich
Torra
Loch Beinn Uraraidh
Claggain Bay
LAGGAN BAY
Glenegedale
Castlehill
Kintour
Ardmore Pt.
KILDALTON CHURCH AND CROSSES
2:20
347 BEINN SHOLUM
Arivoichallum
Eilean Craobhach
Eilean a'Chuirn
Port Alsaig
Rubha Mòr
Machrie Hotel
Kintra
Leorin
ARDBEG DISTILLERY
Eilean Bhride
Kildaton Ho
Kilbride
Cornabus
Lagavulin
Imeraval
Port Ellen
Ardbeg
Eilean Imersay
Lower Cragabus
LAGAVULIN DISTILLERY
Laphroaig
LAPHROAIG DISTILLERY
Texa
Dùn Mór Ghil
THE OA
Risabus
Lower Killeyan
Upper Killeyan
Kinabus
Inerval
AMERICAN MONUMENT
Mull of Oa
Rubha nan Leacan
A R G Y
A N
B U T
Earadale Pt.
N O R T H C H A N N E L
Rubh'a'Mharaiche
Rathlin Island
MULL OF KINTYRE
Bushmills
Ballycastle Bay
Ballycastle
0 2 4 6 miles
0 2 4 6 8 10 km

Kilberry
Tarbert
Gigha Island
Campbeltown
Carradale
Saddell
Sanda Island
Loch Ranza
Lochranza
NORTH ARRAN
ISLE OF ARRAN
Brodick
Brodick Bay
Lamlash
Holy Island
Whiting Bay
Kildonan
Blackwaterfoot
Machrie Bay
KILBRANNAN SOUND
SOUND OF BUTE
ISLAND OF BUTE
Rothesay
Kingarth
Inchmarnock
Ailsa Craig
Sound of Pladda
Pladda
NORTH AYRSHIRE
Southend
Carskiey
Machrihanish
Skipness
Claonaig
Clachan
Tayinloan
Muasdale
Glenbarr
Bellochantuy
Westport
Davaar Island
Ardrossan 0:55

FIRTH OF CLYDE
Isle of Arran
NORTH AYRSHIRE
Brodick
Merkland
Glenrosa
Cladach
Old Quay
BRODICK CASTLE
ARRAN AROMATICS VISITOR CENTRE
ISLE OF ARRAN HERITAGE MUSEUM
Merkland Pt
Wine Port
Port nam Balach
Maol Donn 368
ARDROSSAN 0:55
BRODICK 0:55
Strathwhillan
Corriegills Pt
North Corriegills
South Corriegills
Dun Dubh
Clauchland Hills
Clauchlands Fm
Clauchlands Pt
Hamilton Isle
Clauchlands
Kerr's Port
Margnaheglish
Blairbeg
Lamlash
Cordon
Monamore Br
Monamore Glen
The Ross 311
Benlister Glen
Fairy Glen
Glen Cloy
Glen Ormidale
Glen Shurig
THE STRING
Holy Island
Mullach Beag
Mullach Mor
314
White Pt
Pillar Rock Pt
Gortonallister
The Knowe Fm
Auchencairn
Kingscross Pt
Kingscross
Knockenkelly
Sandbraes
Urie Loch
Glas Choirein
Borrach
North Kiscadale
Whiting Bay
South Kiscadale
Cnoc an Fheidh
Cnoc Mòr
GLENASHDALE FALLS
Glenashdale Burn
Largymore
Auchareoch
Kilmory Water
Largymeanoch
Largybeg
Largybeg Pt
Cnoc Craobhach
Cnoc na Garbad
Cnoc na Comhairle
Port na Gaillin
Dippin Head
Dippin
Margenaish Fm
Levencorroch Hill
Levencorroch
Southbank
East Bennan
West Bennan
STRUEY ROCKS
Bennan Head
Auchenhew
Drumla
Kildonan
Porta Leacach
Port a'Ghillie Ghlais
Porta Buidhe
Port Dearg
Sound of Pladda
Pladda
Saltcoats
ARDROSSAN HARBOUR
South Bay
NORTH AYRSHIRE MUSEUM
Outer Nebbock
Broad Craig
Culzean Bay
CULZEAN CASTLE
Glasson Rock
Barwhin Pt
Maidenhead Bay
Morriston
Balvaird
Birniehill
Port Murray
Kirkoswald
A841
B880
A719
255
266
244

EAST AYRSHIRE
SOUTH AYRSHIRE
Irvine
Kilmarnock
Ayr
Troon
Prestwick
Irvine Bay
Lady Isle
LARNE 1:50 (MARCH-OCT)
Auchenharvie
Stevenston
Cunninghamhead
Kilmaurs
Onthank
Perceton
Girdle Toll
Stanecastle
Springside
Knockentiber
Dreghorn
Fullarton
Crosshouse
Gatehead
Riccarton
Shortlees
Hurlford
Crookedholm
Galston
Shewalton
Drybridge
Dundonald
Lappock Rock
Barassie
Muirhead
Loans
Symington
Hansel Village
Bogend
Craigie
Monkton
Tarbolton
Mauchline
Crossroads
Sornhill
Auchmillan
Mossgiel
Failford
Haugh
Ochiltree
Mossblown
St Quivox
Woodfield
Newton on Ayr
Wallacetown
Whitletts
Annbank
Stair
Trabboch
Coylton
Joppa
Hillhead
Coalhall
Drongan
Belston
Craigie
Holmston
Masonhill
Castlehill
Belmont
Seafield
Doonfoot
Alloway
Kincaidston
Laigh Glengall
Fisherton
Dalrymple
Hollybush
Culroy
Minishant
Maybole
Whitefaulds
Kirkmichael
Aitkenhead
Patna
Polnessan
Littlemill
Rankinston
Hayhill
Sinclairston
Barbieston
Waterside
Burnfoot
Bracken Bay
Longhill Pt
Bell Rock
Meikle Craigs
Black Rocks
Little Craigs
Seal Rock
South Sands
South Bay
Pan Rocks
North Sands
Mill Rock
Stinking Rocks
Barassie Sands
Blackburn Rocks
St Nicholas Rock
Cunning Park
GLASGOW PRESTWICK
PRESTWICK INTERNATIONAL AIRPORT
DUNDONALD CASTLE
SCOTTISH MARITIME MUSEUM
GLASGOW VENNEL MUSEUM
DEAN CASTLE
DICK INSTITUTE
KILMARNOCK SERVICES
BURNS HOUSE MUSEUM
BACHELOR'S CLUB
AYR GORGE WOODLANDS
SOMERSET PARK
MACLAURIN GALLERY & ROZELLE HOUSE
BURNS COTTAGE
TAM O'SHANTER EXPERIENCE & BURNS NATIONAL HERITAGE PARK
HEADS OF AYR FARM PARK
CROSSRAGUEL ABBEY
COLLEGIATE CHURCH
ELECTRIC BRAE
A71
A77
A78
A76
A70
A719
A713
A759
A735
A736
B730
B742
B743
B744
266
267
245
258

Newmilns
Greenholm
Darvel
Galston
Sornhill
Mauchline
Catrine
Sorn
Auchinleck
Cumnock
Netherthird
Ochiltree
Logan
Lugar
Cronberry
Muirkirk
Smallburn
Kames
Glenbuck
Caldermill
Gilmourton
Stobieside
Drumclog
New Cumnock
Pathhead
Mansfield
Connel Park
Bankglen
Burnside
Dalleagles
Kirkconnel
Kirkland
Kelloholm
Skares
Holmhead
Craigens
Dalblair
Netherwood
Auchmillan
Haugh
Whiteflat
Welton
EAST
AYRSHIRE
River Ayr
River Irvine
Glenbuck Loch
Lugar Water
River Nith
Glengavel Reservoir
Bibblon Hill 431
Wedder Hill 434
Tulloch Hill 300
Cairn Table 593
Wardlaw Hill 497
Avisyard Hill 330
Cairnscadden Hill 314
Carnivan Hill 324
Benalt 382
Black Hill 431
Crook Brae 385
Clocklowie Hill 442
Halfmerk Hill 453
Nivistion Hill 460
Stony Hill 562
Priesthill Height 493
Starpet Rig 451
Spirebush Hill 469
Nutberry Hill 522
Goodbush Hill 475
Auchengilloch 462
Dungavel Hill 457
Middlefield Law 466
Fingland Hill 462
Auchtitench Hill 467
Loudoun Hill 316
Distinkhorn 386
Edge Hill 345
Kirkland Hill 507
Todholes Hill 481
Corsencon Hill 475
High Mount 365
Stannery Knowe 363
High Cairn 553
McCrierick's Cairn 556
Polnagrie Hill
Whitehill 406
Peat Hill 385
Blarene Hill 358
Knocknide Hill 344
Chang Hill 463
Benty Cowan Hill 477
Ewe Hill 437
Barbeys Hill 468
Quintin Knowe 554
Corserig Hill 395
Auchinleck
Boswell Museum & Mausoleum
Baird Institute Museum
Kier Hardie Statue
Doon Valley Rare Breeds Centre
Sorn Castle
Loudoun Castle Park
A71
A76
A70
B743
B705
B7036
B7083
B7046
B741
B745
B713
B7037
NS
257
267
268
246
3 miles
5 km

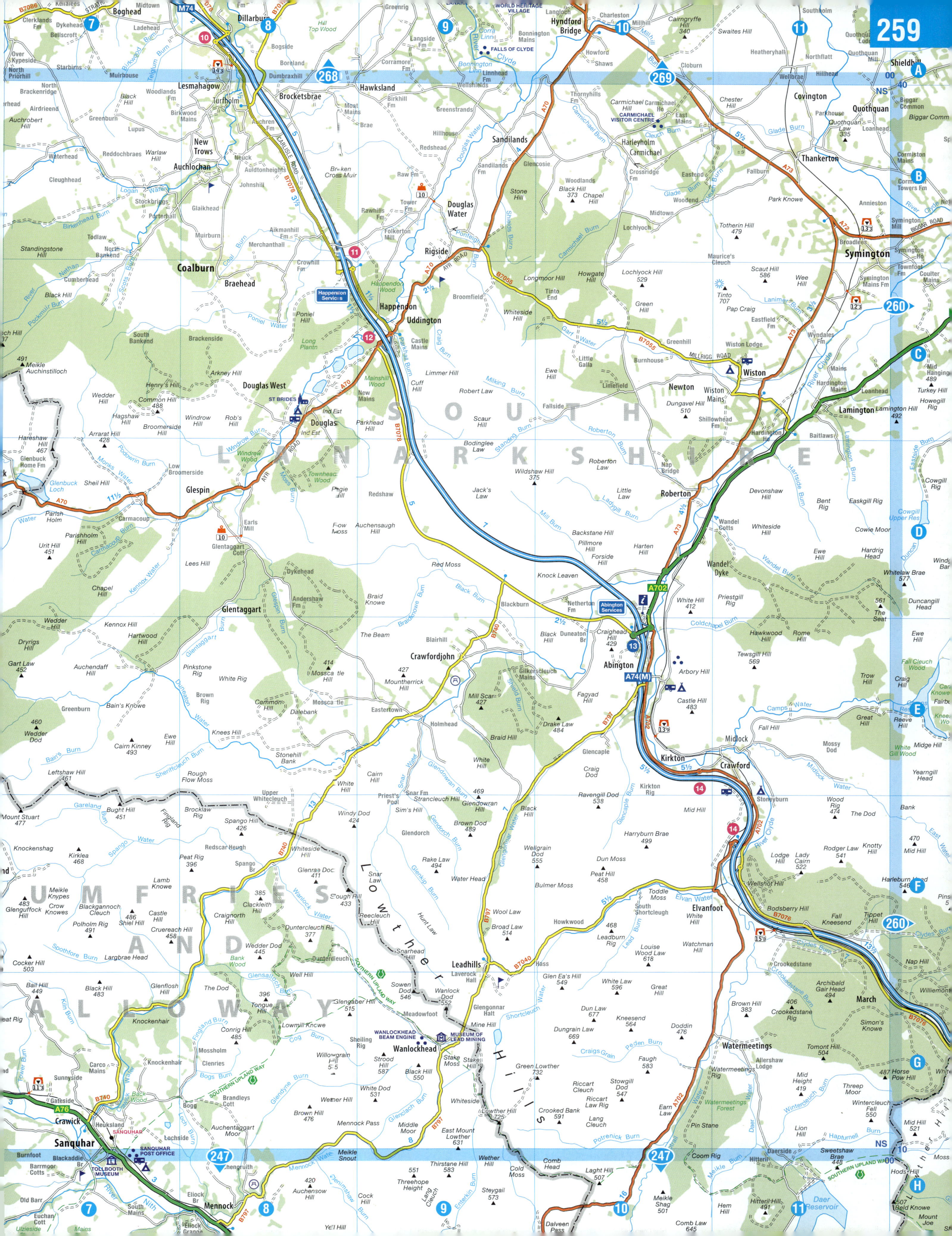

268
269
260
247
SOUTH LANARKSHIRE
DUMFRIES AND GALLOWAY
Lowther Hills
Boghead
Dillarburn
Hyndford Bridge
Lesmahagow
Turfholm
Brocketsbrae
Hawksland
New Trows
Auchlochan
Sandilands
Harleyholm
Carmichael
Thankerton
Covington
Quothquan
Symington
Shieldhill
Douglas Water
Rigside
Coalburn
Braehead
Happendon
Happendon Services
Uddington
Douglas West
Douglas
Glespin
Glentaggart
Wiston
Newton
Lamington
Roberton
Crawfordjohn
Abington
Abington Services
Kirkton
Crawford
Midlock
Elvanfoot
Leadhills
Wanlockhead
Watermeetings
Sanquhar
Crawick
Mennock
March
Wandel Dyke
Tinto
707
Glenbuck Loch
Daer Reservoir
River Clyde
Camps Water
Elvan Water
FALLS OF CLYDE
CARMICHAEL VISITOR CENTRE
WORLD HERITAGE VILLAGE
MUSEUM OF LEAD MINING
WANLOCKHEAD BEAM ENGINE
TOLLBOOTH MUSEUM
SANQUHAR POST OFFICE
SOUTHERN UPLAND WAY
M74
A74(M)
A70
A73
A72
A702
A76
B7078
B7055
B740
B797
B7040
B7076
AYR ROAD
BIGGAR ROAD
MILLRIGG ROAD
CARLISLE ROAD
Green Lowther
732
Lowther Hill
725
East Mount Lowther
631
Mennock Pass
Dalveen Pass
Comb Law
645

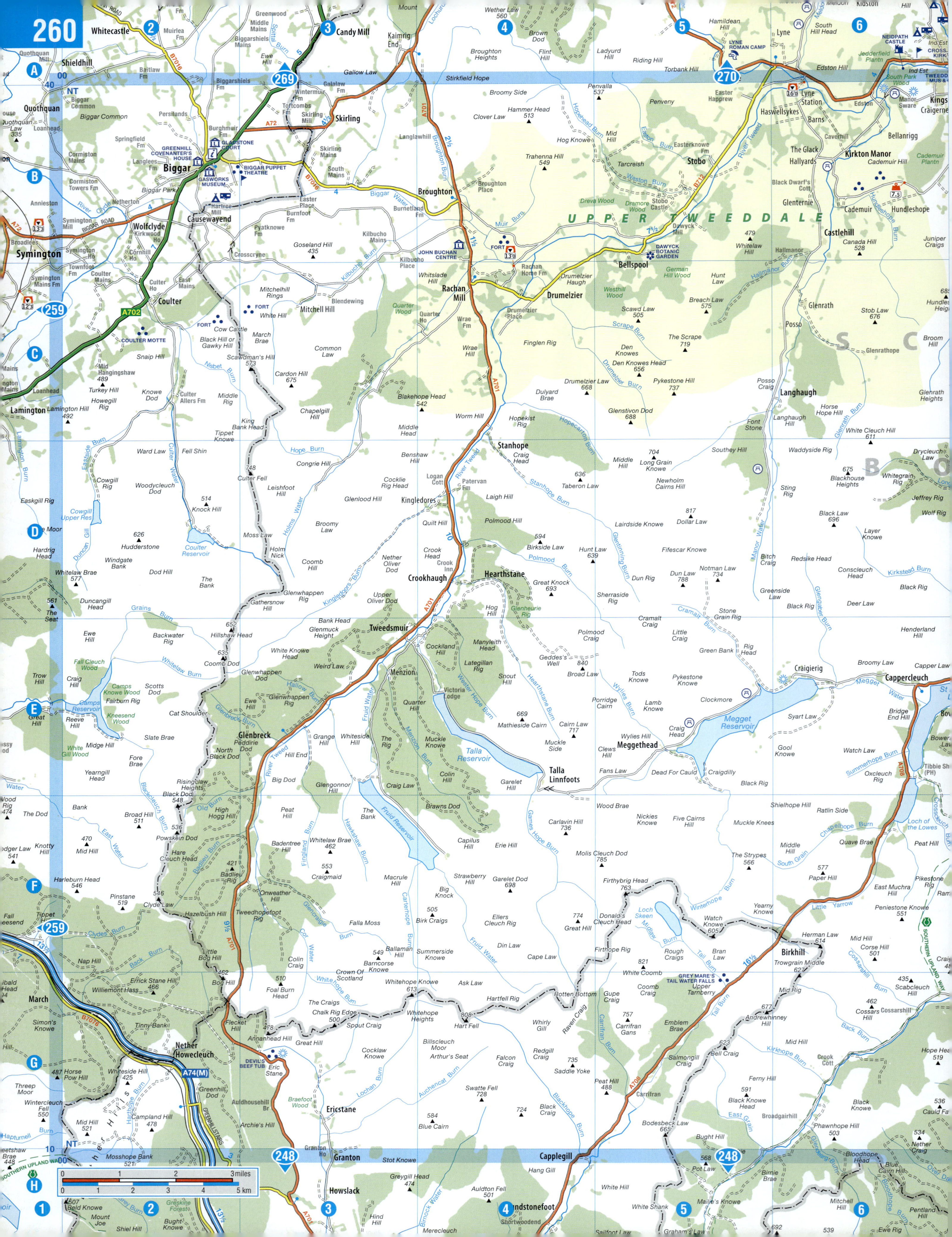
Whitecastle
Candy Mill
Kaimrig End
Mount
Wether Law 560
Brown Dod
Flint Hill
Broughton Heights
Ladyurd Hill
Riding Hill
Hamildean Hill
Lyne
South Hill Head
Lyne Roman Camp
Neidpath Castle
Jedderfield Plantn
Quothquan Mill
Shieldhill
Baitlaw Fm
Muirlea Fm
Greenwood
Middle Mains
Biggarshiels Mains
Ewe Hill
Gallow Law
Stirkfield Hope
Torbank Hill
Edston Hill
Quothquan
Biggar Common
Quothquan Law 335
Biggarshiels Fm
Galalaw Fm
Wintermuir Fm
Toftcombs Fm
Skirling Mill
Skirling
Broomy Side
Hammer Head 513
Clover Law
Penvalla 537
Easter Happrew
Penveny
Haswellsykes
Barns
Lyne Station
Edston
Manor Sware
Kings Craigerne
Cornistone Mains
Springfield Fm
Greenhill Covenanter's House
Gladstone Court
Langlees Fm
Biggar
Biggar Puppet Theatre
Gasworks Museum
Skirling Mains
South Mains
Langlawhill
Broughton
Broughton Place
Trahenna Hill 549
Hog Knowe
Mid Hill
Tarcreish
Stobo
Easterknowe Fm
The Glack
Hallyards
Kirkton Manor
Cademuir Hill
Bellanrigg
Caverhill
Cornistone Towers Fm
Biggar Park
Netherton
Causewayend
Hartree Mill
Easter Place
Burnfoot Fm
Burnetland Fm
Dreva Wood
Dramore Wood
Stobo Castle
Black Dwarf's Cott
Glenternie
Cademuir
Hundleshope
UPPER TWEEDDALE
Annieston
Symington Mill
Wolfclyde
Kirkwood Ho
Pyatknowe Fm
Kilbucho Mains
Dawyck Mill
Dawyck Botanic Garden
Castlehill
Canada Hill 528
Juniper Craigs
Broadlees
Symington
Cornhill Ho
Crosscryne
Goseland Hill 435
John Buchan Centre
Kilbucho Place
Fort
Rachan Home Fm
Bellspool
Whitelaw Hill 479
Hallmanor
German Hill Wood
Townfoot Fm
Coulter Mains
Symington Mains Fm
Culter Ho
East Mains
Coulter
Mitchellhill Rings
Blendewing
Whitslade Hill
Rachan Mill
Drumelzier Haugh
Drumelzier
Westhill Wood
Hunt Law
Breach Law 575
Glenrath
Quarter Wood
Quarter Ho
Wrae Fm
Drumelzier Place
Scawd Law 505
Stob Law 676
Posso
Mitchell Hill
White Hill
Cow Castle
Coulter Motte
Black Hill or Gawky Hill
March Brae
Common Law
Wrae Hill
Finglen Rig
The Scrape 719
Den Knowes
Den Knowes Head 656
Glenrathope
Broom Hill
Snaip Hill
Scawdman's Hill 573
Cardon Hill 675
Pykestone Hill 737
Posso Craig
Langhaugh
Mid Hangingshaw 489
Turkey Hill
Howegill Rig
Knowe Dod
Culter Allers Fm
Middle Rig
Drumelzier Law 668
Dulyard Brae
Blakehope Head 542
Chapelgill Hill
Glenstivon Dod 688
Glenrath Heights
Lamington
Lamington Hill 492
King Bank Head
Tippet Knowe
Middle Head
Worm Hill
Hopekist Rig
Horse Hope Hill
Langhaugh Hill
Font Stone
White Cleuch Hill 611
Ward Law
Fell Shin
Congrie Hill
Benshaw Hill
Stanhope
Craig Head
Middle Hill
Long Grain Knowe 704
Southey Hill
Waddyside Rig
Drycleuch Law
Cowgill Rig
Woodycleuch Dod
Culter Fell 748
Leishfoot Hill
Cockle Rig Head
Logan Cott
Paterwan Fm
Taberon Law 636
Newholm Cairns Hill
Blackhouse Heights 675
Whitegrain Rig
Easkgill Rig
Cowgill Upper Res
Knock Hill 514
Glenlood Hill
Kingledores
Laigh Hill
Sting Rig
Jeffrey Rig
Broomy Law
Quilt Hill
Polmood Hill
Lairdside Knowe
Dollar Law 817
Black Law 696
Wolf Rig
Moor
Hudderstone 626
Coulter Reservoir
Moss Law
Holm Nick
Birkside Law 594
Hunt Law 639
Fifescar Knowe
Layer Knowe
Hardrig Head
Windgate Bank
Dod Hill
Coomb Hill
Nether Oliver Dod
Crook Head
Crook Inn
Crookhaugh
Hearthstane
Great Knock 693
Dun Rig
Dun Law 788
Notman Law 734
Bitch Craig
Redsike Head
Consclcuch Head
Whitelaw Brae 577
The Bank
Glenwhappen Rig
Upper Oliver Dod
Sherraside Rig
Greenside Law
Black Rig
Duncangill Head
The Seat 561
Gathersnow Hill
Bank Head
Glenmuck Height
Tweedsmuir
Hog Hill
Glenheurie Rig
Stone Grain Rig
Cramalt Craig
Deer Law
Ewe Hill
Backwater Rig
Hillshaw Head 652
Polmood Craig
Little Craig
Henderland Hill
White Knowe Head
Cockland Hill
Manyleith Head
Green Bank
Rig Head
Coomb Dod 635
Weird Law
Lategillan Rig
Geddes's Well
Broad Law 840
Tods Knowe
Pykestone Knowe
Broomy Law
Craigierig
Cappercleuch
Capper Law
Trow Hill
Craig Hill
Fall Cleuch Wood
Camps Knowe Wood
Scotts Dod
Glenwhappen Dod
Menzion
Snout Hill
Victoria Lodge
Glenwhappen Rig
Ewe Hill
Quarter Hill
Porridge Cairn
Lamb Knowe
Clockmore
Megget Reservoir
Syart Law
Bridge End Hill
Great Hill
Reeve Hill
Camps Reservoir
Kneesend Wood
Fairburn Rig
Cat Shoulder
Glenbreck
Peddirie Dod
Grange Hill
Whiteside Hill
The Rig
Mathieside Cairn 669
Cairn Law 717
Wylies Hill
Meggethead
Craig Head
Slate Brae
Midge Hill
White Gill Wood
North Black Dod
Hill End
Muckle Knowe
Talla Reservoir
Muckle Side
Clews Hill
Gool Knowe
Watch Law
Fore Brae
Yearngill Head
Risingclaw Heights
Big Dod
Glengonnor Hill
Craig Law
Colin Hill
Garelet Hill
Talla Linnfoots
Fans Law
Dead For Cauld
Craigdilly
Black Rig
Oxcleuch Rig
Tibbie Sh
Bank
The Dod
Broad Hill 511
Black Dod 548
High Hogg Hill
Peat Hill
The Bank
Brawns Dod
Wood Brae
Nickies Knowe
Five Cairns Hill
Muckle Knees
Shielhope Hill
Ratlin Side
Carlavin Hill 736
Loch of the Lowes
Mid Hill 470
Knotty Hill
Powskein Dod 536
Hare Cleuch Head
Badentree Hill
Whitelaw Brae 462
Capilus Hill
Erie Hill
Molis Cleuch Dod 785
The Strypes 566
Middle Hill
Quave Brae
Peat Hill
Badlieu Rig 421
Craigmaid 553
Macrule Hill
Strawberry Hill
Garelet Dod 698
Firthybrig Head 763
Paper Hill 577
Harleburn Head 546
Pinstane 519
Clyde Law 546
Hazelbush Hill
Onweather Hill
Tweedhopefoot Rig
Big Knock
Birk Craigs 505
Falla Moss
Ellers Cleuch Rig
Great Hill 774
Donald's Cleuch Head
Loch Skeen
Yearny Knowe
Watch Knowe 605
East Mucbra Hill
Peniestone Knowe 551
Pikestone Rig
Tippet Hill
Nap Hill
Little Bog Hill
Colin Craig
Ballaman Hill 549
Barncorse Knowe
Summerside Knowe
Din Law
Cape Law
Firthope Rig
Rough Craigs
Bran Law
Birkhill
Herman Law 614
Trowgrain Middle 627
Mid Hill
Corse Hill 501
Errick Stane Hill
Williemont Hass 466
Bog Hill 462
Crown Of Scotland
Foal Burn Head 510
Whitehope Knowe 613
Ask Law
Hartfell Rig
White Coomb 821
Coomb Craig
Gupe Craig
Rotten Bottom
Grey Mare's Tail Water Falls
Upper Tarnberry
Mid Rig
Scabcleuch Hill 435
March
Simon's Knowe
Tinny Bank
The Craigs
Chalk Rig Edge 500
Whitehope Heights
Spout Craig
Hart Fell 808
Whirly Gill
Raven Craig
Carrifran Gans 757
Emblem Brae
Andrewhinney Hill 677
Cossars Hill 462
Cossarshill
Flecket Hill
Annanhead Hill 478
Great Hill
Nether Howecleuch
Devil's Beef Tub
Eric Stane
Cocklaw Knowe
Billscleuch Moor
Arthur's Seat
Falcon Craig
Redgill Craig
Saddle Yoke 735
Bell Craig 623
Salmongill Craig
Mid Hill
Crook Cott
Hope Head 519
Horse Pow Hill 487
Whiteside Hill 425
Threep Moor
Wintercleuch Fell 550
Greenhill Dod
Auldhousehill
Braefoot Wood
Ericstane
Swatte Fell 728
Peat Hill 488
Carrifran
Ferny Hill
Black Knowe Head 591
Black Knowe
Campland Hill 478
Mid Hill 521
Archie's Hill
Blue Cairn 584
Black Craig
Bodesbeck Law 665
Broadgairhill
Phawnhope Hill 503
Nether Craig 534
Cauld Fa 536
Hapturnell
Greenshaw Brae 448
Mosshope Bank 521
Granton Ho
Granton
Stot Knowe
Capplegill
Bught Hill
Pot Law 568
Bloodhope Head
Blue Cairn Hill
Howslack
Greygill Head 474
Hang Gill
Auldton Fell 501
Tindstonefoot
Shortwoodend
White Hill
White Shank
Mailie's Knowe
Birnie Brae
Mitchell Hill
Pentland Hill
Beld Knowe 507
Mount Joe
Shiel Hill
Bught Knowe
Greskine Forest
Hind Hill
Merecleuch
Saifoot Law
Graham's Law
Ewe Rig
Southern Upland Way
A701
A702
A72
B7016
A74(M)
A708
B7076
B712
NT
0 1 2 3 miles
0 1 2 3 4 5 km

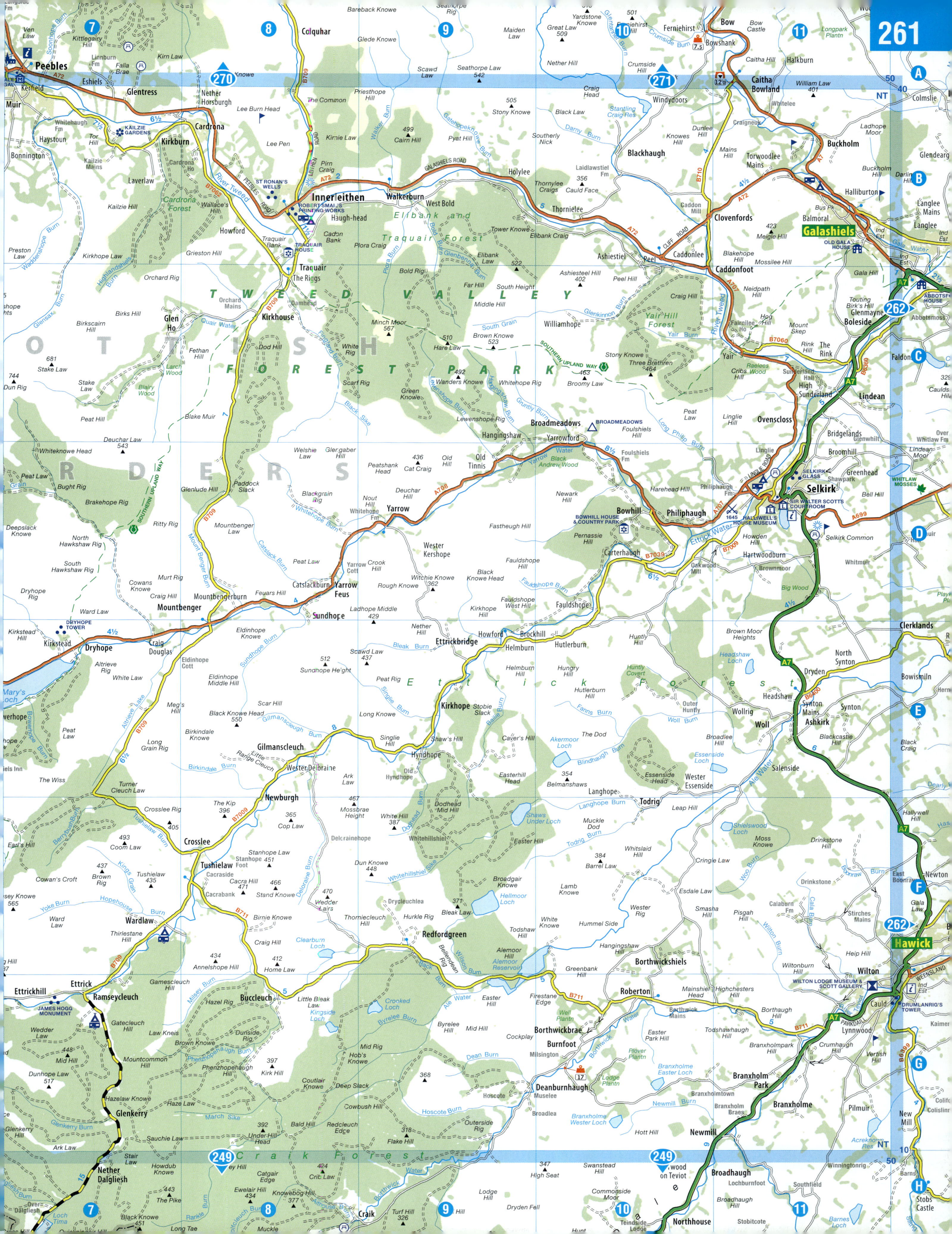

261
Peebles
Innerleithen
Walkerburn
Galashiels
Selkirk
Hawick
Clovenfords
Caddonfoot
Caddonlee
Peel
Ashiestiel
Thornielee
Holylee
Traquair
The Riggs
Kirkhouse
Cardrona
Kirkburn
Glentress
Eshiels
Kerfield
Haystoun
Bonnington
Laverlaw
Colquhar
Blackhaugh
Windydoors
Caitha Bowland
Bow
Bowshank
Buckholm
Halliburton
Boleside
Glenmayne
Yair
Lindean
Ovenscloss
Philiphaugh
Bowhill
Broadmeadows
Yarrowford
Hangingshaw
Yarrow
Yarrow Feus
Catslackburn
Sundhope
Mountbenger
Dryhope
Kirkstead
Ettrickbridge
Howford
Brockhill
Helmburn
Hutlerburn
Kirkhope
Hyndhope
Gilmanscleuch
Newburgh
Crosslee
Tushielaw
Wardlaw
Ramseycleuch
Ettrick
Ettrickhill
Buccleuch
Redfordgreen
Roberton
Borthwickbrae
Burnfoot
Deanburnhaugh
Newmill
Branxholm Park
Branxholme
Broadhaugh
Northhouse
Glenkerry
Nether Dalgliesh
Craik
Borthwickshiels
Todrig
Langhope
Wilton
Cauld
Lynnwood
Ashkirk
Woll
Salenside
Dryden
North Synton
Clerklands
Carterhaugh
Fauldshope
Hartwoodburn
TWEED VALLEY
SCOTTISH FOREST PARK
BORDERS
Elibank and Traquair Forest
Ettrick Forest
Craik Forest
Yair Hill Forest
Cardrona Forest
River Tweed
Ettrick Water
Yarrow Water
Ale Water
SOUTHERN UPLAND WAY
A7
A72
A707
A708
A699
B709
B7009
B711
B710
B7060
270
271
262
249

271
272
261
250
Earlston
Melrose
Newstead
Gattonside
Darnick
Bowden
Newtown St Boswells
St Boswells
Dryburgh
Clintmains
Bemersyde
Redpath
Smailholm
Brotherstone
Nenthorn
Ednam
Stichill
Hume
Falsidehill
Fans
Kelso
Maxwellheugh
Heiton
Roxburgh
Trows
Rutherford
Maxton
Muirhouselaw
Longnewton
Lilliesleaf
Midlem
Belses
Ancrum
Bonjedward
Nisbet
Crailing
Eckford
Cessford
Crailinghall
Jedburgh
Howden
Hundalee
Oxnam
Mossburnford
Camptown
Mervinslaw
Edgerston
Chesters
Southdean
Bonchester Bridge
Hobkirk
Hallrule
Bedrule
Denholm
Minto
Hassendean
Horsleyhill
Clarilaw
Appletreehall
Burnfoot
Hawick
Wilton
Cauldmill
Kirkton
Lanton
Newton
Kalemouth
Ormiston
Lindean
Faldonside
Boleside
Langshaw
Colmslie
Clerklands
Greenhouse
Bloomfield
Swinnie
Bairnkine
Falla
Dolphinston
Newbigging
Swinside Hall
S C O T T I S H
B O R D E R S
E I L D O N   A N D
L E A D F O O T
Teviot Dale
River Tweed
River Teviot
Jedburgh Abbey
Dryburgh Abbey
Melrose Abbey
Abbotsford House
Floors Castle
Mellerstain House
Smailholm Tower
Harestanes Countryside Visitor Centre
Mary Queen of Scots House
Jedburgh Castle Jail Mus
Ferniehirst Castle
Jedforest Deer & Farm Park
Drumlanrig's Tower
Waterloo Monument
Teviot Water Gardens
Mertoun Gardens
Priorwood Gardens
Trimontium Exhibition
Border Union Showground
Ancrum Moor 1545
Lilliardsedge Park
A68
A7
A698
A699
A6091
A6088
B6400
B6397
B6358
B6356
NT
0 1 2 3 miles
0 1 2 3 4 5 km

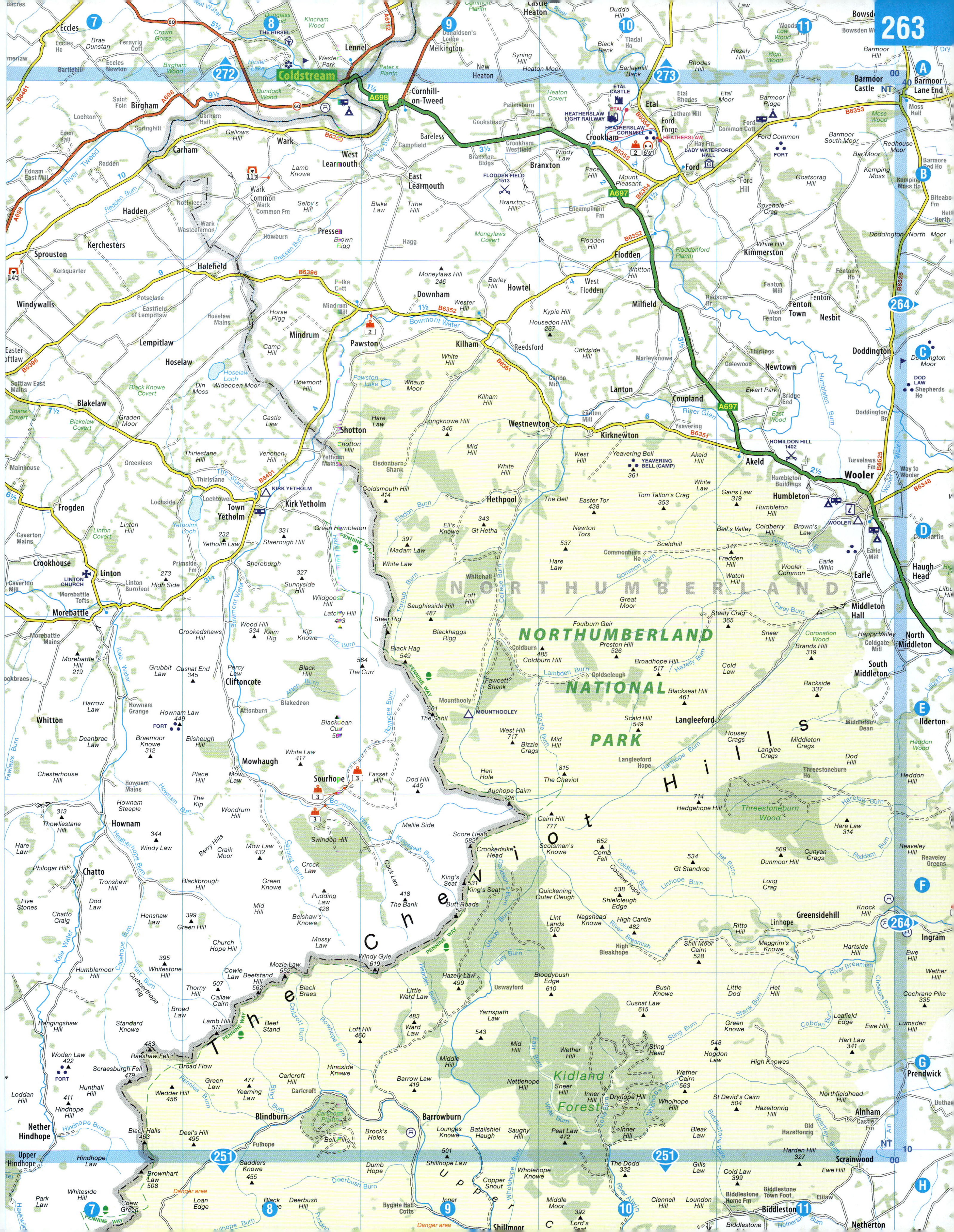
Eccles
Coldstream
THE HIRSEL
Lennel
Cornhill-on-Tweed
Melkington
Donaldson's Lodge
New Heaton
Heaton Moor
Castle Heaton
Duddo
Tindal Ho
Barleymill Bank
Etal
ETAL CASTLE
Rhodes Hill
Bowsden
Barmoor Castle
Barmoor Lane End
Birgham
Carham
Wark
West Learmouth
East Learmouth
Campfield
Bareless
Cookstead
Pallinsburn Ho
Crookham
Branxton
FLODDEN FIELD 1513
HEATHERSLAW LIGHT RAILWAY
HEATHERSLAW CORNMILL
Ford
Ford Forge
LADY WATERFORD HALL
Ford Common
Goatscrag Hill
Kimmerston
Hadden
Kerchesters
Sprouston
Redden
River Tweed
Wark Common
Presson
Hagg
Flodden
Howtel
Milfield
Holefield
Windywalls
Lempitlaw
Hoselaw
Mindrum
Pawston
Kilham
Downham
Moneylaws Hill 246
Bowmont Water
Reedsford
Housedon Hill 267
Lanton
Coupland
Doddington
Nesbit
Fenton Town
Newtown
Blakelaw
Shotton
Kirk Yetholm
Town Yetholm
KIRK YETHOLM
Frogden
Linton
Crookhouse
Morebattle
Westnewton
Kirknewton
River Glen
Akeld
Wooler
Humbleton
Yeavering Bell 361
YEAVERING BELL (CAMP)
HOMILDON HILL 1402
Hethpool
Tom Tallon's Crag 353
Earle
Middleton Hall
North Middleton
South Middleton
NORTHUMBERLAND
NORTHUMBERLAND NATIONAL PARK
The Cheviot Hills
The Cheviot 815
Cairn Hill 777
Hedgehope Hill 714
Langleeford
Threestoneburn Wood
Ilderton
Ingram
Greensidehill
Linhope
Whitton
Hownam
Chatto
Mowhaugh
Sourhope
Cliftoncote
Windy Gyle 619
PENNINE WAY
Blindburn
Barrowburn
Kidland Forest
Alnham
Scrainwood
Netherton
Biddlestone
Upper Coquetdale
Nether Hindhope
Upper Hindhope
Shillmoor
Uswayford
Danger area
A697
A698
B6350
B6352
B6351
B6401
B6396
B6525
B6348
B6353
272
273
251
263
264

Lowick
Barmoor Castle
Barmoor Lane End
West Kyloe
East Kyloe
Fenwick
Fenham
Granary Pt
Holy Island
Lindisfarne Castle
Lindisfarne Priory
Guile Pt
Lowmoor Pt
Teallhole Pt
Ross Pt
Ross Back Sands
Buckton
Smeafield
Elwick
Ross
Detchant
Middleton
Holburn
Hetton Steads
North Hazelrigg
South Hazelrigg
St Cuthbert's Cave
Budle Bay
Budle Pt
Harkess Rocks
RNLI Grace Darling Museum
Bamburgh
Bamburgh Castle
Budle
Waren Mill
Easington
Belford
Glororum
Burton
Spindlestone
Bradford
Mousen
Adderstone
Bellshill
Warenton
Lucker
Seahouses
North Sunderland
Beadnell
Swinhoe
Elford
Newham
Chathill
Ellingham
Preston
Brunton
Christon Bank
Embleton
Farne Islands
Staple Sound
Inner Farne
Longstone Lighthouse
Staple Island
Megstone
Wooler
Doddington
Dod Law
West Horton
East Horton
Chatton
Chillingham
Chillingham Castle
Wild Cattle of Chillingham
Ros Castle
Hepburn
Lilburn Tower
East Lilburn
Newtown
North Middleton
South Middleton
Ilderton
Roseden
Roddam
Wooperton
Old Bewick
New Bewick
Harehope
Eglingham
Beanley
Warenford
Rosebrough
Newstead
Brownieside
North Charlton
South Charlton
Rock
Rennington
Stamford
Dunstan
Dunstan Steads
Littlemill
Howick Hall Gardens
Littlehoughton
High Newton-by-the-Sea
Low Newton-by-the-Sea
Beadnell Bay
Benthall
Craster
N O R T H U M B E R L A N D
Brandon
Branton
Powburn
Ingram
Ingram National Park Centre
Shipley
Titlington
Glanton
Glanton Pike
Shawdon Hall
Bolton
Broome Park
Abberwick
Whittingham
Great Ryle
Little Ryle
Prendwick
Alnham
Thrunton
Yetlington
Callaly
Edlingham
Edlingham Castle
Alnwick
Alnwick Castle
Alnwick Abbey Gatehouse
Alnwick Garden
Hulne Park
Hulne Priory
House of Hardy Museum
Denwick
Hawkhill
Lesbury
Hipsburn
Bilton
Alnmouth
Rugley
Shilbottle
High Buston
NORTHUMBERLAND
A1
A697
A1068
B1340
B1342
B6346
B6347
B6348
B6349
B6341
B1341
B1339
B6353
B6525
273
263
252
0 1 2 3 miles
0 1 2 3 4 5 km

NORTH SEA
Embleton Bay
Castle Pt
DUNSTANBURGH CASTLE
Queen Margaret's Cove
Craster
Cullernose Pt
Howick
Rumbling Kern
Red Stead
Howick Haven
Sugar Sands
Low Stead
Howdiemont Sands
Red Ends
Boulmer
Boulmer Haven
Field Ho
Seaton Pt
Marden Rocks
Alnmouth Bay
NU
253
253

FIRTH OF CLYDE
CLYDE-MUIRSHIEL REGIONAL PARK
NORTH AYRSHIRE
INVERCLYDE
Island of Bute
Great Cumbrae Island
Little Cumbrae Island
Inverkip
Wemyss Bay
Skelmorlie
Upper Skelmorlie
Meigle
Routenburn
Largs
Fairlie
Millport
Hunterston Sands
Portencross
Ardneil
West Kilbride
Seamill
Ardrossan
Saltcoats
Stevenston
Kilwinning
Irvine
Dalry
Kilbirnie
Beith
Glengarnock
Longbar
Lochwinnoch
Giffordland
Dalgarven
Drakemyre
Innellan
Toward
Castle Toward
Craigmore
Montford
Ascog
Kerrycroy
Scoulag
Piperhall
Kingarth
Kilchattan Bay
Garvock
Shielhill
Hill of Stake 522
Misty Law 510
Kelburn Country Centre
Vikingar!
Christian Heritage Museum
Museum of the Cumbraes
Robertson Mus & Aquarium
Hunterston Power Station Visitor Centre
North Ayrshire Museum
Kilwinning Abbey
Eglinton
Scottish Maritime Museum
Ayrshire Museum of Country Life & Costume
Blackshaw Farm Park
Muirshiel
Mount Stuart
Victorian Fernery
Ardencraig Gardens
Brodick 0:55
A78
A760
A737
B780
B781
B782
275
276
255
256
257
0 1 2 3 miles
0 1 2 3 4 5 km

Bearsden
Kilmacolm
Bishopton
Erskine
Clydebank
Renfrew
Dalmuir
Drumchapel
Knightswood
Kessington
Bishopbriggs
Milton
Springburn
Maryhill
Kelvinside
Partick
Scotstoun
Yoker
Whiteinch
Govan
Houston
Crosslee
Craigends
Bridge of Weir
Ranfurly
Linwood
RENFREWSHIRE
Johnstone
Kilbarchan
Brookfield
Paisley
Elderslie
Millarston
Glasgow
Pollokshields
Govanhill
Rutherglen
Cardonald
Hillington
Mosspark
Pollok
Millikenpark
Howwood
Foxbar
Glenburn
Barrhead
Nitshill
Priesthill
Darnley
Thornliebank
Giffnock
Newlands
Cathcart
Castlemilk
Shawlands
Arden
Neilston
Uplawmoor
Newton Mearns
Clarkston
Busby
Sheddens
Thorntonhall
Waterfoot
Eaglesham
Lugton
Burnhouse
Dunlop
Barrmill
Greenhills
Stewarton
Auchentiber
Kingsford
Fenwick
Laigh Fenwick
Waterside
Kilmaurs
Onthank
Moscow
Cunninghamhead
Knockentiber
Springside
Kilmarnock
EAST AYRSHIRE
EAST RENFREWSHIRE
CITY OF GLASGOW
Hairmyres
Newlandsmuir
Peel Park
M77
M8
A737
A726
A77
A736
A761
277
268
257
258

NORTH LANARKSHIRE
Jellyhill
Millerston
Auchinloch
Bridgend
Moodiesburn
Chryston
Muirhead
Stepps
Garnkirk
Gartcosh
Marnock
Glenboig
Riggend
Wattston
Greengairs
Longriggend
Stand
Caldercruix
Plains
Airdrie
Glenmavis
Thrashbush
Clarkston
Craigneuk
Petersburn
Gartness
Chapelhall
Calderbank
Salsburgh
Kirk o' Shotts
Forrestfield
Coatbridge
Old Monkland
Carnbroe
Coatdyke
Gartlea
Greenend
Cairnhill
Easterhouse
Garthamlock
Riddrie
Craigend
Ruchazie
Cranhill
Queenslie
Carntyne
Barlanark
Springboig
Shettleston
Baillieston
Garrowhill
Mount Vernon
Swinton
Bridgeton
Parkhead
Dalmarnock
Fullarton
Broomhouse
Carmyle
Govanhill
Polmadie
Rutherglen
Croftfoot
Burnside
Castlemilk
Cambuslang
Flemington
Westburn
Newton
Uddingston
Bothwell
Birkenshaw
Tannochside
Viewpark
Fallside
Bellshill
Mossend
Holytown
New Stevenston
Newarthill
Newhouse
Hareshaw
Hartwood
Orbiston
Forgewood
Cleland
Motherwell
Wishaw
Newmains
Coltness
Cambusnethan
Morningside
Waterloo
Bogside
Overtown
Wildmanbridge
Larkhall
Dalserf
Rosebank
Ashgill
Netherburn
Crossford
Braidwood
Hazelbank
Blackwood
Kirkmuirhill
Boghead
Dillarburn
Lesmahagow
Brocketsbrae
Stonehouse
Strathaven
Sandford
Glassford
Chapelton
Shawton
Shawtonhill
Quarter
Limekilnburn
Millheugh
Machan
Hareleeshill
Strutherhill
Birkenshaw
Hamilton
Blantyre
High Blantyre
Burnbank
Whitehill
Earnock
Meikle Earnock
Fairhill
Laighstonehall
Low Waters
Cadzow
Eddlewood
Ferniegair
Allanton
Silvertonhill
Knowetop
Carmunnock
Nerston
Kelvin
East Kilbride
Calderwood
Auchentibber
Thorntonhall
Peel Park
College Milton
Hairmyres
West Mains
Westwood
The Murray
Birniehill
Newlandsmuir
Greenhills
Whitehills
Saint Leonards
LANARKSHIRE
SOUTH
Auldhouse
Carnduff
Gilmourton
Caldermill
Stobieside
Whitelaw
Newmains
Udstonhead
Heads
M8
M73
M74
M80
A725
A71
A726
A8
267
278
258
259
0 1 2 3 miles
0 1 2 3 4 5 km

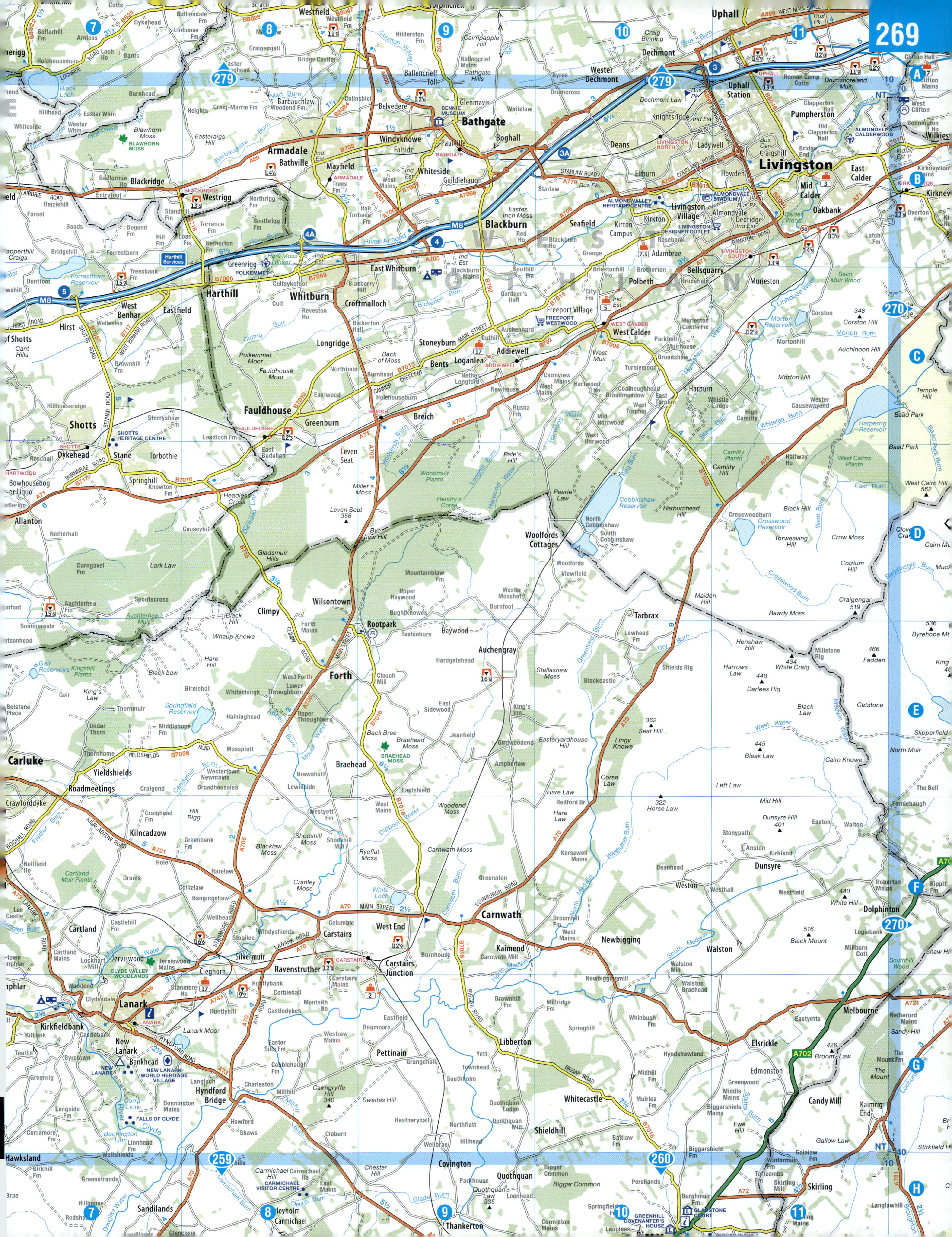

Uphall
Dechmont
Wester Dechmont
Uphall Station
Pumpherston
Livingston
East Calder
Mid Calder
Oakbank
Kirknewton
Bathgate
Boghall
Armadale
Bathville
Mayfield
Whiteside
Blackridge
Westrigg
Blackburn
Seafield
Livingston Village
Kirkton Campus
Adambrae
Polbeth
Bellsquarry
Murieston
West Calder
Harthill
Whitburn
East Whitburn
Croftmalloch
Greenrigg
Eastfield
West Benhar
Hirst
Shotts
Dykehead
Stane
Torbothie
Springhill
Fauldhouse
Greenburn
Breich
Longridge
Stoneyburn
Bents
Loganlea
Addiewell
Harburn
Woolfords Cottages
Cobbinshaw Reservoir
Allanton
Climpy
Wilsontown
Rootpark
Forth
Auchengray
Tarbrax
Carluke
Yieldshields
Roadmeetings
Kilncadzow
Braehead
Carnwath
Newbigging
Walston
Dunsyre
Dolphinton
Cartland
Carstairs
Carstairs Junction
Ravenstruther
Lanark
New Lanark
Kirkfieldbank
Hyndford Bridge
Pettinain
Libberton
Whitecastle
Shieldhill
Elsrickle
Melbourne
Candy Mill
Covington
Quothquan
Thankerton
Skirling
Sandilands
Carmichael
Hawksland
M8
A70
A71
A702
A706
279
259
260
270
LOTHIAN

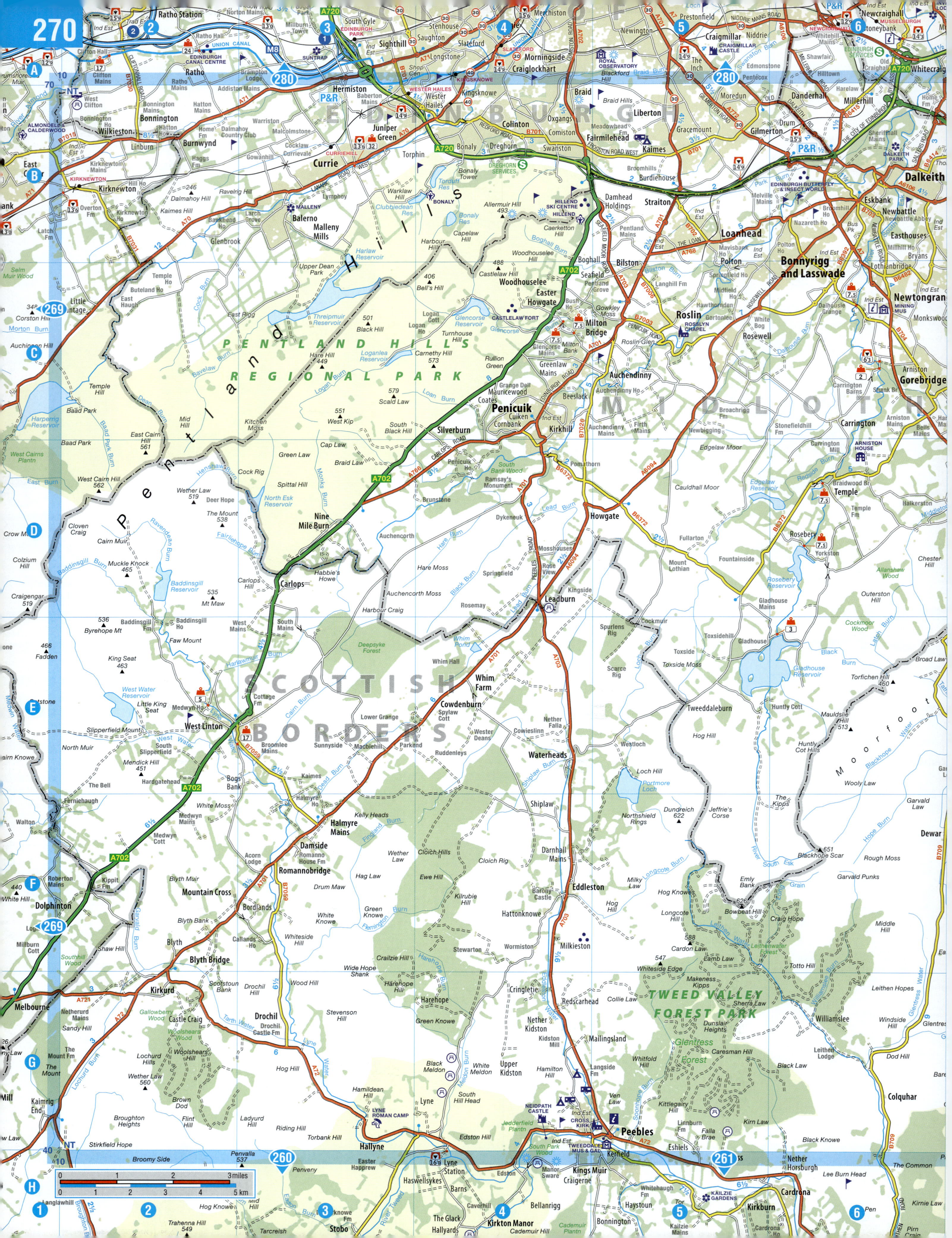
EDINBURGH
PENTLAND HILLS REGIONAL PARK
Pentland Hills
MIDLOTHIAN
SCOTTISH BORDERS
Moorfoot
TWEED VALLEY FOREST PARK
Ratho Station
Ratho
Newbridge
Hermiston
Sighthill
Wester Hailes
Longstone
Slateford
Morningside
Craiglockhart
Merchiston
Newington
Prestonfield
Craigmillar
Niddrie
Newcraighall
Whitecraig
Musselburgh
Bonnington
Wilkieston
Burnwynd
Currie
Juniper Green
Colinton
Bonaly
Dreghorn
Swanston
Oxgangs
Braid
Fairmilehead
Liberton
Kaimes
Gilmerton
Danderhall
Millerhill
Dalkeith
Eskbank
Newbattle
Easthouses
Loanhead
Straiton
Bilston
Bonnyrigg and Lasswade
Newtongrange
Roslin
Rosewell
Kirknewton
Balerno
Malleny Mills
Glenbrook
Little Vantage
Milton Bridge
Auchendinny
Penicuik
Silverburn
Kirkhill
Howgate
Gorebridge
Carrington
Temple
Rosebery
Nine Mile Burn
Carlops
Leadburn
West Linton
Whim Farm
Cowdenburn
Waterheads
Halmyre Mains
Romannobridge
Mountain Cross
Dolphinton
Blyth Bridge
Kirkurd
Melbourne
Drochil
Eddleston
Milkieston
Peebles
Kings Muir
Lyne
Hallyne
Stobo
Kirkton Manor
Cardrona
Kirkburn
Williamslee
Colquhar
Tweeddaleburn
Gladhouse Reservoir
Loganlea Reservoir
Threipmuir Reservoir
Harlaw Reservoir
Glencorse Reservoir
North Esk Reservoir
Baddinsgill Reservoir
West Water Reservoir
Portmore Loch
Rosebery Reservoir
Edgelaw Reservoir
Harperrig Reservoir
A702
A720
A701
A703
A72
A71
A70
A7
A68
A6094
A766
B7059
B6372
280
269
260
261

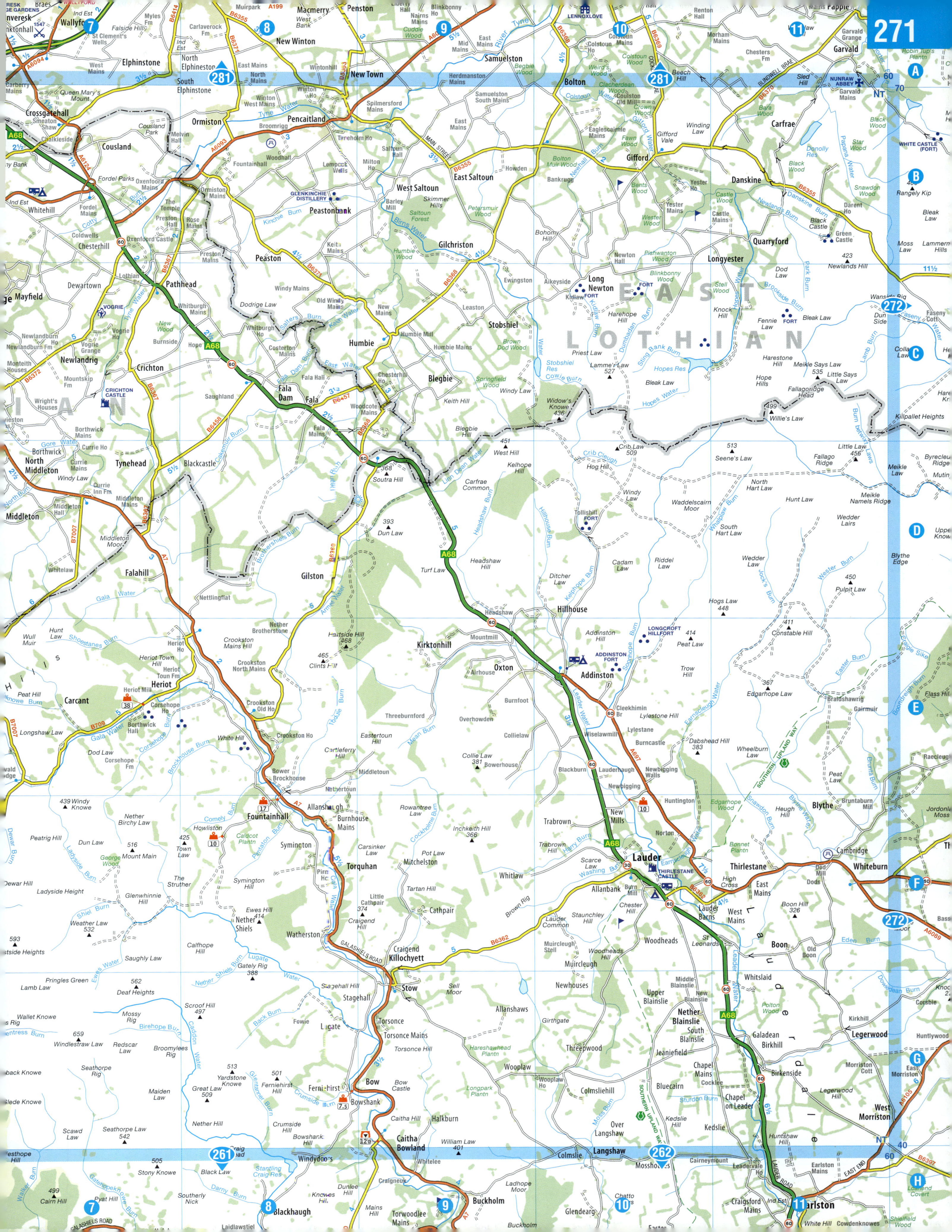

Wallyford
Elphinstone
North Elphinstone
South Elphinstone
Macmerry
Penston
New Winton
New Town
Samuelston
Lennoxlove
Bolton
Garvald
Nunraw Abbey
Crossgatehall
Smeaton Shaw
Chalkieside
Cousland
Ormiston
Pencaitland
Gifford
Carfrae
White Castle (Fort)
Danskine
West Saltoun
East Saltoun
Glenkinchie Distillery
Peastonbank
Whitehill
Chesterhill
Oxenford Castle
Gilchriston
Quarryford
Longyester
Peaston
Pathhead
Dewartown
Mayfield
Vogrie
Long Newton
Kidlaw
East Lothian
Stobshiel
Humbie
Newlandrig
Crichton
Crichton Castle
Fala Dam
Fala
Blegbie
Soutra Hill
Borthwick
North Middleton
Tynehead
Blackcastle
Middleton
Falahill
Gilston
Headshaw
Hillhouse
Kirktonhill
Oxton
Addinston
Longcroft Hillfort
Addinston Fort
Heriot
Carcant
Crookston
Fountainhall
Allanshaugh
Burnhouse Mains
Symington
Torquhan
Mitchelston
Lauder
Thirlestane Castle
Thirlestane
Cambridge
Whiteburn
Blythe
New Mills
Trabrown
Whitlaw
Allanbank
Cathpair
Watherston
Killochyett
Stow
Stagehall
Woodheads
Boon
Muircleugh
Newhouses
Upper Blainslie
Nether Blainslie
Legerwood
Galadean
Birkhill
Birkenside
Torsonce
Torsonce Mains
Bow
Ferniehirst
Bowshank
Caitha Bowland
Halkburn
Colmsliehill
Bluecairn
Chapel on Leader
Over Langshaw
Kedslie
West Morriston
Langshaw
Colmslie
Windydoors
Blackhaugh
Buckholm
Torwoodlee Mains
Glendearg
Earlston
Southern Upland Way
A68
A7
A697
B6362
Lammer Law
Hopes Res
Bleak Law
Willie's Law
Crib Law
Headshaw Hill
Turf Law
Dun Law
Seene's Law
Hogs Law
Peat Law
Edgarhope Law
Windlestraw Law
Mossy Rig
Scroof Hill
Gately Rig
Deaf Heights
Cairn Hill

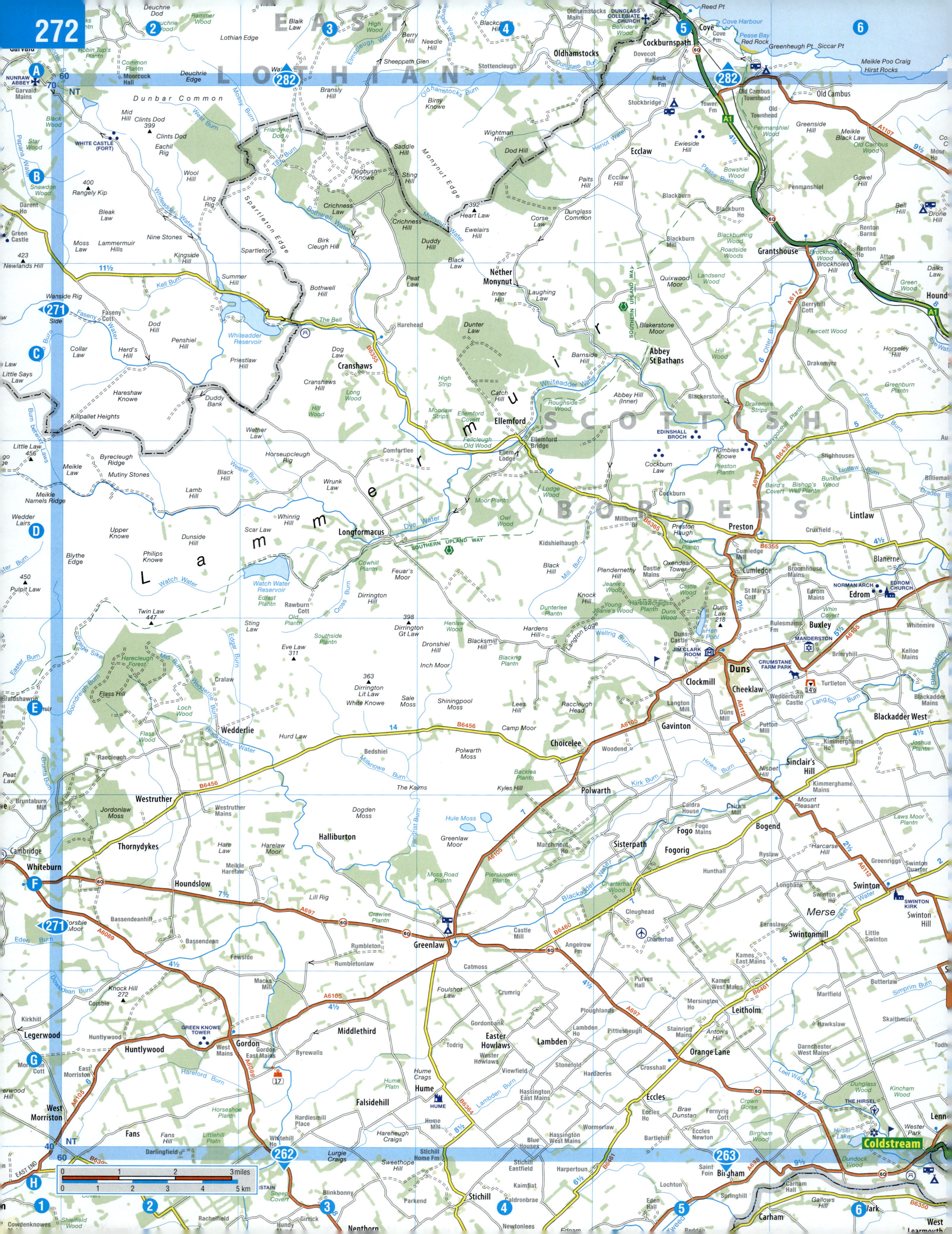
EAST
LOTHIAN
SCOTTISH
BORDERS
Lammermuir
Deuchrie Dod
Rammer Wood
Blaik Law
High Wood
Berry Hill
Needle Hill
Blackcastle Hill
Oldhamstocks Mains
DUNGLASS COLLEGIATE CHURCH
Belvidere Wood
Reed Pt
Coye
Cove
Cove Harbour
Pease Bay
Red Rock
Greenheugh Pt
Siccar Pt
Lothian Edge
Robin Tup's Plantn
Garvald
NUNRAW ABBEY
Garvald Mains
Common Plantn
Moorcock Hall
Deuchrie Edge
Sheeppath Glen
Stottencleugh
Oldhamstocks
Cockburnspath
Dovecot Hall
Meikle Poo Craig
Hirst Rocks
Bransly Hill
Dunbar Common
Birny Knowe
Neuk Fm
Stockbridge
Tower Fm
Old Cambus Townhead
Old Cambus
Old Townhead
Mid Hill
Clints Dod 399
Clints Dod
Black Wood
WHITE CASTLE (FORT)
Eachil Rig
Friardykes Dod
Saddle Hill
Monynut Edge
Wightman Hill
Dod Hill
Ecclaw
Ewieside Hill
Penmanshiel Wood
Greenside Hill
Meikle Black Law
Old Cambus Wood
Star Wood
Snawdon Wood
400 Rangely Kip
Wool Hill
Dogbush Knowe
Sting Hill
Paits Hill
Ecclaw Hill
Bowshiel Wood
Penmanshiel
Gowel Hill
Darent Ho
Bleak Law
Ling Rig
Spartleton Edge
Crichness Law
392 Heart Law
Dunglass Common
Blackburn
Blackburn Ho
Bell Hill
Drone Hill
Green Castle
Moss Law
Lammermuir Hills
Nine Stones
Crichness Hill
Ewelairs Hill
Corse Law
Blackburn Mill
Renton Barns
423 Newlands Hill
Kingside Hill
Spartleton
Birk Cleugh Hill
Duddy Hill
Blackburnrig Wood
Roadside Woods
Grantshouse
Renton Ho
Brockholes Wood
Brockholes Hill
Atton Cott
Dalks Law
Summer Hill
Kell Burn
Black Law
Nether Monynut
Quixwood Moor
Landsend Wood
Green Wood
Hound
Wanside Rig
Faseny Cott
Bothwell Hill
Peat Law
Inner Hill
Laughing Law
Berryhill Cott
Side
Dod Hill
The Bell
Harehead
Dunter Law
SOUTHERN UPLAND WAY
Blakerstone Moor
Fawcett Wood
Collar Law
Herd's Hill
Penshiel Hill
Whiteadder Reservoir
Dog Law
Cranshaws
Barnside Hill
Abbey St Bathans
Hill Wood
Horseley Hill
Drakemyre
Law
Little Says Law
Priestlaw Hill
Cranshaws Hill
High Strip
Catch Hill
Greenburn Plantn
Hareshaw Knowe
Duddy Bank
Long Wood
Roughside Wood
Abbey Hill (Inner)
Blackerstone
Drakemire Strips
Kilpallet Heights
Hill Wood
Moorlaw Strips
Ellemford Covert
Ellemford
EDINSHALL BROCH
Wether Law
Fellcleugh Old Wood
Ellemford Bridge
Ellem Lodge
Little Law 456
Byrecleugh Ridge
Meikle Law
Mutiny Stones
Horseupcleugh Rig
Comfortlee
Humbles Knowe
Cockburn Law
Preston Plantn
Slighhouses
Lintlaw Burn
Bunkle Wood
Baird's Covert
Bishop's Well Plantn
Billiemains
Black Hill
Wrunk Law
Lodge Wood
Cockburn
Meikle Namels Ridge
Lamb Hill
Moor Plantn
Lintlaw
Wedder Lairs
Whinrig Hill
Owl Wood
Millburn
Preston Haugh
Preston
Cruxfield
Upper Knowe
Dunside Hill
Scar Law
Longformacus
Dye Water
Kidshielhaugh
Baramill Plantn
Cumledge Mill
Blanerne
Blythe Edge
Philips Knowe
Black Hill
Plendernethy Hill
Castle Mains
Oxendean Tower
Cumledge
Broomhouse Mains
NORMAN ARCH
EDROM CHURCH
Edrom
450 Pulpit Law
Watch Water
Watch Water Reservoir
Edfast Plantn
Cowhill Plantn
Feuar's Moor
Jeanie's Wood
Castle Wood
St Mary's Cott
Edrom Mains
Rawburn Cott
Dirrington Hill
Knock Hill
Harelawcraigs Plantn
Young Jeanie's Wood
Duns Wood
Duns Law 218
Whin Covert
Twin Law 447
Old Plantn
398 Dirrington Gt Law
Henlaw Wood
Dunterlee Plantn
Buxley
Whitemire
Sting Law
Southside Plantn
Dronshiel Hill
Blacksmill Hill
Hardens Hill
Langton Edge
Wellrig Burn
Hen Poo
Rulesmains Fm
MANDERSTON
Kelloe Mains
Eve Law 311
Duns Castle
JIM CLARK ROOM
Duns
CRUMSTANE FARM PARK
Brieryhill
Harecleugh Forest
Blackrig Plantn
Inch Moor
Clockmill
Cheeklaw
Turtleton
Cralaw
363 Dirrington Lit Law
White Knowe
Sale Moss
Shiningpool Moss
Lees Hill
Raccleugh Head
Langton Mill
Wedderburn Castle
Blackadder Mains
Flass Hill
Loch Wood
Duns Mill
Putton Mill
Blackadder West
Wedderlie
Hurd Law
Camp Moor
Choicelee
Gavinton
Flass Wood
Bedshiel
Polwarth Moss
Woodend
Kimmerghame Ho
Joshua Plantn
Raecleugh
Milknowe Burn
Backlea Plantn
Peat Law
Nisbet Hill
Sinclair's Hill
Bruntaburn Mill
Westruther
Westruther Mains
The Kaims
Kyles Hill
Polwarth
Kirk Burn
Kimmerghame Mains
Jordonlaw Moss
Dogden Moss
Hule Moss
Caldra House
Cairn's Mill
Mount Pleasant
Laws Moor Plantn
Thornydykes
Hare Law
Harelaw Moor
Halliburton
Greenlaw Moor
Marchmont Ho
Sisterpath
Fogo
Fogo Mains
Bogend
Fogorig
Harcarse Hill
Cambridge
Whiteburn
Meikle Harelaw
Moss Road Plantn
Pierscknowe Plantn
Hunthall
Ryslaw
Greenriggs
Swinton Quarter
Houndslow
Lill Rig
Charterhall Wood
Blackadder Water
Longbank
Swinton
Swinton Ho
SWINTON KIRK
Swinton Hill
Bassendeanhill
Crawlee Plantn
Cleughead
Merse
Corsbie Moor
Bassendean
Rumbleton
Greenlaw
Castle Mill
Charterhall
Earnslaw
Swintonmill
Little Swinton
Eden Burn
Fawside
Rumbletonlaw
Catmoss
Angelrow Fm
Kames East Mains
Knock Hill 272
Macks Mill
Foulshot Law
Crumrig
Purves Hall
Kames West Mains
Butterlaw
Simprim Burn
Kirkhill
Corsbie
Mersington Ho
Leitholm
Marlfield
Legerwood
Huntlywood
GREEN KNOWE TOWER
Gordon
Gordon East Mains
Middlethird
Gordonbank
Ploughlands
Stainrigg Mains
Hawkslaw
Skaithmuir
Huntlywood
West Mains
Byrewalls
Easter Howlaws
Lambden
Lambden Ho
Pittlesheugh
Anton's Hill
Orange Lane
Darnchester West Mains
Morton Cott
East Morriston
Hareford Burn
Todrig
Wester Howlaws
Viewfield
Stonefold
Hardacres
Crosshall
Hume Crags
Hume Plantn
Hume
HUME
Hassington East Mains
Lambden Burn
Dunglass Wood
Kincham Wood
THE HIRSEL
Eccles
Eccles Ho
Brae Dunstan
Fernyrig Cott
Crown Gorse
Lennel
West Morriston
Horseshoe Plantn
Falsidehill
Hardiesmill Place
Hume Mill
Hareheugh Craigs
Wormerlaw
Bartlehill
Eccles Newton
Birgham Wood
Hirsel Lake
Wester Park
Fans
Fans Hill
Littlehill Plantn
Whitehill
Blue Houses
Hassington West Mains
Coldstream
Darlingfield
Lurgie Craigs
Stichill Home Fm
Sweethope Hill
Stichill Eastfield
Harpertoun
Saint Foin
Birgham
Dundock Wood
East End
Kaimflat
Lochton
Carham Hall
Sheep Cover
Blinkbonny
Parkend
Stichill
Caldronbrae
Eden Hall
Springhill
Gallows Hill
Wark
Shielfield Wood
Rachelfield
Girrick
Hundy
Carham
West Learmouth
Cowdenknowes
Nenthorn
Newtonlees
NT
0 1 2 3 miles
0 1 2 3 4 5 km

Fast Castle Head
Wheat Stack
Telegraph Hill
FAST CASTLE
283
St Abb's Head
ST ABB'S HEAD
Horsecastle Bay
Oatlee Hill
Lumsdaine
Lumsdaine Moor
Coldingham Loch
Mire Loch
Cross Law
Moorside Plantn
Bell Hill
Northfield
Starney Bay
KITTIWAKE GALLERY
St Abbs
St Abb's Haven
Coldingham Bay
Yellow Craig
COLDINGHAM SANDS
Buskinburn Ho
Coldingham Moor
Huxton
PRIORY
Coldingham
Temple Hall
Press Castle
Press Mains
Abbey Park
Three Burn Grange
Grange Plantn
Gallows Law
Callercove Pt
Hairy Ness
EYEMOUTH MUSEUM
Hallydown
Eyemouth
A1107
Ind Est
Blackhill
Cairncross
North Wood
Ale Water
Scout Pt
Horse Head
Fancove Head
Heugh Head
Howburn
Reston
East Reston Mill
Fox Covert
Littledean Fm
Aytonwood Ho
Ayton Mains
Breeches Rock
Loanside
Auchencrow
Ayton
AYTON CASTLE
Burnmouth
Ross
Ross Pt
Berrybank
Peelwalls Ho
Cockdaw
Peelwalls Fm
Ayton Hill
Chester Hill
Bowie's Plantn
Prenderguest
Millerton Hill
Hilton Bay
Causeway Bank
Horn Burn
Billiemire Burn
Oldcastles
Lamberton Moor
Lamberton Beach
Meg's Dub
Foulden Hill
Westmains Moor
Greenfield Plantns
Edington Covert
North Crowlea Plantn
Hag Wood
Witches' Knowe
Lamberton
Lazybeds Plantn
Chirnside
Deans Hill
Kilbarn
Mordington Holdings
Marshall Meadows Bay
Marshall Meadows
Chirnsidebridge
Nether Mains
Foulden New Mains
FOULDEN TITHE BARN
Foulden
Clappers
Mordington Ho
West Edge Fm
St John's Haven
Stuartslaw
Whiteadder Water
Mains
West Foulden
Foulden Bastle
Edrington Ho
Bus Pk
Allanton
Hutton Hall Barns
Broadmeadows Ho
Clarabad Mill
Edrington Mains
DUNS ROAD
Halidon Hill
Highfields
Sharper's Head
Hutton
Meadow House Mains
Paxton North Mains
Baldersbury Hill
A6105
Bucket Rocks
Broomdykes
Blackadder Mount
Blackadder Bank
Hutton Mains
Paxton Spital Mains
Paxton
New Mills
Grangeburn Mill Fm
Berwick-upon-Tweed
BERWICK-UPON-TWEED
BERWICK BOROUGH MUSEUM
Mill Fm
Tweedmouth
Sandstell Pt
Sunwick Fm
TOWER HOUSE POTTERY
SHIELFIELD PARK
Spittal
Doons Law
Winfield
PAXTON HOUSE
River Tweed
Fishwick
UNION SUSPENSION BRIDGE
Loanend
East Ord
Prior Park
Bear's Head
Whitsome
Hilton
Jardinefield
Ord Mains
ETAL ROAD
Redshin Cove
Fishwick Mains
Horncliffe
Longridge Towers
Tweedmouthmoor
Borewell Fm
Whitsomehill
Horncliffe Mains
Saltpan Rocks
East Newton
Horndean
Green Hill
Norham East Mains
Velvet Hall
Ewe Hill
Murton
Unthank Fm
Unthank Blue Ho
Scremerston
Fellow Hills
Horndean Burn
Norham West Mains
Thornton Park
Murton White Ho
Old Town Covert
New Ladykirk
Thornton Cottage
Cheswick Black Rocks
Ladykirk
NORHAM CASTLE
Emerick Fm
Royalty Fm
Allerdean Mill
Scremerston Hill
Scremerston Town Fm
Greenside Cotts
Norham
Morris Hall
Thornton
Thornton Mains
Allerdeanmill Burn
Camphouses
Shoreswood
Shoreswood Hall
Cheswick Sands
Fox Covert
East Newbiggin
Shoresdean
West Allerdean
Cheswick Ho
Cheswick
Simprim Mains
Ladykirk House
Battle Moor
Grievestead Moor
Allerdean Greens
Coal Burn
Cheswick Buildings
Windmill Hill
Goswick
Milne Graden East Mains
Groat Haugh
Allerdean Grange
Ancroft Northmoor
Ancroft
Widepen Plantn
Grindon
NORTHUMBERLAND
Ancroft Mill
Milne Graden West Mains
Tillmouth Fm
Ancroft Southmoor
Goswick Sands
Holy Island (Lindisfarne)
Westmoor Plantn
Felkington
Bridge Mill
Haggerston
Haydon Dean
Berrington Law
Berrington Ho
Snook Pt
Toft Hill
Mattilees Hill
Berrington Lough
Berrington
New Haggerston
Brockmill Fm
Beal Pt
Tweedmill
Shellacres
Dean Burn
Beal
LINDISFARNE
Brick Kiln Plantn
Bowsden Moor
Twizel Mill
Fadden Hill
Duddo
Mattilees Cott
Jacks Law
Lickar Dean
Beal Sands
Holy Island Sands
Castle Heaton
Common Plantn
Duddo Hill
West Mains Inn
Bowsden
Lowick Mill
Woodside Fm
Bowsden West Fm
Old Dryburn
Granary Pt
Donaldson's Lodge
River Till
Low Wood
The Low
Mount Hooley
Melkington
Black Bank
Tindal Ho
Hazely Hill
High Wood
Barmoor Hill
Dry Burn
Hunting Hall
Syning Hill
Barleymill Bank
Fenham
Peter's Plantn
New Heaton
Heaton Moor
Rhodes Hill
Black Heddon
West Kyloe
Fenwick
Barmoor Castle
Barmoor Lane End
Cornhill-on-Tweed
Heaton Covert
263
ETAL CASTLE
Etal
Etal Rhodes
Etal Moor
Barmoor Ridge
Lowick
264
Kyloe Hills
East Kyloe
Lowmoor Pt
Pallinsburn Ho
Moss Hall
Moss Wood
Fenwick Wood
HEATHERSLAW LIGHT RAILWAY
Letham Hill
Ford Forge
Ford Cott
Kyloe Wood
Tealhole Pt
Cookstead
Common Cott
Shepherdskirk Hill
Bareless
HEATHERSLAW CORNMILL
Crookham
Ford Common
Barmoor South Moor
Redhouse Moor
Buckton
Campfield
Crookham Westfield
HEATHERSLAW
Hay Fm
Crookham Hall
Buckton Moor
Shiellow Wood
Windy Law
LADY WATERFORD HALL
FORT
Bar Moor
Barmoore Red Ho
Brownridge
Smeafield
Branxton Bldgs
Branxton
Ford
Goswick Sands
Snook Pt
Keel Head
Emmanuel Head
Snipe Pt
Beal Pt
Beal Sands
Holy Island Sands
LINDISFARNE HERITAGE CENTRE
Holy Island
LINDISFARNE CASTLE
LINDISFARNE PRIORY
Castle Pt
Granary Pt
Hole Mouth
Fenham
Burrows Hole
Guile Pt
Lowmoor Pt
Tealhole Pt
Ross Pt
Ross Back Sands
Buckton
A1
A698
A6105
B6355
B6437
B6438
B6460
B6461
B6470
B6354
B6525
B6353

IONA ABBEY AND CATHEDRAL
IONA HERITAGE CENTRE
Baile Mor
Iona
Sligneach
Stac an Aoineidh
ST COLUMBA EXHIBITION & WELCOME CENTRE
Fionnphort
Aridhglas
Fidden
Tiraghoil
Eorabus
Achnahard
Knokan
Lower Ardtun
Lee
Bunessan
Loch Assapol
BROLASS
Carsaig
Carsaig Bay
Rubha Dubh
CARSAIG ARCHES
Malcolm's Pt.
Glenbyr
BEINN NA CROISE
CRUACHAN MIN
Erraid
Soa I.
Knockvologan
ROSS OF MULL
Ardalanish
Uisken
Scoor
Ardchiavaig
Rubha nam Braithrean
Eilean a'Chalmain
Rubh Ardalanish
Torran Rocks
OBAN 2:20
Dubh Artach
Rubh'a'Geadha
Balnahard
Kiloran Bay
Uragaig
KILORAN GARDENS
Kiloran
Kilchattan
COLONSAY
Scalasaig
Loch Staosnaig
Ardskenish
Garvard
Rubha Dubh
Balerominhor
PRIORY
Dubh Eilean
Oronsay
Eilean nan Ron
Corpach Bay
Shian Bay
RAINBERG MOR
Loch Righ Mòr
Rubh'an t-Sàilein
1:10 (Summer Only)
Loch Tarbert
Rubha Lang-aoinidh
Rubha a'Mhail
Rubha Bholsa
SGARBH BREAC
Lagg
Loch an Aircil
Loch Lesgamaill
JURA
PAPS OF JURA
Ardmenish
Nave Island
Ardnave Pt.
Gortantaoid
Bunnahabhain
BUNNAHABHAIN DISTILLERY
Loch a Chnuic Bhric
Cnocbreac
JURA FOREST
Corran
Knockrome
Lowlandman's Bay
Carraig Bhan
Ardnave
Kilnave
Killinallan
Learygbreck
Ardfernal
Loch na Mile
Gleann Astaile
An Clachan
Sanaigmore
Garra Eallabus
Caol Ila
CAOL ILA DISTILLERY
Port Askaig
Feolin Ferry
Braigo
Leckgruinart
Loch Gruinart
FINLAGGAN CENTRE
Keills
Keils
Smaull
LOCH GRUINART RSPB
LOCH GRUINART NATURE RESERVE VISITORS CENTRE
Loch Finlaggan
Gleann Ullibh
Craighouse
ISLE OF JURA DISTILLERY
Small Isles
Ballinaby
Carnduncan
Craigens
Loch Cam
Loch Ballygrant
Saligo Bay
Saligo
Aoradh
Tighnacachla
Balole
Ballygrant
Lossit Lodge
BRAT BHEINN
Crackaig
Loch Gorm
Coull
Coul Pt.
Foreland Ho
ISLAY
Lyrabus
Kilmeny
Knockfearoch
Eskrish
Cabrach
Strone
Sunderland
Camas an Staca
JURA HOUSE WALLED GARDEN
Machir Bay
Kilchoman
Blackrock
Redhouses
Daill
BEINN DUBH
Gortan
Conisby
Bridgend
Islay Ho
Am Fraoch Eilean
Rubha na Tràille
Brosdale I.
Bruichladdich
Kilchiaran Bay
Kilchiaran
Bowmore
BOWMORE ROUND CHURCH
ISLAY LIFE MUSEUM
Cattadale
Tormisdale
RHINNS OF ISLAY
Port Charlotte
Mulindry
Gartbreck
LOCH INDAAL
Bridge Ho
Laggan
Kelsay
Nerabus
Lossit Pt.
BEINN BHAN
BEINN BHEIGEIR
Carraig Mhór
A849
A846
A847
B8086
B8087
B8085
B8018
B8017
288
289
254
NM
NR
0 2 4 6 miles
0 2 4 6 8 10 km

Lochbuie Ho
Laggan Lodge
Laggan
Deer Forest
Lord Lovat's Cave
FIRTH OF LORN
Frank Lockwood's Island
Dubh-fheith
Garbh Eileach
Garvellachs
Eileach an Naoimh
SCARBA, LUNGA AND THE GARVELLACHS
Eilean Dubh Mor
Lunga
Scarba
Kilmory Lodge
Rubha Aird Luing
Gulf of Corryvreckan
Seil
Insh I.
Ardencaple Ho
Clachan-Seil
Easdale
An Cala Gardens
Easdale Island Folk Museum
Balvicar
Acha
Cuan
Torsa
Cullipool
Bardrishaig
Luing
Ardinamir
Achafolla
Toberonochy
Shuna
Shuna Ho
Shuna Pt.
Arduaine Garden
Arduaine
Kames
Garraron
Craobh Haven
Lunga
Barravullin
Turnalt
Barbreck Ho
Kintraw
Leragoligan
Ardfern
Craignish Castle
Kirkton
Aird
Rèisa an t-Sruith
Craignish Pt.
Island Macaskin
Eilean Righ
Carnassarie Castle
Kilmartin
Kilmartin House Trust
Glebe Cairn
Kilmartin Stones
Ri Cruin Cairn
Slockavullin
Poltalloch
Duntrune Castle
Loch Crinan
Crinan
Crinan Ferry
Kilmahumaig
Bellanoch
Moine Mhor
Cairnbaan Cup & Ring Marks
Dunadd
Kilmichael Glassary
Cup and Ring Marks
Bridgend
Achnashelloch
Rock Markings
Achnabreck
Cairnbaan
Dounie
Lochgilphead
Kilmory Castle Gardens
Kilmory Woodland Park
Ardrishaig
Loch Gilp
Castleton
Kinuachdrachd
Kinuachdrachd Harbour
Glengarrisdale Bay
ARGYLL AND BUTE
Lealt
Ardlussa
Ardlussa Bay
Inverlussa
Lussagiven
Tarbert
JURA
KNAPDALE
Gallachoille
Carsaig
Tayvallich
Kintallan
Achanamara
Kilmichael of Inverlussa
Ashfield
Barrahormid
Taynish
Keills Chapel
Keillmore
New Ulva
Dunrostan
Castle Sween
Island of Danna
Danna na Cloiche
St Columba's Cave
Lochead
Achahoish
Ellary
Kilmory
Eilean Mòr
St Cormac's Chapel
Kilmory Bay
Clachbreck
Baile Boidheach
Ormsary
Druimdrishaig
Pt. of Knap
Loch Caolisport
Sound of Jura
Miller's Bay
Cretshengan
Coulaghailtro
Kilberry
Kilberry Hd.
Sculptured Stones
Carse Ho.
Dunmore
Loch Stornoway
Ardpatrick
Ardpatrick Ho.
Portachoillan
Ardpatrick Pt.
Eilean Traighe
Ronachan Pt.
Clachan
Port Ellen
Achabraid
Inverneill
Stronachullin Lodge
Erines
Loch Chaorain
Barmore I.
Stonefield Castle Hotel
Tarbert
Barfad
East Loch Tarbert
Dubhchladach
West Tarbert
Corranbuie
Rhu
Kennacraig
Redhouse
Whitehouse
Kilchamaig
West Loch Tarbert
Gartnagrenach
Glenreasdell Mains
Skipness
Skipness Castle
Skipness Pt.
Skipness Bay
Claonaig
Claonaig Bay
Rubha Leathan
Inchmarnock
Sound of Bute
Am Buth
Dunach
Glen Feochan
Glenfeochan House Gardens
Kilbride
Ardentallen
Knipoch
Kilninver
Barochreal
Raera
Glen Euchar
Scammadale
Bragleenmore
Loch Scamadale
Musdale
Braes of Lorn
Innie
Blarian
Loch Tralaig
Melfort
Ardanstur
Kilchoan
Loch Melfort
Kilmelford
Glenmore
Lagalochan
Arichamish
Inverliever Lodge
Torran
Ford
Ederline
Finchairn
Finchairn Castle
Stroneskar
Barmolloch
Kilbride
Leckuary
Loch Gaineamhach
Inverliever Forest
Kilmaha
Loch Avich
Kilmun
Lochavich Ho
Drissaig
Barnaline
Dalavich
Fineglen
Inverinan
Loch Awe
Ballimeanoch
Blarghour
Ardchonnell
Ardchonnell Castle
Portinnisherrich
Eredine
Durran
Braevallich
Eredine Forest
Kilchrenan
Annat
North Port
South Port
Portsonachan
Ardanaiseig
Cladich Steading
Cladich
Lower Achachenna
Tullich
Glen Aray
Three Bridges
Inveraray Castle
Combined Operations Museum
Inveraray Jail
Bell Tower
Inveraray
Newtown
Dalchenna
Auchindrain
Auchindrain Open Air Museum
Brenchoillie
Kenmore
Strachur Bay
Creggans
Clachan Strachur
Strachurmore
Mid Letter
Furnace
Sandhole
Cumlodden
Newton
Glen Sluain
Crarae
Crarae Glen Gardens
Minard
Leanach
Garbhallt
Lachlan Castle
Glenbranter
Glenbranter Forest
Tullochgorm
Birdfield
Lachlan Bay
Minard Castle
Strathlachlan Forest
Lephinmore
Asknish
Lochgair
Lephinchapel
Kilbridemore
Dunans
Bernice
Middle Kames
Largiemore
Loch Fyne
Carrick
Port Ann
Achnaba
Otter Ferry
Ballimore
Auchnaha
Eilean Mòr
Clachan of Glendaruel
Kilmodan Sculptured Stones
Bealach Maim
Conchra
Glendaruel Forest
Achanelid
Garrachra
Glenmassan
Camquhart
Craigendive
Balliemore
Ardtaraig
Auchnagarron
Auchenbreck
Lochead
Loch Riddon
Springfield
Ormidale Lodge
Clachaig
Glenstriven
Kilfinan
Otter Ho
Kilfinan Bay
Acharossan
Drum
Melldalloch
Kyles of Bute
Glachavoil
Ardentraive
Colintraive
Rhubodach
Couston
Algaltraig
Loch Striven
Inverchaolain
Auchenlochan
Tighnabruaich
Port Driseach
Kames
Millhouse
Crispie
Asgog Loch
Portavadie
Blair's Ferry
Kilmichael
Glenmore
Glecknabae
Craig Lodge
Asgog Bay
Kilbride Bay
Ardlamont Ho.
Ardlamont Pt.
Carry
Ardnagowan
Invernoaden
Knockdow
Port Lamont
Newton Park
Castle Toward
Toward Pt.
Ardyne Pt.
Ardamaleish
Ettrickdale
Kames B.
Port Bannatyne
St Colmac
Kildavanan
Ettrick Bay
Westland
Rothesay Castle
Rothesay Bay
Rothesay
Upper Ardroscadale
Craigmore
Montford
Victorian Fernery
Greenan
Town Head
Ascog
Straad
Woodend Ho
Kerrycroy
Northpark
Midpark
Scoulag
Mount Stuart House and Garden
Scalpsie
Scalpsie Bay
Piperhall
Ardscalpsie Pt.
Kingarth
Stravanan Bay
Kilchattan Bay
ISLAND OF BUTE

Lochgoilhead
ARGYLL FOREST PARK
Loch Goil
Loch Long
Loch Eck
Holy Loch
Gare Loch
River Clyde
ARGYLL AND BUTE
INVERCLYDE
Glen Fruin
Glen Douglas
Glen Mollochan
Drimsynie
Corrow
Lettermay
Blairmore
Carrick Castle
Whistlefield
Portincaple
Garelochhead
Rowardennan
Inverbeg
Glenmallan
Edentaggart
Rockville
Mambeg
Faslane Port
Rahane
Shandon
Rhu
Helensburgh
Craigendoran
Colgrain
Clynder
Rosneath
Kilcreggan
Cove
Coulport
Ardpeaton
Gairletter
Ardentinny
Barnacabber
Sligrachan
Drynain
Benmore
Uig
Ardbeg
Inverchaolain
Dalinlongart
Sandhaven
Sandbank
Ardnadam
Kilmun
Strone
Hunter's Quay
Kirn
Dunoon
Bullwood
Innellan
Toward
Glenbranter
Invernoaden
Glen Branter
Garrachra
Clachaig
Glenstriven
Port Lamont
Knockdow
Gourock
Ashton
Levan
Midton
Larkfield
Braeside
Greenock
Port Glasgow
Inverkip
Wemyss Bay
Upper Skelmorlie
Shielhill
Garvock
Devol
Bardrainney
Kilmacolm
Cardross
Ardmore

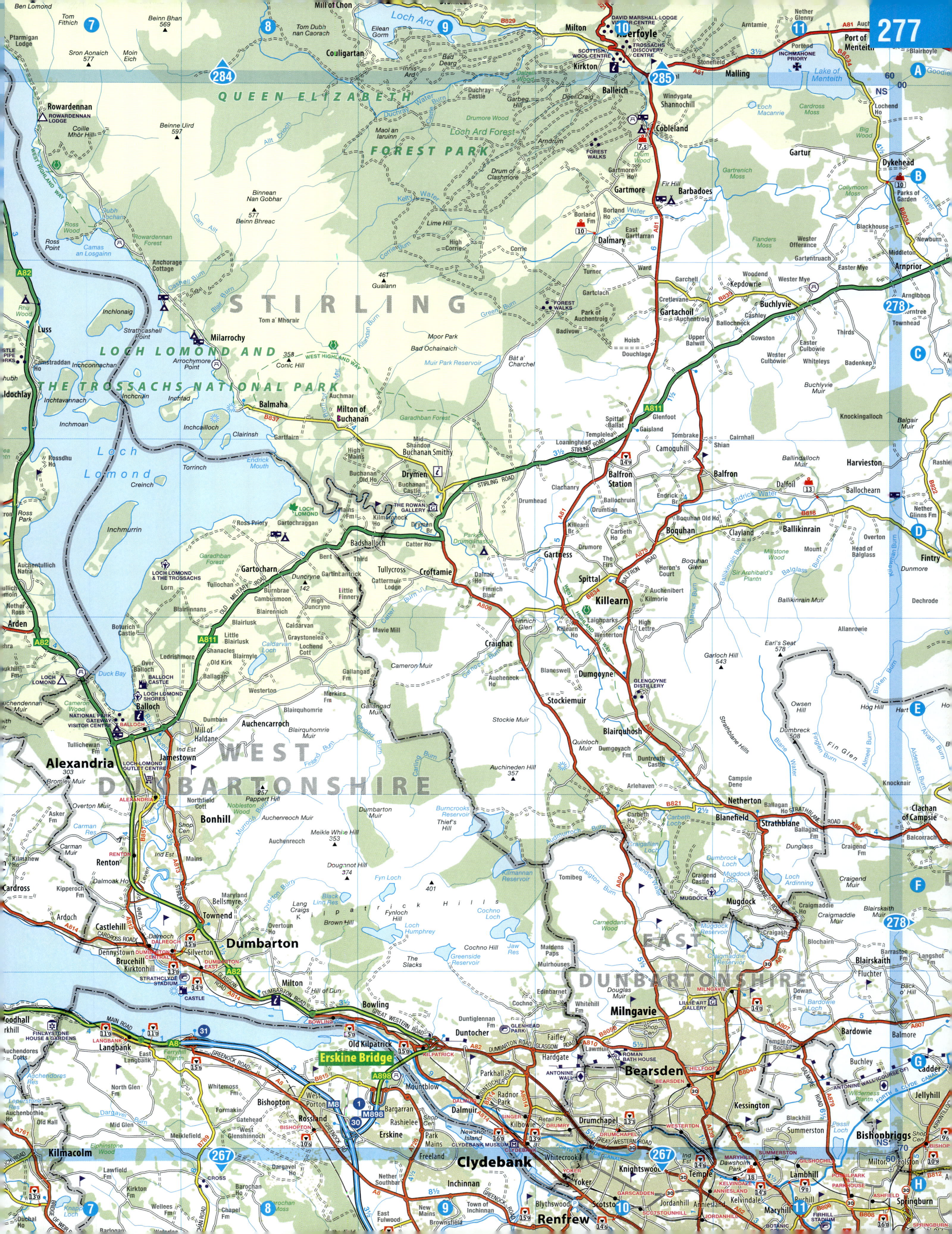
STIRLING
LOCH LOMOND AND THE TROSSACHS NATIONAL PARK
QUEEN ELIZABETH FOREST PARK
WEST DUNBARTONSHIRE
EAST DUNBARTONSHIRE
Aberfoyle
Drymen
Balfron
Killearn
Strathblane
Alexandria
Dumbarton
Milngavie
Bearsden
Clydebank
Erskine Bridge
Loch Lomond

Doune
Dunblane
Deanston
Thornhill
Ruskie
Boquhapple
Flanders Moss
Dykehead
Arnprior
Kippen
Cauldhame
Gargunnock
Old Leckie
Glentirranmuir
Blair Drummond
Blair Drummond Moss
Bridge of Allan
Causewayhead
Raploch
Cambusbarron
Cambuskenneth
Stirling
Torbrex
St Ninians
Bannockburn
Plean
Cowie
Fallin
Menstrie
Blairlogie
Badgergate
Sunnylaw
S T I R L I N G
Gargunnock Hills
Fintry Hills
Fintry
Culcreuch Castle
Carron Valley Reservoir
Carron Valley Forest
Kilsyth Hills
Campsie Fells
Earl's Hill
Easter Cringate
Muirmill
Carron Bridge
Dunipace
Denny
Head of Muir
Dennyloanhead
Bonnybridge
High Bonnybridge
Torwood
Fankerton
Stoneywood
Longcroft
Banknock
Castlecary
Kilsyth
Queenzieburn
Croy
Dullatur
Cumbernauld
Cumbernauld Village
Abronhill
Condorrat
Westfield
Balloch
Luggiebank
Kirkintilloch
Lennoxtown
Clachan of Campsie
Milton of Campsie
Birdston
Torrance
Lenzie
Auchinloch
Bishopbriggs
Lumloch
Muirhead
Chryston
Moodiesburn
Mollinsburn
Riggend
Greengairs
Wattston
Glenboig
Gartcosh
Springburn
Stepps
Caldercruix
Glenmavis
Longriggend
EAST DUNBARTONSHIRE
N O R T H
L A N A R K S H I R E
Blairskaith
Bardowie
Harviestoun
Balgrochan
Twechar
Harestanes
Cadder
Jellyhill
Dunmore
Allanrowie
Blackhouse
Middleton
Knockingalloch
Ballochearn
Balgair Muir
Kippen Muir
0 1 2 3 miles
0 1 2 3 4 5 km

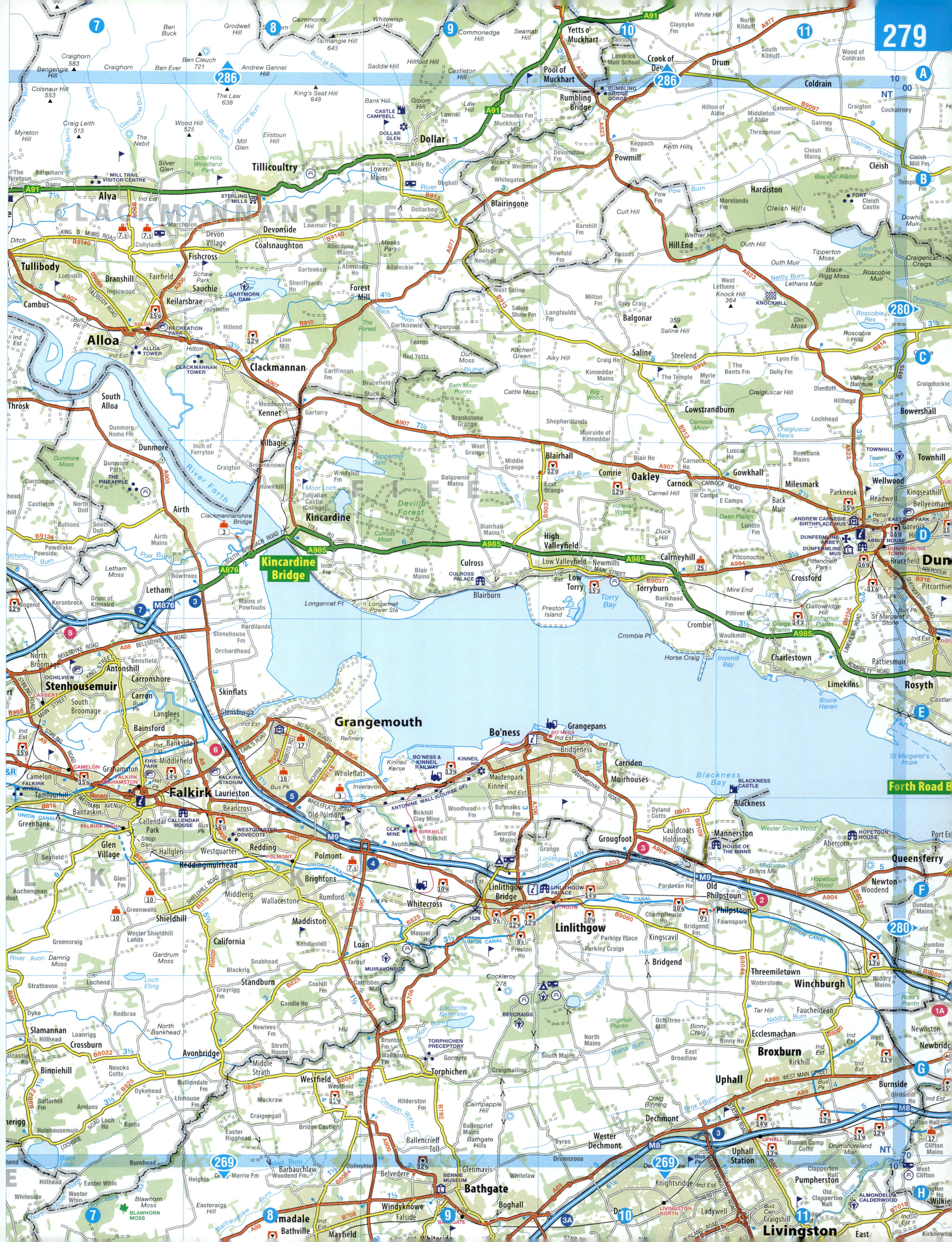
CLACKMANNANSHIRE
FIFE
FALKIRK
Tillicoultry
Dollar
Alva
Menstrie
Tullibody
Alloa
Clackmannan
Kincardine
Kincardine Bridge
Airth
Stenhousemuir
Falkirk
Grangemouth
Bo'ness
Linlithgow
Polmont
Maddiston
Whitecross
Bathgate
Livingston
Broxburn
Uphall
Winchburgh
Philpstoun
Queensferry
Rosyth
Limekilns
Charlestown
Culross
High Valleyfield
Torryburn
Cairneyhill
Crossford
Oakley
Saline
Steelend
Blairhall
Comrie
Carnock
Gowkhall
Milesmark
Crook of Devon
Drum
Powmill
Blairingone
Coaldsnaughton
Devonside
Sauchie
Fishcross
Keilarsbrae
Kennet
Kilbagie
Dunmore
South Alloa
Carronshore
Larbert
Bainsford
Camelon
Laurieston
Redding
Reddingmuirhead
Brightons
Rumford
California
Shieldhill
Avonbridge
Slamannan
Binniehill
Torphichen
Westfield
Armadale
Blackburn
Ecclesmachan
Dechmont
Pumpherston
Muckhart
Rumbling Bridge
Cleish
Hardiston
Hill End
Coldrain
Townhill
Wellwood
Parkneuk
River Forth
Forth Road Bridge
Blackness
Carriden
Muirhouses
Grougfoot
Mannerston
Torry Bay
Crombie
A91
A907
A985
A876
A904
A803
A89
M9
M8
M876
286
280
269

Loch Leven
St Serf's Island
Scotlandwell
Auchmuirbridge
Leslie
Glenrothes
Markinch
Milton of Balgonie
Balcurvie
Windygates
Coaltown of Balgonie
Kinglassie
Thornton
Ballingry
Lochore
FIFE REGIONAL PARK
Crosshill
Woodend
Auchterderran
Cardenden
Cluny
Kelty
Lochgelly
Chapel
Dysart
Gallatown
Pathhead
Kirkcaldy
Linktown
Cowdenbeath
Lumphinnans
Lassodie
Kingseat
Hill of Beath
Townhill
Wellwood
Halbeath
Crossgates
Fordell
Donibristle
Auchtertool
Balmule
Kinghorn
Pettycur
Burntisland
Aberdour
Dunfermline
Pitcorthie
Duloch
Rosyth
Hillend
Dalgety Bay
Inverkeithing
North Queensferry
Forth Road Bridge
FIRTH OF FORTH
Inchkeith
Inchcolm
Cramond Island
Queensferry
Dalmeny
Newbigging
Kirkliston
Newbridge
Ratho Station
Ratho
Cramond
Barnton
Davidson's Mains
Granton
Newhaven
Leith
Edinburgh
Portobello
Joppa
Musselburgh
Corstorphine
Murrayfield
Gorgie
Sighthill
Colinton
Liberton
Gilmerton
Danderhall
Millerhill
Whitecraig
Juniper Green
Currie
Burnwynd
EDINBURGH
CITY OF

Kennoway
Leven
Innerleven
Methil
Buckhaven
Largo Bay
St Monans
Elie
Earlsferry
Grange
Ardross
Kincraig Pt
Chapel Ness
Elie Ness
Sauchar Pt
Fidra
Craigleith
North Berwick
Dirleton
Gullane
Aberlady
Aberlady Bay
Craigielaw
Longniddry
Cockenzie and Port Seton
Prestonpans
Tranent
Elphinstone
Ormiston
Pencaitland
Gladsmuir
Samuelston
Haddington
Athelstaneford
Drem
East Fortune
East Linton
Whitekirk
Tyninghame
Auldhame
Scoughall
Kingston
Ballencrieff
Spittal
Macmerry
Penston
New Winton
New Town
Bolton
Gifford
Garvald
Papple
Traprain
Carfrae
Wallyford
Cousland
EAST
LOTHIAN
A1
A198
A199
A6137
287
282
271

287
North Ness
ISLE OF MAY
Isle of May
South Ness
281
Bass Rock
N O R T H
S E A
Gin Head
TANTALLON CASTLE
Auldhame
Car Rocks
Scoughall
Scoughall Rocks
New Mains
Peffer Sands
Ravensheugh Sands
Frances Craig
Whitekirk
Tyninghame Links
Tyne Sands
Tyninghame
Heckies Hole
JOHN MUIR
Belhaven Bay
Long Craigs
Scart Rock
Meikle Spiker
Dunbar
Belhaven
BELHAVEN ROAD
JOHN MUIR BIRTHPLACE
West Barns
Preston
PRESTON MILL & PHANTASSIE DOVECOT
Knowes
Mill Stone Neuk
White Sands
Barns Ness
Broxburn
East Barns
Hedderwick
South Belton
Old Belton
Wester Broomhouse
Spott
Little Pinkerton
Dunbar Cement Works
Meikle Pinkerton
Skateraw
Chapel Pt
Skateraw Harbour
Torness Pt
Long Craig
Thorntonloch Power Sta
Thorntonloch
Traprain
Grangemuir
Pitcox
Doon Hill
Pinkerton Hill
Crowhill
Luggate Burn
Stenton
THE CHESTERS (FORT)
Thurston Manor
Innerwick
Thurston Mains
Highside Hill
Halls
Blaik Law
Old Branxton
Bilsdean Creek
Reed Pt
Cove Harbour
Cove
Pease Bay
Red Rock
Greenheugh Pt
Siccar Pt
Blackcastle Hill
Oldhamstocks Mains
DUNGLASS COLLEGIATE CHURCH
Cockburnspath
Oldhamstocks
Meikle Poo Craig
Hirst Rocks
E A S T
L O T H I A N
Lothian Edge
Deuchrie Dod
Deuchrie Edge
Watch Law
Bransly Hill
Sheeppath Glen
Needle Hill
Stottencleugh
Dovecot Hall
Neuk Fm
Stockbridge
Old Cambus
Old Cambus Townhead
Garvald
Dunbar Common
272
0 1 2 3 miles
0 1 2 3 4 5 km
WHITE CASTLE (FORT)
Friardykes Dod
Saddle Hill
Birny Knowe
Wightman Hill
Ecclaw
Greenside Hill
Penmanshiel Wood
A1
A199
A1087
A198
B6370
A1107
NT

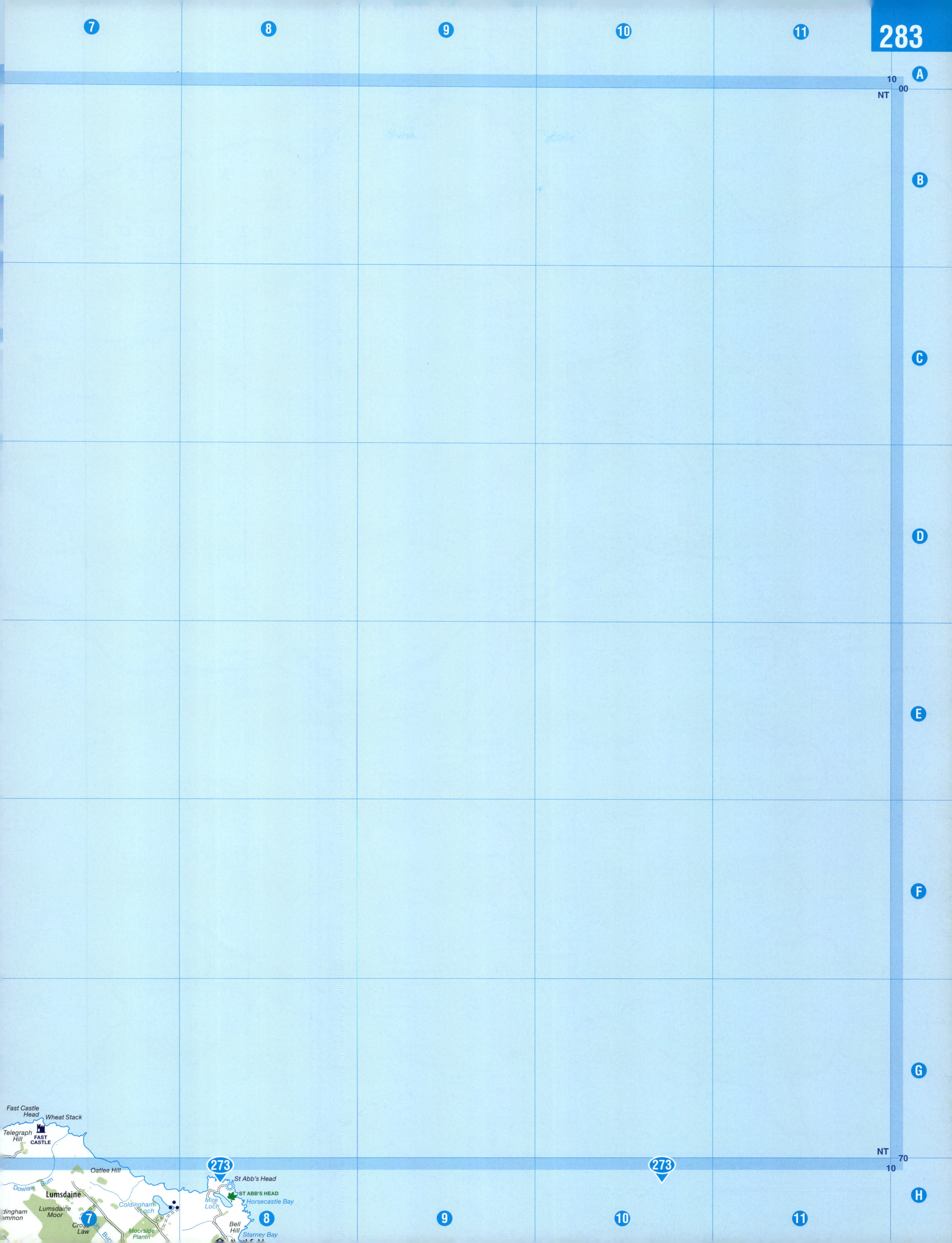
7
8
9
10
11
A
B
C
D
E
F
G
H
10
00
NT
NT
70
10
Fast Castle Head
Wheat Stack
Telegraph Hill
FAST CASTLE
Oatlee Hill
273
St Abb's Head
ST ABB'S HEAD
Horsecastle Bay
Mire Loch
Dowlaw Burn
Lumsdaine
Lumsdaine Moor
Coldingham Loch
Cross Law
Moorside Plantn
Bell Hill
Starney Bay
273

LOCH LINNHE
BEN NEVIS AND GLEN COE
ARGYLL AND BUTE
LORN
LYNN OF LORN
Lismore
Kerrera
Shuna I.
Shuna
Oban
Glencoe
Kinlochleven
Ballachulish
North Ballachulish
South Ballachulish
Kentallen
Duror
Keil
Appin
Portnacroish
Port Appin
Creagan
Glasdrum
Barcaldine
Benderloch
Ledaig
Connel
Dunbeg
Bonawe
Taynuilt
Brochroy
Bridge of Awe
Lochawe
Dalmally
Inveraray
Kilchrenan
Portsonachan
Furnace
Auchindrain
Strachur
Lochgoilhead
Cairndow
Kilmartin
Kilmelford
Kilninver
Loch Etive
Loch Awe
Loch Fyne
Loch Creran
Glen Orchy
Bridge of Orchy
Inveroran Hotel
Kingshouse Hotel
Dalness
Glen Etive
Loch Feochan
Loch Avich
Loch Scamadale
Ardchattan Priory
Dunstaffnage Castle
Kilchurn Castle
Inveraray Castle
Oban Distillery
Scottish Sea Life Sanctuary
Cruachan Reservoir
Ben Lui
Strontian
Inversanda
Kingairloch
Camasnacroise
Glensanda
Eilean Musdile
Lunga
Carnasserie Castle
Kilmartin House Trust
Crarae Glen Gardens
Minard
Tullochgorm
Lachlan Castle
Glenbranter
0 2 4 6 miles
0 2 4 6 8 10 km

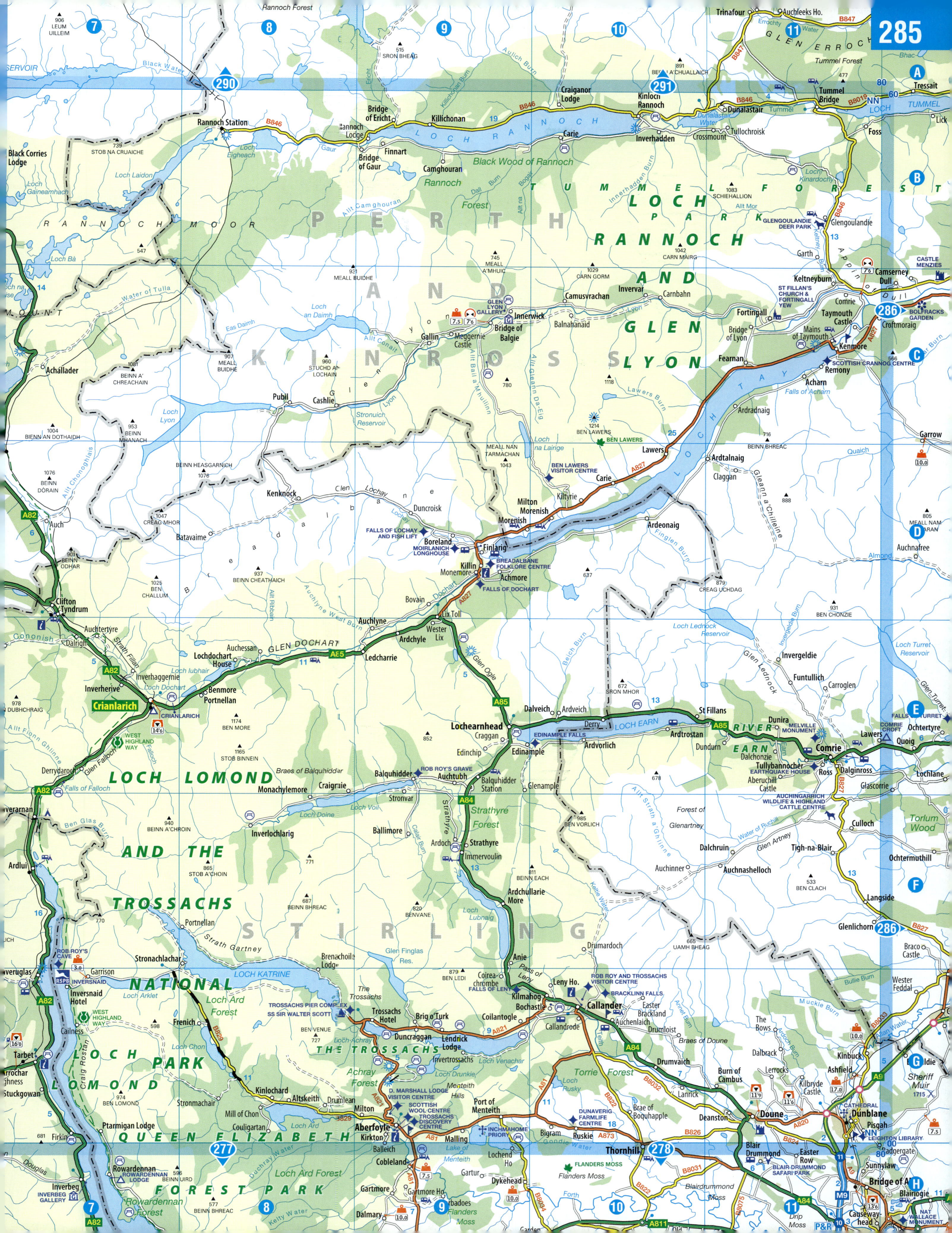
Rannoch Forest
Trinafour
Auchleeks Ho.
GLEN ERROCHTY
Tummel Forest
Tressait
Rannoch Station
Black Corries Lodge
Bridge of Ericht
Killichonan
Kinloch Rannoch
Dunalastair
Tummel Bridge
LOCH TUMMEL
LOCH RANNOCH
Black Wood of Rannoch
Craiganour Lodge
Carie
Inverhadden
Crossmount
Tullochroisk
Foss
Finnart
Bridge of Gaur
Camghouran
Rannoch Forest
Loch Eigheach
Loch Laidon
Loch Gaineamhach
RANNOCH MOOR
Loch Bà
PERTH AND KINROSS
TUMMEL FOREST
LOCH RANNOCH AND GLEN LYON
SCHIEHALLION
GLENGOULANDIE DEER PARK
Glengoulandie
Garth
Keltneyburn
Camserney
Dull
CASTLE MENZIES
Appin of Dull
CARN MAIRG
CARN GORM
MEALL BUIDHE
Inverwick
GLEN LYON GALLERY
Bridge of Balgie
Innerwick
Camusvrachan
Invervar
Carnbahn
Balnahanaid
Fortingall
ST FILLAN'S CHURCH & FORTINGALL YEW
Comrie
Taymouth Castle
Mains of Taymouth
Kenmore
Croftmoraig
BOLFRACKS GARDEN
Bridge of Lyon
Fearnan
SCOTTISH CRANNOG CENTRE
Remony
Acharn
Falls of Acharn
LOCH TAY
Ardradnaig
Gallin
Meggernie Castle
Achallader
Pubil
Cashlie
Glen Lyon
Stronuich Reservoir
Loch Lyon
Loch an Daimh
BEN LAWERS
BEN LAWERS VISITOR CENTRE
Lawers
Carie
Kiltyrie
Ardtalnaig
Claggan
Garrow
Kenknock
Glen Lochay
Duncroisk
Milton Morenish
Morenish
FALLS OF LOCHAY AND FISH LIFT
Boreland
MOIRLANICH LONGHOUSE
Finlarig
Killin
BREADALBANE FOLKLORE CENTRE
Monemore
Achmore
FALLS OF DOCHART
Ardeonaig
CREAG UCHDAG
Batavaime
Auch
Clifton
Tyndrum
Auchtertyre
Dalrigh
Strath Fillan
Bovain
Lix Toll
Wester Lix
Auchlyne
Ardchyle
Ledcharrie
Auchessan
GLEN DOCHART
Lochdochart House
Benmore
Portnellan
Inverhaggernie
Inverherive
Crianlarich
CRIANLARICH
BEN MORE
STOB BINNEIN
WEST HIGHLAND WAY
Glen Ogle
Loch Lednock Reservoir
Glen Lednock
Invergeldie
Funtullich
Carroglen
BEN CHONZIE
SRON MHOR
Dalveich
Ardveich
Derry
LOCH EARN
St Fillans
Lochearnhead
Craggan
Edinchip
Edinample
EDINAMPLE FALLS
Ardvorlich
Ardtrostan
Dundurn
RIVER EARN
Dunira
MELVILLE MONUMENT
Dalchonzie
Tullybannocher
EARTHQUAKE HOUSE
Aberuchill Castle
Comrie
Dalginross
Ross
Lawers
Quoig
Ochtertyre
Lochlane
Glascorrie
AUCHINGARRICH WILDLIFE & HIGHLAND CATTLE CENTRE
Culloch
Torlum Wood
Tigh-na-Blair
Ochtermuthill
Langside
Glenlichorn
Braco Castle
Derrydaroch
Glen Falloch
Falls of Falloch
LOCH LOMOND AND THE TROSSACHS NATIONAL PARK
Braes of Balquhidder
Monachylemore
Craigruie
Balquhidder
ROB ROY'S GRAVE
Auchtubh
Balquhidder Station
Glenample
Stronvar
Loch Doine
Loch Voil
Inverlochlarig
Ballimore
Strathyre Forest
Strathyre
Ardoch
Immervoulin
BEN VORLICH
Forest of Glenartney
Dalchruin
Glen Artney
Auchinner
Auchnashelloch
BEN CLACH
BEN A'CHROIN
STOB A'CHOIN
BEINN BHREAC
BENVANE
Loch Lubnaig
BEINN EACH
Ardchullarie More
Drumardoch
UAMH BHEAG
Ardlui
Portnellan
Strath Gartney
Stronachlachar
Brenachoile Lodge
Glen Finglas Res.
BEN LEDI
Coirea-chrombe
FALLS OF LENY
Pass of Leny
Leny Ho.
Kilmahog
Bochastle
Callander
ROB ROY AND TROSSACHS VISITOR CENTRE
BRACKLINN FALLS
Easter Brackland
Auchenlaich
Drumloist
Braes of Doune
Western Feddal
Bullie Burn
The Bows
Dalbrack
Kinbuck
Ashfield
Muckle Burn
ROB ROY'S CAVE
Garrison
Inversnaid
Inversnaid Hotel
LOCH KATRINE
Loch Arklet
Loch Ard Forest
TROSSACHS PIER COMPLEX
SS SIR WALTER SCOTT
The Trossachs
Trossachs Hotel
Brig o'Turk
Coilantogle
Callandrode
Duncraggan
Lendrick Lodge
BEN VENUE
THE TROSSACHS
Invertrossachs
Loch Venachar
Frenich
Cailness
Tarbet
Craig Rostan
LOCH LOMOND
Loch Chon
Achray Forest
Loch Drunkie
Menteith Hills
D. MARSHALL LODGE VISITOR CENTRE
SCOTTISH WOOL CENTRE
TROSSACHS DISCOVERY CENTRE
Torrie Forest
Loch Rusky
Drumvaich
Burn of Cambus
Lerrocks
Kilbryde Castle
Lanrick
Deanston
Doune
Dunblane
CATHEDRAL
Pisgah
LEIGHTON LIBRARY
Sheriff Muir
Kinlochard
Altskeith
Drumlean
Stronmachair
Mill of Chon
Couligartan
Milton
Aberfoyle
Kirkton
Port of Menteith
DUNAVERIG FARMLIFE CENTRE
Brae of Boquhapple
Ptarmigan Lodge
QUEEN ELIZABETH FOREST PARK
Firkin
BEN LOMOND
INCHMAHOME PRIORY
Malling
Bigram
Ruskie
Thornhill
Blair Drummond
BLAIR DRUMMOND SAFARI PARK
Easter Row
Badgergate
Sunnylaw
Bridge of Allan
Blairlogie
Rowardennan
ROWARDENNAN LODGE
Balleich
Cobleland
Lake of Menteith
Lochend Ho
Gartur
Dykehead
FLANDERS MOSS
Flanders Moss
Blairdrummond Moss
Loch Ard Forest
Inverbeg
INVERBEG GALLERY
Rowardennan Forest
BEINN BHREAC
Gartmore
Gartmore Ho
Dalmary
Drip Moss
Causewayhead
NAT WALLACE MONUMENT

LOCH TUMMEL
TUMMEL FOREST PARK
Tummel Forest
Tummel Bridge
Tressait
Queen's View
Pitlochry
Moulin
Kinnaird
Milton of Edradour
EDRADOUR DISTILLERY
BLAIR ATHOLL DISTILLERY VISITOR CENTRE
SCOTTISH PLANT COLLECTORS GARDEN
THEATRE FESTIVAL
HYDRO-ELECTRIC VISITOR CENTRE
Killiecrankie
KILLIECRANKIE VISITOR CENTRE
Bonskeid House
Fincastle Ho.
Aldclune
Tenandry
Balmore
Milton of Dalcapon
Straloch
Enochdhu
Glen Brerachan
Dalnavaid
Tarvie
Kirkmichael
Ashintully Castle
Middle Balnald
Ballintuim
Blacklunans
Forest of Alyth
Glenisla
Brewlands Bridge
Folda
Freuchies
Bellaty
Dykends
Kirkton of Glenisla
Bridge of Craigisla
Tullymurdoch
Bamff
Netherton
Milton of Drimmie
Blackcraig Forest
Bridge of Cally
Cochrage Muir
Forest of Clunie
Loch Benachally
Loch Ordie
Riemore
Riechip
Butterstone
Forneth
Achalader
Blairgowrie
Rattray
Kinloch
Rosemount
Craighall
Middleton
Lornty
Westfields of Rattray
Tullyfergus
Alyth
New Alyth
Balhary
Meigle
MEIGLE MUSEUM
Kinloch
Longleys
Arthurstone
Ardler
Keillor
Kirkton
West Myreriggs
Coupar Angus
Kettins
Hallyburton House
Leys
Lundie
Pitcur
Campmuir
Markethill
Woodside
Burrelton
Whitefield
Littleton
Saucher
Collace
Kirkton of Collace
Dunsinnan
Kinrossie
KING'S SEAT
Abernyte
Knapp
Rossie
Bandirran
Balbeggie
Ballindean
Inchture
Craigdallie
Kinnaird
Westown
Rait
BRAES OF THE CARSE
Kilspindie
Grange
MEGGINCH CASTLE GARDENS
Errol
Errol Station
Myreside
Glendoick
Glencarse
Kinfauns
Kinfauns Forest
FAIRWAYS HEAVY HORSE CENTRE
Leetown
Chapelhill
St Madoes
Port Allen
Cartagena Bank
Ballinbreich
FIRTH OF TAY
Mugdrum I.
LAING MUSEUM
Newburgh
Higham
Dunbog
Lindores
Lindores Loch
Grange of Lindores
Abernethy
ROUND TOWER
Pitmedden Forest
Colzie
Newhill
Monimail
Collessie
Auchtermuchty
Kinloch
Giffordtown
Dunshalt
Charlottetown
Strathmiglo
Kilgour
FALKLAND PALACE
Falkland
Newton of Falkland
Freuchie
LOMOND HILLS
FIFE REGIONAL PARK
Glasslie
Pitkevy
Cadham
Auchmuirbridge
Leslie
Glenrothes
Kinglassie
Stenton
Thornton
Auchterderran
Cluny
Woodend
Lochore
Crosshill
Glencraig
Ballingry
Scotlandwell
Kinnesswood
Wester Balgedie
Easter Balgedie
MICHAEL BRUCE'S COTTAGE
Glenlomond
Gateside
Nether Urquhart
Burnside
Wester Gospetry
Glenfarg
Duncrievie
Newhill
Arngask
Newton of Balcanquhal
Glentarkie
Glenfoot
Aberargie
Dron
Kintillo
Bridge of Earn
Forgandenny
Forteviot
Glenearn
Path of Condie
Rossie Ochill
Pathstruie
Middle Rigg
Middleton
Athron Hall
Craigow
Upper Tillyrie
Milnathort
BURLEIGH CASTLE
Kinross
KINROSS HOUSE
LOCHLEVEN CASTLE
LOCH LEVEN
St Serf's Island
VANE FARM
Gairney Bank
Dalqueich
Carnbo
Coldrain
Drum
Cleish
Cleish Hills
Hardiston
Blairingone
Powmill
Crook of Devon
RUMBLING BRIDGE GORGE
Rumbling Bridge
Yetts o' Muckhart
Pool o' Muckhart
Glendevon
Burnfoot
Glen Devon
Dollar
CASTLE CAMPBELL
DOLLAR GLEN
Devonside
Coalsnaughton
Fishcross
Alva
Tullibody
NATIONAL WALLACE MONUMENT
Causewayhead
Dunblane
CATHEDRAL
Pisgah
Radgergate
Sunnylaw
Kinbuck
Ashfield
Balhaldie
Sheriff Muir
BLAIRDENON HILL
BEN CLEUCH
KING'S SEAT HILL
STEEL'S KNOWE
Coulshill
Gleneagles House
Glen Eagles
Blackford
Braes of Ogilvie
TULLIBARDINE DISTILLERY
Greenloaning
Dam of Quoiggs
STRATHALLAN
Braco
Braco Castle
ARDOCH ROMAN CAMP
BLACKHALL ROMAN CAMPS
Gleneagles Hotel
GLENEAGLES
Auchterarder
HERITAGE CENTRE
Milton
Muirton
Aberuthven
Tullibardine
TULLIBARDINE CHAPEL
Strathallan Castle
Kinkell Bridge
Dunning
Newton of Pitcairns
Common of Dunning
Dupplin Castle
Aberdalgie
Milltown of Aberdalgie
Craigend
Kirkton
Lawhill
Findo Gask
Clathy
Chapelhill
St David's
Madderty
Balgowan
Tibbermore
Keillour Forest
Mains of Innerpeffray
INNERPEFFRAY LIBRARY
Muthill
DRUMMOND CASTLE GARDENS
OLD CHURCH & TOWER
Middleton
Drumness
Milton of Machany
Machany Water
Garrick
Langside
Ochtermuthill
Culloch
Torlum Wood
Concraig
The Balloch
STUART CRYSTAL VISITOR CENTRE
CRIEFF VISITORS' CENTRE
Crieff
Pittachar
Tomaknock
Milton of Cultoquhey
Lochlane
Glascorrie
AUCHINGARRICH WILDLIFE & HIGHLAND CATTLE CENTRE
Dalginross
Ross
Comrie
Lawers
Quoig
COMRIE CROFT
Ochtertyre
FALLS OF TURRET
GLENTURRET DISTILLERY
Hosh
Monzie Castle
Monzie
Gilmerton
Fowlis Wester
STANDING STONES
Carroglen
Glen Turret
Loch Turret Reservoir
BEN CHONZIE
MEALL TARSUINN
Shaggie Burn
Sma' Glen
Newton
Glenalmond Ho.
Buchanty
Glenalmond College
Drummick
Keillour
Craigend
Methven
Burnside
Braegrum
Almondbank
Lochty
Huntingtower
HUNTINGTOWER CASTLE
Letham
Tulloch
Cherrybank
BELL'S CHERRYBANK GARDENS
BROXDEN SERVICES
Craigie
Friarton
Perth
BRANKLYN GARDEN
FERGUSSON GALLERY
MUSEUM & ART GALLERY
Gannochy
Muirton
PERTH MART VISITOR CENTRE
CAITHNESS GLASS
Pitcairngreen
Scone
SCONE PALACE
East Bonhard
Murrayshall
Balthayock
Walnut Grove
ELCHO
Tarsappie
Rhynd
Echo
Inchyra
Easter Rhynd
Luncarty
Redgorton
Moneydie
Stormontfield
Downhill
Newmiln
St Martins
Guildtown
Wolfhill
Stanley
West Tofts
Airntully
Bankfoot
LIQUEUR CENTRE
Tullybelton
Waterloo
Balquharn
Upper Obney
Meikle Obney
Obney Hills
Murthly
Ardoch
Gellyburn
Caputh
Muir of Thorn
Cargill
Ballathie
Kinclaven
Meikleour
BEECH HEDGE
Delvine
Spittalfield
Kirkton of Lethendy
Wester Essendy
Muirton of Ardblair
Clunie
Craigie
Concraigie
Snaigow House
LOCH OF LOWES NATURE RESERVE
Dunkeld
Little Dunkeld
Birnam
CATHEDRAL
SCOTTISH HORSE REGIONAL MUSEUM
Inver
THE HERMITAGE
Ballinlick
Trochry
WATERFALL
RIVER TAY
Craigvinean Forest
Dalmarnock
Dalguise
Kincraigie
Guay
Dowally
Kindallachan
Balmacneil
Logierait
Ballinluig
Tulliemet
Balnaguard
CHILDHOOD HERITAGE MUSEUM
Ballechin
Strathtay
Grandtully
Little Ballinluig
Edradynate
CLUNY HOUSE GARDENS
Tombuie
ST MARY'S CHURCH
CASTLE MENZIES
Weem
Aberfeldy
ABERFELDY DISTILLERY & DEWAR'S WORLD OF WHISKY
WADE'S BRIDGE
Falls of Moness
BOLFRACKS GARDEN
Camserney
Dull
Comrie
Taymouth Castle
Croftmoraig
Urlar
Kenmore
SCOTTISH CRANNOG CENTRE
Remony
Glengoulandie
Appin of Dull
Bishopric
Loch Skiach
Loch Kennard
Glen Cochill
Scotston
Garrow
Turrerich
Glen Quaich
Loch Freuchie
Glenquaich Lodge
Cablea
Strathbraan
Amulree
Corrymuckloch
MEALL NAN FUARAN
Auchnafree
PERTH AND KINROSS
Little Glenshee
MEALL NAN CAORAICH
Logiealmond Lodge
Logiealmond
Harrietfield
Chapelhill
GLEN ALMOND
CREAG DHUBH
MEALL REAMHAR
CREAG NAM MIAL
DEUCHARY HILL
BEN VRACKIE
FARRAGON HILL
MEALL DEARG
MOUNT BLAIR
CAIRN GIBBS
CREIGH HILL
KNOCK OF BALMYLE
BALDUFF HILL
REEKIE LINN WATERFALL
PEEL FARM TRAIL
Glenhead Farm
Longdrum
Glenmarkie Lodge
Backwater Res.
Loch of Lintrathen
Bridgend of Lintrathen
Auchrannie
Shanzie
Pitcrocknie
Ruthven
SIDLAW HILLS
CARSE OF GOWRIE
POLE HILL
WEST LOMOND
INNERDOUNY HILL
Glenfarg Res.
Glendevon Resrs.
Castlehill Reservoir
Pitlochry
Dalnaglar Castle
Lair
Lettoch
Tulach Hill
ATHOLL COUNTRY COLLECTION
Balavoulin
Foss
Lick
Loch Bhac
Loch Derculich
Loch Broom
Loch Hoil
Loch Kennard
Loch of Clunie
Loch of Drumellie
Lunan Burn
Isla
Tay
Almond
Braan
Earn
Quaich
River Ericht
Black Water
Ardle
Tummel
6 miles
10 km

Glenmoy
Redheugh
Auchnacree
Kirkton of Menmuir
Tigerton
Balrownie
Newtonmill
Logie Pert
North Craigo
Craigo
Logie
Lochside
Morphie
St Cyrus
Milton Ness
Pathhead
Easter Lednathie
CAPTAIN SCOTT & DR. WILSON CAIRN
CAT LAW
Glenogil
Deuchar
Balnamoon
Lochty
Mains of Balhall
Bellichill
Little Brechin
Keithock
Muirton of Ballochy
Hillside
Kirkhill
Maitings
Horniehaugh
Dykehead
Glenquiech
Ogil
Fern
Noranside
Careston
West Muir
Trinity
Brechin
BRECHIN CASTLE CENTRE
CALEDONIAN RAILWAY
HOUSE OF DUN
Dun
Montrose Basin
Montrose
MUSEUM AND ART GALLERY
Scurdie Ness
WILLIAM LAMB MEMORIAL STUDIO
MONTROSE BASIN VISITOR CENTRE
Pearsie
Cortachy
Memus
Tannadice
Careston Castle
PICTAVIA
CATHEDRAL & ROUND TOWER
Bridge of Dun
Barnhead
Inchbraoch
Ferryden
Balloch
Inverquharity
Murthill
Aldbar Castle
Middle Drums
Kinnaird Castle
MILE HILL
Shielhill
Oathlaw
Finavon
Netherton
Mains of Melgund
Farnell
Bonnyton
Maryton
Kirkton of Craig
Kirkton of Kingoldrum
Kinnordy
Northmuir
Crosston
ABERLEMNO SCULPTURED STONES
Aberlemno
Montreathmont Forest
Carcary
Long Craig
Fishtown of Usan
Dunninald
LOCH OF KINNORDY NATURE RESERVE
Kirriemuir
BARRIE'S BIRTHPLACE
Carse Gray
Pitkennedy
Rossie Moor
Westerton
Boddin Pt.
Westmuir
GATEWAY TO THE GLENS MUS
Maryton
Mosside
Blackgate
Turin
Montreathmont Moor
Braehead of Lunan
Kirkton of Airlie
Mains of Ballindarg
Lunanhead
RESTENNETH PRIORY
Rescobie
Dubton
Bolshan
Renmure
Lunan
Craigton
Padanaram
FORFAR LOCH
Forfar
Reswallie
Balgavies
Glasterlaw
LUNAN BAY
Airlie
Roundyhill
Drumgley
Burnside
Milldens
Guthrie
Kinnell
Redcastle
Leys of Cossans
ST ORLAND'S STONE
Kingsmuir
Pitmuies
HOUSE OF PITMUIES GARDEN
Friockheim
Boysack
Inverkeilor
Lang Craig
Ruthven House
EASSIE SCULPTURED STONE
Dean Water
GLAMIS CASTLE
Caldhame
Dunnichen
Letham
Lowrie Moor
Bowriefauld
Idvies
Chapelton
Lunan Water
Ethie Mains
Red Head
Ethie Castle
Glamis
Douglastown
Crachie
Leysmill
Cauldcots
Inchock
Castleton
Thornton
Kirkton
Lour
Tulloes
Cononsyth
Drunkendub
Charleston
ANGUS FOLK MUSEUM
Balkeerie
Eassie
Foffarty
Letham Grange
Kirkinch
Gateside
Inverarity
Whigstreet
Mosston
Colliston
Boysack
Auchmithie
Meg's Craig
Milton
Greystone
Redford
St Vigeans
Marywell
Wester Denoon
ARK HILL
Kirkbuddo
ST VIGEANS MUSEUM
Woodville
Feus
Hayshead
Nether Handwick
Gallowfauld
Carrot
Hayhillock
Carmyllie
Denhead of Arbirlot
Cliffburn
The Deil's Heid
ARBROATH ABBEY
Arbroath
Arbirlot
CROMBIE
SIGNAL TOWER MUSEUM
Coldstream
Husbandtown
Fauldiehill
Easter Knox
Elliot
Bonnyton
Balkemback
Tealing
Todhill
Kirkton of Monikie
Balmirmer
Kirkton of Auchterhouse
Newbigging
Monikie
Craigton
Auchterhouse
North Dronley
Leoch
Muir of Pert
Kirkton
Inveraldie
East March
Wellbank
CARLUNGIE SOUTERRAIN
Monikie Burn
Salmond's Muir
Hatton
Muirdrum
Dronley
Bridgefoot
Burnside of Duntrune
Bucklerheads
Drumsturdy
Newbigging
Clayholes
BARRY MILL
East Haven
Panbride
Muirhead
Birkhill
CLATTO
Kellas
SOUTERRAIN ARDESTIE
Barry
Carnoustie
Fowlis
CAMPERDOWN AND TEMPLETON WOODS
Liff
Downfield
Fintry
Mains of Ardestie
CARNOUSTIE
Barry Links
Denhead of Gray
DUNDEE
VERDANT WORKS
Craigie
Douglas & Angus
Baldovie
West Ferry
Monifieth
Benvie
Lochee
MILLS OBSERVATORY
MUS AND ART GALL
Stannergate
Broughty Ferry
Barnhill
Invergowrie
Dundee
FRIGATE UNICORN
BROUGHTY CASTLE MUSEUM
Buddon Ness
Longforgan
Kingoodie
DUNDEE
Tay Bridge
Castle Huntly
UNIVERSITY BOTANIC GARDENS
DISCOVERY POINT
Scotscraig
Tayport
TENTSMUIR
Newport-on-Tay
Woodhaven
Tay Rail Bridge
Wormit
Shanwell
Fetterdale
Tentsmuir Forest
Balmerino
BALMERINO ABBEY
Kirkton
Bottomcraig
Pickletillem
Coultra
Gauldry
Rhynd
Hazelton Walls
Kilmany
Lucklawhill
Carrick
Creich
Balmullo
Leuchars
LEUCHARS NORMAN CHURCH
Brunton
Rathillet
Logie
Milton
Eden Mouth
Luthrie
Muirhead
Guardbridge
EDEN ESTUARY CENTRE
ST ANDREWS BAY
Moonzie
Dairsie or Osnaburgh
Kincaple
ST ANDREWS
ST ANDREWS AQUARIUM
Lindifferon
Kilmaron Castle
Foodieash
BRITISH GOLF MUS
St Andrews
MOUNT HILL
Newpark
CATH & ST RULE'S TOWER
DOUGLAS BADER GARDENS FOR THE DISABLED
Cupar
Strathkinness
Clatto
Brownhills
Buddo Ness
Kemback
Blebocraigs
ST ANDREWS BOTANIC GARDEN
Boarhills
Babbet Ness
Cupar Muir
CRAIGTOUN
Balone
Denhead
Bow of Fife
Prior Muir
SCOTTISH DEER CENTRE
HILL OF TARVIT MANSIONHOUSE
Pitscottie
Kingsbarns
Cambo Ness
CAMBO GARDENS
Carr Brigs
Bridgend
Baldinnie
Cameron Res.
Stravithie
Tullybothy Craigs
Springfield
Ceres
Dunino
Balcomie
Craighead
Ladybank
SCOTSTARVIT TOWER
FIFE FOLK MUSEUM
Craigrothie
Peat Inn
Beleybridge
Wormiston
Fife Ness
Pitlessie
Chance Inn
Struthers
Radernie
Kingsmuir
CRAIL TOLBOOTH
Kingskettle
Cults
Falfield
Lathones
Lochty
SCOTLAND'S SECRET BUNKER
Crail
CRAIL MUSEUM AND HERITAGE CENTRE
Balmalcolm
Burnturk
Kettlehill
Woodside
New Gilston
Largoward
Carnbee
Pitcorthie
West Ness
Kame
Pitkierie
Rameldry Mill Bank
Montrave
Lathallan Mill
KELLIE CASTLE AND GARDEN
Kilrenny
FIFE COASTAL PATH
Wester Newburn
Langdyke
PRAYTIS FARM PARK
SCOTLAND'S LARDER
Kirkton of Largo
Arncroach
Newton of Balcormo
Grangemuir
Anstruther Easter
SCOTTISH FISHERIES MUSEUM
Star
Baintown
Drumeldrie
Colinsburgh
Lundin Links
Kennoway
Bonnybank
Pittenweem
Anstruther Wester
ROBINSON CRUSOE STATUE
Balchrystie
Abercrombie
Kilconquhar
Scoonie
SILVERBURN ESTATE
Lower Largo
ST FILLAN'S CAVE
ST MONAN'S WINDMILL
BALGONIE CASTLE
Balcurvie
Ardross
St Monans
ST MONAN'S CHURCH
LETHAM GLEN
Leven
Milton of Balgonie
Windygates
Largo Bay
Ruddons Pt.
Innerleven
Earlsferry
Grange
Elie
Sauchar Pt.
ISLE OF MAY
Isle of May
Coaltown of Balgonie
Methil
Chapel Ness
MUSEUM
Buckhaven
FIFE COASTAL PATH
East Wemyss
Coaltown of Wemyss
ANGUS
STRATHMORE
A90
A92
A94
A91
A917
A915
A935
A933
A934
A932
A930
A926
A928
A923
A913
A914
A916
A911
A955
B951
B955
B957
B9134
B9113
B9128
B9127
B978
B961
B965
B9126
B954
B945
B946
B939
B940
B941
B942
B9131
B9171
B932
NO
281
282
292
293
A
B
C
D
E
F
G
H
7
8
9
10
11

294
Eilean nan Each
Gallanach
Port Mor
Muck
SOUND OF EIGG
RÙM 1:10
0:35
AN SGURR
Galmisdale
NM
Sanna Point
Sanna
Sanna Bay
Achnaha
Portuairk
Achosnich
Point of Ardnamurchan
ARDNAMURCHAN LIGHTHOUSE
B8007
Ormsaigmore
Ormsaigbeg
Kilchoan Bay
An Acairseid
Cairns of Coll
Eilean Mor
Rubha Mor
Sorisdale
Bousd
Cornaigmore
B8072
Cliad Bay
Arnabost
Gallanach
Grishipoll
B8071
Clabhach
Ballyhaugh
Loch Cliad
COLL
Hogh Bay
RSPB COLL
Arinagour
B8070
Totronald
Acha
Arileod
Uig
Feall Bay
Loch Eatharna
Eilean Ornsay
Breachacha Castle
Friesland
Calgary Pt.
Gunna
Crossapol Bay
Soa
Loch Breachacha
OBAN 2:40
TIREE
Vaul Bay
Salum
Caolas
Rubha Dubh
Balephetrish Bay
Vaul
B8069
Cornaigmore
Kirkapol
Ruaig
B8068
Cornaigbeg
Kenovay
Gott Bay
Kilmoluaig
Gott
Soa
TIREE
Scarinish
Moss
Baugh
B8065
Heylipol
Crossapol
Heanish
Rubha Traigh an Duin
Balinoe
Hynish Bay
B8067
Balemartine
B8066
Mannal
West Hynish
Hynish
0:55
Ardmore Bay
Ardmore Pt.
Bloody Bay
Quinish Pt.
Glengorm Castle
MULL MUSEUM
Tobermory
Rubha an Aird
Caliach Pt.
Caliach
Sunipol
Croig
Penmore Mill
Mishnish
MULL LITTLE THEATRE
Dervaig
Achnadrish
B8073
Calgary
Calgary Bay
West Ardhu
THE OLD BYRE HERITAGE CENTRE
Druimnacroish
Treshnish Pt.
Ensay
CARN MOR
Kengharair
Haunn
Rubh a'Chaoil
Burg
Kilninian
Achnacraig
Achleck
Fanmore
Treshnish Isles
Fladda
LOCH TUATH
Ballygown
EAS FORS WATERFALL
BEINN NAN DRISE
Eilean Dioghlum
Gometra
Baligortan
Bearnus
Lagganulva
Laggan Bay
Oskamull
Lunga
Gometra Ho
Ulva
Ardalum Ho
Ulva House
Killiechronan
Bac Mor
LOCH NA KEAL
Sound of Ulva
Eorsa
Little Colonsay
Staffa
STAFFA
FINGAL'S CAVE
INCH KENNETH CHAPEL
Inch Kenneth
ISLE OF
Balnahard
Erisgeir
MACKINNON'S CAVE
Balmeanach
(April-Oct) 0:45
BEINN NA SREINE
Glen Seilisdeir
ARDMEANACH
Tiroran
THE BURG
Burg
Kilfinichen Bay
Eilean Annraidh
LOCH SCRIDAIN
MACLEAN'S CROSS
IONA ABBEY AND CATHEDRAL
Rubha nan Cearc
Torrans
IONA HERITAGE CENTRE
Kintra
Loch na Lathaich
Iona
Baile Mor
ST COLUMBA EXHIBITION & WELCOME CENTRE
Achnahard
Knockan
Eorabus
Stac an Aoineidh
Sligneach
Aridhglas
Lower Ardtun
Lee
Fionnphort
A849
Sound of Iona
Fidden
Tiraghoil
Bunessan
CRUACHAN MIN
Loch Assapol
Erraid
Knockvologan
ROSS OF MULL
274
Soa I.
Ardalanish
Uisken
Ardchiavaig
Scoor
Malcolm's Pt.
Eilean a'Chalmain
Rubha nam Braithrean
Rubh Ardalanish
0 2 4 6 miles
0 2 4 6 8 10 km
Gunna
TIREE
Vaul Bay
Salum
Caolas
Rubha Dubh
Sraid Ruadh
Cornaigmore
Balephetrish Bay
Vaul
Kirkapol
B8069
Ruaig
Balevullin
Cornaigbeg
B8068
Gott Bay
Hough
Kilmoluaig
Kenovay
Gott
Soa
Kilkenneth
Moss
TIREE
Scarinish
COLL 0:55
Middleton
Heylipol
Baugh
B8065
Port Mor
Barrapol
Crossapol
Heanish
Rubha Traigh an Duin
Hynish Bay
Loch a'Phuill
B8067
Balinoe
Balemartine
Balephuil
Mannal
Balephuil Bay
West Hynish
Hynish
Port Snoig

MORAR, MOIDART AND ARDNAMURCHAN
HIGHLAND
LOCH SHIEL
LOCH SUNART
MOIDART
SUNART
ARDNAMURCHAN
MORVERN
KINGAIRLOCH
LOCH LINNHE
LYNN OF LORN
LORN
ISLAND OF MULL
ARGYLL AND BUTE
SOUND OF MULL
FIRTH OF LORN
SOUND OF ARISAIG
ARISAIG 1:00 (April-Sept)
Glenfinnan
Glenfinnan Lodge
GLENFINNAN MONUMENT
STATION MUSEUM
LOCH NAN UAMH CAIRN
Lochailort
Inverailort
Arieniskill
Ranochan
Polnish
Druimindarroch
Laggan
Alisary
Roshven
Glenuig
Smirisary
Eilean Shona
Invermoidart
Kinlochmoidart
Ardmolich
Brunery
Glenmoidart Ho
Gaskan
Scamodale
Gorstanvorran
Polloch
Glenhurich
Kinlochan
CASTLE TIORAM
Cul Doirlinn
Newton of Ardtoe
Ardtoe
Shielfoot
Mingarrypark
Kentra
Blain
Moss
Acharacle
Arivegaig
Langal
Dalnabreck
Dalelia
CLAISH MOSS
Arnshealach
Gortenfern
Gorteneorn
Kilmory
Fascadale
Ockle
Branault
MINGARY CASTLE
ARDNAMURCHAN NATURAL HISTORY CENTRE
Glenbeg
GLENBORRODALE
Glenborrodale
Ardslignish
Laga
Salen
Resipole
Ardery
Camuschoirk
Liddesdale
Strontian
Scotstown
Anaheilt
Ariundle
ARIUNDLE CENTRE
ARIUNDLE OAKWOOD
Ardnastang
Archnalea
Glen Tarbert
Sallachan
Corran
Clovullin
Ardgour
Inversanda
Inversanda Bay
Kilmalieu
Kingairloch
Camasnacroise
Lochuisge
Beach
Kinloch
Drimnin
Doirlinn
Carna
Oronsay
Glencripesdale
GLENCRIPESDALE
TOBERMORY
TOBERMORY DISTILLERY
Calve I.
Upper Druimfin
Ardnacross
Killundine
Rhemore
Fiunary
Savary
Lochaline
Larachbeg
KINLOCHALINE CASTLE
ARDTORNISH GARDENS
Achranich
ARDTORNISH CASTLE
Inninbeg
Aros Mains
Salen
Pennygown
Killbeg
Corrynachenchy
Balmeanach
Garmony
Fishnish Bay
Gruline
Kellan
Killiechronan
Knock
Benmore Lodge
Craignure
Java
Scallastle
MULL AND WEST HIGHLAND NARROW GAUGE RAILWAY
TOROSAY CASTLE & GARDEN
DUART CASTLE
Duart Pt.
Duart Bay
Lochdon
Druimsornaig
Grass Pt.
Auchnacraig
Strathcoill
Ishriff
Lochbuie
Kinlochspelve
Barachandroman
Croggan
Portfield
Carsaig
Glenbyre
Laggan Lodge
LORD LOVAT'S CAVE
Pennycross
Lismore
Port Ramsay
Clachan
Balliveolan
Achnacroish
Baligrundle
Achinduin
Kilcheran
Bernera I.
Eilean Musdile
Dalnarrow
Eriska
Port Appin
North Shian
South Shian
Appin
Appin House
Portnacroish
Ardtur
KINLOCHLAICH HOUSE GARDENS
Creagan
South Creagan
SCOTTISH SEA LIFE SANCTUARY
Barcaldine
Barcaldine Forest
Dalrannoch
Ferlochan
Benderloch
Ledaig
Kiel Crofts
Achnacree
Achnaba
ARDCHATTAN PRIORY GARDENS
ARDCHATTAN PRIORY
Bonawe
North Connel
Connel
Black Crofts
FALLS OF LORA
DUNSTAFFNAGE CASTLE
Dunbeg
Ganavan
Dunollie
Achnacloich
ACHNACLOICH GARDEN
Stonefield
Taynuilt
BONAWE IRON FURNACE
INVERAWE FISHERIES & SMOKERIES EXHIBITION
OBAN VISITOR CENTRE
OBAN DISTILLERY
MCCAIG'S TOWER
OBAN RARE BREEDS FARM PARK
BARGUILLEAN GARDENS
Oban
Kerrera
Balliemore
Ardmore
Glenamachrie
Kilbride
Kilmore
Barran
Dunach
GLENFEOCHAN HOUSE GARDENS
Musdale
Kilninver
Barochreal
Raera
Scammadale
Bragleenmore
Seil
AN CALA GARDENS
Easdale
EASDALE ISLAND FOLK MUSEUM
Balvicar
Clachan-Seil
CLACHAN BRIDGE
Clachan
Ardencaple
Blarian
Lochavich
Inverinan
Fineglen
Kilchrenan
Ballimeanoch
CASTLEBAY 4:50 COLL 2:40 TIREE 3:20 LOCHBOISDALE 5:20
COLONSAY 2:20
Sound of Kerrera
Bach I.
Loch Don
Loch Spelve
Loch Buie
Carsaig Bay
Loch Etive
Loch Creran
Shuna I.
Loch Eil
Kinlocheil
Drumsallie
Fassfern
Duisky
South Garvan
Callop
Guesachan
Conaglen House
Aryhoulan
A830
A861
A884
A849
A828
A85
A816
B8043
B8007
B845
B844
295
290
284
274
275

GLEN AFFRIC
KINTAIL
Affric Lodge
Hilton Lodge
Morvich
Cougie
Guisachan Forest
Fort Augustus
Invermoriston
LOCH NESS
Dundreggan Lodge
Dundreggan
Torgyle
Tomchrasky
Ceannacroc Lodge
Cluanie Inn
Cluanie Lodge
LOCH CLUANIE
Lundie
Bun Loyne
LOCH LOYNE
Alltbeithe
LOCH QUOICH
Kingie
Tomdoun
Inchlaggan
LOCH GARRY
Greenfield
Invergarry
Mandally
Munerigie
Aberchalder
Bridge of Oich
Newtown
Laggan
North Laggan
Kilfinnan
LOCH LOCHY
Letterfinlay
LOCH ARKAIG
Murlaggan
Strathan
Ardechvie
Achnasaul
Clunes
Bunarkaig
Achnacarry
Gairlochy
Spean Bridge
Roybridge
Tulloch
Fersit
LOCH TREIG
Corrour Station
LOCH EIL
Kinlocheil
Fassfern
Corpach
Banavie
Caol
Fort William
Glen Nevis
BEN NEVIS AND GLEN COE
MAMORE FOREST
Kinlochleven
Kinlochmore
North Ballachulish
Glencoe
BLACKWATER RESERVOIR
Rannoch Station
Corran
Ardgour
Onich
Altnafeadh
Black Corries Lodge
0 2 4 6 miles
0 2 4 6 8 10 km

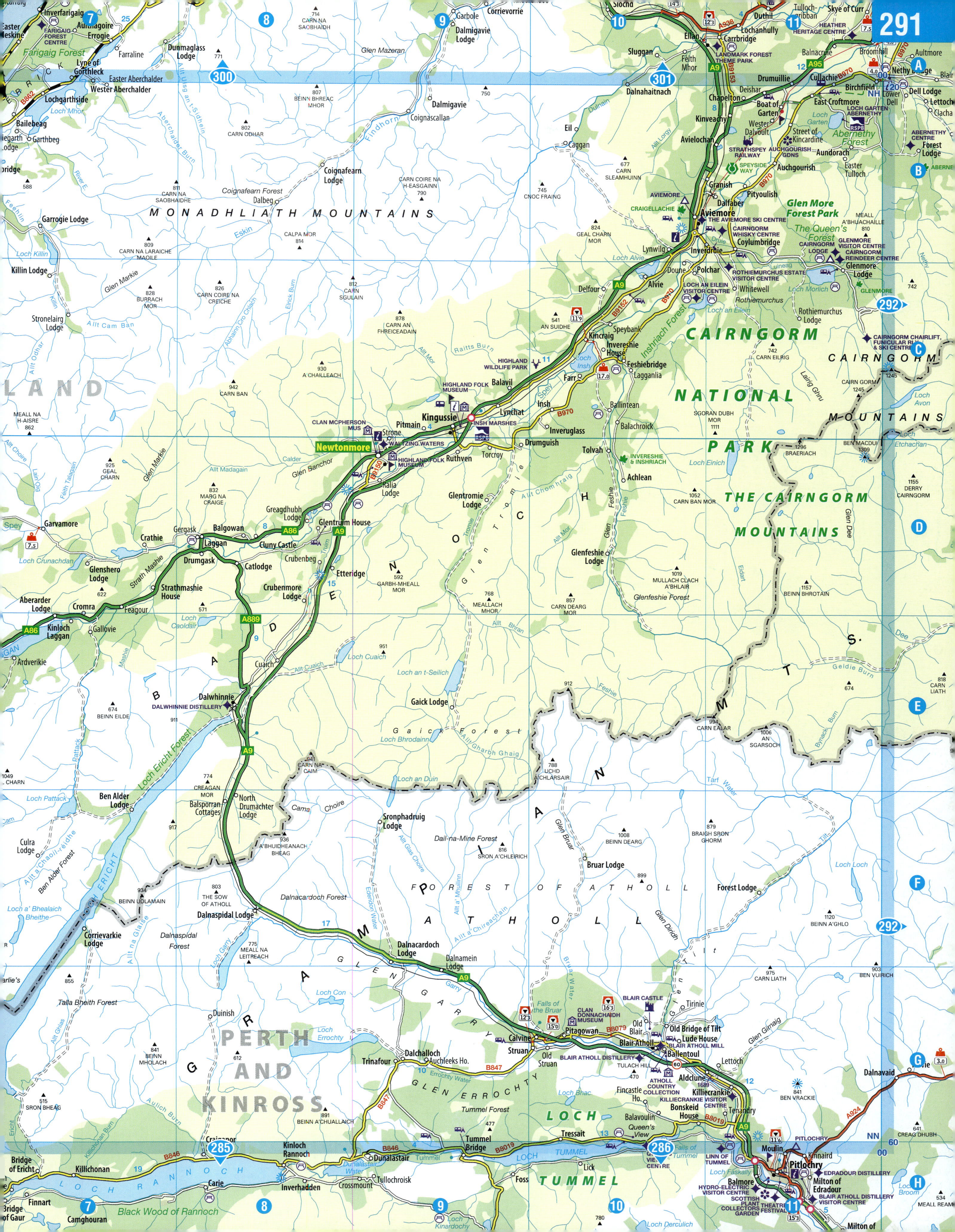

MONADHLIATH MOUNTAINS
CAIRNGORM NATIONAL PARK
CAIRNGORM MOUNTAINS
THE CAIRNGORM MOUNTAINS
PERTH AND KINROSS
FOREST OF ATHOLL
LOCH TUMMEL
LOCH RANNOCH
GLEN GARRY
GLEN ERROCHTY
Glen More Forest Park
The Queen's Forest
Abernethy Forest
Rothiemurchus
Inshriach Forest
Glenfeshie Forest
Gaick Forest
Loch Ericht Forest
Ben Alder Forest
Talla Bheith Forest
Dalnacardoch Forest
Dalnaspidal Forest
Dail-na-Mine Forest
Tummel Forest
Black Wood of Rannoch
Coignafearn Forest
Farigaig Forest
Inverfarigaig
Aultnagoire
Errogie
Farraline
Dunmaglass Lodge
Lyne of Gorthleck
Easter Aberchalder
Wester Aberchalder
Lochgarthside
Loch Mhor
Bailebeag
Garthbeg
Garrogie Lodge
Killin Lodge
Stronelairg Lodge
Garbole
Corrievorrie
Dalmigavie Lodge
Dalmigavie
Coignascallan
Coignafearn Lodge
Dalbeg
Glen Mazeran
Findhorn
Sluggan
Dalnahaitnach
Carrbridge
Duthil
Lochanhully
Ellan
Docharn
Feith Mhor
Drumuillie
Boat of Garten
Deishar
Chapelton
Kinveachy
Avielochan
Aviemore
Granish
Dalfaber
Pityoulish
Coylumbridge
Inverdruie
Lynwilg
Alvie
Doune
Polchar
Whitewell
Rothiemurchus Lodge
Glenmore Lodge
Loch Morlich
Loch an Eilein
Loch Alvie
Delfour
Kincraig
Speybank
Invereshie House
Feshiebridge
Lagganlia
Farr
Insh
Loch Insh
Inveruglass
Drumguish
Tolvah
Achlean
Ballintean
Balachroick
Kingussie
Lynchat
Balavil
Pitmain
Newtonmore
Strone
Ruthven
Torcroy
Ralia Lodge
Glentromie Lodge
Glentruim House
Etteridge
Crubenbeg
Crubenmore Lodge
Catlodge
Cluny Castle
Gergask
Laggan
Balgowan
Drumgask
Crathie
Garvamore
Glenshero Lodge
Strathmashie House
Aberarder Lodge
Cromra
Feagour
Kinloch Laggan
Gallovie
Ardverikie
Loch Crunachdan
Loch Caoldair
Cuaich
Loch Cuaich
Dalwhinnie
DALWHINNIE DISTILLERY
Loch an t-Seilich
Gaick Lodge
Loch Bhrodainn
Loch an Duin
Glenfeshie Lodge
Ben Alder Lodge
North Drumachter Lodge
Balsporran Cottages
Culra Lodge
Corrievarkie Lodge
Dalnaspidal Lodge
Sronphadruig Lodge
Bruar Lodge
Forest Lodge
Dalnacardoch Lodge
Dalnamein Lodge
Calvine
Struan
Old Struan
Pitagowan
Blair Atholl
Old Blair
BLAIR CASTLE
Old Bridge of Tilt
Lude House
Tirinie
Ballentoul
Lettoch
Aldclune
Killiecrankie
Tenandry
Fincastle Ho.
Bonskeid House
Balavoulin
Queen's View
Tressait
Moulin
Pitlochry
Kinnaird
Balmore
Milton of Edradour
Dalnavaid
Dalchalloch
Auchleeks Ho.
Trinafour
Duinish
Loch Errochty
Kinloch Rannoch
Dunalastair
Tummel Bridge
Tullochroisk
Crossmount
Inverhadden
Carie
Killichonan
Bridge of Ericht
Finnart
Camghouran
Foss
Lick
Craiganour
HIGHLAND FOLK MUSEUM
HIGHLAND WILDLIFE PARK
CLAN MCPHERSON MUS
WALTZING WATERS
INSH MARSHES
CAIRNGORM WHISKY CENTRE
ROTHIEMURCHUS ESTATE VISITOR CENTRE
LOCH AN EILEIN VISITOR CENTRE
THE AVIEMORE SKI CENTRE
CAIRNGORM CHAIRLIFT, FUNICULAR RLY & SKI CENTRE
GLENMORE VISITOR CENTRE
CAIRNGORM REINDEER CENTRE
STRATHSPEY RAILWAY
AUCHGOURISH GDNS
LANDMARK FOREST THEME PARK
LOCH GARTEN ABERNETHY
ABERNETHY CENTRE
CRAIGELLACHIE
INVERESHIE & INSHRIACH
CLAN DONNACHAIDH MUSEUM
BLAIR ATHOLL DISTILLERY
BLAIR ATHOLL MILL
ATHOLL COUNTRY COLLECTION
KILLIECRANKIE VISITOR CENTRE
LINN OF TUMMEL
EDRADOUR DISTILLERY
HYDRO-ELECTRIC VISITOR CENTRE
SCOTTISH PLANT COLLECTORS GARDEN
THEATRE FESTIVAL
A9
A86
A889
A95
A924
B9152
B970
B846
B847
B8019
B8079
B9153
B851
B862
300
301
292
285
286

MORAY
CAIRNGORM NATIONAL PARK
CAIRNGORM MOUNTAINS
THE CAIRNGORM MOUNTAINS
GRAMPIAN MTS.
DEESIDE AND LOCHNAGAR
PERTH AND KINROSS
ANGUS
LADDER HILLS
Forest of Mar
Forest of Glen Tanar
BALMORAL FOREST
Glenlivet Forest
Tornashean Forest
Glendoll Forest
Glenisla Forest
Caenlochan Forest
Forest of Alyth
Forest of Glenavon
Abernethy Forest
Ballochbuie Forest
Nethy Bridge
Tomintoul
Braemar
Ballater
Crathie
Inverey
Kirkton of Glenisla
Dinnet
Clova
Glen Clova
Glen Prosen
Glen Muick
Lochnagar
Loch Muick
Dykehead
Memus
Kirriemuir
Bridge of Gairn
Cock Bridge
Corgarff
Gairnshiel Lodge
Spittal of Glenshee
Spittal of Glenmuick
Glenshee Ski Centre
Lecht Ski Centre
Braemar Castle
Balmoral Castle and Gardens
Corgarff Castle
Kildrummy Castle & Garden
Royal Lochnagar Distillery Visitor Centre
Linn of Dee
Linn of Muick Waterfall
Glenmuick Visitor Centre
Cairngorm Chairlift, Funicular Rly & Ski Centre
Captain Scott & Dr Wilson Cairn
1151 Ben Avon
1309 Ben Macdui
1245 Cairn Gorm
1155 Derry Cairngorm
1150 Lochnagar
1068 Glas Maol
933 The Cairnwell

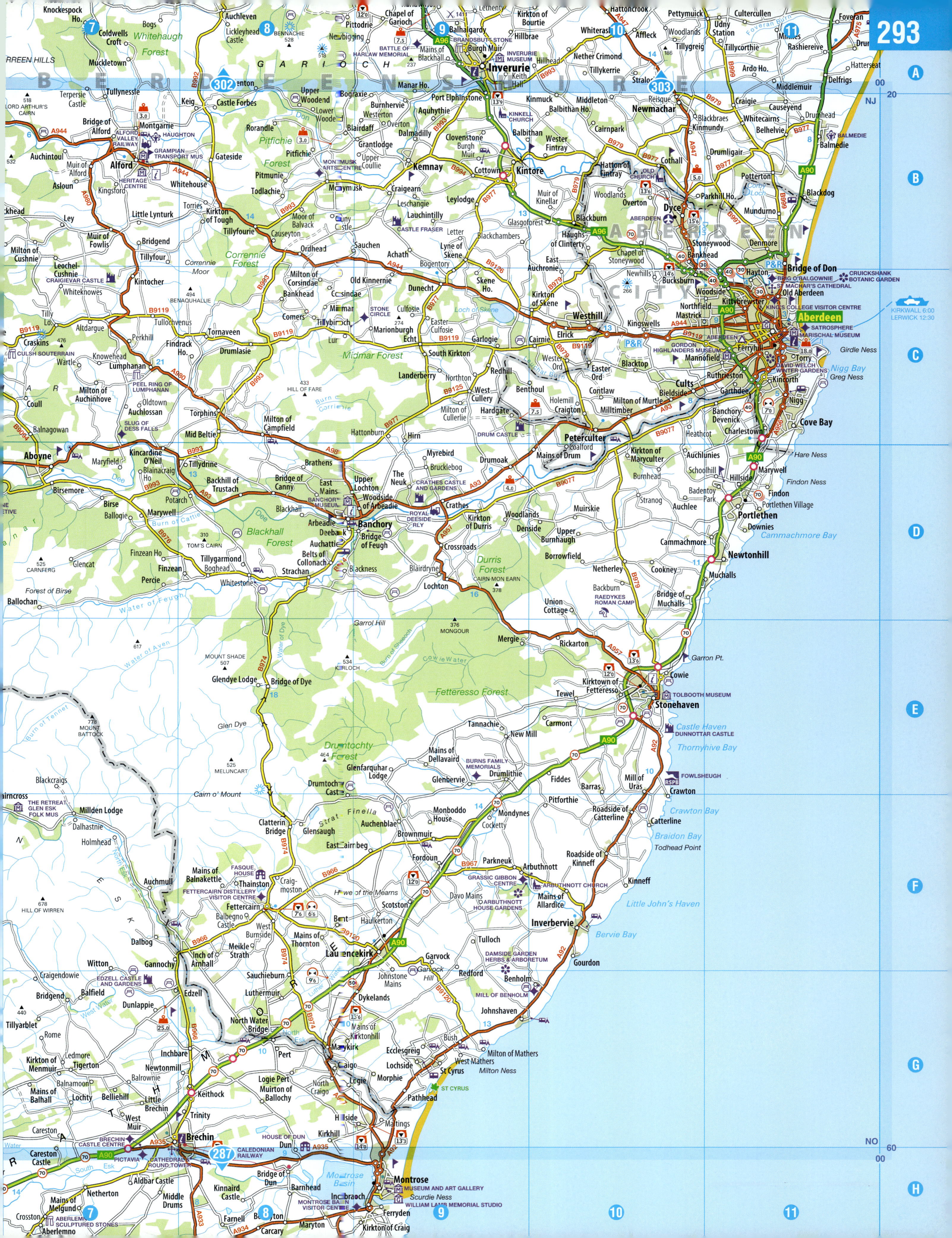
Knockespock Ho.
Bogs
Coldwells Croft
Whitehaugh Forest
Auchleven
Lickleyhead Castle
Bennachie
Pittodrie
Newbigging
Chapel of Garioch
Balhalgardy
Kirkton of Bourtie
Hillbrae
Hattoncrook
Whiterashes
Affleck
Pettymuick
Woodlands
Tillygreig
Udny Station
Culterculen
Tillycorthie
Foveran
Rashiereive
Mucklestown
Tullynessle
Terpersie Castle
Keig
Castle Forbes
Upper Woodend
Lower Woodend
Bograxie
Manar Ho.
Battle of Harlaw Memorial
Mains of Blackhall
Brandsbutt Stone
Burgh Muir
Inverurie
Inverurie Museum
Hillhead
Nether Crimond
Tillykerrie
Keith Hall
Port Elphinstone
Kinmuck
Kinkell Church
Middleton
Balbithan Ho.
Newmachar
Stralock
Reisque
Blackbraes
Kinmundy
Craigie
Whitecairns
Belhelvie
Causeyend
Drumhead
Balmedie
Middlemuir
Ardo Ho.
Hatterseat
Delfrigs
Lord Arthur's Cairn
Bridge of Alford
Montgarrie
Alford Valley Railway
Grampian Transport Mus
Haughton
Alford
Heritage Centre
Muir of Alford
Auchintoul
Asloun
Kingsford
Whitehouse
Gateside
Rorandle
Pitfichie Forest
Pitfichie
Pitmunie
Todlachie
Monymusk Arts Centre
Monymusk
Burnhervie
Westerton
Overton
Blairdaff
Grantlodge
Upper Coullie
Aquhythie
Dalmadilly
Clovenstone
Burgh Muir
Kemnay
Craigearn
Leschangie
Lauchintilly
Leylodge
Cottown
Kintore
Balbithan
Wester Fintray
Cairnpark
Hatton of Fintray
Old Church
Woodlands
Overton
Cothall
Drumligair
Potterton
Blackdog
Parkhill Ho.
Mundurno
Dyce
Muir of Kinellar
Glasgoforest
Blackburn
Haughs of Clinterty
Aberdeen City
Stoneywood
Denmore
Bankhead
Chapel of Stoneywood
Bucksburn
Newhills
Bridge of Don
Brig o' Balgownie
Cruickshank Botanic Garden
St Machar's Cathedral
Old Aberdeen
Hayton
Woodside
Kittybrewster
King's College Visitor Centre
Aberdeen
Satrosphere
Marischal Museum
Kirkwall 6:00
Lerwick 12:30
Torhead
Ley
Little Lynturk
Kirkton of Tough
Tillyfourie
Torries
Moor of Balvack
Causeyton
Lynturk Castle
Castle Fraser
Letter
Blackchambers
Muir of Fowlis
Bridgend
Tillyfour
Milton of Cushnie
Corrennie Moor
Corrennie Forest
Ordhead
Sauchen
Achath
Lyne of Skene
East Auchronie
Leochel-Cushnie
Craigievar Castle
Whiteknowes
Kintocher
Benaquhallie
Milton of Corsindae
Bankhead
Old Kinnernie
Bogentory
Dunecht
Skene Ho.
Kirkton of Skene
Westhill
Kingswells
Northfield
Mastrick
Corsindae
Midmar
Stone Circle
Cullfosie
Tillybirloch
Marionburgh
Easter Cullfosie
Loch of Skene
Comers
Echt
Garlogie
Cairnie
Elrick
Gordon Highlanders Museum
Mannofield
Ferryhill
Torry
Girdle Ness
Duthie
Craskins
Altdargue
Tullochvenus
Perkhill
Findrack Ho.
Tornaveen
Drumlasie
Lumphanan
Knowehead
Wartle
Lur
Midmar Forest
South Kirkton
Westfield
Wester Ord
Easter Ord
Blacktop
Cults
Bieldside
Ruthrieston
Garthdee
Kincorth
David Welch Winter Gardens
Nigg Bay
Greg Ness
Culsh Souterrain
Milton of Auchinhove
Peel Ring of Lumphanan
Oldtown
Auchlossan
Hill of Fare
Northton
Landerberry
Redhill
West Cullery
Benthoul
Holemill
Craigton
Contlaw
Milton of Murtle
Milltimber
Nigg
Coull
Slug of Dess Falls
Torphins
Milton of Campfield
Burn of Corrie
Milton of Cullerie
Hardgate
Drum Castle
Banchory Devenick
Charlestown
Cove Bay
Balnagowan
Mid Beltie
Hattonburn
Hirn
Peterculter
Coalford
Heathcot
Aboyne
Maryfield
Kincardine O'Neil
Blainacraig Ho
Tillydrine
Brathens
Myrebird
Brucklebog
Drumoak
Mains of Drum
Kirkton of Maryculter
Auchlunies
Schoolhill
Hillside
Maryhill
Hare Ness
Findon Ness
Findon
Birsemore
Birse
Ballogie
Backhill of Trustach
Bridge of Canny
East Mains
Upper Lochton
Woodside of Arbeadie
The Neuk
Crathes Castle and Gardens
Crathes
Royal Deeside Rly
Burnhead
Stranog
Badentoy Park
Auchlee
Portlethen
Portlethen Village
Downies
Cammachmore Bay
Potarch
Marywell
Burn of Cattie
Blackhall
Banchory Museum
Arbeadie
Banchory
Deebank
Bridge of Feugh
Auchattie
Kirkton of Durris
Woodlands
Denside
Muirskie
Upper Burnhaugh
Borrowfield
Cammachmore
Newtonhill
Tom's Cairn
Blackhall Forest
Belts of Collonach
Crossroads
Durris Forest
Cairn-mon-earn
Netherley
Cookney
Muchalls
Carnferg
Glencat
Finzean Ho
Finzean
Tillygarmond
Boghead
Strachan
Blackness
Blairdryne
Lochton
Backburn
Raedykes Roman Camp
Bridge of Muchalls
Percie
Whitestone
Forest of Birse
Ballochan
Water of Feugh
Union Cottage
Garrol Hill
Mongour
Mergie
Rickarton
Water of Aven
Mount Shade
Kerloch
Burn of Sheeoch
Cowie Water
Cowie
Garron Pt.
Glendye Lodge
Bridge of Dye
Water of Dye
Fetteresso Forest
Kirktown of Fetteresso
Tewel
Stonehaven
Tolbooth Museum
Mount Battock
Burn of Tennet
Glen Dye
Tannachie
New Mill
Carmont
Castle Haven
Dunnottar Castle
Thornyhive Bay
Melluncart
Drumtochty Forest
Mains of Dellavaird
Burns Family Memorials
Glenbervie
Drumlithie
Fiddes
Barras
Mill of Uras
Fowlsheugh
Crawton
Blackcraigs
Cairn o' Mount
Drumtochty Castle
Glenfarquhar Lodge
Pitforthie
Roadside of Catterline
Catterline
Crawton Bay
The Retreat: Glen Esk Folk Mus
Millden Lodge
Dalhastnie
Holmhead
Clatterin Brig
Glensaugh
Strath Finella
Auchenblae
Monboddo House
Mondynes
Cocketty
Brownmuir
Braidon Bay
Todhead Point
East Fairnbeg
Fordoun
Parkneuk
Arbuthnott
Roadside of Kinneff
Kinneff
Auchmull
Mains of Balnakettle
Fasque House
Thainston
Fettercairn Distillery Visitor Centre
Craigmoston
Howe of the Mearns
Grassic Gibbon Centre
Arbuthnott Church
Davo Mains
Arbuthnott House Gardens
Mains of Allardice
Little John's Haven
Hill of Wirren
Fettercairn
Balbegno Castle
Scotston
Haulkerton
Inverbervie
Bervie Bay
Dalbog
West Burnside
Mains of Thornton
Inch of Arnhall
Meikle Strath
Laurencekirk
Tulloch
Damside Garden Herbs & Arboretum
Gourdon
Witton
Gannochy
Craigendowie
Edzell Castle and Gardens
Sauchieburn
Luthermuir
Johnstone Mains
Garvock
Garvock Hill
Redford
Benholm
Bridgend
Balfield
Edzell
Dunlappie
Dykelands
Mill of Benholm
Johnshaven
Tillyarblet
Rome
North Water Bridge
Mains of Kirktonhill
Marykirk
Ecclesgreig
Bush
Milton of Mathers
West Mathers
Milton Ness
Kirkton of Menmuir
Ledmore
Tigerton
Inchbare
Newtonmill
Pert
Caigo
Lochside
St Cyrus
Balnamoon
Balrownie
Logie Pert
Muirton of Ballochy
North Craigo
Logie
Morphie
Mains of Balhall
Lochty
Belliehill
Little Brechin
Keithock
Pathhead
West Muir
Trinity
Hillside
Kirkhill
Maltings
Careston
Brechin Castle Centre
Brechin
House of Dun
Dun
Caledonian Railway
Careston Castle
Pictavia
Cathedral & Round Tower
Aldbar Castle
Bridge of Dun
Barnhead
Montrose Basin
Montrose
Museum and Art Gallery
Scurdie Ness
William Lamb Memorial Studio
Mains of Melgund
Netherton
Middle Drums
Kinnaird Castle
Montrose Basin Visitor Centre
Inchbraoch
Ferryden
Crosston
Aberlemno Sculptured Stones
Aberlemno
Farnell
Bolshan
Carcary
Maryton
Kirkton of Craig

Ramasaig
Hoe Rape
Hoe Point
Macleod's Tables
HEALABHAL BHEAG
Roag
Orbost
Greep
Vatten
Harlosh
Balmore
Loch Varkasaig
Feorlig
Balmeanach
Loch Caroy
Ose
Eabost West
Eabost
Bracadale
Ullinish
Struan
Coillore
Totardor
Loch Connan
Glengrasco
Sluggans
Torvaig
THE AROS EXPERIENCE
Portree
Heatherfield
Penifiler
Glenmore
Mugeary
Loch Duagrich
Conordan
Geodha Mor
Harlosh I.
Tarner I.
Loch Bracadale
Wiay
Idrigill Point
MACLEOD'S MAIDENS
Oronsay
Gesto
Portnalong
Ardtreck
Fiskavaig
Rubha nan Clach
Loch Harport
Fernilea
TALISKER DISTILLERY
Carbost
ARNAVAL
Drynoch
Merkadale
Satran
Crossal
ROINEVAL
Talisker Bay
Gleann Oraid
Talisker
Sligachan Hotel
Glen Varragill
A863
A87
B885
B8009
I S L A N D
Glen Brittle Forest
Eynort
Grula
BEINN BHREAC
Loch Eynort
Kraiknish
GLENBRITTLE
Glenbrittle House
Bualintur
Culnaneam
CUILLIN HILLS
SGURR NAN GILLEAN
SGURR A'GHREADAIDH
SGURR ALASDAIR
SGURR NAN EAG
Loch Coruisk
Loch Brittle
Glen Sligachan
Rubh an Dunain
Soay Sound
Soay
Mol-chlach
Canna
Garrisdale Pt.
A'Chill
Canna Harbour
Sanday
Sound of Canna
Guirdil Bay
Kilmory
Kilmory Glen
Rubha Shamhnan Insir
Kinloch Glen
Rubha na Roinne
MALLAIG 2:30
A'Bhrideanach
Kinloch
Loch Scresort
KINLOCH CASTLE
Rubha Port na Caranean
ORVAL
R Ù M
Schooner Pt.
Oigh-sgeir
Harris
Glen Harris
ASKIVAL
AINSHVAL
Rubha Sgorr an t-Snidhe
Rubha nam Meirleach
T H E S M A L L I S L E S
SOUND OF RÙM
Bay of Laig
Cleadale
Laig
Rubha an Fhasaidh
Eigg
Sandavore
Kildonnan
Galmisdale
AN SGURR
SOUND OF EIGG
Eilean nan Each
Gallanach
Port Mor
Muck
NG
NM
0 2 4 6 miles
0 2 4 6 8 10 km

297
298
288

ISLAND OF RAASAY
INNER SOUND
Applecross
Milton
Camusteel
Camusterrach
Ard-dhubh
Culduie
Ardban
Toscaig
Uags
Eilean na Bà
Crowlin Islands
Caolas Mòr
Eilean Mòr
An Dubh-aird
BEALACH NA BA
Couldoran
RASSAL ASHWOOD
Rassal
Tornapress
Russel
Sanachan
Glen Mor
Slumbay
Ardarroch
Achintraid
Loch Kishorn
Kishorn I.
Reraig
Cot
STROME CASTLE
Stromemore
Ardaneaskan
Ardnarff
LOCH CARRON
Stromeferry
Plockton
Achmore
CRAIG HIGHLAND FARM
Braeintra
Tullich
New Kelso
Strathcarron
Achintee
Kirkton
Lochcarron
Attadale Ho.
GARDEN
Attadale
Bendronaig Lodge
Attadale Forest
Loch an Iasaich
CARN NAN IOMAIREAN
SGUMAN COINNTICH
Glamaig
Balmeanach
Balachuirn
Camastianavaig
Holoman Bay
Tianavaig Bay
Oskaig
Lower Ollach
RAASAY
Clachan
Raasay Ho.
RAASAY OUTDOOR CENTRE
Inverarish
East Suisnish
Narrows of Raasay
The Braes
Gedintailor
BATTLE OF THE BRAES MONUMENT
Balmeanach
Peinchorran
North Fearns
Eyre Point
Eyre
Caol Mòr
Rubha na'Leac
Loch Sligachan
Sconser
Moll
Scalpay
Corran a Chan Uachdaraich
Scalpay Ho.
Longay
Guillamon I.
Pabay
Caolas Scalpay
Loch Ainort
Luib
Dunan
OLD SKYE CROFTER'S HOUSE
GLAS BHEINN MHOR
MARSCO
RED HILLS
CUILLIN HILLS
BLA BHEINN
SKYE
Loch na Crèitheach
Torrin
Faoilean
Kilbride
Camasunary
Strathaird
Kirkibost
Kilmarie
BEN MEABOST
Loch Slapin
Suisnish
Rubha Suisnish
Elgol
Glasnakille
Eilean na h-Airde
Rubha na h-Easgainne
Tarskavaig
Tarskavaig Pt.
DUNSGIATH CASTLE
Tokavaig
Ord
Loch Eishort
Achnacloich
SLEAT
Teangue
Sasaig
Ferindonald
Kilbeg
Kilmore
KNOCK CASTLE
Knock Bay
Armadale Castle
ARMADALE GARDENS
CLAN DONALD CENTRE
Armadale Bay
Glen Meadhonach
Rubha Charn nan Cearc
Calligarry
Ardvasar
Aird of Sleat
Tormore
Ard Thurinish
Point of Sleat
Corry
BROADFORD
Broadford Bay
Old Corry
Harrapool
WORLD OF WOOD
SKYE SERPENTARIUM
SKYE ENVIRONMENTAL CENTRE
Waterloo
Lower Breakish
Upper Breakish
Ashaig
Skulamus
Strath Suardal
Heast
Kinloch
Drumfearn
Loch na Dal
Duisdalemore
Duisdaleberg
Ornsay
Camuscross
Isleornsay
Kyle of Lochalsh
Black Is.
Erbusaig
Badicaul
Duirinish
Drumbuie
Achnandarach
Port Cam
Balmacara Square
Reraig
Balmacara
Kirkton
Auchertyre
CASTLE MOIL
Kyleakin
BRIGHTWATER VISITOR CENTRE
LOCHALSH WOODLAND GARDEN
LOCH ALSH
Avernish
Nostie
Ardelve
Glas Eilean
Ardintoul Pt.
Ardintoul
Totaig
Dornie
EILEAN DONAN CASTLE
Conchra
Allt-nan-sùgh
Sallachy
Gleann Udalain
Loch Long
Bundalloch
Carndu
Camas-luinie
Glen Elchaig
Loch nan Eun
Killilan
Faddonch
Carnach
Iron Lodge
Loch na Leitreach
FALLS OF GLOMACH
Keppoch
SGURR AN AIRGID
Inverinate
Dorusduain
Loch Duich
Letterfearn
BEINN A'CHUIRN
Ruarach
Morvich
Carn-gorm
Loch a'Bhealaich
Ault a'chruinn
Ratagan
RATAGAN
Inverhiel
Shiel Bridge
KINTAIL
BEINN FADA OR BEN ATTOW
Kintail Forest
Glenlicht Ho.
FIVE SISTERS
Glen Shiel
Glenshiel Forest
THE SADDLE
SGURR AN LOCHAIN
SGURR NA COINNICH
Glen Arroch
Allt Mòr
KYLERHEA OTTER HAVEN
Kyle Rhea
Kylerhea
Bernera
Galltair
Glenelg
BERNERA BARRACKS
Scallasaig
Cnoc Fhionn
Cosag
Glenelg Bay
Eilanreach
GLENELG BROCHS
Balvraid
Glen More
BEN ASLAK
Upper Sandaig
BEINN A'CHAPAILL
Sandaig Is.
SOUND OF SLEAT
BEINN SGRITHEALL
BEINN NAN CAORACH
Arnisdale
Glen Arnisdale
LOCH HOURN
Corran
Kinlochhourn Forest
DRUIM FADA
BUIDHE BHEINN
SGURR THIONAIL
Glen Quoich
Alltbeithe
Kinloch Hourn
SGURR A'MHAORAICH
Barrisdale Bay
Rubha Ard Slisneach
Inverguseran
HIGHLAND
KNOYDART
BEINN NA CAILLICH
LADHAR BHEINN
Glen Barrisdale
Airor
Gleann na Guiserein
Sandaig
Sandaig Bay
Scottas
LUINNE BHEINN
Abhainn Chosaidh
LOCH QUOICH
GAIRICH
Inverie Bay
Inverie
Gleann an Dubh-Lochain
MEALL BUIDHE
SGURR MOR
BEINN BHUIDHE
Camusrory
SGURR NA CICHE
SGURR MHURLAGAIN
Rubha Raonuill
MALLAIG HERITAGE CENTRE
Mallaig
Mallaig Bheag
Glasnacarcoch
LOCH NEVIS
Kylesknoydart
Beoraidbeg
Stoul
Bracara
Bracorina
Tarbet
Kylesmorar
North Morar
Glen Dessarry
Murlaggan
Morar
Glenancross
Garamor
Brinacory
Swordland
Kinlochmorar
Strathan
Traigh Ho.
Scamadale
LOCH MORAR
CARN MOR
Glen Pean
Portnaluchaig
Lettermorar
Oban
AN STAC
Bunacaimb
Back of Keppoch
South Morar
Meoble
Gleann Camgharaidh
Eilean Ighe
Gortenacullish
Kinloid
Arisaig
SGURR NAN COIREACHAN
SGURR THUILM
GAOR BHEINN
Luinga Bheag
Luinga Mhor
Loch nan Ceall
SIDHEAN MOR
MEITH BHEINN
Rumach
LOCH NAN UAMH CAIRN
Druimindarroch
Kinlochbeoraid
Loch Beoraid
Glenfinnan Lodge
Glen Dubh Lighe
Gleann Suileag
Polnish
SGURR AN UTHA
Loch nan Uamh
Arieniskill
Lochailort
Ranochan
STATION MUSEUM
GLENFINNAN MONUMENT
Glen Finnan
Wauchan
Loch Eilt
Ardnish
Inverailort
Glenfinnan
Laggan
MORAR, MOIDART AND ARDNAMURCHAN
SOUND OF ARISAIG
Eilean nan Gobhar
Loch Ailort
Alisary
DRUIM FIACLACH
Kinlocheil
Fassfern
Samalaman I.
Glenuig Bay
Roshven
ROIS-BHEINN
BEINN ODHAR BHEAG
Callop
Drumsallie
LOCH EIL
Smirisary
Glenuig
Guesachan
South Garvan
Duisky

SOUTH LEWIS, HARRIS
AND NORTH UIST
NA HEARADH (HARRIS)
UIBHIST A TUATH (NORTH UIST)
BEINN NA FAOGHLA (BENBECULA)
Tarasaigh (Taransay)
Eilean Scalpaigh (Scalpay)
Eilean Bhearnaraigh (Berneray)
Pabaidh (Pabbay)
Griomasaigh (Grimsay)
Baile Sear (Baleshare)
Na h-eileanan Monach (Heisker or Monach Islands)
CAOLAS NA HEARADH
AN CAOLAS MHONACH
LOCH DUNVEGAN
ST KILDA
St Kilda or Hirta (Hiort)
Boreray
Soay
Tairbeart (Tarbert)
Loch nam Madadh (Lochmaddy)
Roghadal
An t-Ob (Leverburgh)
Sound of Spuir
Sound of Shillay
Caolas Bhearnaraigh
Kirkibost Island
Vallay
Hermetray
Ensay
Killegray
Boreray
Oronsay
Ronay
Flodday

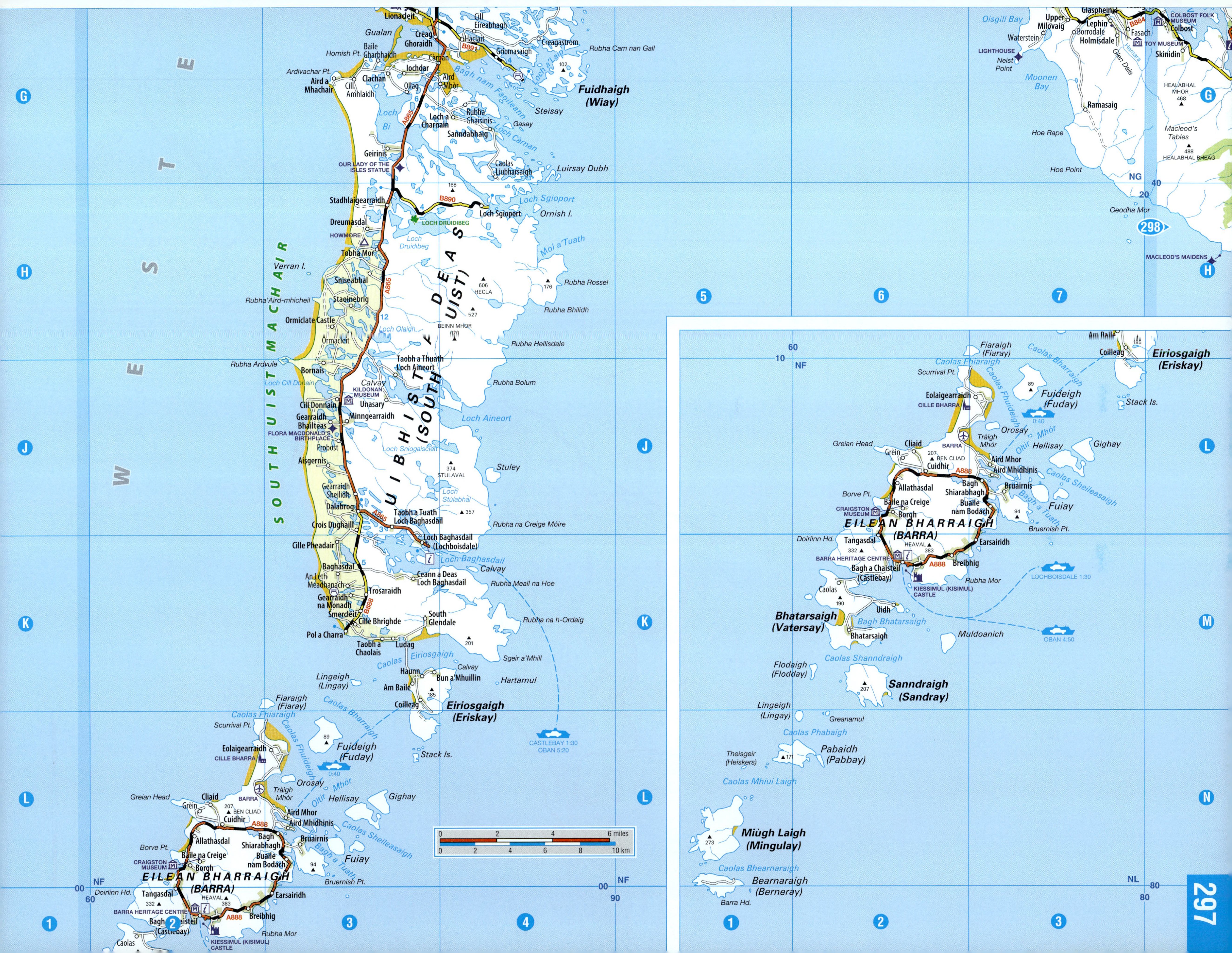
WEST
SOUTH UIST MACHAIR
UIBHIST A DEAS (SOUTH UIST)
Lionacleit
Gualan
Creag Ghoraidh
Cill Eireabhagh
Haclait
Creagastrom
Griomasaigh
Rubha Cam nan Gall
Hornish Pt.
Baile Gharbhaidh
Iochdar
Carnan
Ardivachar Pt.
Aird a Mhachair
Cill Amhlaidh
Clachan
Ollag
Aird Mhòr
Bagh nam Faoileann
Fuidhaigh (Wiay)
Steisay
Loch Bi
Loch a Charnain
Rubha Ghaisinis
Gasay
Sanndabhaig
Loch Carnan
Geirinis
OUR LADY OF THE ISLES STATUE
Caolas Liubharsaigh
Luirsay Dubh
Stadhlaigearraidh
B890
Loch Sgioport
Ornish I.
LOCH DRUIDIBEG
Loch Druidibeg
Dreumasdal
HOWMORE
Tobha Mor
Mol a'Tuath
Verran I.
Sniseabhal
606 HECLA
Rubha Rossel
Rubha Aird-mhicheil
Staoinebrig
Rubha Bhilidh
Ormiclate Castle
BEINN MHOR
Ormacleit
Rubha Hellisdale
Rubha Ardvule
Bornais
Taobh a Tuath Loch Aineort
Loch Cill Donain
Calvay
KILDONAN MUSEUM
Rubha Bolum
Cill Donnain
Unasary
Gearraidh Bhailteas
Minngearraidh
FLORA MACDONALD'S BIRTHPLACE
Loch Aineort
Probost
Loch Sniogaisgeil
Aisgernis
374 STULAVAL
Stuley
Gearraidh Sheilidh
Loch Stùlabhal
Dalabrog
Taobh a Tuath Loch Baghasdail
Crois Dughaill
Rubha na Creige Móire
Loch Baghasdail (Lochboisdale)
Cille Pheadair
Loch Baghasdail
Calvay
Baghasdal
Ceann a Deas Loch Baghasdail
An Leth Meadhanach
Rubha Meall na Hoe
Trosaraidh
Gearraidh na Monadh
Smercleit
Cille Bhrighde
South Glendale
Rubha na h-Ordaig
Pol a Charra
Taobh a Chaolais
Ludag
Sgeir a'Mhill
Caolas Eiriosgaigh
Lingeigh (Lingay)
Haunn
Bun a'Mhuillin
Hartamul
Am Baile
Coilleag
Eiriosgaigh (Eriskay)
Fiaraigh (Fiaray)
Caolas Fhiaraigh
Scurrival Pt.
Caolas Bharraigh
CASTLEBAY 1:30 OBAN 5:20
Eolaigearraidh
CILLE BHARRA
Caolas Fhuideigh
Fuideigh (Fuday)
Stack Is.
Traigh Mhòr
Orosay
Greian Head
Claid
BARRA
Oltir Mhòr
Hellisay
Gighay
Grein
207 BEN CLIAD
Cuidhir
Aird Mhor
Aird Mhidhinis
Caolas Sheileasaigh
Allathasdal
Bagh Shiarabhagh
Bruairnis
Borve Pt.
Baile na Creige
Buaile nam Bodach
Fuiay
Bagh a Tuath
CRAIGSTON MUSEUM
Borgh
EILEAN BHARRAIGH (BARRA)
Bruernish Pt.
Doirlinn Hd.
Tangasdal
HEAVAL
Earsairidh
BARRA HERITAGE CENTRE
Bagh a Chaisteil (Castlebay)
Breibhig
Rubha Mor
KIESSIMUL (KISIMUL) CASTLE
Caolas
0 2 4 6 miles
0 2 4 6 8 10 km
Oisgill Bay
Upper Milovaig
Glaspheinn
Lephin
Borrodale
Holmisdale
Fasach
COLBOST FOLK MUSEUM
Colbost
TOY MUSEUM
Skinidin
Waterstein
LIGHTHOUSE
Neist Point
Glen Dale
Moonen Bay
HEALABHAL MHOR 468
Ramasaig
Hoe Rape
Macleod's Tables
488 HEALABHAL BHEAG
Hoe Point
Geodha Mor
298
MACLEOD'S MAIDENS
Uidh
Bhatarsaigh (Vatersay)
Bagh Bhatarsaigh
Bhatarsaigh
Muldoanich
LOCHBOISDALE 1:30
OBAN 4:50
Flodaigh (Flodday)
Caolas Shanndraigh
Sanndraigh (Sandray)
Greanamul
Caolas Phabaigh
Pabaidh (Pabbay)
Theisgeir (Heiskers)
Caolas Mhiui Laigh
Miùgh Laigh (Mingulay)
Caolas Bhearnaraigh
Bearnaraigh (Berneray)
Barra Hd.

Fladda-chùain
305
307
Eilean Trodday
Rubha Hunish
Rubha na h-Aiseig
TARBERT 1:40
LOCHMADDY 1:45
Shulista
Balmacqueen
DUNTULM CASTLE
Duntulm
Kilmaluag
Connista
Lub Score
Hungladder
MUSEUM OF ISLAND LIFE
Peingown
Borneskеtaig
Heribusta
Eilean Flodigarry
Flodigarry
MEALL NA SUIRAMACH 543
Kilmuir
FLORA MACDONALD'S MEMORIAL
Digg
Staffin I.
Staffin Bay
Glashvin
Kilvaxter
THE QUIRAING
Brogaig
Garafad
Staffin
Balgown
296
Waternish Point
Stenscholl
TROTTERNISH
Linicro
Kilt Rock
KILT ROCK & MEALT FALLS
Clachan
Totscore
Kilbride Point
BIOD BUIDHE 466
Elishader
Scuddaborg
Maligar
Loch Mealt
Ascrib Islands
Idrigill
Uig
Valtos
Uig Bay
Conon
Marishader
Rubha nam Brathairean
Garros
BEN GEARY 284
South Cuil
Balnaknock
Ru Chorachan
BEINN EDRA 611
Breckrey
Culnaknock
Ard Beag
TRUMPAN CHURCH
Trumpan
Geary
Knockbreck
Gillen
Earlish
Lealt
LEALT FALLS
Lower Tote
LOCH
Upper Tote
Ardmore Pt.
Lower Halistra
Upper Halistra
SNIZORT
WATERNISH
A87
Peinlich
Peinaha
Hinnisdal
CREAG A'LAIN 607
Island of Rona
Hallin
Mingay
Lyndale Pt.
Isay
Greshornish Pt.
Stein
Lusta
Kingsburgh
Eilean Garbh
LOCH DUNVEGAN
Loch Bay
Loch Greshornish
Lyndale Ho.
Loch Snizort Beag
Romesdal
Galtrigill
PIPING HERITAGE CENTRE
Claigan
Bay
Greshornish
Kildonan
Romesdal
THE STORR 719
OLD MAN OF STORR
Callakille
Borreraig
BORRERAIG PARK MUSEUM
Uig
BEINN BHREAC 327
Flashader
Treaslane
Suladale
Eyre
TROTTERNISH
Bearreraig Bay
Eilean Tigh
Garbh Eilean
Lonbain
Loch a' Sguirr
An Caol
Husabost
Bay River
Edinbane
The Aird
Kensaleyre
Bernisdale
Keistle
Rhenetra
Annishader
Haultin
Holm I.
Eilean Fladday
Feriniquarrie
Totaig
DUNVEGAN CASTLE
Blackhill
Loch Leathan
SOUND OF RAASAY
254
Glaspheins
COLBOST FOLK MUSEUM
Colbost
Tote
Crepkill
Borve
Loch Fada
Manish Pt.
Loch Arnish
Torran
Lephin
Fasach
TOY MUSEUM
Glen Bernisdale
PRINCE CHARLES'S CAVE
Arnish
Holmisdale
Skinidin
Dunvegan
GIANT ANGUS MACASKILL MUSEUM
Kilmuir
Lonmore
Skeabost
Carbost
CRUACHAN BEINN A'CHEARCAILL 266
Uigshader
Drumuie
Achachork
Rubha na h Airde Glaise
392
BROCHEL CASTLE
Brochel
Glen Dale
HEALABHAL MHOR 468
Roskhill
Upper Feorlig
Glen Vic Askil
Glengrasco
Sluggans
AN TUIREANN ARTS CENTRE
Torvaig
Balmeanach
Roag
Vatten
Loch Connan
Shulishadermor
THE AROS EXPERIENCE
Portree
Inveralivaig
Glame
385
ISLAND OF RAASAY
Applecross Bay
Macleod's Tables
Orbost
Harlosh
Greep
Ose
Heatherfield
Penifiler
BEN TIANAVAIG 413
Balmeanach
Balachuirn
HEALABHAL BHEAG 488
Loch Varkasaig
Balmore
Loch Caroy
Ose
Peinmore
Glenmore
417
Camus
Ard-dhu
Eabost West
Eabost
Bracadale
Totardor
Loch Duagrich
Tungadal
Mugeary
Holoman Bay
Camastianavaig
DUN CAAN 443
Ardban
Geodha Mor
Harlosh I.
Tarner I.
Ullinish
Struan
Coillore
Tianavaig Bay
Oskaig
Rubha na'Leac
Eilean na Bà
Loch Bracadale
Conordan
RAASAY
Lower Ollach
Wiay
Raasay Ho.
Clachan
RAASAY OUTDOOR CENTRE
Idrigill Point
Gesto Ho
Glen Varragill
MACLEOD'S MAIDENS
Oronsay
Portnalong
Upper Ollach
The Braes
Narrows of Raasay
North Fearns
Crowlin Islands
Inverarish
Ardtreck
Loch Harport
ROINEVAL 439
Gedintailor
BATTLE OF THE BRAES MONUMENT
East Suisnish
Eyre Point
Rubha nan Clach
Fiskavaig
Eyre
Balmeanach
297
Fernilea
Caol Mòr
Peinchorran
ARNAVAL 369
TALISKER DISTILLERY
BEN LEE 444
Loch Sligachan
Carbost
Drynoch
Crossal
Longay
Sconser
Gleann Oraid
Merkadale
Satran
Drynoch
GLAMAIG 775
Scalpay
Talisker Bay
Moll
Corran a Chan Uachdaraich
Talisker
Sligachan Hotel
Loch Ainort
Scalpay Ho.
Eynort
Luib
Dunan
Guillamon I.
Pabay
Glen Brittle Forest
OLD SKYE CROFTER'S HOUSE
Caolas Scalpay
BEINN BHREAC 445
Eynort
Grula
SGURR NAN GILLEAN 964
Glen Sligachan
MARSCO 736
GLAS BHEINN MHOR 570
SKYE SERPENTARIUM
459
RED HILLS
Broadford Bay
Corry
Waterloo
MINGINISH
Loch Eynort
SGURR A'GHREADAIDH 973
BROADFORD
Broadford
Lower Breakish
Upper Breakish
Kraiknish
Glen Brittle
THE CUILLIN HILLS
SKYE
732
Old Corry
Harrapool
Skulamus
GLENBRITTLE
CUILLIN HILLS
BLA BHEINN 928
WORLD OF WOOD
SKYE ENVIRONMENTAL CENTRE
Glenbrittle House
Torrin
Bualintur
SGURR ALASDAIR 992
Loch Coruisk
Loch na Crèitheach
Faoilean
Culnaneam
294
295
Kilbride
SGURR NAN EAG 924
Camasunary
Strathaird
Loch Brittle
Kirkibost
Strath
301
Heast
Rubh an Dunain
BEN MEABOST 346
Kilmarie
Loch Slapin
Soay Sound
Soay
Loch Scavaig
Suisnish
Drumfearn
0 2 4 6 miles
0 2 4 6 8 10 km

Peterburn
Brae
Rubha 'Ard na Bà
Naast
Tournaig
Aird Dubh
Loch na Sealga
1062
Clachan
Inverbroom
Auchlunachan
Port Erradale
North Erradale
Rubha Bàn
Big Sand
onga Island
Caolas Beag
INVEREWE GARDEN
Loch Bad a'Chreamh
Poolewe
Londubh
680 BEINN A'CHAISGEIN BEAG
Fisherfield Forest
Dundonnell
Forest
Auchindrean
Braemore
CORRIESHALLOCH GORGE
Fain
Achneigie
Strath na Sealga
908 BEINN DEARG MHOR
CARN DEARG
Lonemore
Mial
Strath
Smithstown
Gairloch
GAIRLOCH HERITAGE MUSEUM
Loch Tollaidh
Inveran
Kernsary
MEALL AN DOIREAN 420
Dubh Loch
914 BEINN A' CHLAIDHEIMH
LOCH GAIRLOCH
Charlestown
Ardlair
791 BEINN AIRIGH CHARR
Braemore Forest
Bodantionail
Port Henderson
Aird
Badachro
Leacnasaide
Opinan
Kerrysdale
Shieldaig
BEINN LAIR 860
1019 MULLACH COIRE MHIC FHEARCHAIR
Letterewe Forest
Loch a'Bhraoin
807
934
Eilean Subhainn
LOCH MAREE ISLANDS
South Erradale
Redpoint
Loch Clàir
Loch Bad an Sgalaig
VICTORIA FALLS
Talladale
Letterewe
Furnace
Lochan Fada
748 GROBAN
999 A'CHAILLEACH
1110 SGURR MOR
Loch Garbhaig
980 SLIOCH
949 MEALL GORM
Fannich
Dubh Loch
Loch Gaineamhach
WESTER ROSS
Abhainn a'Chadh'Bhuidhe
BEINN NAN RAMH 711
Fannich Lodge
Flowerdale
Forest
875 BAOSBHEINN Forest
Loch na h-Oidhche
Glen Grudie
692
Kinlochewe Forest
Heights of Kinlochewe
Leckie
Strath Chrombuill
LOCH FANNICH
Lochrosque Forest
Shieldaig
624 BEINN BHREAC
Loch a'Bhealaich
878
Taagan
Anancaun
Incheril
Kinlochewe Forest
BEINN EIGHE NATIONAL NATURE RESERVE
Kinlochewe
Rubha na Fearn
Fearnmore
Fearnbeg
Arinacrinachd
1010 BEINN EIGHE
985 BEINN ALLIGIN
Upper Diabaig
Lower Diabaig
Cromasaig
933 FIONN BHEINN
Dosmuckeran
Glen Docherty
Badavanich
Achnasheen
Loch a'Chroisg
Reaulay
Cuaig
Kenmore
Loch a' Chracaich
Torridon
Alligin Shuas
Rechullin
Inveralligin
Torridon Ho.
Torridon Forest
1054 LIATHACH
Glen Torridon
Loch Clair
Coulin Lodge
Abhainn Dubh
Ledgowan Forest
Ledgowan
Loch Gowan
550 CARN-BEAG
Ardheslaig
Loch Shieldaig
Upper Loch Torridon
Fasag
TORRIDON VISITOR CENTRE
782 SGURR DUBH
Loch Coulin
Coulin
Coulin Forest
538 CARN MHARTUIN
Shieldaig
Balgy
Annat
Abhainn Thrail
Loch Gaineamhach
493 CROIC-BHEINN
Ben-damph Forest
Loch Damh
678 CARN BREAC
Glen Carron
Loch Sgamhain
Scardroy
Loch Beannacharain
Loch an Eoin
BEN SHIELDAIG 439
902 BEINN DAMH
993 MAOL CHEAN-DEARG
Ashnashellach Lodge
Craig
Glencarron Lodge
928
Glencarron and Glenuig Forest
Glenuaig Lodge
Glen Fhiodhaig
626 BEINN A'CHLACHAIN
Glenshieldaig Forest
Loch Lundie
613
907
Lair
Achnashellach Forest
849 BAC AN EICH
PEL OF RUBHA
Hartfield
Applecross Forest
Kinloch Damph
Loch Coultrie
Loch Dughaill
Balnacra
Loch na Caoidhe
Applecross Ho.
Applecross
896 BEINN BHAN
Coulags
Arineckaig
862
1053
1007 MAOILE LUNNDAIDH
814
Couldoran
RASSAL ASHWOOD
Rassal
Tornapress
Tullich
New Kelso
Strathcarron
Achintee
Kirkton
West Monar Forest
East Monar Forest
Camusterrach
BEALACH NA BA
Loch an Laoigh
LURG MHOR 986
Culduie
710 MEALL GORM
Russel
Sanachan
Glen Mor
Loch Monar
Monar Lodge
Toscaig
Ardarroch
Slumbay
Lochcarron
Attadale Ho. GARDEN
594
Bendronaig Lodge
Loch Calavie
An Gead Loch
Loch an Tachdaidh
Uisge Misgeach
Loch Kishorn
Achintraid
Attadale
Kishorn I.
Reraig
Cot
395
STROME CASTLE
Stromemore
Attadale Forest
Loch Cruoshie
Caolas Mòr
Uags
Ardaneaskan
Ardnarff
706 AN CRUACHAN
LOCH CARRON
Stromeferry
Loch an Iasaich
486 CARN NAN IOMAIREAN
1086 AN RIABHACHAN
1150 SGURR NA LAPAICH
An Dubh-aird
Plockton
Achmore
Killilan Forest
Eilean Mór
CRAIG HIGHLAND FARM
Braeintra
AONACH BUIDHE 899
1069
Port Cam
Duirinish
East Benula Forest
Drumbuie
Achnandarach
Gleann Udalain
SGUMAN COINNTICH 879
Loch Mullardoch
Glen Cannich
Erbusaig
Black Is.
Allt-nan-sugh
Sallachy
Killilan
Iron Lo.
Badicaul
Balmacara Square
Loch Long
Faddonch
Carnach
Kyle of Lochalsh
Reraig
Auchertyre
Conchra
Camas-luinie
West Benula Forest
1005
1053 TOLL CREAGACH
CASTLE MOIL
Balmacara
Kirkton
Nostie
Bundalloch
Glen Elchaig
Loch na Leitreach
Gleann Sithidh
Gleann A'Choilich
1111 TOM A CHOINICH
Kyleakin
LOCHALSH WOODLAND GARDEN
BRIGHTWATER VISITOR CENTRE
Ardelve
Camdu
634
Loch nan Eun
Dornie
EILEAN DONAN CASTLE
FALLS OF GLOMACH
1183 CARN EIGE
Gleann nam Fiadh
LOCH ALSH
Avernish
Glas Eilean
Totaig
Ardintoul Pt.
Ardintoul
Letterfearn
Keppoch
SGURR NAN CEATHREAMHNAN 1151
GLEN AFFRIC
Affric Lodge
Glen Affric
SGURR NA COINNICH 739
KYLERHEA OTTER HAVEN
Kyle Rhea
SGURR AN AIRGID 841
Inverinate
Dorusduain
Loch Duich
Loch a'Bhealaich
Loch Affric
Glen Arroch
603 BEINN A'CHUIRN
Ruarach
Glenaffric Forest
Athnamulloch
539
Bernera
Galltair
Morvich
Carn-gorm
Kylerhea
Glenelg
BARRACKS
Scallasaig
Ratagan
RATAGAN
Aultachruinn
GLEN AFFRIC
BEINN FADA OR BEN ATTOW
Glen Gniomh
Invershiel
Glenelg Bay
610 BEN ASLAK
Eilanreach
Cnoc Fhionn
Shiel Bridge
KINTAIL
1032
888 AONACH SHASUINN
706 CARN A CHAOCHA
GLENELG BROCHS
Glen More
Kintail Forest
Glenlicht Ho.
Balvraid
779
1067 FIVE SISTERS
979 CISTE DHUBH
Allt Garbh
Allt Riabhach

EASTER ROSS
HIGHLAND
Auchindrean
Braemore Forest
Lochdrum
Dirrie More
Loch Droma
Loch Glascarnoch
A835
CORRIESHALLOCH GORGE
Deanich Lodge
Gleann Beag
Lubachlaggan
Loch Vaich
Strathvaich Forest
Strathvaich Lodge
Strathrannoch
Inchbae Forest
Aultguish Inn
Inchbae Lodge
Garbat
Garbat Forest
BEN WYVIS
Wyvis Lodge
Loch Glass
Culzie Lodge
Kildermorie Lodge
Kildermorie Forest
Loch Morie
Braeantra
Inchlumpie
Boath
Ardross Castle
Ardross
Ardross Forest
Stittenham
Dalnavie
Alness
Dalmore
DALMORE DISTILLERY
Alness Bay
Evanton
Novar House
Assynt Ho.
Eileanach Lodge
Glen Glass
Redburn
Swordale
Drummond
Foulis Castle
Ardullie
STOREHOUSE OF FOULIS
Clach Liath Forest
Woodlands
Mountgerald
Heights of Brae
Tulloch Castle
Fodderty
Dingwall
Bottacks
Auchterneed
Knockfarrel
Strathpeffer
Jamestown
Maryburgh
Conon Bridge
Conon House
Marybank
Urray
Muir of Fairburn
Contin
Torrachilty Wood
TORR ACHILTY DAM FISH LIFT
Tarvie
Craigdarroch
Loch Achilty
Comrie
Scatwell Ho
Little Scatwell
Glenmarksie
Loch Luichart
Lochluichart
Corriemoillie
Corriemoillie Forest
Gorstan
Garve
Strathgarve Forest
Strathgarve Lodge
Loch Garve
Grudie
Achanalt
Loch Achanalt
Strath Bran
Dosmuckeran
Loch Fannich
Fannich Lodge
Fannich Forest
Aultdearg
Kinlochluichart Forest
Milton
Porin
Dalnacroich
Bridgend
Achlorachan
Curin
STRATHCONON
STRATH CONON FOREST
Scardroy
Loch Beannacharain
Carnoch
Inverchoran
Glenmeanie
Gleann Meinich
Muirton Mains
Fairburn House
Cabaan Forest
Orrin Reservoir
Glen Orrin
Corriehallie Forest
Muir of Ord
Balvaird
Newmore
GLEN ORD DISTILLERY
Ardnagrask
Corry of Ardnagrask
Aultvaich
Windhill
Tomich House
Clashandorran
Dunmore
Ruilick
Beauly
BEAULY PRIORY
Lovat
Groam
Kirkhill
BEAULY FIRTH
Redcastle
Milton
Kilcoy
Fettes
Tore
Muckernich
Leanaig
Bishop Kinkell
Easter Kinkell
Culbokie
Urquhart
Duncanston
Knockbain
Munlochy
Munlochy Bay
Allangrange Mains
Bogallan
Drumderfit
BLACK ISLE WILDLIFE AND COUNTRY PARK
Arpafeelie
Charlestown
North Kessock
South Kessock
Craigton
Kilmuir
Bunchrew
Lentran
Drumchardine
Easter Moniack
Newtonhill
HIGHLAND WINERIES
Inverness
Leachkin
Culcabock
Dunain Ho.
Dochgarroch
Caledonian Canal
Castleton Village
Leys Castle
Scaniport
Essich
Dochfour Ho.
DOCHFOUR GARDENS
Lochend
Lairgmore
Aldourie Castle
Dores
Drumashie Moor
Abriachan
Balchraggan
ABRIACHAN GARDEN NURSERY
GREAT GLEN WAY
Kiltarlity
Cabrich
Balchraggan
THE AIRD
South Clunes
Kilmorack
Beaufort Castle
AIGAS DAM FISH LADDER
Culburnie
Crask of Aigas
Breakachy
Aigas Forest
Erchless Castle
Erchless Forest
Craigdhu
Inchmore
Struy
Mid Main
Mauld
Eskadale
Camault Muir
Glaichbea
Druimkinnerras
Boblainy
Ardendrain
Boblainy Forest
Cragganvallie
Culligran Falls
GLEN STRATHFARRAR
Glenstrathfarrar Forest
Braulen Lodge
Monar Lodge
Ardchuilk
Loch Beannacharan
Loch a'Mhuillidh
Struy Forest
Carnoch
Glassburn
Balmore
Cannich
STRATHGLASS
Muchrachd
Glencannich Forest
Liatrie
Mullardoch House
Loch Mullardoch
Glen Cannich
Fasnakyle Forest
DOG FALLS
Fasnakyle Ho
Tomich
Guisachan
Balcladaich
Hilton Lodge
PLODDA FALLS
Affric Lodge
Glen Affric
Cougie
Guisachan Forest
Corrimony
CAIRN CORRIMONY
RSPB CORRIMONY
Buntait
Braefield
Shenval
BCC LOCH NESS
Lochletter
Glen Urquhart
Balbeg
Uppertown
Balnain
Balnagrantach
Polmaily
Rychraggan
Milton
Drumnadrochit
LOCH NESS MONSTER EXHIBITIONS
Lewiston
DRUM FARM
Strone
URQUHART CASTLE
Glenurquhart Forest
Balmacaan Forest
Divach
Upper Lenie
Lower Lenie
COBB MEMORIAL
Bunloit
Balbeg
Grotaig
Ancarraig
Easter Boleskine
Inverfarigaig
FARIGAIG FOREST CENTRE
Farigaig Forest
Aultnagoire
Errogie
Foyers
Lyne of Gorthleck
Farraline
Dunmaglass Lodge
Creagnaneun Forest
LOCH NESS
Alltsigh
Lochgarthside
Bailebeag
Garthbeg
Corriegarth Lodge
Whitebridge
Achnaconeran
Levishie
Levishie Forest
Bhlaraidh
FALLS OF MORISTON
Invermoriston
Portclair Forest
Achnabat
Torness
Loch Ruthven
RSPB LOCH RUTHVEN
Aberarder House
East Croachy
Tullich
Brinmore
Mains of Flichity
Dunlichity Lodge
Balnafoich
Farr
STRATHNAIRN
Kilmuir
Culbo
Findon Mains
Shoreton
Cullicudden
Springfield
Newhall House
Resolis
Ferryton
Balblair
Kinbeachie
Braefindon
Killen
Wester Strath
Belmaduthy
Rosehaugh Mains
Suddie
Roskill House
Bennetsfield
Kinkell
Kilmuir
Raddery
Meikle Gluich
Aultnamain Inn
Strathy
Baldoon
Loanreoch
Easter Ardross
Wester Lealty
Achandunie
Nonikiln
Mossfield
Moultavie
FYRISH MONUMENT
ROGIE FALLS
HIGHLAND MUS OF CHILDHOOD
MACDONALD MONUMENT
Lochussie
Tollie
Corntown
Alcaig
Drynie Park
Auchmore
Drumindorsair
Balblair
MADE IN SCOTLAND
INVERNESS CASTLE
CRAIG PHADRIG FORT
BUGHT FLORAL HALL VISITOR CENTRE
BALNAIN HOUSE
GARRISON ENCOUNTER
PRINGLE WEAVERS
DOLPHINS & SEALS OF THE MORAY FIRTH
Longman Pt.
CROMARTY FIRTH
BLACK ISLE
A9
A82
A831
A832
A834
A835
A862
A887
B862
B852
B851
B9169
B9163
B9164
B9176
B861
Chaochain
0 2 4 6 miles
0 2 4 6 8 10 km

MORAY FIRTH
MORAY
BURGHEAD BAY
Cromarty Bay
Nigg Bay
Port an Righ
CAIRNGORMS NATIONAL PARK
HILLS OF CROMDALE
STRATHDEARN
Tain
Portmahomack
Tarbat Discovery Centre
Balintore
Shandwick Stone
Hilton of Cadboll
Fearn
Fearn Abbey
Hill of Fearn
Loans of Tullich
Arabella
Nigg
Nigg Ferry
Castlecraig
King's Cave
Invergordon
Balnapaling
Cromarty
Cromarty Court House
Hugh Miller's Cottage
Sutors of Cromarty
Newhall Point
Udale Bay
Jemimaville
Newton
Navity
Upper Eathie
Craighead
Whitebog
Balmungie
Groam House Museum and Pictish Centre
Rosemarkie
Fortrose
Fortrose Cathedral
Fort George
Chanonry Pt.
Avoch
Whiteness Head
Nairn
Nairn Literary Institute Museum
Nairn Fishertown Museum
Culbin Sands
Culbin Forest
The Bar
Kincorth Ho.
Findhorn
Findhorn Bay
Kinloss
Kinloss Abbey
Burghead
Burghead Museum
Hopeman
Lossiemouth
Lossiemouth Fisheries & Community Museum
Covesea Skerries
Covesea
Duffus
Duffus Castle
Gordonstoun
Cummingston
Roseisle Forest
Roseisle
Palace of Spynie
Spynie
Elgin
Elgin Museum
Elgin Cathedral
Bishopmill
New Elgin
Old Mills
Moray Motor Museum
Lhanbryde
Pittendreich
Miltonduff
Pluscarden Abbey
Monaughty Forest
Forres
Sueno's Stone
Nelson Tower
Falconer Museum
Dallas Dhu Distillery
Rafford
Brodie Castle
Brodie
Boath Dovecote
Auldearn
Kingsteps
Dyke
Moy Ho.
Darnaway Castle
Darnaway Forest
Logie Steading Visitor Centre
Drumine Forest
Altyre Woods
Newtyle Forest
Dallas Forest
Dallas
Kellas
Glenlatterach
Craigroy
Ballachraggan
Elchies Forest
Cardhu Distillery
Archiestown
Knockando
Carron
Speyside Way
Marypark
Glenfarclas Distillery
Bellehiglash
Cragganmore Distillery
Bridge of Avon
Advie
Dalchiarach
Mains of Dalvey
Cromdale
Grantown-on-Spey
Speybridge
Anagach
Revack Estate
Dulnain Bridge
Skye of Curr
Heather Heritage Centre
Clan Grant Mausoleum & Church
Carrbridge
Landmark Forest Theme Park
Lochanhully
Duthil
Sluggan
Slochd
Boat of Garten
Strathspey Railway
Loch Garten Abernethy
Abernethy Forest
Nethy Bridge
Dell Lodge
Forest Lodge
Tomintoul
Tomintoul Museum
Glenlivet Estate Visitor Centre
Glenlivet
Glenlivet Distillery
Tomnavoulin
Knockandhu
Fodderletter
Strathavon Lo.
Lochindorb
Lochindorb Castle
Lochindorb Lodge
Dava
Aitnoch
Ferness
Ardclach Tower
Glenferness House
Dulsie
Relugas
Logie
Cawdor
Cawdor Castle and Gardens
Kilravock Castle
Croy
Culloden
Culloden Forest
Culloden Battlefield
Clava Cairns
Inverness
Castle Stuart
Ardersier
Gollanfield
Tornagrain
Daviot
House of Daviot
Moy
Moy Hall
Strathdearn Forest
Tomatin
Tomatin Distillery
Findhorn Bridge
Kyllachy House
Dalmigavie Lodge
Dalnahaitnach
Cairn Glas-Choire
Carn Kitty
Larig Hill
Carn Uish
Carn na Cailliche
Mill Buie
Carn na Loine
Carn-Ruigh Charrach
Carn Sgumain
Beinn Bhuidhe Mhor
Beinn Bhreac
Carn nan Tri-Tighearnan
Carn na H-Easgainn
Carn an t-Sean-Liathanaich
Carn a' Choire Mhoir
Beinn Mhor
Creagan a' Chaise
Carn Daimh
Carn Meadhonach
Baddoch
Dorback Lodge
Blairnamarrow
Auchgourish
Aviemore

302
311
SPEY BAY
Lossiemouth
Covesea Skerries
Halliman Skerries
LOSSIEMOUTH FISHERIES & COMMUNITY MUSEUM
Stotfield
Branderburgh
Covesea
Gordonstoun
Duffus
DUFFUS CASTLE
Lossie Forest
PALACE OF SPYNIE
Spynie
Elgin
ELGIN MUSEUM
CATHEDRAL
CASHMERE VISITOR CENTRE
Bishopmill
Laverackloch
Quarrywood
OLD MILLS
Pittendreich
MORAY MOTOR MUSEUM
New Elgin
Mayne Ho.
Glassgreen
Moss of Barmuckity
COXTON TOWER
Milltonduff
Muir of Miltonduff
Paddockhaugh
Auchtertyre
Longmorn
BIRNIE CHURCH
Fogwatt
Thomshill
Upper Bogside
MILLBUIES
Coleburn
Glenlatterach
Leuchars Ho.
Lochhill
Stonewells
Burniestrype
Urquhart
Moss of Meft
Darkland
Lhanbryde
Lochs Crofts
Clackmarras
Blackhills
Altonside
Orbliston
Teindland Forest
MORAY
Kingston
DOLPHIN CENTRE
Garmouth
TUGNET ICE HOUSE
Spey Bay
Portannachy
Nether Dallachy
Upper Dallachy
Bogmoor
SPEYSIDE WAY
Auchenhalrig
BAXTERS HIGHLAND VILLAGE
Newlands of Tynet
Mosstodloch
Dipple
Fochabers
FOCHABERS FOLK MUSEUM
Speymouth Forest
Westerton
Ordiequish
Wood of Ordiequish
Inchberry
Mains of Cairnty
Shalloch
Auchinroath
Newlands
Rothes
GLEN GRANT DISTILLERY
Burn of Rothes
Glen of Rothes
Aikenway
Arndilly Ho.
BEN AIGAN
Balnabreich
Mulben
Rosarie
Rosarie Forest
Belnagarrow
Elchies Forest
CARN NA CAILLICHE
CARN UISH
Whiteacen
Dandaleith
LADYCROFT AGRICULTURAL MUSEUM
Archiestown
Ringorm
Craigellachie
SPEYSIDE COOPERAGE VISITOR CENTRE
CARDHU DISTILLERY
Cardow
Knockando
Knockando Ho.
Robertstown
BALVENIE SPEYSIDE WAY VISITOR CENTRE
Speyview
Carron
Charlestown of Aberlour
Daugh of Kinermony
Glenallachie
Milltown of Edinvillie
Sheandow
Maggieknockater
Kininvie Ho.
Midtown of Buchromb
Balvenie
BALVENIE CASTLE
GLENFIDDICH DISTILLERY
Dufftown
Kirktown of Mortlach
Lettoch
MEIKLE CONVAL
Marypark
GLENFARCLAS DISTILLERY
Bellehiglash
Craggenmore
Bridge of Avon
Dalchirach
Haugh of Urnaichie
Cragganmore
BEN RINNES
Favillar
Glen Rinnes
Kirkhill
Glenfiddich Lodge
Glen Fiddich
Glenfiddich Forest
Badeach
Downan
Glenlivet
GLENLIVET DISTILLERY
Shenval
Auchbreck
Castleton
Tomnavoulin
Blackwater Lodge
Blackwater Forest
CARN AN T-SUIDHE
Achdregnie
Knockandhu
Auchnarrow
Glenlivet Forest
Croughly
Clashnoir
Braes of Glenlivet
Chapeltown
Scalan
Milton
TOMINTOUL
Glenchat Lodge
Badenyon
GLENLIVET ESTATE VISITOR CENTRE
Blairnamarrow
292
Buckie
THE BUCKIE DRIFTER
Findochty
Portknockie
Bauds of Cullen
Gordonsburgh
Portessie
Buckpool
Rathven
Portgordon
Slackhead
Cairnfield Ho.
Drybridge
Connage
Slate Haugh
Preshome
ST. GREGORY'S CHAPEL
Clochan
Broadley
Shiel Muir
Braes of Enzie
MILLSTONE HILL
Broadrashes
Forgie
Aultmore
Newmill
Garralburn
Glen of Newmill
Deerhill
Grange Crossroads
Goukstone
Fife Keith
Keith
STRATHISLA DISTILLERY
Mains of Auchoynanie
Blackhillock
KEITH AND DUFFTOWN RLY
Poolside
North Whiteley
Edintore
Towiemore
Drummuir Castle
Drummuir
Little Pitlurg
Coachford
Glen of Coachford
Balloch Wood
Glack of Midthird
Slogan
Midtown of Glass
Milltown of Auchindoun
Daugh of Cairnborrow
Cairnborrow
Torry
Haugh of Glass
AUCHINDOUN CASTLE
THE SCALP
Ballochford
Mill of Lynebain
Tomnaven
Succoth
Bridgend
Inverharroch
Ardwell
Auchmair
Cabrach
ROUND HILL
Aldivalloch
Aldunie
HILL OF THREE STONES
THE BUCK
B9002
ST MARY'S CHURCH
Craig Castle
Wheedlemont
Clova
SCOTTISH SCULPTURE WORKSHOP
Lumsden
CORREEN HILLS
Mossat
Kildrummy
KILDRUMMY CASTLE & GARDEN
Milltown of Kildrummy
Rinmore
Belnacraig
Glenbuchat Castle
Cullen
Cullen Bay
Logie Hd.
Old Cullen
Lintmill
FINDLATER CASTLE
Sandend
Sandend Bay
Redhythe Pt.
Portsoy
Seatown
Boyne Bay
Birkenbog
Bogtown
Fordyce
JOINER'S WORKSHOP
Milton
Kirktown of Deskford
DESKFORD CHURCH
Ardiecow
Berryhillock
Craibstone
Drochedlie
Brodiesord
Little Toux
Mickle Toux
Mains of Edingight
Edingight Ho.
Whitehill
Sillyearn
Crannoch
Bracobrae
Davoch of Grange
KNOCK HILL
Knock
Drumnagorrach
Farmtown
Nethermills
Gordonstown
Ordiquhill
Grantown
Knowes of Elrick
Knowehead
Ramsburn
Crombie Castle
Old Crombie
Knauchland
Moss-side
Marnoch
Tarryblake Ho.
Rothiemay Crossroads
Milltown of Rothiemay
Mains of Mayen
Yonder Bognie
Bogniebrae
FOURMAN HILL
Midplaugh
Ruthven
Haddoch
Cairnie
THE BIN
The Bin Forest
NORDIC SKI CENTRE
Corse of Kinnoir
Milburn
Drumdelgie
Westerton
HUNTLY CASTLE
Huntly
BRANDER MUSEUM
Brideswell
Sliach
Wellheads
Terryhorn
STRATHBOGIE
Bailiesward
Bridgend
Tillathrowie
Coynachie
Clashindarroch Forest
Clashindarroch
Mytice
Kirkney
Gartly
Cults
Culdrain
Kirkstile
Wester Mosshead
Shanquhar
Corncatterach
Blackburn
Clashindarroch Forest
KNOCKANDY HILL
LEITH HALL
Milton of Lesmore
Milton of Noth
Rhynie
Belhinnie
Smithston
Kennethmont
Cottown
Clatt
Auchmenzie
Knockespock Ho.
CORREEN HILLS
LORD ARTHUR'S CAIRN
Terpersie Castle
Tullynessle
Bridge of Alford
ALFORD VALLEY RAILWAY
Montgarrie
HAUGHTON
GRAMPIAN TRANSPORT MUS
Alford
Muir of Alford
Auchintoul
Knock Hd.
Boyndie Bay
Whitehills
Macduff
Inverboyndie
Banff
COLLEONARD SCULPTURE GARDEN & GALLERY
DUFF HOUSE
Boyndie
Easter Whyntie
Tillynaught
Wester Culbeuchly
Auds
Tipperty
Montcoffer Ho.
Alehousehurn
Newmills of Boyne
Oldtown of Ord
Mill of Brydock
Kirktown of Alvah
Cornhill
Upper Culphin
Weachyburn
Loanhead
Fattahead
Greenlaw
The Pole of Itlaw
Drywells
Blacklaw
Finnygaud
Lower Deuchries
Hill of Mountblairy
Hillhead of Blairy
Aberchirder
Cranna
Bogton
Clunie
Carnousie
Auldtown of Carnousie
Mains of Laithers
Deveron
Hillbrae
Inverkeithny
Auchininna
Bogniebrae
Forgue
Fortrie
Brownhill
GLENDRONACH DISTILLERY
Pitfancy
Cobairdy
Drumblair
Carlincraig
Lessendrum
Drumblade
Corse
Denmoss
Balgaveny
Nether Lenshie
Largue
Auchamie
Logie Newton
Badenscoth
Ythanwells
Redhill
Drumdollo
Denend
Thomastown
Milburn
Hillhead
Greystone
Fisherford
Bainshole
Glens of Foudland
HILL OF FOUDLAND
Kirkton of Culsalmond
Rothmaise
Colpy
Cairnhill
Largie
Upper Boddam
Wrangham
Bonnyton
Wraes
Knockenbaird
Auchleven
Newton Ho.
Old Rayne
Insch
Aulton
Oldtown
Ardlair
Towie
Duncanstone
Pitmachie
Ardoyne
ARCHAEOLINK PREHISTORY PARK
Westhall
Oyne
Kirkton of Oyne
Kirkton
Leslie
Bogs
Coldwells Croft
Whitehaugh Forest
Muckletown
Lickleyhead Castle
Auchleven
BENNACHIE
Newbigging
Glenton
GARIOCH
ABERDEEN
Keig
Castle Forbes
Upper Woodend
Lower Woodend
Rorandle
Pitfichie Forest
Pitfichie
Gateside
Pitmunie
MONYMUSK ARTS CENTRE
293
A96
A95
A941
A920
A97
A98
A942
A990
A944
B9040
B9012
B9135
B9103
B9015
B9014
B9016
B9018
B9022
B9023
B9025
B9117
B9118
B9139
B9121
B9024
B9001
B9002
B9008
B9009
B9102
B9136
B9138
B9115
B9113
B992
B9010
B9009
0 2 4 6 miles
0 2 4 6 8 10 km

Fraserburgh
Peterhead
Inverurie
Oldmeldrum
Ellon
Mintlaw
Turriff
Rosehearty
Sandhaven
Pittulie
Broadsea
Kinnaird Head
Kinnaird Castle Lighthouse & Scotland's Lighthouse Museum
Fraserburgh Heritage Museum
Sandhaven Meal Mill
Pitsligo Castle
Fraserburgh Bay
Cairnbulg Pt.
Inverallochy
Maggie's Hoosie
Cairnbulg Castle
St Combs
Inzie Head
Troup Hd.
Pennan Hd.
Quarry Hd.
Head of Garness
Gamrie Bay
Macduff Marine Aquarium
Banff Museum
Crovie
Gardenstown
Pennan
Dubford
Easter Silverford
West Greenskares
Longmanhill
Glenquithlie
Cummerton
Towie
Mill Farm
New Aberdour
Ardlawhill
Windyheads Hill
Upper Boyndlie
Coburty
Peathill
Percyhorner
Pitblae
Easter Cardno
Mid Ardlaw
Tyrie
Memsie
Memsie Burial Cairn
Whitewell
Cairnmuir
Quarryhead
Ladysford
Mountsolie
Hillhead of Auchentumb
Newburgh
Mains of Hatton
Mormond Hill
Fernebrae
Ardglassie
Rathen
Gowanhill
Mosstown
Strathellie
Cairness
Netherton of Lonmay
Loch of Strathbeg
Spillardsford
Lonmay
Crimonmogate
Crimond
Loch of Strathbeg
Old Rattray
Seatown
Rattray Head
Blackhill
Nether Park
Upper Ridinghill
Balearn
North Mosstown
Gas Terminal
St Fergus Moss
St Fergus
Scotstown Hd.
North Kirkton
Kirkton Hd.
Kirktown
Gallowhills
Backfolds
Outhills
Rora Moss
Cuttyhill
Rora
Lunderton
Middleton of Rora
Woodside
Newseat
Inverugie
Inverugie Castle
Ugie Salmon Fish House
Buchanhaven
Torterston
Longside
Flushing
Inverquhomery
Parkhill
Toddlehills
Hillhead of Cocklaw
Arbuthnot Museum & Art Gallery
Keith Inch
Peterhead Maritime
Peterhead Bay
Invernettie
Sandford Bay
Boddam
Buchan Ness
Stirling
Little Dens
Nether Savock
Blackhill
Millbank
Sandfordhill
Nether Kinmundy
Easterton of Lenabo
Moss of Cruden
Croft of Tillymaud
Coldwells
Longhaven
Stoneygate
Teuchan
Auchiries
North Haven
Bullers of Buchan
Twa Havens
Cruden Bay
Slains Castle
Port Erroll
Bay of Cruden
Chapel Hill
Hatton
Bogbrae
Whinnyfold
Eastertown of Auchleuchries
Nether Leask
Blackhill
Artrochie
Auchmacoy
Alehousehill
Kirktown of Slains
Visitor Centre
Collieston
St Catherine's Dub
Waterside
Forvie
Sands of Forvie
Hackley Hd. or Forvie Ness
Newburgh
Newburgh Bar
Foveran
Drums
Hatterseat
Delfrigs
Tarty Burn
Foveran Burn
Knockhall Castle
Meikle Tarty
Kirkton of Logie Buchan
Waterton
Ellon
Ythanbank
Ythan
Newmachar
Reisque
Kinmuck
Craigie
Causeyend
Drumhead
Balmedie
Belhelvie
Whitecairns
Kinmundy
Drumligair
Blackbraes
Potterton
Strichen
Knowhead
Hillfoot
New Leeds
Longhill
Leys
Borrohill
Adziel
North Ugie Water
Little Skillymarno
Denhead
Hythie
Fetterangus
South Auchmachar
Forest of Deer
Deer Abbey
Toux
Dunshillock
Aden
Aberdeenshire Farming Museum
Old Deer
Stuartfield
Millbreck
Clola
Kinknockie
Smallburn
Backhill
Ardallie
Upper Hawkhillock
Muirtack
Arthrath
Toll of Birness
Mains of Birness
Broomfield
Knockothie
Hilton
Inverebrie
Berefold
Upper Ardgrain
Milton Coldwells
Drakemyre
Backhill of Fortree
Mains of Annochie
Auchnagatt
Kinnadie
Skelmuir
Mains of Crichie
Crichie
Bulwark
Clochan
Nethermuir
Drymuir
Backhill of Clackriach
Maud
Maud Railway Museum
Aulton of Atherb
Culsh
New Deer
Mains of Fedderate
Ironside
Eonnykelly
Oldwhat
New Pitsligo
Balnamoon
North Sealmanae
Middlemuir
Craigmaud
Broadleys
Nether Glasslaw
Cowbog
Northside
New Byth
Whitestones
Balthangie
Middle Cairncake
Cuminestown
Slacks of Cairnbanno
Rush-head
Burnside
Cairnbanno Ho.
Middlemuir
North Millbrex
South Millbrex
Crofts of Meikle Ardo
Burnend
Little Gight
Gowanwell
Cairnorrie
Brownhill
Knaven
Barrack
Mains of Inkhorn
Skelmonae
Quilquox
South Cuilquox
Drumwhindle
Cairnleith Crofts
Mains of Arnage
Methlick
Haddo
Haddo House and Gardens
Collynie
Crofts of Haddo
Ardo
Cottown
Stonehouse
Woodhead
Fyvie
Fyvie Castle
Backhill
Tifty
Gourdas
Monkshill
Lethenty
Steinmanhill
Inverythan
Seggat
Biffin
Mains of Towie
Heatherybanks
Bodiechell
Netherton
Crofts of Inverthernie
Dykeside
Kirkton of Auchterless
Burreldales
Gordonstown
Rothiebrisbane
Rothienorman
Petty
Cromblet
St John's Wells
North Flobbets
South Flobbets
Barthol Chapel
Earlsford
St Katherine's
Cross of Jackston
Byebush
Balgove
Tulloch
Redmoss
Wedderlairs
Thornroan
Raxton
Medieval Tomb
Tarves
Ythsie
Craigdam
Auquhorthies
Mains of Cairnbrogie
Tolquhon Castle
Pitmedden Gardens
Pitmedden
Udny Green
Cairnbrogie
Mill of Kingoodie
Hattoncrook
Affleck
Whiterashes
Nether Crimond
Tillykerrie
Straloch
Woodlands
Tillygreig
Udny Station
Pettymuick
Culterculien
Old Craig
Minnes
Rashierieve
Tillycorthie
Ardo Ho.
Middlemuir
Kinknockie
Tipperty
Cross Stone
Ulaw
Esslemont
West Kinharrachie
Meldrum Ho.
Fingask
Whitefield
Barra Castle
Lethenty
Kirkton of Bourtie
Hillbrae
Harlaw Ho.
Balhalgardy
Brandsbutt Stone
Battle of Harlaw Memorial
Mains of Blackhall
Burgh Muir
Inverurie Museum
Keith Hall
Hillhead
Chapel of Garioch
Pitcaple
Whiteford
Daviot
Mounie Castle
Loanhead Stone Circle
Durno
Kirkton of Rayne
Meikle Wartle
Newseat
Jackstown
Westertown
Bethelnie
Folla Rule
Manar Ho.
Port Elphinstone
Burnhervie
Westerton
Aquhythie
Overton
Dalmadilly
Grantlodge
Upper Coullie
Kemnay
Clovenstone
Burgh Muir
Cottown
Kintore
Balbithan
Kinkell Church
Wester Fintray
Cairnpark
Hatton of Fintray
Cothall
Balbithan Ho.
Muiryfold
Delgatie Castle
Little Colp
Delgaty Forest
Hatton Castle
Howe of Teuchar
Darra
Craighill
Birkenhills
Fintry
Upper Cotburn
Lower Cotburn
Craigston Castle
Milltown of Craigston
Plaidy
Milltack
King Edward
Muirton
Fortrie
Gorrachie
Pitgair
Auchlinn
Crudie
Mid Cloch Forbie
Netherbrae
Hill of Overbrae
Brackamore
Hill of Fishrie
Cauldwells
Garmond
Cairnhill
Minnonie
Cushnie
Muirden
Forglen Ho.
Turriff
Formartine
Aberdeenshire
Fraserburgh
Peterhead
Inverurie
NK
A90
A98
A96
A947
A948
A950
A952
A975
A981
A920
B9031
B9032
B9033
B9093
B9105
B9107
B9027
B9029
B9030
B9170
B9005
B9000
B9001
B993
B979
B977
B999
B9106
293

RUBHA ROBHANAIS (BUTT OF LEWIS)
Cunndal
Eòropaidh
Coig Peighinnean
CHURCH OF ST MOULAG
HARBOUR VIEW GALLERY
Port Nis
Lional
Adabroc
Sgiogarstaigh
Tàbost
Suainebost
Cros
Dail bho Thuath
Dail bho Dheas
Aird Dhail
Cross Sands
Glen Cross
Cuiashader
Cellar Head
Gabhsann bho Thuath
Gabhsann bho Dheas
Mealabost Bhuirgh
Bail Àrd Bhuirgh
Coig Peighinnean Bhuirgh
Siadar
Siadar Iarach
Siadar Uarach
Baile an Truiseil
Rubha Leathann
Aird Barvas
TRUSHAL STONE
Barabhas
Barabhas Uarach
Barabhas Iarach
Brù
Arnol
BLACK HOUSE MUSEUM
LOCH NA MUILNE
Labost
Bragar
Siabost bho Thuath
SHAWBOST NORSE MILL
Siabost bho Dheas
SHAWBOST MUSEUM
Pairc Shiaboist
Bàgh Dhail Beag
Dail Beag
Dail Mòr
GEARRANNAN BLACKHOUSE VILLAGE
GARENIN
Na Gearrannan
Borghastan
Carlabhagh
Campay
Loch Chàrlabhaigh
DUN CARLOWAY BROCH
Ciribhig
Dùn Charlabhaigh
Little Bernera
Floday
Harsgeir
Bostadh
IRON AGE HOUSE
Crothair
An Galan Uigeach
AN CAOLAS
Pabay Mor
Tobson
Tolastadh a Chaolais
Loch Ròg an Ear
Vacsay
BERNERA
Bhalasaigh
Breacleit
Great Bernera
Circebost
Barraglom
Tacleit
Iarsiadar
Tobhtarol
Vuia Mòr
Vuia Beag
Loch Ròg
Keava
Eilean Kearstay
Breascleit
CALLANISH VISITOR CENTRE
Calanais
CALLANISH STANDING STONES
Linsiadar
Gearraidh na h-Aibhne
Crulabhig
Lundal
Aird Uig
Bhaltos
Cnip
Riof
Uigen
Cliobh
Miabhig
Bàgh Fhiabhaig
Timsgearraidh
Cradhlastadh
Ard More Mangersta
Càrnais
Mangurstadh
Eadar Dha Fhadhail
SUAINAVAL
Cairisiadar
Floday
Geisiadar
Ungisiadar
Einacleite
Loch Suaineabhal
Aird Fenish
Aird Brenish
Islibhig
Breanais
MEALISVAL
Mealasta
Mealasta Island
Caolas an Eilein
BEINN MHEADHONACH
Loch Chaolartan
Scaliscro
Giosla
Loidse Mhorsgail
Tamanabhagh
Loch Tungabhat
Loch Smuaisaibhal
Loch Lacsabhat Ard
Loch Urabhal
Gleann Mòr Barvas
Gleann Bhruthadail
Glen Bragar
BEINN BRAGAR
BEINN MHOLACH
MUIRNEAG
Loch Mòr Shanndabhat
Loch Langabhat
Gabhsann bho Dheas
Abhainn Ghearadha
Loch Sgeireach Mòr
Loch Scarabhat Mhòr
Loch Mòran Stàir
Loch nan Stearnag
Loch Urghag
Loch Breibhat
Bail' Ur Tholastaidh
Tolastadh bho Thuath
Tolsta Head
Gleann Tholàstaidh
Port Bun a'Ghlinne
Creag Fhraoch
Griais
Cnoc an t-Solais
Bac
Col
Col Uarach
Vatisker Pt.
Breibhig
Coll Sands
BROAD BAY OR LOCH A TUATH
Aird Thunga
Tunga
Sròn Ruadh
Port Nan Giùran
Rubha an t-Siumpain
Cnoc
Amhlaigh
Port Mholair
Aird
Sulaisiadar
Seisiadar
Garrabost
EYE PENINSULA
Pabail Uarach
Pabail Iarach
Bàgh Phabail
Aiginis
An Cnoc
Suardail
A'Chearc
Mealabost
ST COLUMBA'S
STORNOWAY
Grianan
Newmarket
An Gleann Ur
Cnoc Màiri
Lacasdal
MUSEUM NAN EILEAN
LEWIS LOOM CENTRE
Stornoway
Sanndabhaig
AN LANNTAIR GALLERY
Tolm
Holm I.
Airinis
Arnish Moor
ULLAPOOL 2:40
Loch a' Ghainmhich
Acha Mor
Loch Orasaigh
Loch Tobhta Bridein
Loch nam Falcag
Liurbost
Griomsidar
Ben Casgro
Raerinish Pt.
Ranais
Crosbost
Barkin Is.
Tabhaidh Mhor
Soval Lodge
Loch Trealabhal
Loch Airigh na h-Airde
Ceos
Eilean Chaluim Chille
Eilean Orasaidh
Lacasaidh
Baile Ailein
Eireasort
Crobeag
Cromor
Gearraidh Bhaird
Eilean Thoraidh
ISLE OF LEWIS (EILEAN LEÒDHAIS)
306
307
A857
A858
A859
A866
B895
B897
B8011
B8059
B8013
B8014
B8015
NB

SOUTH LEWIS,
HARRIS AND
NORTH UIST
CEANN A TUATH NA HEARADH
NA HEARADH (HARRIS)
HIGHLAND
LOCH A SIAR
LOCH LANGABHAT
CAOLAS NAN EILEAN
CAOLAS NA HEARADH
PARK OR PAIRC
Forest of Harris
South Harris Forest
Morsgail Forest
Scarp
Kearstay
Huisinis
Hushinish Pt.
Beitearsaig
Gobhaig
Horsanish
Abhainn Suidhe
Arda Mòra
Cliasmol
Soay Beag
Soay Mòr
Taransay Glorigs
Camus an t-suithcan
Tarasaigh (Taransay)
Paible
Rubha Sgeirigin
Toe Head
Miabhag
Bun Abhainn Eadarra
OLD WHALING STATION
Aird Asaig
Cul na h-Aird
Isay
Leacainn
Taobh Siar
Oban
Urgha Beag
Urgha
Tairbeart (Tarbert)
Direcleit
Losgaintir
LUSKENTYRE BEACH
BEN LUSKENTYRE
Seilebost
Horgabost
Borve Lodge
Buirgh
Sgarasta Mhor
CHAIPAVAL
Rubha'an Teampuill
Taobh Tuath
SEALLAM!
An t-Ob (Leverburgh)
Fionnsbhagh
ROINEABHAL
BLEAVAL
Loch Langabhat
Loch Steiseabhat
Cairminis
Srannda
Borghasdal
ST CLEMENT'S CHURCH
Roghadal
Renish Pt.
Vallay
Lingarabay I.
Boirseam
Lingreabhagh
Cuidhtinis
Aird Mhighe
Fleoideabhagh
Manais
Beacrabhaic
Caolas Stocinis
Stockinish I.
Geocrab
Liceasto
Collam
Cliuthar
Leac a Li
Aird Mhighe
Greosabhagh
Kennavay
Drinisiadar
Miabhag
Plocrapol
Scadabhagh
Plocropol Pt.
Rubha Bhocaig
Loch Greosabhagh
Loch an Tairbeart
Loch Ceann Dibig
Sgeotasaigh
Carragraich
Caolas Scalpaigh
Carnach
Rudha Crago
Scalpay
Ceann a Bhàigh
Eilean Scalpaigh (Scalpay)
Stulaval
Tirga Mòr
Ullaval
Uisgnaval Mòre
Clisham
Loch Reasort
Loch Tealasabhaigh
Loch Tamanabhaigh
Loch Crabhadail
Loch a' Ghlinne
Bràighe Mòr
Caolas an Scarp
Bàgh Huisinis
Ensay
Killegray
Carminish Is.
Langay
Groay
Gilsay
Lingay
Scaravay
Sursay
Tahay
Opsay
Hermetray
Groatay
Tobha Beag
Bagha Chaise
Weaver's Pt.
Madadh Gruamach
Ceann Tarabhaigh
Airidh a Bhruaich
Aird an Troim
Aline Lodge
Seaforth I.
Aird a' Mhulaidh
Scaladal
Maraig
Eilean Anabaich
Loch Shiphoirt
Rheinigidale
Reinigeadal
Loch Trollamarig
Loch Claidh
Loch Bhrollum
Rubha a'Bhaird
Rubha Bhrollum
Gob Rubh'Uisenis
Mol Truisg
BEINN MHOR
CRIONAIG
Ceann Shiphoirt
Tabost
KERSHADER
Marbhig
Calbost
Grabhair
Taobh a' Ghlinne
Tom an Fhuadain
Orasaigh
Eisgean
Leumrabhagh
Loch Odhairn
Kebock Head
Loch Shell or Loch Sealg
Srianach
Eilean Iubhard
Garbh Eilean
Eilean Mhuire
Na h-Eileanan Mòra (Shiant Islands)
Eilean an Tighe
UIG 1:40
N. UIST 1:00
Fladda-chùain
Eilean Trodday
Rubha Hunish
Rubha na h-Aiseig
Shulista
DUNTULM CASTLE
Duntulm
Balmacqueen
Kilmaluag
Connista
Lub Score
MUSEUM OF ISLAND LIFE
Peingown
Hungladder
Borneskeitaig
Heribusta
Kilmuir
FLORA MACDONALD'S MEMORIAL
Kilvaxter
Balgown
Linicro
Totscore
MEALL NA SUIRAMACH
Digg
Glashvin
Staffin Bay
Staffin I.
Eilean Flodigarry
Flodigarry
THE QUIRAING
Brogaig
Garafad
Staffin
Stenscholl
Clachan
TROTTERNISH
Kilt Rock
Waternish Point
0 2 4 6 miles
0 2 4 6 8 10 km
296
298
307
NG

CAPE WRATH
Kearvaig
SGRIBHIS-BHEINN
Inshore
Geodha Ruadh na Fola
Bay of Keisgaig
Loch Keisgaig
Geodha Ruadh
308
FASHVEN
Am Balg
BEINN DEARG
Sandwood Loch
CREAG RIABHACH
Rubh'an Fhir Léithe
Loch na Gainimh
Strath Shinary
GHLAS BHEINN
Sheigra
Balchrick
Blairmore
Droman
Oldshore Beg
Eilean Roin Mor
Oldshoremore
FARRMHEALL
A838
Gualin Ho.
Kinlochbervie
Loch Clash
Badcall
B801
Inshegra
Achriesgill
Bagh Loch an Roin
Rhivichie
Loch Inchard
Achlyness
Loch Dughaill
Ceathramh Garbh
Ardmore Pt.
Rhiconich
L. na Claise Carnaich
Rubha Ruadh
Ardmore
Portlevorchy
Skerricha
GANU MOR
Foinaven
NORTH-WEST SUTHERLAND
Fanagmore
Tarbet
Loch Laxford
Handa Island
Foindle
Loch a'Garbh-bhaid Mór
Sound of Handa
Loch nam Brac
A894
Laxford Bridge
Badnabay
Loch an Easain Uaine
ARKLE
Scourie Bay
Scourie More
Scourie
Rubh'Aird an t-Sionnaich
Gorm Loch
Laxford
Lochstack Lodge
BEN STACK
Upper Badcall
Lower Badcall
Strath Stack
Airdachuilinn
Badcall Bay
BEINN AUSKAIRD
Eil. a'Bhreitheimh
LOCH A'MHUILINN
Duartbeg
Loch Crocach
REAY FOREST
Achfary
Lochmore Lodge
Rubha a'Mhucard
Meall Mór
Calbha Mór
Duartmore Bridge
Loch More
Loch na Creige Duibhe
Eddrachillis Bay
Calbha Beag
Duartmore Forest
Loch an Leathaid Bhuain
Point of Stoer
Oldany Island
Loch a'Chàirn Bhàin
Cirean Geardail
R. nan Còsan
Loch Nedd
Kylestrome
Glendhu Forest
Eilean Chrona
Culkein Drumbeg
Ardvar
Culkein
Kylesku
Loch Glendhu
Cluas Deas
Rubha Stoer
Clashnessie Bay
Oldany
B869
Unapool
BEINN AIRD DA LOCH
Achnacarnin
Drumbeg
Nedd
Loch Poll
Glenleraig
Newton
Loch Glencoul
Clashmore
Balchladich
Clashnessie
NB
NC

HIGHLAND
ASSYNT-
COIGACH
Loch Assynt
Inchnadamph
Skiag Bridge
Ardvreck Castle
Lochassynt Lodge
Little Assynt
Brackloch
Glencanisp Lodge
Lochinver
Baddidarach
Rhicarn
Clachtoll
Stoer
Rienachait
Achmelvich
Achmelvich Bay
Bay of Stoer
Loch Inver
Badnaban
Strathan
Inverkirkaig
Loch Kirkaig
Stronchrubie
Inverpolly Lodge
Enard Bay
Rhegreanoch
Brae of Achnahaird
Altandhu
Reiff
Polbain
Achiltibuie
Polglass
Badenscallie
Culnacraig
Achduart
Horse Sound
Summer Isles
Isle Ristol
Tanera Mór
Tanera Beg
Horse I.
Priest I.
Bottle I.
Ullapool
Loch Broom
Little Loch Broom
Isle Martin
Annat Bay
Ardmair
Rhue
Morefield
Braes of Ullapool
Leckmelm
Ardcharnich
Inverlael
Lael Forest
Braemore
Corrieshalloch Gorge
Strath More
Auchindrean
Fain
Dundonnell
Dundonnell House
Camusnagaul
Ardessie
Badcaul
Durnamuck
Scoraig
Badluarach
Gruinard House
Gruinard Bay
Gruinard I.
Laide
Mellon Udrigle
Achgarve
Opinan
Aultbea
Mellon Charles
Isle of Ewe
Loch Ewe
Poolewe
Inverewe Garden
Naast
Inverasdale
Cove
Melvaig
Aultgrishan
Peterburn
North Erradale
Big Sand
Gairloch
Strath
Smithstown
Longa Island
Rubha Reidh
Fionn Loch
Kernsary
Ledmore
Elphin
Knockan
Knockan Crag
Altnacealgach Hotel
Benmore Lodge
Drumrunie
Strathcanaird
Langwell Lodge
Cromalt Hills
Rhidorroch Forest
East Rhidorroch Lodge
Glen Achall
A835
A837
A832
A894
B869
B8021
B8057
299
304
305
309
0 2 4 6 miles
0 2 4 6 8 10 km

CAPE WRATH
Kearvaig
Geodha Ruadh na Fola
Bay of Keisgaig
Geodha Ruadh
Am Balg
Rubh'an Fhir Léithe
Sheigra
Balchrick
Blairmore
Oldshore Beg
Oldshoremore
Droman
Eilean Roin Mor
Sandwood Loch
Strath Shinary
Inshore
Achiemore
Balnakeil Bay
Faraid Head
Balnakeil
Durness
Sangomore
Keoldale
Kyle of Durness
DURNESS VISITOR CENTRE
SMOO CAVE
BALNAKEIL CHURCH
BALNAKEIL CRAFT VILLAGE
Leirinmore
Sangobeg
Sarsgrum
Rispond
Eilean Hoan
Eilean Clùimhrig
Geodh'a' Bhrideoin
Whiten Head
LOCH ERIBOLL
Portnancon
Eilean Choraidh
Heilam
Kempie
Eriboll
Polla
Strath Beag
Hope
Lochside
Loch Hope
A'Mhoine
Druim nan Cliar
Rubha Thormaid
BEN HUTIG
Strathan
Achininver
Lubinvullin
Portvasgo
Midfield
Achinahuagh
West Strathan
Talmine
Skinnet
Midtown
Port Vasgo
Rabbit Is.
Tongue Bay
Eilean nan Ron
Caol Raineach
Neave I. or Coombe I.
Torrisdale Bay
Skerray
Strathan Skerray
Clashedy
Achtoty
Airdtorrisdale
Torrisdale
Skullomie
Coldbackie
Blandy
Dallcharn
Rhitongue
Achuvoldrach
TONGUE
Kirkiboll
Tongue
CASTLE VARRICH
Ribigill
KYLE OF TONGUE
Kinloch Lodge
BEN HOPE
BEN LOYAL
Loch Loyal Lodge
CNOC NAN CULLEAN
Alltnacaillich
DUN DORNAIGIL BROCH
Strathmore
Inchkinloch
Borgie
Borgie Forest
Invernaver
Bettyhill
Achina
STRATHNAVER MUSEUM
Farr
Farr Pt.
Crask
Swordly
Kirtomy
Kirtomy Pt.
Ardmore Pt.
Armadale
Armadale Bay
Brawl
Aultiphurst
Strathy
Strathy Point
Totegan
Strath Bay
Lednagullin
Bowside Lodge
Strathy Forest
Leckfurin
ACHANLOCHY CLEARANCE VILLAGE
Skelpick
Achagary
STRATHNAVER
Carnachy
Rhifail
Rough Haugh
BEINN RIFA-GIL
Skail
Langdale
Syre
Dalvina Lo.
ROSAL CLEARANCE TRAIL
BEINN STUMANADH
POLE HILL
Grumbeg
Achadh an Eas
Naver Forest
Loch Naver
Altnaharra
Clebrig
Mudale
Garvault Hotel
BEN GRIAM MORE
BEN GRIAM BEG
Greamchar
Badanloch Forest
Badanloch Lodge
Loch Choire Lodge
BEN KLIBRECK
HIGHLAND
Gobernuisgach Lodge
BEN HEE
Glen Golly
FEINNE-BHEINN MOR
AN LEAN-CHARN
MEALL MEADHONACH
BEINN SPIONNAIDH
CRANSTACKIE
Rhigolter
Strath Dionard
GHLAS BHEINN
FARRMHEALL
Gualin Ho.
Kinlochbervie
Loch Clash
Badcall
Inshegra
Achriesgill
Loch Inchard
Rhivichie
Achlyness
Rhiconich
Loch Dughaill
Ardmore Pt.
Ardmore
Portlevorchy
Skerricha
NORTH-WEST SUTHERLAND
GANU MOR
Foinaven
ARKLE
Loch Dionard
Fanagmore
Tarbet
Foindle
Loch Laxford
Laxford Bridge
Badnabay
Lochstack Lodge
BEN STACK
Strath Stack
Airdachuilinn
Scourie
Badcall Upper
Lower Badcall
BEINN AUSKAIRD
Duartbeg
Calbha Mòr
REAY FOREST
Achfary
Lochmore Lodge
Loch More
Aultanrynie
Kinloch
West Merkland
Duartmore Bridge
Duartmore Forest
Kylestrome
Kylesku
Unapool
Newton
Loch Glendhu
Loch Glencoul
Ardvar
BEINN AIRD DA LOCH
A838
A836
B871
B873
A894
B869
B801

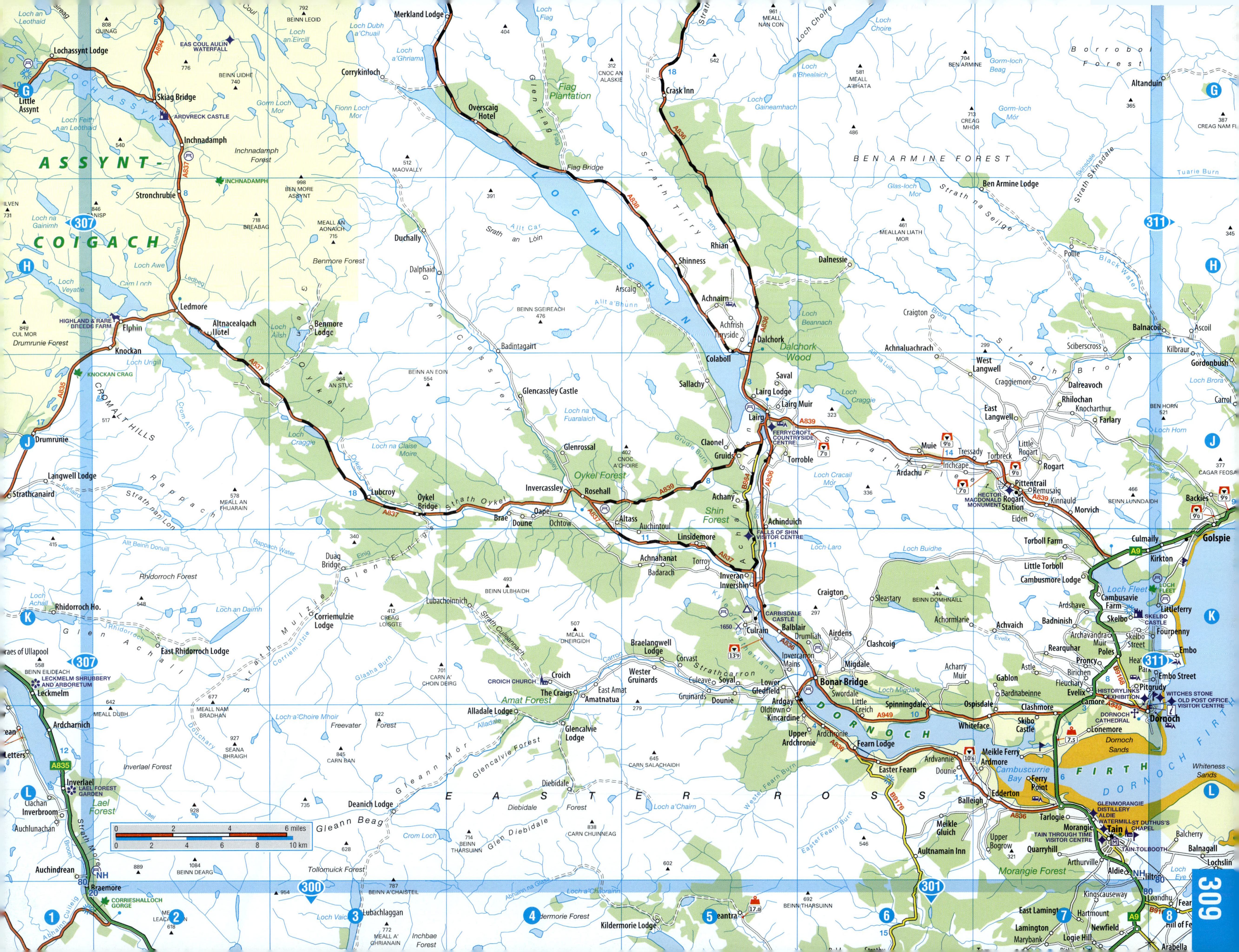

Golspie
Dornoch
Dornoch Firth
Tain
Bonar Bridge
Lairg
Loch Shin
Rogart
Edderton
Ardgay
Invershin
Oykel Bridge
Ledmore
Inchnadamph
Skiag Bridge
Ardvreck Castle
Elphin
Knockan
Overscaig Hotel
Crask Inn
Merkland Lodge
Altnacealgach Hotel
Rosehall
Kildermorie Lodge
Ben Armine Forest
Assynt-Coigach
Shin Forest
Oykel Forest
Amat Forest
Morangie Forest
Cambuscurrie Bay
Dornoch Sands
Skelbo Castle
Loch Fleet
Falls of Shin Visitor Centre
Carbisdale Castle
Croick Church
Glenmorangie Distillery
Knockan Crag
Leckmelm Shrubbery and Arboretum
Corrieshalloch Gorge
Ullapool
A9
A836
A837
A838
A839
A949
B864
B9168
B9176
A835
A894
311
301
300
307

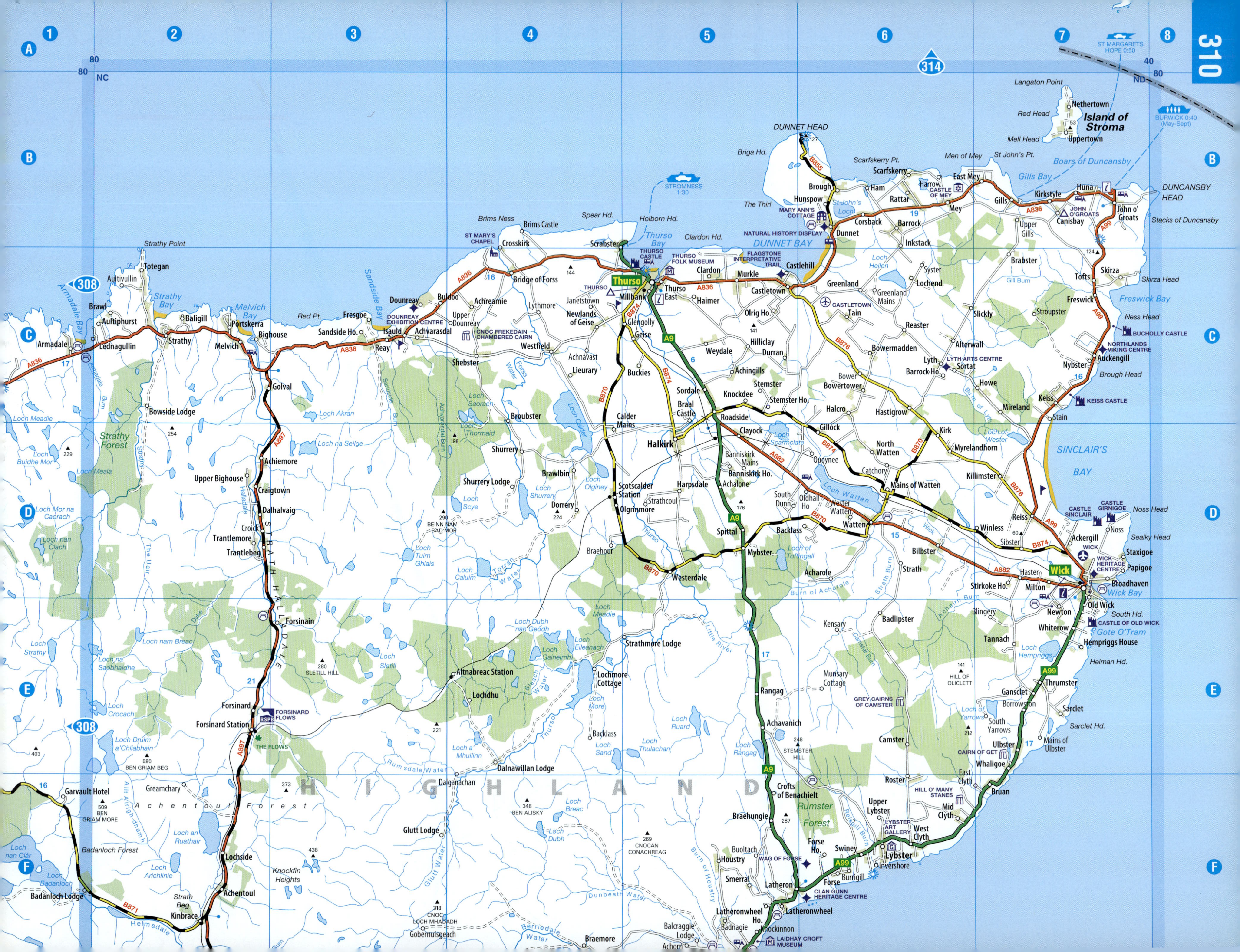

DUNCANSBY HEAD
Stacks of Duncansby
Island of Stroma
Nethertown
Uppertown
John o' Groats
Huna
Canisbay
Freswick Bay
Freswick
Skirza Head
Ness Head
BUCHOLLY CASTLE
NORTHLANDS VIKING CENTRE
Auckengill
Nybster
KEISS CASTLE
Keiss
SINCLAIR'S BAY
Noss Head
CASTLE GIRNIGOE
CASTLE SINCLAIR
Ackergill
WICK HERITAGE CENTRE
Wick
Wick Bay
Papigoe
Staxigoe
Broadhaven
Old Wick
CASTLE OF OLD WICK
Hempriggs House
Thrumster
Sarclet
Mains of Ulbster
Ulbster
Whaligoe
CAIRN OF GET
Bruan
HILL O' MANY STANES
Mid Clyth
East Clyth
West Clyth
Lybster
Swiney
Forse
CLAN GUNN HERITAGE CENTRE
Latheronwheel
Latheron
LAIDHAY CROFT MUSEUM
WAG OF FORSE
Rumster Forest
GREY CAIRNS OF CAMSTER
Camster
Watten
Loch Watten
Mains of Watten
Bilbster
Killimster
Reiss
Gills Bay
Mey
CASTLE OF MEY
East Mey
Barrock
Lyth
Lochend
Dunnet
DUNNET HEAD
DUNNET BAY
Brough
MARY ANN'S COTTAGE
Castlehill
Castletown
CASTLETOWN
Murkle
Thurso
THURSO CASTLE
THURSO FOLK MUSEUM
Thurso Bay
Scrabster
Halkirk
Spittal
Mybster
Westerdale
Strathmore Lodge
Altnabreac Station
Dalnawillan Lodge
Loch Calder
Westfield
Bridge of Forss
Crosskirk
ST MARY'S CHAPEL
Dounreay
DOUNREAY EXHIBITION CENTRE
Reay
Sandside Bay
Fresgoe
Melvich Bay
Melvich
Portskerra
Bighouse
Strathy Point
Strathy
Strathy Bay
Strathy Forest
Armadale
Armadale Bay
Forsinard
Forsinard Station
FORSINARD FLOWS
STRATH HALLADALE
Kinbrace
Achentoul
Badanloch Lodge
Garvault Hotel
Achentoul Forest
H I G H L A N D
A9
A99
A836
A882
A897
B870
B874
B876
B855
B871
308
314

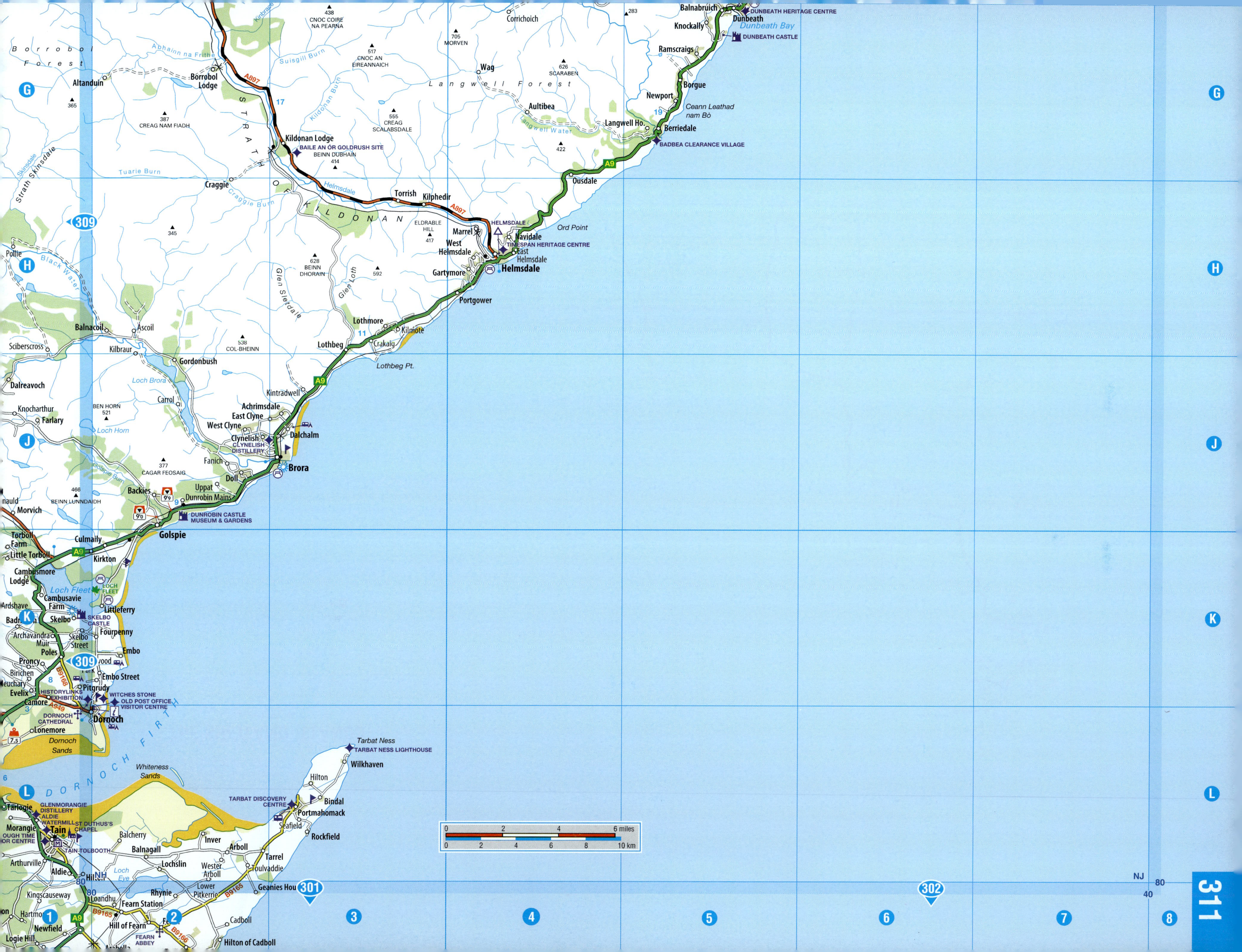

Dunbeath
DUNBEATH HERITAGE CENTRE
DUNBEATH CASTLE
Dunbeath Bay
Balnabruich
Knockally
Ramscraigs
Borgue
Newport
Berriedale
BADBEA CLEARANCE VILLAGE
Ceann Leathad nam Bò
Langwell Ho.
Ousdale
Ord Point
Navidale
TIMESPAN HERITAGE CENTRE
HELMSDALE
East Helmsdale
Helmsdale
West Helmsdale
Marrel
Gartymore
Portgower
Kilphedir
Torrish
Kildonan Lodge
BAILE AN ÒR GOLDRUSH SITE
Borrobol Lodge
Craggie
Altanduin
Langwell Forest
Langwell Water
Corrichoich
Wag
Aultibea
MORVEN
SCARABEN
CNOC AN EIREANNAICH
CNOC COIRE NA FEARNA
CREAG SCALABSDALE
BEINN DUBHAIN
ELDRABLE HILL
BEINN DHORAIN
CREAG NAM FIADH
COL-BHEINN
STRATH OF KILDONAN
Kildonan Burn
Suisgill Burn
Abhainn na Frithe
Craggie Burn
Tuarie Burn
Glen Loth
Glen Sletdale
Strath Skinsdale
Black Water
Borrobol Forest
Kilmote
Crakaig
Lothbeg Pt.
Lothmore
Lothbeg
Kintradwell
Dalchalm
Brora
Achrimsdale
East Clyne
West Clyne
Clynelish
CLYNELISH DISTILLERY
Doll
Fanich
Uppat
Dunrobin Mains
DUNROBIN CASTLE MUSEUM & GARDENS
Golspie
Gordonbush
Carrol
Loch Brora
CAGAR FEOSAIG
BEN HORN
Loch Horn
Backies
Ascoil
Kilbraur
Balnacoil
Sciberscross
Dalreavoch
Knocharthur
Farlary
BEINN LUNNDAIDH
Morvich
Culmaily
Kirkton
Torboll Farm
Little Torboll
Cambusmore Lodge
Cambusavie Farm
Loch Fleet
LOCH FLEET
Littleferry
SKELBO CASTLE
Skelbo
Skelbo Street
Fourpenny
Embo
Embo Street
Muir
Poles
Proncy
Birichen
Evelix
Pitgrudy
Dornoch
DORNOCH CATHEDRAL
WITCHES STONE
OLD POST OFFICE VISITOR CENTRE
HISTORYLINKS EXHIBITION
Lonemore
Dornoch Sands
DORNOCH FIRTH
Whiteness Sands
Tarbat Ness
TARBAT NESS LIGHTHOUSE
Wilkhaven
Hilton
Bindal
Portmahomack
Rockfield
TARBAT DISCOVERY CENTRE
Seafield
Tarrel
Toulvaddie
Arboll
Inver
Cadboll
Hilton of Cadboll
Wester Arboll
Lower Pitkerrie
Balnagall
Lochslin
Balcherry
Rhynie
Fearn Station
Hill of Fearn
FEARN ABBEY
Loandhu
Tain
TAIN TOLBOOTH
GLENMORANGIE DISTILLERY
ALDIE WATERMILL
ST DUTHUS'S CHAPEL
Tarlogie
Morangie
Aldie
Arthurville
Kingscauseway
Newfield
Logie Hill
Loch Eye
A9
A897
B9165
B9166
B9168
A949
301
302
309
NJ
0 2 4 6 miles
0 2 4 6 8 10 km

THE SHETLAND ISLANDS
Scale 1:250,000
Muckle Flugga
Herma Ness
The Noup
Burra Firth
Norwick
Haroldswick
Baltasound
Balta
Huney
UNST
Uyeasound
Bluemull Sound
Gutcher
Cullivoe
Gloup
Uyea
Linga
Mid Yell
Hascosay
COLGRAVE SOUND
FETLAR
Funzie
Burravoe
YELL SOUND
Samphrey
Lunna Holm
Housay
Bruray
Out Skerries
WHALSAY
Sullom Voe Oil Terminal
Brae
Mossbank
Hillswick
ESHA NESS
Ronas Voe
North Roe
Uyea
Ramna Stacks
ST. MAGNUS BAY
SHETLAND
Voe
Vidlin
Lunna

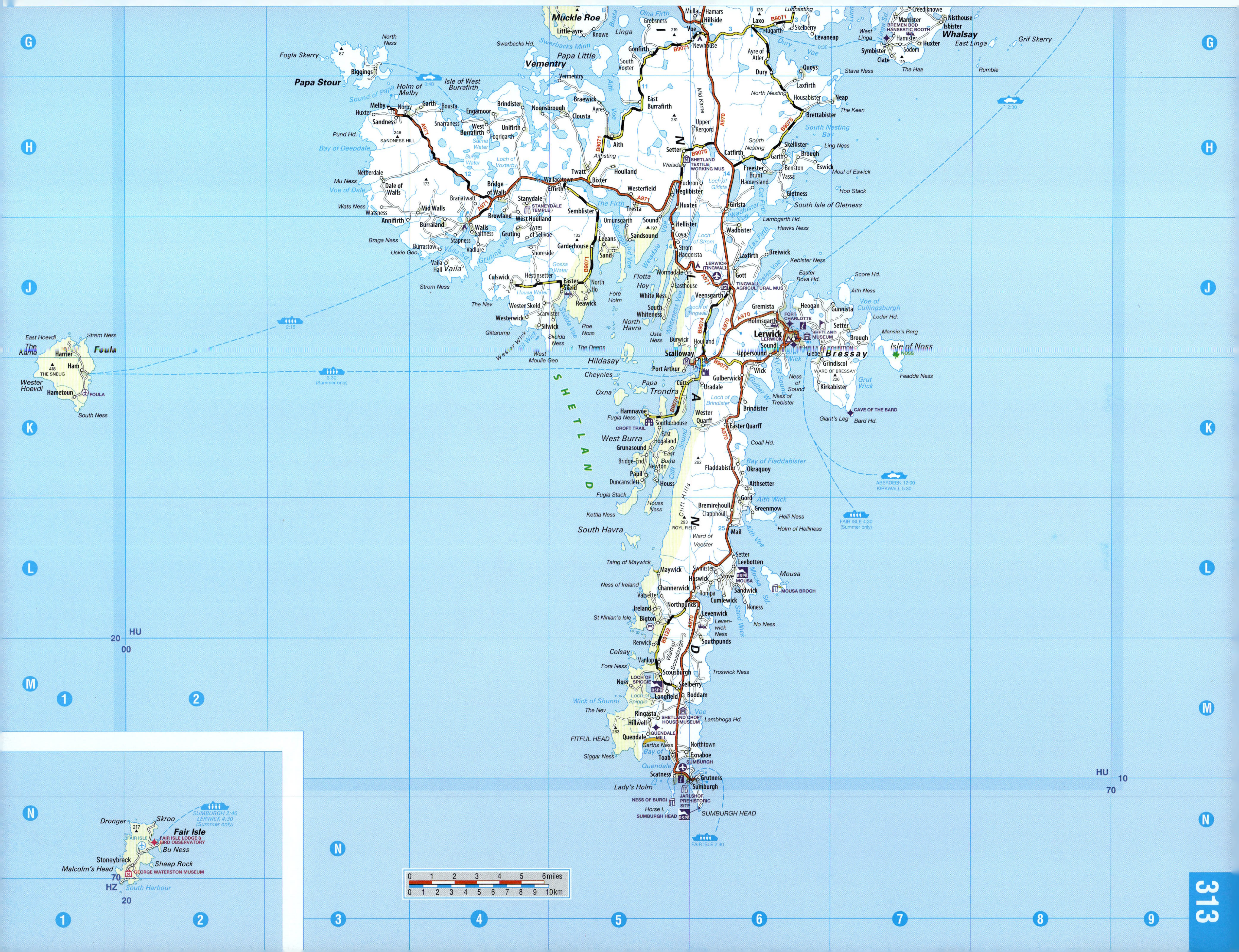
SHETLAND
Lerwick
Bressay
Isle of Noss
Scalloway
Whalsay
Papa Stour
Vementry
Muckle Roe
Papa Little
Trondra
West Burra
South Havra
Mousa
Sumburgh
SUMBURGH HEAD
FITFUL HEAD
Foula
Fair Isle
Vaila
Hildasay
Oxna
Papa
0 1 2 3 4 5 6miles
0 1 2 3 4 5 6 7 8 9 10km

THE ORKNEY ISLANDS
Scale 1:250,000
North Ronaldsay
Papa Westray
Holm of Papa
Westray
Rousay
Egilsay
Wyre
Gairsay
Eday
Sanday
Stronsay
Shapinsay
Auskerry
Copinsay
Mainland
Kirkwall
Stromness
HOY AND
WEST MAINLAND
Hoy
Flotta
Burray
South Ronaldsay
Swona
Island of Stroma
NORTH RONALDSAY FIRTH
THE NORTH SOUND
WESTRAY FIRTH
SANDAY SOUND
STRONSAY FIRTH
SCAPA FLOW
PENTLAND FIRTH
Eynhallow
Faray
Muckle Green Holm
Papa Stronsay
Linga Holm
Cava
Fara
Rysa Little
Calf of Flotta
Graemsay
Switha
Lamb Holm
Glims Holm
Hunda
Muckle Skerry
Pentland Skerries
DUNNET HEAD
DUNCANSBY HEAD
Finstown
St Margaret's Hope
St Mary's
Longhope
Dounby
Lyness
Birsay
Orphir
Deerness
Tankerness
Holland
Balfour
Pierowall
Kettletoft
Whitehall Village
Burwick
Skara Brae
Maes Howe
Ring of Brodgar
Stenness Standing Stones
Old Man of Hoy
Scrabster 1:30
Aberdeen 7:15
Lerwick 7:45
Kirkwall 1:50
Muckle Water
Loch of Harray
Loch of Stenness
Loch of Swannay
Loch of Boardhouse
Ward Hill
Brough Head
Marwick Head
Noup Head
Mull Hd.
Rora Head
Tor Ness
Brough Ness
Cantick Hd.
Halcro Hd.
Old Hd.
1 2 3 4 5 6 7
A B C D E F G H
HY
ND
20
50
70
80
0 1 2 3 4 5 6 miles
0 1 2 3 4 5 6 7 8 9 10 km
310

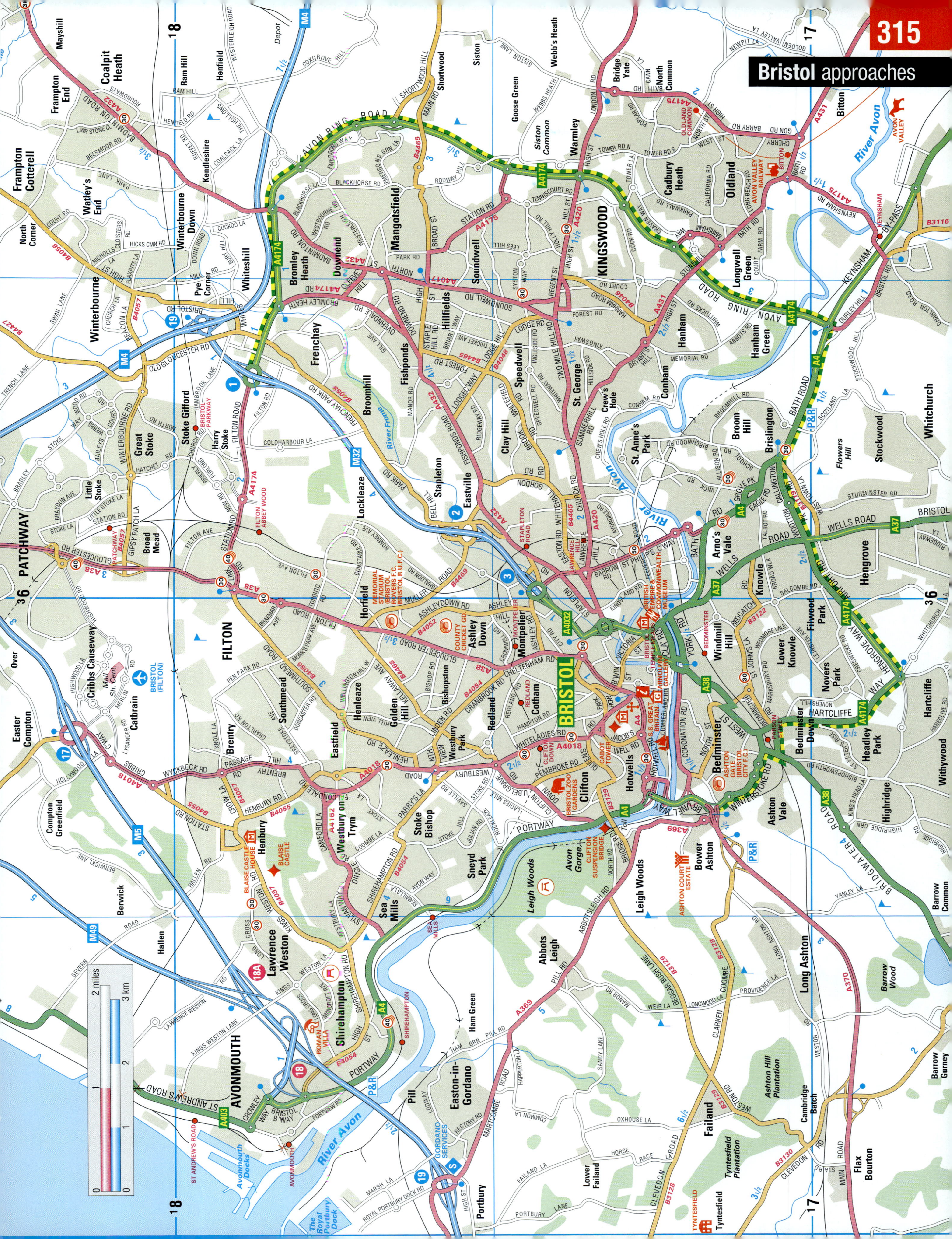
BRISTOL
AVONMOUTH
PATCHWAY
FILTON
KINGSWOOD
Mayshill
Frampton End
Coalpit Heath
Frampton Cotterell
Watley's End
North Corner
Winterbourne
Winterbourne Down
Kendleshire
Henfield
Ram Hill
Pye Corner
Whiteshill
Bromley Heath
Downend
Mangotsfield
Soundwell
Siston
Siston Common
Goose Green
Webb's Heath
Warmley
Bridge Yate
North Common
Oldland Common
Cadbury Heath
Oldland
Bitton
Longwell Green
Hanham
Hanham Green
Conham
Frenchay
Fishponds
Hillfields
Speedwell
St. George
Crew's Hole
Broomhill
Stoke Gifford
Harry Stoke
Great Stoke
Little Stoke
Broad Mead
Stapleton
Eastville
Clay Hill
Lockleaze
St. Anne's Park
Broom Hill
Brislington
Flowers Hill
Stockwood
Whitchurch
Horfield
Ashley Down
Montpelier
Arno's Vale
Knowle
Hengrove
Filwood Park
Lower Knowle
Windmill Hill
Novers Park
Hartcliffe
Headley Park
Withywood
Highridge
Bedminster
Bedminster Down
Ashton Vale
Bower Ashton
Hotwells
Clifton
Cotham
Redland
Bishopston
Golden Hill
Henleaze
Westbury Park
Eastfield
Southmead
Brentry
Henbury
Westbury on Trym
Stoke Bishop
Sneyd Park
Sea Mills
Shirehampton
Lawrence Weston
Cribbs Causeway
Catbrain
Easter Compton
Over
Compton Greenfield
Berwick
Hallen
Leigh Woods
Avon Gorge
Abbots Leigh
Ham Green
Pill
Easton-in-Gordano
Portbury
Long Ashton
Failand
Lower Failand
Tyntesfield
Cambridge Batch
Flax Bourton
Barrow Common
Barrow Wood
Barrow Gurney
Ashton Hill Plantation
Tyntesfield Plantation
Avonmouth Docks
The Royal Portbury Dock
River Avon
River Frome
Avon Ring Road
Gordano Services
Bristol (Filton)
Bristol Parkway
Blaise Castle
Clifton Suspension Bridge
Ashton Court Estate
Memorial Stadium
County Cricket Gd.
Ashton Gate
Avon Valley Railway
M4
M5
M32
M49
A4
A38
A37
A4174
A369
A370
A403
A4018
A432
A420
A431
2 miles
3 km

WOLVERHAMPTON
WALSALL
WILLENHALL
Bentley
BILSTON
DARLASTON
WEDNESBURY
COSELEY
SEDGLEY
TIPTON
Upper Gornal
Gornalwood
DUDLEY
The Black Country
WEST BROMWICH
Lyndon
NEWTON
Sandwell
Oakham
OLDBURY
Pensnett
Netherton
ROWLEY REGIS
Langley Green
SMETHWICK
Wordsley
BRIERLEY HILL
Old Hill
Blackheath
AMBLECOTE
Hurst Green
Lye
Cradley
World's End
STOURBRIDGE
HALESOWEN
Pedmore
Woodgate
Weoley Castle
Wychbury Hill
Hagley Wood
Uffmoor Wood
Twiland Wood
Hagley
Hagley Park
Shenley Fields
Bartley Resr
Himley Park
HIMLEY HALL
BAGGERIDGE COUNTRY PARK
Nature Centre
Merry Hill Centre
BROADFIELD HOUSE GLASS MUSEUM
BLACK COUNTRY MUSEUM
DUDLEY ZOO & CASTLE
SANDWELL VALLEY COUNTRY PARK
WOODGATE VALLEY COUNTRY PARK
HALESOWEN ABBEY (REMS)
FRANKLEY SERVICES
BIRCHILLS CANAL MUSEUM
ARBORETUM
BESCOT STADIUM
THE HAWTHORNS
Warley Woods
Tame Valley Canal
Dudley Canal
M5
M6

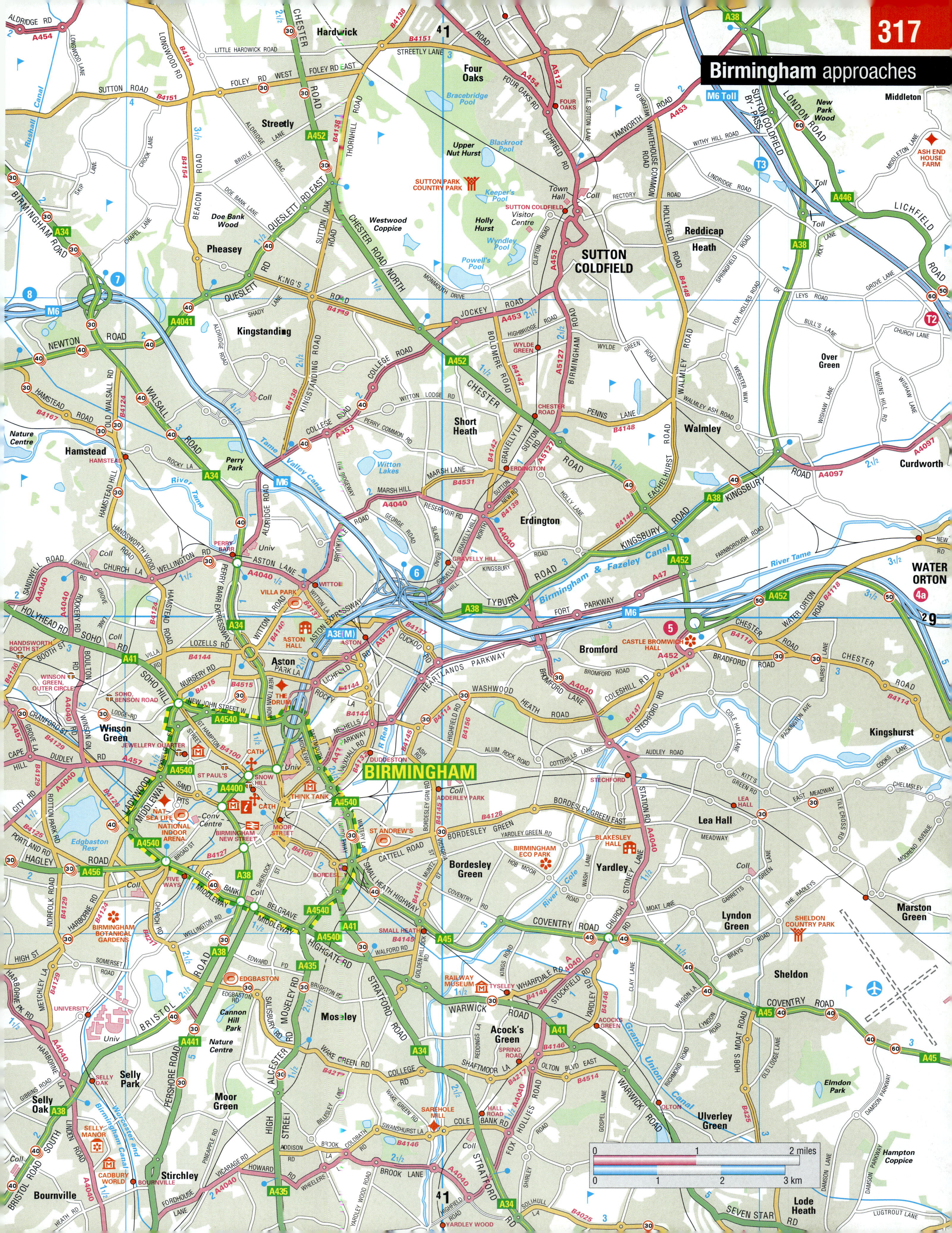
Hardwick
Four Oaks
Bracebridge Pool
Blackroot Pool
Upper Nut Hurst
SUTTON PARK COUNTRY PARK
Keeper's Pool
Holly Hurst
Wyndley Pool
Powell's Pool
Westwood Coppice
Streetly
Doe Bank Wood
Pheasey
SUTTON COLDFIELD
Town Hall
Reddicap Heath
New Park Wood
Middleton
ASH END HOUSE FARM
Kingstanding
Over Green
Short Heath
Witton Lakes
Erdington
Walmley
Curdworth
Hamstead
Nature Centre
Perry Park
River Tame
Tame Valley Canal
WATER ORTON
River Tame
Birmingham & Fazeley Canal
Bromford
CASTLE BROMWICH HALL
Aston
ASTON HALL
VILLA PARK
Winson Green
JEWELLERY QUARTER
BIRMINGHAM
Edgbaston Resr
NAT SEA LIFE
NATIONAL INDOOR ARENA
Conv Centre
BIRMINGHAM NEW STREET
ST ANDREW'S
Bordesley Green
BIRMINGHAM ECO PARK
BLAKESLEY HALL
Yardley
Lea Hall
Kingshurst
Lyndon Green
SHELDON COUNTRY PARK
Marston Green
Sheldon
BIRMINGHAM BOTANICAL GARDENS
EDGBASTON
Cannon Hill Park
Nature Centre
Moseley
RAILWAY MUSEUM
Acock's Green
River Cole
Grand Union Canal
Elmdon Park
Ulverley Green
Selly Park
Selly Oak
UNIVERSITY
Moor Green
SAREHOLE MILL
SELLY MANOR
CADBURY WORLD
Stirchley
Bournville
Worcester and Birmingham Canal
Hampton Coppice
Lode Heath
0 1 2 miles
0 1 2 3 km

CARDIFF
(CAERDYDD)
CAERPHILLY
(CAERFFILI)
Garth
Park Wood
Michaelston-y-Fedw
Castleton
Marshfield
Blacktown
Peterstone Wentlooge
(Llanbedr Gwynllŵg)
St. Mellons
(Llaneirwg)
Trowbridge
Newton
Rumney
(Rhymni)
Pwll-Mawr
Llanrumney
Llanedeyrn
Began
Pentwyn
Tremorfa
Splott
Roath
Roath Park
Cyncoed
Lisvane
(Llys Faen)
Llanishen
Thornhill
Heath
Cathays
Maindy
Gabalfa
Birchgrove
Rhiwbina
Whitchurch
Llandaff North
Llandaff
(Llandaf)
Pontcanna
Canton
Riverside
Grangetown
Butetown
Saltmead
Leckwith
Llandough
Michaelston-le-Pit
Fairwater
Ely
(Tre-lai)
Caerau
St Fagans
(Sain Ffagan)
Michaelston-super-Ely
Culverhouse Cross
Wenvoe
(Gwenfô)
Wallston
Twyn-yr-odyn
Downs
Drope
St Nicholas
(Sain Nicolas)
St Lythans
St Brides-super-Ely
(Llansanffraid-ar-Elai)
St George's
Radyr
Morganstown
Pentyrch
Garth Wood
Taff's Well
(Ffynnon-Taf)
Tongwynlais
Glan-y-llyn
Nantgarw
Groes-wen
Gwaelod-y-garth
Coed y Gedrys
Church Village
Soar
Watford
Caerphilly Common
The Warren
Van
Rudry
Berllangollen
Llwyncelyn
Coed Coesau-Whips
Coed Parc-y-Van
Coed Craig Ruperra
Cefn Mably Woods
Coed-y-Wenallt
Pantmawr
Taff vale
Grath Hill
Llanishen Reservoir
The Lake
Roath Dock
Bute E Dock
Queen Alexandra Dock
Cardiff Bay Barrage
River Taff
River Ely
Rhymney River
WALES MILLENNIUM CENTRE
NATIONAL ASSEMBLY
CARDIFF BAY VISITOR CENTRE
TECHNIQUEST
MILLENNIUM STADIUM
CARDIFF CASTLE
CARDIFF ARMS PARK
NATIONAL MUSEUM CARDIFF
LLANDAFF CATHEDRAL
CASTELL COCH
FOREST FARM
PARC CEFN ONN
ST FAGANS NATIONAL HISTORY MUSEUM
TINKINSWOOD BURIAL CHAMBER
CARDIFF GATE SERVICES
CARDIFF WEST SERVICES
SOPHIA GDNS
M4
A48(M)
A48
A469
A470
A4232
A4119
A4161
A4160
A4055
A4054
A4050
A4234
B4562
B4487
B4239
B4263
B4262
B4267
B4488
P&R
0
1
2
2 miles
3 km

EDINBURGH
MUSSELBURGH
Inveresk
Fisherrow Sands
Portobello
Joppa
Leith
Newhaven
Granton
Wardie
Craigleith
Drylaw
Blackhall
Murrayfield
Corstorphine
Clermiston
Cramond
Cramond Island
Drum Sands
Cammo
Cramond Bridge
Craigiehall
South Gyle
Hermiston
CURRIE
Baberton
Colinton
Bonaly
Longstone
Craiglockhart
Merchiston
Churchill
Nether Liberton
Liberton
Braid Hills
Alnwickhill
Comiston
Torphin
Pentland Hills
Prestonfield
The Inch
Moredun
Gilmerton
Danderhall
Niddrie
Bingham
Northfield
Craigentinny
Holyrood Park
Arthur's Seat
Stoneybank
Old Craighall
Easter Millerhill
DALKEITH
NEWMILLS
Easthouses
BONNYRIGG
Eskgrove
LOANHEAD
Muirhouse
Middle Craigs
Eastern Craigs
Granton Harbour
Port of Leith
EDINBURGH AIRPORT
THE CITY OF EDINBURGH BY-PASS
Water of Leith
River Esk
Duddingston Loch
Dunsapie Loch
Torduff Resr
Clubbiedean Resr
Bonaly Resr
Harlaw Resr

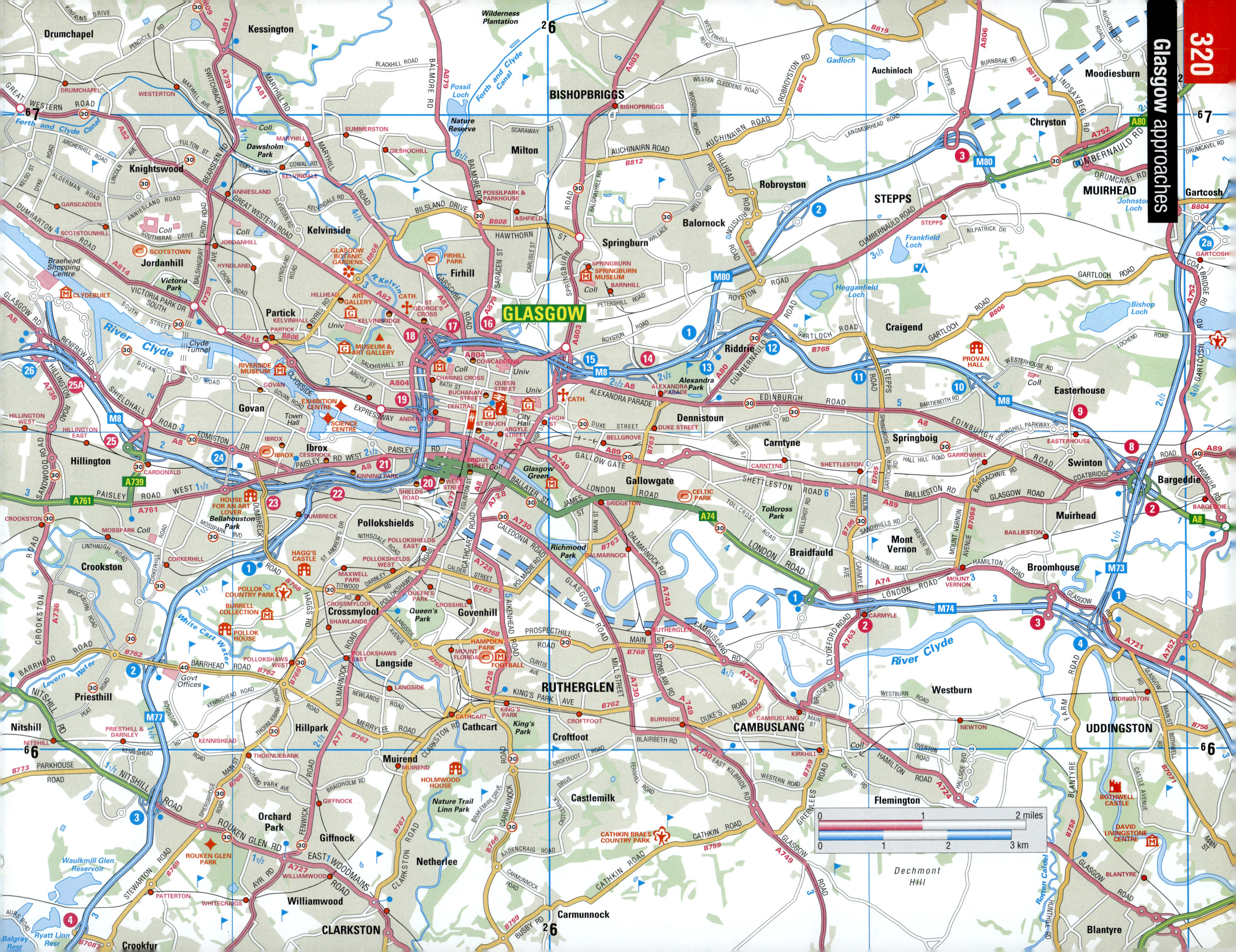
GLASGOW
BISHOPBRIGGS
STEPPS
MUIRHEAD
UDDINGSTON
CAMBUSLANG
RUTHERGLEN
CLARKSTON
Drumchapel
Kessington
Knightswood
Jordanhill
Kelvinside
Partick
Govan
Ibrox
Hillington
Crookston
Priesthill
Nitshill
Pollokshields
Crossmyloof
Langside
Govanhill
Hillpark
Cathcart
Muirend
Orchard Park
Giffnock
Netherlee
Williamwood
Crookfur
Carmunnock
Castlemilk
Croftfoot
Flemington
Westburn
Blantyre
Milton
Firhill
Springburn
Balornock
Robroyston
Auchinloch
Moodiesburn
Chryston
Gartcosh
Craigend
Riddrie
Dennistoun
Carntyne
Springboig
Easterhouse
Swinton
Bargeddie
Muirhead
Broomhouse
Mont Vernon
Braidfauld
Gallowgate
Wilderness Plantation
Dechmont Hill
River Clyde
White Cart Water
Forth and Clyde Canal
2 miles
3 km

LEEDS
Adel
Govt Bldgs
Shadwell
Saw Wood
Arthursdale
Scholes
Univ
HORSFORTH
West End
Cragg Wood
Calverley Wood
Lodge Wood
Horsforth
Cragg Hill
Hawksworth
Ireland Wood
West Park
Meanwood
Park Villas
Tropical World
Roundhay Park
Lady Wood
Monkswood
Oakwood
Brooklands
Swarcliffe
Parklands
Manston
Austhorpe
Cross Gates
Whitekirk
Harehills
Met Univ
Beckett Park
Headingley
Headingley Carnegie
Kirkstall Abbey
Hyde Park
Burley Park
Burley
CALVERLEY
Farsley
Bramley
Hill Top
New Pudsey
Leeds Industrial Museum
University
Civic Hall
Town Hall
Cath
City Art Gall
Coll
Bank
Royal Armouries
Osmondthorpe
East End Park
Halton Moor
Barrowby Park
Swillington Common
Colton
Home Farm
Temple Newsam House
Temple Newsam Country Park
Dawson's Wood
Avenue Wood
PUDSEY
Park Spring
Armley
Upper Wortley
Farnley
Holbeck
Hunslet
Knowsthorpe
Thwaitemills Industrial Museum
River Aire
Aire & Calder Navigation
Newsam Green
Swillington
Little Preston
Fulneck
Black Carr
Park Wood
Tong
New Farnley
Cottingley
Elland Road
Beeston
Cross Flatts Park
Moor Road
South Leeds Stadium
Middleton Railway
Park Halt
Middleton Park
Country Park
Rothwell Haigh
Woodlesford
Cockpit Round
Gildersome
Drighlington
BIRKENSHAW
Churwell
White Rose Centre
Troy Hill
Morley
Town Hall
Bruntcliffe
MORLEY
Topcliffe
Tingley Common
Middleton
ROTHWELL
Oulton
Robin Hood
Carlton
Almshouses Wood
Oakwell Hall Country Park
Birstall
Copley Hill
Sports Centre
Red House
Bagshaw Museum
Thorpe on the Hill
Oulzewell Green
Moss Carr Wood
The Fall
P&R
Moortown Ring Road
Meanwood Ring Rd
Wetwood Ring Rd
Horsforth Ring Rd
Horsforth New Rd
Ring Road Farsley
Stanningley By-Pass
Bramley Ring Rd
Farnley Ring Rd
Beeston Ring Road
Seacroft Ring Road
Wetherby Road
York Road
Selby Road
Pontefract Lane
Leeds and Bradford Road
Stanningley Road
Otley Road
Headingley Lane
Scott Hall Road
Easterly Road
Roundhay Road
Dewsbury Road
Wakefield Road
Gelderd Road
Whitehall Road
Bradford Road
Leeds Road
Aberford Road
Methley Lane
Newmarket Lane
Castle Gate
Middleton Lane
Howden Clough Road
Victoria Rd
Elland Road
A6120
A660
A65
A657
A647
A6110
A58
A58(M)
A64(M)
A64
A61
A63
A643
A62
A653
A639
A642
A654
A650
A652
A6029
A6120
M1
M62
M621
B6157
B6159
B6154
B6155
B6156
B6126
B6135
B6123
B6481
Wellington Hill
43
1 miles 2 miles
1 2 3 km

BARNET
EAST BARNET
Whetstone
North Finchley
Colney Hatch
FINCHLEY
Muswell Hill
East Finchley
Highgate
Hampstead
HAMPSTEAD
Gospel Oak
Belsize Park
Swiss Cottage
Primrose Hill
Regents Park
MARYLEBONE
St. John's Wood
Maida Vale
Kilburn
West Hampstead
Hampstead Garden Suburb
Golders Green
Temple Fortune
Church End
Totteridge
Arkley
Mill Hill
Mill Hill East
HENDON
West Hendon
Brent Cross
Cricklewood
Dollis Hill
Gladstone Park
WILLESDEN
Brondesbury
Kensal Rise
Harlesden
Neasden
Stonebridge
Old Oak Common
Wormwood Scrubs Park
EDGWARE
Burnt Oak
Colindale
Grahame Park
Kingsbury
Queensbury
Edgware Bury
Elstree
Little Stanmore
Canons Park
STANMORE
KENTON
Preston Road
North Wembley
WEMBLEY
Wembley Park
Alperton
Park Royal
West Acton
Hanger Lane
EALING
Perivale
Sudbury
Northwick Park
Belmont
Wealdstone
Greenhill
HARROW
Harrow on the Hill
Headstone
Harrow Weald
Stanmore Common
Bushey Heath
BUSHEY
OXHEY
South Oxhey
Oxhey Woods
Hatch End
Headstone Lane
PINNER
North Harrow
West Harrow
South Harrow
Roxeth
Rayners Lane
Eastcote
Eastcote Village
South Ruislip
Ruislip Manor
NORTHOLT
GREENFORD
Yeading
Brockley Hill
Barnet Gate
Stirling Corner
Underhill
Hilfield Park Reservoir
Brent Reservoir (Welsh Harp)
2 miles
3 km

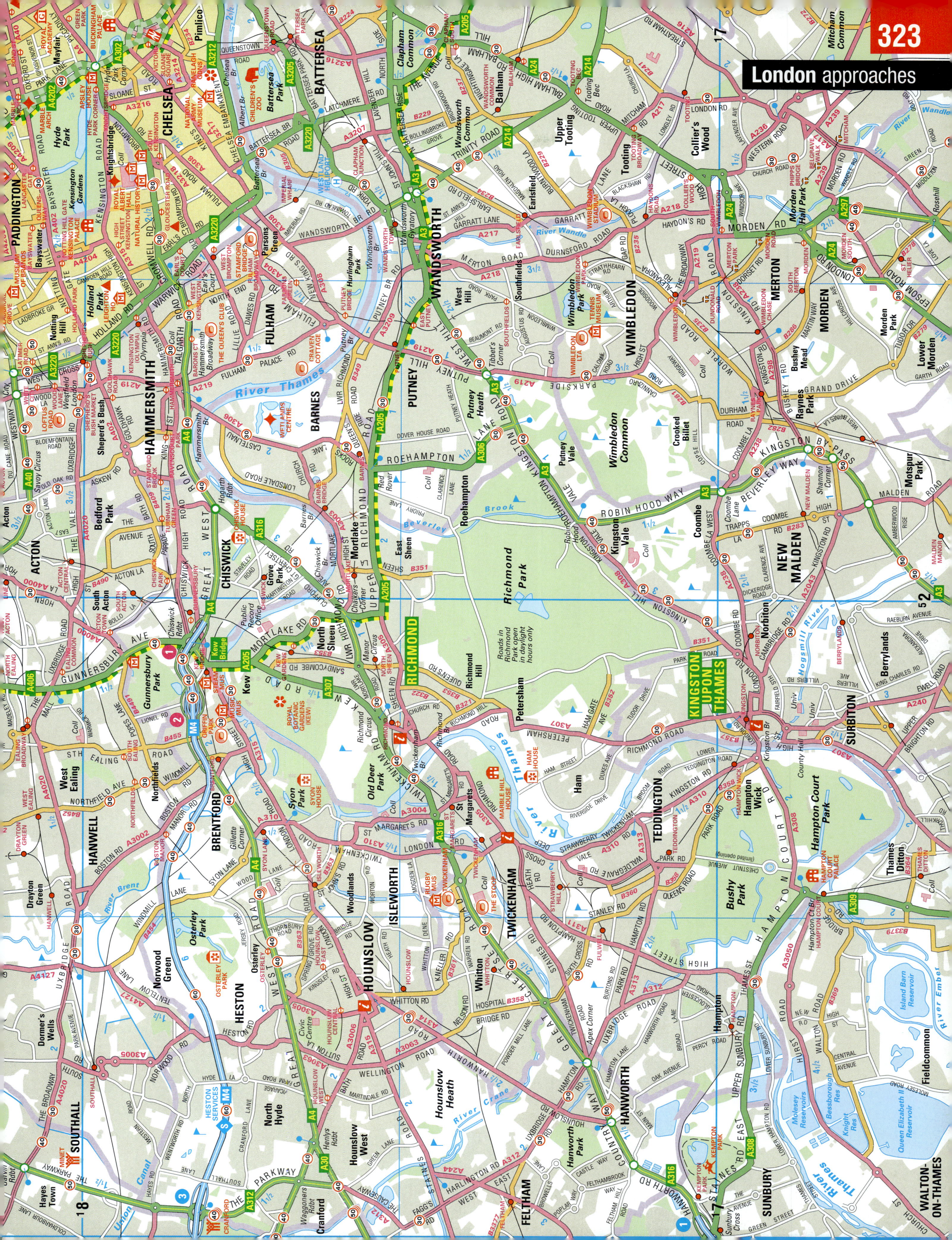
PADDINGTON
Bayswater
Notting Hill
Holland Park
Kensington Gardens
Hyde Park
Knightsbridge
Mayfair
Pimlico
CHELSEA
BATTERSEA
Battersea Park
Parsons Green
FULHAM
HAMMERSMITH
Shepherd's Bush
ACTON
Bedford Park
CHISWICK
BARNES
PUTNEY
WANDSWORTH
Wandsworth Common
Clapham Common
Balham
Upper Tooting
Tooting
Earlsfield
Southfields
West Hill
Wimbledon Park
WIMBLEDON
Wimbledon Common
Collier's Wood
MERTON
MORDEN
Morden Park
Lower Morden
Bushey Mead
Raynes Park
Crooked Billet
Putney Vale
Putney Heath
Roehampton
Kingston Vale
Coombe
NEW MALDEN
Motspur Park
Berrylands
SURBITON
KINGSTON UPON THAMES
Norbiton
Richmond Park
RICHMOND
Richmond Hill
Petersham
Ham
TEDDINGTON
Hampton Wick
Hampton Court Park
Bushy Park
Hampton
Thames Ditton
Mortlake
East Sheen
North Sheen
Kew
Gunnersbury Park
BRENTFORD
Syon Park
Old Deer Park
Isleworth
ISLEWORTH
St Margarets
TWICKENHAM
Whitton
HOUNSLOW
Hounslow Heath
Hounslow West
HESTON
Osterley Park
Norwood Green
HANWELL
West Ealing
Drayton Green
SOUTHALL
Dormer's Wells
North Hyde
Cranford
Hayes Town
FELTHAM
Hanworth Park
HANWORTH
SUNBURY
WALTON-ON-THAMES
Fieldcommon
River Thames
River Wandle
River Brent
River Crane
Hogsmill River
Beverley Brook
Queen Elizabeth II Reservoir
Island Barn Reservoir
Molesey Reservoirs
Bessborough Res
Knight Res
Roads in Richmond Park open in daylight hours only

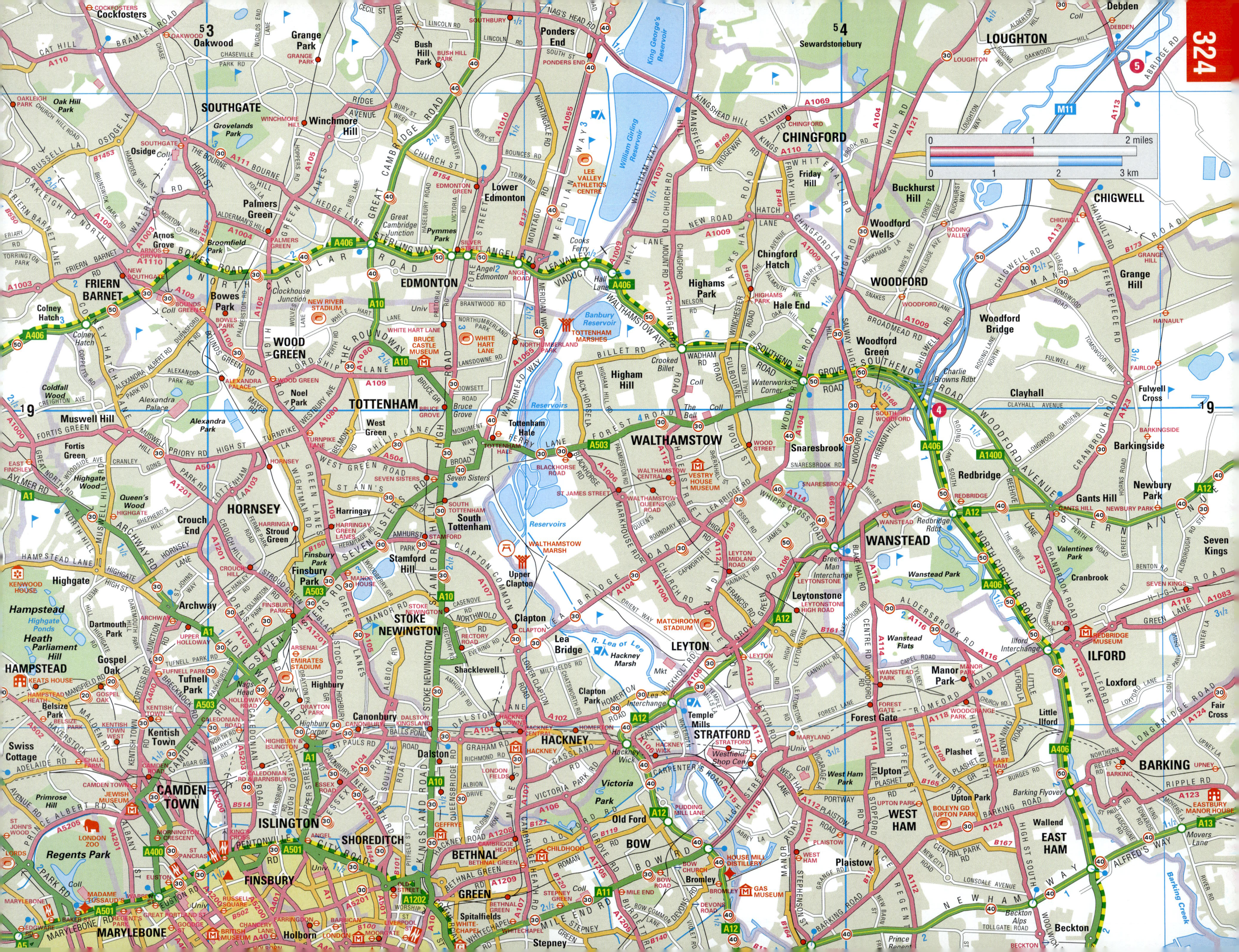
LOUGHTON
Debden
CHIGWELL
Grange Hill
Buckhurst Hill
Woodford Wells
WOODFORD
Woodford Bridge
Woodford Green
CHINGFORD
Chingford Hatch
Highams Park
Hale End
Friday Hill
Sewardstonebury
WALTHAMSTOW
Higham Hill
Clayhall
Barkingside
Redbridge
Gants Hill
Newbury Park
Seven Kings
ILFORD
Cranbrook
Valentines Park
Loxford
BARKING
WANSTEAD
Wanstead Park
Wanstead Flats
Manor Park
Forest Gate
Little Ilford
EAST HAM
Wallend
Beckton
Upton Park
WEST HAM
Plaistow
Upton
Leytonstone
LEYTON
STRATFORD
Temple Mills
Bromley
BOW
Old Ford
Victoria Park
HACKNEY
Hackney Marsh
Clapton Park
Lea Bridge
Clapton
Upper Clapton
Shacklewell
Dalston
BETHNAL GREEN
Stepney
Spitalfields
SHOREDITCH
FINSBURY
ISLINGTON
Canonbury
Highbury
Holborn
CAMDEN TOWN
Kentish Town
Regents Park
Primrose Hill
MARYLEBONE
HAMPSTEAD
Gospel Oak
Belsize Park
Swiss Cottage
Dartmouth Park
Tufnell Park
Archway
Highgate
Crouch End
HORNSEY
Stroud Green
Finsbury Park
Harringay
STOKE NEWINGTON
Stamford Hill
South Tottenham
TOTTENHAM
Tottenham Hale
West Green
WOOD GREEN
Noel Park
Alexandra Palace
Muswell Hill
Fortis Green
Coldfall Wood
Colney Hatch
FRIERN BARNET
Bowes Park
Arnos Grove
Palmers Green
SOUTHGATE
Grovelands Park
Winchmore Hill
Bush Hill Park
Oakwood
Cockfosters
Oak Hill Park
Osidge
EDMONTON
Lower Edmonton
Pymmes Park
Ponders End
King George's Reservoir
William Girling Reservoir
Banbury Reservoir
Walthamstow Marsh
Fulwell Cross
Barking Creek
M11
A406
A12
A13
A10
A1
A501

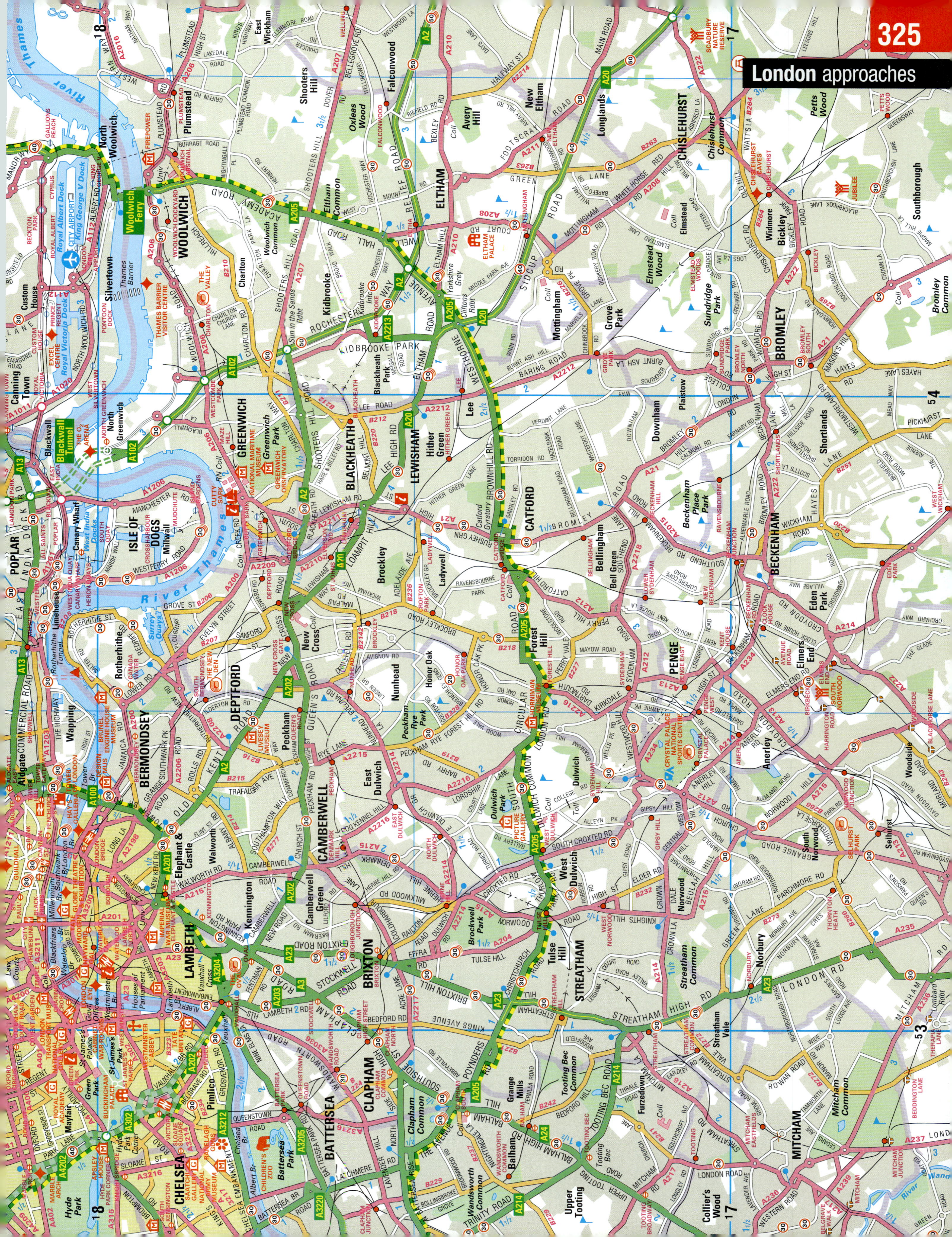
River Thames
North Woolwich
Woolwich Ferry
WOOLWICH
Plumstead
Shooters Hill
Falconwood
Oxleas Wood
New Eltham
Avery Hill
ELTHAM
Longlands
CHISLEHURST
Chislehurst Common
Petts Wood
Silvertown
Thames Barrier
Charlton
Woolwich Common
Kidbrooke
Eltham Common
Mottingham
Grove Park
Elmstead Wood
Sundridge Park
BROMLEY
Bromley Common
Southborough
Canning Town
Custom House
London City Airport
Blackwall Tunnel
North Greenwich
GREENWICH
Greenwich Park
BLACKHEATH
Lee
Hither Green
LEWISHAM
CATFORD
Downham
Plaistow
Shortlands
Beckenham Place Park
BECKENHAM
POPLAR
Canary Wharf
ISLE OF DOGS
Millwall
Brockley
Ladywell
Bellingham
Bell Green
Eden Park
Limehouse
Rotherhithe
Surrey Quays
DEPTFORD
New Cross
Nunhead
Honor Oak
Forest Hill
PENGE
Elmers End
Wapping
BERMONDSEY
Peckham
Peckham Rye Park
East Dulwich
Dulwich
Sydenham
Anerley
Crystal Palace
Woodside
Elephant & Castle
Walworth
CAMBERWELL
Camberwell Green
Kennington
LAMBETH
Vauxhall
BRIXTON
Brockwell Park
Tulse Hill
West Dulwich
West Norwood
South Norwood
Selhurst
Norbury
STREATHAM
Streatham Common
Streatham Vale
Pimlico
Battersea Park
BATTERSEA
CLAPHAM
Clapham Common
Balham
Tooting Bec Common
Furzedown
Upper Tooting
Tooting Bec
Wandsworth Common
MITCHAM
Mitcham Common
Collier's Wood
CHELSEA
Hyde Park
Mayfair
Green Park
St James's
Westminster
Waterloo
Blackfriars
Southwark
Aldgate
Tower Br.

CROSBY
Great Crosby
Waterloo
LITHERLAND
Seaforth
BOOTLE
Ferries to Belfast, Douglas & Dublin
New Brighton
WALLASEY
BIRKENHEAD
Birkenhead Park
Oxton
Tranmere
BEBINGTON
Port Sunlight
Spital
Brimstage
Bromborough
Eastham Ferry
LIVERPOOL
Kingsway
Queensway
Toxteth
Dingle
Edge Hill
Aigburth
Otterspool
Allerton
GARSTON
SPEKE
LIVERPOOL AIRPORT
LIVERPOOL JOHN LENNON AIRPORT
Terminal Building
Woolton
Hunt's Cross
Childwall
Wavertree
Broad Green
Roby
HUYTON-WITH-ROBY
Knowsley
Stockbridge Village
West Derby
Anfield
Walton
Fazakerley
Aintree
AINTREE RACE COURSE
Kirkby Park
Westvale
KIRKBY
Croxteth
CROXTETH COUNTRY PARK
Melling
Sefton Park
Newsham Park
Stanley Park
Wavertree Park
HALEWOOD COUNTRY PARK
EASTHAM COUNTRY PARK
RIMROSE VALLEY COUNTRY PARK
River Mersey
M53
M57
M62
0 1 2 miles
0 1 2 3 km

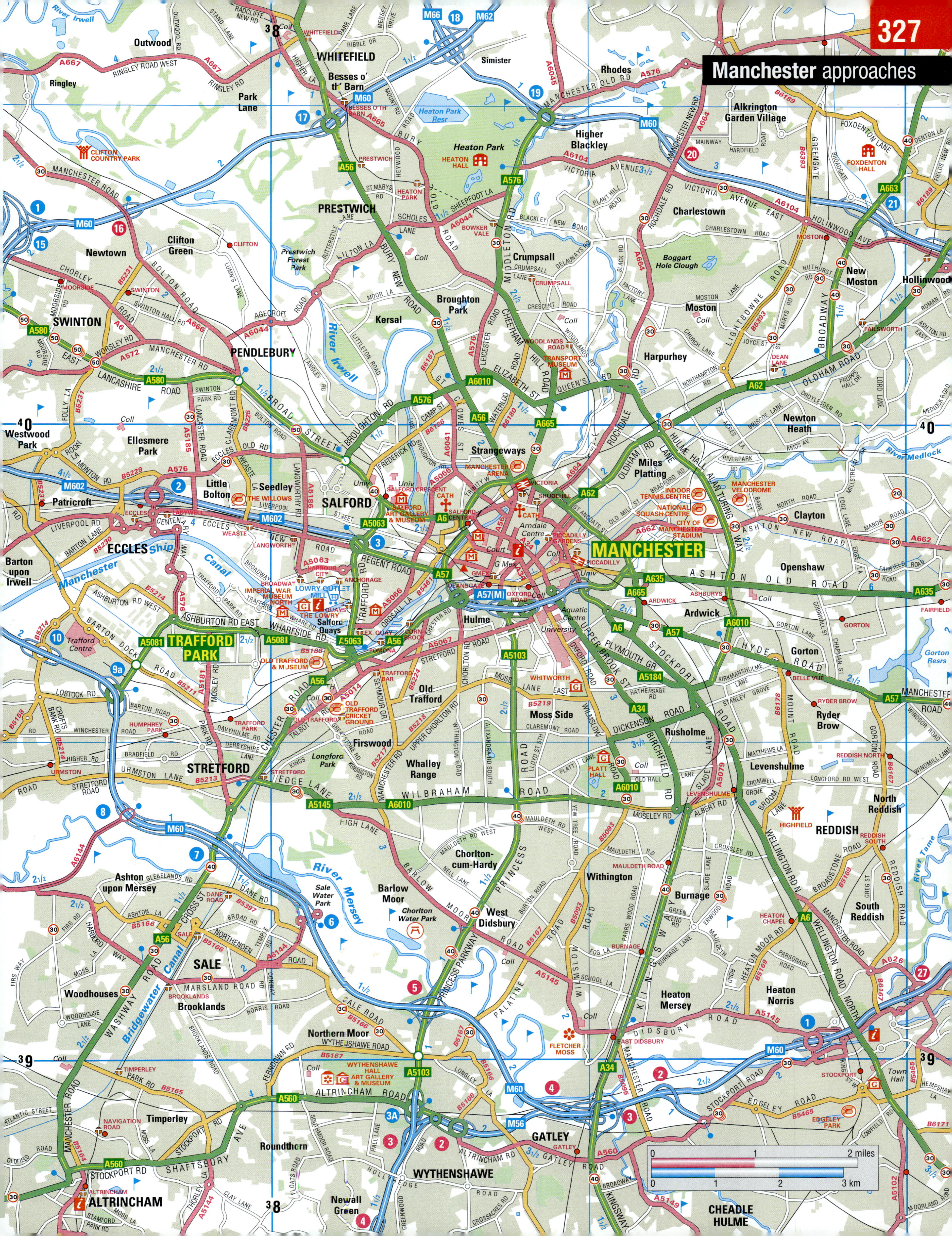
MANCHESTER
SALFORD
TRAFFORD PARK
STRETFORD
SALE
ALTRINCHAM
WYTHENSHAWE
CHEADLE HULME
PRESTWICH
WHITEFIELD
ECCLES
SWINTON
PENDLEBURY
REDDISH
STOCKPORT
GATLEY
0 1 2 miles
0 1 2 3 km

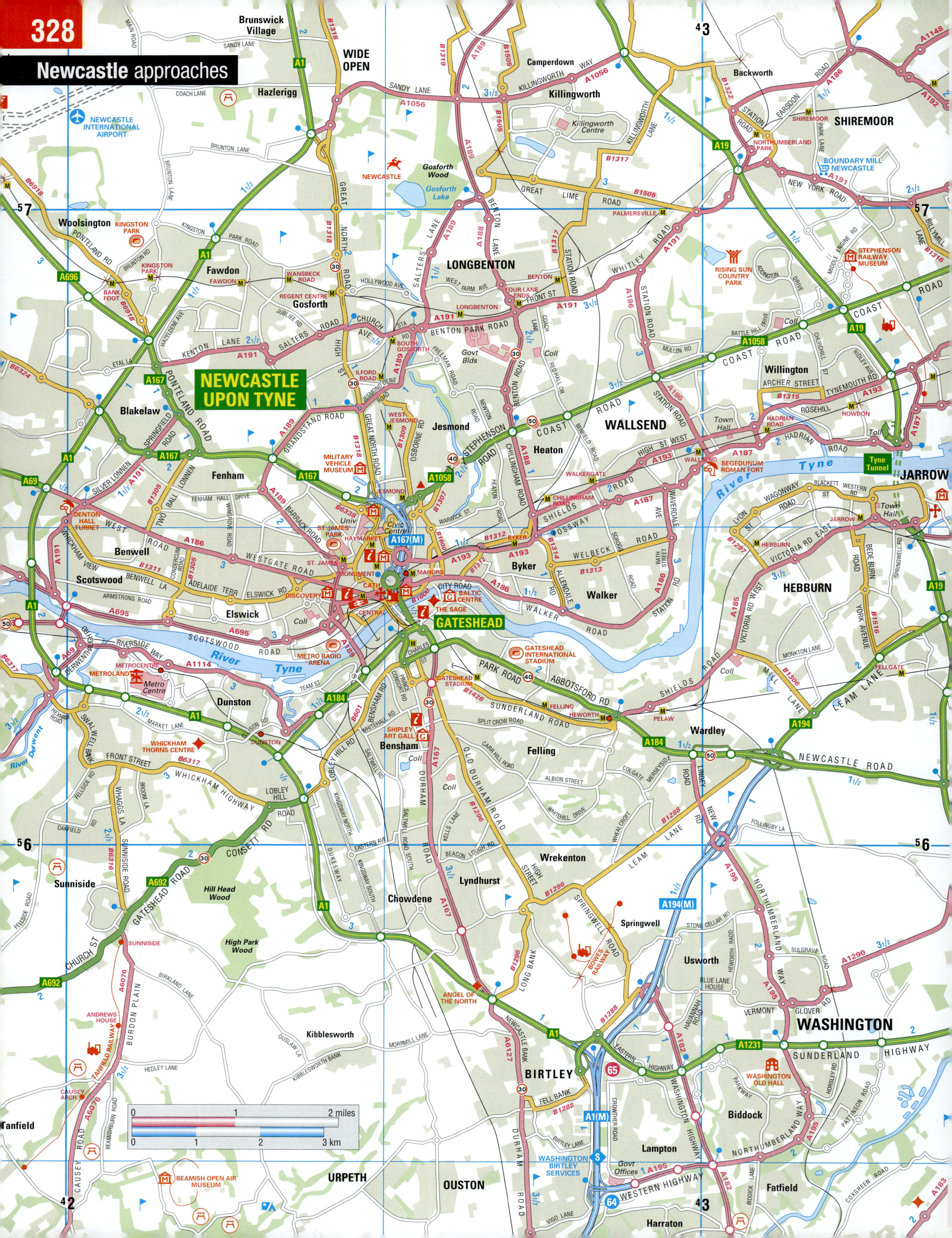
NEWCASTLE UPON TYNE
GATESHEAD
WALLSEND
JARROW
HEBBURN
WASHINGTON
BIRTLEY
SHIREMOOR
LONGBENTON
WIDE OPEN
Brunswick Village
Hazlerigg
Camperdown
Killingworth
Backworth
NEWCASTLE INTERNATIONAL AIRPORT
Woolsington
Fawdon
Gosforth
Blakelaw
Fenham
Jesmond
Heaton
Byker
Walker
Benwell
Scotswood
Elswick
Dunston
Bensham
Felling
Wardley
Wrekenton
Lyndhurst
Chowdene
Springwell
Usworth
Sunniside
Kibblesworth
Willington
Biddick
Lampton
Fatfield
Harraton
URPETH
OUSTON
Tanfield
River Tyne
River Derwent
Hill Head Wood
High Park Wood
Gosforth Wood
Gosforth Lake
Tyne Tunnel
0 1 2 miles
0 1 2 3 km

Dobpark
HUCKNALL
Misk Hill
Long Hill
Watnall Coppice
Beauvale
Ruffs
Hazelgrove
Westville
Broomhill
Butler's Hill
Raceground Hill
Cobbler's Hill
Bestwood Park
Bestwood Village
BESTWOOD
The Warren
Gaunt's Hill
Bogend
Rise Park
Top Valley
Redhill
TUMBLE TOWN
ARNOLD
Bonner Hill
Woodborough
Hunger Hill
PLAYWORLD
Lambley
Crocket Beck
Wicketwood Hill
Nuthall
Bestwood
Highbury Vale
Bulwell
PHOENIX PARK
CINDERHILL
Cinderhill
Old Basford
Wodthorpe
Mapperley
Porchester
Gelding
Broxtowe
COURTAULD'S
Sherwood
Aspley
New Basford
Carrington
Mapperley Park
CARLTON
Strelley
Bilborough
Whitemoor
Catstone Hill
Thorneywood
NOTTINGHAM
St. Ann's
Trowell Moor
Radford
Netherfield
Martin's Pond
WOLLATON
Wollaton
Wollaton Park
WOLLATON HALL
New Lenton
GREENS MILL & SCIENCE CENTRE
Sneinton
Colwick
HOLME PIERREPONT
Stapleford Hill
The Lake
Nottingham University
Nottingham Canal
Meadows
THE CITY GROUND
River Trent
NATIONAL WATER SPORTS CENTRE
Holme Pierrepont
Bramcore Hills
MEADOW LANE
TRENT BRIDGE
Bramcote
Gamston
Bassingfield
Wilford
WEST BRIDGFORD
Chilwell
NOTTINGHAM CITY AIRPORT
Grantham Canal
Rylands
BEESTON
Beeston Canal
Clifton Grove
Edwalton
Wilford Hill
Sharp Hill
ATTENBOROUGH
ATTENBOROUGH NATURE RESERVE
CLIFTON
Tollerton
Clipston
Mill Hill
Brook Hill
Hoe Hill
Barton in Fabis
Manor Park
FRAMEWORK KNITTERS' MUSEUM
RUDDINGTON
Plumtree
Normanton-on-the-Wolds
Clifton Pasture
Fairham Brook
Ruddington Fields
RUSHCLIFFE
GREAT CENTRAL RAILWAY (NOTTINGHAM)
Plumtree Park
Normanton Wolds
Thrumpton
THRUMPTON HALL
Gotham Hill
Ruddington Moor
Bradmore
KEYWORTH
Gotham Moor
Bradmore Moor
M1
A60
A610
A611
A52
A453
A606
A612
A6011
A6514
A6002
0 1 2 miles
0 1 2 3 km

ROTHERHAM
SHEFFIELD
ECCLESFIELD
GRENOSIDE
HILLSBOROUGH
Wickersley
Sunnyside
Listerdale
Brecks
Dalton
Dalton Parva
Dalton Magna
Whiston
Morthen
Ulley
Upper Whiston
Herringthorpe
Moorgate
Eastwood
Clifton
Broom
Canklow
Brinsworth
Masbrough
Bradgate
Greasbrough
Kimberworth
Kimberworth Park
Dropping Well
Blackburn
Tinsley
Jordan
Catcliffe
Treeton
Guilthwaite
Aughton
Aston
Swallownest
Fence
Woodhouse Mill
Woodhouse
Handsworth
Orgreave
Richmond
Darnall
Greenland
Attercliffe
Brightside
Wincobank
Concord Park
Shiregreen
Pismire Hill
Firth Park
Grimesthorpe
Pitsmoor
Fir Vale
Burngreave
Parson Cross
Longley Estate
Shirecliffe
Owlerton
Parkwood Springs
Wadsley Bridge
Birley Carr
Birley Edge
Beeley Wood
Oughtibridge
Wharncliffe Wood
Worrall
Wadsley
Loxley
Malinbridge
Walkley
Steel Bank
Crookesmoor
Crookes
Tapton Hill
Sandygate
Ranmoor
Greystones
Fulwood
Sharrow
Nether Edge
Brincliffe
Ecclesall
Carter Knowle
Millhouses
Whirlow
Abbeydale
Abbeydale Park
Parkhead
Knowle Top
Manor Park
Manor Estate
Sheffield Park
Arbourthorne
Heeley
Lowfield
Newfield Green
Meersbrook
Intake
Gleadless
Gleadless Valley
Hemsworth
Backmoor
Norton Woodseats
Norton
Jordanthorpe
Meadow Head
Greenhill
Lowedges
Totley Rise
Totley Bents
Frecheville
Hollins End
Normanton Spring
Birley
Hackenthorpe
Beighton
Sothall
Drakehouse
Waterthorpe
Westfield
Halfway
Holbrook
Mosborough
Owlthorpe
Moorhole
Highlane
Plumbley
Ridgeway
Ridgeway Moor
Lightwood
Ford
Killamarsh
Norwood
Wales
Woodall
Upperthorpe
Rother Valley Lake
River Rother
River Don
River Sheaf
Limb Brook
River Rivelin
Ulley Reservoir
Dore House Industrial Estate
Sheffield City
Meadowhall
Magna Science Adventure Centre
Ridgeway Cottage Industry Centre
Sheffield Services
Sheffield Station/Sheffield Hallam
Botanical Gardens
Bishop's House
Heeley City Farm
Brampton en le Morthen
Tinsley Park
Sheffield to Keadby Canal
M1
M18
A57
A61
A618
A630
A631
A6102
A6135
A6178
A621
A625
2 miles
3 km

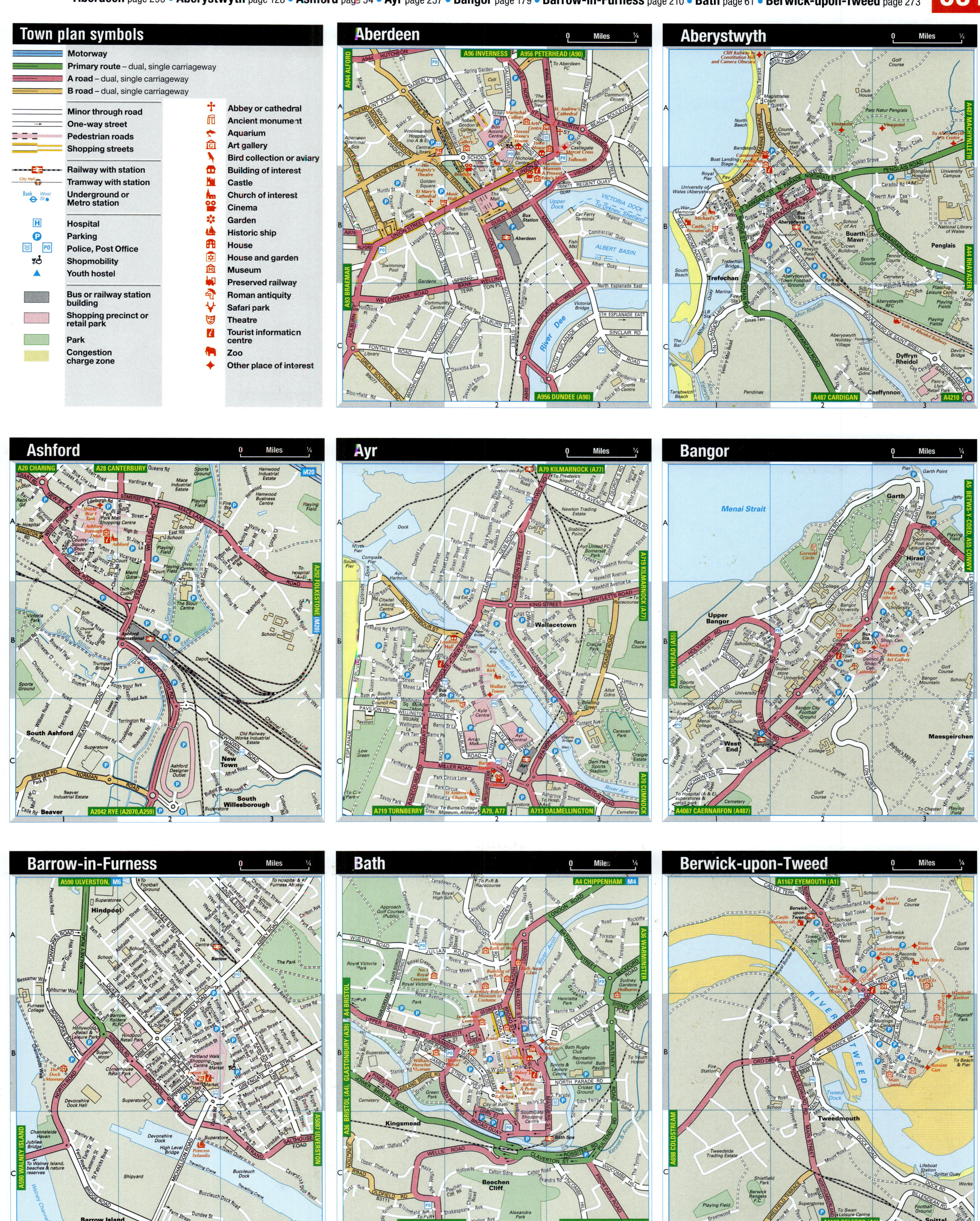

Town plan symbols
Motorway
Primary route – dual, single carriageway
A road – dual, single carriageway
B road – dual, single carriageway
Minor through road
One-way street
Pedestrian roads
Shopping streets
Railway with station
Tramway with station
Underground or Metro station
Hospital
Parking
Police, Post Office
Shopmobility
Youth hostel
Bus or railway station building
Shopping precinct or retail park
Park
Congestion charge zone
Abbey or cathedral
Ancient monument
Aquarium
Art gallery
Bird collection or aviary
Building of interest
Castle
Church of interest
Cinema
Garden
Historic ship
House
House and garden
Museum
Preserved railway
Roman antiquity
Safari park
Theatre
Tourist information centre
Zoo
Other place of interest
Aberdeen
A96 INVERNESS
A956 PETERHEAD (A90)
A93 BRAEMAR
A956 DUNDEE (A90)
Aberystwyth
A487 MACHYNLLETH
A44 RHAYADER
A487 CARDIGAN
A4210
Ashford
A20 CHARING
A28 CANTERBURY
A20 FOLKESTONE (M20)
A2042 RYE (A2070,A259)
Ayr
A79 KILMARNOCK (A77)
A719 KILMARNOCK (A77)
A719 TURNBERRY
A713 DALMELLINGTON
A70 CUMNOCK
Bangor
A5 BETWS-Y-COED, A55 CONWY
A5 HOLYHEAD (A55)
A4087 CAERNARFON (A487)
Menai Strait
Barrow-in-Furness
A590 ULVERSTON, M6
A590 ULVERSTON
A590 WALNEY ISLAND
Bath
A4 CHIPPENHAM M4
A36 WARMINSTER
A4 BRISTOL
A36 BRISTOL (A4), GLASTONBURY (A39)
A367 SHEPTON MALLET
A3062 SHEPTON MALLET (A367)
Berwick-upon-Tweed
A1167 EYEMOUTH (A1)
A698 COLDSTREAM
A1167 BELFORD (A1)

Birmingham

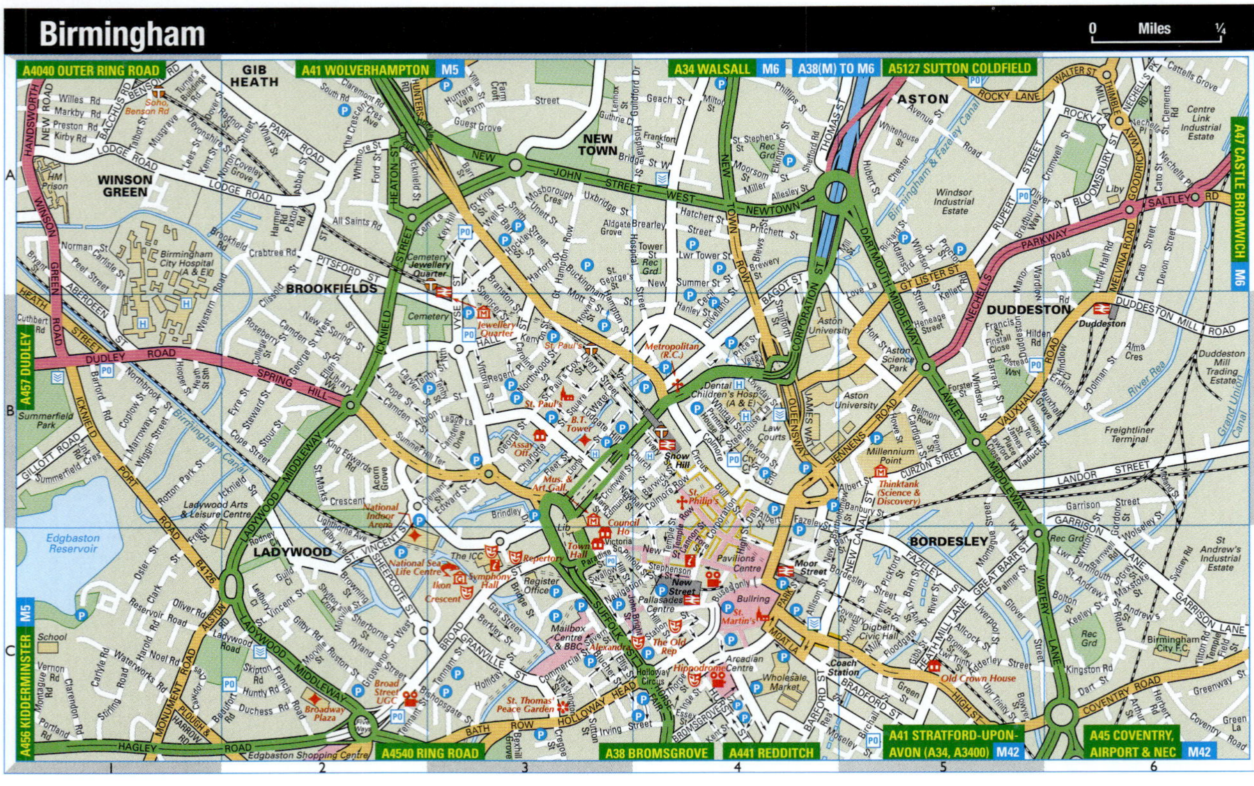

Blackpool

Bournemouth

Bradford

Brighton

Bristol

Bury St Edmunds

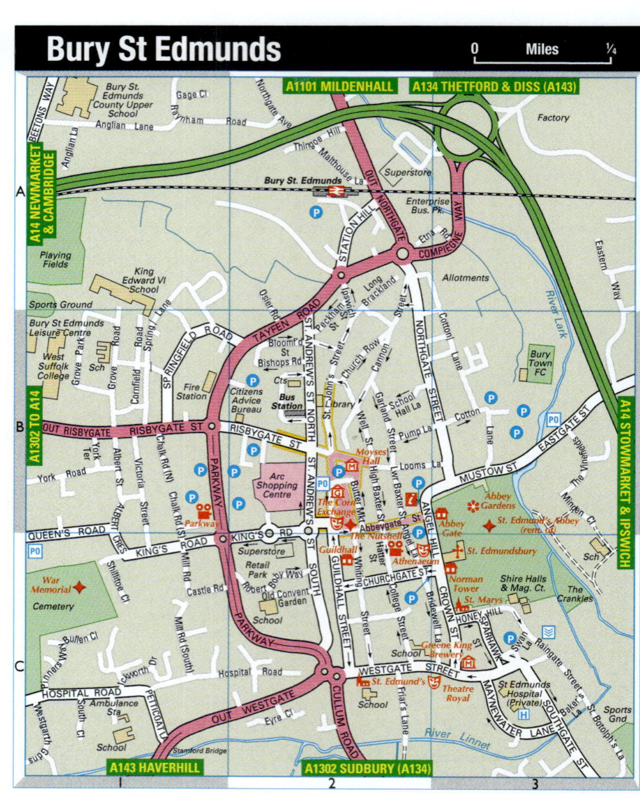

Cambridge
0 Miles ¼
Canterbury
0 Miles ¼
Cardiff / Caerdydd
0 Miles ¼
Carlisle
0 Miles ¼
Chelmsford
0 Miles ¼
Cheltenham
0 Miles ¼
Chester
0 Miles ¼
Chichester
0 Miles ¼
Colchester
0 Miles ¼

Coventry
Miles
B4098 TAMWORTH (A51)
B4113 NUNEATON (A444) M6
A4114 BIRMINGHAM (A45) M42
A4600 LEICESTER M69, M6
B4101 KNOWLE
A4114 TO A46 WARWICK & ROYAL LEAMINGTON SPA
A429 KENILWORTH
Derby
A6 MATLOCK
A61 MANSFIELD (A38) M1
A52 ASHBOURNE
A516 UTTOXETER
A52 NOTTINGHAM M1
A6 to A50 M1
A601 BURTON (A38)
A514 SWADLINCOTE
Dorchester
B3147 YEOVIL (A37)
A3150 POOLE (A35)
B3150 BRIDPORT (A35)
A35 POOLE A352 WAREHAM
B3147 WEYMOUTH (A354) & A35
A35 WEYMOUTH (A354)
Dumfries
A76 KILMARNOCK
A701 MOFFAT
A709 LOCKERBIE
A780 ANNAN (A75)
A711 DALBEATTIE A780 STRANRAER (A75)
A710 DALBEATTIE
Dundee
A929 FORFAR (A90)
A923 COUPAR ANGUS
A92 ABERDEEN (A90)
A85 PERTH (A90)
A92 EDINBURGH
FIRTH OF TAY
Durham
A691 CONSETT
A690 SUNDERLAND A1 (M)
A167 NEWCASTLE A1 (M)
A690 WILLINGTON
A177 STOCKTON A1 (M)
A167 DARLINGTON A1 (M)
A177 TO A167
Edinburgh
A900 LEITH
A90 FORTH ROAD BRIDGE & PERTH (M90)
A1 BERWICK-UPON-TWEED
A8 GLASGOW & STIRLING (M8) (M9)
A71 KILMARNOCK
A702 CARLISLE & M74
TO A7
A7 GALASHIELS & JEDBURGH (A68)
Exeter
B3183 TIVERTON (A396)
A377 BARNSTAPLE
B3212
B3183 HONITON (A30), TAUNTON (M5)
B3212 OKEHAMPTON (A30)
A377 PLYMOUTH (A38)
A3015 EXMOUTH (A376)

Fort William
Miles
Inverlochy
Loch Linnhe
A82 INVERNESS
A82 CRIANLARICH
BELFORD ROAD

Glasgow
Miles
A82 ERSKINE BRIDGE
A81 BEARSDEN
A879 MILNGAVIE (A807)
A803 KIRKINTILLOCH
KELVINGROVE
GARNETHILL
COWCADDENS
FINNIESTON
ANDERSTON
CENTRAL QUAY
RIVER CLYDE
KINNING
LAURIESTON
M8
M77
GREENOCK
A77 KILMARNOCK
A730 RUTHERGLEN
A728
A74
A749
A89 AIRDRIE
M8 EDINBURGH

Gloucester
Miles
A38 CHELTENHAM (A40) & M5
A417 ROSS-ON-WYE (A40)
A4302 TO A38
A430 TO A38 BRISTOL
M5
LONDON ROAD
Gloucestershire Royal Hospital (A & E)

Grimsby
Miles
A180 IMMINGHAM
M180
A180 CLEETHORPES
Alexandra Dock
WESTGATE
CORPORATION ROAD
FREDERICK WARD WAY
A1136 TO A180 & A46 (B1444)
A16 LOUTH & A46

Hanley (Stoke-on-Trent)
Miles
A50 KIDSGROVE
A53 LEEK
Central Forest Park
A53 NEWCASTLE-UNDER-LYME
A52 ASHBOURNE
A5006 STOKE
A50 UTTOXETER
ETRURIA ROAD

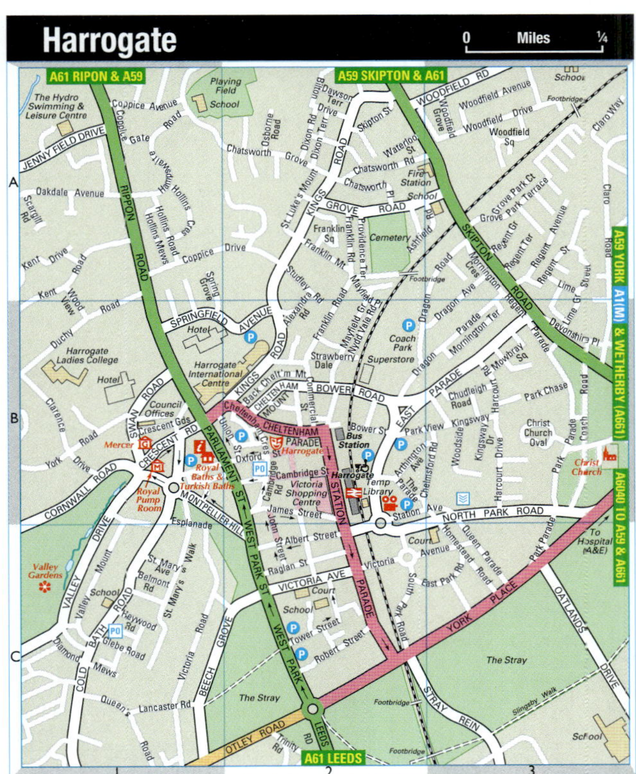
Harrogate
Miles
A61 RIPON & A59
A59 SKIPTON & A61
A59 YORK, A1(M) & WETHERBY (A661)
A61 LEEDS
The Stray

Holyhead / Caergybi
Miles
New Harbour
Old Harbour
Salt Island (Ynys Halen)
Morawelon
Black Bridge
Kingsland
A55 BANGOR
A5 VALLEY

Hull
Miles
A1079 YORK
A165 BRIDLINGTON
A1033 HEDON & CAR FERRY TERMINAL
A1105 HUMBER BRIDGE & SELBY (A63)
A63 SELBY & AIRPORT (A15)
FREETOWN WAY
CASTLE STREET
River Humber

Inverness

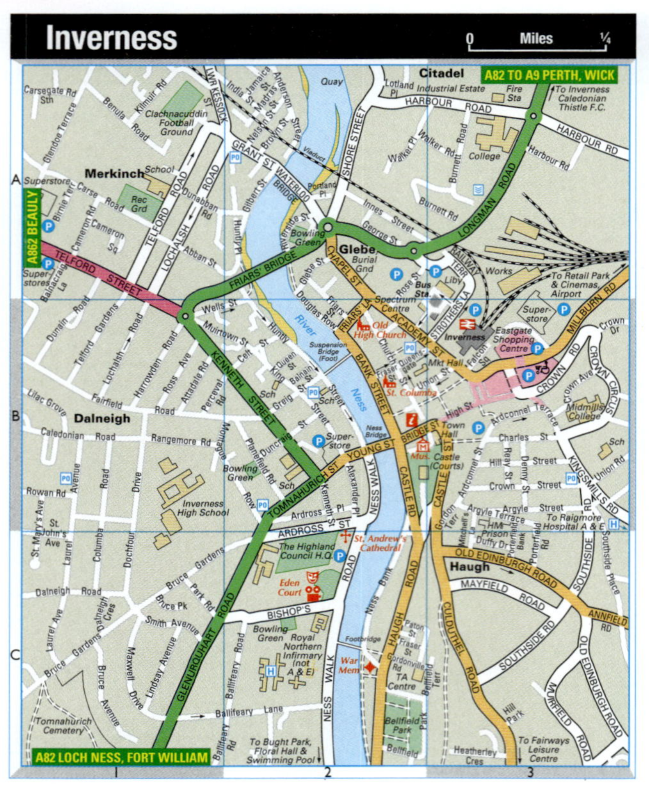

Ipswich

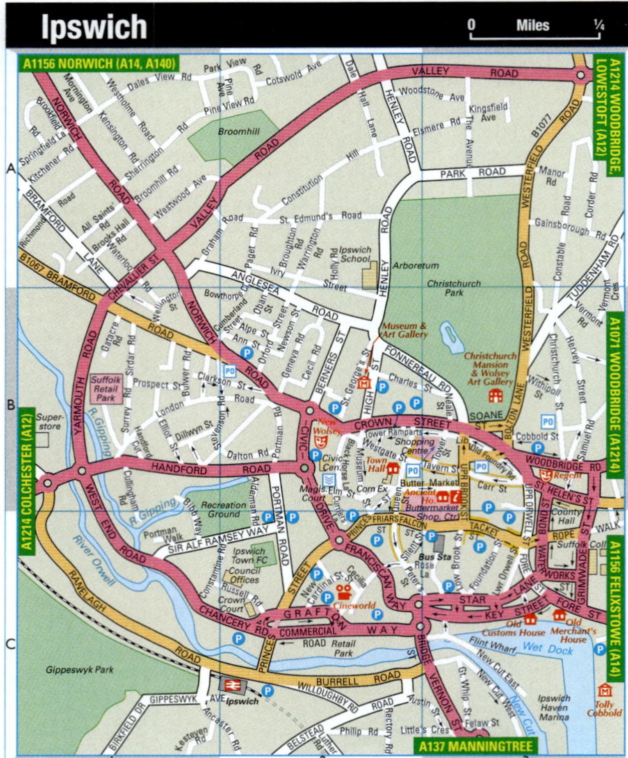

Kendal

King's Lynn

Leeds

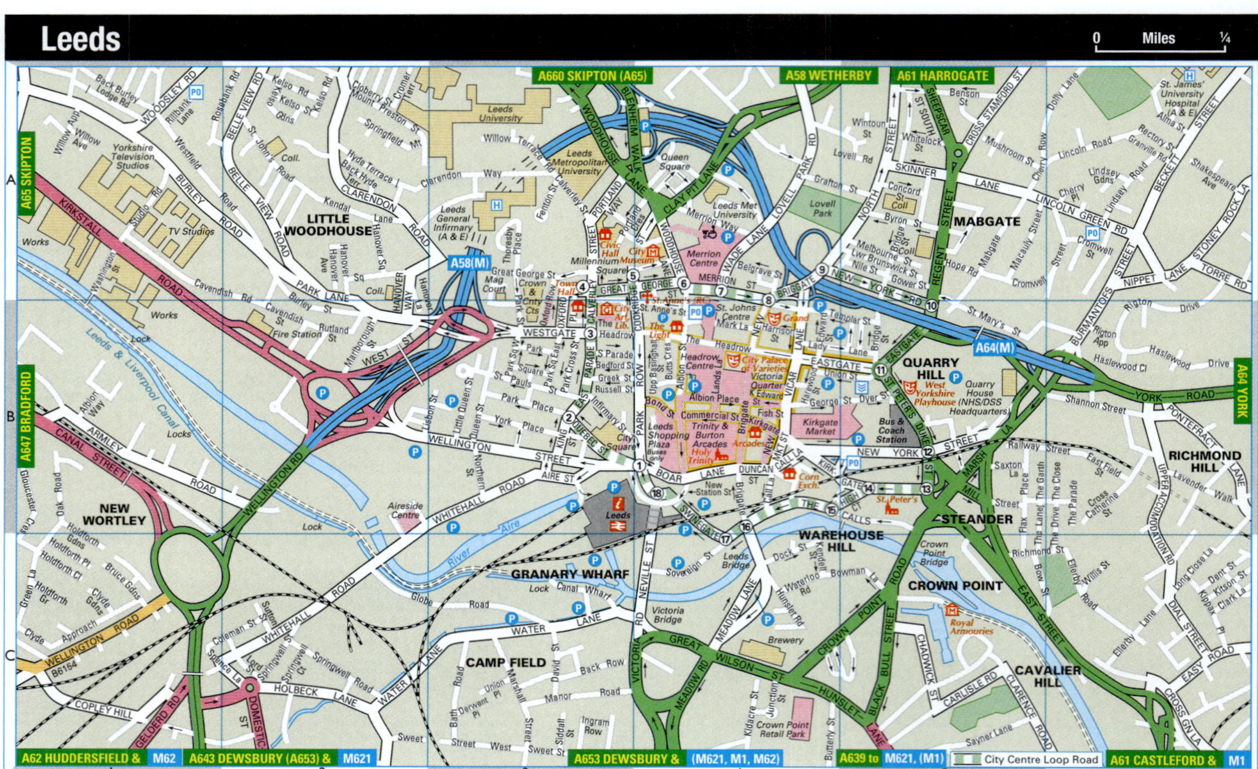

Lancaster

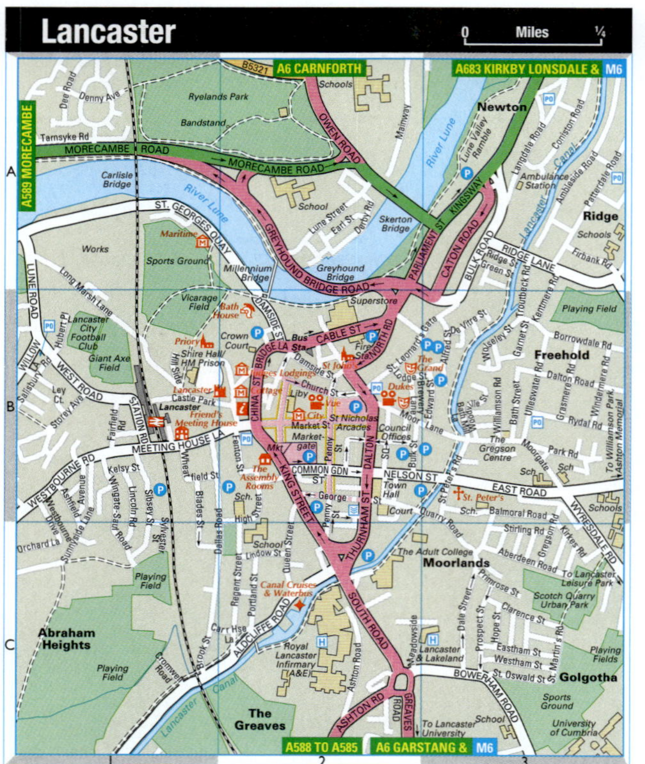

Leicester

Lewes

Lincoln

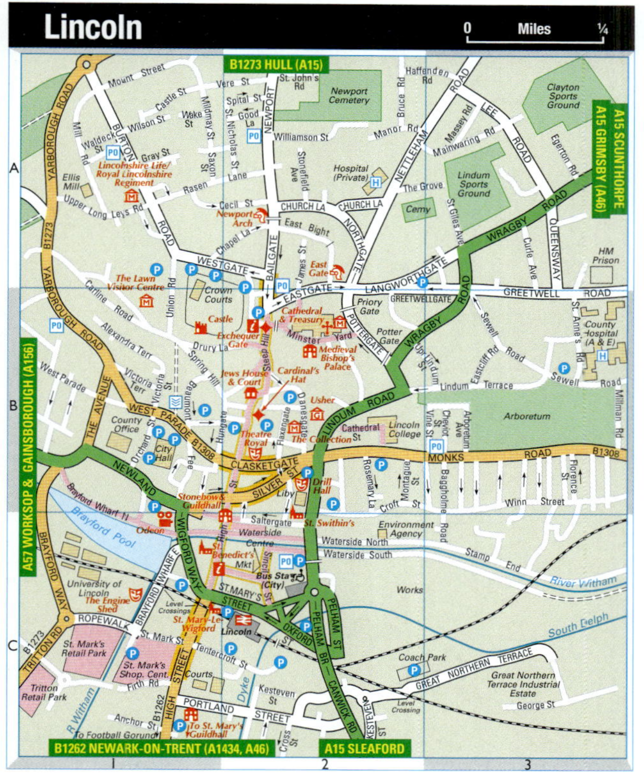

Liverpool

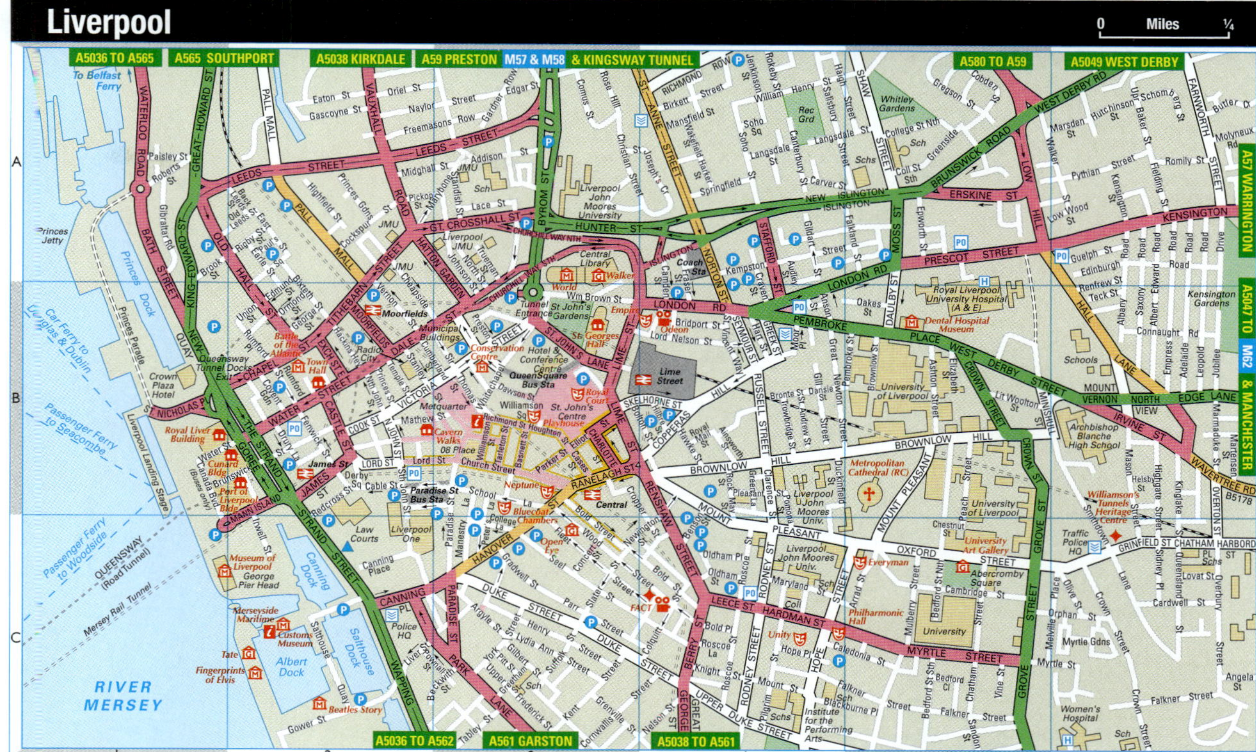

Llandudno

Llanelli

Luton

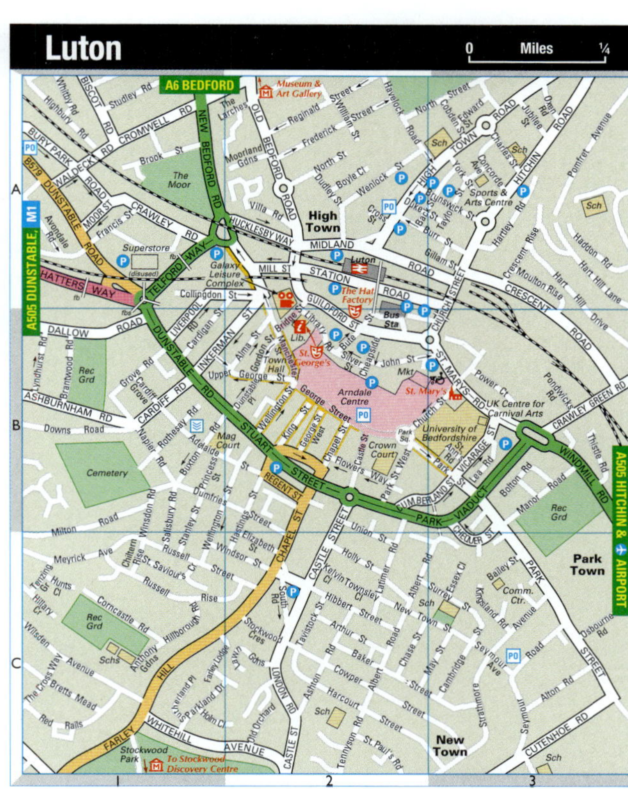

Macclesfield

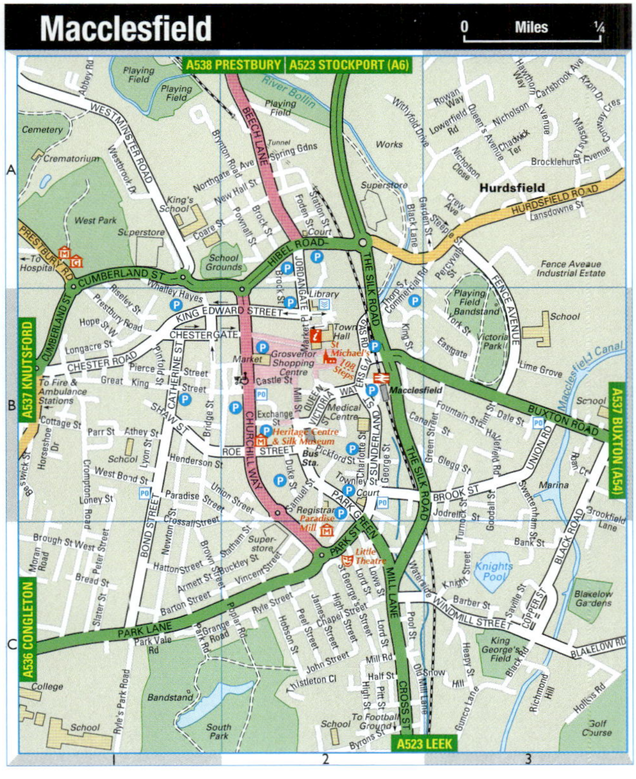

Manchester

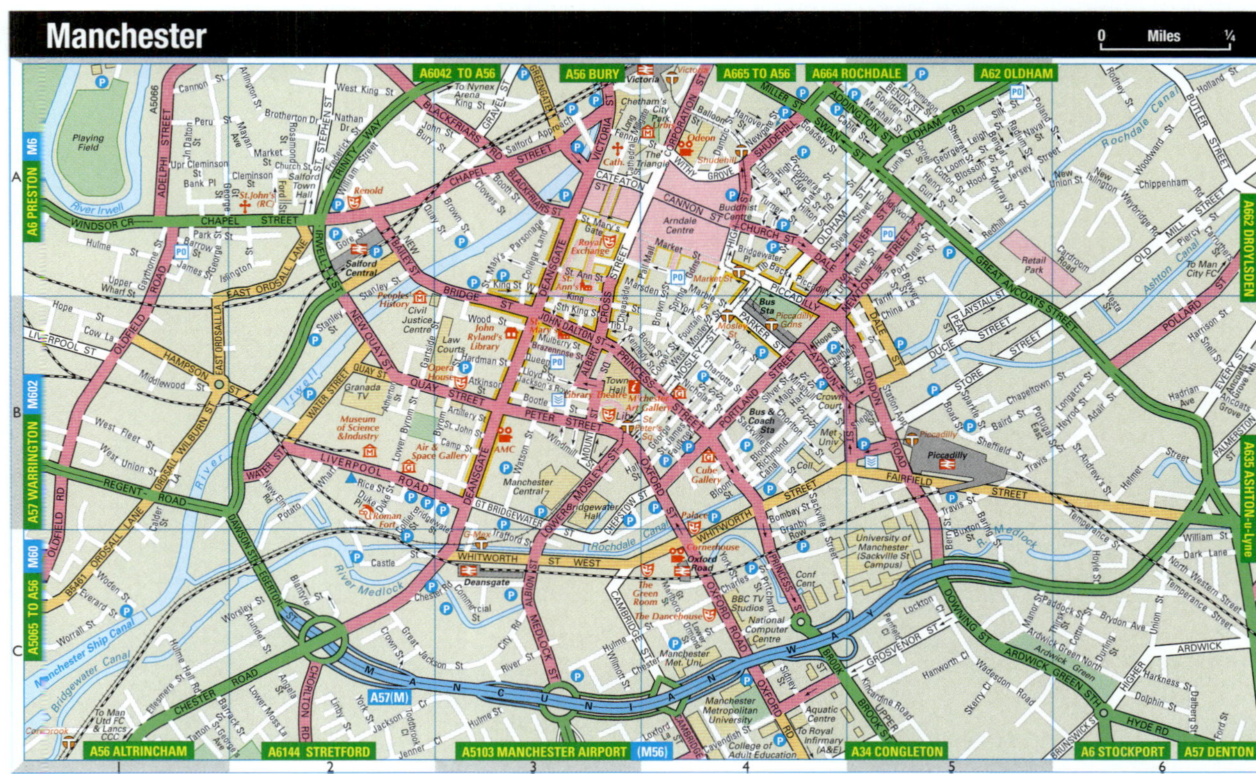

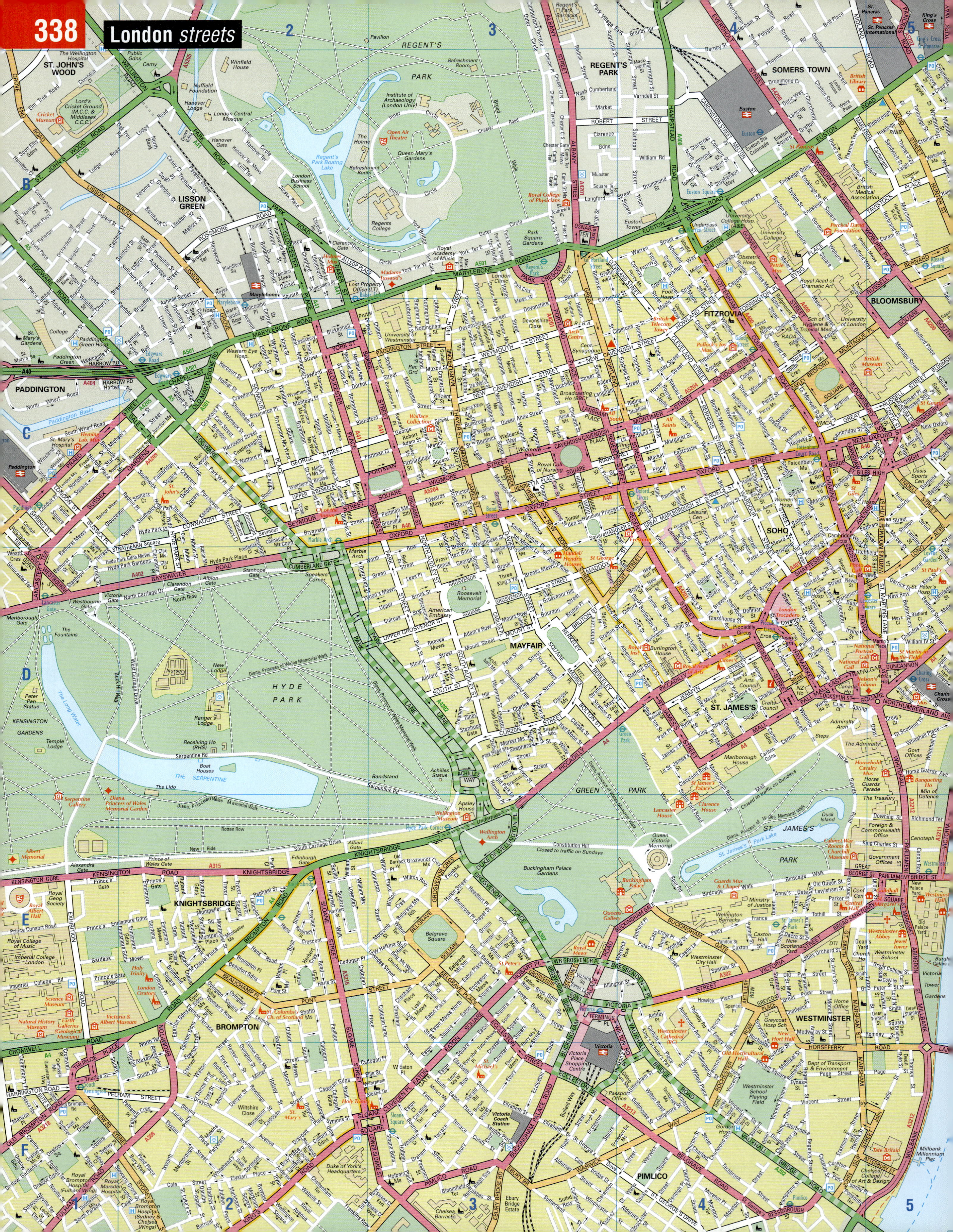

REGENT'S PARK
ST. JOHN'S WOOD
LISSON GREEN
PADDINGTON
MARYLEBONE ROAD
EUSTON ROAD
SOMERS TOWN
FITZROVIA
BLOOMSBURY
OXFORD STREET
SOHO
MAYFAIR
HYDE PARK
KENSINGTON GARDENS
THE SERPENTINE
GREEN PARK
ST. JAMES'S
ST. JAMES'S PARK
KNIGHTSBRIDGE
BROMPTON
WESTMINSTER
PIMLICO
Buckingham Palace Gardens
Victoria
Euston
St. Pancras
King's Cross

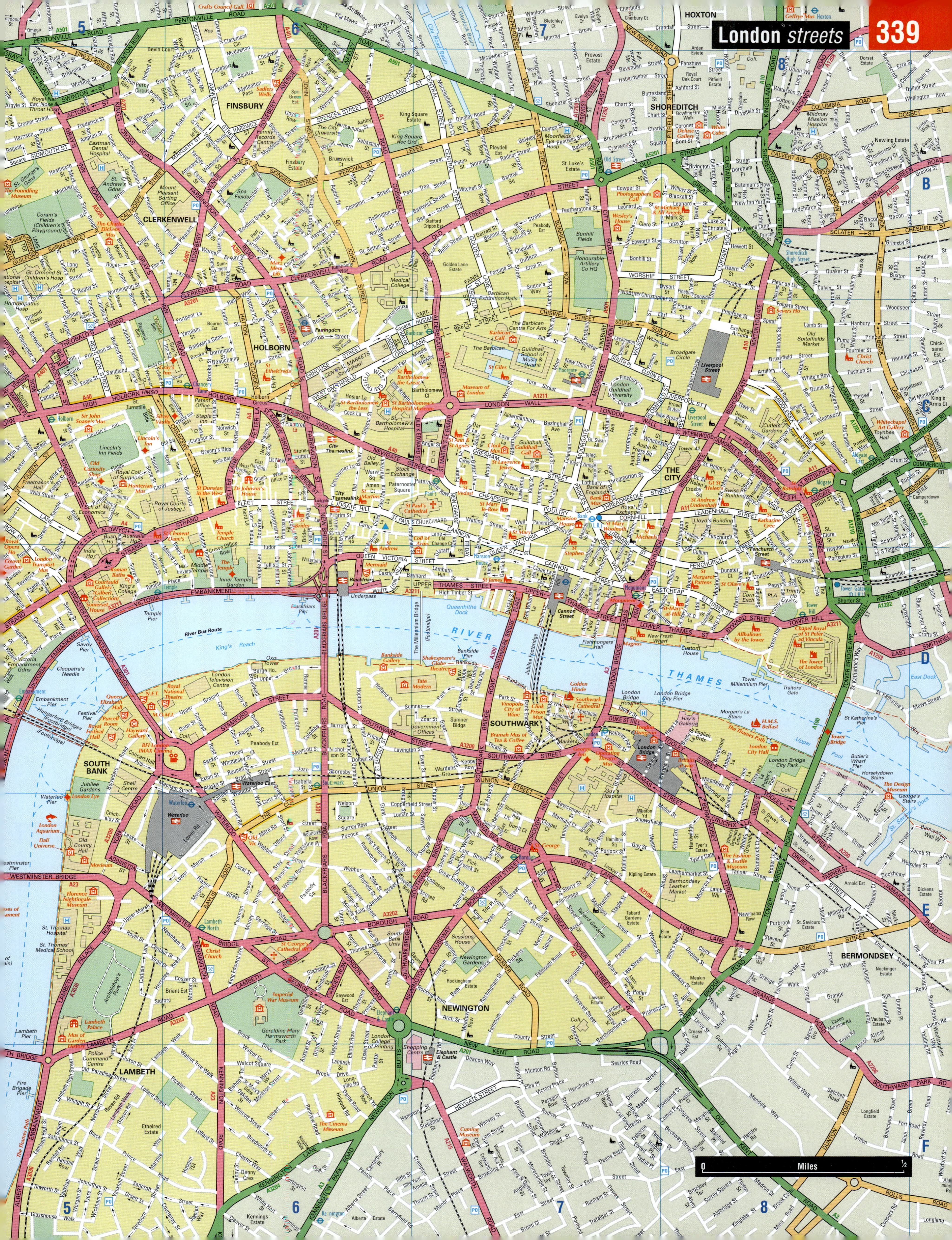
HOXTON
FINSBURY
SHOREDITCH
CLERKENWELL
HOLBORN
THE CITY
SOUTHWARK
SOUTH BANK
LAMBETH
NEWINGTON
BERMONDSEY
RIVER
THAMES
River Bus Route
King's Reach
Waterloo
London Bridge
Liverpool Street
Barbican
Elephant & Castle
Miles
0
½

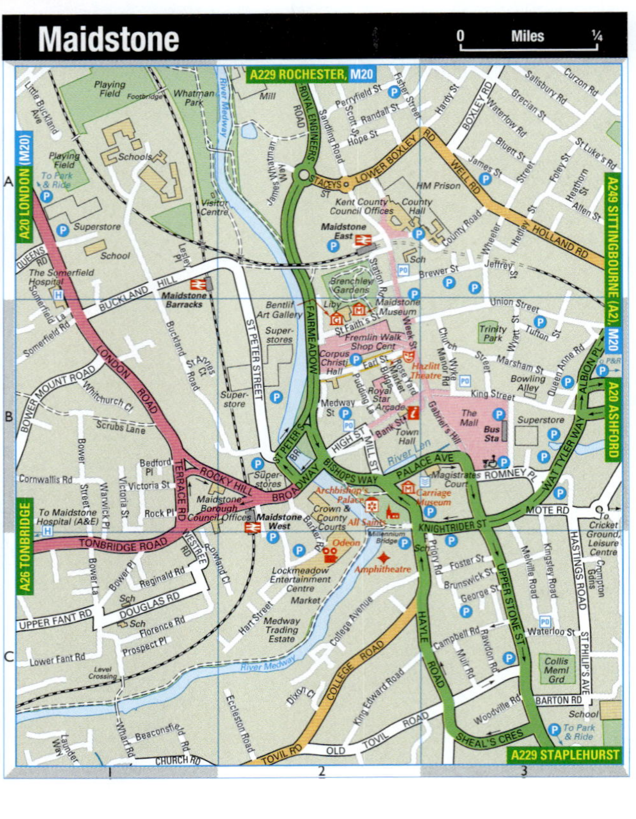
Maidstone
Miles
A229 ROCHESTER, M20
A20 LONDON (M20)
A249 SITTINGBOURNE (A2, M20)
A20 ASHFORD
A26 TONBRIDGE
A229 STAPLEHURST

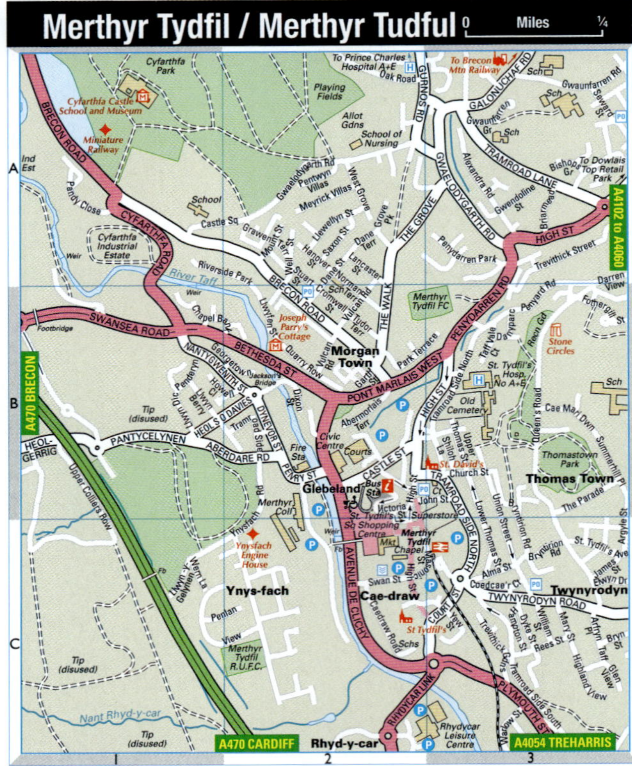
Merthyr Tydfil / Merthyr Tudful
Miles
A470 BRECON
A470 CARDIFF
A4054 TREHARRIS
Morgan Town
Thomas Town
Ynys-fach
Cae-draw
Twynyrodyn
Rhyd-y-car

Middlesbrough
Miles
A1046
A1046 STOCKTON-ON-TEES
A178 HARTLEPOOL
A1046 STOCKTON-ON-TEES
A1032 BILLINGHAM (A19)
A19 SUNDERLAND
A66 REDCAR (A1085)
A1085 REDCAR
A19 THIRSK
A1032 HEMLINGTON & A19
A172 STOKESLEY & THIRSK (A19)
River Tees

Milton Keynes
Miles
A422 BEDFORD
A509 NEWPORT PAGNELL & M1
A422 TO A5
A5 NORTHAMPTON (A508)
A5 DUNSTABLE
Bradville
Neath Hill
Downhead
Linford Wood
Downs Barn
Heelands
Conniburrow
Bradwell Common
Rooksley
Fishermead
Oldbrook
Eaglestone
Winterhill
Loughton

Newcastle upon Tyne
Miles
B1318 GOSFORTH, A1
A1058 TYNEMOUTH
A167 JEDBURGH & (A696), MORPETH (A1)
A193 WALLSEND (A187)
A189 GATESHEAD
A167 DURHAM & A1(M)
BRANDLING VILLAGE
SHIELDFIELD
RIVER TYNE
GATESHEAD
SALT MEADOWS

Newport / Casnewydd
Miles
A4051 M4 JUNC 26
A4042 PONTYPOOL
A4042
Maesglas
Gold Tops
Barnardstown
Clytha Park
Baneswell
St Woolos
Stow Park

Newquay
Miles
Newquay Bay
A3058 ST AUSTELL
A392 INDIAN QUEENS, A30
Towan Beach
Great Western Beach
Tolcarne Beach
Trenance
The Gannel

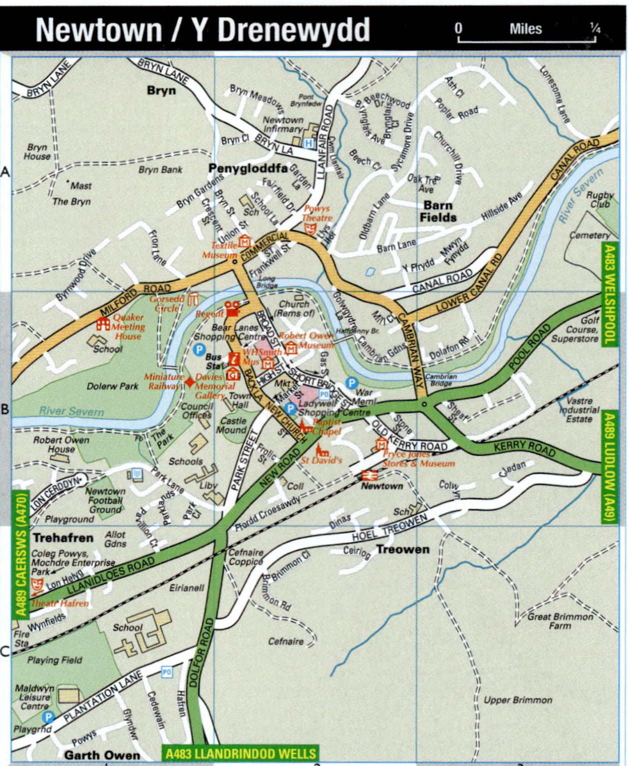
Newtown / Y Drenewydd
Miles
A483 WELSHPOOL
A489 LUDLOW (A49)
A489 CAERSWS (A470)
A483 LLANDRINDOD WELLS
Bryn
Penygloddfa
Barn Fields
River Severn
Trehafren
Treowen
Garth Owen

Northampton
Miles
A5095 LEICESTER (A5199)
A508 MARKET HARBOROUGH
A5123 KETTERING (A43)
A4500 WELLINGBOROUGH (A45)
A428 BEDFORD & WELLINGBOROUGH (A45)
A428 RUGBY
M1 jn 18
M1 jn 16
A4500 COVENTRY (A45)
A5123 OXFORD (A43)
M1 jn 15A
A508 TO M1 jn 15
Semilong
Cotton End

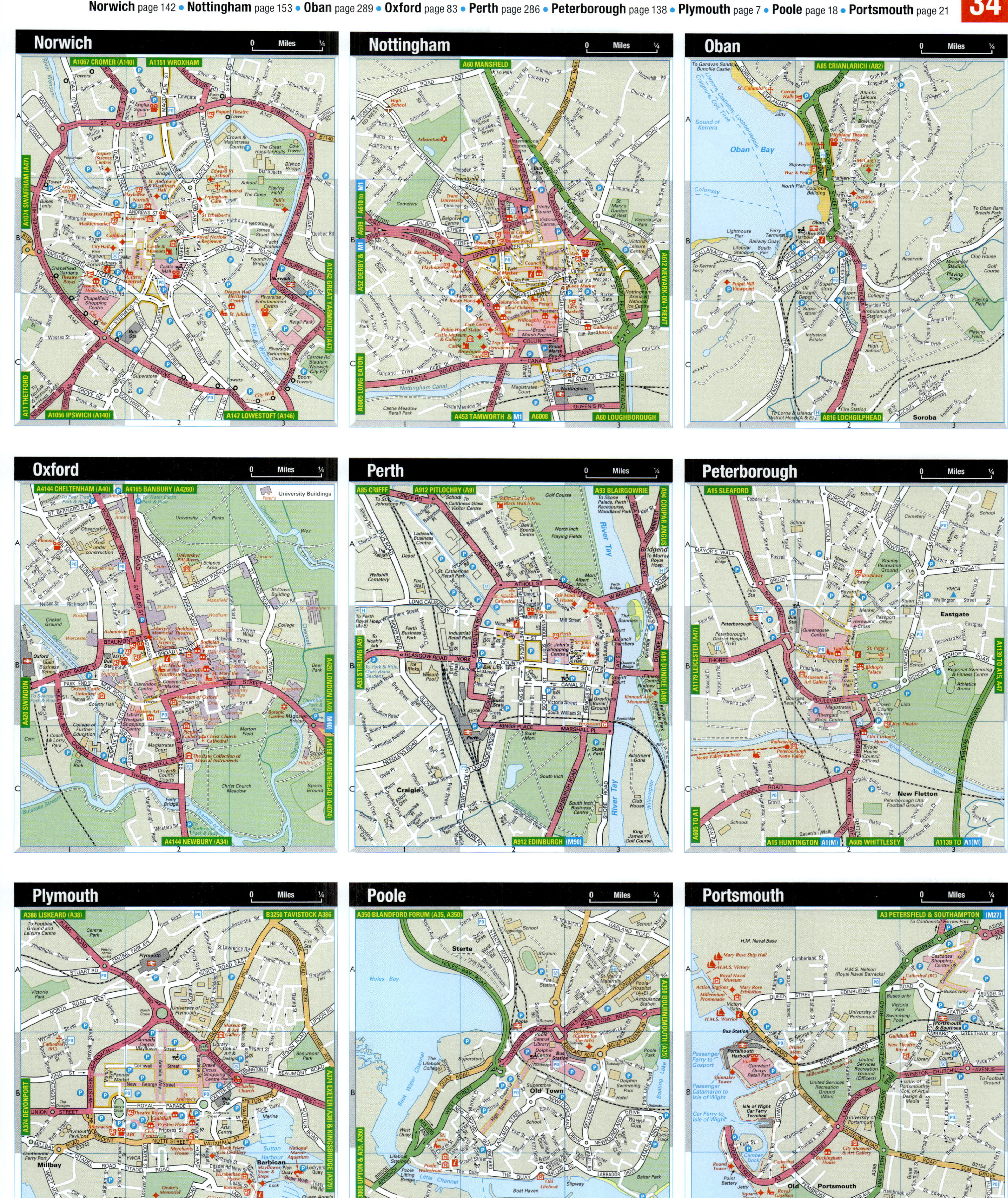
Norwich
A1067 CROMER (A140)
A1151 WROXHAM
A1074 SWAFFHAM (A47)
A11 THETFORD
A1056 IPSWICH (A140)
A147 LOWESTOFT (A146)
A1242 GREAT YARMOUTH (A47)
Nottingham
A60 MANSFIELD
A610 to M1
A52 DERBY & M1
A6005 LONG EATON
A453 TAMWORTH & M1
A6008
A60 LOUGHBOROUGH
A612 NEWARK-ON-TRENT
Oban
A85 CRIANLARICH (A82)
A816 LOCHGILPHEAD
Oban Bay
Soroba
Oxford
A4144 CHELTENHAM (A40)
A4165 BANBURY (A4260)
University Buildings
A420 SWINDON
A4144 NEWBURY (A34)
A420 LONDON (A40, M40)
A4158 MAIDENHEAD (A4074)
Perth
A85 CRIEFF
A912 PITLOCHRY (A9)
A93 BLAIRGOWRIE
A94 COUPAR ANGUS
A93 STIRLING (A9)
A912 EDINBURGH (M90)
A85 DUNDEE (A90)
River Tay
Peterborough
A15 SLEAFORD
A1179 LEICESTER (A47)
A605 TO A1
A15 HUNTINGTON A1(M)
A605 WHITTLESEY
A1139 TO A1(M)
A1139 TO A15, A47
Eastgate
New Fletton
Plymouth
A386 LISKEARD (A38)
B3250 TAVISTOCK A386
A374 DEVONPORT
A374 EXETER (A38) & KINGSBRIDGE (A379)
The Sound
Poole
A350 BLANDFORD FORUM (A35, A350)
A350 BOURNEMOUTH (A35)
B3068 UPTON & A35, A350
Holes Bay
Poole Harbour
Old Town
Portsmouth
A3 PETERSFIELD & SOUTHAMPTON (M27)
A288 SOUTHSEA
Old Portsmouth

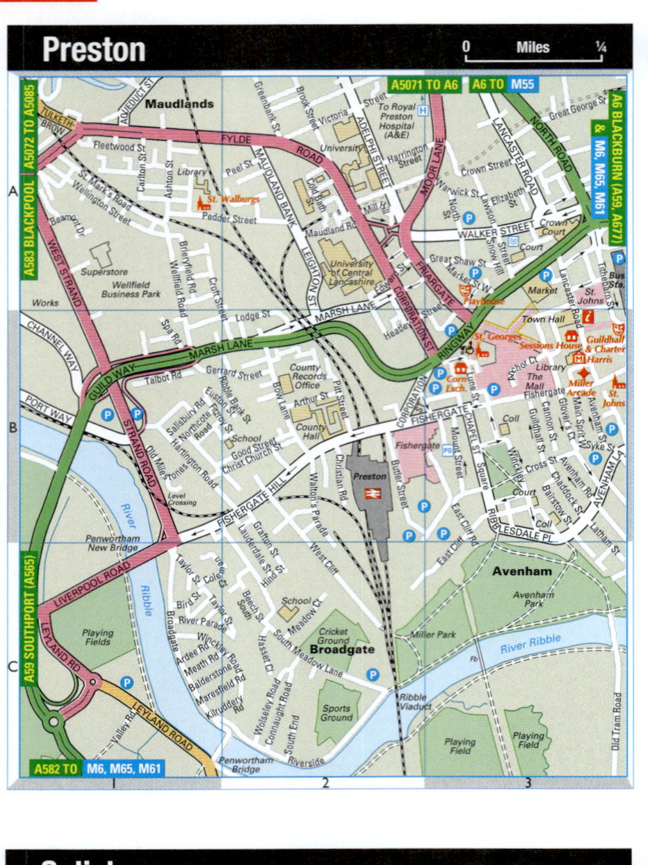
Preston
0 Miles ¼
A5071 TO A6
A6 TO M55
A6 BLACKBURN (A59, A677) & M6, M65, M61
A583 BLACKPOOL
A5072 TO A5085
A59 SOUTHPORT (A565)
A582 TO M6, M65, M61
Maudlands
Avenham
Broadgate
River Ribble

Reading
0 Miles ¼
A4155 HENLEY-ON-THAMES, OXFORD (A4074)
A329 OXFORD
A4155 TO A4
M4 NEWBURY A4
A33 BASINGSTOKE M4
A327 ALDERSHOT
A4 MAIDENHEAD A329 WOKINGHAM M4
River Thames
River Kennet

St Andrews
0 Miles ¼
St Andrews Bay
A91 CUPAR
A915 KIRKCALDY
A917 CRAIL
University and College Buildings

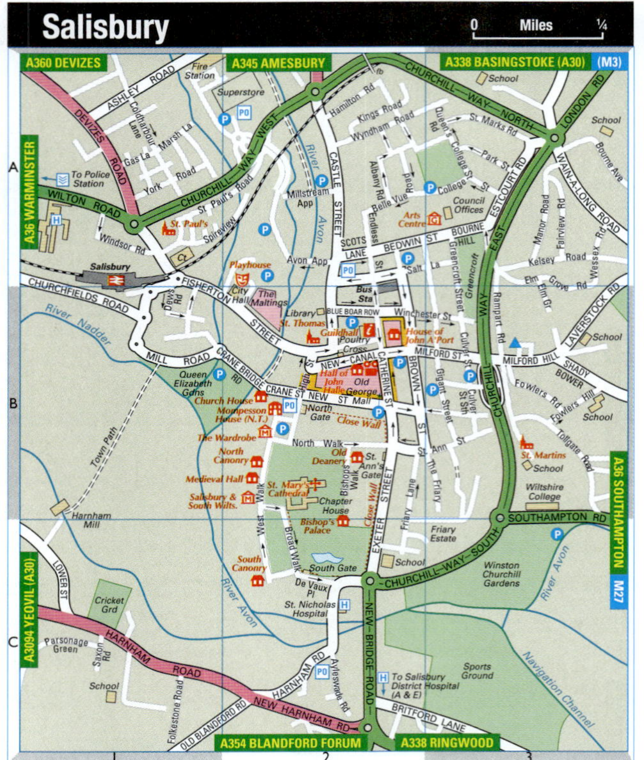
Salisbury
0 Miles ¼
A360 DEVIZES
A345 AMESBURY
A338 BASINGSTOKE (A30)
M3
A36 WARMINSTER
A3094 YEOVIL (A30)
A354 BLANDFORD FORUM
A338 RINGWOOD
A36 SOUTHAMPTON
M27
River Nadder
River Avon

Scarborough
0 Miles ¼
A165 WHITBY (A171)
NORTH BAY
SOUTH BAY
A64 YORK, THIRSK (A170)
A165 BRIDLINGTON, FILEY
North Sands
South Sands

Shrewsbury
0 Miles ¼
A528 ELLESMERE
A5191 WHITCHURCH (A49)
A458 WELSHPOOL
A488 BISHOP'S CASTLE
A5112 WHITCHURCH (A49)
A458 BRIDGNORTH
A5191 LUDLOW (A49)
A5112 LUDLOW (A49)
River Severn
Castle Fields

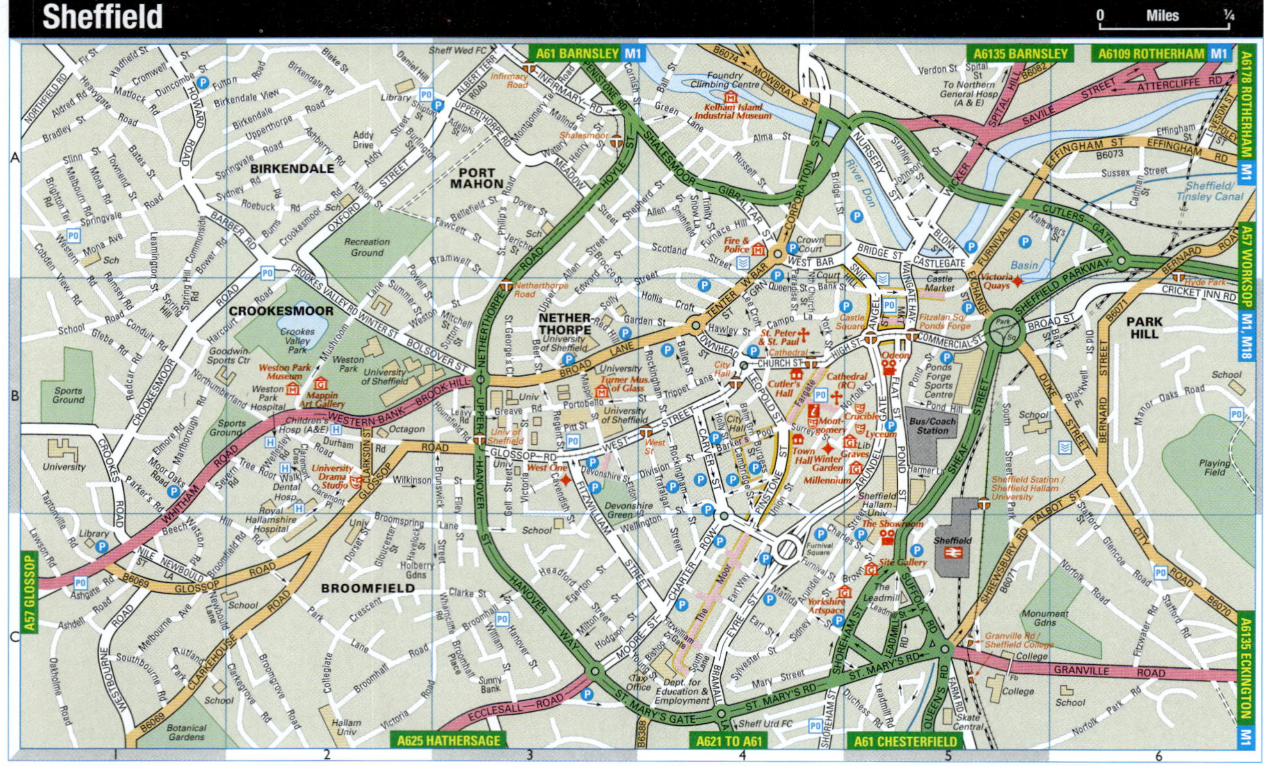
Sheffield
0 Miles ¼
A61 BARNSLEY M1
A6135 BARNSLEY
A6109 ROTHERHAM M1
A6178 ROTHERHAM M1
A57 WORKSOP M1 M18
A6135 ECKINGTON M1
A57 GLOSSOP
A625 HATHERSAGE
A621 TO A61
A61 CHESTERFIELD
BIRKENDALE
PORT MAHON
CROOKESMOOR
NETHER THORPE
PARK HILL
BROOMFIELD

Southampton
0 Miles ¼
A33 WINCHESTER M3
A3024 FAREHAM M27
A3024 SALISBURY (A36) RINGWOOD (A31) M27
A3025 PORTSMOUTH M27
Polygon
Queensland
Chapel
Ocean Village
River Test
River Itchen

Southend-on-Sea
Stirling
Stoke
Stratford-upon-Avon
Sunderland
Swansea / Abertawe
Swindon
Taunton
Telford
Miles

Torquay
Tor Bay
A3022 Exeter (A380, A38)
A379 Teignmouth
A3022 Plymouth (A385, A38)
Truro
B3284 Newquay (A3075) & A30
A390 St Austell
A390 Redruth (A30)
A39 Falmouth
Wick
A99 John O'Groats
A882 Thurso
A99 Inverness (A9)
Wick Bay
Winchester
B3420 Newbury (A34)
B3047 Basingstoke (A33)
B3049 Stockbridge
B3040 Romsey (A3090)
B3404 Alton (A31)
B3335 To M3
B3330 To M3 Sthbnd
Windsor
A332 Slough M4
Heathrow Airport (M4)
A308 Maidenhead
B470 To A4 M4
A308 Staines M25
Wolverhampton
A449 To M54, Stafford
A41 Telford (M54)
A454 Bridgnorth
A449 Stourbridge (A491)
A4123 Dudley
A460 To M54, Cannock
A4124 Brownhills
A454 To M6, Walsall
A41 Birmingham
Worcester
A38 Kidderminster (A449)
A443 Ludlow (A456, A49)
A44 Leominster, Hereford (A4103)
A449 Ross-on-Wye
A38 Tewkesbury M50
A44 M5 Jc 7
Wrexham / Wrecsam
A483 Chester
A5152 Chester (A483)
A541 Mold
A534 Nantwich
A525 Ruthin
A5152 Oswestry (A483)
A525 Whitchurch
York
A19 Thirsk
A1036 Scarborough (A64)
A59 Harrogate
A1036 Leeds (A64)
A19 Selby
A1079 Hull

Town plan indexes

Aberdeen 331

Aberdeen B2
Aberdeen Grammar School A1
Academy, The B2
Albert Basin B3
Albert Quay B3
Albury Rd C1
Alford Pl B1
Art Gallery A2
Arts Centre A2
Back Wynd A2
Baker St A1
Beach Blvd A3
Belmont B2
Belmont St B2
Berry St A2
Blackfriars St A2
Bloomfield Rd C1
Bon Accord Centre A2
Bon-Accord St B1/C1
Bridge St B2
Broad St A2
Bus Station B2
Car Ferry Terminal B3
Castlegate A3
Central Library A1
Chapel St B1
College A2
College St B2
Commerce St A3
Commercial Quay B3
Community Centre A3/C1
Constitution St A3
Cotton St A3
Crown St B2
Denburn Rd A2
Devanha Gdns C2
Devanha Gdns South C2
East North St A3
Esslemont Ave A1
Ferryhill Rd C2
Ferryhill Terr C2
Fish Market B3
Fonthill Rd C1
Galleria, The B1
Gallowgate A2
George St A2
Glenbervie Rd C3
Golden Sq B1
Grampian Rd C3
Great Southern Rd C1
Guild St B2
Hardgate B1/C1
His Majesty's Theatre A1
Holburn St C1
Hollybank Pl C1
Huntly St B1
Hutcheon St A1
Information Ctr B2
John St A2
Justice St A3
King St A2
Langstane Pl B1
Lemon Tree, The A2
Library C1
Loch St A2
Maberly St A1
Marischal College A2
Maritime Museum & Provost Ross's House B2
Market St B2/B3
Menzies Rd C3
Mercat Cross A3
Millburn St C2
Miller St A3
Market B2
Mount St A1
Music Hall B1
North Esp East C3
North Esp West C2
Oscar Rd C3
Palmerston Rd C2
Park St A3
Police Station A2
Polmuir Rd C2
Post Office A1/A2/A3/B1/C3
Provost Skene's House A2
Queen St A2
Regent Quay B3
Regent Road B3
Robert Gordon's College A2
Rose St B1
Rosemount Pl A1
Rosemount Viaduct A1
St Andrew St A2
St Andrew's Cathedral A3
St Mary's Cathedral B1
St Nicholas Centre A2
St Nicholas St A2
School Hill A2
Sinclair Rd C3
Skene Sq A1
Skene St B1
South College St C2
South Crown St C2
South Esp East C3
South Esp West C3
South Mount St A1
Sports Centre C3
Spring Garden A2
Springbank Terr C2
Summer St B1
Swimming Pool B1
The Mall B1
Thistle St B1
Tolbooth A3
Town House A2
Trinity Quay B3
Union Row B1
Union St B1/B2
Upper Dock B3
Upper Kirkgate A2
Victoria Bridge C3
Victoria Dock B3
Victoria Rd C3
Victoria St B2
Virginia St A3
Vue B2
Wellington Pl B2
West North St A2
Whinhill Rd C1
Willowbank Rd C1
Windmill Brae B2
Woolmanhill Hospital A1

Aberystwyth 331

Aberystwyth Holiday Village C2
Aberystwyth RFC C3
Aberystwyth Station B2
Aberystwyth Town Football Ground B2
Alexandra Rd B2
Ambulance Station C3
Baker St B1
Banadl Rd B2
Bandstand A1
Bath St A2
Boat Landing Stage A1
Boulevard de Saint-Brieuc C3
Bridge St B1
Bronglais Hospital B3
Bryn-y-Mor Rd A2
Buarth Rd B2
Bus Station B2
Cae Ceredig C3
Cae Melyn A2
Cae'r-Gog B3
Cambrian St B2
Caradoc Rd B3
Caravan Site C2
Castle (Remains of) B1
Castle St B1
Cemetery B3
Ceredigion Museum A1
Chalybeate St B1
Cliff Terr A2
Club House A2
Commodore A1
County Court A2
Crown Buildings B2
Dan-y-Coed A3
Dinas Terr C1
Eastgate B1
Edge-hill Rd B2
Elm Tree Ave B2
Elysian Gr A2
Felin-y-Mor Rd C1
Fifth Ave C2
Fire Station C1
Glanrafon Terr B1
Glyndŵr Rd B2
Golf Course A3
Gray's Inn Rd B1
Great Darkgate St B1
Greenfield St B2
Heol-y-Bryn A2
High St B1
Infirmary Rd A2
Information Ctr B1
Iorwerth Ave B3
King St B1
Lauraplace B1
Library B1
Lifeboat Station C1
Llanbadarn Rd B3
Loveden Rd A2
Magistrates Court A1
Marina C1
Marine Terr A1
Market B1
Mill St B1
Moor La B2
National Library of Wales B3
New Promenade B1
New St B1
North Beach A1
North Parade B2
North Rd A2
Northgate St B2
Parc Natur Penglais A3
Parc-y-Llyn Retail Park C3
Park & Ride B2
Park Ave B2
Pavillion B1
Pendinas C1
Penglais Rd A3
Penrheidol C2
Pen-y-Craig A2
Pen-yr-angor C1
Pier St B1
Plas Ave B3
Plas Helyg C2
Plascrug Ave B2/C3
Plascrug Leisure Centre C3
Police Station C2
Poplar Row B2
Portland Rd B2
Portland St A2
Post Office B1/B3
Powell St B1
Prospect St B1
Quay Rd B1
Queen St B1
Queen's Ave A2
Queen's Rd A2
Rheidol Retail Park B2
Riverside Terr B1
St Davids Rd B3
St Michael's B1
School of Art B2
Seaview Pl B1
South Beach B1
South Rd B1
Sports Ground B2
Spring Gdns C1
Stanley Terr B2
Swimming Pool & Leisure Centre C3
Superstore B2/C3
Tanybwlch Beach C1
Tennis Courts B3
Terrace Rd B1
The Bar C1
Town Hall A2
Trefechan Bridge B1
Trefechan Rd C2
Trefor Rd A2
Trinity Rd B2
University Campus B3
University of Wales (Aberystwyth) B1
Vale of Rheidol Railway C3
Vaynor St B2
Victoria Terr A1
Viewpoint A2
Viewpoint A3
War Memorial B1
Wharf Quay C1
Y Lanfa C1
Ystwyth Retail Park B2

Ashford 331

Albert Rd A1
Alfred Rd C3
Apsley St A1
Ashford A1
Ashford Borough Museum A1
Ashford International Station B2
Bank St A1
Barrowhill Gdns A1
Beaver Industrial Estate C1
Beaver Rd C1
Beazley Ct C3
Birling Rd B3
Blue Line La A1
Bond Rd C1
Bowens Field B1
Bulleid Pl C1
Cade Rd C1
Chart Rd A1
Chichester Cl B1
Christchurch Rd B1
Chunnel Industrial Estate B1
Church Rd A1
Civic Centre A2
County Square Shopping Centre A1
Court A1
Court A2
Croft Rd A3
Cudworth Rd C3
Curtis Rd C3
Dering Rd A3
Dover Pl B2
Drum La A1
East Hill A1
East St A1
Eastmead Ave B2
Edinburgh Rd A1
Elwick Rd A1
Essella Pk B3
Essella Rd B3
Fire Sta A3
Forge La A1
Francis Rd C1
George St B1
Godfrey Walk B1
Godinton Rd A1
Gordon Cl A3
Hardinge Rd A2
Henwood A3
Henwood Business Centre A3
Henwood Industrial Estate A3
High St A2
Hythe Rd A3
Information Ctr A1
Jemmett Rd B1
Kent Ave A1
Library A1
Linden Rd B3
Lower Denmark Rd C1
Mabledon Ave B3
Mace Industrial Est A2
Mace La A2
Maunsell Pl C3
McArthur Glen Designer Outlet C2
Memorial Gdns A2
Mill Ct A2
Miller Cl A2
Mortimer Cl C1
New St A1
Newtown Green C3
Newtown Rd B2/C3
Norman Rd C1
North St A2
Norwood Gdns A1
Norwood St A1
Old Railway Works Industrial Estate C3
Orion Way C3
Park Mall Shopping Centre A1
Park Pl A1
Park St A1/A2
Pemberton Rd A3
Police Station A1
Post Office A1/A3
Providence St C2
Queen St A1
Queens Rd A2
Regents Pl A1
Riversdale Rd C2
Romney Marsh Rd B2
St John's La A2
Somerset Rd A2
South Stour Ave B2
Star Rd A3
Station Rd B2
Stirling Rd C2
Stour Centre, The B2
Sussex Ave A1
Tannery La A2
Technical College B2
Torrington Rd C2
Trumper Bridge B1
Tufton Rd A3
Tufton St A1
Vicarage La A2
Victoria Cres B1
Victoria Park B1
Victoria Rd B1
Wallis Rd A3
Wellesley Rd A2
West St A1
Whitfeld Rd C1
William Rd C1
World War I Tank A1

Ayr 331

Ailsa Pl B1
Alexandra Terr A3
Allison St B2
Alloway Pk C1
Alloway Pl C1
Alloway St C2
Arran Mall C2
Arran Terr B1
Arthur St B2
Ashgrove St C2
Auld Brig B2
Auld Kirk B2
Ayr C2
Ayr Academy B1
Ayr Central Shopping Centre C2
Ayr Harbour A1
Ayr United FC A3
Back Hawkhill Ave A3
Back Main St B2
Back Peebles St A2
Barns Cres C1
Barns Pk C1
Barns St C1
Barns Street La C1
Bath Pl B1
Bellevue Cres C1
Bellevue La C1
Beresford La C2
Beresford Terr C2
Boswell Pk B2
Britannia Pl A3
Bruce Cres B1
Burns Statue C2
Bus Sta B1
Carrick St C2
Cassillis St B1
Cathcart St B1
Charlotte St B1
Citadel Leisure Centre B1
Citadel Pl B1
Compass Pier A1
Content Ave C3
Content St B2
Craigie Ave B3
Craigie Rd B3
Craigie Way B3
Cromwell Rd B1
Crown St A2
Dalblair Rd C2
Dam Park Sports Stadium C3
Damside A2
Dongola Rd C3
Eglinton Pl B1
Eglinton Terr B1
Elba St B2
Elmbank St A2
Esplanade B1
Farifield Rd C1
Fort St B1
Fothringham Rd C3
Fullarton St C1
Gaiety C2
Garden St B2
George St B2
George's Ave A3
Glebe Cres A2
Glebe Rd A2
Gorden Terr B3
Green St A2
Green Street La A2
Hawkhill Ave A3
Hawkhill Avenue La B3
High St B2
Holmston Rd C3
Information Ctr B1
James St B3
John St B2
King St B2
Kings Ct B2
Kyle Centre C2
Kyle St C2
Library B2
Limekiln Rd A2
Limonds Wynd B2
Loudoun Hall B2
Lymburn Pl B3
Macadam Pl B2
Main St B2
Mcadam's Monument C1
Mccall's Ave A3
Mews La B1
Mill Brae C3
Mill St C2
Mill Wynd C2
Miller Rd C1
Montgomerie Terr B1
New Bridge B2
New Bridge St B2
New Rd A2
Newmarket St B2
Newton-on-Ayr Station A2
North Harbour St B1
North Pier A1
Odeon C2
Oswald La A1
Park Circus C1
Park Circus La C1
Park Terr C1
Pavilion Rd C1
Peebles St A2
Philip Sq B2
Police Station B2
Post Office A2/B2
Prestwick Rd B1
Princes Ct A2
Queen St B3
Queen's Terr B1
Racecourse Rd C1
River St B2
Riverside Pl B2
Russell Dr A2
St Andrews Church C2
St George's Rd A3
Sandgate B1
Savoy Park C1
Seabank Rd B2
Smith St C2
Somerset Rd A3
South Beach Rd B1
South Harbour St B1
South Pier A1
Station Rd C2
Strathaye Pl B2
Taylor St A2
Town Hall B2
Tryfield Pl A3
Turner's Bridge B2
Union Ave A3
Victoria Bridge C3
Victoria St B3
Viewfield Rd A3
Virginia Gdns A2
Waggon Rd A2
Walker Rd A3
Wallace Tower B2
Weaver St A2
Weir Rd A2
Wellington La C1
Wellington Sq C1
West Sanouhar Rd A3
Whitletts Rd B3
Wilson St A3
York St A1
York Street La B1

Bangor 331

Abbey Rd C2
Albert St B1
Ambrose St A3
Ambulance Station A3
Arfon Sports Hall C1
Ashley Rd B3
Bangor City Football Ground C2
Bangor Mountain B3
Bangor Station C1
Bangor University B2
Beach Rd A3
Belmont St C1
Bishop's Mill Rd C3
Boat Yard A3
Brick St B3
Buckley Rd B2
Bus Station B3
Caellepa B2
Caernarfon Rd C1
Cathedral B2
Cemetery C1
Clarence St C1
Clock B3
College B2/C2
College La B2
College Rd B2
Convent La C1
Council Offices B2
Craig y Don Rd B2
Dean St B3
Deiniol Rd B2
Deiniol Shopping Centre B2
Deiniol St B2
Edge Hill A3
Euston Rd C1
Fairview Rd A3
Farrar Rd C2
Ffordd Cynfal C1
Ffordd Elfed C3
Ffordd Islwyn A3
Ffordd y Castell C3
Ffriddoedd Rd B1
Field St B1
Fountain St A3
Friars Ave B3
Friars Rd B3
Friary (site of) B3
Gardd Deman C1
Garth Hill A3
Garth Point A3
Garth Rd A3
Glanrafon B2
Glanrafon Hill B2
Glynne Rd B3
Golf Course B3
Golf Course C2
Gorad Rd B1
Gorsedd Circle A2
Gwern Las C3
Heol Dewi C1
High St B3/C2
Hill St B1
Holyhead Rd B1
Hwfa Rd B1
Information Ctr B2
James St B3
Library B2
Llys Emrys A3
Lon Ogwen C1
Lon-Pobty C2
Lon-y-Felin C3
Lon-y-Glyder C1
Love La B2
Lower Penrallt Rd B2
Lower St B2
Maes-y-Dref B3
Maeshyfryd A3
Meirion La A2
Meirion Rd A2
Menai Ave B1
Menai College C1
Menai Shopping Centre B3
Min-y-Ddol C3
Minafon B2
Mount St B3
Museum & Art Gallery B2
Orme Rd A3
Parc Victoria B1
Penchwintan Rd C1
Penlon Gr B3
Penrhyn Ave C3
Pier A3
Police Station B2
Post Office B2/B3/C1/C3
Prince's Rd B2
Queen's Ave C3
Sackville Rd B2
St Paul's St B2
Seion Rd A3
Seiriol Rd A3
Siliwen Rd A2
Snowdon View B1
Sports Ground B1
Station Rd C1
Strand St B3
Swimming Pool and Leisure Centre A3
Tan-y-Coed C3
Tegid Rd B3
Temple Rd B2
The Crescent B2
Theatr Gwynedd B2
Totton Rd A3
Town Hall B2
Treflan B2
Trem Elidir C1
Upper Garth Rd A3
Victoria Ave B1
Victoria Dr B1
Victoria St B1
Vron St B2
Well St B3
West End C1
William St B3
York Pl B3

Barrow-in-Furness 331

Abbey Rd A3/B2
Adelaide St A2
Ainslie St A3
Albert St C3
Allison St B3
Anson St A2
Argyle St B3
Arthur St B3
Ashburner Way A1
Barrow Raiders RLFC B1
Barrow Station A2
Bath St A1/B2
Bedford Rd A3
Bessamer Way A1
Blake St A1/A2
Bridge Rd C1
Buccleuch Dock C3
Buccleuch Dock Rd C2/C3
Buccleuch St B2/B3
Byron St A2
Calcutta St A1
Cameron St C1
Carlton Ave A3
Cavendish Dock Rd C3
Cavendish St B2/B3
Channelside Walk B1
Channelside Haven C1
Chatsworth St A2
Cheltenham St A3
Church St C3
Clifford St B2
Clive St B1
Collingwood St B2
Cook St A2
Cornerhouse Retail Park B2
Cornwallis St B2
Courts A2
Crellin St B3
Cross St C3
Dalkeith St B2
Dalton Rd B2/C2
Derby St B3
Devonshire Dock C2
Devonshire Dock Hall B1
Dock Museum, The B1
Drake St A2
Dryden St A2
Duke St A1/B2/C3
Duncan St B2
Dundee St C2
Dundonald St B2
Earle St C1
Emlyn St B2
Exmouth St A2
Farm St C2
Fell St B3
Fenton St B3
Ferry Rd C1
Forum 28 B2
Furness College B1
Glasgow St B3
Goldsmith St A2
Greengate St B3
Hardwick St A2
Harrison St B2
Hartington St A2
Hawke St B2
Hibbert Rd A2
High Level Bridge C2
High St B2
Hindpool Retail Park B2
Hindpool Rd B2
Holker St A2
Hollywood Retail & Leisure Park B1
Hood St A2
Howard St B2
Howe St A2
Information Ctr B2
Ironworks Rd A1/B1
James St B3
Jubliee Bridge C1
Keith St B2
Keyes St A2
Lancaster St A3
Lawson St B2
Library B2
Lincoln St A3
Longreins Rd A3
Lonsdale St C3
Lord St B3
Lorne Rd B3
Lyon St A2
Manchester St B2
Market B2
Market St B2
Marsh St B3
Michaelson Rd C2
Milton St A2
Monk St B2
Mount Pleasant B3
Nan Tait Centre B2
Napier St B2
Nelson St B2
North Rd B1
Open Market B2
Parade St B2
Paradise St B3
Park Ave A3
Park Dr A3
Parker St A2
Parry St A2
Peter Green Way A1
Phoenix Rd A1
Police Station B2
Portland Walk Shopping Centre B2
Post Office A3/B2
Princess Selandia C2
Raleigh St A2
Ramsden St B3
Rawlinson St B3
Robert St B3
Rodney St B2
Rutland St A2
St Patricks Rd C1
Salthouse Rd C3
School St B2
Scott St B2
Settle St A3
Shore St C3
Sidney St B2
Silverdale St B3
Slater St B2
Smeaton St B3
Stafford St A3
Stanley Rd C1
Stark St C3
Steel St B1
Storey Sq B3
Strand C3
Superstore A1/B1/C3
Sutherland St B3
TA Centre A2
The Park A3
Thwaite St B3
Town Hall B2
Town Quay C3
Vernon St B2
Vincent St B2
Walney Rd A1
West Gate Rd A2
West View Rd A3
Westmorland St A3
Whitehead St A3
Wordsworth St A2

Bath 331

Alexandra Park C2
Alexandra Rd C2
Approach Golf Courses (Public) A1
Bath Aqua Glass A2
Archway St C3
Assembly Rooms & Museum of Costume A2
Avon St B2
Barton St B2
Bath Abbey B2
Bath City College B2
Bath Pavilion B3
Bath Rugby Club B3
Bath Spa Station C3
Bathwick St A3
Beckford Road A3
Beechen Cliff Rd C2
Bennett St A2
Bloomfield Ave C1
Broad Quay C2
Broad St B2
Brock St A1
Building of Bath Museum A2
Bus Station C2
Calton Gdns C2
Calton Rd C2
Camden Cr A2
Cavendish Rd A1
Cemetery B1
Charlotte St B2
Chaucer Rd C2
Cheap St B2
Circus Mews A2
Claverton St C2
Corn St C2
Cricket Ground B3
Daniel St A3
Edward St B3
Ferry La B3
First Ave C1
Forester Ave A3
Forester Rd A3
Gays Hill A2
George St B2
Great Pulteney St B3
Green Park B1
Green Park Rd B2
Grove St B2
Guildhall B2
Harley St A2
Hayesfield Park C1
Henrietta Gdns A3
Henrietta Mews B3
Henrietta Park B3
Henrietta Rd A3
Henrietta St B3
Henry St B2
Holburne Museum B3
Holloway C2
Information Ctr B2
James St West B1/B2
Jane Austen Centre B2
Julian Rd A1
Junction Rd C1
Kipling Ave C2
Lansdown Cr A1
Lansdown Gr A2
Lansdown Rd A2
Library B2
London Rd A3
London St A2
Lower Bristol Rd B1
Lower Oldfield Park C1
Lyncombe Hill C3
Manvers St B3
Maple Gr C1
Margaret's Hill A2
Marlborough Buildings A1
Marlborough La B1
Midland Bridge Rd B1
Milk St B2
Milsom St B2
Monmouth St B2
Morford St A2
Museum of Bath at Work A2
New King St B1
No. 1 Royal Crescent A1
Norfolk Bldgs B1
Norfolk Cr B1
North Parade Rd B3
Oldfield Rd C1
Paragon A2
Pines Way B1
Podium Shopping Centre B2
Police Station B3
Portland Pl A2
Post Office A1/A3/B2/C1/C2
Postal Museum B2
Powlett Rd A3
Prior Park Rd C3
Pulteney Bridge B2
Pulteney Gdns B3
Pulteney Rd B3/C3
Queen Sq B2
Raby Pl B3
Recreation Ground B3
Rivers St A2
Rockliffe Ave A3
Rockliffe Rd A3
Roman Baths & Pump Room B2
Rossiter Rd C3
Royal Ave A1
Royal Cr A1
Royal High School, The A2
Royal Victoria Park A1
St James Sq A1
St John's Rd A3
Shakespeare Ave C2
Southgate C2
South Pde B3
Sports & Leisure Centre B3
Spring Gdns C3
Stall St B2
Stanier Rd B1
Superstore B1
Sydney Gdns A3
Sydney Pl B3
Sydney Rd B3
Theatre Royal B2
Thermae Bath Spa B2
The Tyning C3
Thomas St A3
Union St B2
Upper Bristol Rd B1
Upper Oldfield Park C1
Victoria Art Gallery B2
Victoria Bridge Rd B1
Walcot St B2
Wells Rd C1
Westgate Buildings B2
Westgate St B2
Weston Rd A1
Widcombe Hill C3
William Herschel Museum B1

Berwick-upon-Tweed 331

Bank Hill B2
Barracks A3
Bell Tower A3
Bell Tower Pl A2
Berwick Br B2
Berwick Infirmary A3
Berwick Rangers F.C. C1
Berwick-upon-Tweed A2
Billendean Rd C3
Blakewell Gdns B2
Blakewell St B2
Brass Bastion A3
Bridge St B3
Brucegate St A2
Castle (Remains of) A2
Castle Terr A2
Castlegate A2
Chapel St A3
Church Rd C2
Church St B3
Court B3
Coxon's La A3
Cumberland Bastion A3
Dean Dr C2
Dock Rd C2/C3
Elizabethan Walls A2/B3
Fire Station B1
Flagstaff Park B3
Football Ground C3
Foul Ford B3
Gallery A3
Golden Sq B2
Golf Course A3
Greenwood C1
Gunpowder Magazine B3
Hide Hill B3
High Greens A2
Holy Trinity A3
Information Ctr A2
Kiln Hill B2
King's Mount B3
Ladywell Rd C2
Library A3
Lifeboat Station C3
Lord's Mount A3
Lovaine Terr A2
Low Greens A3
Main Guard B3
Main St B2/C2
Maltings Art Centre, The B3
Marygate B3
Meg's Mount A2
Middle St C3
Mill St C2
Mount Rd C2
Museum B3
Ness St B3
North Rd A2
Northumberland Ave A2
Northumberland Rd C2
Ord Dr B1
Osborne Cr B2
Osborne Rd B1
Palace Gr B3
Palace St B3
Palace St East B3
Pier Rd B3
Playing Field C1
Police Station B3
Post Office A2/B2/B2
Prince Edward Rd B2
Prior Rd C2
Quay Walls B3
Railway St A2
Ravensdowne B3
Records Office A3
Riverdene B1
Riverside Rd B2
Royal Border Br A2
Royal Tweed Br B2
Russian Gun B3
Scots Gate A2
Scott's Pl A2
Shielfield Park C1
Shielfield Terr C2
Silver St B3
Spittal Quay C3
Superstores C2
The Avenue B3
The Parade A3
Tower Gdns A2
Tower Rd C2
Town Hall B3
Turret Gdns C2
Tweed Dock B2
Tweed St A2
Tweedside Trading Estate C1
Union Brae B2
Union Park Rd B2
Walkergate A3
Wallace Gr A3
War Memorial A2
War Memorial B2
Warkworth Terr A2
Well Close Sq A2
West End B2
West End Pl B1
West End Rd B1
West St B3
West St C3
Windmill Bastion B3
Woolmarket B3
Works C3

Birmingham 332

Abbey St A2
Aberdeen St A1
Acorn Gr B2
Adams St A5
Adderley St C5
Albert St B4/B5
Albion St B2
Alcester St C5
Aldgate Gr A3

Alexandra Theatre C3
All Saint's St A2
All Saints Rd A2
Allcock St C5
Allesley St A4
Allison St C4
Alma Cr B6
Alston Rd C1
Arcadian Centre C4
Arthur St C6
Assay Office B3
Aston Expressway A5
Aston Science Park B5
Aston St B4
Aston University B4/B5
Avenue Rd A5
BT Tower B3
Bacchus Rd A1
Bagot St B4
Banbury St B5
Barford Rd B1
Barford St C4
Barn St C5
Barnwell Rd C6
Barr St A3
Barrack St B5
Bartholomew St C4
Barwick St B4
Bath Row C3
Beaufort Rd C1
Belmont Row B5
Benson Rd A1
Berkley St C3
Bexhill Gr C3
Birchall St C5
Birmingham City F.C. C6
Birmingham City Hospital (A&E) A1
Bishopsgate St C3
Blews St A4
Bloomsbury St A6
Blucher St C3
Bordesley St C4
Bowyer St C5
Bradburne Way A5
Bradford St C5
Branston St A3
Brearley St A4
Brewery St A4
Bridge St A3
Bridge St C3
Bridge St West A4
Brindley Dr B3
Broad St C3
Broad St UGC C2
Broadway Plaza C2
Bromley St C5
Bromsgrove St C4
Brookfield Rd A2
Browning St C2
Bryant St A1
Buckingham St A3
Bullring C4
Bull St B4
Cambridge St C3
Camden Dr B3
Camden St B2
Cannon St C4
Cardigan St B5
Carlisle St A1
Carlyle Rd C1
Caroline St B3
Carver St B2
Cato St A6
Cattell Rd C6
Cattells Gr A6
Cawdor Cr C1
Cecil St B4
Cemetery A2/B2
Cemetery La A2
Centre Link Industrial Estate A6
Charlotte St B3
Cheapside C4
Chester St A5
Children's Hospital (A&E) B4
Church St B4
Claremont Rd A2
Clarendon Rd C1
Clark St C1
Clement St B3
Clissold St B2
Cleveland St B4
Coach Station C5
College St B2
Colmore Circus B4
Colmore Row B4
Commercial St C3
Constitution Hill B3
Convention Centre, The C3
Cope St B2
Coplow St B1
Corporation St B4
Council House B3
County Court B4
Coveley Gr A2
Coventry Rd C6
Coventry St C5
Cox St B3
Crabtree Rd A2
Cregoe St C3
Crescent Ave A2
Crescent Theatre C3
Cromwell St A6
Cromwell St B3
Curzon St B5
Cuthbert Rd B1
Dale End B4
Dart St C6
Dartmouth Circus A4
Dartmouth Middleway A5
Dental Hospital B4
Deritend C5
Devon St A6
Devonshire St A1
Digbeth Civic Hall C5
Digbeth High St C4
Dolman St B6
Dover St A1
Duchess Rd C2
Duddeston B6
Duddeston Manor Rd B5
Duddeston Mill Rd B6
Duddeston Mill Trading Estate B6
Dudley Rd B1
Edgbaston Shopping Centre C2
Edmund St B3
Edward St B3
Elkington St A4
Ellen St B2
Ellis St C3
Erskine St B6
Essex St C4
Eyre St B2
Farm Croft A3
Farm St A3
Fazeley St B4/C5
Felstead Way B5
Finstall Cl B5
Five Ways C2
Fleet St B3
Floodgate St C5
Ford St A2
Fore St C4
Forster St B5
Francis Rd C2
Francis St B5
Frankfort St A4
Frederick St B3
Freeth St C1
Freightliner Terminal B6
Garrison La C6
Garrison St B6
Gas St C3
Geach St A4
George St B3
George St West B2
Gibb St C5
Gillott Rd B1
Gilby Rd C2
Glover St C5
Goode Ave A2
Goodrick Way A6
Gordon St B6
Graham St B3
Granville St C3
Gray St C6
Great Barr St C5
Great Charles St B3
Great Francis St B6
Great Hampton Row A3
Great Hampton St A3
Great King St A3
Great Lister St A5
Great Tindal St C2
Green La C6
Green St C5
Greenway St C6
Grosvenor St West C2
Guest Gr A3
Guild Cl C2
Guildford Dr A4
Guthrie Cl A3
Hagley Rd C1
Hall St B3
Hampton St A3
Handsworth New Rd A1
Hanley St B4
Harford St A3
Harmer Rd A2
Harold Rd C1
Hatchett St A4
Heath Mill La C5
Heath St B1
Heath St South B1
Heaton St A2
Heneage St B5
Henrietta St B4
Herbert Rd C6
High St C4
High St C5
Hilden Rd B5
Hill St C3/C4
Hindlow Cl B6
Hingeston St B2
Hippodrome Theatre C4
HM Prison A1
Hockley Circus A2
Hockley Hill A3
Hockley St A3
Holliday St C3
Holloway Circus C4
Holloway Head C3
Holt St B5
Hooper St B1
Horse Fair C4
Hospital St A4
Howard St B3
Howe St B5
Hubert St A5
Hunters Rd A2
Hunters Vale A3
Huntly Rd C2
Hurst St C4
Icknield Port Rd B1
Icknield Sq B2
Icknield St A2/B2
Ikon Gallery C3
Information Ctr C4
Inge St C4
Irving St C3
Ivy La C5
James Watt Queensway B4
Jennens Rd B5
Jewellery Quarter A3
Jewellery Quarter Museum B3
John Bright St C4
Keeley St C6
Kellett Rd B5
Kent St C4
Kenyon St B3
Key Hill A3
Kilby Ave C2
King Edwards Rd B2
King Edwards Rd C3
Kingston Rd C6
Kirby Rd A1
Ladywood Arts & Leisure Centre B1
Ladywood Middleway C2/C3
Ladywood Rd C1
Lancaster St B4
Landor St B6
Law Courts B4
Lawford Cl B5
Lawley Middleway B5
Ledbury Cl C2
Ledsam St B2
Lees St A1
Legge La B3
Lennox St A3
Library A6/C3
Library Walk B2
Lighthorne Ave B2
Link Rd B1
Lionel St B3
Lister St B5
Little Ann St C5
Little Hall Rd A6
Liverpool St C5
Livery St B3/B4
Lodge Rd A1
Lord St A5
Love La A5
Loveday St B4
Lower Dartmouth St C6
Lower Loveday St B4
Lower Tower St A4
Lower Trinity St C5
Ludgate Hill B3
Mailbox Centre & BBC C3
Margaret St B3
Markby Rd A1
Marroway St B1
Maxstoke St C6
Melvina Rd A6
Meriden St C4
Metropolitan (R.C.) B4
Midland St B6
Milk St C5
Mill St A5
Millennium Point B5
Miller St A4
Milton St A4
Moat La C4
Montague Rd C1
Montague St C5
Monument Rd C1
Moor Street C4
Moor St Queensway C4
Moorsom St A4
Morville St C2
Mosborough Cr A3
Moseley St C4
Mott St B3
Museum & Art Gallery B3
Musgrave Rd A1
National Indoor Arena C2
National Sea Life Centre C3
Navigation St C3
Nechell's Park Rd A6
Nechells Parkway B5
Nechells Pl A6
New Bartholomew St C4
New Canal St C5
New John St West A3
New Spring St B2
New St C4
New Street C4
New Summer St A4
New Town Row A4
Newhall Hill B3
Newhall St B3
Newton St B4
Newtown A4
Noel Rd C1
Norman St A1
Northbrook St B1
Northwood St B3
Norton St A2
Old Crown House C5
Old Rep Theatre, The C4
Old Snow Hill B4
Oliver Rd C1
Oliver St A5
Osler St C1
Oxford St C5
Pallasades Centre C4
Palmer St C5
Paradise Circus C3
Paradise St C3
Park Rd A2
Park St C4
Pavilions Centre C4
Paxton Rd A2
Peel St A1
Penn St B5
Pershore St C4
Phillips St A4
Pickford St C5
Pinfold St C4
Pitsford St A2
Plough & Harrow Rd C1
Police Station A4/B1/B4/C2/C4
Pope St B2
Portland Rd C1
Post Office A3/A5/B1/B3/B4/B5/C2/C3/C5
Preston Rd A1
Price St B4
Princip St B4
Printing House St B4
Priory Queensway B4
Pritchett St A4
Proctor St A5
Queensway B3
Radnor St A2
Rea St C4
Regent Pl B3
Register Office C3
Repertory Theatre C3
Reservoir Rd C1
Richard St A5
River St C5
Rocky La A5/A6
Rodney Cl C2
Roseberry St B2
Rotton Park St B1
Rupert St A5
Ruston St C2
Ryland St C2
St Andrew's Industrial Estate C6
St Andrew's Rd C6
St Andrew's St C6
St Bolton St C6
St Chads Queensway B4
St Clements Rd A6
St George's St A3
St James Pl B5
St Marks Cr B2
St Martin's C4
St Paul's B3
St Paul's Metro Sta B3
St Paul's Sq B3
St Philip's B4
St Stephen's St A4
St Thomas' Peace Garden C3
St Vincent St C2
Saltley Rd A6
Sand Pits Pde B3
Severn St C3
Shadwell St B4
Sheepcote St C2
Shefford Rd A4
Sherborne St C2
Shylton's Croft C2
Skipton Rd C2
Smallbrook Queensway C4
Smith St A3
Snow Hill B4
Snow Hill Queensway B4
Soho, Benson Rd Metro Sta A1
South Rd A2
Spencer St B3
Spring Hill B2
Staniforth St B4
Station St C4
Steelhouse La B4
Stephenson St C4
Steward St B2
Stirling Rd C1
Stour St B2
Suffolk St C3
Summer Hill Rd B2
Summer Hill St B2
Summer Hill Terr B2
Summer La A4
Summer Row B3
Summerfield Cr B1
Summerfield Park B1
Sutton St C3
Swallow St C3
Sydney Rd C6
Symphony Hall C3
Talbot St A1
Temple Row C4
Temple St C4
Templefield St C6
Tenby St B3
Tenby St North B2
Tennant St C2/C3
The Crescent A2
Thimble Mill La A6
Thinktank (Science & Discovery) B5
Thomas St A4
Thorpe St C4
Tilton Rd C6
Tower St A4
Town Hall C3
Trent St C5
Turner's Buildings A1
Unett St A3
Union Terr B5
Upper Trinity St C5
Uxbridge St A3
Vauxhall Gr B5
Vauxhall Rd B5
Vernon Rd C1
Vesey St B4
Viaduct St B5
Victoria Sq C3
Villa St A3
Vittoria St B3
Vyse St B3
Walter St A6
Wardlow Rd A5
Warstone La B2
Washington St C3
Water St B3
Waterworks Rd C1
Watery La C5
Well St A3
Western Rd B1
Wharf St A2
Wheeler St A3
Whitehouse St A5
Whitmore St A2
Whittall St B4
Wholesale Market C4
Wiggin St B1
Willes Rd A1
Windsor Industrial Estate A5
Windsor St A5
Windsor St B5
Winson Green Rd A1
Witton St C6
Wolseley St C6
Woodcock St B5

Blackpool 332

Abingdon St A1
Addison Cr A3
Adelaide St B1
Albert Rd B2
Alfred St B2
Ascot Rd A3
Ashton Rd C2
Auburn Gr C3
Bank Hey St B1
Banks St A1
Beech Ave A3
Bela Gr C2
Belmont Ave B2
Birley St A1
Blackpool & Fleetwood Tram B1
Blackpool & The Fylde College B2
Blackpool F.C. C2
Blackpool North A2
Blackpool Tower B1
Blundell St C1
Bonny St B1
Breck Rd B3
Bryan Rd A3
Buchanan St A2
Cambridge Rd A3
Caunce St A2/A3
Central Dr B1/C2
Central Pier C1
Central Pier C1
Central Pier Theatre C1
Chapel St C1
Charles St A2
Charnley Rd B2
Church St A1/A2
Clinton Ave B2
Coach Station A2/C1
Cocker St A1
Cocker St A1
Coleridge Rd A3
Collingwood Ave A3
Condor Gr C3
Cookson St A2
Coronation St B1
Corporation St A1
Courts B1
Cumberland Ave B3
Cunliffe Rd C3
Dale St C1
Devonshire Rd A3
Devonshire Sq A3
Dickson Rd A1
Elizabeth St A2
Ferguson Rd C3
Forest Gate B3
Foxhall Rd C1
Foxhall Sq C1
Freckleton St C2
George St A2
Gloucester Ave B3
Golden Mile, The C1
Gorse Rd B3
Gorton St A2
Grand Theatre, The B1
Granville Rd A2
Grasmere Rd C2
Grosvenor St A2
Grundy Art Gallery A1
Harvey Rd B3
Hornby Rd B2
Hounds Hill Shopping Centre B1
Hull Rd B1
Ibbison Ct C2
Information Ctr A1
Kent Rd C2
Keswick Rd C2
King St A2
Knox Gr C3
Laycock Gate A3
Layton Rd A3
Leamington Rd B2
Leeds Rd B3
Leicester Rd B2
Levens Gr C2
Library A1
Lifeboat Station B1
Lincoln Rd B2
Liverpool Rd B3
Livingstone Rd B2
London Rd A3
Louis Tussaud's Waxworks B1
Lune Gr C2
Lytham Rd C1
Manchester Sq C1
Manor Rd B3
Maple Ave B3
Market St A1
Marlboro Rd B3
Mere Rd B3
Milbourne St A2
Newcastle Ave B3
Newton Dr A3
North Pier A1
North Pier A1
North Pier Theatre A1
Odeon C2
Olive Gr B3
Palatine Rd B2
Park Rd B2/C3
Peter St A2
Police Station B1
Post Office A1/A3B1/B2/B3
Princess Pde A1
Princess St C1/C2
Promenade A1/C1
Queen St A1
Queen Victoria Rd C2
Raikes Pde B2
Reads Ave B2
Regent Rd B2
Register Office B2
Ribble Rd B2
Rigby Rd C1/C2
Ripon Rd B3
St Albans Rd B3
St Ives Ave C3
St Vincent Ave C3
Salisbury Rd B3
Salthouse Ave C2
Salvation Army Ctr A2
Sands Way C2
Sealife Centre B1
Seasiders Way C1
Selbourne Rd A2
Sharrow Gr C3
Somerset Ave C3
Springfield Rd A1
South King St B2
Sutton Pl B2
Talbot Rd A1/A2
Thornber Gr C2
Topping St A1
Tower B1
Town Hall A1
Tram Depot C1
Tyldesley Rd C1
Vance Rd B1
Victoria St B1
Victory Rd A2
Wayman Rd A3
Westmorland Ave C2/C3
Whitegate Dr B3
Winter Gardens Theatre B1
Woodland Gr B3
Woolman Rd B2

Bournemouth 332

Ascham Rd A3
Avenue Rd B1
Bath Rd C2
Beacon Rd C1
Beach Office C2
Beechey Rd A3
Bodorgan Rd B1
Bourne Ave B1
Bournemouth A3
Bournemouth & Poole College B3
Bournemouth Eye C2
Bournemouth International Ctr C1
Bournemouth Pier C2
Bournemouth Station rbt B3
Braidley Rd A1
Cavendish Place A2
Cavendish Rd A2
Central Drive A1
Christchurch Rd B3
Cliff Lift C1/C3
Coach House Pl A3
Coach Station A3
Commercial Rd B1
Cotlands Rd B3
Courts B3
Cranborne Rd C1
Cricket Ground A2
Cumnor Rd B2
Dean Park A2
Dean Park Cr B2
Dean Park Rd A2
Durrant Rd B1
East Overcliff Dr C3
Exeter Cr C1
Exeter La C2
Exeter Rd C1
Gervis Place B1
Gervis Rd B3
Glen Fern Rd B2
Golf Club B1
Grove Rd B3
Hinton Rd C2
Holdenhurst Rd B3
Horseshoe Common B2
Nuffield Health Bournemouth Hospital (private) A2
Information Ctr B2
Lansdowne rbt B3
Lansdowne Rd A2
Lorne Park Rd B2
Lower Central Gdns B1/C2
Madeira Rd B2
Methuen Rd A3
Meyrick Park A1
Meyrick Rd B3
Milton Rd A2
Oceanarium C2
Odeon Cinema C2
Old Christchurch Rd B2
Ophir Rd A3
Oxford Rd B3
Park Rd A3
Parsonage Rd B2
Pavilion C2
Pier Approach C2
Pier Theatre C2
Police Station A3/B3
Portchester Rd A3
Post Office B1/B3
Priory Rd C1
Quadrant, The B2
Recreation Ground A1
Richmond Hill Rd B1
Russell Cotes Art Gallery & Museum C2
Russell Cotes Rd C2
St Anthony's Rd A1
St Michael's Rd C1
St Paul's rbt A3
St Paul's La B3
St Paul's Rd A3
St Peter's B2
St Peter's rbt B2
St Peter's Rd B2
St Stephen's Rd B1/B2
St Swithun's rbt B3
St Swithun's Rd B3
St Swithun's Rd South B3
St Valerie Rd A2
St Winifred's Rd A2
Stafford Rd B3
Terrace Rd B1
The Square B1
The Triangle B1
Town Hall B1
Tregonwell Rd C1
Trinity Rd B2
Undercliff Drive C3
Upper Central Gdns B1
Upper Hinton Rd B2
Upper Terr Rd C1
Waterfront, The C2
Wellington Rd A2/A3
Wessex Way A3/B1/B2
West Cliff Promenade C1
West Hill Rd C1
West Undercliff Promenade C1
Westover Rd B2
Wimborne Rd A2
Wootton Mount B2
Wychwood Dr A1
Yelverton Rd B2
York Rd B3
Zig-Zag Walks C1/C3

Bradford 332

Alhambra B2
Back Ashgrove B1
Barkerend Rd A3
Barnard Rd C3
Barry St B2
Bolling Rd C3
Bolton Rd A3
Bowland St A1
Bradford College B1
Bradford Forster Sq A2
Bradford Interchange B3
Bridge St B2
Britannia St B2
Broadway B2
Burnett St B3
Bus Station B2
Butler St West A3
Caledonia St C2
Canal Rd A2
Carlton St B1
Cathedral A3
Centenary Sq B2
Chapel St B3
Cheapside A2
Church Bank B3
City Hall B2
City Rd A1
Claremont B1
Colour Museum B1
Croft St B2
Darfield St A1
Darley St A2
Drewton Rd A1
Drummond Trading Estate A1
Dryden St B3
Dyson St A1
Easby Rd C1
East Parade B3
Eldon Pl A1
Filey St B3
Forster Square Retail Park A2
Gallery B3
Garnett St B3
Godwin St B2
Gracechurch St A1
Grattan Rd B1
Great Horton Rd B1/B2
Grove Terr B1
Hall Ings B2
Hall La C3
Hallfield Rd A1
Hammstrasse A2
Harris St B3
Holdsworth St A2
Ice Rink B2
Information Ctr B2
Ivegate B2
Inland Revenue A2
Jacob's Well Municipal Offices B2
James St A2
John St A2
Kirkgate B2
Kirkgate Centre B2
Laisteridge La C1
Law Courts B2/B3
Leeds Rd B3
Library B1/B2
Listerhills Rd B1
Little Horton La C1
Little Horton Gn C1
Longside La B1
Lower Kirkgate B2
Lumb La A1
Manchester Rd C2
Manningham La A1
Manor Row A2
Market C3
Market St B2
Melbourne Place C1
Midland Rd A2
Mill La C2
Morley St B1
Nelson St B2/C2
Nesfield St A2
New Otley Rd A3
Norcroft St B1
North Parade A2
North St A3
North Wing A3
Oastler Shopping Centre A2
Otley Rd A3
Park Ave C1
Park La C1
Park Rd C2
Parma St C2
Peckover St B3
Piccadilly A2
Police Station B2/C2
Post Office A2/B1/B2/C3
Preistley B3
Princes Way B2
Prospect St C3
Radwell Drive C2
Rawson Rd A1
Rebecca St A1
Richmond Rd B1
Russell St C1
St George's Hall B2
St Lukes Hospital C1
St Mary's A3
Shipley Airedale Rd A3/B3
Simes St A1
Smith St B1
Spring Mill St C2
Stott Hill A3
Sunbridge Rd A1/B1/B2
The Leisure Exchange B3
Thornton Rd A1/B1
Trafalgar St A2
Trinity Rd C1
Tumbling Hill St B1
Tyrrel St B2
University of Bradford B1/C1
Usher St C3
Valley Rd A2
Vicar La B3
Wakefield Rd C3
Wapping Rd A3
Westgate A1
White Abbey Rd A1
Wigan Rd A1
Wilton St B1
Wood St A1
Wool Exchange B2
Worthington St A1

Brighton 332

Addison Rd A1
Albert Rd B2
Albion Hill B3
Albion St B3
Ann St A3
Art Gallery & Museum B3
Baker St A3
Black Lion St C2
Brighton A2
Brighton Centre C2
Brighton Fishing Museum C2
Broad St C3
Buckingham Pl A2
Buckingham Rd B2
Cannon Pl C1
Carlton Hill B3
Chatham Pl A1
Cheapside A3
Church St B2
Churchill Square Shopping Centre B2
Clifton Hill B1
Clifton Pl B1
Clifton Rd B1
Clifton St A2
Clifton Terr B1
Clock Tower B2
Clyde Rd A3
Coach Park C3
Compton Ave A2
Davigdor Rd A1
Denmark Terr B1
Ditchling Rd A3
Dome, The B2
Duke St B2
Duke's La C2
Dyke Rd A1/B2
East St C2
Edward St B3
Elmore Rd B3
Frederick St B2
Gardner St B2
Gloucester Pl B3
Gloucester Rd B2
Goldsmid Rd A1
Grand Junction Rd C2
Grand Pde B3
Grove Hill B3
Guildford St A2
Hampton Pl B1
Hanover Terr A3
High St C3
Highdown Rd A1
Information Ctr C2
John St B3
Kemp St B2
Kensington Pl B2
Kings Rd C1
Law Courts B3
Lewes Rd A3
Library B2
London Rd A3
Madeira Dr C3
Marine Pde C3
Middle St C2
Montpelier Pl B1
Montpelier Rd B1
Montpelier St B1
New England Rd A2
New England St A2
New Rd B2
Nizells Ave A1
Norfolk Rd B1
Norfolk Terr B1
North Rd B2
North St B2
Odeon Cinema C2
Old Shoreham Rd A1
Old Steine C3
Osmond Rd A1
Over St B2
Oxford St A3
Palace Pier C3
Park Crescent Terr A3
Phoenix Art Gallery B3
Phoenix Rise A3
Police Station B3
Post Office A1/A2/A3/B1/B2/B3/C3
Preston Rd A2
Preston St B1
Prestonville Rd A1
Queen's Rd B2
Regency Sq C1
Regent St B2
Richmond Pl B3
Richmond St B3
Richmond Terr A3
Rose Hill Terr A3
Royal Alexandra Hospital B1
Royal Pavilion B2
St Bartholomew's A3
St James's St C3
St Nicholas Rd B2
St Nicholas' B2
St Peter's A3
Sea Life Centre C3
Shaftesbury Rd A3
Ship St C2
Silwood Rd B1
Silwood St B1
Southover St A3
Spring Gdns B2
Stanford Rd A1
Stanley Rd A3
Surrey St B2
Sussex St B3
Sussex Terr B3
Swimming Pool B3
Sydney St B3
Temple Gdns B1
Terminus Rd A2
The Lanes C2
Theatre Royal B2
Tidy St B2
Town Hall C2
Toy & Model Museum A2
Trafalgar St B2
Union Rd A3
University of Brighton B3
Upper Lewes Rd A3
Upper North St B1
Viaduct Rd A3
Victoria Gdns B3
Victoria Rd B1
Volk's Electric Railway C3
West Pier (Closed to the Public) C1
West St C2
Western Rd B1
Whitecross St B2
York Ave B1
York Pl B3

Bristol 332

Acramans Rd C4
Albert Rd C6
Alfred Hill A4
All Saint's St A4
All Saints' B4
Allington Rd C3
Alpha Rd C4
Ambra Vale B1
Ambra Vale East B2
Ambrose Rd B2
Amphitheatre C3
Anchor Rd B3
Anvil St B6
Architecture Centre B4
Argyle Pl B2
Arlington Villas A2
Arnolfini Arts Centre, The B4
Art Gallery A3
Ashton Gate Rd C2
Ashton Rd C1
at-Bristol B3
Avon Bridge C1
Avon Cr C1
Avon St B6
Baldwin St B4
Baltic Wharf C2
Baltic Wharf Leisure Centre & Caravan Park C2
Barossa Pl C4
Barton Manor B6
Barton Rd B6
Barton Vale B6
Bath Rd C6
Bathurst Basin C4
Bathurst Parade C4
Beauley Rd C3
Bedminster Bridge C5
Bedminster Parade C4
Bellevue B2
Bellevue Cr B2
Bellevue Rd C6
Berkeley Pl A2
Berkeley Sq A3
Birch Rd C2
Blackfriars A4
Bond St A5
Braggs La A6
Brandon Hill B3
Brandon Steep B3
Bristol Bridge B5
Bristol Cathedral (CE) B3
Bristol Central Library B3
Bristol Eye Hospital (A & E) A4
Bristol Grammar School A3
Bristol Harbour Railway C3
Bristol Marina C2
Bristol Royal Children's Hospital A4

Bristol Royal Infirmary (A & E) ⊞ A4
Bristol Temple Meads Station ⇌ B6
Broad Plain B6
Broad Quay B4
Broad St A4
Broad Weir A5
Broadcasting House A3
Broadmead A5
Brunel Way C1
Brunswick Sq A5
Burton Cl C5
Bus Station A4
Butts Rd B3
Cabot Tower ✦ B3
Caledonia Pl B1
Callowhill Ct A5
Cambridge St C6
Camden Rd C3
Camp Rd A1
Canada Way C2
Cannon St A4
Canon's Rd B3/B4
Canon's Way B3
Cantock's Cl A3
Canynge Rd A1
Canynge Sq A1
Castle Park A5
Castle St A5
Catherine Meade St C4
Cattle Market Rd C6
Charles Pl B1
Charlotte St B3
Charlotte St South B3
Chatterton House ⊞ B5
Chatterton Sq C5
Chatterton St C5
Cheese La B5
Christchurch ⛪ A4
Christchurch Rd A1
Christmas Steps ✦ A4
Church La B2/B5
Church St B5
City Museum ⊞ A3
City of Bristol College B3
Clare St B4
Clarence Rd C5
Cliff Rd C1
Clift House Rd C1
Clifton Cathedral (RC) + A2
Clifton Down A1
Clifton Down Rd A1
Clifton Hill B2
Clifton Park A1/A2
Clifton Park Rd A1
Clifton Rd A2
Cliftonwood Cr B2
Cliftonwood Rd B2
Cliftonwood Terr B2
Clifton Vale B1
Cobblestone Mews A1
College Green B3
College Rd A1
College St B3
Colston Almshouses ⊞ A4
Colston Ave B4
Colston Hall ⊌ B4
Colston Parade C5
Colston St A4
Commercial Rd C4
Commonwealth Museum ⊞ B5
Constitution Hill B2
Cooperage La C2
Corn St B4
Cornwallis Ave B1
Cornwallis Cr B1
Coronation Rd C2/C4
Council House ⊞ B3
Counterslip B5
Courts A4
Create Centre, The ✦ C1
Crosby Row B2
Culver St B3
Cumberland Basin C1
Cumberland Cl C2
Cumberland Rd C2/C3
Dale St A6
David St A6
Dean La C4
Deanery Rd B3
Denmark St B4
Dowry Sq B1
East St A5
Eaton Cr A2
Elmdale Rd A3
Elton Rd A3
Eugene St A4/A6
Exchange, The and St Nicholas' Mkts ⊞ B4
Fairfax St A4
Fire Station B5
Floating Harbour C3
Foster Almshouses ⊞ A4
Frayne Rd C1
Frederick Pl A2
Freeland Pl B1
Frogmore St B3
Fry's Hill B2
Gas La B6
Gasferry Rd C3
General Hospital ⊞ C4
Georgian House ⊞ B3
Glendale B1
Glentworth Rd B2
Gloucester St A1
Goldney Hall B2
Goldney Rd B1
Gordon Rd A2
Granby Hill B1
Grange Rd A1
Great Ann St A6
Great George St A6/B3
Great George Rd B3
Great Western Way B6
Green St North B1
Green St South B1
Greenay Bush La C2
Greenbank Rd C2
Greville Smyth Park C1
Guildhall ⊞ A4
Guinea St C4
Hamilton Rd C3
Hanbury Rd A2
Hanover Pl C2
Harbour Way B3
Harley Pl A1
Haymarket A5
Hensman's Hill B1
High St B4
Highbury Villas A3
Hill St B3
Hill St C6
Hippodrome ⊌ B4
Hopechapel Hill B1
Horfield Rd A4
Horton St B6
Host St A4
Hotwell Rd B1/B2
Houlton St A6
Howard Rd C3
Ice Rink B3
IMAX Cinema ⊌ B4
Information Ctr ℹ B4
Islington Rd C3
Jacob St A5/A6
Jacob's Wells Rd B2
John Carr's Terr B2
John Wesley's Chapel ⛪ A5
Joy Hill B1
Jubilee St B6
Kensington Pl A2
Kilkenny St B6
King St B4
Kingsland Rd B6
Kingston Rd C3
Lamb St A6
Lansdown Rd A2
Lawford St A6
Lawfords Gate A6
Leighton Rd C3
Lewins Mead A4
Lime Rd C2
Little Ann St A6
Little Caroline Pl B1
Little George St A6
Little King St B4
Litfield Rd A1
Llandoger Trow ⊞ B4
Lloyds' Building, The C3
Lodge St A4
Lord Mayor's Chapel, The ⛪ B4
Lower Castle St A5
Lower Church La A4
Lower Clifton Hill B2
Lower Guinea St C4
Lower Lamb St B3
Lower Maudlin St A4
Lower Park Rd A4
Lower Sidney St C2
Lucky La C4
Lydstep Terr C3
Mall (Galleries Shopping Centre), The A5
Manilla Rd A1
Mardyke Ferry Rd C2
Maritime Heritage Centre ⊞ B3
Marlborough Hill A4
Marlborough St A4
Marsh St B4
Mead St C5
Meadow St A5
Merchant Dock B2
Merchant Seamen's Almshouses ⊞ B4
Merchant St A5
Merchants Rd A1
Merchants Rd C1
Meridian Pl A2
Meridian Vale A2
Merrywood Rd C3
Midland Rd A6
Milford St C3
Millennium Sq B3
Mitchell La B5
Mortimer Rd A1
Murray Rd C4
Myrtle Rd A3
Narrow Plain B5
Narrow Quay B4
Nelson St A4
New Charlotte St C4
New Kingsley Rd B6
New Queen St C5
New St A6
Newfoundland St A5
Newgate A5
Newton St A6
Norland Rd A1
North St C2
Oakfield Gr A2
Oakfield Pl A2
Oakfield Rd A2
Old Bread St B6
Old Market St A6
Old Park Hill A4
Oldfield Rd B1
Orchard Ave B4
Orchard La B4
Orchard St B4
Osbourne Rd C3
Oxford St B6
Park Pl A2
Park Rd C3
Park Row A3
Park St A3
Passage St B5
Pembroke Gr A2
Pembroke Rd A2
Pembroke Rd C3
Pembroke St A5
Penn St A5
Pennywell Rd A6
Percival Rd A1
Pero's Bridge B4
Perry Rd A4
Pip & Jay ⛪ A5
Plimsoll Bridge B1
Polygon Rd B1
Portland St A1
Portwall La B5
Post Office ⊠ A1/A3/A4/A5/A6/B1/B4/C4/C5
Prewett St C5
Prince St B4
Prince St Bridge C4
Princess St C5
Princess Victoria St B1
Priory Rd A3
Pump La C5
QEH Theatre ⊌ A2
Queen Charlotte St B4
Quakers Friars A5
Quay St A4
Queen Elizabeth Hospital School B2
Queen Sq B4
Queen St A5
Queen's Ave A3
Queen's Parade B3
Queen's Rd A2/A3
Raleigh Rd C2
Randall Rd B2
Redcliffe Backs B5
Redcliffe Bridge B4
Redcliffe Hill C5
Redcliffe Parade C4
Redcliffe St B5
Redcliffe Way B5
Redcross La A6
Redcross St A6
Redgrave Theatre ⊌ A1
Red Lodge ⊞ A4
Regent St B1
Richmond Hill A2
Richmond Hill Ave A2
Richmond La A2
Richmond Park Rd A2
Richmond St C6
Richmond Terr A2
River St A6
Rownham Mead B2
Royal Fort Rd A3
Royal Park A2
Royal West of England Academy ⊞ A3
Royal York Cr B1
Royal York Villas B1
Rupert St A4
Russ St B6
St Andrew's Walk B2
St George's ⊌ B3
St George's Rd B3
St James ⛪ A4
St John's ⛪ A4
St John's Rd C4
St Luke's Rd C5
St Mary Redcliffe ⛪ C5
St Mary's Hospital ⊞ A3
St Matthias Park A6
St Michael's Hill A3
St Michael's Hospital ⊞ A4
St Michael's Park A3
St Nicholas St B4
St Paul St A5
St Paul's Rd A2
St Peter's (ruin) ⛪ A5
St Philip's Bridge B5
St Philips Rd A6
St Stephen's ⛪ B4
St Stephen's St B4
St Thomas St B5
St Thomas the Martyr ⛪ B5
Sandford Rd B1
Sargent St C5
Saville Pl B1
Ship La C5
Silver St A4
Sion Hill A1
Small St A4
Smeaton Rd C1
Somerset Sq C5
Somerset St C5
Southernhay Ave B2
Southville Rd C4
Spike Island Artspace ⊞ C2
Spring St C5
SS Great Britain and The Matthew ⛴ B2
Stackpool Rd C3
Staight St B6
Stillhouse La C4
Stracey Rd C2
Stratton St A5
Sydney Row C2
Tankard's Cl A3
Temple Back B5
Temple Boulevard B5
Temple Bridge B5
Temple Church ⛪ B5
Temple Circus B5
Temple Gate C5
Temple St B5
Temple Way B5
Terrell St A4
The Arcade A5
The Fosseway A2
The Grove B4
The Horsefair A5
The Mall A1
Theatre Royal ⊌ B4
Thomas La B5
Three Kings of Cologne ⛪ A4
Three Queens La B5
Tobacco Factory, The ⊌ C2
Tower Hill B5
Tower La A4
Trenchard St A4
Triangle South A3
Triangle West A3
Trinity Rd A6
Trinity St A6
Tucker St A5
Tyndall Ave A3
Union St A5
Union St B6
Unity St A6
Unity St B3
University of Bristol A3
University Rd A3
Upper Maudlin St A4
Upper Perry Hill C3
Upper Byron Pl A3
Upton Rd C2
Valentine Bridge B6
Victoria Gr C5
Victoria Rd C6
Victoria Rooms ⊞ A2
Victoria Sq A2
Victoria St B5
Vyvyan Rd A1
Vyvyan Terr A1
Wade St A6
Walter St C2
Wapping Rd C4
Water La B5
Waterloo Rd A6
Waterloo St A1
Waterloo St A6
Watershed, The ✦ B4
Welling Terr B1
Wellington Rd A6
Welsh Back B4
West Mall A1
West St A6
Westfield Pl A1
Wetherell Pl A2
Whitehouse Pl C5
Whitehouse St C5
Whiteladies Rd A2
Whitson St A4
William St C5
Willway St C5
Windsor Pl B1
Windsor Terr B1
Wine St A4
Woodland Rise A3
Woodland Rd A3
Worcester Rd A1
Worcester Terr A1
YHA ▲ B4
York Gdns B1
York Pl A2
York Rd C5

Bury St Edmunds 332

Abbey Gardens ✿ B3
Abbey Gate ⊞ B3
Abbeygate St B2
Albert Cr B1
Albert St B1
Ambulance Sta C1
Angel Hill B2
Angel La B2
Anglian Lane A1
Arc Shopping Centre B2
Athenaeum ⊌ C2
Baker's La C3
Beetons Way A1
Bishops Rd B2
Bloomfield St B2
Bridewell La C2
Bullen Cl C1
Bury St Edmunds ⇌ A2
Bury St Edmunds County Upper School A1
Bury St Edmunds Leisure Centre B1
Bury Town FC B1
Bus Station B2
Butter Mkt B2
Cannon St B2
Castle Rd C1
Cemetery C1
Chalk Rd (N) B1
Chalk Rd (S) B1
Church Row B2
Churchgate St C2
Citizens Advice Bureau B2
College St C2
Compiegne Way A3
Corn Exchange, The ⊌ B2
Cornfield Rd B1
Cotton Lane B3
Courts B2
Covent Garden C2
Crown St C3
Cullum Rd C2
Eastern Way A3
Eastgate St B3
Enterprise Business Park A2
Etna Rd A2
Eyre Cl C2
Fire Station B1
Friar's Lane C2
Gage Cl A1
Garland St B2
Greene King Brewery ⊞ C3
Grove Park B1
Grove Rd B1
Guildhall ⊞ C2
Guildhall St C2
Hatter St C2
High Baxter St B2
Honey Hill C3
Hospital Rd C1/C2
Ickworth Dr C1
Information Ctr ℹ B2
Ipswich St A2
King Edward VI Sch A1
King's Rd C1/B2
Library B2
Long Brackland A2
Looms La B2
Lwr Baxter St B2
Malthouse La A2
Manor House ⊞ C3
Maynewater La C3
Mill Rd C1
Mill Rd (South) C1
Minden Close B3
Moyses Hall ⊞ B2
Mustow St B3
Norman Tower ⊞ C3
Northgate Ave A2
Northgate St B2
Nutshell, The ⊌ B2
Osier Rd A2
Out Northgate A2
Out Risbygate B1
Out Westgate C2
Parkway B1/C2
Parkway ⊌ B1
Peckham St B2
Petticoat La C1
Phoenix Day Hospital ⊞ C1
Pinners Way C1
Police Station ⊠ C3
Post Office ⊠ B2/B3
Pump La B2
Queen's Rd B1
Raingate St C3
Raynham Rd A1
Retail Park C2
Risbygate St B1/B2
Robert Boby Way C2
St Andrew's St North B2
St Andrew's St South B2
St Botolph's La C3
St Edmunds Hospital (private) ⊞ C3
St Edmund's ⛪ C2
St Edmund's Abbey (Remains) ✦ B3
St Edmundsbury + C3
St John's St B2
St Marys ⛪ C3
School Hall La B2
Shillitoe Cl C1
Shire Halls & Magistrates Ct C3
South Cl C1
Southgate St C3
Sparhawk St C3
Spring Lane B1
Springfield Rd B1
Station Hill A2
Swan La C3
Tayfen Rd B2
The Vinefields B3
Theatre Royal ⊌ C3
Thingoe Hill A2
Victoria St B1
War Memorial ✦ C1
Well St B2
West Suffolk College B1
Westgarth Gdns C1
Westgate St C2
Whiting St C2
York Rd B1
York Terr B1

Cambridge 333

Abbey Rd A3
ADC ⊌ A2
Anglia Ruskin University B3
Archaeology & Anthropology ⊞ B2
Art Gallery ⊞ A1
Arts Picture House ⊌ B2
Arts Theatre ⊌ B1
Auckland Rd A3
Bateman St C2
B.B.C. C3
Benet St B1
Bradmore St B3
Bridge St A1
Broad St B3
Brookside C2
Brunswick Terr A3
Burleigh St B3
Bus Station B2
Butt Green A2
Cambridge Contemporary Art Gallery ⊞ B1
Castle Mound ⊞ A1
Castle St A1
Chesterton La A1
Christ's (Coll) B2
Christ's Pieces B2
City Rd B3
Clare (Coll) B1
Clarendon St B2
Coe Fen C2
Coronation St C2
Corpus Christi (Coll) B1
Council Offices C3
Cross St C3
Crusoe Bridge C1
Darwin (Coll) C1
Devonshire Rd C3
Downing (Coll) C2
Downing St B2
Earl St B2
East Rd B3
Eden St B3
Elizabeth Way A3
Elm St B2
Emery St B3
Emmanuel (Coll) B2
Emmanuel Rd B2
Emmanuel St B2
Fair St A3
Fenners Physical Education Centre C3
Fire Station B3
Fitzroy St A3
Fitzwilliam Museum ⊞ C2
Fitzwilliam St C2
Folk Museum ⊞ A1
Glisson Rd C3
Gonville & Caius (Coll) B1
Gonville Place C2
Grafton Centre A3
Grand Arcade B2
Gresham Rd C3
Green St B1
Guest Rd B3
Guildhall ⊞ B2
Harvey Rd C3
Hills Rd C3
Hobson St B2
Hughes Hall (Coll) B3
Information Ctr ℹ B2
James St A3
Jesus (Coll) A2
Jesus Green A2
Jesus La A2
Jesus Terr B3
John St B3
Kelsey Kerridge Sports Centre B3
King St A2
King's (Coll) B1
King's College Chapel ⛪ B1
King's Parade B1
Lammas Land Recreation Ground C1
Lensfield Rd C2
Little St Mary's La B1
Lyndewod Rd C3
Magdalene (Coll) A1
Magdalene St A1
Maid's Causeway A3
Malcolm St A2
Market Hill B1
Market St B1
Mathematical Bridge B1
Mawson Rd C3
Midsummer Common A3
Mill La B1
Mill Rd B3
Mill St C3
Napier St A3
New Square A3
Newmarket Rd A3
Newnham Rd C1
Norfolk St B3
Northampton St A1
Norwich St C2
Orchard St B2
Panton St C2
Paradise Nature Reserve C1
Paradise St B3
Park Parade A1
Park St A2
Park Terr B2
Parker St B2
Parker's Piece B2
Parkside B3
Parkside Pools B3
Parsonage St A3
Pemberton Terr C2
Pembroke (Coll) B2
Pembroke St B1
Perowne St B3
Peterhouse (Coll) C1
Petty Cury B2
Police Station ⊠ B3
Post Office ⊠ A1/A3/B2/B3/C1/C2/C3
Queens' (Coll) B1
Queen's La B1
Queen's Rd B1
Regent St B2
Regent Terr B2
Ridley Hall (Coll) C1
Riverside A3
Round Church, The ⛪ A1
Russell St C2
St Andrew's St B2
St Benet's ⛪ B1
St Catharine's (Coll) B1
St Eligius St C2
St John's (Coll) A1
St Mary's ⛪ B1
St Paul's Rd C3
Saxon St C2
Scott Polar Institute & Museum ⊞ C2
Sedgwick Museum ⊞ B2
Sheep's Green C1
Shire Hall A1
Sidgwick Ave C1
Sidney St A2
Sidney Sussex (Coll) A2
Silver St B1
Station Rd C3
Tenison Ave C3
Tenison Rd C3
Tennis Court Rd B2
The Backs B1
The Fen Causeway C1
Thompson's La A1
Trinity (Coll) A1
Trinity Hall (Coll) B1
Trinity St B1
Trumpington Rd C2
Trumpington St B1
Union Rd C2
University Botanic Gardens ✿ C2
Victoria Ave A2
Victoria St B2
Warkworth St B3
Warkworth Terr B3
Wesley House (Coll) A2
West Rd B1
Westcott House (Coll) A2
Westminster (Coll) A1
Whipple ⊞ B2
Willis Rd B3
Willow Walk A2
Zoology ⊞ B2

Canterbury 333

Artillery St B2
Barton Mill Rd A3
Beaconsfield Rd A1
Beverley Rd A1
Bingley's Island B1
Black Griffin La B1
Broad Oak Rd A2
Broad St B2
Brymore Rd A3
Burgate B2
Bus Station C2
Canterbury College C3
Canterbury East ⇌ C1
Canterbury Tales, The ✦ B2
Canterbury West ⇌ A1
Castle ⊞ C1
Castle Row C1
Castle St C1
Cathedral + B2
Chaucer Rd A3
Christ Church University B3
Christchurch Gate ✦ B2
City Council Offices A3
City Wall B2
Coach Park A2
College Rd B3
Cossington Rd C2
Court B2
Craddock Rd A3
Crown & County Courts B3
Dane John Gdns C2
Dane John Mound ✦ C1
Deanery B2
Dover St C2
Duck La B2
Eastbridge Hosp ⊞ B1
Edgar Rd B3
Ersham Rd C3
Ethelbert Rd C3
Fire Station C2
Forty Acres Rd A1
Gordon Rd C1
Greyfriars ✦ B1
Guildford Rd C1
Havelock St B2
Heaton Rd C1
High St B2
HM Prison B3
Information Ctr ℹ A2/B2
Ivy La B2
Ivy Pl C1
King St B2
King's School B2/B3
King's School Leisure Facilities A2
Kingsmead Leisure Centre A2
Kingsmead Rd A2
Kirby's La B1
Lansdown Rd C2
Lime Kiln Rd C1
Longport B3
Lower Chantry La C3
Mandeville Rd A1
Market Way A2
Marlowe Arcade B2
Marlowe Ave C2
Marlowe Theatre ⊌ B2
Martyrs Field Rd C1
Mead Way B1
Military Rd B2
Monastery St B2
Museum of Canterbury (Rupert Bear Museum) ⊞ B1
New Dover Rd C3
Norman Rd C2
North Holmes Rd B3
North La B1
Northgate A2
Nunnery Fields C2
Nunnery Rd C2
Oaten Hill C2
Odeon Cinema ⊌ C2
Old Dover Rd C2
Old Palace B2
Old Ruttington La B2
Old Weavers ⊞ B2
Orchard St B1
Oxford Rd C1
Palace St B2
Pilgrims Way C3
Pin Hill C1
Pine Tree Ave A1
Police Station ⊠ C2
Post Office ⊠ B2/C1/C2
Pound La B1
Puckle La C2
Raymond Ave C2
Registry Office A2
Rheims Way B1
Rhodaus Cl C2
Rhodaus Town C2
Roman Museum ⊞ B2
Roper Gateway A1
Roper Rd A1
Rose La B2
Royal Museum ⊞ B2
St Augustine's Abbey (remains) + B3
St Augustine's Rd C3
St Dunstan's ⛪ A1
St Dunstan's St A1
St George's Pl C2
St George's St B2
St George's Tower ✦ B2
St Gregory's Rd B3
St John's Hospital ⊞ A2
St Margaret's St B2
St Martin's ⛪ B3
St Martin's Ave B3
St Martin's Rd B3
St Michael's Rd A1
St Mildred's ⛪ C1
St Peter's Gr B1
St Peter's La B2
St Peter's Pl B1
St Peter's St B1
St Radigunds St B2
St Stephen's Ct A1
St Stephen's Path A1
St Stephen's Rd A2
Salisbury Rd A1
Simmonds Rd C1
Spring La C3
Station Rd West B1
Stour St B1
Sturry Rd A3
The Causeway A2
The Friars B2
Tourtel Rd A3
Tudor Rd C1
Union St B2
University for the Creative Arts C3
Vernon Pl C2
Victoria Rd C1
Watling St B2
Westgate Gdns B1
Westgate Towers ⊞ B1
Whitefriars B2
Whitehall Gdns B1
Whitehall Rd B1
Wincheap C1
York Rd C1
Zealand Rd C2

Cardiff *Caerdydd* 333

Adam St B3
Alexandra Gdns A2
Allerton St C1
Arran St A3
ATRiuM (Univ. of Glamorgan) C3
Beauchamp St C1
Bedford St A3
Blackfriars Priory + B1
Boulevard De Nantes B2
Brains Brewery C2
Brook St B1
Bus Station C2
Bute Park B1
Bute St C2
Bute Terr C3
Callaghan Sq C2/C3
Capitol Shopping Centre, The B3
Cardiff Bridge B1
Cardiff Castle ⊞ B2
Cardiff Central Station ⇌ C2
Cardiff Centre Trading Estate C3
Cardiff International Arena ✦ C3
Cardiff Rugby Football Ground B1
Cardiff University A1/A2/B3
Cardiff University Student's Union A2
Caroline St C2
Castle Green B2
Castle Mews A1
Castle St (Heol y Castell) B1
Cathays Station ⇌ A2
Celerity Drive C3
Central Sq C2
Charles St (Heol Siarl) B3
Churchill Way B3
City Hall ⊞ A2
City Rd A3
Clare Rd C1
Clare St C1
Coburn St A3
Coldstream Terr B1
College Rd A1
Colum Rd A1
Court C2
Court Rd C1
Craiglee Drive C3
Cranbrook St A3
Customhouse St C2
Cyfartha St A3
Despenser Place C1
Despenser St C1
Dinas St C1
Duke St (Heol y Dug) B2
Dumfries Place B3
East Grove A3
Ellen St C3
Fire Station B3
Fitzalan Place B3
Fitzhamon Embankment C1
Fitzhamon La C1
Gloucester St C1
Glynrhondda St A2
Gordon Rd A3
Gorsedd Gdns B2
Green St B1
Greyfriars Rd B2
H.M. Prison B3
Hafod St C1
Herbert St C3
High St B2
Industrial Estate C3
John St C3
Jubilee St C1
King Edward VII Ave A1
Kingsway (Ffordd y Brenin) B2
Knox Rd B3
Law Courts B2
Library C2
Llanbleddian Gdns A2
Llantwit St A2
Lloyd George Ave C3
Lower Cathedral Rd B1
Lowther Rd A3
Magistrates Court B3
Mansion House A3
Mardy St C1
Mark St B1
Market B2
Mary Ann St C3
Merches Gdns C1
Mill La C2
Millennium Bridge B1
Millennium Plaza Leisure Complex ⊌ C1
Millennium Stadium C1
Millennium Stadium Tours (Gate 3) ✦ B2
Miskin St A2
Monmouth St C1
Museum Ave A2
Museum Place A2
National Museum of Wales ⊞ A2
National War Memorial ✦ A2
Neville Place C1
New Theatre ⊌ B2
Newport Rd B3
Northcote La A3
Northcote St A3
Park Grove A4
Park Place A2
Park St C2
Penarth Rd C2
Pendyris St C1
Plantaganet St C1
Quay St B2
Queen Anne Sq A1
Queen St (Heol y Frenhines) B2
Queen St Station ⇌ B3
Regimental Museums ⊞ B2
Rhymney St A3
Richmond Rd A3
Royal Welsh College of Music and Drama A1
Russell St A3
Ruthin Gdns A2
St Andrews Place A2
St David's + B2
St David's 2 C2
St David's Centre B2
St David's Hall ✦ B2
St John The Baptist ⛪ B2
St Mary St (Heol Eglwys Fair) B2
St Peter's St A3
Salisbury Rd A3
Sandon St B3
Schooner Way C3
Scott Rd C2
Scott St C2
Senghennydd Rd A2
Sherman Theatre ⊌ A2
Sophia Gardens A1
South Wales Baptist College A3
Stafford Rd C1
Station Terr B3
Stuttgarter Strasse B2
Sussex St C1
Taffs Mead Embankment C1
Talworth St A3
Temple of Peace & Health ✦ A1
The Friary B2
The Hayes B2
The Parade A3
The Walk A3
Treharris St A3
Trinity St B2
Tudor La C1
Tudor St C1
Welsh Assembly Offices A1
Welsh Institute of Sport ✦ A1
West Grove A3
Westgate St (Heol y Porth) B2
Windsor Place B3
Womanby St B2
Wood St C2
Working St B2
Wyeverne Rd A2

Carlisle 333

Abbey St A1
Aglionby St B3
Albion St C3
Alexander St C3
AMF Bowl ✦ C2
Annetwell St A1
Bank St B2
Bitts Park A1
Blackfriars St B2
Blencome St C1
Blunt St C1
Botchergate B2
Boustead's Grassing C2
Bowman St B3
Broad St B3
Bridge St A1
Brook St C3
Brunswick St B2
Bus Station B2
Caldew Bridge A1
Caldew St C1
Carlisle (Citadel) Station ⇌ B2
Castle ⊞ A1
Castle St A1
Castle Way A1
Cathedral + A1
Cecil St B2
Chapel St A2
Charles St B3
Charlotte St B1
Chatsworth Square A2
Chiswick St B2
Citadel, The B2
City Walls A1
Civic Centre A2

Clifton St C1
Close St B3
Collingwood St C1
Colville St C1
Colville Terr C1
Court B2
Court St B2
Crosby St B2
Crown St C2
Currock Rd C2
Dacre Rd A1
Dale St C1
Denton St C1
Devonshire Walk A1
Duke's Rd A2
East Dale St C1
East Norfolk St C1
Eden Bridge A2
Edward St B3
Elm St C2
English St B2
Fire Station A2
Fisher St A1
Flower St B3
Freer St C1
Fusehill St B3
Georgian Way A2
Gloucester Rd C3
Golf Course A2
Graham St C1
Grey St B3
Guildhall Museum A2
Halfey's La B3
Hardwicke Circus A2
Hart St B3
Hewson St C2
Howard Pl A3
Howe St B3
Information Ctr A2
James St B2
Junction St B1
King St B2
Lancaster St B2
Lanes Shopping Centre B2
Laserquest B2
Library A2/B1
Lime St B1
Lindisfarne St C3
Linton St B3
Lismore Pl A3
Lismore St B3
London Rd C3
Lonsdale Rd B2
Lord St C3
Lorne Cres B1
Lorne St B1
Lowther St B2
Market Hall A2
Mary St B2
Memorial Bridge A3
Metcalfe St C1
Milbourne St B1
Myddleton St B3
Nelson St C1
Norfolk St C1
Old Town Hall A2
Oswald St C3
Peter St A2
Petteril St B3
Police Station A2
Portland Pl B2
Portland Sq B2
Post Office A2/B2/B3/C1/C3
Princess St C2
Pugin St B1
Red Bank Terr C2
Regent St C3
Richardson St C1
Rickerby Park A3
Rickergate A2
River St B3
Rome St C2
Rydal St B3
St Cuthbert's B2
St Cuthbert's La B2
St James' Park C1
St James' Rd C1
St Nicholas St C3
Sands Centre A2
Scotch St A2
Shaddongate B1
Sheffield St B1
South Henry St B3
South John St C2
South St B3
Spencer St B2
Sports Centre A2
Strand Rd A2
Swimming Baths B2
Sybil St B3
Tait St B2
Thomas St B1
Thomson St C3
Trafalgar St C1
Tullie House Museum A1
Tyne St C3
Viaduct Estate Rd B1
Victoria Pl A2
Victoria Viaduct B2
Vue B2
Warwick Rd B2
Warwick Sq B3
Water St B2
West Walls B1
Westmorland St C1

Chelmsford 333

Ambulance Station B1
Anchor St C1
Anglia Polytechnic University A2
Arbour La A3
Baddow Rd B2/C3
Baker St C1
Barrack Sq B2
Bellmead B2
Bishop Hall La A2
Bishop Rd A2
Bond St B2
Boswells Dr B3
Boudicca Mews C2
Bouverie Rd C2
Bradford St C1
Braemar Ave C1
Brook St A2
Broomfield Rd A1
Burns Cres C2
Bus Station B2
Can Bridge Way B2
Cedar Ave A1
Cedar Ave West A1
Cemetery A1
Cemetery A2
Cemetery C1
Central Park B1
Chelmsford + B2
Chelmsford ≋ A1
Chichester Dr A3
Chinery Cl A3
Cinema B2
Civic Centre A1
College C1
Cottage Pl A2
County Hall B2
Coval Ave B1
Coval La B1
Coval Wells B1
Cricket Ground B2
Crown Court B2
Duke St B2
Elm Rd C1
Elms Dr A1
Essex Record Office, The B3
Fairfield Rd B1
Falcons Mead B1
George St C2
Glebe Rd A1
Godfrey's Mews C2
Goldlay Ave C3
Goldlay Rd C2
Grove Rd C2
HM Prison A3
Hall St C2
Hamlet Rd C2
Hart St C1
Henry Rd A2
High Bridge Rd B2
High Chelmer Shopping Centre B2
High St B2
Hill Cres B3
Hill Rd Sth B3
Hill Rd B3
Hillview Rd A3
Hoffmans Way A2
Hospital B2
Information Ctr B2
Lady La C2
Langdale Gdns C3
Legg St B2
Library A1
Library B2
Library C2
Lionfield Terr A3
Lower Anchor St C1
Lynmouth Ave C2
Lynmouth Gdns C3
Magistrates Court B2
Maltese Rd A1
Manor Rd C2
Marconi Rd A2
Market B2
Market Rd B2
Marlborough Rd C1
Meadows Shopping Centre, The B2
Meadowside A3
Mews Ct C2
Mildmay Rd C2
Moulsham Dr C2
Moulsham Mill C3
Moulsham St C1/C2
Navigation Rd B3
New London Rd B2/C1
New St A2/B2
New Writtle St C1
Nursery Rd C2
Orchard St C2
Park Rd B1
Parker Rd C2
Parklands Dr A3
Parkway A1/B1/B2
Police Station A2
Post Office A3/B2/C2
Primrose Hill A1
Prykes Dr B1
Queen St C1
Queen's Rd B3
Railway St A1
Rainsford Rd A1
Ransomes Way A2
Rectory La A2
Regina Rd A2
Riverside Leisure Centre B2
Rosebery Rd C2
Rothesay Ave C1
St John's Rd C2
Sandringham Pl B3
Seymour St B1
Shrublands Cl B3
Southborough Rd C1
Springfield Basin B3
Springfield Rd A3/B2/B3
Stapleford Cl C1
Swiss Ave A1
Telford Pl A3
The Meades B1
Tindal St B2
Townfield St A1
Trinity Rd B3
University B1
Upper Bridge Rd C1
Upper Roman Rd C2
Van Dieman's Rd C3
Viaduct Rd B1
Vicarage Rd C1
Victoria Rd A2
Victoria Rd South B2
Vincents Rd C2
Waterloo La B2
Weight Rd B3
Westfield Ave A1
Wharf Rd B3
Writtle Rd C1
YMCA A2
York Rd C1

Cheltenham 333

Albert Rd A3
Albion St B3
All Saints Rd B3
Ambrose St B2
Andover Rd C1
Art Gallery & Museum B2
Axiom Centre B3
Back Montpellier Terr C2
Bandstand C2
Bath Pde B2
Bath Rd C2
Bays Hill Rd C1
Beechwood Place Shopping Centre B3
Bennington St B2
Berkeley St B3
Brewery A2
Brunswick St South A2
Bus Station B2
CAB B2
Carlton St B3
Central Cross Road A3
Cheltenham College C2
Cheltenham F.C. A3
Cheltenham General (A & E) C3
Christchurch Rd B1
Cineworld A2
Clarence Rd A2
Clarence Sq A2
Clarence St B2
Cleeveland St A1
Coach Park A2
College Baths Road C3
College Rd C2
Colletts Dr A1
Corpus St C3
Devonshire St A2
Douro Rd B1
Duke St B3
Dunalley Pde A2
Dunalley St A2
Everyman B2
Evesham Rd A3
Fairview Rd B3
Fairview St B3
Fire Station C3
Folly La A2
Gloucester Rd A1
Grosvenor St B3
Grove St A1
Gustav Holst A3
Hanover St A2
Hatherley St C1
Henrietta St A2
Hewlett Rd B3
High St B2/B3
Hudson St A2
Imperial Gdns C2
Imperial La B2
Imperial Sq C2
Information Ctr B2
Keynsham Rd C3
King St A2
Knapp Rd B2
Ladies College B2
Lansdown Cr C1
Lansdown Rd C1
Leighton Rd B3
London Rd C3
Lypiatt Rd C1
Malvern Rd B1
Manser St A2
Market St A1
Marle Hill Pde A2
Marle Hill Rd A2
Millbrook St A1
Milsom St A2
Montpellier Gdns C2
Montpellier Gr C2
Montpellier Pde C2
Montpellier Spa Rd C2
Montpellier St C1
Montpellier Terr C2
Montpellier Walk C2
New St B2
North Pl B2
Old Bath Rd C3
Oriel Rd B2
Overton Park Rd B1
Overton Rd B1
Oxford St C3
Parabola Rd B1
Park Pl C1
Park St A1
Pittville Circus A3
Pittville Cr A3
Pittville Lawn A3
Playhouse B2
Police Station B1/C1
Portland St B3
Post Office B2/C2
Prestbury Rd A3
Prince's Rd C1
Priory St B3
Promenade B2
Queen St A1
Recreation Ground A2
Regent Arcade B2
Regent St B2
Rodney Rd B2
Royal Cr B2
Royal Wells Rd B2
St George's Pl B2
St Georges Rd B1
St Gregory's B2
St James St B3
St John's Ave B3
St Luke's Rd C2
St Margarets Rd A2
St Mary's B2
St Matthew's B2
St Paul's La A2
St Paul's Rd A2
St Paul's St A2
St Stephen's Rd C1
Sandford Lido C3
Sandford Mill Road C3
Sandford Park C3
Sandford Rd C2
Selkirk St A3
Sherborne Pl B3
Sherborne St B3
Suffolk Pde C2
Suffolk Rd C1
Suffolk Sq C1
Sun St A1
Swindon Rd B2
Sydenham Villas Rd C3
Tewkesbury Rd A1
The Courtyard B1
Thirlstaine Rd C2
Tivoli Rd C1
Tivoli St C1
Town Hall & Theatre B2
Townsend St A1
Trafalgar St C2
Union St B3
University of Gloucestershire (Francis Close Hall) A2
University of Gloucestershire (Hardwick) A1
Victoria Pl B3
Victoria St A2
Vittoria Walk C2
Wel Pl B2
Wellesley Rd A2
Wellington Rd A3
Wellington Sq A3
Wellington St B2
West Drive A3
Western Rd B1
Winchcombe St B3

Chester 333

Abbey Gateway A2
Appleyards La C3
Bedward Row B1
Beeston View C3
Bishop Lloyd's Palace B2
Black Diamond St A2
Bottoms La C3
Boughton B3
Bouverie St A1
Bridge St B2
Bridgegate C2
British Heritage Centre B2
Brook St A3
Brown's La C2
Bus Station B2
Cambrian Rd A1
Canal St A2
Carrick Rd C1
Castle C2
Castle Dr C2
Cathedral B2
Catherine St A1
Chester ≋ A3
Cheyney Rd A1
Chichester St A1
City Rd A3
City Walls B1/B2
City Walls Rd B1
Cornwall St A2
County Hall C2
Cross Hey C3
Cuppin St B2
Curzon Park North C1
Curzon Park South C1
Dee Basin A1
Dee La B3
Delamere St A2
Dewa Roman Experience B2
Duke St B2
Eastgate B2
Eastgate St B2
Eaton Rd C2
Edinburgh Way C3
Elizabeth Cr B3
Fire Station A2
Foregate St B2
Frodsham St B2
Gamul House B2
Garden La A1
Gateway Theatre B2
George St A2
Gladstone Ave A1
God's Providence House B2
Gorse Stacks A2
Greenway St C2
Grosvenor Bridge C1
Grosvenor Museum B2
Grosvenor Park B3
Grosvenor Precinct B2
Grosvenor Rd C1
Grosvenor St B2
Groves Rd B3
Guildhall Museum B1
Handbridge C2
Hartington St C3
Hoole Way A2
Hunter St B2
Information Ctr B2
King Charles' Tower A2
King St A2
Library B2
Lightfoot St A3
Little Roodee C2
Liverpool Rd A2
Love St B3
Lower Bridge St B2
Lower Park Rd B3
Lyon St A2
Magistrates Court B2
Meadows La C3
Military Museum C2
Milton St A3
New Crane St B1
Nicholas St B2
Northgate A2
Northgate Arena A2
Northgate St A2
Nun's Rd B1
Old Dee Bridge C2
Overleigh Rd C2
Park St B2
Police Station B2
Post Office A2/A3/B2/C2
Princess St B2
Queen St B2
Queen's Park Rd C3
Queen's Rd A3
Race Course B1
Raymond St A1
River La C2
Roman Amphitheatre & Gardens B2
Roodee, The (Chester Racecourse) B1
Russell St A3
St Anne St A2
St George's Cr C3
St Martin's Gate A1
St Martin's Way B1
St Oswalds Way A2
Saughall Rd A1
Sealand Rd A1
South View Rd A1
Stanley Palace B1
Station Rd A3
Steven St A3
The Bars B3
The Cross B2
The Groves B3
The Meadows B3
Tower Rd B1
Town Hall B2
Union St B3
Vicar's La B2
Victoria Cr C3
Victoria Rd A2
Walpole St A1
Water Tower St A1
Watergate B1
Watergate St B2
Whipcord La A1
White Friars B2
York St B3

Chichester 333

Adelaide Rd A3
Alexandra Rd A3
Arts Centre B2
Ave de Chartres B1/B2
Barlow Rd A1
Basin Rd C2
Beech Ave B1
Bishops Palace Gardens B2
Bishopsgate Walk A3
Bramber Rd C3
Broyle Rd A2
Bus Station B2
Caledonian Rd B3
Cambrai Ave B3
Canal Wharf C2
Canon La B2
Cathedral B2
Cavendish St A1
Cawley Rd B2
Cedar Dr A1
Chapel St A2
Cherry Orchard Rd C3
Chichester By-Pass C2/C3
Chichester Festival A2
Chichester ≋ B2
Churchside A2
Cinema B3/C1
City Walls B2
Cleveland Rd B3
College La A2
College of Science & Technology B1
Cory Cl C2
Council Offices B2
County Hall B2
Courts B2
District B2
Duncan Rd A1
Durnford Cl A1
East Pallant B2
East Row B2
East St B2
East Walls B3
Eastland Rd C3
Ettrick Cl B3
Ettrick Rd B3
Exton Rd C3
Fire Station A2
Football Ground A2
Franklin Pl A2
Friary (Rems. of) C2
Garland Cl C3
Green La A3
Grove Rd C3
Guilden Rd B3
Guildhall A2
Hawthorn Cl A1
Hay Rd C3
Henty Gdns B1
Herald Dr C3
Information Ctr B2
John's St B2
Joys Croft A3
Jubilee Pk A3
Jubilee Rd A3
Juxon Cl B2
Kent Rd A3
King George Gdns A2
King's Ave C2
Kingsham Ave C3
Kingsham Rd C2
Laburnum Gr B2
Leigh Rd C1
Lennox Rd A3
Lewis Rd A3
Library B2
Lion St B2
Litten Terr A3
Little London B2
Lyndhurst Rd B3
Market B3
Market Ave B2
Market Cross B2
Market Rd B3
Martlet Cl C2
Melbourne Rd A3
Mount La B1
New Park Rd A3
Newlands La A1
North Pallant B2
North St A2
North Walls A2
Northgate A2
Oak Ave A1
Oak Cl A1
Oaklands Park A2
Oaklands Way A2
Orchard Ave A1
Orchard St A1
Ormonde Ave B3
Pallant House B2
Parchment St A1
Parklands Rd A1/B1
Peter Weston Pl B3
Police Station C2
Post Office A1/B2/B3
Priory La A2
Priory Park A2
Priory Rd A2
Queen's Ave C1
Riverside A3
Roman Amphitheatre B3
St Cyriacs A2
St Pancras A3
St Paul's Rd A2
St Richard's Hospital (A+E) A3
Shamrock Cl A3
Sherbourne Rd A1
Somerstown A2
South Bank C2
South Pallant B2
South St B2
Southgate B2
Spitalfield La A3
Stirling Rd B3
Stockbridge Rd C1/C2
Swanfield Dr A3
Terminus Industrial Estate C1
Terminus Rd C1
The Hornet B3
The Litten A3
Tower St A2
Tozer Way A3
Turnbull Rd A3
Upton Rd C1
Velyn Ave B3
Via Ravenna B1
Walnut Ave A1
West St B2
Westgate B1
Westgate Fields B1
Westgate Leisure Centre B1
Weston Ave C1
Whyke Cl C3
Whyke La B3
Whyke Rd C3
Winden Ave B3

Colchester 333

Abbey Gateway C2
Albert St A1
Albion Grove C2
Alexandra Rd C1
Artillery St C3
Arts Centre B1
Balkerne Hill B1
Barrack St C3
Beaconsfield Rd C1
Beche Rd C3
Bergholt Rd A1
Bourne Rd C3
Brick Kiln Rd A1
Bristol Rd B2
Broadlands Way A3
Brook St B3
Bury Cl B2
Butt Rd C1
Camp Folley North C2
Camp Folley South C2
Campion Rd C2
Cannon St C3
Canterbury Rd C2
Castle B2
Castle Park B2
Castle Rd B2
Catchpool Rd A1
Causton Rd B1
Cavalry Barracks C1
Chandlers Row C3
Circular Rd East C2
Circular Rd North C1
Circular Rd West C1
Clarendon Way A1
Claudius Rd C2
Clock B1
Colchester Camp Abbey Field C1
Colchester Institute B1
Colchester ≋ A1
Colchester Town ≋ C2
Colne Bank Ave A1
Colne View Retail Park A2
Compton Rd A3
Cowdray Ave A1/A2
Cowdray Centre, The A2
Crouch St B1
Crowhurst Rd B1
Culver Centre B1
Culver St East B2
Culver St West B1
Dilbridge Rd A3
East Hill B2
East St B3
East Stockwell St B1
Eld La B1
Essex Hall Rd A1
Exeter Dr B2
Fairfax Rd C2
Fire Station A2
Flagstaff Rd C1
George St B2
Gladstone Rd C2
Golden Noble Hill C2
Goring Rd A3
Granville Rd C3
Greenstead Rd B3
Guildford Rd B2
Harsnett Rd C3
Harwich Rd B3
Head St B1
High St B1/B2
High Woods Country Park A2
Hythe Hill C3
Information Ctr B2
Ipswich Rd A3
Kendall Rd C2
Kimberley Rd C3
King Stephen Rd C3
Le Cateau Barracks C1
Leisure World A2
Library B1
Lincoln Way B2
Lion Walk Shopping Centre B1
Lisle Rd C2
Lucas Rd C2
Magdalen Green C3
Magdalen St C2
Maidenburgh St B2
Maldon Rd C1
Manor Rd B1
Margaret Rd A1
Mason Rd A2
Mercers Way A1
Mercury B1
Mersea Rd C2
Meyrick Cr C2
Mile End Rd A1
Military Rd C2
Mill St C2
Minories B2
Moorside B3
Morant Rd C3
Napier Rd C2
Natural History B2
New Town Rd C2
Norfolk Cr A3
North Hill B1
North Station Rd A1
Northgate St B1
Nunns Rd B1
Odeon B1
Old Coach Rd B3
Old Heath Rd C3
Osborne St B2
Petrolea Cl A1
Police Station C1
Popes La B1
Port La C3
Post Office A1/B1/B2/C2/C3
Priory St B2
Queen St B2
Rawstorn Rd B1
Rebon St C3
Recreation Rd C3
Ripple Way A3
Roman Rd B2
Roman Wall B2
Romford Cl A3
Rosebery Ave B2
St Andrews Ave B3
St Andrews Gdns B3
St Botolph St B2
St Botolphs B2
St John's Abbey (site of) C2
St John's St B1
St John's Walk Shopping Centre B1
St Leonards Rd C3
St Marys Fields B1
St Peters B1
St Peter's St B1
Salisbury Ave C1
Serpentine Walk A1
Sheepen Pl B1
Sheepen Rd B1
Sir Isaac's Walk B1
Smythies Ave B2
South St C1
South Way C1
Sports Way A2
Suffolk Cl A3
Town Hall B1
Valentine Dr A3
Victor Rd C3
Wakefield Cl B2
Wellesley Rd C1
Wells Rd B2/B3
West St C1
West Stockwell St B1
Weston Rd C3
Westway A1
Wickham Rd C1
Wimpole Rd C3
Winchester Rd C2
Winnock Rd C2
Wolfe Ave C2
Worcester Rd B2

Coventry 334

Abbots La A1
Albany Rd B1
Alma St B3
Art Faculty B3
Asthill Grove C2
Bablake School A1
Barras La A1/B1
Barrs Hill School A1
Belgrade B2
Bishop Burges St A2
Bond's Hospital B1
Broad Gate B2
Broadway C1
Bus Station A3
Butts Radial B1
Canal Basin A2
Canterbury St A3
Cathedral B3
Chester St A1
Cheylesmore Manor House B2
Christ Church Spire B2
City Walls & Gates A2
Corporation St B2
Council House B2
Coundon Rd A1
Coventry & Warwickshire Hospital (A&E) A2
Coventry Station ≋ C2
Coventry Transport Museum A2
Cox St A3
Croft Rd B1
Dalton Rd C1
Deasy Rd C3
Earl St B2
Eaton Rd C2
Fairfax St B3
Foleshill Rd A2
Ford's Hospital B2
Fowler Rd A1
Friars Rd C2
Gordon St C1
Gosford St B3
Greyfriars Green B2
Greyfriars Rd B2
Gulson Rd B3
Hales St A2
Harnall Lane East A3
Harnall Lane West A2
Herbert Art Gallery & Museum B3
Hertford St B2
Hewitt Ave A1
High St B2
Hill St B1
Holy Trinity B2
Holyhead Rd A1
Howard St A3
Huntingdon Rd C1
Information Ctr B2
Jordan Well B3
King Henry VIII Sch C1
Lady Godiva Statue B2
Lamb St A2
Leicester Row A2
Library B2
Little Park St B2
London Rd C3
Lower Ford St B3
Magistrates & Crown Courts B2
Manor House Drive B2
Manor Rd C2
Market B2
Martyr's Memorial C2
Meadow St B1
Meriden St A1
Michaelmas Rd C2
Middleborough Rd A1
Mile La C3
Millennium Place A2
Much Park St B3
Naul's Mill Park A1
New Union B2
Park Rd C2
Parkside C3
Police HQ B2
Post Office B2
Primrose Hill St A3
Priory Gardens & Visitor Centre B2
Priory St B3
Puma Way C3
Quarryfield La C3
Queen's Rd B1
Quinton Rd C2
Radford Rd A2
Raglan St B3
Retail Park C1
Ringway (Hill Cross) A1
Ringway (Queens) B1
Ringway (Rudge) B1
Ringway (St Johns) B3
Ringway (St Nicholas) A2
Ringway (St Patricks) C2
Ringway (Swanswell) A2
Ringway (Whitefriars) B3
St John St B2
St John The Baptist B2
St Nicholas St A2
Skydome B1
Spencer Ave C1
Spencer Park C1
Spon St B1
Sports Centre B3
Stoney Rd C2
Stoney Stanton Rd A3
Swanswell Pool A3
Sydney Stringer School A3
Technical College B1
Technology Park C3
The Precinct B2
Theatre B1
Thomas Landsdail St C2
Tomson Ave A1
Top Green C1
Toy Museum B3
Trinity St B2
University B3
Upper Hill St A1
Upper Well St A2
Victoria St A3
Vine St A3
Warwick Rd C2
Waveley Rd B1
Westminster Rd C1
White St A3
Windsor St B1

Derby 334

Abbey St C1
Agard St B1
Albert St B2
Albion St B2
Ambulance Station B1
Arthur St A1
Ashlyn Rd B3
Assembly Rooms B2
Babington La C2
Becket St B1
Belper Rd A1
Bold La B1
Bradshaw Way C2
Bradshaw Way Retail Park C2
Bridge St B1
Brook St B1
Burrows Walk C2
Burton Rd C1
Bus Station B2
Caesar St A2
Canal St C3
Carrington St C3
Cathedral B2
Cathedral Rd B1
Charnwood St C2
Chester Green Rd A2
City Rd A2
Clarke St A3
Cock Pitt B3
Council House B2
Courts B1
Cranmer Rd B3
Crompton St C1
Crown & County Courts B2
Crown Walk C2
Curzon St B1
Darley Grove A1
Derby ≋ C3
Derbyshire County Cricket Ground B3
Derwent Business Centre A2
Derwent St B2
Devonshire Walk C2
Drewry La C1
Duffield Rd A1
Duke St A2
Dunton Cl B3
Eagle Market C2
Eastgate B3
East St B2
Exeter St B2
Farm St C1
Ford St B1
Forester St C1
Fox St A2
Friar Gate B1
Friary St B1
Full St B2
Gerard St C1
Gower St C2
Green La C2
Grey St C1
Guildhall B2
Harcourt St C1
Highfield Rd A1
Hill La C2
Information Ctr B2
Iron Gate B2
John St C3
Joseph Wright Centre B1
Kedleston Rd A1
Key St B2
King Alfred St C1
King St A1
Kingston St A1
Leopold St C2
Library B1
Liversage St C3
Lodge La B1
London Rd C2
London Rd Community Hospital C2
Macklin St C1
Mansfield Rd A2
Market B2
Market Pl B2
May St C1
Meadow La B3
Melbourne St C2
Midland Rd C3
Monk St C1
Morledge B2
Mount St C1
Museum & Art Gallery B1
Noble St C3
North Parade A1
North St A1
Nottingham Rd B3
Osmaston Rd C2
Otter St A1
Park St C3
Parker St A1
Pickfords House B1
Playhouse C2
Police HQ A2
Police Station B2

Post Office A1/A2/B1/B2/C2/C3
Prime Enterprise Park A2
Pride Parkway C3
Prime Parkway A2
Queens Leisure Centre B1
Racecourse A3
Railway Terr C3
Register Office B2
Sacheverel St C2
Sadler Gate B1
St Alkmund's Way B1/B2
St Helens House A1
St Mary's A1
St Mary's Bridge A2
St Mary's Bridge Chapel A2
St Mary's Gate B1
St Paul's Rd A2
St Peter's C2
St Peter's St C2
Siddals Rd C3
Silk Mill B2
Sir Frank Whittle Rd A3
Spa La C1
Spring St C1
Stafford St B1
Station Approach C3
Stockbrook St C1
Stores Rd A3
Traffic St C2
Wardwick B1
Werburgh St C1
West Ave A1
Westfield Centre C2
West Meadows Industrial Estate B3
Wharf Rd A2
Wilmot St C2
Wilson St C1
Wood's La C1

Dorchester 334

Ackerman Rd B3
Acland Rd A2
Albert Rd A1
Alexandra Rd B1
Alfred Place B3
Alfred Rd B2
Alington Ave B3
Alington Rd B3
Ambulance Station B3
Ashley Rd B1
Balmoral Cres C3
Barnes Way B2/C2
Borough Gdns A1
Bridport Rd A1
Buckingham Way C3
Caters Place A1
Cemetery A3/C1
Charles St A2
Coburg Rd B1
Colliton St A1
Cornwall Rd A1
Cromwell Rd B1
Culliford Rd B2
Culliford Rd North B2
Dagmar Rd B1
Damer's Rd B1
Diggory Cres C2
Dinosaur Museum A2
Dorchester Bypass C2
Dorchester South Station B1
Dorchester West Station B1
Dorset County Council Offices A1
Dorset County (A+E) B1
Dorset County Museum A1
Duchy Close C3
Duke's Ave B2
Durngate St A2
Durnover Court A3
Eddison Ave B3
Edward Rd B1
Egdon Rd C2
Elizabeth Frink Statue B2
Farfrae Cres B2
Forum Centre, The B1
Friary Hill A2
Friary Lane A2
Frome Terr A2
Garland Cres C3
Glyde Path Rd A1
Government Offices B3
Gt. Western Rd B1
Grosvenor Cres C1
Grosvenor Rd C1
H M Prison A1
Herringston Rd C1
High St East A2
High Street Fordington A2
High Street West A1
Holloway Rd A2
Icen Way A2
Keep Military Museum, The A1
Kings Rd A3/B3
Kingsbere Cres C2
Lancaster Rd B2
Library A1
Lime Cl C1
Linden Ave B2
London Cl A3
London Rd A2/A3
Lubbecke Way A3
Lucetta La B2
Maiden Castle Rd C1
Manor Rd C2
Maumbury Rd B1
Maumbury Rings B1
Mellstock Ave C2
Mill St A3
Miller's Cl A1
Mistover Cl C1
Monmouth Rd B1/B2
Nature Reserve A2
North Sq A2
Northernhay A1
Old Crown Court & Cells A1
Olga Rd B1
Orchard St A2
Police Station B1
Post Office A1/B1
Pound Lane A2
Poundbury Rd A1
Prince of Wales Rd B2
Prince's St A1
Queen's Ave B1
Roman Town House A1
Roman Wall A1
Rothesay Rd C2
St George's Rd B3
Salisbury Field A2
Sandringham Sports Centre B3
Shaston Cres C2
Smokey Hole La B3
South Court Ave C1
South St B1
South Walks Rd B2
Superstore C3
Teddy Bear House A1
Temple Cl C1
Terracotta Warriors & Teddy Bear Museum A2
The Grove A1
Town Hall A2
Town Pump A2
Trinity St A1
Tutankhamun Exhibition A1
Victoria Rd B1
Weatherbury Way C2
Wellbridge Cl C1
West Mills Rd A1
West Walks Rd A1
Weymouth Ave C1
Williams Ave B1
Winterbourne (BMI) C1
Wollaston Rd A2
York Rd B2

Dumfries 334

Academy St A2
Aldermanhill Rd B3
Ambulance Station C3
Annan Rd A3
Ardwall Rd A3
Ashfield Dr A1
Atkinson Rd C1
Averill Cres C1
Balliol Ave C1
Bank St B2
Bankend Rd C3
Barn Slaps B3
Barrie Ave B3
Beech Ave A1
Bowling Green A3
Brewery St B1
Bridge House B1
Brodie Ave C2
Brooke St B2
Broomlands Dr C1
Brooms Rd B3
Buccleuch St A2
Burns House B2
Burns Mausoleum B3
Burns St B2
Burns Statue A2
Bus Station B1
Cardoness St A3
Castle St A2
Catherine St A2
Cattle Market A3
Cemetery B3
Cemetery C2
Church Cres A2
Church St B2
College Rd A1
College St A1
Corbelly Hill B1
Convent, The B1
Corberry Park B1
Cornwall Mt A3
County Offices A2
Court A2
Craigs Rd C3
Cresswell Ave B3
Cresswell Hill B3
Cumberland St B3
David Keswick Athletic Centre A3
David St B1
Dock Park C3
Dockhead B2
Dumfries A3
Dumfries Academy A2
Dumfries Museum & Camera Obscura B2
Dumfries Royal Infirmary (A & E) C3
East Riverside Dr C3
Edinburgh Rd A2
English St B2
Fire Station B3
Friar's Vennel A2
Galloway St B1
George Douglas Dr C1
George St A2
Gladstone Rd C2
Glasgow St A1
Glebe St B3
Glencaple Rd C3
Goldie Ave A1
Goldie Cres A1
Golf Course C3
Greyfriars A2
Grierson Ave B3
HM Prison B1
Hamilton Ave C1
Hamilton Starke Park C2
Hazelrigg Ave C1
Henry St B3
Hermitage Dr C1
High Cemetery C3
High St A2
Hill Ave C2
Hill St B1
Holm Ave C2
Hoods Loaning A3
Howgate St B1
Huntingdon Rd A3
Information Ctr B2
Irish St B2
Irving St A2
King St A1
Kingholm Rd C3
Kirkpatrick Ct C2
Laurieknowe B1
Leafield Rd B3
Library A2
Lochfield Rd A1
Loreburn Pk A3
Loreburn St A2
Loreburne Shopping Centre B2
Lover's Walk A2
Martin Ave B3
Maryholm Dr A1
Mausoleum B3
Maxwell St B1
McKie Ave B3
Mews La A2
Mill Green B2
Mill Rd B1
Moat Rd C2
Moffat Rd A3
Mountainhall Pk C3
Nelson St B1
New Abbey Rd B1/C1
New Bridge B1
Newall Terr A2
Nith Ave A2
Nith Bank C3
Nithbank Hospital C3
Nithside Ave A1
Odeon B2
Old Bridge B1
Palmerston Park (Queen of the South F.C.) A1
Park Rd C1
Pleasance Ave C1
Police H.Q. A3
Police Station A2
Portland Dr A1
Post Office A2/B1/B2/B3/B3
Priestlands Dr C1
Primrose St B1
Queen St B3
Queensberry St A2
Rae St A2
Richmond Ave C2
Robert Burns Centre B2
Roberts Cres C1
Robertson Ave C3
Robinson Dr C1
Rosefield Rd C2
Rosemount St B1
Rotchell Park C1
Rotchell Rd B1
Rugby Football Ground C1
Ryedale Rd C2
St Andrews B2
St John the Evangelist A2
St Josephs College B3
St Mary's Industrial Estate A3
St Mary's St A3
St Michael St B2
St Michael's B2
St Michael's Bridge B2
St Michael's Bridge Rd B2
St Michael's Cemetery B3
Shakespeare St B2
Solway Dr C2
Stakeford St A1
Stark Cres C2
Station Rd A3
Steel Ave A1
Sunderries Ave A1
Sunderries Rd A1
Suspension Brae B2
Swimming Pool A1
Terregles St B1
Theatre Royal B2
Troqueer Rd C2
Union St A1
Wallace St B3
Welldale B2
West Riverside Dr C2
White Sands B2

Dundee 334

Adelaide Pl A1
Airlie Pl C1
Albany Terr A1
Albert St A3
Alexander St A2
Ann St A2
Arthurstone Terr A3
Bank St B2
Barrack Rd A1
Barrack St B2
Bell St B2
Blackscroft A3
Blinshall St B1
Brown St B1
Bus Station B3
Caird Hall B2
Camperdown St B3
Candle La B3
Carmichael St A1
City Churches B2
City Quay B3
City Sq B2
Commercial St B2
Constable St A3
Constitution Ct A1
Constitution Cres A1
Constitution St A1/B2
Cotton Rd A3
Courthouse Sq B1
Cowgate A3
Crescent St A3
Crichton St B2
Dens Brae A3
Dens Rd A3
Discovery Point C2
Douglas St B1
Drummond St A1
Dudhope Castle A1
Dudhope St A2
Dudhope Terr A1
Dundee C2
Dundee College A1
Dundee Contemporary Arts C2
Dundee High School B2
Dura St A3
East Dock St B3
East Whale La B3
East Marketgait B3
Erskine St A3
Euclid Cr B2
Forebank Rd A2
Foundry La A3
Gallagher Retail Park B3
Gellatly St B3
Government Offices C2
Guthrie St B1
Hawkhill B1
Hilltown A2
HMS Unicorn B3
Howff Cemetery, The B2
Information Ctr B2
King St A3
Kinghorne Rd A1
Ladywell Ave A3
Laurel Bank A2
Law Hill, The A1
Law Rd A1
Law St A1
Library A2
Little Theatre A2
Lochee Rd B1
Lower Princes St A3
Lyon St A3
McManus Galleries B2
Meadow Side B2
Meadowside St Pauls B2
Mercat Cross B2
Murraygate B2
Nelson St A2
Nethergate B2/C1
North Marketgait B2
North Lindsay St B2
Old Hawkhill B1
Olympia Leisure Centre C3
Overgate Shopping Centre B2
Park Pl B1
Perth Rd C1
Police Station A2/B1
Post Office B2
Princes St A3
Prospect Pl A2
Reform St B2
Repertory C1
Riverside Dr C2
Roseangle C1
Rosebank St A2
RRS Discovery C2
St Andrew's C2
St Pauls Episcopal B3
Science Centre C2
Seagate B2
Sheriffs Court B1
South George St A2
South Marketgait B3
South Tay St B2
South Ward Rd B2
Steps A2
Tay Road Bridge C3
Tayside House B2
Trades La B3
Union St B2
Union Terr A1
University Library B2
University of Abertay B2
University of Dundee B1
Upper Constitution St A1
Verdant Works B1
Victoria Dock B3
Victoria Rd A2
Victoria St A3
West Marketgait B1/B2
Ward Rd B1
Wellgate B2
West Bell St B1
Westfield Pl C1
William St A3
Wishart Arch A3

Durham 334

Alexander Cr B2
Allergate B2
Archery Rise C1
Assize Courts B3
Back Western Hill A1
Bakehouse La A3
Baths B3
Baths Bridge B3
Boat House B3
Bowling A2
Boyd St C3
Bus Station B2
Castle B2
Castle Chare B2
Cathedral B2
Church St C3
Clay La C1
Claypath B3
College of St Hild & St Bede B3
County Hall A1
County Hospital B1
Crook Hall & Gardens A3
Crossgate B2
Crossgate Peth C1
Darlington Rd C1
Durham A2
Durham Light Infantry Museum & Arts Centre A2
Durham School C2
Ellam Ave C1
Elvet Bridge B3
Elvet Court B3
Farnley Hey B1
Ferens Cl A3
Fieldhouse La A1
Flass St B1
Framwelgate A2
Framewelgate Bridge B2
Framwelgate Peth A2
Framwelgate Waterside A2
Frankland La A3
Freeman's Pl A3
Gala & Sacred Journey B3
Gate Shopping Centre, The B2
Geoffrey Ave C1
Gilesgate B3
Grey College C3
Hallgarth St C3
Hatfield College B3
Hawthorn Terr B1
Heritage Centre B3
HM Prison B3
John St B1
Kingsgate Bridge B3
Laburnum Terr B1
Lawson Terr B1
Leazes Rd B2/B3
Library B2
Margery La B2
Mavin St C3
Millburngate B2
Millburngate Bridge B2
Millennium Bridge (foot/cycle) A2
Mountjoy Research Centre C3
Museum of Archaeology B2
Nevilledale Terr B1
New Elvet B3
New Elvet Bridge B3
North Bailey B3
North End A1
North Rd A1/B2
Observatory C1
Old Elvet B3
Oriental Museum C2
Oswald Court C3
Parkside B2
Passport Office A2
Percy Terr B1
Pimlico C2
Police Station B3
Post Office A1/B2
Potters Bank C1/C2
Prebends Bridge C2
Prebends Walk C2
Prince Bishops Shopping Centre B3
Princes St A1
Providence Row A3
Quarryheads La C2
Redhills La B1
Redhills Terr B1
Saddler St B3
St Chad's College C3
St Cuthbert's Society C2
St John's College C2
St Margaret's B2
St Mary The Less C2
St Mary's College C2
St Monica Grove B1
St Nicholas' B3
St Oswald's C3
Sidegate A2
Silver St B2
Add Sixth Form Centre (Durham Gilesgate) A3
South Bailey C2
South Rd C3
South St B2
Springwell Ave A1
Stockton Rd C3
Students' Rec Centre B3
Sutton St B2
The Avenue B1
The Crescent A1
The Grove A1
The Sands A3
Town Hall B2
Treasury Museum B2
University B2
University Arts Block B3
University Library C3
University Science Site C3
Walkergate Centre A3
Wearside Dr A3
Western Hill A1
Wharton Park A2
Whinney Hill C3

Edinburgh 334

Abbey Strand B6
Abbeyhill B6
Abbeyhill Cr B6
Abbeymount A6
Abercromby Pl A3
Adam St C5
Albany La A4
Albany St A4
Albert Memorial B2
Albyn Pl A2
Alva Pl A6
Alva St B1
Ann St A1
Appleton Tower C4
Archibald Pl C3
Argyle House C3
Assembly Rooms & Musical Hall A3
Atholl Cr B1
Atholl Crescent La C1
Bank St B4
Barony St A4
Beaumont Pl C5
Belford Rd B1
Belgrave Cr A1
Belgrave Crescent La A1
Bell's Brae B1
Blackfriars St B4
Blair St B4
Bread St C2
Bristo Pl C4
Bristo St C4
Brougham St C2
Broughton St A4
Brown St C5
Brunton Terr A6
Buckingham Terr A1
Burial Ground A6
Bus Station A4
Caledonian Cr C1
Caledonian Rd C1
Calton Hill A4
Calton Hill A5
Calton Rd B4
Camera Obscura & Outlook Tower B3
Candlemaker Row C4
Canning St B2
Canongate B5
Canongate B5
Carlton St A1
Carlton Terr A6
Carlton Terrace La A6
Castle St B2
Castle Terr B2
Castlehill B3
Central Library B4
Chalmers Hospital C3
Chalmers St C3
Chambers St C4
Chapel St C4
Charles St C4
Charlotte Sq B2
Chester St B1
Circus La A2
Circus Pl A2
City Art Centre B4
City Chambers B4
City Observatory A5
Clarendon Cr A1
Clerk St C5
Coates Cr B1
Cockburn St B4
College of Art C3
Comely Bank Ave A1
Comely Bank Row A1
Cornwall St C2
Cowans Cl C5
Cowgate B4
Cranston St B5
Crichton St C4
Croft-An-Righ A6
Cumberland St A2
Dalry Pl C1
Dalry Rd C1
Danube St A1
Darnawy St A2
David Hume Tower C4
Davie St C5
Dean Bridge A1
Dean Gdns A1
Dean Park Cr A1
Dean Park Mews A1
Dean Park St A1
Dean Path B1
Dean St A1
Dean Terr A1
Dewar Pl C1
Dewar Place La C1
Doune Terr A2
Drummond Pl A3
Drummond St C5
Drumsheugh Gdns B1
Dublin Mews A3
Dublin St A4
Dublin Street Lane South A4
Dumbiedykes Rd B5
Dundas St A3
Earl Grey St C2
East Crosscauseway C5
East Market St B4
East Norton Pl A6
East Princes St Gdns B3
Easter Rd A6
Edinburgh (Waverley) B4
Edinburgh Castle B3
Edinburgh Dungeon B4
Edinburgh Festival Theatre C4
Edinburgh International Conference Ctr C2
Elder St A4
Esplanade B3
Eton Terr A1
Eye Pavilion C3
Festival Office B3
Filmhouse C2
Fire Station C2
Floral Clock B3
Forres St A2
Forth St A4
Fountainbridge C2
Frederick St A3
Freemasons' Hall B3
Fruit Market B4
Gardner's Cr C2
George Heriot's School C3
George IV Bridge B4
George Sq C4
George Sq La C4
George St B3
Georgian House B2
Gladstone's Land B3
Glen St C3
Gloucester La A2
Gloucester Pl A2
Gloucester St A2
Graham St C3
Grassmarket C3
Great King St A3
Great Stuart B1
Greenside La A5
Greenside Row A5
Greyfriars Kirk C4
Grindlay St C2
Grosvenor St C1
Grove St C1
Gullan's Cl B5
Guthrie St B4
Hanover St A3
Hart St A4
Haymarket C1
Haymarket Station C1
Heriot Pl C3
Heriot Row A2
High School Yard B5
High St B4
Hill Pl C5
Hill St A2
Hillside Cr A5
Holyrood Park C6
Holyrood Rd B5
Home St C2
Hope St B2
Horse Wynd B6
Howden St C5
Howe St A2
India Pl A2
India St A2
Infirmary St B4
Information Ctr B4
Jamaica Mews A2
Jeffrey St B4
John Knox's House B4
Johnston Terr C3
Keir St C3
Kerr St A2
King's Stables Rd B2
Lady Lawson St C3
Lady Stair's House B4
Laserquest C2
Lauriston Gdns C3
Lauriston Park C3
Lauriston Pl C3
Lauriston St C3
Lawnmarket B3
Learmonth Gdns A1
Learmonth Terr A1
Leith St A4
Lennox St A1
Lennox St La A1
Leslie Pl A2
London Rd A5
Lothian Health Board C5
Lothian Rd B2
Lothian St C4
Lower Menz Pl A6
Lynedoch Pl B1
Manor Pl B1
Market St B4
Marshall St C4
Maryfield A6
Maryfield Pl A6
McEwan Hall C4
Medical School C4
Melville St B1
Meuse La B4
Middle Meadow Walk C4
Milton St A6
Montrose Terr A6
Moray House (college) B5
Moray Place A2
Morrison Link C1
Morrison St C1
Mound Pl B3
Museum of Childhood B5
Museum of Edinburgh B5
National Gallery B3
National Library of Scotland B4
National Monument A5
National Museum of Scotland C4
National Portrait Gallery & Museum of Antiquities A3
Nelson Monument A5
Nelson St A3
New St B5
Nicolson Sq C4
Nicolson St C4
Niddry St B4
North Bridge B4
North Meadow Walk C3
North Bank St B3
North Castle St A2
North Charlotte St A2
North St Andrew St A4
North St David St A3
North West Circus Pl A2
Northumberland St A3
Odeon C2
Old Royal High School A5
Old Tolbooth Wynd B5
Our Dynamic Earth B6
Oxford Terr A1
Palace of Holyrood House B6
Palmerston Pl B1
Panmure Pl C3
Parliament House B4
Parliament Sq B4
People's Story, The B5
Playhouse Theatre A4
Pleasance C5
Police Station A4
Ponton St C2
Post Office A3/A4/B5/C1/C2/C4/C5
Potterrow C4
Princes Mall B4
Princes St B3
Queen St A2
Queen Street Gdns A3
Queen's Dr B6/C6
Queensferry Rd A1
Queensferry St B1
Queensferry Street La B2
Radical Rd C6
Randolph Cr B1
Regent Gdns A5
Regent Rd B5
Regent Rd Park A6
Regent Terr A5
Register House A4
Remains of Holyrood Abbey (AD 1128) A6
Richmond La C5
Richmond Pl C5
Rose St B2
Rosemount Bldgs C1
Ross Open Air Theatre B3
Rothesay Pl B1
Rothesay Terr B1
Roxburgh Pl C5
Roxburgh St C5
Royal Bank of Scotland A4
Royal Circus A2
Royal Lyceum C2
Royal Scottish Academy B3
Royal Terr A5
Royal Terrace Gdns A5
Rutland Sq B2
Rutland St B2
St Andrew Sq A3
St Andrew's House A4
St Bernard's Cr A1
St Cecilia's Hall B4
St Colme St A2
St Cuthbert's B2
St Giles' B4
St James Centre A4
St John St B5
St John's B2
St John's Hill B5
St Leonard's Hill C5
St Leonard's La C5
St Leonard's St C5
St Mary's (R.C.) A4
St Mary's Scottish Episcopal B1
St Mary's St B5
St Stephen St A2
Salisbury Crags C6
Saunders St A2
Scott Monument B4
Scottish Arts Council Gallery B2
Scottish Parliament B6
Semple St C2
Shandwick Pl B2
South Bridge B4
South Charlotte St B2
South College St C4
South Learmonth Gdns A1
South St Andrew St A4
South St David St A3
Spittal St C2
Stafford St B1
Student Centre C4
TA Centre C4
Tattoo Office B4
Teviot Pl C4
The Mall B6
The Mound B3
The Royal Mile B5
Thistle St A3
Torphichen Pl C1
Torphichen St C1
Traverse Theatre B2
Tron Sq B4
Tron, The B4
Union St A4
University C4
University Library C4
Upper Grove Pl C1
Usher Hall C2
Vennel C3
Victoria St B3
Viewcraig Gdns B5
Viewcraig St B5
Walker St B1
Waterloo Pl A4
Waverley Bridge B4
Wemyss Pl A2
West Approach Rd C1
West Crosscauseway C5
West Maitland St C1
West of Nicholson St C4
West Port C3
West Princes Street Gdns B3
West Richmond St C5
West Tollcross C2
White Horse Cl B5
William St B1
Windsor St A5
York La A4
York Pl A4
Young St B2

Exeter 334

Alphington St C1
Athelstan Rd B3
Bampfylde St B2
Barnardo Rd C3
Barnfield Hill B3
Barnfield Rd B2/B3
Barnfield Theatre B2
Bartholomew St East B1
Bartholomew St West B1
Bear St B2
Beaufort Rd C1
Bedford St B2
Belgrave Rd A3
Belmont Rd A3
Blackall Rd A2
Blackboy Rd A3
Bonhay Rd B1
Bull Meadow Rd C2
Bus & Coach Sta B3
Castle St B2
Cecil Rd C1
Cheeke St A3
Church Rd C1
Chute St A3
City Industrial Est C2
City Wall B1/B2
Civic Centre B2
Clifton Rd B3
Clifton St B3
Clock Tower A1
College Rd B3
Colleton Cr C2
Commercial Rd C1
Coombe St B2
Cowick St C1
Crown Courts B2
Custom House C2
Danes' Rd A2
Denmark Rd B3
Devon County Hall C3
Devonshire Pl A3
Dinham Rd B1
East Grove Rd C3
Edmund St C1
Elmgrove Rd A1
Exe St B1
Exeter Central Station A1
Exeter City Football Ground A3
Exeter College A1
Exeter Picture House B1
Fire Station A1
Fore St B1
Friars Walk C2
Guildhall B2
Guildhall Shopping Centre B2
Harlequins Shopping Centre B1
Haven Rd C2
Heavitree Rd B3
Hele Rd A1
High St B2
HM Prison A2
Holloway St C2
Hoopern St A2
Horseguards A2
Howell Rd A1
Information Ctr B3
Iron Bridge B1
Isca Rd C1
Jesmond Rd A3
King William St A2
King St B1
Larkbeare Rd C2
Leisure Centre C1
Library B2
Longbrook St A2
Longbrook Terr A2
Lower North St B1
Lucky La C2
Lyndhurst Rd C3
Magdalen Rd B3
Magdalen St B2
Magistrates & Crown Courts A2
Market B2
Market St B2
Marlborough Rd C3
Mary Arches St B1
Matford Ave C3
Matford La C3
Matford Rd C3
May St A3
Mol's Coffee House B2
New Bridge St B1
New North Rd A1/A2
North St B1
Northernhay St B1
Norwood Ave C3
Odeon A3
Okehampton St C1
Old Mill Cl C2
Old Tiverton Rd A3
Oxford Rd A3
Paris St B2
Parr St A3
Paul St B1
Pennsylvania Rd A2
Police H.Q. B3
Portland Street A3
Post Office A3/B1/B3/C1
Powderham Cr A3
Preston St B1
Princesshay Shopping Centre B2
Queen St A1
Queens Rd C1
Queen's Terr A1
Radford Rd C2
Richmond Rd A1
Roberts Rd C2
Rougemont Castle A2
Rougemont House B2
Royal Albert Memorial Museum B2
St David's Hill A1
St James' Park Station A3
St James' Rd A3
St Leonard's Rd C3

St Lukes College B3
St Mary Steps C1
St Nicholas Priory B1
St Peter's Cathedral B2
St Thomas Station C1
Sandford Walk B3
School for the Deaf C2
School Rd C1
Sidwell St A2
Smythen St B1
South St B2
Southernhay East B2
Southernhay West B2
Spacex Gallery B1
Spicer Rd B3
Sports Centre A3
Summerland St A3
Swimming Pool B3
Sydney Rd C1
Tan La C2
The Quay C2
Thornton Hill A2
Topsham Rd C3
Tucker's Hall B1
Tudor St B1
Velwell Rd A1
Verney St A3
Water La C1/C2
Weirfield Rd C2
Well St A3
West Ave A2
West Grove Rd C3
Western Way A3/B1/B2
Wonford Rd B3/C3
York Rd A2

Fort William 335

Abrach Rd A3
Achintore Rd C1
Alma Rd B2
Am Breun Chamas A2
Ambulance Station A3
An Aird A2
Argyll Rd C1
Argyll Terr C1
Bank St B2
Belford Hospital B2
Belford Rd B2/B3
Black Parks A3
Braemore Pl C2
Bruce Pl C2
Bus Station B2
Camanachd Cr A3/B2
Cameron Rd C1
Cameron Sq B1
Carmichael Way A2
Claggan Rd B3
Connochie Rd C1
Cow Hill C3
Creag Dhubh A2
Croft Rd B3
Douglas Pl B2
Dudley Rd B2
Dumbarton Rd C1
Earl of Inverness Rd A3
Fassifern Rd B1
Fort William Station B2
Fort William (Remains) B2
Glasdrum Rd C1
Glen Nevis Pl B3
Gordon Sq B1
Grange Rd C1
Heather Croft Rd C1
Henderson Row C2
High St B1
Highland Visitor Centre B3
Hill Rd B2
Hospital Belhaven Annexe B3
Information Ctr A3
Inverlochy Ct A3
Kennedy Rd B2/C2
Library B2
Linnhe Rd B1
Lochaber College A2
Lochaber Leisure Centre B3
Lochiel Rd A3
Lochy Rd A3
Lundavra Cres C1
Lundavra Rd C1
Lundy Rd A3
Mamore Cr B2
Mary St B2
Middle St B1
Montrose Ave A3
Moray Pl C1
Morven Pl C2
Moss Rd B2
Nairn Cres C1
Nevis Bridge B3
Nevis Rd A3
Nevis Sports Centre A2
Nevis Terr B2
North Rd B3
Obelisk B2
Ocean Frontier Underwater Centre A2
Parade Rd B2
Police Station A3/C1
Post Office A3/B2
Ross Pl C1
St Andrews B2
Shaw Pl B2
Station Brae B1
Studio B1
Treig Rd A3
Union Rd C1
Victoria Rd B2
Wades Rd A3
West Highland B2
Young Pl B2

Glasgow 335

Admiral St C2
Albert Bridge C5
Albion St B5
Anderston C2
Anderston Centre B3
Anderston Quay B3
Arches B4
Argyle St A1/A2/B3/B4/B5
Argyle Street B5
Argyll Arcade B5
Arlington St A3
Art Gallery & Museum A1
Arts Centre B3
Ashley St A3
Bain St C6
Baird St A6
Baliol St A3
Ballater St C5
Barras, The (Market) C6
Bath St A3
BBC Scotland/SMG B1
Bell St C6
Bell's Bridge B1
Bentinck St A2
Berkeley St A3
Bishop La B3
Black St A6
Blackburn St C2
Blackfriars St B6
Blantyre St A1
Blythswood Sq A4
Blythswood St B4
Bothwell St B4
Brand St C1
Breadalbane St A2
Bridge St C4
Bridge St Metro Sta C4
Bridgegate C5
Briggait C5
Broomhill Park A6
Broomielaw B4
Broomielaw Quay Gdns B3
Brown St B4
Brunswick St B5
Buccleuch St A3
Buchanan Bus Station A5
Buchanan Galleries A5
Buchanan St B5
Buchanan St Metro Sta B5
Cadogan St B4
Caledonian University A5
Calgary St A5
Cambridge St A4
Canal St A5
Candleriggs B5
Carlton Pl C4
Carnarvon St A3
Carnoustie St C3
Carrick St B4
Castle St B6
Cathedral Sq B6
Cathedral St B5
Central College of Commerce B5
Centre for Contemporary Arts A4
Centre St C4
Cessnock Metro Sta C1
Cessnock St C1
Charing Cross A3
Charlotte St C6
Cheapside St B3
Citizens' Theatre C5
City Chambers Complex B5
City Halls B5
Clairmont Gdns A2
Claremont St A2
Claremont Terr A2
Claythorne St C6
Cleveland St A3
Clifford La C1
Clifford St C1
Clifton Pl A2
Clifton St A2
Clutha St C1
Clyde Arc B2
Clyde Auditorium B2
Clyde Pl C4
Clyde Place Quay C4
Clyde St C5
Clyde Walkway C3
Clydeside Expressway B2
Coburg St C4
Cochrane St B5
College of Nautical Studies C5
College St B6
Collins St B6
Commerce St C4
Cook St C4
Cornwall St C2
Couper St A5
Cowcaddens Metro Sta A4
Cowcaddens Rd A4
Crimea St B3
Custom House C4
Custom House Quay Gdns C4
Dalhousie St A4
Dental Hospital A4
Derby St A2
Dobbie's Loan A4/A5
Dobbie's Loan Pl A5
Dorset St A3
Douglas St B4
Doulton Fountain C6
Dover St A2
Drury St B4
Drygate B6
Duke St B6
Dunaskin St A1
Dunblane St A4
Dundas St B5
Dunlop St C5
East Campbell St C6
Eastvale Pl A1
Eglinton St C4
Elderslie St A3
Elliot St B2
Elmbank St A3
Esmond St A1
Exhibition Centre B2
Exhibition Way B2
Eye Infirmary A2
Festival Park C1
Film Theatre A4
Finnieston Quay B2
Finnieston Sq B2
Finnieston St B2
Fitzroy Pl A2
Florence St C5
Fox St C5
Gallowgate C6
Garnet St A3
Garnethill St A4
Garscube Rd A4
George Sq B5
George St B5
George V Bridge C4
Gilbert St A1
Glasgow Bridge C4
Glasgow Cathedral B6
Glasgow Central B4
Glasgow Green C6
Glasgow Metropolitan College B5/C5
Glasgow Science Centre B1
Glasgow Science Centre Footbridge B1
Glassford St B5
Glebe St A6
Gloucester St C3
Gorbals Cross C5
Gorbals St C5
Gordon St B4
Govan Rd B1/C1/C2
Grace St B3
Grafton Pl A5
Grant St A3
Granville St A3
Gray St A2
Greendyke St C6
Harley St C1
Harvie St C1
Haugh Rd A1
Heliport B1
Henry Wood Hall A2
High Court C5
High St B6
High Street B6
Hill St A3
Holland St A3
Holm St B4
Hope St A5
Houldsworth St B2
Houston Pl C3
Houston St C3
Howard St C5
Hunter St C6
Hutcheson St B5
Hutchesons Hall B5
Hydepark St B3
Imax Cinema B1
India St A3
Information Ctr B5
Ingram St B5
Jamaica St B4
James Watt St B4
John Knox St B6
John St B5
Kelvin Hall A1
Kelvin Statue A2
Kelvin Way A2
Kelvingrove Park A2
Kelvingrove St A2
Kelvinhaugh St A1
Kennedy St A6
Kent Rd A2
Killermont St A5
King St B5
King's C5
Kingston Bridge C3
Kingston St C4
Kinning Park Metro Sta C2
Kinning St C3
Kyle St A5
Laidlaw St C3
Lancefield Quay B2
Lancefield St B3
Langshot St C1
Lendel Pl C1
Lighthouse B4
Lister St A6
Little St B3
London Rd C6
Lorne St C1
Lower Harbour B1
Lumsden St A1
Lymburn St A1
Lyndoch Cr A3
Lynedoch Pl A3
Lynedoch St A3
Maclellan St C1
Mair St C2
Maitland St A4
Mavisbank Gdns C2
Mcalpine St B3
Mcaslin St A6
McLean Sq C2
McLellan Gallery A4
McPhater St A4
Merchants' House B5
Middlesex St C2
Middleton St C1
Midland St B4
Miller St B5
Millroad St C6
Milnpark St C2
Milton St A4
Minerva St B2
Mitchell Library A3
Mitchell St West B4
Mitchell Theatre A3
Modern Art Gallery B5
Moir St C6
Molendinar St C6
Moncur St C6
Montieth Row C6
Montrose St B5
Morrison St C3
Mosque C5
Museum of Religious Life B6
Nairn St A1
Nelson Mandela Sq B5
Nelson St C4
Nelson's Monument C6
New City Rd A4
Newton St A3
Newton Pl A3
Nicholson St C4
Nile St B5
Norfolk Court C4
Norfolk St C4
North Frederick St B5
North Hanover St B5
North Portland St B6
North St A3
North Wallace St A5
Odeon C3
Old Dumbarton Rd A1
Osborne St B5/C5
Oswald St B4
Overnewton St A1
Oxford St C4
Pacific Dr B1
Paisley Rd C3
Paisley Rd West C1
Park Circus A2
Park Gdns A2
Park St South A2
Park Terr A2
Parkgrove Terr A2
Parnie St C5
Parson St A6
Partick Bridge A1
Passport Office A5
Paterson St C3
Pavilion Theatre A4
Pembroke St A3
People's Palace C6
Pinkston Rd A6
Piping Centre, The National A5
Pitt St A4/B4
Plantation Park C1
Plantation Quay B1
Police Station A4/A6/B5
Port Dundas Rd A5
Port St B2
Portman St C2
Prince's Dock B1
Princes Sq B5
Provand's Lordship B6
Queen St B5
Queen Street B5
Regimental Museum A3
Renfrew St A3/A4
Renton St A5
Richmond St B5
Robertson St B4
Rose St A4
Rottenrow A5
Royal Concert Hall A5
Royal Cr A2
Royal Exchange Sq B5
Royal Hospital For Sick Children A1
Royal Infirmary B6
Royal Scottish Academy of Music & Drama A4
Royal Terr A2
Rutland Cr C2
St Kent St C6
St Andrew's (R.C.) C5
St Andrew's C6
St Andrew's St C5
St Enoch Metro Sta B5
St Enoch Shopping Centre B5
St Enoch Sq B4
St George's Rd A3
St James Rd B6
St Mungo Ave A5/A6
St Mungo Pl A6
St Vincent Cr A2
St Vincent Pl B5
St Vincent St B3/B4
St Vincent Street Church B4
St Vincent Terr B3
Saltmarket C5
Sandyford Pl A3
Sauchiehall St A2/A4
School of Art A4
Scotland St C2/C3
Scott St A4
Scottish Exhibition & Conference Centre B1
Seaward St C2
Shaftesbury St B3
Sheriff Court C5
Shields Rd Metro Sta C3
Shuttle St B6
Somerset Pl A2
South Portland St C4
Springburn Rd A6
Springfield Quay C3
Stanley St C2
Stevenson St C6
Stewart St A4
Stirling Rd B6
Stirling's Library B5
Stobcross Quay B1
Stobcross Rd B1
Stock Exchange B5
Stockwell Pl C5
Stockwell St B5
Stow College A5
Strathclyde University B6
Sussex St C2
Synagogues A3/C4
Tall Ship B1
Taylor Pl A6
Tenement House A3
Teviot St A1
Theatre Royal A4
Tolbooth Steeple & Mercat Cross C6
Tower St C2
Trades House B5
Tradeston St C4
Transport Museum A1
Tron Steeple & Theatre C5
Trongate B5
Tunnel St B2
Turnbull St C5
UGC A5
Union St B4
Victoria Bridge C5
Virginia St B5
West Greenhill Pl B2
West Regent St A4
Wallace St C3
Walls St B6
Walmer Cr C1
Warrock St B3
Washington St B3
Waterloo St B4
Watson St B6
Watt St C3
Wellington St B4
West Campbell St B4
West George St B4
West Graham St A4
West Regent St B4
West St C4
West St Metro Sta C4
Westminster Terr A2
Whitehall St B3
Wilson St B5
Woodlands Gate A3
Woodlands Rd A3
Woodlands Terr A2
Woodside Cr A3
Woodside Pl A3
Woodside Terr A3
York St B4
Yorkhill Pde A1
Yorkhill St A1

Gloucester 335

Albion St C1
Alexandra Rd B3
Alfred St C3
All Saints Rd C2
Alvin St B2
Arthur St C2
Baker St C1
Barton St C2
Blackfriars B1
Blenheim Rd C2
Bristol Rd C1
Brunswick Rd C2
Bruton Way B2
Bus Station B2
Cattle Market A1
City Council Offices B1
City Mus., Art Gall. & Library B2
Clarence St B2
College of Art C2
Commercial Rd B1
Cromwell St C2
Deans Way A2
Denmark Rd A3
Derby Rd C3
Docks C1
Eastgate St B2
Edwy Pde A2
Estcourt Cl A3
Estcourt Rd A3
Falkner St C2
Folk Museum B1
GL1 Leisure Centre C2
Gloucester Cathedral B1
Gloucester Station B2
Gloucestershire Royal Hospital (A & E) B3
Goodyere St C2
Gouda Way A1
Great Western Rd B3
Guildhall B2
Heathville Rd A3
Henry Rd B3
Henry St B2
High Orchard St C1
Hinton Rd A2
India Rd C3
Information Ctr B1
Jersey Rd C3
King's C2
King's Sq B2
Kingsholm Rd A2
Kingsholm Rugby Football Ground A2
Lansdown Rd A3
Library C2
Llanthony Rd C1
London Rd B3
Longsmith St B1
Malvern Rd A3
Market Pde B2
Merchants Rd C1
Mercia Rd A1
Metz Way C3
Midland Rd C2
Millbrook St C3
Market B2
Montpellier C1
Napier St C3
National Waterways C1
Nettleton Rd C2
New Inn B2
New Olympus C3
North Rd A3
Northgate St B2
Oxford Rd A2
Oxford St B2
Park & Ride Gloucester A1
Park Rd C2
Park St B2
Parliament St C1
Pitt St B1
Police Station B1
Post Office B2
Quay St B1
Recreation Gd A1/A2
Regent St C2
Robert Raikes House B1
Royal Oak Rd B1
Russell St B2
Ryecroft St C2
St Aldate St B2
St Ann Way C1
St Catherine St A2
St Mark St A2
St Mary De Crypt B1
St Mary De Lode B1
St Nicholas's B1
St Oswald's Rd A1
St Oswald's Trading Estate A1
St Peter's B2
Seabroke Rd A3
Sebert St A2
Severn Rd C1
Sherborne St B2
Shire Hall B1
Sidney St C3
Soldiers of Gloucestershire B1
Southgate St B1/C1
Spa Field C1
Spa Rd C1
Sports Ground A2/B2
Station Rd B2
Stratton Rd C3
Stroud Rd C1
Superstore A1
Swan Rd A2
Technical College C1
The Mall B1
The Park C2
The Quay B1
Trier Way C1/C2
Union St A2
Vauxhall Rd C3
Victoria St C2
Wellington St C2
Westgate St B1
Widden St C2
Worcester St B2

Grimsby 335

Abbey Drive East C2
Abbey Drive West C2
Abbey Park Rd C2
Abbey Rd C2
Abbey Walk C2
Abbeygate Shopping Centre C2
Abbotsway C2
Adam Smith St A1/A2
Ainslie St C2
Albert St A3
Alexandra Dock A2/B2
Alexandra Retail Park A2
Alexandra Rd A2/B2
Annesley St A2
Armstrong St A1
Arthur St B1
Augusta St C1
Bargate C1
Beeson St A1
Bethlehem St C2
Bodiam Way B3
Bradley St B3
Brighowgate C1/C2
Bus Station B2/C2
Canterbury Dr C1
Cartergate B1/C1
Catherine St C3
Caxton A3
Chantry La B1
Charlton St A1
Church La C2
Church St A3
Cleethorpe Rd A3
College A3
College St C1
Compton Dr C1
Corporation Bridge A2
Corporation Rd A1
Court B3
Crescent St B1
Deansgate C1
Doughty Rd C2
Dover St B1
Duchess St C2
Dudley St C1
Duke of York Gardens B1
Duncombe St B3
Earl La B1
East Marsh St B3
East St B2
Eastgate B3
Eastside Rd A3
Eaton Ct C1
Eleanor St B3
Ellis Way B3
Fisherman's Chapel A3
Fisherman's Wharf B2
Fishing Heritage Centre B2
Flour Sq A3
Frederick St B1
Frederick Ward Way B2
Freeman St A3/B3
Freshney Dr B1
Freshney Pl B2
Garden St C2
Garibaldi St A3
Garth La B2
Grime St B3
Grimsby Docks Station A3
Grimsby Town Station C2
Hainton Ave C3
Har Way B3
Hare St C3
Harrison St B1
Haven Ave B1
Hay Croft Ave B1
Hay Croft St B1
Heneage Rd B3/C3
Henry St B1
Holme St B3
Hume St C1
James St B1
Joseph St B1
Kent St A3
King Edward St A3
Lambert Rd C2
Library B2
Lime St B1
Lister St B1
Littlefield La C1
Lockhill A3
Lord St B1
Ludford St C3
Macaulay St B1
Mallard Mews C3
Manor Ave C2
Market A3
Market Hall B2
Market St B3
Moss Rd C2
Nelson St A3
New St B2
Osbourne St B2
Pasture St B3
Peaks Parkway C3
Pelham Rd C1
Police Station A3
Post Office B1/B2/C2
PS Lincoln Castle B2
Pyewipe Rd A1
Railway Pl A3
Railway St A3
Recreation Ground C2
Rendel St A2
Retail Park B3
Richard St B1
Ripon St B1
Robinson St East B3
Royal St A3
St Hilda's Ave C1
St James C2
Sheepfold St B3/C3
Sixhills St C3
South Park B2
Spring St A3
Superstore B3
Tasburgh St C3
Tennyson St B2
The Close C1
Thesiger St A3
Time Trap C2
Town Hall B2
Veal St B1
Victoria Retail Park A3
Victoria St North A2
Victoria St South B2
Victoria St West B2
Watkin St A1
Welholme Ave C2
Welholme Rd C3
Wellington St B3
Wellowgate C2
Werneth Rd B3
West Coates Rd A1
Westgate A2
Westminster Dr C1
Willingham St C3
Wintringham Rd C2
Wood St B3
Yarborough Dr B1
Yarborough Hotel C2

Hanley 335

Acton St A3
Albion St B2
Argyle St C1
Ashbourne Gr A2
Avoca St A3
Baskerville Rd B3
Bedford Rd C1
Bedford St C1
Bethesda St B2
Bexley St A2
Birches Head Rd A3
Botteslow St C3
Boundary St A2
Broad St C2
Broom St A3
Bryan St A2
Bucknall New Rd B3
Bucknall Old Rd B3
Bus Station B3
Cannon St C2
Castlefield St C1
Hanley Park C2
Cavendish St B1
Central Forest Pk A2
Charles St B3
Cheapside B2
Chell St A3
Clarke St C1
Cleveland Rd C2
Clifford St C3
Clough St B2
Clyde St C1
College Rd C2
Cooper St C2
Corbridge Rd A1
Cutts St C2
Davis St C1
Denbigh St A1
Derby St C3
Dilke St A3
Dundas St A3
Dundee Rd C1
Dyke St B3
Eastwood Rd C3
Eaton St A3
Etruria Park B1
Etruria Rd B1
Etruria Vale Rd C1
Festing St A3
Fire Station C2
Foundry St B2
Franklyn St C3
Garnet St B1
Garth St B3
George St A3
Gilman St B3
Glass St B3
Goodson St B3
Greyhound Way A1
Grove Pl C1
Hampton St C3
Hanley Park C2
Harding Rd C2
Hassall St B3
Havelock Pl C1
Hazlehurst St C3
Hinde St C2
Hope St B2
Houghton St C2
Hulton St A3
Hypermarket A1/B2
Information Ctr B3
Jasper St C2
Jervis St A3
John Bright St A3
John St B2
Keelings Rd A3
Kimberley Rd C1
Ladysmith Rd C1
Lawrence St C2
Leek Rd C3
Library C2
Lichfield St C3
Linfield Rd B3
Loftus St A2
Lower Bedford St C1
Lower Bryan St A2
Lower Mayer St A3
Lowther St A1
Magistrates Court C2
Malham St A2
Marsh St B2
Matlock St C3
Mayer St A3
Milton St C1
Mitchell Memorial Theatre B2
Morley St B2
Moston St A3
Mount Pleasant C1
Mulgrave St A1
Mynors St B3
Nelson Pl B3
New Century St B1
New Forest Industrial Estate A3
Octagon, The Shopping Park B1
Ogden Rd C3
Old Hall St B3
Old Town Rd A3
Pall Mall B2
Palmerston St C3
Park and Ride C2
Parker St B2
Pavilion Dr A1
Pelham St C3
Percy St B2
Piccadilly B2
Picton St B3
Plough St A3
Police Station C2
Portland St A1
Post Office A3/B3/C3
Potteries Museum & Art Gallery B2
Potteries Shopping Centre B2
Potteries Way C2
Powell St A1
Pretoria Rd C1
Quadrant Rd B2
Ranelagh St C2
Raymond St C2
Rectory Rd C1
Regent Rd C2
Regent Theatre B2
Richmond Terr C2
Ridgehouse Dr A1
Robson St C2
St Ann St B3
St Luke St B3
Sampson St B2
Shaw St A1
Sheaf St C2
Shearer St C1
Shelton New Rd C1
Shirley Rd C2
Slippery La B2
Snow Hill C2
Sports Stadium A1
Spur St C3
Stafford St B2
Statham St B2
Stubbs La C3
Sun St C1
Talbot St C3
The Parkway C2
Town Hall B2
Town Rd A3
Trinity St B2
Union St A2
Upper Hillchurch St A3
Upper Huntbach St B3
Victoria Hall Theatre B3
Warner St C2
Warwick St C1
Waterloo Rd A1
Waterloo St B3
Well St B3
Wellesley St C2
Wellington Rd B3
Wellington St B3
Whitehaven Dr A2
Whitmore St C1
Windermere St A1
Woodall St A1
Yates St C2
York St A2

Harrogate 335

Albert St C2
Alexandra Rd B2
Arthington Ave B2
Ashfield Rd A2
Back Cheltenham Mount B2
Beech Grove C1
Belmont Rd C1
Bilton Dr A2
Bower Rd B2
Bower St B2
Bus Station B2
Cambridge Rd B2
Cambridge St B2
Cemetery A2
Chatsworth Pl A2
Chatsworth Grove A2
Chatsworth Rd A2
Chelmsford Rd B3
Cheltenham Cr B2
Cheltenham Mt B2
Cheltenham Pde B2
Christ Church B3
Christ Church Oval B3
Chudleigh Rd B3
Clarence Dr B1
Claro Rd A3
Claro Way A3
Coach Park B2
Coach Rd B3
Cold Bath Rd C1
Commercial St B2
Coppice Ave A1
Coppice Dr A1
Coppice Gate A1
Cornwall Rd B1
Council Offices B1
Court C3
Crescent Gdns B1
Crescent Rd B1
Dawson Terr A2
Devonshire Pl B3
Diamond Mews C1
Dixon Rd A2
Dixon Terr A2
Dragon Ave B3
Dragon Parade B2
Dragon Rd B2
Duchy Rd B1
East Parade B2
East Park Rd C3
Esplanade B1
Fire Station A2
Franklin Mount A2
Franklin Rd B2
Franklin Square A2
Glebe Rd C1
Grove Park Ct A3
Grove Park Terr A3
Grove Rd A2
Hampswaite Rd A1
Harcourt Dr B3
Harcourt Rd B3
Harrogate B2
Harrogate International Centre B1
Harrogate Ladies College B1
Harrogate Theatre B2
Heywood Rd C1
Hollins Cr A1
Hollins Mews A1
Hollins Rd A1
Homestead Rd C3
Hydro Leisure Centre, The A1
Information Ctr B1
James St B2
Jenny Field Dr A1
John St B2
Kent Dr A1
Kent Rd A1
Kings Rd A2
Kingsway B3
Kingsway Dr B3
Lancaster Rd C1
Leeds Rd C2
Lime Grove B3
Lime St A3
Mayfield Grove B2
Mayfield Pl A2
Mercer B1
Montpellier Hill B1
Mornington Cr A3
Mornington Terr B3
Mowbray Sq B3
North Park Rd B3
Nydd Vale Rd B2
Oakdale Ave A1
Oatlands Dr C3
Odeon B2
Osborne Rd A2
Otley Rd C2
Oxford St B2
Park Chase B3
Park Parade B3
Park View B2
Parliament St B1
Police Station B3
Post Office B2/C1
Providence Terr A2
Queen Parade C3
Queen's Rd C1
Raglan St C2
Regent Ave A3
Regent Grove A3
Regent Parade A3
Regent St A3
Regent Terr A3

Rippon Rd A1
Robert St C2
Royal Baths & Turkish Baths B1
Royal Pump Room B1
St Luke's Mount A2
St Mary's Ave C1
St Mary's Walk C1
Scargill Rd A1
Skipton Rd A3
Skipton St A2
Slingsby Walk C3
South Park Rd C2
Spring Grove A1
Springfield Ave B1
Station Ave B2
Station Parade B2
Strawberry Dale B2
Stray Rein C3
Studley Rd A2
Superstore B2
Swan Rd B1
The Parade B2
The Stray C2/C3
Tower St C2
Trinity Rd C2
Union St B2
Valley Dr C1
Valley Gardens C1
Valley Mount C1
Victoria Ave C2
Victoria Rd C1
Victoria Shopping Centre B2
Waterloo St A2
West Park C2
West Park St C2
Wood View A1
Woodfield Ave A3
Woodfield Dr A3
Woodfield Grove A3
Woodfield Rd A3
Woodfield Square A3
Woodside B3
York Pl C3
York Rd B1

Holyhead Caergybi 335

Armenia St A2
Arthur St C2
Beach Rd A1
Boston St B2
Bowling Green C3
Bryn Erw Rd C3
Bryn Glas Cl C3
Bryn Glas Rd C3
Bryn Gwyn Rd C1
Bryn Marchog A1
Bryn Mor Terr A2
Bryngoleu Ave A1
Cae Braenar C3
Cambria St B1
Captain Skinner's Obelisk B2
Cecil St C2
Cemetery C1/C2
Cleveland Ave C2
Coastguard Lookout A2
Court B2
Customs House A3
Cybi Pl A2
Cyttir Rd C3
Edmund St B1
Empire B2
Ferry Terminals B2
Ffordd Beibio B3
Ffordd Feurig C3
Ffordd Hirnos C3
Ffordd Jasper C3
Ffordd Tudur B3
Fire Station C2
Garreglwyd Rd B1
Gilbert St C2
Gorsedd Circle B1
Gwelfor Ave A1
Harbour View B3
Henry St C2
High Terr C1
Hill St B2
Holborn Rd C2
Holland Park Industrial Estate C3
Holyhead Park B1
Holyhead Station B2
Information Ctr B2
King's Rd C2
Kingsland Rd C2
Lewascote C3
Library B2
Lifeboat Station A1
Llanfawr Cl C3
Llanfawr Rd C3
Lligwy St C2
Lon Deg C3
London Rd C3
Longford Rd B1
Longford Terr B1
Maes Cybi B1
Maes Hedd A1
Maes-Hyfryd Rd C1
Maes-y-Dref B1
Maes-yr-Haf A2/B1
Maes-yr-Ysgol C3
Marchog C3
Marina A1
Maritime Museum A1
Market B2
Market St B2
Mill Bank B1
Min-y-Mor Rd A1
Morawelon Industrial Estate B3
Morawelon Rd B3
Moreton Rd C1
New Park Rd B1
Newry St A2
Old Harbour Lighthouse A3
Plas Rd C1
Police Station B2
Porth-y-Felin Rd A1
Post Office A1/B1/B2/B3/C2/C3
Prince of Wales Rd A2
Priory La B3
Pump St C1
Queens Park B1
Reseifion Rd C1
Rock St B1
Roman Fort B2
St Cybi St B2
St Cybi's Church B2
St Seiriol's Cl B1
Salt Island Bridge A2
Seabourne Rd A1
South Stack Rd B1
Sports Ground B1
Stanley St B2
Station St B2
Tan-y-Bryn Rd A1
Tan-yr-Efail C2
Tara St C1
Thomas St B1
Town Hall A2
Treseifion Estate C2
Turkey Shore Rd B2
Ucheldre Arts Centre B1
Ucheldre Ave B1
Upper Baptist St B1
Victoria Rd B2
Victoria Terr B2
Vulcan St B1
Walthew Ave A1
Walthew La A1
Wian St C2

Hull 335

Adelaide St C1
Albert Dock C1
Albion St B2
Alfred Gelder St B2
Anlaby Rd B1
Beverley Rd A1
Blanket Row C2
Bond St B2
Bridlington Ave A2
Brook St B1
Brunswick Ave A1
Bus Station B1
Camilla Cl C3
Canning St B1
Cannon St A2
Cannon's C1
Caroline St A2
Carr La B2
Castle St C2
Central Library B1
Charles St A2
Citadel Way B3
City Hall B2
Clarence St B3
Cleveland St A3
Clifton St A1
Collier St B1
Colonial St B1
Court B2
Deep, The C3
Dock Office Row B3
Dock St B2
Drypool Bridge B3
Egton St A3
English St C1
Ferens Gallery B2
Ferensway B1
Francis St A2
Francis St West A2
Freehold St A1
Freetown Way A2
Garrison Rd B3
George St B2
Gibson St A2
Great Thornton St B1
Great Union St A3
Green La A2
Grey St A1
Grimston St B2
Grosvenor St A1
Guildhall B2
Guildhall Rd B2
Hands-on History B2
Harley St A1
Hessle Rd C1
High St B3
Holy Trinity B2
Hull & East Riding Museum B3
Hull Arena C1
Hull College B3
Hull (Paragon) Station B1
Hull Truck Theatre B1
Humber Dock Marina C2
Humber Dock St C2
Humber St C2
Hyperion St A3
Information Ctr B2
Jameson St B1
Jarratt St B2
Jenning St A3
King Billy Statue C2
King Edward St B2
King St A2
Kingston Retail Park C1
Kingston St C2
Library Theatre B1
Liddell St A1
Lime St A3
Lister St C1
Lockwood St A2
Maister House B3
Maritime Museum B2
Market B2
Market Place B2
Minerva Pier C2
Mulgrave St A3
Myton Bridge C3
Myton St B1
Nelson St C2
New Cleveland St A3
New George St A2
New Theatre A2
Norfolk St A1
North Bridge A3
North St B1
Odeon C1
Old Harbour C3
Osborne St B1
Paragon St B2
Park St B1
Percy St A2
Pier St C2
Police Station B2
Post Office A1/B1/B2
Porter St C1
Portland St B1
Posterngate B2
Prince's Quay B2
Prospect Centre B1
Prospect St B1
Queen's Gdns B2
Railway Dock Marina C2
Railway St C2
Reform St A2
Retail Park B1
Riverside Quay C2
Roper St B2
St James St C1
St Luke's St B1
St Mark St A3
St Mary the Virgin B3
Scott St A2
South Bridge Rd B3
Spring Bank A1
Spring St B1
Spurn Lightship C2
Spyvee St A3
Streetlife Transport Museum B3
Sykes St A2
Tidal Surge Barrier C3
Tower St B3
Trinity House B2
University B2
Vane St A1
Victoria Pier C2
Waterhouse La B2
Waterloo St A1
Waverley St C1
Wellington St C2
Wellington St West C2
West St B1
Whitefriargate B2
Wilberforce Dr B2
Wilberforce House B3
Wilberforce Monument B3
William St C1
Wincolmlee A3
Witham A3
Wright St A1

Inverness 336

Abban St A1
Academy St B2
Alexander Pl B2
Anderson St A2
Annfield Rd C3
Ardconnel St B3
Ardconnel Terr B3
Ardross Pl B2
Ardross St B2
Argyle St B3
Argyle Terr B3
Attadale Rd B1
Ballifeary La C2
Ballifeary Rd C1/C2
Balnacraig La A1
Balnain St B2
Bank St B2
Bellfield Park C2
Bellfield Terr C3
Benula Rd A1
Birnie Terr A1
Bishop's Rd C2
Bowling Green A2
Bowling Green B2
Bowling Green C2
Bridge St B2
Brown St A2
Bruce Ave C1
Bruce Gdns C1
Bruce Pk C1
Burial Ground A2
Burnett Rd A3
Bus Station B3
Caledonian Rd B1
Cameron Rd A1
Cameron Sq A1
Carse Rd A1
Carsegate Rd Sth A1
Castle (Courts) B3
Castle Rd B2
Castle St B3
Celt St B2
Chapel St A2
Charles St B3
Church St B2
Clachnacuddin Football Ground A1
College A3
Columba Rd B1/C1
Crown Ave B3
Crown Circus B3
Crown Dr B3
Crown Rd B3
Crown St B3
Culduthel Rd C3
Dalneigh Cres C1
Dalneigh Rd C1
Denny St B3
Dochfour Dr B1/C1
Douglas Row B2
Duffy Dr C3
Dunabban Rd A1
Dunain Rd B1
Duncraig St B2
Eastgate Shopping Centre B3
Eden Court C2
Fairfield Rd B1
Falcon Sq B3
Fire Station A3
Fraser St B2
Fraser St C2
Friars' Bridge A2
Friars' La B2
Friars' St A2
George St A2
Gilbert St A2
Glebe St A2
Glendoe Terr A1
Glenurquhart Rd C1
Gordon Terr B3
Gordonville Rd C2
Grant St A2
Greig St B2
H.M. Prison B3
Harbour Rd A3
Harrowden Rd B1
Haugh Rd C2
Heatherley Cres C3
High St B3
Highland Council H.Q., The C2
Hill Park C3
Hill St B3
Huntly Pl A2
Huntly St B2
India St A2
Industrial Estate A3
Information Ctr B2
Innes St A2
Inverness B3
Inverness High Sch B1
Jamaica St A2
Kenneth St B2
Kilmuir Rd A1
King St B2
Kingsmills Rd B3
Laurel Ave B1/C1
Library A3
Lilac Gr B1
Lindsay Ave C1
Lochalsh Rd A1/B1
Longman Rd A3
Lotland Pl A2
Lower Kessock St A1
Madras St A2
Market Hall B3
Maxwell Dr C1
Mayfield Rd C3
Midmills College B3
Millburn Rd B3
Mitchell's La C3
Montague Row B2
Muirfield Rd C3
Muirtown St B1
Museum B2
Nelson St A2
Ness Bank C2
Ness Bridge B2
Ness Walk B2/C2
Old Edinburgh Rd C3
Old High Church B2
Park Rd C1
Paton St C2
Perceval Rd B1
Planefield Rd B2
Police Station A3
Porterfield Bank C3
Porterfield Rd C3
Portland Pl A2
Post Office A2/B1/B2/B3
Queen St B2
Queensgate B2
Railway Terr A3
Rangemore Rd B1
Reay St B3
Riverside St A2
Rose St A2
Ross Ave B1
Rowan Rd B1
Royal Northern Infirmary C2
St Andrew's Cathedral C2
St Columba B2
St John's Ave C1
St Mary's Ave C1
Shore St A2
Smith Ave C1
Southside Pl C3
Southside Rd C3
Spectrum Centre B2
Strothers La B3
Superstore B2
TA Centre C2
Telford Gdns B1
Telford Rd A1
Telford St A1
Tomnahurich Cemetery C1
Tomnahurich St B2
Town Hall B3
Union Rd B3
Union St B3
Walker Pl A2
Walker Rd A3
War Memorial C2
Waterloo Bridge A2
Wells St B1
Young St B2

Ipswich 336

Alderman Rd B2
All Saints' Rd A1
Alpe St B2
Ancaster Rd C1
Ancient House B3
Anglesea Rd B2
Ann St B2
Arboretum A2
Austin St C2
Belstead Rd C2
Berners St B2
Bibb Way B1
Birkfield Dr C1
Black Horse La B2
Bolton La B3
Bond St C3
Bowthorpe Cl B2
Bramford La A1
Bramford Rd A1
Bridge St C2
Brookfield Rd A1
Brooks Hall Rd A1
Broomhill A2
Broomhill Rd A1
Broughton Rd A2
Bulwer Rd B1
Burrell Rd C2
Butter Market B2
Butter Market Centre B3
Carr St B3
Cecil Rd B2
Cecilia St C2
Chancery Rd C2
Charles St B2
Chevallier St A1
Christchurch Mansion & Wolsey Art Gallery B3
Christchurch Park A3
Christchurch St B3
Cineworld C2
Civic Centre B2
Civic Dr B2
Clarkson St B1
Cobbold St B3
Commercial Rd C2
Constable Rd A3
Constantine Rd C1
Constitution Hill A2
Corder Rd A3
Corn Exchange B2
Cotswold Ave A2
Council Offices C2
County Hall B3
Crown Court C2
Crown St B2
Cullingham Rd B1
Cumberland St B2
Curriers La B2
Dale Hall La A2
Dales View Rd A1
Dalton Rd B2
Dillwyn St B1
Elliot St B1
Elm St B2
Elsmere Rd A3
Falcon St C2
Felaw St C3
Flint Wharf C3
Fonnereau Rd B2
Fore St C3
Foundation St C3
Franciscan Way C2
Friars St C2
Gainsborough Rd A3
Gatacre Rd B1
Geneva Rd B2
Gippeswyk Ave C1
Gippeswyk Park C1
Grafton Way C2
Graham Rd A1
Grimwade St C3
Great Whip St C3
Handford Cut B1
Handford Rd B1
Henley Rd A2
Hervey St B3
High St B2
Holly Rd A2
Information Ctr B3
Ipswich Haven Marina C3
Ipswich School A2
Ipswich Station C2
Ipswich Town FC (Portman Road) C2
Ivry St A2
Kensington Rd A1
Kesteven Rd C1
Key St C3
Kingsfield Ave A3
Kitchener Rd A1
Magistrates Court B2
Little's Cr C2
London Rd B1
Low Brook St C3
Lower Orwell St C3
Luther Rd C2
Manor Rd A3
Mornington Ave A1
Museum & Art Gallery B2
Museum St B2
Neale St B3
New Cardinal St C2
New Cut East C3
New Cut West C3
New Wolsey B2
Newson St B2
Norwich Rd A1/B1
Oban St B2
Old Customs House C3
Old Foundry Rd B3
Old Merchant's House C3
Orford St B2
Paget Rd A2
Park Rd A3
Park View Rd A2
Peter's St C2
Philip Rd C2
Pine Ave A2
Pine View Rd A2
Police Station B2
Portman Rd B2
Portman Walk C1
Post Office B2/B3
Princes St C2
Prospect St B1
Queen St B2
Ranelagh Rd C1
Recreation Ground B1
Rectory Rd C2
Regent Theatre B3
Retail Park C2
Richmond Rd A1
Rope Walk C3
Rose La C2
Russell Rd C2
St Edmund's Rd A2
St George's St B2
St Helen's St B3
Samuel Rd B3
Sherrington Rd A1
Silent St C2
Sir Alf Ramsey Way C1
Sirdar Rd B1
Soane St B3
Springfield La A1
Star La C3
Stevenson Rd B1
Suffolk College C3
Suffolk Retail Park B1
Superstore B1
Surrey Rd B1
Tacket St C3
Tavern St B3
The Avenue A3
Tolly Cobbold Museum C3
Tower Ramparts B2
Tower Ramparts Shopping Centre B2
Tower St B3
Town Hall B2
Tuddenham Rd A3
Upper Brook St B3
Upper Orwell St B3
Valley Rd A2
Vermont Cr B3
Vermont Rd B3
Vernon St C3
Warrington Rd A2
Waterloo Rd A1
Waterworks St C3
Wellington St B1
West End Rd B1
Westerfield Rd A3
Westgate St B2
Westholme Rd A1
Westwood Ave A1
Willoughby Rd C2
Withipoll St B3
Woodbridge Rd B3
Woodstone Ave A3
Yarmouth Rd B1

Kendal 336

Abbot Hall Art Gallery & Museum of Lakeland Life C2
Ambulance Station A2
Anchorite Fields C2
Anchorite Rd C2
Ann St A3
Appleby Rd A3
Archers Meadow C3
Ashleigh Rd A2
Aynam Rd B2
Bankfield Rd B1
Beast Banks B2
Beezon Fields A2
Beezon Rd A2
Beezon Trad Est A3
Belmont B2
Birchwood Cl C1
Blackhall Rd B2
Brewery Arts Centre B2
Bridge St B2
Brigsteer Rd C1
Burneside Rd A2
Bus Station B2
Buttery Well La C2
Canal Head North B3
Captain French La C2
Caroline St A2
Castle Hill B3
Castle Howe B2
Castle Rd B3
Castle St A3/B3
Cedar Gr C1
Council Offices B2
County Council Offices A2
Cricket Ground A3
Cricket Ground C3
Cross La C2
Dockray Hall Ind Estate A2
Dowker's La B2
Dry Ski Slope B3
East View B1
Echo Barn Hill C1
Elephant Yard Shopping Centre B2
Fairfield La A1
Finkle St B2
Fire Station A2
Fletcher Square C3
Football Ground C3
Fowling La A3
Gillinggate C2
Glebe Rd C2
Golf Course B1
Goose Holme B3
Gooseholme Bridge B3
Green St A1
Greengate C2
Greengate La C1/C2
Greenside B1
Greenwood C1
Gulfs Rd B2
High Tenterfell B1
Highgate B2
Hillswood Ave C1
Horncop La A2
Information Ctr B2
K Village and Heritage Centre B2
Kendal Business Park A3
Kendal Castle (Remains Of) B3
Kendal Fell B1
Kendal Green A1
Kendal A3
Kendal Station A3
Kent Pl B2
Kirkbarrow C2
Kirkland C2
Library B2
Library Rd B2
Little Aynam B3
Little Wood B1
Long Cl C1
Longpool A3
Lound Rd C3
Lound St C3
Low Fellside B2
Lowther St B2
Maple Dr C1
Market Pl B2
Maude St B2
Miller Bridge B2
Milnthorpe Rd C2
Mint St A3
Mintsfeet Rd A3
Mintsfeet Rd South A2
New Rd B2
Noble's Rest C2
Parish Church C2
Park Side Rd C3
Parkside Business Park C3
Parr St B3
Police Station A2
Post Office A3/B2/C2
Quaker Tapestry B2
Queen's Rd B1
Riverside Walk C2
Rydal Mount A2
Sandes Ave A2
Sandgate A3
Sandylands Rd A3
Serpentine Rd B1
Serpentine Wood B1
Shap Rd A3
South Rd C2
Stainbank Rd C1
Station Rd A3
Stramongate B2
Stramongate Bridge B2
Stricklandgate A2/B2
Sunnyside C3
Thorny Hills B3
Town Hall B2
Undercliff Rd B1
Underwood C1
Union St A2
Vicar's Fields C2
Vicarage Dr C1/C2
Wainwright Yard Shopping Centre B2
Wasdale Cl C3
Well Ings C2
Westmorland Shopping Centre & Market Hall B2
Westwood Ave C1
Wildman St A3
Windermere Rd A1
YHA B2
YWCA B2

King's Lynn 336

Albert St A2
Albion St B2
All Saints B2
All Saints St B2
Austin Fields A2
Austin St A2
Avenue Rd B3
Bank Side B1
Beech Rd C2
Birch Tree Cl B2
Birchwood St A2
Blackfriars Rd B2
Blackfriars St B2
Boal St B1
Bridge St B2
Broad St B2
Broad Walk B3
Burkitt St A2
Bus Station A2
Carmelite Terr C2
Chapel St A2
Chase Ave C3
Checker St C2
Church St B2
Clough La B2
Coburg St B2
College of West Anglia A3
Columbia Way A3
Corn Exchange A1
County Court Rd B2
Cresswell St A2
Custom House A1
Eastgate St A2
Edma St A2
Exton's Rd C3
Ferry La B1
Ferry St A1
Framingham's Almshouses B2
Friars St C2
Friars Walk C2
Gaywood Rd A3
George St A2
Gladstone Rd C2
Goodwin's Rd C3
Green Quay B1
Greyfriars' Tower B2
Guanock Terr C2
Guildhall A1
Hansa Rd C3
Hardwick Rd C2
Hextable Rd A2
High St B1
Holcombe Ave C3
Hospital Walk C2
Information Ctr B1
John Kennedy Rd A2
Kettlewell Lane A2
King George V Ave B3
King's Lynn Art Ctr A1
King's Lynn FC B3
King's Lynn Station B2
King St B1
Library B2
Littleport St A2
Loke Rd A2
London Rd B2
Lynn Museum B2
Majestic B2
Magistrates Court B2
Market La A1
Millfleet B2
Milton Ave A3
Nar Valley Walk C2
Nelson St B1
New Conduit St B2
Norfolk St A2
North Lynn Discovery Centre A3
North St A2
Oldsunway B2
Ouse Ave C1
Page Stair Lane A1
Park Ave B3
Police Station B2
Portland Pl C1
Portland St B2
Post Office A3/C2
Purfleet B1
Queen St B1
Raby Ave A3
Railway Rd A2
Red Mount Chapel B3
Regent Way B2
River Walk A1
Robert St C2
Saddlebow Rd C2
St Ann's St A1
St James' Rd B2
St James' Swimming Pool B2
St James St B2
St John's Walk B3
St Margaret's B1
St Nicholas A2
St Nicholas St A1
St Peter's Rd B1
Sir Lewis St A2
Smith Ave A3
South Everard St C2
South Gate C2
South Quay B1
South St B2
Southgate St C2
Stonegate St B2
Surrey St A1
Sydney St C3
Tennyson Ave B3
Tennyson Rd B3
Tower St B2
Town Hall B1
Town House & Tales of The Old Gaol House B1
Town Wall (Remains) B3
True's Yard Museum A2
Valingers Rd C2
Vancouver Ave C2
Waterloo St B2
Wellesley St B2
White Friars Rd C2
Windsor Rd C2
Winfarthing St C2
Wyatt St A2
York Rd C3

Lancaster 336

Aberdeen Rd C3
Adult College, The C3
Aldcliffe Rd C2
Alfred St B3
Ambleside Rd A3
Ambulance Sta A3
Ashfield Ave B1
Ashton Rd C2
Assembly Rooms, The B2
Balmoral Rd B3
Bath House B3
Bath Mill La B3
Bath St B3
Blades St B1
Borrowdale Rd B3
Bowerham Rd C3
Brewery La B2
Bridge La B2
Brook St C1
Bulk Rd A3
Bulk St B2
Bus Station B2
Cable St B2
Canal Cruises & Waterbus C2
Carlisle Bridge A1
Carr House La C2
Castle B1
Castle Park B1
Caton Rd A3
China St B2
Church St B2
City Museum B2
Clarence St C3
Common Gdn St B2
Coniston Rd A3
Cottage Museum B2
Council Offices B2
Court B2
Cromwell Rd C1
Crown Court B1
Dale St C3
Dallas Rd B1/C1
Dalton Rd B3
Dalton Sq B2
Damside St B2
De Vitre St B3
Dee Rd A1
Denny Ave A1
Derby Rd A2
Dukes B2
Earl St A2
East Rd B3
Eastham St C3
Edward St B3
Fairfield Rd B1
Fenton St B2
Firbank Rd A3
Fire Station B2
Friend's Meeting House B1
Garnet St B3
George St B2
Giant Axe Field B1
Grand, The B2
Grasmere Rd B3
Greaves Rd C2
Green St A3
Gregson Centre, The B3
Gregson Rd C3
Greyhound Bridge A2
Greyhound Bridge Rd A2
High St B2
Hill Side B1
Hope St C3
Hubert Pl B1
Information Ctr B2
Judges Lodgings B2
Kelsy St B1
Kentmere Rd B3
King St B2
Kingsway A3
Kirkes Rd C3
Lancaster & Lakeland C3
Lancaster City Football Club B1
Lancaster Station B1
Langdale Rd A3
Ley Ct B1
Library B2
Lincoln Rd B1
Lindow St C2
Lodge St B2
Long Marsh La B1
Lune Rd A1
Lune St A2
Lune Valley Ramble A3
Mainway A2
Maritime Museum A1
Market St B2
Marketgate Shopping Centre B2
Meadowside C2
Meeting House La B1
Millennium Bridge A2
Moor La B2
Moorgate B3
Morecambe Rd A1/A2
Nelson St B2
North Rd B2
Orchard La C1
Owen Rd A2
Park Rd B3
Parliament St A3
Patterdale Rd A3
Penny St B2
Police Station B2
Portland St C2
Post Office A3/B1/B2/B3/C3
Primrose St C3
Priory B1
Prospect St C3
Quarry Rd B3
Queen St C2
Regent St C2
Ridge La A3
Ridge St A3
Royal Lancaster Infirmary (A&E) C2
Rydal Rd B3
Ryelands Park A1
St Georges Quay A1
St John's B2
St Leonard's Gate B2
St Martin's Rd C3
St Nicholas Arcades Shopping Centre B2
St Oswald St C3
St Peter's B3
St Peter's Rd B3
Salisbury Rd B1
Scotch Quarry Urban Park C3
Shire Hall/HM Prison B1
Sibsey St B1
Skerton Bridge A2
South Rd C2
Station Rd B1
Stirling Rd C3
Storey Ave B1
Sunnyside La C1
Sylvester St C1
Tarnsyke Rd A1
Thurnham St C2
Town Hall B2
Troutbeck Rd B3
Ulleswater Rd B3
University of Cumbria C3
Vicarage Field B1
Vue B2
West Rd B1
Westbourne Dr C1
Westbourne Rd B1
Westham St C3
Wheatfield St B1
Williamson Rd B3
Willow La B1
Windermere Rd B3
Wingate-Saul Rd B1
Wolseley St B3
Woodville St B3
Wyresdale Rd C3

Leeds 336

Aire St B3
Aireside Centre B2
Albion Pl B4
Albion St B4
Albion Way B1
Alma St A6
Arcades B4
Armley Rd B1
Back Burley Lodge Rd A1
Back Hyde Terr A2
Back Row C3
Bath Rd C3
Beckett St A6
Bedford St B3
Belgrave St A4
Belle View Rd A2
Benson St A5
Black Bull St C5
Blenheim Walk A3
Boar La B4
Bond St B4
Bow St C5
Bowman La C4
Brewery C4
Bridge St A5/B5
Briggate B4
Bruce Gdns C1
Burley Rd A1
Burley St B1
Burmantofs St B6
Bus & Coach Station B5
Butterly St C4
Butts Cr B4
Byron St A5
Call La B4
Calverley St A3/B3
Canal St B1
Canal Wharf C3
Carlisle Rd C5
Cavendish Rd A1
Cavendish St A2
Chadwick St C5
Cherry Pl A6
Cherry Row A5
City Art Gallery & Library B3
City Museum A4
City Palace of Varieties B4
City Sq B3
Civic Hall A3
Clarence Road C5
Clarendon Rd A2
Clarendon Way A3
Clark La C6
Clay Pit La A4
Cloberry St A2
Clyde Approach C1
Clyde Gdns C1
Coleman St C2
Commercial St B4
Concord St A5
Cookridge St A4
Copley Hill C1
Corn Exchange B4
Cromer Terr A2
Cromwell St A5
Cross Catherine St B6
Cross Green La C6
Cross Stamford St A5
Crown & County Courts A3
Crown Point Bridge C5
Crown Point Retail Park C4
Crown Point Rd C4
David St C3
Dent St C6
Derwent Pl C3
Dial St C6
Dock St C4
Dolly La A6
Domestic St C2
Duke St B5
Duncan St B4
Dyer St B5
East Field St B6
East Pde B3
East St C5
Eastgate B5
Easy Rd C6
Edward St B4
Ellerby La C6
Ellerby Rd C6
Fenton St A3
Fire Station B2
Fish St B4
Flax Pl B5
Gelderd Rd C1
George St B4
Globe Rd C2
Gloucester Cr B1
Gower St A5
Grafton St A4
Grand Theatre B4
Granville Rd A6
Great George St A3
Great Wilson St C4
Greek St B3
Green La C1
Hanover Ave A2
Hanover La A2
Hanover Sq A2
Hanover Way A2
Harewood St B4
Harrison St B4
Haslewood Cl B6
Haslewood Drive B6
Headrow Centre B4
High Court B5
Holbeck La C2
Holdforth Cl C1
Holdforth Gdns B1
Holdforth Gr C1
Holdforth Pl C1
Holy Trinity B4
Hope Rd A5
Hunslet La C4
Hunslet Rd C4
Hyde Terr A2
Infirmary St B3
Information Ctr B3
Ingram Row C3
Junction St C4
Kelso Gdns A2
Kelso Rd A2
Kelso St A2
Kendal La A2
Kendell St C4
Kidacre St C4
King Edward St B4
King St B3
Kippax Pl C6
Kirkgate B4
Kirkgate Market B4
Kirkstall Rd A1
Kitson St C6
Lady La B4
Lands La B4
Lavender Walk B6
Leeds Bridge C4
Leeds General Infirmary (A&E) A3
Leeds Metropolitan University A3/A4
Leeds Shopping Plaza B4
Leeds Station B3
Leeds University A3
Light, The B4
Lincoln Green Rd A6
Lincoln Rd A6
Lindsey Gdns A6
Lindsey Rd A6
Lisbon St B3
Little Queen St B3
Long Close La C6
Lord St C2
Lovell Park A4
Lovell Park Rd A4
Lovell Rd A4
Lower Brunswick St A5
Mabgate A5
Macauly St A5
Magistrates Court A3
Manor Rd C3
Mark La B4
Marlborough St B2
Marsh La B5
Marshall St C3
Meadow La C4
Meadow Rd C4
Melbourne St A5
Merrion Centre A4
Merrion St A4
Merrion Way A4
Mill St B5
Millennium Sq A3
Mount Preston St A2
Mushroom St A5
Neville St C4
New Briggate A4/B4
New Market St B4
New Station St B4
New York Rd A5
New York St B5
Nile St A5
Nippet La A6
North St A4
Northern St B3
Oak Rd B1
Oxford Pl B3
Oxford Row A3
Park Cross St B3
Park La A2
Park Pl B3
Park Row B4
Park Sq B3
Park Sq East B3
Park Sq West B3
Park St B3
Police Station B5
Pontefract La B6
Portland Cr A3
Portland Way A3
Post Office B4/B5
Quarry House (NHS/DSS Headquarters) B5
Quebec St B3
Queen St B3
Railway St B5
Rectory St A6
Regent St A5
Richmond St C5
Rigton Approach B6
Rigton Dr B6
Rillbank La A1
Rosebank Rd A1
Royal Armouries C5
Russell St B3
Rutland St B2
St Anne's Cathedral (RC) A4
St Anne's St B4
St James' Hospital A6
St Johns Centre B4
St John's Rd A2
St Mary's St B5
St Pauls St B3
St Peter's B5
Saxton La B5
Sayner La C5
Shakespeare Ave A6
Shannon St B6
Sheepscar St South A5
Siddall St C3
Skinner La A5
South Pde B3
Sovereign St C4
Spence La C2
Springfield Mount A2
Springwell Ct C2
Springwell Rd C2
Springwell St C2
Stoney Rock La A6
Studio Rd A1
Sutton St C2
Sweet St C3
Sweet St West C3
Swinegate B4
Templar St B5
The Calls B5
The Close B6
The Drive B6
The Garth B5
The Headrow B3/B4
The Lane B5
The Parade B6
Thoresby Pl A3
Torre Rd A6
Town Hall A3
Trinity & Burton Arcades B4
Union Pl C3
Union St B5
Upper Accomodation Rd B6
Upper Basinghall St B4
Vicar La B4
Victoria Bridge C4
Victoria Quarter B4
Victoria Rd C4
Wade La A4
Washington St A1
Water La C3
Waterloo Rd C4
Wellington Rd B2/C1
Wellington St B3
West St B2
West Yorkshire Playhouse B5
Westfield Rd A1
Westgate B3
Whitehall Rd B3/C2
Whitelock St A5
Willis St C6
Willow Approach A1
Willow Ave A1
Willow Terrace Rd A3
Wintoun St A5
Woodhouse La A3/A4
Woodsley Rd A1
York Pl B3
York Rd B6
Yorkshire Television Studios A1

Leicester 336

Abbey St A2
All Saints' A1
Aylestone Rd C2
Bath La A1
Bede Park C1
Bedford St A3
Bedford St South A3
Belgrave Gate A2
Belle Vue A2
Belvoir St B2
Braunstone Gate B1
Burleys Way A2
Burnmoor St C2
Bus Station A2
Canning St A2
Castle B1
Castle Gardens B1
Cathedral B2
Causeway La A2
Charles St B3
Chatham St B2
Christow St A3
Church Gate A2
City Gallery B3
Civic Centre B2
Clock Tower B2
Clyde St A3
Colton St B3
Conduit St B3
Corn Exchange B2
Crafton St A3
Craven St A1
Crown Courts B3
De Montfort Hall C3
De Montfort St C3
De Montfort University C1
Deacon St C2
Dover St B3
Duns La B1
Dunton St A1
East St B3
Eastern Boulevard C1
Edmonton Rd A3
Erskine St A3
Filbert St C1
Filbert St East C2
Fire Station C3
Fleet St A3
Friar La B2
Friday St A2
Gateway St C2
Glebe St B3
Granby St B3
Grange La C2
Grasmere St C1
Great Central St A1
Guildhall B2
Guru Nanak Sikh Museum B1
Halford St B2
Havelock St C2
Haymarket A2
Haymarket Shopping Centre A2
High St A2
Highcross St A1
H.M. Prison C2
Horsefair St B2
Humberstone Gate B2
Humberstone Rd A3
Information Ctr B2
Jarrom St C2
Jewry Wall B1
Kamloops Cr A3
King Richards Rd B1
King St B2
Lancaster Rd C3
Lee St A3
Leicester R.F.C. C2
Leicester Royal Infirmary (A & E) C2
Leicester Station B3
Library B2
Little Theatre B3
London Rd C3
Lower Brown St B2
Magistrates Court B2
Manitoba Rd A3
Mansfield St A2
Market B2
Market St B2
Mill La C2
Montreal Rd A3
Museum & Art Gallery C3
Narborough Rd North B1
Nelson Mandela Park C2
New Park St B1
New St B2
New Walk C3
Newarke Houses B2
Newarke St B2
Northgate St A1
Orchard St A2
Ottawa Rd A3
Oxford St C2
Phoenix B2
Police Station B3
Post Office A1/B2/C2/C3
Prebend St C3
Princess Rd East C3
Princess Rd West C3
Queen St B3
Regent College C3
Regent Rd C2/C3
Repton St A1
Rutland St B3
St George St B3
St Georges Way B3
St John St A2
St Margaret's A2
St Margaret's Way A2
St Martins B2
St Mary de Castro B1
St Matthew's Way A3
St Nicholas B1
St Nicholas Circle B1
St Peter's La A2
Sanvey Gate A2
Shires Shopping Centre A2
Silver St B2
Slater St A1
Soar La A1
South Albion St B3
Southampton St B3
Swain St B3
Swan St A1
The Gateway C2
The Newarke B1
The Rally Community Park A2
Tigers Way C3
Tower St C3
Town Hall B2
Tudor Rd B1
University of Leicester C3
University Rd C3
Upperton Rd C1
Vaughan Way A2
Walnut St C2
Watling St A2
Welford Rd B2
Wellington St B2
West Bridge B1
West St C3
West Walk C3
Western Boulevard C1
Western Rd C1
Wharf St North A3
Wharf St South A3
'Y' Theatre B3
Yeoman St B3
York Rd B2

Lewes 336

Abinger Pl B1
All Saints Centre B2
Ambulance Station B2
Anne of Cleves House C1
Barbican House Mus B1
Brewery B2
Brook St A2
Brooks Rd A2
Bus Station B2
Castle Ditch La B1
Castle Precincts B1
Chapel Hill B3
Church La A1/A2
Cliffe High St B2
Cliffe Industrial Estate C3
Cluny St C1
Cockshut Rd C1
Convent Field C2
Coombe Rd A2
County Hall B1
County Records Office B1
Court B2
Court Rd B2
Crown Court B2
Cuilfail Tunnel B3
Davey's La A3
East St B2
Eastport La C1
Fire Station A2
Fisher St B2
Friars Walk B2
Garden St B1
Government Offices C2
Grange Rd B1
Ham La C2
Harveys Way B2
Hereward Way A2
High St B1/B2
Information Ctr B2
Keere St B1
King Henry's Rd B1
Lancaster St B2
Landport Rd A1
Leisure Centre C3
Lewes Bridge B2
Lewes Castle B1
Lewes Football Ground C2
Lewes Golf Course B3
Lewes Living History Model B1
Lewes Southern By-Pass C2
Lewes Station B2
Library B2
Malling Ind Est A2
Malling Brook Ind Est A3
Malling Hill A3
Malling St A3/B3
Market St B2
Martyr's Monument B3
Mayhew Way A2
Morris Rd B3
Mountfield Rd C2
New Rd B1
Newton Rd A1
North St A2/B2
Offham Rd B1
Old Malling Way A1
Old Needlemakers Craft Centre B2
Orchard Rd A3
Paddock La B1
Paddock Rd B1
Paddock Sports Ground B1
Park Rd B1
Pelham Terr A1
Pells Open Air Swimming Pool A1
Phoenix Causeway B2
Phoenix Ind Est B2
Phoenix Pl B2
Pinwell Rd B2
Police Station B1
Post Office A2/B1/B2/C1
Prince Edward's Rd B1
Priory St C1
Priory of St Pancras (remains of) C1
Railway La B2
Railway Land Nature Reserve B3
Rotten Row B1
Rufus Cl B1
St Pancras Rd C1
St John St B2
St John's Terr B1
St Nicholas La B2
Sewage Works C3
South Downs Business Park A3
South St B3/C3
Southdowns Rd A2
Southerham Junction C3
Southover Grange Gdns C1
Southover High St C1
Southover Rd B1
Spences Field A3
Spences La A2
Stansfield Rd A1
Station Rd B2
Station St B2
Sun St B1
Sussex Downs College C2
Sussex Police HQ A2
Talbot Terr B1
The Avenue B1
The Course C1
The Martlets A2
The Pells A1
Thebes Gallery B2
Toronto Terr B1
Town Hall B2
West St B2
White Hill B1
Willeys Bridge A1

Lincoln 337

Alexandra Terr B1
Anchor St C1
Arboretum B3
Arboretum Ave B3
Baggholme Rd B3
Bailgate A2
Beaumont Fee B1
Brayford Way C1
Brayford Wharf East C1
Brayford Wharf North B1
Bruce Rd A2
Burton Rd A1
Bus Station (City) C2
Canwick Rd C2
Cardinal's Hat B2
Carline Rd B1
Castle B1
Castle St A1
Cathedral & Treasury B2
Cathedral St B2
Cecil St A2
Chapel La A2
Cheviot St B3
Church La A2
City Hall B1
Clasketgate B2
Clayton Sports Ground A3
Coach Park C2
Collection, The B2
County Hospital (A & E) B3
County Office B1
Courts C1
Croft St B2
Cross St C2
Crown Courts B1
Curle Ave A3
Danesgate B2
Drill Hall B2
Drury La B1
East Bight A2
East Gate A2
Eastcliff Rd B3
Eastgate B2
Egerton Rd A3
Ellis Mill A1
Engine Shed, The C1
Environment Agency C2
Exchequer Gate B2
Firth Rd C1
Flaxengate B2
Florence St B3
George St C3
Good La A2
Gray St A1
Great Northern Terr C3
Great Northern Terrace Industrial Estate C3
Greetwell Rd B3
Greetwellgate B3
Haffenden Rd A2
High St B2/C1
HM Prison A3
Hospital (Private) A2
Hungate B2
Information Ctr B2
James St A2
Jews House & Court B2
Kesteven St C2
Langworthgate A2
Lawn Visitor Centre, The B1
Lee Rd A3
Library B2
Lincoln College B2
Lincoln Station C2
Lincolnshire Life/Royal Lincolnshire Regiment Museum A1
Lindum Rd B2
Lindum Sports Grd A3
Lindum Terr B3
Mainwaring Rd A3
Manor Rd A2
Market C2
Massey Rd A3
Medieval Bishop's Palace B2
Mildmay St A1
Mill Rd A1
Millman Rd B3
Minster Yard B2
Monks Rd B3
Montague St B2
Mount St A1
Nettleham Rd A2
Newland B1
Newport A2
Newport Arch A2
Newport Cemetery A2
Northgate A2
Odeon C1
Orchard St B1
Oxford St C2
Pelham Bridge C2
Pelham St C2
Police Station B1
Portland St C2
Post Office A1/A2/B1/B3/C2
Potter Gate B2
Priory Gate B2
Queensway A3
Rasen La A1
Ropewalk C1
Rosemary La B2
St Anne's Rd B3
St Benedict's C1
St Giles Ave A3
St John's Rd A2
St Mark St C1
St Mark's Retail Park C1
St Mark's Shopping Centre C1
St Mary-Le-Wigford C1
St Mary's St C2
St Nicholas St A2
St Swithin's B2
Saltergate C1
Saxon St A1
Sewell Rd B3
Silver St B2
Sincil St C2
Spital St A2
Spring Hill B1
Stamp End C3
Steep Hill B2
Stonebow & Guildhall C2
Stonefield Ave A2
Tentercroft St C1
The Avenue B1
The Grove A3
Theatre Royal B2
Tritton Retail Park C1
Tritton Rd C1
Union Rd B1
University of Lincoln C1
Upper Lindum St B3
Upper Long Leys Rd A1
Usher B2
Vere St A2
Victoria St B1
Victoria Terr B1
Vine St B3
Wake St A1
Waldeck St A1
Waterside Centre C2
Waterside North C2
Waterside South C2
West Pde B1
Westgate A2
Wigford Way C1
Williamson St A2
Wilson St A1
Winn St B3
Wragby Rd A3
Yarborough Rd A1

Liverpool 337

Abercromby Sq C5
Addison St A3
Adelaide Rd B6
Ainsworth St B4
Albany Rd B6
Albert Dock C2
Albert Edward Rd B6
Angela St C6
Anson St B4
Archbishop Blanche High School B6
Argyle St C3
Arrad St C4
Ashton St B5
Audley St A4
Back Leeds St A2
Basnett St B3
Bath St A1
Battle of the Atlantic B2
Beatles Story C2
Beckwith St C3
Bedford Close C5
Bedford St North C5
Bedford St South C5
Benson St C4
Berry St C4
Birkett St A4
Bixteth St B2
Blackburne Place C4
Bluecoat Chambers B3
Bold Place C4
Bold St C4
Bolton St B3
Bridport St B4
Bronte St B4
Brook St A1
Brownlow Hill B4/B5
Brownlow St B5
Brunswick Rd A5
Brunswick St B1
Butler Cr A6
Byrom St A3
Cable St B2
Caledonia St C4
Cambridge St C5
Camden St A4
Canada Blvd B1
Canning Dock C2
Canning Place C2
Canterbury St A4
Cardwell St C6
Carver St A4
Cases St B3
Castle St B2
Cavern Walks B3
Central Library A3
Central Station B3
Chapel St B2
Charlotte St B3
Chatham Place C6
Chatham St C5
Cheapside B2
Chestnut St C5
Christian St A3
Church St B3
Churchill Way North A3
Churchill Way South B3
Clarence St B4
Coach Station A4
Cobden St A5
Cockspur St A2
College La C3
College St North A5
College St South A5
Colquitt St C4
Comus St A3
Concert St C3
Connaught Rd B6
Conservation Centre B3
Cook St B2
Copperas Hill B4
Cornwallis St C3
Covent Garden B2
Craven St A4
Cropper St B3
Crown St B5/C6
Cumberland St B2
Cunard Building B1
Dale St B2
Dansie St B4
Daulby St B5
Dawson St B3
Dental Hospital Museum B5
Derby Sq B2
Drury La B2
Duckinfield St B4
Duke St C3
Earle St A2
East St A2
Eaton St A2
Edgar St A3
Edge La B6
Edinburgh Rd A6
Edmund St B2
Elizabeth St B5
Elliot St B3
Empire Theatre B4
Empress Rd B6
Epworth St A5
Erskine St A5
Everyman Theatre C5
Exchange St East B2
Fact Centre, The C4
Falkland St A5
Falkner St C5/C6
Farnworth St A6
Fenwick St B2
Fielding St A6
Fingerprints of Elvis C2
Fleet St C3
Fraser St A4
Freemasons Row A2
Gardner Row A3
Gascoyne St A2
George Pier Head C1
George St B2
Gibraltar Road A1
Gilbert St C3
Gildart St A4
Gill St B4
Goree B2
Gower St C2
Gradwell St C3
Great Crosshall St A3
Great George St C4
Great Howard St A1
Great Newton St B4
Greek St B4
Green La B4
Greenside A5
Greetham St C3
Gregson St A5
Grenville St C3
Grinfield St C6
Grove St C5
Guelph St A6
Hackins Hey B2
Haigh St A4
Hall La B6
Hanover St B3
Harbord St C6
Hardman St C4
Harker St A4
Hart St B4
Hatton Garden A2
Hawke St B4
Helsby St B6
Henry St C3
HM Customs & Excise National Museum C2
Highfield St A2
Highgate St B6
Hilbre St B4
Hope Place C4
Hope St C4
Houghton St B3
Hunter St A3
Hutchinson St A6
Information Ctr B3/C2
Institute For The Performing Arts C4
Irvine St B6
Irwell St B2
Islington A4
James St B2
James St Station B2
Jenkinson St A4
Johnson St A3
Jubilee Drive B6
Kempston St A4
Kensington A6
Kensington Gdns B6
Kensington St A6
Kent St C3
King Edward St A1
Kinglake St B6
Knight St C4
Lace St A3
Langsdale St A4
Law Courts C2
Leece St C4
Leeds St A2
Leopold Rd B6
Lime St B3
Lime St Station B4
Little Woolton St B5
Liver St C2
Liverpool John Moores University A3/B4/C4
Liverpool Landing Stage B1
Liverpool One C2
London Rd A4/B4
Lord Nelson St B4
Lord St B2
Lovat St C6
Low Hill A5
Low Wood St A6
Lydia Ann St C3
Manestry La C3
Mansfield St A4
Marmaduke St B6
Marsden St A6
Martensen St B6
Marybone A3
Maryland St C4
Mason St B6
Mathew St B2
May St B4
Melville Place C6
Merseyside Maritime Museum C2
Metquarter B3
Metropolitan Cathedral (RC) B5
Midghall St A2
Molyneux Rd A6
Moor Place B4
Moorfields B2
Moorfields Station B2
Moss St A5
Mount Pleasant B4/B5
Mount St C4
Mount Vernon B6
Mulberry St C5
Municipal Buildings B2
Myrtle Gdns C6
Myrtle St C5
Naylor St A2
Nelson St C4
Neptune Theatre B3
New Islington A4
New Quay B1
Newington St C3
North John St B2
North St A3
North View B6
Norton St A4
Oakes St B5
Odeon B4
Old Hall St A1
Old Leeds St A2
Oldham Place C4
Oldham St C4
Olive St C6
Open Eye Gallery C3
Oriel St A2
Ormond St B2
Orphan St C6
Overbury St C6
Overton St B6
Oxford St C5
Paisley St A1
Pall Mall A2
Paradise St C3
Paradise St Bus Station B2
Park La C3
Parker St B3
Parr St C3
Peach St B5
Pembroke Place B4
Pembroke St B5
Peter's La C3
Philharmonic Hall C5
Pickop St A2
Pilgrim St C4
Pitt St C3
Playhouse Theatre B3
Pleasant St B4
Police Headquarters C2
Police Station A4/B4
Pomona St B4
Port of Liverpool Building B2
Post Office A2/A4/A5/A6/B2/B3/B4/C4
Pownall St C2
Prescot St A5
Preston St B3
Princes Dock A1
Princes Gdns A2
Princes Jetty A1
Princes Pde B1
Princes St B2
Pythian St A6
Queen Square Bus Station B3
Queensland St C6
Queensway Tunnel (Docks exit) B1
Queensway Tunnel (Entrance) B3
Radio City B2
Ranelagh St B3
Redcross St B2
Renfrew St B6
Renshaw St B4
Richmond Row A4
Richmond St B3
Rigby St A2
Roberts St A1
Rock St B4
Rodney St C4
Rokeby St A4
Romily St A6
Roscoe La C4
Roscoe St C4
Rose Hill A3
Royal Court Theatre B3
Royal Liver Building B1
Royal Liverpool Hospital (A&E) B5
Royal Mail St B4
Rumford Place B2
Rumford St B2
Russell St B4
St Andrew St B4
St Anne St A4
St Georges Hall B3
St John's Centre B3
St John's Gdns B3
St John's La B3
St Joseph's Cr A4
St Minishull St B5
St Nicholas Place B1
St Paul's Sq A2
St Vincent Way B4
Salisbury St A4
Salthouse Dock C2
Salthouse Quay C2
Sandon St C5
Saxony Rd B6
Schomberg St A6
School La B2
Seel St C3
Seymour St B4
Shaw St A5
Sidney Place C6
Sir Thomas St B3
Skelhorne St B4
Slater St C3
Smithdown La B6
Soho Sq A4
Soho St A4
South John St B2
Springfield A4
Stafford St A4
Standish St A3
Stanley St B2
Strand St C2
Suffolk St C3
Tabley St C3
Tarleton St B3
Tate Gallery C2
Teck St B6
Temple St B2
The Strand B2
Tithebarn St B2
Town Hall B2
Traffic Police Headquarters C6
Trowbridge St B4
Trueman St A3
Union St B2
Unity Theatre C4
University C5
University Art Gallery C5
University of Liverpool B5

Upper Duke St C4
Upper Frederick St C3
Upper Baker St A6
Vauxhall Rd A2
Vernon St B2
Victoria St B2
Vine St C5
Wakefield St A4
Walker Art Gallery A3
Walker St A6
Wapping C2
Water St B1/B2
Waterloo Rd A1
Wavertree Rd B6
West Derby Rd A6
West Derby St B5
Whitechapel B3
Whitley Gdns A5
William Brown St B3
William Henry St A4
Williamson Sq B3
Williamson St B3
Williamson's Tunnels Heritage Centre C6
Women's Hospital C6
Wood St B3
World Museum, Liverpool A3
York St C3

Llandudno 337

Abbey Pl B1
Abbey Rd B1
Adelphi St B3
Alexandra Rd C2
Alice in Wonderland Centre B3
Anglesey Rd A1
Argyll Rd B3
Arvon Ave A2
Atlee Cl C3
Augusta St B3
Back Madoc St B2
Bodafon St B3
Bodhyfryd Rd A2
Bodnant Cr C3
Bodnant Rd C3
Bridge Rd C2
Bryniau Rd C1
Builder St B3
Builder St West C2
Cabin Lift A2
Camera Obscura A3
Caroline Rd B2
Chapel St A2
Charlton St B3
Church Cr C1
Church Walks A2
Claremont Rd B2
Clement Ave A2
Clifton Rd B2
Clonmel St B3
Coach Station B3
Conway Rd B3
Council St West C3
Cricket and Recreation Ground B2
Cwlach Rd A2
Cwlach St A1
Cwm Howard La C3
Cwm Pl C3
Cwm Rd C3
Dale Rd C1
Deganwy Ave B2
Denness Pl C2
Dinas Rd C2
Dolydd B1
Erol Pl B2
Ewloe Dr C3
Fairways C2
Ffordd Dewi C3
Ffordd Dulyn C2
Ffordd Dwyfor C3
Ffordd Elisabeth C3
Ffordd Gwynedd C3
Ffordd Las C3
Ffordd Morfa C3
Ffordd Penrhyn C3
Ffordd Tudno C3
Ffordd yr Orsedd C3
Ffordd Ysbyty C2
Fire & Ambulance Station B3
Garage St B3
George St A2
Gloddaeth Ave B1
Gloddaeth St B2
Gogarth Rd B1
Great Orme Mines A1
Great Ormes Rd B1
Happy Valley A2
Happy Valley Rd A3
Haulfre Gardens A1
Herkomer Cr C1
Hill Terr A2
Hospice B1
Howard Rd B3
Information Ctr B2
Invalids' Walk B1
James St B2
Jubilee St B3
King's Ave C2
King's Rd C2
Knowles Rd C2
Lees Rd C2
Library B2
Lifeboat Station B2
Llandudno A2
Llandudno (A & E) C2
Llandudno Station B3
Llandudno Town Football Ground C2
Llewelyn Ave A2
Lloyd St West B1
Lloyd St B2
Llwynon Rd A1
Llys Maelgwn B1
Madoc St B2
Maelgwn Rd B2
Maesdu Bridge C2
Maesdu Rd C2/C3
Maes-y-Cwm C3
Maes-y-Orsedd C3
Marian Pl C2
Marian Rd C2
Marine Drive (Toll) A3
Market Hall A2
Market St A2
Miniature Golf Course A1
Morfa Rd B1
Mostyn B3
Mostyn Broadway B3
Mostyn St B3
Mowbray Rd C2
New St A2
Norman Rd B3
North Parade A2
North Wales Golf Links C1
Old Rd A2
Oxford Rd B3
Parc Llandudno Shopping Centre B3
Pier A3
Plas Rd A2
Police Station B3
Post Office B3/C2
Promenade A3
Pyllau Rd A1
Rectory La A2
Rhuddlan Ave C3
St Andrew's Ave B2
St Andrew's Pl B2
St Beuno's Rd A1
St David's Pl B2
St David's Rd B2
St George's Pl A3
St Mary's Rd B2
St Seriol's Rd B2
Salisbury Pass B1
Salisbury Rd B2
Somerset St B3
South Parade A2
Stephen St B3
TA Centre B3
Tabor Hill A2
The Oval B1
The Parade B3
Town Hall B2
Trinity Ave B1
Trinity Cres C1
Trinity Sq B3
Tudno St A2
Ty-Coch Rd A2
Ty-Gwyn Rd A1/A2
Ty'n-y-Coed Rd A1
Vaughan St B3
Victoria Shopping Centre B3
Victoria Tram Station A2
War Memorial A2
Werny Wylan C3
West Parade B1
Whiston Pass A2
Winllan Ave C2
Wyddyfd Rd A2
York Rd A2

Llanelli 337

Alban Rd B3
Albert St B1
Als St B3
Amos St C1
Andrew St A3
Ann St C2
Annesley St B2
Arfryn Ave A3
Arthur St B2
Belvedere Rd A1
Bigyn La B3
Bigyn Park Terr C3
Bigyn Rd C3
Bond Ave C3
Brettenham St A1
Bridge St B2
Bryn Pl C1
Bryn Rd C1
Bryn Terr C1
Brynhyfryd Rd A2
Brynmelyn Ave A3
Brynmor Rd B1
Bryn-More Rd C1
Burry St C1
Bus Station B2
Caersalem Terr C2
Cambrian St C1
Caswell St C3
Cedric St B3
Cemetery A2
Chapman St A1
Charles Terr C2
Church St B2
Clos Caer Elms A1
Clos Sant Paul C2
Coastal Link Rd B1/C1
Coldstream St B2
Coleshill Terr B1
College Hill B3
College Sq B3
Copperworks Rd C2
Coronation Rd C3
Corporation Ave A3
Council Offices B2
Court B2
Cowell St B2
Cradock St C2
Craig Ave C3
Cricket Ground A1
Derwent St A1
Dillwyn St C2
Druce St C1
Elizabeth St B2
Emma St C2
Erw Rd B1
Felinfoel Rd A2
Fire Station A3
Firth Rd C3
Fron Terr C3
Furnace Rugby Football Ground A1
Gelli-On B2
George St C2
Gilbert Cres A2
Gilbert Rd A2
Glanmor Rd C2
Glanmor Terr C2
Glasfryn Terr A3
Glenalla Rd B3
Glevering St B3
Goring Rd A2
Gorsedd Circle A2
Grant St C3
Graveyard C2
Great Western Cl C2
Greenway St B1
Hall St B2
Harries Ave A2
Hedley Terr A2
Heol Elli B3
Heol Goffa A3
Heol Nant-y-Felin A3
Heol Siloh B2
Hick St C2
High St C1
Indoor Bowls Centre B1
Inkerman St B2
Island Pl B2
James St B3
John St B2
King George Ave B3
Lake View Cl A2
Lakefield Pl C1
Lakefield Rd C1
Langland Rd C3
Leisure Centre B1
Library B2
Llanelli House B2
Llanelli Parish Church B2
Llanelli R.U.F.C. (Stradey Park) A1
Llanelli Station C2
Llewellyn St C2
Lliedi Cres A3
Lloyd St B2
Llys Alys B2
Llys Fran A3
Llysnewedd C1
Long Row A3
Maes Gors C2
Maesyrhaf A3
Mansel St C3
Marblehall Rd B3
Marborough Rd A2
Margam St C3
Marged St C2
Marine St C1
Market B2
Market St B2
Marsh St C2
Martin Rd C3
Miles St A1
Mill La A3/B2
Mincing La B2
Murray St B2
Myn y Mor B1
Nathan St C1
Nelson Terr C1
Nevill St C2
New Dock Rd C2
New Rd A1
New Zealand St A1
Old Lodge C2
Old Rd A2
Paddock St C2
Palace Ave B3
Parc Howard A2
Parc Howard Museum & Art Gallery A2
Park Cres B1
Park St B2
Parkview Terr B1
Pemberton St C2
Pembrey Rd A1
Peoples Park B1
Police Station B2
Post Office A1/A2/B2/C1/C2
Pottery Pl B3
Pottery St B3
Princess St B1
Prospect Pl A2
Pryce St A1
Queen Mary's Walk C3
Queen Victoria Rd C1
Raby St B1
Railway Terr C2
Ralph St B2
Ralph Terr C1
Regalia Terr B3
Rhydyrafon A3
Richard St B2
Robinson St B2
Roland Ave A1
Russell St C3
St David's Cl C1
St Elli Shopping Centre B2
St Margaret's Dr A1
Spowart Ave A1
Station Rd B2/C2
Stepney Pl B2
Stepney St B2
Stewart St A1
Stradey Park Ave A1
Sunny Hill A2
Swansea Rd A3
TA Centre B2
Talbot St C3
Temple St B3
The Avenue Cilfig A2
The Mariners C1
Theatr Elli B2
Thomas St A2
Toft Pl A3
Town Hall B2
Traeth Ffordd C1
Trinity Rd C3
Trinity Terr C2
Tunnel Rd B3
Tyisha Rd C3
Union Blgs A2
Upper Robinson St B2
Vauxhall Rd B2
Walter's Rd B3
Waun Lanyrafon B2
Waun Rd A3
Wern Rd B3
West End A2
Y Bwthyn C3
Zion Row B3

London 338

Abbey Orchard St E3
Abchurch La D6
Abingdon St E4
Achilles Way D2
Acton St B4
Addington St E4
Air St D3
Albany St B2
Albemarle St D3
Albert Embankment F4
Aldenham St A3
Aldersgate St C6
Aldford St D2
Aldgate C7
Aldgate High St C7
Aldwych C4
Allsop Pl B1
Amwell St B5
Andrew Borde St C3
Angel A5
Appold St C7
Argyle Sq B4
Argyle St B4
Argyll St C3
Arnold Circus B7
Artillery La C7
Artillery Row E3
Ashbridge St B1
Association of Photographers Gallery B6
Baker St B1
Baker St B1
Baldwin's Gdns C5
Baltic St B6
Bank C6
Bank Museum C6
Bank of England C6
Bankside D6
Bankside Gallery D5
Banner St B6
Barbican C6
Barbican Gallery C6
Baroness Rd B7
Basil St E1
Bastwick St B6
Bateman's Row B7
Bath St B6
Bayley St C3
Baylis Rd E5
Beak St D3
Bedford Row C4
Bedford Sq C3
Bedford St D4
Bedford Way B3
Beech St C6
Belgrave Pl E2
Belgrave Sq E2
Bell La C7
Belvedere Rd E4
Berkeley Sq D2
Berkeley St D2
Bernard St B4
Berners Pl C3
Berners St C3
Berwick St C3
Bethnal Green Rd B7
Bevenden St B6
Bevis Marks C7
BFI London IMAX Cinema D5
Bidborough St B4
Binney St C2
Birdcage Walk E3
Bishopsgate C7
Blackfriars D5
Blackfriars Bridge D5
Blackfriars Rd E5
Blandford St C1
Blomfield St C6
Bloomsbury St C3
Bloomsbury Way C4
Bolton St D2
Bond St C2
Borough High St E6
Boswell St C4
Bow St C4
Bowling Green La B5
Brad St D5
Bressenden Pl E3
Brewer St D3
Brick St D2
Bridge St E4
Britain at War D7
Britannia Walk B6
British Library B3
British Museum C4
Britton St B5
Broad Sanctuary E3
Broadway E3
Brook Dr F5
Brook St D2
Brown St C1
Brunswick Pl B6
Brunswick Sq B4
Brushfield St C7
Bruton St D2
Bryanston St C1
Buckingham Gate E3
Buckingham Palace E3
Buckingham Palace Rd F2
Bunhill Row B6
Byward St D7
Cabinet War Rooms & Churchill Mus E3
Cadogan La E2
Cadogan Pl E1
Cadogan Sq F1
Caledonian Rd A4
Calshot St A4
Calthorpe St B4
Calvert Ave B7
Cambridge Circus C3
Camomile St C7
Cannon St D6
Cannon St D6
Carey St C4
Carlisle La E4
Carlisle Pl E3
Carlton House Terr D3
Carmelite St D5
Carnaby St C3
Carter La C5
Carthusian St C6
Cartwright Gdns B4
Castle Baynard St D5
Cavendish Pl C2
Cavendish Sq C2
Caxton Hall E3
Caxton St E3
Central St B6
Chalton St B3
Chancery Lane C5
Chapel Market A5
Chapel St E2
Charing Cross D4
Charing Cross Rd C3
Charles II St D3
Charles Sq B6
Charles St D2
Charlotte Rd B7
Charlotte St C3
Charrington St A3
Chart St B6
Charterhouse Sq C5
Charterhouse St C5
Cheapside C6
Chenies St C3
Chesham St E2
Chester Sq F2
Chesterfield Hill D2
Chiltern St C2
Chiswell St C6
City Garden Row A5
City Rd B6
City Thameslink C5
City University, The B5
Claremont Sq A5
Clarges St D2
Clerkenwell Cl B5
Clerkenwell Green B5
Clerkenwell Rd B5
Cleveland St C3
Clifford St D3
Clink Prison Mus D6
Clock Museum C6
Club Row B7
Cockspur St D3
Coleman St C6
Collier St A4
Columbia Rd B7
Commercial St C7
Compton St B5
Conduit St D2
Constitution Hill E2
Copperfield St E5
Coptic St C4
Cornhill C6
Cornwall Rd D5
Coronet St B7
Courtauld Gallery D4
Covent Garden D4
Covent Garden D4
Cowcross St C5
Cowper St B6
Cranbourn St D3
Craven St D4
Crawford Pl C1
Crawford St C1
Creechurch La C7
Cremer St A7
Cromer St B4
Crondall St A6
Cumberland Gate D1
Cumberland Terr A2
Curtain Rd B7
Curzon St D2
Dalí Universe E4
D'arblay St C3
Davies St C2
Dean St C3
Deluxe Gallery B7
Denmark St C3
Dering St C2
Devonshire St C2
Diana, Princess of Wales Memorial Walk E3
Dingley Rd B6
Donegal St A4
Dorset St C1
Doughty St B4
Dover St D2
Downing St E4
Druid St E7
Drummond St B3
Drury La C4
Drysdale St B7
Duchess St C2
Dufferin St B6
Duke of Wellington Pl E2
Duke St C2
Duke St D3
Duke St Hill D6
Duke's Pl C7
Duncannon St D4
East Rd B6
Eastcastle St C3
Eastcheap D7
Eastman Dental Hospital B4
Eaton Pl E2
Eaton Sq E2
Eccleston Bridge F3
Eccleston St E2
Edgware Rd C1
Eldon St C6
Embankment D4
Endell St C4
Endsleigh Pl B3
Enismore Gdns E1
Euston B3
Euston Rd B3
Euston Square B3
Eversholt St A3
Exmouth Market B5
Fann St B6
Farringdon C5
Farringdon Rd C5
Farringdon St C5
Featherstone St B6
Fenchurch St D7
Fenchurch St D7
Fetter La C5
Finsbury Circus C6
Finsbury Pavement C6
Finsbury Sq B6
Fitzalan St F5
Fitzmaurice Pl D2
Fleet St C5
Floral St D4
Florence Nightingale Museum E4
Folgate St C7
Foot Hospital B3
Fore St C6
Foster La C6
Francis St F3
Frazier St E5
Freemason's Hall C4
Friday St C6
Gainsford St E7
Garden Row E5
Gee St B6
George St C1
Gerrard St D3
Giltspur St C5
Glasshouse St D3
Gloucester Pl C1
Golden Hinde D6
Golden La B6
Golden Sq D3
Goodge St C3
Goodge St C3
Gordon Sq B3
Gosset St B7
Goswell Rd B5
Gough St B4
Goulston St C7
Gower St B3
Gracechurch St D6
Grafton Way B3
Graham St A5
Gray's Inn Rd B4
Great College St E4
Great Cumberland Pl C1
Great Eastern St B7
Great Guildford St D6
Great Marlborough St C3
Great Ormond St B4
Great Ormond Street Children's Hospital B4
Great Percy St B4
Great Peter St E3
Great Portland St B2
Great Portland St C2
Great Queen St C4
Great Russell St C3
Great Scotland Yd D4
Great Smith St E3
Great Suffolk St E5
Great Titchfield St C3
Great Tower St D7
Great Windmill St D3
Greek St C3
Green Park D3
Green St D2
Greencoat Pl F3
Gresham St C6
Greville St B4/C5
Greycoat Hosp Sch E3
Greycoat Pl E3
Grosvenor Cres E2
Grosvenor Gdns E2
Grosvenor Pl E2
Grosvenor Sq D2
Grosvenor St D2
Guards Museum and Chapel E3
Guildhall Art Gallery C6
Guilford St B4
Guy's Hospital D6
Haberdasher St B6
Hackney Rd B7
Half Moon St D2
Halkin St E2
Hall St B5
Hallam St C2
Hampstead Rd B3
Hanover Sq C2
Hans Cres E1
Hanway St C3
Hardwick St B5
Harley St C2
Harrison St B4
Hastings St B4
Hatfields D5
Hayles St F5
Haymarket D3
Hayne St C5
Hay's Galleria D7
Hay's Mews D2
Hayward Gallery D4
Helmet Row B6
Herbrand St B4
Hercules Rd E4
Hertford St D2
High Holborn C4
Hill St D2
HMS Belfast D7
Hobart Pl E2
Holborn C4
Holborn C5
Holborn Viaduct C5
Holland St D5
Holmes Mus B1
Holywell La B7
Horse Guards' Rd D3
Houndsditch C7
Houses of Parliament E4
Howland St C3
Hoxton Sq B7
Hoxton St B7
Hunter St B4
Hunterian Mus C4
Hyde Park D1
Hyde Park Cnr E2
Imperial War Mus E5
Inner Circle B2
Institute of Archaeology (London Univ) B2
Ironmonger Row B6
James St C2
James St D4
Jermyn St D3
Jockey's Fields C4
John Carpenter St D5
John St B4
Judd St B4
Killick St A4
King Charles St E4
King St D3
King St D4
King William St C6
Kingley St C3
King's Cross A4
King's Cross Rd B4
King's Cross St Pancras A4
King's Rd E2
Kingsland Rd B7
Kingsway C4
Kinnerton St E2
Knightsbridge E1
Lamb St C7
Lambeth Bridge F4
Lambeth High St F4
Lambeth North E5
Lambeth Palace E4
Lambeth Palace Rd E4
Lambeth Rd E5
Lambeth Walk F4
Lamb's Conduit St C4
Lancaster Pl D4
Lancaster St E5
Langham Pl C2
Leadenhall St C7
Leake St E4
Leather La C5
Leicester Sq D3
Leicester St D3
Leman St C7
Leonard St B6
Lever St B6
Lexington St D3
Lidlington Pl A3
Lime St D7
Lincoln's Inn Fields C4
Lindsey St C5
Lisle St D3
Liverpool Rd A5
Liverpool St C7
Liverpool St C7
Lloyd Baker St B5
Lloyd Sq B5
Lombard St C6
London Aquarium E4
London Bridge D6
London Bridge Hospital D6
London Canal Museum A4
London City Hall D7
London Dungeon D6
London Guildhall University C6
London Rd E5
London Transport Museum D4
London Wall C6
London-Eye E4
Long Acre D4
Long La C5
Longford St B2
Lower Belgrave St E2
Lower Grosvenor Pl E2
Lower Marsh E5
Lower Thames St D6
Lowndes St E2
Ludgate Circus C5
Ludgate Hill C5
Luxborough St C2
Lyall St E2
Macclesfield Rd B6
Madame Tussaud's B2
Maddox St D2
Malet St C3
Manchester Sq C2
Manchester St C2
Mandeville Pl C2
Mansell St C7
Mansion House C6
Mansion House D6
Maple St C3
Marble Arch C1
Marble Arch D1
Marchmont St B4
Margaret St C3
Margery St B5
Mark La D7
Marlborough Rd D3
Marshall St C3
Marsham St E3
Marylebone High St C2
Marylebone La C2
Marylebone Rd B2
Marylebone St C2
Mecklenburgh Sq B4
Middle Temple La C5
Middlesex St (Petticoat La) C7
Midland Rd A3
Mildmay Mission Hospital B7
Milner St F1
Minories C7
Mintern St A6
Monck St E3
Monmouth St C4
Montagu Pl C1
Montagu Sq C1
Montagu St C4
Montague Pl C3
Monument D6
Monument St D6
Monument, The D6
Moor La C6
Moorfields C6
Moorfields Eye Hospital B6
Moorgate C6
Moorgate C6
Moreland St B5
Morley St E5
Mornington Crescent A3
Mornington Pl A3
Mortimer St C3
Mount Pleasant B5
Mount St D2
Movieum of London E4
Murray Gr A6
Museum of Garden History E4
Museum of London C6
Museum St C4
Myddelton Sq B5
Myddelton St B5
National Film Theatre D4
National Gallery D3
National Hospital B4
National Portrait Gallery D3
Neal St C4
Nelson's Column D4
New Bond St C2/D2
New Bridge St C5
New Cavendish St C2
New Change C6
New Fetter La C5
New Inn Yard B7
New North Rd A6
New Oxford St C3
New Scotland Yard E3
New Sq C4
Newgate St C5
Newton St C4
Nile St B6
Noble St C6
Noel Rd A5
Noel St C3
North Audley St D2
North Cres C3
North Row D2
Northampton Sq B5
Northington St B4
Northumberland Ave D4
Norton Folgate C7
Nottingham Pl C2
Oakley Sq A3
Obstetric Hosp B3
Old Bailey C5
Old Broad St C6
Old Compton St C3
Old County Hall E4
Old Gloucester St C4
Old King Edward St C6
Old Nichol St B7
Old Paradise St F4
Old Spitalfields Mkt C7
Old St B6
Old St B6
Old Vic E5
Open Air Theatre B2
Operating Theatre Museum D6
Orange St D3
Orchard St C2
Ossulston St A3
Outer Circle B1
Oxford Circus C3
Oxford St C2/C3
Paddington St C2
Palace St E3
Pall Mall D3
Pall Mall East D3
Pancras Rd A3/A4
Panton St D3
Paris Gdn D5
Park Cres B2
Park La D2
Park Rd B1
Park St D2
Park St D6
Parker St C4
Parliament Sq E4
Parliament St E4
Paternoster Sq C5
Paul St B6
Pear Tree St B5
Penton Rise B4
Penton St A5
Pentonville Rd A4/A5
Percival St B5
Petticoat La (Middlesex St) C7
Petty France E3
Phoenix Pl B4
Phoenix Rd A3
Photo Gallery D3
Piccadilly D2
Piccadilly Circus D3
Pitfield St B7
Pocock St E5
Pollock's Toy Museum C3
Polygon Rd A3
Pont St E1
Portland Pl C2
Portman Mews C2
Portman Sq C2
Portman St C1
Portugal St C4
Poultry C6
Primrose St C7
Princes St C6
Procter St C4
Provost St B6
Quaker St B7
Queen Anne St C2
Queen Elizabeth Hall D4
Queen Sq B4
Queen St D6
Queen Street Pl D6
Queen Victoria St D5
Queens Gallery E3
Radnor St B6
Rathbone Pl C3
Rawstorne St B5
Red Lion Sq C4
Red Lion St C4
Redchurch St B7
Redcross Way D6
Regency St F3
Regent Sq B4
Regent St C3
Regent's Park B2
Richmond Terr E4
Ridgmount St C3
Rivington St B7
Robert St B2
Rochester Row F3
Rodney St A4
Ropemaker St C6
Rosebery Ave B5
Roupell St D5
Royal Academy of Arts D3
Royal Academy of Dramatic Art B3
Royal Academy of Music B2
Royal College of Nursing C2
Royal College of Surgeons C4
Royal Festival Hall D4
Royal National Theatre D5
Royal National Throat, Nose and Ear Hospital B4
Royal Opera House D4
Russell Sq B3
Russell Square B4
Sackville St D3
Sadlers Wells B5
Saffron Hill C5
Savile Row D3
Savoy Pl D4
Savoy St D4
School of Hygiene & Tropical Medicine C3
Sclater St B7
Scrutton St B7
Sekforde St B5
Serpentine Rd D1
Seven Dials C4
Seward St B5
Seymour St C1
Shad Thames D7
Shaftesbury Ave D3
Shaftesbury St A6
Shakespeare's Globe Theatre D6
Shepherd Market D2
Shepherdess Walk A6
Sherwood St D3
Shoe La C5
Shoreditch High St B7
Shorts Gdns C4
Sidmouth St B4
Silk St C6
Sir John Soane's Museum C4
Skinner St B5
Sloane St E1
Snow Hill C5
Soho Sq C3
Somerset House D4
South Audley St D2
South Carriage Dr E1
South Molton St C2
South Pl C6
South St D2
Southampton Row C4
Southampton St D4
Southwark D5
Southwark Bridge D6
Southwark Bridge Rd D6
Southwark Cath D6
Southwark St D6
Speakers' Corner D1
Spencer St B5
Spital Sq C7
St Alban's St D3
St Andrew St C5
St Bartholomew's Hospital C5
St Botolph St C7
St Bride St C5
St George's Rd E5
St Giles High St C3
St James's Palace D3
St James's Park E3
St James's St D3
St John St B5
St Margaret St E4
St Mark's Hosp B5
St Martin's La D4
St Martin's Le Grand C6
St Mary Axe C7
St Pancras International B4
St Paul's C6
St Paul's Cath C6
St Paul's Churchyard C5
St Peter's Hosp D4
St Thomas' Hosp E4
St Thomas St D6

Stamford St D5
Stanhope St B3
Stephenson Way B3
Stock Exchange C5
Stoney St D6
Strand C5
Stratton St D2
Sumner St D5
Sutton's Way B6
Swanfield St B7
Swinton St B4
Tabernacle St B6
Tate Modern D6
Tavistock Pl B4
Tavistock Sq B3
Tea & Coffee Museum D6
Temple ⊖ D5
Temple Ave D5
Temple Pl D4
Terminus Pl E2
Thayer St C2
The Barbican Centre for Arts C6
The Cut E5
The Mall E3
Theobald's Rd C4
Thorney St F4
Threadneedle St C6
Throgmorton St C6
Tonbridge St B4
Tooley St D7
Torrington Pl B3
Tothill St E3
Tottenham Court Rd B3
Tottenham Court Rd ⊖ C3
Tottenham St C3
Tower Bridge ✦ D7
Tower Bridge App D7
Tower Bridge Rd E7
Tower Hill ⊖ D7
Tower of London, The D7
Toynbee St C7
Trafalgar Square D3
Trinity Sq D7
Trocadero Centre D3
Tudor St D5
Turnmill St C5
Ufford St E5
Union St D5
University College Hospital B3
University of London C3
University of Westminster C2
University St B3
Upper Belgrave St E2
Upper Berkeley St C1
Upper Brook St D2
Upper Grosvenor St D2
Upper Ground D5
Upper Montague St C1
Upper St A5
Upper St Martin's La D4
Upper Thames St D6
Upper Wimpole St C2
Upper Woburn Pl B3
Vere St C2
Vernon Pl C4
Vestry St B6
Victoria ⊖ E2
Victoria Embankment D4
Victoria Place Shopping Centre F2
Victoria St E3
Villiers St D4
Vincent Sq F3
Vinopolis City of Wine D6
Virginia Rd B7
Wakley St B5
Walbrook C6
Wallace Collection C2
Wardour St C3/D3
Warner St B5
Warren St ⊖ B3
Warren St B3
Waterloo ⊖ E4
Waterloo Bridge D4
Waterloo East D5
Waterloo Rd E5
Watling St C6
Webber St E5
Welbeck St C2
Wellington Arch ✦ E2
Wellington Mus E2
Wells St C3
Wenlock Rd A6
Wenlock St A6
Wentworth St C7
Werrington St A3
West Smithfield C5
West Sq E5
Westminster ⊖ E4
Westminster Abbey ✝ E4
Westminster Bridge E4
Westminster Bridge Rd E5
Westminster Cathedral (RC) ✝ E3
Westminster City Hall E3
Westminster Hall E4
Weymouth St C2
Wharf Rd A6
Wharfdale Rd A4
Wharton St B4
Whitcomb St D3
White Cube B7
White Lion Hill D5
White Lion St A5
Whitecross St B6
Whitefriars St C5
Whitehall D4
Whitehall Pl D4
Wigmore Hall C2
Wigmore St C2
William IV St D4
Wilmington Sq B5
Wilson St C6
Wilton Cres E2
Wimpole St C2
Windmill Walk D5
Woburn Pl B4
Woburn Sq B3
Women's Hosp C3
Wood St C6
Woodbridge St B5
Wootton St D5
Wormwood St C7
Worship St B6
Wren St B4
Wynford Rd A4
Wynyatt St B5
York Rd E4
York St C1
York Terrace East B2
York Terrace West B2
York Way A4

Luton 337

Adelaide St B1
Albert Rd C2
Alma St B2
Alton Rd C3
Anthony Gdns C1
Arndale Centre B2
Arthur St C2
Ashburnham Rd B1
Ashton Rd C2
Avondale Rd A1
Back St A2
Bailey St C3
Baker St C2
Biscot Rd A1
Bolton Rd B3
Boyle Cl A2
Brantwood Rd B1
Bretts Mead C1
Bridge St B2
Brook St A1
Brunswick St A3
Burr St A3
Bury Park Rd A1
Bus Station B2
Bute St B2
Buxton Rd B2
Cambridge St C3
Cardiff Grove B1
Cardiff Rd B1
Cardigan St A2
Castle St B2/C2
Chapel St C2
Charles St A3
Chase St C2
Cheapside B2
Chequer St C3
Chiltern Rise C1
Church St B2/B3
Cinema A2
Cobden St A3
Collingdon St A1
Community Centre C3
Concorde Ave A3
Corncastle Rd C1
Cowper St C2
Crawley Green Rd B3
Crawley Rd A1
Crescent Rise A3
Crescent Rd A3
Cromwell Rd A1
Cross St A2
Crown Court B2
Cumberland St B2
Cutenhoe Rd C3
Dallow Rd B1
Downs Rd B1
Dudley St A2
Duke St A2
Dumfries St B1
Dunstable Place B2
Dunstable Rd A1/B1
Edward St A3
Elizabeth St C2
Essex Cl C3
Farley Hill C1
Farley Lodge C1
Flowers Way B2
Francis St A1
Frederick St A2
Galaxy Leisure Complex A2
George St B2
George St West B2
Gillam St A3
Gordon St B2
Grove Rd B1
Guildford St A2
Haddon Rd A3
Harcourt St C2
Hart Hill Drive A3
Hart Hill Lane A3
Hartley Rd A3
Hastings St B2
Hat Factory, The B2
Hatters Way A1
Havelock Rd A2
Hibbert St C2
High Town Rd A3
Highbury Rd A1
Hightown Community Sports & Arts Centre A3
Hillary Cres C1
Hillborough Rd C1
Hitchin Rd A3
Holly St C2
Holm C1
Hucklesby Way A2
Hunts Cl C1
Information Ctr B2
Inkerman St A2
John St B2
Jubilee St A3
Kelvin Cl C2
King St B2
Kingsland Rd C3
Latimer Rd C2
Lawn Gdns C2
Lea Rd B3
Library B2
Library Rd B2
Liverpool Rd B1
London Rd C2
Luton Station A2
Lyndhurst Rd B1
Magistrates Court B2
Manchester St B2
Manor Rd B3
May St C3
Meyrick Ave C1
Midland Rd A2
Mill St A2
Milton Rd B1
Moor St A1
Moor, The A1
Moorland Gdns A2
Moulton Rise A3
Museum & Art Gallery A2
Napier Rd B1
New Bedford Rd A1
New Town St C2
North St A3
Old Bedford Rd A2
Old Orchard C2
Osbourne Rd C3
Oxen Rd A3
Park Sq B2
Park St B3/C3
Park St West B2
Park Viaduct B3
Parkland Drive C1
Police Station B1
Pomfret Ave A3
Pondwicks Rd B3
Post Office A1/A2/B2/C3
Power Court B3
Princess St B1
Red Rails C1
Regent St B2
Reginald St A2
Rothesay Rd B1
Russell Rise C1
Russell St C1
St Ann's Rd B3
St George's B2
St Mary's B3
St Marys Rd B3
St Paul's Rd C2
St Saviour's Cres C1
Salisbury Rd B1
Seymour Ave C3
Seymour Rd C3
Silver St B2
South Rd C2
Stanley St B1
Station Rd A2
Stockwood Cres C2
Stockwood Park C1
Strathmore Ave C2
Stuart St B2
Studley Rd A1
Surrey St C3
Sutherland Place C1
Tavistock St C2
Taylor St A3
Telford Way A1
Tennyson Rd C2
Tenzing Grove C1
The Cross Way C1
The Larches A2
Thistle Rd B3
Town Hall B2
Townsley Cl C2
UK Centre for Carnival Arts ✦ B3
Union St B2
University of Bedfordshire B3
Upper George St B2
Vicarage St B3
Villa Rd A2
Waldeck Rd A1
Wellington St B1/B2
Wenlock St A2
Whitby Rd A1
Whitehill Ave C1
William St A2
Wilsden Ave C1
Windmill Rd B3
Windsor St C2
Winsdon Rd B1
York St A3

Macclesfield 337

108 Steps B2
Abbey Rd A1
Alton Dr A3
Armett St C1
Athey St C1
Bank St C3
Barber St C3
Barton St C1
Beech La A2
Beswick St B1
Black La A2
Black Rd C3
Blakelow Gardens C3
Blakelow Rd C3
Bond St B1/C1
Bread St C1
Bridge St B1
Brock St A2
Brocklehurst Ave A3
Brook St B3
Brookfield La B3
Brough St West C1
Brown St C1
Brynton Rd A2
Buckley St C2
Bus Station B2
Buxton Rd B3
Byrons St C2
Canal St B3
Carlsbrook Ave A3
Castle St B2
Catherine St B1
Cemetery A1
Chadwick Terr A3
Chapel St C2
Charlotte St B2
Chester Rd B1
Chestergate B1
Churchill Way B2
Coare St A1
Commercial Rd B2
Conway Cres A3
Copper St C3
Cottage St B1
Court A2
Court B2
Crematorium A1
Crew Ave A3
Crompton Rd B1/C1
Cross St C2
Crossall St C1
Cumberland St A1/B1
Dale St B3
Duke St B2
Eastgate B3
Exchange St B2
Fence Ave B3
Fence Avenue Industrial Estate A3
Flint St B3
Foden St A2
Fountain St B3
Garden St A3
Gas Rd B2
George St B2
Glegg St B3
Golf Course C3
Goodall St B3
Grange Rd C1
Great King St B1
Green St B3
Grosvenor Shopping Centre B2
Gunco La C3
Half St C2
Hallefield Rd B3
Hatton St C1
Hawthorn Way A3
Heapy St C3
Henderson St B1
Heritage Centre & Silk Museum B2
Hibel Rd A2
High St C2
Hobson St C2
Hollins Rd C3
Hope St West C1
Horseshoe Dr B1
Hurdsfield Rd A3
Information Ctr B2
James St C2
Jodrell St B3
John St C2
Jordangate A2
King Edward St B2
King George's Field C3
King St B2
King's School A1
Knight Pool C3
Knight St C3
Lansdowne St A3
Library B2
Lime Gr B3
Little Theatre C2
Loney St B1
Longacre St B1
Lord St C2
Lowe St C2
Lowerfield Rd A3
Lyon St B1
Macclesfield Station B2
Marina B3
Market B2
Market Pl B2
Masons La A3
Mill La C2
Mill Rd C2
Mill St B2
Moran Rd C1
New Hall St A2
Newton St C1
Nicholson Ave A3
Nicholson Cl A3
Northgate Ave A2
Old Mill La C2
Paradise Mill C2
Paradise St B1
Park Green B2
Park La C1
Park Rd C1
Park St C2
Park Vale Rd C1
Parr St B1
Peel St C2
Percyvale St A3
Peter St C1
Pickford St B2
Pierce St B1
Pinfold St B1
Pitt St C2
Police Station B2
Pool St C2
Poplar Rd C2
Post Office B1/B2/B3
Pownall St A2
Prestbury Rd A1/B1
Queen Victoria St B2
Queen's Ave A3
Registrar B2
Richmond Hill C3
Riseley St B1
Roan Ct B3
Roe St B2
Rowan Way A3
Ryle St C2
Ryle's Park Rd C1
St George's St C2
St Michael's B2
Samuel St B2
Saville St C3
Shaw St B1
Slater St C1
Snow Hill C3
South Park C1
Spring Gdns A2
Statham St C2
Station St A2
Steeple St A3
Sunderland St B2
Superstore A1/A2/C2
Swettenham St B3
The Silk Rd A2/B2
Thistleton Cl C2
Thorp St B2
Town Hall B2
Townley St B2
Turnock St C3
Union Rd B3
Union St B2
Victoria Park B3
Vincent St C2
Waters Green B2
Waterside C2
West Bond St B1
West Park A1
Westbrook Dr A1
Westminster Rd A1
Whalley Hayes B1
Windmill St C3
Withyfold Dr A2
York St B3

Maidstone 340

Albion Pl B3
All Saints B2
Allen St A3
Amphitheatre ✦ C2
Archbishop's Palace B2
Bank St B2
Barker Rd C2
Barton Rd C3
Beaconsfield Rd C1
Bedford Pl B1
Bentlif Art Gallery B2
Bishops Way B2
Bluett St A3
Bower La C1
Bower Mount Rd B1
Bower Pl C1
Bower St B1
Bowling Alley B3
Boxley Rd A3
Brenchley Gardens A2
Brewer St A3
Broadway B2
Brunswick St C3
Buckland Hill A1
Buckland Rd B1
Bus Station B3
Campbell Rd C3
Carriage Museum B2
Church Rd C1
Church St B3
Cinema C2
College Ave C2
College Rd C2
Collis Memorial Garden C3
Cornwallis Rd B1
Corpus Christi Hall B2
County Hall A2
County Rd A3
Crompton Gdns C3
Crown & County Courts B2
Curzon Rd A3
Dixon Cl C2
Douglas Rd C1
Earl St B2
Eccleston Rd C2
Fairmeadow B2
Fisher St A2
Florence Rd C1
Foley St A3
Foster St C3
Fremlin Walk Shopping Centre B2
Gabriel's Hill B3
George St C3
Grecian St A3
Hardy St A3
Hart St C2
Hastings Rd C3
Hayle Rd C3
Hazlitt Theatre B2
Heathorn St A3
Hedley St A3
High St B2
HM Prison A3
Holland Rd A3
Hope St A2
Information Ctr B2
James St A3
James Whatman Way A2
Jeffrey St A3
Kent County Council Offices A2
King Edward Rd C2
King St B3
Kingsley Rd C3
Knightrider St B3
Launder Way C1
Lesley Pl A1
Library B2
Little Buckland Ave A1
Lockmeadow Leisure Complex C2
London Rd B1
Lower Boxley Rd A2
Lower Fant Rd C1
Magistrates Court B3
Maidstone Barracks Station A1
Maidstone Borough Council Offices B1
Maidstone East Station A2
Maidstone Museum B2
Maidstone West Station B2
Market C2
Market Buildings B2
Marsham St B3
Medway St B2
Medway Trading Estate C2
Melville Rd C3
Mill St B2
Millennium Bridge C2
Mote Rd B3
Muir Rd C3
Old Tovil Rd C2
Palace Ave B3
Perryfield St A2
Police Station B3
Post Office A2/B2/B3/C3
Priory Rd C3
Prospect Pl C1
Pudding La B2
Queen Anne Rd B3
Queens Rd A1
Randall St A2
Rawdon Rd C3
Reginald Rd C1
Rock Pl B1
Rocky Hill B1
Romney Pl B3
Rose Yard B2
Rowland Cl C1
Royal Engineers' Rd A2
Royal Star Arcade B2
St Annes Ct B1
St Faith's St B2
St Luke's Rd A3
St Peter's Br B2
St Peter St B2
St Philip's Ave C3
Salisbury Rd A3
Sandling Rd A2
Scott St A2
Scrubs La B1
Sheal's Cres C3
Somerfield La B1
Somerfield Rd B1
Staceys St A2
Station Rd A2
Superstore A1/B2/B3
Terrace Rd B1
The Mall C3
The Somerfield Hospital A1
Tonbridge Rd C1
Tovil Rd C2
Town Hall B2
Trinity Park B3
Tufton St B3
Union St B3
Upper Fant Rd C1
Upper Stone St C3
Victoria St B1
Visitor Centre A2
Warwick Pl B1
Wat Tyler Way B3
Waterloo St C3
Waterlow Rd A3
Week St B2
Well Rd A3
Westree Rd C1
Wharf Rd C1
Whatman Park A1
Wheeler St A3
Whitchurch Cl B1
Woodville Rd C3
Wyatt St B3
Wyke Manor Rd B3

Manchester 337

Adair St B6
Addington St A5
Adelphi St A1
Air & Space Gallery B2
Albert Sq B3
Albion St C3
AMC Great Northern B3
Ancoats Gr B6
Ancoats Gr North B6
Angela St C2
Aquatic Centre C4
Ardwick Green C5
Ardwick Green North C5
Ardwick Green South C5
Arlington St A2
Arndale Centre A4
Artillery St B3
Arundel St C2
Atherton St B2
Atkinson St B3
Aytoun St B4
Back Piccadilly A4
Baird St B5
Balloon St A4
Bank Pl A1
Baring St B5
Barrack St C1
Barrow St A1
BBC TV Studios C4
Bendix St A5
Bengal St A5
Berry St C5
Blackfriars Rd A3
Blackfriars St A3
Blantyre St C2
Bloom St B4
Blossom St A5
Boad St B5
Bombay St B4
Booth St A3
Booth St B4
Bootle St B3
Brazennose St B3
Brewer St A5
Bridge St A3
Bridgewater Hall B3
Bridgewater Pl A4
Bridgewater St B2
Brook St C4
Brotherton Dr A2
Brown St A3
Brown St B4
Brunswick St C6
Brydon Ave C6
Buddhist Centre A4
Bury St A2
Bus & Coach Station B4
Bus Station B4
Butler St A6
Buxton St C5
Byrom St B3
Cable St A5
Calder St B1
Cambridge St C3/C4
Camp St B3
Canal St B4
Cannon St A1
Cannon St A4
Cardroom Rd A6
Carruthers St A6
Castle St C2
Cateaton St A3
Cathedral ✝ A3
Cathedral St A3
Cavendish St C4
Chapel St A1/A3
Chapeltown St B5
Charles St C4
Charlotte St B4
Chatham St B4
Cheapside A3
Chepstow St B3
Chester Rd C1/C2
Chester St C4
Chetham's (Dept Store) A3
China La B5
Chippenham Rd A6
Chorlton Rd C2
Chorlton St B4
Church St A2
Church St A4
City Park A4
City Rd C3
Civil Justice Centre B2
Cleminson St A2
Clowes St A3
College Land A3
College of Adult Education C4
Collier St B2
Commercial St C3
Conference Centre C4
Cooper St B4
Copperas St A4
Cornbrook Metro Sta C1
Cornell St A5
Cornerhouse C4
Corporation St A4
Cotter St C6
Cotton St A5
Cow La B1
Cross St B3
Crown Court B4
Crown St C2
Cube Gallery B4
Dalberg St C6
Dale St A4/B5
Dancehouse, The C4
Dantzic St A4
Dark La C6
Dawson St C2
Dean St A5
Deansgate A3/B3
Deansgate Station C3
Dolphin St C6
Downing St C5
Ducie St B5
Duke Pl B2
Duke St B2
Durling St C6
East Ordsall La A2/B1
Edge St A4
Egerton St C2
Ellesmere St C1
Everard St C1
Every St B6
Fairfield St B5
Faulkner St B4
Fennel St A3
Ford St A2
Ford St C6
Fountain St B4
Frederick St A2
Gartside St B2
Gaythorne St A1
George Leigh St A5
George St A1
George St B4
G-Mex Metro Sta C3
Goadsby St A4
Gore St A2
Goulden St A5
Granada TV Studios B2
Granby Row B4
Gravel St A3
Great Ancoats St A5
Great Bridgewater St B3
Great George St A1
Great Jackson St C2
Great Marlborough St C4
Greengate A3
Green Room, The C4
Grosvenor St C5
Gun St A5
Hadrian Ave B6
Hall St B3
Hampson St B1
Hanover St A4
Hanworth Cl C5
Hardman St B3
Harkness St C6
Harrison St B6
Hart St B4
Helmet St B6
Henry St A5
Heyrod St B6
High St A4
Higher Ardwick C6
Hilton St A4/A5
Holland St A6
Hood St A5
Hope St B1
Hope St B4
Houldsworth St A5
Hoyle St C6
Hulme Hall Rd C1
Hulme St A1
Hulme St C3
Hyde Rd C6
Information Ctr B3
Irwell St A2
Islington St A2
Jackson Cr C2
Jackson's Row B3
James St A1
Jenner Cl C2
Jersey St A5
John Dalton St A1
John Dalton St B3
John Ryland's Library B3
John St A2
Kennedy St B3
Kincardine Rd C5
King St A3
King St West A3
Law Courts B3
Laystall St B5
Lever St A5
Library B3
Library Theatre B3
Linby St C2
Little Lever St A4
Liverpool Rd B2
Liverpool St B1
Lloyd St B3
Lockton Cl C5
London Rd B5
Long Millgate A3
Longacre St B6
Loom St A5
Lower Byrom St B2
Lower Mosley St B3
Lower Moss La C2
Lower Ormond St C4
Loxford La C4
Luna St A5
Major St B4
Manchester Art Gallery B4
Manchester Central B3
Manchester Metropolitan University B4/C4
Mancunian Way C3
Manor St C5
Marble St A4
Market St A2
Market St A4
Market St Metro Sta A4
Marsden St A3
Marshall St A5
Mayan Ave A2
Medlock St C3
Middlewood St B1
Miller St A4
Minshull St B4
Mosley St A4
Mosley St Metro Sta B4
Mount St B3
Mulberry St B3
Murray St A5
Museum of Science & Technology B2
Nathan Dr A2
National Computer Centre C4
Naval St A5
New Bailey St A2
New Elm Rd B1
New Islington A6
New Quay St B2
New Union St A6
Newgate St A4
Newton St A5
Nicholas St B4
North Western St C6
Oak St A4
Odeon A4
Old Mill St A6
Oldfield Rd A1/C1
Oldham Rd A5
Oldham St A4
Opera House B3
Ordsall La C1
Oxford Rd C4
Oxford Rd Station C4
Oxford St B4
Paddock St C6
Palace Theatre B4
Pall Mall A3
Palmerston St B6
Park St A1
Parker St B4
Peak St B5
Penfield Cl C5
Peoples' History Museum B2
Peru St A1
Peter St B3
Piccadilly A4
Piccadilly Metro Sta B5
Piccadilly Gdns Metro Sta B4
Piccadilly Station B5
Piercy St A6
Poland St A5
Police Station B3/B5
Pollard St B6
Port St A5
Portland St B4
Portugal St East B5
Post Office A1/A4/A5/B3
Potato Wharf B2
Princess St B3/C4
Pritchard St C4
Quay St A2
Quay St B2
Queen St B3
Radium St A5
Redhill St A5
Regent Rd B1
Renold Theatre A2
Retail Park A5
Rice St B2
Richmond St B4
River St C3
Roby St B5
Rodney St A6
Roman Fort B2
Rosamond St A2
Royal Exchange A3
Sackville St B4
St Andrew's St B6
St Ann St A3
St Ann's A3
St George's Ave C1
St James St B4
St John St B3
St John's Cathedral (RC) ✝ A2
St Mary's B3
St Mary's Gate A3
St Mary's Parsonage A3
St Peter's Sq Metro Sta B3
St Stephen St A2
Salford Approach A3
Salford Central A2
Sheffield St B5
Shepley St B5
Sherratt St A5
Shudehill A4
Shudehill Metro Sta A4
Sidney St C4
Silk St A5
Silver St B4
Skerry Cl C5
Snell St B6
South King St B3
Sparkle St B5
Spear St A4
Spring Gdns A4
Stanley St A2/B2
Station Approach B5
Store St B5
Swan St A4
Tariff St B5
Tatton St C1
Temperance St B6/C6
The Triangle A4
Thirsk St C6
Thomas St A4
Thompson St A5
Tib La B3
Tib St A4
Town Hall (Manchester) B3
Town Hall (Salford) A2
Trafford St C3
Travis St B5
Trinity Way A2
Turner St A4
Union St C6
University of Manchester (Sackville Street Campus) C5
Upper Brook St C5
Upper Cleminson St A1
Upper Wharf St A1
Urbis Museum A4
Vesta St B6
Victoria Metro Sta A4
Victoria Station A4
Victoria St A3
Wadesdon Rd C5
Water St B2
Watson St B3
West Fleet St B1
West King St A2
West Mosley St B4
West Union St B1
Weybridge Rd A6
Whitworth St B4
Whitworth St West C3
Wilburn St B1
William St A2
William St C6
Wilmott St C3
Windmill St B3
Windsor Cr A1
Withy Gr A4
Woden St C1
Wood St B3
Woodward St A6
Worrall St C1
Worsley St C2
York St B4
York St C2
York St C4

Merthyr Tydfil
Merthyr Tudful 340

Aberdare Rd B2
Abermorlais Terr B2
Alexandra Rd A3
Alma St C3
Arfryn Pl C3
Argyle St C3
Avenue De Clichy C2
Bethesda St B2
Bishops Gr A3
Brecon Rd A1/B2
Briarmead A3
Bryn St C3
Bryntirion Rd B3/C3
Bus Station B2
Caedraw Rd C2
Cae Mari Dwn B3
Castle Sq A1
Castle St B2
Chapel C2
Chapel Bank B1
Church St B3
Civic Centre B2
Coedcae'r Ct C3

Court B3
Courts B2
Court St C3
Cromwell St B2
Cyfarthfa Castle School and Museum A1
Cyfarthfa Industrial Estate A1
Cyfarthfa Park A1
Cyfarthfa Rd A1
Dane St A2
Dane Terr A2
Danyparc B3
Darren View A3
Dixon St B2
Dyke St C3
Dynevor St B2
Elwyn Dr C3
Fire Station B2
Fothergill St B3
Galonuchaf Rd A3
Garth St B2
Georgetown B2
Grawen Terr A2
Grove Pk A2
Gurnos Rd A2
Gwaelodygarth Rd A2/A3
Gwaunfarren Gr A3
Gwaunfarren Rd A3
Gwendoline St A3
Hampton St C3
Hanover St A2
Heol S O Davies B1
Heol-Gerrig B1
Highland View C3
High St A3/B2/B3/C2
Howell Cl B1
Information Ctr B2
Jackson's Bridge B2
James St C3
John St B3
Joseph Parry's Cottage B2
Lancaster St A2
Llewellyn St A2
Llwyfen St B2
Llwyn Berry B1
Llwyn Dic Penderyn B1
Llwyn-y-Gelynen C1
Lower Thomas St B3
Market C2
Mary St C3
Masonic St C2
Merthyr College B2
Merthyr Tydfil FC B3
Merthyr Tydfil R.U.F.C. C2
Merthyr Tydfil Station C3
Meyrick Villas A2
Miniature Railway A1
Mount St A2
Nantygwenith St B1
Norman Terr B2
Oak Rd A2
Old Cemetery B3
Pandy Cl A1
Pantycelynen B1
Park Terr B2
Penlan View C2
Penry St B2
Pentwyn Villas A2
Penyard Rd B3
Penydarren Park A3
Penydarren Rd B3
Plymouth St C3
Police Station C2
Pont Marlais West B2
Post Office A3/B2/C3
Quarry Row B2
Queen's Rd B3
Rees St C3
Rhydycar Leisure Centre C3
Rhydycar Link C2
Riverside Park A1
St David's B3
St Tydfil's C2
St Tydfil's Ave C3
St Tydfil's Hospital (No A+E) B3
St Tydfil's Square Shopping Centre C2
Saxon St A2
School of Nursing A2
Seward St A3
Shiloh La B3
Stone Circles B3
Stuart St A2
Summerhill Pl B3
Superstore B3
Swan St C2
Swansea Rd B1
Taff Glen View C3
Taff Vale Ct B3
The Grove A2
The Parade B3
The Walk B2
Thomastown Park B3
Tramroad La A3
Tramroad Side B2
Tramroad Side North B3
Tramroad Side South C3
Trevithick Gdns C3
Trevithick St A3
Tudor Terr B2
Twynyrodyn Rd C3
Union St B3
Upper Colliers Row B1
Upper Thomas St B3
Victoria St B2
Vulcan Rd B2
Warlow St C3
Well St A2
Wern La C1
West Gr A2
William St C3
Yew St C3
Ynysfach Engine House C2
Ynysfach Rd C2

Middlesbrough 340

Abingdon Rd C3
Acklam Rd C1
Albert Park C2
Albert Rd B2
Albert Terr C2
Aubrey St C3
Ayresome Gdns C2
Ayresome Green La C1
Ayresome St C2
Barton Rd A1
Bilsdale Rd C3
Bishopton Rd C3
Borough Rd B2/B3
Bowes Rd A2
Breckon Hill Rd B3
Bridge St East B3
Bridge St West B2
Brighouse Rd A1
Burlam Rd C1
Bus Station B2
Cannon Park B1
Cannon Park Way B2
Cannon St B1
Captain Cook Sq B2
Carlow St C1
Castle Way C3
Chipchase Rd C2
Cineworld B3
Clairville Sports Stadium C3
Cleveland Centre B2
Clive Rd C2
Commercial St A2
Corporation Rd B2
Costa St C2
Council Offices B3
Crescent Rd C2
Cumberland Rd C2
Depot Rd A2
Derwent St B2
Devonshire Rd C2
Diamond Rd B2
Disabled Driver Test Circuit B1
Dorman Museum C2
Douglas St B3
Eastbourne Rd C2
Eden Rd C3
Enterprise Centre A2
Forty Foot Rd A2
Gilkes St B2
Gosford St A2
Grange Rd B2
Gresham Rd B2
Harehills Rd C1
Harford St C2
Hartington Rd B2
Haverton Hill Rd A1
Hey Wood St B1
Highfield Rd C3
Hill St Centre B2
Holwick Rd B1
Hudson Quay C3
Hutton Rd C3
I.C.I. Works A1
Information Ctr B2
Lambton Rd C3
Lancaster Rd C2
Lansdowne Rd C3
Latham Rd C2
Law Courts B2/B3
Lees Rd B2
Leeway B3
Linthorpe Cemetery C1
Linthorpe Rd B2
Lloyd St B2
Longford St C2
Longlands Rd C3
Lower East St A3
Lower Lake C3
Maldon Rd C1
Manor St B2
Marsh St B2
Marton Rd B3
Middlehaven B3
Middlesbrough By-Pass B2/C1
Middlesbrough F.C. B3
Middlesbrough Leisure Park B3
Middlesbrough Station B2
Middletown Park C2
MIMA B3
Mosque B2
Mosque C3
Mulgrave Rd C2
North Ormesby Rd B3
Newport Bridge B1
Newport Bridge Approach Rd B1
Newport Rd B2
North Rd B2
Northern Rd C1
Outram St B2
Oxford Rd C2
Park La C2
Park Rd North C2
Park Rd South C2
Park Vale Rd C3
Parliament Rd B1
Police Station A2
Port Clarence Rd A3
Portman St B2
Post Office B3/C1/C2/C3
Princes Rd B2
Riverside Business Park A2
Riverside Park Rd A1
Rockliffe Rd C2
Romaldkirk Rd B1
Roman Rd C2
Roseberry Rd C3
St Barnabas' Rd C2
St Paul's Rd B2
Saltwells Rd B3
Scott's Rd A3
Seaton Carew Rd A3
Shepherdson Way B3
Sikh Temple B2
Snowdon Rd A2
South West Ironmasters Park B1
Southfield Rd B2
Southwell Rd C2
Springfield Rd C1
Startforth Rd A2
Stockton Rd C1
Stockton St A2
Surrey St C2
Sycamore Rd C2
Synagogue C2
Tax Offices B3
Tees Viaduct C1
Teessaurus Park A2
Teesside Tertiary College C3
The Avenue C2
The Crescent C2
Thornfield Rd C1
Town Hall B2
Transporter Bridge (Toll) A3
Union St B2
University of Teesside B2
Upper Lake C2
Valley Rd C3
Ventnor Rd C2
Victoria Rd B2
Vulcan St A2
Warwick St C2
Wellesley Rd B3
West Lane Hospital C1
Westminster Rd C2
Wilson St B2
Windward Way B3
Woodlands Rd B2
York Rd C3

Milton Keynes 340

Abbey Way A1
Arbrook Ave B1
Armourer Dr A3
Arncliffe Dr A1
Avebury rbt C2
Avebury Blvd C2
Bankfield rbt B3
Bayard Ave A2
Belvedere rbt A2
Bishopstone A1
Blundells Rd A1
Boycott Ave C2
Bradwell Common Blvd B1
Bradwell Rd C1
Bramble Ave A2
Brearley Ave C2
Breckland A1
Brill Place B1
Burnham Dr B1
Bus Station C1
Campbell Park rbt B3
Cantle Ave A3
Central Milton Keynes Shopping Area B2
Century Ave C2
Chaffron Way C3
Childs Way C1
Christ the Cornerstone B2
Cineworld B2
Civic Offices B2
Cleavers Ave B2
Colesbourne Dr A3
Conniburrow Blvd B2
County Court B2
Currier Dr A2
Dansteed Way A2/A3/B1
Deltic Ave B1
Downs Barn rbt B2
Downs Barn Blvd A3
Eaglestone rbt C3
Eelbrook Ave B1
Elder Gate B1
Evans Gate C2
Fairford Cr A3
Falcon Ave B3
Fennel Dr A2
Fishermead Blvd C3
Food Centre B3
Fulwoods Dr C3
Glazier Dr A2
Glovers La A1
Grafton Gate C1
Grafton St A1/C2
Gurnards Ave B3
Harrier Dr C3
Ibstone Ave B1
Langcliffe Dr A1
Leisure Plaza C1
Leys Rd C1
Library B2
Linford Wood A2
Marlborough Gate B3
Marlborough St A2/B3
Mercers Dr A1
Midsummer rbt C2
Midsummer Blvd B2
Milton Keynes Central C1
Monks Way A1
Mullen Ave A3
Mullion Pl C3
National Hockey Stadium B1
Neath Hill rbt A3
North Elder rbt C1
North Grafton rbt B1
North Overgate rbt A3
North Row B2
North Saxon rbt B2
North Secklow rbt B2
North Skeldon rbt A3
North Witan rbt B1
Oakley Gdns A3
Oldbrook Blvd C2
Open-Air Theatre B3
Overgate A3
Overstreet A3
Patriot Dr B1
Pencarrow Pl B3
Penryn Ave B3
Perran Ave C3
Pitcher La C1
Place Retail Park, The C1
Point Centre, The B2
Police Station B2
Portway rbt B2
Post Office A2/B2/C3
Precedent Dr B1
Quinton Dr B1
Ramsons Ave B2
Rockingham Dr A2
Rooksley rbt B1
Rooksley Retail Park C1
Saxon Gate B2
Saxon St A1/C3
Secklow Gate B2
Shackleton Pl C2
Silbury Blvd B2
Skeldon rbt A3
South Grafton rbt C2
South Row C2
South Saxon rbt C2
South Secklow rbt B3
South Witan rbt C2
Springfield rbt B3
Stanton Wood rbt A1
Stantonbury rbt A1
Stantonbury Leisure Centre A1
Strudwick Dr C2
Sunrise Parkway A2
Telephone Exchange C3
The Boundary C3
Theatre & Art Gallery B3
Tolcarne Ave C3
Towan Ave C3
Trueman Pl C2
Vauxhall A1
Winterhill Retail Park C2
Witan Gate B2
X-Scape B3

Newcastle upon Tyne 340

Albert St B3
Argyle St B3
Back New Bridge St B3
BALTIC The Centre for Contemporary Art C3
Bank Rd C3
Barker St A3
Barrack Rd B1
Bath La B1
Bell's Court B2
Bessie Surtees House C2
Bigg Market C2
Biscuit Factory B3
Black Gate C2
Blackett St B2
Blandford Sq C1
Boating Lake A1
Boyd St B3
Brandling Park A2
Bus Station B2
Buxton St B3
Byron St A3
Camden St B2
Castle Keep C2
Central Metro Sta C1
Central Library B2
Central Motorway B2
Chester St A3
City Hall B2
City Rd B3/C3
City Walls C1
Civic Centre A2
Claremont Rd A1
Clarence St B3
Clarence Walk B3
Clayton St C1/B1
Clayton St West C1
Coach Station C1
College St B2
Collingwood St C2
Copland Terr B3
Coppice Way B3
Corporation St B1
Courts C3
Crawhall Rd B3
Dean St C2
Dinsdale Pl A3
Dinsdale Rd A3
Doncaster Rd A3
Durant Rd B2
Eldon Sq B1
Eldon Sq Shopping Centre B3
Ellison Pl B2
Eskdale Terr A2
Eslington Terr A2
Exhibition Park A1
Falconar St B3
Fenkle St C1
Forth Banks C1
Forth St C1
Gallowgate B1
Gateshead Millennium Bridge C3
Gibson St B3
Goldspink La A3
Grainger Market B2
Grainger St B2
Grantham Rd A3
Granville Rd A3
Grey St B2
Groat Market C2
Guildhall C2
Hancock Museum A2
Hancock St A2
Hanover St C2
Hawks Rd C3
Haymarket Metro Sta B2
Heber St B1
Helmsley Rd A3
High Bridge B2
High Level Bridge C2
Hillgate C3
Howard St B3
Hutton Terr A3
Information Ctr C2
Jesmond Metro Sta A2
Jesmond Rd A2/A3
John Dobson St B2
John George Joicey Museum C2
Jubilee Rd B3
Kelvin Gr A3
Kensington Terr A2
Laing Gallery B2
Lambton Rd A2
Leazes Cr B1
Leazes La B1
Leazes Park B1
Leazes Park Rd B1
Leazes Terr B1
Low Friar St C1
Manor Chare C2
Manors Metro Sta B2
Manors Station B2
Market St B2
Melbourne St B2
Mill Rd C3
Millennium Sq C1
Monument Metro Sta B2
Monument Mall Shopping Centre B2
Morpeth St A2
Mosley St C2
Napier St A3
Nazareth House A3
New Bridge St B2/B3
Newcastle Central Station C1
Newcastle University A1
Newgate Shopping Centre C1
Newgate St B1
Newington Rd A3
Northern Stage Theatre A2
Northumberland Rd B2
Northumberland St B2
Northumbria University A2
Northwest Radial Rd A1
Oakwellgate C3
Odeon B1
Orchard St C2
Osborne Rd A2
Osborne Terr A3
Pandon C3
Pandon Bank B3
Park Terr A1
Percy St B1
Pilgrim St B2
Pipewellgate C2
Pitt St B1
Plummer Tower A2
Police Station B2
Portland Rd A3/B3
Portland Terr A3
Post Office A3/B1/B2/B3
Pottery La C1
Prudhoe Pl B1
Prudhoe St B1
Quayside C2
Queen Elizabeth II Bridge C2
Queen Victoria Rd A1
Richardson Rd A1
Ridley Pl B2
Rock Terr B3
Rosedale Terr A3
Royal Victoria Infirmary A1
Sage Gateshead, The C3
St Andrew's St B1
St James Metro Sta B1
St James' Blvd C1
St James' Park (Newcastle Utd FC) B1
St Mary's (RC) C1
St Mary's Place B2
St Nicholas C2
St Nicholas St C2
St Thomas' St B1
Sandyford Rd A2/A3
Science Park B2
Shield St B3
Shieldfield B3
Simpson Terr B3
South Shore Rd C3
South St C1
Starbeck Ave A3
Stepney Rd B3
Stoddart St B3
Stowell St B1
Strawberry Pl B1
Swing Bridge C2
Temple St C1
Terrace Pl B1
The Close C2
The Side C2
Theatre Royal B2
Tower St B3
Trinity House C2
Tyne Bridge C2
Tyne Bridges C2
Tyneside B2
Victoria Sq A2
Warwick St A3
Waterloo St C1
Wellington St B1
Westgate Rd C1/C2
Windsor Terr A2
Worswick St B2
Wrentham Pl B3

Newport *Casnewydd* 340

Albert Terr B1
Allt-yr-Yn Ave A1
Alma St C2
Ambulance Station C3
Bailey St B2
Barrack Hill A2
Bath St A3
Bedford Rd B3
Belle Vue La C1
Belle Vue Park C1
Bishop St A3
Blewitt St B1
Bolt Cl C3
Bolt St C3
Bond St A2
Bosworth Dr A1
Bridge St B1
Bristol St A3
Bryngwyn Rd B1
Brynhyfryd Ave C1
Brynhyfryd Rd C1
Bus Station B2
Caerau Cres C1
Caerau Rd B1
Caerleon Rd A3
Cambrian Retail Centre B2
Capel Cres C2
Cardiff Rd C2
Caroline St B3
Castle (Remains) A2
Cattle Market and Saturday General Market C3
Cedar Rd B3
Charles St B2
Charlotte Dr C2
Chepstow Rd A3
Church Rd A3
City Cinema B1
Civic Centre B1
Clarence Pl A2
Clifton Pl B1
Clifton Rd C1
Clyffard Cres B1
Clytha Park Rd B1
Clytha Sq C2
Coldra Rd C1
Collier St A3
Colne St B3
Comfrey Cl A1
Commercial Rd C3
Commercial St B2
Corelli St A3
Corn St B2
Corporation Rd B3
Coulson Cl C2
County Court A2
Courts A1
Courts B1
Crawford St A3
Cyril St B3
Dean St A3
Devon Pl B1
Dewsland Park Rd C2
Dolman B2
Dolphin St C3
East Dock Rd C3
East St B1
East Usk Rd A3
Ebbw Vale Wharf B3
Emlyn St B2
Enterprise Way C3
Eton Rd B3
Evans St A2
Factory Rd A2
Fields Rd B1
Francis Dr C2
Frederick St C3
Friars Rd C1
Gaer La C1
George St C3
George Street Bridge C3
Godfrey Rd B1
Gold Tops B1
Gore St A3
Gorsedd Circle C1
Grafton Rd A3
Graham St B1
Granville St C3
Harlequin Dr A1
Harrow Rd B3
Herbert Rd A3
Herbert Walk C2
Hereford St A3
High St B2
Hill St B2
Hoskins St A2
Information Ctr B2
Ivor Sq B2
John Frost Sq B2
Jones St B1
Junction Rd A3
Keynshaw Ave C2
King St C2
Kingsway B2
Kingsway Shopping Centre B2
Ledbury Dr A2
Library A3
Library, Museum & Art Gallery B2
Liverpool Wharf B3
Llanthewy Rd B1
Llanvair Rd A3
Locke St A2
Lower Dock St C3
Lucas St A2
Manchester St A3
Market B2
Marlborough Rd B3
Mellon St C3
Mill St A2
Morgan St A3
Mountjoy Rd C2
Newport Athletic Club Grounds B3
Newport Bridge A2
Newport Leisure and Conference Ctr B2
Newport Station B2
North St B2
Oakfield Rd B1
Park Sq C2
Police Station A3/C2
Post Office B1/B2/C1/C3
Power St A1
Prince St A3
Pugsley St A2
Queen St C2
Queen's Cl A1
Queen's Hill A1
Queen's Hill Cres A1
Queensway B2
Railway St B2
Riverfront Arts Centre B2
Riverside A3
Rodney Rd B2
Royal Gwent (A+E) C2
Rudry St A3
Rugby Rd B3
Ruperra La C3
Ruperra St C3
St Edmund St B1
St Mark's Cres A1
St Mary St B1
St Vincent Rd A3
St Woolos C2
St Woolos General (no A+E) C1
St Woolos Rd B1
School La B2
Serpentine Rd B1
Shaftesbury Park A2
Sheaf La A3
Skinner St B2
Sorrel Dr A1
South Market St C3
Spencer Rd B1
Stow Hill B2/C1/C2
Stow Park Ave C1
Stow Park Dr C1
TA Centre A1
Talbot St B2
Tennis Club A1
Tregare St A3
Trostrey St A3
Tunnel Terr B1
Turner St A3
Usk St A3
Usk Way B3/C3
Victoria Cr B1
War Memorial A3
Waterloo Rd C1
West St B1
Wharves B2
Wheeler St A2
Whitby Pl A3
Windsor Terr B1
York Pl C1

Newquay 340

Agar Rd B2
Alma Pl B1
Ambulance Station B2
Anthony Rd C1
Atlantic Hotel A1
Bank St B1
Barrowfields A3
Bay View Terr B2
Beachfield Ave B1
Beach Rd B1
Beacon Rd A1
Belmont Pl A1
Berry Rd B2
Blue Reef Aquarium B1
Boating Lake C2
Bus Station B1
Chapel Hill B1
Chester Rd A3
Cheviot Rd C1/C2
Chichester Cres C2
Chynance Dr C1
Chyverton Cl C1
Cliff Rd B2
Coach Park B2
Colvreath Rd A3
Council Offices B1
Crantock St B1
Criggar Rocks A3
Dale Cl C3
Dale Rd C3
Dane Rd A1
East St B2
Edgcumbe Ave B3
Edgcumbe Gdns B3
Eliot Gdns B3
Elm Cl C3
Ennor's Rd B1
Fernhill Rd B1
Fire Station B1
Fore St B1
Gannel Rd C2
Golf Driving Range B3
Gover La B1
Great Western Beach A2
Grosvenor Ave B2
Harbour A1
Hawkins Rd C2
Headleigh Rd B2
Hilgrove Rd A3/B3
Holywell Rd B3
Hope Terr B1
Huer's House, The A1
Information Ctr B1
Island Cres B2
Jubilee St B1
Kew Cl C3
Killacourt Cove A2
King Edward Cres A1
Lanhenvor Ave B2
Library B1
Lifeboat Station A1
Linden Ave C2
Listry Rd C2
Lusty Glaze Beach A3
Lusty Glaze Rd A3
Manor Rd B1
Marcus Hill B2
Mayfield Rd B2
Meadowside C3
Mellanvrane La C2
Michell Ave B2
Miniature Golf Course C3
Miniature Railway B3
Mount Wise B1
Mowhay Cl C3
Narrowcliff A3
Newquay B2
Newquay Hospital (no A&E) B2
Newquay Town Football Ground B1
Newquay Zoo B3
North Pier A1
North Quay Hill A1
Oakleigh Terr B2
Pargolla Rd B2
Pendragon Cres C3
Pengannel Cl C1
Penina Ave C3
Police Station & Courts B2
Post Office B1/B2
Quarry Park Rd B3
Rawley La C2
Reeds Way B1
Robartes Rd B2
St Anne's Rd A3
St Aubyn Cres B3
St George's Rd B1
St John's Rd B1
St Mary's Rd B1
St Michael's B1
St Michael's Rd B1
St Thomas' Rd B2
Seymour Ave B2
South Pier A1
South Quay Hill A1
Sweet Briar Cres C3
Sydney Rd A1
The Crescent B1
Tolcarne Beach A2
Tolcarne Point A2
Tolcarne Rd B2
Tor Rd B2
Towan Beach A1
Towan Blystra Rd B3
Tower Rd A1
Trebarwith Cres B2
Tredour Rd C2
Treforda Rd C3
Tregoss Rd B3
Tregunnel Hill B1/C1
Tregunnel Saltings C1
Trelawney Rd B2
Treloggan La C3
Treloggan Rd C3
Trembath Cres C1
Trenance Ave B2
Trenance Gardens B2
Trenance La C2
Trenance Leisure Park B3
Trenance Rd B2
Trenarth Rd B2
Treninnick Hill C3
Tretherras Rd B3
Trethewey Way C1
Trevemper Rd C2
Tunnels Through Time B1
Ulalia Rd B3
Vivian Cl B2
Waterworld B3
Whitegate Rd B3
Wych Hazel Way C3

Newtown *Y Drenewydd* 340

Ash Cl A3
Back La B2
Baptist Chapel B2
Barn La A2
Bear Lanes Shopping Centre B2
Beech Cl A2
Beechwood Dr A2
Brimmon Cl C2
Brimmon Rd C2
Broad St B2
Bryn Bank A1
Bryn Cl A1
Bryn Gdns A1
Bryn House A1
Bryn La A1/A2
Bryn Meadows A2
Bryn St A1
Brynglais Ave A2
Brynglais Cl A2
Bus Station B2
Bynwood Dr A1
Cambrian Bridge B3
Cambrian Gdns B2
Cambrian Way B2
Canal Rd A3
Castle Mound B2
Cedewain C1
Cefnaire A3
Cefnaire Coppice C2
Ceiriog C2
Cemetery A3
Church (Remains of) B2
Churchill Dr A3
Cledan B3
Colwyn B3
Commercial St A2
Council Offices B1
Crescent St A1
Cwm Llanfair B1
Davies Memorial Gallery B2
Dinas B2
Dolafon Rd B3
Dolerw Park B1
Dolfor Rd C1
Eirianell C1
Fairfield Dr A2
Ffordd Croesawdy B2
Fire Station C1
Frankwell St A2
Frolic St B2
Fron La A1
Garden La A2
Gas St B2
Glyndŵr C1
Golwgydre La B2
Gorsedd Circle B1
Great Brimmon Farm C3
Hafren C1
Halfpenny Bridge B2
High St B2
Hillside Ave A3
Hoel Treowen C2
Information Ctr B2
Kerry Rd B3
Ladywell Shopping Centre B2
Library B1
Llanfair Rd A2
Llanidloes Rd C1
Llys Ifor A2
Lon Cerddyn B1
Lonesome La A3
Long Bridge A2
Lon Helyg C1
Lower Canal Rd B3
Maldwyn Leisure Centre C1
Market B2
Market St B2
Milford Rd B1
Mill Cl B2
Miniature Railway B1
Mwyn Fynydd A3
New Church St B2
New Rd B2
Newtown Football Ground B1
Newtown Infirmary A2
Newtown Station B2
Oak Tree Ave A3
Old Kerry Rd B2
Oldbarn La A2
Park Cl B1
Parklands B1
Park La B1
Park St B2
Pavillion Ct C1
Plantation La C1
Police Station B1
Pont Brynfedw A2
Pool Rd B3
Poplar Rd A3
Post Office B2/C1
Powys C1
Powys Theatre A2
Pryce Jones Stores & Museum B2
Quaker Meeting House B1
Regent B2
Robert Owen House B1
Robert Owen Museum B2
Rugby Club A3
St David's B2
School La A2
Sheaf St B3
Short Bridge St B2
Stone St B2
Sycamore Dr A2
Textile Museum A2
The Bryn A1
The Park B1
Town Hall B2
Union St A2
Upper Brimmon C3
Vastre Industrial Estate B3
War Memorial B2
WHSmith Museum B2
Wynfields C1
Y Ffrydd A3

Northampton 340

78 Derngate B3
Abington Sq B3
Abington St B3
Alcombe St A3
All Saints' B2
Ambush St B1
Angel St B2
Arundel St A2
Ash St A2
Auctioneers Way C2
Bailiff St A2
Barrack Rd A2
Beaconsfield Terr A3
Becketts Park C3
Bedford Rd B3
Billing Rd B3
Brecon St A1
Brewery C2
Bridge St C2
Bridge St Depot C3
Broad St B2
Burns St A3
Bus Station B2
Campbell St A2
Castle (Site of) B2
Castle St B2
Cattle Market Rd C2
Central Museum & Art Gallery B2
Charles St A3
Cheyne Walk B3
Church La A2
Clare St A3
Cloutsham St A3
College St B2
Colwyn Rd A3
Cotton End C2

Countess Rd A1
County Hall B2
Court A2
Craven St A3
Crown & County Courts B3
Denmark Rd B3
Derngate B3
Derngate & Royal Theatres B3
Doddridge Church B2
Duke St A3
Dunster St A3
Earl St A3
Euston Rd C2
Fire Station A3
Foot Meadow B2
Gladstone Rd A1
Gold St B2
Grafton St A2
Gray St A3
Green St B1
Greenwood Rd B1
Greyfriars B2
Grosvenor Centre B2
Grove Rd A3
Guildhall B2
Hampton St A2
Harding Terr A2
Hazelwood Rd B3
Herbert St B2
Hervey St A3
Hester St A2
Holy Sepulchre A2
Hood St A3
Horse Market B2
Hunter St A3
Information Ctr B1
Kettering Rd A3
Kingswell St B2
Lady's La B2
Leicester St A2
Leslie Rd A2
Library B3
Lorne Rd A2
Lorry Park A1
Louise Rd A2
Lower Harding St A2
Lower Hester St A2
Lower Mounts B3
Lower Priory St A2
Main Rd C1
Marefair B2
Market Sq B2
Marlboro Rd B1
Marriott St A2
Military Rd A3
Nene Valley Retail Park C1
New South Bridge Rd C2
Northampton General Hospital (A & E) B3
Northampton Station B1
Northcote St A2
Nunn Mills Rd C3
Old Towcester Rd C2
Overstone Rd A3
Peacock Pl B2
Pembroke Rd A1
Penn Court C2
Police Station B3
Post Office A1/A2/B3/C2
Quorn Way A2
Ransome Rd C3
Regent Sq A2
Robert St A2
St Andrew's Rd B1
St Andrew's St A2
St Edmund's Rd B3
St George's St A2
St Giles B3
St Giles St B3
St Giles' Terr B3
St James' Mill Rd B1
St James' Mill Rd East C1
St James Park Rd B1
St James Retail Park C1
St James Rd B1
St Leonard's Rd C2
St Mary's St B2
St Michael's Rd A3
St Peter's B2
St Peter's Square Shopping Precinct B2
St Peter's Way B2
Salisbury St A2
Scarletwell St B2
Semilong Rd A2
Sheep St B2
Sol Central (Leisure Centre) B2
Somerset St A3
South Bridge C2
Southfield Ave C2
Spencer Bridge Rd A1
Spencer Rd A3
Spring Gdns B3
Spring La B2
Swan St B3
TA Centre A3
Tanner St B2
The Drapery B2
The Ridings B3
Tintern Ave A1
Towcester Rd C2
Upper Bath St B2
Upper Mounts A2
Victoria Park A1
Victoria Promenade B2
Victoria Rd B3
Victoria St A2
Wellingborough Rd B3
West Bridge B1
York Rd B3

Norwich 341

Albion Way C3
All Saints Green C2
Anchor Cl A3
Anchor St A3
Anglia Sq A2
Argyle St C3
Arts Centre B1
Ashby St C2
Assembly House B1
Bank Plain B2
Barker St A1
Barn Rd A1
Barrack St A3
Ber St C2
Bethel St B1
Bishop Bridge A3
Bishopbridge Rd A3
Bishopgate A3
Blackfriars St A2
Botolph St A2
Bracondale C3
Brazen Gate C2
Bridewell B2
Brunswick Rd C1
Bull Close Rd A2
Bus Station C2
Calvert St A2
Cannell Green A3
Carrow Rd C3
Castle Mall B2
Castle Meadow B2
Castle & Museum B2
Cathedral B2
Cattlemarket St B2
Chantry Rd C1
Chapel Loke C2
Chapelfield East B1
Chapelfield Gdns B1
Chapelfield North B1
Chapelfield Rd B1
Chapelfield Shopping Centre C1
City Hall B1
City Rd C2
City Wall C1/C3
Colegate A2
Coslany St B1
Cow Hill B1
Cow Tower A3
Cowgate A2
Crown & Magistrates Courts A2
Dragon Hall Heritage Centre C3
Duke St A2
Edward St A2
Elm Hill B2
Erpingham Gate B2
Fire Station B1
Fishergate A2
Foundry Bridge B3
Fye Bridge A2
Garden St C2
Gas Hill B3
Grapes Hill B1
Great Hospital Halls, The A3
Grove Ave C1
Grove Rd C1
Guildhall B1
Gurney Rd A3
Hall Rd C2
Heathgate A3
Heigham St A1
Horn's La C2
Information Ctr B1
Inspire (Science Centre) B1
Ipswich Rd C1
James StuArt Gdns B3
King Edward VI School B2
King St B2
King St C3
Koblenz Ave C3
Library B1
London St B2
Lower Clarence Rd B3
Lower Cl B3
Maddermarket B1
Magdalen St A2
Mariners La C2
Market B2
Market Ave B2
Mountergate B3
Mousehold St A3
Newmarket Rd C1
Norfolk Gallery B2
Norfolk St C1
Norwich City FC C3
Norwich Station B3
Oak St A1
Palace St A2
Pitt St A1
Playhouse B2
Post Office A2/B2/C2
Pottergate B1
Prince of Wales Rd B2
Princes St B2
Pull's Ferry B3
Puppet Theatre A2
Quebec Rd B3
Queen St B2
Queens Rd C2
Recorder Rd B3
Retail Park C3
Riverside Entertainment Centre C3
Riverside Swimming Centre C3
Riverside Rd B3
Rosary Rd B3
Rose La B2
Rouen Rd C2
Royal Norfolk Regiment Museum B2
St Andrew's & Blackfriars Hall B2
St Andrews St B2
St Augustines St A1
St Benedicts St B1
St Ethelbert's Gate B2
St Faiths La B3
St Georges St A2
St Giles St B1
St James Cl A3
St Julians C2
St Martin's La A1
St Peter Mancroft B2
St Peters St B1
St Stephens Rd C1
St Stephens St C1
Silver Rd A2
Silver St A2
Southwell Rd C2
Strangers Hall B1
Superstore C2
Surrey St C2
Sussex St A1
The Close B3
The Forum B1
The Walk B2
Theatre Royal B1
Theatre St B1
Thorn La C2
Thorpe Rd B3
Tombland B2
Union St C1
Vauxhall St B1
Victoria St C1
Walpole St B1
Wensum St A2
Wessex St C1
Westwick St A1
Wherry Rd C3
Whitefriars A2
Willow La B1
Yacht Station B3

Nottingham 341

Abbotsford Dr A3
Addison St A1
Albert Hall B1
Alfred St South A3
Alfreton Rd B1
All Saints Rd A1
Annesley Gr A2
Arboretum A1
Arboretum St A1
Arthur St A1
Arts Theatre B3
Ashforth St A3
Balmoral Rd A1
Barker Gate B3
Bath St B3
Belgrave Centre B1
Bellar Gate B3
Belward St B3
Blue Bell Hill Rd B3
Brewhouse Yard C2
Broad Marsh Bus Station C2
Broad Marsh Precinct C2
Broad St B3
Brook St B3
Burns St A1
Burton St B2
Bus Station A2
Canal St C2
Carlton St B3
Carrington St C2
Castle Blvd C1
Castle C2
Castle Gate C2
Castle Meadow Retail Park C1
Castle Meadow Rd C1
Castle Museum & Gallery C2
Castle Rd C2
Castle Wharf C2
Cavendish Rd East C1
Cemetery B1
Chaucer St B1
Cheapside B2
Church Rd A3
City Link C3
City of Caves C2
Clarendon St B1
Cliff Rd C3
Clumber Rd East C1
Clumber St B2
College St B1
Collin St C2
Conway Cl A2
Council House B2
Court B2
Cranbrook St B3
Cranmer St A2
Cromwell St B1
Curzon St B3
Derby Rd B1
Dryden St A2
Fishpond Dr C1
Fletcher Gate B3
Forest Rd East A1
Forest Rd West A1
Friar La C2
Galleries of Justice C3
Gedling Gr A1
Gedling St B3
George St B3
Gill St A2
Glasshouse St B2
Goldsmith St B2
Goose Gate B3
Great Freeman St A2
Guildhall B2
Hamilton Dr C1
Hampden St A1
Heathcote St B3
High Pavement C3
High School A1
Holles Cr C1
Hope Dr C1
Hungerhill Rd A3
Huntingdon Dr C1
Huntingdon St A2
Information Ctr B2
Instow Rise A3
International Community Centre A2
Kent St B3
King St B2
Lace Centre, The C2
Lace Market B3
Lace Market Theatre C3
Lamartine St B3
Lenton Rd C1
Lewis Cl A3
Lincoln St B2
London Rd C3
Long Row B2
Low Pavement C2
Lower Parliament St B3
Magistrates Court C2
Maid Marian Way B2
Mansfield Rd A2/B2
Middle Hill C2
Milton St B2
Mount St B2
National Ice Centre C3
Newcastle Dr B1
Newstead Gr A2
North Sherwood St A2
Nottingham Arena C3
Nottingham Station C3
Old Market Square B2
Oliver St A1
Park Dr C1
Park Row C1
Park Terr B1
Park Valley C1
Peas Hill Rd A3
Peel St A1
Pelham St B2
Peveril Dr C1
Plantagenet St A3
Playhouse Theatre B1
Plumptre St C3
Police Station B2
Poplar St C3
Portland Rd B1
Post Office B2
Queen's Rd C3
Raleigh St A1
Regent St B1
Rick St B3
Robin Hood Statue C2
Robin Hood St B3
Royal Centre B2
Royal Children Inn C2
Royal Concert Hall B2
St Ann's Hill Rd A2
St Ann's Way A2
St Ann's Well Rd A3
St Barnabas B1
St James' St B2
St Mark's St B3
St Mary's Garden of Rest B3
St Mary's Gate B3
St Nicholas C2
St Peter's C2
St Peter's Gate B2
Salutation Inn C2
Shakespeare St B2
Shelton St A2
South Pde B2
South Rd C1
South Sherwood St B2
Station St C3
Station Street C3
Stoney St B3
Talbot St B1
Tales of Robin Hood C2
Tattershall Dr C1
Tennis Dr B1
Tennyson St A1
The Park C1
The Ropewalk B1
Theatre Royal B2
Trent St C3
Trent University A2/B2
Trent University B2
Trinity Square Shopping Centre B2
Trip To Jerusalem Inn C2
Union Rd B3
Upper Parliament St B2
Victoria Centre B2
Victoria Leisure Centre B3
Victoria Park B3
Victoria St B2
Walter St A1
Warser Gate B3
Watkin St A2
Waverley St A1
Wheeler Gate B2
Wilford Rd C2
Wilford St C2
Willoughby House C2
Wollaton St B1
Woodborough Rd A2
Woolpack La B3
York St A2

Oban 341

Aird's Cres B2
Albany St B2
Albert La A2
Albert Rd A2
Alma Cres B1
Ambulance Station C2
Angus Terr C3
Ardconnel Rd A2
Ardconnel Terr B2
Argyll Sq B2
Argyll St B2
Atlantis Leisure Centre A2
Bayview Rd A1
Benvoulin Rd A2
Bowling Green A2
Breadalbane St A2
Bus Station B2
Campbell St B2
College B2
Colonsay Terr C3
Columba Building B2
Combie St B2
Corran Brae A1
Corran Esplanade A1/A2
Corran Halls A2
Court B2
Crannaig-a-Mhinisteir B1
Crannog La C2
Croft Ave A2
Dalintart Dr C3
Dalriach Rd A2
Drummore Rd C2
Duncraggan Rd A2
Dunollie Rd A2
Dunuaran Rd B1
Feochan Gr C3
Ferry Terminal B1
Gallanach Rd B1
George St A2
Glencruitten Dr C3
Glencruitten Rd B3
Glenmore Rd C1
Glenshellach Rd C1
Glenshellach Terr B2
Harbour Bowl B2
Hazeldean Cres A3
High St B2
Highland Theatre Cinema A2
Hill St B2
Industrial Estate C2
Information Ctr B2
Islay Rd C3
Jacob's Ladder B2
Jura Rd C3
Knipoch Pl C3
Laurel Cres A2
Laurel Rd A2/A3
Library B1
Lifeboat Station B1
Lighthouse Pier B1
Lismore Cres A2
Lochavullin Dr B2
Lochavullin Rd C2
Lochside St B2
Longsdale Cres A3
Longsdale Rd A2/A3
Longsdale Terr A3
Lunga Rd C3
Lynn Rd C2
Market St B2
McCaig Rd C3
McCaig's Tower A2
Mill La C2
Miller Rd C2
Millpark Ave C2
Millpark Rd C2
Mossfield Ave B3
Mossfield Dr B3
Mossfield Stadium B3
Nant Dr C3
Nelson Rd C2
North Pier B2
Nursery La A2
Oban B2
Police Station B2
Polvinister Rd B3
Post Office A2/B2
Pulpit Dr C1
Pulpit Hill C1
Pulpit Hill Viewpoint B1
Quarry Rd C2
Queen's Park Pl B2
Railway Quay B1
Rockfield Rd B2
St Columba's A1
St John's A2
Scalpay Terr C3
Shore St B1
Shuna Terr C3
Sinclair Dr C2
Soroba Rd B2/C2
South Pier B1
Stevenson St B2
Tweedale St B2
Ulva Rd C2
Villa Rd B1
War & Peace A2

Oxford 341

Adelaide St A1
Albert St A1
All Souls (Coll) B2
Ashmolean Museum B2
Balliol (Coll) B2
Banbury Rd A2
Bate Collection of Musical Instruments C2
Beaumont St B1
Becket St B1
Blackhall Rd A2
Blue Boar St B2
Bodleian Library B2
Botanic Garden B3
Brasenose (Coll) B2
Brewer St C2
Broad St B2
Burton-Taylor Theatre B2
Bus Station B1
Canal St A1
Cardigan St A1
Carfax Tower B2
Castle B1
Castle St B1
Catte St B2
Cemetery C1
Christ Church (Coll) B2
Christ Church Cathedral C2
Christ Church Meadow C2
Clarendon Centre B2
Coach & Lorry Park C1
College B3
College of Further Education C1
Cornmarket St B2
Corpus Christi (Coll) B2
County Hall B1
Covered Market B2
Cowley Pl C3
Cranham St A1
Cranham Terr A1
Cricket Ground B1
Crown & County Courts C2
Deer Park B3
Exeter (Coll) B2
Folly Bridge C2
George St B1
Great Clarendon St A1
Hart St A1
Hertford (Coll) B2
High St B3
Hollybush Row B1
Holywell St B2
Hythe Bridge St B1
Ice Rink C1
Information Ctr B2
Jericho St A1
Jesus (Coll) B2
Jowett Walk B3
Juxon St A1
Keble (Coll) A2
Keble Rd A2
Library B2
Linacre (Coll) A3
Lincoln (Coll) B2
Little Clarendon St A1
Longwall St B3
Magdalen (Coll) B3
Magdalen Bridge B2
Magdalen St B2
Magistrate's Court C2
Manchester (Coll) B2
Manor Rd B3
Mansfield (Coll) A2
Mansfield Rd A3
Market B1
Marlborough Rd C2
Martyrs' Memorial B2
Merton Field B3
Merton (Coll) B3
Merton St B2
Museum of Modern Art B2
Museum of Oxford B2
Museum Rd A2
New College (Coll) B3
New Inn Hall St B2
New Rd B1
New Theatre B2
Norfolk St C1
Nuffield (Coll) B1
Observatory A1
Observatory St A1
Odeon B1/B2
Old Fire Station B1
Old Greyfriars St C2
Oriel (Coll) B2
Oxford Station B1
Oxford Story, The B2
Oxford University Research Centres A1
Oxpens Rd C1
Paradise Sq C1
Paradise St B1
Park End St B1
Parks Rd A2/B2
Pembroke (Coll) C2
Phoenix A1
Picture Gallery C2
Plantation Rd A1
Playhouse B2
Police Station C2
Post Office A1/B2
Pusey St B1
Queen's La B3
Queen's (Coll) B3
Radcliffe Camera B2
Rewley Rd B1
Richmond Rd A1
Rose La B3
Ruskin (Coll) B1
Saïd Business School B1
St Aldates C2
St Anne's (Coll) A1
St Antony's (Coll) A1
St Bernard's Rd A1
St Catherine's (Coll) B3
St Cross Building A3
St Cross Rd A3
St Edmund Hall (Coll) B3
St Giles St A2
St Hilda's (Coll) C3
St John St B2
St John's (Coll) B2
St Mary the Virgin B2
St Michael at the Northgate B2
St Peter's (Coll) B1
St Thomas St B1
Science Area A2
Science Museum B2
Sheldonian Theatre B2
Somerville (Coll) A1
South Parks Rd A2
Speedwell St C2
Sports Ground C3
Thames St C2
Town Hall B2
Trinity (Coll) B2
Turl St B2
University College B3
University Museum & Pitt Rivers Museum A2
University Parks A2
Wadham (Coll) B2
Walton Cr A1
Walton St A1
Western Rd C2
Westgate Shopping Centre B2
Woodstock Rd A1
Worcester (Coll) B1

Perth 341

A K Bell Library B2
Abbot Cres C1
Abbot St C1
Albany Terr A1
Albert Monument A3
Alexandra St B2
Art Gallery B3
Atholl St A2
Balhousie Ave A2
Balhousie Castle Black Watch Museum A2
Balhousie St A2
Ballantine Pl A1
Barossa Pl A2
Barossa St A2
Barrack St A2
Bell's Sports Centre A2
Bellwood B3
Blair St B1
Burn Park C1
Bus Station B2
Caledonian Rd B2
Canal Cres B2
Canal St B3
Cavendish Ave C1
Charles St B2
Charlotte Pl A2
Charlotte St A3
Church St A1
City Hall B3
Club House C3
Clyde Pl C1
Commercial St A3
Concert Hall B3
Council Chambers B3
County Pl B2
Court B3
Craigie Pl C2
Crieff Rd A1
Croft Park C2
Cross St B2
Darnhall Cres C1
Darnhall Dr C1
Dundee Rd B3
Dunkeld Rd A1
Earl's Dykes B1
Edinburgh Rd C3
Elibank St B1
Fair Maid's House A3
Fergusson B3
Feus Rd A1
Fire Station A1
Fitness Centre B3
Foundary La A2
Friar St C2
George St B3
Glamis Pl C1
Glasgow Rd B1
Glenearn Rd C2
Glover St B1/C1
Golf Course A2
Gowrie St A3
Gray St B1
Graybank Rd B1
Greyfriars Burial Grnd B3
Hay St A2
High St B2/B3
Hotel B3
Ice Rinks B1
Inchaffray St A1
Industrial/Retail Park B1
Information Ctr B2
Isla Rd A3
James St B2
Keir St A3
King Edward St B3
King James VI Golf Course C3
King St B2
Kings Pl C2
Kinnoull Causeway B2
Kinnoull Aisle 'Monument' B3
Kinnoull St B2
Knowelea Pl C1
Knowelea Terr C1
Ladeside Business Centre A1
Leisure Pool B1
Leonard St B2
Lickley St B2
Lochie Brae A3
Long Causeway A1
Low St A2
Main St A3
Marshall Pl C3
Melville St A2
Mill St B3
Milne St B2
Murray Cres C1
Murray St B2
Needless Rd C1
New Rd B2
North Inch A3
North Methven St A2
Park Pl C1
Perth B2
Perth Bridge A3
Perth Business Park B1
Perth Station C2
Pickletulllum Rd B1
Pitheavlis Cres C1
Playhouse B2
Police Station A2
Pomarium St B2
Post Office A3/B2/C2
Princes St C3
Priory Pl C2
Queen St C1
Queen's Bridge B3
Riggs Rd B1
Riverside B3
Riverside Park B3
Rodney Park B3
Rose Terr A2
St Catherines Retail Park A1
St Catherine's Rd A1/A2
St John St B3
St John's Kirk B3
St John's Shopping Centre B2
St Leonards Bridge C2
St Ninians Cathedral A2
Scott Monument C2
Scott St B2
Sheriff Court B3
Shore Rd C3
Skate Park C3
South Inch C2
South Inch Business Centre C3
South Inch Park C2
South Inch View C2
South Methven St B2
South St B3
South William St B2
Stormont St A2
Strathmore St A3
Stuart Ave C1
Tay St B3
The Stables A1
The Stanners B3
Union La A2
Victoria St B2
Watergate B3
Wellshill Cemetery A1
West Bridge St A3
West Mill St B2
Whitefriers Cres B1
Whitefriers St B1
Wilson St C1
Windsor Terr C1
Woodside Cres C1
York Pl B2
Young St C1

Peterborough 341

Athletics Arena B3
Bishop's Palace B2
Bishop's Rd B2/B3
Boongate A3
Bourges Boulevard A1
Bourges Retail Park B1/B2
Bridge House (Council Offices) C2
Bridge St B2
Bright St A1
Broadway A2
Broadway A2
Brook St A2
Burghley Rd A2
Bus Station B2
Cavendish St A3
Charles St A3
Church St B2
Church Walk A2
Cobden Ave A1
Cobden St A1
Cowgate B2
Craig St A2
Crawthorne Rd A2
Cripple Sidings La C2
Cromwell Rd A1
Dickens St A3
Eastfield Rd A3
Eastgate B3
Fire Station A1
Fletton Ave C2
Frank Perkins Parkway C3
Geneva St A2
George St C1
Gladstone St A1
Glebe Rd C2
Gloucester Rd C3
Granby St B3
Grove St C1
Guildhall B2
Hadrians Ct C3
Henry St A2
Hereward Cross (shopping) B2
Hereward Rd B3
Information Ctr B2
Jubilee St C1
Key Theatre C2
Kent Rd B1
Kirkwood Cl B1
Lea Gdns B1
Library A2
Lincoln Rd A2
London Rd C2
Long Causeway B2
Lower Bridge St C2
Magistrates Court B2
Manor House St A2
Mayor's Walk A1
Midland Rd A1
Monument St A2
Morris St A3
Museum & Art Gallery B2
Nene Valley Railway C1
New Rd A2
New Rd C1
Northminster A2
Old Customs House C2
Oundle Rd C1
Padholme Rd A3
Palmerston Rd C1
Park Rd A2
Passport Office B2
Peterborough District Hospital (A+E) B1
Peterborough Station B1
Peterborough Nene Valley C1
Peterborough United FC C2
Police Station B2
Post Office A3/B1/B2/B3/C1
Priestgate B2
Queen's Walk C2
Queensgate Centre B2
Railworld C1
Regional Swimming & Fitness Centre B3
River La B1
Rivergate Shopping Centre B2
Riverside Mead C3
Russell St A1
St John's B2
St John's St A3
St Marks St A2
St Peter's B2
St Peter's Rd B2
Saxon Rd A3
Spital Bridge A1
Stagshaw Dr C3
Star Rd B3
Thorpe Lea Rd B1
Thorpe Rd B1
Thorpe's Lea Rd B1
Tower St A2
Town Hall B2
Viersen Platz B2
Vineyard Rd B3
Wake Rd B3
Wellington St A3
Wentworth St B2
Westgate B2
Whalley St A3
Wharf Rd C1
Whitsed St A3
YMCA A3

Plymouth 341

ABC B2
Alma Rd A1
Anstis St B1
Armada Centre B2
Armada St A3
Armada Way B2
Arts Centre B2
Athenaeum B1
Athenaeum St C1
Barbican C3
Barbican C3
Baring St A3
Bath St B1
Beaumont Park B3
Beaumont Rd B3
Black Friars Gin Distillery C2
Breton Side B3
Bus Station B2
Castle St C3
Cathedral (RC) B1
Cecil St B1
Central Park A1
Central Park Ave A2
Charles Church B3
Charles Cross rbt B3
Charles St B2
Citadel Rd C2
Citadel Rd East C2
Civic Centre B2
Cliff Rd C1
Clifton Pl A3
Cobourg St A2
College of Art & Design B2
Continental Ferry Port B1
Cornwall St B2
Dale Rd A2
Deptford Pl A3
Derry Ave A2
Derry's Cross rbt B1
Drake Circus B2
Drake Circus Shopping Centre B2
Drake's Memorial C2
Eastlake St B2
Ebrington St B3
Elizabethan House C3
Elliot St C1
Endsleigh Pl A2
Exeter St B3
Fire Station A3
Fish Quay C3
Gibbons St A3
Glen Park Ave A2
Grand Pde C1
Great Western Rd C1
Greenbank Rd A3
Greenbank Terr A3
Guildhall B2
Hampton St B3
Harwell St B1
Hill Park Cr A3
Hoe Approach B2
Hoe Rd C2
Hoegate St C2
Houndiscombe Rd A2
Information Ctr C3
James St A2
Kensington Rd A3
King St B1
Lambhay Hill C3
Leigham St C1
Library B2
Lipson Rd A3/B3
Lockyer St C2
Lockyers Quay C3
Madeira Rd C2
Marina B3
Market Ave B1
Martin St B1
Mayflower Stone & Steps C3
Mayflower St B2

Mayflower Visitor Centre ✦ C3
Merchants House B2
Millbay Rd B1
Museum & Art Gallery B2
National Marine Aquarium C3
Neswick St B1
New George St B2
New St C3
North Cross rbt A2
North Hill A3
North Quay B2
North Rd East A2
North Rd West A1
North St B3
Notte St B2
Octagon St B1
Pannier Market B1
Pennycomequick rbt A1
Pier St C1
Plymouth Pavilions B1
Plymouth Station A2
Police Station B3
Portland Sq A2
Post Office A1/B1/B2
Princess St B2
Prysten House B2
Queen Anne's Battery Seasports Centre C3
Radford Rd C1
Regent St B3
Rope Walk C3
Royal Citadel C2
Royal Pde B2
St Andrew's B2
St Andrew's Cross rbt B2
St Andrew's St B2
St Lawrence Rd A2
Saltash Rd A2
Smeaton's Tower ✦ C2
Southern Terr A3
Southside St C2
Stuart Rd A1
Sutherland Rd A2
Sutton Rd B3
Sydney St A1
Teats Hill Rd C3
The Crescent B1
The Hoe C2
The Octagon rbt B1
The Promenade C2
Theatre Royal B2
Tothill Ave B3
Union St B1
University of Plymouth A2
Vauxhall St B2/3
Victoria Park A1
West Hoe Rd C1
Western Approach B1
Whittington St A1
Wyndham St B1
YMCA B2
YWCA C2

Poole 341

Ambulance Station A3
Baiater Gdns C2
Baiter Park C3
Ballard Cl C2
Ballard Rd C2
Bay Hog La B1
Bridge Approach C1
Bus Station B2
Castle St B2
Catalina Dr B3
Chapel La B2
Church St B1
Cinnamon La B1
Colborne Cl B3
Dear Hay La B2
Denmark La A3
Denmark Rd A3
East St B2
Elizabeth Rd A3
Emerson Rd B2
Ferry Rd C1
Ferry Terminal C1
Fire Station A2
Freightliner Terminal C1
Furnell Rd B3
Garland Rd A3
Green Rd B2
Heckford La A3
Heckford Rd A3
High St B2
High St North A3
Hill St B2
Holes Bay Rd A1
Hospital (A+E) A3
Information Ctr C2
Kingland Rd B3
Kingston Rd A3
Labrador Dr C3
Lagland St B2
Lander Cl C3
Old Lifeboat C2
Lighthouse -Poole Centre for the Arts ✦ B3
Longfleet Rd A3
Maple Rd A3
Market Cl B2
Market St B2
Mount Pleasant Rd B3
New Harbour Rd C1
New Harbour Rd South C1
New Harbour Rd West C1
New Orchard B1
New Quay Rd C1
New St B2
Newfoundland Dr B2
North St B2
Old Orchard B2
Parish Rd A3
Park Lake Rd B3
Parkstone Rd A3
Perry Gdns B2
Pitwines Cl B2
Police Station A2
Poole Central Library B2
Poole Lifting Bridge C1
Poole Park B2
Poole Station A2
Poole Waterfront Museum C1
Post Office A2/B2
St John's Rd A3
St Margaret's Rd A2
St Mary's Maternity Unit A3
St Mary's Rd A3
Seldown Bridge B3
Seldown La B3
Seldown Rd B3
Serpentine Rd A2
Shaftesbury Rd A3
Skinner St B2
Slipway B1
Stanley Rd C2
Sterte Ave A2
Sterte Ave West A1
Sterte Cl A2
Sterte Esplanade A2
Sterte Rd A2
Strand St C2
Swimming Pool B3
Taverner Cl B3
Thames St B1
The Lifeboat College B2
The Quay C2
Towngate Bridge B2
Vallis Cl C3
West Quay B1
West Quay Rd B1
West St B1
West View Rd A2
Whatleigh Cl B2
Wimborne Rd A3

Portsmouth 341

Action Stations ✦ C1
Admiralty Rd A1
Alfred Rd A2
Anglesea Rd B2
Arundel St B3
Bishop St A2
Broad St C1
Buckingham House C2
Burnaby Rd B2
Bus Station B1
Camber Dock C1
Cambridge Rd B2
Car Ferry to Isle of Wight B1
Cascades Shopping Centre A3
Castle Rd C3
Cathedral ✝ C1
Cathedral (RC) ✝ A3
City Museum & Art Gallery B2
Civic Offices B3
Clarence Pier C2
College St B1
Commercial Rd A3
Cottage Gr B3
Cross St A1
Cumberland St A2
Duisburg Way C2
Durham St A3
East St B1
Edinburgh Rd A2
Elm Gr C3
Great Southsea St C3
Green Rd B3
Greetham St B3
Grosvenor St B3
Grove Rd North C3
Grove Rd South C3
Guildhall B3
Guildhall Walk B3
Gunwharf Quays Retail Park B1
Gunwharf Rd B1
Hambrook St C2
Hampshire Terr B2
Hanover St A1
High St C2
HM Naval Base A1
HMS Nelson (Royal Naval Barracks) A2
HMS Victory A1
HMS Warrior A1
Hovercraft Terminal C2
Hyde Park Rd B3
Information Ctr A1/B3
Isambard Brunel Rd B3
Isle of Wight Car Ferry Terminal B1
Kent Rd C3
Kent St A2
King St B3
King's Rd C3
King's Terr C2
Lake Rd A3
Law Courts B3
Library B3
Long Curtain Rd C2
Market Way A3
Marmion Rd C3
Mary Rose Exhibition A1
Mary Rose Ship Hall A1
Middle St B3
Millennium Blvd B2
Millennium Promenade A1/C1
Museum Rd B2
Naval Recreation Ground C2
Nightingale Rd C3
Norfolk St B3
North St A2
Osborne Rd C3
Park Rd B2
Passenger Catamaran to Isle of Wight B1
Passenger Ferry to Gosport B1
Pelham Rd C3
Pembroke Gdns C2
Pier Rd C2
Point Battery C1
Police Station B3
Portsmouth & Southsea A3
Portsmouth Harbour B1
Post Office A2/A3/B1/B3/C3
Queen St A1
Queen's Cr C3
Round Tower ✦ C1
Royal Garrison Church C1
Royal Naval Museum A1
St Edward's Rd C3
St George's Rd B2
St George's Sq B1
St George's Way B2
St James's Rd B3
St James's St A2
St Thomas's St B2
Somers Rd B3
Southsea Common C2
Southsea Terr C2
Spinnaker Tower ✦ B1
Square Tower ✦ C1
Station St A3
Swimming Pool A2
The Hard B1
Town Fortifications ✦ C1
Unicorn Rd A2
United Services Recreation Ground B2
University of Portsmouth A2/B2
University of Portsmouth – College of Art, Design & Media B3
Upper Arundel St A3
Victoria Ave C2
Victoria Park A2
Victory Gate A1
Vue B1
Warblington St B1
Western Pde C2
White Hart Rd C1
Winston Churchill Ave B3

Preston 342

Adelphi St A2
Anchor Ct B3
Aqueduct St A1
Ardee Rd C1
Arthur St B2
Ashton St A1
Avenham La B3
Avenham Park C3
Avenham Rd B3
Avenham St B3
Bairstow St B3
Balderstone Rd C1
Beamont Dr A1
Beech St South C2
Bird St C1
Bow La B2
Brieryfield Rd B1
Broadgate C1
Brook St A2
Bus Station A3
Butler St B2
Cannon St B3
Carlton St A1
Chaddock St B3
Channel Way B1
Chapel St B3
Christ Church St B2
Christian Rd B2
Cold Bath St A2
Coleman Ct C1
Connaught Rd C2
Corporation St A2/B2
County Hall B2
County Records Office B2
Court A3
Court B3
Cricket Ground C2
Croft St A1
Cross St B3
Crown Court A3
Crown St A3
East Cliff C3
East Cliff Rd B3
Edward St A2
Elizabeth St A3
Euston St B1
Fishergate B2/B3
Fishergate Hill B2
Fishergate Shopping Centre B2
Fitzroy St B1
Fleetwood St A1
Friargate A3
Fylde Rd A1/A2
Gerrard St B2
Glover's Ct B3
Good St B2
Grafton St B2
Great George St A3
Great Shaw St A3
Greenbank St A2
Guild Way B1
Guildhall & Charter B3
Guildhall St B3
Harrington St A2
Harris Museum B3
Hartington Rd B1
Hasset Cl C2
Heatley St B2
Hind St C2
Information Ctr B3
Kilruddery Rd C1
Lancaster Rd A3/B3
Latham St B3
Lauderdale St C2
Lawson St A3
Leighton St A2
Leyland Rd C1
Library A1
Library B3
Liverpool Rd C1
Lodge St B2
Lune St B3
Main Sprit West B3
Maresfield Rd C1
Market St West A3
Marsh La B1/B2
Maudland Bank A2
Maudland Rd A2
Meadow Ct C2
Meath Rd C1
Mill Hill A2
Miller Arcade ✦ B3
Miller Park C3
Moor La A3
Mount St B3
North Rd A3
North St A3
Northcote Rd B1
Old Milestones B1
Old Tram Rd C3
Pedder St A1/A2
Peel St A2
Penwortham Bridge C2
Penwortham New Bridge C1
Pitt St B2
Playhouse A3
Police Station A3
Port Way B1
Post Office B3
Preston Station B2
Ribble Bank St B2
Ribble Viaduct C2
Ribblesdale Pl B3
Ringway B3
River Parade C1
Riverside C2
St Georges B3
St Georges Shopping Centre B3
St Johns B3
St Johns Shopping Centre A3
St Mark's Rd A1
St Walburges A1
Salisbury Rd B1
Sessions House B3
Snow Hill A3
South End C2
South Meadow La C2
Spa Rd B1
Sports Ground C2
Strand Rd B1
Syke St B3
Talbot Rd B1
Taylor St C1
Tithebarn St A3
Town Hall B3
Tulketh Brow A1
University of Central Lancashire A2
Valley Rd C1
Victoria St A2
Walker St A3
Walton's Parade B2
Warwick St A3
Wellfield Business Park A1
Wellfield Rd A1
Wellington St A1
West Cliff C2
West Strand A1
Winckley Rd C1
Winckley Square B3
Wolseley Rd C2

Reading 342

Abbey Ruins ✝ B2
Abbey Sq B2
Abbey St B2
Abbot's Walk B2
Acacia Rd C2
Addington Rd C3
Addison Rd A1
Allcroft Rd C3
Alpine St C2
Baker St B1
Berkeley Ave C1
Bridge St B1
Brigham Rd A1
Broad St B1
Broad Street Mall B1
Carey St B1
Castle Hill C1
Castle St B1
Caversham Rd A1
Christchurch Playing Fields A2
Civic Offices & Magistrate's Court B1
Coley Hill C1
Coley Pl C1
Craven Rd C3
Crown St C2
De Montfort Rd A1
Denmark Rd C3
Duke St B2
East St B2
Edgehill St C2
Eldon Rd B3
Eldon Terr B3
Elgar Rd C1
Erleigh Rd C3
Field Rd C1
Fire Station A1
Fobney St B1
Forbury Gdns B2
Forbury Retail Park B2
Forbury Rd B2
Francis St C1
Friar St B1
Garrard St B1
Gas Works Rd B3
George St A2
Great Knollys St B1
Greyfriars B1
Gun St B1
Henry St C1
Hexagon Theatre, The B1
Hill's Meadow A2
HM Prison B2
Howard St B1
Information Ctr B1
Inner Distribution Rd B1
Katesgrove La C1
Kenavon Dr B2
Kendrick Rd C2
King's Meadow Rec Ground A2
King's Rd B2
Library B2
London Rd C3
London St B2
Lynmouth Rd A1
Market Pl B2
Mill La B2
Mill Rd A2
Minster St B1
Morgan Rd C3
Mount Pleasant C2
Museum of English Rural Life C3
Napier Rd A2
Newark St C2
Newport Rd A1
Old Reading University C3
Oracle Shopping Centre, The B1
Orts Rd B3
Pell St C1
Queen Victoria St B1
Queen's Rd A2
Queen's Rd B2
Police Station B1
Post Office B2
Randolph Rd A1
Reading Bridge A2
Reading Station A1
Redlands Rd C3
Renaissance Hotel B1
Riverside Museum B3
Rose Kiln La C1
Royal Berks Hospital (A & E) C3
St Giles C2
St Laurence B2
St Mary's B1
St Mary's Butts B1
St Saviour's Rd C1
Send Rd A3
Sherman Rd C2
Sidmouth St B2
Silver St C2
South St B2
Southampton St C2
Station Hill A1
Station Rd B1
Superstore A3
Swansea Rd A1
Technical College B3
The Causeway A3
The Grove B2
Valpy St B2
Vastern Rd A1
Vue B2
Waldeck St C2
Watlington St B3
West St B1
Whitby Dr C3
Wolseley St C1
York Rd A1
Zinzan St B1

St Andrews 342

Abbey St B2
Abbey Walk B3
Abbotsford Cres A1
Albany Pk C3
Allan Robertson Dr C2
Ambulance Station C1
Anstruther Rd C3
Argyle St B1
Argyll Business Park C1
Auld Burn Rd B2
Bassaguard Industrial Estate B1
Bell St B2
Blackfriars Chapel (Ruins) B2
Boase Ave B2
Braid Cres C3
Brewster Pl C3
Bridge St B1
British Golf Museum A1
Broomfaulds Ave C1
Bruce Embankment A1
Bruce St C2
Bus Station A1
Byre B2
Canongate C1
Cathedral and Priory (Ruins) ✝ B3
Cemetery B3
Chamberlain St C1
Church St B2
Churchill Cres C2
City Rd A1
Claybraes C1
Cockshaugh Public Park B1
Cosmos Community Centre B3
Council Office A2
Crawford Gdns C1
Doubledykes Rd B1
Drumcarrow Rd C1
East Sands B3
East Scores A3
Fire Station C1
Forrest St C1
Fraser Ave C1
Freddie Tait St C2
Gateway Centre A1
Glebe Rd B2
Golf Pl A1
Grange Rd C3
Greenside Pl B2
Greyfriars Gdns A2
Hamilton Ave C2
Hepburn Gdns B1
Horseleys Park C1
Information Ctr B2
Irvine Cres C3
James Robb Ave C1
James St B1
John Knox Rd C1
Kennedy Gdns B1
Kilrymont Cl C3
Kilrymont Pl C3
Kilrymont Rd C3
Kinburn Park B1
Kinkell Terr C3
Kinnesburn Rd B2
Ladebraes Walk B2
Lady Buchan's Cave A3
Lamberton Pl C1
Lamond Dr C2
Langlands Rd B2
Largo Rd C1
Learmonth Pl C1
Library B2
Links Clubhouse A1
Links, The A1
Livingstone Cres B2
Long Rocks A2
Madras College B2
Market St A2
Martyr's Monument A1
Memorial Hospital (No A+E) B3
Murray Pk A2
Murray Pl A2
Nelson St B2
New Course, The A1
New Picture House A2
North Castle St A3
North St A2
Old Course, The A1
Old Station Rd A1
Pends, The B3
Pilmour Links A1
Pipeland Rd B2/C2
Police Station A2
Post Office B2
Preservation Trust B3
Priestden Pk C3
Priestden Pl C3
Priestden Rd C3
Queen's Gdns B2
Queen's Terr B2
Roundhill Rd C2
Royal & Ancient Golf Club A1
St Andrews B1
St Andrews Aquarium A2
St Andrews Botanic Gardens B1
St Andrews Castle (Ruins) & Visitor Centre A2
St Mary St B3
St Mary's College B2
St Nicholas St C3
St Rules Tower B3
St Salvator's College A2
Sandyhill Cres C2
Sandyhill Rd C2
Scooniehill Rd C2
Shields Ave C3
Shoolbraids C2
Sloan St B1
South St B2
Spottiswoode Gdns C1
Station Rd A1
Swilken Bridge A1
The Scores A2
The Shore B3
Tom Morris Dr C2
Tom Stewart La C1
Town Church B2
Town Hall B2
Union St A2
University Chapel A2
University Library A2
University of St Andrews A1
Viaduct Walk B1
War Memorial A3
Wardlaw Gdns B1
Warrack St C3
Watson Ave B2
West Port B2
West Sands A1
Westview B2
Windmill Rd A1
Winram Pl C1
Wishart Gdns C2
Woodburn Pk B3
Woodburn Pl B3
Woodburn Terr B3
Younger Hall A2

Salisbury 342

Albany Rd A2
Arts Centre A3
Ashley Rd A1
Avon Approach A2
Ayleswade Rd C2
Bedwin St A2
Belle Vue A2
Bishop's Palace C2
Bishops Walk B2
Blue Boar Row B2
Bourne Ave A3
Bourne Hill A3
Britford La C2
Broad Walk C2
Brown St B2
Bus Station B2
Castle St A2
Catherine St B2
Chapter House B2
Church House B2
Churchfields Rd B1
Churchill Way East B3
Churchill Way North A3
Churchill Way South C2
Churchill Way West A1
City Hall B2
Close Wall B2
Coldharbour La A1
College St A3
Council Offices A3
Court A1
Crane Bridge Rd B2
Crane St B2
Cricket Ground C1
Culver St South B3
De Vaux Pl C2
Devizes Rd A1
Dews Rd B1
Elm Grove B3
Elm Grove Rd A3
Endless St A2
Estcourt Rd A3
Exeter St C2
Fairview Rd A3
Fire Station A1
Fisherton St A1
Folkestone Rd C1
Fowlers Hill B3
Fowlers Rd B3
Friary Estate C3
Friary La B2
Gas La A1
Gigant St B3
Greencroft B3
Greencroft St A3
Guildhall B2
Hall of John Halle B2
Hamilton Rd A2
Harnham Mill B1
Harnham Rd C1/C2
High St B2
Hospital A1
House of John A'Port B2
Information Ctr B2
Kelsey Rd A3
King's Rd A2
Laverstock Rd B3
Library A2
London Rd A3
Lower St C1
Maltings, The B2
Manor Rd A3
Marsh La A1
Medieval Hall B2
Milford Hill B3
Milford St B3
Mill Rd B1
Millstream Approach A2
Mompesson House (N.T.) B2
New Bridge Rd C2
New Canal B2
New Harnham Rd C2
New St B2
North Canonry B2
North Gate B2
North Walk B2
Old George Hall B2
Old Blandford Rd C1
Old Deanery B2
Park St A3
Parsonage Green C1
Playhouse Theatre A2
Post Office A2/B2/C2
Poultry Cross B2
Queen Elizabeth Gdns B1
Queen's Rd A3
Rampart Rd B3
St Ann's Gate B2
St Ann St B2
St Marks Rd A3
St Martins B3
St Mary's Cathedral ✝ B2
St Nicholas Hospital C2
St Paul's A1
St Paul's Rd A1
St Thomas B2
Salisbury & South Wiltshire Museum B2
Salisbury General Hospital (A & E) C2
Salisbury Station A1
Salt La A3
Saxon Rd C1
Scots La A2
Shady Bower B3
South Canonry C2
South Gate C2
Southampton Rd B2
Spire View A1
Sports Ground C3
The Friary B3
Tollgate Rd B3
Town Path B1
Wain-a-Long Rd A3
Wardrobe, The B2
Wessex Rd A3
West Walk C2
Wilton Rd A1
Wiltshire College B3
Winchester St B3
Windsor Rd A1
Winston Churchill Gdns C3
Wyndham Rd A2
YHA B3
York Rd A1

Scarborough 342

Aberdeen Walk B2
Albert Rd A2
Albion Rd C2
Alexandra Bowling Hall A1
Alexandra Gardens A1
Auborough St B2
Belle Vue St C1
Belmont Rd C2
Brunswick Shopping Centre B2
Castle Dykes B3
Castlegate B3
Castle Holms A3
Castle Hill A3
Castle Rd B2
Castle Walls A3
Cemetery B1
Central Lift ✦ C2
Clarence Gardens A2
Coach Park B1
Columbus Ravine A1
Court B1
Cricket Ground A1
Cross St B2
Crown Terr C2
Dean Rd B1
Devonshire Dr A1
East Harbour B3
East Pier B3
Eastborough B2
Elmville Ave B1
Esplanade C2
Falconers Rd B2
Falsgrave Rd C1
Fire Station B2
Foreshore Rd B2
Friargate B2
Futurist Theatre B2
Gladstone Rd B1
Gladstone St B1
Hoxton Rd B1
Information Ctr B2/B3
King St B2
Londesborough Rd C1
Longwestgate B3
Marine Dr A3
Miniature Railway A1
Nelson St B1
Newborough B2
Nicolas St B2
North Marine Rd A1
North St B2
Northway B1
Old Harbour B3
Peasholm Park A1
Peasholm Rd A1
Plaza A1
Police Station B1
Post Office B2/C1
Princess St B3
Prospect Rd B1
Queen St B2
Queen's Parade A2
Queen's Tower (Remains) A3
Ramshill Rd C2
Roman Signal Station A3
Roscoe St C1
Rotunda Museum C2
Royal Albert Dr A2
St Martin-on-the-Hill C2
St Martin's Ave C2
St Mary's B3
St Nicholas' Lift ✦ C2
St Thomas St B2
Sandside B3
Scarborough Art Gallery and Crescent Art Studio C2
Scarborough Castle A3
Scarborough C1
Somerset Terr C2
South Cliff Lift ✦ C2
Spa, The ✦ C2
Spa Theatre, The C2
Stephen Joseph Theatre B1
Tennyson Ave B1
The Crescent C2
Tollergate B2
Town Hall B2
Trafalgar Rd B1
Trafalgar Square A1
Trafalgar St West B1
Valley Bridge Parade C2
Valley Rd C1
Vernon Rd C2
Victoria Park Mount A1
Victoria Rd B1
West Pier B3
Westborough C1
Westover Rd C1
Westwood C1
Woodall Ave A1
York Pl B2
Yorkshire Coast College (Westwood Campus) C1

Sheffield 342

Addy Dr A2
Addy St A2
Adelphi St A3
Albert Terrace Rd A3
Albion St A2
Aldred Rd A1
Allen St A4
Alma St A4
Angel St B5
Arundel Gate B5
Arundel St C4
Ashberry Rd A2
Ashdell Rd C1
Ashgate Rd C1
Athletics Centre B2
Attercliffe Rd A6
Bailey St B4
Ball St A4
Balm Green B4
Bank St B4
Barber Rd A2
Bard St B5
Barker's Pool B4
Bates St A1
Beech Hill Rd C1
Beet St B3
Bellefield St A3
Bernard Rd A6
Bernard St B6
Birkendale A2
Birkendale Rd A2
Birkendale View A1
Bishop St C4
Blackwell Pl B6
Blake St A2
Blonk St A5
Bolsover St B2
Botanical Gdns C1
Bower Rd A1
Bradley St A1
Bramall La C4
Bramwell St A3
Bridge St A4/A5
Brighton Terrace Rd A1
Broad La B3
Broad St B6
Brocco St A3
Brook Hill B3
Broomfield Rd C1
Broomgrove Rd C2
Broomhall Pl C3
Broomhall Rd C2
Broomhall St C3
Broomspring La C2
Brown St C5
Brunswick St B3
Burgess St B4
Burlington St A2
Burns Rd A2
Bus/Coach Station B5
Cadman St A6
Cambridge St B4
Campo La B4
Carver St B4
Castle Market B5
Castle Square B5
Castlegate A5
Cathedral (RC) ✝ B4
Cathedral ✝ B4
Cavendish St B3
Charles St C4
Charter Row C4
Children's Hospital (A&E) B2
Church St B4
City Hall B4
City Hall B4
City Rd C6
Claremont Cr B2
Claremont Pl B2
Clarke St C3
Clarkegrove Rd C2
Clarkehouse Rd C1
Clarkson St B2
Cobden View Rd A1
Collegiate Cr C2
Commercial St B5
Commonside A1
Conduit Rd B1
Cornish St A3
Corporation St A4
Court B4
Cricket Inn Rd B6
Cromwell St A1
Crookes Rd B1
Crookes Valley Park B2
Crookes Valley Rd B2
Crookesmoor Rd A2
Crown Court A4
Crucible Theatre B5
Cutlers Gate A6
Cutler's Hall B4
Daniel Hill A2
Dental Hospital B2
Dept for Education & Employment C4
Devonshire Green B3
Devonshire St B3
Division St B4
Dorset St C2
Dover St A3
Duchess Rd C5
Duke St B5
Duncombe St A1
Durham Rd B2
Earl St C4
Earl Way C4
Ecclesall Rd C3
Edward St B3
Effingham Rd A6
Effingham St A6
Egerton St C3
Eldon St B3
Elmore Rd B1
Exchange St B5
Eyre St C4
Fargate B4
Farm Rd C5
Fawcett St A3
Filey St B3
Fire & Police Museum A4
Fir St A1
Fitzalan Sq/Ponds Forge B5
Fitzwater Rd C6
Fitzwilliam Gate C4
Fitzwilliam St B3
Flat St B5
Foley St A6
Foundry Climbing Centre A4
Fulton Rd A1
Furnace Hill A4
Furnival Rd A5

Furnival Sq C4
Furnival St C4
Garden St B3
Gell St B3
Gibraltar St A4
Glebe Rd B1
Glencoe Rd C6
Glossop Rd B2/B3/C1
Gloucester St C2
Granville Rd C6
Granville Rd/ Sheffield College C5
Graves Gallery B5
Greave Rd B3
Green La A4
Hadfield St A1
Hanover St C3
Hanover Way C3
Harcourt Rd B1
Harmer La B5
Havelock St C2
Hawley St B4
Haymarket B5
Headford St C3
Heavygate Rd A1
Henry St A3
High St B4
Hodgson St C3
Holberry Gdns C2
Hollis Croft B4
Holly St B4
Hounsfield Rd B3
Howard Rd A1
Hoyle St A3
Hyde Park A6
Infirmary Rd A3
Infirmary Rd A3
Information Ctr B4
Jericho St A3
Johnson St A5
Kelham Island Industrial Museum A4
Lawson Rd C1
Leadmill Rd C5
Leadmill St C5
Leadmill, The C5
Leamington St A1
Leavy Rd B3
Lee Croft B4
Leopold St B4
Leveson St A6
Library A2
Library B5
Library C1
Lyceum Theatre B5
Malinda St A3
Maltravers St A5
Manor Oaks Rd B6
Mappin Art Gallery B2
Mappin St B3
Marlborough Rd B1
Mary St C4
Matilda St C4
Matlock Rd A1
Meadow St A3
Melbourn Rd A1
Melbourne Ave C1
Millennium Galleries B5
Milton St C3
Mitchell St B3
Mona Ave A1
Mona Rd A1
Montgomery Terrace Rd A3
Montgomery Theatre B4
Monument Gdns C6
Moor Oaks Rd B1
Moore St C3
Mowbray St A4
Mushroom La B2
Netherthorpe Rd B3
Netherthorpe Rd B3
Newbould La C1
Nile St C1
Norfolk Park Rd C6
Norfolk Rd C6
Norfolk St B4
North Church St B4
Northfield Rd A1
Northumberland Rd B1
Nursery St A5
Oakholme Rd C1
Octagon B2
Odeon B5
Old St B6
Oxford St A2
Paradise St B4
Park La C2
Park Sq B5
Parker's Rd B1
Pearson Building (Univ) C2
Penistone Rd A3
Pinstone St B4
Pitt St B3
Police Station A4/B5
Pond Hill B5
Pond St B5
Ponds Forge Sports Centre B5
Portobello St B3
Post Office A1/A2/B3/B4/B5/B6/C1/C3/C4/C6
Powell St A2
Queen St B4
Queen's Rd C5
Ramsey Rd B1
Red Hill B3
Redcar Rd B1
Regent St B3
Rockingham St B4
Roebuck Rd A2
Royal Hallamshire Hospital C2
Russell St A4
Rutland Park C1
St George's Cl B3
St Mary's Gate C3
St Mary's Rd C4/C5
St Peter & St Paul Cathedral B4
St Philip's Rd A3
Savile St A5
School Rd B1
Scotland St A4
Severn Rd B1
Shalesmoor A4
Shalesmoor A3
Sheaf St B5
Sheffield Hallam University B5
Sheffield Ice Sports Centre - Skate Central C5
Sheffield Parkway A6
Sheffield Station C5
Sheffield Station/ Sheffield Hallam University B5
Sheffield University B2
Shepherd St A3
Shipton St A2
Shoreham St C4
Showroom, The C5
Shrewsbury Rd C5
Sidney St C4
Site Gallery C5
Slinn St A1
Smithfield A4
Snig Hill A5
Snow La A4
Solly St B3
Southbourne Rd C1
South La C4
South Street Park B5
Spital Hill A5
Spital St A5
Spring Hill B1
Spring Hill Rd B1
Springvale Rd A1
Stafford Rd C6
Stafford St B6
Stanley St A5
Suffolk Rd C5
Summer St B2
Sunny Bank C3
Surrey St B4
Sussex St A6
Sutton St B3
Sydney Rd A2
Sylvester St C4
Talbot St B5
Taptonville Rd B1
Tax Office C4
Tenter St B4
The Moor C4
Town Hall B4
Townend St A1
Townhead St B4
Trafalgar St B4
Tree Root Walk B2
Trinity St A4
Trippet La B4
Turner Museum of Glass B3
Union St B4
University Drama Studio B2
University of Sheffield B3
Upper Allen St A3
Upper Hanover St B3
Upperthorpe Rd A2/A3
Verdon St A5
Victoria Quays B5
Victoria Rd C2
Victoria St B3
Waingate B5
Watery St A3
Watson Rd C1
Wellesley Rd B2
Wellington St C3
West Bar A4
West Bar Green A4
West One B3
West St B3
West St B4
Westbourne Rd C1
Western Bank B2
Western Rd A1
Weston Park B2
Weston Park Hospital B2
Weston Park Museum B2
Weston St B2
Wharncliffe Rd C3
Whitham Rd B1
Wicker A5
Wilkinson St B2
William St C3
Winter Garden B4
Winter St B2
York St B4
Yorkshire Artspace C5
Young St C4

Shrewsbury 342

Abbey Church B3
Abbey Foregate B3
Abbey Lawn Business Park B3
Abbots House B2
Agricultural Show Gd A1
Albert St A2
Alma St B1
Ashley St A3
Ashton Rd C1
Avondale Dr A3
Bage Way C3
Barker St B1
Beacall's La A2
Beeches La C2
Beehive La C1
Belle Vue Gdns C2
Belle Vue Rd C2
Belmont Bank C1
Berwick Ave A1
Berwick Rd A1
Betton St C2
Bishop St B3
Bradford St B3
Bridge St B1
Bus Station B2
Butcher Row B2
Burton St A3
Butler Rd C1
Bynner St C2
Canon St B3
Canonbury C1
Castle Business Park, The A2
Castle Foregate A2
Castle Gates B2
Castle Museum B2
Castle St B2
Cathedral (RC) C1
Chester St A2
Cineworld C3
Claremont Bank B1
Claremont Hill B1
Cleveland St B3
Coleham Head B2
Coleham Pumping Station C2
College Hill B1
Corporation La A1
Coton Cres A1
Coton Hill A2
Coton Mount A1
Crescent La C1
Crewe St A2
Cross Hill B1
Darwin Centre B2
Dingle, The B1
Dogpole B2
Draper's Hall B2
English Bridge B2
Fish St B2
Frankwell B1
Gateway Centre, The A2
Gravel Hill La A1
Greyfriars Rd C2
Guildhall B1
Hampton Rd A3
Haycock Way C3
HM Prison B2
High St B1
Hills La B1
Holywell St B3
Hunter St A1
Information Ctr B1
Ireland's Mansion & Bear Steps B1
John St A3
Kennedy Rd C1
King St B3
Kingsland Bridge (toll) C1
Kingsland Rd C1
Library B2
Lime St C2
Longden Coleham C2
Longden Rd C1
Longner St B1
Luciefelde Rd C1
Mardol B1
Market B1
Marine Terr C2
Monkmoor Rd B3
Moreton Cr C2
Mount St A1
Music Hall B1
New Park Cl A3
New Park Rd A2
New Park St A3
North St A2
Oakley St C1
Old Coleham C2
Old Market Hall B1
Old Potts Way C3
Parade Centre B2
Police Station B1
Post Office A2/B1/B2/B3
Pride Hill B1
Pride Hill Centre B1
Priory Rd B1
Pritchard Way C3
Queen St A3
Raby Cr C2
Rad Brook C1
Rea Brook C3
Riverside B1
Roundhill La A1
Rowley's House B1
St Alkmund's B2
St Chad's B1
St Chad's Terr B1
St John's Hill B1
St Julians Friars C2
St Mary's B2
St Mary's St B2
Salters La C2
Scott St C3
Severn Bank A3
Severn St A2
Shrewsbury B2
Shrewsbury High School for Girls C1
Shrewsbury School C1
Shropshire Wildlife Trust B3
Smithfield Rd B1
South Hermitage C1
Swan Hill B1
Sydney Ave A3
Tankerville St B3
The Dana B2
The Quarry B1
The Square B1
Tilbrook Dr A3
Town Walls C1
Trinity St C2
Underdale Rd B3
Victoria Ave B1
Victoria Quay B1
Victoria St B2
Welsh Bridge B1
Whitehall St B3
Wood St A2
Wyle Cop B2

Southampton 342

Above Bar St A2
Albert Rd North B3
Albert Rd South C3
Anderson's Rd B3
Archaeology Museum C2
Argyle Rd A2
Arundel Tower B1
Bargate, The B2
Bargate Centre B2
BBC Regional Centre A1
Bedford Pl A1
Belvidere Rd A3
Bernard St C2
Blechynden Terr A1
Brazil Rd C3
Brinton's Rd A2
Britannia Rd A3
Briton St C2
Brunswick Pl A2
Bugle St C1
Canute Rd C3
Castle Way C2
Catchcold Tower B1
Central Bridge C3
Central Rd C2
Channel Way C3
Chapel Rd B3
Cineworld C3
City Art Gallery A1
City College B3
Civic Centre A1
Civic Centre Rd A1
Coach Station B1
Commercial Rd A1
Cumberland Pl A1
Cunard Rd C2
Derby Rd A3
Devonshire Rd A1
Dock Gate 4 C2
Dock Gate 8 B1
East Park A2
East Park Terr A2
East St B2
East St Shopping Centre B2
Endle St B3
European Way C2
Fire Station A2
Floating Bridge Rd C3
God's House Tower C2
Golden Gr A3
Graham Rd A2
Guildhall A1
Hanover Bldgs B2
Harbour Lights C3
Harbour Pde B1
Hartington Rd A3
Havelock Rd A1
Henstead Rd A1
Herbert Walker Ave B1
High St B2
Hoglands Park B2
Holy Rood (Rems), Merchant Navy Memorial B2
Hospital A2
Houndwell Park B2
Houndwell Pl B2
Hythe Ferry C2
Information Ctr A1
Isle of Wight Ferry Terminal C1
James St B3
Java Rd C3
Kingsland Market B2
Kingsway A2
Leisure World B1
Library A1
Lime St B2
London Rd A2
Marine Pde B3
Maritime C1
Marsh La B2
Mayflower Memorial C1
Mayflower Park C1
Mayflower Theatre, The A1
Medieval Merchant's House C1
Melbourne St B3
Millais A2
Morris Rd A1
Neptune Way C3
New Rd A2
Nichols Rd A3
Northam Rd A3
Ocean Dock C2
Ocean Village Marina C3
Ocean Way C3
Odeon B1
Ogle Rd B1
Old Northam Rd A2
Orchard La B2
Oxford Ave A2
Oxford St C2
Palmerston Park A2
Palmerston Rd A2
Parsonage Rd A3
Peel St A3
Platform Rd C2
Police Station A1
Portland Terr B1
Post Office A2/A3/B2
Pound Tree Rd B2
Quays Swimming & Diving Complex, The B1
Queen's Park C2
Queen's Peace Fountain A2
Queen's Terr C2
Queen's Way B2
Radcliffe Rd A3
Rochester St A3
Royal Pier C1
St Andrew's Rd A2
St Mary St A2
St Mary's B3
St Mary's Leisure Centre A2
St Mary's Pl B2
St Mary's Rd A2
St Mary's Stadium (Southampton F.C.) A3
St Michael's C1
Solent Sky C3
South Front B2
Southampton Central Station A1
Southampton Solent University A2
Southampton Oceanography Centre C3
SS Shieldhall C2
Terminus Terr C2
The Mall, Marlands A1
The Polygon A1
Threefield La B2
Titanic Engineers' Memorial A2
Town Quay C1
Town Walls B2
Tudor House C1
Vincent's Walk B2
West Gate C1
West Marlands Rd A1
West Park A1
West Park Rd A1
West Quay Rd B1
West Quay Retail Park B1
West Quay Shopping Centre B1
West Rd C2
Western Esplanade B1

Southend-on-Sea 343

Adventure Island C3
Albany Ave A1
Albert Rd C3
Alexandra Rd C2
Alexandra St C2
Art Gallery C1
Ashburnham Rd B2
Ave Rd B1
Avenue Terr B1
Balmoral Rd B1
Baltic Ave B3
Baxter Ave A2/B2
Bircham Rd A2
Boscombe Rd B3
Boston Ave A1/B2
Bournemouth Park Rd A3
Browning Ave A3
Bus Station C3
Byron Ave A3
Cambridge Rd C1/C2
Canewdon Rd B1
Carnarvon Rd A2
Central Ave A3
Chelmsford Ave A1
Chichester Rd B2
Church Rd C3
Civic Centre A2
Clarence Rd C2
Clarence St C2
Cliff Ave B1
Cliffs Pavilion C1
Clifftown Parade C2
Clifftown Rd C2
Colchester Rd A1
College Way B2
County Court B3
Cromer Rd B3
Crowborough Rd A2
Dryden Ave A3
East St A2
Elmer App B2
Elmer Ave B2
Gainsborough Dr A1
Gayton Rd A2
Glenhurst Rd A2
Gordon Pl B2
Gordon Rd B2
Grainger Rd A2
Greyhound Way A3
Guildford Rd B3
Hamlet Ct Rd B1
Hamlet Rd C1
Harcourt Ave A1
Hartington Rd C3
Hastings Rd B3
Herbert Gr C3
Heygate Ave C3
High St B2/C2
Information Ctr C3
Kenway A2
Kilworth Ave B3
Lancaster Gdns B3
Library B2
London Rd B1
Lucy Rd C3
MacDonald Ave A1
Magistrates Court A2
Maldon Rd A2
Marine Parade C3
Milton Rd B1
Milton St B2
Napier Ave B2
Never Never Land C2
North Ave A3
North Rd A1/B1
Odeon B2
Osborne Rd B1
Park Cres B1
Park Rd B1
Park St B1
Park Terr B1
Peter Pan's Playground C2
Pier Hill C3
Pleasant Rd C3
Police Station A2
Post Office B2/B3
Princes St B2
Queens Rd B2
Queensway B2/B3/C3
Rayleigh Ave A1
Redstock Rd A2
Rochford Ave A1
Royal Mews C2
Royal Terr C2
Royals Shopping Precinct, The C3
Ruskin Ave A3
St Ann's Rd B3
St Helen's Rd B1
St John's Rd B1
St Leonard's Rd C3
St Lukes Rd A3
St Vincent's Rd C1
Salisbury Ave A1/B1
Scratton Rd C2
Shakespeare Dr A1
Short St A2
South Ave A3
Southchurch Rd B3
South East Essex College B2
Southend Central B2
Southend Pier Railway C3
Southend United FC A1
Southend Victoria B2
Stadium Rd A2
Stanfield Rd A2
Stanley Rd C3
Sutton Rd A3/B3
Swanage Rd B3
Sweyne Ave A1
Swimming Pool B3
Sycamore Gr A3
Tennyson Ave A3
The Grove A3
Tickfield Ave A2
Tudor Rd A1
Tunbridge Rd A2
Tylers Ave B3
Tyrrel Dr B3
Vale Ave A2
Victoria Ave A2
Victoria Plaza Shopping Precinct B2
Warrior Sq B3
Wesley Rd C3
West Rd A1
West St A1
Westcliff Parade C1
Western Esplanade C1
Weston Rd C2
Whitegate Rd B3
Wilson Rd C1
Wimborne Rd B3
York Rd C3

Stirling 343

Abbey Rd A3
Abbotsford Pl A3
Abercromby Pl C1
Albert Halls B1
Albert Pl B1
Alexandra Pl A3
Allan Park C2
Ambulance Station A2
AMF Ten Pin Bowling B2
Argyll Ave A3
Argyll's Lodging B1
Back O' Hill Industrial Estate A1
Back O' Hill Rd A1
Baker St B2
Ballengeich Pass A1
Balmoral Pl B1
Barn Rd B1
Barnton St B2
Bow St B1
Bruce St A2
Burghmuir Industrial Estate C2
Burghmuir Rd A2/B2/C2
Bus Station B2
Cambuskenneth Bridge A3
Carlton C2
Castle Ct B1
Causewayhead Rd A2
Cemetery A1
Cemetery B1
Church of the Holy Rude B1
Clarendon Pl C1
Club House B1
Colquhoun St C3
Corn Exchange B2
Council Offices C2
Court B2
Cowane A2
Cowane St A2
Cowane's Hospital B1
Crawford Shopping Arcade B2
Crofthead Rd A1
Dean Cres A3
Douglas St A2
Drip Rd A1
Drummond La C1
Drummond Pl C1
Drummond Pl La C1
Dumbarton Rd C2
Eastern Access Rd B2
Edward Ave A3
Edward Rd A2
Forrest Rd A2
Fort A1
Forth Cres B2
Forth St A2
Gladstone Pl C1
Glebe Ave C1
Glebe Cres C1
Glendevon Dr A1
Golf Course C1
Goosecroft Rd B2
Gowanhill A1
Greenwood Ave B1
Harvey Wynd A1
Information Ctr A1/C2
Irvine Pl B2
James St A2
John St B1
Kerse Rd C3
King's Knot B1
King's Park C1
King's Park Rd C1
Laurencecroft Rd A2
Leisure Pool B2
Library B2
Linden Ave C2
Lovers Wk A2
Lower Back Walk B1
Lower Bridge St A2
Lower Castlehill A1
Mar Pl B1
Meadow Pl A3
Meadowforth Rd C3
Middlemuir Rd C3
Millar Pl A3
Morris Terr B2
Mote Hill A1
Murray Pl B2
Nelson Pl C2
Old Town Jail B1
Orchard House Hospital (No A+E) A2
Park Terr C1
Phoenix Industrial Estate C3
Players Rd C3
Port St C2
Princes St B2
Queen St B2
Queen's Rd B1
Queenshaugh Dr A3
Rainbow Slides B2
Ramsay Pl A2
Riverside Dr A3
Ronald Pl A2
Rosebery Pl A2
Royal Gardens B1
Royal Gdns B1
St Mary's Wynd B1
St Ninian's Rd C2
Scott St A2
Seaforth Pl B2
Shore Rd B2
Smith Art Gallery & Museum B1
Snowdon Pl C1
Snowdon Pl La C1
Spittal St B2
Springkerse Industrial Estate C3
Springkerse Rd C3
Stirling Business Centre C2
Stirling Castle A1
Stirling County Rugby Football Club A3
Stirling Enterprise Park B3
Stirling Old Bridge A2
Stirling Station B2
Superstore A2
Sutherland Ave A3
TA Centre C3
Tannery La A2
Thistle Industrial Estate C3
Thistles Shopping Centre, The B2
Tolbooth, The B1
Town Wall B1
Union St A2
Upper Back Walk B1
Upper Bridge St A1
Upper Castlehill B1
Upper Craigs C2
Victoria Pl C1
Victoria Rd B1
Victoria Sq B1/C1
Vue B2
Wallace St A2
Waverley Cres A3
Wellgreen Rd C2
Windsor Pl C1
YHA B1

Stoke-on-Trent 343

Ashford St A3
Avenue Rd A3
Aynsley Rd A2
Barnfield C1
Bath St C2
Beresford St A3
Bilton St C2
Boon Ave C2
Booth St C2
Boothen Rd C2/C3
Boughey St C2
Boughley Rd B3
Brighton St B1
Campbell Rd C2
Carlton Rd B3
Cauldon Rd A3
Cemetery A2
Cemetery Rd A2
Chamberlain Ave C1
Church (RC) B2
Church St C2
City Rd C3
Civic Centre & King's Hall B3
Cliff Vale Pk A1
College Rd A3
Convent Cl B2
Copeland St B2
Cornwallis St C3
Corporation St C2
Crowther St A3
Dominic St B1
Elenora St B2
Elgin St B2
Epworth St A3
Etruscan St A1
Film Theatre B3
Fleming Rd C2
Fletcher Rd C2
Floyd St B2
Foden St C2
Frank St C2
Franklin Rd C1
Frederick Ave B1
Garden St C1
Garner St A1
Gerrard St B2
Glebe St B3
Greatbach Ave B1
Hanley Park A3
Harris St B1
Hartshill Rd B1
Hayward St A2
Hide St B2
Higson Ave B2
Hill St B2
Honeywall B2
Hunters Dr C1
Hunters Way C1
Keary St C2
Kingsway B2
Leek Rd B3
Library C2
Lime St C2
Liverpool Rd B2
London Rd C2
Lonsdale St C2
Lovatt St B2
Lytton St B3
Market C2
Newcastle La C1
Newlands St A2
Norfolk St A2
North St A1/B2
North Staffordshire Royal Infirmary (A&E) B1
Northcote Ave B2
Oldmill St C3
Oriel St B1
Oxford St B1
Penkhull New Rd C1
Penkhull St C1
Police Station C2
Portmeirion Pottery C2
Post Office A3/B1/B3/C1/C2
Prince's Rd B1
Pump St B2
Quarry Ave B1
Quarry Rd B1
Queen Anne St A3
Queen's Rd C1
Queensway A1/B2/C3
Richmond St B1
Rothwell St C1
St Peter's B3
St Thomas Pl C1
Scrivenor Rd A1
Seaford St A3
Selwyn St C3
Shelton New Rd A1
Shelton Old Rd B2
Sheppard St C2
Spark St C2
Spencer Rd B3
Spode St C2
Squires View B3
Staffordshire Univ B3
Stanley Matthews Sports Centre B3
Station Rd B3
Stoke Business Park C3
Stoke Recreation Centre C2
Stoke Rd A2
Stoke-on-Trent College A3
Stoke-on-Trent Station B3
Sturgess St C2
The Villas C1
Thistley Hough C1
Thornton Rd B3
Tolkien Way B1
Trent Valley Rd C1
Vale St B2
Watford St A3
Wellesley St A3
West Ave B1
Westland St B1
Yeaman St C2
Yoxall Ave B1

Stratford-upon-Avon 343

Albany Rd B1
Alcester Rd B1
Ambulance Station B1
Arden St B2
Avenue Farm A1
Avenue Farm Industrial Estate A1
Avenue Rd A3
Avon Industrial Estate A2
Baker Ave A1
Bandstand C3
Benson Rd A3
Birmingham Rd A2
Boat Club B3
Borden Pl C1
Brass Rubbing Centre C2
Bridge St B2
Bridgetown Rd C3
Bridgeway B3
Broad St C2
Broad Walk C2
Brookvale Rd C1
Bull St C2
Bus Station B2
Butterfly Farm & Jungle Safari C3
Cemetery C1
Chapel La B2
Cherry Orchard C1
Chestnut Walk B2
Children's Playground C3
Church St C2
Civic Hall B2
Clarence Rd B1
Clopton Bridge B3
Clopton Rd A2
Coach Terminal and Park B3
College B1
College La C2
College St C2
Community Sports Centre B1
Council Offices (District) B2
Courtyard C2
Cox's Yard B3
Cricket Ground C3
Ely Gdns B2
Ely St B2
Evesham Rd C1
Fire Station B1
Foot Ferry C3
Fordham Ave A2
Gallery, The B3
Garrick Way C1
Gower Memorial B3
Great William St B2
Greenhill St B2
Grove Rd B2
Guild St B2
Guildhall & School B2
Hall's Croft C2
Hartford Rd C1
Harvard House B2
Henley St B2
High St B2
Holton St C2
Holy Trinity C2
Information Ctr B3
Jolyffe Park Rd A2
Judith Shakespeare's House B2
Kipling Rd C3
Leisure & Visitor Centre B3
Library B2
Lodge Rd B1
Maidenhead Rd A3
Mansell St B2
Masons Court B2
Masons Rd A1
Maybird Retail Park A2
Maybrook Rd A1
Mayfield Ave A2
Meer St B2
Mill La C2
Moat House Hotel B3
Narrow La C2
New Place & Nash's House B2
New St C2
Old Town C2
Orchard Way C1
Paddock La C1
Park Rd A1
Payton St B2
Percy St A2
Police Station B2
Post Office B2/B3
Recreation Ground C2
Regal Road A2
Regal Road Trading Estate A2
Rother St B2
Rowley Cr A3
Ryland St C2
Saffron Meadow C2
St Andrew's Cr B1
St Gregory's A3
St Gregory's Rd A3
St Mary's Rd A2
Sanctus Dr C2
Sanctus St C1
Sandfield Rd C2
Scholars La B2
Seven Meadows Rd C2
Shakespeare Centre B2
Shakespeare Institute C2
Shakespeare St B2
Shakespeare's Birthplace B2
Sheep St B2
Shelley Rd C3
Shipston Rd C3
Shottery Rd C1
Slingates Rd A2
Southern La C2
Station Rd B1
Stratford Healthcare B2
Stratford Hospital B2
Stratford Sports Club B1
Stratford-upon-Avon Station B1
Talbot Rd A2
The Greenway C2
The Willows B1
The Willows North B1
Tiddington Rd B3
Timothy's Bridge Rd A1
Town Hall & Council Offices B2
Town Sq B2
Tramway Bridge B3
Trinity St C2
Tyler St B2
War Memorial Gdns B3
Warwick Rd B3
Waterside B3

Welcombe Rd A3
West St C2
Western Rd A2
Wharf Rd A2
Wood St B2

Sunderland 343

Albion Pl C2
Alliance Pl B1
Argyle St C2
Ashwood St C1
Athenaeum St B2
Azalea Terr C2
Beach St A1
Bede Theatre C3
Bedford St B2
Beechwood Terr C1
Belvedere Rd C2
Blandford St B2
Borough Rd B3
Bridge Cr B2
Bridge St B2
Brooke St A2
Brougham St B2
Burdon Rd C2
Burn Park C1
Burn Park Rd C1
Burn Park Tech Park C1
Carol St B1
Charles St A3
Chester Rd C1
Chester Terr B1
Church St A3
Cineworld B2
Civic Centre C2
Cork St B3
Coronation St B3
Cowan Terr C2
Crowtree Rd B2
Dame Dorothy St A2
Deptford Rd B1
Deptford Terr A1
Derby St C2
Derwent St C2
Dock St A3
Dundas St A2
Durham Rd C1
Easington St A2
Egerton St C3
Empire Theatre B2
Farringdon Row B1
Fawcett St B2
Fox St C1
Foyle St B3
Frederick St B3
Gill Rd B2
Hanover Pl A1
Havelock Terr C1
Hay St A2
Headworth Sq B3
Hendon Rd B3
High St East B3
High St West B2/B3
Holmeside B2
Hylton Rd B1
Information Ctr B2
John St B3
Kier Hardie Way A2
Lambton St B3
Laura St C3
Lawrence St B3
Leisure Centre B2
Library & Arts Centre B3
Lily St B1
Lime St B1
Livingstone Rd B2
Low Row B2
Matamba Terr B1
Millburn St B1
Millennium Way A2
Minster B2
Monkwearmouth Station Museum A2
Mowbray Park C3
Mowbray Rd C3
Murton St C3
Museum B3
National Glass Centre A3
New Durham Rd C1
Newcastle Rd A2
Nile St B3
Norfolk St B3
North Bridge St A2
Otto Terr C1
Park La C2
Park Lane Metro Sta C2
Park Rd C2
Paul's Rd B3
Peel St C3
Police Station B2
Post Office B2
Priestly Cr A1
Queen St B2
Railway Row B1
Retail Park B1
Richmond St A2
Roker Ave A2
Royalty Theatre C1
Ryhope Rd C2
St Mary's Way B2
St Michael's Way B2
St Peter's A3
St Peter's Metro Sta A2
St Peter's Way A3
St Vincent St C3
Salem Rd C3
Salem St C3
Salisbury St C3
Sans St B3
Silkworth Row B1
Southwick Rd A2
Stadium of Light (Sunderland AFC) A2
Stadium Way A2
Stobart St A2
Stockton Rd C2
Suffolk St C3
Sunderland Metro Sta B2
Sunderland Station B2
Sunderland St B3
Tatham St C3
Tavistock Pl B3
The Bridges B2
The Place B3
The Royalty C1
Thelma St C1
Thomas St North A2
Thornholme Rd C1
Toward Rd C3
Transport Interchange C2
Trimdon St Way B1
Tunstall Rd C2
University Metro Sta C1
University Library C1
University of Sunderland (City Campus) B1
University of Sunderland (Sir Tom Cowie Campus) A3
Vaux Brewery Way A2
Villiers St B3
Villiers St South B3
Vine Pl C2
Violet St B1
Walton La B3
Waterworks Rd B1
Wearmouth Bridge B2
Wellington La A1
West Sunniside B3
West Wear St B3
Westbourne Rd B1
Western Hill C1
Wharncliffe B1
Whickham St A3
White House Rd C3
Wilson St North A2
Winter Gdns C3
Wreath Quay A1

Swansea
Abertawe 343

Adelaide St C3
Albert Row C3
Alexandra Rd B3
Argyle St C1
Baptist Well Pl A2
Beach St C1
Belle Vue Way B3
Berw Rd A1
Berwick Terr A2
Bond St C1
Brangwyn Concert Hall C1
Bridge St A3
Brookands Terr B1
Brunswick St C1
Bryn-Syfi Terr A2
Bryn-y-Mor Rd C1
Bullins La B1
Burrows Rd C1
Bus/Rail link A3
Bus Station C2
Cadfan Rd A1
Cadrawd Rd A1
Caer St B3
Carig Cr A1
Carlton Terr B2
Carmarthen Rd A3
Castle Square B3
Castle St B3
Catherine St C1
City & County of Swansea Offices (County Hall) C2
City & County of Swansea Offices (Guildhall) C1
Clarence St C2
Colbourne Terr A2
Constitution Hill B1
Court B3
Creidiol Rd A2
Cromwell St B2
Duke St B1
Dunvant Pl C2
Dyfatty Park A3
Dyfatty St A3
Dyfed Ave A1
Dylan Thomas Ctr B3
Dylan Thomas Theatre C3
Eaton Cr C1
Eigen Cr A1
Elfed Rd A1
Emlyn Rd A1
Evans Terr A3
Fairfield Terr B1
Ffynone Dr B1
Ffynone Rd B1
Fire Station B3
Firm St A2
Fleet St C1
Francis St C1
Fullers Row B2
George St B2
Glamorgan St C2
Glyndŵr Pl A1
Glynn Vivian B3
Graig Terr A3
Grand Theatre C2
Granogwen Rd A2
Guildhall Rd South C1
Gwent Rd A1
Gwynedd Ave A1
Hafod St A3
Hanover St B1
Harcourt St B2
Harries St A2
Heathfield B2
Henrietta St B1
Hewson St B2
High St A3/B3
High View A2
Hill St A2
Historic Ships Berth C3
HM Prison C2
Information Ctr C2
Islwyn Rd A1
King Edward's Rd C1
Law Courts C1
Library B3
Long Ridge A2
Madoc St C2
Mansel St B2
Maritime Quarter C3
Market B3
Mayhill Gdns B1
Mayhill Rd A1
Mega Bowl B3
Milton Terr A2
Mission Gallery C3
Montpellier Terr B1
Morfa Rd A3
Mount Pleasant B2
National Waterfront Museum C3
Nelson St C2
New Cut Rd A3
New St A3
Nicander Pde A2
Nicander Pl A2
Nicholl St B2
Norfolk St B2
North Hill Rd A2
Northampton La B2
Orchard St B3
Oxford St B2
Oystermouth Rd C1
Page St B2
Pant-y-Celyn Rd B1
Parc Tawe Link B3
Parc Tawe North B3
Parc Tawe Shopping & Leisure Centre B3
Patti Pavilion C1
Paxton St C2
Penmaen Terr B1
Pen-y-Graig Rd A1
Phillips Pde C1
Picton Terr B2
Plantasia B3
Police Station B2
Post Office A1/A2/B2/C1
Powys Ave A1
Primrose St B2
Princess Way B3
Promenade B2
Pryder Gdns A1
Quadrant Centre C2
Quay Park B3
Rhianfa La B1
Rhondda St B2
Richardson St C2
Rodney St C1
Rose Hill B1
Rosehill Terr B1
Russell St B1
St David's Sq C3
St Helen's Ave C1
St Helen's Cr C1
St Helen's Rd C1
St James Gdns B1
St James's Cr B1
St Mary's B3
Sea View Terr A3
Singleton St C2
South Dock C3
Stanley Pl B2
Strand B3
Swansea Castle B3
Swansea College Arts Centre C1
Swansea Metropolitan University B2
Swansea Museum C3
Swansea Station A3
Taliesyn Rd B1
Tan y Marian Rd A1
Tegid Rd A2
Teilo Cr A1
Terrace Rd B1/B2
The Kingsway B2
The LC C3
Tontine St A3
Tower of Eclipse C3
Townhill Rd A1
Tram Museum C3
Trawler Rd C3
Union St B2
Upper Strand A3
Vernon St A3
Victoria Quay C3
Victoria Rd B3
Vincent St C1
Walter Rd B1
Watkin St A2
Waun-Wen Rd A2
Wellington St C2
Westbury St C1
Western St C1
Westway C2
William St C2
Wind St B3
Woodlands Terr B1
YMCA B2
York St C3

Swindon 343

Albert St C3
Albion St C1
Alfred St A2
Alvescot Rd C3
Art Gallery & Museum C3
Ashford Rd C1
Aylesbury St A2
Bath Rd C2
Bathampton St B1
Bathurst Rd B3
Beatrice St A2
Beckhampton St B3
Bowood Rd C1
Bristol St B1
Broad St A3
Brunel Arcade B2
Brunel Plaza B2
Brunswick St C2
Bus Station B2
Cambria Bridge Rd B1
Cambria Place B1
Canal Walk B2
Carfax St B2
Carr St B1
Cemetery C1/C3
Chandler Cl C3
Chapel C1
Chester St B1
Christ Church C3
Church Place B1
Cirencester Way A3
Clarence St B2
Clifton St C1
Cockleberry Rdbt A2
Colbourne Rdbt A3
Colbourne St A3
College St B2
Commercial Rd B2
Corporation St A2
Council Offices B3
County Rd A3
Courts B2
Cricket Ground A3
Cricklade Street C3
Crombey St B1/C2
Cross St C2
Curtis St B1
Deacon St C1
Designer Outlet (Great Western) B1
Dixon St C2
Dover St C2
Dowling St C2
Drove Rd C3
Dryden St C1
Durham St C3
East St B1
Eastcott Hill C2
Eastcott Rd C2
Edgeware Rd B2
Edmund St C2
Elmina Rd A3
Emlyn Square B1
Euclid St B3
Exeter St B1
Fairview C1
Faringdon Rd B1
Farnsby St B1
Fire Station B3
Fleet St B2
Fleming Way B2/B3
Florence St A2
Gladstone St A3
Gooch St A2
Graham St A3
Great Western Way A1/A2
Groundwell Rd B3
Hawksworth Way A1
Haydon St A2
Henry St B2
Hillside Ave C1
Holbrook Way B2
Hunt St C3
Hydro B1
Hythe Rd C2
Information Ctr B2
Joseph St C1
Kent Rd C2
King William St C2
Kingshill Rd C1
Lansdown Rd C2
Leicester St B3
Library B2
Lincoln St B3
Little London C3
London St B1
Magic Rdbt B3
Maidstone Rd C2
Manchester Rd A3
Maxwell St B1
Milford St B2
Milton Rd B1
Morse St C2
National Monuments Record Centre B1
Newcastle St B3
Newcombe Drive A1
Newcombe Trading Estate A1
Newhall St C2
North St C2
North Star Ave A1
North Star Rdbt A1
Northampton St B3
Oasis Leisure Centre A1
Ocotal Way A3
Okus Rd C1
Old Town C3
Oxford St B1
Park Lane B1
Park Lane Rdbt B1
Pembroke St C2
Plymouth St B3
Polaris House A2
Polaris Way A2
Police Station B2
Ponting St A2
Post Office B1/B2/C1/C3
Poulton St A3
Princes St B2
Prospect Hill C2
Prospect Place C2
Queen St B2
Queen's Park C3
Radnor St C1
Read St C1
Reading St B1
Regent St B2
Retail Park A2/A3/B3
Rosebery St A3
St Mark's B1
Salisbury St A3
Savernake St C2
Shelley St C1
Sheppard St B1
South St C2
Southampton St B3
Spring Gardens B3
Stafford Street C2
Stanier St C2
Station Road A2
Steam B1
Swindon College A2
Swindon Rd C2
Swindon Station A2
Swindon Town Football Club A3
TA Centre B1
Tennyson St B1
The Lawn C3
The Nurseries C1
The Parade B2
The Park B1
Theobald St B1
Town Hall B2
Transfer Bridges Rdbt A3
Union St C2
Upham Rd C3
Victoria Rd C3
Walcot Rd B3
War Memorial B2
Wells St B3
Western St C2
Westmorland Rd B3
Whalebridge Rdbt B2
Whitehead St C1
Whitehouse Rd A2
William St C1
Wood St C3
Wyvern Theatre & Arts Centre B2
York Rd B3

Taunton 343

Addison Gr A1
Albemarle Rd A1
Alfred St B3
Alma St C2
Bath Pl C1
Belvedere Rd A1
Billet St B2
Billetfield C2
Birch Gr A1
Brewhouse Theatre B2
Bridge St B1
Bridgwater & Taunton Canal A2
Broadlands Rd C1
Burton Pl C1
Bus Station B1
Canal Rd A2
Cann St C1
Canon St B2
Castle B1
Castle St B1
Cheddon Rd A2
Chip Lane A1
Clarence St B1
Cleveland St B1
Clifton Terr A2
Coleridge Cres C3
Compass Hill C1
Compton Cl A2
Corporation St B1
Council Offices A1
County Walk Shopping Centre C2
Courtyard B2
Cranmer Rd B2
Critchard Way B3
Cyril St A1
Deller's Wharf B1
Duke St B2
East Reach B3
East St B3
Eastbourne Rd B2
Eastleigh Rd C3
Eaton Cres A2
Elm Gr A1
Elms Cl A1
Fons George C1
Fore St B2
Fowler St A1
French Weir Recreation Grd B1
Geoffrey Farrant Wk A2
Gray's Almshouses B2
Grays Rd B3
Greenway Ave A1
Guildford Pl C1
Hammet St B2
Haydon Rd B3
Heavitree Way A2
Herbert St A1
High St C2
Holway Ave C3
Hugo St B3
Huish's Almshouses B2
Hurdle Way C2
Information Ctr C2
Jubilee St A1
King's College C3
Kings Cl C3
Laburnum St B2
Lambrook Rd B3
Lansdowne Rd A3
Leslie Ave A1
Leycroft Rd B3
Library C2
Linden Gr A1
Magdalene St B2
Magistrates Court B1
Malvern Terr A2
Market House B2
Mary St C2
Middle St B2
Midford Rd B2
Mitre Court B3
Mount Nebo C1
Mount St C2
Mountway C2
Museum of Somerset B1
North St B2
Northfield Ave B1
Northfield Rd B1
Northleigh Rd C3
Obridge Allotments A3
Obridge Lane A3
Obridge Rd A3
Obridge Viaduct A3
Old Market Shopping Centre C2
Osborne Way C1
Park St C1
Paul St C2
Plais St A2
Playing Field C3
Police Station C1
Portland St B1
Post Office B1/B2/C1
Priorswood Industrial Estate A3
Priorswood Rd A2
Priory Ave B2
Priory Bridge Rd B2
Priory Fields Retail Park A3
Priory Park A2
Priory Way A3
Queen St B3
Railway St A1
Records Office A2
Recreation Grd A1
Riverside Place B2
St Augustine St B2
St George's C2
St Georges Sq C2
St James B2
St James St B2
St John's C1
St John's Rd B1
St Josephs Field C2
St Mary Magdalene's B2
Samuels Ct A1
Shire Hall & Law Courts C1
Somerset County Cricket Ground B2
Somerset County Hall C1
Somerset Cricket B2
South Rd C3
South St C3
Staplegrove Rd B1
Station Rd A1
Stephen St B2
Swimming Pool A1
Tancred St B2
Tauntfield Cl C3
Taunton Dean Cricket Club C2
Taunton Station A2
The Avenue A1
The Crescent C1
The Mount C2
Thomas St A1
Toneway A3
Tower St B1
Trevor Smith Pl C3
Trinity Business Centre C3
Trinity Rd C3
Trinity St B3
Trull Rd C1
Tudor House B2
Upper High St C1
Venture Way A3
Victoria Gate B3
Victoria Park B3
Victoria St B3
Viney St B3
Vivary Park C2
Vivary Rd C1
War Memorial C1
Wellesley St A2
Wheatley Cres A3
Whitehall A1
Wilfred Rd B3
William St A1
Wilton Church C1
Wilton Cl C1
Wilton Gr C1
Wilton St C1
Winchester St B2
Winters Field B2
Wood St B1
Yarde Pl B1

Telford 343

Alma Ave C1
Amphitheatre C2
Bowling Alley B2
Brandsfarm Way C3
Brunel Rd B1
Bus Station B2
Buxton Rd C1
Central Park A2
Civic Offices B2
Coach Central B2
Coachwell Cl B1
Colliers Way A1
Courts B2
Dale Acre Way B3
Darliston C3
Deepdale B3
Deercote B2
Dinthill C3
Doddington C3
Dodmoor Grange C3
Downemead B3
Duffryn B3
Dunsheath B3
Euston Way A3
Eyton Mound C1
Eyton Rd C1
Forgegate A2
Grange Central B2
Hall Park Way B1
Hinkshay Rd C2
Hollinsworth Rd A2
Holyhead Rd A3
Housing Trust A1
Ice Rink B2
Information Ctr B2
Ironmasters Way A2
Job Centre B1
Land Registry B1
Lawn Central B2
Lawnswood C1
Library B2
Malinsgate B1
Matlock Ave C1
Moor Rd C1
Mount Rd C1
NFU Offices B1
Odeon B2
Park Lane A1
Police Station B2
Priorslee Ave A3
Queen Elizabeth Ave C3
Queen Elizabeth Way B1
Queensway A2/B3
Rampart Way A2
Randlay Ave C3
Randlay Wood C3
Rhodes Ave C1
Royal Way B1
St Leonards Rd B1
St Quentin Gate B2
Shifnal Rd A3
Sixth Ave A1
Southwater Way B1
Spout Lane C1
Spout Mound B1
Spout Way C1
Stafford Court B3
Stafford Park B3
Stirchley Ave C3
Stone Row C1
Telford Bridge Retail Park A1
Telford Central Station A3
Telford Centre, The B2
Telford Forge Retail Park A1
Telford Hornets RFC C2
Telford International Centre C2
Telford Way A3
Third Ave A2
Town Park C2
Town Park Visitor Centre B2
Walker House B2
Wellswood Ave A2
West Centre Way B1
Withywood Drive C1
Woodhouse Central B2
Yates Way A1

Torquay 344

Abbey Rd B2
Alexandra Rd A2
Alpine Rd B3
Ash Hill Rd A2
Babbacombe Rd B3
Bampfylde Rd B1
Barton Rd A1
Beacon Quay C2
Belgrave Rd A1/B1
Belmont Rd A3
Berea Rd A3
Braddons Hill Rd East B3
Brewery Park A3
Bronshill Rd A2
Castle Rd A2
Cavern Rd A3
Central B2
Chatsworth Rd A2
Chestnut Ave B1
Church St A1
Civic Offices A2
Coach Station A1
Corbyn Head C1
Croft Hill B1
Croft Rd B1
Daddyhole Plain C3
East St A1
Egerton Rd A3
Ellacombe Church Rd A3
Ellacombe Rd A2
Falkland Rd B1
Fleet St B2
Fleet Walk Shopping Centre B2
Grafton Rd B3
Haldon Pier C2
Hatfield Rd A2
Highbury Rd A2
Higher Warberry Rd A3
Hillesdon Rd B3
Hollywood Bowl C3
Hoxton Rd A3
Hunsdon Rd B3
Information Ctr B2
Inner Harbour C3
Kenwyn Rd A3
Laburnum St A1
Law Courts A2
Library A2
Lime Ave B1
Living Coasts C3
Lower Warberry Rd B3
Lucius St B1
Lymington Rd A1
Magdalene Rd A1
Marina C2
Market St B2
Meadfoot Lane C3
Meadfoot Rd C3
Melville St B2
Middle Warberry Rd B3
Mill Lane A1
Montpellier Rd B3
Morgan Ave A1
Museum Rd B3
Newton Rd A1
Oakhill Rd A1
Outer Harbour C2
Parkhill Rd C3
Pavilion C2
Pimlico B2
Police Station A1
Post Office A1/B2
Princes Rd A3
Princes Rd East A3
Princes Rd West A3
Princess Gdns C2
Princess Pier C2
Princess Theatre C2
Rathmore Rd B1
Recreation Grd B1
Riviera Centre International B1
Rock End Ave C3
Rock Rd B2
Rock Walk B2
Rosehill Rd A3
St Efride's Rd A1
St John's B3
St Luke's Rd B2
St Luke's Rd North B2
St Luke's Rd South B2
St Marychurch Rd A2
Scarborough Rd B1
Shedden Hill B2
South Pier C2
South St A1
Spanish Barn B1
Stitchill Rd B3
Strand B3
Sutherland Rd B3
Teignmouth Rd A1
Temperance St B2
The King's Drive B1
The Terrace B3
Thurlow Rd A1
Tor Bay B1
Tor Church Rd A1
Tor Hill Rd A1
Torbay Rd B2
Torquay Museum B3
Torquay Station C1
Torre Abbey Mansion B1
Torre Abbey Meadows B1
Torre Abbey Sands B1
Torwood Gdns B3
Torwood St C3
Union Square A2
Union St A1
Upton Hill A2
Upton Park A1
Upton Rd A1
Vanehill Rd C3
Vansittart Rd A1
Vaughan Parade C2
Victoria Parade C3
Victoria Rd A2
Warberry Rd West B2
Warren Rd B2
Windsor Rd A2/A3
Woodville Rd A3

Truro 344

Adelaide Ter B1
Agar Rd B3
Arch Hill C2
Arundell Pl C2
Avondale Rd B1
Back Quay B3
Barrack La C3
Barton Meadow A1
Benson Rd A2
Bishops Cl A2
Bosvean Gdns B1
Bosvigo Gardens B1
Bosvigo La A1
Bosvigo Rd B2
Broad St A3
Burley Cl C3
Bus Station B3
Calenick St B2
Campfield Hill B3
Carclew St B3
Carew Rd A2
Carey Park B2
Carlyon Rd A2
Carvoza Rd A3
Castle St B2
Cathedral View A2
Chainwalk Dr A2
Chapel Hill B1
Charles St B2
City Hall B3
City Rd B2
Coinage Hall B3
Comprigney Hill A1
Coosebean La A1
Copes Gdns A2
County Hall B1
Courtney Rd A2
Crescent Rd B1
Crescent Rise B1
Daniell Court C2
Daniell Rd C2
Daniell St C2
Daubuz Cl A2
Dobbs La B1
Edward St B2
Eliot Rd A2
Elm Court A3
Enys Cl A1
Enys Rd A1
Fairmantle St B3
Falmouth Rd C2
Ferris Town B2
Fire Station B1
Frances St B2
George St B2
Green Cl C2
Green La C1
Grenville Rd A2
Hall For Cornwall B3
Hendra Rd A2
Hendra Vean A1
High Cross B3
Higher Newham La C3
Higher Trehaverne A2
Hillcrest Ave B1
Hospital B2
Hunkin Cl A2
Hurland Rd C3
Infirmary Hill B2
James Pl B3
Kenwyn Church Rd A1
Kenwyn Hill A1
Kenwyn Rd A2
Kenwyn St B2
Kerris Gdns A1
King St B3
Lemon Quay B3
Lemon Street Gallery B3
Library B1/B3
Malpas Rd B3
Market B3
Memorial Gdns B3
Merrifield Close B1
Mitchell Hill A3
Moresk Cl A3
Moresk Rd A3
Morlaix Ave C3
Nancemere Rd A3
Newham Business Park C3
Newham Industrial Estate C3
Newham Rd C3
Northfield Dr C3
Oak Way A3
Old County Hall B1
Pal's Terr A3
Park View C2
Pendarves Rd A2
Plaza Cinema B3
Police Station B3
Post Office B2/B3
Prince's St B3
Pydar St A2
Quay St B3
Redannick Cres C2
Redannick La B2
Richard Lander Monument C2
Richmond Hill B1
River St B2
Rosedale Rd A2
Royal Cornwall Museum B2
St Aubyn Rd C3
St Clement St B3
St George's Rd A1
School La C2
Station Rd B1
Stokes Rd A2
Strangways Terr C3
Tabernacle St B3
The Avenue A3
The Crescent B1
The Leats B2
The Spires A2
Trehaverne La A2
Tremayne Rd A2
Treseder's Gdns A3
Treworder Rd B1
Treyew Rd B1
Truro Cathedral B3
Truro Harbour Office B3
Truro Station B3
Union St B2
Upper School La C2
Victoria Gdns B2
Waterfall Gdns B2

Wick 344

Ackergill Cres A2
Ackergill St A2
Albert St C2
Ambulance Station A2
Argyle Sq C2
Assembly Rooms C2
Bank Row C2
Bankhead B1
Barons Well B2
Barrogill St C2
Bay View B3
Bexley Terr C3
Bignold Park C2
Bowling Green C2
Breadalbane Terr C2
Bridge of Wick B1
Bridge St B2
Brown Pl C2
Burn St B2
Bus Station B1
Caithness General Hospital (A+E) B1
Cliff Rd B1
Coach Rd B2
Coastguard Station C3
Corner Cres B3
Coronation St C1
Council Offices B2
Court B2
Crane Rock C3
Dempster St C2
Dunnet Ave A2
Fire Station B2
Fish Market C3
Francis St C1
George St A1
Girnigoe St B2
Glamis Rd B2
Gowrie Pl B1
Grant St C2
Green Rd B2
Gunns Terr B3
Harbour Quay B2
Harbour Rd C3
Harbour Terr C2
Harrow Hill C2
Henrietta St A2/B2
Heritage Centre C2
High St B2
Hill Ave A2
Hillhead Rd B3
Hood St C1

Huddart St . . . C2
Information Ctr . . . B2
Kenneth St . . . C1
Kinnaird St . . . C2
Kirk Hill . . . B1
Langwell Cres . . . B3
Leishman Ave . . . B3
Leith Walk . . . A2
Library . . . B2
Lifeboat Station . . . C3
Lighthouse . . . C3
Lindsay Dr . . . B3
Lindsay Pl . . . B3
Loch St . . . C2
Louisburgh St . . . B2
Lower Dunbar St . . . C2
Macleay La . . . B1
Macleod Rd . . . B3
MacRae St . . . C2
Martha Terr . . . B2
Miller Ave . . . B1
Miller La . . . B1
Moray St . . . C2
Mowat Pl . . . B3
Murchison St . . . C3
Newton Ave . . . C1
Newton Rd . . . C1
Nicolson St . . . C3
North Highland College . . . B2
North River Pier . . . B3
Northcote St . . . C2
Owen Pl . . . A2
Police Station . . . B1
Port Dunbar . . . B3
Post Office . . . B2/C2
Pulteney Distillery . . . C2
River St . . . B2
Robert St . . . A1
Rutherford St . . . C2
St John's Episcopal . . . C2
Sandigoe Rd . . . B3
Scalesburn . . . B3
Seaforth Ave . . . C1
Shore La . . . B2
Sinclair Dr . . . B3
Sinclair Terr . . . C2
Smith Terr . . . C3
South Pier . . . C3
South Quay . . . C3
South Rd . . . C1
South River Pier . . . B3
Station Rd . . . B1
Swimming Pool . . . B2
TA Centre . . . C2
Telford St . . . B2
The Shore . . . B2
Thurso Rd . . . B1
Thurso St . . . B1
Town Hall . . . B2
Union St . . . B2
Upper Dunbar St . . . C2
Vansittart St . . . C3
Victoria Pl . . . B2
War Memorial . . . A1
Well of Cairndhuna . . . C3
Wellington Ave . . . C3
Wellington St . . . C3
West Banks Ave . . . C1
West Banks Terr . . . C1
West Park . . . C1
Whitehorse Park . . . B2
Wick Harbour Bridge . . . B2
Wick Industrial Estate . . . A2
Wick Parish Church . . . B1
Wick Station . . . B1
Williamson St . . . B2
Willowbank . . . B2

Winchester 344

Andover Rd . . . A2
Andover Road Retail Park . . . A2
Archery La . . . C2
Arthur Rd . . . A2
Bar End Rd . . . C3
Beaufort Rd . . . C2
Beggar's La . . . B3
Bereweeke Ave . . . A1
Bereweeke Rd . . . A1
Boscobel Rd . . . A2
Brassey Rd . . . A2
Broadway . . . B3
Brooks Shopping Centre, The . . . B3
Bus Station . . . B3
Butter Cross . . . B2
Canon St . . . C2
Castle Wall . . . C2/C3
Castle, King Arthur's Round Table . . . B2
Cathedral . . . C2
Cheriton Rd . . . A1
Chesil St . . . C3
Chesil Theatre . . . C3
Christchurch Rd . . . C1
City Museum . . . B2
City Offices . . . C3
City Rd . . . B2
Clifton Rd . . . B1
Clifton Terr . . . B2
Close Wall . . . C2/C3
Coach Park . . . A2
Colebrook St . . . C3
College St . . . C2
College Walk . . . C3
Compton Rd . . . C2
County Council Offices . . . B2
Cranworth Rd . . . A2
Cromwell Rd . . . C1
Culver Rd . . . C2
Domum Rd . . . C3
Durngate Pl . . . B3
Eastgate St . . . B3
Edgar Rd . . . C2
Egbert Rd . . . A2
Elm Rd . . . B1
Fairfield Rd . . . A1
Fire Station . . . B3
Fordington Ave . . . B1
Fordington Rd . . . A1
Friarsgate . . . B3
Gordon Rd . . . B3
Greenhill Rd . . . B1
Guildhall . . . B3
HM Prison . . . B1
Hatherley Rd . . . A1
High St . . . B2
Hillier Way . . . A3
Hyde Abbey (Remains) . . . A2
Hyde Abbey Rd . . . B2
Hyde Cl . . . A2
Hyde St . . . A2
Information Ctr . . . B3
Jane Austen's House . . . C2
Jewry St . . . B2
John Stripe Theatre . . . C1
King Alfred Pl . . . A2
Kingsgate Arch . . . C2
Kingsgate Park . . . C2
Kingsgate Rd . . . C2
Kingsgate St . . . C2
Lankhills Rd . . . A2
Library . . . B2
Lower Brook St . . . B3
Magdalen Hill . . . B3
Market La . . . B2
Mews La . . . B1
Middle Brook St . . . B3
Middle Rd . . . B1
Military Museums . . . B2
Milland Rd . . . C3
Milverton Rd . . . B1
Monks Rd . . . A3
North Hill Cl . . . A2
North Walls . . . B2
North Walls Rec Gnd . . . A3
Nuns Rd . . . A3
Oram's Arbour . . . B1
Owen's Rd . . . A2
Parchment St . . . B2
Park & Ride . . . C3
Park Ave . . . B3
Playing Field . . . A1
Police H.Q. . . . B1
Police Station . . . B3
Portal Rd . . . C3
Post Office . . . B2/C1
Quarry Rd . . . C3
Ranelagh Rd . . . C1
Regiment Museum . . . B2
River Park Leisure Centre . . . B3
Romans' Rd . . . C2
Romsey Rd . . . B1
Royal Hampshire County Hospital (A & E) . . . B1
St Cross Rd . . . C2
St George's St . . . B2
St Giles Hill . . . C3
St James' La . . . B1
St James' Terr . . . B1
St James Villas . . . C2
St John's . . . B3
St John's St . . . B3
St Michael's Rd . . . C2
St Paul's Hill . . . B1
St Peter St . . . B2
St Swithun St . . . C2
St Thomas St . . . C2
Saxon Rd . . . A2
School of Art . . . B3
Screen . . . C2
Sleepers Hill Rd . . . C1
Southgate St . . . B2
Sparkford Rd . . . C1
Staple Gdns . . . B2
Station Rd . . . B2
Step Terr . . . B1
Stockbridge Rd . . . A1
Stuart Cres . . . C1
Sussex St . . . B2
Swan Lane . . . B2
Tanner St . . . B3
The Square . . . B2
The Weirs . . . C3
The Winchester Gallery . . . B3
Theatre Royal . . . B2
Tower St . . . B2
Town Hall . . . C3
Union St . . . B3
University of Winchester (King Alfred Campus) . . . C1
Upper Brook St . . . B2
Wales St . . . B3
Water Lane . . . B3
West End Terr . . . B1
Western Rd . . . B1
West Gate . . . B2
Wharf Hill . . . C3
Winchester College . . . C2
Winchester Station . . . A2
Wolvesey Castle . . . C3
Worthy Lane . . . A2
Worthy Rd . . . A2

Windsor 344

Adelaide Sq . . . C3
Albany Rd . . . C2
Albert St . . . B1
Alexandra Gdns . . . B2
Alexandra Rd . . . C2
Alma Rd . . . B2
Ambulance Station . . . B1
Arthur Rd . . . B2
Bachelors Acre . . . B3
Barry Ave . . . B2
Beaumont Rd . . . C2
Bexley St . . . B1
Boat House . . . B2
Brocas St . . . B2
Brook St . . . C3
Bulkeley Ave . . . C1
Castle Hill . . . B3
Charles St . . . B2
Claremont Rd . . . C2
Clarence Cr . . . B2
Clarence Rd . . . B1
Clewer Court Rd . . . B1
Coach Park . . . B2
College Cr . . . C1
Courts . . . C2
Cricket Ground . . . C3
Dagmar Rd . . . C2
Datchet Rd . . . B3
Devereux Rd . . . C2
Dorset Rd . . . C2
Duke St . . . B1
Elm Rd . . . C1
Eton College . . . A3
Eton Ct . . . A2
Eton Sq . . . A2
Eton Wick Rd . . . A2
Fire Station . . . C2
Farm Yard . . . B3
Frances Rd . . . C2
Frogmore Dr . . . B3
Gloucester Pl . . . C3
Goslar Way . . . C1
Goswell Hill . . . B2
Goswell Rd . . . B2
Green La . . . C1
Grove Rd . . . C2
Guildhall . . . B3
Helena Rd . . . C2
Helston La . . . B1
High St . . . A2/B3
Holy Trinity . . . C2
Hospital (Private) . . . C2
Household Cavalry . . . C2
Imperial Rd . . . C1
Information Ctr . . . B2/B3
Keats La . . . A2
King Edward Ct . . . B2
King Edward VII Ave . . . A3
King Edward VII Hospital . . . C2
King George V Memorial . . . B3
King's Rd . . . C3
King Stable St . . . A2
Library . . . C2
Maidenhead Rd . . . B1
Meadow La . . . A2
Municipal Offices . . . C3
Nell Gwynne's House . . . B3
Osborne Rd . . . C2
Oxford Rd . . . B1
Park St . . . B3
Peascod St . . . B2
Police Station . . . C2
Post Office . . . A2
Princess Margaret Hospital . . . C2
Queen Victoria's Walk . . . B3
Queen's Rd . . . C2
River St . . . B2
Romney Island . . . A3
Romney Lock . . . A3
Romney Lock Rd . . . A3
Russell St . . . C2
St John's . . . B3
St John's Chapel . . . A2
St Leonards Rd . . . C2
St Mark's Rd . . . C2
Sheet St . . . C3
South Meadow . . . A2
South Meadow La . . . A2
Springfield Rd . . . C1
Stovell Rd . . . B1
Sunbury Rd . . . A2
Tangier La . . . A2
Tangier St . . . A3
Temple Rd . . . C2
Thames St . . . B3
The Brocas . . . A2
The Home Park . . . A3/C3
The Long Walk . . . C3
Theatre Royal . . . B3
Trinity Pl . . . C2
Vansittart Rd . . . B1/C1
Vansittart Rd Gdns . . . C1
Victoria Barracks . . . C2
Victoria St . . . C2
Ward Royal . . . B2
Westmead . . . C1
White Lilies Island . . . A1
William St . . . B2
Windsor Arts Centre . . . C2
Windsor Castle . . . B3
Windsor & Eton Central . . . B2
Windsor & Eton Riverside . . . A3
Windsor Bridge . . . B3
Windsor Great Park . . . C3
Windsor Leisure Centre . . . B1
Windsor Relief Rd . . . A1
Windsor Royal Shopping . . . B2
York Ave . . . C1
York Rd . . . C1

Wolverhampton 344

Albion St . . . B3
Alexandra St . . . C1
Arena . . . B3
Art Gallery . . . B2
Ashland St . . . C1
Austin St . . . A1
Badger Dr . . . A3
Bailey St . . . B3
Bath Ave . . . B1
Bath Rd . . . B1
Bell St . . . C2
Berry St . . . B3
Bilston Rd . . . C3
Bilston St . . . C3
Birmingham Canal . . . A3
Bone Mill La . . . A2
Brewery Rd . . . B1
Bright St . . . A1
Burton Cres . . . B3
Bus Station . . . B3
Cambridge St . . . A3
Camp St . . . B2
Cannock Rd . . . A3
Castle St . . . C2
Chapel Ash . . . C1
Cherry St . . . C1
Chester St . . . A1
Church La . . . C2
Church St . . . C2
Civic Centre . . . B2
Clarence Rd . . . B2
Cleveland St . . . C2
Clifton St . . . C1
Coach Station . . . B2
Compton Rd . . . B1
Corn Hill . . . B3
Coven St . . . A3
Craddock St . . . A1
Cross St North . . . A2
Crown & County Courts . . . C3
Crown St . . . A2
Culwell St . . . B3
Dale St . . . C1
Darlington St . . . C1
Dartmouth St . . . C3
Devon Rd . . . A1
Drummond St . . . B2
Dudley Rd . . . C2
Dudley St . . . B2
Duke St . . . C3
Dunkley St . . . B2
Dunstall Ave . . . A2
Dunstall Hill . . . A2
Dunstall Rd . . . A1/A2
Evans St . . . A1
Fawdry St . . . A1
Field St . . . B3
Fire Station . . . C1
Fiveways rbt . . . A2
Fowler Playing Fields . . . A3
Fox's La . . . A2
Francis St . . . A2
Fryer St . . . B3
Gloucester St . . . A1
Gordon St . . . C3
Graiseley St . . . C1
Grand . . . B3
Granville St . . . C3
Great Brickkiln St . . . C1
Great Hampton St . . . A1
Great Western St . . . A2
Grimstone St . . . B3
Harrow St . . . A1
Hilton St . . . A3
Horseley Fields . . . C3
Humber Rd . . . C1
Jack Hayward Way . . . A2
Jameson St . . . A1
Jenner St . . . C3
Kennedy Rd . . . B3
Kimberley St . . . C1
King St . . . B2
Laburnum St . . . C1
Lansdowne Rd . . . B1
Leicester St . . . A1
Lever St . . . C3
Library . . . C3
Lichfield St . . . B2
Light House . . . B3
Little's La . . . B3
Lock St . . . B3
Lord St . . . C1
Lowe St . . . A1
Lower Stafford St . . . A2
Magistrates Court . . . B2
Mander Centre . . . C2
Mander St . . . C1
Market St . . . B2
Market . . . C2
Melbourne St . . . C3
Merridale St . . . C1
Middlecross . . . C3
Molineux St . . . B2
Mostyn St . . . A1
New Hampton Rd East . . . A1
Nine Elms La . . . A3
North Rd . . . A2
Oaks Cres . . . C1
Oxley St . . . A2
Paget St . . . A1
Park Ave . . . B1
Park Road East . . . A1
Park Road West . . . B1
Paul St . . . C2
Pelham St . . . C1
Penn Rd . . . C2
Piper's Row . . . B3
Pitt St . . . C2
Police Station . . . C3
Pool St . . . C2
Poole St . . . A3
Post Office . . . A1/A2/B2/B2
Powlett St . . . C3
Queen St . . . B2
Raby St . . . C3
Raglan St . . . C1
Railway Dr . . . B3
Red Hill St . . . A2
Red Lion St . . . B2
Retreat St . . . C1
Ring Rd . . . B2
Rugby St . . . A1
Russell St . . . C1
St Andrew's . . . B1
St David's . . . B3
St George's . . . C3
St George's Pde . . . C2
St James St . . . C3
St John's . . . C2
St John's . . . C2
St John's Retail Park . . . C2
St John's Square . . . C2
St Mark's . . . C1
St Marks Rd . . . C1
St Marks St . . . C1
St Patrick's . . . B2
St Peter's . . . B2
St Peter's . . . B2
Salisbury St . . . C1
Salop St . . . C2
School St . . . C2
Sherwood St . . . A2
Smestow St . . . A3
Snowhill . . . C2
Springfield Rd . . . A3
Stafford St . . . B2
Staveley Rd . . . A1
Steelhouse La . . . C3
Stephenson St . . . C1
Stewart St . . . C2
Sun St . . . B3
Sutherland Pl . . . C3
Tempest St . . . C2
Temple St . . . C2
Tettenhall Rd . . . B1
The Maltings . . . B2
The Royal Metro Sta . . . C3
Thomas St . . . C2
Thornley St . . . B2
Tower St . . . C2
Town Hall . . . B2
University . . . B2
Upper Zoar St . . . C1
Vicarage Rd . . . C3
Victoria St . . . C2
Walpole St . . . A1
Walsall St . . . C3
Ward St . . . C3
Warwick St . . . C3
Water St . . . A3
Waterloo Rd . . . B2
Wednesfield Rd . . . B3
West Park (not A&E) . . . B1
West Park Swimming Pool . . . B1
Wharf St . . . C3
Whitmore Hill . . . B2
Wolverhampton . . . B3
Wolverhampton St George's Metro Sta . . . C2
Wolverhampton Wanderers Football Gnd. (Molineux) . . . B2
Worcester St . . . C2
Wulfrun Centre . . . C2
Yarwell Cl . . . A3
York St . . . C3
Zoar St . . . C1

Worcester 344

Albany Terr . . . A1
Alice Otley School . . . A2
Angel Pl . . . B2
Angel St . . . B2
Ashcroft Rd . . . A2
Athelstan Rd . . . C3
Back Lane North . . . A1
Back Lane South . . . A1
Barbourne Rd . . . A2
Bath Rd . . . C2
Battenhall Rd . . . C3
Bridge St . . . B2
Britannia Sq . . . A1
Broad St . . . B2
Bromwich La . . . C1
Bromwich Rd . . . C1
Bromyard Rd . . . C1
Bus Station . . . B2
Carden St . . . B3
Castle St . . . A2
Cathedral . . . C2
Cathedral Plaza . . . B2
Charles St . . . B3
Chequers La . . . B1
Chestnut St . . . A2
Chestnut Walk . . . A2
Citizens' Advice Bureau . . . B2
City Walls Rd . . . B2
Cole Hill . . . C3
College of Technology . . . B2
College St . . . C2
Commandery . . . C3
County Cricket Ground . . . C1
Cripplegate Park . . . B1
Croft Rd . . . B1
Cromwell St . . . B3
Crowngate Centre . . . B2
Deansway . . . B2
Diglis Pde . . . C2
Diglis Rd . . . C2
Edgar Tower . . . C2
Farrier St . . . A2
Fire Station . . . B2
Foregate St . . . B2
Foregate Street . . . B2
Fort Royal Hill . . . C3
Fort Royal Park . . . C3
Foundry St . . . B3
Friar St . . . C2
George St . . . B3
Grand Stand Rd . . . B1
Greenhill . . . C3
Greyfriars . . . B2
Guildhall . . . B2
Henwick Rd . . . B1
High St . . . B2
Hill St . . . B3
Huntingdon Hall . . . B2
Hylton Rd . . . B1
Information Ctr . . . B2
King Charles Place Shopping Centre . . . C1
King's School . . . C2
King's School Playing Field . . . C2
Kleve Walk . . . C2
Lansdowne Cr . . . A3
Lansdowne Rd . . . A3
Lansdowne Walk . . . A3
Laslett St . . . A3
Leisure Centre . . . A3
Library, Museum & Art Gallery . . . A2
Little Chestnut St . . . A2
Little London . . . A2
London Rd . . . C3
Lowell St . . . A3
Lowesmoor . . . B2
Lowesmoor Terr . . . A3
Lowesmoor Wharf . . . A3
Magistrates Court . . . A2
Midland Rd . . . B3
Mill St . . . C2
Moors Severn Terr . . . A1
New Rd . . . C1
New St . . . B2
Northfield St . . . A2
Odeon . . . B2
Padmore St . . . B3
Park St . . . C3
Pheasant St . . . B3
Pitchcroft Racecourse . . . A1
Police Station . . . A2
Portland St . . . C2
Post Office . . . B2
Quay St . . . B2
Queen St . . . B2
Rainbow Hill . . . A3
Recreation Ground . . . A2
Reindeer Court . . . B2
Rogers Hill . . . A3
Sabrina Rd . . . A1
St Dunstan's Cr . . . C3
St John's . . . C1
St Martin's Gate . . . B3
St Oswald's Rd . . . A2
St Paul's St . . . B3
St Swithin's Church . . . B2
St Wulstans Cr . . . C3
Sansome Walk . . . A2
Severn St . . . C2
Shaw St . . . B2
Shire Hall . . . A2
Shrub Hill . . . B3
Shrub Hill Retail Park . . . B3
Shrub Hill Rd . . . B3
Slingpool Walk . . . C1
South Quay . . . B2
Southfield St . . . A2
Sports Ground . . . A2/C1
Stanley Rd . . . B3
Swan, The . . . A1
Swimming Pool . . . A2
Tallow Hill . . . B3
Tennis Walk . . . A2
The Avenue . . . C1
The Butts . . . B2
The Cross . . . B2
The Shambles . . . B2
The Tything . . . A2
Tolladine Rd . . . B3
Tudor House . . . B2
Tybridge St . . . B1
University of Worcester . . . B2
Vincent Rd . . . B3
Vue . . . C2
Washington St . . . A3
Woolhope Rd . . . C3
Worcester Bridge . . . B2
Worcester Library & History Centre . . . B3
Worcester Porcelain Museum . . . C2
Worcester Royal Grammar School . . . A2
Wylds La . . . C3

Wrexham Wrecsam 344

Abbot St . . . B2
Acton Rd . . . A3
Albert St . . . C3
Alexandra Rd . . . C1
Aran Rd . . . A3
Barnfield . . . C3
Bath Rd . . . C2
Beechley Rd . . . C3
Belgrave Rd . . . C2
Belle Vue Park . . . C2
Belle Vue Rd . . . C2
Belvedere Dr . . . A1
Bennion's Rd . . . C3
Berse Rd . . . A1
Bersham Rd . . . C1
Birch St . . . C3
Bodhyfryd . . . B3
Border Retail Park . . . B3
Bradley Rd . . . C2
Bright St . . . B1
Bron-y-Nant . . . B1
Brook St . . . C2
Bryn-y-Cabanau Rd . . . C3
Bury St . . . C3
Bus Station . . . B2
Butchers Market . . . B3
Caia Rd . . . C3
Cambrian Industrial Estate . . . C3
Caxton Pl . . . B2
Cemetery . . . C1
Centenary Rd . . . C1
Chapel St . . . C2
Charles St . . . B3
Chester Rd . . . A3
Chester St . . . B3
Cilcen Gr . . . A3
Citizens Advice Bureau . . . B2
Cobden Rd . . . B1
Council Offices . . . B3
County . . . B2
Crescent Rd . . . B3
Crispin La . . . A2
Croesnewyth Rd . . . B1
Cross St . . . A2
Cunliffe St . . . A2
Derby Rd . . . C3
Dolydd Rd . . . B1
Duke St . . . B2
Eagles Meadow . . . C3
Earle St . . . C2
East Ave . . . A2
Edward St . . . C2
Egerton St . . . B2
Empress Rd . . . C1
Erddig Rd . . . C2
Fairy Rd . . . C2
Fire Station . . . B2
Foster Rd . . . A3
Foxwood Dr . . . C1
Garden Rd . . . A2
General Market . . . B3
Gerald St . . . B2
Gibson St . . . C1
Glyndŵr University Plas Coch Campus . . . A1
Greenbank St . . . C3
Greenfield . . . A2
Grosvenor Rd . . . B2
Grove Park . . . B2
Grove Park Rd . . . B3
Grove Rd . . . A3
Guildhall . . . B2
Haig Rd . . . C3
Hampden Rd . . . C2
Hazel Gr . . . A3
Henblas St . . . B3
High St . . . B3
Hightown Rd . . . C3
Hill St . . . B2
Holt Rd . . . B3
Holt St . . . B3
Hope St . . . B2
Huntroyde Ave . . . C3
Information Ctr . . . B3
Island Green Shopping Centre . . . B2
Job Centre . . . B2
Jubilee Rd . . . B2
King St . . . B2
Kingsmills Rd . . . C3
Lambpit St . . . B3
Law Courts . . . B3
Lawson Cl . . . A3
Lawson Rd . . . A3
Lea Rd . . . C2
Library & Arts Centre . . . C2
Lilac Way . . . B1
Llys David Lord . . . B1
Lorne St . . . A2
Maesgwyn Rd . . . B1
Maesydre Rd . . . A3
Manley Rd . . . B3
Market St . . . B3
Mawddy Ave . . . A2
Mayville Ave . . . A2
Memorial Gallery . . . B2
Memorial Hall . . . B3
Mold Rd . . . A1
Mount St . . . C3
Neville Cres . . . A3
New Rd . . . A2
North Wales Regional Tennis Centre . . . A1
North Wales School of Art & Design . . . B2
Oak Dr . . . A3
Park Ave . . . A3
Park St . . . A2
Peel St . . . C1
Pentre Felin . . . C2
Pen y Bryn . . . C2
Penymaes Ave . . . A3
Peoples Market . . . B3
Percy St . . . C2
Plas Coch Retail Park . . . A1
Plas Coch Rd . . . A1
Police Station . . . B3
Poplar Rd . . . C2
Post Office . . . A2/B2/C2/C3
Powell Rd . . . B3
Poyser St . . . C2
Price's La . . . A2
Primose Way . . . B1
Princess St . . . C1
Queen St . . . B3
Queens Sq . . . B2
Regent St . . . B2
Rhosddu Rd . . . A2/B2
Rhosnesni La . . . A3
Rivulet Rd . . . C3
Ruabon Rd . . . C2
Ruthin Rd . . . C1/C2
St Giles . . . C3
St Giles Way . . . C3
St James Ct . . . A2
St Mary's . . . B2
Salisbury Rd . . . C3
Salop Rd . . . C3
Sontley Rd . . . C2
Spring Rd . . . A2
Stanley St . . . C3
Stansty Rd . . . A2
Station Approach . . . B2
Studio . . . B2
Talbot Rd . . . C2
Techniquest Glyndŵr . . . A2
The Beeches . . . A3
The Pines . . . A3
Town Hill . . . C2
Trevor St . . . C2
Trinity St . . . B2
Tuttle St . . . C3
Vale Park . . . A1
Vernon St . . . B2
Vicarage Hill . . . B2
Victoria Rd . . . C2
Walnut St . . . A2
War Memorial . . . B3
Waterworld Leisure Centre . . . B3
Watery Rd . . . B1/B2
Wellington Rd . . . C2
Westminster Dr . . . A3
William Aston Hall . . . A1
Windsor Rd . . . A1
Wrexham AFC . . . A1
Wrexham Central . . . B2
Wrexham General . . . B2
Wrexham Maelor Hospital (A+E) . . . B1
Wrexham Technology Park . . . B1
Wynn Ave . . . A2
Yale College . . . B3
Yale Gr . . . A3
Yorke St . . . C3

York 344

Aldwark . . . B2
Ambulance Station . . . B3
Arc Museum, The . . . B2
Barbican Rd . . . C3
Barley Hall . . . B2
Bishopgate St . . . C2
Bishopthorpe Rd . . . C2
Blossom St . . . C1
Bootham . . . A1
Bootham Cr . . . A1
Bootham Terr . . . A1
Bridge St . . . B2
Brook St . . . A2
Brownlow St . . . A2
Burton Stone La . . . A1
Castle Museum . . . C2
Castlegate . . . B2
Cemetery Rd . . . C3
Cherry St . . . C2
City Art Gallery . . . A1
City Screen . . . B2
City Wall . . . A2/B1/C3
Clarence St . . . A2
Clementhorpe . . . C2
Clifford St . . . B2
Clifford's Tower . . . B2
Clifton . . . A1
Coach park . . . A2
Coney St . . . B2
Cromwell Rd . . . C2
Crown Court . . . C2
Davygate . . . B2
Deanery Gdns . . . A2
DIG . . . B2
Ebor Industrial Estate . . . B3
Fairfax House . . . B2
Fishergate . . . C3
Foss Islands Rd . . . B3
Fossbank . . . A3
Fossil Island Retail Park . . . B3
Garden St . . . A2
George St . . . C3
Gillygate . . . A2
Goodramgate . . . B2
Grand Opera House . . . B2
Grosvenor Terr . . . A1
Guildhall . . . B2
Hallfield Rd . . . A3
Heslington Rd . . . C3
Heworth Green . . . A3
Holy Trinity . . . B2
Hope St . . . C3
Huntington Rd . . . A3
Information Ctr . . . A2
James St . . . B3
Jorvik Viking Centre . . . B2
Kent St . . . C3
Lawrence St . . . C3
Layerthorpe . . . A3
Leeman Rd . . . B1
Lendal . . . B2
Lendal Bridge . . . B1
Library . . . B1
Longfield Terr . . . A1
Lord Mayor's Walk . . . A2
Lower Eldon St . . . A2
Lowther St . . . A2
Margaret St . . . C3
Marygate . . . A1
Melbourne St . . . C3
Merchant Adventurer's Hall . . . B2
Merchant Taylors' Hall . . . B2
Micklegate . . . B1
Minster, The . . . A2
Monkgate . . . A2
Moss St . . . C1
Museum Gdns . . . B1
Museum St . . . B1
National Railway Museum . . . B1
Navigation Rd . . . B3
Newton Terr . . . C2
North Pde . . . A1
North St . . . B2
Nunnery La . . . C1
Nunthorpe Rd . . . C1
Odeon . . . C1
Ouse Bridge . . . B2
Paragon St . . . C3
Park Gr . . . A3
Park St . . . C1
Parliament St . . . B2
Peasholme Green . . . B3
Penley's Grove St . . . A2
Piccadilly . . . B2
Police Station . . . B2
Post Office . . . B1/B2/C3
Priory St . . . B1
Purey Cust Nuffield Hospital, The . . . A2
Queen Anne's Rd . . . A1
Regimental Museum . . . B2
Rowntree Park . . . C2
St Andrewgate . . . B2
St Benedict Rd . . . C1
St John St . . . A2
St Olave's Rd . . . A1
St Peter's Gr . . . A1
St Saviourgate . . . B2
Scarcroft Hill . . . C1
Scarcroft Rd . . . C1
Skeldergate . . . C2
Skeldergate Bridge . . . C2
Station Rd . . . B1
Stonegate . . . B2
Sycamore Terr . . . A1
Terry Ave . . . C2
The Shambles . . . B2
The Stonebow . . . B2
Theatre Royal . . . B2
Thorpe St . . . C1
Toft Green . . . B1
Tower St . . . C2
Townend St . . . A2
Treasurer's House . . . A2
Trinity La . . . B1
Undercroft Museum . . . A2
Union Terr . . . A2
Victor St . . . C2
Vine St . . . C2
Walmgate . . . B3
Wellington St . . . C3
YMR Exhibition . . . B1
York Dungeon, The . . . B2
York Station . . . B1
Yorkshire Museum . . . B1
Yorkshire Wheel, The . . . B1

Index to road maps of Britain

Abbreviations used in the index

Aberdeen Aberdeen City
Aberds Aberdeenshire
Ald Alderney
Anglesey Isle of Anglesey
Angus Angus
Argyll Argyll and Bute
Bath Bath and North East Somerset
Bedford Bedford
Bl Gwent Blaenau Gwent
Blackburn Blackburn with Darwen
Blackpool Blackpool
Bmouth Bournemouth
Borders Scottish Borders
Brack Bracknell
Bridgend Bridgend
Brighton City of Brighton and Hove
Bristol City and County of Bristol
Bucks Buckinghamshire
C Beds Central Bedfordshire
Caerph Caerphilly
Cambs Cambridgeshire
Cardiff Cardiff
Carms Carmarthenshire
Ceredig Ceredigion
Ches E Cheshire East
Ches W Cheshire West and Chester
Clack Clackmannanshire
Conwy Conwy
Corn Cornwall
Cumb Cumbria
Darl Darlington
Denb Denbighshire
Derby City of Derby
Derbys Derbyshire
Devon Devon
Dorset Dorset
Dumfries Dumfries and Galloway
Dundee Dundee City
Durham Durham
E Ayrs East Ayrshire
E Dunb East Dunbartonshire
E Loth East Lothian
E Renf East Renfrewshire
E Sus East Sussex
E Yorks East Riding of Yorkshire
Edin City of Edinburgh
Essex Essex
Falk Falkirk
Fife Fife
Flint Flintshire
Glasgow City of Glasgow
Glos Gloucestershire
Gtr Man Greater Manchester
Guern Guernsey
Gwyn Gwynedd
Halton Halton
Hants Hampshire
Hereford Herefordshire
Herts Hertfordshire
Highld Highland
Hrtlpl Hartlepool
Hull Hull
IoM Isle of Man
IoW Isle of Wight
Invclyd Inverclyde
Jersey Jersey
Kent Kent
Lancs Lancashire
Leicester City of Leicester
Leics Leicestershire
Lincs Lincolnshire
London Greater London
Luton Luton
M Keynes Milton Keynes
M Tydf Merthyr Tydfil
Mbro Middlesbrough
Medway Medway
Mers Merseyside
Midloth Midlothian
Mon Monmouthshire
Moray Moray
N Ayrs North Ayrshire
N Lincs North Lincolnshire
N Lanark North Lanarkshire
N Som North Somerset
N Yorks North Yorkshire
NE Lincs North East Lincolnshire
Neath Neath Port Talbot
Newport City and County of Newport
Norf Norfolk
Northants Northamptonshire
Northumb Northumberland
Nottingham City of Nottingham
Notts Nottinghamshire
Orkney Orkney
Oxon Oxfordshire
Pboro Peterborough
Pembs Pembrokeshire
Perth Perth and Kinross
Plym Plymouth
Poole Poole
Powys Powys
Ptsmth Portsmouth
Reading Reading
Redcar Redcar and Cleveland
Renfs Renfrewshire
Rhondda Rhondda Cynon Taff
Rutland Rutland
S Ayrs South Ayrshire
S Glos South Gloucestershire
S Lanark South Lanarkshire
S Yorks South Yorkshire
Scilly Scilly
Shetland Shetland
Shrops Shropshire
Slough Slough
Som Somerset
Soton Southampton
Staffs Staffordshire
Southend Southend-on-Sea
Stirling Stirling
Stockton Stockton-on-Tees
Stoke Stoke-on-Trent
Suff Suffolk
Sur Surrey
Swansea Swansea
Swindon Swindon
T&W Tyne and Wear
Telford Telford & Wrekin
Thurrock Thurrock
Torbay Torbay
Torf Torfaen
V Glam The Vale of Glamorgan
W Berks West Berkshire
W Dunb West Dunbartonshire
W Isles Western Isles
W Loth West Lothian
W Mid West Midlands
W Sus West Sussex
W Yorks West Yorkshire
Warks Warwickshire
Warr Warrington
Wilts Wiltshire
Windsor Windsor and Maidenhead
Wokingham Wokingham
Worcs Worcestershire
Wrex Wrexham
York City of York

How to use the index

Example **Blatherwycke** Northants **137** D9

grid square
page number
county or unitary authority

A

Aaron's Hill Sur 50 E3
Aaron's Town Cumb 240 E2
Ab Kettleby Leics 154 E4
Ab Lench Worcs 117 G10
Abbas Combe Som 30 C2
Abberley Worcs 116 D5
Abberton Essex 89 B8
Abberton Worcs 117 G9
Abberwick Northumb 264 G4
Abbess End Essex 87 C9
Abbess Roding Essex 87 C9
Abbey Devon 27 E10
Abbey-cwm-hir Powys 113 C11
Abbey Dore Hereford 97 E7
Abbey Field Essex 107 G9
Abbey Gate Kent 53 B9
Abbey Green Shrops 149 C10
Abbey Green Staffs 169 D7
Abbey Hey Gtr Man 184 B5
Abbey Hulton Stoke 168 F6
Abbey Mead Sur 66 F4
Abbey St Bathans Borders 272 C5
Abbey Town Cumb 238 G5
Abbey Village Lancs 194 C6
Abbey Wood London 68 D3
Abbeycwmhir Powys 113 C11
Abbeydale Glos 80 B5
Abbeydale S Yorks 186 E4
Abbeydale Park S Yorks 186 E4
Abbeyhill Edin 280 G5
Abbeystead Lancs 203 C7
Abbots Bickington Devon 24 E5
Abbots Bromley Staffs 151 E11
Abbots Langley Herts 85 E9
Abbots Leigh N Som 60 E4
Abbot's Meads W Ches 166 B5
Abbots Morton Worcs 117 F10
Abbots Ripton Cambs 122 B4
Abbots Salford Warks 117 G11
Abbots Worthy Hants 48 G3
Abbotsbury Dorset 17 D7
Abbotsford W Sus 36 C4
Abbotsham Devon 24 B6
Abbotskerswell Devon 9 B7
Abbotsleigh Devon 8 F6
Abbotsley Cambs 122 F4
Abbotstone Hants 48 G5
Abbotswood Hants 32 C5
Abbotswood Sur 50 C4
Abbotts Ann Hants 47 E10
Abcott Shrops 115 B7
Abdon Shrops 131 F11
Abdy S Yorks 186 B6
Aber Ceredig 93 B9
Aber-Arad Carms 92 D6
Aber-banc Ceredig 93 C7
Aber Cowarch Gwyn 147 F7
Aber-Giâr Carms 93 C10
Aber-gwynfi Neath 57 B11
Aber-Hirnant Gwyn 147 C9
Aber miwl = Abermule Powys 130 E3
Aber-nant Rhondda 77 E8
Aber-oer Wrex 166 F3
Aber-Rhiwlech Gwyn 147 E8
Aber-Village Powys 96 G2
Aberaeron Ceredig 111 E9
Aberaman Rhondda 77 E8
Aberangell Gwyn 146 G6
Aberarder Highld 290 E6
Aberarder House Highld 300 G6
Aberarder Lodge Highld 291 E7
Aberargie Perth 286 F5
Aberarth Ceredig 111 E9
Aberavon Neath 57 C8
Aberbargoed Caerph 77 E11
Aberbechan Powys 130 E2
Aberbeeg Bl Gwent 78 E2
Aberbran Powys 95 F9
Abercanaid M Tydf 77 E9
Abercarn Caerph 78 G2
Abercastle Pembs 91 E7
Abercegir Powys 128 C6
Aberchalder Highld 290 C5
Aberchirder Aberds 302 D6
Abercorn W Loth 279 F11
Abercraf Powys 76 C4
Abercregan Neath 57 B11
Abercrombie Fife 287 G9
Abercwmboi Rhondda 77 F8
Abercych Pembs 92 C4
Abercynafon Powys 77 B9
Abercynffig = Aberkenfig Bridgend 57 E11
Abercynon Rhondda 77 F9
Aberdâr = Aberdare Rhondda 77 E7
Aberdalgie Perth 286 E4
Aberdare = Aberdâr Rhondda 77 E7
Aberdaron Gwyn 144 D3
Aberdeen Aberdeen 293 C11
Aberdesach Gwyn 162 E6
Aberdour Fife 280 D3
Aberdovey = Aberdyfi Gwyn 128 D2
Aberdulais Neath 76 E3
Aberdyfi = Aberdovey Gwyn 128 D2
Aberedw Powys 95 B11
Abereiddy Pembs 90 E5
Abererch Gwyn 145 B7
Aberfan M Tydf 77 E9
Aberfeldy Perth 286 C2
Aberffraw Anglesey 162 B5
Aberffrwd Ceredig 112 B3
Aberffrwd Mon 78 D5
Aberford W Yorks 206 F4
Aberfoyle Stirl 285 G9
Abergarw Bridgend 58 C2
Abergarwed Neath 76 E4
Abergavenny Mon 78 C3
Abergele Conwy 180 F6
Abergorlech Carms 93 E11
Abergwaun = Fishguard Pembs 91 D9
Abergwesyn Powys 113 G7
Abergwili Carms 93 G8
Abergwynant Gwyn 146 F3
Abergwynfi Neath 57 B11
Abergwyngregyn Gwyn 179 G11
Abergynolwyn Gwyn 128 B3
Aberhosan Powys 128 D6
Aberkenfig = Abercynffig Bridgend 57 E11
Aberlady E Loth 281 E9
Aberlemno Angus 287 B9
Aberlerry Ceredig 128 E2
Aberllefenni Gwyn 128 B5
Aberllydan = Broad Haven Pembs 72 C5
Aberllynfi = Three Cocks Powys 96 D3
Abermagwr Ceredig 112 C3
Abermaw = Barmouth Gwyn 146 F2
Abermeurig Ceredig 111 F11
Abermorddu Flint 166 D4
Abermule = Aber-miwl Powys 130 E3
Abernaint Powys 148 E2
Abernant Carms 92 G6
Abernant Powys 130 D3
Abernethy Perth 286 F5
Abernyte Perth 286 D6
Aberogwr = Ogmore by Sea V Glam 57 F11
Aberpennar = Mountain Ash Rhondda 77 F8
Aberporth Ceredig 110 G5
Aberriw = Berriew Powys 130 C3
Abersoch Gwyn 144 D6
Abersychan Torf 78 E3
Abertawe = Swansea Swansea 56 C6
Aberteifi = Cardigan Ceredig 92 B3
Aberthin V Glam 58 D4
Abertillery Bl Gwent 78 E2
Abertridwr Caerph 58 B6
Abertridwr Powys 147 F10
Abertrinant Gwyn 128 B2
Abertysswg Caerph 77 D10
Aberuchill Castle Perth 285 E11
Aberuthven Perth 286 F3
Aberyscir Powys 95 F9
Aberystwyth Ceredig 111 A11
Abhainn Suidhe W Isles 305 H2
Abingdon Oxon 83 F7
Abinger Common Sur 50 D6
Abinger Hammer Sur 50 D5
Abington Northants 120 E5
Abington S Lnrk 259 E10
Abington Pigotts Cambs 104 C6
Abington Vale Northants 120 E5
Abingworth W Sus 35 D10
Ablington Glos 81 D10
Ablington Wilts 47 D7
Abney Derbys 185 F11
Aboyne Aberds 293 D7
Abraham Heights Lancs 211 G9
Abram Gtr Man 194 G6
Abriachan Highld 300 F5
Abridge Essex 87 F7
Abronhill N Lnrk 278 F5
Abshot Hants 33 F8
Abson S Glos 61 E8
Abthorpe Northants 102 B2
Abune-the-Hill Orkney 314 D2
Aby Lincs 190 F6
Acaster Malbis York 207 D7
Acaster Selby N Yorks 207 E7
Accrington Lancs 195 B9
Acha Argyll 275 B8
Acha Argyll 288 D3
Acha Mor W Isles 304 F5
Achabraid Argyll 275 E9
Achachork Highld 298 E4
Achad nan Darach Highld 284 B4
Achadh an Eas Highld 308 F6
Achadunan Argyll 284 F5
Achafolla Argyll 275 B8
Achagary Highld 308 D7
Achaglass Argyll 255 C8
Achahoish Argyll 275 F8
Achalader Perth 286 C5
Achallader Argyll 285 C7
Achalone Highld 310 D5
Ach'an Todhair Highld 290 F2
Achanalt Highld 300 C2
Achanamara Argyll 275 E8
Achandunie Highld 300 B6
Achanelid Argyll 275 E11
Achany Highld 309 J5
Achaphubuil Highld 290 F2
Acharacle Highld 289 C8
Acharn Highld 289 D9
Acharn Perth 285 C11
Acharole Highld 310 D6
Acharossan Argyll 275 F10
Acharry Muir Highld 309 K6
Achath Aberds 293 B9
Achavanich Highld 310 E5
Achavelgin Highld 301 D9
Achavraat Highld 301 E9
Achddu Carms 74 E6
Achdregnie Moray 302 G2
Achduart Highld 307 J5
Achentoul Highld 310 F2
Achfary Highld 306 F7
Achfrish Highld 309 H5
Achgarve Highld 307 K3
Achiemore Highld 308 C3
Achiemore Highld 310 D2
A'Chill Highld 294 E4
Achiltibuie Highld 307 J5
Achina Highld 308 C7
Achinahuagh Highld 308 C5
Achindaul Highld 290 E3
Achindown Highld 301 E8
Achinduich Highld 309 J5
Achinduin Argyll 289 F10
Achingills Highld 310 C5
Achininver Highld 308 C5
Achintee Highld 290 F3
Achintee Highld 299 E9
Achintraid Highld 295 B10
Achlaven Argyll 289 F11
Achlean Highld 291 D10
Achleck Argyll 288 E6
Achlorachan Highld 300 D3
Achluachrach Highld 290 E4
Achlyness Highld 306 D7
Achmelvich Highld 307 G5
Achmore Highld 295 B10
Achmore Stirl 285 D9
Achnaba Argyll 275 E10
Achnaba Argyll 289 F11
Achnabat Highld 300 F5
Achnabreck Argyll 275 D9
Achnacarnin Highld 306 F5
Achnacarry Highld 290 E3
Achnacloich Argyll 289 F11
Achnacloich Highld 295 E7
Achnaconeran Highld 290 B6
Achnacraig Argyll 288 E6
Achnacree Argyll 289 F11
Achnacree Bay Argyll 289 F11
Achnacroish Argyll 289 E10
Achnadrish Argyll 288 D6
Achnafalnich Argyll 284 E6
Achnagarron Highld 300 C6
Achnaha Highld 288 C6
Achnahanat Highld 309 K5
Achnahannet Highld 301 G9
Achnahard Argyll 288 G5
Achnairn Highld 309 H5
Achnaluachrach Highld 309 J6
Achnandarach Highld 295 B10
Achnanellan Highld 290 E2
Achnasaul Highld 290 E3
Achnasheen Highld 299 D11
Achnashelloch Argyll 275 D9
Achnavast Highld 310 C4
Achneigie Highld 299 B10
Achormlarie Highld 309 K6
Achorn Highld 310 F5
Achosnich Highld 288 C6
Achranich Highld 289 E9
Achreamie Highld 310 C4
Achriabhach Highld 290 G3
Achriesgill Highld 306 D7
Achrimsdale Highld 311 J3
Achtoty Highld 308 C6
Achurch Northants 137 G10
Achuvoldrach Highld 308 D5
Achvaich Highld 309 K7
Achvarasdal Highld 310 C3
Ackenthwaite Cumb 211 C10
Ackergill Highld 310 D7
Acklam M'bro 225 B9
Acklam N Yorks 216 G5
Ackleton Shrops 132 D5
Acklington Northumb 252 C6
Ackton W Yorks 198 C2
Ackworth Moor Top W Yorks 198 D2
Acle Norf 161 G8
Acock's Green W Mid 134 G2
Acol Kent 71 F10
Acomb Northumb 241 D10
Acomb York 207 C7
Acre Gtr Man 196 F2
Acre Lancs 195 C9
Acre Street W Sus 21 B11
Acrefair Wrex 166 G3
Acres Nook Staffs 168 E4
Acton Dorset 18 F5
Acton E Ches 167 E10
Acton London 67 C8
Acton Shrops 130 G6
Acton Staffs 168 G4
Acton Suff 107 C7
Acton Worcs 116 D6
Acton Wrex 166 E4
Acton Beauchamp Hereford 116 G3
Acton Bridge W Ches 183 F9
Acton Burnell Shrops 131 C10
Acton Green Hereford 116 G3
Acton Green London 67 D8
Acton Pigott Shrops 131 C10
Acton Place Suff 107 B7
Acton Reynald Shrops 149 E10
Acton Round Shrops 132 D2
Acton Scott Shrops 131 F9
Acton Trussell Staffs 151 F8
Acton Turville S Glos 61 C10
Adabroc W Isles 304 B7
Adambrae W Loth 269 A10
Adam's Green Dorset 29 F8
Adbaston Staffs 150 D5
Adber Dorset 29 C9
Adbolton Notts 154 B2
Adderbury Oxon 101 D9
Adderley Shrops 150 B3
Adderley Green Stoke 168 G6
Adderstone Northumb 264 C4
Addiewell W Loth 269 C9
Addingham W Yorks 205 D7
Addingham Moorside W Yorks 205 D7
Addington Bucks 102 F4
Addington Corn 6 B5
Addington Kent 53 B7
Addington London 67 G11
Addinston Borders 271 E10
Addiscombe London 67 F10
Addlestone Sur 66 G5
Addlestonemoor Sur 66 F4
Addlethorpe Lincs 175 B8
Adel W Yorks 205 F11
Adeney Telford 150 F4
Adeyfield Herts 85 D9
Adfa Powys 129 C11
Adforton Hereford 115 C8
Adgestone I o W 21 D7
Adisham Kent 55 C8
Adlestrop Glos 100 F4
Adlingfleet E Yorks 199 C10
Adlington E Ches 184 E6
Adlington Lancs 194 E6
Adlington Park Lancs 194 E5
Admaston Staffs 151 E10
Admaston Telford 150 G2
Admington Warks 100 B4
Adpar Ceredig 92 C6
Adsborough Som 28 B3
Adscombe Som 43 F7
Adstock Bucks 102 E4
Adstone Northants 119 G11
Adswood Gtr Man 184 D5
Adversane W Sus 35 C9
Advie Highld 301 F11
Adwalton W Yorks 197 B8
Adwell Oxon 83 F11
Adwick le Street S Yorks 198 F4
Adwick upon Dearne S Yorks 198 G3
Adziel Aberds 303 D9
Ae Dumfries 247 F11
Ae Village Dumfries 247 F11
Affetside Gtr Man 195 E9
Affleck Aberds 303 G8
Affpuddle Dorset 18 C2
Affric Lodge Highld 299 G11
Afon Eitha Wrex 166 F3
Afon-wen Flint 181 G10
Afon Wen Gwyn 145 B8
Afton I o W 20 D2
Agar Nook Leics 153 G9
Agbrigg W Yorks 197 D10
Aggborough Worcs 116 B6
Agglethorpe N Yorks 213 B11
Aglionby Cumb 239 F10
Agneash I o M 192 D5
Aifft Denb 165 B10
Aigburth Mers 182 D5
Aiginis W Isles 304 E6
Aike E Yorks 209 D7
Aikenway Moray 302 E2
Aikerness Orkney 314 A4
Aikers Orkney 314 G4
Aiketgate Cumb 230 B5
Aikton Cumb 239 G7
Ailby Lincs 190 F6
Ailey Hereford 96 B6
Ailstone Warks 118 G4
Ailsworth P'boro 138 D2
Aimes Green Essex 86 E5
Ainderby Quernhow N Yorks 215 C7
Ainderby Steeple N Yorks 224 G6
Aingers Green Essex 108 G2
Ainley Top W Yorks 196 D6
Ainsdale Mers 193 E10
Ainsdale-on-Sea Mers 193 E9
Ainstable Cumb 230 B6
Ainsworth Gtr Man 195 E9
Ainthorpe N Yorks 226 D4
Aintree Mers 182 B5
Aird Argyll 275 C8
Aird Dumfries 236 C2
Aird Highld 299 B7
Aird W Isles 296 F3
Aird W Isles 304 E7
Aird a Mhachair W Isles 297 G3
Aird a' Mhulaidh W Isles 305 G3
Aird Asaig W Isles 305 H3
Aird Dhail W Isles 304 B6
Aird Mhòr W Isles 297 G4
Aird Mhidhinis W Isles 297 L3
Aird Mhighe W Isles 296 C6
Aird Mhighe W Isles 305 J3
Aird Mhor W Isles 297 L3
Aird of Sleat Highld 295 E7
Aird Thunga W Isles 304 E6
Aird Uig W Isles 304 E2
Airdachuilinn Highld 306 E7
Airdens Highld 309 K6
Airdeny Argyll 289 G11
Airdrie N Lnrk 268 B5
Airds of Kells Dumfries 237 B8
Airdtorrisdale Highld 308 C6
Aire View N Yorks 204 D5
Airedale W Yorks 198 B3
Airidh a Bhruaich W Isles 305 G4
Airieland Dumfries 237 D9
Airinis W Isles 304 E6
Airlie Angus 287 B7
Airlies Dumfries 236 D5
Airmyn E Yorks 199 B8
Airntully Perth 286 D4
Airor Highld 295 E9
Airth Falk 279 D7
Airthrey Castle Stirl 278 B6
Airton N Yorks 204 B4
Airy Hill N Yorks 227 D7
Airyhassen Dumfries 236 E5
Airyligg Dumfries 236 C4
Aisby Lincs 155 B10
Aisby Lincs 188 C5
Aisgernis W Isles 297 J3
Aish Devon 8 C3
Aish Devon 8 D6
Aisholt Som 43 F7
Aiskew N Yorks 214 B5
Aislaby N Yorks 216 B5
Aislaby N Yorks 227 D7
Aislaby Stockton 225 C8
Aisthorpe Lincs 188 E6
Aith Orkney 314 E2
Aith Shetland 312 D8
Aith Shetland 313 H5
Aithnen Powys 148 E4
Aithsetter Shetland 313 K6
Aitkenhead S Ayrs 245 B8
Aitnoch Highld 301 F9
Akeld Northumb 263 D11
Akeley Bucks 102 D4
Akenham Suff 108 B2
Albany T & W 243 F7
Albaston Corn 12 G4
Alberbury Shrops 149 G7
Albert Town Pembs 72 B6
Albert Village Leics 152 F6
Albourne W Sus 36 D3
Albourne Green W Sus 36 D3
Albrighton Shrops 132 C6
Albrighton Shrops 149 F9
Albro Castle Ceredig 92 B3
Alburgh Norf 142 F5
Albury Herts 105 G8
Albury Sur 50 D5
Albury End Herts 105 G8
Albury Heath Sur 50 D5
Alby Hill Norf 160 C3
Albyfield Cumb 240 G2
Alcaig Highld 300 D5
Alcaston Shrops 131 F9
Alcester Dorset 30 C5
Alcester Warks 117 F11
Alcester Lane's End W Mid 133 G11
Alciston E Sus 23 D8
Alcombe Som 42 D3
Alcombe Wilts 61 F10
Alconbury Cambs 122 B3
Alconbury Weston Cambs 122 B3
Aldbar Castle Angus 287 B9
Aldborough N Yorks 215 F8
Aldborough Norf 160 C3
Aldborough Hatch London 68 B3
Aldbourne Wilts 63 D9
Aldbrough E Yorks 209 F10
Aldbrough St John N Yorks 224 C4
Aldbury Herts 85 C7
Aldcliffe Lancs 211 G9
Aldclune Perth 291 G11
Aldeburgh Suff 127 F9
Aldeby Norf 143 E8
Aldenham Herts 85 F10
Alder Forest Gtr Man 184 B3
Alder Moor Staffs 152 D4
Alder Row Som 45 E9
Alderbrook E Sus 37 B8
Alderbury Wilts 31 B11
Aldercar Derbys 170 F6
Alderford Norf 160 F2
Alderholt Dorset 31 E10
Alderley Glos 80 G3
Alderley Edge E Ches 184 F4
Alderman's Green W Mid 135 G7
Aldermaston W Berks 64 F5
Aldermaston Soke W Berks 64 G6
Aldermaston Wharf W Berks 64 F6
Alderminster Warks 100 B4
Aldermoor Soton 32 D5
Alderney Poole 18 C6
Alder's End Hereford 98 C2
Aldersbrook London 68 B2
Aldersey Green W Ches 167 D7
Aldershawe Staffs 134 B2
Aldershot Hants 49 C11
Alderton Glos 99 E10
Alderton Northants 102 B4
Alderton Shrops 149 E9
Alderton Suff 108 C6
Alderton Wilts 61 C10
Alderton Fields Glos 99 E10
Alderwasley Derbys 170 E4
Aldfield N Yorks 214 F5
Aldford W Ches 166 D6
Aldgate Rutland 137 C9
Aldham Essex 107 F8
Aldham Suff 107 B10
Aldie Highld 309 L7
Aldingbourne W Sus 22 B6
Aldingham Cumb 210 E5
Aldington Kent 54 F5
Aldington Worcs 99 C11
Aldington Frith Kent 54 F4
Aldivalloch Moray 302 G3
Aldochlay Argyll 277 C7
Aldon Shrops 115 B8
Aldoth Cumb 229 B8
Aldourie Castle Highld 300 F6
Aldreth Cambs 123 C8
Aldridge W Mid 133 C11
Aldringham Suff 127 E8
Aldrington Brighton 36 F3
Aldsworth Glos 81 C11
Aldunie Moray 302 G3
Aldwark Derbys 170 D2
Aldwark N Yorks 215 G9
Aldwarke S Yorks 186 C6
Aldwick W Sus 22 D6
Aldwincle Northants 137 G10
Aldworth W Berks 64 D5
Ale Oak Shrops 130 G4
Alehouseburn Aberds 302 C6
Alehousehill Aberds 303 G10
Alexandria W Dunb 277 F7
Aley Som 43 F7
Aley Green C Beds 85 B9
Alfardisworthy Devon 24 E3
Alfington Devon 15 B8
Alfold Sur 50 G4
Alfold Bars W Sus 50 G4
Alfold Crossways Sur 50 F4
Alford Aberds 293 B7
Alford Lincs 191 F7
Alford Som 44 G6
Alfred's Well Worcs 117 C8
Alfreton Derbys 170 D6
Alfrick Worcs 116 G4
Alfrick Pound Worcs 116 G4
Alfriston E Sus 23 E8
Algakirk Lincs 156 B5
Algaltraig Argyll 275 F11
Algarkirk Lincs 156 B5
Alhampton Som 44 G6
Aline Lodge W Isles 305 G3
Alisary Highld 289 B9
Alkborough N Lincs 199 C11
Alkerton Glos 80 D3
Alkerton Oxon 101 C7
Alkham Kent 55 E9
Alkington Shrops 149 B10
Alkmonton Derbys 152 B3
Alkrington Garden Village Gtr Man 195 G11
All Cannings Wilts 62 G5
All Saints Devon 28 G4
All Saints South Elmham Suff 142 G6
All Stretton Shrops 131 D9
Alladale Lodge Highld 309 L4
Allaleigh Devon 8 E6
Allanaquoich Aberds 292 D3
Allanbank Borders 271 F10
Allanbank N Lnrk 268 D6
Allangrange Mains Highld 300 D6
Allanshaugh Borders 271 F8
Allanshaws Borders 271 G9
Allanton Borders 273 E7
Allanton N Lnrk 269 D7
Allanton S Lnrk 268 E4
Allaston Glos 79 E10
Allathasdal W Isles 297 L2
Allbrook Hants 33 C7
Allen End Warks 134 D3
Allendale Town Northumb 241 F8
Allenheads Northumb 232 B3
Allens Green Herts 87 B7
Allensford Durham 242 G3
Allensmore Hereford 97 D9
Allenton Derby 153 C7
Allenwood Cumb 239 F11
Aller Devon 9 B7
Aller Devon 27 F9
Aller Dorset 30 G3
Aller Som 28 B6
Aller Park Devon 9 B7
Allerby Cumb 229 D7
Allerford Som 27 B11
Allerford Som 42 D2
Allerston N Yorks 217 C7
Allerthorpe E Yorks 207 D11
Allerton Mers 182 D6
Allerton W Yorks 205 G8
Allerton Bywater W Yorks 198 B2
Allerton Mauleverer N Yorks 206 B4
Allesley W Mid 134 G6
Allestree Derby 152 B6
Allet Corn 4 F5
Allexton Leics 136 C6
Allgreave E Ches 169 B7
Allhallows Medway 69 D10
Allhallows-on-Sea Medway 69 D10
Alligin Shuas Highld 299 D8
Allimore Green Staffs 151 F7
Allington Kent 53 B8
Allington Lincs 172 G5
Allington Wilts 47 F8
Allington Wilts 61 D11
Allington Wilts 62 G5
Allington Bar Wilts 61 E11
Allithwaite Cumb 211 D7
Alloa Clack 279 C7
Allonby Cumb 229 C7
Allostock W Ches 184 G2
Alloway S Ayrs 257 F8
Allowenshay Som 28 E5
Allscot Shrops 132 D4
Allscott Telford 150 G2
Allt Carms 75 E9
Allt-na-giubhsaich Aberds 292 E4
Allt na h-Airbhe Highld 307 K6
Allt-nan-sùgh Highld 295 C11
Allt-yr-yn Newport 59 B9
Alltami Flint 166 B3
Alltbeithe Highld 290 C2
Alltchaorunn Highld 284 B5
Alltforgan Powys 147 E9
Alltmawr Powys 95 B11
Alltnacaillich Highld 308 E4
Alltrech Argyll 289 F8
Alltsigh Highld 290 B6
Alltwalis Carms 93 E8
Alltwen Neath 76 E2
Alltyblaca Ceredig 93 B10
Allwood Green Suff 125 C10
Alma Notts 171 E7
Almagill Dumfries 238 B3
Almeley Hereford 114 G6

Almeley Wooton Hereford 114 G6
Almer Dorset 18 B4
Almholme S Yorks 198 F5
Almington Staffs 150 C4
Alminstone Cross Devon 24 C4
Almondbank Perth 286 E4
Almondbury W Yorks 197 D7
Almondsbury S Glos 60 C6
Almondvale W Loth 269 B11
Almshouse Green Essex 106 E5
Alne N Yorks 215 F9
Alne End Warks 118 F2
Alne Hills Warks 118 E2
Alne Station N Yorks 215 F10
Alness Highld 300 C6
Alnessferry Highld 300 C6
Alnham Northumb 263 G11
Alnmouth Northumb 264 G6
Alnwick Northumb 264 G5
Alperton London 67 C7
Alphamstone Essex 107 D7
Alpheton Suff 125 G7
Alphington Devon 14 C4
Alpington Norf 142 C5
Alport Derbys 170 C2
Alport Powys 130 D5
Alpraham E Ches 167 D9
Alresford Essex 107 G11
Alrewas Staffs 152 F3
Alsager E Ches 168 D3
Alsagers Bank Staffs 168 F4
Alscot Bucks 84 E4
Alsop en le Dale Derbys 169 D11
Alston Cumb 231 B10
Alston Devon 28 G4
Alston Sutton Som 44 C2
Alstone Glos 99 E9
Alstone Glos 99 G8
Alstone Som 43 D10
Alstonefield Staffs 169 D10
Alswear Devon 26 C2
Alt Gtr Man 196 G2
Alt Hill Gtr Man 196 G2
Altandhu Highld 307 H4
Altanduin Highld 311 G2
Altarnun Corn 11 E10
Altass Highld 309 J4
Altbough Hereford 97 E10
Altdargue Aberds 293 C7
Alterwall Highld 310 C6
Altham Lancs 203 G11
Althorne Essex 88 F6
Althorpe N Lincs 199 F10
Alticane S Ayrs 244 F6
Alticry Dumfries 236 D4
Altmore Windsor 65 D11
Altnabreac Station Highld 310 E4
Altnacealgach Hotel Highld 307 H7
Altnacraig Argyll 289 G10
Altnafeadh Highld 284 B6
Altnaharra Highld 308 F6
Altofts W Yorks 197 C11
Alton Derbys 170 C5
Alton Hants 49 F8
Alton Staffs 169 G9
Alton Wilts 47 D7
Alton Barnes Wilts 62 G6
Alton Pancras Dorset 30 G2
Alton Priors Wilts 62 G6
Altonhill E Ayrs 257 B10
Altonside Moray 302 D2
Altour Highld 290 E4
Altrincham Gtr Man 184 D3
Altrua Highld 290 E4
Altskeith Stirl 285 G8
Altyre Ho Moray 301 D10
Alum Rock W Mid 134 F2
Alva Clack 279 B7
Alvanley W Ches 183 G7
Alvaston Derby 153 C7
Alvechurch Worcs 117 C10
Alvecote Warks 134 C4
Alvediston Wilts 31 C7
Alveley Shrops 132 G5
Alverdiscott Devon 25 B8
Alverstoke Hants 21 B8
Alverstone I o W 21 D7
Alverthorpe W Yorks 197 C10
Alverton Notts 172 G3
Alves Moray 301 C11
Alvescot Oxon 82 E3
Alveston S Glos 60 B6
Alveston Warks 118 F4
Alveston Down S Glos 60 B6
Alveston Hill Warks 118 G4
Alvie Highld 291 C10
Alvingham Lincs 190 C5
Alvington Glos 79 E10
Alvington Som 29 D8
Alwalton Cambs 138 D2
Alway Newport 59 B10
Alweston Dorset 29 E11
Alwington Devon 24 C6
Alwinton Northumb 251 B10
Alwoodley W Yorks 205 E11
Alwoodley Gates W Yorks 206 E2
Alwoodley Park W Yorks 205 E11
Alyth Perth 286 C6
Am Baile W Isles 297 K3
Am Buth Argyll 289 G10
Amalebra Corn 1 B5
Amalveor Corn 1 B5
Amatnatua Highld 309 K4
Ambaston Derbys 153 C8
Amber Hill Lincs 174 F2
Ambergate Derbys 170 E4
Amberley Glos 80 E5
Amberley Hereford 97 B10
Amberley W Sus 35 E8
Amble Northumb 253 C7
Amblecote W Mid 133 F7
Ambler Thorn W Yorks 196 B5
Ambleside Cumb 221 E7
Ambleston Pembs 91 F10
Ambrosden Oxon 83 B10
Amcotts N Lincs 199 E11
Amen Corner Brack 65 F10
Amersham Bucks 85 F7
Amersham Common Bucks 85 F7
Amersham Old Town Bucks 85 F7
Amersham on the Hill Bucks 85 F7
Amerton Staffs 151 D9
Amesbury Bath 45 B7
Amesbury Wilts 47 E7
Ameysford Dorset 31 G9
Amington Staffs 134 C4
Amisfield Dumfries 247 G11
Amlwch Anglesey 178 C6
Amlwch Port Anglesey 179 C7
Ammanford = Rhydaman Carms 75 C10
Ammerham Som 28 F5
Amod Argyll 255 D8
Amotherby N Yorks 216 E4
Ampfield Hants 32 C6
Ampleforth N Yorks 215 D11
Ampney Crucis Glos 81 E9
Ampney St Mary Glos 81 E9
Ampney St Peter Glos 81 E9
Amport Hants 47 E9
Ampthill C Beds 103 D10
Ampton Suff 125 C7
Amroth Pembs 73 D11
Amulree Perth 286 D2
Amwell Herts 85 C11
An Caol Highld 298 D6
An Cnoc W Isles 304 E6
An Gleann Ur W Isles 304 E6
An Leth Meadhanach W Isles 297 K3
An t-Ob W Isles 296 C6
Anagach Highld 301 G10
Anaheilt Highld 289 C10
Anancaun Highld 299 C10
Ancarraig Highld 300 G4
Ancaster Lincs 173 G7
Anchor Shrops 130 G3
Anchor Corner Norf 141 D10
Anchor Street Norf 160 E6
Anchorage Park Ptsmth 33 G11
Anchorsholme Blkpool 202 E2
Ancoats Gtr Man 184 B5
Ancroft Northumb 273 F9
Ancroft Northmoor Northumb 273 F9
Ancrum Borders 262 E4
Ancton W Sus 35 G7
Ancumtoun Orkney 314 A7
Anderby Lincs 191 F8
Anderby Creek Lincs 191 F8
Andersea Som 43 G10
Andersfield Som 43 G8
Anderson Dorset 18 B3
Anderton Corn 7 E8
Anderton Lancs 194 E6
Anderton W Ches 183 F10
Andertons Mill Lancs 194 E4
Andover Hants 47 D11
Andover Down Hants 47 D11
Andoversford Glos 81 B8
Andreas I o M 192 C5
Andwell Hants 49 C7
Anelog Gwyn 144 D3
Anerley London 67 F10
Anfield Mers 182 C5
Angarrack Corn 2 B3
Angarrick Corn 3 B7
Angelbank Shrops 115 B11
Angersleigh Som 27 D11
Angerton Cumb 238 F6
Angle Pembs 72 E5
Angmering W Sus 35 G9
Angram N Yorks 206 D6
Angram N Yorks 223 F7
Anick Northumb 241 D11
Anie Stirl 285 F9
Ankerdine Hill Worcs 116 F4
Ankerville Highld 301 B8
Anlaby E Yorks 200 B4
Anlaby Park Hull 200 B5
Anmer Norf 158 D4
Anmore Hants 33 E11
Anna Valley Hants 47 E10
Annan Dumfries 238 D5
Annaside Cumb 210 B1
Annat Argyll 284 E4
Annat Highld 290 D5
Annat Highld 299 D8
Annbank S Ayrs 257 E10
Annesley Notts 171 E8
Annesley Woodhouse Notts 171 E7
Annfield Plain Durham 242 G5
Anniesland Glasgow 267 B10
Annifirth Shetland 313 J3
Annis Hill Suff 143 F7
Annishader Highld 298 D4
Annitsford T & W 243 C7
Ann's Hill Hants 33 G9
Annscroft Shrops 131 B9
Annwell Place Derbys 152 F6
Ansdell Lancs 193 B10
Ansells End Herts 85 B11
Ansford Som 44 G6
Ansley Warks 134 E5
Ansley Common Warks 134 E6
Anslow Staffs 152 D4
Anslow Gate Staffs 152 D3
Ansteadbrook Sur 50 G2
Anstey Herts 105 E8
Anstey Leics 135 B10
Anstruther Easter Fife 287 G9
Anstruther Wester Fife 287 G9
Ansty Hants 49 E8
Ansty W Sus 36 C3
Ansty Warks 135 G7
Ansty Wilts 31 B7
Ansty Coombe Wilts 31 B7
Ansty Cross Dorset 30 G3
Anthill Common Hants 33 E10
Anthony Corn 7 E7
Anthony's Cross Glos 98 G4
Anthorn Cumb 238 F5
Antingham Norf 160 C5
Anton's Gowt Lincs 174 F3
Antonshill Falk 279 E7
Antony Corn 7 E7
Antony Passage Corn 7 D8
Antrobus W Ches 183 F10
Anvil Green Kent 54 D6
Anvilles W Berks 63 F10
Anwick Lincs 173 E10
Anwoth Dumfries 237 D7
Aonachan Highld 290 E4
Aoradh Argyll 274 G3
Apedale Staffs 168 F4
Aperfield London 52 B2
Apes Dale Worcs 117 C9
Apes Hall Cambs 139 E11
Apethorpe Northants 137 D10
Apeton Staffs 151 F7
Apley Lincs 189 F10
Apley Forge Shrops 132 D4
Apperknowle Derbys 186 F5
Apperley Glos 99 F7
Apperley Bridge W Yorks 205 F9
Apperley Dene Northumb 242 F3
Appersett N Yorks 223 G7
Appin Argyll 289 E11
Appin House Argyll 289 E11
Appleby N Lincs 200 E3
Appleby-in-Westmorland Cumb 231 G9
Appleby Magna Leics 134 B6
Appleby Parva Leics 134 B6
Applecross Highld 299 E7
Applecross Ho Highld 299 E7
Appledore Devon 27 E9
Appledore Devon 40 G3
Appledore Kent 39 B7
Appledore Heath Kent 54 G3
Appleford Oxon 83 G8
Applegarthtown Dumfries 248 G4
Applehouse Hill Windsor 65 C10
Applemore Hants 32 F5
Appleshaw Hants 47 D10
Applethwaite Cumb 229 F11
Appleton Halton 183 D8
Appleton Oxon 82 E6
Appleton-le-Moors N Yorks 216 B4
Appleton-le-Street N Yorks 216 E4
Appleton Park Warr 183 E10
Appleton Roebuck N Yorks 207 E7
Appleton Thorn Warr 183 E10
Appleton Wiske N Yorks 225 E7
Appletreehall Borders 262 F2
Appletreewick N Yorks 213 G11
Appley I o W 21 C8
Appley Som 27 C9
Appley Bridge Lancs 194 F4
Apse Heath I o W 21 E7
Apsey Green Suff 126 E5
Apsley Herts 85 D9
Apsley End C Beds 104 E2
Apuldram W Sus 22 C4
Aqueduct Telford 132 B3
Aquhythie Aberds 293 B9
Arabella Highld 301 B8
Arbeadie Aberds 293 D8
Arberth = Narberth Pembs 73 C10
Arbirlot Angus 287 C10
Arboll Highld 311 L2
Arborfield Wokingham 65 F9
Arborfield Cross Wokingham 65 F9
Arborfield Garrison Wokingham 65 F9
Arbourthorne S Yorks 186 D5
Arbroath Angus 287 C10
Arbury Cambs 123 E8
Arbuthnott Aberds 293 F9
Archavandra Muir Highld 309 K7
Archdeacon Newton Darl 224 B5
Archenfield Hereford 96 C5
Archiestown Moray 302 E2
Archnalea Highld 289 C10
Arclid E Ches 168 C3
Arclid Green E Ches 168 C3
Ard-dhubh Highld 299 E7
Ardachu Highld 309 J6
Ardailly Argyll 255 B7
Ardalanish Argyll 274 B4
Ardallie Aberds 303 F10
Ardalum Ho Argyll 288 F6
Ardamaleish Argyll 275 G11
Ardanaiseig Argyll 284 E4
Ardaneaskan Highld 295 B10
Ardanstur Argyll 275 B9
Ardargie House Hotel Perth 286 F4
Ardarroch Highld 295 B10
Ardban Highld 295 B9
Ardbeg Argyll 254 C5
Ardbeg Argyll 276 E3
Ardcharnich Highld 307 L6
Ardchiavaig Argyll 274 B4
Ardchonnell Argyll 275 B10
Ardchronie Highld 309 L6
Ardchuilk Highld 300 F2
Ardchullarie More Stirl 285 F9
Ardchyle Stirl 285 E9
Ardclach Highld 301 E9
Arddleen Powys 148 F5
Ardechvie Highld 290 D3
Ardeley Herts 104 F6
Ardelve Highld 295 C10
Arden Argyll 277 E7
Arden E Renf 267 D10
Arden Park Gtr Man 184 C6
Ardencaple Ho Argyll 275 B8
Ardendrain Highld 300 F5
Ardens Grafton Warks 118 G2
Ardentallen Argyll 289 G10
Ardentinny Argyll 276 D3
Ardentraive Argyll 275 F11
Ardeonaig Stirl 285 D10
Ardersier Highld 301 D7
Ardery Highld 289 C9
Ardessie Highld 307 L5
Ardfern Argyll 275 C9
Ardfernal Argyll 274 F6
Ardgartan Argyll 284 G6
Ardgay Highld 309 K5
Ardglassie Aberds 303 C10
Ardgour Highld 290 G2
Ardgye Moray 301 C11
Ardheslaig Highld 299 D7
Ardiecow Moray 302 C5
Ardinamir Argyll 275 B8
Ardindrean Highld 307 L6
Ardingly W Sus 36 B4
Ardington Oxon 64 B2
Ardington Wick Oxon 64 B2
Ardintoul Highld 295 C10
Ardlair Aberds 302 G5
Ardlair Highld 299 B9
Ardlamey Argyll 255 C7
Ardlamont Ho Argyll 275 G10
Ardlawhill Aberds 303 C8
Ardleigh Essex 107 F11
Ardleigh Green London 68 B4
Ardleigh Heath Essex 107 E10
Ardler Perth 286 C6
Ardley Oxon 101 F10
Ardley End Essex 87 C8
Ardlui Argyll 285 F7
Ardlussa Argyll 275 E7
Ardmair Highld 307 K6
Ardmay Argyll 284 G6
Ardmenish Argyll 274 F6
Ardminish Argyll 255 C7
Ardmolich Highld 289 B9
Ardmore Argyll 289 G9
Ardmore Highld 306 D7
Ardmore Highld 309 L7
Ardnacross Argyll 289 E7
Ardnadam Argyll 276 F3
Ardnagowan Argyll 284 G4
Ardnagrask Highld 300 E5
Ardnarff Highld 295 B10
Ardnastang Highld 289 C10
Ardnave Argyll 274 F3
Ardneil N Ayrs 266 F3
Ardno Argyll 284 G5
Ardo Aberds 303 F8
Ardo Ho Aberds 303 G9
Ardoch Argyll 277 F7
Ardoch Perth 286 D4
Ardoch Stirl 285 F9
Ardochy House Highld 290 C4
Ardoyne Aberds 302 G6
Ardpatrick Argyll 275 G8
Ardpatrick Ho Argyll 255 B8
Ardpeaton Argyll 276 D4
Ardradnaig Perth 285 C11
Ardrishaig Argyll 275 E9
Ardross Fife 287 G9
Ardross Highld 300 B6
Ardross Castle Highld 300 B6
Ardrossan N Ayrs 266 G4
Ardshave Highld 309 K7
Ardsheal Highld 289 D11
Ardshealach Highld 289 C8
Ardskenish Argyll 274 D4
Ardsley S Yorks 197 F11
Ardslignish Highld 289 C7
Ardtalla Argyll 254 B5
Ardtalnaig Perth 285 D11
Ardtaraig Argyll 275 E11
Ardtoe Highld 289 B8
Ardtreck Highld 294 B5
Ardtrostan Perth 285 E10
Ardtur Argyll 289 E11
Arduaine Argyll 275 B8
Ardullie Highld 300 C5
Ardvannie Highld 309 L6
Ardvar Highld 306 F6
Ardvasar Highld 295 E8
Ardveich Stirl 285 E10
Ardverikie Highld 291 E7
Ardvorlich Perth 285 E10
Ardwell Dumfries 236 E3
Ardwell Moray 302 F3
Ardwell S Ayrs 244 E5
Ardwell Mains Dumfries 236 E3
Ardwick Gtr Man 184 B5
Areley Kings Worcs 116 C6
Arford Hants 49 F10
Argoed Caerph 77 F11
Argoed Powys 113 E9
Argoed Powys 130 E5
Argoed Shrops 130 G6
Argoed Shrops 148 E6
Argos Hill E Sus 37 B9
Arichamish Argyll 275 C10
Arichastlich Argyll 284 D6
Aridhglas Argyll 288 G5
Arieniskill Highld 295 G9
Arileod Argyll 288 D3
Arinacrinachd Highld 299 D7
Arinagour Argyll 288 D4
Arineckaig Highld 299 E9
Arion Orkney 314 E2
Arisaig Highld 295 G8
Ariundle Highld 289 C10
Arivegaig Highld 289 C8
Arivoichallum Argyll 254 C4
Arkendale N Yorks 215 G7
Arkesden Essex 105 E9
Arkholme Lancs 211 E11
Arkle Town N Yorks 223 E10
Arkleby Cumb 229 D8
Arkleton Dumfries 249 E9
Arkley London 86 F2
Arksey S Yorks 198 F5
Arkwright Town Derbys 186 G6
Arle Glos 99 G8
Arlebrook Glos 80 D4
Arlecdon Cumb 219 B10
Arlescote Warks 101 B7
Arlesey C Beds 104 D3
Arleston Telford 150 G3
Arley E Ches 183 E11
Arley Green E Ches 183 E11
Arlingham Glos 80 C2
Arlington Devon 40 E6
Arlington E Sus 23 D8
Arlington Glos 81 D10
Arlington Beccott Devon 40 E6
Armadale Highld 308 C7
Armadale W Loth 269 B8
Armadale Castle Highld 295 E8
Armathwaite Cumb 230 B6
Armigers Essex 105 F11
Arminghall Norf 142 C5
Armitage Staffs 151 F11
Armitage Bridge W Yorks 196 E6
Armley W Yorks 205 G11
Armscote Warks 100 C4
Armsdale Staffs 150 C5
Armshead Staffs 168 F6
Armston Northants 137 F11
Armthorpe S Yorks 198 F6
Arnabost Argyll 288 D4
Arnaby Cumb 210 C3
Arncliffe N Yorks 213 E8
Arncroach Fife 287 G9
Arndilly Ho Moray 302 E2
Arne Dorset 18 D5
Arnesby Leics 136 E2
Arngask Perth 286 F5
Arnisdale Highld 295 D10
Arnish Highld 298 E5
Arniston Midloth 270 C6
Arnol W Isles 304 D5
Arnold E Yorks 209 E8
Arnold Notts 171 F9
Arno's Vale Bristol 60 E6
Arnprior Stirl 278 C2
Arnside Cumb 211 D9
Aros Mains Argyll 289 E7
Arowry Wrex 149 B9
Arpafeelie Highld 300 D6
Arpinge Kent 55 F7
Arrad Foot Cumb 210 C6
Arram E Yorks 208 E6
Arrathorne N Yorks 224 G4
Arreton I o W 20 D6
Arrington Cambs 122 G6
Arrivain Argyll 284 D6
Arrochar Argyll 284 G6
Arrow Warks 117 F11
Arrow Green Hereford 115 F8
Arrowe Hill Mers 182 D3
Arrowfield Top Worcs 117 C10
Arrunden W Yorks 196 F6
Arscaig Highld 309 H5
Arscott Shrops 131 B8
Arthill E Ches 184 D2
Arthington W Yorks 205 E11
Arthingworth Northants 136 G5
Arthog Gwyn 146 G2
Arthrath Aberds 303 F9
Arthursdale W Yorks 206 F3
Arthurstone Perth 286 C6
Arthurville Highld 309 L7
Artington Sur 50 D3
Artrochie Aberds 303 F10
Arundel W Sus 35 F8
Arwick Orkney 314 D3
Aryhoulan Highld 290 G2
Asby Cumb 229 G7
Ascog Argyll 266 C2
Ascoil Highld 311 H2
Ascot Windsor 66 F2
Ascott Warks 100 E6
Ascott d'Oyley Oxon 82 B4
Ascott Earl Oxon 82 B3
Ascott-under-Wychwood Oxon 82 B4
Asenby N Yorks 215 D7
Asfordby Leics 154 F4
Asfordby Hill Leics 154 F4
Asgarby Lincs 173 F10
Asgarby Lincs 174 B4
Ash Devon 8 F6
Ash Dorset 30 E5
Ash Kent 55 B9
Ash Kent 68 G5
Ash Som 28 C3
Ash Som 29 C7
Ash Sur 49 C11
Ash Bank Staffs 168 F6
Ash Bullayne Devon 26 G3
Ash Green Sur 50 D2
Ash Green Warks 134 F6
Ash Grove Wrex 166 G5
Ash Hill Devon 14 G4
Ash Magna Shrops 149 B11
Ash Mill Devon 26 C3
Ash Moor Devon 26 D3
Ash Parva Shrops 149 B11
Ash Priors Som 27 B11
Ash Street Suff 107 B10
Ash Thomas Devon 27 E8
Ash Vale Sur 49 C11
Ashaig Highld 295 C8
Ashampstead W Berks 64 D5
Ashampstead Green W Berks 64 D5
Ashansworth Hants 48 B2
Ashbank Kent 53 C10
Ashbeer Som 42 F5
Ashbocking Suff 126 G3
Ashbourne Derbys 169 F11
Ashbrittle Som 27 C9
Ashbrook Shrops 131 E9
Ashburnham Forge E Sus 23 B11
Ashburton Devon 8 B5
Ashbury Devon 12 B6
Ashbury Oxon 63 C9
Ashby N Lincs 200 F2
Ashby by Partney Lincs 174 B6
Ashby cum Fenby NE Lincs 201 G9
Ashby de la Launde Lincs 173 D9
Ashby-de-la-Zouch Leics 153 F7
Ashby Folville Leics 154 G4
Ashby Hill NE Lincs 201 G8
Ashby Magna Leics 135 E11
Ashby Parva Leics 135 F10
Ashby Puerorum Lincs 190 G4
Ashby St Ledgers Northants 119 D11
Ashby St Mary Norf 142 C6
Ashchurch Glos 99 E8
Ashcombe Devon 14 F4
Ashcombe Park N Som 59 G10
Ashcott Som 44 F2
Ashcott Corner Som 44 F2
Ashculme Devon 27 E10
Ashdon Essex 105 C11
Ashe Hants 48 D4
Asheldham Essex 89 E7
Ashen Essex 106 C4
Ashendon Bucks 84 C2
Asheridge Bucks 84 E6
Ashey I o W 21 D7
Ashfield Argyll 275 E8
Ashfield Carms 94 F3
Ashfield Hants 32 D5
Ashfield Hereford 97 G11
Ashfield Shrops 148 D6
Ashfield Stirl 285 G11
Ashfield Suff 126 E4
Ashfield Cum Thorpe Suff 126 E4
Ashfield Green Suff 124 F5
Ashfield Green Suff 126 C5
Ashfields Shrops 150 D4
Ashfold Crossways W Sus 36 B2
Ashfold Side N Yorks 214 F2
Ashford Devon 8 F3
Ashford Devon 40 F4
Ashford Hants 31 E10
Ashford Kent 54 E4
Ashford Sur 66 E5
Ashford Bowdler Shrops 115 C10
Ashford Carbonell Shrops 115 C10
Ashford Common Sur 66 E5
Ashford Hill Hants 64 G5
Ashford in the Water Derbys 185 G11
Ashgate Derbys 186 G5
Ashgill S Lnrk 268 F5
Ashgrove Bath 45 B8
Ashiestiel Borders 261 B10
Ashill Devon 27 E9
Ashill Norf 141 C7
Ashill Som 28 D4
Ashingdon Essex 88 G5
Ashington Northumb 253 F7
Ashington Poole 18 B6
Ashington Som 29 C9
Ashington W Sus 35 D10
Ashington End Lincs 175 B8
Ashintully Castle Perth 292 G3
Ashkirk Borders 261 E11
Ashlett Hants 33 G7
Ashleworth Glos 98 F6
Ashley Cambs 124 E3
Ashley Devon 25 E10
Ashley Dorset 31 G10
Ashley E Ches 184 E3
Ashley Glos 80 G6
Ashley Hants 19 B11
Ashley Hants 47 G11
Ashley Kent 55 D10
Ashley Northants 136 E5
Ashley Staffs 150 B5
Ashley Wilts 61 F10
Ashley Dale Staffs 150 B5
Ashley Down Bristol 60 D5
Ashley Green Bucks 85 D7
Ashley Heath Dorset 31 G10
Ashley Heath Staffs 150 B4
Ashley Moor Hereford 115 D9
Ashley Park Sur 66 F6
Ashleyhay Derbys 170 E3
Ashmanhaugh Norf 160 E6
Ashmansworth Hants 48 B2
Ashmansworthy Devon 24 D4
Ashmead Green Glos 80 F3
Ashmill Devon 12 B3
Ashmore Dorset 30 D6
Ashmore Green W Berks 64 F4
Ashmore Lake W Mid 133 D9
Ashmore Park W Mid 133 C9
Ashnashellach Lodge Highld 299 E10
Ashopton Derbys 185 D11
Ashorne Warks 118 F6
Ashover Derbys 170 C4
Ashover Hay Derbys 170 C5
Ashow Warks 118 C6
Ashperton Hereford 98 C2
Ashprington Devon 8 D6
Ashreigney Devon 25 E10
Ashridge Court Devon 25 G11
Ashstead Sur 51 B7
Ashton Corn 2 D4
Ashton Hants 33 D9
Ashton Hereford 115 E10
Ashton Invclyd 276 F4
Ashton Northants 102 B5
Ashton Northants 137 F11
Ashton P'boro 138 B2
Ashton Som 44 D2
Ashton Common Wilts 45 B11
Ashton Gate Bristol 60 E5
Ashton Green E Sus 23 C7
Ashton Hayes W Ches 167 B8
Ashton Heath Halton 183 F9
Ashton-in-Makerfield Gtr Man 183 B9
Ashton Keynes Wilts 81 G8
Ashton under Hill Worcs 99 D9
Ashton-under-Lyne Gtr Man 184 B6
Ashton upon Mersey Gtr Man 184 C3
Ashton Vale Bristol 60 E5
Ashurst Hants 32 E4
Ashurst Kent 52 F4
Ashurst Lancs 194 F3
Ashurst W Sus 35 D11
Ashurst Bridge Hants 32 E4
Ashurst Wood W Sus 52 F2
Ashvale Bl Gwent 77 C10
Ashwater Devon 12 B3
Ashwell Devon 14 G3
Ashwell Herts 104 D5
Ashwell Rutland 155 G7
Ashwell Som 28 D5
Ashwell End Herts 104 C5
Ashwellthorpe Norf 142 D2
Ashwick Som 44 D6
Ashwicken Norf 158 F4
Ashwood Staffs 133 F7
Ashybank Borders 262 F2
Askam in Furness Cumb 210 D4
Askern S Yorks 198 E5
Askerswell Dorset 16 C6
Askerton Hill Lincs 172 F4
Askett Bucks 84 D4
Askham Cumb 230 G6
Askham Notts 188 G2
Askham Bryan York 207 D7
Askham Richard York 206 D6
Asknish Argyll 275 D10
Askrigg N Yorks 223 G8
Askwith N Yorks 205 D9
Aslackby Lincs 155 C11
Aslacton Norf 142 E3
Aslockton Notts 154 B4
Asloun Aberds 293 B7
Asney Som 44 F3
Aspall Suff 126 D3
Aspatria Cumb 229 C8
Aspenden Herts 105 F7
Asperton Lincs 156 B5
Aspley Nottingham 171 G8
Aspley Staffs 150 C6
Aspley Guise C Beds 103 D8
Aspley Heath C Beds 103 D8
Aspley Heath Warks 117 C11
Aspull Gtr Man 194 F6
Aspull Common Gtr Man 183 B10
Assater Shetland 312 F4
Asselby E Yorks 199 B8
Asserby Lincs 191 F7
Asserby Turn Lincs 191 F7
Assington Suff 107 D8
Assington Green Suff 124 G5
Assynt Ho Highld 300 C5
Astbury E Ches 168 C4
Astcote Northants 120 G3
Asterby Lincs 190 F3
Asterley Shrops 131 B7
Asterton Shrops 131 E7
Asthall Oxon 82 C3
Asthall Leigh Oxon 82 C4
Astle E Ches 184 G4
Astle Highld 309 K7
Astley Gtr Man 195 G8
Astley Shrops 149 F10
Astley Warks 134 F6
Astley Worcs 116 D5
Astley Abbotts Shrops 132 D4
Astley Bridge Gtr Man 195 E8
Astley Cross Worcs 116 D6
Astley Green Gtr Man 184 B2
Astmoor Halton 183 E8
Aston Derbys 152 C3
Aston Derbys 185 E11
Aston E Ches 167 F10
Aston Flint 166 B4
Aston Hereford 115 C9
Aston Hereford 115 E9
Aston Herts 104 G5
Aston Oxon 82 E4
Aston Powys 130 E5
Aston S Yorks 187 D7
Aston Shrops 132 E6
Aston Shrops 149 D10
Aston Staffs 151 E7
Aston Staffs 168 G3
Aston Telford 132 B2
Aston W Ches 183 F9
Aston W Mid 133 F11
Aston Wokingham 65 C9
Aston Abbotts Bucks 102 G6
Aston Bank Worcs 116 C2
Aston Botterell Shrops 132 G2
Aston-by-Stone Staffs 151 C8
Aston Cantlow Warks 118 F2
Aston Clinton Bucks 84 C5
Aston Crews Hereford 98 G3
Aston Cross Glos 99 E8
Aston End Herts 104 G5
Aston Eyre Shrops 132 E3
Aston Fields Worcs 117 D9
Aston Flamville Leics 135 E9
Aston Ingham Hereford 98 G3
Aston juxta Mondrum E Ches 167 D11
Aston le Walls Northants 119 G9
Aston Magna Glos 100 D3
Aston Munslow Shrops 131 F10
Aston on Carrant Glos 99 E8
Aston on Clun Shrops 131 G7
Aston-on-Trent Derbys 153 D8
Aston Pigott Shrops 130 B6
Aston Rogers Shrops 130 B6
Aston Rowant Oxon 84 F2
Aston Sandford Bucks 84 D3
Aston Somerville Worcs 99 D10
Aston Square Shrops 148 D6
Aston Subedge Glos 100 C2
Aston Tirrold Oxon 64 B5
Aston Upthorpe Oxon 64 B5
Astrop Northants 101 D10
Astrope Herts 84 C5
Astwick C Beds 104 D4
Astwith Derbys 170 C6
Astwood M Keynes 103 B9
Astwood Worcs 117 B7
Astwood Worcs 117 F7
Astwood Bank Worcs 117 E10
Aswarby Lincs 173 G9
Aswardby Lincs 190 G5
Atch Lench Worcs 117 G10
Atcham Shrops 131 B10
Athelhampton Dorset 17 C11
Athelington Suff 126 C4
Athelney Som 28 B4
Athelstaneford E Loth 281 F10
Atherfield Green I o W 20 F5
Atherington Devon 25 C9
Atherington W Sus 35 G8
Athersley North S Yorks 197 F11
Athersley South S Yorks 197 F11
Atherstone Som 28 D5
Atherstone Warks 134 D6
Atherstone on Stour Warks 118 G4
Atherton Gtr Man 195 G7
Athnamulloch Highld 299 G11
Athron Hall Perth 286 G4
Atley Hill N Yorks 224 E5
Atlow Derbys 170 F2
Attadale Highld 295 B11
Attadale Ho Highld 295 B11
Attenborough Notts 153 B10
Atterbury M Keynes 103 D7
Atterby Lincs 189 C7
Attercliffe S Yorks 186 D5
Atterley Shrops 132 D2
Atterton Leics 135 D7
Attleborough Norf 141 D10
Attleborough Warks 135 E7
Attlebridge Norf 160 F2
Attleton Green Suff 124 G4
Atwick E Yorks 209 C9
Atworth Wilts 61 F11
Auberrow Hereford 97 B9
Aubourn Lincs 172 C6
Auch Argyll 285 D7
Auchagallon N Ayrs 255 D9
Auchallater Aberds 292 E3
Auchareoch N Ayrs 255 E10
Aucharnie Aberds 302 E6
Auchattie Aberds 293 D8
Auchavan Angus 292 G3
Auchbreck Moray 302 G2
Auchenback E Renf 267 D10
Auchenbainzie Dumfries 247 D8
Auchenblae Aberds 293 F9
Auchenbrack Dumfries 247 D7
Auchenbreck Argyll 255 C8
Auchenbreck Argyll 275 E11
Auchencairn Dumfries 237 D9
Auchencairn Dumfries 247 G11
Auchencairn N Ayrs 256 D2
Auchencairn Ho Dumfries 237 D10
Auchencar N Ayrs 255 D9
Auchencarroch W Dunb 277 E8
Auchencrosh S Ayrs 236 B3
Auchencrow Borders 273 C7
Auchendinny Midloth 270 C5
Auchengray S Lnrk 269 E9
Auchenhalrig Moray 302 C3
Auchenharvie N Ayrs 266 G5
Auchenheath S Lnrk 268 G6
Auchenhew N Ayrs 256 E2
Auchenlaich Stirl 285 G10
Auchenleck Dumfries 237 D9
Auchenlochan Argyll 275 F10
Auchenmalg Dumfries 236 D4
Auchenreoch E Dunb 278 F3
Auchensoul S Ayrs 245 E7
Auchentibber S Lnrk 268 E3
Auchentiber N Ayrs 267 F7
Auchertyre Highld 295 C10
Auchessan Stirl 285 E8
Auchgourish Highld 291 B11
Auchinairn E Dunb 268 B2
Auchindrain Argyll 284 G4
Auchindrean Highld 307 L6
Auchininna Aberds 302 E6
Auchinleck Dumfries 236 B6
Auchinleck E Ayrs 258 E3
Auchinloch N Lnrk 278 G3
Auchinner Perth 285 F10
Auchinraith S Lnrk 268 D3
Auchinroath Moray 302 D2
Auchinstarry N Lnrk 278 F4
Auchintoul Aberds 293 B7
Auchintoul Highld 309 K5
Auchiries Aberds 303 F10
Auchlee Aberds 293 D10
Auchleeks Ho Perth 291 G9
Auchleven Aberds 302 G6
Auchlinn Aberds 303 D7
Auchlochan S Lnrk 259 B8
Auchlossan Aberds 293 C7
Auchlunachan Highld 307 L6
Auchlunies Aberds 293 D10
Auchlyne Stirl 285 E9
Auchmacoy Aberds 303 F9
Auchmair Moray 302 G3
Auchmantle Dumfries 236 C3
Auchmenzie Aberds 302 G5
Auchmillan E Ayrs 258 D2
Auchmithie Angus 287 C10
Auchmore Highld 300 D4
Auchmuirbridge Fife 286 G6
Auchmull Angus 293 F7
Auchnacraig Argyll 289 G9
Auchnacree Angus 292 G6
Auchnafree Perth 286 D2
Auchnagallin Highld 301 F10
Auchnagarron Argyll 275 E11
Auchnagatt Aberds 303 E9
Auchnaha Argyll 275 E10
Auchnahillin Highld 301 F7
Auchnarrow Moray 302 G2
Auchnashelloch Perth 285 F11
Auchnotteroch Dumfries 236 C1
Aucholzie Aberds 292 D5
Auchrannie Angus 286 B6
Auchroisk Highld 301 G10
Auchronie Angus 292 E6
Auchterarder Perth 286 F3
Auchteraw Highld 290 C5
Auchterderran Fife 280 B4
Auchterhouse Angus 287 D7
Auchtermuchty Fife 286 F6
Auchterneed Highld 300 D4
Auchtertool Fife 280 C4
Auchtertyre Moray 301 D11
Auchtertyre Stirl 285 E7
Auchtubh Stirl 285 E9
Auckengill Highld 310 C7
Auckley S Yorks 199 G7
Audenshaw Gtr Man 184 B6
Audlem E Ches 167 G11
Audley Staffs 168 E3
Audley End Essex 105 D10
Audley End Essex 106 D6
Audley End Norf 142 G2
Audley End Suff 125 G7
Auds Aberds 302 C6
Aughertree Cumb 229 D11
Aughton E Yorks 207 F10
Aughton Lancs 193 F11
Aughton Lancs 211 F10
Aughton S Yorks 187 D7
Aughton Wilts 47 B8
Aughton Park Lancs 194 F2
Aukside Durham 232 F4
Auldearn Highld 301 D9
Aulden Hereford 115 G9
Auldgirth Dumfries 247 F10
Auldhame E Loth 281 E11
Auldhouse S Lnrk 268 E2
Auldtown of Carnoustie Aberds 302 E6
Ault a'chruinn Highld 295 C11
Ault Hucknall Derbys 171 B7
Aultanrynie Highld 308 F3
Aultbea Highld 307 L3
Aultdearg Highld 300 C2
Aultgrishan Highld 307 L2
Aultguish Inn Highld 300 B3
Aultibea Highld 311 G4
Aultiphurst Highld 310 C2
Aultivullin Highld 310 C2
Aultmore Highld 301 G10
Aultmore Moray 302 D4
Aultnagoire Highld 300 G5
Aultnamain Inn Highld 309 L6
Aultnaslat Highld 290 C3
Aulton Aberds 302 G6
Aulton of Atherb Aberds 303 E9
Aultvaich Highld 300 E5
Aunby Lincs 155 G10
Aundorach Highld 291 B11
Aunk Devon 27 G8
Aunsby Lincs 155 B10
Auquhorthies Aberds 303 G8
Aust S Glos 60 B5
Austen Fen Lincs 190 C5
Austendike Lincs 156 E5
Austenwood Bucks 66 B3
Austerfield S Yorks 187 C11
Austerlands Gtr Man 196 F3
Austhorpe W Yorks 206 G3
Austrey Warks 134 B5
Austwick N Yorks 212 F5
Authorpe Lincs 190 E6
Authorpe Row Lincs 191 G8
Avebury Wilts 62 F6
Avebury Trusloe Wilts 62 F5
Aveley Thurrock 68 C5
Avening Glos 80 F5
Avening Green S Glos 80 G2
Averham Notts 172 E3
Avernish Highld 295 C10
Avery Hill London 68 E2
Aveton Gifford Devon 8 F3
Avielochan Highld 291 B11
Aviemore Highld 291 B10
Avington Hants 48 G4
Avington W Berks 63 F11
Avoch Highld 301 D7
Avon Hants 19 B8
Avon Wilts 62 D3
Avon Dassett Warks 101 B8
Avonbridge Falk 279 G8
Avoncliff Wilts 45 B10
Avonmouth Bristol 60 D4
Avonwick Devon 8 D4
Awbridge Hants 32 C4
Awhirk Dumfries 236 D2
Awkley S Glos 60 B5
Awliscombe Devon 27 G10
Awre Glos 80 D2
Awsworth Notts 171 G7
Axbridge Som 44 C2
Axford Hants 48 E6
Axford Wilts 63 F8
Axmansford Hants 64 G5
Axminster Devon 15 B11
Axmouth Devon 15 C11
Axton Flint 181 E10
Axtown Devon 7 B10
Axwell Park T & W 242 E5
Aycliff Kent 55 E10
Aycliffe Durham 233 G11
Aydon Northumb 242 D2
Aykley Heads Durham 233 C11
Aylburton Glos 79 E10
Aylburton Common Glos 79 E10
Ayle Northumb 231 B10
Aylesbeare Devon 14 C6
Aylesbury Bucks 84 C4
Aylesby NE Lincs 201 F8
Aylesford Kent 53 B8
Aylesham Kent 55 C8
Aylestone Leicester 135 C11
Aylestone Hill Hereford 97 C10
Aylestone Park Leicester 135 C11
Aylmerton Norf 160 B3
Aylsham Norf 160 D3
Aylton Hereford 98 D3
Aylworth Glos 100 G2
Aymestrey Hereford 115 D8
Aynho Northants 101 E10
Ayot Green Herts 86 C2
Ayot St Lawrence Herts 85 B11
Ayot St Peter Herts 86 B2
Ayr S Ayrs 257 E8
Ayre of Atler Shetland 313 G6
Ayres Shetland 313 H5
Ayres End Herts 85 C11
Ayres of Selivoe Shetland 313 J4
Ayres Quay T & W 243 F9
Aysgarth N Yorks 213 B10
Ayshford Devon 27 D8
Ayside Cumb 211 C7
Ayston Rutland 137 C7
Aythorpe Roding Essex 87 B9
Ayton Borders 273 C8
Ayton T & W 243 F7
Ayton Castle Borders 273 C8
Aywick Shetland 312 E7
Azerley N Yorks 214 E5

B

Babbacombe Torbay 9 B8
Babbington Notts 171 G7
Babbinswood Shrops 148 C6
Babbs Green Herts 86 B5
Babcary Som 29 B9
Babel Carms 94 D6

Babel Green Suff 106 B4
Babell Flint 181 G11
Babeny Devon 13 F9
Babingley Norf 158 D3
Babraham Cambs 123 G10
Babworth Notts 187 E11
Bac W Isles 304 D6
Bach-y-gwreiddyn Swansea 75 E10
Bachau Anglesey 178 E6
Bache Shrops 131 G9
Bache Mill Shrops 131 F10
Bacheldre Powys 130 E4
Bachelor's Bump E Sus 38 E4
Back of Keppoch Highld 295 G8
Back o' th' Brook Staffs 169 E9
Back Muir Fife 279 D11
Back Rogerton E Ayrs 258 E3
Back Street Suff 124 F4
Backaland Orkney 314 C5
Backaskaill Orkney 314 A4
Backbarrow Cumb 211 C7
Backbower Gtr Man 185 C7
Backburn Aberds 293 D10
Backe Carms 74 B3
Backfolds Aberds 303 D10
Backford W Ches 182 G6
Backford Cross W Ches 182 G5
Backhill Aberds 303 F10
Backhill Aberds 303 F7
Backhill of Clackriach Aberds 303 E9
Backhill of Fortree Aberds 303 E9
Backhill of Trustach Aberds 293 D8
Backies Highld 311 J2
Backlass Highld 310 D6
Backlass Highld 310 E4
Backwell N Som 60 F3
Backwell Common N Som 60 F3
Backwell Green N Som 60 F3
Backworth T & W 243 C8
Bacon End Essex 87 B10
Baconend Green Essex 87 B10
Bacon's End W Mid 134 F3
Baconsthorpe Norf 160 B2
Bacton Hereford 97 E7
Bacton Norf 160 C6
Bacton Suff 125 D11
Bacton Green Norf 160 C6
Bacton Green Suff 125 D10
Bacup Lancs 195 C11
Badachonacher Highld 300 B6
Badachro Highld 299 B7
Badanloch Lodge Highld 308 F7
Badarach Highld 309 K5
Badavanich Highld 299 D11
Badbea Highld 307 K5
Badbury Swindon 63 C7
Badbury Wick Swindon 63 C7
Badby Northants 119 F11
Badcall Highld 306 D7
Badcaul Highld 307 K5
Baddeley Edge Stoke 168 E6
Baddeley Green Stoke 168 E6
Baddesley Clinton Warks 118 C4
Baddesley Ensor Warks 134 D5
Baddidarach Highld 307 G5
Baddoch Aberds 292 E3
Baddock Highld 301 D7
Baddow Park Essex 88 E2
Badeach Moray 302 F2
Badenscallie Highld 307 J5
Badenscoth Aberds 303 F7
Badentoy Park Aberds 293 D11
Badenyon Aberds 292 B5
Badgall Corn 11 D10
Badgeney Cambs 139 D8
Badger Shrops 132 D5
Badger Street Som 28 D3
Badgergate Stirl 278 B5
Badger's Hill Worcs 99 B10
Badger's Mount Kent 68 G3
Badgeworth Glos 80 B6
Badgworth Som 43 C11
Badharlick Corn 11 D11
Badicaul Highld 295 C9
Badingham Suff 126 D6
Badintagairt Highld 309 H4
Badlesmere Kent 54 C4
Badlipster Highld 310 E6
Badluarach Highld 307 K4
Badminton S Glos 61 C10
Badnaban Highld 307 G5
Badnabay Highld 306 E7
Badnagie Highld 310 F5
Badninish Highld 309 K7
Badrallach Highld 307 K5
Badsey Worcs 99 C11
Badshalloch W Dunb 277 D9
Badshot Lea Sur 49 D11
Badsworth W Yorks 198 E3
Badwell Ash Suff 125 D9
Badwell Green Suff 125 D10
Badworthy Devon 8 C3
Bae Cinmel = Kinmel Bay Conwy 181 E7
Bae Colwyn = Colwyn Bay Conwy 180 F4
Bae Penrhyn = Penrhyn Bay Conwy 180 E4
Baffins Ptsmth 33 G11
Bag Enderby Lincs 190 G5
Bagber Dorset 30 E3
Bagby N Yorks 215 C9
Bagby Grange N Yorks 215 C9
Bagendon Glos 81 D8
Bagginswood Shrops 132 G3
Baggrow Cumb 229 C9
Bàgha Chàise W Isles 296 D5
Bagh a Chaisteil W Isles 297 M2
Bagh Mor W Isles 296 F4
Bagh Shiarabhagh W Isles 297 L3
Bagham Kent 54 C5
Baghasdal W Isles 297 K3
Bagillt Flint 182 F2
Baginton Warks 118 C6
Baglan Neath 57 C8
Bagley Shrops 149 D8
Bagley Som 44 D3
Bagley W Yorks 205 F10
Bagley Green Som 27 D10
Bagley Marsh Shrops 149 D7
Bagmore Hants 49 E7
Bagnall Staffs 168 E6
Bagnor W Berks 64 F3
Bagpath Glos 80 E5
Bagpath Glos 80 G4
Bagshaw Derbys 185 E9
Bagshot Sur 66 G2
Bagshot Wilts 63 F10
Bagshot Heath Sur 66 G2
Bagslate Moor Gtr Man 195 E11
Bagstone S Glos 61 B7
Bagthorpe Norf 158 C5
Bagthorpe Notts 171 E7
Baguley Gtr Man 184 D4
Bagworth Leics 135 B8
Bagwy Llydiart Hereford 97 F8
Bagwyllydiart Hereford 97 F8
Bail Àrd Bhuirgh W Isles 304 C6
Bail' Iochdrach W Isles 296 F4
Bail Uachdraich W Isles 296 E4
Bail' Ur Tholastaidh W Isles 304 D7
Bailbrook Bath 61 F9
Baildon W Yorks 205 F9
Baildon Green W Yorks 205 F8
Baile W Isles 296 C5
Baile a Mhanaich W Isles 296 F3
Baile Ailein W Isles 304 F4
Baile an Truiseil W Isles 304 C5
Baile Boidheach Argyll 275 F8
Baile Gharbhaidh W Isles 297 G3
Baile Glas W Isles 296 F4
Baile Mhartainn W Isles 296 D3
Baile Mhic Phail W Isles 296 D4
Baile Mor Argyll 288 G4
Baile Mor W Isles 296 E3
Baile na Creige W Isles 297 L2
Baile nan Cailleach W Isles 296 F3
Baile Raghaill W Isles 296 D3
Bailebeag Highld 291 B7
Bailey Green Hants 33 B11
Baileyhead Cumb 240 B2
Bailiesward Aberds 302 F4
Bailiff Bridge W Yorks 196 C6
Baillieston Glasgow 268 C3
Bailrigg Lancs 202 B5
Bainbridge N Yorks 223 G8
Bainsford Falk 279 E7
Bainshole Aberds 302 F6
Bainton E Yorks 208 C5
Bainton Oxon 101 F11
Bainton P'boro 137 B11
Baintown Fife 287 G7
Bairnkine Borders 262 F5
Baker Street Thurrock 68 C6
Baker's Cross Kent 53 F9
Bakers End Herts 86 B5
Baker's Hill Glos 79 C9
Baker's Wood Bucks 66 B4
Bakesdown Corn 24 G2
Bakestone Moor Derbys 187 F8
Bakewell Derbys 170 B2
Bala = Y Bala Gwyn 147 B8
Balachroick Highld 291 C10
Balachuirn Highld 298 E5
Balance Hill Staffs 151 C11
Balavil Highld 291 C9
Balavoulin Perth 291 G10
Balbeg Highld 300 F4
Balbeg Highld 300 G4
Balbeggie Perth 286 E5
Balbegno Castle Aberds 293 F8
Balbithan Aberds 293 B9
Balbithan Ho Aberds 293 B10
Balblair Highld 300 E5
Balblair Highld 301 C7
Balblair Highld 309 K5
Balby S Yorks 198 G5
Balcherry Highld 311 L2
Balchladich Highld 306 F5
Balchraggan Highld 300 E5
Balchraggan Highld 300 F5
Balchraggan Highld 300 G5
Balchrick Highld 306 D6
Balchrystie Fife 287 G8
Balcladaich Highld 300 G2
Balcombe W Sus 51 G10
Balcombe Lane W Sus 51 G10
Balcomie Fife 287 F10
Balcraggie Lodge Highld 310 F5
Balcurvie Fife 287 G7
Baldersby N Yorks 215 D7
Baldersby St James N Yorks 215 D7
Balderstone Gtr Man 196 E2
Balderstone Lancs 203 G8
Balderton Notts 172 E4
Balderton W Ches 166 C5
Baldhu Corn 4 G5
Baldingstone Gtr Man 195 E10
Baldinnie Fife 287 F8
Baldock Herts 104 E4
Baldon Row Oxon 83 E9
Baldoon Highld 300 B6
Baldovie Dundee 287 D8
Baldrine I o M 192 D5
Baldslow E Sus 38 E3
Baldwin I o M 192 D4
Baldwinholme Cumb 239 G8
Baldwin's Gate Staffs 168 G3
Baldwins Hill W Sus 51 F11
Bale Norf 159 B10
Balearn Aberds 303 D10
Balemartine Argyll 288 E1
Balephuil Argyll 288 E1
Balerno Edin 270 B3
Balerominhor Argyll 274 D4
Balevullin Argyll 288 E1
Balfield Angus 293 G7
Balfour Orkney 314 E4
Balfour Mains Orkney 314 E4
Balfron Stirl 277 D10
Balfron Station Stirl 277 D10
Balgaveny Aberds 302 E6
Balgavies Angus 287 B9
Balgonar Fife 279 C10
Balgove Aberds 303 F8
Balgowan Highld 291 D8
Balgowan Perth 286 E3
Balgown Highld 298 C3
Balgrennie Aberds 292 C6
Balgrochan E Dunb 278 F2
Balgy Highld 299 D8
Balhaldie Stirl 286 G2
Balhalgardy Aberds 303 G7
Balham London 67 E9
Balhary Perth 286 C6
Baliasta Shetland 312 C8
Baligill Highld 310 C2
Baligortan Argyll 288 E5
Baligrundle Argyll 289 E10
Balindore Argyll 289 F11
Balinoe Argyll 288 E1
Balintore Angus 286 B6
Balintore Highld 301 B8
Balintraid Highld 301 B7
Balintuim Aberds 292 E3
Balk N Yorks 215 C9
Balk Field Notts 188 E2
Balkeerie Angus 287 C7
Balkemback Angus 287 D7
Balkholme E Yorks 199 B9
Balkissock S Ayrs 244 G4
Ball Shrops 148 D6
Ball Corn 10 G6
Ball Shrops 148 D5
Ball Green Stoke 168 E5
Ball Haye Green Staffs 169 D7
Ball Hill Hants 64 G2
Ball o'Ditton Halton 183 D7
Ballabeg I o M 192 E3
Ballacannel I o M 192 D5
Ballachraggan Moray 301 E11
Ballachrochin Highld 301 F8
Ballachulish Highld 284 B4
Balladen Lancs 195 C10
Ballajora I o M 192 C5
Ballaleigh I o M 192 D4
Ballamodha I o M 192 E3
Ballantrae S Ayrs 244 G3
Ballaquine I o M 192 D5
Ballard's Ash Wilts 62 C5
Ballards Gore Essex 88 G5
Ballard's Green Warks 134 E5
Ballasalla I o M 192 C4
Ballasalla I o M 192 E3
Ballater Aberds 292 D5
Ballathie Perth 286 D5
Ballaugh I o M 192 C4
Ballaveare I o M 192 E4
Ballcorach Moray 301 G11
Ballechin Perth 286 B3
Balleich Stirl 277 B10
Balleigh Highld 309 L7
Ballencrieff E Loth 281 F9
Ballencrieff Toll W Loth 279 G9
Ballentoul Perth 291 G10
Ballhill Devon 24 C3
Ballidon Derbys 170 E2
Balliekine N Ayrs 255 D9
Balliemore Argyll 275 E11
Balliemore Argyll 289 G10
Ballikinrain Stirl 277 D11
Ballimeanoch Argyll 284 F4
Ballimore Argyll 275 E10
Ballimore Stirl 285 F9
Ballinaby Argyll 274 G3
Ballinbreich Fife 286 E6
Ballindean Perth 286 E6
Ballingdon Suff 107 C7
Ballingdon Bottom Herts 85 C8
Ballinger Bottom Bucks 84 E6
Ballinger Bottom (South) Bucks 84 E6
Ballinger Common Bucks 84 E6
Ballingham Hereford 97 E11
Ballingham Hill Hereford 97 E11
Ballingry Fife 280 B3
Ballinlick Perth 286 C3
Ballinluig Perth 286 B3
Ballintean Highld 291 C10
Ballintuim Perth 286 B5
Balliveolan Argyll 289 E10
Balloch Angus 287 B7
Balloch Highld 301 E7
Balloch N Lnrk 278 G4
Balloch W Dunb 277 E7
Ballochan Aberds 293 D7
Ballochearn Stirl 277 D11
Ballochford Moray 302 F3
Ballochmorrie S Ayrs 244 G6
Ballogie Aberds 293 D7
Balls Cross W Sus 35 B7
Balls Green E Sus 52 F3
Balls Green Essex 107 G11
Ball's Green Glos 80 F5
Balls Green W Sus 35 C10
Balls Hill W Mid 133 E9
Ballygown Argyll 288 E6
Ballygrant Argyll 274 G4
Ballygroggan Argyll 255 F7
Ballyhaugh Argyll 288 D3
Balmacara Highld 295 C10
Balmacara Square Highld 295 C10
Balmaclellan Dumfries 237 B8
Balmacneil Perth 286 B3
Balmacqueen Highld 298 B4
Balmae Dumfries 237 E8
Balmaha Stirl 277 C8
Balmalcolm Fife 287 G7
Balmalloch N Lnrk 278 F4
Balmeanach Argyll 288 F6
Balmeanach Argyll 289 E8
Balmeanach Highld 295 B7
Balmeanach Highld 298 E3
Balmeanach Highld 298 E5
Balmedie Aberds 293 B11
Balmer Shrops 149 C8
Balmer Heath Shrops 149 C8
Balmerino Fife 287 E7
Balmerlawn Hants 32 G4
Balmesh Dumfries 236 D3
Balmichael N Ayrs 255 D10
Balminnoch Dumfries 236 C4
Balmirmer Angus 287 D9
Balmoral Borders 261 B11
Balmore E Dunb 278 G2
Balmore Highld 298 E2
Balmore Highld 300 F3
Balmore Highld 301 E8
Balmore Perth 286 B3
Balmule Fife 280 D4
Balmullo Fife 287 E8
Balmungie Highld 301 D7
Balmurrie Dumfries 236 C4
Balnaboth Angus 292 G5
Balnabreich Moray 302 D3
Balnabruaich Highld 301 C7
Balnabruich Highld 311 G5
Balnacarn Highld 290 B4
Balnacoil Highld 311 H2
Balnacra Highld 299 E9
Balnacruie Highld 301 G9
Balnafoich Highld 300 F6
Balnagall Highld 311 L2
Balnagowan Aberds 293 C7
Balnagrantach Highld 300 F4
Balnaguard Perth 286 B3
Balnahanaid Perth 285 C10
Balnahard Argyll 274 D5
Balnahard Argyll 288 F6
Balnain Highld 300 F4
Balnakeil Highld 308 C3
Balnakelly Aberds 293 B7
Balnaknock Highld 298 C4
Balnamoon Aberds 303 D9
Balnamoon Angus 293 G7
Balnapaling Highld 301 C7
Balne N Yorks 198 D5
Balnoon Corn 2 B2
Balochroy Argyll 255 B8
Balole Argyll 274 G4
Balone Fife 287 F8
Balornock Glasgow 268 B2
Balquharn Perth 286 D4
Balquhidder Stirl 285 E9
Balquhidder Station Stirl 285 E9
Balquhindachy Aberds 303 E8
Balrownie Angus 293 G7
Balsall W Mid 118 B4
Balsall Common W Mid 118 B4
Balsall Heath W Mid 133 G11
Balsall Street W Mid 118 B4
Balsam Cambs 123 G11
Balscote Oxon 101 C7
Balscott Oxon 101 C7
Balsham Cambs 123 G11
Balsporran Cottages Highld 291 F8
Balstonia Thurrock 69 C7
Baltasound Shetland 312 C8
Balterley Staffs 168 E3
Balterley Green Staffs 168 E3
Balterley Heath Staffs 168 E2
Baltersan Dumfries 236 C6
Balthangie Aberds 303 D8
Balthayock Perth 286 E5
Baltonsborough Som 44 G4
Balure Argyll 289 E11
Balvaird Highld 300 D5
Balvenie Moray 302 E3
Balvicar Argyll 275 B8
Balvraid Highld 295 D10
Balvraid Highld 301 F8
Balwest Corn 2 C3
Bamber Bridge Lancs 194 B5
Bamber's Green Essex 105 G11
Bamburgh Northumb 264 C5
Bamff Perth 286 B6
Bamford Derbys 186 E2
Bamford Gtr Man 195 E11
Bamfurlong Glos 99 G8
Bamfurlong Gtr Man 194 G5
Bampton Cumb 221 B10
Bampton Devon 27 C7
Bampton Oxon 82 E4
Bampton Grange Cumb 221 B10
Banavie Highld 290 F3
Banbury Oxon 101 C9
Banc-y-Darren Ceredig 128 G3
Bancffosfelen Carms 75 C7
Banchor Highld 301 E9
Banchory Aberds 293 D8
Banchory-Devenick Aberds 293 C11
Bancycapel Carms 74 B6
Bancyfelin Carms 74 B4
Bancyffordd Carms 93 D8
Bandirran Perth 286 D6
Bandonhill London 67 G9
Bandrake Head Cumb 210 B6
Banff Aberds 302 C6
Bangor Gwyn 179 G9
Bangor is y coed = Bangor on Dee Wrex 166 F5
Bangor on Dee = Bangor-is-y-coed Wrex 166 F5
Bangor Teifi Ceredig 93 C7
Bangors Corn 11 B10
Banham Norf 141 F11
Bank Hants 32 F3
Bank End Cumb 210 B3
Bank End Cumb 228 D6
Bank Fold Blkburn 195 C8
Bank Hey Blkburn 203 G9
Bank Houses Lancs 202 C4
Bank Lane Gtr Man 195 D9
Bank Newton N Yorks 204 C4
Bank Street Worcs 116 E2
Bank Top Gtr Man 195 E8
Bank Top Lancs 194 F4
Bank Top Stoke 168 E5
Bank Top T & W 242 D4
Bank Top W Yorks 196 C6
Bank Top W Yorks 205 F9
Bankend Dumfries 238 D2
Bankfoot Perth 286 D4
Bankglen E Ayrs 258 G4
Bankhead Aberdeen 293 B10
Bankhead Aberds 293 C8
Bankhead Dumfries 236 C2
Bankhead Falk 278 E6
Bankhead S Lnrk 269 G7
Bankland Som 28 B4
Banknock Falk 278 F5
Banks Cumb 240 E3
Banks Lancs 193 C11
Banks Orkney 314 C4
Bank's Green Worcs 117 D9
Bankshead Shrops 130 F6
Bankshill Dumfries 248 G5
Bankside Falk 279 E7
Banners Gate W Mid 133 D11
Banningham Norf 160 D4
Banniskirk Ho Highld 310 D5
Banniskirk Mains Highld 310 D5
Bannister Green Essex 106 G3
Bannockburn Stirl 278 C5
Banns Corn 4 F4
Banstead Sur 51 B8
Bantam Grove W Yorks 197 B9
Bantaskin Falk 279 F7
Bantham Devon 8 G3
Banton N Lnrk 278 F5
Banwell N Som 43 B11
Banyard's Green Suff 126 C5
Bapchild Kent 70 G2
Baptist End W Mid 133 F8
Bapton Wilts 46 F3
Bar End Hants 33 B7
Bar Hill Cambs 123 E7
Bar Hill Staffs 168 G3
Bar Moor T & W 242 E4
Barabhas W Isles 304 D5
Barabhas Iarach W Isles 304 D5
Barabhas Uarach W Isles 304 C5
Barachandroman Argyll 289 G8
Baramore Highld 289 B8
Barassie S Ayrs 257 C8
Baravullin Argyll 289 F10
Barbadoes Stirl 277 B11
Barbaraville Highld 301 B7
Barbauchlaw W Loth 269 B8
Barber Booth Derbys 185 E10
Barber Green Cumb 211 C7
Barber's Moor Lancs 194 D3
Barbican Plym 7 E9
Barbieston S Ayrs 257 F10
Barbon Cumb 212 C2
Barbourne Worcs 116 F6
Barbreck Ho Argyll 275 C9
Barbridge E Ches 167 D10
Barbrook Devon 41 D8
Barby Northants 119 C10
Barby Nortoft Northants 119 C11
Barcaldine Argyll 289 E11
Barcelona Corn 6 E4
Barcheston Warks 100 D5
Barclose Cumb 239 E10
Barcombe E Sus 36 E6
Barcombe Cross E Sus 36 D6
Barcroft W Yorks 204 F6
Barden N Yorks 224 G2
Barden Park Kent 52 D5
Barden Scale N Yorks 205 B7
Bardennoch Dumfries 246 E3
Bardfield End Green Essex 106 E2
Bardfield Saling Essex 106 F3
Bardister Shetland 312 F5
Bardnabeinne Highld 309 K7
Bardney Lincs 173 B10
Bardon Leics 153 G8
Bardon Mill Northumb 241 E7
Bardowie E Dunb 277 G11
Bardown E Sus 37 B11
Bardrainney Invclyd 276 G6
Bardrishaig Argyll 275 B8
Bardsea Cumb 210 E6
Bardsey W Yorks 206 E3
Bardsley Gtr Man 196 G2
Bardwell Suff 125 C8
Bare Lancs 211 G9
Bare Ash Som 43 F9
Bareless Northumb 263 B9
Barepot Cumb 228 F6
Bareppa Corn 3 D7
Barfad Argyll 275 G9
Barfad Dumfries 236 C5
Barford Norf 142 B2
Barford Sur 49 F11
Barford Warks 118 E5
Barford St John Oxon 101 E8
Barford St Martin Wilts 46 G5
Barford St Michael Oxon 101 E8
Barfrestone Kent 55 C9
Bargaly Dumfries 236 C6
Bargarran Renfs 277 G9
Bargate Derbys 170 F5
Bargeddie N Lnrk 268 C4
Bargoed Caerph 77 F10
Bargrennan Dumfries 236 B5
Barham Cambs 122 B2
Barham Kent 55 C8
Barham Suff 126 G2
Barharrow Dumfries 237 D8
Barhill Dumfries 237 C10
Barholm Dumfries 237 D7
Barholm Lincs 155 G11
Barkby Leics 136 B2
Barkby Thorpe Leics 136 B2
Barkers Green Shrops 149 D10
Barkers Hill Wilts 30 B6
Barkestone-le-Vale Leics 154 C5
Barkham Wokingham 65 F9
Barking London 68 C2
Barking Suff 125 G11
Barking Tye Suff 125 G11
Barkingside London 68 B2
Barkisland W Yorks 196 D5
Barkla Shop Corn 4 E4
Barkston Lincs 172 G6
Barkston N Yorks 206 F5
Barkston Ash N Yorks 206 F5
Barkway Herts 105 D7
Barlake Som 45 D7
Barlanark Glasgow 268 C3
Barland Powys 114 E5
Barland Common Swansea 56 D5
Barlaston Staffs 151 B7
Barlavington W Sus 35 D7
Barlborough Derbys 187 F7
Barlby N Yorks 207 G8
Barlestone Leics 135 B8
Barley Herts 105 D7
Barley Lancs 204 E2
Barley End Bucks 85 C7
Barley Green Lancs 204 E2
Barley Mow T & W 243 G7
Barleycroft End Herts 105 F8
Barleythorpe Rutland 136 B6
Barling Essex 70 B2
Barlings Lincs 189 G9
Barlow Derbys 186 G4
Barlow N Yorks 198 B6
Barlow T & W 242 E5
Barlow Moor Gtr Man 184 C4
Barmby Moor E Yorks 207 D11
Barmby on the Marsh E Yorks 199 B7
Barmer Norf 158 C6
Barming Heath Kent 53 B8
Barmolloch Argyll 275 D9
Barmoor Castle Northumb 263 B11
Barmoor Lane End Northumb 264 B2
Barmouth = Abermaw Gwyn 146 F2
Barmpton Darl 224 B6
Barmston E Yorks 209 B9
Barmston T & W 243 F8
Barmulloch Glasgow 268 B2
Barnaby Green Suff 127 B9
Barnacabber Argyll 276 D3
Barnack P'boro 137 B11
Barnacle Warks 135 G7
Barnaline Argyll 275 B10
Barnard Castle Durham 223 B11
Barnard Gate Oxon 82 C6
Barnardiston Suff 106 B4
Barnard's Green Worcs 98 B5
Barnardtown Newport 59 B10
Barnbarroch Dumfries 237 D10
Barnbow Carr W Yorks 206 F3
Barnburgh S Yorks 198 G3
Barnby Suff 143 F9
Barnby Dun S Yorks 198 F6
Barnby in the Willows Notts 172 E5
Barnby Moor Notts 187 E11
Barncluith S Lnrk 268 E4
Barndennoch Dumfries 247 F9
Barne Barton Plym 7 D8
Barnehurst London 68 D4
Barnes Cray London 68 D4
Barnes Hall S Yorks 186 B4
Barnes Street Kent 52 D6
Barnet London 86 F2
Barnet Gate London 86 F2
Barnetby le Wold N Lincs 200 F5
Barnett Brook E Ches 167 G10
Barnettbrook Worcs 117 B7
Barney Norf 159 C9
Barnfield Kent 54 D2
Barnfields Hereford 97 C9
Barnfields Staffs 169 D7
Barnham Suff 125 B7
Barnham W Sus 35 G7
Barnham Broom Norf 141 B11
Barnhead Angus 287 B10
Barnhill Dundee 287 D8
Barnhill Moray 301 D11
Barnhill W Ches 167 E7
Barnhills Dumfries 236 B1
Barningham Durham 223 C11
Barningham Suff 125 B9
Barningham Green Norf 160 C2
Barnmoor Green Warks 118 E3
Barnoldby le Beck NE Lincs 201 G8
Barnoldswick Lancs 204 D3
Barnoldswick N Yorks 212 E3
Barns Borders 260 B6
Barns Green W Sus 35 B10
Barnsbury London 67 C10
Barnsdale Rutland 137 B8
Barnside W Yorks 197 F7
Barnsley Glos 81 D9
Barnsley S Yorks 197 F10
Barnsley Shrops 132 E5
Barnsole Kent 55 B9
Barnstaple Devon 40 G5
Barnston Essex 87 B10
Barnston Mers 182 E3
Barnstone Notts 154 B4
Barnt Green Worcs 117 C10
Barnton Edin 280 F3
Barnton W Ches 183 F10
Barnwell Northants 137 G10
Barnwell All Saints Northants 137 G10
Barnwell St Andrew Northants 137 F10
Barnwood Glos 80 B5
Barochreal Argyll 289 G10
Barons Cross Hereford 115 F9
Barr Highld 289 D8
Barr S Ayrs 245 E7
Barr Som 27 C11
Barr Common W Mid 133 D11
Barra Castle Aberds 303 G7
Barrachan Dumfries 236 E5
Barrachnie Glasgow 268 C3
Barrack Aberds 303 E8
Barrack Hill Newport 59 B10
Barraer Dumfries 236 C5
Barraglom W Isles 304 E3
Barrahormid Argyll 275 E8
Barran Argyll 289 G10
Barranrioch Argyll 289 G10
Barrapol Argyll 288 E1
Barras Aberds 293 E10
Barras Cumb 222 C6
Barrasford Northumb 241 C10
Barravullin Argyll 275 C9
Barregarrow I o M 192 D4
Barrets Green E Ches 167 D9
Barrhead E Renf 267 D9
Barrhill S Ayrs 244 G6
Barrington Cambs 105 B7
Barrington Som 28 D5
Barripper Corn 2 B4
Barrmill N Ayrs 267 E7
Barrock Highld 310 B6
Barrock Ho Highld 310 C6
Barrow Glos 99 G7
Barrow Lancs 203 F10
Barrow Rutland 155 F7
Barrow S Yorks 186 B5
Barrow Shrops 132 C3
Barrow Som 44 E5
Barrow Suff 124 E5
Barrow Bridge Gtr Man 195 E7
Barrow Burn Northumb 263 G9
Barrow Common N Som 60 F4
Barrow Green Kent 70 G3
Barrow Gurney N Som 60 F4
Barrow Hann N Lincs 200 C5
Barrow Haven N Lincs 200 C5
Barrow Hill Derbys 186 F6
Barrow Hill Dorset 18 B5
Barrow-in-Furness Cumb 210 F4
Barrow Island Cumb 210 F3
Barrow Nook Lancs 194 G2
Barrow Street Wilts 45 G10
Barrow upon Humber N Lincs 200 C5
Barrow upon Soar Leics 153 F11
Barrow upon Trent Derbys 153 D7
Barrow Vale Bath 60 G6
Barrow Wake Glos 80 B6
Barroway Drove Norf 139 C11
Barrowburn Northumb 263 G9
Barrowby Lincs 155 B7
Barrowcliff N Yorks 217 B10
Barrowden Rutland 137 C8
Barrowford Lancs 204 F3
Barrowhill Kent 54 F6
Barrowmore Estate W Ches 167 B7
Barrows Green Cumb 211 B10
Barrows Green E Ches 167 D11
Barrow's Green Mers 183 D8
Barrows Green Notts 171 E7
Barry Angus 287 D9
Barry V Glam 58 F6
Barry Dock V Glam 58 F6
Barry Island V Glam 58 F6
Barsby Leics 154 G3
Barsham Suff 143 F7
Barshare E Ayrs 258 F3
Barstable Essex 69 B8
Barston W Mid 118 B4
Bartestree Hereford 97 C11
Barthol Chapel Aberds 303 F8
Bartholomew Green Essex 106 G4
Barthomley E Ches 168 E3
Bartington W Ches 183 F10
Bartley Hants 32 E4
Bartley Green W Mid 133 G10
Bartlow Cambs 105 B11
Barton Cambs 123 F8
Barton Glos 80 B4
Barton Glos 99 F11
Barton I o W 20 D6
Barton Lancs 193 F11
Barton Lancs 202 F6
Barton N Som 43 B11
Barton N Yorks 224 D4
Barton Oxon 83 D9
Barton Torbay 9 B8
Barton W Ches 166 E6
Barton Warks 118 G2
Barton Abbey Oxon 101 G9
Barton Bendish Norf 140 B4
Barton Court Hereford 98 C4
Barton End Glos 80 F4
Barton Gate Devon 41 E7
Barton Gate Staffs 152 F3
Barton Green Staffs 152 F3
Barton Hartshorn Bucks 102 E2
Barton Hill Bristol 60 E6
Barton Hill N Yorks 216 G4
Barton in Fabis Notts 153 C10
Barton in the Beans Leics 135 B7
Barton-le-Clay C Beds 103 E11
Barton-le-Street N Yorks 216 E4
Barton-le-Willows N Yorks 216 G4
Barton Mills Suff 124 C4
Barton on Sea Hants 19 C10
Barton on the Heath Warks 100 E5
Barton St David Som 44 G4
Barton Seagrave Northants 121 B7
Barton Stacey Hants 48 E2
Barton Town Devon 41 E7
Barton Turf Norf 161 E7
Barton Turn Staffs 152 F4
Barton-under-Needwood Staffs 152 F3
Barton-upon-Humber N Lincs 200 C4
Barton Upon Irwell Gtr Man 184 B3
Barton Waterside N Lincs 200 C4
Bartonsham Hereford 97 D10
Barugh S Yorks 197 F10
Barugh Green S Yorks 197 F10
Barway Cambs 123 B10
Barwell Leics 135 D8
Barwell London 67 G7
Barwick Devon 25 F9
Barwick Herts 86 B5
Barwick Som 29 E9
Barwick in Elmet W Yorks 206 F3
Baschurch Shrops 149 E8
Bascote Warks 119 E8
Bascote Heath Warks 119 E7
Base Green Suff 125 E10
Basford Shrops 131 F7
Basford Staffs 168 F5
Basford Green Staffs 169 E7
Bashall Eaves Lancs 203 E9
Bashley Hants 19 B10
Bashley Park Hants 19 B10
Basildon Essex 69 B8
Basingstoke Hants 48 C6
Baslow Derbys 186 G3
Bason Bridge Som 43 D10
Bassaleg Newport 59 B9
Bassenthwaite Cumb 229 E10
Bassett S Yorks 186 E3
Bassett Soton 32 D6
Bassett Green Soton 32 D6
Bassingbourn Cambs 104 C6
Bassingfield Notts 154 B2
Bassingham Lincs 172 C6
Bassingthorpe Lincs 155 D9
Bassus Green Herts 104 F6
Basta Shetland 312 D7
Basted Kent 52 B6
Baston Lincs 156 G2
Bastonford Worcs 116 G6
Bastwick Norf 161 F8
Baswich Staffs 151 E8
Baswick Steer E Yorks 209 D7
Batavaime Stirl 285 D8
Batch Som 43 B10
Batchcott Shrops 115 C9
Batchfields Hereford 98 B3
Batchley Worcs 117 D10
Batchworth Herts 85 G9
Batchworth Heath Herts 85 G9
Batcombe Dorset 29 G10
Batcombe Som 45 F7
Bate Heath E Ches 183 F11
Bateman's Green Worcs 117 B11
Bateman's Hill Pembs 73 E8
Batemoor S Yorks 186 E5
Batford Herts 85 B10
Bath Bath 61 F8
Bath Side Essex 108 E5
Bath Vale E Ches 168 C5
Bathampton Bath 61 F9
Bathealton Som 27 C9
Batheaston Bath 61 F9
Bathford Bath 61 F9
Bathgate W Loth 269 B9
Bathley Notts 172 D3
Bathpool Corn 11 G11
Bathpool Som 28 B3
Bathville W Loth 269 B8
Bathway Som 44 C5
Bathwick Bath 61 F9
Batlers Green Herts 85 F11
Batley W Yorks 197 C8
Batley Carr W Yorks 197 C8
Batsford Glos 100 E3
Batson Devon 9 G9
Batsworthy Devon 26 D4
Batten's Green Som 28 D3
Battenton Green Worcs 116 D6
Battersby N Yorks 225 D11
Battersea London 67 D9
Battisborough Cross Devon 7 F11
Battisford Suff 125 G11
Battisford Tye Suff 125 G10
Battle E Sus 38 D2
Battle Powys 95 E10
Battle Hill T & W 243 D8
Battledown Glos 99 G9
Battledown Cross Devon 25 F7
Battlefield Shrops 149 F10
Battlesbridge Essex 88 G3
Battlescombe Glos 80 D6
Battlesden C Beds 103 F9
Battlesea Green Suff 126 B4
Battleton Som 26 B6
Battlies Green Suff 125 E7
Battram Leics 135 B8
Battramsley Hants 20 B2
Battramsley Cross Hants 20 B2
Batt's Corner Hants 49 E10
Battyeford W Yorks 197 C7
Batworthy Devon 13 D10
Bauds of Cullen Moray 302 C4
Baugh Argyll 288 E2
Baughton Worcs 99 C7
Baughurst Hants 48 B5
Baulking Oxon 82 G4
Baumber Lincs 190 G2
Baunton Glos 81 E8
Baverstock Wilts 46 G4
Bawburgh Norf 142 B3
Bawdeswell Norf 159 E10
Bawdrip Som 43 F10
Bawdsey Suff 108 C6
Bawsey Norf 158 F3
Bawtry S Yorks 187 C11
Baxenden Lancs 195 B9
Baxterley Warks 134 D5
Baxter's Green Suff 124 F5
Bay Dorset 30 B4
Bay Highld 298 D2
Bay Gate Lancs 203 D11
Bay Horse Lancs 202 C5
Bay View Kent 70 E4
Baybridge Hants 33 C8
Baybridge Northumb 241 G11
Baycliff Cumb 210 E5
Baydon Wilts 63 D9
Bayford Herts 86 D4
Bayford Som 30 B2
Bayles Cumb 231 C10
Bayley's Hill Kent 52 C4
Baylham Suff 126 G2
Baylis Green Worcs 117 C11
Baynard's Green Oxon 101 F10
Baynhall Worcs 99 B7
Baysham Hereford 97 F11
Bayston Hill Shrops 131 B9
Bayswater London 67 C9
Baythorne End Essex 106 C4
Baythorpe Lincs 174 G2
Bayton Worcs 116 C3
Bayton Common Worcs 116 C4
Bayworth Oxon 83 E8
Beach Highld 289 D9
Beach S Glos 61 E8
Beach Hay Worcs 116 C4
Beachampton Bucks 102 D5
Beachamwell Norf 140 B5
Beachans Moray 301 E10
Beacharr Argyll 255 C7
Beachborough Kent 55 F7
Beachlands E Sus 23 E11
Beachley Glos 79 G9
Beacon Corn 2 B5
Beacon Devon 27 F11
Beacon Devon 28 F2
Beacon Down E Sus 37 C9
Beacon End Essex 107 G9
Beacon Hill Bath 61 F9
Beacon Hill Bucks 84 G6
Beacon Hill Cumb 210 E4
Beacon Hill Dorset 18 C5
Beacon Hill Essex 88 C5
Beacon Hill Kent 53 G10
Beacon Hill Notts 172 E4
Beacon Hill Suff 108 B4
Beacon Hill Sur 49 F11
Beacon Lough T & W 243 F7
Beaconhill Northumb 243 B7
Beacon's Bottom Bucks 84 F3
Beaconsfield Bucks 66 B2
Beaconside Staffs 151 E8
Beacrabhaic W Isles 305 J3
Beadlam N Yorks 216 C3
Beadlow C Beds 104 D2
Beadnell Northumb 264 D6
Beaford Devon 25 E9
Beal N Yorks 198 B4
Beal Northumb 273 G11
Bealach Highld 289 D11
Bealach Maim Argyll 275 E10
Bealbury Corn 7 B7
Beal's Green Kent 53 G9
Bealsmill Corn 12 F3
Beam Bridge Som 27 D10
Beam Hill Staffs 152 D4
Beambridge Shrops 131 F10
Beamhurst Staffs 151 B11
Beamhurst Lane Staffs 151 B11
Beaminster Dorset 29 G7
Beamish Durham 242 G6
Beamond End Bucks 84 F6
Beamsley N Yorks 205 C7
Bean Kent 68 E5
Beanacre Wilts 62 F2
Beancross Falk 279 F8
Beanhill M Keynes 103 D7
Beanley Northumb 264 F3
Beansburn E Ayrs 257 B10
Beanthwaite Cumb 210 C4
Beaquoy Orkney 314 D3
Bear Cross Bmouth 19 B7
Beard Hill Som 44 E6
Beardly Batch Som 44 E6
Beardwood Blkburn 195 B7
Beare Devon 27 G7
Beare Green Sur 51 E7
Bearley Warks 118 E3
Bearley Cross Warks 118 E3
Bearnus Argyll 288 E5
Bearpark Durham 233 C10
Bearsbridge Northumb 241 F7
Bearsden E Dunb 277 G10
Bearsted Kent 53 B9
Bearstone Shrops 150 B4
Bearwood Hereford 115 F7
Bearwood Poole 18 B6
Bearwood W Mid 133 F10
Beasley Staffs 168 F4
Beattock Dumfries 248 C3
Beauchamp Roding Essex 87 C9
Beauchief S Yorks 186 E4
Beauclerc Northumb 242 E2
Beaudesert Warks 118 D3
Beaufort Bl Gwent 77 C11
Beaufort Castle Highld 300 E5
Beaulieu Hants 32 G5
Beaulieu Wood Dorset 30 F2
Beauly Highld 300 E5
Beaumaris Anglesey 179 F10
Beaumont Cumb 239 F8
Beaumont Essex 108 G3
Beaumont Windsor 66 E3
Beaumont Hill Darl 224 B5
Beaumont Leys Leicester 135 B11
Beausale Warks 118 C4
Beauvale Notts 171 F8
Beauworth Hants 33 B9
Beavan's Hill Hereford 98 G3
Beaworthy Devon 12 B5
Beazley End Essex 106 F4
Bebington Mers 182 E4
Bebside Northumb 253 G7
Beccles Suff 143 E8
Becconsall Lancs 194 C2
Beck Bottom Cumb 210 C5
Beck Bottom W Yorks 197 C10
Beck Foot Cumb 222 F2
Beck Foot W Yorks 205 F8
Beck Head Cumb 211 C8
Beck Hole N Yorks 226 E6
Beck Houses Cumb 221 F11
Beck Row Suff 124 B3
Beck Side Cumb 210 C4
Beck Side Cumb 211 C7
Beckbury Shrops 132 C5
Beckces Cumb 230 F4
Beckenham London 67 F11
Beckermet Cumb 219 D10
Beckermonds N Yorks 213 C7
Beckery Som 44 F3
Beckett End Norf 140 D5
Beckfoot Cumb 220 E3
Beckfoot Cumb 229 B7
Beckford Worcs 99 D9
Beckhampton Wilts 62 F5
Beckingham Lincs 172 E5

Beckingham Notts 188 D3
Beckington Som 45 C10
Beckjay Shrops 115 B7
Beckley E Sus 38 C5
Beckley Hants 19 B10
Beckley Oxon 83 C9
Beckley Furnace E Sus 38 C4
Beckside Cumb 212 B2
Beckton London 68 C2
Beckwith N Yorks 205 C11
Beckwithshaw N Yorks 205 C11
Becontree London 68 B3
Bed-y-coedwr Gwyn 146 D4
Bedale N Yorks 214 B5
Bedburn Durham 233 E8
Bedchester Dorset 30 D5
Beddau Rhondda 58 B5
Beddgelert Gwyn 163 F9
Beddingham E Sus 36 F6
Beddington London 67 G10
Beddington Corner London 67 F9
Bedfield Suff 126 D4
Bedford Beds 121 G11
Bedford Gtr Man 183 B11
Bedford Park London 67 D8
Bedgebury Cross Kent 53 G8
Bedgrove Bucks 84 C4
Bedham W Sus 35 C8
Bedhampton Hants 22 B2
Bedingfield Suff 126 D3
Bedingham Green Norf 142 E5
Bedlam N Yorks 214 G5
Bedlam Som 45 D9
Bedlam Street W Sus 36 D3
Bedlar's Green Essex 105 G10
Bedlington Northumb 253 G7
Bedlington Station Northumb 253 G7
Bedlinog M Tydf 77 E9
Bedminster Bristol 60 E5
Bedminster Down Bristol 60 F5
Bedmond Herts 85 E9
Bednall Staffs 151 F9
Bednall Head Staffs 151 F9
Bedrule Borders 262 F4
Bedstone Shrops 115 B7
Bedwas Caerph 59 B7
Bedwell Herts 104 G4
Bedwell Wrex 166 F5
Bedwellty Caerph 77 E11
Bedwellty Pits Bl Gwent 77 D11
Bedwlwyn Wrex 148 B4
Bedworth Warks 135 F7
Bedworth Heath Warks 134 F6
Bedworth Woodlands Warks 134 F6
Beeby Leics 136 B3
Beech Hants 49 F7
Beech Staffs 151 B7
Beech Hill Gtr Man 194 F5
Beech Hill W Berks 65 G7
Beech Lanes W Mid 133 F10
Beechcliff Staffs 151 B7
Beechcliffe W Yorks 205 E7
Beechen Cliff Bath 61 G9
Beechingstoke Wilts 46 B5
Beechwood Halton 183 E8
Beechwood Newport 59 B10
Beechwood W Mid 118 B5
Beechwood W Yorks 206 F2
Beecroft C Beds 103 G10
Beedon W Berks 64 D3
Beedon Hill W Berks 64 D3
Beeford E Yorks 209 C8
Beeley Derbys 170 B3
Beelsby NE Lincs 201 G8
Beenham W Berks 64 F5
Beenham Stocks W Berks 64 F5
Beenham's Heath Windsor 65 D10
Beeny Corn 11 C8
Beer Devon 15 D10
Beer Som 44 G2
Beer Hackett Dorset 29 E9
Beercrocombe Som 28 C4
Beesands Devon 8 G6
Beesby Lincs 191 E7
Beeslack Midloth 270 C4
Beeson Devon 8 G6
Beeston C Beds 104 B3
Beeston Norf 159 F8
Beeston Notts 153 B10
Beeston W Ches 167 D8
Beeston W Yorks 205 G11
Beeston Hill W Yorks 205 G11
Beeston Park Side W Yorks 197 B9
Beeston Regis Norf 177 E11
Beeston Royds W Yorks 205 G11
Beeston St Lawrence Norf 160 E6
Beeswing Dumfries 237 C10
Beetham Cumb 211 D9
Beetham Som 28 E3
Beetley Norf 159 F9
Beffcote Staffs 150 F6
Began Cardiff 59 C8
Begbroke Oxon 83 C7
Begdale Cambs 139 B9
Begelly Pembs 73 D10
Beggar Hill Essex 87 E10
Beggarington Hill W Yorks 197 C9
Beggars Ash Hereford 98 D4
Beggar's Bush Powys 114 E5
Beggars Bush W Sus 35 F11
Beggars Pound V Glam 58 F4
Beggearn Huish Som 42 F4
Beguildy Powys 114 B3
Beighton Norf 143 B7
Beighton S Yorks 186 E6
Beighton Hill Derbys 170 E3
Beili-glas Mon 78 C4
Beitearsaig W Isles 305 G1
Beith N Ayrs 266 E6
Bekesbourne Kent 55 B7
Bekesbourne Hill Kent 55 B7
Belah Cumb 239 F9
Belan Powys 130 C4
Belaugh Norf 160 F5
Belbins Hants 32 C5
Belbroughton Worcs 117 B8
Belchalwell Dorset 30 F3
Belchalwell Street Dorset 30 F3
Belchamp Otten Essex 106 C6
Belchamp St Paul Essex 106 C5
Belchamp Walter Essex 106 C6
Belcher's Bar Leics 135 B8
Belchford Lincs 190 F3
Beleybridge Fife 287 F9
Belfield Gtr Man 196 E2
Belford Northumb 264 C4
Belgrano Conwy 181 F7
Belgrave Leicester 135 B11
Belgrave Staffs 134 C4
Belgrave W Ches 166 C5
Belgravia London 67 D9
Belhaven E Loth 282 F3
Belhelvie Aberds 293 B11
Belhinnie Aberds 302 G4
Bell Bar Herts 86 D3
Bell Busk N Yorks 204 B4
Bell Common Essex 86 E6
Bell End Worcs 117 B8
Bell Green London 67 E11
Bell Green W Mid 135 G7
Bell Heath Worcs 117 B9
Bell Hill Hants 34 C2
Bell o' th' Hill W Ches 167 F8
Bellabeg Aberds 292 B5
Bellamore S Ayrs 244 F6
Bellanoch Argyll 275 D8
Bellanrigg Borders 260 B6
Bellasize E Yorks 199 B10
Bellaty Angus 286 B6
Belle Eau Park Notts 171 D11
Belle Green S Yorks 197 F11
Belle Isle W Yorks 197 B10
Belle Vale Mers 182 D6
Belle Vale W Mid 133 G9
Belle Vue Cumb 229 E8
Belle Vue Cumb 239 F9
Belle Vue Gtr Man 184 B5
Belle Vue S Yorks 198 G5
Belle Vue Shrops 149 G9
Belle Vue W Yorks 197 D10
Belleau Lincs 190 F6
Bellehiglash Moray 301 F11
Bellerby N Yorks 224 G2
Bellever Devon 13 F9
Bellevue Worcs 117 C9
Bellfield E Ayrs 257 B10
Bellfields Sur 50 C3
Belliehill Angus 293 G7
Bellingdon Bucks 84 D6
Bellingham London 67 E11
Bellingham Northumb 251 G8
Bellmount Norf 157 E10
Belloch Argyll 255 D7
Bellochantuy Argyll 255 D7
Bell's Close T & W 242 E5
Bell's Corner Suff 107 D9
Bells Yew Green E Sus 52 F6
Bellsbank E Ayrs 245 C11
Bellshill N Lnrk 268 C4
Bellshill Northumb 264 C4
Bellside N Lnrk 268 D6
Bellsmyre W Dunb 277 F8
Bellspool Borders 260 B5
Bellsquarry W Loth 269 C10
Belluton Bath 60 G6
Belmaduthy Highld 300 D6
Belmesthorpe Rutland 155 G10
Belmont Blkburn 195 D7
Belmont Durham 234 C2
Belmont E Sus 38 E4
Belmont London 67 G9
Belmont London 85 G11
Belmont Oxon 63 B11
Belmont S Ayrs 257 E8
Belmont Shetland 312 C7
Belnacraig Aberds 292 B5
Belnagarrow Moray 302 E3
Belnie Lincs 156 C5
Belowda Corn 5 C9
Belper Derbys 170 F4
Belper Lane End Derbys 170 F4
Belph Derbys 187 F8
Belsay Northumb 242 B4
Belses Borders 262 D3
Belsford Devon 8 D5
Belsize Herts 85 E8
Belstead Suff 108 C2
Belston S Ayrs 257 E9
Belstone Devon 13 C8
Belstone Corner Devon 13 B8
Belthorn Blkburn 195 C8
Beltinge Kent 71 F7
Beltingham Northumb 241 E7
Beltoft N Lincs 199 F10
Belton Leics 153 E8
Belton Lincs 155 B8
Belton N Lincs 199 F9
Belton Norf 143 C9
Belton in Rutland Rutland 136 C6
Beltring Kent 53 D7
Belts of Collonach Aberds 293 D8
Belvedere London 68 D3
Belvedere W Loth 269 B9
Belvoir Leics 154 C6
Bembridge I o W 21 D8
Bemersyde Borders 262 C3
Bemerton Wilts 46 G6
Bemerton Heath Wilts 46 G6
Bempton E Yorks 218 E3
Ben Alder Lodge Highld 291 F7
Ben Armine Lodge Highld 309 H7
Ben Casgro W Isles 304 F6
Ben Rhydding W Yorks 205 D8
Benacre Suff 143 G10
Benbuie Dumfries 246 D6
Benchill Gtr Man 184 D4
Bencombe Glos 80 F3
Benderloch Argyll 289 F11
Bendish Herts 104 G3
Bendronaig Lodge Highld 299 F10
Benenden Kent 53 G10
Benfield Dumfries 236 C5
Benfieldside Durham 242 G3
Bengal Pembs 91 E9
Bengate Norf 160 D6
Bengeo Herts 86 C4
Bengeworth Worcs 99 C10
Bengrove Glos 99 E9
Benhall Glos 99 G8
Benhall Green Suff 127 E7
Benhall Street Suff 127 E7
Benhilton London 67 F9
Benholm Aberds 293 G10
Beningbrough N Yorks 206 B6
Benington Herts 104 G5
Benington Lincs 174 F5
Benington Sea End Lincs 174 F6
Benllech Anglesey 179 E8
Benmore Argyll 276 E2
Benmore Stirl 285 E8
Benmore Lodge Argyll 289 F7
Benmore Lodge Highld 309 H3
Bennacott Corn 11 C11
Bennah Devon 14 E2
Bennan N Ayrs 255 E10
Bennane Lea S Ayrs 244 F3
Bennetland E Yorks 199 B10
Bennetsfield Highld 300 D6
Bennett End Bucks 84 F3
Bennetts End Herts 85 D9
Benniworth Lincs 190 E2
Benover Kent 53 D8
Bensham T & W 242 E6
Benslie N Ayrs 266 G6
Benson Oxon 83 G10
Benston Shetland 313 H6
Bent Aberds 293 F8
Bent Gate Lancs 195 C9
Benter Som 44 D6
Bentfield Bury Essex 105 F9
Bentfield Green Essex 105 F10
Bentgate Gtr Man 196 E2
Benthall Northumb 264 D6
Benthall Shrops 132 C3
Bentham Glos 80 B6
Benthoul Aberdeen 293 C10
Bentilee Stoke 168 F6
Bentlass Pembs 73 E7
Bentlawnt Shrops 130 C6
Bentley E Yorks 208 F6
Bentley Essex 87 F9
Bentley Hants 49 E9
Bentley S Yorks 198 F5
Bentley Suff 108 D2
Bentley W Mid 133 D9
Bentley Warks 134 D5
Bentley Worcs 117 D9
Bentley Common Warks 134 D5
Bentley Heath Herts 86 F2
Bentley Heath W Mid 118 B3
Bentley Rise S Yorks 198 G5
Benton Devon 41 F7
Benton Green W Mid 118 B5
Bentpath Dumfries 249 E8
Bents W Loth 269 C9
Bents Head W Yorks 205 F7
Bentwichen Devon 41 G8
Bentworth Hants 49 E7
Benvie Dundee 287 D7
Benville Dorset 29 G8
Benwell T & W 242 E6
Benwick Cambs 138 E6
Beobridge Shrops 132 E5
Beoley Worcs 117 D11
Beoraidbeg Highld 295 F8
Bepton W Sus 34 D5
Berden Essex 105 F9
Bere Alston Devon 7 B8
Bere Ferrers Devon 7 C9
Bere Regis Dorset 18 C2
Berechurch Essex 107 G9
Berefold Aberds 303 F9
Berepper Corn 2 E5
Bergh Apton Norf 142 C6
Berghers Hill Bucks 66 B2
Berhill Som 44 F2
Berinsfield Oxon 83 F9
Berkeley Glos 79 F11
Berkeley Heath Glos 79 F11
Berkeley Road Glos 80 E2
Berkeley Towers E Ches 167 E11
Berkhamsted Herts 85 D7
Berkley Som 45 D10
Berkley Down Som 45 D9
Berkley Marsh Som 45 D10
Berkswell W Mid 118 B4
Bermondsey London 67 D10
Bermuda Warks 135 F7
Bernards Heath Herts 85 D11
Bernera Highld 295 C10
Berner's Cross Devon 25 F10
Berner's Hill E Sus 53 G8
Berners Roding Essex 87 D10
Bernice Argyll 276 C2
Bernisdale Highld 298 D4
Berrick Salome Oxon 83 G10
Berriedale Highld 311 G5
Berrier Cumb 230 F3
Berriew = Aberriw Powys 130 C3
Berrington Northumb 273 G10
Berrington Shrops 131 B10
Berrington Worcs 115 D11
Berrington Green Worcs 115 D11
Berrow Som 43 C10
Berrow Worcs 98 E5
Berrow Green Worcs 116 F4
Berry Swansea 56 D3
Berry Brow W Yorks 196 E6
Berry Cross Devon 25 E7
Berry Down Cross Devon 40 E5
Berry Hill Glos 79 C9
Berry Hill Pembs 91 C11
Berry Hill Stoke 168 F6
Berry Hill Worcs 117 E7
Berry Moor S Yorks 197 G9
Berry Pomeroy Devon 8 C6
Berryfield Wilts 61 G11
Berrygate Hill E Yorks 201 C8
Berryhillock Moray 302 C5
Berrylands London 67 F7
Berrynarbor Devon 40 D5
Berry's Green London 52 B2
Berrysbridge Devon 26 G6
Bersham Wrex 166 F4
Berstane Orkney 314 E4
Berth-ddu Flint 166 B2
Berthengam Flint 181 F10
Berwick E Sus 23 D8
Berwick Kent 54 F6
Berwick S Glos 60 C5
Berwick Bassett Wilts 62 E5
Berwick Hill Northumb 242 B5
Berwick Hills M'bro 225 B10
Berwick St James Wilts 46 F5
Berwick St John Wilts 30 C6
Berwick St Leonard Wilts 46 G2
Berwick-upon-Tweed Northumb 273 E9
Berwick Wharf Shrops 149 G10
Berwyn Denb 165 G11
Bescaby Leics 154 D6
Bescar Lancs 193 E11
Bescot W Mid 133 D10
Besford Shrops 149 E11
Besford Worcs 99 C8
Bessacarr S Yorks 198 G6
Bessels Green Kent 52 B4
Bessels Leigh Oxon 83 E7
Besses o' th' Barn Gtr Man 195 F10
Bessingby E Yorks 218 F3
Bessingham Norf 160 B3
Best Beech Hill E Sus 52 G6
Besthorpe Norf 141 D11
Besthorpe Notts 172 C4
Bestwood Nottingham 171 G9
Bestwood Village Notts 171 F9
Beswick E Yorks 208 D6
Beswick Gtr Man 184 B5
Betchcott Shrops 131 D8
Betchton Heath E Ches 168 C3
Betchworth Sur 51 D8
Bethania Ceredig 111 E11
Bethania Gwyn 163 E10
Bethania Gwyn 164 F2
Bethany Corn 6 D6
Bethel Anglesey 178 G5
Bethel Corn 5 E10
Bethel Gwyn 147 B9
Bethel Gwyn 163 B8
Bethelnie Aberds 303 F7
Bethersden Kent 54 E2
Bethesda Gwyn 163 B10
Bethesda Pembs 73 B9
Bethlehem Carms 94 F3
Bethnal Green London 67 C10
Betley Staffs 168 F3
Betley Common Staffs 168 F2
Betsham Kent 68 E6
Betteshanger Kent 55 C10
Bettiscombe Dorset 16 B3
Bettisfield Wrex 149 B9
Betton Shrops 130 C6
Betton Shrops 150 B3
Betton Strange Shrops 131 B10
Bettws Bridgend 58 B2
Bettws Mon 78 B3
Bettws Newport 78 G3
Bettws Cedewain Powys 130 D2
Bettws Gwerfil Goch Denb 165 F8
Bettws Ifan Ceredig 92 B6
Bettws Newydd Mon 78 D5
Bettws-y-crwyn Shrops 130 G4
Bettyhill Highld 308 C7
Betws Bridgend 57 D11
Betws Carms 75 C10
Betws Bledrws Ceredig 111 G11
Betws-Garmon Gwyn 163 D8
Betws Ifan Ceredig 92 B6
Betws-y-Coed Conwy 164 D4
Betws-yn-Rhos Conwy 180 G6
Beulah Ceredig 92 B5
Beulah Powys 113 G8
Bevendean Brighton 36 F4
Bevercotes Notts 187 G11
Bevere Worcs 116 F6
Beverley E Yorks 208 F6
Beverston Glos 80 G5
Bevington Glos 79 F11
Bewaldeth Cumb 229 E10
Bewbush W Sus 51 F8
Bewcastle Cumb 240 C3
Bewdley Worcs 116 B5
Bewerley N Yorks 214 G3
Bewholme E Yorks 209 C9
Bewley Common Wilts 62 F2
Bewlie Borders 262 D3
Bewlie Mains Borders 262 D3
Bewsey Warr 183 D9
Bexfield Norf 159 D10
Bexhill E Sus 38 F2
Bexley London 68 E3
Bexleyheath London 68 D3
Bexleyhill W Sus 34 B6
Bexon Kent 53 B11
Bexwell Norf 140 C2
Beyton Suff 125 E8
Beyton Green Suff 125 E8
Bhalasaigh W Isles 304 E3
Bhaltos W Isles 304 E2
Bhatarsaigh W Isles 297 M2
Bhlàraidh Highld 290 B5
Bibstone S Glos 79 G11
Bibury Glos 81 D10
Bicester Oxon 101 G11
Bickenhall Som 28 D3
Bickenhill W Mid 134 G3
Bicker Lincs 156 B4
Bicker Bar Lincs 156 B4
Bicker Gauntlet Lincs 156 B4
Bickershaw Gtr Man 194 G6
Bickerstaffe Lancs 194 G2
Bickerton Devon 9 G11
Bickerton E Ches 167 E8
Bickerton N Yorks 206 C5
Bickford Staffs 151 G7
Bickham Som 42 E3
Bickingcott Devon 26 B3
Bickington Devon 13 G11
Bickington Devon 40 G4
Bickleigh Devon 7 C10
Bickleigh Devon 26 F6
Bickleton Devon 40 G4
Bickley London 68 F2
Bickley W Ches 167 F8
Bickley Worcs 116 C2
Bickley Moss W Ches 167 F8
Bickley Town W Ches 167 F8
Bickleywood W Ches 167 F8
Bickmarsh Warks 100 B2
Bicknacre Essex 88 E3
Bicknoller Som 42 F6
Bicknor Kent 53 B11
Bickton Hants 31 E11
Bicton Hereford 115 E9
Bicton Pembs 72 D4
Bicton Shrops 130 G5
Bicton Shrops 149 F8
Bicton Heath Shrops 149 G9
Bidborough Kent 52 E5
Bidden Hants 49 D8
Biddenden Kent 53 F11
Biddenden Green Kent 53 E11
Biddenham Beds 103 B10
Biddestone Wilts 61 E11
Biddick T & W 243 F8
Biddick Hall T & W 243 E9
Biddisham Som 43 C11
Biddlesden Bucks 102 C2
Biddlestone Northumb 251 B11
Biddulph Staffs 168 D5
Biddulph Moor Staffs 168 D6
Bideford Devon 25 B7
Bidford-on-Avon Warks 118 G2
Bidlake Devon 12 D5
Bidston Mers 182 C3
Bidston Hill Mers 182 D3
Bidwell C Beds 103 G10
Bielby E Yorks 207 E11
Bieldside Aberdeen 293 C10
Bierley I o W 20 F6
Bierley W Yorks 205 G9
Bierton Bucks 84 B4
Big Mancot Flint 166 B4
Big Sand Highld 299 B7
Bigbury Devon 8 F3
Bigbury-on-Sea Devon 8 G3
Bigby Lincs 200 F5
Bigfrith Windsor 65 C11
Biggar Cumb 210 F3
Biggar S Lnrk 260 B2
Biggar Road N Lnrk 268 C5
Biggin Derbys 169 D11
Biggin Derbys 170 F3
Biggin N Yorks 206 F6
Biggin Thurrock 69 D7
Biggin Hill London 52 B2
Biggings Shetland 313 G3
Biggleswade C Beds 104 C3
Bighouse Highld 310 C2
Bighton Hants 48 G6
Biglands Cumb 239 G7
Bignall End Staffs 168 E4
Bignor W Sus 35 E7
Bigods Essex 106 G2
Bigram Stirl 285 G10
Bigrigg Cumb 219 C10
Bigswell Orkney 314 E3
Bigton Shetland 313 L5
Bilberry Corn 5 C10
Bilborough Nottingham 171 G8
Bilbrook Som 42 E4
Bilbrook Staffs 133 C7
Bilbrough N Yorks 206 D6
Bilbster Highld 310 D6
Bilby Notts 187 E10
Bildershaw Durham 233 G10
Bildeston Suff 107 B9
Bill Quay T & W 243 E7
Billacombe Plym 7 E10
Billacott Corn 11 C11
Billericay Essex 87 G11
Billesdon Leics 136 C4
Billesley W Mid 133 G11
Billesley Warks 118 F2
Billesley Common W Mid 133 G11
Billingborough Lincs 156 C2
Billinge Mers 194 G4
Billingford Norf 126 B3
Billingford Norf 159 E10
Billingham Stockton 234 G5
Billinghay Lincs 173 E11
Billingley S Yorks 198 G2
Billingshurst W Sus 35 B9
Billingsley Shrops 132 F4
Billington C Beds 103 G8
Billington Lancs 203 F10
Billington Staffs 151 E7
Billockby Norf 161 G8
Billy Mill T & W 243 D8
Billy Row Durham 233 D9
Bilmarsh Shrops 149 D9
Bilsborrow Lancs 202 F6
Bilsby Lincs 191 F7
Bilsby Field Lincs 191 F7
Bilsdon Devon 14 C2
Bilsham W Sus 35 G7
Bilsington Kent 54 G4
Bilson Green Glos 79 C11
Bilsthorpe Notts 171 C10
Bilsthorpe Moor Notts 171 D11
Bilston Midloth 270 C5
Bilston W Mid 133 D9
Bilstone Leics 135 B7
Bilting Kent 54 D5
Bilton E Yorks 209 G9
Bilton N Yorks 206 B2
Bilton Northumb 264 G6
Bilton Warks 119 C9
Bilton Haggs N Yorks 206 D5
Bilton in Ainsty N Yorks 206 D5
Bimbister Orkney 314 E3
Binbrook Lincs 190 C2
Binchester Blocks Durham 233 E10
Bincombe Dorset 17 E9
Bincombe Som 43 F7
Bindal Highld 311 L3
Bindon Som 27 C10
Binegar Som 44 D6
Bines Green W Sus 35 D11
Binfield Brack 65 E10
Binfield Heath Oxon 65 D8
Bingfield Northumb 241 C11
Bingham Edin 280 G6
Bingham Notts 154 B4
Bingley W Yorks 205 F8
Bings Heath Shrops 149 F10
Binham Norf 159 B9
Binley Hants 48 C2
Binley W Mid 119 B7
Binley Woods Warks 119 B7
Binnegar Dorset 18 D3
Binniehill Falk 279 G7
Binscombe Sur 50 D3
Binsey Oxon 83 D7
Binsoe N Yorks 214 D4
Binstead Hants 49 E9
Binstead I o W 21 C7
Binsted Hants 49 E9
Binsted W Sus 35 F7
Binton Warks 118 G2
Bintree Norf 159 E10
Binweston Shrops 130 C6
Birch Essex 88 B6
Birch Gtr Man 195 F11
Birch Acre Worcs 117 C11
Birch Berrow Worcs 116 E4
Birch Cross Staffs 152 C2
Birch Green Essex 88 B6
Birch Green Herts 86 C3
Birch Green Lancs 194 F3
Birch Green Worcs 99 B7
Birch Heath W Ches 167 C8
Birch Hill Brack 65 F11
Birch Hill W Ches 183 G8
Birch Vale Derbys 185 D8
Birchall Hereford 98 D3
Birchall Staffs 169 E7
Bircham Newton Norf 158 C5
Bircham Tofts Norf 158 C5
Birchan Coppice Worcs 116 C6
Birchanger Essex 105 G10
Birchburn N Ayrs 255 E10
Birchden E Sus 52 F4
Birchencliffe W Yorks 196 D6
Birchend Hereford 98 C3
Birchendale Staffs 151 B11
Bircher Hereford 115 D9
Birches Green W Mid 134 E2
Birches Head Stoke 168 F5
Birchett's Green E Sus 53 G7
Birchfield Highld 301 G9
Birchfield W Mid 133 E11
Birchgrove Cardiff 59 D7
Birchgrove E Sus 36 B6
Birchgrove Swansea 57 B8
Birchhall Corner Essex 107 E10
Birchill Devon 28 G4
Birchills W Mid 133 D10
Birchington Kent 71 F9
Birchley Heath Warks 134 E5
Birchmoor Warks 134 C5
Birchmoor Green C Beds 103 E8
Bircholt Forstal Kent 54 E5
Birchover Derbys 170 C2
Birchwood Herts 86 D2
Birchwood Lincs 172 B6
Birchwood Som 28 E2
Birchwood Warr 183 C10
Birchy Hill Hants 19 B11
Bircotes Notts 187 C10
Bird Street Suff 125 G10
Birdbrook Essex 106 C4
Birdbush Wilts 30 C6
Birdfield Argyll 275 D10
Birdforth N Yorks 215 D9
Birdham W Sus 22 D4
Birdholme Derbys 170 B5
Birdingbury Warks 119 D8
Birdlip Glos 80 C6
Birds Edge W Yorks 197 F8
Birds End Suff 124 E5
Birds Green Essex 87 D9
Birdsall N Yorks 216 F6
Birdsgreen Shrops 132 F5
Birdsmoorgate Dorset 28 G5
Birdston E Dunb 278 F3
Birdwell S Yorks 197 G10
Birdwood Glos 80 B2
Birgham Borders 263 B7
Birichen Highld 309 K7
Birkacre Lancs 194 D5
Birkby Cumb 229 D7
Birkby N Yorks 224 E6
Birkby W Yorks 196 D6
Birkdale Mers 193 D10
Birkenbog Aberds 302 C5
Birkenhead Mers 182 D4
Birkenhills Aberds 303 E7
Birkenshaw N Lnrk 268 C3
Birkenshaw S Lnrk 268 F5
Birkenshaw W Yorks 197 B8
Birkenshaw Bottoms W Yorks 197 B8
Birkenside Borders 271 G11
Birkett Mire Cumb 230 G2
Birkhall Aberds 292 D5
Birkhill Angus 287 D7
Birkhill Borders 260 F6
Birkhill Borders 271 G11
Birkholme Lincs 155 E9
Birkhouse W Yorks 197 C7
Birkin N Yorks 198 B4
Birks Cumb 222 G3
Birks W Yorks 197 B9
Birkshaw Northumb 241 D7
Birley Hereford 115 G9
Birley Carr S Yorks 186 C4
Birley Edge S Yorks 186 C4
Birleyhay Derbys 186 E5
Birling Kent 69 G7
Birling Northumb 252 B6
Birling Gap E Sus 23 F9
Birlingham Worcs 99 C8
Birmingham W Mid 133 F11
Birnam Perth 286 C4
Birniehill S Lnrk 268 E2
Birse Aberds 293 D7
Birsemore Aberds 293 D7
Birstall Leics 135 B11
Birstall W Yorks 197 B8
Birstall Smithies W Yorks 197 B8
Birstwith N Yorks 205 B10
Birthorpe Lincs 156 C2
Birtle Gtr Man 195 E10
Birtley Hereford 115 D7
Birtley Northumb 241 B9
Birtley Shrops 131 E9
Birtley T & W 243 F7
Birtley Green Sur 50 E4
Birts Street Worcs 98 D5
Birtsmorton Worcs 98 D6
Bisbrooke Rutland 137 D7
Biscathorpe Lincs 190 D2
Biscombe Som 27 E11
Biscot Luton 103 G11
Biscovey Corn 5 E11
Bish Mill Devon 26 B2
Bisham Windsor 65 C10
Bishampton Worcs 117 G9
Bishon Common Hereford 97 C8
Bishop Auckland Durham 233 F10
Bishop Burton E Yorks 208 F5
Bishop Kinkell Highld 300 D5
Bishop Middleham Durham 234 E2
Bishop Monkton N Yorks 214 F6
Bishop Norton Lincs 189 C7
Bishop Sutton Bath 44 B5
Bishop Thornton N Yorks 214 G5
Bishop Wilton E Yorks 207 B11
Bishopbridge Lincs 189 C8
Bishopbriggs E Dunb 278 G2
Bishopdown Wilts 47 G7
Bishopmill Moray 302 C2
Bishops Cannings Wilts 62 G4
Bishop's Castle Shrops 130 F6
Bishop's Caundle Dorset 29 E11
Bishop's Cleeve Glos 99 F9
Bishop's Down Dorset 29 E11
Bishops Frome Hereford 98 B3
Bishops Green Essex 87 B5
Bishop's Green W Berks 64 G4
Bishop's Hull Som 28 C2
Bishop's Itchington Warks 119 F7
Bishops Lydeard Som 27 B11
Bishop's Norton Glos 98 G6
Bishops Nympton Devon 26 C3
Bishop's Offley Staffs 150 D5
Bishop's Quay Corn 2 D6
Bishop's Stortford Herts 105 G9
Bishop's Sutton Hants 48 G6
Bishop's Tachbrook Warks 118 E6
Bishops Tawton Devon 40 G5
Bishop's Waltham Hants 33 D9
Bishop's Wood Staffs 132 B6
Bishopsbourne Kent 55 C7
Bishopsgarth Stockton 234 G4
Bishopsgate Sur 66 E3
Bishopsteignton Devon 14 G4
Bishopstoke Hants 33 D7
Bishopston Bristol 60 D5
Bishopston Swansea 56 D5
Bishopstone Bucks 84 C4
Bishopstone E Sus 23 E7
Bishopstone Hereford 97 C8
Bishopstone Kent 71 F8
Bishopstone Swindon 63 C8
Bishopstone Wilts 31 B9
Bishopstrow Wilts 45 E11
Bishopswood Som 28 E3
Bishopsworth Bristol 60 F5
Bishopthorpe York 207 D7
Bishopton Darl 234 G3
Bishopton Dumfries 236 E6
Bishopton N Yorks 214 E6
Bishopton Renfs 277 G8
Bishopton Warks 118 F3
Bishopwearmouth T & W 243 F9
Bishpool Newport 59 B10
Bishton Newport 59 B11
Bishton Staffs 151 E10
Bisley Glos 80 D6
Bisley Sur 50 B3
Bisley Camp Sur 50 B2
Bispham Blkpool 202 E2
Bispham Green Lancs 194 E3
Bissoe Corn 4 G5
Bissom Corn 3 C7
Bisterne Hants 31 G10
Bisterne Close Hants 32 G2
Bitchet Green Kent 52 C5
Bitchfield Lincs 155 D9
Bittadon Devon 40 E4
Bittaford Devon 8 D3
Bittering Norf 159 F8
Bitterley Shrops 115 B11
Bitterne Soton 33 E7
Bitterne Park Soton 32 E6
Bitterscote Staffs 134 C4
Bitteswell Leics 135 F10
Bittles Green Dorset 30 C5
Bitton S Glos 61 F7
Bix Oxon 65 B8
Bixter Shetland 313 H5
Blaby Leics 135 D11
Black Bank Cambs 139 F10
Black Bank Warks 135 F7
Black Banks Darl 224 C5
Black Barn Lincs 157 D8
Black Bourton Oxon 82 E3
Black Callerton T & W 242 D5
Black Carr Norf 141 D11
Black Clauchrie S Ayrs 245 G7
Black Corner W Sus 51 F9
Black Corries Lodge Highld 284 B6
Black Crofts Argyll 289 F11
Black Cross Corn 5 C8
Black Dam Hants 48 C6
Black Dog Devon 26 F4
Black Heddon Northumb 242 B3
Black Hill W Yorks 204 E6
Black Horse Drove Cambs 139 E11
Black Lake W Mid 133 E9
Black Lane Gtr Man 195 F9
Black Marsh Shrops 130 D6
Black Moor Lancs 194 E3
Black Moor W Yorks 205 F11
Black Mount Argyll 284 C6
Black Notley Essex 106 G5
Black Park Wrex 166 G4
Black Pill Swansea 56 C6
Black Pole Lancs 202 F5
Black Rock Brighton 36 G4
Black Rock Corn 2 C5
Black Street Suff 143 F10
Black Tar Pembs 73 D7
Black Torrington Devon 25 F7
Black Vein Caerph 78 G2
Blackacre Dumfries 248 E2
Blackadder West Borders 272 E6
Blackawton Devon 8 E6
Blackbeck Cumb 219 D10
Blackbird Leys Oxon 83 E9
Blackborough Devon 27 F9
Blackborough Norf 158 G3
Blackborough End Norf 158 G3
Blackboys E Sus 37 C8
Blackbraes Aberds 293 B10
Blackbrook Derbys 170 F4
Blackbrook Derbys 185 E9
Blackbrook Mers 183 B8
Blackbrook Staffs 150 B5
Blackbrook Sur 51 D7
Blackburn Aberds 293 B10
Blackburn Aberds 302 F5
Blackburn Blkburn 195 B7
Blackburn S Yorks 186 C5
Blackburn W Loth 269 B9
Blackcastle Midloth 271 D8
Blackchambers Aberds 293 B9
Blackcraig Dumfries 246 G6
Blackcraigs Angus 293 E7
Blackden Heath E Ches 184 G3
Blackditch Oxon 82 D6
Blackdog Aberds 293 B11
Blackdown Dorset 28 G5
Blackdown Hants 33 C8
Blackdown Warks 118 D6
Blackdyke Cumb 238 G4
Blackdykes E Loth 281 E11
Blacker Hill S Yorks 197 G11
Blacketts Kent 70 F2
Blackfell T & W 243 F7
Blackfen London 68 E3
Blackfield Hants 32 G6
Blackford Cumb 239 E9
Blackford Dumfries 248 G4
Blackford Perth 286 G2
Blackford Shrops 131 G11
Blackford Som 29 B11
Blackford Som 42 D2
Blackford Som 44 D2
Blackford Bridge Gtr Man 195 F10
Blackfordby Leics 152 F6
Blackfords Staffs 151 G9
Blackgang I o W 20 F5
Blackgate Angus 287 B8
Blackhall Aberds 293 D8
Blackhall Edin 280 G4
Blackhall Renfs 267 C9
Blackhall Colliery Durham 234 D5
Blackhall Mill T & W 242 F4
Blackhall Rocks Durham 234 D5
Blackham E Sus 52 F3
Blackhaugh Borders 261 B10
Blackheath Essex 107 G10
Blackheath London 67 D11
Blackheath Suff 127 C8
Blackheath Sur 50 D4
Blackheath W Mid 133 F9
Blackheath Park London 68 D2
Blackhill Aberds 303 D10
Blackhill Aberds 303 E10
Blackhill Aberds 303 F10
Blackhill Durham 242 G3
Blackhill Hants 32 D4
Blackhill Highld 298 D3
Blackhillock Moray 302 E4
Blackhills Highld 301 D9
Blackhills Moray 302 D2
Blackhills Swansea 56 C5
Blackhorse Devon 14 C5
Blackhorse S Glos 61 D7
Blackjack Lincs 156 B5
Blackland Wilts 62 F4
Blacklands E Sus 38 E4
Blacklands Hereford 98 C2
Blacklaw Aberds 302 D6
Blackleach Lancs 202 G5
Blackley Gtr Man 195 G11
Blackley W Yorks 196 D6
Blacklunans Perth 292 G3
Blackmarstone Hereford 97 D10
Blackmill Bridgend 58 B2
Blackminster Worcs 99 C11
Blackmoor Bath 60 G5
Blackmoor Gtr Man 195 G7
Blackmoor Hants 49 G9
Blackmoor N Som 60 G3
Blackmoor Som 27 D11
Blackmoor Gate Devon 41 E7
Blackmoorfoot W Yorks 196 E5
Blackmore Essex 87 E10
Blackmore Shrops 130 B6
Blackmore End Essex 106 E4
Blackmore End Herts 85 B11
Blackmore End Worcs 98 C6
Blackness Aberds 293 D8
Blackness E Sus 52 G4
Blackness Falk 279 F11
Blacknest Hants 49 E9
Blacknest Windsor 66 F3
Blacknoll Dorset 18 D2
Blacko Lancs 204 E3
Blackoe Shrops 149 B10
Blackpark Dumfries 236 C5
Blackpole Worcs 117 F7
Blackpool Blkpool 202 F2
Blackpool Devon 7 E11
Blackpool Devon 9 F7
Blackpool Devon 14 G2
Blackpool Pembs 73 C9
Blackpool Gate Cumb 240 B2
Blackridge W Loth 269 B7
Blackrock Argyll 274 G4
Blackrock Bath 60 F6
Blackrock Mon 78 C2
Blackrod Gtr Man 194 E6
Blackshaw Dumfries 238 D2
Blackshaw Head W Yorks 196 B3
Blackshaw Moor Staffs 169 D8
Blacksmith's Corner Suff 108 C2
Blacksmith's Green Suff 126 D2
Blacksnape Blkburn 195 C8
Blackstone W Sus 36 D2
Blackstone Worcs 116 C5
Blackthorn Oxon 83 B10
Blackthorpe Suff 125 E8
Blacktoft E Yorks 199 C10
Blacktop Aberdeen 293 C10
Blacktown Newport 59 C9
Blackwall Derbys 170 F3
Blackwall London 67 C11
Blackwall Tunnel London 67 C11
Blackwater Corn 4 F4
Blackwater Dorset 19 B8
Blackwater Hants 49 B11
Blackwater I o W 20 D6
Blackwater Norf 159 E11
Blackwater Som 28 D3
Blackwater Lodge Moray 302 G3
Blackwaterfoot N Ayrs 255 E9
Blackweir Cardiff 59 D7
Blackwell Cumb 239 G10
Blackwell Darl 224 C5
Blackwell Derbys 170 D6
Blackwell Derbys 185 G10
Blackwell Devon 27 B8
Blackwell W Sus 51 F11
Blackwell Warks 100 C4
Blackwell Worcs 117 C9
Blackwood Caerph 77 F11
Blackwood S Lnrk 268 G5
Blackwood Warr 183 C10
Blackwood Hill Staffs 168 D6
Blacon W Ches 166 B5
Bladbean Kent 55 D7
Blades N Yorks 223 F9
Bladnoch Dumfries 236 D6
Bladon Oxon 82 C6
Blaen-Cil-Llech Ceredig 92 C6
Blaen Clydach Rhondda 77 G7
Blaen-gwynfi Neath 57 B11
Blaen-pant Ceredig 92 C5
Blaen-waun Carms 92 F4
Blaen-waun Ceredig 111 G7
Blaen-y-coed Carms 92 F6
Blaen-y-cwm Bl Gwent 77 C10
Blaen-y-Cwm Denb 147 C10
Blaen-y-cwm Gwyn 146 E4
Blaen-y-cwm Powys 147 E11
Blaenannerch Ceredig 92 B4
Blaenau Carms 75 C10
Blaenau Flint 166 D2
Blaenau Dolwyddelan Conwy 164 E2
Blaenau Ffestiniog Gwyn 164 F2
Blaenau-Gwent Bl Gwent 78 E2
Blaenavon Torf 78 D3
Blaenbedw Fawr Ceredig 111 G7
Blaencaerau Bridgend 57 C11
Blaencelyn Ceredig 111 G7
Blaencwm Rhondda 76 F6
Blaendulais = Seven Sisters Neath 76 D4
Blaendyryn Powys 95 D8
Blaenffos Pembs 92 D3
Blaengarw Bridgend 76 G6
Blaengwrach Neath 76 D5
Blaengwynfi Neath 57 B11
Blaenllechau Rhondda 77 F8
Blaenpennal Ceredig 112 E2
Blaenplwyf Ceredig 111 B11
Blaenporth Ceredig 92 B5
Blaenrhondda Rhondda 76 E6
Blaenwaun Carms 92 F4
Blaenycwm Ceredig 112 B6
Blagdon N Som 44 B4
Blagdon Torbay 9 C7
Blagdon Hill Som 28 D2
Blagill Cumb 231 B10
Blaguegate Lancs 194 F3
Blaich Highld 290 F2
Blain Highld 289 C8
Blaina Bl Gwent 78 D2
Blainacraig Ho Aberds 293 D7
Blair Fife 280 C6
Blair Atholl Perth 291 G10
Blair Drummond Stirl 278 B4
Blairbeg N Ayrs 256 C2
Blairburn Fife 279 D9
Blairdaff Aberds 293 B8
Blairdryne Aberds 293 D9
Blairglas Argyll 276 D6
Blairgorm Highld 301 G10
Blairgowrie Perth 286 C5
Blairhall Fife 279 D10
Blairhill N Lnrk 268 B4
Blairingone Perth 279 B9
Blairland N Ayrs 266 F6
Blairlinn N Lnrk 278 G5
Blairlogie Stirl 278 B6
Blairlomond Argyll 276 B3
Blairmore Argyll 276 E3
Blairmore Highld 306 D6
Blairnamarrow Moray 292 B4
Blairquhosh Stirl 277 E10
Blair's Ferry Argyll 275 G10
Blairskaith E Dunb 277 F11
Blaisdon Glos 80 B2
Blaise Hamlet Bristol 60 D5
Blake End Essex 106 G4
Blakebrook Worcs 116 B6
Blakedown Worcs 117 B7
Blakelands M Keynes 103 C7
Blakelaw Borders 263 C7

Blakelaw T & W 242 D6
Blakeley Staffs 133 E7
Blakeley Lane Staffs 169 F7
Blakelow E Ches 167 E11
Blakemere Hereford 97 C7
Blakenall Heath W Mid 133 C10
Blakeney Glos 79 D11
Blakeney Norf 177 E8
Blakenhall E Ches 168 F2
Blakenhall W Mid 133 D8
Blakeshall Worcs 132 G6
Blakesley Northants 120 G2
Blanchland Northumb 241 G11
Bland Hill N Yorks 205 C10
Blandford Camp Dorset 30 F6
Blandford Forum Dorset 30 F5
Blandford St Mary Dorset 30 F5
Blandy Highld 308 D6
Blanefield Stirl 277 F11
Blanerne Borders 272 D6
Blank Bank Staffs 168 F4
Blankney Lincs 173 C9
Blantyre S Lnrk 268 D3
Blar a'Chaorainn Highld 290 G3
Blaran Argyll 275 B9
Blarghour Argyll 275 B10
Blarmachfoldach Highld 290 G2
Blarnalearoch Highld 307 K6
Blasford Hill Essex 88 C2
Blashford Hants 31 F11
Blaston Leics 136 D6
Blatchbridge Som 45 D9
Blatherwycke Northants 137 D9
Blawith Cumb 210 B5
Blaxhall Suff 127 F7
Blaxton S Yorks 199 G7
Blaydon T & W 242 E5
Blaydon Burn T & W 242 E5
Blaydon Haughs T & W 242 E5
Bleach Green Cumb 219 B9
Bleach Green Suff 126 B4
Bleadney Som 44 D3
Bleadon N Som 43 B10
Bleak Acre Hereford 98 B2
Bleak Hall M Keynes 103 D7
Bleak Hey Nook Gtr Man 196 F4
Bleak Hill Hants 31 E10
Blean Kent 70 G6
Bleasby Lincs 189 E10
Bleasby Notts 172 F2
Bleasby Moor Lincs 189 E10
Bleasdale Lancs 203 D7
Bleatarn Cumb 222 C4
Blebocraigs Fife 287 F8
Bleddfa Powys 114 D4
Bledington Glos 100 G4
Bledlow Bucks 84 E3
Bledlow Ridge Bucks 84 F3
Bleet Wilts 45 B11
Blegbie E Loth 271 C9
Blegbury Devon 24 B2
Blencarn Cumb 231 E8
Blencogo Cumb 229 B9
Blendworth Hants 34 E2
Blenheim Oxon 83 D9
Blenheim Oxon 83 E9
Blenheim Park Norf 158 C6
Blenkinsopp Hall Northumb 240 E5
Blennerhasset Cumb 229 C9
Blervie Castle Moray 301 D10
Bletchingdon Oxon 83 B8
Bletchingley Sur 51 C10
Bletchley M Keynes 103 E7
Bletchley Shrops 150 C2
Bletherston Pembs 91 G11
Bletsoe Beds 121 F10
Blewbury Oxon 64 B4
Bliby Kent 54 F4
Blickling Norf 160 D3
Blidworth Notts 171 D9
Blidworth Bottoms Notts 171 E9
Blidworth Dale Notts 171 E9
Blindburn Northumb 263 G8
Blindcrake Cumb 229 E8
Blindley Heath Sur 51 D11
Blindmoor Som 28 E3
Blingery Highld 310 E7
Blisland Corn 11 G8
Bliss Gate Worcs 116 C4
Blissford Hants 31 E11
Blisworth Northants 120 G4
Blithbury Staffs 151 E11
Blitterlees Cumb 238 G4
Blo' Norton Norf 125 B10
Blockley Glos 100 D3
Blofield Norf 142 B6
Blofield Heath Norf 160 G6
Bloodman's Corner Suff 143 D10
Bloomfield Bath 45 B7
Bloomfield Bath 61 G8
Bloomfield Borders 262 E3
Bloomfield W Mid 133 E9
Bloomsbury London 67 C10
Blore Staffs 150 C4
Blore Staffs 169 F10
Bloreheath Staffs 150 B4
Blossomfield W Mid 118 B2
Blount's Green Staffs 151 C11
Blowick Mers 193 D11
Blowinghouse Corn 4 E4
Bloxham Oxon 101 D8
Bloxholm Lincs 173 E9
Bloxwich W Mid 133 C9
Bloxworth Dorset 18 C3
Blubberhouses N Yorks 205 B9
Blue Anchor Corn 5 D8
Blue Anchor Som 42 E4
Blue Anchor Swansea 56 B4
Blue Bell Hill Kent 69 G8
Blue Hill Herts 104 G5
Blue Row Essex 89 C8
Blue Town Kent 70 D2
Blue Vein Wilts 61 F10
Bluebell Telford 149 G11
Bluecairn Borders 271 G10
Bluetown Kent 54 B2
Bluewater Kent 68 E5
Blughasary Highld 307 J6
Blundellsands Mers 182 B4
Blundeston Suff 143 D10
Blunham C Beds 122 G3
Blunsdon St Andrew Swindon 62 B6
Bluntington Worcs 117 C7
Bluntisham Cambs 123 C7
Blunts Corn 6 C6
Blunt's Green Warks 118 D2
Blurton Stoke 168 G5
Blyborough Lincs 188 C6
Blyford Suff 127 B8
Blymhill Staffs 150 G6
Blymhill Lawns Staffs 150 G6
Blyth Borders 270 F2
Blyth Northumb 253 G8
Blyth Notts 187 D10
Blyth Bridge Borders 270 F2
Blyth End Warks 134 E4
Blythburgh Suff 127 B9
Blythe Borders 271 F11
Blythe Bridge Staffs 169 G7
Blythe Marsh Staffs 169 G7
Blythswood Renfs 267 B10
Blyton Lincs 188 C5
Boarhills Fife 287 F9
Boarhunt Hants 33 F10
Boars Hill Oxon 83 E7
Boarsgreave Lancs 195 C10
Boarshead E Sus 52 G4
Boarstall Bucks 83 C10
Boasley Cross Devon 12 C5
Boat of Garten Highld 291 B11
Boath Highld 300 B5
Bobbing Kent 69 F11
Bobbington Staffs 132 E6
Bobbingworth Essex 87 D8
Bobby Hill Suff 125 C10
Boblainy Highld 300 F4
Bocaddon Corn 6 D3
Bochastle Stirl 285 G10
Bockhanger Kent 54 E4
Bocking Essex 106 G5
Bocking Churchstreet Essex 106 F5
Bocking's Elm Essex 89 B11
Bockleton Worcs 115 E11
Bockmer End Bucks 65 B10
Bocombe Devon 24 C5
Bodantionail Highld 299 B7
Boddam Aberds 303 E11
Boddam Shetland 313 M5
Bodden Som 44 E6
Boddington Glos 99 F7
Bodedern Anglesey 178 E4
Bodellick Corn 10 G5
Bodelva Corn 5 E11
Bodelwyddan Denb 181 F8
Bodenham Hereford 115 G10
Bodenham Wilts 31 B11
Bodenham Bank Hereford 98 E2
Bodenham Moor Hereford 115 G10
Bodermid Gwyn 144 D3
Bodewryd Anglesey 178 C5
Bodfari Denb 181 G9
Bodffordd Anglesey 178 F6
Bodham Norf 177 E10
Bodiam E Sus 38 B3
Bodicote Oxon 101 D9
Bodiechell Aberds 303 E7
Bodieve Corn 10 G5
Bodiggo Corn 5 D10
Bodilly Corn 2 C5
Bodinnick Corn 6 E2
Bodle Street Green E Sus 23 C11
Bodley Devon 41 D7
Bodmin Corn 5 B11
Bodmiscombe Devon 27 F10
Bodney Norf 140 D6
Bodorgan Anglesey 162 B5
Bodsham Kent 54 D6
Boduan Gwyn 144 B6
Boduel Corn 6 C4
Bodwen Corn 5 C10
Bodymoor Heath Warks 134 D4
Bofarnel Corn 6 C2
Bogallan Highld 300 D6
Bogbrae Aberds 303 F10
Bogend Borders 272 F5
Bogend Notts 171 F7
Bogend S Ayrs 257 C9
Bogentory Aberds 293 C9
Boghall Midloth 270 B4
Boghall W Loth 269 B9
Boghead Aberds 293 D8
Boghead S Lnrk 268 G5
Bogmoor Moray 302 C3
Bogniebrae Aberds 302 E5
Bogniebrae Aberds 302 E6
Bognor Regis W Sus 22 D6
Bograxie Aberds 293 B9
Bogs Aberds 302 G5
Bogs Bank Borders 270 E3
Bogside N Lnrk 268 E6
Bogthorn W Yorks 204 F6
Bogton Aberds 302 D6
Bogtown Aberds 302 C5
Bogue Dumfries 246 G4
Bohemia E Sus 38 E4
Bohemia Wilts 32 D2
Bohenie Highld 290 E4
Bohetherick Corn 7 B8
Bohortha Corn 3 C9
Bohuntine Highld 290 E4
Bohuntinville Highld 290 E4
Boirseam W Isles 296 C6
Bojewyan Corn 1 C3
Bokiddick Corn 5 C11
Bolahaul Fm Carms 74 B6
Bolam Durham 233 G9
Bolam Northumb 252 G3
Bolam West Houses Northumb 252 G3
Bolas Heath Telford 150 E3
Bolberry Devon 9 G8
Bold Heath Mers 183 D8
Boldmere W Mid 134 E2
Boldon T & W 243 E9
Boldon Colliery T & W 243 E8
Boldre Hants 20 B2
Boldron Durham 223 C10
Bole Notts 188 D3
Bole Hill Derbys 186 G4
Bolehall Staffs 134 C4
Bolehill Derbys 170 E3
Bolehill Derbys 186 G6
Bolehill S Yorks 186 E5
Bolenowe Corn 2 B5
Boleside Borders 261 C11
Boley Park Staffs 134 B2
Bolham Devon 27 E7
Bolham Notts 188 E2
Bolham Water Devon 27 E11
Bolholt Gtr Man 195 E9
Bolingey Corn 4 E5
Bolitho Corn 2 C5
Bollihope Durham 232 E6
Bollington E Ches 184 F6
Bollington Cross E Ches 184 F6
Bolney W Sus 36 C3
Bolnhurst Beds 121 F11
Bolnore W Sus 36 C4
Bolshan Angus 287 B10
Bolsover Derbys 187 G7
Bolsterstone S Yorks 186 B3
Bolstone Hereford 97 E11
Boltby N Yorks 215 B9
Bolter End Bucks 84 G3
Bolton Cumb 231 G8
Bolton E Loth 281 G10
Bolton E Yorks 207 C11
Bolton Gtr Man 195 F8
Bolton Northumb 264 G4
Bolton W Yorks 205 F9
Bolton Abbey N Yorks 205 C7
Bolton Bridge N Yorks 205 C7
Bolton-by-Bowland Lancs 203 D11
Bolton Green Lancs 194 D5
Bolton Houses Lancs 202 G4
Bolton-le-Sands Lancs 211 F9
Bolton Low Houses Cumb 229 C10
Bolton New Houses Cumb 229 C10
Bolton-on-Swale N Yorks 224 F5
Bolton Percy N Yorks 206 E6
Bolton Town End Lancs 211 F9
Bolton upon Dearne S Yorks 198 G3
Bolton Wood Lane Cumb 229 C11
Bolton Woods W Yorks 205 F9
Boltonfellend Cumb 239 D11
Boltongate Cumb 229 C10
Boltshope Park Durham 232 E4
Bolventor Corn 11 F9
Bomarsund Northumb 253 G7
Bombie Dumfries 237 D9
Bomby Cumb 221 B10
Bomere Heath Shrops 149 F9
Bon-y-maen Swansea 57 E7
Bonaly Edin 270 E4
Bonar Bridge Highld 309 K6
Bonawe Argyll 284 D4
Bonby N Lincs 200 D4
Boncath Pembs 92 D4
Bonchester Bridge Borders 262 G3
Bonchurch I o W 21 F7
Bond End Staffs 152 F2
Bondend Glos 80 B5
Bondleigh Devon 25 G11
Bondman Hays Leics 135 B9
Bonds Lancs 202 E5
Bondstones Devon 25 F9
Bonehill Devon 13 F10
Bonehill Staffs 134 C3
Bo'ness Falk 279 E9
Bonhill W Dunb 277 F7
Boningale Shrops 132 C6
Bonjedward Borders 262 E5
Bonkle N Lnrk 268 D6
Bonnavoulin Highld 289 D7
Bonning Gate Cumb 221 F9
Bonnington Borders 261 E7
Bonnington Edin 270 B2
Bonnington Kent 54 F5
Bonnybank Fife 287 G7
Bonnybridge Falk 278 E6
Bonnykelly Aberds 303 D8
Bonnyrigg and Lasswade Midloth 270 B6
Bonnyton Aberds 302 F6
Bonnyton Angus 287 B10
Bonnyton Angus 287 D7
Bonnyton E Ayrs 257 B10
Bonsall Derbys 170 D3
Bonskeid House Perth 291 G10
Bonson Som 43 E8
Bont Mon 78 B5
Bont-Dolgadfan Powys 129 C7
Bont Fawr Carms 94 F4
Bont goch = Elerch Ceredig 128 F3
Bont-newydd Conwy 181 G8
Bont Newydd Gwyn 146 E5
Bont Newydd Gwyn 164 G2
Bontddu Gwyn 146 F3
Bonthorpe Lincs 191 G7
Bontnewydd Ceredig 112 D2
Bontnewydd Gwyn 163 D7
Bontuchel Denb 165 D9
Bonvilston = Tresimwn V Glam 58 E5
Boode Devon 40 F4
Booker Bucks 84 G4
Bookham Dorset 30 G2
Boolebank Shrops 149 D11
Boon Borders 271 F11
Boon Hill Staffs 168 E4
Boorley Green Hants 33 E8
Boosbeck Redcar 226 B3
Boose's Green Essex 106 E6
Boot Cumb 220 E3
Boot Street Suff 108 B4
Booth Staffs 151 D10
Booth W Yorks 196 B4
Booth Bank E Ches 184 D2
Booth Bridge Lancs 204 D4
Booth Green E Ches 184 E6
Booth Wood W Yorks 196 D4
Boothby Graffoe Lincs 173 D7
Boothby Pagnell Lincs 155 C9
Boothen Stoke 168 G5
Boothferry E Yorks 199 B8
Boothgate Derbys 170 F5
Boothroyd W Yorks 197 C8
Boothsdale W Ches 167 B8
Boothstown Gtr Man 195 G8
Boothtown W Yorks 196 B5
Boothville Northants 120 E5
Bootle Cumb 210 B2
Bootle Mers 182 B4
Booton Norf 160 E2
Boots Green W Ches 184 G3
Booze N Yorks 223 E10
Boquhan Stirl 277 D10
Boquio Corn 2 C5
Boraston Shrops 116 D2
Boraston Dale Shrops 116 C2
Borden Kent 69 G11
Border Cumb 238 G5
Bordesley W Mid 133 F11
Bordesley Green W Mid 134 F2
Bordlands Borders 270 F3
Bordley N Yorks 213 G8
Bordon Hants 49 F10
Bordon Camp Hants 49 F9
Boreham Essex 88 D3
Boreham Wilts 45 E11
Boreham Street E Sus 23 C11
Borehamwood Herts 85 F11
Boreland Dumfries 236 C5
Boreland Dumfries 248 E5
Boreland Fife 280 C6
Boreland Stirl 285 D9
Boreland of Southwick Dumfries 237 C11
Boreley Worcs 116 D6
Borestone Stirl 278 C5
Borgh W Isles 296 C5
Borgh W Isles 297 L2
Borghasdal W Isles 296 C6
Borghastan W Isles 304 D4
Borgie Highld 308 D6
Borgue Dumfries 237 E8
Borgue Highld 311 G5
Borley Essex 106 C6
Borley Green Essex 106 C6
Borley Green Suff 125 E9
Bornais W Isles 297 J3
Bornesketaig Highld 298 B3
Borness Dumfries 237 E8
Borough Scilly 1 G3
Borough Green Kent 52 B6
Borough Marsh Wokingham 65 D9
Borough Park Staffs 134 B4
Borough Post Som 28 C4
Boroughbridge N Yorks 215 F7
Borras Wrex 166 E4
Borras Head Wrex 166 E5
Borreraig Highld 296 F7
Borrobol Lodge Highld 311 G2
Borrodale Highld 297 G7
Borrohill Aberds 303 D9
Borrowash Derbys 153 C8
Borrowby N Yorks 215 B8
Borrowby N Yorks 226 B5
Borrowdale Cumb 220 C5
Borrowfield Aberds 293 D10
Borrowston Highld 310 E7
Borrowstoun Mains Falk 279 E9
Borstal Medway 69 F8
Borth = Y Borth Ceredig 128 E2
Borth-y-Gest Gwyn 145 B11
Borthwick Midloth 271 D7
Borthwickbrae Borders 261 G10
Borthwickshiels Borders 261 F10
Borve Highld 298 E4
Borve Lodge W Isles 305 J2
Borwick Lancs 211 E10
Borwick Rails Cumb 210 D3
Bosavern Corn 1 C3
Bosbury Hereford 98 C3
Boscadjack Corn 2 C5
Boscastle Corn 11 C8
Boscean Corn 1 C3
Boscombe Bmouth 19 C8
Boscombe Wilts 47 F8
Boscomoor Staffs 151 G8
Boscoppa Corn 5 E10
Boscreege Corn 2 C3
Bosham W Sus 22 C4
Bosham Hoe W Sus 22 C4
Bosherston Pembs 73 G7
Boskednan Corn 1 C4
Boskenna Corn 1 E4
Bosleake Corn 4 G3
Bosley E Ches 168 B6
Boslowick Corn 3 C7
Boslymon Corn 5 C11
Bosoughan Corn 5 C7
Bosporthennis Corn 1 B4
Bossall N Yorks 216 G4
Bossiney Corn 11 D7
Bossingham Kent 54 D6
Bossington Hants 47 G10
Bossington Kent 55 B8
Bossington Som 41 D11
Bostadh W Isles 304 D3
Bostock Green W Ches 167 B11
Boston Lincs 174 G4
Boston Long Hedges Lincs 174 F5
Boston Spa W Yorks 206 D4
Boston West Lincs 174 F3
Boswednack Corn 1 B4
Boswin Corn 2 C5
Boswinger Corn 5 G9
Boswyn Corn 2 B5
Botallack Corn 1 C3
Botany Bay London 86 F3
Botany Bay Mon 79 E8
Botcherby Cumb 239 F10
Botcheston Leics 135 B9
Botesdale Suff 125 B10
Bothal Northumb 252 F6
Bothampstead W Berks 64 D4
Bothamsall Notts 187 G11
Bothel Cumb 229 D9
Bothenhampton Dorset 16 C5
Bothwell S Lnrk 268 D4
Bothy Highld 290 F4
Botley Bucks 85 E7
Botley Hants 33 E8
Botley Oxon 83 D7
Botloe's Green Glos 98 F4
Botolph Claydon Bucks 102 G4
Botolphs W Sus 35 F11
Bottacks Highld 300 C4
Botternell Corn 11 G11
Bottesford Leics 154 B6
Bottesford N Lincs 199 F11
Bottisham Cambs 123 E10
Bottlesford Wilts 46 B6
Bottom Boat W Yorks 197 C11
Bottom House Staffs 169 E8
Bottom o' th' Moor Gtr Man 195 E7
Bottom of Hutton Lancs 194 B3
Bottom Pond Kent 53 B11
Bottomcraig Fife 287 E7
Bottomley W Yorks 196 D5
Bottoms Corn 1 E3
Bottreaux Mill Devon 26 B4
Bottrells Close Bucks 85 G7
Botts Green Warks 134 E4
Botusfleming Corn 7 C8
Botwnnog Gwyn 144 C5
Bough Beech Kent 52 D3
Boughrood Powys 96 D2
Boughrood Brest Powys 96 D2
Boughspring Glos 79 F9
Boughton Lincs 173 F10
Boughton Norf 140 C3
Boughton Northants 120 D5
Boughton Notts 171 B11
Boughton W Ches 166 B6
Boughton Aluph Kent 54 D4
Boughton Corner Kent 54 D4
Boughton Green Kent 53 C9
Boughton Heath W Ches 166 B6
Boughton Lees Kent 54 D4
Boughton Malherbe Kent 53 D11
Boughton Monchelsea Kent 53 C9
Boughton Street Kent 54 B5
Bougton End C Beds 103 D9
Boulby Redcar 226 B5
Bould Oxon 100 G4
Boulden Shrops 131 F10
Boulder Clough W Yorks 196 C4
Bouldnor I o W 20 D3
Bouldon Shrops 131 F10
Boulmer Northumb 265 G7
Boulsdon Glos 98 G4
Boulston Pembs 73 C7
Boultenstone Aberds 292 B6
Boultham Lincs 173 B7
Boultham Moor Lincs 173 B7
Boulton Derbys 153 C7
Boulton Moor Derbys 153 C7
Boundary Leics 152 F6
Boundary Staffs 169 G7
Boundstone Sur 49 E10
Bountis Thorne Devon 24 D5
Bourn Cambs 122 F6
Bournbrook W Mid 133 G10
Bourne Lincs 155 E11
Bourne N Som 44 B3
Bourne End Beds 121 E10
Bourne End Bucks 65 B11
Bourne End C Beds 103 C9
Bourne End Herts 85 D8
Bourne Vale W Mid 133 D11
Bourne Valley Poole 19 C7
Bournemouth Bmouth 19 C7
Bournes Green Glos 80 E6
Bournes Green Sthend 70 B2
Bournes Green Worcs 117 C8
Bournheath Worcs 117 C9
Bournmoor Durham 243 G8
Bournside Glos 99 G8
Bournstream Glos 80 G2
Bournville W Mid 133 G10
Bourton Bucks 102 E4
Bourton Dorset 45 G9
Bourton N Som 59 G11
Bourton Oxon 63 B8
Bourton Shrops 131 D11
Bourton Wilts 62 G4
Bourton on Dunsmore Warks 119 C8
Bourton-on-the-Hill Glos 100 E3
Bourton-on-the-Water Glos 100 G3
Bourtreehill N Ayrs 257 B8
Bousd Argyll 288 C4
Bousta Shetland 313 H4
Boustead Hill Cumb 239 F7
Bouth Cumb 210 B6
Bouthwaite N Yorks 214 E2
Bouts Worcs 117 F10
Bovain Stirl 285 D9
Boveney Bucks 66 D2
Boveridge Dorset 31 E9
Boverton V Glam 58 F3
Bovey Tracey Devon 14 F2
Bovingdon Herts 85 E8
Bovingdon Green Bucks 65 B10
Bovingdon Green Herts 85 E8
Bovinger Essex 87 D8
Bovington Camp Dorset 18 D2
Bow Borders 271 G9
Bow Devon 8 D6
Bow Devon 26 G2
Bow Orkney 314 G3
Bow Oxon 82 G4
Bow Brickhill M Keynes 103 E8
Bow Broom S Yorks 187 B7
Bow Common London 67 C11
Bow of Fife Fife 287 F7
Bow Street Ceredig 128 G2
Bow Street Norf 141 D10
Bowbank Durham 232 G4
Bowbeck Suff 125 B8
Bowbridge Glos 80 E5
Bowbrook Shrops 149 G9
Bowburn Durham 234 D2
Bowcombe I o W 20 D5
Bowd Devon 15 C8
Bowden Borders 262 C3
Bowden Devon 8 F6
Bowden Dorset 30 C3
Bowden Hill Wilts 62 F2
Bowdens Som 28 B6
Bowderdale Cumb 222 E3
Bowdon Gtr Man 184 D3
Bower Highld 310 C6
Bower Northumb 251 G7
Bower Ashton Bristol 60 E5
Bower Heath Herts 85 B10
Bower Hinton Som 29 D7
Bower House Tye Suff 107 C9
Bowerchalke Wilts 31 C8
Bowerhill Wilts 62 G2
Bowerhope Borders 261 E7
Bowermadden Highld 310 C6
Bowers Staffs 150 B6
Bowers Gifford Essex 69 B9
Bowershall Fife 279 C11
Bowertower Highld 310 C6
Bowes Durham 223 C9
Bowes Park London 86 G4
Bowgreave Lancs 202 E5
Bowgreen Gtr Man 184 D3
Bowhill Borders 261 D10
Bowhill Fife 280 B4
Bowhouse Dumfries 238 D2
Bowhousebog or Liquo N Lnrk 269 D7
Bowing Park Mers 182 D6
Bowismiln Borders 262 E2
Bowithick Corn 11 E9
Bowker's Green Lancs 194 G2
Bowland Bridge Cumb 211 B8
Bowldown Wilts 62 D2
Bowlee Gtr Man 195 F10
Bowlees Durham 232 F4
Bowler's Town E Sus 38 C6
Bowley Hereford 115 G10
Bowley Lane Hereford 98 C3
Bowley Town Hereford 115 G10
Bowlhead Green Sur 50 F2
Bowling W Dunb 277 G9
Bowling W Yorks 205 G9
Bowling Alley Hants 49 D9
Bowling Bank Wrex 166 F5
Bowling Green Corn 5 D10
Bowling Green Corn 12 G3
Bowling Green Glos 81 E8
Bowling Green Hants 19 B11
Bowling Green Shrops 150 D2
Bowling Green W Mid 133 F8
Bowling Green Worcs 116 G6
Bowlish Som 44 E6
Bowmans Kent 68 E4
Bowmanstead Cumb 220 F6
Bowmore Argyll 254 B4
Bowness-on-Solway Cumb 238 E6
Bowness-on-Windermere Cumb 221 F8
Bowridge Hill Dorset 30 B4
Bowrie-fauld Angus 287 C9
Bowsden Northumb 273 G9
Bowsey Hill Windsor 65 C10
Bowshank Borders 271 G9
Bowside Lodge Highld 310 C2
Bowston Cumb 221 F9
Bowthorpe Norf 142 B3
Bowyer's Common Hants 34 B3
Box Glos 80 E5
Box Wilts 61 F10
Box End Beds 103 B10
Box Hill Sur 51 C7
Box Hill Wilts 61 F10
Box Trees W Mid 118 C2
Boxbush Glos 80 C2
Boxbush Glos 98 G3
Boxford Suff 107 C9
Boxford W Berks 64 E2
Boxgrove W Sus 22 B6
Boxley Kent 53 B9
Boxmoor Herts 85 D9
Box's Shop Corn 24 G2
Boxted Essex 107 E9
Boxted Suff 124 G6
Boxted Cross Essex 107 E10
Boxted Heath Essex 107 E10
Boxwell Glos 80 G4
Boxworth Cambs 122 E6
Boxworth End Cambs 123 D7
Boyatt Wood Hants 32 C6
Boyden End Suff 124 F4
Boyden Gate Kent 71 F8
Boyland Common Norf 141 G11
Boylestone Derbys 152 B3
Boylestonfield Derbys 152 B3
Boyn Hill Windsor 65 C11
Boyndie Aberds 302 C6
Boynton E Yorks 218 F2
Boys Hill Dorset 29 E11
Boys Village V Glam 58 F4
Boysack Angus 287 C10
Boysack Angus 287 C9
Boythorpe Derbys 186 G5
Boyton Corn 12 C2
Boyton Suff 109 B7
Boyton Wilts 46 F3
Boyton Cross Essex 87 D10
Boyton End Essex 106 E2
Boyton End Suff 106 C4
Bozeat Northants 121 F8
Bozen Green Herts 105 F8
Brù W Isles 304 D5
Braaid I o M 192 E4
Braal Castle Highld 310 C5
Brabling Green Suff 126 E5
Brabourne Kent 54 E5
Brabourne Lees Kent 54 E5
Brabster Highld 310 C7
Bracadale Highld 294 B5
Bracara Highld 295 F9
Braceborough Lincs 155 G11
Bracebridge Lincs 173 B7
Bracebridge Heath Lincs 173 B7
Bracebridge Low Fields Lincs 173 B7
Braceby Lincs 155 B10
Bracewell Lancs 204 D3
Bracken Bank W Yorks 204 F6
Bracken Hill W Yorks 197 C7
Bracken Park W Yorks 206 E3
Brackenber Cumb 222 B4
Brackenbottom N Yorks 212 E6
Brackenfield Derbys 170 D5
Brackenhall W Yorks 197 D7
Brackenlands Cumb 229 B11
Brackenthwaite Cumb 229 B11
Brackenthwaite Cumb 229 G9
Brackenthwaite N Yorks 205 C11
Brackla = Bragle Bridgend 58 D2
Bracklamore Aberds 303 D8
Bracklesham W Sus 22 D4
Brackletter Highld 290 E3
Brackley Argyll 255 C8
Brackley Northants 101 D11
Brackloch Highld 307 G6
Bracknell Brack 65 F11
Braco Perth 286 G2
Braco Castle Perth 286 F2
Braco Park Aberds 303 C9
Bracobrae Moray 302 D5
Bracon N Lincs 199 F9
Bracon Ash Norf 142 D3
Bracorina Highld 295 F9
Bradaford Devon 12 C3
Bradbourne Derbys 170 E2
Bradbury Durham 234 F2
Bradda I o M 192 F2
Bradden Northants 102 B2
Braddock Corn 6 C3
Braddocks Hay Staffs 168 D5
Bradeley Stoke 168 E5
Bradeley Green E Ches 167 G8
Bradenham Bucks 84 F4
Bradenham Norf 141 B8
Bradenstoke Wilts 62 D4
Brades Village W Mid 133 E9
Bradfield Devon 27 F9
Bradfield Essex 108 E2
Bradfield Norf 160 C5
Bradfield W Berks 64 E6
Bradfield Combust Suff 125 F7
Bradfield Green E Ches 167 D11
Bradfield Heath Essex 108 F2
Bradfield St Clare Suff 125 F8
Bradfield St George Suff 125 E8
Bradford Corn 11 F8
Bradford Derbys 170 C2
Bradford Devon 24 F6
Bradford Gtr Man 184 B5
Bradford Northumb 264 C5
Bradford W Yorks 205 G9
Bradford Abbas Dorset 29 E9
Bradford Leigh Wilts 61 G10
Bradford-on-Avon Wilts 61 G10
Bradford-on-Tone Som 27 C11
Bradford Peverell Dorset 17 C9
Bradgate S Yorks 186 C6
Bradiford Devon 40 G5
Brading I o W 21 D8
Bradley Derbys 170 F2
Bradley Glos 80 G3
Bradley Hants 48 E6
Bradley NE Lincs 201 F8
Bradley Staffs 151 F7
Bradley W Mid 133 D8
Bradley W Yorks 197 C7
Bradley Wrex 166 E4
Bradley Cross Som 44 C3
Bradley Fold Gtr Man 195 F9
Bradley Green Glos 80 G2
Bradley Green Som 43 F9
Bradley Green W Ches 167 F8
Bradley Green Warks 134 C5
Bradley Green Worcs 117 E9
Bradley in the Moors Staffs 169 G9
Bradley Mills W Yorks 197 D7
Bradley Mount E Ches 184 F6
Bradley Stoke S Glos 60 C6
Bradlow Hereford 98 D4
Bradmore Notts 153 C11
Bradmore W Mid 133 D7
Bradney Shrops 132 D5
Bradney Som 43 F10
Bradninch Devon 27 G8
Bradnock's Marsh W Mid 118 B4
Bradnop Staffs 169 D8
Bradnor Green Hereford 114 F5
Bradpole Dorset 16 C5
Bradshaw Gtr Man 195 E8
Bradshaw Staffs 168 D6
Bradshaw W Yorks 196 E5
Bradshaw W Yorks 205 G7
Bradstone Devon 12 E3
Bradville M Keynes 102 C6
Bradwall Green E Ches 168 C3
Bradway S Yorks 186 E4
Bradwell Derbys 185 E11
Bradwell Devon 40 E3
Bradwell Essex 106 G6
Bradwell M Keynes 102 D6
Bradwell Norf 143 C10
Bradwell Staffs 168 F4
Bradwell Common M Keynes 102 D6
Bradwell Grove Oxon 82 D2
Bradwell Hills Derbys 185 E11
Bradwell on Sea Essex 89 D8
Bradwell Waterside Essex 89 D7
Bradworthy Devon 24 E4
Bradworthy Cross Devon 24 E4
Brae Dumfries 237 B10
Brae Highld 307 L3
Brae Highld 309 J4
Brae Shetland 312 G5
Brae of Achnahaird Highld 307 H5
Brae of Boquhapple Stirl 285 G10
Brae Roy Lodge Highld 290 D5
Braeantra Highld 300 B5
Braebuster Orkney 314 F5
Braedownie Angus 292 F4
Braeface Falk 278 E5
Braefield Highld 300 F4
Braefindon Highld 300 D6
Braegrum Perth 286 E4
Braehead Dumfries 236 D6
Braehead Orkney 314 B4
Braehead Orkney 314 F5
Braehead S Ayrs 257 E8
Braehead S Lnrk 259 C8
Braehead S Lnrk 267 D11
Braehead S Lnrk 269 E9
Braehead Stirl 278 C6
Braehead of Lunan Angus 287 B10
Braehoulland Shetland 312 F4
Braehour Highld 310 D4
Braehungie Highld 310 F5
Braeintra Highld 295 B10
Braelangwell Lodge Highld 309 K5
Braemar Aberds 292 D3
Braemore Highld 299 B11
Braemore Highld 310 F4
Braepark Edin 280 F3
Braes of Enzie Moray 302 D3
Braes of Ullapool Highld 307 K6
Braeside Invclyd 276 F4
Braeswick Orkney 314 C6
Braevallich Argyll 275 C10
Braewick Shetland 312 F4
Braewick Shetland 313 H5
Brafferton Darl 233 G11
Brafferton N Yorks 215 E8
Brafield-on-the-Green Northants 120 F6
Bragar W Isles 304 D4
Bragbury End Herts 104 G5
Bragenham Bucks 103 F8
Bragle = Brackla Bridgend 58 D2
Bragleenmore Argyll 289 G11
Braichmelyn Gwyn 163 B10
Braichyfedw Powys 129 E7
Braid Edin 280 G4
Braides Lancs 202 C4
Braidfauld Glasgow 268 C2
Braidley N Yorks 213 C10
Braids Argyll 255 C8
Braidwood S Lnrk 268 F6
Braigh Chalasaigh W Isles 296 D5
Braigo Argyll 274 G3
Brailsford Derbys 170 G3
Brailsford Green Derbys 170 G3
Braingortan Argyll 275 F11
Brain's Green Glos 79 D11
Brainshaugh Northumb 252 C6
Braintree Essex 106 G5
Braiseworth Suff 126 C2
Braishfield Hants 32 B5
Braiswick Essex 107 F9
Braithwaite Cumb 229 G10
Braithwaite S Yorks 198 E6
Braithwaite W Yorks 204 E6
Braithwell S Yorks 187 C8
Brakefield Green Norf 141 B10
Brakenhill W Yorks 198 D2
Bramber W Sus 35 E11
Bramblecombe Dorset 30 G3
Brambridge Hants 33 C7
Bramcote Notts 153 B10
Bramcote Warks 135 F8
Bramcote Hills Notts 153 B10
Bramcote Mains Warks 135 F8
Bramdean Hants 33 B10
Bramerton Norf 142 C5
Bramfield Herts 86 B3
Bramfield Suff 127 C7
Bramford Suff 108 B2
Bramford W Mid 133 E8
Bramhall Gtr Man 184 D5
Bramhall Moor Gtr Man 184 D6
Bramhall Park Gtr Man 184 D5
Bramham W Yorks 206 E4
Bramhope W Yorks 205 E11
Bramley Derbys 186 F6
Bramley Hants 48 B6
Bramley S Yorks 187 C7
Bramley Sur 50 D4
Bramley W Yorks 205 F10
Bramley Corner Hants 48 B6
Bramley Green Hants 49 B7
Bramley Head N Yorks 205 B8
Bramley Vale Derbys 171 B7
Bramling Kent 55 B8
Brampford Speke Devon 14 B4
Brampton Cambs 122 C4
Brampton Cumb 231 G9
Brampton Cumb 240 E2
Brampton Derbys 186 G5
Brampton Hereford 97 D8
Brampton Lincs 188 F4
Brampton Norf 160 E4
Brampton S Yorks 198 G2
Brampton Suff 143 G8
Brampton Abbotts Hereford 98 F2
Brampton Ash Northants 136 F5
Brampton Bryan Hereford 115 C7
Brampton en le Morthen S Yorks 187 D7
Brampton Park Cambs 122 C4
Brampton Street Suff 143 G8
Bramshall Staffs 151 C11
Bramshaw Hants 32 D3
Bramshill Hants 65 G8
Bramshott Hants 49 G10
Bramwell Som 28 B6
Bran End Essex 106 F3
Branatwatt Shetland 313 H4
Branault Highld 289 C7
Branbridges Kent 53 D7
Brancaster Norf 176 E3
Brancaster Staithe Norf 176 E3
Brancepeth Durham 233 D10
Branch End Northumb 242 E3
Branchill Moray 301 D10
Branchton Invclyd 276 F4
Brand End Lincs 174 F5
Brand Green Glos 98 F4
Brand Green Hereford 98 C5
Branderburgh Moray 302 B2
Brandesburton E Yorks 209 D8
Brandeston Suff 126 E4
Brandhill Shrops 115 B8
Brandis Corner Devon 24 G6
Brandish Street Som 42 D2
Brandiston Norf 160 E2
Brandlingill Cumb 229 F8
Brandon Durham 233 D10
Brandon Lincs 172 F6
Brandon Northumb 264 F2
Brandon Suff 140 F5
Brandon Warks 119 B8
Brandon Bank Cambs 140 F2
Brandon Creek Norf 140 E2
Brandon Parva Norf 141 B11
Brands Hill Windsor 66 D4
Brandsby N Yorks 215 E11
Brandwood Shrops 149 D9
Brandwood End W Mid 117 B11
Brandy Carr W Yorks 197 C10
Brandy Hole Essex 88 F4
Brandy Wharf Lincs 189 B8
Brandyquoy Orkney 314 G4
Brane Corn 1 D4
Branksome Darl 224 B5
Branksome Poole 18 C6
Branksome Park Poole 19 C7
Bransbury Hants 48 E2
Bransby Lincs 188 F5
Branscombe Devon 15 D9
Bransford Worcs 116 G5
Bransgore Hants 19 B9
Branshill Clack 279 C7
Bransholme Hull 209 G8
Branson's Cross Worcs 117 C11
Branston Leics 154 D6
Branston Lincs 173 B8
Branston Staffs 152 E4
Branston Booths Lincs 173 B9
Branstone I o W 21 E7
Bransty Cumb 219 B9
Brant Broughton Lincs 172 E6
Brantham Suff 108 E2
Branthwaite Cumb 229 D11
Branthwaite Cumb 229 G7
Branthwaite Edge Cumb 229 G7
Brantingham E Yorks 200 B2
Branton Northumb 264 F2
Branton S Yorks 198 G6
Branton Green N Yorks 215 G8
Branxholm Park Borders 261 G11
Branxholme Borders 261 G11
Branxton Northumb 263 B9
Brascote Leics 135 C8
Brassey Green W Ches 167 C8
Brassington Derbys 170 E2
Brasted Kent 52 C3
Brasted Chart Kent 52 C3
Brathens Aberds 293 D8
Bratoft Lincs 175 B7
Brattle Kent 54 G2
Brattleby Lincs 188 E6
Bratton Som 42 D2
Bratton Telford 150 G2
Bratton Wilts 46 C2
Bratton Clovelly Devon 12 C5
Bratton Fleming Devon 40 F6
Bratton Seymour Som 29 B11
Braughing Herts 105 F7
Braughing Friars Herts 105 G8
Braulen Lodge Highld 300 F2
Braunston Northants 119 D10
Braunston-in-Rutland Rutland 136 B6
Braunstone Leics 135 C11
Braunstone Town Leicester 135 C11
Braunton Devon 40 F3
Brawby N Yorks 216 D4
Brawith N Yorks 225 D10
Brawl Highld 310 C2
Brawlbin Highld 310 D4
Bray Windsor 66 D2
Bray Shop Corn 12 G2
Bray Wick Windsor 65 D11
Braybrooke Northants 136 G5
Braydon Side Wilts 62 B4
Brayford Devon 41 G7
Brayfordhill Devon 41 G7
Brays Grove Essex 87 D7
Braystones Cumb 219 D10
Brayswick Worcs 98 B6
Braythorn N Yorks 205 D10
Brayton N Yorks 207 G8
Braytown Dorset 18 D2
Braywoodside Windsor 65 D11
Brazacott Corn 11 C11
Brazenhill Staffs 151 E7
Brea Corn 4 G3
Breach Bath 60 G6
Breach Kent 69 F10
Breach W Sus 22 B3
Breachacha Castle Argyll 288 D3
Breachwood Green Herts 104 G2
Breacleit W Isles 304 E3
Bread Street Glos 80 D4
Breaden Heath Shrops 149 B8
Breadsall Derbys 153 B7
Breadsall Hilltop Derby 153 B7
Breadstone Glos 80 E2
Breage Corn 2 D4
Breakachy Highld 300 E4
Brealeys Devon 25 D8
Bream Glos 79 D10
Breamore Hants 31 D11
Bream's Meend Glos 79 D9
Brean Som 43 B9
Breanais W Isles 304 F1
Brearley W Yorks 196 B4
Brearton N Yorks 214 G6
Breascleit W Isles 304 E4
Breaston Derbys 153 C9
Brechfa Carms 93 E10
Brechin Angus 293 G7
Breck of Cruan Orkney 314 E3
Breckan Orkney 314 F2
Breckles Norf 141 E9
Breckrey Highld 298 C5
Brecks S Yorks 187 C7
Brecon Powys 95 F10
Bredbury Gtr Man 184 C6
Bredbury Green Gtr Man 184 C6
Brede E Sus 38 D4
Bredenbury Hereford 116 F2
Bredfield Suff 126 G5
Bredgar Kent 69 G11

Bredhurst Kent 69 G9
Bredicot Worcs 117 G8
Bredon Worcs 99 D8
Bredon's Hardwick Worcs 99 D8
Bredon's Norton Worcs 99 D8
Bredwardine Hereford 96 C6
Breedon on the Hill Leics 153 E8
Breeds Essex 87 C11
Breedy Butts Lancs 202 E2
Breibhig W Isles 297 M2
Breibhig W Isles 304 E6
Breich W Loth 269 C9
Breightmet Gtr Man 195 F8
Breighton E Yorks 207 G10
Breinton Hereford 97 D9
Breinton Common Hereford 97 C9
Breiwick Shetland 313 J6
Brelston Green Hereford 97 G11
Bremhill Wilts 62 E3
Bremhill Wick Wilts 62 E3
Bremirehoull Shetland 313 L6
Brenachoile Lodge Stirl 285 G8
Brenchley Kent 53 E7
Brenchoillie Argyll 284 G4
Brendon Devon 24 E5
Brendon Devon 24 F5
Brendon Devon 41 D9
Brenkley T & W 242 B6
Brent Corn 6 E4
Brent Eleigh Suff 107 B8
Brent Knoll Som 43 C10
Brent Mill Devon 8 D3
Brent Pelham Herts 105 E8
Brentford London 67 D7
Brentford End London 67 D7
Brentingby Leics 154 F5
Brentry Bristol 60 D5
Brentwood Essex 87 G9
Brenzett Kent 39 B8
Brenzett Green Kent 39 B8
Brereton Staffs 151 F11
Brereton Cross Staffs 151 F11
Brereton Green E Ches 168 C3
Brereton Heath E Ches 168 C4
Breretonhill Staffs 151 F11
Bressingham Norf 141 G11
Bressingham Common Norf 141 G11
Bretby Derbys 152 E5
Bretford Warks 119 B8
Bretforton Worcs 99 C11
Bretherdale Head Cumb 221 E11
Bretherton Lancs 194 C3
Brettabister Shetland 313 H6
Brettenham Norf 141 G8
Brettenham Suff 125 G9
Bretton Derbys 186 F2
Bretton Flint 166 C5
Bretton P'boro 138 C3
Brewer Street Sur 51 C10
Brewer's End Essex 105 G11
Brewers Green Norf 142 G2
Brewlands Bridge Angus 292 G3
Brewood Staffs 133 B7
Briach Moray 301 D10
Briants Puddle Dorset 18 C2
Briar Hill Northants 120 F4
Brick End Essex 105 F11
Brick Hill Sur 66 G3
Brick House End Essex 105 F9
Brick Houses S Yorks 186 E4
Brick-kiln End Notts 171 D9
Brickendon Herts 86 D4
Bricket Wood Herts 85 E10
Brickfields Worcs 117 F7
Brickhill Beds 121 G11
Brickhouses E Ches 168 C3
Brickkiln Green Essex 106 E4
Bricklehampton Worcs 99 C9
Bride I o M 192 B5
Bridekirk Cumb 229 E8
Bridell Pembs 92 C3
Bridestowe Devon 12 D6
Brideswell Aberds 302 F5
Bridford Devon 14 D2
Bridfordmills Devon 14 D2
Bridge Corn 2 D6
Bridge Corn 4 G3
Bridge Kent 55 C7
Bridge Som 28 F5
Bridge Ball Devon 41 D8
Bridge End Beds 121 G10
Bridge End Cumb 230 B3
Bridge End Devon 8 F3
Bridge End Durham 232 D6
Bridge End Essex 106 E3
Bridge End Flint 166 D4
Bridge End Hereford 98 B2
Bridge End Lincs 156 B2
Bridge End Northumb 241 D10
Bridge End Northumb 241 E10
Bridge End Oxon 83 G9
Bridge-End Shetland 313 K5
Bridge End Sur 50 B5
Bridge End Warks 118 E5
Bridge End Worcs 98 E6
Bridge Green Essex 105 D9
Bridge Green Norf 142 G2
Bridge Hewick N Yorks 214 E6
Bridge Ho Argyll 254 B4
Bridge of Alford Aberds 293 B7
Bridge of Allan Stirl 278 B5
Bridge of Avon Moray 301 F11
Bridge of Avon Moray 301 G11
Bridge of Awe Argyll 284 E4
Bridge of Balgie Perth 285 C9
Bridge of Cally Perth 286 B5
Bridge of Canny Aberds 293 D8
Bridge of Craigisla Angus 286 B6
Bridge of Dee Dumfries 237 D9
Bridge of Don Aberdeen 293 B11
Bridge of Dun Angus 287 B10
Bridge of Dye Aberds 293 E8
Bridge of Ericht Perth 285 B9
Bridge of Feugh Aberds 293 D9
Bridge of Forss Highld 310 C4
Bridge of Gairn Aberds 292 D5
Bridge of Gaur Perth 285 B9
Bridge of Lyon Perth 285 C11
Bridge of Muchalls Aberds 293 D10
Bridge of Muick Aberds 292 D5
Bridge of Oich Highld 290 C5
Bridge of Orchy Argyll 284 D6
Bridge of Waith Orkney 314 E2
Bridge of Walls Shetland 313 H4
Bridge of Weir Renfs 267 B7
Bridge Reeve Devon 25 E11
Bridge Sollers Hereford 97 C8
Bridge Street Suff 107 B7
Bridge Town Warks 118 G4
Bridge Trafford W Ches 183 G7
Bridge Yate S Glos 61 E7
Bridgefoot Aberds 292 C6
Bridgefoot Angus 287 D7
Bridgefoot Cumb 229 F7
Bridgehampton Som 29 C9
Bridgehill Durham 242 G3
Bridgeholm Green Derbys 185 E8
Bridgehouse Gate N Yorks 214 F3
Bridgelands Borders 261 C11
Bridgemary Hants 33 G9
Bridgemere E Ches 168 F2
Bridgemont Derbys 185 E8
Bridgend Aberds 293 B7
Bridgend Aberds 302 F5
Bridgend Angus 293 G7
Bridgend Argyll 255 D8
Bridgend Argyll 274 G4
Bridgend Argyll 275 D9
Bridgend Corn 6 D2
Bridgend Cumb 221 C7
Bridgend Devon 7 F11
Bridgend Fife 287 F7
Bridgend Glos 80 E4
Bridgend Highld 300 D3
Bridgend Invclyd 276 F5
Bridgend Moray 302 F3
Bridgend N Lnrk 278 G3
Bridgend Pembs 92 B3
Bridgend W Loth 279 F10
Bridgend = Pen-y-Bont ar-ogwr Bridgend 58 C2
Bridgend of Lintrathen Angus 286 B6
Bridgeness Falk 279 E10
Bridgerule Devon 24 G3
Bridges Corn 5 D10
Bridges Shrops 131 D7
Bridgeton Glasgow 268 C2
Bridgetown Corn 12 D2
Bridgetown Devon 8 C6
Bridgetown Som 42 G2
Bridgetown Staffs 133 B9
Bridgham Norf 141 F9
Bridgnorth Shrops 132 E4
Bridgtown Staffs 133 B9
Bridgwater Som 43 F10
Bridlington E Yorks 218 F3
Bridport Dorset 16 C5
Bridstow Hereford 97 G11
Brierfield Lancs 204 F2
Brierholme Carr S Yorks 199 E7
Brierley Glos 79 B10
Brierley Hereford 115 F9
Brierley S Yorks 198 E2
Brierley Hill W Mid 133 F8
Brierton Hrtlpl 234 E5
Briery Cumb 229 G11
Briery Hill Bl Gwent 77 D11
Briestfield W Yorks 197 D8
Brig o'Turk Stirl 285 G9
Brigflatts Cumb 222 G2
Brigg N Lincs 200 F3
Brigg N Lincs 200 F4
Briggate Norf 160 D6
Briggswath N Yorks 227 D7
Brigham Cumb 229 E7
Brigham Cumb 229 G11
Brigham E Yorks 209 C7
Brighouse W Yorks 196 C6
Brighstone I o W 20 E4
Brightgate Derbys 170 D3
Brighthampton Oxon 82 E5
Brightholmlee S Yorks 186 B3
Brightley Devon 13 B7
Brightling E Sus 37 C11
Brightlingsea Essex 89 B9
Brighton Brighton 36 G4
Brighton Corn 5 E8
Brighton Hill Hants 48 D6
Brighton le Sands Mers 182 B4
Brightons Falk 279 F8
Brightside S Yorks 186 D5
Brightwalton W Berks 64 D2
Brightwalton Green W Berks 64 D2
Brightwalton Holt W Berks 64 D2
Brightwell Suff 108 C4
Brightwell Baldwin Oxon 83 F11
Brightwell cum Sotwell Oxon 83 G9
Brigmerston Wilts 47 D7
Brignall Durham 223 C11
Brigsley NE Lincs 201 G9
Brigsteer Cumb 211 B9
Brigstock Northants 137 F8
Brill Bucks 83 C11
Brill Corn 2 D6
Brilley Hereford 96 B5
Brilley Mountain Powys 114 G5
Brimaston Pembs 91 G8
Brimfield Hereford 115 D10
Brimington Derbys 186 G6
Brimington Common Derbys 186 G6
Brimley Devon 13 F11
Brimley Devon 28 G4
Brimps Hill Glos 79 B11
Brimpsfield Glos 80 C6
Brimpton W Berks 64 G5
Brimpton Common W Berks 64 G5
Brims Orkney 314 H2
Brims Castle Highld 310 B4
Brimscombe Glos 80 E5
Brimsdown London 86 F5
Brimstage Mers 182 E4
Brinacory Highld 295 F9
Brincliffe S Yorks 186 D4
Brind E Yorks 207 G10
Brindham Som 44 E4
Brindister Shetland 313 H4
Brindister Shetland 313 K6
Brindle Lancs 194 C6
Brindle Heath Gtr Man 195 G10
Brindley E Ches 167 E9
Brindley Ford Stoke 168 E5
Brindwoodgate Derbys 186 F4
Brineton Staffs 150 G6
Bringewood Forge Hereford 115 C9
Bringhurst Leics 136 E6
Bringsty Common Hereford 116 F4
Brington Cambs 121 B11
Brinian Orkney 314 D4
Briningham Norf 159 C10
Brinkhill Lincs 190 G5
Brinkley Cambs 124 G2
Brinkley Notts 172 E2
Brinkley Hill Hereford 97 E11
Brinklow M Keynes 103 D8
Brinklow Warks 119 B8
Brinkworth Wilts 62 C4
Brinmore Highld 300 G6
Brinnington Gtr Man 184 C6
Brinscall Lancs 194 C6
Brinsea N Som 60 G2
Brinsford Staffs 133 B8
Brinsley Notts 171 F7
Brinsop Hereford 97 C8
Brinsop Common Hereford 97 C8
Brinsworth S Yorks 186 D6
Brinsworthy Devon 41 G9
Brinton Norf 159 B10
Brisco Cumb 239 G10
Briscoe Cumb 219 C10
Briscoerigg N Yorks 205 C11
Brisley Norf 159 E8
Brislington Bristol 60 E6
Brissenden Green Kent 54 F2
Bristnall Fields W Mid 133 F9
Bristol Bristol 60 E5
Briston Norf 159 C11
Britain Bottom S Glos 61 B9
Britannia Lancs 195 C11
Britford Wilts 31 B11
Brithdir Caerph 77 E11
Brithdir Ceredig 92 B6
Brithdir Gwyn 146 F5
Brithem Bottom Devon 27 E8
British Torf 78 E3
Briton Ferry = Llansawel Neath 57 C8
Britten's Bath 45 B7
Britwell Slough 66 C3
Britwell Salome Oxon 83 G11
Brixham Torbay 9 D8
Brixton Devon 7 E11
Brixton London 67 D10
Brixton Deverill Wilts 45 F11
Brixworth Northants 120 C4
Brize Norton Oxon 82 D4
Broad Alley Worcs 117 D7
Broad Blunsdon Swindon 81 G11
Broad Campden Glos 100 D3
Broad Carr W Yorks 196 D5
Broad Chalke Wilts 31 B8
Broad Clough Lancs 195 C11
Broad Colney Herts 85 E11
Broad Common Worcs 117 D7
Broad Ford Kent 53 F8
Broad Green C Beds 103 C9
Broad Green Cambs 124 F3
Broad Green Essex 105 D8
Broad Green Essex 107 G7
Broad Green London 67 F10
Broad Green Mers 182 C6
Broad Green Suff 124 F5
Broad Green Suff 125 F11
Broad Green Worcs 116 F5
Broad Green Worcs 117 C9
Broad Haven = Aberllydan Pembs 72 C5
Broad Heath Powys 114 E6
Broad Heath Staffs 151 D7
Broad Heath Worcs 116 D3
Broad Hill Cambs 123 B11
Broad Hinton Wilts 62 D6
Broad Ings E Yorks 208 C2
Broad Lane Corn 4 G3
Broad Lanes Shrops 132 F5
Broad Laying Hants 64 G2
Broad Layings Hants 64 G2
Broad Marston Worcs 100 B2
Broad Meadow Staffs 168 F4
Broad Oak Carms 93 G11
Broad Oak Cumb 220 G2
Broad Oak Dorset 30 E3
Broad Oak E Sus 37 C10
Broad Oak E Sus 38 D4
Broad Oak Hants 49 C9
Broad Oak Hereford 97 G9
Broad Oak Kent 54 F4
Broad Oak Kent 71 G7
Broad Oak Mers 183 B8
Broad Oak Shrops 132 F5
Broad Parkham Devon 24 C5
Broad Street E Sus 38 D5
Broad Street Kent 53 B10
Broad Street Kent 54 E6
Broad Street Kent 55 E7
Broad Street Medway 69 E9
Broad Street Suff 107 C9
Broad Street Wilts 46 B6
Broad Street Green Essex 88 D5
Broad Tenterden Kent 53 G11
Broad Town Wilts 62 D5
Broadbottom Gtr Man 185 C7
Broadbridge W Sus 22 B4
Broadbridge Heath W Sus 50 G6
Broadbury Devon 12 B5
Broadbush Swindon 81 G11
Broadclyst Devon 14 B5
Broadfield Gtr Man 195 E10
Broadfield Invclyd 276 G6
Broadfield Lancs 194 C4
Broadfield Lancs 195 B8
Broadfield Pembs 73 E10
Broadfield W Sus 51 G9
Broadford Highld 295 C8
Broadford Sur 50 D3
Broadford Bridge W Sus 35 C9
Broadgate Hants 32 C6
Broadgrass Green Suff 125 E9
Broadgreen Wood Herts 86 D4
Broadhalgh Gtr Man 195 E11
Broadham Green Sur 51 C11
Broadhaugh Borders 249 B10
Broadhaven Highld 310 D7
Broadheath Gtr Man 184 D3
Broadhembury Devon 27 G10
Broadhempston Devon 8 B6
Broadholm Derbys 170 F4
Broadholme Derbys 170 F5
Broadholme Lincs 188 G5
Broadland Row E Sus 38 D4
Broadlands Devon 14 G3
Broadlane Corn 2 C4
Broadlay Carms 74 D5
Broadley Lancs 195 D11
Broadley Moray 302 C3
Broadley Common Essex 86 D6
Broadleys Aberds 303 C8
Broadmayne Dorset 17 D10
Broadmeadows Borders 261 C10
Broadmere Hants 48 D6
Broadmoor Pembs 73 D9
Broadmoor Sur 50 D6
Broadmoor Common Hereford 98 D2
Broadmore Green Worcs 116 G6
Broadoak Dorset 16 B4
Broadoak Glos 80 C2
Broadoak Hants 33 E8
Broadoak Shrops 149 F9
Broadoak Wrex 166 D5
Broadoak End Herts 86 C4
Broadoak Park Gtr Man 195 G9
Broadplat Oxon 65 C8
Broadrashes Moray 302 D4
Broadrock Glos 79 F8
Broad's Green Essex 87 C11
Broad's Green Wilts 62 F3
Broadsands Torbay 9 D7
Broadsea Aberds 303 C9
Broadshard Som 28 E6
Broadstairs Kent 71 F11
Broadstone Kent 53 D11
Broadstone Mon 79 E8
Broadstone Poole 18 B6
Broadstone Shrops 131 F10
Broadstreet Common Newport 59 C11
Broadwas Worcs 116 F5
Broadwater Herts 104 G4
Broadwater W Sus 35 G11
Broadwater Down Kent 52 F5
Broadwaters Worcs 116 B6
Broadwath Cumb 239 F11
Broadway Carms 74 D3
Broadway Carms 74 D5
Broadway Pembs 72 C5
Broadway Som 28 D4
Broadway Suff 127 B7
Broadway Worcs 99 D11
Broadway Lands Hereford 97 E11
Broadwell Glos 79 C9
Broadwell Glos 100 F4
Broadwell Oxon 82 E3
Broadwell Warks 119 D9
Broadwey Dorset 17 E9
Broadwindsor Dorset 28 G6
Broadwood Kelly Devon 25 F10
Broadwoodwidger Devon 12 D4
Brobury Hereford 96 C6
Brochel Highld 298 E5
Brochroy Argyll 284 D4
Brock Lancs 202 E6
Brock Hill Essex 88 F2
Brockamin Worcs 116 G5
Brockbridge Hants 33 D10
Brockdish Norf 126 B4
Brockencote Worcs 117 C7
Brockenhurst Hants 32 G4
Brocketsbrae S Lnrk 259 B8
Brockfield Devon 28 F4
Brockford Green Suff 126 D2
Brockford Street Suff 126 D2
Brockhall Northants 120 E2
Brockhall Village Lancs 203 F10
Brockham Sur 51 D7
Brockham End Bath 61 F8
Brockham Park Sur 51 D8
Brockhampton Glos 99 F8
Brockhampton Glos 99 G10
Brockhampton Hants 22 B2
Brockhampton Hereford 97 E11
Brockhampton Green Dorset 30 F2
Brockhill Borders 261 E9
Brockholes W Yorks 197 E7
Brockhollands Glos 79 D10
Brockhurst Derbys 170 C4
Brockhurst Hants 33 G10
Brockhurst Warks 135 G9
Brocklebank Cumb 230 C2
Brocklehirst Dumfries 238 C3
Brocklesby Lincs 200 E6
Brockley London 67 E11
Brockley N Som 60 F3
Brockley Corner Suff 124 C6
Brockley Green Suff 106 B4
Brockley Green Suff 124 G6
Brockleymoor Cumb 230 D5
Brockloch Dumfries 246 D2
Brockmanton Hereford 115 F10
Brockmoor W Mid 133 F8
Brock's Green Hants 64 G4
Brock's Watering Norf 142 E2
Brockscombe Devon 12 C5
Brockton Shrops 130 C6
Brockton Shrops 130 F6
Brockton Shrops 131 E11
Brockton Shrops 132 C4
Brockton Staffs 150 C6
Brockton Telford 150 F4
Brockweir Glos 79 E8
Brockwell Som 42 E2
Brockwood Hants 33 B10
Brockworth Glos 80 B5
Brocton Corn 5 B10
Brocton Staffs 151 F9
Brodick N Ayrs 256 B2
Brodie Moray 301 D9
Brodiesord Aberds 302 C5
Brodsworth S Yorks 198 F4
Brogaig Highld 298 C4
Brogborough C Beds 103 D9
Broke Hall Suff 108 C3
Broken Cross E Ches 184 G5
Broken Cross W Ches 183 G11
Broken Green Herts 105 G8
Brokenborough Wilts 62 B2
Brokerswood Wilts 45 C10
Brokes N Yorks 224 F3
Brombil Neath 57 D9
Bromborough Mers 182 E4
Bromborough Pool Mers 182 E4
Bromdon Shrops 132 G2
Brome Suff 126 B2
Brome Street Suff 126 B3
Bromesberrow Glos 98 E4
Bromesberrow Heath Glos 98 E4
Bromeswell Suff 126 G6
Bromfield Cumb 229 B9
Bromfield Shrops 115 B9
Bromford W Mid 134 E2
Bromham Beds 121 G10
Bromham Wilts 62 F3
Bromley Herts 105 G8
Bromley London 67 C11
Bromley London 68 F2
Bromley S Yorks 186 B4
Bromley Shrops 132 D4
Bromley Shrops 149 D8
Bromley W Mid 133 F8
Bromley Common London 68 F2
Bromley Cross Essex 107 F11
Bromley Cross Gtr Man 195 E8
Bromley Green Kent 54 F3
Bromley Hall Staffs 150 C5
Bromley Heath S Glos 61 D7
Bromley Park London 67 F11
Bromley Wood Staffs 152 E2
Bromlow Shrops 130 C6
Brompton London 67 D9
Brompton Medway 69 F9
Brompton N Yorks 217 C8
Brompton N Yorks 225 F7
Brompton Shrops 131 B10
Brompton-by-Sawdon N Yorks 217 C8
Brompton-on-Swale N Yorks 224 F4
Brompton Ralph Som 42 G5
Brompton Regis Som 42 G3
Bromsash Hereford 98 G2
Bromsberrow Heath Glos 98 E4
Bromsgrove Worcs 117 C9
Bromstead Common Staffs 150 F6
Bromstead Heath Staffs 150 F6
Bromstone Kent 71 F11
Bromyard Hereford 116 G3
Bromyard Downs Hereford 116 F3
Bronaber Gwyn 146 C4
Broncroft Shrops 131 F10
Brondesbury London 67 C8
Brondesbury Park London 67 C8
Broneirion Powys 129 F10
Brongest Ceredig 92 B6
Brongwyn Ceredig 92 C5
Bronington Wrex 149 B9
Bronllys Powys 96 D2
Bronnant Ceredig 112 D2
Bronwydd Ceredig 93 C7
Bronwydd Arms Carms 93 G8
Bronydd Powys 96 B4
Bronygarth Shrops 148 B5
Brook Carms 74 D3
Brook Devon 12 G5
Brook Devon 14 C2
Brook Hants 32 B4
Brook Hants 32 E3
Brook I o W 20 E3
Brook Kent 54 E5
Brook Sur 50 D5
Brook Sur 50 F2
Brook Bottom Gtr Man 185 D7
Brook Bottom Gtr Man 196 G3
Brook Bottom Lancs 202 F6
Brook End Beds 121 E11
Brook End C Beds 104 B3
Brook End Cambs 121 C11
Brook End Herts 104 F6
Brook End M Keynes 103 C8
Brook End Wilts 61 C10
Brook End Worcs 99 B7
Brook Green London 67 D8
Brook Green Suff 125 F7
Brook Hill Hants 32 E3
Brook Hill Notts 153 C11
Brook Place Sur 66 G3
Brook Street Essex 87 G9
Brook Street Kent 52 D5
Brook Street Kent 54 G2
Brook Street Suff 106 B6
Brook Street W Sus 36 B4
Brook Waters Wilts 30 C6
Brooke Norf 142 D5
Brooke Rutland 136 B6
Brookenby Lincs 190 B2
Brookend Glos 79 E11
Brookend Glos 79 F9
Brookend Oxon 100 G6
Brookfield Derbys 185 B8
Brookfield Lancs 203 G7
Brookfield M'bro 225 B9
Brookfield Renfs 267 C8
Brookfoot W Yorks 196 C6
Brookgreen I o W 20 E3
Brookhampton Oxon 83 F10
Brookhampton Shrops 131 E11
Brookhampton Som 29 B10
Brookhouse Blkburn 195 B7
Brookhouse Denb 165 B9
Brookhouse E Ches 184 F6
Brookhouse Lancs 211 G10
Brookhouse S Yorks 187 D8
Brookhouse W Yorks 196 B5
Brookhouse Green E Ches 168 C4
Brookhouses Derbys 185 D8
Brookhouses Staffs 169 G7
Brookhurst Mers 182 E4
Brookland Kent 39 B7
Brooklands Dumfries 237 B10
Brooklands Gtr Man 184 C3
Brooklands Shrops 167 G8
Brooklands Sur 66 G5
Brooklands W Yorks 206 F2
Brookleigh Devon 14 B5
Brookmans Park Herts 86 E2
Brookpits W Sus 35 G8
Brookrow Shrops 116 C2
Brooks Corn 6 C3
Brooks Powys 130 D2
Brooks End Kent 71 F9
Brooks Green W Sus 35 C10
Brooksbottoms Gtr Man 195 D9
Brooksby Leics 154 F3
Brookside Brack 66 E2
Brookside Derbys 186 G5
Brookside Telford 132 B3
Brookthorpe Glos 80 C4
Brookvale Halton 183 E8
Brookville Norf 140 D4
Brookwood Sur 50 B2
Broom C Beds 104 C3
Broom Cumb 231 G9
Broom Devon 28 G4
Broom E Renf 267 D10
Broom Pembs 73 D10
Broom S Yorks 186 C5
Broom Warks 117 G11
Broom Green Norf 159 E9
Broom Hill Bristol 60 E6
Broom Hill Dorset 31 G8
Broom Hill Durham 242 G4
Broom Hill London 68 F3
Broom Hill Suff 108 B5
Broom Hill Worcs 117 B8
Broom Street Kent 70 G4
Broombank Worcs 116 C3
Broome Norf 143 E7
Broome Shrops 131 D10
Broome Shrops 131 G8
Broome Worcs 117 B8
Broome Park Northumb 264 G4
Broomedge Warr 184 D2
Broomer's Corner W Sus 35 C10
Broomershill W Sus 35 D9
Broomfield Aberds 303 F9
Broomfield Cumb 230 B2
Broomfield Essex 88 D2
Broomfield Kent 53 C10
Broomfield Kent 71 F7
Broomfield Som 43 G8
Broomfield Wilts 61 D11
Broomfields Shrops 149 F8
Broomfleet E Yorks 199 B11
Broomhall E Ches 167 F10
Broomhall Windsor 66 F3
Broomhall Green E Ches 167 F10
Broomham E Sus 23 C8
Broomhaugh Northumb 242 E2
Broomhill Borders 261 D11
Broomhill Bristol 60 D6
Broomhill Highld 301 G9
Broomhill Kent 55 B8
Broomhill Norf 140 C2
Broomhill Northumb 252 C6
Broomhill Notts 171 F8
Broomhill S Yorks 198 G2
Broomhill W Ches 167 B7
Broomhill Bank Kent 52 E5
Broomholm Norf 160 C6
Broomhouse Glasgow 268 C3
Broomlands N Ayrs 257 B8
Broomley Northumb 242 E2
Broompark Durham 233 C10
Broomridge Stirl 278 C6
Broom's Barn Suff 124 D5
Broom's Green Glos 98 E4
Broomsgrove E Sus 38 E4
Broomsthorpe Norf 158 D6
Broomton Highld 301 B8
Broomy Hill Hereford 97 D9
Broomy Lodge Hants 32 E2
Broomyshaw Staffs 169 F9
Brora Highld 311 J3
Broseley Shrops 132 C3
Brotherhouse Bar Lincs 156 G5
Brotheridge Green Worcs 98 C6
Brotherlee Durham 232 D4
Brotherstone Borders 262 B4
Brothertoft Lincs 174 F3
Brotherton N Yorks 198 B3
Brothybeck Cumb 230 C2
Brotton Redcar 226 B3
Broubster Highld 310 C4
Brough Cumb 222 C5
Brough Derbys 185 E11
Brough E Yorks 200 B2
Brough Highld 310 B6
Brough Notts 172 D4
Brough Orkney 314 H4
Brough Orkney 314 E3
Brough Shetland 312 C7
Brough Shetland 312 F6
Brough Shetland 312 F7
Brough Shetland 313 G7
Brough Shetland 313 H6
Brough Shetland 313 J7
Brough Lodge Shetland 312 D7
Brough Sowerby Cumb 222 C5
Broughall Shrops 167 G9
Broughton Borders 260 B4
Broughton Bucks 84 C4
Broughton Cambs 122 B5
Broughton Edin 280 F5
Broughton Flint 166 C4
Broughton Hants 47 G10
Broughton Lancs 202 F6
Broughton M Keynes 103 C7
Broughton N Lincs 200 F3
Broughton N Yorks 204 C4
Broughton N Yorks 216 E5
Broughton Northants 120 B6
Broughton Orkney 314 B4
Broughton Oxon 101 D8
Broughton Shrops 132 E6
Broughton Staffs 150 C5
Broughton V Glam 58 E2
Broughton Astley Leics 135 E10
Broughton Beck Cumb 210 C5
Broughton Common N Lincs 200 E3
Broughton Common Wilts 61 G11
Broughton Cross Cumb 229 E7
Broughton Gifford Wilts 61 G11
Broughton Green Worcs 117 E9
Broughton Hackett Worcs 117 G8
Broughton in Furness Cumb 210 B4
Broughton Lodges Leics 154 E4
Broughton Mills Cumb 220 G4
Broughton Moor Cumb 228 E6
Broughton Park Gtr Man 195 G10
Broughton Poggs Oxon 82 E2
Broughtown Orkney 314 B6
Broughty Ferry Dundee 287 D8
Brow Edge Cumb 211 C7
Browhouses Dumfries 239 D7
Browland Shetland 313 H4
Brown Bank N Yorks 205 C10
Brown Candover Hants 48 F5
Brown Edge Lancs 193 E11
Brown Edge Mers 183 C8
Brown Edge Staffs 168 E6
Brown Heath Hants 33 D8
Brown Heath W Ches 167 B7
Brown Knowl W Ches 167 E7
Brown Lees Staffs 168 D5
Brown Moor W Yorks 206 G3
Brown Street Suff 125 E11
Brownber Cumb 222 D4
Brownbread Street E Sus 23 B11
Brownedge E Ches 168 C3
Brownheath Devon 27 D10
Brownheath Shrops 149 D9
Brownheath Common Worcs 117 E7
Brownhill Aberds 302 E6
Brownhill Aberds 303 E8
Brownhill Blkburn 203 G9
Brownhill Shrops 149 E8
Brownhills Fife 287 F9
Brownhills Shrops 150 B3
Brownhills W Mid 133 B10
Brownieside Northumb 264 E5
Browninghill Green Hants 48 B5
Brownlow E Ches 168 C4
Brownlow Mers 194 G4
Brownlow Fold Gtr Man 195 E8
Brownlow Heath E Ches 168 C4
Brownmuir Aberds 293 F9
Brown's Bank E Ches 167 G10
Brown's End Glos 98 E4
Brown's Green W Mid 133 E10
Browns Wood M Keynes 103 D8
Brownshill Glos 80 E5
Brownshill Green W Mid 134 G6
Brownside Lancs 204 G3
Brownsover Warks 119 B10
Brownston Devon 8 E3
Brownston Green Norf 143 C9
Browntop Cumb 229 G7
Broxa N Yorks 227 G8
Broxbourne Herts 86 D5
Broxburn E Loth 282 F3
Broxburn W Loth 279 G11
Broxfield Northumb 264 F6
Broxholme Lincs 188 F6
Broxted Essex 105 F11
Broxton W Ches 167 E7
Broxtowe Nottingham 171 G8
Broxwood Hereford 115 G7
Broyle Side E Sus 23 C7
Bruairnis W Isles 297 L3
Bruan Highld 310 F7
Bruar Lodge Perth 291 F10
Brucefield Fife 280 D2
Brucehill W Dunb 277 F7
Bruche Warr 183 D10
Brucklebog Aberds 293 D9
Bruera W Ches 166 C6
Bruern Abbey Oxon 100 G5
Bruichladdich Argyll 274 G3
Bruisyard Suff 126 D6
Brumby N Lincs 199 F11
Brunant Powys 130 B5
Brund Staffs 169 C10
Brundall Norf 142 B6
Brundish Norf 143 D7
Brundish Suff 126 D5
Brundish Street Suff 126 C5
Brundon Suff 107 C7
Brunery Highld 289 B9
Brunnion Corn 2 B2
Brunshaw Lancs 204 G3
Brunstane Edin 280 G6
Brunstock Cumb 239 F10
Brunswick Gtr Man 184 B4
Brunswick Park London 86 G3
Brunswick Village T & W 242 C6
Brunt Hamersland Shetland 313 H6
Bruntcliffe W Yorks 197 B9
Brunthwaite W Yorks 205 D7
Bruntingthorpe Leics 136 F2
Brunton Fife 287 E7
Brunton Northumb 264 E6
Brunton Wilts 47 B8
Brushes Gtr Man 185 B7
Brushfield Derbys 185 G11
Brushford Devon 25 F11
Brushford Som 26 B6
Bruton Som 45 G7
Bryans Midloth 270 C6
Bryan's Green Worcs 117 D7
Bryanston Dorset 30 F5
Bryant's Bottom Bucks 84 F5
Brydekirk Dumfries 238 C5
Bryher Scilly 1 G3
Brymbo Conwy 180 G4
Brymbo Wrex 166 E3
Brympton Som 29 D8
Brympton D'Evercy Som 29 D8
Bryn Caerph 77 F11
Bryn Carms 75 E8
Bryn Gtr Man 194 G5
Bryn Gwyn 179 G9
Bryn Neath 57 C10
Bryn Powys 130 C3
Bryn Rhondda 76 D6
Bryn Shrops 130 F5
Bryn Swansea 56 C4
Bryn W Ches 183 G10
Bryn Bwbach Gwyn 146 B2
Bryn-coch Neath 57 B8
Bryn Celyn Anglesey 179 F10
Bryn Celyn Flint 181 F11
Bryn Common Flint 166 D3
Bryn Du Anglesey 178 G4
Bryn Dulas Conwy 180 F6
Bryn Eglwys Gwyn 163 B10
Bryn Gates Gtr Man 194 G5
Bryn-glas Conwy 164 B4
Bryn Golau Rhondda 58 B3
Bryn-henllan Pembs 91 D10
Bryn-Iwan Carms 92 E6
Bryn-mawr Gwyn 144 C4
Bryn Mawr Powys 148 F5
Bryn Myrddin Carms 93 G8
Bryn-nantllech Conwy 164 B6
Bryn-newydd Denb 165 G11
Bryn Offa Wrex 166 F4
Bryn Pen-y-lan Wrex 166 G4
Bryn-penarth Powys 130 C2
Bryn Pydew Conwy 180 F4
Bryn Rhyd-yr-Arian Conwy 165 B7
Bryn-rhys Conwy 180 F4
Bryn Saith Marchog Denb 165 E9
Bryn Sion Gwyn 147 F7
Bryn Tanat Powys 148 E4
Bryn-y-cochin Shrops 149 B7
Bryn-y-gwenin Mon 78 B4
Bryn-y-maen Conwy 180 F4
Bryn-yr-Eos Wrex 166 G3
Bryn-yr-eryr Gwyn 162 F5
Bryn-yr-ogof Denb 165 D11
Brynafan Ceredig 112 C4
Brynamman Carms 76 C2
Brynawel Caerph 77 G11
Brynberian Pembs 92 D2
Brynbryddan Neath 57 C9
Bryncae Rhondda 58 C3
Bryncethin Bridgend 58 C2
Brynbryddan Neath 57 C9
Bryncae Rhondda 58 C3
Bryncethin Bridgend 58 C2
Bryncir Gwyn 163 G7
Bryncoch Bridgend 58 C2
Bryncroes Gwyn 144 C4
Bryncrug Gwyn 128 C2
Brynderwen Powys 130 D3
Bryndu Carms 75 D8
Bryneglwys Denb 165 F10
Brynford Flint 181 G11
Bryngias Newport 59 B10
Brynglas Sta Gwyn 128 C2
Bryngwran Anglesey 178 F4
Bryngwyn Ceredig 92 B5
Bryngwyn Mon 78 D5
Bryngwyn Powys 96 B3
Brynhenllan Pembs 91 D10
Brynheulog Bridgend 57 C11
Brynhoffnant Ceredig 110 G6
Bryniau Denb 181 E9
Bryning Lancs 194 B2
Brynithel Bl Gwent 78 E2
Brynllywarch Powys 130 F3
Brynmawr Bl Gwent 77 C11
Brynmenyn Bridgend 58 B2
Brynmill Swansea 56 C6
Brynmorfudd Conwy 164 C4
Brynna Rhondda 58 C3
Brynnau Gwynion Rhondda 58 C3
Brynore Shrops 149 B7
Brynrefail Anglesey 179 D7
Brynrefail Gwyn 163 C9
Brynsadler Rhondda 58 C4
Brynsiencyn Anglesey 163 B7
Brynsworthy Devon 40 G4
Brynteg Anglesey 179 E7
Brynteg Ceredig 93 C9
Brynteg Wrex 166 E4
Bryntirion Bridgend 57 E11
Buaile nam Bodach W Isles 297 L3
Bualintur Highld 294 C6
Bualnaluib Highld 307 K3
Buarthmeini Gwyn 146 C6
Bubbenhall Warks 119 C7
Bubblewell Glos 80 E5
Bubnell Derbys 186 G2
Bubwith E Yorks 207 F10
Buccleuch Borders 261 G8
Buchan Hill W Sus 51 G9
Buchanan Smithy Stirl 277 D9
Buchanhaven Aberds 303 E11
Buchanty Perth 286 E3
Buchley E Dunb 277 G11
Buchlyvie Stirl 277 C11
Buck Hill Wilts 62 E3
Buckabank Cumb 230 B3
Buckbury Worcs 98 E6
Buckden Cambs 122 D3
Buckden N Yorks 213 D8
Buckenham Norf 143 B7
Buckerell Devon 27 G10
Bucket Corner Hants 32 C6
Buckfast Devon 8 B4
Buckfastleigh Devon 8 B4
Buckham Dorset 29 G7
Buckhaven Fife 281 B7
Buckholm Borders 261 B11
Buckholt Mon 79 B8
Buckhorn Devon 12 B3
Buckhorn Weston Dorset 30 C3
Buckhurst Kent 53 E10
Buckhurst Hill Essex 86 G6
Buckie Moray 302 C4
Buckies Highld 310 C5
Buckingham Bucks 102 E3
Buckland Bucks 84 C5
Buckland Devon 8 G3
Buckland Devon 14 G3
Buckland Glos 99 D11
Buckland Hants 20 B2
Buckland Herts 105 E7
Buckland Kent 55 E10
Buckland Oxon 82 F4
Buckland Sur 51 C8
Buckland Brewer Devon 24 C6
Buckland Common Bucks 84 D6
Buckland Dinham Som 45 C9
Buckland Down Som 45 C8
Buckland End W Mid 134 F2
Buckland Filleigh Devon 25 F7
Buckland in the Moor Devon 13 G10
Buckland Marsh Oxon 82 F4
Buckland Monachorum Devon 7 B9
Buckland Newton Dorset 29 F11
Buckland Ripers Dorset 17 E8
Buckland St Mary Som 28 E3
Buckland Valley Kent 55 E10
Bucklands Borders 262 F2
Bucklandwharf Bucks 84 C5
Bucklebury W Berks 64 E5
Bucklebury Alley W Berks 64 E4
Bucklegate Lincs 156 B6
Buckleigh Devon 24 B6
Bucklerheads Angus 287 D8
Bucklers Hard Hants 20 B4
Bucklesham Suff 108 C4
Buckley = Bwcle Flint 166 C3
Buckley Green Warks 118 D3
Buckley Hill Mers 182 B4
Bucklow Hill E Ches 184 E2
Buckminster Leics 155 E7
Buckmoorend Bucks 84 E4
Bucknall Lincs 173 B11
Bucknall Stoke 168 F6
Bucknell Oxon 101 F11
Bucknell Shrops 115 C7
Buckoak W Ches 183 G8
Buckover S Glos 79 G11
Buckpool Moray 302 C4
Buckpool W Mid 133 F7
Buckridge Worcs 116 C4
Buck's Cross Devon 24 C4
Bucks Green W Sus 50 G5
Bucks Hill Herts 85 E9
Bucks Horn Oak Hants 49 E10
Buck's Mills Devon 24 C5
Bucksburn Aberdeen 293 C10
Buckshaw Village Lancs 194 C5
Buckskin Hants 48 C6
Buckton E Yorks 218 E3
Buckton Hereford 115 C7
Buckton Northumb 264 B3
Buckton Vale Gtr Man 196 G3
Buckworth Cambs 122 B2
Budbrooke Warks 118 D5
Budby Notts 171 B10
Buddgbrake Shetland 312 B8
Budd's Titson Corn 24 G2
Bude Corn 24 F2
Budge's Shop Corn 6 D6
Budlake Devon 14 B5
Budle Northumb 264 B5
Budleigh Som 27 D11
Budleigh Salterton Devon 15 E7
Budlett's Common E Sus 37 C7
Budock Water Corn 3 C7
Budworth Heath W Ches 183 F11
Buersil Head Gtr Man 196 E2
Buerton E Ches 167 G11
Buffler's Holt Bucks 102 D3
Bufton Leics 135 B8
Bugbrooke Northants 120 F3
Bugford Devon 40 E6
Bughtlin Edin 280 G3
Buglawton E Ches 168 C5
Bugle Corn 5 D10
Bugle Gate Worcs 116 D6
Bugley Dorset 30 C3
Bugley Wilts 45 E11
Bugthorpe E Yorks 207 B11
Building End Essex 105 D8
Buildwas Shrops 132 C2
Builth Road Powys 113 G10
Builth Wells Powys 113 G10
Buirgh W Isles 305 J2
Bulbourne Herts 84 C6
Bulbridge Wilts 46 G5
Bulby Lincs 155 D11
Bulcote Notts 171 G11
Buldoo Highld 310 C3
Bulford Wilts 47 E7
Bulford Camp Wilts 47 E7
Bulkeley E Ches 167 E8
Bulkeley Hall Shrops 168 G2
Bulkington Warks 135 F7
Bulkington Wilts 46 B2
Bulkworthy Devon 24 E5
Bull Bay = Porthllechog Anglesey 178 C6
Bull Hill Hants 20 B2
Bullamoor N Yorks 225 G7
Bullbridge Derbys 170 E5
Bullbrook Brack 65 F11
Bulleign Kent 53 G11
Bullenhill Wilts 45 B11
Bullen's Green Herts 86 D2
Bulley Glos 80 B3
Bullgill Cumb 229 D7
Bullhurst Hill Derbys 170 G3
Bullinghope Hereford 97 D10
Bullington Hants 48 E3
Bullington Lincs 189 F9

Bullo Glos 79 D11
Bullock's Horn Wilts 81 G7
Bullockstone Kent 71 F7
Bulls Cross London 86 F4
Bull's Green Herts 86 B3
Bull's Green Norf 143 E8
Bulls Green Som 45 D8
Bull's Hill Hereford 97 G11
Bullwood Argyll 276 G3
Bullyhole Bottom Mon 79 F7
Bulmer Essex 106 C6
Bulmer N Yorks 216 F3
Bulmer Tye Essex 106 D6
Bulphan Thurrock 68 B6
Bulstrode Herts 85 E8
Bulthy Shrops 148 G6
Bulverhythe E Sus 38 F3
Bulwark Aberds 303 E9
Bulwark Mon 79 G8
Bulwell Nottingham 171 F8
Bulwell Forest Nottingham 171 F8
Bulwick Leics 136 E3
Bulwick Northants 137 E9
Bumble's Green Essex 86 D6
Bumwell Hill Norf 142 E2
Bun Abhainn Eadarra W Isles 305 H3
Bun a'Mhuillin W Isles 297 K3
Bun Loyne Highld 290 C4
Bunacaimb Highld 295 G8
Bunarkaig Highld 290 E3
Bunbury E Ches 167 D9
Bunbury Heath E Ches 167 D9
Bunce Common Sur 51 D8
Bunchrew Highld 300 E6
Bundalloch Highld 295 C10
Buness Shetland 312 C8
Bunessan Argyll 288 G5
Bungay Suff 142 F6
Bunker's Hill Cambs 139 B8
Bunkers Hill Gtr Man 184 D6
Bunker's Hill Lincs 174 E3
Bunker's Hill Lincs 189 G7
Bunker's Hill Norf 142 B3
Bunkers Hill Oxon 83 B7
Bunker's Hill Suff 143 C10
Bunloit Highld 300 G5
Bunnahabhain Argyll 274 F5
Bunny Notts 153 D11
Bunny Hill Notts 153 D11
Bunree Highld 290 G2
Bunroy Highld 290 E4
Bunsley Bank E Ches 167 G11
Bunstead Hants 32 C6
Buntait Highld 300 F3
Buntingford Herts 105 F7
Bunting's Green Essex 106 E6
Bunwell Norf 142 E2
Bunwell Bottom Norf 142 D2
Buoltach Highld 310 F5
Burbage Derbys 185 G8
Burbage Leics 135 E8
Burbage Wilts 63 G8
Burcher Hereford 114 E6
Burchett's Green Windsor 65 C10
Burcombe Wilts 46 G5
Burcot Oxon 83 F9
Burcot Worcs 117 C9
Burcote Shrops 132 D4
Burcott Bucks 84 B4
Burcott Bucks 103 G7
Burcott Som 44 D4
Burdiehouse Edin 270 B5
Burdon T & W 243 G9
Burdonshill V Glam 58 E6
Burdrop Oxon 101 D7
Bures Suff 107 E8
Bures Green Suff 107 D8
Burford Devon 24 C4
Burford E Ches 167 E10
Burford Oxon 82 C3
Burford Shrops 115 D11
Burford Som 44 E5
Burg Argyll 288 E5
Burg Argyll 288 G6
Burgar Orkney 314 D3
Burgate Hants 31 D11
Burgate Suff 125 B11
Burgates Hants 34 B3
Burge End Herts 104 E2
Burgedin Powys 148 G4
Burgess Hill W Sus 36 D4
Burgh Suff 126 G4
Burgh by Sands Cumb 239 F8
Burgh Castle Norf 143 B9
Burgh Common Norf 141 E11
Burgh Heath Sur 51 B8
Burgh Hill E Sus 23 C8
Burgh Hill E Sus 38 B2
Burgh le Marsh Lincs 175 B8
Burgh Muir Aberds 293 B9
Burgh Muir Aberds 303 G7
Burgh next Aylsham Norf 160 D4
Burgh on Bain Lincs 190 D2
Burgh St Margaret = Fleggburgh Norf 161 G8
Burgh St Peter Norf 143 E9
Burgh Stubbs Norf 159 C10
Burghclere Hants 64 G3
Burghclere Common Hants 64 G3
Burghead Moray 301 C11
Burghfield W Berks 65 F7
Burghfield Common W Berks 64 F6
Burghfield Hill W Berks 65 F7
Burghill Hereford 97 C9
Burghwallis S Yorks 198 E4
Burgois Corn 10 G4
Burham Kent 69 G8
Burham Court Kent 69 G8
Buriton Hants 34 C2
Burland E Ches 167 E10
Burlawn Corn 10 G5
Burleigh Brack 65 E11
Burleigh Glos 80 E5
Burlescombe Devon 27 D9
Burleston Dorset 17 C11
Burlestone Devon 8 F6
Burley Hants 32 G2
Burley Rutland 155 G7
Burley Shrops 131 G9
Burley W Yorks 205 G11
Burley Beacon Hants 32 G2
Burley Gate Hereford 97 B11
Burley in Wharfedale W Yorks 205 D9
Burley Lawn Hants 32 G2
Burley Lodge Hants 32 F2
Burley Street Hants 32 G2
Burley Woodhead W Yorks 205 E9
Burleydam E Ches 167 G10
Burlinch Som 28 B3
Burlingham Green Norf 161 G7
Burlingjobb Powys 114 F5
Burlish Park Worcs 116 C6
Burlorne Tregoose Corn 5 B10
Burlow E Sus 23 B9
Burlton Shrops 149 D9
Burmantofts W Yorks 206 G2
Burmarsh Hereford 97 B10
Burmarsh Kent 54 G5
Burmington Warks 100 D5
Burn N Yorks 198 B5
Burn Bridge N Yorks 206 C2
Burn Naze Lancs 202 E2
Burn of Cambus Stirl 285 G11
Burnage Gtr Man 184 C5
Burnard's Ho Devon 24 G4
Burnaston Derbys 152 C5
Burnbank S Lnrk 268 D4
Burnby E Yorks 208 D2
Burncross S Yorks 186 B4
Burndell W Sus 35 G7
Burnden Gtr Man 195 F8
Burnedge Gtr Man 196 E2
Burnend Aberds 303 E8
Burneside Cumb 221 F10
Burness Orkney 314 B6
Burneston N Yorks 214 B6
Burnett Bath 61 F7
Burnfoot Borders 261 G10
Burnfoot Borders 262 F2
Burnfoot Dumfries 239 C7
Burnfoot Dumfries 247 E11
Burnfoot E Ayrs 245 B10
Burnfoot N Lnrk 268 B5
Burnfoot Perth 286 G3
Burngreave S Yorks 186 D5
Burnham Bucks 66 C2
Burnham N Lincs 200 D5
Burnham Deepdale Norf 176 E4
Burnham Green Herts 86 B3
Burnham Market Norf 176 E4
Burnham Norton Norf 176 E4
Burnham-on-Crouch Essex 88 F6
Burnham-on-Sea Som 43 D10
Burnham Overy Staithe Norf 176 E4
Burnham Overy Town Norf 176 E4
Burnham Thorpe Norf 176 E5
Burnhead Aberds 293 D10
Burnhead Borders 262 F2
Burnhead Dumfries 247 D9
Burnhead Dumfries 247 G10
Burnhead S Ayrs 244 C6
Burnhervie Aberds 293 B9
Burnhill Green Staffs 132 C5
Burnhope Durham 233 B9
Burnhouse N Ayrs 267 E7
Burnhouse Mains Borders 271 F8
Burniere Corn 10 G5
Burniestrype Moray 302 C3
Burniston N Yorks 227 G10
Burnlee W Yorks 196 F6
Burnley Lancs 204 G2
Burnley Lane Lancs 204 G2
Burnley Wood Lancs 204 G2
Burnmouth Borders 273 C9
Burnopfield Durham 242 F5
Burnrigg Cumb 239 F11
Burn's Green Herts 104 G6
Burnsall N Yorks 213 G10
Burnside Aberds 303 E8
Burnside Angus 287 B9
Burnside E Ayrs 258 G3
Burnside Fife 286 G5
Burnside Perth 286 E4
Burnside S Lnrk 268 C2
Burnside Shetland 312 F4
Burnside T & W 243 G8
Burnside W Loth 279 G11
Burnside of Duntrune Angus 287 D8
Burnstone Devon 24 C4
Burnswark Dumfries 238 B5
Burnt Ash Glos 80 E5
Burnt Heath Derbys 186 F2
Burnt Heath Essex 107 F11
Burnt Hill W Berks 64 E5
Burnt Houses Durham 233 G8
Burnt Mills Essex 88 G3
Burnt Oak E Sus 37 B8
Burnt Oak London 86 G2
Burnt Tree W Mid 133 E9
Burnt Yates N Yorks 214 G5
Burntcommon Sur 50 C4
Burntheath Derbys 152 C4
Burnthouse Corn 3 B7
Burntisland Fife 280 D4
Burnton E Ayrs 245 B11
Burnturk Fife 287 G7
Burntwood Staffs 133 B11
Burntwood Green Staffs 133 B11
Burntwood Pentre Flint 166 C3
Burnworthy Som 27 D11
Burnwynd Edin 270 B2
Burpham Sur 50 C4
Burpham W Sus 35 F8
Burradon Northumb 251 B11
Burradon T & W 243 C7
Burrafirth Shetland 312 B8
Burraland Shetland 312 F5
Burraland Shetland 313 J4
Burras Corn 2 C5
Burrastow Shetland 313 J4
Burraton Corn 7 D8
Burraton Coombe Corn 7 D8
Burravoe Shetland 312 F7
Burravoe Shetland 312 G5
Burray Village Orkney 314 G4
Burreldales Aberds 303 F7
Burrells Cumb 222 B3
Burrelton Perth 286 D6
Burridge Devon 28 F4
Burridge Devon 40 F5
Burridge Hants 33 E8
Burrigill Highld 310 F6
Burrill N Yorks 214 B4
Burringham N Lincs 199 F10
Burrington Devon 25 D10
Burrington Hereford 115 C8
Burrington N Som 44 B3
Burrough End Cambs 124 F2
Burrough Green Cambs 124 F2
Burrough on the Hill Leics 154 G5
Burroughs Grove Bucks 65 B11
Burroughston Orkney 314 D5
Burrow Devon 14 B5
Burrow Som 28 C6
Burrow Som 42 E2
Burrow-bridge Som 28 B5
Burrowbridge Som 43 G11
Burrowhill Sur 66 G3
Burrows Cross Sur 50 D5
Burrowsmoor Holt Notts 172 G2
Burrsville Park Essex 89 B11
Burrswood Kent 52 F4
Burry Swansea 56 C3
Burry Green Swansea 56 C3
Burry Port = Porth Tywyn Carms 74 E6
Burscott Devon 24 C4
Burscough Lancs 194 E2
Burscough Bridge Lancs 194 E2
Bursdon Devon 24 D3
Bursea E Yorks 208 G2
Burshill E Yorks 209 D7
Bursledon Hants 33 F7
Burslem Stoke 168 F5
Burstall Suff 107 C11
Burstallhill Suff 107 B11
Burstock Dorset 28 G6
Burston Devon 26 G2
Burston Norf 142 G2
Burston Staffs 151 C8
Burstow Sur 51 E10
Burstwick E Yorks 201 B8
Burtersett N Yorks 213 B7
Burtholme Cumb 240 E2
Burthorpe Suff 124 E5
Burthwaite Cumb 230 B1
Burtle Som 43 E11
Burtle Hill Som 43 E11
Burtoft Lincs 156 B5
Burton Dorset 17 C9
Burton Dorset 19 C9
Burton Lincs 189 G7
Burton Northumb 264 C5
Burton Pembs 73 D7
Burton Som 29 E8
Burton Som 43 E7
Burton V Glam 58 F4
Burton W Ches 167 C8
Burton W Ches 182 G4
Burton Wilts 45 G10
Burton Wilts 61 D10
Burton Wrex 166 D5
Burton Agnes E Yorks 218 G2
Burton Bradstock Dorset 16 D5
Burton Corner Lincs 174 F4
Burton Dassett Warks 119 G7
Burton End Cambs 106 B2
Burton End Essex 105 G10
Burton Ferry Pembs 73 D7
Burton Fleming E Yorks 217 E11
Burton Green Essex 106 F6
Burton Green W Mid 118 B5
Burton Green Wrex 166 D4
Burton Hastings Warks 135 E8
Burton-in-Kendal Cumb 211 D10
Burton in Lonsdale N Yorks 212 E3
Burton Joyce Notts 171 G10
Burton Latimer Northants 121 C8
Burton Lazars Leics 154 F5
Burton-le-Coggles Lincs 155 D9
Burton Leonard N Yorks 214 G6
Burton Manor Staffs 151 E8
Burton on the Wolds Leics 153 E11
Burton Overy Leics 136 D3
Burton Pedwardine Lincs 173 G10
Burton Pidsea E Yorks 209 G10
Burton Salmon N Yorks 198 B3
Burton Stather N Lincs 199 D11
Burton upon Stather N Lincs 199 D11
Burton upon Trent Staffs 152 E5
Burton Westwood Shrops 132 D2
Burtonwood Warr 183 C9
Burwardsley W Ches 167 D8
Burwarton Shrops 132 F2
Burwash E Sus 37 C11
Burwash Common E Sus 37 C10
Burwash Weald E Sus 37 C10
Burwell Cambs 123 D11
Burwell Lincs 190 F5
Burwen Anglesey 178 C6
Burwick Orkney 314 H4
Burwick Shetland 313 J5
Burwood Shrops 131 F9
Burwood Park Sur 66 G6
Bury Cambs 138 G5
Bury Gtr Man 195 E10
Bury Som 26 B6
Bury W Sus 35 E8
Bury End Beds 121 G9
Bury End C Beds 104 E2
Bury End Worcs 99 D11
Bury Green Herts 85 E4
Bury Green Herts 105 G8
Bury Hollow W Sus 35 E8
Bury Park Luton 103 G11
Bury St Edmunds Suff 125 E7
Buryas Br Corn 1 D4
Burybank Staffs 151 B7
Bury's Bank W Berks 64 F3
Burythorpe N Yorks 216 G5
Busbiehill N Ayrs 257 B9
Busbridge Sur 50 E3
Busby E Renf 267 D11
Buscot Oxon 82 F2
Buscott Som 44 F2
Bush Aberds 293 G9
Bush Corn 24 F2
Bush Bank Hereford 115 G9
Bush Crathie Aberds 292 D4
Bush End Essex 87 B9
Bush Estate Norf 161 D8
Bush Green Norf 141 D10
Bush Green Norf 142 F4
Bush Green Suff 125 F8
Bush Hill Park London 86 F4
Bushbury Sur 50 D7
Bushbury W Mid 133 C8
Bushby Leics 136 C3
Bushey Dorset 18 E5
Bushey Herts 85 G10
Bushey Ground Oxon 82 D4
Bushey Heath Herts 85 G11
Bushey Mead London 67 F8
Bushfield Cumb 249 G11
Bushley Worcs 99 E7
Bushley Green Worcs 99 E7
Bushmead Beds 122 E2
Bushmoor Shrops 131 F8
Bushton Wilts 62 D5
Bushy Common Norf 159 G9
Bushy Hill Sur 50 C4
Busk Cumb 231 C8
Busk N Yorks 196 F2
Buslingthorpe Lincs 189 D9
Bussage Glos 80 E5
Bussex Som 43 F11
Busta Shetland 312 G5
Bustard Green Essex 106 F2
Bustard's Green Norf 142 E3
Bustatoun Orkney 314 A7
Busveal Corn 4 G4
Butcher's Common Norf 160 E6
Butcher's Cross E Sus 37 B9
Butcombe N Som 60 G4
Bute Town Caerph 77 D10
Butetown Cardiff 59 D7
Butlane Head Shrops 149 G8
Butleigh Som 44 G4
Butleigh Wootton Som 44 G4
Butler's Cross Bucks 84 D4
Butlers Cross Bucks 85 G7
Butler's End Warks 134 G4
Butler's Hill Notts 171 F8
Butlers Marston Warks 118 G6
Butlersbank Shrops 149 E11
Butley Suff 127 G7
Butley High Corner Suff 109 B7
Butley Low Corner Suff 109 B7
Butley Town E Ches 184 F6
Butlocks Heath Hants 33 F7
Butt Green E Ches 167 E11
Butt Lane Staffs 168 E4
Butt Yeats Lancs 211 F11
Butter Bank Staffs 151 E7
Butterburn Cumb 240 C5
Buttercrambe N Yorks 207 B10
Butteriss Gate Corn 2 C6
Butterknowle Durham 233 F8
Butterleigh Devon 27 F7
Butterley Derbys 170 C4
Butterley Derbys 170 E6
Buttermere Cumb 220 B3
Buttermere Wilts 63 G10
Butterrow Glos 80 E5
Butters Green Staffs 168 E4
Buttershaw W Yorks 196 B6
Butterstone Perth 286 C4
Butterton Staffs 168 G4
Butterton Staffs 169 D9
Butterwick Cumb 221 B10
Butterwick Durham 234 F3
Butterwick Lincs 174 G5
Butterwick N Yorks 216 D4
Butterwick N Yorks 217 E9
Butteryhaugh Northumb 250 E4
Buttington Powys 130 B5
Button Haugh Green Suff 125 D9
Buttonbridge Shrops 116 B4
Buttonoak Worcs 116 B5
Button's Green Suff 125 G8
Butts Devon 14 D2
Butt's Green Essex 88 E3
Butts Green Essex 105 E9
Butt's Green Hants 32 B4
Buttsash Hants 32 F6
Buttsbear Cross Corn 24 G3
Buttsbury Essex 87 F11
Buttsole Kent 55 C10
Buxhall Suff 125 F10
Buxhall Fen Street Suff 125 F10
Buxley Borders 272 E6
Buxted E Sus 37 C7
Buxton Derbys 185 G9
Buxton Norf 160 E4
Buxworth Derbys 185 E8
Bwcle = Buckley Flint 166 C3
Bwlch Flint 181 G11
Bwlch Powys 96 G2
Bwlch-derwin Gwyn 163 F7
Bwlch-Llan Ceredig 111 F11
Bwlch-newydd Carms 93 G7
Bwlch-y-cibau Powys 148 F3
Bwlch-y-cwm Cardiff 58 C6
Bwlch-y-fadfa Ceredig 93 B8
Bwlch-y-ffridd Powys 129 D11
Bwlch-y-Plain Powys 114 B4
Bwlch-y-sarnau Powys 113 C10
Bwlchgwyn Wrex 166 E3
Bwlchnewydd Carms 93 G7
Bwlchtocyn Gwyn 144 D6
Bwlchyddar Powys 148 E3
Bwlchygroes Pembs 92 D4
Bwlchyllyn Gwyn 163 D8
Bybrook Kent 54 E4
Bycross Hereford 97 C7
Bye Green Bucks 84 C5
Byeastwood Bridgend 58 C2
Byebush Aberds 303 F7
Byerhope Northumb 232 B3
Byermoor T & W 242 F5
Byers Green Durham 233 E10
Byfield Northants 119 G10
Byfleet Sur 66 G5
Byford Hereford 97 C7
Byford Common Hereford 97 C7
Bygrave Herts 104 D5
Byker T & W 243 E7
Byland Abbey N Yorks 215 D10
Bylchau Conwy 165 C7
Byley W Ches 168 B2
Bynea Carms 56 B4
Byram N Yorks 198 B3
Byrness Northumb 251 C7
Bythorn Cambs 121 B11
Byton Hereford 115 E7
Byton Hand Hereford 115 E7
Bywell Northumb 242 E2
Byworth W Sus 35 C7

C

Cabbacott Devon 24 C6
Cabbage Hill Brack 65 E11
Cabharstadh W Isles 304 F5
Cabin Shrops 130 F6
Cablea Perth 286 D3
Cabourne Lincs 200 G6
Cabrach Argyll 274 G5
Cabrach Moray 302 G3
Cabrich Highld 300 E5
Cabus Lancs 202 D5
Cackle Hill Lincs 157 D7
Cackle Street E Sus 23 B11
Cackle Street E Sus 37 B7
Cackle Street E Sus 38 D4
Cackleshaw W Yorks 204 F6
Cad Green Som 28 D4
Cadboll Highld 301 B8
Cadbury Devon 26 G6
Cadbury Barton Devon 25 D11
Cadbury Heath S Glos 61 E7
Cadder E Dunb 278 G2
Cadderlie Argyll 284 D4
Caddington C Beds 85 B9
Caddleton Argyll 275 B8
Caddonfoot Borders 261 C10
Caddonlee Borders 261 B10
Cade Street E Sus 37 C10
Cadeby Leics 135 C8
Cadeby S Yorks 198 G4
Cadeleigh Devon 26 F6
Cademuir Borders 260 B6
Cader Denb 165 C8
Cadger Path Angus 287 B8
Cadgwith Corn 2 G6
Cadham Fife 286 G6
Cadishead Gtr Man 184 C2
Cadle Swansea 56 B6
Cadley Lancs 202 G6
Cadley Wilts 47 C8
Cadley Wilts 63 F8
Cadmore End Bucks 84 G3
Cadnam Hants 32 E3
Cadney N Lincs 200 G4
Cadney Bank Wrex 149 C9
Cadole Flint 166 C2
Cadoxton V Glam 58 F6
Cadoxton-Juxta-Neath Neath 57 B9
Cadshaw Blkburn 195 D8
Cadwell Herts 104 E3
Cadzow S Lnrk 268 E4
Cae Clyd Gwyn 164 G2
Cae-gors Carms 75 E9
Caeathro Gwyn 163 C7
Caehopkin Powys 76 C4
Caemorgan Ceredig 92 B3
Caenby Lincs 189 D8
Caenby Corner Lincs 189 D7
Cae'r-bont Powys 76 C4
Cae'r-bryn Carms 75 C9
Caer-Estyn Wrex 166 D4
Caer-Farchell Pembs 90 F5
Cae'r-Lan Powys 76 C4
Caer Llan Mon 79 D7
Caerau Bridgend 57 C11
Caerau Cardiff 58 D6
Caerau Park Newport 59 B9
Caerdeon Gwyn 146 F2
Caerfarchell Pembs 90 F5
Caerffili = Caerphilly Caerph 59 B7
Caerfyrddin = Carmarthen Carms 93 G8
Caergeiliog Anglesey 178 F4
Caergwrle Flint 166 D4
Caergybi = Holyhead Anglesey 178 E2
Caerhendy Neath 57 C9
Caerhun Gwyn 163 B9
Caerleon Newport 78 G4
Caermead V Glam 58 F3
Caermeini Pembs 92 E2
Caernarfon Gwyn 163 C7
Caerphilly = Caerffili Caerph 59 B7
Caersws Powys 129 E10
Caerwedros Ceredig 111 F7
Caerwent Mon 79 G7
Caerwent Brook Mon 60 B3
Caerwych Gwyn 146 B2
Caerwys Flint 181 G10
Caethle Gwyn 128 D2
Cage Green Kent 52 D5
Caggan Highld 291 B10
Caggle Street Mon 78 B5
Cailness Stirl 285 G7
Caim Anglesey 179 E10
Cainscross Glos 80 D4
Caio Carms 94 D3
Cairinis W Isles 296 E4
Cairisiadar W Isles 304 E2
Cairminis W Isles 296 C6
Cairnbaan Argyll 275 D9
Cairnbanno Ho Aberds 303 E8
Cairnborrow Aberds 302 E4
Cairnbrogie Aberds 303 G8
Cairnbulg Castle Aberds 303 C10
Cairncross Angus 292 F6
Cairncross Borders 273 C7
Cairnderry Dumfries 236 B5
Cairndow Argyll 284 F5
Cairness Aberds 303 C10
Cairneyhill Fife 279 D10
Cairnfield Ho Moray 302 C4
Cairngaan Dumfries 236 F3
Cairngarroch Dumfries 236 E2
Cairnhill Aberds 302 F6
Cairnhill Aberds 303 D7
Cairnhill N Lnrk 268 C5
Cairnie Aberds 293 C10
Cairnie Aberds 302 E4
Cairnlea S Ayrs 244 G6
Cairnleith Crofts Aberds 303 F9
Cairnmuir Aberds 303 C9
Cairnorrie Aberds 303 E8
Cairnpark Aberds 293 B10
Cairnpark Dumfries 247 D9
Cairnryan Dumfries 236 C2
Cairnton Orkney 314 F3
Caister-on-Sea Norf 161 G10
Caistor Lincs 200 G6
Caistor St Edmund Norf 142 C4
Caistron Northumb 251 C11
Caitha Bowland Borders 271 G9
Cakebole Worcs 117 C7
Calais Street Suff 107 D9
Calanais W Isles 304 E4
Calbost W Isles 305 G6
Calbourne I o W 20 D4
Calceby Lincs 190 F5
Calcoed Flint 181 G11
Calcot Glos 81 C9
Calcot W Berks 65 E7
Calcot Row W Berks 65 E7
Calcott Kent 71 G7
Calcott Shrops 149 G8
Calcott's Green Glos 80 B3
Calcutt N Yorks 206 B2
Calcutt Wilts 81 G10
Caldback Shetland 312 C8
Caldbeck Cumb 230 D2
Caldbergh N Yorks 213 B11
Caldecote Cambs 122 F6
Caldecote Cambs 138 F2
Caldecote Herts 104 D4
Caldecote Northants 120 G3
Caldecote Warks 135 E7
Caldecote Hill Herts 85 G11
Caldecott Northants 121 D9
Caldecott Oxon 83 F7
Caldecott Rutland 137 E7
Caldecotte M Keynes 103 D7
Calder Cumb 219 E10
Calder Bridge Cumb 219 D10
Calder Grove W Yorks 197 D10
Calder Hall Cumb 219 E10
Calder Mains Highld 310 D4
Calder Vale Lancs 202 D6
Calderbank N Lnrk 268 C5
Calderbrook Gtr Man 196 D2
Caldercruix N Lnrk 268 B6
Caldermill S Lnrk 268 G3
Caldermoor Gtr Man 196 D2
Calderstones Mers 182 D6
Calderwood S Lnrk 268 D2
Caldhame Angus 287 C8
Caldicot = Cil-y-coed Mon 60 B3
Caldmore W Mid 133 D10
Caldwell Derbys 152 F5
Caldwell N Yorks 224 C3
Caldy Mers 182 D2
Cale Green Gtr Man 184 D5
Caledrhydiau Ceredig 111 G9
Calenick Corn 4 G6
Caleys Fields Worcs 100 C4
Calf Heath Staffs 133 B8
Calford Green Suff 106 B3
Calfsound Orkney 314 C5
Calgary Argyll 288 D5
Caliach Argyll 288 D5
Califer Moray 301 D10
California Cambs 139 G10
California Falk 279 F8
California Norf 161 G10
California Suff 108 C3
California W Mid 133 G10
Calke Derbys 153 E7
Callakille Highld 298 D6
Callaly Northumb 252 B3
Callander Stirl 285 G10
Callandrode Stirl 285 G10
Callands Warr 183 C9
Callaughton Shrops 132 D2
Callendar Park Falk 279 F7
Callert Ho Highld 290 G2
Callerton T & W 242 D5
Callerton Lane End T & W 242 D5
Callestick Corn 4 E5
Calligarry Highld 295 E8
Callington Corn 7 B7
Callingwood Staffs 152 E3
Calloose Corn 2 B3
Callop Highld 289 B11
Callow Derbys 170 E3
Callow Hereford 97 E9
Callow End Worcs 98 B6
Callow Hill Mon 79 B8
Callow Hill Som 44 E4
Callow Hill Wilts 62 C4
Callow Hill Worcs 116 C4
Callow Marsh Hereford 98 B3
Callows Grave Worcs 115 D11
Calmore Hants 32 E4
Calmsden Glos 81 D8
Calne Wilts 62 E4
Calne Marsh Wilts 62 E4
Calow Derbys 186 G6
Calow Green Derbys 170 B6
Calrofold E Ches 184 G6
Calshot Hants 33 G7
Calstock Corn 7 B8
Calstone Wellington Wilts 62 F4
Calthorpe Norf 160 C3
Calthorpe Oxon 101 D9
Calthwaite Cumb 230 C5
Calton Glasgow 268 C2
Calton N Yorks 204 B4
Calton Staffs 169 E10
Calton Lees Derbys 170 B3
Calvadnack Corn 2 B5
Calveley E Ches 167 D9
Calver Derbys 186 G2
Calver Hill Hereford 97 B7
Calver Sough Derbys 186 F2
Calverhall Shrops 150 B2
Calverleigh Devon 26 E6
Calverley W Yorks 205 F10
Calvert Bucks 102 G3
Calverton M Keynes 102 D5
Calverton Notts 171 F10
Calvine Perth 291 G10
Calvo Cumb 238 G4
Cam Glos 80 F3
Camaghael Highld 290 F3
Camas an Staca Argyll 274 G5
Camas-luinie Highld 295 C11
Camas Salach Highld 289 C8
Camasnacroise Highld 289 D10
Camastianavaig Highld 295 B7
Camasunary Highld 295 D7
Camault Muir Highld 300 E5
Camb Shetland 312 D7
Camber E Sus 39 D7
Camberley Sur 65 G11
Camberwell London 67 D10
Camblesforth N Yorks 199 B7
Cambo Northumb 252 F2
Cambois Northumb 253 G8
Camborne Corn 4 G3
Cambourne Cambs 122 F6
Cambridge Borders 271 F11
Cambridge Cambs 123 F9
Cambridge Glos 80 E3
Cambridge W Yorks 205 D10
Cambridge Batch N Som 60 F4
Cambridge Town Sthend 70 C2
Cambrose Corn 4 F3
Cambus Clack 279 C7
Cambusavie Farm Highld 309 K7
Cambusbarron Stirl 278 C5
Cambusdrenny Stirl 278 C5
Cambuskenneth Stirl 278 C6
Cambuslang S Lnrk 268 D2
Cambusmore Lodge Highld 309 K7
Cambusnethan N Lnrk 268 D6
Camden London 67 C9
Camden Hill Kent 53 F9
Camden Park Kent 52 F5
Cameley Bath 44 B6
Camelford Corn 11 E8
Camelon Falk 279 E7
Camelsdale Sur 49 G11
Camer Kent 69 F7
Cameron Fife 280 B6
Cameron Bridge Fife 280 B6
Camerory Highld 301 F10
Camer's Green Worcs 98 D5
Camerton Bath 45 B7
Camerton Cumb 228 E6
Camerton E Yorks 201 B8
Camghouran Perth 285 B9
Cammachmore Aberds 293 D11
Cammeringham Lincs 188 E6
Camnant Powys 113 F11
Camoquhill Stirl 277 D10
Camore Highld 309 K7
Camp Lincs 172 E5
Camp Corner Oxon 83 E10
Camp Hill N Yorks 214 C6
Camp Hill Pembs 73 C10
Camp Hill W Yorks 196 D5
Camp Hill Warks 134 E6
Camp Town W Yorks 206 F2
Campbeltown Argyll 255 E8
Camperdown T & W 243 C7
Camphill Derbys 185 F11
Campion Hills Warks 118 D6
Campions Essex 87 C7
Campile Dumfries 247 E9
Campmuir Perth 286 D6
Camps End Cambs 106 C2
Camps Heath Suff 143 E10
Campsall S Yorks 198 E4
Campsea Ashe Suff 126 F6
Campsey Ash Suff 126 F6
Campsfield Oxon 83 B7
Campton C Beds 104 D2
Camptoun E Loth 281 F10
Camptown Borders 262 G5
Camquhart Argyll 275 E10
Camrose Pembs 91 G8
Camserney Perth 286 C2
Camster Highld 310 E6
Camuschoirk Highld 289 C9
Camuscross Highld 295 D8
Camusnagaul Highld 290 F2
Camusnagaul Highld 307 L5
Camusrory Highld 295 F10
Camusteel Highld 299 E7
Camusterrach Highld 299 E7
Camusvrachan Perth 285 C10
Canada Hants 32 D3
Canada Lincs 200 G6
Canadia E Sus 38 D2
Canal Foot Cumb 210 D6
Canal Side S Yorks 199 E7
Canbus Clack 279 C7
Candacraig Ho Aberds 292 B5
Candle Street Suff 125 C10
Candlesby Lincs 175 B7
Candy Mill S Lnrk 269 G11
Cane End Oxon 65 D7
Caneheath E Sus 23 D9
Canewdon Essex 88 G5
Canford Bottom Dorset 31 G8
Canford Cliffs Poole 19 D7
Canford Heath Poole 18 C6
Canford Magna Poole 18 B6
Cangate Norf 160 F6
Canham's Green Suff 125 D11
Canholes Derbys 185 G8
Canisbay Highld 310 B7
Canklow S Yorks 186 C6
Canley W Mid 118 B6
Cann Dorset 30 C5
Cann Common Dorset 30 C5
Cannalidgey Corn 5 B8
Cannard's Grave Som 44 E6
Cannich Highld 300 F3
Canning Town London 68 C2
Cannington Som 43 F9
Cannock Staffs 133 B9
Cannock Wood Staffs 151 G10
Cannon's Green Essex 87 D9
Cannop Glos 79 C10
Canon Bridge Hereford 97 C8
Canon Frome Hereford 98 C3
Canon Pyon Hereford 97 B9
Canonbie Dumfries 239 B9
Canonbury London 67 C10
Canons Ashby Northants 119 G11
Canons Park London 85 G11
Canon's Town Corn 2 B2
Canonsgrove Som 28 C2
Canterbury Kent 54 B6
Cantley Norf 143 C7
Cantley S Yorks 198 G6
Cantlop Shrops 131 B10
Canton Cardiff 59 D7
Cantraybruich Highld 301 E7
Cantraydoune Highld 301 E7
Cantraywood Highld 301 E7
Cantsfield Lancs 212 E2
Canvey Island Essex 69 C9
Canwick Lincs 173 B7
Canworthy Water Corn 11 C10
Caol Highld 290 F3
Caol Ila Argyll 274 F5
Caolas Argyll 288 E2
Caolas W Isles 297 M2
Caolas Fhlodaigh W Isles 296 F4
Caolas Liubharsaigh W Isles 297 G4
Caolas Scalpaigh W Isles 305 J4
Caolas Stocinis W Isles 305 J3
Caoslasnacon Highld 290 G3
Capel Carms 75 E8
Capel Kent 52 E6
Capel Sur 51 E7
Capel Bangor Ceredig 128 G3
Capel Betws Lleucu Ceredig 112 F2
Capel Carmel Gwyn 144 D3
Capel Coch Anglesey 179 E7
Capel Cross Kent 53 E8
Capel Curig Conwy 164 D2
Capel Cynon Ceredig 93 B7
Capel Dewi Carms 93 G9
Capel Dewi Ceredig 93 C9
Capel Dewi Ceredig 128 G2
Capel Garmon Conwy 164 D4
Capel Green Suff 109 B7
Capel-gwyn Anglesey 178 F4
Capel Gwyn Carms 93 G9
Capel Gwynfe Carms 94 G4
Capel Hendre Carms 75 C9
Capel Hermon Gwyn 146 D4
Capel Isaac Carms 93 F11
Capel Iwan Carms 92 D5
Capel-le-Ferne Kent 55 F8
Capel Llanilltern Cardiff 58 C5
Capel Mawr Anglesey 178 G6
Capel Newydd = Newchapel Pembs 92 D4
Capel Parc Anglesey 178 D6
Capel St Andrew Suff 109 B7
Capel St Mary Suff 107 D11
Capel Seion Carms 75 C8
Capel Seion Ceredig 112 B2
Capel Siloam Conwy 164 E4
Capel Tygwydd Ceredig 92 C5
Capel Uchaf Gwyn 162 F6
Capel-y-ffin Powys 96 E5
Capel-y-graig Gwyn 163 B8
Capelulo Conwy 180 F2
Capenhurst W Ches 182 G5
Capernwray Lancs 211 E10
Capheaton Northumb 252 G2
Capland Som 28 D4
Cappercleuch Borders 260 E6
Capplegill Dumfries 248 B4
Capstone Medway 69 F9
Captain Fold Gtr Man 195 E11
Capton Devon 8 E6
Capton Som 42 F5
Caputh Perth 286 D4
Car Colston Notts 172 G2
Caradon Town Corn 11 G11
Carbis Corn 5 D10
Carbis Bay Corn 2 B2
Carbost Highld 294 B5
Carbost Highld 298 E4
Carbrain N Lnrk 278 G5
Carbrook S Yorks 186 D5
Carbrooke Norf 141 C9
Carburton Notts 187 G10
Carcant Borders 271 E7
Carcary Angus 287 B10
Carclaze Corn 5 E10
Carclew Corn 3 B7
Carcroft S Yorks 198 E4
Cardenden Fife 280 C4
Cardeston Shrops 149 G7
Cardew Cumb 230 B2
Cardewlees Cumb 239 G8
Cardiff Cardiff 59 D7
Cardigan = Aberteifi Ceredig 92 B3
Cardinal's Green Cambs 106 B2
Cardington Beds 103 B11
Cardington Shrops 131 D10
Cardinham Corn 6 B2
Cardonald Glasgow 267 C10
Cardow Moray 301 E11
Cardrona Borders 261 B8
Cardross Argyll 276 F6
Cardurnock Cumb 238 F5
Care Village Leics 136 D4
Careby Lincs 155 F10
Careston Angus 293 G7
Careston Castle Angus 287 B9
Carew Pembs 73 E8
Carew Cheriton Pembs 73 E8
Carew Newton Pembs 73 E8
Carey Hereford 97 E11
Carey Park Corn 6 E4
Carfin N Lnrk 268 D5
Carfrae E Loth 271 B11
Carfury Corn 1 C4
Cargate Common Norf 142 E2
Cargenbridge Dumfries 237 B11
Cargill Perth 286 D5
Cargo Cumb 239 F9
Cargo Fleet M'bro 234 G6
Cargreen Corn 7 C8
Carham Northumb 263 B8
Carhampton Som 42 E4
Carharrack Corn 4 G4
Carie Perth 285 B10
Carie Perth 285 D10
Carines Corn 4 D5
Carisbrooke I o W 20 D5
Cark Cumb 211 D7
Carkeel Corn 7 C8
Carlabhagh W Isles 304 D4
Carland Cross Corn 5 E7
Carlbury N Yorks 224 B4
Carlby Lincs 155 G11
Carlecotes S Yorks 197 G7
Carleen Corn 2 C4
Carlenrig Borders 249 C9
Carlesmoor N Yorks 214 E3
Carleton Cumb 219 D10
Carleton Cumb 230 F6
Carleton Cumb 239 G10
Carleton Lancs 202 F2
Carleton N Yorks 204 D5
Carleton W Yorks 198 C3
Carleton Forehoe Norf 141 B11
Carleton Hall Cumb 219 F11
Carleton-in-Craven N Yorks 204 D5
Carleton Rode Norf 142 E2
Carleton St Peter Norf 142 C6
Carley Hill T & W 243 F9
Carlidnack Corn 3 D7
Carlin How Redcar 226 B4
Carlincraig Aberds 302 E6
Carlingcott Bath 45 B7
Carlinghow W Yorks 197 C8
Carlingwark Devon 27 E11
Carlisle Cumb 239 F10
Carloggas Corn 5 B7
Carloggas Corn 5 E9
Carloonan Argyll 284 F4
Carlops Borders 270 D3
Carlton Beds 121 F9
Carlton Cambs 124 G2
Carlton Leics 135 C7
Carlton N Yorks 198 C6
Carlton N Yorks 213 C11
Carlton N Yorks 216 B2
Carlton N Yorks 224 C3
Carlton Notts 171 G10
Carlton S Yorks 197 E11
Carlton Stockton 234 G3
Carlton Suff 127 E7
Carlton W Yorks 197 B10
Carlton Colville Suff 143 F10
Carlton Curlieu Leics 136 D3
Carlton Green Cambs 124 G2
Carlton Husthwaite N Yorks 215 D9
Carlton in Cleveland N Yorks 225 E10
Carlton in Lindrick Notts 187 E9
Carlton le Moorland Lincs 172 D6
Carlton Miniott N Yorks 215 C7
Carlton on Trent Notts 172 C3
Carlton Purlieus Northants 136 F6
Carlton Scroop Lincs 172 G6
Carluddon Corn 5 D10
Carluke S Lnrk 268 E6
Carlyon Bay Corn 5 E11
Carmarthen = Caerfyrddin Carms 93 G8
Carmel Anglesey 178 E5
Carmel Carms 75 B9
Carmel Flint 181 F11
Carmel Gwyn 163 E7
Carmel Powys 113 D11
Carmichael S Lnrk 259 B10
Carminow Cross Corn 5 B11
Carmont Aberds 293 E10
Carmunnock Glasgow 268 D2
Carmyle Glasgow 268 C2
Carmyllie Angus 287 C9
Carn Arthen Corn 2 B5
Carn Brea Village Corn 4 G3
Carn-gorm Highld 295 C11
Carn Towan Corn 1 D3
Carnaby E Yorks 218 F2
Carnach Highld 299 G10
Carnach Highld 307 K5
Carnach W Isles 305 J4
Carnachy Highld 308 D7
Càrnais W Isles 304 E2
Càrnan W Isles 297 G3
Carnbahn Perth 285 C10
Carnbee Fife 287 G9
Carnbo Perth 286 G4
Carnbrea Corn 4 G3
Carnbroe N Lnrk 268 C4
Carndu Highld 295 C10
Carnduff S Lnrk 268 F3
Carnduncan Argyll 274 G3
Carne Corn 3 B10
Carne Corn 3 E7
Carne Corn 5 D9
Carnebone Corn 2 C6
Carnedd Powys 129 E10
Carnetown Rhondda 77 G9
Carnforth Lancs 211 E10
Carnforth Lancs 211 E9
Carnglas Swansea 56 C6

Carnhedryn Pembs 90 F6
Carnhedryn Uchaf Pembs 90 F5
Carnhell Green Corn 2 B4
Carnhot Corn 4 F4
Carnkie Corn 2 B5
Carnkie Corn 2 C6
Carnkief Corn 4 E5
Carno Powys 129 D9
Carnoch Highld 300 D2
Carnoch Highld 300 F3
Carnock Fife 279 D10
Carnon Downs Corn 4 G5
Carnousie Aberds 302 D6
Carnoustie Angus 287 D9
Carnsmerry Corn 5 D10
Carntyne Glasgow 268 B2
Carnwadric E Renf 267 D10
Carnwath S Lnrk 269 F9
Carnyorth Corn 1 C3
Caroe Corn 11 C9
Carol Green W Mid 118 B5
Carpalla Corn 5 E9
Carpenders Park Herts 85 G10
Carpenter's Hill Worcs 117 C11
Carperby N Yorks 213 B10
Carpley Green N Yorks 213 B8
Carr Gtr Man 195 D9
Carr S Yorks 187 C8
Carr Bank Cumb 211 D9
Carr Cross Lancs 193 E11
Carr Gate W Yorks 197 C10
Carr Green Gtr Man 184 D2
Carr Hill T & W 243 E7
Carr Houses Mers 193 G10
Carr Vale Derbys 171 B7
Carradale Argyll 255 D9
Carragraich W Isles 305 J3
Carrbridge Highld 301 G9
Carrbrook Gtr Man 196 G3
Carreg-wen Pembs 92 C4
Carreg y Garth Gwyn 163 B9
Carreglefn Anglesey 178 D5
Carrhouse Devon 26 F3
Carrick Argyll 275 E10
Carrick Dumfries 237 D7
Carrick Fife 287 E8
Carrick Castle Argyll 276 C3
Carrick Ho Orkney 314 C5
Carriden Falk 279 E10
Carrington Gtr Man 184 C2
Carrington Lincs 174 E4
Carrington Midloth 270 C6
Carrington Nottingham 171 G9
Carroch Dumfries 246 E5
Carrog Conwy 164 F3
Carrog Denb 165 G10
Carroglen Perth 285 E11
Carrol Highld 311 J2
Carron Falk 279 E7
Carron Moray 302 E2
Carron Bridge Stirl 278 E4
Carronbridge Dumfries 247 D9
Carronshore Falk 279 E7
Carrot Angus 287 C8
Carrow Hill Mon 78 G6
Carroway Head Staffs 134 D3
Carrshield Northumb 232 B2
Carrutherstown Dumfries 238 C4
Carrville Durham 234 C2
Carry Argyll 275 G10
Carsaig Argyll 275 E8
Carsaig Argyll 289 G7
Carscreugh Dumfries 236 D4
Carse Gray Angus 287 B8
Carse Ho Argyll 275 G8
Carsegowan Dumfries 236 D6
Carseriggan Dumfries 236 C5
Carsethorn Dumfries 237 D11
Carshalton London 67 G9
Carshalton Beeches London 67 G9
Carshalton on the Hill London 67 G9
Carsington Derbys 170 E3
Carskiey Argyll 255 G7
Carsluith Dumfries 236 D6
Carsphairn Dumfries 246 E3
Carstairs S Lnrk 269 F8
Carstairs Junction S Lnrk 269 F9
Carswell Marsh Oxon 82 F4
Cartbridge Sur 50 B4
Carter Knowle S Yorks 186 E4
Carterhaugh Borders 261 D10
Carter's Clay Hants 32 C4
Carter's Green Essex 87 C8
Carter's Hill Wokingham 65 F9
Carterspiece Glos 79 C9
Carterton Oxon 82 D3
Carterway Heads Northumb 242 G2
Carthamartha Corn 12 F3
Carthew Corn 2 B5
Carthew Corn 5 D10
Carthorpe N Yorks 214 C6
Cartington Northumb 252 C2
Cartland S Lnrk 269 F7
Cartledge Derbys 186 F4
Cartmel Cumb 211 D7
Cartmel Fell Cumb 211 B8
Cartsdyke Invclyd 276 F5
Cartworth W Yorks 196 F6
Carty Port Dumfries 236 C6
Carway Carms 75 D7
Carwinley Cumb 239 C10
Carwynnen Corn 2 B5
Cary Fitzpaine Som 29 B9
Carzantic Corn 12 E3
Carzield Dumfries 247 G11
Carzise Corn 2 C3
Cas Mael = Puncheston Pembs 91 F10
Cascob Powys 114 D4
Cashes Green Glos 80 D4
Cashlie Perth 285 C8
Cashmoor Dorset 31 E7
Cassey Compton Glos 81 C9
Cassington Oxon 83 C7
Cassop Durham 234 D2
Castallack Corn 1 D5
Castell Conwy 164 B3
Castell Denb 165 B10
Castell-Howell Ceredig 93 B8
Castell nedd = Neath Neath 57 B8
Castell Newydd Emlyn = Newcastle Emlyn Carms 92 C6
Castell-y-bwch Torf 78 G3
Castell-y-rhingyll Carms 75 C9
Castellau Rhondda 58 B5
Casterton Cumb 212 D2
Castle Devon 28 G4
Castle Som 27 B9
Castle Acre Norf 158 F6
Castle Ashby Northants 121 F7
Castle Bolton N Yorks 223 G10
Castle Bromwich W Mid 134 F2
Castle Bytham Lincs 155 F9
Castle Caereinion Powys 130 B3
Castle Camps Cambs 106 C2
Castle Carlton Lincs 190 E5
Castle Carrock Cumb 240 F2
Castle Cary Som 44 G6
Castle Combe Wilts 61 D10
Castle Donington Leics 153 D8
Castle Douglas Dumfries 237 C9
Castle Eaton Swindon 81 F10
Castle Eden Durham 234 D4
Castle End P'boro 138 B2
Castle Fields Shrops 149 G10
Castle Forbes Aberds 293 B8
Castle Frome Hereford 98 B3
Castle Gate Corn 1 C5
Castle Green London 68 C3
Castle Green S Yorks 197 G9
Castle Green Sur 66 G3
Castle Gresley Derbys 152 F5
Castle Heaton Northumb 273 G8
Castle Hedingham Essex 106 D5
Castle Hill E Sus 37 B9
Castle Hill Gtr Man 184 C6
Castle Hill Kent 53 E7
Castle Hill Suff 108 B3
Castle Hill Worcs 116 F5
Castle Huntly Perth 287 E7
Castle Kennedy Dumfries 236 D3
Castle O'er Dumfries 248 E6
Castle Rising Norf 158 E3
Castle Street W Yorks 196 C3
Castle Stuart Highld 301 E7
Castle Toward Argyll 266 B2
Castle Town W Sus 36 E2
Castle-upon-Alun V Glam 58 E2
Castle Vale W Mid 134 E2
Castlebythe Pembs 91 F10
Castlecary N Lnrk 278 F5
Castle Craig Borders 270 G2
Castlecraig Highld 301 C8
Castlecroft Staffs 133 D7
Castlefairn Dumfries 246 F6
Castlefields Halton 183 E8
Castleford W Yorks 198 B2
Castlegreen Shrops 130 F6
Castlehead Renfs 267 C9
Castlehill Argyll 254 B4
Castlehill Borders 260 B6
Castlehill Highld 310 C5
Castlehill S Ayrs 257 E9
Castlehill W Dunb 277 F7
Castlemaddy Dumfries 246 F3
Castlemartin Pembs 72 F6
Castlemilk Dumfries 238 B5
Castlemilk Glasgow 268 D2
Castlemorris Pembs 91 E8
Castlemorton Worcs 98 D5
Castlerigg Cumb 229 G11
Castleside Durham 233 B7
Castlethorpe M Keynes 102 C6
Castlethorpe N Lincs 200 F3
Castleton Angus 287 C7
Castleton Argyll 275 E9
Castleton Derbys 185 E11
Castleton Gtr Man 195 E11
Castleton Moray 301 G11
Castleton N Yorks 226 D3
Castleton Newport 59 C9
Castleton Village Highld 300 E6
Castletown Cumb 230 E6
Castletown Dorset 17 G9
Castletown Highld 301 E7
Castletown Highld 310 C5
Castletown I o M 192 F3
Castletown Staffs 151 E8
Castletown T & W 243 F9
Castletown W Ches 166 E6
Castletump Glos 98 F4
Castleweary Borders 249 C10
Castlewigg Dumfries 236 E6
Castley N Yorks 205 D11
Castling's Heath Suff 107 C9
Caston Norf 141 D9
Castor P'boro 138 D2
Caswell Swansea 56 D5
Cat Bank Cumb 220 F6
Cat Hill S Yorks 197 F8
Catacol N Ayrs 255 C10
Catbrain S Glos 60 C5
Catbrook Mon 79 E8
Catch Flint 182 G2
Catchall Corn 1 D4
Catchems Corner W Mid 118 B4
Catchems End Worcs 116 B5
Catchgate Durham 242 G5
Catchory Highld 310 D6
Catcleugh Northumb 250 C6
Catcliffe S Yorks 186 D6
Catcomb Wilts 62 D4
Catcott Som 43 F11
Caterham Sur 51 B10
Catfirth Shetland 313 H6
Catford London 67 E11
Catforth Lancs 202 F5
Cathays Cardiff 59 D7
Cathays Park Cardiff 59 D7
Cathcart Glasgow 267 C11
Cathedine Powys 96 F2
Catherine-de-Barnes W Mid 134 G3
Catherine Slack W Yorks 196 B5
Catherington Hants 33 E11
Catherton Shrops 116 B3
Cathiron Warks 119 B9
Catholes Cumb 222 G3
Cathpair Borders 271 F9
Catisfield Hants 33 F8
Catley Lane Head Gtr Man 195 D11
Catley Southfield Hereford 98 C3
Catlodge Highld 291 D8
Catlowdy Cumb 239 B11
Catmere End Essex 105 D9
Catmore W Berks 64 C3
Caton Devon 13 G11
Caton Lancs 211 G10
Caton Green Lancs 211 F10
Catrine E Ayrs 258 D2
Cat's Ash Newport 78 G5
Cat's Common Norf 160 E6
Cats Edge Staffs 169 E7
Cat's Hill Cross Staffs 150 C6
Catsfield E Sus 38 E2
Catsfield Stream E Sus 38 E2
Catsgore Som 29 B8
Catsham Som 44 G5
Catshaw S Yorks 197 G8
Catshill W Mid 133 B11
Catshill Worcs 117 C9
Catslackburn Borders 261 D8
Catslip Oxon 65 B8
Catstree Shrops 132 D4
Cattadale Argyll 274 G4
Cattal N Yorks 206 C4
Cattawade Suff 108 E2
Cattedown Plym 7 E9
Catterall Lancs 202 E5
Catterick N Yorks 224 F4
Catterick Bridge N Yorks 224 F4
Catterick Garrison N Yorks 224 F3
Catterlen Cumb 230 E5
Catterline Aberds 293 F10
Catterton N Yorks 206 D6
Catteshall Sur 50 E3
Catthorpe Leics 119 B11
Cattistock Dorset 17 B7
Cattle End Northants 102 C3
Catton N Yorks 215 D7
Catton Northumb 241 F8
Catwick E Yorks 209 D8
Catworth Cambs 121 C11
Caudle Green Glos 80 C6
Caudlesprings Norf 141 C8
Caulcott C Beds 103 C9
Caulcott Oxon 101 G10
Cauld Borders 261 G11
Cauldcoats Holdings Falk 279 F10
Cauldcots Angus 287 C10
Cauldhame Stirl 278 C2
Cauldmill Borders 262 G2
Cauldon Staffs 169 F9
Cauldon Lowe Staffs 169 F9
Cauldwells Aberds 303 D7
Caulkerbush Dumfries 237 D11
Caulside Dumfries 249 G10
Caundle Marsh Dorset 29 E11
Caunsall Worcs 132 G6
Caunton Notts 172 D2
Causeway Hants 33 E11
Causeway Hants 34 C2
Causeway Mon 60 B2
Causeway End Cumb 210 C6
Causeway End Cumb 211 B9
Causeway End Dumfries 236 C6
Causeway End Essex 87 B11
Causeway End Wilts 62 C4
Causeway Foot W Yorks 197 E7
Causeway Foot W Yorks 205 G7
Causeway Green W Mid 133 F9
Causewayend S Lnrk 260 B2
Causewayhead Cumb 238 G4
Causewayhead Stirl 278 B6
Causewaywood Shrops 131 D10
Causey Durham 242 F6
Causey Park Bridge Northumb 252 E5
Causeyend Aberds 293 B11
Causeyton Aberds 293 B8
Caute Devon 24 E6
Cautley Cumb 222 G3
Cavendish Suff 106 B6
Cavendish Bridge Leics 153 D8
Cavenham Suff 124 D5
Cavers Carre Borders 262 D3
Caversfield Oxon 101 F11
Caversham Reading 65 E8
Caversham Heights Reading 65 D8
Caverswall Staffs 169 G7
Cavil E Yorks 207 G11
Cawdor Highld 301 D8
Cawkeld E Yorks 208 C5
Cawkwell Lincs 190 F3
Cawood N Yorks 207 F7
Cawsand Corn 7 E8
Cawston Norf 160 E2
Cawston Warks 119 C9
Cawthorne N Yorks 216 B5
Cawthorne S Yorks 197 F9
Cawthorpe Lincs 155 E11
Cawton N Yorks 216 D2
Caxton Cambs 122 F6
Caynham Shrops 115 C11
Caythorpe Lincs 172 F6
Caythorpe Notts 171 F11
Cayton N Yorks 217 C11
Ceallan W Isles 296 F4
Ceann a Bhàigh W Isles 305 J4
Ceann a Bhaigh W Isles 296 E3
Ceann a Deas Loch Baghasdail W Isles 297 K3
Ceann Shiphoirt W Isles 305 G4
Ceann Tarabhaigh W Isles 305 G4
Ceannacroc Lodge Highld 290 B4
Cearsiadair W Isles 304 F5
Ceathramh Meadhanach W Isles 296 D4
Cefn Newport 59 B9
Cefn Powys 148 G5
Cefn Berain Conwy 165 B7
Cefn-brith Conwy 164 E6
Cefn-bryn-brain Carms 76 C2
Cefn-bychan Swansea 56 B4
Cefn-bychan Wrex 166 G3
Cefn Canol Powys 148 C4
Cefn-coch Conwy 164 B5
Cefn Coch Powys 129 C10
Cefn Coch Powys 148 D2
Cefn-coed-y-cymmer M Tydf 77 D9
Cefn Cribbwr Bridgend 57 E11
Cefn Cross Bridgend 57 E11
Cefn-ddwysarn Gwyn 147 B9
Cefn Einion Shrops 130 F5
Cefn-eurgain Flint 166 B2
Cefn Fforest Caerph 77 F11
Cefn Glas Bridgend 57 E11
Cefn Golau Bl Gwent 77 D10
Cefn-gorwydd Powys 95 B8
Cefn Hengoed Caerph 77 F10
Cefn-hengoed Swansea 57 B7
Cefn Llwyd Ceredig 128 G2
Cefn-mawr Wrex 166 G3
Cefn Rhigos Rhondda 76 D6
Cefn-y-bedd Flint 166 D4
Cefn-y-crib Torf 78 F2
Cefn-y-Garth Swansea 76 E2
Cefn-y-pant Carms 92 F3
Cefncaeau Carms 56 B4
Cefneithin Carms 75 C9
Cefnpennar Rhondda 77 E8
Cegidfa = Guilsfield Powys 148 G4
Cei-bach Ceredig 111 F8
Ceinewydd = New Quay Ceredig 111 F7
Ceint Anglesey 179 F7
Ceinws Powys 128 B5
Cellan Ceredig 94 B2
Cellarhead Staffs 169 F7
Cellarhill Kent 70 G3
Celyn-Mali Flint 165 B11
Cemaes Anglesey 178 C5
Cemmaes Powys 128 B6
Cemmaes Road = Glantwymyn Powys 128 C6
Cenarth Carms 92 C5
Cenin Gwyn 163 F7
Central Invclyd 276 F5
Central Milton Keynes M Keynes 102 D6
Ceos W Isles 304 F5
Ceres Fife 287 F8
Ceri = Kerry Powys 130 F2
Cerne Abbas Dorset 29 G11
Cerney Wick Glos 81 F9
Cerrig Llwydion Neath 57 C9
Cerrig-mân Anglesey 179 C7
Cerrigceinwen Anglesey 178 G6
Cerrigydrudion Conwy 165 F7
Cess Norf 161 F8
Cessford Borders 262 E6
Ceunant Gwyn 163 C8
Chaceley Glos 99 E7
Chaceley Hole Glos 98 E6
Chaceley Stock Glos 99 F7
Chacewater Corn 4 G4
Chackmore Bucks 102 D3
Chacombe Northants 101 C9
Chad Valley W Mid 133 F10
Chadbury Worcs 99 B10
Chadderton Gtr Man 196 F2
Chadderton Fold Gtr Man 195 F11
Chaddesden Derby 153 B7
Chaddesley Corbett Worcs 117 C7
Chaddlehanger Devon 12 F5
Chaddlewood Plym 7 D11
Chaddleworth W Berks 64 D2
Chadkirk Gtr Man 184 D6
Chadlington Oxon 100 G6
Chadshunt Warks 118 G6
Chadsmoor Staffs 151 G9
Chadstone Northants 121 F7
Chadwell Leics 154 E5
Chadwell Shrops 150 G5
Chadwell End Beds 121 D11
Chadwell Heath London 68 B3
Chadwell St Mary Thurrock 68 D6
Chadwick Worcs 116 D6
Chadwick End W Mid 118 C4
Chadwick Green Mers 183 B8
Chaffcombe Som 28 E5
Chafford Hundred Thurrock 68 D5
Chagford Devon 13 D10
Chailey E Sus 36 D5
Chain Bridge Lincs 174 G4
Chainbridge Cambs 139 C8
Chainhurst Kent 53 D8
Chalbury Dorset 31 F8
Chalbury Common Dorset 31 F8
Chaldon Sur 51 B10
Chaldon Herring or East Chaldon Dorset 17 E11
Chale I o W 20 F5
Chale Green I o W 20 F5
Chalfont Common Bucks 85 G8
Chalfont Grove Bucks 85 G7
Chalfont St Giles Bucks 85 G7
Chalfont St Peter Bucks 85 G8
Chalford Glos 80 E5
Chalford Oxon 84 E2
Chalford Wilts 45 C11
Chalford Hill Glos 80 E5
Chalgrave C Beds 103 F10
Chalgrove Oxon 83 F10
Chalk Kent 69 E7
Chalk End Essex 87 C10
Chalkfoot Cumb 230 B2
Chalkhill Norf 141 C7
Chalkhouse Green Oxon 65 D8
Chalkshire Bucks 84 D4
Chalksole Kent 55 E9
Chalkway Som 28 F5
Chalkwell Kent 69 G11
Chalkwell Sthend 69 B11
Challaborough Devon 8 G3
Challacombe Devon 41 E7
Challister Shetland 312 G7
Challoch Dumfries 236 C5
Challock Kent 54 C4
Chalmington Dorset 29 G9
Chalton C Beds 103 F10
Chalton Hants 34 D2
Chalvedon Essex 69 B8
Chalvey Slough 66 D3
Chalvington E Sus 23 D8
Chambercombe Devon 40 D4
Chamber's Green Kent 54 E2
Champson Devon 26 B4
Chance Inn Fife 287 F7
Chancery = Rhydgaled Ceredig 111 B11
Chance's Pitch Hereford 98 C4
Chandler's Cross Herts 85 F9
Chandler's Cross Worcs 98 D5
Chandler's Ford Hants 32 C6
Chandlers Green Hants 49 B8
Channel Tunnel Kent 55 F7
Channel's End Beds 122 F2
Channerwick Shetland 313 L6
Chantry Devon 25 C9
Chantry Som 45 D8
Chantry Suff 108 C2
Chapel Corn 4 C6
Chapel Cumb 229 E10
Chapel Fife 280 C5
Chapel Allerton Som 44 C2
Chapel Allerton W Yorks 206 F2
Chapel Amble Corn 10 F5
Chapel Brampton Northants 120 D4
Chapel Chorlton Staffs 150 B6
Chapel Cleeve Som 42 E4
Chapel Cross E Sus 37 C10
Chapel-en-le-Frith Derbys 185 E9
Chapel End Beds 103 B11
Chapel End Beds 122 F2
Chapel End C Beds 103 C11
Chapel End Cambs 138 G2
Chapel End E Ches 167 G11
Chapel End Essex 105 G11
Chapel End Northants 138 F2
Chapel End Warks 134 E6
Chapel Field Gtr Man 195 F9
Chapel Field Norf 161 E7
Chapel Fields W Mid 118 B6
Chapel Fields York 207 C7
Chapel Green Herts 104 D6
Chapel Green Warks 119 E9
Chapel Green Warks 134 F5
Chapel Haddlesey N Yorks 198 B5
Chapel Head Cambs 138 G6
Chapel Hill Aberds 303 F10
Chapel Hill Glos 79 E10
Chapel Hill Lincs 174 E2
Chapel Hill Mon 79 F8
Chapel Hill N Yorks 206 D2
Chapel House Lancs 194 F3
Chapel Knapp Wilts 61 F11
Chapel Lawn Shrops 114 B6
Chapel-le-Dale N Yorks 212 D4
Chapel Leigh Som 27 B10
Chapel Mains Borders 271 G11
Chapel Milton Derbys 185 E9
Chapel of Garioch Aberds 303 G7
Chapel of Stoneywood Aberdeen 293 B10
Chapel on Leader Borders 271 G11
Chapel Outon Dumfries 236 E6
Chapel Plaister Wilts 61 F10
Chapel Row E Sus 23 C10
Chapel Row Essex 88 E3
Chapel Row W Berks 64 F5
Chapel St Leonards Lincs 191 G9
Chapel Stile Cumb 220 D6
Chapel Town Corn 5 D7
Chapelgate Lincs 157 E8
Chapelhall N Lnrk 268 C5
Chapelhill Dumfries 248 E3
Chapelhill Highld 301 B8
Chapelhill N Ayrs 266 G4
Chapelhill Perth 286 D4
Chapelhill Perth 286 E3
Chapelhill Perth 286 E6
Chapelknowe Dumfries 239 C8
Chapels Blkburn 195 C7
Chapels Cumb 210 C4
Chapelthorpe W Yorks 197 D10
Chapelton Angus 287 C10
Chapelton Devon 25 B9
Chapelton Highld 291 B11
Chapelton S Lnrk 268 F3
Chapelton Row Dumfries 237 E8
Chapeltown Blkburn 195 D8
Chapeltown Moray 302 G2
Chapeltown S Yorks 186 B5
Chapeltown W Yorks 206 F2
Chapman's Hill Worcs 117 B9
Chapmans Town E Sus 23 B10
Chapmans Well Devon 12 C3
Chapmanslade Wilts 45 D10
Chapmore End Herts 86 B4
Chappel Essex 107 F7
Charaton Cross Corn 6 B6
Charcott Kent 52 D4
Chard Som 28 F4
Chard Junction Dorset 28 G4
Chardleigh Green Som 28 E4
Chardstock Devon 28 G4
Charfield S Glos 80 G2
Charfield Green S Glos 80 G2
Charfield Hill S Glos 80 G2
Charford Worcs 117 D9
Chargrove Glos 80 B6
Charing Kent 54 D3
Charing Cross Dorset 31 E10
Charing Heath Kent 54 D2
Charing Hill Kent 54 C3
Charingworth Glos 100 D4
Charlbury Oxon 82 B5
Charlcombe Bath 61 F8
Charlcutt Wilts 62 D3
Charlecote Warks 118 F5
Charlemont W Mid 133 E10
Charles Devon 41 G7
Charles Bottom Devon 41 G7
Charles Tye Suff 125 G10
Charlesfield Borders 262 D3
Charlesfield Dumfries 238 D5
Charleshill Sur 49 E11
Charleston Angus 287 C7
Charleston Renfs 267 C9
Charlestown Aberdeen 293 C11
Charlestown Corn 5 E10
Charlestown Derbys 185 C8
Charlestown Dorset 17 F9
Charlestown Fife 279 E11
Charlestown Gtr Man 195 G10
Charlestown Gtr Man 195 G11
Charlestown Highld 299 B8
Charlestown Highld 300 E6
Charlestown W Yorks 196 B3
Charlestown W Yorks 205 F9
Charlestown of Aberlour Moray 302 E2
Charlesworth Derbys 185 C8
Charleton Devon 8 G5
Charlinch Som 43 F8
Charlottetown Fife 286 F6
Charlton Hants 47 D11
Charlton Herts 104 F3
Charlton London 68 D2
Charlton Northants 101 D10
Charlton Northumb 251 F8
Charlton Oxon 64 B2
Charlton Redcar 226 B2
Charlton Som 28 B3
Charlton Som 44 E6
Charlton Som 45 C7
Charlton Sur 66 F5
Charlton Telford 149 G11
Charlton W Sus 34 E5
Charlton Wilts 30 C6
Charlton Wilts 46 B6
Charlton Wilts 62 B3
Charlton Worcs 99 B10
Charlton Worcs 116 C6
Charlton Abbots Glos 99 G10
Charlton Adam Som 29 B8
Charlton-All-Saints Wilts 31 C11
Charlton Down Dorset 17 C9
Charlton Horethorne Som 29 C11
Charlton Kings Glos 99 G9
Charlton Mackrell Som 29 B8
Charlton Marshall Dorset 30 G5
Charlton Musgrove Som 30 B2
Charlton on Otmoor Oxon 83 B9
Charlton on the Hill Dorset 30 G5
Charlton Park Glos 99 G9
Charlton St Peter Wilts 46 B6
Charltonbrook S Yorks 186 B4
Charlwood E Sus 51 G11
Charlwood Hants 49 G7
Charlwood Sur 51 E8
Charlynch Som 43 F8
Charminster Bmouth 19 C8
Charminster Dorset 17 C9
Charmouth Dorset 16 C3
Charnage Wilts 45 G10
Charndon Bucks 102 G3
Charnes Staffs 150 C5
Charney Bassett Oxon 82 G5
Charnock Green Lancs 194 D5
Charnock Hall S Yorks 186 E5
Charnock Richard Lancs 194 D5
Charsfield Suff 126 F5
Chart Corner Kent 53 C9
Chart Hill Kent 53 D9
Chart Sutton Kent 53 D10
Charter Alley Hants 48 B5
Charterhouse Som 44 B3
Chartershall Stirl 278 C6
Charterville Allotments Oxon 82 C4
Chartham Kent 54 C6
Chartham Hatch Kent 54 B6
Chartridge Bucks 84 E6
Charvil Wokingham 65 D9
Charwelton Northants 119 F10
Chase Cross London 87 G8
Chase End Street Worcs 98 D5
Chase Hill S Glos 61 B8
Chase Terrace Staffs 133 B10
Chasetown Staffs 133 B10
Chastleton Oxon 100 F4
Chasty Devon 24 G4
Chat Hill W Yorks 205 G8
Chatburn Lancs 203 E11
Chatcull Staffs 150 C5
Chatford Shrops 131 B9
Chatham Caerph 59 B8
Chatham Medway 69 F9
Chatham Green Essex 88 B2
Chathill Northumb 264 D5
Chatley Worcs 117 E7
Chattenden Medway 69 E9
Chatter End Essex 105 F9
Chatteris Cambs 139 F7
Chatterley Staffs 168 E4
Chattern Hill Sur 66 E5
Chatterton Lancs 195 D9
Chattisham Suff 107 C11
Chattle Hill Warks 134 E3
Chatto Borders 263 F7
Chatton Northumb 264 D3
Chaul End C Beds 103 G11
Chaulden Herts 85 D8
Chavel Shrops 149 G8
Chavenage Green Glos 80 F5
Chavey Down Brack 65 F11
Chawleigh Devon 26 E2
Chawley Oxon 83 E7
Chawson Worcs 117 E7
Chawston Beds 122 F3
Chawton Hants 49 F8
Chaxhill Glos 80 C2
Chazey Heath Oxon 65 D7
Cheadle Gtr Man 184 D5
Cheadle Staffs 169 G8
Cheadle Heath Gtr Man 184 D5
Cheadle Hulme Gtr Man 184 D5
Cheadle Park Staffs 169 G8
Cheam London 67 G8
Cheapside Herts 105 E8
Cheapside Sur 50 B4
Cheapside Windsor 66 F2
Chearsley Bucks 84 C2
Chebsey Staffs 151 D7
Checkendon Oxon 65 C7
Checkley E Ches 168 F2
Checkley Hereford 97 D11
Checkley Staffs 151 B10
Checkley Green E Ches 168 F2
Chedburgh Suff 124 F5
Cheddar Som 44 C3
Cheddington Bucks 84 B6
Cheddleton Staffs 169 E7
Cheddleton Heath Staffs 169 E7
Cheddon Fitzpaine Som 28 B2
Chedglow Wilts 80 G6
Chedgrave Norf 143 D7
Chedington Dorset 29 F7
Chediston Suff 127 B7
Chediston Green Suff 127 B7
Chedworth Glos 81 C9
Chedworth Laines Glos 81 C8
Chedzoy Som 43 F10
Cheeklaw Borders 272 E5
Cheeseman's Green Kent 54 F4
Cheetham Hill Gtr Man 195 G10
Cheglinch Devon 40 E4
Chegworth Kent 53 C10
Cheldon Devon 26 E2
Chelfham Devon 40 F6
Chelford E Ches 184 G4
Chell Heath Stoke 168 E5
Chellaston Derby 153 C7
Chellington Beds 121 F9
Chells Herts 104 F5
Chelmarsh Shrops 132 F4
Chelmer Village Essex 88 D2
Chelmick Shrops 131 E9
Chelmondiston Suff 108 D4
Chelmorton Derbys 169 B10
Chelmsford Essex 88 D2
Chelmsine Som 27 D11
Chelmsley Wood W Mid 134 F3
Chelsea London 67 D9
Chelsfield London 68 G3
Chelsham Sur 51 B11
Chelston Som 27 C11
Chelston Torbay 9 C7
Chelston Heathfield Som 27 C11
Chelsworth Suff 107 B9
Chelsworth Common Suff 107 B9
Cheltenham Glos 99 G8
Chelveston Northants 121 D9
Chelvey N Som 60 F3
Chelvey Batch N Som 60 F3
Chelwood Bath 60 G6
Chelwood Common E Sus 36 B6
Chelwood Gate E Sus 36 B6
Chelworth Wilts 81 G7
Chelworth Lower Green Wilts 81 G9
Chelworth Upper Green Wilts 81 G9
Chelynch Som 45 E7
Chemistry Shrops 167 G8
Cheney Longville Shrops 131 G8
Chenhalls Corn 2 B3
Chenies Bucks 85 F8
Cheny Longville Shrops 131 G8
Chepstow Mon 79 G8
Chequerbent Gtr Man 195 F7
Chequerfield W Yorks 198 C3
Chequers Corner Norf 139 B9
Chequertree Kent 54 F4
Cherhill Wilts 62 E4
Cherington Glos 80 F6
Cherington Warks 100 D5
Cheristow Devon 24 B3
Cheriton Devon 41 D8
Cheriton Hants 33 B9
Cheriton Kent 55 F7
Cheriton Swansea 56 C3
Cheriton Bishop Devon 13 C11
Cheriton Cross Devon 13 C11
Cheriton Fitzpaine Devon 26 F5
Cheriton or Stackpole Elidor Pembs 73 F7
Cherrington Telford 150 E3
Cherry Burton E Yorks 208 E5
Cherry Green Essex 105 F11
Cherry Green Herts 105 F7
Cherry Hinton Cambs 123 F9
Cherry Orchard Shrops 149 G9
Cherry Orchard Worcs 117 G7
Cherry Tree Blkburn 195 B7
Cherry Tree Gtr Man 185 C7
Cherry Willingham Lincs 189 G8
Cherrybank Perth 286 E5
Cherrytree Hill Derby 153 B7
Chertsey Sur 66 F4
Chertsey Meads Sur 66 F5
Cheselbourne Dorset 17 B11
Chesham Bucks 85 E7
Chesham Gtr Man 195 E10
Chesham Bois Bucks 85 F7
Cheshunt Herts 86 E5
Chesley Kent 69 G11
Cheslyn Hay Staffs 133 B9
Chessetts Wood Warks 118 C3
Chessington London 67 G7
Chessmount Bucks 85 E7
Chestall Staffs 151 G11
Chester W Ches 166 B6
Chester-le-Street Durham 243 G7
Chester Moor Durham 233 B11
Chesterblade Som 45 E7
Chesterfield Derbys 186 G5
Chesterfield Staffs 134 B2
Chesterhill Midloth 271 B7
Chesterhope Northumb 251 F9
Chesterknowes Borders 262 D2
Chesters Borders 262 E4
Chesters Borders 262 G4
Chesterton Cambs 123 E9
Chesterton Cambs 138 D2
Chesterton Glos 81 E8
Chesterton Oxon 101 G11
Chesterton Shrops 132 D5
Chesterton Staffs 168 F4
Chesterton Warks 118 F6
Chesterton Green Warks 118 F6
Chesterwood Northumb 241 D8
Chestfield Kent 70 F6
Chestnut Hill Cumb 229 G11
Chestnut Street Kent 69 G11
Cheston Devon 8 D3
Cheswardine Shrops 150 D4
Cheswell Telford 150 F4
Cheswick Northumb 273 F10
Cheswick Buildings Northumb 273 F10
Cheswick Green W Mid 118 B2
Chetnole Dorset 29 F10
Chettiscombe Devon 27 E7
Chettisham Cambs 139 G10
Chettle Dorset 31 E7
Chetton Shrops 132 E3
Chetwode Bucks 102 F2
Chetwynd Aston Telford 150 F5
Cheveley Cambs 124 E3
Chevening Kent 52 B3
Cheverell's Green Herts 85 B9
Chevin End W Yorks 205 E9
Chevington Suff 124 F5
Chevithorne Devon 27 D7
Chew Magna Bath 60 G5
Chew Moor Gtr Man 195 F7
Chew Stoke Bath 60 G5
Chewton Keynsham Bath 61 F7
Chewton Mendip Som 44 C5
Cheylesmore W Mid 118 B6
Chichacott Devon 13 B8
Chicheley M Keynes 103 B8
Chichester W Sus 22 C5
Chickenley W Yorks 197 C9
Chickerell Dorset 17 E8
Chicklade Wilts 46 G2
Chickney Essex 105 F11
Chicksands C Beds 104 D2
Chicksgrove Wilts 46 G3
Chickward Hereford 114 G5
Chidden Hants 33 D11
Chiddingfold Sur 50 F3
Chiddingly E Sus 23 C8
Chiddingstone Kent 52 D3
Chiddingstone Causeway Kent 52 D4
Chiddingstone Hoath Kent 52 E3
Chideock Dorset 16 C4
Chidgley Som 42 F4
Chidham W Sus 22 C3
Chidswell W Yorks 197 C9
Chieveley W Berks 64 E3
Chignall St James Essex 87 D11
Chignall Smealy Essex 87 C11
Chigwell Essex 86 G6
Chigwell Row Essex 87 G7
Chilbolton Hants 47 F11
Chilbolton Down Hants 48 F2
Chilbridge Dorset 31 G7
Chilcomb Hants 33 B8
Chilcombe Dorset 16 C6
Chilcombe Som 42 F6
Chilcompton Som 44 C6
Chilcote Leics 152 G5
Child Okeford Dorset 30 E4
Childer Thornton W Ches 182 F5
Childerditch Essex 68 B6
Childerley Gate Cambs 123 F7
Childrey Oxon 63 B11
Child's Ercall Shrops 150 E3
Child's Hill London 67 B8
Childsbridge Kent 52 B5
Childswickham Worcs 99 D11
Childwall Mers 182 D6
Childwick Bury Herts 85 C10
Childwick Green Herts 85 C10
Chilfrome Dorset 17 B7
Chilgrove W Sus 34 E4
Chilham Kent 54 C5
Chilhampton Wilts 46 G5
Chilla Devon 24 G6
Chillaton Devon 12 E4
Chillenden Kent 55 C9
Chillerton I o W 20 E5
Chillesford Suff 127 G7
Chillingham Northumb 264 D3
Chillington Devon 8 G5
Chillington Som 28 E5
Chillmill Kent 53 E7
Chilmark Wilts 46 G3
Chilmington Green Kent 54 E3
Chilsham E Sus 23 C10
Chilson Oxon 82 B4
Chilson Som 28 G4
Chilson Common Som 28 G4
Chilsworthy Corn 12 G4
Chilsworthy Devon 24 F4
Chiltern Green Beds 85 B10
Chilthorne Domer Som 29 D8
Chiltington E Sus 36 D5
Chilton Bucks 83 C11
Chilton Durham 233 F11
Chilton Kent 71 G11
Chilton Oxon 64 B3
Chilton Suff 107 C7
Chilton Candover Hants 48 E5
Chilton Cantelo Som 29 C9
Chilton Foliat Wilts 63 E10
Chilton Lane Durham 234 E2
Chilton Moor T & W 234 B2
Chilton Polden Som 43 F11
Chilton Street Suff 106 B5
Chilton Trinity Som 43 F9
Chilvers Coton Warks 135 E7
Chilwell Notts 153 B10
Chilworth Hants 32 D6
Chilworth Sur 50 D4
Chilworth Old Village Hants 32 D6
Chimney Oxon 82 E5
Chimney-end Oxon 82 B4
Chimney Street Suff 106 B4
Chineham Hants 49 C7
Chingford London 86 G5
Chingford Green London 86 G5
Chingford Hatch London 86 G5
Chinley Derbys 185 E8
Chinley Head Derbys 185 E9
Chinnor Oxon 84 E3
Chipley Som 27 C10
Chipmans Platt Glos 80 D3
Chipnall Shrops 150 C4
Chippenhall Green Suff 126 B5
Chippenham Cambs 124 D3
Chippenham Wilts 62 E2
Chipperfield Herts 85 E8
Chipping Herts 105 E7
Chipping Lancs 203 E8
Chipping Barnet London 86 F2
Chipping Campden Glos 100 D3
Chipping Hill Essex 88 B4
Chipping Norton Oxon 100 F6
Chipping Ongar Essex 87 E8
Chipping Sodbury S Glos 61 C8
Chipping Warden Northants 101 B9
Chipstable Som 27 B8
Chipstead Kent 52 B3
Chipstead Sur 51 B9
Chirbury Shrops 130 D5
Chirk = Y Waun Wrex 148 B5
Chirk Bank Shrops 148 B5
Chirk Green Wrex 148 B5
Chirmorrie S Ayrs 236 B4
Chirnside Borders 273 D7
Chirnsidebridge Borders 273 D7
Chirton T & W 243 D8
Chirton Wilts 46 B5
Chisbridge Cross Bucks 65 B10
Chisbury Wilts 63 F9
Chiselborough Som 29 E7
Chiseldon Swindon 63 D7
Chiserley W Yorks 196 B4
Chislehampton Oxon 83 F9
Chislehurst London 68 E2
Chislehurst West London 68 E2
Chislet Kent 71 G8
Chislet Forstal Kent 71 G8
Chiswell Dorset 17 G9
Chiswell Green Herts 85 E10
Chiswick London 67 D8
Chiswick End Cambs 105 B7
Chisworth Derbys 185 C7
Chitcombe E Sus 38 C4
Chithurst W Sus 34 C4
Chittering Cambs 123 C9
Chitterley Devon 26 G6
Chitterne Wilts 46 E3
Chittlehamholt Devon 25 C11
Chittlehampton Devon 25 B10
Chittoe Wilts 62 F3
Chitts Hills Essex 107 F9
Chitty Kent 71 G8
Chivelstone Devon 9 G10
Chivenor Devon 40 G4
Chivery Bucks 84 D6
Chobham Sur 66 G3
Choicelee Borders 272 E4
Cholderton Wilts 47 E8
Cholesbury Bucks 84 D6
Chollerford Northumb 241 C10
Chollerton Northumb 241 C10
Cholmondeston E Ches 167 C10
Cholsey Oxon 64 B5
Cholstrey Hereford 115 F9
Cholwell Bath 44 B6
Chop Gate N Yorks 225 F11
Choppington Northumb 253 G7
Chopwell T & W 242 F4
Chorley E Ches 167 E9
Chorley Lancs 194 D5
Chorley Shrops 132 G3
Chorley Staffs 151 G11
Chorley Common W Sus 34 B4
Chorleywood Herts 85 F8
Chorleywood Bottom Herts 85 F8
Chorleywood West Herts 85 F8
Chorlton E Ches 168 E2
Chorlton-cum-Hardy Gtr Man 184 C4
Chorlton Lane W Ches 167 F7
Choulton Shrops 131 F7
Chowdene T & W 243 F7
Chowley W Ches 167 D7
Chownes Mead W Sus 36 C4
Chreagain Highld 289 C10
Chrishall Essex 105 D8
Christchurch Cambs 139 D9
Christchurch Dorset 19 C9
Christchurch Glos 79 C9
Christchurch Newport 59 B10
Christian Malford Wilts 62 D3
Christleton W Ches 166 B6
Christmas Common Oxon 84 G2
Christon N Som 43 B11
Christon Bank Northumb 264 E6
Christow Devon 14 D2
Chryston N Lnrk 278 G3
Chub Tor Devon 7 B10
Chuck Hatch E Sus 52 G3
Chudleigh Devon 14 F3
Chudleigh Knighton Devon 14 F2
Chulmleigh Devon 25 E11

Chunal Derbys 185 C8
Church Lancs 195 B8
Church Aston Telford 150 F4
Church Brampton Northants 120 D4
Church Brough Cumb 222 C5
Church Broughton Derbys 152 C4
Church Charwelton Northants 119 F10
Church Clough Lancs 204 F3
Church Common Hants 34 B2
Church Coombe Corn 4 G3
Church Cove Corn 2 G6
Church Crookham Hants 49 C10
Church Eaton Staffs 150 F6
Church End Beds 122 F2
Church End Bucks 84 B6
Church End Bucks 84 D2
Church End C Beds 85 B8
Church End C Beds 103 E9
Church End C Beds 103 G9
Church End C Beds 104 D3
Church End C Beds 122 G3
Church End Cambs 121 C11
Church End Cambs 123 C7
Church End Cambs 123 D7
Church End Cambs 138 G4
Church End Cambs 139 B7
Church End E Yorks 209 C7
Church End Essex 88 B2
Church End Essex 105 C11
Church End Essex 105 F11
Church End Essex 106 F4
Church End Glos 80 D2
Church End Glos 99 D7
Church End Hants 49 B7
Church End Herts 85 C10
Church End Herts 85 F8
Church End Herts 104 E5
Church End Herts 105 G8
Church End Lincs 156 C4
Church End Lincs 190 B6
Church End London 67 C8
Church End London 86 G2
Church End Norf 157 F10
Church End Oxon 82 D5
Church End Oxon 100 E6
Church End Suff 108 D4
Church End Sur 50 B5
Church End W Mid 119 B7
Church End Warks 134 E4
Church End Warks 134 E5
Church End Wilts 62 D4
Church End Worcs 98 C6
Church Enstone Oxon 101 G7
Church Fenton N Yorks 206 F6
Church Green Devon 15 B9
Church Green Norf 141 E11
Church Gresley Derbys 152 F5
Church Hanborough Oxon 82 C6
Church Hill Pembs 73 C7
Church Hill Staffs 151 G10
Church Hill W Ches 167 C10
Church Hill W Mid 133 D9
Church Hill Worcs 117 D11
Church Hougham Kent 55 E9
Church Houses N Yorks 226 F3
Church Knowle Dorset 18 E4
Church Laneham Notts 188 F4
Church Langton Leics 136 E4
Church Lawford Warks 119 B8
Church Lawton E Ches 168 D4
Church Leigh Staffs 151 B10
Church Lench Worcs 117 G10
Church Mayfield Staffs 169 G11
Church Minshull E Ches 167 C11
Church Norton W Sus 22 D5
Church Oakley Hants 48 C5
Church Preen Shrops 131 D10
Church Pulverbatch Shrops 131 C8
Church Stowe Northants 120 F2
Church Street Essex 106 C5
Church Street Kent 69 E8
Church Stretton Shrops 131 E9
Church Town Corn 4 G3
Church Town Leics 153 F7
Church Town N Lincs 199 F9
Church Town Sur 51 C11
Church Village Rhondda 58 B5
Church Warsop Notts 171 B9
Church Westcote Glos 100 G4
Church Whitfield Kent 55 D10
Church Wilne Derbys 153 C8
Churcham Glos 80 B3
Churchbank Shrops 114 B6
Churchbridge Corn 6 D4
Churchbridge Staffs 133 B9
Churchdown Glos 80 B5
Churchend Essex 89 G8
Churchend Essex 106 G2
Churchend Glos 80 D3
Churchend Reading 65 E7
Churchend S Glos 80 G2
Churches Green E Sus 23 B10
Churchfield Hereford 98 B4
Churchfield W Mid 133 E10
Churchfields Wilts 31 B10
Churchgate Herts 86 E4
Churchgate Street Essex 87 C7
Churchill Devon 28 G4
Churchill Devon 40 E5
Churchill N Som 44 B2
Churchill Oxon 100 G5
Churchill Worcs 117 B7
Churchill Worcs 117 G8
Churchill Green N Som 60 G2
Churchinford Som 28 E2
Churchmoor Rough Shrops 131 F8
Churchover Warks 135 G10
Churchstanton Som 27 E11
Churchstoke Powys 130 E5
Churchstow Devon 8 F4
Churchton Pembs 73 D10
Churchtown Corn 11 F7
Churchtown Cumb 230 C3
Churchtown Derbys 170 C3
Churchtown Devon 24 G3
Churchtown Devon 41 E7
Churchtown I o M 192 C5
Churchtown Lancs 202 E5
Churchtown Mers 193 D11
Churchtown Shrops 130 F5
Churchtown Som 42 F3
Churchwood W Sus 35 D8
Churnet Grange Staffs 169 E7
Churnsike Lodge Northumb 240 B5
Churscombe Torbay 9 C7
Churston Ferrers Torbay 9 D8
Churt Sur 49 F11
Churton W Ches 166 D6
Churwell W Yorks 197 B9
Chute Cadley Wilts 47 C10
Chute Standen Wilts 47 C10
Chwefford Conwy 180 G4
Chwilog Gwyn 145 B8
Chwitffordd = Whitford Flint 181 F10
Chyandour Corn 1 C5
Chyanvounder Corn 2 E5
Chycoose Corn 3 B8
Chynhale Corn 2 C4
Chynoweth Corn 2 C2
Chyvarloe Corn 2 E5
Cicelyford Mon 79 E8
Cil y coed = Caldicot Mon 60 B3
Cilan Uchaf Gwyn 144 E5
Cilau Pembs 91 D8
Cilcain Flint 165 B11
Cilcennin Ceredig 111 E10
Cilcewydd Powys 130 C4
Cilfor Gwyn 146 B2
Cilfrew Neath 76 E3
Cilfynydd Rhondda 77 G9
Cilgerran Pembs 92 C3
Cilgwyn Carms 94 F4
Cilgwyn Ceredig 92 C6
Cilgwyn Gwyn 163 E7
Cilgwyn Pembs 91 D11
Ciliau Aeron Ceredig 111 F9
Cill Amhlaidh W Isles 297 G3
Cill Donnain W Isles 297 J3
Cill Eireabhagh W Isles 297 G4
Cille Bhrighde W Isles 297 K3
Cille Pheadair W Isles 297 K3
Cilmaengwyn Neath 76 D2
Cilmery Powys 113 G10
Cilsan Carms 93 G11
Ciltalgarth Gwyn 164 G5
Ciltwrch Powys 96 C3
Cilwendeg Pembs 92 D4
Cilybebyll Neath 76 E2
Cilycwm Carms 94 D5
Cimla Neath 57 B9
Cinder Hill Gtr Man 195 F9
Cinder Hill Kent 52 D4
Cinder Hill W Mid 133 E8
Cinder Hill W Sus 36 B5
Cinderford Glos 79 C11
Cinderhill Derbys 170 F5
Cinderhill Nottingham 171 G8
Cinnamon Brow Warr 183 C10
Cippenham Slough 66 C2
Cippyn Pembs 92 B2
Circebost W Isles 304 E3
Cirencester Glos 81 E8
Ciribhig W Isles 304 D3
City London 67 C10
City Powys 130 F4
City V Glam 58 D3
City Dulas Anglesey 179 D7
Clabhach Argyll 288 D3
Clachaig Argyll 276 E2
Clachaig Highld 292 B2
Clachaig N Ayrs 255 E10
Clachan Argyll 255 B8
Clachan Argyll 275 B8
Clachan Argyll 284 F5
Clachan Argyll 289 E10
Clachan Highld 295 B7
Clachan Highld 298 C4
Clachan Highld 307 L6
Clachan W Isles 297 G3
Clachan na Luib W Isles 296 E4
Clachan of Campsie E Dunb 278 F2
Clachan of Glendaruel Argyll 275 E10
Clachan-Seil Argyll 275 B8
Clachan Strachur Argyll 284 G4
Clachaneasy Dumfries 236 B5
Clachanmore Dumfries 236 E2
Clachbreck Argyll 275 F8
Clachnabrain Angus 292 G5
Clachtoll Highld 307 G5
Clackmannan Clack 279 C8
Clackmarras Moray 302 D2
Clacton-on-Sea Essex 89 B11
Cladach N Ayrs 256 B2
Cladach Chairinis W Isles 296 F4
Cladach Chireboist W Isles 296 E3
Claddach Argyll 254 B2
Claddach-knockline W Isles 296 E3
Cladich Argyll 284 E4
Cladich Steading Argyll 284 E4
Cladswell Worcs 117 F10
Claggan Highld 289 E8
Claggan Highld 290 F3
Claggan Perth 285 D11
Claigan Highld 298 D2
Claines Worcs 117 F7
Clandown Bath 45 B7
Clanfield Hants 33 D11
Clanfield Oxon 82 E3
Clanking Bucks 84 D4
Clanville Hants 47 D10
Clanville Som 44 G6
Clanville Wilts 62 D2
Claonaig Argyll 255 B9
Claonel Highld 309 J5
Clap Hill Kent 54 F5
Clapgate Dorset 31 G8
Clapgate Herts 105 G8
Clapham Beds 121 G10
Clapham Devon 14 D3
Clapham London 67 D9
Clapham N Yorks 212 F4
Clapham W Sus 35 F9
Clapham Green Beds 121 G10
Clapham Green N Yorks 205 B10
Clapham Hill Kent 70 G6
Clapham Park London 67 E9
Clapper Corn 10 G6
Clapper Hill Kent 53 F10
Clappers Borders 273 D8
Clappersgate Cumb 221 E7
Clapphoull Shetland 313 L6
Clapton Som 28 F6
Clapton Som 44 C6
Clapton W Berks 63 E11
Clapton in Gordano N Som 60 E3
Clapton-on-the-Hill Glos 81 B11
Clapton Park London 67 B11
Clapworthy Devon 25 C11
Clara Vale T & W 242 E4
Clarach Ceredig 128 G2
Clarack Aberds 292 D6
Clarbeston Pembs 91 G10
Clarbeston Road Pembs 91 G10
Clarborough Notts 188 E2
Clardon Highld 310 C5
Clare Oxon 83 F11
Clare Suff 106 B5
Clarebrand Dumfries 237 C9
Claregate W Mid 133 C7
Claremont Park Sur 66 G6
Claremount W Yorks 196 B5
Clarence Park N Som 59 G10
Clarendon Park Leicester 135 C11
Clareston Pembs 73 C7
Clarilaw Borders 262 D2
Clarilaw Borders 262 F2
Clark Green E Ches 184 F6
Clarken Green Hants 48 C5
Clark's Green Sur 51 F7
Clark's Hill Lincs 157 E7
Clarksfield Gtr Man 196 G2
Clarkston E Renf 267 D11
Clarkston N Lnrk 268 B5
Clase Swansea 57 B7
Clashandorran Highld 300 E5
Clashcoig Highld 309 K6
Clasheddy Highld 308 C6
Clashgour Argyll 284 C6
Clashindarroch Aberds 302 F4
Clashmore Highld 306 F5
Clashmore Highld 309 L7
Clashnessie Highld 306 F5
Clashnoir Moray 302 G2
Clate Shetland 313 G7
Clatford Wilts 63 F7
Clatford Oakcuts Hants 47 F10
Clathy Perth 286 F3
Clatt Aberds 302 G5
Clatter Powys 129 E9
Clatterford I o W 20 D5
Clatterford End Essex 87 C10
Clatterford End Essex 87 D9
Clatterford End Essex 87 E8
Clatterin Bridge Aberds 293 F8
Clatto Fife 287 F8
Clatworthy Som 42 G5
Clauchlands N Ayrs 256 C2
Claughton Lancs 202 E6
Claughton Lancs 211 F11
Claughton Mers 182 D4
Clavelshay Som 43 G9
Claverdon Warks 118 E3
Claverham N Som 60 F2
Claverhambury Essex 86 E6
Clavering Essex 105 E9
Claverley Shrops 132 E5
Claverton Bath 61 G9
Claverton Down Bath 61 G9
Clawdd-côch V Glam 58 D5
Clawdd-newydd Denb 165 E9
Clawdd Poncen Denb 165 G9
Clawthorpe Cumb 211 D10
Clawton Devon 12 B3
Claxby Lincs 189 C10
Claxby Lincs 191 G7
Claxby St Andrew Lincs 191 G7
Claxton N Yorks 216 G3
Claxton Norf 142 C6
Clay Common Suff 143 G9
Clay Coton Northants 119 B11
Clay Cross Derbys 170 C5
Clay End Herts 104 F6
Clay Hill Bristol 60 E6
Clay Hill London 86 F4
Clay Hill W Berks 64 E5
Clay Lake Lincs 156 E5
Clay Mills Derbys 152 D5
Claybokie Aberds 292 D2
Claybrooke Magna Leics 135 F9
Claybrooke Parva Leics 135 F9
Claydon Glos 99 E8
Claydon Oxon 119 G9
Claydon Suff 126 G2
Claygate Dumfries 239 B9
Claygate Kent 52 C6
Claygate Kent 53 E8
Claygate Sur 67 G7
Claygate Cross Kent 52 B6
Clayhall Hants 21 B8
Clayhall London 86 G6
Clayhanger Devon 27 C8
Clayhanger Som 28 E4
Clayhanger W Mid 133 C10
Clayhidon Devon 27 D11
Clayhill E Sus 38 C4
Clayhill Hants 32 F4
Clayhithe Cambs 123 E10
Clayholes Angus 287 D9
Clayland Stirl 277 D11
Claylock Highld 310 D5
Claypit Hill Cambs 123 G7
Claypits Devon 27 B7
Claypits Glos 80 D3
Claypits Kent 55 B9
Claypits Suff 140 G4
Claypole Lincs 172 F5
Clays End Bath 61 G8
Claythorpe Lincs 190 F6
Clayton Gtr Man 184 B5
Clayton S Yorks 198 F3
Clayton Staffs 168 G5
Clayton W Sus 36 E3
Clayton W Yorks 205 G8
Clayton Brook Lancs 194 C5
Clayton Green Lancs 194 C5
Clayton Heights W Yorks 205 G8
Clayton-le-Dale Lancs 203 G9
Clayton-le-Moors Lancs 203 G10
Clayton-le-Woods Lancs 194 C5
Clayton West W Yorks 197 E9
Clayworth Notts 188 D2
Cleadale Highld 294 G6
Cleadon T & W 243 E9
Cleadon Park T & W 243 E9
Clearbrook Devon 7 B10
Clearwell Glos 79 D9
Clearwell Newport 59 B9
Clearwood Wilts 45 D10
Cleasby N Yorks 224 C5
Cleat Orkney 314 B4
Cleat Orkney 314 H4
Cleatlam Durham 224 B2
Cleator Cumb 219 C10
Cleator Moor Cumb 219 B10
Cleave Devon 28 G2
Clebrig Highld 308 F5
Cleckheaton W Yorks 197 B7
Cleddon Mon 79 E8
Clee St Margaret Shrops 131 G11
Cleedownton Shrops 131 G11
Cleehill Shrops 115 B11
Cleekhimin N Lnrk 268 D5
Cleemarsh Shrops 131 G11
Cleestanton Shrops 115 B11
Cleethorpes NE Lincs 201 F10
Cleeton St Mary Shrops 116 B2
Cleeve Glos 80 C2
Cleeve N Som 60 F3
Cleeve Oxon 64 C6
Cleeve Hill Glos 99 F9
Cleeve Prior Worcs 99 B11
Cleghorn S Lnrk 269 F8
Clegyrnant Powys 129 B8
Clehonger Hereford 97 D9
Cleirwy = Clyro Powys 96 C4
Cleish Perth 279 B11
Cleland N Lnrk 268 D5
Clement Street Kent 68 E4
Clement's End C Beds 85 B8
Clements End Glos 79 D9
Clench Wilts 63 G7
Clench Common Wilts 63 F7
Clencher's Mill Hereford 98 D4
Clenchwarton Norf 157 E11
Clennell Northumb 251 B10
Clent Worcs 117 B8
Cleobury Mortimer Shrops 116 B3
Cleobury North Shrops 132 F2
Cleongart Argyll 255 D7
Clephanton Highld 301 D8
Clerk Green W Yorks 197 C8
Clerkenwater Corn 5 B11
Clerkenwell London 67 C10
Clerklands Borders 262 E2
Clermiston Edin 280 G3
Clestrain Orkney 314 F3
Cleuch Head Borders 262 G3
Cleughbrae Dumfries 238 C3
Clevancy Wilts 62 D5
Clevans Renfs 267 B7
Clevedon N Som 60 E2
Cleveley Oxon 101 G7
Cleveleys Lancs 202 E2
Cleverton Wilts 62 B3
Clevis Bridgend 57 F10
Clewer Som 44 C2
Clewer Green Windsor 66 D2
Clewer New Town Windsor 66 D3
Clewer Village Windsor 66 D3
Cley next the Sea Norf 177 E8
Cliaid W Isles 297 L2
Cliasmol W Isles 305 H2
Cliburn Cumb 231 G7
Click Mill Orkney 314 D3
Cliddesden Hants 48 D6
Cliff Derbys 185 D8
Cliff Warks 134 D4
Cliff End E Sus 38 E5
Cliff End W Yorks 196 D6
Cliffburn Angus 287 C10
Cliffe Lancs 203 G10
Cliffe Medway 69 D8
Cliffe N Yorks 207 G9
Cliffe N Yorks 224 B4
Cliffe Woods Medway 69 E8
Clifford Devon 24 C4
Clifford Hereford 96 B4
Clifford W Yorks 206 E4
Clifford Chambers Warks 118 G3
Clifford's Mesne Glos 98 G4
Cliffs End Kent 71 G10
Clifftown Sthend 69 B11
Clifton Bristol 60 E5
Clifton C Beds 104 D3
Clifton Cumb 230 F6
Clifton Derbys 169 G11
Clifton Devon 40 E5
Clifton Gtr Man 195 G9
Clifton Lancs 202 G5
Clifton N Yorks 205 D9
Clifton Northumb 252 G6
Clifton Nottingham 153 C11
Clifton Oxon 101 E9
Clifton S Yorks 186 C6
Clifton S Yorks 187 B8
Clifton Stirl 285 D7
Clifton W Ches 183 F8
Clifton W Yorks 197 C7
Clifton Worcs 98 B6
Clifton York 207 C7
Clifton Campville Staffs 152 G5
Clifton Green Gtr Man 195 G9
Clifton Hampden Oxon 83 F8
Clifton Junction Gtr Man 195 G9
Clifton Manor C Beds 104 D3
Clifton Maybank Dorset 29 E9
Clifton Moor York 207 B7
Clifton Reynes M Keynes 121 G8
Clifton upon Dunsmore Warks 119 B10
Clifton upon Teme Worcs 116 E4
Cliftoncote Borders 263 E8
Cliftonville Kent 71 E11
Cliftonville N Lnrk 268 B4
Cliftonville Norf 160 B6
Climping W Sus 35 G8
Climpy S Lnrk 269 D8
Clink Som 45 D9
Clinkham Wood Mers 183 B8
Clint N Yorks 205 B11
Clint Green Norf 159 G10
Clintmains Borders 262 C4
Clints N Yorks 224 E2
Cliobh W Isles 304 E2
Clipiau Gwyn 146 G6
Clippesby Norf 161 G8
Clippings Green Norf 159 G10
Clipsham Rutland 155 F9
Clipston Northants 136 G4
Clipston Notts 154 C2
Clipstone C Beds 103 F8
Clitheroe Lancs 203 E10
Cliuthar W Isles 305 J3
Clive Shrops 149 E10
Clive W Ches 167 B11
Clive Green W Ches 167 C11
Clive Vale E Sus 38 E4
Clivocast Shetland 312 C8
Clixby Lincs 200 G6
Cloatley Wilts 81 G7
Cloatley End Wilts 81 G7
Clocaenog Denb 165 E9
Clochan Aberds 303 E9
Clochan Moray 302 C4
Clock Face Mers 183 C8
Clock House London 67 G9
Clock Mills Hereford 96 B5
Clockmill Borders 272 E5
Cloddiau Powys 130 B4
Cloddymoss Moray 301 D9
Clodock Hereford 96 F6
Cloford Som 45 E8
Cloford Common Som 45 E8
Cloigyn Carms 74 C6
Clola Aberds 303 E10
Clophill C Beds 103 D11
Clopton Northants 137 G11
Clopton Suff 126 G4
Clopton Corner Suff 126 G4
Clopton Green Suff 124 G5
Clopton Green Suff 125 E9
Close Clark I o M 192 E3
Close House Durham 233 F10
Closeburn Dumfries 247 E9
Closworth Som 29 E9
Clothall Herts 104 E5
Clothall Common Herts 104 E5
Clotton W Ches 167 C8
Clotton Common W Ches 167 C8
Cloud Side Staffs 168 C6
Cloudesley Bush Warks 135 F9
Clouds Hereford 97 D11
Clough Gtr Man 196 D2
Clough Gtr Man 196 F2
Clough W Yorks 196 E5
Clough Dene Durham 242 F5
Clough Foot W Yorks 196 C2
Clough Hall Staffs 168 E4
Clough Head W Yorks 196 C5
Cloughfold Lancs 195 C10
Cloughton N Yorks 227 G10
Cloughton Newlands N Yorks 227 F10
Clounlaid Highld 289 D9
Clousta Shetland 313 H5
Clouston Orkney 314 E2
Clova Aberds 302 G4
Clova Angus 292 F5
Clove Lodge Durham 223 B8
Clovelly Devon 24 C4
Clovenfords Borders 261 B10
Clovenstone Aberds 293 B9
Cloves Moray 301 C11
Clovullin Highld 290 G2
Clow Bridge Lancs 195 B10
Clowance Wood Corn 2 C4
Clowne Derbys 187 F7
Clows Top Worcs 116 C4
Cloy Wrex 166 G5
Cluanie Inn Highld 290 B2
Cluanie Lodge Highld 290 B2
Clubmoor Mers 182 C5
Clubworthy Corn 11 C11
Cluddley Telford 150 G2
Clun Shrops 130 G6
Clunbury Shrops 131 G7
Clunderwen Carms 73 B10
Clune Highld 301 D9
Clune Highld 301 G7
Clunes Highld 290 E4
Clungunford Shrops 115 B7
Clunie Aberds 302 D6
Clunie Perth 286 C5
Clunton Shrops 130 G6
Cluny Fife 280 B4
Cluny Castle Aberds 293 B8
Cluny Castle Highld 291 D8
Clutton Bath 44 B6
Clutton W Ches 167 E7
Clutton Hill Bath 44 B6
Clwt-grugoer Conwy 165 C7
Clwt-y-bont Gwyn 163 C9
Clwydyfagwyr M Tydf 77 D8
Clydach Mon 78 C2
Clydach Swansea 75 E11
Clydach Terrace Powys 77 C11
Clydach Vale Rhondda 77 G7
Clydebank W Dunb 277 G9
Clyffe Pypard Wilts 62 D5
Clynder Argyll 276 E4
Clyne Neath 76 E4
Clynelish Highld 311 J2
Clynnog-fawr Gwyn 162 F6
Clyro = Cleirwy Powys 96 C4
Clyst Honiton Devon 14 C5
Clyst Hydon Devon 27 G8
Clyst St George Devon 14 D5
Clyst St Lawrence Devon 27 G8
Clyst St Mary Devon 14 C5
Cnip W Isles 304 E2
Cnoc Amhlaigh W Isles 304 E7
Cnoc an t-Solais W Isles 304 D6
Cnoc Fhionn Highld 295 D10
Cnoc Màiri W Isles 304 E6
Cnoc Rolum W Isles 296 F3
Cnocbreac Argyll 274 F5
Cnwch-coch Ceredig 112 B3
Coachford Aberds 302 E4
Coad's Green Corn 11 F11
Coal Aston Derbys 186 F5
Coal Bank Darl 234 G3
Coal Pool W Mid 133 C10
Coalbrookdale Telford 132 C3
Coalbrookvale Bl Gwent 77 D11
Coalburn S Lnrk 259 C8
Coalburns T & W 242 E4
Coalcleugh Northumb 232 B2
Coaley Glos 80 E3
Coaley Peak Glos 80 E3
Coalford Aberds 293 D10
Coalhall E Ayrs 257 F10
Coalhill Essex 88 F3
Coalmoor Telford 132 B3
Coalpit Field Warks 135 F7
Coalpit Heath S Glos 61 C7
Coalpit Hill Staffs 168 E4
Coalport Telford 132 C3
Coalsnaughton Clack 279 B8
Coaltown of Balgonie Fife 280 B5
Coaltown of Wemyss Fife 280 B6
Coalville Leics 153 G8
Coalway Glos 79 C9
Coanwood Northumb 240 F5
Coarsewell Devon 8 E4
Coat Som 29 C7
Coatbridge N Lnrk 268 C4
Coatdyke N Lnrk 268 C5
Coate Swindon 63 C7
Coate Wilts 62 G4
Coates Cambs 138 D6
Coates Glos 81 E7
Coates Lancs 204 D3
Coates Lincs 188 E6
Coates Midloth 270 C4
Coates Notts 188 E4
Coates W Sus 35 D7
Coatham Redcar 235 F7
Coatham Mundeville Darl 233 G11
Coatsgate Dumfries 248 B3
Cobairdy Aberds 302 E5
Cobbaton Devon 25 B10
Cobbler's Corner Worcs 116 F5
Cobbler's Green Norf 142 E5
Cobbler's Plain Mon 79 E7
Cobbs Warr 183 D10
Cobb's Cross Glos 98 E5
Cobbs Fenn Essex 106 E5
Cobby Syke N Yorks 205 B9
Coberley Glos 81 B7
Cobhall Common Hereford 97 D9
Cobham Kent 69 F7
Cobham Sur 66 G6
Cobleland Stirl 277 B10
Cobler's Green Essex 87 B11
Cobley Dorset 31 C8
Cobley Hill Worcs 117 C10
Cobnash Hereford 115 E9
Cobridge Stoke 168 F5
Cobscot Shrops 150 B3
Coburty Aberds 303 C9
Cock and End Suff 124 G4
Cock Alley Derbys 186 G6
Cock Bank Wrex 166 F5
Cock Bevington Warks 117 G11
Cock Bridge Aberds 292 C4
Cock Clarks Essex 88 E4
Cock Gate Hereford 115 D9
Cock Green Essex 87 B11
Cock Hill N Yorks 206 B6
Cock Marling E Sus 38 D5
Cock Street Kent 53 C9
Cock Street Suff 107 D9
Cockadilly Glos 80 E4
Cockayne N Yorks 226 F2
Cockburnspath Borders 282 G5
Cockden Lancs 204 G3
Cockenzie and Port Seton E Loth 281 F8
Cocker Bar Lancs 194 C4
Cockerham Lancs 202 C5
Cockermouth Cumb 229 E8
Cockernhoe Herts 104 G2
Cockernhoe Green Herts 104 G2
Cockersdale W Yorks 197 B8
Cockerton Darl 224 B5
Cockett Swansea 56 C6
Cocketty Aberds 293 F9
Cockfield Durham 233 G8
Cockfield Suff 125 G8
Cockfosters London 86 F3
Cockhill Som 44 G6
Cocking W Sus 34 D5
Cocking Causeway W Sus 34 D5
Cockington Torbay 9 C7
Cocklake Som 44 D2
Cocklaw Northumb 241 C10
Cockleford Glos 81 C7
Cockley Beck Cumb 220 E4
Cockley Cley Norf 140 C5
Cockley Hill W Yorks 197 D7
Cocknowle Dorset 18 E4
Cockpole Green Wokingham 65 C9
Cocks Corn 4 E5
Cocks Green Suff 125 F7
Cockshead Ceredig 112 F2
Cockshoot Hereford 97 D11
Cockshutford Shrops 131 F11
Cockshutt Shrops 132 G4
Cockshutt Shrops 149 D8
Cockthorpe Norf 177 E7
Cockwells Corn 2 C2
Cockwood Devon 14 E5
Cockwood Som 43 E8
Cockyard Derbys 185 F8
Cockyard Hereford 97 E8
Codda Corn 11 F9
Coddenham Suff 126 G2
Coddenham Green Suff 126 F2
Coddington Hereford 98 C4
Coddington Notts 172 E4
Coddington W Ches 167 D7
Codford St Mary Wilts 46 F3
Codford St Peter Wilts 46 F3
Codicote Herts 86 B2
Codicote Bottom Herts 86 B2
Codmore Bucks 85 E7
Codmore Hill W Sus 35 C9
Codnor Derbys 170 F6
Codnor Breach Derbys 170 F6
Codnor Gate Derbys 170 E6
Codnor Park Derbys 170 E6
Codrington S Glos 61 D8
Codsall Staffs 133 C7
Codsall Wood Staffs 132 B6
Codsend Som 41 F11
Coed Cwnwr Mon 78 F6
Coed Eva Torf 78 G3
Coed Llai = Leeswood Flint 166 D3
Coed Mawr Gwyn 179 G9
Coed Morgan Mon 78 C5
Coed-Talon Flint 166 D3
Coed-y-bryn Ceredig 93 C7
Coed-y-caerau Newport 78 G5
Coed-y-fedw Mon 78 D6
Coed y Garth Ceredig 128 E3
Coed y go Shrops 148 D5
Coed-y-paen Mon 78 F4
Coed-y-parc Gwyn 163 B10
Coed-y-wlad Powys 130 B4
Coed-yr-ynys Powys 96 G3
Coed Ystumgwern Gwyn 145 E11
Coedcae Bl Gwent 77 D11
Coedcae Torf 78 D3
Coedely Rhondda 58 B4
Coedkernew Newport 59 C9
Coedpoeth Wrex 166 E3
Coedway Powys 148 G6
Coelbren Powys 76 C4
Coffee Hall M Keynes 103 D7
Coffinswell Devon 9 B7
Cofton Devon 14 E5
Cofton Common W Mid 117 B10
Cofton Hackett Worcs 117 B10
Cog V Glam 59 F7
Cogan V Glam 59 E7
Cogenhoe Northants 120 E6
Cogges Oxon 82 D5
Coggeshall Essex 106 G6
Coggeshall Hamlet Essex 107 G7
Coggins Mill E Sus 37 B9
Coig Peighinnean W Isles 304 B7
Coig Peighinnean Bhuirgh W Isles 304 C6
Coignafearn Lodge Highld 291 B8
Coignascallan Highld 291 B9
Coilacriech Aberds 292 D5
Coilantogle Stirl 285 G9
Coilessan Argyll 284 G6
Coilleag W Isles 297 K3
Coillemore Highld 300 B6
Coillore Highld 294 B5
Coirea-chrombe Stirl 285 G9
Coisley Hill S Yorks 186 E6
Coity Bridgend 58 C2
Cokenach Herts 105 D7
Cokhay Green Derbys 152 D5
Col W Isles 304 D6
Col Uarach W Isles 304 E6
Colaboll Highld 309 H5
Colan Corn 5 C7
Colaton Raleigh Devon 15 D7
Colbost Highld 298 E2
Colburn N Yorks 224 F3
Colby Cumb 231 G9
Colby I o M 192 E3
Colby Norf 160 C4
Colchester Essex 107 F10
Colchester Green Suff 125 F8
Colcot V Glam 58 F6
Cold Ash W Berks 64 F4
Cold Ash Hill Hants 49 G10
Cold Ashby Northants 120 B3
Cold Ashton S Glos 61 E9
Cold Aston Glos 81 B10
Cold Blow Pembs 73 C10
Cold Brayfield M Keynes 121 G8
Cold Christmas Herts 86 B5
Cold Cotes N Yorks 212 E4
Cold Elm Glos 98 E6
Cold Hanworth Lincs 189 E8
Cold Harbour Dorset 18 D4
Cold Harbour Herts 85 B10
Cold Harbour Kent 69 G11
Cold Harbour Lincs 155 C9
Cold Harbour Oxon 64 D6
Cold Harbour Wilts 45 B11
Cold Harbour Wilts 45 D11
Cold Harbour Windsor 65 D10
Cold Hatton Telford 150 E2
Cold Hatton Heath Telford 150 E2
Cold Hesledon Durham 234 B4
Cold Hiendley W Yorks 197 E11
Cold Higham Northants 120 G3
Cold Inn Pembs 73 D10
Cold Kirby N Yorks 215 C10
Cold Moss Heath E Ches 168 C3
Cold Newton Leics 136 B4
Cold Northcott Corn 11 D10
Cold Norton Essex 88 E4
Cold Overton Leics 154 G6
Cold Row Lancs 202 E3
Cold Well Staffs 151 G11
Coldbackie Highld 308 D6
Coldbeck Cumb 222 E4
Coldblow London 68 E4
Coldbrook Powys 96 D3
Coldean Brighton 36 F4
Coldeast Devon 14 G2
Coldeaton Derbys 169 D10
Colden W Yorks 196 B3
Colden Common Hants 33 C7
Coldfair Green Suff 127 E8
Coldham Cambs 139 C8
Coldham Staffs 133 B7
Coldham's Common Cambs 123 F9
Coldharbour Corn 4 F5
Coldharbour Devon 27 E9
Coldharbour Dorset 17 E9
Coldharbour Glos 79 E9
Coldharbour Kent 52 C5
Coldharbour London 68 D4
Coldharbour Sur 50 E6
Coldingham Borders 273 B8
Coldmeece Staffs 151 C7
Coldoch Stirl 278 B3
Coldra Newport 59 B11
Coldrain Perth 286 G4
Coldred Kent 55 D9
Coldridge Devon 25 F11
Coldstream Angus 287 D7
Coldstream Borders 263 B8
Coldvreath Corn 5 D9
Coldwaltham W Sus 35 D8
Coldwells Aberds 303 E11
Coldwells Croft Aberds 302 G5
Cole Som 45 G7
Cole End Essex 105 D11
Cole End Warks 134 F3
Cole Green Herts 86 C3
Cole Green Herts 105 E8
Cole Henley Hants 48 C3
Cole Park London 67 E7
Colebatch Shrops 130 F6
Colebrook Devon 27 F8
Colebrooke Devon 13 B11
Coleburn Moray 302 D2
Coleby Lincs 173 C7
Coleby N Lincs 199 D11
Coleford Devon 26 G3
Coleford Glos 79 C9
Coleford Som 45 D7
Coleford Water Som 42 G6
Colegate End Norf 142 F3
Colehall W Mid 134 F2
Colehill Dorset 31 G8
Coleman Green Herts 85 C11
Coleman's Hatch E Sus 52 G3
Colemere Shrops 149 C8
Colemore Hants 49 G8
Colemore Green Shrops 132 D4
Coleorton Leics 153 F8
Coleorton Moor Leics 153 F8
Colerne Wilts 61 E10
Cole's Cross Dorset 28 G5
Coles Green Suff 107 C11
Cole's Green Suff 126 E5
Coles Green Worcs 116 G5
Coles Meads Sur 51 C9
Colesbourne Glos 81 C7
Colesbrook Dorset 30 B4
Colesden Beds 122 F2
Coleshill Bucks 85 F7
Coleshill Oxon 82 G2
Coleshill Warks 134 F4
Colestocks Devon 27 G9
Colethrop Glos 80 C4
Coley Bath 44 B5
Coley Reading 65 E8
Coley W Yorks 196 B6
Colfin Dumfries 236 D2
Colgate W Sus 51 G8
Colgrain Argyll 276 E6
Colham Green London 66 C5
Colindale London 67 B8
Colinsburgh Fife 287 G8
Colinton Edin 270 B4
Colintraive Argyll 275 F11
Colkirk Norf 159 D8
Collace Perth 286 D6
Collafield Glos 79 C11
Collafirth Shetland 312 G6
Collam W Isles 305 J3
Collamoor Head Corn 11 C9
Collaton Devon 9 G9
Collaton St Mary Torbay 9 D7
College Milton S Lnrk 268 D2
College of Roseisle Moray 301 C11
College Park London 67 C8
College Town Brack 65 G11
Collennan S Ayrs 257 C8
Collessie Fife 286 F6
Colleton Mills Devon 25 D11
Collett's Br Norf 139 B9
Collett's Green Worcs 116 G6
Collier Row London 87 G8
Collier Street Kent 53 D8
Collier's End Herts 105 G7
Collier's Green E Sus 38 C3
Colliers Green Kent 53 F9
Colliers Hatch Essex 87 E8
Collier's Wood London 67 E9
Colliery Row T & W 234 B2
Collieston Aberds 303 G10
Collin Dumfries 238 B2
Collingbourne Ducis Wilts 47 C8
Collingbourne Kingston Wilts 47 B8
Collingham Notts 172 C4
Collingham W Yorks 206 D3
Collington Hereford 116 E2
Collingtree Northants 120 F5
Collingwood Northumb 243 B7
Collins End Oxon 65 D7
Collins Green Warr 183 C9
Collins Green Worcs 116 F4
Collipriest Devon 27 E7
Colliston Angus 287 C10
Colliton Devon 27 G9
Collycroft Warks 135 F7
Collyhurst Gtr Man 195 G11
Collynie Aberds 303 F8
Collyweston Northants 137 C9
Colmonell S Ayrs 244 F4
Colmslie Borders 262 B2
Colmsliehill Borders 271 G10
Colmworth Beds 122 F2
Coln Rogers Glos 81 D9
Coln St Aldwyns Glos 81 D10
Coln St Dennis Glos 81 C9
Colnabaichin Aberds 292 C4
Colnbrook Slough 66 D4
Colne Cambs 123 B7
Colne Lancs 204 E3
Colne Bridge W Yorks 197 C7
Colne Edge Lancs 204 E3
Colne Engaine Essex 107 E7
Colnefields Cambs 123 B7
Colney Norf 142 B3
Colney Hatch London 86 G3
Colney Heath Herts 86 D2
Colney Street Herts 85 E11
Cologin Argyll 289 G10
Colpitts Grange Northumb 241 F11
Colpy Aberds 302 F6
Colquhar Borders 270 G6
Colscott Devon 24 E5
Colshaw Staffs 169 B8
Colsterdale N Yorks 214 C2
Colsterworth Lincs 155 E8
Colston E Dunb 268 B2
Colston Pembs 91 F9
Colston Bassett Notts 154 C3
Colstrope Bucks 65 B9
Colt Hill Hants 49 C8
Colt Park Cumb 210 E5
Coltfield Moray 301 C11
Colthouse Cumb 221 F7
Colthrop W Berks 64 F4
Coltishall Norf 160 F5
Coltness N Lnrk 268 D6
Colton Cumb 210 B6
Colton N Yorks 206 E6
Colton Norf 142 B2
Colton Staffs 151 E11
Colton Suff 125 D7
Colton W Yorks 206 G3
Colton Hills Staffs 133 D8
Colt's Green S Glos 61 C8
Colt's Hill Kent 52 E6
Columbia T & W 243 F8
Columbjohn Devon 14 B5
Colva Powys 114 G4
Colvend Dumfries 237 D10
Colvister Shetland 312 D7
Colwall Hereford 98 C4
Colwall Green Hereford 98 C4
Colwall Stone Hereford 98 C5
Colwell I o W 20 D2
Colwell Northumb 241 B11
Colwich Staffs 151 E10
Colwick Notts 171 G10
Colwinston = Tregolwyn V Glam 58 D2
Colworth W Sus 22 C6
Colwyn Bay = Bae Colwyn Conwy 180 F4
Colychurch Bridgend 58 D2
Colyford Devon 15 C10
Colyton Devon 15 C10
Colzie Fife 286 F6
Combe Devon 8 B4
Combe Devon 8 G4
Combe Devon 9 G9
Combe E Sus 37 B10
Combe Hereford 114 E6
Combe Oxon 82 B6
Combe Som 28 B6
Combe W Berks 63 G11
Combe Almer Dorset 18 B5
Combe Common Sur 50 F3
Combe Down Bath 61 G9
Combe Fishacre Devon 8 C6
Combe Florey Som 43 G7
Combe Hay Bath 45 B8
Combe Martin Devon 40 D5
Combe Moor Hereford 115 E7
Combe Pafford Torbay 9 B8
Combe Raleigh Devon 27 G11
Combe St Nicholas Som 28 E4
Combe Throop Som 30 C2
Combebow Devon 12 D5
Combeinteignhead Devon 14 G4
Comberbach W Ches 183 F10
Comberford Staffs 134 B3
Comberton Cambs 123 F7
Comberton Hereford 115 D9
Combpyne Devon 15 C11
Combrew Devon 40 G4
Combridge Staffs 151 B11
Combrook Warks 118 G6
Combs Derbys 185 F8
Combs Suff 125 F10
Combs W Yorks 197 D8
Combs Ford Suff 125 F11
Combwich Som 43 E9
Come-to-Good Corn 4 G6
Comers Aberds 293 C8
Comeytrowe Som 28 C2
Comford Corn 2 B6
Comfort Corn 2 D6
Commins Coch Ceredig 128 G2
Comiston Edin 270 B4
Comley Shrops 131 D9
Commercial End Cambs 123 E11
Commins Denb 165 C10
Commins Capel Betws Ceredig 112 F2
Commins Coch Powys 128 C6
Common Cefn-llwyn Mon 78 G4
Common Edge Blkpool 202 G2
Common End Cumb 228 G6
Common End Derbys 170 C6
Common Hill Hereford 97 E11
Common Moor Corn 6 B4
Common Platt Wilts 62 B6
Common Side Derbys 170 F6
Common Side Derbys 186 F4
Common Side W Ches 167 B9
Common-y-coed Mon 60 B2

Commondale N Yorks 226 C3
Commonmoor Corn 6 B4
Commonside Derbys 170 G2
Commonside Notts 171 D7
Commonside W Ches 183 G8
Commonwood Herts 85 E8
Commonwood Shrops 149 D9
Commonwood Wrex 166 E5
Comp Kent 52 B6
Compass Som 43 G9
Compstall Gtr Man 185 C7
Compton Derbys 169 F11
Compton Devon 9 C7
Compton Hants 32 B4
Compton Hants 33 B7
Compton Plym 7 D9
Compton Staffs 132 G6
Compton Sur 49 D11
Compton Sur 50 D3
Compton W Berks 64 D4
Compton W Mid 133 D7
Compton W Sus 34 E3
Compton W Yorks 206 E3
Compton Wilts 46 C6
Compton Abbas Dorset 30 D5
Compton Abdale Glos 81 B9
Compton Bassett Wilts 62 E4
Compton Beauchamp Oxon 63 B9
Compton Bishop Som 43 B11
Compton Chamberlayne Wilts 31 B8
Compton Common Bath 60 G6
Compton Dando Bath 60 G6
Compton Dundon Som 44 G3
Compton Durville Som 28 D6
Compton End Hants 33 B7
Compton Green Glos 98 F4
Compton Greenfield S Glos 60 C5
Compton Martin Bath 44 B4
Compton Pauncefoot Som 29 B10
Compton Valence Dorset 17 C7
Comrie Fife 279 D10
Comrie Highld 300 D4
Comrie Perth 285 C11
Comrie Perth 285 E11
Comrue Dumfries 248 F3
Conaglen House Highld 290 G2
Conanby S Yorks 187 B7
Conchra Argyll 275 E11
Conchra Highld 295 C10
Concord T & W 243 F8
Concraig Perth 286 F2
Concraigie Perth 286 C5
Conder Green Lancs 202 B5
Conderton Worcs 99 D9
Condicote Glos 100 F3
Condorrat N Lnrk 278 G4
Condover Shrops 131 B9
Coney Hall London 67 G11
Coney Hill Glos 80 B5
Coney Weston Suff 125 B9
Coneyhurst W Sus 35 C10
Coneysthorpe N Yorks 216 E4
Coneythorpe N Yorks 206 B3
Conford Hants 49 G10
Congash Highld 301 G10
Congdon's Shop Corn 11 F11
Congeith Dumfries 237 C10
Congelow Kent 53 D7
Congerstone Leics 135 B7
Congham Norf 158 E4
Congl-y-wal Gwyn 164 G2
Congleton E Ches 168 C5
Congleton Edge E Ches 168 C5
Congresbury N Som 60 G2
Congreve Staffs 151 G8
Conham Bristol 60 E6
Conicavel Moray 301 D9
Coningsby Lincs 174 D2
Conington Cambs 122 D6
Conington Cambs 138 F3
Conisbrough S Yorks 187 B8
Conisby Argyll 274 G3
Conisholme Lincs 190 B6
Coniston Cumb 220 F6
Coniston E Yorks 209 F9
Coniston Cold N Yorks 204 B4
Conistone N Yorks 213 F9
Conkwell Wilts 61 G9
Connage Moray 302 C4
Connah's Quay Flint 166 B3
Connel Argyll 289 F11
Connel Park E Ayrs 258 G4
Conniburrow M Keynes 103 D7
Connista Highld 298 B4
Connon Corn 6 C3
Connor Downs Corn 2 B3
Conock Wilts 46 B5
Conon Bridge Highld 300 D5
Conon House Highld 300 D5
Cononish Stirl 285 E7
Cononley N Yorks 204 D5
Cononley Woodside N Yorks 204 D5
Cononsyth Angus 287 C9
Conordan Highld 295 B7
Conquermoor Heath Telford 150 F3
Consall Staffs 169 F7
Consett Durham 242 G4
Constable Burton N Yorks 224 G3
Constable Lee Lancs 195 C10
Constantine Corn 2 D6
Constantine Bay Corn 10 G3
Contin Highld 300 D4
Contlaw Aberdeen 293 C10
Conwy Conwy 180 F3
Conyer Kent 70 G3
Conyers Green Suff 125 D7
Cooden E Sus 38 F2
Cooil I o M 192 E4
Cookbury Devon 24 F6
Cookbury Wick Devon 24 F5
Cookham Windsor 65 B11
Cookham Dean Windsor 65 C11
Cookham Rise Windsor 65 C11
Cookhill Worcs 117 F11
Cookley Suff 126 B6
Cookley Worcs 132 G6
Cookley Green Oxon 83 G11
Cookney Aberds 293 D10
Cookridge W Yorks 205 E11
Cook's Green Essex 89 B11
Cook's Green Suff 125 G9
Cooksbridge E Sus 36 E6
Cooksey Corner Worcs 117 D8
Cooksey Green Worcs 117 D8
Cookshill Staffs 168 G6
Cooksland Corn 5 B11
Cooksmill Green Essex 87 D10
Cooksongreen W Ches 183 G9
Coolham W Sus 35 C10
Cooling Medway 69 D9
Cooling Street Medway 69 E8
Coolinge Kent 55 F8
Coomb Hill Kent 69 G7
Coombe Bucks 84 D4
Coombe Corn 4 G2
Coombe Corn 4 G5
Coombe Corn 5 E9
Coombe Corn 6 C4
Coombe Corn 24 E2
Coombe Devon 14 G4
Coombe Devon 27 D8
Coombe Glos 80 G3
Coombe Hants 33 C11
Coombe Kent 55 B9
Coombe London 67 E8
Coombe Som 28 B3
Coombe Som 28 F6
Coombe Wilts 30 C5
Coombe Wilts 47 C7
Coombe Bissett Wilts 31 B10
Coombe Dingle Bristol 60 D5
Coombe Hill Glos 99 F7
Coombe Keynes Dorset 18 E2
Coombes W Sus 35 F11
Coombesdale Staffs 150 B6
Coombeswood W Mid 133 F9
Coombs End S Glos 61 C9
Coombses Som 28 F4
Cooper Street Kent 55 B10
Cooper Turning Gtr Man 194 F6
Cooper's Corner Kent 52 D3
Cooper's Green E Sus 37 C7
Cooper's Green Herts 85 D11
Cooper's Hill C Beds 103 D10
Cooper's Hill Sur 66 E3
Coopersale Common Essex 87 E7
Coopersale Street Essex 87 E7
Cootham W Sus 35 E9
Cop Street Kent 55 B9
Copcut Worcs 117 E7
Copdock Suff 108 C2
Coped Hall Wilts 62 C5
Copenhagen Denb 165 B8
Copford Essex 107 G8
Copford Green Essex 107 G8
Copgrove N Yorks 214 G6
Copister Shetland 312 F6
Cople Beds 104 B2
Copley Durham 233 F7
Copley Gtr Man 185 B7
Copley W Yorks 196 C5
Copley Hill W Yorks 197 B8
Coplow Dale Derbys 185 F11
Copmanthorpe York 207 D7
Copmere End Staffs 150 D6
Copnor Ptsmth 33 G11
Copp Lancs 202 F4
Coppathorne Corn 24 G2
Coppenhall E Ches 168 D2
Coppenhall Staffs 151 F8
Coppenhall Moss E Ches 168 D2
Copperhouse Corn 2 B3
Coppice Gtr Man 196 G2
Coppicegate Shrops 132 G4
Coppingford Cambs 138 G3
Coppins Corner Kent 54 D2
Coppleham Som 42 G2
Copplestone Devon 26 G3
Coppull Lancs 194 E5
Coppull Moor Lancs 194 E5
Copsale W Sus 35 C11
Copse Hill London 67 E8
Copster Green Lancs 203 G9
Copster Hill Gtr Man 196 G2
Copston Magna Warks 135 F9
Copt Green Warks 118 D3
Copt Heath W Mid 118 B3
Copt Hewick N Yorks 214 E6
Copt Oak Leics 153 G9
Copthall Green Essex 86 E6
Copthill Durham 232 C3
Copthorne Corn 11 C11
Copthorne E Ches 167 G11
Copthorne Shrops 149 G9
Copthorne Sur 51 F10
Coptiviney Shrops 149 B8
Copton Kent 54 B4
Copy's Green Norf 159 B8
Copythorne Hants 32 E4
Corbets Tey London 68 B5
Corbridge Northumb 241 E11
Corbriggs Derbys 170 B6
Corby Northants 137 F7
Corby Glen Lincs 155 E9
Corby Hill Cumb 239 F11
Cordon N Ayrs 256 C2
Cordwell Norf 142 E2
Coreley Shrops 116 C2
Cores End Bucks 66 B2
Corfe Som 28 D2
Corfe Castle Dorset 18 E5
Corfe Mullen Dorset 18 B5
Corfton Shrops 131 F9
Corfton Bache Shrops 131 F9
Corgarff Aberds 292 C4
Corgee Corn 5 C10
Corhampton Hants 33 C10
Corlae Dumfries 246 D5
Corlannau Neath 57 C9
Corley Warks 134 F6
Corley Ash Warks 134 F5
Corley Moor Warks 134 F5
Cornaa I o M 192 D5
Cornabus Argyll 254 C4
Cornaigbeg Argyll 288 E1
Cornaigmore Argyll 288 C4
Cornaigmore Argyll 288 E1
Cornard Tye Suff 107 C8
Cornbank Midloth 270 C4
Cornbrook Shrops 116 B2
Corncatterach Aberds 302 F5
Cornel Conwy 164 C2
Corner Row Lancs 202 F4
Cornett Hereford 97 B11
Corney Cumb 220 G2
Cornforth Durham 234 E2
Cornharrow Dumfries 246 E5
Cornhill Aberds 302 D5
Cornhill Powys 96 C2
Cornhill Stoke 168 E5
Cornhill-on-Tweed Northumb 263 B9
Cornholme W Yorks 196 B2
Cornish Hall End Essex 106 D3
Cornquoy Orkney 314 G5
Cornriggs Durham 232 C2
Cornsay Durham 233 C8
Cornsay Colliery Durham 233 C9
Cornton Stirl 278 B5
Corntown Highld 300 D5
Corntown V Glam 58 D2
Cornwell Oxon 100 F5
Cornwood Devon 8 D2
Cornworthy Devon 8 D6
Corpach Highld 290 F2
Corpusty Norf 160 C2
Corran Highld 290 G2
Corran Highld 295 E10
Corran a Chan Uachdaraich Highld 295 C7
Corranbuie Argyll 275 G9
Corrany I o M 192 D5
Corrichoich Highld 311 G4
Corrie N Ayrs 255 C11
Corrie Common Dumfries 248 F6
Corriecravie N Ayrs 255 E10
Corriecravie Moor N Ayrs 255 E10
Corriedoo Dumfries 246 G5
Corriegarth Lodge Highld 291 B7
Corriemoillie Highld 300 C3
Corriemulzie Lodge Highld 309 K3
Corrievarkie Lodge Perth 291 F7
Corrievorrie Highld 301 G7
Corrigall Orkney 314 E3
Corrimony Highld 300 F3
Corringham Lincs 188 C5
Corringham Thurrock 69 C8
Corris Gwyn 128 B5
Corris Uchaf Gwyn 128 B4
Corrour Highld 290 G5
Corrour Shooting Lodge Highld 290 G6
Corrow Argyll 284 G5
Corry Highld 295 C8
Corry of Ardnagrask Highld 300 E5
Corrybrough Highld 301 G8
Corrydon Perth 292 G3
Corryghoil Argyll 284 E5
Corrykinloch Highld 309 G3
Corrylach Argyll 255 D8
Corrymuckloch Perth 286 D2
Corrynachenchy Argyll 289 E8
Cors-y-Gedol Gwyn 145 E11
Corsback Highld 310 B6
Corscombe Dorset 29 F8
Corse Aberds 302 E6
Corse Glos 98 F5
Corse Lawn Worcs 98 E6
Corse of Kinnoir Aberds 302 E5
Corsewall Dumfries 236 C2
Corsham Wilts 61 E11
Corsindae Aberds 293 C8
Corsley Wilts 45 D10
Corsley Heath Wilts 45 D10
Corsock Dumfries 237 B9
Corston Bath 61 F7
Corston Orkney 314 E3
Corston Wilts 62 C2
Corstorphine Edin 280 G3
Cortachy Angus 287 B7
Corton Suff 143 D10
Corton Wilts 46 E2
Corton Denham Som 29 C10
Cortworth S Yorks 186 B6
Coruanan Lodge Highld 290 G2
Corunna W Isles 296 E4
Corvast Highld 309 K5
Corwen Denb 165 G9
Cory Devon 24 D5
Coryates Dorset 17 D8
Coryton Cardiff 58 C6
Coryton Devon 12 E5
Coryton Thurrock 69 C8
Còsag Highld 295 D10
Cosby Leics 135 E10
Coscote Oxon 64 B4
Coseley W Mid 133 E8
Cosford Warks 119 B9
Cosgrove Northants 102 C5
Cosham Ptsmth 33 F11
Cosheston Pembs 73 E8
Cosmeston V Glam 59 F7
Cosmore Dorset 29 F11
Cossall Notts 171 G7
Cossall Marsh Notts 171 G7
Cosses S Ayrs 244 G4
Cossington Leics 154 G2
Cossington Som 43 E11
Costa Orkney 314 D3
Costessey Norf 160 G3
Costessey Park Norf 160 G3
Costhorpe Notts 187 D9
Costislost Corn 10 G6
Costock Notts 153 D11
Coston Leics 154 E6
Coston Norf 141 B11
Coswinsawsin Corn 2 B4
Cote Oxon 82 E4
Cote Som 43 E10
Cote W Sus 35 F10
Cotebrook W Ches 167 B9
Cotehill Cumb 239 G11
Cotes Cumb 211 B9
Cotes Leics 153 E11
Cotes Staffs 150 C6
Cotes Heath Staffs 150 C6
Cotes Park Derbys 170 E6
Cotesbach Leics 135 G10
Cotford St Lukes Som 27 B11
Cotgrave Notts 154 B2
Cothall Aberds 293 B10
Cotham Bristol 60 E5
Cotham Notts 172 F3
Cothelstone Som 43 G7
Cotheridge Worcs 116 G5
Cotherstone Durham 223 B10
Cothill Oxon 83 F7
Cotland Mon 79 E8
Cotleigh Devon 28 G2
Cotmanhay Derbys 171 G7
Cotmarsh Wilts 62 D5
Cotmaton Devon 15 D8
Coton Cambs 123 F8
Coton Northants 120 C3
Coton Shrops 149 C10
Coton Staffs 134 B3
Coton Staffs 150 E6
Coton Staffs 151 C9
Coton Staffs 151 D8
Coton Clanford Staffs 151 E7
Coton Hayes Staffs 151 C9
Coton Hill Shrops 149 G9
Coton Hill Staffs 151 C9
Coton in the Clay Staffs 152 D3
Coton in the Elms Derbys 152 F4
Coton Park Derbys 152 F5
Cotonwood Shrops 149 B10
Cotonwood Staffs 150 E6
Cotswold Community Wilts 81 F8
Cott Devon 8 C5
Cottam E Yorks 217 F9
Cottam Lancs 202 G6
Cottam Notts 188 F4
Cottartown Highld 301 F10
Cottenham Cambs 123 D8
Cottenham Park London 67 F8
Cotterdale N Yorks 222 G6
Cottered Herts 104 F6
Cotterhill Woods S Yorks 187 E9
Cotteridge W Mid 117 B10
Cotterstock Northants 137 E10
Cottesbrooke Northants 120 C4
Cottesmore Rutland 155 G8
Cotteylands Devon 26 E6
Cottingham E Yorks 208 G6
Cottingham Northants 136 E6
Cottingley W Yorks 205 F8
Cottisford Oxon 101 E11
Cotton Staffs 169 F9
Cotton Suff 125 D11
Cotton End Beds 103 B11
Cotton End Northants 120 F5
Cotton Stones W Yorks 196 C4
Cotton Tree Lancs 204 F4
Cottonworth Hants 47 F11
Cottown Aberds 293 B9
Cottown Aberds 302 G5
Cottown Aberds 303 E8
Cotts Devon 7 B8
Cottwood Devon 25 E10
Cotwall Telford 150 F2
Cotwalton Staffs 151 B8
Coubister Orkney 314 E3
Couch Green Hants 48 G4
Couch's Mill Corn 6 D2
Coughton Hereford 97 G11
Coughton Warks 117 E11
Coughton Fields Warks 117 F11
Cougie Highld 300 G2
Coulaghailtro Argyll 275 G8
Coulags Highld 299 E9
Coulby Newham M'bro 225 B10
Coulderton Cumb 219 D9
Couldoran Highld 299 E8
Couligartan Stirl 285 G8
Coulin Highld 299 D10
Coulin Lodge Highld 299 D10
Coull Aberds 293 C7
Coull Argyll 274 G3
Coulmony Ho Highld 301 E9
Coulport Argyll 276 D4
Coulsdon London 51 B9
Coulshill Perth 286 G3
Coulston Wilts 46 C3
Coulter S Lnrk 260 C2
Coultings Som 43 E8
Coulton N Yorks 216 E2
Coultra Fife 287 E7
Cound Shrops 131 C11
Coundlane Shrops 131 B11
Coundmoor Shrops 131 C11
Coundon Durham 233 F10
Coundon W Mid 134 G6
Coundon Grange Durham 233 F10
Coundongate Durham 233 F10
Counters End Herts 85 D8
Countersett N Yorks 213 B8
Countess Wilts 47 E7
Countess Cross Essex 107 E7
Countess Wear Devon 14 D4
Countesthorpe Leics 135 D11
Countisbury Devon 41 D8
County Oak W Sus 51 F9
Coup Green Lancs 194 B5
Coupar Angus Perth 286 C6
Coupland Cumb 222 B4
Coupland Northumb 263 C10
Cour Argyll 255 C9
Courance Dumfries 248 E3
Coursley Som 42 G6
Court-at-Street Kent 54 F5
Court Barton Devon 14 D2
Court Colman Bridgend 57 E11
Court Corner Hants 48 B6
Court Henry Carms 93 G11
Court House Green W Mid 135 G7
Courteenhall Northants 120 G5
Courthill Perth 286 C5
Courtsend Essex 89 G8
Courtway Som 43 G8
Cousland Midloth 271 B7
Cousley Wood E Sus 53 G7
Couston Argyll 275 F11
Cova Shetland 313 J5
Cove Argyll 276 E4
Cove Borders 282 G5
Cove Devon 27 D7
Cove Hants 49 B11
Cove Highld 307 K3
Cove Bay Aberdeen 293 C11
Cove Bottom Suff 127 B9
Covehithe Suff 143 G10
Coven Staffs 133 B8
Coven Heath Staffs 133 C8
Coven Lawn Staffs 133 C8
Covender Hereford 98 C2
Coveney Cambs 139 G9
Covenham St Bartholomew Lincs 190 C4
Covenham St Mary Lincs 190 C4
Coventry W Mid 118 B6
Coverack Corn 3 F7
Coverack Bridges Corn 2 C5
Coverham N Yorks 214 B2
Covesea Moray 301 B11
Covingham Swindon 63 B7
Covington Cambs 121 C11
Covington S Lnrk 259 B11
Cow Ark Lancs 203 D9
Cow Green Suff 125 D11
Cow Hill Lancs 203 G7
Cow Roast Herts 85 C7
Cowan Bridge Lancs 212 D2
Cowbar Redcar 226 B5
Cowbeech E Sus 23 C10
Cowbeech Hill E Sus 23 C10
Cowbit Lincs 156 F5
Cowbog Aberds 303 D8
Cowbridge Lincs 174 F4
Cowbridge Som 42 E3
Cowbridge = Y Bont-Faen V Glam 58 E3
Cowcliffe W Yorks 196 D6
Cowdale Derbys 185 G9
Cowden Kent 52 E3
Cowdenbeath Fife 280 C3
Cowdenburn Borders 270 E4
Cowen Head Cumb 221 F9
Cowers Lane Derbys 170 F4
Cowes I o W 20 B5
Cowesby N Yorks 215 B9
Cowesfield Green Wilts 32 C3
Cowfold W Sus 36 C2
Cowgill Cumb 212 B5
Cowgrove Dorset 18 B5
Cowhill Derbys 170 F5
Cowhill S Glos 79 G10
Cowhorn Hill S Glos 61 E7
Cowie Aberds 293 E10
Cowie Stirl 278 D6
Cowlands Corn 4 G6
Cowleaze Corner Oxon 82 E4
Cowley Derbys 186 F4
Cowley Devon 14 B4
Cowley Glos 81 C7
Cowley London 66 C5
Cowley Oxon 83 E8
Cowley Peachy London 66 C5
Cowleymoor Devon 27 E7
Cowling Lancs 194 D5
Cowling N Yorks 204 E5
Cowling N Yorks 214 B4
Cowlinge Suff 124 G4
Cowlow Derbys 185 G9
Cowmes W Yorks 197 D7
Cowpe Lancs 195 C10
Cowpen Northumb 253 G7
Cowpen Bewley Stockton 234 G5
Cowplain Hants 33 E11
Cowshill Durham 232 C3
Cowslip Green N Som 60 G3
Cowstrandburn Fife 279 C10
Cowthorpe N Yorks 206 C4
Cox Common Suff 143 G8
Cox Green Gtr Man 195 E8
Cox Green Sur 50 G5
Cox Green Windsor 65 D11
Cox Hill Corn 4 G4
Cox Moor Notts 171 D8
Coxall Hereford 115 C7
Coxbank E Ches 167 G11
Coxbench Derbys 170 G5
Coxbridge Som 44 F4
Coxford Corn 11 B9
Coxford Norf 158 D6
Coxford Soton 32 D5
Coxgreen Staffs 132 F6
Coxheath Kent 53 C8
Coxhill Kent 55 D8
Coxhoe Durham 234 D2
Coxley Som 44 E4
Coxley W Yorks 197 D9
Coxley Wick Som 44 E4
Coxlodge T & W 242 D6
Coxpark Corn 12 G4
Coxtie Green Essex 87 F9
Coxwold N Yorks 215 D10
Coychurch Bridgend 58 D2
Coylton S Ayrs 257 E10
Coylumbridge Highld 291 B11
Coynach Aberds 292 C6
Coynachie Aberds 302 F4
Coytrahen Bridgend 57 D11
CoytrahÛn Bridgend 57 D11
Crab Orchard Dorset 31 F9
Crabadon Devon 8 E5
Crabbet Park W Sus 51 F10
Crabble Kent 55 E9
Crabbs Cross Worcs 117 E10
Crabbs Green Herts 105 F9
Crabgate Norf 159 D11
Crabtree Plym 7 D10
Crabtree W Sus 36 B2
Crabtree Green Wrex 166 G4
Crackaig Argyll 274 G6
Crackenedge W Yorks 197 C8
Crackenthorpe Cumb 231 G9
Crackington Haven Corn 11 B8
Crackley Staffs 168 E4
Crackley Warks 118 C5
Crackleybank Shrops 150 G5
Crackpot N Yorks 223 F9
Crackthorn Corner Suff 125 B10
Cracoe N Yorks 213 G9
Cracow Moss E Ches 168 F2
Cracow Moss Staffs 168 F3
Craddock Devon 27 E9
Cradhlastadh W Isles 304 E2
Cradle Edge W Yorks 205 F7
Cradle End Herts 105 G9
Cradley Hereford 98 B4
Cradley W Mid 133 G8
Cradley Heath W Mid 133 F8
Cradoc Powys 95 E10
Crafthole Corn 7 E7
Crafton Bucks 84 B5
Crag Bank Lancs 211 E9
Crag Foot Lancs 211 E9
Cragg Hill W Yorks 205 F10
Cragg Vale W Yorks 196 C4
Craggan Highld 301 G10
Craggan Moray 301 F11
Craggan Stirl 285 E9
Craggaunallie Highld 300 F5
Craggenmore Moray 301 F11
Craggie Highld 301 F7
Craggie Highld 311 H2
Craggiemore Highld 309 J7
Craghead Durham 242 G6
Crahan Corn 2 C5
Crai Powys 95 G7
Craibstone Moray 302 D4
Craichie Angus 287 C9
Craig Dumfries 237 B8
Craig Dumfries 237 C8
Craig Highld 299 E10
Craig Berthlwyd M Tydf 77 F9
Craig Castle Aberds 302 G4
Craig-cefn-parc Swansea 75 E11
Craig Douglas Borders 261 E7
Craig Llangiwg Neath 76 D2
Craig-llwyn Shrops 148 D4
Craig Lodge Argyll 275 G10
Craig-moston Aberds 293 F8
Craig Penllyn V Glam 58 D3
Craig-y-don Conwy 180 E3
Craig-y-Duke Swansea 76 E2
Craig-y-nos Powys 76 B4
Craig-y-penrhyn Ceredig 128 E3
Craig-y-Rhacca Caerph 59 B7
Craiganor Lodge Perth 285 B10
Craigdallie Perth 286 E6
Craigdam Aberds 303 F8
Craigdarroch Dumfries 246 E6
Craigdarroch Highld 300 D4
Craigdhu Highld 300 E4
Craigearn Aberds 293 B9
Craigellachie Moray 302 E2
Craigencallie Ho Dumfries 237 B7
Craigencross Dumfries 236 C2
Craigend Borders 271 F9
Craigend Glasgow 268 B3
Craigend Perth 286 E3
Craigend Perth 286 E5
Craigend Stirl 278 D5
Craigendive Argyll 275 E11
Craigendoran Argyll 276 E6
Craigendowie Angus 293 G7
Craigends Renfs 267 B8
Craigens Argyll 274 G3
Craigens E Ayrs 258 F3
Craigentinny Edin 280 G5
Craigerne Borders 261 B7
Craighall Perth 286 C5
Craighat Stirl 277 E9
Craighead Fife 287 G10
Craighead Highld 301 C7
Craighill Aberds 303 E7
Craighlaw Mains Dumfries 236 C5
Craighouse Argyll 274 G6
Craigie Aberds 293 B11
Craigie Dundee 287 D8
Craigie Perth 286 C5
Craigie Perth 286 E5
Craigie S Ayrs 257 C10
Craigie S Ayrs 257 E8
Craigiefield Orkney 314 E4
Craigiehall Edin 280 F3
Craigielaw E Loth 281 F9
Craigierig Borders 260 E6
Craigleith Edin 280 G4
Craiglockhart Edin 280 G4
Craigmalloch E Ayrs 245 E11
Craigmaud Aberds 303 D8
Craigmill Stirl 278 B6
Craigmillar Edin 280 G5
Craigmore Argyll 266 B2
Craignant Shrops 148 B5
Craignell Dumfries 237 B7
Craigneuk N Lnrk 268 C5
Craigneuk N Lnrk 268 D5
Craignish Castle Argyll 275 C8
Craignure Argyll 289 F9
Craigo Angus 293 G8
Craigow Perth 286 G4
Craigrory Highld 300 E6
Craigrothie Fife 287 F7
Craigroy Moray 301 D11
Craigruie Stirl 285 E8
Craig's End Essex 106 D4
Craigsford Mains Borders 262 B3
Craigshall Dumfries 237 D10
Craigshill W Loth 269 B11
Craigside Durham 233 D8
Craigston Castle Aberds 303 D7
Craigton Aberdeen 293 C10
Craigton Angus 287 B7
Craigton Angus 287 D9
Craigton Glasgow 267 C10
Craigton Highld 300 E6
Craigton Highld 309 H6
Craigton Highld 309 K6
Craigtown Highld 310 D2
Craik Borders 249 B8
Crail Fife 287 G10
Crailing Borders 262 E5
Crailinghall Borders 262 E5
Crakaig Highld 311 H3
Crakehill N Yorks 215 E8
Crakemarsh Staffs 151 B11
Crambe N Yorks 216 G4
Crambeck N Yorks 216 F4
Cramhurst Sur 50 E2
Cramlington Northumb 243 B7
Cramond Edin 280 F3
Cramond Bridge Edin 280 F3
Crampmoor Hants 32 C5
Cranage E Ches 168 B3
Cranberry Staffs 150 B6
Cranborne Dorset 31 E9
Cranbourne Brack 66 E2
Cranbourne Hants 48 C6
Cranbrook Kent 53 F9
Cranbrook London 68 B2
Cranbrook Common Kent 53 F9
Crane Moor S Yorks 197 G10
Crane's Corner Norf 159 G8
Cranfield C Beds 103 C9
Cranford Devon 24 C4
Cranford London 66 D6
Cranford St Andrew Northants 121 B8
Cranford St John Northants 121 B8
Cranham Glos 80 C5
Cranham London 68 B5
Cranhill Glasgow 268 B2
Cranhill Warks 118 G2
Crank Mers 183 B8
Crank Wood Gtr Man 194 G6
Crankwood Gtr Man 194 G6
Cranleigh Sur 50 F5
Cranley Suff 126 C3
Cranley Gardens London 67 B9
Cranmer Green Suff 125 C10
Cranmore I o W 20 D3
Cranmore Som 45 E7
Cranna Aberds 302 D6
Crannich Argyll 289 E7
Crannoch Moray 302 D4
Cranoe Leics 136 D5
Cransford Suff 126 E6
Cranshaws Borders 272 C3
Cranstal I o M 192 B5
Cranswick E Yorks 208 C6
Crantock Corn 4 C5
Cranwell Lincs 173 F8
Cranwich Norf 140 E5
Cranworth Norf 141 C9
Craobh Haven Argyll 275 C8
Crapstone Devon 7 B10
Crarae Argyll 275 D10
Crask Highld 308 C7
Crask Inn Highld 309 G5
Crask of Aigas Highld 300 E4
Craskins Aberds 293 C7
Craster Northumb 265 F7
Craswall Hereford 96 D5
Crateford Shrops 132 F4
Crateford Staffs 133 B8
Cratfield Suff 126 B6
Crathes Aberds 293 D9
Crathie Aberds 292 D4
Crathie Highld 291 D7
Crathorne N Yorks 225 D8
Craven Arms Shrops 131 G8
Crawcrook T & W 242 E4
Crawford Lancs 194 G3
Crawford S Lnrk 259 E11
Crawforddyke S Lnrk 269 F7
Crawfordjohn S Lnrk 259 E9
Crawick Dumfries 259 G7
Crawley Devon 28 F3
Crawley Hants 48 G2
Crawley Oxon 82 C4
Crawley W Sus 51 F9
Crawley Down W Sus 51 F10
Crawley End Essex 105 C8
Crawley Hill Sur 65 G11
Crawleyside Durham 232 C5
Crawshaw W Yorks 197 E8
Crawshawbooth Lancs 195 B10
Crawton Aberds 293 F10
Cray N Yorks 213 D8
Cray Perth 292 G3
Crayford London 68 E4
Crayke N Yorks 215 E11
Craymere Beck Norf 159 C11
Crays Hill Essex 88 G2
Cray's Pond Oxon 64 C6
Crazies Hill Wokingham 65 C9
Creacombe Devon 26 D4
Creag Aoil Highld 290 F3
Creag Ghoraidh W Isles 297 G3
Creagan Argyll 289 E11
Creagan Sithe Argyll 284 G6
Creagastrom W Isles 297 G4
Creaguaineach Lodge Highld 290 G5
Creaksea Essex 88 F6
Creamore Bank Shrops 149 C10
Crean Corn 1 E3
Creaton Northants 120 C4
Creca Dumfries 238 C6
Credenhill Hereford 97 C9
Crediton Devon 26 G4
Creebridge Dumfries 236 C6
Creech Dorset 18 E4
Creech Bottom Dorset 18 E4
Creech Heathfield Som 28 B3
Creech St Michael Som 28 B3
Creed Corn 5 F8
Creediknowe Shetland 312 G7
Creegbrawse Corn 4 G4
Creekmoor Poole 18 C6
Creekmouth London 68 C3
Creeksea Essex 88 F6
Creeting Bottoms Suff 126 F2
Creeting St Mary Suff 125 F11
Creeton Lincs 155 E10
Creetown Dumfries 236 D6
Creg-ny-Baa I o M 192 D4
Creggans Argyll 284 G4
Cregneash I o M 192 F2
Cregrina Powys 114 G2
Creich Fife 287 E7
Creigau Mon 79 F7
Creighton Staffs 151 B11
Creigiau Cardiff 58 C5
Crelly Corn 2 C5
Cremyll Corn 7 E9
Crendell Dorset 31 E9
Crepkill Highld 298 E4
Creslow Bucks 102 G6
Cress Green Glos 80 E3
Cressage Shrops 131 C11
Cressbrook Derbys 185 G11
Cresselly Pembs 73 D9
Cressex Bucks 84 G4
Cressing Essex 106 G5
Cresswell Northumb 253 E7
Cresswell Staffs 151 B9
Cresswell Quay Pembs 73 D9
Creswell Derbys 187 G8
Creswell Staffs 151 D7
Creswell Staffs 151 D8
Creswell Green Staffs 151 G11
Cretingham Suff 126 E4
Cretshengan Argyll 275 G8
Creunant = Crynant Neath 76 E3
Crewe E Ches 168 D2
Crewe W Ches 166 E6
Crewe-by-Farndon W Ches 166 E6
Crewgarth Cumb 231 E8
Crewgreen Powys 148 F6
Crewkerne Som 28 F6
Crews Hill London 86 F4
Crew's Hole Bristol 60 E6
Crewton Derby 153 C7
Crianlarich Stirl 285 E7
Cribbs Causeway S Glos 60 C5
Cribden Side Lancs 195 C9
Cribyn Ceredig 111 G10
Criccieth Gwyn 145 B9
Crich Derbys 170 E5
Crich Carr Derbys 170 E4
Crichie Aberds 303 E9
Crichton Midloth 271 C7
Crick Mon 79 G7
Crick Northants 119 C11
Crickadarn Powys 95 C11
Cricket Hill Hants 65 G10
Cricket Malherbie Som 28 E5
Cricket St Thomas Som 28 F5
Crickham Som 44 D2
Crickheath Shrops 148 E5
Crickheath Wharf Shrops 148 E5
Crickhowell Powys 78 B2
Cricklade Wilts 81 G10
Cricklewood London 67 B8
Crickmery Shrops 150 D3
Crick's Green Hereford 116 G2
Criddlestyle Hants 31 E11
Cridling Stubbs N Yorks 198 C4
Cridmore I o W 20 E5
Crieff Perth 286 E2
Criggan Corn 5 C10
Criggion Powys 148 F5
Crigglestone W Yorks 197 D10
Crimble Gtr Man 195 E11
Crimchard Som 28 F4
Crimdon Park Durham 234 D5
Crimond Aberds 303 D10
Crimonmogate Aberds 303 D10
Crimp Corn 24 D3
Crimplesham Norf 140 C3
Crimscote Warks 100 B4
Crinan Argyll 275 D8
Crinan Ferry Argyll 275 D8
Crindau Newport 59 B10
Crindledyke N Lnrk 268 D6
Cringleford Norf 142 B3
Cringles W Yorks 204 D6
Cringletie Borders 270 G4
Crinow Pembs 73 C10
Cripple Corner Essex 107 E7
Cripplesease Corn 2 B2
Cripplestyle Dorset 31 E9
Cripp's Corner E Sus 38 C3
Crispie Argyll 275 F10
Crist Derbys 185 E8
Crit Hall Kent 53 G9
Critchell's Green Hants 32 B3
Critchill Som 45 D9
Critchmere Sur 49 G11
Crizeley Hereford 97 E8
Croanford Corn 10 G6
Croasdale Cumb 219 B11
Crobeag W Isles 304 F5
Crock Street Som 28 E4
Crockenhill Kent 68 F4
Crocker End Oxon 65 B8
Crockerhill Hants 33 F9
Crockerhill W Sus 22 B6
Crockernwell Devon 13 C11
Crockers Devon 40 F5
Crocker's Ash Hereford 79 B8
Crockerton Wilts 45 E11
Crockerton Green Wilts 45 E11
Crocketford or Ninemile Bar Dumfries 237 B10
Crockey Hill York 207 D8
Crockham Heath W Berks 64 G2
Crockham Hill Kent 52 C2
Crockhurst Street Kent 52 E6
Crockleford Heath Essex 107 F10
Crockleford Hill Essex 107 F10
Crockness Orkney 314 G3
Croes-goch Pembs 87 E11
Croes-Hywel Mon 78 C4
Croes-lan Ceredig 93 C7
Croes Llanfair Mon 78 D4
Croes-wian Flint 181 G10
Croes-y-mwyalch Torf 78 G4
Croes y pant Mon 78 E4
Croesau Bach Shrops 148 D4
Croeserw Neath 57 B11
Croesor Gwyn 163 G10
Croespenmaen Caerph 77 F11
Croesyceiliog Carms 74 B6
Croesyceiliog Torf 78 F4
Croesywaun Gwyn 163 D8
Croft Hereford 115 D9
Croft Leics 135 D10
Croft Lincs 175 C8
Croft Pembs 92 C3
Croft Warr 183 C10
Croft Mitchell Corn 2 B5
Croft of Tillymaud Aberds 303 F11
Croft-on-Tees N Yorks 224 D5
Croftamie Stirl 277 D9
Croftfoot S Lnrk 268 C2
Crofthandy Corn 4 G4
Croftlands Cumb 210 D5
Croftmalloch W Loth 269 C8
Croftmoraig Perth 285 C11
Crofton Cumb 239 G8
Crofton London 68 F2
Crofton W Yorks 197 D11
Crofton Wilts 63 G9
Crofts Dumfries 237 B9
Crofts Bank Gtr Man 184 B3
Crofts of Benachielt Highld 310 F5
Crofts of Haddo Aberds 303 F8
Crofts of Inverthernie Aberds 303 E7
Crofts of Meikle Ardo Aberds 303 E8
Crofty Swansea 56 B4
Croggan Argyll 289 G9
Croglin Cumb 231 B7
Croich Highld 309 K4
Croick Highld 310 D2
Croig Argyll 288 D5
Crois Dughaill W Isles 297 J3
Cromarty Highld 301 C7
Cromasaig Highld 299 C10
Crombie Fife 279 D10
Crombie Castle Aberds 302 D5
Cromblet Aberds 303 F7
Cromdale Highld 301 G10
Cromer Herts 104 F5
Cromer Norf 160 A4
Cromer-Hyde Herts 86 C2
Cromford Derbys 170 D3
Cromhall S Glos 79 G11
Cromhall Common S Glos 61 B7
Cromor W Isles 304 F6
Crompton Fold Gtr Man 196 F2
Cromra Highld 291 D7
Cromwell Notts 172 C3
Cromwell Bottom W Yorks 196 C6
Cronberry E Ayrs 258 E4
Crondall Hants 49 D9
Cronk-y-Voddy I o M 192 D4
Cronton Mers 183 D7
Crook Cumb 221 G9
Crook Devon 27 G11
Crook Durham 233 D9
Crook of Devon Perth 286 G4
Crookdake Cumb 229 C9
Crooke Gtr Man 194 F5
Crooked Billet London 67 E8
Crooked Soley Wilts 63 E10
Crooked Withies Dorset 31 F9
Crookedholm E Ayrs 257 B11
Crookes S Yorks 186 D4
Crookesmoor S Yorks 186 D4
Crookfur E Renf 267 D10
Crookgate Bank Durham 242 F5
Crookhall Durham 242 G4
Crookham Northumb 263 B10
Crookham W Berks 64 G4
Crookham Village Hants 49 C9
Crookhaugh Borders 260 D4
Crookhill T & W 242 E5
Crookhouse Borders 263 D7
Crooklands Cumb 211 C10
Crookston Glasgow 267 C10
Cropredy Oxon 101 B9
Cropston Leics 153 G11
Cropthorne Worcs 99 C9
Cropton N Yorks 216 B5
Cropwell Bishop Notts 154 B3
Cropwell Butler Notts 154 B3
Cros W Isles 304 B7
Crosben Highld 289 D9
Crosbost W Isles 304 F5
Crosby Cumb 229 D7
Crosby I o M 192 E4
Crosby Mers 182 B4
Crosby N Lincs 199 E11
Crosby Court N Yorks 225 G7
Crosby Garrett Cumb 222 D4
Crosby-on-Eden Cumb 239 E11
Crosby Ravensworth Cumb 222 C2
Crosby Villa Cumb 229 D7
Croscombe Som 44 E5
Crosemere Shrops 149 D8
Crosland Edge W Yorks 196 E6
Crosland Hill W Yorks 196 E6
Crosland Moor W Yorks 196 D6
Croslands Park Cumb 210 E4
Cross Devon 40 F3
Cross Devon 40 G6
Cross Shrops 149 B7
Cross Som 44 C2
Cross Ash Mon 78 B6
Cross-at-Hand Kent 53 D9
Cross Bank Worcs 116 C4
Cross Coombe Corn 4 E4
Cross End Beds 121 F11
Cross End Essex 107 E7
Cross End M Keynes 103 D8
Cross Gate W Sus 35 E8
Cross Gates W Yorks 205 F7
Cross Gates W Yorks 206 G3
Cross Green Devon 12 D3
Cross Green Staffs 133 B8
Cross Green Suff 124 G6
Cross Green Suff 125 F7
Cross Green Suff 125 G9
Cross Green Telford 150 G2
Cross Green W Yorks 206 G2
Cross Green Warks 119 F7
Cross Hands Carms 75 C9
Cross-hands Carms 92 G3
Cross Hands Pembs 73 C9
Cross Heath Staffs 168 F4
Cross Hill Corn 10 G6
Cross Hill Derbys 170 F6
Cross Hill Glos 79 F9
Cross Hills N Yorks 204 E6
Cross Holme N Yorks 225 F11
Cross Houses Shrops 131 B10
Cross Houses Shrops 132 E3
Cross in Hand E Sus 37 C9
Cross in Hand Leics 135 G10
Cross Inn Carms 74 C3
Cross Inn Ceredig 111 E10
Cross Inn Ceredig 111 F7
Cross Inn Rhondda 58 C5
Cross Keys Kent 52 C4
Cross Keys Wilts 61 E11
Cross Lane E Ches 167 C11
Cross Lane Head Shrops 132 D4
Cross Lanes Corn 2 E5
Cross Lanes Dorset 30 G3
Cross Lanes N Yorks 215 F10
Cross Lanes Oxon 65 D7

Cross Lanes Wrex 166 F5
Cross Llyde Hereford 97 F8
Cross o' th' hands Derbys 170 F3
Cross o' th' Hill W Ches 167 F7
Cross Oak Powys 96 G2
Cross of Jackston Aberds 303 F7
Cross Roads Devon 12 D5
Cross Roads W Yorks 204 F6
Cross Stone Aberds 303 G9
Cross Street Suff 126 B3
Cross Town E Ches 184 F3
Crossaig Argyll 255 B9
Crossal Highld 294 B6
Crossapol Argyll 288 E1
Crossbrae Aberds 302 D6
Crossburn Falk 279 G7
Crossbush W Sus 35 F8
Crosscanonby Cumb 229 D7
Crossdale Street Norf 160 B4
Crossens Mers 193 D11
Crossflatts W Yorks 205 E8
Crossford Fife 279 D11
Crossford S Lnrk 268 F6
Crossgate Lincs 156 D4
Crossgate Orkney 314 E4
Crossgate Staffs 151 B8
Crossgatehall E Loth 271 B7
Crossgates Cumb 229 G7
Crossgates Fife 280 D2
Crossgates N Yorks 217 C10
Crossgates Powys 113 E11
Crossgill Cumb 231 C10
Crossgill Lancs 211 G11
Crossgreen Shrops 149 F9
Crosshands Carms 92 G3
Crosshill E Ayrs 257 E11
Crosshill Fife 280 B3
Crosshill S Ayrs 245 B8
Crosshouse E Ayrs 257 B9
Crossings Cumb 240 B2
Crosskeys Caerph 78 G2
Crosskirk Highld 310 B4
Crosslands Cumb 210 B6
Crosslanes Shrops 148 F6
Crosslee Borders 261 F8
Crosslee Renfs 267 B8
Crossley Hall W Yorks 205 G8
Crossmichael Dumfries 237 C9
Crossmill E Renf 267 D10
Crossmoor Lancs 202 F4
Crossmount Perth 285 B11
Crosspost W Sus 36 C3
Crossroads Aberds 293 D9
Crossroads E Ayrs 257 C11
Crossroads Fife 281 B7
Crosston Angus 287 B9
Crosstown Corn 24 D2
Crosstown V Glam 58 F4
Crosswater Sur 49 F11
Crossway Hereford 98 E2
Crossway Mon 78 B6
Crossway Powys 113 F11
Crossway Green Mon 79 G8
Crossway Green Worcs 116 D6
Crossways Dorset 17 D11
Crossways Kent 68 D5
Crossways Mon 96 G6
Crossways S Glos 79 G11
Crossways Sur 49 F11
Crosswell = Ffynnongroes Pembs 92 D2
Crosswood Ceredig 112 C3
Crosthwaite Cumb 221 G8
Croston Lancs 194 D3
Crostwick Norf 160 F5
Crostwight Norf 160 D6
Crothair W Isles 304 E3
Crouch Kent 52 B6
Crouch Kent 54 B5
Crouch End London 67 B9
Crouch Hill Dorset 30 E2
Crouch House Green Kent 52 D2
Crouchers W Sus 22 C4
Croucheston Wilts 31 B9
Croughly Moray 301 G11
Croughton Northants 101 E10
Crovie Aberds 303 C8
Crow Hants 31 G11
Crow Edge S Yorks 197 G7
Crow Green Essex 87 F9
Crow Hill Hereford 98 F2
Crow Nest W Yorks 205 F8
Crow Wood Halton 183 D8
Crowan Corn 2 C4
Crowborough E Sus 52 G4
Crowborough Staffs 168 D6
Crowborough Warren E Sus 52 G4
Crowcombe Som 42 F6
Crowcroft Worcs 116 G5
Crowden Derbys 185 B9
Crowden Devon 12 B5
Crowder Park Devon 8 D4
Crowdhill Hants 33 C7
Crowdicote Derbys 169 B10
Crowdleham Kent 52 B5
Crowdon N Yorks 227 F9
Crowell Oxon 84 F2
Crowell Hill Oxon 84 F3
Crowfield Northants 102 C2
Crowfield Suff 126 F2
Crowgate Street Norf 160 E6
Crowgreaves Shrops 132 D4
Crowhill E Loth 282 F4
Crowhill Gtr Man 184 B6
Crowhill M Keynes 102 D6
Crowhole Derbys 186 F4
Crowhurst E Sus 38 E3
Crowhurst Sur 51 D11
Crowhurst Lane End Sur 51 D11
Crowland Lincs 156 G4
Crowlas Corn 2 C2
Crowle N Lincs 199 E9
Crowle Worcs 117 F8
Crowle Green Worcs 117 F8
Crowle Hill N Lincs 199 E9
Crowle Park N Lincs 199 E9
Crowmarsh Gifford Oxon 64 B6
Crown Corner Suff 126 C5
Crown East Worcs 116 G6
Crown Hills Leicester 136 C2
Crown Wood Brack 65 F11
Crownfield Bucks 84 F4
Crownhill Plym 7 D9
Crownland Suff 125 D10
Crownpits Sur 50 E3
Crownthorpe Norf 141 C11
Crowntown Corn 2 C4
Crows-an-wra Corn 1 D3
Crow's Green Essex 106 F3
Crow's Nest Corn 6 B5
Crowshill Norf 141 B8
Crowsley Oxon 65 D8
Crowsnest Shrops 131 C7
Crowther's Pool Powys 96 B4
Crowthorne Brack 65 G10
Crowton W Ches 183 G9
Croxall Staffs 152 G3
Croxby Lincs 189 B11
Croxby Top Lincs 189 B11
Croxdale Durham 233 D11
Croxden Staffs 151 B11
Croxley Green Herts 85 F9
Croxteth Mers 182 B6
Croxton Cambs 122 E4
Croxton N Lincs 200 E5
Croxton Norf 141 F7
Croxton Norf 159 C9
Croxton Staffs 150 C5
Croxton Green E Ches 167 E8
Croxton Kerrial Leics 154 D6
Croxtonbank Staffs 150 C5
Croy Highld 301 E7
Croy N Lnrk 278 F4
Croyde Devon 40 F2
Croyde Bay Devon 40 F2
Croydon Cambs 104 B6
Croydon London 67 F10
Crozen Hereford 97 B11
Crubenbeg Highld 291 D8
Crubenmore Lodge Highld 291 D8
Cruckmeole Shrops 131 B8
Cruckton Shrops 149 G8
Cruden Bay Aberds 303 F10
Crudgington Telford 150 F2
Crudie Aberds 303 D7
Crudwell Wilts 81 G7
Crug Powys 114 C3
Crugmeer Corn 10 F4
Crugybar Carms 94 D3
Cruise Hill Worcs 117 E10
Crulabhig W Isles 304 E3
Crumlin Caerph 78 F2
Crumplehorn Corn 6 E4
Crumpsall Gtr Man 195 G10
Crumpsbrook Shrops 116 B2
Crumpton Hill Worcs 98 B5
Crundale Kent 54 D5
Crundale Pembs 73 B7
Crurys Morchard Devon 26 E5
Crux Easton Hants 48 B2
Cruxton Dorset 17 B8
Crwbin Carms 75 C7
Crya Orkney 314 F3
Cryers Hill Bucks 84 F5
Crymlyn Gwyn 179 G10
Crymych Pembs 92 E3
Crynant = Creunant Neath 76 E3
Crynfryn Ceredig 111 E11
Cuaich Highld 291 E8
Cuaig Highld 299 D7
Cuan Argyll 275 B8
Cubbington Warks 118 D6
Cubeck N Yorks 213 B9
Cubert Corn 4 D5
Cubitt Town London 67 D11
Cubley S Yorks 197 G8
Cubley Common Derbys 152 B3
Cublington Bucks 102 G6
Cublington Hereford 97 D8
Cuckfield W Sus 36 B4
Cucklington Som 30 B3
Cuckney Notts 187 G9
Cuckold's Green Suff 143 G9
Cuckold's Green Wilts 46 B3
Cuckoo Green Suff 143 D10
Cuckoo Hill Notts 188 C2
Cuckoo Tye Suff 107 C7
Cuckoo's Corner Hants 49 E8
Cuckoo's Corner Wilts 46 B4
Cuckoo's Knob Wilts 63 G7
Cuckron Shetland 313 H6
Cucumber Corner Norf 143 B7
Cuddesdon Oxon 83 E10
Cuddington Bucks 84 C2
Cuddington W Ches 183 G10
Cuddington Heath W Ches 167 F7
Cuddy Hill Lancs 202 F5
Cudham London 52 B2
Cudliptown Devon 12 F6
Cudworth Som 28 E5
Cudworth S Yorks 197 F11
Cudworth Som 28 E5
Cudworth Sur 51 E8
Cudworth Common S Yorks 197 F11
Cuerden Green Lancs 194 C5
Cuerdley Cross Warr 183 D8
Cufaude Hants 48 B6
Cuffern Pembs 91 G7
Cuffley Herts 86 E4
Cuiashader W Isles 304 C7
Cuidhir W Isles 297 L2
Cuidhtinis W Isles 296 C6
Cuiken Midloth 270 C4
Cuilcheanna Ho Highld 290 G2
Cuin Argyll 288 D6
Cul na h-Aird W Isles 305 H3
Cùl Doirlinn Highld 289 B8
Culbo Highld 300 C6
Culbokie Highld 300 D6
Culburnie Highld 300 E4
Culcabock Highld 300 E6
Culcairn Highld 300 C6
Culcharry Highld 301 D8
Culcheth Warr 183 B11
Culcronchie Dumfries 237 C7
Culdrain Aberds 302 F5
Culduie Highld 299 E7
Culeave Highld 309 K5
Culford Suff 124 D6
Culfordheath Suff 125 C7
Culfosie Aberds 293 C9
Culgaith Cumb 231 F8
Culham Oxon 83 F8
Culkein Highld 306 F5
Culkein Drumbeg Highld 306 F6
Culkerton Glos 80 F6
Cullachie Highld 301 G9
Cullen Moray 302 C5
Cullercoats T & W 243 C9
Cullicudden Highld 300 C6
Cullingworth W Yorks 205 F7
Cullipool Argyll 275 B8
Cullivoe Shetland 312 C7
Culloch Perth 285 F11
Culloden Highld 301 E7
Cullompton Devon 27 F8
Culm Davy Devon 27 D10
Culmaily Highld 311 K2
Culmazie Dumfries 236 D5
Culmer Sur 50 F2
Culmers Kent 70 G5
Culmington Shrops 131 G9
Culmore Stirl 278 B3
Culmstock Devon 27 E10
Culnacraig Highld 307 J5
Culnaightrie Dumfries 237 D9
Culnaknock Highld 298 C5
Culnaneam Highld 294 C6
Culpho Suff 108 B4
Culra Lodge Highld 291 F7
Culrain Highld 309 K5
Culross Fife 279 D9
Culroy S Ayrs 257 G8
Culscadden Dumfries 236 E6
Culsh Aberds 292 C5
Culsh Aberds 303 E8
Culshabbin Dumfries 236 D5
Culswick Shetland 313 J4
Cultercullen Aberds 303 G9
Cults Aberdeen 293 C10
Cults Aberds 302 F5
Cults Dumfries 236 E6
Cults Fife 287 G7
Culverlane Devon 8 C4
Culverstone Green Kent 68 G6
Culverthorpe Lincs 173 G8
Culworth Northants 101 B10
Culzie Lodge Highld 300 B5
Cumberlow Green Herts 104 E6
Cumbernauld N Lnrk 278 G5
Cumbernauld Village N Lnrk 278 F5
Cumber's Bank Wrex 149 B8
Cumberworth Lincs 191 G8
Cumdivock Cumb 230 B2
Cumeragh Village Lancs 203 F7
Cuminestown Aberds 303 D8
Cumledge Borders 272 D5
Cumlewick Shetland 313 L6
Cumlodden Argyll 275 D11
Cumloden Dumfries 236 C6
Cummersdale Cumb 239 G9
Cummerton Aberds 303 C8
Cummertrees Dumfries 238 D4
Cummingston Moray 301 C11
Cumnock E Ayrs 258 E3
Cumnor Oxon 83 E7
Cumnor Hill Oxon 83 D7
Cumrew Cumb 240 G2
Cumwhinton Cumb 239 G10
Cumwhitton Cumb 240 G2
Cundall N Yorks 215 E8
Cundy Cross S Yorks 197 F11
Cundy Hos S Yorks 186 B4
Cunninghamhead N Ayrs 257 G[illegible]
Cunnister Shetland 312 D[illegible]
Cupar Fife 287 F[illegible]
Cupar Muir Fife 287 F[illegible]
Cupernham Hants 32 C[illegible]
Cupid Green Herts 85 D[illegible]
Cupid's Hill Mon 97 F[illegible]
Curbar Derbys 186 G[illegible]
Curborough Staffs 152 G[illegible]
Curbridge Hants 33 E[illegible]
Curbridge Oxon 82 D[illegible]
Curdridge Hants 33 E[illegible]
Curdworth Warks 134 E[illegible]
Curgurrell Corn 3 B[illegible]
Curin Highld 300 D[illegible]
Curland Som 28 D[illegible]
Curland Common Som 28 D[illegible]
Curlew Green Suff 127 D[illegible]
Curling Tye Green Essex 88 D[illegible]
Curload Som 28 B[illegible]
Currarie S Ayrs 244 E[illegible]
Currian Vale Corn 5 D[illegible]
Curridge W Berks 64 E[illegible]
Currie Edin 270 B[illegible]
Currock Cumb 239 G10
Curry Lane Corn 11 C1[illegible]
Curry Mallet Som 28 C[illegible]
Curry Rivel Som 28 B[illegible]
Cursiter Orkney 314 E[illegible]
Curteis' Corner Kent 53 F1[illegible]
Curtisden Green Kent 53 E[illegible]
Curtisknowle Devon 8 E[illegible]
Curtismill Green Essex 87 F[illegible]
Cury Corn 2 E[illegible]
Cusbay Orkney 314 C[illegible]
Cusgarne Corn 4 G[illegible]
Cushnie Aberds 303 C7
Cushuish Som 43 G7
Cusop Hereford 96 C4
Custards Hants 32 F3
Custom House London 68 C2
Cusveorth Coombe Corn 4 G5
Cusworth S Yorks 198 G4
Cutcloy Dumfries 236 F6
Cutcombe Som 42 F2
Cutgate Gtr Man 195 E11
Cuthill E Loth 281 G7
Cutiau Gwyn 146 F2
Cutlers Green Essex 105 E11
Cutler's Green Som 44 C5
Cutmadoc Corn 5 C11
Cutmere Corn 6 C6
Cutnall Green Worcs 117 D7
Cutsdean Glos 99 E11
Cutsyke W Yorks 198 C2
Cutteslowe Oxon 83 C8
Cutthorpe Derbys 186 G4
Cuttiford's Door Som 28 E4
Cutts Shetland 313 K6
Cuttybridge Pembs 72 B6
Cuttyhill Aberds 303 D10
Cuxham Oxon 83 F11
Cuxton Medway 69 F8
Cuxwold Lincs 201 G7
Cwm Bl Gwent 77 D11
Cwm Denb 181 F9
Cwm Neath 57 C10
Cwm Powys 129 D11
Cwm Powys 130 E5
Cwm Shrops 114 B6
Cwm Swansea 57 B7
Cwm-byr Carms 94 E2
Cwm Capel Carms 75 E7
Cwm-celyn Bl Gwent 78 D2
Cwm-Cewydd Gwyn 147 G7
Cwm-cou Ceredig 92 C5
Cwm Dows Caerph 78 F2
Cwm-Dulais Swansea 75 E10
Cwm felin fach Caerph 77 G11
Cwm Ffrwd-oer Torf 78 E3
Cwm-Fields Torf 78 E3
Cwm Gelli Caerph 77 F11
Cwm Gwyn Swansea 56 C6
Cwm Head Shrops 131 F8
Cwm-hesgen Gwyn 146 D5
Cwm-hwnt Rhondda 76 D6
Cwm Irfon Powys 95 B7
Cwm-Llinau Powys 128 B6
Cwm-mawr Carms 75 C8
Cwm-miles Carms 92 G3
Cwm Nant-gam Bl Gwent 78 C2
Cwm-parc Rhondda 77 F7
Cwm Penmachno Conwy 164 F3
Cwm Plysgog Ceredig 92 C3
Cwm-twrch Isaf Powys 76 C3
Cwm-twrch Uchaf Powys 76 C3
Cwm-y-glo Carms 75 C9
Cwm-y-glo Gwyn 163 C8
Cwmafan Neath 57 C9
Cwmaman Rhondda 77 F8
Cwmann Carms 93 B11
Cwmavon Torf 78 D3
Cwmbâch Rhondda 77 E8
Cwmbach Carms 75 E7
Cwmbach Carms 92 F5
Cwmbach Powys 96 D3
Cwmbach Rhondda 77 E8
Cwmbach Llechrhyd Powys 113 G10
Cwmbelan Powys 129 G8
Cwmbran Torf 78 G3
Cwmbrwyno Ceredig 128 G4
Cwmcarn Caerph 78 G2
Cwmcarvan Mon 79 D7
Cwmcoednerth Ceredig 92 B6
Cwmcrawnon Powys 77 B10
Cwmcych Carms 92 D5
Cwmdare Rhondda 77 E7
Cwmdu Carms 94 E2
Cwmdu Powys 96 G3
Cwmdu Swansea 56 C6
Cwmduad Carms 93 E7
Cwmdwr Carms 94 E4
Cwmerfyn Ceredig 128 G3
Cwmfelin Bridgend 57 D11
Cwmfelin M Tydf 77 E9
Cwmfelin Boeth Carms 73 B11
Cwmfelin Mynach Carms 92 G4
Cwmffrwd Carms 74 B6
Cwmgiedd Powys 76 C3
Cwmgors Neath 76 C2
Cwmgwili Carms 75 C9
Cwmgwrach Neath 76 E5
Cwmhiraeth Carms 92 D6
Cwmifor Carms 94 F3
Cwmisfael Carms 75 B7
Cwmllynfell Neath 76 C2
Cwmnantyrodyn Caerph 77 F11
Cwmorgan Pembs 92 E5
Cwmparc Rhondda 77 F7
Cwmpengraig Carms 92 D6
Cwmpennar Rhondda 77 E8
Cwmrhos Powys 96 G3
Cwmrhydyceirw Swansea 57 B7
Cwmsychpant Ceredig 93 B9
Cwmsyfiog Caerph 77 E11
Cwmsymlog Ceredig 128 G4
Cwmtillery Bl Gwent 78 D2
Cwmwdig Water Pembs 90 E6
Cwmwysg Powys 95 F7
Cwmyoy Mon 96 G5
Cwmystwyth Ceredig 112 C5
Cwrt Gwyn 128 C3
Cwrt-newydd Ceredig 93 B9
Cwrt-y-cadno Carms 94 C3
Cwrt-y-gollen Powys 78 B2
Cydweli = Kidwelly Carms 74 D6
Cyffordd Llandudno = Llandudno Junction Conwy 180 F3
Cyffylliog Denb 165 D9
Cyfronydd Powys 130 B2
Cymau Flint 166 D3
Cymdda Bridgend 58 C2
Cymer Neath 57 B11
Cymmer Rhondda 77 G8
Cyncoed Cardiff 59 C7
Cynghordy Carms 94 C6
Cynheidre Carms 75 D7
Cynonville Neath 57 B10
Cyntwell Cardiff 58 D6
Cynwyd Denb 165 G9
Cynwyl Elfed Carms 93 F7
Cywarch Gwyn 147 F7

D

Daccombe Devon 9 B8
Dacre Cumb 230 F5
Dacre N Yorks 214 G3
Dacre Banks N Yorks 214 G3
Daddry Shield Durham 232 D3
Dadford Bucks 102 D3
Dadlington Leics 135 D8
Dafarn Faig Gwyn 163 F7
Dafen Carms 75 E8
Daffy Green Norf 141 B9
Dagdale Staffs 151 C11
Dagenham London 68 C3
Daggons Dorset 31 E10
Daglingworth Glos 81 D7
Dagnall Bucks 85 B7
Dagtail End Worcs 117 E10
Dagworth Suff 125 E10
Dail Beag W Isles 304 D4
Dail bho Dheas W Isles 304 B6
Dail bho Thuath W Isles 304 B6
Dail Mor W Isles 304 D4
Daill Argyll 274 G4
Dailly S Ayrs 245 C7
Dainton Devon 9 B7
Dairsie or Osnaburgh Fife 287 F8
Daisy Green Suff 125 D10
Daisy Green Suff 125 D11
Daisy Hill Gtr Man 195 G7
Daisy Hill W Yorks 197 B9
Daisy Hill W Yorks 205 G8
Daisy Nook Gtr Man 196 G2
Dalabrog W Isles 297 J3
Dalavich Argyll 275 B10
Dalbeattie Dumfries 237 C10
Dalbeg Highld 291 B8
Dalblair E Ayrs 258 F4
Dalbog Angus 293 F7
Dalbrack Stirl 285 G11
Dalbury Derbys 152 C5
Dalby I o M 192 E3
Dalby Lincs 190 G6
Dalby N Yorks 216 E2
Dalchalloch Perth 291 G9
Dalchalm Highld 311 J3
Dalchenna Argyll 284 G4
Dalchirach Moray 301 F11
Dalchonzie Perth 285 E11
Dalchork Highld 309 H5
Dalchreichart Highld 290 B4
Dalchruin Perth 285 F11
Dalderby Lincs 174 B2
Dale Cumb 230 C6
Dale Gtr Man 196 F3
Dale Pembs 72 D4
Dale Shetland 312 G6
Dale Abbey Derbys 153 B8
Dale Bottom Cumb 229 G11
Dale Brow E Ches 184 F6
Dale End Derbys 170 C2
Dale End N Yorks 204 D5
Dale Head Cumb 221 B8
Dale Hill E Sus 53 G7
Dale Hill E Sus 53 G8
Dale Moor Derbys 153 B8
Dale of Walls Shetland 313 H3
Dalebank Derbys 170 C5
Dalelia Highld 289 C9
Dales Brow Gtr Man 195 G9
Dales Green Staffs 168 D5
Daless Highld 301 F8
Dalestie Moray 292 B3
Dalestorth Notts 171 C8
Dalfaber Highld 291 B11
Dalfoil Stirl 277 D11
Dalganachan Highld 310 E4
Dalgarven N Ayrs 266 F5
Dalgety Bay Fife 280 E3
Dalginross Perth 285 E11
Dalguise Perth 286 C3
Dalhalvaig Highld 310 D2
Dalham Suff 124 E4
Dalhastnie Angus 293 F7
Dalhenzean Perth 292 G3
Dalinlongart Argyll 276 E2
Dalkeith Midloth 270 B6
Dallam Warr 183 C9
Dallas Moray 301 D11
Dallas Lodge Moray 301 D11
Dallcharn Highld 308 D6
Dalleagles E Ayrs 258 G3
Dallicott Shrops 132 E5
Dallimores I o W 20 C6
Dallinghoo Suff 126 G5
Dallington E Sus 23 B11
Dallington Northants 120 E4
Dallow N Yorks 214 E3
Dalmadilly Aberds 293 B9
Dalmally Argyll 284 E5
Dalmarnock Glasgow 268 C2
Dalmarnock Perth 286 C3
Dalmary Stirl 277 B10
Dalmellington E Ayrs 245 B11
Dalmeny Edin 280 F2
Dalmigavie Highld 291 B9
Dalmigavie Lodge Highld 301 G7
Dalmilling S Ayrs 257 E9
Dalmore Highld 300 C6
Dalmuir W Dunb 277 G9
Dalnabreck Highld 289 C8
Dalnacardoch Lodge Perth 291 F9
Dalnacroich Highld 300 D3
Dalnaglar Castle Perth 292 G3
Dalnahaitnach Highld 301 G8
Dalnamein Lodge Perth 291 G9
Dalnarrow Argyll 289 F9
Dalnaspidal Lodge Perth 291 F8
Dalnavaid Perth 292 G2
Dalnavie Highld 300 B6
Dalnaw Dumfries 236 B5
Dalnawillan Lodge Highld 310 E4
Dalness Highld 284 B5
Dalnessie Highld 309 H6
Dalphaid Highld 309 H3
Dalqueich Perth 286 G4
Dalrannoch Argyll 289 E11
Dalreavoch Highld 309 J7
Dalriach Highld 301 F10
Dalrigh Stirl 285 E7
Dalry Edin 280 G4
Dalry N Ayrs 266 F5
Dalrymple E Ayrs 257 G9
Dalscote Northants 120 G3
Dalserf S Lnrk 268 E6
Dalshannon N Lnrk 278 G4
Dalston Cumb 239 G9
Dalston London 67 C10
Dalswinton Dumfries 247 F10
Dalton Cumb 211 D10
Dalton Dumfries 238 C4
Dalton Lancs 194 F3
Dalton N Yorks 215 D8
Dalton N Yorks 224 D2
Dalton Northumb 241 F10
Dalton Northumb 242 C4
Dalton S Lnrk 268 D3
Dalton S Yorks 187 C7
Dalton W Yorks 197 D7
Dalton-in-Furness Cumb 210 E4
Dalton-le-Dale Durham 234 B4
Dalton Magna S Yorks 187 C7
Dalton-on-Tees N Yorks 224 D5
Dalton Parva S Yorks 187 C7
Dalton Piercy Hrtlpl 234 E5
Dalveallan Highld 300 F6
Dalveich Stirl 285 E10
Dalvina Lo Highld 308 E6
Dalwey Telford 132 B3
Dalwhinnie Highld 291 E8
Dalwood Devon 28 G3
Dalwyne S Ayrs 245 D8
Dam Green Norf 141 F11
Dam Head W Yorks 196 B6
Dam Mill Staffs 133 C7
Dam of Quoiggs Perth 286 G2
Dam Side Lancs 202 D4
Damask Green Herts 104 F5
Damems W Yorks 204 F6
Damerham Hants 31 D10
Damery Glos 80 G2
Damgate Norf 143 B8
Damgate Norf 161 F9
Damhead Moray 301 D10
Damhead Holdings Midloth 270 B5
Damnaglaur Dumfries 236 F3
Damside Borders 270 F3
Dan Caerlan Rhondda 58 C5
Danaway Kent 69 G11
Danbury Essex 88 E3
Danbury Common Essex 88 E3
Danby N Yorks 226 D4
Danby Wiske N Yorks 224 F6
Dancers Hill Herts 86 F2
Dancing Green Hereford 98 G2
Dandaleith Moray 302 E2
Danderhall Midloth 270 B6
Dandy Corner Suff 125 D11
Dane Bank Gtr Man 184 B6
Dane End Herts 104 G6
Dane in Shaw E Ches 168 C5
Dane Street Kent 54 C5
Danebank E Ches 185 E7
Danebridge E Ches 169 B7
Danegate E Sus 52 G5
Danehill E Sus 36 B6
Danemoor Green Norf 141 B11
Danes Moss E Ches 184 G6
Danesbury Herts 86 B2
Danesfield Bucks 65 C10
Danesford Shrops 132 E4
Daneshill Hants 49 C7
Danesmoor Derbys 170 C6
Daneway Glos 80 E6
Dangerous Corner Gtr Man 195 G7
Dangerous Corner Lancs 194 E4
Daniel's Water Kent 54 E3
Danna na Cloiche Argyll 275 F7
Dannonchapel Corn 10 E6
Danskine E Loth 271 B11
Danthorpe E Yorks 209 G10
Danygraig Caerph 78 G2
Danzey Green Warks 118 D2
Dapple Heath Staffs 151 D10
Darby End W Mid 133 F9
Darby Green Hants 65 G10
Darbys Green Worcs 116 F4
Darby's Hill W Mid 133 F9
Darcy Lever Gtr Man 195 F8
Dardy Powys 78 B2
Darenth Kent 68 E5
Daresbury Halton 183 E9
Daresbury Delph Halton 183 E9
Darfield S Yorks 198 G2
Darfoulds Notts 187 F9
Dargate Kent 70 G5
Dargate Common Kent 70 G5
Darite Corn 6 B5
Darkland Moray 302 C2
Darland Wrex 166 D5
Darlaston W Mid 133 D9
Darlaston Green W Mid 133 D9
Darley N Yorks 205 B10
Darley Shrops 132 D3
Darley Abbey Derby 153 B7
Darley Bridge Derbys 170 C3
Darley Dale Derbys 170 C3
Darley Green Warks 118 C3
Darley Head N Yorks 205 B9
Darley Hillside Derbys 170 C3
Darleyford Corn 11 G11
Darleyhall Herts 104 G2
Darlingscott Warks 100 C4
Darlington Darl 224 C5
Darliston Shrops 149 C11
Darlton Notts 188 G3
Darmsden Suff 125 G11
Darn Hill Gtr Man 195 E10
Darnall S Yorks 186 D5
Darnaway Castle Moray 301 D9
Darnford Staffs 134 B2
Darnhall W Ches 167 C10
Darnhall Mains Borders 270 F4
Darnick Borders 262 C2
Darowen Powys 128 C6
Darra Aberds 303 E7
Darracott Devon 24 D2
Darracott Devon 40 F3
Darras Hall Northumb 242 C5
Darrington W Yorks 198 D3
Darrow Green Norf 142 F5
Darsham Suff 127 D8
Darshill Som 44 E6
Dartford Kent 68 E4
Dartford Crossing Kent 68 D5
Dartington Devon 8 C5
Dartmeet Devon 13 G9
Dartmouth Devon 9 E7
Dartmouth Park London 67 B9
Darton S Yorks 197 F10
Darvel E Ayrs 258 B3
Darvillshill Bucks 84 F4
Darwell Hole E Sus 23 B11
Darwen Blkburn 195 C7
Dassels Herts 105 F7
Datchet Windsor 66 D3
Datchet Common Windsor 66 D3
Datchworth Herts 86 B3
Datchworth Green Herts 86 B3
Daubhill Gtr Man 195 F8
Daugh of Kinermony Moray 302 E2
Dauntsey Wilts 62 C3
Dauntsey Lock Wilts 62 C3
Dava Moray 301 F10
Davenham W Ches 183 G11
Davenport E Ches 168 B4
Davenport Gtr Man 184 D5
Davenport Green E Ches 184 F4
Davenport Green Gtr Man 184 D4
Daventry Northants 119 E11
David Street Kent 68 G6
David's Well Powys 113 B11
Davidson's Mains Edin 280 F4
Davidston Highld 301 C7
Davidstow Corn 11 D9
Davington Dumfries 248 C6
Davington Kent 70 G4
Daviot Aberds 303 G7
Daviot Highld 301 F7
Davis's Town E Sus 23 B8
Davo Mains Aberds 293 F9
Davoch of Grange Moray 302 D4
Davyhulme Gtr Man 184 B3
Daw Cross N Yorks 205 C11
Daw End W Mid 133 C10
Dawdon Durham 234 B4
Dawesgreen Sur 51 D8
Dawker Hill N Yorks 207 F7
Dawley Telford 132 B3
Dawley Bank Telford 132 B3
Dawlish Devon 14 F5
Dawlish Warren Devon 14 F5
Dawn Conwy 180 G5
Daw's Cross Essex 107 E7
Daw's Green Som 27 C11
Daws Heath Essex 69 B10
Daw's House Corn 12 E2
Dawshill Worcs 116 G6
Dawsmere Lincs 157 C8
Day Green E Ches 168 D3
Daybrook Notts 171 F9
Dayhills Staffs 151 C9
Dayhouse Bank Worcs 117 B9
Daylesford Glos 100 F4
Daywall Shrops 148 C5
Ddol Flint 181 G10
Ddôl Cownwy Powys 147 F10
Ddrydwy Anglesey 178 G5
De Beauvoir Town London 67 C10
Deacons Hill Herts 85 F11
Deadman's Cross C Beds 104 C2
Deadman's Green Staffs 151 B10
Deadwater Hants 49 F10
Deadwater Northumb 250 D4
Deaf Hill Durham 234 D3
Deal Kent 55 C11
Deal Hall Essex 89 F8
Dean Cumb 229 F7
Dean Devon 8 C4
Dean Devon 40 D6
Dean Devon 40 E4
Dean Devon 41 D8
Dean Dorset 31 D7
Dean Edin 280 G4
Dean Hants 33 D9
Dean Hants 48 G2
Dean Lancs 195 B11
Dean Oxon 100 G6
Dean Som 45 E7
Dean Bank Durham 233 E11
Dean Court Oxon 83 D7
Dean Cross Devon 40 E4
Dean Head S Yorks 197 G9
Dean Lane Head W Yorks 205 G7
Dean Park Renfs 267 B10
Dean Prior Devon 8 C4
Dean Row E Ches 184 E5
Dean Street Kent 53 C8
Deanburnhaugh Borders 261 G9
Deane Gtr Man 195 F7
Deane Hants 48 C4
Deanend Dorset 31 D7
Deanich Lodge Highld 309 L3
Deanland Dorset 31 D7
Deanlane End W Sus 34 E2
Deans W Loth 269 B10
Deans Bottom Kent 69 G11
Dean's Green Warks 118 D2
Deans Hill Kent 69 G11
Deanscales Cumb 229 F7
Deansgreen E Ches 183 D11
Deanshanger Northants 102 D5
Deanston Stirl 285 G11
Dearham Cumb 229 D7
Dearnley Gtr Man 196 D2
Debach Suff 126 G4
Debdale Gtr Man 184 B5
Debden Essex 86 F6
Debden Essex 105 E11
Debden Cross Essex 105 E11
Debden Green Essex 86 F6
Debden Green Essex 105 E11
Debenham Suff 126 E3
Deblin's Green Worcs 98 B6
Dechmont W Loth 279 G10
Deckham T & W 243 E7
Deddington Oxon 101 E9
Dedham Essex 107 E11
Dedham Heath Essex 107 E11
Dedridge W Loth 269 B11
Dedworth Windsor 66 D2
Deebank Aberds 293 D8
Deecastle Aberds 292 D6
Deene Northants 137 E8
Deenethorpe Northants 137 E9
Deepcar S Yorks 186 B3
Deepclough Derbys 185 B8
Deepcut Sur 50 B2
Deepdale C Beds 104 B4
Deepdale Cumb 212 C4
Deepdale N Yorks 213 D7
Deepdene Sur 51 D7
Deepfields W Mid 133 E8
Deeping Gate Lincs 138 B2
Deeping St James Lincs 138 B3
Deeping St Nicholas Lincs 156 F4
Deepthwaite Cumb 211 C10
Deepweir Mon 60 B3
Deerhill Moray 302 D4
Deerhurst Glos 99 F7
Deerhurst Walton Glos 99 F7
Deerland Pembs 73 C7
Deerness Orkney 314 F5
Deer's Green Essex 105 E9
Deerstones N Yorks 205 C7
Deerton Street Kent 70 G3
Defford Worcs 99 C8
Defynnog Powys 95 F8
Deganwy Conwy 180 F3
Degar V Glam 58 D4
Degibna Corn 2 D5
Deighton N Yorks 225 E7
Deighton W Yorks 197 D7
Deighton York 207 E8
Deiniolen Gwyn 163 C9
Deishar Highld 291 B11
Delabole Corn 11 E7
Delamere W Ches 167 B9
Delfour Highld 291 C10
Delfrigs Aberds 303 G9
Dell Lodge Highld 292 B2
Dell Quay W Sus 22 C4
Delliefure Highld 301 F10
Delly End Oxon 82 C5
Delnabo Moray 292 B3
Delnadamph Aberds 292 C4
Delnamer Angus 292 G3
Delph Gtr Man 196 F3
Delves Durham 233 B8
Delvin End Essex 106 D5
Delvine Perth 286 C5
Dembleby Lincs 155 B10
Demelza Corn 5 C9
Denaby Main S Yorks 187 B7
Denbeath Fife 281 B7
Denbigh Denb 165 B9
Denbury Devon 8 B6
Denby Derbys 170 F5
Denby Bottles Derbys 170 F5
Denby Common Derbys 170 F6
Denby Dale W Yorks 197 F8
Denchworth Oxon 82 G5
Dendron Cumb 210 E4
Dene Park Kent 52 C5
Denel End C Beds 103 D10
Denend Aberds 302 F6
Deneside Durham 234 B4
Denford Northants 121 B9
Denford Staffs 169 E7
Dengie Essex 89 E7
Denham Bucks 66 B4
Denham Bucks 102 G5
Denham Suff 124 E5
Denham Suff 126 C3
Denham Corner Suff 126 C3
Denham End Suff 124 E5
Denham Green Bucks 66 B4
Denham Street Suff 126 C3
Denhead Aberds 303 D9
Denhead Fife 287 F8
Denhead of Arbilot Angus 287 C9
Denhead of Gray Dundee 287 D7
Denholm Borders 262 F3
Denholme W Yorks 205 G7
Denholme Clough W Yorks 205 G7
Denholme Edge W Yorks 205 G7
Denholme Gate W Yorks 205 G7
Denholmhill Borders 262 F3
Denio Gwyn 145 B7
Denmead Hants 33 E11
Denmore Aberdeen 293 B11
Denmoss Aberds 302 E6
Dennington Suff 126 D5
Dennington Corner Suff 126 D5
Dennington Hall Suff 126 D5
Denny Falk 278 E6
Denny Bottom Kent 52 F5
Denny End Cambs 123 D9
Denny Lodge Hants 32 F4
Dennyloanhead Falk 278 E6
Dennystown W Dunb 277 F7
Denshaw Gtr Man 196 E3
Denside Aberds 293 D10
Densole Kent 55 E8
Denston Suff 124 G5
Denstone Staffs 169 G9
Denstroude Kent 70 G6
Dent Cumb 212 B4
Dent Bank Durham 232 F4
Denton Cambs 138 F2
Denton Darl 224 B4
Denton E Sus 23 E7
Denton Gtr Man 184 B6
Denton Kent 55 D8
Denton Kent 69 E7
Denton Lincs 155 C7
Denton N Yorks 205 D8
Denton Norf 142 F5
Denton Northants 120 F6
Denton Oxon 83 E9
Denton Burn T & W 242 D5
Denton Holme Cumb 239 G10
Denton's Green Mers 183 B7
Denver Norf 140 C2
Denvilles Hants 22 B2
Denwick Northumb 264 G6
Deopham Norf 141 C11
Deopham Green Norf 141 D10
Deopham Stalland Norf 141 D10
Depden Suff 124 F5
Depden Green Suff 124 F5
Deppers Bridge Warks 119 F7
Deptford London 67 D11
Deptford T & W 243 F9
Deptford Wilts 46 F4
Derby Derbys 153 B7
Derby Devon 40 G5
Derbyhaven I o M 192 F3
Derbyshire Hill Mers 183 C8
Dereham Norf 159 G9
Dergoals Dumfries 236 D4
Deri Caerph 77 E10
Derriford Plym 7 D9
Derril Devon 24 G4
Derringstone Kent 55 D8
Derrington Shrops 132 E2
Derrington Staffs 151 E7
Derriton Devon 24 G4
Derry Stirl 285 E10
Derry Downs London 68 F3
Derry Fields Wilts 81 G8
Derry Hill Wilts 62 E3
Derry Lodge Aberds 292 D2
Derrydaroch Stirl 285 E7
Derryguaig Argyll 288 F6
Derrythorpe N Lincs 199 F10
Dersingham Norf 158 C3
Dertfords Wilts 45 D10
Dervaig Argyll 288 D6
Derwen Bridgend 58 C2
Derwen Denb 165 E9
Derwenlas Powys 128 D4
Desborough Northants 136 G6
Desford Leics 135 C9
Deskryshiel Aberds 292 B6
Detchant Northumb 264 B3
Detling Kent 53 B9
Deuchar Angus 292 G6
Deuddwr Powys 148 F4
Deuxhill Shrops 132 F3
Devauden Mon 79 F7
Deveral Corn 2 B3
Devil's Bridge = Pontarfynach Ceredig 112 B4
Devitts Green Warks 134 E5
Devizes Wilts 62 G4
Devol Invclyd 276 G6
Devon Village Clack 279 B8
Devonport Plym 7 E9
Devonside Clack 279 B8
Devoran Corn 3 B7
Dewar Borders 270 F6
Dewartown Midloth 271 C7
Dewes Green Essex 105 E9
Dewlands Common Dorset 31 F9
Dewlish Dorset 17 B11
Dewsbury W Yorks 197 C8
Dewsbury Moor W Yorks 197 C8
Dewshall Court Hereford 97 E9
Dhoon I o M 192 D5
Dhoor I o M 192 C5
Dhowin I o M 192 B5
Dhustone Shrops 115 B11
Dial Green W Sus 34 B6
Dial Post W Sus 35 D11
Dibberford Dorset 29 G7
Dibden Hants 32 F6
Dibden Purlieu Hants 32 F6
Dickens Heath W Mid 118 B2
Dickleburgh Norf 142 G3
Dickleburgh Moor Norf 142 G3
Dickon Hills Lincs 174 D6
Didbrook Glos 99 E11
Didcot Oxon 64 B4
Diddington Cambs 122 D3
Diddlebury Shrops 131 F10
Diddywell Devon 25 B7
Didley Hereford 97 E9
Didling W Sus 34 D4
Didlington Norf 140 D5
Didmarton Glos 61 B10
Didsbury Gtr Man 184 C4
Didworthy Devon 8 C3
Diebidale Highld 309 L4
Digbeth W Mid 133 F11
Digby Lincs 173 E9
Digg Highld 298 C4
Diggle Gtr Man 196 F4
Diglis Worcs 116 G6
Digmoor Lancs 194 F3
Digswell Herts 86 B3
Digswell Park Herts 86 C2
Digswell Water Herts 86 C3
Dihewyd Ceredig 111 F9
Dilham Norf 160 D6
Dilhorne Staffs 169 G7
Dill Hall Lancs 195 B8
Dillarburn S Lnrk 268 G6
Dillington Cambs 122 D2
Dillington Som 28 D5
Dilston Northumb 241 E11
Dilton Marsh Wilts 45 D11
Dilwyn Hereford 115 G8
Dimlands V Glam 58 F3
Dimmer Som 44 G6
Dimple Derbys 170 C3
Dimple Gtr Man 195 D8
Dimsdale Staffs 168 F4
Dimson Corn 12 G4
Dinas Carms 92 E5
Dinas Corn 10 G4
Dinas Gwyn 144 B5
Dinas Gwyn 163 D7
Dinas Cross Pembs 91 D10
Dinas Dinlle Gwyn 162 D6
Dinas-Mawddwy Gwyn 147 G7
Dinas Mawr Conwy 164 E4
Dinas Powys V Glam 59 E7
Dinbych y Pysgod = Tenby Pembs 73 E10
Dinckley Lancs 203 F9
Dinder Som 44 E5
Dinedor Hereford 97 D10
Dinedor Cross Hereford 97 D10
Dines Green Worcs 116 F6
Dingestow Mon 79 C7
Dinghurst N Som 44 B2
Dingle Mers 182 D5
Dingleden Kent 53 G10
Dingleton Borders 262 C2
Dingley Northants 136 F5
Dingwall Highld 300 D5
Dinlabyre Borders 250 E2
Dinmael Conwy 165 G8
Dinnet Aberds 292 D6
Dinnington S Yorks 187 D8
Dinnington Som 28 E6
Dinnington T & W 242 C6
Dinorwic Gwyn 163 C9
Dinton Bucks 84 C3
Dinton Wilts 46 G4
Dinwoodie Mains Dumfries 248 E4

Dinworthy Devon 24 D4
Dipford Som 28 C2
Dipley Hants 49 B8
Dippenhall Sur 49 D10
Dippertown Devon 12 E4
Dippin N Ayrs 256 E2
Dipple Devon 24 D4
Dipple Moray 302 D3
Dipple S Ayrs 244 C6
Diptford Devon 8 D4
Dipton Durham 242 G5
Diptonmill Northumb 241 E10
Dirdhu Highld 301 G10
Direcleit W Isles 305 J3
Dirleton E Loth 281 E10
Dirt Pot Northumb 232 B3
Discoed Powys 114 E5
Discove Som 45 G7
Diseworth Leics 153 E9
Dishes Orkney 314 D6
Dishforth N Yorks 215 E7
Dishley Leics 153 E10
Disley E Ches 185 E7
Diss Norf 126 B2
Disserth Powys 113 F10
Distington Cumb 228 G6
Ditchampton Wilts 46 G5
Ditcheat Som 44 F6
Ditchfield Bucks 84 G4
Ditchford Hill Worcs 100 D4
Ditchingham Norf 142 E6
Ditchling E Sus 36 D4
Ditherington Shrops 149 G10
Ditteridge Wilts 61 F10
Dittisham Devon 9 E7
Ditton Halton 183 D7
Ditton Kent 53 B8
Ditton Green Cambs 124 F3
Ditton Priors Shrops 132 F2
Dittons E Sus 23 E10
Divach Highld 300 G4
Divlyn Carms 94 D5
Dixton Glos 99 E9
Dixton Mon 79 C8
Dizzard Corn 11 B9
Dobcross Gtr Man 196 F3
Dobs Hill Flint 166 C4
Dobson's Bridge Shrops 149 C9
Dobwalls Corn 6 C4
Doc Penfro = Pembroke Dock Pembs 73 E7
Doccombe Devon 13 D11
Dochfour Ho Highld 300 F6
Dochgarroch Highld 300 E6
Dockeney Norf 143 E7
Dockenfield Sur 49 E10
Docker Lancs 211 E11
Docking Norf 158 B5
Docklow Hereford 115 F11
Dockray Cumb 230 G3
Dockroyd W Yorks 204 F6
Docton Devon 24 C2
Dodbrooke Devon 8 G4
Dodburn Borders 249 B11
Doddenham Worcs 116 F5
Doddinghurst Essex 87 F9
Doddington Cambs 139 E7
Doddington Kent 54 B2
Doddington Lincs 188 G6
Doddington Northumb 263 C11
Doddington Shrops 116 B2
Doddiscombsleigh Devon 14 D3
Doddshill Norf 158 C3
Doddycross Corn 6 C6
Dodford Northants 120 E2
Dodford Worcs 117 C8
Dodington S Glos 61 C9
Dodington Som 43 E7
Dodleston W Ches 166 C5
Dodmarsh Hereford 97 C11
Dods Leigh Staffs 151 C10
Dodscott Devon 25 D8
Dodworth S Yorks 197 F10
Dodworth Bottom S Yorks 197 G10
Dodworth Green S Yorks 197 G10
Doe Bank W Mid 134 D2
Doe Green Warr 183 D9
Doe Lea Derbys 171 B7
Doehole Derbys 170 D5
Doffcocker Gtr Man 195 E7
Dog & Gun Mers 182 B5
Dog Hill Gtr Man 196 F3
Dog Village Devon 14 B5
Dogdyke Lincs 174 D2
Dogingtree Estate Staffs 151 G9
Dogley Lane W Yorks 197 E7
Dogmersfield Hants 49 C9
Dogridge Wilts 62 B5
Dogsthorpe P'boro 138 C3
Doirlinn Highld 289 D8
Dol-fôr Powys 128 B6
Dol-ffanog Gwyn 146 G4
Dôl-y-Bont Ceredig 128 F2
Dol-y-cannau Powys 96 B3
Dolanog Powys 147 G11
Dolau Powys 114 D2
Dolau Rhondda 58 C3
Dolbenmaen Gwyn 163 G8
Dole Ceredig 128 F2
Dolemeads Bath 61 G9
Doley Staffs 150 D4
Dolfach Powys 129 C8
Dolfor Powys 130 F2
Dolgarrog Conwy 164 B3
Dolgellau Gwyn 146 F4
Dolgerdd Ceredig 111 G8
Dolgoch Gwyn 128 C3
Dolgran Carms 93 E8
Dolhelfa Powys 113 C8
Dolhendre Gwyn 147 C7
Doll Highld 311 J2
Dollar Clack 279 B9
Dolley Green Powys 114 D5
Dollis Hill London 67 B8
Dollwen Ceredig 128 G3
Dolphin Flint 181 G11
Dolphingstone E Loth 281 G7
Dolphinholme Lancs 202 C6
Dolphinstone Borders 262 F5
Dolphinton S Lnrk 270 F2
Dolton Devon 25 E9
Dolwen Conwy 180 G5
Dolwen Powys 129 B9
Dolwyd Conwy 180 F4
Dolwyddelan Conwy 164 E2
Dolydd Gwyn 163 D7
Dolyhir Powys 114 F4
Dolymelinau Powys 129 D11
Dolywern Wrex 148 B4
Domewood Sur 51 E10
Domgay Powys 148 F5
Dommett Som 28 E3
Don Johns Essex 106 F6
Doncaster S Yorks 198 G5
Doncaster Common S Yorks 198 G6
Dones Green W Ches 183 F10
Donhead St Andrew Wilts 30 C6
Donhead St Mary Wilts 30 C6
Donibristle Fife 280 D3
Doniford Som 42 E5
Donington Lincs 156 B4
Donington Shrops 132 C6
Donington Eaudike Lincs 156 B4
Donington le Heath Leics 153 G8
Donington on Bain Lincs 190 E2
Donington South Ing Lincs 156 C4
Donisthorpe Leics 152 G6
Donkey Street Kent 54 G6
Donkey Town Sur 66 G2
Donna Nook Lincs 190 B6
Donnington Glos 100 F3
Donnington Hereford 98 E4
Donnington Shrops 131 B11
Donnington Telford 150 G4
Donnington W Berks 64 F3
Donnington W Sus 22 C5
Donnington Wood Telford 150 G4
Donwell T & W 243 F7
Donyatt Som 28 E4
Doomsday Green W Sus 35 B11
Doonfoot S Ayrs 257 F8
Dora's Green Hants 49 D10
Dorback Lodge Highld 292 B2
Dorcan Swindon 63 C7
Dorchester Dorset 17 C9
Dorchester Oxon 83 G9
Dordale Worcs 117 C8
Dordon Warks 134 C5
Dore S Yorks 186 E4
Dores Highld 300 F5
Dorking Sur 51 D7
Dorking Tye Suff 107 D8
Dorley's Corner Suff 127 D7
Dormans Park Sur 51 E11
Dormansland Sur 52 E2
Dormanstown Redcar 235 G7
Dormer's Wells London 66 C6
Dormington Hereford 97 C11
Dormston Worcs 117 F9
Dorn Glos 100 E4
Dorn Hill Worcs 100 E3
Dornal S Ayrs 236 B4
Dorney Bucks 66 D2
Dorney Reach Bucks 66 D2
Dornie Highld 295 C10
Dornoch Highld 309 L7
Dornock Dumfries 238 D6
Dorrery Highld 310 D4
Dorridge W Mid 118 B3
Dorrington Lincs 173 E9
Dorrington Shrops 131 C9
Dorsington Warks 100 B2
Dorstone Hereford 96 C6
Dorton Bucks 83 C11
Dorusduain Highld 295 C11
Doseley Telford 132 B3
Dosmuckeran Highld 300 C2
Dosthill Staffs 134 C4
Dosthill Staffs 134 D4
Dothan Anglesey 178 G5
Dothill Telford 150 G2
Dottery Dorset 16 B5
Double Hill Bath 45 B8
Doublebois Corn 6 C3
Dougarie N Ayrs 255 D9
Doughton Glos 80 G5
Doughton Norf 159 D7
Douglas I o M 192 E4
Douglas S Lnrk 259 C8
Douglas & Angus Dundee 287 D8
Douglas Water S Lnrk 259 B9
Douglas West S Lnrk 259 C8
Douglastown Angus 287 C8
Doulting Som 44 E6
Dounby Orkney 314 D2
Doune Highld 291 C10
Doune Highld 309 J4
Doune Stirl 285 G11
Doune Park Aberds 303 C7
Douneside Aberds 292 C6
Dounie Argyll 275 D8
Dounie Highld 309 K5
Dounie Highld 309 L6
Dounreay Highld 310 C3
Doura N Ayrs 266 G6
Dousland Devon 7 B10
Dovaston Shrops 149 E7
Dove Green Notts 171 E7
Dove Holes Derbys 185 F9
Dove Point Mers 182 C2
Dovecot Mers 182 C6
Dovecothall Glasgow 267 D10
Dovenby Cumb 229 E7
Dovendale Lincs 190 E4
Dover Gtr Man 194 G6
Dover Kent 55 E10
Dovercourt Essex 108 E5
Doverdale Worcs 117 D7
Doverhay Som 41 D11
Doveridge Derbys 152 C2
Doversgreen Sur 51 D9
Dowally Perth 286 C4
Dowanhill Glasgow 267 B11
Dowbridge Lancs 202 G4
Dowdeswell Glos 81 B7
Dowe Hill Norf 161 F10
Dowlais M Tydf 77 D10
Dowlais Top M Tydf 77 D9
Dowland Devon 25 E9
Dowles Worcs 116 B5
Dowlesgreen Wokingham 65 F10
Dowlish Ford Som 28 E5
Dowlish Wake Som 28 E5
Down Ampney Glos 81 F10
Down End Som 43 E10
Down Field Cambs 124 C2
Down Hall Cumb 239 G7
Down Hatherley Glos 99 G7
Down Park W Sus 51 F10
Down St Mary Devon 26 G2
Down Street E Sus 36 C6
Down Thomas Devon 7 E10
Downall Green Gtr Man 194 G5
Downan Moray 301 F11
Downan S Ayrs 244 G3
Downcraig Ferry N Ayrs 266 E3
Downderry Corn 6 E6
Downe London 68 G2
Downend Glos 80 F4
Downend I o W 20 D6
Downend S Glos 60 D6
Downend W Berks 64 D3
Downfield Dundee 287 D7
Downgate Corn 11 G11
Downgate Corn 12 G3
Downham Essex 88 F2
Downham Lancs 203 E11
Downham London 67 E11
Downham Northumb 263 C9
Downham Market Norf 140 C2
Downhead Som 29 B9
Downhead Som 45 D7
Downhead Park M Keynes 103 C7
Downhill Corn 5 B7
Downhill Perth 286 D4
Downhill T & W 243 F9
Downholland Cross Lancs 193 F11
Downholme N Yorks 224 F2
Downicary Devon 12 C3
Downies Aberds 293 D11
Downinney Corn 11 C10
Downley Bucks 84 G4
Downs V Glam 58 E6
Downside C Beds 103 G10
Downside E Sus 23 E9
Downside N Som 60 F3
Downside Som 44 D6
Downside Sur 50 B6
Downside Sur 51 B7
Downton Hants 19 C11
Downton Powys 114 E4
Downton Shrops 149 G10
Downton Wilts 31 C11
Downton on the Rock Hereford 115 C8
Dowsby Lincs 156 D2
Dowsdale Lincs 156 G5
Dowslands Som 28 C2
Dowthwaitehead Cumb 230 G3
Doxey Staffs 151 E8
Doxford Park T & W 243 G9
Doynton S Glos 61 E8
Drabblegate Norf 160 D4
Draethen Newport 59 B8
Draffan S Lnrk 268 F5
Dragley Beck Cumb 210 D5
Dragonby N Lincs 200 E2
Dragons Green W Sus 35 C10
Drakehouse S Yorks 186 E6
Drakeland Corner Devon 7 D11
Drakelow Worcs 132 G6
Drakemyre Aberds 303 F9
Drakemyre N Ayrs 266 E5
Drake's Broughton Worcs 99 B8
Drakes Cross Worcs 117 B11
Drakestone Green Suff 107 B9
Drakewalls Corn 12 G4
Draughton N Yorks 204 C6
Draughton Northants 120 B5
Drawbridge Corn 6 B3
Drax N Yorks 199 B7
Draycot Oxon 83 D10
Draycot Cerne Wilts 62 D2
Draycot Fitz Payne Wilts 62 G6
Draycot Foliat Swindon 63 D7
Draycote Warks 119 C8
Draycott Derbys 153 C8
Draycott Glos 80 E2
Draycott Glos 100 D3
Draycott Shrops 132 E6
Draycott Som 29 C8
Draycott Som 44 C3
Draycott Worcs 99 B7
Draycott in the Clay Staffs 152 D3
Draycott in the Moors Staffs 169 G7
Drayford Devon 26 E3
Drayton Leics 136 E6
Drayton Lincs 156 B4
Drayton Norf 160 G3
Drayton Northants 119 E11
Drayton Oxon 83 G7
Drayton Oxon 101 C8
Drayton Ptsmth 33 F11
Drayton Som 28 C6
Drayton Som 29 D7
Drayton Warks 118 F3
Drayton Worcs 117 B8
Drayton Bassett Staffs 134 C3
Drayton Beauchamp Bucks 84 C6
Drayton Parslow Bucks 102 F6
Drayton St Leonard Oxon 83 F10
Dre-fach Carms 75 B11
Dre-fach Ceredig 93 B10
Dre-gôch Denb 165 B10
Drebley N Yorks 205 B7
Dreemskerry I o M 192 C5
Dreenhill Pembs 72 C6
Drefach Carms 75 C8
Drefach Carms 92 G5
Drefach Carms 93 D7
Drefelin Carms 93 D7
Dreggie Highld 301 G10
Dreghorn Edin 270 B4
Dreghorn N Ayrs 257 B9
Drellingore Kent 55 E8
Drem E Loth 281 F10
Dresden Stoke 168 G6
Dreumasdal W Isles 297 H3
Drewsteignton Devon 13 C10
Driby Lincs 190 G5
Driffield E Yorks 208 B6
Driffield Glos 81 F9
Drift Corn 1 D4
Drigg Cumb 219 F11
Drighlington W Yorks 197 B8
Drimnin Highld 289 D7
Drimnin Ho Highld 289 D7
Drimpton Dorset 28 F6
Drimsynie Argyll 284 G5
Dringhoe E Yorks 209 C9
Dringhouses York 207 D7
Drinisiadar W Isles 305 J3
Drinkstone Suff 125 E9
Drinkstone Green Suff 125 E9
Drishaig Argyll 284 F5
Drissaig Argyll 275 B10
Drive End Dorset 29 F9
Driver's End Herts 86 B2
Drochedlie Aberds 302 C5
Drochil Borders 270 G3
Drointon Staffs 151 D10
Droitwich Spa Worcs 117 E7
Droman Highld 306 D6
Dromore Dumfries 237 C7
Dron Perth 286 F5
Dronfield Derbys 186 F5
Dronfield Woodhouse Derbys 186 F4
Drongan E Ayrs 257 F10
Dronley Angus 287 D7
Droop Dorset 30 F3
Drope Cardiff 58 D6
Dropping Well S Yorks 186 C5
Droughduil Dumfries 236 D3
Droxford Hants 33 D10
Droylsden Gtr Man 184 B6
Drub W Yorks 197 B7
Druggers End Worcs 98 D5
Druid Denb 165 G8
Druidston Pembs 72 B5
Druim Highld 301 D9
Druimarbin Highld 290 F2
Druimavuic Argyll 284 C4
Druimdrishaig Argyll 275 F8
Druimindarroch Highld 295 G8
Druimkinnerras Highld 300 F4
Druimnacroish Argyll 288 E6
Druimsornaig Argyll 289 F9
Druimyeon More Argyll 255 B7
Drum Argyll 275 F10
Drum Edin 270 B6
Drum Perth 286 G4
Drumardoch Stirl 285 F10
Drumbeg Highld 306 F6
Drumblade Aberds 302 E5
Drumblair Aberds 302 E6
Drumbuie Dumfries 246 G3
Drumbuie Highld 295 B9
Drumburgh Cumb 239 F7
Drumburn Dumfries 237 C11
Drumchapel Glasgow 277 G10
Drumchardine Highld 300 E5
Drumchork Highld 307 L3
Drumclog S Lnrk 258 B4
Drumderfit Highld 300 D6
Drumeldrie Fife 287 G8
Drumelzier Borders 260 C4
Drumfearn Highld 295 D8
Drumgask Highld 291 D8
Drumgelloch N Lnrk 268 B5
Drumgley Angus 287 B8
Drumgreen Angus 292 F6
Drumguish Highld 291 D9
Drumhead Aberds 293 B11
Drumin Moray 301 F11
Drumindorsair Highld 300 E4
Drumlasie Aberds 293 C8
Drumlean Stirl 285 G8
Drumlemble Argyll 255 F7
Drumliah Highld 309 K6
Drumligair Aberds 293 B11
Drumlithie Aberds 293 E9
Drumloist Stirl 285 G10
Drummersdale Lancs 193 E11
Drummick Perth 286 E3
Drummoddie Dumfries 236 E5
Drummond Highld 300 C6
Drummore Dumfries 236 F3
Drummuir Moray 302 E3
Drummuir Castle Moray 302 E3
Drumnadrochit Highld 300 G5
Drumnagorrach Moray 302 D5
Drumness Perth 286 F2
Drumoak Aberds 293 D9
Drumore Argyll 255 E8
Drumpark Dumfries 247 G9
Drumpellier N Lnrk 268 B4
Drumphail Dumfries 236 C4
Drumrash Dumfries 237 B8
Drumrunie Highld 307 J6
Drumry W Dunb 277 G10
Drums Aberds 303 G9
Drumsallie Highld 289 B11
Drumsmittal Highld 300 E6
Drumstinchall Dumfries 237 D10
Drumsturdy Angus 287 D8
Drumtochty Castle Aberds 293 F8
Drumtroddan Dumfries 236 E5
Drumuie Highld 298 E4
Drumuillie Highld 301 G9
Drumvaich Stirl 285 G10
Drumwalt Dumfries 236 D5
Drumwhindle Aberds 303 F9
Drunkendub Angus 287 C10
Drury Flint 166 C3
Drury Lane Wrex 167 G7
Drury Square Norf 159 F8
Drurylane Norf 141 C8
Dry Doddington Lincs 172 F4
Dry Drayton Cambs 123 E7
Dry Hill Hants 49 F7
Dry Sandford Oxon 83 E7
Dry Street Essex 69 B7
Drybeck Cumb 222 B3
Drybridge Moray 302 C4
Drybridge N Ayrs 257 B9
Drybrook Glos 79 B10
Dryburgh Borders 262 C3
Dryden Borders 261 E11
Dryhill Kent 52 B3
Dryhope Borders 261 E7
Drylaw Edin 280 F4
Drym Corn 2 C4
Drymen Stirl 277 D9
Drymere Norf 140 B5
Drymuir Aberds 303 E9
Drynachan Lodge Highld 301 F8
Drynie Park Highld 300 D5
Drynoch Highld 294 B6
Dryslwyn Carms 93 G11
Dryton Shrops 131 B11
Drywells Aberds 302 D6
Duag Bridge Highld 309 K3
Duartbeg Highld 306 F6
Duartmore Bridge Highld 306 F6
Dubbs Cross Devon 12 C3
Dubford Aberds 303 C8
Dubhchladach Argyll 275 G9
Dublin Suff 126 D3
Dubton Angus 287 B9
Dubwath Cumb 229 E9
Duchally Highld 309 H3
Duchlage Argyll 276 D6
Duchrae Dumfries 246 G5
Duck Corner Suff 109 C7
Duck End Beds 103 C11
Duck End Beds 121 G9
Duck End Bucks 102 F5
Duck End Cambs 122 E4
Duck End Essex 105 G10
Duck End Essex 106 E3
Duck End Essex 106 F3
Duckend Green Essex 106 G4
Duckhole S Glos 79 G10
Ducklington Oxon 82 D5
Duckmanton Derbys 186 G6
Duck's Cross Beds 122 F2
Ducks Island London 86 F2
Duckswich Worcs 98 D6
Dudbridge Glos 80 E4
Dudden Hill London 67 B8
Duddenhoe End Essex 105 D9
Duddingston Edin 280 G5
Duddington Northants 137 C9
Duddlestone Som 28 C2
Duddleswell E Sus 37 B7
Duddlewick Shrops 132 G3
Duddo Northumb 273 G8
Duddon W Ches 167 C8
Duddon Bridge Cumb 210 B3
Duddon Common W Ches 167 B8
Dudleston Shrops 148 B6
Dudleston Grove Shrops 149 B7
Dudleston Heath (Criftins) Shrops 149 B7
Dudley T & W 243 C7
Dudley W Mid 133 E8
Dudley Hill W Yorks 205 G9
Dudley Port W Mid 133 E9
Dudley Wood W Mid 133 F8
Dudley's Fields W Mid 133 C9
Dudlows Green Warr 183 E10
Dudsbury Dorset 19 B7
Dudswell Herts 85 D7
Dudwells Pembs 91 G8
Duerdon Devon 24 D4
Duffield Derbys 170 G4
Duffieldbank Derbys 170 G5
Duffryn Neath 57 B10
Duffryn Newport 59 B9
Duffryn Shrops 130 G4
Dufftown Moray 302 F3
Duffus Moray 301 C11
Dufton Cumb 231 F9
Duggleby N Yorks 217 F7
Duich Argyll 254 B4
Duiletter Argyll 284 D5
Duinish Perth 291 G8
Duirinish Highld 295 B9
Duisdalebeg Highld 295 D8
Duisdalemore Highld 295 D9
Duisky Highld 290 F2
Duke End Warks 134 F4
Dukesfield Northumb 241 F10
Dukestown Bl Gwent 77 C10
Dukinfield Gtr Man 184 B6
Dulas Anglesey 179 D7
Dulcote Som 44 E5
Dulford Devon 27 F9
Dull Perth 286 C2
Dullatur N Lnrk 278 F4
Dullingham Cambs 124 F2
Dullingham Ley Cambs 124 F2
Dulnain Bridge Highld 301 G9
Duloch Fife 280 D2
Duloe Beds 122 E3
Duloe Corn 6 D4
Dulsie Highld 301 E9
Dulverton Som 26 B6
Dulwich London 67 E10
Dulwich Village London 67 E10
Dumbarton W Dunb 277 F7
Dumbleton Glos 99 D10
Dumcrieff Dumfries 248 C4
Dumfries Dumfries 237 B11
Dumgoyne Stirl 277 E10
Dummer Hants 48 D5
Dumpford W Sus 34 C4
Dumpinghill Devon 24 F6
Dumpling Green Norf 159 G10
Dumplington Gtr Man 184 B3
Dumpton Kent 71 F11
Dun Angus 287 B10
Dun Charlabhaigh W Isles 304 D3
Dunach Argyll 289 G10
Dunadd Argyll 275 D9
Dunain Ho Highld 300 E6
Dunalastair Perth 285 B11
Dunan Highld 295 C7
Dunans Argyll 275 D11
Dunans Argyll 275 D9
Dunball Som 43 E10
Dunbar E Loth 282 F3
Dunbeath Highld 311 G5
Dunbeg Argyll 289 F10
Dunblane Stirl 285 G11
Dunbog Fife 286 F6
Dunbridge Hants 32 B4
Duncansclett Shetland 313 K5
Duncanston Highld 300 D5
Duncanstone Aberds 302 G5
Dunchideock Devon 14 D3
Dunchurch Warks 119 C9
Duncombe Lancs 202 F6
Duncote Northants 120 G3
Duncow Dumfries 247 G11
Duncraggan Stirl 285 G9
Duncrievie Perth 286 G5
Duncroisk Stirl 285 D9
Duncton W Sus 35 D7
Dundas Ho Orkney 314 H4
Dundee Dundee 287 D8
Dundeugh Dumfries 246 F3
Dundon Som 44 G3
Dundon Hayes Som 44 G3
Dundonald Fife 280 C4
Dundonald S Ayrs 257 C9
Dundonnell Highld 307 L5
Dundonnell Hotel Highld 307 L5
Dundonnell House Highld 307 L6
Dundraw Cumb 229 B10
Dundreggan Highld 290 B5
Dundreggan Lodge Highld 290 B5
Dundrennan Dumfries 237 E9
Dundridge Hants 33 D9
Dundry N Som 60 F5
Dundurn Perth 285 E11
Dundyvan N Lnrk 268 C4
Dunecht Aberds 293 C9
Dunfermline Fife 279 D11
Dunfield Glos 81 F10
Dunford Bridge S Yorks 197 G7
Dungate Kent 54 B2
Dunge Wilts 45 C11
Dungeness Kent 39 D9
Dungworth S Yorks 186 D3
Dunham Notts 188 G4
Dunham-on-the-Hill W Ches 183 G7
Dunham on Trent Notts 188 G4
Dunham Town Gtr Man 184 D2
Dunham Woodhouses Gtr Man 184 D2
Dunhampstead Worcs 117 E8
Dunhampton Worcs 116 D6
Dunholme Lincs 189 F8
Dunino Fife 287 F9
Dunipace Falk 278 E6
Dunira Perth 285 E11
Dunkeld Perth 286 C4
Dunkerton Bath 45 B8
Dunkeswell Devon 27 F10
Dunkeswick N Yorks 206 D2
Dunkirk Cambs 139 F10
Dunkirk Kent 54 B5
Dunkirk Norf 160 D4
Dunkirk Nottingham 153 B11
Dunkirk S Glos 61 B9
Dunkirk Staffs 168 E4
Dunkirk W Ches 182 G5
Dunkirk Wilts 62 G3
Dunk's Green Kent 52 C6
Dunlappie Angus 293 G7
Dunley Hants 48 C3
Dunley Worcs 116 D5
Dunlichity Lodge Highld 300 F6
Dunlop E Ayrs 267 F8
Dunmaglass Lodge Highld 300 G5
Dunmere Corn 5 B10
Dunmore Argyll 275 G8
Dunmore Falk 279 D7
Dunmore Highld 300 E5
Dunn Street Kent 54 D3
Dunn Street Kent 69 G9
Dunnet Highld 310 B6
Dunnichen Angus 287 C9
Dunnikier Fife 280 C5
Dunninald Angus 287 B11
Dunning Perth 286 F4
Dunnington E Yorks 209 C9
Dunnington Warks 117 G11
Dunnington York 207 C9
Dunningwell Cumb 210 C3
Dunnockshaw Lancs 195 B10
Dunnose I o W 21 F7
Dunnsheath Shrops 149 F9
Dunollie Argyll 289 F10
Dunoon Argyll 276 F3
Dunragit Dumfries 236 D3
Dunrobin Mains Highld 311 J2
Dunrostan Argyll 275 E8
Duns Borders 272 E5
Duns Tew Oxon 101 F9
Dunsa Derbys 186 G2
Dunsby Lincs 156 D2
Dunscar Gtr Man 195 E8
Dunscore Dumfries 247 G9
Dunscroft S Yorks 199 F7
Dunsdale Redcar 226 B2
Dunsden Green Oxon 65 D8
Dunsfold Sur 50 F4
Dunsfold Common Sur 50 F4
Dunsfold Green Sur 50 F4
Dunsford Devon 14 D2
Dunsford Sur 50 F4
Dunshalt Fife 286 F6
Dunshillock Aberds 303 E9
Dunsill Notts 171 C7
Dunsinnan Perth 286 D5
Dunskey Ho Dumfries 236 D2
Dunslea Corn 11 G11
Dunsley N Yorks 227 C7
Dunsley Staffs 133 G7
Dunsmore Bucks 84 D5
Dunsmore Warks 119 B10
Dunsop Bridge Lancs 203 C9
Dunstable C Beds 103 G10
Dunstal Staffs 151 D11
Dunstall Staffs 152 E3
Dunstall Common Worcs 99 C7
Dunstall Green Suff 124 E4
Dunstall Hill W Mid 133 C8
Dunstan Northumb 265 F7
Dunstan Steads Northumb 264 E6
Dunster Som 42 E3
Dunston Derbys 186 G5
Dunston Lincs 173 C9
Dunston Norf 142 C4
Dunston Staffs 151 F8
Dunston T & W 242 E6
Dunston Heath Staffs 151 F8
Dunston Hill T & W 242 E6
Dunstone Devon 7 E11
Dunstone Devon 8 G5
Dunsville S Yorks 198 F6
Dunswell E Yorks 209 F7
Dunsyre S Lnrk 269 F11
Dunterton Devon 12 F3
Dunthrop Oxon 101 F7
Duntisbourne Abbots Glos 81 D7
Duntisbourne Leer Glos 81 D7
Duntisbourne Rouse Glos 81 D7
Duntish Dorset 29 F11
Duntocher W Dunb 277 G9
Dunton Bucks 102 G6
Dunton C Beds 104 C4
Dunton Norf 159 C7
Dunton Bassett Leics 135 E10
Dunton Green Kent 52 B4
Dunton Patch Norf 159 C7
Dunton Wayletts Essex 87 G11
Duntrune Castle Argyll 275 D8
Duntulm Highld 298 B4
Dunure S Ayrs 257 F7
Dunvant = Dynfant Swansea 56 C5
Dunvegan Highld 298 E2
Dunveth Corn 10 G5
Dunwear Som 43 F10
Dunwich Suff 127 C9
Dunwood Staffs 168 D6
Dupplin Castle Perth 286 F4
Durdar Cumb 239 G10
Durgan Corn 3 D7
Durgates E Sus 52 G6
Durham Durham 233 C11
Durisdeer Dumfries 247 C9
Durisdeermill Dumfries 247 C9
Durkar W Yorks 197 D10
Durleigh Som 43 F9
Durleighmarsh W Sus 34 C3
Durley Hants 33 D8
Durley Wilts 63 G8
Durley Street Hants 33 D8
Durlock Kent 55 B9
Durlow Common Hereford 98 D2
Durn Gtr Man 196 D2
Durnamuck Highld 307 K5
Durness Highld 308 C4
Durnfield Som 29 C7
Durno Aberds 303 G7
Durns Town Hants 19 B11
Duror Highld 289 D11
Durran Argyll 275 C10
Durran Highld 310 C5
Durrant Green Kent 53 F11
Durrants Hants 22 B2
Durrington W Sus 35 F10
Durrington Wilts 47 E7
Durrisdale Orkney 314 D3
Dursley Glos 80 F3
Dursley Wilts 45 C11
Dursley Cross Glos 98 G3
Durston Som 28 B3
Durweston Dorset 30 F5
Dury Shetland 313 G6
Duryard Devon 14 C4
Duston Northants 120 E4
Dutch Village Essex 69 C9
Duthil Highld 301 G9
Dutlas Powys 114 B4
Duton Hill Essex 106 F2
Dutson Corn 12 D2
Dutton Ches 183 F9
Duxford Cambs 105 B9
Duxford Oxon 82 F5
Duxmoor Shrops 115 B8
Dwygyfylchi Conwy 180 F2
Dwyran Anglesey 162 B6
Dwyrhiw Powys 129 C11
Dyce Aberdeen 293 B10
Dyche Som 43 E7
Dye House Northumb 241 F10
Dyer's Common S Glos 60 C5
Dyer's Green Cambs 105 B7
Dyffryn Bridgend 57 C11
Dyffryn Carms 92 G6
Dyffryn Ceredig 110 G5
Dyffryn Pembs 91 D8
Dyffryn Ardudwy Gwyn 145 E11
Dyffryn-bern Ceredig 110 G5
Dyffryn Castell Ceredig 128 G5
Dyffryn Ceidrych Carms 94 F4
Dyffryn Cellwen Neath 76 D5
Dyke Lincs 156 E2
Dyke Moray 301 D9
Dykehead Angus 292 G5
Dykehead N Lnrk 269 D7
Dykehead Stirl 277 B11
Dykelands Aberds 293 G9
Dykends Angus 286 B6
Dykeside Aberds 303 E7
Dykesmains N Ayrs 266 G5
Dylife Powys 129 E7
Dymchurch Kent 39 B9
Dymock Glos 98 E4
Dynfant = Dunvant Swansea 56 C5
Dyrham S Glos 61 D8
Dysart Fife 280 C6
Dyserth Denb 181 F9

E

Eabost Highld 294 B5
Eabost West Highld 298 E3
Each End Kent 55 B10
Eachway Worcs 117 B9
Eachwick Northumb 242 C4
Eadar Dha Fhadhail W Isles 304 E2
Eagland Hill Lancs 202 D4
Eagle Lincs 172 B5
Eagle Barnsdale Lincs 172 B5
Eagle Moor Lincs 172 B5
Eagle Tor Derbys 170 C2
Eaglesfield Cumb 229 F7
Eaglesfield Dumfries 238 C6
Eaglesham E Renf 267 E11
Eaglestone M Keynes 103 D7
Eaglethorpe Northants 137 E11
Eagley Gtr Man 195 E8
Eairy I o M 192 E3
Eakley Lanes M Keynes 120 G6
Eakring Notts 171 C11
Ealand N Lincs 199 E9
Ealing London 67 C7
Eals Northumb 240 F5
Eamont Bridge Cumb 230 F6
Earby Lancs 204 D3
Earcroft Blkburn 195 C7
Eardington Shrops 132 E4
Eardisland Hereford 115 F8
Eardisley Hereford 96 B6
Eardiston Shrops 149 D7
Eardiston Worcs 116 D3
Earith Cambs 123 C7
Earl Shilton Leics 135 D9
Earl Soham Suff 126 E4
Earl Sterndale Derbys 169 B9
Earl Stoneham Suff 126 F2
Earl Stonham Suff 126 F2
Earle Northumb 263 D11
Earlestown Mers 183 B9
Earley Wokingham 65 E9
Earlham Norf 142 B4
Earlish Highld 298 C3
Earls Barton Northants 121 E7
Earls Colne Essex 107 F7
Earl's Common Worcs 117 F9
Earl's Court London 67 D9
Earl's Croome Worcs 99 C7
Earl's Down E Sus 23 B10
Earl's Green Suff 125 D10
Earlsdon W Mid 118 B6
Earlsferry Fife 281 B9
Earlsfield Lincs 155 B8
Earlsfield London 67 E9
Earlsford Aberds 303 F8
Earlsheaton W Yorks 197 C9
Earlsmill Moray 301 D9
Earlston Borders 262 B3
Earlston E Ayrs 257 B10
Earlstone Common Hants 64 G3
Earlstoun Dumfries 246 G4
Earlswood Mon 79 F7
Earlswood Sur 51 D9
Earlswood Warks 118 C2
Earnley W Sus 22 D4
Earnock S Lnrk 268 E3
Earnshaw Bridge Lancs 194 C4
Earsairidh W Isles 297 M3
Earsdon T & W 243 C8
Earsham Norf 142 F6
Earsham Street Suff 126 B4
Earswick York 207 B8
Eartham W Sus 22 B6
Earthcott Green S Glos 61 B7
Easby N Yorks 224 E3
Easby N Yorks 225 D11
Easdale Argyll 275 B8
Easebourne W Sus 34 C5
Easenhall Warks 119 B9
Eashing Sur 50 E2
Easington Bucks 83 C11
Easington Durham 234 C4
Easington E Yorks 201 D11
Easington Lancs 203 C10
Easington Northumb 264 C4
Easington Oxon 83 F11
Easington Oxon 101 D9
Easington Redcar 226 B4
Easington Colliery Durham 234 C4
Easington Lane T & W 234 B3
Easingwold N Yorks 215 F10
Easole Street Kent 55 C9
Eason's Green E Sus 23 B8
Eassie Angus 287 C7
East Aberthaw V Glam 58 F4
East Acton London 67 C8
East Adderbury Oxon 101 D9
East Allington Devon 8 F5
East Amat Highld 309 K4
East Anstey Devon 26 B5
East Anton Hants 47 D11
East Appleton N Yorks 224 F4
East Ardsley W Yorks 197 B10
East Ashling W Sus 22 B4
East Aston Hants 48 D2
East Auchronie Aberds 293 C10
East Ayton N Yorks 217 B9
East Bank Bl Gwent 78 D2
East Barkwith Lincs 189 E11
East Barming Kent 53 C8
East Barnby N Yorks 226 C6
East Barnet London 86 F3
East Barns E Loth 282 F4
East Barsham Norf 159 C8
East Barton Suff 125 D8
East Beach W Sus 22 E5
East Beckham Norf 177 E11
East Bedfont London 66 E5
East Bergholt Suff 107 D11
East Bierley W Yorks 197 B7
East Bilney Norf 159 F9
East Blackdene Durham 232 D3
East Blatchington E Sus 23 E7
East Bloxworth Dorset 18 C3
East Boldon T & W 243 E9
East Boldre Hants 32 G5
East Bower Som 43 F10
East Bonhard Perth 286 E5
East Brent Som 43 C11
East Bridgford Notts 171 G11
East Briscoe Durham 223 B9
East Buckland Devon 41 G7
East Budleigh Devon 15 E7
East Burnham Bucks 66 C3
East Burrafirth Shetland 313 H5
East Burton Dorset 18 D2
East Butsfield Durham 233 B8
East Butterleigh Devon 27 F7
East Butterwick N Lincs 199 F10
East Cairnbeg Aberds 293 F9
East Calder W Loth 269 B11
East Carleton Norf 142 C3
East Carlton Northants 136 F6
East Carlton W Yorks 205 E10
East Chaldon or Chaldon Herring Dorset 17 E11
East Challow Oxon 63 B11
East Charleton Devon 8 G5
East Chelborough Dorset 29 F9
East Chiltington E Sus 36 D5
East Chinnock Som 29 E7
East Chisenbury Wilts 46 C6
East Cholderton Hants 47 D9
East Clandon Sur 50 C5
East Claydon Bucks 102 F4
East Clevedon N Som 60 E2
East Clyne Highld 311 J3
East Clyth Highld 310 F7
East Coker Som 29 E8
East Combe Som 43 G7
East Common N Yorks 207 G8
East Compton Dorset 30 D5
East Compton Som 44 E6
East Cornworthy Devon 8 D6
East Cottingwith E Yorks 207 E10
East Cowes I o W 20 B6
East Cowick E Yorks 199 C7
East Cowton N Yorks 224 E6
East Cramlington Northumb 243 B7
East Cranmore Som 45 E7
East Creech Dorset 18 E4
East Croachy Highld 300 G6
East Croftmore Highld 291 B11
East Curthwaite Cumb 230 B2
East Dean E Sus 23 F9
East Dean Glos 98 G3
East Dean Hants 32 B3
East Dean W Sus 34 E6
East Dene S Yorks 186 C6
East Denton T & W 242 D6
East Didsbury Gtr Man 184 C5
East Down Devon 40 E6
East Drayton Notts 188 F3
East Dulwich London 67 E10
East Dundry N Som 60 F5
East Ella Hull 200 B5
East End Beds 122 F2
East End Bucks 84 B4
East End C Beds 103 C9
East End Dorset 18 B5
East End E Yorks 201 B9
East End E Yorks 209 B9
East End E Yorks 209 G9
East End Essex 89 D8
East End Glos 81 E11
East End Hants 20 B3
East End Hants 33 C11
East End Hants 64 G2
East End Herts 105 F9
East End Kent 38 B2
East End Kent 53 F10
East End Kent 70 E3
East End M Keynes 103 C8
East End N Som 60 E3
East End Oxon 82 C5
East End Oxon 101 D9
East End Oxon 101 E7
East End S Glos 61 E9
East End Som 29 B10
East End Som 44 C5
East End Som 45 D7
East End Suff 108 E2
East End Suff 126 F3
East End Green Herts 86 C3
East Everleigh Wilts 47 C8
East Ewell Sur 67 G8
East Farleigh Kent 53 C8
East Farndon Northants 136 F4
East Fen Common Cambs 124 C2
East Ferry Lincs 188 B4
East Fields W Berks 64 F3
East Finchley London 67 B9
East Finglassie Fife 280 B5
East Firsby Lincs 189 D8
East Fleet Dorset 17 E8
East Fortune E Loth 281 F10
East Garforth W Yorks 206 G4
East Garston W Berks 63 D11
East Gateshead T & W 243 E7
East Ginge Oxon 64 B2
East Gores Essex 107 G7
East Goscote Leics 154 G2
East Grafton Wilts 63 G9
East Grange Moray 301 C10
East Green Hants 49 E9
East Green Suff 124 G3
East Green Suff 127 D8
East Grimstead Wilts 32 B2
East Grinstead W Sus 51 F11
East Guldeford E Sus 38 C6
East Haddon Northants 120 D3
East Hagbourne Oxon 64 B4
East Halton N Lincs 200 D6
East Ham London 68 C2
East Hampnett W Sus 22 B6
East Hanney Oxon 82 G6
East Hanningfield Essex 88 E3
East Hardwick W Yorks 198 D3
East Harling Norf 141 F9
East Harlsey N Yorks 225 F8
East Harnham Wilts 31 B10
East Harptree Bath 44 B5
East Hartford Northumb 243 B7
East Harting W Sus 34 D3
East Hatch Wilts 30 B6
East Hatley Cambs 122 G5
East Hauxwell N Yorks 224 G3
East Heckington Lincs 173 G11
East Haven Angus 287 D9
East Hedleyhope Durham 233 C9
East Helmsdale Highld 311 H4
East Hendred Oxon 64 B3
East Herringthorpe S Yorks 187 C7
East Herrington T & W 243 G9
East Heslerton N Yorks 217 D8
East Hewish N Som 59 G11
East Hill Kent 68 G5

East Hoathly E Sus 23 B8
East Hogaland Shetland 313 K5
East Holme Dorset 18 D3
East Holton Dorset 18 C5
East Holywell Northumb 243 C8
East Horndon Essex 68 B6
East Horrington Som 44 D5
East Horsley Sur 50 C5
East Horton Northumb 264 C2
East Howdon T & W 243 D8
East Howe Bmouth 19 B7
East Huntspill Som 43 E10
East Hyde C Beds 85 B10
East Ilkerton Devon 41 D8
East Ilsley W Berks 64 C3
East Keal Lincs 174 C5
East Kennett Wilts 62 F6
East Keswick W Yorks 206 E3
East Kilbride S Lnrk 268 E2
East Kimber Devon 12 B5
East Kingston W Sus 35 G9
East Kirkby Lincs 174 C4
East Knapton N Yorks 217 D7
East Knighton Dorset 18 D2
East Knowstone Devon 26 C4
East Knoyle Wilts 45 G11
East Kyloe Northumb 264 B3
East Kyo Durham 242 G5
East Lambrook Som 28 D6
East Lamington Highld 301 B7
East Langdon Kent 55 D10
East Langton Leics 136 E4
East Langwell Highld 309 J7
East Lavant W Sus 22 B5
East Lavington W Sus 34 D6
East Law Northumb 242 G3
East Layton N Yorks 224 D3
East Leake Notts 153 D11
East Learmouth Northumb 263 B9
East Leigh Devon 8 E3
East Leigh Devon 25 F11
East Lexham Norf 159 F7
East Lilburn Northumb 264 E2
East Linton E Loth 281 F11
East Liss Hants 34 B3
East Lockinge Oxon 64 B2
East Loftus Redcar 226 B4
East Looe Corn 6 E5
East Lound N Lincs 188 B3
East Lulworth Dorset 18 E3
East Lutton N Yorks 217 F8
East Lydeard Som 27 B11
East Lydford Som 44 G5
East Lyng Som 28 B4
East Mains Aberds 293 D8
East Mains Borders 271 F11
East Mains S Lnrk 268 E2
East Malling Kent 53 B8
East Malling Heath Kent 53 B7
East March Angus 287 D8
East Marden W Sus 34 E4
East Markham Notts 188 G2
East Marsh NE Lincs 201 E9
East Martin Hants 31 D9
East Marton N Yorks 204 C4
East Melbury Dorset 30 C5
East Meon Hants 33 C11
East Mere Devon 27 D7
East Mersea Essex 89 C9
East Mey Highld 310 B7
East Molesey Sur 67 F7
East Moor W Yorks 197 C10
East Moors Cardiff 59 D8
East Morden Dorset 18 B4
East Morton W Yorks 205 E7
East Moulsecoomb Brighton 36 F4
East Ness N Yorks 216 D3
East Newton E Yorks 209 F11
East Newton N Yorks 216 D2
East Norton Leics 136 C5
East Nynehead Som 27 C11
East Oakley Hants 48 C5
East Ogwell Devon 14 G2
East Orchard Dorset 30 D4
East Ord Northumb 273 E9
East Panson Devon 12 C3
East Parley Dorset 19 B8
East Peckham Kent 53 D7
East Pennard Som 44 F5
East Perry Cambs 122 D3
East Portholland Corn 5 G9
East Portlemouth Devon 9 G9
East Prawle Devon 9 G10
East Preston W Sus 35 G9
East Pulham Dorset 30 F2
East Putford Devon 24 D5
East Quantoxhead Som 42 E6
East Rainton T & W 234 B2
East Ravendale NE Lincs 190 B2
East Raynham Norf 159 D7
East Rhidorroch Lodge Highld 307 K7
East Rigton W Yorks 206 E3
East Rolstone N Som 59 G11
East Rounton N Yorks 225 E8
East Row N Yorks 227 C7
East Rudham Norf 158 D6
East Runton Norf 177 E11
East Ruston Norf 160 D6
East Saltoun E Loth 271 B9
East Sheen London 67 D8
East Skelston Dumfries 247 F8
East Sleekburn Northumb 253 G7
East Somerton Norf 161 F9
East Stanley Durham 242 G6
East Stockwith Lincs 188 C3
East Stoke Dorset 18 D3
East Stoke Notts 172 F3
East Stoke Som 29 D7
East Stour Dorset 30 C4
East Stour Common Dorset 30 C4
East Stourmouth Kent 71 G9
East Stowford Devon 25 B10
East Stratton Hants 48 F4
East Street Kent 55 B10
East Street Som 44 F4
East Studdal Kent 55 D10
East Suisnish Highld 295 B7
East Taphouse Corn 6 C3
East-the-Water Devon 25 B7
East Third Borders 262 B4
East Thirston Northumb 252 D5
East Tilbury Thurrock 69 D7
East Tisted Hants 49 G8
East Torrington Lincs 189 E10
East Town Som 42 G6
East Town Som 44 E6
East Town Wilts 45 B11
East Trewent Pembs 73 F8
East Tuddenham Norf 159 G11
East Tuelmenna Corn 6 B4
East Tytherley Hants 32 B3
East Tytherton Wilts 62 E3
East Village Devon 26 F4
East Village V Glam 58 E3
East Wall Shrops 131 E10
East Walton Norf 158 F4
East Water Som 44 C4
East Week Devon 13 C9
East Wellow Hants 32 C4
East Wemyss Fife 280 B6
East Whitburn W Loth 269 B9
East Wickham London 68 D3
East Williamston Pembs 73 E9
East Winch Norf 158 F3
East Winterslow Wilts 47 G8
East Wittering W Sus 21 B11
East Witton N Yorks 214 B2
East Woodburn Northumb 251 F10
East Woodhay Hants 64 G2
East Woodlands Som 45 E9
East Worldham Hants 49 F8
East Worlington Devon 26 E3
East Worthing W Sus 35 G11
East Wretham Norf 141 E8
East Youlstone Devon 24 D3
Eastacombe Devon 25 B8
Eastacombe Devon 25 C9
Eastacott Devon 25 C10
Eastbourne Darl 224 C6
Eastbourne E Sus 23 F10
Eastbridge Suff 127 D9
Eastbrook Som 28 C2
Eastbrook V Glam 59 E7
Eastburn E Yorks 208 B5
Eastburn W Yorks 204 E6
Eastburn Br W Yorks 204 E6
Eastbury London 85 G9
Eastbury W Berks 63 D10
Eastby N Yorks 204 C6
Eastchurch Kent 70 E3
Eastcombe Glos 80 E5
Eastcote London 66 B6
Eastcote Northants 120 G3
Eastcote W Mid 118 B3
Eastcote Village London 66 B6
Eastcott Corn 24 D3
Eastcott Wilts 46 B4
Eastcotts Beds 103 B11
Eastcourt Wilts 63 G8
Eastcourt Wilts 81 G7
Eastdon Devon 14 F5
Eastdown Devon 8 F6
Eastend Essex 86 C6
Eastend Oxon 100 G6
Easter Aberchalder Highld 291 B7
Easter Ardross Highld 300 B6
Easter Balgedie Perth 286 G5
Easter Balmoral Aberds 292 D4
Easter Boleskine Highld 300 G5
Easter Brackland Stirl 285 G10
Easter Brae Highld 300 C6
Easter Cardno Aberds 303 C9
Easter Compton S Glos 60 C5
Easter Cringate Stirl 278 D4
Easter Culfosie Aberds 293 C9
Easter Davoch Aberds 292 C6
Easter Earshaig Dumfries 248 C2
Easter Ellister Argyll 254 B3
Easter Fearn Highld 309 L6
Easter Galcantray Highld 301 E8
Easter Housebyres Borders 262 B2
Easter Howgate Midloth 270 C4
Easter Howlaws Borders 272 G4
Easter Kinkell Highld 300 D5
Easter Knox Angus 287 D9
Easter Langlee Borders 262 B2
Easter Lednathie Angus 292 G5
Easter Milton Highld 301 D9
Easter Moniack Highld 300 E5
Easter Ord Aberdeen 293 C10
Easter Quarff Shetland 313 K6
Easter Rhynd Perth 286 F5
Easter Row Stirl 278 B5
Easter Silverford Aberds 303 C7
Easter Skeld Shetland 313 J5
Easter Softlaw Borders 263 C7
Easter Tulloch Highld 291 B11
Easter Whyntie Aberds 302 C6
Eastergate W Sus 22 B6
Easterhouse Glasgow 268 B3
Eastern Green W Mid 134 G5
Easterside Mbro 225 B10
Easterton Wilts 46 C4
Easterton of Lenabo Aberds 303 E10
Easterton Sands Wilts 46 B4
Eastertown Som 43 C10
Eastertown of Auchleuchries Aberds 303 F10
Eastfield Borders 262 D2
Eastfield Bristol 60 D5
Eastfield N Lnrk 269 C7
Eastfield N Lnrk 278 G4
Eastfield N Yorks 217 C10
Eastfield Northumb 243 B7
Eastfield P'boro 138 D4
Eastfield S Lnrk 268 C2
Eastfield S Yorks 197 G9
Eastfield Hall Northumb 252 B6
Eastgate Durham 232 D5
Eastgate Norf 160 E2
Eastgate P'boro 138 D4
Easthall Herts 104 G3
Eastham Mers 182 E5
Eastham Worcs 116 D3
Eastham Ferry Mers 182 E5
Easthampstead Brack 65 F11
Easthampton Hereford 115 E8
Easthaugh Norf 159 F11
Eastheath Wokingham 65 F10
Easthope Shrops 131 D11
Easthopewood Shrops 131 D11
Easthorpe Essex 107 G8
Easthorpe Leics 154 B6
Easthorpe Notts 172 E2
Easthouse Shetland 313 J5
Easthouses Midloth 270 B6
Easting Orkney 314 A7
Eastington Devon 26 F2
Eastington Glos 80 D3
Eastington Glos 81 C10
Eastland Gate Hants 33 E11
Eastleach Martin Glos 82 D2
Eastleach Turville Glos 81 D11
Eastleigh Devon 25 B7
Eastleigh Hants 32 D6
Eastling Kent 54 B3
Eastmoor Derbys 186 G4
Eastmoor Norf 140 C4
Eastney Ptsmth 21 B9
Eastnor Hereford 98 D4
Eastoft N Lincs 199 D10
Eastoke Hants 21 B10
Easton Bristol 60 E6
Easton Cambs 122 C2
Easton Cumb 239 C10
Easton Cumb 239 F7
Easton Devon 8 F3
Easton Devon 13 D10
Easton Dorset 17 G9
Easton Hants 48 G4
Easton I o W 20 D2
Easton Lincs 155 E8
Easton Norf 160 C2
Easton Som 44 E4
Easton Suff 126 E5
Easton W Berks 64 E2
Easton Wilts 61 E11
Easton Grey Wilts 61 B11
Easton in Gordano N Som 60 C4
Easton Maudit Northants 121 E7
Easton on the Hill Northants 137 C10
Easton Royal Wilts 63 G8
Easton Town Som 44 G5
Easton Town Wilts 61 B11
Eastover Som 43 F10
Eastpark Dumfries 238 D2
Eastrea Cambs 138 D5
Eastriggs Dumfries 238 D6
Eastrington E Yorks 199 B9
Eastrip Wilts 61 E10
Eastrop Hants 48 C6
Eastry Kent 55 C10
Eastville Bristol 60 E6
Eastville Lincs 174 D6
Eastwell Leics 154 D5
Eastwell Park Kent 54 D4
Eastwick Herts 86 C6
Eastwick Shetland 312 F5
Eastwood Hereford 98 C2
Eastwood Notts 171 F7
Eastwood S Yorks 186 C5
Eastwood Sthend 69 B10
Eastwood W Yorks 196 B3
Eastwood End Cambs 139 E8
Eastwood Hall Notts 171 F7
Eathorpe Warks 119 D7
Eaton E Ches 168 B5
Eaton Hereford 115 F10
Eaton Leics 154 D5
Eaton Norf 142 B4
Eaton Notts 188 F2
Eaton Oxon 82 E6
Eaton Shrops 131 F10
Eaton Shrops 131 F7
Eaton W Ches 167 C9
Eaton Bishop Hereford 97 D9
Eaton Bray C Beds 103 G9
Eaton Constantine Shrops 131 B11
Eaton Ford Cambs 122 E3
Eaton Green C Beds 103 G9
Eaton Hastings Oxon 82 F3
Eaton Mascott Shrops 131 B10
Eaton on Tern Shrops 150 E3
Eaton Socon Cambs 122 F3
Eaton upon Tern Shrops 150 E3
Eau Brink Norf 157 F11
Eau Withington Hereford 97 C10
Eaves Green W Mid 134 G5
Eavestone N Yorks 214 F4
Ebberly Hill Devon 25 D9
Ebberston N Yorks 217 C7
Ebbesbourne Wake Wilts 31 C7
Ebblake Hants 31 F10
Ebbw Vale Bl Gwent 77 D11
Ebchester Durham 242 F4
Ebdon N Som 59 G11
Ebernoe W Sus 35 B7
Ebford Devon 14 D5
Ebley Glos 80 D4
Ebnal W Ches 167 F7
Ebnall Hereford 115 F9
Ebreywood Shrops 149 F10
Ebrington Glos 100 C3
Ecchinswell Hants 48 B4
Ecclaw Borders 272 B5
Eccle Riggs Cumb 210 B4
Ecclefechan Dumfries 238 C5
Eccles Borders 272 G5
Eccles Gtr Man 184 B3
Eccles Kent 69 G8
Eccles on Sea Norf 161 D8
Eccles Road Norf 141 E10
Ecclesall S Yorks 186 E4
Ecclesfield S Yorks 186 C5
Ecclesgreig Aberds 293 G9
Eccleshall Staffs 150 D6
Eccleshill W Yorks 205 F9
Ecclesmachan W Loth 279 G11
Eccleston Lancs 194 D4
Eccleston Mers 183 B7
Eccleston W Ches 166 C6
Eccleston Park Mers 183 C7
Eccliffe Dorset 30 B3
Eccup W Yorks 205 E11
Echt Aberds 293 C9
Eckford Borders 262 D6
Eckfordmoss Borders 262 D6
Eckington Derbys 186 F6
Eckington Worcs 99 C8
Eckington Corner E Sus 23 D8
Ecklands S Yorks 197 G8
Eckworthy Devon 24 D6
Ecton Northants 120 E6
Ecton Staffs 169 D9
Ecton Brook Northants 120 E6
Edale Derbys 185 D10
Edale End Derbys 185 D11
Edbrook Som 43 E8
Edburton W Sus 36 E2
Edderside Cumb 229 B7
Edderton Highld 309 L7
Eddington Kent 71 F7
Eddington W Berks 63 F10
Eddistone Devon 24 C3
Eddleston Borders 270 F4
Eddlewood S Lnrk 268 E4
Eden Mount Cumb 211 D8
Eden Park London 67 F11
Eden Vale Durham 234 D4
Eden Vale Wilts 45 C11
Edenbridge Kent 52 D2
Edenfield Lancs 195 D9
Edenhall Cumb 231 E7
Edenham Lincs 155 E11
Edensor Derbys 170 B2
Edentaggart Argyll 276 C6
Edenthorpe S Yorks 198 F6
Edentown Cumb 239 F9
Ederline Argyll 275 C9
Edern Gwyn 144 B5
Edford Som 45 D7
Edgarley Som 44 F4
Edgbaston W Mid 133 G11
Edgcote Northants 101 B10
Edgcott Bucks 102 G3
Edgcott Som 41 F10
Edgcumbe Corn 2 C6
Edge Glos 80 D4
Edge Shrops 131 B7
Edge End Glos 79 C9
Edge End Lancs 203 G10
Edge Fold Blkburn 195 D8
Edge Fold Gtr Man 195 F8
Edge Green Gtr Man 183 B9
Edge Green Norf 141 F10
Edge Green W Ches 167 E7
Edge Hill Mers 182 C5
Edge Hill Warks 134 D4
Edge Mount S Yorks 186 C3
Edgebolton Shrops 149 E11
Edgefield Norf 159 C11
Edgefield Street Norf 159 C11
Edgehill Warks 101 B7
Edgeley Gtr Man 184 D5
Edgerley Shrops 148 F6
Edgerston Borders 262 G5
Edgerton W Yorks 196 D6
Edgeside Lancs 195 C10
Edgeworth Glos 80 D6
Edginswell Devon 9 B7
Edgiock Worcs 117 E10
Edgmond Telford 150 F4
Edgmond Marsh Telford 150 E4
Edgton Shrops 131 F7
Edgware London 85 G11
Edgwick W Mid 134 G6
Edgworth Blkburn 195 D8
Edham Borders 262 B6
Edial Staffs 133 B11
Edinample Stirl 285 E9
Edinbane Highld 298 D3
Edinburgh Edin 280 G5
Edinchip Stirl 285 E9
Edingale Staffs 152 G4
Edingight Ho Moray 302 D5
Edinglassie Ho Aberds 292 B5
Edingley Notts 171 D11
Edingthorpe Norf 160 C6
Edingthorpe Green Norf 160 C6
Edington Som 43 E11
Edington Wilts 46 C2
Edingworth Som 43 C11
Edintore Moray 302 E4
Edistone Devon 24 C2
Edith Weston Rutland 137 B8
Edithmead Som 43 D10
Edlaston Derbys 169 G11
Edlesborough Bucks 85 B7
Edlingham Northumb 252 B4
Edlington Lincs 190 G2
Edmondsham Dorset 31 E9
Edmondsley Durham 233 B10
Edmondstown Rhondda 77 G8
Edmondthorpe Leics 155 F7
Edmonston S Lnrk 269 G11
Edmonstone Orkney 314 D5
Edmonton Corn 10 G5
Edmonton London 86 G4
Edmundbyers Durham 242 G2
Ednam Borders 262 B6
Ednaston Derbys 170 G2
Edney Common Essex 87 E11
Edradynate Perth 286 B2
Edrom Borders 272 D6
Edstaston Shrops 149 C10
Edstone Warks 118 E3
Edvin Loach Hereford 116 F3
Edwalton Notts 153 B11
Edwardstone Suff 107 C8
Edwardsville M Tydf 77 F9
Edwinsford Carms 94 E2
Edwinstowe Notts 171 B10
Edworth C Beds 104 C4
Edwyn Ralph Hereford 116 F2
Edzell Angus 293 G7
Efail-fôch Neath 57 B9
Efail Isaf Rhondda 58 C5
Efailnewydd Gwyn 145 B7
Efailwen Carms 92 F2
Efenechtyd Denb 165 D10
Effingham Sur 50 C6
Effirth Shetland 313 H5
Effledge Borders 262 F3
Efflinch Staffs 152 F3
Efford Devon 26 G5
Efford Plym 7 D10
Egbury Hants 48 C2
Egdon Worcs 117 G8
Egerton Gtr Man 195 E8
Egerton Kent 54 D2
Egerton Forstal Kent 53 D11
Egerton Green E Ches 167 E8
Egford Som 45 D9
Eggbeare Corn 12 D2
Eggborough N Yorks 198 C5
Eggbuckland Plym 7 D10
Eggesford Station Devon 25 E11
Eggington C Beds 103 F9
Egginton Derbys 152 D5
Egginton Common Derbys 152 D5
Egglesburn Durham 232 G5
Egglescliffe Stockton 225 C8
Eggleston Durham 232 G5
Egham Sur 66 E4
Egham Hythe Sur 66 E4
Egham Wick Sur 66 E3
Egleton Rutland 137 B7
Eglingham Northumb 264 F4
Egloshayle Corn 10 G5
Egloskerry Corn 11 D11
Eglwys-Brewis V Glam 58 F4
Eglwys Cross Wrex 167 G7
Eglwys Fach Ceredig 128 D3
Eglwysbach Conwy 180 G4
Eglwyswen Pembs 92 D3
Eglwyswrw Pembs 92 D2
Egmanton Notts 172 B2
Egmere Norf 159 B8
Egremont Cumb 219 C10
Egremont Mers 182 C4
Egton N Yorks 226 D6
Egton Bridge N Yorks 226 D6
Egypt Bucks 66 B3
Egypt Hants 48 E3
Egypt W Berks 64 D2
Egypt W Yorks 205 G7
Eiden Highld 309 J7
Eight Ash Green Essex 107 F8
Eighton Banks T & W 243 F7
Eign Hill Hereford 97 D10
Eignaig Highld 289 E9
Eil Highld 291 B10
Eilanreach Highld 295 D10
Eildon Borders 262 C3
Eilean Anabaich W Isles 305 H4
Eilean Darach Highld 307 L6
Eilean Shona Ho Highld 289 B8
Eileanach Lodge Highld 300 C5
Einacleite W Isles 304 F3
Einsiob = Evenjobb Powys 114 E5
Eisgean W Isles 305 G5
Eisingrug Gwyn 146 C2
Elan Village Powys 113 D8
Eland Green Northumb 242 C5
Elberton S Glos 60 B6
Elborough N Som 43 B11
Elbridge Shrops 149 E7
Elbridge W Sus 22 C6
Elburton Plym 7 E10
Elcho Perth 286 E5
Elcock's Brook Worcs 117 E10
Elcombe Glos 80 F3
Elcombe Swindon 62 C6
Elcot W Berks 63 F11
Eldene Swindon 63 C7
Elder Street Essex 105 E11
Eldernell Cambs 138 D6
Eldersfield Worcs 98 E6
Elderslie Renfs 267 C8
Eldon Durham 233 F10
Eldon Lane Durham 233 F10
Eldrick S Ayrs 245 G7
Eldroth N Yorks 212 F5
Eldwick W Yorks 205 E8
Elemore Vale T & W 234 B3
Elerch = Bont-goch Ceredig 128 F3
Elfhowe Cumb 221 F9
Elford Northumb 264 C5
Elford Staffs 152 G3
Elford Closes Cambs 123 C10
Elgin Moray 302 C2
Elgol Highld 295 D7
Elham Kent 55 E7
Eliburn W Loth 269 B10
Elie Fife 287 G8
Elim Anglesey 178 D5
Eling Hants 32 E5
Eling W Berks 64 D4
Elishader Highld 298 C5
Elishaw Northumb 251 D9
Elizafield Dumfries 238 C2
Elkesley Notts 187 F11
Elkington Northants 120 B2
Elkins Green Essex 87 E10
Elkstone Glos 81 C7
Ellacombe Torbay 9 C8
Ellan Highld 301 G8
Elland W Yorks 196 C6
Elland Lower Edge W Yorks 196 C6
Elland Upper Edge W Yorks 196 C6
Ellary Argyll 275 F8
Ellastone Staffs 169 G10
Ellel Lancs 202 B5
Ellemford Borders 272 C4
Ellenborough Cumb 228 D6
Ellenbrook Herts 86 D2
Ellenbrook I o M 192 E4
Ellenglaze Corn 4 D5
Ellenhall Staffs 150 D6
Ellen's Green Sur 50 F5
Ellerbeck N Yorks 225 F8
Ellerburn N Yorks 216 C6
Ellerby N Yorks 226 C5
Ellerdine Telford 150 E2
Ellerdine Heath Telford 150 E2
Ellerhayes Devon 27 G7
Elleric Argyll 284 C4
Ellerker E Yorks 200 B2
Ellerton E Yorks 207 F10
Ellerton N Yorks 224 F5
Ellerton Shrops 150 D4
Ellesborough Bucks 84 D4
Ellesmere Shrops 149 C8
Ellesmere Park Gtr Man 184 B3
Ellesmere Port W Ches 182 F6
Ellicombe Som 42 E3
Ellingham Hants 31 F10
Ellingham Norf 143 E7
Ellingham Northumb 264 D5
Ellingstring N Yorks 214 C3
Ellington Cambs 122 C3
Ellington Northumb 253 E7
Ellington Thorpe Cambs 122 C3
Elliot Angus 287 D10
Elliots Green Som 45 D9
Elliot's Town Caerph 77 E10
Ellisfield Hants 48 D6
Elliston Borders 262 D3
Ellistown Leics 153 G8
Ellon Aberds 303 F9
Ellonby Cumb 230 D4
Ellough Suff 143 F8
Elloughton E Yorks 200 B2
Ellwood Glos 79 D9
Elm Cambs 139 B9
Elm Corner Sur 50 B5
Elm Cross Wilts 62 D6
Elm Hill Dorset 30 B4
Elm Park London 68 B4
Elmbridge Glos 80 B5
Elmbridge Worcs 117 D8
Elmdon Essex 105 D9
Elmdon W Mid 134 G3
Elmdon Heath W Mid 134 G3
Elmer W Sus 35 G7
Elmers End London 67 F11
Elmers Green Lancs 194 F3
Elmers Marsh W Sus 34 B5
Elmesthorpe Leics 135 D9
Elmfield I o W 21 C8
Elmhurst Bucks 84 B4
Elmhurst Staffs 152 G2
Elmley Castle Worcs 99 C9
Elmley Lovett Worcs 117 D7
Elmore Glos 80 B3
Elmore Back Glos 80 B3
Elms Green Hereford 115 F10
Elms Green Worcs 116 D4
Elmscott Devon 24 C2
Elmsett Suff 107 B11
Elmslack Lancs 211 D9
Elmstead Essex 107 F11
Elmstead London 68 E2
Elmstead Heath Essex 107 G11
Elmstead Market Essex 107 G11
Elmsted Kent 54 E6
Elmsthorpe Leics 135 D9
Elmstone Kent 71 G9
Elmstone Hardwicke Glos 99 F8
Elmswell E Yorks 208 B5
Elmswell Suff 125 E9
Elmton Derbys 187 G8
Elness Orkney 314 C6
Elphin Highld 307 H7
Elphinstone E Loth 281 G7
Elrick Aberds 293 C10
Elrig Dumfries 236 E5
Elrigbeag Argyll 284 F5
Elrington Northumb 241 E9
Elscar S Yorks 197 G11
Elsdon Hereford 114 G6
Elsdon Northumb 251 E10
Elsecar S Yorks 186 B5
Elsecar S Yorks 197 G11
Elsenham Essex 105 F10
Elsenham Sta Essex 105 F10
Elsfield Oxon 83 C8
Elsham N Lincs 200 E4
Elsing Norf 159 F11
Elslack N Yorks 204 D4
Elson Hants 33 G10
Elson Shrops 149 B7
Elsrickle S Lnrk 269 G11
Elstead Sur 50 E2
Elsted W Sus 34 D4
Elsthorpe Lincs 155 E11
Elstob Durham 234 G2
Elston Devon 26 G3
Elston Notts 172 F3
Elston Wilts 46 E5
Elstone Devon 25 D11
Elstow Beds 103 B11
Elstree Herts 85 F11
Elstronwick E Yorks 209 G10
Elswick Lancs 202 F4
Elswick T & W 242 E6
Elswick Leys Lancs 202 F4
Elsworth Cambs 122 E6
Elterwater Cumb 220 E6
Eltham London 68 E2
Eltisley Cambs 122 F5
Elton Cambs 137 E11
Elton Derbys 170 C2
Elton Glos 80 C2
Elton Gtr Man 195 E9
Elton Hereford 115 C9
Elton Notts 154 B5
Elton Stockton 225 B8
Elton W Ches 183 F7
Elton Green W Ches 183 G7
Elton's Marsh Hereford 97 C9
Eltringham Northumb 242 E3
Elvanfoot S Lnrk 259 F11
Elvaston Derbys 153 C8
Elveden Suff 124 B6
Elvet Hill Durham 233 C11
Elvingston E Loth 281 G9
Elvington Kent 55 C9
Elvington York 207 D9
Elwell Devon 41 G7
Elwell Dorset 17 E9
Elwick Hrtlpl 234 E5
Elwick Northumb 264 B4
Elworth E Ches 168 C2
Elworthy Som 42 G5
Ely Cambs 139 G10
Ely Cardiff 58 D6
Emberton M Keynes 103 B7
Embleton Cumb 229 E9
Embleton Durham 234 F4
Embleton Northumb 264 E6
Embo Highld 311 K2
Embo Street Highld 311 K2
Emborough Som 44 C6
Embsay N Yorks 204 C6
Emerson Park London 68 B4
Emerson Valley M Keynes 102 E6
Emerson's Green S Glos 61 D7
Emery Down Hants 32 F3
Emley W Yorks 197 E8
Emmbrook Wokingham 65 F9
Emmer Green Reading 65 D8
Emmett Carr Derbys 187 F7
Emmington Oxon 84 E2
Emneth Norf 139 B9
Emneth Hungate Norf 139 B10
Emorsgate Norf 157 E10
Empingham Rutland 137 B8
Empshott Hants 49 G9
Empshott Green Hants 49 G8
Emscote Warks 118 D5
Emstrey Shrops 149 G10
Emsworth Hants 22 B2
Enborne W Berks 64 F2
Enborne Row W Berks 64 G2
Enchmarsh Shrops 131 D10
Enderby Leics 135 D10
Endmoor Cumb 211 C10
Endon Staffs 168 E6
Endon Bank Staffs 168 E6
Energlyn Caerph 58 B6
Enfield London 86 F4
Enfield Worcs 117 D10
Enfield Highway London 86 F5
Enfield Lock London 86 F5
Enfield Town London 86 F4
Enfield Wash London 86 F5
Enford Wilts 46 C6
Engamoor Shetland 313 H4
Engedi Anglesey 178 F5
Engine Common S Glos 61 C7
Englefield W Berks 64 E6
Englefield Green Sur 66 E3
Englesea-brook E Ches 168 E3
English Bicknor Glos 79 B9
English Frankton Shrops 149 D9
Englishcombe Bath 61 G8
Engollan Corn 10 G3
Enham Alamein Hants 47 D11
Enis Devon 25 B9
Enisfirth Shetland 312 F5
Enmore Som 43 G8
Enmore Field Hereford 115 E9
Enmore Green Dorset 30 C5
Ennerdale Bridge Cumb 219 B11
Enniscaven Corn 5 D9
Enoch Dumfries 247 C9
Enochdhu Perth 292 G2
Ensay Argyll 288 E5
Ensbury Bmouth 19 B7
Ensbury Park Bmouth 19 C7
Ensdon Shrops 149 F8
Ensis Devon 25 B9
Enslow Oxon 83 B7
Enstone Oxon 101 G7
Enterkinfoot Dumfries 247 C9
Enterpen N Yorks 225 D9
Enton Green Sur 50 E3
Enville Staffs 132 F6
Eolaigearraidh W Isles 297 L3
Eorabus Argyll 288 G5
Eòropaidh W Isles 304 B7
Epney Glos 80 C3
Epperstone Notts 171 F11
Epping Essex 87 E7
Epping Green Essex 86 D6
Epping Green Herts 86 D3
Epping Upland Essex 86 E6
Eppleby N Yorks 224 C3
Eppleworth E Yorks 208 G6
Epsom Sur 67 G8
Epwell Oxon 101 C7
Epworth N Lincs 199 G9
Epworth Turbary N Lincs 199 G9
Erbistock Wrex 166 G5
Erbusaig Highld 295 C9
Erchless Castle Highld 300 E4
Erdington W Mid 134 E2
Eredine Argyll 275 C10
Eriboll Highld 308 D4
Ericstane Dumfries 260 G3
Eridge Green E Sus 52 F5
Erines Argyll 275 F9
Eriswell Suff 124 B4
Erith London 68 D4
Erlestoke Wilts 46 C3
Ermine Lincs 189 G7
Ermington Devon 8 E2
Ernesettle Plym 7 D8
Erpingham Norf 160 C3
Erriottwood Kent 54 B2
Errogie Highld 300 G5
Errol Perth 286 E6
Errol Station Perth 286 E6
Erskine Renfs 277 G9
Erskine Bridge Renfs 277 G9
Ervie Dumfries 236 C2
Erwarton Suff 108 E4
Erwood Powys 95 C11
Eryholme N Yorks 224 D6
Eryrys Denb 166 D2
Escomb Durham 233 E9
Escott Som 42 F5
Escrick N Yorks 207 E8
Esgairdawe Carms 94 C2
Esgairgeiliog Powys 128 B5
Esgyryn Conwy 180 F4
Esh Durham 233 C9
Esh Winning Durham 233 C9
Esher Sur 66 G6
Eshiels Borders 261 B7
Esholt W Yorks 205 E9
Eshott Northumb 252 D6
Eshton N Yorks 204 B4
Esk Valley N Yorks 226 E6
Eskadale Highld 300 F4
Eskbank Midloth 270 B6
Eskdale Green Cumb 220 E2
Eskdalemuir Dumfries 249 D7
Eske E Yorks 209 E7
Eskham Lincs 190 B5
Eskholme S Yorks 198 D6
Esknish Argyll 274 G4
Eslington Park Northumb 264 G2
Esperley Lane Ends Durham 233 G8
Esprick Lancs 202 F4
Essendine Rutland 155 G10
Essendon Herts 86 D3
Essich Highld 300 F6
Essington Staffs 133 C9
Esslemont Aberds 303 G9
Eston Redcar 225 B11
Estover Plym 7 D10
Eswick Shetland 313 H6
Etal Northumb 263 B10
Etchilhampton Wilts 62 G4
Etchingham E Sus 38 B2
Etchinghill Kent 55 F7
Etchinghill Staffs 151 F10
Etchingwood E Sus 37 C8
Etherley Dene Durham 233 F9
Ethie Castle Angus 287 C10
Ethie Mains Angus 287 C10
Etling Green Norf 159 G10
Etloe Glos 79 D11
Eton Windsor 66 D3
Eton Wick Windsor 66 D2
Etruria Stoke 168 F5
Etsell Shrops 131 C7
Etterby Cumb 239 F9
Etteridge Highld 291 D8
Ettersgill Durham 232 F3
Ettiley Heath E Ches 168 C2
Ettingshall W Mid 133 D8
Ettingshall Park W Mid 133 D8
Ettington Warks 100 B5
Etton E Yorks 208 E5
Etton P'boro 138 B2
Ettrick Borders 261 G7
Ettrickbridge Borders 261 E9
Ettrickdale Argyll 275 G11
Ettrickhill Borders 261 G7
Etwall Derbys 152 C5
Etwall Common Derbys 152 C5
Eudon Burnell Shrops 132 F3
Eudon George Shrops 132 F3
Euston Suff 125 B7
Euximoor Drove Cambs 139 D9
Euxton Lancs 194 D5
Evanstown Bridgend 58 B3
Evanton Highld 300 C6
Eve Hill W Mid 133 E8
Evedon Lincs 173 F9
Evelix Highld 309 K7
Even Pits Hereford 97 D11
Even Swindon Swindon 62 B6
Evendine Hereford 98 C5
Evenjobb = Einsiob Powys 114 E5
Evenley Northants 101 E11
Evenlode Glos 100 F4
Evenwood Durham 233 G9
Evenwood Gate Durham 233 G9
Everbay Orkney 314 D6
Evercreech Som 44 F6
Everdon Northants 119 F11
Everingham E Yorks 208 E2
Everland Shetland 312 D8
Everleigh Wilts 47 C8
Everley N Yorks 217 B9
Eversholt C Beds 103 E9
Evershot Dorset 29 G9
Eversley Essex 69 B8
Eversley Hants 65 G9
Eversley Centre Hants 65 G9
Eversley Cross Hants 65 G9
Everthorpe E Yorks 208 G4
Everton C Beds 122 G4
Everton Hants 19 C11
Everton Mers 182 C5
Everton Notts 187 C11
Evertown Dumfries 239 B9
Evesbatch Hereford 98 B3
Evesham Worcs 99 C10
Evington Kent 54 D6
Evington Leicester 136 C2
Ewanrigg Cumb 228 D6
Ewden Village S Yorks 186 B3
Ewell Sur 67 G8
Ewell Minnis Kent 55 E9
Ewelme Oxon 83 G10
Ewen Glos 81 F8
Ewenny V Glam 58 D2
Ewerby Lincs 173 F10
Ewerby Thorpe Lincs 173 F10
Ewes Dumfries 249 E9
Ewesley Northumb 252 E3
Ewhurst Sur 50 E5
Ewhurst Green E Sus 38 C3
Ewhurst Green Sur 50 F5
Ewloe Flint 166 B4
Ewloe Green Flint 166 B3
Ewood Blkburn 195 B7
Ewood Bridge Lancs 195 C9
Eworthy Devon 12 C5
Ewshot Hants 49 D10
Ewyas Harold Hereford 97 F7
Exbourne Devon 25 G10
Exbury Hants 20 B4
Exceat E Sus 23 F8
Exebridge Devon 26 C6
Exelby N Yorks 214 B5
Exeter Devon 14 C4
Exford Som 41 F11
Exfords Green Shrops 131 B9
Exhall Warks 118 F2
Exhall Warks 135 F7
Exlade Street Oxon 65 C7
Exley W Yorks 196 C5
Exley Head W Yorks 204 F6
Exminster Devon 14 D4
Exmouth Devon 14 E6
Exnaboe Shetland 313 M5
Exning Suff 124 D2
Exted Kent 55 E7
Exton Devon 14 D5
Exton Hants 33 C10
Exton Rutland 155 G8
Exton Som 42 G2
Exwick Devon 14 C4
Eyam Derbys 186 F2
Eydon Northants 119 G10
Eye Hereford 115 E9
Eye P'boro 138 C4
Eye Suff 126 C2
Eye Green P'boro 138 C4
Eyemouth Borders 273 C8
Eyeworth C Beds 104 B4
Eyhorne Street Kent 53 C10
Eyke Suff 126 G6
Eynesbury Cambs 122 F3
Eynort Highld 294 C5
Eynsford Kent 68 F4
Eynsham Oxon 82 D6
Eype Dorset 16 C5
Eyre Highld 295 B7
Eyre Highld 298 D4
Eyres Monsell Leicester 135 D11
Eythorne Kent 55 D9
Eyton Hereford 115 E9
Eyton Shrops 131 F7
Eyton Shrops 149 G7
Eyton Wrex 166 G4
Eyton on Severn Shrops 131 B11
Eyton upon the Weald Moors Telford 150 G3

F

Faberstown Wilts 47 C9
Faccombe Hants 47 B11
Faceby N Yorks 225 E9
Fachell Gwyn 163 B8
Fachwen Gwyn 163 C9
Facit Lancs 195 D11
Fackley Notts 171 C7
Faddiley E Ches 167 E9
Faddonch Highld 295 C11
Fadmoor N Yorks 216 B3
Faerdre Swansea 75 E11
Fagley W Yorks 205 G9
Fagwyr Swansea 75 E11
Faichem Highld 290 C4
Faifley W Dunb 277 G10
Failand N Som 60 E4
Failford S Ayrs 257 D11
Failsworth Gtr Man 195 G11
Fain Highld 299 B11
Faindouran Lodge Moray 292 C2
Fair Cross London 68 B3
Fair Green Norf 158 F3
Fair Hill Cumb 230 E6
Fair Moor Northumb 252 F5
Fair Oak Hants 33 D7
Fair Oak Hants 64 G5
Fair Oak Lancs 203 D8
Fair Oak Green Hants 65 G7
Fairbourne Gwyn 146 G2
Fairbourne Heath Kent 53 C11
Fairburn N Yorks 198 B3
Fairburn House Highld 300 D4
Fairfield Clack 279 C7
Fairfield Derbys 185 G9
Fairfield Gtr Man 184 B6
Fairfield Gtr Man 195 E10
Fairfield Kent 39 B7
Fairfield Mers 182 C5
Fairfield Stockton 225 B8
Fairfield Worcs 99 C10
Fairfield Worcs 117 B8
Fairfield Park Bath 61 F9
Fairfields Glos 98 E4
Fairford Glos 81 E11
Fairhaven Lancs 193 B10
Fairhaven N Ayrs 255 C10
Fairhill S Lnrk 268 E4
Fairlands Sur 50 C3
Fairlee I o W 20 C6
Fairlie N Ayrs 266 D4
Fairlight E Sus 38 E5
Fairlight Cove E Sus 38 E5
Fairlop London 87 G7
Fairmile Devon 15 B7
Fairmile Dorset 19 C9
Fairmile Sur 66 G6
Fairmilehead Edin 270 B4
Fairoak Caerph 77 F11
Fairoak Staffs 150 C5
Fairseat Kent 68 G6
Fairstead Essex 88 B3
Fairstead Norf 158 F2
Fairview Glos 99 G9
Fairwarp E Sus 37 B7
Fairwater Cardiff 58 D6
Fairwater Torf 78 G3
Fairwood Wilts 45 C10
Fairy Cottage I o M 192 D5
Fairy Cross Devon 24 C6
Fakenham Norf 159 D8
Fakenham Magna Suff 125 B8
Fala Midloth 271 C8
Fala Dam Midloth 271 C8
Falahill Borders 271 D7
Falcon Hereford 98 E2
Falcon Lodge W Mid 134 D2
Falconwood London 68 D3
Falcutt Northants 101 C11
Faldingworth Lincs 189 E9
Faldonside Borders 262 C2
Falfield Fife 287 G8
Falfield S Glos 79 G11
Falkenham Suff 108 D5
Falkenham Sink Suff 108 D5
Falkirk Falk 279 F7
Falkland Fife 286 G6
Falla Borders 262 G6
Fallgate Derbys 170 C5
Fallin Stirl 278 C6
Fallinge Derbys 170 B3
Fallings Heath W Mid 133 D9
Fallowfield Gtr Man 184 C5
Fallside N Lnrk 268 C4
Falmer E Sus 36 F5
Falmouth Corn 3 C8
Falnash Borders 249 B9
Falsgrave N Yorks 217 B10
Falside W Loth 269 B9
Falsidehill Borders 272 G3
Falstone Northumb 250 F6
Fanagmore Highld 306 E6
Fancott C Beds 103 F10
Fangdale Beck N Yorks 225 G11
Fangfoss E Yorks 207 C11
Fanich Highld 311 J2
Fankerton Falk 278 E5
Fanmore Argyll 288 E6
Fanner's Green Essex 87 C11
Fannich Lodge Highld 300 C2
Fans Borders 272 G2
Fanshowe E Ches 184 G5
Fant Kent 53 B8
Faoilean Highld 295 C7
Far Arnside Cumb 211 D8
Far Bank S Yorks 198 E6
Far Banks Lancs 194 C2
Far Bletchley M Keynes 103 E7

Far Coton Leics 135 C7
Far Cotton Northants 120 F4
Far End Cumb 220 F6
Far Forest Worcs 116 C4
Far Green Glos 80 E3
Far Hoarcross Staffs 152 E2
Far Laund Derbys 170 F5
Far Ley Staffs 132 D5
Far Moor Gtr Man 194 G4
Far Oakridge Glos 80 E6
Far Royds W Yorks 205 G11
Far Sawrey Cumb 221 F7
Far Thrupp Glos 80 E5
Farcet Cambs 138 E4
Farden Shrops 115 B11
Fareham Hants 33 F9
Farewell Staffs 151 G11
Farforth Lincs 190 F4
Farhill Derbys 170 C5
Faringdon Oxon 82 F3
Farington Lancs 194 B4
Farington Moss Lancs 194 C4
Farlam Cumb 240 F3
Farlands Booth Derbys 185 D9
Farlary Highld 309 J7
Farleigh N Som 60 F3
Farleigh Sur 67 G11
Farleigh Court Sur 67 G11
Farleigh Green Kent 53 C8
Farleigh Hungerford Som 45 B10
Farleigh Wallop Hants 48 D6
Farleigh Wick Wilts 61 G10
Farlesthorpe Lincs 191 G7
Farleton Cumb 211 C10
Farleton Lancs 211 F11
Farley Bristol 60 E2
Farley Derbys 170 C3
Farley Shrops 131 B7
Farley Shrops 132 C2
Farley Staffs 169 G9
Farley Wilts 32 B2
Farley Green Suff 124 G4
Farley Green Sur 50 D5
Farley Hill Luton 103 G11
Farley Hill Wokingham 65 G8
Farleys End Glos 80 B3
Farlington N Yorks 216 F2
Farlington Ptsmth 33 F11
Farlow Shrops 132 G2
Farm Town Leics 153 F7
Farmborough Bath 61 G7
Farmbridge End Essex 87 C10
Farmcote Glos 99 F11
Farmcote Shrops 132 E5
Farmington Glos 81 B10
Farmoor Oxon 82 D6
Farms Common Corn 2 C5
Farmtown Moray 302 D5
Farnah Green Derbys 170 F4
Farnborough Hants 49 C11
Farnborough London 68 G2
Farnborough W Berks 64 C2
Farnborough Warks 101 B8
Farnborough Green Hants 49 B11
Farnborough Park Hants 49 B11
Farnborough Street Hants 49 B11
Farncombe Sur 50 E3
Farndish Beds 121 E8
Farndon Notts 172 E3
Farndon W Ches 166 E6
Farnell Angus 287 B10
Farnham Dorset 31 D7
Farnham Essex 105 G9
Farnham N Yorks 215 G7
Farnham Suff 127 E7
Farnham Sur 49 D10
Farnham Common Bucks 66 C3
Farnham Green Essex 105 F9
Farnham Park Bucks 66 C3
Farnham Royal Bucks 66 C3
Farnhill N Yorks 204 D6
Farningham Kent 68 F4
Farnley N Yorks 205 D10
Farnley W Yorks 205 G11
Farnley Bank W Yorks 197 E7
Farnley Tyas W Yorks 197 E7
Farnsfield Notts 171 D10
Farnworth Gtr Man 195 F8
Farnworth Halton 183 D8
Farr Highld 291 C10
Farr Highld 300 F6
Farr Highld 308 C7
Farr House Highld 300 F6
Farraline Highld 300 G5
Farringdon Devon 14 C6
Farringdon T & W 243 G9
Farrington Dorset 30 D4
Farrington Gurney Bath 44 B6
Farsley W Yorks 205 F10
Farsley Beck Bottom W Yorks 205 F10
Farther Howegreen Essex 88 E4
Farthing Corner Medway 69 G10
Farthing Green Kent 53 D10
Farthinghoe Northants 101 D10
Farthingloe Kent 55 E9
Farthingstone Northants 120 F2
Fartown W Yorks 196 D6
Farway Devon 15 B9
Farway Marsh Devon 28 G4
Fasach Highld 297 G7
Fasag Highld 299 D8
Fascadale Highld 289 B7
Faslane Port Argyll 276 D4
Fasnacloich Argyll 284 C4
Fasnakyle Ho Highld 300 G3
Fassfern Highld 290 F2
Fatfield T & W 243 G8
Fattahead Aberds 302 D6
Faucheldean W Loth 279 G11
Faugh Cumb 240 G2
Faughill Borders 262 C2
Fauld Staffs 152 D3
Fauldhouse W Loth 269 C8
Fauldiehill Angus 287 D9
Fauldshope Borders 261 D10
Faulkbourne Essex 88 B3
Faulkland Som 45 C8
Fauls Shrops 149 C11
Faverdale Darl 224 B5
Faversham Kent 70 G4
Favillar Moray 302 F2
Fawdington N Yorks 215 E8
Fawdon Northumb 264 F2
Fawdon T & W 242 D6
Fawfieldhead Staffs 169 C9
Fawkham Green Kent 68 F5
Fawler Oxon 63 B10
Fawler Oxon 82 B5
Fawley Bucks 65 B9
Fawley Hants 33 G7
Fawley W Berks 63 C11
Fawley Bottom Bucks 65 B8
Fawley Chapel Hereford 97 F11
Faxfleet E Yorks 199 C11
Faygate W Sus 51 G8
Fazakerley Mers 182 B5
Fazeley Staffs 134 C4
Feagour Highld 291 D7
Fearby N Yorks 214 C3
Fearn Highld 301 B8
Fearn Lodge Highld 309 L6
Fearn Station Highld 301 B8
Fearnan Perth 285 C11
Fearnbeg Highld 299 D7
Fearnhead Warr 183 C10
Fearnmore Highld 299 C7
Fearnville W Yorks 206 F2
Featherstone Staffs 133 B8
Featherstone W Yorks 198 C2
Featherwood Northumb 251 C8
Feckenham Worcs 117 E10
Fedw Fawr Anglesey 179 E10
Feering Essex 107 G7
Feetham N Yorks 223 F9
Fegg Hayes Stoke 168 E5
Fèith Mhor Highld 301 G8
Feizor N Yorks 212 F5
Felbridge Sur 51 F11
Felbrigg Norf 160 B4
Felcourt Sur 51 E11
Felden Herts 85 E8
Felderland Kent 55 B10
Feldy E Ches 183 F11
Felhampton Shrops 131 F8
Felin-Crai Powys 95 G7
Felin Newydd Carms 94 D3
Felin-newydd Powys 96 D2
Felin Newydd = New Mills Powys 129 C11
Felin Puleston Wrex 166 F4
Felin-Wnda Ceredig 92 B6
Felindre Carms 75 C7
Felindre Carms 93 D7
Felindre Carms 93 G11
Felindre Carms 94 E3
Felindre Carms 94 F4
Felindre Ceredig 111 F10
Felindre Powys 96 D3
Felindre Powys 96 G3
Felindre Powys 130 C3
Felindre Powys 130 G3
Felindre Rhondda 58 C3
Felindre Swansea 75 E10
Felindre Farchog Pembs 92 D2
Felinfach Ceredig 111 F10
Felinfach Powys 95 E11
Felinfoel Carms 75 E8
Felingwmisaf Carms 93 G10
Felingwmuchaf Carms 93 G10
Felinwynt Ceredig 110 G4
Felixkirk N Yorks 215 C9
Felixstowe Suff 108 E5
Felixstowe Ferry Suff 108 D6
Felkington Northumb 273 G8
Felkirk W Yorks 197 E11
Fell End Cumb 222 F4
Fell Lane W Yorks 204 E6
Fell Side Cumb 230 D2
Felldyke Cumb 219 B11
Fellgate T & W 243 E8
Felling T & W 243 E7
Felling Shore T & W 243 E7
Fellside T & W 242 E5
Felmersham Beds 121 F9
Felmingham Norf 160 D5
Felmore Essex 69 B8
Felpham W Sus 35 H7
Felsham Suff 125 F8
Felsted Essex 106 G3
Feltham London 66 E6
Feltham Som 28 D2
Felthamhill London 66 E5
Felthorpe Norf 160 F3
Felton Hereford 97 B11
Felton N Som 60 F4
Felton Northumb 252 C5
Felton Butler Shrops 149 F7
Feltwell Norf 140 E4
Fen Ditton Cambs 123 E9
Fen Drayton Cambs 122 D6
Fen End Lincs 156 E4
Fen End W Mid 118 B4
Fen Side Lincs 174 D4
Fen Street Norf 141 G11
Fen Street Suff 125 B11
Fen Street Suff 125 B9
Fenay Bridge W Yorks 197 D7
Fence Lancs 204 F2
Fence Houses T & W 243 G8
Fencott Oxon 83 B9
Fengate Norf 160 E3
Fengate P'boro 138 D4
Fenham Northumb 273 G11
Fenham T & W 242 D6
Fenhouses Lincs 174 G3
Feniscliffe Blkburn 195 B7
Feniscowles Blkburn 194 B6
Feniton Devon 15 B8
Fenlake Beds 103 B11
Fenn Green Shrops 132 G5
Fenn Street Medway 69 D9
Fennington Som 27 B11
Fenn's Bank Wrex 149 B10
Fenny Bentley Derbys 169 E11
Fenny Bridges Devon 15 B8
Fenny Castle Som 44 E4
Fenny Compton Warks 119 G8
Fenny Drayton Leics 134 D6
Fenny Stratford M Keynes 103 E7
Fenrother Northumb 252 E5
Fenstanton Cambs 122 D6
Fenstead End Suff 124 G6
Fenton Cambs 122 B6
Fenton Cumb 240 F2
Fenton Lincs 172 E5
Fenton Lincs 188 F4
Fenton Northumb 263 C11
Fenton Stoke 168 G5
Fenton Barns E Loth 281 E10
Fenton Low Stoke 168 F5
Fenton Pits Corn 5 C11
Fenton Town Northumb 263 C11
Fentonadle Corn 11 F7
Fenwick E Ayrs 267 G9
Fenwick Northumb 242 C3
Fenwick Northumb 273 G11
Fenwick S Yorks 198 D5
Feochaig Argyll 255 F8
Feock Corn 3 B8
Feolin Ferry Argyll 274 G5
Ferguslie Park Renfs 267 C9
Ferindonald Highld 295 E8
Feriniquarrie Highld 296 F7
Ferlochan Argyll 289 E11
Fern Angus 292 G6
Fern Bucks 65 B11
Fern Bank Gtr Man 185 B7
Fern Hill Suff 106 B6
Ferndale Kent 52 E5
Ferndale Rhondda 77 F7
Ferndown Dorset 31 G9
Ferne Wilts 30 C6
Ferness Highld 301 E9
Ferney Green Cumb 221 F8
Fernham Oxon 82 G3
Fernhill Gtr Man 195 E10
Fernhill Rhondda 77 F8
Fernhill W Sus 51 E10
Fernhill Gate Gtr Man 195 F7
Fernhill Heath Worcs 117 F7
Fernhurst W Sus 34 B5
Fernie Fife 287 F7
Ferniebrae Aberds 303 D9
Ferniegair S Lnrk 268 E4
Ferniehirst Borders 271 G8
Fernilea Highld 294 B5
Fernilee Derbys 185 F8
Fernsplatt Corn 4 G5
Ferrensby N Yorks 215 G7
Ferring W Sus 35 G9
Ferry Hill Cambs 139 G7
Ferry Point Highld 309 L7
Ferrybridge W Yorks 198 C3
Ferryden Angus 287 B11
Ferryhill Aberdeen 293 C11
Ferryhill Durham 233 E11
Ferryhill Station Durham 234 E2
Ferryside = Glan-y-Ffer Carms 74 C5
Ferryton Highld 300 C6
Fersfield Norf 141 G11
Fersit Highld 290 F5
Feshiebridge Highld 291 C10
Fetcham Sur 50 B6
Fetterangus Aberds 303 D9
Fettercairn Aberds 293 F8
Fetterdale Fife 287 E8
Fettes Highld 300 D5
Fewcott Oxon 101 F10
Fewston N Yorks 205 C9
Fewston Bents N Yorks 205 C9
Ffair-Rhos Ceredig 112 D4
Ffairfach Carms 94 G2
Ffaldybrenin Carms 94 C2
Ffarmers Carms 94 C3
Ffawyddog Powys 78 B2
Ffodun = Forden Powys 130 C4
Ffont y gari = Font y gary V Glam 58 F5
Fford-las Denb 165 C10
Ffordd-y-Gyfraith Bridgend 57 E11
Ffordd-las Powys 96 D4
Fforest Carms 75 E9
Fforest-fach Swansea 56 B6
Fforest Gôch Neath 76 E2
Ffos-y-ffin Ceredig 111 E8
Ffos-y-go Wrex 166 E4
Ffostrasol Ceredig 93 B7
Ffridd Powys 130 D3
Ffrith Wrex 166 D3
Ffrwd Gwyn 163 D7
Ffwl y mwn = Fonmon V Glam 58 F4
Ffynnon Carms 74 B5
Ffynnon ddrain Carms 93 G8
Ffynnon Gron Pembs 91 F8
Ffynnon Gynydd Powys 96 C3
Ffynnon-oer Ceredig 111 G10
Ffynnongroes = Crosswell Pembs 92 D2
Ffynnongroyw Flint 181 E10
Fickleshole Sur 67 G11
Fidden Argyll 288 G5
Fiddes Aberds 293 E10
Fiddington Glos 99 E8
Fiddington Som 43 E8
Fiddington Sands Wilts 46 C4
Fiddleford Dorset 30 E4
Fiddler' Green Norf 141 D10
Fiddler's Ferry Mers 193 C11
Fiddler's Ferry Warr 183 D9
Fiddler's Green Glos 99 G8
Fiddler's Green Hereford 97 D11
Fiddlers Hamlet Essex 87 E7
Field Hereford 114 G6
Field Som 44 E6
Field Staffs 151 C10
Field Assarts Oxon 82 C4
Field Broughton Cumb 211 C7
Field Common Sur 66 F6
Field Dalling Norf 159 B10
Field Green Kent 38 B3
Field Head Leics 135 B9
Fields End Herts 85 D8
Field's Place Hereford 115 G8
Fife Keith Moray 302 D4
Fifehead Magdalen Dorset 30 C3
Fifehead Neville Dorset 30 E3
Fifehead St Quintin Dorset 30 E3
Fifield Oxon 82 B2
Fifield Wilts 46 C6
Fifield Windsor 66 D2
Fifield Bavant Wilts 31 B8
Figheldean Wilts 47 D7
Filands Wilts 62 B2
Filby Norf 161 G9
Filby Heath Norf 161 G9
Filchampstead Oxon 83 D7
Filey N Yorks 218 C2
Filgrave M Keynes 103 B7
Filham Devon 8 D2
Filkins Oxon 82 E2
Filleigh Devon 25 B11
Filleigh Devon 26 E2
Fillingham Lincs 188 D6
Fillongley Warks 134 F5
Filmore Hill Hants 33 B11
Filton S Glos 60 D6
Filwood Park Bristol 60 F5
Fimber E Yorks 217 G7
Finavon Angus 287 B8
Fincastle Ho Perth 291 G10
Finchairn Argyll 275 C10
Fincham Mers 182 C6
Fincham Norf 140 B3
Finchampstead Wokingham 65 G9
Finchdean Hants 34 E2
Finchingfield Essex 106 E3
Finchley London 86 G3
Findern Derbys 152 C6
Findhorn Moray 301 C10
Findhorn Bridge Highld 301 G8
Findo Gask Perth 286 E4
Findochty Moray 302 C4
Findon Aberds 293 D11
Findon W Sus 35 F10
Findon Mains Highld 300 C6
Findon Valley W Sus 35 F10
Findrack Ho Aberds 293 C8
Fine Street Hereford 96 D6
Finedon Northants 121 C8
Fineglen Argyll 275 B10
Fingal Street Suff 126 D4
Fingask Aberds 303 G7
Fingerpost Worcs 116 C4
Fingest Bucks 84 G3
Finghall N Yorks 214 B3
Fingland Cumb 239 F7
Fingland Dumfries 259 F7
Finglesham Kent 55 C10
Fingringhoe Essex 107 G10
Finham W Mid 118 B6
Finkle Street S Yorks 186 B4
Finlarig Stirl 285 D9
Finmere Oxon 102 E2
Finnart Perth 285 B9
Finney Green E Ches 184 E5
Finney Green Staffs 168 F3
Finningham Suff 125 D11
Finningley S Yorks 187 B11
Finnygaud Aberds 302 D5
Finsbury London 67 C10
Finsbury Park London 67 B10
Finstall Worcs 117 D9
Finsthwaite Cumb 211 B7
Finstock Oxon 82 B5
Finstown Orkney 314 E3
Fintry Aberds 303 D7
Fintry Dundee 287 D8
Fintry Stirl 278 D2
Finwood Warks 118 D3
Finzean Aberds 293 D8
Finzean Ho Aberds 293 D7
Fionnphort Argyll 288 G5
Fionnsbhagh W Isles 296 C6
Fir Toll Kent 54 E2
Fir Tree Durham 233 E8
Fir Vale S Yorks 186 C5
Firbank Cumb 222 G2
Firbeck S Yorks 187 D9
Firby N Yorks 214 B5
Firby N Yorks 216 F4
Firemore Highld 307 L3
Firgrove Gtr Man 196 E2
Firkin Argyll 285 G7
Firle E Sus 23 D7
Firs Lane Gtr Man 194 G6
First Coast Highld 307 K4
Firsby Lincs 175 C7
Firsdown Wilts 47 G8
Firswood Gtr Man 184 B4
Firth Borders 262 E2
Firth Moor Darl 224 C6
Firth Park S Yorks 186 C5
Firwood Fold Gtr Man 195 E8
Fishbourne I o W 21 C7
Fishbourne W Sus 22 C4
Fishburn Durham 234 E3
Fishcross Clack 279 B7
Fisher Place Cumb 220 B6
Fisherford Aberds 302 F6
Fishermead M Keynes 103 D7
Fisherrow E Loth 280 G6
Fishers Green Herts 104 F4
Fisher's Pond Hants 33 C7
Fishersgate Brighton 36 F3
Fisherstreet W Sus 50 G3
Fisherton Highld 301 D7
Fisherton S Ayrs 257 F7
Fisherton de la Mere Wilts 46 F4
Fisherwick Staffs 134 B3
Fishery Windsor 65 C11
Fishguard = Abergwaun Pembs 91 D9
Fishlake S Yorks 199 E7
Fishleigh Devon 25 F8
Fishleigh Barton Devon 25 C9
Fishleigh Castle Devon 25 F8
Fishley Norf 161 G8
Fishley W Mid 133 C10
Fishmere End Lincs 156 B5
Fishponds Bristol 60 D6
Fishpool Glos 98 F3
Fishpool Gtr Man 195 F10
Fishpool N Yorks 205 D10
Fishpools Powys 114 D3
Fishtoft Lincs 174 G5
Fishtoft Drove Lincs 174 F4
Fishtown of Usan Angus 287 B11
Fishwick Borders 273 E8
Fishwick Lancs 194 B5
Fiskavaig Highld 294 B5
Fiskerton Lincs 189 G8
Fiskerton Notts 172 E2
Fitling E Yorks 209 G11
Fittleton Wilts 46 D6
Fittleworth W Sus 35 D8
Fitton End Cambs 157 G8
Fitton Hill Gtr Man 196 G2
Fitz Shrops 149 F8
Fitzhead Som 27 B10
Fitzwilliam W Yorks 198 D2
Fiunary Highld 289 E8
Five Acres Glos 79 C9
Five Ash Down E Sus 37 C7
Five Ashes E Sus 37 C9
Five Bells Som 42 E5
Five Bridges Hereford 98 B3
Five Houses I o W 20 D4
Five Lane Ends Lancs 202 C6
Five Lanes Mon 78 G6
Five Oak Green Kent 52 D6
Five Oaks W Sus 35 B9
Five Roads Carms 75 D7
Five Ways Warks 118 D4
Five Wents Kent 53 C10
Fivecrosses W Ches 183 F8
Fivehead Som 28 C5
Fivelanes Corn 11 E10
Fixby W Yorks 196 C6
Flackley Ash E Sus 38 C5
Flack's Green Essex 88 B3
Flackwell Heath Bucks 65 B11
Fladbury Worcs 99 B9
Fladbury Cross Worcs 99 B9
Fladda Shetland 312 E5
Fladdabister Shetland 313 K6
Flagg Derbys 169 B10
Flaggoners Green Hereford 116 G2
Flamborough E Yorks 218 E4
Flamstead Herts 85 C9
Flamstead End Herts 86 E4
Flansham W Sus 35 G7
Flanshaw W Yorks 197 C10
Flappit Spring W Yorks 205 F7
Flasby N Yorks 204 B4
Flash Staffs 169 B8
Flashader Highld 298 D3
Flask Inn N Yorks 227 E8
Flathurst W Sus 35 C7
Flaunden Herts 85 E8
Flawborough Notts 172 G3
Flawith N Yorks 215 F9
Flax Bourton N Som 60 F4
Flax Moss Lancs 195 C9
Flaxby N Yorks 206 B3
Flaxholme Derbys 170 G4
Flaxlands Norf 142 E2
Flaxley Glos 79 B11
Flaxpool Som 42 F6
Flaxton N Yorks 216 G3
Fleckney Leics 136 E2
Flecknoe Warks 119 E10
Fledborough Notts 188 G4
Fleet Dorset 17 E8
Fleet Hants 22 C2
Fleet Hants 49 C10
Fleet Lincs 157 E7
Fleet Downs Kent 68 E5
Fleet Hargate Lincs 157 E7
Fleetend Hants 33 F8
Fleetlands Hants 33 G9
Fleets N Yorks 213 G9
Fleetville Herts 85 D11
Fleetwood Lancs 202 D2
Fleggburgh = Burgh St Margaret Norf 161 G8
Fleming Field Durham 234 C3
Flemings Kent 55 B9
Flemingston V Glam 58 E4
Flemington S Lnrk 268 D3
Flemington S Lnrk 268 G4
Flempton Suff 124 D6
Fleoideabhagh W Isles 296 C6
Fletcher's Green Kent 52 C4
Fletchersbridge Corn 6 B2
Fletchertown Cumb 229 C10
Fletching E Sus 36 C6
Fletching Common E Sus 36 C6
Fleuchary Highld 309 K7
Fleuchlang Dumfries 237 D8
Fleur-de-lis Caerph 77 F11
Flexbury Corn 24 F2
Flexford Hants 32 C6
Flexford Sur 50 D2
Flimby Cumb 228 E6
Flimwell E Sus 53 G8
Flint Flint 182 G2
Flint Cross Cambs 105 C8
Flint Hill Durham 242 G5
Flint Mountain = Mynydd Fflint Flint 182 G2
Flintham Notts 172 F2
Flinton E Yorks 209 F10
Flint's Green W Mid 134 G5
Flintsham Hereford 114 F6
Flishinghurst Kent 53 F9
Flitcham Norf 158 D4
Flitholme Cumb 222 B5
Flitton C Beds 103 D11
Flitwick C Beds 103 D10
Flixborough N Lincs 199 D11
Flixborough Stather N Lincs 199 E11
Flixton Gtr Man 184 C2
Flixton N Yorks 217 D10
Flixton Suff 142 F6
Flockton W Yorks 197 E8
Flockton Green W Yorks 197 D8
Flockton Moor W Yorks 197 E8
Flodaigh W Isles 296 F4
Flodden Northumb 263 B10
Flodigarry Highld 298 B4
Flood Street Hants 31 D10
Floodgates Hereford 114 F5
Flood's Ferry Cambs 139 E7
Flookburgh Cumb 211 D7
Flordon Norf 142 D3
Flore Northants 120 E2
Florence Stoke 168 G6
Flotterton Northumb 251 C11
Flowers Bottom Bucks 84 F4
Flowers Green E Sus 23 C10
Flowery Field Gtr Man 184 B6
Flowton Suff 107 B11
Fluchter E Dunb 277 G11
Flugarth Shetland 313 G6
Flush House W Yorks 196 F6
Flushdyke W Yorks 197 C9
Flushing Aberds 303 E10
Flushing Corn 3 C8
Flushing Corn 3 D7
Flyford Flavell Worcs 117 G9
Foals Green Suff 126 C5
Fobbing Thurrock 69 C8
Fochabers Moray 302 D3
Fochriw Caerph 77 D10
Fockerby N Lincs 199 D10
Fodderletter Moray 301 G11
Fodderstone Gap Norf 140 B3
Fodderty Highld 300 D5
Foddington Som 29 B9
Foel Powys 147 G9
Foel-gastell Carms 75 C8
Foffarty Angus 287 C8
Foggathorpe E Yorks 207 F11
Foggbrook Gtr Man 184 D6
Fogo Borders 272 F5
Fogorig Borders 272 F5
Fogrigarth Shetland 313 H4
Fogwatt Moray 302 D2
Foindle Highld 306 E6
Fold Head Lancs 195 D11
Fold Hill Lincs 175 E7
Folda Angus 292 G3
Foldrings S Yorks 186 C3
Fole Staffs 151 B10
Foleshill W Mid 135 G7
Foley Park Worcs 116 B6
Folke Dorset 29 E11
Folkestone Kent 55 F8
Folkingham Lincs 155 C11
Folkington E Sus 23 E9
Folksworth Cambs 138 F2
Folkton N Yorks 217 D11
Folla Rule Aberds 303 F7
Folley Shrops 132 D5
Follifoot N Yorks 206 C2
Follingsby T & W 243 E8
Folly Dorset 30 G2
Folly Pembs 91 G8
Folly Cross Devon 25 F7
Folly Gate Devon 13 B7
Folly Green Essex 106 F6
Fonmon = Ffwl-y-mwn V Glam 58 F4
Fonston Corn 11 C10
Font-y-gary = Ffont-y-gari V Glam 58 F5
Fonthill Bishop Wilts 46 G2
Fonthill Gifford Wilts 46 G2
Fontmell Magna Dorset 30 D5
Fontmell Parva Dorset 30 E4
Fontwell W Sus 35 F7
Foodieash Fife 287 F7
Foolow Derbys 185 F11
Footbridge Glos 99 F10
Footherley Staffs 134 C2
Footrid Worcs 116 C3
Foots Cray London 68 E3
Forbestown Aberds 292 B5
Force Forge Cumb 220 G6
Force Green Kent 52 B2
Force Mills Cumb 220 G6
Forcett N Yorks 224 C3
Ford Argyll 275 C9
Ford Bucks 84 D3
Ford Derbys 186 E6
Ford Devon 8 E2
Ford Devon 8 G5
Ford Devon 24 C6
Ford Devon 28 G2
Ford Glos 99 F11
Ford Hereford 115 F10
Ford Kent 71 F8
Ford Mers 182 B4
Ford Northumb 263 B10
Ford Pembs 91 F9
Ford Plym 7 D9
Ford Shrops 149 G8
Ford Som 27 B9
Ford Som 44 C5
Ford Staffs 169 E9
Ford W Sus 35 G7
Ford Wilts 47 G7
Ford Wilts 61 E10
Ford End Essex 87 B11
Ford End Essex 105 E9
Ford Forge Northumb 263 B10
Ford Green Lancs 202 D5
Ford Green Stoke 168 E5
Ford Heath Shrops 149 G8
Ford Hill Northumb 263 B11
Ford Street Som 27 D11
Forda Devon 12 C6
Forda Devon 40 F3
Fordbridge W Mid 134 F3
Fordcombe Kent 52 E4
Fordell Fife 280 D3
Forden = Ffodun Powys 130 C4
Forder Corn 7 D8
Forder Green Devon 8 B5
Fordgate Som 43 G10
Fordham Cambs 124 C2
Fordham Essex 107 F8
Fordham Norf 140 D2
Fordham Heath Essex 107 F8
Fordhouses W Mid 133 C8
Fordingbridge Hants 31 E10
Fordington Lincs 190 G6
Fordley T & W 243 C7
Fordon E Yorks 217 D10
Fordoun Aberds 293 F9
Ford's Green E Sus 36 B6
Ford's Green Suff 125 D11
Fordstreet Essex 107 F8
Fordton Devon 14 B2
Fordwater Devon 28 G4
Fordwells Oxon 82 C4
Fordwich Kent 55 B7
Fordyce Aberds 302 C5
Forebridge Staffs 151 E8
Foredale N Yorks 212 F6
Forehill S Ayrs 257 E8
Foreland Fields I o W 21 D9
Foreland Ho Argyll 274 G3
Foremark Derbys 152 D6
Forest Lodge Argyll 284 C6
Forest Lodge Highld 292 B2
Forest Lodge Perth 291 F11
Forest Becks Lancs 203 C11
Forest Coal Pit Mon 96 G5
Forest Gate Hants 33 E10
Forest Gate London 68 C2
Forest Green Glos 80 E4
Forest Green Sur 50 E6
Forest Hall Cumb 221 E10
Forest Hall T & W 243 D7
Forest Head Cumb 240 F3
Forest Hill London 67 E11
Forest Hill Oxon 83 D9
Forest Hill Wilts 63 F8
Forest Holme Lancs 195 B10
Forest-in-Teesdale Durham 232 F3
Forest Lane Head N Yorks 206 B2
Forest Mill Clack 279 C9
Forest Moor N Yorks 206 B2
Forest Row E Sus 52 G2
Forest Side I o W 20 D5
Forest Town Notts 171 C9
Forestburn Gate Northumb 252 D3
Forestdale London 67 G11
Foresterseat Moray 301 D11
Forestreet Devon 24 E5
Forestside W Sus 34 E3
Forewoods Common Wilts 61 G10
Forfar Angus 287 B8
Forgandenny Perth 286 F4
Forge Corn 4 F3
Forge Powys 128 D5
Forge Hammer Torf 78 F3
Forge Side Torf 78 D2
Forgewood N Lnrk 268 D4
Forgie Moray 302 D3
Forglen Ho Aberds 302 D6
Forgue Aberds 302 E6
Forhill Worcs 117 B11
Formby Mers 193 F10
Forncett End Norf 142 E2
Forncett St Mary Norf 142 E3
Forncett St Peter Norf 142 E3
Forneth Perth 286 C4
Fornham All Saints Suff 124 D6
Fornham St Genevieve Suff 124 D6
Fornham St Martin Suff 125 D7
Fornighty Highld 301 D9
Forrabury Corn 11 C7
Forres Moray 301 D10
Forrest Lodge Dumfries 246 F3
Forrestfield N Lnrk 269 B7
Forry's Green Essex 106 E5
Forsbrook Staffs 169 G7
Forse Highld 310 F6
Forse Ho Highld 310 F6
Forshaw Heath Warks 117 C11
Forsinain Highld 310 E3
Forsinard Highld 310 E2
Forsinard Station Highld 310 E2
Forstal Kent 53 B8
Forston Dorset 17 B9
Fort Augustus Highld 290 C5
Fort George Highld 301 D7
Fort Matilda Invclyd 276 F5
Fort William Highld 290 F3
Forteviot Perth 286 F4
Forth S Lnrk 269 E8
Forth Road Bridge Edin 280 F2
Forthampton Glos 99 E7
Forthay Glos 80 F2
Fortingall Perth 285 C11
Fortis Green London 67 B9
Forton Hants 33 G10
Forton Hants 48 E2
Forton Lancs 202 C5
Forton Shrops 149 F8
Forton Som 28 F4
Forton Staffs 150 E5
Forton Heath Shrops 149 F8
Fortrie Aberds 302 E6
Fortrie Aberds 303 E7
Fortrose Highld 301 D7
Fortuneswell Dorset 17 G9
Forty Green Bucks 84 E3
Forty Green Bucks 84 G6
Forty Hill London 86 F4
Forward Green Suff 125 F11
Forwood Glos 80 E5
Fosbury Wilts 47 B10
Foscot Oxon 100 G4
Foscote Bucks 102 D4
Foscote Northants 102 B3
Foscote Wilts 61 D11
Fosdyke Lincs 156 C6
Fosdyke Bridge Lincs 156 C6
Foss Perth 285 B11
Foss Cross Glos 81 D9
Fossebridge Glos 81 C9
Fostall Kent 70 G5
Fosten Green Kent 53 F10
Foster Street Essex 87 D7
Fosterhouses S Yorks 199 E7
Foster's Booth Northants 120 G3
Foster's Green Worcs 117 D9
Foston Derbys 152 C3
Foston Leics 136 D2
Foston Lincs 172 G5
Foston N Yorks 216 F3
Foston on the Wolds E Yorks 209 B8
Fotherby Lincs 190 C4
Fothergill Cumb 228 E6
Fotheringhay Northants 137 E11
Foubister Orkney 314 F5
Foul Anchor Cambs 157 F9
Foul End Warks 134 E4
Foul Mile E Sus 23 B10
Foulbridge Cumb 230 B4
Foulby W Yorks 197 D11
Foulden Borders 273 D8
Foulden Norf 140 D5
Foulford Hants 31 F11
Foulis Castle Highld 300 C5
Foulride Green E Sus 23 E9
Foulridge Lancs 204 E3
Foulsham Norf 159 E10
Foundry Corn 2 B3
Foundry Hill Norf 159 D11
Fountain Bridgend 57 E11
Fountainhall Borders 271 F8
Four Ashes Bucks 84 F5
Four Ashes Staffs 132 F6
Four Ashes Staffs 133 B8
Four Ashes Suff 125 C10
Four Ashes W Mid 118 B3
Four Crosses Powys 129 B11
Four Crosses Powys 148 F5
Four Crosses Staffs 133 B9
Four Crosses Wrex 166 E3
Four Elms Devon 28 F3
Four Elms Kent 52 D3
Four Foot Som 44 G5
Four Forks Som 43 F8
Four Gates Gtr Man 194 F6
Four Gotes Cambs 157 F9
Four Houses Corner W Berks 64 F6
Four Lane End S Yorks 197 G9
Four Lane Ends Blkburn 195 B7
Four Lane Ends Gtr Man 195 E9
Four Lane Ends W Ches 167 C9
Four Lane Ends W Yorks 205 G8
Four Lanes Corn 2 B5
Four Marks Hants 49 G7
Four Mile Bridge Anglesey 178 F3
Four Mile Elm Glos 80 C4
Four Oaks E Sus 38 C5
Four Oaks Glos 98 F3
Four Oaks Kent 70 G3
Four Oaks W Mid 134 D2
Four Oaks W Mid 134 G4
Four Oaks Park W Mid 134 D2
Four Points W Berks 64 D5
Four Pools Worcs 99 C10
Four Roads Carms 74 D6
Four Roads I o M 192 F3
Four Throws Kent 38 B3
Four Wantz Essex 87 C10
Four Wents Kent 53 F9
Fourlane Ends Derbys 170 D5
Fourlanes End E Ches 168 D4
Fourpenny Highld 311 K2
Fourstones Northumb 241 D9
Fovant Wilts 31 B8
Foveran Aberds 303 G9
Fowey Corn 6 E2
Fowler's Plot Som 43 F10
Fowley Common Warr 183 B11
Fowlis Angus 287 D7
Fowlis Wester Perth 286 E3
Fowlmere Cambs 105 B8
Fownhope Hereford 97 E11
Fox Corner C Beds 103 F8
Fox Corner Sur 50 C3
Fox Hatch Essex 87 F9
Fox Hill Bath 61 G9
Fox Hill Hereford 98 B3
Fox Hole Swansea 56 D5
Fox Holes Wilts 45 E11
Fox Lane Hants 49 B11
Fox Royd W Yorks 197 D8
Fox Street Essex 107 F10
Foxash Estate Essex 107 E11
Foxbar Renfs 267 C9
Foxbury London 68 E2
Foxcombe Hill Oxon 83 E7
Foxcote Glos 81 B8
Foxcote Som 45 B8
Foxdale I o M 192 E3
Foxdown Hants 48 C4
Foxearth Essex 106 C6
Foxendown Kent 69 F7
Foxfield Cumb 210 B4
Foxford W Mid 135 G7
Foxham Wilts 62 D3
Foxhills Hants 32 E4
Foxhole Corn 5 E9
Foxhole Norf 142 D4
Foxhole Swansea 57 C7
Foxholes N Yorks 217 E10
Foxhunt Green E Sus 23 B8
Foxley Hereford 97 B8
Foxley Norf 159 E10
Foxley Staffs 168 E3
Foxley Wilts 61 B11
Foxlydiate Worcs 117 D10
Foxt Staffs 169 F8
Foxton Cambs 105 B8
Foxton Durham 234 F3
Foxton Leics 136 E4
Foxton N Yorks 225 F8
Foxup N Yorks 213 D7
Foxwist Green W Ches 167 B10
Foxwood Shrops 116 B2
Foy Hereford 97 F11
Foyers Highld 300 G4
Foynesfield Highld 301 D8
Fraddam Corn 2 C3
Fraddon Corn 5 D8
Fradley Staffs 152 G3
Fradley Junction Staffs 152 G2
Fradswell Staffs 151 C9
Fraisthorpe E Yorks 218 G3
Framfield E Sus 37 C7
Framingham Earl Norf 142 C5
Framingham Pigot Norf 142 C5
Framlingham Suff 126 E5
Frampton Dorset 17 B8
Frampton Lincs 156 B6
Frampton Cotterell S Glos 61 C7
Frampton Court Glos 99 E10
Frampton End S Glos 61 C7
Frampton Mansell Glos 80 E6
Frampton on Severn Glos 80 D2
Frampton West End Lincs 174 G3
Framsden Suff 126 F3
Framwellgate Moor Durham 233 C11
France Lynch Glos 80 E6
Franche Worcs 116 B6
Frandley W Ches 183 F10
Frankby Mers 182 D2
Frankfort Norf 160 E6
Franklands Gate Hereford 97 B10
Frankley Worcs 133 G9
Frankley Green Worcs 133 G9
Frankley Hill Worcs 117 B9
Frank's Bridge Powys 114 F2
Frankton Warks 119 C8
Frankwell Shrops 149 G9
Frans Green Norf 160 G2
Frant E Sus 52 F5
Fraserburgh Aberds 303 C9
Frating Essex 107 G11
Frating Green Essex 107 G11
Fratton Ptsmth 21 B9
Freasley Warks 134 D4
Freathy Corn 7 E8
Freckenham Suff 124 C3
Freckleton Lancs 194 B2
Fredley Sur 51 C7
Free Town Gtr Man 195 E10
Freebirch Derbys 186 G4
Freeby Leics 154 E6
Freefolk Hants 48 D3
Freehay Staffs 169 G8
Freeland Oxon 82 C6
Freeland Renfs 267 B9
Freeland Corner Norf 160 F3
Freemantle Soton 32 E6
Freeport Village W Loth 269 C10
Freester Shetland 313 H6
Freethorpe Norf 143 B8
Freezy Water London 86 F5
Freiston Lincs 174 G5
Freiston Shore Lincs 174 G5
Fremington Devon 40 G4
Fremington N Yorks 223 F10
French Street Kent 52 C3
Frenchay S Glos 60 D6
Frenchbeer Devon 13 D9
Frenches Green Essex 106 G4
Frenchmoor Hants 32 B3
Frenchwood Lancs 194 B4
Frenich Stirl 285 G8
Frensham Sur 49 E10
Frenze Norf 142 G2
Fresgoe Highld 310 C3
Freshbrook Swindon 62 C6
Freshfield Mers 193 F9
Freshford Bath 61 G9
Freshwater I o W 20 D2
Freshwater Bay I o W 20 D2
Freshwater East Pembs 73 F8
Fressingfield Suff 126 B5
Freston Suff 108 D3
Freswick Highld 310 C7
Fretherne Glos 80 D2
Frettenham Norf 160 F4
Freuchie Fife 286 G6
Freuchies Angus 292 G4
Freystrop Pembs 73 C7
Friar Park W Mid 133 E10
Friarn Som 43 F7
Friars Cliff Dorset 19 C9
Friar's Gate E Sus 52 G3
Friar's Hill E Sus 38 E5
Friarton Perth 286 E5
Friday Bridge Cambs 139 C9
Friday Hill London 86 G5
Friday Street E Sus 23 E10
Friday Street Suff 126 G6
Friday Street Suff 127 E7
Friday Street Sur 50 D6
Fridaythorpe E Yorks 208 B3
Friendly W Yorks 196 C5
Friern Barnet London 86 G3
Friesland Argyll 288 D3
Friesthorpe Lincs 189 E9
Frieston Lincs 172 F6
Frieth Bucks 84 G3
Frieze Hill Som 28 B2
Friezeland Notts 171 E7
Frilford Oxon 82 F6
Frilford Heath Oxon 82 F6
Frilsham W Berks 64 E4
Frimley Sur 49 B11
Frimley Green Sur 49 B11
Frimley Ridge Sur 49 B11
Frindsbury Medway 69 E8
Fring Norf 158 C4
Fringford Oxon 102 F2
Friningham Kent 53 B10
Frinkle Green Essex 106 C4
Frinsted Kent 53 B11
Frinton-on-Sea Essex 108 G4
Friockheim Angus 287 C9
Friog Gwyn 146 G2
Frisby Leics 136 C4
Frisby on the Wreake Leics 154 F3
Friskney Lincs 175 D7
Friskney Eaudyke Lincs 175 D7
Friskney Tofts Lincs 175 E7
Friston E Sus 23 F8
Friston Suff 127 E8
Fritchley Derbys 170 E5
Frith Kent 54 B2
Frith Bank Lincs 174 F4
Frith Common Worcs 116 D3
Frith-hill Bucks 84 E6
Frith Hill Sur 50 E3
Fritham Hants 32 E2
Frithelstock Devon 25 D7
Frithelstock Stone Devon 25 D7
Frithend Hants 49 F10
Frithsden Herts 85 D8
Frithville Lincs 174 E4
Frittenden Kent 53 E10
Frittiscombe Devon 8 G6
Fritton Norf 142 E4
Fritton Norf 143 C9
Fritton Norf 161 F8
Fritwell Oxon 101 F10
Frizinghall W Yorks 205 F8
Frizington Cumb 219 B10
Frizzeler's Green Suff 124 E5
Frobost W Isles 297 J3
Frocester Glos 80 E3
Frochas Powys 148 G5
Frodesley Shrops 131 C10
Frodingham N Lincs 199 E11
Frodsham W Ches 183 F8
Frog End Cambs 123 F10
Frog End Cambs 123 G8
Frog Moor Swansea 56 C3
Frog Pool Worcs 116 E5
Frogden Borders 263 D7
Froggatt Derbys 186 F2

Froghall Staffs 169 F8
Frogham Hants 31 E11
Frogham Kent 55 C9
Froghole Kent 52 C2
Frogholt Kent 55 F7
Frogland Cross S Glos 60 C6
Frogmore Devon 8 G5
Frogmore Hants 33 C11
Frogmore Hants 49 B10
Frogmore Herts 85 E11
Frognal S Ayrs 257 D8
Frognall Lincs 156 G3
Frogpool Corn 4 G5
Frogs' Green Essex 105 D11
Frogshail Norf 160 B5
Frogwell Corn 6 B6
Frolesworth Leics 135 E10
Frome Som 45 D9
Frome St Quintin Dorset 29 G9
Fromebridge Glos 80 D3
Fromefield Som 45 D9
Fromes Hill Hereford 98 B3
Fromington Hereford 97 B10
Fron Denb 165 B9
Fron Gwyn 145 B7
Fron Gwyn 163 D8
Fron Powys 113 D11
Fron Powys 129 C8
Fron Powys 130 C4
Fron Powys 130 D3
Fron Shrops 148 B5
Fron-Bache Denb 166 G2
Fron-dêg Wrex 166 F3
Fron Isaf Wrex 166 G3
Froncysyllte Wrex 166 G3
Frongoch Gwyn 147 B8
Frost Devon 26 F3
Frost Hill N Som 60 G2
Frost Row Norf 141 C10
Frostenden Suff 143 G9
Frostenden Corner Suff 143 G9
Frosterley Durham 232 D6
Frostlane Hants 32 F6
Frotoft Orkney 314 D4
Froxfield C Beds 103 E9
Froxfield Wilts 63 F9
Froxfield Green Hants 34 B2
Froyle Hants 49 E9
Fryern Hill Hants 32 C6
Fryerning Essex 87 E10
Fryerns Essex 69 B8
Fryton N Yorks 216 E3
Fugglestone St Peter Wilts 46 G6
Fulbeck Lincs 172 E6
Fulbeck Northumb 252 F5
Fulbourn Cambs 123 F10
Fulbrook Oxon 82 C3
Fulflood Hants 33 B7
Fulford Som 28 B2
Fulford Staffs 151 B9
Fulford York 207 D8
Fulham London 67 D8
Fulking W Sus 36 E2
Full Sutton E Yorks 207 B10
Fullabrook Devon 40 E4
Fullarton Glasgow 268 C2
Fullarton N Ayrs 257 B8
Fuller Street Essex 88 B2
Fuller's End Essex 105 F10
Fuller's Moor W Ches 167 E7
Fullerton Hants 47 F11
Fulletby Lincs 190 G3
Fullshaw S Yorks 197 G8
Fullwell Cross London 86 G6
Fullwood E Ayrs 267 E8
Fullwood Gtr Man 196 F2
Fulmer Bucks 66 B3
Fulmodestone Norf 159 C9
Fulneck W Yorks 205 G10
Fulnetby Lincs 189 F9
Fulney Lincs 156 E5
Fulready Warks 100 B5
Fulshaw Park E Ches 184 E4
Fulstone W Yorks 197 F7
Fulstow Lincs 190 B4
Fulthorpe Stockton 234 G4
Fulwell Oxon 101 G7
Fulwell T & W 243 F9
Fulwood Lancs 202 G6
Fulwood S Yorks 186 D4
Fulwood Som 28 C2
Fundenhall Norf 142 D3
Fundenhall Street Norf 142 D2
Funtington W Sus 22 B3
Funtley Hants 33 F9
Funtullich Perth 285 E11
Funzie Shetland 312 D8
Furley Devon 28 G3
Furnace Argyll 284 G4
Furnace Carms 74 E6
Furnace Carms 75 E8
Furnace Ceredig 128 D3
Furnace Highld 299 B9
Furnace End Warks 134 E4
Furnace Green W Sus 51 F9
Furnace Wood W Sus 51 F11
Furneaux Pelham Herts 105 F8
Furner's Green E Sus 36 B6
Furness Vale Derbys 185 E8
Furneux Pelham Herts 105 F8
Furnham Som 28 F4
Further Ford End Essex 105 E9
Further Quarter Kent 53 F11
Furtho Northants 102 C5
Furze Devon 25 B10
Furze Hill Hants 31 E11
Furze Platt Windsor 65 C11
Furzebrook Dorset 18 E4
Furzedown Hants 32 B5
Furzedown London 67 E9
Furzehill Devon 41 D8
Furzehill Dorset 31 G8
Furzeley Corner Hants 33 E11
Furzey Lodge Hants 32 G5
Furzley Hants 32 D3
Furzton M Keynes 102 D6
Fyfett Som 28 E2
Fyfield Essex 87 D9
Fyfield Glos 82 E2
Fyfield Hants 47 D9
Fyfield Oxon 82 F6
Fyfield Wilts 63 F7
Fyfield Wilts 63 G7
Fylingthorpe N Yorks 227 D8
Fyning W Sus 34 C4
Fyvie Aberds 303 F7

G

Gabalfa Cardiff 59 D7
Gabhsann bho Dheas W Isles 304 C6
Gabhsann bho Thuath W Isles 304 C6
Gable Head Hants 21 B10
Gablon Highld 309 K7
Gabroc Hill E Ayrs 267 E9
Gadbrook Sur 51 D8
Gaddesby Leics 154 G3
Gadebridge Herts 85 D8
Gadfa Anglesey 179 D7
Gadfield Elm Worcs 98 E5
Gadlas Shrops 149 B7
Gadlys Rhondda 77 E7
Gadshill Kent 69 E8
Gaer Newport 59 B9
Gaer Powys 96 G3
Gaer-fawr Mon 78 F6
Gaerllwyd Mon 78 F6
Gaerwen Anglesey 179 G7
Gagingwell Oxon 101 F8
Gaick Lodge Highld 291 E9
Gailey Staffs 151 G8
Gailey Wharf Staffs 151 G8
Gain Hill Kent 53 D8
Gainfield Oxon 82 F4
Gainford Durham 224 B3
Gainsborough Lincs 188 C4
Gainsborough Suff 108 C3
Gainsford End Essex 106 D4
Gairletter Argyll 276 E3
Gairloch Highld 299 B8
Gairlochy Highld 290 E3
Gairney Bank Perth 280 B2
Gairnshiel Lodge Aberds 292 C4
Gaisgill Cumb 222 D2
Gaitsgill Cumb 230 B3
Galadean Borders 271 G11
Galashiels Borders 261 B11
Galdlys Flint 182 G2
Gale Gtr Man 196 D2
Galgate Lancs 202 B5
Galhampton Som 29 B10
Gallaberry Dumfries 247 G11
Gallachoille Argyll 275 E8
Gallanach Argyll 288 C4
Gallanach Argyll 289 G10
Gallanach Highld 294 G6
Gallantry Bank E Ches 167 E8
Gallatown Fife 280 C5
Galley Common Warks 134 E6
Galley Hill Cambs 122 D6
Galley Hill Lincs 190 F6
Galleyend Essex 88 E2
Galleywood Essex 88 E2
Galligill Cumb 231 B11
Gallin Perth 285 C9
Gallovie Highld 291 E7
Gallowfauld Angus 287 C8
Gallowhill Glasgow 267 D11
Gallowhill Renfs 267 B9
Gallowhills Aberds 303 D10
Gallows Corner London 87 G8
Gallows Green Essex 106 F2
Gallows Green Essex 107 F8
Gallows Green Staffs 169 G9
Gallows Green Worcs 117 E8
Gallows Inn Derbys 171 G7
Gallowsgreen Torf 78 D3
Gallowstree Common Oxon 65 C7
Gallt Melyd = Meliden Denb 181 E9
Gallt-y-foel Gwyn 163 C9
Galltair Highld 295 C10
Galltegfa Denb 165 D10
Gallypot Street E Sus 52 F3
Galmington Som 28 C2
Galmisdale Highld 294 G6
Galmpton Devon 8 G3
Galmpton Torbay 9 D7
Galon Uchaf M Tydf 77 D9
Galphay N Yorks 214 E5
Galston E Ayrs 258 B2
Galtrigill Highld 296 F7
Gam Corn 11 F7
Gamble Hill W Yorks 205 G11
Gamble's Green Essex 88 C3
Gamblesby Cumb 231 D8
Gamelsby Cumb 239 G7
Gamesley Derbys 185 C8
Gamlingay Cambs 122 G4
Gamlingay Cinques Cambs 122 G4
Gamlingay Great Heath Cambs 122 G4
Gammaton Devon 25 B7
Gammaton Moor Devon 25 C7
Gammersgill N Yorks 213 C11
Gamston Notts 154 B2
Gamston Notts 188 F2
Ganarew Hereford 79 B8
Ganavan Argyll 289 F10
Ganders Green Glos 98 G4
Gang Corn 6 B6
Ganllwyd Gwyn 146 E4
Gannetts Dorset 30 D3
Gannochy Angus 293 F7
Gannochy Perth 286 E5
Ganstead E Yorks 209 G9
Gansclet Highld 310 E7
Ganthorpe N Yorks 216 E3
Ganton N Yorks 217 D9
Gants Hill London 68 B2
Ganwick Corner Herts 86 F3
Gaodhail Argyll 289 F8
Gappah Devon 14 F3
Garafad Highld 298 C4
Garamor Highld 295 F8
Garbat Highld 300 C4
Garbhallt Argyll 275 D11
Garboldisham Norf 141 G10
Garbole Highld 301 G7
Garden City Bl Gwent 77 D11
Garden City Flint 166 B4
Garden Village S Yorks 186 B3
Garden Village Swansea 56 B5
Garden Village W Yorks 206 G4
Garden Village Wrex 166 E4
Gardeners Green Wokingham 65 F10
Gardenstown Aberds 303 C7
Garderhouse Shetland 313 J5
Gardham E Yorks 208 E5
Gardie Shetland 312 D7
Gardin Shetland 312 G6
Gare Hill Som 45 E9
Garelochhead Argyll 276 C4
Garford Oxon 82 F6
Garforth W Yorks 206 G4
Gargrave N Yorks 204 C4
Gargunnock Stirl 278 C4
Garizim Conwy 179 F11
Garker Corn 5 E10
Garlandhayes Devon 27 D11
Garlands Cumb 239 G10
Garleffin S Ayrs 244 G3
Garlic Street Norf 142 G4
Garlieston Dumfries 236 E6
Garliford Devon 26 B3
Garlinge Kent 71 F10
Garlinge Green Kent 54 C6
Garlogie Aberds 293 C9
Garmelow Staffs 150 D5
Garmond Aberds 303 D8
Garmondsway Durham 234 E2
Garmony Argyll 289 E8
Garmouth Moray 302 C3
Garmston Shrops 132 B2
Garn Powys 130 G2
Garn-swllt Swansea 75 D10
Garn-yr-erw Torf 78 C2
Garnant Carms 75 C11
Garndiffaith Torf 78 E3
Garndolbenmaen Gwyn 163 G7
Garnedd Conwy 164 E2
Garnett Bridge Cumb 221 F10
Garnetts Essex 87 B10
Garnfadryn Gwyn 144 C5
Garnkirk N Lnrk 268 B3
Garnlydan Bl Gwent 77 C11
Garnsgate Lincs 157 E8
Garnswllt Swansea 75 D10
Garra Eallabus Argyll 274 F3
Garrabost W Isles 304 E7
Garrachra Argyll 275 E11
Garralburn Moray 302 D4
Garraron Argyll 275 C9
Garras Corn 2 E6
Garreg Flint 181 F10
Garreg Gwyn 163 G10
Garrets Green W Mid 134 F2
Garrick Perth 286 F2
Garrigill Cumb 231 C10
Garrison Stirl 285 G7
Garriston N Yorks 224 G3
Garroch Dumfries 246 G3
Garrogie Lodge Highld 291 B7
Garros Highld 298 C4
Garrow Perth 286 C2
Garrowhill Glasgow 268 C3
Garrygualach Highld 290 C3
Garryhorn Dumfries 246 E2
Garsdale Cumb 212 B4
Garsdale Head Cumb 222 G5
Garsdon Wilts 62 B3
Garshall Green Staffs 151 C9
Garsington Oxon 83 E9
Garstang Lancs 202 D5
Garston Herts 85 F10
Garston Mers 182 E6
Garswood Mers 183 B9
Gartachoil Stirl 277 C10
Gartbreck Argyll 254 B3
Gartcosh N Lnrk 268 B3
Garth Bridgend 57 C11
Garth Ceredig 128 G2
Garth Flint 181 E10
Garth Gwyn 179 G9
Garth Newport 59 B9
Garth Newport 78 G4
Garth Perth 285 B11
Garth Powys 95 B9
Garth Powys 114 C5
Garth Shetland 313 H4
Garth Shetland 313 H5
Garth Wrex 156 G3
Garth Owen Powys 130 E2
Garth Row Cumb 221 F10
Garth Trevor Wrex 156 G3
Garthamlock Glasgow 258 B3
Garthbeg Highld 291 B7
Garthbrengy Powys 95 E10
Garthdee Aberdeen 293 C11
Gartheli Ceredig 111 F11
Garthmyl Powys 130 D3
Garthorpe Leics 154 E6
Garthorpe N Lincs 199 D10
Gartlea N Lnrk 268 C5
Gartloch Glasgow 268 B3
Gartly Aberds 302 F5
Gartmore Stirl 277 B10
Gartmore Ho Stirl 277 B10
Gartnagrenach Argyll 255 B8
Gartness N Lnrk 268 C5
Gartness Stirl 277 D10
Gartocharn W Dunb 277 D8
Garton E Yorks 209 F11
Garton-on-the-Wolds E Yorks 208 B5
Gartsherrie N Lnrk 268 B4
Gartur Stirl 277 B11
Gartymore Highld 311 H4
Garvald E Loth 281 G11
Garvamore Highld 291 D7
Garvard Argyll 274 D4
Garvault Hotel Highld 308 F7
Garve Highld 300 C3
Garvestone Norf 141 B10
Garvock Aberds 293 F9
Garvock Invclyd 276 G5
Garvock Hill Fife 280 D2
Garway Hereford 97 G9
Garway Hill Hereford 97 F8
Gaskan Highld 289 B9
Gasper Wilts 45 G9
Gastard Wilts 61 F11
Gasthorpe Norf 141 G9
Gaston Green Essex 87 B7
Gatacre Park Shrops 132 F5
Gatcombe I o W 20 D5
Gate Burton Lincs 188 E4
Gate Helmsley N Yorks 207 B9
Gateacre Mers 182 D6
Gatebeck Cumb 211 B10
Gateford Notts 187 E9
Gateford Common Notts 187 E9
Gateforth N Yorks 198 B5
Gatehead E Ayrs 257 B9
Gatehouse Northumb 251 F7
Gatehouse of Fleet Dumfries 237 D8
Gatelawbridge Dumfries 247 D10
Gateley Norf 159 E9
Gatenby N Yorks 214 B6
Gatesgarth Cumb 220 B3
Gateshead T & W 243 E7
Gatesheath W Ches 167 C7
Gateside Aberds 293 B8
Gateside Angus 287 C8
Gateside Dumfries 248 E4
Gateside E Renf 267 D9
Gateside Fife 286 G5
Gateside N Ayrs 267 E7
Gateside Shetland 312 F4
Gatewen Wrex 166 E4
Gatherley Devon 12 E3
Gathurst Gtr Man 194 F4
Gatlas Newport 78 G4
Gatley Gtr Man 184 D4
Gatley End Cambs 104 C5
Gatton Sur 51 C9
Gattonside Borders 262 B2
Gatwick Glos 80 C2
Gatwick Airport W Sus 51 E9
Gaufron Powys 113 D9
Gaulby Leics 136 C3
Gauldry Fife 287 E7
Gauntons Bank E Ches 167 F9
Gaunt's Common Dorset 31 F8
Gaunt's Earthcott S Glos 60 C6
Gaunt's End Essex 105 F10
Gautby Lincs 189 G11
Gavinton Borders 272 E5
Gawber S Yorks 197 F10
Gawcott Bucks 102 E3
Gawsworth E Ches 168 B5
Gawthorpe W Yorks 197 C9
Gawthorpe W Yorks 197 D7
Gawthrop Cumb 212 B3
Gawthwaite Cumb 210 C5
Gay Bowers Essex 88 E3
Gay Street W Sus 35 C9
Gaydon Warks 119 G7
Gayfield Orkney 314 A4
Gayhurst M Keynes 103 B7
Gayle N Yorks 213 B7
Gayles N Yorks 224 D2
Gayton Mers 182 E3
Gayton Norf 158 F4
Gayton Northants 120 G4
Gayton Staffs 151 D9
Gayton Engine Lincs 191 D7
Gayton le Marsh Lincs 190 E6
Gayton le Wold Lincs 190 D2
Gayton Thorpe Norf 158 F4
Gaywood Norf 158 E2
Gaza Shetland 312 F5
Gazeley Suff 124 E4
Gearraidh Sheilidh W Isles 297 J3
Geanies House Highld 301 B8
Gearraidh Bhailteas W Isles 297 J3
Gearraidh Bhaird W Isles 304 F5
Gearraidh Dubh W Isles 296 F4
Gearraidh na h-Aibhne W Isles 304 E4
Gearraidh na Monadh W Isles 297 K3
Geary Highld 298 C2
Geat Wolford Warks 100 E4
Geddes House Highld 301 D8
Gedding Suff 125 F9
Geddington Northants 137 G7
Gedgrave Hall Suff 109 B8
Gedintailor Highld 295 B7
Gedling Notts 171 G10
Gedney Lincs 157 E8
Gedney Broadgate Lincs 157 E8
Gedney Drove End Lincs 157 D9
Gedney Dyke Lincs 157 D8
Gedney Hill Lincs 156 G6
Gee Cross Gtr Man 185 C7
Geeston Rutland 137 C9
Gegin Wrex 166 E3
Geilston Argyll 276 F6
Geinas Denb 165 B9
Geirinis W Isles 297 G3
Geise Highld 310 C5
Geisiadar W Isles 304 E3
Geldeston Norf 143 E7
Gell Conwy 164 B5
Gelli Pembs 73 B9
Gelli Rhondda 77 G7
Gelli-gaer Neath 57 C9
Gelli-hôf Caerph 77 F11
Gellideg M Tydf 77 D8
Gellifor Denb 165 C10
Gelligaer Caerph 77 F10
Gelligroes Caerph 77 G11
Gellilydan Gwyn 146 B3
Gellinud Neath 76 E2
Gellinudd Neath 76 E2
Gellyburn Perth 286 D4
Gellygron Neath 76 E2
Gellywen Carms 92 G5
Gelsmoor Leics 153 F8
Gelston Dumfries 237 D9
Gelston Lincs 172 G6
Gembling E Yorks 209 B8
Gemini Warr 183 C9
Gendros Swansea 56 B6
Genesis Green Suff 124 F4
Gentleshaw Staffs 151 G11
Geocrab W Isles 305 J3
George Green Bucks 66 C4
George Nympton Devon 26 C2
Georgefield Dumfries 249 E7
Georgeham Devon 40 F3
Georgetown Bl Gwent 77 D10
Georgia Corn 1 B5
Gergask Highld 291 D8
Gerlan Gwyn 163 B10
Germansweek Devon 12 C4
Germiston Glasgow 268 B2
Germoe Corn 2 D3
Gernon Bushes Essex 87 E7
Gerrans Corn 3 B9
Gerrard's Bromley Staffs 150 C5
Gerrards Cross Bucks 66 B4
Gerrick Redcar 226 C4
Geseilfa Powys 129 E8
Gestingthorpe Essex 106 D6
Gesto Ho Highld 294 B5
Geuffordd Powys 148 G4
Geufron Denb 166 G2
Gib Heath W Mid 133 F11
Gibb Hill W Ches 183 F10
Gibbet Hill W Mid 118 C6
Gibbet Hill Warks 135 G10
Gibbshill Dumfries 237 B9
Gibraltar Beds 103 B10
Gibraltar Bucks 84 C3
Gibraltar Kent 55 F8
Gibraltar Oxon 83 B7
Gibshill Invclyd 276 G6
Gibsmere Notts 172 F2
Giddeahall Wilts 61 E11
Giddy Green Dorset 18 D2
Gidea Park London 68 B4
Gidleigh Devon 13 D9
Giffard Park M Keynes 103 C7
Giffnock E Renf 267 D11
Gifford E Loth 271 B10
Giffordland N Ayrs 266 F5
Giffordtown Fife 286 F6
Gigg Gtr Man 195 F10
Giggetty Staffs 133 E7
Giggleswick N Yorks 212 G6
Giggshill Sur 67 F7
Gignog Pembs 91 G7
Gilberdyke E Yorks 199 B10
Gilbert Street Hants 49 G7
Gilbert's Coombe Corn 4 G3
Gilbert's End Worcs 98 C6
Gilbert's Green Warks 118 C2
Gilbertstone W Mid 134 G2
Gilchriston E Loth 271 B9
Gilcrux Cumb 229 D8
Gildersome W Yorks 197 B8
Gildersome Street W Yorks 197 B8
Gildingwells S Yorks 187 D9
Gilesgate Durham 233 C11
Gilesgate Moor Durham 233 C11
Gileston V Glam 58 F4
Gilfach Caerph 77 F11
Gilfach Hereford 96 E6
Gilfach Goch Rhondda 58 B3
Gilfachrheda Ceredig 111 F8
Gilgarran Cumb 228 G6
Gill N Yorks 204 E5
Gillamoor N Yorks 216 B3
Gillan Corn 3 E7
Gillar's Green Mers 183 B7
Gillbank Cumb 221 F7
Gillbent Gtr Man 184 E5
Gillen Highld 298 D2
Gillesbie Dumfries 248 E5
Gilling East N Yorks 216 D2
Gilling West N Yorks 224 D3
Gillingham Dorset 30 B4
Gillingham Medway 69 F9
Gillingham Norf 143 E8
Gillock Highld 310 D6
Gillow Heath Staffs 168 D5
Gills Highld 310 B7
Gill's Green Kent 53 G9
Gillway Staffs 134 C4
Gilmanscleuch Borders 261 E8
Gilmerton Edin 270 B5
Gilmerton Perth 286 E2
Gilmonby Durham 223 C9
Gilmorton Leics 135 F11
Gilmourton S Lnrk 268 G3
Gilnow Gtr Man 195 F8
Gilroyd S Yorks 197 G10
Gilsland Northumb 240 D4
Gilsland Spa Cumb 240 D4
Gilson Warks 134 E3
Gilstead W Yorks 205 F8
Gilston Borders 271 D8
Gilston Herts 86 C6
Gilston Park Herts 86 C6
Giltbrook Notts 171 F7
Gilver's Lane Worcs 98 C6
Gilwell Park Essex 86 F5
Gilwern Mon 78 C2
Gimingham Norf 160 B5
Gincliough E Ches 185 F7
Ginger's Green E Sus 23 C10
Giosla W Isles 304 F3
Gipping Suff 125 E11
Gipsey Bridge Lincs 174 F3
Gipsy Row Suff 107 D11
Gipsyville Hull 200 B5
Gipton W Yorks 206 F2
Gipton Wood W Yorks 206 F2
Girdle Toll N Ayrs 266 G6
Girlington W Yorks 205 G8
Girlsta Shetland 313 H6
Girsby Lincs 190 D2
Girsby N Yorks 225 D7
Girt Som 29 C10
Girtford C Beds 104 B3
Girtford C Beds 122 G3
Girthon Dumfries 237 D8
Girton Cambs 123 E8
Girton Notts 172 B4
Girvan S Ayrs 244 D5
Gisburn Lancs 204 D2
Gisleham Suff 143 F10
Gislingham Suff 125 C11
Gissing Norf 142 F2
Gittisham Devon 15 B8
Givons Grove Sur 51 C7
Glachavoil Argyll 275 F11
Glack of Midthird Moray 302 E3
Glackmore Highld 300 D6
Gladestry Powys 114 F4
Gladsmuir E Loth 281 G9
Glaichbea Highld 300 F5
Glais Swansea 76 E2
Glaisdale N Yorks 226 D5
Glame Highld 298 E5
Glamis Angus 287 C7
Glan Adda Gwyn 179 G9
Glan-Conwy Conwy 164 E4
Glan-Duar Carms 93 C10
Glan-Dwyfach Gwyn 163 G7
Glan Gors Anglesey 179 F7
Glan-rhyd Gwyn 163 D7
Glan-rhyd Powys 76 D3
Glan-traeth Anglesey 178 F3
Glan-y-don Flint 181 F11
Glan y Ffer = Glanyferi Carms 74 C5
Glan-y-llyn Rhondda 58 C6
Glan-y-môr Carms 74 C4
Glan-y-nant Caerph 77 F10
Glan-y-nant Powys 129 G8
Glan-y-wern Gwyn 146 C2
Glan-yr-afon Anglesey 179 E10
Glan-yr-afon Flint 181 E10
Glan-yr-afon Gwyn 164 G6
Glan-yr-afon Gwyn 165 G8
Glan-yr-afon Shrops 148 E4
Glanafon Pembs 73 B7
Glanaman Carms 75 C11
Glandford Norf 177 E8
Glandwr Caerph 78 E2
Glandwr Pembs 92 F3
Glandy Cross Carms 92 F2
Glandyfi Ceredig 128 D3
Glangrwyney Powys 78 B2
Glanhanog Powys 129 D8
Glanmule Powys 130 E3
Glanrafon Ceredig 128 G2
Glanrhyd Gwyn 144 B5
Glanrhyd Pembs 92 C2
Glantlees Northumb 252 B4
Glanton Northumb 264 G3
Glanton Pike Northumb 264 G3
Glantwymyn = Cemmaes Road Powys 128 C6
Glanvilles Wootton Dorset 29 F11
Glanwern Ceredig 128 F2
Glanwydden Conwy 180 E4
Glapthorn Northants 137 E10
Glapwell Derbys 171 B7
Glas-allt Shiel Aberds 292 E4
Glasbury Powys 96 D3
Glaschoil Highld 301 F10
Glascoed Denb 181 G7
Glascoed Mon 78 E4
Glascoed Powys 129 F11
Glascoed Powys 148 G2
Glascoed Wrex 166 E3
Glascorrie Aberds 292 D5
Glascorrie Perth 286 E2
Glascote Staffs 134 C4
Glascwm Powys 114 G3
Glasdir Flint 181 E10
Glasdrum Argyll 284 C4
Glasfryn Conwy 164 E6
Glasgoed Ceredig 92 B6
Glasgoforest Aberds 293 B10
Glasgow Glasgow 267 B11
Glashvin Highld 298 C4
Glasinfryn Gwyn 163 B9
Glasllwch Newport 59 B9
Glasnacardoch Highld 295 F8
Glasnakille Highld 295 D7
Glasphein Highld 297 G7
Glaspwll Powys 128 D4
Glass Houghton W Yorks 198 C2
Glassburn Highld 300 F3
Glassenbury Kent 53 F8
Glasserton Dumfries 236 F6
Glassford S Lnrk 268 F4
Glassgreen Moray 302 C2
Glasshouse Glos 98 G4
Glasshouse Hill Glos 98 G4
Glasshouses N Yorks 214 G3
Glasslie Fife 286 G6
Glasson Cumb 228 D6
Glasson Cumb 239 E7
Glasson Lancs 202 B4
Glassonby Cumb 231 D7
Glasterlaw Angus 287 B9
Glaston Rutland 137 C7
Glastonbury Som 44 F4
Glatton Cambs 138 F3
Glazebrook Warr 183 C11
Glazebury Warr 183 B11
Glazeley Shrops 132 F4
Gleadless S Yorks 186 E5
Gleadless Valley S Yorks 186 E5
Gleadmoss E Ches 168 B4
Gleadsmoss E Ches 168 B4
Gleanhead Dumfries 245 G10
Gleann Tholàstaidh W Isles 304 D7
Gleaston Cumb 210 E5
Glebe Hants 33 D9
Glebe Shetland 313 J6
Glebe T & W 243 F8
Glecknabae Argyll 275 G11
Gledhow W Yorks 206 F2
Gledrid Shrops 148 B5
Gelaintiant Powys 129 E9
Glemsford Suff 106 B6
Glen Dumfries 237 B10
Glen Dumfries 237 D7
Glen Auldyn I o M 192 C5
Glen Bernisdale Highld 298 E4
Glen Ho Borders 261 C7
Glen Mona I o M 192 D5
Glen Mor Highld 295 B10
Glen Nevis House Highld 290 F3
Glen of Newmill Moray 302 D4
Glen Parva Leics 135 D11
Glen Sluain Argyll 275 D11
Glen Tanar House Aberds 292 D6
Glen Trool Lodge Dumfries 245 G10
Glen Vic Askil Highld 298 E3
Glen Village Falk 279 F7
Glen Vine I o M 192 E4
Glenallachie Moray 302 E2
Glenalmond College Perth 286 E3
Glenalmond Ho Perth 286 E3
Glenamachrie Argyll 289 G11
Glenample Stirl 285 E9
Glenancross Highld 295 F8
Glenapp Castle S Ayrs 244 G3
Glenaros Ho Argyll 289 E7
Glenbarr Argyll 255 D7
Glenbeg Highld 289 C7
Glenbeg Highld 301 G10
Glenbervie Aberds 293 E9
Glenboig N Lnrk 268 B4
Glenborrodale Highld 289 C8
Glenbranter Argyll 276 B2
Glenbreck Borders 260 E3
Glenbrein Lodge Highld 290 B6
Glenbrittle House Highld 294 C6
Glenbrook Edin 270 B2
Glenbuchat Castle Aberds 292 B5
Glenbuchat Lodge Aberds 292 B5
Glenbuck E Ayrs 259 D7
Glenburn Renfs 267 C9
Glenbyre Argyll 289 G7
Glencalvie Lodge Highld 309 L4
Glencanisp Lodge Highld 307 G6
Glencaple Dumfries 237 C11
Glencarron Lodge Highld 299 D10
Glencarse Perth 286 E5
Glencassley Castle Highld 309 J4
Glencat Aberds 293 D7
Glenceitlein Highld 284 C5
Glencoe Highld 284 B4
Glencraig Fife 280 B3
Glencripesdale Highld 289 D8
Glencrosh Dumfries 247 F7
Glendavan Ho Aberds 292 C6
Glendearg Borders 262 B2
Glendevon Perth 286 G3
Glendoe Lodge Highld 290 C6
Glendoebeg Highld 290 C6
Glendoick Perth 286 E6
Glendoll Lodge Angus 292 F4
Glendoune S Ayrs 244 D5
Glenduckie Fife 286 F6
Glendye Lodge Aberds 293 E8
Gleneagles Hotel Perth 286 F3
Gleneagles House Perth 286 G3
Glenearn Perth 286 F5
Glenegedale Argyll 254 B4
Glenelg Highld 295 D10
Glenernie Moray 301 E10
Glenfarg Perth 286 F5
Glenfarquhar Lodge Aberds 293 E9
Glenferness House Highld 301 E9
Glenfeshie Lodge Highld 291 D10
Glenfiddich Lodge Moray 302 F3
Glenfield Leics 135 B10
Glenfinnan Highld 295 G10
Glenfinnan Lodge Highld 295 G11
Glenfintaig Ho Highld 290 E4
Glenfoot Perth 286 F5
Glenfyne Lodge Argyll 284 F6
Glengap Dumfries 237 D8
Glengarnock N Ayrs 266 E6
Glengolly Highld 310 C5
Glengorm Castle Argyll 288 D6
Glengoulandie Perth 285 B11
Glengrasco Highld 298 E4
Glenhead Farm Angus 292 G4
Glenholt Plym 7 C10
Glenhoul Dumfries 246 F4
Glenhurich Highld 289 C10
Glenkerry Borders 261 G7
Glenkiln Dumfries 237 B10
Glenkindie Aberds 292 B6
Glenlair Dumfries 237 B9
Glenlatterach Moray 301 D11
Glenlee Dumfries 246 G4
Glenleigh Park E Sus 38 F2
Glenleraig Highld 306 F6
Glenlichorn Perth 285 F11
Glenlivet Moray 301 G11
Glenlochar Dumfries 237 C9
Glenlochsie Perth 292 F2
Glenlocksie Lodge Perth 292 F2
Glenloig N Ayrs 255 D10
Glenlomond Perth 286 G5
Glenluce Dumfries 236 D4
Glenlussa Ho Argyll 255 E8
Glenmallan Argyll 276 B5
Glenmarkie Argyll 292 E6
Glenmarkie Lodge Angus 292 G4
Glenmarksie Highld 300 D3
Glenmavis N Lnrk 268 B4
Glenmavis W Loth 269 B9
Glenmaye I o M 192 E3
Glenmayne Borders 261 C11
Glenmeanie Highld 300 D2
Glenmidge Dumfries 247 F9
Glenmoidart Ho Highld 289 B9
Glenmore Argyll 275 B9
Glenmore Argyll 275 G11
Glenmore Highld 298 E4
Glenmore Lodge Highld 291 C11
Glenmoy Angus 292 G6
Glennoe Argyll 284 D4
Glenogil Angus 292 G6
Glenowen Pembs 73 D7
Glenprosen Lodge Angus 292 G4
Glenprosen Village Angus 292 G5
Glenquaich Lodge Perth 286 D2
Glenquiech Angus 292 G6
Glenquithlie Aberds 303 C8
Glenrath Borders 260 C6
Glenrazie Dumfries 236 C5
Glenreasdell Mains Argyll 255 B9
Glenree N Ayrs 255 E10
Glenridding Cumb 221 B7
Glenrosa N Ayrs 256 B2
Glenrossal Highld 309 J4
Glenrothes Fife 286 G6
Glensanda Highld 289 E10
Glensaugh Aberds 293 F8
Glensburgh Falk 279 E8
Glenshero Lodge Highld 291 D7
Glenshoe Lodge Perth 292 G3
Glenstockadale Dumfries 236 C2
Glenstriven Argyll 275 F11
Glentaggart S Lnrk 259 D8
Glentarkie Perth 286 F5
Glenternie Borders 260 B6
Glentham Lincs 189 C8
Glentirranmuir Stirl 278 C3
Glenton Aberds 302 G6
Glentress Borders 261 B7
Glentromie Lodge Highld 291 D9
Glentrool Village Dumfries 236 B5
Glentruan I o M 192 B5
Glentruim House Highld 291 D8
Glentworth Lincs 188 D6
Glenuaig Lodge Highld 299 E11
Glenuig Highld 289 B8
Glenure Argyll 284 C4
Glenurquhart Highld 301 C7
Glenview Argyll 284 E5
Glespin S Lnrk 259 D8
Gletness Shetland 313 H6
Glewstone Hereford 97 G11
Glinton P'boro 138 B3
Globe Town London 67 C11
Glodwick Gtr Man 196 G2
Glogue Pembs 92 E4
Glooston Leics 136 D4
Glororum Northumb 264 C5
Glossop Derbys 185 C8
Gloster Hill Northumb 253 C7
Gloucester Glos 80 B4
Gloup Shetland 312 C7
Gloweth Corn 4 G5
Glusburn N Yorks 204 E6
Glutt Lodge Highld 310 F3
Glutton Bridge Staffs 169 B9
Gluvian Corn 5 C8
Glympton Oxon 101 G8
Glyn Mon 79 F7
Glyn Powys 129 F8
Glyn Castle Neath 76 E4
Glyn-Ceiriog Wrex 148 B4
Glyn-cywarch Gwyn 146 C2
Glyn Etwy Bl Gwent 77 D11
Glyn-neath = Glynedd Neath 76 D5
Glynarthen Ceredig 92 B6
Glynbrochan Powys 129 G8
Glyncoch Rhondda 77 G9
Glyncoed Bl Gwent 77 C11
Glyncorrwg Neath 57 B11
Glynde E Sus 23 D7
Glyndebourne E Sus 23 C7
Glyndyfrdwy Denb 165 G10
Glyne Gap E Sus 38 F3
Glynedd = Glyn neath Neath 76 D5
Glynhafren Powys 129 G7
Glynllan Bridgend 58 B2
Glynmorlas Shrops 148 B6
Glynogwr Bridgend 58 B3
Glyntaff Rhondda 58 B5
Glyntawe Powys 76 B4
Gnosall Staffs 150 E6
Gnosall Heath Staffs 150 E6
Goadby Leics 136 D4
Goadby Marwood Leics 154 D5
Goat Lees Kent 54 D4
Goatacre Wilts 62 D4
Goatham Green E Sus 38 C4
Goathill Dorset 29 D11
Goathland N Yorks 226 E6
Goathurst Som 43 G9
Goathurst Common Kent 52 C3
Gobernuisgach Lodge Highld 308 E4
Gobernuisgeach Highld 310 F3
Gobhaig W Isles 305 H2
Gobley Hole Hants 48 D6
Gobowen Shrops 148 C6
Godalming Sur 50 E3
Goddards Bucks 84 G3
Goddard's Corner Suff 126 D5
Goddard's Green Kent 53 G10
Goddards' Green W Berks 65 F7
Goddards' Green W Sus 36 C3
Godden Green Kent 52 B5
Goddington London 68 F3
Godley Gtr Man 185 B7
Godley Hill Gtr Man 185 C7
Godleybrook Staffs 169 G7
Godleys Green E Sus 36 D5
Godmanchester Cambs 122 C4
Godmanstone Dorset 17 B9
Godmersham Kent 54 C5
Godney Som 44 E3
Godolphin Cross Corn 2 C4
Godre'r-graig Neath 76 D3
God's Blessing Green Dorset 31 G8
Godshill Hants 31 E11
Godshill I o W 20 E6
Godstone Sur 51 C10
Godswinscroft Hants 19 B9
Godwell Devon 8 D2
Godwick Norf 159 E8
Godwinscroft Hants 19 B9
Goetre Mon 78 D4
Goferydd Anglesey 178 E2
Goff's Oak Herts 86 E4
Gogar Edin 280 G3
Goginan Ceredig 128 G3
Goirtean a'Chladaich Highld 290 F2
Golan Gwyn 163 G8
Golant Corn 6 E2
Golberdon Corn 12 G2
Golborne Gtr Man 183 B10
Golcar W Yorks 196 D6
Golch Flint 181 F11
Gold Hill Dorset 30 E4
Gold Hill Norf 139 E10
Goldcliff Newport 59 C11
Golden Balls Oxon 83 F9
Golden Cross E Sus 23 C10
Golden Cross E Sus 23 C8
Golden Green Kent 52 D6
Golden Grove Carms 75 B9
Golden Hill Bristol 60 D5
Golden Hill Hants 19 B11
Golden Hill Pembs 73 E7
Golden Hill Pembs 91 G9
Golden Park Devon 24 C2
Golden Pot Hants 49 E8
Golden Valley Derbys 170 E6
Golden Valley Glos 99 G8
Golden Valley Hereford 98 B3
Goldenhill Stoke 168 E5
Golder Field Hereford 115 E11
Golders Green London 67 B9
Goldfinch Bottom W Berks 64 G4
Goldhanger Essex 88 D6
Golding Shrops 131 C10
Goldington Beds 121 G11
Goldsborough N Yorks 206 B3
Goldsborough N Yorks 226 C6
Goldsithney Corn 2 C2
Goldstone Shrops 150 D4
Goldthorn Park W Mid 133 D8
Goldthorpe S Yorks 198 G3
Goldworthy Devon 24 C5
Golford Kent 53 F9
Golftyn Flint 182 G3
Golgotha Kent 55 D9
Gollanfield Highld 301 D8
Gollawater Corn 4 E5
Gollinglith Foot N Yorks 214 C3
Golly Wrex 166 D4
Golsoncott Som 42 F4
Golspie Highld 311 J2
Golval Highld 310 C2
Golynos Torf 78 E3
Gomeldon Wilts 47 F7
Gomersal W Yorks 197 B8
Gometra Ho Argyll 288 E5
Gomshall Sur 50 D5
Gonalston Notts 171 F11
Gonamena Corn 11 G11
Gonerby Hill Foot Lincs 155 B8
Gonfirth Shetland 313 G5
Good Easter Essex 87 C10
Gooderstone Norf 140 C5
Goodleigh Devon 40 G6
Goodley Stock Kent 52 C2
Goodmanham E Yorks 208 E3
Goodmayes London 68 B3
Goodnestone Kent 55 C9
Goodnestone Kent 70 G4
Goodrich Hereford 79 B9
Goodrington Torbay 9 D7
Good's Green Worcs 132 G5
Goodshaw Lancs 195 B10
Goodshaw Chapel Lancs 195 B10
Goodshaw Fold Lancs 195 B10
Goodstone Devon 13 G11
Goodwick = Wdig Pembs 91 D8
Goodworth Clatford Hants 47 E11
Goodyers End Warks 134 F6
Goodyhills Cumb 229 B8
Goom's Hill Worcs 117 G10
Goon Gumpas Corn 4 G4
Goon Piper Corn 3 B8
Goonabarn Corn 5 E9
Goonbell Corn 4 F4
Goonhavern Corn 4 E5
Goonhusband Corn 2 D5
Goonlaze Corn 2 B6
Goonown Corn 4 E4
Goonvrea Corn 4 F4
Goose Eye W Yorks 204 E6
Goose Green Cumb 211 C10
Goose Green Essex 108 F2
Goose Green Gtr Man 194 G5
Goose Green Hants 32 F4
Goose Green Herts 86 D5
Goose Green Kent 52 C6
Goose Green Kent 53 F10
Goose Green Lancs 150 E6
Goose Green Norf 142 F2
Goose Green S Glos 61 C8
Goose Green W Sus 34 C3
Goose Green W Sus 35 D10
Goose Hill Hants 64 G4
Goose Pool Hereford 97 D9
Gooseberry Green Essex 87 F11
Gooseford Devon 13 C9
Goosehill Green Worcs 117 E8
Gooseham Mill Devon 24 D2
Goosehill W Yorks 197 C11
Goosemoor Staffs 150 F6
Goosemoor Green Staffs 151 G11
Goosenford Som 28 B2
Goosewell Devon 40 D5
Goosey Oxon 82 G5
Goosnargh Lancs 203 F7
Goostrey E Ches 184 G3
Gorbals Glasgow 267 C11
Gorcott Hill Warks 117 D11
Gord Shetland 313 L6
Gorddinog Conwy 179 G11
Gordon Borders 272 G2
Gordonbush Highld 311 J2
Gordonsburgh Moray 302 C4
Gordonstoun Moray 301 C11
Gordonstown Aberds 302 D5
Gordonstown Aberds 303 F7
Gore Dorset 29 D9
Gore Kent 55 B10
Gore Cross Wilts 46 C4
Gore End Hants 64 G2
Gore Pit Essex 88 B5
Gore Street Kent 71 F9
Gorebridge Midloth 270 C6
Gorefield Cambs 157 G8
Gorehill W Sus 35 C7
Gorgie Edin 280 G4

Gorhambury Herts 85 D10
Goring Oxon 64 C6
Goring-by-Sea W Sus 35 G10
Goring Heath Oxon 65 D7
Gorleston-on-Sea Norf 143 C10
Gornalwood W Mid 133 E8
Gorrachie Aberds 303 D7
Gorran Churchtown Corn 5 G9
Gorran Haven Corn 5 G10
Gorran High Lanes Corn 5 G9
Gorrenberry Borders 249 D11
Gorrig Ceredig 93 C8
Gorse Covert Warr 183 C11
Gorse Hill Gtr Man 184 B4
Gorse Hill Swindon 63 B7
Gorsedd Flint 181 F11
Gorseinon Swansea 56 B5
Gorseness Orkney 314 E4
Gorsethorpe Notts 171 B9
Gorseybank Derbys 170 E3
Gorsgoch Ceredig 111 G9
Gorslas Carms 75 C9
Gorsley Glos 98 F3
Gorsley Common Hereford 98 F3
Gorsley Ley Staffs 133 B11
Gorst Hill Worcs 116 C4
Gorstage W Ches 183 G10
Gorstan Highld 300 C3
Gorstanvorran Highld 289 B10
Gorstella W Ches 166 C5
Gorsteyhill E Ches 168 E2
Gorsty Hill Staffs 151 D11
Gorstyhill Staffs 168 E2
Gortan Argyll 274 G3
Gortantaoid Argyll 274 F4
Gortenacullish Highld 295 G8
Gorteneorn Highld 289 C8
Gortenfern Highld 289 C8
Gortinanane Argyll 255 C8
Gorton Gtr Man 184 B5
Gortonallister N Ayrs 256 D2
Gosbeck Suff 126 F3
Gosberton Lincs 156 C4
Gosberton Cheal Lincs 156 D4
Gosberton Clough Lincs 156 D3
Goscote W Mid 133 C10
Goseley Dale Derbys 152 E6
Gosfield Essex 106 F5
Gosford Hereford 115 D10
Gosford Oxon 83 C7
Gosford Green W Mid 118 B6
Gosforth Cumb 219 E11
Gosforth T & W 242 D6
Gosforth Valley Derbys 186 F4
Gosland Green Suff 124 G5
Gosling Green Suff 107 C9
Gosmere Kent 54 B4
Gosmore Herts 104 F3
Gospel Ash Staffs 132 E6
Gospel End Village Staffs 133 E7
Gospel Green W Sus 50 G2
Gospel Oak London 67 B9
Gosport Hants 21 B8
Gosport Hants 32 C5
Gossabrough Shetland 312 E7
Gossard's Green C Beds 103 C9
Gossington Glos 80 E2
Gossops Green W Sus 51 F9
Goswick Northumb 273 F11
Gotham Dorset 31 E9
Gotham E Sus 38 F2
Gotham Notts 153 C10
Gothelney Green Som 43 F9
Gotherington Glos 99 F9
Gothers Corn 5 D9
Gott Argyll 288 E2
Gott Shetland 313 J6
Gotton Som 28 B2
Goudhurst Kent 53 F8
Goukstone Moray 302 D4
Goulceby Lincs 190 F3
Goulton N Yorks 225 E9
Gourdas Aberds 303 E7
Gourdon Aberds 293 F10
Gourock Invclyd 276 F4
Govan Glasgow 267 B11
Govanhill Glasgow 267 C11
Gover Hill Kent 52 C6
Goverton Notts 172 E2
Goveton Devon 8 F5
Govilon Mon 78 C3
Gowanhill Aberds 303 C10
Gowanwell Aberds 303 E8
Gowdall E Yorks 198 C6
Gowerton = Tre-Gwyr Swansea 56 B5
Gowhole Derbys 185 E8
Gowkhall Fife 279 D11
Gowkthrapple N Lnrk 268 E5
Gowthorpe E Yorks 207 C11
Goxhill E Yorks 209 E9
Goxhill N Lincs 200 C6
Goxhill Haven N Lincs 200 B6
Goybre Neath 57 D9
Goytre Neath 57 D9
Gozzard's Ford Oxon 83 F7
Grabhair W Isles 305 G5
Graby Lincs 155 D11
Gracca Corn 5 D10
Gracemount Edin 270 B5
Grade Corn 2 G6
Graffham W Sus 34 D6
Grafham Cambs 122 D3
Grafham Sur 50 E4
Grafton Hereford 97 D9
Grafton N Yorks 215 G8
Grafton Oxon 82 E3
Grafton Shrops 149 F8
Grafton Worcs 99 D9
Grafton Worcs 115 E11
Grafton Flyford Worcs 117 F9
Grafton Regis Northants 102 B5
Grafton Underwood Northants 137 G8
Grafty Green Kent 53 D11
Grahamston Falk 279 E7
Graianrhyd Denb 166 D2
Graig Carms 74 E6
Graig Conwy 180 G4
Graig Denb 181 G9
Graig Rhondda 58 B5
Graig Wrex 148 B4
Graig-Fawr Swansea 75 E10
Graig-fechan Denb 165 E10
Graig Felen Swansea 75 E11
Graig Penllyn V Glam 58 D3
Graig Trewyddfa Swansea 57 B7
Grain Medway 69 D11
Grains Bar Gtr Man 196 F3
Grainsby Lincs 190 B3
Grainthorpe Lincs 190 B5
Grainthorpe Fen Lincs 190 B5
Graiselound N Lincs 188 B3
Grampound Corn 5 F8
Grampound Road Corn 5 E8
Gramsdal W Isles 296 F4
Granborough Bucks 102 F5
Granby Notts 154 B5
Grandborough Warks 119 D9
Grandpont Oxon 83 D8
Grandtully Perth 286 B3
Grange Cumb 220 B5
Grange Dorset 31 G8
Grange E Ayrs 257 B10
Grange Fife 287 G8
Grange Halton 183 E8
Grange Lancs 203 G7
Grange Medway 69 F9
Grange Mers 182 D2
Grange N Yorks 223 G8
Grange NE Lincs 201 F9
Grange Perth 286 E6
Grange Warr 183 C10
Grange Crossroads Moray 302 D4
Grange Estate Dorset 31 G10
Grange Hall Moray 301 C10
Grange Hill Durham 233 F10
Grange Hill Essex 86 G6
Grange Moor W Yorks 197 D8
Grange of Cree Dumfries 236 D6
Grange of Lindores Fife 286 F6
Grange-over-Sands Cumb 211 D8
Grange Park London 86 F4
Grange Park Mers 183 C7
Grange Park Northants 120 F5
Grange Park Swindon 62 C6
Grange Villa Durham 242 G6
Grange Village Glos 79 C11
Grangemill Derbys 170 D2
Grangemouth Falk 279 E8
Grangemuir Fife 287 G9
Grangepans Falk 279 E10
Grangetown Cardiff 59 E7
Grangetown Redcar 235 G7
Grangetown T & W 243 G10
Granish Highld 291 B11
Gransmoor E Yorks 209 B8
Gransmore Green Essex 106 G3
Granston = Treopert Pembs 91 E7
Grant Thorold NE Lincs 201 F9
Grantchester Cambs 123 F8
Grantham Lincs 155 B8
Grantley N Yorks 214 F4
Grantley Hall N Yorks 214 F4
Grantlodge Aberds 293 B9
Granton Dumfries 248 B3
Granton Edin 280 F4
Grantown Aberds 302 D5
Grantown-on-Spey Highld 301 G10
Grantsfield Hereford 115 E10
Grantshouse Borders 272 B6
Graplin Dumfries 237 E8
Grappenhall Warr 183 D10
Grasby Lincs 200 G5
Grasmere Cumb 220 D6
Grass Green Essex 106 D4
Grasscroft Gtr Man 196 G3
Grassendale Mers 182 D5
Grassgarth Cumb 221 F8
Grassgarth Cumb 230 C2
Grassholme Durham 232 G4
Grassington N Yorks 213 G10
Grassmoor Derbys 170 B6
Grassthorpe Notts 172 B3
Grasswell T & W 243 G8
Grateley Hants 47 E9
Gratton Devon 24 E5
Gratwich Staffs 151 C10
Gravel W Ches 167 B11
Gravel Castle Kent 55 D8
Gravel Hill Bucks 85 G8
Gravel Hole Gtr Man 196 F2
Gravel Hole Shrops 149 B7
Graveley Cambs 122 E4
Graveley Herts 104 F4
Gravelhill Shrops 149 G9
Gravelly Hill W Mid 134 E2
Gravels Shrops 130 D6
Gravelsbank Shrops 130 C6
Graven Shetland 312 F6
Graveney Kent 70 G5
Gravenhunger Moss Shrops 168 G2
Gravesend Herts 105 F8
Gravesend Kent 68 E6
Grayingham Lincs 188 B6
Grayrigg Cumb 221 F11
Grays Thurrock 68 D6
Grayshott Hants 49 F11
Grayson Green Cumb 228 F5
Grayswood Sur 50 G2
Graythorp Hrtlpl 234 F6
Grazeley Wokingham 65 F7
Grazeley Green W Berks 65 F7
Greagdhubh Lodge Highld 291 D8
Greamchary Highld 310 F2
Greasbrough S Yorks 186 B6
Greasby Mers 182 D3
Greasley Notts 171 F7
Great Abington Cambs 105 B10
Great Addington Northants 121 B9
Great Alne Warks 118 F2
Great Altcar Lancs 193 F10
Great Amwell Herts 86 C5
Great Asby Cumb 222 C3
Great Ashfield Suff 125 D9
Great Ashley Wilts 61 G10
Great Ayton N Yorks 225 C11
Great Baddow Essex 88 E2
Great Bardfield Essex 106 E3
Great Barford Beds 122 G2
Great Barrington Glos 82 C2
Great Barrow W Ches 167 B7
Great Barton Suff 125 D7
Great Barugh N Yorks 216 D4
Great Bavington Northumb 251 G11
Great Bealings Suff 108 B4
Great Bedwyn Wilts 63 G9
Great Bentley Essex 108 G2
Great Berry Essex 69 B7
Great Billing Northants 120 E6
Great Bircham Norf 158 C5
Great Blakenham Suff 126 G2
Great Blencow Cumb 230 E5
Great Bolas Telford 150 E2
Great Bookham Sur 50 C6
Great Bosullow Corn 1 C4
Great Bourton Oxon 101 B9
Great Bowden Leics 136 F4
Great Bower Kent 54 C4
Great Bradley Suff 124 G3
Great Braxted Essex 88 C5
Great Bricett Suff 125 G10
Great Brickhill Bucks 103 E8
Great Bridge W Mid 133 E9
Great Bridgeford Staffs 151 D7
Great Brington Northants 120 D3
Great Bromley Essex 107 F11
Great Broughton Cumb 229 E7
Great Broughton N Yorks 225 D10
Great Buckland Kent 69 G7
Great Budworth W Ches 183 F11
Great Burdon Darl 224 B6
Great Burgh Sur 51 B8
Great Burstead Essex 87 G11
Great Busby N Yorks 225 D10
Great Canfield Essex 87 B9
Great Carlton Lincs 190 D6
Great Casterton Rutland 137 B10
Great Casterton Rutland 137 B9
Great Cellws Powys 113 E11
Great Chalfield Wilts 61 G11
Great Chart Kent 54 E3
Great Chatwell Staffs 150 G5
Great Chell Stoke 168 E5
Great Chesterford Essex 105 C10
Great Cheveney Kent 53 E8
Great Cheverell Wilts 46 C3
Great Chilton Durham 233 E11
Great Chishill Cambs 105 D8
Great Clacton Essex 89 B11
Great Claydons Essex 88 E3
Great Cliff W Yorks 197 D10
Great Clifton Cumb 228 F6
Great Coates NE Lincs 201 F8
Great Comberton Worcs 99 C9
Great Common Suff 143 F7
Great Common W Sus 35 B8
Great Corby Cumb 239 G11
Great Cornard Suff 107 C7
Great Cowden E Yorks 209 E10
Great Coxwell Oxon 82 G3
Great Crakehall N Yorks 224 G4
Great Cransley Northants 120 B6
Great Cressingham Norf 141 C7
Great Crosby Mers 182 B4
Great Crosthwaite Cumb 229 G11
Great Cubley Derbys 152 B3
Great Dalby Leics 154 G4
Great Denham Beds 103 B10
Great Doddington Northants 121 E7
Great Doward Hereford 79 B9
Great Dunham Norf 159 G7
Great Dunmow Essex 106 G2
Great Durnford Wilts 46 F6
Great Easton Essex 106 F2
Great Easton Leics 136 E6
Great Eccleston Lancs 202 E4
Great Edstone N Yorks 216 C4
Great Ellingham Norf 141 D10
Great Elm Som 45 D8
Great Eppleton T & W 234 B3
Great Eversden Cambs 123 G7
Great Fencote N Yorks 224 G5
Great Finborough Suff 125 F10
Great Fransham Norf 159 G7
Great Gaddesden Herts 85 C8
Great Gate Staffs 169 G9
Great Gidding Cambs 138 G2
Great Givendale E Yorks 208 C2
Great Glemham Suff 126 E6
Great Glen Leics 136 D3
Great Gonerby Lincs 155 B7
Great Gransden Cambs 122 F5
Great Green Cambs 104 C5
Great Green Norf 142 F5
Great Green Suff 125 B11
Great Green Suff 125 F8
Great Green Suff 126 B2
Great Habton N Yorks 216 D5
Great Hale Lincs 173 G10
Great Hallingbury Essex 87 B8
Great Hampden Bucks 84 E4
Great Harrowden Northants 121 C7
Great Harwood Lancs 203 G10
Great Haseley Oxon 83 E10
Great Hatfield E Yorks 209 E9
Great Haywood Staffs 151 E10
Great Haywood Staffs 151 E9
Great Heath W Mid 134 G6
Great Heck N Yorks 198 C5
Great Henny Essex 107 D7
Great Hinton Wilts 46 B2
Great Hivings Bucks 85 E7
Great Hockham Norf 141 E9
Great Holcombe Oxon 83 F10
Great Holland Essex 89 B12
Great Hollands Brack 65 F11
Great Holm M Keynes 102 D6
Great Honeyborough Pembs 73 D7
Great Horkesley Essex 107 E9
Great Hormead Herts 105 F7
Great Horton W Yorks 205 G8
Great Horwood Bucks 102 E5
Great Houghton Northants 120 F5
Great Houghton S Yorks 198 F2
Great Howarth Gtr Man 196 D2
Great Hucklow Derbys 185 F11
Great Job's Cross Kent 38 B4
Great Kelk E Yorks 209 B8
Great Kendale E Yorks 217 G10
Great Kimble Bucks 84 D4
Great Kingshill Bucks 84 F5
Great Langton N Yorks 224 F5
Great Lea Common Reading 65 F8
Great Leighs Essex 88 B2
Great Lever Gtr Man 195 F8
Great Limber Lincs 200 F6
Great Linford M Keynes 103 C7
Great Livermere Suff 125 C7
Great Longstone Derbys 186 G2
Great Lumley Durham 233 B11
Great Lyth Shrops 131 B9
Great Malgraves Thurrock 69 C7
Great Malvern Worcs 98 B5
Great Maplestead Essex 106 E6
Great Marton Blkpool 202 F2
Great Marton Moss Blkpool 202 G2
Great Massingham Norf 158 E5
Great Melton Norf 142 B2
Great Milton Oxon 83 E10
Great Missenden Bucks 84 E5
Great Mitton Lancs 203 F10
Great Mongeham Kent 55 C10
Great Moor Gtr Man 184 D6
Great Moor Staffs 132 D6
Great Moulton Norf 142 E3
Great Munden Herts 105 G7
Great Musgrave Cumb 222 C5
Great Ness Shrops 149 F7
Great Notley Essex 106 G4
Great Oak Mon 78 D5
Great Oakley Essex 108 F3
Great Oakley Northants 137 F7
Great Offley Herts 104 F2
Great Ormside Cumb 222 B4
Great Orton Cumb 239 G8
Great Ouseburn N Yorks 215 G8
Great Oxendon Northants 136 G4
Great Oxney Green Essex 87 D11
Great Palgrave Norf 158 G6
Great Parndon Essex 86 D6
Great Pattenden Kent 53 E8
Great Paxton Cambs 122 E4
Great Plumpton Lancs 202 G3
Great Plumstead Norf 160 G6
Great Ponton Lincs 155 C8
Great Preston W Yorks 198 B2
Great Purston Northants 101 D10
Great Raveley Cambs 138 G5
Great Rissington Glos 81 B11
Great Rollright Oxon 100 E6
Great Ryburgh Norf 159 D9
Great Ryle Northumb 264 G2
Great Ryton Shrops 131 C9
Great Saling Essex 106 F4
Great Salkeld Cumb 231 D7
Great Sampford Essex 106 D2
Great Sankey Warr 183 D9
Great Saredon Staffs 133 B9
Great Saxham Suff 124 E5
Great Shefford W Berks 63 E11
Great Shelford Cambs 123 G9
Great Shoddesden Hants 47 D9
Great Smeaton N Yorks 224 E6
Great Snoring Norf 159 C8
Great Somerford Wilts 62 C3
Great Stainton Darl 234 G2
Great Stambridge Essex 88 G5
Great Staughton Cambs 122 E2
Great Steeping Lincs 174 C6
Great Stoke S Glos 60 C6
Great Stonar Kent 55 B10
Great Stretton Leics 136 C3
Great Strickland Cumb 231 G7
Great Stukeley Cambs 122 C4
Great Sturton Lincs 190 F2
Great Sutton Shrops 131 G10
Great Sutton W Ches 182 F5
Great Swinburne Northumb 241 B10
Great Tew Oxon 101 F7
Great Tey Essex 107 F7
Great Thirkleby N Yorks 215 D9
Great Thurlow Suff 124 G3
Great Torrington Devon 25 D7
Great Tosson Northumb 252 C2
Great Totham Essex 88 C5
Great Tows Lincs 190 C2
Great Tree Corn 6 D5
Great Urswick Cumb 210 E5
Great Wakering Essex 70 B2
Great Waldingfield Suff 107 C8
Great Walsingham Norf 159 B8
Great Waltham Essex 87 C11
Great Warley Essex 87 G9
Great Washbourne Glos 99 E9
Great Weeke Devon 13 D10
Great Weldon Northants 137 F8
Great Welnetham Suff 125 F7
Great Wenham Suff 107 D11
Great Whittington Northumb 242 C2
Great Wigborough Essex 89 C7
Great Wilbraham Cambs 123 F10
Great Wilne Derbys 153 C8
Great Wishford Wilts 46 F5
Great Witchingham Norf 160 E2
Great Witcombe Glos 80 C6
Great Witley Worcs 116 D5
Great Wolford Warks 100 E4
Great Wratting Suff 106 B3
Great Wymondley Herts 104 F4
Great Wyrley Staffs 133 B9
Great Wytheford Shrops 149 F11
Great Yarmouth Norf 143 B10
Great Yeldham Essex 106 D5
Greater Doward Hereford 79 B9
Greatfield Wilts 62 B5
Greatford Lincs 155 G11
Greatgap Bucks 84 B6
Greatgate Staffs 169 G9
Greatham Hants 49 G9
Greatham Hrtlpl 234 F5
Greatham W Sus 35 D8
Greatmoor Bucks 102 G4
Greatness Kent 52 B4
Greatstone-on-Sea Kent 39 C9
Greatworth Northants 101 C11
Greave Gtr Man 184 C6
Greave Lancs 195 C11
Grebby Lincs 174 B6
Greeba I o M 192 D4
Green Denb 165 B9
Green Pembs 73 E7
Green Powys 130 E5
Green Bank Cumb 211 C7
Green Bottom Corn 4 F5
Green Bottom Glos 79 B11
Green Close N Yorks 212 F4
Green Clough W Yorks 205 G7
Green Crize Hereford 97 D10
Green Cross Sur 49 F11
Green Down Devon 28 G3
Green End Beds 103 B10
Green End Beds 121 E11
Green End Beds 122 E2
Green End Beds 122 G2
Green End Bucks 84 F4
Green End Bucks 103 E8
Green End C Beds 103 D11
Green End Cambs 122 C4
Green End Cambs 123 F7
Green End Herts 85 D8
Green End Herts 104 E5
Green End Herts 104 E6
Green End Herts 104 G6
Green End Herts 105 F7
Green End Lancs 204 D4
Green End N Yorks 226 E6
Green End Warks 134 F5
Green Gate Devon 27 D8
Green Hailey Bucks 84 E4
Green Hammerton N Yorks 206 B5
Green Haworth Lancs 195 B9
Green Head Cumb 230 B3
Green Heath Staffs 151 G9
Green Hill Kent 53 C9
Green Hill Lincs 155 B8
Green Hill W Yorks 206 F4
Green Hill Wilts 62 B5
Green Lane Devon 13 F11
Green Lane Hereford 98 B2
Green Lane Powys 130 D3
Green Lane Warks 117 E11
Green Lane Worcs 118 B6
Green Moor S Yorks 186 B3
Green Ore Som 44 C5
Green Parlour Bath 45 C8
Green Quarter Cumb 221 E9
Green St Green London 68 G3
Green Side W Yorks 197 E7
Green Side W Yorks 205 G11
Green Street E Sus 38 E3
Green Street Essex 87 F10
Green Street Glos 80 B5
Green Street Glos 80 E3
Green Street Herts 85 F11
Green Street Herts 105 G9
Green Street W Sus 35 C10
Green Street Worcs 99 B7
Green Street Worcs 99 C8
Green Street Green Kent 68 E5
Green Street Green London 68 G3
Green Tye Herts 86 B6
Greenacres Gtr Man 196 F2
Greenan Argyll 275 G11
Greenbank Falk 279 F7
Greenbank Shetland 312 C7
Greenbank W Ches 183 G10
Greenburn W Loth 269 C8
Greencroft Durham 242 G5
Greendale E Ches 184 F5
Greendikes Northumb 264 D3
Greendown Som 44 C5
Greendykes Northumb 264 D3
Greenend N Lnrk 268 C4
Greenend Oxon 100 G6
Greenfaulds N Lnrk 278 G5
Greenfield C Beds 103 E11
Greenfield Glasgow 268 C2
Greenfield Gtr Man 196 G3
Greenfield Highld 289 D11
Greenfield Highld 290 C4
Greenfield Oxon 84 G2
Greenfield = Maes-Glas Flint 181 F11
Greenfoot N Lnrk 268 B4
Greenford London 66 C6
Greengairs N Lnrk 278 G5
Greengarth Hall Cumb 219 E11
Greengate Gtr Man 196 D2
Greengate Norf 159 F10
Greengates W Yorks 205 F9
Greengill Cumb 229 D8
Greenhalgh Lancs 202 F4
Greenhall S Lnrk 268 D3
Greenham Dorset 28 G6
Greenham Som 27 C9
Greenham W Berks 64 F3
Greenhaugh Northumb 251 F7
Greenhead Borders 261 D11
Greenhead Dumfries 247 D9
Greenhead N Lnrk 268 E6
Greenhead Northumb 240 D5
Greenhead Staffs 169 F7
Greenheys Gtr Man 195 G8
Greenhill Dumfries 238 B4
Greenhill Durham 234 B3
Greenhill Falk 278 F6
Greenhill Hereford 98 B4
Greenhill Kent 71 F7
Greenhill Leics 153 G8
Greenhill London 67 B7
Greenhill S Yorks 186 E4
Greenhill Worcs 99 B10
Greenhill Worcs 116 B6
Greenhill Bank Shrops 149 B7
Greenhillocks Derbys 170 F6
Greenhills N Ayrs 267 E7
Greenhills S Lnrk 268 E2
Greenhithe Kent 68 E5
Greenholm E Ayrs 258 B2
Greenholme Cumb 221 D11
Greenhouse Borders 262 E3
Greenhow N Yorks 214 G2
Greenhow Hill N Yorks 214 G2
Greenigoe Orkney 314 F4
Greenland Highld 310 C6
Greenland S Yorks 186 D5
Greenland Mains Highld 310 C6
Greenlands Bucks 65 B9
Greenlands Worcs 117 D11
Greenlaw Aberds 302 D6
Greenlaw Borders 272 F4
Greenlaw Mains Midloth 270 C4
Greenlea Dumfries 238 B2
Greenley M Keynes 102 C6
Greenloaning Perth 286 G2
Greenlooms W Ches 167 C7
Greenman's Lane Wilts 62 C3
Greenmeadow Swindon 62 B6
Greenmeadow Torf 78 F3
Greenmount Gtr Man 195 E9
Greenmow Shetland 313 L6
Greenock Invclyd 276 F5
Greenock West Invclyd 276 F5
Greenodd Cumb 210 C6
Greenrigg W Loth 269 C8
Greenrow Cumb 238 G4
Greens Borders 249 F11
Greens Norton Northants 102 B3
Greensforge Staffs 133 F7
Greensgate Norf 160 F2
Greenside Cumb 222 E4
Greenside Derbys 186 F5
Greenside Gtr Man 184 B6
Greenside T & W 242 E4
Greenside W Yorks 197 D7
Greensidehill Northumb 263 F11
Greensplat Corn 5 D9
Greenstead Essex 107 F10
Greenstead Green Essex 106 F6
Greensted Essex 87 E8
Greensted Green Essex 87 E8
Greenstreet Green Suff 107 B10
Greenway Hereford 98 E4
Greenway Pembs 91 E11
Greenway Som 27 B11
Greenway V Glam 58 E5
Greenway Worcs 116 C4
Greenwell Cumb 240 F2
Greenwells Borders 262 C3
Greenwich London 67 D11
Greenwich Suff 108 C3
Greenwich Wilts 46 G2
Greenwith Common Corn 4 G5
Greenwoods Essex 87 F11
Greeny Orkney 314 D2
Greep Highld 298 E2
Greet Glos 99 E10
Greet Kent 54 B2
Greete Shrops 115 C11
Greetham Lincs 190 G4
Greetham Rutland 155 G8
Greetland W Yorks 196 C5
Greetland Wall Nook W Yorks 196 C5
Greetwell N Lincs 200 G2
Gregg Hall Cumb 221 G9
Gregson Lane Lancs 194 B5
Gregynog Powys 129 D11
Greinetobht W Isles 296 D4
Grèin W Isles 297 L2
Greinton Som 44 F2
Gremista Shetland 313 J6
Grenaby I o M 192 E3
Grendon Northants 121 E7
Grendon Warks 134 C5
Grendon Bishop Hereford 115 F11
Grendon Common Warks 134 D5
Grendon Green Hereford 115 F11
Grendon Underwood Bucks 102 G3
Grenofen Devon 12 G5
Grenoside S Yorks 186 C4
Greosabhagh W Isles 305 J3
Gresford Wrex 166 E5
Gresham Norf 160 B3
Greshornish Highld 298 D3
Gressenhall Norf 159 F9
Gressingham Lancs 211 F11
Gresty Green E Ches 168 E2
Greta Bridge Durham 223 C11
Gretna Dumfries 239 D8
Gretna Green Dumfries 239 D8
Gretton Glos 99 E10
Gretton Northants 137 E7
Gretton Shrops 131 D10
Gretton Fields Glos 99 E10
Grewelthorpe N Yorks 214 D4
Grey Green N Lincs 199 F9
Greyfield Bath 44 B6
Greygarth N Yorks 214 E3
Greylake Som 43 G11
Greylake Fosse Som 44 F2
Greynor Carms 75 D9
Greynor-isaf Carms 75 D9
Greyrigg Dumfries 248 F3
Greys Green Oxon 65 C8
Greysouthen Cumb 229 F7
Greystead Northumb 251 F7
Greystoke Cumb 230 E4
Greystoke Gill Cumb 230 F4
Greystone Aberds 292 D6
Greystone Aberds 302 F6
Greystone Angus 287 C9
Greystone Cumb 211 D10
Greystone Dumfries 237 B11
Greystonegill N Yorks 212 F3
Greystones S Yorks 186 D4
Greystones Warks 99 B11
Greytree Hereford 97 F11
Greywell Hants 49 C8
Griais W Isles 304 D6
Grianan W Isles 304 E6
Gribb Dorset 28 G5
Gribthorpe E Yorks 207 F11
Gridley Corner Devon 12 C3
Griff Warks 135 F7
Griffins Hill W Mid 133 G10
Griffithstown Torf 78 F3
Griffydam Leics 153 F8
Grigg Kent 53 E11
Griggs Green Hants 49 G10
Grillis Corn 2 B5
Grilstone Devon 26 C2
Grimbister Orkney 314 E3
Grimblethorpe Lincs 190 D2
Grimeford Village Lancs 194 E6
Grimes Hill Worcs 117 B11
Grimesthorpe S Yorks 186 C5
Grimethorpe S Yorks 198 F2
Griminis W Isles 296 D3
Griminis W Isles 296 F3
Grimister Shetland 312 D6
Grimley Worcs 116 E6
Grimness Orkney 314 G4
Grimoldby Lincs 190 D5
Grimpo Shrops 149 D7
Grimsargh Lancs 203 G7
Grimsbury Oxon 101 C9
Grimsby NE Lincs 201 E9
Grimscote Northants 120 G3
Grimscott Corn 24 F3
Grimshaw Blkburn 195 C8
Grimshaw Green Lancs 194 E3
Grimsthorpe Lincs 155 E10
Grimston E Yorks 209 F11
Grimston Leics 154 E3
Grimston Norf 158 E4
Grimston York 207 C8
Grimstone Dorset 17 C8
Grimstone End Suff 125 D8
Grinacombe Moor Devon 12 C4
Grindale E Yorks 218 E2
Grindigar Orkney 314 F5
Grindiscol Shetland 313 K6
Grindle Shrops 132 C5
Grindleford Derbys 186 F2
Grindleton Lancs 203 D11
Grindley Staffs 151 D10
Grindley Brook Shrops 167 G8
Grindlow Derbys 185 F11
Grindon Northumb 273 G8
Grindon Staffs 169 E9
Grindon Stockton 234 F3
Grindon T & W 243 G9
Grindonmoor Gate Staffs 169 E9
Grindsbrook Booth Derbys 185 D10
Gringley on the Hill Notts 188 C2
Grinsdale Cumb 239 F9
Grinshill Shrops 149 E10
Grinstead Hill Suff 125 G11
Grinton N Yorks 223 F10
Griomasaigh W Isles 297 G4
Griomsidar W Isles 304 F5
Grisdale Cumb 222 G5
Grishipoll Argyll 288 D3
Grisling Common E Sus 36 C6
Gristhorpe N Yorks 217 C11
Griston Norf 141 D8
Gritley Orkney 314 F5
Grittenham Wilts 62 C4
Grittlesend Hereford 98 B4
Grittleton Wilts 61 C11
Grizebeck Cumb 210 C4
Grizedale Cumb 220 G6
Groam Highld 300 E5
Grobister Orkney 314 D6
Grobsness Shetland 313 G5
Groby Leics 135 B10
Groes Conwy 165 C8
Groes Neath 57 D9
Groes Efa Denb 165 B10
Groes-faen Rhondda 58 C5
Groes-fawr Denb 165 B10
Groes-lwyd Mon 96 G6
Groes-lwyd Powys 148 G4
Groes-wen Caerph 58 B6
Groesffordd Gwyn 144 B5
Groesffordd Powys 95 F11
Groesffordd Marli Denb 181 G8
Groeslon Gwyn 163 C8
Groeslon Gwyn 163 D7
Groespluan Powys 130 B4
Grogport Argyll 255 C9
Gromford Suff 127 F7
Gronant Flint 181 E9
Gronwen Shrops 148 D5
Groombridge E Sus 52 F4
Grosmont Mon 97 G8
Grosmont N Yorks 226 D6
Gross Green Warks 119 F7
Grotaig Highld 300 G4
Groton Suff 107 C9
Grotton Gtr Man 196 G3
Grougfoot Falk 279 F10
Grove Bucks 103 G8
Grove Dorset 17 G10
Grove Hereford 98 C2
Grove Kent 71 G8
Grove Notts 188 F2
Grove Oxon 82 G6
Grove Pembs 73 E7
Grove End Kent 69 G11
Grove End Warks 100 D6
Grove End Warks 134 D3
Grove Green Kent 53 B9
Grove Hill E Sus 23 C10
Grove Hill Kent 71 G8
Grove Park London 67 D8
Grove Park London 68 E2
Grove Town W Yorks 198 C3
Grove Vale W Mid 133 E10
Grovehill E Yorks 208 F6
Grovehill Herts 85 D9
Groves Kent 55 B9
Grovesend Swansea 75 E9
Grub Street Staffs 150 D5
Grubb Street Kent 68 F5
Grudie Highld 300 C3
Gruids Highld 309 J5
Gruinard House Highld 307 K4
Gruinards Highld 309 K5
Grula Highld 294 C5
Gruline Argyll 289 E7
Gruline Ho Argyll 289 F7
Grumbeg Highld 308 F6
Grumbla Corn 1 D4
Grunasound Shetland 313 K5
Grundisburgh Suff 126 G4
Grunsagill Lancs 203 C11
Gruting Shetland 313 J4
Grutness Shetland 313 N6
Gryn Goch Gwyn 162 F6
Gualachulain Highld 284 C5
Gualin Ho Highld 308 D3
Guard House W Yorks 204 E6
Guardbridge Fife 287 F8
Guarlford Worcs 98 B6
Guay Perth 286 C4
Gubbion's Green Essex 88 B2
Gubblecote Herts 84 C6
Guesachan Highld 289 B10
Guestling Green E Sus 38 E5
Guestling Thorn E Sus 38 D4
Guestwick Norf 159 D11
Guestwick Green Norf 159 D11
Guide Blkburn 195 B8
Guide Bridge Gtr Man 184 B6
Guide Post Northumb 253 F7
Guilden Morden Cambs 104 C5
Guilden Sutton W Ches 166 B6
Guildford Sur 50 D3
Guildford Park Sur 50 D3
Guildiehaugh W Loth 269 B9
Guildtown Perth 286 D5
Guilford Pembs 73 D7
Guilsborough Northants 120 C3
Guilsfield = Cegidfa Powys 148 G4
Guilthwaite S Yorks 187 D7
Guilton Kent 55 B9
Guineaford Devon 40 F5
Guisachan Highld 300 G3
Guisborough Redcar 226 B2
Guiseley W Yorks 205 E9
Guist Norf 159 D9
Guith Orkney 314 C5
Guiting Power Glos 99 G11
Gulberwick Shetland 313 K6
Gullane E Loth 281 E9
Guller's End Worcs 99 D7
Gulling Green Suff 124 F6
Gullom Holme Cumb 231 F9
Gulval Corn 1 C5
Gulworthy Devon 12 G4
Gumfreston Pembs 73 E10
Gumley Leics 136 E3
Gummow's Shop Corn 5 D7
Gun Green Kent 53 G9
Gun Hill E Sus 23 C9
Gunby E Yorks 207 F10
Gunby Lincs 155 E8
Gunby Lincs 175 B7
Gundenham Som 27 C10
Gundleton Hants 48 G6
Gunn Devon 40 G6
Gunnersbury London 67 D7
Gunnerside N Yorks 223 F9
Gunnerton Northumb 241 C10
Gunness N Lincs 199 E10
Gunnislake Corn 12 G4
Gunnista Shetland 313 J7
Guns Village W Mid 133 E9
Gunstone Staffs 133 C7
Gunter's Bridge W Sus 35 C7
Gunthorpe Lincs 188 B4
Gunthorpe Norf 159 C10
Gunthorpe Notts 171 G11
Gunthorpe P'boro 138 C3
Gunthorpe Rutland 137 B7
Gunton Suff 143 D10
Gunville I o W 20 D5
Gunwalloe Corn 2 E5
Gunwalloe Fishing Cove Corn 2 E5
Gupworthy Som 42 F3
Gurnard I o W 20 B5
Gurnett E Ches 184 G6
Gurney Slade Som 44 D6
Gurnos M Tydf 77 D8
Gurnos Powys 76 D3
Gushmere Kent 54 B4
Gussage All Saints Dorset 31 E8
Gussage St Andrew Dorset 31 E7
Gussage St Michael Dorset 31 E7
Gustard Wood Herts 85 B11
Guston Kent 55 E10
Gutcher Shetland 312 D7
Guthram Gowt Lincs 156 E3
Guthrie Angus 287 B9
Guyhirn Cambs 139 C7
Guyhirn Gull Cambs 139 C7
Guy's Cliffe Warks 118 D5
Guy's Head Lincs 157 D9
Guy's Marsh Dorset 30 C4
Guyzance Northumb 252 C6
Gwaelod-y-garth Cardiff 58 C6
Gwaenysgor Flint 181 E9
Gwalchmai Anglesey 178 F5
Gwalchmai Uchaf Anglesey 178 F5
Gwallon Corn 2 C2
Gwastad Pembs 91 G10
Gwastadgoed Gwyn 145 G11
Gwastadnant Gwyn 163 D10
Gwaun-Cae-Gurwen Neath 76 C2
Gwaun-Leision Neath 76 C2
Gwavas Corn 2 D5
Gwavas Corn 2 G6
Gwbert Ceredig 92 B3
Gwedna Corn 2 C4
Gweek Corn 2 D6
Gwehelog Mon 78 E5
Gwenddwr Powys 95 C11
Gwennap Corn 2 B6
Gwenter Corn 2 F6
Gwern y brenin Shrops 148 D6
Gwern-y-Steeple V Glam 58 D5
Gwernaffield-y-Waun Flint 166 C2
Gwernafon Powys 129 E8
Gwerneirin Powys 129 F10
Gwerneirin Powys 130 G3
Gwernesney Mon 78 E6
Gwernogle Carms 93 E10
Gwernol Denb 166 E2
Gwernydd Powys 129 C11
Gwernymynydd Flint 166 C2
Gwersyllt Wrex 166 E4
Gwespyr Flint 181 E10
Gwills Corn 4 D6
Gwinear Corn 2 B3
Gwinear Downs Corn 2 C4
Gwithian Corn 2 A3
Gwredog Anglesey 178 D6
Gwrhay Caerph 77 F11
Gwyddelwern Denb 165 F9
Gwyddgrug Carms 93 D9
Gwynfryn Wrex 166 E3
Gwystre Powys 113 D11
Gwytherin Conwy 164 C5
Gyfelia Wrex 166 F4
Gyffin Conwy 180 F3
Gylen Park Argyll 289 G10
Gyre Orkney 314 F3
Gyrn Denb 165 D11
Gyrn-goch Gwyn 162 F6

H

Habberley Shrops 131 C7
Habberley Worcs 116 B6
Habergham Lancs 204 G2
Habertoft Lincs 175 B8
Habin W Sus 34 C4
Habrough NE Lincs 200 E6
Haccombe Devon 14 G3
Haceby Lincs 155 B10
Hacheston Suff 126 F6
Hack Green E Ches 167 F10
Hackbridge London 67 F9
Hackenthorpe S Yorks 186 E6
Hackford Norf 141 C11
Hackforth N Yorks 224 G4
Hackland Orkney 314 D3
Hackleton Northants 120 F6
Hacklinge Kent 55 C10
Hackman's Gate Worcs 117 B7
Hackness N Yorks 227 G9
Hackness Orkney 314 G3
Hackness Som 43 D10
Hackney London 67 C10
Hackney Wick London 67 C11
Hackthorn Lincs 189 E7
Hackthorn Wilts 47 D7
Hackthorpe Cumb 230 G6
Haclait W Isles 297 G4
Haconby Lincs 156 D2
Hacton London 68 B4
Haddacott Devon 25 C8
Hadden Borders 263 B7
Haddenham Bucks 84 D2
Haddenham Cambs 123 B9
Haddenham End Field Cambs 123 B9
Haddington E Loth 281 G10
Haddington Lincs 172 C6
Haddiscoe Norf 143 D8
Haddoch Aberds 302 E5
Haddon Cambs 138 E2
Haddon E Ches 169 B7
Hade Edge W Yorks 196 F6
Hademore Staffs 134 B3
Haden Cross W Mid 133 F9
Hadfield Derbys 185 B8
Hadham Cross Herts 86 B6
Hadham Ford Herts 105 G8
Hadleigh Essex 69 B10
Hadleigh Suff 107 C10
Hadleigh Heath Suff 107 C9
Hadley London 86 F2
Hadley Telford 150 G3
Hadley Worcs 117 E7
Hadley Castle Telford 150 G3
Hadley End Staffs 152 E2
Hadley Wood London 86 F3
Hadlow Kent 52 D6
Hadlow Down E Sus 37 C8
Hadlow Stair Kent 52 D6
Hadnall Shrops 149 F10
Hadspen Som 45 G7
Hadstock Essex 105 C11
Hadston Northumb 253 D7
Hady Derbys 186 G5
Hadzor Worcs 117 E8
Haffenden Quarter Kent 53 E11
Hafod Swansea 57 C7
Hafod-Dinbych Conwy 164 E5
Hafod Grove Pembs 92 C2
Hafod-lom Conwy 180 G5
Hafod-y-Green Denb 181 G8
Hafodiwan Ceredig 111 G7
Hafodyrynys Bl Gwent 78 F2
Hag Fold Gtr Man 195 G7
Haggate Gtr Man 196 F2
Haggate Lancs 204 F3
Haggbeck Cumb 239 C11
Haggersta Shetland 313 J5
Haggerston London 67 C10
Haggerston Northumb 273 G10
Hagginton Hill Devon 40 D5
Haggrister Shetland 312 F5
Haggs Falk 278 F5
Haghill Glasgow 268 B2
Hagley Hereford 97 C11
Hagley Worcs 133 G8
Hagloe Glos 79 D11
Hagmore Green Suff 107 D9
Hagnaby Lincs 174 C4
Hagnaby Lincs 191 F7
Hagnaby Lock Lincs 174 D4
Hague Bar Derbys 185 D7
Hagworthingham Lincs 174 B4
Haigh Gtr Man 194 F6
Haigh S Yorks 197 E9
Haigh Moor W Yorks 197 C9
Haighton Green Lancs 203 G7
Haighton Top Lancs 203 G7
Hail Weston Cambs 122 E3
Haile Cumb 219 D10
Hailes Glos 99 E10
Hailey Herts 86 C5
Hailey Oxon 64 B6
Hailey Oxon 82 C5
Hailsham E Sus 23 D9
Hailstone Hill Wilts 81 G9
Haimer Highld 310 C5
Haimwood Powys 148 F6
Hainault London 87 G7
Haine Kent 71 F11
Hainford Norf 160 F4

Hains Dorset 30 D3
Hainton Lincs 189 E11
Hainworth W Yorks 205 F7
Hainworth Shaw W Yorks 205 F7
Hairmyres S Lnrk 268 E2
Haisthorpe E Yorks 218 G2
Hakeford Devon 40 F6
Hakin Pembs 72 D5
Halabezack Corn 2 C6
Halam Notts 171 E11
Halamanning Corn 2 C3
Halbeath Fife 280 D2
Halberton Devon 27 E8
Halcon Som 28 B2
Halcro Highld 310 C6
Haldens Herts 86 C2
Hale Cumb 211 D10
Hale Gtr Man 184 D3
Hale Halton 183 E7
Hale Hants 31 D11
Hale Kent 71 F9
Hale Medway 69 F9
Hale Som 30 B3
Hale Sur 49 D10
Hale Bank Halton 183 E7
Hale Barns Gtr Man 184 D3
Hale Coombe N Som 44 B2
Hale End London 86 G5
Hale Green E Sus 23 C9
Hale Mills Corn 4 G5
Hale Nook Lancs 202 E3
Hale Street Kent 53 D7
Halecommon W Sus 34 C4
Hales Norf 143 D7
Hales Staffs 150 C4
Hales Bank Hereford 116 G2
Hales Green Derbys 169 G11
Hales Green Norf 143 D7
Hales Park Worcs 116 B5
Hales Place Kent 54 B6
Hales Street Norf 142 F3
Hales Wood Hereford 98 E2
Halesfield Telford 132 C4
Halesgate Lincs 156 D6
Halesowen W Mid 133 G9
Halesworth Suff 127 B7
Halewood Mers 183 D7
Half Moon Village Devon 14 B3
Halford Shrops 131 G8
Halford Warks 100 B5
Halfpenny Cumb 211 B10
Halfpenny Furze Carms 74 C3
Halfpenny Green Staffs 132 E6
Halfway Carms 75 E8
Halfway Carms 94 E2
Halfway Carms 94 E6
Halfway S Yorks 186 E6
Halfway W Berks 64 F2
Halfway Wilts 45 D11
Halfway Bridge W Sus 34 C6
Halfway House Shrops 148 G6
Halfway Houses Gtr Man 195 F9
Halfway Houses Kent 70 E2
Halfway Street Kent 55 D9
Halgabron Corn 11 D7
Halifax W Yorks 196 B5
Halkburn Borders 271 G9
Halket E Ayrs 267 E8
Halkirk Highld 310 D5
Halkyn = Helygain Flint 182 G2
Halkyn Mountain Flint 182 G2
Hall Bower W Yorks 196 E6
Hall Broom S Yorks 186 D3
Hall Cross Lancs 202 G4
Hall Dunnerdale Cumb 220 F4
Hall End Beds 103 B10
Hall End C Beds 103 D11
Hall End Lincs 174 E6
Hall End S Glos 61 B8
Hall End Warks 134 C5
Hall Flat Worcs 117 C9
Hall Garth York 207 C9
Hall Green E Ches 168 D4
Hall Green Essex 106 D5
Hall Green Lancs 194 C3
Hall Green Lancs 194 F4
Hall Green W Mid 133 E10
Hall Green W Mid 134 G2
Hall Green W Mid 135 G7
Hall Green W Yorks 197 D10
Hall Green Wrex 167 G7
Hall Grove Herts 89 C8
Hall i' th' Wood Gtr Man 195 E8
Hall of Clestrain Orkney 314 F2
Hall of Tankerness Orkney 314 F5
Hall of the Forest Shrops 130 G4
Hall Santon Cumb 220 E2
Hall Waberthwaite Cumb 220 F2
Hallam Fields Derbys 153 B9
Halland E Sus 23 B8
Hallaton Leics 136 D5
Hallatrow Bath 44 B6
Hallbankgate Cumb 240 F3
Halleaths Dumfries 248 G3
Hallen S Glos 60 C5
Hallend Warks 118 D2
Hallew Corn 5 D10
Hallfield Gate Derbys 170 D5
Hallgarth Durham 234 C2
Hallglen Falk 279 F7
Halliburton Borders 261 B11
Halliburton Borders 272 F3
Hallin Highld 298 D2
Halling Medway 69 G8
Hallingbury Street Essex 87 B8
Hallington Lincs 190 D4
Hallington Northumb 241 B11
Halliwell Gtr Man 195 E8
Hallon Shrops 132 D5
Hallonsford Shrops 132 D5
Halloughton Notts 171 E11
Hallow Worcs 116 F6
Hallow Heath Worcs 116 F6
Hallowes Derbys 186 F5
Hallowsgate W Ches 167 B8
Hallrule Borders 262 G3
Halls E Loth 282 G3
Hall's Cross E Sus 23 D11
Halls Green Essex 86 D6
Hall's Green Herts 104 F5
Hall's Green Kent 52 D4
Hallsands Devon 9 G11
Hallsford Bridge Essex 87 E9
Hallspill Devon 25 C7
Hallthwaites Cumb 210 B3
Hallwood Green Glos 98 E3
Hallworthy Corn 11 D9
Hallyards Borders 260 B6
Hallyburton House Perth 286 D6
Hallyne Borders 270 G3
Halmer End Staffs 168 F3
Halmond's Frome Hereford 98 B3
Halmore Glos 79 E11
Halmyre Mains Borders 270 F3
Halnaker W Sus 22 B6
Halsall Lancs 193 E11
Halse Northants 101 C11
Halse Som 27 B10
Halsetown Corn 2 B2
Halsfordwood Devon 14 C3
Halsham E Yorks 201 B9
Halsinger Devon 40 F4
Halstead Essex 106 E6
Halstead Kent 68 G3
Halstead Leics 136 B4
Halstock Dorset 29 F8
Halsway Som 42 F6
Haltcliff Bridge Cumb 230 D3
Halterworth Hants 32 C5
Haltham Lincs 174 C2
Haltoft End Lincs 174 F5
Halton Bucks 84 C5
Halton Halton 183 E8
Halton Lancs 211 G10
Halton Northumb 241 D11
Halton W Yorks 206 G2
Halton Wrex 148 B6
Halton Barton Corn 7 B8
Halton Brook Halton 183 E8
Halton East N Yorks 204 C6
Halton Fenside Lincs 174 C6
Halton Gill N Yorks 213 D7
Halton Green Lancs 211 F10
Halton Holegate Lincs 174 B6
Halton Lea Gate Northumb 240 F5
Halton Moor W Yorks 206 G2
Halton Shields Northumb 242 D2
Halton View Halton 183 D8
Halton West N Yorks 204 C2
Haltwhistle Northumb 240 E6
Halvergate Norf 143 B8
Halvosso Corn 2 C6
Halwell Devon 8 E5
Halwill Devon 12 B4
Halwill Junction Devon 24 G6
Halwin Corn 2 C5
Ham Devon 28 G2
Ham Glos 79 F11
Ham Glos 99 G9
Ham Highld 310 B6
Ham Kent 55 C10
Ham London 67 E7
Ham Plym 7 D9
Ham Shetland 313 K1
Ham Som 27 C11
Ham Som 28 B3
Ham Som 28 E3
Ham Som 45 D7
Ham Wilts 63 G10
Ham Common Dorset 30 B4
Ham Green Bucks 83 B11
Ham Green Hants 48 G2
Ham Green Hereford 98 C4
Ham Green Kent 38 B5
Ham Green Kent 69 F10
Ham Green N Som 60 D4
Ham Green Wilts 61 G11
Ham Green Worcs 117 E10
Ham Hill Kent 69 G8
Ham Moor Sur 66 G5
Ham Street Som 44 G5
Hamar Shetland 312 F5
Hamarhill Orkney 314 C5
Hamars Shetland 313 G6
Hamble-le-Rice Hants 33 F7
Hambleden Bucks 65 B9
Hambledon Hants 33 E10
Hambledon Sur 50 F3
Hambleton Lancs 202 E3
Hambleton N Yorks 205 C7
Hambleton N Yorks 207 G7
Hambleton Moss Side Lancs 202 E3
Hambridge Som 28 C5
Hambrook S Glos 60 D6
Hambrook W Sus 22 B3
Hameringham Lincs 174 B4
Hamerton Cambs 122 B2
Hametoun Shetland 313 K1
Hamilton S Lnrk 268 D3
Hamister Shetland 313 G7
Hamlet Dorset 29 F9
Hammer W Sus 49 G11
Hammer Bottom Hants 49 G11
Hammerfield Herts 85 D8
Hammerpot W Sus 35 F9
Hammersmith Derbys 170 E5
Hammersmith London 67 D8
Hammerwich Staffs 133 B11
Hammerwood E Sus 52 F2
Hammill Kent 55 B9
Hammond Street Herts 86 E4
Hammoon Dorset 30 E4
Hamnavoe Shetland 312 E4
Hamnavoe Shetland 312 E6
Hamnavoe Shetland 312 F6
Hamnavoe Shetland 313 K5
Hamnish Clifford Hereford 115 F10
Hamp Som 43 F10
Hampden Park E Sus 23 E10
Hampen Glos 81 B9
Hamperden End Essex 105 E11
Hamperley Shrops 131 F8
Hampers Green W Sus 35 C7
Hampeth Northumb 252 B5
Hampnett Glos 81 B10
Hampole S Yorks 198 E4
Hampreston Dorset 19 B7
Hampsfield Cumb 211 C8
Hampson Green Lancs 202 C5
Hampstead London 67 B9
Hampstead Garden Suburb London 67 B9
Hampstead Norreys W Berks 64 D4
Hampsthwaite N Yorks 205 B11
Hampton Kent 71 F7
Hampton London 66 F6
Hampton Shrops 132 F4
Hampton Swindon 81 G11
Hampton Worcs 99 C10
Hampton Bank Shrops 149 C9
Hampton Beech Shrops 130 B6
Hampton Bishop Hereford 97 D11
Hampton Fields Glos 80 F5
Hampton Gay Oxon 83 B7
Hampton Green Glos 80 E5
Hampton Green W Ches 167 F8
Hampton Hargate P'boro 138 E3
Hampton Heath W Ches 167 F7
Hampton Hill London 66 E6
Hampton in Arden W Mid 134 G4
Hampton Loade Shrops 132 F5
Hampton Lovett Worcs 117 D7
Hampton Lucy Warks 118 F5
Hampton Magna Warks 118 D5
Hampton on the Hill Warks 118 E5
Hampton Park Hereford 97 D10
Hampton Park Soton 32 D6
Hampton Poyle Oxon 83 B8
Hampton Wick London 67 F7
Hamptons Kent 52 C6
Hamptworth Wilts 32 D2
Hamrow Norf 159 E8
Hamsey E Sus 36 E6
Hamsey Green London 51 E10
Hamshill Glos 80 E3
Hamstall Ridware Staffs 152 F2
Hamstead I o W 20 C4
Hamstead W Mid 133 E10
Hamstead Marshall W Berks 64 F2
Hamsterley Durham 233 E8
Hamsterley Durham 242 F4
Hamstreet Kent 54 G4
Hamworthy Poole 18 C5
Hanbury Staffs 152 D3
Hanbury Worcs 117 E9
Hanbury Woodend Staffs 152 D3
Hanby Lincs 155 C10
Hanchett Village Suff 106 B3
Hanchurch Staffs 168 G4
Hand Green W Ches 167 C8
Handbridge W Ches 166 B6
Handcross W Sus 36 B3
Handforth E Ches 184 E5
Handless Shrops 131 E7
Handley Derbys 170 C5
Handley W Ches 167 D7
Handley Green Essex 87 E11
Handsacre Staffs 151 F11
Handside Herts 86 C2
Handsworth S Yorks 186 D6
Handsworth W Mid 133 E10
Handsworth Wood W Mid 133 E11
Handy Cross Bucks 84 G5
Handy Cross Devon 24 B6
Handy Cross Som 42 G6
Hanford Dorset 30 E4
Hanford Stoke 168 G5
Hangersley Hants 31 F11
Hanging Bank Kent 52 C3
Hanging Heaton W Yorks 197 C9
Hanging Houghton Northants 120 C5
Hanging Langford Wilts 46 F4
Hangingshaw Borders 261 C9
Hangingshaw Dumfries 248 F4
Hangleton Brighton 36 F3
Hangleton W Sus 35 G9
Hangsman Hill S Yorks 199 E7
Hanham S Glos 60 E6
Hanham Green S Glos 60 E6
Hankelow E Ches 167 F11
Hankerton Wilts 81 G7
Hankham E Sus 23 D10
Hanley Stoke 168 F5
Hanley Castle Worcs 98 C6
Hanley Child Worcs 116 E3
Hanley Swan Worcs 98 C6
Hanley William Worcs 116 E3
Hanlith N Yorks 213 G8
Hanmer Wrex 149 B9
Hannaford Devon 25 B10
Hannafore Corn 6 E5
Hannah Lincs 191 F8
Hanningfields Green Suff 125 G7
Hannington Hants 48 B4
Hannington Northants 120 C6
Hannington Swindon 81 G11
Hannington Wick Swindon 81 F11
Hanscombe End C Beds 104 E2
Hansel Devon 8 F6
Hansel Village S Ayrs 257 C9
Hansley Cross Staffs 169 G9
Hanslope M Keynes 102 B6
Hanthorpe Lincs 155 E11
Hanwell London 67 C7
Hanwell Oxon 101 C8
Hanwood Shrops 131 B8
Hanwood Bank Shrops 149 G8
Hanworth Brack 65 F11
Hanworth London 66 E5
Hanworth Norf 160 B3
Happendon S Lnrk 259 C9
Happisburgh Norf 161 C7
Happisburgh Common Norf 161 D7
Hapsford Som 45 D9
Hapsford W Ches 183 G7
Hapton Lancs 203 G11
Hapton Norf 142 D3
Hararden = Penarlâg Flint 166 B4
Harberton Devon 8 D5
Harbertonford Devon 8 D5
Harbledown Kent 54 B5
Harborne W Mid 133 G10
Harborough Magna Warks 119 B9
Harborough Parva Warks 119 B9
Harbottle Northumb 251 C10
Harbour Heights E Sus 36 G5
Harbour Village Pembs 91 D3
Harbourland Kent 53 B9
Harbourneford Devon 8 C4
Harbours Hill Worcs 117 D9
Harbridge Hants 31 E10
Harbridge Green Hants 31 E10
Harburn W Loth 269 C10
Harbury Warks 119 F7
Harby Leics 154 C4
Harby Notts 188 G5
Harcombe Devon 14 E3
Harcombe Devon 15 C9
Harcourt Corn 3 B8
Harcourt Hill Oxon 83 E7
Hardbreck Orkney 314 F4
Hardeicke Glos 80 C4
Harden S Yorks 197 G7
Harden W Mid 133 C10
Harden W Yorks 205 F7
Harden Park E Ches 184 F4
Hardendale Cumb 221 C11
Hardenhuish Wilts 62 E2
Hardgate Aberds 293 C9
Hardgate Dumfries 237 C10
Hardgate N Yorks 214 G5
Hardgate W Dunb 277 G10
Hardham W Sus 35 D8
Hardhorn Lancs 202 F3
Hardingham Norf 141 C10
Hardings Booth Staffs 169 C9
Hardings Wood Staffs 168 E4
Hardingstone Northants 120 F5
Hardington Som 45 C8
Hardington Mandeville Som 29 E8
Hardington Marsh Som 29 F8
Hardington Moor Som 29 E8
Hardiston Perth 279 B11
Hardisworthy Devon 24 C2
Hardley Hants 32 G6
Hardley Street Norf 143 C7
Hardmead M Keynes 103 B8
Hardrow N Yorks 223 G7
Hardstoft Derbys 170 C6
Hardstoft Common Derbys 170 C6
Hardway Hants 33 G10
Hardway Som 45 G8
Hardwick Bucks 84 B4
Hardwick Cambs 122 D3
Hardwick Cambs 123 F7
Hardwick Norf 142 F4
Hardwick Norf 158 F2
Hardwick Northants 121 D7
Hardwick Oxon 82 D5
Hardwick Oxon 101 C8
Hardwick Oxon 101 F11
Hardwick S Yorks 187 D7
Hardwick Shrops 131 E7
Hardwick Stockton 234 G4
Hardwick W Mid 133 D11
Hardwick Green Worcs 98 E6
Hardwick Village Notts 187 F10
Hardwicke Glos 80 C3
Hardwicke Glos 99 F8
Hardwicke Hereford 96 C5
Hardy's Green Essex 107 G8
Hare Som 28 D3
Hare Appletree Lancs 202 B6
Hare Edge Derbys 186 G4
Hare Green Essex 107 G11
Hare Hatch Wokingham 65 D10
Hare Street Essex 86 D6
Hare Street Herts 104 F6
Hare Street Herts 105 F7
Hareby Lincs 174 B4
Harecroft W Yorks 205 F7
Hareden Lancs 203 C8
Harefield London 85 G9
Harefield Soton 33 E7
Harefield Grove London 85 G9
Haregate Staffs 169 D7
Harehill Derbys 152 B3
Harehills W Yorks 206 G2
Harehope Borders 270 G4
Harehope Northumb 264 E3
Harelaw Durham 242 G5
Hareleeshill S Lnrk 268 E5
Hareplain Kent 53 F10
Haresceugh Cumb 231 C8
Harescombe Glos 80 C4
Haresfield Glos 80 C4
Haresfield Swindon 82 G2
Haresfinch Mers 183 B8
Hareshaw N Lnrk 268 C6
Hareshaw Head Northumb 251 F9
Harestanes E Dunb 278 G3
Harestock Hants 48 G3
Harewood W Yorks 206 D2
Harewood End Hereford 97 F10
Harewood Hill W Yorks 204 F6
Harford Carms 94 C2
Harford Devon 8 D2
Harford Devon 40 G6
Hargate Norf 142 E2
Hargate Hill Derbys 185 C8
Hargatewall Derbys 185 F10
Hargrave Northants 121 C10
Hargrave Suff 124 F5
Hargrave W Ches 167 C7
Harker Cumb 239 E9
Harker Marsh Cumb 229 E7
Harkland Shetland 312 E6
Harknett's Gate Essex 86 D6
Harkstead Suff 108 E3
Harlaston Staffs 152 G4
Harlaw Ho Aberds 303 G7
Harlaxton Lincs 155 C7
Harle Syke Lancs 204 F3
Harlech Gwyn 145 C11
Harlequin Notts 154 B3
Harlescott Shrops 149 F10
Harlesden London 67 C8
Harlesthorpe Derbys 187 F7
Harleston Devon 8 F5
Harleston Norf 142 G4
Harleston Suff 125 F10
Harlestone Northants 120 E4
Harley S Yorks 186 B5
Harley Shrops 131 C11
Harley Shute E Sus 38 F3
Harleyholm S Lnrk 259 B10
Harleywood Glos 80 F4
Harling Road Norf 141 F9
Harlington C Beds 103 E10
Harlington London 66 D5
Harlington S Yorks 198 G3
Harlosh Highld 298 E2
Harlow Essex 86 C6
Harlow Carr N Yorks 205 C11
Harlow Green T & W 243 F7
Harlow Hill N Yorks 205 C11
Harlow Hill Northumb 242 D3
Harlthorpe E Yorks 207 F10
Harlton Cambs 123 G7
Harlyn Corn 10 F3
Harman's Corner Kent 69 G11
Harman's Cross Dorset 18 E5
Harmans Water Brack 65 F11
Harmby N Yorks 214 B2
Harmer Green Herts 86 B3
Harmer Hill Shrops 149 E9
Harmondsworth London 66 D5
Harmston Lincs 173 C7
Harnage Shrops 131 C11
Harnham Northumb 242 B3
Harnham Wilts 31 B10
Harnham Wilts 31 B11
Harnhill Glos 81 E9
Harold Hill London 87 G8
Harold Park London 87 G9
Harold Wood London 87 G8
Haroldston West Pembs 72 B5
Haroldswick Shetland 312 B8
Harome N Yorks 216 C2
Harpenden Herts 85 C10
Harpenden Common Herts 85 C10
Harper Green Gtr Man 195 F8
Harperley Durham 242 G5
Harper's Gate Staffs 169 D7
Harper's Green Norf 159 E8
Harpford Devon 15 C7
Harpham E Yorks 217 G11
Harpley Norf 158 D5
Harpley Worcs 116 E3
Harpole Northants 120 E3
Harpsdale Highld 310 D5
Harpsden Oxon 65 C9
Harpsden Bottom Oxon 65 C9
Harpswell Lincs 188 D6
Harpton Powys 114 F4
Harpur Hill Derbys 185 G9
Harpurhey Gtr Man 195 G11
Harraby Cumb 239 G10
Harracott Devon 25 B9
Harrapool Highld 295 C8
Harras Cumb 219 B9
Harraton T & W 243 G7
Harrier Shetland 313 J1
Harrietfield Perth 286 E3
Harrietsham Kent 53 C11
Harringay London 67 B10
Harrington Cumb 228 F5
Harrington Lincs 190 G5
Harrington Northants 136 G5
Harringworth Northants 137 D8
Harris Highld 294 F5
Harriseahead Staffs 168 D5
Harriston Cumb 229 C9
Harrogate N Yorks 206 C2
Harrold Beds 121 F8
Harrop Dale Gtr Man 196 F4
Harrow Highld 310 B6
Harrow London 67 B7
Harrow Green Suff 125 G7
Harrow Hill Glos 79 B10
Harrow on the Hill London 67 B7
Harrow Street Suff 107 D9
Harrow Weald London 85 G11
Harrowbarrow Corn 7 B7
Harrowbeer Devon 7 B10
Harrowby Lincs 155 B8
Harrowden Beds 103 B11
Harrowgate Hill Darl 224 B5
Harrowgate Village Darl 224 B5
Harry Stoke S Glos 60 D6
Harston Cambs 123 G8
Harston Leics 154 C6
Harswell E Yorks 208 E2
Hart Hrtlpl 234 E5
Hart Common Gtr Man 194 F6
Hart Hill Luton 104 G2
Hart Station Hrtlpl 234 D5
Hartbarrow Cumb 221 G8
Hartburn Northumb 252 F3
Hartburn Stockton 225 B8
Hartcliffe Bristol 60 F5
Hartest Suff 124 G6
Hartest Hill Suff 124 G6
Hartfield E Sus 52 F3
Hartfield Highld 299 E7
Hartford Cambs 122 C5
Hartford Som 27 B7
Hartford W Ches 183 G10
Hartford End Essex 87 B11
Hartfordbeach W Ches 183 G10
Hartfordbridge Hants 49 B9
Hartforth N Yorks 224 D3
Hartgrove Dorset 30 D4
Hartham Herts 86 C4
Harthill N Lnrk 269 C8
Harthill S Yorks 187 E7
Harthill W Ches 167 D8
Hartington Derbys 169 C10
Hartland Devon 24 C3
Hartle Worcs 117 B8
Hartlebury Shrops 132 D4
Hartlebury Worcs 116 C6
Hartlebury Common Worcs 116 C6
Hartlepool Hrtlpl 234 E6
Hartley Cumb 222 D5
Hartley Kent 53 G9
Hartley Kent 68 F6
Hartley Northumb 243 B8
Hartley Plym 7 D9
Hartley Green Kent 68 F6
Hartley Green Staffs 151 D9
Hartley Mauditt Hants 49 F8
Hartley Westpall Hants 49 B7
Hartley Wintney Hants 49 B9
Hartlington N Yorks 213 G10
Hartlip Kent 69 G10
Hartmoor Dorset 30 C3
Hartmount Highld 301 B7
Hartoft End N Yorks 226 G5
Harton N Yorks 216 G4
Harton Shrops 131 F9
Harton T & W 243 E9
Hartpury Glos 98 F5
Hart's Green Suff 125 F7
Hart's Hill W Mid 133 F8
Hartsgreen Shrops 132 G5
Hartshead W Yorks 197 C7
Hartshead Green Gtr Man 196 G3
Hartshead Moor Side W Yorks 197 C7
Hartshead Moor Top W Yorks 197 B7
Hartshead Pike Gtr Man 196 G3
Hartshill Stoke 168 F5
Hartshill Warks 134 E6
Hartshill Green Warks 134 E6
Hartshorne Derbys 152 E6
Hartsop Cumb 221 C8
Hartswell Som 27 B9
Hartwell Northants 120 G5
Hartwell Staffs 151 B8
Hartwith N Yorks 214 G4
Hartwood Lancs 194 D5
Hartwood N Lnrk 268 D6
Hartwoodburn Borders 261 D11
Harvel Kent 68 G6
Harvest Hill W Mid 134 G5
Harvieston Stirl 277 D11
Harvills Hawthorn W Mid 133 E9
Harvington Worcs 99 B11
Harvington Worcs 117 C7
Harvington Cross Worcs 99 B11
Harvington Hill Worcs 99 B10
Harwell Notts 187 C11
Harwell Oxon 64 B3
Harwich Essex 108 E5
Harwood Durham 232 E2
Harwood Gtr Man 195 E8
Harwood Dale N Yorks 227 F9
Harwood Lee Gtr Man 195 E8
Harwood on Teviot Borders 249 B10
Harworth Notts 187 C10
Hasbury W Mid 133 G9
Hascombe Sur 50 E3
Haselbech Northants 120 B4
Haselbury Plucknett Som 29 E7
Haseley Warks 118 D4
Haseley Green Warks 118 D4
Haseley Knob Warks 118 C4
Haselor Warks 118 F2
Hasfield Glos 98 F6
Hasguard Pembs 72 D5
Haskayne Lancs 193 F11
Hasketon Suff 126 G4
Hasland Derbys 170 B5
Haslemere Sur 50 G2
Haslingbourne W Sus 35 C7
Haslingden Lancs 195 C9
Haslingfield Cambs 123 G8
Haslington E Ches 168 D2
Hasluck's Green W Mid 118 B2
Hassall E Ches 168 D3
Hassall Green E Ches 168 D3
Hassell Street Kent 54 D5
Hassendean Borders 262 E2
Hassingham Norf 143 B7
Hassocks W Sus 36 D3
Hassop Derbys 186 G2
Haster Highld 310 D7
Hasthorpe Lincs 175 B7
Hastigrow Highld 310 C6
Hasting Hill T & W 243 G9
Hastingleigh Kent 54 E5
Hastings E Sus 38 F4
Hastings Som 28 D4
Hastingwood Essex 87 D7
Hastoe Herts 84 D6
Haston Shrops 149 E10
Haswell Durham 234 C3
Haswell Moor Durham 234 C3
Haswell Plough Durham 234 C3
Haswellsykes Borders 260 B6
Hatch C Beds 104 B3
Hatch Devon 8 F4
Hatch Hants 49 C7
Hatch Wilts 30 B6
Hatch Beauchamp Som 28 C4
Hatch Bottom Hants 33 E7
Hatch End Beds 121 E11
Hatch End London 85 G10
Hatch Farm Hill W Sus 34 B6
Hatch Green Som 28 D4
Hatch Warren Hants 48 D6
Hatchet Gate Hants 32 G5
Hatchet Green Hants 31 D11
Hatching Green Herts 85 C10
Hatchmere W Ches 183 G9
Hatcliffe NE Lincs 201 G8
Hateley Heath W Mid 133 E10
Hatfield Hereford 115 F11
Hatfield Herts 86 D2
Hatfield S Yorks 199 F7
Hatfield Worcs 117 G7
Hatfield Broad Oak Essex 87 B8
Hatfield Chase S Yorks 199 E8
Hatfield Garden Village Herts 86 D2
Hatfield Heath Essex 87 B8
Hatfield Hyde Herts 86 C2
Hatfield Peverel Essex 88 C3
Hatfield Woodhouse S Yorks 199 F7
Hatford Oxon 82 G4
Hatherden Hants 47 C10
Hatherleigh Devon 25 G8
Hatherley Glos 99 G8
Hathern Leics 153 E9
Hatherop Glos 81 D11
Hathersage Derbys 186 E2
Hathersage Booths Derbys 186 E2
Hathershaw Gtr Man 196 G2
Hatherton E Ches 167 F11
Hatherton Staffs 151 G9
Hatley St George Cambs 122 G5
Hatston Orkney 314 E4
Hatt Corn 7 C7
Hatt Hill Hants 32 B4
Hatterseat Aberds 303 G9
Hattersley Gtr Man 185 C7
Hattingley Hants 48 F6
Hatton Aberds 303 F10
Hatton Angus 287 D9
Hatton Derbys 152 D4
Hatton Lincs 189 F11
Hatton London 66 D5
Hatton Moray 301 D11
Hatton Shrops 131 E9
Hatton Warks 118 D4
Hatton Warr 183 E9
Hatton Castle Aberds 303 E7
Hatton Grange Shrops 132 C5
Hatton Heath W Ches 167 C7
Hatton Hill Sur 66 G2
Hatton of Fintray Aberds 293 B10
Hatton Park Northants 121 D7
Hattonburn Aberds 293 C9
Hattoncrook Aberds 303 G8
Hattonknowe Borders 270 F4
Haugh E Ayrs 257 D11
Haugh Gtr Man 196 E2
Haugh Lincs 190 F6
Haugh-head Borders 261 B8
Haugh Head Northumb 264 D2
Haugh of Glass Moray 302 F4
Haugh of Kilnmaichlie Moray 301 F11
Haugh of Urr Dumfries 237 C10
Haugham Lincs 190 E4
Haughland Orkney 314 E5
Haughley Suff 125 E10
Haughley Green Suff 125 E10
Haughley New Street Suff 125 E10
Haughs of Clinterty Aberdeen 293 B10
Haughton E Ches 167 D9
Haughton Notts 187 G11
Haughton Powys 148 F6
Haughton Shrops 132 B4
Haughton Shrops 132 D3
Haughton Shrops 149 D7
Haughton Shrops 149 F11
Haughton Staffs 151 E7
Haughton Castle Northumb 241 C10
Haughton Green Gtr Man 184 C6
Haughton Le Skerne Darl 224 B6
Haughurst Hill W Berks 64 G5
Haulkerton Aberds 293 F9
Haultwick Herts 104 G6
Haunn Argyll 288 E5
Haunn W Isles 297 K3
Haunton Staffs 152 G4
Hauxton Cambs 123 G8
Havannah E Ches 168 C5
Havant Hants 22 B2
Haven Hereford 97 B11
Haven Hereford 115 G8
Haven Bank Lincs 174 E2
Haven Side E Yorks 201 B7
Havenstreet I o W 21 C7
Havercroft W Yorks 197 E11
Haverfordwest = Hwlffordd Pembs 73 B7
Haverhill Suff 106 B3
Haverigg Cumb 210 D3
Havering-atte-Bower London 87 G8
Haveringland Norf 160 E3
Haversham M Keynes 102 C6
Haverthwaite Cumb 210 C6
Haverton Hill Stockton 234 G5
Haviker Street Kent 53 D8
Havyatt Som 44 F4
Havyatt Green N Som 60 G3
Hawarden = Penarlâg Flint 166 B4
Hawbridge Worcs 99 B8
Hawbush Green Essex 106 G5
Hawcoat Cumb 210 E4
Hawcross Glos 98 E5
Hawddamor Gwyn 146 F3
Hawen Ceredig 92 B6
Hawes N Yorks 213 B7
Hawes' Green Norf 142 D4
Hawes Side Blkpool 202 G2
Hawford Worcs 116 E6
Hawgreen Shrops 150 D2
Hawick Borders 262 F2
Hawk Green Gtr Man 185 D7
Hawk Hill Cumb 228 F6
Hawkchurch Devon 28 G4
Hawkcombe Som 41 D11
Hawkedon Suff 124 G5
Hawkenbury Kent 52 F5
Hawkenbury Kent 53 E10
Hawkeridge Wilts 45 C11
Hawkerland Devon 15 D7
Hawkersland Cross Hereford 97 B10
Hawkes End W Mid 134 G6
Hawkesbury S Glos 61 B9
Hawkesbury Warks 135 G7
Hawkesbury Upton S Glos 61 B9
Hawkesley W Mid 117 B10
Hawkhill Northumb 264 G6
Hawkhope Northumb 250 F6
Hawkhurst Kent 53 G9
Hawkhurst Common E Sus 23 B8
Hawkinge Kent 55 F8
Hawkin's Hill Essex 106 E3
Hawkley Gtr Man 194 G5
Hawkley Hants 34 B2
Hawkridge Som 41 G11
Hawks Green Staffs 151 G9
Hawks Hill Bucks 66 B2
Hawk's Hill Sur 51 B7
Hawks Stones W Yorks 196 B2
Hawksdale Cumb 230 B3
Hawkshaw Blkburn 195 D9
Hawkshead Cumb 221 F7
Hawkshead Hill Cumb 220 F6
Hawksland S Lnrk 259 B8
Hawkspur Green Essex 106 E3
Hawkswick N Yorks 213 E9
Hawksworth Notts 172 G3
Hawksworth W Yorks 205 E9
Hawksworth W Yorks 205 F11
Hawkwell Essex 88 G4
Hawley Hants 49 B11
Hawley Kent 68 E4
Hawley Bottom Devon 28 G2
Hawley Lane Hants 49 B11
Hawling Glos 99 G11
Hawn Orkney 314 D4
Hawnby N Yorks 215 B10
Hawne W Mid 133 G9
Haworth W Yorks 204 F6
Haws Bank Cumb 220 F6
Hawstead Suff 125 F7
Hawstead Green Suff 125 F7
Hawthorn Durham 234 B4
Hawthorn Hants 49 G7
Hawthorn Rhondda 58 B6
Hawthorn Wilts 61 F11
Hawthorn Corner Kent 71 F8
Hawthorn Hill Brack 65 E11
Hawthorn Hill Lincs 174 D2
Hawthorns Staffs 168 F4
Hawthorpe Lincs 155 D10
Hawton Notts 172 E3
Haxby York 207 B8
Haxey N Lincs 188 B3
Haxey N Lincs 199 G9
Haxey Carr N Lincs 199 G9
Haxted Sur 52 D2
Haxton Wilts 46 D6
Hay Corn 10 G5
Hay Field S Yorks 187 B10
Hay Green Essex 87 E10
Hay Green Herts 104 D6
Hay Green Norf 157 F10
Hay Mills W Mid 134 G2
Hay-on-Wye Powys 96 C4
Hay Street Herts 105 F7
Haybridge Shrops 116 C2
Haybridge Som 44 D4
Haybridge Telford 150 G3
Hayden Glos 99 G8
Haydock Mers 183 B9
Haydon Bath 45 C7
Haydon Dorset 29 D11
Haydon Som 28 C3
Haydon Som 44 D5
Haydon Swindon 62 B6
Haydon Bridge Northumb 241 E8
Haydon Wick Swindon 62 B6
Haye Corn 7 B7
Haye Fm Corn 6 B6
Hayes London 66 C6
Hayes London 68 F2
Hayes Staffs 169 C9
Hayes End London 66 C5
Hayes Knoll Wilts 81 G10
Hayes Town London 66 C6
Hayfield Derbys 185 D8
Hayfield Fife 280 C5
Hayfield Green S Yorks 187 B11
Haygate Telford 150 G2
Haygrass Som 28 C2
Hayhill E Ayrs 257 F11
Hayhillock Angus 287 C9
Haylands I o W 21 C7
Hayle Corn 2 B3
Hayley Green W Mid 133 G8
Haymoor End Som 28 B4
Haymoor Green E Ches 167 E11
Hayne Devon 26 F5
Haynes C Beds 103 C11
Haynes Church End C Beds 103 C11
Haynes West End C Beds 103 C11
Hayscastle Pembs 91 F7
Hayscastle Cross Pembs 91 G8
Haysford Pembs 91 G8
Hayshead Angus 287 C10
Hayston E Dunb 278 G2
Haystoun Borders 261 B7
Haythorne Dorset 31 F8
Hayton Aberdeen 293 C11
Hayton Cumb 229 C8
Hayton Cumb 240 F2
Hayton E Yorks 208 D2
Hayton Notts 188 E2
Hayton's Bent Shrops 131 G10
Haytor Vale Devon 13 F11
Haytown Devon 24 E5
Haywards Heath W Sus 36 C4
Haywood S Lnrk 269 E9
Haywood S Yorks 198 E5
Haywood Oaks Notts 171 D10
Hazard's Green E Sus 23 C11
Hazel End Essex 105 G9
Hazel Grove Gtr Man 184 D6
Hazel Street Kent 53 B11
Hazel Street Kent 53 F7
Hazel Stub Suff 106 C3
Hazelbank S Lnrk 268 F6
Hazelbeach Pembs 72 E6
Hazelbury Bryan Dorset 30 F2
Hazeleigh Essex 88 E4
Hazeley Hants 49 B8
Hazeley Bottom Hants 49 B9
Hazeley Heath Hants 49 B9
Hazeley Lea Hants 49 B8
Hazelgrove Notts 171 F8
Hazelhurst Gtr Man 195 D9
Hazelhurst Gtr Man 195 G9
Hazelhurst Gtr Man 196 G3
Hazelslack Cumb 211 D9
Hazelslade Staffs 151 G10
Hazelton Glos 81 B9
Hazelton Walls Fife 287 E7
Hazelwood Derbys 170 F4
Hazelwood Devon 8 E4
Hazelwood London 68 G2
Hazlehead S Yorks 197 G7
Hazlemere Bucks 84 F5
Hazler Shrops 131 E9
Hazlerigg T & W 242 C6
Hazles Staffs 169 F8
Hazlescross Staffs 169 F8
Hazleton Glos 81 B9
Hazlewood N Yorks 205 C7
Hazon Northumb 252 C5
Heacham Norf 158 B3
Head of Muir Falk 278 E6
Headbourne Worthy Hants 48 G3
Headbrook Hereford 114 F6
Headcorn Kent 53 E10
Headingley W Yorks 205 F11
Headington Oxon 83 D8
Headington Hill Oxon 83 D8
Headlam Durham 224 B3
Headless Cross Cumb 211 D7
Headless Cross Worcs 117 D10
Headley Hants 49 F10
Headley Hants 64 G4
Headley Sur 51 C8
Headley Down Hants 49 F10
Headley Heath Worcs 117 B11
Headley Park Bristol 60 F5
Headon Devon 24 G5
Headon Notts 188 F2
Heads S Lnrk 268 F4
Heads Nook Cumb 239 F11
Headshaw Borders 261 E11
Headstone London 66 B6
Headwell Fife 279 D11
Heady Hill Gtr Man 195 E10
Heage Derbys 170 E5
Healaugh N Yorks 206 D5
Healaugh N Yorks 223 F10
Heald Green Gtr Man 184 D4
Healds Green Gtr Man 195 F11
Heale Devon 40 D6
Heale Som 28 B5
Heale Som 28 D2
Heale Som 45 E7
Healey Gtr Man 195 D11
Healey N Yorks 214 C3
Healey Northumb 242 F2
Healey W Yorks 197 C8
Healey W Yorks 197 D9
Healey Cote Northumb 252 C4
Healey Hall Northumb 242 F2
Healeyfield Durham 233 B7
Healing NE Lincs 201 E8
Heamoor Corn 1 C5
Heaning Cumb 221 F8
Heanish Argyll 288 E2
Heanor Derbys 170 F6
Heanor Gate Derbys 170 F6
Heanton Punchardon Devon 40 F4
Heap Bridge Gtr Man 195 E10
Heapham Lincs 188 D5
Hearn Hants 49 F10
Hearnden Green Kent 53 D10
Hearthstane Borders 260 D4
Hearthstone Derbys 170 D4
Hearts Delight Kent 69 G11
Heasley Mill Devon 41 G8
Heast Highld 295 D8
Heath Cardiff 59 D7
Heath Derbys 170 B6
Heath Halton 183 E8
Heath and Reach C Beds 103 F8
Heath Charnock Lancs 194 E5
Heath Common W Sus 35 E10
Heath Common W Yorks 197 D11
Heath Cross Devon 13 B10
Heath Cross Devon 14 C2
Heath End Bucks 84 F5
Heath End Bucks 85 D7
Heath End Derbys 153 E7
Heath End Hants 64 G2
Heath End Hants 64 G5
Heath End S Glos 61 B7
Heath End Sur 49 D10
Heath End W Mid 133 C10
Heath End W Sus 35 D7
Heath End Warks 118 E4
Heath Green Hants 48 F6
Heath Green Worcs 117 C11
Heath Hayes Staffs 151 G10
Heath Hill Shrops 150 G5
Heath House Som 44 D2
Heath Lanes Telford 150 E2
Heath Park London 68 B4
Heath Side Kent 68 E4
Heath Town W Mid 133 D8
Heathcot Aberds 293 C10
Heathcote Derbys 169 C10
Heathcote Shrops 150 D3
Heathcote Warks 118 E6
Heather Leics 153 G7
Heather Row Hants 49 C8
Heathercombe Devon 13 E10
Heatherfield Highld 298 E4
Heatherside Sur 50 B2
Heatherwood Park Highld 311 K2
Heatherybanks Aberds 303 E7
Heathfield Cambs 105 B9
Heathfield Devon 14 F2
Heathfield E Sus 37 C9
Heathfield Glos 80 F2
Heathfield Hants 33 F9
Heathfield Lincs 189 C10
Heathfield N Yorks 214 F2
Heathfield S Ayrs 257 E9
Heathfield Som 27 B11
Heathfield Som 43 G7
Heathfield Village Oxon 83 B8
Heathhall Dumfries 237 B11
Heathlands Wokingham 65 F10
Heathrow Airport London 66 D5
Heathstock Devon 28 G2
Heathton Shrops 132 E6
Heathtop Derbys 152 C4
Heathwaite Cumb 221 F8
Heathwaite N Yorks 225 E9
Heatley Staffs 151 D11
Heatley Warr 184 D2
Heaton Gtr Man 195 F7
Heaton Lancs 211 G8
Heaton Staffs 169 C7
Heaton T & W 243 D7
Heaton W Yorks 205 G8
Heaton Chapel Gtr Man 184 C5

Heaton Mersey Gtr Man 184 C5
Heaton Moor Gtr Man 184 C5
Heaton Norris Gtr Man 184 C5
Heaton Royds W Yorks 205 F8
Heaton Shay W Yorks 205 F8
Heaton's Bridge Lancs 194 E2
Heaven's Door Som 29 C10
Heaverham Kent 52 B5
Heaviley Gtr Man 184 D6
Heavitree Devon 14 C4
Hebburn T & W 243 E8
Hebburn Colliery T & W 243 D8
Hebburn New Town T & W 243 E8
Hebden N Yorks 213 G10
Hebden Bridge W Yorks 196 B3
Hebden Green W Ches 167 B10
Hebing End Herts 104 G6
Hebron Anglesey 179 E7
Hebron Carms 92 F3
Hebron Northumb 252 F5
Heck Dumfries 248 G3
Heckdyke N Lincs 188 B3
Heckfield Hants 65 G8
Heckfield Green Suff 126 B3
Heckfordbridge Essex 107 G8
Heckingham Norf 143 D7
Heckington Lincs 173 G10
Heckmondwike W Yorks 197 C8
Heddington Wilts 62 F3
Heddington Wick Wilts 62 F3
Heddle Orkney 314 E3
Heddon Devon 25 B11
Heddon-on-the-Wall Northumb 242 D4
Hedenham Norf 142 E6
Hedge End Dorset 30 F4
Hedge End Hants 33 E7
Hedgehog Bridge Lincs 174 F3
Hedgerley Bucks 66 B3
Hedgerley Green Bucks 66 B3
Hedgerley Hill Bucks 66 B3
Hedging Som 28 B4
Hedley Hill Durham 233 C9
Hedley on the Hill Northumb 242 F3
Hednesford Staffs 151 G9
Hedon E Yorks 201 B7
Hedsor Bucks 66 B2
Hedworth T & W 243 E8
Heelands M Keynes 102 D6
Heeley S Yorks 186 E5
Hegdon Hill Hereford 115 G11
Heggerscales Cumb 222 C6
Heggle Lane Cumb 230 D3
Heglibister Shetland 313 H5
Heighington Darl 233 G11
Heighington Lincs 173 B8
Heighley Staffs 168 F3
Height End Lancs 195 C9
Heightington Worcs 116 C5
Heights Gtr Man 196 F3
Heights of Brae Highld 300 C5
Heights of Kinlochewe Highld 299 C10
Heilam Highld 308 C4
Heiton Borders 262 C6
Helbeck Cumb 222 B5
Hele Devon 12 C2
Hele Devon 13 G10
Hele Devon 27 G7
Hele Devon 40 D4
Hele Som 27 C11
Hele Torbay 9 B8
Helebridge Corn 24 G2
Helensburgh Argyll 276 E5
Helford Corn 3 D7
Helford Passage Corn 3 D7
Helham Green Herts 86 B5
Helhoughton Norf 159 D7
Helions Bumpstead Essex 106 C3
Hell Corner W Berks 63 G11
Hellaby S Yorks 187 C8
Helland Corn 11 G7
Helland Som 28 C4
Hellandbridge Corn 11 G7
Hellesdon Norf 160 G4
Hellesveor Corn 2 A2
Hellidon Northants 119 F10
Hellifield N Yorks 204 B3
Hellifield Green N Yorks 204 B3
Hellingly E Sus 23 C9
Hellington Norf 142 C6
Hellister Shetland 313 J5
Hellman's Cross Essex 87 B9
Helm N Yorks 223 G8
Helm Northumb 252 D5
Helmburn Borders 261 E9
Helmdon Northants 101 C11
Helme W Yorks 196 E5
Helmingham Suff 126 F3
Helmington Row Durham 233 D9
Helmsdale Highld 311 H4
Helmshore Lancs 195 C9
Helmside Cumb 212 B3
Helmsley N Yorks 216 C2
Helperby N Yorks 215 F8
Helperthorpe N Yorks 217 E9
Helpringham Lincs 173 G10
Helpston P'boro 138 B2
Helsby W Ches 183 F7
Helscott Corn 24 G2
Helsey Lincs 191 G8
Helston Corn 2 D5
Helston Water Corn 4 G5
Helstone Corn 11 E7
Helton Cumb 230 G6
Helwith Bridge N Yorks 212 F6
Helygain = Halkyn Flint 182 G2
Hem Heath Stoke 168 G5
Hemblington Norf 160 G6
Hemblington Corner Norf 160 G6
Hembridge Som 44 F5
Hemel Hempstead Herts 85 D9
Hemerdon Devon 7 D11
Hemford Shrops 130 C6
Hemingbrough N Yorks 207 G9
Hemingby Lincs 190 G2
Hemingfield S Yorks 197 G11
Hemingford Abbots Cambs 122 C5
Hemingford Grey Cambs 122 C5
Hemingstone Suff 126 G3
Hemington Leics 153 D9
Hemington Northants 137 F11
Hemington Som 45 C8
Hemley Suff 108 C5
Hemlington M'bro 225 C10
Hemp Green Suff 127 D7
Hempholme E Yorks 209 C7
Hempnall Norf 142 E4
Hempnall Green Norf 142 E4
Hempriggs House Highld 310 E7
Hemp's Green Essex 107 F8
Hempshill Vale Notts 171 G8
Hempstead Essex 106 D2
Hempstead Medway 69 G9
Hempstead Norf 160 B2
Hempstead Norf 161 D8
Hempsted Glos 80 B4
Hempton Norf 159 D8
Hempton Oxon 101 E8
Hempton Wainhill Oxon 84 E3
Hemsby Norf 161 F9
Hemsted Kent 54 E6
Hemswell Lincs 188 C6
Hemswell Cliff Lincs 188 D6
Hemsworth Dorset 31 F7
Hemsworth S Yorks 186 E5
Hemsworth W Yorks 198 E2
Hemyock Devon 27 E10
Hen Bentref Llandegfan Anglesey 179 G9
Hên-efail Denb 165 C9
Hen-feddau fawr Pembs 92 E4
Henaford Devon 24 D2
Henbrook Worcs 117 D8
Henbury Bristol 60 D5
Henbury Dorset 18 B5
Henbury E Ches 184 G5
Hendomen Powys 130 D4
Hendon London 67 B8
Hendon T & W 243 F10
Hendra Corn 2 B6
Hendra Corn 2 C5
Hendra Corn 2 D3
Hendra Corn 2 F6
Hendra Corn 5 C9
Hendra Corn 5 D9
Hendra Corn 11 E7
Hendra Croft Corn 4 D5
Hendrabridge Corn 6 B5
Hendraburnick Corn 11 D8
Hendre Flint 165 B11
Hendre Gwyn 110 B2
Hendre Powys 129 D9
Hendre-ddu Conwy 164 B5
Hendredenny Park Caerph 58 B6
Hendreforgan Rhondda 58 B3
Hendrerwydd Denb 165 C10
Hendrewen Swansea 75 D10
Hendy Carms 75 E9
Hendy-Gwyn Carms 74 B2
Hendy Gwyn = Whitland Carms 73 B11
Heneglwys Anglesey 178 F6
Henfield S Glos 61 D7
Henfield W Sus 36 D2
Henford Devon 12 C3
Henfords Marsh Wilts 45 E11
Henghurst Kent 54 F3
Hengoed Caerph 77 F10
Hengoed Denb 165 D9
Hengoed Powys 114 G4
Hengoed Shrops 148 C5
Hengrave Norf 160 F2
Hengrave Suff 124 D6
Hengrove Bristol 60 F6
Hengrove Park Bristol 60 F5
Henham Essex 105 F10
Heniarth Powys 130 B2
Henlade Som 28 C3
Henleaze Bristol 60 D5
Henley Dorset 29 G11
Henley Glos 80 B6
Henley Shrops 115 B10
Henley Shrops 131 F9
Henley Som 44 G2
Henley Suff 126 G3
Henley W Sus 34 B5
Henley Wilts 47 B10
Henley Wilts 61 F10
Henley Common W Sus 34 B5
Henley Green W Mid 135 G7
Henley-in-Arden Warks 118 D3
Henley-on-Thames Oxon 65 C9
Henley Street Kent 69 F7
Henley's Down E Sus 38 E2
Henllan Ceredig 93 C7
Henllan Denb 165 B8
Henllan Amgoed Carms 92 G3
Henllys Torf 78 G3
Henllys Vale Torf 78 G3
Henlow C Beds 104 D3
Hennock Devon 14 E2
Henny Street Essex 107 D7
Henryd Conwy 180 G3
Henry's Moat Pembs 91 F10
Hensall N Yorks 198 C5
Henshaw Northumb 241 E7
Henshaw W Yorks 205 E10
Hensingham Cumb 219 B9
Hensington Oxon 83 B7
Henstead Suff 143 F9
Hensting Hants 33 C7
Henstridge Devon 40 E5
Henstridge Som 30 D2
Henstridge Ash Som 30 C2
Henstridge Bowden Som 29 C11
Henstridge Marsh Som 30 C2
Henton Oxon 84 E3
Henton Som 44 D3
Henwood Corn 11 G11
Henwood Oxon 83 E7
Henwood Green Kent 52 E6
Heogan Shetland 313 J6
Heol-ddu Carms 75 E7
Heol-ddu Swansea 56 B6
Heol-laethog Bridgend 58 C2
Heol-las Bridgend 58 C2
Heol-las Swansea 57 B7
Heol Senni Powys 95 G8
Heol-y-gaer Powys 96 D3
Heol-y-mynydd V Glam 57 G11
Heolgerrig M Tydf 77 D8
Hepburn Northumb 264 E3
Hepple Northumb 251 C11
Hepscott Northumb 252 G6
Hepthorne Lane Derbys 170 C6
Heptonstall W Yorks 196 B3
Hepworth Suff 125 C9
Hepworth W Yorks 197 F7
Herbrandston Pembs 72 D5
Hereford Hereford 97 C10
Heribusta Highld 298 B4
Heriot Borders 271 E7
Hermiston Edin 280 G3
Hermit Hill S Yorks 197 G10
Hermit Hole W Yorks 205 F7
Hermitage Borders 250 D2
Hermitage Dorset 29 F10
Hermitage W Berks 64 E4
Hermitage W Sus 22 B3
Hermitage Green Mers 183 C10
Hermon Anglesey 162 B5
Hermon Carms 93 E7
Hermon Carms 94 F3
Hermon Pembs 92 E4
Herne Kent 71 F7
Herne Bay Kent 71 F7
Herne Common Kent 71 F7
Herne Hill London 67 E10
Herne Pound Kent 53 C7
Herner Devon 25 B9
Hernhill Kent 70 G5
Herniss Corn 2 C6
Herodsfoot Corn 6 C4
Heron Cross Stoke 168 G5
Heronden Kent 55 C9
Herongate Essex 87 G10
Heron's Ghyll E Sus 37 B7
Herons Green Bath 44 B5
Heronsford S Ayrs 244 G4
Heronsgate Herts 85 G8
Heronston Bridgend 58 D2
Herra Shetland 312 D8
Herriard Hants 49 D7
Herringfleet Suff 143 D9
Herring's Green Beds 103 C11
Herringswell Suff 124 C4
Herringthorpe S Yorks 186 C6
Hersden Kent 71 G8
Hersham Corn 24 F3
Hersham Sur 66 G6
Herstmonceux E Sus 23 C10
Herston Dorset 18 F6
Herston Orkney 314 G4
Hertford Herts 86 C4
Hertford Heath Herts 86 C4
Hertingfordbury Herts 86 C4
Hesket Newmarket Cumb 230 D2
Hesketh Bank Lancs 194 C2
Hesketh Lane Lancs 203 E8
Hesketh Moss Lancs 194 C2
Heskin Green Lancs 194 D4
Hesleden Durham 234 D4
Hesleyside Northumb 251 G8
Heslington York 207 C8
Hessay York 206 C6
Hessenford Corn 6 D6
Hessett Suff 125 E8
Hessle E Yorks 200 B4
Hessle W Yorks 198 D2
Hest Bank Lancs 211 F9
Hester's Way Glos 99 G8
Hestinsetter Shetland 313 J4
Heston London 66 D6
Hestwall Orkney 314 E2
Heswall Mers 182 E3
Hethe Oxon 101 F11
Hethel Norf 142 C3
Hethelpit Cross Glos 98 F5
Hethersett Norf 142 C3
Hethersgill Cumb 239 D11
Hetherside Cumb 239 D10
Hetherson Green W Ches 167 F8
Hethpool Northumb 263 D9
Hett Durham 233 D11
Hetton N Yorks 204 B5
Hetton Downs T & W 234 B3
Hetton-le-Hill T & W 234 B3
Hetton-le-Hole T & W 234 B3
Hetton Steads Northumb 264 B2
Heugh Northumb 242 C3
Heugh-head Aberds 292 B5
Heveningham Suff 126 C6
Hever Kent 52 E3
Heversham Cumb 211 C9
Hevingham Norf 160 E3
Hew Green N Yorks 205 B10
Hewas Water Corn 5 F9
Hewelsfield Glos 79 E9
Hewelsfield Common Glos 79 E8
Hewer Hill Cumb 230 D3
Hewish N Som 60 G2
Hewish Som 28 F6
Hewood Dorset 28 G5
Heworth T & W 243 E7
Heworth York 207 C8
Hexham Northumb 241 E10
Hextable Kent 68 E4
Hexthorpe S Yorks 198 G5
Hexton Herts 104 E2
Hexworthy Devon 13 G9
Hey Lancs 204 E3
Hey Green W Yorks 196 E4
Hey Houses Lancs 193 B10
Heybridge Essex 88 D5
Heybridge Essex 87 F10
Heybridge Basin Essex 88 D5
Heybrook Bay Devon 7 F10
Heydon Cambs 105 C8
Heydon Norf 160 D2
Heydour Lincs 155 B10
Heyheads Gtr Man 196 G3
Heylipol Argyll 288 E1
Heylor Shetland 312 E4
Heyope Powys 114 C4
Heyrod Gtr Man 185 B7
Heysham Lancs 211 G8
Heyshaw N Yorks 214 G3
Heyshott W Sus 34 D5
Heyshott Green W Sus 34 D5
Heyside Gtr Man 196 F2
Heytesbury Wilts 46 E2
Heythrop Oxon 101 F7
Heywood Gtr Man 195 E11
Heywood Wilts 45 C11
Hibaldstow N Lincs 200 G3
Hibb's Green Suff 125 G7
Hickford Hill Essex 106 C5
Hickleton S Yorks 198 F3
Hickling Norf 161 E8
Hickling Notts 154 D3
Hickling Green Norf 161 E8
Hickling Heath Norf 161 E8
Hickling Pastures Notts 154 D3
Hickmans Green Kent 54 B5
Hicks Forstal Kent 71 G7
Hicks Gate Bath 60 F6
Hick's Mill Corn 4 G5
Hickstead W Sus 36 C3
Hidcote Bartrim Glos 100 C3
Hidcote Boyce Glos 100 C3
Hifnal Shrops 132 D4
Higginshaw Gtr Man 196 F2
High Ackworth W Yorks 198 D2
High Angerton Northumb 252 F3
High Bankhill Cumb 231 C7
High Banton N Lnrk 278 E4
High Barn Lincs 174 C5
High Barnes T & W 243 F9
High Barnet London 86 F2
High Beach Essex 86 F6
High Bentham N Yorks 212 F3
High Bickington Devon 25 C10
High Biggins Cumb 212 D2
High Birkwith N Yorks 212 D5
High Birstwith N Yorks 205 B10
High Blantyre S Lnrk 268 D3
High Bonnybridge Falk 278 F6
High Bradfield S Yorks 186 C3
High Bradley N Yorks 204 D6
High Bray Devon 41 G7
High Brooms Kent 52 E5
High Brotheridge Glos 80 C5
High Bullen Devon 25 C8
High Buston Northumb 252 B6
High Callerton Northumb 242 C5
High Cark Cumb 211 C7
High Casterton Cumb 212 D2
High Catton E Yorks 207 C10
High Church Northumb 252 F5
High Cogges Oxon 82 D5
High Common Norf 141 B9
High Coniscliffe Darl 224 B4
High Crompton Gtr Man 196 F2
High Cross Cambs 123 F8
High Cross Corn 2 D6
High Cross E Sus 37 B9
High Cross Hants 34 B2
High Cross Herts 85 F10
High Cross Herts 86 B5
High Cross Newport 59 B9
High Cross W Sus 36 D2
High Cross Warks 118 D3
High Crosshill S Lnrk 268 C2
High Cunsey Cumb 221 G7
High Dubmire T & W 234 B2
High Dyke Durham 232 F5
High Easter Essex 87 C10
High Eggborough N Yorks 198 C5
High Eldrig Dumfries 236 C4
High Ellington N Yorks 214 C3
High Ercall Telford 149 F11
High Etherley Durham 233 F9
High Ferry Lincs 174 F5
High Field Lancs 203 C10
High Flatts W Yorks 197 F8
High Forge Durham 242 G6
High Friarside Durham 242 F5
High Gallowhill E Dunb 278 G2
High Garrett Essex 106 F5
High Grange Durham 233 E9
High Grantley N Yorks 214 F4
High Green Cumb 221 E8
High Green Norf 141 B8
High Green Norf 142 B2
High Green Norf 159 G8
High Green S Yorks 186 B4
High Green Shrops 132 G4
High Green Suff 125 E7
High Green W Yorks 197 E7
High Green Worcs 99 B7
High Halden Kent 53 F11
High Halstow Medway 69 D9
High Ham Som 44 G2
High Handenhold Durham 242 G6
High Harrington Cumb 228 F6
High Harrogate N Yorks 206 B2
High Haswell Durham 234 C3
High Hatton Shrops 150 E2
High Hauxley Northumb 253 C7
High Hawsker N Yorks 227 D8
High Heath Shrops 150 D3
High Heath W Mid 133 C10
High Hesket Cumb 230 C5
High Hesleden Durham 234 D5
High Hill Cumb 229 G11
High Houses Essex 87 C11
High Hoyland S Yorks 197 E9
High Hunsley E Yorks 208 F4
High Hurstwood E Sus 37 B7
High Hutton N Yorks 216 F5
High Ireby Cumb 229 D10
High Kelling Norf 177 E10
High Kilburn N Yorks 215 D10
High Lands Durham 233 F8
High Lane Gtr Man 185 D7
High Lane Worcs 116 E3
High Lanes Corn 2 B3
High Laver Essex 87 D8
High Legh E Ches 184 E2
High Leven Stockton 225 C8
High Littleton Bath 44 B6
High Longthwaite Cumb 229 B11
High Lorton Cumb 229 F9
High Marishes N Yorks 216 D6
High Marnham Notts 188 G4
High Melton S Yorks 198 G4
High Mickley Northumb 242 E3
High Mindork Dumfries 236 D5
High Moor Derbys 187 E7
High Moor Lancs 194 E4
High Moorsley T & W 234 B2
High Nash Glos 79 C9
High Newton Cumb 211 C8
High Newton-by-the-Sea Northumb 264 D6
High Nibthwaite Cumb 210 B5
High Oaks Cumb 222 G2
High Offley Staffs 150 D5
High Ongar Essex 87 E9
High Onn Staffs 150 F6
High Onn Wharf Staffs 150 F6
High Park Cumb 221 G10
High Park Mers 193 D11
High Risby N Lincs 200 E2
High Roding Essex 87 B10
High Rougham Suff 125 E8
High Row Cumb 230 D3
High Row Cumb 230 G3
High Salvington W Sus 35 F10
High Scales Cumb 229 B9
High Sellafield Cumb 219 E10
High Shaw N Yorks 223 G7
High Shields T & W 243 D9
High Shincliffe Durham 233 C11
High Side Cumb 229 E10
High Southwick T & W 243 F9
High Spen T & W 242 F4
High Stakesby N Yorks 227 C7
High Stoop Durham 233 C8
High Street Corn 5 E9
High Street Kent 53 G8
High Street Pembs 73 B11
High Street Suff 107 B7
High Street Suff 127 C8
High Street Suff 127 F8
High Street Suff 143 G7
High Street Green Suff 125 F10
High Sunderland Borders 261 C11
High Throston Hrtlpl 234 E5
High Tirfergus Argyll 255 F7
High Town Luton 103 G11
High Town Shrops 132 E4
High Town Staffs 151 G9
High Toynton Lincs 174 B3
High Trewhitt Northumb 252 B2
High Urpeth Durham 242 G6
High Valleyfield Fife 279 D10
High Walton Cumb 219 C9
High Warden Northumb 241 D10
High Water Head Cumb 220 F6
High Westwood Durham 242 F4
High Whinnow Cumb 239 G8
High Woolaston Glos 79 F9
High Worsall N Yorks 225 D7
High Wray Cumb 221 F7
High Wych Herts 87 C7
High Wycombe Bucks 84 G5
Higham Derbys 170 D5
Higham Fife 286 F6
Higham Kent 69 E8
Higham Lancs 204 F2
Higham S Yorks 197 F10
Higham Suff 107 D10
Higham Suff 124 E4
Higham Common S Yorks 197 F10
Higham Dykes Northumb 242 B4
Higham Ferrers Northants 121 D9
Higham Gobion C Beds 104 E2
Higham Hill London 86 G5
Higham on the Hill Leics 135 D7
Higham Wood Kent 52 D5
Highampton Devon 25 G7
Highams Park London 86 G5
Highbridge Cumb 230 C3
Highbridge Hants 33 C7
Highbridge Highld 290 E3
Highbridge Som 43 D10
Highbridge W Mid 133 C10
Highbrook W Sus 51 G11
Highburton W Yorks 197 E7
Highbury London 67 B10
Highbury Ptsmth 33 G11
Highbury Som 45 D7
Highbury Vale Nottingham 171 G8
Highclere Hants 64 G2
Highcliffe Derbys 186 F2
Highcliffe Dorset 19 C10
Higher Alham Som 45 E7
Higher Ansty Dorset 30 G3
Higher Ashton Devon 14 E3
Higher Audley Blkburn 195 B7
Higher Bal Corn 4 E4
Higher Ballam Lancs 202 G3
Higher Bartle Lancs 202 G6
Higher Bebington Mers 182 D4
Higher Berry End C Beds 103 E9
Higher Blackley Gtr Man 195 G10
Higher Boarshaw Gtr Man 195 F11
Higher Bockhampton Dorset 17 C10
Higher Bojewyan Corn 1 C3
Higher Boscaswell Corn 1 C3
Higher Brixham Torbay 9 D8
Higher Broughton Gtr Man 195 G10
Higher Burrow Som 28 C6
Higher Burwardsley W Ches 167 D8
Higher Chalmington Dorset 29 G9
Higher Cheriton Devon 27 G10
Higher Chillington Som 28 E5
Higher Chiswick Derbys 185 C7
Higher Clovelly Devon 24 C4
Higher Condurrow Corn 2 B5
Higher Crackington Corn 11 B9
Higher Cransworth Corn 5 B9
Higher Croft Blkburn 195 B7
Higher Denham Bucks 66 B4
Higher Dinting Derbys 185 C8
Higher Disley E Ches 185 E7
Higher Downs Corn 2 C3
Higher Durston Som 28 B3
Higher End Gtr Man 194 G4
Higher Folds Gtr Man 195 G7
Higher Gabwell Torbay 9 B8
Higher Green Gtr Man 195 G8
Higher Halstock Leigh Dorset 29 F8
Higher Heysham Lancs 211 G8
Higher Hogshead Lancs 195 C11
Higher Holton Som 29 B11
Higher Hurdsfield E Ches 184 G6
Higher Kingcombe Dorset 16 B6
Higher Kinnerton Flint 166 C4
Higher Land Corn 12 G3
Higher Marsh Som 30 C2
Higher Melcombe Dorset 30 G2
Higher Menadew Corn 5 D10
Higher Molland Devon 41 G8
Higher Muddiford Devon 40 F5
Higher Nyland Dorset 30 C2
Higher Penwortham Lancs 194 B4
Higher Pertwood Wilts 45 F11
Higher Porthpean Corn 5 E10
Higher Poynton E Ches 184 E6
Higher Prestacott Devon 12 B3
Higher Rads End C Beds 103 E9
Higher Ridge Shrops 149 C7
Higher Rocombe Barton Devon 9 B8
Higher Row Dorset 31 G8
Higher Runcorn Halton 183 E8
Higher Sandford Dorset 29 C10
Higher Shotton Flint 166 B4
Higher Shurlach W Ches 183 G11
Higher Slade Devon 40 D4
Higher Street Som 42 E6
Higher Tale Devon 27 G9
Higher Tolcarne Corn 5 B7
Higher Totnell Dorset 29 F10
Higher Town Corn 5 C10
Higher Town Scilly 1 F4
Higher Town Som 42 D3
Higher Tremarcoombe Corn 6 B5
Higher Vexford Som 42 F6
Higher Walreddon Devon 12 G5
Higher Walton Lancs 194 B5
Higher Walton Warr 183 D9
Higher Wambrook Som 28 F3
Higher Warcombe Devon 40 D3
Higher Weaver Devon 27 G9
Higher Whatcombe Dorset 30 G4
Higher Wheelton Lancs 194 C6
Higher Whitley W Ches 183 E10
Higher Wincham W Ches 183 F11
Higher Woodsford Dorset 17 D11
Higher Wraxall Dorset 29 G9
Higher Wych W Ches 167 G7
Highercliff Corn 6 D4
Higherford Lancs 204 E3
Highertown Corn 4 G6
Highertown Corn 11 E8
Highfield E Yorks 207 F10
Highfield Glos 79 E10
Highfield Gtr Man 194 G5
Highfield Gtr Man 195 F8
Highfield Herts 85 D9
Highfield N Ayrs 266 E6
Highfield Oxon 101 G11
Highfield S Yorks 186 D5
Highfield Soton 32 E6
Highfield T & W 242 F4
Highfields Cambs 123 F7
Highfields Derbys 170 B6
Highfields Essex 88 B5
Highfields Glos 80 F3
Highfields Leicester 136 C2
Highfields Northumb 273 E9
Highfields S Yorks 198 F4
Highfields Staffs 151 E8
Highgate E Sus 52 G2
Highgate Kent 53 G9
Highgate London 67 B9
Highgate Powys 130 D2
Highgate S Yorks 198 G3
Highgate W Mid 133 F11
Highlane Derbys 186 E6
Highlane E Ches 168 B5
Highlanes Corn 10 G4
Highlanes Staffs 150 C5
Highlaws Cumb 229 B8
Highleadon Glos 98 G5
Highleigh W Sus 22 D4
Highley Shrops 132 G4
Highmoor Cumb 229 B11
Highmoor Oxon 65 B8
Highmoor Cross Oxon 65 C8
Highmoor Hill Mon 60 B3
Highnam Glos 80 B3
Highnam Green Glos 98 G5
Highoak Norf 141 C11
Highridge Bristol 60 F5
Highroad Well Moor W Yorks 196 B5
Highstead Kent 71 F8
Highsted Kent 70 G2
Highstreet Kent 70 G5
Highstreet Green Essex 106 E5
Highstreet Green Sur 50 F3
Hightae Dumfries 238 B3
Highter's Heath W Mid 117 B11
Hightown E Ches 168 C5
Hightown Hants 31 G11
Hightown Mers 193 G10
Hightown Soton 33 E7
Hightown W Yorks 197 C7
Hightown Wrex 166 F4
Hightown Green Suff 125 F9
Hightown Heights W Yorks 197 C7
Highway Corn 4 G4
Highway Hereford 97 B9
Highway Som 29 C7
Highway Wilts 62 E4
Highway Windsor 65 C11
Highweek Devon 14 G2
Highwood Devon 27 F10
Highwood Dorset 18 D3
Highwood Essex 87 E10
Highwood Hants 31 F11
Highwood Worcs 116 D3
Highwood Hill London 86 G2
Highworth Swindon 82 G2
Highworthy Devon 24 F6
Hilborough Norf 140 C6
Hilborough Ho Norf 140 C6
Hilcot Glos 81 B7
Hilcot End Glos 81 E9
Hilcote Derbys 171 D7
Hilcott Wilts 46 B6
Hilden Park Kent 52 D5
Hildenborough Kent 52 D5
Hildersham Cambs 105 B10
Hildersley Hereford 98 G2
Hilderstone Staffs 151 C8
Hilderthorpe E Yorks 218 F3
Hilfield Dorset 29 F10
Hilgay Norf 140 D2
Hill S Glos 79 G10
Hill W Mid 134 D2
Hill Warks 119 D9
Hill Bottom Oxon 64 D6
Hill Brow W Sus 34 B3
Hill Chorlton Staffs 150 B5
Hill Common Norf 161 E8
Hill Corner Som 45 D10
Hill Croome Worcs 99 C7
Hill Dale Lancs 194 E3
Hill Deverill Wilts 45 E11
Hill Dyke Lincs 174 F4
Hill End Durham 232 D6
Hill End Fife 279 B10
Hill End Glos 99 D8
Hill End London 85 G8
Hill End N Yorks 205 C7
Hill End Som 29 E8
Hill End Worcs 117 E8
Hill Furze Worcs 99 B9
Hill Gate Hereford 97 F9
Hill Green Essex 105 E9
Hill Green Kent 69 G10
Hill Head Hants 33 G8
Hill Head Northumb 241 D10
Hill Hoath Kent 52 E3
Hill Hook W Mid 134 C2
Hill Houses Shrops 116 B2
Hill Mountain Pembs 73 D7
Hill of Beath Fife 280 C3
Hill of Drip Stirl 278 B5
Hill of Fearn Highld 301 B8
Hill of Keillor Angus 286 C6
Hill of Mountblairy Aberds 302 D6
Hill of Overbrae Aberds 303 C8
Hill Park Hants 33 F9
Hill Park Kent 52 B2
Hill Ridware Staffs 151 F11
Hill Side Hants 34 B3
Hill Side S Yorks 197 G8
Hill Side W Yorks 197 D7
Hill Side Worcs 116 E5
Hill Somersal Derbys 152 C2
Hill Street Kent 54 D6
Hill Top Derbys 186 F5
Hill Top Durham 232 G5
Hill Top Durham 233 C10
Hill Top Durham 242 G5
Hill Top Gtr Man 195 G8
Hill Top Hants 32 G6
Hill Top N Yorks 214 G3
Hill Top N Yorks 214 G5
Hill Top Notts 171 F7
Hill Top S Yorks 186 C5
Hill Top S Yorks 186 D3
Hill Top S Yorks 187 B7
Hill Top S Yorks 197 F9
Hill Top Staffs 133 E7
Hill Top W Mid 133 E9
Hill Top W Sus 34 C5
Hill Top W Yorks 196 E5
Hill Top W Yorks 197 D10
Hill Top W Yorks 205 G11
Hill Top W Yorks 205 G7
Hill Top W Yorks 134 E5
Hill Top Warks 135 E7
Hill View Dorset 18 B5
Hill Wood W Mid 134 C2
Hill Wootton Warks 118 D6
Hillam N Yorks 198 B4
Hillbeck Cumb 222 B5
Hillblock Pembs 73 B8
Hillborough Kent 71 F8
Hillbourne Poole 18 C6
Hillbrae Aberds 302 E6
Hillbrae Aberds 303 G7
Hillbutts Dorset 31 G7
Hillcliffe Warr 183 D10
Hillclifflane Derbys 170 F3
Hillcommon Som 27 B11
Hillcross Derbys 152 C6
Hilldyke Lincs 174 F4
Hillend Fife 280 E2
Hillend N Lnrk 268 B6
Hillend N Som 43 B11
Hillend Shrops 132 E6
Hillend Swansea 56 C2
Hillend Green Glos 98 F4
Hillersland Glos 79 C9
Hillerton Devon 13 B10
Hillesden Bucks 102 F3
Hillesden Hamlet Bucks 102 E3
Hillesley Glos 61 B9
Hillfarance Som 27 C11
Hillfarrance Som 27 C11
Hillfield Devon 8 E6
Hillfield W Mid 118 B2
Hillfields S Glos 60 D6
Hillfields W Mid 118 B6
Hillfoot Aberds 303 D9
Hillfoot W Yorks 205 G10
Hillfoot End C Beds 104 E2
Hillgreen W Berks 64 D3
Hillgrove W Sus 34 B6
Hillhampton Hereford 97 B11
Hillhead Aberds 302 F5
Hillhead Aberds 303 G8
Hillhead Corn 5 C11
Hillhead Devon 9 E8
Hillhead E Ayrs 257 B10
Hillhead S Ayrs 257 F10
Hillhead of Auchentumb Aberds 303 D9
Hillhead of Blairy Aberds 302 D6
Hillhead of Cocklaw Aberds 303 E10
Hillhouse Borders 271 D10
Hilliard's Cross Staffs 152 G3
Hilliclay Highld 310 C5
Hillingdon London 66 C5
Hillingdon Heath London 66 C5
Hillington Glasgow 267 C10
Hillington Norf 158 D4
Hillis Corner I o W 20 C5
Hillmoor Devon 27 E10
Hillmorton Warks 119 C10
Hillock Vale Lancs 195 B9
Hillockhead Aberds 292 B6
Hillockhead Aberds 292 C5
Hillpool Worcs 117 B7
Hillpound Hants 33 D9
Hills Town Derbys 171 B7
Hillsborough S Yorks 186 C4
Hillside Aberds 293 D11
Hillside Angus 293 G9
Hillside Devon 8 C4
Hillside Devon 27 F11
Hillside Hants 49 C9
Hillside Mers 193 E10
Hillside Orkney 314 G4
Hillside Orkney 314 D3
Hillside Shetland 313 G6
Hillside Shrops 131 F11
Hillside Wilts 81 G9
Hillstreet Hants 32 D4
Hillswick Shetland 312 F4
Hilltop Bl Gwent 77 D11
Hilltop Bucks 85 E7
Hilltop Derbys 170 C4
Hillview T & W 243 G9
Hillway I o W 21 D8
Hillwell Shetland 313 M5
Hillyfields Hants 32 D5
Hilmarton Wilts 62 D4
Hilperton Wilts 45 B11
Hilperton Marsh Wilts 45 B11
Hilsea Ptsmth 33 G11
Hilston E Yorks 209 G11
Hiltingbury Hants 32 C6
Hilton Aberds 303 F9
Hilton Borders 273 E7
Hilton Cambs 122 D5
Hilton Cumb 231 G10
Hilton Derbys 152 C4
Hilton Dorset 30 G3
Hilton Durham 233 G9
Hilton Highld 309 L7
Hilton Highld 311 L3
Hilton Shrops 132 D5
Hilton Staffs 133 B11
Hilton Stockton 225 C9
Hilton House Gtr Man 194 F6
Hilton Lodge Highld 300 G2
Hilton of Cadboll Highld 301 B8
Hilton Park Gtr Man 195 G10
Himbleton Worcs 117 F8
Himley Staffs 133 E7
Hincaster Cumb 211 C10
Hinchley Wood Sur 67 F7
Hinchliffe Mill W Yorks 196 F6
Hinchwick Glos 100 E2
Hinckley Leics 135 E8
Hinderclay Suff 125 B10
Hinderton W Ches 182 F4
Hinderwell N Yorks 226 B5
Hindford Shrops 148 C6
Hindhead Sur 49 F11
Hindle Fold Lancs 203 G10
Hindley Gtr Man 194 G6
Hindley Northumb 242 F2
Hindley Green Gtr Man 194 G6
Hindlip Worcs 117 F7
Hindolveston Norf 159 D10
Hindon Wilts 46 G2
Hindpool Cumb 210 F3
Hindringham Norf 159 B9
Hindsford Gtr Man 195 G7
Hingham Norf 141 C10
Hinstock Shrops 150 D3
Hintlesham Suff 107 C11
Hinton Glos 79 E11
Hinton Hants 19 B10
Hinton Hereford 96 D6
Hinton Northants 119 G10
Hinton S Glos 61 D8
Hinton Shrops 131 B8
Hinton Som 29 C9
Hinton Ampner Hants 33 B9
Hinton Blewett Bath 44 B5
Hinton Charterhouse Bath 45 B9
Hinton Cross Worcs 99 C10
Hinton-in-the-Hedges Northants 101 D11
Hinton Martell Dorset 31 F8
Hinton on the Green Worcs 99 C10
Hinton Parva Dorset 31 G7
Hinton Parva Swindon 63 C8
Hinton St George Som 28 E6
Hinton St Mary Dorset 30 D3
Hinton Waldrist Oxon 82 F5
Hints Shrops 116 C2
Hints Staffs 134 C3
Hinwick Beds 121 E8
Hinwood Shrops 131 B7
Hinxhill Kent 54 E5
Hinxton Cambs 105 B9
Hinxworth Herts 104 C4
Hipperholme W Yorks 196 B6
Hipplecote Worcs 116 F4
Hipsburn Northumb 264 G6
Hipswell N Yorks 224 F3
Hirael Gwyn 179 G9
Hiraeth Carms 92 G3
Hirn Aberds 293 C9
Hirnant Powys 147 E11
Hirst N Lnrk 269 C7
Hirst Northumb 253 F7
Hirst Courtney N Yorks 198 C6
Hirwaen Denb 165 C10
Hirwaun Rhondda 77 D7
Hirwaun Common Bridgend 58 C2
Hiscott Devon 25 B8
Hislop Borders 249 C9
Hisomley Wilts 45 D11
Histon Cambs 123 E8
Hitcham Suff 125 G9
Hitchill Dumfries 238 D4
Hitchin Herts 104 F3
Hitchin Hill Herts 104 F3
Hitcombe Bottom Wilts 45 E10
Hither Green London 67 E11
Hittisleigh Devon 13 C10
Hittisleigh Barton Devon 13 B10
Hive E Yorks 208 G2
Hixon Staffs 151 D10
Hoaden Kent 55 B9
Hoar Cross Staffs 152 E2
Hoarwithy Hereford 97 F10
Hoath Kent 71 G8
Hoath Corner Kent 52 E3
Hob Hill W Ches 167 E7
Hobarris Shrops 114 B6
Hobbister Orkney 314 F3
Hobble End Staffs 133 B10
Hobbles Green Suff 124 G4
Hobbs Cross Essex 87 C7
Hobbs Cross Essex 87 F7
Hobbs Wall Bath 61 G7
Hobkirk Borders 262 G3
Hobroyd Derbys 185 C8
Hobson Durham 242 F5
Hoby Leics 154 F3
Hoccombe Som 27 B10
Hockenden London 68 F3
Hockerill Herts 105 G9
Hockering Norf 159 G11
Hockering Heath Norf 159 G11
Hockerton Notts 172 D2
Hockholler Som 27 C11
Hockholler Green Som 27 C11
Hockley E Ches 184 E6
Hockley Essex 88 G4
Hockley Kent 54 B3
Hockley Staffs 134 C4
Hockley W Mid 118 B5
Hockley Heath W Mid 118 C3
Hockliffe C Beds 103 F9
Hockwold cum Wilton Norf 140 F4
Hockworthy Devon 27 D8
Hocombe Hants 32 C6
Hoddesdon Herts 86 D5
Hoddlesden Blkburn 195 C8
Hoddom Mains Dumfries 238 C5
Hoddomcross Dumfries 238 C5
Hoden Worcs 99 B11
Hodgefield Staffs 168 E6
Hodgehill E Ches 168 B4
Hodgehill W Mid 134 F2
Hodgeston Pembs 73 F8
Hodley Powys 130 E3
Hodnet Shrops 150 D2
Hodnetheath Shrops 150 D2
Hodsock Notts 187 D10
Hodsoll Street Kent 68 G6
Hodson Swindon 63 C7
Hodthorpe Derbys 187 F8
Hoe Hants 33 D9
Hoe Norf 159 F9
Hoe Sur 50 D5
Hoe Benham W Berks 64 F2
Hoe Gate Hants 33 E10
Hoff Cumb 222 B3
Hoffleet Stow Lincs 156 B4
Hog Hatch Sur 49 D10
Hogaland Shetland 312 F5
Hogben's Hill Kent 54 B4
Hogganfield Glasgow 268 B2
Hoggard's Green Suff 125 F7
Hoggeston Bucks 102 G6
Hoggington Wilts 45 B10
Hoggrill's End Warks 134 E4
Hogha Gearraidh W Isles 296 D3
Hoghton Lancs 194 B6
Hoghton Bottoms Lancs 194 B6
Hogley Green W Yorks 196 F6
Hognaston Derbys 170 E2
Hogpits Bottom Herts 85 E8
Hogsthorpe Lincs 191 G8
Hogstock Dorset 31 F7
Holbeach Lincs 157 E7
Holbeach Bank Lincs 157 D7
Holbeach Clough Lincs 156 D6
Holbeach Drove Lincs 156 G6
Holbeach Hurn Lincs 157 D7
Holbeach St Johns Lincs 156 F6
Holbeach St Marks Lincs 157 C7
Holbeach St Matthew Lincs 157 C8
Holbeache Worcs 116 B5
Holbeck Notts 187 G8
Holbeck W Yorks 205 G11
Holbeck Woodhouse Notts 187 G8
Holberrow Green Worcs 117 F10
Holbeton Devon 8 E2
Holborn London 67 C10
Holborough Kent 69 G8
Holbrook Derbys 170 G5
Holbrook S Yorks 186 E6
Holbrook Suff 108 D3
Holbrook Common S Glos 61 E7
Holbrook Moor Derbys 170 F5
Holbrooks W Mid 134 G6
Holburn Northumb 264 B2
Holbury Hants 32 G6
Holcombe Devon 14 G5
Holcombe Gtr Man 195 D9
Holcombe Som 45 D7
Holcombe Brook Gtr Man 195 E9
Holcombe Rogus Devon 27 D9
Holcot Northants 120 D5
Holdbrook London 86 F5
Holden Lancs 203 D11
Holden Fold Gtr Man 196 F2
Holdenby Northants 120 D3
Holdenhurst Bmouth 19 B8
Holder's Green Essex 106 F2
Holders Hill London 86 G2
Holdfast Worcs 99 D7
Holdgate Shrops 131 F11
Holdingham Lincs 173 F9

Holditch Dorset 28 G4
Holdsworth W Yorks 196 B5
Holdworth S Yorks 186 C3
Hole Devon 24 D4
Hole W Yorks 204 F6
Hole Bottom W Yorks 196 C2
Hole-in-the-Wall Hereford 98 F2
Hole Street W Sus 35 E10
Holefield Borders 263 C8
Holehills N Lnrk 268 B5
Holehouse Derbys 185 C8
Holehouses E Ches 184 F2
Holemill Aberdeen 293 C10
Holemoor Devon 24 F6
Hole's Hole Devon 7 B8
Holestane Dumfries 247 D9
Holestone Derbys 170 C4
Holewater Devon 41 F8
Holford Som 43 E7
Holgate York 207 C7
Holker Cumb 211 D7
Holkham Norf 176 E5
Hollacombe Devon 24 G5
Hollacombe Devon 26 G4
Hollacombe Hill Devon 7 E10
Holland Orkney 314 D6
Holland Orkney 314 A4
Holland Sur 52 C2
Holland Fen Lincs 174 F2
Holland Lees Lancs 194 F4
Holland-on-Sea Essex 89 B12
Hollands Som 29 D9
Hollandstoun Orkney 314 A7
Hollee Dumfries 239 D7
Hollesley Suff 109 C7
Hollicombe Torbay 9 C7
Hollies Common Staffs 150 E6
Hollin Hall Lancs 204 F4
Hollin Park W Yorks 206 F2
Hollinfare Warr 183 C11
Hollingbourne Kent 53 B10
Hollingbury Brighton 36 F4
Hollingdean Brighton 36 F4
Hollingdon Bucks 103 F7
Hollingrove E Sus 37 C11
Hollingthorpe W Yorks 197 D10
Hollington Derbys 152 B4
Hollington E Sus 38 E3
Hollington Hants 48 B2
Hollington Staffs 151 B11
Hollington Cross Hants 48 B2
Hollington Grove Derbys 152 B4
Hollingwood Derbys 186 G6
Hollingworth Gtr Man 185 B8
Hollins Cumb 222 G3
Hollins Derbys 186 G4
Hollins Gtr Man 195 F10
Hollins Gtr Man 195 F11
Hollins Gtr Man 195 F8
Hollins Staffs 168 D6
Hollins Staffs 168 E4
Hollins Staffs 169 F7
Hollins End S Yorks 186 E5
Hollins Green Warr 183 C11
Hollins Lane Lancs 202 C5
Hollins Lane Shrops 149 B10
Hollinsclough Staffs 169 B9
Hollinsgreen E Ches 168 C2
Hollinswood Telford 132 B4
Hollinthorpe W Yorks 206 G3
Hollinwood Gtr Man 196 G2
Hollinwood Shrops 149 B10
Hollis Green Devon 27 F9
Hollis Head Devon 27 G7
Hollocombe Devon 25 E10
Hollocombe Town Devon 25 E10
Hollow Brook Bath 60 G5
Hollow Meadows S Yorks 186 D2
Hollow Oak Dorset 18 C2
Hollow Street Kent 71 G8
Holloway Derbys 170 D4
Holloway Wilts 45 G11
Holloway Windsor 65 C10
Holloway Hill Sur 50 E3
Hollowell Northants 120 C3
Hollowmoor Heath W Ches 167 B7
Hollows Dumfries 239 B9
Holly Bank W Mid 133 C11
Holly Brook Som 44 D4
Holly Bush Wrex 166 G6
Holly Cross Windsor 65 C10
Holly End Norf 139 B9
Holly Green Bucks 84 E3
Holly Green Worcs 99 C7
Holly Hill N Yorks 224 E3
Hollyberry End W Mid 134 G5
Hollybush Caerph 77 E11
Hollybush E Ayrs 257 G9
Hollybush Stoke 168 G5
Hollybush Torf 78 G3
Hollybush Worcs 98 D5
Hollybush Corner Bucks 66 B3
Hollybush Corner Suff 125 F8
Hollybush Hill Bucks 66 C3
Hollybush Hill Essex 89 B10
Hollybushes Kent 54 B2
Hollycroft Leics 135 E8
Hollyhurst Shrops 131 D9
Hollyhurst Warks 135 F7
Hollym E Yorks 201 B10
Hollywaste Shrops 116 B2
Hollywater Hants 49 G10
Hollywood Worcs 117 B11
Holmacott Devon 25 B8
Holman Clavel Som 28 D2
Holmbridge W Yorks 196 F6
Holmbury St Mary Sur 50 E6
Holmbush Corn 5 E10
Holmbush Dorset 28 G5
Holmcroft Staffs 151 D8
Holme C Beds 104 C3
Holme Cambs 138 F3
Holme Cumb 211 D10
Holme N Lincs 200 F2
Holme N Yorks 215 C7
Holme Notts 172 D4
Holme W Yorks 196 F6
Holme W Yorks 205 G9
Holme Chapel Lancs 195 B11
Holme Green C Beds 104 C3
Holme Green N Yorks 207 E7
Holme Green Wokingham 65 F10
Holme Hale Norf 141 B7
Holme Hill NE Lincs 201 F9
Holme Lacy Hereford 97 D11
Holme Lane Notts 154 B2
Holme Marsh Hereford 114 G6
Holme Mills Cumb 211 D10
Holme next the Sea Norf 176 E2
Holme-on-Spalding-Moor E Yorks 208 F2
Holme on the Wolds E Yorks 208 D5
Holme Pierrepont Notts 154 B2
Holme St Cuthbert Cumb 229 B8
Holme Slack Lancs 203 G7
Holme Wood W Yorks 205 G9
Holmebridge Dorset 18 D3
Holmer Hereford 97 C10
Holmer Green Bucks 84 F6
Holmes Lancs 194 D2
Holmes Chapel E Ches 168 B3
Holmesdale Derbys 186 F5
Holmesfield Derbys 186 F4
Holmes's Hill E Sus 23 C8
Holmeswood Lancs 194 D2
Holmethorpe Sur 51 C9
Holmewood Derbys 170 B6
Holmfield W Yorks 196 B5
Holmfirth W Yorks 196 F6
Holmhead Angus 293 F7
Holmhead Dumfries 246 F6
Holmhead E Ayrs 258 E3
Holmhill Dumfries 247 D9
Holmisdale Highld 297 G7
Holmley Common Derbys 186 F5
Holmpton E Yorks 201 C11
Holmrook Cumb 219 F11
Holmsgarth Shetland 313 J6
Holmside Durham 233 B10
Holmsleigh Green Devon 28 G2
Holmston S Ayrs 257 E9
Holmwood Corner Sur 51 E7
Holmwrangle Cumb 230 B6
Holne Devon 8 B4
Holnest Dorset 29 E11
Holnicote Som 42 D2
Holsworthy Devon 24 G4
Holsworthy Beacon Devon 24 F5
Holt Dorset 31 G8
Holt Hants 49 C8
Holt Mers 183 C7
Holt Norf 159 B11
Holt Wilts 61 G11
Holt Worcs 116 E6
Holt Wrex 166 E6
Holt End Hants 49 F7
Holt End Worcs 117 D11
Holt Fleet Worcs 116 E6
Holt Green Lancs 193 G11
Holt Head W Yorks 196 E5
Holt Heath Dorset 31 G9
Holt Heath Worcs 116 E6
Holt Hill Kent 53 B8
Holt Hill Staffs 152 D2
Holt Park W Yorks 205 E11
Holt Pound Hants 49 E10
Holt Wood Dorset 31 F8
Holtby York 207 C9
Holton Oxon 83 D10
Holton Som 30 B2
Holton Suff 127 B7
Holton cum Beckering Lincs 189 E10
Holton Heath Dorset 18 C4
Holton le Clay Lincs 201 G9
Holton le Moor Lincs 189 B9
Holton St Mary Suff 107 D11
Holts Gtr Man 196 G3
Holtspur Bucks 84 G6
Holtye E Sus 52 F3
Holway Dorset 28 G5
Holway Dorset 29 C10
Holway Flint 181 F11
Holway Som 28 C2
Holwell Dorset 30 E2
Holwell Herts 104 E3
Holwell Leics 154 E4
Holwell Oxon 82 D2
Holwell Som 45 D8
Holwellbury C Beds 104 E3
Holwick Durham 232 F4
Holworth Dorset 17 E11
Holy City Devon 28 G3
Holy Cross T & W 243 D8
Holy Cross Worcs 117 B8
Holy Island Northumb 273 B11
Holy Vale Scilly 1 G4
Holybourne Hants 49 E8
Holyfield Essex 86 E5
Holyhead = Caergybi Anglesey 178 E2
Holylee Borders 261 B9
Holymoorside Derbys 170 B4
Holyport Windsor 65 D11
Holystone Northumb 251 C11
Holytown N Lnrk 268 C5
Holywell C Beds 85 B8
Holywell Cambs 122 C6
Holywell Corn 4 D5
Holywell Dorset 29 G9
Holywell E Sus 23 F9
Holywell Glos 80 G3
Holywell Hereford 97 C7
Holywell Herts 85 F9
Holywell Northumb 243 C8
Holywell Som 29 E8
Holywell Warks 118 D3
Holywell = Treffynnon Flint 181 F11
Holywell Green W Yorks 196 D5
Holywell Lake Som 27 C10
Holywell Row Suff 124 B4
Holywood Dumfries 247 G10
Hom Green Hereford 97 G11
Homedowns Glos 99 E8
Homer Shrops 132 C2
Homer Green Mers 193 G10
Homersfield Suff 142 F5
Homerton London 67 B11
Homington Wilts 31 B10
Honey Hall N Som 60 G2
Honey Hill Kent 70 G6
Honey Street Wilts 62 G6
Honey Tye Suff 107 D9
Honeyborne Worcs 100 C2
Honeychurch Devon 25 G10
Honeydon Beds 122 F2
Honeyhill Wokingham 65 F10
Honeystreet Wilts 62 G6
Honeywick C Beds 103 G9
Honicknowle Plym 7 D9
Honiley Warks 118 C4
Honing Norf 160 D6
Honingham Norf 160 G2
Honington Lincs 172 G6
Honington Suff 125 C8
Honington Warks 100 C5
Honiton Devon 27 G11
Honkley Wrex 166 D4
Honley W Yorks 196 E6
Honley Moor W Yorks 196 E6
Honnington Telford 150 F4
Honor Oak London 67 E11
Honor Oak Park London 67 E11
Honresfeld Gtr Man 196 D2
Hoo Kent 71 G9
Hoo End Herts 85 B11
Hoo Green E Ches 184 E2
Hoo Hole W Yorks 196 B4
Hoo Meavy Devon 7 B10
Hoo St Werburgh Medway 69 E9
Hoober S Yorks 186 B6
Hoobrook Worcs 116 C6
Hood Green S Yorks 197 G10
Hood Hill S Yorks 186 B5
Hood Manor Warr 183 D9
Hooe E Sus 23 D11
Hooe Plym 7 E10
Hooe Common E Sus 23 C11
Hoofield W Ches 167 C8
Hoohill Blkpool 202 F2
Hook Cambs 139 E8
Hook Devon 28 F4
Hook E Yorks 199 B9
Hook Hants 33 F8
Hook Hants 49 C8
Hook London 67 G7
Hook Pembs 73 C7
Hook Wilts 62 C5
Hook-a-gate Shrops 131 B9
Hook Bank Worcs 98 C6
Hook End Essex 87 F9
Hook End Oxon 65 C7
Hook End W Mid 134 G4
Hook Green Kent 53 F7
Hook Green Kent 68 F6
Hook Heath Sur 50 B3
Hook Norton Oxon 101 E7
Hook Park Hants 33 G7
Hook Street Glos 79 F11
Hook Street Wilts 62 C5
Hooke Dorset 16 B6
Hooker Gate T & W 242 F4
Hookgate Staffs 150 B4
Hook's Cross Herts 104 G5
Hooksway W Sus 34 D4
Hookway Devon 14 B3
Hookwood Sur 51 E9
Hoole W Ches 166 B6
Hoole Bank W Ches 166 B6
Hooley Sur 51 B9
Hooley Bridge Gtr Man 195 E11
Hooley Brow Gtr Man 195 E11
Hooley Hill Gtr Man 184 B6
Hoop Mon 79 D8
Hoopers Pool Wilts 45 C10
Hoops Devon 24 C5
Hooton W Ches 182 F5
Hooton Levitt S Yorks 187 C8
Hooton Pagnell S Yorks 198 F3
Hooton Roberts S Yorks 187 B7
Hop Pole Lincs 156 G3
Hopcroft's Holt Oxon 101 F9
Hope Derbys 185 E11
Hope Devon 9 G8
Hope Highld 308 D4
Hope Powys 130 B5
Hope Shrops 130 C6
Hope Staffs 169 D10
Hope = Yr Hôb Flint 166 D4
Hope Bagot Shrops 115 C11
Hope Bowdler Shrops 131 E9
Hope End Green Essex 105 G11
Hope Green E Ches 184 E6
Hope Mansell Hereford 79 B10
Hope Park Shrops 130 C6
Hope under Dinmore Hereford 115 G10
Hopebeck Cumb 229 G9
Hopedale Staffs 169 D10
Hopeman Moray 301 C11
Hope's Green Essex 69 B9
Hope's Rough Hereford 98 B2
Hopesay Shrops 131 G7
Hopesgate Shrops 130 C6
Hopetown W Yorks 197 C11
Hopgoods Green W Berks 64 F4
Hopkinstown Rhondda 77 G9
Hopley's Green Hereford 114 G6
Hopperton N Yorks 206 B4
Hopsford Warks 135 G8
Hopstone Shrops 132 E5
Hopton Derbys 170 E3
Hopton Shrops 149 D11
Hopton Shrops 149 E7
Hopton Staffs 151 D8
Hopton Suff 125 B9
Hopton Cangeford Shrops 131 G10
Hopton Castle Shrops 115 B7
Hopton Heath Staffs 151 D9
Hopton on Sea Norf 143 D10
Hopton Wafers Shrops 116 B2
Hoptonbank Shrops 116 B2
Hoptongate Shrops 131 G10
Hoptonheath Shrops 115 B7
Hopwas Staffs 134 B3
Hopwood Gtr Man 195 F11
Hopwood Worcs 117 B10
Hopworthy Devon 24 G4
Horam E Sus 23 B9
Horbling Lincs 156 B2
Horbury W Yorks 197 D9
Horbury Bridge W Yorks 197 D9
Horbury Junction W Yorks 197 D10
Horcott Glos 81 E11
Horden Durham 234 C4
Horderley Shrops 131 F8
Hordle Hants 19 B11
Hordley Shrops 149 C7
Horeb Carms 75 D7
Horeb Carms 93 F10
Horeb Ceredig 93 C7
Horeb Flint 166 D3
Horfield Bristol 60 D6
Horgabost W Isles 305 J2
Horham Suff 126 C4
Horkesley Heath Essex 107 F9
Horkstow N Lincs 200 D3
Horkstow Wolds N Lincs 200 D3
Horley Oxon 101 C8
Horley Sur 51 E9
Horn Ash Dorset 28 G5
Horn Hill Som 43 E8
Horn Street Kent 55 F7
Horn Street Kent 69 G7
Hornblotton Som 44 G5
Hornblotton Green Som 44 G5
Hornby Lancs 211 F11
Hornby N Yorks 224 G4
Hornby N Yorks 225 D7
Horncastle Lincs 174 B3
Horncastle Reading 65 E7
Hornchurch London 68 B4
Horncliffe Northumb 273 F8
Horndean Borders 273 F7
Horndean Hants 34 E2
Horndon Devon 12 F6
Horndon on the Hill Thurrock 69 C7
Horne Sur 51 E10
Horne Row Essex 88 E3
Horner Som 41 D11
Horner's Green Suff 107 C9
Hornestreet Essex 107 E10
Horney Common E Sus 37 B7
Hornick Corn 5 E9
Horniehaugh Angus 292 G6
Horning Norf 160 F6
Horninghold Leics 136 D6
Horninglow Staffs 152 D4
Horningsea Cambs 123 E9
Horningsham Wilts 45 E10
Horningtoft Norf 159 E8
Horningtops Corn 6 C5
Horns Corner Kent 38 B2
Horns Cross Devon 24 C5
Horns Cross E Sus 38 C4
Horns Green Kent 52 B3
Hornsbury Som 28 E4
Hornsby Cumb 240 G2
Hornsea E Yorks 209 D10
Hornsea Bridge E Yorks 209 D10
Hornsea Burton E Yorks 209 D10
Hornsey London 67 B10
Hornsey Vale London 67 B10
Hornton Oxon 101 B7
Horpit Swindon 63 C8
Horrabridge Devon 7 B10
Horringer Suff 124 E6
Horringford I o W 20 D6
Horrocks Fold Gtr Man 195 E8
Horrocksford Lancs 203 E10
Horsalls Kent 53 C11
Horse Bridge Staffs 169 E7
Horsebridge Devon 12 G4
Horsebridge Hants 47 G10
Horsebridge Shrops 131 B7
Horsebrook Devon 8 D4
Horsebrook Staffs 151 G7
Horsecastle N Som 60 F2
Horsedown Wilts 61 D10
Horsedowns Corn 2 C4
Horsehay Telford 132 B3
Horseheath Cambs 106 B2
Horseholm Dumfries 238 C2
Horsehouse N Yorks 213 C10
Horseley Heath W Mid 133 E9
Horsell Sur 50 B3
Horsell Birch Sur 50 B3
Horseman Side Essex 87 F8
Horseman's Green Wrex 166 G6
Horsemere Green W Sus 35 G7
Horsenden Bucks 84 E3
Horsepools Glos 80 C4
Horseway Cambs 139 F8
Horseway Head Hereford 114 E6
Horsey Norf 161 E9
Horsey Som 43 F10
Horsey Corner Norf 161 E9
Horsey Down Wilts 81 G9
Horsford Norf 160 F3
Horsforth W Yorks 205 F10
Horsforth Woodside W Yorks 205 F10
Horsham W Sus 51 G7
Horsham Worcs 116 F4
Horsham St Faith Norf 160 F4
Horshoe Green Kent 52 E3
Horsington Lincs 173 B11
Horsington Som 30 C2
Horsley Derbys 170 G5
Horsley Glos 80 F4
Horsley Northumb 242 D3
Horsley Northumb 251 D8
Horsley Cross Essex 108 F2
Horsley Hill T & W 243 D9
Horsley Woodhouse Derbys 170 G5
Horsleycross Street Essex 108 F2
Horsleyhill Borders 262 F2
Horsleyhope Durham 233 B7
Horsleys Green Bucks 84 F3
Horsmonden Kent 53 E7
Horspath Oxon 83 E9
Horstead Norf 160 F5
Horsted Green E Sus 23 B7
Horsted Keynes W Sus 36 B5
Horton Bucks 84 B6
Horton Dorset 31 F8
Horton Kent 54 B6
Horton Lancs 204 C3
Horton Northants 120 G6
Horton S Glos 61 C9
Horton Shrops 149 D9
Horton Som 28 E4
Horton Staffs 168 D6
Horton Swansea 56 D3
Horton Telford 150 G3
Horton Wilts 62 G5
Horton Windsor 66 D4
Horton Common Dorset 31 F9
Horton Cross Som 28 D4
Horton-cum-Studley Oxon 83 C9
Horton Green W Ches 167 F7
Horton Heath Dorset 31 F9
Horton Heath Hants 33 D7
Horton in Ribblesdale N Yorks 212 E6
Horton Kirby Kent 68 F5
Horton Wharf Bucks 84 B6
Hortonlane Shrops 149 G8
Hortonwood Telford 150 G3
Horwich Gtr Man 194 E6
Horwich End Derbys 185 E8
Horwood Devon 25 B8
Horwood Riding S Glos 61 B8
Hoscar Lancs 194 E3
Hose Leics 154 D4
Hoselaw Borders 263 C8
Hoses Cumb 220 G4
Hosey Hill Kent 52 C3
Hosh Perth 286 E2
Hosta W Isles 296 D3
Hoswick Shetland 313 L6
Hotham E Yorks 208 G3
Hothfield Kent 54 D3
Hotley Bottom Bucks 84 E5
Hoton Leics 153 E11
Hotwells Bristol 60 E5
Houbans Shetland 312 F5
Houbie Shetland 312 D8
Houdston S Ayrs 244 D5
Hough Argyll 288 E1
Hough E Ches 168 E2
Hough E Ches 184 F5
Hough Green Halton 183 D7
Hough-on-the-Hill Lincs 172 F6
Hough Side W Yorks 205 G10
Hougham Lincs 172 G5
Houghton Cambs 122 C5
Houghton Cumb 239 F10
Houghton Hants 47 G10
Houghton Northumb 242 D4
Houghton Pembs 73 D7
Houghton W Sus 35 E8
Houghton Bank Darl 233 G10
Houghton Conquest C Beds 103 C10
Houghton Green E Sus 38 C6
Houghton Green Warr 183 C10
Houghton-le-Side Darl 233 G10
Houghton-le-Spring T & W 234 B2
Houghton on the Hill Leics 136 C3
Houghton Regis C Beds 103 G10
Houghton St Giles Norf 159 B8
Houghwood Mers 194 G4
Houlland Shetland 312 B7
Houlland Shetland 312 F7
Houlland Shetland 313 H5
Houlland Shetland 313 J6
Houlsyke N Yorks 226 D4
Hound Hants 33 F7
Hound Green Hants 49 B8
Hound Hill Dorset 31 G7
Houndmills Hants 48 C6
Houndscroft Glos 80 E5
Houndslow Borders 272 F2
Houndsmoor Som 27 B10
Houndstone Som 29 D8
Houndwood Borders 272 C6
Hounsdown Hants 32 E5
Hounsley Batch N Som 60 G4
Hounslow London 66 D6
Hounslow Green Essex 87 B11
Hounslow West London 66 D6
Hourston Orkney 314 E2
Housabister Shetland 313 H6
Housay Shetland 312 F8
House of Daviot Highld 301 E7
House of Glenmuick Aberds 292 D5
Househill Highld 301 D8
Houses Hill W Yorks 197 D7
Housetter Shetland 312 E5
Housham Tye Essex 87 C8
Houss Shetland 313 K5
Houston Renfs 267 B8
Houstry Highld 310 F5
Houton Orkney 314 F3
Hove Brighton 36 G3
Hove Edge W Yorks 196 C6
Hoveringham Notts 171 F11
Hoveton Norf 160 F6
Hovingham N Yorks 216 D3
How Cumb 240 F2
How Caple Hereford 98 E2
How End C Beds 103 C10
How Green Kent 52 D3
How Hill Norf 161 F7
How Wood Herts 85 E10
Howbeck Bank E Ches 167 F11
Howbrook S Yorks 186 B4
Howden Borders 262 E5
Howden E Yorks 199 B8
Howden W Loth 269 B11
Howden Clough W Yorks 197 B8
Howden-le-Wear Durham 233 E9
Howdon T & W 243 D8
Howdon Pans T & W 243 D8
Howe Highld 310 C7
Howe N Yorks 214 C6
Howe Norf 142 C5
Howe Bridge Gtr Man 195 G7
Howe Green Essex 88 E2
Howe Green Essex 87 B8
Howe Green Warks 134 F6
Howe of Teuchar Aberds 303 E7
Howe Street Essex 87 C11
Howe Street Essex 106 E3
Howegreen Essex 88 E4
Howell Lincs 173 F10
Howey Powys 113 F11
Howford Borders 261 B8
Howford Borders 261 E9
Howgate Cumb 228 G5
Howgate Midloth 270 D4
Howgill Cumb 222 F2
Howgill Lancs 204 D2
Howgill N Yorks 205 B7
Howick Mon 79 F8
Howick Northumb 265 F7
Howick Cross Lancs 194 B4
Howle Durham 233 F7
Howle Telford 150 E3
Howleigh Som 28 D2
Howlett End Essex 105 E11
Howley Glos 80 G2
Howley Som 28 F3
Howley Warr 183 D10
Hownam Borders 263 F7
Hownam Mains Borders 263 E7
Howpasley Borders 249 B8
Howsen Worcs 116 G5
Howsham N Lincs 200 G4
Howsham N Yorks 216 G4
Howslack Dumfries 248 B3
Howt Green Kent 69 F11
Howtel Northumb 263 C9
Howton Hereford 97 F8
Howtown Cumb 221 B8
Howwood Renfs 267 C7
Hoxne Suff 126 B3
Hoxton London 67 C10
Hoy Orkney 314 F2
Hoylake Mers 182 D2
Hoyland S Yorks 197 G11
Hoyland Common S Yorks 197 G11
Hoylandswaine S Yorks 197 G9
Hoyle W Sus 34 D6
Hoyle Mill S Yorks 197 F11
Hubbard's Hill Kent 52 C4
Hubberholme N Yorks 213 D8
Hubberston Pembs 72 D5
Hubbersty Head Cumb 221 G8
Hubberton Green W Yorks 196 C4
Hubbert's Bridge Lincs 174 G3
Huby N Yorks 205 D11
Huby N Yorks 215 F11
Hucclecote Glos 80 B5
Hucking Kent 53 B10
Hucknall Notts 171 F8
Hud Hey Lancs 195 C9
Huddersfield W Yorks 196 D6
Huddington Worcs 117 F8
Huddisford Devon 24 D4
Huddlesford Staffs 134 B3
Hudnall Herts 85 C8
Hudnalls Glos 79 E8
Hudswell N Yorks 224 E3
Hudswell Wilts 61 F11
Huggate E Yorks 208 B3
Hugglepit Devon 24 C4
Hugglescote Leics 153 G8
Hugh Mill Lancs 195 C10
Hugh Town Scilly 1 G4
Hughenden Valley Bucks 84 F5
Hughley Shrops 131 D11
Hugus Corn 4 G5
Huish Devon 25 B7
Huish Devon 25 E8
Huish Wilts 62 G6
Huish Champflower Som 27 B9
Huish Episcopi Som 28 B6
Huisinis W Isles 305 G1
Hulcote C Beds 103 D8
Hulcote Northants 102 B4
Hulcott Bucks 84 B5
Hull End Derbys 185 E9
Hulland Derbys 170 F2
Hulland Moss Derbys 170 F3
Hulland Ward Derbys 170 F3
Hullavington Wilts 61 C11
Hullbridge Essex 88 G4
Hulme Gtr Man 184 B4
Hulme Staffs 168 F6
Hulme Warr 183 C10
Hulme End Staffs 169 D10
Hulme Walfield E Ches 168 B4
Hulseheath E Ches 184 E2
Hulver Street Suff 143 F9
Hulverstone I o W 20 E3
Humber Devon 14 G3
Humber Hereford 115 F10
Humber Bridge N Lincs 200 C4
Humberston NE Lincs 201 F10
Humberston Fitties NE Lincs 201 F10
Humberstone Leicester 136 B2
Humbie E Loth 271 C9
Humble Green Suff 107 B8
Humbleton E Yorks 209 G10
Humbleton Northumb 263 D11
Humby Lincs 155 C10
Hume Borders 272 G4
Hummersknott Darl 224 C5
Humshaugh Northumb 241 C10
Huna Highld 310 B7
Huncoat Lancs 203 G11
Huncote Leics 135 D10
Hundalee Borders 262 F4
Hundall Derbys 186 F5
Hunderthwaite Durham 232 G5
Hundle Houses Lincs 174 E3
Hundleby Lincs 174 B5
Hundleshope Borders 260 B6
Hundleton Pembs 73 E7
Hundon Suff 106 B4
Hundred Acres Hants 33 E9
Hundred End Lancs 194 C2
Hundred House Powys 114 G2
Hungarton Leics 136 B3
Hungate W Yorks 197 B11
Hunger Hill Gtr Man 195 F7
Hunger Hill Lancs 194 E4
Hungerford Hants 31 E11
Hungerford Shrops 131 F10
Hungerford Som 42 E4
Hungerford W Berks 63 F10
Hungerford Windsor 65 E10
Hungerford Green W Berks 64 D5
Hungerford Newtown W Berks 63 E11
Hungershall Park Kent 52 F5
Hungerstone Hereford 97 D8
Hungerton Lincs 155 D7
Hungladder Highld 298 B3
Hungreyhatton Shrops 150 D3
Hunmanby N Yorks 217 D11
Hunmanby Moor N Yorks 218 D2
Hunningham Warks 119 D7
Hunningham Hill Warks 119 D7
Hunnington Worcs 133 G9
Hunny Hill I o W 20 D5
Hunsdon Herts 86 C6
Hunsdonbury Herts 86 C6
Hunsingore N Yorks 206 C4
Hunslet W Yorks 206 G2
Hunslet Carr W Yorks 206 G2
Hunsonby Cumb 231 D7
Hunspow Highld 310 B6
Hunstanton Norf 175 G11
Hunstanworth Durham 232 B5
Hunsterson E Ches 167 F11
Hunston Suff 125 D9
Hunston W Sus 22 C5
Hunston Green Suff 125 D9
Hunstrete Bath 60 G6
Hunsworth W Yorks 197 B7
Hunt End Worcs 117 E10
Huntenhull Green Wilts 45 D10
Huntercombe End Oxon 65 B7
Hunters Forstal Kent 71 F7
Hunter's Quay Argyll 276 F3
Huntham Som 28 B4
Hunthill Lodge Angus 292 F6
Hunting-tower Perth 286 E4
Huntingdon Cambs 122 C4
Huntingfield Suff 126 C6
Huntingford Dorset 45 G10
Huntingford S Glos 80 G2
Huntington E Loth 281 F9
Huntington Hereford 97 C9
Huntington Hereford 114 G5
Huntington Staffs 151 G9
Huntington Telford 132 B3
Huntington W Ches 166 C6
Huntington York 207 B8
Huntley Glos 80 B2
Huntley Staffs 169 G8
Huntly Aberds 302 F5
Huntlywood Borders 272 G2
Hunton Hants 48 F3
Hunton Kent 53 D8
Hunton N Yorks 224 G3
Hunton Bridge Herts 85 E9
Hunt's Corner Norf 141 F11
Hunt's Cross Mers 182 D6
Hunt's Green Bucks 84 E5
Hunt's Green W Berks 64 E2
Hunts Green Warks 134 D3
Hunt's Hill Bucks 84 F4
Hunt's Lane Leics 135 C9
Huntscott Som 42 E2
Huntsham Devon 27 C8
Huntshaw Devon 25 C8
Huntshaw Water Devon 25 C8
Huntspill Som 43 D10
Huntstile Som 43 G9
Huntworth Som 43 G10
Hunwick Durham 233 E9
Hunworth Norf 159 B11
Hurcott Som 28 D5
Hurcott Som 29 B8
Hurcott Worcs 117 B7
Hurdcott Wilts 47 G7
Hurdley Powys 130 E5
Hurdsfield E Ches 184 G6
Hurgill N Yorks 224 E3
Hurlet Glasgow 267 C10
Hurley Warks 134 D4
Hurley Windsor 65 C10
Hurley Bottom Windsor 65 C10
Hurley Common Warks 134 D4
Hurlford E Ayrs 257 B11
Hurliness Orkney 314 H2
Hurlston Lancs 194 E2
Hurlston Green Lancs 193 E11
Hurn Dorset 19 B8
Hurn E Yorks 208 E6
Hurn's End Lincs 174 F6
Hursey Dorset 28 G6
Hursley Hants 32 B6
Hurst Cumb 230 C4
Hurst Dorset 17 C11
Hurst Gtr Man 196 G2
Hurst N Yorks 223 E10
Hurst Som 29 D7
Hurst Wokingham 65 E9
Hurst Green E Sus 38 B2
Hurst Green Essex 89 B9
Hurst Green Lancs 203 F9
Hurst Green Sur 51 C11
Hurst Green W Mid 133 F9
Hurst Hill W Mid 133 E8
Hurst Park Sur 66 F6
Hurst Wickham W Sus 36 D3
Hurstbourne Priors Hants 48 D2
Hurstbourne Tarrant Hants 47 C11
Hurstead Gtr Man 196 D2
Hurstley Hereford 97 B7
Hurstpierpoint W Sus 36 D3
Hurstwood Lancs 204 G3
Hurtmore Sur 50 D3
Hurworth-on-Tees Darl 224 C6
Hurworth Place Darl 224 D5
Hury Durham 223 B9
Husabost Highld 298 D2
Husbands Bosworth Leics 136 G2
Husbandtown Angus 287 D8
Husborne Crawley C Beds 103 D9
Husthwaite N Yorks 215 D10
Hut Green N Yorks 198 C5
Hutcherleigh Devon 8 E5
Hutchesontown Glasgow 267 C11
Hutchwns Bridgend 57 F10
Huthwaite Notts 171 D7
Hutlerburn Borders 261 E10
Huttock Top Lancs 195 C11
Huttoft Lincs 191 F8
Hutton Borders 273 E9
Hutton Cumb 230 F4
Hutton E Yorks 208 C6
Hutton Essex 87 F10
Hutton Lancs 194 B3
Hutton N Som 43 B11
Hutton Bonville N Yorks 224 E6
Hutton Buscel N Yorks 217 C9
Hutton Conyers N Yorks 214 E6
Hutton Cranswick E Yorks 208 C6
Hutton End Cumb 230 D4
Hutton Gate Redcar 225 B11
Hutton Hang N Yorks 214 B3
Hutton Henry Durham 234 D4
Hutton-le-Hole N Yorks 226 G4
Hutton Magna Durham 224 C2
Hutton Mount Essex 87 G10
Hutton Roof Cumb 211 D11
Hutton Roof Cumb 230 E3
Hutton Rudby N Yorks 225 D9
Hutton Sessay N Yorks 215 D9
Hutton Village Redcar 225 C11
Hutton Wandesley N Yorks 206 C6
Huttons Ambo N Yorks 216 F5
Huxham Devon 14 B4
Huxham Green Som 44 F5
Huxley W Ches 167 C8
Huxter Shetland 313 G7
Huxter Shetland 313 H3
Huxter Shetland 313 H5
Huxton Borders 273 B7
Huyton Mers 182 C6
Huyton Park Mers 182 C6
Huyton Quarry Mers 183 C7
Hwlffordd = Haverfordwest Pembs 73 B7
Hycemoor Cumb 210 B1
Hyde Glos 80 E5
Hyde Glos 99 F11
Hyde Gtr Man 184 B6
Hyde Hants 31 E11
Hyde Hants 48 G3
Hyde Hereford 115 F9
Hyde Chase Essex 88 E4
Hyde End W Berks 64 G5
Hyde Heath Bucks 84 E6
Hyde Lea Staffs 151 E8
Hyde Park S Yorks 198 G5
Hydestile Sur 50 E3
Hylton Castle T & W 243 F9
Hylton Red House T & W 243 F9
Hyltons Crossways Norf 160 D4
Hyndburn Bridge Lancs 203 G10
Hyndford Bridge S Lnrk 269 G8
Hyndhope Borders 261 E9
Hynish Argyll 288 F1
Hyssington Powys 130 E6
Hystfield Glos 79 F11
Hythe Hants 32 F6
Hythe Kent 55 G7
Hythe Som 44 C2
Hythe Sur 66 E4
Hythe End Windsor 66 E4
Hythie Aberds 303 D10
Hyton Cumb 210 B1

I

Iarsiadar W Isles 304 E3
Ibberton Dorset 30 F3
Ible Derbys 170 D2
Ibslev Hants 31 F11
Ibstock Leics 153 G8
Ibstone Bucks 84 G3
Ibthorpe Hants 47 C11
Iburndale N Yorks 227 D7
Ibworth Hants 48 C5
Icelton N Som 59 F11
Ichrachan Argyll 284 D4
Ickburgh Norf 140 E6
Ickenham London 66 B5
Ickenthwaite Cumb 210 B6
Ickford Bucks 83 D11
Ickham Kent 55 B8
Ickleford Herts 104 E3
Icklesham E Sus 38 D5
Ickleton Cambs 105 C9
Icklingham Suff 124 C5
Ickornshaw N Yorks 204 E5
Ickwell C Beds 104 B3
Ickwell Green C Beds 104 B3
Icomb Glos 100 G4
Icy Park Devon 8 F3
Idbury Oxon 82 B2
Iddesleigh Devon 25 F9
Ide Devon 14 C3
Ide Hill Kent 52 C3
Ideford Devon 14 F3
Iden E Sus 38 C6
Iden Green Kent 53 F8
Iden Green Kent 53 G10
Idle W Yorks 205 F9
Idle Moor W Yorks 205 F9
Idless Corn 4 F6
Idlicote Warks 100 C5
Idmiston Wilts 47 F7
Idole Carms 74 B6
Idridgehay Derbys 170 F3
Idridgehay Green Derbys 170 F3
Idrigill Highld 298 C3
Idstone Oxon 63 C9
Idvies Angus 287 C9
Iet-y-bwlch Carms 92 F3
Iffley Oxon 83 E8
Ifield W Sus 51 F8
Ifield Green W Sus 51 F9
Ifieldwood W Sus 51 F8
Ifold W Sus 50 G4
Iford Bmouth 19 C8
Iford E Sus 36 F6
Ifton Heath Shrops 148 B6
Ightfield Shrops 149 B11
Ightfield Heath Shrops 149 B11
Ightham Kent 52 B5
Igtham Common Kent 52 B5
Iken Suff 127 F8
Ilam Staffs 169 E10
Ilchester Som 29 C8
Ilchester Mead Som 29 C8
Ilderton Northumb 264 E2
Ileden Kent 55 C8
Ilford London 68 B2
Ilford Som 28 D5
Ilfracombe Devon 40 D4
Ilkeston Derbys 171 G7
Ilketshall St Andrew Suff 143 F7
Ilketshall St Lawrence Suff 143 G7
Ilketshall St Margaret Suff 142 F6
Ilkley W Yorks 205 D8
Illand Corn 11 F11
Illey W Mid 133 G9
Illidge Green E Ches 168 C3
Illington Norf 141 F8
Illingworth W Yorks 196 B5
Illogan Corn 4 G3
Illogan Highway Corn 4 G3
Illshaw Heath W Mid 118 C2
Illston on the Hill Leics 136 D4
Ilmer Bucks 84 D3
Ilmington Warks 100 C4
Ilminster Som 28 E5
Ilsington Devon 13 F11
Ilston Swansea 56 C5
Ilton N Yorks 214 D3
Ilton Som 28 D5
Imachar N Ayrs 255 C9
Imber Wilts 46 D3
Imeraval Argyll 254 C4
Immervoulin Stirl 285 F9
Immingham NE Lincs 201 E7
Impington Cambs 123 E8
Ince W Ches 183 F7
Ince Blundell Mers 193 G10
Ince in Makerfield Gtr Man 194 G5
Inch of Arnhall Aberds 293 F8
Inchbae Lodge Highld 300 C4
Inchbare Angus 293 G8
Inchberry Moray 302 D3
Inchbraoch Angus 287 B11
Inchbrook Glos 80 E4
Incheril Highld 299 C10
Inchgrundle Angus 292 F6
Inchina Highld 307 K4
Inchinnan Renfs 267 B9
Inchkinloch Highld 308 E5
Inchlaggan Highld 290 C3
Inchlumpie Highld 300 B5
Inchmore Highld 300 E3
Inchmore Highld 300 E5
Inchnacardoch Hotel Highld 290 B5
Inchnadamph Highld 307 G7
Inchock Angus 287 C10
Inchree Highld 290 G2
Inchrory Moray 292 C3
Inchs Corn 5 C9
Inchture Perth 286 E6
Inchyra Perth 286 E5
Indian Queens Corn 5 D8
Inerval Argyll 254 C4
Ingatestone Essex 87 F11
Ingbirchworth S Yorks 197 F8
Ingerthorpe N Yorks 214 F5
Ingestre Staffs 151 E9
Ingham Lincs 188 E6
Ingham Norf 161 D7
Ingham Suff 125 C7
Ingham Corner Norf 161 D7
Ingleborough Norf 157 F9
Ingleby Derbys 152 D6
Ingleby Lincs 188 F5
Ingleby Arncliffe N Yorks 225 E8
Ingleby Barwick Stockton 225 C9
Ingleby Cross N Yorks 225 E9
Ingleby Greenhow N Yorks 225 D11
Ingleigh Green Devon 25 F10
Inglemire Hull 209 G7
Inglesbatch Bath 61 G8
Inglesham Swindon 82 F2
Ingleton Durham 233 G9
Ingleton N Yorks 212 E3
Inglewhite Lancs 202 E6
Ingmanthorpe N Yorks 206 C4
Ingoe Northumb 242 C2
Ingol Lancs 202 G6
Ingoldisthorpe Norf 158 C3
Ingoldmells Lincs 175 B9
Ingoldsby Lincs 155 C10
Ingon Warks 118 F4
Ingram Northumb 264 F2
Ingrams Green W Sus 34 C4
Ingrave Essex 87 G10
Ingrow W Yorks 205 F7
Ings Cumb 221 F8
Ingst S Glos 60 B5
Ingthorpe Rutland 137 B9
Ingworth Norf 160 D3
Inham's End Cambs 138 D5
Inhurst Hants 64 G5
Inkberrow Worcs 117 F10
Inkerman Durham 233 D8
Inkersall Derbys 186 G6
Inkersall Green Derbys 186 G6
Inkford Worcs 117 C11
Inkpen W Berks 63 G11
Inkpen Common W Berks 63 G11
Inkstack Highld 310 B6
Inlands W Sus 22 B3
Inmarsh Wilts 62 G2
Inn Cumb 221 D8
Innellan Argyll 276 G3
Inner Hope Devon 9 G8
Innerleithen Borders 261 B8
Innerleven Fife 287 G7
Innermessan Dumfries 236 C2
Innerwick E Loth 282 G4
Innerwick Perth 285 C9
Innie Argyll 275 B9
Inninbeg Highld 289 E8
Innis Chonain Argyll 284 E5
Innistrynich Argyll 284 E5
Innox Hill Som 45 D9
Innsworth Glos 99 G7

Insch Aberds 302 G6
Insh Highld 291 C10
Inshegra Highld 306 D7
Inshore Highld 308 C3
Inskip Lancs 202 F5
Inskip Moss Side Lancs 202 F5
Instoneville S Yorks 198 E5
Instow Devon 40 G3
Insworke Corn 7 E8
Intack Blkburn 195 B8
Intake S Yorks 186 E5
Intake S Yorks 198 G5
Intake W Yorks 205 F10
Interfield Worcs 98 B5
Intwood Norf 142 C3
Inver Aberds 292 D4
Inver Highld 311 L2
Inver Perth 286 C4
Inver Mallie Highld 290 E3
Inverailort Highld 295 G9
Inveraldie Angus 287 D8
Inveralivaig Highld 298 E4
Inveralligin Highld 299 D8
Inverallochy Aberds 303 C10
Inveran Highld 299 B8
Inveran Highld 309 K5
Inveraray Argyll 284 G4
Inverarish Highld 295 B7
Inverarity Angus 287 C8
Inverarnan Stirl 285 F7
Inverasdale Highld 307 L3
Inverawe Ho Argyll 284 D4
Inverbeg Argyll 276 B6
Inverbervie Aberds 293 F10
Inverboyndie Aberds 302 C6
Inverbroom Highld 307 L6
Invercarron Mains Highld 309 K5
Invercassley Highld 309 J4
Invercauld House Aberds 292 D3
Inverchaolain Argyll 275 F11
Invercharnan Highld 284 C5
Inverchoran Highld 300 D2
Invercreran Argyll 284 C4
Inverdruie Highld 291 B11
Inverebrie Aberds 303 F9
Invereck Argyll 276 E2
Invererernan Ho Aberds 292 B5
Invereshie House Highld 291 C10
Inveresk E Loth 280 G6
Inverey Aberds 292 E2
Inverfarigaig Highld 300 G5
Invergarry Highld 290 C5
Invergelder Aberds 292 D4
Invergeldie Perth 285 E11
Invergordon Highld 301 C7
Invergowrie Perth 287 D7
Inverguseran Highld 295 E9
Inverhadden Perth 285 B10
Inverhaggernie Stirl 285 E7
Inverharroch Moray 302 F3
Inverherive Stirl 285 E7
Inverie Highld 295 F9
Inverinan Argyll 275 B10
Inverinate Highld 295 C11
Inverkeilor Angus 287 C10
Inverkeithing Fife 280 E2
Inverkeithny Aberds 302 E6
Inverkip Invclyd 276 G4
Inverkirkaig Highld 307 H5
Inverlael Highld 307 L6
Inverleith Edin 280 F4
Inverliever Lodge Argyll 275 C9
Inverliver Argyll 284 D4
Inverlochlarig Stirl 285 F8
Inverlochy Argyll 284 E5
Inverlochy Highld 290 F3
Inverlochy Moray 301 G11
Inverlounin Argyll 276 B4
Inverlussa Argyll 275 E7
Invermark Lodge Angus 292 E6
Invermoidart Highld 289 B8
Invermoriston Highld 290 B6
Invernaver Highld 308 C7
Inverneill Argyll 275 E9
Inverness Highld 300 E6
Invernettie Aberds 303 E11
Invernoaden Argyll 276 B2
Inveronich Argyll 284 G6
Inveroran Hotel Argyll 284 C6
Inverpolly Lodge Highld 307 H5
Inverquharity Angus 287 B8
Inverquhomery Aberds 303 E10
Inverroy Highld 290 E4
Inversanda Highld 289 D11
Invershiel Highld 295 D11
Invershin Highld 309 K5
Invershore Highld 310 F6
Inversnaid Hotel Stirl 285 G7
Invertrossachs Stirl 285 G9
Inverugie Aberds 303 E11
Inveruglas Argyll 285 G7
Inveruglass Highld 291 C10
Inverurie Aberds 303 G7
Invervar Perth 285 C10
Inverythan Aberds 303 E7
Inwardleigh Devon 13 B7
Inwood Shrops 131 D9
Inworth Essex 88 B5
Iochdar W Isles 297 G3
Iping W Sus 34 C5
Ipplepen Devon 8 B6
Ipsden Oxon 64 B6
Ipsley Worcs 117 D11
Ipstones Staffs 169 F8
Ipswich Suff 108 C3
Irby Mers 182 E3
Irby in the Marsh Lincs 175 C7
Irby upon Humber NE Lincs 201 G7
Irchester Northants 121 D8
Ireby Cumb 229 D10
Ireby Lancs 212 D3
Ireland C Beds 104 C2
Ireland Orkney 314 F3
Ireland Shetland 313 L5
Ireland Wilts 45 C10
Ireland Wood W Yorks 205 F11
Ireland's Cross Shrops 168 G2
Ireleth Cumb 210 D4
Ireshopeburn Durham 232 D3
Ireton Wood Derbys 170 F3
Irlam Gtr Man 184 C2
Irlams o' th' Height Gtr Man 195 G9
Irnham Lincs 155 D10
Iron Acton S Glos 61 C7
Iron Bridge Cambs 139 D9
Iron Cross Warks 117 G11
Iron Lo Highld 299 G10
Ironbridge Telford 132 C3
Irongray Dumfries 237 B11
Ironmacannie Dumfries 237 B8
Irons Bottom Sur 51 D9
Ironside Aberds 303 D8
Ironville Derbys 170 E6
Irstead Norf 161 E7
Irstead Street Norf 161 F7
Irthington Cumb 239 E11
Irthlingborough Northants 121 C8
Irton N Yorks 217 C10
Irvine N Ayrs 257 B8
Irwell Vale Lancs 195 C9
Isabella Pit Northumb 253 G8
Isallt Bach Anglesey 178 F3
Isauld Highld 310 C3
Isbister Orkney 314 E3
Isbister Orkney 314 D2
Isbister Shetland 312 D5
Isbister Shetland 313 G7
Isel Cumb 229 E9
Isfield E Sus 36 D6
Isham Northants 121 C7
Ishriff Argyll 289 F8
Isington Hants 49 E9
Island Carr N Lincs 200 F3
Islands Common Cambs 122 E3
Islay Ho Argyll 274 G4
Isle Abbotts Som 28 C5
Isle Brewers Som 28 C5
Isle of Axholme N Lincs 199 F9
Isle of Dogs London 67 D11
Isle of Man Dumfries 238 B2
Isle of Whithorn Dumfries 236 F6
Isleham Cambs 124 C2
Isleornsay Highld 295 D9
Islesburgh Shetland 312 G5
Islesteps Dumfries 237 B11
Isleworth London 67 D7
Isley Walton Leics 153 D8
Islibhig W Isles 304 F1
Islington London 67 C10
Islington Telford 150 E4
Islip Northants 121 B9
Islip Oxon 83 C8
Isombridge Telford 150 G2
Istead Rise Kent 68 F6
Isycoed Wrex 166 E6
Itchen Soton 32 E6
Itchen Abbas Hants 48 G4
Itchen Stoke Hants 48 G5
Itchingfield W Sus 35 B10
Itchington S Glos 61 B7
Itteringham Norf 160 C2
Itteringham Common Norf 160 D3
Itton Devon 13 B9
Itton Mon 79 F7
Itton Common Mon 79 F7
Ivegill Cumb 230 C4
Ivelet N Yorks 223 F8
Iver Bucks 66 C4
Iver Heath Bucks 66 C4
Iverley Staffs 133 G7
Iveston Durham 242 G4
Ivinghoe Bucks 84 B6
Ivinghoe Aston Bucks 85 B7
Ivington Hereford 115 F9
Ivington Green Hereford 115 F9
Ivy Chimneys Essex 86 E6
Ivy Cross Dorset 30 C5
Ivy Hatch Kent 52 C5
Ivy Todd Norf 141 B7
Ivybridge Devon 8 D2
Ivychurch Kent 39 B8
Iwade Kent 69 F11
Iwerne Courtney or Shroton Dorset 30 E5
Iwerne Minster Dorset 30 E5
Iwood N Som 60 G3
Ixworth Suff 125 C8
Ixworth Thorpe Suff 125 C8

J

Jack Green Lancs 194 B5
Jack Hayes Staffs 168 F6
Jack Hill N Yorks 205 C10
Jack in the Green Devon 14 B6
Jackfield Telford 132 C3
Jack's Green Essex 105 G11
Jack's Green Glos 80 D5
Jack's Hatch Essex 86 D6
Jacksdale Notts 170 E6
Jackson Bridge W Yorks 197 F7
Jackstown Aberds 303 F7
Jacobs Well Sur 50 C3
Jacobstow Corn 11 B9
Jacobstowe Devon 25 G9
Jagger Green W Yorks 196 D5
Jameston Pembs 73 F9
Jamestown Dumfries 249 D8
Jamestown Highld 300 D4
Jamestown W Dunb 277 E7
Jamphlars Fife 280 B4
Janetstown Highld 310 C4
Janke's Green Essex 107 F8
Jarrow T & W 243 D8
Jarvis Brook E Sus 37 B8
Jasper's Green Essex 106 F4
Java Argyll 289 F9
Jaw Hill W Yorks 197 C9
Jawcraig Falk 278 F6
Jaywick Essex 89 C11
Jealott's Hill Brack 65 E11
Jeaniefield Borders 271 G10
Jedburgh Borders 262 E5
Jedurgh Borders 262 F5
Jeffreyston Pembs 73 D9
Jellyhill E Dunb 278 G2
Jemimaville Highld 301 C7
Jennetts Hill W Berks 64 E5
Jennyfield N Yorks 205 B11
Jericho Gtr Man 195 E10
Jersey Farm Herts 85 D11
Jersey Marine Neath 57 C8
Jerviswood S Lnrk 269 F7
Jesmond T & W 243 D7
Jevington E Sus 23 E9
Jewell's Cross Corn 24 G3
Jingle Street Mon 79 C7
Jockey End Herts 85 C8
Jodrell Bank E Ches 184 G3
John O'Gaunt Leics 136 B4
John O'Gaunts W Yorks 197 B11
John o'Groats Highld 310 B7
Johnby Cumb 230 E4
John's Cross E Sus 38 C2
Johnshaven Aberds 293 G9
Johnson Fold Gtr Man 195 E7
Johnson Street Norf 161 F7
Johnson's Hillock Lancs 194 C5
Johnston Pembs 72 C6
Johnstone Renfs 267 C8
Johnstone Mains Aberds 293 F9
Johnstonebridge Dumfries 248 E3
Johnstown Carms 74 B6
Johnstown Wrex 166 F4
Jolly's Bottom Corn 4 F5
Joppa Corn 2 B3
Joppa Edin 280 G6
Joppa S Ayrs 257 F10
Jordan Green Norf 159 E11
Jordanhill Glasgow 267 B10
Jordans Bucks 85 G7
Jordanston Pembs 91 E8
Jordanthorpe S Yorks 186 E5
Jordon S Yorks 186 C6
Joyford Glos 79 C9
Joy's Green Glos 79 B10
Jubilee Gtr Man 196 E2
Jubilee Notts 170 E6
Jugbank Staffs 150 B5
Jump S Yorks 197 G11
Jumpers Common Dorset 19 C8
Jumpers Green Dorset 19 C8
Jumper's Town E Sus 52 G3
Junction N Yorks 204 D6
Juniper Northumb 241 F10
Juniper Green Edin 270 B3
Jurby East I o M 192 C4
Jurby West I o M 192 C4
Jurston Devon 13 E9
Jury's Gap E Sus 39 D7

K

Kaber Cumb 222 C5
Kaimend S Lnrk 269 F9
Kaimes Edin 270 B5
Kaimrig End Borders 269 G11
Kalemouth Borders 262 D6
Kame Fife 287 G7
Kames Argyll 275 B9
Kames Argyll 275 F10
Kames E Ayrs 258 D5
Kates Hill W Mid 133 E9
Kea Corn 4 G6
Keadby N Lincs 199 E10
Keal Cotes Lincs 174 C5
Kearby Town End N Yorks 206 D2
Kearnsey Kent 55 E9
Kearsley Gtr Man 195 F9
Kearstwick Cumb 212 C2
Kearton N Yorks 223 F9
Kearvaig Highld 306 B7
Keasden N Yorks 212 F4
Kebroyd W Yorks 196 C4
Keckwick Halton 183 E9
Keddington Lincs 190 D4
Keddington Corner Lincs 190 D5
Kedington Suff 106 B4
Kedleston Derbys 170 G4
Kedslie Borders 271 G11
Keekle Cumb 219 B10
Keelars Tye Essex 107 G11
Keelby Lincs 201 E7
Keele Staffs 168 F4
Keeley Green Beds 103 B10
Keelham W Yorks 205 G7
Keenley Northumb 241 F7
Keenthorne Som 43 F8
Keeres Green Essex 87 C9
Keeston Pembs 72 B6
Keevil Wilts 46 B2
Kegworth Leics 153 D9
Kehelland Corn 4 G2
Keig Aberds 293 B8
Keighley W Yorks 205 E7
Keil Highld 289 D11
Keilarsbrae Clack 279 C7
Keilhill Aberds 303 D7
Keillmore Argyll 275 E7
Keillor Perth 286 C6
Keillour Perth 286 E3
Keills Argyll 274 G5
Keils Argyll 274 G6
Keinton Mandeville Som 44 G4
Keir Mill Dumfries 247 E9
Keisby Lincs 155 D10
Keislo Highld 310 C7
Keiss Highld 310 C7
Keistle Highld 298 D4
Keith Moray 302 D4
Keith Hall Aberds 303 G7
Keith Inch Aberds 303 E11
Keithock Angus 293 G8
Kelbrook Lancs 204 E4
Kelby Lincs 173 G8
Kelcliffe W Yorks 205 E9
Keld Cumb 221 C11
Keld N Yorks 223 E7
Keld Houses N Yorks 214 G2
Keldholme N Yorks 216 B4
Kelfield N Lincs 199 G10
Kelfield N Yorks 207 F7
Kelham Notts 172 D3
Kelhurn Argyll 276 F6
Kellacott Devon 12 D4
Kellamergh Lancs 194 B2
Kellan Argyll 289 E7
Kellas Angus 287 D8
Kellas Moray 301 D11
Kellaton Devon 9 G11
Kellaways Wilts 62 D3
Kelleth Cumb 222 D3
Kelleythorpe E Yorks 208 B5
Kelleythorpe E Yorks 208 B6
Kelling Norf 177 E9
Kellingley N Yorks 198 C4
Kellington N Yorks 198 C5
Kelloe Durham 234 D2
Kelloholm Dumfries 258 G6
Kells Cumb 219 B9
Kelly Corn 10 G6
Kelly Devon 12 E3
Kelly Bray Corn 12 G3
Kelmarsh Northants 120 B4
Kelmscott Oxon 82 F3
Kelsale Suff 127 D7
Kelsall W Ches 167 B8
Kelsall Hill W Ches 167 B8
Kelsay Argyll 254 B2
Kelshall Herts 104 D6
Kelsick Cumb 238 G5
Kelso Borders 262 C6
Kelstedge Derbys 170 C4
Kelstern Lincs 190 C3
Kelsterton Flint 182 G3
Kelston Bath 61 F8
Keltneyburn Perth 285 C11
Kelton Dumfries 237 B11
Kelton Durham 232 G4
Kelty Fife 280 C2
Keltybridge Fife 280 B2
Kelvedon Essex 88 B5
Kelvedon Hatch Essex 87 F9
Kelvin S Lnrk 268 E2
Kelvindale Glasgow 267 B11
Kelvinside Glasgow 267 B11
Kelynack Corn 1 D3
Kemacott Devon 41 D7
Kemback Fife 287 F8
Kemberton Shrops 132 C4
Kemble Glos 81 F7
Kemble Wick Glos 81 F7
Kemerton Worcs 99 D8
Kemeys Commander Mon 78 E4
Keminchan E Ches 168 B4
Kemnay Aberds 293 B9
Kemp Town Brighton 36 G4
Kempe's Corner Kent 54 D4
Kempie Highld 308 D4
Kempley Glos 98 F3
Kempley Green Glos 98 F3
Kemps Green Warks 118 C2
Kempsey Worcs 99 B7
Kempsford Glos 81 F11
Kempshott Hants 48 C6
Kempston Beds 103 B10
Kempston Church End Beds 103 B10
Kempston Hardwick Beds 103 C10
Kempston West End Beds 103 B9
Kempton Shrops 131 G7
Kemsing Kent 52 B4
Kemsley Kent 70 F2
Kemsley Street Kent 69 G10
Kenardington Kent 54 G3
Kenchester Hereford 97 C8
Kencot Oxon 82 E3
Kendal Cumb 221 G10
Kendal End Worcs 117 C10
Kendleshire S Glos 61 D7
Kendon Caerph 77 F11
Kendoon Dumfries 246 F4
Kendray S Yorks 197 F11
Kenfig Bridgend 57 E10
Kenfig Hill Bridgend 57 E10
Kenghsarair Argyll 288 E6
Kenilworth Warks 118 C5
Kenknock Stirl 285 D8
Kenley London 51 B10
Kenley Shrops 131 C11
Kenmore Argyll 284 G4
Kenmore Highld 299 D7
Kenmore Perth 285 C11
Kenmure Dumfries 237 B8
Kenn Devon 14 D4
Kenn N Som 60 F2
Kenn Moor Gate N Som 60 F2
Kennacley W Isles 305 J3
Kennacraig Argyll 275 G9
Kennards House Corn 11 E11
Kenneggy Corn 2 D3
Kenneggy Downs Corn 2 D3
Kennerleigh Devon 26 F4
Kennet Clack 279 C8
Kennet End Suff 124 D4
Kennethmont Aberds 302 G5
Kennett Cambs 124 D3
Kennford Devon 14 D4
Kenninghall Norf 141 F10
Kenninghall Heath Norf 141 G10
Kennington Kent 54 E4
Kennington London 67 D10
Kennington Oxon 83 E8
Kennoway Fife 287 G7
Kenny Som 28 D4
Kenny Hill Suff 124 B3
Kennythorpe N Yorks 216 F5
Kenovay Argyll 288 E1
Kensal Green London 67 C8
Kensal Rise London 67 C8
Kensal Town London 67 C8
Kensaleyre Highld 298 D4
Kensary Highld 310 E6
Kensington London 67 D9
Kensington Mers 182 C5
Kensworth C Beds 85 B8
Kent Street E Sus 38 D3
Kent Street Kent 53 C7
Kent Street W Sus 36 C2
Kentallen Highld 284 B4
Kentchurch Hereford 97 F8
Kentford Suff 124 D4
Kentisbeare Devon 27 F9
Kentisbury Devon 40 E6
Kentisbury Ford Devon 40 E6
Kentish Town London 67 C9
Kentmere Cumb 221 E9
Kenton Devon 14 E5
Kenton London 67 B7
Kenton Suff 126 D3
Kenton T & W 242 D6
Kenton Bankfoot T & W 242 D6
Kenton Bar T & W 242 D6
Kenton Corner Suff 126 D4
Kenton Green Glos 80 C3
Kentra Highld 289 C8
Kentrigg Cumb 221 G10
Kents Corn 11 B9
Kents Bank Cumb 211 D7
Kent's Green Glos 98 G4
Kents Hill M Keynes 103 D7
Kent's Oak Hants 32 C4
Kenwick Shrops 149 C8
Kenwick Park Shrops 149 D8
Kenwyn Corn 4 F6
Kenyon Warr 183 B10
Keoldale Highld 308 C3
Keonchulish Ho Highld 307 K6
Kepdowrie Stirl 277 C11
Kepnal Wilts 63 G7
Keppanach Highld 290 G2
Keppoch Highld 295 C11
Keprigan Argyll 255 F7
Kepwick N Yorks 225 G9
Kerchesters Borders 263 B7
Kerdiston Norf 159 E11
Keresforth Hill S Yorks 197 F10
Keresley W Mid 134 G6
Keresley Newlands Warks 134 G6
Kerfield Borders 270 G5
Kerley Downs Corn 4 G5
Kernborough Devon 8 G5
Kerne Bridge Hereford 79 B9
Kernsary Highld 299 B8
Kerridge E Ches 184 F6
Kerridge-end E Ches 184 F6
Kerris Corn 1 D4
Kerry = Ceri Powys 130 F2
Kerry Hill Staffs 168 F6
Kerrycroy Argyll 266 C2
Kerry's Gate Hereford 97 E7
Kerrysdale Highld 299 B8
Kersal Gtr Man 195 G10
Kersall Notts 172 C2
Kersbrook Cross Corn 12 F2
Kerscott Devon 25 B10
Kersey Suff 107 C10
Kersey Tye Suff 107 C9
Kersey Upland Suff 107 C9
Kershopefoot Cumb 249 G11
Kersoe Worcs 99 C9
Kerswell Devon 27 F9
Kerswell Green Worcs 99 B7
Kerthen Wood Corn 2 C3
Kesgrave Suff 108 B4
Kessingland Suff 143 F10
Kessingland Beach Suff 143 F10
Kessington E Dunb 277 G11
Kestle Corn 5 F9
Kestle Mill Corn 5 D7
Keston London 68 G2
Keston Mark London 68 F2
Keswick Cumb 229 G11
Keswick Norf 142 C4
Keswick Norf 161 C7
Kete Pembs 72 E4
Ketford Glos 98 E4
Ketley Telford 150 G3
Ketley Bank Telford 150 G3
Ketsby Lincs 190 F5
Kettering Northants 121 B7
Ketteringham Norf 142 C3
Kettins Perth 286 D6
Kettle Corner Kent 53 C8
Kettle Green Herts 86 B6
Kettlebaston Suff 125 G9
Kettlebridge Fife 287 G7
Kettlebrook Staffs 134 C4
Kettleburgh Suff 126 E5
Kettlehill Fife 287 G7
Kettleholm Dumfries 238 B4
Kettleness N Yorks 226 B6
Kettleshulme E Ches 185 F7
Kettleshume E Ches 185 F7
Kettlesing N Yorks 205 B10
Kettlesing Bottom N Yorks 205 B10
Kettlesing Head N Yorks 205 B10
Kettlestone Norf 159 C9
Kettlethorpe Lincs 188 F4
Kettlethorpe W Yorks 197 D10
Kettletoft Orkney 314 C6
Kettlewell N Yorks 213 E9
Ketton Rutland 137 C9
Kevingtown London 68 F3
Kew London 67 D7
Kew Bridge London 67 D7
Keward Som 44 E4
Kewstoke N Som 59 G10
Kexbrough S Yorks 197 F9
Kexby Lincs 188 D5
Kexby York 207 C10
Key Green E Ches 168 C5
Key Green N Yorks 226 E6
Key Street Kent 69 G11
Keybridge Corn 11 G7
Keycol Kent 69 G11
Keyford Som 45 D9
Keyham Leics 136 B3
Keyhaven Hants 20 C2
Keyingham E Yorks 201 B8
Keymer W Sus 36 D4
Keynsham Bath 61 F7
Key's Green Kent 53 F7
Keysers Estate Essex 86 D5
Keysoe Beds 121 E11
Keysoe Row Beds 121 E11
Keyston Cambs 121 B10
Keyworth Notts 154 C2
Khantore Aberds 292 D4
Kibbear Som 28 C2
Kibblesworth T & W 242 F6
Kibworth Beauchamp Leics 136 E3
Kibworth Harcourt Leics 136 E3
Kidbrooke London 68 D2
Kidburngill Cumb 229 G7
Kiddal Lane End W Yorks 206 F4
Kiddemore Green Staffs 133 B7
Kidderminster Worcs 116 B6
Kiddington Oxon 101 G8
Kidd's Moor Norf 142 C2
Kidlington Oxon 83 C7
Kidmore End Oxon 65 D7
Kidnal W Ches 167 F7
Kidsdale Dumfries 236 F6
Kidsgrove Staffs 168 E4
Kidstones N Yorks 213 C9
Kidwelly = Cydweli Carms 74 D6
Kiel Crofts Argyll 289 F11
Kielder Northumb 250 E4
Kierfiold Ho Orkney 314 E2
Kiff Green W Berks 64 F5
Kilbagie Fife 279 D8
Kilbarchan Renfs 267 C8
Kilbeg Highld 295 E8
Kilberry Argyll 275 G8
Kilbirnie N Ayrs 266 E6
Kilbowie W Dunb 277 G10
Kilbraur Highld 311 H2
Kilbride Argyll 254 C4
Kilbride Argyll 275 D9
Kilbride Argyll 289 G10
Kilbride Argyll 289 G11
Kilbride Highld 295 C7
Kilbridemore Argyll 275 D11
Kilbryde Castle Stirl 285 G11
Kilburn Angus 292 G5
Kilburn Derbys 170 F5
Kilburn London 67 C9
Kilburn N Yorks 215 D10
Kilby Leics 136 D2
Kilby Bridge Leics 136 D2
Kilchamaig Argyll 275 G9
Kilchattan Argyll 274 D4
Kilchattan Bay Argyll 266 E2
Kilchenzie Argyll 255 E7
Kilcheran Argyll 289 F10
Kilchiaran Argyll 274 G3
Kilchoan Argyll 275 B8
Kilchoan Highld 288 C6
Kilchoman Argyll 274 G3
Kilchrenan Argyll 284 E4
Kilconquhar Fife 287 G8
Kilcot Glos 98 F3
Kilcoy Highld 300 D5
Kilcreggan Argyll 276 E4
Kildale N Yorks 226 D2
Kildalloig Argyll 255 F8
Kildary Highld 301 B7
Kildaton Ho Argyll 254 C5
Kildavanan Argyll 275 G11
Kildermorie Lodge Highld 300 B5
Kildonan Dumfries 236 D2
Kildonan Highld 298 D3
Kildonan N Ayrs 256 E2
Kildonan Lodge Highld 311 G3
Kildonnan Highld 294 G6
Kildrum N Lnrk 278 F5
Kildrummy Aberds 292 B6
Kildwick N Yorks 204 D6
Kilfinan Argyll 275 F10
Kilfinnan Highld 290 D4
Kilgetty Pembs 73 D10
Kilgour Fife 286 G6
Kilgrammie S Ayrs 245 C7
Kilgwrrwg Common Mon 79 F7
Kilhallon Corn 5 E11
Kilham E Yorks 217 G11
Kilham Northumb 263 C9
Kilkeddan Argyll 255 E8
Kilkenneth Argyll 288 E1
Kilkenny Glos 81 B8
Kilkerran Argyll 255 F8
Kilkhampton Corn 24 E3
Killamarsh Derbys 187 E7
Killaworgey Corn 5 C8
Killay Swansea 56 C6
Killbeg Argyll 289 E8
Killean Argyll 255 C7
Killearn Stirl 277 D10
Killegruer Argyll 255 D7
Killen Highld 300 D6
Killerby Darl 224 B3
Killichonan Perth 285 B9
Killiechoinich Argyll 289 G10
Killiechonate Highld 290 E4
Killiechronan Argyll 289 E7
Killiecrankie Perth 291 G11
Killiemor Argyll 288 F6
Killiemore House Argyll 288 G6
Killilan Highld 295 B11
Killimster Highld 310 D7
Killin Stirl 285 D9
Killin Lodge Highld 291 C7
Killinallan Argyll 274 F4
Killingbeck W Yorks 206 G2
Killinghall N Yorks 205 B11
Killington Cumb 212 B2
Killington Devon 41 D7
Killingworth T & W 243 C7
Killingworth Moor T & W 243 C7
Killingworth Village T & W 243 C7
Killivose Corn 2 B4
Killmahumaig Argyll 275 D8
Killochyett Borders 271 F9
Killocraw Argyll 255 D7
Killundine Highld 289 E7
Killylung Dumfries 247 G11
Kilmacolm Invclyd 267 B7
Kilmaha Argyll 275 C10
Kilmahog Stirl 285 G10
Kilmalieu Highld 289 D10
Kilmaluag Highld 298 B4
Kilmany Fife 287 E7
Kilmarie Highld 295 D7
Kilmarnock E Ayrs 257 B10
Kilmaron Castle Fife 287 F7
Kilmartin Argyll 275 D9
Kilmaurs E Ayrs 267 G8
Kilmelford Argyll 275 B9
Kilmeny Argyll 274 G4
Kilmersdon Som 45 C7
Kilmeston Hants 33 B9
Kilmichael Argyll 255 E7
Kilmichael Argyll 275 F10
Kilmichael Glassary Argyll 275 D9
Kilmichael of Inverlussa Argyll 275 E8
Kilmington Devon 15 B11
Kilmington Wilts 45 F9
Kilmington Common Wilts 45 F9
Kilmoluaig Argyll 288 E1
Kilmonivaig Highld 290 E3
Kilmorack Highld 300 E4
Kilmore Argyll 289 G10
Kilmore Highld 295 E8
Kilmory Argyll 275 F8
Kilmory Highld 289 B7
Kilmory Highld 294 E5
Kilmory N Ayrs 255 E10
Kilmory Lodge Argyll 275 C8
Kilmote Highld 311 H3
Kilmuir Highld 298 E3
Kilmuir Highld 298 E2
Kilmuir Highld 300 E6
Kilmuir Highld 301 B7
Kilmun Argyll 275 B10
Kilmun Argyll 276 E3
Kiln Green Hereford 79 B10
Kiln Green Wokingham 65 D10
Kiln Pit Hill Northumb 242 G2
Kilnave Argyll 274 F3
Kilncadzow S Lnrk 269 F7
Kilndown Kent 53 G8
Kilnhill Cumb 229 E10
Kilnhurst S Yorks 187 B7
Kilninian Argyll 288 E5
Kilninver Argyll 289 G10
Kilnsea E Yorks 201 D12
Kilnsey N Yorks 213 F9
Kilnwick E Yorks 208 D5
Kilnwick Percy E Yorks 208 C2
Kiloran Argyll 274 D4
Kilpatrick N Ayrs 255 E10
Kilpeck Hereford 97 E8
Kilphedir Highld 311 H3
Kilpin E Yorks 199 B9
Kilpin Pike E Yorks 199 B9
Kilrenny Fife 287 G9
Kilsby Northants 119 C11
Kilspindie Perth 286 E6
Kilsyth N Lnrk 278 F4
Kiltarlity Highld 300 E5
Kilton Notts 187 F9
Kilton Redcar 226 B4
Kilton Som 43 E7
Kilton Thorpe Redcar 226 B3
Kiltyrie Perth 285 D10
Kilvaxter Highld 298 C3
Kilve Som 43 E7
Kilvington Notts 172 G3
Kilwinning N Ayrs 266 G6
Kimberley Norf 141 C11
Kimberley Notts 171 G8
Kimberworth S Yorks 186 C6
Kimberworth Park S Yorks 186 C6
Kimble Wick Bucks 84 D4
Kimblesworth Durham 233 B11
Kimbolton Cambs 121 D11
Kimbolton Hereford 115 E10
Kimbridge Hants 32 B4
Kimcote Leics 135 F11
Kimmeridge Dorset 18 F4
Kimmerston Northumb 263 B11
Kimpton Hants 47 D9
Kimpton Herts 85 B11
Kimworthy Devon 24 E4
Kinabus Argyll 254 C3
Kinbeachie Highld 300 C6
Kinbrace Highld 310 F2
Kinbuck Stirl 285 G11
Kincaidston S Ayrs 257 F9
Kincaple Fife 287 F8
Kincardine Fife 279 D8
Kincardine Highld 309 L6
Kincardine Bridge Falk 279 D8
Kincardine O'Neil Aberds 293 D7
Kinclaven Perth 286 D5
Kincorth Aberdeen 293 C11
Kincorth Ho Moray 301 C10
Kincraig Highld 291 C10
Kincraigie Perth 286 C3
Kindallachan Perth 286 C3
Kine Moor S Yorks 197 G9
Kineton Glos 99 F11
Kineton Warks 118 G6
Kineton Green W Mid 134 G2
Kinfauns Perth 286 E5
King Edward Aberds 303 D7
King Sterndale Derbys 185 G9
King Street Essex 87 E9
Kingairloch Highld 289 D10
Kingarth Argyll 255 B11
Kingates I o W 20 F6
Kingbeare Corn 11 G11
Kingcoed Mon 78 D6
Kingdown N Som 60 G4
Kingerby Lincs 189 C9
Kingfield Sur 50 B4
Kingford Devon 24 F3
Kingford Devon 25 D10
Kingham Oxon 100 G5
Kinghay Wilts 30 B5
Kingholm Quay Dumfries 237 B11
Kinghorn Fife 280 D5
Kingie Highld 290 C3
Kinglassie Fife 280 B4
Kingledores Borders 260 D4
Kingoodie Perth 287 E7
King's Acre Hereford 97 C9
King's Bromley Staffs 152 F2
Kings Caple Hereford 97 F11
King's Cliffe Northants 137 D10
Kings Clipstone Notts 171 C10
King's Coughton Warks 117 F11
King's Dyke Cambs 138 D4
King's End Oxon 101 G11
King's End Worcs 116 G6
King's Furlong Hants 48 C6
King's Green Glos 98 E5
King's Green Worcs 116 E5
King's Heath Northants 120 E4
King's Heath W Mid 133 G11
Kings Hedges Cambs 123 E9
Kings Hill Glos 81 E8
King's Hill Kent 53 C7
King's Hill W Mid 133 D9
Kings Langley Herts 85 E9
King's Lynn Norf 158 E2
King's Meaburn Cumb 231 G8
King's Mills Derbys 153 D8
King's Mills Wrex 166 F4
Kings Moss Mers 194 G4
Kings Muir Borders 261 B7
King's Newnham Warks 119 B9
King's Newton Derbys 153 D7
King's Norton Leics 136 C3
King's Norton W Mid 117 B11
King's Nympton Devon 25 D11
King's Pyon Hereford 115 G8
Kings Ripton Cambs 122 B5
King's Somborne Hants 47 G11
King's Stag Dorset 30 E2
King's Stanley Glos 80 E4
King's Sutton Northants 101 D9
King's Tamerton Plym 7 D9
King's Thorn Hereford 97 E10
King's Walden Herts 104 G3
Kings Worthy Hants 48 G3
Kingsand Corn 7 E8
Kingsash Bucks 84 D5
Kingsbarns Fife 287 F9
Kingsbridge Devon 8 G4
Kingsbridge Som 42 F3
Kingsburgh Highld 298 D3
Kingsbury London 67 B8
Kingsbury Warks 134 D4
Kingsbury Episcopi Som 28 C6
Kingsbury Regis Som 29 D11
Kingscauseway Highld 301 B7
Kingscavil W Loth 279 F10
Kingsclere Hants 48 B4
Kingsclere Woodlands Hants 64 G4
Kingscote Glos 80 F4
Kingscott Devon 25 D8
Kingscourt Glos 80 E4
Kingscross N Ayrs 256 D2
Kingsditch Glos 99 G8
Kingsdon Som 29 B8
Kingsdown Kent 54 B2
Kingsdown Kent 55 D11
Kingsdown Swindon 63 B7
Kingsdown Wilts 61 F10
Kingseat Fife 280 C2
Kingseathill Fife 280 D2
Kingsett Devon 12 E6
Kingsey Bucks 84 D2
Kingsfield Hereford 97 B10
Kingsfold Lancs 194 B4
Kingsfold W Sus 51 F7
Kingsford Aberds 293 B7
Kingsford E Ayrs 267 F8
Kingsford Worcs 132 G6
Kingsforth N Lincs 200 D4
Kingsgate Kent 71 E11
Kingshall Green Suff 125 E8
Kingshall Street Suff 125 E8
Kingsheanton Devon 40 F5
Kingshill Glos 80 F3
Kingshill Swindon 62 C6
Kingsholm Glos 80 B4
Kingshouse Hotel Highld 284 B6
Kingshurst W Mid 134 F3
Kingside Hill Cumb 238 G5
Kingskerswell Devon 9 B7
Kingskettle Fife 287 G7
Kingsknowe Edin 280 G4
Kingsland Anglesey 178 E2
Kingsland Hereford 115 E8
Kingsland London 67 C10
Kingsland Shrops 149 G9
Kingsley Hants 49 F9
Kingsley Staffs 169 F8
Kingsley W Ches 183 G9
Kingsley Green W Sus 49 G11
Kingsley Holt Staffs 169 F8
Kingsley Moor Staffs 169 F7
Kingsley Park Northants 120 E5
Kingslow Shrops 132 D5
Kingsmead Hants 33 E9
Kingsmoor Essex 86 D6
Kingsmuir Angus 287 C8
Kingsmuir Fife 287 G9
Kingsnordley Shrops 132 F5
Kingsnorth Kent 54 F4
Kingstanding W Mid 133 E11
Kingsteignton Devon 14 G3
Kingsteps Highld 301 D9
Kingsthorpe Northants 120 E5
Kingsthorpe Hollow Northants 120 E5
Kingston Cambs 122 F6
Kingston Devon 8 F2
Kingston Devon 9 E8
Kingston Dorset 18 F5
Kingston Dorset 30 F3
Kingston E Loth 281 E10
Kingston Gtr Man 184 B6
Kingston Hants 31 G11
Kingston I o W 20 E5
Kingston Kent 55 C7
Kingston M Keynes 103 D8
Kingston Moray 302 C3
Kingston Ptsmth 33 G11
Kingston Suff 108 B5
Kingston Bagpuize Oxon 82 F6
Kingston Blount Oxon 84 F2
Kingston by Sea W Sus 36 G2
Kingston Deverill Wilts 45 F10
Kingston Gorse W Sus 35 G9
Kingston Lisle Oxon 63 B10
Kingston Maurward Dorset 17 C10
Kingston near Lewes E Sus 36 F5
Kingston on Soar Notts 153 D10
Kingston Park T & W 242 D6
Kingston Russell Dorset 17 C7
Kingston St Mary Som 28 B2
Kingston Seymour N Som 60 F2
Kingston Stert Oxon 84 E2
Kingston upon Hull Hull 200 B5
Kingston upon Thames London 67 F7
Kingston Vale London 67 E8
Kingstone Hereford 97 D8
Kingstone Hereford 98 G2
Kingstone S Yorks 197 F10
Kingstone Som 28 E5
Kingstone Staffs 151 D11
Kingstone Winslow Oxon 63 B9
Kingstown Cumb 239 F9
Kingsway Bath 61 G8
Kingsway Halton 183 D8
Kingswear Devon 9 E7
Kingswells Aberdeen 293 C10
Kingswinford W Mid 133 F7
Kingswood Bucks 83 B11
Kingswood Essex 69 B8
Kingswood Glos 80 G2
Kingswood Hereford 114 G5
Kingswood Herts 85 E10
Kingswood Kent 53 C10
Kingswood Powys 130 C4
Kingswood S Glos 60 E6
Kingswood Som 42 F6
Kingswood Sur 51 B8
Kingswood Warks 118 C3
Kingswood Warr 183 C9
Kingswood Brook Warks 118 C3
Kingswood Common Staffs 132 C6
Kingswood Common Worcs 116 D4
Kingthorpe Lincs 189 F10
Kington Hereford 114 F5
Kington S Glos 79 G10
Kington Worcs 117 F9
Kington Langley Wilts 62 D2
Kington Magna Dorset 30 C3
Kington St Michael Wilts 62 D2
Kingussie Highld 291 C9
Kingweston Som 44 G4
Kinhrive Highld 301 B7
Kininvie Ho Moray 302 E3
Kinkell Bridge Perth 286 F3
Kinknockie Aberds 303 E10
Kinknockie Aberds 303 G9
Kinkry Hill Cumb 240 B2
Kinlet Shrops 132 G4
Kinloch Fife 286 F6
Kinloch Highld 289 D8
Kinloch Highld 294 F5
Kinloch Highld 295 D8
Kinloch Highld 308 F3
Kinloch Perth 286 C5
Kinloch Perth 286 C6
Kinloch Damph Highld 299 E8
Kinloch Hourn Highld 295 E11
Kinloch Laggan Highld 291 E7
Kinloch Lodge Highld 308 D5
Kinloch Rannoch Perth 285 B10
Kinlochan Highld 289 C10
Kinlochard Stirl 285 G8
Kinlochbeoraid Highld 295 G10
Kinlochbervie Highld 306 D7
Kinlocheil Highld 289 B11
Kinlochewe Highld 299 C10
Kinlochleven Highld 290 G3
Kinlochmoidart Highld 289 B9
Kinlochmorar Highld 295 F10
Kinlochmore Highld 290 G3
Kinlochspelve Argyll 289 G8
Kinloid Highld 295 G8
Kinloss Moray 301 C10
Kinmel Bay = Bae Cinmel Conwy 181 E7
Kinmuck Aberds 293 B10
Kinmundy Aberds 293 B10
Kinnadie Aberds 303 E9
Kinnaird Perth 286 B3
Kinnaird Perth 286 E6
Kinnaird Castle Angus 287 B10
Kinnauld Highld 309 J7
Kinneff Aberds 293 F10
Kinneil Falk 279 E9
Kinnelhead Dumfries 248 C2
Kinnell Angus 287 B10
Kinnerley Shrops 148 E6
Kinnersley Hereford 96 B6
Kinnersley Worcs 99 C7
Kinnerton Powys 114 E4
Kinnerton Green Flint 166 C4
Kinnesswood Perth 286 G5
Kinninvie Durham 233 G7
Kinnordy Angus 287 B7
Kinoulton Notts 154 C3
Kinross Perth 286 G5
Kinrossie Perth 286 D5
Kinsbourne Green Herts 85 B10
Kinsey Heath E Ches 167 G11
Kinsham Hereford 115 E7
Kinsham Worcs 99 D8
Kinsley W Yorks 198 E2
Kinson Bmouth 19 B7
Kintallan Argyll 275 E8
Kintbury W Berks 63 F11
Kintessack Moray 301 C9
Kintillo Perth 286 F5
Kintocher Aberds 293 C7
Kinton Hereford 115 C8
Kinton Shrops 149 F7
Kintore Aberds 293 B9
Kintour Argyll 254 B5
Kintra Argyll 254 C4
Kintra Argyll 288 G5
Kintradwell Highld 311 J3
Kintraw Argyll 275 C9
Kinuachdrachd Argyll 275 D8
Kinveachy Highld 291 B11
Kinver Staffs 132 G6
Kinwalsey Warks 134 F5
Kip Hill Durham 242 G6
Kiplin N Yorks 224 F5
Kippax W Yorks 206 G4
Kippen Stirl 278 C3
Kippford or Scaur Dumfries 237 D10

Kippilaw Borders 262 D2
Kippilaw Mains Borders 262 D2
Kipping's Cross Kent 52 F6
Kippington Kent 52 C4
Kirbister Orkney 314 E2
Kirbister Orkney 314 D6
Kirbister Orkney 314 F3
Kirbuster Orkney 314 D2
Kirby Bedon Norf 142 B5
Kirby Bellars Leics 154 F4
Kirby Cane Norf 143 E7
Kirby Corner W Mid 118 B5
Kirby Cross Essex 108 G4
Kirby Fields Leics 135 C10
Kirby Green Norf 143 E7
Kirby Grindalythe N Yorks 217 F8
Kirby Hill N Yorks 215 F7
Kirby Hill N Yorks 224 D2
Kirby Knowle N Yorks 215 B9
Kirby-le-Soken Essex 108 G4
Kirby Misperton N Yorks 216 D5
Kirby Moor Cumb 240 E2
Kirby Muxloe Leics 135 C10
Kirby Row Norf 143 E7
Kirby Sigston N Yorks 225 G8
Kirby Underdale E Yorks 208 B2
Kirby Wiske N Yorks 215 C7
Kirdford W Sus 35 B8
Kirk Highld 310 D6
Kirk Bramwith S Yorks 198 E6
Kirk Deighton N Yorks 206 C3
Kirk Ella E Yorks 200 B4
Kirk Hallam Derbys 171 G7
Kirk Hammerton N Yorks 206 B5
Kirk Ireton Derbys 170 E3
Kirk Langley Derbys 152 B5
Kirk Merrington Durham 233 E11
Kirk Michael I o M 192 C4
Kirk of Shotts N Lnrk 268 C6
Kirk Sandall S Yorks 198 F6
Kirk Smeaton N Yorks 198 D4
Kirk Yetholm Borders 263 D8
Kirkabister Shetland 312 G6
Kirkabister Shetland 313 K6
Kirkandrews Dumfries 237 E8
Kirkandrews-on-Eden Cumb 239 F9
Kirkapol Argyll 288 E2
Kirkbampton Cumb 239 F8
Kirkbean Dumfries 237 D11
Kirkborough Cumb 229 D7
Kirkbrae Orkney 314 B4
Kirkbride Cumb 238 F6
Kirkbridge N Yorks 224 G5
Kirkbuddo Angus 287 C9
Kirkburn Borders 261 B7
Kirkburn E Yorks 208 B5
Kirkburton W Yorks 197 E7
Kirkby Lincs 189 C9
Kirkby Mers 182 B6
Kirkby N Yorks 225 D10
Kirkby Fenside Lincs 174 C4
Kirkby Fleetham N Yorks 224 G5
Kirkby Green Lincs 173 D9
Kirkby Hill N Yorks 215 F7
Kirkby in Ashfield Notts 171 D8
Kirkby-in-Furness Cumb 210 C4
Kirkby la Thorpe Lincs 173 F10
Kirkby Lonsdale Cumb 212 D2
Kirkby Malham N Yorks 213 G7
Kirkby Mallory Leics 135 C9
Kirkby Malzeard N Yorks 214 E4
Kirkby Mills N Yorks 216 B4
Kirkby on Bain Lincs 174 C2
Kirkby Overblow N Yorks 206 D2
Kirkby Stephen Cumb 222 D5
Kirkby Thore Cumb 231 F8
Kirkby Underwood Lincs 155 D11
Kirkby Wharfe N Yorks 206 E6
Kirkby Woodhouse Notts 171 E7
Kirkbymoorside N Yorks 216 B3
Kirkcaldy Fife 280 C5
Kirkcambeck Cumb 240 D2
Kirkcarswell Dumfries 237 E9
Kirkcolm Dumfries 236 C2
Kirkconnel Dumfries 258 G6
Kirkconnell Dumfries 237 C11
Kirkcowan Dumfries 236 C5
Kirkcudbright Dumfries 237 D8
Kirkdale Mers 182 C4
Kirkfieldbank S Lnrk 269 G7
Kirkforthar Feus Fife 286 G6
Kirkgunzeon Dumfries 237 C10
Kirkham Lancs 202 G4
Kirkham N Yorks 216 F4
Kirkhamgate W Yorks 197 C9
Kirkhams Gtr Man 195 F10
Kirkharle Northumb 252 G2
Kirkheaton Northumb 242 B2
Kirkheaton W Yorks 197 D7
Kirkhill Angus 293 G8
Kirkhill E Renf 267 D11
Kirkhill Highld 300 E5
Kirkhill Midloth 270 C4
Kirkhill Moray 302 F2
Kirkhill W Loth 279 G11
Kirkholt Gtr Man 195 E11
Kirkhope Borders 261 E9
Kirkhouse Borders 261 C8
Kirkhouse Cumb 240 F3
Kirkiboll Highld 308 D5
Kirkibost Highld 295 D7
Kirkinch Angus 287 C7
Kirkinner Dumfries 236 D6
Kirkintilloch E Dunb 278 G3
Kirkland Cumb 219 B11
Kirkland Cumb 229 B11
Kirkland Cumb 231 E8
Kirkland Dumfries 247 E8
Kirkland Dumfries 258 G6
Kirkland S Ayrs 244 E6
Kirkland Guards Cumb 229 C9
Kirkleatham Redcar 235 G7
Kirklees Gtr Man 195 E9
Kirklevington Stockton 225 D8
Kirkley Suff 143 E10
Kirklington N Yorks 214 C6
Kirklington Notts 171 D11
Kirklinton Cumb 239 D10
Kirkliston Edin 280 G2
Kirkmaiden Dumfries 236 F3
Kirkmichael Perth 286 B4
Kirkmichael S Ayrs 245 B8
Kirkmichael Mains Dumfries 248 F2
Kirkmuirhill S Lnrk 268 G5
Kirknewton Northumb 263 C10
Kirknewton W Loth 270 B2
Kirkney Aberds 302 F5
Kirkoswald Cumb 231 C7
Kirkoswald S Ayrs 244 B6
Kirkpatrick Dumfries 247 E10
Kirkpatrick Durham Dumfries 237 B9
Kirkpatrick-Fleming Dumfries 239 C7
Kirksanton Cumb 210 C2
Kirkshaw N Lnrk 268 C4
Kirkstall W Yorks 205 F11
Kirkstead Borders 261 E7
Kirkstead Lincs 173 C11
Kirkstile Aberds 302 F5
Kirkstyle Highld 310 B7
Kirkthorpe W Yorks 197 C11
Kirkton Aberds 302 E6
Kirkton Aberds 302 G6
Kirkton Angus 286 C6
Kirkton Angus 287 C8
Kirkton Angus 287 D8
Kirkton Argyll 275 C8
Kirkton Borders 262 G2
Kirkton Dumfries 247 G11
Kirkton Fife 280 D4
Kirkton Fife 287 E7
Kirkton Highld 295 C10
Kirkton Highld 299 E9
Kirkton Highld 301 D7
Kirkton Highld 309 K7
Kirkton Perth 286 F3
Kirkton S Lnrk 259 E10
Kirkton Stirl 285 G9
Kirkton W Loth 269 B10
Kirkton Manor Borders 260 B6
Kirkton of Airlie Angus 287 B7
Kirkton of Auchterhouse Angus 287 D7
Kirkton of Auchterless Aberds 303 E7
Kirkton of Barevan Highld 301 E8
Kirkton of Bourtie Aberds 303 G8
Kirkton of Collace Perth 286 D5
Kirkton of Craig Angus 287 B11
Kirkton of Culsalmond Aberds 302 F6
Kirkton of Durris Aberds 293 D9
Kirkton of Glenbuchat Aberds 292 B5
Kirkton of Glenisla Angus 292 G4
Kirkton of Kingoldrum Angus 287 B7
Kirkton of Largo Fife 287 G8
Kirkton of Lethendy Perth 286 C5
Kirkton of Logie Buchan Aberds 303 G9
Kirkton of Maryculter Aberds 293 D10
Kirkton of Menmuir Angus 293 G7
Kirkton of Monikie Angus 287 D9
Kirkton of Oyne Aberds 302 G6
Kirkton of Rayne Aberds 302 G6
Kirkton of Skene Aberds 293 C10
Kirkton of Tough Aberds 293 B8
Kirktonhill Borders 271 E9
Kirktonhill W Dunb 277 G7
Kirktoun E Ayrs 267 G8
Kirktown Aberds 303 D10
Kirktown of Alvah Aberds 302 C6
Kirktown of Deskford Moray 302 C5
Kirktown of Fetteresso Aberds 293 E10
Kirktown of Mortlach Moray 302 F3
Kirktown of Slains Aberds 303 G10
Kirkurd Borders 270 G2
Kirkwall Orkney 314 E4
Kirkwhelpington Northumb 251 G11
Kirkwood Dumfries 238 B4
Kirkwood N Lnrk 268 C4
Kirmington N Lincs 200 E6
Kirmond le Mire Lincs 189 C11
Kirn Argyll 276 F3
Kirriemuir Angus 287 B7
Kirstead Green Norf 142 D5
Kirtlebridge Dumfries 238 C6
Kirtleton Dumfries 249 G7
Kirtling Cambs 124 F3
Kirtling Green Cambs 124 F3
Kirtlington Oxon 83 B7
Kirtomy Highld 308 C7
Kirton Lincs 156 B6
Kirton Notts 171 B11
Kirton Suff 108 D5
Kirton Campus W Loth 269 B10
Kirton End Lincs 174 G3
Kirton Holme Lincs 174 G3
Kirton in Lindsey N Lincs 188 B6
Kiskin Cumb 210 B1
Kislingbury Northants 120 F3
Kit Hill Dorset 30 D4
Kitbridge Devon 28 G4
Kitchenroyd W Yorks 197 F8
Kite Green Warks 118 D3
Kite Hill I o W 21 C7
Kitebrook Warks 100 E4
Kites Hardwick Warks 119 D9
Kitlye Glos 80 E5
Kit's Coty Kent 69 G8
Kitt Green Gtr Man 194 F5
Kittisford Som 27 C9
Kittle Swansea 56 D5
Kitts End Herts 86 F2
Kitt's Green W Mid 134 F3
Kitt's Moss Gtr Man 184 E5
Kittwhistle Dorset 28 G5
Kittybrewster Aberdeen 293 C11
Kitwell W Mid 133 G9
Kitwood Hants 49 G7
Kivernoll Hereford 97 E9
Kiveton Park S Yorks 187 E7
Knaith Lincs 188 E4
Knaith Park Lincs 188 D4
Knap Corner Dorset 30 C4
Knaphill Sur 50 B3
Knapp Hants 32 C6
Knapp Perth 286 D6
Knapp Som 28 B4
Knapp Wilts 31 B8
Knapp Hill Wilts 30 B5
Knapthorpe Notts 172 D2
Knaptoft Leics 136 F2
Knapton Norf 160 C6
Knapton York 207 C7
Knapton Green Hereford 115 G8
Knapwell Cambs 122 E6
Knaresborough N Yorks 206 B3
Knarsdale Northumb 240 G5
Knatts Valley Kent 68 G5
Knauchland Moray 302 D5
Knaven Aberds 303 E8
Knave's Ash Kent 71 G7
Knaves Green Suff 126 D2
Knavesmire York 207 D7
Knayton N Yorks 215 B8
Knebworth Herts 104 G5
Knedlington E Yorks 199 B8
Kneesall Notts 172 C2
Kneesworth Cambs 104 C6
Kneeton Notts 172 F2
Knelston Swansea 56 D3
Knenhall Staffs 151 B8
Knettishall Suff 141 G9
Knightacott Devon 41 F7
Knightcote Warks 119 G7
Knightcott N Som 43 B11
Knightley Staffs 150 D6
Knightley Dale Staffs 150 E6
Knighton Devon 7 F10
Knighton Dorset 29 E10
Knighton Leicester 135 C11
Knighton Oxon 63 B9
Knighton Poole 18 B6
Knighton Som 43 E7
Knighton Staffs 150 D4
Knighton Staffs 168 G2
Knighton Wilts 63 E9
Knighton Worcs 117 F10
Knighton = Tref-y-Clawdd Powys 114 C5
Knighton Fields Leicester 135 C11
Knighton on Teme Worcs 116 C2
Knightor Corn 5 D10
Knight's End Cambs 139 E8
Knights Enham Hants 47 D11
Knight's Hill London 67 E10
Knightsbridge Glos 99 F7
Knightsbridge London 67 D9
Knightsmill Corn 11 E7
Knightsridge W Loth 269 B10
Knightswood Glasgow 267 B10
Knightwick Worcs 116 F4
Knill Hereford 114 E5
Knipe Fold Cumb 220 F5
Knipoch Argyll 289 G10
Knipton Leics 154 C6
Knitsley Durham 233 B8
Kniveton Derbys 170 E2
Knocharthur Highld 309 J7
Knock Argyll 289 F7
Knock Cumb 231 F9
Knock Moray 302 D5
Knockally Highld 311 G5
Knockan Highld 307 H7
Knockandhu Moray 302 G2
Knockando Moray 301 E11
Knockando Ho Moray 302 E2
Knockandoo Highld 301 G7
Knockbain Highld 300 D6
Knockbreck Highld 298 C2
Knockbrex Dumfries 237 E7
Knockcarrach Highld 290 B6
Knockdee Highld 310 C5
Knockdolian S Ayrs 244 F4
Knockdow Argyll 276 G2
Knockdown Glos 61 B10
Knockenbaird Aberds 302 G6
Knockenkelly N Ayrs 256 D2
Knockentiber E Ayrs 257 B9
Knockerdown Derbys 170 E2
Knockespock Ho Aberds 302 G5
Knockfarrel Highld 300 D5
Knockglass Dumfries 236 D2
Knockhall Kent 68 E5
Knockhall Castle Aberds 303 G9
Knockholt Kent 52 B3
Knockholt Pound Kent 52 B3
Knockie Lodge Highld 290 B6
Knockin Shrops 148 E6
Knockin Heath Shrops 149 E7
Knockinlaw E Ayrs 257 B10
Knockinnon Highld 310 F5
Knocklaw Northumb 252 C3
Knocklearn Dumfries 237 B9
Knocklearoch Argyll 274 G4
Knockmill Kent 68 G5
Knocknaha Argyll 255 F7
Knocknain Dumfries 236 C1
Knockothie Aberds 303 F9
Knockrome Argyll 274 F6
Knocksharry I o M 192 D3
Knockstapplemore Argyll 255 F7
Knockvologan Argyll 274 B4
Knodishall Suff 127 E8
Knokan Argyll 288 G6
Knole Som 29 B7
Knoll Green Som 43 F8
Knoll Top N Yorks 214 F3
Knollbury Mon 60 B2
Knolls Green E Ches 184 F4
Knolton Wrex 149 B7
Knolton Bryn Wrex 149 B7
Knook Wilts 46 E2
Knossington Leics 136 B6
Knotbury Staffs 169 B8
Knott End-on-Sea Lancs 202 D3
Knotting Beds 121 E10
Knotting Green Beds 121 E10
Knottingley W Yorks 198 C4
Knotts Cumb 230 G4
Knotts Lancs 203 C11
Knotty Ash Mers 182 C6
Knotty Corner Devon 24 B6
Knotty Green Bucks 84 G6
Knowbury Shrops 115 C11
Knowe Dumfries 236 B5
Knowe Shetland 313 G5
Knowefield Cumb 239 F10
Knowehead Aberds 293 C7
Knowehead Aberds 302 D5
Knowehead Dumfries 246 E4
Knowes E Loth 282 F3
Knowes of Elrick Aberds 302 D6
Knowesgate Northumb 251 F11
Knoweton N Lnrk 268 D5
Knowetop N Lnrk 268 D5
Knowhead Aberds 303 D9
Knowl Bank Staffs 168 F3
Knowl Green Essex 106 C5
Knowl Hill Windsor 65 D10
Knowl Wall Staffs 151 B7
Knowl Wood W Yorks 196 C2
Knowle Bristol 60 E6
Knowle Devon 15 E7
Knowle Devon 26 G3
Knowle Devon 27 F8
Knowle Devon 40 F3
Knowle Hants 33 F9
Knowle Shrops 115 C11
Knowle Som 43 F10
Knowle W Mid 118 B3
Knowle Wilts 63 G7
Knowle Fields Worcs 117 F10
Knowle Green Lancs 203 F8
Knowle Green Sur 65 E4
Knowle Grove W Mid 118 B3
Knowle Hill Sur 66 F3
Knowle Park W Yorks 205 E7
Knowle St Giles Som 28 E4
Knowlegate Shrops 115 C11
Knowles Hill Devon 14 G3
Knowlesands Shrops 132 E4
Knowlton Dorset 31 E8
Knowlton Kent 55 C9
Knowsley Mers 182 B6
Knowsthorpe W Yorks 206 G2
Knowstone Devon 26 C4
Knox Bridge Kent 53 E9
Knucklas Powys 114 C5
Knuston Northants 121 D8
Knutsford E Ches 184 F3
Knutton Staffs 168 F4
Knuzden Brook Lancs 195 B8
Knypersley Staffs 168 D5
Kraiknish Highld 294 C5
Krumlin W Yorks 196 D5
Kuggar Corn 2 F6
Kyle of Lochalsh Highld 295 C9
Kyleakin Highld 295 C9
Kylepark N Lnrk 268 C3
Kylerhea Highld 295 C9
Kylesknoydart Highld 295 F10
Kylesku Highld 306 F7
Kylesmorar Highld 295 F10
Kylestrome Highld 306 F7
Kyllachy House Highld 301 G7
Kymin Hereford 97 B11
Kymin Mon 79 C8
Kynaston Hereford 97 E10
Kynaston Shrops 149 E7
Kynnersley Telford 150 F3
Kyre Worcs 116 E2
Kyre Green Worcs 116 E2
Kyre Magna Worcs 116 E2
Kyre Park Worcs 116 E2
Kyrewood Worcs 116 D2

L

Labost W Isles 304 D4
Lacasaidh W Isles 304 F5
Lacasdal W Isles 304 E6
Laceby NE Lincs 201 F8
Laceby Acres NE Lincs 201 F8
Lacey Green Bucks 84 F4
Lacey Green E Ches 184 E4
Lach Dennis W Ches 184 G2
Lache W Ches 166 C5
Lackenby Redcar 225 B11
Lackford Suff 124 C5
Lacock Wilts 62 F2
Ladbroke Warks 119 F8
Laddenvean Corn 3 E7
Laddingford Kent 53 D7
Lade Kent 39 C9
Lade Bank Lincs 174 E5
Ladies Riggs N Yorks 214 F2
Ladmanlow Derbys 185 G8
Ladock Corn 5 E7
Ladwell Hants 32 C6
Lady Orkney 314 B6
Lady Green Mers 193 G10
Lady Hall Cumb 210 B3
Lady Halton Shrops 115 B9
Lady House Gtr Man 196 E2
Lady Park T & W 242 F6
Lady Wood W Yorks 206 F2
Ladybank Fife 287 F7
Ladybrook Notts 171 C8
Ladyburn Invclyd 276 F6
Ladycross Corn 12 D2
Ladyes Hill Warks 118 C5
Ladykirk Borders 273 F7
Ladyoak Shrops 131 C7
Ladyridge Hereford 97 E11
Lady's Green Suff 124 F5
Ladysford Aberds 303 C9
Ladywell London 67 E11
Ladywell Shrops 149 C9
Ladywell W Loth 269 B10
Ladywood Telford 132 C3
Ladywood W Mid 133 F11
Ladywood Worcs 117 E7
Laffak Mers 183 B8
Laga Highld 289 C8
Lagafater Lodge Dumfries 236 B3
Lagalochan Argyll 275 B9
Lagavulin Argyll 254 C5
Lagg Argyll 274 F6
Lagg N Ayrs 255 E10
Laggan Argyll 254 B3
Laggan Highld 289 B9
Laggan Highld 290 D4
Laggan Highld 291 D8
Laggan S Ayrs 245 G7
Laggan Lodge Argyll 289 G8
Lagganlia Highld 291 C10
Lagganmullan Dumfries 237 D7
Lagganulva Argyll 288 E6
Lagness W Sus 22 C5
Laide Highld 307 K3
Laig Highld 294 G6
Laigh Carnduff S Lnrk 268 F3
Laigh Fenwick E Ayrs 267 G9
Laigh Glengall S Ayrs 257 F8
Laighmuir E Ayrs 267 F9
Laighstonehall S Lnrk 268 E4
Laindon Essex 69 B7
Lair Highld 299 E10
Lair Perth 292 G3
Laira Plym 7 D10
Lairg Highld 309 J5
Lairg Lodge Highld 309 J5
Lairg Muir Highld 309 J5
Lairgmore Highld 300 F5
Laisterdyke W Yorks 205 G9
Laithes Cumb 230 E5
Laithkirk Durham 232 G5
Laity Moor Corn 4 F3
Lake Devon 12 D6
Lake Devon 24 F6
Lake Devon 40 G5
Lake I o W 21 E7
Lake Poole 18 C5
Lake Wilts 46 F6
Lake End Bucks 66 D2
Lakenham Norf 142 B4
Lakenheath Suff 140 G4
Laker's Green Sur 50 F4
Lakesend Norf 139 D10
Lakeside Cumb 211 B7
Lakeside Thurrock 68 D5
Lakeside Worcs 117 D11
Laleham Sur 66 F5
Laleston = Trelales Bridgend 57 F11
Lamanva Corn 3 C7
Lamarsh Essex 107 D7
Lamas Norf 160 E4
Lamb Corner Essex 107 E10
Lambden Borders 272 G4
Lamberden Kent 38 B4
Lamberhead Green Gtr Man 194 G4
Lamberhurst Kent 53 F7
Lamberhurst Quarter Kent 53 F7
Lamberton Borders 273 D9
Lambert's End W Mid 133 E9
Lambeth London 67 D10
Lambfair Green Suff 124 G4
Lambfoot Cumb 229 E9
Lambhill Glasgow 267 B11
Lambley Northumb 240 F5
Lambley Notts 171 F10
Lambourn W Berks 63 D10
Lambourn Woodlands W Berks 63 D10
Lambourne Corn 4 E5
Lambourne Essex 87 F7
Lambourne End Essex 87 G7
Lambridge Bath 61 F9
Lamb's Cross Kent 53 D9
Lambs' Green Dorset 18 B5
Lambs Green W Sus 51 F8
Lambston Pembs 72 B6
Lambton T & W 243 G7
Lamellion Corn 6 C4
Lamerton Devon 12 F5
Lamesley T & W 243 F7
Laminess Orkney 314 C6
Lamington Highld 301 B7
Lamington S Lnrk 259 C11
Lamlash N Ayrs 256 C2
Lamledra Corn 5 G10
Lamloch Dumfries 246 D2
Lammack Blkburn 195 B7
Lamonby Cumb 230 D4
Lamorick Corn 5 C10
Lamorna Corn 1 E4
Lamorran Corn 5 G7
Lampardbrook Suff 126 E5
Lampeter = Llanbedr Pont Steffan Ceredig 93 B11
Lampeter Velfrey Pembs 73 C11
Lamphey Pembs 73 E8
Lamplugh Cumb 229 G7
Lamport Northants 120 C5
Lampton London 66 D6
Lamyatt Som 45 F7
Lana Devon 12 B2
Lana Devon 24 F4
Lanark S Lnrk 269 G7
Lancaster Lancs 211 G9
Lanchester Durham 233 B9
Lancing W Sus 35 G11
Land Gate Gtr Man 194 G5
Landbeach Cambs 123 D9
Landcross Devon 25 C7
Landerberry Aberds 293 C9
Landewednack Corn 2 G6
Landford Wilts 32 D3
Landford Manor Wilts 32 C3
Landfordwood Wilts 32 C3
Landguard Manor I o W 21 E7
Landhill Devon 12 B4
Landican Mers 182 D3
Landimore Swansea 56 C3
Landkey Devon 40 G5
Landkey Newland Devon 40 G5
Landore Swansea 57 B7
Landport E Sus 36 E6
Landport Ptsmth 33 G10
Landrake Corn 7 C7
Landscove Devon 8 B5
Landshipping Pembs 73 C8
Landshipping Quay Pembs 73 C8
Landslow Green Gtr Man 185 B7
Landulph Corn 7 C8
Landwade Suff 124 D2
Landywood Staffs 133 B9
Lane Corn 4 C6
Lane Bottom Lancs 204 F3
Lane Bottom W Yorks 205 F7
Lane End Bucks 84 G4
Lane-end Corn 5 B10
Lane End Cumb 220 G2
Lane End Derbys 170 C6
Lane End Devon 24 G6
Lane End Dorset 18 C3
Lane End Flint 166 C3
Lane End Glos 79 B10
Lane End Hants 33 B11
Lane End Hants 33 B9
Lane End I o W 21 D9
Lane End Kent 68 E5
Lane End Lancs 204 D3
Lane End S Yorks 186 B5
Lane End Sur 49 E10
Lane End W Yorks 204 F6
Lane End Wilts 45 D10
Lane Ends Corn 210 C6
Lane Ends Derbys 152 C4
Lane Ends E Ches 168 D2
Lane Ends E Ches 185 E7
Lane Ends Gtr Man 185 C7
Lane Ends Lancs 194 D6
Lane Ends Lancs 203 C11
Lane Ends Lancs 203 G11
Lane Ends N Yorks 204 E5
Lane Ends Stoke 168 E5
Lane Ends W Yorks 205 F10
Lane Green Staffs 133 C7
Lane Head Derbys 185 F11
Lane Head Durham 224 C2
Lane Head Gtr Man 183 B10
Lane Head Pembs 73 E8
Lane Head W Mid 133 C9
Lane Head W Yorks 197 F7
Lane Heads Lancs 202 F4
Lane Side Lancs 195 C9
Laneast Corn 11 E10
Laneham Notts 188 F4
Lanehead Durham 232 C2
Lanehead Northumb 251 F7
Lanehouse Dorset 17 F9
Lanercost Cumb 240 E3
Lanes End Bucks 84 D6
Lane's End Shrops 132 G2
Lanescot Corn 5 D11
Lanesend Pembs 73 D9
Lanesfield W Mid 133 D8
Laneshaw Bridge Lancs 204 E4
Laney Green Staffs 133 B9
Lanfach Caerph 78 F2
Langaford Devon 12 B4
Langage Devon 7 E11
Langal Highld 289 C9
Langaller Som 28 B3
Langar Notts 154 C4
Langbank Renfs 277 G7
Langbar N Yorks 205 C7
Langbaurgh N Yorks 225 C11
Langburnshiels Borders 250 C2
Langcliffe N Yorks 212 G6
Langdale Highld 308 E6
Langdale End N Yorks 227 G8
Langdon Corn 12 D2
Langdon Beck Durham 232 E3
Langdon Hills Essex 69 B7
Langdown Hants 32 F6
Langdyke Dumfries 238 C3
Langdyke Fife 287 G7
Langenhoe Essex 89 B8
Langford C Beds 104 C3
Langford Devon 14 B4
Langford Devon 27 G8
Langford Essex 88 D4
Langford Notts 172 D4
Langford Oxon 82 E2
Langford Som 28 B2
Langford Budville Som 27 C10
Langford Green Devon 27 G8
Langford Green N Som 44 B3
Langham Dorset 30 B3
Langham Essex 107 E10
Langham Norf 177 E8
Langham Rutland 154 G6
Langham Som 28 E4
Langham Suff 125 C9
Langhaugh Borders 260 C6
Langho Lancs 203 G10
Langholm Dumfries 249 G9
Langholme N Lincs 188 B3
Langhope Borders 261 E10
Langland Swansea 56 D6
Langlee Borders 262 B2
Langlee Mains Borders 262 B2
Langleeford Northumb 263 E10
Langlees Falk 279 E7
Langley Derbys 170 F6
Langley E Ches 184 G6
Langley Essex 105 D8
Langley Glos 99 F10
Langley Gtr Man 195 F11
Langley Hants 32 G6
Langley Herts 104 G4
Langley Kent 53 C10
Langley Northumb 241 E8
Langley Oxon 82 B4
Langley Slough 66 D4
Langley Som 27 B9
Langley W Mid 133 F9
Langley W Sus 34 B4
Langley Warks 118 E3
Langley Burrell Wilts 62 D2
Langley Common Derbys 152 B5
Langley Common Wokingham 65 F9
Langley Corner Bucks 66 C4
Langley Green Derbys 152 B5
Langley Green Essex 107 G7
Langley Green W Mid 133 F9
Langley Green W Sus 51 F9
Langley Green Warks 118 E3
Langley Heath Kent 53 C10
Langley Marsh Som 27 B9
Langley Mill Derbys 170 F6
Langley Moor Durham 233 C11
Langley Park Durham 233 C10
Langley Street Norf 143 C7
Langley Vale Sur 51 B8
Langleybury Herts 85 E9
Langloan N Lnrk 268 C4
Langney E Sus 23 E10
Langold Notts 187 D9
Langore Corn 12 D2
Langport Som 28 B6
Langrick Lincs 174 F3
Langrick Bridge Lincs 174 F3
Langridge Bath 61 F8
Langridge Ford Devon 25 C9
Langrigg Cumb 222 C5
Langrigg Cumb 229 B9
Langrish Hants 34 C2
Langsett S Yorks 197 G8
Langshaw Borders 262 B2
Langside Glasgow 267 C11
Langside Perth 285 F11
Langskaill Orkney 314 B4
Langstone Devon 13 E10
Langstone Hants 22 C2
Langstone Newport 78 G5
Langthorne N Yorks 224 G5
Langthorpe N Yorks 215 F7
Langthwaite N Yorks 223 E10
Langtoft E Yorks 217 F10
Langtoft Lincs 156 G2
Langton Durham 224 B3
Langton Lincs 174 B2
Langton Lincs 190 G5
Langton N Yorks 216 F5
Langton by Wragby Lincs 189 F11
Langton Green Kent 52 F4
Langton Green Suff 126 C2
Langton Herring Dorset 17 E8
Langton Long Blandford Dorset 30 F5
Langton Matravers Dorset 18 F6
Langtree Devon 25 D7
Langtree Week Devon 25 D7
Langwathby Cumb 231 E7
Langwell Ho Highld 311 G5
Langwell Lodge Highld 307 J6
Langwith Derbys 171 B8
Langwith Junction Derbys 171 B8
Langworth Lincs 189 F9
Lanham Green Essex 106 G5
Lanivet Corn 5 C10
Lanjeth Corn 5 E9
Lanjew Corn 5 C9
Lank Corn 11 F7
Lanlivery Corn 5 D11
Lanner Corn 2 B6
Lanreath Corn 6 D3
Lanrick Stirl 285 G10
Lansallos Corn 6 E3
Lansbury Park Caerph 59 B7
Lansdown Bath 61 F8
Lansdown Glos 99 G8
Lanstephan Corn 12 D2
Lanteglos Corn 11 E7
Lanteglos Highway Corn 6 E3
Lanton Borders 262 E4
Lanton Northumb 263 C10
Lantuel Corn 5 B9
Lantyan Corn 6 D2
Lapal W Mid 133 G9
Lapford Devon 26 F2
Lapford Cross Devon 26 F2
Laphroaig Argyll 254 C4
Lapley Staffs 151 G7
Lapworth Warks 118 C3
Larachbeg Highld 289 E8
Larbert Falk 279 E7
Larbreck Lancs 202 E4
Larches Lancs 202 G6
Larden Green E Ches 167 E9
Larg Highld 292 B2
Largie Aberds 302 F6
Largiebaan Argyll 255 F7
Largiemore Argyll 275 E10
Largoward Fife 287 G8
Largs N Ayrs 266 D4
Largue Aberds 302 E6
Largybeg N Ayrs 256 E3
Largymeanoch N Ayrs 256 E2
Largymore N Ayrs 256 E2
Lark Hill Gtr Man 195 G7
Larkfield Invclyd 276 F4
Larkfield Kent 53 B8
Larkfield W Yorks 205 F10
Larkhall Bath 61 F9
Larkhall S Lnrk 268 E5
Larkhill Wilts 46 E6
Larklands Derbys 171 G7
Larks' Hill Suff 108 B3
Larling Norf 141 F9
Larport Hereford 97 D11
Larrick Corn 12 F2
Larriston Borders 250 E2
Lartington Durham 223 B10
Lary Aberds 292 C5
Lasborough Glos 80 G4
Lasham Hants 49 E7
Lashenden Kent 53 E10
Lask Edge Staffs 168 D6
Lassington Glos 98 G5
Lassodie Fife 280 C2
Lastingham N Yorks 226 G4
Latcham Som 44 D2
Latchbrook Corn 7 D8
Latchford Herts 105 G7
Latchford Oxon 83 E11
Latchford Warr 183 D10
Latchingdon Essex 88 E5
Latchley Corn 12 G4
Latchmere Green Hants 64 G6
Latchmore Bank Essex 87 B7
Lately Common Warr 183 B11
Lathallan Mill Fife 287 G8
Lathbury M Keynes 103 B7
Latheron Highld 310 F5
Latheronwheel Highld 310 F5
Latheronwheel Ho Highld 310 F5
Lathom Lancs 194 E3
Lathones Fife 287 G8
Latimer Bucks 85 F8
Latteridge S Glos 61 C7
Lattiford Som 29 B11
Lattinford Hill Suff 107 D11
Latton Wilts 81 F9
Latton Bush Essex 87 D7
Lauchintilly Aberds 293 B9
Laudale Ho Highld 289 D9
Lauder Borders 271 F10
Lauder Barns Borders 271 F10
Laugharne = Talacharn Carms 74 C4
Laughern Hill Worcs 116 F5
Laughterton Lincs 188 F4
Laughton E Sus 23 C8
Laughton Leics 136 F3
Laughton Lincs 155 C11
Laughton Lincs 188 B4
Laughton Common E Sus 23 C7
Laughton Common S Yorks 187 D8
Laughton en le Morthen S Yorks 187 D8
Launcells Corn 24 F2
Launcells Cross Corn 24 F3
Launceston Corn 12 D2
Launcherley Som 44 E4
Laund Lancs 195 C10
Launton Oxon 102 G2
Laurencekirk Aberds 293 F9
Laurieston Dumfries 237 C8
Laurieston Falk 279 F8
Lavendon M Keynes 121 G8
Lavenham Suff 107 B8
Laverackloch Moray 301 C11
Laverhay Dumfries 248 D4
Laverlaw Borders 261 B7
Laverley Som 44 F5
Lavernock V Glam 59 F7
Laversdale Cumb 239 E11
Laverstock Wilts 47 G7
Laverstoke Hants 48 D3
Laverton Glos 99 D11
Laverton N Yorks 214 E4
Laverton Som 45 C9
Lavington Sands Wilts 46 B4
Lavister Wrex 166 D5
Law S Lnrk 268 E6
Law Hill S Lnrk 268 E6
Lawers Perth 285 D10
Lawers Perth 285 E11
Lawford Essex 107 E11
Lawford Som 42 F6
Lawford Heath Warks 119 C9
Lawhill Perth 286 F3
Lawhitton Corn 12 E3
Lawkland N Yorks 212 F5
Lawkland Green N Yorks 212 F5
Lawley Telford 132 B3
Lawn Swindon 63 C7
Lawnhead Staffs 150 E6
Lawns W Yorks 197 C10
Lawnswood W Yorks 205 F11
Lawnt Denb 165 B8
Lawrence Hill Newport 59 B10
Lawrence Weston Bristol 60 D4
Lawrenny Pembs 73 D8
Lawrenny Quay Pembs 73 D8
Lawshall Suff 125 G7
Lawshall Green Suff 125 G7
Lawton Hereford 115 F8
Lawton Shrops 131 G10
Lawton-gate E Ches 168 D4
Lawton Heath End E Ches 168 D3
Laxey I o M 192 D5
Laxfield Suff 126 C5
Laxfirth Shetland 313 H6
Laxfirth Shetland 313 J6
Laxford Bridge Highld 306 E7
Laxo Shetland 313 G6
Laxobigging Shetland 312 F6
Laxton E Yorks 199 B9
Laxton Northants 137 D8
Laxton Notts 172 B2
Laycock W Yorks 204 E6
Layer Breton Essex 88 B6
Layer de la Haye Essex 89 B7
Layer Marney Essex 88 B6
Layerthorpe York 207 C8
Layham Suff 107 C10
Laymore Dorset 28 G5
Layters Green Bucks 85 G7
Laytham E Yorks 207 F10
Layton Blkpool 202 F2
Lazenby Redcar 225 B11
Lazonby Cumb 230 D6
Le Skerne Haughton Darl 224 B6
Lea Derbys 170 D4
Lea Hereford 98 G3
Lea Lancs 202 G5
Lea Lincs 188 D4
Lea Shrops 131 B8
Lea Shrops 131 F7
Lea Wilts 62 B3
Lea Bridge London 67 B11
Lea by Backford W Ches 182 G5
Lea End Worcs 117 B10
Lea Forge E Ches 168 F2
Lea Green Mers 183 C8
Lea Hall W Mid 134 F2
Lea Heath Staffs 151 D10
Lea Line Hereford 98 G3
Lea Marston Warks 134 E4
Lea Town Lancs 202 G5
Lea Valley Herts 85 B11
Lea Yeat Cumb 212 B5
Leabrooks Derbys 170 E6
Leac a Li W Isles 305 J3
Leacainn W Isles 305 H3
Leachkin Highld 300 E6
Leacnasaide Highld 299 B7
Leadburn Midloth 270 D4
Leaden Roding Essex 87 C9
Leadendale Staffs 151 B8
Leadenham Lincs 173 E7
Leadgate Cumb 231 C10
Leadgate Durham 242 G4
Leadgate T & W 242 F4
Leadhills S Lnrk 259 G9
Leadingcross Green Kent 53 C11
Leadmill Derbys 186 E2
Leadmill Flint 166 C2
Leafield Oxon 82 B4
Leafield Wilts 61 F11
Leagrave Luton 103 G10
Leagreen Hants 19 C11
Leake Lincs 174 F6
Leake N Yorks 225 G8
Leake Commonside Lincs 174 E5
Leake Fold Hill Lincs 174 E6
Lealholm N Yorks 226 D5
Lealholm Side N Yorks 226 D5
Lealt Argyll 275 D7
Lealt Highld 298 C5
Leam Derbys 186 F2
Leamington Hastings Warks 119 D8
Leamoor Common Shrops 131 F8
Leamore W Mid 133 C9
Leamside Durham 234 B2
Leanach Argyll 275 D11
Leanachan Highld 290 F4
Leanaig Highld 300 D5
Leapgate Worcs 116 C6
Leargybreck Argyll 274 F6
Lease Rigg N Yorks 226 E6
Leasey Bridge Herts 85 C11
Leasgill Cumb 211 C9
Leasingham Lincs 173 F9
Leasingthorne Durham 233 F11
Leason Swansea 56 C3
Leasowe Mers 182 C3
Leatherhead Sur 51 B7
Leatherhead Common Sur 51 B7
Leathern Bottle Glos 80 E2
Leathley N Yorks 205 D10
Leaths Dumfries 237 C9
Leaton Shrops 149 F9
Leaton Telford 150 G2
Leaton Heath Shrops 149 F9
Leaveland Kent 54 C4
Leavenheath Suff 107 D9
Leavening N Yorks 216 G5
Leaves Green London 68 G2
Leavesden Green Herts 85 E9
Leazes Durham 242 F5
Lebberston N Yorks 217 C11
Leburnick Corn 12 E3
Lechlade-on-Thames Glos 82 F2
Leck Lancs 212 D2
Leckford Hants 47 F11
Leckfurin Highld 308 D7
Leckgruinart Argyll 274 G3
Leckhampstead Bucks 102 D4
Leckhampstead W Berks 64 D2
Leckhampstead Thicket W Berks 64 D2
Leckhampton Glos 80 B6
Leckie Highld 299 C10
Leckmelm Highld 307 K6
Leckuary Argyll 275 D9
Leckwith V Glam 59 E7
Leconfield E Yorks 208 E6
Ledaig Argyll 289 F11
Ledburn Bucks 103 G8
Ledbury Hereford 98 D4
Ledcharrie Stirl 285 E9
Leddington Glos 98 E3
Ledgemoor Hereford 115 G8
Ledgowan Highld 299 D11
Ledicot Hereford 115 E8
Ledmore Angus 293 G7
Ledmore Highld 307 H7
Lednagullin Highld 308 C7
Ledsham W Ches 182 G5
Ledsham W Yorks 198 B3
Ledston W Yorks 198 B2
Ledston Luck W Yorks 206 G4
Ledstone Devon 8 F4
Ledwell Oxon 101 F8
Lee Argyll 288 G6
Lee Devon 40 D3
Lee Devon 40 D5
Lee Hants 32 D5
Lee Lancs 203 B7
Lee London 67 E11
Lee Northumb 241 F10
Lee Shrops 149 C8
Lee Bank W Mid 133 F11
Lee Brockhurst Shrops 149 D10
Lee Chapel Essex 69 B7
Lee Clump Bucks 84 E6
Lee Common Bucks 84 E6
Lee Gate Bucks 84 D5
Lee Ground Hants 33 F8
Lee Head Derbys 185 C8
Lee Mill Devon 8 D2
Lee Moor Devon 7 C11
Lee Moor W Yorks 197 B10
Lee-on-the-Solent Hants 33 G9
Lee-over-Sands Essex 89 C10
Leeans Shetland 313 J5
Leebotten Shetland 313 L6
Leebotwood Shrops 131 D9
Leece Cumb 210 F4
Leechpool Mon 60 B4
Leeds Kent 53 C10
Leeds W Yorks 205 G11
Leedstown Corn 2 C4
Leeford Devon 41 D9
Leegomery Telford 150 G3
Leeholme Durham 233 E10
Leek Staffs 169 D7
Leek Wootton Warks 118 D5
Leekbrook Staffs 169 E7
Leeming N Yorks 214 B5
Leeming W Yorks 204 G6
Leeming Bar N Yorks 224 G5
Leemings Lancs 203 D10
Lees Derbys 152 B5
Lees Gtr Man 196 G3

Lees W Yorks 204 F6
Leesthorpe Leics 154 G5
Leeswood = Coed-Llai Flint 166 D3
Leetown Perth 286 E6
Leftwich W Ches 183 G11
Legar Powys 78 B2
Legbourne Lincs 190 E5
Legburthwaite Cumb 220 B6
Legerwood Borders 271 G11
Leggatt Hill W Sus 34 C6
Legsby Lincs 189 D10
Leicester Leicester 135 C11
Leicester Forest East Leics 135 C10
Leicester Grange Warks 135 E8
Leigh Devon 26 E2
Leigh Dorset 18 B6
Leigh Dorset 29 F10
Leigh Dorset 30 F3
Leigh Glos 99 F7
Leigh Gtr Man 195 G7
Leigh Kent 52 D4
Leigh Shrops 130 C6
Leigh Sur 51 D8
Leigh Wilts 81 G9
Leigh Worcs 116 G5
Leigh Beck Essex 69 C10
Leigh Common Som 30 B2
Leigh Delamere Wilts 61 D11
Leigh Green Kent 54 G2
Leigh-on-Sea Sthend 69 B10
Leigh Park Hants 22 B2
Leigh Sinton Worcs 116 G5
Leigh upon Mendip Som 45 D7
Leigh Woods N Som 60 E5
Leigham Plym 7 D10
Leighland Chapel Som 42 F4
Leighswood W Mid 133 C11
Leighterton Glos 80 G4
Leighton N Yorks 214 D3
Leighton Shrops 132 B2
Leighton Som 45 E8
Leighton = Tre'r llai Powys 130 B4
Leighton Bromswold Cambs 122 B2
Leighton Buzzard C Beds 103 F8
Leinthall Earls Hereford 115 D8
Leinthall Starkes Hereford 115 D8
Leintwardine Hereford 115 C8
Leire Leics 135 E10
Leirinmore Highld 308 C4
Leiston Suff 127 E8
Leitfie Perth 286 C6
Leith Edin 280 F5
Leithenhall Dumfries 248 D4
Leitholm Borders 272 G5
Lelant Corn 2 B2
Lelant Downs Corn 2 B2
Lelley E Yorks 209 G10
Lem Hill Worcs 116 C4
Lemington T & W 242 E5
Lemmington Hall Northumb 264 G4
Lempitlaw Borders 263 C7
Lemsford Herts 86 C2
Lenacre Cumb 212 B3
Lenborough Bucks 102 E3
Lenchwick Worcs 99 B10
Lendalfoot S Ayrs 244 F4
Lendrick Lodge Stirl 285 G9
Lenham Kent 53 C11
Lenham Forstal Kent 54 C2
Lenham Heath Kent 54 D2
Lennel Borders 273 G7
Lennoxtown E Dunb 278 F2
Lent Bucks 66 C2
Lent Rise Bucks 66 C2
Lenten Pool Denb 165 B8
Lenton Lincs 155 C10
Lenton Nottingham 153 B11
Lenton Abbey Nottingham 153 B10
Lentran Highld 300 E5
Lenwade Norf 159 F11
Leny Ho Stirl 285 G10
Lenzie E Dunb 278 G3
Lenziemill N Lnrk 278 G5
Leoch Angus 287 D7
Leochel-Cushnie Aberds 293 B7
Leominster Hereford 115 F9
Leomonsley Staffs 134 B2
Leonard Stanley Glos 80 E4
Leonardston Pembs 72 D6
Leorin Argyll 254 C4
Lepe Hants 20 B5
Lephin Highld 297 G7
Lephinchapel Argyll 275 D10
Lephinmore Argyll 275 D10
Leppington N Yorks 216 G5
Lepton W Yorks 197 D8
Lepton Edge W Yorks 197 D8
Lerigoligan Argyll 275 C9
Lerrocks Stirl 285 G11
Lerryn Corn 6 D2
Lerwick Shetland 313 J6
Lesbury Northumb 264 G6
Leschangie Aberds 293 B9
Leslie Aberds 302 G5
Leslie Fife 286 G6
Lesmahagow S Lnrk 259 B8
Lesnewth Corn 11 C8
Lessendrum Aberds 302 E5
Lessingham Norf 161 D7
Lessness Heath London 68 D3
Lessonhall Cumb 238 G6
Leswalt Dumfries 236 C2
Letchmore Heath Herts 85 F11
Letchworth Herts 104 E4
Letcombe Bassett Oxon 63 B11
Letcombe Regis Oxon 63 B11
Letham Angus 287 C9
Letham Falk 279 D7
Letham Fife 287 F7
Letham Perth 286 E4
Letham Grange Angus 287 C10
Lethem Borders 250 B5
Lethen Ho Highld 301 D9
Lethenty Aberds 303 E8
Lethenty Aberds 303 G7
Letheringham Suff 126 F5
Letheringsett Norf 159 B11
Lettaford Devon 13 E10
Lettan Orkney 314 B7
Letter Aberds 293 B9
Letterewe Highld 299 B9
Letterfearn Highld 295 C10
Letterfinlay Highld 290 D4
Lettermay Argyll 284 G5
Lettermorar Highld 295 G9
Lettermore Argyll 288 E6
Letters Highld 307 L6
Letterston = Treletert Pembs 91 F8
Lettoch Highld 292 B2
Lettoch Highld 301 F10
Lettoch Moray 302 F3
Lettoch Perth 291 G11
Letton Hereford 96 B6
Letton Hereford 115 C7
Letton Green Norf 141 B9
Lett's Green Kent 52 B3
Letty Brongu Bridgend 57 D11
Letty Green Herts 86 C3
Letwell S Yorks 187 D9
Leuchars Fife 287 E8
Leuchars Ho Moray 302 C2
Leumrabhagh W Isles 305 G5
Levalsa Meor Corn 5 F10
Levan Invclyd 276 F4
Levaneap Shetland 313 G6
Levedale Staffs 151 F7
Level of Mendalgief Newport 59 B10
Level's Green Essex 105 G9
Leven E Yorks 209 D8
Leven Fife 287 G7
Leven Seat W Loth 269 D8
Levencorroch N Ayrs 256 E2
Levenhall E Loth 281 G7
Levens Cumb 211 B9
Levens Green Herts 105 G7
Levenshulme Gtr Man 184 C5
Leventhorpe W Yorks 205 G8
Levenwick Shetland 313 L6
Lever-Edge Gtr Man 195 F8
Leverington Cambs 157 G8
Leverington Common Cambs 157 G8
Leverstock Green Herts 85 D9
Leverton Lincs 174 F6
Leverton W Berks 63 E10
Leverton Highgate Lincs 174 F6
Leverton Lucasgate Lincs 174 F6
Leverton Outgate Lincs 174 F6
Levington Suff 108 D4
Levisham N Yorks 226 G6
Levishie Highld 290 B6
Lew Oxon 82 D4
Lewannick Corn 11 E11
Lewcombe Dorset 29 F9
Lewdown Devon 12 D4
Lewes E Sus 36 E6
Leweston Pembs 91 G8
Lewisham London 67 D11
Lewiston Highld 300 G5
Lewistown Bridgend 58 B2
Lewknor Oxon 84 F2
Leworthy Devon 24 G4
Leworthy Devon 41 F7
Lewson Street Kent 70 G3
Lewth Lancs 202 F5
Lewthorn Cross Devon 13 F11
Lewtrenchard Devon 12 D5
Lexden Essex 107 G9
Ley Aberds 293 B7
Ley Corn 6 B3
Ley Som 41 F10
Ley Green Herts 104 G3
Ley Hey Park Gtr Man 185 D7
Ley Hill W Mid 134 D2
Leybourne Kent 53 B7
Leyburn N Yorks 224 G2
Leycett Staffs 168 F3
Leyfields Staffs 134 B4
Leyhill Bucks 85 E7
Leyhill S Glos 79 G11
Leyland Lancs 194 C4
Leylodge Aberds 293 B9
Leymoor W Yorks 196 D6
Leys Aberds 292 C6
Leys Aberds 303 D10
Leys Cumb 219 B11
Leys Perth 286 D6
Leys Staffs 169 F8
Leys Castle Highld 300 E6
Leys Hill Hereford 79 B9
Leys of Cossans Angus 287 C7
Leysdown-on-Sea Kent 70 E4
Leysmill Angus 287 C10
Leysters Hereford 115 E11
Leysters Pole Hereford 115 E11
Leyton London 67 B11
Leytonstone London 67 B11
Lezant Corn 12 F2
Lezerea Corn 2 C5
Leziate Norf 158 F3
Lhanbryde Moray 302 C2
Liatrie Highld 300 F2
Libanus Powys 95 F9
Libberton S Lnrk 269 G9
Libbery Worcs 117 F9
Liberton Edin 270 B5
Liceasto W Isles 305 J3
Lichfield Staffs 134 B2
Lick Perth 286 B2
Lickey Worcs 117 B9
Lickey End Worcs 117 C9
Lickfold W Sus 34 B6
Lickhill Worcs 116 C6
Lickleyhead Castle Aberds 302 G6
Liddaton Devon 12 E5
Liddel Orkney 314 H4
Liddesdale Highld 289 D9
Liddeston Pembs 72 D5
Liddington Swindon 63 C8
Liden Swindon 63 C7
Lidgate Suff 124 F4
Lidget S Yorks 199 G7
Lidget Green W Yorks 205 G8
Lidgett Notts 171 B10
Lidgett Park W Yorks 206 F2
Lidham Hill E Sus 38 D4
Lidlington C Beds 103 D9
Lidsey W Sus 22 C6
Lidsing Kent 69 G9
Lidstone Oxon 101 G7
Lieurary Highld 310 C4
Liff Angus 287 D7
Lifford W Mid 117 B11
Lifton Devon 12 D3
Liftondown Devon 12 D3
Light Oaks Staffs 168 E6
Lightcliffe W Yorks 196 B6
Lighteach Shrops 149 C10
Lightfoot Green Lancs 202 G6
Lighthorne Warks 118 F6
Lighthorne Heath Warks 119 F7
Lighthorne Rough Warks 118 F6
Lightmoor Telford 132 B3
Lightpill Glos 80 E4
Lightwater Sur 66 G2
Lightwood S Yorks 186 E5
Lightwood Shrops 132 E2
Lightwood Shrops 150 D3
Lightwood Staffs 169 G8
Lightwood Stoke 168 G6
Lightwood Green E Ches 167 G10
Lightwood Green Wrex 166 G5
Liglartrie S Ayrs 244 F6
Lilbourne Northants 119 B11
Lilburn Tower Northumb 264 E2
Lilford Gtr Man 195 G7
Lillesdon Som 28 C4
Lilleshall Telford 150 F4
Lilley Herts 104 F2
Lilley W Berks 64 D2
Lilliesleaf Borders 262 E2
Lillingstone Dayrell Bucks 102 D4
Lillingstone Lovell Bucks 102 C4
Lillington Dorset 29 E10
Lillington Warks 118 D6
Lilliput Poole 18 C6
Lilstock Som 43 E7
Lilybank Invclyd 276 G6
Lilyhurst Shrops 150 G4
Lilyvale Kent 54 F5
Limbrick Lancs 194 D6
Limbury Luton 103 G11
Lime Side Gtr Man 196 G2
Lime Street Worcs 98 E6
Lime Tree Park W Mid 118 B5
Limebrook Hereford 115 D7
Limefield Gtr Man 195 E10
Limehouse London 67 C11
Limehurst Gtr Man 196 G2
Limekiln Field Derbys 187 G7
Limekilnburn S Lnrk 268 E4
Limekilns Fife 279 E11
Limerigg Falk 279 G7
Limerstone I o W 20 E4
Limestone Brae Northumb 231 B11
Limington Som 29 C8
Limpenhoe Norf 143 C7
Limpenhoe Hill Norf 143 C8
Limpers Hill Wilts 45 G10
Limpley Stoke Wilts 61 G9
Limpsfield Sur 52 C2
Limpsfield Chart Sur 52 C2
Limpsfield Common Sur 52 C2
Linbriggs Northumb 251 B9
Linburn W Loth 270 B2
Linby Notts 171 E8
Linchmere W Sus 49 G11
Lincluden Dumfries 237 B11
Lincoln Lincs 189 G7
Lincomb Worcs 116 D6
Lincombe Devon 8 D4
Lincombe Devon 40 D3
Lindal in Furness Cumb 210 D5
Lindale Cumb 211 C8
Lindean Borders 261 C11
Linden Glos 80 B4
Lindfield W Sus 36 B4
Lindford Hants 49 F10
Lindifferon Fife 287 F7
Lindley N Yorks 205 D10
Lindley W Yorks 196 D6
Lindley Green N Yorks 205 D10
Lindores Fife 286 F6
Lindow End E Ches 184 F4
Lindrick Dale S Yorks 187 E8
Lindridge Worcs 116 D3
Lindsell Essex 106 F2
Lindsey Suff 107 C9
Lindsey Tye Suff 107 B9
Lindwell W Yorks 196 C5
Lineholt Worcs 116 D6
Lineholt Common Worcs 116 D6
Liney Som 43 F11
Linfitts Gtr Man 196 F3
Linford Hants 31 F11
Linford Thurrock 69 D7
Lingague I o M 192 E3
Lingards Wood W Yorks 196 E5
Lingbob W Yorks 205 F7
Lingdale Redcar 226 B3
Lingen Hereford 115 D7
Lingfield Darl 224 C6
Lingfield Sur 51 E11
Lingfield Common Sur 51 E11
Lingley Green Warr 183 D9
Lingley Mere Warr 183 D10
Lingreabhagh W Isles 296 C6
Lingwood Norf 143 B7
Linhope Borders 249 C10
Linhope Northumb 263 F11
Linicro Highld 298 C3
Link N Som 44 B3
Linkend Worcs 98 E6
Linkenholt Hants 47 B11
Linkhill Kent 38 B4
Linkinhorne Corn 12 G2
Linklater Orkney 314 H4
Linklet Orkney 314 A7
Linksness Orkney 314 E5
Linksness Orkney 314 F2
Linktown Fife 280 C5
Linley Shrops 131 E7
Linley Shrops 132 D3
Linley Brook Shrops 132 D3
Linley Green Hereford 116 G3
Linleygreen Shrops 132 D3
Linlithgow W Loth 279 F10
Linlithgow Bridge W Loth 279 F9
Linndhu Ho Argyll 289 D7
Linneraineach Highld 307 J6
Linns Angus 292 F3
Linnyshaw Gtr Man 195 G8
Linshiels Northumb 251 B9
Linsiadar W Isles 304 E4
Linsidemore Highld 309 K5
Linslade C Beds 103 F8
Linstead Parva Suff 126 B6
Linstock Cumb 239 F10
Linthorpe M'bro 225 B9
Linthurst Worcs 117 C9
Linthwaite W Yorks 196 E6
Lintlaw Borders 272 D6
Lintmill Moray 302 C5
Linton Borders 263 D7
Linton Cambs 105 B11
Linton Derbys 152 F5
Linton Hereford 98 F3
Linton Kent 53 D9
Linton N Yorks 213 G9
Linton Northumb 253 E7
Linton W Yorks 206 D3
Linton Heath Derbys 152 F5
Linton Hill Hereford 98 G3
Linton-on-Ouse N Yorks 215 G9
Lintridge Glos 98 E4
Lintz Durham 242 F5
Lintzford T & W 242 F4
Lintzgarth Durham 232 C4
Linwood Hants 31 F11
Linwood Lincs 189 D10
Linwood Renfs 267 C8
Lionacleit W Isles 297 G3
Lional W Isles 304 B7
Lions Green E Sus 23 B9
Liphook Hants 49 G10
Lipley Shrops 150 C4
Lippitts Hill Essex 86 F5
Liquo or Bowhousebog N Lnrk 269 D7
Liscard Mers 182 C4
Liscombe Som 41 G11
Liskeard Corn 6 C5
Liss Hants 34 B3
Liss Forest Hants 34 B3
Lissett E Yorks 209 B8
Lissington Lincs 189 E10
Lisson Grove London 67 C9
Listerdale S Yorks 187 C7
Listock Som 28 C4
Listoft Lincs 191 G8
Liston Essex 107 C7
Liston Garden Essex 106 B6
Lisvane Cardiff 59 C7
Liswerry Newport 59 B10
Litcham Norf 159 F7
Litchard Bridgend 58 C2
Litchborough Northants 120 G2
Litchfield Hants 48 C3
Litchurch Derbys 153 B7
Litherland Mers 182 B4
Litlington Cambs 104 C6
Litlington E Sus 23 E8
Litmarsh Hereford 97 B10
Little Abington Cambs 105 B10
Little Addington Northants 121 C9
Little Airmyn E Yorks 199 B8
Little Almshoe Herts 104 F3
Little Alne Warks 118 E2
Little Altcar Mers 193 F10
Little Ann Hants 47 E10
Little Arowry Wrex 167 G7
Little Asby Cumb 222 D3
Little Ashley Wilts 61 G10
Little Assynt Highld 307 G6
Little Aston Staffs 133 C11
Little Atherfield I o W 20 E5
Little Ayre Orkney 314 G3
Little-ayre Shetland 313 G5
Little Ayton N Yorks 225 C11
Little Baddow Essex 88 D3
Little Badminton S Glos 61 C10
Little Ballinluig Perth 286 B3
Little Bampton Cumb 239 F7
Little Bardfield Essex 106 E3
Little Barford Beds 122 F3
Little Barningham Norf 160 C3
Little Barrington Glos 82 C2
Little Barrow W Ches 183 G7
Little Barugh N Yorks 216 D5
Little Bavington Northumb 241 B11
Little Bayham E Sus 52 F6
Little Bealings Suff 108 B4
Little Beckford Glos 99 E9
Little Bedwyn Wilts 63 F9
Little Bentley Essex 108 F2
Little Berkhamsted Herts 86 D3
Little Billing Northants 120 E6
Little Billington C Beds 103 G8
Little Birch Hereford 97 E10
Little Bispham Blkpool 202 E2
Little Blakenham Suff 108 B2
Little Blencow Cumb 230 E5
Little Bloxwich W Mid 133 C10
Little Bognor W Sus 35 C8
Little Bolehill Derbys 170 E3
Little Bollington E Ches 184 D2
Little Bolton Gtr Man 184 B3
Little Bookham Sur 50 C6
Little Bosullow Corn 1 C4
Little Bourton Oxon 101 C9
Little Bowden Leics 136 F4
Little Boys Heath Bucks 84 F6
Little Bradley Suff 124 G3
Little Braithwaite Cumb 229 G10
Little Brampton Shrops 131 G7
Little Braxted Essex 88 C4
Little Bray Devon 41 F7
Little Brechin Angus 293 G7
Little Brickhill M Keynes 103 E8
Little Bridgeford Staffs 151 D7
Little Brington Northants 120 E3
Little Bristol S Glos 80 G2
Little Britain Warks 118 G2
Little Bromley Essex 107 F11
Little Bromwich W Mid 134 F2
Little Broughton Cumb 229 E7
Little Budworth W Ches 167 B9
Little Burstead Essex 87 G11
Little Bytham Lincs 155 F10
Little Cambridge Essex 106 F2
Little Canfield Essex 105 G11
Little Canford Poole 18 B6
Little Carleton Lancs 202 F2
Little Carlton Lincs 190 D5
Little Carlton Notts 172 D3
Little Casterton Rutland 137 B10
Little Catwick E Yorks 209 E8
Little Catworth Cambs 122 C2
Little Cawthorpe Lincs 190 E5
Little Chalfield Wilts 61 G11
Little Chalfont Bucks 85 F7
Little Chart Kent 54 D2
Little Chart Forstal Kent 54 D3
Little Chell Stoke 168 E5
Little Chester Derby 153 B7
Little Chesterford Essex 105 C10
Little Chesterton Oxon 101 G11
Little Cheverell Wilts 46 C3
Little Chishill Cambs 105 D8
Little Clacton Essex 89 B11
Little Clanfield Oxon 82 E3
Little Clegg Gtr Man 196 E2
Little Clifton Cumb 229 F7
Little Coates NE Lincs 201 F8
Little Colp Aberds 303 E7
Little Comberton Worcs 99 C9
Little Comfort Corn 12 E2
Little Common E Sus 38 F2
Little Common Lincs 156 D6
Little Common Shrops 115 B7
Little Common W Sus 34 C6
Little Compton Warks 100 E5
Little Corby Cumb 239 F11
Little Cornard Suff 107 D7
Little Cowarne Hereford 116 G2
Little Coxwell Oxon 82 G3
Little Crakehall N Yorks 224 G4
Little Cransley Northants 120 B6
Little Crawley M Keynes 103 B8
Little Creaton Northants 120 C4
Little Creich Highld 309 L6
Little Cressingham Norf 141 D7
Little Crosby Mers 193 G10
Little Cubley Derbys 152 B3
Little Dalby Leics 154 G5
Little Dawley Telford 132 B3
Little Dens Aberds 303 E10
Little Dewchurch Hereford 97 E10
Little Ditton Cambs 124 F3
Little Doward Hereford 79 B8
Little Downham Cambs 139 G10
Little Drayton Shrops 150 C3
Little Driffield E Yorks 208 B6
Little Drybrook Glos 79 D9
Little Dunham Norf 159 G7
Little Dunkeld Perth 286 C4
Little Dunmow Essex 106 G3
Little Durnford Wilts 46 G6
Little Eastbury Worcs 116 F6
Little Easton Essex 106 G2
Little Eaton Derbys 170 G5
Little Eccleston Lancs 202 E4
Little Ellingham Norf 141 D10
Little End Cambs 122 F3
Little End E Yorks 208 F2
Little End Essex 87 E8
Little Everdon Northants 119 F11
Little Eversden Cambs 123 G7
Little Faringdon Oxon 82 E2
Little Fencote N Yorks 224 G5
Little Fenton N Yorks 206 F6
Little Finborough Suff 125 G10
Little Fransham Norf 159 G8
Little Frith Kent 54 B2
Little Gaddesden Herts 85 C7
Little Gidding Cambs 138 G2
Little Gight Aberds 303 F8
Little Glemham Suff 126 F6
Little Glenshee Perth 286 D3
Little Gorsley Glos 98 F3
Little Gransden Cambs 122 F5
Little Green Cambs 104 B5
Little Green Notts 172 G2
Little Green Som 45 D8
Little Green Suff 125 C11
Little Green Wrex 167 G7
Little Grimsby Lincs 190 C4
Little Gringley Notts 188 E2
Little Gruinard Highld 307 L4
Little Habton N Yorks 216 D4
Little Hadham Herts 105 G8
Little Hale Lincs 173 G10
Little Hale Norf 141 B8
Little Hallam Derbys 171 G7
Little Hallingbury Essex 87 B7
Little Hampden Bucks 84 E5
Little Haresfield Glos 80 D4
Little Harrowden Northants 121 C7
Little Harwood Blkburn 195 B7
Little Haseley Oxon 83 E10
Little Hatfield E Yorks 209 E9
Little Hautbois Norf 160 E5
Little Haven Pembs 72 C5
Little Haven W Sus 51 G7
Little Hay Staffs 134 C2
Little Hayfield Derbys 185 E8
Little Haywood Staffs 151 E10
Little Heath E Ches 167 G11
Little Heath Herts 85 D8
Little Heath Herts 86 E3
Little Heath London 68 B3
Little Heath Staffs 151 F8
Little Heath Sur 66 G6
Little Heath W Berks 65 E7
Little Heath W Ches 166 B6
Little Heath W Mid 134 G6
Little Heck N Yorks 198 C5
Little Henham Essex 105 E10
Little Henny Essex 107 D7
Little Herbert's Glos 81 B7
Little Hereford Hereford 115 D11
Little Hill Hereford 97 F9
Little Hill Som 28 E3
Little Holbury Hants 32 G6
Little Honeyborough Pembs 73 D7
Little Hoole Moss Houses Lancs 194 C3
Little Horkesley Essex 107 E9
Little Hormead Herts 105 F8
Little Horsted E Sus 23 B7
Little Horton W Yorks 205 G9
Little Horton Wilts 62 G4
Little Horwood Bucks 102 E5
Little Houghton Northants 120 F6
Little Houghton S Yorks 198 G2
Little Hucklow Derbys 185 F11
Little Hulton Gtr Man 195 G8
Little Humber E Yorks 201 C8
Little Hungerford W Berks 64 E4
Little Ilford London 68 B2
Little Ingestre Staffs 151 E9
Little Inkberrow Worcs 117 F10
Little Irchester Northants 121 D8
Little Keyford Som 45 D9
Little Kimble Bucks 84 D4
Little Kineton Warks 118 G6
Little Kingshill Bucks 84 F5
Little Knowles Green Suff 124 F5
Little Langdale Cumb 220 E6
Little Langford Wilts 46 F4
Little Laver Essex 87 D8
Little Lawford Warks 119 B9
Little Layton Blkpool 202 F2
Little Leigh W Ches 183 F10
Little Leighs Essex 88 B2
Little Lepton W Yorks 197 E8
Little Leven E Yorks 209 D8
Little Lever Gtr Man 195 F9
Little Limber Lincs 200 E6
Little Linford M Keynes 102 C6
Little Load Som 29 C7
Little London Bucks 83 C10
Little London Bucks 84 D5
Little London Cambs 139 D8
Little London E Sus 23 B9
Little London Essex 105 F9
Little London Essex 106 D3
Little London Glos 80 B2
Little London Hants 47 D11
Little London Hants 48 B6
Little London Lincs 156 E4
Little London Lincs 157 E8
Little London Lincs 189 D10
Little London Lincs 190 F4
Little London Norf 140 D5
Little London Norf 160 C2
Little London Norf 160 C5
Little London Norf 160 E3
Little London Oxon 83 E8
Little London Powys 129 F10
Little London Shrops 131 F10
Little London Som 44 D6
Little London Suff 125 F10
Little London W Yorks 205 F10
Little London Worcs 116 C2
Little Longstone Derbys 185 G11
Little Lynturk Aberds 293 B7
Little Lyth Shrops 131 B9
Little Madeley Staffs 168 F3
Little Malvern Worcs 98 C5
Little Mancot Flint 166 B4
Little Maplestead Essex 106 E6
Little Marcle Hereford 98 D3
Little Marlow Bucks 65 B11
Little Marsden Lancs 204 F3
Little Marsh Bucks 102 G3
Little Marsh Norf 159 B10
Little Marsh Wilts 45 B11
Little Marton Blkpool 202 G2
Little Mascalls Essex 88 E2
Little Massingham Norf 158 E5
Little Melton Norf 142 B3
Little Merthyr Hereford 96 B5
Little Milford Pembs 73 C7
Little Mill Kent 53 D7
Little Mill Mon 78 E4
Little Milton Newport 59 B11
Little Milton Oxon 83 E10
Little Minster Oxon 82 C4
Little Missenden Bucks 84 F6
Little Mongeham Kent 55 C10
Little Moor Gtr Man 184 D6
Little Moor Lancs 203 E10
Little Moor End Lancs 195 B8
Little Morrell Warks 118 F6
Little Mountain Flint 166 C3
Little Musgrave Cumb 222 C5
Little Ness Shrops 149 F8
Little Neston W Ches 182 F3
Little Newcastle Pembs 91 F9
Little Newsham Durham 224 B2
Little Norlington E Sus 23 C7
Little Norton Som 29 D7
Little Oakley Essex 108 F4
Little Oakley Northants 137 F7
Little Odell Beds 121 F9
Little Offley Herts 104 F2
Little Onn Staffs 150 F6
Little Ormside Cumb 222 B4
Little Orton Cumb 239 F9
Little Orton Leics 134 B6
Little Ouse Norf 140 F2
Little Ouseburn N Yorks 215 G8
Little Overton Wrex 166 G5
Little Oxney Green Essex 87 D11
Little Packington Warks 134 G4
Little Parndon Essex 86 C6
Little Paxton Cambs 122 E3
Little Petherick Corn 10 G4
Little Pitlurg Moray 302 E4
Little Plumpton Lancs 202 G3
Little Plumstead Norf 160 G6
Little Ponton Lincs 155 C8
Little Posbrook Hants 33 G8
Little Poulton Lancs 202 F3
Little Preston Kent 53 B8
Little Preston W Yorks 206 G3
Little Raveley Cambs 122 B5
Little Reedness E Yorks 199 C10
Little Reynoldston Swansea 56 D3
Little Ribston N Yorks 206 C3
Little Rissington Glos 81 B11
Little Rogart Highld 309 J7
Little Rollright Oxon 100 E5
Little Ryburgh Norf 159 D9
Little Ryle Northumb 264 G2
Little Ryton Shrops 131 C9
Little Salisbury Wilts 63 G7
Little Salkeld Cumb 231 D7
Little Sampford Essex 106 E3
Little Sandhurst Brack 65 G10
Little Saredon Staffs 133 B8
Little Saxham Suff 124 E5
Little Scatwell Highld 300 D3
Little Scotland Gtr Man 194 E6
Little Sessay N Yorks 215 E9
Little Shelford Cambs 123 G9
Little Shoddesden Hants 47 D9
Little Shrewley Warks 118 D4
Little Shurdington Glos 80 B6
Little Silver Devon 26 F6
Little Silver Devon 40 E4
Little Singleton Lancs 202 F3
Little Skillymarno Aberds 303 D9
Little Skipwith N Yorks 207 F9
Little Smeaton N Yorks 198 D4
Little Smeaton N Yorks 224 E6
Little Snoring Norf 159 C9
Little Sodbury S Glos 61 C9
Little Sodbury End S Glos 61 C8
Little Somborne Hants 47 G11
Little Somerford Wilts 62 C3
Little Soudley Shrops 150 D4
Little Stainforth N Yorks 212 F6
Little Stainton Darl 234 G2
Little Stanmore London 85 G11
Little Stanney W Ches 182 G6
Little Staughton Beds 122 E2
Little Steeping Lincs 174 C6
Little Stoke S Glos 60 C6
Little Stoke Staffs 151 C8
Little Stonham Suff 126 E2
Little Stretton Leics 136 C3
Little Stretton Shrops 131 E8
Little Strickland Cumb 221 B11
Little Studley N Yorks 214 E6
Little Stukeley Cambs 122 B4
Little Sugnall Staffs 150 C6
Little Sutton Lincs 157 E9
Little Sutton Shrops 131 G10
Little Sutton W Ches 182 F5
Little Swinburne Northumb 241 B10
Little Tarrington Hereford 98 C2
Little Tew Oxon 101 F7
Little Tey Essex 107 G7
Little Thetford Cambs 123 B10
Little Thirkleby N Yorks 215 D9
Little Thornage Norf 159 B11
Little Thornton Lancs 202 E3
Little Thorpe Durham 234 C4
Little Thorpe W Yorks 197 C7
Little Thurlow Suff 124 G3
Little Thurlow Green Suff 124 G3
Little Thurrock Thurrock 68 D6
Little Torboll Highld 309 K7
Little Torrington Devon 25 D7
Little Totham Essex 88 C5
Little Toux Aberds 302 D5
Little Town Cumb 220 B4
Little Town Lancs 203 F9
Little Town Warr 183 C10
Little Tring Herts 84 C6
Little Twycross Leics 134 B6
Little Urswick Cumb 210 E5
Little Vantage W Loth 270 C2
Little Wakering Essex 70 B2
Little Walden Essex 105 C10
Little Waldingfield Suff 107 B8
Little Walsingham Norf 159 B8
Little Waltham Essex 88 C2
Little Walton Warks 135 G9
Little Warley Essex 87 G10
Little Warton Warks 134 C5
Little Washbourne Glos 99 E9
Little Weighton E Yorks 208 G5
Little Weldon Northants 137 F8
Little Welland Worcs 98 D6
Little Welnetham Suff 125 E7
Little Welton Lincs 190 D4
Little Wenham Suff 107 D11
Little Wenlock Telford 132 B2
Little Weston Som 29 B10
Little Whitehouse I o W 20 C5
Little Whittingham Green Suff 126 B5
Little Wigborough Essex 89 B7
Little Wilbraham Cambs 123 F10
Little Wisbeach Lincs 156 C2
Little Wishford Wilts 46 F5
Little Witcombe Glos 80 B6
Little Witley Worcs 116 E5
Little Wittenham Oxon 83 G9
Little Wolford Warks 100 D5
Little Wood Corner Bucks 84 E6
Little Woodcote London 67 G9
Little Woolgarston Dorset 18 E5
Little Worthen Shrops 130 B6
Little Wratting Suff 106 B3
Little Wymington Beds 121 D9
Little Wymondley Herts 104 F4
Little Wyrley Staffs 133 B10
Little Wytheford Shrops 149 F11
Little Yeldham Essex 106 D5
Littlebeck N Yorks 227 D7
Littleborough Devon 26 E4
Littleborough Gtr Man 196 D2
Littleborough Notts 188 E4
Littlebourne Kent 55 B8
Littlebredy Dorset 17 D7
Littlebury Essex 105 D10
Littlebury Green Essex 105 D9
Littlecote Bucks 102 G6
Littlecott Wilts 46 C6
Littledean Glos 79 C11
Littledean Hill Glos 79 C11
Littledown Bmouth 19 C8
Littledown Hants 47 B10
Littleferry Highld 311 K2
Littlefield NE Lincs 201 F9
Littlefield Common Sur 50 C3
Littlefield Green Windsor 65 D11
Littlegain Shrops 132 D5
Littleham Devon 14 E6
Littleham Devon 24 C6
Littlehampton W Sus 35 G8
Littlehempston Devon 8 C6
Littlehoughton Northumb 264 F6
Littlemill Aberds 292 D5
Littlemill E Ayrs 257 F11
Littlemill Highld 301 D9
Littlemill Northumb 264 F6
Littlemoor Derbys 170 C5
Littlemoor Dorset 17 E9
Littlemore Oxon 83 E8
Littlemoss Gtr Man 184 B6
Littleover Derby 152 C6
Littleport Cambs 139 F11
Littler W Ches 167 B10
Littlestead Green Oxon 65 D8
Littlestone-on-Sea Kent 39 C9
Littlethorpe Leics 135 D10
Littlethorpe N Yorks 214 F6
Littleton Bath 60 G5
Littleton Dorset 30 G5
Littleton Hants 48 G3
Littleton Perth 286 D6
Littleton Som 44 G3
Littleton Sur 50 D3
Littleton Sur 66 F5
Littleton W Ches 166 B6
Littleton Wilts 62 G2
Littleton Common Sur 66 E5
Littleton Drew Wilts 61 C10
Littleton-on-Severn S Glos 60 B5
Littleton Panell Wilts 46 C4
Littleton-upon-Severn S Glos 79 G9
Littletown Durham 234 C2
Littletown I o W 20 C6
Littletown W Yorks 197 C8
Littlewick Green Windsor 65 D10
Littlewindsor Dorset 28 G6
Littlewood Staffs 133 B9
Littlewood Green Warks 117 E11
Littleworth Beds 103 C11
Littleworth Glos 80 E4
Littleworth Glos 100 D2
Littleworth Oxon 82 F4
Littleworth Oxon 83 D9
Littleworth Oxon 83 G10
Littleworth S Yorks 187 B10
Littleworth Staffs 151 E8
Littleworth Staffs 151 G10
Littleworth W Sus 35 C11
Littleworth Warks 118 E4
Littleworth Wilts 63 G7
Littleworth Worcs 117 E9
Littleworth Worcs 117 G7
Littleworth Common Bucks 66 B2
Littleworth End Warks 134 D3
Littley Green Essex 87 B11
Litton Derbys 185 F11
Litton N Yorks 213 E8
Litton Som 44 C5
Litton Cheney Dorset 17 C7
Litton Mill Derbys 185 G11
Liurbost W Isles 304 F5
Livermead Torbay 9 C8
Liverpool Mers 182 C4
Liverpool Airport Mers 182 E6
Liversedge W Yorks 197 C8
Liverton Devon 14 F2
Liverton Redcar 226 B4
Liverton Mines Redcar 226 B4
Liverton Street Kent 53 C11
Livesey Street Kent 53 C8
Livingshayes Devon 27 G7
Livingston W Loth 269 B11
Livingston Village W Loth 269 B10
Lix Toll Stirl 285 D9
Lixwm Flint 181 G11
Lizard Corn 2 G6
Llaingoch Anglesey 178 E2
Llaithddu Powys 129 G11
Llampha V Glam 58 D2
Llan Powys 129 C7
Llan-dafal Bl Gwent 77 E11
Llan eurgain = Northop Flint 166 B2
Llan Ffestiniog Gwyn 164 G2
Llan-mill Pembs 73 C10
Llan-non = Llanon Ceredig 111 D10
Llan-y-pwll Wrex 166 E5
Llanaber Gwyn 146 F2
Llanaelhaearn Gwyn 162 G5
Llanafan Ceredig 112 C3
Llanafan-fawr Powys 113 F9
Llanallgo Anglesey 179 D7
Llananno Powys 113 C11
Llanarmon Gwyn 145 B8
Llanarmon Dyffryn Ceiriog Wrex 148 C3
Llanarmon Mynydd-mawr Powys 148 D2
Llanarmon-yn-Ial Denb 165 D11
Llanarth Ceredig 111 F8
Llanarth Mon 78 C5
Llanarthne Carms 93 G10
Llanasa Flint 181 E10
Llanbabo Anglesey 178 D5
Llanbad Rhondda 58 C3
Llanbadarn Fawr Ceredig 128 G2
Llanbadarn Fynydd Powys 114 B2
Llanbadarn-y-Garreg Powys 96 B2
Llanbadoc Mon 78 E5
Llanbadrig Anglesey 178 C5
Llanbeder Newport 78 G5
Llanbedr Gwyn 145 D11
Llanbedr Powys 96 B2
Llanbedr Powys 96 G4
Llanbedr-Dyffryn-Clwyd Denb 165 D10
Llanbedr Pont Steffan = Lampeter Ceredig 93 B11
Llanbedr-y-cennin Conwy 164 B3
Llanbedrgoch Anglesey 179 E8
Llanbedrog Gwyn 144 C6
Llanberis Gwyn 163 C9
Llanbethery V Glam 58 F4
Llanbister Powys 114 C2
Llanblethian = Llanfleiddan V Glam 58 E3
Llanboidy Carms 92 G4
Llanbradach Caerph 77 G10
Llanbrynmair Powys 129 C7
Llancadle = Llancatal V Glam 58 F4
Llancaiach Caerph 77 F10
Llancarfan V Glam 58 E5
Llancatal = Llancadle V Glam 58 F4
Llancayo Mon 78 E5
Llancloudy Hereford 97 G9
Llancowrid Powys 130 E3
Llancynfelyn Ceredig 128 E2
Llandaff Cardiff 59 D7
Llandaff North Cardiff 59 D7
Llandanwg Gwyn 145 D11
Llandarcy Neath 57 B8
Llandawke Carms 74 C3
Llanddaniel Fab Anglesey 179 G7
Llanddarog Carms 75 B8
Llanddeiniol Ceredig 111 C11
Llanddeiniolen Gwyn 163 B8
Llandderfel Gwyn 147 B9
Llanddeusant Anglesey 178 D4
Llanddeusant Carms 94 G5
Llanddew Powys 95 E11
Llanddewi Swansea 56 D3
Llanddewi-Brefi Ceredig 112 F3
Llanddewi Fach Mon 78 F4
Llanddewi Rhydderch Mon 78 C5
Llanddewi Skirrid Mon 78 B4
Llanddewi Velfrey Pembs 73 B10
Llanddewi Ystradenni Powys 114 D2
Llanddewi'r Cwm Powys 95 B10
Llanddoged Conwy 164 C4
Llanddona Anglesey 179 F9
Llanddowror Carms 74 C3
Llanddulas Conwy 180 F6
Llanddwywe Gwyn 145 E11
Llanddyfynan Anglesey 179 F8
Llandecwyn Gwyn 146 B2
Llandefaelog Powys 95 E10
Llandefaelog Fach Powys 95 E10
Llandefaelog-tre'r-graig Powys 96 F2
Llandefalle Powys 96 D2
Llandegai Gwyn 179 G9
Llandegfan Anglesey 179 G9
Llandegla Denb 165 E11
Llandegley Powys 114 E2
Llandegveth Mon 78 F4
Llandegwning Gwyn 144 C5
Llandeilo Carms 94 G2
Llandeilo Graban Powys 95 C11
Llandeilo'r Fan Powys 95 E7
Llandeloy Pembs 91 F7
Llandenny Mon 78 E6
Llandenny Walks Mon 78 E5
Llandevaud Newport 78 G6
Llandevenny Mon 60 B2
Llandewi Ystradenny Powys 114 D2
Llandilo Pembs 92 F2
Llandilo-yr-ynys Carms 93 G9
Llandinabo Hereford 97 F10
Llandinam Powys 129 F10
Llandissilio Pembs 92 G2
Llandogo Mon 79 E8
Llandough V Glam 58 E3
Llandough V Glam 59 E7
Llandovery = Llanymddyfri Carms 94 E5
Llandow = Llandw V Glam 58 E2
Llandre Carms 94 C3
Llandre Ceredig 128 F2
Llandrillo Denb 147 B10
Llandrillo-yn-Rhôs Conwy 180 E4
Llandrindod Wells Powys 113 E11
Llandrinio Powys 148 F5
Llandruidion Pembs 90 G5
Llandudno Conwy 180 E3
Llandudno Junction = Cyffordd Llandudno Conwy 180 F3
Llandudoch = St Dogmaels Pembs 92 B3
Llandw = Llandow V Glam 58 E2
Llandwrog Gwyn 163 D7
Llandybie Carms 75 B10
Llandyfaelog Carms 74 C6
Llandyfan Carms 75 B10
Llandyfriog Ceredig 92 C6
Llandyfrydog Anglesey 178 D6
Llandygwydd Ceredig 92 C4
Llandynan Denb 165 F11
Llandyrnog Denb 165 B10
Llandysilio Powys 148 F5
Llandyssil Powys 130 D3
Llandysul Ceredig 93 C8

Llanedeyrn Cardiff 59 C8
Llanedi Carms 75 D9
Llanedwen Anglesey 163 B8
Llaneglwys Powys 95 D11
Llanegryn Gwyn 110 B2
Llanegwad Carms 93 G10
Llaneilian Anglesey 179 C7
Llanelian yn-Rhôs Conwy 180 F5
Llanelidan Denb 165 E10
Llanelieu Powys 96 E3
Llanellen Mon 78 C4
Llanelli Carms 56 B4
Llanelltyd Gwyn 146 F4
Llanelly Mon 78 C2
Llanelly Hill Mon 78 C2
Llanelwedd Powys 113 G10
Llanelwy = St Asaph Denb 181 G8
Llanenddwyn Gwyn 145 E11
Llanengan Gwyn 144 D5
Llanerch Powys 130 E6
Llanerch Emrys Powys 148 E4
Llanerchymedd Anglesey 178 E6
Llanerfyl Powys 129 B10
Llaneuddog Anglesey 179 D7
Llanfabon Caerph 77 G10
Llanfachraeth Anglesey 178 E4
Llanfachreth Gwyn 146 E5
Llanfaelog Anglesey 178 G4
Llanfaelrhys Gwyn 144 D4
Llanfaenor Mon 78 B6
Llanfaes Anglesey 179 F10
Llanfaes Powys 95 F10
Llanfaethlu Anglesey 178 D4
Llanfaglan Gwyn 163 C7
Llanfair Gwyn 145 D11
Llanfair Caereinion Powys 130 B2
Llanfair Clydogau Ceredig 112 G2
Llanfair-Dyffryn-Clwyd Denb 165 D10
Llanfair Kilgeddin Mon 78 D4
Llanfair Kilgheddin Mon 78 D4
Llanfair-Nant-Gwyn Pembs 92 D3
Llanfair Talhaiarn Conwy 180 G6
Llanfair Waterdine Shrops 114 B4
Llanfairfechan Conwy 179 F11
Llanfairpwll-gwyngyll Anglesey 179 G8
Llanfairyneubwll Anglesey 178 F3
Llanfairynghornwy Anglesey 178 C4
Llanfallteg Carms 73 B11
Llanfallteg West Carms 73 B10
Llanfaredd Powys 113 G11
Llanfarian Ceredig 111 B11
Llanfechain Powys 148 E3
Llanfechan Powys 113 G9
Llanfechell Anglesey 178 C5
Llanferres Denb 165 C11
Llanfflewyn Anglesey 178 D5
Llanfigael Anglesey 178 E4
Llanfihangel-ar-arth Carms 93 D9
Llanfihangel-Crucorney Mon 96 G6
Llanfihangel Glyn Myfyr Conwy 165 F7
Llanfihangel-helygen Powys 113 E10
Llanfihangel Nant Bran Powys 95 E8
Llanfihangel-nant-Melan Powys 114 F3
Llanfihangel Rhydithon Powys 114 D3
Llanfihangel Rogiet Mon 60 B2
Llanfihangel Tal-y-llyn Powys 96 F2
Llanfihangel Tor y Mynydd Mon 79 E7
Llanfihangel-uwch-Gwili Carms 93 G9
Llanfihangel-y-Creuddyn Ceredig 112 B3
Llanfihangel-y-pennant Gwyn 128 B3
Llanfihangel-y-pennant Gwyn 163 F8
Llanfihangel-yng-Ngwynfa Powys 147 F11
Llanfihangel yn Nhowyn Anglesey 178 F4
Llanfilo Powys 96 E2
Llanfleiddan = Llanblethian V Glam 58 E3
Llanfoist Mon 78 C3
Llanfor Gwyn 147 B8
Llanfrechfa Torf 78 G4
Llanfrothen Gwyn 163 G10
Llanfrynach Powys 95 F11
Llanfwrog Anglesey 178 E4
Llanfwrog Denb 165 D10
Llanfyllin Powys 148 F2
Llanfynydd Carms 93 F11
Llanfynydd Flint 166 D3
Llanfyrnach Pembs 92 E4
Llangadfan Powys 147 G10
Llangadog Carms 74 D6
Llangadog Carms 94 F4
Llangadwaladr Anglesey 162 B5
Llangadwaladr Powys 148 C3
Llangaffo Anglesey 162 B6
Llangain Carms 74 B5
Llangammarch Wells Powys 95 B8
Llangan V Glam 58 D3
Llangarron Hereford 97 G10
Llangasty Talyllyn Powys 96 F2
Llangathen Carms 93 G11
Llangattock Powys 78 B2
Llangattock Lingoed Mon 97 G7
Llangattock nigh Usk Mon 78 D4
Llangattock-Vibon-Avel Mon 79 B7
Llangedwyn Powys 148 E3
Llangefni Anglesey 179 F7
Llangeinor Bridgend 58 B2
Llangeitho Ceredig 112 F2
Llangeler Carms 93 D7
Llangendeirne Carms 75 C7
Llangennech Carms 75 E9
Llangennith Swansea 56 C2
Llangenny Powys 78 B2
Llangernyw Conwy 164 B5
Llangeview Mon 78 E5
Llangewydd Court Bridgend 57 E11
Llangian Gwyn 144 D5
Llangloffan Pembs 91 E8
Llanglydwen Carms 92 F3
Llangoed Anglesey 179 F10
Llangoedmor Ceredig 92 B3
Llangollen Denb 166 G2
Llangolman Pembs 92 F2
Llangors Powys 96 F2
Llangorwen Ceredig 128 G2
Llangovan Mon 79 D7
Llangower Gwyn 147 C8
Llangrannog Ceredig 110 G6
Llangristiolus Anglesey 178 G6
Llangrove Hereford 79 B8
Llangua Mon 97 F7
Llangunllo Powys 114 C4
Llangunnor Carms 74 B6
Llangurig Powys 113 B8
Llangwm Conwy 165 G7
Llangwm Mon 78 E6
Llangwm Pembs 73 D7
Llangwnnadl Gwyn 144 C4
Llangwyfan Denb 165 B10
Llangwyfan-isaf Anglesey 162 B4
Llangwyllog Anglesey 178 F6
Llangwyryfon Ceredig 111 C11
Llangybi Ceredig 112 G2
Llangybi Gwyn 162 G6
Llangybi Mon 78 F5
Llangyfelach Swansea 56 B6
Llangyndeyrn Carms 75 C7
Llangynhafal Denb 165 C10
Llangynidr Powys 77 B11
Llangyniew Powys 130 B2
Llangynin Carms 74 B2
Llangynog Carms 74 B4
Llangynog Powys 147 D11
Llangynwyd Bridgend 57 D11
Llanhamlach Powys 95 F11
Llanharan Rhondda 58 C4
Llanharry Rhondda 58 C4
Llanhennock Mon 78 G5
Llanhilleth Bl Gwent 78 E2
Llanhowel Pembs 90 F6
Llanidloes Powys 129 G9
Llaniestyn Gwyn 144 C5
Llanifyny Powys 129 G7
Llanigon Powys 96 D4
Llanilar Ceredig 112 C2
Llanilid Rhondda 58 C3
Llanilltud Fawr = Llantwit Major V Glam 58 F3
Llanio Ceredig 112 F2
Llanion Pembs 73 E7
Llanishen Cardiff 59 C7
Llanishen Mon 79 E7
Llanllawddog Carms 93 F9
Llanllechid Gwyn 163 B10
Llanllowell Mon 78 F5
Llanllugan Powys 129 C11
Llanllwch Carms 74 B5
Llanllwchaiarn Powys 130 E2
Llanllwni Carms 93 D9
Llanllwyd Shrops 130 G3
Llanllyfni Gwyn 163 E7
Llanmadoc Swansea 56 C2
Llanmaes Cardiff 58 D6
Llanmaes V Glam 58 F3
Llanmartin Newport 59 B11
Llanmerwig Powys 130 E3
Llanmihangel V Glam 58 E3
Llanmiloe Carms 74 D3
Llanmorlais Swansea 56 C4
Llannefydd Conwy 181 G7
Llannerch-y-môr Flint 181 F11
Llannon Carms 75 D8
Llannor Gwyn 145 B7
Llanon Pembs 90 E6
Llanon = Llan-non Ceredig 111 D10
Llanover Mon 78 D4
Llanpumsaint Carms 93 F8
Llanreath Pembs 73 E7
Llanreithan Pembs 91 F7
Llanrhaeadr Denb 165 C9
Llanrhaeadr-ym-Mochnant Powys 148 D2
Llanrhian Pembs 90 E6
Llanrhidian Swansea 56 C3
Llanrhos Conwy 180 E3
Llanrhyddlad Anglesey 178 D4
Llanrhystud Ceredig 111 D10
Llanrosser Hereford 96 D5
Llanrothal Hereford 79 B7
Llanrug Gwyn 163 C8
Llanrumney Cardiff 59 C8
Llanrwst Conwy 164 C4
Llansadurnen Carms 74 C3
Llansadwrn Anglesey 179 F9
Llansadwrn Carms 94 E3
Llansaint Carms 74 D5
Llansamlet Swansea 57 B7
Llansanffraid Glan Conwy Conwy 180 F4
Llansannan Conwy 164 B6
Llansannor V Glam 58 D3
Llansantffraed Ceredig 111 D10
Llansantffraed Powys 96 G2
Llansantffraed Cwmdeuddwr Powys 113 D9
Llansantffraed-in-Elwel Powys 113 G11
Llansantffraid-ym-Mechain Powys 148 E4
Llansawel Carms 94 D2
Llansawel = Briton Ferry Neath 57 C8
Llansilin Powys 148 D4
Llansoy Mon 78 E6
Llanspyddid Powys 95 F10
Llanstadwell Pembs 72 D6
Llansteffan Carms 74 C5
Llanstephan Powys 96 C2
Llantarnam Torf 78 G4
Llanteems Mon 96 G6
Llanteg Pembs 73 C11
Llanthony Mon 96 F5
Llantilio Crossenny Mon 78 C5
Llantilio Pertholey Mon 78 B4
Llantood Pembs 92 C3
Llantrisant Anglesey 178 E5
Llantrisant Mon 78 F5
Llantrisant Rhondda 58 C4
Llantrithyd V Glam 58 E4
Llantwit Neath 57 B9
Llantwit Fardre Rhondda 58 B5
Llantwit Major = Llanilltud Fawr V Glam 58 F3
Llanuwchllyn Gwyn 147 C7
Llanvaches Newport 78 G6
Llanvair Discoed Mon 78 G6
Llanvapley Mon 78 C5
Llanvetherine Mon 78 B5
Llanveynoe Hereford 96 E6
Llanvihangel Crucorney Mon 96 G6
Llanvihangel Gobion Mon 78 D4
Llanvihangel-Ystern-Llewern Mon 78 C6
Llanwarne Hereford 97 F10
Llanwddyn Powys 147 F10
Llanwenarth Mon 78 C3
Llanwenog Ceredig 93 B9
Llanwern Newport 59 B11
Llanwinio Carms 92 F5
Llanwnda Gwyn 163 D7
Llanwnda Pembs 91 D8
Llanwnnen Ceredig 93 B10
Llanwnog Powys 129 E10
Llanwrda Carms 94 E4
Llanwrin Powys 128 C5
Llanwrthwl Powys 113 E9
Llanwrtud = Llanwrtyd Wells Powys 95 B7
Llanwrtyd Wells = Llanwrtud Powys 95 B7
Llanwyddelan Powys 129 C11
Llanyblodwel Shrops 148 E4
Llanybri Carms 74 C4
Llanybydder Carms 93 C10
Llanycefn Pembs 91 G11
Llanychaer Pembs 91 D9
Llanycil Gwyn 147 C8
Llanycrwys Carms 94 B2
Llanymawddwy Gwyn 147 F8
Llanymddyfri = Llandovery Carms 94 E5
Llanymynech Powys 148 E5
Llanynghenedl Anglesey 178 E4
Llanynys Denb 165 C10
Llanyrafon Torf 78 G4
Llanyre Powys 113 E10
Llanystumdwy Gwyn 145 B9
Llanywern Powys 96 F2
Llawhaden Pembs 73 B9
Llawnt Shrops 148 C5
Llawrdref Bellaf Gwyn 144 D5
Llawr-y-glyn Powys 129 E8
Llay Wrex 166 D4
Llechcynfarwy Anglesey 178 E5
Llecheiddior Gwyn 163 G7
Llechfaen Powys 95 F11
Llechfraith Gwyn 146 F3
Llechryd Caerph 77 D10
Llechryd Ceredig 92 C4
Llechrydau Powys 148 C4
Llechwedd Conwy 180 F3
Lledrod Ceredig 112 C2
Llenmerewig Powys 130 E3
Llethrid Swansea 56 C4
Llettyrychen Carms 75 E7
Llidiad Nenog Carms 93 D10
Llidiardau Gwyn 147 B7
Llidiart-y-parc Denb 165 G10
Llithfaen Gwyn 162 G5
Lloc Flint 181 F10
Llong Flint 166 C3
Llowes Powys 96 C3
Lloyney Powys 114 B4
Llugwy Powys 128 D4
Llundain-fach Ceredig 111 F11
Llwydarth Bridgend 57 C11
Llwydcoed Rhondda 77 E7
Llwyn Denb 165 C9
Llwyn Shrops 130 G5
Llwyn-derw Powys 129 G8
Llwyn-du Mon 78 B3
Llwyn-hendy Carms 56 B4
Llwyn-on Village M Tydf 77 C8
Llwyn-têg Carms 75 D9
Llwyn-y-brain Carms 73 C11
Llwyn-y-go Shrops 148 E6
Llwyn-y-groes Ceredig 111 F11
Llwyn-yr-hwrdd Pembs 92 E4
Llwyncelyn Ceredig 111 F8
Llwyndafydd Ceredig 111 F7
Llwynderw Powys 130 C4
Llwynduris Ceredig 92 C4
Llwyndyrys Gwyn 162 G5
Llwyneinion Wrex 166 F3
Llwyngwril Gwyn 110 B2
Llwynhendy Carms 56 B4
Llwynmawr Wrex 148 B4
Llwynygog Powys 129 E7
Llwynypia Rhondda 77 G7
Llynclys Shrops 148 E5
Llynfaes Anglesey 178 F6
Llys-y-frôn Pembs 91 G10
Llysfaen Conwy 180 F5
Llyswen Powys 96 D2
Llysworney V Glam 58 E3
Llywel Powys 95 E7
Llywernog Ceredig 128 G4
Load Brook S Yorks 186 D3
Loan Falk 279 F9
Loanend Northumb 273 E8
Loanhead Aberds 302 D6
Loanhead Midloth 270 B5
Loanhead Perth 286 D5
Loanreoch Highld 300 B6
Loans S Ayrs 257 C8
Loans of Tullich Highld 301 B8
Loansdean Northumb 252 G5
Lobb Devon 40 F3
Lobhillcross Devon 12 D5
Lobley Hill T & W 242 E6
Lobthorpe Lincs 155 E9
Loch a Charnain W Isles 297 G4
Loch a' Ghainmhich W Isles 304 F4
Loch Baghasdail W Isles 297 K3
Loch Choire Lodge Highld 308 F6
Loch Eil Highld 290 F2
Loch Euphoirt W Isles 296 E4
Loch Head Dumfries 236 E5
Loch Head Dumfries 245 E11
Loch Loyal Lodge Highld 308 E6
Loch nam Madadh W Isles 296 E5
Loch Sgioport W Isles 297 H4
Lochailort Highld 295 G9
Lochaline Highld 289 E8
Lochanhully Highld 301 G9
Lochans Dumfries 236 D2
Locharbriggs Dumfries 247 G11
Lochassynt Lodge Highld 307 G6
Lochavich Ho Argyll 275 B10
Lochawe Argyll 284 E5
Lochbuie Argyll 289 G8
Lochbuie Ho Argyll 289 G8
Lochcallater Lodge Aberds 292 E3
Lochcarron Highld 295 B10
Lochdhu Highld 310 E4
Lochdochart House Stirl 285 E8
Lochdon Argyll 289 F9
Lochdrum Highld 300 B2
Lochead Argyll 275 E11
Lochead Argyll 275 F8
Lochearnhead Stirl 285 E9
Lochee Dundee 287 D7
Lochend Edin 280 G5
Lochend Highld 300 F5
Lochend Highld 310 C6
Lochend Ho Stirl 277 B11
Locherben Dumfries 247 D11
Lochetive Ho Highld 284 C5
Lochfoot Dumfries 237 B10
Lochgair Argyll 275 D10
Lochgarthside Highld 291 B7
Lochgelly Fife 280 C3
Lochgilphead Argyll 275 E9
Lochgoilhead Argyll 284 G6
Lochhill Moray 302 C2
Lochhussie Highld 300 D4
Lochinch Castle Dumfries 236 C3
Lochindorb Lodge Highld 301 F9
Lochinver Highld 307 G5
Lochlane Perth 286 E2
Lochletter Highld 300 G4
Lochluichart Highld 300 C3
Lochmaben Dumfries 248 G3
Lochmore Cottage Highld 310 E4
Lochmore Lodge Highld 306 F7
Lochnell Ho Argyll 289 F10
Lochore Fife 280 B3
Lochorodale Argyll 255 F7
Lochportain W Isles 296 D5
Lochran Perth 280 B2
Lochranza N Ayrs 255 B10
Lochs Crofts Moray 302 C3
Lochside Aberds 293 G9
Lochside Highld 301 D8
Lochside Highld 308 D4
Lochside Highld 310 F2
Lochside S Ayrs 257 E8
Lochslin Highld 311 L2
Lochstack Lodge Highld 306 E7
Lochton Aberds 293 D9
Lochty Angus 293 G7
Lochty Fife 287 G9
Lochty Perth 286 E4
Lochuisge Highld 289 D9
Lochurr Dumfries 247 F7
Lochwinnoch Renfs 267 D7
Lochwood Dumfries 248 D3
Lochwood Glasgow 268 B3
Lochyside Highld 290 F3
Lockengate Corn 5 C10
Lockerbie Dumfries 248 G4
Lockeridge Wilts 62 F6
Lockeridge Dene Wilts 62 F6
Lockerley Hants 32 B3
Lockhills Cumb 230 B6
Locking N Som 43 B11
Locking Stumps Warr 183 C10
Lockinge Oxon 64 B2
Lockington E Yorks 208 D5
Lockington Leics 153 D9
Locklease Bristol 60 D6
Lockleywood Shrops 150 D3
Locks Heath Hants 33 F8
Locksbottom London 68 F2
Locksbrook Bath 61 G8
Locksgreen I o W 20 C4
Lockton N Yorks 226 G6
Lockwood W Yorks 196 D6
Loddington Leics 136 C5
Loddington Northants 120 B6
Loddiswell Devon 8 F4
Loddon Norf 143 D7
Loddon Ingloss Norf 142 D6
Lode Cambs 123 E10
Lode Heath W Mid 134 G3
Loders Dorset 16 C5
Lodge Green N Yorks 223 F9
Lodge Green W Mid 134 G5
Lodge Hill Corn 6 C4
Lodge Hill W Mid 133 G10
Lodge Hill W Yorks 197 C9
Lodge Lees Kent 55 D8
Lodge Moor S Yorks 186 D3
Lodge Park Worcs 117 D10
Lodgebank Shrops 149 D11
Lodsworth W Sus 34 C6
Lodsworth Common W Sus 34 C6
Lodway Bristol 60 D4
Lofthouse N Yorks 214 E2
Lofthouse W Yorks 197 B10
Lofthouse Gate W Yorks 197 C10
Loftus Redcar 226 B4
Logan E Ayrs 258 E3
Logan Mains Dumfries 236 E2
Loganlea W Loth 269 C9
Loggerheads Denb 165 C11
Loggerheads Staffs 150 B4
Logie Angus 293 G8
Logie Fife 287 E8
Logie Moray 301 D10
Logie Coldstone Aberds 292 C6
Logie Hill Highld 301 B7
Logie Newton Aberds 302 F6
Logie Pert Angus 293 G8
Logiealmond Lodge Perth 286 D3
Logierait Perth 286 B3
Login Carms 92 G3
Logmore Green Sur 50 D6
Loidse Mhorsgail W Isles 304 F3
Lolworth Cambs 123 E7
Lomeshaye Lancs 204 F2
Lôn Gwyn 147 C7
Lon-las Swansea 57 B8
Lonbain Highld 298 D6
Londesborough E Yorks 208 D3
London Apprentice Corn 5 E10
London Beach Kent 53 F11
London Colney Herts 85 E11
London End Cambs 121 D11
London Fields W Mid 133 E8
London Minstead Hants 32 E3
Londonderry N Yorks 214 B6
Londonderry W Mid 133 F10
Londonthorpe Lincs 155 B9
Londubh Highld 307 L3
Lonemore Highld 299 B7
Lonemore Highld 309 L7
Long Ashton N Som 60 E5
Long Bank Worcs 116 C5
Long Bennington Lincs 172 G4
Long Bredy Dorset 17 C7
Long Buckby Northants 120 D2
Long Buckby Wharf Northants 120 D2
Long Clawson Leics 154 D4
Long Common Hants 33 E8
Long Compton Staffs 151 E7
Long Compton Warks 100 E5
Long Crendon Bucks 83 D11
Long Crichel Dorset 31 E7
Long Cross Wilts 45 G9
Long Dean Wilts 61 D11
Long Ditton Sur 67 F7
Long Drax N Yorks 199 B7
Long Duckmanton Derbys 186 G6
Long Eaton Derbys 153 C9
Long Gardens Essex 106 D6
Long Green Suff 125 B11
Long Green W Ches 183 G7
Long Green Worcs 98 E6
Long Hanborough Oxon 82 C6
Long Itchington Warks 119 D8
Long John's Hill Norf 142 B4
Long Lane Telford 150 F2
Long Lawford Warks 119 B9
Long Lee W Yorks 205 E7
Long Load Som 29 C7
Long Marston Herts 84 B5
Long Marston N Yorks 206 C6
Long Marston Warks 100 B3
Long Marton Cumb 231 G9
Long Meadow Cambs 123 E10
Long Meadowend Shrops 131 G8
Long Melford Suff 107 B7
Long Newnton Glos 80 G6
Long Newton E Loth 271 C10
Long Oak Shrops 149 E7
Long Park Hants 48 G2
Long Preston N Yorks 204 B2
Long Riston E Yorks 209 E8
Long Sandall S Yorks 198 F6
Long Sight Gtr Man 196 F2
Long Stratton Norf 142 E3
Long Street M Keynes 102 B5
Long Sutton Hants 49 D8
Long Sutton Lincs 157 E8
Long Sutton Som 29 B7
Long Thurlow Suff 125 D10
Long Whatton Leics 153 E9
Long Wittenham Oxon 83 G8
Longbar N Ayrs 266 E6
Longbarn Warr 183 C10
Longbenton T & W 243 D7
Longborough Glos 100 F3
Longbridge Plym 7 D10
Longbridge W Mid 117 B10
Longbridge Warks 118 E5
Longbridge Deverill Wilts 45 E11
Longbridge Hayes Stoke 168 E5
Longbridgemuir Dumfries 238 D3
Longburgh Cumb 239 F8
Longburton Dorset 29 E11
Longcause Devon 8 C5
Longcliffe Derbys 170 D2
Longcot Oxon 82 G3
Longcroft Cumb 238 F6
Longcroft Falk 278 F5
Longcross Devon 12 F4
Longcross Sur 66 F3
Longdale Cumb 222 D2
Longdales Cumb 230 C6
Longden Shrops 131 B8
Longden Common Shrops 131 C8
Longdon Staffs 151 G11
Longdon Worcs 98 D6
Longdon Green Staffs 151 G11
Longdon Heath Worcs 98 D6
Longdon Hill End Worcs 98 D6
Longdon on Tern Telford 150 F2
Longdown Devon 14 C3
Longdowns Corn 2 C6
Longdrum Angus 292 G4
Longfield Kent 68 F6
Longfield Shetland 313 M5
Longfield Wilts 45 B11
Longfield Hill Kent 68 F6
Longfleet Poole 18 C6
Longford Derbys 152 B4
Longford Glos 98 G6
Longford Kent 52 B4
Longford London 66 D4
Longford Shrops 150 C2
Longford Telford 150 F4
Longford W Mid 135 G7
Longford Warr 183 C10
Longfordlane Derbys 152 B4
Longforgan Perth 287 E7
Longformacus Borders 272 D3
Longframlington Northumb 252 C4
Longham Dorset 19 B7
Longham Norf 159 F8
Longhaven Aberds 303 F11
Longhedge Wilts 45 E10
Longhill Aberds 303 D9
Longhirst Northumb 252 F6
Longhope Glos 79 B11
Longhope Orkney 314 G3
Longhorsley Northumb 252 E4
Longhoughton Northumb 264 G6
Longhouse Bath 61 G8
Longlands Cumb 229 D11
Longlands London 68 E2
Longlane Derbys 152 B5
Longlane W Berks 64 E3
Longlevens Glos 99 G7
Longley W Yorks 196 C5
Longley W Yorks 196 F6
Longley Estate S Yorks 186 C5
Longley Green Worcs 116 G4
Longleys Perth 286 C6
Longmanhill Aberds 303 C7
Longmoor Camp Hants 49 G9
Longmorn Moray 302 D2
Longmoss E Ches 184 G5
Longnewton Borders 262 D3
Longnewton Stockton 225 B7
Longney Glos 80 C3
Longniddry E Loth 281 F8
Longnor Shrops 131 C9
Longnor Staffs 151 G7
Longnor Staffs 169 C9
Longnor Park Shrops 131 C9
Longparish Hants 48 E2
Longpark Cumb 239 E10
Longpark E Ayrs 257 B10
Longport Stoke 168 F5
Longridge Glos 80 D5
Longridge Lancs 203 F8
Longridge Staffs 151 F8
Longridge W Loth 269 C9
Longridge End Glos 98 G6
Longrigg N Lnrk 278 G6
Longriggend N Lnrk 278 G6
Longrock Corn 1 C5
Longscales N Yorks 205 B10
Longsdon Staffs 169 E7
Longshaw Gtr Man 194 G4
Longshaw Staffs 169 F9
Longside Aberds 303 E10
Longsight Gtr Man 184 B5
Longslow Shrops 150 B3
Longsowerby Cumb 239 G9
Longstanton Cambs 123 D7
Longstock Hants 47 F11
Longstone Corn 2 B2
Longstone Corn 11 G7
Longstone Edin 280 G4
Longstone Pembs 73 D10
Longstone Som 28 B5
Longstowe Cambs 122 G6
Longstreet Wilts 46 C6
Longthorpe P'boro 138 D3
Longthwaite Cumb 230 G4
Longton Lancs 194 B3
Longton Stoke 168 G6
Longtown Cumb 239 D9
Longtown Hereford 96 F6
Longtownmail Orkney 314 F4
Longview Mers 182 C6
Longville in the Dale Shrops 131 E10
Longway Bank Derbys 170 E4
Longwell Green S Glos 61 E7
Longwick Bucks 84 D3
Longwitton Northumb 252 F3
Longwood Shrops 132 B2
Longwood W Yorks 196 D6
Longwood Edge W Yorks 196 D6
Longworth Oxon 82 F5
Longyester E Loth 271 B10
Lonmay Aberds 303 D10
Lonmore Highld 298 E2
Looe Corn 6 E5
Looe Mills Corn 6 C4
Loose Kent 53 C9
Loose Hill Kent 53 C9
Loosegate Lincs 156 D6
Loosley Row Bucks 84 E4
Lopcombe Corner Wilts 47 F9
Lopen Som 28 E6
Lopen Head Som 28 E6
Loppergarth Cumb 210 D5
Loppington Shrops 149 D9
Lopwell Devon 7 C9
Lorbottle Northumb 252 B2
Lorbottle Hall Northumb 252 B2
Lord's Hill Soton 32 D5
Lords Wood Medway 69 G9
Lordsbridge Norf 157 G11
Lordshill Common Sur 50 E4
Lordswood Soton 32 D6
Lornty Perth 286 C5
Loscoe Derbys 170 F6
Loscoe W Yorks 198 C2
Loscombe Dorset 16 B6
Losgaintir W Isles 305 J2
Lossiemouth Moray 302 B2
Lossit Argyll 254 B2
Lossit Lodge Argyll 274 G5
Lostford Shrops 150 C2
Lostock Gtr Man 195 F7
Lostock Gralam W Ches 183 F11
Lostock Green W Ches 183 G11
Lostock Hall Lancs 194 B4
Lostock Junction Gtr Man 195 F7
Lostwithiel Corn 6 D2
Loth Orkney 314 C6
Lothbeg Highld 311 H3
Lothersdale N Yorks 204 D5
Lothianbridge Midloth 270 C6
Lothmore Highld 311 H3
Lottisham Som 44 G5
Loudwater Bucks 84 G6
Loudwater Herts 85 F9
Loughborough Leics 153 F10
Loughor Swansea 56 B5
Loughton Essex 86 F6
Loughton M Keynes 102 D6
Loughton Shrops 132 G2
Lound Lincs 155 F11
Lound Notts 187 D11
Lound Suff 143 D10
Loundsley Green Derbys 186 G5
Lount Leics 153 F7
Lour Angus 287 C8
Louth Lincs 190 D4
Lovat Highld 300 E5
Lovaton Devon 7 B10
Love Clough Lancs 195 B10
Love Green Bucks 66 C4
Lovedean Hants 33 E11
Lover Wilts 32 C2
Loversall S Yorks 187 B9
Loves Green Essex 87 E10
Lovesome Hill N Yorks 225 F7
Loveston Pembs 73 D9
Lovington Som 44 G5
Low Ackworth W Yorks 198 D3
Low Alwinton Northumb 251 B10
Low Angerton Northumb 252 G3
Low Barlings Lincs 189 G9
Low Barugh S Yorks 197 F10
Low Bentham N Yorks 212 F2
Low Biggins Cumb 212 D2
Low Blantyre S Lnrk 268 D3
Low Borrowbridge Cumb 222 E2
Low Bradfield S Yorks 186 C3
Low Bradley N Yorks 204 D6
Low Braithwaite Cumb 230 C4
Low Bridge Wilts 62 E3
Low Brunton Northumb 241 C10
Low Burnham N Lincs 199 G9
Low Burton N Yorks 214 C4
Low Buston Northumb 252 B6
Low Catton E Yorks 207 C10
Low Clanyard Dumfries 236 F3
Low Common Norf 142 E2
Low Compton Gtr Man 196 F2
Low Coniscliffe Darl 224 C5
Low Cotehill Cumb 239 G11
Low Coylton S Ayrs 257 F10
Low Crosby Cumb 239 F10
Low Dalby N Yorks 217 B7
Low Dinsdale Darl 224 C6
Low Eighton T & W 243 F7
Low Ellington N Yorks 214 C4
Low Etherley Durham 233 F9
Low Fell T & W 243 F7
Low Fold W Yorks 205 F10
Low Fulney Lincs 156 E5
Low Garth N Yorks 226 D4
Low Gate N Yorks 214 F5
Low Gate Northumb 241 E10
Low Geltbridge Cumb 240 E2
Low Grantley N Yorks 214 E4
Low Green N Yorks 205 B10
Low Green Suff 125 E7
Low Green W Yorks 205 F10
Low Greenside T & W 242 E4
Low Habberley Worcs 116 B6
Low Ham Som 28 B6
Low Hauxley Northumb 253 C7
Low Hawsker N Yorks 227 D8
Low Hesket Cumb 230 B5
Low Hesleyhurst Northumb 252 D3
Low Hill W Mid 133 C8
Low Hutton N Yorks 216 F5
Low Knipe Cumb 230 G6
Low Laithe N Yorks 214 G3
Low Laithes S Yorks 197 G11
Low Leighton Derbys 185 D8
Low Lorton Cumb 229 F9
Low Marishes N Yorks 216 D6
Low Marnham Notts 172 B4
Low Mill N Yorks 226 F3
Low Moor Lancs 203 E10
Low Moor W Yorks 197 B7
Low Moorsley T & W 234 B2
Low Moresby Cumb 228 G5
Low Newton Cumb 211 C8
Low Newton-by-the-Sea Northumb 264 E6
Low Prudhoe Northumb 242 E4
Low Risby N Lincs 200 E2
Low Row Cumb 229 C9
Low Row Cumb 240 E3
Low Row N Yorks 223 F9
Low Salchrie Dumfries 236 C2
Low Smerby Argyll 255 E8
Low Snaygill N Yorks 204 D5
Low Street Norf 141 B10
Low Street Thurrock 69 D7
Low Tharston Norf 142 D3
Low Thornley T & W 242 E5
Low Torry Fife 279 D10
Low Town Shrops 132 E4
Low Toynton Lincs 190 G3
Low Valley S Yorks 198 G2
Low Valleyfield Fife 279 D10
Low Walton Cumb 219 C9
Low Waters S Lnrk 268 E4
Low Westwood Durham 242 F4
Low Whinnow Cumb 239 G8
Low Whita N Yorks 223 F10
Low Wood Cumb 210 C6
Low Worsall N Yorks 225 C7
Low Wray Cumb 221 E7
Lowbands Glos 98 E5
Lowbridge House Cumb 221 E10
Lowca Cumb 228 G5
Lowcross Hill W Ches 167 E7
Lowdham Notts 171 F11
Lowe Shrops 149 C10
Lowe Hill Staffs 169 D7
Lowedges S Yorks 186 E4
Lower Achachenna Argyll 284 E4
Lower Aisholt Som 43 F8
Lower Allscot Shrops 132 D4
Lower Altofts W Yorks 197 C11
Lower Amble Corn 10 G5
Lower Ansty Dorset 30 G3
Lower Ardtun Argyll 288 G5
Lower Arncott Oxon 83 B10
Lower Ashtead Sur 51 B7
Lower Ashton Devon 14 E2
Lower Assendon Oxon 65 C8
Lower Badcall Highld 306 E6
Lower Ballam Lancs 202 G3
Lower Bartle Lancs 202 G5
Lower Basildon W Berks 64 D6
Lower Bassingthorpe Lincs 155 D9
Lower Bearwood Hereford 115 F7
Lower Bebington Mers 182 E4
Lower Beeding W Sus 36 B2
Lower Benefield Northants 137 F9
Lower Bentley Worcs 117 D9
Lower Beobridge Shrops 132 E5
Lower Berry Hill Glos 79 C9
Lower Binton Warks 118 G2
Lower Birchwood Derbys 170 E6
Lower Bitchet Kent 52 C5
Lower Blandford St Mary Dorset 30 F5
Lower Blunsdon Swindon 81 G10
Lower Bobbingworth Green Essex 87 D8
Lower Bockhampton Dorset 17 C10
Lower Boddington Northants 119 G9
Lower Bodham Norf 160 B2
Lower Bodinnar Corn 1 C4
Lower Bois Bucks 85 E7
Lower Bordean Hants 33 C11
Lower Boscaswell Corn 1 C3
Lower Bourne Sur 49 E10
Lower Bradley W Mid 133 D9
Lower Brailes Warks 100 D6
Lower Breakish Highld 295 C8
Lower Bredbury Gtr Man 184 C6
Lower Breinton Hereford 97 D9
Lower Broadheath Worcs 116 F6
Lower Brook Hants 32 B4
Lower Broughton Gtr Man 184 B4
Lower Brynamman Neath 76 C2
Lower Brynn Corn 5 C9
Lower Buckenhill Hereford 98 E2
Lower Buckland Hants 20 B2
Lower Bullingham Hereford 97 D10
Lower Bullington Hants 48 E3
Lower Bunbury E Ches 167 D9
Lower Burgate Hants 31 D11
Lower Burrow Som 28 C6
Lower Burton Hereford 115 F8
Lower Bush Medway 69 F7
Lower Cadsden Bucks 84 E4
Lower Caldecote C Beds 104 B3
Lower Cam Glos 80 E2
Lower Canada N Som 43 B11
Lower Carden W Ches 167 E7
Lower Catesby Northants 119 F10
Lower Cator Devon 13 F9
Lower Caversham Reading 65 E8
Lower Chapel Powys 95 D10
Lower Chedworth Glos 81 C9
Lower Cheriton Devon 27 G10
Lower Chicksgrove Wilts 46 G3
Lower Chute Wilts 47 C10
Lower Clapton London 67 B11
Lower Clent Worcs 117 B8
Lower Clicker Corn 6 C5
Lower Clopton Warks 118 F3
Lower Common Hants 48 E6
Lower Common Hants 65 G9
Lower Common Mon 78 B2
Lower Common Mon 78 E4
Lower Common Shrops 131 B9
Lower Copthurst Lancs 194 C5
Lower Cotburn Aberds 303 D7
Lower Cousley Wood E Sus 53 G7
Lower Cox Street Kent 69 G10
Lower Cragabus Argyll 254 C4
Lower Creedy Devon 26 G4
Lower Croan Corn 10 G6
Lower Crossings Derbys 185 E8
Lower Cumberworth W Yorks 197 F8
Lower Cwm-twrch Powys 76 C3
Lower Daggons Hants 31 E9
Lower Darwen Blkburn 195 B7
Lower Dean Beds 121 D11
Lower Dean Devon 8 C4
Lower Dell Highld 292 B2
Lower Denby W Yorks 197 F8
Lower Denzell Corn 5 B7
Lower Deuchries Aberds 302 D6
Lower Diabaig Highld 299 C7
Lower Dicker E Sus 23 C9
Lower Dinchope Shrops 131 G9
Lower Dowdeswell Glos 81 B8
Lower Down Shrops 130 G6
Lower Drift Corn 1 D4
Lower Dunsforth N Yorks 215 G8
Lower Durston Som 28 B3
Lower Earley Wokingham 65 E9
Lower East Carleton Norf 142 C3
Lower Eastern Green W Mid 118 B5
Lower Edmonton London 86 G4
Lower Egleton Hereford 98 B2
Lower Elkstone Staffs 169 D9
Lower Ellastone Staffs 169 G10
Lower End Bucks 83 D11
Lower End Bucks 102 E4
Lower End C Beds 103 D8
Lower End C Beds 103 G9
Lower End Glos 81 E7
Lower End Northants 120 F6
Lower End Northants 120 G5
Lower End Northants 121 E7
Lower End Oxon 82 B4
Lower Everleigh Wilts 47 C7
Lower Eythorne Kent 55 D9
Lower Failand N Som 60 E4
Lower Faintree Shrops 132 F3
Lower Falkenham Suff 108 D5
Lower Farringdon Hants 49 F8
Lower Feltham London 66 E5
Lower Fittleworth W Sus 35 D8
Lower Forge Shrops 132 F4
Lower Foxdale I o M 192 E3
Lower Frankton Shrops 149 C7
Lower Freystrop Pembs 73 C7
Lower Froyle Hants 49 E9
Lower Gabwell Devon 9 B8
Lower Gledfield Highld 309 K5
Lower Godney Som 44 E3
Lower Goldstone Kent 71 G9
Lower Gornal W Mid 133 E8
Lower Grange W Yorks 205 G8
Lower Gravenhurst C Beds 104 D2
Lower Green Essex 88 E2
Lower Green Essex 105 E8
Lower Green Essex 106 E4
Lower Green Gtr Man 184 B2
Lower Green Herts 104 E3
Lower Green Kent 52 E5
Lower Green Kent 52 E6
Lower Green Norf 159 B9
Lower Green Staffs 133 B8
Lower Green Suff 124 D4
Lower Green Sur 66 F6
Lower Green W Berks 63 G11
Lower Green Warks 119 D10
Lower Grove Common Hereford 97 F11
Lower Hacheston Suff 126 F6
Lower Halistra Highld 298 D2
Lower Halliford Sur 66 F5
Lower Halstock Leigh Dorset 29 F8
Lower Halstow Kent 69 F11
Lower Hamswell S Glos 61 E8
Lower Hamworthy Poole 18 C6
Lower Hardres Kent 55 C7
Lower Hardwick Hereford 115 F8
Lower Harpton Powys 114 E5
Lower Hartlip Kent 69 G10
Lower Hartshay Derbys 170 E5
Lower Hartwell Bucks 84 C3
Lower Hatton Staffs 150 B6
Lower Hawthwaite Cumb 210 B4
Lower Haysden Kent 52 D5
Lower Hayton Shrops 131 G10
Lower Hazel S Glos 60 B6
Lower Heath E Ches 168 C5
Lower Hempriggs Moray 301 C11
Lower Heppington Kent 54 C6
Lower Hergest Hereford 114 F5
Lower Herne Kent 71 F7
Lower Heyford Oxon 101 G9
Lower Heysham Lancs 211 G8
Lower Higham Kent 69 E8
Lower Highmoor Oxon 65 B8
Lower Holbrook Suff 108 E3
Lower Holditch Dorset 28 G4
Lower Holloway London 67 B10
Lower Holwell Dorset 31 E9
Lower Hook Worcs 98 C6
Lower Hookner Devon 13 E10
Lower Hopton Shrops 149 E7
Lower Hopton W Yorks 197 D7
Lower Hordley Shrops 149 D7
Lower Horncroft W Sus 35 E8
Lower Horsebridge E Sus 23 C9
Lower House Halton 183 D8
Lower Houses W Yorks 197 D7
Lower Howsell Worcs 98 B5
Lower Illey Worcs 133 G9
Lower Island Kent 70 F6
Lower Kersal Gtr Man 195 G10
Lower Kilburn Derbys 170 F5
Lower Kilcott Glos 61 B9
Lower Killeyan Argyll 254 C3
Lower Kingcombe Dorset 17 B7
Lower Kingswood Sur 51 C8
Lower Kinnerton W Ches 166 C4
Lower Kinsham Hereford 115 E7
Lower Knapp Som 28 B4
Lower Knightley Staffs 150 E6
Lower Knowle Bristol 60 E5
Lower Langford N Som 60 G3
Lower Largo Fife 287 G8
Lower Layham Suff 107 C10
Lower Ledwyche Shrops 115 C10
Lower Leigh Staffs 151 B10
Lower Lemington Glos 100 E4
Lower Lenie Highld 300 G5
Lower Lode Glos 99 E7
Lower Lovacott Devon 25 B8
Lower Loxhore Devon 40 F6
Lower Lydbrook Glos 79 B9
Lower Lye Hereford 115 D8
Lower Machen Newport 59 B8
Lower Maes-coed Hereford 96 E6
Lower Mains Clack 279 B9
Lower Mannington Dorset 31 F9
Lower Marsh Som 30 C2
Lower Marston Som 45 E9
Lower Meend Glos 79 E9

Lower Menadue Corn 5 D10
Lower Merridge Som 43 G8
Lower Mickletown W Yorks 198 B2
Lower Middleton Cheney Northants 101 C10
Lower Midway Derbys 152 E6
Lower Mill Corn 3 B10
Lower Milovaig Highld 296 F7
Lower Milton Som 44 D4
Lower Moor Wilts 81 G8
Lower Moor Worcs 99 B9
Lower Morton S Glos 79 G10
Lower Mountain Flint 166 D4
Lower Nazeing Essex 86 D5
Lower Netchwood Shrops 132 E2
Lower Netherton Devon 14 G3
Lower New Inn Torf 78 F4
Lower Ninnes Corn 1 C5
Lower Nobut Staffs 151 C10
Lower North Dean Bucks 84 F5
Lower Norton Warks 118 E4
Lower Nyland Dorset 30 C2
Lower Ochrwyth Caerph 59 B8
Lower Odcombe Som 29 D8
Lower Oddington Glos 100 F4
Lower Ollach Highld 295 B7
Lower Padworth W Berks 64 F6
Lower Penarth V Glam 59 F7
Lower Penn Staffs 133 D7
Lower Pennington Hants 20 C2
Lower Penwortham Lancs 194 B4
Lower Peover W Ches 184 G2
Lower Pexhill E Ches 184 G5
Lower Pilsley Derbys 170 C6
Lower Pitkerrie Highld 311 L2
Lower Place Gtr Man 196 E2
Lower Place London 67 C8
Lower Pollicot Bucks 84 C2
Lower Porthkerry V Glam 58 F5
Lower Porthpean Corn 5 E10
Lower Quinton Warks 100 B3
Lower Rabber Hereford 114 G5
Lower Race Torf 78 E3
Lower Radley Oxon 83 F8
Lower Rainham Medway 69 F10
Lower Ratley Hants 32 C4
Lower Raydon Suff 107 D10
Lower Rea Glos 80 B4
Lower Ridge Devon 28 G2
Lower Ridge Shrops 148 C6
Lower Roadwater Som 42 F4
Lower Rochford Worcs 116 D2
Lower Rose Corn 4 E5
Lower Row Dorset 31 G8
Lower Sapey Worcs 116 E3
Lower Seagry Wilts 62 C3
Lower Sheering Essex 87 C7
Lower Shelton C Beds 103 C9
Lower Shiplake Oxon 65 D9
Lower Shuckburgh Warks 119 E9
Lower Skelty Swansea 56 C6
Lower Slackstead Hants 32 B5
Lower Slade Devon 40 D4
Lower Slaughter Glos 100 G3
Lower Solva Pembs 87 G11
Lower Soothill W Yorks 197 C9
Lower Soudley Glos 79 D11
Lower Southfield Hereford 98 C3
Lower Stanton St Quintin Wilts 62 C2
Lower Stoke Medway 69 D10
Lower Stoke W Mid 119 B7
Lower Stondon C Beds 104 D3
Lower Stone Glos 79 G11
Lower Stonnall Staffs 133 C11
Lower Stow Bedon Norf 141 E9
Lower Stratton Som 28 D6
Lower Stratton Swindon 63 B7
Lower Street E Sus 38 E2
Lower Street Norf 160 B5
Lower Street Norf 160 C3
Lower Street Norf 160 F6
Lower Street Suff 108 E3
Lower Street Suff 124 G5
Lower Strensham Worcs 99 C8
Lower Stretton Warr 183 E10
Lower Studley Wilts 45 B11
Lower Sundon C Beds 103 F10
Lower Swainswick Bath 61 F9
Lower Swanwick Hants 33 F7
Lower Swell Glos 100 F3
Lower Sydenham London 67 E11
Lower Tadmarton Oxon 101 D8
Lower Tale Devon 27 G9
Lower Tasburgh Norf 142 D3
Lower Tean Staffs 151 B10
Lower Thorpe Northants 101 B10
Lower Threapwood Wrex 166 G6
Lower Thurlton Norf 143 D8
Lower Thurnham Lancs 202 C5
Lower Thurvaston Derbys 152 B4
Lower Todding Hereford 115 B8
Lower Tote Highld 298 C5
Lower Town Devon 27 E8
Lower Town Hereford 98 C2
Lower Town Pembs 91 D9
Lower Town W Yorks 204 G6
Lower Town Worcs 117 F7
Lower Trebullett Corn 12 F2
Lower Tregunnon Corn 11 E10
Lower Treworrick Corn 6 B4
Lower Tuffley Glos 80 C4
Lower Turmer Hants 31 F10
Lower Twitchen Devon 24 D5
Lower Twydall Medway 69 F10
Lower Tysoe Warks 100 B6
Lower Upham Hants 33 D8
Lower Upnor Medway 69 E9
Lower Vexford Som 42 F6
Lower Wainhill Oxon 84 E3
Lower Walton Warr 183 D10
Lower Wanborough Swindon 63 C8
Lower Weacombe Som 42 E6
Lower Weald M Keynes 102 D5
Lower Wear Devon 14 D4
Lower Weare Som 44 C2
Lower Weedon Northants 120 F2
Lower Welson Hereford 114 G5
Lower Westholme Som 44 E5
Lower Westhouse N Yorks 212 E3
Lower Westmancote Worcs 99 D8
Lower Weston Bath 61 F8
Lower Whatcombe Dorset 30 G4
Lower Whatley Som 45 D8
Lower Whitley W Ches 183 F10
Lower Wick Glos 80 F2
Lower Wick Worcs 116 G6
Lower Wield Hants 48 E6
Lower Willingdon E Sus 23 E9
Lower Winchendon or Nether Winchendon Bucks 84 C2
Lower Withington E Ches 168 B4
Lower Wolverton Worcs 117 G8
Lower Woodend Aberds 293 B8
Lower Woodend Bucks 65 B10
Lower Woodford Wilts 46 G6
Lower Woodley Corn 5 B10
Lower Woodside Herts 86 D2
Lower Woolston Som 29 B11
Lower Woon Corn 5 C10
Lower Wraxall Dorset 29 G9
Lower Wraxall Som 44 F6
Lower Wraxall Wilts 61 G10
Lower Wych W Ches 167 G7
Lower Wyche Worcs 98 C5
Lower Wyke W Yorks 197 B7
Lower Yelland Devon 40 G3
Lower Zeals Wilts 45 G9
Lowerford Lancs 204 F3
Lowerhouse E Ches 184 F6
Lowerhouse Lancs 204 G2
Lowertown Corn 2 D5
Lowertown Corn 5 C11
Lowertown Devon 12 E5
Lowes Barn Durham 233 C11
Lowesby Leics 136 B4
Lowestoft Suff 143 E10
Loweswater Cumb 229 G8
Lowfield S Yorks 186 D5
Lowfield Heath W Sus 51 E9
Lowford Hants 33 E7
Lowgill Cumb 222 F2
Lowgill Lancs 212 G3
Lowick Cumb 210 B5
Lowick Northants 137 G9
Lowick Northumb 264 B2
Lowick Bridge Cumb 210 B5
Lowick Green Cumb 210 B5
Lowlands Torf 78 F3
Lowmoor Row Cumb 231 F8
Lowna N Yorks 226 G3
Lownie Moor Angus 287 C8
Lowood Borders 262 B2
Lowsonford Warks 118 D3
Lowther Cumb 230 G6
Lowthertown Dumfries 238 D6
Lowtherville I o W 21 F7
Lowthorpe E Yorks 217 G11
Lowton Gtr Man 183 B10
Lowton Som 27 D11
Lowton Common Gtr Man 183 B10
Lowton Heath Gtr Man 183 B10
Lowton St Mary's Gtr Man 183 B10
Loxbeare Devon 26 D6
Loxford London 68 B2
Loxhill Sur 50 F4
Loxhore Devon 40 F6
Loxhore Cott Devon 40 F6
Loxley S Yorks 186 D4
Loxley Warks 118 G5
Loxley Green Staffs 151 C11
Loxter Hereford 98 C4
Loxton N Som 43 B11
Loxwood W Sus 50 G4
Loyter's Green Essex 87 C8
Loyterton Kent 70 G3
Lozells W Mid 133 F11
Lubachlaggan Highld 300 B3
Lubachoinnich Highld 309 K4
Lubberland Shrops 116 B2
Lubcroy Highld 309 J3
Lubenham Leics 136 F4
Lubinvullin Highld 308 C5
Lucas End Herts 86 E4
Lucas Green Lancs 194 C5
Lucas Green Sur 50 B2
Luccombe Som 42 E2
Luccombe Village I o W 21 F7
Lucker Northumb 264 C5
Luckett Corn 12 G3
Lucking Street Essex 106 E6
Luckington Wilts 61 C10
Lucklawhill Fife 287 E8
Luckwell Bridge Som 42 F2
Lucton Hereford 115 E8
Ludag W Isles 297 K3
Ludborough Lincs 190 B3
Ludbrook Devon 8 E3
Ludchurch Pembs 73 C10
Luddenden W Yorks 196 B4
Luddenden Foot W Yorks 196 C4
Ludderburn Cumb 221 G8
Luddesdown Kent 69 F7
Luddington N Lincs 199 D10
Luddington Warks 118 G3
Luddington in the Brook Northants 138 G2
Lude House Perth 291 G10
Ludford Lincs 190 D2
Ludford Shrops 115 C10
Ludgershall Bucks 83 B11
Ludgershall Wilts 47 C9
Ludgvan Corn 2 C2
Ludham Norf 161 F7
Ludlow Shrops 115 C10
Ludney Lincs 190 B5
Ludney Som 28 E5
Ludstock Hereford 98 D3
Ludstone Shrops 132 E6
Ludwell Wilts 30 C6
Ludworth Durham 234 C3
Luffenhall Herts 104 F5
Luffincott Devon 12 C2
Lufton Som 29 D8
Lugar E Ayrs 258 E3
Lugate Borders 271 G8
Lugg Green Hereford 115 E9
Luggate Burn E Loth 282 G2
Luggiebank N Lnrk 278 G5
Lugsdale Halton 183 D8
Lugton E Ayrs 267 E8
Lugwardine Hereford 97 C11
Luib Highld 295 C7
Luibeilt Highld 290 G4
Lulham Hereford 97 C8
Lullenden Sur 52 E2
Lullington Derbys 152 G5
Lullington Som 45 C9
Lulsgate Bottom N Som 60 F4
Lulsley Worcs 116 F4
Lulworth Camp Dorset 18 E2
Lumb Lancs 195 C10
Lumb Lancs 195 D9
Lumb W Yorks 196 C4
Lumb W Yorks 197 E7
Lumb Foot W Yorks 204 F6
Lumburn Devon 12 G5
Lumbutts W Yorks 196 C3
Lumby N Yorks 206 G5
Lumley W Sus 22 B3
Lumley Thicks Durham 243 G8
Lumloch E Dunb 268 B2
Lumphanan Aberds 293 C7
Lumphinnans Fife 280 C3
Lumsdaine Borders 273 B7
Lumsden Aberds 302 G4
Lunan Angus 287 B10
Lunanhead Angus 287 B8
Luncarty Perth 286 E4
Lund E Yorks 208 D5
Lund N Yorks 207 G9
Lund Shetland 312 C7
Lundal W Isles 304 E3
Lundavra Highld 290 G2
Lunderton Aberds 303 E11
Lundie Angus 286 D6
Lundie Highld 290 B3
Lundin Links Fife 287 G8
Lundwood S Yorks 197 F11
Lundy Green Norf 142 E4
Lunga Argyll 275 C8
Lunna Shetland 312 G6
Lunning Shetland 312 G7
Lunnister Shetland 312 F5
Lunnon Swansea 56 D4
Lunsford Kent 53 B7
Lunsford's Cross E Sus 38 E2
Lunt Mers 193 G10
Luntley Hereford 115 F7
Lunts Heath Halton 183 D8
Lupin Staffs 152 F2
Luppitt Devon 27 F11
Lupridge Devon 8 E4
Lupset W Yorks 197 D10
Lupton Cumb 211 C11
Lurg Aberds 293 C8
Lurgashall W Sus 34 B6
Lurignich Argyll 289 D11
Lurley Devon 26 E6
Lusby Lincs 174 B4
Luscott Shrops 131 D11
Luson Devon 8 F2
Luss Argyll 277 C7
Lussagiven Argyll 275 E7
Lusta Highld 298 D2
Lustleigh Devon 13 E11
Lustleigh Cleave Devon 13 E11
Luston Hereford 115 E9
Lusty Som 45 G7
Luthermuir Aberds 293 G8
Luthrie Fife 287 F7
Lutley W Mid 133 G8
Luton Devon 14 F4
Luton Devon 27 G9
Luton Luton 103 G11
Luton Medway 69 F9
Lutsford Devon 24 D3
Lutterworth Leics 135 G10
Lutton Devon 7 D11
Lutton Devon 8 C3
Lutton Lincs 157 D8
Lutton Northants 138 F2
Lutton Gowts Lincs 157 E8
Lutworthy Devon 26 D3
Luxborough Som 42 F3
Luxley Glos 98 G3
Luxted London 68 G2
Luxton Devon 28 E2
Luxulyan Corn 5 D11
Luzley Gtr Man 196 G3
Luzley Brook Gtr Man 196 F2
Lyatts Som 29 E8
Lybster Highld 310 F6
Lydbury North Shrops 131 F7
Lydcott Devon 41 F7
Lydd Kent 39 C8
Lydd on Sea Kent 39 C9
Lydden Kent 55 D9
Lydden Kent 71 F11
Lyddington Rutland 137 D7
Lyde Orkney 314 E3
Lyde Shrops 130 C6
Lyde Cross Hereford 97 C10
Lyde Green Hants 49 B8
Lyde Green S Glos 61 D7
Lydeard St Lawrence Som 42 G6
Lydford Devon 12 E6
Lydford Fair Place Som 44 G5
Lydford-on-Fosse Som 44 G5
Lydgate Derbys 186 F4
Lydgate Gtr Man 196 G3
Lydgate W Yorks 196 B2
Lydham Shrops 130 E6
Lydiard Green Wilts 62 B5
Lydiard Millicent Wilts 62 B5
Lydiard Plain Wilts 62 B5
Lydiard Tregoze Swindon 62 C6
Lydiate Mers 193 G11
Lydiate Ash Worcs 117 B9
Lydlinch Dorset 30 E2
Lydmarsh Som 28 F5
Lydney Glos 79 E10
Lydstep Pembs 73 F9
Lye W Mid 133 G8
Lye Cross N Som 60 G3
Lye Green Bucks 85 E7
Lye Green E Sus 52 G4
Lye Green Warks 118 D3
Lye Green Wilts 45 B10
Lye Head Worcs 116 C5
Lye Hole N Som 60 G4
Lyewood Common E Sus 52 F4
Lyford Oxon 82 G5
Lymbridge Green Kent 54 E6
Lyme Green E Ches 184 G6
Lyme Regis Dorset 16 C2
Lymiecleuch Borders 249 C9
Lyminge Kent 55 E7
Lymington Hants 20 B2
Lyminster W Sus 35 G8
Lymm Warr 183 D11
Lymore Hants 19 C11
Lympne Kent 54 F6
Lympsham Som 43 C10
Lympstone Devon 14 E5
Lynbridge Devon 41 D8
Lynch Hants 48 D4
Lynch Som 42 D2
Lynch Hill Hants 48 D3
Lynch Hill Slough 66 C2
Lynchat Highld 291 C9
Lynchgate Shrops 131 F7
Lyndale Ho Highld 298 D3
Lyndhurst Hants 32 F4
Lyndon Rutland 137 C8
Lyndon Green W Mid 134 F2
Lyne Borders 270 G4
Lyne Sur 66 F4
Lyne Down Hereford 98 E2
Lyne of Gorthleck Highld 300 G5
Lyne of Skene Aberds 293 B9
Lyne Station Borders 260 B6
Lyneal Shrops 149 C8
Lyneal Mill Shrops 149 C9
Lyneal Wood Shrops 149 C9
Lyneham Oxon 100 G5
Lyneham Wilts 62 D4
Lynemore Highld 301 G10
Lynemouth Northumb 253 E7
Lyness Orkney 314 G3
Lynford Norf 140 E6
Lyng Norf 159 F11
Lyng Som 28 B4
Lyngate Norf 160 C5
Lyngford Som 28 B2
Lynmore Highld 301 F10
Lynmouth Devon 41 D8
Lynn Staffs 133 C11
Lynn Telford 150 F5
Lynnwood Borders 261 G11
Lynsore Bottom Kent 55 D7
Lynsted Kent 70 G2
Lynstone Corn 24 F2
Lynton Devon 41 D8
Lynwilg Highld 291 B10
Lynworth Glos 99 G9
Lyons T & W 234 B3
Lyon's Gate Dorset 29 F11
Lyon's Green Norf 159 G8
Lyons Hall Essex 88 B2
Lyonshall Hereford 114 F6
Lypiatt Glos 80 D6
Lyrabus Argyll 274 G3
Lytchett Matravers Dorset 18 B4
Lytchett Minster Dorset 18 C5
Lyth Highld 310 C6
Lytham Lancs 193 B11
Lytham St Anne's Lancs 193 B10
Lythbank Shrops 131 B9
Lythe N Yorks 226 C6
Lythes Orkney 314 H4
Lythmore Highld 310 C4

M

Maam Argyll 284 F5
Mabe Burnthouse Corn 3 C7
Mabie Dumfries 237 B11
Mableden Kent 52 E5
Mablethorpe Lincs 191 D8
Macclesfield E Ches 184 G6
Macclesfield Forest E Ches 185 G7
Macduff Aberds 303 C7
Mace Green Suff 108 C2
Machan S Lnrk 268 E5
Macharioch Argyll 255 G8
Machen Caerph 59 B8
Machrie N Ayrs 255 D9
Machrie Hotel Argyll 254 C4
Machrihanish Argyll 255 E7
Machroes Gwyn 144 D6
Machynlleth Powys 128 C4
Machynys Carms 56 B4
Mackerel's Common W Sus 35 B8
Mackerye End Herts 85 B11
Mackham Devon 27 F11
Mackney Oxon 64 B5
Mackside Borders 262 G4
Mackworth Derbys 152 B6
Macmerry E Loth 281 G8
Madderty Perth 286 E3
Maddington Wilts 46 E5
Maddiston Falk 279 F8
Maddox Moor Pembs 73 C7
Madehurst W Sus 35 E7
Madeley Staffs 168 G3
Madeley Telford 132 C3
Madeley Heath Staffs 168 F3
Madeley Heath Worcs 117 B9
Madeley Park Staffs 168 G3
Madeleywood Telford 132 C3
Maders Corn 12 G2
Madford Devon 27 E10
Madingley Cambs 123 E7
Madjeston Dorset 30 B4
Madley Hereford 97 D8
Madresfield Worcs 98 B6
Madron Corn 1 C5
Maen-y-groes Ceredig 111 F7
Maenaddwyn Anglesey 179 E7
Maenclochog Pembs 91 F11
Maendy V Glam 58 D4
Maenporth Corn 3 D7
Maentwrog Gwyn 163 G11
Maer Corn 24 F2
Maer Staffs 150 B5
Maerdy Carms 94 G2
Maerdy Conwy 165 G8
Maerdy Rhondda 77 F7
Maes-bangor Ceredig 128 G3
Maes-glas Newport 59 B9
Maes Glas = Greenfield Flint 181 F11
Maes llyn Ceredig 93 C7
Maes Pennant Flint 181 F11
Maes-Treylow Powys 114 D5
Maes-y-dre Flint 166 C2
Maesbrook Shrops 148 E5
Maesbury Shrops 148 D6
Maesbury Marsh Shrops 148 D6
Maesgeirchen Gwyn 179 G9
Maesgwyn-Isaf Powys 148 G3
Maesgwynne Carms 92 G4
Maeshafn Denb 166 C2
Maesllyn Ceredig 93 C7
Maesmynis Powys 95 B10
Maesteg Bridgend 57 C10
Maesybont Carms 75 B9
Maesycoed Rhondda 58 B5
Maesycrugiau Carms 93 C9
Maesycwmmer Caerph 77 G11
Maesygwartha Mon 78 C2
Maesymeillion Ceredig 93 B8
Maesypandy Powys 129 D9
Maesyrhandir Powys 129 E11
Magdalen Laver Essex 87 D8
Maggieknockater Moray 302 E3
Maggots End Essex 105 F9
Magham Down E Sus 23 C10
Maghull Mers 193 G11
Magor Mon 60 B2
Magpie Green Suff 125 B11
Mahaar Dumfries 236 B2
Maida Vale London 67 C9
Maiden Bradley Wilts 45 F10
Maiden Head N Som 60 F5
Maiden Law Durham 233 B9
Maiden Newton Dorset 17 B7
Maiden Wells Pembs 73 F7
Maidenbower W Sus 51 F9
Maidencombe Torbay 9 B8
Maidenhall Suff 108 C3
Maidenhead Windsor 65 C11
Maidenhead Court Windsor 66 C2
Maidenpark Falk 279 E9
Maidens S Ayrs 244 B6
Maiden's Green Brack 65 E11
Maiden's Hall Northumb 252 D6
Maidensgrave Suff 108 B5
Maidensgrove Oxon 65 B8
Maidenwell Corn 11 G8
Maidenwell Lincs 190 F4
Maidford Northants 120 G2
Maids Moreton Bucks 102 D4
Maidstone Kent 53 B9
Maidwell Northants 120 B4
Mail Shetland 313 L6
Mailand Shetland 312 C8
Mailingsland Borders 270 G4
Main Powys 148 F3
Maindee Newport 59 B10
Maindy Cardiff 59 D7
Mainholm S Ayrs 257 E9
Mains Cumb 229 G7
Mains of Airies Dumfries 236 C1
Mains of Allardice Aberds 293 F10
Mains of Annochie Aberds 303 E9
Mains of Ardestie Angus 287 D9
Mains of Arnage Aberds 303 F9
Mains of Auchoynanie Moray 302 E4
Mains of Baldoon Dumfries 236 D6
Mains of Balhall Angus 293 G7
Mains of Ballindarg Angus 287 B8
Mains of Balnakettle Aberds 293 F8
Mains of Birness Aberds 303 F9
Mains of Blackhall Aberds 303 G7
Mains of Burgie Moray 301 D10
Mains of Cairnbrogie Aberds 303 G8
Mains of Cairnty Moray 302 D3
Mains of Clunas Highld 301 E8
Mains of Crichie Aberds 303 E9
Mains of Daltulich Highld 301 E7
Mains of Dalvey Highld 301 F11
Mains of Dellavaird Aberds 293 E9
Mains of Drum Aberds 293 D10
Mains of Edingight Moray 302 D5
Mains of Fedderate Aberds 303 E8
Mains of Flichity Highld 300 G6
Mains of Hatton Aberds 303 D9
Mains of Hatton Aberds 303 E7
Mains of Inkhorn Aberds 303 F9
Mains of Innerpeffray Perth 286 F3
Mains of Kirktonhill Aberds 293 G8
Mains of Laithers Aberds 302 E6
Mains of Mayen Moray 302 E5
Mains of Melgund Angus 287 B9
Mains of Taymouth Perth 285 C11
Mains of Thornton Aberds 293 F8
Mains of Towie Aberds 303 E7
Mains of Ulbster Highld 310 E7
Mains of Watten Highld 310 D6
Mainsforth Durham 234 E2
Mainsriddle Dumfries 237 D11
Mainstone Shrops 130 F5
Maisemore Glos 98 G6
Maitland Park London 67 C9
Major's Green W Mid 118 B2
Makeney Derbys 170 G5
Malacleit W Isles 296 D3
Malborough Devon 9 G9
Malcoff Derbys 185 E9
Malden Rushett London 67 G7
Maldon Essex 88 D4
Malehurst Shrops 131 B7
Malham N Yorks 213 G8
Maligar Highld 298 C4
Malinbridge S Yorks 186 D4
Malinslee Telford 132 B3
Malkin's Bank E Ches 168 D3
Mallaig Highld 295 F8
Mallaig Bheag Highld 295 F8
Malleny Mills Edin 270 B3
Malling Stirl 285 G9
Mallows Green Essex 105 F9
Malltraeth Anglesey 162 B6
Mallwyd Gwyn 147 G7
Malmesbury Wilts 62 B2
Malmsmead Devon 41 D9
Malpas Corn 4 G6
Malpas Newport 78 G4
Malpas W Berks 64 E6
Malpas W Ches 167 F7
Malswick Glos 98 F4
Maltby Lincs 190 E4
Maltby S Yorks 187 C8
Maltby Stockton 225 C9
Maltby le Marsh Lincs 191 E7
Malting End Suff 124 G4
Malting Green Essex 107 G9
Maltings Angus 293 G9
Maltman's Hill Kent 54 E2
Malton N Yorks 216 E5
Malvern Common Worcs 98 C5
Malvern Link Worcs 98 B5
Malvern Wells Worcs 98 C5
Mambeg Argyll 276 D4
Mamble Worcs 116 C3
Mamhilad Mon 78 E4
Man-moel Caerph 77 E11
Manaccan Corn 3 E7
Manadon Plym 7 D9
Manafon Powys 130 C2
Manais W Isles 296 C7
Manar Ho Aberds 303 G7
Manaton Devon 13 E11
Manby Lincs 190 D5
Mancetter Warks 134 D6
Manchester Gtr Man 184 B4
Manchester Airport Gtr Man 184 D4
Mancot Flint 166 B4
Mancot Royal Flint 166 B4
Mandally Highld 290 C4
Manea Cambs 139 F9
Maney W Mid 134 D2
Manfield N Yorks 224 C4
Mangaster Shetland 312 F5
Mangotsfield S Glos 61 D7
Mangrove Green Herts 104 G2
Mangurstadh W Isles 304 E2
Manhay Corn 2 C5
Manian-fawr Pembs 92 B3
Mankinholes W Yorks 196 C3
Manley Devon 27 E7
Manley W Ches 183 G8
Manley Common W Ches 183 G8
Manmoel Caerph 77 E11
Mannal Argyll 288 E1
Mannamead Plym 7 D9
Mannerston W Loth 279 F10
Manningford Abbots Wilts 46 B6
Manningford Bohune Wilts 46 B6
Manningford Bruce Wilts 46 B6
Manningham W Yorks 205 G9
Mannings Heath W Sus 36 B2
Mannington Dorset 31 F9
Manningtree Essex 107 E11
Mannofield Aberdeen 293 C11
Manor London 68 B2
Manor Bourne Devon 7 F9
Manor Estate S Yorks 186 D5
Manor Hill Corner Lincs 157 F8
Manor House W Mid 135 G7
Manor Park Bucks 84 C4
Manor Park E Sus 37 C7
Manor Park London 68 B2
Manor Park Notts 153 C11
Manor Park S Yorks 186 D5
Manor Park Slough 66 C3
Manor Park W Ches 167 B11
Manor Park W Yorks 205 D9
Manor Parsley Corn 4 F4
Manor Royal W Sus 51 F9
Manorbier Pembs 73 F9
Manorbier Newton Pembs 73 F8
Manordeilo Carms 94 F3
Manorhill Borders 262 C5
Manorowen Pembs 91 D8
Man's Cross Essex 106 D5
Mansegate Dumfries 247 G9
Mansel Lacy Hereford 97 B8
Manselfield Swansea 56 D5
Mansell Gamage Hereford 97 C7
Manselton Swansea 57 B7
Mansergh Cumb 212 C2
Mansewood Glasgow 267 C11
Mansfield E Ayrs 258 G4
Mansfield Notts 171 C8
Mansfield Woodhouse Notts 171 C8
Manson Green Norf 141 C10
Mansriggs Cumb 210 C5
Manston Dorset 30 D4
Manston Kent 71 F10
Manston W Yorks 206 F3
Manswood Dorset 31 F7
Manthorpe Lincs 155 B8
Manthorpe Lincs 155 F11
Mantles Green Bucks 85 F7
Manton N Lincs 200 G2
Manton Notts 187 F9
Manton Rutland 137 C7
Manton Wilts 63 F7
Manton Warren N Lincs 200 F2
Manuden Essex 105 F9
Manwood Green Essex 87 C8
Manywells Height W Yorks 205 F7
Maperton Som 29 B11
Maple Cross Herts 85 G8
Maple End Essex 105 D11
Maplebeck Notts 172 C2
Mapledurham Oxon 65 D7
Mapledurwell Hants 49 C7
Maplehurst W Sus 35 C11
Maplescombe Kent 68 G5
Mapleton Derbys 169 F11
Mapperley Derbys 170 G6
Mapperley Nottingham 171 G9
Mapperley Park Nottingham 171 G9
Mapperton Dorset 16 B6
Mapperton Dorset 18 B4
Mappleborough Green Warks 117 D11
Mappleton E Yorks 209 E10
Mapplewell S Yorks 197 F10
Mappowder Dorset 30 F2
Mar Lodge Aberds 292 D2
Maraig W Isles 305 H3
Marazanvose Corn 4 E6
Marazion Corn 2 C2
Marbhig W Isles 305 G6
Marbrack Dumfries 246 E3
Marbury E Ches 167 F9
March Cambs 139 D8
March S Lnrk 259 G11
Marcham Oxon 83 F7
Marchamley Shrops 149 D11
Marchamley Wood Shrops 149 C11
Marchington Staffs 152 C2
Marchington Woodlands Staffs 152 D2
Marchroes Gwyn 144 D6
Marchwiel Wrex 166 F5
Marchwood Hants 32 E5
Marcross V Glam 58 F2
Marden Hereford 97 B10
Marden Kent 53 E8
Marden T & W 243 C9
Marden Wilts 46 B5
Marden Ash Essex 87 E9
Marden Beech Kent 53 E8
Marden Thorn Kent 53 E9
Marden's Hill E Sus 52 G3
Mardleybury Herts 86 B3
Mardu Shrops 130 G5
Mardy Mon 78 B4
Mardy Shrops 148 C5
Marefield Leics 136 B4
Mareham le Fen Lincs 174 C3
Mareham on the Hill Lincs 174 B3
Marehay Derbys 170 F5
Marehill W Sus 35 D9
Maresfield E Sus 37 C7
Maresfield Park E Sus 37 C7
Marfleet Hull 200 B6
Marford Wrex 166 D5
Margam Neath 57 D9
Margaret Marsh Dorset 30 D4
Margaret Roding Essex 87 C9
Margaretting Essex 87 E11
Margaretting Tye Essex 87 E11
Margate Kent 71 E11
Margery Sur 51 C9
Margnaheglish N Ayrs 256 C2
Margreig Dumfries 237 B10
Margrove Park Redcar 226 B3
Marham Norf 158 G4
Marhamchurch Corn 24 G2
Marholm P'boro 138 C2
Marian Flint 181 F9
Marian Cwm Denb 181 F9
Marian-glas Anglesey 179 E8
Marian y de = South Beach Gwyn 145 C7
Marian y mor = West End Gwyn 145 C7
Mariandyrys Anglesey 179 E10
Marianglas Anglesey 179 E8
Mariansleigh Devon 26 C2
Marine Town Kent 70 E2
Marionburgh Aberds 293 C9
Marishader Highld 298 C4
Marjoriebanks Dumfries 248 G3
Mark Dumfries 236 D3
Mark Dumfries 237 C7
Mark S Ayrs 236 B2
Mark Som 43 D11
Mark Causeway Som 43 D11
Mark Cross E Sus 23 C7
Mark Cross E Sus 52 G5
Mark Hall North Essex 87 C7
Mark Hall South Essex 87 C7
Markbeech Kent 52 E3
Markby Lincs 191 F7
Markeaton Derbys 152 B6
Market Bosworth Leics 135 C8
Market Deeping Lincs 138 B2
Market Drayton Shrops 150 C3
Market Harborough Leics 136 F4
Market Lavington Wilts 46 C4
Market Overton Rutland 155 F7
Market Rasen Lincs 189 D10
Market Stainton Lincs 190 F2
Market Warsop Notts 171 B9
Market Weighton E Yorks 208 E3
Market Weston Suff 125 B9
Markethill Perth 286 D6
Markfield Leics 153 G9
Markham Caerph 77 E11
Markham Moor Notts 188 G2
Markinch Fife 286 G6
Markington N Yorks 214 F5
Markland Hill Gtr Man 195 F7
Markle E Loth 281 F11
Mark's Corner I o W 20 C5
Marks Gate London 87 G7
Marks Tey Essex 107 G8
Marksbury Bath 61 G7
Markyate Herts 85 B9
Marl Bank Worcs 98 C5
Marland Gtr Man 195 E11
Marlas Hereford 97 F8
Marlborough Wilts 63 F7
Marlbrook Hereford 115 G10
Marlbrook Worcs 117 C9
Marlcliff Warks 117 G11
Marldon Devon 9 C7
Marle Green E Sus 23 B9
Marle Hill Glos 99 G9
Marlesford Suff 126 F6
Marley Kent 55 C10
Marley Kent 55 C7
Marley Green E Ches 167 F9
Marley Heights W Sus 49 G11
Marley Hill T & W 242 F6
Marley Pots T & W 243 F9
Marlingford Norf 142 B2
Marloes Pembs 72 D3
Marlow Bucks 65 B10
Marlow Hereford 115 B8
Marlow Bottom Bucks 65 B11
Marlow Common Bucks 65 B10
Marlpit Hill Kent 52 D2
Marlpits E Sus 38 E2
Marlpool Derbys 170 F6
Marnhull Dorset 30 D3
Marnoch Aberds 302 D5
Marnock N Lnrk 268 B4
Marple Gtr Man 185 D7
Marple Bridge Gtr Man 185 D7
Marpleridge Gtr Man 185 D7
Marr S Yorks 198 F4
Marr Green Wilts 63 G8
Marrel Highld 311 H4
Marrick N Yorks 223 F11
Marrister Shetland 313 G7
Marros Carms 74 D2
Marsden T & W 243 E9
Marsden W Yorks 196 E4
Marsden Height Lancs 204 F3
Marsett N Yorks 213 B8
Marsh Bucks 84 D4
Marsh Devon 28 E3
Marsh W Yorks 196 D6
Marsh W Yorks 204 F6
Marsh Baldon Oxon 83 F9
Marsh Benham W Berks 64 F2
Marsh Common S Glos 60 C5
Marsh End Worcs 98 D6
Marsh Gate W Berks 63 F10
Marsh Gibbon Bucks 102 G2
Marsh Green Devon 14 C6
Marsh Green Gtr Man 194 F5
Marsh Green Kent 52 E2
Marsh Green Staffs 168 D5
Marsh Green Telford 150 G2
Marsh Green W Ches 183 F8
Marsh Houses Lancs 202 C5
Marsh Lane Derbys 186 F6
Marsh Lane Glos 79 D9
Marsh Mills Som 43 F7
Marsh Side Norf 176 E3
Marsh Street Som 42 E3
Marshall Meadows Northumb 273 D9
Marshall's Cross Mers 183 C8
Marshall's Elm Som 44 G3
Marshall's Heath Herts 85 B11
Marshalsea Dorset 28 G5
Marshalswick Herts 85 D11
Marsham Norf 160 E3
Marshaw Lancs 203 C7
Marshborough Kent 55 B10
Marshbrook Shrops 131 F8
Marshchapel Lincs 190 B5
Marshfield Newport 59 C9
Marshfield S Glos 61 E9
Marshfield Bank E Ches 167 D11
Marshgate Corn 11 C9
Marshland St James Norf 139 B10
Marshmoor Herts 86 D2
Marshside Kent 71 F8
Marshside Mers 193 D11
Marshwood Dorset 16 B3
Marske N Yorks 224 E2
Marske-by-the-Sea Redcar 235 G8
Marston Hereford 115 F7
Marston Lincs 172 G5
Marston Oxon 83 D8
Marston Staffs 150 G6
Marston Staffs 151 D8
Marston W Ches 183 F11
Marston Warks 119 B8
Marston Warks 134 E4
Marston Wilts 46 B3
Marston Bigot Som 45 E9
Marston Doles Warks 119 F9
Marston Gate Som 45 D9
Marston Green W Mid 134 F3
Marston Hill Glos 81 F10
Marston Jabbett Warks 135 F7
Marston Magna Som 29 C9
Marston Meysey Wilts 81 F10
Marston Montgomery Derbys 152 B2
Marston Moretaine C Beds 103 C9
Marston on Dove Derbys 152 D4
Marston St Lawrence Northants 101 C10
Marston Stannett Hereford 115 F11
Marston Trussell Northants 136 F3
Marstow Hereford 79 B9
Marsworth Bucks 84 C6
Marten Wilts 47 B9
Marthall E Ches 184 F4
Martham Norf 161 F9
Marthwaite Cumb 222 G2
Martin Hants 31 D9
Martin Kent 55 D10
Martin Lincs 173 D10
Martin Lincs 174 B2
Martin Dales Lincs 173 C11
Martin Drove End Hants 31 C9
Martin Hussingtree Worcs 117 E7
Martin Mill Kent 55 D10
Martin Moor Lincs 174 C2
Martindale Cumb 221 B8
Martinhoe Devon 41 D7
Martinhoe Cross Devon 41 D7
Martin's Moss E Ches 168 C4
Martinscroft Warr 183 D11
Martinstown Dorset 17 D8
Martinstown or Winterbourne St Martin Dorset 17 D8
Martlesham Suff 108 B4
Martlesham Heath Suff 108 B4
Martletwy Pembs 73 C8
Martley Worcs 116 E5
Martock Som 29 D7
Marton Cumb 210 D4
Marton E Ches 168 B5
Marton E Yorks 209 F9
Marton Lincs 188 E4
Marton M'bro 225 B10
Marton N Yorks 215 G8
Marton N Yorks 216 C4
Marton Shrops 130 C5
Marton Shrops 149 E8
Marton W Ches 167 B10
Marton Warks 119 D8
Marton Green W Ches 167 B10
Marton Grove M'bro 225 B9
Marton-in-the-Forest N Yorks 215 F11
Marton-le-Moor N Yorks 215 E7
Marton Moor Warks 119 D8
Marton Moss Side Blkpool 202 G2
Martyr Worthy Hants 48 G4
Martyr's Green Sur 50 B5
Marwick Orkney 314 D2
Marwood Devon 40 F4
Mary Tavy Devon 12 F6
Marybank Highld 300 D4
Marybank Highld 301 B7
Maryburgh Highld 300 D5
Maryfield Aberds 293 D7
Maryfield Corn 7 D8
Maryhill Glasgow 267 B11
Marykirk Aberds 293 G8
Maryland Mon 79 D8
Marylebone Gtr Man 194 F5
Marypark Moray 301 F11
Maryport Cumb 228 D6
Maryport Dumfries 236 F3
Maryton Angus 287 B10
Maryton Angus 287 B7
Marywell Aberds 293 D11
Marywell Aberds 293 D7
Marywell Angus 287 C10
Masbrough S Yorks 186 C6
Mascle Bridge Pembs 73 D7
Masham N Yorks 214 C4
Mashbury Essex 87 C11
Masongill N Yorks 212 D3
Masonhill S Ayrs 257 E9
Mastin Moor Derbys 187 F7
Mastrick Aberdeen 293 C10
Matchborough Worcs 117 D11
Matching Essex 87 C8
Matching Green Essex 87 C8
Matching Tye Essex 87 C8
Matfen Northumb 242 C2
Matfield Kent 53 E7
Mathern Mon 79 G8
Mathon Hereford 98 B4
Mathry Pembs 91 E7
Matlaske Norf 160 C3
Matley Gtr Man 185 B7
Matlock Derbys 170 C3
Matlock Bank Derbys 170 C3
Matlock Bath Derbys 170 D3
Matlock Bridge Derbys 170 D3
Matlock Cliff Derbys 170 D4
Matlock Dale Derbys 170 D3
Matshead Lancs 202 E6
Matson Glos 80 B4
Matterdale End Cumb 230 G3
Mattersey Notts 187 D11
Mattersey Thorpe Notts 187 D11
Matthewsgreen Wokingham 65 F10
Mattingley Hants 49 B8
Mattishall Norf 159 G11
Mattishall Burgh Norf 159 G11
Mauchline E Ayrs 257 D11
Maud Aberds 303 E9
Maudlin Corn 5 C11
Maudlin Dorset 28 F5
Maudlin W Sus 22 B5
Maudlin Cross Dorset 28 F5
Maugersbury Glos 100 F4
Maughold I o M 192 C5
Mauld Highld 300 F4
Maulden C Beds 103 D11
Maulds Meaburn Cumb 222 B2
Maunby N Yorks 215 B7
Maund Bryan Hereford 115 G11
Maundown Som 27 B9
Mauricewood Midloth 270 C4
Mautby Norf 161 G9
Mavesyn Ridware Staffs 151 F11
Mavis Enderby Lincs 174 B5
Maviston Highld 301 D9
Maw Green E Ches 168 D2
Mawbray Cumb 229 B7
Mawdesley Lancs 194 E3
Mawdlam Bridgend 57 E10
Mawgan Corn 2 D6
Mawgan Porth Corn 5 B7
Mawla Corn 4 F4
Mawnan Corn 3 D7
Mawnan Smith Corn 3 D7
Mawsley Northants 120 B6
Mawson Green S Yorks 198 D6
Mawthorpe Lincs 191 G7
Maxey P'boro 138 B2
Maxstoke Warks 134 F4

Maxted Street Kent 54 E6
Maxton Borders 262 C4
Maxton Kent 55 E10
Maxwellheugh Borders 262 C6
Maxwelltown Dumfries 237 B11
Maxworthy Corn 11 C11
May Bank Staffs 168 F5
May Hill Mon 79 C8
May Hill Village Glos 98 G4
Mayals Swansea 56 C6
Maybole S Ayrs 257 G8
Maybury Sur 50 B4
Maybush Soton 32 E5
Mayer's Green W Mid 133 E10
Mayes Green Sur 50 F6
Mayeston Pembs 73 E8
Mayfair London 67 C9
Mayfield E Sus 37 B9
Mayfield Midloth 271 C7
Mayfield Northumb 243 B7
Mayfield Staffs 169 F11
Mayfield W Loth 269 B8
Mayford Sur 50 B3
Mayhill Swansea 56 C6
Mayland Essex 88 E6
Maylandsea Essex 88 E6
Maynard's Green E Sus 23 B9
Mayne Ho Moray 302 C2
Mayon Corn 1 D3
Maypole Kent 68 E4
Maypole Kent 68 G3
Maypole Kent 71 G8
Maypole Mon 79 B7
Maypole Scilly 1 G4
Maypole Green Essex 107 G9
Maypole Green Norf 143 D8
Maypole Green Suff 125 F8
Maypole Green Suff 126 D5
May's Green N Som 59 G11
Mays Green Oxon 65 C8
May's Green Sur 50 B5
Mayshill S Glos 61 C7
Maythorn S Yorks 197 F7
Maythorne Notts 171 D11
Maywick Shetland 313 L5
Mead Devon 13 G11
Mead Devon 24 D2
Mead End Hants 19 B11
Mead End Hants 33 E11
Mead End Wilts 31 C8
Mead Vale Sur 51 D9
Meadgate Bath 45 B7
Meadle Bucks 84 D4
Meadow Green Hereford 116 F4
Meadow Hall S Yorks 186 C5
Meadow Head S Yorks 186 E4
Meadowbank Edin 280 G5
Meadowbank W Ches 167 B11
Meadowend Essex 106 C4
Meadowfield Durham 233 D10
Meadowfoot N Ayrs 266 F4
Meadowley Shrops 132 E3
Meadowmill E Loth 281 G8
Meadows Nottingham 153 B11
Meadowtown Shrops 130 C6
Meads E Sus 23 F10
Meadside Oxon 83 G9
Meadwell Devon 12 E4
Meaford Staffs 151 B7
Meagill N Yorks 205 B9
Meal Bank Cumb 221 F10
Meal Hill W Yorks 197 F7
Mealabost Bhuirgh W Isles 304 C6
Mealabost W Isles 304 E6
Mealasta W Isles 304 F1
Mealrigg Cumb 229 B8
Mealsgate Cumb 229 C10
Meanwood W Yorks 205 F11
Mearbeck N Yorks 212 G6
Meare Som 44 E3
Meare Green Som 28 B4
Meare Green Som 28 C3
Mearns Bath 45 B7
Mearns E Renf 267 D10
Mears Ashby Northants 120 D6
Measborough Dike S Yorks 197 F11
Measham Leics 152 G6
Meath Green Sur 51 E9
Meathop Cumb 211 C8
Meaux E Yorks 209 F7
Meaver Corn 2 F5
Meavy Devon 7 B10
Medbourne Leics 136 E5
Medbourne M Keynes 102 D6
Medburn Northumb 242 C4
Meddon Devon 24 D3
Meden Vale Notts 171 B9
Medhurst Row Kent 52 D3
Medlam Lincs 174 D4
Medlar Lancs 202 F4
Medlicott Shrops 131 E8
Medlyn Corn 2 C6
Medmenham Bucks 65 C10
Medomsley Durham 242 G4
Medstead Hants 49 F7
Meer Common Hereford 115 G7
Meer End W Mid 118 C4
Meerbrook Staffs 169 C7
Meerhay Dorset 29 G7
Meers Bridge Lincs 191 D7
Meersbrook S Yorks 186 E5
Meesden Herts 105 E8
Meeson Telford 150 E3
Meeson Heath Telford 150 E3
Meeth Devon 25 F8
Meethe Devon 25 C11
Meeting Green Suff 124 F4
Meeting House Hill Norf 160 D6
Meggernie Castle Perth 285 C9
Meggethead Borders 260 E5
Meidrim Carms 92 G5
Meifod Denb 165 D8
Meifod Powys 148 G3
Meigle N Ayrs 266 B3
Meigle Perth 286 C6
Meikle Earnock S Lnrk 268 E4
Meikle Ferry Highld 309 L7
Meikle Forter Angus 292 G3
Meikle Gluich Highld 309 L6
Meikle Obney Perth 286 D4
Meikle Pinkerton E Loth 282 F4
Meikle Strath Aberds 293 F8
Meikle Tarty Aberds 303 G9
Meikle Wartle Aberds 303 F7
Meikleour Perth 286 D5
Meinciau Carms 75 C7
Meir Stoke 168 G6
Meir Heath Staffs 168 G6
Melbourn Cambs 105 C7
Melbourne Derbys 153 D7
Melbourne E Yorks 207 E11
Melbourne S Lnrk 269 G11
Melbury Abbas Dorset 30 D5
Melbury Bubb Dorset 29 F9
Melbury Osmond Dorset 29 F9
Melbury Sampford Dorset 29 F9
Melby Shetland 313 H3
Melchbourne Beds 121 D10
Melcombe Som 43 G9
Melcombe Bingham Dorset 30 G3
Melcombe Regis Dorset 17 E9
Meldon Devon 13 C7
Meldon Northumb 252 G4
Meldreth Cambs 105 B7
Meldrum Ho Aberds 303 G8
Melfort Argyll 275 B9
Melgarve Highld 290 D6
Meliden = Gallt Melyd Denb 181 E9
Melin Caiach Caerph 77 F10
Melin-y-coed Conwy 164 C4
Melin-y-ddôl Powys 129 B11
Melin-y-grug Powys 129 B11
Melin-y-Wig Denb 165 F8
Melinbyrhedyn Powys 128 D6
Melincourt Neath 76 E4
Melincryddan Neath 57 B8
Melinsey Corn 3 B10
Melkington Northumb 273 G7
Melkinthorpe Cumb 231 F7
Melkridge Northumb 240 E6
Melksham Wilts 62 G2
Melksham Forest Wilts 62 G2
Mell Green W Berks 64 D3
Mellangaun Highld 307 L3
Mellangoose Corn 2 D5
Melldalloch Argyll 275 F10
Mellguards Cumb 230 B4
Melling Lancs 211 E11
Melling Mers 193 G11
Melling Mount Mers 194 G2
Mellingey Corn 10 G4
Mellis Suff 126 C2
Mellis Green Suff 125 C11
Mellon Charles Highld 307 K3
Mellon Udrigle Highld 307 K3
Mellor Gtr Man 185 D7
Mellor Lancs 203 G9
Mellor Brook Lancs 203 G8
Mells Som 45 D8
Mells Suff 127 B8
Mells Green Som 45 D8
Melmerby Cumb 231 D8
Melmerby N Yorks 213 B11
Melmerby N Yorks 214 D6
Melon Green Suff 124 F6
Melplash Dorset 16 B5
Melrose Borders 262 C2
Melsetter Orkney 314 H2
Melsonby N Yorks 224 D3
Meltham W Yorks 196 E6
Meltham Mills W Yorks 196 E6
Melton E Yorks 200 B3
Melton Suff 126 G5
Melton Constable Norf 159 C10
Melton Mowbray Leics 154 F5
Melton Ross N Lincs 200 E5
Meltonby E Yorks 207 C11
Melvaig Highld 307 L2
Melverley Shrops 148 F6
Melverley Green Shrops 148 F6
Melvich Highld 310 C2
Membland Devon 7 F11
Membury Devon 28 G3
Memsie Aberds 303 C9
Memus Angus 287 B8
Mena Corn 5 C10
Menabilly Corn 5 E11
Menadarva Corn 4 G2
Menagissey Corn 4 F4
Menai Bridge = Porthaethwy Anglesey 179 G9
Mendham Suff 142 G5
Mendlesham Suff 126 D2
Mendlesham Green Suff 125 E11
Menethorpe N Yorks 216 F5
Mengham Hants 21 B10
Menheniot Corn 6 C5
Menherion Corn 2 B6
Menithwood Worcs 116 D4
Menna Corn 5 E8
Mennock Dumfries 247 B8
Menston W Yorks 205 E9
Menstrie Clack 278 B6
Mentmore Bucks 84 B6
Menzion Borders 260 E3
Meoble Highld 295 G9
Meole Brace Shrops 149 G9
Meols Mers 182 C2
Meon Hants 33 G8
Meonstoke Hants 33 D10
Meopham Kent 68 F6
Meopham Green Kent 68 F6
Meopham Station Kent 68 F6
Mepal Cambs 139 G8
Meppershall C Beds 104 D2
Merbach Hereford 96 B6
Mercaton Derbys 170 G3
Merchant Fields W Yorks 197 B7
Merchiston Edin 280 G4
Mere E Ches 184 E2
Mere Wilts 45 G10
Mere Brow Lancs 194 D2
Mere Green W Mid 134 D2
Mere Green Worcs 117 E9
Mere Heath W Ches 183 G11
Mereclough Lancs 204 G3
Merehead Wrex 149 B9
Meresborough Medway 69 G10
Mereside Blkpool 202 G2
Meretown Staffs 150 E5
Mereworth Kent 53 C7
Mergie Aberds 293 E9
Meriden Herts 85 F10
Meriden W Mid 134 G4
Merkadale Highld 294 B5
Merkland Dumfries 237 B9
Merkland N Ayrs 256 B2
Merkland S Ayrs 244 E6
Merkland Lodge Highld 309 G4
Merle Common Sur 52 D2
Merley Poole 18 B6
Merlin's Bridge Pembs 72 C6
Merlin's Cross Pembs 73 E7
Merridale W Mid 133 D7
Merridge Som 43 G8
Merrie Gardens I o W 21 E7
Merrifield Devon 8 F6
Merrifield Devon 24 G3
Merrington Shrops 149 E9
Merrion Pembs 72 F6
Merriott Dorset 16 B6
Merriott Som 28 E6
Merriottsford Som 28 E6
Merritown Dorset 19 B8
Merrivale Devon 12 F6
Merrivale Hereford 98 G2
Merrow Sur 50 C4
Merry Field Hill Dorset 31 G8
Merry Hill Herts 85 G10
Merry Hill W Mid 133 D7
Merry Lees Leics 135 B9
Merry Meeting Corn 11 G7
Merry Oak Soton 32 E6
Merrybent Darl 224 C4
Merryhill Green Wokingham 65 E9
Merrylee E Renf 267 D11
Merrymeet Corn 6 B5
Mersham Kent 54 F5
Merstham Sur 51 C9
Merston W Sus 22 C5
Merstone I o W 20 E6
Merther Corn 5 G7
Merther Lane Corn 5 G7
Merthyr Carms 93 G7
Merthyr Cynog Powys 95 D9
Merthyr-Dyfan V Glam 58 F6
Merthyr Mawr Bridgend 57 F11
Merthyr Tydfil M Tydf 77 D8
Merthyr Vale M Tydf 77 F9
Merton Devon 25 E8
Merton London 67 F9
Merton Norf 141 D8
Merton Oxon 83 B9
Merton Park London 67 F9
Mervinslaw Borders 262 G5
Meshaw Devon 26 D3
Messing Essex 88 B5
Messingham N Lincs 199 G11
Mesty Croft W Mid 133 E9
Mesur-y-dorth Pembs 87 E11
Metal Bridge Durham 233 E11
Metfield Suff 142 G5
Metherell Corn 7 B8
Metheringham Lincs 173 C9
Methersgate Suff 108 B5
Methil Fife 281 B7
Methilhill Fife 281 B7
Methlem Gwyn 144 C3
Methley W Yorks 197 B11
Methley Junction W Yorks 197 B11
Methley Lanes W Yorks 197 B11
Methlick Aberds 303 F8
Methven Perth 286 E4
Methwold Norf 140 E4
Methwold Hythe Norf 140 E4
Mettingham Suff 143 F7
Metton Norf 160 B3
Mevagissey Corn 5 G10
Mewith Head N Yorks 212 F4
Mexborough S Yorks 187 B7
Mey Highld 310 B6
Meyrick Park Bmouth 19 C7
Meysey Hampton Glos 81 F10
Miabhag W Isles 305 H2
Miabhag W Isles 305 J3
Miabhig W Isles 304 E2
Mial Highld 299 B7
Michaelchurch Hereford 97 F10
Michaelchurch Escley Hereford 96 E6
Michaelchurch on Arrow Powys 114 G4
Michaelston-le-Pit V Glam 59 E7
Michaelston-y-Fedw Newport 59 C8
Michaelstow Corn 11 F7
Michealston-super-Ely Cardiff 58 D6
Michelcombe Devon 8 B3
Micheldever Hants 48 F4
Michelmersh Hants 32 B4
Mickfield Suff 126 E2
Mickle Trafford W Ches 166 B6
Micklebring S Yorks 187 C8
Mickleby N Yorks 226 C6
Micklefield Bucks 84 G5
Micklefield W Yorks 206 G4
Micklefield Green Herts 85 F8
Mickleham Sur 51 C7
Micklehurst Gtr Man 196 G3
Mickleover Derby 152 C6
Micklethwaite Cumb 239 G7
Micklethwaite W Yorks 205 E8
Mickleton Durham 232 G5
Mickleton Glos 100 C3
Mickletown W Yorks 197 B11
Mickley Derbys 186 F4
Mickley N Yorks 214 D5
Mickley Shrops 150 C2
Mickley Green Suff 124 F6
Mickley Square Northumb 242 E3
Mid Ardlaw Aberds 303 C9
Mid Auchinleck Invclyd 276 G6
Mid Beltie Aberds 293 C8
Mid Calder W Loth 269 B11
Mid Cloch Forbie Aberds 303 D7
Mid Clyth Highld 310 F6
Mid Garrary Dumfries 237 B7
Mid Holmwood Sur 51 D7
Mid Lambrook Som 28 D6
Mid Lavant W Sus 22 B5
Mid Letter Argyll 284 G4
Mid Main Highld 300 F4
Mid Murthat Dumfries 248 D3
Mid Urchany Highld 301 E8
Mid Walls Shetland 313 H4
Mid Yell Shetland 312 D7
Midanbury Hants 33 E7
Midbea Orkney 314 B4
Middle Assendon Oxon 65 B8
Middle Aston Oxon 101 F9
Middle Balnald Perth 286 B4
Middle Barton Oxon 101 F8
Middle Bickenhill W Mid 134 G4
Middle Bockhampton Dorset 19 B9
Middle Bourne Sur 49 E10
Middle Bridge N Som 60 D3
Middle Burnham Som 43 D10
Middle Cairncake Aberds 303 E8
Middle Chinnock Som 29 E7
Middle Claydon Bucks 102 F4
Middle Cliff Staffs 169 E8
Middle Crackington Corn 11 B9
Middle Drums Angus 287 B9
Middle Duntisbourne Glos 81 D7
Middle Green Bucks 66 C4
Middle Green Som 27 D10
Middle Green Suff 124 D4
Middle Handley Derbys 186 F6
Middle Harling Norf 141 F9
Middle Herrington T & W 243 G9
Middle Hill Pembs 73 C7
Middle Hill Staffs 133 B9
Middle Kames Argyll 275 E10
Middle Littleton Worcs 99 B11
Middle Luxton Devon 28 E2
Middle Madeley Staffs 168 F3
Middle Maes-coed Hereford 96 E6
Middle Marwood Devon 40 F4
Middle Mayfield Staffs 169 G10
Middle Mill Pembs 87 F11
Middle Quarter Kent 53 F11
Middle Rainton T & W 234 B2
Middle Rasen Lincs 189 D9
Middle Rigg Perth 286 G4
Middle Rocombe Devon 9 B8
Middle Side Durham 232 F4
Middle Stoford Som 27 C11
Middle Stoke Devon 13 G9
Middle Stoke Medway 69 D10
Middle Stoke W Mid 119 B7
Middle Stoughton Som 44 D2
Middle Strath W Loth 279 G8
Middle Street Glos 80 E3
Middle Taphouse Corn 6 C3
Middle Town Scilly 1 F4
Middle Tysoe Warks 100 C6
Middle Wallop Hants 47 F9
Middle Weald M Keynes 102 D5
Middle Wick Glos 80 F2
Middle Winterslow Wilts 47 G8
Middle Woodford Wilts 46 F6
Middlebie Dumfries 238 B6
Middlecave N Yorks 216 E5
Middlecliffe S Yorks 198 F2
Middlecott Devon 13 D10
Middlecott Devon 24 F6
Middlecott Devon 26 F3
Middlecroft Derbys 186 G6
Middlefield Falk 279 E7
Middleforth Green Lancs 194 B4
Middleham N Yorks 214 B2
Middlehill Corn 6 B5
Middlehill Wilts 61 F10
Middlehope Shrops 131 F9
Middlemarsh Dorset 29 F11
Middlemoor Devon 12 G5
Middlemuir Aberds 303 D9
Middlemuir Aberds 303 E8
Middlemuir Aberds 303 G9
Middleport Stoke 168 F5
Middlerig Falk 279 F8
Middlesbrough M'bro 234 G5
Middlesceugh Cumb 230 C4
Middleshaw Cumb 211 B11
Middlesmoor N Yorks 213 E11
Middlestone Durham 233 E11
Middlestone Moor Durham 233 E10
Middlestown W Yorks 197 D9
Middlethird Borders 272 G3
Middlethorpe York 207 D7
Middleton Aberds 293 B10
Middleton Argyll 288 E1
Middleton Cumb 212 B2
Middleton Derbys 169 C11
Middleton Derbys 170 D3
Middleton Essex 107 D7
Middleton Gtr Man 195 F11
Middleton Hants 48 E2
Middleton Hereford 115 D10
Middleton Hrtlpl 234 E6
Middleton I o W 20 D2
Middleton Lancs 202 B4
Middleton Midloth 271 D7
Middleton N Yorks 204 E5
Middleton N Yorks 205 D8
Middleton N Yorks 216 B5
Middleton Norf 158 F3
Middleton Northants 136 F6
Middleton Northumb 252 F3
Middleton Northumb 264 B4
Middleton Perth 286 C5
Middleton Perth 286 F2
Middleton Perth 286 G5
Middleton Shrops 115 B10
Middleton Shrops 130 D5
Middleton Shrops 148 D6
Middleton Suff 127 D8
Middleton Swansea 56 D2
Middleton W Yorks 197 B10
Middleton Warks 134 D3
Middleton Baggot Shrops 132 E2
Middleton Cheney Northants 101 C9
Middleton Green Staffs 151 B9
Middleton Hall Northumb 263 D11
Middleton-in-Teesdale Durham 232 F4
Middleton Junction Gtr Man 195 G11
Middleton Moor Suff 127 D8
Middleton of Rora Aberds 303 E10
Middleton-on-Leven N Yorks 225 D9
Middleton-on-Sea W Sus 35 G7
Middleton on the Hill Hereford 115 E10
Middleton-on-the-Wolds E Yorks 208 D4
Middleton One Row Darl 225 C7
Middleton Place Cumb 219 G11
Middleton Priors Shrops 132 E2
Middleton Quernhow N Yorks 214 D6
Middleton St George Darl 224 C6
Middleton Scriven Shrops 132 F3
Middleton Stoney Oxon 101 G10
Middleton Tyas N Yorks 224 D4
Middletown Cumb 219 D9
Middletown N Som 60 E3
Middletown Powys 148 G6
Middletown Warks 117 E11
Middlewich E Ches 167 B11
Middlewick Wilts 61 E11
Middlewood Corn 11 F11
Middlewood E Ches 184 E6
Middlewood S Yorks 186 C4
Middlewood Green Suff 125 E11
Middleyard Glos 80 E4
Middlezoy Som 43 G11
Middridge Durham 233 F11
Midelney Som 28 C6
Midfield Highld 308 C5
Midford Bath 61 G9
Midgard Borders 262 F3
Midge Hall Lancs 194 C4
Midgeholme Cumb 240 F4
Midgham W Berks 64 F5
Midgham Green W Berks 64 F5
Midgley W Yorks 196 B4
Midgley W Yorks 197 E9
Midhopestones S Yorks 186 B2
Midhurst W Sus 34 C5
Midland Orkney 314 F3
Midlem Borders 262 D2
Midlock S Lnrk 259 E11
Midmar Aberds 293 C8
Midmuir Argyll 289 G11
Midpark Argyll 255 B11
Midplaugh Aberds 302 E5
Midsomer Norton Bath 45 C7
Midton Invclyd 276 F4
Midtown Highld 307 L3
Midtown Highld 308 C5
Midtown of Buchromb Moray 302 E3
Midtown of Glass Aberds 302 E4
Midville Lincs 174 D5
Midway E Ches 184 E6
Midway Som 45 D7
Miekle Toux Aberds 302 D5
Migdale Highld 309 K6
Migvie Aberds 292 C6
Milarrochy Stirl 277 C8
Milber Devon 14 G3
Milborne Port Som 29 D11
Milborne St Andrew Dorset 18 B2
Milborne Wick Som 29 C11
Milbourne Northumb 242 B4
Milbourne Wilts 62 B2
Milburn Aberds 302 E5
Milburn Aberds 302 F6
Milburn Cumb 231 F9
Milbury Heath S Glos 79 G11
Milby N Yorks 215 F8
Milch Hill Essex 106 G4
Milcombe Corn 6 D4
Milcombe Oxon 101 E8
Milden Suff 107 B9
Mildenhall Suff 124 C4
Mildenhall Wilts 63 F8
Mile Cross Norf 160 G4
Mile Elm Wilts 62 F3
Mile End Cambs 140 G2
Mile End Devon 14 G2
Mile End Essex 107 F9
Mile End Glos 79 C9
Mile End London 67 C11
Mile End Suff 124 G6
Mile Oak Brighton 36 F2
Mile Oak Kent 53 E7
Mile Oak Shrops 148 D6
Mile Oak Staffs 134 C3
Mile Town Kent 70 E2
Milebrook Powys 114 C6
Milebush Kent 53 D9
Mileham Norf 159 F8
Miles Green Staffs 168 F4
Miles Green Sur 50 B3
Miles Hill W Yorks 205 F11
Miles Platting Gtr Man 184 B5
Milesmark Fife 279 D11
Miles's Green W Berks 64 F4
Milfield Northumb 263 C10
Milford Derbys 170 F5
Milford Devon 24 C2
Milford Powys 129 E11
Milford Shrops 149 E8
Milford Staffs 151 E9
Milford Sur 50 E2
Milford Wilts 31 B11
Milford Haven Pembs 72 D6
Milford on Sea Hants 19 C11
Milkhouse Water Wilts 63 G7
Milkieston Borders 270 F4
Milkwall Glos 79 D9
Milkwell Wilts 30 C6
Mill Bank W Yorks 196 C4
Mill Brow Gtr Man 185 D7
Mill Common Norf 142 C6
Mill Common Suff 143 G8
Mill Corner E Sus 38 C4
Mill Dam N Yorks 212 F3
Mill End Bucks 65 C9
Mill End Cambs 124 F3
Mill End Glos 81 C10
Mill End Herts 85 G8
Mill End Herts 104 E6
Mill End Green Essex 106 F2
Mill Farm Aberds 303 C8
Mill Green Cambs 106 B2
Mill Green Essex 87 E10
Mill Green Hants 64 G4
Mill Green Herts 86 D2
Mill Green Norf 142 G2
Mill Green Shrops 150 D3
Mill Green Suff 107 C9
Mill Green Suff 125 F9
Mill Green Suff 126 E2
Mill Green W Mid 133 C11
Mill Hall Dumfries 237 E8
Mill Hill Blkburn 195 B7
Mill Hill E Sus 23 D10
Mill Hill Essex 88 G4
Mill Hill Glos 79 D10
Mill Hill Gtr Man 195 F8
Mill Hill Kent 55 C11
Mill Hill Lincs 175 B8
Mill Hill London 86 G2
Mill Hill Suff 125 G7
Mill Hills Suff 108 B5
Mill Hirst N Yorks 214 G3
Mill Lane Hants 49 C9
Mill Meads London 67 C11
Mill of Brydock Aberds 302 D6
Mill of Chon Stirl 285 G8
Mill of Haldane W Dunb 277 E8
Mill of Kingoodie Aberds 303 G8
Mill of Lynebain Aberds 302 F4
Mill of Muiresk Aberds 302 E6
Mill of Rango Orkney 314 E2
Mill of Sterin Aberds 292 D5
Mill of Uras Aberds 293 E10
Mill Park Argyll 255 G8
Mill Place N Lincs 200 F3
Mill Shaw W Yorks 205 G11
Mill Side Cumb 211 C8
Mill Street Kent 53 B7
Mill Street Norf 159 F11
Mill Street Suff 107 D9
Mill Throp Bmouth 19 B8
Milland W Sus 34 B4
Millarston Renfs 267 C9
Millbank Aberds 303 E11
Millbank Highld 310 C5
Millbank Kent 71 F8
Millbeck Cumb 229 F11
Millbounds Orkney 314 C5
Millbreck Aberds 303 E10
Millbridge Sur 49 E10
Millbrook C Beds 103 D10
Millbrook Corn 7 E8
Millbrook Devon 41 G9
Millbrook Gtr Man 185 B7
Millbrook Soton 32 E5
Millburn S Ayrs 257 D10
Millcombe Devon 8 F6
Milldale Staffs 169 E10
Millden Lodge Angus 293 F7
Milldens Angus 287 B9
Millend Glos 80 D3
Millend Glos 80 F3
Millendreath Corn 6 E5
Millerhill Midloth 270 B6
Miller's Dale Derbys 185 G10
Miller's Green Derbys 170 E3
Miller's Green Essex 87 D9
Millersneuk E Dunb 278 G3
Millerston Glasgow 268 B2
Millfield P'boro 138 C3
Millfield T & W 243 F9
Millgate Lancs 195 D11
Millgate Norf 160 D3
Millgillhead Cumb 229 G7
Millhalf Hereford 96 B5
Millhall Kent 53 B8
Millhayes Devon 27 E10
Millhayes Devon 28 G2
Millhead Lancs 211 E9
Millheugh S Lnrk 268 E5
Millhill Devon 12 G5
Millholme Cumb 221 G11
Millhouse Argyll 275 F10
Millhouse Cumb 230 D3
Millhouse Green S Yorks 197 G8
Millhousebridge Dumfries 248 F4
Millhouses S Yorks 186 E4
Millhouses S Yorks 198 G2
Millikenpark Renfs 267 C8
Millin Cross Pembs 73 C7
Millington E Yorks 208 C2
Millington Green Derbys 170 F3
Millmeece Staffs 150 C6
Millmoor Devon 27 E10
Millness Cumb 211 C10
Millom Cumb 210 C3
Millook Corn 11 B9
Millow C Beds 104 C4
Millpool Corn 2 C3
Millpool Corn 11 G8
Millport N Ayrs 266 E3
Millquarter Dumfries 246 G4
Milltack Aberds 303 D7
Millthorpe Derbys 186 F4
Millthorpe Lincs 156 C2
Millthrop Cumb 222 G3
Milltimber Aberdeen 293 C10
Milltown Aberds 292 C4
Milltown Corn 6 B2
Milltown Corn 6 D2
Milltown Derbys 170 C5
Milltown Devon 40 F5
Milltown Highld 301 E9
Milltown of Aberdalgie Perth 286 E4
Milltown of Auchindoun Moray 302 E3
Milltown of Craigston Aberds 303 D7
Milltown of Edinvillie Moray 302 E2
Milltown of Kildrummy Aberds 292 B6
Milltown of Rothiemay Moray 302 E5
Milltown of Towie Aberds 292 B6
Millwall London 67 D11
Milnathort Perth 286 G5
Milner's Heath W Ches 167 C7
Milngavie E Dunb 277 G11
Milnquarter Falk 278 F6
Milnrow Gtr Man 196 E2
Milnsbridge W Yorks 196 D6
Milnshaw Lancs 195 B9
Milnthorpe Cumb 211 C9
Milnthorpe W Yorks 197 D10
Milnwood N Lnrk 268 D4
Milo Carms 75 B9
Milson Shrops 116 C2
Milstead Kent 54 B2
Milston Wilts 47 D7
Milthorpe Northants 101 B11
Milton Angus 287 C7
Milton Angus 292 E6
Milton Cambs 123 E9
Milton Cumb 211 C10
Milton Cumb 240 E3
Milton Derbys 152 D6
Milton Dumfries 236 D4
Milton Dumfries 237 B10
Milton Dumfries 247 G8
Milton Fife 287 E8
Milton Glasgow 267 B11
Milton Highld 299 E7
Milton Highld 300 D3
Milton Highld 300 E5
Milton Highld 300 F4
Milton Highld 301 B7
Milton Highld 301 E7
Milton Highld 311 D7
Milton Kent 69 E7
Milton Moray 292 B3
Milton Moray 302 C5
Milton N Som 59 G10
Milton Notts 188 G2
Milton Oxon 83 G7
Milton Oxon 101 E8
Milton Pembs 73 E8
Milton Perth 286 F3
Milton Powys 114 G4
Milton Ptsmth 21 B9
Milton S Yorks 197 G11
Milton Som 29 C7
Milton Stirl 285 G9
Milton Stoke 168 E6
Milton W Dunb 277 G8
Milton Wilts 45 G11
Milton Abbas Dorset 30 G4
Milton Bridge Midloth 270 C4
Milton Bryan C Beds 103 E9
Milton Clevedon Som 45 F7
Milton Coldwells Aberds 303 F9
Milton Combe Devon 7 B9
Milton Common Oxon 83 E11
Milton Coombe Devon 7 B9
Milton Damerel Devon 24 E5
Milton End Glos 80 C2
Milton End Glos 81 E10
Milton Ernest Beds 121 F10
Milton Green Devon 12 F4
Milton Green W Ches 167 D7
Milton Heights Oxon 83 G7
Milton Hill Devon 14 F4
Milton Hill Oxon 83 G7
Milton Keynes M Keynes 103 D7
Milton Keynes Village M Keynes 103 D7
Milton Lilbourne Wilts 63 G7
Milton Malsor Northants 120 F4
Milton Morenish Perth 285 D10
Milton of Auchinhove Aberds 293 C7
Milton of Balgonie Fife 287 G7
Milton of Buchanan Stirl 277 C8
Milton of Campfield Aberds 293 C8
Milton of Campsie E Dunb 278 F3
Milton of Corsindae Aberds 293 C8
Milton of Cullerlie Aberds 293 C9
Milton of Cultoquhey Perth 286 E2
Milton of Cushnie Aberds 293 B7
Milton of Dalcapon Perth 286 B3
Milton of Drimmie Perth 286 B5
Milton of Edradour Perth 286 B3
Milton of Gollanfield Highld 301 D7
Milton of Lesmore Aberds 302 G4
Milton of Logie Aberds 292 C6
Milton of Machany Perth 286 F3
Milton of Mathers Aberds 293 G9
Milton of Murtle Aberdeen 293 C10
Milton of Noth Aberds 302 G5
Milton of Tullich Aberds 292 D5
Milton on Stour Dorset 30 B3
Milton Regis Kent 70 F2
Milton Street E Sus 23 E8
Milton under Wychwood Oxon 82 B3
Miltonduff Moray 301 C11
Miltonhill Moray 301 C10
Miltonise Dumfries 236 B3
Milverton Som 27 B10
Milverton Warks 118 D6
Milwich Staffs 151 C9
Milwr Flint 181 G11
Mimbridge Sur 66 G3
Minard Argyll 275 D10
Minard Castle Argyll 275 D10
Minchington Dorset 31 E7
Minchinhampton Glos 80 E5
Mindrum Northumb 263 C8
Minehead Som 42 D3
Minera Wrex 166 E3
Mineshope Corn 11 B9
Minety Wilts 81 G8
Minffordd Gwyn 145 B11
Minffordd Gwyn 146 G4
Minffordd Gwyn 179 G9
Mingarrypark Highld 289 C8
Mingoose Corn 4 F4
Miningsby Lincs 174 C4
Minions Corn 11 G11
Minishant S Ayrs 257 G8
Minllyn Gwyn 147 G7
Minnes Aberds 303 G9
Minngearraidh W Isles 297 J3
Minnigaff Dumfries 236 C6
Minnonie Aberds 303 C7
Minnow End Essex 88 C2
Minnygap Dumfries 248 D2
Minshull Vernon E Ches 167 C11
Minskip N Yorks 215 G7
Minstead Hants 32 E3
Minsted W Sus 34 C5
Minster Kent 70 E3
Minster Kent 71 G10
Minster Lovell Oxon 82 C4
Minsterley Shrops 131 C7
Minsterworth Glos 80 B3
Minterne Magna Dorset 29 G11
Minterne Parva Dorset 29 G11
Minting Lincs 189 G11
Mintlaw Aberds 303 E10
Minto Borders 262 E3
Minto Kames Borders 262 E3
Minton Shrops 131 E8
Mintsfeet Cumb 221 G10
Minwear Pembs 73 C8
Minworth W Mid 134 E3
Mirbister Orkney 314 E2
Mirehouse Cumb 219 B9
Mireland Highld 310 C7
Mirfield W Yorks 197 D8
Miserden Glos 80 D6
Misery Corner Norf 142 F5
Miskin Rhondda 58 C4
Miskin Rhondda 77 F8
Misselfore Wilts 31 C8
Misson Notts 187 C11
Misterton Leics 135 G11
Misterton Notts 188 C3
Misterton Som 29 F7
Misterton Soss Notts 188 B3
Mistley Essex 108 E2
Mistley Heath Essex 108 E2
Mitcham London 67 F9
Mitchel Troy Mon 79 C7
Mitcheldean Glos 79 B11
Mitchell Corn 5 E7
Mitchell Hill Borders 260 C3
Mitchellslacks Dumfries 247 D11
Mitchelston Borders 271 F9
Mitcheltroy Common Mon 79 D7
Mite Houses Cumb 219 F11
Mitford Northumb 252 F5
Mithian Corn 4 E4
Mithian Downs Corn 4 F4
Mitton Staffs 151 F7
Mitton Worcs 99 E8
Mixbury Oxon 102 E2
Mixenden W Yorks 196 B5
Mixtow Corn 6 E2
Moat Cumb 239 C10
Moats Tye Suff 125 F10
Mobberley E Ches 184 F3
Mobberley Staffs 169 G8
Moblake E Ches 167 G11
Mobwell Bucks 84 E5
Moccas Hereford 97 C7
Mochdre Conwy 180 F4
Mochdre Powys 129 F11
Mochrum Dumfries 236 E5
Mockbeggar Hants 31 F11
Mockbeggar Kent 54 E6
Mockbeggar Medway 69 E8
Mockerkin Cumb 229 G7
Moclett Orkney 314 B4
Modbury Devon 8 E3
Moddershall Staffs 151 B8
Model Village Derbys 187 G8
Model Village Warks 119 E8
Modest Corner Kent 52 E5
Moel Tryfan Gwyn 163 D8
Moel-y-crio Flint 165 B11
Moelfre Anglesey 179 D8
Moelfre Conwy 181 G7
Moelfre Powys 148 D3
Moffat Dumfries 248 B3
Moffat Mills N Lnrk 268 C5
Mogador Sur 51 C8
Moggerhanger C Beds 104 B2
Mogworthy Devon 26 D5
Moira Leics 152 F6
Moity Powys 96 C3
Mol-chlach Highld 294 D6
Molash Kent 54 C4
Mold Flint 166 C2
Moldgreen W Yorks 197 D7
Molehill Green Essex 105 G11
Molehill Green Essex 106 G4
Molescroft E Yorks 208 E6
Molesden Northumb 252 G4
Molesworth Cambs 121 B11
Monnis Corn 5 D10
Moll Highld 295 B7
Molland Devon 26 B4
Mollington Oxon 101 B8
Mollington W Ches 182 G5
Mollinsburn N Lnrk 278 G4
Monachty Ceredig 111 E10
Monachylemore Stirl 285 F8
Monar Lodge Highld 300 E2
Monaughty Powys 114 D4
Monboddo House Aberds 293 F9
Mondaytown Shrops 130 B6
Mondynes Aberds 293 F9
Monemore Stirl 285 D9
Monevechadan Argyll 284 G5
Monewden Suff 126 F4
Money Hill Leics 153 F7
Moneyacres E Ayrs 267 E8
Moneydie Perth 286 E4
Moneyhill Herts 85 G8
Moneyrow Green Windsor 65 D11
Moneystone Staffs 169 F9
Mongleath Corn 3 C7
Moniaive Dumfries 247 E7
Monifieth Angus 287 D8
Monikie Angus 287 D8
Monimail Fife 286 F6
Monington Pembs 92 C2
Monk Bretton S Yorks 197 F11
Monk End N Yorks 224 D5
Monk Fryston N Yorks 198 B4
Monk Hesleden Durham 234 D5
Monk Sherborne Hants 48 B6
Monk Soham Suff 126 D4
Monk Street Essex 106 F2
Monken Hadley London 86 F3
Monkerton Devon 14 C5
Monkhide Hereford 98 C2
Monkhill Cumb 239 F8
Monkhill W Yorks 198 C3
Monkhopton Shrops 132 E2
Monkland Hereford 115 F9
Monkleigh Devon 25 C7
Monkmoor Shrops 149 G10
Monknash V Glam 58 E2
Monkokehampton Devon 25 F9
Monks Eleigh Suff 107 B9
Monk's Gate W Sus 36 B2
Monks Heath E Ches 184 G4
Monks Hill Kent 53 E11
Monks Kirby Warks 135 G9
Monks Orchard London 67 F11
Monk's Park Wilts 61 F11
Monks Risbrough Bucks 84 E4
Monkscross Corn 12 G3
Monkseaton T & W 243 C8
Monkshill Aberds 303 E7
Monksilver Som 42 F5
Monkspath W Mid 118 B2
Monksthorpe Lincs 174 B6
Monkston Park M Keynes 103 D7
Monkswood Midloth 270 C6
Monkswood Mon 78 E4
Monkswood W Yorks 206 F2
Monkton Devon 27 G11
Monkton Kent 71 G9
Monkton Pembs 73 E7
Monkton S Ayrs 257 D9
Monkton T & W 243 E8
Monkton V Glam 58 E2
Monkton Combe Bath 61 G9
Monkton Deverill Wilts 45 F11
Monkton Farleigh Wilts 61 F10
Monkton Heathfield Som 28 B3
Monkton Up Wimborne Dorset 31 E8
Monktonhall E Loth 280 G6
Monkwearmouth T & W 243 F9
Monkwood Hants 49 G7
Monkwood Green Worcs 116 E6
Monmarsh Hereford 97 B10
Monmore Green W Mid 133 D8
Monmouth = Trefynwy Mon 79 C8
Monmouth Cap Mon 97 F7
Monnington on Wye Hereford 97 C7
Monreith Dumfries 236 E5
Monreith Mains Dumfries 236 E5
Montacute Som 29 D7
Montcliffe Gtr Man 195 E7
Montcoffer Ho Aberds 302 C6
Montford Argyll 266 C2
Montford Shrops 149 G8
Montford Bridge Shrops 149 F8
Montgarrie Aberds 293 B7
Montgomery Powys 130 D4
Montgomery Lines Hants 49 C11
Monton Gtr Man 184 B3
Montpelier Bristol 60 E5
Montrave Fife 287 G7
Montrose Angus 287 B11
Montsale Essex 89 F8
Monwode Lea Warks 134 E5
Monxton Hants 47 E10
Monyash Derbys 169 B11
Monymusk Aberds 293 B8
Monzie Perth 286 E2
Monzie Castle Perth 286 E2
Moodiesburn N Lnrk 278 G3
Moolham Som 28 E5
Moon's Green Kent 38 B5
Moon's Moat Worcs 117 D11
Moonzie Fife 287 F7
Moor Som 28 D6
Moor Allerton W Yorks 205 F11
Moor Common Bucks 84 G4
Moor Crichel Dorset 31 F7
Moor Cross Devon 8 D2
Moor Edge W Yorks 205 F7
Moor End Bucks 84 G4
Moor End C Beds 103 G9
Moor End Cambs 105 B7
Moor End Durham 234 C2
Moor End E Yorks 208 F2
Moor End Glos 99 G9
Moor End Lancs 202 E3
Moor End N Yorks 207 F7
Moor End N Yorks 215 G7
Moor End S Yorks 197 G9
Moor End W Yorks 196 B5
Moor End W Yorks 206 D4
Moor End Worcs 116 E2
Moor End York 207 B9
Moor End Field N Yorks 215 F8
Moor Green Herts 104 F6
Moor Green Staffs 169 G7
Moor Green W Mid 133 G11
Moor Green Wilts 61 F11
Moor Hall W Mid 134 D2
Moor Head W Yorks 197 B8
Moor Head W Yorks 197 E8
Moor Monkton N Yorks 206 B6
Moor Monkton Moor N Yorks 206 B6

Moor of Balvack Aberds 293 B8
Moor of Granary Moray 301 D10
Moor of Ravenstone Dumfries 236 E5
Moor Park Cumb 229 D7
Moor Park Hereford 97 C9
Moor Park Herts 85 G9
Moor Park Sur 49 D11
Moor Row Cumb 219 C10
Moor Row Cumb 229 B10
Moor Side Lancs 202 F5
Moor Side Lancs 202 G4
Moor Side Lincs 174 D2
Moor Side W Yorks 197 B7
Moor Side W Yorks 204 F6
Moor Street Kent 69 F10
Moor Top W Yorks 197 C7
Mooradale Shetland 312 F6
Mooray Wilts 46 G3
Moorby Lincs 174 C3
Moorclose Cumb 228 F5
Moorclose Gtr Man 195 F11
Moorcot Hereford 115 F7
Moordown Bmouth 19 C7
Moore Halton 183 E9
Moorend Cumb 239 G8
Moorend Derbys 170 F2
Moorend Dumfries 239 C7
Moorend Glos 80 C5
Moorend Glos 80 E2
Moorend Gtr Man 185 D7
Moorend S Glos 61 D7
Moorend Cross Hereford 98 B4
Moorends S Yorks 199 D7
Moorfield Derbys 185 C8
Moorgate Norf 160 C3
Moorgate S Yorks 186 C6
Moorgreen Hants 33 D7
Moorgreen Notts 171 F7
Moorhaigh Notts 171 C8
Moorhall Derbys 186 G4
Moorhampton Hereford 97 B7
Moorhaven Village Devon 8 D3
Moorhayne Devon 28 F2
Moorhead W Yorks 205 F8
Moorhey Gtr Man 196 G2
Moorhole S Yorks 186 E6
Moorhouse Cumb 239 F8
Moorhouse Cumb 239 G7
Moorhouse Notts 172 B3
Moorhouse S Yorks 198 E3
Moorhouse Bank Sur 52 C2
Moorhouses Lincs 174 D3
Moorland or Northmoor Green Som 43 G10
Moorledge Bath 60 G5
Moorlinch Som 43 F11
Moorsholm Redcar 226 C3
Moorside Dorset 30 D3
Moorside Durham 233 B7
Moorside Gtr Man 195 G9
Moorside Gtr Man 196 F3
Moorside W Ches 182 F3
Moorside W Yorks 197 B8
Moorside W Yorks 205 F10
Moorstock Kent 54 F6
Moorswater Corn 6 C4
Moorthorpe W Yorks 198 E3
Moortown Devon 12 B2
Moortown Devon 12 G6
Moortown Devon 25 C8
Moortown Hants 31 G11
Moortown I o W 20 E4
Moortown Lincs 189 B9
Moortown Telford 150 F2
Moortown W Yorks 206 F2
Morangie Highld 309 L7
Morar Highld 295 F8
Moravian Settlement Derbys 153 B8
Môrawelon Anglesey 178 E3
Morayhill Highld 301 E7
Morborne Cambs 138 E2
Morchard Bishop Devon 26 F3
Morchard Road Devon 26 G3
Morcombelake Dorset 16 C4
Morcott Rutland 137 C8
Morda Shrops 148 D5
Morden Dorset 18 B4
Morden London 67 F9
Morden Green Cambs 104 C5
Morden Park London 67 F8
Mordiford Hereford 97 D11
Mordington Holdings Borders 273 D8
Mordon Durham 234 F2
More Shrops 130 E6
More Crichel Dorset 31 F7
Morebath Devon 27 C7
Morebattle Borders 263 E7
Morecambe Lancs 211 G8
Moredon Swindon 62 B6
Moredun Edin 270 B5
Morefield Highld 307 K6
Morehall Kent 55 F8
Moreleggan Argyll 284 G6
Moreleigh Devon 8 E5
Morenish Perth 285 D9
Moresby Cumb 228 G5
Moresby Parks Cumb 219 B9
Morestead Hants 33 B8
Moreton Dorset 18 D2
Moreton Essex 87 D8
Moreton Hereford 115 E10
Moreton Mers 182 C3
Moreton Oxon 82 E6
Moreton Oxon 83 E11
Moreton Staffs 150 F5
Moreton Staffs 152 D2
Moreton Corbet Shrops 149 E11
Moreton-in-Marsh Glos 100 E4
Moreton Jeffries Hereford 98 B2
Moreton Morrell Warks 118 F6
Moreton on Lugg Hereford 97 B10
Moreton Paddox Warks 118 G6
Moreton Pinkney Northants 101 B11
Moreton Say Shrops 150 C2
Moreton Valence Glos 80 D3
Moretonhampstead Devon 13 D11
Moretonwood Shrops 150 C2
Morfa Carms 56 B4
Morfa Carms 75 C9
Morfa Ceredig 110 G6
Morfa Gwyn 144 C3
Morfa Bach Carms 74 C5
Morfa Bychan Gwyn 145 B10
Morfa Dinlle Gwyn 162 D6
Morfa Glas Neath 76 D5
Morfa Nefyn Gwyn 162 G3
Morfydd Denb 165 F10
Morgan's Vale Wilts 31 C11
Morganstown Cardiff 58 C6
Moriah Ceredig 112 B2
Mork Glos 79 D9
Morland Cumb 231 G7
Morley Derbys 170 G5
Morley Durham 233 F8
Morley E Ches 184 E4
Morley W Yorks 197 B9
Morley Green E Ches 184 E4
Morley Park Derbys 170 F5
Morley St Botolph Norf 141 D11
Morley Smithy Derbys 170 G5
Morleymoor Derbys 170 G5
Mornick Corn 12 G2
Morningside Edin 280 G3
Morningside N Lnrk 268 D6
Morningthorpe Norf 142 E4
Morpeth Northumb 252 F6
Morphie Aberds 293 G9
Morrey Staffs 152 F2
Morridge Side Staffs 169 E8
Morrilow Heath Staffs 151 B9
Morris Green Essex 106 E4
Morriston = Treforys Swansea 57 B7
Morristown V Glam 59 E7
Morston Norf 177 E8
Mortehoe Devon 40 D3
Morthen S Yorks 187 D7
Mortimer W Berks 65 G7
Mortimer West End Hants 64 G6
Mortimer's Cross Hereford 115 E8
Mortlake London 67 D8
Mortomley S Yorks 186 B4
Morton Cumb 230 D4
Morton Cumb 239 G9
Morton Derbys 170 C6
Morton I o W 21 D8
Morton Lincs 155 E11
Morton Lincs 172 C5
Morton Lincs 188 C4
Morton Norf 160 F2
Morton Notts 172 E2
Morton S Glos 79 G10
Morton Shrops 148 E5
Morton Bagot Warks 118 E2
Morton Common Shrops 148 E5
Morton Mains Dumfries 247 D9
Morton Mill Shrops 149 E11
Morton-on-Swale N Yorks 224 G6
Morton Spirt Warks 117 G10
Morton Tinmouth Durham 233 G9
Morton Underhill Worcs 117 F10
Morvah Corn 1 B4
Morval Corn 6 D5
Morven Lodge Aberds 292 C5
Morvich Highld 295 C11
Morvich Highld 309 J7
Morville Shrops 132 E3
Morville Heath Shrops 132 E3
Morwellham Quay Devon 7 B8
Morwenstow Corn 24 E2
Mosborough S Yorks 186 E6
Moscow E Ayrs 267 G9
Mose Shrops 132 E5
Mosedale Cumb 230 E3
Moseley W Mid 133 D8
Moseley W Mid 133 G11
Moseley Worcs 116 F6
Moses Gate Gtr Man 195 F8
Mosley Common Gtr Man 195 G8
Moss Argyll 288 E1
Moss Highld 289 C8
Moss S Yorks 198 E5
Moss Wrex 166 E4
Moss Bank Halton 183 D8
Moss Bank Mers 183 B8
Moss Edge Lancs 202 D4
Moss Edge Lancs 202 E4
Moss End Brack 65 E11
Moss End E Ches 183 F11
Moss Houses E Ches 184 G5
Moss Lane E Ches 184 G6
Moss Nook Gtr Man 184 D4
Moss Nook Mers 183 C8
Moss of Barmuckity Moray 302 C2
Moss of Meft Moray 302 C2
Moss Pit Staffs 151 E8
Moss Side Cumb 238 G5
Moss Side Gtr Man 184 B4
Moss-side Highld 301 D8
Moss Side Lancs 193 G11
Moss Side Lancs 194 C4
Moss Side Lancs 202 G3
Moss Side Mers 182 B6
Moss-side Moray 302 D5
Mossat Aberds 292 B6
Mossbank Shetland 312 F6
Mossbay Cumb 228 F5
Mossblown S Ayrs 257 E10
Mossbrow Gtr Man 184 D2
Mossburnford Borders 262 F5
Mossdale Dumfries 237 B8
Mossedge Cumb 239 D11
Mossend N Lnrk 268 C4
Mosser Mains Cumb 229 F8
Mossfield Highld 300 B6
Mossgate Staffs 151 B8
Mossgiel E Ayrs 257 D11
Mosshouses Borders 262 B2
Mossley E Ches 168 C5
Mossley Gtr Man 196 G3
Mossley Brow Gtr Man 196 G3
Mossley Hill Mers 182 D5
Mosspark Glasgow 267 C10
Mosstodloch Moray 302 D3
Mosston Angus 287 C9
Mosstown Aberds 303 C10
Mossy Lea Lancs 194 E4
Mosterton Dorset 29 F7
Moston E Ches 168 C2
Moston Gtr Man 195 G11
Moston Shrops 149 D11
Moston W Ches 182 G6
Moston Green E Ches 168 C2
Mostyn Flint 181 E11
Mostyn Quay Flint 181 E11
Motcombe Dorset 30 B5
Mothecombe Devon 8 F2
Motherby Cumb 230 F4
Motherwell N Lnrk 268 D5
Motspur Park London 67 F8
Mottingham London 68 E2
Mottisfont Hants 32 B4
Mottistone I o W 20 E4
Mottram in Longdendale Gtr Man 185 B7
Mottram Rise Gtr Man 185 B7
Mottram St Andrew E Ches 184 F5
Mott's Green Essex 87 B8
Mott's Mill E Sus 52 F4
Mouldsworth W Ches 183 G8
Moulin Perth 286 B3
Moulsecoomb Brighton 36 F4
Moulsford Oxon 64 C5
Moulsham Essex 88 D2
Moulsoe M Keynes 103 C8
Moultavie Highld 300 B6
Moulton Lincs 156 E6
Moulton N Yorks 224 E4
Moulton Northants 120 D5
Moulton Suff 124 E3
Moulton V Glam 58 E5
Moulton W Ches 167 B11
Moulton Chapel Lincs 156 F5
Moulton Eaugate Lincs 156 F6
Moulton Park Northants 120 E5
Moulton St Mary Norf 143 B7
Moulton Seas End Lincs 156 D6
Moulzie Angus 292 F4
Mounie Castle Aberds 303 G7
Mount Corn 4 D5
Mount Corn 6 B2
Mount Highld 301 E9
Mount W Yorks 196 D5
Mount Ambrose Corn 4 G4
Mount Ballan Mon 60 B3
Mount Batten Plym 7 E9
Mount Bovers Essex 88 G4
Mount Bures Essex 107 E8
Mount Canisp Highld 301 B7
Mount Charles Corn 5 B10
Mount Charles Corn 5 E10
Mount Cowdown Wilts 47 C9
Mount End Essex 87 E7
Mount Ephraim E Sus 23 B7
Mount Gould Plym 7 D9
Mount Hawke Corn 4 F4
Mount Hermon Corn 2 F6
Mount Hermon Sur 50 B4
Mount Hill S Glos 61 E7
Mount Lane Devon 12 D3
Mount Pleasant Bucks 102 E3
Mount Pleasant Corn 5 C10
Mount Pleasant Derbys 152 D6
Mount Pleasant Derbys 152 F5
Mount Pleasant Derbys 170 F4
Mount Pleasant Devon 27 G11
Mount Pleasant Durham 233 E11
Mount Pleasant E Ches 168 D4
Mount Pleasant E Sus 23 E7
Mount Pleasant E Sus 36 D6
Mount Pleasant Flint 182 G2
Mount Pleasant Hants 19 B11
Mount Pleasant Kent 71 F10
Mount Pleasant London 85 G8
Mount Pleasant M Tydf 77 F9
Mount Pleasant Neath 57 B9
Mount Pleasant Norf 141 E9
Mount Pleasant Pembs 73 D8
Mount Pleasant Shrops 149 G9
Mount Pleasant Stockton 234 G4
Mount Pleasant Stoke 168 G5
Mount Pleasant Suff 106 B4
Mount Pleasant T & W 243 E7
Mount Pleasant W Yorks 197 C8
Mount Pleasant Warks 135 F7
Mount Pleasant Worcs 99 D10
Mount Pleasant Worcs 117 E10
Mount Sion Wrex 166 E3
Mount Skippett Oxon 82 B5
Mount Sorrel Wilts 31 C8
Mount Tabor W Yorks 196 B5
Mount Vernon Glasgow 268 C3
Mount Wise Corn 7 E9
Mountain Anglesey 178 E2
Mountain W Yorks 205 G7
Mountain Air Bl Gwent 77 D11
Mountain Ash = Aberpennar Rhondda 77 F8
Mountain Bower Wilts 61 D10
Mountain Cross Borders 270 F2
Mountain Street Kent 54 C5
Mountain Water Pembs 91 G8
Mountbenger Borders 261 D8
Mountbengerburn Borders 261 D8
Mountblow W Dunb 277 G9
Mounters Dorset 30 D3
Mountfield E Sus 38 C2
Mountgerald Highld 300 C5
Mountjoy Corn 5 C7
Mountnessing Essex 87 F10
Mounton Mon 79 G8
Mountsolie Aberds 303 D9
Mountsorrel Leics 153 F11
Mousehill Sur 50 E2
Mousehole Corn 1 D5
Mousen Northumb 264 C4
Mousley End Warks 118 D4
Mouswald Dumfries 238 C3
Mouth Mill Devon 24 B3
Mow Cop E Ches 168 D5
Mowbreck Lancs 202 G4
Mowden Darl 224 B5
Mowden Essex 88 C3
Mowhaugh Borders 263 E8
Mowmacre Hill Leicester 135 B11
Mowshurst Kent 52 D3
Mowsley Leics 136 F2
Moxby N Yorks 215 F11
Moxley W Mid 133 D9
Moy Argyll 255 E8
Moy Highld 290 E6
Moy Highld 301 F7
Moy Hall Highld 301 F7
Moy Ho Moray 301 C10
Moy Lodge Highld 290 E6
Moyles Court Hants 31 F11
Moylgrove = Trewyddel Pembs 92 C2
Muasdale Argyll 255 C7
Much Birch Hereford 97 E10
Much Cowardne Hereford 98 B2
Much Cowarne Hereford 98 B2
Much Dewchurch Hereford 97 E9
Much Hadham Herts 86 B5
Much Hoole Lancs 194 C3
Much Hoole Moss Houses Lancs 194 C3
Much Hoole Town Lancs 194 C3
Much Marcle Hereford 98 E3
Much Wenlock Shrops 132 C2
Muchalls Aberds 293 D11
Muchelney Som 28 C6
Muchelney Ham Som 28 C6
Muchlarnick Corn 6 D4
Muchrachd Highld 300 F2
Muckairn Argyll 289 F11
Muckernich Highld 300 D5
Mucking Thurrock 69 C7
Muckle Breck Shetland 312 G7
Muckleford Dorset 17 C8
Mucklestone Staffs 150 B4
Muckleton Norf 158 B6
Muckleton Shrops 149 E11
Muckletown Aberds 302 G5
Muckley Shrops 132 D2
Muckley Corner Staffs 133 B11
Muckley Cross Shrops 132 D2
Muckton Lincs 190 E5
Muckton Bottom Lincs 190 E5
Mudale Highld 308 F5
Mudd Gtr Man 185 C7
Muddiford Devon 40 F5
Muddlebridge Devon 40 G4
Muddles Green E Sus 23 C8
Mudeford Dorset 19 C9
Mudford Som 29 D9
Mudford Sock Som 29 D9
Mudgley Som 44 D2
Mugdock Stirl 277 F11
Mugeary Highld 294 B6
Mugginton Derbys 170 G3
Muggintonlane End Derbys 170 G3
Muggleswick Durham 232 B6
Mugswell Sur 51 C9
Muie Highld 309 J6
Muir Aberds 292 E2
Muir of Alford Aberds 293 B7
Muir of Fairburn Highld 300 D4
Muir of Fowlis Aberds 293 B7
Muir of Kinellar Aberds 293 B10
Muir of Miltonduff Moray 301 D11
Muir of Ord Highld 300 D5
Muir of Pert Angus 287 D8
Muircleugh Borders 271 F10
Muirden Aberds 303 D7
Muirdrum Angus 287 D9
Muiredge Fife 281 B7
Muirend Glasgow 267 C11
Muirhead Angus 287 D7
Muirhead Fife 286 G6
Muirhead Fife 287 F8
Muirhead N Lnrk 268 B3
Muirhead S Ayrs 257 C8
Muirhouse Edin 280 F4
Muirhouse N Lnrk 268 D5
Muirhouselaw Borders 262 D4
Muirhouses Falk 279 E10
Muirkirk E Ayrs 258 D5
Muirmill Stirl 278 E4
Muirshearlich Highld 290 E3
Muirskie Aberds 293 D10
Muirtack Aberds 303 F9
Muirton Aberds 303 D7
Muirton Highld 301 C7
Muirton Perth 286 E5
Muirton Perth 286 F3
Muirton Mains Highld 300 D4
Muirton of Ardblair Perth 286 C5
Muirton of Ballochy Angus 293 G8
Muiryfold Aberds 303 D7
Muker N Yorks 223 F8
Mulbarton Norf 142 C3
Mulben Moray 302 D3
Mulberry Corn 5 B10
Mulfra Corn 1 C5
Mulindry Argyll 254 B4
Mulla Shetland 313 G6
Mullardoch House Highld 300 F2
Mullenspond Hants 47 D9
Mullion Corn 2 F5
Mullion Cove Corn 2 F5
Mumbles Hill Swansea 56 D6
Mumby Lincs 191 G8
Mumps Gtr Man 196 F2
Mundale Moray 301 D10
Munderfield Row Hereford 116 G2
Munderfield Stocks Hereford 116 G2
Mundesley Norf 160 B6
Mundford Norf 140 E6
Mundham Norf 142 D6
Mundon Essex 88 E5
Mundurno Aberdeen 293 B11
Mundy Bois Kent 54 D2
Munerigie Highld 290 C4
Muness Shetland 312 C8
Mungasdale Highld 307 K4
Mungrisdale Cumb 230 E3
Munlochy Highld 300 D6
Munsary Cottage Highld 310 E6
Munsley Hereford 98 C3
Munslow Shrops 131 F10
Munstone Hereford 97 C10
Murch V Glam 59 E7
Murchington Devon 13 D9
Murcot Worcs 99 C11
Murcott Oxon 83 B9
Murcott Wilts 81 G7
Murdieston Stirl 278 B3
Murdishaw Halton 183 E9
Murieston W Loth 269 C11
Murkle Highld 310 C5
Murlaggan Highld 290 D2
Murlaggan Highld 290 E5
Murra Orkney 314 F2
Murrayfield Edin 280 G4
Murrayshall Perth 286 E5
Murraythwaite Dumfries 238 C4
Murrell Green Hants 49 B8
Murrell's End Glos 98 E4
Murrell's End Glos 98 G5
Murrion Shetland 312 F4
Murrow Cambs 139 B7
Mursley Bucks 102 F6
Murston Kent 70 G2
Murthill Angus 287 B8
Murthly Perth 286 D4
Murton Cumb 231 G10
Murton Durham 234 B3
Murton Northumb 273 F9
Murton Swansea 56 D5
Murton T & W 243 C8
Murton York 207 C8
Murton Grange N Yorks 215 B10
Murtwell Devon 8 D5
Musbury Devon 15 C11
Muscliff Bmouth 19 B7
Muscoates N Yorks 216 C3
Muscott Northants 120 E2
Musdale Argyll 289 G11
Mushroom Green W Mid 133 F8
Musselburgh E Loth 280 G6
Musselwick Pembs 72 D4
Mustard Hyrn Norf 161 F8
Muston Leics 154 B6
Muston N Yorks 217 D11
Mustow Green Worcs 117 C7
Muswell Hill London 86 G3
Mutehill Dumfries 237 E8
Mutford Suff 143 F9
Muthill Perth 286 F2
Mutley Plym 7 D9
Mutterton Devon 27 G8
Mutton Hall E Sus 37 C9
Muxton Telford 150 G4
Mwdwl-eithin Flint 181 F11
Mwynbwll Flint 165 B11
Mybster Highld 310 D5
Myddfai Carms 94 F5
Myddle Shrops 149 E9
Myddlewood Shrops 149 E9
Myddyn-fych Carms 75 C10
Mydroilyn Ceredig 111 F9
Myerscough Lancs 202 F5
Myerscough Smithy Lancs 203 G8
Mylor Bridge Corn 3 B8
Mylor Churchtown Corn 3 B8
Mynachdy Cardiff 59 D7
Mynachdy Rhondda 77 F8
Mynachlog-ddu Pembs 92 E2
Mynd Shrops 115 C7
Myndd Llandegai Gwyn 163 B10
Myndtown Shrops 131 F7
Mynydd Bach Ceredig 112 B4
Mynydd-bach Mon 79 G7
Mynydd-Bach Swansea 57 B7
Mynydd-bach-y-glo Swansea 56 B6
Mynydd Bodafon Anglesey 179 D7
Mynydd Fflint = Flint Mountain Flint 182 G2
Mynydd Gilan Gwyn 144 E5
Mynydd-isa Flint 166 C3
Mynydd-llan Flint 181 G11
Mynydd Marian Conwy 180 F5
Mynydd Mechell Anglesey 178 D5
Mynyddislwyn Caerph 77 G11
Mynyddygarreg Carms 74 D6
Mynytho Gwyn 144 C6
Myrebird Aberds 293 D9
Myrelandhorn Highld 310 D6
Myreside Perth 286 E6
Myrtle Hill Carms 94 E5
Mytchett Sur 49 B11
Mytchett Place Sur 49 C11
Mytholm W Yorks 196 B3
Mytholmes W Yorks 204 F6
Mytholmroyd W Yorks 196 B4
Mythop Lancs 202 G3
Mytice Aberds 302 F4
Myton Warks 118 E6
Myton Hall N Yorks 215 F8
Myton-on-Swale N Yorks 215 F8
Mytton Shrops 149 F8

N

Na Gearrannan W Isles 304 D3
Naast Highld 307 L3
Nab Hill W Yorks 197 D7
Nab Wood W Yorks 205 F8
Nab's Head Lancs 194 B6
Naburn York 207 D7
Naccolt Kent 54 E4
Nackington Kent 55 C7
Nacton Suff 108 C4
Nadderwater Devon 14 C3
Nafferton E Yorks 209 B7
Nag's Head Glos 80 F5
Naid-y-march Flint 181 F11
Nailbridge Glos 79 B10
Nailsbourne Som 28 B2
Nailsea N Som 60 D3
Nailstone Leics 135 B8
Nailsworth Glos 80 F5
Nailwell Bath 61 G8
Nairn Highld 301 D8
Nalderswood Sur 51 D8
Nance Corn 4 G3
Nanceddan Corn 2 C2
Nancegollan Corn 2 C4
Nancemellin Corn 4 G2
Nancenoy Corn 2 D6
Nancledra Corn 1 B5
Nangreaves Lancs 195 D10
Nanhoron Gwyn 144 C5
Nanhyfer = Nevern Pembs 91 D11
Nannau Gwyn 146 E4
Nannerch Flint 165 B11
Nanpantan Leics 153 F10
Nanpean Corn 5 D9
Nanquidno Corn 1 D3
Nanstallon Corn 5 B10
Nant Carms 74 B6
Nant Denb 165 D11
Nant Alyn Flint 165 B11
Nant-ddu Powys 77 B8
Nant-glas Powys 113 E9
Nant Mawr Flint 166 C3
Nant Peris = Old Llanberis Gwyn 163 D10
Nant Uchaf Denb 165 D8
Nant-y-Bai Carms 94 C5
Nant-y-Bwch Bl Gwent 77 C10
Nant-y-cafn Neath 76 D4
Nant y Caws Shrops 148 D5
Nant-y-ceisiad Caerph 59 B8
Nant-y-derry Mon 78 D4
Nant-y-felin Conwy 179 G11
Nant-y-ffin Carms 93 E11
Nant-y-gollen Shrops 148 D4
Nant-y-moel Bridgend 76 G6
Nant-y-pandy Conwy 179 G11
Nant-y-Rhiw Conwy 164 D4
Nanternis Ceredig 111 F7
Nantgaredig Carms 93 G9
Nantgarw Rhondda 58 B6
Nantglyn Denb 165 C8
Nantgwyn Powys 113 B9
Nantithet Corn 2 E5
Nantlle Gwyn 163 E8
Nantmawr Shrops 148 E5
Nantmel Powys 113 D10
Nantmor Gwyn 163 F10
Nantserth Powys 113 C9
Nantwich E Ches 167 E11
Nantycaws Carms 75 B7
Nantyffyllon Bridgend 57 C11
Nantyglo Bl Gwent 77 C11
Nantyronen Station Ceredig 112 B3
Napchester Kent 55 D10
Naphill Bucks 84 F4
Napleton Worcs 99 B7
Napley Staffs 150 B4
Napley Heath Staffs 150 B4
Nappa N Yorks 204 C3
Nappa Scar N Yorks 223 G9
Napton on the Hill Warks 119 E9
Narberth = Arberth Pembs 73 C10
Narberth Bridge Pembs 73 C10
Narborough Leics 135 D10
Narborough Norf 158 G4
Narfords Som 28 F3
Narkurs Corn 6 D6
Narracott Devon 24 D5
Narrowgate Corner Norf 161 F8
Nasareth Gwyn 163 E7
Naseby Northants 120 B3
Nash Bucks 102 E5
Nash Hereford 114 E6
Nash Kent 55 B9
Nash London 68 G2
Nash Newport 59 C10
Nash Shrops 116 C2
Nash Som 29 E8
Nash End Worcs 132 G5
Nash Lee Bucks 84 D4
Nash Mills Herts 85 E9
Nash Street E Sus 23 C8
Nash Street Kent 68 F6
Nashend Glos 80 D5
Nashes Green Hants 49 D7
Nassington Northants 137 D11
Nast Hyde Herts 86 D2
Nastend Glos 80 D3
Nasty Herts 105 G7
Natcott Devon 24 C3
Nateby Cumb 222 D5
Nateby Lancs 202 E5
Nately Scures Hants 49 C8
Natland Cumb 211 B10
Natton Glos 99 E8
Naughton Suff 107 B10
Naunton Glos 100 G2
Naunton Worcs 99 D7
Naunton Beauchamp Worcs 117 G9
Navant Hill W Sus 34 B6
Navenby Lincs 173 D7
Navestock Heath Essex 87 F8
Navestock Side Essex 87 F9
Navidale Highld 311 H4
Navity Highld 301 C7
Nawton N Yorks 216 C3
Nayland Suff 107 E9
Nazeing Essex 86 D6
Nazeing Gate Essex 86 D6
Nazeing Long Green Essex 86 E6
Nazeing Mead Essex 86 D5
Neacroft Hants 19 B9
Nealhouse Cumb 239 G8
Neal's Green Warks 134 G6
Neames Forstal Kent 54 B5
Neap Shetland 313 H7
Near Hardcastle N Yorks 214 F2
Near Sawrey Cumb 221 F7
Nearton End Bucks 102 F6
Neasden London 67 B8
Neasham Darl 224 C6
Neat Enstone Oxon 101 G7
Neat Marsh E Yorks 209 G9
Neath = Castell-nedd Neath 57 B8
Neath Abbey Neath 57 B8
Neath Hill M Keynes 103 C7
Neatham Hants 49 E8
Neatishead Norf 160 E6
Neaton Norf 141 C8
Nebo Anglesey 179 C7
Nebo Ceredig 111 D10
Nebo Conwy 164 D4
Nebo Gwyn 163 E7
Nebsworth Warks 100 C3
Nechells W Mid 133 F11
Necton Norf 141 B7
Nedd Highld 306 F6
Nedderton Northumb 252 G6
Nedge Hill Som 44 C5
Nedge Hill Telford 132 B4
Nedging Suff 107 B9
Nedging Tye Suff 107 B10
Needham Norf 142 G4
Needham Green Essex 87 B9
Needham Market Suff 125 G11
Needham Street Suff 124 D4
Needingworth Cambs 122 C6
Needwood Staffs 152 E3
Neen Savage Shrops 116 B3
Neen Sollars Shrops 116 C3
Neenton Shrops 132 F2
Nefod Shrops 148 B6
Nefyn Gwyn 162 G4
Neighbourne Som 44 D6
Neight Hill Worcs 117 F8
Neilston E Renf 267 D9
Neinthirion Powys 129 B9
Neithrop Oxon 101 C8
Nelly Andrews Green Powys 130 B5
Nelson Caerph 77 F10
Nelson Lancs 204 F3
Nelson Village Northumb 243 B7
Nemphlar S Lnrk 269 G7
Nempnett Thrubwell N Som 60 G4
Nene Terrace Lincs 138 B5
Nenthall Cumb 231 B11
Nenthead Cumb 231 C11
Nenthorn Borders 262 B5
Neopardy Devon 13 B11
Nep Town W Sus 36 D2
Nepcote W Sus 35 F10
Nepgill Cumb 229 F7
Nerabus Argyll 254 B3
Nercwys Flint 166 C2
Nerston S Lnrk 268 D2
Nesbit Northumb 263 C11
Ness Orkney 314 C4
Ness W Ches 182 F4
Nesscliffe Shrops 149 F7
Nessholt W Ches 182 F4
Nesstoun Orkney 314 A7
Neston W Ches 182 F3
Neston Wilts 61 F11
Netchells Green W Mid 133 F11
Netham Bristol 60 E6
Nethanfoot S Lnrk 268 F6
Nether Alderley E Ches 184 F4
Nether Blainslie Borders 271 G10
Nether Booth Derbys 185 D10
Nether Broughton Leics 154 D3
Nether Burrow Lancs 212 D2
Nether Burrows Derbys 152 B5
Nether Cassock Dumfries 248 C6
Nether Cerne Dorset 17 B9
Nether Chanderhill Derbys 186 G4
Nether Compton Dorset 29 D9
Nether Crimond Aberds 303 G8
Nether Dalgliesh Borders 249 B7
Nether Dallachy Moray 302 C3
Nether Edge S Yorks 186 E4
Nether End Derbys 186 G3
Nether End Leics 154 G4
Nether End W Yorks 197 F8
Nether Exe Devon 26 G6
Nether Glasslaw Aberds 303 D8
Nether Hall Leicester 136 B2
Nether Handley Derbys 186 F6
Nether Handwick Angus 287 C7
Nether Haugh S Yorks 186 B6
Nether Headon Notts 188 F2
Nether Heage Derbys 170 E5
Nether Heyford Northants 120 F3
Nether Hindhope Borders 263 G7
Nether Horsburgh Borders 261 B8
Nether Howecleuch S Lnrk 260 G2
Nether Kellet Lancs 211 F10
Nether Kidston Borders 270 G4
Nether Kinmundy Aberds 303 E10
Nether Kirton E Renf 267 D9
Nether Langwith Notts 187 G8
Nether Leask Aberds 303 F10
Nether Lenshie Aberds 302 E6
Nether Loads Derbys 170 B4
Nether Monynut Borders 272 C4
Nether Moor Derbys 170 B5
Nether Padley Derbys 186 F3
Nether Park Aberds 303 D10
Nether Poppleton York 207 B7
Nether Row Cumb 230 D2
Nether Savock Aberds 303 E10
Nether Shiels Borders 271 F8
Nether Silton N Yorks 225 G9
Nether Skyborry Shrops 114 C5
Nether St Suff 125 E8
Nether Stowe Staffs 152 G2
Nether Stowey Som 43 F7
Nether Street Essex 87 C9
Nether Street Herts 86 B6
Nether Urquhart Fife 286 G5
Nether Wallop Hants 47 F10
Nether Warden Northumb 241 D10
Nether Wasdale Cumb 220 E2
Nether Welton Cumb 230 B3
Nether Westcote Glos 100 G4
Nether Whitacre Warks 134 E4
Nether Winchendon or Lower Winchendon Bucks 84 C2
Nether Worton Oxon 101 E8
Nether Yeadon W Yorks 205 E10
Netheravon Wilts 46 D6
Netherbrae Aberds 303 D7
Netherbrough Orkney 314 E3
Netherburn S Lnrk 268 F6
Netherbury Dorset 16 B5
Netherby Cumb 239 C9
Netherby N Yorks 206 D2
Netherclay Som 28 C3
Nethercote Oxon 101 C9
Nethercote Warks 119 E10
Nethercott Devon 12 B3
Nethercott Devon 40 F3
Nethercott Oxon 101 G9
Nethercott Som 42 G6
Netherdale Shetland 313 H3
Netherend Glos 79 E9
Netherfield E Sus 38 D2
Netherfield M Keynes 103 D7
Netherfield Notts 171 G10
Nethergate Norf 159 D11
Netherhampton Wilts 31 B10
Netherland Green Staffs 152 C2
Netherlaw Dumfries 237 E9
Netherley Dorset 28 F6
Netherlee E Renf 267 D11
Netherley Aberds 293 D10
Netherley Mers 182 D6
Netherley W Yorks 196 E4
Nethermill Dumfries 248 F2
Nethermills Moray 302 D5
Nethermuir Aberds 303 E9
Netherne on-the-Hill Sur 51 B9
Netheroyd Hill W Yorks 196 D6
Netherplace E Renf 267 D10
Netherraw Borders 262 E3
Netherseal Derbys 152 G5
Netherstoke Dorset 29 F8
Netherthird E Ayrs 258 F3
Netherthong W Yorks 196 F6
Netherthorpe Derbys 186 G6
Netherthorpe S Yorks 187 E8
Netherton Aberds 303 E8
Netherton Angus 287 B9
Netherton Corn 11 G11
Netherton Cumb 228 D6
Netherton Devon 14 G3
Netherton Glos 81 E11
Netherton Hants 47 B11
Netherton Hereford 97 F10
Netherton Mers 193 G11
Netherton N Lnrk 268 E5
Netherton Northumb 251 B11
Netherton Oxon 82 F6
Netherton Perth 286 B5
Netherton Shrops 132 G4
Netherton Stirl 277 F11
Netherton W Ches 183 F8
Netherton W Mid 133 F8
Netherton W Yorks 196 E6
Netherton W Yorks 197 D9
Netherton Worcs 99 C9
Netherton of Lonmay Aberds 303 C10
Nethertown Cumb 219 D9
Nethertown Highld 310 B7
Nethertown Lancs 203 F10
Nethertown Staffs 152 F2
Netherwitton Northumb 252 E4
Netherwood E Ayrs 258 D5
Nethy Bridge Highld 301 G10
Netley Hants 33 F7
Netley Hill Soton 33 E7
Netley Marsh Hants 32 E4
Nettacott Devon 14 B4
Netteswell Essex 87 C7
Nettlebed Oxon 65 B8
Nettlebridge Som 44 D6
Nettlecombe Dorset 16 B6
Nettlecombe I o W 20 F6
Nettleden Herts 85 C8
Nettleham Lincs 189 F8
Nettlestead Kent 53 C7
Nettlestead Suff 107 B11
Nettlestead Green Kent 53 C7
Nettlestone I o W 21 C8
Nettlesworth Durham 233 B11
Nettleton Glos 80 C6
Nettleton Lincs 200 G6
Nettleton Wilts 61 D10
Nettleton Green Wilts 61 D10
Nettleton Hill W Yorks 196 D5
Nettleton Shrub Wilts 61 D10
Nettleton Top Lincs 189 B10
Netton Wilts 46 F6
Neuadd Carms 94 G3
Nevendon Essex 88 G2
Nevern = Nanhyfer Pembs 91 D11
Nevilles Cross Durham 233 C11
New Abbey Dumfries 237 C11
New Aberdour Aberds 303 C8
New Addington London 67 G11
New Alresford Hants 48 G5
New Alyth Perth 286 C6
New Arley Warks 134 F5
New Arram E Yorks 208 E6
New Ash Green Kent 68 F6
New Balderton Notts 172 E4
New Barn Kent 68 F6
New Barnet London 86 F3
New Barnetby N Lincs 200 E5
New Barton Northants 121 E7
New Basford Nottingham 171 G9
New Beaupre V Glam 58 E4
New Beckenham London 67 E11
New Bewick Northumb 264 E3
New-bigging Angus 286 C6
New Bilton Warks 119 B9
New Bolingbroke Lincs 174 D4
New Bolsover Derbys 187 G7
New Boston Mers 183 B9
New Botley Oxon 83 D7
New Boultham Lincs 189 G7
New Bradwell M Keynes 102 C6
New Brancepeth Durham 233 C10
New Bridge Wrex 166 G3
New Brighton Flint 166 B3
New Brighton Mers 182 C4
New Brighton W Sus 22 B3
New Brighton W Yorks 197 B9
New Brighton W Yorks 205 F8
New Brighton Wrex 166 E3
New Brimington Derbys 186 G6
New Brinsley Notts 171 E7
New Brotton Redcar 235 G9
New Broughton Wrex 166 E4
New Buckenham Norf 141 E11
New Buildings Bath 45 B7
New Buildings Dorset 18 E5
New Bury Gtr Man 195 F8
New Byth Aberds 303 D8
New Catton Norf 160 G4
New Charlton London 68 D2
New Cheltenham S Glos 61 E7
New Cheriton Hants 33 B9
New Clipstone Notts 171 C9
New Costessey Norf 160 G3
New Coundon Durham 233 E10
New Cowper Cumb 229 B8
New Crofton W Yorks 197 D11
New Cross Ceredig 112 B2
New Cross London 67 D11
New Cross Oxon 65 D9
New Cross Som 28 D6
New Cross Gate London 67 D11
New Cumnock E Ayrs 258 G4
New Deer Aberds 303 E8
New Delaval Northumb 243 B7
New Delph Gtr Man 196 F3
New Denham Bucks 66 C4
New Downs Corn 1 C3
New Downs Corn 4 E4
New Duston Northants 120 E4
New Earswick York 207 B8
New Eastwood Notts 171 F7
New Edlington S Yorks 187 B8
New Elgin Moray 302 C2
New Ellerby E Yorks 209 F9
New Eltham London 68 E2
New End Lincs 190 G2
New End Warks 118 E2
New End Worcs 117 F11
New England Essex 106 C4
New England Lincs 175 D8
New England P'boro 138 C3
New England Som 28 E4
New Farnley W Yorks 205 G10
New Ferry Mers 182 D4
New Fletton P'boro 138 D3
New Fryston W Yorks 198 B3
New Galloway Dumfries 237 B8
New Gilston Fife 287 G8
New Greens Herts 85 D10
New Grimsby Scilly 1 F3
New Ground Herts 85 C7
New Hainford Norf 160 F4
New Hall Hey Lancs 195 C10
New Hartley Northumb 243 B8
New Haw Sur 66 G5
New Headington Oxon 83 D9
New Heaton Northumb 273 G7
New Hedges Pembs 73 E10
New Herrington T & W 243 G8
New Hinksey Oxon 83 E8
New Ho Durham 232 D3
New Holkham Norf 159 B7
New Holland N Lincs 200 C5
New Holland W Yorks 205 F7
New Horwich Derbys 185 E8
New Houghton Derbys 171 B7
New Houghton Norf 158 D5
New House Kent 68 E6
New Houses Gtr Man 194 G5
New Houses N Yorks 212 E6
New Humberstone Leicester 136 B2
New Hunwick Durham 233 E9
New Hutton Cumb 221 G11
New Hythe Kent 53 B8
New Inn Carms 93 D9
New Inn Devon 24 F6
New Inn Mon 79 E7
New Inn Pembs 91 E11
New Inn Torf 78 F4
New Invention Shrops 114 B5
New Invention W Mid 133 C9
New Kelso Highld 299 E9
New Kingston Notts 153 D10
New Kyo Durham 242 G5
New Ladykirk Borders 273 F7
New Lanark S Lnrk 269 G7
New Lane Lancs 194 E2
New Lane End Warr 183 B10
New Langholm Dumfries 249 G9
New Leake Lincs 174 D6
New Leeds Aberds 303 D9
New Lodge S Yorks 197 F10
New Longton Lancs 194 B4
New Luce Dumfries 236 C3
New Malden London 67 F8
New Marske Redcar 235 G8
New Marston Oxon 83 D8
New Marton Shrops 148 C6
New Micklefield W Yorks 206 G4
New Mill Aberds 293 E9
New Mill Borders 262 G2
New Mill Corn 1 C5
New Mill Corn 4 F6
New Mill Cumb 219 E11
New Mill Herts 84 C6
New Mill W Yorks 197 F7
New Mill Wilts 63 G7
New Mills Borders 271 F10
New Mills Corn 5 E7
New Mills Derbys 185 D7
New Mills E Ches 184 E3
New Mills Glos 79 E10
New Mills Hereford 98 D4
New Mills = Felin Newydd Powys 129 C11
New Milton Hants 19 B10
New Mistley Essex 108 E2
New Moat Pembs 91 F11
New Moston Gtr Man 195 G11
New Ollerton Notts 171 B11
New Oscott W Mid 133 E11
New Pale W Ches 183 G8
New Park N Yorks 205 B11

New Parks Leicester 135 B11
New Passage S Glos 60 B4
New Pitsligo Aberds 303 D8
New Polzeath Corn 10 F4
New Quay =
Ceinewydd Ceredig 111 F7
New Rackheath Norf 160 G5
New Radnor Powys 114 E4
New Rent Cumb 230 D5
New Ridley Northumb 242 F3
New Road Side N Yorks 204 E5
New Road Side W Yorks 197 B7
New Romney Kent 39 C9
New Rossington
S Yorks 187 B10
New Row Ceredig 112 C4
New Row Lancs 203 F8
New Row N Yorks 226 C2
New Sarum Wilts 46 G6
New Sawley Derbys 153 C9
New Scarbro W Yorks 205 G10
New Sharlston W Yorks 197 C11
New Silksworth T & W 243 G9
New Skelton Redcar 226 B3
New Smithy Derbys 185 E9
New Southgate London 86 G3
New Springs Gtr Man 194 F6
New Sprowston Norf 160 G4
New Stanton Derbys 153 B9
New Stevenston N Lnrk 268 D5
New Street Kent 68 G6
New Street Staffs 169 E9
New Swanage Dorset 18 E6
New Swannington Leics 153 F8
New Thirsk N Yorks 215 C8
New Thundersley Essex 69 B9
New Totley S Yorks 186 F4
New Town Bath 45 B9
New Town Bath 60 G5
New Town Dorset 30 C3
New Town Dorset 30 D6
New Town Dorset 31 D7
New Town Dorset 31 F7
New Town E Loth 281 G8
New Town E Sus 37 C7
New Town Edin 280 G4
New Town Edin 280 G5
New Town Glos 99 E10
New Town Kent 53 B7
New Town Kent 68 E4
New Town Lancs 203 F8
New Town Luton 103 G11
New Town Medway 69 G8
New Town Oxon 100 F5
New Town Reading 65 E8
New Town Shetland 312 E6
New Town Som 29 D11
New Town Som 29 D9
New Town Som 44 D3
New Town Soton 33 E7
New Town Swindon 63 C7
New Town T & W 234 B2
New Town T & W 243 E8
New Town W Berks 64 D6
New Town W Mid 133 B11
New Town W Mid 133 E9
New Town W Sus 35 B11
New Town W Yorks 198 C3
New Town Wilts 46 C6
New Town Wilts 63 E9
New Tredegar Caerph 77 E10
New Trows S Lnrk 259 B8
New Ulva Argyll 275 E8
New Village E Yorks 209 G7
New Village S Yorks 198 F5
New Walsoken Cambs 139 B9
New Waltham NE Lincs 201 G9
New Well Powys 113 B11
New Wells Powys 130 D3
New Whittington Derbys 186 F5
New Wimpole
Cambs 104 B6
New Winton E Loth 281 G8
New Woodhouses
Shrops 167 G9
New Works Telford 132 B3
New Wortley W Yorks 205 G11
New Yatt Oxon 82 C5
New York Lincs 174 D2
New York N Yorks 214 G3
New York T & W 243 C8
New Zealand Wilts 62 D4
Newall W Yorks 205 D10
Newall Green Gtr Man 184 D4
Newark Orkney 314 B7
Newark P'boro 138 C4
Newark-on-Trent Notts 172 E3
Newarthill N Lnrk 268 D5
Newball Lincs 189 F9
Newbarn Kent 55 F7
Newbarns Cumb 210 E4
Newbattle Midloth 270 B6
Newbie Dumfries 238 D5
Newbiggin Cumb 210 F5
Newbiggin Cumb 211 D11
Newbiggin Cumb 219 G11
Newbiggin Cumb 230 F5
Newbiggin Cumb 231 B7
Newbiggin Cumb 231 F8
Newbiggin Durham 232 B5
Newbiggin Durham 232 F4
Newbiggin Durham 233 B8
Newbiggin N Yorks 213 B9
Newbiggin N Yorks 223 G9
Newbiggin-by-the-Sea
Northumb 253 F8
Newbiggin Hall Estate
T & W 242 D6
Newbiggin-on-Lune
Cumb 222 D4
Newbigging Aberds 303 G7
Newbigging Angus 287 D8
Newbigging Borders 262 F6
Newbigging Edin 280 F2
Newbigging S Lnrk 269 F10
Newbiggings Orkney 314 B6
Newbold Derbys 186 G5
Newbold Gtr Man 196 E2
Newbold Leics 136 B5
Newbold Leics 153 F8
Newbold Heath Leics 135 B8
Newbold on Avon Warks 119 B9
Newbold on Stour
Warks 100 B4
Newbold Pacey Warks 118 F5
Newbold Verdon Leics 135 C8
Newbolds W Mid 133 C8
Newborough P'boro 138 B4
Newborough Staffs 152 D2
Newbottle Northumb 101 D10
Newbottle T & W 243 G8
Newbourne Suff 108 C5
Newbridge Bath 61 F8
Newbridge Caerph 78 F2
Newbridge Ceredig 111 F10
Newbridge Corn 1 C4
Newbridge Corn 4 G5
Newbridge Corn 7 B7
Newbridge Dumfries 237 B11
Newbridge E Sus 52 G3
Newbridge Edin 280 G2
Newbridge Hants 32 D3
Newbridge I o W 20 D4
Newbridge Lancs 204 F3
Newbridge N Yorks 216 B6
Newbridge Oxon 82 E6
Newbridge Pembs 91 E8
Newbridge Shrops 148 D6
Newbridge W Mid 133 D7
Newbridge Wrex 166 G3
Newbridge Green Worcs 98 D6
Newbridge-on-Usk Mon 78 G5
Newbridge-on-Wye
Powys 113 F10
Newbrough Northumb 241 D9
Newbuildings Devon 26 G3
Newburgh Aberds 303 D9
Newburgh Aberds 303 G9
Newburgh Borders 261 F8
Newburgh Fife 286 F6
Newburgh Lancs 194 E3
Newburn T & W 242 D5
Newbury Kent 54 B2
Newbury W Berks 64 F3
Newbury Wilts 45 E10
Newbury Park London 68 B2
Newby Cumb 231 G7
Newby Lancs 204 D2
Newby N Yorks 205 D11
Newby N Yorks 212 E4
Newby N Yorks 215 F7
Newby N Yorks 225 C10
Newby N Yorks 227 G10
Newby Bridge Cumb 211 B7
Newby Cote N Yorks 212 E4
Newby East Cumb 239 F11
Newby Head Cumb 231 G7
Newby West Cumb 239 G9
Newby Wiske N Yorks 215 B7
Newcastle Bridgend 58 D2
Newcastle Mon 78 B6
Newcastle Shrops 130 G4
Newcastle Emlyn =
Castell Newydd Emlyn
Carms 92 C6
Newcastle-under-Lyme
Staffs 168 F4
Newcastle upon Tyne
T & W 242 E6
Newcastleton or
Copshaw Holm
Borders 249 F11
Newchapel Powys 129 G9
Newchapel Staffs 168 E5
Newchapel Sur 51 E11
Newchapel =
Capel Newydd Pembs 92 D4
Newchurch Bl Gwent 77 C11
Newchurch Carms 93 G7
Newchurch Hereford 115 G7
Newchurch I o W 21 D7
Newchurch Kent 54 G5
Newchurch Lancs 195 C10
Newchurch Mon 79 F7
Newchurch Powys 114 G4
Newchurch Staffs 152 E2
Newchurch in Pendle
Lancs 204 F2
Newcott Devon 28 F2
Newcraighall Edin 280 G6
Newdigate Sur 51 E7
Newell Green Brack 65 E11
Newenden Kent 38 B4
Newent Glos 98 F4
Newerne Glos 79 E10
Newfield Durham 233 E10
Newfield Durham 242 G6
Newfield Highld 301 B7
Newfield Stoke 168 E6
Newford Scilly 1 G4
Newfound Hants 48 C5
Newgale Pembs 90 G6
Newgarth Orkney 314 E2
Newgate Lancs 194 F4
Newgate Norf 177 E9
Newgate Corner Norf 161 G8
Newgate Street Herts 86 D4
Newgrounds Hants 31 E11
Newhailes Edin 280 G6
Newhall Derbys 152 E5
Newhall E Ches 167 F10
Newhall Green Warks 134 F5
Newhall House Highld 300 C6
Newhall Point Highld 301 C7
Newham Lincs 174 E3
Newham Northumb 264 D5
Newhaven Derbys 169 C11
Newhaven Devon 24 C5
Newhaven E Sus 36 G6
Newhaven Edin 280 F5
Newhey Gtr Man 196 E2
Newhill Fife 286 F6
Newhill Perth 286 G5
Newhill S Yorks 186 B6
Newhills Aberdeen 293 C10
Newholm N Yorks 227 C7
Newhouse Borders 262 E2
Newhouse N Lnrk 268 C5
Newhouse Shetland 313 G6
Newhouses Borders 271 G10
Newick E Sus 36 C6
Newingreen Kent 54 F6
Newington Edin 280 G5
Newington Kent 55 F7
Newington Kent 69 G11
Newington Kent 71 F11
Newington London 67 D10
Newington Notts 187 C11
Newington Oxon 83 F10
Newington Shrops 131 G8
Newington Bagpath
Glos 80 G4
Newland Cumb 210 D6
Newland E Yorks 199 B10
Newland Glos 79 D9
Newland Hull 209 G7
Newland N Yorks 199 C7
Newland Oxon 82 C5
Newland Worcs 98 B5
Newland Bottom Cumb 210 C5
Newland Common
Worcs 117 E8
Newland Green Kent 54 D2
Newlandrig Midloth 271 C7
Newlands Borders 250 E2
Newlands Borders 262 E2
Newlands Cumb 229 G10
Newlands Cumb 230 D2
Newlands Derbys 170 F6
Newlands Dumfries 247 F11
Newlands Glasgow 267 C11
Newlands Highld 301 E7
Newlands Moray 302 D3
Newlands Northumb 242 F3
Newlands Notts 171 C9
Newlands Staffs 151 E11
Newlands Corner Sur 50 D4
Newlands of Geise
Highld 310 C4
Newlands of Tynet
Moray 302 C3
Newlands Park Anglesey 178 E3
Newlandsmuir S Lnrk 268 E2
Newliston Edin 280 G2
Newliston Fife 280 C5
Newlot Orkney 314 E5
Newlyn Corn 1 D5
Newmachar Aberds 293 B10
Newmains N Lnrk 268 D6
Newman's End Essex 87 C8
Newman's Green Suff 107 C7
Newman's Place
Hereford 96 B5
Newmarket Glos 80 F4
Newmarket Suff 124 E2
Newmarket W Isles 304 E6
Newmill Borders 261 G11
Newmill Corn 1 C5
Newmill Moray 302 D4
Newmill of Inshewan
Angus 292 G6
Newmillerdam W Yorks 197 D10
Newmills Corn 11 D11
Newmills Fife 279 D10
Newmills Highld 300 C6
Newmills of Boyne
Aberds 302 D5
Newmiln Perth 286 D5
Newmilns E Ayrs 258 B2
Newmore Highld 300 B6
Newmore Highld 300 D5
Newnes Shrops 149 C7
Newney Green Essex 87 D11
Newnham Cambs 123 F8
Newnham Glos 79 C11
Newnham Hants 49 C8
Newnham Herts 104 D4
Newnham Kent 54 B3
Newnham Northants 119 F11
Newnham Warks 118 E3
Newnham Bridge
Worcs 116 D2
Newpark Fife 287 F8
Newpool Staffs 168 D5
Newport Corn 12 D2
Newport Devon 40 G5
Newport Dorset 18 C3
Newport E Yorks 208 G3
Newport Essex 105 E10
Newport Glos 79 F11
Newport Highld 311 G5
Newport I o W 20 D6
Newport Newport 59 B10
Newport Norf 161 F10
Newport Som 28 C4
Newport Telford 150 F4
Newport = Trefdraeth
Pembs 91 D11
Newport-on-Tay Fife 287 E8
Newport Pagnell
M Keynes 103 C7
Newpound Common
W Sus 35 B9
Newquay Corn 4 C6
Newsam Green W Yorks 206 G3
Newsbank E Ches 168 B4
Newseat Aberds 303 E10
Newseat Aberds 303 F7
Newsells Herts 105 D7
Newsham Lancs 202 F6
Newsham N Yorks 215 C7
Newsham N Yorks 224 C2
Newsham Northumb 243 B8
Newsholme E Yorks 199 B8
Newsholme Lancs 204 C2
Newsholme W Yorks 204 F6
Newsome W Yorks 196 E6
Newstead Borders 262 C3
Newstead Northumb 264 D5
Newstead Notts 171 E8
Newstead Staffs 168 G5
Newstead W Yorks 197 E11
Newstreet Lane Shrops 150 B2
Newtake Devon 14 G3
Newthorpe N Yorks 206 G5
Newthorpe Notts 171 F7
Newthorpe Common
Notts 171 F7
Newtoft Lincs 189 D8
Newton Argyll 275 D11
Newton Borders 262 E3
Newton Borders 262 F2
Newton Bridgend 57 F10
Newton C Beds 104 C4
Newton Cambs 105 B8
Newton Cambs 157 G8
Newton Cardiff 59 D8
Newton Corn 5 C11
Newton Cumb 210 E4
Newton Derbys 170 D6
Newton Dorset 30 E3
Newton Dumfries 239 C7
Newton Dumfries 248 E4
Newton Gtr Man 185 B7
Newton Hereford 96 C5
Newton Hereford 96 E6
Newton Hereford 115 D7
Newton Hereford 115 G10
Newton Highld 301 C7
Newton Highld 301 E7
Newton Highld 306 F7
Newton Highld 310 E7
Newton Lancs 202 F2
Newton Lancs 202 G4
Newton Lancs 203 C9
Newton Lancs 211 E11
Newton Lincs 155 B10
Newton Mers 182 D2
Newton Moray 301 C11
Newton Norf 158 F6
Newton Northants 137 G7
Newton Northumb 242 E2
Newton Notts 171 G11
Newton Perth 286 D2
Newton S Glos 79 G10
Newton S Lnrk 259 C10
Newton S Lnrk 268 C3
Newton S Yorks 198 G5
Newton Shetland 312 E5
Newton Shetland 313 K5
Newton Shrops 132 D4
Newton Shrops 149 C8
Newton Som 42 F6
Newton Staffs 151 D10
Newton Suff 107 C8
Newton Swansea 56 D6
Newton W Ches 166 B6
Newton W Ches 167 B8
Newton W Ches 183 F8
Newton W Loth 279 F11
Newton W Mid 133 E10
Newton Warks 119 B10
Newton Wilts 32 C2
Newton Abbot Devon 14 G3
Newton Arlosh Cumb 238 F5
Newton Aycliffe
Durham 233 G11
Newton Bewley Hrtlpl 234 F5
Newton Blossomville
M Keynes 121 G8
Newton Bromswold
Northants 121 D9
Newton Burgoland
Leics 135 B7
Newton by Toft Lincs 189 D9
Newton Cross Pembs 91 F7
Newton Ferrers Devon 7 F10
Newton Flotman Norf 142 D4
Newton Green Mon 79 G8
Newton Hall Durham 233 B11
Newton Hall Northumb 242 D2
Newton Harcourt Leics 136 D2
Newton Heath Gtr Man 195 G11
Newton Hill W Yorks 197 C10
Newton Ho Aberds 302 G6
Newton Hurst Staffs 151 D11
Newton Ketton Darl 234 G2
Newton Kyme N Yorks 206 E5
Newton-le-Willows
Mers 183 B9
Newton-le-Willows
N Yorks 214 B4
Newton Longville Bucks 102 E6
Newton Mearns E Renf 267 D10
Newton Morrell N Yorks 224 D4
Newton Morrell Oxon 102 F2
Newton Mulgrave
N Yorks 226 B5
Newton of Ardtoe
Highld 289 B8
Newton of Balcanqual
Perth 286 F5
Newton of Balcormo
Fife 287 G9
Newton of Falkland
Fife 286 G6
Newton of Mountblairy
Aberds 302 D6
Newton of Pitcairns
Perth 286 F4
Newton on Ayr S Ayrs 257 E8
Newton on Ouse N Yorks 206 B6
Newton-on-Rawcliffe
N Yorks 226 G6
Newton on the Hill
Shrops 149 E9
Newton on the Moor
Northumb 252 B5
Newton on Trent Lincs 188 G4
Newton Park Argyll 266 B2
Newton Park Mers 183 C9
Newton Peveril Dorset 18 B4
Newton Poppleford
Devon 15 D7
Newton Purcell Oxon 102 E2
Newton Regis Warks 134 B5
Newton Reigny Cumb 230 E5
Newton Rigg Cumb 230 E5
Newton St Boswells
Borders 262 C3
Newton St Cyres Devon 14 B3
Newton St Faith Norf 160 F4
Newton St Loe Bath 61 G8
Newton St Petrock Devon 24 E6
Newton Solney Derbys 152 D5
Newton Stacey Hants 48 E2
Newton Stewart
Dumfries 236 C6
Newton Tony Wilts 47 E8
Newton Tracey Devon 25 B8
Newton under
Roseberry Redcar 225 C11
Newton Underwood
Northumb 252 F4
Newton upon
Derwent E Yorks 207 D10
Newton Valence Hants 49 G8
Newton with Scales
Lancs 202 G4
Newton Wood Gtr Man 184 B6
Newtonairds Dumfries 247 G9
Newtongrange Midloth 270 C6
Newtonhill Aberds 293 D11
Newtonhill Highld 300 E5
Newtonia E Ches 167 B11
Newtonmill Angus 293 G8
Newtonmore Highld 291 D9
Newtown Argyll 284 G4
Newtown Bl Gwent 77 C11
Newtown Bucks 85 E7
Newtown Caerph 78 G2
Newtown Cambs 121 D11
Newtown Corn 2 D3
Newtown Corn 11 F11
Newtown Cumb 229 B7
Newtown Cumb 239 F9
Newtown Cumb 240 E2
Newtown Derbys 185 E7
Newtown Devon 26 B3
Newtown Devon 29 G7
Newtown E Ches 184 E6
Newtown Falk 279 E9
Newtown Glos 79 E11
Newtown Glos 80 D3
Newtown Glos 99 E8
Newtown Gtr Man 194 F5
Newtown Gtr Man 195 G9
Newtown Hants 21 B8
Newtown Hants 32 C4
Newtown Hants 32 E3
Newtown Hants 33 D8
Newtown Hants 33 E10
Newtown Hants 33 F7
Newtown Hants 49 F8
Newtown Hants 49 G10
Newtown Hants 64 G3
Newtown Hereford 97 E10
Newtown Hereford 98 C2
Newtown Highld 290 C5
Newtown I o W 20 C4
Newtown I o M 192 E4
Newtown Mers 183 B7
Newtown Norf 143 B10
Newtown Northumb 252 C2
Newtown Northumb 263 C11
Newtown Northumb 264 D2
Newtown Oxon 65 C9
Newtown Poole 18 C6
Newtown Powys 130 E2
Newtown Rhondda 77 F9
Newtown Shrops 132 C2
Newtown Shrops 149 C9
Newtown Shrops 149 E8
Newtown Som 28 E3
Newtown Som 43 F9
Newtown Staffs 133 C9
Newtown Staffs 168 C6
Newtown Staffs 169 C9
Newtown W Ches 183 F8
Newtown W Mid 133 F11
Newtown Wilts 30 B6
Newtown Wilts 63 G10
Newtown Worcs 116 F5
Newtown Worcs 117 E7
Newtown-in-St Martin
Corn 2 E6
Newtown Linford Leics 135 B10
Newtown St Boswells
Borders 262 C3
Newtown Unthank Leics 135 C9
Newtyle Angus 286 C6
Newyears Green London 66 B5
Nextend Hereford 114 F6
Neyland Pembs 73 D7
Niarbyl I o M 192 E3
Nib Heath Shrops 149 F8
Nibley Glos 79 D11
Nibley S Glos 61 C7
Nibley Green Glos 80 F2
Nibon Shetland 312 F5
Nicholashayne Devon 27 D10
Nicholaston Swansea 56 D4
Nidd N Yorks 214 G6
Niddrie Edin 280 G5
Nigg Aberdeen 293 C11
Nigg Highld 301 B8
Nigg Ferry Highld 301 C7
Nightcott Som 26 B5
Nilig Denb 165 D8
Nimble Nook Gtr Man 196 G2
Nimlet S Glos 61 E8
Nimmer Som 28 E4
Nine Ashes Essex 87 E9
Nine Elms London 67 D9
Nine Elms Swindon 62 B6
Nine Maidens Downs Corn 2 B5
Nine Mile Burn Midloth 270 D3
Nine Wells Pembs 90 G5
Ninebanks Northumb 241 G7
Nineveh Worcs 116 C3
Nineveh Worcs 116 E2
Ninewells Glos 79 C9
Ninfield E Sus 38 E2
Ningwood I o W 20 D3
Ningwood Common
I o W 20 D3
Ninnes Bridge Corn 2 B2
Nisbet Borders 262 D5
Nisthouse Orkney 314 E3
Nisthouse Shetland 313 G7
Nithbank Dumfries 247 D9
Niton I o W 20 F6
Nitshill Glasgow 267 C10
No Man's Heath W Ches 167 F8
No Man's Heath Warks 134 B5
No Man's Land Corn 6 D5
No Man's Land Hants 33 B8
Noah's Arks Kent 52 B5
Noah's Green Worcs 117 E10
Noak Bridge Essex 87 G11
Noak Hill Essex 87 G11
Noak Hill London 87 G8
Nob End Gtr Man 195 F9
Nobland Green Herts 86 B5
Noblethorpe S Yorks 197 F9
Nobold Shrops 149 G9
Nobottle Northants 120 E3
Nob's Crook Hants 33 C7
Nocton Lincs 173 C9
Noctorum Mers 182 D3
Nodmore W Berks 64 D2
Noel Park London 86 G4
Nog Tow Lancs 202 G6
Nogdam End Norf 143 C7
Noke Oxon 83 C8
Noke Street Medway 69 E8
Nolton Pembs 72 B5
Nolton Haven Pembs 72 B5
Nomansland Devon 26 E4
Nomansland Herts 85 C11
Nomansland Wilts 32 D3
Noneley Shrops 149 D9
Noness Shetland 313 L6
Nonikiln Highld 300 B6
Nonington Kent 55 C9
Nook Cumb 211 C10
Noon Nick W Yorks 205 F8
Noonsbrough Shetland 313 H4
Noonsun E Ches 184 F4
Noonvares Corn 2 C3
Noranside Angus 292 G6
Norbiton London 67 F7
Norbreck Blkpool 202 E2
Norbridge Hereford 98 C4
Norbury Derbys 169 G10
Norbury E Ches 167 F9
Norbury London 67 F10
Norbury Shrops 131 E7
Norbury Staffs 150 E5
Norbury Common
E Ches 167 F9
Norbury Junction
Staffs 150 E5
Norbury Moor Gtr Man 184 D6
Norby N Yorks 215 C8
Norby Shetland 313 H3
Norchard Worcs 116 D6
Norcote Glos 81 E8
Norcott Brook W Ches 183 E10
Norcross Blkpool 202 E2
Nordelph Norf 139 C11
Nordelph Corner Norf 141 C10
Norden Dorset 18 E4
Norden Gtr Man 195 E11
Norden Heath Dorset 18 E4
Nordley Shrops 132 D3
Norham Northumb 273 F8
Norham West Mains
Northumb 273 F8
Nork Sur 51 B8
Nork Sur 67 G8
Norland Town W Yorks 196 C5
Norleaze Wilts 45 C11
Norley Devon 25 G8
Norley W Ches 183 G9
Norley Common Sur 50 E4
Norleywood Hants 20 B3
Norlington E Sus 36 E6
Normacot Stoke 168 G6
Norman Cross Cambs 138 E3
Norman Hill Glos 80 F3
Normanby N Lincs 199 D11
Normanby N Yorks 216 C4
Normanby Redcar 225 B10
Normanby-by-Spital
Lincs 189 D7
Normanby by Stow
Lincs 188 E5
Normanby le Wold
Lincs 189 B10
Normandy Sur 50 C2
Norman's Bay E Sus 23 D11
Norman's Green Devon 27 G9
Normanston Suff 143 E10
Normanton Derby 152 C6
Normanton Leics 172 G4
Normanton Lincs 172 F6
Normanton Notts 172 E2
Normanton Rutland 137 B8
Normanton W Yorks 197 C11
Normanton Wilts 46 E6
Normanton le Heath
Leics 153 G7
Normanton on Soar
Notts 153 E10
Normanton-on-the-
Wolds Notts 154 C2
Normanton on Trent
Notts 172 B3
Normanton Spring
S Yorks 186 E6
Normanton Turville
Leics 135 D9
Normoss Lancs 202 F2
Norney Sur 50 E2
Norr W Yorks 205 F7
Norrington Common
Wilts 61 G11
Norris Green Corn 7 B8
Norris Green Mers 182 C5
Norris Hill Leics 152 F6
Norristhorpe W Yorks 197 C8
Norseman Orkney 314 E3
North Acton London 67 C8
North Anston S Yorks 187 E8
North Ascot Brack 66 F2
North Aston Oxon 101 F9
North Ayre Shetland 312 F6
North Baddesley Hants 32 D5
North Ballachulish
Highld 290 G2
North Barrow Som 29 B10
North Barsham Norf 159 C8
North Batsom Som 41 G10
North Beer Corn 12 C2
North Benfleet Essex 69 B9
North Bersted W Sus 22 C6
North Berwick E Loth 281 D11
North Bitchburn
Durham 233 E9
North Blyth Northumb 253 G8
North Boarhunt Hants 33 E10
North Bockhampton
Dorset 19 B9
North Bovey Devon 13 E10
North Bradley Wilts 45 C11
North Brentor Devon 12 E5
North Brewham Som 45 F8
North Brook End
Cambs 104 C5
North Broomage Falk 279 E7
North Buckland Devon 40 E3
North Burlingham Norf 161 G7
North Cadbury Som 29 B10
North Cairn Dumfries 236 B1
North Camp Hants 49 C11
North Carlton Lincs 188 F6
North Carlton Notts 187 E9
North Carrine Argyll 255 G7
North Cave E Yorks 208 G3
North Cerney Glos 81 D8
North Chailey E Sus 36 C5
North Charford Wilts 31 D11
North Charlton
Northumb 264 E5
North Cheam London 67 F8
North Cheriton Som 29 B11
North Cliff E Yorks 209 D10
North Cliffe E Yorks 208 F3
North Clifton Notts 188 G4
North Close Durham 233 E11
North Cockerington
Lincs 190 C5
North Coker Som 29 E8
North Collafirth
Shetland 312 E5
North Common S Glos 61 E7
North Common Suff 125 B9
North Connel Argyll 289 F11
North Cornelly Bridgend 57 E10
North Corner Corn 3 F7
North Corner S Glos 61 C7
North Corriegills N Ayrs 256 C2
North Corry Highld 289 D10
North Cotes Lincs 201 G11
North Country Corn 4 G3
North Court Som 41 F11
North Cove Suff 143 F9
North Cowton N Yorks 224 E5
North Craigo Angus 293 G8
North Crawley M Keynes 103 C8
North Cray London 68 E3
North Creake Norf 159 B7
North Curry Som 28 B4
North Dalton E Yorks 208 C4
North Darley Corn 11 G11
North Dawn Orkney 314 F4
North Deighton N Yorks 206 C3
North Denes Norf 161 G10
North Dronley Angus 287 D7
North Drumachter
Lodge Highld 291 F8
North Duffield N Yorks 207 F9
North Dykes Cumb 230 D6
North Eastling Kent 54 B3
North Elham Kent 55 E7
North Elkington Lincs 190 C3
North Elmham Norf 159 E9
North Elmsall W Yorks 198 E3
North Elmshall W Yorks 198 E3
North Elphinestone
E Loth 281 G7
North End Bath 60 G6
North End Beds 103 B9
North End Beds 121 F10
North End Bucks 102 F4
North End Bucks 102 F6
North End Cumb 239 F8
North End Devon 27 D10
North End Dorset 30 B4
North End Durham 233 C11
North End E Yorks 209 C9
North End E Yorks 209 E9
North End E Yorks 209 G11
North End Essex 87 B11
North End Essex 106 D5
North End Hants 31 D10
North End Hants 33 B9
North End Hants 64 G2
North End Leics 153 F11
North End Lincs 174 G2
North End Lincs 189 B8
North End Lincs 190 C5
North End Lincs 190 D6
North End Lincs 191 D7
North End Lincs 201 G10
North End London 67 B9
North End London 68 D4
North End N Lincs 200 C6
North End N Som 60 F2
North End Norf 141 E10
North End Northumb 252 C4
North End Ptsmth 33 G11
North End Som 28 B3
North End W Sus 35 F10
North End W Sus 35 G7
North End W Sus 51 F11
North End Wilts 81 G8
North Erradale Highld 307 L2
North Evington
Leicester 136 C2
North Ewster N Lincs 199 G10
North Fambridge Essex 88 F5
North Fearns Highld 295 B7
North Featherstone
W Yorks 198 C2
North Feltham London 66 E6
North Feorline N Ayrs 255 E10
North Ferriby E Yorks 200 B3
North Finchley London 86 G3
North Flobbets Aberds 303 F7
North Frodingham
E Yorks 209 C8
North Gluss Shetland 312 F5
North Gorley Hants 31 E11
North Green Norf 141 B10
North Green Norf 142 F4
North Green Suff 126 B6
North Green Suff 126 E6
North Green Suff 127 D7
North Greetwell Lincs 189 G8
North Grimston N Yorks 216 F6
North Halley Orkney 314 F5
North Halling Medway 69 F8
North Harrow London 66 B6
North Hayling Hants 22 C2
North Hazelrigg
Northumb 264 C3
North Heasley Devon 41 G8
North Heath W Berks 64 E3
North Heath W Sus 35 C9
North Hill Corn 11 F11
North Hillingdon London 66 C5
North Hinksey Oxon 83 D7
North Hinksey Village
Oxon 83 D7
North Ho Shetland 313 J5
North Holmwood Sur 51 D7
North Houghton Hants 47 G10
North Howden E Yorks 207 G11
North Huish Devon 8 D4
North Hyde London 66 D6
North Hykeham Lincs 172 B6
North Hylton T & W 243 F8
North Kelsey Lincs 200 G4
North Kelsey Moor
Lincs 200 G5
North Kensington London 67 C8
North Kessock Highld 300 E6
North Killingholme
N Lincs 200 D6
North Kilvington
N Yorks 215 B8
North Kilworth Leics 136 G2
North Kingston Hants 31 G11
North Kirkton Aberds 303 D11
North Kiscadale N Ayrs 256 D2
North Kyme Lincs 173 E11
North Laggan Highld 290 D4
North Lancing W Sus 35 F11
North Landing E Yorks 218 E4
North Lee Bucks 84 D4
North Lees N Yorks 214 E6
North Leigh Kent 54 D6
North Leigh Oxon 82 C5
North Leverton with
Habblesthorpe Notts 188 E3
North Littleton Worcs 99 B11
North Looe Sur 67 G8
North Lopham Norf 141 G10
North Luffenham
Rutland 137 C8
North Marden W Sus 34 D4
North Marston Bucks 102 G5
North Middleton
Midloth 271 D7
North Middleton
Northumb 264 E2
North Millbrex Aberds 303 E8
North Molton Devon 26 B2
North Moreton Oxon 64 B5
North Mosstown
Aberds 303 D10
North Motherwell
N Lnrk 268 D4
North Moulsecoomb
Brighton 36 F4
North Mundham W Sus 22 C5
North Muskham Notts 172 D3
North Newbald E Yorks 208 F4
North Newington Oxon 101 D8
North Newnton Wilts 46 B6
North Newton Som 43 G9
North Nibley Glos 80 F2
North Oakley Hants 48 C4
North Ockendon London 68 C5
North Ormesby M'bro 234 G6
North Ormsby Lincs 190 C3
North Otterington
N Yorks 215 B7
North Owersby Lincs 189 C9
North Perrott Som 29 F7
North Petherton Som 43 G9
North Petherwin Corn 11 D11
North Pickenham Norf 141 B7
North Piddle Worcs 117 G9
North Poorton Dorset 16 B6
North Port Argyll 284 E4
North Poulner Hants 31 F11
North Queensferry Fife 280 E2
North Radworthy Devon 41 G9
North Rauceby Lincs 173 F8
North Reddish Gtr Man 184 C5
North Reston Lincs 190 E5
North Rigton N Yorks 205 D11
North Ripley Hants 19 B9
North Rode E Ches 168 B5
North Roe Shetland 312 E5
North Row Cumb 229 E10
North Runcton Norf 158 F2
North Sandwick
Shetland 312 D7
North Scale Cumb 210 F3
North Scarle Lincs 172 B5
North Seaton Northumb 253 F7
North Seaton Colliery
Northumb 253 F7
North Sheen London 67 D7
North Shian Argyll 289 E11
North Shields T & W 243 D9
North Shoebury Shend 70 B2
North Shore Blkpool 202 F2
North Side Cumb 228 F6
North Side P'boro 138 D5
North Skelmanae
Aberds 303 D9
North Skelton Redcar 226 B3
North Somercotes Lincs 190 B6
North Stainley N Yorks 214 D5
North Stainmore Cumb 222 B6
North Stifford Thurrock 68 C6
North Stoke Bath 61 F8
North Stoke Oxon 64 B6
North Stoke W Sus 35 E8
North Stoneham Hants 32 D6
North Street Hants 31 D11
North Street Hants 48 G6
North Street Kent 54 B4
North Street Medway 69 E10
North Street W Berks 64 E6
North Sunderland
Northumb 264 C6
North Synton Borders 261 E11
North Tamerton Corn 12 B2
North Tawton Devon 25 G11
North Thoresby Lincs 190 B3
North Tidworth Wilts 47 D8
North Togston Northumb 252 C6
North Town Devon 25 F8
North Town Hants 49 C11
North Town Som 29 B10
North Town Som 44 E5
North Town Windsor 65 C11
North Tuddenham
Norf 159 G10
North Walbottle T & W 242 D5
North Walney Cumb 210 F3
North Walsham Norf 160 C5
North Waltham Hants 48 D5
North Warnborough
Hants 49 C8
North Water Bridge
Angus 293 G8
North Waterhayne
Devon 28 F3
North Watford Herts 85 F10
North Watten Highld 310 D6
North Weald Bassett
Essex 87 E7
North Weirs Hants 32 G3
North Wembley London 67 B7
North Weston N Som 60 D3
North Weston Oxon 83 D11
North Wheatley Notts 188 D3
North Whilborough
Devon 9 B7
North Whiteley Moray 302 E4
North Wick Bath 60 F5
North Widcombe Bath 44 B5
North Willingham
Lincs 189 D11
North Wingfield Derbys 170 B6
North Witham Lincs 155 E8
North Woolwich London 68 D2
North Wootton Dorset 29 E11
North Wootton Norf 158 E2
North Wootton Som 44 E5
North Wraxall Wilts 61 D10
North Wroughton
Swindon 63 C7
Northacre Norf 141 D9
Northall Bucks 103 G9
Northall Green Norf 159 G9
Northallerton N Yorks 225 G7
Northam Devon 24 B6
Northam Soton 32 E6
Northampton Northants 120 E5
Northaw Herts 86 E3
Northay Devon 28 G5
Northay Som 28 E3
Northbeck Lincs 173 G9
Northborough P'boro 138 B3
Northbourne Bmouth 19 B7
Northbourne Kent 55 C10
Northbridge Street E Sus 38 C2
Northbrook Dorset 17 C11
Northbrook Hants 33 D9
Northbrook Hants 48 F4
Northbrook Oxon 101 G9
Northbrook Wilts 46 C4
Northchapel W Sus 35 B7
Northchurch Herts 85 D7
Northcote Devon 27 G11
Northcott Corn 24 F2
Northcott Devon 12 C2
Northcott Devon 27 E9
Northcott Devon 27 F10
Northcourt Oxon 83 F8
Northdown Kent 71 E11
Northdyke Orkney 314 D2
Northedge Derbys 170 B5
Northend Bath 61 F9
Northend Bucks 84 G2
Northend Essex 89 E7
Northend Essex 105 D10
Northend Warks 119 G7
Northenden Gtr Man 184 C4
Northern Moor Gtr Man 184 C4
Northfield Aberdeen 293 C11
Northfield Borders 262 D3
Northfield Borders 273 B8
Northfield E Yorks 200 B4
Northfield Edin 280 G5
Northfield Highld 301 B7
Northfield M Keynes 103 C7
Northfield Northants 137 E7
Northfield Som 43 F9
Northfield W Mid 117 B10
Northfields Hants 33 B7
Northfields Lincs 137 B10
Northfleet Kent 68 E6
Northfleet Green Kent 68 E6
Northgate Lincs 156 D3
Northgate Som 27 B9
Northgate W Sus 51 F9
Northhouse Borders 249 B10
Northiam E Sus 38 B4
Northill C Beds 104 B2
Northington Glos 80 D2
Northington Hants 48 F5
Northlands Lincs 174 E4
Northlea Durham 243 G10
Northleach Glos 81 C10
Northleigh Devon 15 B9
Northleigh Devon 40 G6
Northlew Devon 12 B6
Northmoor Devon 24 D4
Northmoor Oxon 82 E6
Northmoor Corner Som 43 G10
Northmoor Green or
Moorland Som 43 G10
Northmuir Angus 287 B7
Northney Hants 22 C2
Northolt London 66 C6
Northop =
Llan-eurgain Flint 166 B2
Northop Hall Flint 166 B3
Northorpe Lincs 155 F11
Northorpe Lincs 156 B4
Northorpe Lincs 188 B5
Northorpe W Yorks 197 C8
Northover Som 29 C8
Northover Som 44 F3
Northowram W Yorks 196 B6
Northpark Argyll 275 G11
Northport Dorset 18 D4
Northpunds Shetland 313 L6
Northrepps Norf 160 B4
Northside Aberds 303 D8
Northside Orkney 314 D2
Northton Aberds 293 C9
Northtown Orkney 314 G4
Northtown Shetland 313 M5
Northumberland Heath
London 68 D4
Northville Torf 78 F3
Northway Devon 24 C5
Northway Glos 99 E8
Northway Som 27 B10
Northway Swansea 56 D5
Northwich W Ches 183 G11
Northwick S Glos 60 B5
Northwick Som 43 D11
Northwick Worcs 116 F6
Northwold Norf 140 D5
Northwood Derbys 170 C3
Northwood I o W 20 C5
Northwood Kent 71 F11
Northwood London 85 G9
Northwood Mers 182 B6
Northwood Shrops 149 C9
Northwood Staffs 168 G5
Northwood Stoke 168 F5
Northwood Green Glos 80 B2
Northwood Hills London 85 G9
Norton Devon 9 E7
Norton Devon 24 B3
Norton E Sus 23 E7
Norton Glos 99 G7
Norton Halton 183 E9
Norton Herts 104 E4
Norton I o W 20 D2

Norton Mon 78 B6
Norton N Som 59 G10
Norton Northants 120 E2
Norton Notts 187 G9
Norton Powys 114 D6
Norton S Yorks 186 E5
Norton S Yorks 198 D4
Norton Shrops 131 B11
Norton Shrops 131 G9
Norton Shrops 132 C4
Norton Stockton 234 G4
Norton Suff 125 D9
Norton Swansea 56 D3
Norton Swansea 56 D6
Norton W Mid 133 G7
Norton W Sus 22 B6
Norton W Sus 22 D5
Norton Wilts 61 C11
Norton Worcs 99 B10
Norton Worcs 117 G7
Norton Ash Kent 70 G3
Norton Bavant Wilts 46 E2
Norton Bridge Staffs 151 C7
Norton Canes Staffs 133 B10
Norton Canon Hereford 97 B7
Norton Corner Norf 159 D11
Norton Disney Lincs 172 D5
Norton East Staffs 133 B10
Norton Ferris Wilts 45 F9
Norton Fitzwarren Som 27 B11
Norton Green Herts 104 G4
Norton Green I o W 20 D2
Norton Green Staffs 168 E6
Norton Green W Mid 118 C3
Norton Hawkfield Bath 60 G5
Norton Heath Essex 87 E10
Norton in Hales Shrops 150 B4
Norton-in-the-Moors Stoke 168 E5
Norton-Juxta-Twycross Leics 134 B6
Norton-le-Clay N Yorks 215 E8
Norton Lindsey Warks 118 E4
Norton Little Green Suff 125 D9
Norton Malreward Bath 60 F6
Norton Mandeville Essex 87 E9
Norton-on-Derwent N Yorks 216 E5
Norton St Philip Som 45 B9
Norton sub Hamdon Som 29 D7
Norton Subcourse Norf 143 D8
Norton Woodseats S Yorks 186 E5
Norton's Wood N Som 60 E2
Norwell Notts 172 C3
Norwell Woodhouse Notts 172 C2
Norwich Norf 142 B4
Norwick Shetland 312 B8
Norwood Derbys 187 E7
Norwood Dorset 29 F8
Norwood End Essex 87 D9
Norwood Green London 66 D6
Norwood Green W Yorks 196 B6
Norwood Hill Sur 51 E8
Norwood New Town London 67 E10
Norwoodside Cambs 139 D8
Noseley Leics 136 D4
Noss Highld 310 D7
Noss Shetland 313 M5
Noss Mayo Devon 7 F11
Nosterfield N Yorks 214 C5
Nosterfield End Cambs 106 C2
Nostie Highld 295 C10
Notgrove Glos 100 G2
Nottage Bridgend 57 F10
Notter Corn 7 C7
Notting Hill London 67 C8
Nottingham Nottingham 153 B11
Nottington Dorset 17 E9
Notton W Yorks 197 E10
Notton Wilts 62 F2
Nounsley Essex 88 C3
Noutard's Green Worcs 116 D5
Nova Scotia W Ches 167 B10
Novar House Highld 300 C6
Novers Park Bristol 60 F5
Noverton Glos 99 G9
Nowton Suff 125 E7
Nox Shrops 149 G8
Noyadd Trefawr Ceredig 92 B5
Noyadd Wilym Ceredig 92 C4
Nuffield Oxon 65 B7
Nun Appleton N Yorks 207 F7
Nun Hills Lancs 195 C11
Nun Monkton N Yorks 206 B6
Nunburnholme E Yorks 208 D2
Nuncargate Notts 171 E8
Nunclose Cumb 230 B5
Nuneaton Warks 135 E7
Nuneham Courtenay Oxon 83 F9
Nuney Green Oxon 65 D7
Nunhead London 67 D11
Nunney Som 45 D8
Nunney Catch Som 45 E8
Nunnington N Yorks 216 D3
Nunnykirk Northumb 252 E3
Nunsthorpe NE Lincs 201 F9
Nunthorpe M'bro 225 C10
Nunthorpe York 207 C8
Nunton Wilts 31 B11
Nunwick N Yorks 214 E6
Nup End Bucks 84 B5
Nup End Herts 86 B2
Nupdown S Glos 79 F10
Nupend Glos 80 D3
Nupend Glos 80 F4
Nuper's Hatch Essex 87 G8
Nuppend Glos 79 E10
Nuptown Brack 65 E11
Nursling Hants 32 D5
Nursted Hants 34 C3
Nursteed Wilts 62 G4
Nurston V Glam 58 F5
Nurton Staffs 132 D6
Nurton Hill Staffs 132 D6
Nut Grove Mers 183 C7
Nutbourne W Sus 22 B3
Nutbourne W Sus 35 D9
Nutbourne Common W Sus 35 D9
Nutburn Hants 32 C5
Nutcombe Sur 49 G11
Nutfield Sur 51 C10
Nuthall Notts 171 G8
Nuthampstead Herts 105 E8
Nuthurst W Sus 35 B11
Nuthurst Warks 118 C3
Nutley E Sus 36 B6
Nutley Hants 48 E6
Nuttall Gtr Man 195 D9
Nutwell S Yorks 198 G6
Nybster Highld 310 C7
Nye N Som 60 G2
Nyetimber W Sus 22 D5
Nyewood W Sus 34 C4
Nyland Som 44 C3
Nymet Rowland Devon 26 F2
Nymet Tracey Devon 26 G2
Nympsfield Glos 80 E4
Nynehead Som 27 C10
Nythe Som 44 G2
Nythe Swindon 63 B7
Nyton W Sus 22 B6

O

Oad Street Kent 69 G11
Oadby Leics 136 C2
Oak Bank Gtr Man 195 F10
Oak Cross Devon 12 B6
Oak Hill Stoke 168 G5
Oak Hill Suff 109 B7
Oak Tree Darl 225 C7
Oakall Green Worcs 116 E6
Oakamoor Staffs 169 G9
Oakbank W Loth 269 B11
Oakdale Caerph 77 F11
Oakdale N Yorks 205 B11
Oakdale Poole 18 C6
Oake Som 27 B11
Oake Green Som 27 B11
Oaken Staffs 133 C7
Oakenclough Lancs 202 D6
Oakengates Telford 150 G4
Oakenholt Flint 182 G3
Oakenshaw Durham 233 D10
Oakenshaw Lancs 203 G10
Oakenshaw W Yorks 197 B7
Oakerthorpe Derbys 170 E5
Oakes W Yorks 196 D6
Oakfield Herts 104 F3
Oakfield I o W 21 C7
Oakfield Torf 78 G4
Oakford Ceredig 111 F9
Oakford Devon 26 C6
Oakfordbridge Devon 26 C6
Oakgrove E Ches 168 B6
Oakgrove M Keynes 103 D7
Oakham Rutland 137 B7
Oakham W Mid 133 F9
Oakhanger E Ches 168 E3
Oakhanger Hants 49 F9
Oakhill Som 44 D6
Oakhill W Sus 51 G7
Oakhurst Kent 52 C4
Oakington Cambs 123 E8
Oaklands Carms 74 B6
Oaklands Herts 86 B2
Oaklands Powys 113 G10
Oakle Street Glos 80 B3
Oakleigh Park London 86 G3
Oakley Beds 121 G10
Oakley Bucks 83 C10
Oakley Fife 279 D10
Oakley Glos 99 G9
Oakley Hants 48 C5
Oakley Oxon 84 E3
Oakley Poole 18 B6
Oakley Staffs 150 B4
Oakley Suff 126 B3
Oakley Court Oxon 64 B6
Oakley Green Windsor 66 D2
Oakley Park Powys 129 F9
Oakley Park Suff 126 B3
Oakley Wood Oxon 64 B6
Oakmere W Ches 167 B9
Oakridge Glos 80 E6
Oakridge Hants 48 C6
Oakridge Lynch Glos 80 E6
Oaks Shrops 131 C8
Oaks Green Derbys 152 C3
Oaks in Charnwood Leics 153 F9
Oaksey Wilts 81 G7
Oakshaw Ford Cumb 240 B2
Oakshott Hants 34 B2
Oakthorpe Leics 152 G6
Oakwell W Yorks 197 B8
Oakwood Derby 153 B7
Oakwood London 86 F3
Oakwood Northumb 241 D10
Oakwood W Yorks 206 F2
Oakwood Warr 183 C11
Oakwoodhill Sur 50 F6
Oakworth W Yorks 204 F6
Oape Highld 309 J4
Oare Kent 70 G4
Oare Som 41 D10
Oare W Berks 64 E4
Oare Wilts 63 G7
Oareford Som 41 D10
Oasby Lincs 155 B10
Oath Som 28 B5
Oathill Dorset 28 F6
Oathlaw Angus 287 B8
Oatlands Glasgow 267 C11
Oatlands N Yorks 205 C11
Oatlands Park Sur 66 G5
Oban Argyll 289 G10
Oban Highld 295 G10
Oban W Isles 305 H3
Obley Shrops 114 B6
Oborne Dorset 29 D11
Obthorpe Lincs 155 F11
Obthorpe Lodge Lincs 156 F2
Occlestone Green W Ches 167 C11
Occold Suff 126 C3
Ocean Village Soton 32 E6
Ochiltree E Ayrs 258 E2
Ochr-y-foel Denb 181 F9
Ochtermuthill Perth 286 F2
Ochtertyre Perth 286 E2
Ochtow Highld 309 J4
Ockbrook Derbys 153 B8
Ocker Hill W Mid 133 E9
Ockeridge Worcs 116 E5
Ockford Ridge Sur 50 E3
Ockham Sur 50 B5
Ockle Highld 289 B7
Ockley Sur 50 F6
Ocle Pychard Hereford 97 B11
Octon E Yorks 217 F10
Octon Cross Roads E Yorks 217 F10
Odam Barton Devon 26 D2
Odcombe Som 29 D8
Odd Down Bath 61 G8
Oddendale Cumb 221 C11
Odder Lincs 188 G6
Oddingley Worcs 117 F8
Oddington Glos 100 F4
Oddington Oxon 83 C9
Odell Beds 121 F9
Odham Devon 25 G7
Odie Orkney 314 D6
Odiham Hants 49 C8
Odsal W Yorks 197 B7
Odsey Cambs 104 D5
Odstock Wilts 31 B10
Odstone Leics 135 B7
Offchurch Warks 119 D7
Offenham Worcs 99 B11
Offenham Cross Worcs 99 B11
Offerton Gtr Man 184 D6
Offerton T & W 243 F8
Offerton Green Gtr Man 184 D6
Offham E Sus 36 E5
Offham Kent 53 B7
Offham W Sus 35 F8
Offleyhay Staffs 150 D5
Offleymarsh Staffs 150 D5
Offleyrock Staffs 150 D5
Offord Cluny Cambs 122 D4
Offord D'Arcy Cambs 122 D4
Offton Suff 107 B11
Offwell Devon 15 B9
Ogbourne Maizey Wilts 63 E7
Ogbourne St Andrew Wilts 63 E7
Ogbourne St George Wilts 63 E8
Ogden W Yorks 205 G7
Ogdens Hants 31 E11
Ogil Angus 292 G6
Ogle Northumb 242 B4
Ogmore V Glam 57 F11
Ogmore-by-Sea = Aberogwr V Glam 57 F11
Ogmore Vale Bridgend 76 G6
Okeford Fitzpaine Dorset 30 E4
Okehampton Devon 13 B7
Okehampton Camp Devon 13 C7
Oker Derbys 170 C3
Okewood Hill Sur 50 F6
Okle Green Glos 98 F5
Okraquoy Shetland 313 K6
Okus Swindon 62 C6
Olchard Devon 14 F3
Old Northants 120 C5
Old Aberdeen Aberdeen 293 C11
Old Alresford Hants 48 G5
Old Arley Warks 134 E5
Old Balornock Glasgow 268 B2
Old Basford Nottingham 171 G8
Old Basing Hants 49 C7
Old Belses Borders 262 E3
Old Bewick Northumb 264 E3
Old Bexley London 68 E3
Old Blair Perth 291 G10
Old Bolingbroke Lincs 174 B4
Old Boston Mers 183 B9
Old Bramhope W Yorks 205 E10
Old Brampton Derbys 186 G4
Old Bridge of Tilt Perth 291 G10
Old Bridge of Urr Dumfries 237 C9
Old Buckenham Norf 141 E11
Old Burdon T & W 243 G9
Old Burghclere Hants 48 B3
Old Byland N Yorks 215 B11
Old Cambus Borders 272 B6
Old Cardinham Castle Corn 6 B2
Old Carlisle Cumb 229 B11
Old Cassop Durham 234 D2
Old Castleton Borders 250 E2
Old Catton Norf 160 G4
Old Chalford Oxon 100 F6
Old Church Stoke Powys 130 E5
Old Clee NE Lincs 201 F9
Old Cleeve Som 42 E4
Old Colwyn Conwy 180 F5
Old Coppice Shrops 131 B9
Old Corry Highld 295 C8
Old Coulsdon London 51 B10
Old Country Hereford 98 C4
Old Craig Aberds 303 G9
Old Craig Angus 292 G4
Old Crombie Aberds 302 D5
Old Cryals Kent 53 E7
Old Cullen Moray 302 C5
Old Dailly S Ayrs 244 D6
Old Dalby Leics 154 E3
Old Dam Derbys 185 F10
Old Deer Aberds 303 E9
Old Denaby S Yorks 187 B7
Old Ditch Som 44 D4
Old Dolphin W Yorks 205 G8
Old Down S Glos 60 B6
Old Duffus Moray 301 C11
Old Edlington S Yorks 187 B8
Old Eldon Durham 233 F10
Old Ellerby E Yorks 209 F9
Old Fallings W Mid 133 C8
Old Farm Park M Keynes 103 D8
Old Felixstowe Suff 108 D6
Old Field Shrops 115 B9
Old Fletton P'boro 138 D3
Old Fold T & W 243 E7
Old Ford London 67 C11
Old Forge Hereford 79 B9
Old Furnace Torf 78 E3
Old Gate Lincs 157 E8
Old Glossop Derbys 185 C8
Old Goginan Ceredig 128 G3
Old Goole E Yorks 199 C8
Old Gore Hereford 98 F2
Old Graitney Dumfries 239 D8
Old Grimsby Scilly 1 F3
Old Hall Powys 129 G8
Old Hall Green Herts 105 G7
Old Hall Street Norf 160 C6
Old Harlow Essex 87 C7
Old Hatfield Herts 86 D2
Old Heath Essex 107 G10
Old Heathfield E Sus 37 C9
Old Hill W Mid 133 F9
Old Hills Worcs 98 B6
Old Hunstanton Norf 175 G11
Old Hurst Cambs 122 B5
Old Hutton Cumb 211 B11
Old Johnstone Dumfries 248 D6
Old Kea Corn 4 G6
Old Kilpatrick W Dunb 277 G9
Old Kinnernie Aberds 293 C9
Old Knebworth Herts 104 G4
Old Langho Lancs 203 F10
Old Laxey I o M 192 D5
Old Leake Lincs 174 E6
Old Leckie Stirl 278 C3
Old Lindley W Yorks 196 D5
Old Linslade C Beds 103 F8
Old Llanberis = Nant Peris Gwyn 163 D10
Old Malden London 67 F8
Old Malton N Yorks 216 E5
Old Marton Shrops 148 C6
Old Mead Essex 105 F10
Old Micklefield W Yorks 206 G4
Old Mill Corn 12 G3
Old Milton Hants 19 C10
Old Milverton Warks 118 D5
Old Monkland N Lnrk 268 C4
Old Nenthorn Borders 262 B5
Old Netley Hants 33 F7
Old Neuadd Powys 129 F11
Old Newton Suff 125 E11
Old Oak Common London 67 C8
Old Park Corn 6 B4
Old Park Telford 132 B3
Old Passage S Glos 60 B5
Old Perton Staffs 133 D7
Old Philpstoun W Loth 279 F11
Old Polmont Falk 279 F8
Old Portmouth Ptsmth 21 B8
Old Quarrington Durham 234 D2
Old Radnor Powys 114 F5
Old Rattray Aberds 303 D10
Old Rayne Aberds 302 G6
Old Romney Kent 39 B8
Old Shirley Soton 32 E5
Old Shoreham W Sus 36 F2
Old Snydale W Yorks 198 C2
Old Sodbury S Glos 61 C9
Old Somerby Lincs 155 C9
Old Stillington Stockton 234 G3
Old Storridge Common Worcs 116 G4
Old Stratford Northants 102 C5
Old Struan Perth 291 G10
Old Swan Mers 182 C5
Old Swarland Northumb 252 C5
Old Swinford W Mid 133 G8
Old Tame Gtr Man 196 F3
Old Tebay Cumb 222 D2
Old Thirsk N Yorks 215 C8
Old Tinnis Borders 261 D9
Old Toll S Ayrs 257 E9
Old Town Cumb 211 C11
Old Town Cumb 230 C5
Old Town E Sus 23 F9
Old Town E Sus 38 F2
Old Town E Sus 38 F4
Old Town E Yorks 218 F3
Old Town Edin 280 G5
Old Town Herts 104 F4
Old Town Scilly 1 G4
Old Town Swindon 63 C7
Old Town W Yorks 196 B3
Old Trafford Gtr Man 184 B4
Old Tree Kent 71 G8
Old Tupton Derbys 170 B5
Old Warden C Beds 104 C2
Old Warren Flint 166 C4
Old Way Som 28 D5
Old Weston Cambs 121 B11
Old Wharf Hereford 98 D4
Old Whittington Derbys 186 G5
Old Wick Highld 310 D7
Old Wimpole Cambs 122 G6
Old Windsor Windsor 66 E3
Old Wingate Durham 234 D3
Old Wives Lees Kent 54 C5
Old Woking Sur 50 B4
Old Wolverton M Keynes 102 C6
Old Woodhall Lincs 174 B2
Old Woodhouses Shrops 167 G9
Old Woodstock Oxon 82 B6
Oldany Highld 306 F6
Oldberrow Warks 118 D2
Oldborough Devon 26 F3
Oldbrook M Keynes 103 D7
Oldbury Kent 52 B5
Oldbury Shrops 132 E4
Oldbury W Mid 133 F9
Oldbury Warks 134 E6
Oldbury Naite S Glos 79 G10
Oldbury-on-Severn S Glos 79 G10
Oldbury on the Hill Glos 61 B10
Oldcastle Mon 96 G6
Oldcastle Heath W Ches 167 F7
Oldcotes Notts 187 D9
Oldcroft Glos 79 D10
Oldend Glos 80 D3
Oldfallow Staffs 151 G9
Oldfield Cumb 229 F7
Oldfield Shrops 132 F3
Oldfield W Yorks 196 E6
Oldfield W Yorks 204 F6
Oldfield Worcs 116 E6
Oldfield Brow Gtr Man 184 D3
Oldfield Park Bath 61 G8
Oldford Som 45 C9
Oldfurnace Staffs 169 G8
Oldhall Renfs 267 C10
Oldhall Green Suff 125 F7
Oldhall Ho Highld 310 D6
Oldham Gtr Man 196 F2
Oldham Edge Gtr Man 196 F2
Oldhamstocks E Loth 282 G4
Oldhurst Cambs 122 B6
Oldington Shrops 132 D4
Oldland S Glos 61 E7
Oldland Common S Glos 61 E7
Oldmeldrum Aberds 303 G8
Oldmixon N Som 43 B10
Oldshore Beg Highld 306 D6
Oldshoremore Highld 306 D7
Oldstead N Yorks 215 C10
Oldtown Aberds 293 C7
Oldtown Aberds 302 G5
Oldtown Highld 309 L5
Oldtown of Ord Aberds 302 D6
Oldwalls Swansea 56 C3
Oldway Swansea 56 D5
Oldway Torbay 9 C7
Oldways End Devon 26 B5
Oldwhat Aberds 303 D8
Oldwich Lane W Mid 118 C4
Oldwood Worcs 115 D11
Olgrinmore Highld 310 D4
Olive Green Staffs 152 F2
Oliver's Battery Hants 33 B7
Ollaberry Shetland 312 E5
Ollag W Isles 297 G3
Ollerbrook Booth Derbys 185 D10
Ollerton E Ches 184 F3
Ollerton Notts 171 B11
Ollerton Shrops 150 D2
Ollerton Fold Lancs 194 C6
Ollerton Lane Shrops 150 D3
Olmarch Ceredig 112 F2
Olmstead Green Essex 106 C2
Olney M Keynes 121 G7
Olrig Ho Highld 310 C5
Olton W Mid 134 G2
Olveston S Glos 60 B6
Olwen Ceredig 93 B11
Ombersley Worcs 116 E6
Ompton Notts 171 B11
Omunsgarth Shetland 313 J5
Onchan I o M 192 E4
Onecote Staffs 169 D9
Onehouse Suff 125 F10
Onen Mon 78 C6
Onesacre S Yorks 186 C3
Ongar Hill Norf 157 E11
Ongar Street Hereford 115 D7
Onibury Shrops 115 B9
Onich Highld 290 G2
Onllwyn Neath 76 C4
Onneley Staffs 168 G3
Onslow Village Sur 50 D3
Onthank E Ayrs 267 G8
Onziebust Orkney 314 D4
Openshaw Gtr Man 184 B5
Openwoodgate Derbys 170 F5
Opinan Highld 299 B7
Opinan Highld 307 K3
Orange Lane Borders 272 G5
Orange Row Norf 157 E10
Orasaigh W Isles 305 G5
Orbiston N Lnrk 268 D4
Orbliston Moray 302 D3
Orbost Highld 298 E2
Orby Lincs 175 B7
Orchard Hill Devon 24 B6
Orchard Leigh Bucks 85 E7
Orchard Portman Som 28 C2
Orcheston Wilts 46 D5
Orcop Hereford 97 F9
Orcop Hill Hereford 97 F9
Ord Highld 295 D8
Ordale Shetland 312 C8
Ordhead Aberds 293 B8
Ordie Aberds 292 C6
Ordiequish Moray 302 D3
Ordiqhill Aberds 302 D5
Ordley Northumb 241 F10
Ordsall Gtr Man 184 B4
Ordsall Notts 187 E11
Ore E Sus 38 E4
Oreston Plym 7 E10
Oreton Shrops 132 G3
Orford Suff 109 B8
Orford Warr 183 C10
Organford Dorset 18 C4
Orgreave S Yorks 186 D6
Orgreave Staffs 152 F3
Oridge Street Glos 98 F5
Orlandon Pembs 72 D4
Orlestone Kent 54 G3
Orleton Hereford 115 D9
Orleton Worcs 116 D3
Orleton Common Hereford 115 D9
Orlingbury Northants 121 C7
Ormacleit W Isles 297 H3
Ormathwaite Cumb 229 F11
Ormesby Redcar 225 B10
Ormesby St Margaret Norf 161 G9
Ormesby St Michael Norf 161 G9
Ormiclate Castle W Isles 297 H3
Ormidale Lodge Argyll 275 F11
Ormiscaig Highld 307 K3
Ormiston Borders 262 G2
Ormiston E Loth 271 B8
Ormsaigbeg Highld 288 C6
Ormsaigmore Highld 288 C6
Ormsary Argyll 275 F8
Ormsgill Cumb 210 E3
Ormskirk Lancs 194 F2
Ornsby Hill Durham 233 B9
Orpington London 68 F3
Orrell Gtr Man 194 F4
Orrell Mers 182 B4
Orrell Post Gtr Man 194 G4
Orrisdale I o M 192 C4
Orrock Fife 280 D4
Orroland Dumfries 237 E9
Orsett Thurrock 68 C6
Orsett Heath Thurrock 68 D6
Orslow Staffs 150 F6
Orston Notts 172 G3
Orthwaite Cumb 229 E11
Ortner Lancs 202 C6
Orton Cumb 222 D2
Orton Northants 120 B6
Orton Staffs 133 D7
Orton Brimbles P'boro 138 D3
Orton Goldhay P'boro 138 D3
Orton Longueville P'boro 138 D3
Orton Malborne P'boro 138 D3
Orton-on-the-Hill Leics 134 C6
Orton Rigg Cumb 239 G8
Orton Southgate P'boro 138 E2
Orton Waterville P'boro 138 D3
Orton Wistow P'boro 138 D2
Orwell Cambs 123 G7
Osbaldeston Lancs 203 G8
Osbaldeston Green Lancs 203 G8
Osbaldwick York 207 C8
Osbaston Leics 135 C8
Osbaston Shrops 148 E6
Osbaston Telford 149 F11
Osbaston Hollow Leics 135 B8
Osbournby Lincs 155 B11
Oscroft W Ches 167 B8
Ose Highld 298 E3
Osea Island Essex 88 D6
Osehill Green Dorset 29 F11
Osgathorpe Leics 153 F8
Osgodby Lincs 189 C9
Osgodby N Yorks 207 G8
Osgodby N Yorks 217 C11
Osgodby Common N Yorks 207 F8
Osidge London 86 G3
Oskaig Highld 295 B7
Oskamull Argyll 288 E6
Osleston Derbys 152 B4
Osmaston Derby 153 C7
Osmaston Derbys 170 G2
Osmington Dorset 17 E10
Osmington Mills Dorset 17 E10
Osmondthorpe W Yorks 206 G2
Osmotherley N Yorks 225 F9
Osney Oxon 83 D8
Ospisdale Highld 309 L7
Ospringe Kent 70 G4
Ossaborough Devon 40 E3
Ossemsley Hants 19 B10
Osset Spa W Yorks 197 D9
Ossett W Yorks 197 C9
Ossett Street Side W Yorks 197 C9
Ossington Notts 172 C3
Ostend Essex 88 F6
Ostend Norf 161 C7
Osterley London 66 D6
Oswaldkirk N Yorks 216 D2
Oswaldtwistle Lancs 195 B8
Oswestry Shrops 148 D5
Otby Lincs 189 C10
Oteley Shrops 149 C8
Otford Kent 52 B4
Otham Kent 53 C9
Otham Hole Kent 53 C10
Othery Som 43 G11
Otley Suff 126 F4
Otley W Yorks 205 D10
Otter Ferry Argyll 275 E10
Otter Ho Argyll 275 F10
Otterbourne Hants 33 C7
Otterburn N Yorks 204 B3
Otterburn Northumb 251 E9
Otterburn Camp Northumb 251 D9
Otterden Place Kent 54 C2
Otterford Som 28 E2
Otterham Corn 11 C9
Otterham Quay Kent 69 F10
Otterham Station Corn 11 D9
Otterhampton Som 43 E8
Ottershaw Sur 66 G4
Otterspool Mers 182 D5
Otterswick Shetland 312 E7
Otterton Devon 15 D7
Otterwood Hants 32 G6
Ottery St Mary Devon 15 B8
Ottinge Kent 55 E7
Ottringham E Yorks 201 C9
Oughterby Cumb 239 F7
Oughtershaw N Yorks 213 C7
Oughterside Cumb 229 C8
Oughtibridge S Yorks 186 C4
Oughtrington Warr 183 D11
Oulston N Yorks 215 E10
Oulton Cumb 238 G6
Oulton Norf 160 D2
Oulton Staffs 150 E5
Oulton Staffs 151 B8
Oulton Suff 143 D10
Oulton W Yorks 197 B11
Oulton Broad Suff 143 E10
Oulton Grange Staffs 151 B8
Oulton Heath Staffs 151 B8
Oulton Street Norf 160 D3
Oultoncross Staffs 151 C8
Oundle Northants 137 F10
Ounsdale Staffs 133 E7
Ousby Cumb 231 E8
Ousdale Highld 311 G4
Ousden Suff 124 F4
Ousefleet E Yorks 199 C10
Ousel Hole W Yorks 205 E8
Ouston Durham 243 G7
Ouston Northumb 241 G7
Ouston Northumb 242 C3
Out Elmstead Kent 55 C8
Out Newton E Yorks 201 C11
Out Rawcliffe Lancs 202 E4
Outcast Cumb 210 D6
Outer Hope Devon 8 G3
Outertown Orkney 314 E2
Outgate Cumb 221 F7
Outhgill Cumb 222 E5
Outhill Warks 118 D2
Outhills Aberds 303 D10
Outlands Staffs 150 C5
Outlane W Yorks 196 D5
Outlane Moor W Yorks 196 D5
Outlet Village W Ches 182 G6
Outmarsh Wilts 61 G11
Outwell Norf 139 C10
Outwick Hants 31 D10
Outwood Gtr Man 195 F9
Outwood Som 28 B4
Outwood Sur 51 D10
Outwood W Yorks 197 C10
Outwoods Leics 153 F8
Outwoods Staffs 150 F5
Outwoods Staffs 152 E4
Outwoods Warks 134 G4
Ouzlewell Green W Yorks 197 B10
Ovenden W Yorks 196 B5
Ovenscloss Borders 261 C11
Over Cambs 123 C7
Over Glos 80 B4
Over S Glos 60 C5
Over W Ches 167 B10
Over Burrow Lancs 212 D2
Over Burrows Derbys 152 B5
Over Compton Dorset 29 D9
Over End Cambs 137 E11
Over End Derbys 186 G3
Over Green W Mid 134 E3
Over Haddon Derbys 170 B2
Over Hulton Gtr Man 195 F7
Over Kellet Lancs 211 E10
Over Kiddington Oxon 101 G8
Over Knutsford E Ches 184 F3
Over Langshaw Borders 271 G10
Over Monnow Mon 79 C8
Over Norton Oxon 100 F6
Over Peover E Ches 184 G3
Over Silton N Yorks 225 G9
Over Stowey Som 43 F7
Over Stratton Som 28 D6
Over Tabley E Ches 184 E2
Over Town Lancs 195 B11
Over Wallop Hants 47 F9
Over Whitacre Warks 134 E5
Over Worton Oxon 101 F8
Overa Farm Stud Norf 141 F9
Overbister Orkney 314 B6
Overbury Worcs 99 D9
Overcombe Dorset 17 E9
Overend W Mid 133 G9
Overgreen Derbys 186 G4
Overleigh Som 44 F3
Overley Staffs 152 F3
Overley Green Warks 117 F11
Overmoor Staffs 169 F7
Overpool W Ches 182 F5
Overs Shrops 131 D7
Overscaig Hotel Highld 309 G4
Overseal Derbys 152 F5
Overslade Warks 119 C9
Oversland Kent 54 B5
Oversley Green Warks 117 F11
Overstone Northants 120 D6
Overstrand Norf 160 A4
Overthorpe Northants 101 C9
Overthorpe W Yorks 197 D8
Overton Aberdeen 293 B10
Overton Aberds 293 B9
Overton Dumfries 237 C11
Overton Glos 80 C2
Overton Hants 48 D4
Overton Involyd 276 G5
Overton Lancs 202 B4
Overton N Yorks 207 B7
Overton Shrops 115 C10
Overton Shrops 131 B10
Overton Staffs 151 B10
Overton Swansea 56 D3
Overton W Ches 183 F8
Overton W Yorks 197 D9
Overton = Owrtyn Wrex 166 G5
Overton Bridge Wrex 166 G5
Overtown Lancs 212 D2
Overtown N Lnrk 268 E6
Overtown Swindon 63 D7
Overtown W Yorks 197 D11
Oving Bucks 102 G5
Oving W Sus 22 C6
Ovingdean Brighton 36 G5
Ovingham Northumb 242 E3
Ovington Durham 224 C2
Ovington Essex 106 C5
Ovington Hants 48 G5
Ovington Norf 141 C8
Ovington Northumb 242 E3
Owen's Bank Staffs 152 D4
Ower Hants 32 D4
Owermoigne Dorset 17 D11
Owl End Cambs 122 B4
Owlbury Shrops 130 E6
Owlcotes Derbys 170 B6
Owler Bar Derbys 186 F3
Owlerton S Yorks 186 D4
Owlet W Yorks 205 F9
Owlpen Glos 80 F4
Owl's Green Suff 126 D5
Owlsmoor Brack 65 G11
Owlswick Bucks 84 D3
Owlthorpe S Yorks 186 E6
Owmby Lincs 200 G5
Owmby-by-Spital Lincs 189 D8
Ownham W Berks 64 E2
Owrtyn = Overton Wrex 166 G5
Owslebury Hants 33 C8
Owston Leics 136 B5
Owston S Yorks 198 E5
Owston Ferry N Lincs 199 G10
Owstwick E Yorks 209 G11
Owthorne E Yorks 201 B10
Owthorpe Notts 154 C3
Owton Manor Hrtlpl 234 F5
Oxborough Norf 140 C4
Oxclose S Yorks 186 E6
Oxclose T & W 243 F7
Oxcombe Lincs 190 F4
Oxcroft Derbys 187 G7
Oxcroft Estate Derbys 187 G7
Oxen End Essex 106 F3
Oxen Park Cumb 210 B6
Oxenhall Glos 98 F4
Oxenholme Cumb 211 B10
Oxenhope W Yorks 204 F6
Oxenpill Som 44 E2
Oxenton Glos 99 E9
Oxenwood Wilts 47 B10
Oxford Oxon 83 D8
Oxford Stoke 168 E5
Oxgang E Dunb 278 G3
Oxgangs Edin 270 B4
Oxhey Herts 85 F10
Oxhill Durham 242 G5
Oxhill Warks 100 B6
Oxlease Herts 86 D2
Oxley W Mid 133 C8
Oxley Green Essex 88 C6
Oxley's Green E Sus 37 C11
Oxlode Cambs 139 F9
Oxnam Borders 262 F5
Oxnead Norf 160 E4
Oxshott Sur 66 G6
Oxspring S Yorks 197 G9
Oxted Sur 51 C11
Oxton Borders 271 E9
Oxton Mers 182 D3
Oxton N Yorks 206 E5
Oxton Notts 171 E10
Oxton Rakes Derbys 186 G4
Oxwich Swansea 56 D3
Oxwich Green Swansea 56 D3
Oxwick Norf 159 D8
Oykel Bridge Highld 309 J3
Oyne Aberds 302 G6
Oystermouth Swansea 56 D6
Ozleworth Glos 80 G3

P

Pabail Iarach W Isles 304 E7
Pabail Uarach W Isles 304 E7
Pabo Conwy 180 F4
Pace Gate N Yorks 205 C8
Pachesham Park Sur 51 B7
Packers Hill Dorset 30 E2
Packington Leics 153 G7
Packmoor Staffs 168 E5
Packmores Warks 118 D5
Packwood W Mid 118 C3
Packwood Gullet W Mid 118 C3
Padanaram Angus 287 B8
Padbury Bucks 102 E4
Paddington London 67 C9
Paddington Warr 183 D10
Paddlesworth Kent 55 F7
Paddlesworth Kent 69 G7
Paddock Kent 54 C3
Paddock W Yorks 196 D6
Paddock Wood Kent 53 E7
Paddockhaugh Moray 302 D2
Paddockhill E Ches 184 F4
Paddockhole Dumfries 248 G6
Paddolgreen Shrops 149 C10
Padfield Derbys 185 B8
Padgate Warr 183 D10
Padham's Green Essex 87 F10
Padiham Lancs 203 G11
Padney Cambs 123 C10
Padog Conwy 164 E4
Padside N Yorks 205 B9
Padside Green N Yorks 205 B9
Padson Devon 13 B7
Padstow Corn 10 F4
Padworth W Berks 64 F6
Padworth Common Hants 64 G6
Paganhill Glos 80 D4
Page Bank Durham 233 D10
Page Moss Mers 182 C6
Pagelsham Eastend Essex 88 G6
Page's Green Suff 126 D2
Pagham W Sus 22 D5
Paglesham Churchend Essex 88 G6
Paglesham Eastend Essex 88 G6
Paibeil W Isles 296 E3
Paible W Isles 305 J2
Paignton Torbay 9 C7
Pailton Warks 135 G9
Painleyhill Staffs 151 C10
Pains Hill Sur 52 C2
Painscastle Powys 96 B3
Painshawfield Northumb 242 E3
Painsthorpe E Yorks 208 B2
Painswick Glos 80 D5
Painter's Forstal Kent 54 B3
Painter's Green Herts 86 B3
Painters Green Wrex 167 G8
Painthorpe W Yorks 197 D10
Paintmoor Som 28 F4
Pairc Shiaboist W Isles 304 D4
Paisley Renfs 267 C9
Pakefield Suff 143 E10
Pakenham Suff 125 D8
Pale Gwyn 147 B9
Pale Green Essex 106 C3
Palehouse Common E Sus 23 B7
Palestine Hants 47 E9
Paley Street Windsor 65 D11
Palfrey W Mid 133 D10
Palgowan Dumfries 245 G9
Palgrave Suff 126 B2
Pallaflat Cumb 219 C9
Pallington Dorset 17 C11
Pallion T & W 243 F9
Pallister M'bro 225 B10
Palmarsh Kent 54 G6
Palmer Moor Derbys 152 C2
Palmers Cross Staffs 133 C7
Palmers Cross Sur 50 E4
Palmer's Flat Glos 79 D9
Palmer's Green Kent 53 E7
Palmers Green London 86 G4
Palmersbridge Corn 11 F9
Palmerstown V Glam 58 F6
Palmersville T & W 243 C7
Palmstead Kent 55 D7
Palnackie Dumfries 237 D10
Palnure Dumfries 236 C6
Palterton Derbys 171 B7
Pamber End Hants 48 B6
Pamber Green Hants 48 B6
Pamber Heath Hants 64 G6
Pamington Glos 99 E8
Pamphill Dorset 31 G7
Pampisford Cambs 105 B9
Pan I o W 20 D6
Pan Orkney 314 G3
Panborough Som 44 D3
Panbride Angus 287 D9
Pancakehill Glos 81 C9
Pancrasweek Devon 24 F3
Pancross V Glam 58 F4
Pandy Gwyn 128 C2
Pandy Gwyn 146 F4
Pandy Gwyn 147 D7
Pandy Mon 96 G6
Pandy Powys 129 C8
Pandy Wrex 148 B3
Pandy Wrex 166 G6
Pandy Tudur Conwy 164 C5
Pandy'r Capel Denb 165 E9
Panfield Essex 106 F4
Pangbourne W Berks 64 D6
Panhall Fife 280 C6
Panks Bridge Hereford 98 B2
Pannal N Yorks 206 C2
Pannal Ash N Yorks 205 C11
Pannel's Ash Essex 106 C5
Panpunton Powys 114 C5
Panshanger Herts 86 C3
Pant Denb 166 E2
Pant Flint 181 G10
Pant Gwyn 144 C4
Pant M Tydf 77 D9
Pant Powys 129 C11
Pant Shrops 148 E5
Pant Wrex 166 D5
Pant Wrex 166 F3
Pant-glas Powys 128 D5
Pant-glas Gwyn 163 F7
Pant-glas Shrops 148 C5
Pant-lasau Swansea 57 B7
Pant Mawr Powys 129 G7
Pant-pastynog Denb 165 C8
Pant-teg Carms 93 F9
Pant-y-Caws Carms 92 F3
Pant-y-crûg Ceredig 112 B3
Pant-y-dwr Powys 113 B9
Pant-y-ffridd Powys 130 C3
Pant-y-pyllau Bridgend 58 C2
Pant-y-Wacco Flint 181 F10
Pant-yr-awel Bridgend 58 B2
Pantasaph Flint 181 F11
Pantdu Neath 57 C9
Panteg Ceredig 111 E9
Panteg Torf 78 F4
Pantersbridge Corn 6 B3
Pantgwyn Carms 93 F11
Pantgwyn Ceredig 92 B4
Pantmawr Cardiff 58 C6
Panton Lincs 189 F11
Pantperthog Gwyn 128 C4
Pantside Caerph 78 F2
Pantyffynnon Carms 75 C10
Pantygasseg Torf 78 F3
Pantymwyn Flint 165 C11
Panxworth Norf 161 G7
Papcastle Cumb 229 E8
Papermill Bank Shrops 149 D11
Papigoe Highld 310 D7
Papil Shetland 313 K5
Papley Northants 138 F2
Papley Orkney 314 G4
Papple E Loth 281 G11
Papplewick Notts 171 E8
Papworth Everard Cambs 122 E5
Papworth St Agnes Cambs 122 E5
Papworth Village Settlement Cambs 122 E5
Par Corn 5 E11
Paradise Glos 80 C5
Paradise Green Hereford 97 B10
Paramoor Corn 5 F9
Paramour Street Kent 71 G9
Parbold Lancs 194 E3
Parbrook Som 44 F5
Parbrook W Sus 35 B9
Parc Gwyn 147 C7
Parc Erissey Corn 4 G3
Parc-hendy Swansea 56 B4
Parc Mawr Caerph 77 G10
Parc-Seymour Newport 78 G6
Parc-y-rhôs Carms 93 B11
Parchey Som 43 F10
Parciau Anglesey 179 E7
Parcllyn Ceredig 110 G4
Pardown Hants 48 D5
Pardshaw Cumb 229 G7
Pardshaw Hall Cumb 229 F8
Parham Suff 126 E6
Park Corn 10 G6
Park Devon 14 B2
Park Dumfries 247 E10
Park Som 44 G3
Park Swindon 63 C7
Park Barn Sur 50 C3
Park Bottom Corn 4 G3
Park Bridge Gtr Man 196 G2
Park Broom Cumb 239 F10
Park Close Lancs 204 E3
Park Corner Bath 45 B9
Park Corner E Sus 23 C8
Park Corner E Sus 52 F4
Park Corner Oxon 65 B7
Park Corner Windsor 65 C11
Park End Beds 121 G9
Park End Cambs 123 E11
Park End M'bro 225 B10
Park End Northumb 241 B9
Park End Som 43 G7
Park End Staffs 168 E3
Park End Worcs 116 C5
Park Gate Dorset 30 F2
Park Gate Hants 33 F8
Park Gate Kent 55 D7
Park Gate Suff 124 F4
Park Gate W Yorks 197 E8
Park Gate Worcs 117 C8
Park Green Essex 105 F9
Park Hall Shrops 148 C6
Park Head Cumb 231 C7
Park Head Derbys 170 E5
Park Head W Yorks 197 F7
Park Hill Glos 79 F9
Park Hill Kent 54 G3
Park Hill Mers 194 G3
Park Hill N Yorks 214 F6
Park Hill Notts 171 E11
Park Hill S Yorks 186 D5
Park Lane Staffs 133 B7
Park Lane Wrex 149 B8
Park Langley London 67 F11

Park Mains Renfs 277 G9
Park Mill W Yorks 197 E9
Park Royal London 67 C7
Park Street Herts 85 E10
Park Street W Sus 50 G6
Park Town Luton 103 G11
Park Town Oxon 83 D8
Park Village Northumb 240 E5
Park Village W Mid 133 C8
Park Villas W Yorks 206 F2
Park Wood Kent 53 C9
Park Wood Medway 69 G10
Parkend Glos 79 D10
Parkend Glos 80 C3
Parkengear Corn 5 F8
Parker's Corner W Berks 64 E6
Parker's Green Herts 104 F6
Parker's Green Kent 52 D6
Parkeston Essex 108 E4
Parkfield Corn 6 B6
Parkfield S Glos 61 D7
Parkfield W Mid 133 D8
Parkfoot Falk 278 F6
Parkgate Cumb 229 B10
Parkgate Dumfries 248 F2
Parkgate E Ches 184 G3
Parkgate Essex 87 B11
Parkgate Kent 53 G11
Parkgate S Yorks 186 B6
Parkgate Sur 51 E8
Parkgate W Ches 182 F3
Parkhall W Dunb 277 G9
Parkham Devon 24 C5
Parkham Ash Devon 24 C5
Parkhead Cumb 230 C2
Parkhead Glasgow 268 C2
Parkhead S Yorks 186 E4
Parkhill Aberds 303 E10
Parkhill Invclyd 277 G7
Parkhill Ho Aberds 293 B10
Parkhouse Mon 79 E7
Parkhouse Green Derbys 170 C6
Parkhurst I o W 20 C5
Parklands W Yorks 206 F3
Parkmill Swansea 56 D4
Parkneuk Aberds 293 F9
Parkneuk Fife 279 D11
Parkside C Beds 103 G10
Parkside Cumb 219 B10
Parkside Durham 234 B4
Parkside N Lnrk 268 D6
Parkside Staffs 151 D8
Parkside Wrex 166 D5
Parkstone Poole 18 C6
Parkway Hereford 98 D4
Parkway Som 29 C9
Parkwood Springs S Yorks 186 D4
Parley Cross Dorset 19 B7
Parley Green Dorset 19 B7
Parliament Heath Suff 107 C9
Parlington W Yorks 206 F4
Parmoor Bucks 65 B9
Parnacott Devon 24 F4
Parney Heath Essex 107 E10
Parr Mers 183 C8
Parr Brow Gtr Man 195 G8
Parracombe Devon 41 E7
Parrog Pembs 91 D10
Parsley Hay Derbys 169 C10
Parslow's Hillock Bucks 84 E4
Parson Cross S Yorks 186 C5
Parson Drove Cambs 139 B7
Parsonage Green Essex 88 D2
Parsonby Cumb 229 D8
Parsons Green London 67 D9
Parson's Heath Essex 107 F10
Partick Glasgow 267 B11
Partington Gtr Man 184 C2
Partney Lincs 174 B6
Parton Cumb 228 G5
Parton Cumb 239 G7
Parton Dumfries 237 B8
Parton Glos 99 G7
Parton Hereford 96 B6
Partridge Green W Sus 35 D11
Partrishow Powys 96 G5
Parwich Derbys 169 E11
Pasford Staffs 132 D6
Passenham Northants 102 D5
Passfield Hants 49 G10
Passingford Bridge Essex 87 F8
Passmores Essex 86 D6
Paston Norf 160 C6
Paston P'boro 138 C3
Paston Green Norf 160 C6
Pasturefields Staffs 151 D9
Patchacott Devon 12 B5
Patcham Brighton 36 F4
Patchetts Green Herts 85 F10
Patching W Sus 35 F9
Patchole Devon 40 E6
Patchway S Glos 60 C6
Pategill Cumb 230 F6
Pateley Bridge N Yorks 214 F3
Paternoster Heath Essex 88 C6
Path Head T & W 242 E5
Path of Condie Perth 286 F4
Pathe Som 43 G11
Pather N Lnrk 268 E5
Pathfinder Village Devon 14 C2
Pathhead Aberds 293 G9
Pathhead E Ayrs 258 G4
Pathhead Fife 280 C5
Pathhead Midloth 271 C7
Pathlow Warks 118 F3
Pathstruie Perth 286 F4
Patient End Herts 105 F8
Patmore Heath Herts 105 F8
Patna E Ayrs 257 G10
Patney Wilts 46 B5
Patrick I o M 192 D3
Patrick Brompton N Yorks 224 G4
Patricroft Gtr Man 184 B3
Patrington E Yorks 201 C10
Patrington Haven E Yorks 201 C10
Patrixbourne Kent 55 B7
Patsford Devon 40 F4
Patterdale Cumb 221 B7
Pattiesmuir Fife 279 E11
Pattingham Staffs 132 D6
Pattishall Northants 120 G3
Pattiswick Essex 106 G6
Patton Shrops 131 E11
Patton Bridge Cumb 221 F11
Paul Corn 1 D5
Paulerspury Northants 102 B4
Paull E Yorks 201 B7
Paul's Green Corn 2 C4
Paulsgrove Ptsmth 33 F10
Paulton Bath 45 B7
Paulville W Loth 269 B9
Pave Lane Telford 150 F5
Pavenham Beds 121 F9
Pawlett Som 43 E10
Pawlett Hill Som 43 E9
Pawston Northumb 263 C9
Paxford Glos 100 D3
Paxton Borders 273 E8
Payden Street Kent 54 C2
Payhembury Devon 27 G9
Paynes Green Sur 50 F6
Paynter's Cross Corn 7 C7
Paynter's Lane End Corn 4 G3
Paythorne Lancs 204 C2
Payton Som 27 C10
Peacehaven E Sus 36 G6
Peacehaven Heights E Sus 36 G6
Peacemarsh Dorset 30 B4
Peak Dale Derbys 185 F9
Peak Forest Derbys 185 F10
Peak Hill Lincs 156 F5
Peakirk P'boro 138 B3
Pean Hill Kent 70 G6
Pear Ash Som 45 G9
Pear Tree Derby 153 C7
Pearsie Angus 287 B7
Pearson's Green Kent 53 E7
Peartree Herts 86 C2
Peartree Green Essex 87 F9
Peartree Green Hereford 97 E11
Peartree Green Soton 32 E6
Peartree Green Sur 50 F3
Peas Acre W Yorks 205 E8
Peas Hill Cambs 139 D8
Pease Pottage W Sus 51 G9
Peasedown St John Bath 45 B8
Peasehill Derbys 170 F6
Peaseland Green Norf 159 F11
Peasemore W Berks 64 D3
Peasenhall Suff 127 D7
Peaslake Sur 50 E5
Peasley Cross Mers 183 C8
Peasmarsh E Sus 38 C5
Peasmarsh Som 28 E4
Peasmarsh Sur 50 D3
Peaston E Loth 271 B8
Peastonbank E Loth 271 B8
Peat Inn Fife 287 G8
Peathill Aberds 303 C9
Peatling Magna Leics 135 E11
Peatling Parva Leics 135 F11
Peaton Shrops 131 G10
Peatonstrand Shrops 131 G10
Peats Corner Suff 126 E3
Pebmarsh Essex 107 E7
Pebsham E Sus 38 F3
Pebworth Worcs 100 B2
Pecket Well W Yorks 196 B3
Peckforton E Ches 167 D8
Peckham London 67 D10
Peckham Bush Kent 53 D7
Pecking Mill Som 44 F6
Peckingell Wilts 62 E2
Peckleton Leics 135 C9
Pedair-ffordd Powys 148 E2
Pedham Norf 160 G6
Pedlars End Essex 87 D8
Pedlar's Rest Shrops 131 G9
Pedlinge Kent 54 F6
Pedmore W Mid 133 G8
Pednor Bottom Bucks 84 E6
Pednormead End Bucks 85 E7
Pedwell Som 44 F2
Peebles Borders 270 G5
Peel Borders 261 B10
Peel I o M 192 D3
Peel Lancs 202 G3
Peel Common Hants 33 G9
Peel Green Gtr Man 184 B2
Peel Hall Gtr Man 184 D4
Peel Hill Lancs 202 G3
Peel Park S Lnrk 268 E2
Peene Kent 55 F7
Peening Quarter Kent 38 B5
Peggs Green Leics 153 F8
Pegsdon C Beds 104 E2
Pegswood Northumb 252 F6
Pegwell Kent 71 G11
Peinaha Highld 298 D4
Peinchorran Highld 295 B7
Peingown Highld 298 B4
Peinlich Highld 298 D4
Peinmore Highld 298 E4
Pelaw T & W 243 E7
Pelcomb Pembs 72 B6
Pelcomb Bridge Pembs 72 B6
Pelcomb Cross Pembs 72 B6
Peldon Essex 89 B7
Pelhamfield I o W 21 C7
Pell Green E Sus 52 G6
Pellon W Yorks 196 B5
Pelsall W Mid 133 C10
Pelsall Wood W Mid 133 C10
Pelton Durham 243 G7
Pelton Fell Durham 243 G7
Pelutho Cumb 229 B8
Pelynt Corn 6 D4
Pemberton Carms 75 E8
Pemberton Gtr Man 194 G5
Pembles Cross Kent 53 D11
Pembre = Pembrey Carms 74 E6
Pembrey = Pembre Carms 74 E6
Pembridge Hereford 115 F7
Pembroke Pembs 73 E7
Pembroke Dock = Doc Penfro Pembs 73 E7
Pembroke Ferry Pembs 73 E7
Pembury Kent 52 E6
Pempwell Corn 12 F3
Pen-allt Hereford 97 F11
Pen-bedw Pembs 92 D4
Pen-bont Rhydybeddau Ceredig 128 G3
Pen-caer-fenny Swansea 56 B4
Pen-clawdd Swansea 56 B4
Pen-common Powys 76 D6
Pen-gilfach Gwyn 163 C9
Pen-groes-oped Mon 78 D4
Pen-lan Swansea 56 B6
Pen-Lan-mabws Pembs 91 F7
Pen-llyn Anglesey 178 E4
Pen-lon Anglesey 162 B6
Pen Mill Som 29 D9
Pen-onn V Glam 58 F5
Pen-Rhiw-fawr Neath 76 C2
Pen-rhos Wrex 166 E3
Pen-sarn Gwyn 145 D11
Pen-sarn Gwyn 162 G6
Pen-twyn Caerph 78 E2
Pen-twyn Carms 75 C9
Pen-twyn Mon 79 D8
Pen-twyn Torf 78 E3
Pen-Uchar Plwyf Flint 181 G11
Pen-y-Ball Top Flint 181 F11
Pen-y-banc Carms 93 G8
Pen-y-banc Carms 94 G2
Pen-y-bank Caerph 77 E10
Pen-y-Bont Bl Gwent 78 D2
Pen-y-bont Carms 92 F6
Pen-y-bont Gwyn 128 C4
Pen-y-bont Gwyn 146 D2
Pen-y-bont Powys 148 E4
Pen y Bont ar ogwr = Bridgend Bridgend 58 C2
Pen-y-Bryn Gwyn 145 E9
Pen-y-bryn Gwyn 146 F3
Pen-y-bryn Pembs 92 C3
Pen-y-bryn Powys 130 C3
Pen-y-Bryn Shrops 148 B6
Pen-y-bryn Wrex 166 E3
Pen-y-cae Bridgend 58 C2
Pen-y-cae Neath 57 D9
Pen-y-cae Powys 76 C4
Pen-y-cae-mawr Mon 78 F6
Pen-y-cefn Flint 181 F10
Pen-y-clawdd Mon 79 D7
Pen-y-coed Shrops 148 E5
Pen-y-coedcae Rhondda 58 B5
Pen-y-Darren M Tydf 77 D9
Pen-y-fai Bridgend 57 E11
Pen-y-fai Carms 75 E7
Pen-y-fan Carms 56 B4
Pen-y-fan Mon 79 D8
Pen-y-felin Flint 165 B11
Pen-y-ffordd Denb 181 F8
Pen-y-ffordd Flint 181 E10
Pen y Foel Shrops 148 E5
Pen-y-garn Carms 93 E11
Pen-y-garn Ceredig 128 F2
Pen-y-garnedd Anglesey 179 F8
Pen-y-gop Conwy 164 G5
Pen-y-graig Gwyn 144 C3
Pen-y-groes Carms 75 C9
Pen-y-groeslon Gwyn 144 C4
Pen-y-Gwryd Hotel Gwyn 163 D11
Pen-y-lan Cardiff 59 D7
Pen-y-lan Newport 59 B9
Pen-y-lan V Glam 58 D3
Pen-y-maes Flint 181 F11
Pen-y-Myndd Carms 75 E7
Pen-y-Park Hereford 96 C5
Pen-y-rhiw Rhondda 58 B5
Pen-y-stryt Denb 165 E11
Pen-y-wern Shrops 114 B6
Pen-yr-englyn Rhondda 76 F6
Pen-yr-heol Bridgend 58 C2
Pen-yr-heol Mon 78 C6
Pen-yr-Heolgerrig M Tydf 77 D8
Penallt Mon 79 C8
Penally = Penalun Pembs 73 F10
Penalt Hereford 97 F11
Penalun = Penally Pembs 73 F10
Penare Corn 5 C9
Penarlâg = Hawarden Flint 166 E4
Penarron Powys 130 F2
Penarth V Glam 59 E7
Penarth Moors Cardiff 59 E7
Penbeagle Corn 2 E2
Penbedw Flint 165 B11
Penberth Corn 1 E4
Penbidwal Mon 96 G6
Penbodlas Gwyn 144 C5
Penboyr Carms 93 D7
Penbryn Ceredig 110 G5
Pencader Carms 93 D8
Pencaenewydd Gwyn 162 G6
Pencaerau Neath 57 B8
Pencaitland E Loth 271 B8
Pencarnisiog Anglesey 178 G5
Pencarreg Carms 93 B10
Pencarrow Corn 11 E8
Penceiliogi Carms 75 E8
Pencelli Powys 95 F11
Pencoed Bridgend 58 C3
Pencombe Hereford 115 G11
Pencoyd Hereford 97 F10
Pencoys Corn 2 B5
Pencraig Anglesey 179 F7
Pencraig Hereford 97 G11
Pencraig Powys 147 D10
Pencroesoped Mon 78 D4
Pencuke Corn 11 C9
Pendas Fields W Yorks 206 F3
Pendeen Corn 1 C3
Pendeford W Mid 133 C7
Penderyn Rhondda 77 D7
Pendine = Pentwyn Carms 74 D2
Pendlebury Gtr Man 195 G9
Pendleton Gtr Man 184 B4
Pendleton Lancs 203 F11
Pendock Worcs 98 E5
Pendoggett Corn 10 F5
Pendomer Som 29 E8
Pendoylan V Glam 58 D5
Pendre Bridgend 58 C2
Pendre Gwyn 110 C2
Pendre Powys 95 F10
Pendrift Corn 11 G8
Penegoes Powys 128 C5
Penelewey Corn 4 G5
Penenden Heath Kent 53 B9
Penffordd Pembs 91 G11
Penffordd Lâs = Staylittle Powys 129 E7
Pengam Caerph 77 F11
Penge London 67 E11
Pengegon Corn 2 B5
Pengelly Corn 11 E7
Pengenffordd Powys 96 E3
Pengersick Corn 2 D3
Pengold Corn 11 C8
Pengorffwysfa Anglesey 179 C7
Pengover Green Corn 6 B5
Penguithal Hereford 97 G10
Pengwern Denb 181 F8
Penhale Corn 2 F5
Penhale Corn 5 D8
Penhale Jakes Corn 2 D4
Penhallick Corn 3 F7
Penhallick Corn 4 G3
Penhallow Corn 4 E5
Penhalurick Corn 2 B6
Penhalvean Corn 2 B6
Penhelig Gwyn 128 D2
Penhill Devon 40 G4
Penhill Swindon 63 B7
Penhow Newport 78 G6
Penhurst E Sus 23 B11
Peniarth Gwyn 128 B2
Penicuik Midloth 270 C4
Peniel Carms 93 G8
Peniel Denb 165 C8
Penifiler Highld 298 E4
Peninver Argyll 255 E8
Penisa'r Waun Gwyn 163 C9
Penistone S Yorks 197 G8
Penjerrick Corn 3 C7
Penketh Warr 183 D9
Penkhull Stoke 168 G5
Penkill S Ayrs 244 D6
Penknap Wilts 45 D11
Penkridge Staffs 151 G8
Penleigh Wilts 45 C11
Penley Wrex 149 B8
Penllech Gwyn 144 C4
Penllergaer Swansea 56 B6
Penllwyn Caerph 77 F11
Penllwyn Ceredig 128 G3
Penllyn V Glam 58 D3
Penmachno Conwy 164 E3
Penmaen Caerph 77 F11
Penmaen Swansea 56 D4
Penmaen Rhôs Conwy 180 F5
Penmaenan Conwy 180 F2
Penmaenmawr Conwy 180 F2
Penmaenpool Gwyn 146 F3
Penmark V Glam 58 F5
Penmarth Corn 2 B6
Penmayne Corn 10 F4
Penmon Anglesey 179 E10
Penmore Mill Argyll 288 D6
Penmorfa Ceredig 110 G6
Penmorfa Gwyn 163 G8
Penmynydd Anglesey 179 G8
Penn Bucks 84 G6
Penn W Mid 133 D7
Penn Bottom Bucks 84 G6
Penn Street Bucks 84 F6
Pennal Gwyn 128 C4
Pennan Aberds 303 C8
Pennance Corn 4 G4
Pennant Ceredig 111 E10
Pennant Conwy 164 D5
Pennant Denb 147 C10
Pennant Denb 165 E8
Pennant Powys 129 D7
Pennant Melangell Powys 147 D10
Pennar Pembs 73 E7
Pennar Park Pembs 72 E6
Pennard Swansea 56 D5
Pennerley Shrops 131 D7
Pennington Cumb 210 D5
Pennington Gtr Man 183 B11
Pennington Hants 20 C2
Pennington Green Gtr Man 194 F6
Pennorth Powys 96 F2
Pennsylvania Devon 14 C4
Pennsylvania S Glos 61 E8
Penny Bridge Cumb 210 C6
Penny Green Derbys 187 F8
Penny Hill Lincs 157 D7
Penny Hill W Yorks 196 D5
Pennycross Argyll 289 G7
Pennycross Plym 7 D9
Pennygate Norf 160 E6
Pennygown Argyll 289 E7
Pennylands Lancs 194 F3
Pennymoor Devon 26 E5
Pennypot Kent 54 G6
Penny's Green Norf 142 D3
Pennytinney Corn 10 F6
Pennywell T & W 243 F9
Penparc Ceredig 92 B4
Penparc Pembs 91 E7
Penparcau Ceredig 111 B11
Penpedairheol Caerph 77 F10
Penpedairheol Mon 78 E4
Penpergym Mon 78 C4
Penperlleni Mon 78 E4
Penpethy Corn 11 D7
Penpillick Corn 5 D11
Penplas Carms 74 B5
Penpol Corn 3 B8
Penpoll Corn 6 E2
Penponds Corn 2 B4
Penpont Corn 11 G7
Penpont Dumfries 247 E8
Penpont Powys 95 F9
Penprysg Bridgend 58 C3
Penquit Devon 8 E2
Penrallt Gwyn 145 B7
Penrallt Powys 129 F9
Penrhôs Mon 78 C6
Penrherber Carms 92 D5
Penrhiw Caerph 78 G2
Penrhiw-llan Ceredig 93 C7
Penrhiw-pal Ceredig 92 B6
Penrhiwceiber Rhondda 77 F8
Penrhiwgarreg Bl Gwent 78 E2
Penrhiwtyn Neath 57 B8
Penrhos Anglesey 178 E3
Penrhos Gwyn 144 C6
Penrhos Hereford 114 F6
Penrhos Mon 78 C6
Penrhos Powys 76 C3
Penrhos-garnedd Gwyn 179 G9
Penrhosfeilw Anglesey 178 E2
Penrhyd Lastra Anglesey 178 C6
Penrhyn Bay = Bae-Penrhyn Conwy 180 E4
Penrhyn Castle Pembs 92 B2
Penrhyn-coch Ceredig 128 G2
Penrhyn side Conwy 180 E4
Penrhyndeudraeth Gwyn 146 B2
Penrhynside Conwy 180 E4
Penrhys Rhondda 77 F8
Penrice Swansea 56 D3
Penrith Cumb 230 E6
Penrose Corn 10 G3
Penrose Corn 11 F7
Penrose Hill Corn 2 D4
Penruddock Cumb 230 F4
Penryn Corn 3 C7
Pensarn Carms 74 B6
Pensarn Conwy 181 F7
Pensax Worcs 116 D4
Pensby Mers 182 E3
Penselwood Som 45 G9
Pensford Bath 60 G6
Pensham Worcs 99 C8
Penshaw T & W 243 G8
Penshurst Kent 52 E4
Pensilva Corn 6 B5
Pensnett W Mid 133 F8
Penston E Loth 281 G8
Penstone Devon 26 G3
Penstraze Corn 4 F5
Pentewan Corn 5 F10
Pentiken Shrops 130 G4
Pentir Gwyn 163 B9
Pentire Corn 4 C5
Pentirvin Shrops 130 C6
Pentlepoir Pembs 73 D10
Pentlow Essex 106 B6
Pentlow Street Essex 106 B6
Pentney Norf 158 G4
Penton Corner Hants 47 D10
Penton Grafton Hants 47 D10
Penton Mewsey Hants 47 D10
Pentonville London 67 C10
Pentowin Carms 74 B3
Pentraeth Anglesey 179 F8
Pentrapeod Caerph 77 E11
Pentre Carms 75 C8
Pentre Denb 165 D10
Pentre Flint 165 C11
Pentre Flint 166 B4
Pentre Flint 166 C2
Pentre Flint 166 D2
Pentre Powys 129 F11
Pentre Powys 130 B4
Pentre Powys 130 E5
Pentre Powys 147 D11
Pentre Powys 148 G3
Pentre Rhondda 77 F7
Pentre Shrops 148 D4
Pentre Shrops 149 F7
Pentre Wrex 148 B2
Pentre Wrex 166 G3
Pentre-bâch Ceredig 93 B11
Pentre-bach Powys 95 E8
Pentre Berw Anglesey 179 G7
Pentre-bont Conwy 164 E2
Pentre Broughton Wrex 166 E4
Pentre Bychan Wrex 166 F4
Pentre-cefn Shrops 148 D4
Pentre-celyn Denb 165 E11
Pentre-Celyn Powys 129 B7
Pentre-chwyth Swansea 57 B7
Pentre Cilgwyn Wrex 148 B4
Pentre-clawdd Shrops 148 C5
Pentre-coed Shrops 149 B7
Pentre-cwrt Carms 93 D7
Pentre Dolau-Honddu Powys 95 C9
Pentre-dwr Swansea 57 B7
Pentre-Ffwrndan Flint 182 G3
Pentre-galar Pembs 92 E3
Pentre-Gwenlais Carms 75 B10
Pentre Gwynfryn Gwyn 145 D11
Pentre Halkyn Flint 182 G2
Pentre Hodre Shrops 114 B6
Pentre Isaf Conwy 164 B5
Pentre Llanrhaeadr Denb 165 C9
Pentre Llifior Powys 130 D2
Pentre-llwyn-llwyd Powys 113 G9
Pentre-llyn Ceredig 112 C2
Pentre-llyn cymmer Conwy 165 E7
Pentre Maelor Wrex 166 F5
Pentre Meyrick V Glam 58 D3
Pentre-newydd Shrops 148 B5
Pentre-Piod Torf 78 E3
Pentre-Poeth Carms 75 E8
Pentre-poeth Newport 59 B9
Pentre-rhew Ceredig 112 G3
Pentre-tafarn-y-fedw Conwy 164 C4
Pentre-ty-gwyn Carms 94 D6
Pentre-uchaf Conwy 180 F5
Pentrebach Carms 94 E6
Pentrebach M Tydf 77 E9
Pentrebach Rhondda 58 B5
Pentrebach Swansea 75 D10
Pentrebane Cardiff 58 D6
Pentrebeirdd Powys 148 G3
Pentrecagal Carms 92 C6
Pentredwr Denb 165 F11
Pentref-y-groes Caerph 77 F11
Pentrefelin Anglesey 178 C6
Pentrefelin Carms 93 G11
Pentrefelin Ceredig 94 B2
Pentrefelin Conwy 180 G4
Pentrefelin Denb 166 G2
Pentrefelin Gwyn 145 B10
Pentrefoelas Conwy 164 E5
Pentregat Ceredig 111 G7
Pentreheyling Shrops 130 E4
Pentrellwyn Ceredig 93 C8
Pentre'r beirdd Powys 148 G3
Pentre'r Felin Conwy 164 B4
Pentre'r-felin Denb 165 B10
Pentre'r-felin Powys 95 E8
Pentreuchaf Gwyn 145 B7
Pentrich Derbys 170 E5
Pentridge Dorset 31 D8
Pentrisil Pembs 91 E11
Pentwyn Caerph 77 E10
Pentwyn Cardiff 59 C8
Pentwyn = Pendine Carms 74 D2
Pentwyn Berthlwyd Caerph 77 F10
Pentwyn-mawr Caerph 77 F11
Pentyrch Cardiff 58 C6
Penuchadre V Glam 57 G11
Penuwch Ceredig 111 E11
Penwartha Corn 4 E5
Penwartha Coombe Corn 4 E5
Penweathers Corn 4 G6
Penwithick Corn 5 D10
Penwood Hants 64 G2
Penwortham Lane Lancs 194 B4
Penwyllt Powys 76 B5
Penybanc Carms 75 C10
Penybedd Carms 74 E6
Penybont Ceredig 128 F2
Penybont Powys 114 E2
Penybontfawr Powys 147 E11
Penybryn Caerph 77 F10
Penycae Wrex 166 F3
Penycaerau Gwyn 144 D3
Penycwm Pembs 90 G6
Penydre Swansea 75 E11
Penyfeidr Pembs 91 F7
Penyffordd Flint 166 C4
Penyffridd Gwyn 163 D8
Penygarn Torf 78 E3
Penygarnedd Powys 148 E2
Penygelli Powys 130 E2
Penygraig Rhondda 77 G7
Penygraigwen Anglesey 178 D6
Penygroes Gwyn 163 E7
Penygroes Pembs 92 D3
Penymynydd Flint 166 C4
Penyraber Pembs 91 D9
Penyrheol Caerph 58 B6
Penyrheol Swansea 56 B5
Penyrheol Torf 78 F3
Penysarn Anglesey 179 C7
Penywaun Rhondda 77 E7
Penzance Corn 1 C5
Peopleton Worcs 117 G8
Peover Heath E Ches 184 G3
Peper Harow Sur 50 E2
Pepper Hill Som 43 F7
Pepper Hill W Yorks 196 B6
Peppercombe Devon 24 C5
Peppermoor Northumb 264 F6
Pepper's Green Essex 87 C10
Pepperstock C Beds 85 B9
Per-ffordd-llan Flint 181 F10
Perceton N Ayrs 267 G7
Percie Aberds 293 D7
Percuil Corn 3 C9
Percy Main T & W 243 D8
Percyhorner Aberds 303 C9
Perham Down Wilts 47 D9
Periton Som 42 D3
Perivale London 67 C7
Perkhill Aberds 293 C7
Perkinsville Durham 243 G7
Perlethorpe Notts 187 G11
Perran Downs Corn 2 C3
Perran Wharf Corn 3 B7
Perranarworthal Corn 3 B7
Perrancoombe Corn 4 E5
Perranporth Corn 4 E5
Perranuthnoe Corn 2 D2
Perranwell Corn 3 B7
Perranwell Corn 4 E5
Perranwell Station Corn 3 B7
Perranzabuloe Corn 4 E5
Perrott's Brook Glos 81 D8
Perry Devon 26 F5
Perry Kent 55 B9
Perry W Mid 133 E11
Perry Barr W Mid 133 E11
Perry Beeches W Mid 133 E11
Perry Common W Mid 133 E11
Perry Crofts Staffs 134 C4
Perry Green Essex 106 G6
Perry Green Herts 86 B6
Perry Green Som 43 F9
Perry Green Wilts 62 B3
Perry Street Kent 68 E6
Perry Street Som 28 F4
Perryfields Worcs 117 C8
Perryfoot Derbys 185 E10
Perrymead Bath 61 G9
Perrystone Hill Hereford 98 F2
Perrywood Kent 54 B4
Pershall Staffs 150 C6
Pershore Worcs 99 B8
Pert Angus 293 G8
Pertenhall Beds 121 D11
Perth Perth 286 E5
Perthcelyn Rhondda 77 F9
Perthy Shrops 149 C7
Perton Hereford 97 C11
Perton Staffs 133 D7
Pertwood Wilts 45 F11
Pested Kent 54 C4
Peter Tavy Devon 12 F6
Peterborough P'boro 138 D3
Peterburn Highld 307 L2
Peterchurch Hereford 96 D6
Peterculter Aberdeen 293 C10
Peterhead Aberds 303 E11
Peterlee Durham 234 C4
Peter's Finger Devon 12 D3
Peter's Green Herts 85 B10
Peters Marland Devon 25 E7
Petersburn N Lnrk 268 C5
Petersfield Hants 34 C2
Petersham London 67 E7
Peterston-super-Ely V Glam 58 D5
Peterstone Wentlooge Newport 59 C9
Peterstow Hereford 97 G11
Petertown Orkney 314 F3
Peterville Corn 4 E4
Petham Kent 54 C6
Petherwin Gate Corn 11 D11
Petrockstow Devon 25 F8
Petsoe End M Keynes 103 B7
Pett E Sus 38 E5
Pett Bottom Kent 54 E6
Pett Bottom Kent 55 C7
Pett Level E Sus 38 E5
Pettaugh Suff 126 F3
Petteridge Kent 53 E7
Pettinain S Lnrk 269 G9
Pettings Kent 68 G6
Pettistree Suff 126 G5
Petton Devon 27 C8
Petton Shrops 149 D8
Petts Wood London 68 F2
Petty Aberds 303 F7
Petty France S Glos 61 B9
Pettycur Fife 280 D5
Pettymuick Aberds 303 G9
Pettywell Norf 159 E11
Petworth W Sus 35 C7
Pevensey E Sus 23 E10
Pevensey Bay E Sus 23 E11
Peverell Plym 7 D9
Pewsey Wilts 63 G7
Pewsey Wharf Wilts 63 G7
Pewterspear Warr 183 E10
Phantassie E Loth 281 F11
Pharisee Green Essex 106 G2
Pheasants Bucks 65 B9
Pheasant's Hill Bucks 65 B9
Pheasey W Mid 133 D11
Pheonix Green Hants 49 B9
Phepson Worcs 117 F8
Philadelphia T & W 243 G8
Philham Devon 24 C3
Philiphaugh Borders 261 D10
Phillack Corn 2 B3
Philleigh Corn 3 B9
Phillip's Town Caerph 77 E10
Philpot End Essex 87 B10
Philpstoun W Loth 279 F10
Phocle Green Hereford 98 F2
Phoenix Green Hants 49 B9
Phoenix Row Durham 233 F9
Phorp Moray 301 D10
Pibsbury Som 28 B6
Pibwrlwyd Carms 74 B6
Pica Cumb 228 G6
Piccadilly S Yorks 187 B7
Piccadilly Warks 134 D4
Piccadilly Corner Norf 142 F5
Piccotts End Herts 85 D9
Pickburn S Yorks 198 F4
Picken End Worcs 98 C6
Pickering N Yorks 216 C5
Pickering Nook Durham 242 F5
Picket Hill Hants 31 F11
Picket Piece Hants 47 D11
Picket Post Hants 31 F11
Pickford W Mid 134 G5
Pickford Green W Mid 134 G5
Pickhill N Yorks 214 C6
Picklenash Glos 98 F4
Pickles Hill W Yorks 204 F6
Picklescott Shrops 131 D8
Pickletillem Fife 287 E8
Pickley Green Gtr Man 195 G7
Pickmere E Ches 183 F11
Pickney Som 27 B11
Pickstock Telford 150 E4
Pickup Bank Blkburn 195 C8
Pickwell Devon 40 E3
Pickwell Leics 154 G5
Pickwick Wilts 61 E11
Pickwood Scar W Yorks 196 C5
Pickworth Lincs 155 C10
Pickworth Rutland 155 G9
Picton Flint 181 E10
Picton N Yorks 225 D8
Picton W Ches 182 G6
Pict's Hill Som 28 B6
Piddinghoe E Sus 36 G6
Piddington Bucks 84 G4
Piddington Northants 120 G6
Piddington Oxon 83 B10
Piddlehinton Dorset 17 B10
Piddletrenthide Dorset 17 B10
Pidley Cambs 122 B6
Pidney Dorset 30 F2
Piece Corn 2 B5
Piercebridge Darl 224 B4
Piercing Hill Essex 86 F6
Pierowall Orkney 314 B4
Piff's Elm Glos 99 F8
Pig Oak Dorset 31 G8
Pigdon Northumb 252 F5
Pightley Som 43 F8
Pigstye Green Essex 87 D10
Pike End W Yorks 196 D4
Pike Hill Lancs 204 G3
Pike Law W Yorks 196 D4
Pikehall Derbys 169 D11
Pikeshill Hants 32 F3
Pikestye Hereford 97 B10
Pilford Dorset 31 G8
Pilgrims Hatch Essex 87 F9
Pilham Lincs 188 C5
Pilhough Derbys 170 C3
Pill N Som 60 D4
Pill Pembs 72 D6
Pillaton Corn 7 C7
Pillaton Staffs 151 G8
Pillerton Hersey Warks 100 B6
Pillerton Priors Warks 100 B5
Pilleth Powys 114 D5
Pilley Glos 81 B7
Pilley Hants 20 B2
Pilley S Yorks 197 G10
Pilley Bailey Hants 20 B2
Pillgwenlly Newport 59 B10
Pilling Lancs 202 D4
Pilling Lane Lancs 202 D3
Pillmouth Devon 25 C7
Pillowell Glos 79 D10
Pillows Green Glos 98 F5
Pillwell Dorset 30 D3
Pilmuir Borders 261 G11
Pilning S Glos 60 B5
Pilrig Edin 280 F5
Pilsbury Derbys 169 C10
Pilsdon Dorset 16 B4
Pilsgate P'boro 137 B11
Pilsley Derbys 170 C6
Pilsley Derbys 186 G2
Pilsley Green Derbys 170 C6
Pilson Green Norf 161 G7
Piltdown E Sus 36 C6
Pilton Devon 40 G5
Pilton Edin 280 F4
Pilton Northants 137 G10
Pilton Rutland 137 C8
Pilton Som 44 E5
Pilton Green Swansea 56 D2
Piltown Som 44 F5
Pimhole Gtr Man 195 E10
Pimlico Herts 85 D9
Pimlico Lancs 203 E10
Pimlico London 67 D9
Pimlico Northants 102 C2
Pimperne Dorset 29 F9
Pimperne Dorset 30 F6
Pin Green Herts 104 F4
Pin Mill Suff 108 D4
Pinchbeck Lincs 156 D4
Pinchbeck Bars Lincs 156 D3
Pinchbeck West Lincs 156 E4
Pincheon Green S Yorks 199 D7
Pinckney Green Wilts 61 G10
Pincock Lancs 194 D5
Pineham Kent 55 D10
Pineham M Keynes 103 C7
Pinehurst Swindon 63 B7
Pinfarthings Glos 80 E5
Pinfold Lancs 193 E11
Pinfold Hill S Yorks 197 G9
Pinfoldpond C Beds 103 E8
Pinford End Suff 124 F6
Pinged Carms 74 E6
Pingewood W Berks 65 F7
Pinhoe Devon 14 C5
Pink Green Worcs 117 D11
Pinkett's Booth W Mid 134 G5
Pinkie Braes E Loth 281 G7
Pinkney Wilts 61 B11
Pinkneys Green Windsor 65 C11
Pinksmoor Som 27 D10
Pinley W Mid 119 B7
Pinley Green Warks 118 D4
Pinminnoch Dumfries 236 D2
Pinminnoch S Ayrs 244 E5
Pinmore S Ayrs 244 E6
Pinmore Mains S Ayrs 244 E6
Pinnacles Essex 86 D6
Pinner London 66 B6
Pinner Green London 85 G10
Pinnerwood Park London 85 G10
Pin's Green Worcs 98 B6
Pinsley Green E Ches 167 F9
Pinstones Shrops 131 F9
Pinvin Worcs 99 B9
Pinwall Leics 134 C6
Pinwherry S Ayrs 244 F5
Pinxton Derbys 171 E7
Pipe and Lyde Hereford 97 C10
Pipe Aston Hereford 115 C9
Pipe Gate Shrops 168 G2
Pipe Ridware Staffs 151 F11
Pipehill Staffs 133 B11
Pipehouse Bath 45 B9
Piperhall Argyll 266 D2
Piperhill Highld 301 D8
Piper's Ash W Ches 166 B6
Piper's End Worcs 98 E6
Piper's Hill Worcs 117 D9
Piper's Pool Corn 11 E11
Pipewell Northants 136 F6
Pippacott Devon 40 F4
Pippin Street Lancs 194 C5
Pipps Hill Essex 69 B7
Pipsden Kent 53 G9
Pipton Powys 96 D3
Pirbright Sur 50 B2
Pirbright Camp Sur 50 B2
Pirnmill N Ayrs 255 C9
Pirton Herts 104 E2
Pirton Worcs 99 B7
Pisgah Ceredig 112 B3
Pisgah Stirl 285 G11
Pishill Oxon 65 B8
Pishill Bank Oxon 84 G2
Pismire Hill S Yorks 186 C5
Pistyll Gwyn 162 G4
Pit Mon 78 D5
Pitagowan Perth 291 G10
Pitblae Aberds 303 C9
Pitcairngreen Perth 286 E4
Pitcalnie Highld 301 B8
Pitcaple Aberds 303 G7
Pitch Green Bucks 84 E3
Pitch Place Sur 49 F11
Pitch Place Sur 50 C3
Pitchcombe Glos 80 D5
Pitchcott Bucks 102 G5
Pitcher's Green Suff 125 F8
Pitchford Shrops 131 C10
Pitcombe Som 45 G7
Pitcorthie Fife 280 D2
Pitcorthie Fife 287 G9
Pitcot Som 45 D7
Pitcot V Glam 57 G11
Pitcox E Loth 282 F2
Pitcur Perth 286 D6
Pitfancy Aberds 302 E5
Pitfichie Aberds 293 B8
Pitforthie Aberds 293 F10
Pitgair Aberds 303 D7
Pitgrudy Highld 309 K7
Pithmaduthy Highld 301 B7
Pitkennedy Angus 287 B9
Pitkevy Fife 286 G6
Pitkierie Fife 287 G9
Pitlessie Fife 287 G7
Pitlochry Perth 286 B3
Pitmachie Aberds 302 G6
Pitmain Highld 291 C9
Pitmedden Aberds 303 G8
Pitminster Som 28 D2
Pitmuies Angus 287 C9
Pitmunie Aberds 293 B8
Pitney Som 29 B7
Pitrocknie Perth 286 C6
Pitscottie Fife 287 F8
Pitsea Essex 69 B8
Pitses Gtr Man 196 G2
Pitsford Northants 120 D5
Pitsford Hill Som 42 G6
Pitsmoor S Yorks 186 D5
Pitstone Bucks 84 B6
Pitstone Green Bucks 84 B6
Pitstone Hill Bucks 85 C7
Pitt Hants 33 B7
Pitt Court Glos 80 F3
Pittachar Perth 286 E2
Pittendreich Moray 301 C11
Pittentrail Highld 309 J7
Pittenweem Fife 287 G9
Pitteuchar Fife 280 B5
Pittington Durham 234 C2
Pittodrie Aberds 302 G6
Pitton Swansea 56 D2
Pitton Wilts 47 G8
Pitts Hill Stoke 168 E5
Pittswood Kent 52 D6
Pittulie Aberds 303 C9
Pittville Glos 99 G9
Pity Me Durham 233 B11
Pityme Corn 10 F5
Pityoulish Highld 291 B11
Pixey Green Suff 126 B4
Pixham Sur 51 C7
Pixham Worcs 98 B6
Pixley Hereford 98 D3
Pixley Shrops 150 D3
Pizien Well Kent 53 C7
Place Newton N Yorks 217 E7
Plaidy Aberds 303 D7
Plaidy Corn 6 E5
Plain-an-Gwarry Corn 4 G3
Plain Dealings Pembs 73 B9
Plain Spot Notts 171 E7
Plain Street Corn 10 F5
Plains N Lnrk 268 B5
Plainsfield Som 43 F7
Plaish Shrops 131 D10
Plaistow London 68 C2
Plaistow London 68 E2
Plaistow W Sus 50 G4
Plaistow Green Essex 106 F6
Plaitford Wilts 32 D3
Plaitford Green Hants 32 C3
Plank Lane Gtr Man 194 G6
Plans Dumfries 238 D3
Plantation Bridge Cumb 221 F9
Plantationfoot Dumfries 248 E4
Plardiwick Staffs 150 E6
Plas-canol Gwyn 145 F11
Plas Coch Wrex 166 E4
Plas Dinam Powys 129 F10
Plas Gogerddan Ceredig 128 G2
Plas Llwyngwern Powys 128 C5
Plas Meredydd Powys 130 D3
Plas Nantyr Wrex 148 B3
Plas-yn-Cefn Denb 181 G8
Plasau Shrops 149 E7
Plashet London 68 C2
Plashett Carms 74 D3
Plasiolyn Powys 129 C11
Plasnewydd Powys 129 D9
Plaster's Green Bath 60 G4
Plastow Green Hants 64 G4
Platt Kent 52 B6
Platt Bridge Gtr Man 194 G6
Platt Lane Shrops 149 B10
Platts Common S Yorks 197 G11
Platt's Heath Kent 53 C11
Plawsworth Durham 233 B11
Plaxtol Kent 52 C6
Play Hatch Oxon 65 D8
Playden E Sus 38 C6
Playford Suff 108 B4
Playing Place Corn 4 G6
Playley Green Glos 98 E5
Plealey Shrops 131 B8
Pleamore Cross Som 27 D10
Plean Stirl 278 D6
Pleasant Valley Pembs 73 D10
Pleasington Blkburn 194 B6
Pleasley Derbys 171 C8
Pleasleyhill Notts 171 C8
Pleck Dorset 30 D3
Pleck Dorset 30 E2
Pleck W Mid 133 D9
Pleck or Little Ansty Dorset 30 G3
Pleckgate Blkburn 203 G9
Pledgdon Green Essex 105 F11
Pledwick W Yorks 197 D10
Plemstall W Ches 183 G7
Plenmeller Northumb 240 E6
Plâs Berwyn Denb 165 G11
Pleshey Essex 87 C11
Plockton Highld 295 B10
Plocrapol W Isles 305 J3
Plot Gate Som 44 G4
Plot Street Som 44 F5
Plough Hill Warks 134 E6
Ploughfield Hereford 97 C7
Plowden Shrops 131 F7
Ploxgreen Shrops 131 C7
Pluckley Kent 54 D2
Pluckley Thorne Kent 54 E2
Plucks Gutter Kent 71 G9
Plumbland Cumb 229 D9
Plumbley S Yorks 186 E6
Plumford Kent 54 B4
Plumley E Ches 184 F2
Plump Hill Glos 79 B11
Plumpton Cumb 230 D5
Plumpton E Sus 36 E5
Plumpton Northants 101 B11
Plumpton End Northants 102 B4
Plumpton Foot Cumb 230 D5
Plumpton Green E Sus 36 D5
Plumpton Head Cumb 230 E6
Plumstead London 68 D3
Plumstead Norf 160 C2
Plumstead Common London 68 D3
Plumstead Green Norf 160 C2
Plumtree Notts 154 C2
Plumtree Green Kent 53 D10
Plumtree Park Notts 154 C2
Plungar Leics 154 C5
Plush Dorset 30 G2
Plusha Corn 11 E11
Plushabridge Corn 12 G2
Plusterwine Glos 79 F9
Plwmp Ceredig 111 G7
Plymouth Plym 7 E9

Plympton Plym 7 D10
Plymstock Plym 7 E10
Plymtree Devon 27 G9
Pobgreen Gtr Man 196 F4
Pochin Houses Caerph 77 E11
Pocket Nook Gtr Man 183 B10
Pockley N Yorks 216 B2
Pocklington E Yorks 208 D2
Pockthorpe Norf 141 D8
Pockthorpe Norf 158 D6
Pockthorpe Norf 159 E10
Pockthorpe Norf 159 F11
Pode Hole Lincs 156 E4
Podimore Som 29 C8
Podington Beds 121 E8
Podmoor Worcs 117 C7
Podmore Norf 159 G9
Podmore Staffs 150 B5
Podsmead Glos 80 B4
Poffley End Oxon 82 C5
Pogmoor S Yorks 197 F10
Point Corn 3 B8
Point Clear Essex 89 C9
Pointon Lincs 156 C2
Pokesdown Bmouth 19 C8
Pol a Charra W Isles 297 K3
Polbae Dumfries 236 B4
Polbain Highld 307 H4
Polbathic Corn 7 D7
Polbeth W Loth 269 C10
Polborder Corn 7 C7
Polbrock Corn 5 B10
Polchar Highld 291 C10
Pole Elm Worcs 98 B6
Pole Moor W Yorks 196 D5
Polebrook Northants 137 F11
Polegate E Sus 23 D9
Poles Highld 309 K7
Pole's Hole Wilts 45 C10
Polesden Lacey Sur 50 C6
Poleshill Som 27 C9
Polesworth Warks 134 C5
Polgear Corn 2 B5
Polgigga Corn 1 E3
Polglass Highld 307 J5
Polgooth Corn 5 E9
Poling W Sus 35 G8
Poling Corner W Sus 35 F8
Polkerris Corn 5 E11
Poll Hill Mers 182 E3
Polla Highld 308 D3
Polladras Corn 2 C4
Pollard Street Norf 160 C6
Pollhill Kent 53 C11
Pollie Highld 309 H7
Pollington E Yorks 198 D6
Polliwilline Argyll 255 G8
Polloch Highld 289 C9
Pollok Glasgow 267 C10
Pollokshields Glasgow 267 C11
Polmadie Glasgow 267 C11
Polmarth Corn 2 B6
Polmassick Corn 5 F9
Polmear Corn 5 E11
Polmont Falk 279 F8
Polmorla Corn 10 G5
Polnessan E Ayrs 257 G10
Polnish Highld 295 G9
Polopit Northants 121 B10
Polpenwith Corn 2 D6
Polpeor Corn 2 B2
Polperro Corn 6 E4
Polruan Corn 6 E2
Polsham Som 44 E4
Polsloe Devon 14 C4
Polstead Suff 107 D9
Polstead Heath Suff 107 C9
Poltalloch Argyll 275 D9
Poltesco Corn 2 F6
Poltimore Devon 14 B5
Polton Midloth 270 C5
Polwarth Borders 272 E4
Polwheveral Corn 2 D6
Polyphant Corn 11 E11
Polzeath Corn 10 F4
Pomeroy Derbys 169 B10
Pomphlett Plym 7 E10
Ponciau Wrex 166 F3
Pond Close Som 27 B10
Pond Park Bucks 85 E7
Pond Street Essex 105 D9
Ponde Powys 96 D2
Ponders End London 86 F5
Pondersbridge Cambs 138 E5
Pondtail Hants 49 C10
Pondwell I o W 21 C8
Poniou Corn 1 B4
Ponjeravah Corn 2 D6
Ponsanooth Corn 3 B7
Ponsford Devon 27 F8
Ponsonby Cumb 219 D11
Ponsongath Corn 3 F7
Ponsworthy Devon 13 G10
Pont Corn 6 E2
Pont Aber Carms 94 G4
Pont Aber-Geirw Gwyn 146 D5
Pont-ar-gothi Carms 93 G10
Pont ar Hydfer Powys 95 F7
Pont-ar-llechau Carms 94 G4
Pont Cyfyng Conwy 164 D2
Pont Cysyllte Wrex 166 G3
Pont Dolydd Prysor Gwyn 146 B4
Pont-faen Powys 95 E9
Pont-Faen Shrops 148 B5
Pont Fronwydd Gwyn 146 E6
Pont-gareg Pembs 92 C2
Pont-Henri Carms 75 D7
Pont Hwfa Anglesey 178 E2
Pont iets = Pontyates Carms 75 D7
Pont-Llogel Powys 147 F10
Pont-newydd Carms 74 D6
Pont-newydd Flint 165 B11
Pont Pen-y-benglog Gwyn 163 C10
Pont Rhyd-goch Conwy 163 C11
Pont-Rhyd-sarn Gwyn 147 D7
Pont Rhyd-y-berry Powys 95 D9
Pont Rhyd-y-cyff Bridgend 57 D11
Pont-rhyd-y-groes Ceredig 112 C4
Pont Rhydgaled Powys 128 G6
Pont-Rhythallt Gwyn 163 C8
Pont-rug Gwyn 163 C8
Pont Senni = Sennybridge Powys 95 F8
Pont Siôn Norton Rhondda 77 G9
Pont-siôn Ceredig 93 B8
Pont-Walby Neath 76 D5
Pont-y-blew Shrops 148 B6
Pont-y-gwaith Rhondda 77 G8
Pont-y-pant Conwy 164 E3
Pont y Pennant Gwyn 147 E8
Pont-y-rhyl Bridgend 58 B2
Pont-y-wal Powys 96 D2
Pont yr Afon-Gam Gwyn 164 G2
Pont-yr-hafod Pembs 91 F8
Pont-Ystrad Denb 165 C9
Pontamman Carms 75 C10
Pontantwn Carms 74 C6
Pontardawe Neath 76 E2
Pontarddulais Swansea 75 E9
Pontarfynach = Devils Bridge Ceredig 112 B4
Pontarsais Carms 93 F8
Pontblyddyn Flint 166 C3
Pontbren Araeth Carms 94 G3
Pontbren Llwyd Rhondda 76 D6
Pontcanna Cardiff 59 D7
Pontdolgoch Powys 129 E10
Pontefract W Yorks 198 C3
Ponteland Northumb 242 C5
Ponterwyd Ceredig 128 G4
Pontesbury Shrops 131 B7
Pontesbury Hill Shrops 131 B7
Pontesford Shrops 131 B8
Pontfadog Wrex 148 B4
Pontfaen Pembs 91 E10
Pontgarreg Ceredig 110 G6
Ponthen Shrops 148 F6
Ponthir Torf 78 G4
Ponthirwaun Ceredig 92 B5
Pontiago Pembs 91 D8
Pontithel Powys 96 D3
Pontllanfraith Caerph 77 F11
Pontlliw Swansea 75 E10
Pontllyfni Gwyn 162 E6
Pontlottyn Caerph 77 D10
Pontneddfechan Powys 76 D6
Pontnewydd Torf 78 F3
Pontnewynydd Torf 78 E3
Pontrhydfendigaid Ceredig 112 D4
Pontrhydyfen Neath 57 C9
Pontrhydyrun Torf 78 F3
Pontrilas Hereford 97 F7
Pontrobert Powys 148 G2
Ponts Green E Sus 23 B11
Pontshill Hereford 98 G2
Pontsticill M Tydf 77 C9
Pontwgan Conwy 180 G3
Pontyates = Pont-iets Carms 75 D7
Pontyberem Carms 75 C8
Pontyclun Rhondda 58 C4
Pontycymer Bridgend 76 G6
Pontyglasier Pembs 92 D2
Pontymister Caerph 78 G2
Pontymoel Torf 78 E3
Pontypool Torf 78 E3
Pontypridd Rhondda 58 B5
Pontywaun Caerph 78 G2
Pooksgreen Hants 32 E5
Pool Corn 4 G3
Pool W Yorks 205 D10
Pool Head Hereford 115 G11
Pool Hey Lancs 193 D11
Pool o' Muckhart Clack 286 G4
Pool Quay Powys 148 G5
Poolbrook Worcs 98 C5
Poole N Yorks 198 B3
Poole Poole 18 C6
Poole Som 27 C10
Poole Keynes Glos 81 F8
Poolend Staffs 169 D7
Poolestown Dorset 30 D2
Poolewe Highld 307 L3
Pooley Bridge Cumb 230 G5
Pooley Street Norf 141 G11
Poolfold Staffs 168 D5
Poolhead Shrops 149 C9
Poolhill Glos 98 F4
Poolmill Hereford 97 G11
Poolsbrook Derbys 186 G6
Poolside Moray 302 E4
Poolstock Gtr Man 194 G5
Pooltown Som 42 F3
Pootings Kent 52 D3
Pope Hill Pembs 72 C6
Pope's Hill Glos 79 C11
Popeswood Brack 65 F10
Popham Devon 41 G8
Popham Hants 48 E5
Poplar London 67 C11
Poplar Grove Lincs 190 B6
Poplars Herts 104 G5
Popley Hants 48 C6
Porchester Nottingham 171 G9
Porchfield I o W 20 C4
Porin Highld 300 D3
Poringland Norf 142 C5
Porkellis Corn 2 C5
Porlock Som 41 D11
Porlock Weir Som 41 D11
Porlockford Som 41 D11
Port Allen Perth 286 E6
Port Ann Argyll 275 E10
Port Appin Argyll 289 E11
Port Arthur Shetland 313 K5
Port Askaig Argyll 274 G5
Port Bannatyne Argyll 275 G11
Port Brae Fife 280 C5
Port Bridge Devon 9 D7
Port Carlisle Cumb 238 E6
Port Charlotte Argyll 254 B3
Port Clarence Stockton 234 G5
Port Dinorwic = Y Felinheli Gwyn 163 B8
Port Driseach Argyll 275 F10
Port Dundas Glasgow 267 B11
Port e Vullen I o M 192 C5
Port Edgar Edin 280 F2
Port Ellen Argyll 254 C4
Port Elphinstone Aberds 293 B9
Port Erin I o M 192 F2
Port Erroll Aberds 303 F10
Port-Eynon Swansea 56 D3
Port Gaverne Corn 10 E6
Port Glasgow Invclyd 276 G6
Port Henderson Highld 299 B7
Port Hill Oxon 65 B7
Port Isaac Corn 10 E5
Port Lamont Argyll 275 F11
Port Lion Pembs 73 D7
Port Logan Dumfries 236 E2
Port Mead Swansea 56 B6
Port Mholair W Isles 304 E7
Port Mor Highld 288 B6
Port Mulgrave N Yorks 226 B5
Port Nan Giùran W Isles 304 E7
Port nan Long W Isles 296 D4
Port Nis W Isles 304 B7
Port of Menteith Stirl 285 G9
Port Quin Corn 10 E5
Port Ramsay Argyll 289 E10
Port St Mary I o M 192 F3
Port Solent Ptsmth 33 F10
Port Sunlight Mers 182 E4
Port Sutton Bridge Lincs 157 E9
Port Talbot Neath 57 D9
Port Tennant Swansea 57 C7
Port Wemyss Argyll 254 B2
Port William Dumfries 236 E5
Portachoillan Argyll 255 B8
Portash Wilts 46 G3
Portavadie Argyll 275 G10
Portbury N Som 60 D4
Portchester Hants 33 F10
Portclair Highld 290 B6
Porteath Corn 10 F5
Portencalzie Dumfries 236 B2
Portencross N Ayrs 266 F3
Porterfield Renfs 267 B9
Porter's End Herts 85 B11
Portesham Dorset 17 D8
Portessie Moray 302 C4
Portfield Argyll 289 G9
Portfield Som 28 B6
Portfield W Sus 22 B5
Portfield Gate Pembs 72 B6
Portgate Devon 12 D4
Portgordon Moray 302 C3
Portgower Highld 311 H4
Porth Corn 4 C6
Porth Rhondda 77 G8
Porth Colmon Gwyn 144 C3
Porth Kea Corn 4 G6
Porth Navas Corn 3 D7
Porth Tywyn = Burry Port Carms 74 E6
Porth-y-felin Anglesey 178 E2
Porth-y-waen Shrops 148 E5
Porthallow Corn 3 E7
Porthallow Corn 6 E4
Porthcawl Bridgend 57 F10
Porthcothan Corn 10 G3
Porthcurno Corn 1 E3
Portheiddy Pembs 90 E6
Porthgain Pembs 90 E6
Porthgwarra Corn 1 E3
Porthhallow Corn 3 E7
Porthill Shrops 149 G9
Porthill Staffs 168 F5
Porthilly Corn 10 F4
Porthkerry V Glam 58 F5
Porthleven Corn 2 D4
Porthllechog = Bull Bay Anglesey 178 C6
Porthloo Scilly 1 G4
Porthmadog Gwyn 145 B11
Porthmeor Corn 1 B4
Porthoustock Corn 3 E8
Porthpean Corn 5 E10
Porthtowan Corn 4 F3
Porthyrhyd Carms 75 B8
Porthyrhyd Carms 94 D4
Portico Mers 183 C7
Portincaple Argyll 276 C4
Portington Devon 12 F4
Portington E Yorks 207 G11
Portinnisherrich Argyll 275 B10
Portinscale Cumb 229 G11
Portishead N Som 60 D3
Portkil Argyll 276 E5
Portknockie Moray 302 C4
Portland Som 44 F3
Portlethen Aberds 293 D11
Portlethen Village Aberds 293 D11
Portlevorchy Highld 306 D7
Portling Dumfries 237 D10
Portloe Corn 3 B10
Portlooe Corn 6 E4
Portmahomack Highld 311 L3
Portmeirion Gwyn 145 B11
Portmellon Corn 5 G10
Portmore Hants 20 B2
Portnacroish Argyll 289 E11
Portnahaven Argyll 254 B2
Portnalong Highld 294 B5
Portnaluchaig Highld 295 G8
Portnancon Highld 308 C4
Portnellan Stirl 285 E8
Portnellan Stirl 285 F8
Portobello Edin 280 G6
Portobello T & W 243 F7
Portobello W Mid 133 D9
Portobello W Yorks 197 D10
Porton Wilts 47 F7
Portpatrick Dumfries 236 D2
Portrack Stockton 225 B9
Portreath Corn 4 F3
Portree Highld 298 E4
Portscatho Corn 3 B9
Portsea Ptsmth 33 G10
Portsea Island Ptsmth 21 B9
Portskerra Highld 310 C2
Portskewett Mon 60 B4
Portslade Brighton 36 F3
Portslade-by-Sea Brighton 36 G3
Portslade Village Brighton 36 F3
Portsmouth Ptsmth 21 B9
Portsmouth W Yorks 196 B2
Portsonachan Argyll 284 E4
Portsoy Aberds 302 C5
Portswood Soton 32 E6
Porttanachy Moray 302 C3
Portuairk Highld 288 C6
Portvasgo Highld 308 C5
Portway Dorset 18 D2
Portway Glos 98 E5
Portway Hereford 97 B9
Portway Hereford 97 D9
Portway Som 28 B6
Portway Som 44 F3
Portway W Mid 133 F9
Portway Worcs 117 C11
Portwood Gtr Man 184 C6
Portwrinkle Corn 7 E7
Posenhall Shrops 132 C3
Poslingford Suff 106 B5
Posso Borders 260 C6
Post Green Dorset 18 C5
Post Mawr = Synod Inn Ceredig 111 G8
Postbridge Devon 13 F9
Postcombe Oxon 84 F2
Postling Kent 54 F6
Postlip Glos 99 F10
Postwick Norf 142 B5
Pot Common Sur 50 E2
Potarch Aberds 293 D8
Potash Suff 108 D2
Potbridge Hants 49 C8
Potholm Dumfries 249 F9
Potmaily Highld 300 F4
Potman's Heath Kent 38 B5
Potsgrove C Beds 103 F9
Pott Row Norf 158 E4
Pott Shrigley E Ches 184 F6
Potten End Herts 85 D8
Potten Street Kent 71 F9
Potter Brompton N Yorks 217 D9
Potter Heigham Norf 161 F8
Potter Hill Leics 154 E4
Potter Hill S Yorks 186 B4
Potter Somersal Derbys 152 B2
Potter Street Essex 87 D7
Pottergate Street Norf 142 E3
Potterhanworth Lincs 173 B9
Potterhanworth Booths Lincs 173 B9
Potterne Wilts 46 B3
Potterne Wick Wilts 46 B4
Potternewton W Yorks 206 F2
Potters Bar Herts 86 E3
Potters Brook Lancs 202 C5
Potters Corner Kent 54 E3
Potter's Cross Staffs 132 G6
Potters Crouch Herts 85 D10
Potter's Forstal Kent 53 D11
Potter's Green E Sus 37 C8
Potter's Green W Mid 135 G7
Potters Hill N Som 60 F4
Potters Marston Leics 135 D9
Pottersheath Herts 86 B2
Potterspury Northants 102 C5
Potterton Aberds 293 B11
Potterton W Yorks 206 F4
Pottery Field W Yorks 206 G2
Potthorpe Norf 159 E8
Pottington Devon 40 G5
Potto N Yorks 225 E9
Potton C Beds 104 B4
Pouchen End Herts 85 D8
Poughill Corn 24 F2
Poughill Devon 26 F5
Poulner Hants 31 F11
Poulshot Wilts 46 B3
Poulton Glos 81 E10
Poulton Mers 182 C4
Poulton Mers 182 E4
Poulton W Ches 166 D5
Poulton-le-Fylde Lancs 202 F2
Pound Som 28 D6
Pound Bank Worcs 98 B5
Pound Bank Worcs 116 C4
Pound Green E Sus 37 C8
Pound Green Hants 48 B5
Pound Green I o W 20 D2
Pound Green Suff 124 G4
Pound Green Worcs 116 B5
Pound Hill W Sus 51 F9
Pound Street Hants 64 G3
Poundbury Dorset 17 C9
Poundffald Swansea 56 C5
Poundfield E Sus 52 G4
Poundford E Sus 37 C9
Poundgate E Sus 37 B7
Poundgreen Wokingham 65 F7
Poundland S Ayrs 244 F5
Poundon Bucks 102 F2
Poundsbridge Kent 52 E4
Poundsgate Devon 13 G10
Poundstock Corn 11 B10
Pounsley E Sus 37 C8
Poverest London 68 F3
Povey Cross Sur 51 E9
Pow Green Hereford 98 C4
Powburn Northumb 264 F3
Powder Mills Kent 52 D5
Powderham Devon 14 E5
Powers Hall End Essex 88 B4
Powerstock Dorset 16 B6
Powfoot Dumfries 238 D4
Powhill Cumb 238 F6
Powick Worcs 116 G6
Powler's Piece Devon 24 D5
Powmill Perth 279 B10
Pownall Park E Ches 184 E4
Powntley Copse Hants 49 E8
Poxwell Dorset 17 E10
Poyle Slough 66 D4
Poynings W Sus 36 E3
Poyntington Dorset 29 D11
Poynton E Ches 184 E6
Poynton Telford 149 F11
Poynton Green Telford 149 F11
Poyntzfield Highld 301 C7
Poyston Pembs 73 B7
Poyston Cross Pembs 73 B7
Poystreet Green Suff 125 F9
Praa Sands Corn 2 D3
Pratling Street Kent 53 B8
Pratt's Bottom London 68 G3
Praze Corn 2 B3
Praze-an-Beeble Corn 2 B4
Predannack Wollas Corn 2 F5
Prees Shrops 149 C11
Prees Green Shrops 149 C11
Prees Heath Shrops 149 B11
Prees Higher Heath Shrops 149 B11
Prees Lower Heath Shrops 149 C11
Prees Wood Shrops 149 C11
Preesall Lancs 202 D3
Preesall Park Lancs 202 D3
Preesgweene Shrops 148 B5
Preeshenlle Shrops 148 C6
Pren-gwyn Ceredig 93 C8
Prenbrigog Flint 166 C3
Prendergast Pembs 73 B7
Prendergast Pembs 90 G6
Prenderguest Borders 273 D8
Prendwick Northumb 264 G2
Prenteg Gwyn 163 G9
Prenton Mers 182 D4
Prescot Mers 183 C7
Prescott Devon 27 E9
Prescott Glos 99 F9
Prescott Shrops 132 G3
Prescott Shrops 149 E8
Presdales Herts 86 C5
Preshome Moray 302 C4
Press Derbys 170 B5
Pressen Northumb 263 B8
Prestatyn Denb 181 E9
Prestbury E Ches 184 F6
Prestbury Glos 99 G9
Presteigne Powys 114 E6
Presthope Shrops 131 D11
Prestleigh Som 44 E6
Prestolee Gtr Man 195 F9
Preston Borders 272 D5
Preston Brighton 36 F4
Preston Devon 14 G3
Preston Dorset 17 E10
Preston E Loth 281 F11
Preston E Loth 281 G7
Preston E Yorks 209 G9
Preston Glos 81 E8
Preston Glos 98 E3
Preston Herts 104 G3
Preston Kent 70 G4
Preston Kent 71 G8
Preston Lancs 194 B4
Preston London 67 B7
Preston Northumb 264 D5
Preston Rutland 137 C7
Preston Shrops 149 G10
Preston T & W 243 D8
Preston Torbay 9 C7
Preston Wilts 62 D4
Preston Wilts 63 E9
Preston Bagot Warks 118 D3
Preston Bissett Bucks 102 F3
Preston Bowyer Som 27 B10
Preston Brockhurst Shrops 149 E10
Preston Brook Halton 183 E9
Preston Candover Hants 48 E6
Preston Capes Northants 119 G11
Preston Crowmarsh Oxon 83 G10
Preston Deanery Northants 120 F5
Preston Fields Warks 118 D3
Preston Grange T & W 243 C8
Preston Green Warks 118 D3
Preston Gubbals Shrops 149 F9
Preston-le-Skerne Durham 234 G2
Preston Marsh Hereford 97 B11
Preston Montford Shrops 149 G8
Preston on Stour Warks 118 G4
Preston-on-Tees Stockton 225 B8
Preston on the Hill Halton 183 E9
Preston on Wye Hereford 97 C7
Preston Pastures Worcs 100 B3
Preston Plucknett Som 29 D8
Preston St Mary Suff 125 G8
Preston-under-Scar N Yorks 223 G11
Preston upon the Weald Moors Telford 150 F3
Preston Wynne Hereford 97 B11
Prestonfield Edin 280 G5
Prestonmill Dumfries 237 D11
Prestonpans E Loth 281 G7
Prestwich Gtr Man 195 G10
Prestwick Northumb 242 C5
Prestwick S Ayrs 257 D9
Prestwold Leics 153 E11
Prestwood Bucks 84 E5
Prestwood Staffs 133 F7
Prestwood Staffs 169 G10
Prey Heath Sur 50 B3
Price Town Bridgend 76 G6
Prickwillow Cambs 139 G11
Priddy Som 44 C4
Pride Park Derbys 153 B7
Priest Down Bath 60 G6
Priest Hutton Lancs 211 E10
Priest Weston Shrops 130 D5
Priestacott Devon 24 F6
Priestcliffe Derbys 185 G10
Priestcliffe Ditch Derbys 185 G10
Priestfield W Mid 133 D8
Priestfield Worcs 98 C6
Priesthaugh Borders 249 C11
Priesthill Glasgow 267 C10
Priesthorpe W Yorks 205 F10
Priestley Green W Yorks 196 B6
Prieston Borders 262 D2
Priestside Dumfries 238 D4
Priestthorpe W Yorks 205 F8
Priestwood Brack 65 F11
Priestwood Kent 69 G7
Priestwood Green Kent 69 G7
Primethorpe Leics 135 E10
Primrose T & W 243 E8
Primrose Corner Norf 160 G6
Primrose Green Norf 159 F11
Primrose Hill Bath 61 F8
Primrose Hill Lancs 193 F11
Primrose Hill London 67 C9
Primrose Hill W Mid 133 F8
Primrose Valley N Yorks 218 D2
Primrosehill Herts 85 E9
Primsland Worcs 117 E8
Prince Hill E Ches 168 G2
Prince Royd W Yorks 196 D6
Princes End W Mid 133 E9
Princes Gate Pembs 73 C10
Prince's Marsh Hants 34 B3
Princes Park Mers 182 D5
Princes Risborough Bucks 84 E4
Princethorpe Warks 119 C8
Princetown Caerph 77 C10
Princetown Devon 13 G7
Prinsted W Sus 22 B3
Printstile Kent 52 E5
Prion Denb 165 C9
Prior Muir Fife 287 F9
Prior Park Northumb 273 E9
Prior Rigg Cumb 239 D11
Priors Frome Hereford 97 D11
Priors Halton Shrops 115 B9
Priors Hardwick Warks 119 F9
Priors Marston Warks 119 F9
Prior's Norton Glos 99 G7
Priors Park Glos 99 E7
Priorslee Telford 150 G4
Priorswood Som 28 B2
Priory Pembs 72 D6
Priory Green Suff 107 C8
Priory Heath Suff 108 C3
Priory Wood Hereford 96 B5
Prisk V Glam 58 D4
Pristacott Devon 25 B8
Priston Bath 61 G7
Pristow Green Norf 142 F2
Prittlewell Sthend 69 B11
Privett Hants 21 B7
Privett Hants 33 B11
Prixford Devon 40 F4
Probus Corn 5 F7
Proncy Highld 309 K7
Prospect Cumb 229 C8
Prospect Village Staffs 151 G10
Prospidnick Corn 2 C4
Provanmill Glasgow 268 B2
Prowse Devon 26 F4
Prudhoe Northumb 242 E3
Prussia Cove Corn 2 D3
Ptarmigan Lodge Stirl 285 G7
Pubil Perth 285 C8
Publow Bath 60 G6
Puckeridge Herts 105 G7
Puckington Som 28 D5
Pucklechurch S Glos 61 D7
Pucknall Hants 32 B5
Puckrup Glos 99 D7
Puckshole Glos 80 D4
Puddaven Devon 8 C5
Pudding Pie Nook Lancs 202 F6
Puddinglake W Ches 168 B2
Puddington Devon 26 E4
Puddington W Ches 182 G4
Puddle Corn 5 D11
Puddlebridge Som 28 E4
Puddledock Kent 52 C4
Puddledock Kent 68 E4
Puddledock Norf 141 E11
Puddletown Dorset 17 C11
Pudleigh Som 28 E4
Pudleston Hereford 115 F11
Pudsey W Yorks 196 B2
Pudsey W Yorks 205 G10
Pulborough W Sus 35 D8
Pulcree Dumfries 237 D7
Pule Hill W Yorks 196 B5
Puleston Telford 150 E4
Pulford W Ches 166 D5
Pulham Dorset 30 F2
Pulham Market Norf 142 F3
Pulham St Mary Norf 142 F4
Pullens Green S Glos 79 G10
Pulley Shrops 131 B9
Pullington Kent 53 G10
Pulloxhill C Beds 103 E11
Pulverbatch Shrops 131 C8
Pumpherston W Loth 269 B11
Pumsaint Carms 94 C3
Puncheston = Cas -Mael Pembs 91 F10
Puncknowle Dorset 16 D6
Punnett's Town E Sus 37 C10
Purbrook Hants 33 F11
Purewell Dorset 19 C9
Purfleet Thurrock 68 D5
Puriton Som 43 E10
Purleigh Essex 88 E4
Purley London 67 G10
Purley on Thames W Berks 65 D7
Purlogue Shrops 114 B5
Purlpit Wilts 61 F11
Purls Bridge Cambs 139 F9
Purn N Som 43 B10
Purse Caundle Dorset 29 D11
Purslow Shrops 131 G7
Purston Jaglin W Yorks 198 D2
Purtington Som 28 F5
Purton Glos 79 E11
Purton W Berks 64 D3
Purton Wilts 62 B5
Purton Common Wilts 62 B5
Purton Stoke Wilts 81 G9
Purwell Herts 104 F4
Pury End Northants 102 B4
Pusey Oxon 82 F5
Putley Hereford 98 D2
Putley Common Hereford 98 D2
Putley Green Hereford 98 D3
Putloe Glos 80 D3
Putney London 67 D8
Putney Heath London 67 E8
Putney Vale London 67 E8
Putnoe Beds 121 G11
Putsborough Devon 40 E3
Putson Hereford 97 D10
Puttenham Herts 84 C5
Puttenham Sur 50 D2
Puttock End Essex 106 C6
Puttock's End Essex 87 B9
Putton Dorset 17 E9
Puxey Dorset 30 E3
Puxley Northants 102 C5
Puxton N Som 60 G2
Pwll Carms 75 E7
Pwll Powys 130 C3
Pwll-clai Flint 181 G11
Pwll-glas Denb 165 E10
Pwll-Mawr Cardiff 59 D8
Pwll-melyn Flint 181 G11
Pwll-trap Carms 74 B3
Pwll-y-glaw Neath 57 C9
Pwllcrochan Pembs 72 E6
Pwllgloyw Powys 95 E10
Pwllheli Gwyn 145 B7
Pwllmeyric Mon 79 G8
Pwllypant Caerph 59 B7
Pye Bridge Derbys 170 E6
Pye Corner Devon 14 B4
Pye Corner Kent 53 D11
Pye Corner Newport 59 B9
Pye Corner S Glos 60 D6
Pye Green Staffs 151 G9
Pye Hill Notts 170 E6
Pyecombe W Sus 36 E3
Pyewipe NE Lincs 201 E9
Pyle I o W 20 F5
Pyle Swansea 56 D5
Pyle = Y Pîl Bridgend 57 E10
Pyle Hill Sur 50 B3
Pylehill Hants 33 D7
Pyleigh Som 42 G6
Pylle Som 44 F6
Pymore or Pymoor Cambs 139 F9
Pype Hayes W Mid 134 E2
Pyrford Sur 50 B4
Pyrford Green Sur 50 B4
Pyrford Village Sur 50 B4
Pyrland Som 28 B2
Pyrton Oxon 83 F11
Pytchley Northants 121 C7
Pyworthy Devon 24 G4

Q

Quabbs Shrops 130 G4
Quabrook E Sus 52 G2
Quadring Lincs 156 C4
Quadring Eaudike Lincs 156 C4
Quags Corner W Sus 34 C5
Quainton Bucks 84 B2
Quaker's Yard M Tydf 77 F9
Quaking Houses Durham 242 G5
Quality Corner Cumb 219 B9
Quarhouse Glos 80 E5
Quarley Hants 47 E9
Quarmby W Yorks 196 D6
Quarndon Derbys 170 G4
Quarndon Common Derbys 170 G4
Quarr Hill I o W 21 C7
Quarrelton Renfs 267 C8
Quarrendon Bucks 84 B4
Quarriers Village Invclyd 267 B7
Quarrington Lincs 173 G9
Quarrington Hill Durham 234 D2
Quarry Bank W Mid 133 F8
Quarry Heath Staffs 151 G8
Quarry Hill Staffs 134 C4
Quarrybank W Ches 167 B8
Quarryford E Loth 271 B11
Quarryhead Aberds 303 C9
Quarryhill Highld 309 L7
Quarrywood Moray 301 C11
Quarter S Lnrk 268 E4
Quartley Devon 27 B7
Quatford Shrops 132 E4
Quatquoy Orkney 314 E3
Quatt Shrops 132 F5
Quebec Durham 233 C9
Quebec W Sus 34 C3
Quedgeley Glos 80 C4
Queen Adelaide Cambs 139 G11
Queen Camel Som 29 C9
Queen Charlton Bath 60 F6
Queen Dart Devon 26 D4
Queen Oak Dorset 45 G9
Queen Street Kent 53 D7
Queen Street Wilts 62 B4
Queenborough Kent 70 E2
Queenhill Worcs 99 D7
Queen's Bower I o W 21 E7
Queen's Corner W Sus 34 B5
Queen's Head Shrops 148 D6
Queen's Park Beds 103 B10
Queen's Park Blkburn 195 B7
Queen's Park Essex 87 F11
Queen's Park Northants 120 E5
Queen's Park W Ches 166 B6
Queensbury London 67 B7
Queensbury W Yorks 205 G8
Queensferry Edin 280 F2
Queensferry Flint 166 B4
Queenslie Glasgow 268 B3
Queenstown Blkpool 202 F2
Queensville Staffs 151 E8
Queenzieburn N Lnrk 278 F3
Quemerford Wilts 62 F4
Quendale Shetland 313 M5
Quendon Essex 105 E10
Queniborough Leics 154 G2
Quenington Glos 81 E10
Quernmore Lancs 202 B6
Queslett W Mid 133 E11
Quethiock Corn 6 C6
Quhamm Shetland 312 G6
Quholm Orkney 314 E2
Quick Gtr Man 196 G3
Quick Edge Gtr Man 196 G3
Quicks Green W Berks 64 D5
Quidenham Norf 141 F10
Quidhampton Hants 48 C4
Quidhampton Wilts 46 G6
Quilquox Aberds 303 F9
Quina Brook Shrops 149 C10
Quinbury End Northants 120 G2
Quindry Orkney 314 G4
Quinton Northants 120 G5
Quinton W Mid 133 G9
Quintrell Downs Corn 5 C7
Quixhill Staffs 169 G10
Quoditch Devon 12 B4
Quoig Perth 286 E2
Quoisley E Ches 167 F8
Quoit Corn 5 C8
Quorndon or Quorn Leics 153 F11
Quothquan S Lnrk 259 B11
Quoyloo Orkney 314 D2
Quoynee Highld 310 D6
Quoyness Orkney 314 F2
Quoys Shetland 312 B8
Quoys Shetland 313 G6

R

Raasay Ho Highld 295 B7
Rabbit's Cross Kent 53 D9
Rableyheath Herts 86 B2
Raby Cumb 238 G5
Raby Mers 182 F4
Racecourse Suff 108 C3
Racedown Hants 47 E9
Rachan Mill Borders 260 C4
Rachub Gwyn 163 B10
Rack End Oxon 82 E6
Rackenford Devon 26 D5
Rackham W Sus 35 E9
Rackheath Norf 160 G5
Rackley Som 43 C11
Racks Dumfries 238 C2
Rackwick Orkney 314 B4
Rackwick Orkney 314 G2
Radbourne Derbys 152 B5
Radcliffe Gtr Man 195 F9
Radcliffe Northumb 253 C7
Radcliffe on Trent Notts 154 B2
Radclive Bucks 102 E3
Radcot Oxon 82 F3
Raddery Highld 301 D7
Raddington Som 27 B8
Raddon Devon 26 G6
Radernie Fife 287 G8
Radfall Kent 70 G6
Radfield Kent 70 G2
Radford Bath 45 B7
Radford Nottingham 171 G9
Radford Oxon 101 G8
Radford W Mid 134 G6
Radford Worcs 117 F10
Radford Semele Warks 118 E6
Radfordbridge Oxon 101 G8
Radipole Dorset 17 E9
Radlet Som 43 F8
Radlett Herts 85 F11
Radley Oxon 83 F8
Radley Green Essex 87 D10
Radley Park Oxon 83 F8
Radlith Shrops 131 B8
Radmanthwaite Notts 171 C8
Radmoor Shrops 150 E2
Radmore Green E Ches 167 D9
Radmore Wood Staffs 151 D11
Radnage Bucks 84 F3
Radnor Corn 4 G4
Radnor Park W Dunb 277 G9
Radstock Bath 45 C7
Radstone Northants 101 C11
Radway Warks 101 B7
Radway Green E Ches 168 E3
Radwell Beds 121 F10
Radwell Herts 104 D4
Radwinter Essex 106 D2
Radwinter End Essex 106 D2
Radyr Cardiff 58 C6
Raehills Dumfries 248 E3
Raera Argyll 289 G10
Rafborough Hants 49 B11
Rafford Moray 301 D10
Raga Shetland 312 D6
Ragdale Leics 154 F3
Ragdon Shrops 131 E9
Raggalds W Yorks 205 G7
Ragged Appleshaw Hants 47 D10
Raginnis Corn 1 D5
Raglan Mon 78 D6
Ragmere Norf 141 E11
Ragnal Wilts 63 E10
Ragnall Notts 188 G4
Rahane Argyll 276 D4
Rahoy Highld 289 D8
Raigbeg Highld 301 G8
Rails S Yorks 186 D3
Rain Shore Gtr Man 195 D11
Rainbow Hill Worcs 117 F7
Rainford Mers 194 G3
Rainford Junction Mers 194 G3
Rainham London 68 C4
Rainham Medway 69 F10
Rainhill Mers 183 C7
Rainhill Stoops Mers 183 C8
Rainow E Ches 185 F7
Rainowlow E Ches 185 F7
Rainsough Gtr Man 195 G10
Rainton Dumfries 237 D8
Rainton N Yorks 215 D7
Rainton Bridge T & W 234 B2
Rainton Gate Durham 234 B2
Rainworth Notts 171 D9
Raisbeck Cumb 222 D2
Raise Cumb 231 B10
Rait Perth 286 E6
Raithby Lincs 190 E4
Raithby by Spilsby Lincs 174 B5
Rake W Sus 34 B4
Rake Common Hants 34 B3
Rake End Staffs 151 F11
Rake Head Lancs 195 C10
Rakes Dale Staffs 169 G9
Rakeway Staffs 169 G8
Rakewood Gtr Man 196 E2
Raleigh Devon 40 G5
Ralia Lodge Highld 291 D9
Rallt Swansea 56 C4
Ram Carms 93 B11
Ram Alley Wilts 63 G8
Ram Hill S Glos 61 D7
Ram Lane Kent 54 D3
Ramasaig Highld 297 G7
Rame Corn 2 C6
Rame Corn 7 F8
Rameldry Mill Bank Fife 287 G7
Ramnageo Shetland 312 C8
Rampisham Dorset 29 G9
Rampside Cumb 210 F4
Rampton Cambs 123 D8
Rampton Notts 188 F3
Ramsbottom Gtr Man 195 D9
Ramsburn Moray 302 D5
Ramsbury Wilts 63 E9
Ramscraigs Highld 311 G5
Ramsdean Hants 34 C2
Ramsdell Hants 48 B5
Ramsden London 68 F3
Ramsden Oxon 82 B5
Ramsden Worcs 99 B8
Ramsden Bellhouse Essex 88 G2
Ramsden Heath Essex 88 F2
Ramsden Wood W Yorks 196 C2
Ramsey Cambs 138 F5
Ramsey Essex 108 E4
Ramsey I o M 192 C5
Ramsey Forty Foot Cambs 138 F6
Ramsey Heights Cambs 138 F5
Ramsey Island Essex 89 D7
Ramsey Mereside Cambs 138 F5
Ramsey St Mary's Cambs 138 F5
Ramseycleuch Borders 261 G7
Ramsgate Kent 71 G11
Ramsgill N Yorks 214 E2
Ramshaw Durham 232 B5
Ramshaw Durham 233 F8
Ramsholt Suff 108 C6
Ramshorn Staffs 169 F9
Ramsley Devon 13 C8
Ramslye Kent 52 F5
Ramsnest Common Sur 50 G2
Ranais W Isles 304 F6
Ranby Lincs 190 F2
Ranby Notts 187 E11
Rand Lincs 189 F10
Randwick Glos 80 D4
Ranfurly Renfs 267 C7
Rangag Highld 310 E5
Rangemore Staffs 152 E3
Rangeworthy S Glos 61 B7
Rankinston E Ayrs 257 G11
Rank's Green Essex 88 B3
Ranmoor S Yorks 186 D4
Ranmore Common Sur 50 C6
Rannerdale Cumb 220 B3
Rannoch Lodge Perth 285 B9
Rannoch Station Perth 285 B8
Ranochan Highld 295 G10
Ranskill Notts 187 D11
Ranton Staffs 151 E7
Ranton Green Staffs 150 E6
Ranworth Norf 161 G7
Rapkyns W Sus 50 G6
Raploch Stirl 278 C5
Rapness Orkney 314 B5
Rapps Som 28 D4
Rascal Moor E Yorks 208 F2
Rascarrel Dumfries 237 E9
Rashielee Renfs 277 G9
Rashiereive Aberds 303 G9
Rashwood Worcs 117 D8
Raskelf N Yorks 215 E9
Rassal Highld 299 E8
Rassau Bl Gwent 77 C11
Rastrick W Yorks 196 C6
Ratagan Highld 295 D11
Ratby Leics 135 B10
Ratcliff London 67 C11
Ratcliffe Culey Leics 134 D6
Ratcliffe on Soar Leics 153 D9
Ratcliffe on the Wreake Leics 154 G2
Ratford Wilts 62 E3
Ratfyn Wilts 47 E7
Rathen Aberds 303 C10
Rathillet Fife 287 E7
Rathmell N Yorks 204 B2
Ratho Edin 280 G2
Ratho Station Edin 280 G2
Rathven Moray 302 C4
Ratlake Hants 32 C6
Ratley Warks 101 B7
Ratling Kent 55 C8
Ratlinghope Shrops 131 D8
Ratsloe Devon 14 B5
Rattar Highld 310 B6
Ratten Row Cumb 230 B3
Ratten Row Cumb 230 C2
Ratten Row Lancs 202 E4
Ratten Row Norf 157 G10
Rattery Devon 8 C4
Rattlesden Suff 125 F9
Rattray Perth 286 C5
Raughton Cumb 230 B3
Raughton Head Cumb 230 B3
Raunds Northants 121 C9
Ravelston Edin 280 G4
Ravenfield S Yorks 187 B7
Ravenglass Cumb 219 F11
Ravenhead Mers 183 C8
Ravenhills Green Worcs 116 G4
Raveningham Norf 143 D7
Raven's Green Essex 108 G2
Ravenscar N Yorks 227 E9
Ravenscliffe Stoke 168 E4
Ravenscliffe W Yorks 205 F9
Ravenscraig Invclyd 276 F5
Ravensdale I o M 192 C4
Ravensden Beds 121 G11
Ravenseat N Yorks 223 E7
Ravenshall Staffs 168 F3
Ravenshead Notts 171 E9
Ravensmoor E Ches 167 E10
Ravensthorpe Northants 120 C3
Ravensthorpe P'boro 138 C3
Ravensthorpe W Yorks 197 C8
Ravenstone Leics 153 G8
Ravenstone M Keynes 120 G6
Ravenstonedale Cumb 222 E4
Ravenstown Cumb 211 D7
Ravenstruther S Lnrk 269 F8
Ravenswood Village Settlement Wokingham 65 G10
Ravensworth N Yorks 224 D2
Raw N Yorks 227 D8
Raw Green S Yorks 197 F9

Rawcliffe E Yorks 199 C7
Rawcliffe York 207 C7
Rawcliffe Bridge E Yorks 199 C7
Rawdon W Yorks 205 F10
Rawdon Carrs W Yorks 205 F10
Rawfolds W Yorks 197 C7
Rawgreen Northumb 241 F10
Rawmarsh S Yorks 186 B6
Rawnsley Staffs 151 G10
Rawreth Essex 88 G3
Rawreth Shot Essex 88 G3
Rawridge Devon 28 F2
Rawson Green Derbys 170 F5
Rawtenstall Lancs 195 C10
Rawthorpe W Yorks 197 D7
Rawyards N Lnrk 268 B5
Raxton Aberds 303 F8
Raydon Suff 107 D11
Raygill N Yorks 204 D4
Raylees Northumb 251 E10
Rayleigh Essex 88 G4
Rayne Essex 106 G4
Rayners Lane London 66 B6
Raynes Park London 67 F8
Reabrook Shrops 131 C7
Reach Cambs 123 D11
Read Lancs 203 G11
Reader's Corner Essex 88 E2
Reading Reading 65 E8
Reading Street Kent 54 G2
Reading Street Kent 71 F11
Readings Glos 79 B10
Ready Token Glos 81 E10
Readymoney Corn 6 E2
Reagill Cumb 222 B2
Rearquhar Highld 309 K7
Rearsby Leics 154 G3
Reasby Lincs 189 F9
Rease Heath E Ches 167 E10
Reaster Highld 310 C6
Reaulay Highld 299 D7
Reawick Shetland 313 J5
Reawla Corn 2 B4
Reay Highld 310 C3
Rechullin Highld 299 D8
Reculver Kent 71 F8
Red Ball Devon 27 D9
Red Bridge Lancs 211 D9
Red Bull E Ches 168 D4
Red Bull Staffs 150 B4
Red Dial Cumb 229 B11
Red Hill Bmouth 19 B7
Red Hill Hants 34 E2
Red Hill Hereford 97 D10
Red Hill Kent 53 C7
Red Hill Leics 135 D10
Red Hill Pembs 72 B6
Red Hill W Yorks 198 B2
Red Hill Warks 118 F2
Red Hill Worcs 117 G7
Red House Common E Sus 36 C5
Red Lake Telford 150 G3
Red Lodge Suff 124 C3
Red Lumb Gtr Man 195 D10
Red Pits Norf 159 D11
Red Post Corn 24 F3
Red Rail Hereford 97 F10
Red Rice Hants 47 E10
Red Rock Gtr Man 194 F5
Red Roses Carms 74 C2
Red Row Northumb 253 D7
Red Scar Lancs 203 G7
Red Street Staffs 168 E4
Red Wharf Bay Anglesey 179 E8
Redberth Pembs 73 E9
Redbourn Herts 85 C10
Redbournbury Herts 85 C10
Redbourne N Lincs 189 B7
Redbourne N Lincs 200 G3
Redbridge Dorset 17 D11
Redbridge London 68 B2
Redbridge Soton 32 E5
Redbrook Mon 79 C8
Redbrook Wrex 167 G8
Redburn Highld 300 C5
Redburn Highld 301 E9
Redburn Northumb 241 E7
Redcar Redcar 235 G8
Redcastle Angus 287 B10
Redcastle Highld 300 E5
Redcliff Bay N Som 60 D2
Redcroft Dumfries 237 B9
Redcross Worcs 117 C7
Reddicap Heath W Mid 134 D2
Redding Falk 279 F8
Reddingmuirhead Falk 279 F8
Reddish Gtr Man 184 C5
Reddish Warr 183 D11
Redditch Worcs 117 D10
Rede Suff 124 F6
Redenhall Norf 142 G5
Redenham Hants 47 D10
Redesdale Camp Northumb 251 D8
Redesmouth Northumb 251 G9
Redford Aberds 293 F9
Redford Angus 287 C9
Redford Dorset 29 F10
Redford Durham 233 E7
Redford W Sus 34 B5
Redfordgreen Borders 261 F9
Redgorton Perth 286 E4
Redgrave Suff 125 B10
Redheugh Angus 292 G6
Redhill Aberds 293 C9
Redhill Aberds 302 F6
Redhill Herts 104 E6
Redhill N Som 60 G4
Redhill Notts 171 F9
Redhill Shrops 131 B9
Redhill Shrops 150 G4
Redhill Staffs 150 D6
Redhill Sur 51 C9
Redhill Telford 150 G4
Redhills Cumb 230 F6
Redhills Devon 14 C4
Redhouse Argyll 275 G9
Redhouses Argyll 274 G4
Redisham Suff 143 G8
Redland Bristol 60 D5
Redland Orkney 314 D3
Redland End Bucks 84 E4
Redlands Dorset 17 E9
Redlands Som 44 G3
Redlands Swindon 81 G11
Redlane Som 28 E2
Redlingfield Suff 126 C3
Redlynch Som 45 G8
Redlynch Wilts 32 C2
Redmain Cumb 229 E8
Redmarley D'Abitot Glos 98 E5
Redmarshall Stockton 234 G3
Redmile Leics 154 B5
Redmire N Yorks 223 G10
Redmonsford Devon 24 D4
Redmoor Corn 5 C11
Rednal Shrops 149 D7
Rednal W Mid 117 B10
Redpath Borders 262 B3
Redpoint Highld 299 C7
Redruth Corn 4 G3
Redscarhead Borders 270 G4
Redstocks Wilts 62 G2
Redtye Corn 5 C10
Redvales Gtr Man 195 F10
Redwick Newport 60 C2
Redwick S Glos 60 B4
Redwith Shrops 148 E6
Redworth Darl 233 G10
Reed Herts 105 D7
Reed End Herts 104 D6
Reed Point Lincs 174 E2
Reedham Lincs 174 D2
Reedham Norf 143 C8
Reedley Lancs 204 F2
Reedness E Yorks 199 C9
Reeds Beck Lincs 174 B2
Reeds Holme Lancs 195 C10
Reedsford Northumb 263 C9
Reedy Devon 14 D2
Reen Manor Corn 4 E5
Reepham Lincs 189 G8
Reepham Norf 159 E11
Reeth N Yorks 223 F10
Reeves Green W Mid 118 B5
Refail Powys 130 C3
Regaby I o M 192 C5
Regil Bath 60 G4
Regoul Highld 301 D8
Reiff Highld 307 H4
Reigate Sur 51 C9
Reigate Heath Sur 51 C8
Reighton N Yorks 218 D2
Reighton Gap N Yorks 218 D2
Reinigeadal W Isles 305 H4
Reisque Aberds 293 B10
Reiss Highld 310 D7
Rejerrah Corn 4 D5
Releath Corn 2 C5
Relubbus Corn 2 C3
Relugas Moray 301 E9
Remenham Wokingham 65 C9
Remenham Hill Wokingham 65 C9
Remony Perth 285 C11
Rempstone Notts 153 E11
Remusaig Highld 309 J7
Rendcomb Glos 81 D8
Rendham Suff 126 E6
Rendlesham Suff 126 G6
Renfrew Renfs 267 B10
Renhold Beds 121 G11
Renishaw Derbys 186 F6
Renmure Angus 287 B10
Rennington Northumb 264 F6
Renshaw Wood Shrops 132 B6
Renton W Dunb 277 F7
Renwick Cumb 231 C7
Repps Norf 161 F8
Repton Derbys 152 D6
Reraig Highld 295 C10
Reraig Cot Highld 295 B10
Rerwick Shetland 313 M5
Rescassa Corn 5 G9
Rescobie Angus 287 B9
Rescorla Corn 5 D10
Resipole Highld 289 C9
Reskadinnick Corn 4 G2
Resolfen = Resolven Neath 76 E4
Resolis Highld 300 C6
Resolven = Resolfen Neath 76 E4
Restalrig Edin 280 G5
Reston Borders 273 C7
Reston Cumb 221 F9
Restronguet Passage Corn 3 B8
Restrop Wilts 62 B5
Resugga Green Corn 5 D10
Reswallie Angus 287 B9
Retallack Corn 5 B8
Retew Corn 5 D8
Retford Notts 188 E2
Retire Corn 5 C10
Rettendon Essex 88 F3
Rettendon Place Essex 88 F3
Revesby Lincs 174 C3
Revesby Bridge Lincs 174 C4
Revidge Blkburn 195 B7
Rew Devon 9 G9
Rew Devon 13 G11
Rew Dorset 29 F11
Rew Street I o W 20 C5
Rewe Devon 14 B4
Rexon Devon 12 D4
Rexon Cross Devon 12 D4
Reybridge Wilts 62 F2
Reydon Suff 127 B9
Reydon Smear Suff 127 B9
Reymerston Norf 141 B10
Reynalton Pembs 73 D9
Reynoldston Swansea 56 C3
Rezare Corn 12 F3
Rhadyr Mon 78 E5
Rhaeadr Gwy = Rhayader Powys 113 D9
Rhandir Conwy 180 G4
Rhandirmwyn Carms 94 C5
Rhayader = Rhaeadr Gwy Powys 113 D9
Rhedyn Gwyn 144 C5
Rhegreanoch Highld 307 H5
Rhemore Highld 289 D7
Rhencullen I o M 192 C4
Rhenetra Highld 298 D4
Rhes-y-cae Flint 181 G11
Rhewl Denb 165 C10
Rhewl Denb 165 F11
Rhewl Shrops 148 C6
Rhewl Wrex 149 E7
Rhewl-fawr Flint 181 E10
Rhewl-Mostyn Flint 181 E11
Rhian Highld 309 H5
Rhicarn Highld 307 G5
Rhiconich Highld 306 D7
Rhicullen Highld 300 B6
Rhidorroch Ho Highld 307 K6
Rhiews Shrops 150 B2
Rhifail Highld 308 E7
Rhigolter Highld 308 D3
Rhigos Rhondda 76 D6
Rhilochan Highld 309 J7
Rhippinllwyd Ceredig 92 C5
Rhippinllwyd Ceredig 110 G6
Rhiroy Highld 307 L6
Rhitongue Highld 308 D6
Rhivichie Highld 306 D7
Rhiw Gwyn 144 D4
Rhiwabon = Ruabon Wrex 166 G4
Rhiwbebyll Denb 165 B10
Rhiwbina Cardiff 59 C7
Rhiwbryfdir Gwyn 163 F11
Rhiwceiliog Bridgend 58 C3
Rhiwderin Newport 59 B9
Rhiwen Gwyn 163 C9
Rhiwfawr Neath 76 C2
Rhiwinder Rhondda 58 B4
Rhiwlas Gwyn 147 B8
Rhiwlas Gwyn 163 B9
Rhiwlas Powys 148 C3
Rhode Som 43 G9
Rhode Common Kent 54 B5
Rhodes Gtr Man 195 E11
Rhodes Minnis Kent 55 E7
Rhodesia Notts 187 F9
Rhodiad Pembs 90 F5
Rhonadale Argyll 255 D8
Rhondda Rhondda 77 F7
Rhonehouse or Kelton Hill Dumfries 237 D9
Rhoose V Glam 58 F5
Rhos Carms 93 D7
Rhôs Neath 76 E2
Rhôs Denb 165 C10
Rhos Powys 148 F5
Rhôs Common Powys 148 F5
Rhos-ddû Gwyn 144 B5
Rhos-fawr Gwyn 145 B7
Rhos-goch Powys 96 B3
Rhos Haminiog Ceredig 111 E10
Rhos-hill Pembs 92 C3
Rhos Isaf Gwyn 163 D7
Rhôs Lligwy Anglesey 179 D7
Rhôs-on-Sea Conwy 180 E4
Rhos-y-brithdir Powys 148 E2
Rhos-y-garth Ceredig 112 C2
Rhos-y-gwaliau Gwyn 147 C8
Rhos-y-llan Gwyn 144 B4
Rhos-y-Madoc Wrex 166 G4
Rhos-y-meirch Powys 114 D5
Rhosaman Carms 76 C2
Rhosbeirio Anglesey 178 C5
Rhoscefnhir Anglesey 179 F8
Rhoscolyn Anglesey 178 F3
Rhoscrowther Pembs 72 E6
Rhosddu Wrex 166 E4
Rhosdylluan Gwyn 147 B7
Rhosesmor Flint 166 B2
Rhosfach Pembs 92 F2
Rhosgadfan Gwyn 163 D8
Rhosgoch Anglesey 178 D6
Rhosgoch Powys 96 B3
Rhosgyll Gwyn 163 G7
Rhoshirwaun Gwyn 144 D3
Rhoslan Gwyn 163 G7
Rhoslefain Gwyn 110 B2
Rhosllanerchrugog Wrex 166 F3
Rhosmaen Carms 94 G2
Rhosmeirch Anglesey 179 F7
Rhosneigr Anglesey 178 G4
Rhosnesni Wrex 166 E5
Rhosrobin Wrex 166 E4
Rhossili Swansea 56 D2
Rhosson Pembs 90 F4
Rhostrehwfa Anglesey 178 G6
Rhostryfan Gwyn 163 D7
Rhostyllen Wrex 166 F4
Rhoswiel Shrops 148 B5
Rhosybol Anglesey 178 D6
Rhosycaerau Pembs 91 D8
Rhosygadair Newydd Ceredig 92 B4
Rhosygadfa Shrops 148 C6
Rhosygilwen Pembs 92 C4
Rhosymedre Wrex 166 G3
Rhosyn-coch Carms 92 G5
Rhu Argyll 275 G9
Rhu Argyll 276 E5
Rhuallt Denb 181 F9
Rhubodach Argyll 275 F11
Rhuddall Heath W Ches 167 C9
Rhuddlan Ceredig 93 C9
Rhuddlan Denb 181 F8
Rhue Highld 307 K5
Rhulen Powys 96 B2
Rhunahaorine Argyll 255 C8
Rhyd Ceredig 92 C5
Rhyd Gwyn 163 G10
Rhyd Powys 129 C9
Rhyd-Ddu Gwyn 163 E9
Rhyd-Rosser Ceredig 111 D11
Rhyd-uchaf Gwyn 147 B8
Rhyd-y-Brown Pembs 91 G11
Rhyd-y-clafdy Gwyn 144 B6
Rhyd-y-cwm Shrops 130 G3
Rhyd-y-foel Conwy 180 F6
Rhyd-y-fro Neath 76 D2
Rhyd-y-gwin Swansea 75 E11
Rhyd-y-gwystl Gwyn 145 B8
Rhyd-y-meirch Mon 78 D4
Rhyd-y-meudwy Denb 165 E10
Rhyd-y-pandy Swansea 75 E11
Rhyd-y-sarn Gwyn 163 G11
Rhyd-yr-onen Gwyn 128 C2
Rhydaman = Ammanford Carms 75 C10
Rhydargaeau Carms 93 F8
Rhydcymerau Carms 93 D11
Rhydd Worcs 98 B6
Rhydd Green Worcs 98 B6
Rhydding Neath 57 B8
Rhydfudr Ceredig 111 D11
Rhydgaled Conwy 165 C7
Rhydgaled = Chancery Ceredig 111 B11
Rhydlewis Ceredig 92 B6
Rhydlios Gwyn 144 C3
Rhydlydan Conwy 164 E5
Rhydlydan Powys 129 E11
Rhydmoelddu Powys 113 B11
Rhydness Powys 96 C2
Rhydowen Carms 92 F3
Rhydowen Ceredig 93 B8
Rhydspence Hereford 96 B4
Rhydtalog Flint 166 D2
Rhydwen Gwyn 146 F4
Rhydwyn Anglesey 178 D4
Rhydycroesau Powys 148 C4
Rhydyfelin Carms 92 D5
Rhydyfelin Ceredig 111 B11
Rhydyfelin Powys 129 E11
Rhydyfelin Rhondda 58 B5
Rhydygele Pembs 91 G7
Rhydymain Gwyn 146 E6
Rhydymwyn Flint 166 B2
Rhydywrach Carms 73 B11
Rhyl Denb 181 E8
Rhymney Caerph 77 D10
Rhyn Wrex 148 B6
Rhynd Fife 287 E8
Rhynd Perth 286 E5
Rhynie Aberds 302 G4
Rhynie Highld 301 B8
Ribbesford Worcs 116 C5
Ribble Head N Yorks 212 D5
Ribblehead N Yorks 212 D5
Ribbleton Lancs 203 G7
Ribby Lancs 202 G4
Ribchester Lancs 203 F8
Riber Derbys 170 D4
Ribigill Highld 308 D5
Riby Lincs 201 F7
Riby Cross Roads Lincs 201 F8
Riccall N Yorks 207 F8
Riccarton E Ayrs 257 B10
Richards Castle Hereford 115 D9
Richborough Port Kent 71 G10
Richings Park Bucks 66 D4
Richmond London 67 E7
Richmond N Yorks 224 E3
Richmond S Yorks 186 D6
Richmond Hill W Yorks 206 G2
Richmond's Green Essex 106 F2
Rich's Holford Som 42 G6
Rickard's Down Devon 24 B6
Rickarton Aberds 293 E10
Rickerby Cumb 239 F10
Rickerscote Staffs 151 E8
Rickford N Som 44 B3
Rickinghall Suff 125 B10
Rickleton T & W 243 G7
Rickling Essex 105 E9
Rickling Green Essex 105 F10
Rickmansworth Herts 85 G9
Rickney E Sus 23 D10
Riddell Borders 262 E2
Riddings Derbys 170 E6
Riddlecombe Devon 25 E10
Riddlesden W Yorks 205 E7
Riddrie Glasgow 268 B2
Ridgacre W Mid 133 G10
Ridge Bath 44 B5
Ridge Dorset 18 D4
Ridge Hants 32 D4
Ridge Herts 86 E2
Ridge Lancs 211 G9
Ridge Som 28 F3
Ridge Wilts 46 G3
Ridge Common Hants 34 C2
Ridge Green Sur 51 D10
Ridge Hill Gtr Man 185 B7
Ridge Lane Warks 134 E5
Ridge Row Kent 55 E8
Ridgebourne Powys 113 E11
Ridgehill N Som 60 G4
Ridgemarsh Herts 85 G8
Ridgeway Bristol 60 D6
Ridgeway Derbys 170 E5
Ridgeway Derbys 186 E6
Ridgeway Kent 54 E5
Ridgeway Newport 59 B9
Ridgeway Pembs 73 D10
Ridgeway Som 45 D8
Ridgeway Staffs 168 E5
Ridgeway Cross Hereford 98 B4
Ridgeway Moor Derbys 186 E6
Ridgewell Essex 106 C4
Ridgewood E Sus 23 B7
Ridgmont C Beds 103 D9
Ridgway Shrops 131 F7
Ridgway Sur 50 B4
Riding Gate Som 30 B2
Riding Mill Northumb 242 E2
Ridley Kent 68 G6
Ridley Northumb 241 E7
Ridley Stokoe Northumb 250 F6
Ridleywood Wrex 166 E6
Ridlington Norf 160 C6
Ridlington Rutland 136 C6
Ridlington Street Norf 160 C6
Ridsdale Northumb 251 G10
Riechip Perth 286 C4
Riemore Perth 286 C4
Rienachait Highld 306 F5
Rievaulx N Yorks 215 B11
Riff Orkney 314 E4
Riffin Aberds 303 E7
Rifle Green Torf 78 D3
Rift House Hrtlpl 234 E5
Rigg Dumfries 239 D7
Riggend N Lnrk 278 G5
Rigsby Lincs 190 F6
Rigside S Lnrk 259 B9
Riley Green Lancs 194 B6
Rileyhill Staffs 152 F2
Rilla Mill Corn 11 G11
Rillaton Corn 11 G11
Rillington N Yorks 217 E7
Rimac Lincs 191 C7
Rimington Lancs 204 D2
Rimpton Som 29 C10
Rimswell E Yorks 201 B10
Rimswell Valley E Yorks 201 B10
Rinaston Pembs 91 F9
Rindleford Shrops 132 D4
Ring o' Bells Lancs 194 E3
Ringasta Shetland 313 M5
Ringford Dumfries 237 D8
Ringing Hill Leics 153 F9
Ringinglow S Yorks 186 E3
Ringland Newport 59 B11
Ringland Norf 160 G2
Ringles Cross E Sus 37 C7
Ringlestone Kent 53 B11
Ringlestone Kent 53 B9
Ringley Gtr Man 195 F9
Ringmer E Sus 36 E6
Ringmore Devon 8 F3
Ringorm Moray 302 E2
Ring's End Cambs 139 C7
Ringsfield Suff 143 F8
Ringsfield Corner Suff 143 F8
Ringshall Herts 85 C7
Ringshall Suff 125 G10
Ringshall Stocks Suff 125 G10
Ringstead Norf 176 E2
Ringstead Northants 121 B9
Ringtail Green Essex 87 B11
Ringwood Hants 31 F11
Ringwould Kent 55 D11
Rinmore Aberds 292 B6
Rinnigill Orkney 314 G3
Rinsey Corn 2 D3
Rinsey Croft Corn 2 D4
Riof W Isles 304 E3
Ripe E Sus 23 C8
Ripley Derbys 170 E5
Ripley Hants 19 B9
Ripley N Yorks 214 G5
Ripley Sur 50 B5
Riplingham E Yorks 208 G5
Ripon N Yorks 214 E6
Ripper's Cross Kent 54 E3
Rippingale Lincs 155 D11
Ripple Kent 55 D10
Ripple Worcs 99 D7
Ripponden W Yorks 196 D4
Rireavach Highld 307 K5
Risabus Argyll 254 C4
Risbury Hereford 115 G10
Risby E Yorks 208 G6
Risby Lincs 189 C10
Risby Suff 124 D5
Risca Caerph 78 G2
Rise E Yorks 209 E9
Rise Carr Darl 224 B5
Rise End Derbys 170 D3
Rise Park London 87 G8
Rise Park Nottingham 171 F9
Riseden E Sus 52 G6
Riseden Kent 53 F8
Risegate Lincs 156 D4
Riseholme Lincs 189 F7
Risehow Cumb 228 E6
Riseley Beds 121 E10
Riseley Wokingham 65 G8
Rishangles Suff 126 D3
Rishton Lancs 203 G10
Rishworth W Yorks 196 D4
Rising Bridge Lancs 195 B9
Rising Sun Corn 12 G3
Risingbrook Staffs 151 E8
Risinghurst Oxon 83 D9
Risley Derbys 153 B9
Risley Warr 183 C11
Risplith N Yorks 214 F4
Rispond Highld 308 C4
Rivar Wilts 63 G10
Rivenhall Essex 88 B4
Rivenhall End Essex 88 B4
River Kent 55 E9
River W Sus 34 C6
River Bank Cambs 123 D10
Riverhead Kent 52 B4
Rivers' Corner Dorset 30 E3
Riverside Cardiff 59 D7
Riverside Plym 7 D8
Riverside Stirl 278 C6
Riverside Worcs 117 D10
Riverside Docklands Lancs 194 B4
Riverton Devon 40 G6
Riverview Park Kent 69 E7
Rivington Lancs 194 E6
Rixon Dorset 30 E3
Rixton Warr 183 C11
Roa Island Cumb 210 G4
Roach Bridge Lancs 194 B5
Roaches Gtr Man 196 G3
Roachill Devon 26 C4
Road Green Norf 142 E5
Road Weedon Northants 120 F2
Roade Northants 120 G5
Roadhead Cumb 240 C2
Roadmeetings S Lnrk 269 F7
Roadside Highld 310 C5
Roadside of Catterline Aberds 293 F10
Roadside of Kinneff Aberds 293 F10
Roadwater Som 42 F4
Roag Highld 298 E2
Roast Green Essex 105 E9
Roath Cardiff 59 D7
Roath Park Cardiff 59 D7
Rob Roy's House Argyll 284 F5
Roberton Borders 261 G10
Roberton S Lnrk 259 D10
Robertsbridge E Sus 38 C2
Robertstown Moray 302 E2
Robertstown Rhondda 77 E8
Roberttown W Yorks 197 C7
Robeston Back Pembs 73 B9
Robeston Cross Pembs 72 D5
Robeston Wathen Pembs 73 B9
Robeston West Pembs 72 D5
Robhurst Kent 54 G2
Robin Hill Staffs 168 D6
Robin Hood Derbys 186 G3
Robin Hood Lancs 194 E4
Robin Hood W Yorks 197 B10
Robin Hood's Bay N Yorks 227 D9
Robinhood End Essex 106 D4
Robins W Sus 34 B4
Robinson's End Warks 134 E6
Roborough Devon 7 C10
Roborough Devon 25 D9
Robroyston Glasgow 268 B2
Roby Mers 182 C6
Roby Mill Lancs 194 F4
Rocester Staffs 152 B2
Roch Pembs 91 G7
Roch Gate Pembs 91 G7
Rochdale Gtr Man 195 E11
Roche Corn 5 C9
Roche Grange Staffs 169 C7
Rochester Medway 69 F8
Rochester Northumb 251 D8
Rochford Essex 88 G5
Rochford Worcs 116 D2
Rock Caerph 77 F11
Rock Corn 10 F4
Rock Devon 28 G3
Rock Neath 57 C9
Rock Northumb 264 E6
Rock Som 28 C4
Rock W Sus 35 E10
Rock Worcs 116 C4
Rock End Staffs 168 D5
Rock Ferry Mers 182 D4
Rockbeare Devon 14 C6
Rockbourne Hants 31 D10
Rockcliffe Cumb 239 E9
Rockcliffe Dumfries 237 D10
Rockcliffe Flint 182 G3
Rockcliffe Lancs 195 C11
Rockcliffe Cross Cumb 239 E8
Rockfield Highld 311 L3
Rockfield Mon 79 C7
Rockford Devon 41 D9
Rockford Hants 31 F11
Rockgreen Shrops 115 B10
Rockhampton S Glos 79 G11
Rockhead Corn 11 E7
Rockhill Shrops 114 B5
Rockingham Northants 137 E7
Rockland All Saints Norf 141 D9
Rockland St Mary Norf 142 C6
Rockland St Peter Norf 141 D9
Rockley Notts 188 G2
Rockley Wilts 63 E7
Rockley Ford Som 45 C8
Rockness Glos 80 F4
Rockrobin E Sus 52 G6
Rocks Park E Sus 37 C7
Rocksavage Halton 183 E8
Rockstowes Glos 80 F3
Rockville Argyll 276 C4
Rockwell End Bucks 65 B9
Rockwell Green Som 27 C10
Rocky Hill Scilly 1 G4
Rodbaston Staffs 151 G8
Rodborough Glos 80 E4
Rodbourne Swindon 62 C6
Rodbourne Wilts 62 C2
Rodbourne Bottom Wilts 62 C2
Rodbourne Cheney Swindon 62 B6
Rodbridge Corner Essex 107 C7
Rodd Hereford 114 E6
Rodd Hurst Hereford 114 E6
Roddam Northumb 264 E2
Rodden Dorset 17 E8
Roddymoor Durham 233 D9
Rode Som 45 C10
Rode Heath E Ches 168 D3
Rode Hill Som 45 C10
Rodeheath E Ches 168 B5
Roden Telford 149 F11
Rodford S Glos 61 C7
Rodgrove Som 30 C2
Rodhuish Som 42 F4
Rodington Telford 149 G11
Rodington Heath Telford 149 G11
Rodley Glos 80 C2
Rodley W Yorks 205 F10
Rodmarton Glos 80 F6
Rodmell E Sus 36 F6
Rodmer Clough W Yorks 196 B3
Rodmersham Kent 70 G2
Rodmersham Green Kent 70 G2
Rodney Stoke Som 44 C3
Rodsley Derbys 170 G2
Rodway Som 43 F9
Rodway Telford 150 F3
Rodwell Dorset 17 F9
Roe Cross Gtr Man 185 B7
Roe End Herts 85 B8
Roe Green Gtr Man 195 G9
Roe Green Herts 86 D2
Roe Green Herts 104 E6
Roe Lee Blkburn 203 G9
Roebuck Low Gtr Man 196 F3
Roecliffe N Yorks 215 F7
Roedean Brighton 36 G4
Roehampton London 67 E8
Roesound Shetland 312 G5
Roestock Herts 86 D2
Roffey W Sus 51 G7
Rogart Highld 309 J7
Rogart Station Highld 309 J7
Rogate W Sus 34 C4
Roger Ground Cumb 221 F7
Rogerstone Newport 59 B9
Rogerton S Lnrk 268 D2
Roghadal W Isles 296 C6
Rogiet Mon 60 B3
Rogue's Alley Cambs 139 B7
Roke Oxon 83 G10
Rokemarsh Oxon 83 G10
Roker T & W 243 F10
Rollesby Norf 161 F8
Rolleston Leics 136 C4
Rolleston Notts 172 E2
Rolleston-on-Dove Staffs 152 D4
Rollestone S Yorks 186 E5
Rollestone Wilts 46 E5
Rollestone Camp Wilts 46 E5
Rolls Mill Dorset 30 E3
Rolston E Yorks 209 D10
Rolstone N Som 59 G11
Rolvenden Kent 53 G10
Rolvenden Layne Kent 53 G11
Romaldkirk Durham 232 G5
Roman Hill Suff 143 E10
Romanby N Yorks 225 G7
Romannobridge Borders 270 F3
Romansleigh Devon 26 C2
Rome Angus 293 G7
Romesdal Highld 298 D4
Romford Dorset 31 F9
Romford Kent 52 E6
Romford London 68 B4
Romiley Gtr Man 184 C6
Romney Street Kent 68 G4
Rompa Shetland 313 L6
Romsey Hants 32 C5
Romsey Town Cambs 123 F9
Romsley Shrops 132 G5
Romsley Worcs 117 B9
Romsley Hill Worcs 117 B9
Ronachan Ho Argyll 255 B8
Ronague I o M 192 E3
Rondlay Telford 132 B4
Ronkswood Worcs 117 G7
Rood End W Mid 133 F10
Rook End Essex 105 E11
Rook Street Wilts 45 G10
Rookby Cumb 222 C6
Rookhope Durham 232 C4
Rooking Cumb 221 B8
Rookley I o W 20 E6
Rookley Green I o W 20 E6
Rooks Bridge Som 43 C11
Rooks Hill Kent 52 C5
Rook's Nest Som 42 G5
Rooksey Green Suff 125 G8
Rooksmoor Glos 80 E4
Rookwith N Yorks 214 B4
Rookwood W Sus 21 B11
Roos E Yorks 209 G11
Roose Cumb 210 F4
Roosebeck Cumb 210 F5
Roosecote Cumb 210 F4
Roost End Essex 106 C4
Rootham's Green Beds 122 F2
Rooting Street Kent 54 D3
Rootpark S Lnrk 269 E9
Ropley Hants 48 G6
Ropley Dean Hants 48 G6
Ropley Soke Hants 49 G7
Ropsley Lincs 155 C9
Rora Aberds 303 D10
Rorandle Aberds 293 B8
Rorrington Shrops 130 C5
Rosarie Moray 302 E3
Roscroggan Corn 4 G3
Rose Corn 4 E5
Rose-an-Grouse Corn 2 B2
Rose Ash Devon 26 C3
Rose Green Essex 107 F8
Rose Green Suff 107 C9
Rose Green Suff 107 D8
Rose Green W Sus 22 D6
Rose Grove Lancs 204 G2
Rose Hill Bucks 66 C2
Rose Hill Derbys 153 B7
Rose Hill E Sus 23 B7
Rose Hill Gtr Man 195 F8
Rose Hill Lancs 204 G2
Rose Hill Oxon 83 E8
Rose Hill Suff 108 C3
Rose Hill Sur 51 D7
Rose Valley Pembs 73 E8
Roseacre Kent 53 B9
Roseacre Lancs 202 F4
Rosebank E Dunb 278 G3
Rosebank S Lnrk 268 F6
Rosebery Midloth 270 D6
Rosebrae Moray 301 C11
Rosebrough Northumb 264 D4
Rosebush Pembs 91 F11
Rosecare Corn 11 B9
Rosedale Herts 86 E4
Rosedale Abbey N Yorks 226 F4
Roseden Northumb 264 E2
Rosedinnick Corn 5 B8
Rosedown Devon 24 C3
Rosefield Highld 301 D8
Rosehall Highld 309 J4
Rosehall N Lnrk 268 C4
Rosehaugh Mains Highld 300 D6
Rosehearty Aberds 303 C9
Rosehill Blkburn 195 C8
Rosehill Corn 4 E5
Rosehill Corn 5 C10
Rosehill Gtr Man 184 D3
Rosehill London 67 F9
Rosehill Pembs 72 B5
Rosehill Shrops 150 C3
Rosehill T & W 243 D8
Roseisle Moray 301 C11
Roseland Corn 6 C5
Roselands E Sus 23 E10
Rosemarket Pembs 73 D7
Rosemarkie Highld 301 D7
Rosemary Lane Devon 27 E11
Rosemelling Corn 5 D10
Rosemergy Corn 1 B4
Rosemount Perth 286 C5
Rosenannon Corn 5 B9
Rosenithon Corn 3 E8
Roser's Cross E Sus 37 C9
Rosevean Corn 5 D10
Rosevear Corn 2 E5
Roseville W Mid 133 E8
Rosevine Corn 3 B9
Rosewarne Corn 2 B4
Rosewarne Corn 4 G2
Rosewell Midloth 270 C5
Roseworth Stockton 234 G4
Roseworthy Corn 2 B4
Roseworthy Corn 4 F5
Roseworthy Barton Corn 2 B4
Rosgill Cumb 221 B10
Rosherville Kent 68 E6
Roshven Highld 289 B9
Roskear Croft Corn 4 G3
Roskhill Highld 298 E2
Roskill House Highld 300 D6
Roskorwell Corn 3 E7
Rosley Cumb 230 B2
Roslin Midloth 270 C5
Rosliston Derbys 152 F4
Rosneath Argyll 276 E5
Ross Borders 273 C9
Ross Dumfries 237 E8
Ross Northumb 264 B4
Ross Perth 285 E11
Ross Green Worcs 116 E5
Ross-on-Wye Hereford 98 G2
Rossett Wrex 166 D5
Rossett Green N Yorks 206 C2
Rossie Ochill Perth 286 F4
Rossie Priory Perth 286 D6
Rossington S Yorks 187 B10
Rosskeen Highld 300 C6
Rossland Renfs 277 G8
Rossmore Poole 19 C7
Roster Highld 310 F6
Rostherne E Ches 184 E2
Rostholme S Yorks 198 F5
Rosthwaite Cumb 220 C5
Rosthwaite Cumb 220 G4
Roston Derbys 169 G10
Rosudgeon Corn 2 D3
Rosyth Fife 280 E2
Rotchfords Essex 107 E8
Rotcombe Bath 44 B6
Rothbury Northumb 252 C3
Rotherbridge W Sus 35 C7
Rotherby Leics 154 F3
Rotherfield E Sus 37 B9
Rotherfield Greys Oxon 65 C8
Rotherfield Peppard Oxon 65 C8
Rotherham S Yorks 186 C6
Rotherhithe London 67 D11
Rothersthorpe Northants 120 F4
Rotherwas Hereford 97 D10
Rotherwick Hants 49 B8
Rothes Moray 302 E2
Rothesay Argyll 275 G11
Rothiebrisbane Aberds 303 F7
Rothiemay Crossroads Moray 302 E5
Rothiemurchus Lodge Highld 291 C11
Rothienorman Aberds 303 F7
Rothiesholm Orkney 314 D6
Rothley Leics 153 G11
Rothley Northumb 252 F2
Rothley Plain Leics 153 G11
Rothley Shield East Northumb 252 E2
Rothmaise Aberds 302 F6
Rothwell Lincs 189 B11
Rothwell Northants 136 G6
Rothwell W Yorks 197 B10
Rothwell Haigh W Yorks 197 B10
Rotsea E Yorks 209 C7
Rottal Angus 292 G5
Rotten End Essex 106 F4
Rotten End Suff 127 D7
Rotten Green Hants 49 B9
Rotten Row W Berks 64 E5
Rotten Row W Mid 118 B3
Rottingdean Brighton 36 G5
Rottington Cumb 219 C9
Rotton Park W Mid 133 F10
Roud I o W 20 E6
Rough Bank Gtr Man 196 E2
Rough Close Staffs 151 B8
Rough Common Kent 54 B6
Rough Haugh Highld 308 E7
Rough Hay Staffs 152 E4
Rougham Norf 158 E6
Rougham Suff 125 E8
Rougham Green Suff 125 E8
Roughbirchworth S Yorks 197 G9
Roughburn Highld 290 E5
Roughcote Staffs 168 G6
Roughlee Lancs 204 E2
Roughley W Mid 134 D2
Roughmoor Som 28 B2
Roughmoor Swindon 62 B6
Roughrigg N Lnrk 278 G6
Roughsike Cumb 240 B2
Roughton Lincs 174 C2
Roughton Norf 160 B4
Roughton Shrops 132 E5
Roughton Moor Lincs 174 C2
Roughway Kent 52 C6
Round Bush Herts 85 F10
Round Green Luton 103 G11
Round Maple Suff 107 C9
Round Oak Shrops 131 G7
Round Oak W Mid 133 F8
Round Spinney Northants 120 D5
Round Street Kent 69 F7
Roundbush Essex 88 E5
Roundbush Green Essex 87 C9
Roundham Som 28 F6
Roundhay W Yorks 206 F2
Round's Green W Mid 133 F9
Roundshaw London 67 G10
Roundstonefoot Dumfries 248 B4
Roundstreet Common W Sus 35 B9
Roundswell Devon 40 G4
Roundthorn Gtr Man 184 D4
Roundthwaite Cumb 222 E2
Roundway Wilts 62 G4
Roundyhill Angus 287 B7
Rous Lench Worcs 117 G10
Rousdon Devon 15 C11
Rousham Oxon 101 G9
Routenburn N Ayrs 266 C3
Routh E Yorks 209 E7
Rout's Green Bucks 84 F3
Row Corn 11 F7
Row Cumb 211 B8
Row Cumb 231 E8
Row Ash Hants 33 E8
Row Brow Cumb 229 D7
Row Green Essex 106 G4
Row Heath Essex 89 B10
Row-of-trees E Ches 184 F4
Row Town Sur 66 G4
Rowanburn Dumfries 239 B10
Rowanfield Glos 99 G8
Rowardennan Stirl 277 B7
Rowarth Derbys 185 D8
Rowbarton Som 28 B2
Rowberrow Som 44 B3
Rowde Wilts 62 G3
Rowden Devon 13 B8
Rowden N Yorks 205 B11
Rowe Head Cumb 210 D5
Rowen Conwy 180 G3
Rowfoot Northumb 240 E5
Rowford Som 28 B2
Rowhedge Essex 107 G10
Rowhill Sur 66 G4
Rowhook W Sus 50 G6
Rowington Warks 118 D4
Rowington Green Warks 118 D4
Rowland Derbys 186 G2
Rowlands Castle Hants 34 E2
Rowlands Gill T & W 242 F5
Rowland's Green Hereford 98 D3
Rowledge Sur 49 E10
Rowlestone Hereford 97 F7
Rowley E Yorks 208 G5
Rowley Shrops 130 B6
Rowley Green London 86 F2
Rowley Hill W Yorks 197 E7
Rowley Park Staffs 151 E8
Rowley Regis W Mid 133 F9
Rowley's Green W Mid 134 G6
Rowling Kent 55 C9
Rowly Sur 50 E4
Rownall Staffs 169 F7
Rowner Hants 33 G9
Rowney Green Worcs 117 C10
Rownhams Hants 32 D5
Rowrah Cumb 219 B11
Rowsham Bucks 84 B4
Rowsley Derbys 170 B3
Rowstock Oxon 64 B3
Rowston Lincs 173 D9
Rowthorne Derbys 171 C7
Rowton Shrops 149 G7
Rowton Telford 150 F2
Rowton W Ches 166 C6
Rowton Moor W Ches 166 C6
Roxburgh Borders 262 C5
Roxburgh Mains Borders 262 D5
Roxby N Lincs 200 D2
Roxby N Yorks 226 B5
Roxeth London 66 B6
Roxton Beds 122 G3
Roxwell Essex 87 D10
Royal British Legion Village Kent 53 B8
Royal Leamington Spa Warks 118 D6
Royal Oak Darl 233 G10
Royal Oak Lancs 194 G2
Royal Oak N Yorks 218 D2
Royal Tunbridge Wells = Tunbridge Wells Kent 52 F5
Royal's Green E Ches 167 G10
Roybridge Highld 290 E4
Royd S Yorks 197 G8
Royd Moor S Yorks 197 G8
Royd Moor W Yorks 198 E2
Roydhouse W Yorks 197 E8
Roydon Essex 86 D6
Roydon Norf 141 G11
Roydon Norf 158 E4
Roydon Hamlet Essex 86 D6
Royds Green W Yorks 197 B11
Roystone Glasgow 268 B2
Royston Herts 105 C7
Royston S Yorks 197 E11
Royston Water Som 28 E2
Royton Gtr Man 196 F2
Ruabon = Rhiwabon Wrex 166 G4
Ruaig Argyll 288 E2
Ruan High Lanes Corn 3 B10
Ruan Lanihorne Corn 5 G7
Ruan Major Corn 2 F6
Ruan Minor Corn 2 F6
Ruarach Highld 295 C11
Ruardean Glos 79 B10
Ruardean Hill Glos 79 B10
Ruardean Woodside Glos 79 B10
Rubery Worcs 117 B9
Rubha Ghaisinis W Isles 297 G4
Rubha Stoer Highld 306 F5
Ruchazie Glasgow 268 B3
Ruchill Glasgow 267 B11
Ruckcroft Cumb 230 C6
Ruckhall Hereford 97 D9
Ruckinge Kent 54 G4
Ruckland Lincs 190 F4
Rucklers Lane Herts 85 E9
Ruckley Shrops 131 C10
Rudbaxton Pembs 91 G9
Rudby N Yorks 225 D9
Ruddington Notts 153 C11
Ruddle Glos 79 C11
Rudford Glos 98 G5
Rudge Shrops 132 D6
Rudge Som 45 C10
Rudge Heath Shrops 132 D5
Rudgeway S Glos 60 B6
Rudgwick W Sus 50 G5
Rudhall Hereford 98 F2
Rudheath W Ches 183 G11
Rudheath Woods W Ches 184 G2
Rudhja Garbh Argyll 289 E11
Rudley Green Essex 88 E4
Rudloe Wilts 61 E10
Rudry Caerph 59 B7
Rudston E Yorks 217 F11
Rudyard Staffs 169 D7
Rueswood Shrops 149 D9
Rufford Lancs 194 D3
Rufforth York 206 C6
Ruffs Notts 171 F8
Rugby Warks 119 B10
Rugeley Staffs 151 F10
Ruggin Som 27 D11
Ruglen S Ayrs 245 C7
Rugley Northumb 264 G5
Ruilick Highld 300 E5
Ruishton Som 28 C3
Ruisigearraidh W Isles 296 C5
Ruislip London 66 B5
Ruislip Common London 66 B5
Ruislip Gardens London 66 B5
Ruislip Manor London 66 B6
Ruiton W Mid 133 E8
Ruloe W Ches 183 G9
Rumach Highld 295 G8

Rumbling Bridge Perth 279 B10
Rumbow Cottages Worcs 117 B8
Rumburgh Suff 142 G6
Rumbush W Mid 118 B2
Rumer Hill Staffs 133 B9
Rumford Corn 10 G3
Rumford Falk 279 F8
Rumney Cardiff 59 D8
Rumsam Devon 40 G5
Rumwell Som 27 C11
Runcorn Halton 183 E8
Runcton W Sus 22 C5
Runcton Holme Norf 140 B2
Rundlestone Devon 13 G7
Runfold Sur 49 D11
Runhall Norf 141 B11
Runham Norf 143 B10
Runham Norf 161 G9
Runham Vauxhall Norf 143 B10
Running Hill Head Gtr Man 196 F4
Running Waters Durham 234 C2
Runnington Som 27 C10
Russell Green Essex 88 D3
Runshaw Moor Lancs 194 D4
Runswick Bay N Yorks 226 B6
Runwell Essex 88 G2
Ruscombe Glos 80 D4
Ruscombe Wokingham 65 D9
Ruscote Oxon 101 C8
Rush Green Essex 89 B11
Rush Green Herts 86 C5
Rush Green Herts 104 G4
Rush Green London 68 B4
Rush Green Norf 141 B11
Rush-head Aberds 303 E8
Rush Hill Bath 61 G8
Rushall Hereford 98 E2
Rushall Norf 142 G3
Rushall W Mid 133 C10
Rushall Wilts 46 B6
Rushbrooke Suff 125 E7
Rushbury Shrops 131 E10
Rushcombe Bottom Poole 18 B5
Rushden Herts 104 E6
Rushden Northants 121 D9
Rushenden Kent 70 E2
Rusher's Cross E Sus 37 B10
Rushey Mead Leicester 136 B2
Rushford Devon 12 F4
Rushford Norf 141 G8
Rushgreen Warr 183 D11
Rushington Hants 32 E5
Rushlake Green E Sus 23 B10
Rushland Cross Cumb 210 B6
Rushley Green Essex 106 D5
Rushmere C Beds 103 F8
Rushmere Suff 143 F9
Rushmere St Andrew Suff 108 B4
Rushmere Street Suff 108 B4
Rushmoor Sur 49 E11
Rushmoor Telford 150 G2
Rushmore Hants 33 E11
Rushmore Hill London 68 G3
Rushock Hereford 114 F6
Rushock Worcs 117 C7
Rusholme Gtr Man 184 B5
Rushton Dorset 18 D3
Rushton N Yorks 217 C9
Rushton Northants 136 G6
Rushton Shrops 132 B2
Rushton W Ches 167 C9
Rushton Spencer Staffs 168 C6
Rushwick Worcs 116 G6
Rushy Green E Sus 23 C7
Rushyford Durham 233 F11
Ruskie Stirl 285 G10
Ruskington Lincs 173 E9
Rusland Cumb 210 B6
Rusling End Herts 104 G4
Rusper W Sus 51 F8
Ruspidge Glos 79 C11
Russ Hill Sur 51 E8
Russel Highld 299 E8
Russell Hill London 67 G10
Russell's Green E Sus 38 E2
Russell's Hall W Mid 133 F8
Russell's Water Oxon 65 B8
Russel's Green Suff 126 C5
Rusthall Kent 52 F5
Rustington W Sus 35 G9
Ruston N Yorks 217 C9
Ruston Parva E Yorks 217 G11
Ruswarp N Yorks 227 D7
Ruthall Shrops 131 F11
Rutherford Borders 262 C4
Rutherglen S Lnrk 268 C2
Ruthernbridge Corn 5 B10
Ruthin Denb 165 D10
Ruthin V Glam 58 D3
Ruthrieston Aberdeen 293 C11
Ruthven Aberds 302 E5
Ruthven Angus 286 C6
Ruthven Highld 291 D9
Ruthven Highld 301 F8
Ruthven House Angus 287 C7
Ruthvoes Corn 5 C8
Ruthwaite Cumb 229 D10
Ruthwell Dumfries 238 D3
Ruxley London 68 E3
Ruxton Hereford 97 F11
Ruxton Green Hereford 79 B8
Ruyton-XI-Towns Shrops 149 E7
Ryal Northumb 242 C2
Ryal Fold Blkburn 195 C7
Ryall Dorset 16 C4
Ryall Worcs 99 C7
Ryarsh Kent 53 B7
Rychraggan Highld 300 F4
Rydal Cumb 221 D7
Ryde I o W 21 C7
Rydens Sur 66 F6
Rydeshill Sur 50 C3
Rydon Devon 14 G3
Rye E Sus 38 C6
Rye Common Hants 49 C9
Rye Foreign E Sus 38 C5
Rye Harbour E Sus 38 D6
Rye Park Herts 86 C5
Rye Street Worcs 98 D5
Ryebank Shrops 149 C10
Ryecroft S Yorks 186 B6
Ryecroft W Yorks 205 F7
Ryecroft Gate Staffs 168 C6
Ryeford Glos 80 E4
Ryehill E Yorks 201 B8
Ryeish Green Wokingham 65 F8
Ryelands Hereford 115 F9
Ryeworth Glos 99 G9
Ryhall Rutland 155 G10
Ryhill W Yorks 197 E11
Ryhope T & W 243 G10
Rylah Derbys 171 B7
Rylands Notts 153 B10
Rylstone N Yorks 204 B5
Ryme Intrinseca Dorset 29 E9
Ryther N Yorks 207 F7
Ryton Glos 98 E4
Ryton N Yorks 216 D5
Ryton Shrops 132 C5
Ryton T & W 242 E5
Ryton Warks 135 F7
Ryton-on-Dunsmore Warks 119 C7
Ryton Woodside T & W 242 E4

S

Sabden Lancs 203 F11
Sabine's Green Essex 87 F8
Sackers Green Suff 107 D8
Sacombe Herts 86 B4
Sacombe Green Herts 86 B4
Sacriston Durham 233 B10
Sadberge Darl 224 B6
Saddell Argyll 255 D8
Saddell Ho Argyll 255 D8
Saddington Leics 136 E3
Saddle Bow Norf 158 F2
Saddle Street Dorset 28 G5
Saddlescombe W Sus 36 E3
Sadgill Cumb 221 D9
Saffron Walden Essex 105 D10
Saffron's Cross Hereford 115 G10
Sageston Pembs 73 E9
Saham Hills Norf 141 C8
Saham Toney Norf 141 C8
Saighdinis W Isles 296 E4
Saighton W Ches 166 C6
Sain Dunwyd = St Donats V Glam 58 F2
Sain Tathon = St Athan V Glam 58 F4
St Abbs Borders 273 B8
St Abb's Haven Borders 273 B8
St Agnes Corn 4 E4
St Agnes Scilly 1 H3
St Albans Herts 85 D10
St Allen Corn 4 E6
St Andrews Fife 287 F9
St Andrew's Major V Glam 58 E6
St Andrew's Wood Devon 27 F9
St Annes Lancs 193 B10
St Anne's Park Bristol 60 E6
St Ann's Dumfries 248 E3
St Ann's Nottingham 171 G9
St Ann's Chapel Corn 12 G4
St Ann's Chapel Devon 8 F3
St Anthony Corn 3 C9
St Anthony-in-Meneage Corn 3 D7
St Anthony's T & W 243 E7
St Anthony's Hill E Sus 23 E10
St Arvans Mon 79 F8
St Asaph = Llanelwy Denb 181 G8
St Athan = Sain Tathon V Glam 58 F4
St Augustine's Kent 54 C6
St Austell Corn 5 E10
St Austins Hants 20 B2
St Bees Cumb 219 C9
St Blazey Corn 5 E11
St Blazey Gate Corn 5 E11
St Boswells Borders 262 C3
St Breock Corn 10 G5
St Breward Corn 11 F7
St Briavels Glos 79 E9
St Briavels Common Glos 79 E8
St Bride's Pembs 72 C4
St Brides Major = Saint-y-Brid V Glam 57 G11
St Bride's Netherwent Mon 60 B2
St Brides-super-Ely V Glam 58 D5
St Brides Wentlooge Newport 59 C9
St Budeaux Plym 7 D8
St Buryan Corn 1 D4
St Catherine Bath 61 E9
St Catherine's Argyll 284 G5
St Catherine's Hill Dorset 19 B8
St Chloe Glos 80 E4
St Clears = Sanclêr Carms 74 B3
St Cleer Corn 6 B5
St Clement Corn 4 G6
St Clether Corn 11 E10
St Colmac Argyll 275 G11
St Columb Major Corn 5 C8
St Columb Minor Corn 4 C6
St Columb Road Corn 5 D8
St Combs Aberds 303 C10
St Cross Hants 33 B7
St Cross South Elmham Suff 142 G5
St Cyrus Aberds 293 G9
St David's Perth 286 E3
St David's = Tyddewi Pembs 90 F5
St Day Corn 4 G4
St Decumans Som 42 E5
St Dennis Corn 5 D9
St Denys Soton 32 E6
St Devereux Hereford 97 E8
St Dials Torf 78 G3
St Dogmaels = Llandudoch Pembs 92 B3
St Dominick Corn 7 B8
St Donat's = Sain Dunwyd V Glam 58 F2
St Edith's Wilts 62 G3
St Endellion Corn 10 F5
St Enoder Corn 5 D7
St Erme Corn 4 E6
St Erney Corn 7 D7
St Erth Corn 2 B3
St Erth Praze Corn 2 B3
St Ervan Corn 10 G3
St Eval Corn 5 B7
St Ewe Corn 5 F9
St Fagans Cardiff 58 D6
St Fergus Aberds 303 D10
St Fillans Perth 285 E10
St Florence Pembs 73 E9
St Gennys Corn 11 B8
St George Bristol 60 E6
St George Conwy 181 F7
St George in the East London 67 C10
St George's Gtr Man 184 B4
St Georges N Som 59 G11
St George's Telford 150 G4
St George's V Glam 58 D5
St George's Hill Sur 66 G5
St George's Well Devon 27 F8
St Germans Corn 7 D7
St Giles Lincs 189 G7
St Giles London 67 C10
St Giles in the Wood Devon 25 D8
St Giles on the Heath Devon 12 C3
St Giles's Hill Hants 33 B7
St Gluvias Corn 3 C7
St Godwalds Worcs 117 D9
St Harmon Powys 113 C9
St Helen Auckland Durham 233 F9
St Helena Warks 134 C5
St Helens Cumb 228 E6
St Helen's E Sus 38 E4
St Helens I o W 21 D8
St Helens Mers 183 B8
St Helen's S Yorks 197 F11
St Helen's Wood E Sus 38 E4
St Helier London 67 F9
St Hilary Corn 2 C3
St Hilary V Glam 58 E4
St Ibbs Herts 104 F3
St Illtyd Bl Gwent 78 E2
St Ippollytts Herts 104 F3
St Ishmael's Pembs 72 D4
St Issey Corn 10 G4
St Ive Corn 6 B6
St Ive Cross Corn 6 B6
St Ives Cambs 122 C6
St Ives Corn 2 A2
St Ives Dorset 31 G10
St James Dorset 30 C5
St James London 67 C9
St James Norf 160 E5
St James South Elmham Suff 142 G6
St James's End Northants 120 E4
St Jidgey Corn 5 B8
St John Corn 7 E8
St John's E Sus 52 G4
St John's I o M 192 D3
St John's Kent 52 B4
St John's Kent 52 E4
St Johns London 67 D11
St John's Sur 50 B3
St John's W Yorks 206 F4
St Johns Warks 118 C5
St John's Worcs 116 G6
St John's Chapel Devon 25 B8
St John's Chapel Durham 232 D3
St John's Fen End Norf 157 G10
St John's Highway Norf 157 G10
St John's Park I o W 21 C8
St John's Town of Dalry Dumfries 246 G4
St John's Wells Aberds 303 F7
St John's Wood London 67 C9
St Judes I o M 192 C4
St Julians Herts 85 D10
St Julians Newport 59 B10
St Just in Roseland Corn 3 B9
St Just Corn 1 C3
St Justinian Pembs 90 F4
St Katharines Wilts 63 G9
St Katherine's Aberds 303 F7
St Keverne Corn 3 E7
St Kew Corn 10 F6
St Kew Highway Corn 10 F6
St Keyne Corn 6 C4
St Lawrence Corn 5 B10
St Lawrence Essex 89 E7
St Lawrence I o W 20 F6
St Lawrence Kent 71 F11
St Leonard's Bucks 84 D6
St Leonards Dorset 31 G10
St Leonards E Sus 38 F3
St Leonards S Lnrk 268 E2
St Leonard's Street Kent 53 B7
St Levan Corn 1 E3
St Luke's Derby 152 B6
St Luke's London 67 C10
St Lythans V Glam 58 E6
St Mabyn Corn 10 G6
St Madoes Perth 286 E5
St Margaret South Elmham Suff 142 G6
St Margaret's Hereford 97 E7
St Margarets Herts 86 C5
St Margarets London 67 E7
St Margaret's at Cliffe Kent 55 E11
St Margaret's Hope Orkney 314 G4
St Mark's Glos 99 G8
St Mark's I o M 192 E3
St Martin Corn 2 E6
St Martin Corn 6 E5
St Martins Perth 286 D5
St Martin's Shrops 148 B6
St Martin's Moor Shrops 148 B6
St Mary Bourne Hants 48 C2
St Mary Church V Glam 58 E4
St Mary Cray London 68 F3
St Mary Hill V Glam 58 D3
St Mary Hoo Medway 69 D10
St Mary in the Marsh Kent 39 B9
St Marychurch Torbay 9 B8
St Mary's Orkney 314 F4
St Mary's Bay Kent 39 B9
St Maughans Mon 79 B7
St Maughans Green Mon 79 B7
St Mawes Corn 3 C8
St Mawgan Corn 5 B7
St Mellion Corn 7 B7
St Mellons Cardiff 59 C8
St Merryn Corn 10 G3
St Mewan Corn 5 E9
St Michael Caerhays Corn 5 G9
St Michael Church Som 43 G10
St Michael Penkevil Corn 5 G7
St Michael South Elmham Suff 142 G6
St Michaels Kent 53 F11
St Michaels Torbay 9 C7
St Michaels Worcs 115 D11
St Michael's Hamlet Mers 182 D5
St Michael's on Wyre Lancs 202 E5
St Minver Corn 10 F5
St Monans Fife 287 G9
St Neot Corn 6 B3
St Neots Cambs 122 E3
St Newlyn East Corn 4 D6
St Nicholas Herts 104 F5
St Nicholas Pembs 91 D7
St Nicholas V Glam 58 E5
St Nicholas at Wade Kent 71 F9
St Nicholas South Elmham Suff 142 G6
St Nicolas Park Warks 135 E7
St Ninians Stirl 278 C5
St Olaves Norf 143 D9
St Osyth Essex 89 B10
St Osyth Heath Essex 89 B10
St Owens Cross Hereford 97 G10
St Pancras London 67 C10
St Paul's Glos 80 B4
St Paul's Cray London 68 F3
St Paul's Walden Herts 104 G3
St Peter South Elmham Suff 142 G6
St Peter The Great Worcs 117 G7
St Peter's Glos 99 G8
St Peters Kent 71 F11
St Peter's T & W 243 E7
St Petrox Pembs 73 F7
St Pinnock Corn 6 C4
St Quivox S Ayrs 257 E9
St Ruan Corn 2 F6
St Stephen Corn 5 E8
St Stephens Corn 7 D8
St Stephen's Corn 12 D2
St Stephens Herts 85 D10
St Teath Corn 11 E7
St Thomas Devon 14 C4
St Thomas Swansea 57 C7
St Tudy Corn 11 F7
St Twynnells Pembs 73 F7
St Veep Corn 6 E2
St Vigeans Angus 287 C10
St Vincent's Hamlet Essex 87 G9
St Wenn Corn 5 C9
St Weonards Hereford 97 G9
St Winnow Corn 6 D2
St y-Nyll V Glam 58 D5
Saint Hill Devon 27 F9
Saint Hill W Sus 51 F11
Saint y Brid = St Brides Major V Glam 57 G11
Saintbridge Glos 80 B5
Saintbury Glos 100 D2
Saint's Hill Kent 52 E4
Saith ffynnon Flint 181 F11
Salcombe Devon 9 G9
Salcombe Regis Devon 15 D9
Salcott-cum-Virley Essex 88 C6
Salden Bucks 102 F6
Sale Gtr Man 184 C3
Sale Green Worcs 117 F8
Saleby Lincs 191 F7
Salehurst E Sus 38 C2
Salem Carms 94 F2
Salem Ceredig 128 G3
Salem Corn 4 G4
Salen Argyll 289 E7
Salen Highld 289 C8
Salendine Nook W Yorks 196 D6
Salenside Borders 261 E11
Salesbury Lancs 203 G9
Saleway Worcs 117 F8
Salford C Beds 103 D8
Salford Gtr Man 184 B4
Salford Oxon 100 F5
Salford Ford C Beds 103 D8
Salford Priors Warks 117 G11
Salfords Sur 51 D9
Salhouse Norf 160 G6
Saligo Argyll 274 G3
Salisbury Wilts 31 B10
Salkeld Dykes Cumb 230 D6
Sallachan Highld 289 C11
Sallachy Highld 295 B11
Sallachy Highld 309 J5
Salle Norf 160 E2
Salmans Kent 52 E4
Salmonby Lincs 190 G4
Salmond's Muir Angus 287 D9
Salmonhutch Devon 14 B2
Salperton Glos 99 G11
Salperton Park Glos 81 B9
Salph End Beds 121 G11
Salsburgh N Lnrk 268 C6
Salt Staffs 151 D9
Salt Coates Cumb 238 G5
Salt End E Yorks 201 B7
Salt Hill Slough 66 C3
Salta Cumb 229 B7
Saltaire W Yorks 205 F8
Saltash Corn 7 D8
Saltburn Highld 301 C7
Saltburn-by-the-Sea Redcar 235 G9
Saltby Leics 155 D7
Saltcoats Cumb 219 F11
Saltcoats E Loth 281 E9
Saltcoats N Ayrs 266 G4
Saltcotes Lancs 193 B11
Saltdean Brighton 36 G5
Salter Lancs 212 G2
Salter Street W Mid 118 C2
Salterbeck Cumb 228 F5
Salterforth Lancs 204 D3
Salters Heath Hants 48 B6
Salters Lode Norf 139 C11
Saltershill Shrops 150 D2
Salterswall W Ches 167 B10
Salterton Wilts 46 F6
Saltfleet Lincs 191 C7
Saltfleetby All Saints Lincs 191 C7
Saltfleetby St Clement Lincs 191 C7
Saltfleetby St Peter Lincs 190 D6
Saltford Bath 61 F7
Salthouse Cumb 210 F4
Salthouse Norf 177 E9
Saltley W Mid 133 F11
Saltmarsh Newport 59 C11
Saltmarshe E Yorks 199 C9
Saltness Orkney 314 G2
Saltness Shetland 313 J4
Saltney Flint 166 B5
Salton N Yorks 216 D4
Saltrens Devon 25 C7
Saltwell T & W 243 E7
Saltwick Northumb 242 B5
Saltwood Kent 55 F7
Salum Argyll 288 E2
Salvington W Sus 35 F10
Salwarpe Worcs 117 E7
Salwayash Dorset 16 B5
Sambourne Warks 117 E11
Sambourne Wilts 45 E11
Sambrook Telford 150 E4
Samhla W Isles 296 E3
Samlesbury Lancs 203 G7
Samlesbury Bottoms Lancs 194 B6
Sampford Arundel Som 27 D10
Sampford Brett Som 42 E5
Sampford Chapple Devon 25 G10
Sampford Courtenay Devon 25 G10
Sampford Moor Som 27 D10
Sampford Peverell Devon 27 E8
Sampford Spiney Devon 12 G6
Sampool Bridge Cumb 211 B9
Samuel's Corner Essex 70 B3
Samuelston E Loth 281 G9
Sanachan Highld 299 E8
Sanaigmore Argyll 274 F3
Sanclêr = St Clears Carms 74 B3
Sancreed Corn 1 D4
Sancton E Yorks 208 F4
Sand Highld 307 K4
Sand Shetland 313 J5
Sand Som 44 D2
Sand Gate Cumb 211 D7
Sand Hole E Yorks 208 F2
Sand Hutton N Yorks 207 B9
Sand Side Cumb 210 C4
Sand Side Lancs 202 C4
Sandaig Highld 295 E9
Sandal W Yorks 197 D10
Sandal Magna W Yorks 197 D10
Sandale Cumb 229 C10
Sandavore Highld 294 G6
Sandbach E Ches 168 C3
Sandbach Heath E Ches 168 C3
Sandbank Argyll 276 E3
Sandbanks Kent 70 G4
Sandbanks Poole 18 D6
Sandborough Staffs 152 F2
Sandbraes Lincs 200 G6
Sandend Aberds 302 C5
Sanderstead London 67 G10
Sandfields Glos 99 G8
Sandfields Neath 57 C8
Sandfields Staffs 134 B2
Sandford Cumb 222 B4
Sandford Devon 26 G4
Sandford Dorset 18 D4
Sandford Hants 31 G11
Sandford I o W 20 E6
Sandford N Som 44 B2
Sandford S Lnrk 268 G4
Sandford Shrops 148 E6
Sandford Shrops 149 C11
Sandford W Yorks 205 F11
Sandford Worcs 99 B7
Sandford Batch N Som 44 B2
Sandford Hill Stoke 168 G6
Sandford on Thames Oxon 83 E8
Sandford Orcas Dorset 29 C10
Sandford St Martin Oxon 101 F8
Sandfordhill Aberds 303 E11
Sandgate Kent 55 G7
Sandgreen Dumfries 237 D7
Sandhaven Aberds 303 C9
Sandhaven Argyll 276 E3
Sandhead Dumfries 236 E2
Sandhill Bucks 102 F4
Sandhill Cambs 139 F11
Sandhill S Yorks 198 F2
Sandhills Dorset 29 E11
Sandhills Dorset 29 G9
Sandhills Mers 182 C4
Sandhills Oxon 83 D9
Sandhills Sur 50 F2
Sandhills W Yorks 206 F3
Sandhoe Northumb 241 D11
Sandhole Argyll 275 D11
Sandholme E Yorks 208 G2
Sandholme Lincs 156 B6
Sandhurst Brack 65 G10
Sandhurst Glos 98 G6
Sandhurst Kent 38 B3
Sandhurst Cross Kent 38 B3
Sandhutton N Yorks 215 C7
Sandiacre Derbys 153 B9
Sandilands Lincs 191 E8
Sandilands S Lnrk 259 B9
Sandiway W Ches 183 G10
Sandleheath Hants 31 E10
Sandling Kent 53 B9
Sandlow Green E Ches 168 B3
Sandness Shetland 313 H3
Sandon Essex 88 E2
Sandon Herts 104 E6
Sandon Staffs 151 C8
Sandonbank Staffs 151 D8
Sandown I o W 21 E7
Sandown Park Kent 52 E6
Sandpit Dorset 28 G6
Sandpits Glos 98 F6
Sandplace Corn 6 D5
Sandridge Herts 85 C11
Sandridge Wilts 62 F2
Sandringham Norf 158 D3
Sands Bucks 84 G4
Sands End London 67 D9
Sandsend N Yorks 227 C7
Sandside Cumb 210 D6
Sandside Cumb 211 C9
Sandside Orkney 314 F2
Sandside Ho Highld 310 C3
Sandsound Shetland 313 J5
Sandtoft N Lincs 199 F8
Sandvoe Shetland 312 D5
Sandway Kent 53 C11
Sandwell W Mid 133 F10
Sandwich Kent 55 B10
Sandwich Bay Estate Kent 55 B11
Sandwick Cumb 221 B8
Sandwick Orkney 314 H4
Sandwick Shetland 313 L6
Sandwith Cumb 219 C9
Sandwith Newtown Cumb 219 C9
Sandy C Beds 104 B3
Sandy Carms 75 E7
Sandy Bank Lincs 174 E3
Sandy Carrs Durham 234 C3
Sandy Cross E Sus 37 C9
Sandy Cross Sur 49 D11
Sandy Down Hants 20 B2
Sandy Gate Devon 14 C5
Sandy Haven Pembs 72 D5
Sandy Lane W Yorks 205 F8
Sandy Lane Wilts 62 F3
Sandy Lane Wrex 166 G5
Sandy Way I o W 20 E5
Sandybank Orkney 314 C5
Sandycroft Flint 166 B4
Sandyford Dumfries 248 E6
Sandyford Stoke 168 E5
Sandygate Devon 14 G3
Sandygate I o M 192 C4
Sandygate S Yorks 186 D4
Sandyhills Dumfries 237 D10
Sandylake Corn 6 C2
Sandylands Lancs 211 G8
Sandylands Som 27 C10
Sandylane Swansea 56 D5
Sandypark Devon 13 D10
Sandysike Cumb 239 D9
Sangobeg Highld 308 C4
Sangomore Highld 308 C4
Sanham Green W Berks 63 F10
Sankey Bridges Warr 183 D9
Sankyns Green Worcs 116 E5
Sanna Highld 288 C6
Sanndabhaig W Isles 297 G4
Sanndabhaig W Isles 304 E6
Sannox N Ayrs 255 C11
Sanquhar Dumfries 247 B7
Sansaw Heath Shrops 149 E10
Santon Cumb 220 E2
Santon N Lincs 200 E2
Santon Bridge Cumb 220 E2
Santon Downham Suff 140 F6
Sapcote Leics 135 E9
Sapey Bridge Worcs 116 F4
Sapey Common Hereford 116 E4
Sapiston Suff 125 B8
Sapley Cambs 122 C4
Sapperton Derbys 152 C3
Sapperton Glos 80 E6
Sapperton Lincs 155 C10
Saracen's Head Lincs 156 D6
Sarclet Highld 310 E7
Sardis Carms 75 D9
Sardis Pembs 73 D10
Sarisbury Hants 33 F8
Sarn Bridgend 58 C2
Sarn Flint 181 F10
Sarn Powys 130 E4
Sarn Bach Gwyn 144 D6
Sarn Meyllteyrn Gwyn 144 C4
Sarnau Carms 74 B4
Sarnau Ceredig 110 G6
Sarnau Gwyn 147 B9
Sarnau Powys 95 E10
Sarnau Powys 148 F4
Sarnesfield Hereford 115 G7
Saron Carms 75 C10
Saron Carms 93 D7
Saron Denb 165 C8
Saron Gwyn 163 B8
Saron Gwyn 163 D7
Sarratt Herts 85 F8
Sarratt Bottom Herts 85 F8
Sarre Kent 71 G9
Sarsden Oxon 100 G5
Sarsden Halt Oxon 100 G5
Sarsgrum Highld 308 C3
Sasaig Highld 295 E8
Sascott Shrops 149 G8
Satley Durham 233 C8
Satmar Kent 55 F9
Satran Highld 294 B6
Satron N Yorks 223 F8
Satterleigh Devon 25 C11
Satterthwaite Cumb 220 G6
Satwell Oxon 65 C8
Sauchen Aberds 293 B8
Saucher Perth 286 D5
Sauchie Clack 279 C7
Sauchieburn Aberds 293 G8
Saughall W Ches 182 G5
Saughall Massie Mers 182 D3
Saughton Edin 280 G4
Saughtree Borders 250 D3
Saul Glos 80 D2
Saundby Notts 188 D3
Saundersfoot Pembs 73 E10
Saunderton Bucks 84 E3
Saunderton Lee Bucks 84 F4
Saunton Devon 40 F3
Sausthorpe Lincs 174 B5
Saval Highld 309 J5
Savary Highld 289 E8
Saveock Corn 4 F5
Saverley Green Staffs 151 B9
Savile Park W Yorks 196 C5
Savile Town W Yorks 197 C8
Sawbridge Warks 119 D10
Sawbridgeworth Herts 87 B7
Sawdon N Yorks 217 B8
Sawley Derbys 153 C9
Sawley Lancs 203 D11
Sawley N Yorks 214 F4
Sawood W Yorks 204 G6
Sawston Cambs 105 B9
Sawtry Cambs 138 G3
Sawyer's Hill Som 27 C11
Sawyers Hill Wilts 81 G8
Saxby Leics 154 F6
Saxby Lincs 189 D8
Saxby W Sus 35 G7
Saxby All Saints N Lincs 200 D3
Saxelbye Leics 154 E4
Saxham Street Suff 125 E11
Saxilby Lincs 188 F5
Saxlingham Norf 159 B10
Saxlingham Green Norf 142 D4
Saxlingham Nethergate Norf 142 D4
Saxlingham Thorpe Norf 142 D4
Saxmundham Suff 127 E7
Saxon Street Cambs 124 F3
Saxondale Notts 154 B3
Saxtead Suff 126 D5
Saxtead Green Suff 126 E5
Saxtead Little Green Suff 126 D5
Saxthorpe Norf 160 C2
Saxton N Yorks 206 F5
Sayers Common W Sus 36 D3
Scackleton N Yorks 216 E2
Scadabhagh W Isles 305 J3
Scaftworth Notts 187 C11
Scagglethorpe N Yorks 216 E6
Scaitcliffe Lancs 195 B9
Scaladal W Isles 305 G3
Scalan Moray 292 B4
Scalasaig Argyll 274 D4
Scalby E Yorks 199 B10
Scalby N Yorks 227 G10
Scald End Beds 121 F10
Scaldwell Northants 120 C5
Scale Hall Lancs 211 G9
Scale Houses Cumb 231 B7
Scaleby Cumb 239 E11
Scalebyhill Cumb 239 E10
Scales Cumb 210 E5
Scales Cumb 230 F2
Scales Cumb 231 C7
Scales Lancs 202 G5
Scalford Leics 154 E5
Scaling Redcar 226 C4
Scallasaig Highld 295 D10
Scallastle Argyll 289 F8
Scalloway Shetland 313 K6
Scalpay W Isles 305 J4
Scalpay Ho Highld 295 C8
Scalpsie Argyll 255 B11
Scamadale Highld 295 F9
Scamblesby Lincs 190 F3
Scamland E Yorks 207 E11
Scammadale Argyll 289 G10
Scamodale Highld 289 B10
Scampston N Yorks 217 D7
Scampton Lincs 189 F7
Scaniport Highld 300 F6
Scapa Orkney 314 F4
Scapegoat Hill W Yorks 196 D5
Scar Orkney 314 B6
Scar Head Cumb 220 G5
Scarborough N Yorks 217 B10
Scarcewater Corn 5 E8
Scarcliffe Derbys 171 B7
Scarcroft W Yorks 206 E3
Scarcroft Hill W Yorks 206 E3
Scardroy Highld 300 D2
Scarff Shetland 312 E4
Scarfskerry Highld 310 B6
Scargill Durham 223 C11
Scarinish Argyll 288 E2
Scarisbrick Lancs 193 E11
Scarness Cumb 229 E10
Scarning Norf 159 G9
Scarrington Notts 172 G2
Scarth Hill Lancs 194 F2
Scarthingwell N Yorks 206 F5
Scartho NE Lincs 201 F9
Scarvister Shetland 313 J5
Scarwell Orkney 314 D2
Scatness Shetland 313 M5
Scatraig Highld 301 F7
Scatwell Ho Highld 300 D3
Scawby N Lincs 200 F3
Scawsby S Yorks 198 G5
Scawthorpe S Yorks 198 F5
Scawton N Yorks 215 C10
Scayne's Hill W Sus 36 C5
Scethrog Powys 96 F2
Scholar Green E Ches 168 D4
Scholemoor W Yorks 205 G8
Scholes Gtr Man 194 F5
Scholes S Yorks 186 B5
Scholes W Yorks 197 B7
Scholes W Yorks 197 F7
Scholes W Yorks 204 F6
Scholes W Yorks 206 F3
Scholey Hill W Yorks 197 B11
School Aycliffe Durham 233 G11
School Green Essex 106 E4
School Green I o W 20 D2
School Green W Ches 167 C10
School Green W Yorks 205 G8
School House Dorset 28 G5
Schoolgreen Wokingham 65 F8
Schoolhill Aberds 293 D11
Sciberscross Highld 309 H7
Scilly Bank Cumb 219 B9
Scissett W Yorks 197 E8
Scleddau Pembs 91 E8
Sco Ruston Norf 160 E5
Scofton Notts 187 E10
Scole Norf 126 B2
Scole Common Norf 142 G2
Scolpaig W Isles 296 D3
Scone Perth 286 E5
Sconser Highld 295 B7
Scoonie Fife 287 G7
Scoor Argyll 274 B5
Scopwick Lincs 173 D9
Scoraig Highld 307 K5
Scorborough E Yorks 208 D6
Scorrier Corn 4 G4
Scorriton Devon 8 B4
Scorton Lancs 202 D6
Scorton N Yorks 224 E5
Scot Hay Staffs 168 F4
Scot Lane End Gtr Man 194 F6
Scotbheinn W Isles 296 F4
Scotby Cumb 239 G10
Scotch Corner N Yorks 224 E4
Scotches Derbys 170 F4
Scotforth Lancs 202 B5
Scotgate W Yorks 196 E6
Scothern Lincs 189 F8
Scotland Leics 135 D8
Scotland Leics 136 D3
Scotland Leics 153 E7
Scotland Lincs 155 C10
Scotland W Berks 64 F5
Scotland End Oxon 100 E6
Scotland Gate Northumb 253 G7
Scotland Street Suff 107 D9
Scotlands W Mid 133 C8
Scotlandwell Perth 286 G5
Scots' Gap Northumb 252 F2
Scotsburn Highld 301 B7
Scotscalder Station Highld 310 D4
Scotscraig Fife 287 E8
Scotston Aberds 293 F9
Scotston Perth 286 C3
Scotstoun Glasgow 267 B10
Scotstown Highld 289 C10
Scotswood T & W 242 E5
Scotswood Windsor 66 F2
Scott Willoughby Lincs 155 B11
Scottas Highld 295 E9
Scotter Lincs 199 G11
Scotterthorpe Lincs 199 G11
Scottlethorpe Lincs 155 E11
Scotton Lincs 188 B5
Scotton N Yorks 206 B2
Scotton N Yorks 224 F3
Scottow Norf 160 E5
Scoughall E Loth 282 E2
Scoulag Argyll 266 D2
Scoulton Norf 141 C9
Scounslow Green Staffs 151 D11
Scourie Highld 306 E6
Scourie More Highld 306 E6
Scousburgh Shetland 313 M5
Scout Dike S Yorks 197 G8
Scout Green Cumb 221 D11
Scouthead Gtr Man 196 F3
Scowles Glos 79 C9
Scrabster Highld 310 B4
Scraesburgh Borders 262 F5
Scrafield Lincs 174 B4
Scragged Oak Kent 69 G10
Scrainwood Northumb 251 B11
Scrane End Lincs 174 G5
Scrapsgate Kent 70 E2
Scraptoft Leics 136 B2
Scrapton Som 28 E3
Scratby Norf 161 F10
Scrayingham N Yorks 216 G4
Scredda Corn 5 E10
Scredington Lincs 173 G9
Screedy Som 27 B9
Scremby Lincs 174 B6
Scremerston Northumb 273 F10
Screveton Notts 172 G2
Scrivelsby Lincs 174 B3
Scriven N Yorks 206 B2
Scronkey Lancs 202 D4
Scrooby Notts 187 C11
Scropton Derbys 152 C3
Scrub Hill Lincs 174 D2
Scruton N Yorks 224 G5
Scrwgan Powys 148 E3
Scuddaborg Highld 298 C3
Scuggate Cumb 239 C10
Sculcoates Hull 209 G7
Sculthorpe Norf 159 C7
Scunthorpe N Lincs 199 E11
Scurlage Swansea 56 D3
Sea Som 28 E4
Sea Mill Cumb 210 F5
Sea Mills Bristol 60 D5
Sea Mills Corn 10 G4
Sea Palling Norf 161 D8
Seaborough Dorset 28 F6
Seabridge Staffs 168 G4
Seabrook Kent 55 G7
Seaburn T & W 243 F10
Seacombe Mers 182 C4
Seacox Heath Kent 53 G8
Seacroft Lincs 175 C9
Seacroft W Yorks 206 F3
Seadyke Lincs 156 B6
Seafar N Lnrk 278 G5
Seafield Highld 311 L3
Seafield Midloth 270 C5
Seafield S Ayrs 257 E8
Seafield W Loth 269 B10
Seaford E Sus 23 F7
Seaforth Mers 182 B4
Seagrave Leics 154 F2
Seagry Heath Wilts 62 C3
Seaham Durham 234 B4
Seahouses Northumb 264 C6
Seal Kent 52 B4
Sealand Flint 166 B5
Seale Sur 49 D11
Seamer N Yorks 217 C10
Seamer N Yorks 225 C9
Seamill N Ayrs 266 F4
Searby Lincs 200 F5
Seasalter Kent 70 F5
Seascale Cumb 219 E10
Seathorne Lincs 175 B9
Seathwaite Cumb 220 C4
Seathwaite Cumb 220 F4
Seatle Cumb 211 C7
Seatoller Cumb 220 C4
Seaton Corn 6 E6
Seaton Cumb 228 E6
Seaton Devon 15 C10
Seaton Durham 243 G9
Seaton E Yorks 209 D9
Seaton Kent 55 B8
Seaton Northumb 243 B8
Seaton Rutland 137 D8
Seaton Burn T & W 242 C6
Seaton Carew Hrtlpl 234 F6
Seaton Delaval Northumb 243 B8
Seaton Ross E Yorks 207 E11
Seaton Sluice Northumb 243 B8
Seatown Aberds 302 C5
Seatown Aberds 303 D11
Seatown Dorset 16 C4
Seaureaugh Moor Corn 2 B6
Seave Green N Yorks 225 E11
Seaview I o W 21 C8
Seaville Cumb 238 G5
Seavington St Mary Som 28 E6
Seavington St Michael Som 28 D6
Seawick Essex 89 C10
Sebastopol Torf 78 F3
Sebay Orkney 314 F5
Sebergham Cumb 230 C3
Sebiston Velzian Orkney 314 D2
Seckington Warks 134 B5
Second Coast Highld 307 K4
Second Drove Cambs 139 F10
Sedbergh Cumb 222 G3
Sedbury Glos 79 G8
Sedbusk N Yorks 223 G7
Seddington C Beds 104 B3
Sedgeberrow Worcs 99 D10
Sedgebrook Lincs 155 B7
Sedgefield Durham 234 F3
Sedgeford Norf 158 B4
Sedgehill Wilts 30 B5
Sedgemere W Mid 118 B4
Sedgley W Mid 133 E8
Sedgley Park Gtr Man 195 G10
Sedgwick Cumb 211 B10
Sedlescombe E Sus 38 D3
Sedlescombe Street E Sus 38 D3
Sedrup Bucks 84 C3
Seed Kent 54 B2
Seed Lee Lancs 194 C5
Seedley Gtr Man 184 B4
Seend Wilts 62 G2
Seend Cleeve Wilts 62 G2
Seend Head Wilts 62 G2
Seer Green Bucks 85 G7
Seething Norf 142 D6
Seething Wells London 67 F7
Sefton Mers 193 G11
Segensworth Hants 33 F8
Seggat Aberds 303 E7
Seghill Northumb 243 C7
Seifton Shrops 131 G9
Seighford Staffs 151 D7
Seilebost W Isles 305 J2
Seion Gwyn 163 B8
Seisdon Staffs 132 E6
Seisiadar W Isles 304 E7
Selattyn Shrops 148 C5
Selborne Hants 49 G8
Selby N Yorks 207 G8
Selgrove Kent 54 B4
Selham W Sus 34 C6
Selhurst London 67 F10
Selkirk Borders 261 D11
Sellack Hereford 97 F11
Sellack Boat Hereford 97 F11
Sellafirth Shetland 312 D7
Sellan Corn 1 C4
Sellibister Orkney 314 B7
Sellick's Green Som 28 D2
Sellindge Kent 54 F6
Selling Kent 54 B4
Sells Green Wilts 62 G3
Selly Hill N Yorks 227 D7
Selly Oak W Mid 133 G10
Selly Park W Mid 133 G11
Selmeston E Sus 23 D8
Selsdon London 67 G10
Selsey W Sus 22 E5
Selsfield Common W Sus 51 G10
Selside Cumb 221 F10
Selside N Yorks 212 D5
Selsley Glos 80 E4
Selsmore Hants 21 B10
Selson Kent 55 B10
Selsted Kent 55 E8
Selston Notts 171 E7
Selston Common Notts 171 E7
Selston Green Notts 171 E7
Selwick Orkney 314 F2
Selworthy Som 42 D2
Sem Hill Wilts 30 B5
Semblister Shetland 313 H5
Semer Suff 107 B9
Semington Wilts 61 G11
Semley Wilts 30 B5
Sempringham Lincs 156 C2
Send Sur 50 B4
Send Grove Sur 50 C4
Send Marsh Sur 50 B4
Senghenydd Caerph 77 G10
Sennen Corn 1 D3
Sennen Cove Corn 1 D3

Sennybridge =
Pont Senni Powys 95 F8
Serlby Notts 187 D10
Serrington Wilts 46 F5
Sessay N Yorks 215 D9
Setchey Norf 158 G2
Setley Hants 32 G4
Seton E Loth 281 G8
Seton Mains E Loth 281 F8
Setter Shetland 312 E6
Setter Shetland 313 H5
Setter Shetland 313 J7
Setter Shetland 313 L6
Settiscarth Orkney 314 E3
Settle N Yorks 212 G6
Settrington N Yorks 216 E6
Seven Ash Som 43 G7
Seven Kings London 68 B3
Seven Sisters =
Blaendulais Neath 76 D4
Seven Springs Glos 81 B7
Seven Star Green Essex 107 F8
Sevenhampton Glos 99 G10
Sevenhampton Swindon 82 G2
Sevenoaks Kent 52 C4
Sevenoaks Common
Kent 52 C4
Sevenoaks Weald Kent 52 C4
Severn Beach S Glos 60 B4
Severn Stoke Worcs 99 C7
Severnhampton Swindon 82 G2
Sevick End Beds 121 G11
Sevington Kent 54 E4
Sewards End Essex 105 D11
Sewardstone Essex 86 F5
Sewardstonebury Essex 86 F5
Sewell C Beds 103 G9
Sewerby E Yorks 218 F3
Seworgan Corn 2 C6
Sewstern Leics 155 E7
Sexhow N Yorks 225 D9
Sezincote Glos 100 E3
Sgarasta Mhor W Isles 305 J2
Sgiogarstaigh W Isles 304 B7
Sgiwen = Skewen Neath 57 B8
Shab Hill Glos 80 B6
Shabbington Bucks 83 D11
Shackerley Shrops 132 B6
Shackerstone Leics 135 B7
Shacklecross Derbys 153 C8
Shackleford Sur 50 D2
Shackleton W Yorks 196 B3
Shacklewell London 67 B10
Shackloford Sur 50 D2
Shade W Yorks 196 C2
Shadforth Durham 234 C2
Shadingfield Suff 143 G8
Shadoxhurst Kent 54 F3
Shadsworth Blkburn 195 B8
Shadwell Glos 80 F3
Shadwell London 67 C11
Shadwell Norf 141 G8
Shadwell W Yorks 206 F2
Shaffalong Staffs 169 E7
Shaftenhoe End Herts 105 D8
Shaftesbury Dorset 30 C5
Shafton S Yorks 197 E11
Shafton Two Gates
S Yorks 197 E11
Shaggs Dorset 18 E3
Shakeford Shrops 150 D3
Shakerley Gtr Man 195 G7
Shakesfield Glos 98 E3
Shalbourne Wilts 63 G10
Shalcombe I o W 20 D3
Shalden Hants 49 E7
Shalden Green Hants 49 E7
Shaldon Devon 14 G4
Shalfleet I o W 20 D4
Shalford Essex 106 F4
Shalford Som 45 G8
Shalford Sur 50 D4
Shalford Green Essex 106 F4
Shalloch Moray 302 D3
Shallowford Devon 25 B11
Shallowford Devon 41 E8
Shallowford Staffs 151 D7
Shalmsford Street Kent 54 C5
Shalstone Bucks 102 D2
Shamley Green Sur 50 E4
Shandon Argyll 276 D5
Shandwick Highld 301 B8
Shangton Leics 136 D4
Shankhouse Northumb 243 B7
Shanklin I o W 21 E7
Shannochie N Ayrs 255 E10
Shannochill Stirl 277 B10
Shanquhar Aberds 302 F5
Shanwell Fife 287 E8
Shanzie Perth 286 B6
Shap Cumb 221 B11
Shapridge Glos 79 B11
Shapwick Dorset 30 G6
Shapwick Som 44 F2
Sharcott Wilts 46 B6
Shard End W Mid 134 F3
Shardlow Derbys 153 C8
Shareshill Staffs 133 B8
Sharlston W Yorks 197 D11
Sharlston Common
W Yorks 197 D11
Sharmans Cross W Mid 118 B2
Sharnal Street Medway 69 E9
Sharnbrook Beds 121 F9
Sharneyford Lancs 195 C11
Sharnford Leics 135 E9
Sharnhill Green Dorset 30 F2
Sharoe Green Lancs 202 G6
Sharow N Yorks 214 E6
Sharp Street Norf 161 E7
Sharpenhoe C Beds 103 E11
Sharperton Northumb 251 C11
Sharples Gtr Man 195 E8
Sharpley Heath Staffs 151 B9
Sharpness Glos 79 E11
Sharp's Corner E Sus 23 B9
Sharpsbridge E Sus 36 C6
Sharpstone Bath 45 B9
Sharpthorne W Sus 51 G11
Sharptor Corn 11 G11
Sharpway Gate Worcs 117 D9
Sharrington Norf 159 B10
Sharrow S Yorks 186 D4
Sharston Gtr Man 184 D4
Shatterford Worcs 132 G5
Shatterling Kent 55 B9
Shatton Derbys 185 E11
Shaugh Prior Devon 7 C10
Shavington E Ches 168 E2
Shaw Gtr Man 196 F2
Shaw Swindon 62 B6
Shaw W Berks 64 F3
Shaw W Yorks 204 F6
Shaw Wilts 61 F11
Shaw Common Glos 98 F3
Shaw Green Herts 104 E5
Shaw Green Lancs 194 D4
Shaw Green N Yorks 205 C11
Shaw Heath E Ches 184 F3
Shaw Heath Gtr Man 184 D5
Shaw Lands S Yorks 197 F10
Shaw Mills N Yorks 214 G5
Shaw Side Gtr Man 196 F2
Shawbank Shrops 131 G9
Shawbirch Telford 150 G2
Shawbury Shrops 149 E11
Shawclough Gtr Man 195 E11
Shawdon Hall Northumb 264 G3
Shawell Leics 135 G10
Shawfield Gtr Man 195 E11
Shawfield Staffs 169 C9
Shawfield Head
N Yorks 205 C11
Shawford Hants 33 C7
Shawford Som 45 C9
Shawforth Lancs 195 C11
Shawhead Dumfries 237 B10
Shawhead N Lnrk 268 C4
Shawhill Dumfries 238 D6
Shawlands Glasgow 267 C11
Shawsburn S Lnrk 268 E5
Shawton S Lnrk 268 F3
Shawtonhill S Lnrk 268 F3
Shay Gate W Yorks 205 F8
Sheandow Moray 302 F2
Shear Cross Wilts 45 E11
Shearington Dumfries 238 D2
Shearsby Leics 136 E2
Shearston Som 43 G9
Shebbear Devon 24 F6
Shebdon Staffs 150 D5
Shebster Highld 310 C4
Sheddens E Renf 267 D11
Shedfield Hants 33 E9
Sheen Staffs 169 C10
Sheep Hill Durham 242 F5
Sheepbridge Derbys 186 G5
Sheepdrove W Berks 63 D10
Sheeplane C Beds 103 E8
Sheepridge Bucks 65 B11
Sheepridge W Yorks 197 D7
Sheepscar W Yorks 206 G2
Sheepscombe Glos 80 C5
Sheepstor Devon 7 B11
Sheepwash Devon 25 F7
Sheepwash Northumb 253 F7
Sheepway N Som 60 D3
Sheepy Magna Leics 134 C6
Sheepy Parva Leics 134 C6
Sheering Essex 87 C8
Sheerness Kent 70 E2
Sheerwater Sur 66 G4
Sheet Hants 34 C3
Sheet Shrops 115 C10
Sheets Heath Sur 50 B2
Sheffield Corn 1 D5
Sheffield S Yorks 186 D5
Sheffield Bottom
W Berks 65 F7
Sheffield Green E Sus 36 C6
Sheffield Park S Yorks 186 D5
Shefford C Beds 104 D2
Shefford Woodlands
W Berks 63 E11
Sheigra Highld 306 C6
Sheildmuir N Lnrk 268 D5
Sheinton Shrops 132 C2
Shelderton Shrops 115 B8
Sheldon Derbys 169 B11
Sheldon Devon 27 F10
Sheldon W Mid 134 G3
Sheldwich Kent 54 B4
Sheldwich Lees Kent 54 B4
Shelf Bridgend 58 C2
Shelf W Yorks 196 B6
Shelfanger Norf 142 G2
Shelfield W Mid 133 C10
Shelfield Warks 118 E2
Shelfield Green Warks 118 E2
Shelfleys Northants 120 F4
Shelford Notts 171 G11
Shelford Warks 135 F8
Shell Worcs 117 F9
Shell Green Halton 183 D8
Shelland Suff 125 E10
Shellbrook Leics 152 F6
Shelley Essex 87 E9
Shelley Suff 107 D10
Shelley W Yorks 197 E8
Shelley Woodhouse
W Yorks 197 E8
Shellingford Oxon 82 G4
Shellow Bowells Essex 87 D10
Shellwood Cross Sur 51 D8
Shelsley Beauchamp
Worcs 116 E4
Shelsley Walsh Worcs 116 E4
Shelthorpe Leics 153 F10
Shelton Beds 121 D10
Shelton Norf 142 E4
Shelton Notts 172 G3
Shelton Shrops 149 G9
Shelton Stoke 168 F5
Shelton Green Norf 142 E4
Shelton Lock Derby 153 C7
Shelton under Harley
Staffs 150 B6
Shelve Shrops 130 D6
Shelvin Devon 27 G11
Shelvingford Kent 71 F8
Shelwick Hereford 97 C10
Shelwick Green
Hereford 97 C10
Shenfield Essex 87 G10
Shenington Oxon 101 C7
Shenley Herts 85 E11
Shenley Brook End
M Keynes 102 D6
Shenley Church End
M Keynes 102 D6
Shenley Fields W Mid 133 G10
Shenley Lodge
M Keynes 102 D6
Shenley Wood M Keynes 102 D6
Shenleybury Herts 85 E11
Shenmore Hereford 97 D7
Shennanton Dumfries 236 C5
Shennanton Ho
Dumfries 236 C5
Shenstone Staffs 134 C2
Shenstone Worcs 117 C7
Shenstone Woodend
Staffs 134 C2
Shenton Leics 135 C7
Shenval Highld 300 G4
Shenval Moray 302 G2
Shenvault Moray 301 E10
Shepeau Stow Lincs 156 G6
Shephall Herts 104 G5
Shepherd Hill W Yorks 197 C9
Shepherd's Bush London 67 C8
Shepherd's Gate Norf 157 F11
Shepherd's Green Oxon 65 C8
Shepherd's Hill Sur 50 G2
Shepherd's Patch Glos 80 E2
Shepherd's Port Norf 158 C3
Shepherdswell or
Sibertswold Kent 55 D9
Shepley W Yorks 197 F7
Shepperdine S Glos 79 F10
Shepperton Sur 66 F5
Shepperton Green Sur 66 F5
Shepreth Cambs 105 B7
Shepshed Leics 153 F9
Shepton Beauchamp
Som 28 D6
Shepton Mallet Som 44 E6
Shepton Montague Som 45 G7
Shepway Kent 53 C9
Sheraton Durham 234 D4
Sherberton Devon 13 G8
Sherborne Bath 44 B5
Sherborne Dorset 29 D10
Sherborne Glos 81 C11
Sherborne St John
Hants 48 B6
Sherbourne Warks 118 E5
Sherbourne Street Suff 107 C9
Sherburn Durham 234 C2
Sherburn N Yorks 217 D9
Sherburn Grange
Durham 234 C2
Sherburn Hill Durham 234 C2
Sherburn in Elmet
N Yorks 206 G5
Shere Sur 50 D5
Shereford Norf 159 D7
Sherfield English Hants 32 C3
Sherfield on Loddon
Hants 49 B7
Sherfin Lancs 195 B9
Sherford Devon 8 G5
Sherford Dorset 18 C4
Sherford Som 28 C2
Sheriff Hill T & W 243 E7
Sheriff Hutton N Yorks 216 F3
Sheriffhales Shrops 150 G5
Sheriff's Lench Worcs 99 B10
Sheringham Norf 177 E11
Sherington M Keynes 103 B7
Sheringwood Norf 177 E11
Shermanbury W Sus 36 D2
Shernal Green Worcs 117 E8
Shernborne Norf 158 C4
Sherrard's Green Worcs 98 B5
Sherrardspark Herts 86 C2
Sherriffhales Shrops 150 G5
Sherrington Wilts 46 F3
Sherston Wilts 61 B11
Sherwood Nottingham 171 G9
Sherwood Green Devon 25 C9
Sherwood Park Kent 52 E6
Shettleston Glasgow 268 C2
Shevington Gtr Man 194 F4
Shevington Moor
Gtr Man 194 E4
Shevington Vale
Gtr Man 194 F4
Sheviock Corn 7 D7
Shewalton N Ayrs 257 B8
Shibden Head W Yorks 196 B5
Shide I o W 20 D5
Shiel Aberds 292 B4
Shiel Bridge Highld 295 D11
Shieldaig Highld 299 B8
Shieldaig Highld 299 D8
Shieldhall Glasgow 267 B10
Shieldhill Dumfries 248 F2
Shieldhill Falk 279 F7
Shieldhill S Lnrk 269 G10
Shielfoot Highld 289 C8
Shielhill Angus 287 B8
Shielhill Invclyd 276 G4
Shifford Oxon 82 E5
Shifnal Shrops 132 B4
Shilbottle Northumb 252 B5
Shilbottle Grange
Northumb 252 B6
Shildon Durham 233 F10
Shillford E Renf 267 D8
Shillingford Devon 27 C7
Shillingford Oxon 83 G9
Shillingford Abbot Devon 14 D4
Shillingford St George
Devon 14 D4
Shillingstone Dorset 30 E4
Shillington C Beds 104 E2
Shillmoor Northumb 251 B9
Shilton Oxon 82 D3
Shilton Warks 135 G8
Shilvington Northumb 252 G5
Shimpling Norf 142 G3
Shimpling Suff 125 G7
Shimpling Street Suff 125 G7
Shincliffe Durham 233 C11
Shiney Row T & W 243 G8
Shinfield Wokingham 65 F8
Shingay Cambs 104 B6
Shingham Norf 140 C5
Shingle Street Suff 109 C7
Shinner's Bridge Devon 8 C5
Shinness Highld 309 H5
Shipbourne Kent 52 C5
Shipdham Norf 141 B9
Shipdham Airfield Norf 141 B9
Shipham Som 44 B2
Shiphay Torbay 9 B7
Shiplake Oxon 65 D9
Shiplake Bottom Oxon 65 C8
Shiplake Row Oxon 65 D9
Shiplate N Som 43 B11
Shiplaw Borders 270 E4
Shipley Derbys 170 G6
Shipley Northumb 264 F4
Shipley Shrops 132 D6
Shipley W Sus 35 C10
Shipley W Yorks 205 F8
Shipley Bridge Sur 51 E10
Shipley Common
Derbys 171 G7
Shipley Shiels Northumb 251 E7
Shipmeadow Suff 143 F7
Shipping Pembs 73 D10
Shippon Oxon 83 F7
Shipston-on-Stour
Warks 100 C5
Shipton Bucks 102 F5
Shipton Glos 81 B8
Shipton N Yorks 207 B7
Shipton Shrops 131 E11
Shipton Bellinger Hants 47 D8
Shipton Gorge Dorset 16 C5
Shipton Green W Sus 22 C4
Shipton Lee Bucks 102 G4
Shipton Moyne Glos 61 B11
Shipton Oliffe Glos 81 B8
Shipton on Cherwell
Oxon 83 B7
Shipton Solers Glos 81 B8
Shipton-under-
Wychwood Oxon 82 B3
Shiptonthorpe E Yorks 208 E3
Shirburn Oxon 83 F11
Shirdley Hill Lancs 193 E11
Shire Oak W Mid 133 C11
Shirebrook Derbys 171 B8
Shirecliffe S Yorks 186 C4
Shiregreen S Yorks 186 C5
Shirehampton Bristol 60 D4
Shiremoor T & W 243 C8
Shirenewton Mon 79 G7
Shireoaks Derbys 185 E9
Shireoaks Notts 187 E9
Shires Mill Fife 279 D10
Shirkoak Kent 54 F2
Shirl Heath Hereford 115 F8
Shirland Derbys 170 D6
Shirlett Shrops 132 D3
Shirley Derbys 170 G2
Shirley Hants 19 B9
Shirley London 67 F11
Shirley Soton 32 E6
Shirley W Mid 118 B2
Shirley Heath W Mid 118 B2
Shirley holms Hants 19 B11
Shirley Warren Soton 32 E5
Shirrell Heath Hants 33 E9
Shirwell Devon 40 F5
Shirwell Cross Devon 40 F5
Shiskine N Ayrs 255 E10
Shitterton Dorset 18 C2
Shobdon Hereford 115 E8
Shobley Hants 31 F11
Shobnall Staffs 152 E4
Shobrooke Devon 26 G5
Shoby Leics 154 F3
Shocklach W Ches 166 F6
Shocklach Green
W Ches 166 F6
Shoeburyness Sthend 70 C2
Sholden Kent 55 C11
Sholing Soton 32 E6
Sholing Common Soton 33 E7
Sholver Gtr Man 196 F3
Shoot Hill Shrops 149 G8
Shootash Hants 32 C4
Shooters Hill London 68 D2
Shootersway Herts 85 D7
Shop Corn 10 G3
Shop Corn 24 E2
Shop Devon 24 E5
Shop Corner Suff 108 E4
Shopford Cumb 240 C3
Shopnoller Som 43 G7
Shopp Hill W Sus 34 B6
Shopwyke W Sus 22 B5
Shore Gtr Man 196 D2
Shore W Yorks 196 B2
Shore Bottom Devon 28 G2
Shore Mill Highld 301 C7
Shoreditch London 67 C10
Shoreditch Som 28 C2
Shoregill Cumb 222 E5
Shoreham Kent 68 G4
Shoreham Beach W Sus 36 G2
Shoreham-by-Sea
W Sus 36 F2
Shores Green Oxon 82 D5
Shoresdean Northumb 273 F9
Shoreside Shetland 313 J4
Shoreswood Northumb 273 F8
Shoreton Highld 300 C6
Shorley Hants 33 B9
Shorncliffe Camp Kent 55 F7
Shorncote Glos 81 F8
Shorne Kent 69 E7
Shorne Ridgeway Kent 69 E7
Shorne West Kent 69 E7
Short Cross W Mid 133 G9
Short Green Norf 141 F11
Short Heath Derbys 152 G6
Short Heath W Mid 133 C9
Short Heath W Mid 133 E11
Short Street Wilts 45 D10
Shortacombe Devon 12 D6
Shortacross Corn 6 D5
Shortbridge E Sus 37 C7
Shortfield Common Sur 49 E10
Shortgate E Sus 23 B7
Shorthampton Oxon 100 G6
Shortheath Hants 49 F9
Shortheath Sur 49 E10
Shorthill Shrops 131 B8
Shortlands London 67 F11
Shortlanesend Corn 4 F6
Shortlees E Ayrs 257 B10
Shortmoor Devon 28 G2
Shortmoor Dorset 29 G7
Shorton Torbay 9 C7
Shortroods Renfs 267 B9
Shortstanding Glos 79 C9
Shortstown Beds 103 B11
Shortwood Glos 80 F4
Shortwood S Glos 61 D7
Shorwell I o W 20 E5
Shoscombe Bath 45 B8
Shoscombe Vale Bath 45 B8
Shotatton Shrops 149 E7
Shotesham Norf 142 D5
Shotgate Essex 88 G3
Shotley Northants 137 D8
Shotley Suff 108 D4
Shotley Bridge Durham 242 G3
Shotley Gate Suff 108 E4
Shotleyfield Northumb 242 G3
Shottenden Kent 54 C4
Shottermill Sur 49 G11
Shottery Warks 118 G3
Shotteswell Warks 101 B8
Shottisham Suff 108 C6
Shottle Derbys 170 F4
Shottlegate Derbys 170 F4
Shotton Durham 234 D4
Shotton Durham 234 F3
Shotton Flint 166 B4
Shotton Northumb 242 B6
Shotton Northumb 263 C8
Shotton Colliery
Durham 234 C3
Shotts N Lnrk 269 C7
Shotwick W Ches 182 G4
Shouldham Norf 140 B3
Shouldham Thorpe
Norf 140 B3
Shoulton Worcs 116 F6
Shover's Green E Sus 53 G7
Shraleybrook Staffs 168 F3
Shrawardine Shrops 149 F8
Shrawley Worcs 116 E6
Shreding Green Bucks 66 C4
Shrewley Warks 118 D4
Shrewley Common
Warks 118 D4
Shrewsbury Shrops 149 G9
Shrewton Wilts 46 E5
Shripney W Sus 22 C6
Shrivenham Oxon 63 B8
Shropham Norf 141 E9
Shroton or
Iwerne Courtney Dorset 30 E5
Shrub End Essex 107 G9
Shrubs Hill Sur 66 F3
Shruthernhill S Lnrk 268 F5
Shucknall Hereford 97 C11
Shudy Camps Cambs 106 C2
Shulishadermor Highld 298 E4
Shulista Highld 298 B4
Shuna Ho Argyll 275 C8
Shurdington Glos 80 B6
Shurlock Row Windsor 65 E10
Shurnock Worcs 117 E10
Shurrery Highld 310 D4
Shurrery Lodge Highld 310 D4
Shurton Som 43 E8
Shustoke Warks 134 E4
Shut Heath Staffs 151 E7
Shute Devon 15 B11
Shute Devon 26 G5
Shute End Wilts 31 B11
Shutford Oxon 101 C7
Shuthonger Glos 99 D7
Shutlanger Northants 120 G4
Shutt Green Staffs 133 B7
Shutta Corn 6 E5
Shuttington Warks 134 B5
Shuttlesfield Kent 55 E7
Shuttlewood Derbys 187 G7
Shuttleworth Gtr Man 195 D10
Shutton Hereford 98 F3
Shwt Bridgend 57 D11
Siabost bho Dheas
W Isles 304 D4
Siabost bho Thuath
W Isles 304 D4
Siadar W Isles 304 C5
Siadar Iarach W Isles 304 C5
Siadar Uarach W Isles 304 C5
Sibbaldbie Dumfries 248 F4
Sibbertoft Northants 136 G3
Sibdon Carwood Shrops 131 G8
Sibford Ferris Oxon 101 D7
Sibford Gower Oxon 101 D7
Sible Hedingham Essex 106 E5
Sibley's Green Essex 106 F2
Sibsey Lincs 174 E5
Sibsey Fen Side Lincs 174 E4
Sibson Cambs 137 D11
Sibson Leics 135 C7
Sibster Highld 310 D7
Sibthorpe Notts 172 F3
Sibthorpe Notts 188 G2
Sibton Suff 127 D7
Sibton Green Suff 127 C7
Sicklesmere Suff 125 E7
Sicklinghall N Yorks 206 D3
Sid Devon 15 D8
Sidbrook Som 28 B3
Sidbury Devon 15 C8
Sidbury Shrops 132 F3
Sidcot N Som 44 B2
Sidcup London 68 E3
Siddal W Yorks 196 C6
Siddick Cumb 228 E6
Siddington E Ches 184 G4
Siddington Glos 81 F8
Siddington Heath
E Ches 184 G4
Side of the Moor
Gtr Man 195 E8
Sidemoor Worcs 117 C9
Sidestrand Norf 160 B5
Sideway Stoke 168 G5
Sidford Devon 15 C8
Sidlesham W Sus 22 D5
Sidlesham Common
W Sus 22 C5
Sidley E Sus 38 F2
Sidlow Sur 51 D9
Sidmouth Devon 15 D8
Sidway Staffs 150 B5
Sigford Devon 13 G11
Sigglesthorne E Yorks 209 D9
Sighthill Edin 280 G3
Sighthill Glasgow 268 B2
Sigingstone = Tresigin
V Glam 58 E3
Signet Oxon 82 C2
Sigwells Som 29 C10
Silchester Hants 64 G6
Sildinis W Isles 305 G4
Sileby Leics 153 F11
Silecroft Cumb 210 C2
Silfield Norf 142 D2
Silford Devon 24 B6
Silian Ceredig 111 G11
Silk Willoughby Lincs 173 G9
Silkstead Hants 32 C6
Silkstone S Yorks 197 F9
Silkstone Common
S Yorks 197 G9
Silloth Cumb 238 G4
Sills Northumb 251 C8
Sillyearn Moray 302 D5
Siloh Carms 94 D4
Silpho N Yorks 227 G9
Silsden W Yorks 204 D6
Silsoe C Beds 103 D11
Silton Dorset 30 B3
Silver End Essex 88 B4
Silver End W Mid 133 F8
Silver Green Norf 142 E5
Silver Hill E Sus 38 B2
Silver Knap Som 29 C11
Silver Street Glos 80 E3
Silver Street Kent 69 G11
Silver Street Som 27 C11
Silver Street Som 44 G4
Silver Street Worcs 117 B11
Silverburn Midloth 270 C4
Silverdale Lancs 211 E9
Silverdale Staffs 168 F4
Silverdale Green Lancs 211 E9
Silvergate Norf 160 D3
Silverhill E Sus 38 E3
Silverhill Park E Sus 38 E4
Silverknowes Edin 280 F4
Silverley's Green Suff 126 B5
Silvermuir S Lnrk 269 F8
Silverstone Northants 102 C3
Silverton Devon 27 G7
Silverton W Dunb 277 F8
Silvertonhill S Lnrk 268 E4
Silvertown London 68 D2
Silverwell Corn 4 F4
Silvington Shrops 116 B2
Silwick Shetland 313 J4
Sim Hill S Yorks 197 G9
Simister Gtr Man 195 F10
Simmondley Derbys 185 C8
Simm's Cross Halton 183 D8
Simm's Lane End Mers 194 G4
Simonburn Northumb 241 C9
Simonsbath Som 41 F9
Simonsburrow Devon 27 D10
Simonside T & W 243 E8
Simonstone Lancs 203 G11
Simonstone N Yorks 223 G7
Simprim Borders 272 F6
Simpson M Keynes 103 D7
Simpson Pembs 72 B5
Simpson Cross Pembs 72 B5
Simpson Green W Yorks 205 F9
Sinclair's Hill Borders 272 E6
Sinclairston E Ayrs 257 F11
Sinclairtown Fife 280 C5
Sinderby N Yorks 214 C6
Sinderhope Northumb 241 G8
Sinderland Green
Gtr Man 184 C2
Sindlesham Wokingham 65 F9
Sinfin Derby 152 C6
Sinfin Moor Derby 153 C7
Singdean Borders 250 C3
Single Hill Bath 45 B8
Singleborough Bucks 102 E5
Singleton Lancs 202 F3
Singleton W Sus 34 E5
Singlewell Kent 69 E7
Singret Wrex 166 D4
Sinkhurst Green Kent 53 E10
Sinnahard Aberds 292 B6
Sinnington N Yorks 216 B4
Sinton Worcs 116 E6
Sinton Green Worcs 116 E6
Sion Hill Bath 61 F8
Sipson London 66 D5
Sirhowy Bl Gwent 77 C11
Sisland Norf 142 D6
Sissinghurst Kent 53 F9
Sisterpath Borders 272 F5
Siston S Glos 61 D7
Sithney Corn 2 D4
Sithney Common Corn 2 D4
Sithney Green Corn 2 D4
Sittingbourne Kent 70 G2
Six Ashes Staffs 132 F5
Six Bells Bl Gwent 78 E2
Six Hills Leics 154 E2
Six Mile Bottom Cambs 123 F11
Sixhills Lincs 189 D11
Sixmile Kent 54 E6
Sixpenny Handley
Dorset 31 D7
Sizewell Suff 127 E9
Skaigh Devon 13 C8
Skail Highld 308 E7
Skaill Orkney 314 F5
Skaill Orkney 314 C4
Skaill Orkney 314 F2
Skares E Ayrs 258 F2
Skateraw E Loth 282 F4
Skaw Shetland 312 B8
Skaw Shetland 312 G7
Skeabost Highld 298 E4
Skeabrae Orkney 314 D2
Skeeby N Yorks 224 E4
Skeete Kent 54 E6
Skeffington Leics 136 C4
Skeffling E Yorks 201 D11
Skegby Notts 171 C7
Skegby Notts 188 G3
Skegness Lincs 175 C9
Skelberry Shetland 313 G6
Skelberry Shetland 313 M5
Skelbo Highld 309 K7
Skelbo Street Highld 309 K7
Skelbrooke S Yorks 198 E4
Skeldyke Lincs 156 B6
Skelfhill Borders 249 C11
Skellingthorpe Lincs 188 G6
Skellister Shetland 313 H6
Skellorn Green E Ches 184 E6
Skellow S Yorks 198 E4
Skelmanthorpe W Yorks 197 E8
Skelmersdale Lancs 194 F3
Skelmonae Aberds 303 F8
Skelmorlie N Ayrs 266 B3
Skelmuir Aberds 303 E9
Skelpick Highld 308 D7
Skelton Cumb 230 D4
Skelton E Yorks 199 B9
Skelton N Yorks 223 E11
Skelton Redcar 226 B3
Skelton York 207 B7
Skelton-on-Ure N Yorks 215 F7
Skelwick Orkney 314 B4
Skelwith Bridge Cumb 220 E6
Skendleby Lincs 174 B6
Skendleby Psalter Lincs 190 G6
Skene Ho Aberds 293 C9
Skenfrith Mon 97 G9
Skerne E Yorks 208 B6
Skerne Park Darl 224 C5
Skeroblingarry Argyll 255 E8
Skerray Highld 308 C6
Skerricha Highld 306 D7
Skerryford Pembs 72 C6
Skerton Lancs 211 G9
Sketchley Leics 135 E8
Sketty Swansea 56 C6
Skewen = Sgiwen Neath 57 B8
Skewes Corn 5 B9
Skewsby N Yorks 216 E2
Skeyton Norf 160 D4
Skeyton Corner Norf 160 D5
Skiag Bridge Highld 307 G7
Skibo Castle Highld 309 L7
Skidbrooke Lincs 190 C6
Skidbrooke North End
Lincs 190 B6
Skidby E Yorks 208 G6
Skilgate Som 27 B7
Skillington Lincs 155 D7
Skinburness Cumb 238 F4
Skinflats Falk 279 E8
Skinidin Highld 298 E2
Skinner's Bottom Corn 4 F4
Skinners Green W Berks 64 F2
Skinnet Highld 308 C5
Skinningrove Redcar 226 B4
Skipness Argyll 255 B9
Skippool Lancs 202 E3
Skiprigg Cumb 230 B3
Skipsea E Yorks 209 B9
Skipsea Brough E Yorks 209 C9
Skipton N Yorks 204 C5
Skipton-on-Swale
N Yorks 215 D7
Skipwith N Yorks 207 F9
Skirbeck Lincs 174 G4
Skirbeck Quarter Lincs 174 G4
Skirethorns N Yorks 213 G9
Skirlaugh E Yorks 209 F8
Skirling Borders 260 B3
Skirmett Bucks 65 B9
Skirpenbeck E Yorks 207 B10
Skirwith Cumb 231 E8
Skirza Highld 310 C7
Skitby Cumb 239 D10
Skitham Lancs 202 E4
Skittle Green Bucks 84 E3
Skulamus Highld 295 C8
Skullomie Highld 308 C6
Skyborry Green Shrops 114 C5
Skye Green Essex 107 G7
Skye of Curr Highld 301 G9
Skyfog Pembs 90 F6
Skyreholme N Yorks 213 G11
Slack Derbys 170 C4
Slack W Yorks 196 B3
Slack W Yorks 196 D5
Slack Head Cumb 211 D9
Slackcote Gtr Man 196 F3
Slackhall Derbys 185 E9
Slackhead Moray 302 C4
Slackholme End Lincs 191 G8
Slacks of Cairnbanno
Aberds 303 E8
Slad Glos 80 D5
Sladbrook Glos 98 F5
Slade Devon 27 F10
Slade Devon 40 D4
Slade Kent 54 C2
Slade Pembs 72 B6
Slade Swansea 56 D3
Slade End Oxon 83 G9
Slade Green London 68 D4
Slade Heath Staffs 133 B8
Slade Hooton S Yorks 187 D8
Sladen Green Hants 48 C2
Slades Green Worcs 99 E7
Sladesbridge Corn 10 G6
Slaggyford Northumb 240 G5
Slaidburn Lancs 203 C10
Slaithwaite W Yorks 196 E5
Slaley Derbys 170 D3
Slaley Northumb 241 F11
Slamannan Falk 279 G7
Slape Cross Som 43 F10
Slapewath Redcar 226 B2
Slapton Bucks 103 G8
Slapton Devon 8 G6
Slapton Northants 102 B2
Slate Haugh Moray 302 C4
Slateford Edin 280 G4
Slatepit Dale Derbys 170 B4
Slattocks Gtr Man 195 F11
Slaugham W Sus 36 B3
Slaughter Hill E Ches 168 D2
Slaughterbridge Corn 11 D8
Slaughterford Wilts 61 E10
Slawston Leics 136 E5
Slay Pits S Yorks 199 F7
Sleaford Hants 49 F10
Sleaford Lincs 173 F9
Sleagill Cumb 221 B11
Sleap Shrops 149 D9
Sleapford Telford 150 F2
Sleapshyde Herts 86 D2
Sleastary Highld 309 K6
Slebech Pembs 73 B8
Sledge Green Worcs 98 E6
Sledmere E Yorks 217 G8
Sleeches Cross E Sus 52 G5
Sleepers Hill Hants 33 B7
Sleet Moor Derbys 170 E6
Sleetbeck Cumb 240 B2
Sleight Dorset 18 B5
Sleights N Yorks 227 D7
Slepe Dorset 18 C4
Sliabhna h-Airde
W Isles 296 F3
Slickly Highld 310 C6
Sliddery N Ayrs 255 E10
Slideslow Worcs 117 C9
Sligachan Hotel Highld 294 C6
Sligneach Argyll 288 G4
Sligrachan Argyll 276 C3
Slimbridge Glos 80 E2
Slindon Staffs 150 C6
Slindon W Sus 35 F7
Slinfold W Sus 50 G6
Sling Glos 79 D9
Sling Gwyn 163 B10
Slingsby N Yorks 216 E3
Slioch Aberds 302 F5
Slip End C Beds 85 B9
Slip End Herts 104 D5
Slippery Ford W Yorks 204 E6
Slipton Northants 121 B9
Slitting Mill Staffs 151 F10
Slochd Highld 301 G8
Slockavullin Argyll 275 D9
Slogan Moray 302 E3
Sloley Norf 160 E5
Sloncombe Devon 13 D10
Sloothby Lincs 191 G7
Slough Slough 66 D3
Slough Green Som 28 C3
Slough Green W Sus 36 B3
Slough Hill Suff 125 G7
Sluggan Highld 301 G8
Sluggans Highld 298 E4
Slumbay Highld 295 B10
Sly Corner Kent 54 G3
Slyfield Sur 50 C3
Slyne Lancs 211 F9
Smailholm Borders 262 B4
Small Dole W Sus 36 E2
Small End Lincs 174 D6
Small Heath W Mid 134 F2
Small Hythe Kent 53 G11
Small Way Som 44 G6
Smallbridge Gtr Man 196 D2
Smallbrook Devon 14 B3
Smallbrook Glos 79 E9
Smallburgh Norf 160 E6
Smallburn Aberds 303 E10
Smallburn E Ayrs 258 D5
Smalldale Derbys 185 E11
Smalldale Derbys 185 F9
Smalley Derbys 170 G6
Smalley Common
Derbys 170 G6
Smalley Green Derbys 170 G6
Smallfield Sur 51 E10
Smallford Herts 85 D11
Smallholm Dumfries 238 B4
Smallmarsh Devon 25 C10
Smallrice Staffs 151 C9
Smallridge Devon 28 G4
Smallshaw Gtr Man 196 G2
Smallthorne Stoke 168 E5
Smallwood E Ches 168 C4
Smallwood Worcs 117 D10
Smallwood Green Suff 125 F8
Smallwood Hey Lancs 202 D3
Smallworth Norf 141 G10
Smannell Hants 47 D11
Smardale Cumb 222 D4
Smarden Kent 53 E11
Smarden Bell Kent 53 E11
Smart's Hill Kent 52 E4
Smaull Argyll 274 G3
Smeatharpe Devon 27 E11
Smeaton Fife 280 C5
Smeeth Kent 54 F5
Smeeton Westerby
Leics 136 E3
Smelthouses N Yorks 214 G3
Smerclait W Isles 297 K3
Smerral Highld 310 F5
Smestow Staffs 133 E7
Smethcott Shrops 131 D9
Smethwick W Mid 133 F10
Smethwick Green
E Ches 168 C4
Smirisary Highld 289 B8
Smisby Derbys 152 F6
Smite Hill Worcs 117 F7
Smith End Green Worcs 116 G5
Smith Green Lancs 202 C5
Smithaleigh Devon 7 D11
Smithbrook W Sus 34 C6
Smithfield Cumb 239 D10
Smithies S Yorks 197 F11
Smithincott Devon 27 E9
Smithley S Yorks 197 G11
Smith's End Herts 105 D8
Smith's Green E Ches 168 E3
Smith's Green Essex 105 G11
Smith's Green Essex 106 C3
Smithston Aberds 302 G5
Smithstown Highld 299 B7
Smithton Highld 301 E7
Smithwood Green Suff 125 G8
Smithy Bridge Gtr Man 196 D2
Smithy Gate Flint 181 F11
Smithy Green E Ches 184 G2
Smithy Green Gtr Man 184 D5
Smithy Houses Derbys 170 F5
Smithy Lane Ends Lancs 194 E2
Smock Alley W Sus 35 D9
Smockington Leics 135 F9
Smoky Row Bucks 84 D4
Smoogro Orkney 314 F3
Smug Oak Herts 85 E10
Smyrton S Ayrs 244 G4
Smythe's Green Essex 88 B6
Snagshall E Sus 38 C3
Snaigow House Perth 286 C4
Snailbeach Shrops 131 C7
Snails Hill Som 29 E7
Snailswell Herts 104 E3
Snailwell Cambs 124 D2
Snainton N Yorks 217 C8
Snaisgill Durham 232 F5
Snaith E Yorks 198 C6
Snape N Yorks 214 C5
Snape Suff 127 F7
Snape Green Lancs 193 E11
Snape Hill Derbys 186 F5
Snape Hill S Yorks 198 G2
Snapper Devon 40 G5
Snaresbrook London 67 B11
Snarestone Leics 134 B6
Snarford Lincs 189 E9
Snargate Kent 39 B7
Snarraness Shetland 313 H4
Snatchwood Torf 78 E3
Snave Kent 39 B8
Sneachill Worcs 117 G8
Snead Powys 130 E6
Snead Common Worcs 116 D4
Sneads Green Worcs 117 D7
Sneath Common Norf 142 F3
Sneaton N Yorks 227 D7
Sneatonthorpe N Yorks 227 D8
Snedshill Telford 132 B4
Sneinton Nottingham 153 B11
Snelland Lincs 189 E9
Snelston Derbys 169 G11
Snetterton Norf 141 E9
Snettisham Norf 158 C3
Sneyd Green Stoke 168 F5
Sneyd Park Bristol 60 D5
Snibston Leics 153 G8
Snig's End Glos 98 F5
Snipeshill Kent 70 G2
Sniseabhal W Isles 297 H3
Snitter Northumb 252 C2
Snitterby Lincs 189 C7
Snitterfield Warks 118 F4
Snitterton Derbys 170 C3
Snitton Shrops 115 B11
Snittongate Shrops 115 B11
Snodhill Hereford 96 C6
Snodland Kent 69 G7
Snods Edge Northumb 242 G3
Snow End Herts 105 E8
Snow Hill E Ches 167 E10
Snow Hill W Yorks 197 C10
Snow Lea W Yorks 196 D5
Snow Street Norf 141 G11
Snowden Hill S Yorks 197 G9
Snowdown Kent 55 C8
Snowshill Glos 99 E11
Snydale W Yorks 198 D2
Soake Hants 33 E11
Soar Anglesey 178 G5
Soar Carms 94 F2
Soar Devon 9 G9
Soar Gwyn 146 B2
Soar Powys 95 E9
Soar-y-Mynydd Ceredig 112 G5
Soberton Hants 33 D10
Soberton Heath Hants 33 E10
Sockbridge Cumb 230 F6
Sockburn Darl 224 D6
Sockety Dorset 29 F7
Sodom Denb 181 G9
Sodom Shetland 313 G7
Sodom Wilts 62 C4
Sodylt Bank Shrops 148 B6
Soham Cambs 123 C11
Soham Cotes Cambs 123 B11
Soho London 67 C9
Soho Som 45 D7
Soho W Mid 133 F10
Solas W Isles 296 D4
Soldon Cross Devon 24 E4
Soldridge Hants 49 G7
Sole Street Kent 54 D5
Sole Street Kent 69 F7
Solfach = Solva Pembs 90 G5
Solihull W Mid 118 B2
Solihull Lodge W Mid 117 B11
Sollers Dilwyn Hereford 115 F8
Sollers Hope Hereford 98 E2
Sollom Lancs 194 D3
Solva = Solfach Pembs 90 G5
Somerby Leics 154 G5
Somerby Lincs 200 F5
Somercotes Derbys 170 E6
Somerdale Bath 61 F7
Someford Dorset 19 C9
Somerford E Ches 168 B4
Somerford Staffs 133 B7
Somerford Keynes Glos 81 G8
Somerley W Sus 22 D4
Somerleyton Suff 143 D9
Somers Town London 67 C9
Somers Town Ptsmth 21 B8
Somersal Herbert
Derbys 152 B2
Somersby Lincs 190 G4
Somersham Cambs 123 B7
Somersham Suff 107 B11
Somerton Newport 59 B10
Somerton Oxon 101 F9
Somerton Som 29 B7
Somerton Suff 124 G6
Somerton Hill Som 29 B7
Somerwood Shrops 149 G11
Sompting W Sus 35 G11
Sompting Abbotts
W Sus 35 F11
Sonning Wokingham 65 D9
Sonning Common Oxon 65 C8
Sonning Eye Oxon 65 D9
Sontley Wrex 166 F4
Sookholme Notts 171 B8
Sopley Hants 19 B9
Sopwell Herts 85 D11
Sopworth Wilts 61 B10
Sorbie Dumfries 236 E6
Sordale Highld 310 C5
Sorisdale Argyll 288 C4
Sorley Devon 8 F4
Sorn E Ayrs 258 D3
Sornhill E Ayrs 258 C2
Sortat Highld 310 C6
Sotby Lincs 190 F2
Sots Hole Lincs 173 C10
Sotterley Suff 143 G9
Soudley Shrops 131 E9
Soudley Shrops 150 D4
Soughley S Yorks 197 G7
Soughton = Sychdyn
Flint 166 B2
Soulbury Bucks 103 F7

Soulby Cumb 222 C4
Soulby Cumb 230 F5
Souldern Oxon 101 E10
Souldrop Beds 121 E9
Sound E Ches 167 F10
Sound Shetland 313 H5
Sound Shetland 313 J6
Sound Heath E Ches 167 F10
Soundwell S Glos 60 D6
Sour Nook Cumb 230 C3
Sourhope Borders 263 E8
Sourin Orkney 314 C4
Sourlie N Ayrs 266 G6
Sourton Devon 12 C6
Soutergate Cumb 210 C4
South Acre Norf 158 G6
South Acton London 67 D7
South Alkham Kent 55 E8
South Allington Devon 9 G10
South Alloa Falk 279 C7
South Ambersham W Sus 34 C6
South Anston S Yorks 187 E8
South Ascot Windsor 66 F2
South Ashford Kent 54 E4
South Auchmachar Aberds 303 E9
South Baddesley Hants 20 B3
South Ballachulish Highld 284 B4
South Balloch S Ayrs 245 D8
South Bank Redcar 234 G6
South Bank York 207 C7
South Barrow Som 29 B10
South Beach Northumb 243 B8
South Beach = Marian-y-de Gwyn 145 C7
South Beddington London 67 G9
South Benfleet Essex 69 B9
South Bents T & W 243 E10
South Bersted W Sus 22 C6
South Blainslie Borders 271 G10
South Bockhampton Dorset 19 B9
South Bramwith S Yorks 198 E6
South Brent Devon 8 D3
South Brewham Som 45 F8
South Bromley London 67 C11
South Broomage Falk 279 E7
South Broomhill Northumb 252 D6
South Burlingham Norf 143 B7
South Cadbury Som 29 B10
South Cairn Dumfries 236 C1
South Carlton Lincs 189 F7
South Carlton Notts 187 E9
South Carne Corn 11 E10
South Cave E Yorks 208 G4
South Cerney Glos 81 F8
South Chailey E Sus 36 D5
South Chard Som 28 F4
South Charlton Northumb 264 E5
South Cheriton Som 29 C11
South Church Durham 233 F10
South Cliffe E Yorks 208 F3
South Clifton Notts 188 G4
South Clunes Highld 300 E5
South Cockerington Lincs 190 D5
South Common Devon 28 G4
South Cornelly Bridgend 57 E10
South Corriegills N Ayrs 256 C2
South Corrielaw Dumfries 248 G5
South Cove Suff 143 G9
South Creagan Argyll 289 E11
South Creake Norf 159 B7
South Crosland W Yorks 196 E6
South Croxton Leics 154 G3
South Croydon London 67 G10
South Cuil Highld 298 C3
South Dalton E Yorks 208 D5
South Darenth Kent 68 F5
South Denes Norf 143 C10
South Down Hants 33 C7
South Down Som 28 E2
South Duffield N Yorks 207 G9
South Dunn Highld 310 D5
South Earlswood Sur 51 D9
South Elkington Lincs 190 D3
South Elmsall W Yorks 198 E3
South Elphinstone E Loth 281 G7
South End Beds 103 B10
South End Bucks 103 F7
South End Cumb 210 G4
South End E Yorks 209 E9
South End Hants 31 D10
South-end Herts 86 B6
South End N Lincs 200 C6
South End Norf 141 E9
South Erradale Highld 299 B7
South Fambridge Essex 88 F5
South Farnborough Hants 49 C11
South Fawley W Berks 63 C11
South Ferriby N Lincs 200 C3
South Field E Yorks 200 B4
South Field Windsor 66 D3
South Flobbets Aberds 303 F7
South Garth Shetland 312 D7
South Garvan Highld 289 B11
South Glendale W Isles 297 K3
South Gluss Shetland 312 F5
South Godstone Sur 51 D11
South Gorley Hants 31 E11
South Gosforth T & W 242 D6
South Green Essex 89 B8
South Green Essex 87 G11
South Green Kent 69 G11
South Green Norf 157 F10
South Green Norf 159 G11
South Green Suff 126 B3
South Gyle Edin 280 G3
South-haa Shetland 312 E5
South Hackney London 67 C11
South Ham Hants 48 C6
South Hampstead London 67 C9
South Hanningfield Essex 88 F2
South Harefield London 66 B5
South Harrow London 66 B6
South Harting W Sus 34 D3
South Hatfield Herts 86 D2
South Hayling Hants 21 B10
South Hazelrigg Northumb 264 C3
South Heath Bucks 84 E6
South Heath Essex 89 B10
South Heighton E Sus 23 E7
South-heog Shetland 312 E5
South Hetton Durham 234 B3
South Hiendley W Yorks 197 E11
South Hill Corn 12 G2
South Hill N Som 43 B10
South Hill Pembs 72 C4
South Hinksey Oxon 83 E8
South Hole Devon 24 C2
South Holme N Yorks 216 D3
South Holmwood Sur 51 D7
South Hornchurch London 68 C4
South Huish Devon 8 G3
South Hykeham Lincs 172 C6
South Hylton T & W 243 F9
South Kelsey Lincs 189 B8
South Kensington London 67 D9
South Kessock Highld 300 E6
South Killingholme N Lincs 201 D7
South Kilvington N Yorks 215 C8
South Kilworth Leics 136 G2
South Kirkby W Yorks 198 E2
South Kirkton Aberds 293 C9
South Kiscadale N Ayrs 256 D2
South Knighton Devon 14 G2
South Knighton Leicester 136 C2
South Kyme Lincs 173 F11
South Lambeth London 67 D10
South Lancing W Sus 35 G11
South Lane S Yorks 197 F9
South Leigh Oxon 82 D5
South Leverton Notts 188 E3
South Littleton Worcs 99 B11
South Lopham Norf 141 G10
South Luffenham Rutland 137 C8
South Malling E Sus 36 E6
South Marston Swindon 63 B7
South Merstham Sur 51 C9
South Middleton Northumb 263 E11
South Milford N Yorks 206 G5
South Millbrex Aberds 303 E8
South Milton Devon 8 G4
South Mimms Herts 86 E2
South Molton Devon 26 B2
South Moor Durham 242 G5
South Moreton Oxon 64 B5
South Mundham W Sus 22 C5
South Muskham Notts 172 D3
South Newbald E Yorks 208 F4
South Newbarns Cumb 210 F4
South Newington Oxon 101 E8
South Newsham Northumb 243 B8
South Newton Wilts 46 G5
South Normanton Derbys 170 D6
South Norwood London 67 F10
South Nutfield Sur 51 D10
South Ockendon Thurrock 68 C5
South Ormsby Lincs 190 F5
South Ossett W Yorks 197 D9
South Otterington N Yorks 215 B7
South Owersby Lincs 189 C9
South Oxhey Herts 85 G10
South Park Sur 51 D8
South Pelaw Durham 243 G7
South Perrott Dorset 29 F7
South Petherton Som 28 D6
South Petherwin Corn 12 E2
South Pickenham Norf 141 C7
South Pill Corn 7 D8
South Pool Devon 8 G5
South Poorton Dorset 16 B6
South Port Argyll 284 E4
South Quilquox Aberds 303 F8
South Radworthy Devon 41 G9
South Rauceby Lincs 173 F8
South Raynham Norf 159 E7
South Reddish Gtr Man 184 C5
South Reston Lincs 190 E6
South Ruislip London 66 B6
South Runcton Norf 140 B2
South Scarle Notts 172 C4
South Shian Argyll 289 E11
South Shields T & W 243 D9
South Shore Blkpool 202 G2
South Side Durham 233 F8
South Side Orkney 314 D5
South Somercotes Lincs 190 C6
South Stainley N Yorks 214 G6
South Stainmore Cumb 222 C6
South Stanley Durham 242 G5
South Stifford Thurrock 68 D6
South Stoke Oxon 64 C6
South Stoke W Sus 35 F8
South Stour Kent 54 F4
South Street E Sus 36 D5
South Street Kent 54 B5
South Street Kent 68 G6
South Street Kent 69 G10
South Street Kent 70 F6
South Street London 52 B2
South Tawton Devon 13 C9
South Tehidy Corn 4 G3
South Thoresby Lincs 190 F6
South Tidworth Wilts 47 D8
South Tottenham London 67 B10
South Town Devon 14 E5
South Town Hants 49 F7
South Twerton Bath 61 G8
South Ulverston Cumb 210 D6
South View Hants 48 C6
South Voxter Shetland 313 G5
South Walsham Norf 161 G7
South Warnborough Hants 49 D8
South Weald Essex 87 G9
South Weirs Hants 32 G3
South Weston Oxon 84 F2
South Wheatley Corn 11 C10
South Wheatley Notts 188 D3
South Whiteness Shetland 313 J5
South Widcombe Bath 44 B5
South Wigston Leics 135 D11
South Willesborough Kent 54 E4
South Willingham Lincs 189 E11
South Wimbledon London 67 E9
South Wingate Durham 234 E4
South Wingfield Derbys 170 D5
South Witham Lincs 155 F8
South Wonford Devon 24 F5
South Wonston Hants 48 F3
South Woodford London 86 G6
South Woodham Ferrers Essex 88 F4
South Wootton Norf 158 E2
South Wraxall Wilts 61 G10
South Yardley W Mid 134 G2
South Yarrows Highld 310 E7
South Yeo Devon 25 G8
South Zeal Devon 13 C9
Southall London 66 C6
Southam Cumb 219 C9
Southam Glos 99 F9
Southam Warks 119 E8
Southampton Soton 32 E6
Southay Som 28 D6
Southborough Kent 52 E5
Southborough London 67 F7
Southborough London 68 F2
Southbourne Bmouth 19 C8
Southbourne W Sus 22 B3
Southbrook Wilts 45 G10
Southburgh Norf 141 C9
Southburn E Yorks 208 C5
Southchurch Sthend 70 B2
Southcoombe Oxon 100 F6
Southcote Reading 65 E7
Southcott Corn 11 B9
Southcott Devon 24 D6
Southcott Wilts 47 B7
Southcourt Bucks 84 C4
Southcrest Worcs 117 D10
Southdean Borders 250 B4
Southdene Mers 182 B6
Southdown Bath 61 G8
Southdown Corn 7 E8
Southease E Sus 36 F6
Southend Argyll 255 G7
Southend Bucks 65 B9
Southend Glos 80 F2
Southend London 67 E11
Southend Oxon 83 E9
Southend W Berks 64 D2
Southend W Berks 64 E5
Southend Wilts 63 E7
Southend-on-Sea Sthend 69 B11
Southerhouse Shetland 313 K5
Southerly Devon 12 D6
Southern Cross Brighton 36 F3
Southern Green Herts 104 E6
Southernby Cumb 230 D3
Southernden Kent 53 D11
Southerndown V Glam 57 G11
Southerness Dumfries 237 D11
Southery Norf 140 E2
Southey Green Essex 106 E5
Southfield Northumb 243 B7
Southfields London 67 E9
Southfields Thurrock 69 C7
Southfleet Kent 68 E6
Southford I o W 20 F6
Southgate Ceredig 111 A11
Southgate London 86 G3
Southgate Norf 159 C7
Southgate Norf 160 E2
Southgate Swansea 56 D5
Southgate W Sus 51 F9
Southill C Beds 104 C3
Southill Dorset 17 E9
Southington Hants 48 D4
Southlands Dorset 17 F9
Southleigh Devon 15 C10
Southmarsh Som 45 G8
Southmead Bristol 60 D5
Southminster Essex 89 F7
Southmoor Oxon 82 F5
Southoe Cambs 122 E3
Southolt Suff 126 D3
Southorpe P'boro 137 C11
Southover Dorset 17 C8
Southover E Sus 36 F6
Southover E Sus 37 B11
Southowram W Yorks 196 C6
Southport Mers 193 D10
Southpunds Shetland 313 L6
Southrepps Norf 160 B5
Southrey Lincs 173 B10
Southrop Glos 81 E11
Southrop Oxon 101 E7
Southrope Hants 49 E7
Southsea Ptsmth 21 B8
Southsea Wrex 166 E4
Southstoke Bath 61 G8
Southtown Norf 143 B10
Southtown Orkney 314 G4
Southtown Som 28 D4
Southtown Som 44 F5
Southville Devon 8 G4
Southville Torf 78 F3
Southwaite Cumb 230 C4
Southwark London 67 D10
Southwater W Sus 35 B11
Southwater Street W Sus 35 B11
Southway Plym 7 C9
Southway Som 44 E4
Southwell Dorset 17 G9
Southwell Notts 172 E2
Southwick Hants 33 F10
Southwick Northants 137 E10
Southwick Som 43 D11
Southwick T & W 243 F9
Southwick W Sus 36 F2
Southwick Wilts 45 B10
Southwold Suff 127 B10
Southwood Derbys 153 E7
Southwood Hants 49 B10
Southwood Norf 143 B7
Southwood Som 44 G5
Southwood Worcs 116 E4
Soval Lodge W Isles 304 F5
Sowber Gate N Yorks 215 B7
Sower Carr Lancs 202 E3
Sowerby N Yorks 215 C8
Sowerby W Yorks 196 C4
Sowerby Bridge W Yorks 196 C5
Sowerby Row Cumb 230 D3
Sowley Green Suff 124 G4
Sowood W Yorks 196 D5
Sowood Green W Yorks 196 D5
Sowton Devon 14 C5
Sowton Barton Devon 14 D2
Soyal Highld 309 K5
Soyland Town W Yorks 196 C4
Spa Common Norf 160 C5
Spacey Houses N Yorks 206 C2
Spalding Lincs 156 E4
Spaldington E Yorks 207 G11
Spaldwick Cambs 122 C2
Spalford Notts 172 B4
Spanby Lincs 155 B11
Spango Invclyd 276 G4
Spanish Green Hants 49 B7
Sparham Norf 159 F11
Sparhamhill Norf 159 F11
Spark Bridge Cumb 210 C6
Sparkbrook W Mid 133 G11
Sparkford Som 29 B10
Sparkhill W Mid 133 G11
Sparkwell Devon 7 D11
Sparl Shetland 312 G5
Sparnon Corn 1 E3
Sparnon Gate Corn 4 G3
Sparrow Green Norf 159 G9
Sparrow Hill Som 44 C2
Sparrowpit Derbys 185 E9
Sparrow's Green E Sus 52 G6
Sparsholt Hants 48 G2
Sparsholt Oxon 63 B10
Spartylea Northumb 232 B3
Spath Staffs 151 B11
Spaunton N Yorks 226 G4
Spaxton Som 43 F8
Spean Bridge Highld 290 E4
Spear Hill W Sus 35 D10
Spearywell Hants 32 B4
Speckington Som 29 C9
Speed Gate Kent 68 F5
Speedwell Bristol 60 E6
Speen Bucks 84 F4
Speen W Berks 64 F3
Speeton N Yorks 218 E2
Speke Mers 182 E6
Speldhurst Kent 52 E5
Spellbrook Herts 87 B7
Spelsbury Oxon 101 G7
Spelter Bridgend 57 C11
Spen W Yorks 197 B7
Spen Green E Ches 168 C4
Spencers Wood Wokingham 65 F8
Spennells Worcs 116 C6
Spennithorne N Yorks 214 B2
Spennymoor Durham 233 E11
Spernall Warks 117 E11
Spetchley Worcs 117 G7
Spetisbury Dorset 30 G6
Spexhall Suff 143 G7
Spey Bay Moray 302 C3
Speybank Highld 291 C10
Speybridge Highld 301 G10
Speyview Moray 302 E2
Spillardsford Aberds 303 D10
Spilsby Lincs 174 B6
Spindlestone Northumb 264 C5
Spinkhill Derbys 187 F7
Spinney Hill Northants 120 E5
Spinney Hills Leicester 136 C2
Spinningdale Highld 309 L6
Spion Kop Notts 171 B9
Spirthill Wilts 62 D3
Spital Mers 182 E4
Spital Windsor 66 D3
Spital Hill S Yorks 187 C10
Spital in the Street Lincs 189 D7
Spital Tongues T & W 242 D6
Spitalbrook Herts 86 D5
Spitalfields London 67 C10
Spitalhill Derbys 169 F11
Spithurst E Sus 36 D6
Spittal Dumfries 236 D5
Spittal E Loth 281 F9
Spittal E Yorks 207 C11
Spittal Highld 310 D5
Spittal Northumb 273 E10
Spittal Pembs 91 G9
Spittal Stirl 277 D10
Spittal Houses S Yorks 186 B5
Spittal of Glenmuick Aberds 292 E5
Spittal of Glenshee Perth 292 F3
Spittalfield Perth 286 C5
Spittlegate Lincs 155 C8
Spixworth Norf 160 F4
Splatt Corn 10 F4
Splatt Corn 11 D10
Splatt Devon 25 F10
Splatt Som 43 F8
Splayne's Green E Sus 36 C6
Splott Cardiff 59 D7
Spofforth N Yorks 206 C3
Spon End W Mid 118 B6
Spon Green Flint 166 C3
Spondon Derby 153 B8
Spooner Row Norf 141 D11
Spoonleygate Shrops 132 D6
Sporle Norf 158 G6
Spotland Bridge Gtr Man 195 E11
Spott E Loth 282 F3
Spratton Northants 120 C4
Spreakley Sur 49 E10
Spreyton Devon 13 B9
Spriddlestone Devon 7 E10
Spridlington Lincs 189 E8
Sprig's Alley Oxon 84 F3
Spring Bank Cumb 229 G10
Spring Cottage Leics 152 F6
Spring End N Yorks 223 F9
Spring Gardens Som 45 D9
Spring Green Lancs 204 E4
Spring Grove London 67 D7
Spring Hill Gtr Man 196 F2
Spring Hill Lancs 195 B8
Spring Hill W Mid 133 D7
Spring Park London 67 F11
Spring Vale S Yorks 197 G9
Spring Valley I o M 192 E4
Springbank Glos 99 G8
Springboig Glasgow 268 C3
Springbourne Bmouth 19 C8
Springburn Glasgow 268 B2
Springfield Argyll 275 F11
Springfield Caerph 77 F11
Springfield Dumfries 239 D8
Springfield Essex 88 D2
Springfield Fife 287 F7
Springfield Gtr Man 194 F5
Springfield Highld 300 C6
Springfield M Keynes 103 D7
Springfield Moray 301 D10
Springfield W Mid 133 D8
Springfield W Mid 133 F9
Springfield W Mid 133 G11
Springfields Stoke 168 G5
Springhead Gtr Man 196 G3
Springhill E Renf 267 D10
Springhill I o W 20 B6
Springhill N Lnrk 269 D7
Springhill Staffs 133 B11
Springhill Staffs 133 C9
Springholm Dumfries 237 C10
Springkell Dumfries 239 B7
Springside N Ayrs 257 B9
Springthorpe Lincs 188 D5
Springwell Essex 105 C10
Springwell T & W 243 F7
Springwell T & W 243 F9
Springwells Dumfries 248 E3
Sproatley E Yorks 209 G9
Sproston Green W Ches 168 B2
Sprotbrough S Yorks 198 G4
Sproughton Suff 108 C2
Sprouston Borders 263 B7
Sprowston Norf 160 G4
Sproxton Leics 155 E7
Sproxton N Yorks 216 C2
Sprunston Cumb 230 B3
Spunhill Shrops 149 C8
Spurlands End Bucks 84 F5
Spurstow E Ches 167 D9
Spurtree Shrops 116 D2
Spynie Moray 302 C2
Spyway Dorset 16 C6
Square and Compass Pembs 91 E7
Squires Gate Blkpool 202 G2
Sraid Ruadh Argyll 288 E1
Srannda W Isles 296 C6
Sronphadruig Lodge Perth 291 F9
Stableford Shrops 132 D5
Stableford Staffs 150 B6
Stacey Bank S Yorks 186 C3
Stackhouse N Yorks 212 F6
Stackpole Pembs 73 F7
Stackpole Quay Pembs 73 F7
Stacksford Norf 141 E11
Stacksteads Lancs 195 C10
Stackyard Green Suff 107 B9
Staddiscombe Plym 7 E10
Staddlethorpe E Yorks 199 B10
Staddon Devon 24 C3
Staddon Devon 24 G5
Staden Derbys 185 G9
Stadhampton Oxon 83 F10
Stadhlaigearraidh W Isles 297 H3
Stadmorslow Staffs 168 D5
Staffield Cumb 230 C6
Staffin Highld 298 C4
Stafford Staffs 151 E8
Stafford Park Telford 132 B4
Stafford's Corner Essex 89 B7
Stafford's Green Dorset 29 C10
Stagbatch Hereford 115 F9
Stagden Cross Essex 87 C10
Stagehall Borders 271 G9
Stag's Head Devon 25 B11
Stagsden Beds 103 B9
Stagsden West End Beds 103 B9
Stain Highld 310 C7
Stainburn Cumb 228 F6
Stainburn N Yorks 205 D10
Stainby Lincs 155 E8
Staincliffe W Yorks 197 C8
Staincross S Yorks 197 E10
Staindrop Durham 233 G8
Staines Sur 66 E4
Staines Green Herts 86 C3
Stainfield Lincs 155 D11
Stainfield Lincs 189 G10
Stainforth N Yorks 212 F6
Stainforth S Yorks 198 E6
Staining Lancs 202 F3
Stainland W Yorks 196 D5
Stainsacre N Yorks 227 D8
Stainsby Derbys 170 B6
Stainsby Lincs 190 G4
Stainton Cumb 211 B10
Stainton Cumb 230 F5
Stainton Cumb 239 F9
Stainton Durham 223 B11
Stainton M'bro 225 C9
Stainton N Yorks 224 F2
Stainton S Yorks 187 C9
Stainton by Langworth Lincs 189 F9
Stainton le Vale Lincs 189 C11
Stainton with Adgarley Cumb 210 E5
Staintondale N Yorks 227 F9
Stair Cumb 229 G10
Stair E Ayrs 257 E10
Stairfoot S Yorks 197 F11
Stairhaven Dumfries 236 D4
Staithes N Yorks 226 B5
Stake Hill Gtr Man 195 F11
Stake Pool Lancs 202 D4
Stakeford Northumb 253 F7
Stalbridge Dorset 30 D2
Stalbridge Weston Dorset 30 D2
Stalham Norf 161 D7
Stalham Green Norf 161 E7
Stalisfield Green Kent 54 C3
Stallen Dorset 29 D10
Stalling Busk N Yorks 213 B8
Stallingborough NE Lincs 201 E7
Stallington Staffs 151 B8
Stalmine Lancs 202 D3
Stalmine Moss Side Lancs 202 D3
Stalybridge Gtr Man 185 B7
Stambermill W Mid 133 G8
Stamborough Som 42 F4
Stambourne Essex 106 D4
Stambourne Green Essex 106 D4
Stamford Lincs 137 B10
Stamford Bridge E Yorks 207 B10
Stamford Bridge W Ches 167 B7
Stamford Hill London 67 B10
Stamfordham Northumb 242 C3
Stamperland E Renf 267 D11
Stamshaw Ptsmth 33 G10
Stanah Cumb 220 B6
Stanah Lancs 202 E3
Stanborough Herts 86 C2
Stanbridge C Beds 103 G9
Stanbridge Dorset 31 G8
Stanbridgeford C Beds 103 G9
Stanbrook Essex 106 F2
Stanbrook Worcs 98 B6
Stanbury W Yorks 204 F6
Stand Gtr Man 195 F9
Stand N Lnrk 268 B5
Standburn Falk 279 G8
Standeford Staffs 133 B8
Standen Kent 53 E11
Standen Hall Lancs 203 E10
Standen Street Kent 53 G10
Standerwick Som 45 C10
Standford Hants 49 G10
Standford Bridge Telford 150 E4
Standingstone Cumb 229 B11
Standingstone Cumb 229 E7
Standish Glos 80 D4
Standish Gtr Man 194 E5
Standish Lower Ground Gtr Man 194 F5
Standlake Oxon 82 E5
Standon Hants 32 B6
Standon Herts 105 G7
Standon Staffs 150 B6
Standon Green End Herts 86 B5
Stane N Lnrk 269 D7
Stanecastle N Ayrs 257 B8
Stanfield Norf 159 E8
Stanfield Stoke 168 E5
Stanford C Beds 104 C3
Stanford Kent 54 F6
Stanford Norf 141 E7
Stanford Shrops 148 G6
Stanford Bishop Hereford 116 G3
Stanford Bridge Worcs 116 D4
Stanford Dingley W Berks 64 E5
Stanford End Wokingham 65 G8
Stanford Hills Notts 153 E10
Stanford in the Vale Oxon 82 G4
Stanford-le-Hope Thurrock 69 C7
Stanford on Avon Northants 119 B11
Stanford on Soar Notts 153 E10
Stanford on Teme Worcs 116 D4
Stanford Rivers Essex 87 E8
Stanfree Derbys 187 G7
Stanground P'boro 138 D4
Stanhill Lancs 195 B8
Stanhoe Norf 158 B6
Stanhope Borders 260 D4
Stanhope Durham 232 D5
Stanhope Kent 54 E3
Stanion Northants 137 F8
Stank Cumb 210 E4
Stanklyn Worcs 117 C7
Stanks W Yorks 206 F3
Stanley Derbys 170 G6
Stanley Durham 242 G5
Stanley Lancs 194 F3
Stanley Notts 171 C7
Stanley Perth 286 D5
Stanley Shrops 132 G3
Stanley Shrops 132 G5
Stanley Staffs 168 E6
Stanley W Yorks 197 C10
Stanley Wilts 62 E3
Stanley Common Derbys 170 G6
Stanley Crook Durham 233 D9
Stanley Downton Glos 80 E4
Stanley Ferry W Yorks 197 C11
Stanley Gate Lancs 194 G2
Stanley Green E Ches 184 E5
Stanley Green Poole 18 C6
Stanley Green Shrops 149 B10
Stanley Hill Hereford 98 C3
Stanley Moor Staffs 168 E6
Stanley Pontlarge Glos 99 E9
Stanleytown Rhondda 77 G8
Stanlow Staffs 132 D5
Stanlow W Ches 182 F6
Stanmer Brighton 36 F4
Stanmore Hants 33 B7
Stanmore London 85 G11
Stanmore Shrops 132 E4
Stanmore W Berks 64 D3
Stanner Powys 114 F5
Stannergate Dundee 287 D8
Stanners Hill Sur 66 G3
Stannersburn Northumb 250 F6
Stanningfield Suff 125 F7
Stanningley W Yorks 205 G10
Stannington Northumb 242 B6
Stannington S Yorks 186 D4
Stanpit Dorset 19 C9
Stansbatch Hereford 114 E6
Stansfield Suff 124 G5
Stanshope Staffs 169 E10
Stanstead Suff 106 B6
Stanstead Abbotts Herts 86 C5
Stansted Kent 68 G6
Stansted Airport Essex 105 G11
Stansted Mountfitchet Essex 105 G10
Stanthorne W Ches 167 B11
Stanton Glos 99 E11
Stanton Mon 96 G6
Stanton Northumb 252 F4
Stanton Staffs 169 F10
Stanton Suff 125 C9
Stanton by Bridge Derbys 153 D7
Stanton-by-Dale Derbys 153 B9
Stanton Chare Suff 125 C9
Stanton Drew Bath 60 G5
Stanton Fitzwarren Swindon 81 G11
Stanton Gate Notts 153 B9
Stanton Harcourt Oxon 82 D6
Stanton Hill Notts 171 C7
Stanton in Peak Derbys 170 C2
Stanton Lacy Shrops 115 B9
Stanton Lees Derbys 170 C3
Stanton Long Shrops 131 E11
Stanton-on-the-Wolds Notts 154 C2
Stanton Prior Bath 61 G7
Stanton St Bernard Wilts 62 G5
Stanton St John Oxon 83 D9
Stanton St Quintin Wilts 62 D2
Stanton Street Suff 125 D9
Stanton under Bardon Leics 153 G9
Stanton upon Hine Heath Shrops 149 E11
Stanton Wick Bath 60 G6
Stantonbury M Keynes 102 C6
Stantway Glos 80 C2
Stanwardine in the Fields Shrops 149 E8
Stanwardine in the Wood Shrops 149 D8
Stanway Essex 107 G8
Stanway Glos 99 E11
Stanway Green Essex 107 G9
Stanway Green Suff 126 C4
Stanwell Sur 66 E5
Stanwell Moor Sur 66 E4
Stanwick Northants 121 C9
Stanwick-St-John N Yorks 224 C3
Stanwix Cumb 239 F10
Stanycliffe Gtr Man 195 F11
Stanydale Shetland 313 H4
Staoinebrig W Isles 297 H3
Stape N Yorks 226 G5
Stapehill Dorset 31 G9
Stapeley E Ches 167 F11
Stapenhill Staffs 152 E5
Staple Kent 55 B9
Staple Som 42 E6
Staple Cross Devon 27 C8
Staple Fitzpaine Som 28 D3
Staple Hill S Glos 61 D7
Staple Hill Worcs 117 C9
Staple Lawns Som 28 D3
Staplecross E Sus 38 C3
Stapleford Cambs 123 G9
Stapleford Herts 86 B4
Stapleford Leics 154 F6
Stapleford Lincs 172 D5
Stapleford Notts 153 B9
Stapleford Wilts 46 F5
Stapleford Abbotts Essex 87 G8
Stapleford Tawney Essex 87 F8
Staplegrove Som 28 B2
Staplehay Som 28 C2
Staplehurst Kent 53 E9
Staplers I o W 20 D6
Staples Hill W Sus 35 B8
Staplestreet Kent 70 G5
Stapleton Bristol 60 D6
Stapleton Cumb 240 C2
Stapleton Hereford 114 D6
Stapleton Leics 135 D8
Stapleton N Yorks 198 D4
Stapleton N Yorks 224 C5
Stapleton Shrops 131 C9
Stapleton Som 29 C7
Stapley Som 27 E11
Staploe Beds 122 E2
Staplow Hereford 98 C3
Stapness Shetland 313 J4
Star Fife 287 G7
Star Pembs 92 E4
Star Som 44 B2
Star Hill Mon 79 E7
Stara Orkney 314 D2
Starbeck N Yorks 206 B2
Starbotton N Yorks 213 E9
Starcross Devon 14 E5
Stareton Warks 118 C6
Stargate T & W 242 E5
Starkholmes Derbys 170 D4
Starling Gtr Man 195 E9
Starlings Green Essex 105 E9
Starr's Green E Sus 38 D3
Starston Norf 142 G4
Start Devon 8 G6
Start Hill Essex 105 G10
Startforth Durham 223 B10
Startley Wilts 62 C2
Startop's End Bucks 84 C6
Starveall S Glos 61 B9
Starvecrow Kent 52 D5
Statenborough Kent 55 B10
Statford St Andrew Suff 127 E7
Statham Warr 183 D11
Stathe Som 28 B5
Stathern Leics 154 C5
Station Hill Cumb 229 B11
Station Town Durham 234 D4
Statland Common Norf 141 D10
Staughton Green Cambs 122 D2
Staughton Highway Cambs 122 E2
Staughton Moor Beds 122 E2
Staunton Glos 79 C8
Staunton Glos 98 F5
Staunton in the Vale Notts 172 G4
Staunton on Arrow Hereford 115 E7
Staunton on Wye Hereford 97 B7
Staupes N Yorks 205 B10
Staveley Cumb 211 B7
Staveley Cumb 221 F9
Staveley Derbys 186 G6
Staveley N Yorks 215 G7
Staveley-in-Cartmel Cumb 211 B7
Staverton Devon 8 C5
Staverton Glos 99 G7
Staverton Northants 119 E10
Staverton Wilts 61 G11
Staverton Bridge Glos 99 G7
Stawell Som 43 F11
Stawley Som 27 C9
Staxigoe Highld 310 D7
Staxton N Yorks 217 D10
Staylittle Ceredig 128 F2
Staylittle = Penffordd-Lâs Powys 129 E7
Staynall Lancs 202 E3
Staythorpe Notts 172 E3
Stead W Yorks 205 D8
Steam Mills Glos 79 B10
Stean N Yorks 213 E11
Steanbow Som 44 F5
Stearsby N Yorks 216 E2
Steart Som 29 B9
Steart Som 43 D9
Stebbing Essex 106 G3
Stebbing Green Essex 106 G3
Stechford W Mid 134 F2
Stede Quarter Kent 53 F11
Stedham W Sus 34 C5
Steel Northumb 241 F10
Steel Northumb 251 G9
Steel Bank S Yorks 186 D4
Steel Cross E Sus 52 G4
Steel Green Cumb 210 D3
Steel Heath Shrops 149 B10
Steele Road Borders 250 E2
Steelend Fife 279 C10
Steeleroad-end Borders 250 E2
Steen's Bridge Hereford 115 F10
Steep Hants 34 B2
Steep Lane W Yorks 196 C4
Steep Marsh Hants 34 B3
Steephill I o W 21 F7
Steeple Dorset 18 E4
Steeple Essex 88 E6
Steeple Ashton Wilts 46 B2
Steeple Aston Oxon 101 F9
Steeple Barton Oxon 101 G8
Steeple Bumpstead Essex 106 C3
Steeple Claydon Bucks 102 F3
Steeple Gidding Cambs 138 G2
Steeple Langford Wilts 46 F4
Steeple Morden Cambs 104 C5
Steeraway Telford 132 B3
Steeton W Yorks 204 E6
Stein Highld 298 D2
Steinmanhill Aberds 303 E7
Stella T & W 242 E5
Stelling Minnis Kent 54 D6
Stelvio Newport 59 B9
Stembridge Som 28 C6
Stembridge Swansea 56 C3
Stemster Highld 310 C5
Stemster Ho Highld 310 C5
Stenalees Corn 5 D10
Stenaquoy Orkney 314 C5
Stencoose Corn 4 F4
Stenhill Devon 27 E9
Stenhouse Dumfries 247 E8
Stenhouse Edin 280 G4
Stenhousemuir Falk 279 E7
Stenigot Lincs 190 E3
Stennack Corn 2 B5
Stenness Shetland 312 F4
Stenscholl Highld 298 C4
Stenso Orkney 314 D3
Stenson Derbys 152 D6
Stenton E Loth 282 G2
Stenton Fife 280 B5
Stentwood Devon 27 F10
Stenwith Lincs 154 B6
Stepaside Corn 5 F9
Stepaside Pembs 73 D10
Stepaside Powys 129 F11
Stepping Hill Gtr Man 184 D6
Steppingley C Beds 103 D10
Stepps N Lnrk 268 B3
Sterndale Moor Derbys 169 B10
Sternfield Suff 127 E7
Sterridge Devon 40 D5
Stert Wilts 46 B4
Sterte Poole 18 C6
Stetchworth Cambs 124 F2
Stevenage Herts 104 G4
Steven's Crouch E Sus 38 D2
Stevenston N Ayrs 266 G5
Stevenstone Devon 25 D8
Steventon Hants 48 D4
Steventon Oxon 83 G7
Steventon Shrops 115 C10
Steventon End Essex 105 C11
Stevington Beds 121 G9
Stewards Essex 87 D7
Steward's Green Essex 87 E7
Stewartby Beds 103 C10
Stewarton Argyll 255 F7
Stewarton E Ayrs 267 F8
Stewkley Bucks 103 F7
Stewkley Dean Bucks 102 F6
Stewley Som 28 D4
Stewton Lincs 190 D5
Steyne Cross I o W 21 D8
Steyning W Sus 35 E11
Steynton Pembs 72 D6
Stibb Corn 24 E2
Stibb Cross Devon 24 E6
Stibb Green Wilts 63 G8
Stibbard Norf 159 D9
Stibbington Cambs 137 D11
Stichill Borders 262 B6
Stick Hill Kent 52 E3
Sticker Corn 5 E9
Stickford Lincs 174 D5
Sticklepath Devon 13 C8
Sticklepath Devon 40 G5
Sticklepath Som 28 E4
Sticklepath Som 42 F4
Sticklinch Som 44 F5
Stickling Green Essex 105 E9
Stickney Lincs 174 D4
Stiff Street Kent 69 G11
Stiffkey Norf 177 E7
Stifford's Bridge Hereford 98 B4
Stileway Som 44 E3
Stillingfleet N Yorks 207 E7
Stillington N Yorks 215 F11
Stillington Stockton 234 G3
Stilton Cambs 138 F3
Stinchcombe Glos 80 F2
Stinsford Dorset 17 C10
Stiperstones Shrops 131 C7
Stirchley Telford 132 B4
Stirchley W Mid 133 G11
Stirkoke Ho Highld 310 D7
Stirling Aberds 303 E11
Stirling Stirl 278 C5
Stirtloe Cambs 122 D3
Stirton N Yorks 204 C5
Stisted Essex 106 G5
Stitchcombe Wilts 63 F8
Stitchin's Hill Worcs 116 G5
Stithians Corn 2 B6
Stittenham Highld 300 B6
Stivichall W Mid 118 B6
Stixwould Lincs 173 B11
Stoak W Ches 182 G6
Stobhill Northumb 252 G6
Stobhillgate Northumb 252 F6
Stobieside S Lnrk 258 B4
Stobo Borders 260 B5
Stoborough Dorset 18 D4
Stoborough Green Dorset 18 D4
Stobs Castle Borders 250 B2
Stobshiel E Loth 271 C9
Stobswood Northumb 252 E6
Stock Essex 87 F11
Stock Lancs 204 D3
Stock N Som 60 G3
Stock Green Worcs 117 F9
Stock Hill Suff 125 D9
Stock Wood Worcs 117 F10
Stockbridge Hants 47 G11
Stockbridge S Yorks 198 F5
Stockbridge W Sus 22 C5
Stockbridge W Yorks 205 E7
Stockbridge Village Mers 182 C6
Stockbury Kent 69 G10
Stockcross W Berks 64 F2
Stockend Glos 80 D4
Stocker's Head Kent 54 C3
Stockerston Leics 136 D6
Stockfield W Mid 134 G2
Stockheath Hants 22 B2
Stockholes Turbary N Lincs 199 F9
Stockiemuir Stirl 277 E10
Stocking Hereford 98 E2
Stocking Green Essex 105 D11
Stocking Pelham Herts 105 F9
Stockingford Warks 134 E6
Stockland Devon 28 G2
Stockland Bristol Som 43 E8
Stockland Green Kent 52 E5
Stockland Green W Mid 133 E11
Stockleigh English Devon 26 F5
Stockleigh Pomeroy Devon 26 G5
Stockley Wilts 62 F4
Stocklinch Som 28 D5
Stockport Gtr Man 184 C5
Stocks Green Kent 52 D5
Stocksbridge S Yorks 186 B3
Stocksfield Northumb 242 E3
Stockstreet Essex 106 G6
Stockton Hereford 115 E10
Stockton Norf 143 E7
Stockton Shrops 130 C5
Stockton Shrops 132 D4
Stockton Telford 150 F5
Stockton Warks 119 E8
Stockton Wilts 46 F3
Stockton Brook Staffs 168 E6
Stockton Heath Warr 183 D10
Stockton-on-Tees Stockton 225 B8
Stockton on Teme Worcs 116 D4
Stockton on the Forest York 207 B9
Stocktonwood Shrops 130 C5
Stockwell Devon 27 G7
Stockwell Glos 80 C6
Stockwell London 67 D10
Stockwell End W Mid 133 C7
Stockwell Heath Staffs 151 E11
Stockwitch Cross Som 29 C9
Stockwood Bristol 60 F6
Stockwood Dorset 29 F9
Stockwood Vale Bath 60 F6
Stodday Lancs 202 B5
Stodmarsh Kent 71 G8
Stody Norf 159 C11
Stoer Highld 307 G5
Stoford Som 29 E9
Stoford Wilts 46 F5
Stoford Water Devon 27 F9
Stogumber Som 42 F5

Stogursey Som 43 E8
Stoke Devon 24 C2
Stoke Hants 22 C2
Stoke Hants 48 C2
Stoke Medway 69 D10
Stoke Plym 7 D9
Stoke Suff 108 C3
Stoke W Mid 119 B7
Stoke Abbott Dorset 29 G7
Stoke Albany Northants 136 F6
Stoke Aldermoor W Mid 119 B7
Stoke Ash Suff 126 C2
Stoke Bardolph Notts 171 G10
Stoke Bishop Bristol 60 D5
Stoke Bliss Worcs 116 E3
Stoke Bruerne Northants 102 B4
Stoke by Clare Suff 106 C4
Stoke-by-Nayland Suff 107 D9
Stoke Canon Devon 14 B4
Stoke Charity Hants 48 F3
Stoke Climsland Corn 12 G3
Stoke Common Hants 33 C7
Stoke Cross Hereford 116 G2
Stoke D'Abernon Sur 50 B6
Stoke Doyle Northants 137 F10
Stoke Dry Rutland 137 D7
Stoke Edith Hereford 98 C2
Stoke End Warks 134 D3
Stoke Farthing Wilts 31 B9
Stoke Ferry Norf 140 D4
Stoke Fleming Devon 9 F7
Stoke Gabriel Devon 8 D6
Stoke Gifford S Glos 60 D6
Stoke Golding Leics 135 D7
Stoke Goldington M Keynes 102 B6
Stoke Green Bucks 66 C3
Stoke Hammond Bucks 103 F7
Stoke Heath Shrops 150 D3
Stoke Heath W Mid 135 G7
Stoke Heath Worcs 117 D8
Stoke Hill Devon 14 C4
Stoke Hill Hereford 98 B2
Stoke Holy Cross Norf 142 C4
Stoke Lacy Hereford 98 B2
Stoke Lane Hereford 116 G2
Stoke Lyne Oxon 101 F11
Stoke Mandeville Bucks 84 C4
Stoke Newington London 67 B10
Stoke on Tern Shrops 150 D2
Stoke-on-Trent Stoke 168 F5
Stoke Orchard Glos 99 F8
Stoke Park Suff 108 C3
Stoke Poges Bucks 66 C3
Stoke Pound Worcs 117 D9
Stoke Prior Hereford 115 F10
Stoke Prior Worcs 117 D8
Stoke Rivers Devon 40 F6
Stoke Rochford Lincs 155 D8
Stoke Row Oxon 65 C7
Stoke St Gregory Som 28 B4
Stoke St Mary Som 28 C3
Stoke St Michael Som 45 D7
Stoke St Milborough Shrops 131 G11
Stoke sub Hamdon Som 29 D7
Stoke Talmage Oxon 83 F11
Stoke Trister Som 30 B2
Stoke Wake Dorset 30 F3
Stoke Water Dorset 29 G7
Stoke Wharf Worcs 117 D9
Stokeford Dorset 18 D3
Stokegorse Shrops 131 G11
Stokeham Notts 188 F3
Stokeinteignhead Devon 14 G4
Stokenchurch Bucks 84 F3
Stokenham Devon 8 G6
Stokesay Shrops 131 G8
Stokesby Norf 161 G8
Stokesley N Yorks 225 D10
Stokoe Northumb 250 F6
Stolford Som 43 D8
Ston Easton Som 44 C6
Stondon Massey Essex 87 E9
Stone Bucks 84 C3
Stone Glos 79 F11
Stone Kent 38 B6
Stone Kent 68 E5
Stone S Yorks 187 D9
Stone Som 44 G5
Stone Staffs 151 C8
Stone Worcs 117 B7
Stone Allerton Som 44 C2
Stone Bridge Corner P'boro 138 C5
Stone Chair W Yorks 196 B6
Stone Cross E Sus 23 E10
Stone Cross E Sus 37 B8
Stone Cross E Sus 52 G6
Stone Cross Kent 52 F4
Stone Cross Kent 54 F4
Stone Cross Kent 55 B10
Stone Cross W Mid 133 E10
Stone-edge Batch N Som 60 E3
Stone Head N Yorks 204 E4
Stone Heath Staffs 151 B9
Stone Hill Kent 54 D2
Stone Hill Kent 54 F5
Stone Hill S Glos 60 E6
Stone Hill S Yorks 199 F7
Stone House Cumb 212 B5
Stone in Oxney Kent 38 B6
Stone Raise Cumb 230 B4
Stone Street Kent 52 C5
Stone Street Suff 107 D9
Stone Street Suff 143 G7
Stonea Cambs 139 E9
Stoneacton Shrops 131 E10
Stonebow Worcs 99 B8
Stonebridge Essex 70 B2
Stonebridge London 67 C8
Stonebridge N Som 43 B11
Stonebridge Norf 141 E8
Stonebridge Sur 51 D7
Stonebridge W Mid 134 G4
Stonebridge Green Kent 54 D2
Stonebroom Derbys 170 D6
Stonebyres Holdings S Lnrk 268 G6
Stoneclough Gtr Man 195 F9
Stonecombe Devon 40 E6
Stonecrouch Kent 53 G7
Stonedge Borders 250 B3
Stoneferry Hull 209 G8
Stonefield Argyll 289 F11
Stonefield S Lnrk 268 D3
Stonefield Staffs 151 C7
Stonefield Castle Hotel Argyll 275 F9
Stonegate E Sus 37 B11
Stonegate N Yorks 226 D5
Stonegrave N Yorks 216 D3
Stonegravels Derbys 186 G5
Stonehall Kent 55 D9
Stonehall Worcs 99 B7
Stonehaugh Northumb 241 B7
Stonehaven Aberds 293 E10
Stonehill Sur 66 G4
Stonehills Hants 33 G7
Stonehouse Aberds 303 F8
Stonehouse Glos 80 D4
Stonehouse Northumb 240 F5
Stonehouse Plym 7 E9
Stonehouse S Lnrk 268 F5
Stonehouses Staffs 169 G7
Stoneleigh London 67 G8
Stoneleigh Warks 118 C6
Stoneley Green E Ches 167 E10
Stonely Cambs 122 D2
Stonepits Worcs 117 F10
Stonequarry W Sus 52 F2
Stoner Hill Hants 34 B2
Stones Green Essex 108 F3
Stonesby Leics 154 E6
Stonesfield Oxon 82 B5
Stonestreet Green Kent 54 F5
Stonethwaite Cumb 220 C5
Stoneton Warks 119 G9
Stonewells Moray 302 C2
Stonewood Kent 68 E5
Stoney Cross Hants 32 E3
Stoney Hill Worcs 117 C9
Stoney Middleton Derbys 186 F2
Stoney Royd W Yorks 196 C5
Stoney Stanton Leics 135 E9
Stoney Stoke Som 45 G8
Stoney Stratton Som 45 F7
Stoney Stretton Shrops 131 B7
Stoneyard Green Hereford 98 C4
Stoneybank E Loth 280 G6
Stoneybreck Shetland 313 N2
Stoneyburn W Loth 269 C9
Stoneycombe Devon 9 B7
Stoneycroft Mers 182 C5
Stoneyfield Gtr Man 195 E11
Stoneyfield Moray 301 D11
Stoneyford Derbys 170 F6
Stoneyford Devon 27 F8
Stoneygate Aberds 303 F10
Stoneygate Leicester 136 C2
Stoneyhills Essex 88 F6
Stoneykirk Dumfries 236 D2
Stoneylane Shrops 115 B11
Stoneywood Aberdeen 293 B10
Stoneywood Falk 278 E5
Stonganess Shetland 312 C7
Stonham Aspal Suff 126 F2
Stonnall Staffs 133 C11
Stonor Oxon 65 B8
Stonton Wyville Leics 136 D4
Stony Batter Hants 32 B3
Stony Cross Devon 25 B8
Stony Cross Hereford 98 B4
Stony Cross Hereford 115 D10
Stony Dale Notts 172 G2
Stony Gate T & W 243 G9
Stony Green Bucks 84 F5
Stony Heap Durham 242 G4
Stony Heath Hants 48 B5
Stony Houghton Derbys 171 B7
Stony Knaps Dorset 28 G5
Stony Littleton Bath 45 B8
Stony Stratford M Keynes 102 C5
Stonyfield Highld 300 B6
Stonyford Hants 32 D4
Stonyland Devon 25 B8
Stonymarsh Hants 32 B4
Stoodleigh Devon 26 D6
Stop-and-Call Pembs 91 D8
Stopes S Yorks 186 D3
Stopgate Devon 28 F2
Stopham W Sus 35 D8
Stopper Lane Lancs 204 D2
Stopsley Luton 104 G2
Stoptide Corn 10 F4
Stores Corner Suff 109 B7
Storeton Mers 182 E4
Storiths N Yorks 205 C7
Stormontfield Perth 286 E5
Stormore Wilts 45 D10
Stornoway W Isles 304 E6
Storridge Hereford 98 B4
Storrington W Sus 35 E9
Storrs Cumb 221 G7
Storrs S Yorks 186 D3
Storth Cumb 211 C9
Storwood E Yorks 207 E10
Stotfield Moray 302 B2
Stotfold C Beds 104 D4
Stottesdon Shrops 132 G3
Stoughton Leics 136 C2
Stoughton Sur 50 C3
Stoughton W Sus 34 E4
Stoughton Cross Som 44 D2
Stoul Highld 295 F9
Stoulton Worcs 99 B8
Stour Provost Dorset 30 C3
Stour Row Dorset 30 C4
Stourbridge W Mid 133 G8
Stourpaine Dorset 30 F5
Stourport on Severn Worcs 116 C6
Stourton Staffs 133 F7
Stourton W Yorks 206 G2
Stourton Warks 100 D5
Stourton Wilts 45 G9
Stourton Caundle Dorset 30 D2
Stourton Hill Warks 100 D6
Stout Som 44 G2
Stove Orkney 314 C6
Stove Shetland 313 L6
Stoven Suff 143 G8
Stow Borders 271 G9
Stow Lincs 155 B11
Stow Lincs 188 E5
Stow Bardolph Norf 140 B2
Stow Bedon Norf 141 D9
Stow cum Quy Cambs 123 E10
Stow Lawn W Mid 133 D8
Stow Longa Cambs 122 C2
Stow Maries Essex 88 F4
Stow-on-the-Wold Glos 100 F3
Stow Park Newport 59 B10
Stowbridge Norf 140 B2
Stowe Glos 79 D9
Stowe Hereford 96 B5
Stowe Lincs 156 G2
Stowe Shrops 114 C6
Stowe Staffs 152 G2
Stowe-by-Chartley Staffs 151 D10
Stowe Green Glos 79 D9
Stowell Glos 81 C9
Stowell Som 29 C11
Stowey Bath 44 B5
Stowfield Hereford 79 B9
Stowford Devon 12 B4
Stowford Devon 12 D4
Stowford Devon 24 E3
Stowford Devon 25 B10
Stowford Devon 41 E7
Stowgate Lincs 156 G3
Stowlangtoft Suff 125 D9
Stowmarket Suff 125 F10
Stowting Kent 54 E6
Stowting Common Kent 54 E6
Stowting Court Kent 54 E6
Stowupland Suff 125 F11
Straad Argyll 275 G11
Strachan Aberds 293 D8
Strachurmore Argyll 284 G5
Stradbroke Suff 126 C4
Stradishall Suff 124 G4
Stradsett Norf 140 C3
Stragglethorpe Lincs 172 E6
Straid S Ayrs 244 E4
Straight Soley Wilts 63 E10
Straith Dumfries 247 F8
Straiton Edin 270 B5
Straiton S Ayrs 245 C9
Straloch Aberds 303 G8
Straloch Perth 292 G2
Stramshall Staffs 151 B11
Strand Glos 80 C2
Strand London 67 C10
Strands Cumb 210 C3
Strang I o M 192 E4
Strangeways Gtr Man 184 B4
Strangford Hereford 97 F11
Stranghow Redcar 226 B3
Strangways Wilts 46 E6
Stranog Aberds 293 D10
Stranraer Dumfries 236 C2
Strata Florida Ceredig 112 D4
Stratfield Mortimer W Berks 65 G7
Stratfield Saye Hants 65 G7
Stratfield Turgis Hants 49 B7
Stratford C Beds 104 B3
Stratford Glos 99 D7
Stratford London 67 C11
Stratford Marsh London 67 C11
Stratford New Town London 67 C11
Stratford St Andrew Suff 127 E7
Stratford St Mary Suff 107 E10
Stratford Sub Castle Wilts 46 G6
Stratford Tony Wilts 31 B9
Stratford-upon-Avon Warks 118 F3
Strath Highld 299 B7
Strath Highld 310 D6
Strathallan Castle Perth 286 F3
Strathan Highld 295 F11
Strathan Highld 307 G5
Strathan Highld 308 C5
Strathan Skerray Highld 308 C6
Strathaven S Lnrk 268 G4
Strathblane Stirl 277 F11
Strathcanaird Highld 307 J6
Strathcarron Highld 299 E9
Strathcoil Argyll 289 F8
Strathcoul Highld 310 D5
Strathdon Aberds 292 B5
Strathellie Aberds 303 C10
Strathgarve Lodge Highld 300 C4
Strathkinness Fife 287 F8
Strathmashie House Highld 291 D7
Strathmiglo Fife 286 F6
Strathmore Lodge Highld 310 E5
Strathpeffer Highld 300 D4
Strathrannoch Highld 300 B3
Strathtay Perth 286 B3
Strathvaich Lodge Highld 300 B3
Strathwhillan N Ayrs 256 B2
Strathy Highld 300 B6
Strathy Highld 310 C2
Strathyre Stirl 285 F9
Stratton Corn 24 F2
Stratton Dorset 17 C9
Stratton Glos 81 E8
Stratton Audley Oxon 102 F2
Stratton Chase Bucks 85 G7
Stratton-on-the-Fosse Som 45 C7
Stratton St Margaret Swindon 63 B7
Stratton St Michael Norf 142 E4
Stratton Strawless Norf 160 E4
Stravithie Fife 287 F9
Strawberry Bank Cumb 211 B8
Strawberry Hill E Sus 52 F5
Strawberry Hill London 67 E7
Strawberry Hill W Yorks 198 C2
Streat E Sus 36 D5
Streatham London 67 E10
Streatham Hill London 67 E10
Streatham Park London 67 E9
Streatham Vale London 67 E9
Streatley C Beds 103 F11
Streatley W Berks 64 C5
Street Cumb 222 D2
Street Lancs 202 C6
Street N Yorks 226 E4
Street Som 28 F5
Street Som 44 F3
Street Ash Som 28 E3
Street Ashton Warks 135 G9
Street Dinas Shrops 148 B6
Street End Hants 33 D9
Street End Kent 54 C6
Street End W Sus 22 D5
Street Gate T & W 242 F6
Street Houses N Yorks 206 D6
Street Lane Derbys 170 F5
Street Lydan Wrex 149 B8
Street of Kincardine Highld 291 B11
Street on the Fosse Som 44 F6
Streethay Staffs 152 G2
Streethouse W Yorks 197 C11
Streetlam N Yorks 224 F6
Streetly W Mid 133 D11
Streetly End Cambs 106 B2
Strefford Shrops 131 F8
Strelley Notts 171 G8
Strensall York 215 G2
Strensham Worcs 99 C8
Stretch Down Devon 26 E4
Stretcholt Som 43 E9
Strete Devon 8 F6
Stretford Gtr Man 184 C4
Stretford Hereford 115 F10
Stretford Court Hereford 115 F8
Strethall Essex 105 D9
Stretham Cambs 123 C10
Strettington W Sus 22 B5
Stretton Derbys 170 C5
Stretton Rutland 155 F8
Stretton Staffs 151 G7
Stretton Staffs 152 D5
Stretton W Ches 166 E6
Stretton Warr 183 E10
Stretton en le Field Leics 152 G6
Stretton Grandison Hereford 98 C2
Stretton-on-Dunsmore Warks 119 C8
Stretton-on-Fosse Warks 100 D4
Stretton Sugwas Hereford 97 C9
Stretton under Fosse Warks 135 G8
Stretton Westwood Shrops 131 D11
Strichen Aberds 303 D9
Strines Gtr Man 185 D7
Stringston Som 43 E7
Strixton Northants 121 E8
Stroat Glos 79 F9
Strode N Som 60 G4
Strom Shetland 313 J5
Stromeferry Highld 295 B10
Stromemore Highld 295 B10
Stromness Orkney 314 F2
Stronaba Highld 290 E4
Stronachlachar Stirl 285 F8
Stronachullin Lodge Argyll 275 F9
Stronchreggan Highld 290 F2
Stronchrubie Highld 307 H7
Strone Argyll 255 F7
Strone Argyll 274 G6
Strone Argyll 276 E3
Strone Highld 290 E3
Strone Highld 291 D9
Strone Highld 300 G5
Strone Invclyd 276 G5
Stronelairg Lodge Highld 291 C7
Stroneskar Argyll 275 C9
Stronmachair Stirl 285 G8
Stronmilchan Argyll 284 E5
Stronord Dumfries 236 C6
Stronsaul Argyll 276 F2
Strontian Highld 289 C10
Stronvar Stirl 285 E9
Strood Kent 53 G11
Strood Medway 69 F8
Strood Green Sur 51 D8
Strood Green W Sus 35 C8
Strood Green W Sus 50 G6
Strothers Dale Northumb 241 F11
Stroud Glos 80 D4
Stroud Hants 34 C2
Stroud Sur 50 F2
Stroud Green Essex 88 G5
Stroud Green Glos 80 D4
Stroud Green London 67 B10
Stroude Sur 66 F4
Strouden Bmouth 19 C8
Stroul Argyll 276 E4
Stroupster Highld 310 C7
Stroxton Lincs 155 C8
Stroxworthy Devon 24 D4
Struan Highld 294 B5
Struan Perth 291 G10
Strubby Lincs 191 E7
Structon's Heath Worcs 116 D5
Strugg's Hill Lincs 156 B5
Strumpshaw Norf 142 B6
Strutherhill S Lnrk 268 F5
Struthers Fife 287 G7
Struy Highld 300 F3
Stryd Anglesey 178 E2
Stryd y Facsen Anglesey 178 E4
Stryt-issa Wrex 166 F3
Stuartfield Aberds 303 E9
Stub Place Cumb 219 G11
Stubb Norf 161 E8
Stubbermere W Sus 22 B3
Stubber's Green W Mid 133 C10
Stubbings Windsor 65 C10
Stubbing's Green Suff 125 C11
Stubbington Hants 33 G9
Stubbins Lancs 195 D9
Stubbins Lancs 202 E6
Stubble Green Cumb 219 F11
Stubbles W Berks 64 D5
Stubbs Cross Kent 54 F3
Stubb's Green Norf 142 D5
Stubbs Green Norf 143 D7
Stubhampton Dorset 30 E6
Stubshaw Cross Gtr Man 194 G5
Stubton Lincs 172 F5
Stubwood Staffs 151 B11
Stuckgowan Argyll 285 G7
Stuckton Hants 31 E11
Stud Green E Ches 168 C2
Stud Green Windsor 65 D11
Studd Hill Kent 71 F7
Studdal Kent 55 D10
Studfold N Yorks 212 E6
Studham C Beds 85 B8
Studland Dorset 18 E6
Studley Warks 117 E11
Studley Wilts 62 E3
Studley Green Bucks 84 F3
Studley Green Wilts 45 B10
Studley Roger N Yorks 214 E5
Studley Royal N Yorks 214 E5
Stump Cross Essex 105 C10
Stumps Cross Glos 99 E11
Stuntney Cambs 123 B11
Stunts Green E Sus 23 C10
Sturbridge Staffs 150 C6
Sturford Wilts 45 E10
Sturgate Lincs 188 D5
Sturmer Essex 106 C3
Sturminster Common Dorset 30 E3
Sturminster Marshall Dorset 31 G7
Sturminster Newton Dorset 30 E3
Sturry Kent 71 G7
Stursdon Corn 24 E2
Sturton N Lincs 200 G3
Sturton by Stow Lincs 188 E5
Sturton le Steeple Notts 188 E3
Stuston Suff 126 B2
Stutton N Yorks 206 E5
Stutton Suff 108 E3
Styal E Ches 184 E4
Styants Bottom Kent 52 B5
Stydd Lancs 203 F9
Styrrup Notts 187 C10
Suainebost W Isles 304 B7
Suardail W Isles 304 E6
Succoth Aberds 302 F4
Succoth Argyll 284 G6
Suckley Worcs 116 G4
Suckley Green Worcs 116 G4
Suckley Knowl Worcs 116 G4
Sucksquoy Orkney 314 H4
Sucksted Green Essex 105 F11
Sudborough Northants 137 G9
Sudbourne Suff 127 G8
Sudbrook Lincs 173 G7
Sudbrook Mon 60 B4
Sudbrooke Lincs 189 F8
Sudbury Derbys 152 C3
Sudbury London 67 C7
Sudbury Suff 107 C7
Sudden Gtr Man 195 E11
Suddie Highld 300 D6
Sudgrove Glos 80 D6
Suffield N Yorks 227 G9
Suffield Norf 160 C4
Sugnall Staffs 150 C5
Sugwas Pool Hereford 97 C9
Suisnish Highld 295 D7
Suladale Highld 298 D3
Sulaisiadar W Isles 304 E7
Sulby I o M 192 C4
Sulgrave Northants 101 B11
Sulgrave T & W 243 F8
Sulham W Berks 64 E6
Sulhampstead W Berks 64 F6
Sulhampstead Abbots W Berks 64 F6
Sulhampstead Bannister Upper End W Berks 64 F6
Sulhamstead W Berks 64 F6
Sulland Orkney 314 B5
Sullington W Sus 35 E9
Sullington Warren W Sus 35 E9
Sullom Shetland 312 F5
Sullom Voe Oil Terminal Shetland 312 F5
Sully V Glam 59 F7
Sumburgh Shetland 313 N6
Summer Bridge N Yorks 214 G4
Summer Heath Bucks 84 G2
Summer Hill E Sus 23 D9
Summer Hill W Mid 133 E9
Summerbridge N Yorks 214 G4
Summercourt Corn 5 D7
Summerfield Kent 55 B9
Summerfield Norf 158 B5
Summerfield Worcs 116 C6
Summerfield Park W Mid 133 F10
Summergangs Hull 209 G8
Summerhill Newport 59 B10
Summerhill Pembs 73 D11
Summerhill Staffs 133 B11
Summerhill Telford 150 F4
Summerhill Worcs 116 B6
Summerhill Wrex 166 E4
Summerhouse Darl 224 B4
Summerlands Cumb 211 B10
Summerlands Som 29 D8
Summerleaze Mon 60 B2
Summerley Derbys 186 F5
Summerscales N Yorks 205 C8
Summersdale W Sus 22 B5
Summerseat Gtr Man 195 E9
Summerston Glasgow 277 G11
Summerstown Bucks 102 G3
Summerstown London 67 E9
Summertown Oxon 83 D8
Summit Gtr Man 195 E10
Summit Gtr Man 196 D2
Summit Gtr Man 196 F2
Sun Green Gtr Man 185 B7
Sunbrick Cumb 210 E5
Sunbury Common Sur 66 E5
Sunbury-on-Thames Sur 66 F6
Sundayshill S Glos 79 G11
Sundaywell Dumfries 247 G8
Sunderland Argyll 274 G3
Sunderland Cumb 229 D9
Sunderland Lancs 202 B4
Sunderland T & W 243 F9
Sunderland Bridge Durham 233 D11
Sundhope Borders 261 D8
Sundon Park Luton 103 F11
Sundridge Kent 52 B3
Sundridge London 68 E2
Sunhill Glos 81 E10
Sunipol Argyll 288 D5
Sunk Island E Yorks 201 D9
Sunken Marsh Essex 69 C10
Sunningdale Windsor 66 F3
Sunninghill Windsor 66 F2
Sunningwell Oxon 83 E7
Sunniside Durham 233 D8
Sunniside T & W 242 F6
Sunniside T & W 243 G8
Sunny Bank Gtr Man 195 F10
Sunny Bower Lancs 203 G10
Sunny Hill Derby 152 C6
Sunnybrow Durham 233 E9
Sunnyfields S Yorks 198 F4
Sunnyhurst Blkburn 195 C7
Sunnylaw Stirl 278 B5
Sunnymead Oxon 83 D8
Sunnymeads Windsor 66 D4
Sunnymede Essex 87 G11
Sunnyside S Yorks 187 C7
Sunnyside W Sus 51 F11
Sunset Hereford 114 F6
Sunton Wilts 47 C8
Surbiton London 67 F7
Surby I o M 192 E3
Surfleet Lincs 156 D5
Surfleet Seas End Lincs 156 D5
Surlingham Norf 142 B6
Surrex Essex 107 G7
Suspension Bridge Norf 139 E10
Sustead Norf 160 B3
Susworth Lincs 199 G10
Sutcombe Devon 24 E4
Sutcombemill Devon 24 E4
Sutherland Grove Argyll 289 E11
Suton Norf 141 D11
Sutors of Cromarty Highld 301 C8
Sutterby Lincs 190 G5
Sutterton Lincs 156 B5
Sutterton Dowdyke Lincs 156 C5
Sutton Bucks 66 D4
Sutton C Beds 104 B4
Sutton Cambs 123 B8
Sutton Devon 8 G4
Sutton E Sus 23 F7
Sutton Kent 55 D10
Sutton Lincs 172 E5
Sutton London 67 G9
Sutton Mers 183 C8
Sutton N Yorks 198 B3
Sutton Norf 161 E7
Sutton Notts 154 B5
Sutton Notts 187 E11
Sutton Oxon 82 D6
Sutton P'boro 137 D11
Sutton Pembs 72 B6
Sutton S Yorks 198 E5
Sutton Shrops 132 F4
Sutton Shrops 149 D7
Sutton Shrops 149 G10
Sutton Shrops 150 C3
Sutton Som 44 G6
Sutton Staffs 150 E5
Sutton Suff 108 B6
Sutton Sur 50 D5
Sutton W Sus 35 D7
Sutton Abinger Sur 50 D6
Sutton at Hone Kent 68 E5
Sutton Bassett Northants 136 E5
Sutton Benger Wilts 62 D2
Sutton Bingham Som 29 E8
Sutton Bonington Notts 153 E10
Sutton Bridge Lincs 157 E9
Sutton Cheney Leics 135 C8
Sutton Coldfield W Mid 134 D2
Sutton Corner Lincs 157 D8
Sutton Courtenay Oxon 83 G8
Sutton Crosses Lincs 157 E8
Sutton Cum Lound Notts 187 E11
Sutton End W Sus 35 D7
Sutton Forest Side Notts 171 D8
Sutton Gault Cambs 123 B8
Sutton Green Sur 50 C4
Sutton Green W Ches 182 F5
Sutton Green Wrex 166 F6
Sutton Hall Shrops 132 C4
Sutton Heath Mers 183 C8
Sutton Hill Telford 132 C4
Sutton Holms Dorset 31 F9
Sutton Howgrave N Yorks 214 D6
Sutton in Ashfield Notts 171 D7
Sutton-in-Craven N Yorks 204 E6
Sutton in the Elms Leics 135 E10
Sutton Ings Hull 209 G8
Sutton Lakes Hereford 97 B10
Sutton Lane Ends E Ches 184 G6
Sutton Leach Mers 183 C8
Sutton Maddock Shrops 132 C4
Sutton Mallet Som 43 F11
Sutton Mandeville Wilts 31 B7
Sutton Manor Mers 183 C8
Sutton Marsh Hereford 97 C10
Sutton Mill N Yorks 204 E6
Sutton Montis Som 29 C10
Sutton on Hull Hull 209 G8
Sutton on Sea Lincs 191 E8
Sutton-on-the-Forest N Yorks 215 G11
Sutton on the Hill Derbys 152 C4
Sutton on Trent Notts 172 B3
Sutton Poyntz Dorset 17 E10
Sutton Row Wilts 31 B7
Sutton St Edmund Lincs 157 G7
Sutton St James Lincs 157 F7
Sutton St Michael Hereford 97 B10
Sutton St Nicholas Hereford 97 B10
Sutton Scarsdale Derbys 170 B6
Sutton Scotney Hants 48 F3
Sutton Street Suff 108 C6
Sutton under Brailes Warks 100 D6
Sutton-under-Whitestonecliffe N Yorks 215 C9
Sutton upon Derwent E Yorks 207 D10
Sutton Valence Kent 53 D10
Sutton Veny Wilts 45 E11
Sutton Waldron Dorset 30 D5
Sutton Weaver W Ches 183 F8
Sutton Wick Bath 44 B5
Sutton Wick Oxon 83 G7
Swaby Lincs 190 F5
Swadlincote Derbys 152 F6
Swaffham Norf 140 B6
Swaffham Bulbeck Cambs 123 E11
Swaffham Prior Cambs 123 E11
Swafield Norf 160 C5
Swaile's Green E Sus 38 C3
Swain House W Yorks 205 F9
Swainby N Yorks 225 E9
Swainshill Hereford 97 C9
Swainsthorpe Norf 142 C4
Swainswick Bath 61 F9
Swaithe S Yorks 197 G11
Swalcliffe Oxon 101 D7
Swalecliffe Kent 70 F6
Swallow Lincs 201 G7
Swallow Beck Lincs 173 B7
Swallowcliffe Wilts 31 B7
Swallowfield Wokingham 65 G8
Swallowfields Devon 8 C5
Swallowhurst Cumb 220 G2
Swallownest S Yorks 187 D7
Swallows Cross Essex 87 F10
Swalwell T & W 242 E6
Swampton Hants 48 C2
Swan Bottom Bucks 84 D6
Swan Green Suff 126 C5
Swan Green W Ches 184 G2
Swan Street Essex 107 F7
Swan Village W Mid 133 E9
Swanage Dorset 18 F6
Swanbach E Ches 167 G11
Swanbister Orkney 314 F3
Swanborough Swindon 81 G11
Swanbourne Bucks 102 F6
Swanbridge V Glam 59 F7
Swanland E Yorks 200 B3
Swanley Glos 80 F2
Swanley Kent 68 F4
Swanley Bar Herts 86 E3
Swanley Village Kent 68 F4
Swanmore Hants 33 D9
Swanmore I o W 21 C7
Swannay Orkney 314 D2
Swannington Leics 153 F8
Swannington Norf 160 F2
Swanpool Lincs 189 G7
Swanscombe Kent 68 E6
Swansea = Abertawe Swansea 56 C6
Swanside Mers 182 C6
Swanston Edin 270 B4
Swanton Abbott Norf 160 D5
Swanton Hill Norf 160 D5
Swanton Morley Norf 159 F10
Swanton Novers Norf 159 C10
Swanton Street Kent 53 B11
Swanwick Derbys 170 E6
Swanwick Hants 33 F8
Swanwick Green E Ches 167 F9
Swarby Lincs 173 G8
Swarcliffe W Yorks 206 F3
Swardeston Norf 142 C4
Swarister Shetland 312 E7
Swarkestone Derbys 153 D7
Swarland Northumb 252 C5
Swarraton Hants 48 F5
Swartha W Yorks 205 D7
Swarthmoor Cumb 210 D5
Swartland Orkney 314 D2
Swathwick Derbys 170 B5
Swaton Lincs 156 B2
Swavesey Cambs 123 D7
Sway Hants 19 B11
Swayfield Lincs 155 E9
Swaythling Soton 32 D6
Sweet Green Worcs 116 E2
Sweetham Devon 14 B3
Sweethaws E Sus 37 B8
Sweethay Som 28 C2
Sweetholme Cumb 221 B11
Sweethouse Corn 5 C11
Sweets Corn 11 B9
Sweetshouse Corn 5 C11
Sweffling Suff 126 E6
Swell Som 28 C5
Swelling Hill Hants 49 G7
Swepstone Leics 153 G7
Swerford Oxon 101 E7
Swettenham E Ches 168 B4
Swetton N Yorks 214 E3
Swffryd Caerph 78 F2
Swift's Green Kent 53 E11
Swiftsden E Sus 38 B2
Swilland Suff 126 G3
Swillbrook Lancs 202 G5
Swillington W Yorks 206 G3
Swillington Common W Yorks 206 G3
Swimbridge Devon 25 B10
Swimbridge Newland Devon 40 G6
Swinbrook Oxon 82 C3
Swincliffe N Yorks 205 B10
Swincliffe W Yorks 197 B8
Swincombe Devon 41 E7
Swinden N Yorks 204 C3
Swinderby Lincs 172 C5
Swindon Glos 99 G8
Swindon Staffs 133 E7
Swindon Swindon 63 C7
Swine E Yorks 209 F8
Swinefleet E Yorks 199 C9
Swineford S Glos 61 F7
Swineshead Beds 121 D11
Swineshead Lincs 174 G2
Swineshead Bridge Lincs 174 G2
Swinethorpe Lincs 172 B5
Swiney Highld 310 F6
Swinford Leics 119 B11
Swinford Oxon 82 D6
Swingate Notts 171 G8
Swingbrow Cambs 139 F7
Swingfield Minnis Kent 55 E8
Swingfield Street Kent 55 E8
Swingleton Green Suff 107 B9
Swinhoe Northumb 264 D6
Swinhope Lincs 190 B2
Swining Shetland 312 G6
Swinister Shetland 312 E5
Swinister Shetland 313 L6
Swinithwaite N Yorks 213 B10
Swinmore Common Hereford 98 C3
Swinnie Borders 262 F4
Swinnow Moor W Yorks 205 G10
Swinscoe Staffs 169 F10
Swinside Cumb 229 G10
Swinside Townfoot Borders 262 F6
Swinstead Lincs 155 E10
Swinton Borders 272 F6
Swinton Glasgow 268 C3
Swinton Gtr Man 195 G9
Swinton N Yorks 214 D4
Swinton N Yorks 216 E5
Swinton S Yorks 186 B6
Swinton Bridge S Yorks 187 B7
Swinton Hill Borders 272 F6
Swinton Park Gtr Man 195 G9
Swintonmill Borders 272 F6
Swiss Valley Carms 75 E8
Swithland Leics 153 G10
Swordale Highld 300 C5
Swordale Highld 309 K6
Swordland Highld 295 F9
Swordly Highld 308 C7
Sworton Heath E Ches 183 E11
Swyddffynnon Ceredig 112 D3
Swynnerton Staffs 151 B7
Swyre Dorset 16 D6
Sycamore Devon 28 F3
Sychdyn = Soughton Flint 166 B2
Sychtyn Powys 129 B9
Sydallt Wrex 166 D4
Syde Glos 81 C7
Sydenham London 67 E11
Sydenham Oxon 84 E2
Sydenham Som 43 F10
Sydenham Damerel Devon 12 F4
Syderstone Norf 158 C6
Sydling St Nicholas Dorset 17 B8
Sydmonton Hants 48 B3
Sydney E Ches 168 D2
Syerston Notts 172 F2
Syke Gtr Man 195 D11
Sykehouse S Yorks 198 D6
Sykes Lancs 203 C8
Syleham Suff 126 B4
Sylen Carms 75 D8
Symbister Shetland 313 G7
Symington Borders 271 F8
Symington S Ayrs 257 C9
Symington S Lnrk 259 B11
Symonds Green Herts 104 F4
Symonds Yat Hereford 79 B9
Symondsbury Dorset 16 C4
Synderford Dorset 28 G5
Synod Inn = Post Mawr Ceredig 111 G8
Synton Borders 261 E11
Synton Mains Borders 261 E11
Synwell Glos 80 G3
Syre Highld 308 E6
Syreford Glos 99 G10
Syresham Northants 102 C2
Syster Highld 310 C6
Syston Leics 154 G2
Syston Lincs 172 G6
Sytch Ho Green Shrops 132 E5
Sytch Lane Telford 150 E2
Sytchampton Worcs 116 D6
Sywell Northants 120 D6

T

Taagan Highld 299 C10
Tabley Hill E Ches 184 F2
Tabor Gwyn 146 F5
Tabost W Isles 305 G5
Tabost W Isles 304 B7
Tachbrook Mallory Warks 118 E6
Tacker Street Som 42 F4
Tackley Oxon 101 G9
Tacleit W Isles 304 E3
Tacolneston Norf 142 D2
Tadcaster N Yorks 206 E5
Tadden Dorset 31 G7
Taddington Derbys 185 G10
Taddington Glos 99 E11
Taddiport Devon 25 D7
Tadhill Som 45 D7
Tadley Hants 64 G6
Tadley Oxon 64 B4
Tadlow Beds 104 B5
Tadlow C Beds 104 B5
Tadmarton Oxon 101 D7
Tadnoll Dorset 17 D11
Tadwick Bath 61 E8
Tadworth Sur 51 B8
Tafarn-y-bwlch Pembs 91 E11
Tafarn-y-gelyn Denb 165 C11
Tafarnau-bach Bl Gwent 77 C10
Taff Merthyr Garden Village M Tydf 77 F10
Taff's Well Rhondda 58 C6
Tafolwern Powys 129 C7
Tai Conwy 164 C3
Tai-bach Powys 148 D3
Tai-mawr Conwy 165 G7
Tai-morfa Gwyn 144 D5
Tai-nant Wrex 166 F3
Tai-Ucha Denb 165 D8
Taibach Neath 57 D9
Taigh a Ghearraidh W Isles 296 D3
Taigh Bhalaigh W Isles 296 D3
Tain Highld 309 L7
Tain Highld 310 C6
Tainlon Gwyn 162 E6
Tai'r-Bull Powys 95 F9
Tai'r-heol Caerph 77 G10
Tai'r-ysgol Swansea 57 B7
Tairbeart W Isles 305 H3
Tairgwaith Neath 76 C2
Takeley Essex 105 G11
Takeley Street Essex 105 G10
Tal-sarn Ceredig 111 F10
Tal-y-bont Ceredig 128 F3
Tal-y-Bont Conwy 164 B3
Tal-y-bont Gwyn 145 E11
Tal-y-bont Gwyn 179 G10
Tal-y-cafn Conwy 180 G3
Tal-y-coed Mon 78 B6
Tal-y-llyn Gwyn 128 B4
Tal-y-waenydd Gwyn 163 F11
Tal-y-wern Powys 128 C6
Talachar = Laugharne Carms 74 C4
Talachddu Powys 95 E11
Talacre Flint 181 E10
Talardd Gwyn 147 D7
Talaton Devon 15 B7
Talbenny Pembs 72 C4
Talbot Green Rhondda 58 C4
Talbot Heath Poole 19 C7
Talbot Village Poole 19 C7
Talbot Woods Bmouth 19 C7
Talbot's End S Glos 80 G2
Tale Devon 27 G9
Talerddig Powys 129 C8
Talgarreg Ceredig 111 G8
Talgarth Powys 96 E3
Talgarth's Well Swansea 56 D2
Talisker Highld 294 B5
Talke Staffs 168 E4
Talke Pits Staffs 168 E4
Talkin Cumb 240 F3
Talla Linnfoots Borders 260 E4
Talladale Highld 299 B9
Tallaminnoch S Ayrs 245 D10
Talland Corn 6 E4
Tallarn Green Wrex 166 G6
Tallentire Cumb 229 D8
Talley Carms 94 E2
Tallington Lincs 137 B11
Talmine Highld 308 C5
Talog Carms 92 F6
Talsarn Carms 94 F5
Talsarnau Gwyn 146 B2
Talskiddy Corn 5 B8
Talwrn Anglesey 179 F7
Talwrn Wrex 166 F3
Talybont-on-Usk Powys 96 G2
Talygarn Rhondda 58 C4
Talyllyn Powys 96 F2
Talysarn Gwyn 163 E7
Talywain Torf 78 E3
Tamanabhagh W Isles 304 F2
Tame Bridge N Yorks 225 D10
Tame Water Gtr Man 196 F3
Tamer Lane End Gtr Man 194 G6
Tamerton Foliot Plym 7 C9
Tamfourhill Falk 279 E7
Tamworth Staffs 134 C4
Tamworth Green Lincs 174 G5
Tan Hills Durham 233 B11
Tan Hinon Powys 129 F7
Tan-lan Conwy 164 C3
Tan-lan Gwyn 163 G10
Tan Office Suff 126 E2
Tan Office Green Suff 124 F5
Tan-y-bwlch Gwyn 163 G11
Tan-y-fron Conwy 165 C7
Tan-y-graig Anglesey 179 F8
Tan-y-graig Gwyn 144 B6
Tan-y-groes Ceredig 92 B5
Tan-y-mynydd Gwyn 144 C6
Tan-y-pistyll Powys 147 D11
Tan-yr-allt Denb 181 E9
Tan-yr-allt Gwyn 163 E7
Tancred N Yorks 206 B5
Tandem W Yorks 197 D7
Tanden Kent 54 F2
Tandlehill Renfs 267 C8
Tandridge Sur 51 C11
Tanerdy Carms 93 G8
Tanfield Durham 242 F5
Tanfield Lea Durham 242 G5
Tang N Yorks 205 B10
Tang Hall York 207 C8
Tangasdal W Isles 297 M2
Tangiers Pembs 73 B7
Tangley Hants 47 C10
Tanglwst Carms 92 E6
Tangmere W Sus 22 B6
Tangwick Shetland 312 F4
Tangy Argyll 255 E7
Tankersley S Yorks 197 G10
Tankerton Kent 70 F6
Tanlan Flint 181 E10
Tanlan Banks Flint 181 E10
Tannach Highld 310 E7
Tannachie Aberds 293 E9
Tannadice Angus 287 B8
Tanner's Green Worcs 117 C11
Tannington Suff 126 D4
Tannington Place Suff 126 D4
Tannochside N Lnrk 268 C4
Tansley Derbys 170 D4
Tansley Hill W Mid 133 F9
Tanis Wilts 62 G3
Tanhouse Lancs 194 F3

Tansley Knoll Derbys 170 C4
Tansor Northants 137 E11
Tanterton Lancs 202 G6
Tantobie Durham 242 G5
Tanton N Yorks 225 C10
Tanwood Worcs 117 C8
Tanworth-in-Arden Warks 118 C2
Tanyfron Wrex 166 E3
Tanygrisiau Gwyn 163 F11
Tanyrhydiau Ceredig 112 D4
Tanysgafell Gwyn 163 B10
Taobh a Chaolais W Isles 297 K3
Taobh a' Ghlinne W Isles 305 G5
Taobh a Thuath Loch Aineort W Isles 297 J3
Taobh a Tuath Loch Baghasdail W Isles 297 J3
Taobh Siar W Isles 305 H3
Taobh Tuath W Isles 296 C5
Taplow Bucks 66 C2
Tapnage Hants 33 E9
Tapton Derbys 186 G5
Tapton Hill S Yorks 186 D4
Tarbat Ho Highld 301 B7
Tarbert Argyll 255 B7
Tarbert Argyll 275 E7
Tarbert Argyll 275 G9
Tarbet Argyll 285 G7
Tarbet Highld 295 F9
Tarbet Highld 306 E6
Tarbock Green Mers 183 D7
Tarbolton S Ayrs 257 D10
Tarbrax S Lnrk 269 D10
Tardebigge Worcs 117 D10
Tardy Gate Lancs 194 B4
Tarfside Angus 292 F6
Tarland Aberds 292 C6
Tarleton Lancs 194 C3
Tarleton Moss Lancs 194 C2
Tarlogie Highld 309 L7
Tarlscough Lancs 194 E2
Tarlton Glos 81 F7
Tarn W Yorks 205 F9
Tarnbrook Lancs 203 B7
Tarnock Som 43 C11
Tarns Cumb 229 B8
Tarnside Cumb 221 G8
Tarporley W Ches 167 C9
Tarpots Essex 69 B9
Tarr Som 42 G6
Tarraby Cumb 239 F10
Tarrant Crawford Dorset 30 G6
Tarrant Gunville Dorset 30 E6
Tarrant Hinton Dorset 30 E6
Tarrant Keyneston Dorset 30 G6
Tarrant Launceston Dorset 30 F6
Tarrant Monkton Dorset 30 F6
Tarrant Rawston Dorset 30 F6
Tarrant Rushton Dorset 30 F6
Tarrel Highld 311 L2
Tarring Neville E Sus 36 G6
Tarrington Hereford 98 C2
Tarrington Common Hereford 98 D2
Tarryblake Ho Moray 302 E5
Tarsappie Perth 286 E5
Tarskavaig Highld 295 E7
Tarts Hill Shrops 149 B8
Tarves Aberds 303 F8
Tarvie Highld 300 D4
Tarvie Perth 292 G2
Tarvin W Ches 167 B7
Tarvin Sands W Ches 167 B7
Tasburgh Norf 142 D4
Tasley Shrops 132 E3
Taston Oxon 101 G7
Tat Bank W Mid 133 F9
Tatenhill Staffs 152 E4
Tatenhill Common Staffs 152 E3
Tathall End M Keynes 102 B6
Tatham Lancs 212 F2
Tathwell Lincs 190 E4
Tatling End Bucks 66 B4
Tatsfield Sur 52 B2
Tattenhall W Ches 167 D7
Tattenhoe M Keynes 102 E6
Tatterford Norf 159 D7
Tattersett Norf 158 C6
Tattershall Lincs 174 D2
Tattershall Bridge Lincs 173 D11
Tattershall Thorpe Lincs 174 D2
Tattingstone Suff 108 D2
Tattingstone White Horse Suff 108 D2
Tattle Bank Warks 118 E3
Tatton Dale E Ches 184 E2
Tatworth Som 28 F4
Taunton Gtr Man 196 G2
Taunton Som 28 C2
Taverham Norf 160 G3
Taverners Green Essex 87 B9
Tavernspite Pembs 73 C11
Tavistock Devon 12 G5
Taw Green Devon 13 B9
Tawstock Devon 25 B9
Taxal Derbys 185 F8
Tay Bridge Dundee 287 E8
Tayinloan Argyll 255 C7
Taymouth Castle Perth 285 C11
Taynish Argyll 275 E8
Taynton Glos 98 G4
Taynton Oxon 82 C2
Taynuilt Argyll 284 D4
Tayport Fife 287 E8
Tayvallich Argyll 275 E8
Tea Green Herts 104 G2
Tealby Lincs 189 C11
Tealing Angus 287 D8
Team Valley T & W 242 E6
Teams T & W 242 E6
Teanford Staffs 169 G8
Teangue Highld 295 E8
Teanna Mhachair W Isles 296 E3
Teasley Mead E Sus 52 F4
Tebay Cumb 222 E2
Tebworth C Beds 103 F9
Tedburn St Mary Devon 14 C2
Teddington Glos 99 E9
Teddington London 67 E7
Teddington Hands Worcs 99 E9
Tedsmore Shrops 149 D7
Tedstone Delamere Hereford 116 F3
Tedstone Wafer Hereford 116 F3
Teesville Redcar 225 B10
Teeton Northants 120 C3
Teffont Evias Wilts 46 G3
Teffont Magna Wilts 46 G3
Tegryn Pembs 92 E4
Teigh Rutland 155 F7
Teign Village Devon 14 E2
Teigncombe Devon 13 D9
Teigngrace Devon 14 G2
Teignmouth Devon 14 G4
Telford Telford 132 B3
Telham E Sus 38 E3
Tellisford Som 45 B10
Telscombe E Sus 36 G6
Telscombe Cliffs E Sus 36 G5
Templand Dumfries 248 F3
Temple Corn 11 G8
Temple Glasgow 267 B10
Temple Midloth 270 D6
Temple Wilts 45 E10
Temple Windsor 65 C10
Temple Balsall W Mid 118 B4
Temple Bar Carms 75 B9
Temple Bar Ceredig 111 G10
Temple Bar W Sus 22 B5
Temple Cloud Bath 44 B6
Temple Cowley Oxon 83 E8
Temple End Essex 106 C6
Temple End Suff 124 G3
Temple Ewell Kent 55 E9
Temple Fields Essex 87 C7
Temple Grafton Warks 118 G2
Temple Guiting Glos 99 F11
Temple Herdewyke Warks 119 G7
Temple Hill Kent 68 D5
Temple Hirst N Yorks 198 C6
Temple Normanton Derbys 170 B6
Temple Sowerby Cumb 231 F8
Templeborough S Yorks 186 C6
Templecombe Som 30 C2
Templehall Fife 280 C5
Templeman's Ash Dorset 28 G6
Templeton Devon 26 E5
Templeton Pembs 73 C10
Templeton W Berks 63 F11
Templeton Bridge Devon 26 E5
Templetown Durham 242 G4
Tempsford C Beds 122 G3
Ten Acres W Mid 133 G11
Ten Mile Bank Norf 140 D2
Tenandry Perth 291 G11
Tenbury Wells Worcs 115 D11
Tenby = Dinbych-y-Pysgod Pembs 73 E10
Tencreek Corn 6 E4
Tendring Essex 108 G2
Tendring Green Essex 108 F2
Tendring Heath Essex 108 F2
Tenston Orkney 314 E2
Tenterden Kent 53 G11
Terfyn Conwy 180 F6
Terfyn Gwyn 163 C9
Terhill Som 43 G7
Terling Essex 88 B3
Ternhill Shrops 150 C2
Terpersie Castle Aberds 302 G5
Terras Corn 5 E8
Terregles Banks Dumfries 237 B11
Terrible Down E Sus 23 B7
Terrick Bucks 84 D4
Terriers Bucks 84 G5
Terrington N Yorks 216 E3
Terrington St Clement Norf 157 E10
Terrington St John Norf 157 G10
Terryhorn Aberds 302 F4
Terry's Green Warks 118 C2
Terwick Common W Sus 34 C4
Teston Kent 53 C8
Testwood Hants 32 E5
Tetbury Glos 80 G5
Tetbury Upton Glos 80 F5
Tetchill Shrops 149 C7
Tetchwick Bucks 83 B11
Tetcott Devon 12 B2
Tetford Lincs 190 G4
Tetley N Lincs 199 E9
Tetney Lincs 201 G10
Tetney Lock Lincs 201 G10
Tetsworth Oxon 83 E11
Tettenhall W Mid 133 D7
Tettenhall Wood W Mid 133 D7
Tetworth Cambs 122 G4
Teuchan Aberds 303 F10
Teversal Notts 171 C7
Teversham Cambs 123 F9
Teviothead Borders 249 B10
Tewel Aberds 293 E10
Tewin Herts 86 C3
Tewin Wood Herts 86 B3
Tewitfield Lancs 211 E10
Tewkesbury Glos 99 E7
Teynham Kent 70 G3
Teynham Street Kent 70 G3
Thackley W Yorks 205 F9
Thackley End W Yorks 205 F9
Thackthwaite Cumb 229 G8
Thainston Aberds 293 F8
Thakeham W Sus 35 D10
Thame Oxon 84 D2
Thames Ditton Sur 67 F7
Thames Haven Thurrock 69 C8
Thames Head Glos 81 F7
Thamesmead London 68 C3
Thanington Kent 54 B6
Thankerton S Lnrk 259 B11
Tharston Norf 142 E3
Thatcham W Berks 64 F4
Thatto Heath Mers 183 C8
Thaxted Essex 106 E2
The Aird Highld 298 D4
The Alders Staffs 134 C3
The Arms Norf 141 D7
The Bage Hereford 96 C5
The Balloch Perth 286 F2
The Bank E Ches 168 D4
The Banks Gtr Man 185 D7
The Banks Wilts 62 D4
The Barony E Ches 167 E11
The Barony Orkney 314 D2
The Barton Wilts 62 D5
The Batch S Glos 61 E7
The Beeches Glos 81 E8
The Bell Gtr Man 194 F4
The Bents Staffs 151 C10
The Blythe Staffs 151 D10
The Bog Shrops 131 D7
The Borough Dorset 30 E2
The Borough London 67 D10
The Bourne Sur 49 E10
The Bourne Worcs 117 F9
The Bows Stirl 285 G11
The Braes Highld 295 B7
The Brampton Staffs 168 F4
The Brand Leics 153 G10
The Bratch Staffs 133 E7
The Breck Orkney 314 F3
The Brents Kent 70 G4
The Bridge Dorset 30 E3
The Broad Hereford 115 E9
The Brook Suff 125 B11
The Brushes Derbys 186 F5
The Bryn Mon 78 D4
The Burf Worcs 116 D6
The Butts Hants 49 F8
The Butts Som 45 D9
The Camp Glos 80 D6
The Camp Herts 85 D11
The Cape Warks 118 D5
The Chart Kent 52 C3
The Chequer Wrex 167 G7
The Chuckery W Mid 133 D10
The City Bucks 84 F3
The Cleaver Hereford 97 F10
The Close W Sus 22 C5
The Colony Oxon 100 D6
The Common Bath 60 G6
The Common Bucks 102 E5
The Common Dorset 30 E3
The Common Shrops 150 D3
The Common Suff 108 B2
The Common Swansea 56 C4
The Common W Sus 51 G7
The Common Wilts 47 G8
The Common Wilts 61 G11
The Common Wilts 62 C4
The Corner Kent 53 E8
The Corner Shrops 131 F8
The Cot Mon 79 F8
The Craigs Highld 309 K4
The Crofts E Yorks 218 E4
The Cronk I o M 192 C4
The Cross Hands Leics 134 C6
The Cwm Mon 79 G7
The Dell Suff 143 D9
The Delves W Mid 133 D10
The Den N Ayrs 266 E6
The Dene Durham 242 G4
The Dene Hants 47 C11
The Down Kent 53 F7
The Down Shrops 132 E3
The Downs Sur 50 F3
The Dunks Wrex 166 E4
The Eals Northumb 251 F7
The Eaves Glos 79 D10
The Fall W Yorks 197 B10
The Fence Glos 79 D8
The Flat Glos 80 B3
The Flatt Cumb 240 B3
The Flourish Derbys 153 B8
The Folly Herts 85 C11
The Folly S Glos 61 E8
The Fording Hereford 98 F3
The Forge Hereford 114 F6
The Forstal Kent 54 F4
The Forties Derbys 152 F6
The Four Alls Shrops 150 C3
The Fox Wilts 62 B6
The Foxholes Shrops 132 G2
The Frenches Hants 32 C4
The Frythe Herts 86 C2
The Garths Shetland 312 B8
The Gibb Wilts 61 D10
The Glack Borders 260 B6
The Gore Shrops 131 G11
The Grange N Yorks 225 F11
The Grange Norf 160 E2
The Green C Beds 85 B8
The Green Cambs 122 D5
The Green Cumb 210 C3
The Green Cumb 211 D7
The Green Essex 88 B3
The Green Hants 32 B3
The Green M Keynes 103 C7
The Green Norf 141 C11
The Green Norf 159 B11
The Green Northants 102 C5
The Green Oxon 101 F9
The Green S Yorks 197 G8
The Green Shrops 130 G6
The Green Warks 118 F4
The Green Wilts 45 G11
The Grove Dumfries 237 B11
The Grove Durham 242 G3
The Grove Herts 85 F9
The Grove Shrops 131 B7
The Grove Shrops 131 G8
The Grove Worcs 99 C7
The Gutter Derbys 170 F5
The Gutter Worcs 117 B9
The Hacket S Glos 61 B7
The Hague Derbys 185 C8
The Hall Shetland 312 D8
The Hallands N Lincs 200 C5
The Ham Wilts 45 C11
The Handfords Staffs 151 E7
The Harbour Kent 53 D10
The Haven W Sus 50 G5
The Headland Hrtlpl 234 E6
The Heath Norf 159 D8
The Heath Norf 160 E3
The Heath Norf 160 E4
The Heath Staffs 151 C11
The Heath Suff 108 D2
The Hem Shrops 132 B4
The Hendre Mon 79 C7
The Herberts V Glam 58 E3
The Hermitage Cambs 123 C7
The High Essex 86 C6
The Highlands E Sus 38 F2
The Hill Cumb 210 C3
The Hobbins Shrops 132 E4
The Hollands Staffs 168 D6
The Hollies Notts 172 E4
The Holmes Derbys 153 B7
The Holt Wokingham 65 D10
The Hook Worcs 98 C6
The Hope Shrops 115 B10
The Howe Cumb 211 B9
The Howe I o M 192 F2
The Humbers Telford 150 G3
The Hundred Hereford 115 E10
The Hyde London 67 B8
The Hyde Worcs 98 C6
The Hythe Essex 107 G10
The Inch Edin 280 G5
The Knab Swansea 56 D6
The Knap V Glam 58 F5
The Knapp Hereford 116 G3
The Knapp S Glos 79 G11
The Knowle W Mid 133 F9
The Laches Staffs 133 B8
The Lake Dumfries 237 E8
The Lakes Worcs 116 B5
The Lawe T & W 243 D9
The Lawns E Yorks 208 G6
The Leacon Kent 54 G3
The Leath Shrops 131 F11
The Lee Bucks 84 E6
The Lees Kent 54 C4
The Leigh Glos 99 F7
The Leys Staffs 134 C4
The Lhen I o M 192 B4
The Ling Norf 142 D6
The Lings Norf 141 B10
The Lings S Yorks 199 F7
The Linleys Wilts 61 F11
The Lunt W Mid 133 D9
The Manor W Sus 22 C4
The Marsh E Ches 168 C4
The Marsh Hereford 115 F9
The Marsh Powys 130 D6
The Marsh Shrops 150 D3
The Marsh Staffs 150 D6
The Marsh Suff 125 B11
The Marsh Suff 126 B2
The Marsh Wilts 62 C5
The Middles Durham 242 G6
The Mint Hants 34 B3
The Moor Flint 166 B4
The Moor Kent 38 B3
The Moors Hereford 97 E10
The Mount Hants 64 G2
The Mount Reading 65 E8
The Mumbles = Y Mwmbwls Swansea 56 D6
The Murray S Lnrk 268 E2
The Mythe Glos 99 E7
The Nant Wrex 166 E3
The Narth Mon 79 D8
The Neuk Aberds 293 D9
The Node Herts 104 G4
The Nook Shrops 149 C11
The Nook Shrops 150 E3
The North Mon 79 D8
The Oval Bath 61 G8
The Park Glos 99 G8
The Parks S Yorks 198 F6
The Pitts Wilts 31 B9
The Platt Oxon 83 E9
The Pludds Glos 79 B10
The Point Devon 14 E5
The Pole of Itlaw Aberds 302 D6
The Port of Felixstowe Suff 108 E5
The Potteries Stoke 168 F5
The Pound Glos 98 E4
The Quarry Glos 80 F2
The Quarry Shrops 149 G9
The Quarter Kent 53 E11
The Quarter Kent 53 G11
The Rampings Worcs 99 E7
The Rectory Lincs 156 G2
The Reddings Glos 99 G8
The Rhos Pembs 73 C8
The Rhydd Hereford 97 E9
The Riddle Hereford 115 E9
The Ridge Wilts 61 F11
The Ridges Wokingham 65 G10
The Ridgeway Herts 86 E3
The Riding Northumb 241 D10
The Riggs Borders 261 C8
The Rink Borders 261 C11
The Rise Windsor 66 F2
The Rock Telford 132 B3
The Rocks Kent 53 B8
The Rocks S Glos 61 C8
The Roe Denb 181 G8
The Rookery Herts 85 G10
The Rookery Staffs 168 D5
The Row Lancs 211 D9
The Rowe Staffs 150 B6
The Ryde Herts 86 D2
The Sands Sur 49 D11
The Scarr Glos 98 F4
The Shoe Wilts 61 E10
The Shruggs Staffs 151 C8
The Slack Durham 233 F8
The Slade W Berks 64 F4
The Smeeth Norf 157 G10
The Smithies Shrops 132 D3
The Spa Wilts 62 G2
The Spring Warks 118 C5
The Square Torf 78 F3
The Stocks Kent 38 B6
The Stocks Wilts 62 G2
The Straits Hants 49 F9
The Straits W Mid 133 E8
The Strand Wilts 46 B2
The Swillett Herts 85 F8
The Sydnall Shrops 150 C3
The Thrift Cambs 104 D6
The Throat Wokingham 65 F10
The Toft Staffs 151 F8
The Towans Corn 2 B3
The Town Scilly 1 F3
The Twittocks Glos 99 D7
The Tynings Glos 80 B6
The Vale W Mid 133 G11
The Valley E Ches 167 D11
The Valley Kent 54 C3
The Valley Leics 154 F4
The Valley Pembs 73 D10
The Vauld Hereford 97 B10
The Village Newport 78 G4
The Village W Mid 133 F7
The Village Windsor 66 E3
The Walshes Worcs 116 C6
The Warren Kent 54 E3
The Warren Wilts 63 F8
The Waterwheel Shrops 131 C7
The Weaven Hereford 97 E10
The Wells Sur 67 G7
The Wern Wrex 166 E3
The Willows NE Lincs 201 F8
The Wood Shrops 148 E6
The Wood Shrops 149 D9
The Woodlands Leics 136 D3
The Woodlands Suff 107 C11
The Woodlands Suff 108 D3
The Woods W Mid 133 D10
The Wrangle Bath 44 B4
The Wrythe London 67 F9
The Wyke Shrops 132 B4
The Wymm Hereford 97 B10
The Yeld Shrops 131 G11
Theakston N Yorks 214 B6
Thealby N Lincs 199 D11
Theale Som 44 D3
Theale W Berks 64 E6
Thearne E Yorks 209 F7
Theberton Suff 127 D8
Theddingworth Leics 136 F3
Theddlethorpe All Saints Lincs 191 D7
Theddlethorpe St Helen Lincs 191 D7
Thelbridge Barton Devon 26 E3
Thelnetham Suff 125 B10
Thelveton Norf 142 G3
Thelwall Warr 183 D10
Themelthorpe Norf 159 E11
Thenford Northants 101 C10
Theobald's Green Wilts 62 F4
Therfield Herts 104 D6
Thetford Lincs 156 F2
Thetford Norf 141 G7
Theydon Bois Essex 86 F6
Theydon Garnon Essex 87 F7
Theydon Mount Essex 87 F7
Thick Hollins W Yorks 196 E6
Thicket Mead Bath 45 B7
Thickthorn Hall Norf 142 B3
Thickwood Wilts 61 E10
Thimble End W Mid 134 E2
Thimbleby Lincs 190 G2
Thimbleby N Yorks 225 F9
Thinford Durham 233 E11
Thingley Wilts 61 E11
Thingwall Mers 182 E3
Thirdpart N Ayrs 266 F3
Thirlby N Yorks 215 C9
Thirlestane Borders 271 F11
Thirn N Yorks 214 B4
Thirsk N Yorks 215 C8
Thirtleby E Yorks 209 G9
Thistleton Lancs 202 F4
Thistleton Rutland 155 F8
Thistley Green Essex 88 B2
Thistley Green Suff 124 B3
Thixendale N Yorks 216 G6
Thockrington Northumb 241 B11
Tholomas Drove Cambs 139 B7
Tholthorpe N Yorks 215 F9
Thomas Chapel Pembs 73 D10
Thomas Close Cumb 230 C4
Thomastown Aberds 302 F5
Thomastown Rhondda 58 B4
Thompson Norf 141 D8
Thomshill Moray 302 D2
Thong Kent 69 E7
Thongsbridge W Yorks 196 F6
Thoralby N Yorks 213 B10
Thoresby Notts 187 G10
Thoresthorpe Lincs 191 F7
Thoresway Lincs 189 B11
Thorganby Lincs 190 B2
Thorganby N Yorks 207 E9
Thorgill N Yorks 226 F4
Thorington Suff 127 C8
Thorington Street Suff 107 D10
Thorlby N Yorks 204 C5
Thorley Herts 87 B7
Thorley I o W 20 D3
Thorley Houses Herts 105 G9
Thorley Street Herts 87 B7
Thorley Street I o W 20 D3
Thormanby N Yorks 215 E9
Thorn Devon 13 D9
Thorn Powys 114 E5
Thorn Hill S Yorks 186 C6
Thornaby on Tees Stockton 225 B9
Thornage Norf 159 B11
Thornborough Bucks 102 E4
Thornborough N Yorks 214 D5
Thornbury Devon 24 F6
Thornbury Hereford 116 F2
Thornbury S Glos 79 G10
Thornbury W Yorks 205 G9
Thornby Cumb 239 G7
Thornby Northants 120 B3
Thorncliff W Yorks 197 E8
Thorncliffe Staffs 169 D8
Thorncombe Dorset 28 G5
Thorncombe Dorset 30 G5
Thorncombe Street Sur 50 E4
Thorncote Green C Beds 104 B3
Thorncross I o W 20 E4
Thorndon Suff 126 D2
Thorndon Cross Devon 12 C6
Thorne Corn 24 G2
Thorne S Yorks 199 E7
Thorne Coffin Som 29 D8
Thorne Moor Devon 12 D3
Thorne St Margaret Som 27 C9
Thornehillhead Devon 24 D6
Thornend Wilts 62 D3
Thorner W Yorks 206 E3
Thornes Staffs 133 C11
Thornes W Yorks 197 D10
Thorney Bucks 66 D4
Thorney Notts 188 G5
Thorney P'boro 138 C5
Thorney Som 28 C6
Thorney Close T & W 243 G9
Thorney Crofts E Yorks 201 C8
Thorney Green Suff 125 E11
Thorney Hill Hants 19 B9
Thorney Island W Sus 22 C3
Thorney Toll P'boro 138 C6
Thorneywood Notts 171 G9
Thornfalcon Som 28 C3
Thornford Dorset 29 E10
Thorngrafton Northumb 241 D7
Thorngrove Som 43 G11
Thorngumbald E Yorks 201 B8
Thornham Norf 176 E2
Thornham Fold Gtr Man 195 F11
Thornham Magna Suff 126 C2
Thornham Parva Suff 126 C2
Thornhaugh P'boro 137 C11
Thornhill Cardiff 59 C7
Thornhill Cumb 219 D10
Thornhill Derbys 185 E11
Thornhill Dumfries 247 D9
Thornhill Soton 33 E7
Thornhill Stirl 278 B3
Thornhill Torf 78 F3
Thornhill W Yorks 197 D9
Thornhill Wilts 62 D5
Thornhill Edge W Yorks 197 D8
Thornhill Lees W Yorks 197 D8
Thornhill Park Hants 33 E7
Thornhills W Yorks 197 C7
Thornholme E Yorks 218 G2
Thornicombe Dorset 30 G5
Thornielee Borders 261 B10
Thornley Durham 233 D8
Thornley Durham 234 D3
Thornliebank E Renf 267 D10
Thornly Park Renfs 267 C9
Thornroan Aberds 303 F8
Thorns N Yorks 223 E7
Thorns Suff 124 F4
Thorns Green Ches 184 E3
Thornseat S Yorks 186 C2
Thornsett Derbys 185 D8
Thornthwaite Cumb 229 F10
Thornthwaite N Yorks 205 B9
Thornton Angus 287 C7
Thornton Bucks 102 D5
Thornton E Yorks 207 D11
Thornton Fife 280 B5
Thornton Lancs 202 E2
Thornton Leics 135 B9
Thornton Lincs 174 B2
Thornton M'bro 225 C9
Thornton Mers 193 G10
Thornton Northumb 273 F9
Thornton Pembs 72 D6
Thornton W Yorks 205 G8
Thornton Curtis N Lincs 200 D5
Thornton Heath London 67 F10
Thornton Hough Mers 182 E4
Thornton in Craven N Yorks 204 D4
Thornton in Lonsdale N Yorks 212 E3
Thornton-le-Beans N Yorks 225 G7
Thornton-le-Clay N Yorks 216 F3
Thornton-le-Dale N Yorks 216 C6
Thornton le Moor Lincs 189 B9
Thornton-le-Moor N Yorks 215 B7
Thornton-le-Moors W Ches 182 G6
Thornton-le-Street N Yorks 215 B8
Thornton Rust N Yorks 213 B9
Thornton Steward N Yorks 214 B3
Thornton Watlass N Yorks 214 B4
Thorntonhall S Lnrk 267 D11
Thorntonloch E Loth 282 G4
Thornton Park Northumb 273 F8
Thornwood Common Essex 87 D7
Thornydykes Borders 272 F2
Thoroton Notts 172 G3
Thorp Gtr Man 196 F2
Thorp Arch W Yorks 206 D4
Thorpe Cumb 230 F5
Thorpe Derbys 169 E11
Thorpe E Yorks 208 D5
Thorpe Lincs 191 E7
Thorpe N Yorks 213 G10
Thorpe Norf 143 D8
Thorpe Notts 172 F3
Thorpe Sur 66 F4
Thorpe Abbotts Norf 126 B3
Thorpe Acre Leics 153 E10
Thorpe Arnold Leics 154 E5
Thorpe Audlin W Yorks 198 D3
Thorpe Bassett N Yorks 217 E7
Thorpe Bay Sthend 70 B2
Thorpe by Water Rutland 137 D7
Thorpe Common Suff 108 D5
Thorpe Constantine Staffs 134 B5
Thorpe Culvert Lincs 175 C7
Thorpe Edge W Yorks 205 F9
Thorpe End Norf 160 G5
Thorpe Fendykes Lincs 175 C7
Thorpe Green Essex 108 G3
Thorpe Green Lancs 194 C5
Thorpe Green Suff 125 G8
Thorpe Green Sur 66 F4
Thorpe Hamlet Norf 142 B4
Thorpe Hesley S Yorks 186 B5
Thorpe in Balne S Yorks 198 E5
Thorpe in the Fallows Lincs 188 E6
Thorpe Langton Leics 136 E4
Thorpe Larches Durham 234 F3
Thorpe Latimer Lincs 156 B2
Thorpe-le-Soken Essex 108 G3
Thorpe le Street E Yorks 208 E2
Thorpe le Vale Lincs 190 C2
Thorpe Lea Sur 66 E4
Thorpe Malsor Northants 120 B6
Thorpe Mandeville Northants 101 B10
Thorpe Market Norf 160 B4
Thorpe Marriot Norf 160 F3
Thorpe Morieux Suff 125 G8
Thorpe on the Hill Lincs 172 B6
Thorpe on the Hill W Yorks 197 B10
Thorpe Row Norf 141 B9
Thorpe St Andrew Norf 142 B5
Thorpe St Peter Lincs 175 C7
Thorpe Salvin S Yorks 187 E8
Thorpe Satchville Leics 154 G4
Thorpe Street Suff 125 B10
Thorpe Thewles Stockton 234 G4
Thorpe Tilney Lincs 173 D10
Thorpe Underwood N Yorks 206 B5
Thorpe Underwood Northants 136 G5
Thorpe Waterville Northants 137 G10
Thorpe Willoughby N Yorks 207 G7
Thorpe Wood N Yorks 207 G7
Thorpeness Suff 127 F9
Thorpland Norf 140 B2
Thorrington Essex 89 B9
Thorverton Devon 26 G6
Thoulstone Wilts 45 D10
Thrandeston Suff 126 B2
Thrapston Northants 121 B9
Thrashbush N Lnrk 268 B5
Threapland Cumb 229 D9
Threapland N Yorks 213 G9
Threapwood Staffs 169 G8
Threapwood W Ches 166 F6
Three Ashes Hants 64 G6
Three Ashes Hereford 97 G10
Three Ashes Shrops 115 B7
Three Ashes Som 45 D7
Three Bridges Argyll 284 F4
Three Bridges Lincs 190 D6
Three Bridges W Sus 51 F9
Three Burrows Corn 4 F4
Three Chimneys Kent 53 F10
Three Cocked Hat Norf 143 D8
Three Cocks = Aberllynfi Powys 96 D3
Three Crosses Swansea 56 C5
Three Cups Corner E Sus 37 C10
Three Fingers Wrex 167 G7
Three Gates Dorset 29 F10
Three Hammers Corn 11 D10
Three Holes Norf 139 C10
Three Holes Cross Corn 10 G6
Three Leg Cross E Sus 53 G7
Three Legged Cross Dorset 31 F9
Three Maypoles W Mid 118 B2
Three Mile Cross Wokingham 65 F8
Three Oaks E Sus 38 E4
Three Sisters Denb 165 C9
Threehammer Common Norf 160 E6
Threekingham Lincs 155 B11
Threelows Staffs 169 F9
Threemile Cross Wokingham 65 F8
Threemilestone Corn 4 G5
Threemiletown W Loth 279 F11
Threepwood Borders 271 G10
Threewaters Corn 5 B10
Threlkeld Cumb 230 F2
Threshers Bush Essex 87 D7
Threshfield N Yorks 213 G9
Thrigby Norf 161 G9
Thringarth Durham 232 G4
Thringstone Leics 153 F8
Thrintoft N Yorks 224 G6
Thriplow Cambs 105 B8
Throapham S Yorks 187 D8
Throckenholt Lincs 139 B7
Throcking Herts 104 E6
Throckley T & W 242 D5
Throckmorton Worcs 99 B9
Throop Dorset 18 C2
Throphill Northumb 252 F4
Thropton Northumb 252 C2
Throsk Stirl 279 C7
Througham Glos 80 D6
Throughgate Dumfries 247 G9
Throwleigh Devon 13 C9
Throwley Kent 54 B3
Throwley Forstal Kent 54 C3
Throxenby N Yorks 217 B10
Thrumpton Notts 153 C10
Thrumpton Notts 188 E2
Thrumster Highld 310 E7
Thrunton Northumb 264 G3
Thrupe Som 44 D6
Thrupp Glos 80 E5
Thrupp Oxon 82 F3
Thrupp Oxon 83 B7
Thruscross N Yorks 205 B9
Thrushelton Devon 12 D4
Thrussington Leics 154 F2
Thruxton Hants 47 D9
Thruxton Hereford 97 E8
Thrybergh S Yorks 187 B7
Thulston Derbys 153 C8
Thunder Bridge W Yorks 197 E7
Thunder Hill Norf 161 F8
Thundergay N Ayrs 255 C9
Thunder's Hill E Sus 23 C9
Thundersley Essex 69 B9
Thundridge Herts 86 B5
Thurcaston Leics 153 G11
Thurcroft S Yorks 187 D7
Thurdon Corn 24 E3
Thurgarton Norf 160 C3
Thurgarton Notts 171 F11
Thurgoland S Yorks 197 G9
Thurlaston Leics 135 D10
Thurlaston Warks 119 C9
Thurlbear Som 28 C3
Thurlby Lincs 156 F2
Thurlby Lincs 172 C6
Thurlby Lincs 191 F7
Thurleigh Beds 121 F11
Thurlestone Devon 8 G3
Thurloxton Som 43 G9
Thurlstone S Yorks 197 G8
Thurlton Norf 143 D8
Thurlton Links Norf 143 D8
Thurlwood E Ches 168 D4
Thurmaston Leics 136 B2
Thurnby Leics 136 C2
Thurne Norf 161 F8
Thurnham Kent 53 B10
Thurnham Lancs 202 C5
Thurning Norf 159 D11
Thurning Northants 137 G11
Thurnscoe S Yorks 198 F3
Thurnscoe East S Yorks 198 F3
Thursby Cumb 239 G8
Thursford Norf 159 C9
Thursford Green Norf 159 C9
Thursley Sur 50 F2
Thurso Highld 310 C5
Thurso East Highld 310 C5
Thurstaston Mers 182 E2
Thurston Suff 125 D8
Thurston Clough Gtr Man 196 F3
Thurston End Suff 124 G5
Thurstonfield Cumb 239 F8
Thurstonland W Yorks 197 E7
Thurton Norf 142 C6
Thurvaston Derbys 152 B2
Thurvaston Derbys 152 B4
Thuxton Norf 141 B10
Thwaite N Yorks 223 F7
Thwaite Suff 126 D2
Thwaite Flat Cumb 210 E4
Thwaite Head Cumb 220 G6
Thwaite St Mary Norf 142 E6
Thwaites W Yorks 205 E7
Thwaites Brow W Yorks 205 E7
Thwing E Yorks 217 E11
Tibbermore Perth 286 E4
Tibberton Glos 98 G5
Tibberton Telford 150 E3
Tibberton Worcs 117 F8
Tibenham Norf 142 F2
Tibshelf Derbys 170 C6
Tibshelf Wharf Notts 171 C7
Tibthorpe E Yorks 208 B5
Ticehurst E Sus 53 G7
Tichborne Hants 48 G5
Tickencote Rutland 137 B9
Tickenham N Som 60 E3
Ticket Wood Devon 8 G4
Tickford End M Keynes 103 C7
Tickhill S Yorks 187 C9
Tickleback Row Brack 65 E11
Ticklerton Shrops 131 E9
Tickmorend Glos 80 F4
Ticknall Derbys 153 E7
Tickton E Yorks 209 E7
Tidbury Green W Mid 117 B11
Tidcombe Wilts 47 B9
Tiddington Oxon 83 E11
Tiddington Warks 118 F4
Tidebrook E Sus 37 B10
Tideford Corn 6 D6
Tideford Cross Corn 6 C6
Tidenham Glos 79 F9
Tidenham Chase Glos 79 F9
Tideswell Derbys 185 F11
Tidmarsh W Berks 64 E6
Tidmington Warks 100 D5
Tidnor Hereford 97 D11
Tidpit Hants 31 D9
Tidworth Wilts 47 D8
Tiers Cross Pembs 72 C6
Tiffield Northants 120 G3
Tifty Aberds 303 E7
Tigerton Angus 293 G7
Tigh-na-Blair Perth 285 F11
Tighnabruaich Argyll 275 F10
Tighnacachla Argyll 274 G3
Tighnafiline Highld 307 L3
Tighness Argyll 284 G6
Tigley Devon 8 C5
Tilbrook Cambs 121 D11
Tilbury Thurrock 68 D6
Tilbury Green Essex 106 C4
Tilbury Juxta Clare Essex 106 C5
Tile Cross W Mid 134 F3
Tile Hill W Mid 118 B5
Tilegate Green Essex 87 D8
Tilehouse Green W Mid 118 B3
Tilehurst Reading 65 E7
Tilekiln Green Essex 105 G10
Tiley Dorset 29 F11
Tilford Sur 49 E11
Tilford Common Sur 49 E11
Tilford Reeds Sur 49 E11
Tilgate W Sus 51 G9
Tilgate Forest Row W Sus 51 G9
Tilkey Essex 106 G6
Tilland Corn 6 C6
Tillathrowie Aberds 302 F4
Tillers' Green Glos 98 E3
Tilley Shrops 149 D10
Tilley Green Shrops 149 D10
Tillicoultry Clack 279 B8
Tillietudlem S Lnrk 268 F6
Tillingham Essex 89 E7
Tillington Hereford 97 B9
Tillington Staffs 151 E8
Tillington W Sus 35 C7
Tillington Common Hereford 97 B9
Tillislow Devon 12 C3
Tillworth Devon 28 G4
Tilly Down Hants 47 D10
Tilly Lo Aberds 293 C7
Tillyarblet Angus 293 G7
Tillybirloch Aberds 293 C8
Tillycorthie Aberds 303 G9
Tillydrine Aberds 293 D8
Tillyfour Aberds 293 B7
Tillyfourie Aberds 293 B8
Tillygarmond Aberds 293 D8
Tillygreig Aberds 303 G8
Tillykerrie Aberds 303 G8
Tillynaught Aberds 302 C5
Tilmanstone Kent 55 C10
Tilney All Saints Norf 157 F11
Tilney cum Islington Norf 157 G11
Tilney Fen End Norf 157 G10
Tilney High End Norf 157 F11
Tilney St Lawrence Norf 157 G10
Tilsdown Glos 80 F2
Tilshead Wilts 46 D4
Tilsmore E Sus 37 C9
Tilsop Shrops 116 C2
Tilstock Shrops 149 B10
Tilston W Ches 167 E7
Tilstone Bank W Ches 167 D9
Tilstone Fearnall W Ches 167 C9
Tilsworth C Beds 103 G9
Tilton on the Hill Leics 136 B4
Tilts S Yorks 198 F5
Tiltups End Glos 80 F4
Tilty Essex 105 F11
Timberden Bottom Kent 68 G4
Timberhonger Worcs 117 C8
Timberland Lincs 173 D10
Timbersbrook E Ches 168 C5
Timberscombe Som 42 E3
Timble N Yorks 205 C9
Timbold Hill Kent 54 B2
Timbrelham Corn 12 E3
Timperley Gtr Man 184 D3
Timsbury Bath 45 B7
Timsbury Hants 32 C4
Timsgearraidh W Isles 304 E2
Timworth Suff 125 D7
Timworth Green Suff 125 D7
Tincleton Dorset 17 C11
Tindale Cumb 240 F4
Tindale Crescent Durham 233 F9
Tindon End Essex 106 E2
Tingewick Bucks 102 E3
Tingley W Yorks 197 B9
Tingon Shetland 312 E4
Tingrith C Beds 103 E10
Tingwall Orkney 314 D3
Tinhay Devon 12 E3
Tinkers End Bucks 102 F5
Tinshill W Yorks 205 F11
Tinsley S Yorks 186 C6
Tinsley Green W Sus 51 F9
Tintagel Corn 11 D7
Tintern Parva Mon 79 E8
Tintinhull Som 29 D8
Tintwistle Derbys 185 B8
Tinwald Dumfries 248 G2
Tinwell Rutland 137 B10
Tipner Ptsmth 33 G10
Tippacott Devon 41 D9
Tipper's Hill Warks 134 F5
Tipperty Aberds 302 C6
Tipperty Aberds 303 G9
Tipple Cross Devon 12 D3
Tipps End Norf 139 D10
Tip's Cross Essex 87 E9
Tiptoe Hants 19 B11
Tipton W Mid 133 E8
Tipton Green W Mid 133 E9
Tipton St John Devon 15 C7
Tiptree Essex 88 B5
Tiptree Heath Essex 88 B5
Tir-y-berth Caerph 77 F11
Tir-y-dail Carms 75 C10
Tirabad Powys 95 C7
Tiraghoil Argyll 288 G5
Tircanol Swansea 57 B7
Tirdeunaw Swansea 57 B7
Tirinie Perth 291 G10
Tirley Glos 98 F6
Tirley Knowle Glos 98 F6
Tiroran Argyll 288 G6
Tirphil Caerph 77 E10
Tirril Cumb 230 F6
Tirryside Highld 309 H5
Tisbury Wilts 30 B6
Tisman's Common W Sus 50 G5
Tissington Derbys 169 E11
Titchberry Devon 24 B2
Titchfield Hants 33 F8
Titchfield Common Hants 33 F8
Titchfield Park Hants 33 F8
Titchmarsh Northants 121 B10
Titchwell Norf 176 E3
Titcomb W Berks 63 F11
Tithby Notts 154 B3
Tithe Barn Hillock Mers 183 B9
Tithebarn Staffs 169 G9
Titley Hereford 114 E6
Titlington Northumb 264 F4
Titmore Green Herts 104 F4
Titsey Sur 52 C2
Titson Corn 24 G2
Tittenhurst Windsor 66 F3
Tittensor Staffs 151 B7
Titterhill Shrops 131 G10
Tittle Row Windsor 65 C11
Tittleshall Norf 159 E7
Titton Worcs 116 D6
Titty Hill W Sus 34 B5
Tiverton Devon 27 E7
Tiverton W Ches 167 C9
Tivetshall St Margaret Norf 142 F3
Tivetshall St Mary Norf 142 F3
Tividale W Mid 133 E9
Tivington Som 42 D2
Tivington Knowle Som 42 E2
Tivoli Cumb 228 G5
Tivy Dale S Yorks 197 F9
Tixall Staffs 151 E9

Tixover Rutland 137 C9
Toab Orkney 314 F5
Toab Shetland 313 M5
Toad Row Suff 143 F10
Toadmoor Derbys 170 E4
Tobermory Argyll 289 D7
Toberonochy Argyll 275 C8
Tobha Beag W Isles 296 D5
Tobha Mor W Isles 297 H3
Tobhtarol W Isles 304 E3
Tobson W Isles 304 E3
Toby's Hill Lincs 191 C7
Tocher Aberds 302 F6
Tockenham Wilts 62 D4
Tockenham Wick Wilts 62 C4
Tockholes Blkburn 195 C7
Tockington S Glos 60 B6
Tockwith N Yorks 206 C5
Todber Dorset 30 C4
Todding Hereford 115 B8
Toddington C Beds 103 F10
Toddington Glos 99 E10
Toddington W Sus 35 G8
Toddlehills Aberds 303 E10
Todd's Green Herts 104 F4
Todenham Glos 100 D4
Todhill Angus 287 D8
Todhills Cumb 239 E9
Todhills Durham 233 E10
Todlachie Aberds 293 B8
Todmorden W Yorks 196 C2
Todpool Corn 4 G4
Todrig Borders 261 F10
Todwick S Yorks 187 E7
Toft Cambs 123 F7
Toft Lincs 155 F11
Toft Shetland 312 F6
Toft Warks 119 C9
Toft Hill Durham 233 F9
Toft Hill Lincs 174 C2
Toft Monks Norf 143 E8
Toft next Newton Lincs 189 D8
Toftrees Norf 159 D7
Tofts Highld 310 C7
Toftshaw W Yorks 197 B7
Toftwood Norf 159 G9
Togston Northumb 252 C6
Tokavaig Highld 295 D8
Tokers Green Oxon 65 D8
Tokyngton London 67 C7
Tolastadh a Chaolais W Isles 304 E3
Tolastadh bho Thuath W Isles 304 D7
Tolborough Corn 11 F9
Tolcarne Corn 2 B5
Tolcarne Corn 2 C5
Tolcarne Wartha Corn 2 B5
Toldish Corn 5 D8
Tolgus Mount Corn 4 G3
Tolhurst E Sus 53 G7
Toll Bar Mers 183 C7
Toll Bar Rutland 137 B10
Toll Bar S Yorks 198 F5
Toll End W Mid 133 E9
Toll of Birness Aberds 303 F10
Tolladine Worcs 117 F7
Tolland Som 42 G6
Tollard Farnham Dorset 30 D6
Tollard Royal Wilts 30 D6
Tollbar End W Mid 119 B7
Toller Fratrum Dorset 17 B7
Toller Porcorum Dorset 17 B7
Toller Whelme Dorset 29 G8
Tollerford Dorset 17 B7
Tollerton N Yorks 215 G10
Tollerton Notts 154 C2
Tollesbury Essex 89 C7
Tollesby M'bro 225 B10
Tolleshunt D'Arcy Essex 88 C6
Tolleshunt Knights Essex 88 C6
Tolleshunt Major Essex 88 C5
Tollie Highld 300 D5
Tolm W Isles 304 E6
Tolmers Herts 86 E4
Tolpuddle Dorset 17 C11
Tolskithy Corn 4 G3
Tolvaddon Downs Corn 4 G3
Tolvah Highld 291 D10
Tolworth London 67 F7
Tom an Fhuadain W Isles 305 G5
Tomaknock Perth 286 E2
Tomatin Highld 301 G8
Tombreck Highld 300 F6
Tombui Perth 286 B2
Tomchrasky Highld 290 B4
Tomdoun Highld 290 C3
Tomich Highld 300 B6
Tomich Highld 300 G3
Tomich House Highld 300 E5
Tomintoul Aberds 292 D3
Tomintoul Moray 292 B3
Tomlow Warks 119 E9
Tomnaven Moray 302 F4
Tomnavoulin Moray 302 G2
Tomperrow Corn 4 G5
Tompkin Staffs 168 E6
Tompset's Bank E Sus 52 G2
Tomsleibhe Argyll 289 F8
Tomthorn Derbys 185 F9
Ton Mon 78 F5
Ton Breigam V Glam 58 D3
Ton-Pentre Rhondda 77 F7
Ton-teg Rhondda 58 B5
Ton-y-pistyll Caerph 77 F11
Tonbridge Kent 52 D5
Tonderghie Dumfries 236 F6
Tondu Bridgend 57 E11
Tone Som 27 C10
Tone Green Som 27 C11
Tonedale Som 27 C10
Tong Kent 53 D10
Tong Shrops 132 B5
Tong W Yorks 205 G10
Tong Forge Shrops 132 B5
Tong Green Kent 54 C3
Tong Norton Shrops 132 B5
Tong Park W Yorks 205 F9
Tong Street W Yorks 205 G9
Tonge Leics 153 E8
Tonge Corner Kent 70 F2
Tonge Fold Gtr Man 195 F8
Tonge Moor Gtr Man 195 E8
Tongham Sur 49 D11
Tongland Dumfries 237 D8
Tongue Highld 308 D5
Tongue End Lincs 156 F3
Tongwell M Keynes 103 C7
Tongwynlais Cardiff 58 C6
Tonmawr Neath 57 B10
Tonna = Tonnau Neath 57 B9
Tonnau = Tonna Neath 57 B9
Tontine Lancs 194 G4
Tonwell Herts 86 B4
Tonypandy Rhondda 77 G7
Tonyrefail Rhondda 58 B4
Toot Baldon Oxon 83 E9
Toot Hill Essex 87 E8
Toot Hill Staffs 169 G9
Toothill Hants 32 D5
Toothill Swindon 62 C6
Toothill W Yorks 196 C6
Tooting Graveney London 67 E9
Top End Beds 121 E10
Top Green Notts 172 F3
Top Lock Gtr Man 194 F6
Top o' th' Lane Lancs 194 C5
Top o' th' Meadows Gtr Man 196 F3
Top of Hebers Gtr Man 195 F11
Top Valley Nottingham 171 F9
Topcliffe N Yorks 215 D8
Topcliffe W Yorks 197 B9
Topcroft Norf 142 E5
Topcroft Street Norf 142 E5
Topham S Yorks 198 D6
Topleigh W Sus 34 D6
Toppesfield Essex 106 D4
Toppings Gtr Man 195 E8
Toprow Norf 142 D3
Topsham Devon 14 D5
Torbeg N Ayrs 255 E10
Torboll Farm Highld 309 K7
Torbothie N Lnrk 269 D7
Torbreck Highld 309 J7
Torbrex Stirl 278 C5
Torbryan Devon 8 B6
Torbush N Lnrk 268 D6
Torcross Devon 8 G6
Torcroy Highld 291 D9
Tore Highld 300 D6
Torfrey Corn 6 E2
Torgyle Highld 290 B5
Torinturk Argyll 275 G9
Torkington Gtr Man 184 D6
Torksey Lincs 188 F4
Torlum W Isles 296 F3
Torlundy Highld 290 F3
Tormarton S Glos 61 D9
Tormisdale Argyll 254 B2
Tormitchell S Ayrs 244 E6
Tormore Highld 295 E8
Tormore N Ayrs 255 D9
Tornagrain Highld 301 E7
Tornahaish Aberds 292 C4
Tornapress Highld 299 E8
Tornaveen Aberds 293 C8
Torness Highld 300 G5
Toronto Durham 233 E9
Torpenhow Cumb 229 D10
Torphichen W Loth 279 G9
Torphin Edin 270 B4
Torphins Aberds 293 C8
Torpoint Corn 7 E8
Torquay Torbay 9 C8
Torquhan Borders 271 F8
Torr Devon 7 E11
Torr Devon 8 C2
Torra Argyll 254 B4
Torran Argyll 275 C9
Torran Highld 298 E5
Torran Highld 301 B7
Torrance E Dunb 278 G2
Torrans Argyll 288 G6
Torranyard N Ayrs 267 G7
Torre Som 42 E4
Torre Torbay 9 C8
Torridon Highld 299 D9
Torridon Ho Highld 299 D8
Torries Aberds 293 B8
Torrin Highld 295 C7
Torrisdale Highld 308 C6
Torrisdale Castle Argyll 255 D8
Torrisdale-Square Argyll 255 D8
Torrish Highld 311 H3
Torrisholme Lancs 211 G9
Torroble Highld 309 J5
Torroy Highld 309 K5
Torrpark Corn 11 D10
Torry Aberdeen 293 C11
Torry Aberds 302 F4
Torryburn Fife 279 D10
Torsonce Borders 271 G9
Torsonce Mains Borders 271 G9
Torterston Aberds 303 E10
Torthorwald Dumfries 238 B2
Tortington W Sus 35 F8
Torton Worcs 116 C6
Tortworth S Glos 80 G2
Torvaig Highld 298 E4
Torver Cumb 220 G5
Torwood Falk 278 E6
Torwoodlee Mains Borders 261 B11
Torworth Notts 187 D11
Tosberry Devon 24 C3
Toscaig Highld 295 B9
Toseland Cambs 122 E4
Tosside N Yorks 203 B11
Tostock Suff 125 E9
Tot Hill Hants 64 G3
Totaig Highld 295 C10
Totaig Highld 298 D2
Totardor Highld 294 B5
Tote Highld 298 E4
Tote Hill Hants 32 C4
Tote Hill W Sus 34 C5
Totegan Highld 310 C2
Totford Hants 48 F5
Totham Hill Essex 88 C5
Totham Plains Essex 88 C5
Tothill Lincs 190 E6
Totland I o W 20 D2
Totley S Yorks 186 F4
Totley Brook S Yorks 186 E4
Totley Rise S Yorks 186 E4
Totmonslow Staffs 151 B9
Totnell Dorset 29 F10
Totnes Devon 8 C6
Totnor Hereford 97 E11
Toton Notts 153 C10
Totronald Argyll 288 D3
Totscore Highld 298 C3
Tottenham London 86 G4
Tottenham Hale London 67 B10
Tottenhill Norf 158 G2
Tottenhill Row Norf 158 G2
Totteridge Bucks 84 G5
Totteridge London 86 G2
Totternhoe C Beds 103 G9
Totteroak S Glos 61 C8
Totterton Shrops 131 F7
Totties W Yorks 197 F7
Tottington Gtr Man 195 E9
Tottington Norf 141 D7
Tottlebank Cumb 210 C6
Tottleworth Lancs 203 G10
Totton Hants 32 E5
Touchen End Windsor 65 D11
Toulston N Yorks 206 E5
Toulton Som 43 G7
Toulvaddie Highld 311 L2
Tournaig Highld 307 L3
Toux Aberds 303 D9
Tovil Kent 53 C9
Tow House Northumb 241 E7
Tow Law Durham 233 D8
Towan Corn 10 G3
Towan Cross Corn 4 F4
Toward Argyll 266 B2
Towcester Northants 102 B3
Towednack Corn 1 B5
Tower End Norf 158 F3
Tower Hamlets Kent 55 E10
Tower Hill Devon 12 C3
Tower Hill E Ches 184 F6
Tower Hill Essex 108 E5
Tower Hill Herts 85 E8
Tower Hill Mers 194 G2
Tower Hill Sur 51 D7
Tower Hill W Mid 133 E11
Tower Hill W Sus 35 B11
Towerage Bucks 84 G4
Towerhead N Som 44 B2
Towersey Oxon 84 D2
Towie Aberds 292 B6
Towie Aberds 302 G5
Towie Aberds 303 C8
Towiemore Moray 302 E3
Town Barton Devon 14 C2
Town End Bucks 84 F3
Town End Cambs 139 D8
Town End Cumb 211 B7
Town End Cumb 211 C8
Town End Cumb 212 C2
Town End Cumb 220 D6
Town End Cumb 221 E8
Town End Cumb 221 F7
Town End Cumb 231 F8
Town End Derbys 185 F11
Town End E Yorks 207 C10
Town End Mers 183 D7
Town End W Yorks 196 D5
Town Fields W Ches 167 B10
Town Green Gtr Man 183 B9
Town Green Lancs 194 F2
Town Green Norf 161 F7
Town Head Cumb 220 D6
Town Head Cumb 221 B8
Town Head Cumb 222 C2
Town Head Cumb 222 G3
Town Head Cumb 231 F7
Town Head Cumb 231 F8
Town Head Cumb 231 G8
Town Head Cumb 231 G9
Town Head Derbys 185 F11
Town Head N Yorks 204 B2
Town Head N Yorks 212 F5
Town Head Staffs 169 F8
Town Head W Yorks 204 E6
Town Kelloe Durham 234 E3
Town Lane Gtr Man 183 B11
Town Littleworth E Sus 36 D6
Town of Lowton Mers 183 B10
Town Park Telford 132 B3
Town Row E Sus 52 G5
Town Street Glos 98 F6
Town Yetholm Borders 263 D8
Townend Derbys 185 E9
Townend Staffs 151 B9
Townend W Dunb 277 F8
Towney Durham 232 B3
Townfield Durham 232 B5
Towngate Cumb 230 B6
Towngate Lincs 156 G2
Townhead Argyll 275 G11
Townhead Cumb 229 D7
Townhead Cumb 230 D6
Townhead Cumb 231 E8
Townhead Dumfries 237 E8
Townhead N Lnrk 268 B4
Townhead Northumb 231 E6
Townhead S Ayrs 244 C6
Townhead S Yorks 186 E4
Townhead S Yorks 197 G7
Townhead of Greenlaw Dumfries 237 C9
Townhill Fife 280 D2
Townhill Swansea 56 C6
Townhill Park Hants 33 E7
Townlake Devon 12 G4
Townland Green Kent 54 G2
Town's End Bucks 102 G2
Town's End Dorset 18 B3
Town's End Dorset 18 E5
Town's End Dorset 29 F9
Towns End Hants 48 B5
Towns End Som 30 D2
Town's End Som 45 D7
Townsend Bath 44 B5
Townsend Bucks 84 D2
Townsend Devon 25 B10
Townsend Herts 85 D10
Townsend Oxon 63 B11
Townsend Pembs 72 D4
Townsend Som 44 C4
Townsend Stoke 168 F6
Townsend Wilts 45 B3
Townsend Wilts 45 B4
Townsend Fold Lancs 195 C10
Townshend Corn 2 C3
Townwell S Glos 79 G11
Towthorpe E Yorks 217 G8
Towthorpe York 207 B8
Towton N Yorks 206 F5
Towyn Conwy 181 F7
Toxteth Mers 182 D5
Toynton All Saints Lincs 174 C5
Toynton Fen Side Lincs 174 C5
Toynton St Peter Lincs 174 C6
Toy's Hill Kent 52 C3
Trabboch E Ayrs 257 E10
Traboe Corn 2 E6
Trabrown Borders 271 F10
Tracebridge Som 27 C9
Tradespark Highld 301 D8
Tradespark Orkney 314 F4
Trafford Park Gtr Man 184 B3
Traigh Ho Highld 295 F8
Trallong Powys 95 F9
Trallwn Rhondda 77 G9
Trallwn Swansea 57 B7
Tram Inn Hereford 97 E9
Tramagenna Corn 11 E7
Tranch Torf 78 E3
Tranent E Loth 281 G8
Tranmere Mers 182 D4
Trantlebeg Highld 310 D2
Trantlemore Highld 310 D2
Tranwell Northumb 252 G5
Trapp Carms 75 B11
Traprain E Loth 281 F11
Trap's Green Warks 118 D2
Trapshill W Berks 63 G11
Traquair Borders 261 C8
Trash Green W Berks 65 F7
Travellers' Rest Carms 74 B5
Trawden Lancs 204 F4
Trawscoed Powys 95 E11
Trawsfynydd Gwyn 146 B4
Trawsnant Ceredig 111 D11
Tre-Aubrey V Glam 58 E4
Tre-Beferad V Glam 58 F3
Tre-boeth Swansea 57 B7
Tre-derwen Powys 148 F4
Tre-Forgan Neath 76 D3
Tre-gagle Mon 79 D8
Tre-Gibbon Rhondda 77 E7
Tre Gwyr = Gowerton Swansea 56 B5
Tre-gynwr Carms 74 B6
Tre-hill V Glam 58 E5
Tre-Ifor Rhondda 77 D7
Tre-lan Flint 165 B11
Tre-Mostyn Flint 181 F10
Tre-pit V Glam 58 E2
Tre-Taliesin Ceredig 128 E3
Tre-vaughan Carms 93 G8
Tre-wyn Mon 96 G6
Treadam Mon 78 B5
Treaddow Hereford 97 G10
Treal Corn 2 F6
Trealaw Rhondda 77 G8
Treales Lancs 202 G4
Trearddur Anglesey 178 F3
Treaslane Highld 298 D3
Treath Corn 3 D7
Treator Corn 10 F4
Trebanog Rhondda 77 G8
Trebanos Neath 76 E2
Trebarber Corn 5 C7
Trebartha Corn 11 F11
Trebarwith Corn 11 D7
Trebarwith Strand Corn 10 D6
Trebeath Corn 11 D11
Trebell Green Corn 5 C11
Treberfydd Powys 96 F2
Trebetherick Corn 10 F4
Trebilcock Corn 5 C9
Treble's Holford Som 43 G7
Treborough Som 42 F4
Trebudannon Corn 5 C7
Trebullett Corn 12 F2
Treburgett Corn 11 F7
Treburgie Corn 6 C4
Treburley Corn 12 F3
Treburrick Corn 10 G3
Trebyan Corn 5 C11
Trecastle Powys 95 F7
Trecenydd Caerph 58 B6
Trecott Devon 25 G10
Trecwn Pembs 91 E9
Trecynon Rhondda 77 E7
Tredannick Corn 10 G6
Tredaule Corn 11 E10
Tredavoe Corn 1 D5
Treddiog Pembs 91 F7
Tredegar Bl Gwent 77 D10
Trederwen Powys 148 F5
Tredethy Corn 11 G7
Tredington Glos 99 F8
Tredington Warks 100 C5
Tredinnick Corn 1 C4
Tredinnick Corn 5 D10
Tredinnick Corn 6 B3
Tredinnick Corn 6 D4
Tredinnick Corn 10 G4
Tredogan V Glam 58 F5
Tredomen Caerph 77 G10
Tredomen Powys 96 E2
Tredown Devon 24 D2
Tredrizzick Corn 10 F5
Tredunnock Mon 78 G5
Tredustan Powys 96 E2
Tredworth Glos 80 B4
Treen Corn 1 B4
Treen Corn 1 E3
Treesmill Corn 5 D11
Treeton S Yorks 186 D6
Tref y Clawdd = Knighton Powys 114 C5
Trefaes Gwyn 144 C5
Trefanny Hill Corn 6 D4
Trefasser Pembs 91 D7
Trefdraeth Anglesey 178 G6
Trefdraeth = Newport Pembs 91 D11
Trefecca Powys 96 E2
Trefechan Ceredig 111 A11
Trefechan M Tydf 77 D8
Trefechan Wrex 166 F3
Trefeglwys Powys 129 E9
Trefeitha Powys 96 E2
Trefenter Ceredig 112 D2
Treffgarne Pembs 91 G9
Treffynnon Pembs 90 F6
Treffynnon = Holywell Flint 181 F11
Trefgarn Owen Pembs 91 F7
Trefil Bl Gwent 77 C10
Trefilan Ceredig 111 F11
Trefin = Trevine Pembs 90 E6
Treflach Shrops 148 D5
Trefnanney Powys 148 F4
Trefnant Denb 181 G9
Trefonen Shrops 148 D5
Trefor Anglesey 178 E5
Trefor Gwyn 162 F5
Treforda Corn 11 E7
Treforest Rhondda 58 B5
Treforgan Ceredig 92 B4
Treforys = Morriston Swansea 57 B7
Trefriw Conwy 164 C3
Trefrize Corn 12 F2
Trefynwy = Monmouth Mon 79 C8
Tregada Corn 12 E2
Tregadgwith Corn 1 D4
Tregadillett Corn 12 E2
Tregaian Anglesey 178 F6
Tregajorran Corn 4 G3
Tregamere Corn 5 C8
Tregardock Corn 10 E6
Tregare Mon 78 C6
Tregarland Corn 6 D5
Tregarlandbridge Corn 6 D4
Tregarne Corn 3 E7
Tregaron Ceredig 112 F3
Tregarrick Mill Corn 6 D4
Tregarth Gwyn 163 B10
Tregath Gwyn 163 B10
Tregatillian Corn 5 C8
Tregatta Corn 11 D7
Tregavarah Corn 1 D4
Tregear Corn 2 E5
Tregeare Corn 11 D10
Tregeiriog Wrex 148 C3
Tregele Anglesey 178 C5
Tregellist Corn 10 F6
Tregeseal Corn 1 C3
Tregew Corn 3 C8
Tregidden Corn 2 E6
Treginnis Pembs 90 G4
Tregla Powys 90 F6
Tregole Corn 11 B9
Tregolls Corn 2 B6
Tregolwyn = Colwinston V Glam 58 D2
Tregona Corn 5 B7
Tregonce Corn 10 G4
Tregonetha Corn 5 C9
Tregonna Corn 10 G4
Tregonning Corn 5 D7
Tregony Corn 5 G8
Tregoodwell Corn 11 E8
Tregorden Corn 10 G6
Tregorrick Corn 5 E10
Tregoss Corn 5 C9
Tregoyd Powys 96 D4
Tregoyd Mill Powys 96 D3
Tregreenwell Corn 11 E7
Tregrehan Mills Corn 5 E10
Tregroes Ceredig 93 C8
Tregullon Corn 5 C11
Tregunna Corn 10 G5
Tregurrian Corn 5 B7
Tregurtha Downs Corn 2 C2
Tregyddulan Pembs 91 D7
Tregynon Powys 129 D11
Trehafod Rhondda 77 G8
Trehafren Powys 129 E11
Trehan Corn 7 D8
Treharris M Tydf 77 F9
Trehemborne Corn 10 G3
Treherbert Rhondda 76 F6
Trehunist Corn 6 C6
Trekeivesteps Corn 11 G10
Trekenner Corn 12 F2
Trekenning Corn 5 C8
Treknow Corn 11 D7
Trelales = Laleston Bridgend 57 F11
Trelan Corn 2 F6
Trelash Corn 11 C9
Trelassick Corn 5 E7
Trelawnyd Flint 181 F9
Trelech Carms 92 E5
Treleddyd-fawr Pembs 90 F5
Treleigh Corn 4 G4
Treletert = Letterston Pembs 91 F8
Trelew Corn 3 B8
Trelewis M Tydf 77 F10
Treligga Corn 11 E7
Trelights Corn 10 F5
Trelill Corn 10 F6
Trelion Corn 5 E8
Treliske Corn 4 F6
Trelissick Corn 3 B8
Treliver Corn 5 B9
Trellech Mon 79 E7
Trelleck Grange Mon 79 E7
Trelogan Flint 181 E10
Treloquithack Corn 2 D5
Trelowia Corn 6 D5
Trelowth Corn 5 E9
Trelystan Powys 130 C5
Tremadog Gwyn 163 G9
Tremail Corn 11 D9
Tremain Ceredig 92 B4
Tremaine Corn 11 D10
Tremains Bridgend 58 D2
Tremar Corn 6 B5
Trematon Corn 7 D7
Trematon Castle Corn 7 D8
Tremayne Corn 2 B4
Trembraze Corn 6 B5
Tremedda Corn 1 B5
Tremeirchion Denb 181 G9
Tremethick Cross Corn 1 C4
Tremore Corn 5 C10
Tremorebridge Corn 5 C10
Tremorfa Cardiff 59 D8
Trenance Corn 4 C6
Trenance Corn 5 B7
Trenance Corn 5 C9
Trenance Corn 10 G4
Trenant Corn 6 B4
Trenant Corn 10 G5
Trenarren Corn 5 F10
Trenay Corn 6 B3
Trench Telford 150 G3
Trench Green Oxon 65 D7
Trench Wood Kent 52 D5
Trencreek Corn 4 C6
Trencrom Corn 2 B2
Trendeal Corn 5 E7
Trenear Corn 2 C5
Treneglos Corn 11 D10
Trenerth Corn 2 B4
Trenewan Corn 6 E3
Trengune Corn 11 C9
Trenhorne Corn 11 F11
Treningle Corn 5 B10
Treninnick Corn 4 C6
Trenoon Corn 2 F6
Trenoweth Corn 3 C7
Trent Dorset 29 D9
Trent Vale Stoke 168 G5
Trentham Stoke 168 G5
Trentishoe Devon 40 D6
Trentlock Derbys 153 C9
Trenwheal Corn 2 C4
Treoes V Glam 58 D2
Treopert = Granston Pembs 91 E7
Treorchy = Treorci Rhondda 77 F7
Treorci = Treorchy Rhondda 77 F7
Treowen Caerph 78 F2
Treowen Powys 130 E2
Trequite Corn 10 F6
Tre'r-ddôl Ceredig 128 E3
Trer Ilai = Leighton Powys 130 B4
Trerhyngyll V Glam 58 D4
Trerise Corn 2 F6
Trerose Corn 3 D7
Trerulefoot Corn 6 D6
Tresaith Ceredig 110 G5
Tresamble Corn 3 B7
Tresarrett Corn 11 G7
Tresavean Corn 2 B6
Tresawle Corn 5 F7
Tresawsen Corn 4 F5
Trescoll Corn 5 C10
Trescott Staffs 132 D6
Trescowe Corn 2 C3
Tresean Corn 4 D5
Tresevern Croft Corn 2 B6
Tresham Glos 80 G3
Tresigin = Sigingstone V Glam 58 E3
Tresillian Corn 5 F7
Tresimwn = Bonvilston V Glam 58 E5
Tresinney Corn 11 E8
Tresinwen Pembs 91 C7
Treskerby Corn 4 G4
Treskillard Corn 2 B5
Treskilling Corn 5 D10
Treskinnick Cross Corn 11 B10
Treslothan Corn 2 B5
Tresmeer Corn 11 D10
Tresowes Green Corn 2 D3
Tresoweshill Corn 2 D4
Tresparrett Corn 11 C8
Tresparrett Posts Corn 11 C8
Tressady Highld 309 J7
Tressait Perth 291 G10
Tresta Shetland 312 D8
Tresta Shetland 313 H5
Treswell Notts 188 F3
Treswithian Corn 4 G2
Treswithian Downs Corn 4 G2
Trethellan Water Corn 2 B6
Trethevy Corn 11 D7
Trethewell Corn 3 B9
Trethewey Corn 1 E3
Trethillick Corn 10 F4
Trethomas Caerph 59 B7
Trethosa Corn 5 E8
Trethowel Corn 5 E10
Trethurgy Corn 5 D10
Tretio Pembs 90 F5
Tretire Hereford 97 G10
Tretower Powys 96 G3
Treuddyn Flint 166 D3
Trevadlock Corn 11 F11
Trevail Corn 4 D5
Trevalga Corn 11 D7
Trevalgan Corn 1 A5
Trevalyn Wrex 166 D5
Trevance Corn 10 G4
Trevanger Corn 10 F5
Trevanson Corn 10 G5
Trevarrack Corn 1 C5
Trevarren Corn 5 C8
Trevarrian Corn 4 B6
Trevarrick Corn 5 G9
Trevarth Corn 4 G4
Trevaughan Carms 73 B11
Trevaughan Carms 93 G7
Treveal Corn 1 A5
Trevegean Corn 1 D3
Treveighan Corn 11 F7
Trevellas Corn 4 E4
Trevelmond Corn 6 C4
Trevelver Corn 10 G5
Trevemper Corn 4 D6
Treven Corn 11 D7
Trevena Corn 2 D4
Trevenen Corn 2 D5
Trevenen Bal Corn 2 D5
Trevenning Corn 11 F7
Treveor Corn 5 G9
Treverbyn Corn 5 D10
Treverbyn Corn 6 B4
Treverva Corn 3 C7
Trevescan Corn 1 E3
Trevethin Torf 78 E3
Trevia Corn 11 E7
Trevigro Corn 6 B6
Trevilder Corn 10 G6
Trevilla Corn 3 B8
Trevilson Corn 4 D6
Trevine Corn 10 F5
Trevine = Trefin Pembs 90 E6
Treviscoe Corn 5 D8
Treviskey Corn 2 B6
Trevithal Corn 1 D5
Trevoll Corn 4 D6
Trevone Corn 10 F3
Trevor Wrex 166 G3
Trevor Uchaf Denb 166 G2
Trevorrick Corn 10 G4
Trevowah Corn 4 D5
Trevowhan Corn 1 B4
Trew Corn 2 D4
Trewalder Corn 11 E7
Trewarmett Corn 11 D7
Trewartha Corn 2 B2
Trewartha Corn 3 B10
Trewassa Corn 11 D8
Treween Corn 11 E10
Trewellard Corn 1 C3
Trewen Corn 11 E11
Trewen Corn 11 E7
Trewen Mon 79 G7
Trewennack Corn 2 D5
Trewennan Corn 11 E7
Trewern Powys 148 G5
Trewetha Corn 10 E6
Trewethen Corn 10 F6
Trewethern Corn 10 F6
Trewey Corn 1 B5
Trewidland Corn 6 D5
Trewindle Corn 6 C2
Trewint Corn 6 C5
Trewint Corn 11 B9
Trewint Corn 11 E10
Trewithian Corn 3 B9
Trewithick Corn 11 D11
Trewollock Corn 5 G10
Trewoodloe Corn 12 G2
Trewoofe Corn 1 D4
Trewoon Corn 2 F5
Trewoon Corn 5 E9
Treworga Corn 5 G7
Treworgan Common Mon 78 D6
Treworlas Corn 3 B9
Treworld Corn 11 C8
Trewornan Corn 10 G5
Treworthal Corn 3 B9
Trewyddel = Moylgrove Pembs 92 C2
Trewyn Devon 24 G4
Treyarnon Corn 10 G3
Treyford W Sus 34 D4
Trezaise Corn 5 D9
Trezelah Corn 1 C5
Triangle Glos 79 E8
Triangle Staffs 133 B11
Triangle W Yorks 196 C4
Trickett's Cross Dorset 31 G9
Triffleton Pembs 91 G9
Trillacott Corn 11 D11
Trimdon Durham 234 E3
Trimdon Colliery Durham 234 D3
Trimdon Grange Durham 234 D3
Trimingham Norf 160 B5
Trimley Lower Street Suff 108 D5
Trimley St Martin Suff 108 D5
Trimley St Mary Suff 108 D5
Trimpley Worcs 116 B5
Trims Green Herts 87 B7
Trimsaran Carms 75 E7
Trimstone Devon 40 E3
Trinafour Perth 291 G9
Trinant Caerph 78 E2
Tring Herts 84 C6
Tring Wharf Herts 84 C6
Tringford Herts 84 C6
Trinity Angus 293 G8
Trinity Devon 27 F7
Trinity Edin 280 F4
Trinity Fields Staffs 151 D8
Trisant Ceredig 112 B4
Triscombe Som 43 F7
Trislaig Highld 290 F2
Trispen Corn 4 E6
Tritlington Northumb 252 E6
Troan Corn 5 D7
Trochry Perth 286 C3
Trodigal Argyll 255 E7
Troearhiwgwair Bl Gwent 77 D11
Troedrhiwdalar Powys 113 G9
Troedrhiwffenyd Ceredig 93 C8
Troedrhiwfuwch Caerph 77 E10
Troedyraur Ceredig 92 B6
Troedyrhiw M Tydf 77 E9
Trofarth Conwy 180 G5
Trolliloes E Sus 23 C10
Tromode I o M 192 E4
Trondavoe Shetland 312 F5
Troon Corn 2 B5
Troon S Ayrs 257 C8
Trooper's Inn Pembs 73 C7
Trosaraidh W Isles 297 K3
Trossachs Hotel Stirl 285 G9
Troston Suff 125 C7
Trostre Carms 56 B4
Trostrey Common Mon 78 E5
Troswell Corn 11 C11
Trotshill Worcs 117 F7
Trotten Marsh W Sus 34 B4
Trottiscliffe Kent 68 G6
Trotton W Sus 34 C4
Trough Gate Lancs 195 C11
Troutbeck Cumb 221 E8
Troutbeck Cumb 230 F3
Troutbeck Bridge Cumb 221 F8
Trow Green Glos 79 D9
Troway Derbys 186 F5
Trowbridge Cardiff 59 C8
Trowbridge Wilts 45 B11
Trowell Notts 153 B9
Trowle Common Wilts 45 B10
Trowley Bottom Herts 85 C9
Trows Borders 262 C5
Trowse Newton Norf 142 B4
Troy Town Kent 52 D2
Troy Town Kent 54 E5
Troy Town Medway 69 F8
Troydale W Yorks 205 G10
Truas Corn 11 D7
Trub Gtr Man 195 F11
Trudoxhill Som 45 E8
True Street Devon 8 C6
Trueman's Heath Worcs 117 B11
Trull Som 28 C2
Trumaisgearraidh W Isles 296 D4
Trumfleet S Yorks 198 E6
Trumpan Highld 298 C2
Trumpet Hereford 98 D3
Trumpington Cambs 123 F8
Trumpsgreen Sur 66 F3
Trunch Norf 160 C5
Trunnah Lancs 202 E2
Truro Corn 4 G6
Truscott Corn 12 D2
Trusham Devon 14 E3
Trusley Derbys 152 B5
Trussall Corn 2 D5
Trussell Corn 11 D10
Trusthorpe Lincs 191 E8
Truthan Corn 4 E6
Truthwall Corn 2 C2
Trwstllewelyn Powys 130 D3
Tryfil Anglesey 178 E6
Trysull Staffs 133 E7
Trythogga Corn 1 C5
Tubbs Mill Corn 5 G9
Tubney Oxon 82 F6
Tubslake Kent 53 G9
Tuckenhay Devon 8 D6
Tuckermarsh Devon 7 B8
Tuckerton Som 28 B3
Tuckhill Shrops 132 F5
Tucking Mill Bath 61 G9
Tuckingmill Corn 4 G3
Tuckingmill Corn 11 F7
Tuckingmill Wilts 30 B6
Tuckton Bmouth 19 C8
Tuddenham Suff 108 B3
Tuddenham Suff 124 C4
Tuddenham St Martin Suff 108 B3
Tudeley Kent 52 D6
Tudeley Hale Kent 52 D6
Tudhay Devon 28 G4
Tudhoe Durham 233 D11
Tudhoe Grange Durham 233 E11
Tudor Hill W Mid 134 D2
Tudorville Hereford 97 G11
Tudweiliog Gwyn 144 B4
Tuebrook Mers 182 C5
Tuesley Sur 50 E3
Tuesnoad Kent 54 E2
Tuffley Glos 80 C4
Tufnell Park London 67 B9
Tufton Hants 48 D3
Tufton Pembs 91 F10
Tugby Leics 136 C5
Tugford Shrops 131 F11
Tughall Northumb 264 D6
Tulchan Lodge Angus 292 F3
Tullecombe W Sus 34 B4
Tullibardine Perth 286 F3
Tullibody Clack 279 B7
Tullich Argyll 284 F4
Tullich Highld 299 E9
Tullich Highld 300 G6
Tullich Muir Highld 301 B7
Tulliemet Perth 286 B3
Tulloch Aberds 293 F9
Tulloch Aberds 303 F8
Tulloch Highld 290 E5
Tulloch Perth 286 E4
Tulloch Castle Highld 300 C5
Tulloch-gribban Highld 301 G9
Tullochgorm Argyll 275 D10
Tullochroisk Perth 285 B11
Tullochvenus Aberds 293 C7
Tulloes Angus 287 C9
Tullybannocher Perth 285 E11
Tullybelton Perth 286 D4
Tullycross Stirl 277 D9
Tullyfergus Perth 286 C6
Tullymurdoch Perth 286 B5
Tullynessle Aberds 293 B7
Tulse Hill London 67 E10
Tumble = Y Tymbl Carms 75 C8
Tumbler's Green Essex 106 F6
Tumby Lincs 174 D2
Tumby Woodside Lincs 174 D3
Tummel Bridge Perth 285 B11
Tumpy Green Glos 80 E2
Tumpy Lakes Hereford 97 B10
Tunbridge Hill Medway 69 E10
Tunbridge Wells = Royal Tunbridge Wells Kent 52 F5
Tunga W Isles 304 E6
Tungate Norf 160 D5
Tunley Bath 45 B7
Tunley Glos 80 E6
Tunnel Hill Worcs 98 C6
Tunnel Pits N Lincs 199 G8
Tunshill Gtr Man 196 E2
Tunstall E Yorks 209 G12
Tunstall Kent 69 G11
Tunstall Lancs 212 E2
Tunstall N Yorks 224 F4
Tunstall Norf 143 B8
Tunstall Staffs 150 D5
Tunstall Stoke 168 E5
Tunstall Suff 127 G7
Tunstall T & W 243 G9
Tunstead Derbys 185 G10
Tunstead Gtr Man 196 G4
Tunstead Norf 160 E5
Tunworth Hants 49 D7
Tupsley Hereford 97 C10
Tupton Derbys 170 B5
Tur Langton Leics 136 E4
Turbary Common Poole 19 C7
Turf Hill Gtr Man 196 E2
Turfdown Corn 5 B11
Turfholm S Lnrk 259 B8
Turfmoor Devon 28 G3
Turfmoor Shrops 149 F7
Turgis Green Hants 49 B7
Turin Angus 287 B9
Turkdean Glos 81 B10
Turkey Island Hants 33 E9
Turkey Island W Sus 34 D3
Turkey Tump Hereford 97 F10
Turleigh Wilts 61 G10
Turleygreen Shrops 132 F5
Turlin Moor Poole 18 C5
Turmer Hants 31 F10
Turn Lancs 195 D10
Turnalt Argyll 275 C9
Turnastone Hereford 97 D7
Turnberry S Ayrs 244 B6
Turnchapel Plym 7 E9
Turnditch Derbys 170 F3
Turner Green Lancs 203 G8
Turner's Green E Sus 23 B10
Turner's Green E Sus 52 G6
Turner's Green W Berks 64 F4
Turner's Green Warks 118 D3
Turners Hill W Sus 51 F10
Turners Puddle Dorset 18 C2
Turnerwood S Yorks 187 E8
Turnford Herts 86 E5
Turnhouse Edin 280 G3
Turnhurst Stoke 168 E5
Turnstead Milton Derbys 185 E8
Turnworth Dorset 30 F4
Turrerich Perth 286 D2
Turriff Aberds 303 D7
Tursdale Durham 234 D2
Turton Bottoms Blkburn 195 D8
Turves Cambs 138 D6
Turves Green W Mid 117 B10
Turvey Beds 121 G8
Turville Bucks 84 G3
Turville Heath Bucks 84 G2
Turweston Bucks 102 D2
Tushielaw Borders 261 F8
Tutbury Staffs 152 D4
Tutnall Worcs 117 C9
Tutnalls Glos 79 E10
Tutshill Glos 79 G8
Tutt Hill Kent 54 D3
Tuttington Norf 160 D4
Tutts Clump W Berks 64 E5
Tutwell Corn 12 F3
Tuxford Notts 188 G2
Twatt Orkney 314 D2
Twatt Shetland 313 H5
Twechar E Dunb 278 F4
Tweedale Telford 132 C4
Tweeddaleburn Borders 270 E5
Tweedmouth Northumb 273 E9
Tweedsmuir Borders 260 E3
Twelve Heads Corn 4 G5
Twelve Oaks E Sus 37 C11
Twelvewoods Corn 6 B4
Twemlow Green E Ches 168 B3
Twenties Kent 71 F10
Twenty Lincs 156 E3
Twerton Bath 61 G8
Twickenham London 67 E7
Twigworth Glos 98 G6
Twineham W Sus 36 D3
Twineham Green W Sus 36 C3
Twinhoe Bath 45 B8
Twinstead Essex 107 D7
Twinstead Green Essex 106 D6
Twiss Green Warr 183 B11
Twist Devon 28 G3
Twiston Lancs 204 E2
Twitchen Devon 41 G9
Twitchen Shrops 115 B7
Twitchen Mill Devon 41 G9
Twitham Kent 55 B9
Twitton Kent 52 B4
Two Bridges Devon 13 G8
Two Bridges Glos 79 D11
Two Burrows Corn 4 F4
Two Dales Derbys 170 C3
Two Gates Staffs 134 C4
Two Mile Ash M Keynes 102 D6
Two Mile Ash W Sus 35 B10
Two Mile Hill Bristol 60 E6
Two Mile Oak Cross Devon 8 B6
Two Mills W Ches 182 G5
Two Pots Devon 40 E4
Two Waters Herts 85 D9
Twr Anglesey 178 E2
Twycross Leics 134 C6
Twydall Medway 69 F9
Twyford Bucks 102 F3
Twyford Derbys 152 D6
Twyford Dorset 30 D5
Twyford Hants 33 C7
Twyford Leics 154 G4
Twyford Lincs 155 E8
Twyford Norf 159 E10
Twyford Oxon 101 D9
Twyford Shrops 148 D6
Twyford Wokingham 65 D9
Twyford Worcs 99 B10
Twyford Common Hereford 97 D10
Twyn-Allws Mon 78 C3
Twyn Shôn-Ifan Caerph 77 G11
Twyn-y-Sheriff Mon 78 D6
Twyn-yr-odyn V Glam 58 E6
Twynholm Dumfries 237 D8
Twyning Glos 99 D7
Twyning Green Glos 99 D8
Twynllanan Carms 94 G5
Twynmynydd Carms 75 C11
Twynyrodyn M Tydf 77 D9
Twywell Northants 121 B9
Ty-coch Swansea 56 C6
Ty-draw Conwy 164 D5
Ty-draw Swansea 57 C7
Ty-fry Mon 78 F6
Ty-hen Carms 92 G6
Ty-hen Gwyn 144 C3
Ty-isaf Carms 56 B4
Ty Llwyn Bl Gwent 77 D11
Ty-mawr Anglesey 179 D7
Ty Mawr Carms 93 C10
Ty-mawr Conwy 181 F7
Ty Mawr Cwm Conwy 164 F6
Ty-nant Conwy 165 G7
Ty-nant Gwyn 147 D8
Ty-Newydd Ceredig 111 D10
Ty Rhiw Rhondda 58 C6
Ty-Sign Caerph 78 G2
Ty-uchaf Powys 147 E10
Tyberton Hereford 97 D7
Tyburn W Mid 134 E2

Tyby Norf 159 D11
Tycroes Carms 75 C10
Tycrwyn Powys 148 F2
Tydd Gote Lincs 157 F9
Tydd St Giles Cambs 157 F8
Tydd St Mary Lincs 157 F8
Tyddewi = St Davids Pembs 90 F5
Tyddyn Powys 129 F9
Tyddyn Angharad Denb 165 F9
Tyddyn Dai Anglesey 178 C6
Tyddyn-mawr Gwyn 163 G9
Tye Hants 22 C2
Tye Common Essex 87 G11
Tye Green Essex 87 C10
Tye Green Essex 87 D7
Tye Green Essex 88 B2
Tye Green Essex 105 D11
Tye Green Essex 105 G10
Tye Green Essex 106 G5
Tyegate Green Norf 161 G7
Tyersal W Yorks 205 G9
Tyganol V Glam 58 E4
Tyla Mon 78 C2
Tylagwyn Bridgend 58 B2
Tyldesley Gtr Man 195 G7
Tyle Carms 94 F3
Tyle-garw Rhondda 58 C4
Tyler Hill Kent 70 G6
Tylers Causeway Herts 86 D3
Tylers Green Bucks 84 G6
Tyler's Green Essex 87 D8
Tyler's Green Sur 51 C11
Tyler's Hill Bucks 85 E7
Tylorstown Rhondda 77 F8
Tylwch Powys 129 G9
Tyn-lon Gwyn 163 D7
Ty'n-y-bryn Rhondda 58 B4
Ty'n-y-celyn Wrex 148 B3
Ty'n-y-coed Shrops 148 D4
Ty'n-y-coedcae Caerph 59 B7
Ty'n-y-cwm Swansea 75 E10
Tyn-y-fedwen Powys 148 C2
Ty'n-y-ffordd Denb 181 G8
Tyn-y-ffridd Powys 148 C2
Ty'n-y-garn Bridgend 57 E11
Tyn-y-graig Powys 113 G10
Ty'n-y-groes Conwy 180 G3
Ty'n-y-maes Gwyn 163 C10
Tyn-y-pwll Anglesey 178 D6
Ty'n-yr-eithin Ceredig 112 E3
Tynant Rhondda 58 B5
Tyncelyn Ceredig 112 E2
Tyndrum Stirl 285 D7
Tyne Dock T & W 243 D9
Tyne Tunnel T & W 243 D8
Tyneham Dorset 18 E3
Tynehead Midloth 271 D7
Tynemouth T & W 243 D9
Tynewydd Ceredig 92 B4
Tynewydd Neath 76 D4
Tynewydd Rhondda 76 F6
Tyning Bath 45 B7
Tyninghame E Loth 282 F2
Tynron Dumfries 247 E8
Tyntesfield N Som 60 E4
Tyntetown Rhondda 77 F9
Tynyfedw Conwy 165 B7
Tynygongl Anglesey 179 E8
Tynygraig Ceredig 112 D3
Tynyrwtra Powys 129 F7
Ty'r-felin-isaf Conwy 164 C5
Tyrells End C Beds 103 E9
Tyrell's Wood Sur 51 B7
Tyrie Aberds 303 C9
Tyringham M Keynes 103 B7
Tyseley W Mid 134 G2
Tythecott Devon 24 D6
Tythegston Bridgend 57 F11
Tytherington E Ches 184 F6
Tytherington S Glos 61 B7
Tytherington Som 45 D9
Tytherington Wilts 46 E2
Tytherleigh Devon 28 G4
Tytherton Lucas Wilts 62 E2
Tyttenhanger Herts 85 D11
Tywardreath Corn 5 E11
Tywardreath Highway Corn 5 D11
Tywyn Conwy 180 F3
Tywyn Gwyn 110 C2

U

Uachdar W Isles 296 F3
Uags Highld 295 B9
Ubberley Stoke 168 F6
Ubbeston Green Suff 126 C6
Ubley Bath 44 B4
Uckerby N Yorks 224 E4
Uckfield E Sus 37 C7
Uckinghall Worcs 99 D7
Uckington Glos 99 G8
Uckington Shrops 131 B11
Uddingston S Lnrk 268 C3
Uddington S Lnrk 259 C9
Udimore E Sus 38 D5
Udley N Som 60 G3
Udny Green Aberds 303 G8
Udny Station Aberds 303 G9
Udston S Lnrk 268 D3
Udstonhead S Lnrk 268 F4
Uffcott Wilts 62 D6
Uffculme Devon 27 E9
Uffington Lincs 137 B11
Uffington Oxon 63 B10
Uffington Shrops 149 G10
Ufford P'boro 137 C11
Ufford Suff 126 G5
Ufton Warks 119 E7
Ufton Green W Berks 64 F6
Ufton Nervet W Berks 64 F6
Ugadale Argyll 255 E8
Ugborough Devon 8 D3
Ugford Wilts 46 G5
Uggeshall Suff 143 G8
Ugglebarnby N Yorks 227 D7
Ughill S Yorks 186 C3
Ugley Essex 105 F10
Ugley Green Essex 105 F10
Ugthorpe N Yorks 226 C5
Uidh W Isles 297 M2
Uig Argyll 276 E2
Uig Argyll 288 D3
Uig Highld 296 F7
Uig Highld 298 C3
Uigen W Isles 304 E2
Uigshader Highld 298 E4
Uisken Argyll 274 B4
Ulaw Aberds 303 G9
Ulbster Highld 310 E7
Ulcat Row Cumb 230 G4
Ulceby Lincs 190 G6
Ulceby N Lincs 200 E6
Ulceby Skitter N Lincs 200 E6
Ulcombe Kent 53 D10
Uldale Cumb 229 D10
Uley Glos 80 F3
Ulgham Northumb 252 E6
Ullapool Highld 307 K6
Ullcombe Devon 28 F2
Ullenhall Warks 118 D2
Ullenwood Glos 80 B6
Ulleskelf N Yorks 206 E6
Ullesthorpe Leics 135 F10
Ulley S Yorks 187 D7
Ullingswick Hereford 97 B11
Ullington Worcs 100 B2
Ullinish Highld 294 B5
Ullock Cumb 229 G10
Ullock Cumb 229 G7
Ulnes Walton Lancs 194 D4
Ulpha Cumb 220 G3
Ulrome E Yorks 209 B9
Ulshaw N Yorks 214 B2
Ulsta Shetland 312 E6
Ulva House Argyll 288 F6
Ulverley Green W Mid 134 G2
Ulverston Cumb 210 D5
Ulwell Dorset 18 E6
Umberleigh Devon 25 C10
Unapool Highld 306 F7
Unasary W Isles 297 J3
Under Bank W Yorks 196 F6
Under the Wood Kent 71 F8
Under Tofts S Yorks 186 D4
Underbarrow Cumb 221 G9
Undercliffe W Yorks 205 G9
Underdale Shrops 149 G10
Underdown Devon 14 D3
Underhill London 86 F3
Underhill Wilts 45 G11
Underhoull Shetland 312 C7
Underling Green Kent 53 D9
Underriver Kent 52 C5
Underriver Ho Kent 52 C5
Underton Shrops 132 E3
Underwood Newport 59 B11
Underwood Notts 171 E7
Underwood Pembs 73 C7
Underwood Plym 7 D10
Undley Suff 140 G3
Undy Mon 60 B2
Ungisiadar W Isles 304 F3
Unifirth Shetland 313 H4
Union Cottage Aberds 293 D10
Union Mills I o M 192 E4
Union Street E Sus 53 G8
United Downs Corn 4 G4
Unstone Derbys 186 F5
Unstone Green Derbys 186 F5
Unsworth Gtr Man 195 F10
Unthank Cumb 230 B3
Unthank Cumb 230 D5
Unthank Cumb 231 C8
Unthank Derbys 186 F4
Unthank End Cumb 230 D5
Up Cerne Dorset 29 G11
Up End M Keynes 103 B8
Up Exe Devon 26 G6
Up Green Hants 65 G9
Up Hatherley Glos 99 G8
Up Holland Lancs 194 F4
Up Marden W Sus 34 E3
Up Mudford Som 29 D9
Up Nately Hants 49 C7
Up Somborne Hants 47 G11
Up Sydling Dorset 29 G10
Upavon Wilts 46 C6
Upchurch Kent 69 F10
Upcott Devon 24 D2
Upcott Devon 25 F11
Upcott Devon 25 F9
Upcott Devon 40 F3
Upcott Hereford 114 G6
Upcott Som 27 C11
Upend Cambs 124 F3
Upgate Norf 160 F2
Upgate Street Norf 141 E11
Upgate Street Norf 142 E5
Uphall Dorset 29 G9
Uphall W Loth 279 G11
Uphall Station W Loth 279 G11
Upham Devon 26 F5
Upham Hants 33 C8
Uphampton Hereford 115 E7
Uphampton Worcs 116 E6
Uphempston Devon 8 C6
Uphill N Som 43 B10
Uphill Manor N Som 43 B10
Uplands Glos 80 D5
Uplands Swansea 56 C6
Uplawmoor E Renf 267 D8
Upleadon Glos 98 F5
Upleadon Court Glos 98 F5
Upleatham Redcar 226 B2
Uplees Kent 70 G3
Uploders Dorset 16 C6
Uplowman Devon 27 D8
Uplyme Devon 16 C2
Upminster London 68 B5
Upnor Medway 69 E9
Upottery Devon 28 F2
Uppacott Devon 25 B9
Uppat Highld 311 J2
Uppend Essex 105 F9
Upper Affcot Shrops 131 F8
Upper Ardchronie Highld 309 L6
Upper Ardgrain Aberds 303 F9
Upper Ardroscadale Argyll 275 G11
Upper Arley Worcs 132 G5
Upper Armley W Yorks 205 G11
Upper Arncott Oxon 83 B10
Upper Astley Shrops 149 F10
Upper Aston Shrops 132 E6
Upper Astrop Northants 101 D10
Upper Badcall Highld 306 E6
Upper Bangor Gwyn 179 G9
Upper Basildon W Berks 64 D5
Upper Batley W Yorks 197 B8
Upper Battlefield Shrops 149 F10
Upper Beeding W Sus 35 E11
Upper Benefield Northants 137 F9
Upper Bentley Worcs 117 D9
Upper Bighouse Highld 310 D2
Upper Birchwood Derbys 170 E6
Upper Blainslie Borders 271 G10
Upper Boat Rhondda 58 B6
Upper Boddam Aberds 302 F6
Upper Boddington Northants 119 G9
Upper Bogrow Highld 309 L7
Upper Bogside Moray 302 D2
Upper Bonchurch I o W 21 F7
Upper Booth Derbys 185 D10
Upper Borth Ceredig 128 F2
Upper Boyndlie Aberds 303 C9
Upper Brailes Warks 100 D6
Upper Brandon Parva Norf 141 B10
Upper Breakish Highld 295 C8
Upper Breinton Hereford 97 C9
Upper Broadheath Worcs 116 F6
Upper Brockholes W Yorks 196 B5
Upper Broughton Notts 154 D3
Upper Broxwood Hereford 115 G7
Upper Bruntingthorpe Leics 136 F2
Upper Brynamman Carms 76 C2
Upper Buckenhill Hereford 97 E11
Upper Bucklebury W Berks 64 F4
Upper Bullington Hants 48 E3
Upper Burgate Hants 31 D11
Upper Burnhaugh Aberds 293 D10
Upper Bush Medway 69 F7
Upper Caldecote C Beds 104 B3
Upper Cam Glos 80 F3
Upper Canada N Som 43 B11
Upper Canterton Hants 32 E3
Upper Catesby Northants 119 F10
Upper Catshill Worcs 117 C9
Upper Chapel Powys 95 C10
Upper Cheddon Som 28 B2
Upper Chicksgrove Wilts 31 B7
Upper Church Village Rhondda 58 B5
Upper Chute Wilts 47 C9
Upper Clapton London 67 B10
Upper Clatford Hants 47 E11
Upper Coberley Glos 81 B7
Upper College Shrops 149 C11
Upper Colwall Hereford 98 C5
Upper Common Hants 48 D6
Upper Cotburn Aberds 303 D7
Upper Cotton Staffs 169 F9
Upper Coullie Aberds 293 B9
Upper Cound Shrops 131 C11
Upper Coxley Som 44 E4
Upper Cudworth S Yorks 197 F11
Upper Culphin Aberds 302 D6
Upper Cumberworth W Yorks 197 F8
Upper Cwm-twrch Powys 76 C3
Upper Cwmbran Torf 78 F3
Upper Dallachy Moray 302 C3
Upper Deal Kent 55 C11
Upper Dean Beds 121 D10
Upper Dean Devon 8 C4
Upper Denby W Yorks 197 D8
Upper Denby W Yorks 197 F8
Upper Denton Cumb 240 D4
Upper Derraid Highld 301 F10
Upper Diabaig Highld 299 C8
Upper Dicker E Sus 23 D9
Upper Dinchope Shrops 131 G9
Upper Dormington Hereford 97 D11
Upper Dounreay Highld 310 C4
Upper Dovercourt Essex 108 E4
Upper Dowdeswell Glos 81 B8
Upper Druimfin Argyll 289 D7
Upper Dunsforth N Yorks 215 G8
Upper Dunsley Herts 84 C6
Upper Eashing Sur 50 E3
Upper Eastern Green W Mid 134 G5
Upper Eathie Highld 301 C7
Upper Edmonton London 86 G4
Upper Egleton Hereford 98 C2
Upper Elkstone Staffs 169 D9
Upper Ellastone Staffs 169 G10
Upper Elmers End London 67 F11
Upper End Derbys 185 F9
Upper End Glos 81 C10
Upper End Glos 81 D8
Upper End Leics 154 G4
Upper Enham Hants 47 D11
Upper Farmcote Shrops 132 E5
Upper Farringdon Hants 49 F8
Upper Feorlig Highld 298 E2
Upper Fivehead Som 28 C4
Upper Forge Shrops 132 F4
Upper Framilode Glos 80 C3
Upper Froyle Hants 49 E9
Upper Gambolds Worcs 117 D9
Upper Gills Highld 310 B7
Upper Glenfintaig Highld 290 E4
Upper Godney Som 44 E3
Upper Goldstone Kent 71 G9
Upper Gornal W Mid 133 E8
Upper Gravenhurst C Beds 104 D2
Upper Green Essex 105 E8
Upper Green Mon 78 B5
Upper Green Suff 124 E4
Upper Green W Berks 63 G11
Upper Green W Yorks 197 B9
Upper Grove Common Hereford 97 F11
Upper Guist Norf 159 D10
Upper Hackney Derbys 170 C3
Upper Hale Sur 49 D10
Upper Halistra Highld 298 D2
Upper Halliford Sur 66 F5
Upper Halling Medway 69 G7
Upper Ham Worcs 99 D7
Upper Hambleton Rutland 137 B8
Upper Hamnish Hereford 115 F10
Upper Harbledown Kent 54 B6
Upper Hardres Court Kent 55 C7
Upper Hardwick Hereford 115 F8
Upper Hartfield E Sus 52 G3
Upper Hartshay Derbys 170 E5
Upper Haselor Worcs 99 C10
Upper Hatton Staffs 150 B6
Upper Haugh S Yorks 186 B6
Upper Hawkhillock Aberds 303 F10
Upper Hayesden Kent 52 E5
Upper Hayton Shrops 131 G10
Upper Heath Shrops 131 F11
Upper Heaton W Yorks 197 D7
Upper Hellesdon Norf 160 G4
Upper Helmsley N Yorks 207 B9
Upper Hengoed Shrops 148 C5
Upper Hergest Hereford 114 G5
Upper Heyford Northants 120 F3
Upper Heyford Oxon 101 F9
Upper Hill Glos 79 F11
Upper Hill Hereford 115 G9
Upper Hindhope Borders 251 B7
Upper Holloway London 67 B9
Upper Holton Suff 127 B8
Upper Hopton W Yorks 197 D7
Upper Horsebridge E Sus 23 C9
Upper Howsell Worcs 98 B5
Upper Hoyland S Yorks 197 G11
Upper Hulme Staffs 169 C8
Upper Hyde I o W 21 E7
Upper Ifold Sur 50 G4
Upper Inglesham Swindon 82 F2
Upper Inverbrough Highld 301 F8
Upper Kergord Shetland 313 H6
Upper Kidston Borders 270 G4
Upper Kilcott Glos 61 B9
Upper Killay Swansea 56 C5
Upper Killeyan Argyll 254 C3
Upper Kinsham Hereford 115 D7
Upper Knockando Moray 301 E11
Upper Lambourn W Berks 63 C10
Upper Landywood Staffs 133 B9
Upper Langford N Som 44 B3
Upper Langwith Derbys 171 B8
Upper Layham Suff 107 C10
Upper Leigh Staffs 151 B10
Upper Lenie Highld 300 G5
Upper Littleton N Som 60 G5
Upper Loads Derbys 170 B4
Upper Lochton Aberds 293 D8
Upper Lode Worcs 99 E7
Upper Longdon Staffs 151 G11
Upper Longwood Shrops 132 B2
Upper Ludstone Shrops 132 D6
Upper Lybster Highld 310 F6
Upper Lydbrook Glos 79 B10
Upper Lyde Hereford 97 C9
Upper Lye Hereford 115 D7
Upper Maes-coed Hereford 96 D6
Upper Marsh W Yorks 204 F6
Upper Midhope S Yorks 186 B2
Upper Midway Derbys 152 E5
Upper Milovaig Highld 297 G7
Upper Milton Oxon 82 B3
Upper Milton Som 44 D4
Upper Minety Wilts 81 G8
Upper Mitton Worcs 116 C6
Upper Moor Worcs 99 B9
Upper Moor Side W Yorks 205 G10
Upper Morton S Glos 79 G11
Upper Nash Pembs 73 E8
Upper Netchwood Shrops 132 E2
Upper Newbold Derbys 186 G5
Upper Nobut Staffs 151 B10
Upper North Dean Bucks 84 F4
Upper Norwood London 67 F10
Upper Norwood W Sus 34 D6
Upper Obney Perth 286 D4
Upper Ochrwyth Caerph 59 B8
Upper Oddington Glos 100 F4
Upper Ollach Highld 295 B7
Upper Padley Derbys 186 F2
Upper Pickwick Wilts 61 E11
Upper Pollicott Bucks 84 C2
Upper Poppleton York 207 C7
Upper Port Highld 301 G10
Upper Postern Kent 52 D6
Upper Quinton Warks 100 B3
Upper Race Torf 78 F3
Upper Ratley Hants 32 C4
Upper Ridinghill Aberds 303 D10
Upper Rissington Glos 82 B2
Upper Rochford Worcs 116 D2
Upper Rodmersham Kent 70 G2
Upper Sandaig Highld 295 D9
Upper Sanday Orkney 314 F5
Upper Sapey Hereford 116 E3
Upper Saxondale Notts 154 B3
Upper Seagry Wilts 62 C2
Upper Shelton C Beds 103 C9
Upper Sheringham Norf 177 E10
Upper Shirley London 67 G11
Upper Shirley Soton 32 E6
Upper Siddington Glos 81 F8
Upper Skelmorlie N Ayrs 266 B4
Upper Slackstead Hants 32 B5
Upper Slaughter Glos 100 G3
Upper Solva Pembs 90 G5
Upper Soudley Glos 79 C11
Upper Stanton Drew Bath 60 G6
Upper Staploe Beds 122 F2
Upper Stoke Norf 142 C5
Upper Stoke W Mid 135 G7
Upper Stondon C Beds 104 D2
Upper Stowe Northants 120 F2
Upper Stratton Swindon 63 B7
Upper Street Hants 31 D11
Upper Street Norf 142 G3
Upper Street Norf 160 B5
Upper Street Norf 160 E6
Upper Street Norf 160 F6
Upper Street Norf 161 F7
Upper Street Suff 108 E2
Upper Street Suff 124 G5
Upper Street Suff 126 G2
Upper Strensham Worcs 99 D8
Upper Studley Wilts 45 B10
Upper Sundon C Beds 103 F10
Upper Swainswick Bath 61 F9
Upper Swanmore Hants 33 D9
Upper Swell Glos 100 F3
Upper Sydenham London 67 E10
Upper Tankersley S Yorks 186 B4
Upper Tean Staffs 151 B10
Upper Threapwood W Ches 166 F6
Upper Thurnham Lancs 202 C5
Upper Tillyrie Perth 286 G5
Upper Tooting London 67 E9
Upper Tote Highld 298 D5
Upper Town Derbys 170 D3
Upper Town Derbys 170 E2
Upper Town Durham 233 D7
Upper Town Hereford 97 B11
Upper Town N Som 60 F4
Upper Town Suff 125 D8
Upper Town W Yorks 204 G6
Upper Town Wilts 62 D3
Upper Treverward Shrops 114 B5
Upper Tysoe Warks 100 C6
Upper Up Glos 81 F8
Upper Upham Wilts 63 D8
Upper Upnor Medway 69 E9
Upper Vobster Som 45 D8
Upper Walthamstow London 67 B11
Upper Wardington Oxon 101 B9
Upper Wardley W Sus 34 B4
Upper Weald M Keynes 102 D5
Upper Weedon Northants 120 F2
Upper Welland Worcs 98 C5
Upper Wellingham E Sus 36 E6
Upper Welson Hereford 114 G5
Upper Westholme Som 44 E5
Upper Weston Bath 61 F8
Upper Weybread Suff 126 B4
Upper Whiston S Yorks 187 D7
Upper Wick Glos 80 F2
Upper Wick Worcs 116 G6
Upper Wield Hants 48 F6
Upper Wigginton Shrops 148 B6
Upper Winchendon Bucks 84 C2
Upper Witton W Mid 133 E11
Upper Wolvercote Oxon 83 D7
Upper Wolverton Worcs 117 G8
Upper Woodend Aberds 293 B8
Upper Woodford Wilts 46 F6
Upper Woolhampton W Berks 64 F5
Upper Wootton Hants 48 C5
Upper Wraxall Wilts 61 E10
Upper Wyche Hereford 98 C5
Upperby Cumb 239 G10
Upperdale Derbys 185 G11
Uppermill Gtr Man 196 F3
Uppersound Shetland 313 J6
Upperthong W Yorks 196 F6
Upperthorpe Derbys 187 E7
Upperthorpe N Lincs 199 G9
Upperton E Sus 23 E10
Upperton Oxon 83 G11
Upperton W Sus 35 C7
Uppertown Derbys 170 C4
Uppertown Highld 300 F4
Uppertown Highld 310 B7
Uppertown Northumb 241 C9
Uppertown Orkney 314 G4
Upperwood Derbys 170 D3
Uppincott Devon 26 G5
Uppingham Rutland 137 D7
Uppington Dorset 31 F8
Uppington Shrops 132 B2
Upsall N Yorks 215 B9
Upsher Green Suff 107 C8
Upshire Essex 86 E6
Upstreet Kent 71 G8
Upthorpe Glos 80 E3
Upthorpe Suff 125 C9
Upton Bucks 84 C3
Upton Cambs 122 B3
Upton Corn 11 G11
Upton Corn 24 G2
Upton Cumb 230 D2
Upton Devon 8 G4
Upton Devon 27 G9
Upton Dorset 17 E10
Upton Dorset 18 C5
Upton E Yorks 209 C8
Upton Hants 32 D5
Upton Hants 47 B11
Upton I o W 21 C7
Upton Kent 71 F11
Upton Leics 135 D7
Upton Lincs 188 D5
Upton London 68 C2
Upton Mers 182 D3
Upton Mers 183 D7
Upton Norf 161 G7
Upton Northants 120 E4
Upton Notts 172 E2
Upton Notts 188 F2
Upton Oxon 64 B4
Upton Oxon 82 C2
Upton P'boro 138 C2
Upton Slough 66 D3
Upton Som 27 B7
Upton Som 29 B7
Upton W Ches 166 B6
Upton W Yorks 198 E3
Upton Warks 118 F2
Upton Wilts 45 G11
Upton Bishop Hereford 98 F2
Upton Cheyney S Glos 61 F7
Upton Cressett Shrops 132 E3
Upton Crews Hereford 98 F2
Upton Cross Corn 11 G11
Upton End C Beds 104 E2
Upton Field Notts 172 E2
Upton Green Norf 161 G7
Upton Grey Hants 49 D7
Upton Heath W Ches 166 B6
Upton Hellions Devon 26 G4
Upton Lea Bucks 66 C3
Upton Lovell Wilts 46 E2
Upton Magna Shrops 149 G11
Upton Noble Som 45 F8
Upton Park London 68 C2
Upton Pyne Devon 14 B4
Upton Rocks Halton 183 D8
Upton St Leonards Glos 80 C5
Upton Scudamore Wilts 45 D11
Upton Snodsbury Worcs 117 G8
Upton upon Severn Worcs 99 C7
Upton Warren Worcs 117 D8
Upwaltham W Sus 34 E6
Upware Cambs 123 C10
Upwell Norf 139 C9
Upwey Dorset 17 E9
Upwick Green Herts 105 G9
Upwood Cambs 138 G5
Uradale Shetland 313 K6
Urafirth Shetland 312 F5
Uragaig Argyll 274 D4
Urchfont Wilts 46 B4
Urdimarsh Hereford 97 B10
Ure Shetland 312 F4
Ure Bank N Yorks 214 E6
Urgashay Som 29 C9
Urgha W Isles 305 J3
Urgha Beag W Isles 305 H3
Urishay Common Hereford 96 D6
Urlar Perth 286 C2
Urlay Nook Stockton 225 C7
Urmston Gtr Man 184 C3
Urpeth Durham 242 G6
Urquhart Highld 300 D5
Urquhart Moray 302 C2
Urra N Yorks 225 E11
Urray Highld 300 D5
Ushaw Moor Durham 233 C10
Usk Mon 78 E5
Usselby Lincs 189 C9
Usworth T & W 243 F8
Utkinton W Ches 167 B8
Utley W Yorks 204 E6
Uton Devon 14 B2
Utterby Lincs 190 C4
Uttoxeter Staffs 151 C11
Uwchmynydd Gwyn 144 D3
Uxbridge London 66 C5
Uxbridge Moor London 66 C5
Uyea Shetland 312 D5
Uyeasound Shetland 312 C7
Uzmaston Pembs 73 C7

V

Vachelich Pembs 90 F5
Vadlure Shetland 313 J4
Vagg Som 29 D8
Vaila Hall Shetland 313 J4
Vaivoe Shetland 312 G7
Vale W Yorks 196 B2
Vale Down Devon 12 D6
Vale of Health London 67 B9
Valeswood Shrops 149 E7
Valley = Y Fali Anglesey 178 F3
Valley Park Hants 32 C6
Valley Truckle Corn 11 E7
Valleyfield Dumfries 237 D8
Valsgarth Shetland 312 B8
Valtos Highld 298 C5
Van Caerph 59 B7
Van Powys 129 F9
Vange Essex 69 B8
Vanlop Shetland 313 M5
Varchoel Powys 148 G4
Varfell Corn 2 C2
Varteg Torf 78 D3
Vassa Shetland 313 H6
Vastern Wilts 62 C5
Vatsetter Shetland 312 E7
Vatsetter Shetland 313 L5
Vatten Highld 298 E2
Vaul Argyll 288 E2
Vauxhall London 67 D10
Vauxhall Mers 182 C4
Vauxhall W Mid 133 F11
Vaynol Hall Gwyn 163 B8
Vaynor M Tydf 77 C8
Veensgarth Shetland 313 J6
Velator Devon 40 F3
Veldo Hereford 97 C11
Velindre Powys 96 D3
Vellanoweth Corn 2 C2
Vellow Som 42 F5
Velly Devon 24 C3
Veness Orkney 314 D5
Venn Devon 8 F4
Venn Green Devon 24 E5
Venn Ottery Devon 15 C7
Venngreen Devon 24 E5
Vennington Shrops 130 B6
Venn's Green Hereford 97 B10
Venny Tedburn Devon 14 B2
Venterdon Corn 12 G3
Vention Devon 40 E3
Ventnor I o W 21 F7
Venton Devon 7 D11
Ventongimps Corn 4 E5
Ventonleague Corn 2 B3
Venus Hill Herts 85 E8
Veraby Devon 26 B3
Vermentry Shetland 313 H5
Vernham Bank Hants 47 B10
Vernham Dean Hants 47 B10
Vernham Row Hants 47 B10
Vernham Street Hants 47 B11
Vernolds Common Shrops 131 G9
Verwood Dorset 31 F9
Veryan Corn 3 B10
Veryan Green Corn 5 G8
Vicarage Devon 15 D10
Vicarscross W Ches 166 B6
Vickerstown Cumb 210 F3
Victoria Corn 5 C9
Victoria S Yorks 197 F7
Victoria Dock Village Hull 200 B6
Victoria Park Bucks 84 C4
Victory Gardens Renfs 267 B10
Vidlin Shetland 312 G6
Viewpark N Lnrk 268 C4
Vigo W Mid 133 C10
Vigo Village Kent 68 G6
Vinegar Hill Mon 60 B2
Vinehall Street E Sus 38 C3
Vines Cross E Sus 23 B9
Viney Hill Glos 79 D11
Vinney Green S Glos 61 D7
Virginia Water Sur 66 F3
Virginstow Devon 12 C3
Viscar Corn 2 C6
Vobster Som 45 D8
Voe Shetland 312 E5
Voe Shetland 313 G6
Vogue Corn 4 G4
Vole Som 43 D11
Vowchurch Hereford 97 D7
Vowchurch Common Hereford 97 D7
Voxmoor Som 27 D10
Voxter Shetland 312 F5
Voy Orkney 314 E2
Vron Gate Shrops 130 B6
Vulcan Village Mers 183 C9

W

Waberthwaite Cumb 220 G2
Wackerfield Durham 233 G9
Wacton Hereford 116 F2
Wacton Norf 142 E3
Wacton Common Norf 142 F3
Wadbister Shetland 313 J6
Wadborough Worcs 99 B8
Wadbrook Devon 28 G4
Waddesdon Bucks 84 B2
Waddeton Devon 9 D7
Waddicar Mers 182 B5
Waddicombe Devon 26 B5
Waddingham Lincs 189 B7
Waddington Lancs 203 E10
Waddington Lincs 173 C7
Waddingworth Lincs 189 G11
Waddon Devon 14 F3
Waddon London 67 G10
Wade Hall Lancs 194 C4
Wadebridge Corn 10 G5
Wadeford Som 28 E4
Wadenhoe Northants 137 G10
Wades Green E Ches 167 C11
Wadesmill Herts 86 B5
Wadhurst E Sus 52 G6
Wadshelf Derbys 186 G4
Wadsley S Yorks 186 C4
Wadsley Bridge S Yorks 186 C4
Wadswick Wilts 61 F10
Wadwick Hants 48 C2
Wadworth S Yorks 187 B9
Waen Denb 165 B10
Waen Denb 165 C7
Waen Flint 181 G11
Waen Powys 129 E9
Waen Aberwheeler Denb 165 B9
Waen-fâch Powys 148 F4
Waen Goleugoed Denb 181 G9
Waen-pentir Gwyn 163 B9
Waen-wen Gwyn 163 B9
Wag Highld 311 G4
Wagbeach Shrops 131 C7
Wagg Som 28 B6
Waggersley Staffs 151 B7
Waggs Plot Devon 28 G4
Wain Lee Staffs 168 D5
Wainfelin Torf 78 E3
Wainfleet All Saints Lincs 175 D7
Wainfleet Bank Lincs 175 D7
Wainfleet St Mary Lincs 175 D8
Wainfleet Tofts Lincs 175 D7
Wainford Norf 142 E6
Waingroves Derbys 170 F6
Wainhouse Corner Corn 11 B9
Wainscott Medway 69 E8
Wainstalls W Yorks 196 B4
Waitby Cumb 222 D5
Waithe Lincs 201 G9
Wake Green W Mid 133 G11
Wake Hill N Yorks 214 E3
Wake Lady Green N Yorks 226 F3
Wakefield W Yorks 197 C10
Wakeley Herts 104 F6
Wakerley Northants 137 D9
Wakes Colne Essex 107 F7
Wakes Colne Green Essex 107 E7
Walberswick Suff 127 C9
Walberton W Sus 35 F7
Walbottle T & W 242 D5
Walby Cumb 239 E10
Walcombe Som 44 D5
Walcot Bath 61 F9
Walcot Lincs 155 B11
Walcot N Lincs 199 C11
Walcot Oxon 82 B4
Walcot Shrops 130 F6
Walcot Swindon 63 C7
Walcot Telford 149 G11
Walcot Worcs 99 B8
Walcot Green Norf 142 G2
Walcote Leics 135 G11
Walcote Warks 118 F2
Walcott Lincs 173 D10
Walcott Norf 161 C7
Walden N Yorks 213 C10
Walden Head N Yorks 213 C9
Walden Stubbs N Yorks 198 D5
Waldersey Cambs 139 C8
Waldershaigh S Yorks 186 B3
Walderslade Medway 69 G9
Walderton W Sus 34 E3
Walditch Dorset 16 C5
Waldley Derbys 152 B2
Waldridge Durham 243 G7
Waldringfield Suff 108 C5
Waldringfield Heath Suff 108 B5
Waldron E Sus 23 B8
Waldron Down E Sus 37 C8
Wales S Yorks 187 E7
Wales Som 29 C9
Wales Bar S Yorks 187 E7
Wales End Suff 106 B5
Walesby Lincs 189 C10
Walesby Notts 187 G11
Walesby Grange Lincs 189 C10
Waleswood S Yorks 187 E7
Walford Hereford 97 G11
Walford Hereford 115 C7
Walford Shrops 149 E8
Walford Som 28 B3
Walford Staffs 150 C6
Walford Heath Shrops 149 F8
Walgherton E Ches 167 F11
Walgrave Northants 120 C6
Walham Glos 98 G6
Walham Green London 67 D9
Walhampton Hants 20 B2
Walk Mill Lancs 204 G3
Walkden Gtr Man 195 G8
Walker T & W 243 E7
Walker Barn E Ches 185 G7
Walker Fold Lancs 203 E9
Walkerburn Borders 261 B9
Walkeringham Notts 188 C3
Walkerith Lincs 188 C3
Walkern Herts 104 F5
Walker's Green Hereford 97 B10
Walker's Heath W Mid 117 B11
Walkerville N Yorks 224 F4
Walkford Dorset 19 C10
Walkhampton Devon 7 B10
Walkington E Yorks 208 F5
Walkley S Yorks 186 D4
Walkmill Shrops 131 E7
Walkmills Shrops 131 D9
Wall Corn 2 B4
Wall Northumb 241 D10
Wall Staffs 134 B2
Wall Bank Shrops 131 E10
Wall End Cumb 210 C4
Wall End Kent 71 G8
Wall Heath W Mid 133 F7
Wall Hill Gtr Man 196 F3
Wall Mead Bath 45 B7
Wall Nook Durham 233 B10
Wall under Heywood Shrops 131 E10
Wallacestone Falk 279 F8
Wallaceton Dumfries 247 F8
Wallacetown S Ayrs 245 C7
Wallacetown S Ayrs 257 E8
Wallacetown Shetland 313 H5
Wallands Park E Sus 36 E6
Wallasey Mers 182 C4
Wallbank Lancs 195 D11
Wallbrook W Mid 133 E8
Wallcrouch E Sus 53 G7
Wallend London 68 C2
Waller's Green Hereford 98 D3
Walley's Green E Ches 167 C11
Wallingford Oxon 64 B6
Wallington Hants 33 F9
Wallington Herts 104 E5
Wallington London 67 G9
Wallington Heath W Mid 133 C9
Wallingwells Notts 187 E9
Wallis Pembs 91 F10
Wallisdown Poole 19 C7
Walliswood Sur 50 F6
Wallow Green Glos 80 F4
Wallridge Northumb 242 B3
Walls Shetland 313 J4
Wallsend T & W 243 D7
Wallston V Glam 58 E6
Wallsuches Gtr Man 195 E7
Wallsworth Glos 98 G6
Wallyford E Loth 281 G7
Walmer Kent 55 C11
Walmer Bridge Lancs 194 C3
Walmersley Gtr Man 195 E10
Walmgate Stray York 207 C8
Walmley W Mid 134 E2
Walmsgate Lincs 190 F5
Walnut Grove Perth 286 E5
Walnut Tree M Keynes 103 D7
Walnuttree Green Herts 105 G9
Walpole Som 43 E10
Walpole Suff 127 C7
Walpole Cross Keys Norf 157 F10
Walpole Highway Norf 157 G10
Walpole Marsh Norf 157 F9
Walpole St Andrew Norf 157 F10
Walpole St Peter Norf 157 F10
Walrow Som 43 D10
Walsal End W Mid 118 B4
Walsall W Mid 133 D10
Walsall Wood W Mid 133 C10
Walsden W Yorks 196 C2
Walsgrave on Sowe W Mid 135 G7
Walsham le Willows Suff 125 C9
Walshaw Gtr Man 195 E9
Walshford N Yorks 206 C4
Walsoken Cambs 157 G9
Walson Mon 97 G8
Walston S Lnrk 269 F11
Walsworth Herts 104 E4
Walters Ash Bucks 84 F4
Walter's Green Kent 52 E4
Walterston V Glam 58 E5
Walterstone Hereford 96 F6
Waltham Kent 54 D6
Waltham NE Lincs 201 G9
Waltham Abbey Essex 86 E5
Waltham Chase Hants 33 D9
Waltham Cross Herts 86 E5
Waltham on the Wolds Leics 154 D6
Waltham St Lawrence Windsor 65 D10
Waltham's Cross Essex 106 E3
Walthamstow London 67 B11
Walton Bucks 84 C4
Walton Cumb 240 E2
Walton Derbys 170 B5
Walton Leics 135 F11
Walton M Keynes 103 D7
Walton Mers 182 C5
Walton P'boro 138 C3
Walton Powys 114 F5
Walton Shrops 115 B9
Walton Som 44 F3
Walton Staffs 151 C7
Walton Staffs 151 D7
Walton Suff 108 D5
Walton Telford 149 F11
Walton W Yorks 197 D11
Walton W Yorks 206 D4
Walton Warks 118 G5
Walton Cardiff Glos 99 E8
Walton Court Bucks 84 C4
Walton East Pembs 91 G10
Walton Elm Dorset 30 D3
Walton Grounds Northants 101 E10
Walton Heath Hants 33 F10
Walton Highway Norf 157 G9
Walton in Gordano N Som 60 E2
Walton-le-Dale Lancs 194 B5
Walton Manor Oxon 83 D8
Walton-on-Thames Sur 66 F6
Walton on the Hill Staffs 151 E9
Walton on the Hill Sur 51 B8
Walton-on-the-Naze Essex 108 G5
Walton on the Wolds Leics 153 F11
Walton-on-Trent Derbys 152 F4
Walton Pool Worcs 117 B8
Walton St Mary N Som 60 E2
Walton Summit Lancs 194 B5
Walton Warren Norf 158 F4
Walton West Pembs 72 C5
Walwen Flint 181 F10
Walwen Flint 181 G11
Walwen Flint 182 F2
Walwick Northumb 241 C10
Walworth Darl 224 B4
Walworth London 67 D10
Walworth Gate Darl 233 G10
Walwyn's Castle Pembs 72 C5
Wambrook Som 28 F3
Wampool Cumb 238 G6
Wanborough Sur 50 D2
Wanborough Swindon 63 C8
Wandel Dyke S Lnrk 259 D11
Wandle Park London 67 F10
Wandon End Herts 104 G2
Wandsworth London 67 E9
Wangford Suff 127 B9
Wanlip Leics 154 G2
Wanlockhead Dumfries 259 G9
Wannock E Sus 23 E9
Wansford E Yorks 209 B7
Wansford P'boro 137 D11
Wanshurst Green Kent 53 D9
Wanson Corn 24 G1
Wanstead London 68 B2
Wanstrow Som 45 E8
Wanswell Glos 79 E11
Wantage Oxon 63 B11
Wants Green Worcs 116 F5
Wapley S Glos 61 D8
Wappenbury Warks 119 D7
Wappenham Northants 102 B2
Wapping London 67 C10
Warbleton E Sus 23 B10
Warblington Hants 22 B2
Warborough Oxon 83 G9
Warboys Cambs 138 G6
Warbreck Blkpool 202 F2
Warbstow Corn 11 C10
Warbstow Cross Corn 11 C10
Warburton Gtr Man 184 D2
Warburton Green Gtr Man 184 E3
Warcop Cumb 222 B4
Ward End W Mid 134 F2
Ward Green S Yorks 197 G10
Ward Green Suff 125 E10
Ward Green Cross Lancs 203 F8
Warden Kent 70 E4
Warden Northumb 241 D10
Warden Powys 114 E6
Warden Hill Glos 99 G8
Warden Point I o W 20 D2
Warden Street C Beds 104 C2
Wardhedges C Beds 103 D11
Wardhill Orkney 314 D6
Wardington Oxon 101 B9
Wardlaw Borders 261 F7
Wardle E Ches 167 D10

Wardle Gtr Man 196 D2
Wardle Bank E Ches 167 D10
Wardley Gtr Man 195 G9
Wardley Rutland 136 C6
Wardley T & W 243 E7
Wardley W Sus 34 B4
Wardlow Derbys 185 G11
Wardour Wilts 30 B6
Wardpark N Lnrk 278 F5
Wardrobes Bucks 84 E4
Wardsend E Ches 184 E6
Wardy Hill Cambs 139 G9
Ware Herts 86 C5
Ware Kent 71 G9
Ware Street Kent 53 B9
Wareham Dorset 18 D4
Warehorne Kent 54 G3
Waren Mill Northumb 264 C4
Warenford Northumb 264 D4
Warenton Northumb 264 C4
Wareside Herts 86 B5
Waresley Cambs 122 G4
Waresley Worcs 116 C6
Warfield Brack 65 E11
Warfleet Devon 9 E7
Wargate Lincs 156 C4
Wargrave Mers 183 C9
Wargrave Wokingham 65 D9
Warham Hereford 97 D9
Warham Norf 176 E6
Warhill Gtr Man 185 B7
Waring's Green W Mid 118 C2
Wark Northumb 241 B9
Wark Northumb 263 B8
Wark Common Northumb 263 B8
Warkleigh Devon 25 C10
Warkton Northants 121 B7
Warkworth Northants 101 C9
Warkworth Northumb 252 B6
Warlaby N Yorks 224 G6
Warland W Yorks 196 C2
Warleggan Corn 6 B3
Warleigh Bath 61 G9
Warley Essex 87 G9
Warley Town W Yorks 196 B5
Warley Woods W Mid 133 F10
Warlingham Sur 51 B11
Warmbrook Derbys 170 E3
Warmfield W Yorks 197 C11
Warmingham E Ches 168 C2
Warminghurst W Sus 35 D10
Warmington Northants 137 E11
Warmington Warks 101 B8
Warminster Wilts 45 D11
Warminster Common Wilts 45 E11
Warmlake Kent 53 C10
Warmley S Glos 61 E7
Warmley Hill S Glos 61 E7
Warmley Tower S Glos 61 E7
Warmonds Hill Northants 121 D9
Warmsworth S Yorks 198 G4
Warmwell Dorset 17 D11
Warnborough Green Hants 49 C8
Warndon Worcs 117 F7
Warners End Herts 85 D8
Warnford Hants 33 C10
Warnham W Sus 51 G7
Warningcamp W Sus 35 F8
Warninglid W Sus 36 B2
Warpsgrove Oxon 83 F10
Warren Dorset 18 C3
Warren E Ches 184 G5
Warren Pembs 72 F6
Warren S Yorks 186 B5
Warren Corner Hants 34 B2
Warren Corner Hants 49 D10
Warren Heath Suff 108 C4
Warren Row Windsor 65 C10
Warren Street Kent 54 C2
Warrenby Redcar 235 F7
Warren's Green Herts 104 F5
Warrington M Keynes 121 G7
Warrington Warr 183 D10
Warriston Edin 280 F5
Warsash Hants 33 F7
Warsill N Yorks 214 F4
Warslow Staffs 169 D9
Warsop Vale Notts 171 B8
Warstock W Mid 117 B11
Warstone Staffs 133 B9
Warter E Yorks 208 C3
Warthermarske N Yorks 214 D4
Warthill N Yorks 207 B9
Wartle Aberds 293 C7
Wartling E Sus 23 D11
Wartnaby Leics 154 E4
Warton Lancs 194 B2
Warton Lancs 211 E9
Warton Northumb 252 C2
Warton Warks 134 C5
Warton Bank Lancs 194 B2
Warwick Warks 118 E5
Warwick Bridge Cumb 239 F11
Warwick on Eden Cumb 239 F11
Warwick Wold Sur 51 C10
Warwicksland Cumb 239 B10
Wasbister Orkney 314 C3
Wasdale Head Cumb 220 D3
Wash Derbys 185 E9
Wash Common W Berks 64 G3
Wash Dyke Norf 157 F10
Wash Water W Berks 64 G3
Washall Green Herts 105 E8
Washaway Corn 5 B10
Washbourne Devon 8 E5
Washbrook Som 44 C2
Washbrook Suff 108 C2
Washbrook Street Suff 108 C2
Washerwall Staffs 168 F6
Washfield Devon 26 D6
Washfold N Yorks 223 E11
Washford Som 42 E5
Washford Worcs 117 D11
Washford Pyne Devon 26 E4
Washingborough Lincs 189 G8
Washingley Cambs 138 F2
Washington T & W 243 F8
Washington W Sus 35 E10
Washington Village T & W 243 F8
Washmere Green Suff 107 B8
Washpit W Yorks 196 F6
Washwood Heath W Mid 134 F2
Wasing W Berks 64 G5
Waskerley Durham 233 B7
Wasp Green Sur 51 D10
Wasperton Warks 118 F5
Wasps Nest Lincs 173 C9
Wass N Yorks 215 D11
Waste Green Warks 118 D4
Wastor Devon 8 F2
Watch House Green Essex 106 G3
Watchet Som 42 E5
Watchfield Oxon 63 B8
Watchfield Som 43 D10
Watchgate Cumb 221 F10
Watchhill Cumb 229 C9
Watchill Dumfries 238 D6
Watchill Dumfries 248 G3
Watcombe Torbay 9 B8
Watendlath Cumb 220 B5
Water Devon 13 E11
Water Lancs 195 B10
Water Eaton M Keynes 103 E7
Water Eaton Oxon 83 C8
Water End Beds 104 B2
Water End C Beds 103 D11
Water End C Beds 104 B5
Water End E Yorks 207 F11
Water End Essex 105 C11
Water End Hants 49 C7
Water End Herts 85 C8
Water End Herts 86 E2
Water Fryston W Yorks 198 B3
Water Garth Nook Cumb 210 F3
Water Houses N Yorks 213 F7
Water Newton Cambs 138 D2
Water Orton Warks 134 E3
Water Stratford Bucks 102 E3
Water Yeat Cumb 210 B5
Waterbeach Cambs 123 D9
Waterbeach W Sus 22 B5
Waterbeck Dumfries 238 B6
Waterdale Herts 85 E10
Waterden Norf 159 B7
Waterditch Hants 19 B9
Waterend Bucks 84 F3
Waterend Cumb 229 G8
Waterend Glos 80 C3
Waterend Herts 86 C2
Waterfall Staffs 169 E9
Waterfoot Argyll 255 D9
Waterfoot Cumb 230 G5
Waterfoot E Renf 267 D11
Waterfoot Lancs 195 C10
Waterford Hants 20 B2
Waterford Herts 86 C4
Watergate Corn 6 E4
Watergate Corn 11 E8
Watergore Som 28 D6
Waterhales Essex 87 F8
Waterham Kent 70 G5
Waterhay Wilts 81 G9
Waterhead Angus 292 F6
Waterhead Cumb 221 E7
Waterhead Devon 8 F3
Waterhead Dumfries 248 E5
Waterhead on Minnoch S Ayrs 245 E9
Waterheads Borders 270 E4
Waterheath Norf 143 E8
Waterhouses Durham 233 C9
Waterhouses Staffs 169 E9
Wateringbury Kent 53 C7
Waterlane Glos 80 E6
Waterlip Som 45 E7
Waterloo Blkburn 195 B7
Waterloo Corn 11 G8
Waterloo Derbys 170 C6
Waterloo Gtr Man 196 G2
Waterloo Highld 295 C8
Waterloo Mers 182 B4
Waterloo N Lnrk 268 E6
Waterloo Norf 126 B2
Waterloo Norf 143 E8
Waterloo Norf 160 F4
Waterloo Pembs 73 E7
Waterloo Perth 286 D4
Waterloo Poole 18 C6
Waterloo Shrops 149 C9
Waterloo Park Mers 182 B4
Waterloo Port Gwyn 163 C7
Waterlooville Hants 33 F11
Waterman Quarter Kent 53 E10
Watermead Glos 80 B5
Watermeetings S Lnrk 259 G11
Watermill E Sus 38 E2
Watermillock Cumb 230 G4
Watermoor Glos 81 E8
Waterperry Oxon 83 D10
Waterrow Som 27 B9
Water's Nook Gtr Man 195 F7
Waters Upton Telford 150 F2
Watersfield W Sus 35 D8
Watersheddings Gtr Man 196 F2
Waterside Aberds 292 B5
Waterside Aberds 303 G10
Waterside Blkburn 195 C8
Waterside Bucks 85 E7
Waterside Cumb 229 B10
Waterside Derbys 185 E8
Waterside E Ayrs 245 B10
Waterside E Ayrs 267 G9
Waterside E Dunb 278 G3
Waterside E Renf 267 D10
Waterside S Yorks 199 E7
Waterside Sur 51 D11
Waterside Telford 150 F2
Waterslack Lancs 211 D9
Waterstein Highld 297 G7
Waterstock Oxon 83 D10
Waterston Pembs 72 D6
Waterthorpe S Yorks 186 E6
Waterton Aberds 303 F9
Waterton Bridgend 58 D2
Watford Herts 85 F10
Watford Northants 120 D2
Watford Gap Staffs 134 C2
Watford Heath Herts 85 G10
Watford Park Caerph 58 B6
Wath Cumb 222 D3
Wath N Yorks 214 D6
Wath N Yorks 214 F2
Wath N Yorks 216 D3
Wath Brow Cumb 219 C10
Wath upon Dearne S Yorks 198 G2
Watherston Borders 271 F8
Watledge Glos 80 E4
Watley's End S Glos 61 C7
Watlington Norf 158 G2
Watlington Oxon 83 G11
Watnall Notts 171 F8
Watsness Shetland 313 H3
Watten Highld 310 D6
Wattisfield Suff 125 C10
Wattisham Suff 125 G10
Wattisham Stone Suff 125 G10
Wattlefield Norf 142 D2
Wattlesborough Heath Shrops 149 G7
Watton E Yorks 208 C6
Watton Norf 141 C8
Watton at Stone Herts 86 B4
Watton Green Norf 141 C8
Watton's Green Essex 87 F8
Wattston N Lnrk 268 B5
Wattstown Rhondda 77 G8
Wattsville Caerph 78 G2
Wauchan Highld 295 G11
Waulkmill Lodge Orkney 314 F3
Waun Gwyn 163 C9
Waun Powys 143 F4
Waun Beddau Pembs 90 F5
Waun Fawr Ceredig 128 G2
Waun-Lwyd Bl Gwent 77 D11
Waun-y-clyn Carms 75 E7
Waun y Gilfach Bridgend 57 D10
Waunarlwydd Swansea 56 B6
Waunclunda Carms 94 E3
Waunfawr Gwyn 163 D8
Waungilwen Carms 92 D6
Waungron Swansea 75 E9
Waunlwyd Bl Gwent 77 D11
Wavendon M Keynes 103 D8
Wavendon Gate M Keynes 103 D8
Waverbridge Cumb 229 B10
Waverton Cumb 229 B10
Waverton W Ches 167 C7
Wavertree Mers 182 D5
Wawcott W Berks 63 F11
Wawne E Yorks 209 F7
Waxham Norf 161 D8
Waxholme E Yorks 201 B10
Way Kent 71 F10
Way Village Devon 26 E5
Way Wick N Som 59 G11
Waye Devon 13 G11
Wayend Street Hereford 98 D4
Wayfield Medway 69 F9
Wayford Som 28 F6
Waymills Shrops 167 G9
Wayne Green Mon 78 B6
Way's Green W Ches 167 B10
Waytown Devon 24 C5
Waytown Devon 40 G5
Wdig = Goodwick Pembs 91 D8
Weachyburn Aberds 302 D6
Weacombe Som 42 E6
Weald Oxon 82 E4
Wealdstone London 67 B7
Wearde Corn 7 D8
Weardley W Yorks 205 E11
Weare Som 44 C2
Weare Giffard Devon 25 C7
Wearhead Durham 232 D3
Wearne Som 28 B6
Weasdale Cumb 222 E3
Weasenham All Saints Norf 158 E6
Weasenham St Peter Norf 159 E7
Weaste Gtr Man 184 B4
Weatherhill Sur 51 E10
Weatheroak Hill Worcs 117 C11
Weaverham W Ches 183 G10
Weavering Street Kent 53 B9
Weaverslake Staffs 152 F2
Weaverthorpe N Yorks 217 E9
Webbington Som 43 B11
Webheath Worcs 117 D10
Webscott Shrops 149 E9
Wecock Hants 33 E11
Wedderlairs Aberds 303 F8
Wedderlie Borders 272 E2
Weddington Kent 55 B9
Weddington Warks 135 E7
Wedhampton Wilts 46 B5
Wedmore Som 44 D2
Wednesbury W Mid 133 D9
Wednesbury Oak W Mid 133 E9
Wednesfield W Mid 133 C8
Weecar Notts 172 B4
Weedon Bucks 84 B4
Weedon Bec Northants 120 F2
Weedon Lois Northants 102 B2
Weeford Staffs 134 C2
Week Devon 8 C5
Week Devon 12 E5
Week Devon 25 B9
Week Devon 26 D2
Week Green Corn 11 B10
Week St Mary Corn 11 B10
Weeke Devon 26 E3
Weeke Hants 48 G3
Weekley Northants 137 G7
Weekmoor Som 27 B10
Weeks I o W 21 C7
Weel E Yorks 209 E7
Weeley Essex 108 C2
Weeley Heath Essex 108 C3
Weelsby NE Lincs 201 F9
Weem Perth 286 C2
Weeping Cross Staffs 151 E8
Weethley Warks 117 F11
Weethley Bank Warks 117 G11
Weethley Gate Warks 117 G11
Weeting Norf 140 F5
Weeton E Yorks 201 C11
Weeton Lancs 202 G3
Weeton N Yorks 205 D11
Weetwood W Yorks 205 F11
Weetwood Common W Ches 167 B8
Weetwood Hall Northumb 264 E2
Weir Essex 69 B10
Weir Lancs 195 B11
Weir Quay Devon 7 B8
Weirbrook Shrops 148 E6
Welborne Norf 159 G11
Welborne Common Norf 141 B11
Welbourn Lincs 173 E7
Welburn N Yorks 216 C3
Welburn N Yorks 216 F4
Welbury N Yorks 225 E7
Welby Lincs 155 B9
Welches Dam Cambs 139 F9
Welcombe Devon 24 D2
Weld Bank Lancs 194 D5
Weldon Northants 137 F8
Weldon Northumb 252 D4
Welford Northants 136 G2
Welford W Berks 64 E2
Welford-on-Avon Warks 118 G3
Welham Leics 136 E5
Welham Notts 188 E2
Welham Som 45 G7
Welham Green Herts 86 D2
Welhambridge E Yorks 207 G11
Well Hants 49 D9
Well Lincs 190 G6
Well N Yorks 214 C5
Well Bottom Dorset 30 D6
Well End Bucks 65 B11
Well End Herts 86 F2
Well Green Gtr Man 184 D3
Well Heads W Yorks 205 G7
Well Hill Kent 68 G3
Well Place Oxon 65 B7
Well Street Kent 53 B7
Well Town Devon 26 F5
Welland Worcs 98 C5
Welland Stone Worcs 98 D5
Wellbank Angus 287 D8
Wellbrook E Sus 37 B9
Welldale Dumfries 238 D5
Weller's Town Kent 52 E4
Wellesbourne Warks 118 F5
Wellheads Aberds 302 F4
Wellhouse W Berks 64 E4
Wellhouse W Yorks 196 E5
Welling London 68 D3
Wellingborough Northants 121 D7
Wellingham Norf 159 E7
Wellingore Lincs 173 D7
Wellington Cumb 219 E11
Wellington Hereford 97 B9
Wellington Som 27 C10
Wellington Telford 150 G3
Wellington Heath Hereford 98 C4
Wellington Hill W Yorks 206 F2
Wellisford Som 27 C9
Wellow Bath 45 B8
Wellow I o W 20 D3
Wellow NE Lincs 201 F9
Wellow Notts 171 B11
Wellow Wood Hants 32 C3
Wellpond Green Herts 105 G8
Wellroyd W Yorks 205 F10
Wells Som 44 D5
Wells Green E Ches 167 E11
Wells-Next-The-Sea Norf 176 E6
Wellsborough Leics 135 C7
Wellsprings Som 28 B2
Wellstye Green Essex 87 B10
Wellswood Torbay 9 C8
Welltown Corn 6 B2
Wellwood Fife 279 D11
Welney Norf 139 E10
Welsford Devon 24 C3
Welsh Bicknor Hereford 79 B9
Welsh End Shrops 149 B10
Welsh Frankton Shrops 149 C7
Welsh Harp London 67 B8
Welsh Hook Pembs 91 F8
Welsh Newton Hereford 79 B7
Welsh Newton Common Hereford 79 B8
Welsh St Donats V Glam 58 D4
Welshampton Shrops 149 B8
Welshpool Powys 130 B4
Welshwood Park Essex 107 F10
Welstor Devon 13 G10
Welton Bath 45 C7
Welton Cumb 230 C3
Welton E Ayrs 258 D2
Welton E Yorks 200 B3
Welton Lincs 189 F8
Welton Northants 119 D11
Welton Hill Lincs 189 E8
Welton le Marsh Lincs 175 B7
Welton le Wold Lincs 190 D3
Welwick E Yorks 201 C10
Welwyn Herts 86 B2
Welwyn Garden City Herts 86 C2
Wem Shrops 149 D10
Wembdon Som 43 F9
Wembley London 67 B7
Wembley Park London 67 B7
Wembury Devon 7 F10
Wembworthy Devon 25 F11
Wemyss Bay Invclyd 266 B3
Wenallt Ceredig 112 C3
Wenallt Gwyn 146 F4
Wenallt Gwyn 165 G7
Wendens Ambo Essex 105 D10
Wendlebury Oxon 83 B9
Wendling Norf 159 G8
Wendover Bucks 84 D5
Wendover Dean Bucks 84 E5
Wendron Corn 2 C5
Wendy Cambs 104 B6
Wenfordbridge Corn 11 F7
Wenhaston Suff 127 B8
Wenhaston Black Heath Suff 127 C8
Wennington Cambs 122 B4
Wennington Lancs 212 E2
Wennington London 68 C4
Wensley Derbys 170 C3
Wensley N Yorks 213 B11
Wentbridge W Yorks 198 D3
Wentnor Shrops 131 E7
Wentworth Cambs 123 B9
Wentworth S Yorks 186 B5
Wenvoe V Glam 58 E6
Weobley Hereford 115 G8
Weobley Marsh Hereford 115 G8
Weoley Castle W Mid 133 G10
Wepham W Sus 35 F8
Wepre Flint 166 B3
Wereham Norf 140 C3
Wereham Row Norf 140 C3
Wereton Staffs 168 E3
Wergs W Mid 133 C7
Wern Gwyn 145 B10
Wern Powys 77 B10
Wern Powys 147 G9
Wern Powys 148 E5
Wern Powys 148 G5
Wern Shrops 148 C5
Wern Swansea 56 C4
Wern ddu Shrops 148 D4
Wern-Gifford Mon 96 G6
Wern-olau Swansea 56 B5
Wern Tarw Bridgend 58 C3
Wern-y-cwrt Mon 78 D5
Wern-y-gaer Flint 166 B2
Werneth Gtr Man 196 G2
Werneth Low Gtr Man 185 C7
Wernffrwd Swansea 56 C4
Wernlas Shrops 148 E6
Wernrheolydd Mon 78 C5
Wernyrheolydd Mon 78 C5
Werrington Corn 12 D2
Werrington P'boro 138 C3
Werrington Staffs 168 F6
Wervin W Ches 182 G6
Wescoe Hill N Yorks 205 D11
Wesham Lancs 202 G4
Wessington Derbys 170 D5
West Aberthaw V Glam 58 F4
West Acre Norf 158 F5
West Acton London 67 C7
West Adderbury Oxon 101 D9
West Allerdean Northumb 273 F9
West Allotment T & W 243 C8
West Alvington Devon 8 G4
West Amesbury Wilts 46 E6
West Anstey Devon 26 B5
West Appleton N Yorks 224 G4
West Ardhu Argyll 288 D6
West Ardsley W Yorks 197 B9
West Ardwell Dumfries 236 E2
West Arthurlie E Renf 267 D9
West Ashby Lincs 190 G3
West Ashford Devon 40 F4
West Ashling W Sus 22 B4
West Ashton Wilts 45 B11
West Auckland Durham 233 F9
West Ayton N Yorks 217 C9
West Bagborough Som 43 G7
West Bank Bl Gwent 78 D2
West Bank Halton 183 E8
West Barkwith Lincs 189 E11
West Barnby N Yorks 226 C6
West Barnes London 67 F8
West Barns E Loth 282 F3
West Barsham Norf 159 C8
West Bay Dorset 16 C5
West Beckham Norf 160 B2
West Bedfont Sur 66 E5
West Benhar N Lnrk 269 C7
West Bergholt Essex 107 F9
West Bexington Dorset 16 D6
West Bilney Norf 158 F4
West Blackdene Durham 232 D3
West Blackdown Devon 12 E5
West Blatchington Brighton 36 F3
West Bold Borders 261 B9
West Boldon T & W 243 E9
West Bourton Dorset 30 B3
West Bowling W Yorks 205 G9
West Bradford Lancs 203 E10
West Bradley Som 44 F5
West Bretton W Yorks 197 E9
West Bridgford Notts 153 B11
West Brompton London 67 D9
West Bromwich W Mid 133 E10
West Broughton Derbys 152 C2
West Buckland Devon 41 G7
West Buckland Som 27 C11
West Burnside Aberds 293 F8
West Burrafirth Shetland 313 H4
West Burton N Yorks 213 B10
West Burton W Sus 35 E7
West Butsfield Durham 233 C8
West Butterwick N Lincs 199 F10
West Byfleet Sur 66 G4
West Caister Norf 161 G10
West Calder W Loth 269 C10
West Camel Som 29 C9
West Carlton W Yorks 205 E10
West Carr Hull 209 G7
West Carr N Lincs 199 F8
West Chadsmoor Staffs 151 G9
West Challow Oxon 63 B11
West Charleton Devon 8 G5
West Chelborough Dorset 29 F8
West Chevington Northumb 252 D6
West Chiltington W Sus 35 D9
West Chiltington Common W Sus 35 D9
West Chinnock Som 29 E7
West Chirton T & W 243 D8
West Chisenbury Wilts 46 C6
West Clandon Sur 50 C4
West Cliff Bmouth 19 C7
West Cliff N Yorks 227 C7
West Cliffe Kent 55 E10
West Clyne Highld 311 J2
West Clyth Highld 310 F6
West Coker Som 29 E8
West Common Hants 32 G6
West Compton Dorset 17 C7
West Compton Som 44 E5
West Cornforth Durham 234 E2
West Cowick E Yorks 199 C7
West Cranmore Som 45 E7
West Cross Kent 53 G10
West Cross Swansea 56 D6
West Crudwell Wilts 80 G6
West Clyne Highld 311 J2
West Clyth Highld 310 F6
West Cullery Aberds 293 C9
West Curry Corn 11 C11
West Curthwaite Cumb 230 B2
West Darlochan Argyll 255 E7
West Dean W Sus 34 E5
West Dean Wilts 32 B3
West Deeping Lincs 138 B2
West Denant Pembs 72 C6
West Denton T & W 242 D5
West Derby Mers 182 C5
West Dereham Norf 140 C3
West Didsbury Gtr Man 184 C4
West Down Devon 40 E4
West Down Hants 47 F11
West Downs Corn 5 C10
West Drayton London 66 D5
West Drayton Notts 188 G2
West Dulwich London 67 E10
West Ealing London 67 C7
West Edge Derbys 170 C4
West Ella E Yorks 200 B4
West End Beds 121 E11
West End Beds 121 G9
West End Brack 65 E11
West End Caerph 78 F2
West End Cumb 239 F8
West End Dorset 30 G6
West End E Yorks 201 B9
West End E Yorks 208 G4
West End E Yorks 209 B9
West End E Yorks 209 G9
West End E Yorks 217 G11
West End Glos 80 F5
West End Hants 33 E7
West End Hants 33 F9
West End Hants 48 F6
West End Herts 86 D3
West End Kent 54 B2
West End Kent 71 F7
West End Lancs 195 B8
West End Lancs 211 G8
West End Leics 153 F8
West End Lincs 174 F5
West End Lincs 190 B5
West End Mon 78 F6
West End N Som 60 F3
West End N Yorks 205 B8
West End N Yorks 206 E6
West End N Yorks 207 F7
West End Norf 141 B8
West End Norf 161 G10
West End Oxon 64 B5
West End Oxon 82 E6
West End S Glos 61 B8
West End S Lnrk 269 F9
West End S Yorks 199 F7
West End Som 44 C5
West End Som 45 G7
West End Suff 143 G9
West End Sur 49 E10
West End Sur 66 G2
West End Sur 66 G6
West End W Sus 36 D2
West End W Yorks 197 B7
West End W Yorks 205 F10
West End Wilts 30 C6
West End Wilts 31 C7
West End Wilts 62 D3
West End Windsor 65 D10
West End Worcs 99 D11
West End = Marian-y-mor Gwyn 145 C7
West End Green Hants 65 G7
West-end Town V Glam 58 F3
West Ewell Sur 67 G8
West Farleigh Kent 53 C8
West Farndon Northants 119 G10
West Felton Shrops 148 D6
West Fenton E Loth 281 E9
West Ferry Dundee 287 D8
West Field N Lincs 200 D6
West Field York 207 C7
West Fields W Berks 64 F3
West Firle E Sus 23 D7
West Fleetham Northumb 264 D5
West Flodden Northumb 263 C10
West Garforth W Yorks 206 G3
West Ginge Oxon 64 B2
West Gorton Gtr Man 184 B5
West Grafton Wilts 63 G8
West Green Hants 49 B8
West Green London 67 B10
West Green S Yorks 197 F11
West Green W Sus 51 F9
West Greenskares Aberds 303 C7
West Grimstead Wilts 32 B2
West Grinstead W Sus 35 C11
West Haddlesey N Yorks 198 B5
West Haddon Northants 120 C2
West Hagbourne Oxon 64 B4
West Hagley Worcs 133 G8
West Hall Cumb 240 D3
West Hallam Derbys 170 G6
West Halton N Lincs 200 C2
West Ham Hants 48 C6
West Ham London 68 C2
West Hampstead London 67 B9
West Handley Derbys 186 F5
West Hanney Oxon 82 G6
West Hanningfield Essex 88 F2
West Hardwick W Yorks 198 D2
West Harling Norf 141 G9
West Harlsey N Yorks 225 F8
West Harnham Wilts 31 B10
West Harptree Bath 44 B5
West Harrow London 66 B6
West Harting W Sus 34 C3
West Harton T & W 243 E9
West Hatch Som 28 C3
West Hatch Wilts 30 B6
West Head Norf 139 B11
West Heath E Ches 168 C4
West Heath Hants 48 B5
West Heath Hants 49 B11
West Heath London 68 D3
West Heath W Mid 117 B10
West Helmsdale Highld 311 H4
West Hendon London 67 B8
West Hendred Oxon 64 B2
West Herrington T & W 243 G8
West Heslerton N Yorks 217 D8
West Hewish N Som 59 G11
West Hill Devon 15 C7
West Hill E Sus 38 E4
West Hill E Yorks 218 F3
West Hill London 67 E8
West Hill N Som 60 D3
West Hill Som 30 B2
West Hill Staffs 151 G9
West Hill W Sus 51 G10
West Hill Wilts 61 F11
West Hoathly W Sus 51 G11
West Holme Dorset 18 D3
West Holywell T & W 243 C8
West Horndon Essex 68 B6
West Horrington Som 44 D5
West Horsley Sur 50 C5
West Horton Northumb 264 C2
West Hougham Kent 55 E9
West Houlland Shetland 313 H4
West Houses Lincs 174 E4
West Howe Bmouth 19 B7
West Howetown Som 42 G2
West Huntington York 207 B8
West Huntspill Som 43 E10
West Hurn Dorset 19 B8
West Hyde Herts 85 G8
West Hynish Argyll 288 F1
West Hythe Kent 54 G6
West Ilkerton Devon 41 D8
West Ilsley W Berks 64 C3
West Itchenor W Sus 22 C3
West Jesmond T & W 243 D7
West Keal Lincs 174 C5
West Kennett Wilts 62 F6
West Kensington London 67 D8
West Kilbride N Ayrs 266 F4
West Kilburn London 67 C8
West Kingsdown Kent 68 G5
West Kingston W Sus 35 G9
West Kington Wilts 61 D10
West Kington Wick Wilts 61 D10
West Kinharrachie Aberds 303 F9
West Kirby Mers 182 D2
West Kirkby Mers 182 D2
West Knapton N Yorks 217 D7
West Knighton Dorset 17 D10
West Knoyle Wilts 45 G11
West Kyloe Northumb 273 G11
West Kyo Durham 242 G5
West Lambrook Som 28 D6
West Langdon Kent 55 D10
West Langwell Highld 309 J6
West Lavington W Sus 34 C5
West Lavington Wilts 46 C4
West Layton N Yorks 224 D2
West Lea Durham 234 B4
West Leake Notts 153 D10
West Learmouth Northumb 263 B9
West Leigh Devon 25 F11
West Leigh Hants 22 B2
West Leigh Som 42 G6
West Lexham Norf 158 F6
West Lilling N Yorks 216 F2
West Linton Borders 270 E2
West Liss Hants 34 B3
West Littleton S Glos 61 D9
West Lockinge Oxon 64 B2
West Looe Corn 6 E5
West Luccombe Som 41 D11
West Lulworth Dorset 18 E2
West Lutton N Yorks 217 F8
West Lydford Som 44 G5
West Lydiatt Hereford 97 C11
West Lyn Devon 41 D8
West Lyng Som 28 B4
West Lynn Norf 158 E2
West Mains Borders 271 F11
West Mains S Lnrk 268 E2
West Malling Kent 53 B7
West Malvern Worcs 98 B5
West Marden W Sus 34 E3
West Marina E Sus 38 F3
West Markham Notts 188 G2
West Marsh NE Lincs 201 E9
West Marton N Yorks 204 C3
West Mathers Aberds 293 G9
West Melbury Dorset 30 C5
West Melton S Yorks 198 G2
West Meon Hants 33 C10
West Meon Woodlands Hants 33 B10
West Merkland Highld 308 F3
West Mersea Essex 89 C8
West Milton Dorset 16 B6
West Minster Kent 70 E2
West Molesey Sur 66 F6
West Monkseaton T & W 243 C8
West Monkton Som 28 B3
West Moor T & W 243 C7
West Moors Dorset 31 G9
West Morden Dorset 18 B4
West Morriston Borders 272 G2
West Morton W Yorks 205 E7
West Mudford Som 29 C9
West Muir Angus 293 G7
West Myreriggs Perth 286 C6
West Ness N Yorks 216 D3
West Newham Northumb 242 B3
West Newton E Yorks 209 F9
West Newton Norf 158 D3
West Newton Som 28 B3
West Norwood London 67 E10
West Ogwell Devon 14 G2
West Orchard Dorset 30 D4
West Overton Wilts 62 F6
West Panson Devon 12 C2
West Park Hrtlpl 234 E5
West Park Hull 200 B5
West Park Mers 183 B7
West Park T & W 243 D9
West Park W Yorks 205 F11
West Parley Dorset 19 B7
West Pasture Durham 232 G4
West Peckham Kent 52 C6
West Pelton Durham 242 G6
West Pennard Som 44 F4
West Pentire Corn 4 C5
West Perry Cambs 122 D2
West Pontnewydd Torf 78 F3
West Poringland Norf 142 C5
West Porlock Som 41 D11
West Portholland Corn 5 G9
West Porton Renfs 277 G8
West Pulham Dorset 30 F2
West Putford Devon 24 D5
West Quantoxhead Som 42 E6
West Rainton Durham 234 B2
West Rasen Lincs 189 D9
West Ravendale NE Lincs 190 B2
West Raynham Norf 159 D7
West Retford Notts 187 E11
West Rounton N Yorks 225 E8
West Row Suff 124 B3
West Royd W Yorks 205 F9
West Rudham Norf 158 D6
West Ruislip London 66 B5
West Runton Norf 177 E11
West Saltoun E Loth 271 B9
West Sandford Devon 26 G4
West Sandwick Shetland 312 E6
West Scholes W Yorks 205 G7
West Scrafton N Yorks 213 C11
West Shepton Som 44 E6
West Side Bl Gwent 77 D11
West Side Orkney 314 C5
West Skelston Dumfries 247 F8
West Sleekburn Northumb 253 G7
West Somerton Norf 161 F9
West Southbourne Bmouth 19 C8
West Stafford Dorset 17 D10
West Stockwith Notts 188 C3
West Stoke Devon 13 G9
West Stoke Som 29 D7
West Stoke W Sus 22 B4
West Stonesdale N Yorks 223 E7
West Stoughton Som 44 D2
West Stour Dorset 30 C3
West Stourmouth Kent 71 G9
West Stow Suff 124 C6
West Stowell Wilts 62 G6
West Strathan Highld 308 C5
West Stratton Hants 48 E4
West Street Kent 54 C2
West Street Kent 55 C10
West Street Medway 69 D8
West Street Suff 125 C9
West Tanfield N Yorks 214 D5
West Taphouse Corn 6 C3
West Tarbert Argyll 275 G9
West Tarring W Sus 35 G10
West Third Borders 262 B4
West Thirston Northumb 252 D5
West Thorney W Sus 22 C3
West Thurrock Thurrock 68 D5
West Tilbury Thurrock 69 D7
West Tisted Hants 33 B11
West Tofts Norf 140 E6
West Tofts Perth 286 D5
West Tolgus Corn 4 G3
West Torrington Lincs 189 E10
West Town Bath 60 G4
West Town Devon 14 B3
West Town Devon 24 C4
West Town Hants 21 B10
West Town Hereford 115 E8
West Town N Som 60 F3
West Town Som 44 F4
West Town W Sus 36 D3
West Tytherley Hants 32 B3
West Tytherton Wilts 62 E2
West Vale W Yorks 196 C5
West View Hrtlpl 234 D5
West Village V Glam 58 E3
West Walton Norf 157 G9
West Walton Highway Norf 157 G9
West Watergate Corn 6 E4
West Watford Herts 85 F10
West Wellow Hants 32 D3
West Wemyss Fife 280 C6
West Wick N Som 59 G11
West Wickham Cambs 106 B2
West Wickham London 67 F11
West Williamston Pembs 73 D8
West Willoughby Lincs 173 G7
West Winch Norf 158 F2
West Winterslow Wilts 47 G8
West Wittering W Sus 21 B11
West Witton N Yorks 213 B11
West Woodburn Northumb 251 F9
West Woodhay W Berks 63 G11
West Woodlands Som 45 E9
West Worldham Hants 49 F8
West Worlington Devon 26 E3
West Worthing W Sus 35 G10
West Wratting Cambs 124 G2
West Wycombe Bucks 84 G4
West Wylam Northumb 242 E4
West Yatton Wilts 61 E11
West Yell Shetland 312 E6
West Yeo Som 43 G10
West Yoke Kent 68 F5
West Youlstone Corn 24 D3
Westacott Devon 40 G5
Westbere Kent 71 G7
Westborough Lincs 172 G5
Westbourne Bmouth 19 C7
Westbourne Suff 108 B2
Westbourne W Sus 22 B3
Westbourne Green London 67 C9
Westbrook Hereford 96 C5
Westbrook Kent 71 E10
Westbrook Sur 50 E3
Westbrook W Berks 64 E2
Westbrook Warr 183 C9
Westbrook Wilts 62 F3
Westbrook Green Norf 142 G2
Westbrook Hay Herts 85 D8
Westburn S Lnrk 268 C3
Westbury Bucks 102 D2
Westbury Shrops 131 B7
Westbury Wilts 45 C11
Westbury Leigh Wilts 45 C11
Westbury-on-Severn Glos 80 C2
Westbury on Trym Bristol 60 D5
Westbury Park Bristol 60 D5
Westbury-sub-Mendip Som 44 D4
Westby Lancs 202 G3
Westby Lincs 155 D9
Westcliff-on-Sea Sthend 69 B11
Westcombe Som 29 B7
Westcombe Som 45 F7
Westcot Oxon 63 B10
Westcote Glos 100 G4
Westcotes Leicester 135 C11
Westcott Bucks 84 B2
Westcott Devon 27 G8
Westcott Shrops 131 C8
Westcott Sur 50 D6
Westcott Barton Oxon 101 F8
Westcourt Wilts 63 G8
Westcroft M Keynes 102 E6
Westcroft W Mid 133 C8
Westdean E Sus 23 F8
Westdene Brighton 36 F3
Westdown Camp Wilts 46 D4
Westdowns Corn 11 E7
Westend Oxon 100 G6
Westend S Glos 79 G10
Westend Town Northumb 241 D7
Westenhanger Kent 54 F6
Wester Aberchalder Highld 300 G5
Wester Arboll Highld 311 L2
Wester Auchinloch N Lnrk 278 G3
Wester Auchnagallin Highld 301 F10
Wester Balgedie Perth 286 G5
Wester Brae Highld 300 C6
Wester Broomhouse E Loth 282 F3
Wester Craiglands Highld 301 D7
Wester Culbeuchly Aberds 302 C6
Wester Dalvoult Highld 291 B11
Wester Dechmont W Loth 269 B10
Wester Deloraine Borders 261 E8
Wester Denoon Angus 287 C7
Wester Ellister Argyll 254 B2
Wester Essendy Perth 286 C5
Wester Essenside Borders 261 E10
Wester Feddal Perth 286 G2
Wester Fintray Aberds 293 B10
Wester Galgantray Highld 301 E8
Wester Gospetry Fife 286 G5
Wester Gruinards Highld 309 K5
Wester Hailes Edin 270 B4
Wester Housebyres Borders 262 B2
Wester Kershope Borders 261 D9
Wester Lealty Highld 300 B6
Wester Lix Stirl 285 E9
Wester Milton Highld 301 D9
Wester Mosshead Aberds 302 F5
Wester Newburn Fife 287 G8
Wester Ord Aberds 293 C10
Wester Parkgate Dumfries 248 F2
Wester Quarff Shetland 313 K6
Wester Skeld Shetland 313 J4
Wester Strath Highld 300 D6
Wester Watten Highld 310 D6
Westerdale Highld 310 D5
Westerdale N Yorks 226 D3
Westerfield Shetland 313 H5
Westerfield Suff 108 B3
Westerfolds Moray 301 C11
Westergate W Sus 22 C6
Westerham Kent 52 C2
Westerhope T & W 242 D5
Westerleigh S Glos 61 D8
Westerleigh Hill S Glos 61 D8
Western Bank Cumb 229 B10
Western Downs Staffs 151 E8
Western Heights Kent 55 E10
Western Hill Durham 233 C11
Western Park Leicester 135 C11
Westerton Aberds 293 B9
Westerton Aberds 302 E5
Westerton Angus 287 B10
Westerton Durham 233 E10
Westerton Moray 302 D3
Westerton W Sus 22 B5
Westertown Aberds 303 F7
Westerwick Shetland 313 J4
Westfield Bath 45 C7
Westfield Cumb 228 F5
Westfield E Sus 38 D4
Westfield Hants 21 B10
Westfield Hereford 98 B4
Westfield Highld 310 C4
Westfield N Lnrk 278 G4
Westfield Norf 141 B9

Westfield Redcar 235 G7
Westfield S Yorks 186 E6
Westfield Sur 50 B4
Westfield W Loth 279 G8
Westfield W Yorks 197 C8
Westfield W Yorks 205 E9
Westfield Sole Kent 69 G9
Westfields Dorset 30 F2
Westfields Hereford 97 C9
Westfields of Rattray Perth 286 C5
Westford Som 27 C10
Westgate Durham 232 D4
Westgate N Lincs 199 F9
Westgate Norf 176 E4
Westgate Norf 177 E7
Westgate Hill W Yorks 197 B8
Westgate on Sea Kent 71 E10
Westgate Street Norf 160 E3
Westhall Aberds 302 G6
Westhall Suff 143 G8
Westhall Hill Oxon 82 C3
Westham Dorset 17 F9
Westham E Sus 23 E10
Westham Som 44 D2
Westhampnett W Sus 22 B5
Westhay Som 44 E2
Westhead Lancs 194 F2
Westhide Hereford 97 C11
Westhill Aberds 293 C10
Westhill E Yorks 209 F10
Westhill Highld 301 E7
Westhope Hereford 115 G9
Westhope Shrops 131 F9
Westhorp Northants 119 G10
Westhorpe Lincs 156 C4
Westhorpe Notts 171 E11
Westhorpe Suff 125 D10
Westhoughton Gtr Man 195 F7
Westhouse N Yorks 212 E3
Westhouses Derbys 170 D6
Westhumble Sur 51 C7
Westing Shetland 312 C7
Westington Glos 100 D2
Westlake Devon 8 E2
Westland Argyll 275 G11
Westland Green Herts 105 G8
Westlands Staffs 168 G4
Westlands Worcs 117 E7
Westlea Northumb 252 G6
Westlea Swindon 62 C6
Westleigh Devon 25 B7
Westleigh Devon 27 D9
Westleigh Gtr Man 194 G6
Westleton Suff 127 D8
Westley Shrops 131 B7
Westley Suff 124 E6
Westley Heights Essex 69 B7
Westley Waterless Cambs 124 F2
Westlington Bucks 84 C3
Westlinton Cumb 239 E9
Westmancote Worcs 99 D8
Westmarsh Kent 71 G9
Westmeston E Sus 36 E4
Westmill Herts 104 E3
Westmill Herts 105 F7
Westminster London 67 D10
Westmoor End Cumb 229 D8
Westmuir Angus 287 B7
Westness Orkney 314 D3
Westnewton Cumb 229 C8
Westnewton Northumb 263 C10
Westoe T & W 243 D9
Weston Bath 61 F8
Weston Devon 15 D9
Weston Devon 27 G10
Weston Dorset 17 G9
Weston Dorset 29 F8
Weston E Ches 168 E2
Weston E Ches 184 G5
Weston Halton 183 E8
Weston Hants 34 C2
Weston Hereford 115 F7
Weston Herts 104 E5
Weston Lincs 156 D5
Weston N Yorks 205 D9
Weston Northants 101 B11
Weston Notts 172 B3
Weston Pembs 73 C8
Weston S Lnrk 269 F10
Weston Shrops 114 C6
Weston Shrops 131 E11
Weston Shrops 148 D5
Weston Shrops 149 D11
Weston Soton 32 E6
Weston Staffs 151 D9
Weston Suff 143 F8
Weston W Berks 63 E11
Weston Bampfylde Som 29 C10
Weston Beggard Hereford 97 C11
Weston by Welland Northants 136 E5
Weston Colley Hants 48 F4
Weston Colville Cambs 124 G2
Weston Common Soton 33 E7
Weston Corbett Hants 49 D7
Weston Coyney Stoke 168 G6
Weston Ditch Suff 124 B3
Weston Favell Northants 120 E5
Weston Green Cambs 124 G2
Weston Green Norf 160 G2
Weston Green Sur 67 F7
Weston Heath Shrops 150 G5
Weston Hills Lincs 156 E5
Weston in Arden Warks 135 F7
Weston-in-Gordano N Som 60 E2
Weston Jones Staffs 150 E5
Weston Longville Norf 160 F2
Weston Lullingfields Shrops 149 E8
Weston Manor I o W 20 D2
Weston Mill Plym 7 D9
Weston-on-Avon Warks 118 G3
Weston-on-the-Green Oxon 83 B8
Weston-on-Trent Derbys 153 D8
Weston Park Bath 61 F8
Weston Patrick Hants 49 D7
Weston Point Halton 183 E7
Weston Rhyn Shrops 148 B5
Weston-sub-Edge Glos 100 C2
Weston-super-Mare N Som 59 G10
Weston Town Som 45 E8
Weston Turville Bucks 84 C5
Weston under Lizard Staffs 150 G6
Weston under Penyard Hereford 98 G2
Weston under Wetherley Warks 119 D7
Weston Underwood Derbys 170 G3
Weston Underwood M Keynes 121 G7
Westonbirt Glos 61 B11
Westoncommon Shrops 149 D8
Westoning C Beds 103 E10
Westonwharf Shrops 149 D8
Westonzoyland Som 43 G11
Westow N Yorks 216 F5
Westowe Som 42 G6
Westown Devon 27 E10
Westown Perth 286 E6
Westport Argyll 255 E7
Westport Som 28 D5
Westquarter Falk 279 F8
Westra V Glam 58 E6
Westridge Green W Berks 64 D5
Westrigg W Loth 269 B8
Westrip Glos 80 D4
Westrop Wilts 61 E11
Westrop Green W Berks 64 E4
Westrum N Lincs 200 F4
Westruther Borders 272 F2
Westry Cambs 139 D7
Westthorpe Derbys 187 F7
Westvale Mers 182 B6
Westville Devon 8 G4
Westville Notts 171 F8
Westward Cumb 229 C11
Westward Ho! Devon 24 B6
Westweekmoor Devon 12 C4
Westwell Kent 54 D3
Westwell Oxon 82 D2
Westwell Leacon Kent 54 D3
Westwells Wilts 61 F11
Westwick Cambs 123 D8
Westwick Durham 223 B11
Westwick Norf 160 D5
Westwick Row Herts 85 D9
Westwood Devon 14 B6
Westwood Devon 14 E5
Westwood Kent 55 D7
Westwood Kent 71 F11
Westwood Notts 171 E7
Westwood P'boro 138 D3
Westwood S Lnrk 268 E2
Westwood Wilts 45 B10
Westwood Wilts 46 G6
Westwood Heath W Mid 118 B5
Westwood Park Essex 107 E9
Westwood Park Gtr Man 184 B3
Westwoodside N Lincs 188 B3
Westy Warr 183 D10
Wetham Green Kent 69 F10
Wetheral Cumb 239 G11
Wetheral Plain Cumb 239 F11
Wetherby W Yorks 206 D4
Wetherden Suff 125 E10
Wetherden Upper Town Suff 125 D10
Wetheringsett Suff 126 D2
Wethersfield Essex 106 E4
Wethersta Shetland 312 G5
Wetherup Street Suff 126 E2
Wetley Rocks Staffs 169 F7
Wetmore Staffs 152 E5
Wettenhall E Ches 167 C10
Wettenhall Green E Ches 167 C10
Wettles Shrops 131 F8
Wetton Staffs 169 D10
Wetwang E Yorks 208 B4
Wetwood Staffs 150 C5
Wexcombe Wilts 47 B9
Wexham Street Bucks 66 C3
Weybourne Norf 177 E10
Weybourne Sur 49 D11
Weybread Suff 142 G4
Weybridge Sur 66 G5
Weycroft Devon 16 B2
Weydale Highld 310 C5
Weyhill Hants 47 D10
Weymouth Dorset 17 F9
Weythel Powys 114 F4
Whaddon Bucks 102 E6
Whaddon Cambs 104 B6
Whaddon Glos 80 C4
Whaddon Glos 99 G9
Whaddon Wilts 31 B11
Whaddon Wilts 61 G11
Whaddon Gap Cambs 104 B6
Whale Cumb 230 G6
Whaley Derbys 187 G8
Whaley Bridge Derbys 185 E8
Whaley Thorns Derbys 187 G8
Whaligoe Highld 310 E7
Whalley Lancs 203 F10
Whalley Banks Lancs 203 F10
Whalley Range Gtr Man 184 C4
Whalleys Lancs 194 F3
Whalton Northumb 252 G4
Wham N Yorks 212 G5
Whaplode Lincs 156 E6
Whaplode Drove Lincs 156 G6
Whaplode St Catherine Lincs 156 E6
Wharf Warks 119 G8
Wharfe N Yorks 212 F5
Wharles Lancs 202 F4
Wharley End C Beds 103 C8
Wharmley Northumb 241 D9
Wharncliffe Side S Yorks 186 C3
Wharram le Street N Yorks 217 F7
Wharram Percy N Yorks 217 G7
Wharton Hereford 115 F10
Wharton Lincs 188 C4
Wharton W Ches 167 B11
Wharton Green W Ches 167 B11
Whashton N Yorks 224 D3
Whasset Cumb 211 C10
Whatcombe Dorset 30 G4
Whatcote Warks 100 C6
Whatcroft W Ches 167 B11
Whateley Staffs 134 D4
Whatfield Suff 107 B10
Whatley Som 28 F5
Whatley Som 45 D8
Whatlington E Sus 38 D3
Whatmore Shrops 116 C2
Whatsole Street Kent 54 E6
Whatstandwell Derbys 170 E4
Whatton Notts 154 B4
Whauphill Dumfries 236 E6
Whaw N Yorks 223 E9
Wheal Alfred Corn 2 B3
Wheal Baddon Corn 4 G5
Wheal Busy Corn 4 G4
Wheal Frances Corn 4 E5
Wheal Kitty Corn 4 E4
Wheal Rose Corn 4 G4
Wheat Hold Hants 64 G5
Wheatacre Norf 143 E9
Wheatcroft Derbys 170 D5
Wheatenhurst Glos 80 D3
Wheathall Shrops 131 C9
Wheathampstead Herts 85 C11
Wheathill Shrops 132 G2
Wheathill Som 44 G5
Wheatley Devon 14 C4
Wheatley Hants 49 E9
Wheatley Oxon 83 D9
Wheatley S Yorks 198 G5
Wheatley W Yorks 196 B5
Wheatley Hill Durham 234 D3
Wheatley Hills S Yorks 198 G6
Wheatley Lane Lancs 204 F2
Wheatley Park S Yorks 198 F5
Wheaton Aston Staffs 151 G7
Wheddon Cross Som 42 F2
Wheedlemont Aberds 302 G4
Wheelbarrow Town Kent 55 D7
Wheeler End Bucks 84 G4
Wheelerstreet Sur 50 E2
Wheelock E Ches 168 D3
Wheelock Heath E Ches 168 D2
Wheelton Lancs 194 C5
Wheelton Lancs 194 C6
Wheelton Lancs 194 D5
Wheelton Lancs 194 D6
Wheen Angus 292 F5
Wheldale W Yorks 198 B3
Wheldrake York 207 D9
Whelford Glos 81 F11
Whelley Gtr Man 194 F5
Whelp Street Suff 107 B8
Whelpley Hill Herts 85 E7
Whelpo Cumb 230 D2
Whelston Flint 182 F2
Whempstead Herts 104 G6
Whenby N Yorks 216 F2
Whepstead Suff 124 F6
Wherry Town Corn 1 D5
Wherstead Suff 108 C3
Wherwell Hants 47 E11
Wheston Derbys 185 F10
Whetley Cross Dorset 29 G7
Whetsted Kent 53 D7
Whetstone Leics 135 D11
Whetstone London 86 G3
Whettleton Shrops 131 G8
Whicham Cumb 210 C2
Whichford Warks 100 E6
Whickham T & W 242 E6
Whickham Fell T & W 242 F6
Whiddon Devon 40 F5
Whiddon Down Devon 13 C9
Whifflet N Lnrk 268 C4
Whigstreet Angus 287 C8
Whilton Northants 120 E2
Whilton Locks Northants 120 E2
Whim Farm Borders 270 E4
Whimble Devon 24 G5
Whimple Devon 14 B6
Whimpwell Green Norf 161 D7
Whinburgh Norf 141 B10
Whin Lane End Lancs 202 E3
Whinfield Darl 224 B6
Whinhall N Lnrk 268 B5
Whinmoor W Yorks 206 F3
Whinney Hill S Yorks 187 C7
Whinney Hill Stockton 225 B7
Whinnieliggate Dumfries 237 D9
Whinny Heights Blkburn 195 B7
Whinnyfold Aberds 303 F10
Whins of Milton Stirl 278 C5
Whins Wood W Yorks 205 F7
Whipcott Devon 27 D9
Whippendell Botton Herts 85 E9
Whippingham I o W 20 C6
Whipsiderry Corn 4 C6
Whipsnade C Beds 85 B8
Whipton Devon 14 C5
Whirley Grove E Ches 184 F5
Whirlow S Yorks 186 E4
Whisby Lincs 172 B6
Whissendine Rutland 154 G6
Whissonsett Norf 159 E8
Whisterfield E Ches 184 G4
Whistlefield Argyll 276 C2
Whistlefield Argyll 276 C4
Whistley Green Wokingham 65 E9
Whistlow Oxon 101 F9
Whiston Mers 183 C7
Whiston Northants 120 E6
Whiston S Yorks 186 D6
Whiston Staffs 151 G7
Whiston Staffs 169 F8
Whiston Cross Mers 183 C7
Whiston Cross Shrops 132 C5
Whitacre Heath Warks 134 E4
Whitbarrow Village Cumb 230 F4
Whitbeck Cumb 210 C2
Whitbourne Hereford 116 F4
Whitbourne Moor Wilts 45 D10
Whitburn T & W 243 E10
Whitburn W Loth 269 C8
Whitburn Colliery T & W 243 E10
Whitby N Yorks 227 C7
Whitby W Ches 182 F5
Whitbyheath W Ches 182 F5
Whitchurch Bath 60 F6
Whitchurch Bucks 102 G5
Whitchurch Cardiff 59 C7
Whitchurch Devon 12 G5
Whitchurch Hants 48 D3
Whitchurch Hereford 79 B9
Whitchurch Pembs 90 F5
Whitchurch Shrops 167 G8
Whitchurch Som 30 C2
Whitchurch Canonicorum Dorset 16 B3
Whitchurch Hill Oxon 64 D6
Whitchurch-on-Thames Oxon 64 D6
Whitcombe Dorset 17 D10
Whitcombe Som 29 C10
Whitcot Shrops 131 E7
Whitcott Keysett Shrops 130 G5
White Ball Som 27 D9
White Colne Essex 107 F7
White Coppice Lancs 194 D6
White Cross Bath 44 B5
White Cross Bath 44 B6
White Cross Corn 2 E5
White Cross Corn 5 D7
White Cross Hereford 97 C9
White Cross Som 43 D10
White Cross Wilts 45 G9
White Cross Hill Cambs 123 B9
White End Worcs 98 E5
White Gate Gtr Man 195 G11
White Gate Som 28 F4
White Grit Shrops 130 D6
White Hall Herts 104 G5
White Hill Bath 45 B8
White Hill W Yorks 204 E6
White Hill Wilts 45 G10
White Hills Northants 120 E4
White House Suff 108 B2
White Houses Notts 188 F2
White Lackington Dorset 17 B10
White Ladies Aston Worcs 117 G8
White-le-Head Durham 242 G5
White Lee W Yorks 197 B8
White Lund Lancs 211 G8
White Mill Carms 93 G9
White Moor Derbys 170 F5
White Ness Shetland 313 J5
White Notley Essex 88 B3
White Oak Kent 68 F4
White Ox Mead Bath 45 B8
White Pit Lincs 190 F5
White Post Kent 52 E4
White Post Notts 171 D10
White Rocks Hereford 97 G8
White Roding or White Roothing Essex 87 C9
White Stake Lancs 194 B4
White Stone Hereford 97 C11
White Waltham Windsor 65 D10
Whiteacen Moray 302 E2
Whiteacre Kent 54 D6
Whiteacre Heath Warks 134 E4
Whiteash Green Essex 106 E5
Whitebirk Blkburn 195 B8
Whitebog Highld 301 C7
Whitebridge Highld 290 B6
Whitebrook Mon 79 D8
Whiteburn Borders 271 F11
Whitebushes Sur 51 D9
Whitecairn Dumfries 236 D4
Whitecairns Aberds 293 B11
Whitecastle S Lnrk 269 G10
Whitechapel Lancs 203 E7
Whitechapel London 67 C10
Whitechurch Maund Hereford 97 B11
Whitecleat Orkney 314 F5
Whitecliff Glos 79 C9
Whiteclosegate Cumb 239 F10
Whitecote W Yorks 205 F10
Whitecraig E Loth 281 G7
Whitecraigs E Renf 267 D11
Whitecroft Glos 79 D10
Whitecrook W Dunb 267 B10
Whitecross Corn 2 C2
Whitecross Corn 6 E2
Whitecross Corn 10 G5
Whitecross Falk 279 F9
Whitecross Som 28 C6
Whitecross Staffs 151 E7
Whiteface Highld 309 L7
Whitefarland N Ayrs 255 C9
Whitefaulds S Ayrs 245 B7
Whitefield Aberds 303 G7
Whitefield Dorset 18 C4
Whitefield Gtr Man 195 F10
Whitefield Perth 286 D5
Whitefield Som 27 B9
Whitefield Lane End Mers 183 D7
Whiteflat E Ayrs 258 D2
Whiteford Aberds 303 G7
Whitegate W Ches 167 B10
Whitehall Blkburn 195 C7
Whitehall Bristol 60 E6
Whitehall Devon 27 E10
Whitehall Devon 40 F4
Whitehall Hants 49 C8
Whitehall Herts 104 E6
Whitehall W Sus 35 C10
Whitehall Village Orkney 314 D6
Whitehaven Cumb 219 B9
Whitehaven Shrops 148 E5
Whitehawk Brighton 36 G4
Whiteheath Gate W Mid 133 F9
Whitehill E Sus 37 B8
Whitehill Hants 49 G9
Whitehill Kent 54 B4
Whitehill Midloth 271 B7
Whitehill Moray 302 D5
Whitehill S Lnrk 268 D4
Whitehill Staffs 168 E3
Whitehills Aberds 302 C6
Whitehills S Lnrk 268 E2
Whitehills T & W 243 E7
Whiteholme Blkpool 202 E2
Whitehough Derbys 185 E8
Whitehouse Aberds 293 B8
Whitehouse Argyll 275 G9
Whitehouse Common W Mid 134 D2
Whitehouse Green W Berks 65 F7
Whiteinch Glasgow 267 B10
Whitekirk E Loth 281 E10
Whiteknights Reading 65 E8
Whiteknowes Aberds 293 C7
Whitelackington Som 28 D5
Whitelaw S Lnrk 268 G2
Whiteleaf Bucks 84 E4
Whiteleas T & W 243 E9
Whiteleaved Oak Glos 98 D5
Whitelee Borders 262 C3
Whitelee Northumb 250 B6
Whitelees S Ayrs 257 C9
Whiteley Bank I o W 21 E7
Whiteley Green E Ches 184 F6
Whiteley Village Sur 66 G5
Whitelye Mon 79 E8
Whitemans Green W Sus 36 B4
Whitemire Moray 301 D9
Whitemoor Corn 5 D9
Whitemoor Nottingham 171 G8
Whitemoor Warks 118 C5
Whitemore Staffs 168 C5
Whitenap Hants 32 C5
Whiteoak Green Oxon 82 C4
Whiteparish Wilts 32 C2
Whitepits Wilts 45 F10
Whiterashes Aberds 303 G8
Whiterigg Borders 262 C3
Whiterock Bridgend 58 D2
Whiterow Highld 310 E7
White's Green W Sus 34 B6
Whiteshill Glos 80 D4
Whiteshill S Glos 60 D6
Whiteside Northumb 240 D6
Whiteside W Loth 269 B9
Whitesmith E Sus 23 C8
Whitespots Dumfries 247 F10
Whitestaunton Som 28 E3
Whitestone Aberds 293 D8
Whitestone Devon 14 C3
Whitestone Devon 40 D3
Whitestone Warks 135 F7
Whitestones Aberds 303 D8
Whitestreet Green Suff 107 D9
Whitewall Common Mon 60 B2
Whitewall Corner N Yorks 216 E5
Whiteway Bath 61 G8
Whiteway Dorset 18 E3
Whiteway Glos 80 C6
Whiteway Glos 80 F4
Whitewell Aberds 303 C9
Whitewell Corn 11 E7
Whitewell Highld 291 C11
Whitewell Lancs 203 D9
Whitewell Wrex 167 G7
Whitewell Bottom Lancs 195 C10
Whiteworks Devon 13 G8
Whitfield Kent 55 D10
Whitfield Northants 102 D2
Whitfield Northumb 241 F7
Whitfield S Glos 79 G11
Whitfield Court Sur 50 B3
Whitfield Hall Northumb 241 F7
Whitford Devon 15 B11
Whitford = Chwitffordd Flint 181 F10
Whitgift E Yorks 199 C10
Whitgreave Staffs 151 D7
Whithaugh Borders 249 F11
Whithebeir Orkney 314 C5
Whithorn Dumfries 236 E6
Whiting Bay N Ayrs 256 D2
Whitington Norf 140 D4
Whitkirk W Yorks 206 G3
Whitland = Hendy-Gwyn Carms 73 B11
Whitlaw Borders 271 F9
Whitleigh Plym 7 C9
Whitletts S Ayrs 257 E9
Whitley Gtr Man 194 F5
Whitley N Yorks 198 C5
Whitley Reading 65 E8
Whitley S Yorks 186 C4
Whitley W Mid 119 B7
Whitley Wilts 61 F11
Whitley Bay T & W 243 C9
Whitley Bridge N Yorks 198 C5
Whitley Chapel Northumb 241 F10
Whitley Head W Yorks 204 E6
Whitley Heath Staffs 150 D6
Whitley Lower W Yorks 197 D8
Whitley Reed W Ches 183 E10
Whitley Row Kent 52 C3
Whitley Sands T & W 243 C9
Whitley Thorpe N Yorks 198 C5
Whitley Wood Reading 65 F8
Whitlock's End W Mid 118 B2
Whitminster Glos 80 D3
Whitmoor Devon 27 E9
Whitmore Dorset 31 F9
Whitmore Staffs 168 G4
Whitmore Park W Mid 134 G6
Whitnage Devon 27 D8
Whitnash Warks 118 E6
Whitnell Som 43 F8
Whitney Bottom Som 28 E4
Whitney-on-Wye Hereford 96 B5
Whitrigg Cumb 229 D10
Whitrigg Cumb 238 F6
Whitriggs Borders 262 F3
Whitsbury Hants 31 D10
Whitslaid Borders 271 G11
Whitsome Borders 273 E7
Whitsomehill Borders 273 F7
Whitson Newport 59 C11
Whitstable Kent 70 F6
Whitstone Corn 11 B11
Whittingham Northumb 264 G3
Whittingslow Shrops 131 F8
Whittington Glos 99 G10
Whittington Lancs 212 D2
Whittington Norf 140 D4
Whittington Shrops 148 C6
Whittington Staffs 133 G7
Whittington Staffs 134 B3
Whittington Staffs 150 C5
Whittington Warks 134 D5
Whittington Worcs 117 G7
Whittington Moor Derbys 186 G5
Whittleford Warks 134 E6
Whittle-le-Woods Lancs 194 C5
Whittlebury Northants 102 C3
Whittlesey Cambs 138 D5
Whittlesford Cambs 105 B9
Whittlestone Head Blkburn 195 D8
Whitton Borders 263 E7
Whitton Hereford 115 C8
Whitton London 66 E6
Whitton N Lincs 200 C2
Whitton Northumb 252 C3
Whitton Powys 114 D5
Whitton Shrops 115 C11
Whitton Stockton 234 G3
Whitton Suff 108 B2
Whittonditch Wilts 63 E9
Whittonstall Northumb 242 F3
Whittytree Shrops 115 B8
Whitway Hants 48 B3
Whitwell Derbys 187 F8
Whitwell Herts 104 G3
Whitwell I o W 20 F6
Whitwell N Yorks 224 F5
Whitwell Rutland 137 B8
Whitwell-on-the-Hill N Yorks 216 F4
Whitwell Street Norf 160 E2
Whitwick Leics 153 F8
Whitwood W Yorks 198 C2
Whitworth Lancs 195 D11
Whixall Shrops 149 C10
Whixley N Yorks 206 B4
Whoberley W Mid 118 B6
Wholeflats Falk 279 E8
Whome Orkney 314 G3
Whorlton Durham 224 C2
Whorlton N Yorks 225 E9
Whydown E Sus 38 F2
Whygate Northumb 241 B7
Whyke W Sus 22 C5
Whyle Hereford 115 E11
Whyteleafe Sur 51 B10
Wibdon Glos 79 F9
Wibsey W Yorks 205 G8
Wibtoft Leics 135 F9
Wichenford Worcs 116 E5
Wichling Kent 54 B2
Wick Bmouth 19 C8
Wick Devon 27 G11
Wick Highld 310 D7
Wick S Glos 61 E8
Wick Shetland 313 K6
Wick Som 28 B6
Wick Som 43 C10
Wick Som 43 E8
Wick V Glam 58 E2
Wick W Sus 35 G8
Wick Wilts 31 C11
Wick Worcs 99 B9
Wick Episcopi Worcs 116 G6
Wick Hill Brack 65 E11
Wick Hill Kent 53 E10
Wick Hill Wokingham 65 G9
Wick Rocks S Glos 61 E8
Wick St Lawrence N Som 59 F11
Wick Street Glos 80 D5
Wicken Cambs 123 C11
Wicken Northants 102 D4
Wicken Bonhunt Essex 105 E9
Wicken Green Village Norf 158 C6
Wickenby Lincs 189 E9
Wicker Street Green Suff 107 C9
Wickersley S Yorks 187 C7
Wickford Essex 88 G3
Wickham Hants 33 E9
Wickham W Berks 63 E11
Wickham Bishops Essex 88 C4
Wickham Green Suff 125 D11
Wickham Green W Berks 63 E11
Wickham Heath W Berks 64 F2
Wickham Market Suff 126 F6
Wickham St Paul Essex 106 D6
Wickham Skeith Suff 125 D11
Wickham Street Suff 124 G5
Wickham Street Suff 125 D11
Wickhambreaux Kent 55 B8
Wickhambrook Suff 124 G4
Wickhamford Worcs 99 C11
Wickhampton Norf 143 B8
Wickham's Cross Som 44 G4
Wickhurst Kent 52 D4
Wicklane Bath 45 B7
Wicklewood Norf 141 C11
Wickmere Norf 160 C3
Wickridge Street Glos 98 F6
Wickstreet E Sus 23 D8
Wickwar S Glos 61 B8
Widbrook Wilts 45 B10
Widcombe Bath 61 G9
Widdington Essex 105 E10
Widdrington Northumb 253 D7
Widdrington Station Northumb 252 E6
Widecombe in the Moor Devon 13 F10
Widegates Corn 6 D5
Widemarsh Hereford 97 C10
Widemouth Bay Corn 24 G2
Wideopen T & W 242 C6
Widewall Orkney 314 G4
Widewell Plym 7 C9
Widford Essex 87 D11
Widford Herts 86 B6
Widgham Green Cambs 124 F3
Widham Wilts 62 B5
Widley Hants 33 F11
Widmer End Bucks 84 F5
Widmerpool Notts 154 D2
Widmoor Bucks 66 B2
Widmore London 68 F2
Widnes Halton 183 D8
Wierton Kent 53 D9
Wig Powys 130 F2
Wig Fach Bridgend 57 F10
Wigan Gtr Man 194 F5
Wiganthorpe N Yorks 216 E3
Wigbeth Dorset 31 F8
Wigborough Som 28 D6
Wiggaton Devon 15 C8
Wiggenhall St Germans Norf 157 G11
Wiggenhall St Mary Magdalen Norf 157 G11
Wiggenhall St Mary the Virgin Norf 157 G11
Wiggenhall St Peter Norf 158 G2
Wiggens Green Essex 106 C3
Wiggington Staffs 134 B4
Wigginstall Staffs 169 C9
Wigginton Herts 84 C6
Wigginton Oxon 101 E7
Wigginton Shrops 148 B6
Wigginton Staffs 134 B4
Wigginton York 207 B7
Wigginton Bottom Herts 84 D6
Wigginton Heath Oxon 101 D7
Wigglesworth N Yorks 204 B2
Wiggonby Cumb 239 G7
Wiggonholt W Sus 35 D9
Wighill N Yorks 206 D5
Wighton Norf 159 B8
Wightwick Manor Staffs 133 D7
Wigley Derbys 186 G4
Wigley Hants 32 D4
Wigmarsh Shrops 149 D7
Wigmore Hereford 115 D8
Wigmore Medway 69 G10
Wigsley Notts 188 G5
Wigsthorpe Northants 137 G10
Wigston Leics 136 D2
Wigston Magna Leics 136 D2
Wigston Parva Leics 135 F9
Wigthorpe Notts 187 E9
Wigtoft Lincs 156 B5
Wigton Cumb 229 B11
Wigtown Dumfries 236 D6
Wigtwizzle S Yorks 186 B2
Wike W Yorks 206 E2
Wike Well End S Yorks 199 E7
Wilbarston Northants 136 F6
Wilberfoss E Yorks 207 C10
Wilberlee W Yorks 196 E5
Wilburton Cambs 123 C9
Wilby Norf 141 F10
Wilby Northants 121 D7
Wilby Suff 126 C4
Wilcot Wilts 62 G6
Wilcott Shrops 149 F7
Wilcott Marsh Shrops 149 F7
Wilcove Corn 7 D8
Wilcrick Newport 60 B2
Wild Mill Bridgend 58 C2
Wilday Green Derbys 186 G4
Wildboarclough E Ches 169 B7
Wilde Street Suff 124 B4
Wilden Beds 121 F11
Wilden Worcs 116 C6
Wildern Hants 33 E7
Wildernesse Kent 52 B4
Wilderspool Warr 183 D10
Wilderswood Gtr Man 194 E6
Wildhern Hants 47 C11
Wildhill Herts 86 D3
Wildmanbridge S Lnrk 268 E6
Wildmoor Hants 49 B7
Wildmoor Oxon 83 F7
Wildmoor Worcs 117 B9
Wildridings Brack 65 F11
Wildsworth Lincs 188 B4
Wildwood Staffs 151 E8
Wilford Nottingham 153 B11
Wilgate Green Kent 54 B3
Wilkesley E Ches 167 G10
Wilkhaven Highld 311 L3
Wilkieston W Loth 270 B2
Wilkin Throop Som 29 C11
Wilkinthroop Som 29 C11
Wilksby Lincs 174 C3
Will Row Lincs 191 D7
Willacy Lane End Lancs 202 F5
Willand Devon 27 E8
Willand Som 27 E11
Willand Moor Devon 27 E8
Willard's Hill E Sus 38 C2
Willaston E Ches 167 E11
Willaston Shrops 149 B11
Willaston W Ches 182 F4
Willen M Keynes 103 C7
Willenhall W Mid 119 B7
Willenhall W Mid 133 D9
Willerby E Yorks 208 G6
Willerby N Yorks 217 D10
Willersey Glos 100 D2
Willersley Hereford 96 B6
Willesborough Kent 54 E4
Willesborough Lees Kent 54 E4
Willesden London 67 C8
Willesden Green London 67 C8
Willesley Leics 152 G6
Willesley Wilts 61 B11
Willett Som 42 G6
Willey Shrops 132 D3
Willey Warks 135 G9
Willey Green Sur 50 C2
Willhayne Som 28 E4
Williamhope Borders 261 C10
William's Green Suff 107 C9
Williamscott Oxon 101 B9
Williamslee Borders 270 G6
Williamstown Rhondda 77 G8
Williamthorpe Derbys 170 B6
Willian Herts 104 E4
Willicote Pastures Worcs 100 B3
Willingale Essex 87 D9
Willingcott Devon 40 E3
Willingdon E Sus 23 E9
Willingham Cambs 123 C8
Willingham Suff 143 G8
Willingham by Stow Lincs 188 E5
Willingham Green Cambs 124 G2
Willington Beds 104 B2
Willington Derbys 152 D5
Willington Durham 233 D9
Willington Kent 53 C9
Willington T & W 243 D8
Willington Warks 100 D5
Willington Corner W Ches 167 B8
Willington Quay T & W 243 D8
Willisham Suff 125 G11
Willisham Tye Suff 125 G11
Willitoft E Yorks 207 F10
Williton Som 42 E5
Willoughbridge Staffs 168 G3
Willoughby Lincs 191 G7
Willoughby Warks 119 D10
Willoughby Hills Lincs 174 F4
Willoughby-on-the-Wolds Notts 154 D2
Willoughby Waterleys Leics 135 E11
Willoughton Lincs 188 C6
Willow Green W Ches 183 F10
Willow Green Worcs 116 F5
Willow Holme Cumb 239 F9
Willowbank Bucks 66 B5
Willows Gtr Man 195 F8
Willows Green Essex 88 B2
Willowtown Bl Gwent 77 C11
Willsbridge S Glos 61 E7
Willslock Staffs 151 C11
Willstone Shrops 131 D9
Willsworthy Devon 12 E6
Wilmcote Warks 118 F3
Wilmington Bath 61 G7
Wilmington Devon 15 B10
Wilmington E Sus 23 E8
Wilmington Kent 68 E4
Wilmington Green E Sus 23 D8
Wilminstone Devon 12 F5
Wilmslow E Ches 184 E4
Wilmslow Park E Ches 184 E5
Wilnecote Staffs 134 C4
Wilney Green Norf 141 G11
Wilpshire Lancs 203 G9
Wilsden W Yorks 205 F7
Wilsden Hill W Yorks 205 F7
Wilsford Lincs 173 G8
Wilsford Wilts 46 B6
Wilsford Wilts 46 F6
Wilsham Devon 41 D9
Wilshaw W Yorks 196 F6
Wilsic S Yorks 187 B9
Wilsill N Yorks 214 G3
Wilsley Green Kent 53 F9
Wilsley Pound Kent 53 F9
Wilsom Hants 49 F8
Wilson Hereford 97 G11
Wilson Leics 153 E8
Wilsontown S Lnrk 269 D9
Wilstead Beds 103 C11
Wilsthorpe Derbys 153 C9
Wilsthorpe Lincs 155 G11
Wilstone Herts 84 C6
Wilstone Green Herts 84 C6
Wilthorpe S Yorks 197 F10
Wilton Borders 261 G11
Wilton Cumb 219 C10
Wilton Hereford 97 G11
Wilton N Yorks 217 C7
Wilton Redcar 225 B11
Wilton Som 28 C2
Wilton Wilts 46 G5
Wilton Wilts 63 G9
Wilton Park Bucks 85 G7
Wiltown Devon 27 D11
Wiltown Som 28 C5
Wimbish Essex 105 D11
Wimbish Green Essex 106 D2
Wimble Hill Hants 49 D10
Wimblebury Staffs 151 G10
Wimbledon London 67 E8
Wimblington Cambs 139 E8
Wimbolds Trafford W Ches 182 G6
Wimboldsley W Ches 167 C11
Wimborne Minster Dorset 18 B6
Wimborne St Giles Dorset 31 E8
Wimbotsham Norf 140 B2
Wimpole Cambs 104 B6
Wimpson Soton 32 E5
Wimpstone Warks 100 B4
Wincanton Som 30 B2
Winceby Lincs 174 B4
Wincham W Ches 183 F11
Winchburgh W Loth 279 G11
Winchcombe Glos 99 F10
Winchelsea E Sus 38 D6
Winchelsea Beach E Sus 38 D6
Winchester Hants 33 B7
Winchestown Bl Gwent 77 C11
Winchet Hill Kent 53 E8
Winchfield Hants 49 C9
Winchfield Hurst Hants 49 C9
Winchmore Hill Bucks 84 G6
Winchmore Hill London 86 G4
Wincle E Ches 169 B7
Wincobank S Yorks 186 C5
Wind Hill Cumb 210 E5
Wind Hill Pembs 73 F9
Winder Cumb 219 B10
Windermere Cumb 221 F8
Winderton Warks 100 C6
Windhill Highld 300 E5
Windhill S Yorks 198 G3
Windhill W Yorks 205 F9
Windhouse Shetland 312 D6
Winding Wood W Berks 63 E11
Windle Hill W Ches 182 F4
Windlehurst Gtr Man 185 D7
Windlesham Sur 66 G2
Windley Derbys 170 F4
Windmill Corn 10 G3
Windmill Flint 181 G11
Windmill Flint 182 G2
Windmill Hill Bristol 60 E5
Windmill Hill E Sus 23 C10
Windmill Hill Halton 183 E9
Windmill Hill Kent 69 F10
Windmill Hill Som 28 D4
Windmill Hill W Yorks 197 D11
Windmill Hill Worcs 99 B8
Windrush Glos 81 C11
Windsor N Lincs 199 E9
Windsor Windsor 66 D3
Windsor Green Suff 125 G7
Windsoredge Glos 80 E4
Windwhistle Som 28 F5
Windy Arbor Mers 183 D7
Windy Arbour Warks 118 C5
Windy Hill Wrex 166 E4
Windy Nook T & W 243 E7
Windy-Yett E Ayrs 267 E9
Windydoors Borders 261 B10
Windygates Fife 287 G7
Windyharbour E Ches 184 G4
Windyknowe W Loth 269 B9
Windywalls Borders 263 C7
Wineham W Sus 36 C2
Winestead E Yorks 201 C9
Winewall Lancs 204 E4
Winfarthing Norf 142 F2
Winford I o W 21 E7
Winford N Som 60 F4
Winforton Hereford 96 B5
Winfrith Newburgh Dorset 18 E2
Wing Bucks 103 G7
Wing Rutland 137 C7
Wingate Durham 234 D4
Wingates Gtr Man 195 F7
Wingates Northumb 252 D4
Wingerworth Derbys 170 B5
Wingfield C Beds 103 F10
Wingfield S Yorks 186 B6
Wingfield Suff 126 B4
Wingfield Wilts 45 B10
Wingfield Green Suff 126 B4
Wingfield Park Derbys 170 E5
Wingham Kent 55 B8
Wingham Green Kent 55 B8
Wingham Well Kent 55 B8
Wingmore Kent 55 D7
Wingrave Bucks 84 B5
Winkburn Notts 172 D2
Winkfield Brack 66 E2
Winkfield Place Brack 66 E2
Winkfield Row Brack 65 E11
Winkhill Staffs 169 E9
Winkhurst Green Kent 52 D3
Winklebury Hants 48 C6
Winkleigh Devon 25 F10
Winksley N Yorks 214 E5
Winkton Dorset 19 B9
Winlaton T & W 242 E5
Winlaton Mill T & W 242 E5
Winless Highld 310 D7
Winllan Powys 148 E4
Winmarleigh Lancs 202 D5
Winmarleigh Moss Lancs 202 D4
Winnal Hereford 97 E9
Winnal Common Hereford 97 E9
Winnall Hants 33 B7
Winnall Worcs 116 D6
Winnard's Perch Corn 5 B8
Winnersh Wokingham 65 E9
Winnington Staffs 150 B4
Winnington W Ches 183 G10
Winnington Green Shrops 148 G6
Winnothdale Staffs 169 G8
Winscales Cumb 228 F6
Winscombe N Som 44 B2
Winsdon Hill Luton 103 G11
Winsford Som 42 G2
Winsford W Ches 167 B10
Winsh-wen Swansea 57 B7
Winsham Devon 40 F3
Winsham Som 28 F5
Winshill Staffs 152 E5
Winsick Derbys 170 B6
Winskill Cumb 231 D7
Winslade Devon 24 D5
Winslade Hants 49 D7
Winsley N Yorks 214 G4
Winsley Wilts 61 G10
Winslow Bucks 102 F5
Winslow Mill Hereford 98 D2
Winson Glos 81 D9
Winson Green W Mid 133 F10
Winsor Hants 32 E4
Winstanley Gtr Man 194 G5
Winstanleys Gtr Man 194 F5
Winster Cumb 221 G8
Winster Derbys 170 C2
Winston Durham 224 B2
Winston Suff 126 E3
Winston Green Suff 126 E3
Winstone Glos 81 D7
Winswell Devon 25 E7
Winter Gardens Essex 69 C9
Winter Well Som 28 C3
Winterborne Bassett Wilts 62 E6
Winterborne Came Dorset 17 D10
Winterborne Clenston Dorset 30 G4
Winterborne Herringston Dorset 17 D9
Winterborne Houghton Dorset 30 G4
Winterborne Kingston Dorset 18 B3
Winterborne Monkton Dorset 17 D9